D1552352

Audi A6

Service Manual

1998, 1999, 2000, 2001, 2002, 2003, 2004
including S6, allroad quattro, RS6

C5 Platform

B® BentleyPublishers .com

Bentley Publishers, a division of Robert Bentley, Inc.
1734 Massachusetts Avenue
Cambridge, MA 02138 USA
800-423-4595 / 617-547-4170

Information that makes the difference®

BentleyPublishers™ .com

Technical contact information

We welcome your feedback. Please submit corrections and additions to our Audi technical discussion forum at:

http://www.BentleyPublishers.com

Updates and corrections

We will evaluate submissions and post appropriate editorial changes online as updates or tech discussion. Appropriate updates and corrections will be added to the book in future printings. Check for updates and corrections for this book before beginning work on your vehicle. See the following web address for additional information:

http://www.BentleyPublishers.com/updates/

WARNING—IMPORTANT SAFETY NOTICE

Do not use this manual unless you are familiar with automotive repair procedures and safe workshop practices. This manual illustrates the workshop procedures required for most service work. It is not a substitute for full and up-to-date information from the vehicle manufacturer or for proper training as an automotive technician. Note that it is not possible for us to anticipate all of the ways or conditions under which vehicles may be serviced or to provide cautions as to all of the possible hazards that may result.

We have endeavored to ensure the accuracy of the information in this manual. Please note, however, that considering the vast quantity and the complexity of the service information involved, we cannot warrant the accuracy or completeness of the information contained in this manual.

FOR THESE REASONS, NEITHER THE PUBLISHER NOR THE AUTHOR MAKES ANY WARRANTIES, EXPRESS OR IMPLIED, THAT THE INFORMATION IN THIS BOOK IS FREE OF ERRORS OR OMISSIONS, AND WE EXPRESSLY DISCLAIM THE IMPLIED WARRANTIES OF MERCHANTABILITY AND OF FITNESS FOR A PARTICULAR PURPOSE, EVEN IF THE PUBLISHER OR AUTHOR HAVE BEEN ADVISED OF A PARTICULAR PURPOSE, AND EVEN IF A PARTICULAR PURPOSE IS INDICATED IN THE MANUAL. THE PUBLISHER AND AUTHOR ALSO DISCLAIM ALL LIABILITY FOR DIRECT, INDIRECT, INCIDENTAL OR CONSEQUENTIAL DAMAGES THAT RESULT FROM ANY USE OF THE EXAMPLES, INSTRUCTIONS OR OTHER INFORMATION IN THIS BOOK. IN NO EVENT SHALL OUR LIABILITY WHETHER IN TORT, CONTRACT OR OTHERWISE EXCEED THE COST OF THIS MANUAL.

Before attempting any work on your Audi, read **00 WARNINGS and CAUTIONS** and any **WARNING** or **CAUTION** that accompanies a procedure in the service manual. Review the **WARNINGS** and **CAUTIONS** each time you prepare to work on your Audi.

Your common sense and good judgment are crucial to safe and successful service work. Read procedures through before starting them. Think about whether the condition of your car, your level of mechanical skill, or your level of reading comprehension might result in or contribute in some way to an occurrence which might cause you injury, damage your car, or result in an unsafe repair. If you have doubts for these or other reasons about your ability to perform safe repair work on your car, have the work done at an authorized Audi dealer or other qualified shop.

Part numbers listed in this manual are for identification purposes only, not for ordering. Always check with your authorized Audi dealer to verify part numbers and availability before beginning service work that may require new parts.

Special tools required to perform certain service operations are identified in the manual and are recommended for use. Use of tools other than those recommended in this manual may be detrimental to the car's safe operation as well as the safety of the person servicing the car.

The vehicle manufacturer will continue to issue service information updates and parts retrofits after the editorial closing of this manual. Some of these updates and retrofits will apply to procedures and specifications in this manual. We regret that we cannot supply updates to purchasers of this manual.

This manual is prepared, published and distributed by Bentley, Publishers, 1734 Massachusetts Avenue, Cambridge, Massachusetts 02138. All information contained in this manual is based on information available to the publisher from Audi at the time of editorial closing. Audi has not reviewed and does not vouch for the accuracy or completeness of the technical specifications and work procedures described and given in this manual.

© 2006 Robert Bentley, Inc. Bentley Publishers is a trademark of Robert Bentley, Inc

All rights reserved. The right is reserved to make changes at any time without notice. No part of this publication may be reproduced, stored in a retrieval system, or transmitted in any form or by any means, electronic, mechanical, photocopying, recording, or otherwise, without the prior written consent of the publisher. This includes text, figures, and tables. All rights reserved under Berne and Pan-American Copyright conventions. This manual is simultaneously published in Canada.

ISBN 978-0-8376-1670-4 **Bentley Stock No. A604** Mfg. code: A604-06-1310, editorial closing 08/2006·

Library of Congress Cataloging-in-Publication Data for 2006 edition

Audi A6. Service manual : 1998, 1999, 2000, 2001, 2002, 2003, 2004 including S6, allroad quattro, rs6, c5 platform.
p. cm.
Includes index.
ISBN: 978-0-8376-1499-1
1. Audi A6 automobile--Maintenance and repair--Handbooks, manuals, etc. I. Robert Bentley, inc.

TL215.A8A8947 2006
629.28'722--dc22

2006022610

Cover design and use of blue band on spine and back cover are trade dress and are trademarks of Bentley Publishers™. All rights reserved.

The paper used in this publication is acid free and meets the requirements of the National Standard for Information Sciences-Permanence of Paper for Printed Library Materials. ♾

Manufactured in the United States of America.

BentleyPublishers.com
BentleyPublishers.com—All Rights Reserved

Foreword

For the Audi owner with automotive mechanical skills, and for independent service professionals, this manual includes maintenance and repair procedures the Audi A6 is likely to need during its service life, including many procedures and specifications that were available in an authorized Audi dealer service department at the time this manual was prepared. The Audi owner with no intention of working on his or her car will find that owning and referring to this manual will make it possible to be better informed and to more knowledgeably discuss repairs with a professional automotive technician.

For those intending to do maintenance and repair on their Audi, it is essential that safety equipment be used and safety precautions observed when working on your vehicle. A minimum safety equipment list includes hand protection, eye protection, and a fire extinguisher. A selection of good quality hand tools is also needed. This includes a torque wrench to ensure that fasteners are tightened in accordance with specifications. In some cases, the text refers to special tools that are recommended or required to accomplish adjustments or repairs. These tools are often identified by their Volkswagen special tool number and illustrated.

Disclaimer

We have endeavored to ensure the accuracy of the information in this manual. When the vast array of data presented in the manual is taken into account, however, no claim to infallibility can be made. We therefore cannot be responsible for the result of any errors that may have crept into the text. Please also read **WARNING–Important Safety Notice** on the copyright page at the beginning of this book.

Prior to starting a repair procedure, read the procedure, **00 WARNINIG and CAUTIONS** and the **WARNINGS** and **CAUTIONS** that accompany the procedure. Reading a procedure before beginning work helps you determine in advance the need for specific skills, identify hazards, prepare for appropriate capture and handling of hazardous materials, and the need for particular tools and replacement parts such as gaskets.

Bentley Publishers encourages comments from the readers of this manual with regard to errors, and suggestions for improvement of our product. These communications have been and will be carefully considered in the preparation of this and other manuals. If you identify inconsistencies in the manual, you may have found an error. Please contact the publisher and we will endeavor to post applicable corrections on our website. Review corrections (errata) that we have posted before beginning work. Please see the following web address:

`http://www.BentleyPublishers.com/updates/`

Audi continues to issue service information and parts retrofits after the editorial closing of this manual. Some of this updated information may apply to procedures and specifications in this manual. For the latest information, please see the following web address:

`https://erwin.audi.com`

Audi offers extensive warranties, especially on components of the fuel delivery and emission control systems. Therefore, before deciding to repair a Audi that may be covered wholly or in part by any warranties issued by Audi of America, Inc., consult your authorized Audi dealer. You may find that the dealer can make the repair either free or at minimum cost. Regardless of its age, or whether it is under warranty, your Audi is both an easy car to service and an easy car to get serviced. So if at any time a repair is needed that you feel is too difficult to do yourself, a trained Audi technician is ready to do the job for you.

Bentley Publishers

00 Warnings and Cautions

Please read these warnings and cautions before proceeding with maintenance and repair work.

WARNINGS—
See also ***CAUTIONS*** *on next page.*

- Read the important safety notice on the copyright page at the beginning of the book.
- Some repairs may be beyond your capability. If you lack the skills, tools and equipment, or a suitable workplace for any procedure described in this manual, we suggest you leave such repairs to an authorized Audi dealer service department or other qualified shop.
- Thoroughly read each procedure and the **WARNINGS** and **CAUTIONS** that accompany the procedure. Also review posted corrections at `www.BentleyPublishers.com/updates/` before beginning work.
- If any procedure, tightening torque, wear limit, specification or data presented in this manual does not appear to be appropriate for a specific application, contact the publisher or the vehicle manufacturer for clarification before using the information in question.
- Audi is constantly improving its cars. Sometimes these changes, both in parts and specifications, are made applicable to earlier models. Therefore, before starting any major jobs or repairs to components on which passenger safety may depend, consult your authorized Audi dealer about technical bulletins that may have been issued.
- Do not reuse any fasteners that are worn or deformed in normal use. Many fasteners are designed to be used only once and become unreliable and may fail when used a second time. This includes, but is not limited to, nuts, bolts, washers, self-locking nuts or bolts, circlips and cotter pins. Replace these fasteners with new parts.
- Do not work under a lifted car unless it is solidly supported on stands designed for the purpose. Do not support a car on cinder blocks, hollow tiles or other props that may crumble under continuous load. Do not work under a car that is supported solely by a jack. Do not work under the car while the engine is running.
- If you are going to work under a car on the ground, make sure that the ground is level. Block the wheels to keep the car from rolling. Disconnect the battery negative (–) terminal (ground strap) to prevent others from starting the car while you are under it.
- Do not run the engine unless the work area is well ventilated. Carbon monoxide kills.
- Remove finger rings, bracelets and other jewelry so that they cannot cause electrical shorts, get caught in running machinery or be crushed by heavy parts.
- Tie long hair behind your head. Do not wear a necktie, a scarf, loose clothing, or a necklace when you work near machine tools or running engines. If your hair, clothing, or jewelry were to get caught in the machinery, severe injury could result.
- Do not attempt to work on your car if you do not feel well. You increase the danger of injury to yourself and others if you are tired, upset or have taken medication or any other substance that may keep you from being fully alert.
- Illuminate your work area adequately but safely. Use a portable safety light for working inside or under the car. Make sure the bulb is enclosed by a wire cage. The hot filament of an accidentally broken bulb can ignite spilled fuel, vapors or oil.
- Catch draining fuel, oil or brake fluid in suitable containers. Do not use food or beverage containers that might mislead someone into drinking from them. Store flammable fluids away from fire hazards. Wipe up spills at once but do not store the oily rags, which can ignite and burn spontaneously.
- Observe good workshop practices. Wear goggles when you operate machine tools or work with battery acid. Wear gloves or other protective clothing whenever the job requires working with harmful substances.
- Greases, lubricants and other automotive chemicals contain toxic substances, many of which are absorbed directly through the skin. Read the manufacturer's instructions and warnings carefully. Use hand and eye protection. Avoid direct skin contact.
- Disconnect the battery negative (–) terminal (ground strap) whenever you work on the fuel system or the electrical system. Do not smoke or work near heaters or other fire hazards. Keep an approved fire extinguisher handy.
- Friction components (such as brake pads or shoes or clutch discs) contain asbestos fibers or other friction materials. Do not create dust by grinding, sanding, or by cleaning with compressed air. Avoid breathing dust. Breathing any friction material dust can lead to serious diseases and may result in death.
- Batteries give off explosive hydrogen gas during charging. Keep sparks, lighted matches and open flame away from the top of the battery. If hydrogen gas escaping from the cap vents is ignited, it will ignite gas trapped in the cells and cause the battery to explode.
- Battery acid (electrolyte) can cause severe burns. Flush contact area with water and seek medical attention.
- Do not quick-charge the battery (for boost starting) for longer than one minute. Wait at least one minute before boosting the battery a second time.
- Connect and disconnect a battery charger only with the battery charger switched OFF.
- Connect and disconnect battery cables, jumper cables or a battery charger only with the ignition switched OFF. Do not disconnect the battery while the engine is running.
- Do not allow the battery charging voltage to exceed 16.5 volts. If the battery begins producing gas or boiling violently, reduce the charging rate. Boosting a sulfated battery at a high rate can cause an explosion.

continued on next page

BentleyPublishers.com
BentleyPublishers.com—All Rights Reserved

Please read these warnings and cautions before proceeding with maintenance and repair work.

WARNINGS— (continued)

- The air-conditioning system is filled with a hazardous refrigerant. Make sure the system is serviced only by a trained technician using approved refrigerant recovery/recycling equipment, trained in related safety precautions, and familiar with regulations governing the discharging and disposal of automotive chemical refrigerants.
- Do not expose any part of the air-conditioning system to high temperatures such as open flame. Excessive heat increases system pressure and may cause the system to burst.
- Some aerosol tire inflators are highly flammable. Be extremely cautious when repairing a tire that may have been inflated using an aerosol tire inflator. Keep sparks, open flame or other sources of ignition away from the tire repair area. Inflate and deflate the tire at least four times before breaking the bead from the rim. Completely remove the tire from the rim before attempting any repair.
- Use extreme care when draining and disposing of engine coolant. Coolant is poisonous and lethal to humans and pets. Pets are attracted to coolant because of its sweet smell and taste. Seek medical attention immediately if coolant is ingested.
- Cars covered by this manual are equipped with a supplemental restraint system, that automatically deploys airbags and pyrotechnic seat belt reels in case of a frontal or side impact. These are explosive devices. Handled improperly or without adequate safeguards, they can be accidently activated and cause serious injury.
- The ignition system produces high voltages that can be fatal. Avoid contact with exposed terminals and use extreme care when working on a car with the engine running or the ignition switched ON.
- Place jack stands only at locations specified by the manufacturer. The vehicle lifting jack supplied with the vehicle is intended for tire changes only. Use a heavy duty floor jack to lift the vehicle before installing jack stands. See **03 Maintenance**.
- Aerosol cleaners and solvents may contain hazardous or deadly vapors and are highly flammable. Use only in a well ventilated area. Do not use on hot surfaces (engines, brakes, etc.).
- Due to risk of personal injury, be sure the engine is cold before beginning work on the cooling system.

CAUTIONS—
*See also **WARNINGS** on previous page.*

- If you lack the skills, tools and equipment, or a suitable workshop for any procedure described in this manual, we suggest you leave such repairs to an authorized Audi dealer or other qualified shop.
- Audi is constantly improving its cars and sometimes these changes, both in parts and specifications, are made applicable to earlier models. Therefore, part numbers listed in this manual are for reference only. Always check with your authorized Audi dealer parts department for the latest information.
- Before starting a job, make certain that you have the necessary tools and parts on hand. Read the instructions thoroughly and do not attempt shortcuts. Use tools appropriate to the work and use replacement parts meeting Audi specifications. Makeshift tools, parts and procedures do not make good repairs
- Do not use pneumatic and electric tools to tighten fasteners, especially on light alloy parts. Use a torque wrench to tighten fasteners to the tightening torque specification listed.
- Be mindful of the environment. Before you drain the crankcase, find out the proper way to dispose of the oil. Do not pour oil onto the ground, down a drain, or into a stream, pond or lake. Dispose of in accordance with Federal, State and Local laws.
- Do not expose electronic control modules to temperatures from a paint-drying booth or a heat lamp in excess of 203°F (95°C). Do not expose control modules to temperatures in excess of 185°F (85°C) for more than two hours.
- Before doing any electrical welding on cars equipped with ABS, disconnect the battery negative (–) terminal (ground strap) and the ABS control module connector.
- Make sure ignition is OFF before disconnecting the battery.
- Label battery cables before disconnecting. On some models, battery cables are not color coded.
- Disconnecting the battery may erase fault code(s) stored in control module memory. Use Audi diagnostic equipment or equivalent to check for fault codes prior to disconnecting the battery cables. If the Check Engine light (malfunction indicator light or MIL) is illuminated or any other system faults are detected (indicated by an illuminated warning light), see an authorized Audi dealer.
- If a normal or rapid charger is used to charge battery, disconnect the battery and remove from the vehicle in order to avoid damaging paint and upholstery.
- Do not quick-charge the battery (for boost starting) for longer than one minute. Wait at least one minute before boosting the battery a second time.
- Slow-charge a sealed or "maintenance free" battery at an amperage rate that is approximately 10% of the battery's ampere-hour (Ah) rating.

BentleyPublishers.com
BentleyPublishers.com—All Rights Reserved

01 Vehicle Identification and VIN Decoder

Vehicle identification number (VIN), decoding

Some of the information in this manual applies only to cars of a particular model year or range of years. For example, 2003 refers to the 2003 model year but does not necessarily match the calendar year in which the car was manufactured or sold. To be sure of the model year of a particular car, check the vehicle identification number (VIN) on the car.

The VIN is a unique sequence of 17 characters assigned by Audi to identify each individual car. When decoded, the VIN tells the country and year of manufacture; make, model and serial number; assembly plant and even some equipment specifications.

The Audi VIN is on a plate mounted on the top of the dashboard, on the driver's side where the number can be seen through the windshield. The 10th character is the model year code. Examples: W for 1998, X for 1999, Y for 2000, 1 for 2001, 2 for 2002, etc. The table below explains some of the codes in the VIN for 1998 - 2004 Audi A6 models covered by this manual.

Sample VIN: WAUDA24B1XN001612

position: 1 2 3 4 5 6 7 8 9 10 11 12 - 17

VIN position	Description	Decoding information	
1	Manufacturing country	T W	Hungary Germany
2	Manufacturer	A R U	Audi germany Audi Hungary quattro GmbH
3	Vehicle Type	U	Passenger vehicle
4	Series		Varies with model year. See BentleyPublishers.com for more information
5	Engine		
6	Restraint system		
7-8	Model	4A 4B	A6 Avant A6, allroad quattro, S6, RS6
9	Check digit		0 - 9 or X, calculated by NHTSA
10	Model year	W X Y 1 2 3 4	1998 1999 2000 2001 2002 2003 2004
11	Assembly plant	A K N	Ingolstadt Karman-Rheine Neckarsulm
12-17	Serial number		Sequential production number for specific vehicle

BentleyPublishers.com
BentleyPublishers.com—All Rights Reserved

02 Product Familiarization

GENERAL

The information included in this section, based on introductory material for 1998 through 2004 Audi A6 vehicles sold in the USA and Canada, is intended to serve as a product familiarization guide.

CAUTION—

- *The information in this section is subject to change. Use it as a general reference only. Check Audi factory repair information or the publisher's website at www.bentleypublishers.com for information that may supersede material in this section.*

A6 06 2

The fifth generation Audi A6, referred to as the C5 platform, was introduced in 1998 and continued through 2004.

The Audi A6 combines the latest engineering expertise with a future-oriented automobile design. The primary goals were to increase body rigidity, reduce body weight, and improve crash properties. Despite being lighter, the body is 50% more rigid than its predecessor. This results in better driving dynamics and increased ride comfort.

The engine hood, window frame modules and side impact bars, for example, are manufactured using aluminum.

BentleyPublishers.com
BentleyPublishers.com—All Rights Reserved

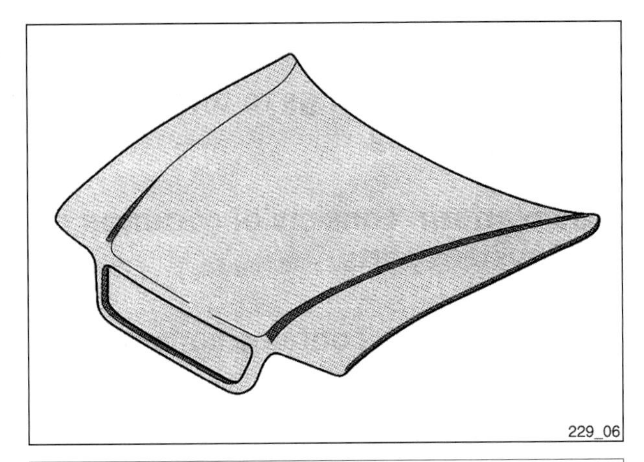
229_06

The aluminum engine hood is the first instance in which Audi used this material for a large exterior part on a high volume production vehicle. It is 36 percent lighter than a conventional engine hood, yet 20 percent more rigid.

The use of magnesium and aluminum components in the running gear such as the transmission housing, suspension components, brake calipers and wheels is another weight-reducing measure.

229_09

The four-link front suspension represents a breakthrough in handling and performance. Additionally, antilock brakes, electronic differential lock and traction control are combined to give the Audi A6 a standard of driving safety unrivalled in its class.

All A6 models feature a host of standard equipment features aimed at improving active and passive safety. Included are:

- ABS (antilock braking)
- EDL (electronic differential lock)
- EBD (electronic brake pressure distribution)
- ASR (antislip regulation)
- Ellipsoidal-principle headlights
- Electrically adjustable and heated outside mirrors
- Outside temperature display
- Driver and passenger airbags
- Side impact airbags for front
- Three-point automatic seat belts and head restraints for all seats
- Pyrotechnic seat-belt tensioners for front and rear seats

In addition, the A6 features extensive anti-theft measures:

- Anti-theft alarm system with ultrasonic interior monitoring
- Button on inside of driver's door for central locking
- Break-in protection for door and luggage compartment locks (lock mechanism cannot be opened even if the car is broken into)
- Remote (keyless) central locking

The A6 was introduced with an all-new engine, the 2.8 liter, thirty-valve V6. This engine has variable intake manifold runners and variable camshaft timing. The engine delivers 200 hp and a peak torque of 207 lb-ft at 3200 rpm. An impressive feature of the 2.8 liter engine is that more than 188 lb-ft (91%) of torque is available at engine speeds between 2500 and 5000 rpm.

Additional engines were introduced throughout the production of the A6, including the 450 hp, 4.2 liter V8 used in the limited edition Audi RS6. See **1 Engine** for engine data.

BentleyPublishers.com
BentleyPublishers.com—All Rights Reserved

Body dimensions

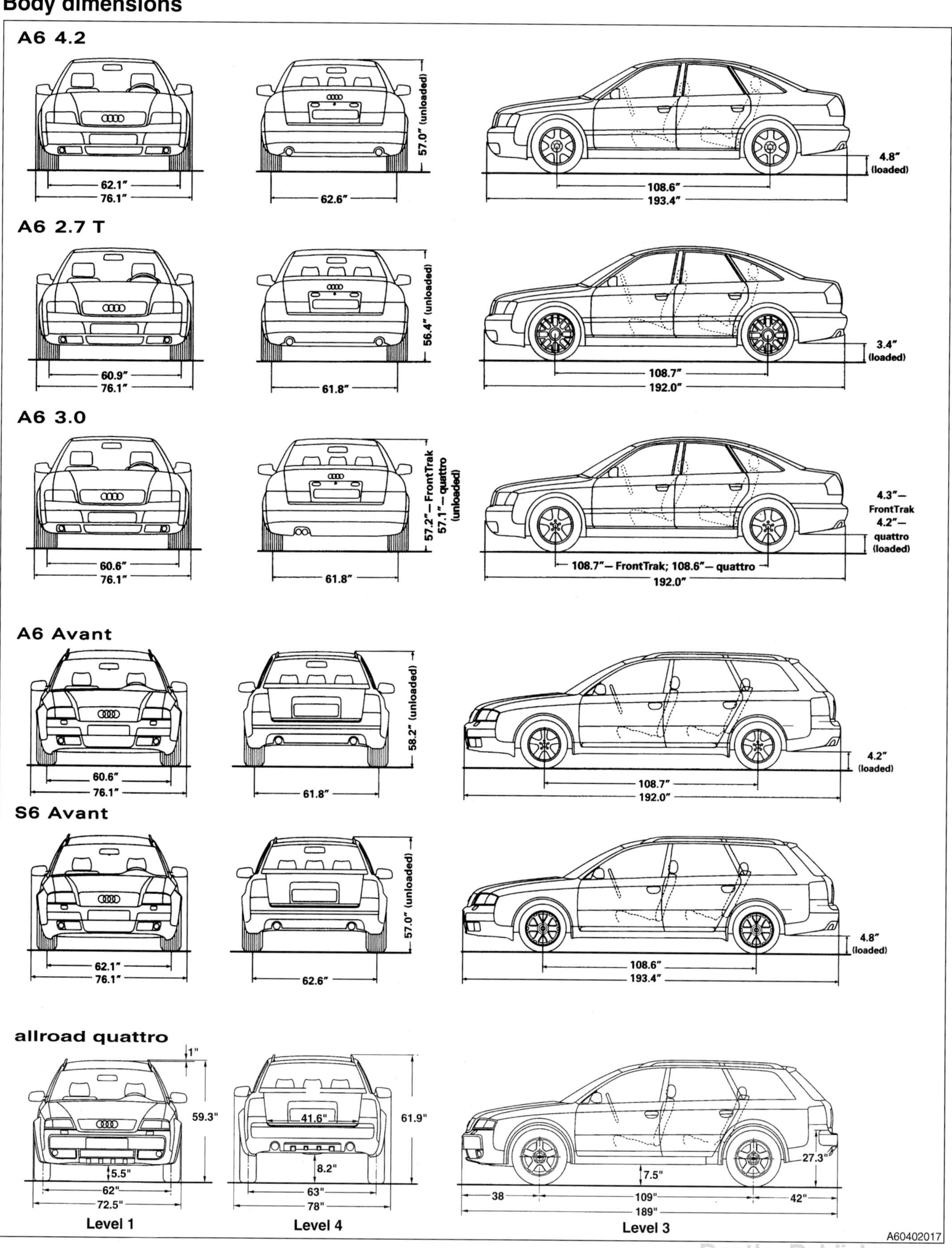

BentleyPublishers.com
BentleyPublishers.com—All Rights Reserved

allroad quattro

The allroad quattro, introduced in 2000, shares its platform with the second generation C5 A6 Avant (station wagon), although an advanced air suspension, larger wheels with all-terrain tires and flared and unpainted bumpers give it a distinct appearance and more overall flexibility over varying terrain. Audi's quattro (all-wheel drive) system is standard equipment.

The allroad quattro was not created by simply changing shock absorbers and springs to give an A6 more ground clearance. There are more than 1,100 new parts that make the allroad a true all-terrain vehicle.

The 4-level suspension system has two operating modes:

- Automatic mode adjusts vehicle height according to vehicle speed and needs no driver input.
- Manual mode allows driver to select vehicle height, depending on speed and terrain.

Other functional and visual modifications to the allroad quattro include:

- Larger exterior mirrors
- Front and rear underbody protection
- Matte paint finish on the bumpers, roof and door sills
- Redesigned headlights

BentleyPublishers.com
BentleyPublishers.com—All Rights Reserved

Audi RS6

The Audi RS6, a limited production powerhouse, is the perfect synthesis of extraordinary power, practicality and poise. The company's most powerful model to date, the RS6 features an engine producing an amazing 450 horsepower at 6,500 revolutions per minute, making it a world-class luxury sedan with the heart of a sports car.

Tuned by quattro GmbH, the high-performance arm of Audi AG and well known for the remarkable RS4 model sold exclusively in Europe, the RS6 is the first vehicle quattro GmbH has produced for the North American market.

1 Engine

Both V6 and V8 engines were available for the A6. V6 power plants included a 2.7 liter biturbo engine and 2.8 liter and 3.0 liter normally-aspirated engines.

The three variants of the V8 engine all displace 4.2 liters, with power output ranging from 299 hp to 450 hp.

V6 engines

2.7 liter V6 biturbo, engine codes APB, BEL

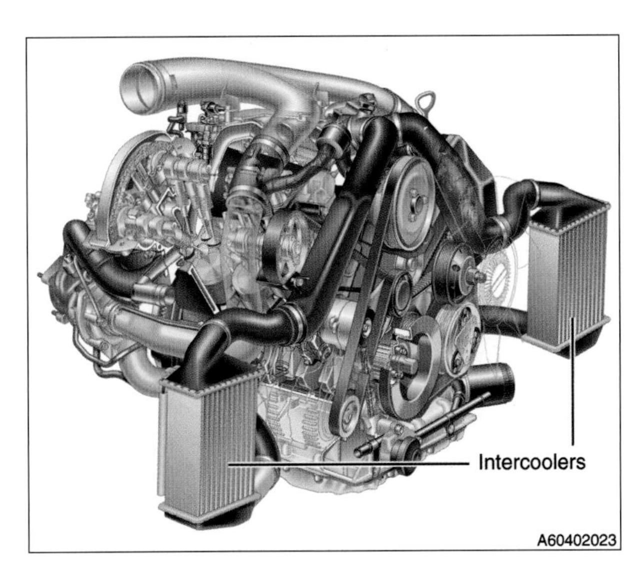

The 2.7 liter V6 biturbo engine, introduced in model year 2000, uses two small turbochargers, one fitted at each bank of cylinders. Due to their lower inertia, the dual turbochargers respond more quickly than a single larger unit.

Each turbocharger has its own intercooler for more efficient cooling of the intake air.

2.7 liter engine, general specifications	
Engine codes	APB (2000 - 2003) BEL (2003 - 2004)
Engine type	dual overhead belt and chain-driven camshafts, 30-valve V6
Engine size	2.7 liters / 163 cu. in.
Horsepower	
APB	258 @ 5800 rpm
BEL	250 @ 5800 rpm
Torque (lb-ft)	258 @ 1800 - 4500 rpm

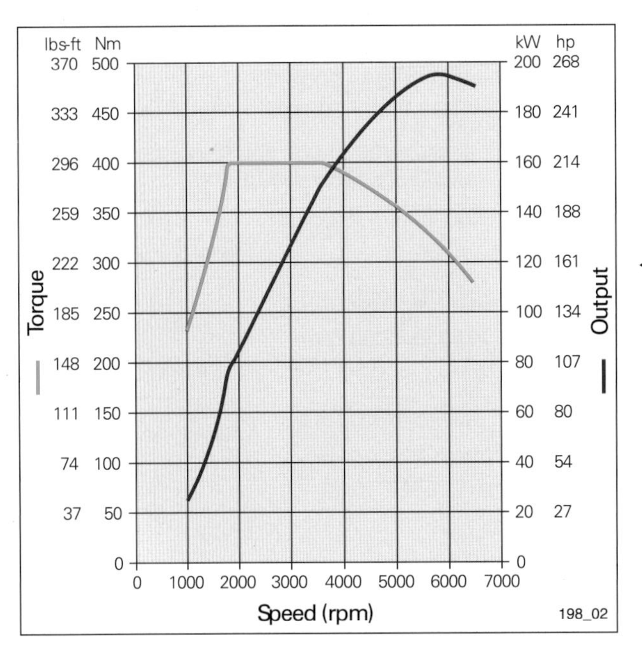

2.7 liter engine horsepower and torque graph.

BentleyPublishers.com
BentleyPublishers.com—All Rights Reserved

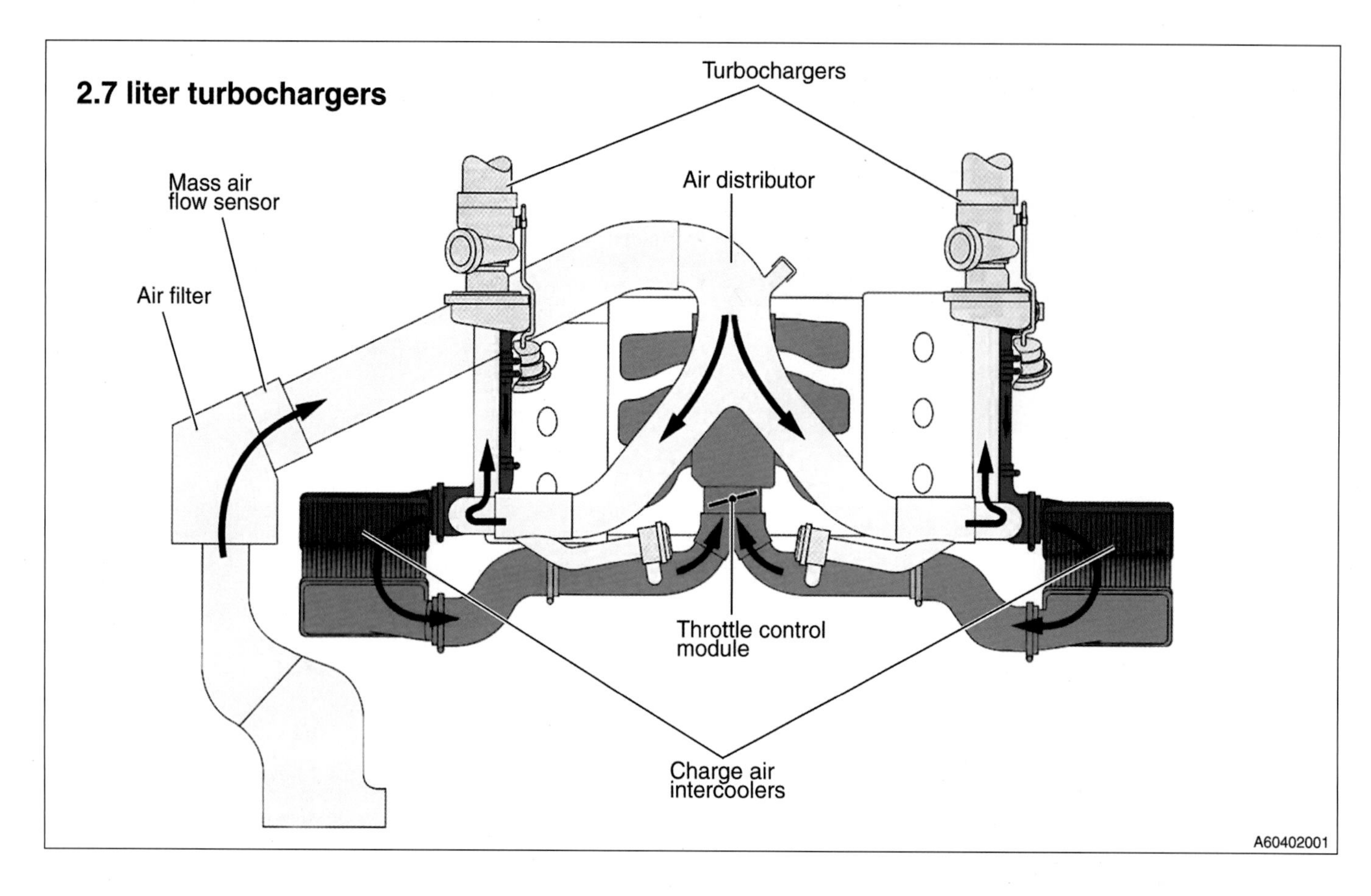

2.8 liter V6 engine, engine codes AHA, ATQ

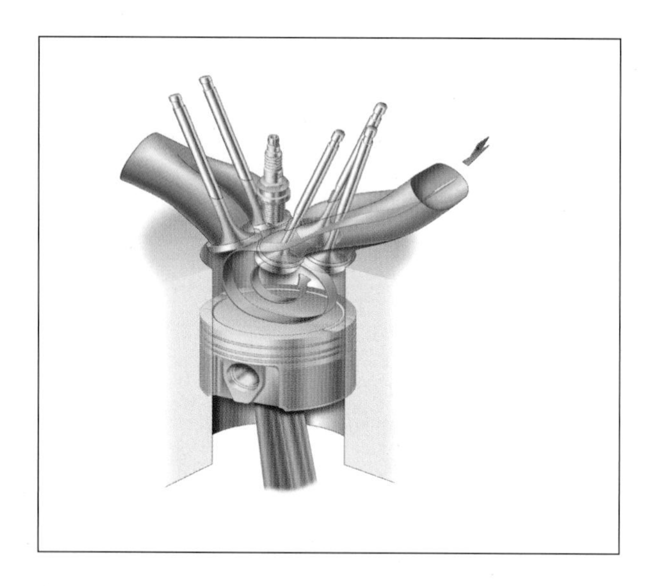

The initial V6 offering was the 2.8 liter naturally-aspirated engine with the cylinder banks positioned 90° apart. Other features include 5 valves per cylinder, distributorless ignition and variable valve timing. Two variants of this engine were available.

A redesigned combustion chamber and centrally-located spark plug are responsible for improving combustion efficiency, reducing fuel consumption and lowering exhaust emissions.

2.8 liter engine, general specifications	
Engine codes	AHA (1998 - 2000) ATQ (2000 - 2001)
Engine type	dual overhead belt and chain-driven camshafts, 30-valve V6
Engine size	2.8 liters / 169 cu. in.
Horsepower	
AHA	200 @ 6000 rpm
ATQ	193 @ 6000 rpm
Torque (lb-ft)	
AHA	207 @ 3200 rpm
ATQ	207 @ 3200 rpm

BentleyPublishers.com
BentleyPublishers.com—All Rights Reserved

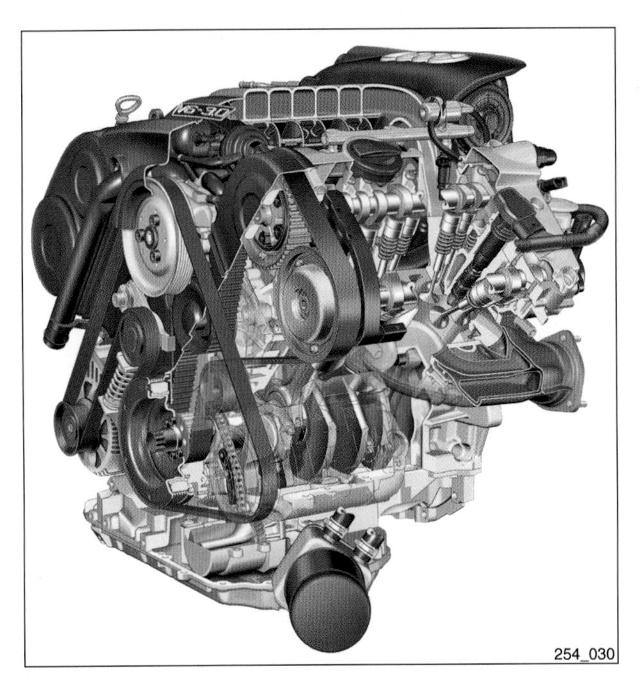

3.0 liter V6 engine, engine code AVK

The 3.0 liter V6, which replaced the 2.8 liter base engine in 2002, features:

- Aluminum cylinder block manufactured using patented Cosworth rollover casting process—a technique used for race car engines.
- Plastic two-position variable intake manifold reduces engine mass and provides torque over a broad range with maximum power available at high rpm.
- Dual overhead camshafts with continuously variable intake camshaft adjustment and two-position exhaust camshaft adjustment boost power output and torque and ensure compliance with exhaust emissions standards.
- Five-valve-per-cylinder technology ensures optimum flow of fuel-air mixture and exhaust gases to keep fuel consumption and exhaust emissions low.

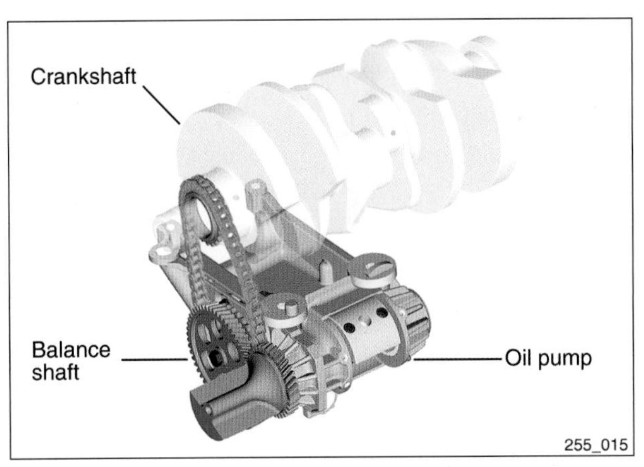

- Balance shaft integrated with oil pump into single module located underneath engine block in sump provides for exceptional running smoothness.
- Motronic ME 7.1.1 engine management system with “drive by wire” electronic throttle control for immediate response to driver input: Accelerator pedal movement is transmitted to the engine management system instantly and without loss.
- Mapped-characteristic ignition and solid-state spark distribution are exceptionally reliable and improve fuel mixture combustion.

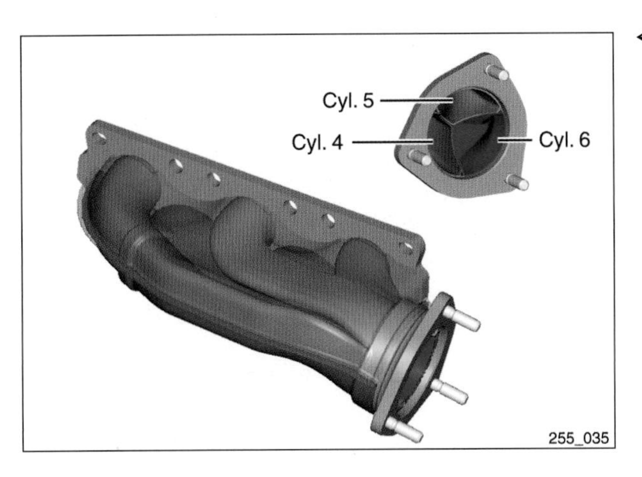

- Tubular air-gap-insulated exhaust manifolds lower weight, improve noise pattern, quickly heat catalytic converters to light-off temperature, and reduce heat transfer to engine compartment.
- Cylinder-bank-selective oxygen sensing, with two pre-converters close to engine and two main catalytic converters farther back ensure long-term stability in exhaust emissions.
- Ultra-low emissions vehicle (ULEV) certification.

3.0 liter engine, general specifications	
Engine code	AVK (2002 - 2004)
Engine type	dual overhead belt-driven camshafts, 30-valve V6
Engine size	3.0 liters / 182 cu. in.
Horsepower	220 @ 6300 rpm
Torque (lb-ft)	221 @ 3200 rpm

BentleyPublishers.com
BentleyPublishers.com—All Rights Reserved

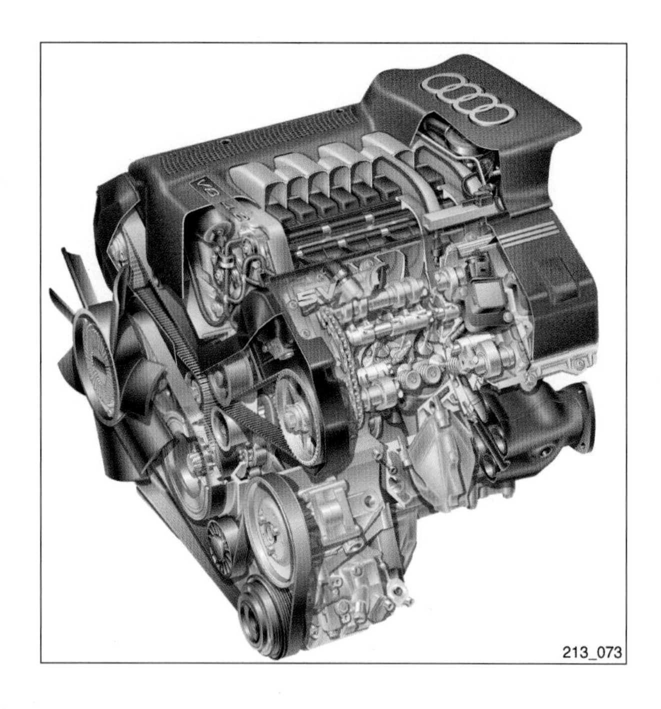
213_073

V8 engines

V8 engines, engine codes ART, AWN, BBD

Starting in 1999, A6 models were optionally equipped with a normally aspirated V8 engine. Features of the 4.2 liter V8 5-valve engine include the following:

- Aluminum engine block with no cylinder liners
- Aluminum five-valve-per-cylinder head with roller rockers
- Belt-driven dual camshafts
- Three-stage variable intake manifold
- Bosch Motronic ME 7.1 engine management system with non-return fuel supply system
- Water-cooled 190-ampere generator
- Oil filter module with integrated oil cooler
- Meets Low-Emission Vehicle (LEV) exhaust emission standards
- Two hot film mass air flow sensors with integrated air temperature sensors
- One ignition coil with integrated output stage per cylinder, each mounted directly over the spark plug
- Two electronically controlled fuel pumps
- Two electric engine cooling fans

The base V8 engines (2 versions, engine codes ART and AWN) produce 299 hp. The Audi S6, introduced in 2002, was equipped with a 340 hp version of the 4.2 liter V8 (engine code BBD).

V8 basic engines, general specifications	
Engine code	ART (1999 - 2001) AWN (2002 - 2004) BBD (2002 - 2004)
Engine type	dual overhead belt-driven camshafts, 40-valve V8
Engine size	4.2 liters / 255 cu. in.
Horsepower	
ART, AWN	299 @ 6000 rpm
BBD	340 @ 6000
Torque (lb-ft)	
ART	302 @ 3300
AWN	305 @ 3300
BBD	295 @ 3500

allroad quattro V8 engine, engine code BAS

For 2003 allroad quattro models, a new 4.2 liter V8 engine was optional. Redesigned to fit into the slim engine compartment, this engine uses chain drive to operate the dual overhead camshafts. The single-row chain is on the clutch end of the engine, allowing the engine to be shortened by 52 mm (approx. 2 in).

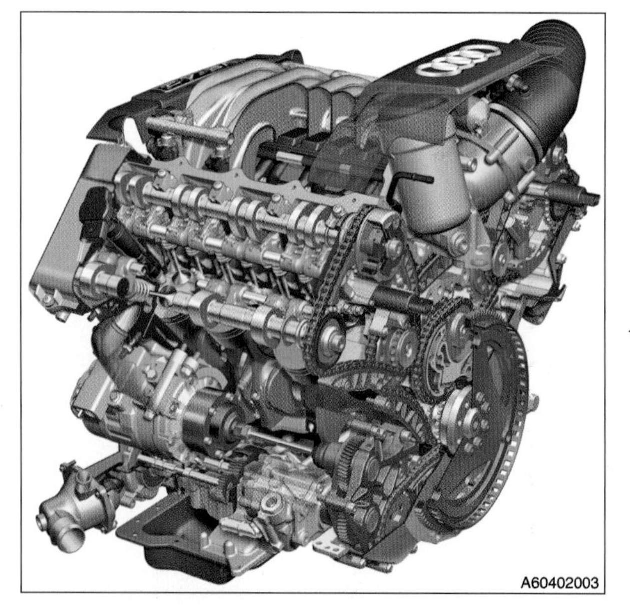
A60402003

BentleyPublishers.com
BentleyPublishers.com—All Rights Reserved

Four chains are used to operate the camshafts and engine accessories. The main chain drives two intermediate shafts, one per cylinder bank. From each of these a short chain drives the two camshafts in each cylinder head. (Camshaft chains are not shown in illustration.)

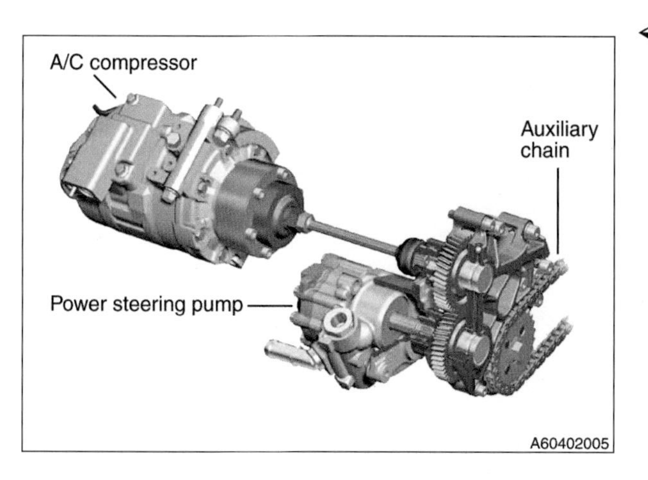

The auxiliary chain, driven off a second crankshaft sprocket, operates engine accessories through driven sprockets or intermediate shafts:

- Oil pump
- Coolant pump
- Power steering pump
- A/C compressor

A two-stage variable intake manifold made from magnesium allows higher engine power output at higher speeds in addition to high low-end torque. Continuous camshaft timing control also contributes to increased performance over a wide range of engine speeds.

V8 allroad quattro engine, general specifications	
Engine code	BAS
Engine type	dual overhead chain-driven camshafts, 40-valve V8
Engine size	4.2 liters / 255 cu. in.
Horsepower	299 @ 6200 rpm
Torque (lb-ft)	280 @ 2700 - 4600 rpm

BentleyPublishers.com
BentleyPublishers.com—All Rights Reserved

RS6 biturbo V8 engine, engine code BCY

At the heart of the high performance RS6 model is the 4.2 liter V8 biturbo engine. Inspired by Audi's victorious LeMans-winning R8 racing car, this power plant makes the sleek RS6 among the fastest five-passenger sport sedans ever to enter the U.S. market, with a 0 - 60 mph time of just 4.6 seconds.

The V8 engine has twin turbochargers, one for each cylinder bank, to provide increased low-end torque, more power at the upper end and overall improved engine response. Electronically controlled waste gates and twin intercoolers optimize engine performance. An intercooler mated with each cylinder bank prevents loss of turbo pressure while cooling the intake air. This force-fed, temperature-optimized "breathing" results in minimal turbo lag and peak engine performance.

Engine architecture includes 90° cylinder banks, a lightweight aluminum block and five valves per cylinder—two of which are sodium-filled for maximum valve temperature control—that help feed and ventilate the cylinders faster, producing more torque and power with less fuel consumption and lower emissions.

RS6 engine, general specifications	
Engine code	BCY
Engine type	dual overhead belt and chain-driven camshafts, 40-valve V8
Engine size	4.2 liters / 255 cu. in.
Horsepower	450 @ 5700 - 6400 rpm
Torque (lb-ft)	413 @ 1950 - 5600 rpm

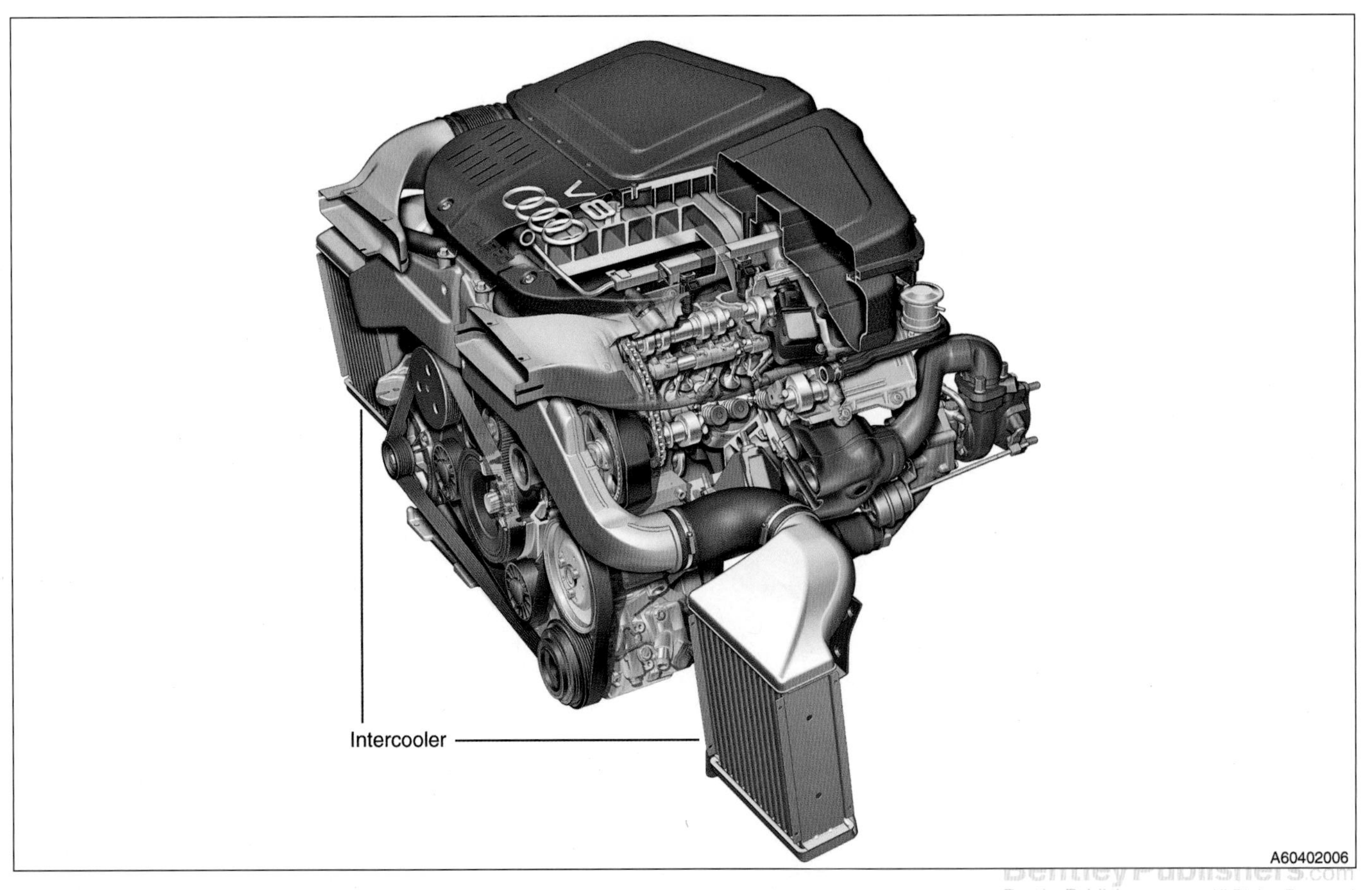

BentleyPublishers.com
BentleyPublishers.com—All Rights Reserved

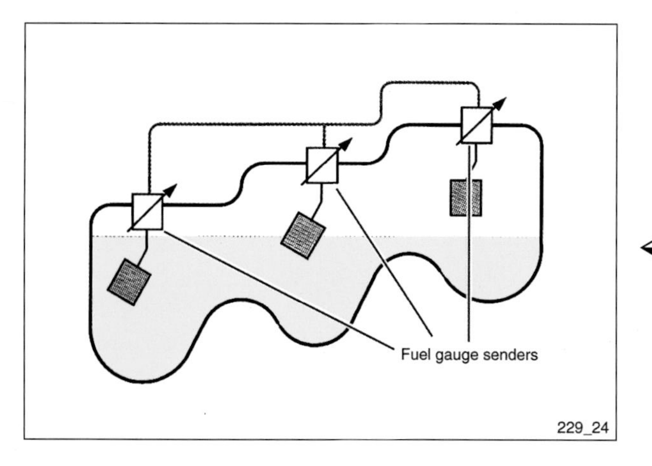

2 Engine Management

Fuel tank

The fuel tank is plastic and placed near the rear axle for crash safety.

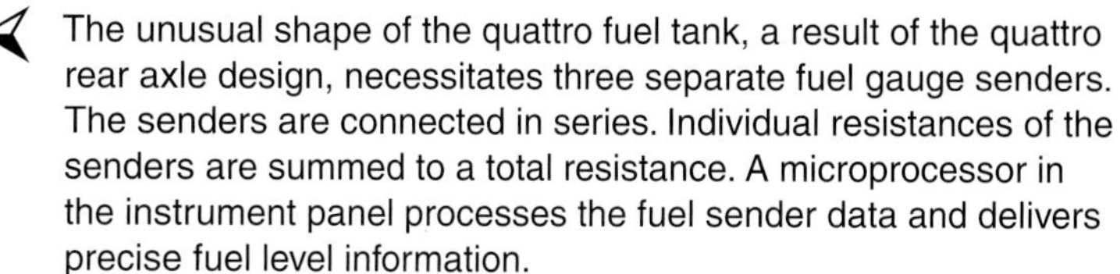

The unusual shape of the quattro fuel tank, a result of the quattro rear axle design, necessitates three separate fuel gauge senders. The senders are connected in series. Individual resistances of the senders are summed to a total resistance. A microprocessor in the instrument panel processes the fuel sender data and delivers precise fuel level information.

The fuel pump and its baffled enclosure are inside the fuel tank on the right. Two siphon pumps, driven by the fuel return feed fuel into the fuel pump baffle. Access to the fuel pump is through a cover under the rear seat cushion.

Fuel tank removal requires disassembly of the rear axle. The filler neck is a separate part from the fuel tank.

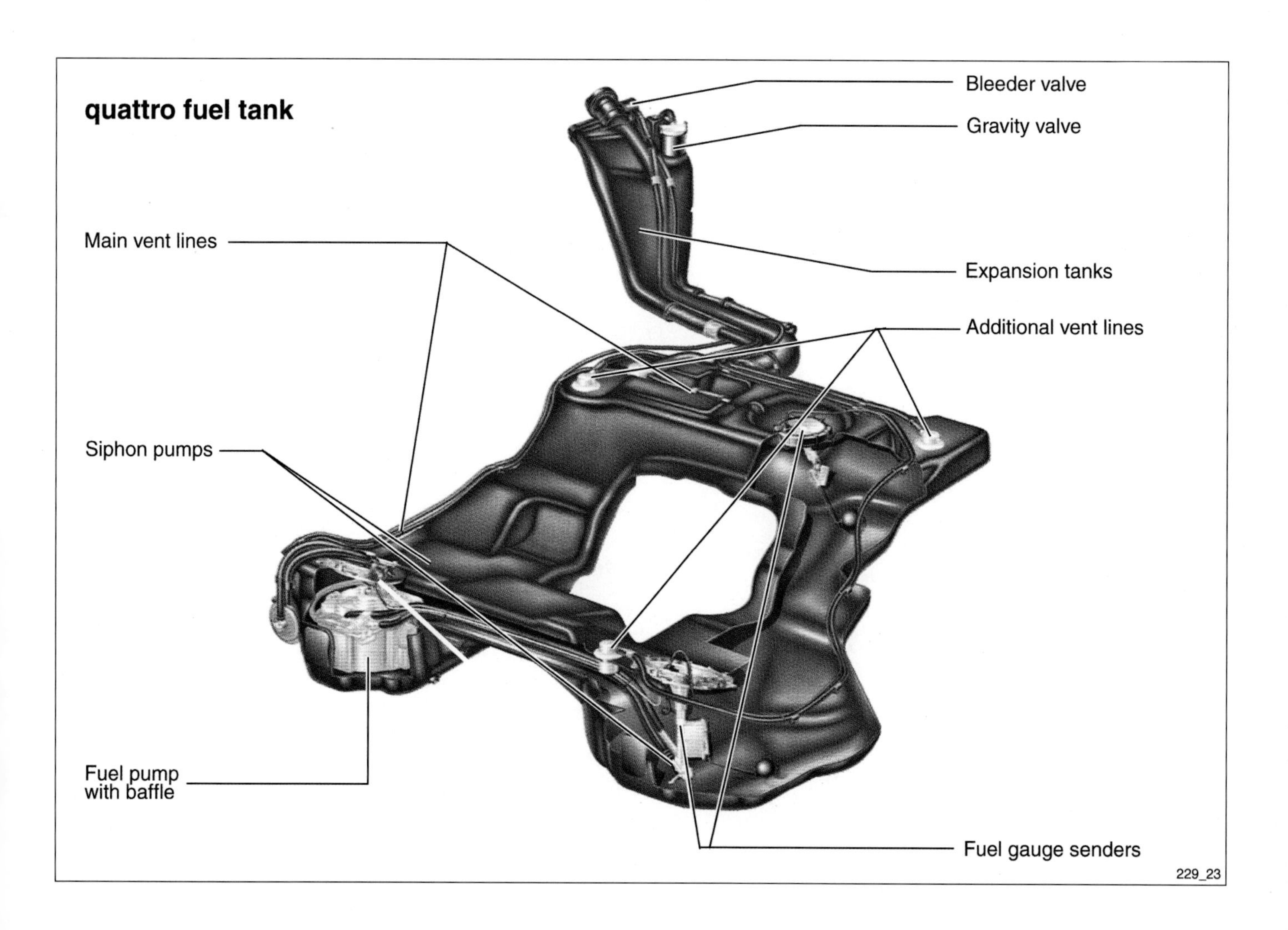

BentleyPublishers.com
BentleyPublishers.com—All Rights Reserved

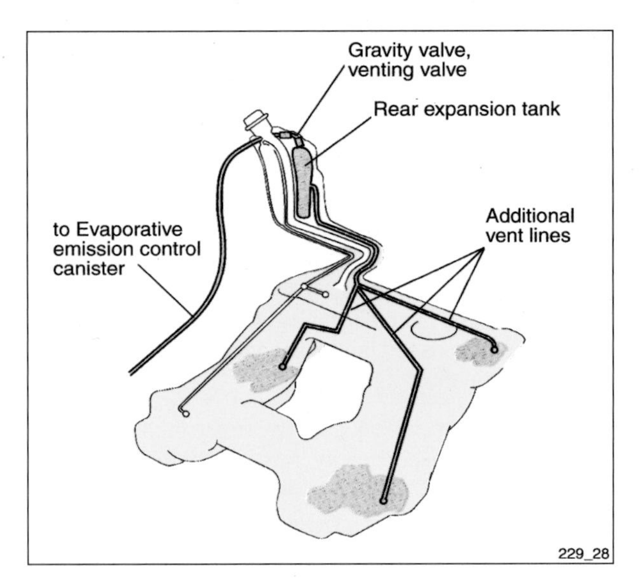

Evaporative emissions control

The fuel tank evaporative emissions control components are two separate expansion tanks at the filer neck and a complex of fuel vent lines.

When the fuel tank is filled, air escapes through the two main vent lines into the front expansion tank at the filler neck.

When the fuel heats up due to an external cause, such as heat from the exhaust system or return fuel flow from the engine, the resulting fumes enter the rear expansion tank through the three auxiliary vent lines. Fumes are then fed to the evaporative emissions control canister (carbon canister) via the gravity valve and the venting valve.

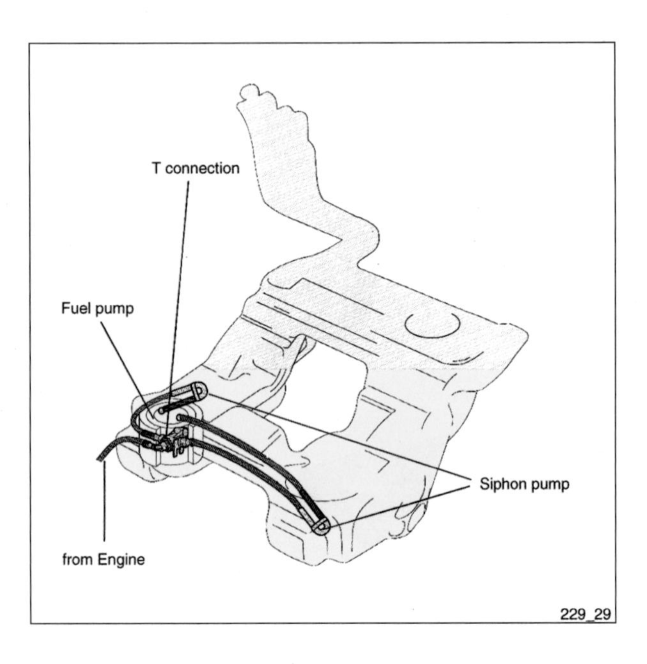

Fuel return system

Fuel returning from the engine flows through the return line to a T-fitting mounted on the fuel pump. From there the fuel is fed to the siphon pumps. If the return volume is excessive, the fuel backs up at the siphon pumps, increasing back-pressure in the return line, and in turn affecting the fuel pressure regulator.

A pressure control valve in the T-fitting limits pressure in the fuel return line to a maximum of 14.5 psi (1 bar). When the valve opens, fuel flows directly into the fuel pump baffle.

Motronic engine management

Bosch Motronic engine management systems in A6 cars combine fuel injection, ignition and on-board diagnostic capabilities. The Motronic engine control module (ECM) coordinates a variety of environmental sensors, drivetrain monitors and driver wishes to determine performance parameters for the engine. It also stores, for later access, diagnostic trouble codes (DTCs). Motronic systems are fully compliant with federal and state mandated second generation on-board diagnostic (OBD II) standards.

The first A6 models in 1998 (2.8 liter engine, engine code AHA) were fitted with Bosch Motronic engine management M5.9.2. The throttle valve in this system is operated via traditional throttle cable.

Starting with 2000 models, A6 models were fitted with Bosch Motronic ME 7 engine management which uses a new internal circuit design to calculate engine torque. The ECM receives relevant signal data from the input sensors and other sources and computes the required amount of engine torque needed based on pre-established priorities.

BentleyPublishers.com
BentleyPublishers.com—All Rights Reserved

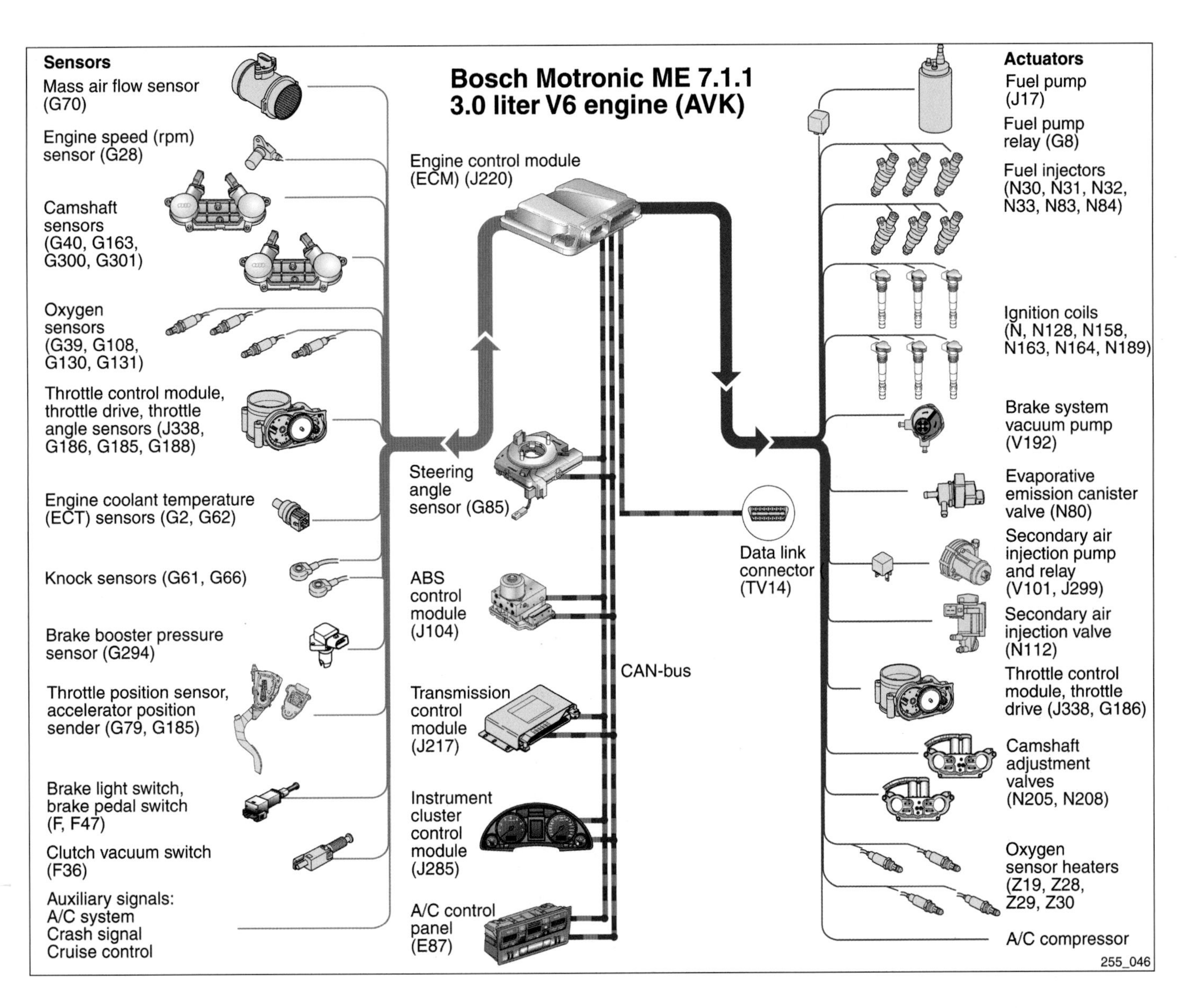

The ME 7 ECM obtains the calculated engine torque through precise coordination of the following output components:

- Throttle valve control module (throttle angle)
- Where applicable, turbo waste gate regulator solenoid valve (boost control)
- Fuel injectors (injection time and fuel cut-off)
- Power output stages with ignition coils (ignition timing)

NOTE—

- *Audi identifies electrical components by a letter and / or a number in the electrical schematics. See* **EWD Electrical Wiring Diagrams**. *These electrical identifiers are listed in parenthesis as an aid to electrical troubleshooting.*

BentleyPublishers.com
BentleyPublishers.com—All Rights Reserved

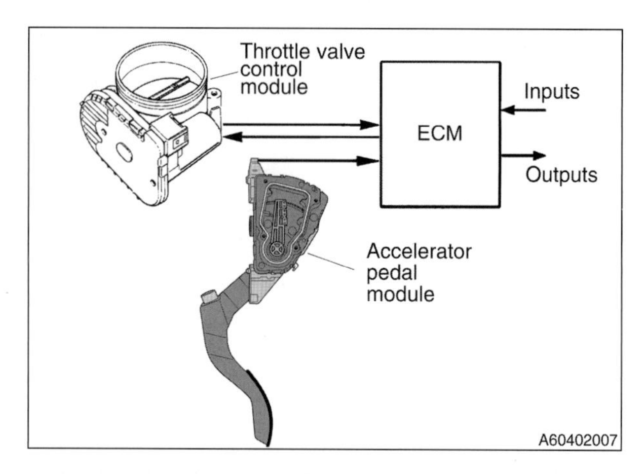

With ME 7, instead of a throttle cable between the accelerator pedal and the throttle valve there is electronic throttle control with the following components:

- Throttle valve control module
- Accelerator pedal module

The electronic throttle valve control system is used to reduce and increase torque without adversely affecting exhaust emissions.

The ideal combination of throttle opening and charge pressure can be controlled for greatest efficiency. Under acceleration, throttle movement is programmed to best increase intake charge velocity. On deceleration, the throttle valve can be held open to reduce emissions.

RS6 engine management

The Motronic adaptive engine management system controls turbo boost pressure, engine knock and exhaust-gas temperature for enhanced engine performance and efficiency.

The optimized dual-branch exhaust system with middle and rear mufflers, larger pipe sections and metal-base catalytic converters adds to performance and sound.

To save weight, the upper section of the air cleaner is made from a carbon fiber composite. A carbon-fiber cover over the front section of the engine adds to the striking under hood appearance.

3 Clutch and Transmission

Self-adjusting clutch (SAC)

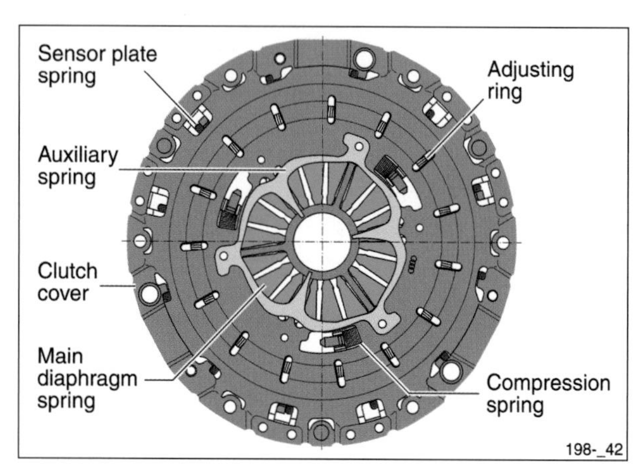

The 2.7 liter biturbo engine is equipped with the self-adjusting clutch (SAC).

The SAC is designed to maintain consistent clutch action throughout the life of the clutch disc. If the clutch disc is replaced without pressure plate replacement, follow the correct procedure to reset the pressure plate adjusting ring. See **30 Clutch**.

Manual transmission

A6 manual transmissions are all-wheel drive (quattro). Manual transmission applications are as follows:

- **5-speed (01A)**:
  2.8 liter V6 models, engine codes AHA, ATQ
- **6-speed (01E)**:
  2.7 liter biturbo V6, engine codes APB, BEL
  V8 models, engine codes ART, AWN, BBD
- For transmission identification and additional application information, see **03 Maintenance**.

BentleyPublishers.com
BentleyPublishers.com—All Rights Reserved

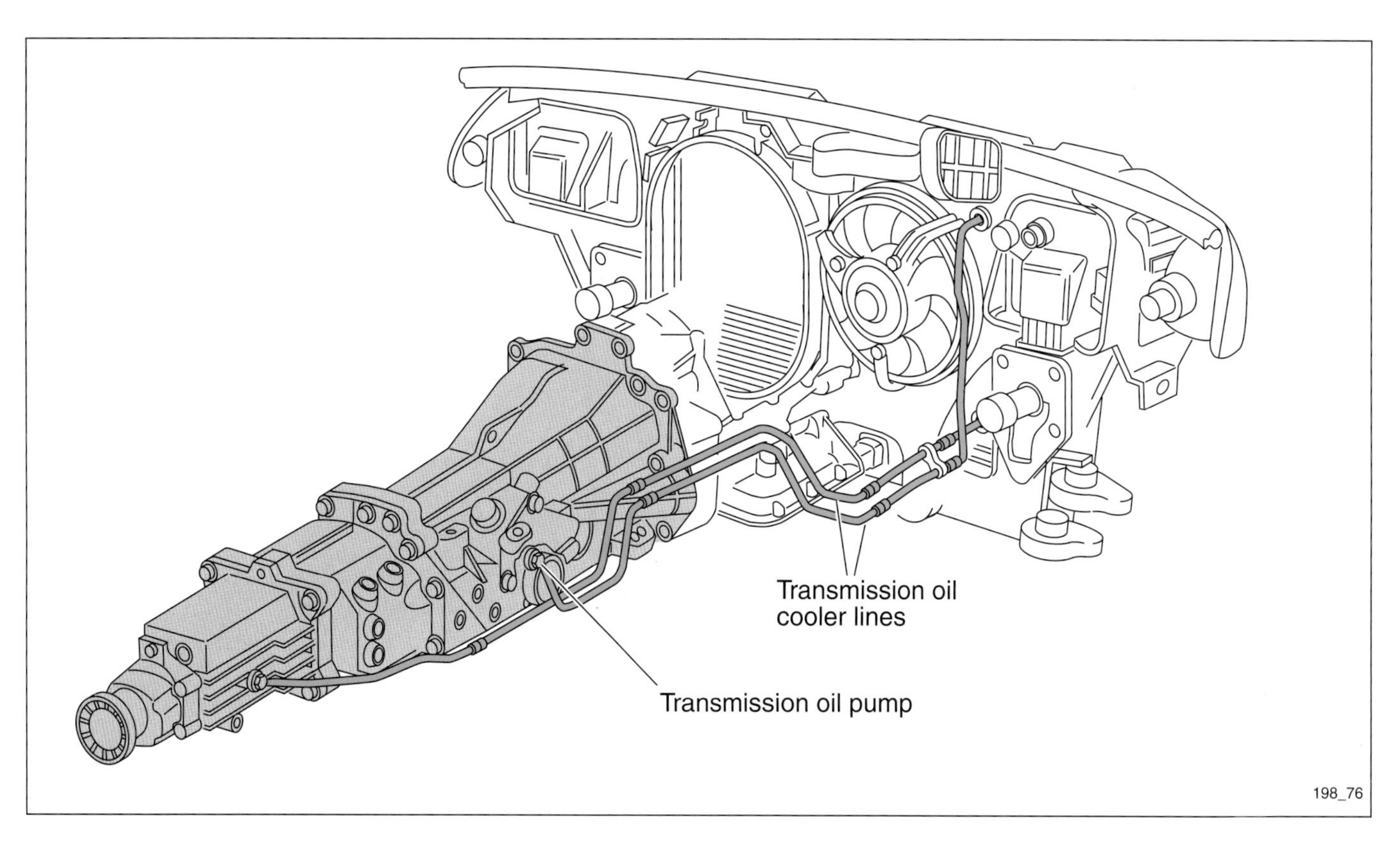

Transmission oil cooling

The six speed manual transmission oil is cooled by circulating through a radiator heat exchanger.

Automatic transmission

Automatic transmission versions

A6 automatic transmission applications are as follows:

- **5-speed front-wheel drive (01V)**:
  2.8 liter V6 models, engine codes AHA, ATQ
- **5-speed all-wheel drive (quattro) (01V)**:
  2.8 liter V6 models, engine codes AHA, ATQ
- **5-speed all-wheel drive (quattro) (01L)**:
  2.7 liter biturbo V6 engine (allroad quattro)
  V8 models
- **Multitronic® (continuously variable transmission or CVT) front-wheel drive (01J)**:
  3.0 liter V6, engine code AVK

01V automatic transmission

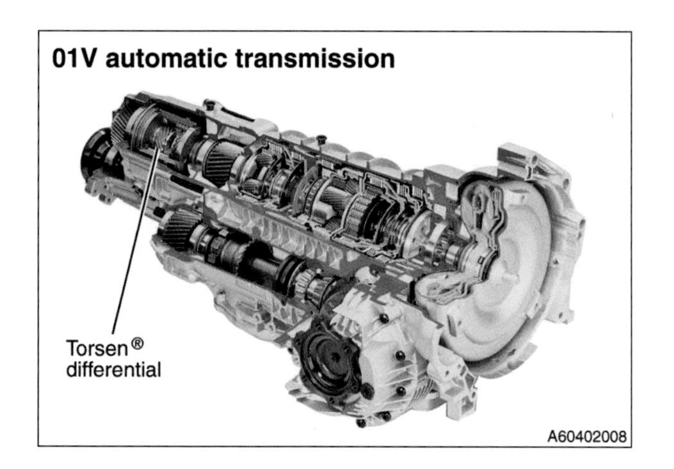

BentleyPublishers.com
BentleyPublishers.com—All Rights Reserved

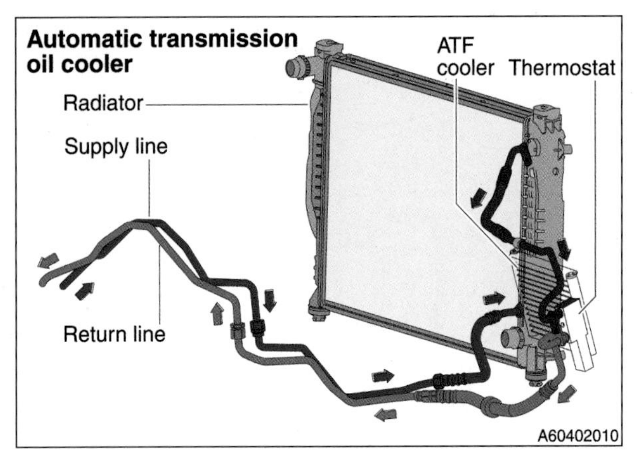

Thermostatically controlled ATF cooler in front of the engine radiator.

Automatic transmission controls

The automatic transmission control module (J217) receives signals from a number of sensors. These signals are used to control when the transmission shifts and how the shift feels.

The on-board diagnostics system also monitors the transmission for faults and stores diagnostic trouble codes (DTCs) which can be accessed via the data link connector.

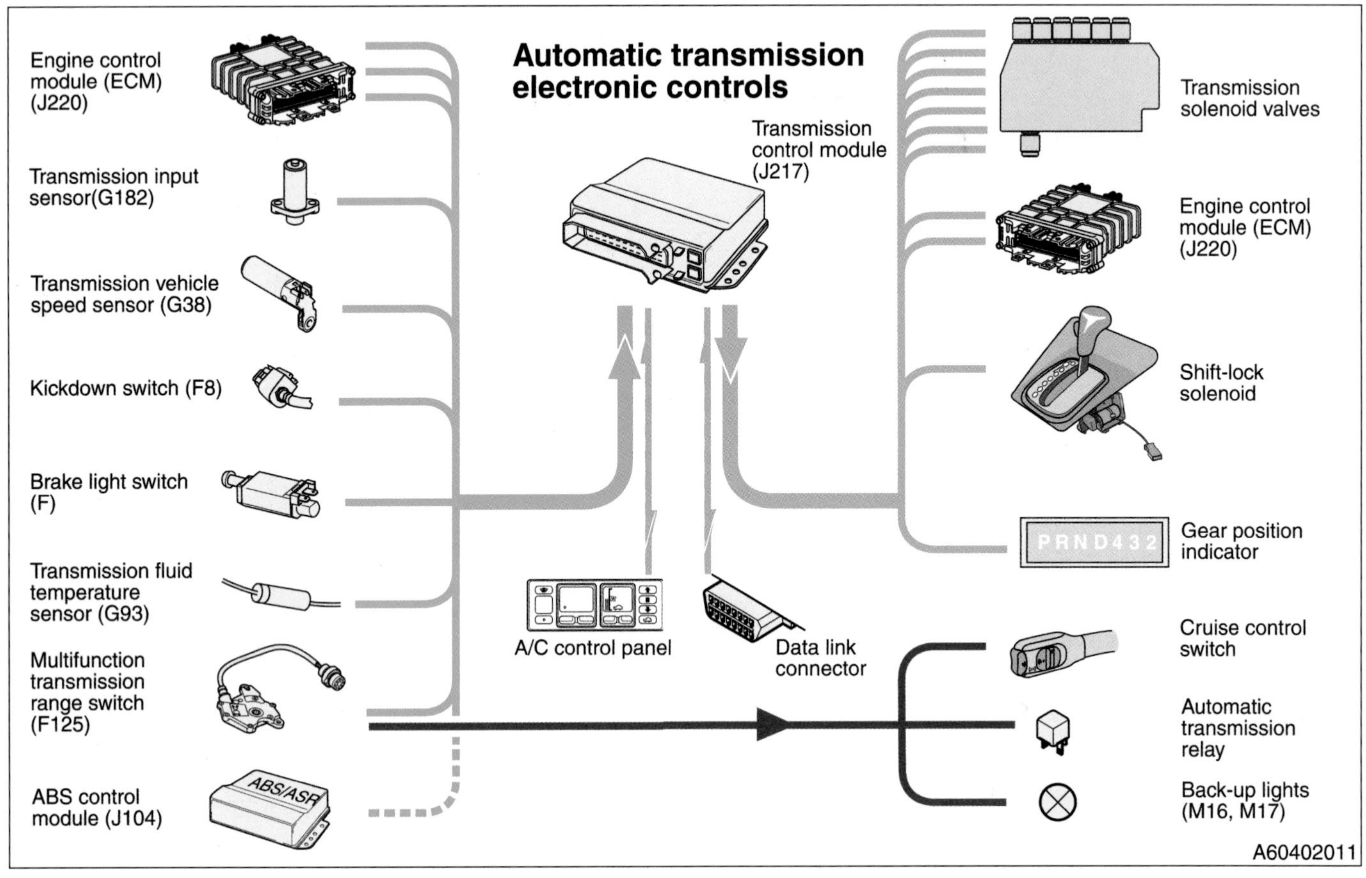

Multitronic® (CVT) transmission

Models with 3.0 liter engine (engine code AVK) were available with the continuously variable transmission (CVT) known as the Multitronic®.

Reverse clutch
Flywheel
Auxiliary reduction gear
Variator and chain
Planetary gears
Forward clutch
Hydraulic control module
Transmision control module (J217)
228_003

The key component of the Multitronic® transmission is the variator. It allows gear ratios to be adjusted continuously between the starting torque multiplication ratio and the final torque multiplication ratio. As a result, a suitable gear ratio is always available. The engine can operate within the optimum rpm range regardless of whether it is optimized for performance or fuel economy.

BentleyPublishers.com
BentleyPublishers.com—All Rights Reserved

The variator has two tapered disc pairs as well as a special chain which runs in the V-shaped gap between the two tapered pulley pairs. The chain acts as a power transmission element.

- Pulley set 1 is driven by the engine through an auxiliary reduction gear step.
- Engine torque is transmitted via a special chain to pulley set 2 and from there to the final drive.

One of the tapered pulley halves in each pulley can be shifted on the shaft for variable adjustment of the chain track diameter and transmission ratio. The two sets of pulleys are adjusted simultaneously so that the chain is always taut and the disc contact pressure is sufficient for power transmission purposes.

Torque is transmitted by the frictional force between the ends of the cradle type pressure pieces and the contact faces of the tapered pulleys.

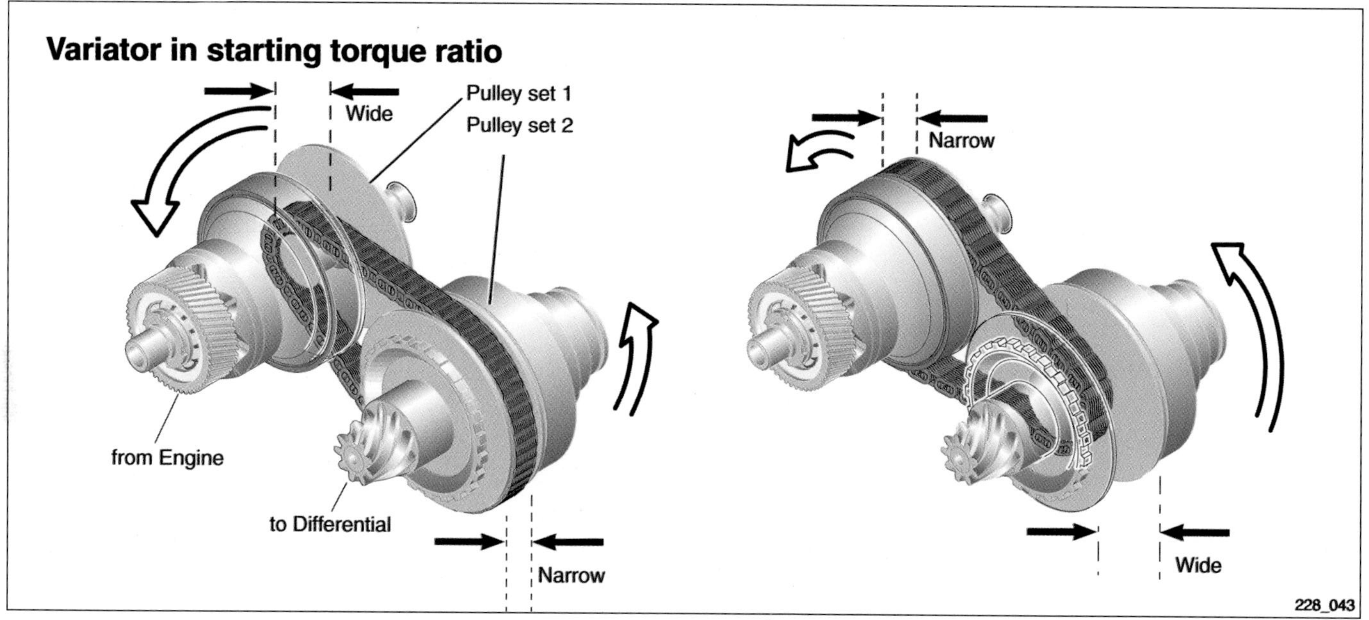

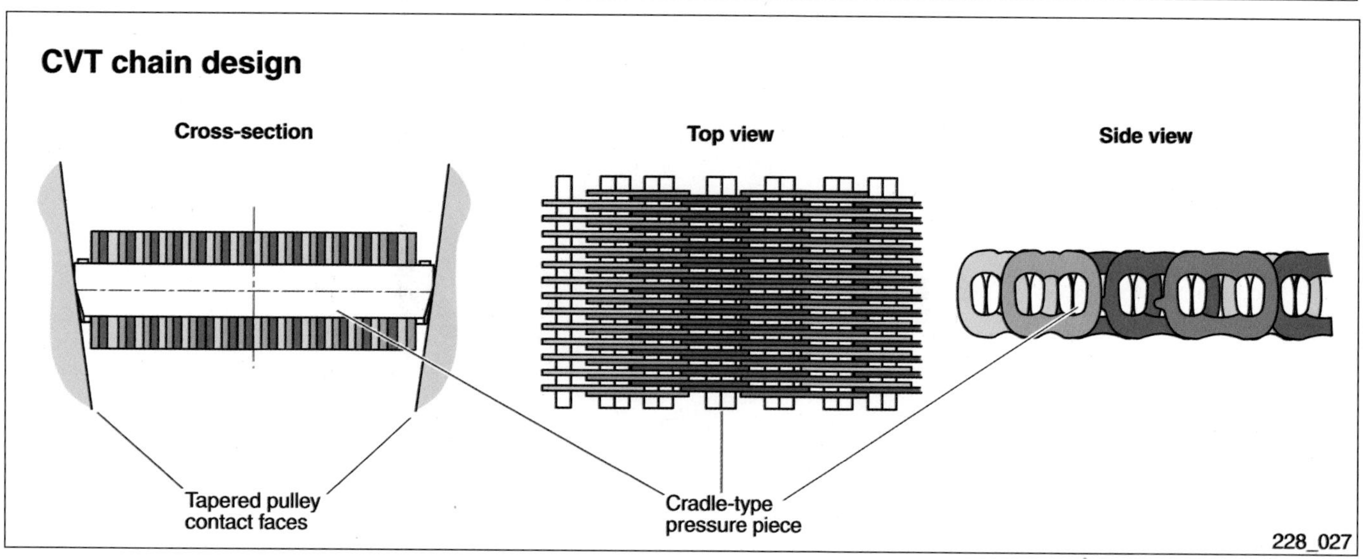

BentleyPublishers.com
BentleyPublishers.com—All Rights Reserved

RS6 transmission

Emphasizing sports car-like driving, engineers specifically tuned a five-speed Tiptronic automatic transmission for the RS6. In Sport setting (gear selector in S position), gears are held longer on upshifts and downshifts are advanced to increase performance. Depending on the level of lateral acceleration, the transmission management system can even adjust to avoid undesirable shifts during cornering.

The RS6 driver can also operate the transmission manually in either of two ways: by briefly pressing the gear selector front or back in the Tiptronic shift gate, or via race-inspired shift paddles behind the steering wheel.

The specially tuned RS6 Tiptronic transmission also incorporates a Dynamic Shift Program (DSP) with hill detection capability. DSP automatically selects from more than 200 shift patterns to match driver characteristics with driving conditions, while hill detection capability prevents gear hunt on inclines.

Torsen® differential

All-wheel drive (quattro) systems rely on a torque sensing (Torsen®) center differential to distribute power between the front and rear axles. This type of differential works without electrical connections or computer controls. There are no driver inputs. The Torsen differential is a sealed unit in the rear of either the manual or the automatic transmission housing.

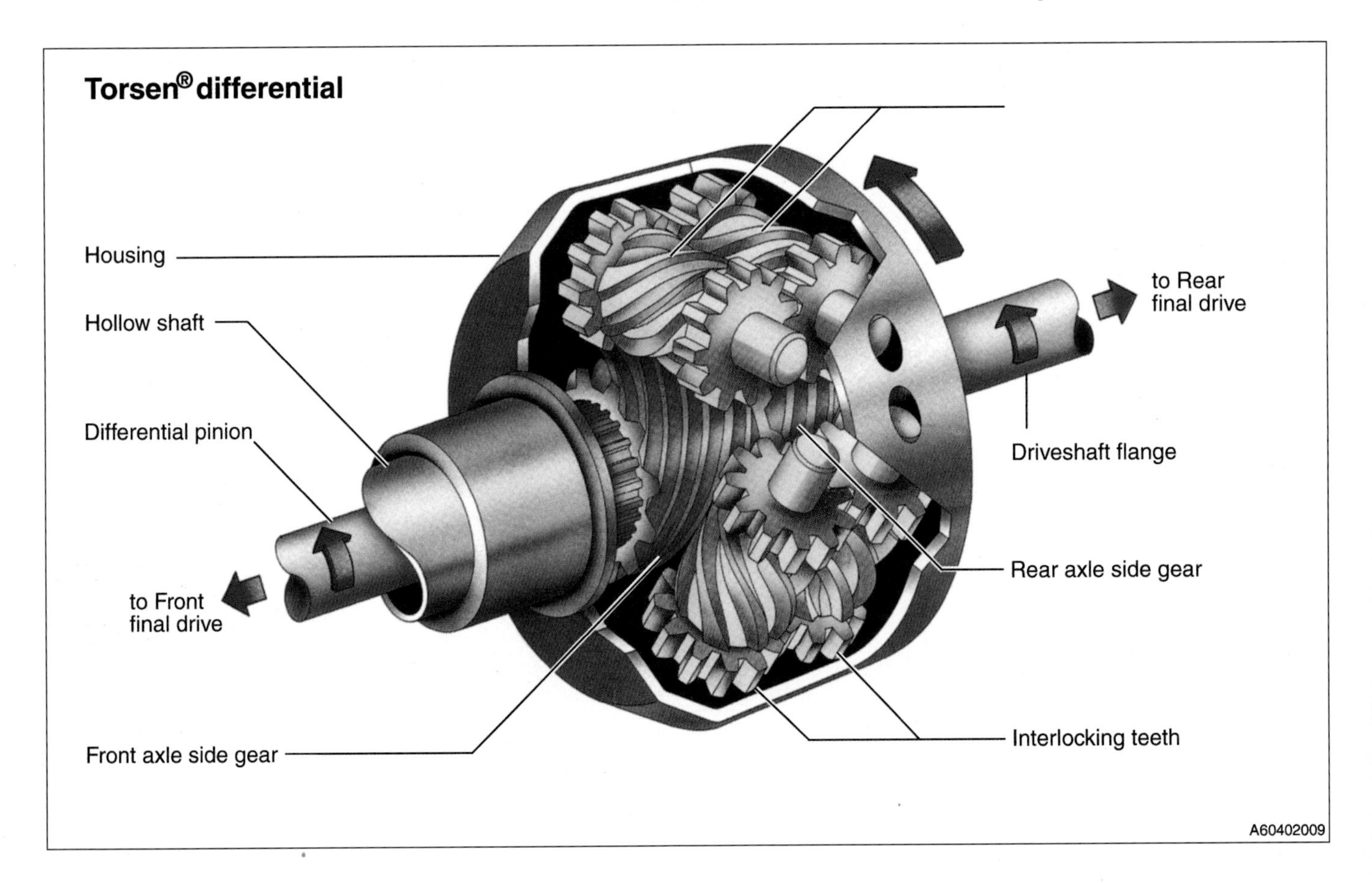

BentleyPublishers.com
BentleyPublishers.com—All Rights Reserved

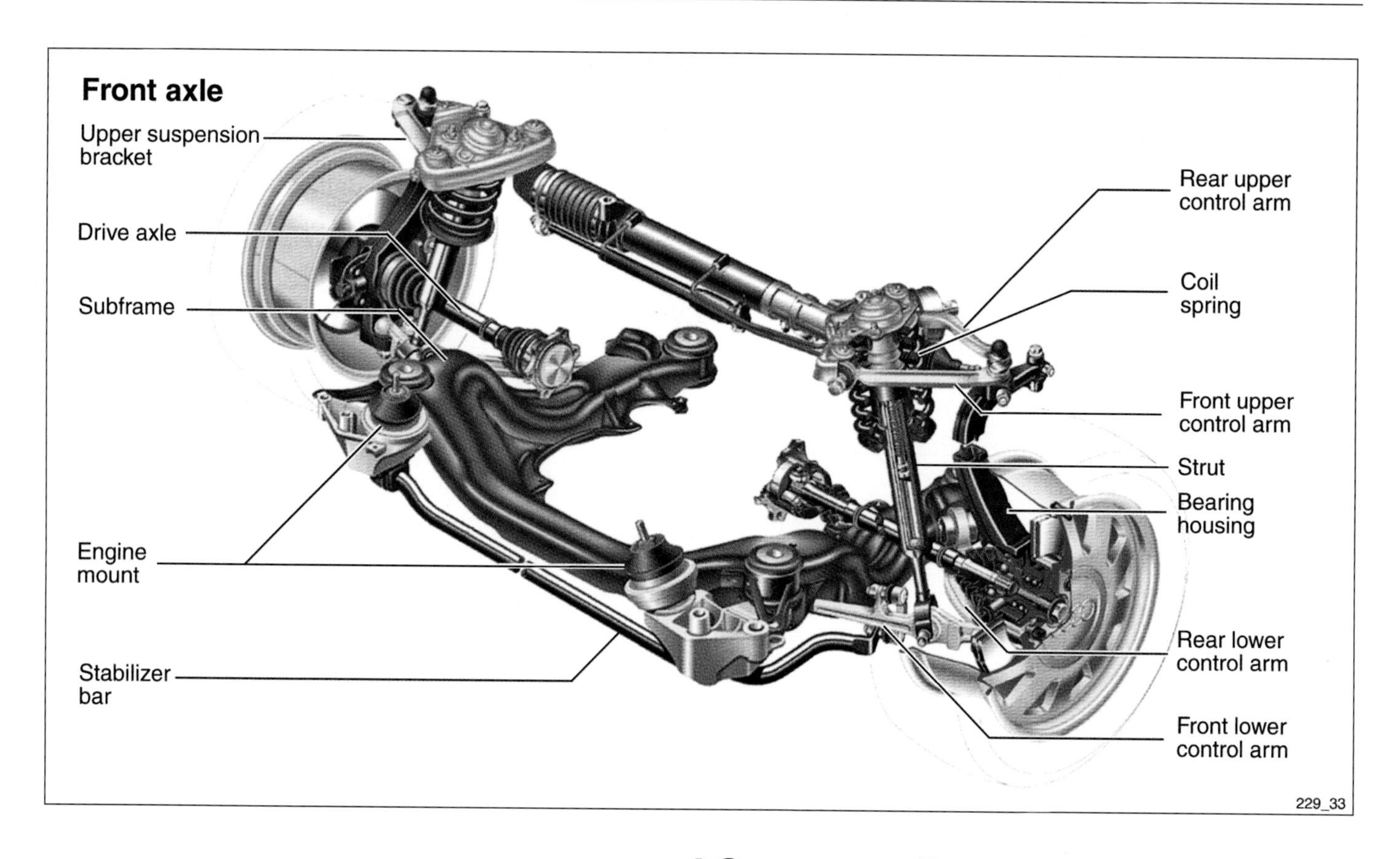

4 Suspension, Brakes and Steering

Suspension

Front suspension

A light and stable subframe rigidly holds the front axle components together and supports the front of the engine- transmission assembly.

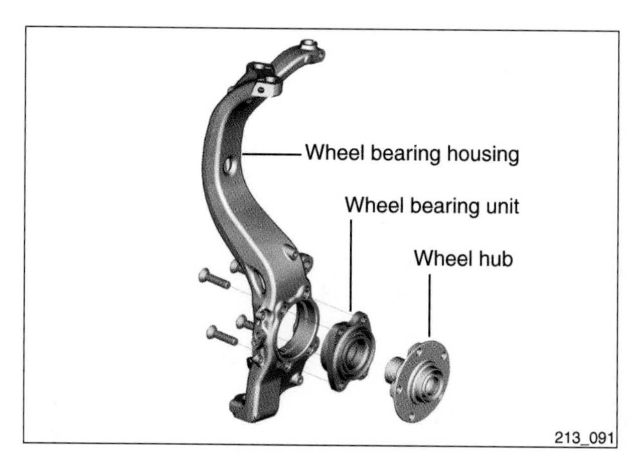

In order to save weight, many suspension components, including front wheel bearing housing, wheel bearing assembly and wheel hub are made of light-weight aluminum (Al-Mg-Si) alloy.

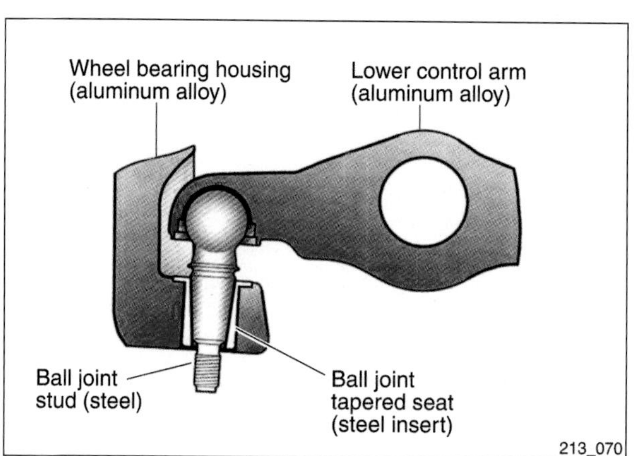

Due to the softness of the aluminum alloy, tapered ball joint stud seats are reinforced with press-fit steel bushings.

BentleyPublishers.com
BentleyPublishers.com—All Rights Reserved

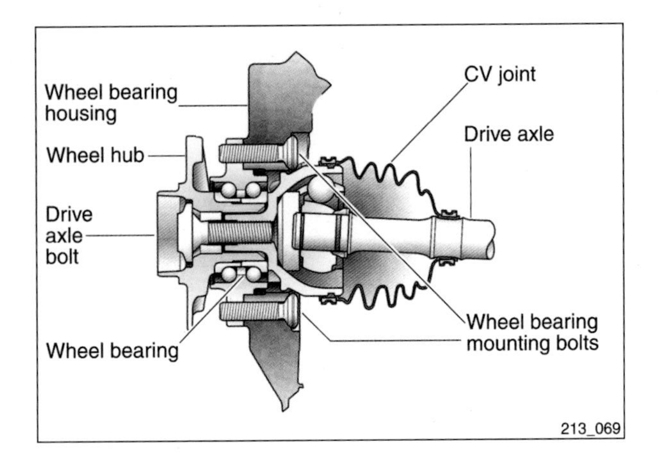

The front wheel bearing is bolted to the wheel bearing housing. This makes it possible to replace the wheel bearing without having to remove the wheel bearing housing or the drive axle.

Rear suspension (front-wheel drive)

The front-wheel drive rear suspension is referred to as the compound-link axle. The support brackets with large rubber bushings are placed on the outside of the rear axle, thus reducing rocking of the assembly. The placement of the shock absorbers and coil springs allows a maximum cargo width of 39.4 inches (1 meter).

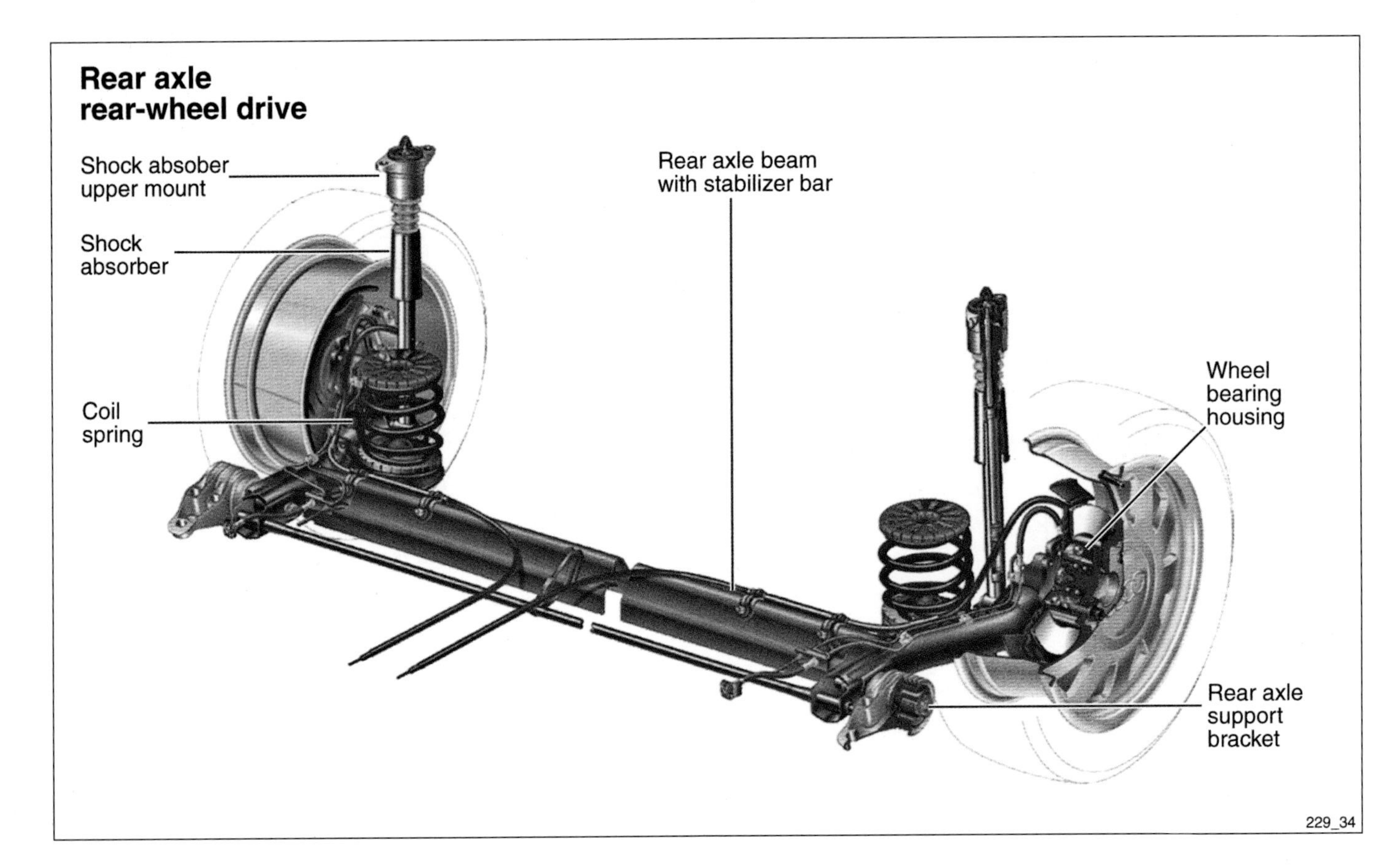

BentleyPublishers.com
BentleyPublishers.com—All Rights Reserved

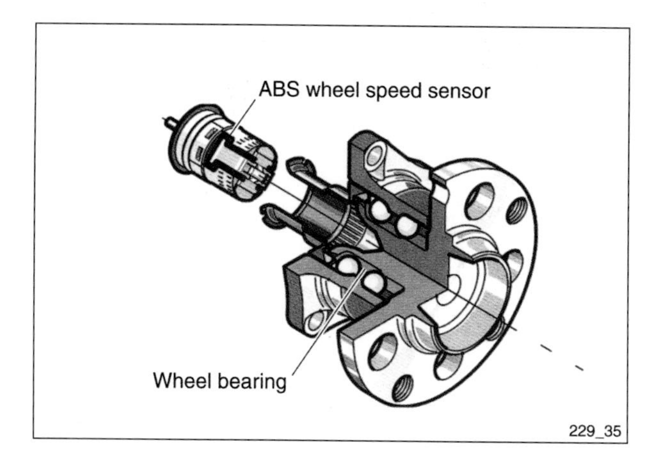

229_35

The double groove ball bearing wheel bearing and hub bolt as a unit to the fwd rear axle.

The ABS wheel speed sensor is in the center of the wheel bearing hub.

Rear suspension (quattro)

The all-wheel drive (quattro) rear suspension is referred to as double transverse link rear axle. Upper and lower transverse control arms are anchored to a tubular subframe.

- The placement of the struts allows a maximum cargo width of 39.4 inches (1 meter).
- The subframe is suspended from the body on rubber and metal bushings, providing good acoustical isolation for the passenger cabin.
- Wheel bearings are double grooved contact bearings.

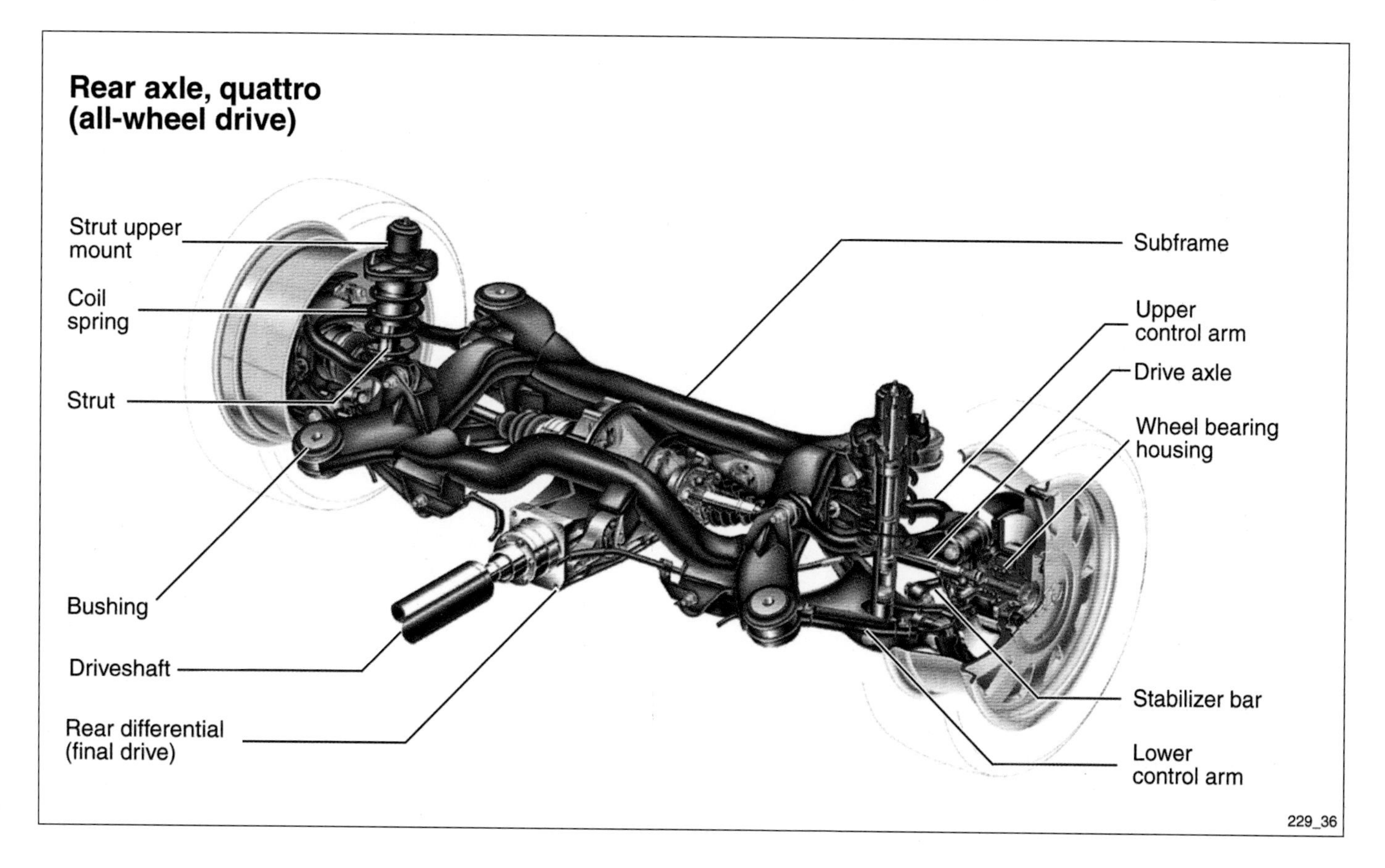

229_36

BentleyPublishers.com
BentleyPublishers.com—All Rights Reserved

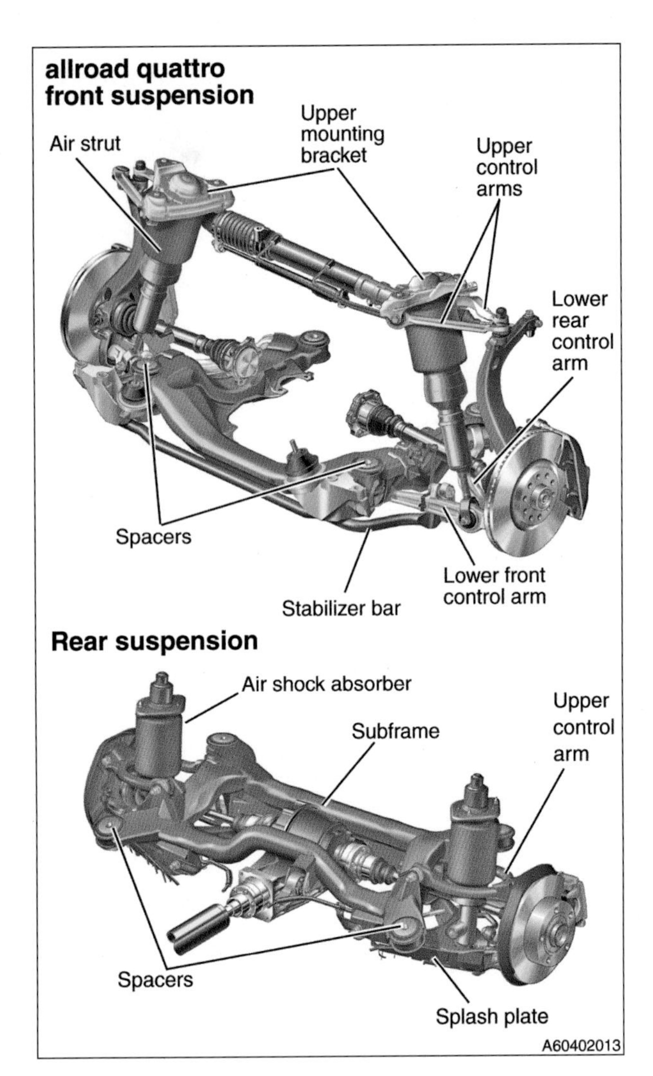

allroad quattro air suspension

> ***CAUTION—***
>
> • *Before jacking up or lifting allroad quattro vehicle, place the suspension in service mode. See* **allroad quattro "jack mode"** *in* **03 Maintenance**.

Although derived from A6 suspension, the allroad quattro suspension differs in a number of ways:

Front suspension

- Subframe strengthened
- Drive axle cutouts deepened for additional suspension travel
- Suspension level sensor brackets welded to subframe
- Suspension strut eye lowered
- Suspension level sensor bracket and coupling rod moved further inboard for clearance reasons
- Rear lower control arm ball joint increased in size
- Rear upper control arm angle changed
- Upper suspension bracket modified to accommodate air struts
- Front subframe bushing spacers 25 mm (1 in) thick
- Front drive axle shafts redesigned due to changed suspension characteristics

Rear suspension

- Upper control arm modified to accommodate air struts
- Upper control arm pivot point set higher to optimize wheel clearance
- Track rod made of aluminum for increased rigidity and improved tracking
- Splash plate under lower control arm to deflect debris away from brakes
- Stabilizer bar shape modified to make room for suspension air pump
- Rear drive axle shaft diameter increased
- Rear outer CV joints reinforced
- Driveshaft center bearing adapted to higher body

BentleyPublishers.com
BentleyPublishers.com—All Rights Reserved

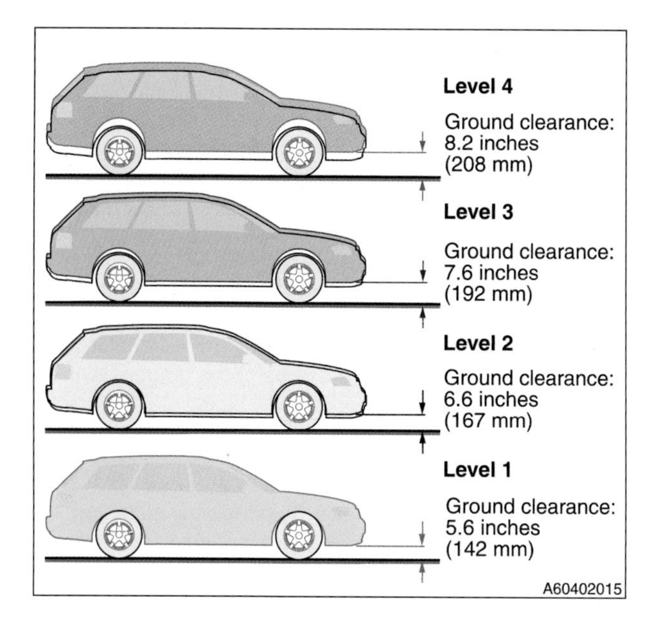

Four-level air suspension

The allroad quattro four-level air suspension system alters ground clearance by 2.6 inches (66 mm) in four stages. Vehicle level is automatically controlled as a function of road speed. Automatic level control overrides manual settings.

The four-level air suspension system is composed of:

- Dashboard display
- Warning light
- Air springs
- Solenoid valves
- Pressure sensor
- Suspension level sensors
- Air compressor
- Temperature sensor
- Pressure accumulator

The air suspension system is designed for fast response and low noise.

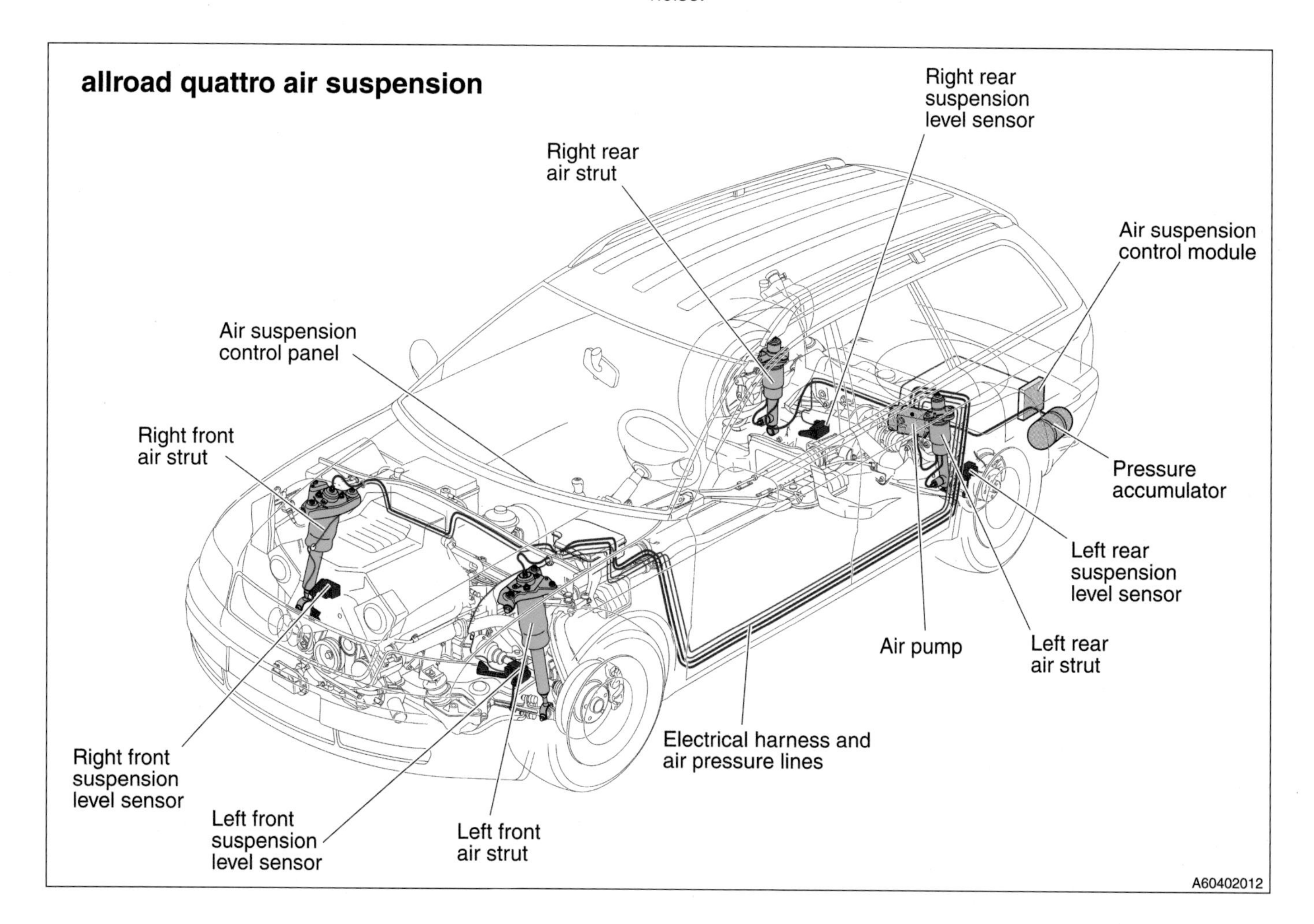

BentleyPublishers.com
BentleyPublishers.com—All Rights Reserved

RS6 suspension

The RS6 suspension system, known as Dynamic Ride Control (DRC), uses diagonally connected, hydraulically dampened shock absorbers. By continually adjusting the hydraulic pressure at each shock absorber, DRC limits body roll and pitch during spirited driving.

DRC connects the shock absorbers at the opposite corners of the vehicle via a hydraulic system incorporating a central reservoir. During sharp cornering, as the hydraulic pressure on the shock absorber on the inside is reduced, hydraulic fluid is transferred through the reservoir to increase pressure to the diagonally linked shock absorber on the outside.

The central reservoir of the DRC also works to balance hydraulic pressure when front or rear shock absorbers are under pressure, such as during hard starts and stops. By effectively stiffening the suspension under pressure, DRC helps maintain the overall stability of the RS6.

Antilock brakes (ABS)

A6 vehicles are equipped with antilock braking (ABS). ABS uses electronic control of brakes and throttle to prevent wheels from locking during hard braking, thus helping to increase vehicle directional control and decrease stopping distance.

Basic ABS is supplemented by software and hardware to achieve additional safety features. The electronic components have self-diagnostic capabilities.

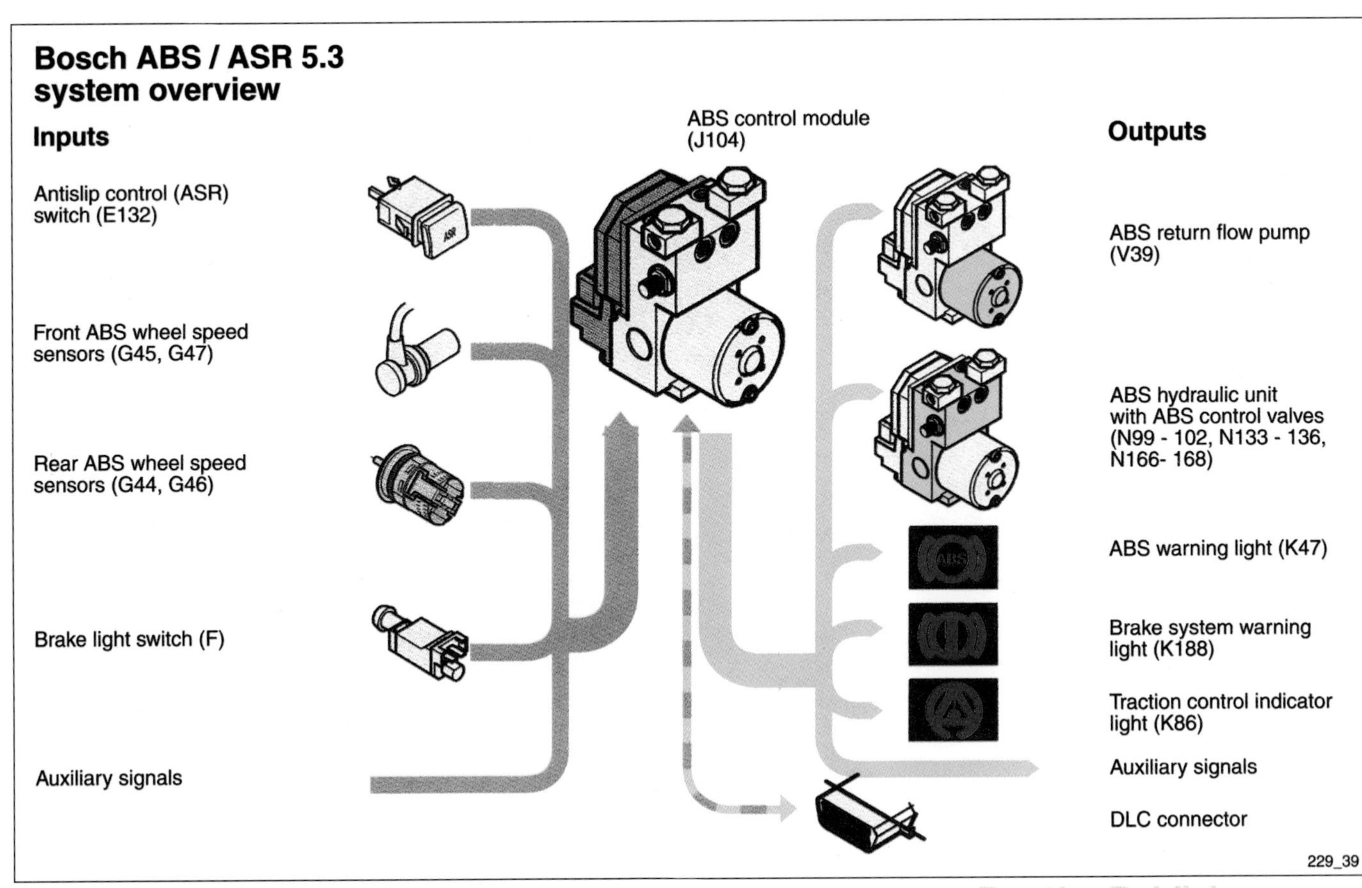

BentleyPublishers.com
BentleyPublishers.com—All Rights Reserved

Bosch ABS / ASR 5.3

Bosch ABS 5.3 includes the following vehicle stability refinements:

- **Electronic brake pressure distribution (EBD)**. ABS control module regulates brake pressure to eliminate rear wheel lock-up. EBD control ends as soon as ABS control is applied.
- **Electronic differential lock (EDL)**. As an aid to starting on a slick surface, EDL automatically brakes the spinning wheel and shunts driving torque to the wheel with traction. EDL controls slip up to 25 mph (40 kph) in a front-wheel drive vehicle and up to 50 mph (80 kph) in a quattro (all-wheel drive) vehicle.
- **Antislip regulation (ASR)**. On front-wheel drive vehicles, if driving wheels spin during acceleration, ASR regulates wheel spin by reducing engine torque. This is done through retarding ignition timing or cyclically shutting down fuel injectors. ASR is effective across the entire range of speed.

Bosch ABS / ESP 5.7

Beginning in 2001, A6 models are equipped with ABS supplemented with electronic stability program (ESP). This traction control system uses acceleration sensors (rotation rate or yaw sensor and lateral acceleration sensor), steering angle sensor and hydraulic pump to maintain precise control of the vehicle under difficult traction conditions. ABS / ESP incorporates ABS / ASR features (EDL, EBD) in addition to:

- **Electronic brake control (EBC)**. This feature prevents driven wheels from locking (and skidding) due to engine braking.

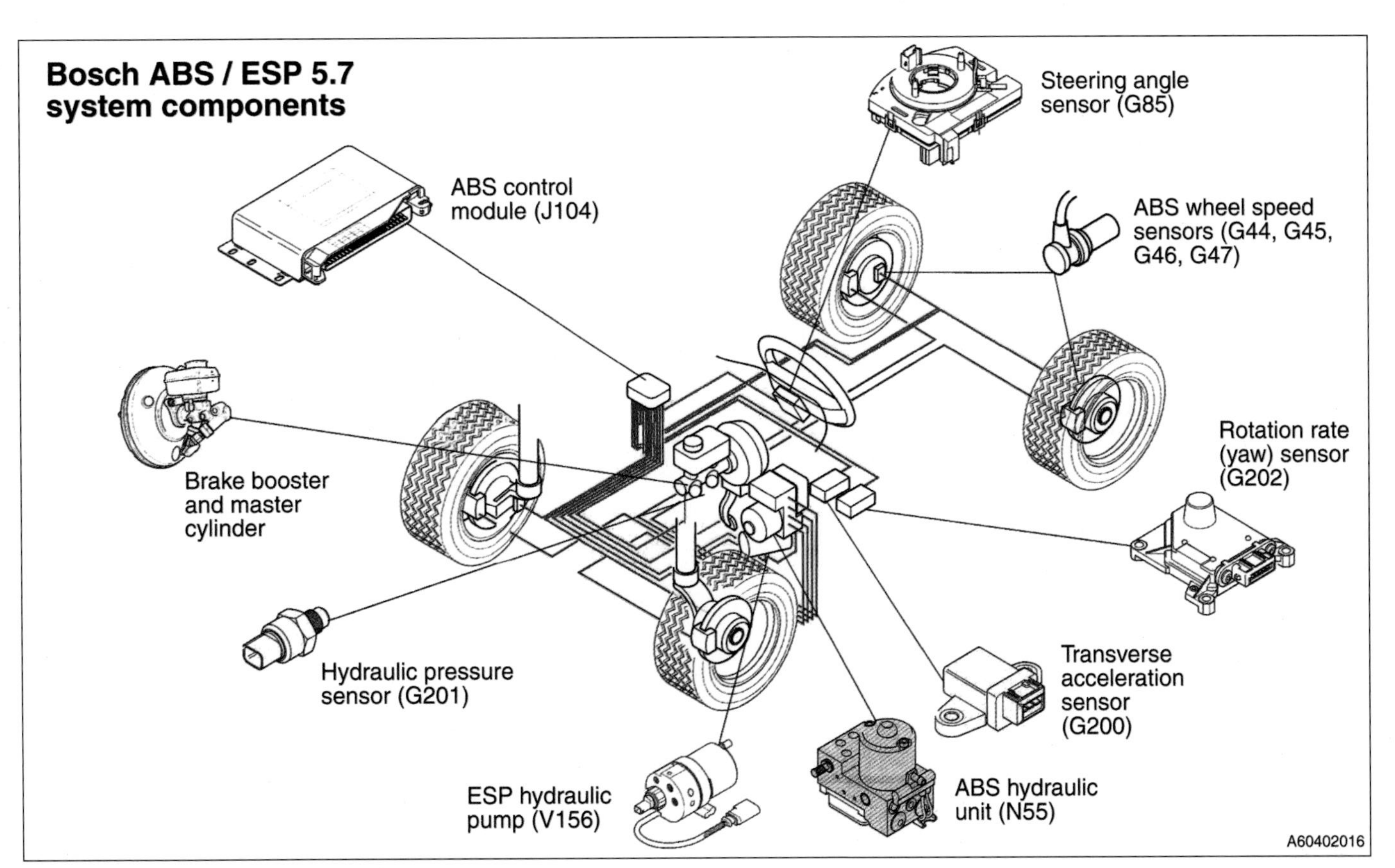

BentleyPublishers.com
BentleyPublishers.com—All Rights Reserved

Brakes

Front brake calipers

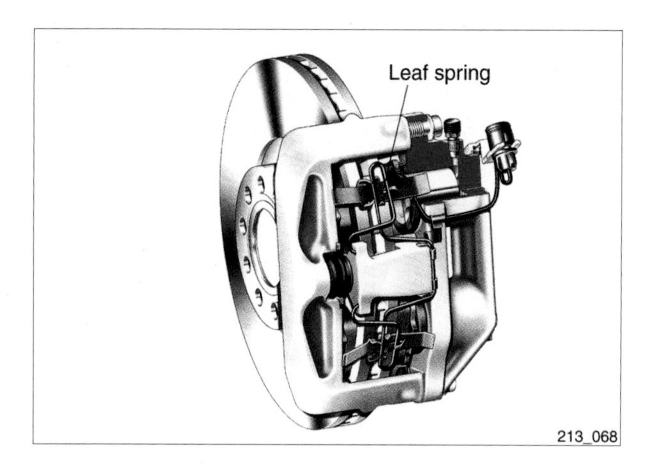

213_068

The A6 has a diagonally partitioned dual circuit brake system with front and rear disc brakes. The front brake disks are ventilated.

Beginning with 2000 models, the high performance HP-2 front brake caliper was fitted (except allroad quattro and RS6). This caliper, partly made of aluminum, makes it possible to reduce the weight of the front axle by 4.9 lb (2.2 kg) despite the fact that the front discs are larger.

A60402024

High performance Brembo front brakes are fitted to RS6 models.

There are four pads per caliper. A stainless steel central leaf spring allows brake pads to be changed with no special tools and without the need to remove the caliper.

To prevent contact corrosion between aluminum and steel components, the caliper carrier and outer (floating) caliper are coated with zinc-cobalt.

Rear brake calipers

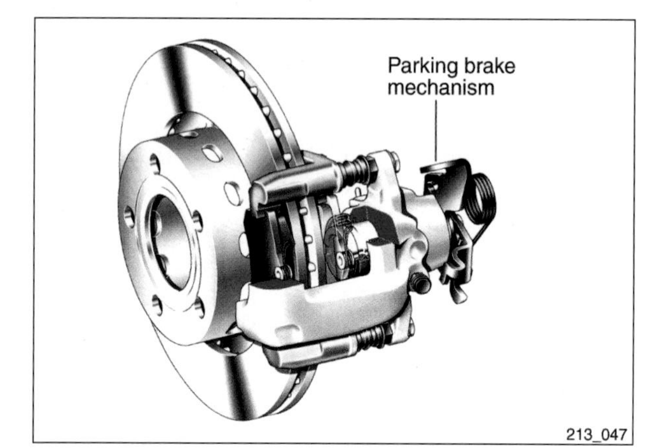

213_047

The aluminum rear brake caliper reduces weight by approximately 1.1 lb (0.5 kg) per side. The parking brake cable attaches to the rear caliper and operates the pads.

BentleyPublishers.com
BentleyPublishers.com—All Rights Reserved

Steering column

The manually adjustable steering column has a range in length of 2 inches (50 mm) and in height of 1.6 inches (40 mm). The steering column adjustment lock system consists of eight plates on either side of the steering shaft. Of these four are for length adjustment and four are for height adjustment. The lock plates have large surfaces, providing for good clamping and ease of use.

The magnesium slide and bearing block supporting the steering column are designed to collapse in case of an accident.

An electrically powered adjustable steering column is an available option with some models.

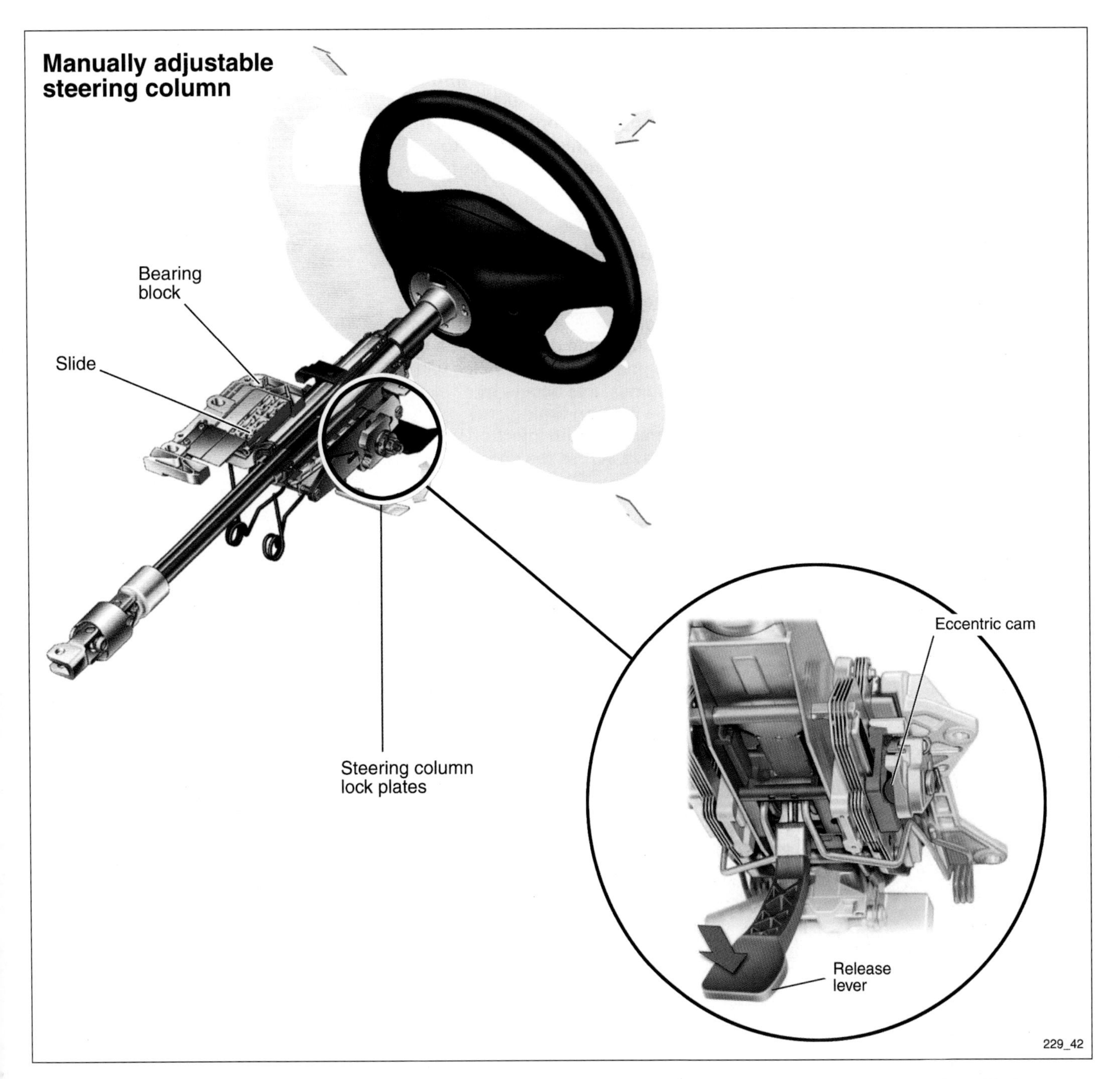

BentleyPublishers.com
BentleyPublishers.com—All Rights Reserved

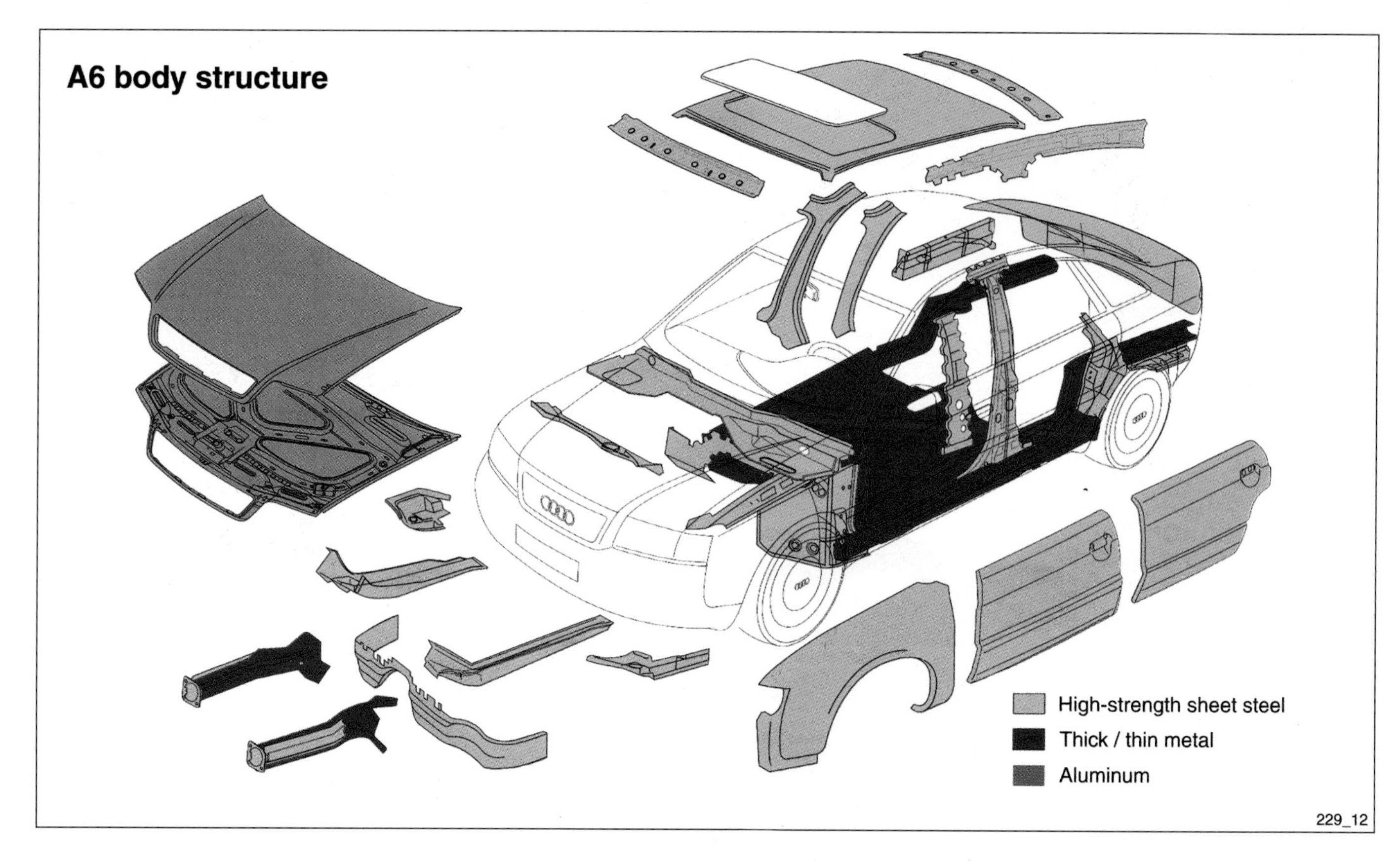

5 Body–Assembly
6 Body–Components and Accessories
7 Body–Interior Trim

Body structure

The A6 body sets high standards for crash safety and meets demands for lighter structure as well.

- Thin high-strength steel structure and sheet metal keeps down vehicle weight without reducing rigidity.
- Continuous laser welds connect the roof and side members, increasing the overall rigidity of the body.
- Aluminum, used in the engine hood, is as strong as steel when utilized correctly.
- Deformation elements in the front and rear insure that the body structure remains intact in minor accidents.

Service position

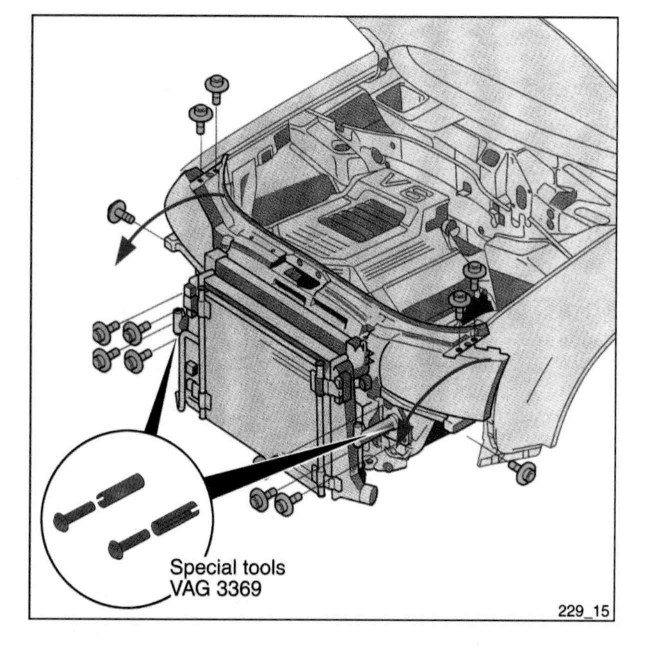

About 2.8 inches (70 mm) of clearance is gained for working in the engine compartment by placing the front lock carrier (radiator support panel) in service position. Use special tools VAG 3369 to support lock carrier panel. See **50 Body–Front**.

BentleyPublishers.com
BentleyPublishers.com—All Rights Reserved

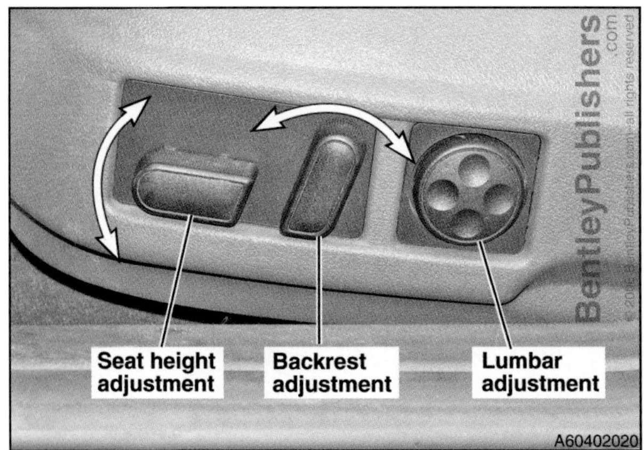

Seats

The power seat is adjusted with convenient self-explanatory button-levers on the side of the seat cushion.

M 1 2 3 STOP

A60402021

Automatic driver and passenger seat adjustment with memory function is an option. The system stores seat and outside rear-view mirror settings for three drivers.

The passenger side outside mirror swivels downward when the vehicle is in reverse, allowing a view of the right rear wheel and the curb.

Passenger safety

A6 passenger safety features include:

- Driver and passenger airbags
- Side (thorax) airbags in seat back bolsters
- Optional side curtain airbags
- Pyrotechnic seat belt retractor reels

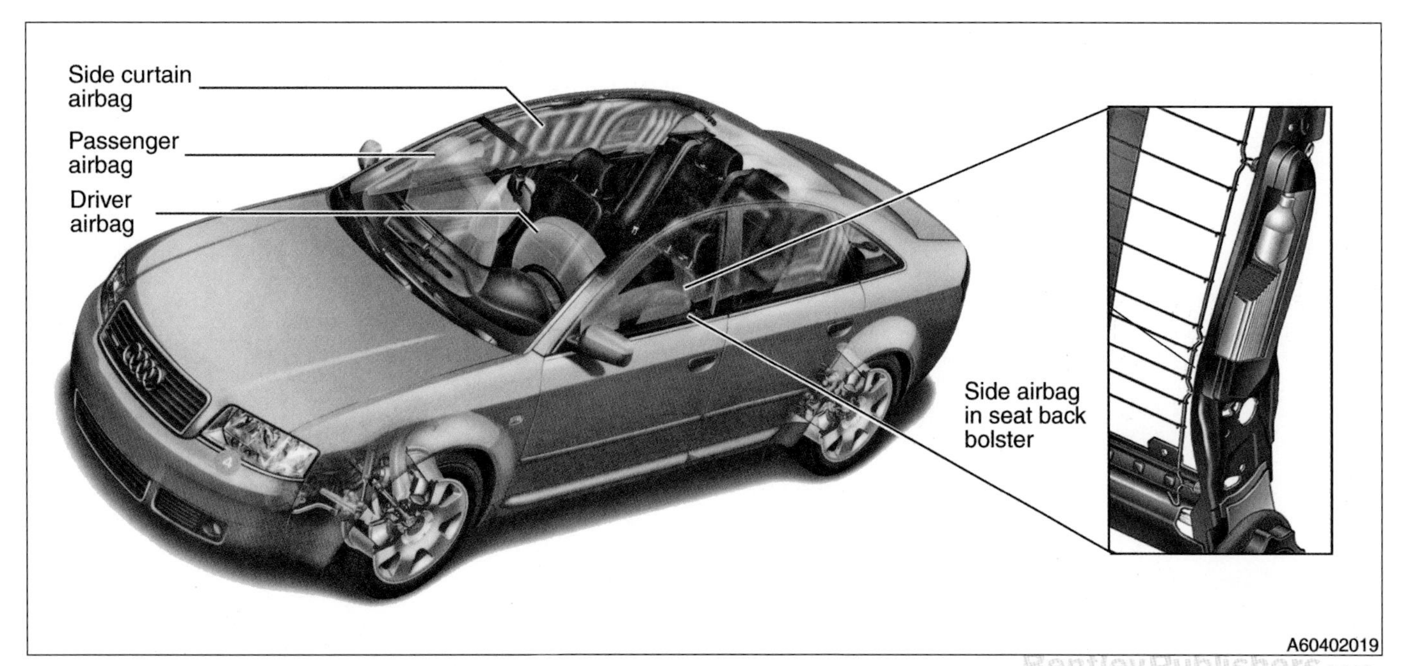

BentleyPublishers.com
BentleyPublishers.com—All Rights Reserved

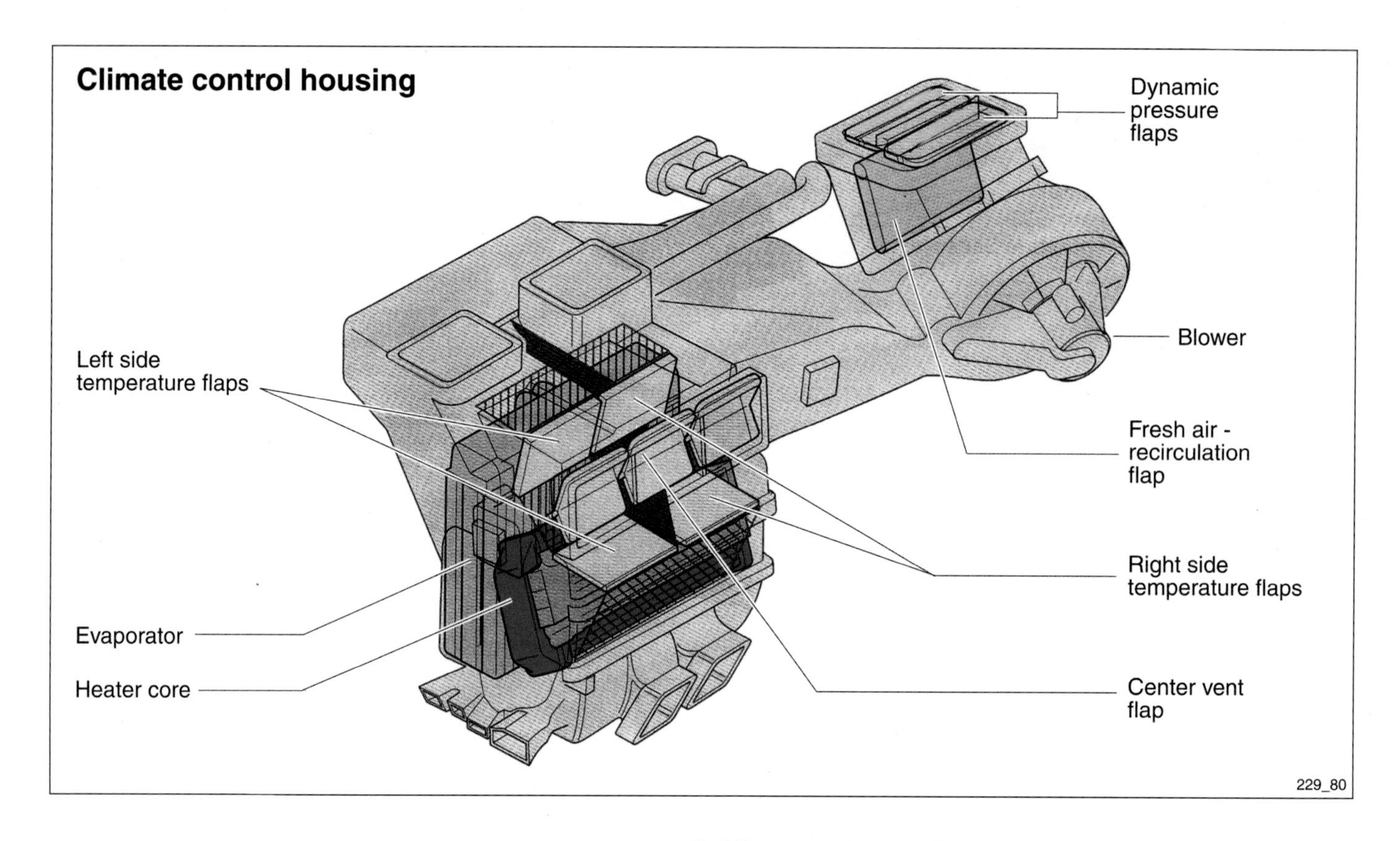

8 Heating and Air-conditioning

The climate control housing under the dashboard contains the evaporator, heater core (heat exchanger) and blower.

Electric motors operate heating and air-conditioning vent flaps:

- Fresh air-recirculation flap is combined with the dynamic pressure flap.
- One temperature flap for each side of the passenger cabin.
- Three-part flap for the center vent and the foot well vents.

Climate control function is fully automatic, with separate temperature controls for left and right.

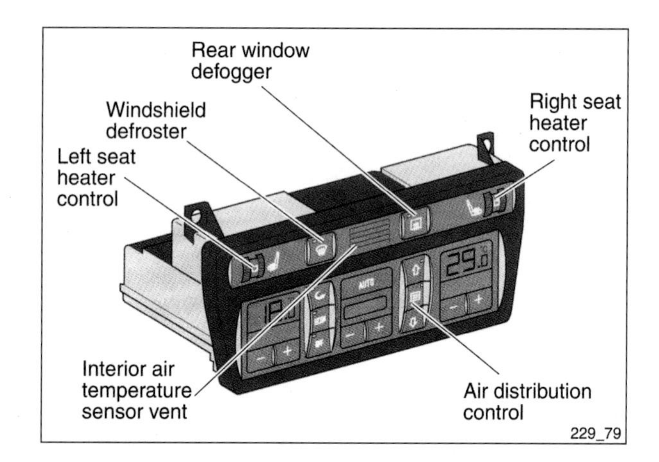

The climate control panel on the dashboard features:

- Rear window heater and windshield defroster control integrated with control panel.
- Left and right side cabin temperature independently adjustable between 64° and 86°F (18° to 30°C).
- Interior air temperature vent and sensor integrated with panel.

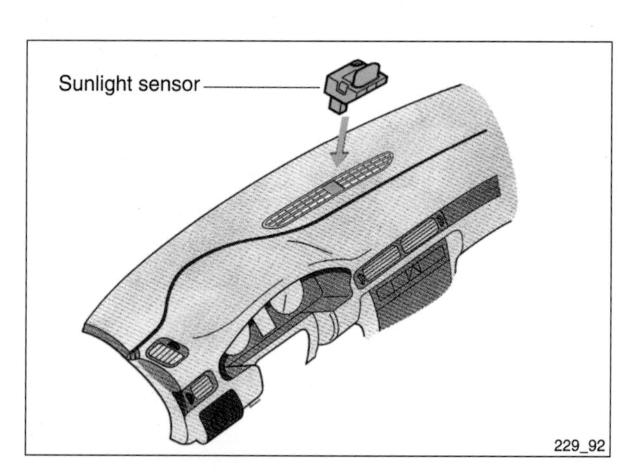

The sunlight sensor, located in the top of the dashboard, detects which side of the vehicle the sunlight is coming from and signals the climate control system to heat or cool that side to compensate.

BentleyPublishers.com
BentleyPublishers.com—All Rights Reserved

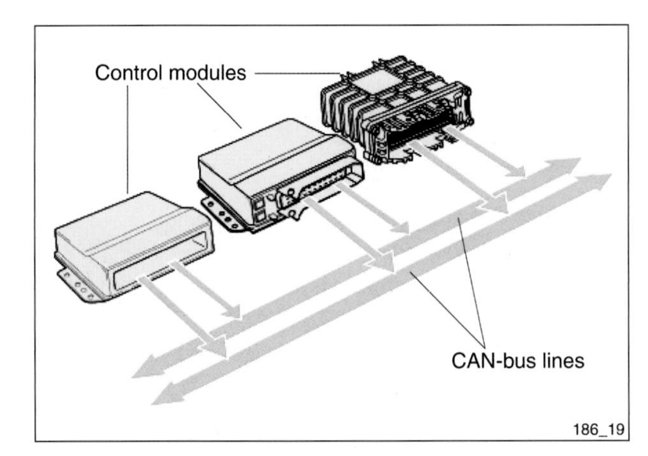

186_19

9 Electrical System

CAN-bus

In response to the growing number of electronic modules and controls, later A6 models were equipped with bus data transfer systems. In these systems, digital signals are shared among the modules. The controller area network (CAN) bus is the most frequently used of these systems. See **9 Electrical System–General**.

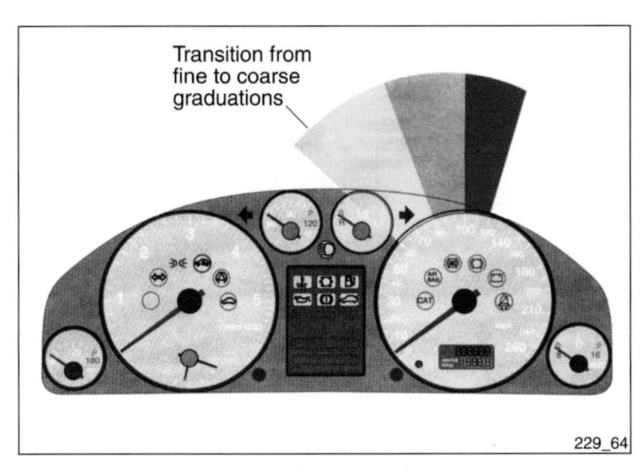

229_64

Graduated speed display

Vehicle speed display in the instrument cluster is not proportional. The graduations are wider in the lower speed up to 50 mph (80 kph), making it easier to maintain a steady rate at speeds that are used most often.

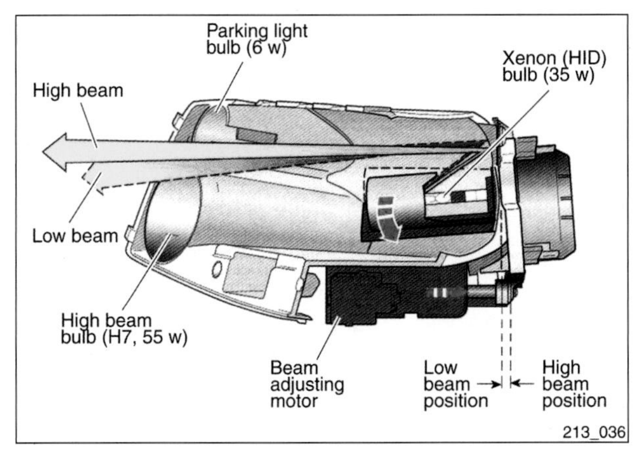

213_036

Xenon (HID) headlights

Optional high intensity discharge (HID) or xenon headlights are designed to use one xenon bulb for both high and low beam. An electric beam adjusting motor switches bulb position in relation to the reflector. An additional halogen (H7) high beam bulb provides long-range illumination and acts as a headlight flasher when the lights are not switched on.

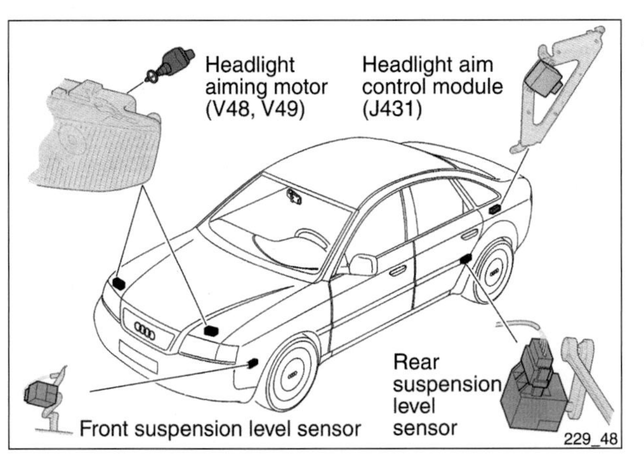

229_48

Vehicles with xenon headlights are additionally equipped with a headlight cleaning system and an automatic headlight aiming system. The aiming system adapts headlight aim to vehicle load.

NOTE—

- *Headlight aim control module position depicted for models up to 2000. In 2000 - 2004 models, module is behind the glove compartment on a bracket.*

BentleyPublishers.com
BentleyPublishers.com—All Rights Reserved

Instrument cluster, dashboard and controls

1. Dashboard vent
2. Engine oil temperature gauge
3. Tachometer with analog clock and warning lights
4. Engine coolant temperature gauge
5. Fuel gauge
6. Speedometer with warning lights
7. Voltmeter
8. Wiper and washer stalk switch
   Trip computer function switch
9. Storage compartment or
   ASR / ESP switch
10. Emergency flasher switch
11. Cup holder
12. Glove compartment lock
13. Passenger airbag
14. Light switch
15. Engine hood release lever
16. Instrument cluster dimmer
17. Horn button
    Driver airbag
18. Steering column adjustment lever
19. Concert radio
20. Climate control panel
21. Seat heater control
22. Windshield defroster switch
23. Gear selector
24. Interior air temperature sampling vent
25. Rear window defogger switch

BentleyPublishers.com
BentleyPublishers.com—All Rights Reserved

03 Maintenance

TABLES

MAINTENANCE TABLES

BentleyPublishers.com
BentleyPublishers.com—All Rights Reserved

GENERAL

This repair group explains the structure of this repair manual and details basic information regarding your vehicle and repair procedures for it. Included are service and maintenance procedures.

Carry out the maintenance work described in this repair group at the factory specified time or mileage interval shown in **Maintenance Schedules**. Following these intervals helps ensure safe and dependable operation.

The owner's manual, maintenance record and warranty booklet originally supplied with the vehicle contain maintenance schedules that apply to your Audi. Many of the maintenance procedures are necessary to maintain warranty protection.

Audi is constantly updating their recommended maintenance procedures and requirements. The information contained here may not include updates or revisions made by Audi since the publication of the documents supplied with the car. If there is any doubt about what procedures apply to a specific model or model year, or what intervals to follow, consult an authorized Audi dealer.

How to use this manual

This manual is divided into 12 main sections, or partitions:

0 General, Maintenance
1 Engine
2 Engine Management, Exhaust and Engine Electrical
3 Clutch and Transmission
4 Suspension, Brakes and Steering
5 Body–Assembly
6 Body–Components and Accessories
7 Body–Interior Trim
8 Heating and Air-conditioning
9 Electrical System
EWD Electrical Wiring Diagrams
OBD On Board Diagnostics

A master listing of the 12 partitions and the corresponding specific repair groups can be found on the inside front cover.

Thumb tabs are used on the first page of each repair group to help locate the groups quickly. Page numbers throughout the manual are organized according to the repair group system. A comprehensive **Index** is at the end of the manual.

BentleyPublishers.com
BentleyPublishers.com—All Rights Reserved

Warnings, Cautions and Notes

Throughout this manual, there are numerous paragraphs with the headings **WARNING, CAUTION** or **NOTE**. These headings have different meanings.

WARNING—
- *Text under this heading warns of unsafe practices that are very likely to cause injury, either by direct threat to the person(s) doing the work or by increased risk of accident or mechanical failure while driving.*

CAUTION—
- *Text under this heading also calls attention to important precautions to be observed during the repair work to help prevent accidentally damaging the car or its parts.*

NOTE—
- *A note contains information, tips, or pointers which help in doing a better job and completing it more easily.*

Read **WARNING, CAUTION** and **NOTE** headings before you begin repair work. See also **00 Warnings and Cautions**.

Towing and Transport

Vehicle transport and towing eyes

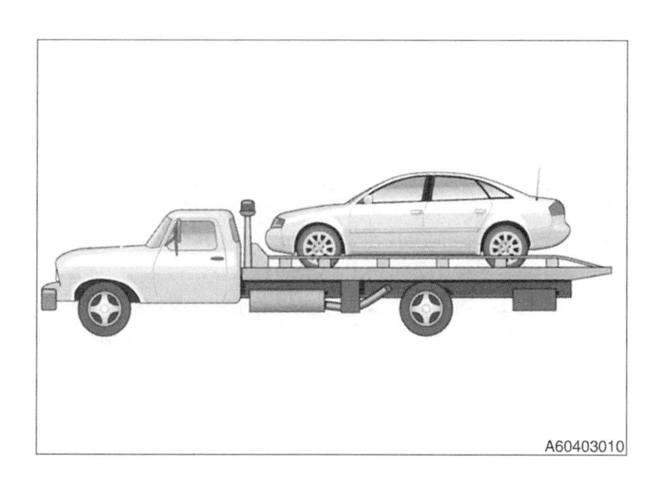

A60403010

The following information is to be used by a commercial flatbed tow truck operator who knows how to operate equipment safely. To prevent damage to the vehicle, read and understand all of the following information before proceeding.

WARNING—
- *Do not allow passengers to ride in a transported vehicle.*
- *To reduce the risk of accident and serious personal injury, stay within manufacturer's rated equipment capacities. Exceeding manufacturer's design specifications is dangerous.*

CAUTION—
- *To prevent damage to the vehicle, transport the Audi A6 with a flatbed carrier, not by towing.*
- *Do not use conventional sling-type equipment or wheel dollies.*
- *Improper attachment of vehicle to the flatbed carrier may cause damage to underbody components.*

Towing eye mounts are located at right front and right rear of the vehicle. A detachable towing eye is stored in the vehicle tool kit.

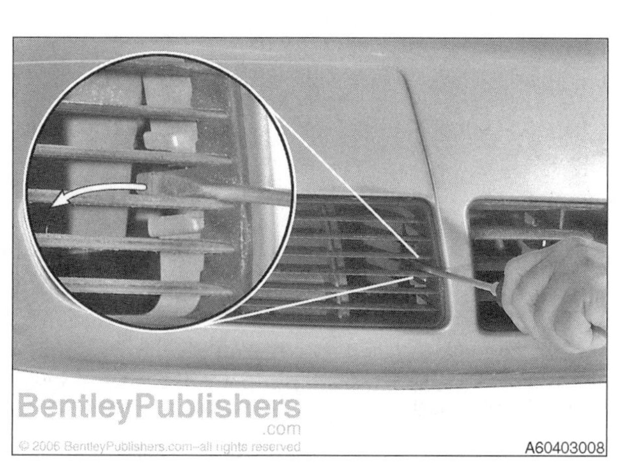

BentleyPublishers.com
© 2006 BentleyPublishers.com–all rights reserved

A60403008

Front towing eye mount: Insert screwdriver in lower right grille slot. Pry gently on lock tab (**arrow**), then pull grille off.

– Locate towing eye in tool kit.

BentleyPublishers.com
BentleyPublishers.com—All Rights Reserved

Thread towing eye (**arrow**) fully into mounting hole and tighten with wheel lug wrench (also from tool kit).

– After use, unscrew towing eye, return to vehicle tool kit, and stow tool kit in trunk.

– Reinstall cover.

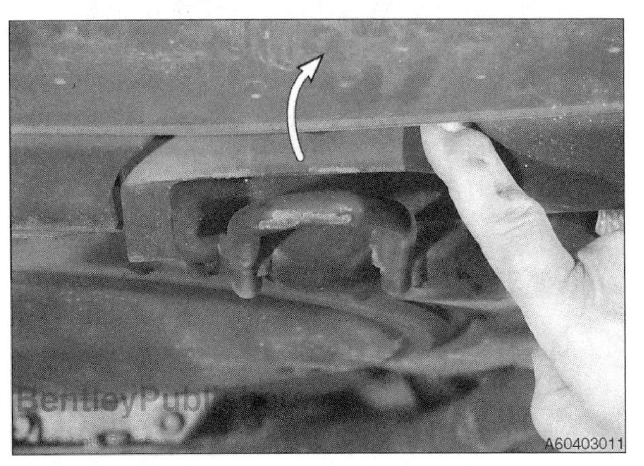

Rear towing hook: Press cover up (**arrow**) to gain access to hook.

RAISING VEHICLE

> ***CAUTION—***
> - *Do not lift or jack the vehicle underneath the engine oil pan, transmission housing, front or rear axle or the body side members. This could lead to serious damage.*
> - *To avoid damage to the underbody or chassis frame, insert a rubber pad between shop jack and lift points.*
> - *Allroad quattro: To prevent damaging the shock absorbers when lifting the vehicle with a lift or jack, set the ride height selection to "jack mode". See* **allroad quattro "jack mode"** *in this repair group.*

Raising car safely using car jack

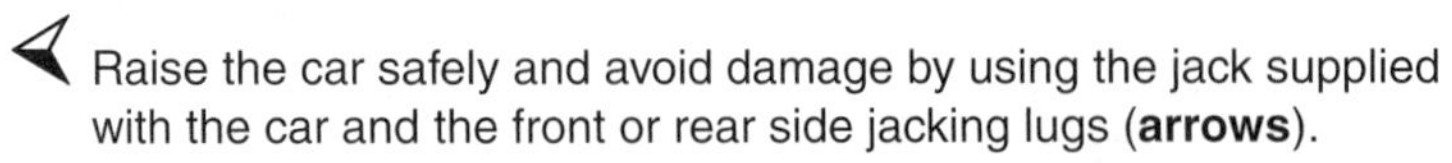

Raise the car safely and avoid damage by using the jack supplied with the car and the front or rear side jacking lugs (**arrows**).

– Park car on flat, level surface. Use chocks to block wheel opposite to one being raised.

> ***WARNING —***
> - *Do not rely on the transmission or the parking brake to keep the car from rolling.*

– If changing a tire, loosen lug bolts before raising car.

– Remove jack and jack handle from jack compartment in trunk.

BentleyPublishers.com
BentleyPublishers.com—All Rights Reserved

If changing a tire, remove large knurled nut at center of spare tire in trunk and lift spare out.

Place jack in position and raise jack arm until jack claw fits under jacking lug (**arrow**) securely. Make sure jack is resting on flat, solid ground. Use a board or other support to provide a firm surface for jack, if necessary.

Correct distance between jacking lug and wheel opening is as follows:

- Audi A6 or S6 sedan; allroad quattro
  A = approx. 15 cm (6 in)
  B = approx. 15 cm (6 in)
- Audi A6 or S6 Avant (station wagon)
  A = approx. 15 cm (6 in)
  B = approx. 25 cm (10 in)

> ***WARNING —***
> - *Watch the jack closely. Make sure it stays stable and does not shift or tilt.*

– Raise car slowly while constantly checking position of jack and car.

– Allroad quattro: See **allroad quattro "jack mode"** in this repair group.

BentleyPublishers.com
BentleyPublishers.com—All Rights Reserved

Raising car safely with lift

> ***WARNING —***
> - *Before driving a vehicle on a lift, confirm that vehicle weight does not exceed the allowable lifting capacity.*
> - *Lift vehicle only at points indicated in order to avoid damaging vehicle floor pan and to prevent tipping.*
> - *Do not start engine and engage a gear with vehicle lifted if any drive wheel has contact with the floor. There is danger of an accident due to possible vehicle movement.*
> - *Exercise care while running the engine with the vehicle on a lift. Engine vibration and vehicle movement could cause the vehicle to slip off the lift.*
> - *Before driving on a lift, be sure there is enough clearance under the vehicle so that it does not drag the lift arms. Pay particular attention to front spoilers and aprons.*
> - *Observe the lift manufacturer's safety instructions and guidelines before raising vehicle.*

Use correct jacking points (**arrows**) to lift car safely and avoid damage.

Front lifting point at longitudinal reinforcement of floor pan at area marked for vehicle jack.

> ***CAUTION—***
> - *Do not raise vehicle at the vertical support on the rocker panel at front of vehicle.*

Rear lifting point at vertical reinforcement of lower sill.

BentleyPublishers.com
BentleyPublishers.com—All Rights Reserved

Raising car safely using floor jack

– Park car on flat, level surface. Use chocks to block wheel opposite to one being raised.

> ***WARNING —***
> *• Do not rely on the transmission or the parking brake to keep the car from rolling.*

– If changing a tire, loosen lug bolts before raising car.

Raising front wheel

< Place floor jack under reinforced floor pan inboard of front jacking point.

– Place jack stand under front jacking point. Lower car slowly until it rests firmly on jack stand.

Raising rear wheel

< Place floor jack under rear trailing arm.

– Place jack stand under rear jacking point. Lower car slowly until it rests firmly on jack stand.

allroad quattro "jack mode"

Prior to jacking or lifting allroad quattro vehicle, set ride height control to "jack mode" to protect suspension shock absorbers and struts.

> ***WARNING —***
> *• Make sure that no one is lying under the vehicle or has head or hands in the wheel housing while the ride height is changing.*

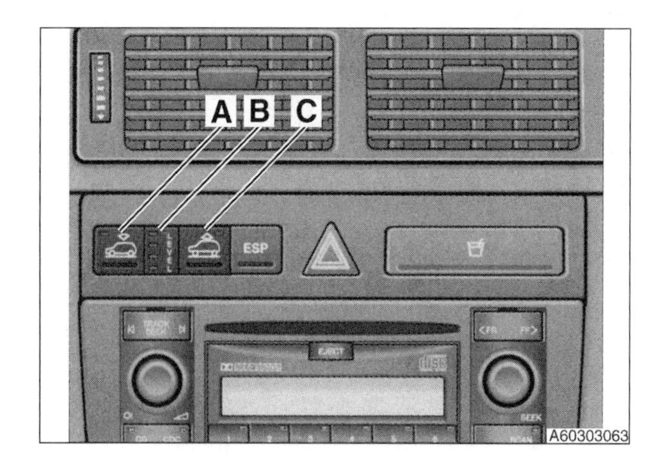

Activating

< Before lifting vehicle with jack or lift:

- Switch ignition ON.
- Press ride height control buttons **A** and **C** in center dashboard for at least 5 seconds.
- When system is in "jack mode", LEDs on control buttons (**A** and **C**), yellow LED for manual mode on level indicator (**B**) and warning light for level control in instrument cluster all illuminate.

– Switch ignition OFF and lift or jack vehicle.

Deactivating

– Press control buttons **1** and **3** for at least 5 seconds. Warning light in instrument cluster, LED for manual mode and control button LEDs turn OFF.

"Jack mode" is automatically deactivated when vehicle speed exceeds 3 mph (5 kph).

BentleyPublishers.com
BentleyPublishers.com—All Rights Reserved

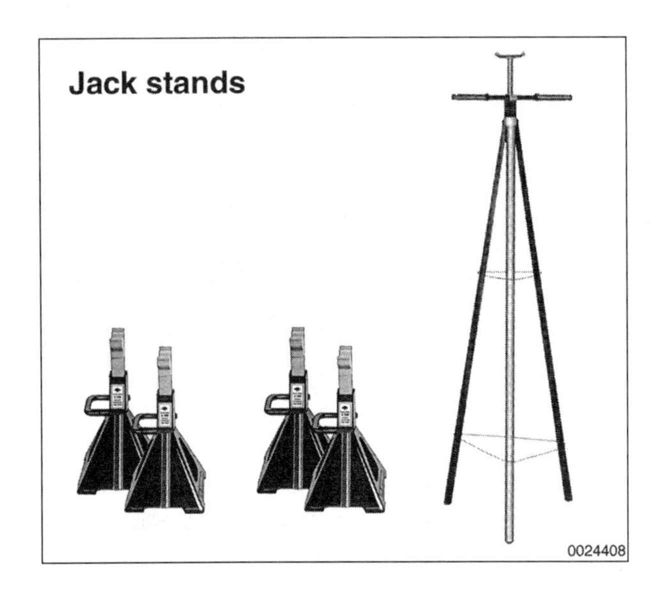

Working under car safely

When working under the car, observe the following points as a matter of safety and good practice:

- Disconnect negative battery cable so that no one else can start vehicle. Let others know that you will be under vehicle.
- Place at least two jack stands under vehicle. A jack is a temporary lifting device. Do not use the jack alone to support vehicle while you are under it. Use positively locking jack stands that are designed for the purpose of supporting a vehicle.
- If you are using a lift, be sure that safety locks are engaged and that vehicle weight is resting on the locks, not the hydraulic system.
- Lower vehicle slowly until its weight is fully supported by jack stands or safety locks. Watch to make sure that jack stands do not tip or lean as vehicle settles on them, and that jack stands and lift arms are placed solidly and will not move.
- Check to make sure vehicle is stable before working under car.
- Observe all jacking precautions again when raising vehicle to remove jack stands.

> ***WARNING —***
> - *Use care when removing major (heavy) components from one end of the vehicle. The sudden change in weight and balance can cause vehicle to tip off lift or jack stands.*
> - *Do not support vehicle at engine oil pan, transmission, fuel tank, or on front or rear axle. Serious damage may result.*

Basic Service Information

Diagnostic trouble codes (DTCs), accessing

Use Volkswagen / Audi diagnostic scan tool or equivalent to retrieve diagnostic trouble codes.

– Place transmission selector lever in PARK or NEUTRAL. Engage parking brake. Make sure ignition is OFF.

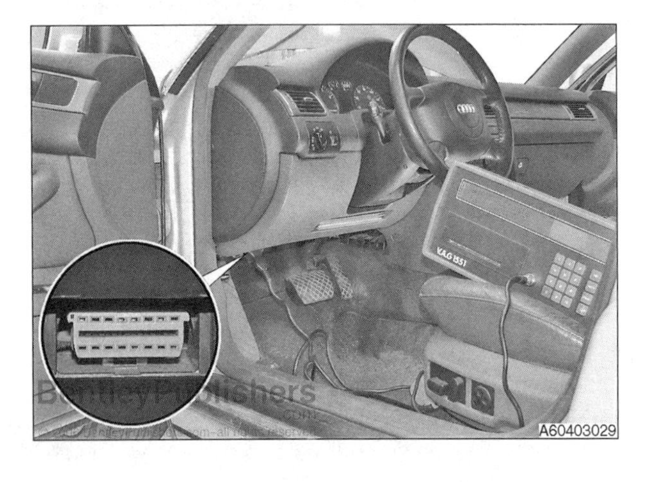

– Connect scan tool (VAG 1551 or equivalent) to data link connector (DLC, **inset**), located under left side dashboard, to left of steering column.

– Start engine and let idle.

– Follow scan tool instructions as they appear on scan tool screen.

– For more information, see **OBD On Board Diagnostics**.

Service reminder indicator (SRI), resetting

When the time for a service is reached, the needed service appears in the trip odometer display when the ignition is switched ON. The display blinks for about 60 seconds after the engine is started:

- Service OIL indicates that oil change service is due.
- Service INSP indicates that inspection service is due.

A service that is due is displayed 1,000 km (625 mi) or 10 days in advance.

BentleyPublishers.com
BentleyPublishers.com—All Rights Reserved

Reset the SRI after each oil change or inspection service. The SRI can be reset in two different ways:

- Use the Volkswagen / Audi scan tool or equivalent.
- Use buttons on the instrument cluster.

SRI, resetting using scan tool

Connect scan tool (VAG 1551 or equivalent) to data link connector (DLC, **inset**), located under left side dashboard, to left of steering column.

– Switch ignition ON.

– Follow scan tool instructions as they appear on scan tool screen.

SRI, resetting with instrument cluster buttons

– Switch ignition OFF.

Press and hold down trip odometer reset button (**B**).

– Switch ignition ON while keeping button depressed:
  - Service OIL appears in trip odometer display (**C**).
  - Release trip odometer reset button.

– Pull clock reset button (**A**) until Service OIL indicator is reset.
  - - - - appears in trip odometer display (**C**).

– Press trip odometer reset button (**B**) again to advance to next service event.
  - Release trip odometer reset button.
  - Service INSP appears in trip odometer display (**C**).

– Pull clock reset button (**A**) until Service INSP indicator is reset.
  - - - - appears in trip odometer display (**C**).

– Switch ignition OFF.

Non-reusable fasteners

Many fasteners used on the cars covered by this manual must be replaced with new ones once they are removed. These include but are not limited to: bolts, nuts (self-locking, nylock, etc.), cotter pins, studs, brake fittings, roll pins, clips and washers. Use genuine Audi parts for this purpose.

Some bolts are designed to stretch during assembly and are permanently altered, rendering them unreliable once removed. These are known as torque-to-yield fasteners. Replace fasteners where instructed to do so. Failure to replace these fasteners could cause vehicle damage and personal injury. See an authorized Audi dealer for applications and ordering information.

BentleyPublishers.com
BentleyPublishers.com—All Rights Reserved

Tightening fasteners

Tighten fasteners gradually and evenly to avoid misalignment or over-stressing any one portion of the component. For components sealed with gaskets, this method helps to ensure that the gasket seals properly.

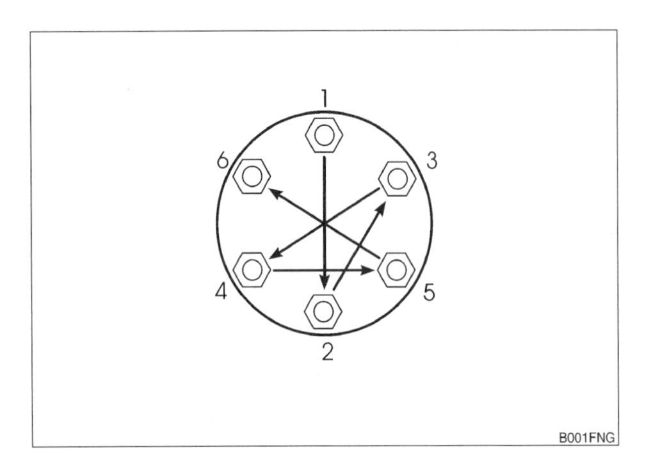

Where there are several fasteners, tighten them in a sequence alternating between opposite sides of the component. Repeat the sequence until all the bolts are evenly tightened to the proper specification.

For some repairs a specific tightening sequence is necessary, or a particular order of assembly is required. Such special conditions are noted in the text, and the necessary sequence is described or illustrated. Where no specific torque is listed, use **Table a** as a general guide for tightening fasteners.

NOTE—
- *Metric bolt classes or grades are marked on the bolt head.*
- *Do not confuse wrench size with bolt diameter.*

Table a. Bolt tightening torques–general (in Nm)

Bolt diameter	Bolt class (according to DIN 267)					
	5.6	5.8	6.8	8.8	10.9	12.9
M5	2.5	3.5	4.5	6	8	10
M6	4.5	6	7.5	10	14	17
M8	11	15	18	24	34	40
M10	23	30	36	47	66	79
M12	39	52	62	82	115	140
M14	62	82	98	130	180	220
M16	94	126	150	200	280	340
M18	130	174	210	280	390	470

CAUTION—
- **Table a** *is a general reference only. The values listed are not intended to be used as a substitute for torques specifically called out in the text.*

Buying parts

Many of the maintenance and repair tasks in this manual call for the installation of new parts, or the use of new gaskets and other materials when reinstalling parts. In most cases, make sure needed parts are on hand before beginning the job. Read the introductory text and the complete procedure to determine which parts are needed.

For bigger jobs, partial disassembly and inspection are required to determine a complete parts list. Read the procedure carefully and, if necessary, make arrangements to get the necessary parts while your car is disassembled.

BentleyPublishers.com
BentleyPublishers.com—All Rights Reserved

Genuine Audi parts

Genuine replacement parts from an authorized Audi dealer are designed and manufactured to the same standards as the original parts. They are guaranteed to fit and work as intended by the engineers who designed the car.

Many independent repair shops make a point of using genuine Audi parts, even though they may at times be more expensive. They know the value of doing the job right with the right parts. Parts from other sources can be as good, particularly if manufactured by one of Audis original equipment suppliers, but it is often difficult to know.

Audi is constantly updating and improving their cars, often making improvements during a given model year. Audi may recommend a newer, improved part as a replacement, and your authorized dealer's parts department will know about it and provide it. The Audi parts organization is best equipped to deal with any Audi parts needs.

Non-returnable parts

Some parts cannot be returned. The best example is electrical parts. Buy electrical parts carefully, and be as sure as possible that a replacement is needed, especially for expensive parts such as electronic control modules. It may be wise to let an authorized Audi dealer or other qualified shop confirm your diagnosis before replacing a non-returnable part.

Audi service

Audi dealers are uniquely qualified to provide service for Audi cars. Their authorized relationship with the large Audi service organization means that they are constantly receiving special tools and equipment, together with the latest and most accurate repair information.

The Audi dealer's service technicians are highly trained and very capable. Authorized Audi dealers are committed to supporting the Audi product. On the other hand, there are many independent shops that specialize in Audi service and are capable of doing high quality repair work. Checking with other Audi owners for recommendations on service facilities is a good way to learn of reputable Audi shops in your area.

Tools

Most maintenance can be accomplished with a small selection of the right tools. Tools range in quality from inexpensive junk, which may break at first use, to very expensive and well-made tools for the professional. The best tools for most do-it-yourself Audi owners lie somewhere in between.

Many reputable tool manufacturers offer good quality, moderately priced tools with a lifetime guarantee. These are your best buy. They cost a little more, but they are good quality tools that will do what is expected of them. Sears Craftsman® line is one such source of good quality tools.

Some of the repairs covered in this manual require the use of special tools, such as a custom puller or specialized electrical test

BentleyPublishers.com
BentleyPublishers.com—All Rights Reserved

equipment. These special tools are called out in the text and can be purchased through an authorized Audi dealer. As an alternative, some special tools mentioned may be purchased from the following tool manufacturers and/or distributors:

Assenmacher Specialty Tools, Inc.
6440 Odell Place
Boulder, CO 80301
(800) 525-2943
www.asttool.com

Baum Tools Unlimited, Inc.
PO Box 5867
Sarasota, FL 34277
(800) 848-6657
www.baumtools.com

Equipment Solutions
P.O. Box 1450
Kenosha, WI 53141-1450
(800) 892-9650

Mac Tools
4635 Hilton Corporate Drive
Columbus, OH 43232
(800) 622-8665
www.mactools.com

Metalnerd
509 Crestview Drive, Suite B
Greensburg, PA, 15601 USA
(412) 601-4270
www.metalnerd.com

Ross-Tech
920 South Broad Street
Lansdale, PA 19446
215-361-8942
www.ross-tech.com

Samstag Sales
115 Main St. N., Suite 216
Carthage, TN 37030
(615) 735-3388
www.samstagsales.com

Shade Tree Software
4186 Culebra Ct.
Boulder, CO 80301
(303) 449-1664
(303) 940-2468
www.shadetreesoftware.com

Snap-On Technologies, Inc.
2801 80th St.
Kenosha, WI 53141-1410
(262) 656-5200
www.snapon.com

Zelenda Automotive, Inc.
66-02 Austin St.
Forest Hills, NY 11374
(888) 892-8348
www.zelenda.com

IDENTIFICATION PLATES AND LABELS

Information you need to know

Model. When ordering parts, it is important that you know the correct model designation for your car. Audi A6 models include the following:

- A6 and S6 (sedan)
- A6 and S6 Avant (station wagon)
- allroad quattro
- RS6

Model year. This is not necessarily the same as date of manufacture or date of sale. A 1999 model may have been manufactured in late 1998, and perhaps not sold until early 2000. It is still a 1999 model. Model years covered by this manual are 1998 to 2004.

Date of manufacture. This information may be necessary when ordering replacement parts or determining if any of the warranty recalls are applicable to your car. The label on the driver's door below the door latch specifies the month and year that the car was built.

BentleyPublishers.com
BentleyPublishers.com—All Rights Reserved

Vehicle identification number (VIN). This is a combination of letters and numbers that identify the particular car. See **Vehicle identification number (VIN)** in this repair group. Also see **01 Vehicle Identification and VIN Decoder**.

Engine code. Cars covered in this manual are powered by various V6 and V8 engines. For information on engine codes and engine applications, see **Engine identification** in this repair group.

Transmission code. The transmission type with its identifying code may be important when buying clutch parts, seals, gaskets, and other transmission-related parts. For information on transmission codes and applications, see **02 Audi A6 Familiarization**. Transmission identification code plates are illustrated in **Transmission identification** in this repair group.

Vehicle identification number (VIN)

Vehicle year, model and engine type, along with other pertinent data can be decoded from the vehicle identification number (VIN). Data plates and stickers containing the VIN are attached to the vehicle at several locations.

United States federal law requires the VIN to be visible from outside the vehicle and at a standard location. A plate with the complete 17 digit VIN is attached to the padded dashboard at the base of the windshield. See **01 VIN Decoder** for detailed VIN decoding information.

Other VIN locations include:
- Stamped on engine compartment bulkhead.
- Stamped into body in rear floor area under rear seat. Access to this area requires lifting rear seat bottom (cushion).

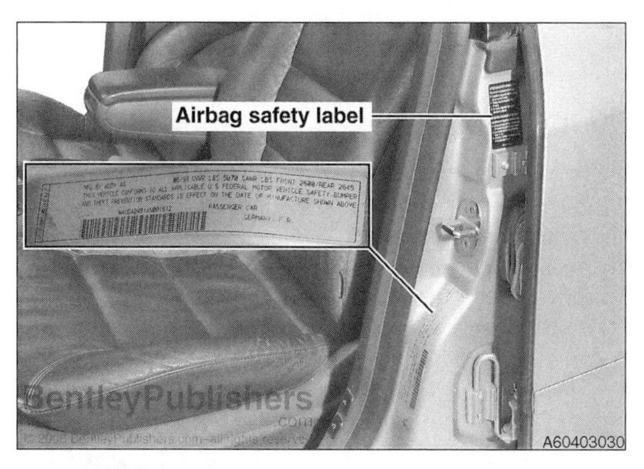

B pillar certification sticker (**inset**) lists full VIN, vehicle weight ratings, production date and emission compliance certification. Airbag safety label is also attached to vehicle at this location.

Data labels

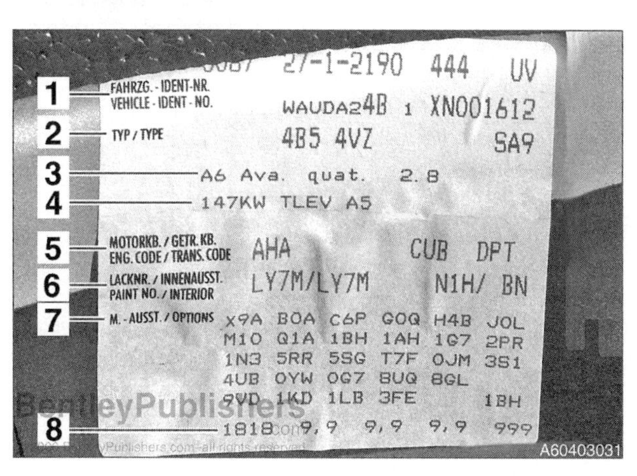

Vehicle data sticker is located in the trunk in the spare tire well. The sticker contains the following vehicle information:

1. Vehicle identification number (VIN)
2. Production control number
3. Model identification number
4. Engine power, exhaust emissions classification, transmission type
5. Engine and transmission code letters
6. Paint code, interior code
7. Option codes (optional equipment identification numbers)
8. Curb weight, fuel consumption, CO_2 emissions

BentleyPublishers.com
BentleyPublishers.com—All Rights Reserved

Identification Plates and Labels

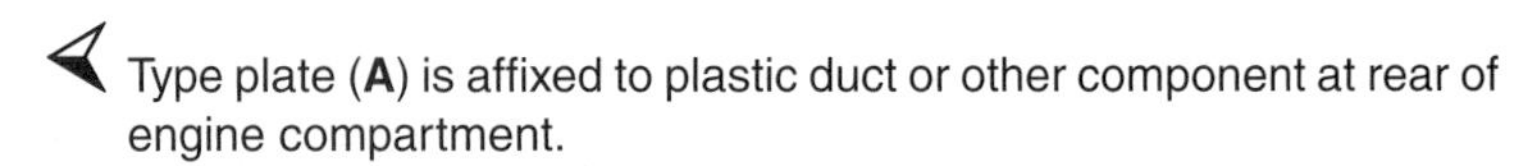

Type plate (**A**) is affixed to plastic duct or other component at rear of engine compartment.

B is engine compartment bulkhead VIN stamping.

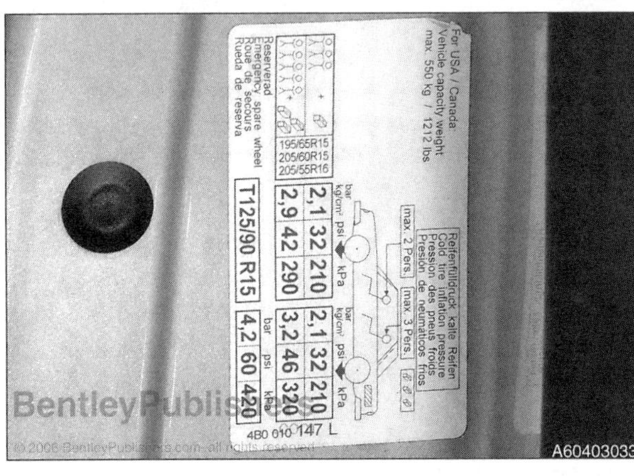

Tire pressure inflation sticker is attached to B or C pillar. Recommended cold tire inflation pressures vary from car to car due to differing equipment levels. Usually, tire pressure information does not appear in the owner's manual.

Engine identification

An Audi engine is identified by a three letter code (example: AHA) followed by a 6-digit serial number. Engine code and engine serial number are located on the engine block. A sticker with engine code and serial number is often attached to the upper timing belt cover or cylinder head cover. In addition, engine code may be stamped on the cylinder head to allow for quick identification. Engine application information is in **Table b**.

Engine type	Engine codes
V6 5-valves per cylinder	AHA, ATQ, APB, BEL, AVK
V8 5-valves per cylinder	ART, AWN, BBD, BAS, BCY

BentleyPublishers.com
BentleyPublishers.com—All Rights Reserved

Table b. Engine applications and specifications

Year model	Engine displacement, type	Engine code	Engine displacement in cc (cu in)	Power in kW @ rpm (hp @ rpm)	Torque in Nm @ rpm (ft-lb @ rpm)	Bore x stroke in mm (in)	Compression ratio	Engine management, emissions category
V6 5-valves per cylinder								
1998- 1999 A6	2.8 liter	AHA	2771 (169)	147 @ 6000 (200 @ 6000)	280 @ 3200 (207 @ 3200)	82.5 x 86.4 (3.25 x 3.40)	10.1 : 1	Motronic M5.9.2
2000 - 2001 A6		ATQ	2771 (169)	142 @ 6000 (193 @ 6000)	280 @ 3200 (207 @ 3200)	82.5 x 86.4 (3.25 x 3.40)	9.9 : 1	Motronic ME 7.1
2000 - 2003 A6, allroad quattro	2.7 liter biturbo	APB	2671 (163)	187@ 5800 (254 @ 5800)	350@ 1800- 4500 (258@ 1800 - 4500)	81.0 x 86.4 (3.19 x 3.40)	9.3 : 1	Motronic ME 7.1 TLEV
2003 - 2004 A6, allroad quattro		BEL	2671 (163)	184 @ 5800 (250 @ 5800)	350@ 1800- 4500 (258@ 1800 - 4500)	81.0 x 86.4 (3.19 x 3.40)	9.3 : 1	Motronic ME 7.1 LEV
2002 - 2004 A6	3.0 liter	AVK	2976 (182)	162 @ 6300 (220 @ 6300)	300 Nm @ 3200 (221@ 3200)	82.5 x 92.8 (3.25 x 3.65)	10.5 : 1	Motronic ME 7.1.1 ULEV
V8 5-valves per cylinder								
1999 - 2001 A6	4.2 liter	ART	4172 (255)	221 @ 6000 (299 @ 6000)	410 @ 3300 (302 @ 3300)	84.5 x 93.0 (3.33 x 3.66)	11 : 1	Motronic ME 7.1
2002 - 2004 A6		AWN	4172 (255)	228 @ 6000 (299 @ 6000)	413 @ 3300 (305 @ 3300)	84.5 x 93.0 (3.33 x 3.66)		Motronic
2002 - 2004 S6		BBD	4172 (255)	250 @ 6000 (340 @ 6000)	400 @ 3500 (295 @ 3500)	84.5 x 93.0 (3.33 x 3.66)		Motronic
2003 - 2004 allroad quattro		BAS	4163 (254)	220 @ 6200 (299 @ 6200)	380 @ 2600- 4700 (280 @ 2600- 4700)	84.5 x 92.8 (3.33 x 3.65)		Motronic ME 7.1.1
2003 - 2004 RS6	4.2 liter biturbo	BCY	4172 (255)	331 @ 5700 - 6400 (450 @ 5700 - 6400)	560 @ 1950 - 5600 (413 @ 1950 - 5600)	84.5 x 93.0 (3.33 x 3.66)	9.3 : 1	Motronic ME 7.1.1

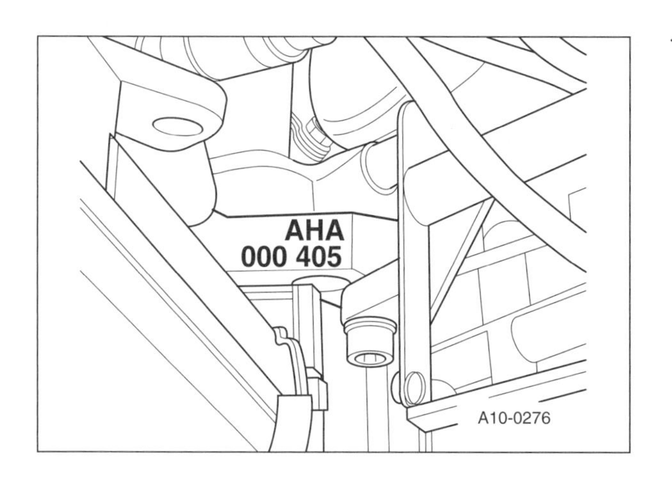

2.7 liter and 2.8 liter V6 engine

Engine codes AHA, ATQ, APB, BEL: Engine code letters and serial number are on flat surface of cylinder block in front of right cylinder head.

Engine code is also included on vehicle data sticker in trunk.

BentleyPublishers.com
BentleyPublishers.com—All Rights Reserved

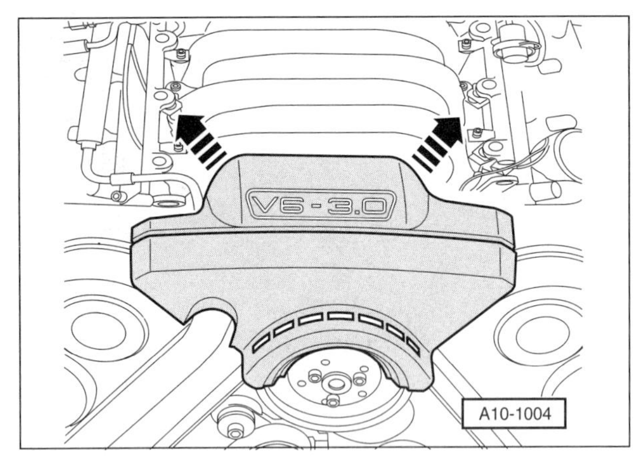

3.0 liter V6 engine

- To gain access to engine identification sticker, remove front engine cover by pulling up (**arrows**).

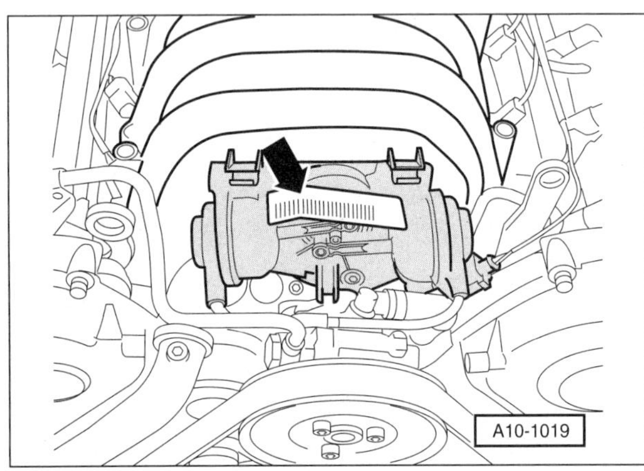

- Engine code AVK: Engine code, serial number and production number are on sticker (**arrow**) on vacuum diaphragm housing of intake manifold.

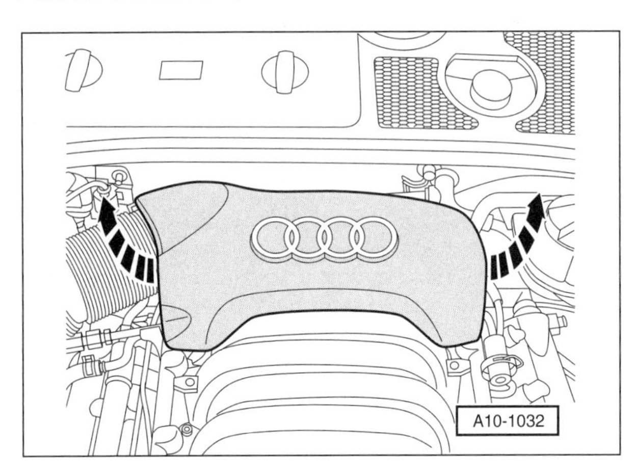

- If sticker is not present, remove rear engine cover by pulling up (**arrows**).

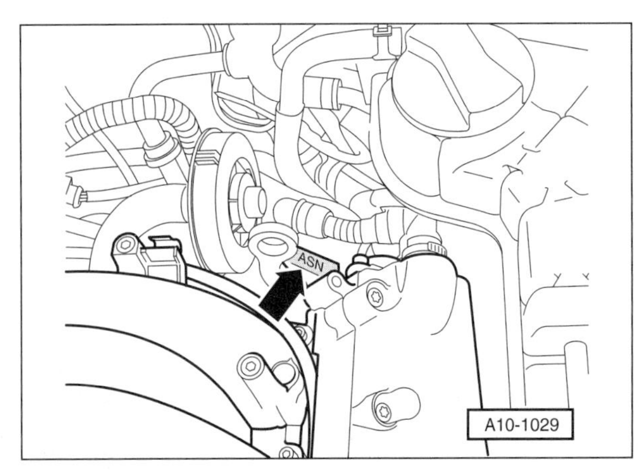

- Engine code AVK: Engine code and serial number are stamped on left rear of cylinder block (**arrow**).

Engine code is also included on vehicle data sticker in trunk.

BentleyPublishers.com
BentleyPublishers.com—All Rights Reserved

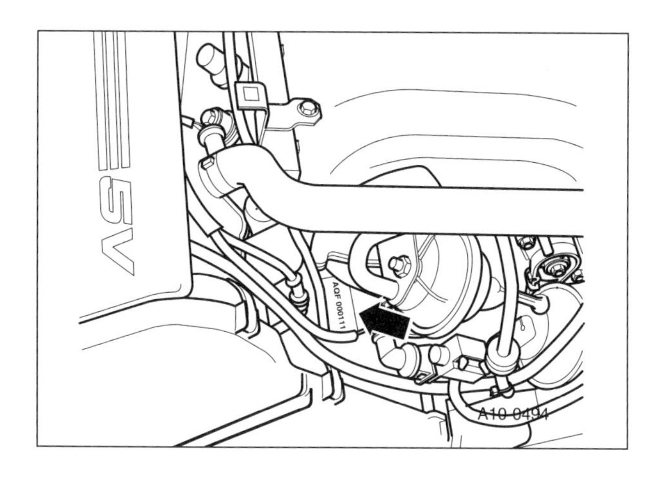

4.2 liter V8 engine (A6, S6)

Engine codes ART, AWN, BBD: Engine code and serial number are on left side of cylinder block (**arrow**).

Engine number is also listed on label on belt cover.

Engine code is also included on vehicle data sticker in trunk.

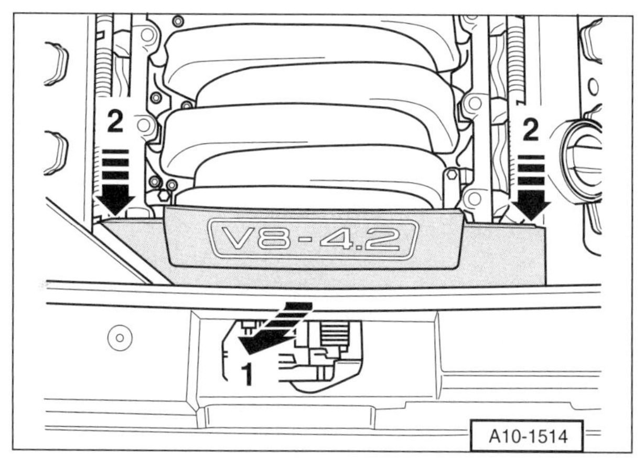

4.2 liter V8 engine (allroad quattro)

To gain access to engine identification stamping, remove front engine cover by pulling up (**2**) and forward (**1**).

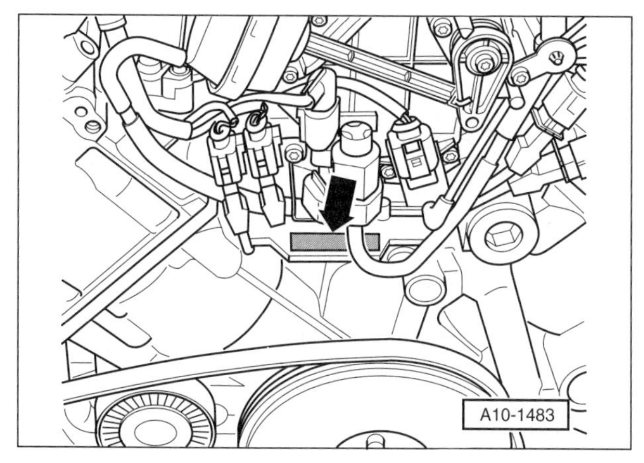

Engine code BAS: Engine code and serial number are stamped on front top of cylinder block (**arrow**).

Engine code and serial number is also on a sticker on right cylinder head cover.

Engine code is also included on vehicle data sticker in trunk.

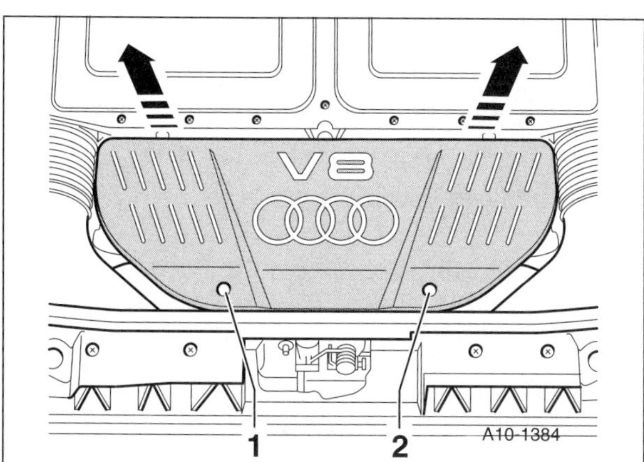

4.2 liter V8 engine (RS6)

Remove front engine cover:

- Undo quick-release fasteners (**1**, **2**).

BentleyPublishers.com
BentleyPublishers.com—All Rights Reserved

Engine code BCY: Engine code and serial number is on top of cylinder block (**arrow**).

An engine code and serial number sticker is on the right timing belt guard.

Engine code is also included on vehicle data sticker in trunk.

Transmission identification

A6 models were equipped with one of five different transmission types. See **Table c**. Each transmission type comes in numerous versions, depending on engine type and front-wheel-drive (FWD) or all-wheel-drive (AWD) requirements. For transmission applications, specifications and lubricants, see **02 Product Familiarization**, **34 Manual Transmission**, **37 Automatic Transmission** or **39 Final Drive**.

Table c. Transmission codes

Description T	**ype**
5-speed manual quattro (AWD)	01A
6-speed manual quattro (AWD)	01E
5-speed automatic with Tiptronic® control quattro (AWD)	01L
5-speed automatic with Tiptronic® control FWD or quattro	01V
Multitronic® or continuously variable transmission (CVT) FWD	01J

Transmissions are identified by type, code, serial number, manufacturer identification code and build date:

- **Type** denotes a family of transmissions that have generally similar characteristics.
- **Code** is a specific group within the type and identifies individual traits such as gear ratios.
- **Serial number** is a multi-digit string such as 0019967.
- **Manufacturer identification code** may be a combination of letters and numbers.
- **Build date** is a five digit number such as 15079. The digits indicate day (15), month (07 = July) and year (9 = 1999) transmission was built.

Transmission identification plates containing some or all of this information are affixed to the transmission housing.

BentleyPublishers.com
BentleyPublishers.com—All Rights Reserved

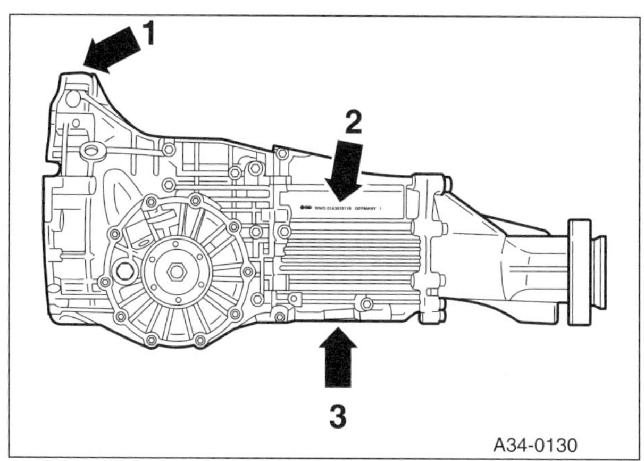

5-speed manual transmission quattro (AWD)

Transmission type 1A: Code, serial number and date of manufacture are on transmission as follows:

1. Top of bellhousing: Code letters and date of manufacture
2. Top of gear housing: Type and serial number
3. Bottom of gear housing: Code letters and date of manufacture

Transmission code letters are also included on vehicle data sticker in trunk.

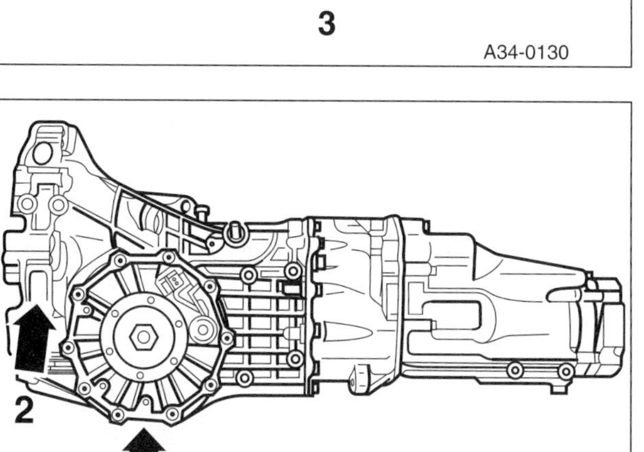

6-speed manual transmission quattro (AWD)

Transmission type 1E: Code, serial number and date of manufacture are on transmission as follows:

1. Bottom of front final drive housing: Code letters and serial number
2. Left side of bellhousing: Type and serial number

Transmission code letters are also included on vehicle data sticker in trunk.

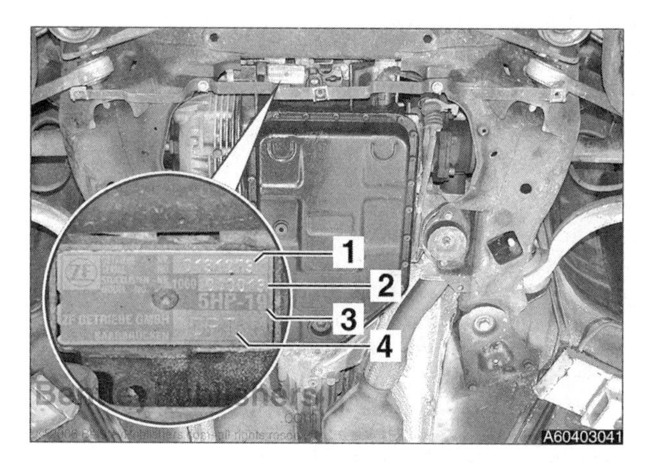

5-speed automatic transmission quattro (AWD)

Transmission type 01L: Identification plate (**inset**), on bottom of transmission at right front of transmission oil pan, contains information as follows:

1. Transmission serial number
2. Part list number
3. Manufacturer identification code
4. Code letters

Transmission code letters are also included on vehicle data sticker in trunk.

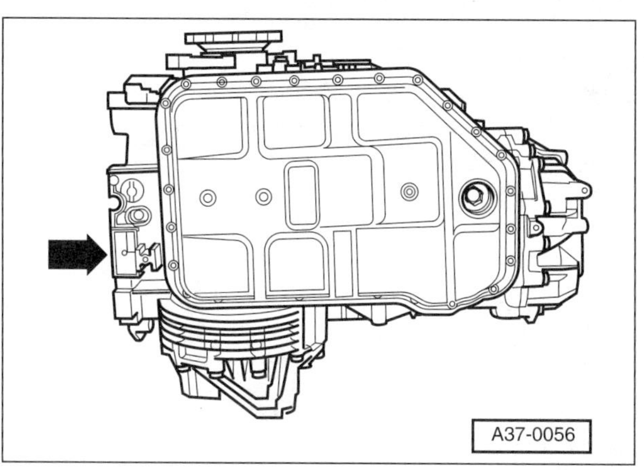

5-speed automatic transmission FWD or quattro (AWD)

Transmission type 01V: Identification plate (**arrow**) is on bottom of transmission at right front of transmission oil pan.

Another identification label (identical to the first one) is on the side of the transmission. It is not accessible when the transmission is installed.

BentleyPublishers.com
BentleyPublishers.com—All Rights Reserved

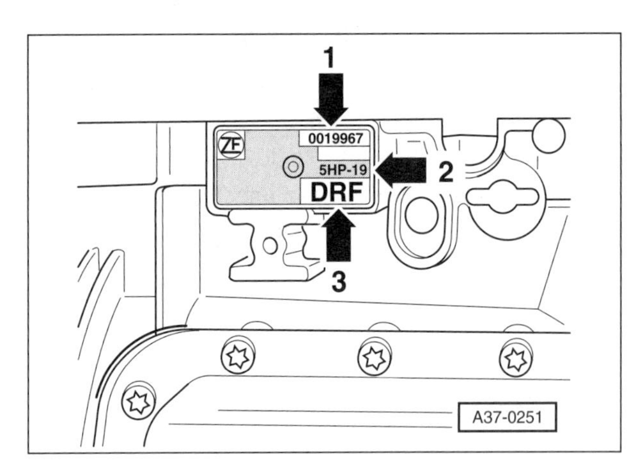

 Transmission identification plate contains information as follows:

1. Transmission serial number
2. Manufacturer identification code
3. Code letters

Transmission code letters are also included on vehicle data sticker in trunk.

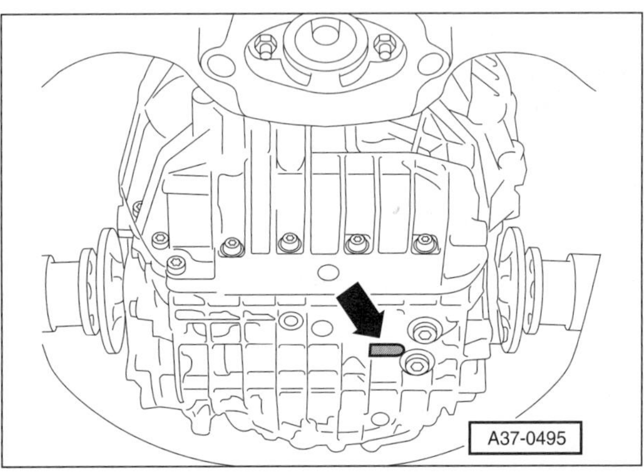

Multitronic® (CVT) transmission FWD

Transmission type 01J: Identification plate (**arrow**) is on bottom of transmission.

Transmission code letters are also included on vehicle data sticker in trunk.

ENGINE COMPARTMENT

Fluid leaks, visual inspection

Remove upper and lower engine covers as required and check engine compartment for signs of fluid leaks. Fluid leaks attract dust making them easier to spot. Many expensive repairs can be avoided by prompt repair of minor fluid leaks.

Visually inspect for oil and ATF leaks at engine and transmission. Also inspect cooling, fuel, heating and air-conditioning systems for leakage. Visually inspect hoses and hose connections for leaks, worn areas, porosity and brittleness. Also see **Fluid and exhaust leaks, visual inspection** in this repair group.

Check that fluid levels are between MIN and MAX marks.

The following illustrations show representative engine compartment layouts for the four major engine types. There may be minor differences among model years, such as engine cover configuration.

BentleyPublishers.com
BentleyPublishers.com—All Rights Reserved

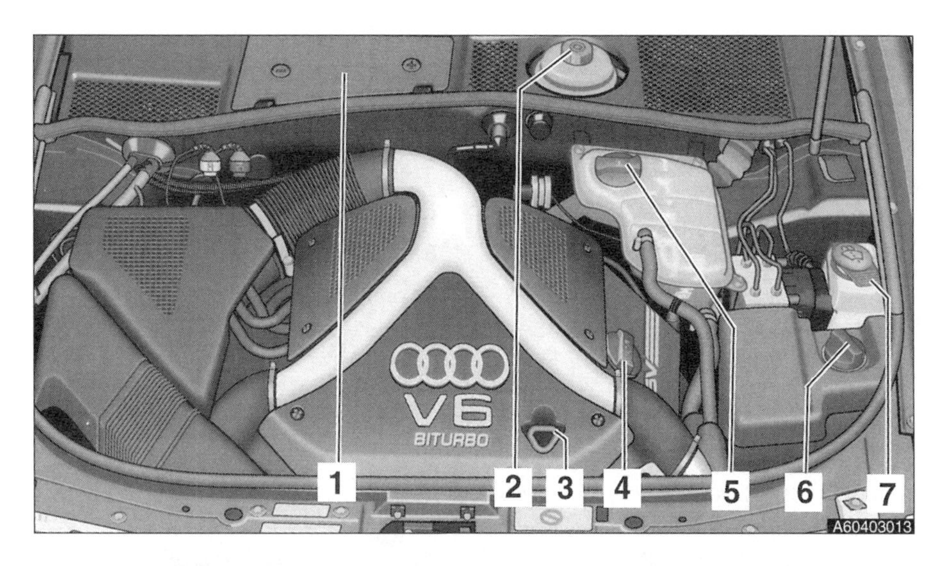

2.7 liter V6 biturbo engine compartment

1. Battery (under plastic cover)
2. Brake fluid reservoir
3. Engine oil dipstick
4. Engine oil filler cap
5. Coolant expansion reservoir
6. Power steering fluid reservoir
7. Washer fluid reservoir

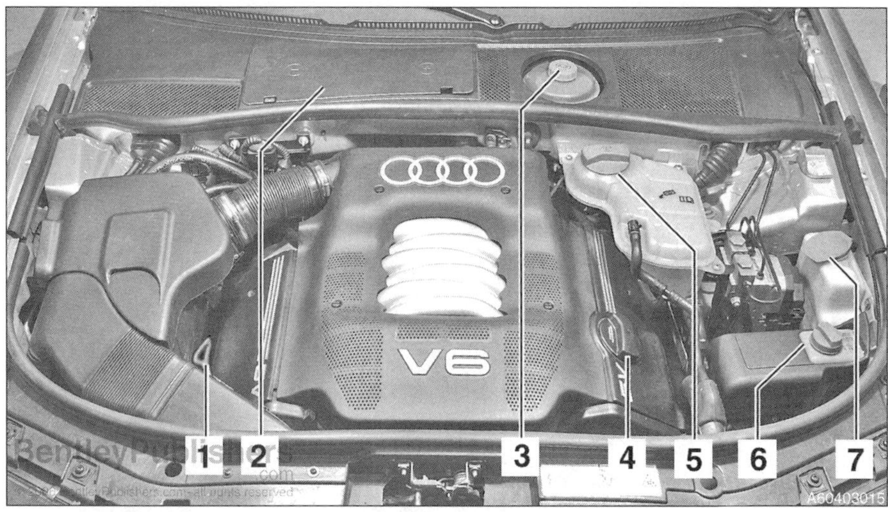

2.8 liter V6 engine compartment

1. Engine oil dipstick
2. Battery (under plastic cover)
3. Brake fluid reservoir
4. Engine oil filler cap
5. Coolant expansion reservoir
6. Power steering fluid reservoir
7. Washer fluid reservoir

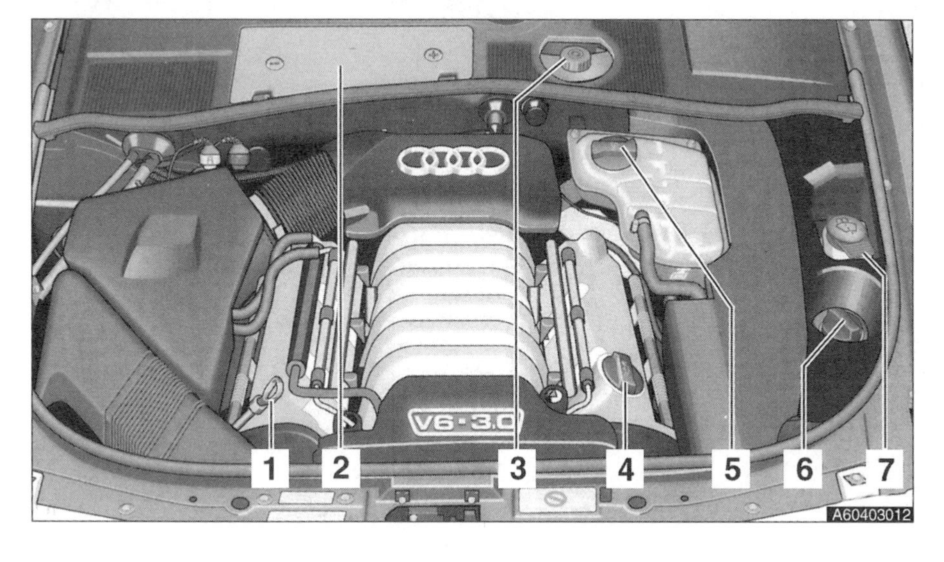

3.0 liter V6 engine compartment

1. Engine oil dipstick
2. Battery (under plastic cover)
3. Brake fluid reservoir
4. Engine oil filler cap
5. Coolant expansion reservoir
6. Power steering fluid reservoir
7. Washer fluid reservoir

BentleyPublishers.com
BentleyPublishers.com—All Rights Reserved

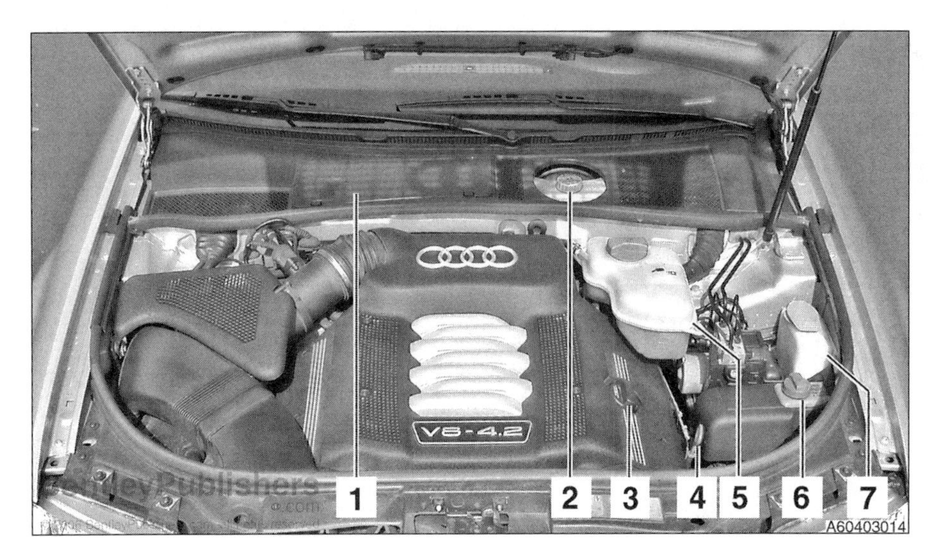

4.2 liter V8 engine compartment

1. **Battery (under plastic cover)**
2. **Brake fluid reservoir**
3. **Engine oil filler cap**
4. **Engine oil dipstick**
5. **Coolant expansion reservoir**
6. **Power steering fluid reservoir**
7. **Windshield washer fluid reservoir**

Engine covers, removing

A6 and S6 models are equipped with a variety of plastic noise absorber panels in the engine compartment, both above and underneath the engine. Following are some of the common types and methods for removing them.

Engine upper cover

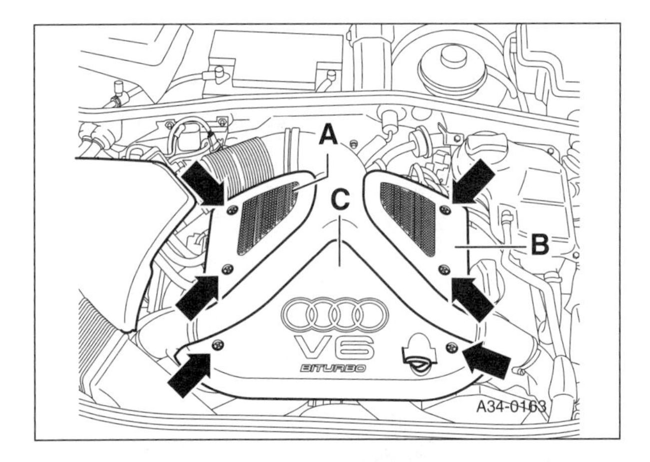

2.7 liter V6 engine top covers:

- Loosen 6 fasteners (**arrows**) by turning each 90°.
- Lift off covers **A**, **B** and **C**.

2.8 liter V6 engine top covers:

- Loosen 8 fasteners (**arrows**) by turning each 90°.
- Lift off top (center) cover, then cylinder head covers.

BentleyPublishers.com
BentleyPublishers.com—All Rights Reserved

3.0 liter V6 rear engine cover:

- Lift cover at corners (**arrows**) and remove.

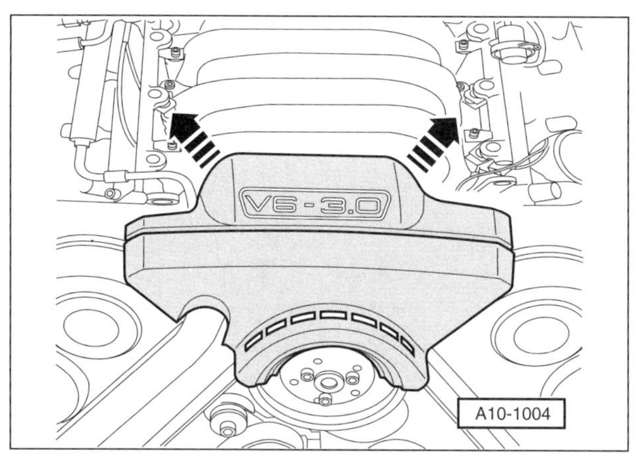

3.0 liter V6 front engine cover:

- Lift cover at corners (**arrows**) and remove.

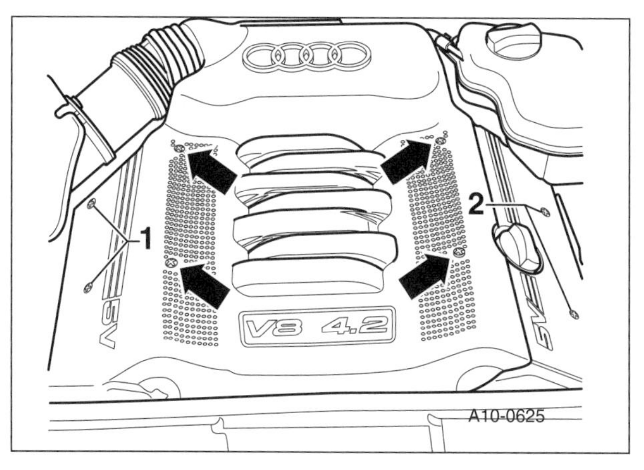

4.2 liter V8 engine top cover:

- Remove air filter hose.
- Loosen 4 fasteners (**arrows**) and lift off cover.

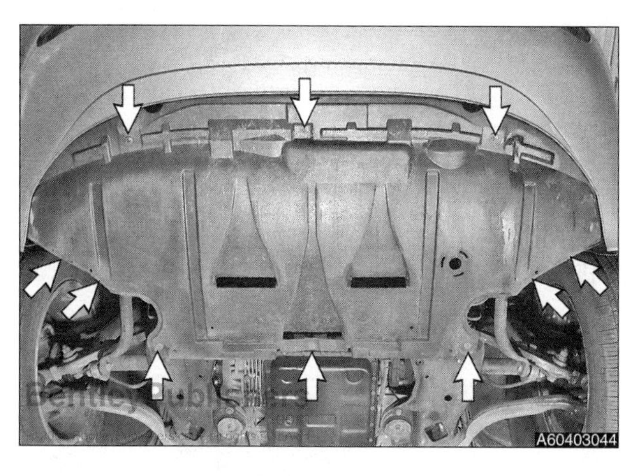

Engine lower cover (splash shield)

V6 models: Loosen or remove quick-release fasteners (**arrows**). Pull engine cover backward away from bumper cover to remove.

– V8 models are similar.

BentleyPublishers.com
BentleyPublishers.com—All Rights Reserved

ENGINE OIL

Lubricating oil for engines covered by this manual are required to meet quality and performance standards specified by Audi. These standards are sufficiently high so that, generally, only synthetic oils meet them.

BentleyPublishers.com
© 2006 BentleyPublishers.com—all rights reserved

Engine oils meeting specified performance standards are available from Audi and other sources.

Current specifications for engine oil are shown in **Table d**. If an oil container is not marked with appropriate specification(s), assume that the oil is not suitable. Use of oils other than those specified may cause engine damage.

Table d. Oil specifications	
Type: • American Petroleum Institute • European industry standard	 API service SJ ACEA A2 or ACEA A3
Viscosity grade: • New engine • At oil change • High temperature service (above 40°C or 100°F)	 SAE 0W30 SAE 5W30 SAE 5W40

Approximate refill capacities in **Table e** are listed by engine code. Engine application information is in **Table b** in this repair group.

Table e. Oil capacities (approximate)	
Engine (engine code)	**Oil capacity (includes oil filter) in liters (US qt)**
2.7 liter V6 (A6, S6) (APB)	6.5 (6.9)
2.7 liter V6 (allroad quattro)(BEL)	6.9 (7.3)
2.8 liter V6 (AHA, ATQ)	5.7 (6.0)
3.0 liter V6 (AVK)	6.4 (6.8)
4.2 liter V8 (A6, S6, RS6) (ART, AWN, BBD, BCY)	7.6 (8.0)
4.2 liter V8 (allroad quattro)(BAS)	8.5 (8.9)

Engine oil level, checking

It is normal for any engine to consume a small amount of oil. The rate of consumption depends on oil quality, viscosity, and operating conditions. Check oil level at every fuel filling and especially before starting out on a trip.

– Switch ignition OFF, then wait at least 3 minutes to allow oil to flow back into oil pan.

– Remove dipstick, wipe with a clean cloth and fully reinsert into tube.

BentleyPublishers.com
BentleyPublishers.com—All Rights Reserved

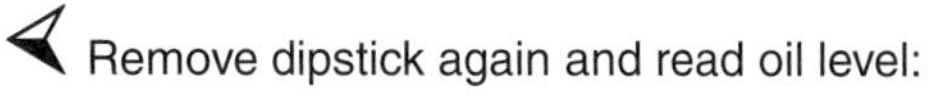

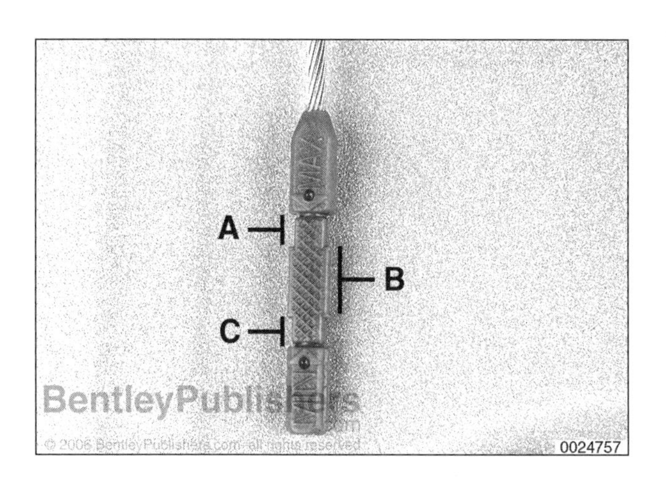

Remove dipstick again and read oil level:

- Oil at **MIN** (**C**). Top off oil. After topping off, level is correct when it is within normal area (**B**).
- Oil at **Normal** operating range (**B**). Oil does not have to be topped off, but can be as long as level does not exceed MAX mark (**A**).
- **MAX** mark (**A**). Do not top off.

> ***WARNING —***
> - *Do not overfill engine oil. Overfilling can cause misfire diagnostic trouble codes (DTCs) to be stored in engine control module (ECM).*
> - *If engine oil level is above maximum mark, catalytic converter may be damaged.*

— When filling engine, note oil specification and capacity information in **Table d** and in **Table e** in this repair group.

Engine oil and filter, changing

Procedures differ slightly by engine due to filter placement and design. 1999 A6 Quattro Avant with 2.8 liter V8 engine is illustrated.

— Warm car to normal operating temperature.

— Shut engine OFF and apply parking brake.

— Raise car and support safely. See **Raising Vehicle** in this repair group.

> ***WARNING —***
> - *Make sure the car is stable and well supported at all times. Use a professional automotive lift or jack stands designed for the purpose. A floor jack is not adequate support.*

— Remove engine lower cover. See **Engine covers, removing** in this repair group.

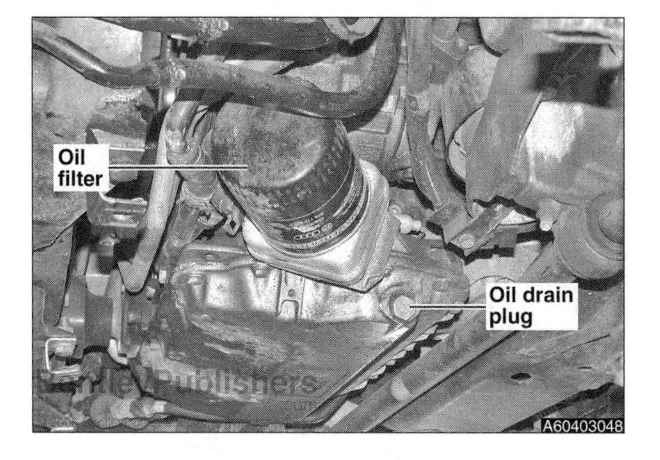

Remove oil drain plug and allow oil to drain into waste oil pan.

> ***WARNING —***
> - *Hot oil scalds. Wear protective clothing, gloves and eye protection.*
> - *Used oil is hazardous waste. Dispose of properly.*

— When oil has completely drained, reinstall drain plug with new gasket and tighten.

Tightening torque	
Drain plug to oil pan (replace gasket)	30 Nm (22 ft-lb)

— Loosen and remove oil filter.

> ***WARNING —***
> - *Dispose of oil filter as hazardous waste.*

— RS6 model: Remove front stabilizer bar to subframe mounting bolts for access to oil filter.

BentleyPublishers.com
BentleyPublishers.com—All Rights Reserved

- When installing new filter:
  - Clean oil filter flange on engine.
  - Lightly lubricate rubber seal on new oil filter.
  - Thread filter on and tighten by hand.

- Fill engine with oil of correct viscosity and rating. See **Table d** in this repair group.
  - See **Table e** for approximate engine capacities.
  - To prevent overfilling, fill to approximately ½ liter (½ US qt) less than listed capacity. After vehicle is lowered and is sitting on a level area, start engine, allow to run until operating temperature is approximately 60°C (140°F) and switch OFF ignition. Wait approximately 3 minutes, check dipstick, then carefully fill to MAX mark.

- Start engine and inspect oil filter housing and drain plug area for leaks. Switch engine OFF.

- Install lower engine cover.

Engine oil may also be changed using an extraction system. This method involves siphoning oil out of engine through the dipstick tube. In some instances this can save considerable time. When using this type of device, follow the manufacturer's instructions.

COOLING SYSTEM

Cooling system maintenance consists of maintaining coolant level, checking coolant freezing point, and inspecting hoses. Coolant flushing is not part of Audi scheduled maintenance.

The coolant used in vehicles covered by this manual, called G12 and identified by its purple color, is phosphate and silicate free. Advantages of G12 over earlier types of coolant include improvements in corrosion protection, thermal stability, heat transfer and control, hard water tolerance and environmental protection.

BentleyPublishers.com
© 2006 BentleyPublishers.com–all rights reserved

0024690g

> ***WARNING —***
> - *Hot coolant can scald. Do not work on cooling system until it has fully cooled.*
> - *Use extreme care when draining and disposing of coolant. Coolant is poisonous and lethal. Children and pets are attracted to it because of its color, sweet smell and taste. See a doctor or veterinarian immediately if any amount is ingested.*

BentleyPublishers.com
BentleyPublishers.com—All Rights Reserved

CAUTION—
- *Use Audi G12 original antifreeze or equivalent when filling cooling system. Use of other antifreeze types may be harmful to the cooling system. Do not use antifreeze containing phosphates.*
- *Do not mix green coolant (G11) with purple coolant (G12). Mixing can cause serious engine damage.*
- *Contamination of G12 with other colored coolants is identifiable by discoloration (brown, gray, etc.). This mixture may cause a foamy deposit to appear in the expansion tank and radiator. Drain and flush cooling system completely before refilling with the correct type of antifreeze.*
- *Make sure the cooling system is filled year-round with a mixture of 40% antifreeze minimum. The antifreeze mixture provides corrosion control and also raises the boiling point of the coolant.*
- *Do not use tap water in the cooling system. Use distilled water to mix antifreeze.*

Table f. Cooling system capacities	
Engine	**Cooling system capacity in liters (US qt)**
2.7 liter V6 (APB, BEL)	6.0 (6.3)
2.8 liter V6 (AHA, ATQ)	8.0 (8.5)
3.0 liter V6 (AVK)	8.0 (8.5)
4.2 liter V8 (ART, AWN, BBD)	9.0 (9.5)
4.2 liter V8 biturbo (RS6) (BCY)	12.0 (12.7)

Coolant level, checking

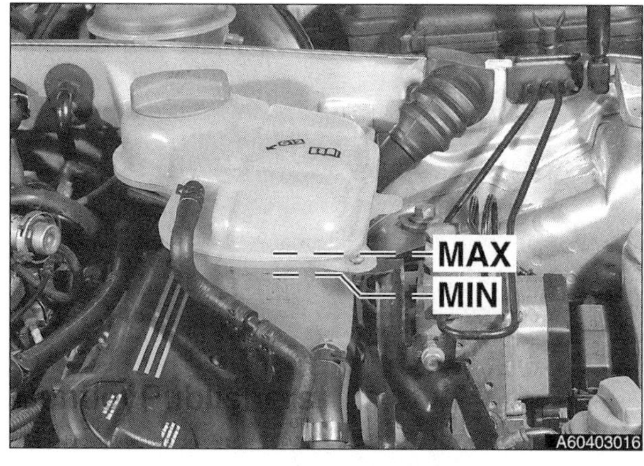

A translucent expansion tank on left side of engine compartment provides easy monitoring of coolant level without opening the system. Check coolant level when engine is cold because level rises as engine warms up. Make sure coolant level is between **MIN** and **MAX** marks.

Coolant / antifreeze concentration, checking

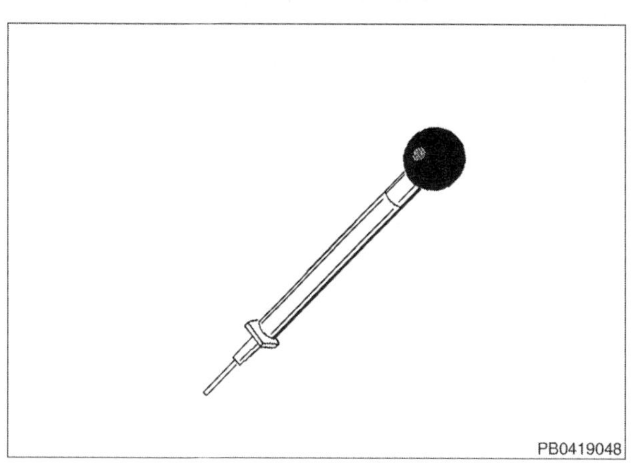

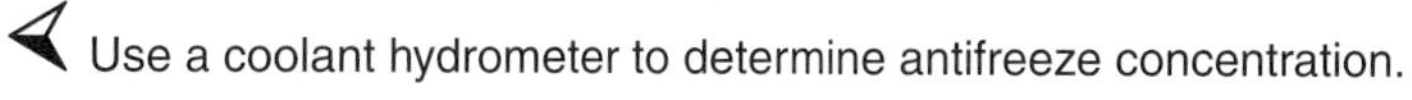

Use a coolant hydrometer to determine antifreeze concentration.

Coolant mixture recommendations	
Concentration	**Cold protection**
50% antifreeze	-35°C (-31°F)
60% antifreeze	-40°C (-40°F)

Do not use a higher concentration of antifreeze than a 60% mixture, as the heat transfer quality of the coolant decreases with higher antifreeze concentrations.

BentleyPublishers.com
BentleyPublishers.com—All Rights Reserved

Cooling system hoses, inspecting

Inspect hoses by first checking that all connections are tight and dry. Coolant seepage indicates that either the hose clamp is loose, hose is damaged, or connection is dirty or corroded. Dried coolant has a chalky appearance.

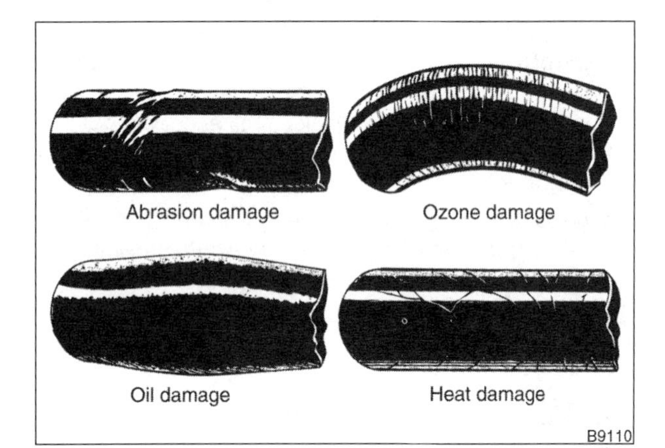

Check hose condition by a visual and tactile inspection, making sure they are firm and springy. Replace hoses that exhibit conditions noted below.

- Leakage: Dripping, moisture, or seepage near clamps or connectors.
- Electromechanical degradation: Difficult to see, but detectable by squeezing hose and feeling for cracks, weak areas, and voids.
- Oil damage: Soft and spongy to touch, visible bulges and swelling.
- Abrasion damage: Wear, abrasion, or scuffing, often due to contact with components in engine compartment.
- Heat damage: Internal and external damage generally due to high underhood temperatures or overheating. Internal heat damage is often indicated by swelling with external damage marked by hardened and cracked areas.
- Ozone damage: Small, parallel cracks in outer layers, but without hardening. Due to exposure to atmospheric conditions.

When installing or reinstalling hoses, use clamps specified by Audi. Some clamp types do not allow for heat expansion and contraction and often leak when hot or cold.

For additional cooling system information and repair procedures, see **19 Engine—Cooling System**.

BRAKE FLUID

Brake fluid level, checking

WARNING —
- *Brake fluid is poisonous. Do not ingest brake fluid. Wash thoroughly with soap and water if brake fluid comes into contact with skin.*

CAUTION—
- *Use only new, previously unopened brake fluid conforming to US Standard FMVSS 116 DOT 4.*
- *Do not let brake fluid come in contact with paint. Wash immediately with soap and water.*
- *Brake fluid absorbs moisture from the air. Store in an airtight container.*
- *Do not use DOT 5 (silicone) brake fluid.*
- *Do not fill brake fluid above* **MAX** *in fluid reservoir.*
- *Do not mix mineral oil products such as gasoline or engine oil with brake fluid. Mineral oil damages rubber seals in the brake system.*
- *Observe hazardous waste regulations when disposing brake fluid as a hazardous waste.*

BentleyPublishers.com
BentleyPublishers.com—All Rights Reserved

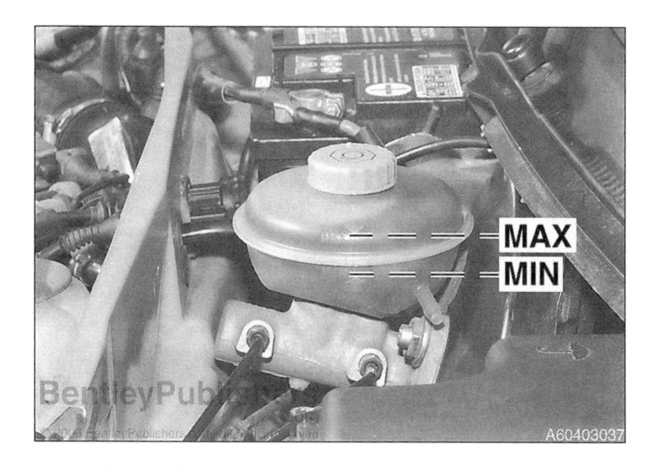

Routine maintenance of the brake system includes maintaining an adequate level of brake fluid in reservoir.

< Check that fluid level in brake fluid reservoir at left rear of engine compartment is between **MIN** and **MAX** marks. Fluid level drops slightly as brake pad material wears.

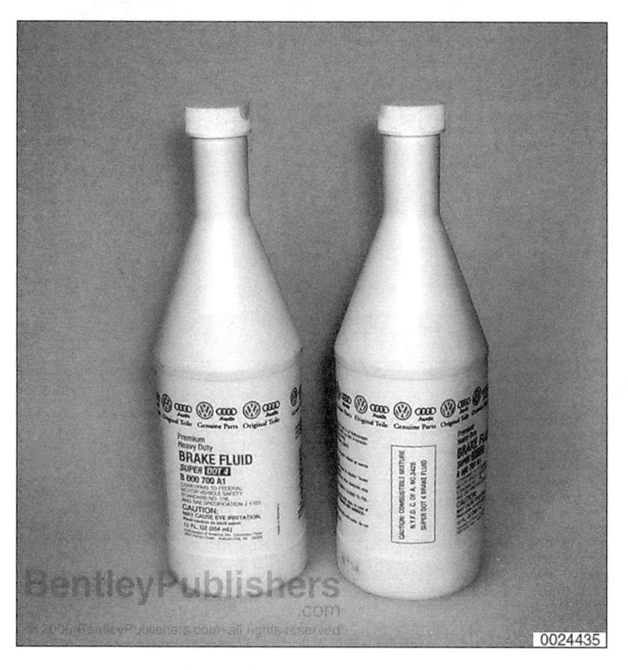

< Audi specifies use of brake fluid that meets Department of Transportation (DOT) 4 Super specification in addition to Federal Motor Vehicle Safety Standard (FMVSS) 116 and Society of Automotive Engineers (SAE) standard J1703. DOT 4 Super is also known as DOT 4+. Brake fluid meeting this specification is also available in 30 and 50 liter containers for use in pressure bleeders.

Brake fluid specifications	
Audi specification	DOT 4 Super (DOT 4+)
Minimum wet boiling point	329°F (165°C)
Dry boiling point (exceeds)	500°F (260°C)

Additional service items include replacing brake fluid every 2 years, checking brake pads for wear, checking parking brake function and inspecting system for fluid leaks or other damage. See also:

- **Under Car Maintenance** in this repair group for pad inspection and brake system inspection.
- **46 Brakes–Mechanical** for brake pad and disc service.
- **47 Brakes–Hydraulic** for brake system bleeding and fluid change.

Power Steering

> ***CAUTION—***
> - *Use only Audi hydraulic oil in the power steering system. Do not use ATF or other non-approved types of power steering fluid. If the wrong fluid is used, power steering components may fail.*
> - *Do not reuse drained power steering fluid.*

Power steering fluid	
Audi specification	G 002 000

Power steering fluid level, checking

Power steering hydraulic fluid can be checked when either hot or cold.

– Cold power steering fluid: Do not start engine. Turn front wheels to straight-ahead position.

BentleyPublishers.com
BentleyPublishers.com—All Rights Reserved

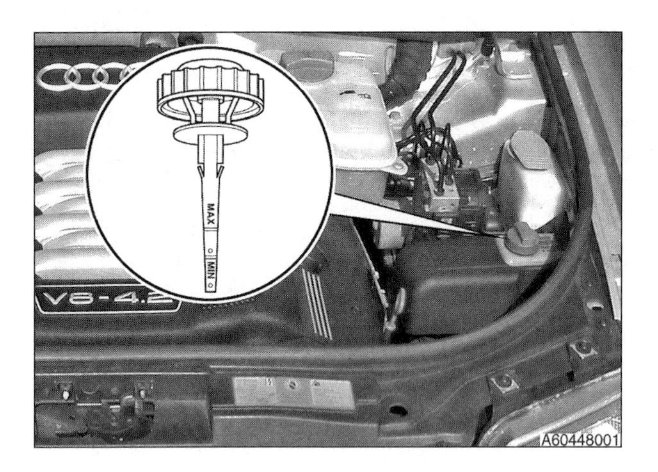

– Power steering fluid at operating temperature (over 50°C or 122°F): Start engine and steer front wheels to straight-ahead position.

< Remove power steering fluid reservoir filler cap (**inset**). Dipstick is attached to cap.

– Wipe dipstick with a clean lint-free shop cloth.

– Screw on cap hand tight and remove.

– Read level on stick:
  - If fluid is cold, correct level is ± 2 mm (± 0.08 in) of MIN mark
  - If fluid is at operating temperature, correct level is between MIN and MAX marks.

– If cold fluid level is more than 2 mm (0.08 in) above MIN mark, extract some fluid.

– If fluid level is more than 2 mm (0.08 in) below MIN mark, check power steering hydraulic system for leaks. It is not sufficient to top up with fluid.

FUEL FILTER

The fuel filter does not have a specified replacement interval. Replace it as conditions or the situation dictates.

> ***WARNING —***
> - *The fuel system is designed to retain pressure even when the ignition is OFF. When working with the fuel system, loosen the fuel lines slowly to allow residual fuel pressure to dissipate. Avoid spraying fuel. Use shop towels to capture leaking fuel.*
> - *Before beginning work on the fuel system, place a fire extinguisher in the vicinity of the work area.*
> - *Fuel is highly flammable. When working around fuel, do not disconnect wires that could cause electrical sparks. Do not smoke or work near heaters or other fire hazards.*
> - *Wear eye protection and protective clothing to avoid injuries from contact with fuel.*
> - *Unscrew the fuel tank cap to release pressure in the tank before working on fuel lines.*
> - *Do not use a work light with an incandescent bulb near fuel. Fuel may spray on the hot bulb causing a fire.*
> - *Make sure the work area is properly ventilated.*

BentleyPublishers.com
BentleyPublishers.com—All Rights Reserved

Fuel filter, replacing

The illustration in this procedure applies to all V6 engines and V8 engines for A6, S6 and allroad quattro vehicles. Also see **Fuel filter, replacing (RS6)**.

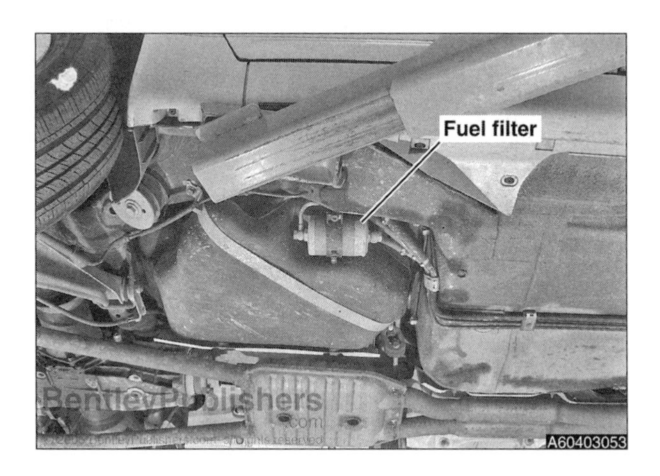

In vehicles covered by this procedure, the fuel filter is underneath the vehicle in front of the right rear wheel.

– Raise car and support safely.

> ***WARNING —***
> - *Make sure the car is stable and well supported at all times.*

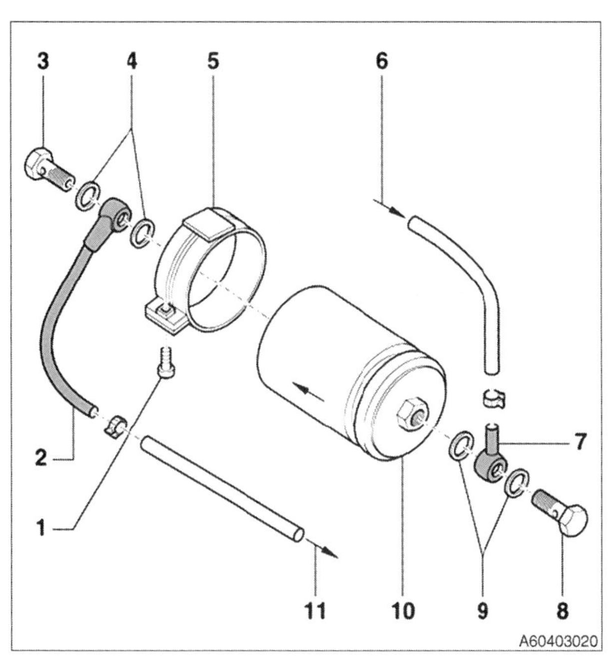

Fuel filter mounting and layout for V6 and V8 engines except V8 biturbo (RS6 models):

1. Clamping screw, tighten to 2 Nm (18 in-lb)
2. Fuel supply to engine
3. Banjo bolt, tighten to 23 Nm (17 ft-lb)
4. Sealing O-rings, replace
5. Fuel filter mounting bracket
6. Fuel line from fuel pump in tank
7. Fuel line flange
8. Banjo bolt, tighten to 23 Nm (17 ft-lb)
9. Sealing O-rings, replace
10. Fuel filter, underneath vehicle, in front of right rear wheel
11. Fuel supply to engine

– Before loosening fuel line banjo bolt, wrap a shop towel around connection.

– Counterhold fuel filter, then loosen banjo bolts.

> ***CAUTION—***
> - *Place fuel-resistant container underneath work area to catch spilling fuel.*

– Remove fuel filter bracket clamping screw. Slide fuel filter out of clamp.

– When installing:
- Note that arrow on fuel filter shows direction of fuel flow.
- Use new sealing O-rings.
- Counterhold fuel filter when tightening banjo bolts.

Tightening torques	
Fuel filter clamping screw	2 Nm (18 in-lb)
Fuel line to fuel filter	23 Nm (17 ft-lb)

– Check for fuel leaks upon completion of repair.

BentleyPublishers.com
BentleyPublishers.com—All Rights Reserved

Fuel filter, replacing (RS6)

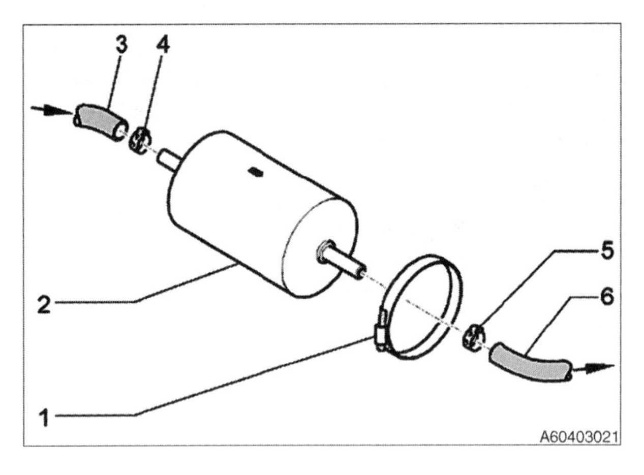

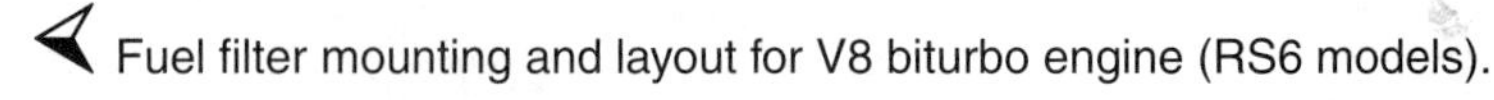

Fuel filter mounting and layout for V8 biturbo engine (RS6 models).

1. Hose clamp
   - Attaches fuel filter to fuel tank
   - Tighten to 0.8 Nm (7 in-lb)
2. Fuel filter
   - Underneath vehicle, at right front of fuel tank, above fuel pump
3. Fuel line from fuel pump
4. Hose clamp
5. Hose clamp
6. Fuel supply line to engine

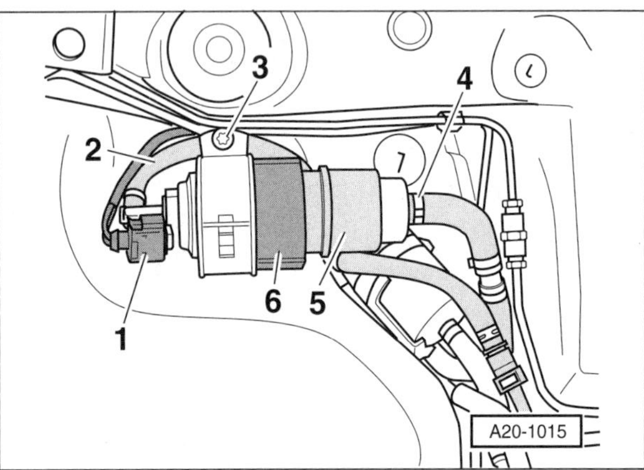

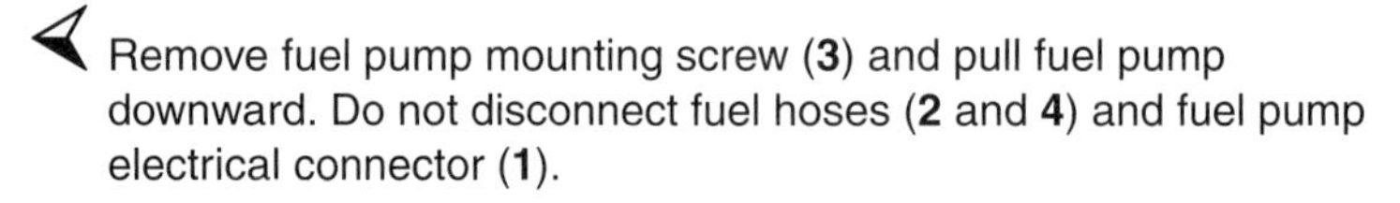

Remove fuel pump mounting screw (**3**) and pull fuel pump downward. Do not disconnect fuel hoses (**2** and **4**) and fuel pump electrical connector (**1**).

– Before loosening fuel line clamps, wrap a shop towel around connection.

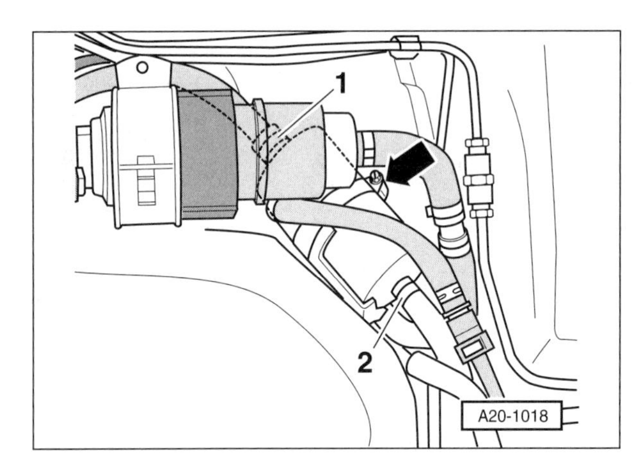

Loosen hose clamps (**1** and **2**) and separate lines from fuel filter carefully.

> ***CAUTION—***
> - *Place fuel-resistant container underneath work area to catch spilling fuel.*

– Loosen hose clamp (**arrow**). Slide fuel filter out of clamp.

– When installing:
- Note that arrow on fuel filter shows direction of fuel flow.
- Use new hose clamps.

Tightening torques	
Fuel filter mounting clamp	0.8 Nm (7 in-lb)
Fuel pump mounting bracket	1.2 Nm (11 in-lb)

– Check for fuel leaks upon completion of repair.

BentleyPublishers.com
BentleyPublishers.com—All Rights Reserved

Battery Service

The battery is in the plenum chamber, behind the engine.

Open engine hood, release battery cover latches (**arrow**s) and flip cover up to gain access to battery.

Routine maintenance of the battery consists of visual checks of battery hold-down bracket, cable clamps, and fluid (electrolyte) level.

Some batteries have a charge indicator (magic eye) on top that displays electrolyte level and charge condition.

- Green: Charge and electrolyte level are OK.
- Black: There is insufficient or no charge.
- Yellow or colorless: Electrolyte level is critically low. Top off with distilled water immediately.

NOTE—

- *If the battery is more than 5 years old and the charge indicators are colorless, replace battery.*
- *Air bubbles that occur normally during battery charging or during vehicle operation may adversely affect charge indicator reading. To obtain an accurate reading, gently tap the charge indicator with a screwdriver handle to displace air bubbles that have formed.*
- *Maintenance-free batteries are equipped with sealing plugs with plastic foil.*
- *When replacing battery, use a replacement battery with central gas venting.*

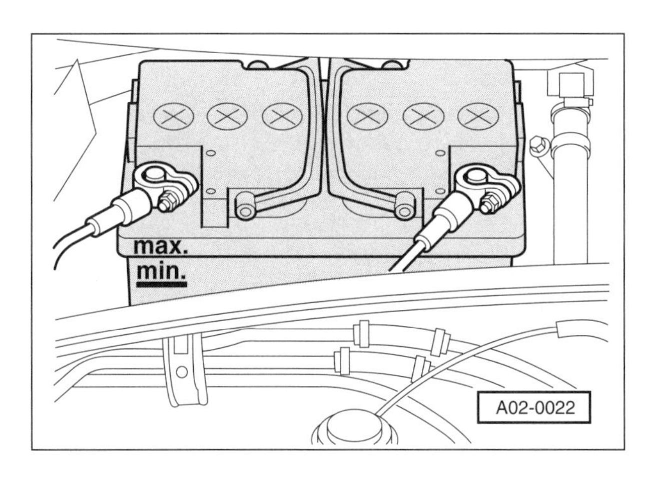

At high outside temperatures, check electrolyte level through translucent battery housing. Make sure electrolyte level is just above battery plates and separators.

- Remove filler caps to see battery plates. Make sure electrolyte level aligns with internal electrolyte level indicator (lip). This equates to external **min** and **max** markings on battery case.

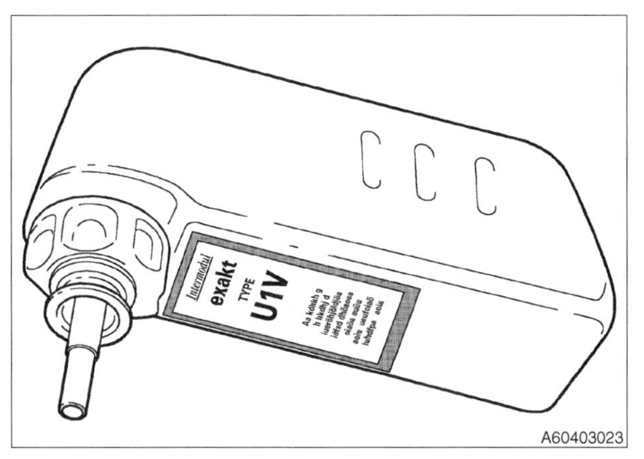

For a battery with removable electrolyte caps:

- If electrolyte level is low, use battery filling bottle VAS 5045 or equivalent to replenish it.
- Add distilled water only. Distilled water prevents electrolyte impurities which cause self-discharging.
- Make sure battery cell caps are equipped with O-ring seals.

CAUTION—

- *Do not overfill battery. Overfilled batteries can boil over.*
- *Too little electrolyte reduces the service life of the battery.*
- *Extract excess electrolyte using a hydrometer.*
- *Dispose of electrolyte (sulfuric acid) as hazardous waste. Refer to local regulations pertaining to electrolyte disposal.*

BentleyPublishers.com
BentleyPublishers.com—All Rights Reserved

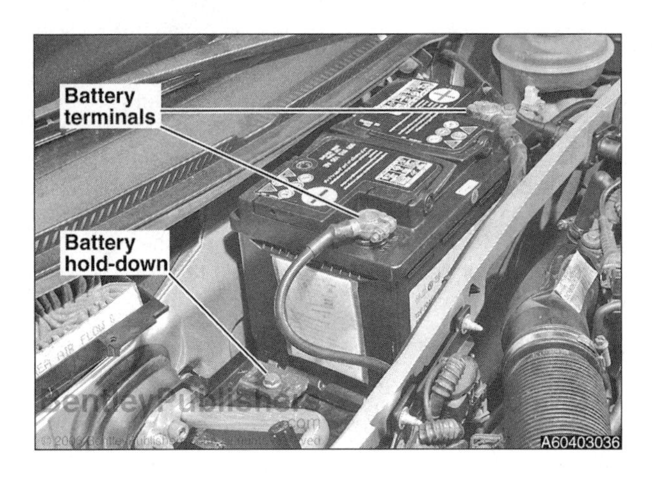

During battery service, check the following:

- Battery securely bolted down.
- Battery terminal clamping bolts tight.

Tightening torques	
Bolts, positive and negative cable clamps	9 Nm (80 in-lb)
Bolt, battery hold-down bracket	20 Nm (15 ft-lb)

– Make sure battery vent hose is routed correctly.

See **27 Battery, Alternator, Starter** for battery testing and installation procedures.

Battery notes

If the battery is disconnected and reconnected, reset or reinitialize the following systems or components:

- Clock
- Power window motor limit position (one-touch operation)
- Seat and outside rear view memory
- Heating and A/C automatic setting
- Radio

See **Battery reconnection notes** in **27 Battery, Alternator, Starter** for procedures necessary to reinitialize components.

ENGINE ACCESSORY BELT

The engine uses one multi-ribbed belt (serpentine belt) to drive engine accessories. Although belts of this design are known for their smooth running, high efficiency and long life, periodic inspection is required.

Accessory belt, checking

– Remove upper and lower engine covers. See **Engine covers, removing** in this repair group.

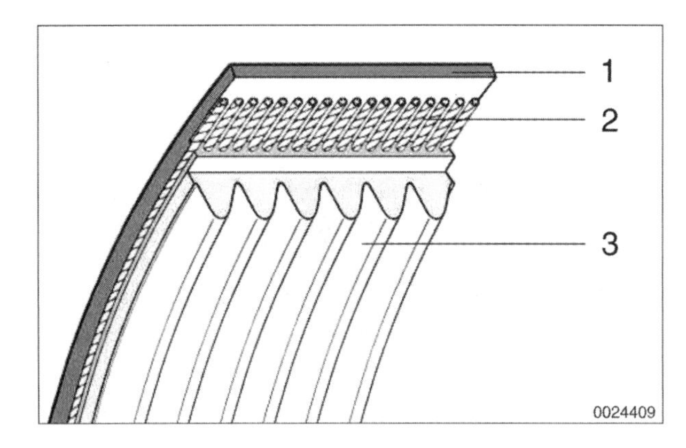

Cross section of accessory belt.

1. Top layer, cover
2. Tension cords
3. Base material (substrata)

– Check belt periodically. Manually turn engine at vibration damper or pulley with a suitable wrench. Inspect belt from above or below for the following conditions:

- Splits, core fractures and cross section fractures.
- Separation between layers, especially between top cover and tension cord layers.
- Chunking or breaking of base or lower layer material.
- Fraying of tension cords.
- Flank (side) wear including flaking, fraying, chunking, glazing, hardening, and surface cracking.
- Oil and grease contamination.

BentleyPublishers.com
BentleyPublishers.com—All Rights Reserved

– If any of above conditions exist, replace belt. Also, consult maintenance tables at end of this repair group for appropriate routine replacement intervals.

Accessory belt, replacing (2.7 liter or 2.8 liter V6 engine)

– Remove lower engine cover. See **Engine covers, removing** in this repair group.

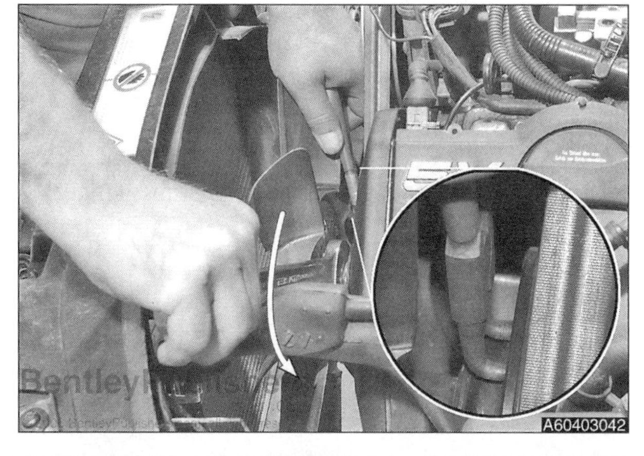

Use dual pin tool (Audi special tool 3212 or equivalent) (**inset**) to counterhold viscous fan pulley. Use 32 mm open end wrench (Audi special tool 3312) to loosen (**arrow**) viscous fan.

> ***CAUTION—***
> *• Viscous fan threads are left-handed. Turn clockwise to remove.*

– Lift out viscous fan.

Use 17 mm box wrench to rotate tensioner in direction of **arrow**. When hole in tensioner lines up with front engine cover, insert steel drift to lock tensioner in released mode.

– If belt is to be reused, such as for an emergency spare, mark direction of rotation prior to removal.

– Remove belt.

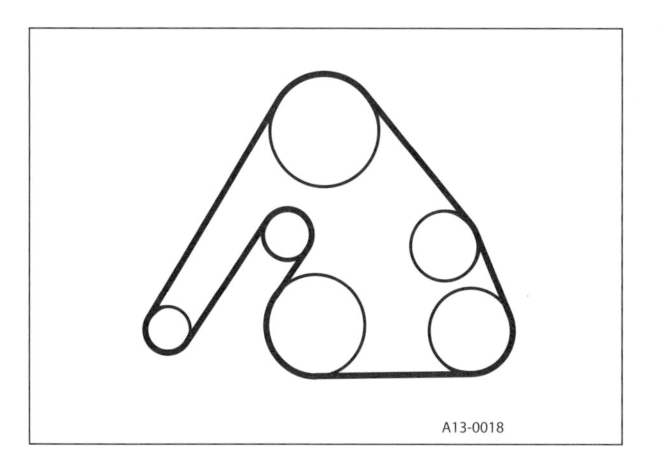

When reinstalling belt:

- Note correct belt routing as illustrated.
- If reusing belt, install in direction of rotation marked previously.
- Be sure belt is seated correctly in pulley grooves.
- Position belt on A/C compressor pulley last.

– Use 17 mm box wrench to release tension and remove steel locking drift.

– Reinstall viscous fan. Note that viscous fan threads are left-handed.

Tightening torque	
Viscous fan to fan pulley: • Using 3312 open end wrench • Without 3312 open end wrench	 37 Nm (27 ft-lb) 70 Nm (52 ft-lb)

– Run engine briefly to make sure belt routing is correct. Then reinstall engine covers.

BentleyPublishers.com
BentleyPublishers.com—All Rights Reserved

Accessory belt, replacing (3.0 liter V6 engine)

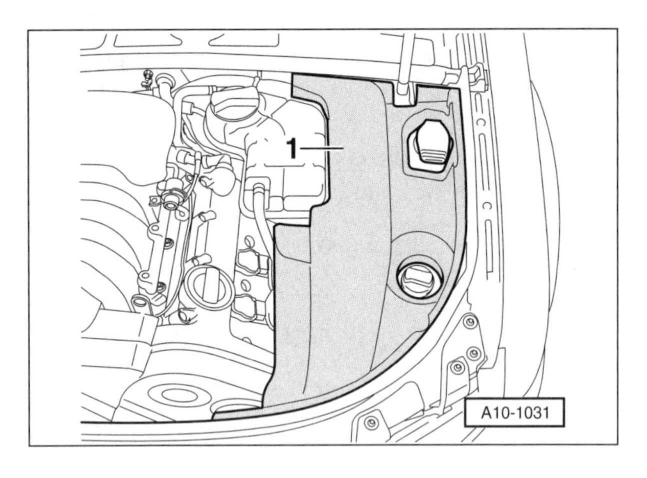

◀ Remove cover (**1**) in engine compartment (left side).

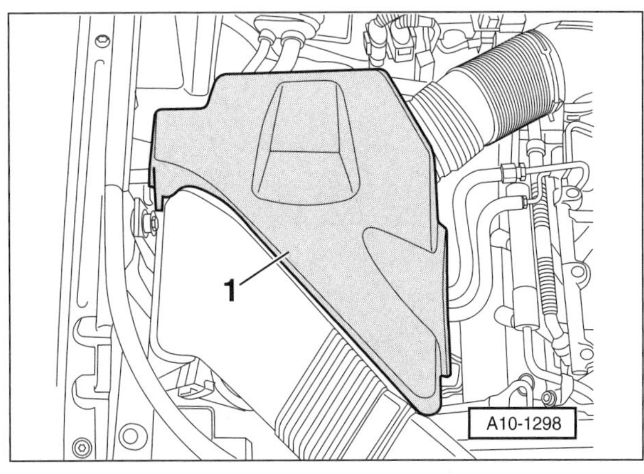

◀ Remove air filter housing cover (**1**) in engine compartment (right side).

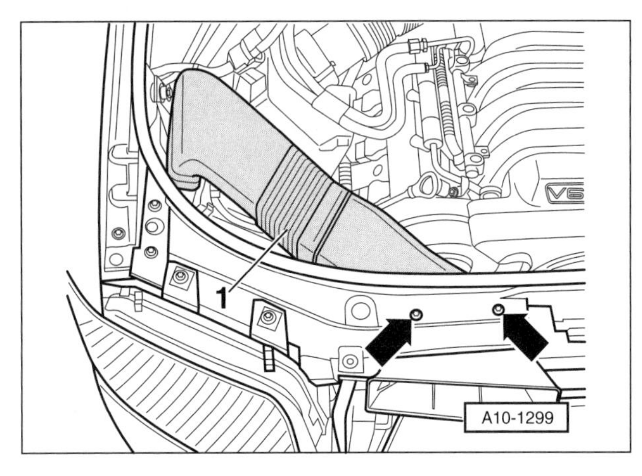

◀ Remove air duct mounting bolts (**arrows**) at lock carrier (radiator support frame). Remove air duct (**1**).

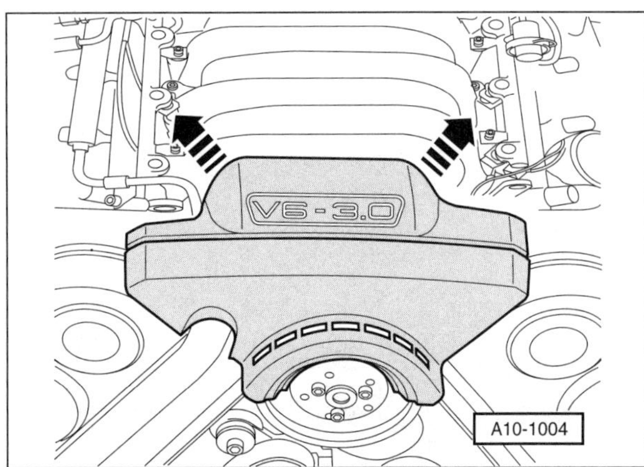

◀ Remove front engine cover (**arrows**).

BentleyPublishers.com
BentleyPublishers.com—All Rights Reserved

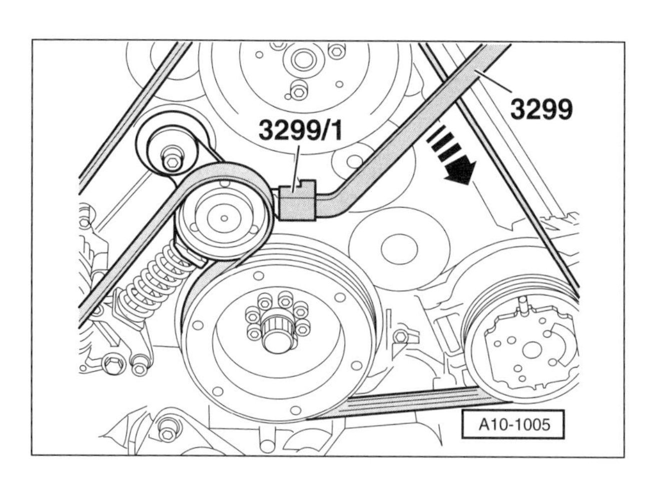

< Use special tools 3299 and 3299/1 to swing belt tensioner in direction of **arrow** to release belt tension.

– If belt is to be reused, such as for an emergency spare, mark direction of rotation prior to removal.

– Remove belt from pulleys.

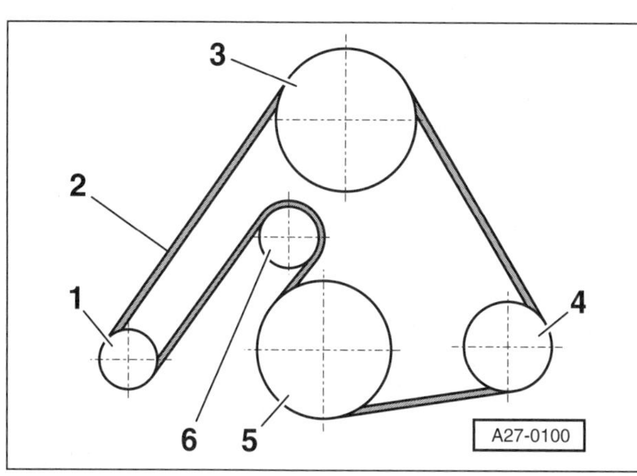

< When installing belt:

- First install belt on crankshaft pulley (**5**).
- Slide belt on tensioner (**6**) last.
- Note correct belt routing as illustrated.
- Be sure belt is seated correctly in pulley grooves.

1. Alternator (generator)
2. Accessory belt
3. Power steering pump
4. A/C compressor
5. Crankshaft pulley
6. Tensioner

– Run engine briefly to make sure belt routing is correct. Then reinstall engine covers.

BentleyPublishers.com
BentleyPublishers.com—All Rights Reserved

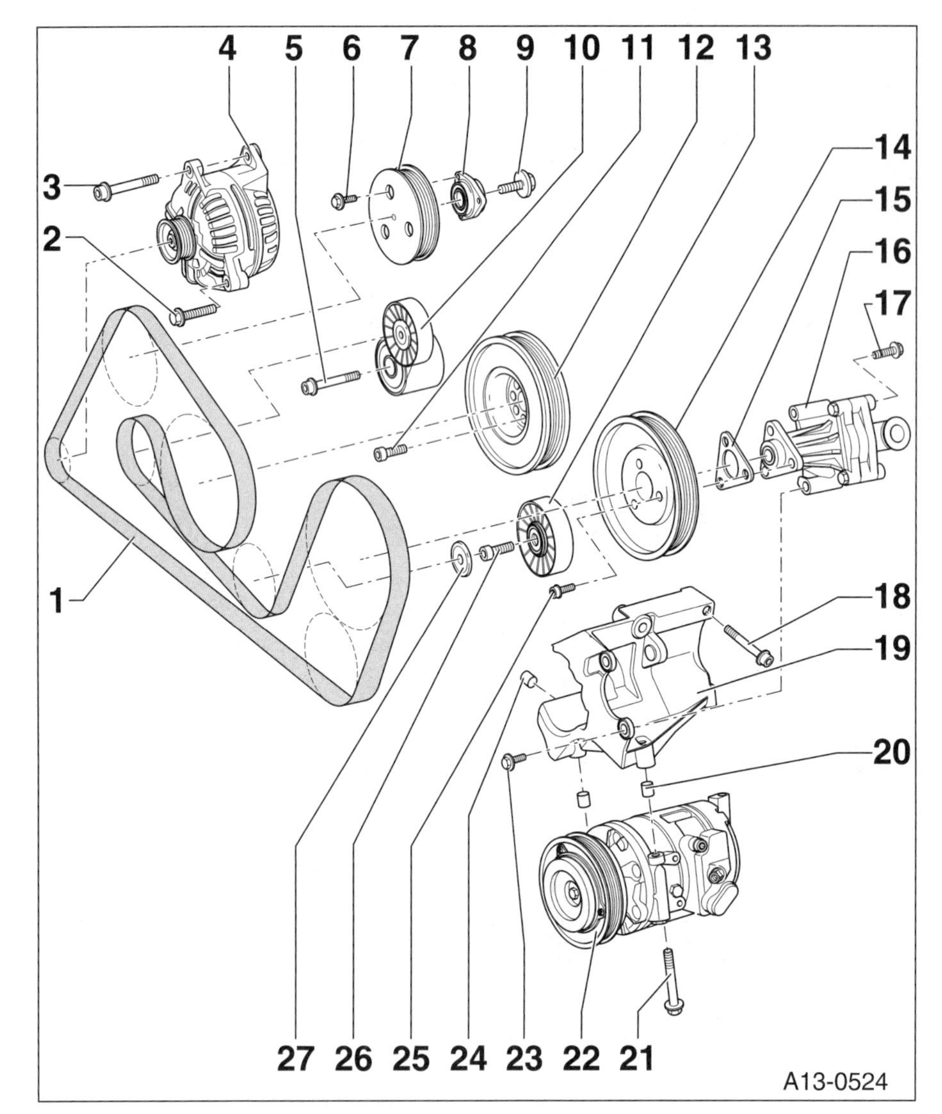

Accessory belt and components (V8 engine, *not* allroad quattro)

1. **Engine accessory belt (serpentine belt), mark direction of rotation before removing**
2. **Alternator mounting bolt, 22 Nm (16 ft-lb)**
3. **Alternator mounting bolt, 45 Nm (33 ft-lb)**
4. **Alternator**
5. **Belt tensioner mounting bolt, 43 Nm (32 ft-lb)**
6. **Upper idler mounting bolt, 10 Nm (7 ft-lb)**
7. **Upper idler pulley**
8. **Upper idler bearing**
9. **Upper idler bearing mounting bolt**
   - Apply thread locking compound when installing. Tighten to 22 Nm (16 ft-lb).
10. **Belt tensioner**
    - Rotate clockwise with 19 mm wrench to release tension
11. **Vibration damper mounting bolt, 22 Nm (16 ft-lb)**
12. **Front pulley (vibration damper)**
13. **Lower idler pulley**
14. **Power steering pump pulley**
15. **Spacer (shim)**
16. **Power steering pump**
17. **Power steering pump mounting bolt**
    - 22 Nm (16 ft-lb)
18. **Power steering pump mounting bracket bolt, 40 Nm (30 ft-lb)**
19. **Power steering pump bracket**
20. **Dowel sleeve**
21. **A/C compressor mounting bolt, 22 Nm (16 ft-lb)**
22. **A/C compressor**
23. **Mounting bolt, 22 Nm (16 ft-lb)**
24. **Dowel sleeve**
25. **Power steering pump pulley mounting bolt, 22 Nm (16 ft-lb)**
26. **Lower tensioner pulley mounting bolt, 23 Nm (17 ft-lb)**
27. **Trim cover**

BentleyPublishers.com
BentleyPublishers.com—All Rights Reserved

Accessory belt, replacing (V8 engine, *not* allroad quattro)

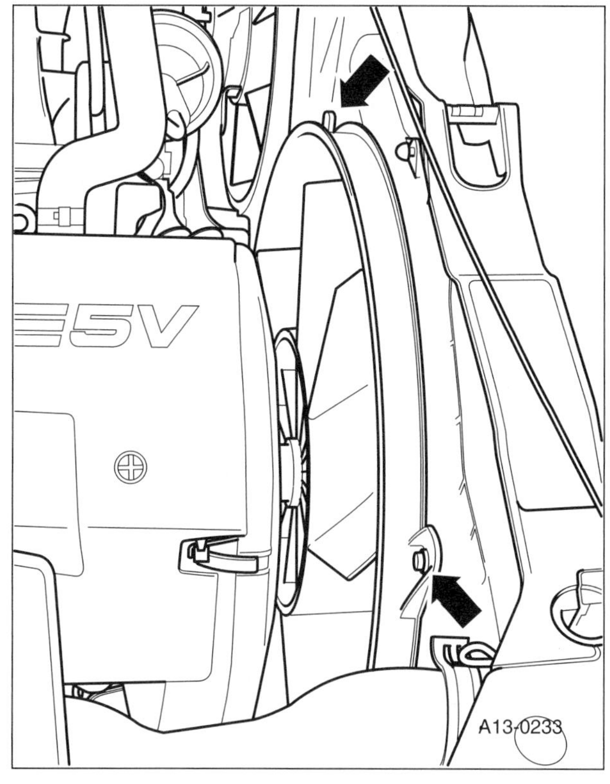

To gain access to engine accessory belt (serpentine belt), remove front bumper cover and place lock carrier (radiator mounting frame in front of engine) in service position

- Raise car and support safely.

> ***WARNING —***
> - *Make sure the car is stable and well supported at all times. Use a professional automotive lift or jack stands designed for the purpose. A floor jack is not adequate support.*

- Remove lower engine cover. See **Engine covers, removing** in this repair group.
- Remove front bumper and bumper cover. See **63 Bumpers**.
- Place lock carrier in service position. See **50 Body-Front**.

◄ Remove mounting screws (**arrows**) for viscous fan outer housing. Remove housing.

- Remove electric fan.

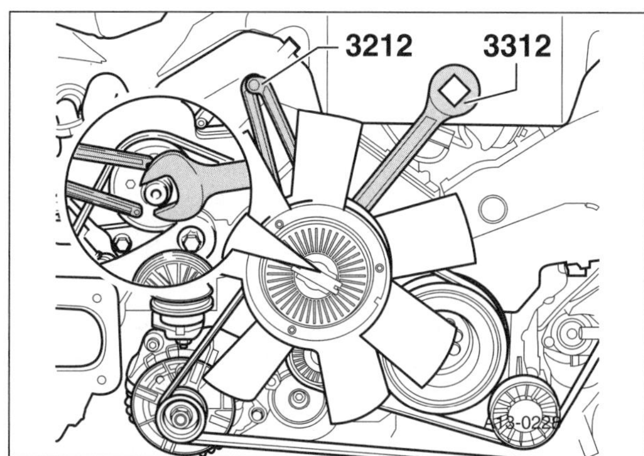

◄ use Audi special tool 3212 (pin wrench) to counterhold viscous fan pulley. Use open end wrench (special tool 3312) to loosen viscous fan.

> ***CAUTION—***
> - *Viscous fan threads are left-handed. Turn clockwise to remove.*

- Lift out viscous fan.

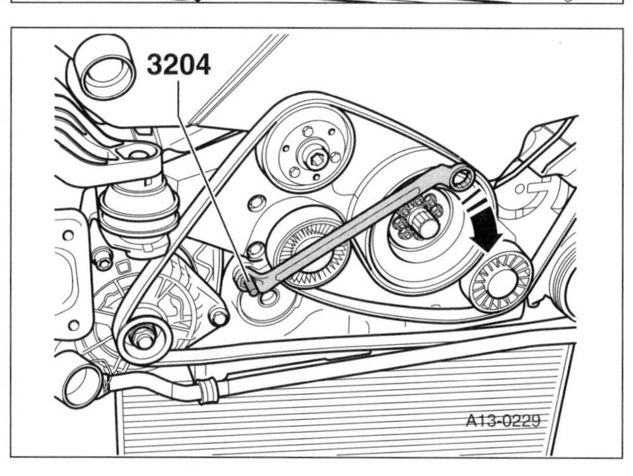

◄ Use 19 mm box wrench to rotate tensioner in direction of **arrow**. When hole in tensioner lines up with hole in front engine cover, insert steel drift (special tool 3204) to lock tensioner in released mode.

- If belt is to be reused, such as for an emergency spare, mark direction of rotation prior to removal.
- Remove belt.

BentleyPublishers.com
BentleyPublishers.com—All Rights Reserved

Engine Accessory Belt

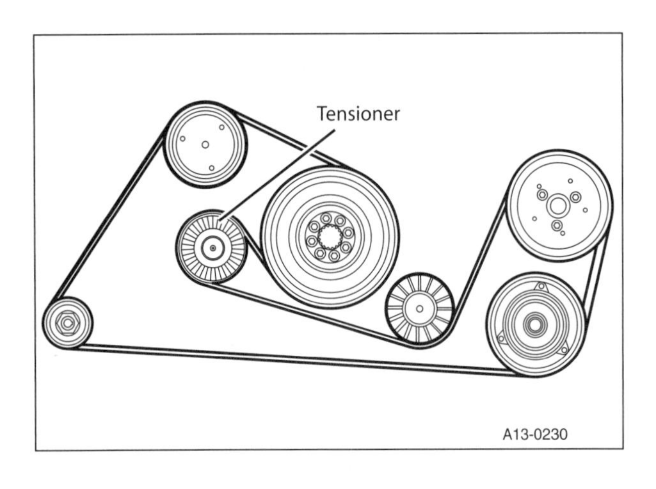

When reinstalling belt:

- Note correct belt routing in illustration.
- If reusing belt, install in direction of rotation marked previously.
- Be sure belt is seated correctly in pulley grooves.
- Place belt on vibration damper and idler pulleys first. Position belt on tensioner pulley last.

– Use 19 mm box wrench to release tension and remove steel drift.

– Reinstall viscous fan. Note that viscous fan threads are left-handed.

Tightening torques	
Fan shroud to radiator	10 Nm (7 ft-lb)
Viscous fan to fan pulley: • Using 3312 open end wrench • Without 3312 open end wrench	 37 Nm (27 ft-lb) 70 Nm (52 ft-lb)

– Run engine briefly to make sure belt routing is correct. Then reinstall lock carrier, front bumper and engine covers.

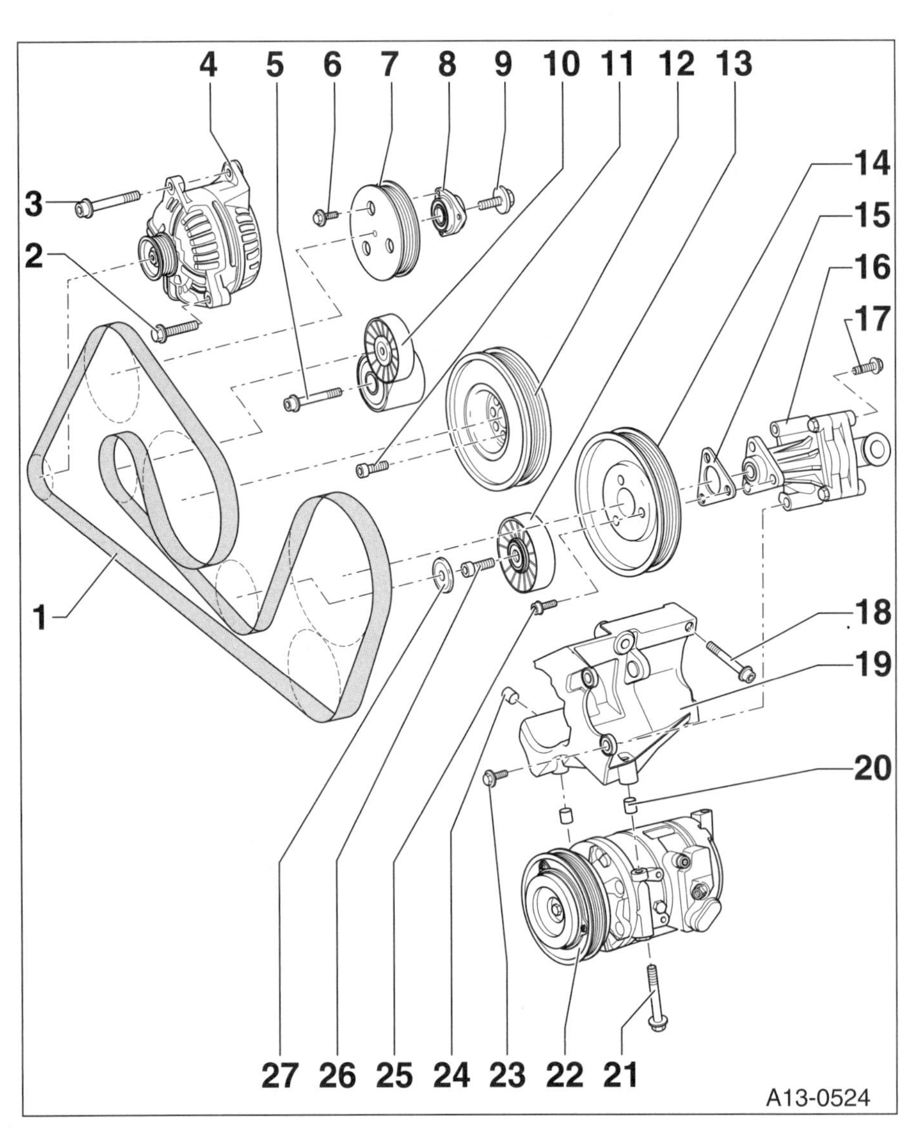

Alternator belt and components (V8 engine, allroad quattro)

1. **Belt tensioner mounting bolt, 22 Nm (16 ft-lb)**
2. **Belt tensioner lock bolt, 22 Nm (16 ft-lb)**
3. **Belt tensioner**
4. **Alternator mounting bolt, 22 Nm (16 ft-lb)**
5. **Alternator**
6. **Alternator bracket mounting bolt, M8: 22 Nm (16 ft-lb), M10: 46 Nm (34 ft-lb)**
7. **Bushing**
8. **Alternator bracket**
9. **Idler pulley mounting bracket bolt, 10 Nm (7 ft-lb)**
10. **Idler pulley mounting bracket**
11. **Front pulley (vibration damper)**
12. **Vibration damper mounting bolt, bolt class 12.9, 42 Nm (31 ft-lb)**
13. **Idler pulley**
14. **Idler pulley mounting bolt, 22 Nm (16 ft-lb)**
15. **Trim cover**
16. **Alternator belt (serpentine belt), mark direction of rotation before removing**
17. **Threaded bushing**

BentleyPublishers.com
BentleyPublishers.com—All Rights Reserved

Alternator belt, removing and installing (V8 engine, allroad quattro)

To gain access to alternator accessory belt (serpentine belt), remove front bumper cover and place lock carrier (radiator mounting frame in front of engine) in service position.

– Raise car and support safely.

> ***WARNING —***
> - *Make sure the car is stable and well supported at all times. Use a professional automotive lift or jack stands designed for the purpose. A floor jack is not adequate support.*

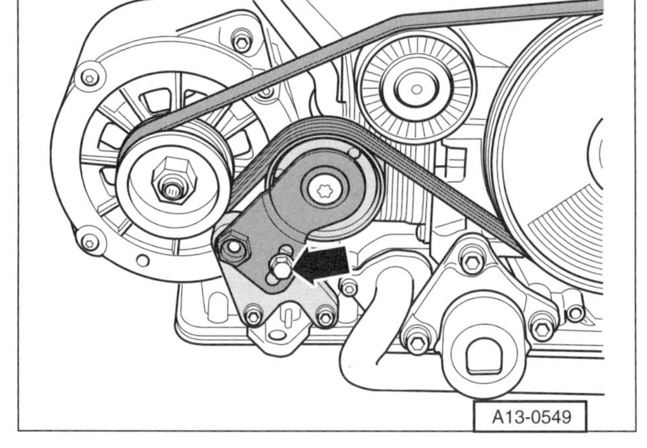

– Remove lower engine cover. See **Engine covers, removing** in this repair group.

– Remove front bumper and bumper cover. See **63 Bumpers**.

– Place lock carrier in service position. See **50 Body-Front**.

◄ Loosen tensioner lock bolt (**arrow**) and remove belt.

– If reusing belt, install in direction of rotation marked previously.

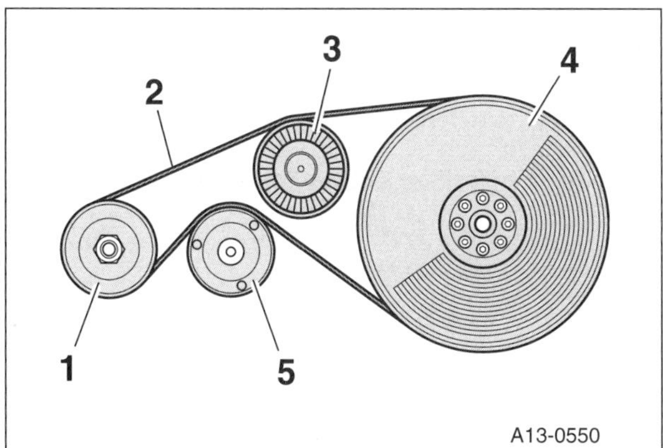

◄ Place belt over pulley in specified sequence:

1. Alternator
2. Belt
3. Idler pulley
4. Vibration damper
5. Tensioner pulley

– Be sure belt is seated correctly in pulley grooves.

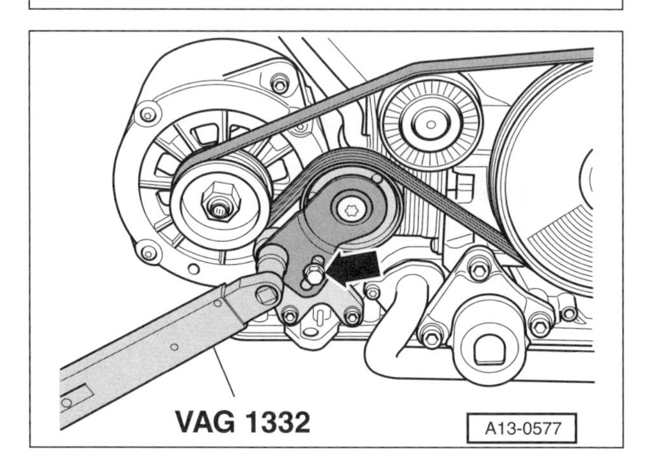

◄ Use torque wrench (special tool VAG 1332 or equivalent) to rotate tensioner and pretension alternator belt. Then tighten tensioner lock bolt.

Tightening torques	
Alternator belt pretension	72 ± 2 Nm (54 ± 2 ft-lb)
Tensioner lock bolt	22 Nm (16 ft-lb)

– Run engine for a short time and make sure belt routing is correct.

– Reset belt pretension. Then reinstall lock carrier, front bumper and engine covers.

BentleyPublishers.com
BentleyPublishers.com—All Rights Reserved

TIMING BELT (TOOTHED BELT)

For timing belt replacement interval, see **Maintenance Schedules** in this repair group.

Timing belt replacement is covered in **13 Timing Belt, Engine Pulley, Rear Main Seal**.

> ***CAUTION—***
> - *To avoid costly damage caused to engine mechanical components if the timing belt breaks, periodically inspect the timing belt for wear, cracks or other damage every 50,000 miles.*

SPARK PLUGS

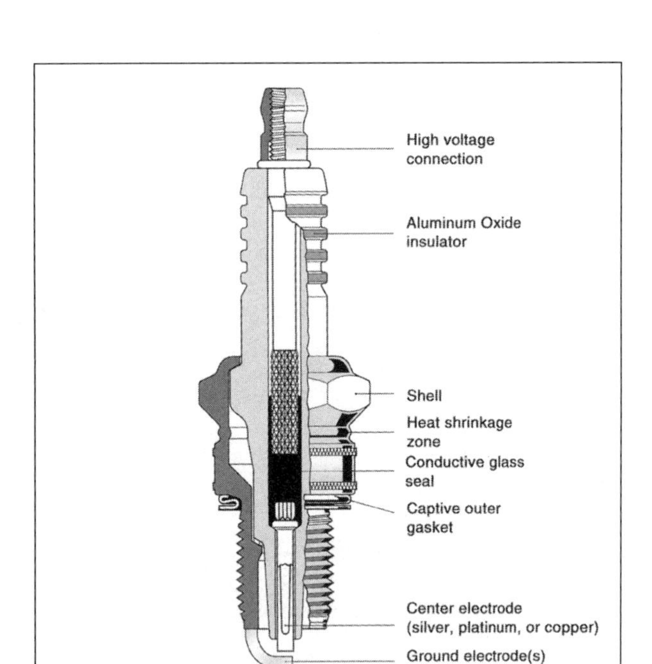

Audi specifies spark plugs made by several different manufacturers. **Table g** is a listing of current spark plug recommendations. Engines are referred to by configuration (V6 or V8), displacement and engine code. For a listing of engine codes and applications, see **Table b** in this repair group.

Because specifications and part numbers can change, check with an authorized Audi dealer parts department or aftermarket parts specialist for the latest application information.

Spark plugs, replacing

– Remove upper engine cover. See **Engine covers, removing** in this repair group.

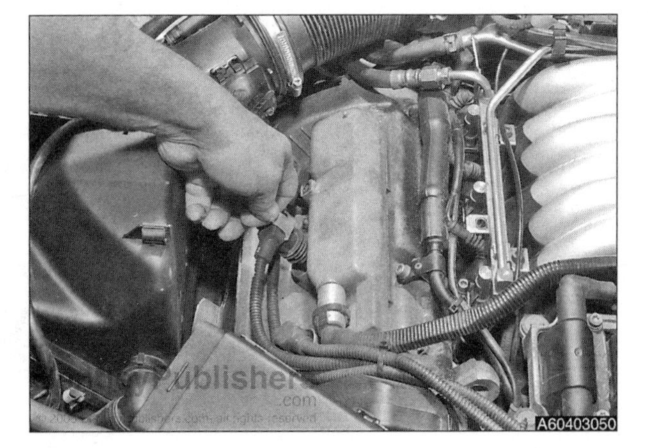

◀ Engine with spark plug wires: Use special plastic puller (stored under cylinder head plastic cover) to pull spark plug wire ends off spark plugs.

BentleyPublishers.com
BentleyPublishers.com—All Rights Reserved

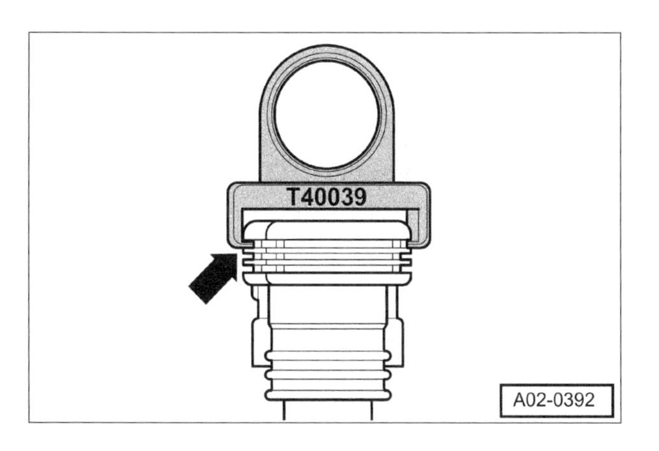

◀ Engine with coils over spark plugs: Attach special tool T40039 (puller) to uppermost thick rib (**arrow**) of each coil. Pull coils off spark plugs.

> ***CAUTION—***
> *• Lower ribs of coil may be damaged if they are pried for removal.*

– If special coil removal tool is not available, use 2 screwdrivers to gently pry each coil off spark plug.

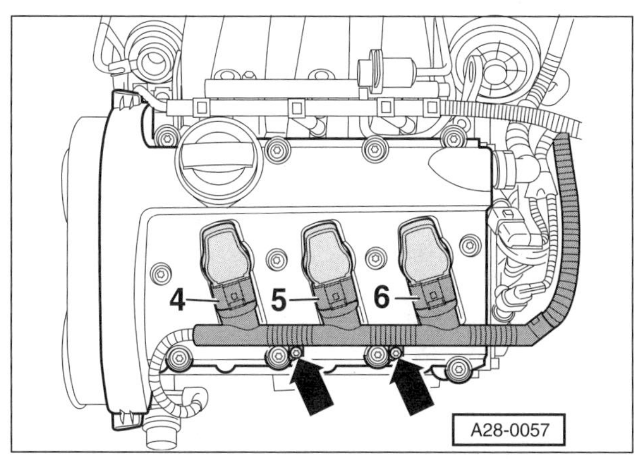

◀ Disconnect harness connectors (**4**, **5**, **6**) and pull all connectors out of ignition coils simultaneously.

– Pull coils out of spark plug holes and set aside.

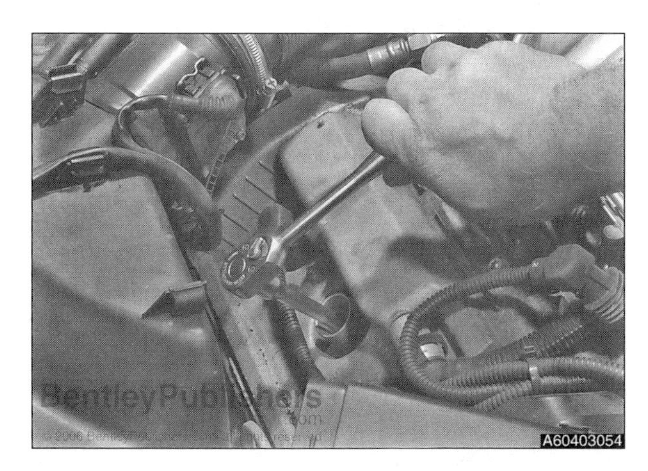

◀ Remove and install spark plugs with 16 mm or 5/8 inch spark plug socket (special tool 3122B or equivalent). See **Table g** for spark plug applications.

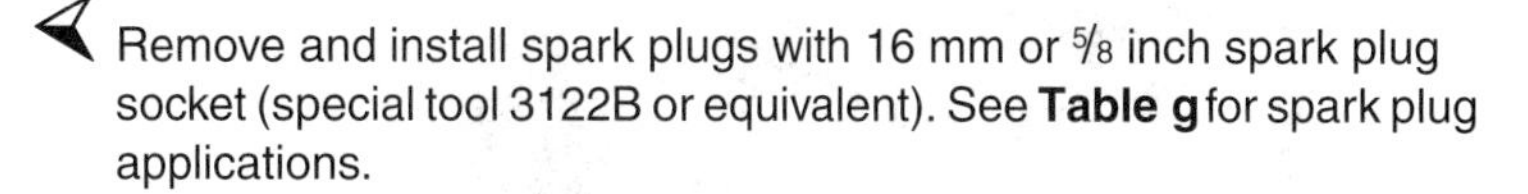

Tightening torque	
Spark plug to cylinder head	30 Nm (22 ft-lb)

– Engine with spark plug wires: Reinstall wires on spark plugs. Press firmly for a positive connection.

– Engine with coils over spark plugs: Insert ignition coils loosely into spark plug shafts.

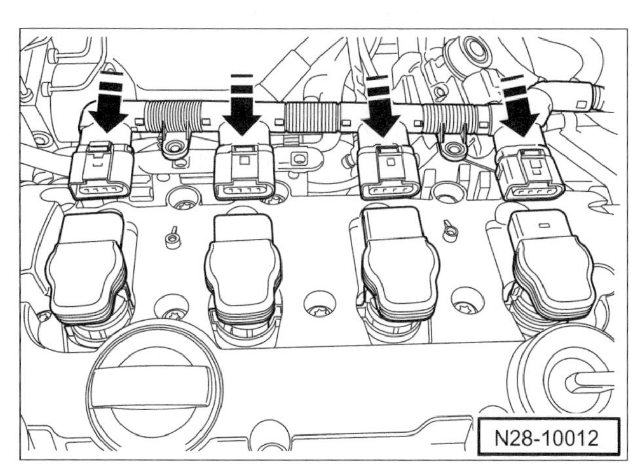

◀ Align ignition coils to connectors (**arrows**) and connect all connectors to ignition coils simultaneously.

– Align ignition coils to recesses in cylinder head cover.

– Press ignition coils evenly on spark plugs by hand.

– Install upper engine cover.

BentleyPublishers.com
BentleyPublishers.com—All Rights Reserved

Table g. Spark plug applications			
Year, engine	**Bosch spark plug**	**NGK spark plug**	**Gap**[1]
1998 - 2001 V6 2.8 liter (AHA, ATQ)	F7LTCR, FGR7DQE0[2]	BKR6EKUB	AHA: 1.6 mm (0.064 in) ATQ: 0.4 - 0.6 mm (0.016 - 0.024 in)
2000 - 2002 V6 2.7 liter (APB)	-	BKR6ES	0.8 mm (0.032 in)
2002 - 2003 V6 2.7 liter (BEL)	FR7DPP222T	BKR6ES[3]	0.8 mm (0.032 in)
2004 V6 2.7 liter (BEL)	FR7DPP33	BKR6ES	0.8 mm (0.032 in)
2002 - 2003 V6 3.0 liter (AVK)	FR7DPP332	BKR6ES-11[4]	1.1 mm (0.044 in)
2004 V6 3.0 liter (AVK)	FR7DPP22U	BKR6ES-11	1.1 mm (0.044 in)
2000 V8 (ART)	HGR7KQC	BKR6EKUB	1.6 mm (0.064 in)
2001 - 2003 V8 (AWN, BAS)	FGR7KQE	BKR6ES	Not adjustable
2002 - 2004 V8 (BBD)	FGR6KQE	BKR6ES	Not adjustable
2004 V8 (AWN, BBD, BAS)	FGR7KQE0	BKR6ES	Not adjustable
2003 - 2004 V8 biturbo (BCY)	-	BKR7E[3]	0.8 mm (0.032 in)

Notes:
1. Spark plugs gapped by the manufacturer during production; However, check gaps before installation.
2. Part numbers current at time of publication, but can change due to supersessions, etc.
3. -11 at end of part number indicates that spark plug is pregapped at factory to 1.1 mm (0.044 in).
4. Standard NGK spark plug listed. See NGK web site for additional listings.

AIR FILTER

Air filter element, replacing

The air filter housing is in the right front corner of the engine compartment. In the procedure that follows, 1999 2.8 liter V6 engine compartment is illustrated. Layout of other engine compartments is similar.

- Where applicable, pull up cover (**A**).
- Detach intake air duct (**2**).

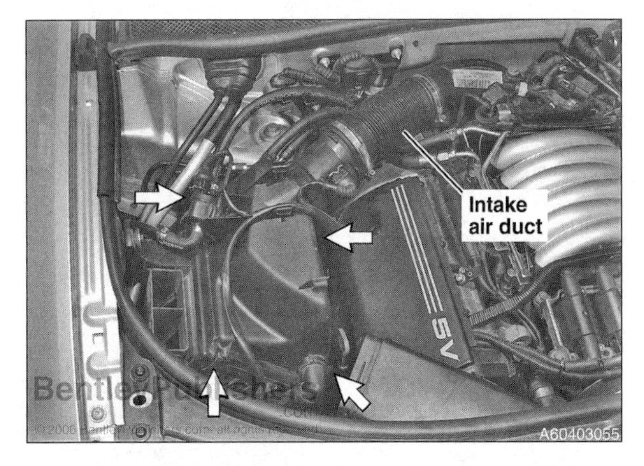

- Unsnap air filter housing cover retaining clips (**arrows**) and swing cover to side. If necessary remove intake air duct.
- Remove old air filter element and discard.
- Clean air filter housing and install new filter element.

NOTE—
- *Make sure sealing surfaces on inner and outer parts of air filter housing are positioned flush and seated correctly.*

- Snap filter housing halves together using retaining clips.

BentleyPublishers.com
BentleyPublishers.com—All Rights Reserved

UNDER CAR MAINTENANCE

Transmission and final drive oil

For transmission and final drive lubricant information see the following repair groups:

- **34 Manual Transmission**
- **37 Automatic Transmission**
- **39 Final Drive**

Brake system, visual inspection

– Brake system items to check:
  - Check power brake servo (booster) and brake master cylinder.
  - Check ABS hydraulic unit.
  - Check front and rear brake calipers, hoses and brake lines for leaks and damage.
  - Inspect parking brake cables and handle for proper operation and damage.
  - Check that brake hoses are not twisted.
  - Check that brake hoses do not touch any part of the vehicle when steering is at full lock and suspension is at limits.
  - Check hoses and lines for porosity, deterioration and chafing.
  - Check brake connections and attachments for correct seating, leaks and corrosion.

Correct problems noted during the visual inspection immediately. For parking brake adjustment, see **46 Brakes–Mechanical**.

Brake pads, checking

Periodic maintenance of the brake system includes inspection of front and rear brake pads for thickness.

Outer brake pad thickness can usually be estimated by looking through wheel openings. For inner brake pad thickness, use flashlight and mirror to inspect. If in doubt, remove wheel for more thorough inspection.

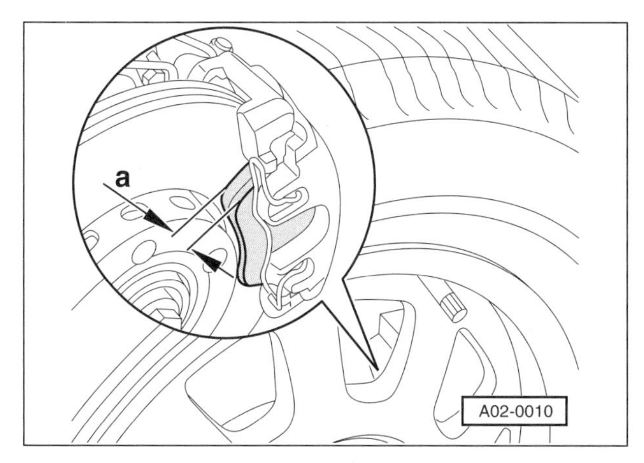

 Front disc brake outer pad.

Brake pad wear limit (incl. pad backing plate)	
Minimum pad thickness (**a**)	7 mm (0.28 in)

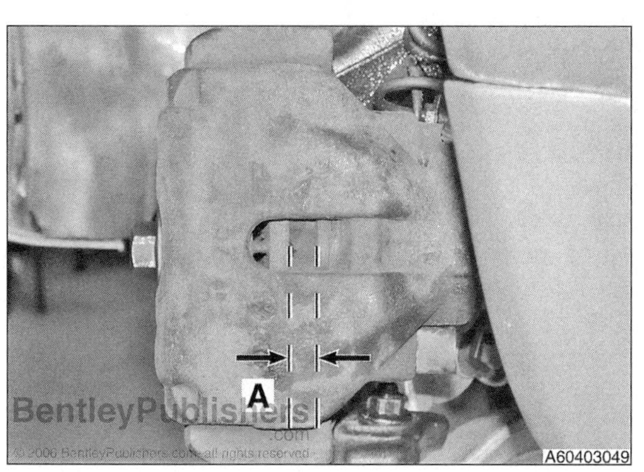

 With front wheel off, check inner front pad thickness (**A**) through caliper opening.

Brake pad wear limit (friction material only)	
Minimum pad thickness (**A**) • A6, S6, allroad quattro • RS6	 approx. 2 mm (0.08 in) approx.3 mm (0.12 in)

BentleyPublishers.com
BentleyPublishers.com—All Rights Reserved

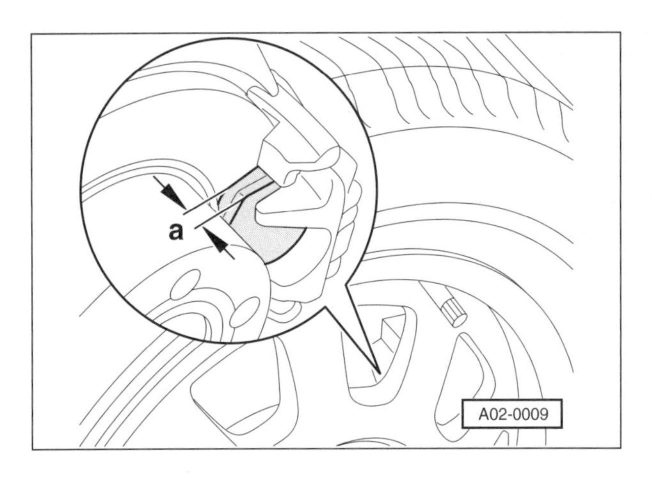

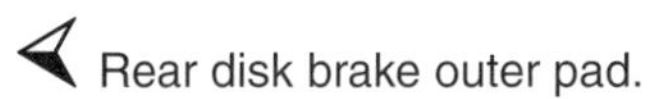
Rear disk brake outer pad.

Brake pad wear limit (incl. pad backing plate)	
Minimum pad thickness (**a**)	7 mm (0.28 in)

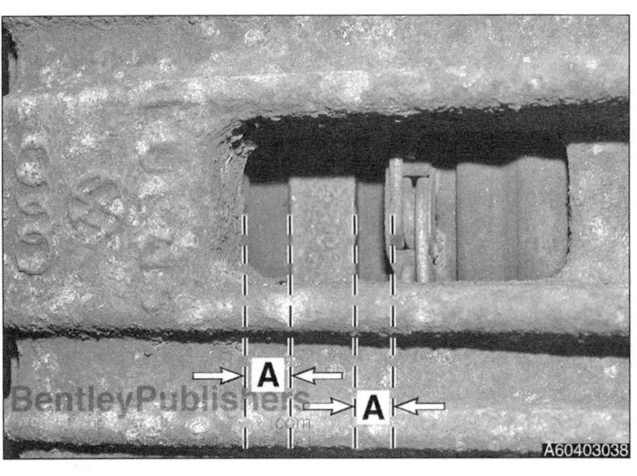

With rear wheel off, check rear pad thickness (**A**) through rear caliper opening.

Brake pad wear limit (friction material only)	
Minimum pad thickness (**A**) • A6, S6, allroad quattro • RS6	 approx. 2 mm (0.08 in) approx.3 mm (0.12 in)

– Replace pads if they are below minimum thickness. See **46 Brakes–Mechanical**.

Tire and wheel service

Tire maintenance

Inspect tires at least as often as every maintenance service. Check tire pressures more often. Remember that tire pressures change with temperature changes. Check tire pressures when the tires are cold due to the normal pressure rise associated with heat generated by driving. Tire pressures vary among vehicles with different engine types and between different model years. Refer to data label on B or C pillar for proper inflation pressures. See **Identification Plates and Labels** at the beginning of this section. Be sure to also check spare tire pressure.

Replace dust caps on tire valves after checking tire pressures. If valve extensions are used (as on steel wheels and wheel covers), inspect extensions for damage and replace as required.

Tire inspection includes the following:

- Check tire tread surface and side walls for signs of damage. Remove foreign objects from tread.
- Check that tires are the same type, size, and tread pattern.
- Measure tread depth. If tread wear has exceeded minimum specification listed, replace tire.
- Check for scuffing, cupping, feathering, irregular tread wear, sidewall checking, dry rot, cuts and fractures.
- Also check rims (both steel and alloy) for damage.

BentleyPublishers.com
BentleyPublishers.com—All Rights Reserved

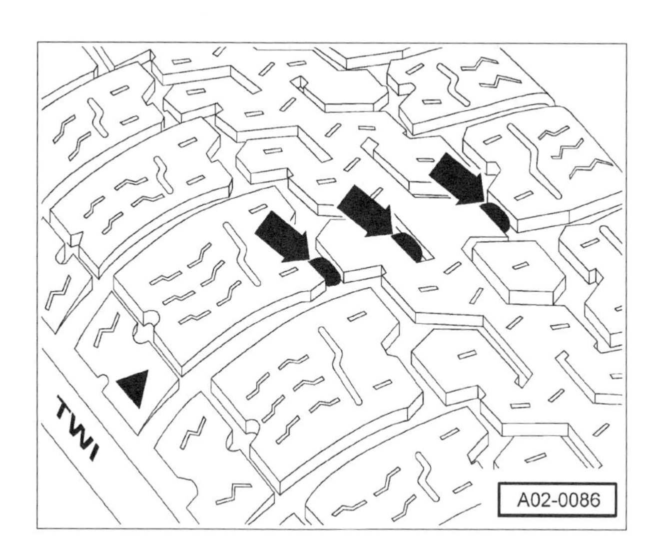

Tires have tread wear indicators (TWI) spaced around the tire, perpendicular to the tread, at marked intervals (**arrows**). Note bold **TWI** lettering and location **triangle** on sidewall at location of TWI bands.

When the TWIs start to become visible, the tire is reaching the wear limit. When the tops of the tread wear indicators are even with the tread, replace the tire.

Some tire manufacturers use symbols in place of TWI lettering on sidewalls; however, TWI bands are still at that location on tire.

New summer and all-season tires generally have $^{10}/_{32}$ of an inch tread depth while snow tires generally have $^{14}/_{32}$ of an inch (tread depth is measured in 32nds of an inch). Tread depth can, however, vary considerably among different designs and manufacturers.

Specifications below are minimum as specified by vehicle and tire manufacturers. State, provincial, and local laws may specify a different tread depth minimum.

Tire wear limit	
Tire tread depth, minimum remaining: • Summer & all-season • Snow	 1.6 mm (0.063 or $^{2}/_{32}$ in) 4.0 mm (0.157 or $^{5}/_{32}$ in)

Snow tires (or winter tires) have a deep lug tread pattern. Be sure to use on all 4 wheels. They are not generally suitable for warm weather use and have different speed ratings.

Because winter tires improve vehicle handling on snow and ice, use them when temperatures below 45°F (7°C) are expected. Also, increase tire pressure by 3 psi (0.2 bar). Consult the tire manufacturer and owner's manual for additional information and applications.

If irregular wear patterns are identified, determine causes by checking wheel alignment (front and rear) and inspecting suspension for damage and worn components.

Vibrations felt through the steering wheel may be an indication of tire out of balance condition. This can shorten tire life, cause wear on steering and suspension components, and often results in an uncomfortable ride. Tire balancing is the usual solution.

Old tires (6 years or more) may experience internal tread separation and not function or perform as originally intended. Use such old tire with caution. Audi specifies replacement after this time period regardless of tread wear.

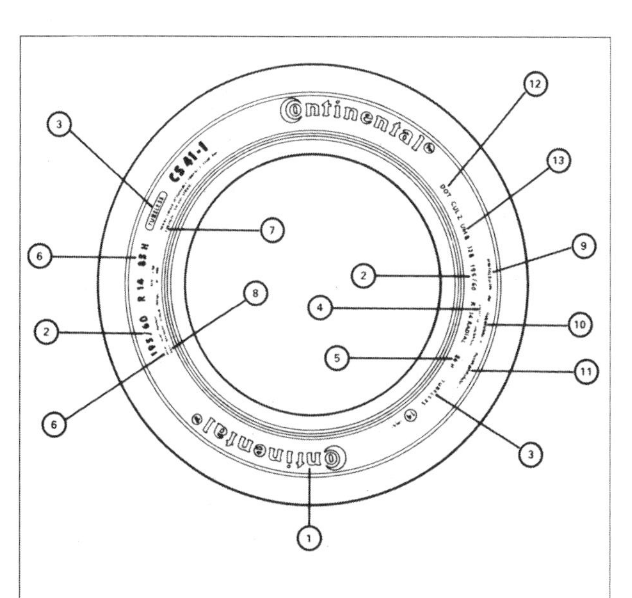

1. Manufacturers brand
2. Tire dimensions (size)
3. Type (tubeless)
4. Country of origin
5. Load index and speed rating
6. Load range (A, B, C)
7. Tread & sidewall ply and construction
8. Maximum cold inflation pressure
9. Treadwear index number
10. Traction index number (AA, A, B, C)
11. Temperature index number (A, B, C)
12. USA DOT compliant
13. Manufacturers production coding

0024416

Overview of typical information found on tire sidewall that complies with United States Federal Motor Vehicle Safety Standard (FMVSS) 109.

When installing new tires, observe the following:

- Audi specifies that only radial tires be used.
- For safety reasons, replace tires in pairs and on the same axle. However, replacement of all tires at the same time is preferred.
- Balance new tires and fit rims (wheels) with new valve stems.
- New tires tend to be slippery and require a break-in period of at least 350 miles (560 km).

BentleyPublishers.com
BentleyPublishers.com—All Rights Reserved

- Acquire new tires of the same size as originally installed. If you change sizes, be aware that gear ratios and speedometer readings will change.
- For safety reasons, install tires of the same type and tread pattern on a vehicle.
- On quattro (AWD) vehicles, be sure to use the same size tires on all four wheels.
- Do not use tires of a lesser speed rating or load range than originally installed.
- Rebalance wheel and tires as they accumulate mileage due to settling and wear.
- Make sure that wheel covers are installed correctly so as to not cut valve stems and extensions.

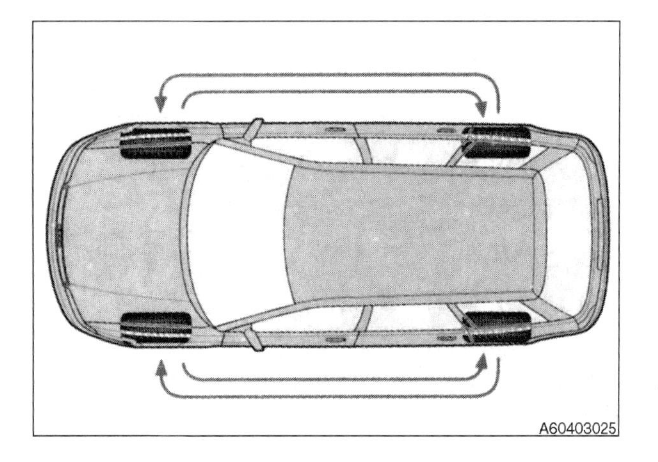

Tire rotation

Rotate tires periodically for maximum service life and quiet running. Tire rotation also equalizes wear between tires. This keeps rolling circumference the same which is important to some vehicle systems, especially in quattro (AWD) models.

Audi specifies tire rotation at the first 5000 mile service and then defers to the tire manufacturers recommendations from that point on. These recommendations can vary slightly among manufacturers, but generally, rotation at 5000 to 7500 mile intervals results in optimal wear.

Audi specifies that under most conditions, tires remain on the same side of vehicle when they are rotated (front to back).

Wheel maintenance

Visually inspect steel and alloy wheels for damage such as dents and missing balance weights. Ensure that all wheel bolts are present and tight.

When removing and installing road wheels (alloy and steel), note the following points to be sure that wheels remain secure at all times:

- Tighten wheel bolts in crisscross pattern and to specified torque. Excessive torque can distort wheel bolt seat in wheel.
- When installing a wheel, do not torque down first wheel bolt immediately (for instance, with an impact wrench). This may prevent other bolts from centering the wheel properly when they are threaded in. The rim could then come loose, even though bolts have been tightened to specified torque.
- Do not apply grease or oil to wheel bolt threads.
- Before installing wheels, examine seats of wheel bolts and contact surfaces between rims and hubs for rust, corrosion and paint. Clean off these parts if necessary.

> ***WARNING —***
> - *If the procedures above are not observed, wheel mountings can come loose and wheel bolt seats can become distorted. This distortion may be very slight and not visible to the naked eye. Even this slight distortion of wheel bolt seats can prevent rims from being held firmly and cause the wheels to come loose.*

BentleyPublishers.com
BentleyPublishers.com—All Rights Reserved

Suspension components, checking

– Raise car and support safely.

> ***WARNING —***
> *• Make sure the car is stable and well supported at all times. Use a professional automotive lift or jack stands designed for the purpose. A floor jack is not adequate support.*

– Check wear in front suspension components:
  - Move wheels and tie-rods and check for free play.
  - Any play indicates wear. Repair as necessary.

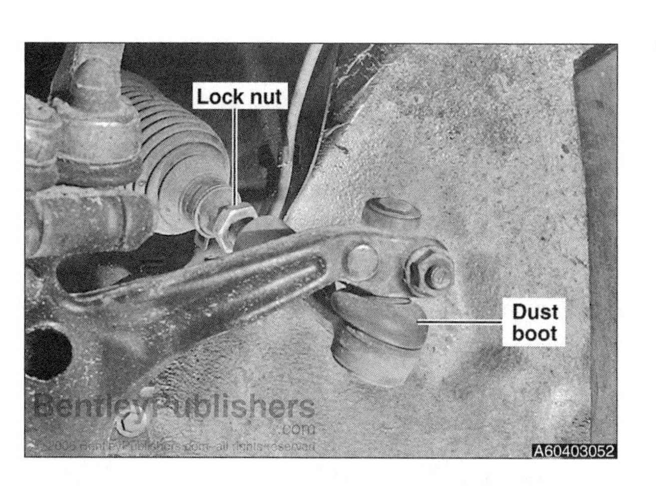

 Check tightness of tie-rod lock nut.

Tightening torque	
Lock nut to tie-rod end	40 Nm (30 ft lb)

– Check dust boot for damage and correct installation.

– Inspect shock absorbers for excessive oil leakage past seals.

– Inspect coil springs for cracks and damage.

– Inspect front and rear suspension bushings for cracks or deterioration.

– Inspect front and rear inner and outer constant velocity (CV) joint boots for leakage, cuts and damage. Ensure that boots are correctly seated.

See **40 Front Suspension** and **42 Rear Suspension** for suspension component replacement.

Underbody visual inspection

– Inspect the following for leaks or damage:
  - Engine
  - Transmission (manual and automatic)
  - Fuel system
  - Cooling and heating systems
  - Brake system
  - Exhaust system

A small amount of dampness is considered normal in some cases, especially around axle and pulley seals since the leaking fluid helps the seal work properly. On the other hand, expensive repairs can be avoided by prompt repair of minor fluid leaks. Judgement and experience are required to distinguish among the different kinds of fluid leaks.

– Inspect underside of vehicle for damage caused by normal wear and tear or by driving over road debris. Whenever vehicle is raised on a lift, inspect underbody, wheel wells and sill or rocker panels for damage to underbody sealants and coatings. Also inspect after major repairs to vehicle systems.

BentleyPublishers.com
BentleyPublishers.com—All Rights Reserved

– Repair damage or defects found. Only use wax-based or tar-based anti-corrosion compounds as specified. Do not use oil-based anti-corrosion sprays due to possible incompatibility with factory applied protection. Check with an authorized Audi dealer's parts department.

Body and Interior Maintenance

Airbags, visual inspection

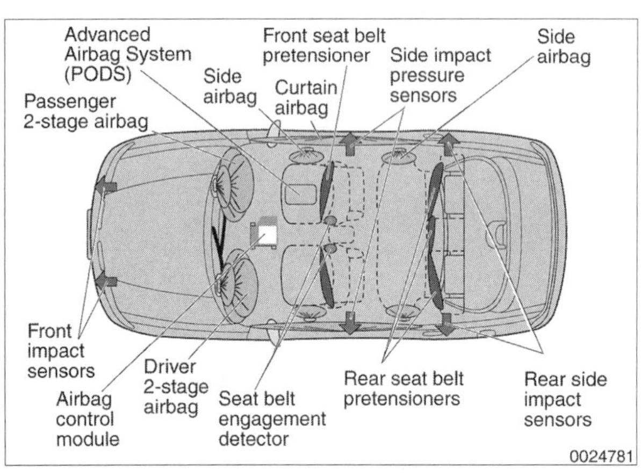
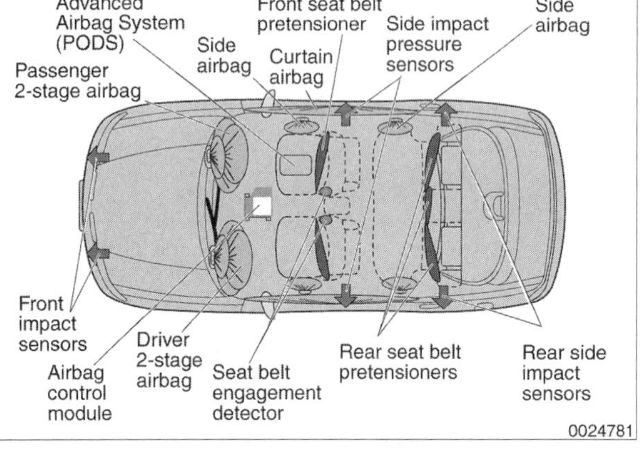

Airbags are installed in several locations in passenger compartment:

- Driver and passenger side front airbags.
- Side airbags built into front seat backrests.
- Side curtain airbags.
- Side airbags built into rear seat backrest bolster (optional).

Inspect padded airbag covers on steering wheel and instrument panel for signs of external damage. Ensure that they are not covered over or have any objects attached to them such as stickers, etc.

Inspect seats front and rear. Do not use seat covers that obstruct or hinder proper deployment of side airbags.

Inspect covers along roof line for side curtain airbags for objects that may hinder proper deployment of side airbag.

Do not apply any chemical treatment to airbag unit covers. Clean with a dry or water moistened cloth only.

Observe airbag warning light in instrument cluster when starting vehicle. It should come on for a few seconds when engine is started, as an electronic systems self-test, and immediately go off and stay off. It should only illuminate during this self-test or if there is a system malfunction.

Door check strap and hinges, lubricating

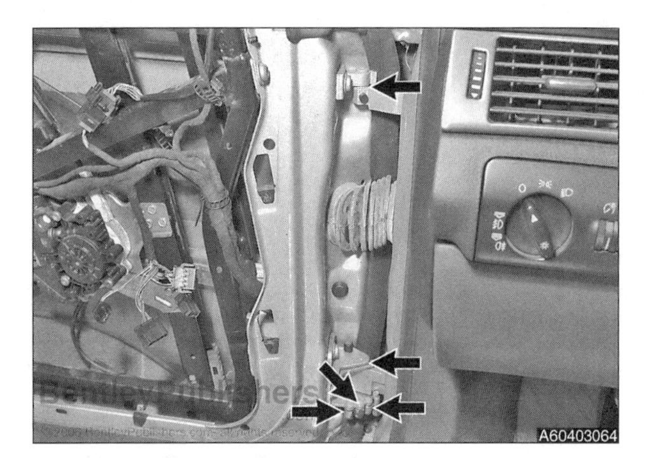

Lubricate door check strap and hinges at places indicated by **arrows** using G 000 150 lubricant.

BentleyPublishers.com
BentleyPublishers.com—All Rights Reserved

Door lock service

– Lock and unlock left and right front doors. Make sure lock buttons move up when unlocking and down when locking each door.

– Press lock buttons down on right front and rear doors and close doors. Make sure doors are locked.

– Make sure that lock button on driver door cannot be pressed down while door is open.

– Move child safety lever on each rear door lock down (in direction of arrow on door lock). Check to make sure that inside door lever is inoperative while lock button is in unlocked position.

– Lubricate lock cylinders using G 000 400 01 lubricant.

Sunroof service

– Check sunroof for leaks.

– Clean guide rails, then spray with D 007 000 00 silicone lubricant.

> ***CAUTION—***
> *• Do not allow silicone spray lubricant to contact painted areas.*

Dust and pollen filter element, replacing

The dust and pollen filter (interior ventilation microfilter) is in the plenum chamber on the right-side.

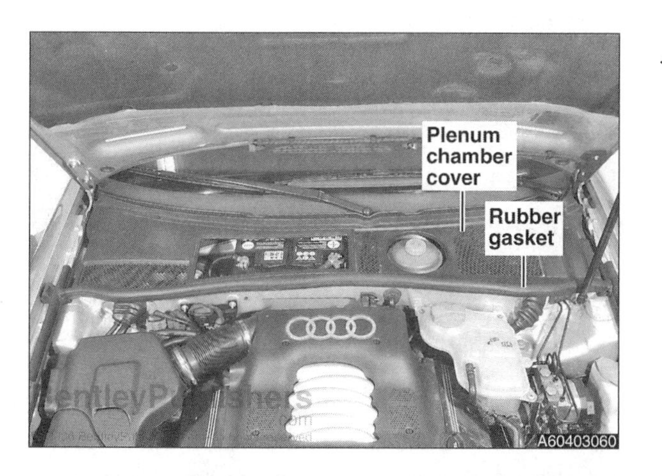

Remove plenum chamber cover:

- Pull rubber gasket forward and off.
- Remove plenum chamber cover toward front.

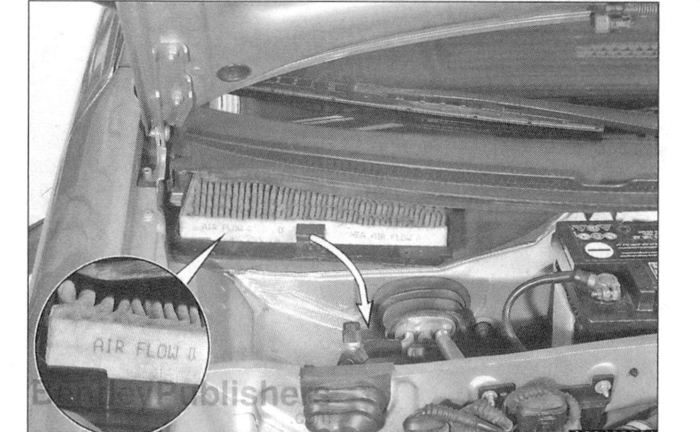

Working inside plenum chamber:

- Press filter retaining clip forward (**arrow**).
- Remove filter from housing by pulling forward and up.
- Insert new filter element with air flow arrow (**inset**) aligned correctly.

– Reinstall plenum chamber cover and rubber gasket.

> ***CAUTION—***
> *• When installing, place the plenum chamber cover carefully into the water deflector below the windshield so that no water can run into the dust and pollen filter and the climate control system.*

BentleyPublishers.com
BentleyPublishers.com—All Rights Reserved

Interior motion detector, checking

Test function of ultra-sonic interior motion detector:

– Open door window approx. 10 cm (4 in).

– Lock vehicle to set alarm and activate interior monitoring.

– Wait 30 seconds until indicator light blinks steadily, approx. once every two seconds.

– Reach through window opening and place hand directly on sensor cover in center of head liner.

– If motion detector is OK, alarm is triggered.

– Unlock vehicle to switch alarm OFF.

Windshield wiper blade, replacing

– Fold wiper arm away from windshield.

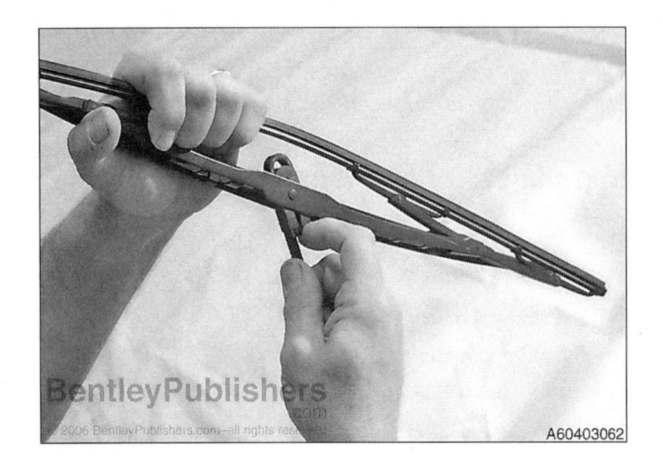

◄ Remove wiper blade:
- Rotate blade on arm approximately ¼ turn.
- Squeeze plastic retainer and slide blade off arm.

> ***CAUTION—***
> *• Do not allow wiper arm to snap back against windshield*

– Installation is reverse of removal.

For windshield wiper troubleshooting or wiper arm replacement, see **92 Wiper and Washers**.

Washer fluid, topping off

◄ Fill windshield washer reservoir.
- When adding windshield washer fluid, be sure freeze protection is adequate for your climate.
- Some windshield washer fluid brands are used as-is, straight from container. Some are in concentrate form and require mixing with water. Refer to product label for proper freeze protection levels.
- Most windshield washer fluids contain additives that help remove insect residue.

BentleyPublishers.com
BentleyPublishers.com—All Rights Reserved

Headlights, adjusting

– For proper adjustment of headlights:
  - Tires inflated to correct pressures.
  - Vehicle on level surface.
  - Headlight lenses and reflectors clean and undamaged with correct type and wattage bulbs.
  - Vehicle loaded with approximately 75 Kg (165 lb) on driver's seat.
  - Vehicle at curb weight with tool kit, jack, spare tire, etc. and 90% fuel load. If fuel tank is not at least 90% full, add weight to compensate.

– Before manually adjusting xenon (HID) headlights, use VAG scan tool or equivalent to check DTC memory, then erase DTC memory.

Left headlight:
  - Height adjustment screw (**1**).
  - Lateral and height adjustment screw (**2**).

– Right headlight adjustment is the reverse.

MAINTENANCE SCHEDULES

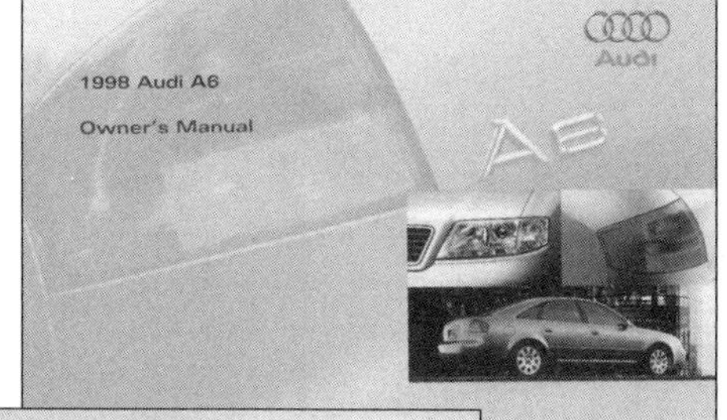

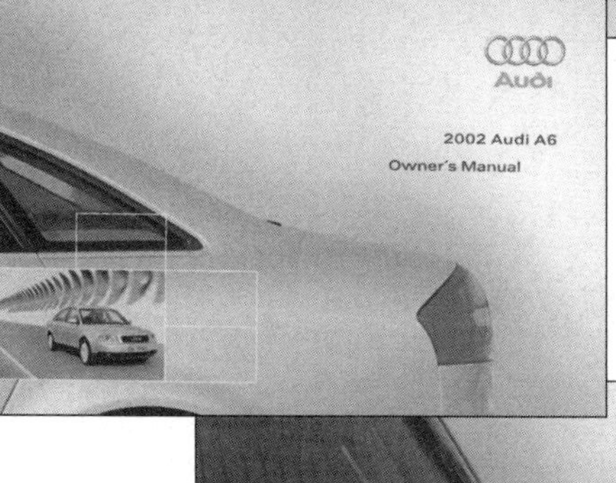

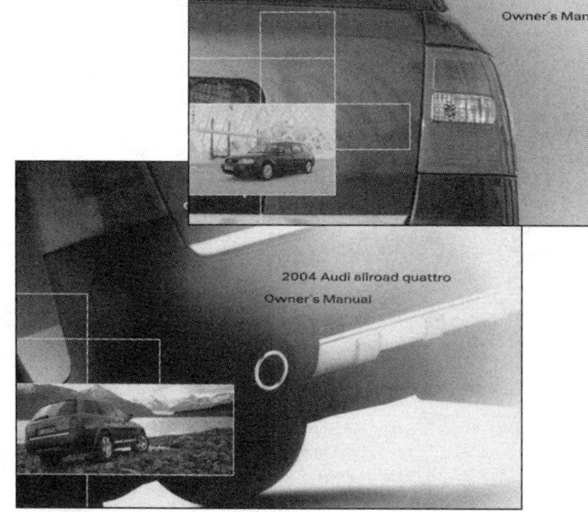

Aside from keeping your Audi in the best possible condition, scheduled maintenance plays a role in maintaining full coverage under Audi's extensive warranties. If in doubt about terms and conditions of your vehicle warranty, consult an authorized Audi dealer.

Maintenance schedules list routine maintenance specified by Audi, as well as time and/or mileage intervals.

Audi continually updates maintenance schedules to suit changing conditions through the issuance of Maintenance Schedule Service Circulars. If in doubt about any of the requirements for your vehicle, consult an authorized Audi dealer.

Maintenance tables on the following pages are year specific.

Maintenance tables may only list service intervals through 100,000 miles (160,000 km). For continued service, repeat intervals listed for the first 100,000 (160,000 km) miles for remaining life of vehicle.

Routine maintenance operations within each engine group can vary depending on vehicle equipment level; not all operations apply to all vehicles.

> ***CAUTION—***
> - *To avoid costly damage caused to engine mechanicai components if the timing belt breaks, periodically inspect the timing belt for wear, cracks or other damage every 50,000 miles.*

BentleyPublishers.com
BentleyPublishers.com—All Rights Reserved

Audi 1998 Maintenance Schedule

1.8T - A4 Miles / Kilometers	5 / 8	10 / 16	15 / 24	20 / 32	25 / 40	30 / 48	35 / 56	40 / 64	45 / 72	50 / 80	55 / 88	60 / 96	65 / 104	70 / 112	75 / 120	80 / 128	85 / 136	90 / 144	95 / 152	100 / 160
V6 & V8 - A4, A6 & A8 Miles / Kilometers	7.5 / 12	15 / 24	22.5 / 36	30 / 48	37.5 / 60	45 / 72	52.5 / 84	60 / 96	67.5 / 108	75 / 120	82.5 / 132	90 / 144	97.5 / 156	105 / 168	112.5 / 180	120 / 192	127.5 / 204	135 / 216	142.5 / 228	150 / 240
Engine Oil – change	•	•	•	•	•	•	•	•	•	•	•	•	•	•	•	•	•	•	•	•
Engine Filter – replace	•	•		•		•		•		•		•		•		•		•		•
Fluid Levels – check	•	•	•	•	•	•	•	•	•	•	•	•	•	•	•	•	•	•	•	•
Auto-Shift Lock – check	•	•	•	•	•	•	•	•	•	•	•	•	•	•	•	•	•	•	•	•
Manual Trans – check shift and clutch interlock	•	•	•	•	•	•	•	•	•	•	•	•	•	•	•	•	•	•	•	•
Wheels – rotate*	•																			
Plenum Water Valve – clean**	•	•	•	•	•	•	•	•	•	•	•	•	•	•	•	•	•	•	•	•
Service Reminder – reset	•	•	•	•	•	•	•	•	•	•	•	•	•	•	•	•	•	•	•	•
Battery – check level		•		•		•		•		•		•		•		•		•		•
Dust/Pollen Filter – replace		•		•		•		•		•		•		•		•		•		•
Cooling System – check level		•		•		•		•		•		•		•		•		•		•
W/W System – check, add fluid		•		•		•		•		•		•		•		•		•		•
Sliding Roof – clean, lubricate rails		•		•		•		•		•		•		•		•		•		•
ATF – replace***						•						•						•		
Transmission – check for leaks		•		•		•		•		•		•		•		•		•		•
Trans Final Drive(s) – check lubricant		•		•		•		•		•		•		•		•		•		•
Brake System – check		•		•		•		•		•		•		•		•		•		•
Drive Shafts – check boots		•		•		•		•		•		•		•		•		•		•
Wheels/Tires/Spare – check condition, pressure		•		•		•		•		•		•		•		•		•		•
OBD – check DTC memory		•		•		•		•		•		•		•		•		•		•
Exterior Lights – check, adjust		•		•		•		•		•		•		•		•		•		•
Door Hinge Mechanism – lubricate		•		•		•		•		•		•		•		•		•		•
Roof Mechanism – lubricate****		•		•		•		•		•		•		•		•		•		•
During Road Test		•		•		•		•		•		•		•		•		•		•
After Road Test		•		•		•		•		•		•		•		•		•		•
Air Cleaner – replace				•				•				•				•				•
Front End – check dust seals, ball joints, tie rod ends				•				•				•				•				•
Brake Fluid – replace†				•				•				•				•				•
1.8T																				
Spark Plugs – replace						•						•						•		
V-Belt – replace																		•		
Ribbed Belt – replace																		•		
Timing Belt – replace																		•		
V6, V8																				
Spark Plugs – replace				•				•				•				•				•
Ribbed Belt – replace												•								
Timing Belt – replace††												•								

*First service only **A6 Wagon only ***Cabriolet, A6 Wagon only ****Cabriolet only
†Every 2 years, regardless of mileage ††Except Cabriolet

> **CAUTION—**
> - *To avoid costly damage caused to engine mechanical components if the timing belt breaks, periodically inspect the timing belt for wear, cracks or other damage every 50,000 miles.*

BentleyPublishers.com
BentleyPublishers.com—All Rights Reserved

1999 Maintenance Schedule

Audi

V6 & V8 – A4, A6 & A8 — Miles	7.5	15	22.5	30	37.5	45	52.5	60	67.5	75	82.5	90	97.5	105	112.5	120	127.5	135	142.5	150
Kilometers	12	24	36	48	60	72	84	96	108	120	132	144	156	168	180	192	204	216	228	240
Engine Oil – change	•	•	•	•	•	•	•	•	•	•	•	•	•	•	•	•	•	•	•	•
Engine Filter – replace	•	•		•		•		•		•		•		•		•		•		•
Fluid Levels – check	•	•	•	•	•	•	•	•	•	•	•	•	•	•	•	•	•	•	•	•
Auto-Shift Lock – check	•	•	•	•	•	•	•	•	•	•	•	•	•	•	•	•	•	•	•	•
Manual Trans – check shift and clutch interlock	•	•	•	•	•	•	•	•	•	•	•	•	•	•	•	•	•	•	•	•
Wheels – rotate*	•																			
Service Reminder – reset	•	•	•	•	•	•	•	•	•	•	•	•	•	•	•	•	•	•	•	•
Battery – check level		•		•		•		•		•		•		•		•		•		•
Dust/Pollen Filter – replace		•		•		•		•		•		•		•		•		•		•
Cooling System – check level		•		•		•		•		•		•		•		•		•		•
W/W System – check, add fluid		•		•		•		•		•		•		•		•		•		•
Sliding Roof – clean, lubricate rails		•		•		•		•		•		•		•		•		•		•
Transmission – check for leaks		•		•		•		•		•		•		•		•		•		•
Trans Final Drive(s) – check lubricant		•		•		•		•		•		•		•		•		•		•
Brake System – check		•		•		•		•		•		•		•		•		•		•
Drive Shafts – check boots		•		•		•		•		•		•		•		•		•		•
Wheels/Tires/Spare – check condition, pressure		•		•		•		•		•		•		•		•		•		•
OBD – check DTC memory		•		•		•		•		•		•		•		•		•		•
Exterior Lights – check, adjust		•		•		•		•		•		•		•		•		•		•
Door Hinge Mechanism – lubricate		•		•		•		•		•		•		•		•		•		•
During Road Test		•		•		•		•		•		•		•		•		•		•
After Road Test		•		•		•		•		•		•		•		•		•		•
Air Cleaner – replace				•				•				•				•				•
Front End – check dust seals, ball joints, tie rod ends				•				•				•				•				•
Brake Fluid – replace**				•				•				•				•				•
1.8T																				
Spark Plugs – replace						•						•						•		
V-Belt – replace																•				
Ribbed Belt – replace																•				
Timing Belt – replace***																				
V6, V8																				
Spark Plugs – replace				•				•				•				•				•
Ribbed Belt – replace										•										
Timing Belt – replace														•						
Timing Belt Tensioner Roller – replace****										•										•

*First service only **Every 2 years, regardless of mileage ***105,000 miles ****A4, A6 only

CAUTION—
- *To avoid costly damage caused to engine mechanical components if the timing belt breaks, periodically inspect the timing belt for wear, cracks or other damage every 50,000 miles.*

BentleyPublishers.com

BentleyPublishers.com—All Rights Reserved

2000 Maintenance Schedule

Audi

1.8T – A4 & TT Miles	5	10	15	20	25	30	35	40	45	50	55	60	65	70	75	80	85	90	95	100
Kilometers	8	16	24	32	40	48	56	64	72	80	88	96	104	112	120	128	136	144	152	160
V6 & V8 – A4, A6 & A8 Miles	8	16	24	32	40	48	56	64	72	80	88	96	104	112	120	128	136	144	152	160
Kilometers	12	24	36	48	60	72	84	96	108	120	132	144	156	168	180	192	204	216	228	240
Engine Oil – change	•	•	•	•	•	•	•	•	•	•	•	•	•	•	•	•	•	•	•	•
Engine Filter – replace	•	•		•		•		•		•		•		•		•		•		•
Fluid Levels – check	•	•	•	•	•	•	•	•	•	•	•	•	•	•	•	•	•	•	•	•
Auto-Shift Lock – check	•	•	•	•	•	•	•	•	•	•	•	•	•	•	•	•	•	•	•	•
Manual Trans – check shift and clutch interlock	•	•	•	•	•	•	•	•	•	•	•	•	•	•	•	•	•	•	•	•
Wheels – rotate*	•																			
Service Reminder – reset	•	•	•	•	•	•	•	•	•	•	•	•	•	•	•	•	•	•	•	•
Battery – check level		•		•		•		•		•		•		•		•		•		•
Dust/Pollen Filter – replace		•		•		•		•		•		•		•		•		•		•
Cooling System – check level		•		•		•		•		•		•		•		•		•		•
W/W System – check, add fluid		•		•		•		•		•		•		•		•		•		•
Sliding Roof – clean, lubricate rails		•		•		•		•		•		•		•		•		•		•
Transmission – check for leaks		•		•		•		•		•		•		•		•		•		•
Trans Final Drive(s) – check lubricant		•		•		•		•		•		•		•		•		•		•
Brake System – check		•		•		•		•		•		•		•		•		•		•
Drive Shafts – check boots		•		•		•		•		•		•		•		•		•		•
Wheels/Tires/Spare – check condition, pressure		•		•		•		•		•		•		•		•		•		•
OBD – check DTC memory		•		•		•		•		•		•		•		•		•		•
Exterior Lights – check, adjust		•		•		•		•		•		•		•		•		•		•
Door Hinge Mechanism – lubricate		•		•		•		•		•		•		•		•		•		•
During Road Test		•		•		•		•		•		•		•		•		•		•
After Road Test		•		•		•		•		•		•		•		•		•		•
Air Cleaner – replace				•				•				•				•				•
Front End – check dust seals, ball joints, tie rod ends				•				•				•				•				•
Brake Fluid – replace**				•				•				•				•				•
Haldex Clutch – change oil***				•				•				•				•				•
1.8T																				
Spark Plugs – replace						•						•						•		
V-Belt – replace																•				
Ribbed Belt – replace																•				
Timing Belt – replace****																				
V6, V8																				
Spark Plugs – replace				•				•				•				•				•
Ribbed Belt – replace									•									•		
Timing Belt – replace													•							
Timing Belt Tensioner Roller – replace†													•							

*First service only **Replace every 2 years, regardless of mileage ***TT quattro only ****105,000 miles
†A4, A6 only

> ***CAUTION—***
> - *To avoid costly damage caused to engine mechanical components if the timing belt breaks, periodically inspect the timing belt for wear, cracks or other damage every 50,000 miles.*

BentleyPublishers.com
BentleyPublishers.com—All Rights Reserved

Audi 2001 Maintenance Schedule

ALL MODELS Miles / Kilometers	5 / 8	10 / 16	20 / 32	30 / 48	40 / 64	50 / 80	60 / 96	70 / 112	80 / 128	90 / 144	100 / 160	105 / 170	110 / 180	120 / 192
Engine Oil – change	•	•	•	•	•	•	•	•	•	•	•		•	•
Engine Filter – replace	•	•	•	•	•	•	•	•	•	•	•		•	•
Service Reminder – reset	•	•	•	•	•	•	•	•	•	•	•		•	•
Wheels – rotate	•											a		
Fluid Levels – check	•	•	•	•	•	•	•	•	•	•	•	d	•	•
Auto-Shift Lock – check	•	•	•	•	•	•	•	•	•	•	•	d	•	•
Brake System – check	•	•	•	•	•	•	•	•	•	•	•	i	•	•
Manual Trans – check shift and clutch interlock	•	•	•	•	•	•	•	•	•	•	•	t	•	•
Cooling System – check level		•	•	•	•	•	•	•	•	•	•	i	•	•
Exhaust System – check		•	•	•	•	•	•	•	•	•	•	o	•	•
OBD – check DTC memory		•	•	•	•	•	•	•	•	•	•	n	•	•
Door Hinge Mechanism – lubricate		•	•	•	•	•	•	•	•	•	•	a	•	•
Battery – check level		•	•	•	•	•	•	•	•	•	•	l	•	•
W/W System – check, add fluid		•	•	•	•	•	•	•	•	•	•		•	•
Wheels/Tires/Spare – check condition, pressure		•	•	•	•	•	•	•	•	•	•	s	•	•
Drive Shafts – check boots		•	•	•	•	•	•	•	•	•	•	e	•	•
Road Test		•	•	•	•	•	•	•	•	•	•	r	•	•
Lights – check		•	•	•	•	•	•	•	•	•	•	v	•	•
Engine – check for leaks			•		•		•		•		•	i		•
Front End – check dust seals, ball joints, tie rod ends			•		•		•		•		•	c		•
Haldex Clutch – change oil & filter*			•		•		•		•		•	e		•
Transmission – check for leaks			•		•		•		•		•			•
Trans Final Drive (man/auto) – check lubricant			• m		• m/a		• m		• m/a		• m			• m/a
Dust/Pollen Filter – replace			•		•		•		•		•			•
Head Lights – adjust			•		•		•		•		•			•
Sliding Roof – clean, lubricate rails			•		•		•		•		•			•
Power Steering Fluid – check					•				•					•
Air Cleaner – replace element					•				•					•
Spark Plugs – replace					•				•					•
Ribbed Belt – replace									•					
V-Belt – replace**									•					
Timing Belt – replace**														•
Timing Belt – replace***												•		
Timing Belt and Tensioner Roller – replace****												•		
Brake Fluid – replace	every 2 years regardless of mileage													

*TT quattro **1.8L Turbo ***2.7L V6 and 4.2L V8 ****2.8L V6

CAUTION—

- *To avoid costly damage caused to engine mechanical components if the timing belt breaks, periodically inspect the timing belt for wear, cracks or other damage every 50,000 miles.*

BentleyPublishers.com
BentleyPublishers.com—All Rights Reserved

2002 Maintenance Schedule A6/S6/allroad

Miles Kilometers	10 15	20 30	30 45	40 60	50 75	60 90	70 105	80 120	90 135	100 150	105 155
Engine Oil – change	•	•	•	•	•	•	•	•	•	•	
Engine Oil Filter – replace	•	•	•	•	•	•	•	•	•	•	
Engine – check for leaks		•		•		•		•		•	
Cooling System – check level, add if necessary	•	•	•	•	•	•	•	•	•	•	
Exhaust System – check for damage, leaks	•	•	•	•	•	•	•	•	•	•	
Engine OBD – check memory, purge	•	•	•	•	•	•	•	•	•	•	
Door Hinge Mechanisms – lubricate	•	•	•	•	•	•	•	•	•	•	
Battery – check electrolyte, add if necessary	•	•	•	•	•	•	•	•	•	•	
Wiper/Washer – check fluid, add if necessary	•	•	•	•	•	•	•	•	•	•	
Automatic Shift Lock – check	•	•	•	•	•	•	•	•	•	•	
Manual Transmission – check shift and interlock	•	•	•	•	•	•	•	•	•	•	
Tires and Spare – check condition, pressure	•	•	•	•	•	•	•	•	•	•	
Service Reminder Display – reset	•	•	•	•	•	•	•	•	•	•	
Brake System – check	•	•	•	•	•	•	•	•	•	•	
Drive Shaft Boots – check	•	•	•	•	•	•	•	•	•	•	
Road Test	•	•	•	•	•	•	•	•	•	•	
Lights – check	•	•	•	•	•	•	•	•	•	•	
Front Axle – check dust seals, ball joints, tie-rod ends		•		•		•		•		•	
Automatic Transmission – check for leaks		•		•		•		•		•	
Automatic Transmission Final Drive – check for leaks				•				•			
Manual Transmission – check for leaks		•		•		•		•		•	
Manual Transmission Final Drive – check for leaks		•		•		•		•		•	
Multitronic™ Transmission – change ATF				•				•			
Power Steering Fluid – check, add if necessary				•				•			
Dust and Pollen Filter – replace		•		•		•		•		•	
Headlights – inspect, adjust if necessary		•		•		•		•		•	
Sliding Roof – clean, lubricate rails		•		•		•		•		•	
Air Cleaner – replace filter element				•				•			
Spark Plugs – replace				•				•			
Ribbed Belt – replace								•			
Ribbed Belt – check**											•
Timing Belt – replace** ***											•
Timing Belt and Tensioner Roller – replace*											•

Brake Fluid – replace every 2 years, regardless of mileage. *2.7L **3.0L ***4.2L

W42AUMAINT2002A4S4A6S6 Printed October 2001

> **CAUTION—**
> - *To avoid costly damage caused to engine mechanical components if the timing belt breaks, periodically inspect the timing belt for wear, cracks or other damage every 50,000 miles.*

BentleyPublishers.com

BentleyPublishers.com—All Rights Reserved

2003 Maintenance Schedule A6/S6/allroad

Miles Kilometers	10 15	20 30	30 45	40 60	50 75	60 90	70 105	80 120	90 135	100 150	105 155
Engine Oil – change	•	•	•	•	•	•	•	•	•	•	
Engine Oil Filter – replace	•	•	•	•	•	•	•	•	•	•	
Engine – check for leaks		•		•		•		•		•	
Cooling System – check level, add if necessary	•	•	•	•	•	•	•	•	•	•	
Exhaust System – check for damage, leaks	•	•	•	•	•	•	•	•	•	•	
Engine OBD – check memory, purge	•	•	•	•	•	•	•	•	•	•	
Door Hinge Mechanisms – lubricate	•	•	•	•	•	•	•	•	•	•	
Battery – check electrolyte, add if necessary	•	•	•	•	•	•	•	•	•	•	
Wiper/Washer – check fluid, add if necessary	•	•	•	•	•	•	•	•	•	•	
Automatic Shift Lock – check	•	•	•	•	•	•	•	•	•	•	
Manual Transmission – check shift and interlock	•	•	•	•	•	•	•	•	•	•	
Tires and Spare – check condition, pressure	•	•	•	•	•	•	•	•	•	•	
Service Reminder Display – reset	•	•	•	•	•	•	•	•	•	•	
Brake System – check	•	•	•	•	•	•	•	•	•	•	
Drive Shaft Boots – check	•	•	•	•	•	•	•	•	•	•	
Road Test	•	•	•	•	•	•	•	•	•	•	
Lights – check	•	•	•	•	•	•	•	•	•	•	
Front Axle – check dust seals, ball joints, tie-rod ends		•		•		•		•		•	
Automatic Transmission – check for leaks		•		•		•		•		•	
Automatic Transmission Final Drive – check for leaks		•		•		•		•		•	
Manual Transmission – check for leaks		•		•		•		•		•	
Manual Transmission Final Drive – check for leaks		•		•		•		•		•	
Multitronic™ Transmission – change ATF				•				•			
Power Steering Fluid – check, add if necessary				•				•			
Dust and Pollen Filter – replace		•		•		•		•		•	
Headlights – inspect, adjust if necessary		•		•		•		•		•	
Sliding Roof – clean, lubricate rails		•		•		•		•		•	
Air Cleaner – replace filter element				•				•			
Spark Plugs – replace (except allroad 4.2L)				•				•			
Spark Plugs – replace (allroad 4.2L)						•					
Ribbed Belt – replace								•			
Ribbed Belt – check**											•
Timing Belt – replace** ***											•
Timing Belt and Tensioner Roller – replace*											•

Brake Fluid – replace every 2 years, regardless of mileage. *2.7L **3.0L ***4.2L

> ***CAUTION—***
> - *To avoid costly damage caused to engine mechanical components if the timing belt breaks, periodically inspect the timing belt for wear, cracks or other damage every 50,000 miles.*

BentleyPublishers.com
BentleyPublishers.com—All Rights Reserved

2004 Maintenance Schedule A6/allroad

Miles	5	15	25	35	45	55	65	75	85	95
Kilometers	8	25	40	55	70	85	100	115	130	145
Engine Oil – change oil and replace filter	•	•	•	•	•	•	•	•	•	•
Wiper/Washer/Headlight washer – check adjust. and function	•	•	•	•	•	•	•	•	•	•
Tires and Spare – check condition, pressure	•	•	•	•	•	•	•	•	•	•
Tires – rotate	•									
Service Reminder Display – reset	•	•	•	•	•	•	•	•	•	•
Road Test	•	•	•	•	•	•	•	•	•	•
Cooling System – check level, add if necessary		•		•		•		•		•
Exhaust System – check for damage, leaks		•		•		•		•		•
Engine OBD – check memory, purge		•		•		•		•		•
Engine – check for leaks		•		•		•		•		•
Battery – check electrolyte, add if necessary		•		•		•		•		•
Dust and Pollen Filter – replace		•		•		•		•		•
Automatic Transmission and Final Drive – check for leaks		•		•		•		•		•
Manual Transmission and Final Drive – check for leaks		•		•		•		•		•
Sunroof – clean, lubricate rails		•		•		•		•		•
Front Axle – check dust seals, ball joints, tie rods, tie rod ends		•		•		•		•		•
Tire Repair Set – check renewal date		•		•		•		•		•
Lights – check		•		•		•		•		•
Headlights – inspect, adjust if necessary		•		•		•		•		•
Brake System – check		•		•		•		•		•
Drive Shaft Boots – check		•		•		•		•		•
Door Hinge Mechanisms – lubricate		•		•		•		•		•
Spark Plugs – replace (except allroad 4.2L)				•				•		
Spark Plugs – replace (allroad 4.2L)						•				
Multitronic™ Transmission – change ATF				•				•		
Power Steering Fluid – check, add if necessary				•				•		
Air Cleaner – clean housing, replace filter element				•				•		
Ribbed Belt – replace (except 3.0)								•		
Ribbed Belt – check (3.0)								•		
Timing Belt – replace (3.0 and 4.2L)								•		
Timing Belt and tensioner – replace (2.7T)								•		

Brake Fluid – replace every 2 years, regardless of mileage

CAUTION—

- *To avoid costly damage caused to engine mechanical components if the timing belt breaks, periodically inspect the timing belt for wear, cracks or other damage every 50,000 miles.*

BentleyPublishers.com
BentleyPublishers.com—All Rights Reserved

13 Timing Belt, Crankshaft Pulley, Rear Main Seal

General

This repair group covers camshaft timing belt (toothed belt), front crankshaft pulley (vibration damper) and rear main (crankshaft) seal replacement.

Engine accessory belt service is covered in **03 Maintenance**.

Timing Belt

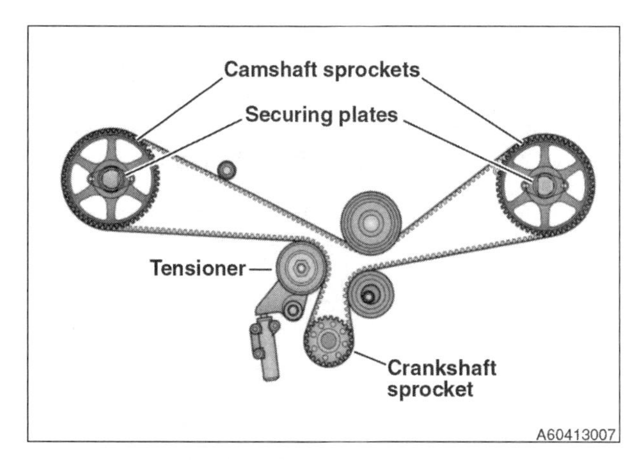

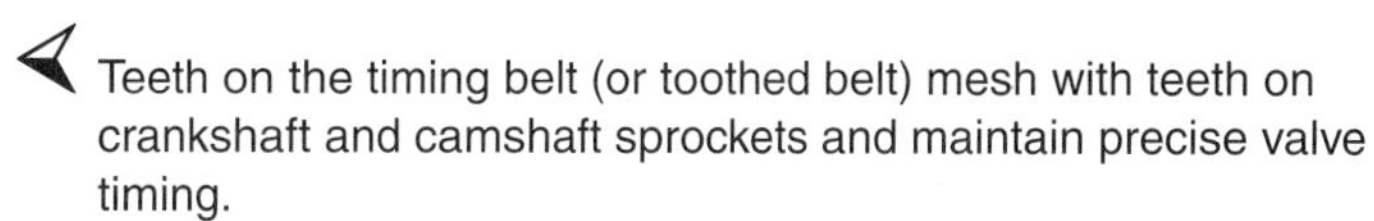

Teeth on the timing belt (or toothed belt) mesh with teeth on crankshaft and camshaft sprockets and maintain precise valve timing.

Each camshaft sprocket is pressed on the camshaft. When loosened, the sprocket spins freely on the camshaft. An oval securing plate is keyed to the camshaft. The sprocket is securely attached to the camshaft only when the securing plate is torqued.

Precise camshaft timing is achieved by lining up the oval securing plates on the camshaft ends with each other, using a special tool. Camshaft sprockets are then tightening in that position.

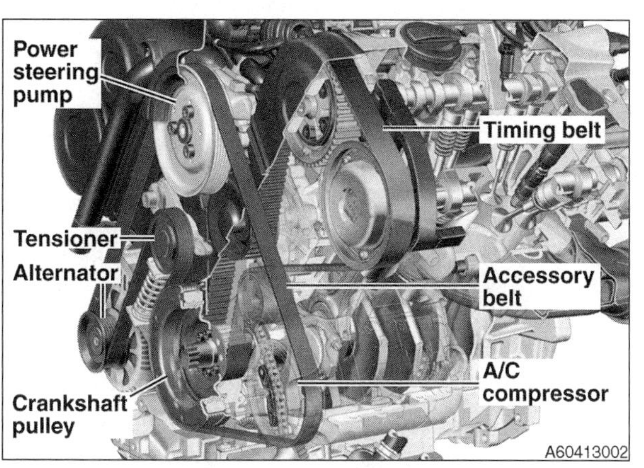

Prior to removing the timing belt, proceed as follows:

- Remove engine cooling fan and viscous clutch.
- Remove front bumper and place front lock carrier (radiator mounting frame in front of engine) in service position. See **50 Body–Front**.
- Remove accessory belt. See **03 Maintenance**.
- Remove accessory belt tensioner.
- Remove timing belt covers.
- Remove front crankshaft pulley. See **Crankshaft pulley, removing and installing (2.7 liter or 2.8 liter V6 engine)** in this repair group.

BentleyPublishers.com
BentleyPublishers.com—All Rights Reserved

CAUTION—
- *To avoid costly damage caused to engine mechanical components if the timing belt breaks, periodically inspect the timing belt for wear, cracks or other damage every 50,000 miles.*
- *When replacing the timing belt, be sure to use crankshaft locking pin and camshaft locking bar as described in the procedure.*
- *When the timing belt is off, any rotation of the camshafts or crankshaft may lead to valve damage.*

Timing belt, removing

In the procedure that follows, timing belt replacement is described for 2.7 liter or 2.8 liter V6 engine. Timing belt replacement requires several special tools and procedures. Read the procedure through before starting the job.

– Raise car and support safely.

WARNING —
- *Make sure the car is stable and well supported at all times. Use a professional automotive lift or jack stands designed for the purpose. A floor jack is not adequate support.*

– Remove upper and lower engine covers. See **03 Maintenance**.

– Remove viscous fan and accessory belt. See **03 Maintenance**.

NOTE —
- *If belt is to be reused, mark direction of rotation prior to removal.*

– Remove front bumper. Place lock carrier (radiator mounting frame in front of engine) in service position. See **50 Body–Front**.

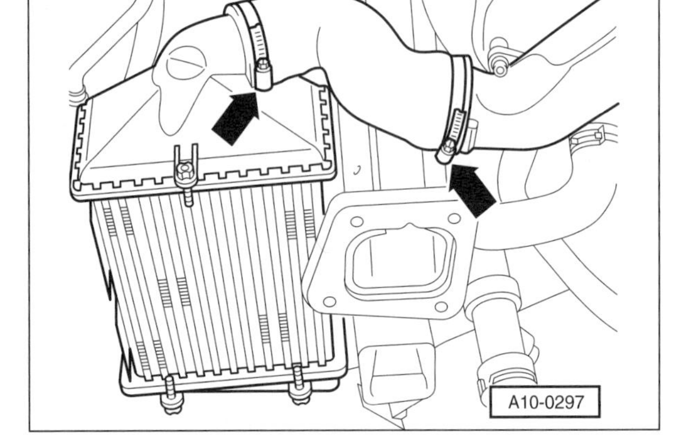

◀ 2.7 liter turbo engine: Loosen hose clamps (**arrows**) and remove pressure hoses between intercoolers and turbo pressure lines on left and right sides.

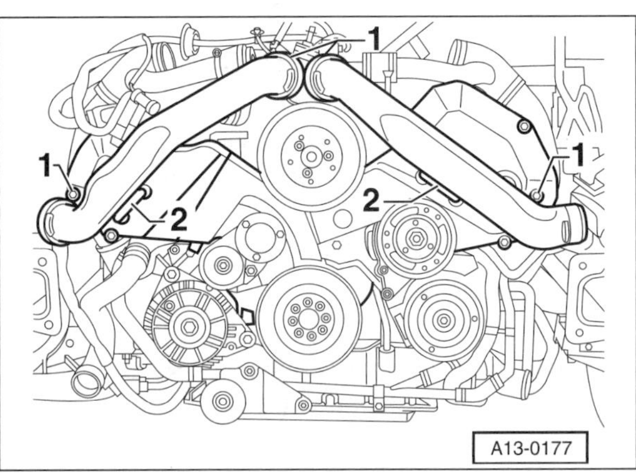

◀ 2.7 liter turbo engine: Remove turbo pressure lines (**1**). Note position of retaining strips (**2**).

BentleyPublishers.com
BentleyPublishers.com—All Rights Reserved

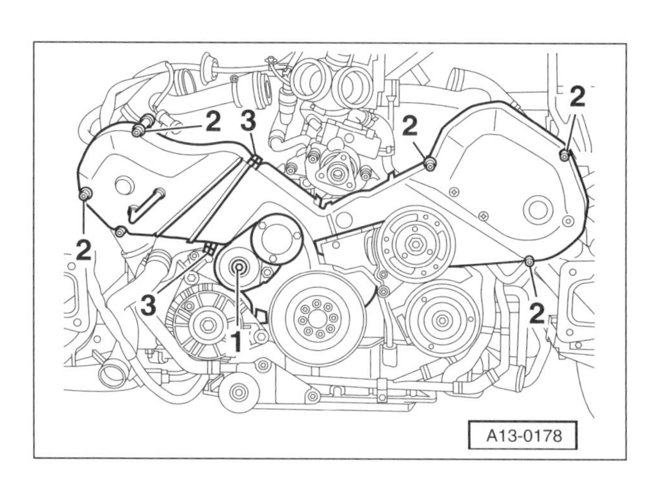

Working at front of engine:

- Remove accessory belt tensioner (**1**).
- Unclip (**2**) timing belt covers on both sides and remove.
- Remove center timing belt cover (**3**).

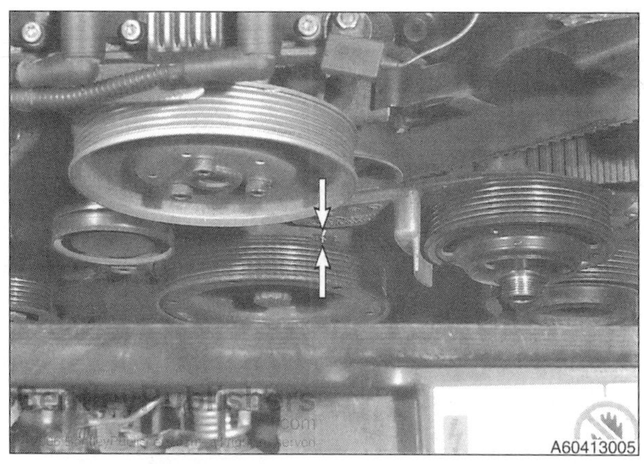

Using wrench on crankshaft pulley, rotate crankshaft until marks (**arrows**) on crankshaft pulley and lower crankcase cover align. This is crankshaft top dead center (TDC).

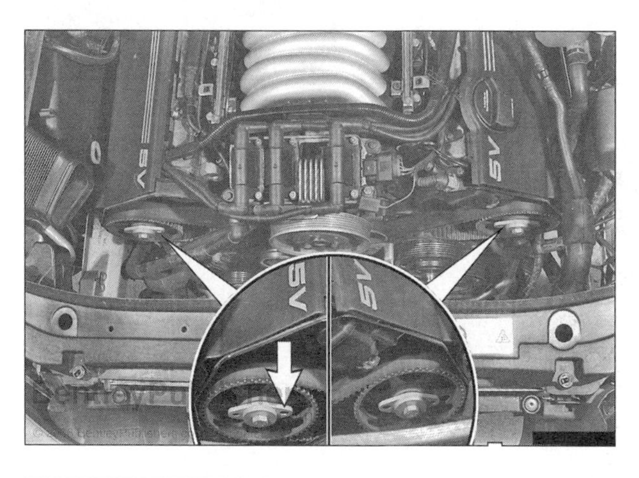

Check position of camshafts:

- Make sure larger holes (**arrows**) of oval securing plates on camshaft sprockets face each other.
- If not, turn crankshaft another full rotation until crankshaft pulley TDC marks line up and larger holes on securing plates face each other.

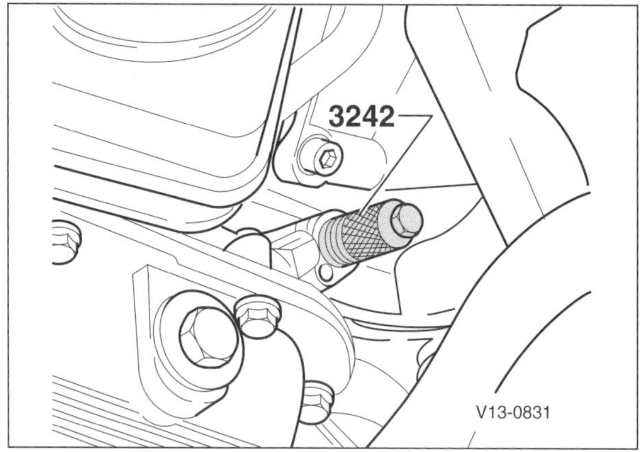

Working underneath engine, above oil drain plug, remove sealing plug from left side of cylinder block and install special tool VAG 3242 (crankshaft locking pin) into opening. If necessary, rock crankshaft back and forth gently to facilitate insertion of lock tool.

– Remove crankshaft pulley (vibration dampener). See **Crankshaft pulley, removing and installing (2.7 liter or 2.8 liter V6 engine)** in this repair group. Do not loosen or remove pulley center bolt.

BentleyPublishers.com
BentleyPublishers.com—All Rights Reserved

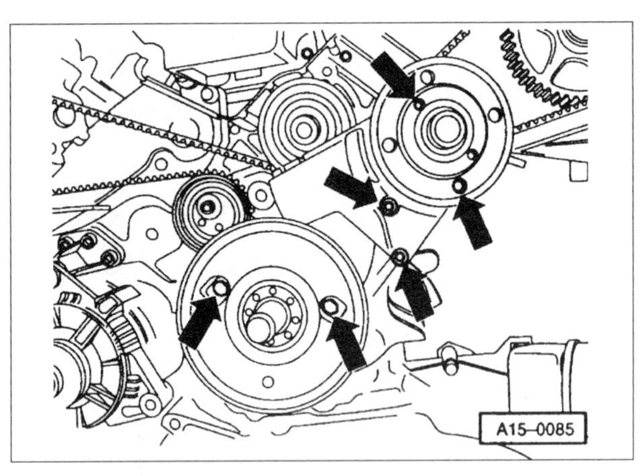

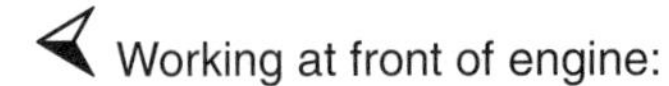

Working at front of engine:

- Remove accessory belt idler mounting bolts (**top arrows**) and lift out idler pulley bracket.
- Remove timing belt cover bolts (**lower arrows**) and take off cover.

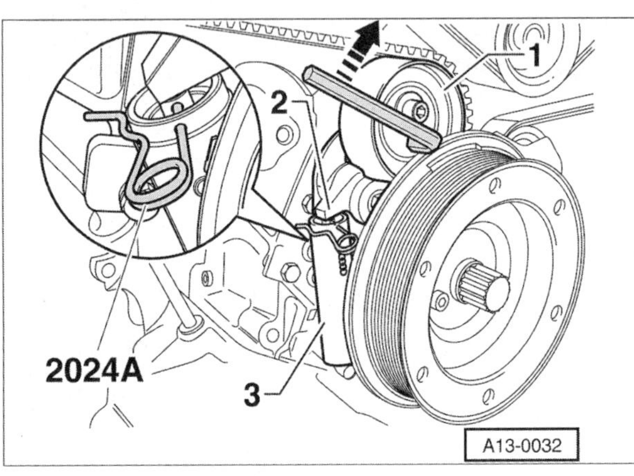

Remove timing belt tensioner.

- Use 8 mm Allen wrench to rotate belt tensioner roller (**1**) clockwise (direction of broken **arrow)**. Apply slow pressure to oil-dampened tensioner.
- When tensioning lever (**2**) compresses tensioner (**3**), insert pin with 2 mm diameter (VAG special tool 2024A or equivalent) into bore. This releases timing belt tension.

NOTE—

- *If belt is to be reused, mark direction of rotation prior to removal.*
- *Crankshaft pulley is shown installed.*

– Remove timing belt.

CAUTION—
- *Do not allow camshafts to rotate once timing belt is off.*

Timing belt, installing

– Prior to reinstalling timing belt, spin timing belt tensioner and listen for worn bearing. Replace if necessary.

Tightening torque	
Timing belt tensioner to crankcase	20 Nm (15 ft-lb)

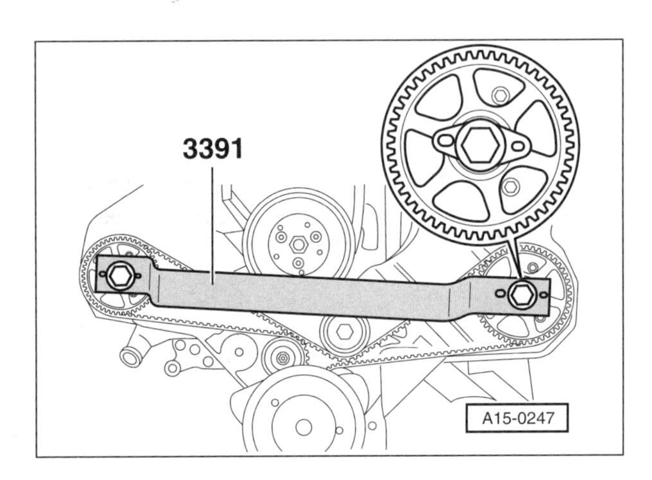

Working at camshafts:

- Install camshaft locking bar, special tool VAG 3391, into oval securing plates.
- Loosen camshaft sprocket bolts and back out approx. 5 turns.
- Remove camshaft locking bar.

BentleyPublishers.com
BentleyPublishers.com—All Rights Reserved

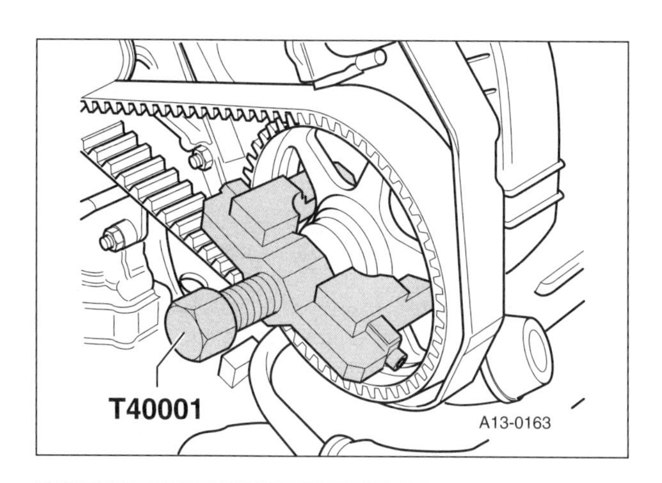

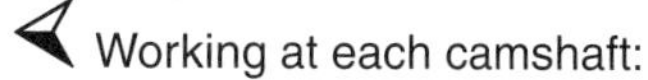

Working at each camshaft:

- Remove camshaft sprocket bolt.
- Use puller VAG T40001 to pull off camshaft sprocket.
- Reinstall camshaft sprocket together with oval securing plate and hand-tighten.
- Make sure camshaft sprocket is just loose enough to be rotated but not so loose as to wobble on camshaft.

NOTE—

- *Once the camshaft sprocket bolt is loosened sufficiently, as an alternative to using puller VAG T40001, tap the sprocket gently with a soft-faced hammer to break free press fit at camshaft.*

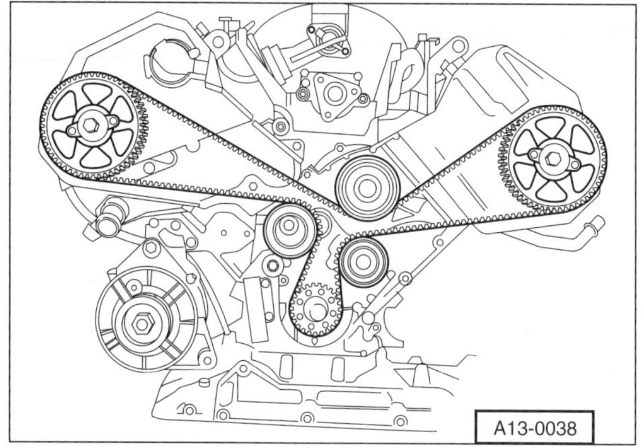

Install timing belt on all sprockets, as shown. If reusing belt, install in direction marked previously.

– Reinstall camshaft locking bar, special tool VAG 3391, into oval securing plates.

– Use 8 mm Allen wrench to back off timing belt tensioner, then remove tensioner locking pin. Release tensioner slowly.

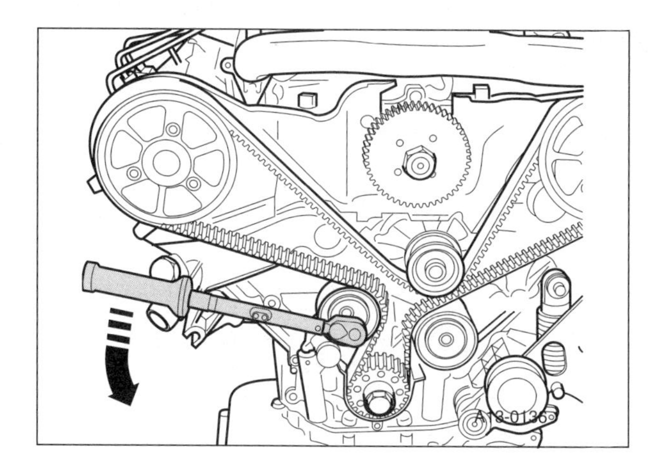

Pretension timing belt as follows: Using torque wrench with 8 mm Allen bit, apply 15 Nm (11 ft-lb) counterclockwise tension to belt (broken **arrow**). This ensures tensioner expands completely and engages timing belt correctly.

– Tighten camshaft sprocket bolts.

Tightening torque	
Camshaft sprocket to camshaft	55 Nm (40 ft-lb)

– Remove camshaft locking bar and crankshaft locking pin.

– Remainder of installation is reverse of removal.

Tightening torques	
Accessory belt idler to crankcase	45 Nm (33 ft-lb)
Crankshaft pulley to timing belt sprocket	25 Nm (18 ft-lb)
Timing belt tensioner element to bracket	10 Nm (7 ft-lb)

BentleyPublishers.com
BentleyPublishers.com—All Rights Reserved

FRONT CRANKSHAFT PULLEY

The front crankshaft pulley is also referred to as vibration damper or harmonic balancer. It is bolted to front of timing belt crankshaft sprocket.

V6 models with 2.7 liter or 2.8 liter engine: Remove lower engine cover, viscous fan and accessory belt to gain access to front pulley.

V6 models with 3.0 liter engine: Remove engine covers and front bumper. Place lock carrier (radiator mounting frame in front of engine) in service position. Remove accessory belt to gain access to engine pulley.

V8 models: Remove lower engine cover and front bumper. Place lock carrier (radiator mounting frame in front of engine) in service position. Remove viscous fan and accessory belt to gain access to engine pulley.

> ***CAUTION—***
> *• Do not loosen or remove the timing belt sprocket mounting bolt in the center of the pulley.*

Crankshaft pulley, removing and installing (2.7 liter or 2.8 liter V6 engine)

– Raise car and support safely.

> ***WARNING —***
> *• Make sure the car is stable and well supported at all times. Use a professional automotive lift or jack stands designed for the purpose. A floor jack is not adequate support.*

– Remove lower engine cover. See **03 Maintenance**.

– Remove viscous fan and accessory belt. See **03 Maintenance**.

NOTE —
• If belt is to be reused, mark direction of rotation prior to removal.

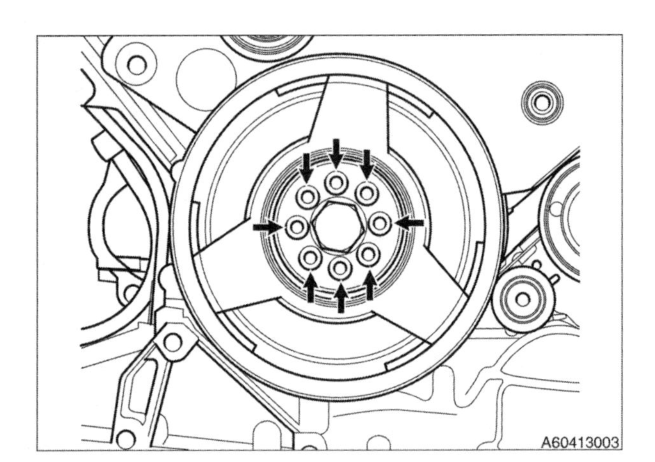

◄ Remove pulley mounting bolts (**arrows**). Do not loosen or remove center bolt.

– Lift pulley off timing belt sprocket.

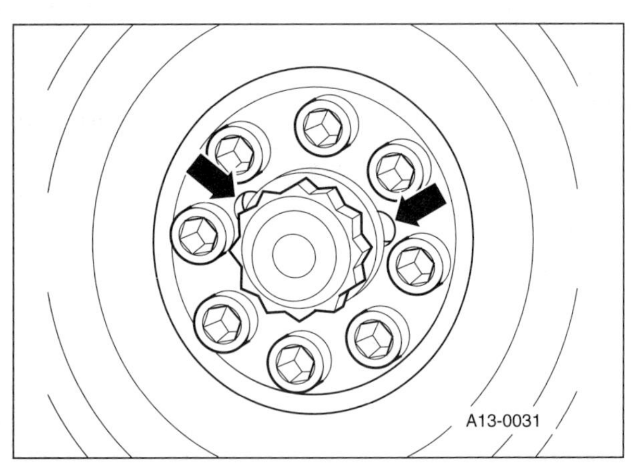

◄ When installing pulley, make sure notches (**arrows**) in pulley are aligned with locating lugs on timing belt sprocket.

Tightening torque	
Crankshaft pulley to timing belt sprocket	25 Nm (18 ft-lb)

– Reinstall belt in direction of rotation marked previously. Reinstall viscous fan. See **03 Maintenance**.

– Run engine briefly to make sure belt routing is correct. Then reinstall engine cover.

BentleyPublishers.com
BentleyPublishers.com—All Rights Reserved

Crankshaft pulley, removing and installing (3.0 liter V6 engine)

– Raise car and support safely.

> ***WARNING—***
> *• Make sure the car is stable and well supported at all times. Use a professional automotive lift or jack stands designed for the purpose. A floor jack is not adequate support.*

– Remove engine covers. See **03 Maintenance**.

– Remove front bumper cover and bumper. See **63 Bumpers**.

– Place lock carrier in service position. See **50 Body-Front**.

– Remove accessory belt. See **03 Maintenance**.

NOTE—

• If belt is to be reused, mark direction of rotation prior to removal.

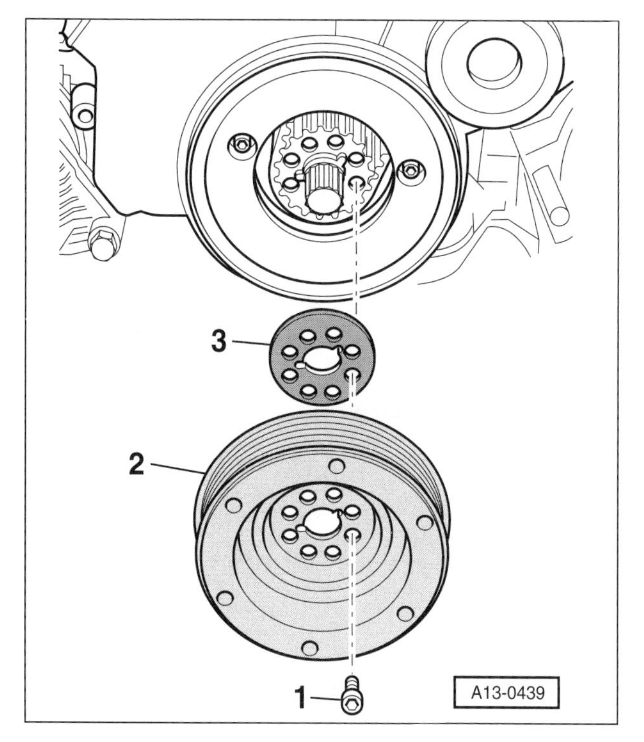

◄ Remove front pulley mounting bolts (**1**). Do not loosen or remove center bolt.

– Lift pulley (**2**) and thrust washer (**3**) off timing belt crankshaft sprocket.

NOTE—

• Thrust washer is only installed on sprocket with part no. 06C 105 063. If sprocket part no. is 06C 105 063 B, do not install thrust washer.

– When installing, make sure convex side of thrust washer rests against pulley.

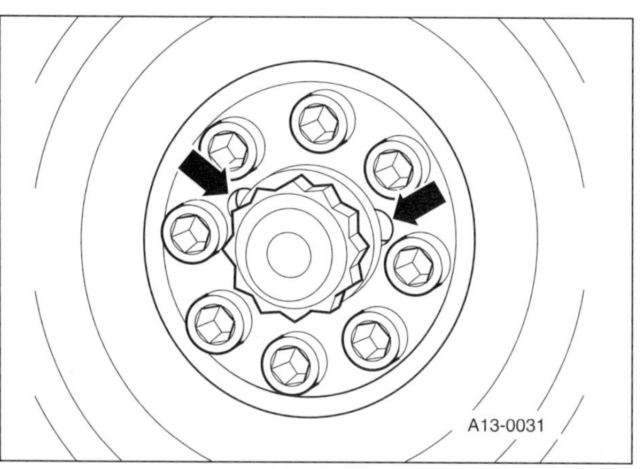

◄ When installing pulley, make sure notches (**arrows**) in pulley are aligned with locating lugs on timing belt sprocket.

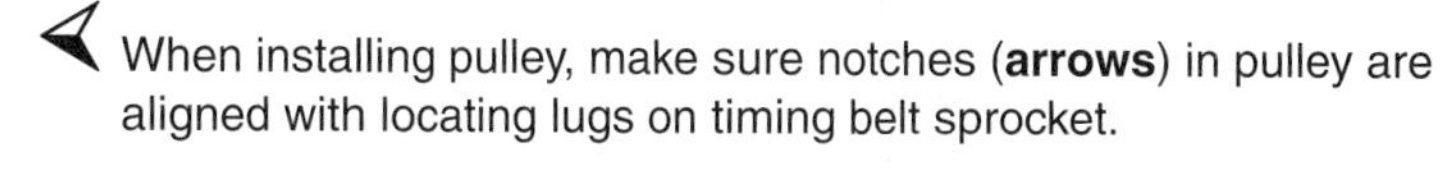

Tightening torque	
Crankshaft pulley to timing belt sprocket	25 Nm (18 ft-lb)

– Reinstall belt in direction of rotation marked previously.

– Run engine briefly to make sure belt routing is correct. Then reinstall lock carrier, front bumper and engine covers.

BentleyPublishers.com
BentleyPublishers.com—All Rights Reserved

Crankshaft pulley, removing and installing (V8 engine)

– Raise car and support safely.

> ***WARNING —***
> • *Make sure the car is stable and well supported at all times. Use a professional automotive lift or jack stands designed for the purpose. A floor jack is not adequate support.*

– Remove lower engine cover. See **03 Maintenance**.

– Remove front bumper and bumper cover. See **63 Bumpers**.

– Place lock carrier in service position. See **50 Body-Front**.

– Remove viscous fan (if equipped) and accessory belt. See **03 Maintenance**.

NOTE —
• *If belt is to be reused, mark direction of rotation prior to removal.*

– Remove pulley mounting bolts. Do not loosen or remove center bolt.

– Lift pulley off timing belt sprocket or crankshaft.

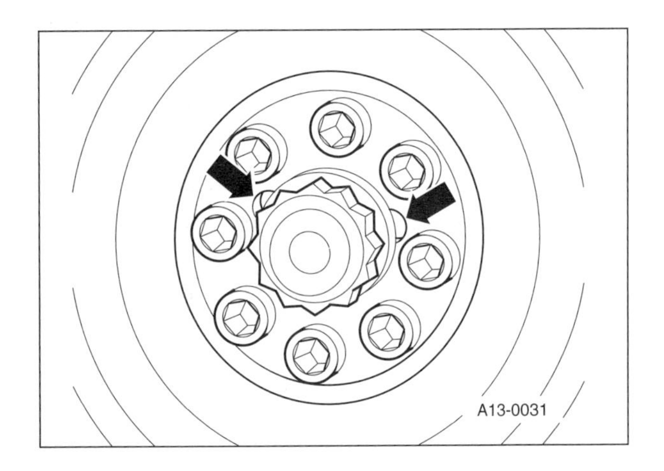

◄ When installing pulley, make sure notches (**arrows**) in pulley are aligned with locating lugs on timing belt sprocket.

Tightening torque	
Crankshaft pulley to timing belt sprocket (*not* allroad quattro)	25 Nm (18 ft-lb)
Crankshaft pulley to crankshaft (allroad quattro V8, engine code BAS)	42 Nm (31 ft-lb)

– Reinstall belt in direction of rotation marked previously. Reinstall viscous fan (if equipped). See **03 Maintenance**.

– Run engine briefly to make sure belt routing is correct. Then reinstall lock carrier, front bumper and engine covers.

REAR MAIN SEAL

Oil drips at the engine-to-transmission seam are usually evidence of crankshaft rear main seal leak. The rear main seal is accessible after removal of transmission and dual-mass flywheel or torque plate.

V6 models with 2.8 liter engine: Replace seal by prying out of seal flange.

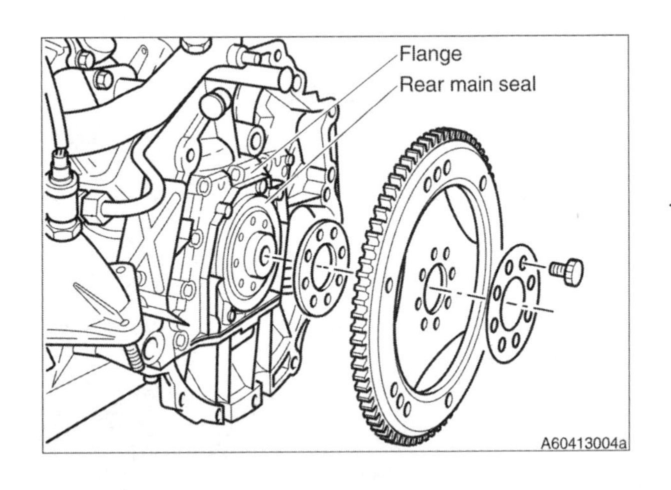

◄ **V6 models with 2.7 liter or 3.0 liter engine, V8 models *not* allroad quattro:** Audi recommends replacing rear main seal with seal flange. When removing flange, sealing of flange against oil pan gasket is disturbed. Remove upper oil pan and replace gasket if it is damaged.

Allroad quattro models with V8 engine (engine code BAS): Replace seal by prying out of timing chain cover.

BentleyPublishers.com
BentleyPublishers.com—All Rights Reserved

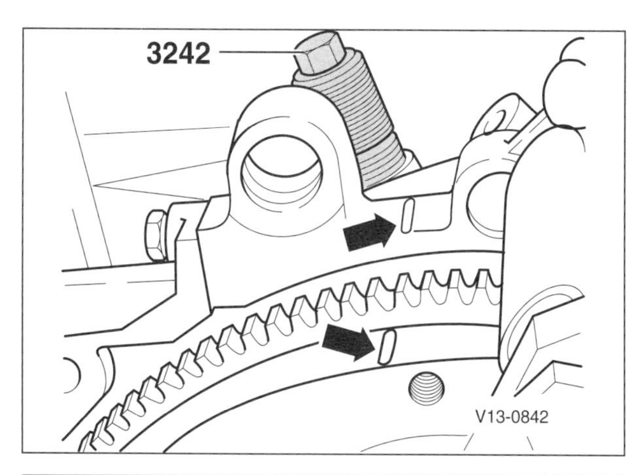

Rear main seal, removing and installing (2.8 liter V6 engine)

With transmission removed, set crankshaft at top dead center (TDC). Install special tool 3242 (crankshaft lock). Mark position of flywheel or torque plate opposite crankcase flange (**arrows**).

– Remove flywheel or torque plate and set aside. Discard mounting bolts.

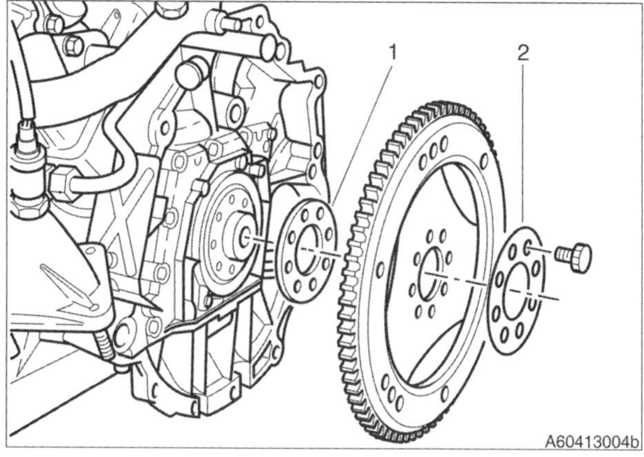

Automatic transmission: Mark positions of shim (**1**) and washer (**2**) in front of and behind torque plate as plate is removed.

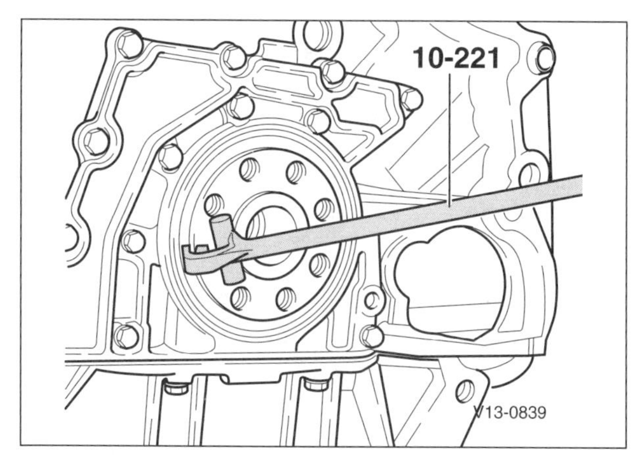

Pry out rear main seal using special tool 10-221 or equivalent.

> ***CAUTION—***
> • *Work carefully to avoid damaging aluminum seal flange or flywheel sealing surface.*

– Clean sealing surfaces.

– Install new seal over end of crankshaft. Use installation sleeve supplied with new seal.

> ***CAUTION—***
> • *Do not lubricate seal lip or seal outer edge before installation.*

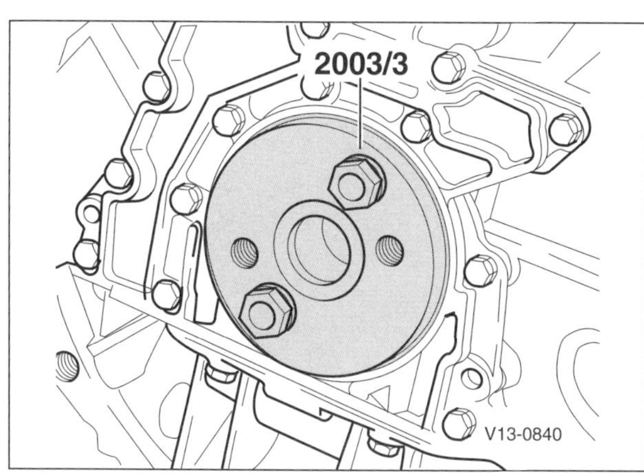

Use special tool 2003 / 3 (seal installer) to press seal in up to stop. Use flywheel or torque plate installation bolts for pressing.

– Using new bolts, install flywheel or torque plate. Use previously made marks to line up with crankcase.

Tightening torque	
Flywheel to crankshaft • Stage 1 • Stage 2	 60 Nm (44 ft-lb) +½ turn (180°)
Torque plate to crankshaft • Stage 1 • Stage 2	 60 Nm (44 ft-lb) +¼ turn (90°)

BentleyPublishers.com
BentleyPublishers.com—All Rights Reserved

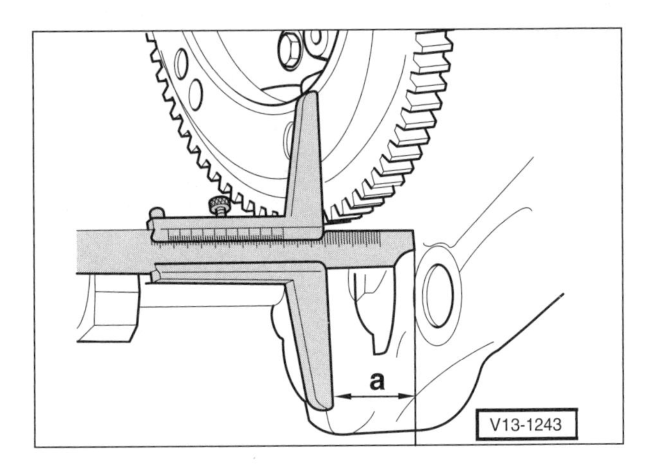

After installing torque plate, measure torque plate offset (**a**) from crankcase bellhousing flange. Measure at three different spots and calculate average value.

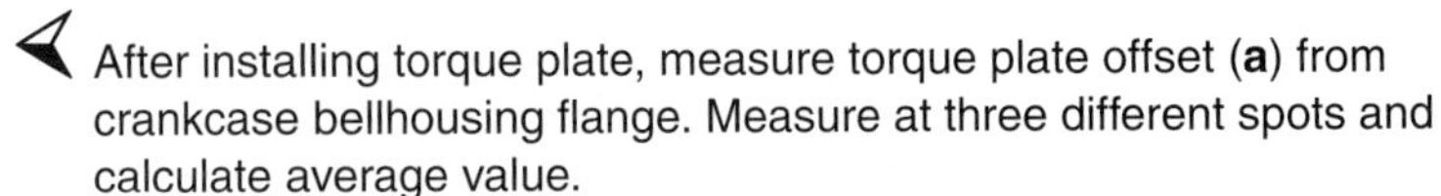

Torque plate offset	
a = offset of torque plate from engine block	approx. 12.3 mm (0.484 in)

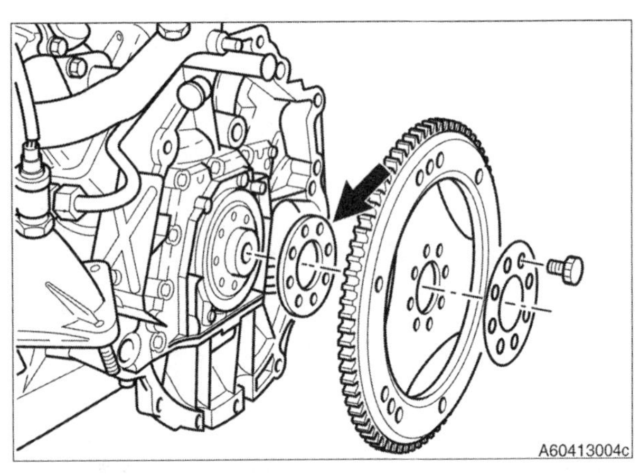

If offset is incorrect, install different shim (**arrow**) between flywheel and crankshaft. Shim thickness is as follows:

- Part no. 054 105 30 = 3.00 mm (0.12 in)
- Part no. 054 105 20 = 4.00 mm (0.16 in)

– Remainder of installation is reverse of removal. Bear in mind:

- Automatic transmission: Check ATF level and top off.
- Cooling system: Check coolant level and top off.
- Engine: Check oil level and top off.

Rear main seal, removing and installing (2.7 liter or 3.0 liter V6 engine; V8 engine *not* allroad quattro)

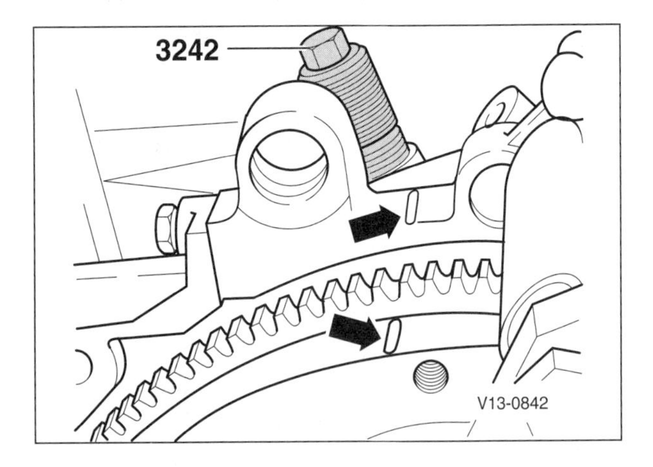

With transmission removed, set crankshaft at top dead center (TDC). Install special tool 3242 (crankshaft lock). Mark position of flywheel or torque plate opposite crankcase flange (**arrows**).

– Manual transmission: Remove flywheel and set aside. Discard mounting bolts.

BentleyPublishers.com
BentleyPublishers.com—All Rights Reserved

◀ Automatic transmission: Using wrench on crankshaft pulley, rotate crankshaft until marks (**arrows**) on crankshaft pulley and lower crankcase cover align. This is crankshaft top dead center (TDC).

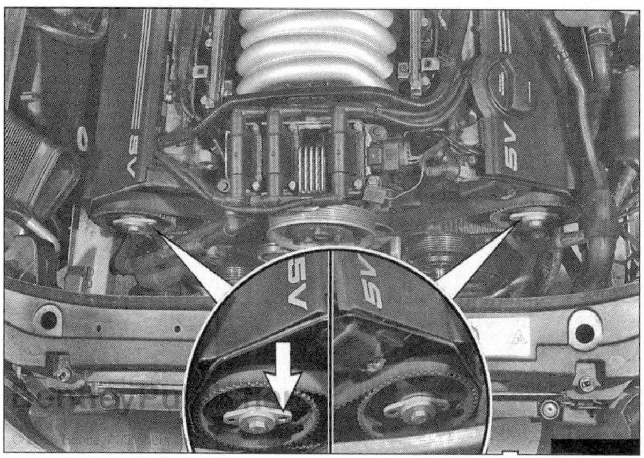

◀ Check position of camshafts:

- Make sure larger holes (**arrows**) of oval securing plates on camshaft sprockets face each other.
- If not, turn crankshaft another full rotation until crankshaft pulley TDC marks line up and larger holes on securing plates face each other.

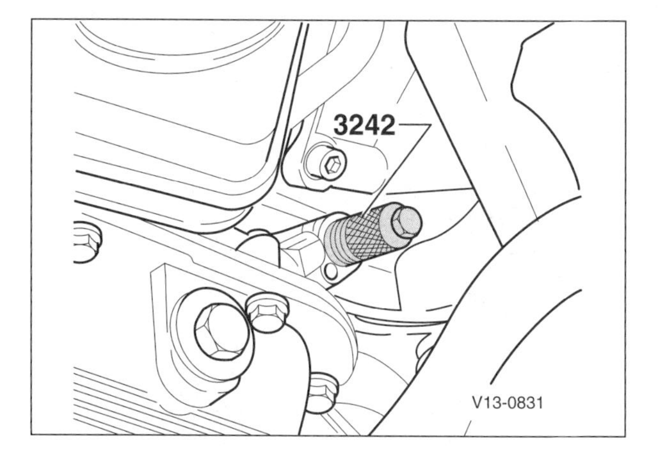

◀ Working underneath engine, above oil drain plug, remove sealing plug from left side of cylinder block and install special tool VAG 3242 (crankshaft locking pin) into opening. If necessary, rock crankshaft back and forth gently to facilitate insertion of lock tool.

– Remove torque plate mounting bolts and discard.

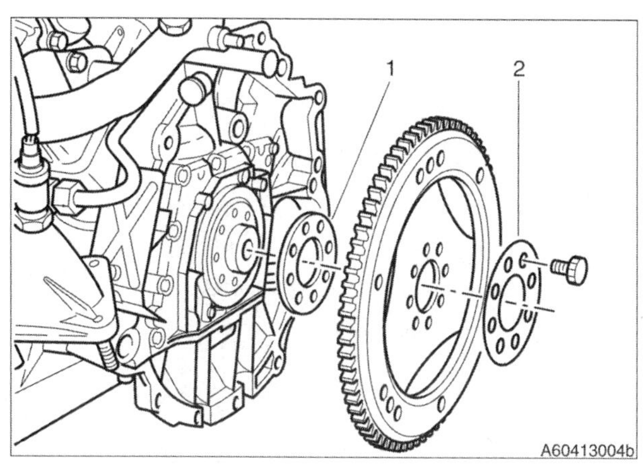

◀ Automatic transmission: Mark positions of shim (**1**) and washer (**2**) in front of and behind torque plate as plate is removed.

BentleyPublishers.com
BentleyPublishers.com—All Rights Reserved

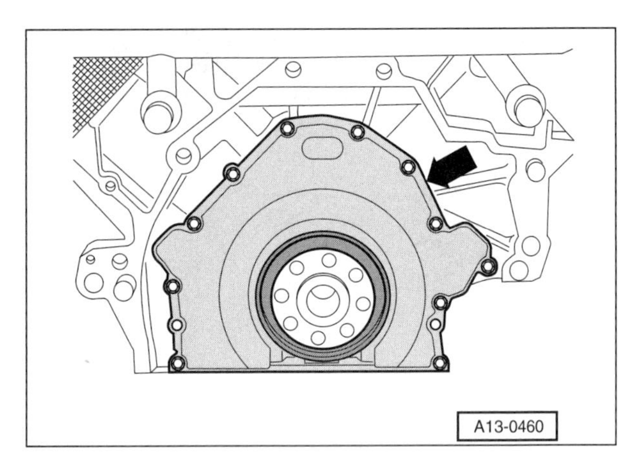

Remove seal flange (**arrow**) with seal. Separate carefully from oil pan gasket.

> ***CAUTION—***
> - *If oil pan gasket is damaged during this procedure, remove oil pan and replace gasket.*

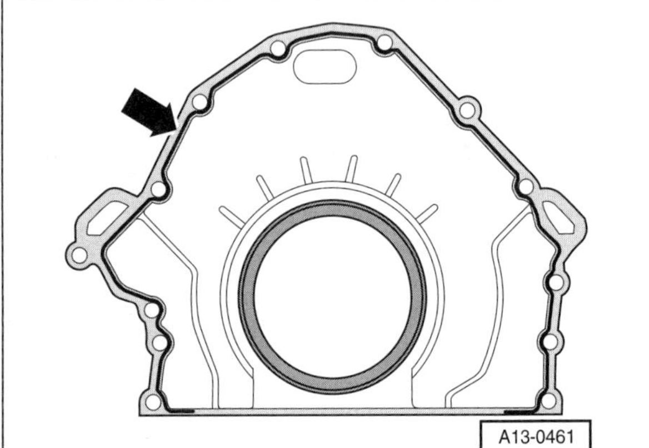
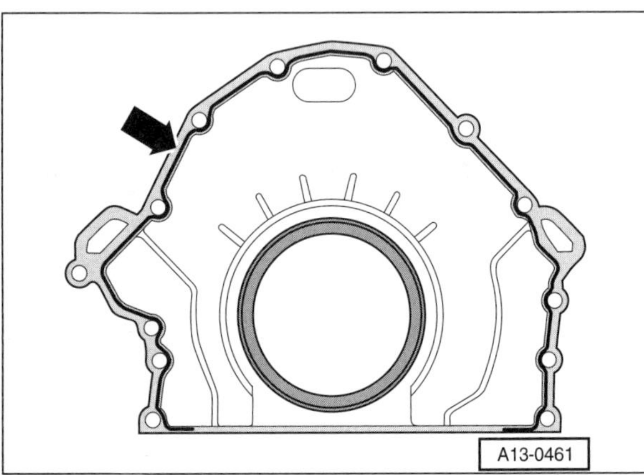

Apply bead of silicone sealant approx. 1.5 mm (0.06 in) to sealing surface of new flange. Install new seal and seal flange over end of crankshaft. Use installation sleeve supplied with new seal.

> ***CAUTION—***
> - *Be sure to install seal flange within 5 minutes after application of silicone sealant.*

– Using new bolts, install flywheel or torque plate. Use previously made marks to line up with crankcase.

Tightening torque	
Flywheel to crankshaft (use new bolts): • Stage 1 • Stage 2	 60 Nm (44 ft-lb) +½ turn (180°)
Torque plate to crankshaft (use new bolts): • Stage 1 • Stage 2	 60 Nm (44 ft-lb) +¼ turn (90°)

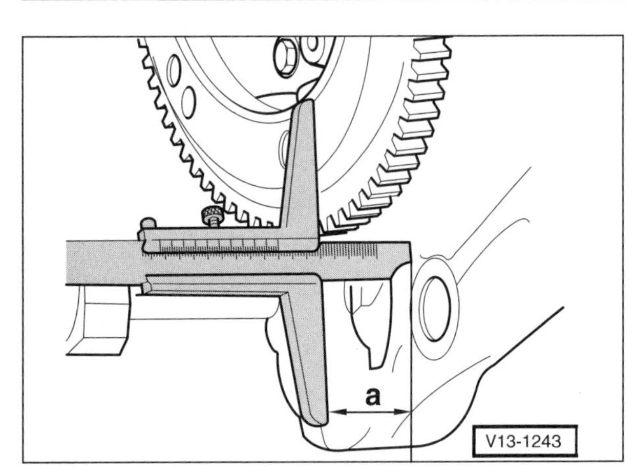

After installing torque plate, measure torque plate offset (**a**) from crankcase bellhousing flange. Measure at three different spots and calculate average value.

Torque plate offset	
a = offset of torque plate from engine block	approx. 12.3 mm (0.484 in)

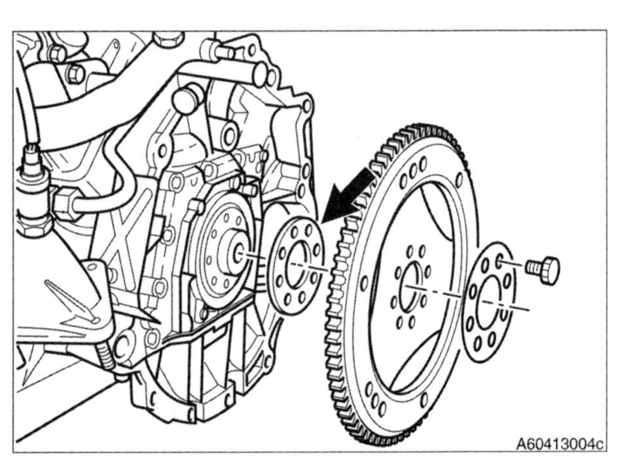

If offset is incorrect, install different shim (**arrow**) between flywheel and crankshaft. Shim thickness is as follows:

- Part no. 054 105 30 = 3.00 mm (0.12 in)
- Part no. 054 105 20 = 4.00 mm (0.16 in)

– Remainder of installation is reverse of removal. Bear in mind:

- Automatic transmission: Check ATF level and top off.
- Cooling system: Check coolant level and top off.
- Engine: Check oil level and top off.

BentleyPublishers.com
BentleyPublishers.com—All Rights Reserved

Rear main seal, removing and installing (V8 engine allroad quattro)

– Raise car and support safely.

> ***WARNING —***
> - *Make sure the car is stable and well supported at all times. Use a professional automotive lift or jack stands designed for the purpose. A floor jack is not adequate support.*

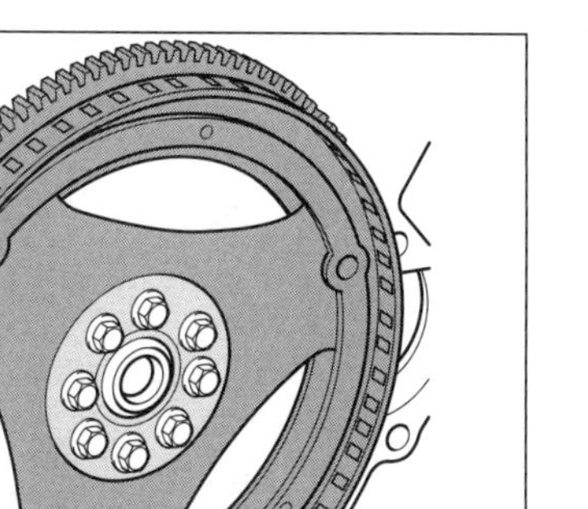

◀ With transmission removed, use special tool 10-201 to counterhold torque plate while loosening mounting bolts. Mark position of torque plate and washer for reinstallation. Set torque plate aside and discard bolts.

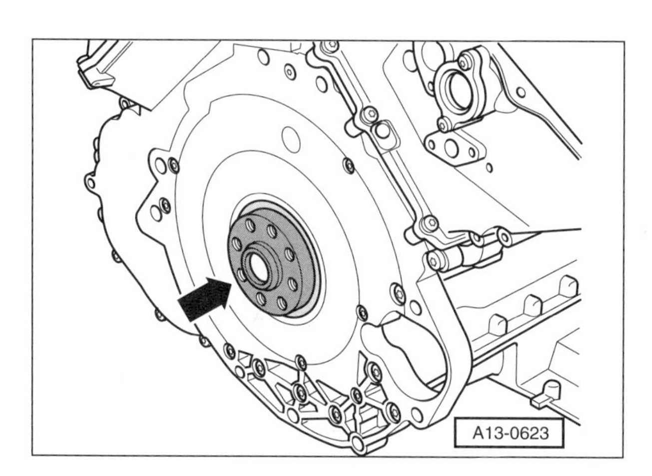

◀ Pull shim (**arrow**) off end of crankshaft. Mark its position for reinstallation.

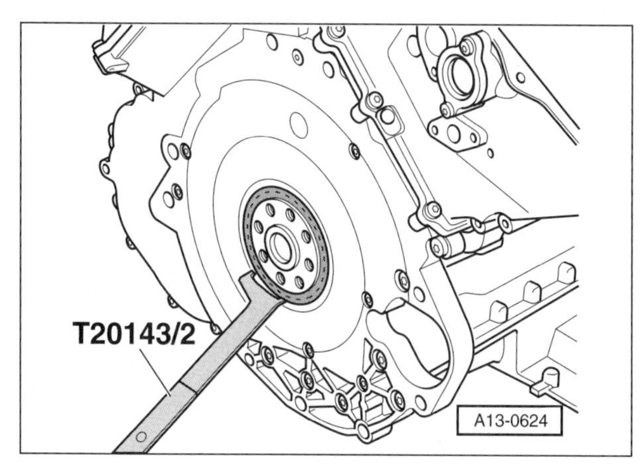

◀ Pry out seal using special prying tool T20143/2.

BentleyPublishers.com
BentleyPublishers.com—All Rights Reserved

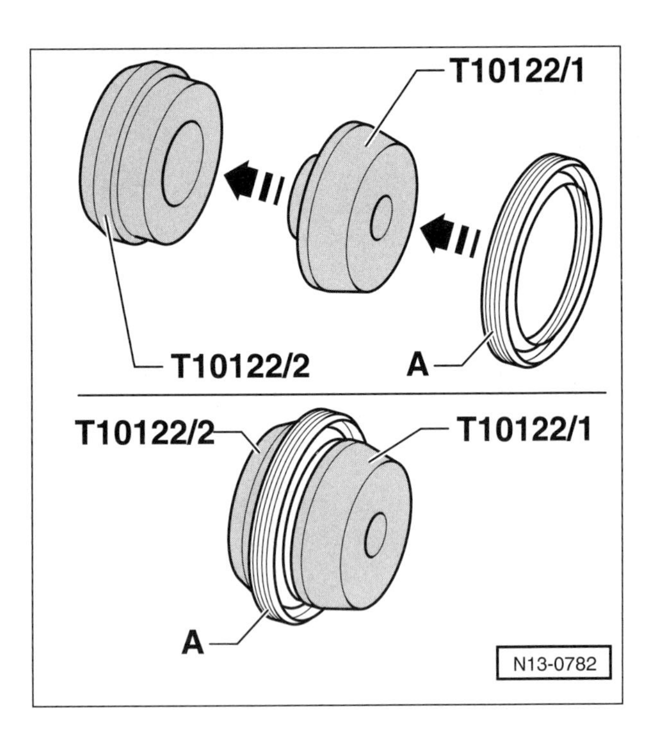

- Insert assembly sleeve T10122/1 on pull sleeve T10122/2. Slide new seal on pull sleeve.
- Remove assembly sleeve.

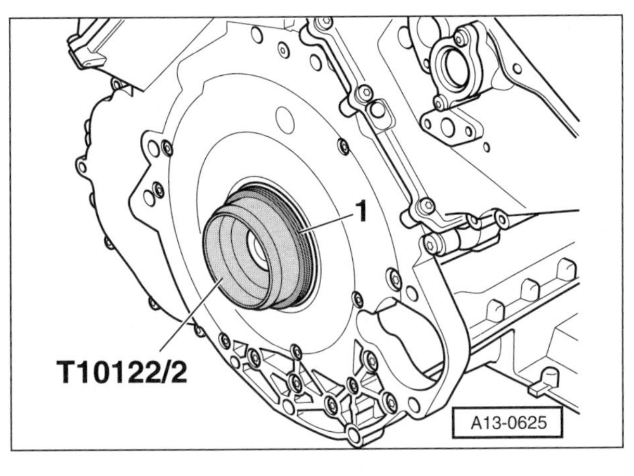

- Install pull sleeve with seal on end of crankshaft.

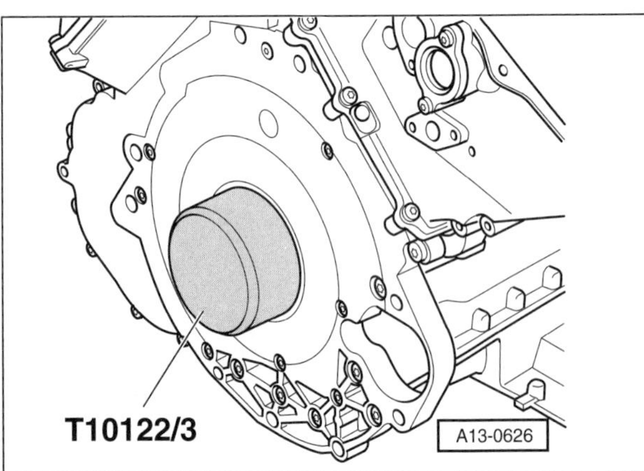

- Use special tool T10122/3 to press in seal evenly until it is flush with surface of cover. Remove pull sleeve.
- Reinstall torque plate, washer and shim according to marks made previously. Use new bolts.

Tightening torque	
Torque plate to crankshaft (use new bolts): • Stage 1 • Stage 2	 30 Nm (22 ft-lb) + ¼ turn (90°)

- Remainder of installation is reverse of removal. Bear in mind:
  - Automatic transmission: Check ATF level and top off.
  - Cooling system: Check coolant level and top off.
  - Engine: Check oil level and top off.

BentleyPublishers.com
BentleyPublishers.com—All Rights Reserved

15 Cylinder Head Cover

GENERAL

This repair group contains information on cylinder head cover removal and installation.

If engine oil leaks are detected or if spark plug wells are filling with oil, remove the cylinder head cover and replace both the outer gasket and the inner spark plug well gasket.

Warnings and Cautions

WARNING—
- *To avoid personal injury, be sure the engine is cold before beginning any procedure in this repair group.*
- *The fuel system is designed to retain pressure even when the ignition is OFF. When working with the fuel system, loosen the fuel lines slowly to allow residual fuel pressure to dissipate. Avoid spraying fuel. Use shop towels to capture leaking fuel.*
- *Before beginning work on the fuel system, place a fire extinguisher in the vicinity of the work area.*
- *Fuel is highly flammable. When working around fuel, do not disconnect wires that could cause electrical sparks. Do not smoke or work near heaters or other fire hazards.*
- *Wear eye protection and protective clothing to avoid injuries from contact with fuel.*
- *Unscrew the fuel tank cap to release pressure in the tank before working on fuel lines.*
- *Do not use a work light with an incandescent bulb near fuel. Fuel may spray on the hot bulb causing a fire.*
- *Make sure the work area is properly ventilated.*

BentleyPublishers.com
BentleyPublishers.com—All Rights Reserved

> **CAUTION—**
> - *When working on internal engine components, maintain absolute cleanliness.*
> - *To prevent damage to vehicle body or paint, use protective body covers.*
> - *Lay removed engine parts on a clean surface and cover immediately. Even small dirt particles can block oil passages.*
> - *Place matching marks on harness connectors, hardware and other components for ease of assembly.*

CYLINDER HEAD COVER (V6 MODELS)

The illustrations in this section illustrate work on a 2000 A6 with 2.7 liter turbocharged engine. The procedures are similar on other models.

Left cylinder head cover, removing and installing (2.7 liter or 2.8 liter engine)

– Remove top engine cover(s). See **03 Maintenance**.

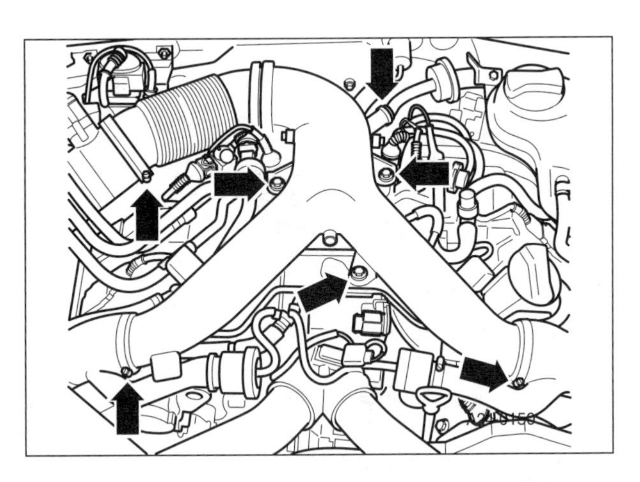

◄ Loosen or remove turbo ram air duct clamps and remove air duct mounting fasteners (**arrows**). Lift off air duct.

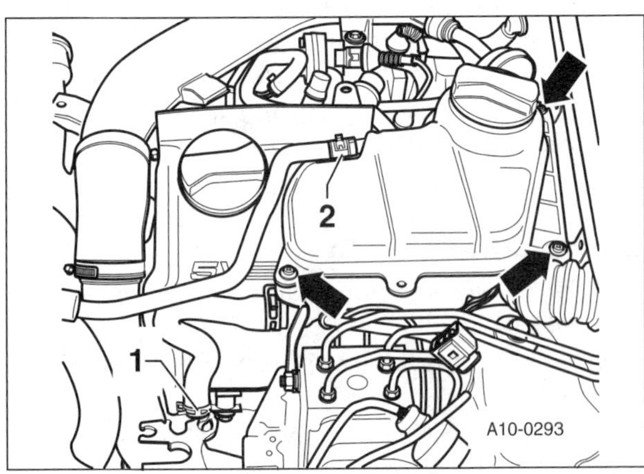

◄ Remove coolant reservoir mounting fasteners (**arrows**).

- Disconnect electrical connector from coolant level sensor at bottom of reservoir.
- Move reservoir aside. Leave coolant hoses connected.

– Remove cover panel from left cylinder head cover.

BentleyPublishers.com
BentleyPublishers.com—All Rights Reserved

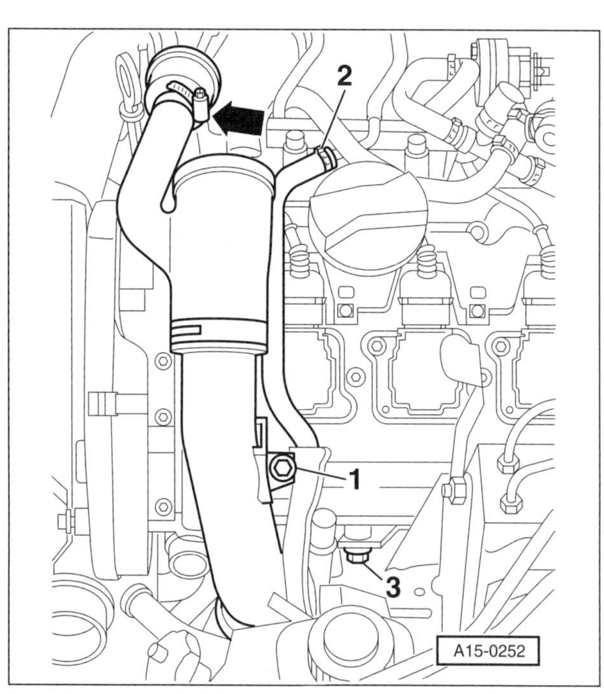

Working at left side of engine:

- Loosen hose clamp (**arrow**).
- Remove intake line (**1**).
- Disconnect hose (**2**).
- Detach coolant line (**3**).
- Plug lower section of intake line.

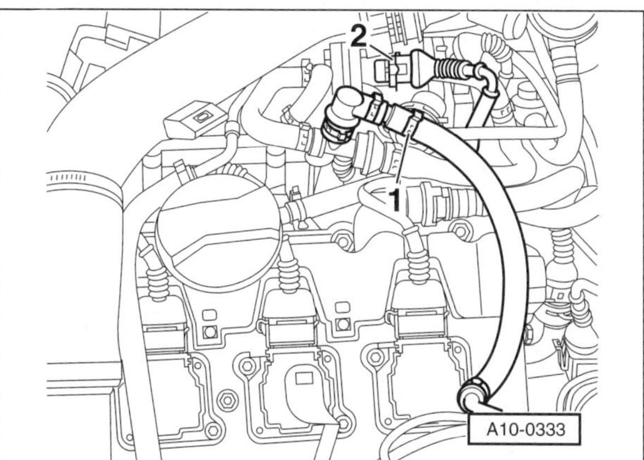

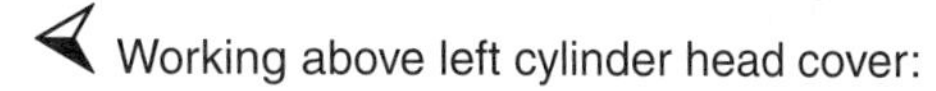

Working above left cylinder head cover:

- Disconnect crankcase breather (**1**) from cylinder head cover.
- Disconnect electrical connectors from ignition coils and remove coils. See **28 Ignition System**.

NOTE —

- *2.8 liter engine: Remove spark plug wires.*

– Remove cylinder head cover.

Installation is reverse of removal; note the following:

- Replace cylinder head cover gasket and spark plug well gasket.
- Seal end points of joints (**arrows**) between bearing caps and cylinder head. Apply small quantity of sealant (D 454 300 A2 or equivalent) at four end points.
- Tighten cylinder head cover fasteners in crisscross pattern, starting from middle fasteners.

Tightening torque	
Cylinder head cover to cylinder head (M6)	10 Nm (7 ft-lb)

BentleyPublishers.com
BentleyPublishers.com—All Rights Reserved

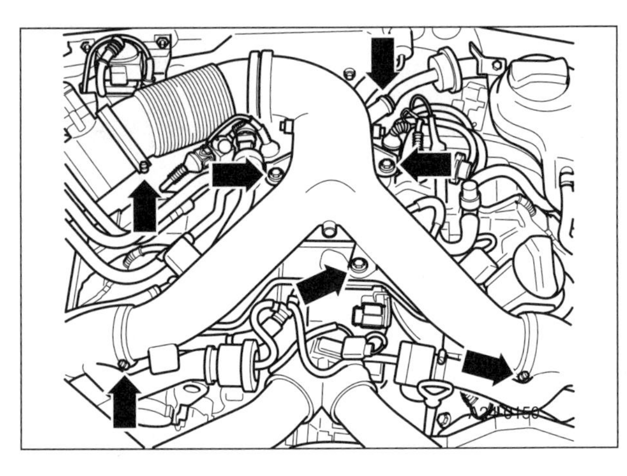

Right cylinder head cover, removing and installing (2.7 liter or 2.8 liter engine)

– Remove top engine cover(s). See **03 Maintenance**.

– Remove cover above air filter housing.

◄ Loosen or remove turbo ram air duct clamps and remove air duct mounting fasteners (**arrows**). Lift off air duct.

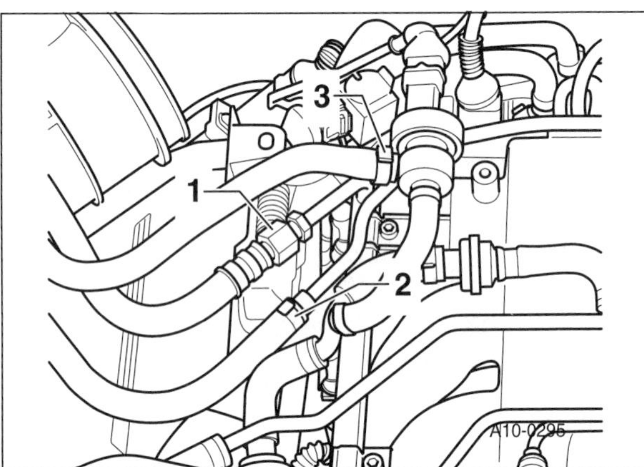

◄ Wrap shop towel around fuel connections, then disconnect fuel supply and return lines (**1** and **2**) and move fuel lines clear.

> ***WARNING—***
>
> - *The fuel system is designed to retain pressure even when the ignition is OFF. When working with the fuel system, loosen the fuel lines slowly to allow residual fuel pressure to dissipate. Avoid spraying fuel. Use shop towels to capture leaking fuel.*

– Disconnect hose from fuel tank evaporative control (EVAP) valve (**3**).

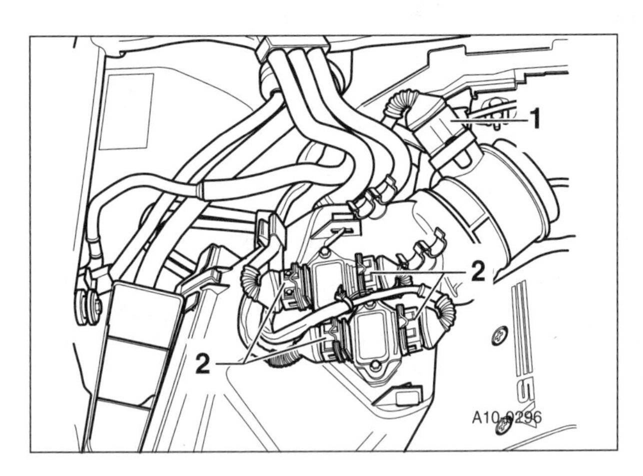

◄ Working at right side of engine:

- Disconnect electrical connector (**1**) from mass air flow sensor.
- Disconnect electrical connectors (**2**) from ignition output stages and move wiring clear.

– Remove air filter housing.

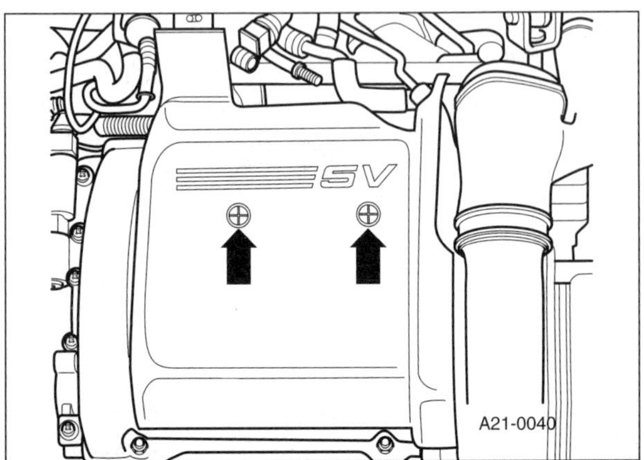

◄ Loosen fasteners (**arrows**) and remove cover panel from right cylinder head cover.

BentleyPublishers.com
BentleyPublishers.com—All Rights Reserved

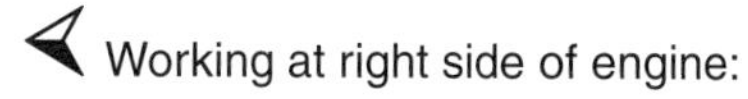

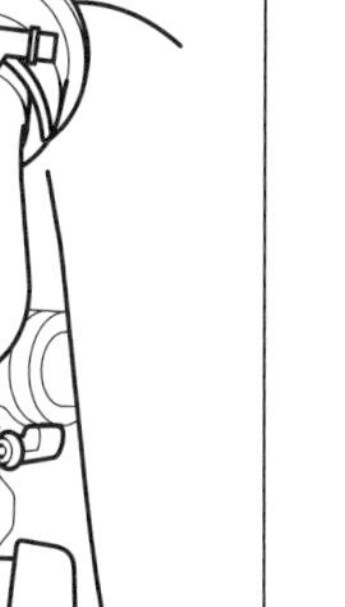

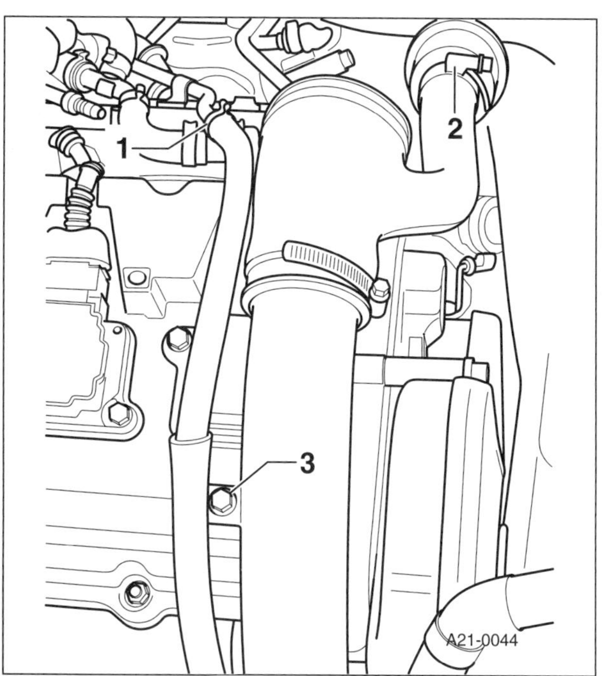

◀ Working at right side of engine:

- Disconnect hose (**1**).
- Loosen hose clamp (**2**).
- Detach upper section of intake line (**3**).
- Plug lower section of intake line.

– Remove bolt securing timing belt guard from cylinder head cover.

– Disconnect electrical connectors from ignition coils and remove coils. See **28 Ignition System**.

NOTE —

• 2.8 liter engine: Remove spark plug wires.

– Detach crankcase breather from cylinder head cover.

– Remove cylinder head cover.

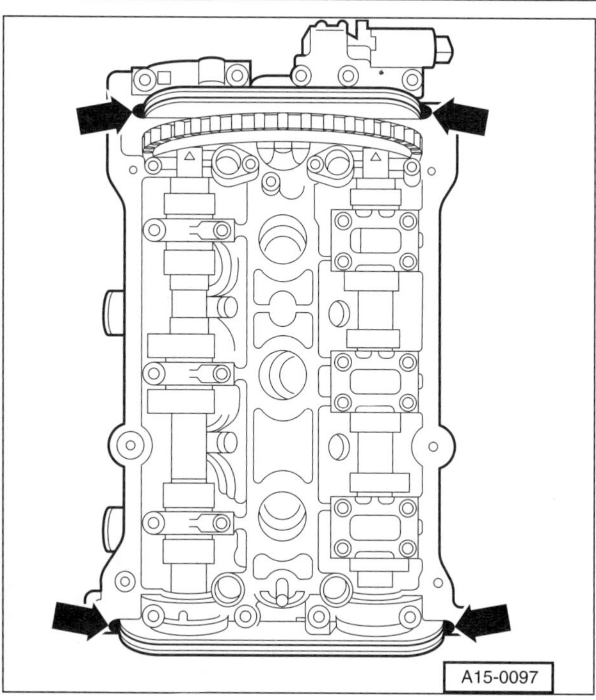

◀ Installation is reverse of removal; note the following:

- Replace cylinder head cover gasket and spark plug well gasket.
- Seal end points of joints (**arrows**) between bearing caps and cylinder head. Apply small quantity of sealant (D 454 300 A2 or equivalent) at four end points.
- Tighten cylinder head cover fasteners in crisscross pattern, starting from middle fasteners.

Tightening torque	
Cylinder head cover to cylinder head (M6)	10 Nm (7 ft-lb)

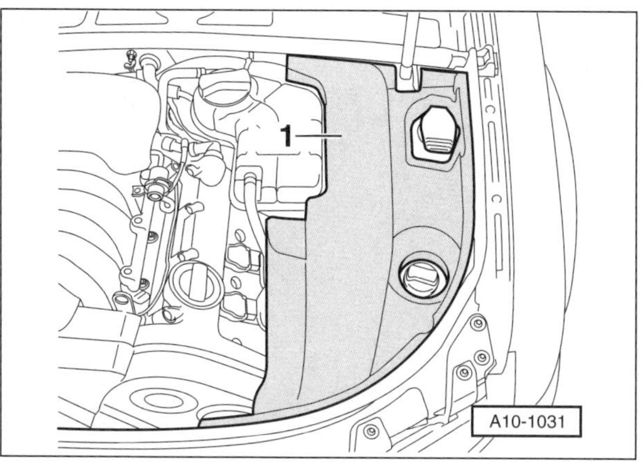

Left cylinder head cover, removing and installing (3.0 liter engine)

– Remove front and rear engine upper covers. See **03 Maintenance**.

◀ Remove cover (**1**) in engine compartment (left side).

BentleyPublishers.com
BentleyPublishers.com—All Rights Reserved

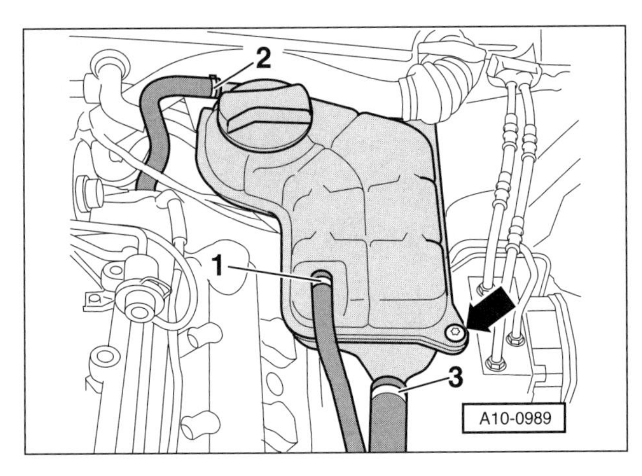

◀ Remove coolant reservoir mounting fastener (**arrow**).

- Disconnect electrical connector from coolant level sensor at bottom of reservoir.
- Move reservoir aside. Leave coolant hoses connected.

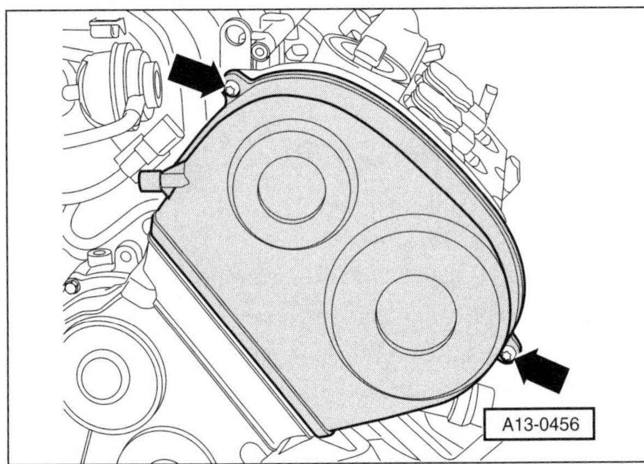

◀ Remove timing belt guard mounting fasteners (**arrows**) from left cylinder head cover.

– Disconnect electrical connectors from ignition coils and remove coils. See **28 Ignition System**.

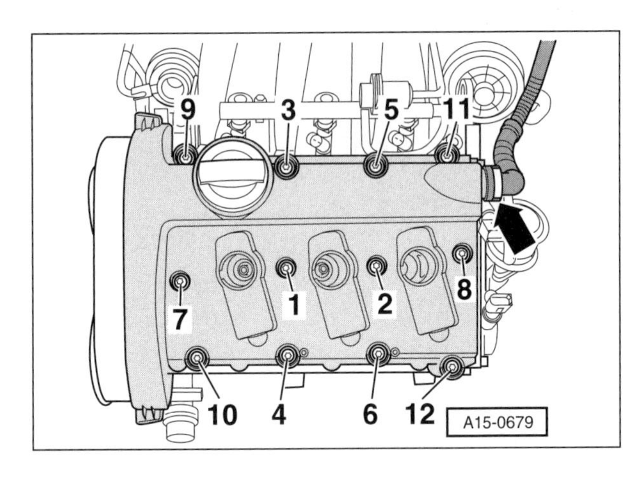

◀ Working above left cylinder head:

- Detach crankcase breather (**arrow**) from cylinder head cover.
- Unscrew cylinder head cover bolts in sequence, **12** to **1**.
- Remove cylinder head cover.

– Installation is reverse of removal; note the following:

- Replace cylinder head cover gasket and spark plug well gasket.
- Use new self-locking fasteners.
- Tighten cylinder head cover fasteners in sequence, **1** to **12**.

Tightening torques	
Cylinder head cover to cylinder head (M6) (use new fasteners)	10 Nm (7 ft-lb)
Timing belt guard to cylinder head cover (use new fasteners)	6 Nm (53 in-lb)

BentleyPublishers.com
BentleyPublishers.com—All Rights Reserved

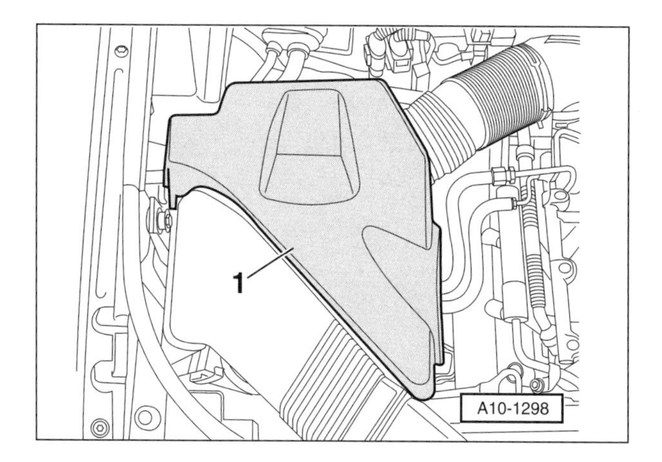

Right cylinder head cover, removing and installing (3.0 liter engine)

– Remove front and rear engine upper cover(s). See **03 Maintenance**.

◄ Remove air filter housing cover (**1**) in engine compartment (right side).

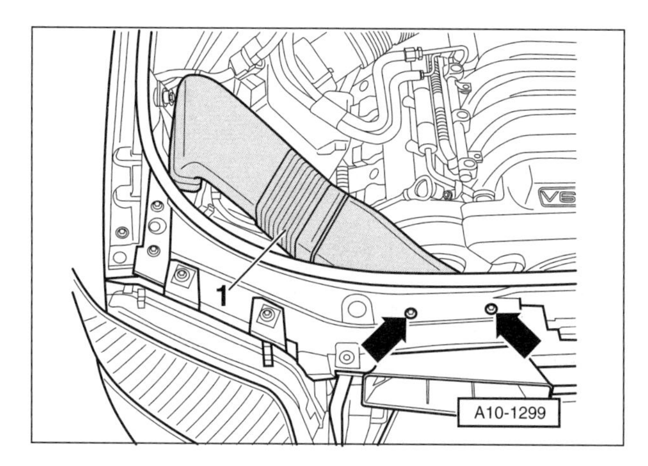

◄ Remove air duct mounting bolts (**arrows**) at lock carrier (radiator support frame). Remove air duct (**1**).

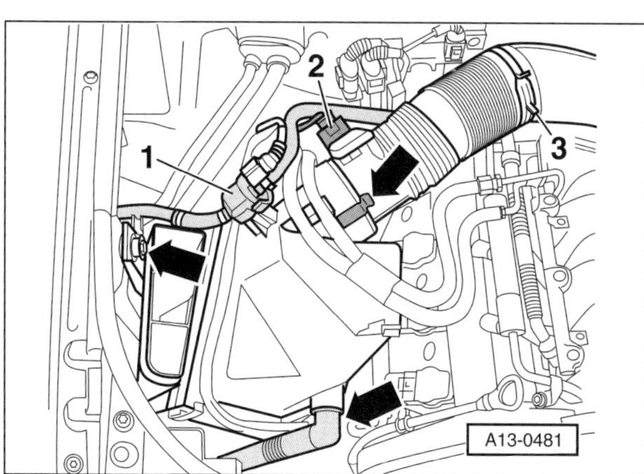

◄ Working in right side of engine compartment:

- Disconnect fuel tank evaporative control (EVAP) valve (**1**) at air filter housing.
- Disconnect electrical harness connector (**2**) at mass air flow (MAF) sensor.
- Remove engine intake air duct (**3**) together with MAF sensor.
- Unclip air filter housing clips (**arrows**). Lift air filter housing out.

BentleyPublishers.com
BentleyPublishers.com—All Rights Reserved

Cylinder Head Cover (V6 Models)

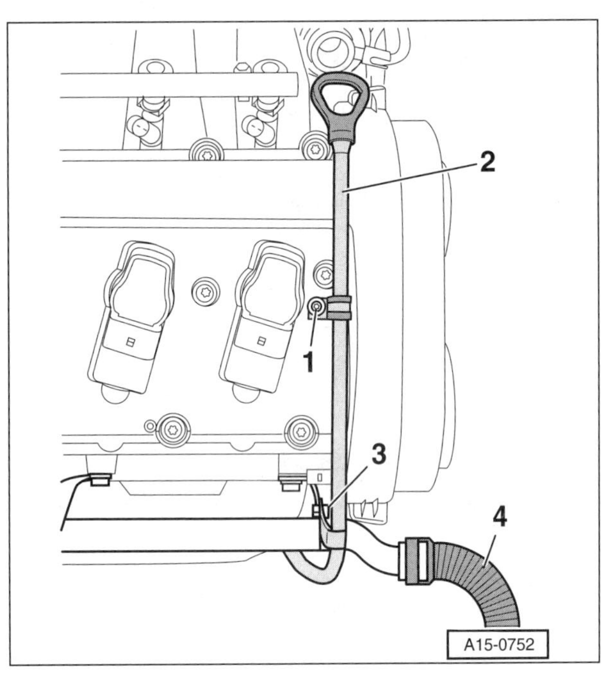

Working above right cylinder head:

- Remove dipstick mounting bolts (**1** and **3**).
- Disconnect secondary air injector hose (**4**).
- Pull dipstick guide tube (**2**) out of oil pan. Swing forward for removal.

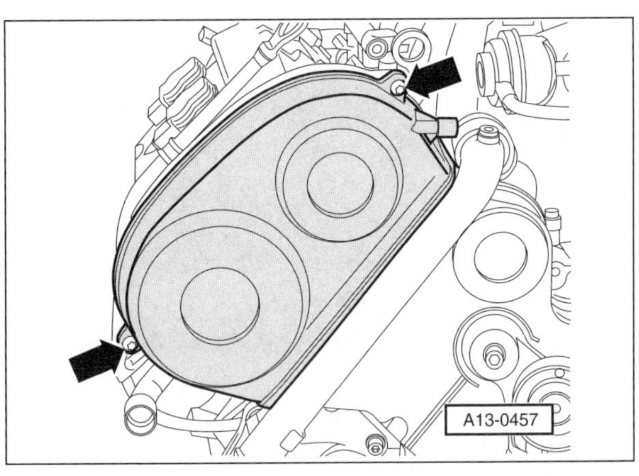

Remove timing belt guard mounting fasteners (**arrows**) from right cylinder head cover.

– Disconnect electrical connectors from ignition coils and remove coils. See **28 Ignition System**.

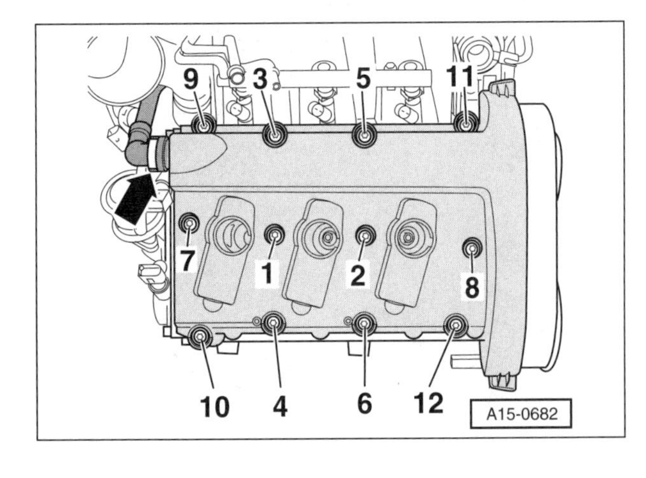

Working above right cylinder head:

- Detach crankcase breather (**arrow**) from cylinder head cover.
- Unscrew cylinder head cover bolts in sequence, **12** to **1**.
- Remove cylinder head cover.

– When reinstalling cylinder head cover, note the following:

- Replace cylinder head cover gasket and spark plug well gasket.
- Use new self-locking fasteners.
- Tighten cylinder head cover fasteners in sequence, **1** to **12**.

Tightening torques	
Cylinder head cover to cylinder head (M6) (use new fasteners)	10 Nm (7 ft-lb)
Timing belt guard to cylinder head cover (use new fasteners)	6 Nm (53 in-lb)

BentleyPublishers.com
BentleyPublishers.com—All Rights Reserved

– Raise car and support safely.

> ***WARNING —***
> • *Make sure the car is stable and well supported at all times. Use a professional automotive lift or jack stands designed for the purpose. A floor jack is not adequate support.*

– Remove engine lower cover. See **03 Maintenance**.

– Replace dipstick guide tube O-ring gasket. Insert guide tube into oil pan, then secure fasteners to cylinder head cover.

Tightening torque	
Dipstick guide tube to cylinder head cover	10 Nm (7 ft-lb)

Cylinder Head Cover (V8 Models)

The illustrations in this section illustrate work on a 2000 A6 with V8 engine. The procedures are similar on other normally aspirated V8 models.

On the biturbo RS6 engine (engine code BCY), the left cylinder head cover can be removed with the engine installed. Right cylinder head cover removal, however, requires engine removal.

Left cylinder head cover, removing and installing (V8 models)

– Remove top engine cover(s). See **03 Maintenance**.

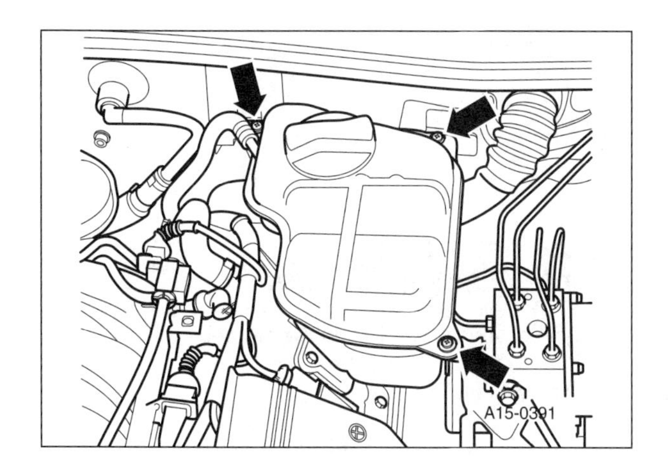

◀ Remove coolant reservoir mounting fasteners (**arrows**).

- Disconnect electrical connector from coolant level sensor at bottom of reservoir.
- Move reservoir aside. Leave coolant hoses connected.

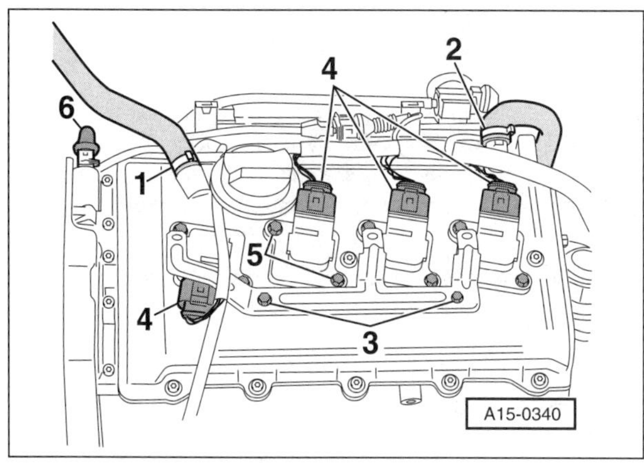

◀ Working above left cylinder head:

- Detach breather hoses (**1** and **2**).
- Remove ignition coil bracket mounting bolts (**3**).
- Disconnect ignition coil connectors (**4**).
- Remove ignition coil mounting bolts (**5**) and remove coils.

– Remove cylinder head cover.

BentleyPublishers.com
BentleyPublishers.com—All Rights Reserved

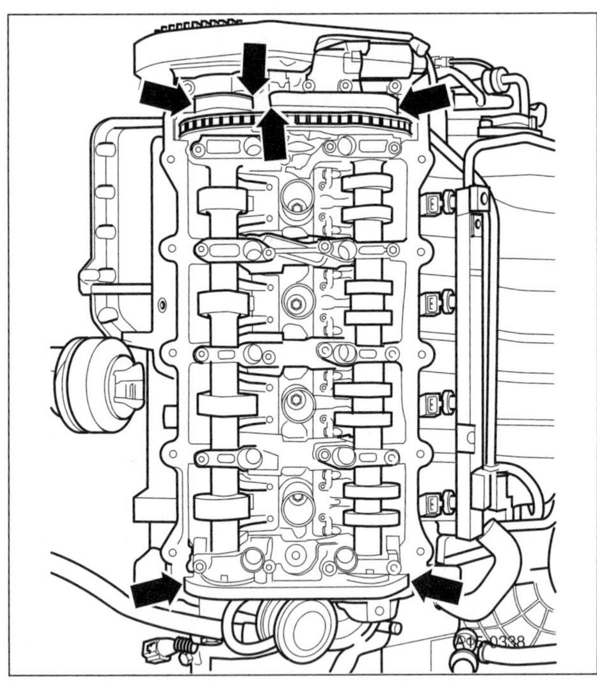

Installation is reverse of removal; note the following:

- Replace cylinder head cover gasket and spark plug well gasket.
- Apply small quantity of sealant (D 454 300 A2 or equivalent) to cylinder head cover sealing points (**arrows**).

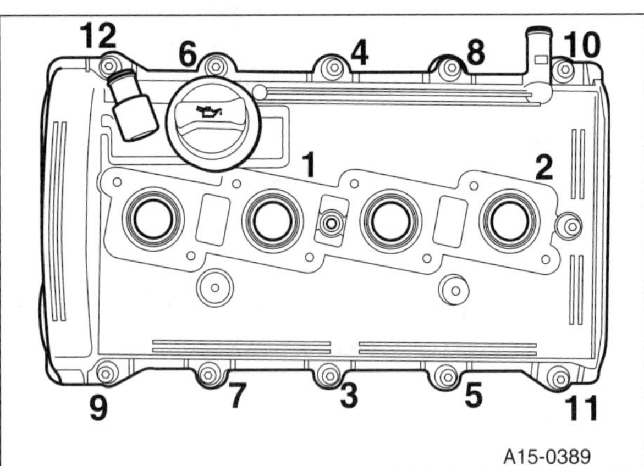

Tighten cylinder head cover fasteners in sequence, **1** to **12**.

Tightening torques	
Cylinder head cover to cylinder head	10 Nm (7 ft-lb)
Ignition coil to cylinder head cover	10 Nm (7 ft-lb)

Right cylinder head cover, removing and installing (V8 models)

– Remove top engine cover(s). See **03 Maintenance**.

– Remove intake air duct for air cleaner.

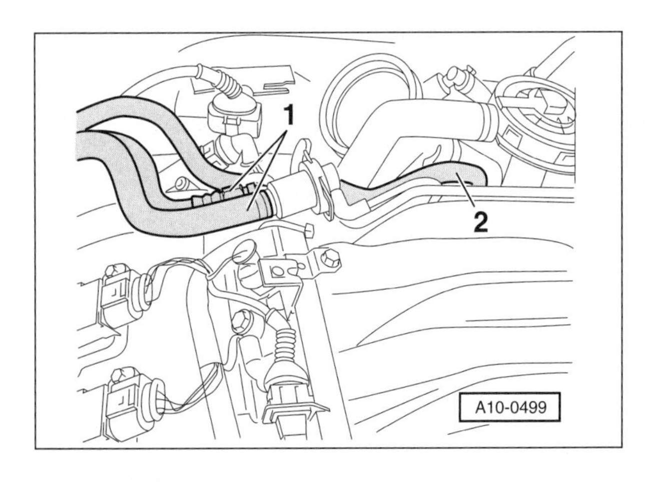

Wrap a shop towel around fuel connections, then disconnect fuel supply and return lines (**1**) and move fuel lines clear.

> ***WARNING —***
> - *The fuel system is designed to retain pressure even when the ignition is OFF. When working with the fuel system, loosen the fuel lines slowly to allow residual fuel pressure to dissipate. Avoid spraying fuel. Use shop towels to capture leaking fuel.*

– Disconnect hose from fuel tank evaporative control (EVAP) valve (**2**).

BentleyPublishers.com
BentleyPublishers.com—All Rights Reserved

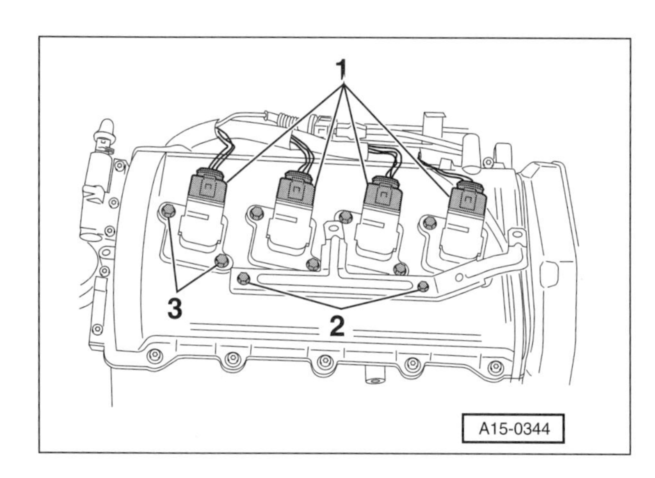

Working above right cylinder head:

- Disconnect ignition coil connectors (**1**).
- Remove ignition coil bracket mounting bolts (**2**).
- Remove ignition coil mounting bolts (**3**) and remove coils.

– Disconnect crankcase breather hose.

– Remove cylinder head cover.

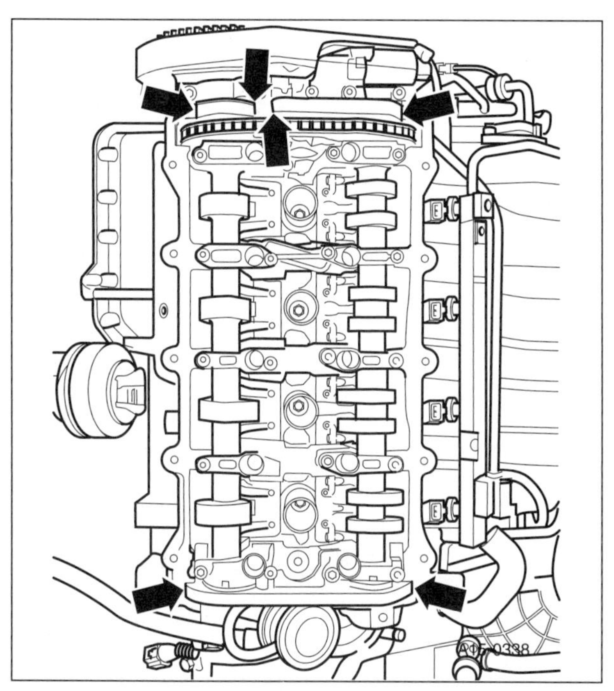

Installation is reverse of removal; note the following:

- Replace cylinder head cover gasket and spark plug well gasket.
- Apply small quantity of sealant (D 454 300 A2 or equivalent) to cylinder head cover sealing points (**arrows**).

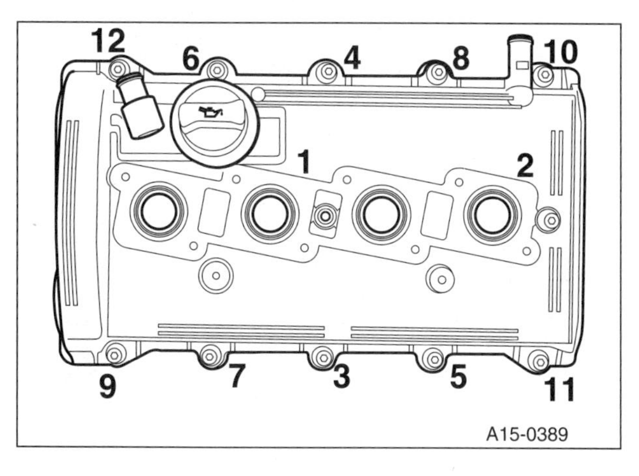

Tighten cylinder head cover fasteners in sequence, **1** to **12**.

Tightening torques	
Cylinder head cover to cylinder head	10 Nm (7 ft-lb)
Ignition coil to cylinder head cover	10 Nm (7 ft-lb)

BentleyPublishers.com
BentleyPublishers.com—All Rights Reserved

17 Engine–Lubrication System

TABLES

General

The repair group covers the A6 engine lubrication system.

Engine Oil

Lubricating oil for engines covered by this manual are required to meet quality and performance standards specified by Audi. These standards are sufficiently high so that, generally, only synthetic oils meet them. Current specifications for engine oil are shown in **Table a**. For oil change procedures, see **03 Maintenance**.

Table a. Oil specifications	
Type: • American Petroleum Institute • European industry standard	 API service SJ ACEA A2 or ACEA A3
Viscosity grade: • New engine • At oil change • High temperature service (above 40°C or 100°F)	 SAE 0W30 SAE 5W30 SAE 5W40

Approximate refill capacities in **Table b** are listed by engine code. For engine application and code information, see **03 Maintenance**.

Table b. Oil capacities (approximate)	
Engine (engine code)	**Oil capacity (includes oil filter) in liters (US qt)**
2.7 liter V6 (A6, S6) (APB)	6.5 (6.9)
2.7 liter V6 (allroad quattro)(BEL)	6.9 (7.3)
2.8 liter V6 (AHA, ATQ)	5.7 (6.0)
3.0 liter V6 (AVK)	6.4 (6.8)

BentleyPublishers.com
BentleyPublishers.com—All Rights Reserved

Table b. Oil capacities (approximate)	
Engine (engine code)	Oil capacity (includes oil filter) in liters (US qt)
4.2 liter V8 (A6, S6, RS6) (ART, AWN, BBD, BCY)	7.6 (8.0)
4.2 liter V8 (allroad quattro)(BAS)	8.5 (8.9)

CAUTION—
- *Lubricant specifications are subject to change. Consult your authorized dealer for the latest information regarding lubricant applications and specifications.*

Oil Pressure

◀ The oil pressure switch (F1) is in the oil pan at the base of the oil filter housing.

Oil pressure and oil pressure switch, testing

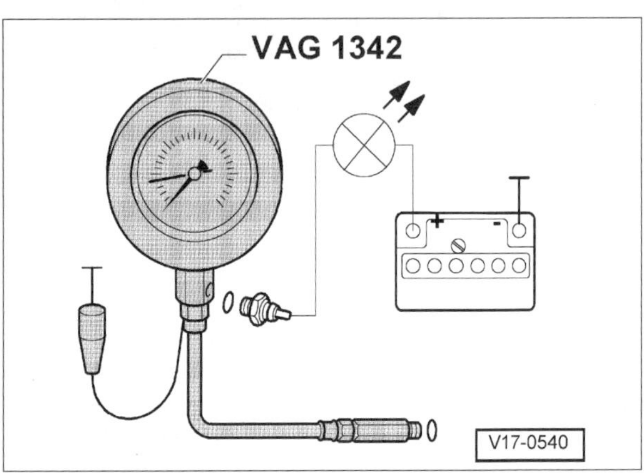

◀ Remove oil pressure switch and screw into oil pressure tester (special tool VAG 1324).

- Screw tester into oil pan in place of oil pressure switch.
- Connect brown wire of tester to ground.
- Connect LED voltage tester (special tool VAG 1527B) to B+ (battery positive terminal) and to oil pressure switch.
- Start engine and gradually increase engine speed.
  - Make sure LED lights up at 1.2 - 1.6 bar (17.4 - 23.2 psi). If not, replace oil pressure switch.
- Warm up engine so that oil pressure is at least 80°C (176°F). Read off oil pressure at 2000 rpm and compare to specification.

Oil pressure specification	
Oil pressure at 2000 rpm, oil at 80°C (176°F)	min 2.0 bar (29 psi)

BentleyPublishers.com
BentleyPublishers.com—All Rights Reserved

LUBRICATION SYSTEM

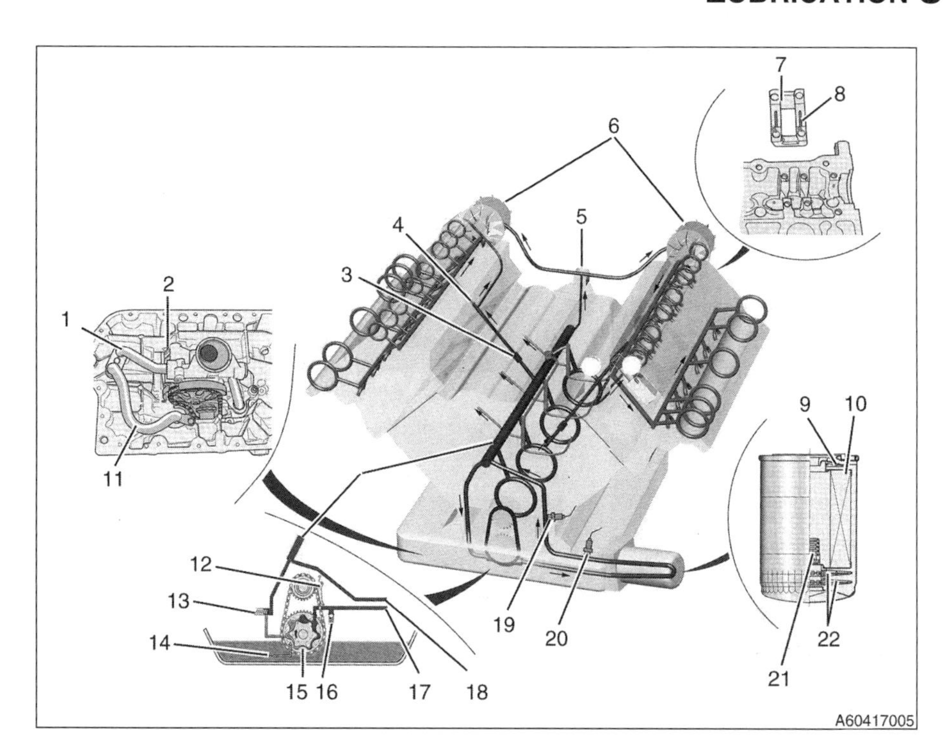

2.7 liter engine lubrication system

1. to Oil filter and oil cooler
2. Oil pressure relief valve
3. Oil retention valve
4. Restrictor
5. Oil distributor
6. Turbochargers
7. Camshaft bearing cap
8. Oil groove
9. Oil retention valve
10. Oil filter element
11. from Oil filter and oil cooler
12. Chain tensioner
13. Oil pressure control valve
14. Oil pickup filter
15. Duocentric oil pump
16. Oil pressure relief valve
17. to Oil filter and oil cooler
18. from Oil filter and oil cooler
19. Oil temperature sender
20. Oil pressure switch
21. Bypass valve
22. Bypass filter

NOTE—

- *2.8 liter engine lubrication system is similar.*

BentleyPublishers.com
BentleyPublishers.com—All Rights Reserved

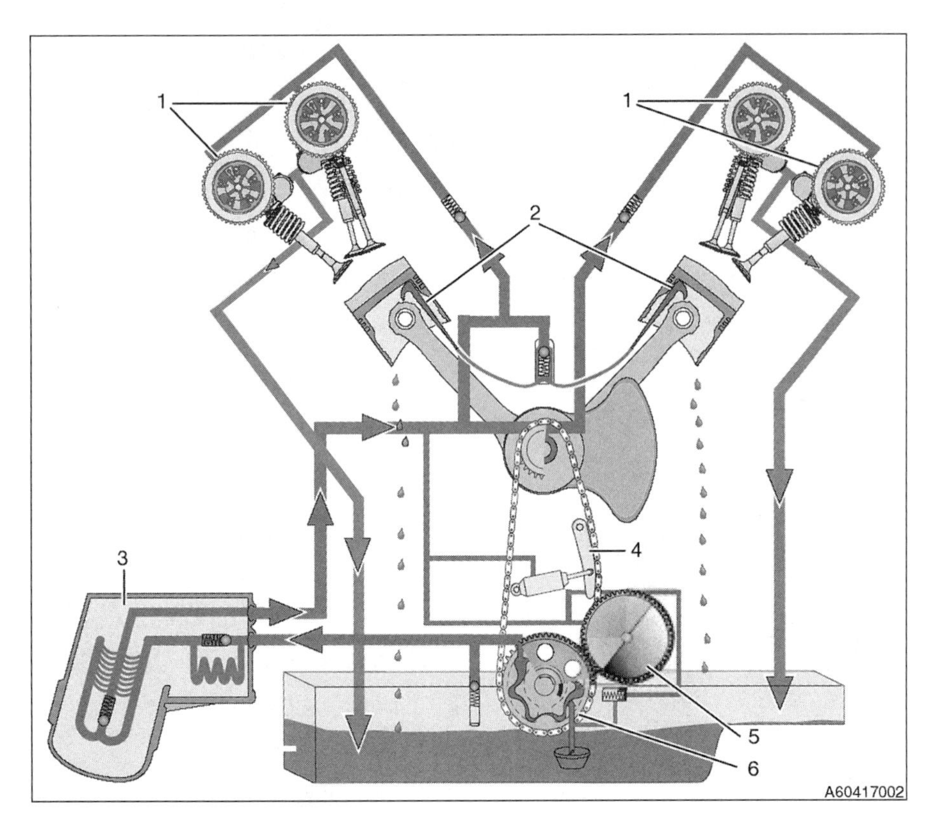

3.0 liter engine lubrication system

1. **Camshaft bearing caps**
2. **Piston cooling jets**
3. **Oil filter housing**
4. **Oil pump drive chain tensioner**
5. **Balance shaft**
   - Driven by oil pump gear
6. **Duocentric oil pump**

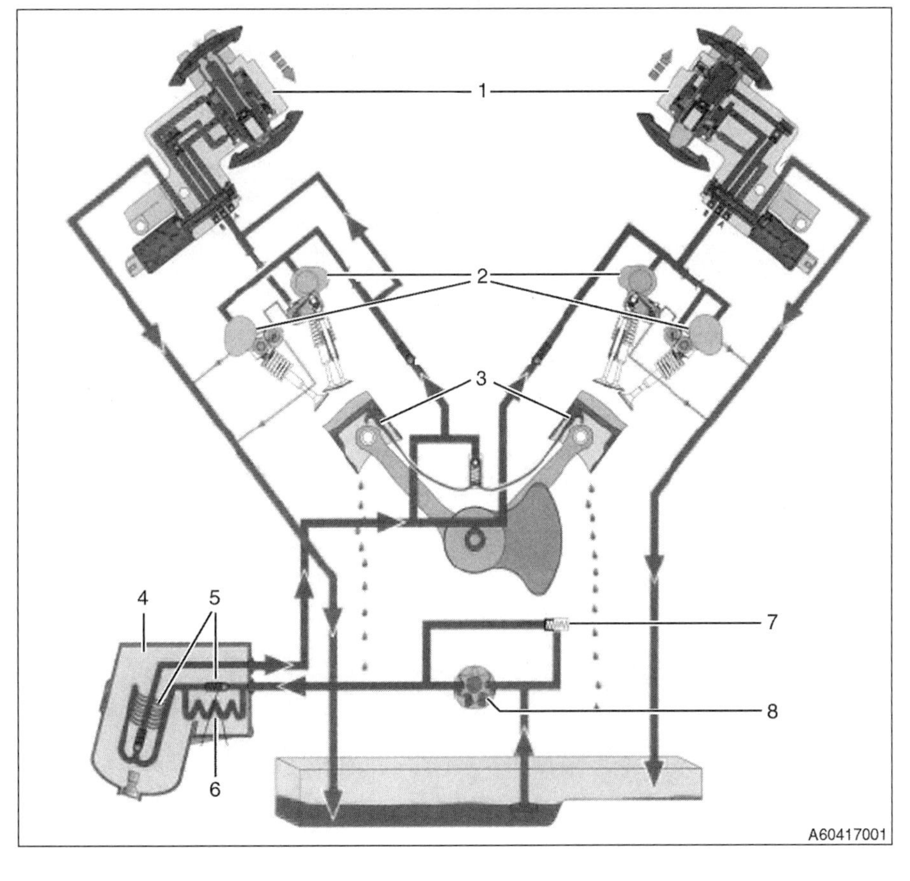

V8 engine lubrication system

1. **Variable camshaft timing assemblies**
2. **Camshafts**
3. **Piston cooling jets**
4. **Oil filter**
5. **Bypass valves**
6. **Oil cooler**
7. **Oil pressure control valve**
8. **Duocentric oil pump**

BentleyPublishers.com
BentleyPublishers.com—All Rights Reserved

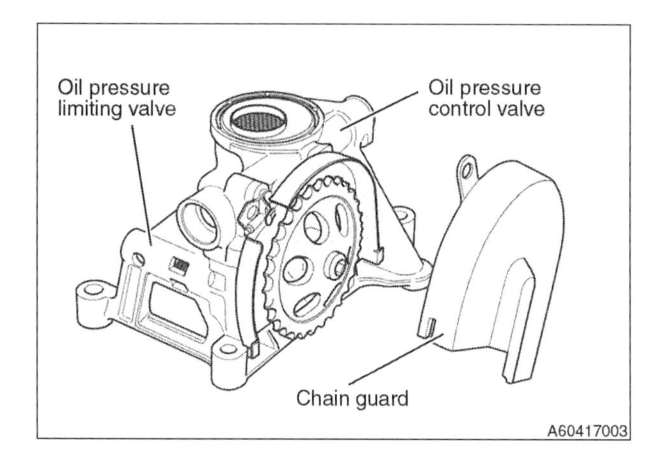

Oil pump

V6 engine: Duocentric oil pump, chain-driven from the crankshaft, is attached to the bottom of the crankcase and projects into the oil sump. It is immersed completely in engine oil when oil level is correct. This prevents the pump from running dry.

This oil pump design, in combination with the extremely short intake path, assures that oil pressure builds up quickly, particularly during cold starts.

The sheet metal chain guard encapsulates the oil pump drive chain and sprocket and sharply reduces oil frothing.

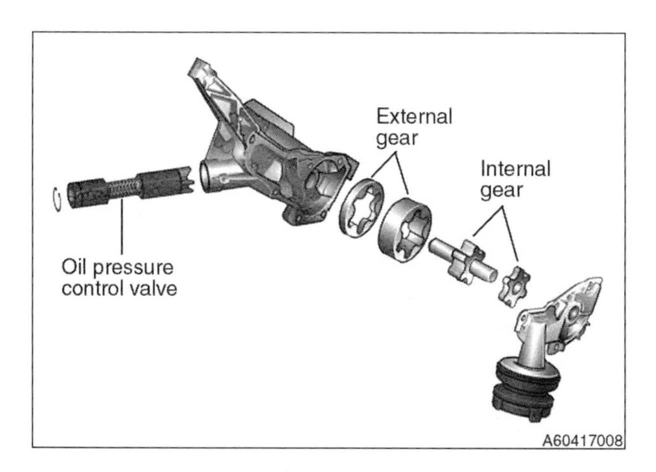

V8 engine: Oil pump is similar in design and placement.

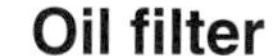

Oil filter

V6 engine: Oil filter unit contains oil retention valve, filter element, bypass filter and filter bypass valve. If filter element becomes clogged or if oil is too thick (high viscosity), bypass valve maintains engine lubrication.

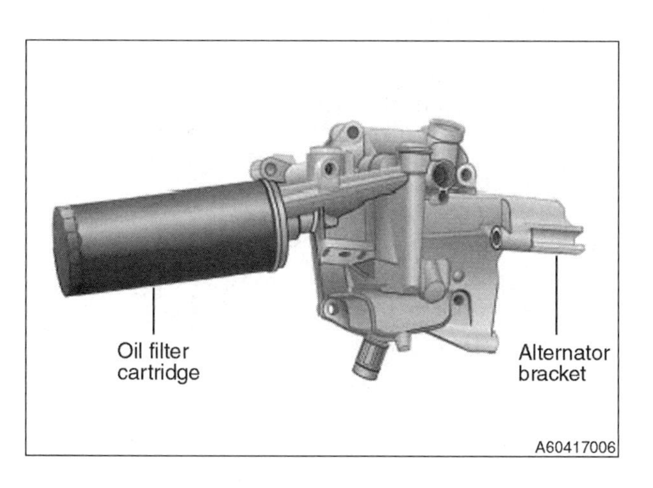

V8 engine: Oil filter housing and function is similar to V6 engine, but for space reasons a filter cartridge is used instead of paper insert.

Oil cooler

2.7 liter turbocharged engine: Oil cooler unit is integrated into the primary oil flow. The capacity and flow resistance are optimized to allow all the oil to flow through the cooler, unlike in the normally aspirated engine, where an oil bypass line is required.

BentleyPublishers.com
BentleyPublishers.com—All Rights Reserved

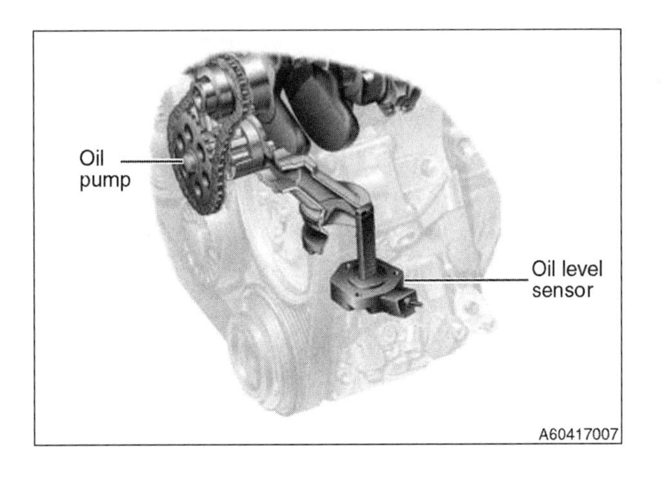

Oil level sensor

V8 engine: Oil level sensor functions as oil quality sensor as well. This allows oil service interval to be calculated. An oil level warning is displayed in the instrument cluster.

Oil passages and valves

Oil pressure control valve is integrated in the oil pump housing. The oil diverted by the control valve is fed to the suction side of the pump, thus optimizing oil pump efficiency.

Oil pressure limiting valve is another internal pressure control valve which opens when oil pressure rises too high, as during a cold start.

Oil retention valves prevent oil running out of the oil filter and the cylinder heads and back into the oil sump while the engine is not running.

Oil passage restrictorsprevent oil flooding of the cylinder heads at high engine speeds.

Control valve for piston cooling oilopens once the oil pressure is above 26 psi (1.8 bar). This allows the low-viscosity engine oil to maintain high pressure. Also, at low engine speeds piston cooling is not necessary.

Integrated oil supply

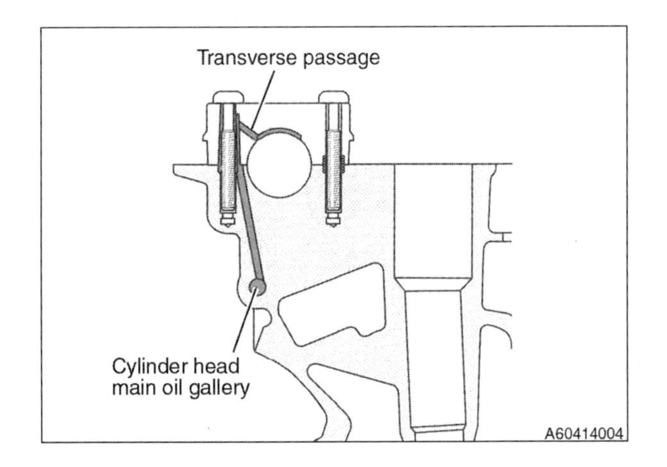

V6 engines installed in A6 models feature the integrated oil supply concept. Each camshaft bearing is supplied via a passage stemming from the cylinder head main oil gallery. The oil is fed along a bolt bore to a transverse passage. A lubrication groove distributes the oil throughout the camshaft bearing.

BentleyPublishers.com
BentleyPublishers.com—All Rights Reserved

19 Engine–Cooling System

19

TABLE

General

This repair group covers the engine cooling system.

See **87 Heating and Air-conditioning** for cooling system relay and fuse locations.

Coolant and antifreeze

Cooling system capacity depends on a variety of options and configurations. If a repair procedure includes opening the sealed cooling system, make sure to fill and bleed the system after repairs. See **Cooling system draining, filling and bleeding** in this repair group.

Table a. Cooling system capacities

Engine	Cooling system capacity in liters (US qt)
2.7 liter V6 (APB, BEL)	6.0 (6.3)
2.8 liter V6 (AHA, ATQ)	8.0 (8.5)
3.0 liter V6 (AVK)	8.0 (8.5)
4.2 liter V8 (ART, AWN, BBD)	9.0 (9.5)
4.2 liter V8 biturbo (RS6) (BCY)	12.0 (12.7)

BentleyPublishers.com
BentleyPublishers.com—All Rights Reserved

The cooling system is factory filled with a lifetime coolant mixture of 50% silicate-free antifreeze and 50% water. If topping up or replacing coolant, use fresh antifreeze mixture. Audi recommends using G12 low-phosphate silicate-free coolant year-round. In an emergency, use clear water.

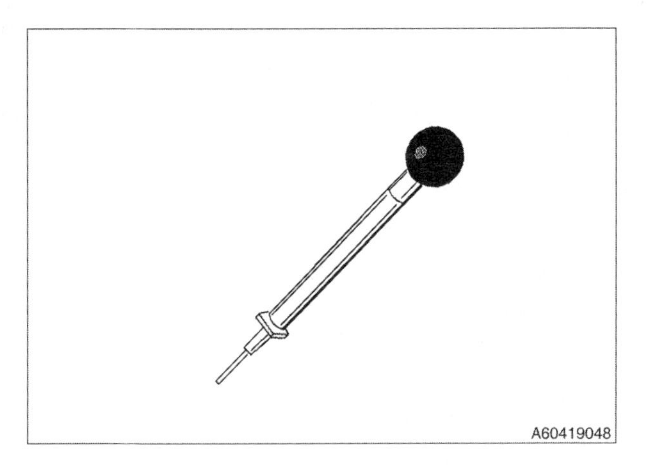

Use a coolant hydrometer to determine antifreeze concentration.

Coolant mixture recommendations	
Concentration	**Cold protection**
50% antifreeze	-35°C (-31°F)
60% antifreeze	-40°C (-40°F)

Antifreeze concentration may also be tested with a refractometer, special tool T10007

Do not use a higher concentration of antifreeze than a 60% mixture, as the heat transfer quality of the coolant decreases with higher antifreeze concentrations.

Do not allow antifreeze concentration to drop below 40%.

Warnings and Cautions

Observe the following warnings and cautions when working on the cooling system.

WARNING —

- *At normal operating temperature the cooling system is pressurized. Allow the engine to cool thoroughly (a minimum of one hour), then cover the overflow tank pressure cap with a cloth and open carefully to relieve system pressure slowly.*
- *Releasing cooling system pressure lowers the boiling point of coolant and it may boil suddenly. Use heavy gloves and wear eye and face protection to guard against scalding.*
- *Use extreme care when draining and disposing of engine coolant. Coolant is poisonous and lethal to humans and pets. Pets are attracted to coolant because of its sweet smell and taste. Seek medical attention immediately if coolant is ingested.*

BentleyPublishers.com
BentleyPublishers.com—All Rights Reserved

CAUTION—

- *Replace sealing O-rings and gaskets when working on cooling system.*
- *When installing coolant hoses, make sure they are free of stress and do not come into contact with other components. Make sure arrows embossed on coolant pipes and coolant hoses face each other.*
- *Avoid adding cold water to the cooling system while the engine is hot or overheated. If it is necessary to add coolant to a hot system, do so only with the engine running and coolant pump turning.*
- *To avoid excess silicate gel precipitation in the cooling system and loss of cooling capacity, use Audi coolant G12 or equivalent silicate-free antifreeze.*
- *G12 coolant is red in color. Do not mix with other coolant types.*
- *If the fluid in the expansion tank is brown, G12 coolant has been mixed with another coolant. Flush cooling system and change coolant. See* **Cooling system draining, filling and bleeding** *in this repair group.*
- *Dispose of coolant in an environmentally safe manner.*
- *If oil enters the cooling system, flush with cleaning agent.*
- *Fill the cooling system year round with frost and corrosion protection additives included in approved coolant.*
- *Be sure to protect the engine against freezing to approx. -25°C (-13°F). In an arctic climate, protect to -35°C (-31°F).*
- *In summer, do not reduce antifreeze concentration below 40%.*
- *In extreme arctic conditions, do not increase the antifreeze concentration above 60%.*
- *Mix antifreeze with clean drinking water.*
- *If the radiator, heater core, cylinder head, cylinder head gasket or engine block is replaced, completely replace the engine coolant. Do not reuse old coolant.*
- *Prior to disconnecting the battery, read the battery disconnection precautions given in* **00 Warnings and Cautions**.

BentleyPublishers.com
BentleyPublishers.com—All Rights Reserved

2.7 Liter Engine Cooling System

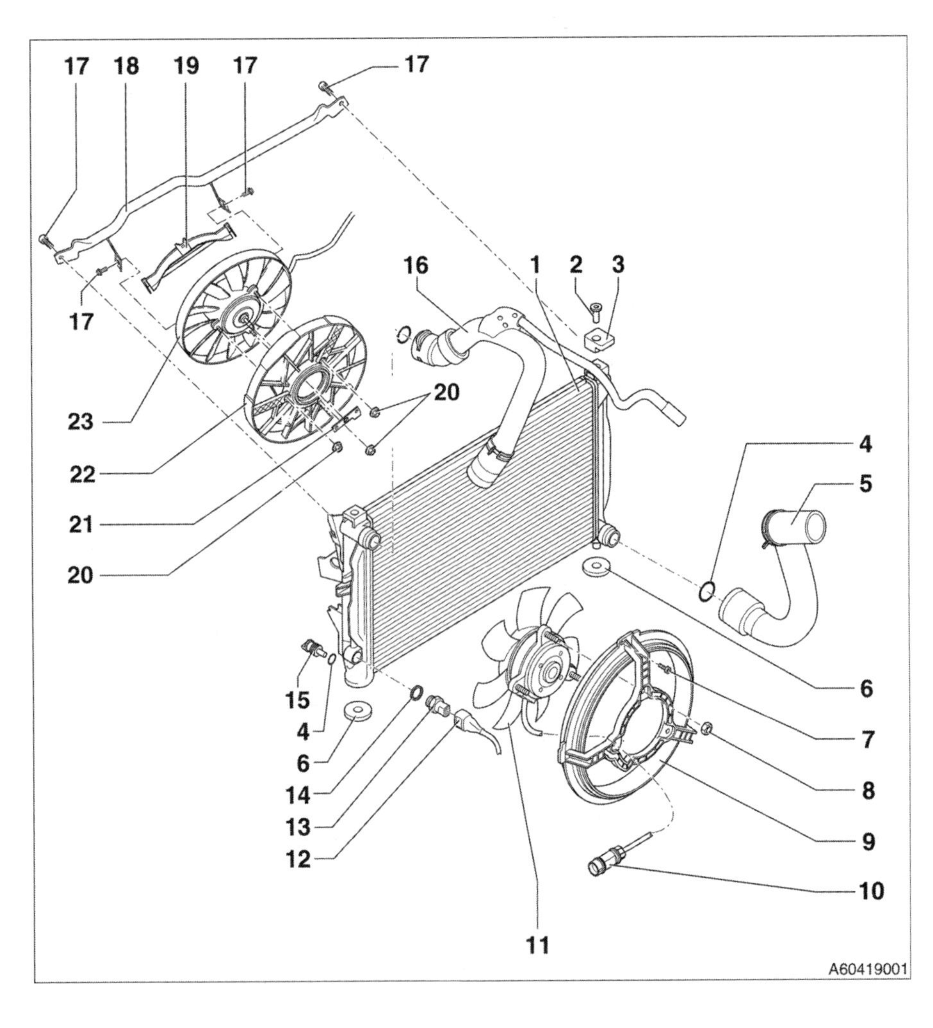

Cooling system components on body (2.7 liter engine)

1. **Radiator**
2. **Mounting clip**
3. **Rubber insulator**
4. **Sealing O-ring**
5. **Lower radiator hose**
6. **Rubber washer**
7. **Self-tapping screw**
8. **Nut, 10 Nm (7 ft-lb)**
9. **Fan shroud**
10. **2-pin harness connector**
11. **Engine cooling fan (V7)**
12. **Fan switch 4-pin connector (F18)**
13. **Fan switch (F54), 35 Nm (26 ft-lb)**
    - Stage 1 switching temperatures:
      on: 92° - 97°C (198° - 207°F)
      off: 84° - 91°C (183° - 196°F)
    - Stage 2 switching temperatures:
      on: 99° - 105°C (210° - 221°F)
      off: 91° - 98°C (196° - 208°F)
14. **Sealing O-ring**
15. **Radiator drain screw, 10 Nm (7 ft-lb)**
16. **Upper radiator hose**
17. **Bolt, 5 Nm (44 in-lb)**
18. **Mounting bracket**
19. **Guard plate**
20. **Nut, 5 Nm (44 in-lb)**
21. **Bracket**
22. **Electric fan shroud**
23. **Electric cooling fan**

BentleyPublishers.com
BentleyPublishers.com—All Rights Reserved

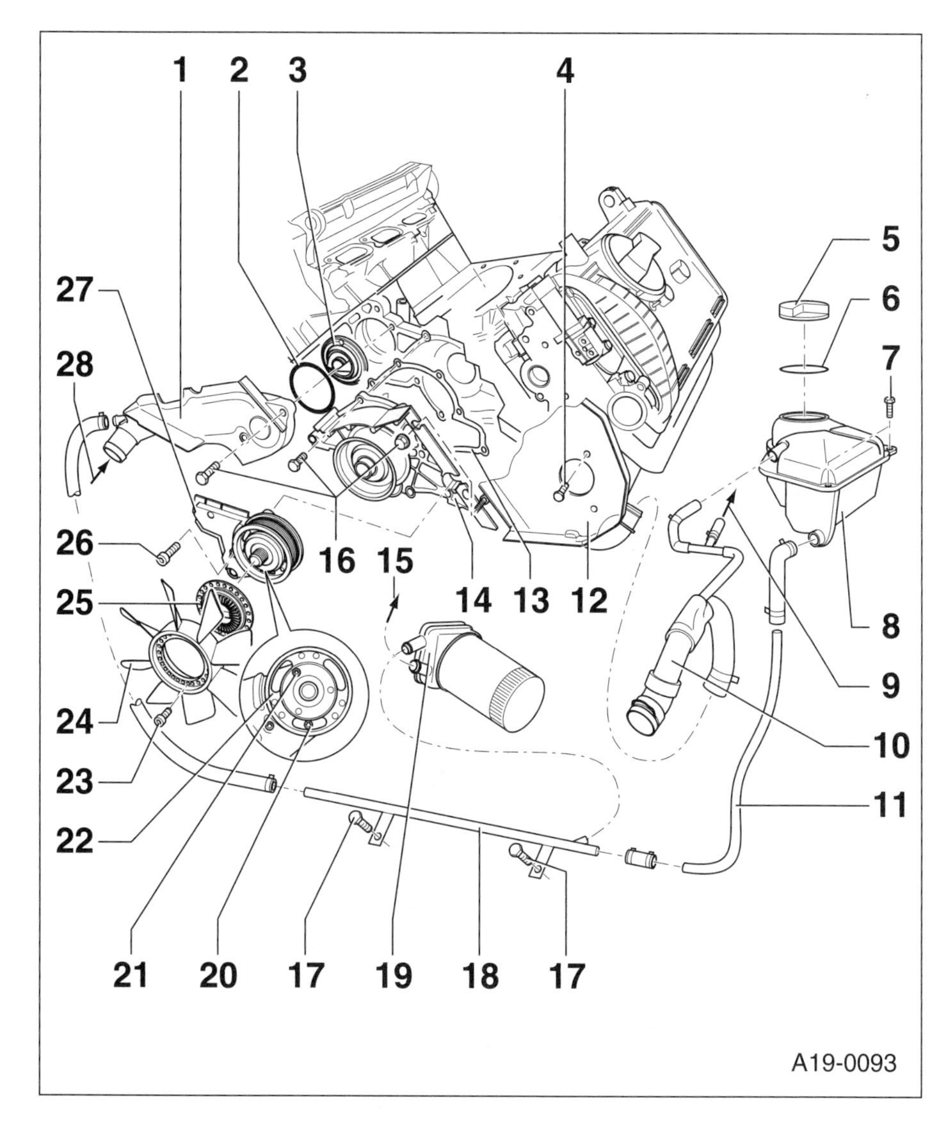

Cooling system components, front of engine (2.7 liter engine)

1. Thermostat cover and coolant duct
2. Sealing O-ring
3. Thermostat
    - Starts to open at 86°C (187°F)
4. Bolt, 10 Nm (7 ft-lb)
5. Filler cap, 1.4 - 1.6 bar (20 - 23 psi)
6. Sealing O-ring
7. Bolt, 10 Nm (7 ft-lb)
8. Coolant expansion tank
9. to Rear coolant line
10. Top radiator hose
11. Coolant line
12. Timing belt guard
13. Coolant pump gasket
14. Coolant pump
15. to Front coolant line
16. Bolt, 10 Nm (7 ft-lb)
17. Bolt, 25 Nm (18 ft-lb)
18. Lower coolant line
19. Oil cooler
20. Bolt, 25 Nm (18 ft-lb)
21. Bolt, 10 Nm (7 ft-lb)
22. Accessory belt pulley
23. Bolt, 10 Nm (7 ft-lb)
24. Viscous fan
25. Viscous fan clutch
    - Left-hand threads
    - 37 Nm (27 ft-lb) using special spanner 3312
26. Bolt, 25 Nm (18 ft-lb)
27. Viscous fan bracket
28. Lower radiator hose connection

BentleyPublishers.com
BentleyPublishers.com—All Rights Reserved

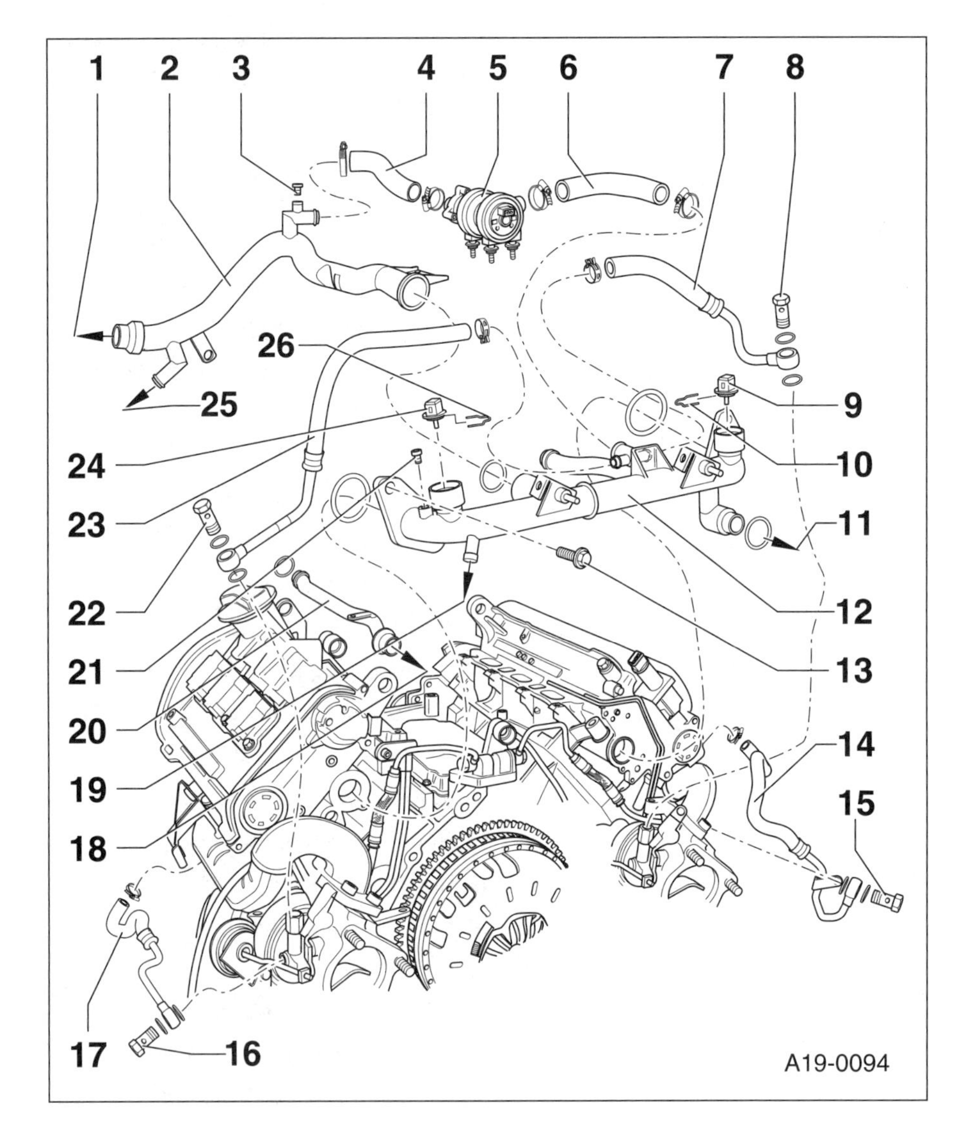

Cooling system components, rear of engine (2.7 liter engine)

1. to Front coolant hose
2. Front coolant hose
3. Bleeder, 20 Nm (15 ft-lb)
4. Coolant line
5. Auxiliary coolant pump
6. Coolant hose
7. Right coolant line
8. Banjo bolt, 35 Nm (26 ft-lb)
   - Replace copper sealing O-rings
9. Engine coolant temperature (ECT) sensor (G62)
10. Retaining clip
11. to Heater core
12. Rear coolant line
13. Bolt, 10 Nm (7 ft-lb)
14. Right coolant line
15. Banjo bolt, 35 Nm (26 ft-lb)
    - Replace copper sealing O-rings
16. Banjo bolt, 35 Nm (26 ft-lb)
    - Replace copper sealing O-rings
17. Left coolant line
18. from Heater core
19. to Front coolant hose
20. Coolant line
21. Bleeder, 20 Nm (15 ft-lb)
22. Banjo bolt, 35 Nm (26 ft-lb)
    - Replace copper sealing O-rings
23. Left coolant line
24. Auxiliary coolant pump sensor, 2-pin
25. to Oil cooler
26. Retaining clip

BentleyPublishers.com
BentleyPublishers.com—All Rights Reserved

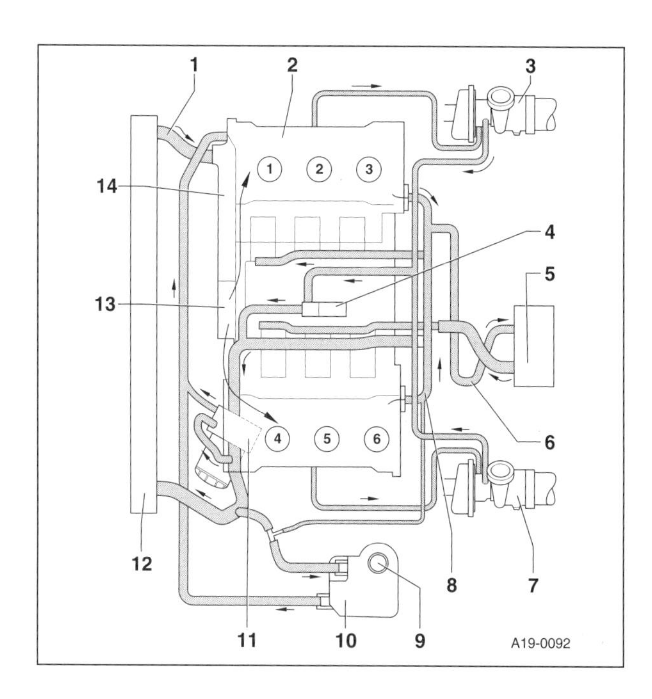

Coolant hose schematic (2.7 liter engine)

1. Lower radiator hose
2. Cylinder head and engine block
3. Right turbocharger
4. Auxiliary coolant pump
5. Heater core
6. to Heater core
7. Left turbocharger
8. Rear coolant line
9. Coolant expansion tank cap
10. Coolant expansion tank
11. Oil cooler
12. Radiator
13. Coolant pump
14. Thermostat cover and duct

2.8 LITER ENGINE COOLING SYSTEM

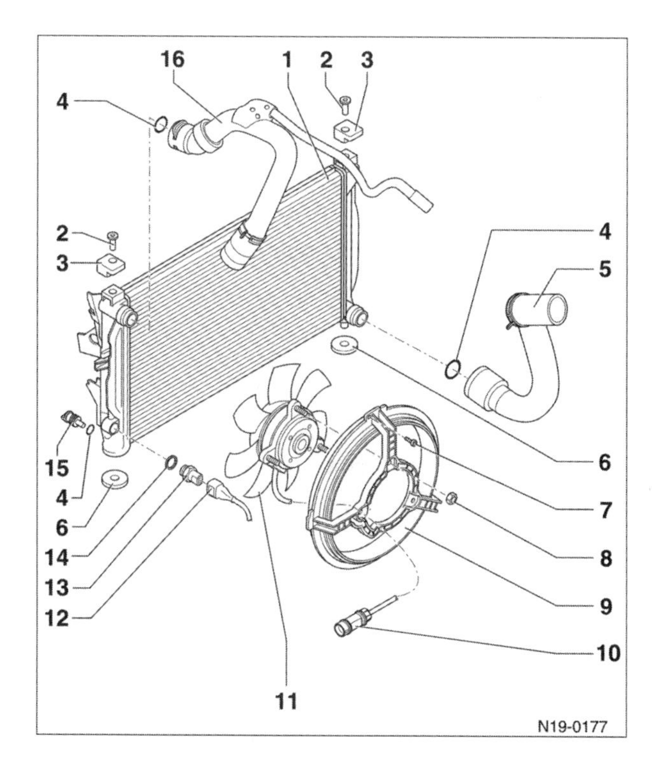

Cooling system components on body (2.8 liter engine)

1. Radiator
2. Mounting clip
3. Rubber insulator
4. Sealing O-ring
5. Lower radiator hose
6. Rubber washer
7. Self-tapping screw
8. Nut, 10 Nm (7 ft-lb)
9. Fan shroud
10. 2-pin harness connector
11. Engine cooling fan (V7)
12. Fan switch 4-pin connector (F18)
13. 2-stage fan switch (F54), 35 Nm (26 ft-lb)
    - Stage 1 switching temperatures:
      on: 92° - 97°C (198° - 207°F)
      off: 84° - 91°C (183° - 196°F)
    - Stage 2 switching temperatures:
      on: 99° - 105°C (210° - 221°F)
      off: 91° - 98°C (196° - 208°F)
14. Sealing O-ring
15. Radiator drain screw, 10 Nm (7 ft-lb)
16. Upper radiator hose

BentleyPublishers.com
BentleyPublishers.com—All Rights Reserved

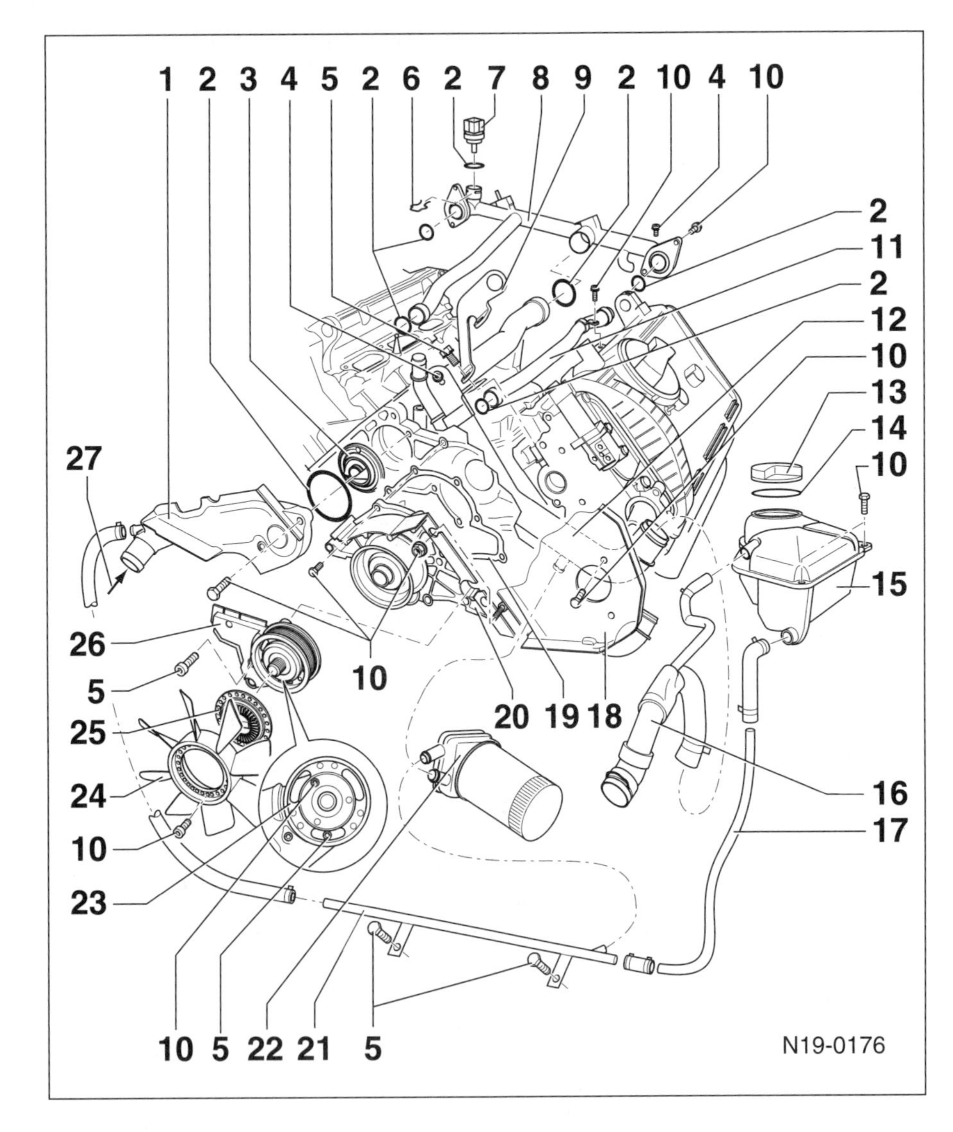

Cooling system components on engine (2.8 liter engine)

1. **Thermostat cover and coolant duct**
2. **Sealing O-ring**
3. **Thermostat**
   - Starts to open at 86°C (187°F)
   - Opening travel: min. 7 mm (¼ in)
4. **Cooling system bleeder**
5. **Bolt, 25 Nm (18 ft-lb)**
6. **Retaining clip**
7. **Engine coolant temperature (ECT) sensor (G62)**
8. **Rear coolant distributor pipe**
9. **Engine lifting eye**
10. **Bolt, 10 Nm (7 ft-lb)**
11. **Coolant pipe**
12. **Front coolant pipe**
13. **Cooling expansion tank pressure cap**
14. **Sealing O-ring**
15. **Coolant expansion tank**
16. **Upper radiator hose**
17. **Coolant hose**
18. **Timing belt guard**
19. **Coolant pump gasket**
20. **Coolant pump**
21. **Lower coolant pipe**
22. **Oil cooler**
23. **Belt pulley**
24. **Viscous fan**
25. **Viscous fan clutch**
    - Left-hand threads
    - 37 Nm (27 ft-lb) using special spanner, VAG 3312
26. **Viscous fan bracket**
27. **Lower radiator hose connection**

BentleyPublishers.com
BentleyPublishers.com—All Rights Reserved

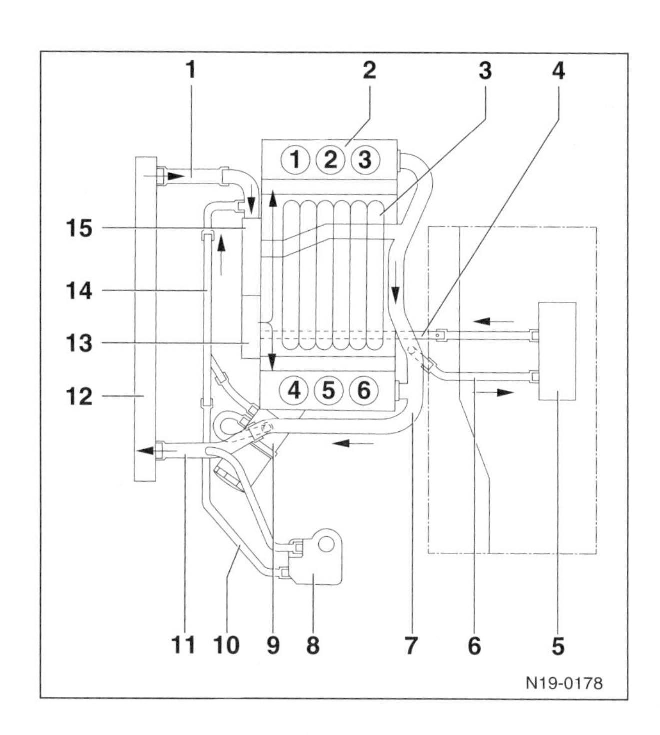

Coolant hose schematic (2.8 liter engine)

1. Lower radiator hose
2. Cylinder head and engine block
3. Intake manifold
4. from Heater core
5. Heater core
6. to Heater core
7. Rear coolant pipe
8. Coolant expansion tank
9. Oil cooler
10. Coolant line
11. Upper radiator hose
12. Radiator
13. Coolant pump
14. Lower coolant pipe
15. Thermostat cover and duct

3.0 LITER ENGINE COOLING SYSTEM

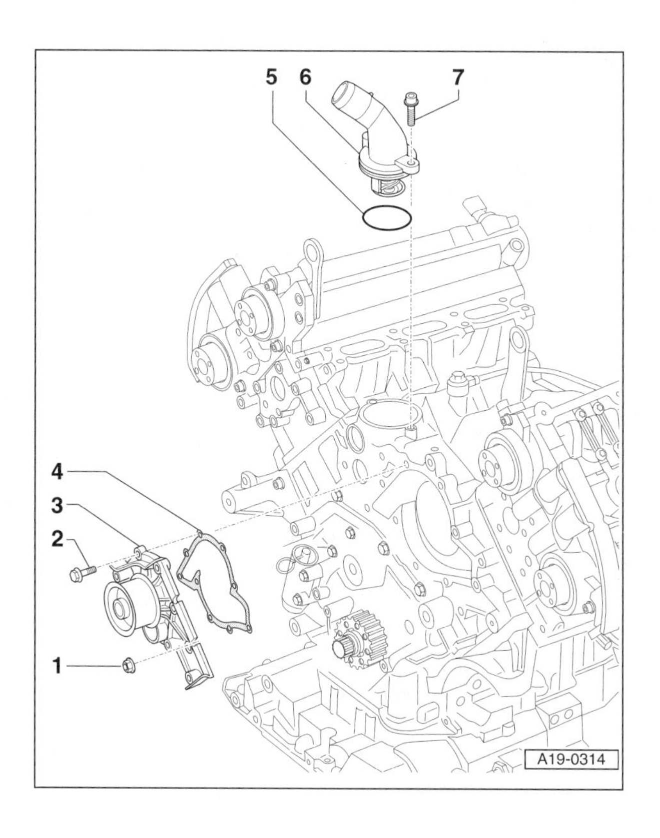

Coolant pump and thermostat (3.0 liter engine)

1. Nut, 10 Nm (7 ft-lb)
2. Bolt, 10 Nm (7 ft-lb)
3. Coolant pump
4. Coolant pump gasket
5. Sealing O-ring
6. Coolant thermostat integral with housing
7. Bolt, 10 Nm (7 ft-lb)

BentleyPublishers.com
BentleyPublishers.com—All Rights Reserved

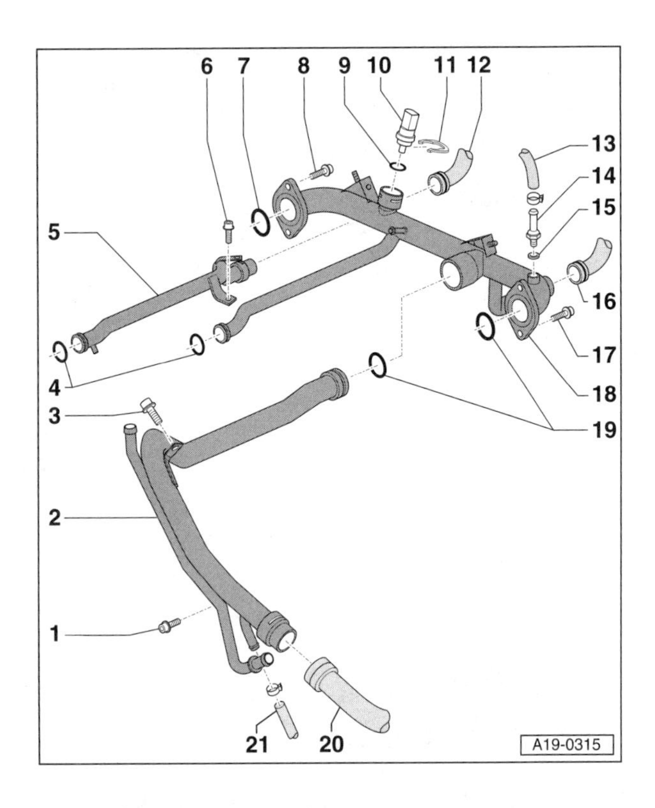

Coolant hoses and pipes (3.0 liter engine)

1. Bolt, 10 Nm (7 ft-lb)
2. Front coolant pipe
3. Bolt, 22 Nm (16 ft-lb)
4. Sealing O-ring
5. Right coolant pipe
6. Bolt, 10 Nm (7 ft-lb)
7. Sealing O-ring
8. Bolt, 10 Nm (7 ft-lb)
9. Sealing O-ring
10. Engine coolant temperature (ETC) sensor (G2, G62)
11. Retaining clip
12. to Heater core
13. to Coolant expansion tank
14. Connector, 20 Nm (14 ft-lb)
15. Seal
16. to Heater core
17. Bolt, 10 Nm (7 ft-lb)
18. Rear coolant pipe
19. Sealing O-rings
20. Top radiator hose
21. to Oil cooler

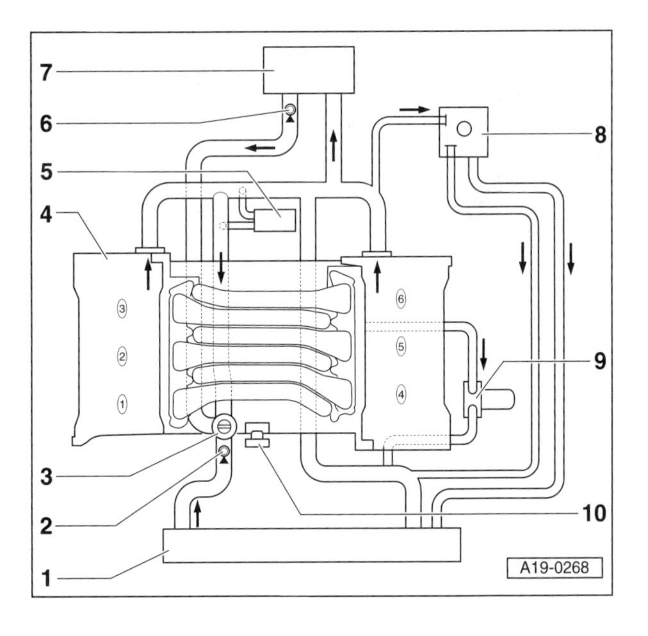

Coolant hose schematic (3.0 liter engine)

1. Radiator
2. Bleeder, 15 Nm (10 ft-lb)
3. Coolant thermostat
4. Cylinder head and engine block
5. Throttle valve control module (J338)
6. Bleeder hole
7. Heater core
8. Coolant expansion tank
9. Oil cooler
10. Coolant pump

BentleyPublishers.com
BentleyPublishers.com—All Rights Reserved

V8 Engine Cooling System

Coolant hose schematic (4.2 liter engine, codes ART, AWN, BBD)

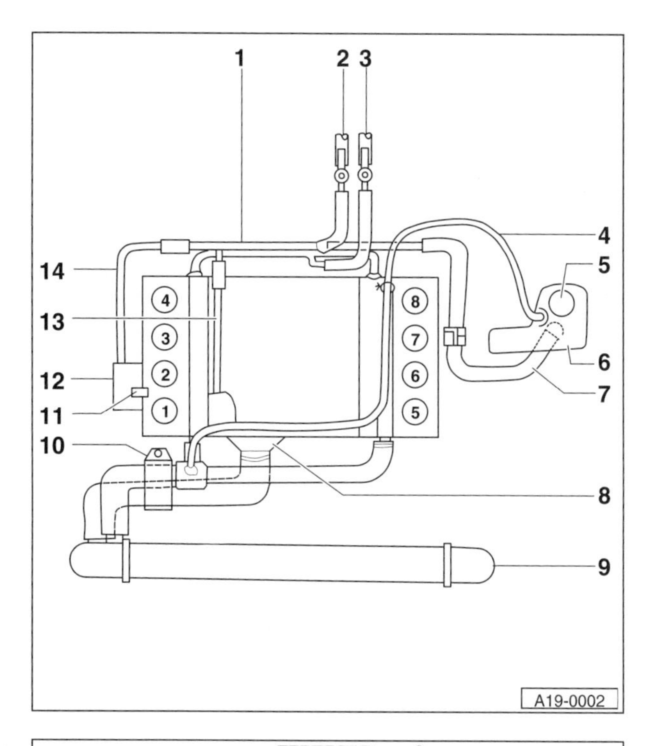

1. Rear coolant pipe
2. from Heater core (with bleeder valve)
3. to Heater core (with bleeder valve)
4. Return hose
5. Coolant expansion tank cap
6. Coolant expansion tank
7. Filler hose
8. Coolant thermostat
9. Radiator
10. Coolant hoses bracket
11. Coolant pipe between oil cooler and engine block
12. Oil cooler
13. Coolant pipe
14. Coolant pipe, to oil cooler

Coolant hose schematic (4.2 liter RS6 biturbo engine, code BCY)

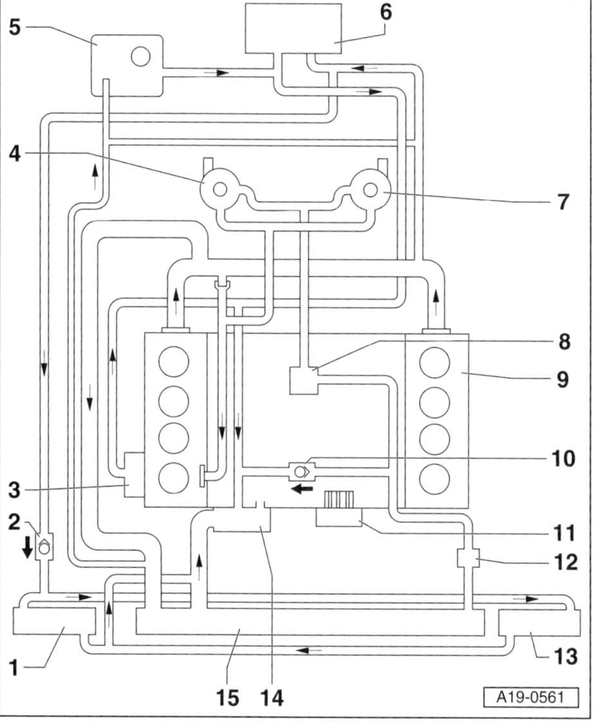

1. Right intercooler
2. Check valve
3. Oil filter
4. Right turbocharger
5. Coolant overflow tank
6. Heater core
7. Left turbocharger
8. After-run coolant pump (V51)
9. Engine
10. Check valve
11. Coolant pump
12. Coolant circulation valve
13. Left intercooler
14. Coolant thermostat
15. Radiator

BentleyPublishers.com
BentleyPublishers.com—All Rights Reserved

Cooling System Service

In the procedures that follow, the 2.8 liter V6 engine is illustrated. Other models are similar.

Cooling system draining, filling and bleeding

> ***WARNING —***
> - *To avoid personal injury, be sure the engine is cold before opening the cooling system.*

– Remove coolant expansion tank filler cap.

– Raise car and support safely.

> ***WARNING —***
> - *Make sure the car is stable and well supported at all times. Use a professional automotive lift or jack stands designed for the purpose. A floor jack is not adequate support.*

– Remove lower engine cover (splash shield). See **03 Maintenance**.

– Place 5 gallon pail underneath engine.

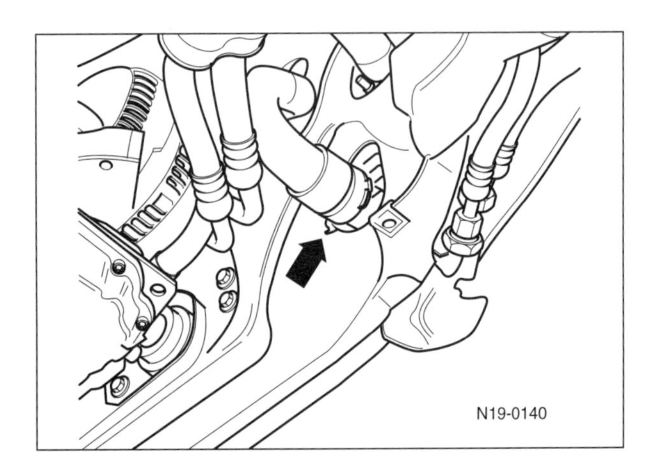

◄ Loosen hose clamp (**arrow**) at bottom radiator hose. Detach hose from radiator and allow coolant to drain into pail.

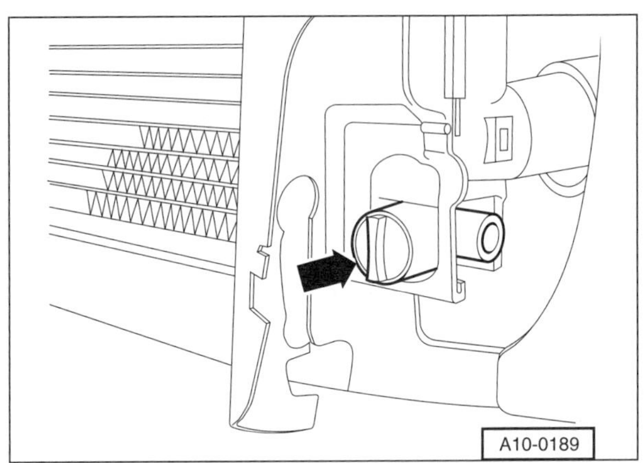

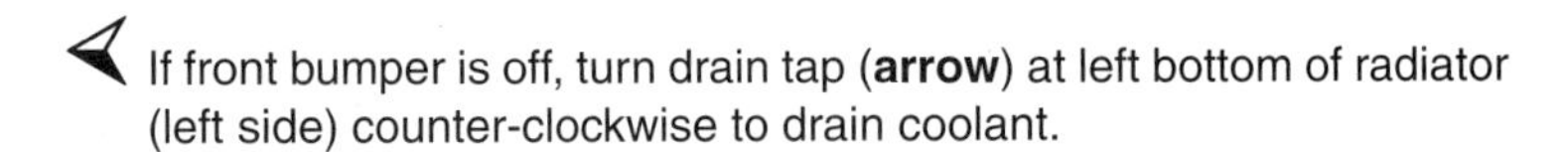

◄ If front bumper is off, turn drain tap (**arrow**) at left bottom of radiator (left side) counter-clockwise to drain coolant.

◄ Drain coolant from engine block at block drain screw (**arrow**).

– Using new sealing O-ring, replace and tighten drain screw.

Tightening torque	
Block drain screw to engine block	20 Nm (15 ft-lb)

– Reattach lower hose to radiator. Alternatively, shut off coolant drain tap on radiator.

BentleyPublishers.com
BentleyPublishers.com—All Rights Reserved

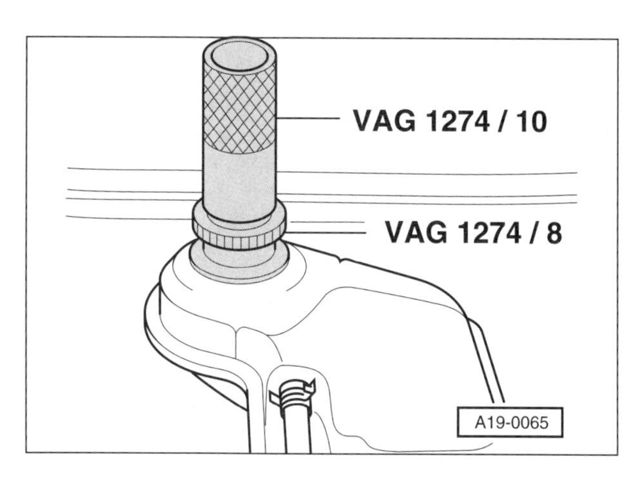

Thread adapter (VAG 1274/8) to coolant expansion tank and insert extension (VAG 1274/10).

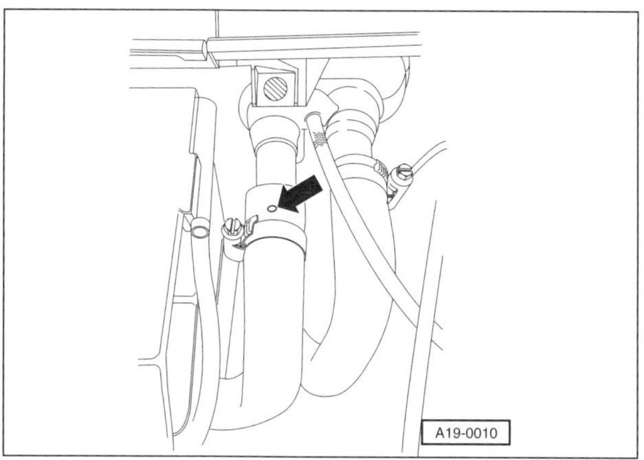

Working at heater core coolant connection:

- Loosen coolant hose clamp at heater core
- Pull hose far enough so that bleeder hole (**arrow**) in coolant line is not covered by hose.
- Fill coolant at tank until it comes out at bleeder.
- Push coolant hose on connection and tighten clamp.

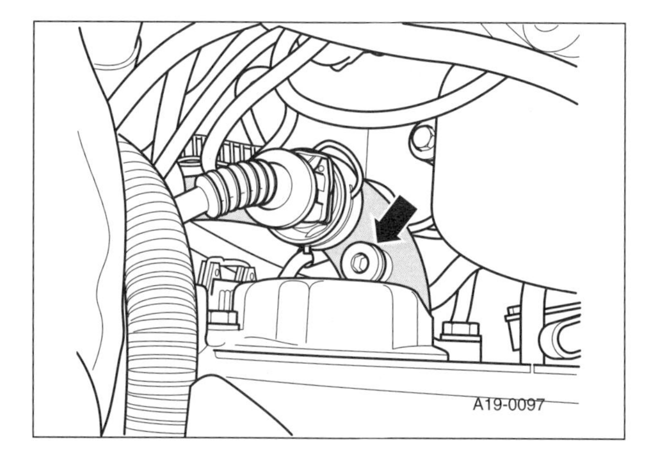

Loosen bleeder screw on rear coolant line (**arrow**) below expansion tank.

- Fill coolant at tank until it comes out at bleeder.
- Tighten bleeder screw.

Tightening torque	
Coolant bleeder screw	20 Nm (15 ft-lb)

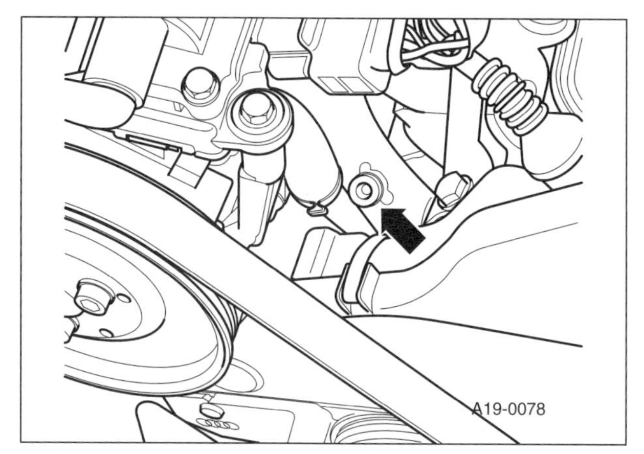

Loosen bleeder screw on front coolant line (**arrow**) between power steering pump and left cylinder head.

- Fill coolant at tank until it comes out at bleeder screw.
- Tighten bleeder screw.

Tightening torque	
Coolant bleeder screw	20 Nm (15 ft-lb)

BentleyPublishers.com
BentleyPublishers.com—All Rights Reserved

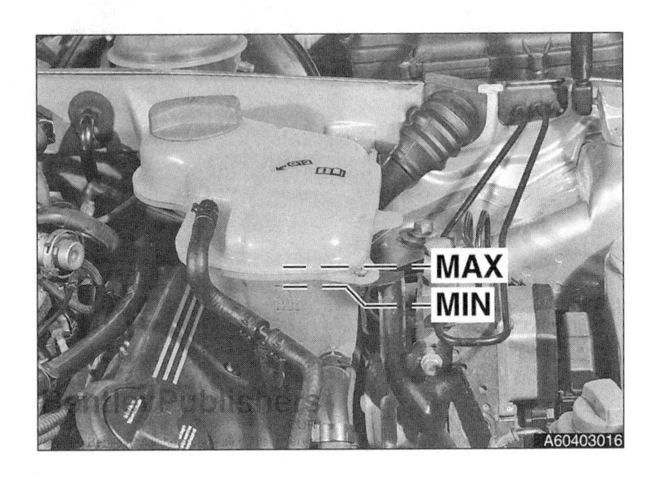

Remove special tools from coolant expansion tank and fill coolant to MAX mark. Close filler cap on expansion tank.

– Set heater controls to maximum heat setting.

– Start engine and let it idle for about 10 minutes.
  - Maintain engine speed of about 2000 rpm for about 5 minutes.
  - Allow engine to run at idling speed until lower radiator hose becomes hot.

– Check coolant level and top off if necessary.

> ***WARNING —***
> - *When the engine is hot, the cooling system is under pressure. Releasing cooling system pressure lowers the boiling point of coolant and it may boil suddenly.*
> - *Cover the filler cap with a cloth and remove it carefully.*
> - *Use heavy gloves and wear eye and face protection to guard against scalding.*

Normal coolant level	
Engine at operating temperature	MAX
Engine cold	between MIN and MAX

Radiator, removing and installing

– Raise car and support safely.

> ***WARNING —***
> - *Make sure the car is stable and well supported at all times. Use a professional automotive lift or jack stands designed for the purpose. A floor jack is not adequate support.*

– Remove front bumper and bumper cover. See **63 Bumpers**.

– Drain coolant. See **Cooling system draining, filling and bleeding** in this repair group.

> ***WARNING —***
> - *To avoid personal injury, be sure the engine is cold before opening the cooling system.*

Loosen hose clamps (**arrow**) and detach upper and lower hoses from radiator.

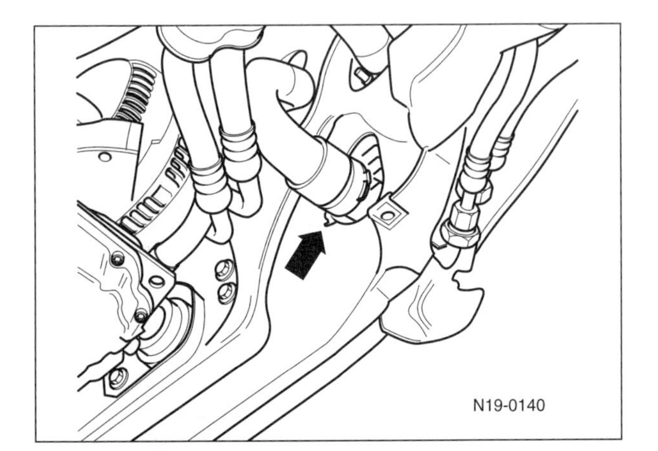

BentleyPublishers.com
BentleyPublishers.com—All Rights Reserved

Disconnect electric cooling fan switch connector (**arrow**) at bottom right of radiator.

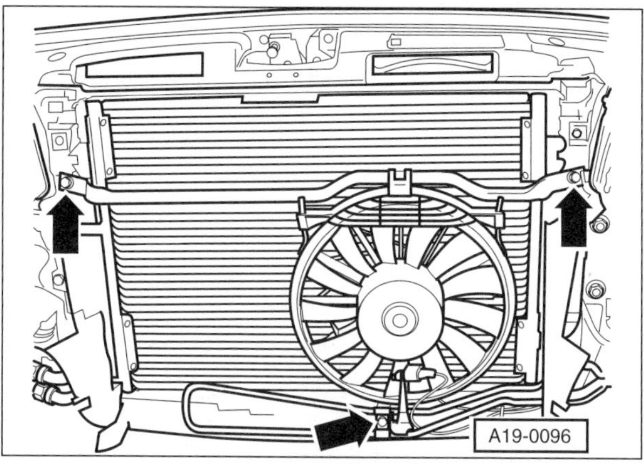

Working in front of radiator:

- Remove electric cooling fan mounting fasteners (**upper arrows**), if applicable.
- Detach power steering fluid cooler mounting screw (**lower arrow**).
- Use stiff wire to suspend these components in front of engine compartment.

> ***CAUTION—***
> - *Do not kink steering fluid line.*
> - *Do not disconnect steering fluid line.*

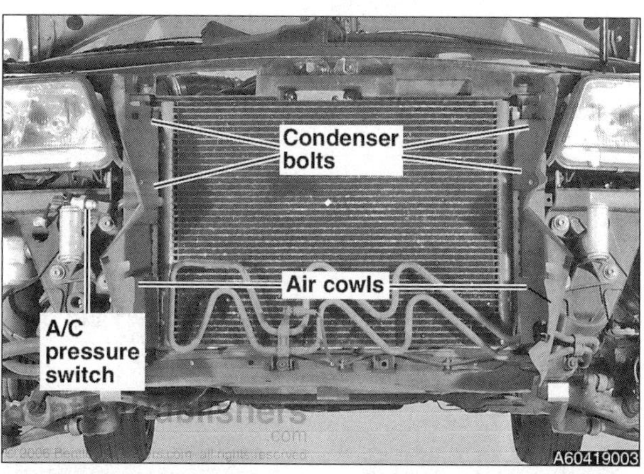

Working at A/C condenser:

- Detach air cowls from radiator on left and right sides.
- Remove condenser securing bolts (hidden by cowl in photo)
- Disconnect A/C pressure switch connector.
- Lift condenser out of bracket, turn sideways and secure with stiff wire.

> ***CAUTION—***
> - *Do not kink or stretch A/C refrigerant lines.*
> - *Do not disconnect refrigerant lines.*

Use small screwdriver tip to release two radiator retaining pins (**arrows**).

– Tip radiator forward and out.

– Installation is reverse of removal.

Tightening torques	
A/C condenser to radiator	10 Nm (7 ft-lb)
Electric cooling fan to radiator	10 Nm (7 ft-lb)
Power steering cooling line to radiator	10 Nm (7 ft-lb)

– After completing assembly, fill cooling system and bleed. See **Cooling system draining, filling and bleeding** in this repair group.

BentleyPublishers.com
BentleyPublishers.com—All Rights Reserved

Coolant pump, removing and installing

This procedure applies to engines with camshaft timing belt. The timing belt drives the coolant pump. When replacing the pump, mark both accessory belt and timing belt with direction of rotation before removal for correct reinstallation.

> ***CAUTION—***
> • *Do not allow coolant to contaminate belts.*

– Raise car and support safely. Remove front bumper. Place lock carrier (radiator mounting frame in front of engine) in service position. See **50 Body–Front**.

> ***WARNING—***
> • *Make sure the car is stable and well supported at all times. Use a professional automotive lift or jack stands designed for the purpose. A floor jack is not adequate support.*

– Remove viscous fan and accessory belt. See **03 Maintenance**.

NOTE—
• *If belt is to be reused, mark direction of rotation prior to removal.*

– Remove timing belt. See **13 Timing Belt, Crankshaft Pulley, Rear Main Seal**.

NOTE—
• *If belt is to be reused, mark direction of rotation prior to removal.*

– Drain coolant. See **Cooling system draining, filling and bleeding** in this repair group.

> ***WARNING—***
> • *To avoid personal injury, be sure the engine is cold before opening the cooling system.*

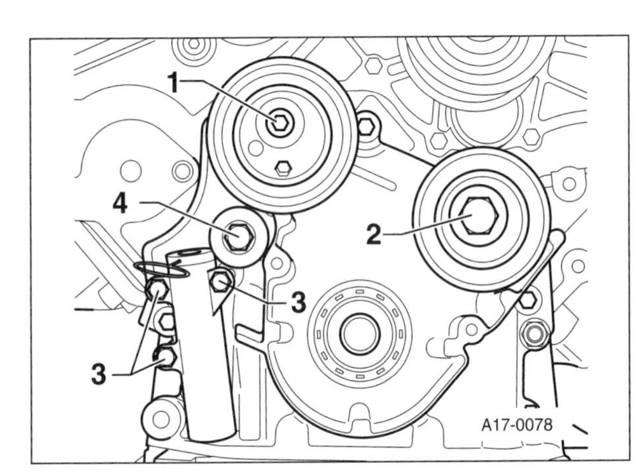

◄ Working at front of engine, remove timing belt tensioner (**1**) and idler pulley (**2**).

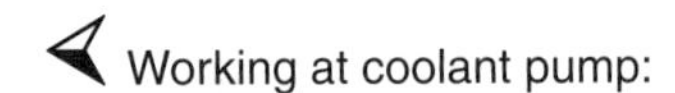

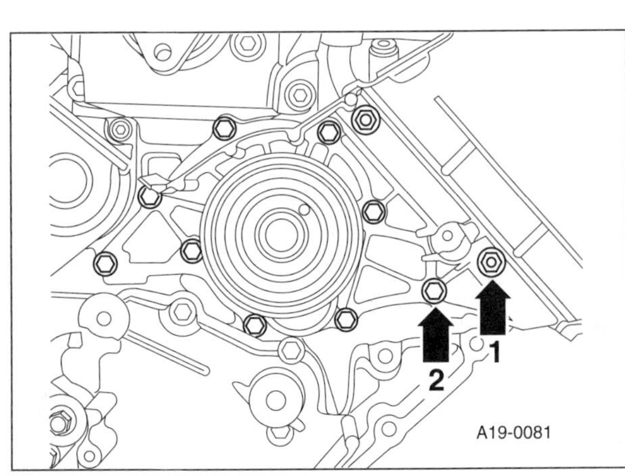

◄ Working at coolant pump:

- Remove timing belt guard mounting nuts (**1**) (2 nuts).
- Remove coolant pump mounting bolts (**2**) (9 bolts).
- Remove coolant pump.

– Prior to installing new pump, clean gasket and sealant residue off sealing surface of engine block.

– Installation is reverse of removal. Use new coolant pump gasket.

Tightening torques	
Timing belt guard to coolant pump	10 Nm (7 ft-lb)
Timing belt idler to engine block	40 Nm (28 ft-lb)
Timing belt tensioner to engine block	20 Nm (14 ft-lb)

BentleyPublishers.com
BentleyPublishers.com—All Rights Reserved

Coolant thermostat, removing and installing

Mark both accessory belt and timing belt with direction of rotation before removal for correct reinstallation.

> ***CAUTION—***
> *• Do not allow coolant to contaminate belts.*

– Raise car and support safely.

> ***WARNING —***
> *• Make sure the car is stable and well supported at all times. Use a professional automotive lift or jack stands designed for the purpose. A floor jack is not adequate support.*

– Remove front bumper. Place lock carrier (radiator mounting frame in front of engine) in service position. See **50 Body–Front**.

– Remove viscous fan and accessory belt. See **03 Maintenance**.

NOTE —

• If belt is to be reused, mark direction of rotation prior to removal.

– Remove timing belt. See **13 Timing Belt, Crankshaft Pulley, Rear Main Seal**.

NOTE —

• If belt is to be reused, mark direction of rotation prior to removal.

– Drain coolant. See **Cooling system draining, filling and bleeding** in this repair group.

> ***WARNING —***
> *• To avoid personal injury, be sure the engine is cold before opening the cooling system.*

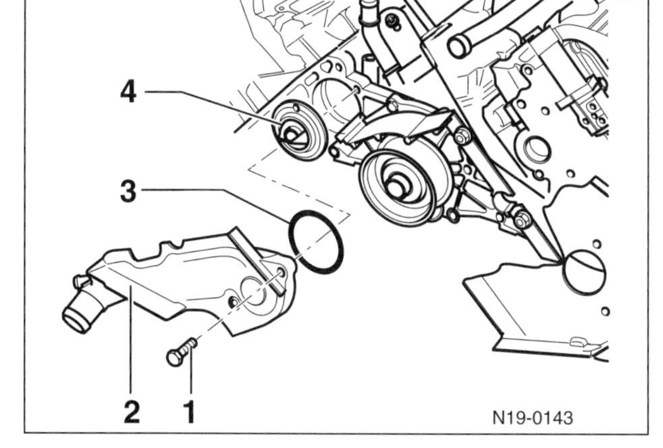

◄ Remove thermostat cover mounting bolts (**1**) and remove thermostat (**4**).

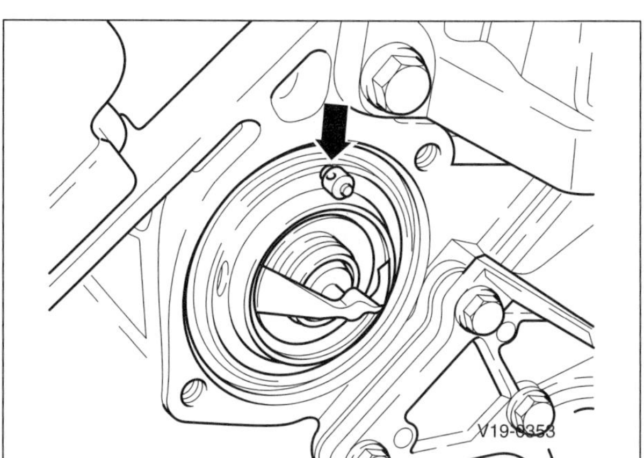

◄ Installation is reverse of removal. Make sure of the following:

- Lubricate new thermostat sealing O-ring with coolant.
- Install thermostat with bleeder valve (**arrow**) at top.

Tightening torque	
Thermostat cover to engine block	10 Nm (7 ft-lb)

– After completing assembly, fill cooling system and bleed. See **Cooling system draining, filling and bleeding** in this repair group.

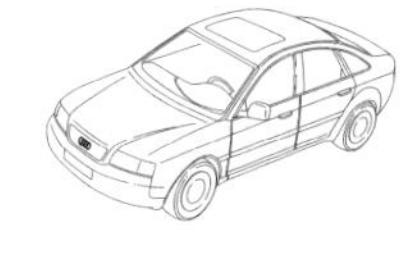

BentleyPublishers.com
BentleyPublishers.com—All Rights Reserved

20 Fuel Storage and Supply

General

This repair group covers repair information for the fuel pump and related fuel storage and supply components.

For additional information, see:

- **03 Maintenance** for fuel filter replacement
- **24 Fuel Injection**
- **OBD On-Board Diagnostics**

Fuel tank capacity

Audi A6 models vary in body style and running gear. This affects fuel tank capacity. The accompanying table gives approximate fuel capacity figures for most models.

Fuel tank capacity	
V6 engine, allroad quattro	18.5 US gal (70 liters)
V8 engine (*not* allroad quattro)	21.6 US gal (82 liters)

Fuel pump and fuel level sender(s)

Front-wheel drive model. Fuel pump and fuel level sender are combined in one unit in fuel tank.

Quattro model. There are three fuel level senders. Right fuel level sender and fuel pump are combined in one unit in fuel tank.

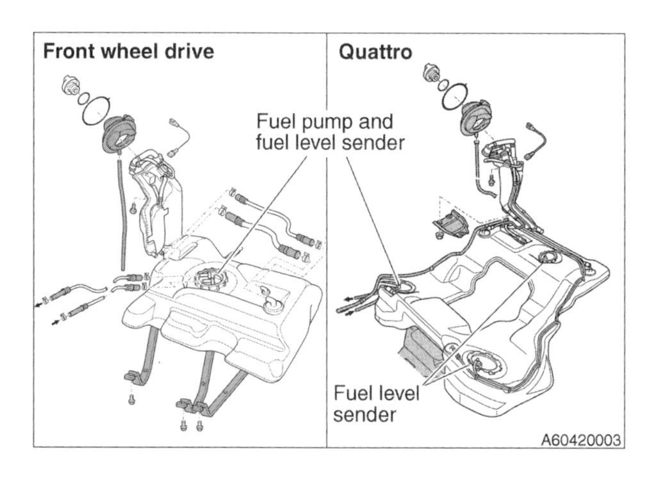

BentleyPublishers.com
BentleyPublishers.com—All Rights Reserved

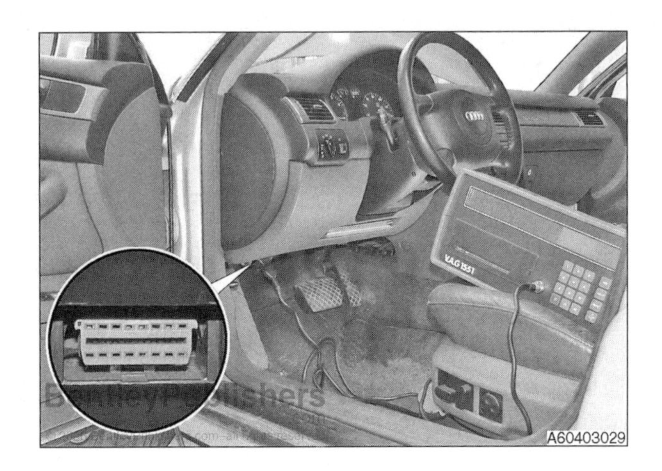

On-Board Diagnostics

The fuel supply system is an integral part of the operation of the fuel injection system. Problems such as a no-start condition, hesitation, or stalling may be due to poor fuel delivery. The fuel pump itself is not directly monitored by on-board diagnostics, but the fuel pump relay is monitored.

In case of poor fuel delivery, the fuel injection system's attempt to adjust for the resultant lean running condition may exceed the system's ability to compensate. This stores diagnostic trouble codes (DTCs) in the engine control module (ECM) memory.

< Investigate suspected fuel pump problems with Volkswagen scan tool or equivalent plugged into to DLC plug (**inset**) under left side dashboard. See **OBD On-Board Diagnostics** for more information.

Warnings and Cautions

Read the following warnings and cautions before starting work on your fuel system.

WARNING —

- *Disconnect negative (-) battery cable and cover terminal with insulated material when working on fuel related components.*
- *Gasoline is dangerous to your health. Wear suitable hand, skin and eye protection when working on fuel system. Do not breathe fuel vapors. Work in a well-ventilated area.*
- *Fuel and fuel vapors may leak during many operations described in this repair group. Do not smoke or create sparks. Be aware of pilot lights in gas operated equipment (heating systems, water heaters, etc.). Have an approved fire extinguisher handy.*
- *The fuel system is designed to maintain pressure in the system after the engine is turned off. Fuel is expelled under pressure when fuel lines are disconnected. This is a fire hazard, especially if the engine is warm. To prevent fuel from spraying, wrap a clean shop rag around fuel line fitting before loosening or disconnecting it.*
- *Exercise extreme caution when using spray-type cleaners on a warm engine. Observe all manufacturer recommendations.*
- *Prior to working on fuel-related components, unscrew the fuel filler cap to release pressure in the tank.*

BentleyPublishers.com
BentleyPublishers.com—All Rights Reserved

CAUTION—

- *Cleanliness is essential when working on any part of the fuel system. Thoroughly clean fuel line unions and hose fittings before disconnecting them. Use only clean tools.*
- *Keep removed components clean. Seal or cover them with plastic or paper, especially if repair cannot be completed immediately. Seal open fuel supply and return lines to prevent contamination.*
- *When replacing parts, install only new, clean components. Replace seals and O-rings.*
- *Prior to disconnecting the battery, read the battery disconnection cautions in* **00 Warnings and Cautions**.
- *Before making any electrical tests with the ignition turned ON, disable the ignition system. Be sure the battery is disconnected when replacing components.*
- *To prevent damage to the ignition system or other DME components, including the engine control module (ECM), always connect and disconnect wires and test equipment with the ignition OFF.*
- *Only use digital multimeter for electrical tests.*

NOTE—

- *Audi identifies electrical components by a letter and/or a number in the electrical schematics. See* **EWD Electrical Wiring Diagrams**. *These electrical identifiers are listed in parentheses as an aid to electrical troubleshooting.*

FUEL DELIVERY TESTS

The fuel pump delivers fuel at high pressure to the fuel injection system. During starting, the fuel pump runs as long as the ignition switch is in the START position and continues to run once the engine starts. If an electrical system fault interrupts power to the fuel pump, the engine does not run.

Checking fuel delivery is a fundamental part of troubleshooting and diagnosing the DME system. Fuel pressure directly influences fuel delivery. An accurate fuel pressure gauge is needed to make the tests.

There are three significant fuel delivery values to be measured:

- **Fuel delivery volume**—created by the fuel pump and affected by restrictions, such as clogged fuel filter.
- **Fuel pressure**—created by the fuel pump and maintained by the pressure regulator.
- **Residual pressure**—the pressure maintained in the closed system after the engine and fuel pump are shut off.

BentleyPublishers.com
BentleyPublishers.com—All Rights Reserved

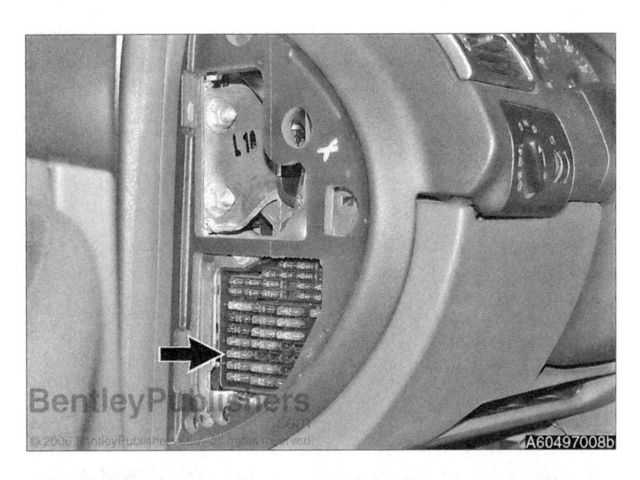

Fuel pump power supply

The fuel pump is powered by fuse 28 (**arrow**) in fuse panel at left end of dashboard.

Fuel pump fuse ratings	
1998 - 2000 fuel pump	15 A
2001 - 2004 fuel pump	20 A

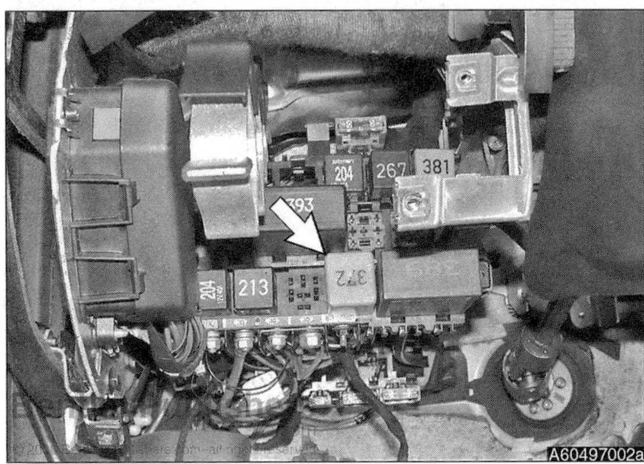

The fuel pump relay (**arrow**) is on micro central electric panel, under left side dashboard. See **97 Fuses, Relays, Component Locations** for additional information.

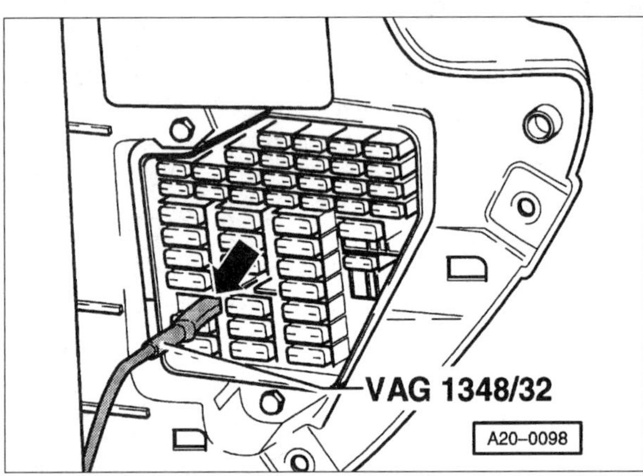

Fuel pump, electrical testing

– Make sure battery has at least 12 volts charge.

– Visually check fuel pump fuse and relay. See **Fuel pump power supply** in this repair group.

Remove fuse 28 (**arrow**). Connect VAG1348/3A remote switch and VAG1348/32 adapter cable to opening for fuse 28. Connect other end of adapter to battery.

- Listen carefully for fuel pump to start running. If necessary, lift up rear seat bottom to hear fuel pump.

– Alternatively, crank engine and listen carefully for fuel pump to start running.

- Fuel pump runs quietly. If necessary, lift up rear seat bottom to hear fuel pump.

– If fuel pump does not run, remove rear seat bottom, see **72 Seats**.

Remove fuel pump cover. Pull off (**arrow**) fuel pump electrical connector.

BentleyPublishers.com
BentleyPublishers.com—All Rights Reserved

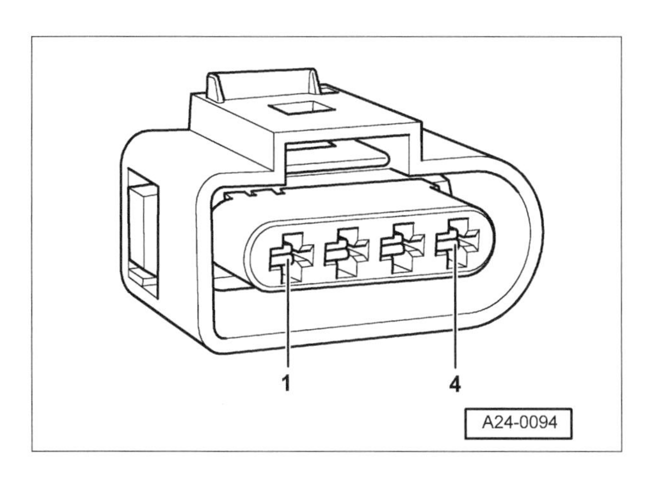

◀ Connect digital multimeter across terminals **1** and **4** of harness connector.

– Crank engine and check voltage at fuel pump connector. If battery voltage (12 volts) is not present, locate and repair open circuit in wiring. See **EWD Electrical Wiring Diagrams**.

– If battery voltage is present, fuel pump power supply is OK. To continue, testing, see **Fuel pump wiring (in tank), checking**in this repair group.

Relieving system fuel pressure

The fuel system retains fuel pressure in the system when the engine is OFF. To prevent fuel from spraying on a hot engine, relieve system fuel pressure before disconnecting fuel lines. One method is to tightly wrap a shop towel around fuel line fitting and loosen or disconnect the fitting.

Fuel pump wiring (in tank), checking

– Loosen fuel tank filler cap.

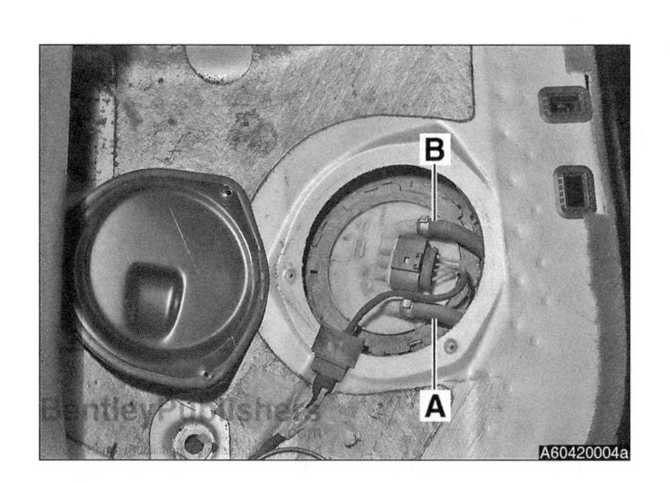

◀ Remove fuel pump cover. Mark fuel supply (**A**) and return (**B**) lines.

– Loosen or cut hose clamps and detach fuel lines from pump assembly.

> ***WARNING—***
> *• Fuel will be expelled under pressure. Wrap a cloth around fuel line to absorb any leaking fuel.*

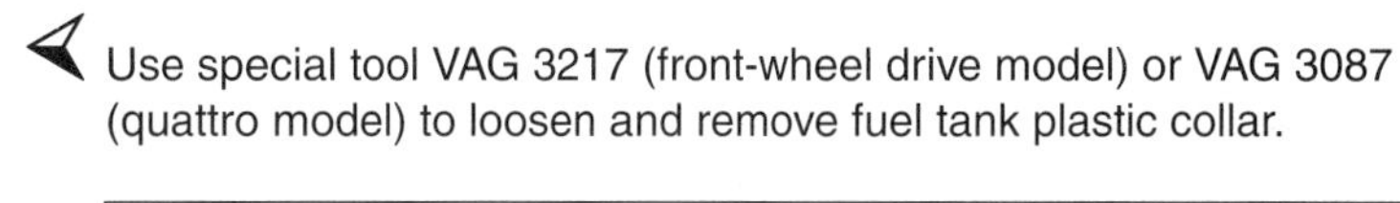

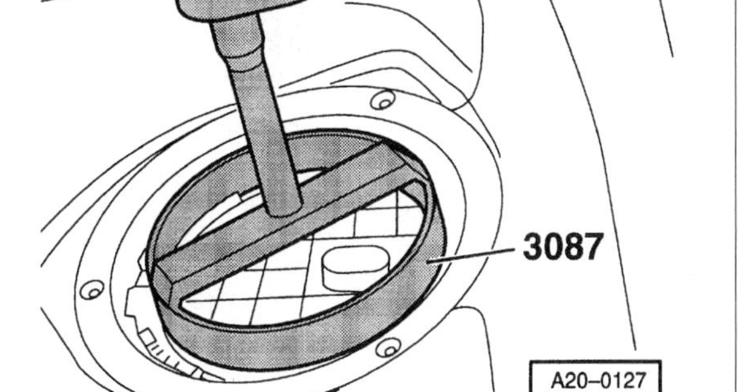

◀ Use special tool VAG 3217 (front-wheel drive model) or VAG 3087 (quattro model) to loosen and remove fuel tank plastic collar.

> ***WARNING—***
> *• Open fuel tank cap only if the tank is below ¼ full. Otherwise large quantities of fuel escape.*

BentleyPublishers.com
BentleyPublishers.com—All Rights Reserved

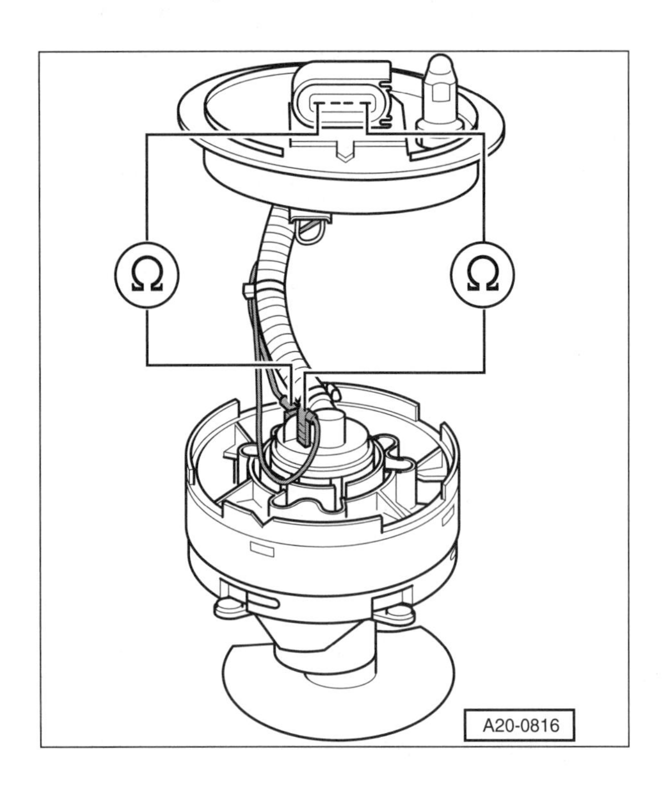

- Pull up fuel pump assembly flange and make sure electrical connections between flange and fuel pump are intact.

– Reattach fuel lines using new hose clamps.

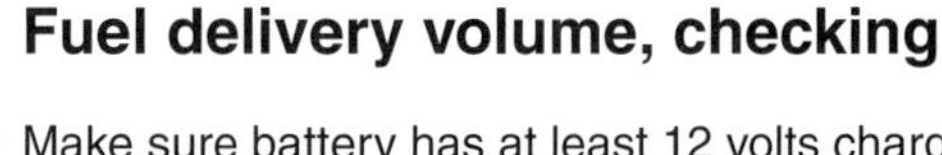

Fuel delivery volume, checking

– Make sure battery has at least 12 volts charge and fuel tank is at least ¼ full. Make sure fuel filter is not plugged. Fuel filter replacement is covered in **03 Maintenance**.

– Loosen fuel tank filler cap.

– Remove upper engine cover. See **03 Maintenance**.

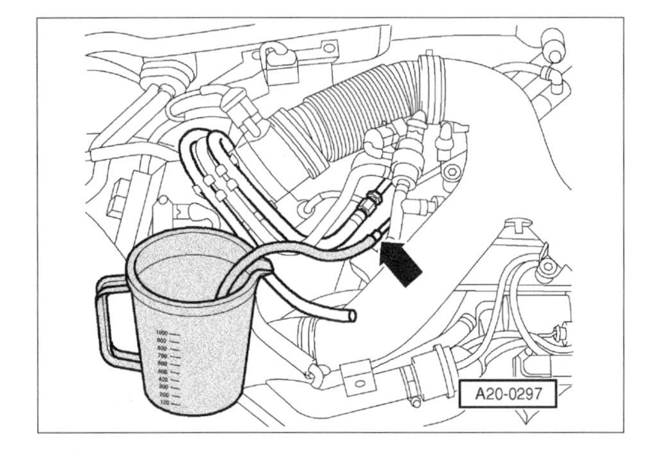

- 2.7 liter V6 turbo engine: Disconnect fuel return line from fuel rail. Connect test hose (**arrow**) and hold end of hose in measuring container.

> ***WARNING —***
> *• Fuel will be expelled under pressure. Wrap a cloth around fuel line to absorb any leaking fuel.*

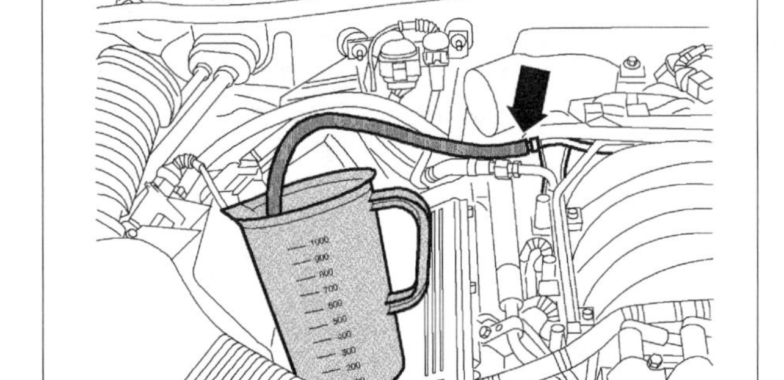

- 2.8 liter V6 engine: Disconnect fuel return line from fuel rail. Connect test hose (**arrow**) and hold end of hose in measuring container.

> ***WARNING —***
> *• Fuel will be expelled under pressure. Wrap a cloth around fuel line to absorb any leaking fuel.*

BentleyPublishers.com
BentleyPublishers.com—All Rights Reserved

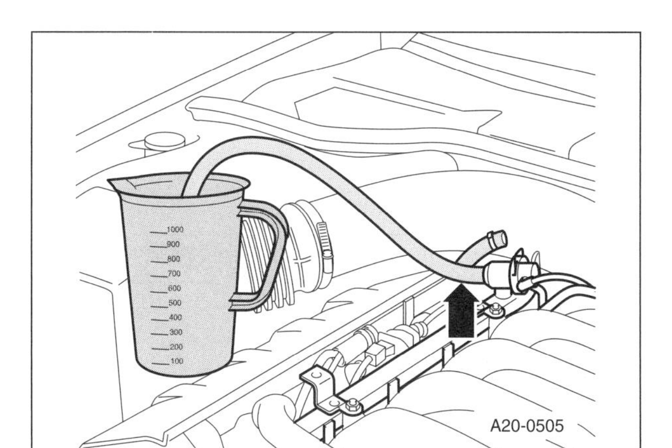

V8 engine with timing belt: Disconnect fuel return line from fuel rail. Connect test hose (**arrow**) and hold end of hose in measuring container.

> ***WARNING —***
> - *Fuel will be expelled under pressure. Wrap a cloth around fuel line to absorb any leaking fuel.*

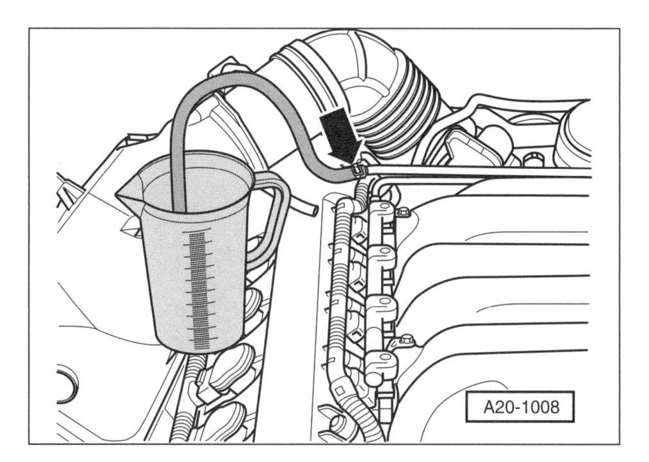

V8 engine with non-return fuel loop: Disconnect fuel return line from fuel rail. Connect test hose (**arrow**) and hold end of hose in measuring container.

> ***WARNING —***
> - *Fuel will be expelled under pressure. Wrap a cloth around fuel line to absorb any leaking fuel.*

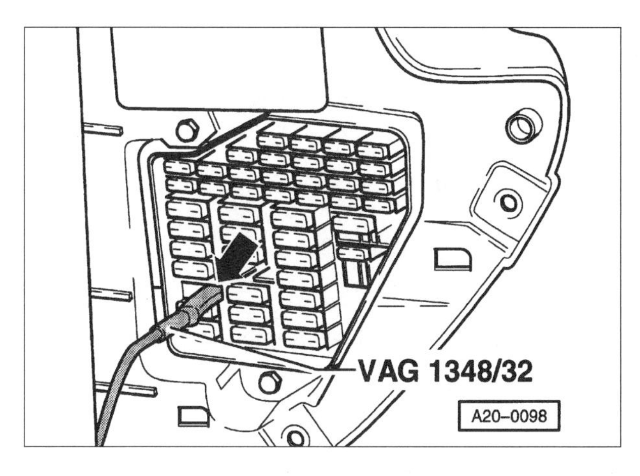

Remove fuse 28 (**arrow**). Connect VAG1348/3A remote switch and VAG1348/32 adapter cable to opening for fuse 28. Connect other end of adapter to battery. Operate fuel pump in this manner for 15 seconds.

– Compare quantity of fuel delivered with correct quantity indicated in graphs below.

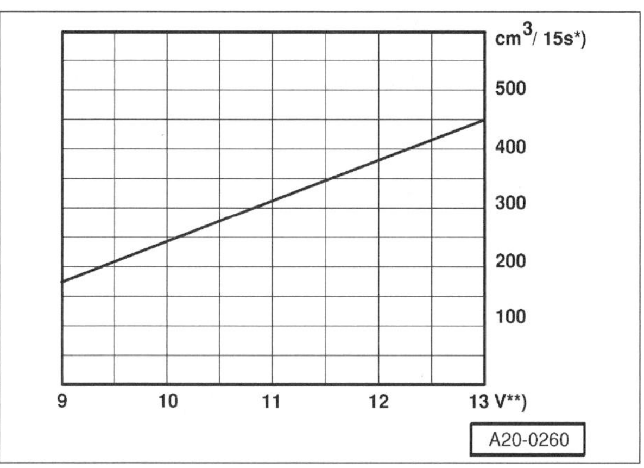

2.8 liter V6 engine: Fuel delivery plotted vs. battery voltage.

- Minimum delivery in milliliters / 15 seconds.
- Engine stationery during test.

Fuel delivery volume (battery @ 12 volts)	
2.8 liter V6 engine	approx. 375 ml (13 oz) / 15 sec

BentleyPublishers.com
BentleyPublishers.com—All Rights Reserved

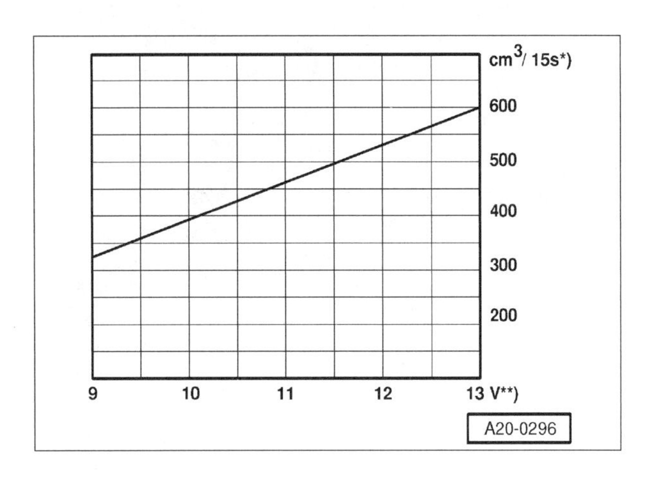

2.7 liter V6 engine, V8 engine: Fuel delivery plotted vs. battery voltage.

- Minimum delivery in milliliters / 15 seconds.
- Engine stationery during test.

Fuel delivery volume (battery @ 12 volts)	
2.7 liter, 3.0 liter, V8 engine	approx. 525 ml (18 oz) / 15 sec

– If fuel delivery volume is below specifications:

- Check for fuel line obstructions.
- Replace fuel pump.

Fuel pressure, checking

– Make sure battery has at least 12 volts charge and fuel tank is at least ¼ full. Make sure fuel filter is not plugged. Fuel filter replacement is covered in **03 Maintenance**.

– Loosen fuel tank filler cap.

– Remove upper engine cover. See **03 Maintenance**.

– Make sure fuel delivery volume is correct. See **Fuel delivery volume, checking** in this repair group.

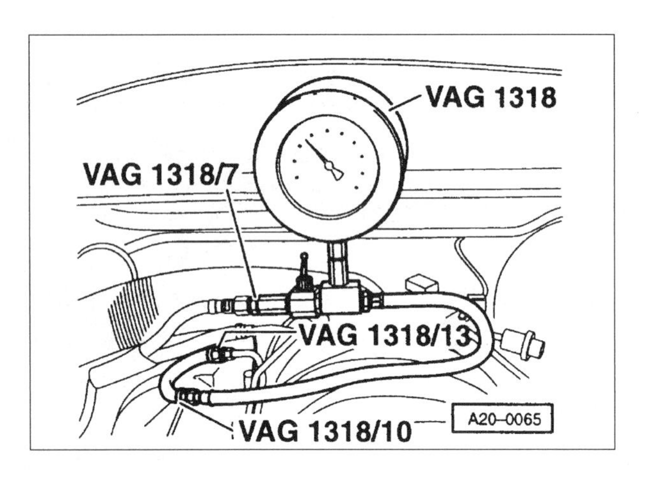

Connect fuel pressure tester (VAG 1318 or equivalent) between fuel supply line and fuel rail using adapters 1318/7, 1318/10 and 1318/13.

- Make sure VAG 1318 shut off valve is between fuel supply line and gauge.

> ***WARNING—***
> - *Fuel will be expelled under pressure. Wrap a cloth around fuel line to absorb any leaking fuel.*

– Open pressure tester shut-off valve (lever parallel to direction of flow).

– Detach and plug vacuum line from fuel pressure regulator to intake manifold.

> ***CAUTION—***
> - *During the following test, do not let the engine run unnecessarily with the vacuum hose removed, since the higher fuel pressure causes enrichment of the fuel air mixture. The richer mixture could exceed the oxygen sensor control limits and set a DTC.*

– Switch off electrical consumers (rear defrost, A/C, etc.).

– Start engine and allow it to idle. Measure fuel pressure.

NOTE—
- *If, during the pressure test, fuel leaks from the fuel pressure regulator vacuum connection, replace the regulator.*

BentleyPublishers.com
BentleyPublishers.com—All Rights Reserved

Fuel pressure specifications	
2.7 liter biturbo V6, all V8 idling: • Fuel pressure regulator vacuum line connected • Fuel pressure regulator vacuum line disconnected	 approx. 3.5 bar (51 psi) approx. 4.0 bar (58 psi)
2.8 liter or 3.0 liter V6 idling: • Fuel pressure regulator vacuum line connected • Fuel pressure regulator vacuum line disconnected	 approx. 3.2 - 3.8 bar (46 - 55 psi) approx. 3.8 - 4.2 bar (55 - 61 psi)

- If specified value is not obtained, replace fuel pressure regulator and repeat pressure test.
- If specified value is still not attained, check the following and replace if necessary:
  - Fuel pump strainer in fuel tank
  - Fuel pump
  - Fuel filter
  - Fuel supply lines (check for damage or kinking)
- If specified value is exceeded, check return line for damage or kinking; replace if necessary.
- Reattach vacuum hose to fuel pressure regulator and check for fuel pressure drop when vacuum is applied.
  - Check that fuel pressure drops by about 0.5 bar (7 psi).
- If pressure does not change, proceed as follows:
  - Check vacuum line for leaks, cracks or damage.
  - Check vacuum line for obstruction; remove hose at fuel pressure regulator and blow into it.
  - If there is no leak and vacuum connection has no obstruction, replace fuel pressure regulator.

Residual pressure, checking

- After fuel pressure test is finished, leave pressure gauge attached.
- Check residual pressure 10 minutes after shutting off engine.

Residual pressure specifications	
Warm engine	approx. 3.0 bar (44 psi)
Cold engine • 2.7 liter biturbo V6, all V8 • 2.8 liter or 3.0 liter V6	 approx. 2.5 bar (36 psi) approx. 2.2 bar (32 psi)

NOTE—

- *The higher pressure with a warm engine, caused by fuel expansion, is normal.*

BentleyPublishers.com
BentleyPublishers.com—All Rights Reserved

– If specified value is not attained:
  - Start engine and, after fuel pressure stabilizes, switch ignition OFF.
  - Close shut-off valve on VAG 1318 pressure tester.

– If residual pressure does not drop, check for the following:
  - Leak at connection between pressure gauge and fuel lines
  - Leak at fuel reservoir line
  - Fuel pump malfunctioning

– If the residual pressure drops excessively, check for the following:
  - Fuel pressure regulator malfunction
  - Injectors leaking
  - Pressure gauge leaking behind shut-off valve

– To remove pressure gauge, close shut-off valve, loosen VAG 1318/7 adapter and drain excess fuel by opening shut-off valve into fuel-resistant container.

Fuel System Components

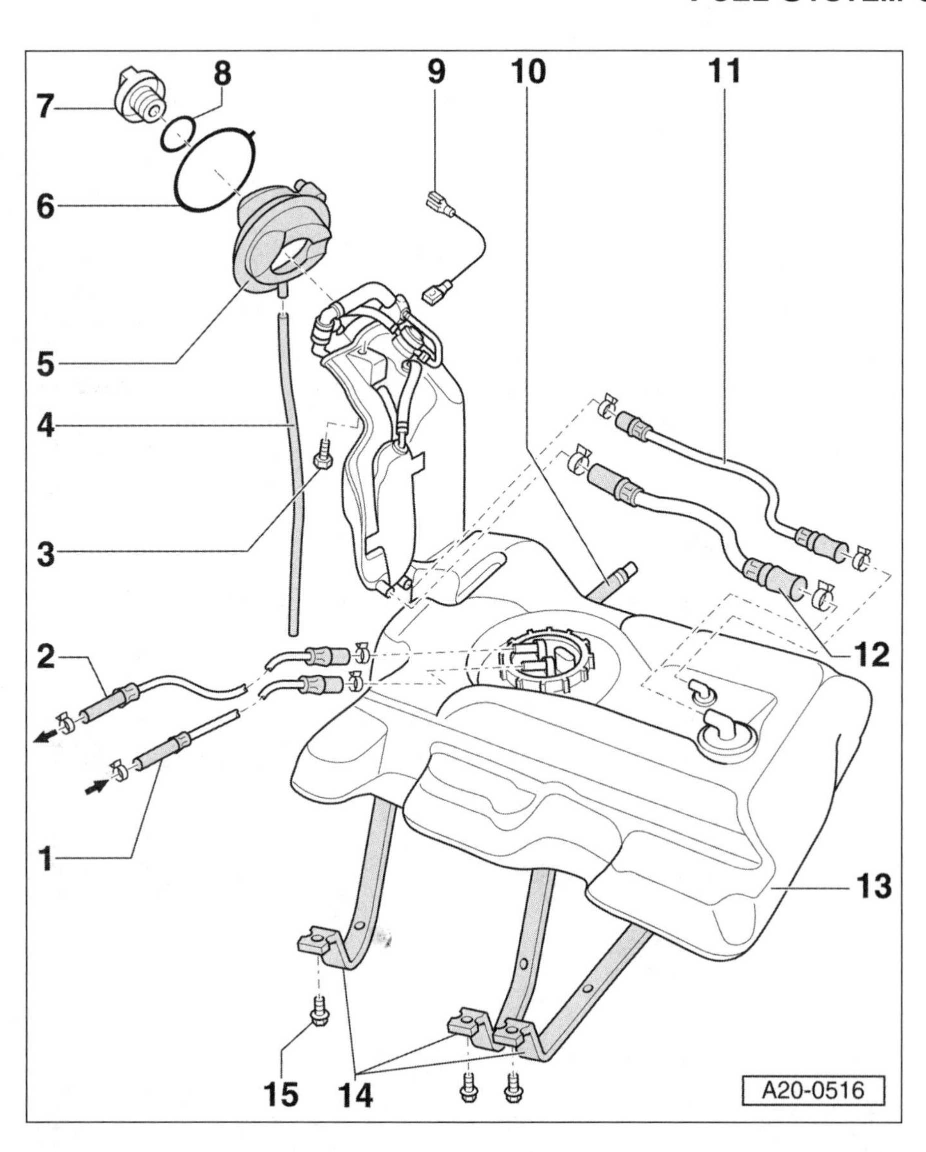

Fuel tank components (front-wheel drive)

1. **Fuel return line**
2. **Fuel supply line**
3. **M8 x 35 mm bolt with washer**
   - Secures fuel tank filler neck and ground wire
4. **Overflow hose**
5. **Rubber cup**
6. **Retaining ring**
7. **Fuel tank filler cap**
8. **Seal**
9. **Filler neck ground wire**
   - Make sure wire end contacts bare metal of body
   - Resistance to ground approx. 0Ω
10. **Vent line**
11. **Vent line**
12. **Vent line**
13. **Fuel tank**
14. **Mounting straps**
15. **M8 x 28 mm bolt with washer**
    - 23 Nm (17 ft-lb)

BentleyPublishers.com
BentleyPublishers.com—All Rights Reserved

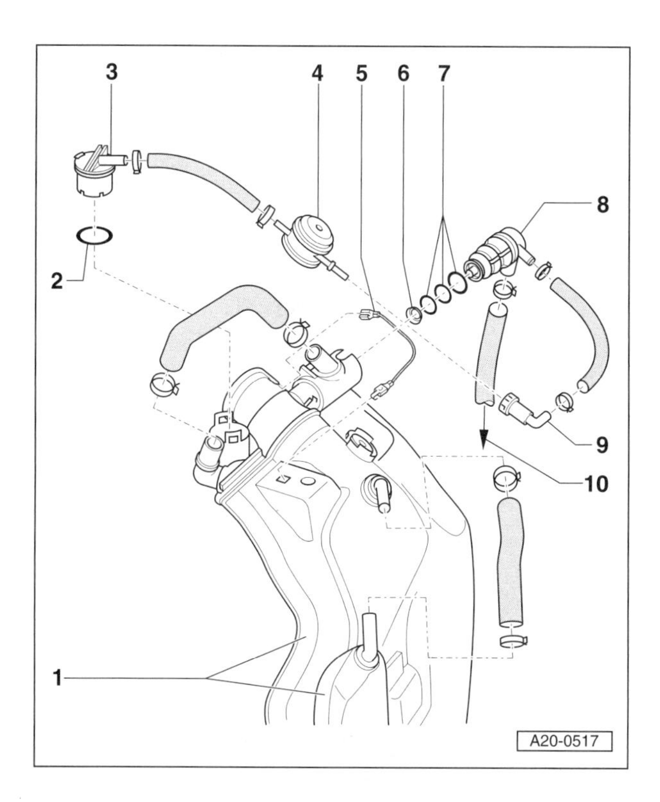

Fuel filler neck components (front-wheel drive)

1. Filler neck and expansion tank
2. O-ring
3. Gravity valve
4. Tank protection valve
5. Filler neck ground wire
6. Sealing ring
7. O-ring
8. Vent valve
9. Hose coupling
10. to Fuel tank evaporative canister

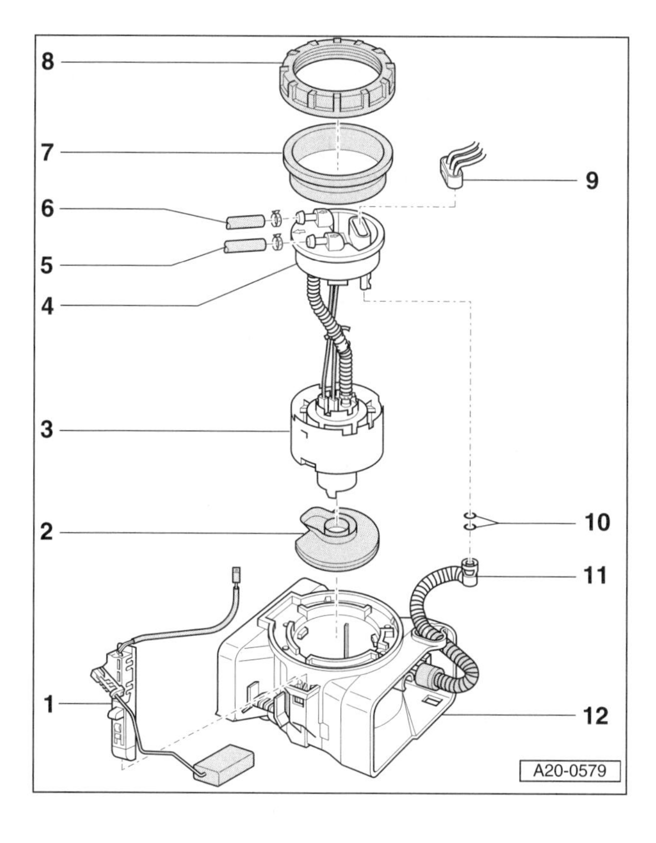

Fuel pump and fuel level sender components (front-wheel drive)

1. Fuel level sender
2. Fuel strainer
3. Fuel pump module
4. Fuel tank flange
5. Fuel return line
6. Fuel supply line
7. Rubber seal
8. Fuel tank flange collar nut
9. Harness connector
10. Sealing O-rings
11. Fuel return line
12. Accumulator housing

BentleyPublishers.com
BentleyPublishers.com—All Rights Reserved

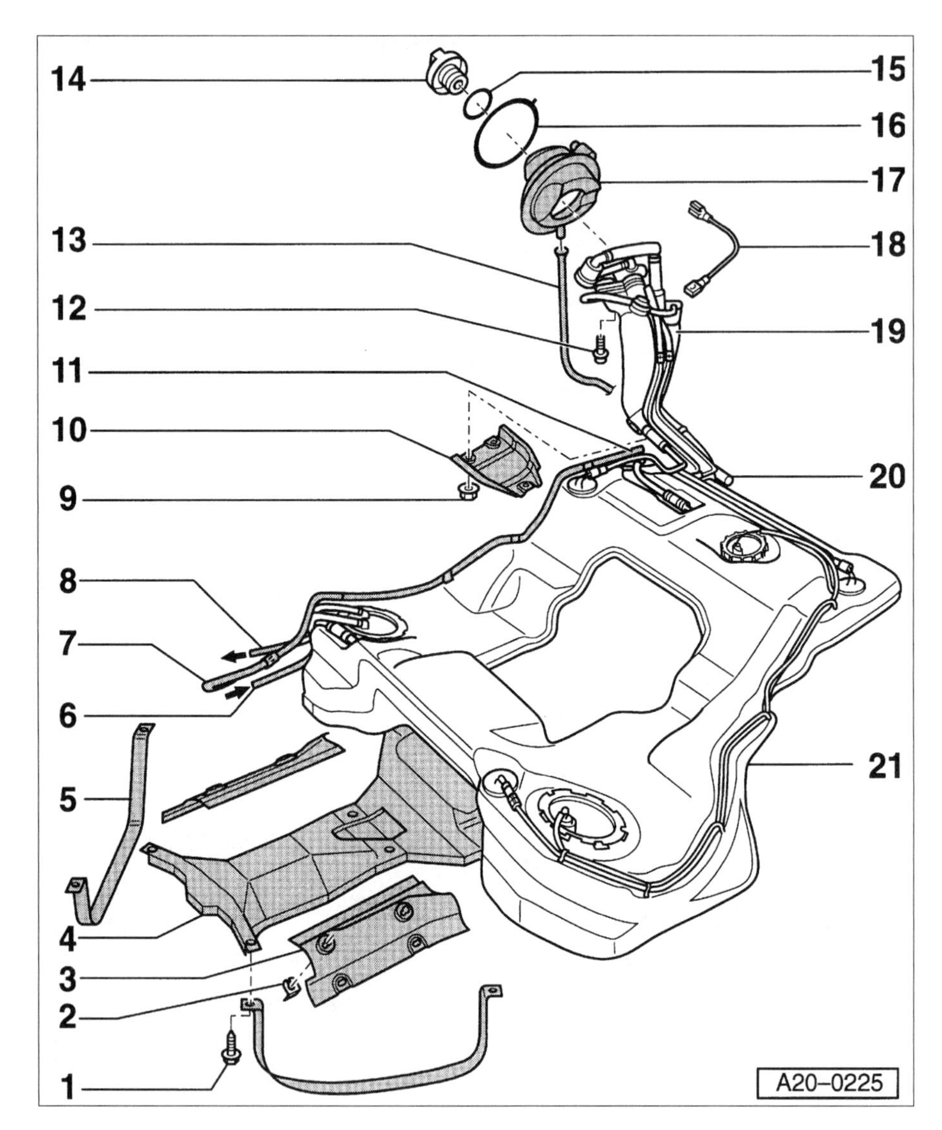

Fuel tank components (V6 quattro, allroad quattro)

1. **M8 x 28 mm bolt with washer**
   - 23 Nm (17 ft-lb)
2. **Speed nut**
3. **Heat shield**
4. **Support bracket**
5. **Mounting strap**
6. **Fuel return line**
7. **Vent line**
   - to Fuel tank evaporative control canister valve (N80)
   - Remains attached when tank is removed
8. **Fuel supply line**
9. **Nut**
   - 2 Nm (18 in-lb)
10. **Filler neck heat shield**
11. **Vent line**
    - to Fuel tank evaporative control canister
    - Remains attached when tank is removed
12. **M8 x 28 mm bolt with washer**
    - 25 Nm (18 ft-lb)
    - Secures fuel tank filler neck and ground wire
13. **Overflow return hose**
14. **Fuel tank filler cap**
15. **Seal**
16. **Retaining ring**
17. **Rubber cup**
18. **Filler neck ground wire**
    - Make sure wire end contacts bare metal of body
    - Resistance to ground approx. 0Ω
19. **Filler neck and expansion tank**
20. **Vent line**
21. **Fuel tank**

BentleyPublishers.com
BentleyPublishers.com—All Rights Reserved

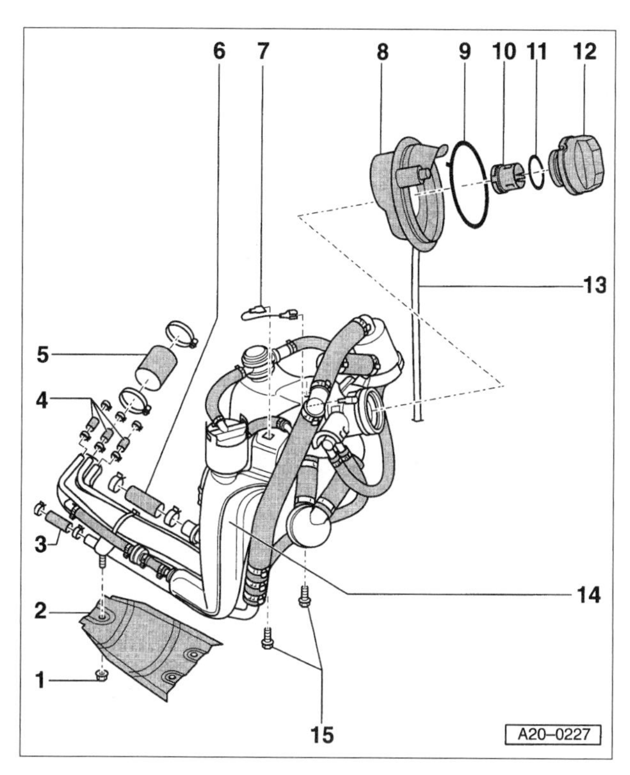

Fuel filler neck components (V6 quattro, allroad quattro)

1. Nut, 2 Nm (18 in-lb)
2. Heat shield
3. Hose
4. Hose connections for vent lines
5. Hose connection
6. Hose connection for fuel tank vent
7. Ground connection
8. Rubber cup
9. Retaining ring
10. Seal insert
11. Seal
12. Fuel tank filler cap
13. Overflow return line
14. Filler neck and expansion tank
15. M8 x 30 mm bolt with washer, 23 Nm (17 ft-lb)

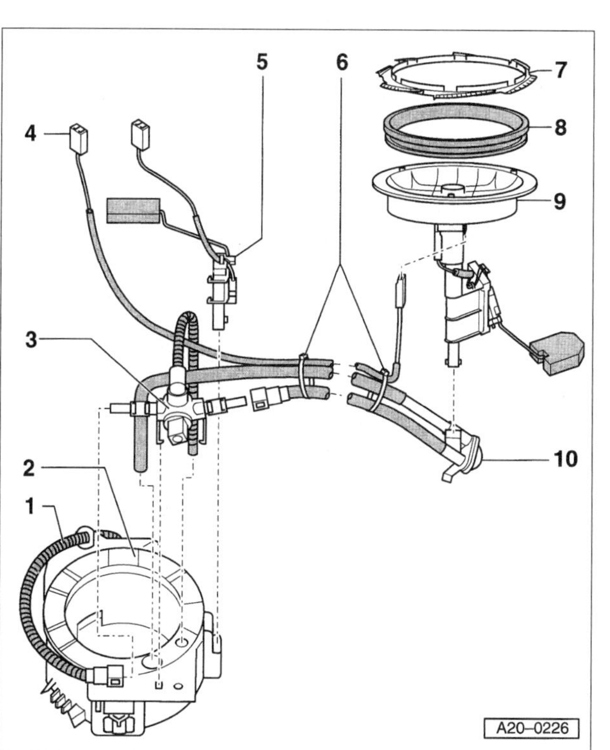

Fuel level sender components (quattro)

1. Fuel supply line to accumulator
2. Accumulator
3. Fuel distributor
   - Attached to accumulator with 3 tabs
   - Push toward left to release
4. Left side fuel level sender wire
5. Right side fuel level sender (G)
   - Attached to accumulator with tabs
6. Wire ties
7. Fuel tank flange collar lock
8. Rubber seal
9. Fuel tank flange and left side fuel level sender (G169)
10. Suction jet pump

BentleyPublishers.com
BentleyPublishers.com—All Rights Reserved

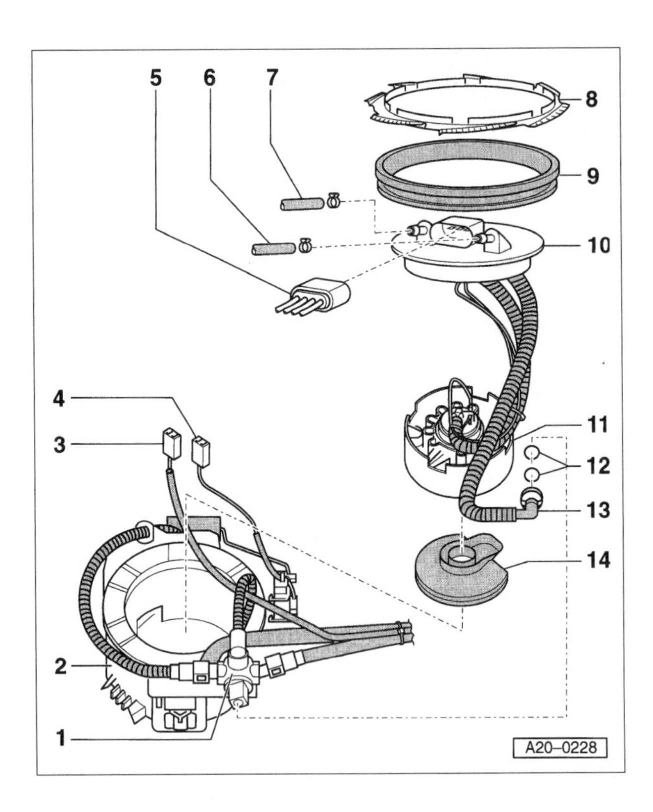

Fuel pump components (quattro)

1. Fuel distributor
   - Attached to accumulator with 3 tabs
   - Push toward left to release
2. Accumulator
3. Left side fuel level sensor wire
4. Right side fuel level sensor wire
5. Harness connector
6. Fuel return line
7. Fuel supply line
8. Fuel tank flange collar lock
9. Rubber seal
10. Fuel tank flange
11. Fuel pump module (G6)
    - To remove:
      -Release fuel distributor
      -Disconnect return line from distributor
12. Sealing O-rings
13. Fuel return line
14. Fuel strainer

BentleyPublishers.com
BentleyPublishers.com—All Rights Reserved

21 Turbocharger and Intercooler

GENERAL

This repair group provides basic turbocharger and intercooler information.

Turbocharger boost pressure is not adjustable and separate parts for the turbocharger are usually not available from Audi.

Turbocharger system

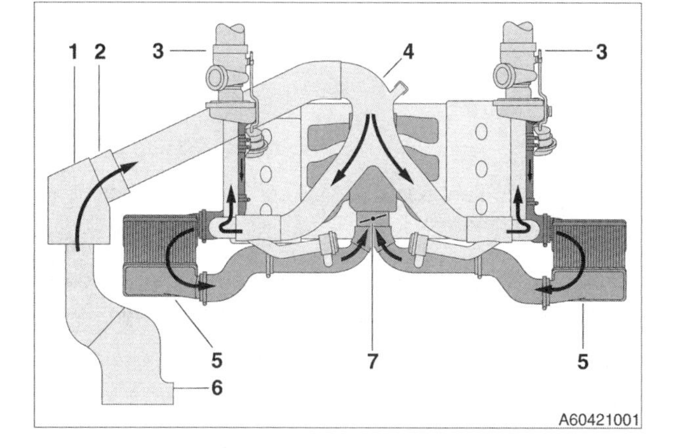

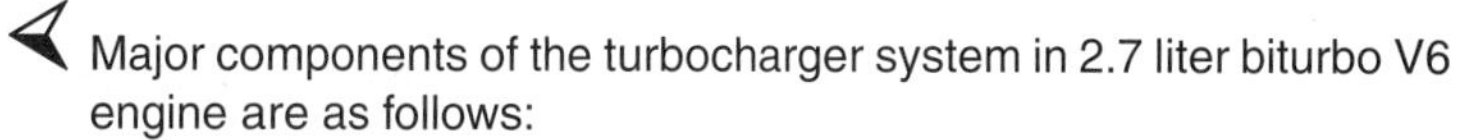

Major components of the turbocharger system in 2.7 liter biturbo V6 engine are as follows:

1. Air filter housing
2. Mass airflow sensor
3. Turbocharger
4. Air distributor
5. Charge air intercooler
6. Fresh air intake
7. Throttle valve control module

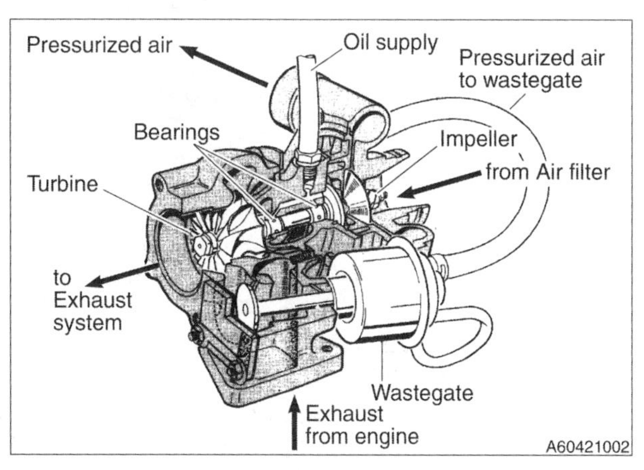

The exhaust-driven turbocharger turbine spins at very high speed and is precisely balanced. The turbine bearings are cooled and lubricated by engine oil with additional cooling supplied by a connection to the engine cooling system.

Turbo boost pressure is controlled by the wastegate bypass regulator valve operated by the engine control module (ECM). When boost pressure exceeds programmed values, the ECM opens the wastegate to bypass some of the exhaust gases around the turbine.

BentleyPublishers.com
BentleyPublishers.com—All Rights Reserved

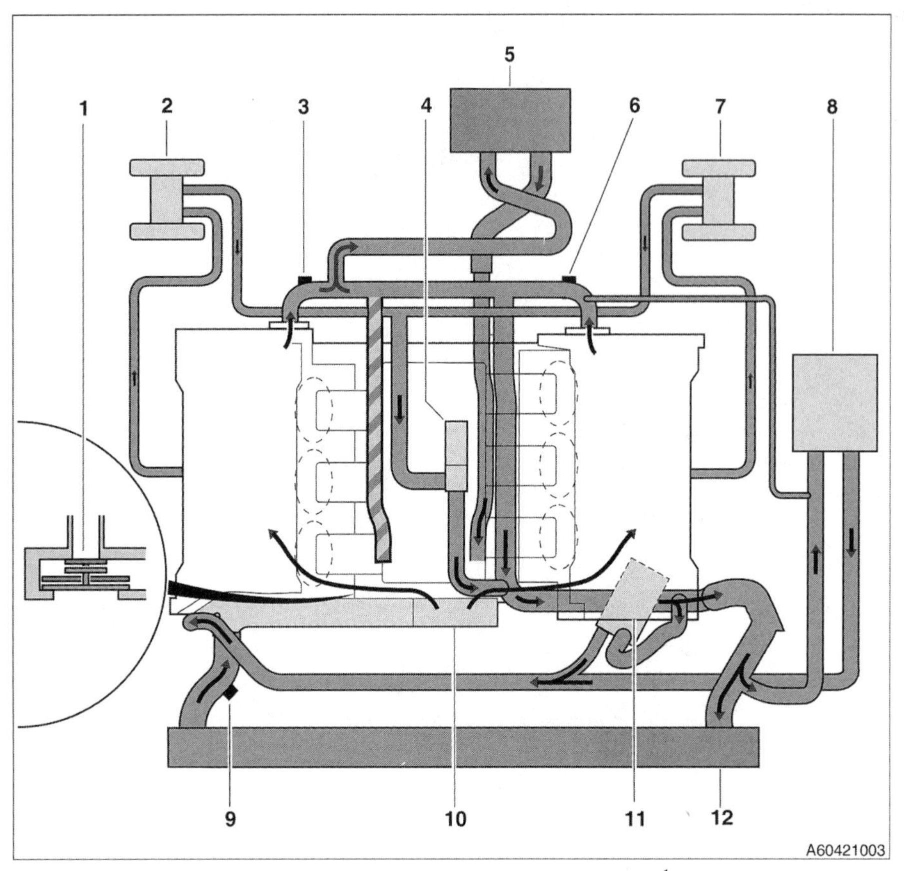

Turbocharger cooling circuit

1. **Coolant thermostat**
2. **Right turbocharger**
3. **Engine coolant temperature (ECT) sensors (G2, G62)**
4. **After-run coolant pump (V51)**
5. **Heater core**
6. **After-run coolant pump switch (F95)**
7. **Left turbocharger**
8. **Coolant overflow reservoir**
9. **Coolant fan switch (F18, F54)**
10. **Coolant pump**
11. **Engine oil cooler**
12. **Radiator**

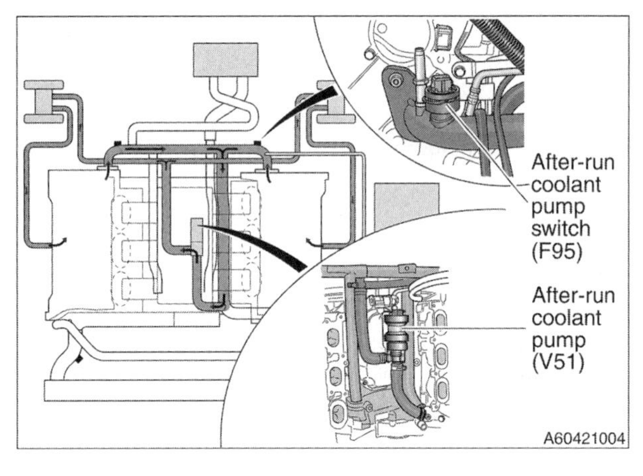

The electric after-run coolant pump (V51) keeps coolant from overheating under high thermal load.

The after-run pump is in the engine V-angle. The after-run function is activated only if the engine is not running. It cycles OFF after 10 minutes.

Warnings and Cautions

WARNING —

- *The turbocharger and related components operate at very high temperature. Prior to working on these components, allow the system to cool thoroughly.*

CAUTION—

- *To prevent dust and dirt contamination, cover turbocharger components with dust-free paper or seal them in plastic bags. Do not use cloth material. Avoid nearby use of compressed air.*
- *Do not move the car or work in dusty conditions while the turbocharger is open or removed.*
- *Make sure turbocharger air lines, hoses and connections are free of oil and grease before reinstalling. When reinstalling air connection O-rings, use water as an assembly lubricant.*

BentleyPublishers.com
BentleyPublishers.com—All Rights Reserved

24 Fuel Injection

General

This repair group covers basic engine management information.

See also the following:

- **03 Maintenance** for engine air filter and fuel filter replacement
- **20 Fuel Storage and Supply** for fuel pressure and delivery tests
- **26 Exhaust System for** oxygen sensor and secondary air injection information
- **28 Ignition** for ignition system repairs
- **OBD On-Board Diagnostics** for information on OBD II system and diagnostic trouble codes (DTCs)

Engine management systems

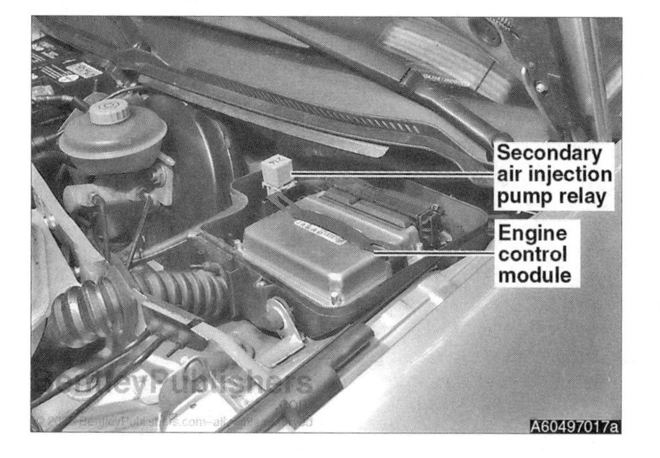

The Motronic (or DME) engine management system uses an electronic engine control module (ECM) to control fuel injection and ignition functions. Audi A6 models are equipped with versions of the Motronic system.

BentleyPublishers.com
BentleyPublishers.com—All Rights Reserved

Bosch Motronic M5.9.2 is used in 1998 - 1999 A6 models with 2.8 liter V6 engine. The main features of the M5.9.2 system are:

- Cylinder sequential fuel injection (separate fuel mixture control for each cylinder)
- Electronically mapped ignition system
- Adaptive heated oxygen sensor control, two sensors per catalytic converter (4 in all)
- Camshaft sensors and variable exhaust camshaft timing
- Hot film mass air flow sensor with integral intake air temperature sensing
- Idle air control
- Adaptive fuel metering
- Adaptive throttle position sensor
- Adaptive ignition knock control
- Check Engine light (malfunction indicator light or MIL) for emissions related and catalytic converter damaging faults
- Secondary air injection
- Evaporative emissions control (EVAP) with fuel tank leak detection
- On-board diagnostics

Bosch Motronic ME 7.1 is used in 1999 and later V8 models and 2000 and later V6 models. The major differences between Bosch ME 7.1 and Bosch M5.9.2 are:

- Electronic throttle control (EPC)
- Individual ignition coils at each cylinder (direct ignition)
- Turbocharger control (if equipped)
- Integrated cruise control
- A/C compressor control
- Control module communication using CAN-bus

Bosch Motronic ME 7.1.1 is an upgrade over the older system.

Engine management systems and major features are summarized in **Table a**. ME 7.1 is shown schematically in **Bosch Motronic 7.1 sensors and actuators (2.7 liter turbo V6 engine)**

Table a. Engine management system (Motronic) applications

Engine	Code	Year, model	Motronic version	Emissions category	Intake manifold tuning	Variable camshaft timing	Secondary air injection	Electronic throttle control (EPC)
2.7 liter V6 biturbo	APB	2000 - 2003 A6, allroad quattro	Motronic ME 7.1	TLEV	Yes	Exhaust	Yes	Yes
	BEL	2003 - 2004 A6, allroad quattro	Motronic ME 7.1	LEV	Yes	Exhaust	Yes	Yes
2.8 liter V6	AHA	1998 - 1999 A6	Motronic M5.9.2	TLEV	Yes	Exhaust	Yes	No
	ATQ	2000 - 2002 A6	Motronic ME 7.1	LEV	Yes	Exhaust	Yes	Yes
3.0 liter V6	AVK	2002 - 2004 A6	Motronic ME 7.1.1	ULEV	Yes	Intake Exhaust	Yes	Yes

BentleyPublishers.com
BentleyPublishers.com—All Rights Reserved

Table a. Engine management system (Motronic) applications								
Engine	Code	Year, model	Motronic version	Emissions category	Intake manifold tuning	Variable camshaft timing	Secondary air injection	Electronic throttle control (EPC)
4.2 liter V8	ART, AWN, BBD	1999 - 2004 A6, S6	Motronic ME 7.1	LEV or TLEV	Yes	Yes	Yes	Yes
4.2 liter V8	BAS	2003 - 2004 allroad quattro	Motronic ME 7.1.1	LEV	Yes	Yes	Yes	Yes
4.2 liter V8 biturbo	BCY	2003 - 2004 RS6	Motronic ME 7.1.1	TLEV	No	Yes	Yes	Yes

NOTE—

- *Audi identifies electrical components by a letter and/or a number in the electrical schematics. See* **EWD Electrical Wiring Diagrams**. *These electrical identifiers are listed in parentheses as an aid to electrical troubleshooting.*
- *Electronic throttle control (EPC) is also referred to as E-gas.*

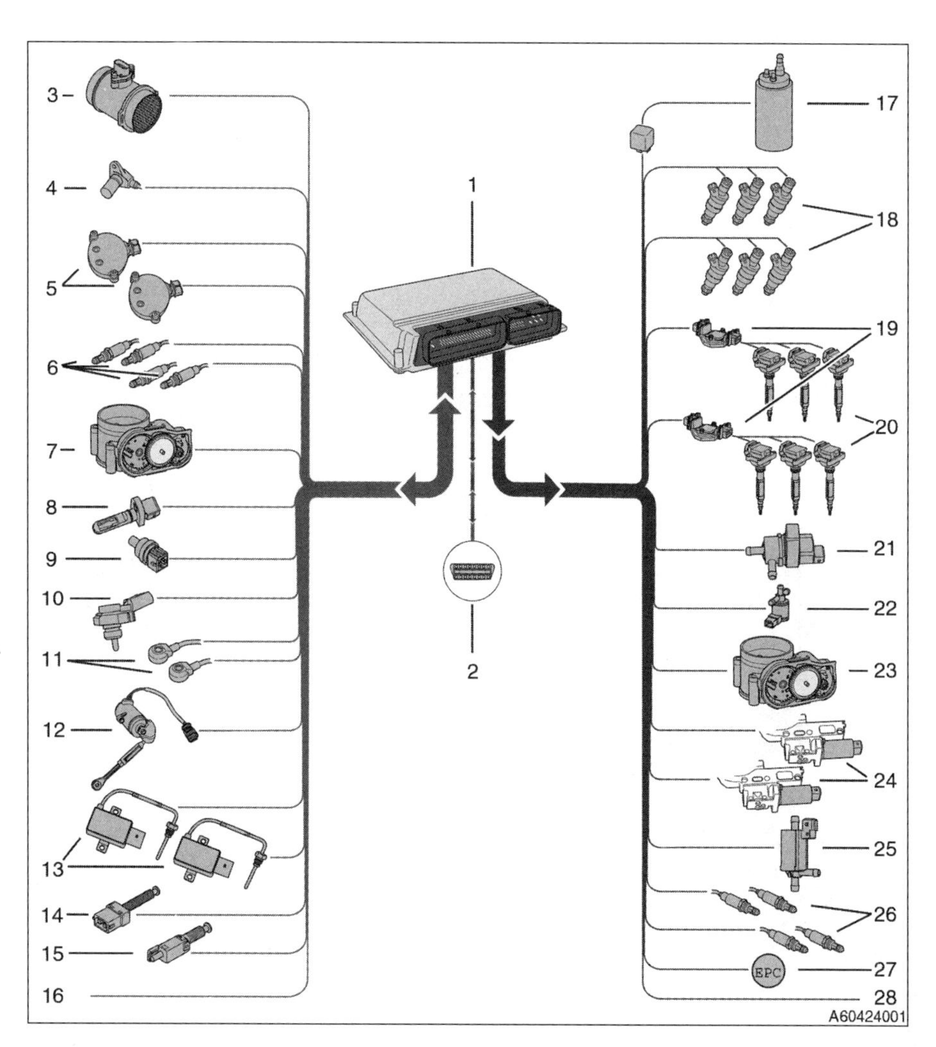

Bosch Motronic 7.1 sensors and actuators (2.7 liter turbo V6 engine)

1. **Engine control module (ECM)(J220)**
2. **DLC (OBD II) plug**

Sensors

3. **Mass airflow sensor (G70)**
4. **Engine speed sensor (G28)**
5. **Exhaust camshaft sensors (G40, G163)**
6. **Oxygen sensors (G39, G108, G130, G131)**
7. **Throttle valve control module (J338) Throttle angle sensors (G187, G188) Throttle drive (G186)**
8. **Intake air temperature sensor (G42)**
9. **Engine coolant temperature (ECT) sensor (G2, G62)**
10. **Charge air pressure sensor (G31)**
11. **Knock sensors (G61, G66)**
12. **Throttle position sensor (G79) Accelerator pedal position sender (G185)**
13. **Exhaust temperature sensors (G235, G236)**
14. **Brake light switch (F) Cruise control vacuum switch (F47)**
15. **Clutch vacuum switch (F36)**
16. **Auxiliary signals**

BentleyPublishers.com
BentleyPublishers.com—All Rights Reserved

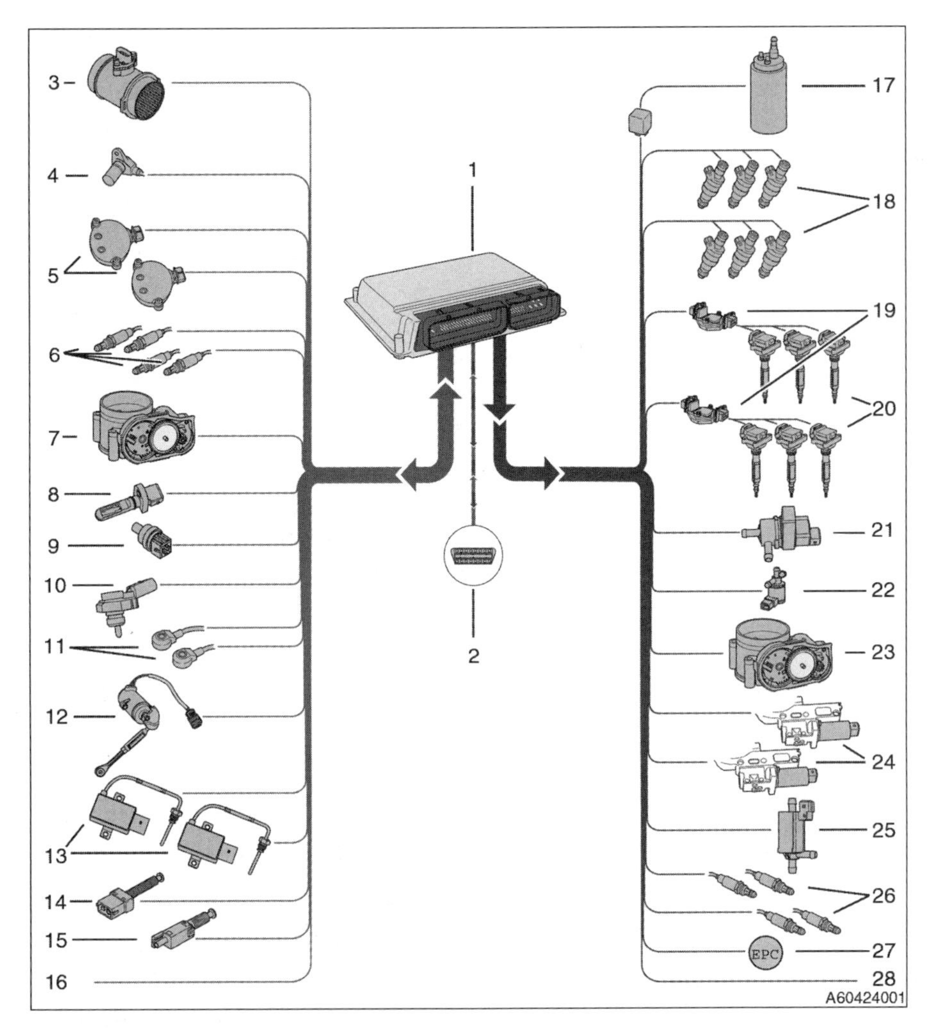

Bosch Motronic 7.1 sensors and actuators (2.7 liter turbo V6 engine) *(continued)*

Actuators

17. **Fuel pump (G6)**
    Fuel pump relay (J17)
18. **Fuel injectors (N30 - N33, N83 - N84)**
19. **Ignition power output stage (N122, N192)**
20. **Ignition coils (N, N128, N158, N163, N164, N189)**
21. **Fuel tank evaporative control (EVAP) canister purge regulator valve (N80)**
22. **Wastegate bypass regulator (N75)**
23. **Throttle valve control module (J338)**
    Throttle drive (G186)
24. **Camshaft adjustment valves (N205, N208)**
25. **Turbocharger recirculating valve (N249)**
26. **Oxygen sensor heaters (Z19, Z28, Z29, Z30)**
27. **Electronic throttle (EPC) warning light (K132)**
28. **Auxiliary signals**

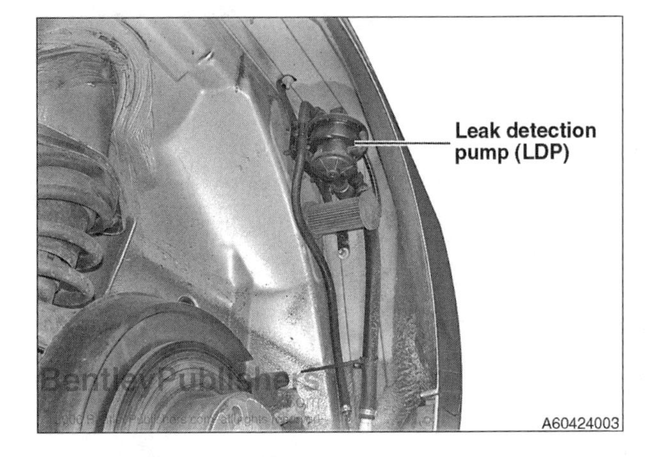
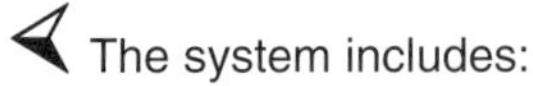

Evaporative control system (EVAP)

The evaporative control system is designed to prevent fuel system evaporative losses from venting into the atmosphere. The evaporative system allows control and monitoring of evaporative losses by the on-board diagnostic (OBD II) software incorporated in the ECM.

The system includes:

- Carbon canister to store evaporated fuel.
- Plumbing to duct vapor from the fuel tank overflow reservoir to the carbon canister and from the canister to the intake manifold.
- Leak detection pump (LDP) (illustrated for 1998 2.8 liter A6) to monitor the integrity of the evaporative control system.

The fuel overflow reservoir and associated valves are in the rear of the right rear fender, behind the wheel housing liner. The leak detection pump is under the front of the same fender.

BentleyPublishers.com
BentleyPublishers.com—All Rights Reserved

On-board diagnostics

Most components and functions of the Motronic system are monitored by on-board diagnostics (OBD II) software in the engine control module (ECM). If faults and malfunctions are detected, one or more diagnostic trouble codes (DTCs) may be set.

In case of an emissions related fault or malfunction considered catalyst-damaging, the Check Engine light (malfunction indicator light or MIL) or EPC light is illuminated.

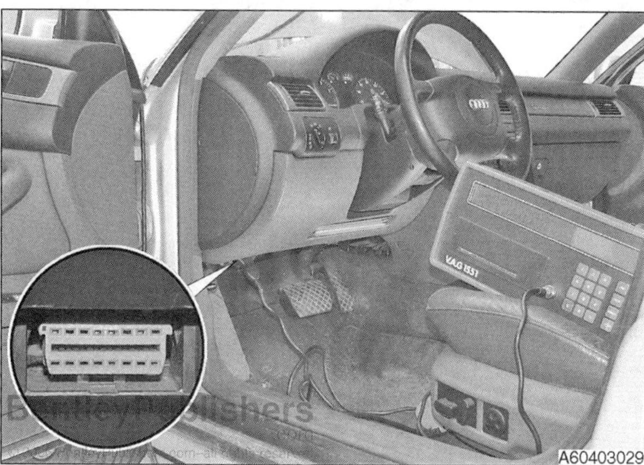

Access fault memory with VAG scan tool or equivalent plugged into the DLC plug (**inset**) under left side dashboard. See **OBD On-Board Diagnostics** for more information.

NOTE—
- *After engine management system tests or repairs, the ECM may recognize a malfunction and store a DTC. Therefore perform the following after ending tests and repairs:*
  -Use VAG scan tool or equivalent to check DTC memory.
  -Erase DTC memory.
  -Reset or erase diagnostic data.
  -Generate readiness code.
- *After engine management system tests or repairs, the engine may start, run for a short period and then cut out because the electronic anti-theft immobilizer disables the ECM. In such cases:*
  -Use VAG scan tool or equivalent to check DTC memory.
  -Adapt ECM.

Warnings and Cautions

Read the following warnings and cautions before starting work on the fuel injection system.

WARNING—
- *Disconnect negative (-) battery cable and cover terminal with insulated material when working on fuel related components.*
- *Gasoline is dangerous to your health. Wear suitable hand, skin and eye protection when working on the fuel system. Do not breathe fuel vapors. Work in a well-ventilated area.*
- *Do not smoke or create sparks around fuel. Be aware of pilot lights in gas operated equipment (heating systems, water heaters, etc.). Have an approved fire extinguisher handy.*
- *The fuel system is designed to maintain pressure in the system after the engine is turned off. Fuel is expelled under pressure when fuel lines are disconnected. This is a fire hazard, especially if the engine is warm. To prevent fuel from spraying, wrap a clean shop rag around fuel line fitting before loosening or disconnecting it.*
- *Exercise extreme caution when using spray-type cleaners on a warm engine. Observe all manufacturer recommendations.*
- *Prior to working on fuel-related components, unscrew the fuel filler cap to release pressure in the tank.*

BentleyPublishers.com
BentleyPublishers.com—All Rights Reserved

> ***CAUTION—***
> - *Cleanliness is essential when working on any part of the fuel system. Thoroughly clean fuel line unions and hose fittings before disconnecting them. Use only clean tools.*
> - *Keep removed components clean. Seal or cover them with plastic or paper, especially if repair cannot be completed immediately. Seal open fuel supply and return lines to prevent contamination.*
> - *When replacing parts, install only new, clean components. Replace seals and O-rings.*
> - *Prior to disconnecting the battery, read the battery disconnection cautions in* **00 Warnings and Cautions**.
> - *Before making any electrical tests with the ignition turned ON, disable the ignition system. Be sure the battery is disconnected when replacing components.*
> - *To prevent damage to the ignition system or other electronic components, including the engine control module (ECM), always connect and disconnect wires and test equipment with the ignition OFF.*
> - *Only use a digital multimeter for electrical tests.*

ENGINE MANAGEMENT FUSES AND RELAYS

To gain access to fuse and relay panels, see **97 Fuses, Relays, Component Locations**.

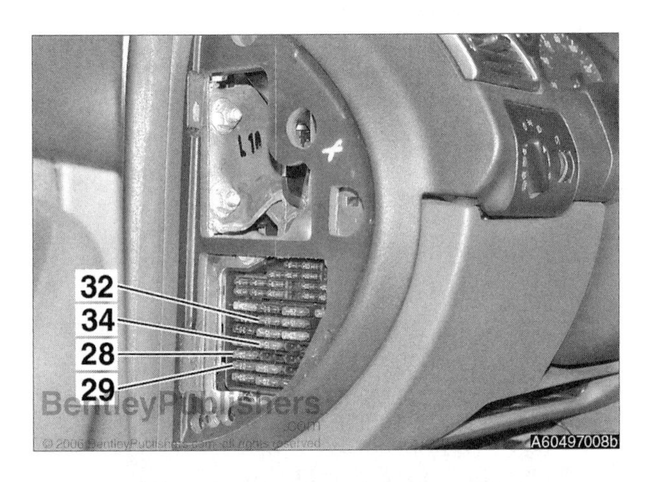

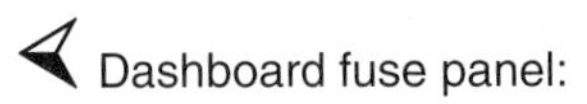

Dashboard fuse panel:

- Fuse **28**: Fuel pump
  1998 - 2000: 15A
  2001 - 2004: 20A
- Fuse **29**: Engine control module (ECM)
  1998 - 2000: 30A
  2001 - 2004: 20A
- Fuse **32**: Engine control module (ECM) 20A
- Fuse **34**: Fuel injectors 15A

Under left dashboard:

- 13-fold relay and fuse panel:

1. Secondary air pump fuse 50A
2. Starter interlock relay (J207), manual transmission Park / neutral position (PNP) relay (J226), multitronic (CVT) transmission

- Micro central electric panel:

3. Fuel pump relay (J17)

NOTE—
- *1998 - 2000 models: Fuel pump relay also powers fuel injectors.*

BentleyPublishers.com
BentleyPublishers.com—All Rights Reserved

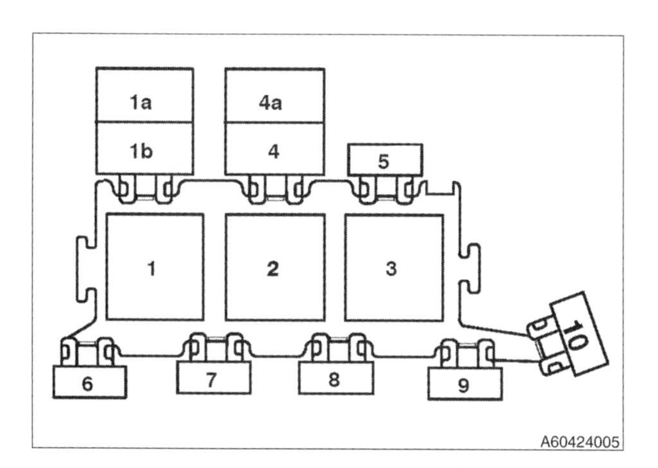

E-box, left rear engine compartment in plenum chamber:

- Fuse **5**: Engine control module (ECM) 15A
- Fuse **7**: Secondary air pump 40A
- Relay **1b**: Coolant circulation pump relay (J151)
- Relay **2**: Secondary air pump relay (J299)
- Relay **3**: Engine control module (ECM) relay (J271)

NOTE —

- *2004 3.0 liter V6 engine illustrated.*

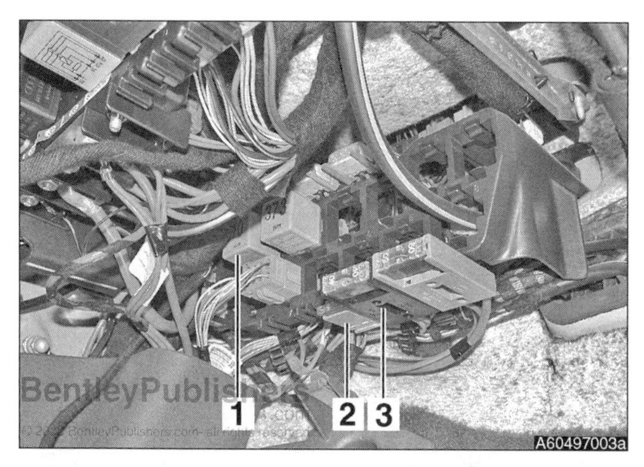

8-fold relay and fuse panel, under left dashboard:

1. Engine cooling fan relay (high speed)
2. Engine cooling fan control module fuse 5A
3. Engine cooling fan fuse 60A

ENGINE CONTROL MODULE (ECM)

ECM, accessing

– Open engine hood and remove plenum chamber cover.

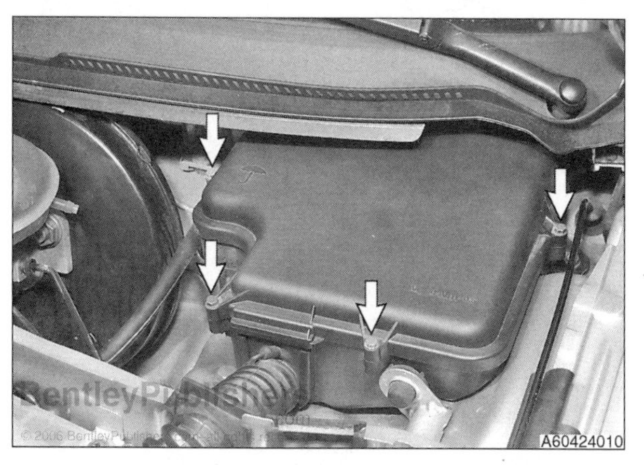

Remove electronics box (E-box) cover retaining bolts (**arrows**). Lift off cover.

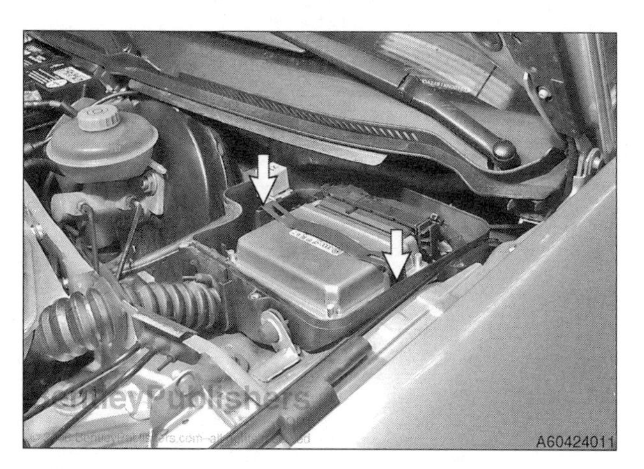

Use small screwdriver to pry off ECM hold-down clip at sides (**arrows**).

BentleyPublishers.com
BentleyPublishers.com—All Rights Reserved

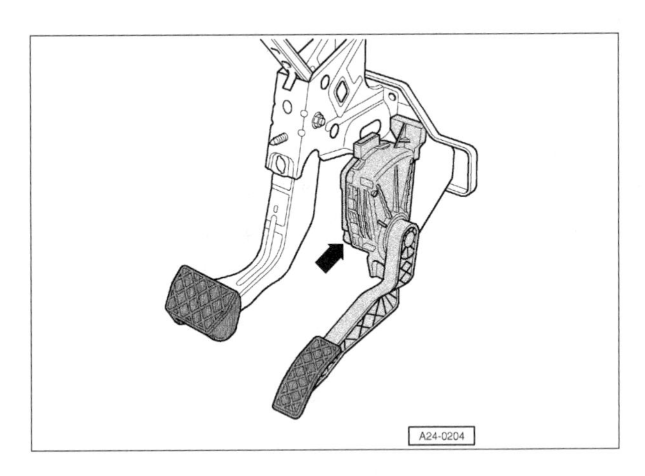

PEDAL SENSORS AND SWITCHES

Accelerator pedal sensors

Throttle position sensor (G79) and accelerator position sensor (G185) (**arrow**)

NOTE—
- *These components are shared by models with Bosch Motronic ME 7.1 and later (engine management with electronic throttle control or EPC).*

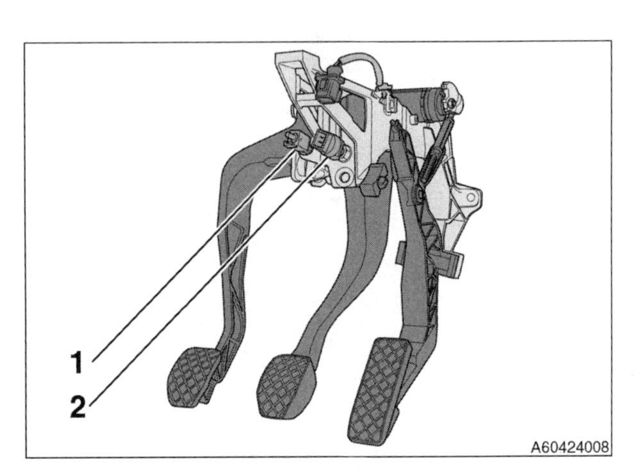

Cruise control switches

Pedal switches:

1. Clutch vacuum switch (F36)
2. Brake light switch (F)
   Cruise control vacuum switch (F47)

Cruise control stalk switch replacement is covered in **48 Steering**.

BentleyPublishers.com
BentleyPublishers.com—All Rights Reserved

2.7 Liter Engine Management Components

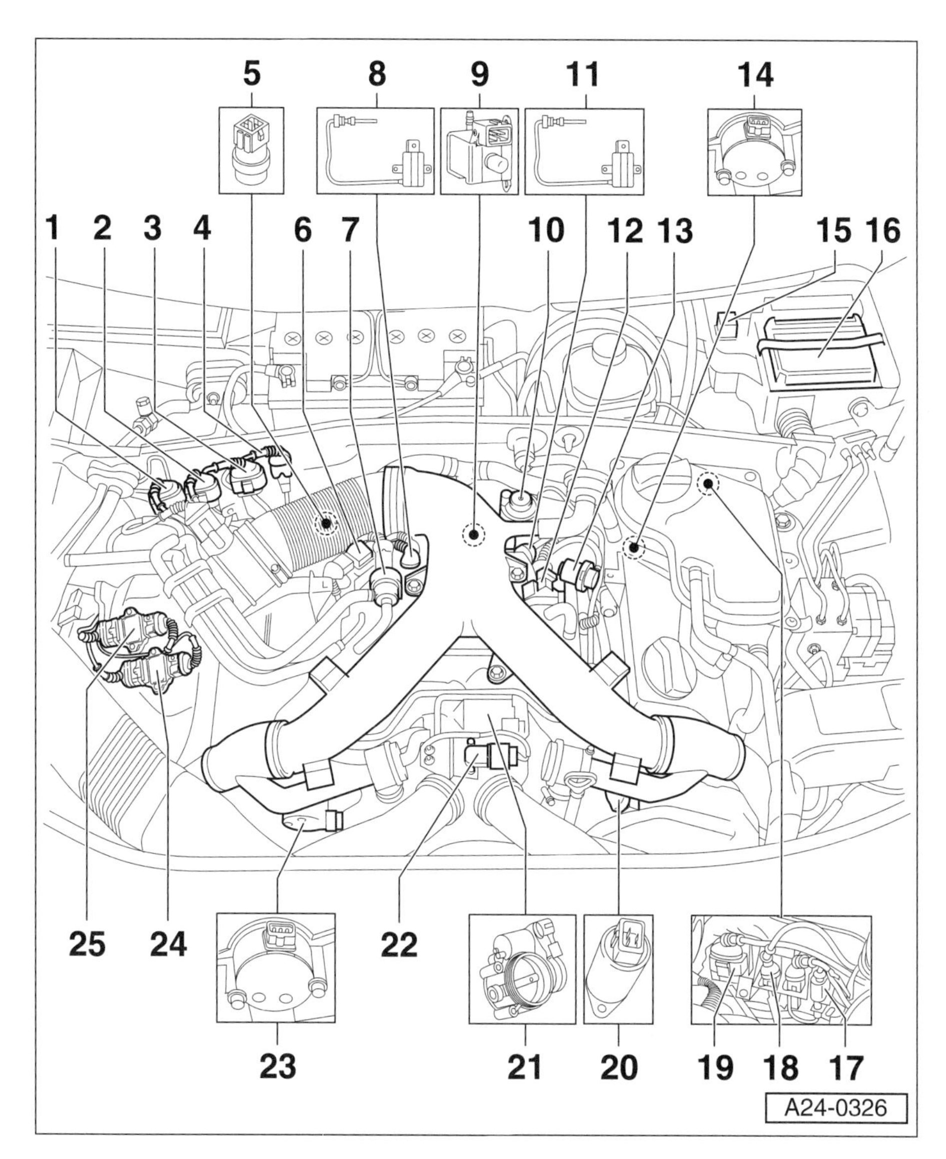

Engine compartment (2.7 liter turbo V6 engine)

1. **Oxygen sensor harness connector**
   - Right bank oxygen sensor (G130) and heater (Z29) behind catalytic converter
   - 4-pin green
2. **Oxygen sensor harness connector**
   - Left bank oxygen sensor (G131) and heater (Z30) behind catalytic converter
   - 4-pin green
3. **Oxygen sensor harness connector**
   - Right bank oxygen sensor (G39) and heater (Z19) in front of catalytic converter.
   - 4-pin black
4. **Right knock sensor (G61) harness connector**
   - 3-pin
5. **Engine coolant temperature (ECT) sensor (G62)**
   - On coolant pipe behind right cylinder bank
6. **Wastegate bypass regulator valve (N75)**
7. **Fuel tank evaporative control (EVAP) canister purge regulator valve (N80)**
8. **Exhaust temperature sensor, right (G235)**
   - Right rear of intake manifold
   - Automatic transmission: Ganged with left exhaust temperature sensor (G236) at this location
9. **Secondary air injection solenoid valve (N112)**
   - Only on vehicle with automatic transmission
10. **Secondary air injection pump motor (V101)**
    - Only on vehicle with automatic transmission
11. **Exhaust temperature sensor, left (G236)**
    - Left rear of intake manifold
    - Automatic transmission: Ganged with right exhaust temperature sensor (G235) at right rear of intake manifold

BentleyPublishers.com
BentleyPublishers.com—All Rights Reserved

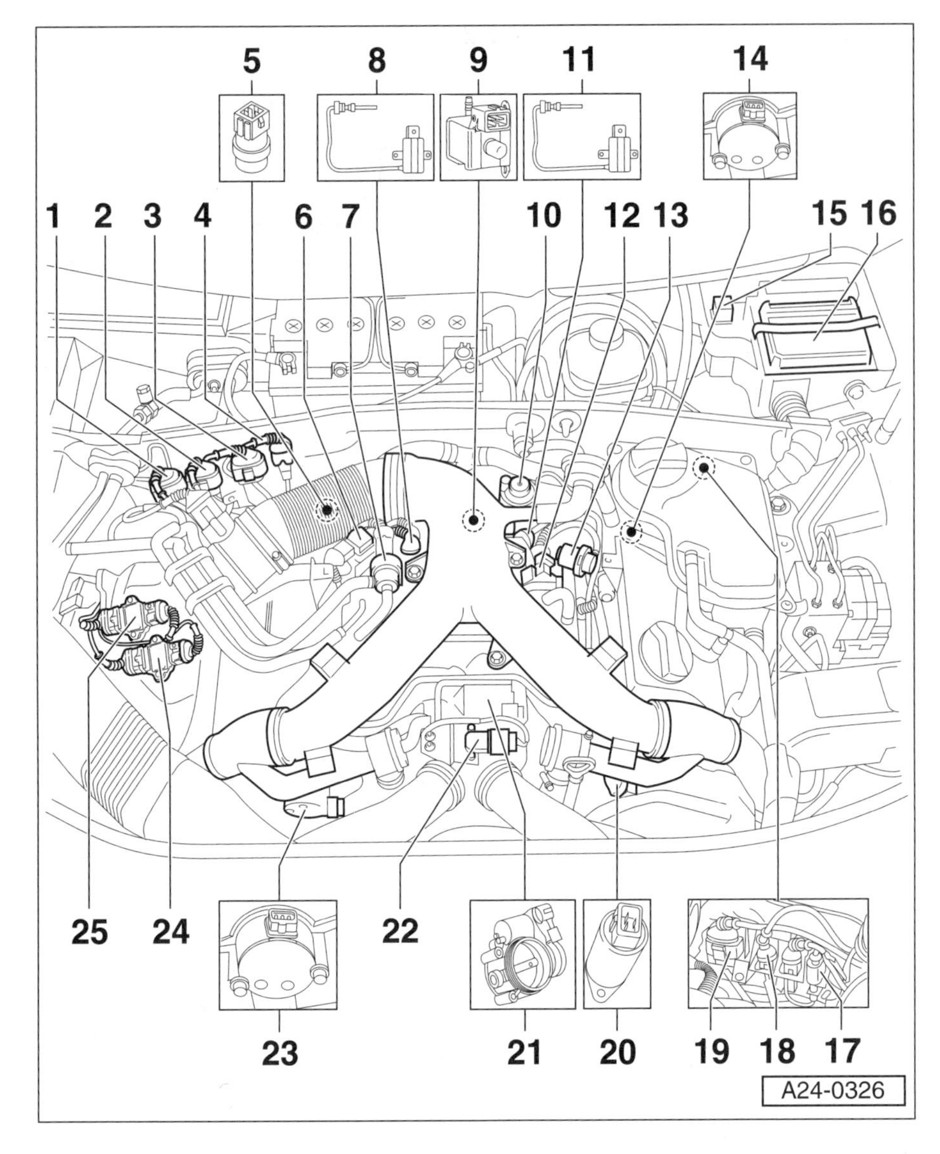

Engine compartment (2.7 liter turbo V6 engine)

(continued)

12. **Turbocharger recirculating valve (N249)**
13. **Fuel pressure regulator**
14. **Left camshaft sensor (G163)**
15. **Secondary air injection pump relay (J299)**
    - Only on vehicle with automatic transmission
16. **Engine control module (ECM) (J220)**
17. **Engine speed sensor harness connector**
    - 3-pin grey
18. **Left knock sensor (G66) harness connector**
    - 3-pin
19. **Oxygen sensor harness connector**
    - Left bank oxygen sensor (G108) and heater (Z28) in front of catalytic converter
    - 4-pin black
20. **Left camshaft adjustment valve (N208)**
21. **Throttle valve control module (J338)**
    - Throttle drive (G186)
    - Throttle valve angle sensors (G187, G188)
22. **Charge air pressure sensor (G31)**
    - In rubber elbow before throttle valve control module
23. **Right camshaft sensor (G40)**
24. **Left bank ignition coils power output stage (N192)**
25. **Right bank ignition coils power output stage (N122)**

BentleyPublishers.com
BentleyPublishers.com—All Rights Reserved

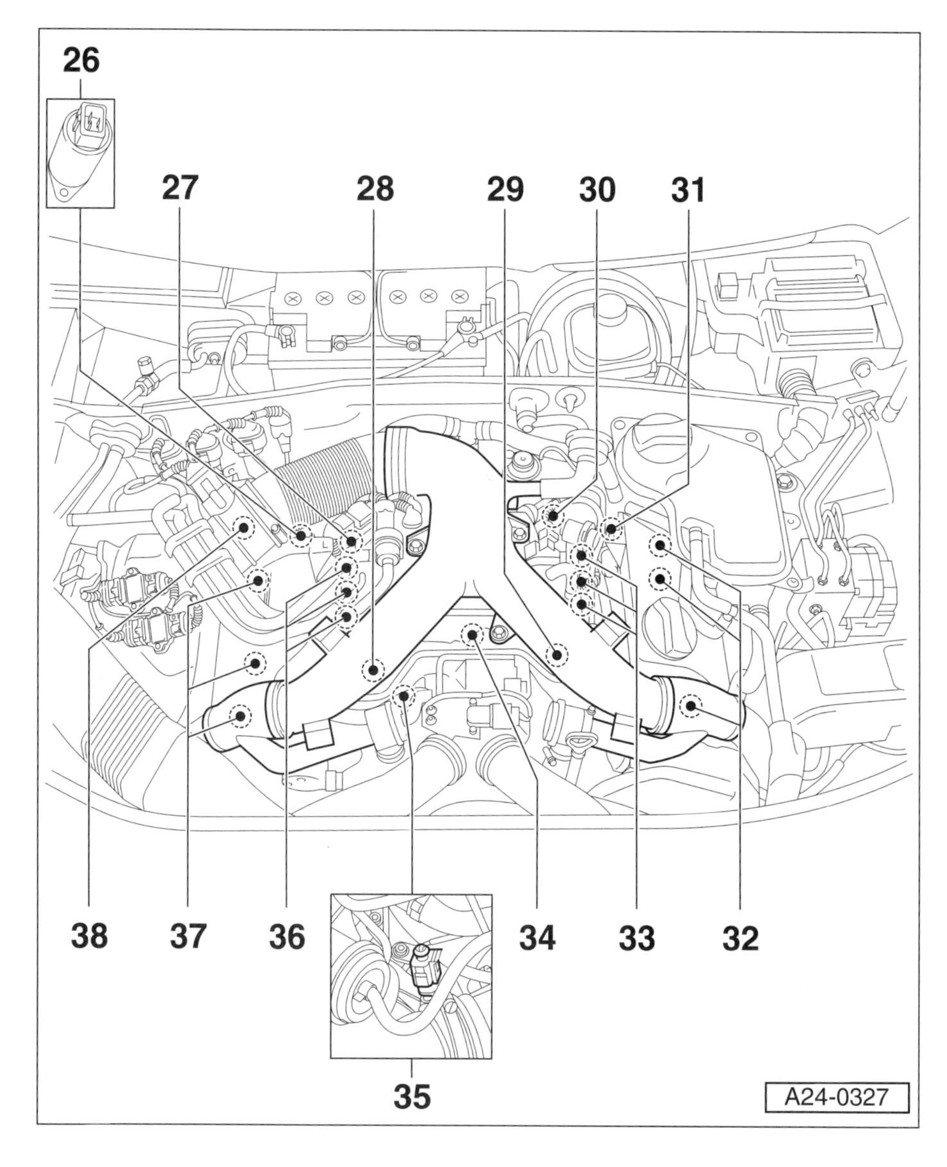

26. **Right camshaft adjustment valve (N205)**
27. **Right bank oxygen sensor (G39) and heater (Z19)**
    - In front of catalytic converter
28. **Knock sensor (G61)**
29. **Knock sensor (G66)**
30. **Engine speed sensor (G28)**
    - In transmission bellhousing, above flywheel or torque plate
31. **Left bank oxygen sensor (G108) and heater (Z28)**
    - In front of catalytic converter
32. **Left bank ignition coils (N163, N164, N189)**
33. **Left bank fuel injectors (N33, N83, N84)**
34. **Intake air temperature (IAT) sensor (G42)**
    - Under front of intake manifold, near throttle valve control module
35. **Intake air temperature harness connector**
    - 2-pin
36. **Right bank fuel injectors (N30 - N32)**
37. **Right bank ignition coils (N, N128, N158)**
38. **Mass airflow sensor (G70)**

Engine management component details (2.7 liter turbo V6 engine)

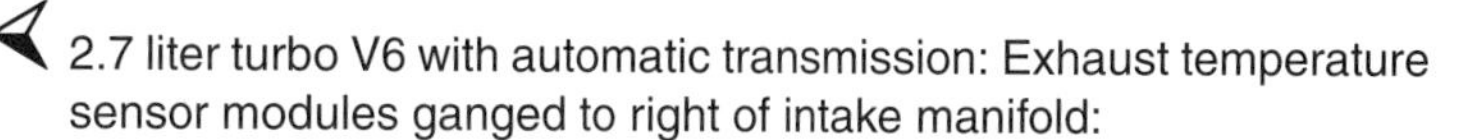

2.7 liter turbo V6 with automatic transmission: Exhaust temperature sensor modules ganged to right of intake manifold:

1. Right bank exhaust temperature sensor module (G235)
2. Left bank exhaust temperature sensor module (G236)

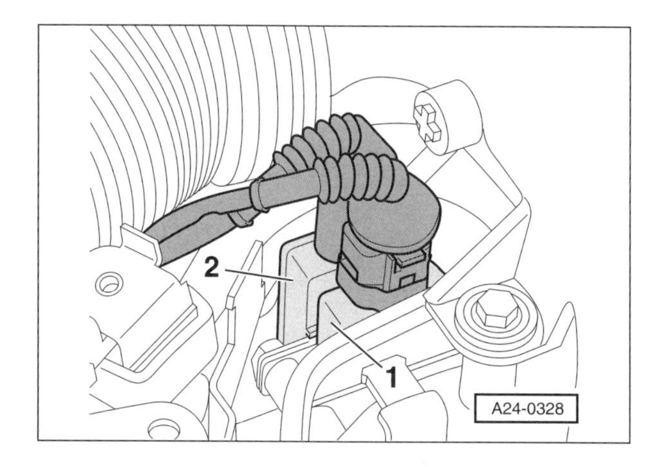

BentleyPublishers.com
BentleyPublishers.com—All Rights Reserved

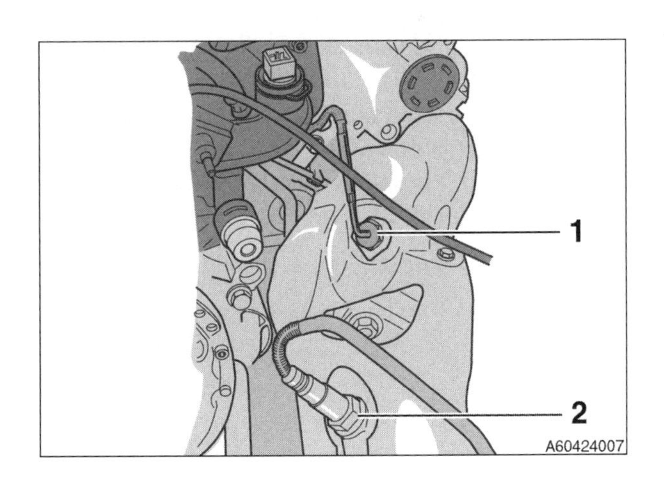

Exhaust manifold:

1. Exhaust temperature sensor
2. Oxygen sensor before catalytic converter

2.8 Liter Engine Management Components

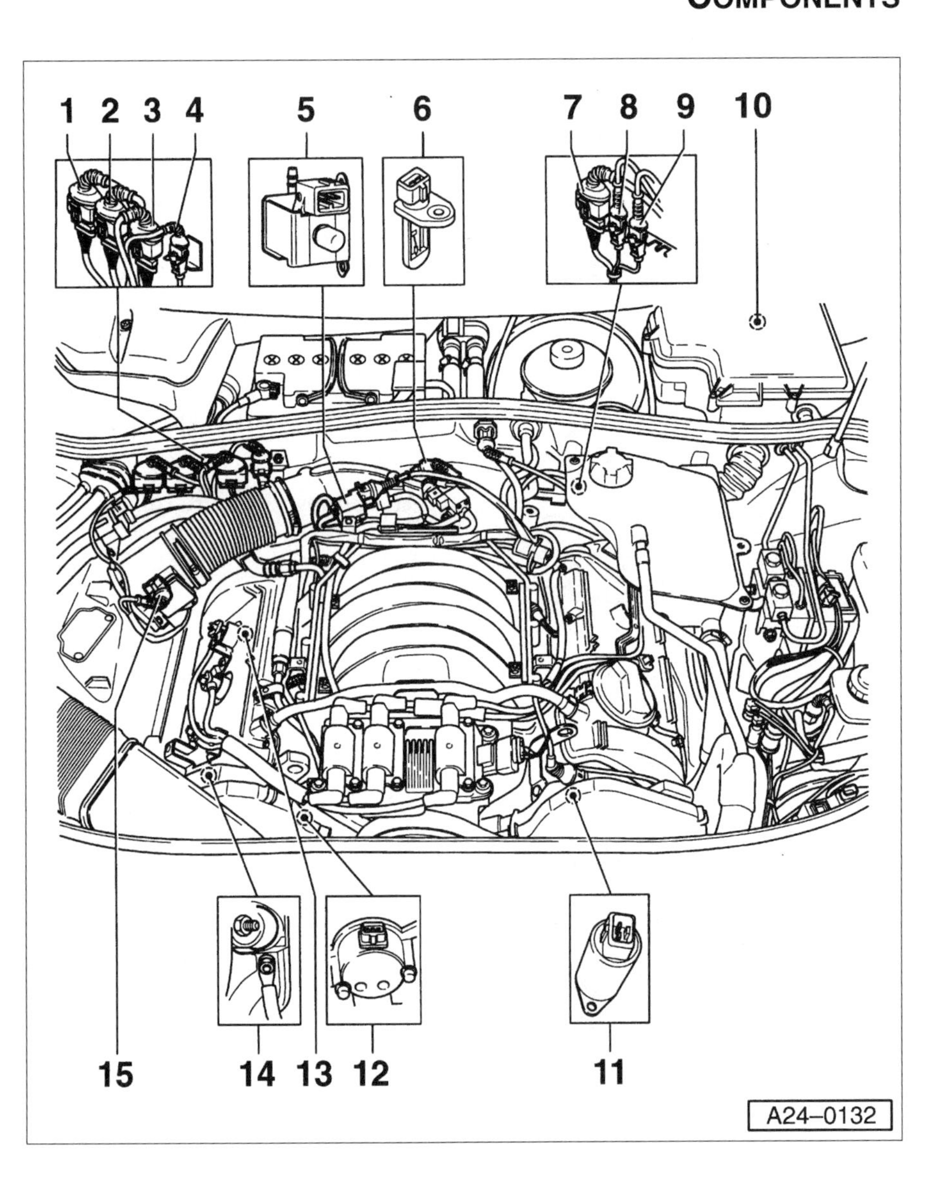

Engine compartment (2.8 liter V6 engine)

1. **Oxygen sensor harness connector**
   - Right bank oxygen sensor (G130) and heater (Z29) behind catalytic converter
   - 4-pin green
2. **Oxygen sensor harness connector**
   - Left bank oxygen sensor (G131) and heater (Z30) behind catalytic converter
   - 4-pin green
3. **Oxygen sensor harness connector**
   - Right bank oxygen sensor (G39) and heater (Z19) in front of catalytic converter.
   - 4-pin black
4. **Knock sensor (G61) harness connector**
5. **Secondary air injection solenoid valve (N112)**
6. **Intake air temperature (IAT) sensor (G42)**
7. **Oxygen sensor harness connector**
   - Left bank oxygen sensor (G108) and heater (Z28) in front of catalytic converter
   - 4-pin black
8. **Engine speed sensor harness connector**
   - 3-pin grey
9. **Left knock sensor (G66) harness connector**

BentleyPublishers.com
BentleyPublishers.com—All Rights Reserved

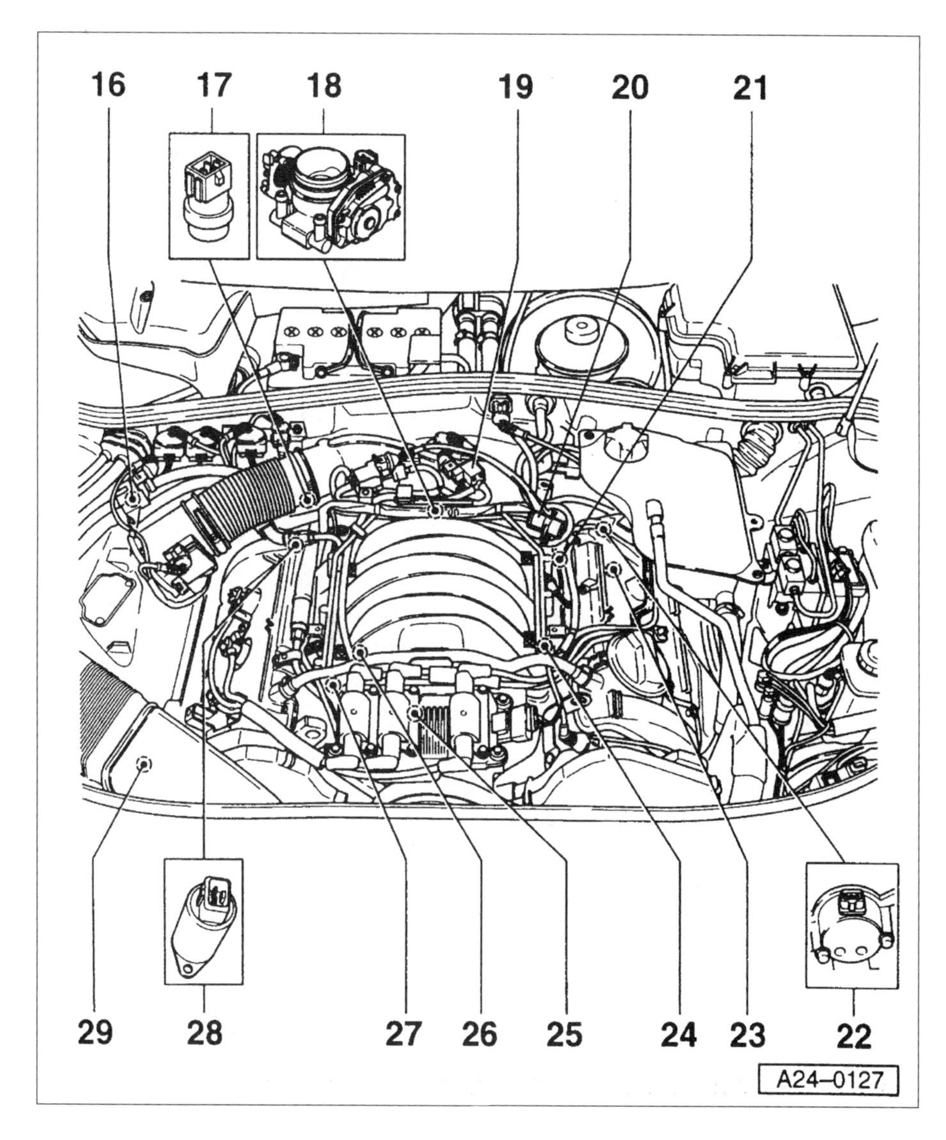

10. **Electronics box (E-box)**
    - Engine control module (ECM) (J220). See **ECM, accessing** in this repair group.
    - Secondary air injection pump relay (J299). See **Engine Management Fuses and Relays** in this repair group.
11. **Left camshaft adjustment valve (N208)**
12. **Right camshaft sensor (G163)**
13. **Right bank oxygen sensor (G39) and heater (Z19)**
    - In front of catalytic converter
14. **Ground cable**
    - to Right engine mount
15. **Mass airflow sensor (G70)**
16. **Fuel tank evaporative control (EVAP) canister purge regulator valve (N80)**
17. **Engine coolant temperature (ECT) sensor (G62)**
    - On coolant pipe behind right cylinder bank
18. **Throttle valve control module (J338)**
    - Throttle drive (G186)
    - Throttle plate angle sensor (G187)
    - Throttle position sensor (G69)
    - Closed throttle position switch (F60)
19. **Intake manifold tuning valve (change-over valve) (N156)**
20. **Fuel pressure regulator**
21. **Engine speed sensor (G28)**
    - In transmission bellhousing, above flywheel or torque plate
22. **Left camshaft sensor (G40)**
23. **Left bank oxygen sensor (G108) and heater (Z28)**
    - In front of catalytic converter
24. **Left knock sensor (G66)**
25. **Ignition coils and power output stage**
26. **Right knock sensor (G61)**
27. **Fuel injectors (N30 - N33, N83, N84)**
28. **Right camshaft adjustment valve (N205)**
29. **Secondary air injection pump motor (V101)**

BentleyPublishers.com
BentleyPublishers.com—All Rights Reserved

3.0 Liter Engine Management Components

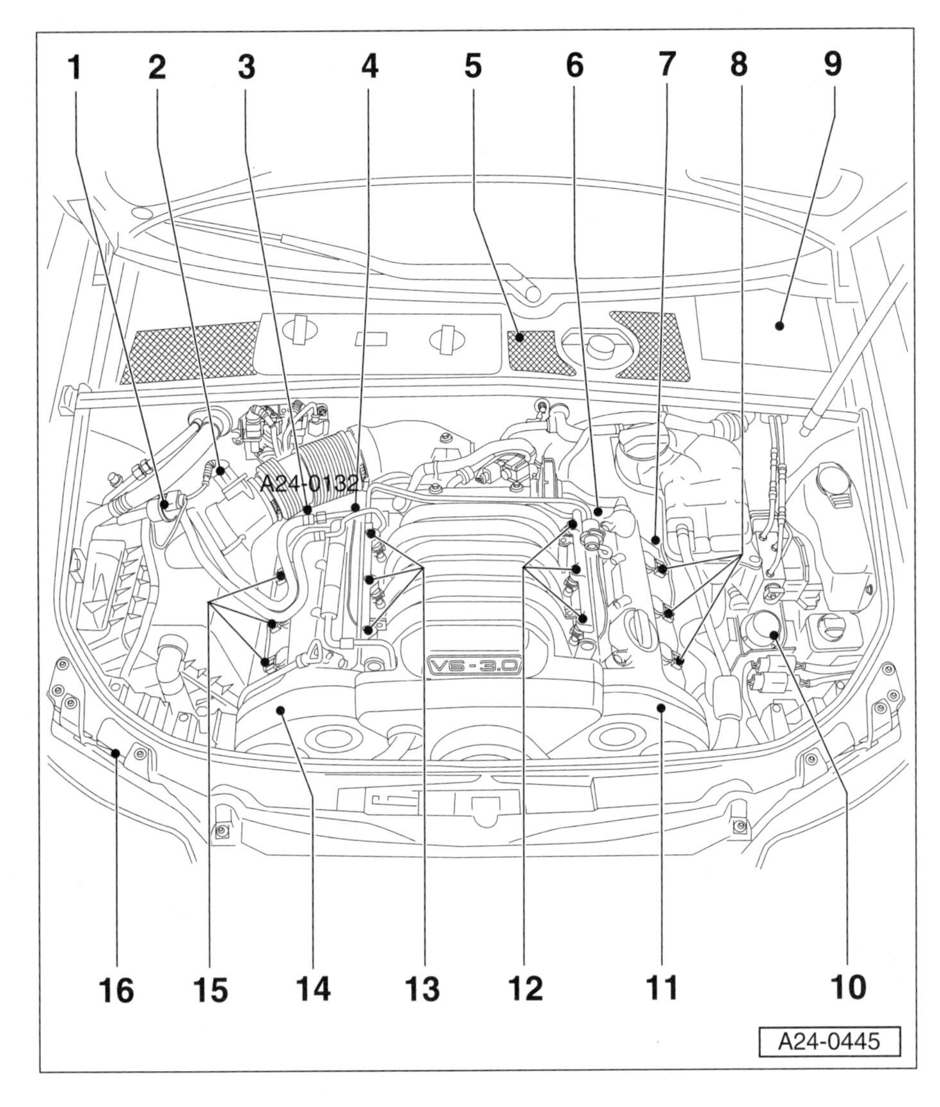

Engine compartment (3.0 liter V6 engine)

1. **Fuel tank evaporative control (EVAP) canister purge regulator valve (N80)**
2. **Mass airflow sensor (G70)**
   - Includes intake air temperature (IAT) sensor (G42).
3. **Right exhaust camshaft sensor (G300)**
4. **Right intake camshaft sensor (G40)**
5. **Brake booster pressure sensor (G294)**
   - Automatic transmission vehicle
6. **Left intake camshaft sensor (G163)**
7. **Left exhaust camshaft sensor (G301)**
8. **Left bank ignition coils with power output stage (N292, N323, N324)**
9. **Electronics box (E-box)**
   - Engine control module (ECM) (J220). See **ECM, accessing** in this repair group.
   - Secondary air injection pump relay (J299). See **Engine Management Fuses and Relays** in this repair group.
   - Secondary air injection pump fuse (S130)
10. **Brake system vacuum pump (V192)**
11. **Left intake camshaft adjustment valve (N208), exhaust camshaft adjustment valve (N319)**
12. **Left bank fuel injectors (N33, N83, N84)**
13. **Right bank fuel injectors (N30 - N32)**
14. **Right intake camshaft adjustment valve (N205), exhaust camshaft adjustment valve (N318)**
15. **Right bank ignition coils with power output stage (N70, N127, N291)**
16. **Secondary air injection pump (V101)**

BentleyPublishers.com
BentleyPublishers.com—All Rights Reserved

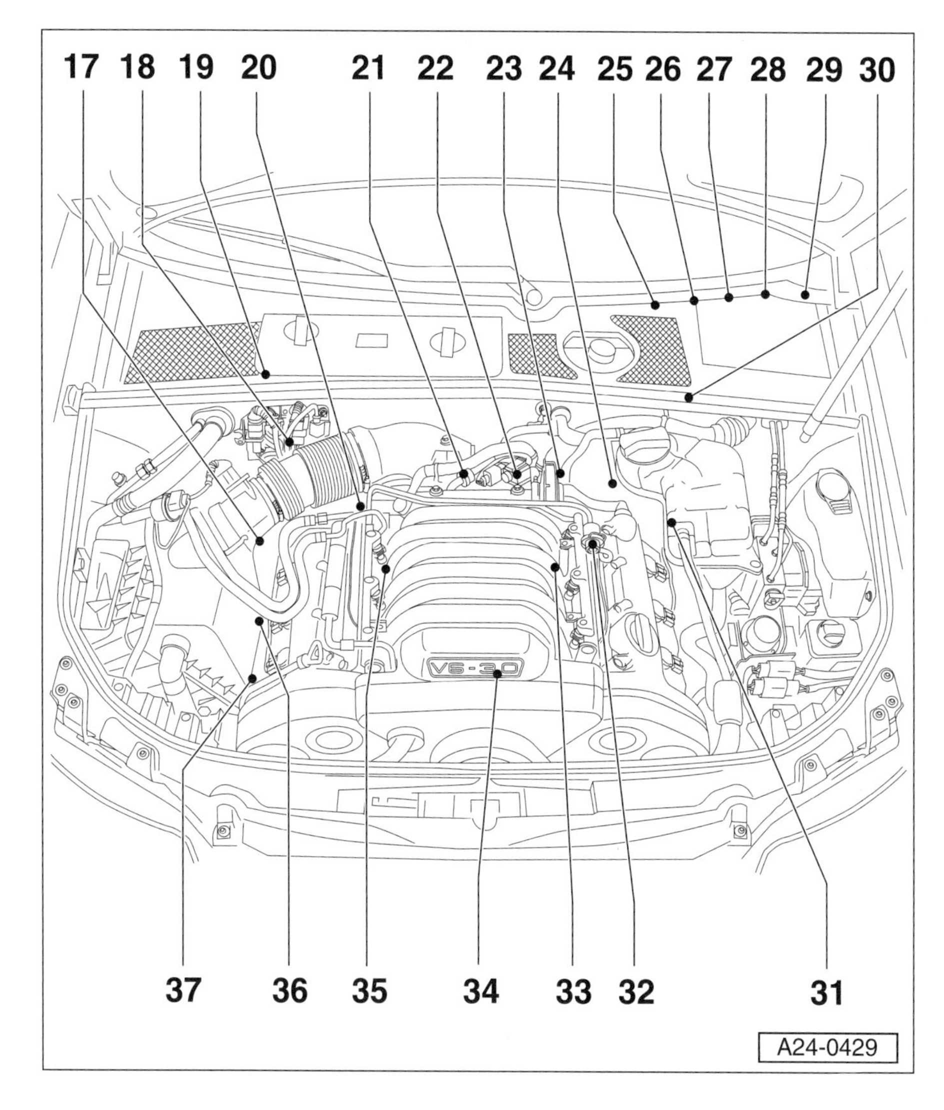

17. **Right bank oxygen sensor (G39) and heater (Z19)**
    - In front of catalytic converter
18. **Harness connectors**
    - Right rear engine compartment
    - See **Electrical connectors (3.0 liter V6 engine compartment)** in this repair group.
19. **Right bank oxygen sensor (G130) and heater (Z29)**
    - Behind catalytic converter
20. **Engine coolant temperature (ECT) sensor (G62)**
    - On coolant pipe behind right cylinder bank
21. **Throttle valve control module (J338)**
    - Throttle drive (G186)
    - Throttle plate angle sensors (G187, G188)
22. **Secondary air injection solenoid valve (N112)**
23. **Engine speed sensor (G28)**
24. **Harness connectors**
    - Left rear engine compartment, underneath coolant overflow tank
    - See **Electrical connectors (3.0 liter V6 engine compartment)** in this repair group.
25. **Accelerator pedal switches**
    - See **Pedal Sensors and Switches** in this repair group.
26. **Fuel pump relay (J17) under dashboard**
    - See **Engine Management Fuses and Relays** in this repair group.

BentleyPublishers.com
BentleyPublishers.com—All Rights Reserved

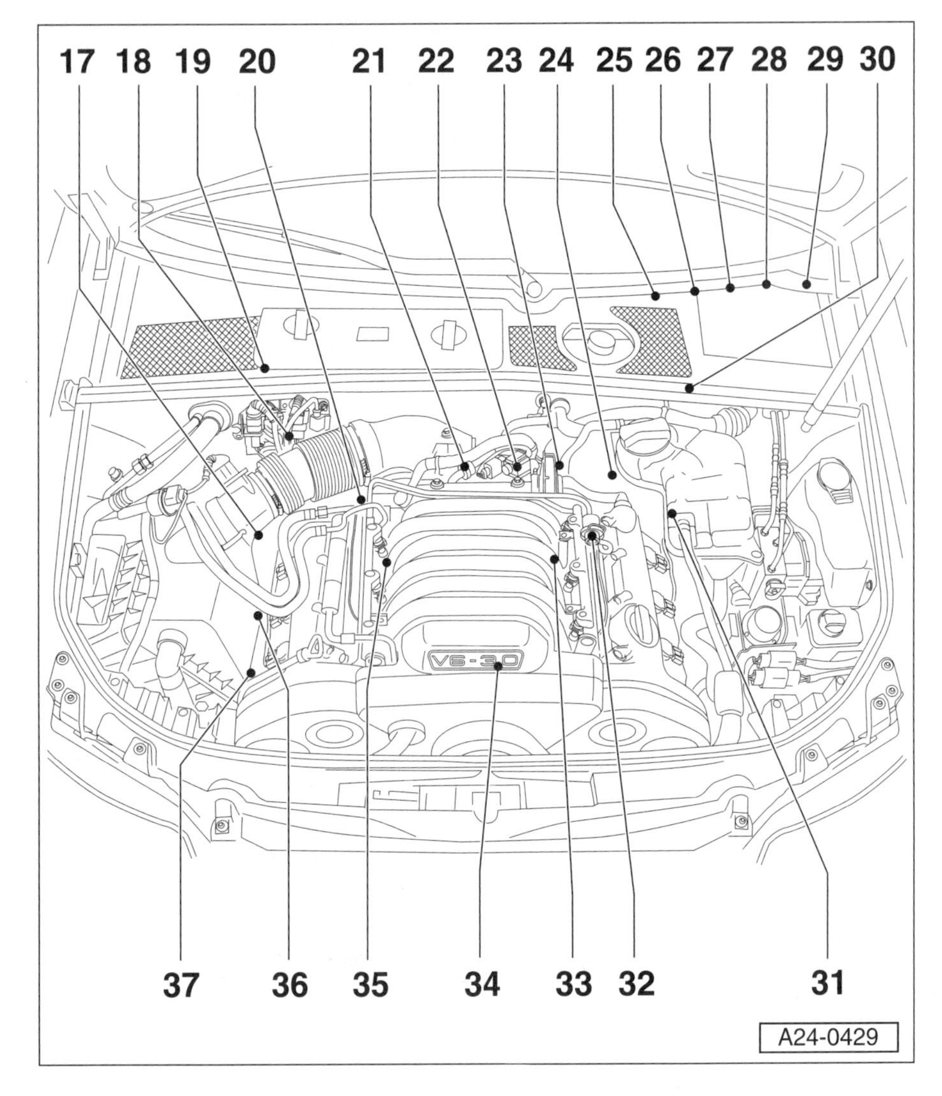

Engine compartment (3.0 liter V6 engine) *(continued)*

27. Brake pedal switches
- See **Pedal Sensors and Switches** in this repair group.

28. Clutch pedal switch
- See **Pedal Sensors and Switches** in this repair group.

29. Instrument cluster warning lights
- See **On-board diagnostics** in this repair group.

30. Left bank oxygen sensor (G131) and heater (Z30)
- Behind catalytic converter

31. Left bank oxygen sensor (G108) and heater (Z28)
- In front of catalytic converter

32. Fuel pressure regulator

33. Left knock sensor (G66)
- Underneath intake manifold

34. Intake manifold tuning valve (change-over valve) (N156)

35. Right knock sensor (G61)
- Underneath intake manifold

36. Ground cable to right engine mount

Electrical connectors (3.0 liter V6 engine compartment)

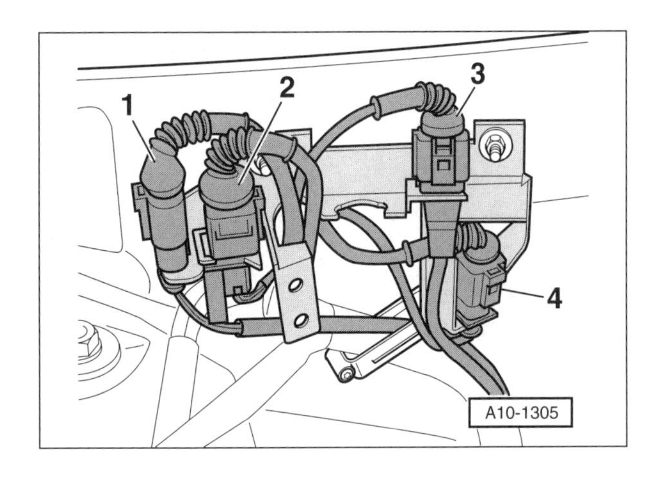

Right rear engine compartment:

1. 4-pin harness connector, green
   Right bank oxygen sensor (G130) and heater (Z29) behind catalytic converter
2. 6-pin harness connector, black
   Right bank oxygen sensor (G39) and heater (Z19) in front of catalytic converter
3. 3-pin harness connector, blue
   Knock sensor (G66)
4. 3-pin harness connector, black
   Terminal 50

BentleyPublishers.com
BentleyPublishers.com—All Rights Reserved

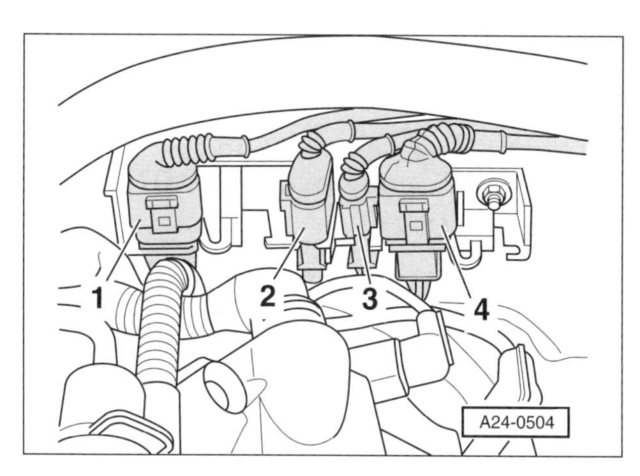

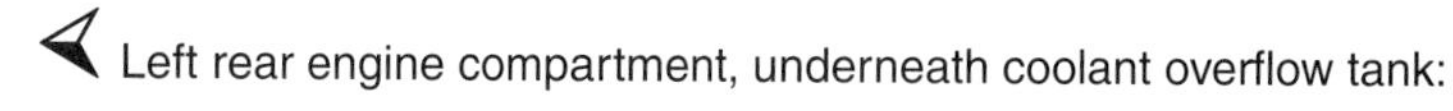

Left rear engine compartment, underneath coolant overflow tank:

1. 6-pin harness connector, black
   Left bank oxygen sensor (G108) and heater (Z28) in front of catalytic converter
2. 3-pin harness connector, grey
   Engine speed sensor (G28)
3. 3-pin harness connector, blue
   Knock sensor (G66)
4. 2-pin connector
   Right camshaft adjustment

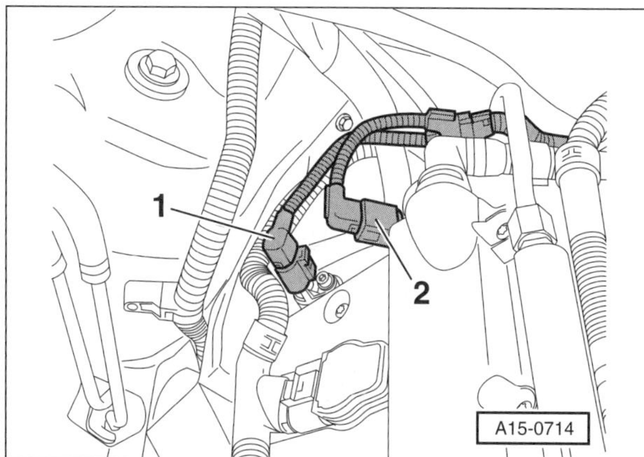

Rear of right cylinder head:

1. Exhaust camshaft sensor (G300) harness connector
2. Intake camshaft sensor (G40) harness connector

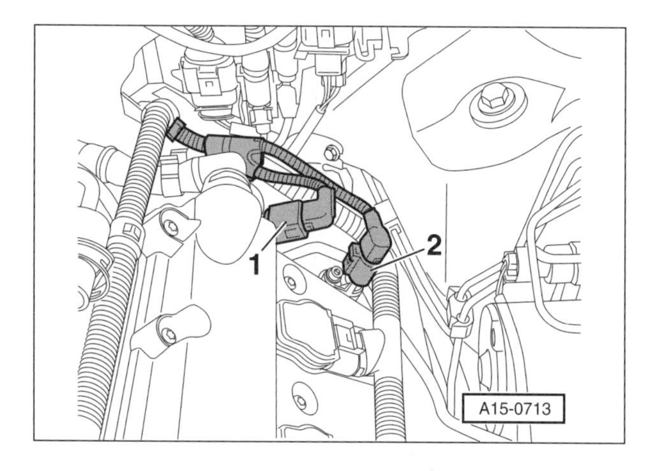

Rear of left cylinder head:

1. Intake camshaft sensor (G163) harness connector
2. Exhaust camshaft sensor (G301) harness connector

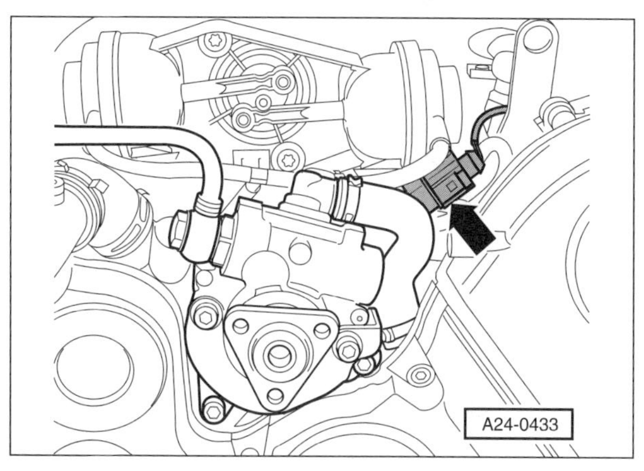

Intake manifold tuning valve (change-over valve) (N156) connector (**arrow**)

BentleyPublishers.com
BentleyPublishers.com—All Rights Reserved

V8 Engine Management Components (Engine Codes ART, AWN, BBD)

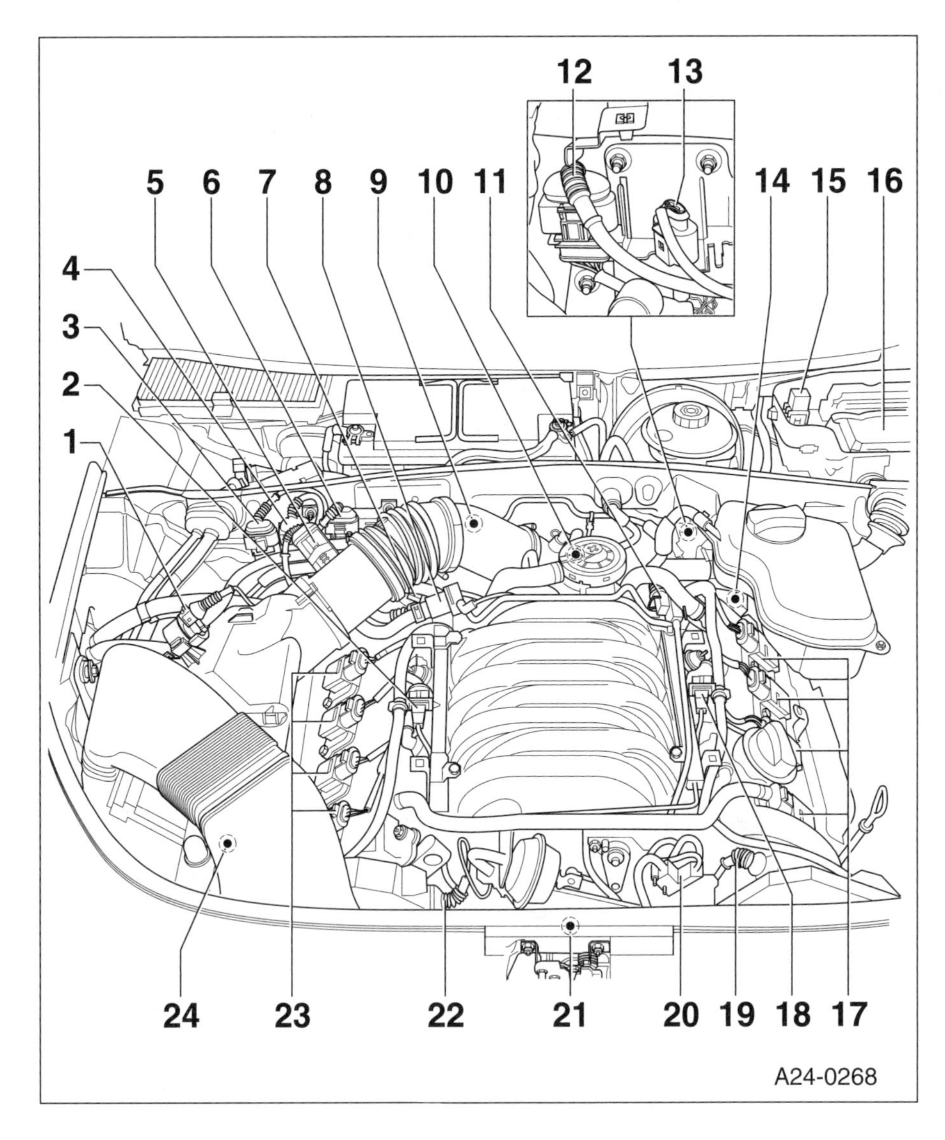

Engine compartment (V8 engine, codes ART, AWN, BBD)

1. **Fuel tank evaporative control (EVAP) canister purge regulator valve (N80)**
2. **Right knock sensor (G61) harness connector**
   - 3-pin
3. **Oxygen sensor harness connector (B1S2)**
   - Post-catalytic converter oxygen sensor (G130) and heater (Z29)
   - 4-pin green
4. **Oxygen sensor harness connector (B2S2)**
   - Post-catalytic converter oxygen sensor (G131) and heater (Z30)
   - 4-pin green
5. **Mass airflow sensor (G70)**
   - Includes intake air temperature (IAT) sensor (G42).
6. **Oxygen sensor harness connector (B1S1)**
   - Pre-catalytic converter oxygen sensor (G39) and heater (Z19)
   - 4-pin
7. **Camshaft adjustment valve (N205)**
8. **Fuel pressure regulator**
9. **Engine coolant temperature (ECT) sensor (G62)**
   - On coolant pipe behind right cylinder bank

BentleyPublishers.com
BentleyPublishers.com—All Rights Reserved

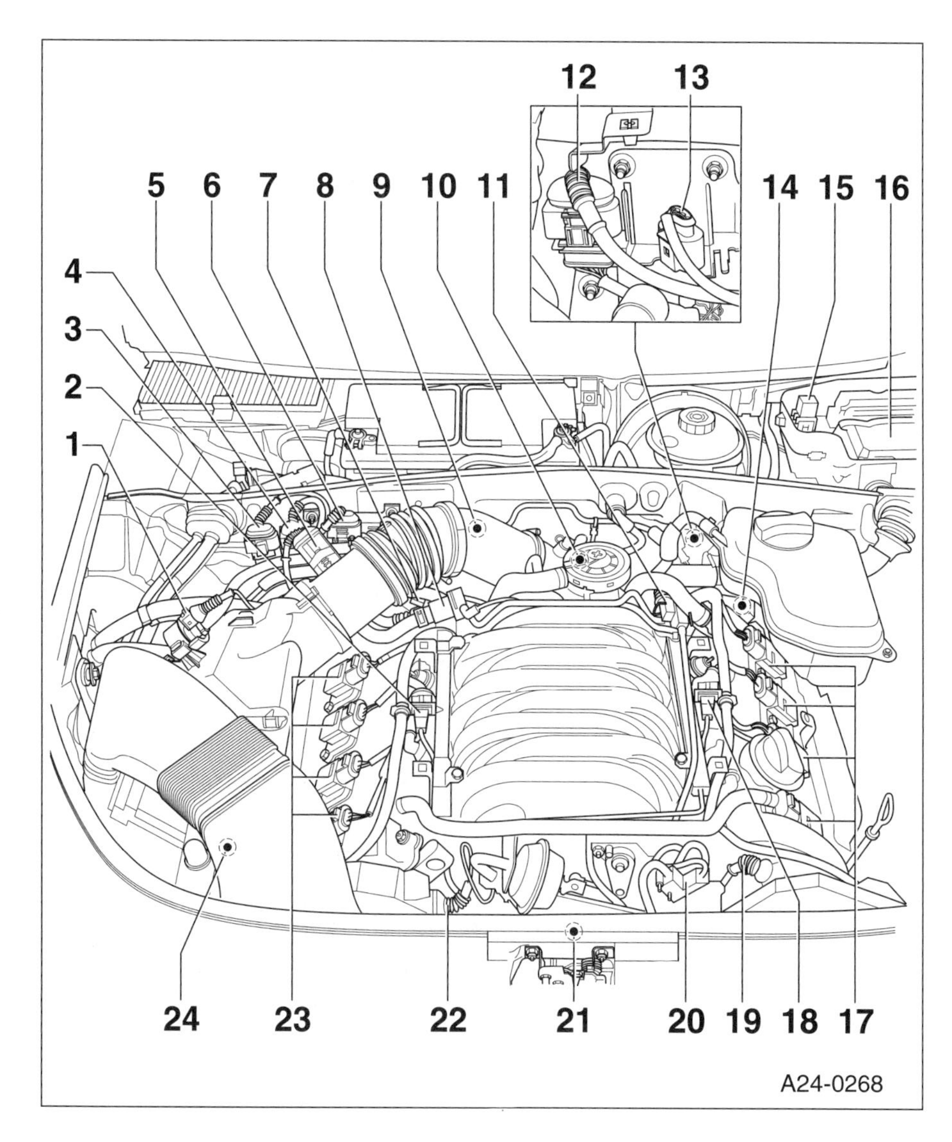

10. **Throttle valve control module (J338)**
    - Throttle drive (G186)
    - Throttle plate angle sensors (G187, G188)
11. **Secondary air injection solenoid valve (N112)**
12. **Oxygen sensor harness connector (B2S1)**
    - Pre-catalytic converter oxygen sensor (G108) and heater (Z28)
    - 4-pin
13. **Engine speed sensor (G28) harness connector**
    - 3-pin
14. **Left camshaft sensor (G163)**
15. **Secondary air injection pump relay (J299)**
    - E-box, left rear engine compartment in air plenum. See **Engine Management Fuses and Relays** in this repair group.
16. **Engine control module (ECM) (J220)**
    - E-box, left rear engine compartment in air plenum. See **ECM, accessing** in this repair group.
17. **Left bank ignition coils with power output stages**
18. **Left knock sensor (G61) harness connector**
    - 3-pin
19. **Left camshaft adjustment valve (N208)**
20. **Intake manifold tuning valve**
21. **Intake manifold tuning valve**
22. **Right camshaft sensor (G40)**
23. **Right bank ignition coils with power output stages**
24. **Secondary air injection pump motor (V101)**
    - Underneath engine, behind right corner of front bumper

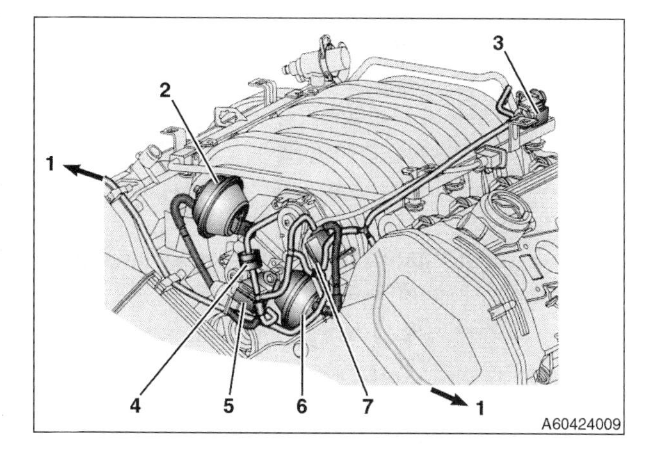

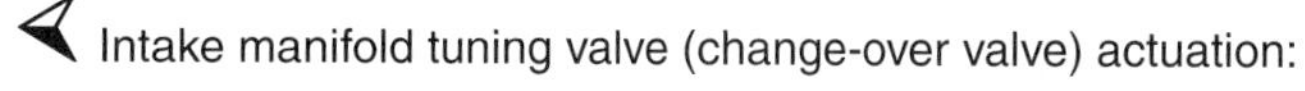
Intake manifold tuning valve (change-over valve) actuation:

1. to Vacuum reservoir
2. Vacuum unit, stage 3
3. Secondary air injection solenoid (N112)
4. Check-valve
5. Intake manifold tuning valve (N261)
6. Vacuum unit, stage 2
7. Intake manifold tuning valve (N156)

BentleyPublishers.com
BentleyPublishers.com—All Rights Reserved

V8 Engine Management Components (Engine Code BAS, allroad quattro)

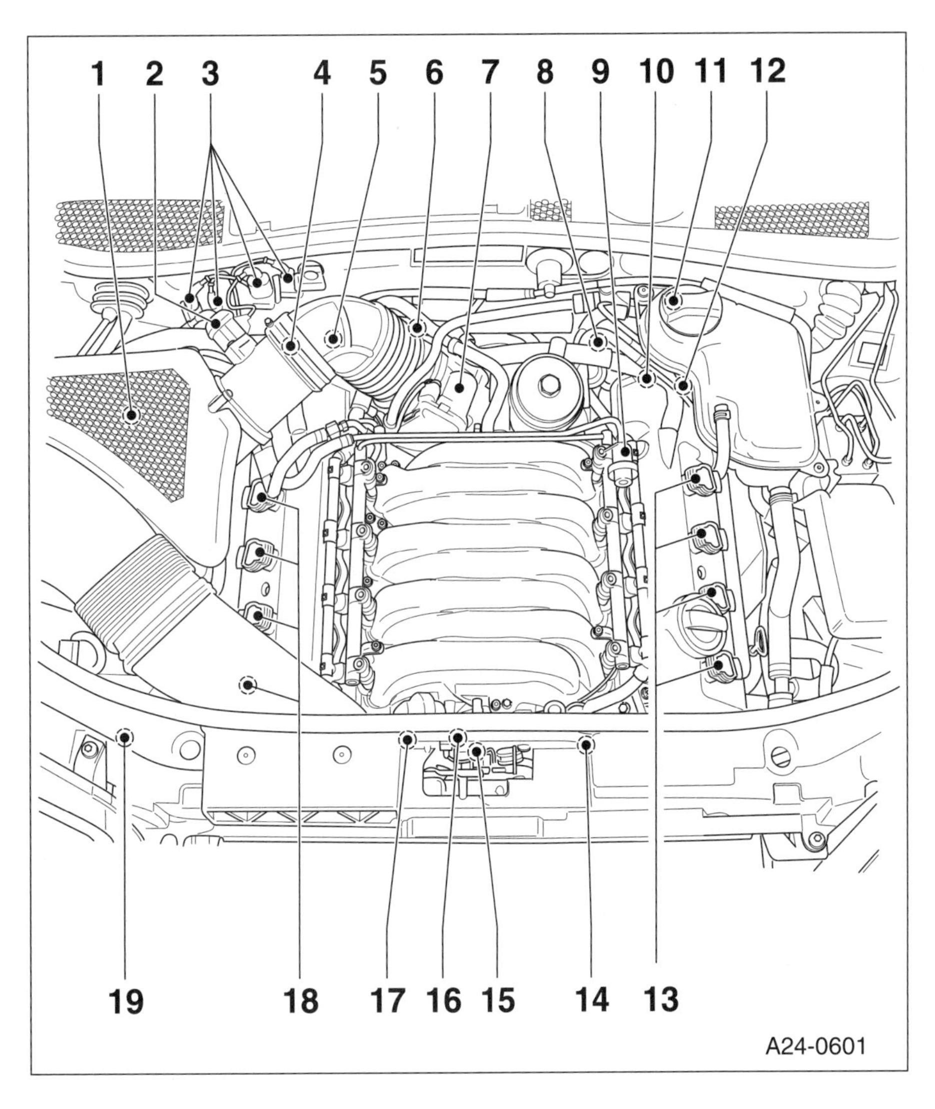

Engine compartment (V8 engine, code BAS)

1. **Fuel tank evaporative control (EVAP) canister purge regulator valve (N80)**
2. **Mass airflow sensor (G70)**
   - Includes intake air temperature (IAT) sensor (G42).
3. **Right rear engine compartment harness connectors**
   - See **Engine management component details (V8 engine, code BAS)** in this repair group.
4. **Right camshaft adjustment valve (N205)**
5. **Right camshaft sensor (G40)**
6. **Engine coolant temperature (ECT) sensor (G62)**
7. **Throttle valve control module (J338)**
   - Throttle drive (G186)
   - Throttle plate angle sensors (G187, G188)
8. **Oil pressure switch (F1)**
9. **Fuel pressure regulator**
10. **Left camshaft sensor (G163)**
11. **Left rear engine compartment harness connectors**
    - See **Engine management component details (V8 engine, code BAS)** in this repair group.
12. **Left camshaft adjustment valve (N208)**
13. **Left bank ignition coils with power output stages**
14. **Left knock sensor (G198, G199) harness 3-pin connector**
15. **Secondary air injection solenoid valve (N112)**
16. **Intake manifold tuning valve (change-over valve) (N156)**
17. **Right knock sensor (G61, G66) harness 3-pin connector**
18. **Right bank ignition coils with power output stages**
19. **Secondary air injection pump motor (V101)**

BentleyPublishers.com
BentleyPublishers.com—All Rights Reserved

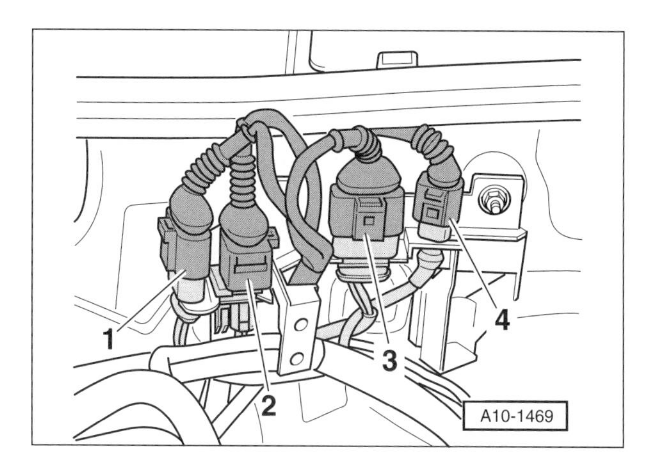

Engine management component details (V8 engine, code BAS)

◀ Right rear engine compartment harness connectors:

1. 4-pin harness connector
   Right bank oxygen sensor (G130) and heater (Z29) behind catalytic converter
2. 4-pin harness connector
   Right bank oxygen sensor (G39) and heater (Z19) in front of catalytic converter
3. Starter and alternator connector
4. 4-pin harness connector, brown
   Left bank oxygen sensor (G131) and heater (Z30) behind catalytic converter

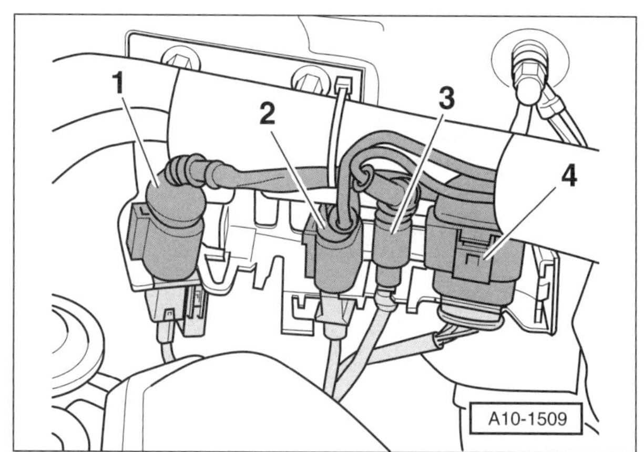

◀ Left rear engine compartment harness connectors, behind coolant expansion tank:

1. 4-pin harness connector
   Left bank oxygen sensor (G108) and heater (Z28) in front of catalytic converter
2. 3-pin harness connector
   Engine speed sensor (G28)
3. 2-pin harness connector
   Left camshaft adjustment
4. 4-pin harness connector
   Left bank oxygen sensor (G131) and heater (Z30) behind catalytic converter

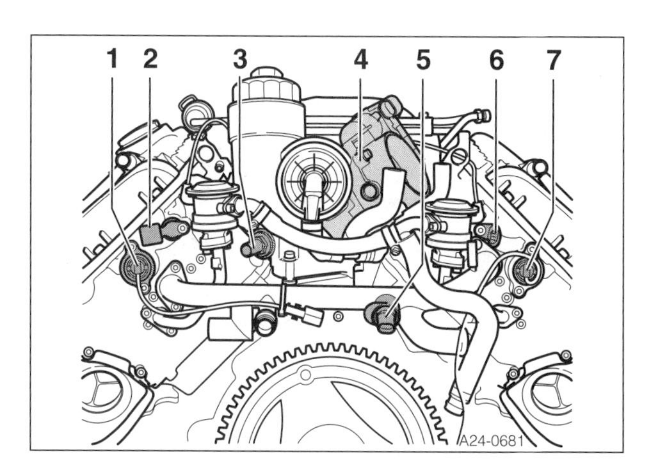

◀ Rear of engine:

1. Left camshaft adjustment valve (N208)
2. Left camshaft sensor (G163)
3. Oil pressure switch (F1)
4. Throttle valve control module (J338)
5. Engine coolant temperature (ECT) sensor (G62)
6. Right camshaft sensor (G40)
7. Right camshaft adjustment valve (N205)

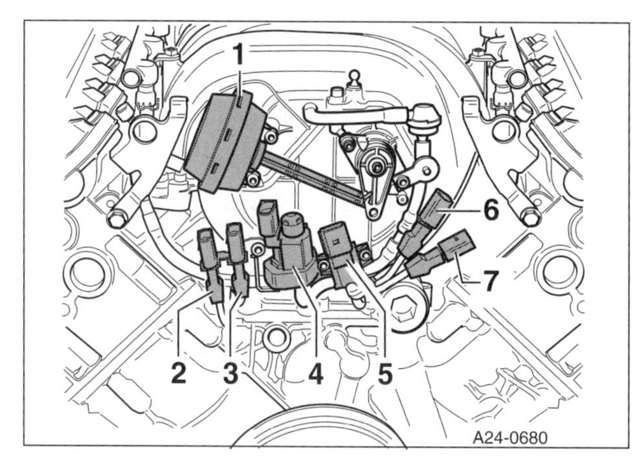

◀ Front of engine:

1. Intake manifold tuning valve (change-over valve) vacuum unit
2. Knock sensor (G61)
3. Knock sensor (G66)
4. Intake manifold tuning valve (change-over valve) (N156)
5. Secondary air injection solenoid (N112)
6. Knock sensor (G198)
7. Knock sensor (G199)

BentleyPublishers.com
BentleyPublishers.com—All Rights Reserved

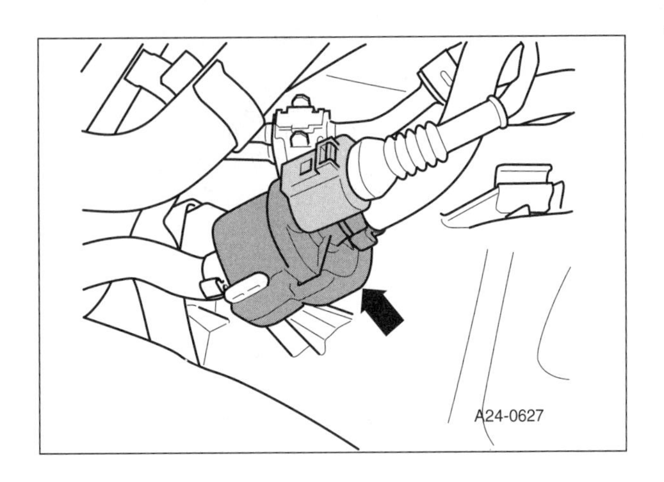

Fuel tank evaporative emission control (EVAP) canister purge regulator valve (N80) (**arrow**)

V8 Engine Management Components (Engine Code BCY, RS6)

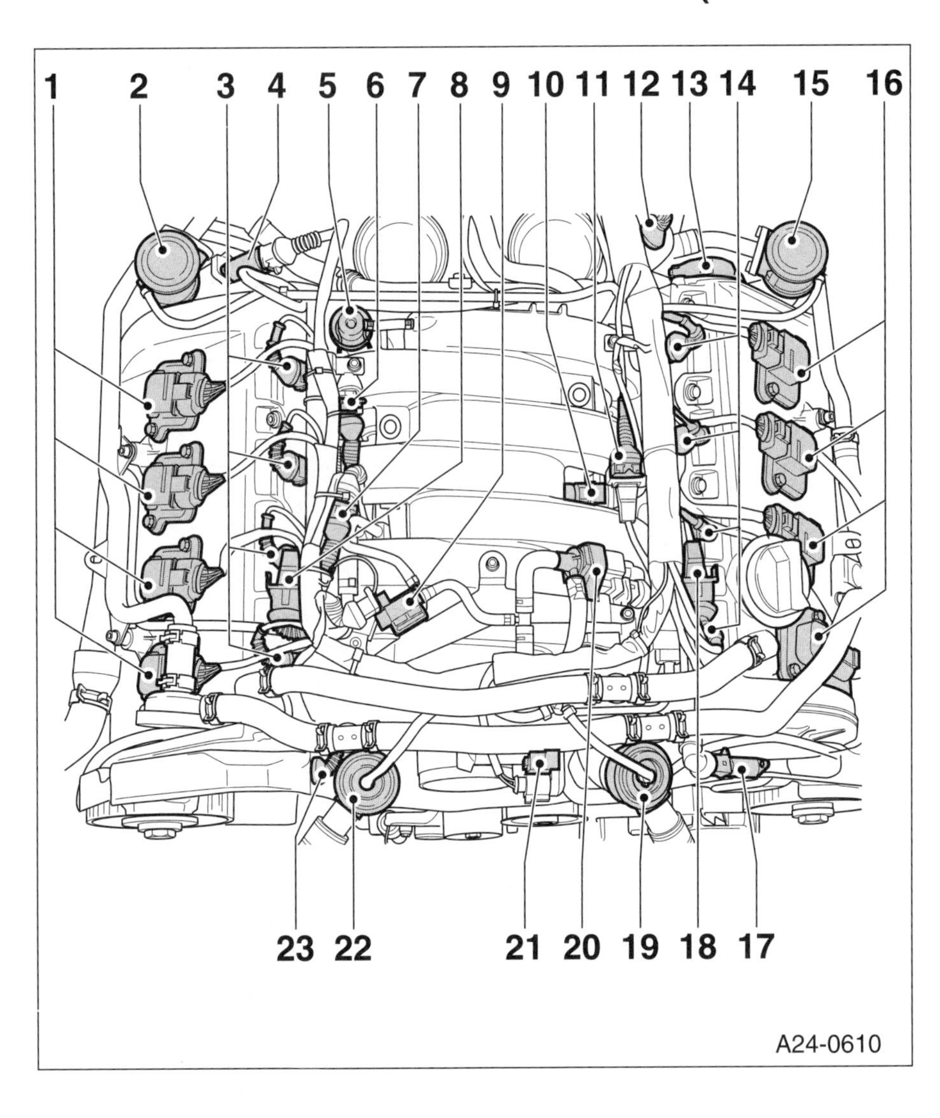

Engine compartment (V8 engine, code BCY)

1. **Right bank ignition coils with power output stages (N70, N127, N291, N292)**
2. **Right bank secondary air injection mechanical combination valve**
3. **Right bank fuel injectors (N30 - N33)**
4. **Right camshaft adjustment valve (N205)**
5. **Fuel pressure regulator**
6. **After-run coolant pump (V51) connector**
7. **Turbocharger recirculating valve (N249) connector**
8. **Knock sensor 1 (G61) connector**
   - 3-pin
9. **Secondary air injection solenoid (N112)**
10. **Intake air temperature (IAT) sensor (G42)**
11. **Knock sensor 3 (G198) connector**
    - 3-pin
12. **Engine coolant temperature (ECT) sensor (G62)**
13. **Left camshaft sensor (G163)**
14. **Left bank fuel injectors (N83 - N86)**
15. **Left bank secondary air injection mechanical combination valve**
16. **Left bank ignition coils with power output stages (N323 - N326)**

BentleyPublishers.com
BentleyPublishers.com—All Rights Reserved

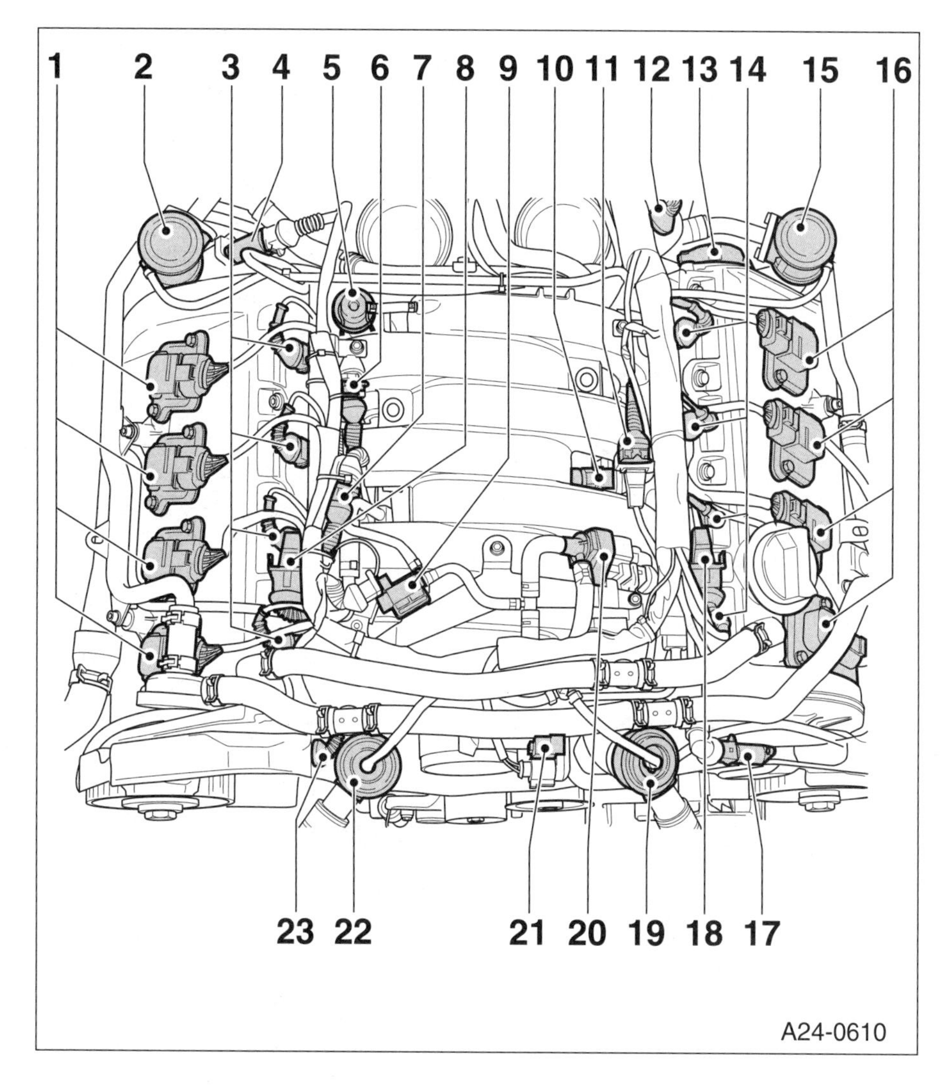

17. **Left camshaft adjustment valve (N208)**
18. **Knock sensor 2 (G66) connector**
    - 3-pin
19. **Left bank turbocharger recirculation valve (mechanical)**
20. **Turbocharger recirculation valve (N249)**
21. **Throttle valve control module (J338)**
    - Throttle drive (G186)
    - Throttle plate angle sensors (G187, G188)
22. **Right bank turbocharger recirculation valve (mechanical)**
23. **Right camshaft sensor (G40)**

BentleyPublishers.com
BentleyPublishers.com—All Rights Reserved

V8 Engine Management Components (Engine Code BCY, RS6)

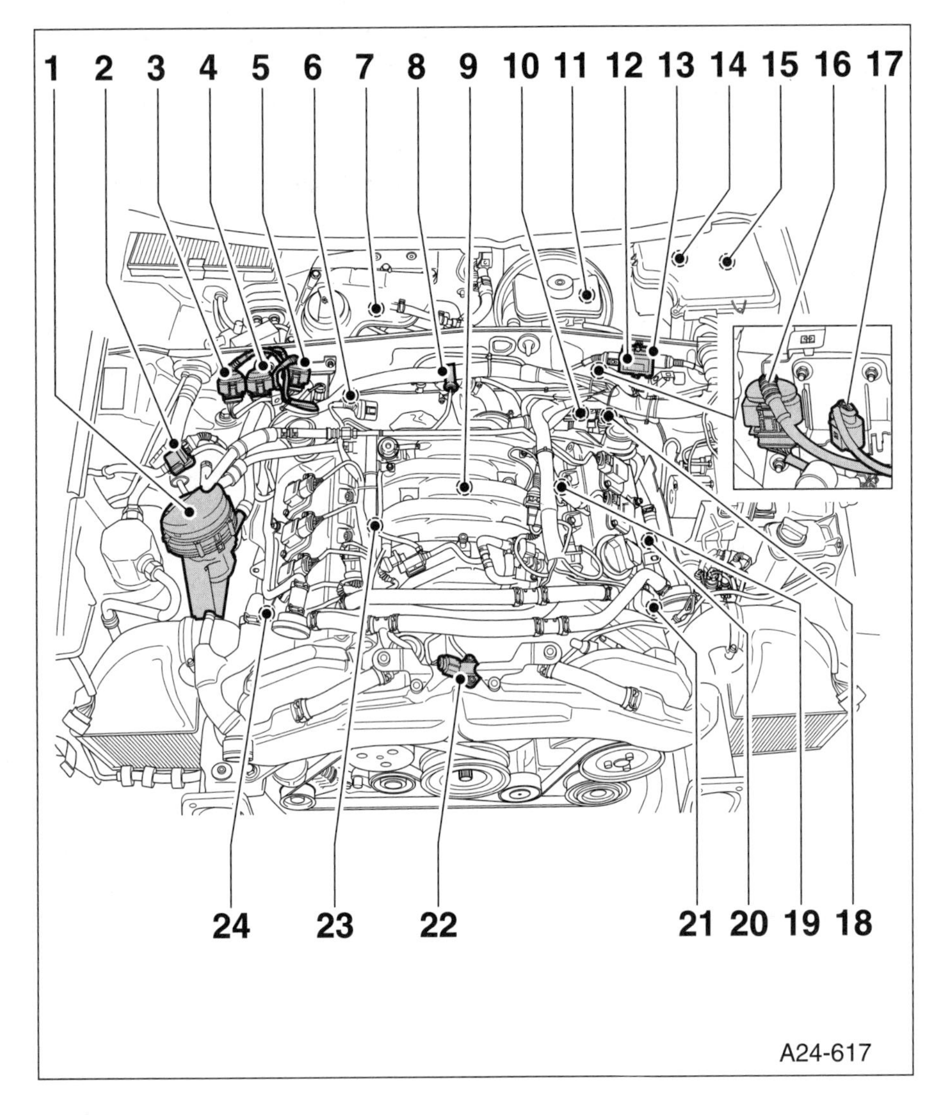

Additional engine compartment components (V8 engine, code BCY)

1. **Secondary air injection pump motor (V101)**
2. **Fuel tank evaporative control (EVAP) canister purge regulator valve (N80)**
3. **Oxygen sensor harness connector**
   - Right bank oxygen sensor (G130) and heater (Z29) behind catalytic converter
   - 4-pin green
4. **Oxygen sensor harness connector**
   - Left bank oxygen sensor (G131) and heater (Z30) behind catalytic converter
   - 4-pin brown
5. **Oxygen sensor harness connector**
   - Right bank oxygen sensor (G39) and heater (Z19) in front of catalytic converter.
   - 4-pin black
6. **Right bank oxygen sensor (G39) and heater (Z19)**
   - In front of catalytic converter
7. **Right bank oxygen sensor (G130) and heater (Z29)**
   - Behind catalytic converter
8. **Wastegate bypass regulator valve (N75)**
9. **After-run coolant pump (V51)**
10. **Left bank oxygen sensor (G108) and heater (Z30)**
    - In front of catalytic converter
11. **Left bank oxygen sensor (G131) and heater (Z30)**
    - Behind catalytic converter
12. **Right exhaust temperature sensor (G235)**
    - Brown
13. **Left exhaust temperature sensor (G236)**
    - Black
14. **Electronics box (E-box)**
    - Secondary air injection pump relay (J299)
    - Coolant circulation pump relay (J151). See **Engine Management Fuses and Relays** in this repair group.

BentleyPublishers.com
BentleyPublishers.com—All Rights Reserved

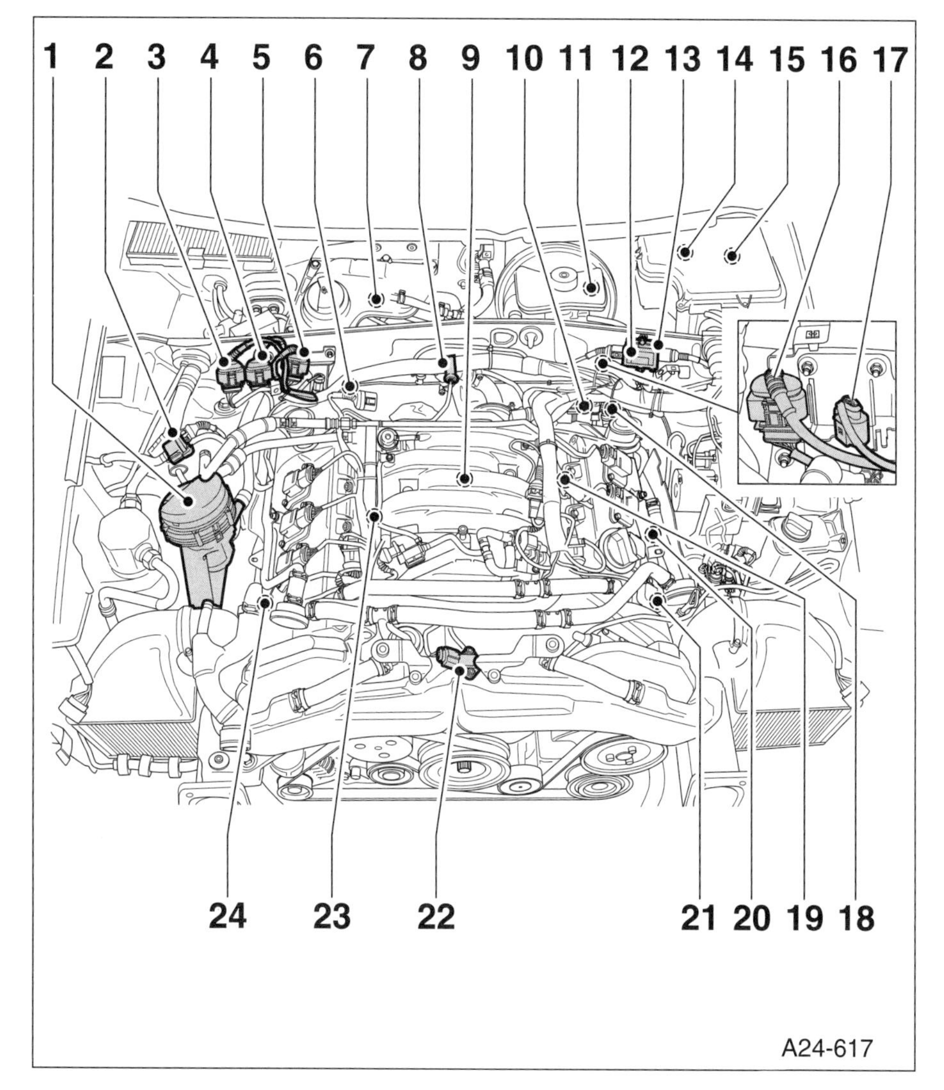

15. **Engine control module (ECM) (J220)**
    - See **ECM, accessing** in this repair group.
16. **Oxygen sensor harness connector**
    - Left bank oxygen sensor (G108) and heater (Z28) in front of catalytic converter
    - 4-pin
17. **Engine speed sensor (G28) connector**
    - 3-pin grey
18. **Engine speed sensor (G28)**
19. **Knock sensor 3 (G198)**
    - Underneath intake manifold at cylinder 7 water jacket
20. **Knock sensor 2 (G66)**
    - Underneath left exhaust manifold at cylinder 6 water jacket
21. **Left electrohydraulic engine mount solenoid (N145)**
22. **Charge air pressure sensor (G31)**
23. **Knock sensor 1 (G61)**
    - Underneath intake manifold between cylinders 2 and 3
24. **Right electrohydraulic engine mount solenoid (N144)**

BentleyPublishers.com
BentleyPublishers.com—All Rights Reserved

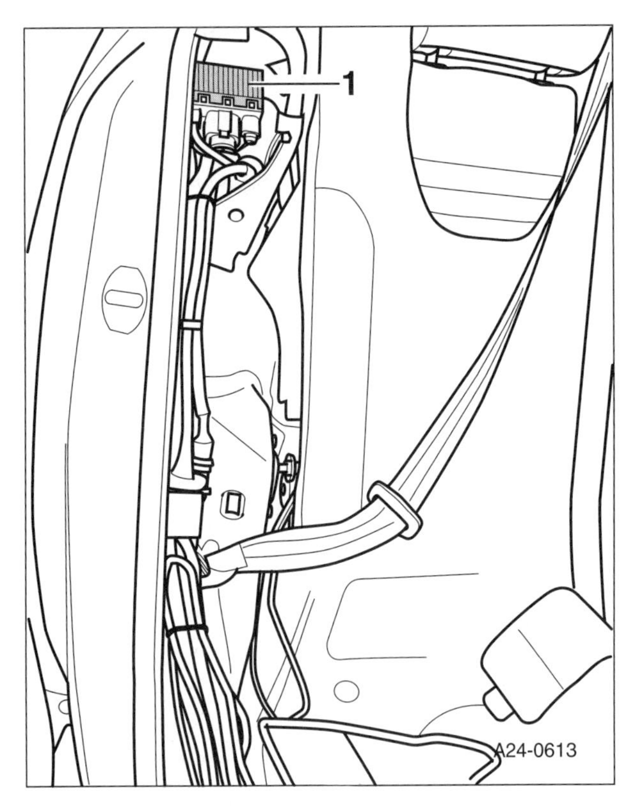

Engine management component details (V8 engine, code BCY)

◀ Fuel pump control module (J534) (**1**)

- Right rear of vehicle
- In a cavity on a level with C-pillar underneath cover next to rear seat backrest (at right rear seat belt reel)

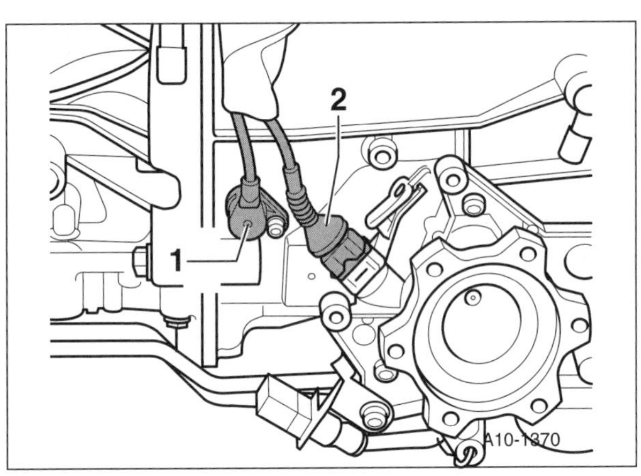

◀ Engine speed sensor (G28) (**1**) and vehicle speed sensor (**2**)

BentleyPublishers.com
BentleyPublishers.com—All Rights Reserved

26 Exhaust System

GENERAL

This repair group covers muffler, catalytic converter and oxygen sensor replacement. Also included is a description of the secondary air injection system.

See **24 Fuel Injection** for engine management power supply fuse and relay locations.

Exhaust manifolds, catalytic converters and oxygen sensors

The 2.7 liter turbocharged engine exhaust manifolds are insulated double-walled elbows.This design leads to lower engine compartment temperature and quicker exhaust system heat-up.

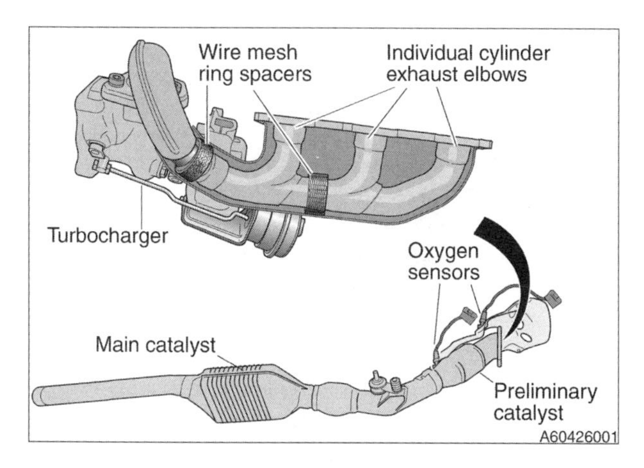

In 2.7 liter, 3.0 liter, V8 allroad quattro and RS6 models, a preliminary catalytic converter is installed in the exhaust manifold just downstream of each turbocharger. This catalyst reaches operating temperature ("light-off") very quickly after a cold start. The main catalysts are further downstream, under the vehicle floor.

BentleyPublishers.com
BentleyPublishers.com—All Rights Reserved

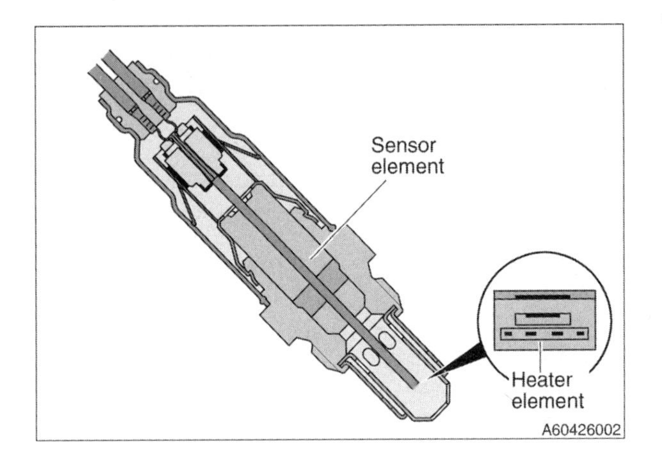

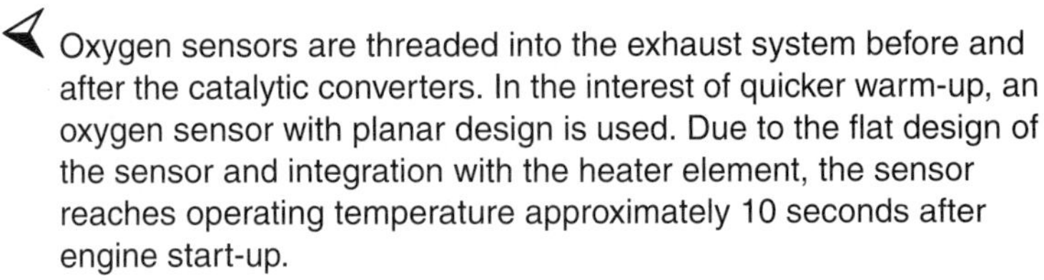

Oxygen sensors are threaded into the exhaust system before and after the catalytic converters. In the interest of quicker warm-up, an oxygen sensor with planar design is used. Due to the flat design of the sensor and integration with the heater element, the sensor reaches operating temperature approximately 10 seconds after engine start-up.

On-Board Diagnostics

Oxygen sensors and other components of the Motronic engine management system are monitored by on-board diagnostics (OBD II) software in the engine control module (ECM). If engine managements faults and malfunctions are detected, one or more diagnostic trouble codes (DTCs) are set.

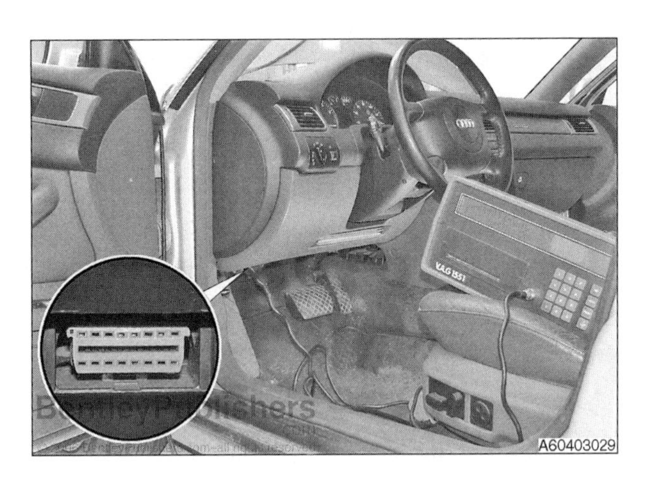

Access fault memory with Volkswagen scan tool or equivalent plugged into the DLC plug (**inset**) under left side dashboard. See **OBD On-Board Diagnostics** for more information.

Warnings and Cautions

WARNING —
- *The exhaust system and catalytic converters operate at very high temperatures. Allow components to cool before servicing. Wear protective gloves to prevent burns. Do not use flammable chemicals near a hot catalytic converter.*
- *Exhaust gases are colorless, odorless and very toxic. Run the engine only in a well-ventilated area. Immediately repair any leaks in the exhaust system or structural damage to the car body that might allow exhaust gases to enter the passenger compartment.*
- *Corroded exhaust system components crumble easily and often have exposed sharp edges. To avoid injury, wear eye protection and heavy gloves when working with exhaust parts.*
- *Heat shields protect the car occupants, undercoating, and various other components from excessive heat. Damaged or missing shields, particularly those above the catalytic converter, increase interior temperatures and create a fire hazard. Repair or replace as necessary.*

CAUTION—
- *Do not continue to operate the starter if the engine fails to start promptly. Extended cranking may allow excess fuel to enter catalytic converter(s), creating a fire hazard and possibly damaging the converter(s).*
- *Do not drag or bang the oxygen sensors.*
- *Do not bend the flexible pipe connection in the front exhaust pipe more than 10°. Otherwise it can be damaged.*
- *When replacing exhaust parts, replace gaskets and fasteners.*

BentleyPublishers.com
BentleyPublishers.com—All Rights Reserved

Exhaust System Diagrams

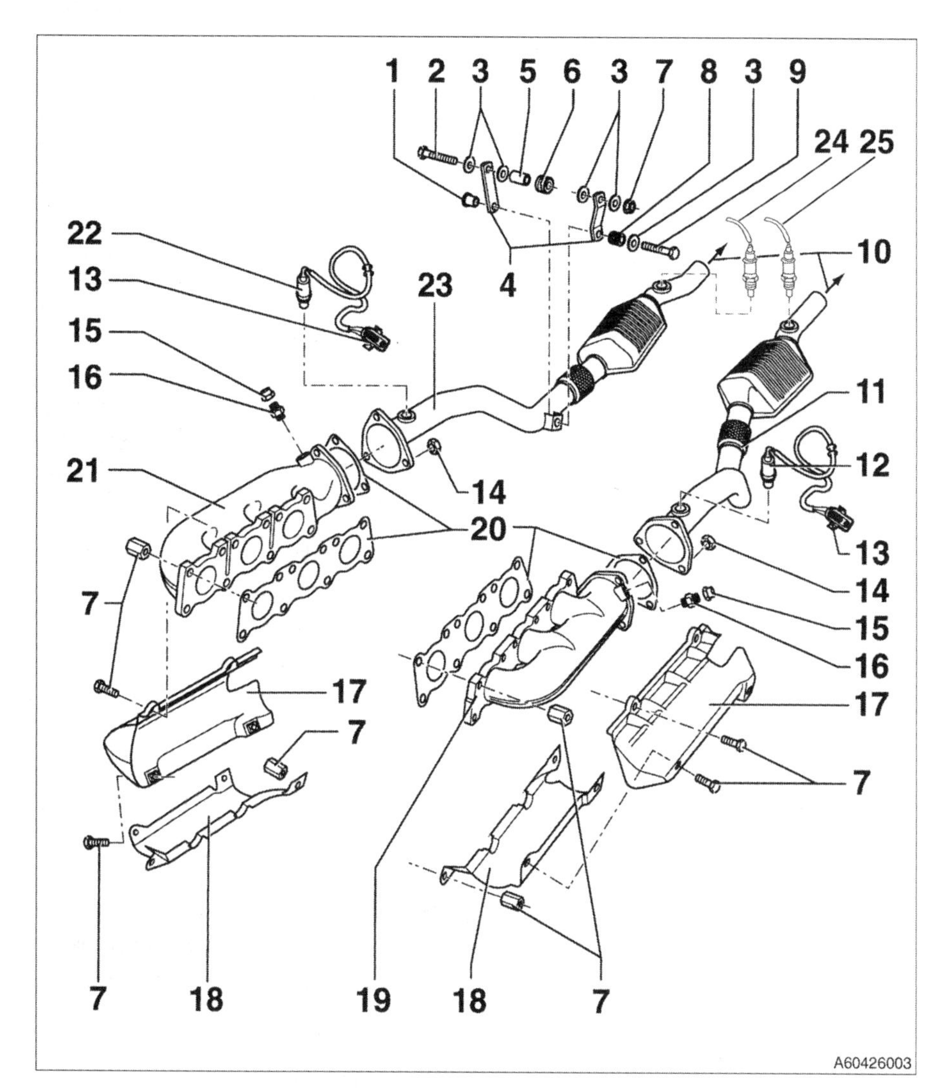

2.8 liter engine exhaust system components (front)

1. Spacer sleeve
2. Bolt
3. Washers
4. Bracket
5. Spacer bushing
6. Rubber bushing
7. Nut, 25 Nm (18 ft-lb)
8. Spring
9. Bolt, 25 Nm (18 ft-lb)
10. to Center muffler
11. Left front pipe with catalytic converter
12. Left bank oxygen sensor (G108) and heater (Z28) in front of catalytic converter
13. Oxygen sensor harness connector
    - 4-pin black
14. Nut, 25 Nm (18 ft-lb)
15. Cap, 10 Nm (7 ft-lb)
16. Plug
17. Upper heat shield
18. Lower heat shield
19. Left exhaust manifold
20. Gaskets
21. Right exhaust manifold
22. Right bank oxygen sensor (G39) and heater (Z19) in front of catalytic converter
23. Right front pipe with catalytic converter
24. Right bank oxygen sensor (G130) and heater (Z29) behind catalytic converter
25. Left bank oxygen sensor (G131)and heater (Z30) behind catalytic converter

NOTE—

- *Audi identifies electrical components by a letter and/or a number in the electrical schematics. See* **EWD Electrical Wiring Diagrams**. *These electrical identifiers are listed in parentheses as an aid to electrical troubleshooting.*

BentleyPublishers.com
BentleyPublishers.com—All Rights Reserved

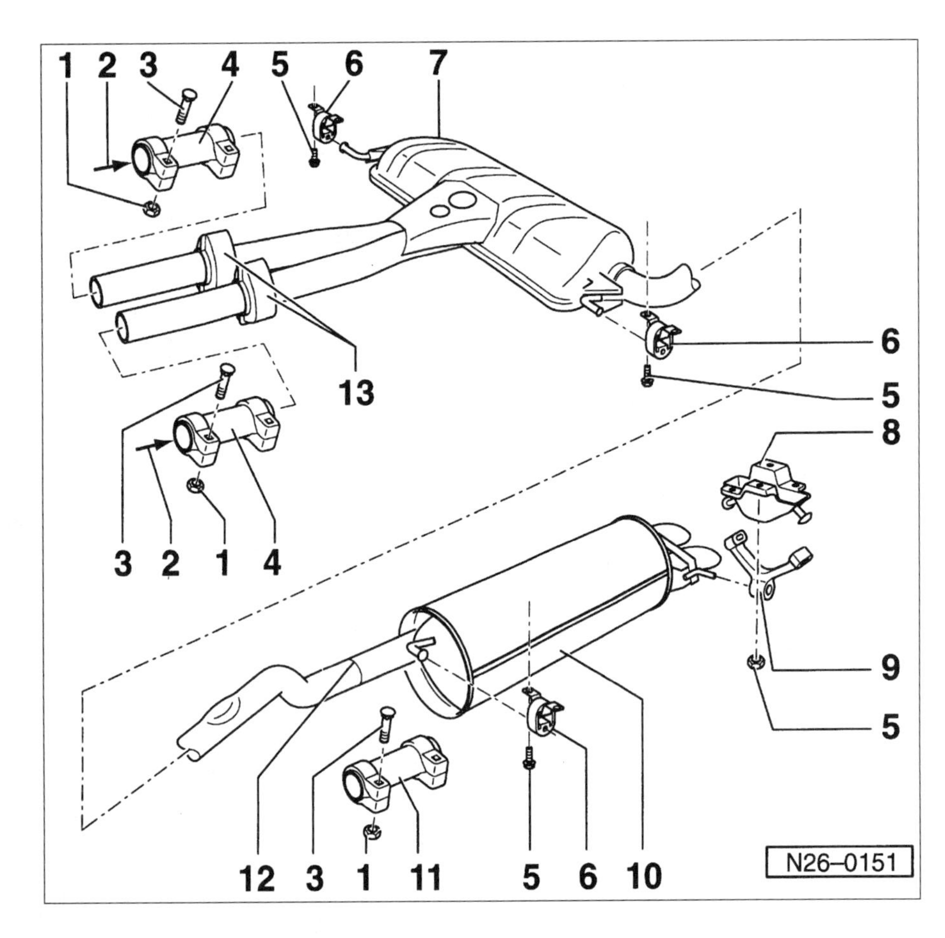

2.7 liter or 2.8 liter engine exhaust system components (rear, front-wheel drive)

1. Clamping nut, 40 Nm (30 ft-lb)
2. from Catalytic converter
3. Carriage bolt
4. Double clamp
5. Exhaust hanger mounting bolt, 25 Nm (18 ft-lb)
6. Exhaust hanger
7. Front muffler
8. Bracket
9. Exhaust hanger
10. Rear muffler
11. Double clamp
12. Cut pipe here
13. Resonators

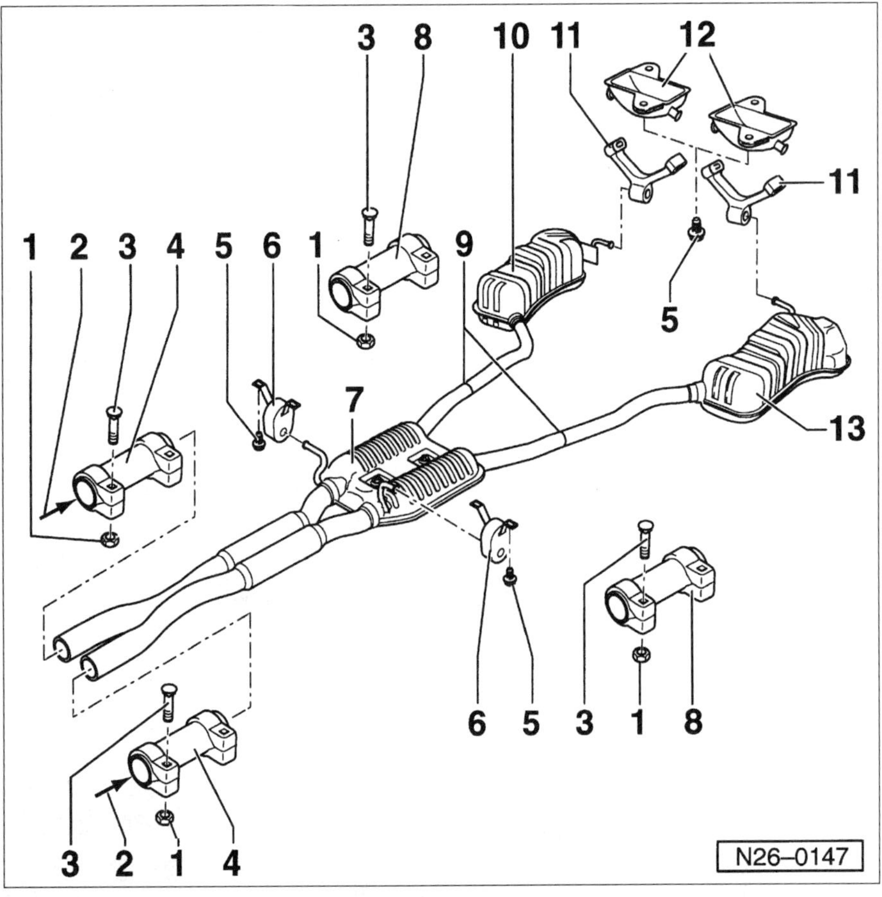

2.7 liter or 2.8 liter engine, V8 engine (engine codes ART, AWD, BBD) exhaust system components (rear, quattro)

1. Clamping nut, 40 Nm (30 ft-lb)
2. from Catalytic converter
3. Carriage bolt
4. Double clamp
5. Exhaust hanger mounting bolt, 25 Nm (18 ft-lb)
6. Exhaust hanger
7. Front muffler
8. Double clamp
9. Cut pipe here
10. Right rear muffler
11. Exhaust hanger
12. Bracket
13. Left rear muffler

BentleyPublishers.com
BentleyPublishers.com—All Rights Reserved

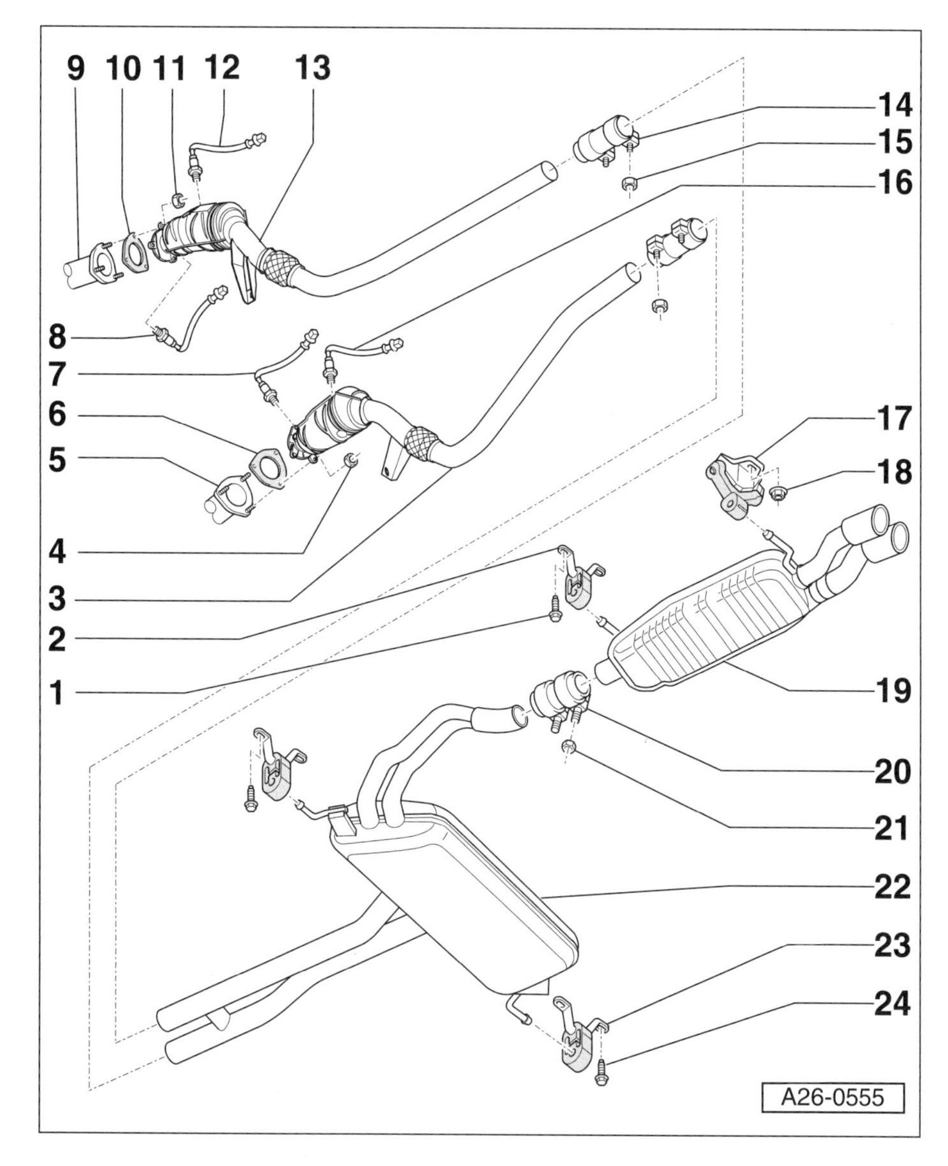

3.0 liter engine exhaust system components (front-wheel drive)

1. **Exhaust hanger mounting bolt, 23 Nm (17 ft-lb)**
2. **Exhaust hanger**
3. **Right front pipe with catalytic converter**
4. **Nut, 27 Nm (20 ft-lb)**
5. **Right exhaust manifold**
6. **Gasket**
7. **Right bank oxygen sensor before catalytic converter**
   - 55 Nm (40 ft-lb)
8. **Left bank oxygen sensor before catalytic converter**
   - 55 Nm (40 ft-lb)
9. **Left exhaust manifold**
10. **Gasket**
11. **Nut, 27 Nm (20 ft-lb)**
12. **Left bank oxygen sensor after catalytic converter**
    - 55 Nm (40 ft-lb)
13. **Left front pipe with catalytic converter**
14. **Double clamp**
15. **Nut, 40 Nm (30 ft-lb)**
16. **Right bank oxygen sensor after catalytic converter**
17. **Exhaust hanger**
18. **Nut, 18 Nm (13 ft-lb)**
19. **Rear muffler**
20. **Double clamp**
21. **Nut, 40 Nm (30 ft-lb)**
22. **Center muffler**
23. **Exhaust hanger**
24. **Exhaust hanger mounting bolt, 23 Nm (17 ft-lb)**

BentleyPublishers.com
BentleyPublishers.com—All Rights Reserved

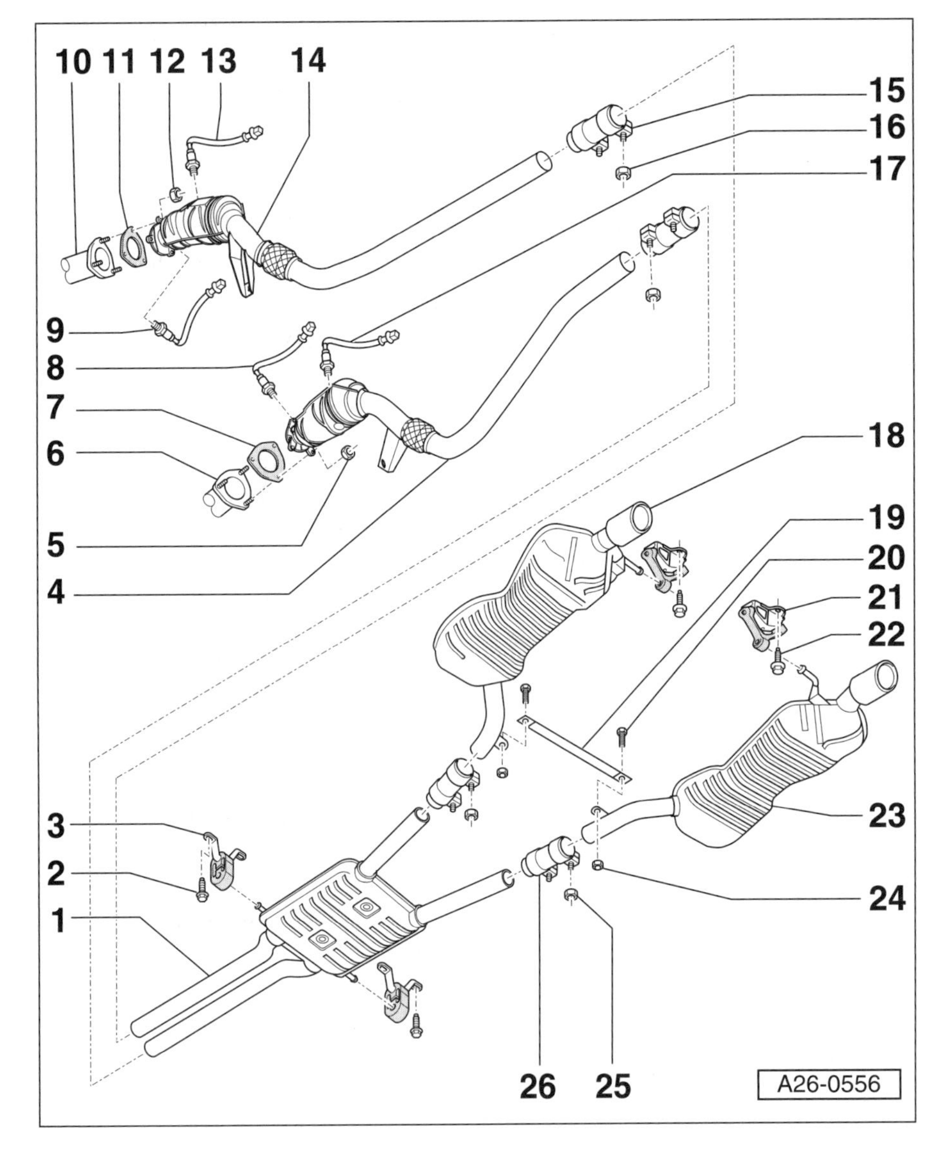

3.0 liter engine exhaust system components (quattro)

1. **Center muffler**
2. **Exhaust hanger mounting bolt, 23 Nm (17 ft-lb)**
3. **Exhaust hanger**
4. **Right front pipe with catalytic converter**
5. **Nut, 27 Nm (20 ft-lb)**
6. **Right exhaust manifold**
7. **Gasket**
8. **Right bank oxygen sensor before catalytic converter**
   - 55 Nm (40 ft-lb)
9. **Left bank oxygen sensor before catalytic converter**
   - 55 Nm (40 ft-lb)
10. **Left exhaust manifold**
11. **Gasket**
12. **Nut, 27 Nm (20 ft-lb)**
13. **Left bank oxygen sensor behind catalytic converter**
    - 55 Nm (40 ft-lb)
14. **Left front pipe with catalytic converter**
15. **Double clamp**
16. **Nut, 40 Nm (30 ft-lb)**
17. **Right bank oxygen sensor behind catalytic converter**
18. **Left rear muffler**
19. **Brace**
20. **Bolt**
21. **Exhaust hanger**
22. **Exhaust hanger mounting bolt, 23 Nm (17 ft-lb)**
23. **Right rear muffler**
24. **Nut, 25 Nm (18 ft-lb)**
25. **Nut, 40 Nm (30 ft-lb)**
26. **Double clamp**

BentleyPublishers.com

BentleyPublishers.com—All Rights Reserved

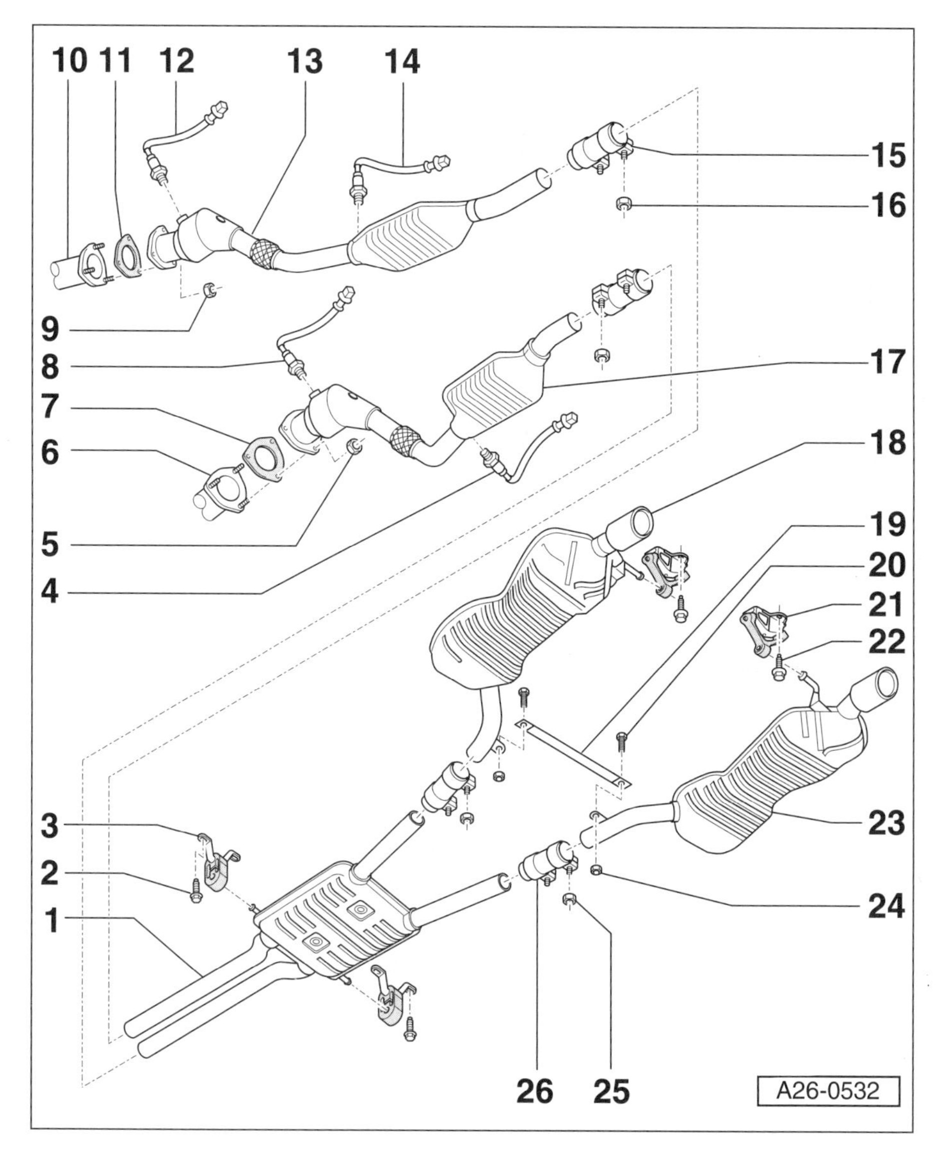

V-8 engine exhaust system components (allroad quattro, engine code BAS)

1. **Center muffler**
2. **Exhaust hanger mounting bolt, 23 Nm (17 ft-lb)**
3. **Exhaust hanger**
4. **Right bank oxygen sensor behind catalytic converter**
   - 55 Nm (40 ft-lb)
5. **Nut, 27 Nm (20 ft-lb)**
6. **Right exhaust manifold**
7. **Gasket**
8. **Right bank oxygen sensor before catalytic converter**
   - 55 Nm (40 ft-lb)
9. **Nut, 27 Nm (20 ft-lb)**
10. **Left exhaust manifold**
11. **Gasket**
12. **Left bank oxygen sensor before catalytic converter**
    - 55 Nm (40 ft-lb)
13. **Left front pipe with preliminary and main catalytic converters**
14. **Left bank oxygen sensor behind catalytic converter**
    - 55 Nm (40 ft-lb)
15. **Double clamp**
16. **Nut, 40 Nm (30 ft-lb)**
17. **Right front pipe with preliminary and main catalytic converters**
18. **Left rear muffler**
19. **Brace**
20. **Bolt**
21. **Exhaust hanger**
22. **Exhaust hanger mounting bolt, 23 Nm (17 ft-lb)**
23. **Right rear muffler**
24. **Nut, 25 Nm (18 ft-lb)**
25. **Nut, 40 Nm (30 ft-lb)**
26. **Double clamp**

BentleyPublishers.com
BentleyPublishers.com—All Rights Reserved

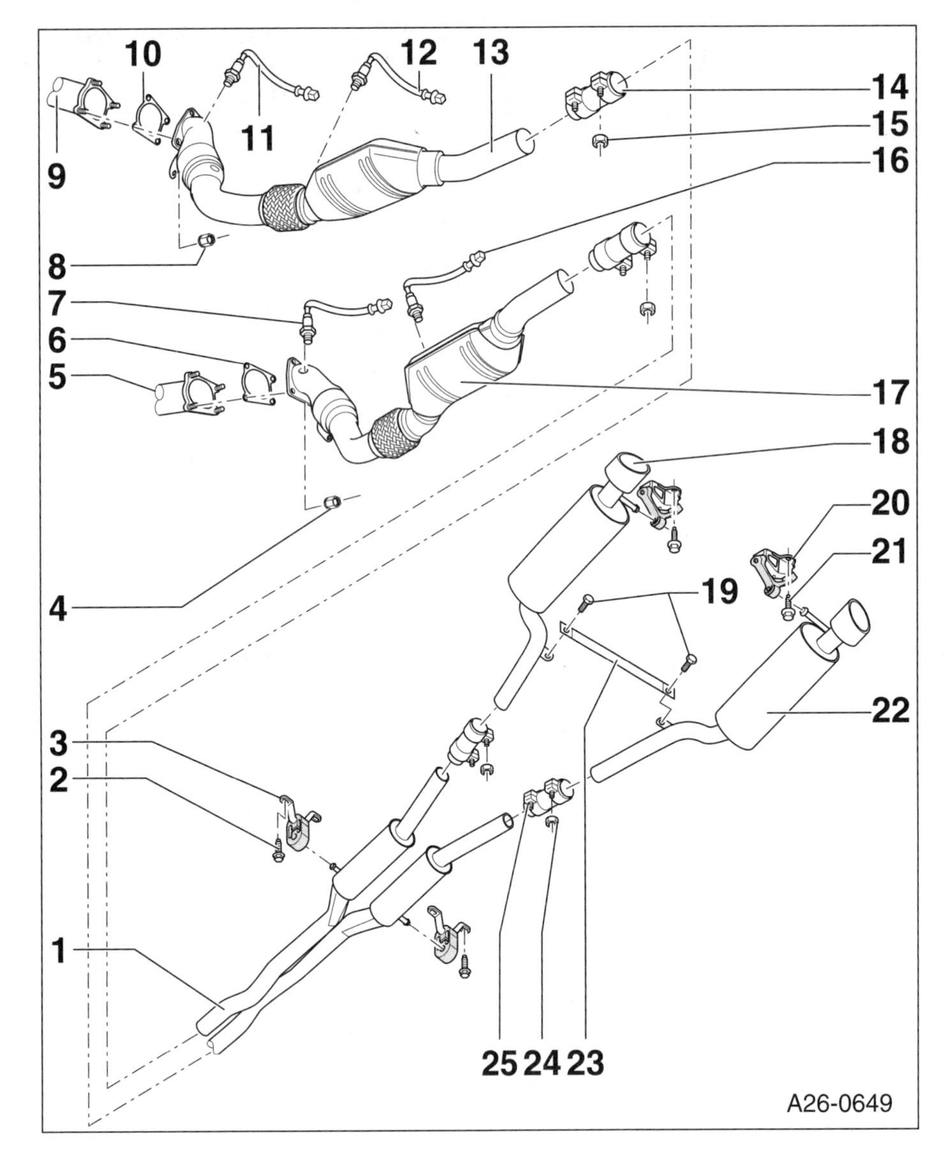

V-8 engine exhaust system components (RS6, engine code BCY)

1. **Center muffler**
2. **Exhaust hanger mounting bolt, 25 Nm (18 ft-lb)**
3. **Exhaust hanger**
4. **Nut, 50 Nm (37 ft-lb)**
5. **Right turbocharger**
6. **Gasket**
7. **Right bank oxygen sensor before catalytic converter**
   - 55 Nm (40 ft-lb)
8. **Nut, 50 Nm (37 ft-lb)**
9. **Left turbocharger**
10. **Gasket**
11. **Left bank oxygen sensor before catalytic converter**
    - 55 Nm (40 ft-lb)
12. **Left bank oxygen sensor behind catalytic converter**
    - 55 Nm (40 ft-lb)
13. **Left front pipe with preliminary and main catalytic converters**
14. **Double clamp**
15. **Nut, 40 Nm (30 ft-lb)**
16. **Right bank oxygen sensor behind catalytic converter**
    - 55 Nm (40 ft-lb)
17. **Right front pipe with preliminary and main catalytic converters**
18. **Left rear muffler**
19. **Bolt, 25 Nm (18 ft-lb)**
20. **Exhaust hanger**
21. **Exhaust hanger mounting bolt, 23 Nm (17 ft-lb)**
22. **Right rear muffler**
23. **Brace**
24. **Nut, 40 Nm (30 ft-lb)**
25. **Double clamp**

BentleyPublishers.com
BentleyPublishers.com—All Rights Reserved

Exhaust System Component Replacement

The exhaust system is designed to be maintenance free, although regular inspection is warranted due to the harsh operating conditions. Under normal conditions, catalytic converters do not require replacement unless they are damaged.

Exhaust system, installation details

The exhaust systems on most of the vehicles covered by this manual are one-piece. One pipe starts at each exhaust manifold or turbocharger and incorporates preliminary catalytic converter (where equipped), main catalytic converter, mufflers and tailpipe.

– Liberally apply penetrating oil to exhaust system fasteners and wait several minutes before attempting removal.

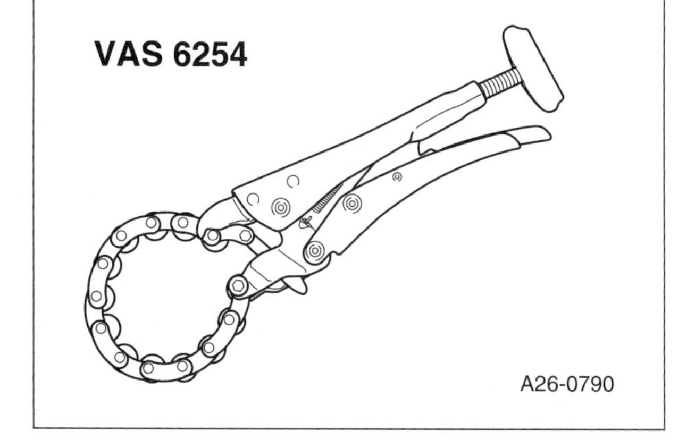

To replace damaged or rusted sections of the exhaust system, use special tool VAS 6254 or equivalent pipe cutter.

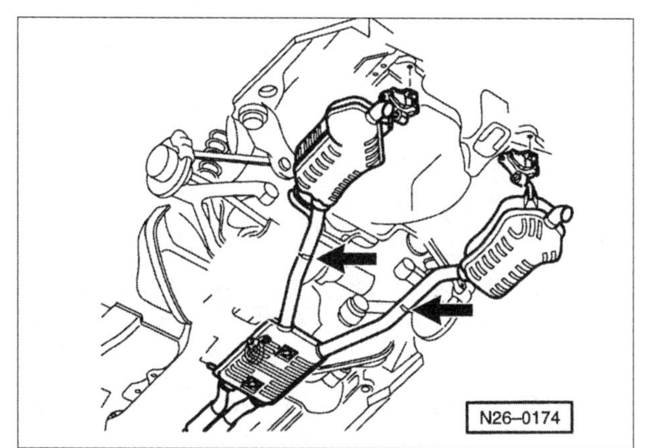

Cut pipe at separation point (**arrows**) marked by depression around circumference.

– Attach new pipe to old section using double clamp.

– Use new fasteners, clamps, mounts and gaskets when replacing exhaust system components.

– Use high-temperature anti-seize compound on threaded fasteners to make future replacement easier.

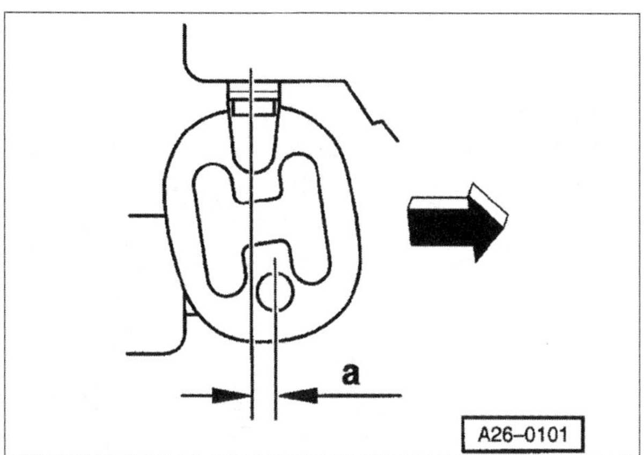

Before tightening clamps and fasteners, push exhaust system forward (in direction of vehicle motion, **arrow**) to preload exhaust hangers.

Exhaust hanger preload (in direction of vehicle motion)	
Distance **a**	5 - 11 mm (¼ - ¾ in)

– Make sure exhaust system is not under stress and has sufficient clearance from body.

BentleyPublishers.com
BentleyPublishers.com—All Rights Reserved

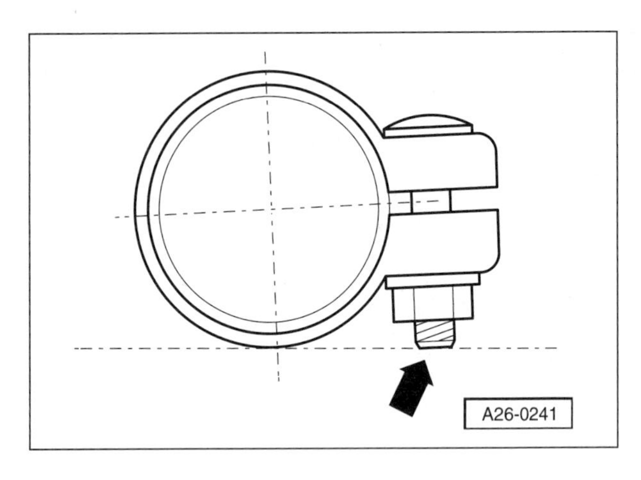

Make sure double clamp bolt threads point down (**arrow**). To prevent bolt threads from becoming snagged and damaged by road debris, make sure threads do not project below bottom edge of clamp.

Tightening torque	
Double clamp pinch fastener	40 Nm (30 ft-lb)

Oxygen sensor, replacing

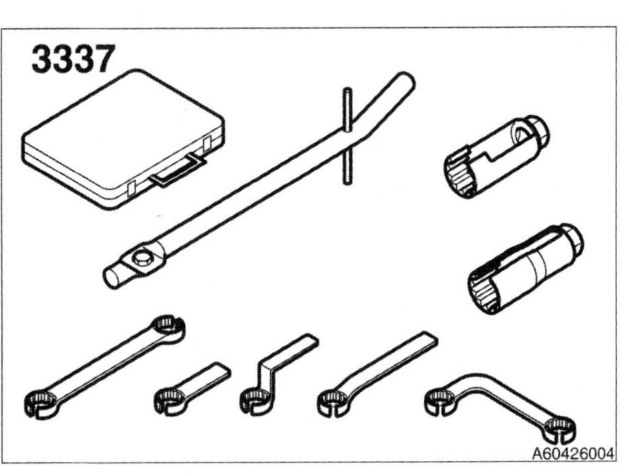

Use special tool VAG 3337 or equivalent to remove and install oxygen sensor requires.

> ***CAUTION—***
> - *When installing oxygen sensor be sure to use high temperature anti-seize compound that contains no silicone and is marked "safe for oxygen sensors" (VAG part no. G 052 112 A3).*
> - *Do not allow anti-seize compound to contaminate oxygen sensor tip.*

– To locate correct oxygen sensor harness connector, see engine compartment diagrams in **24 Fuel Injection**.

Tightening torque	
Oxygen sensor to exhaust system	55 Nm (40 ft-lb)

SECONDARY AIR INJECTION

The secondary air system pumps ambient air into the exhaust stream after a cold engine start to reduce the warm-up time of the catalytic converters and to reduce HC and CO emissions. The engine control module (ECM) controls and monitors the secondary air injection system.

Secondary air injection is used in normally aspirated engines, and in turbocharged engines with automatic transmission.

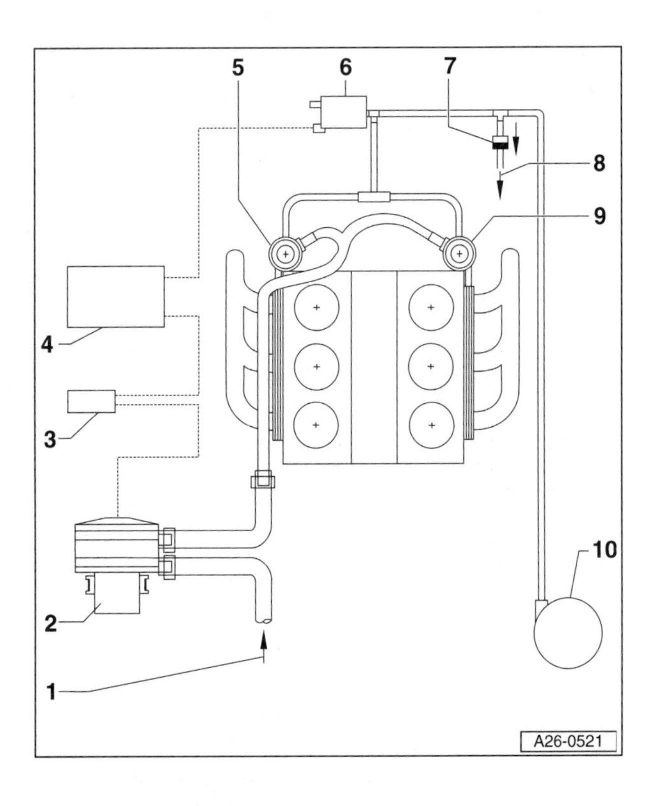

Secondary air injection system schematic

1. from Air filter housing
2. Secondary air injection pump (V101)
3. Secondary air injection pump relay (J299)
4. Engine control module (ECM) (J220)
5. Combination valve on right cylinder head
6. Secondary air injection solenoid (N112)
7. Check valve
8. to Intake manifold
9. Combination valve on left cylinder head
10. Vacuum reservoir

BentleyPublishers.com
BentleyPublishers.com—All Rights Reserved

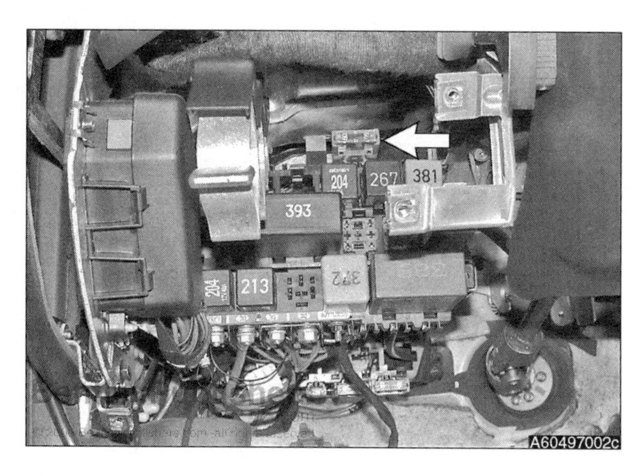

Secondary air injection fuse and relay

Under left dashboard in 13-fold relay and fuse panel: Secondary air pump fuse 50A (**arrow**).

For access information, see **97 Fuses, Relays, Component Locations**.

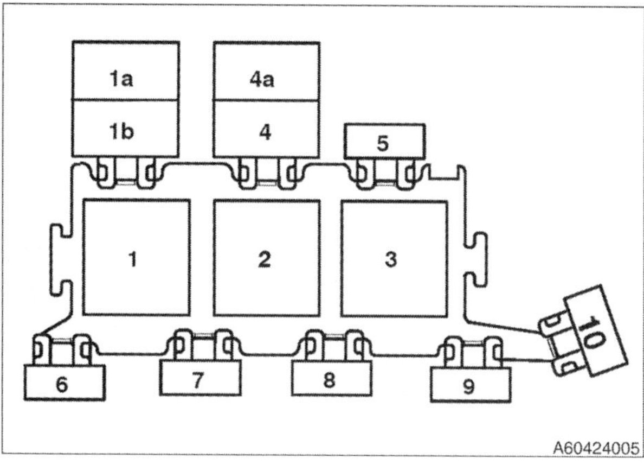

E-box, left rear engine compartment in plenum chamber:

- Fuse **7**: Secondary air pump 40A fuse
- Relay **2**: Secondary air pump relay (J299)

Secondary air injection pump (V101)

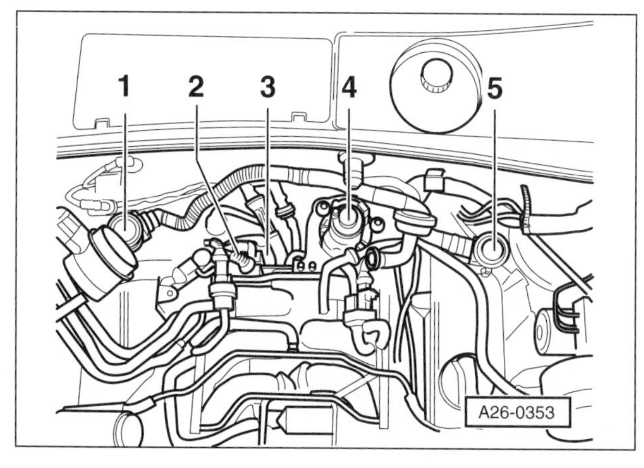

2.7 liter V6 twin-turbo engine: At left rear of intake manifold (**4**).

2.8 liter V6, 3.0 liter V6, V8 engine(allroad quattro, engine code BAS): Underneath right front of engine, behind right end of front bumper.

BentleyPublishers.com
BentleyPublishers.com—All Rights Reserved

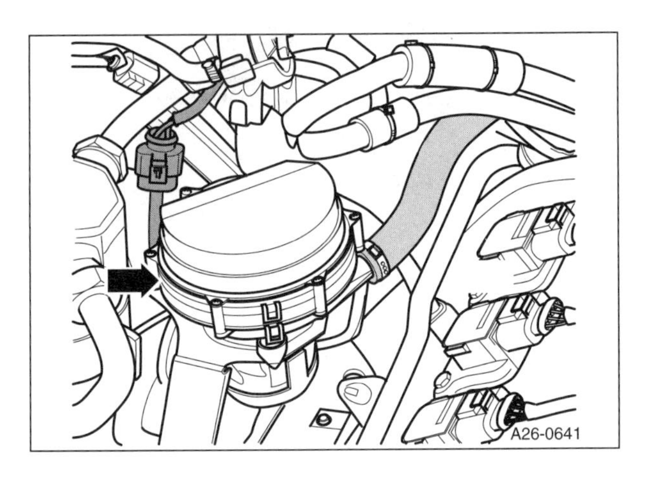

◀ **V8 twin-turbo engine (RS6, engine code BCY)**: Right rear engine compartment (**arrow**).

Combination valve

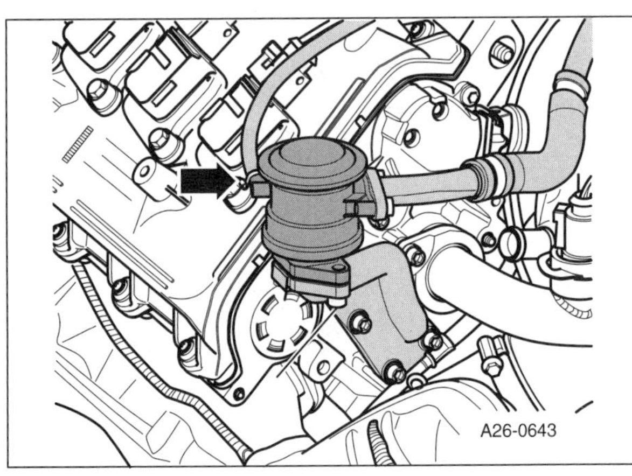

◀ **All engines**: Rear of cylinder head (**arrow**).

Secondary air injection solenoid (N112)

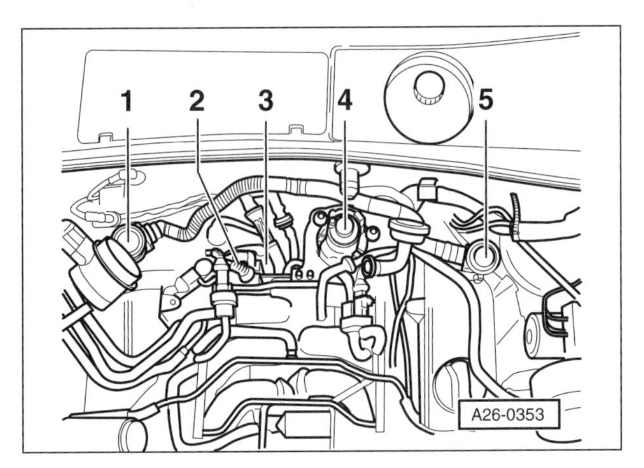

◀ **2.7 liter V6 twin-turbo engine**: At right rear of intake manifold (**3**).

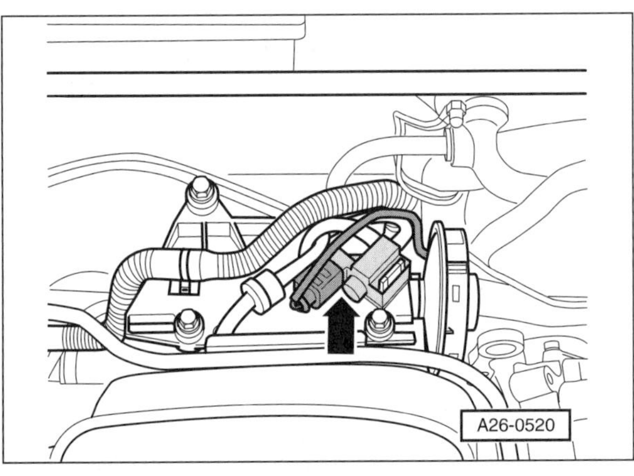

◀ **3.0 liter V6 engine**: At engine compartment bulkhead behind engine (**arrow**).

BentleyPublishers.com
BentleyPublishers.com—All Rights Reserved

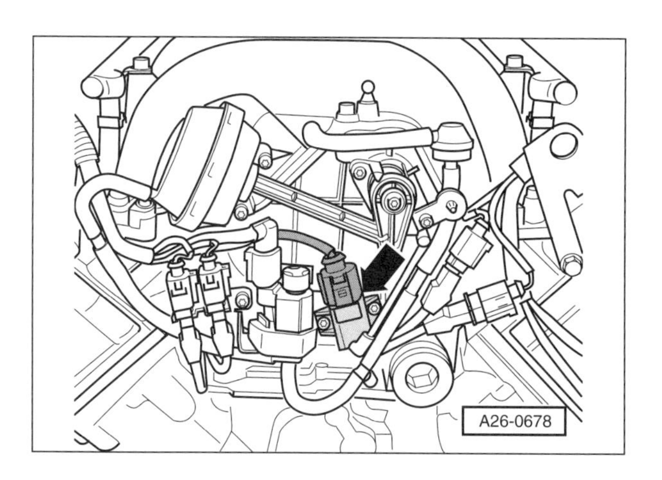

V8 engine (allroad quattro, engine code BAS): Underneath front engine cover, at front of intake manifold (**arrow**).

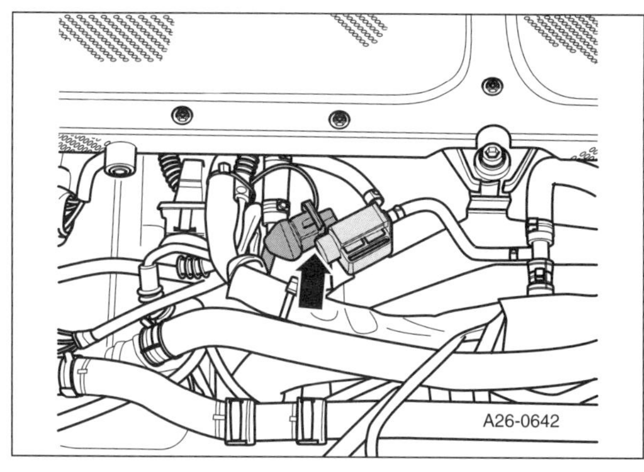

V8 engine (RS6, engine code BCY): Underneath front engine cover, at right front of intake manifold (**arrow**).

Vacuum reservoir

All engines: Under left front wheel housing liner.

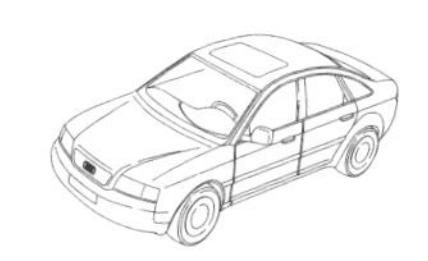

BentleyPublishers.com
BentleyPublishers.com—All Rights Reserved

27 Battery, Alternator, Starter

TABLES

GENERAL

This repair group covers battery, alternator and starter service and repairs.

The alternator and starter are wired directly to the battery. To prevent accidental shorts that might blow a fuse or damage wires and electrical components, disconnect the negative (–) battery lead before working on the electrical system.

Various versions of alternators, starters, and batteries have been used in Audi A6 models. Replace components according to the original equipment specification. When in doubt, consult an authorized Audi parts department.

Also see the following:

- **03 Maintenance** for engine accessory belt replacement
- **9 Electrical System–General** for bus information and electrical diagnostics
- **EWD Electrical Wiring Diagrams** for battery, charging system and starter schematics

NOTE—

- *The alternator is identified as generator by the vehicle manufacturer.*

BentleyPublishers.com
BentleyPublishers.com—All Rights Reserved

Cruise control

Cruise control functions are controlled by the engine control module (ECM). The only serviceable cruise control components are the steering wheel, clutch and brake pedal switches and related wiring. See the following repair groups:

- **30 Clutch** for clutch pedal position switch
- **46 Brakes–Mechanical** for brake light switch
- **96 Interior Lights, Switches, Anti-theft**for steering column stalk switches

Battery reconnection notes

In addition to battery / power supply warnings and cautions in this repair group and in **00 Warnings and Cautions,** observe the following whenever the battery is disconnected or accidentally discharged.

- Reset climate control.
- Reset clock.
- Reset radio presets.
- Disconnect and reconnect Audi Telematics (telephone) back-up battery.
- Reinitialize window regulator motors.

Climate control, resetting

For climate control to operate in automatic mode, set to 75°F (23°C).

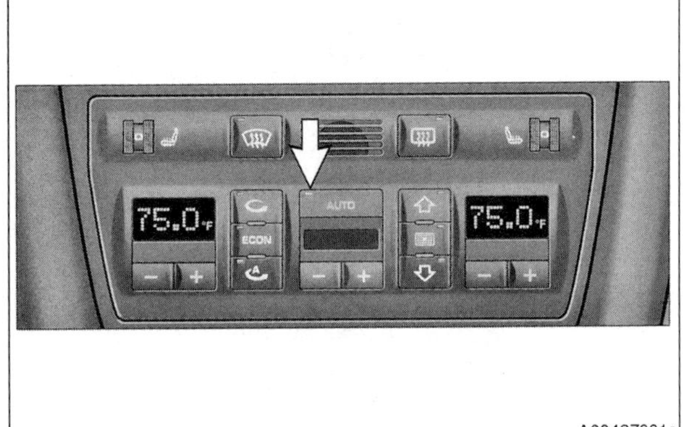

A60487001a

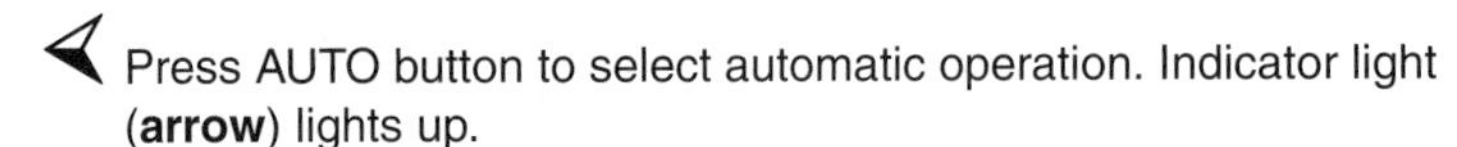

Press AUTO button to select automatic operation. Indicator light (**arrow**) lights up.

– Press - or + buttons on each side to set temperature to 75°F (23°C).

Clock, resetting

– See owner's manual.

Radio presets, resetting

– Prior to disconnecting battery, note down radio presets. To reset, see owner's manual.

Telematics back-up battery, reconnecting

– On vehicle equipped with Audi Telematics by OnStar®, switch OFF emergency (back-up) battery for the Telematic control module prior to disconnecting vehicle battery.

– After reconnecting vehicle battery, switch emergency (back-up) battery ON.

BentleyPublishers.com
BentleyPublishers.com—All Rights Reserved

Window regulator motors, reinitializing

If the battery is disconnected, the window control module loses its memory of window current position and disables one-touch automatic up / down function. To restore one-touch operation:

- Switch ignition ON.
- Use window switches to raise windows to top.
- Operate each window switch in CLOSE direction for 1 second to reinitialize one-touch operation.
- Test each window by briefly pressing switch in OPEN direction. Make sure window opens fully.

Troubleshooting

Tests for individual electrical system components are described under component headings in this repair group. **Table a** gives some general troubleshooting ideas.

Table a. Battery, starter and alternator troubleshooting

Symptom	Probable cause	Corrective action
Engine does not crank.	Fault in immobilizer system.	Try another ignition key. If problem persists, contact your authorized dealer.
	Faulty automatic transmission range switch.	Check, and if necessary, replace automatic transmission range switch.
Engine cranks slowly or not at all, solenoid clicks when starter is operated.	Battery cables loose, dirty or corroded.	See **03 Maintenance** for battery service.
	Battery discharged.	Charge battery and test. Replace if necessary.
	Battery to body ground cable loose, dirty or corroded.	Inspect ground cable. Clean, tighten or replace if necessary.
	Poor connection at starter motor terminal 30.	Check connections, test for voltage at starter. Test for voltage at clutch switch or automatic transmission range switch. See **97 Fuses, Relays, Component Locations**.
	Starter motor or solenoid faulty.	Test starter.
Battery does not stay charged more than a few days.	Short circuit draining battery.	Test for excessive current drain with everything electrical off.
	Short driving trips and high electrical drain on charging system.	Evaluate driving style. Where possible, reduce electrical consumption when making short trips.
	Engine accessory belt loose, worn, damaged.	See **03 Maintenance** for belt service.
	Battery faulty.	Test battery and replace if necessary.
	Battery cables loose, dirty or corroded.	See **03 Maintenance** for battery service.
	Alternator faulty.	Test alternator and voltage regulator.

BentleyPublishers.com
BentleyPublishers.com—All Rights Reserved

Warnings and Cautions

WARNING —

- *Wear goggles, rubber gloves, and a rubber apron when working around the battery and battery acid (electrolyte). Battery acid contains sulfuric acid and can cause skin irritation and burning.*
- *If battery acid is spilled on your skin or clothing, flush the area at once with large quantities of water. If electrolyte gets into your eyes, bathe them with large quantities of clean water for several minutes and call a physician.*
- *Allow a frozen battery to thaw before attempting to recharge it.*
- *Gases given off by the battery during charging are explosive. Do not smoke. Keep open flames away from the battery top, and prevent electrical sparks by turning off the battery charger before connecting or disconnecting it.*
- *If the battery begins gassing (boiling) violently when charging, reduce the charging rate immediately.*
- *Charge the battery in a well ventilated area.*
- *When removing the battery, disconnect negative (–) terminal first.*
- *Before beginning repairs on the electrical system:*
  -Switch electrical consumers OFF.
  -Switch ignition OFF and remove ignition key.

CAUTION—

- *Do not disconnect the battery cables while the engine is running. The alternator will be damaged.*
- *Do not operate the alternator with its output terminal (B+ or 30) disconnected and the other terminals connected. Do not short, bridge, or ground any terminals of the charging system.*
- *Disconnect the negative (–) battery cable when working at or near the alternator. Battery voltage is always present at the rear of the alternator, even with ignition key OFF.*
- *Disconnect the negative (–) battery cable first and reconnect it last. Cover the battery post with an insulating material whenever the cable is removed.*
- *Disconnect the battery cables during battery charging. This prevents damage to the alternator and other solid-state components.*
- *Loosen battery cell caps before charging the battery but leave them on the battery to prevent battery acid splatter.*
- *Do not reverse battery terminals. Even a momentary wrong connection can damage the alternator or electrical components.*
- *Replace the battery if the case is cracked or leaking. Leaking electrolyte can damage the car. If electrolyte is spilled, clean the area with a solution of baking soda and water.*
- *Batteries are damaged by quick charging. Use quick charging as a last alternative when slow charging is not possible.*
- *Do not quick charge a totally discharged battery.*
- *Do not allow the battery charging rate to exceed 16.5 volts.*
- *Prior to disconnecting the battery, read the battery disconnection precautions in* **00 Warnings and Cautions**.

BentleyPublishers.com
BentleyPublishers.com—All Rights Reserved

> ***CAUTION—***
> - *Do not store precision tools in same room where batteries are being charged. Tools may corrode due to chemical reactions.*
> - *Only use a digital multimeter when testing automotive electrical components.*
> - *Disconnecting the battery may erase the radio code and radio presets. Note radio code and stored stations and restore them after reconnecting the battery.*
> - *On-board computer and clock stored settings may be lost when the battery is disconnected.*
> - *Do not depend on the color of insulation to tell battery positive and negative cables apart. Label cables before removing.*
> - *If a quick charger is used to charge the battery, disconnect the battery from vehicle electrical system and remove. This prevents damage to paint work and upholstery.*

BATTERY

The battery is in the plenum chamber, behind the engine.

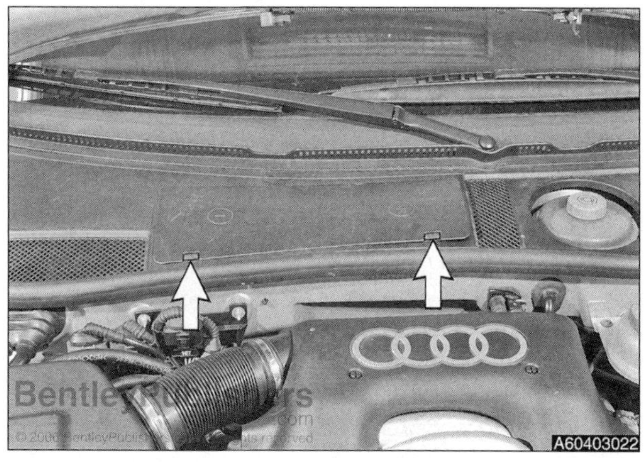

- Open engine hood, release battery cover latches (**arrow**s) and flip cover up to gain access to battery.

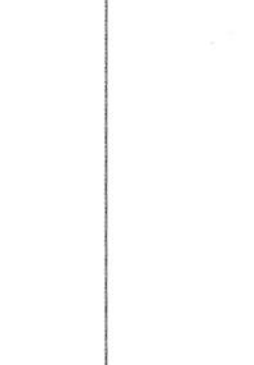

- Battery components:

1. Filler cap
2. Positive (B+) terminal
3. Electrolyte level indicator
4. Negative plate (grey)
5. Separator (insulator)
6. Positive plate (dark brown)
7. Negative (–) terminal

The six-cell, 12-volt lead-acid battery is rated in ampere / hours (Ah) and cold cranking amps (CCA). The Ah rating is determined by the average amount of current the battery can deliver over time without dropping below a specified voltage. The CCA rating is determined by the battery's ability to deliver starting current at 0°F (–18°C).

If the battery discharges when vehicle is not driven, there may be a constant drain or current draw causing it to discharge when ignition is OFF. Depending on the amount of draw and battery condition, a full discharge can happen overnight or it may take a few weeks. Although a small static drain on the battery is normal (example: clock or radio memory), a large drain such as a relay sticking on or

BentleyPublishers.com
BentleyPublishers.com—All Rights Reserved

a faulty switch causes the battery to discharge quickly. Perform a static current draw test first when experiencing battery discharge. See **Static current draw, checking** in this repair group.

If current draw on the battery is not excessive but it still discharges, test the battery charge using open circuit or load voltage test. See **Open circuit voltage test** and **Load voltage, testing** in this repair group. Batteries with removable filler caps can also be tested by checking electrolyte specific gravity. See **Electrolyte specific gravity, testing** in this repair group. Inexpensive specific gravity testers are available at most auto supply stores.

Some vehicles are equipped with batteries that have a central gas venting system with anti-flash protection. The anti-flash protection consists of a small round fiberglass mat. Its purpose is to vent gases that form in the battery through a vent opening in the battery cover. Anti-flash protection is also installed to prevent ignition of flammable gases in the battery.

Electrolyte level, checking

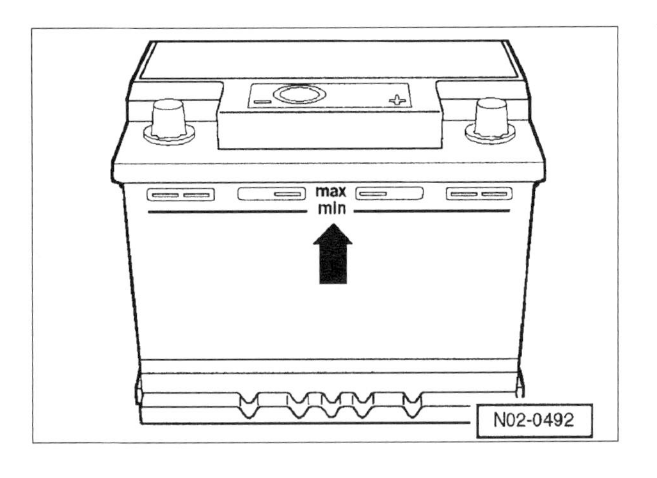

< Visually check battery electrolyte level on batteries with recognizable MAX and MIN markings (**arrow**).

– If there are no markings on battery housing, or if electrolyte level cannot be read, remove battery filler caps to view electrolyte level. Check that level is at same height as visible plastic peg.

– Top off each battery cell as necessary with distilled water using a battery filling bottle. Use a hydrometer to remove excess electrolyte.

CAUTION—
- *Use genuine Audi battery filler caps fitted with O-ring seals.*

NOTE—
- *Use of distilled water to fill battery cells prevents contamination of battery electrolyte and decreases the likelihood of self-discharge.*

Battery residual voltage, checking

– Make sure vehicle is not started or driven for at least 2 hours before taking measurements.

– Make sure no loads are applied to battery for at least 2 hours before taking measurements.

– Make sure battery is not charged for at least 2 hours before taking measurements.

– With ignition switched OFF, use digital multimeter to measure voltage between battery terminals.
- If multimeter displays 12.5 volts or more, battery is OK.
- If voltage is below 12.5 volts, continue with tests.

Static current draw, checking

– Make sure ignition and all electrical accessories are switched OFF.

– Disconnect negative (–) cable from battery.

BentleyPublishers.com
BentleyPublishers.com—All Rights Reserved

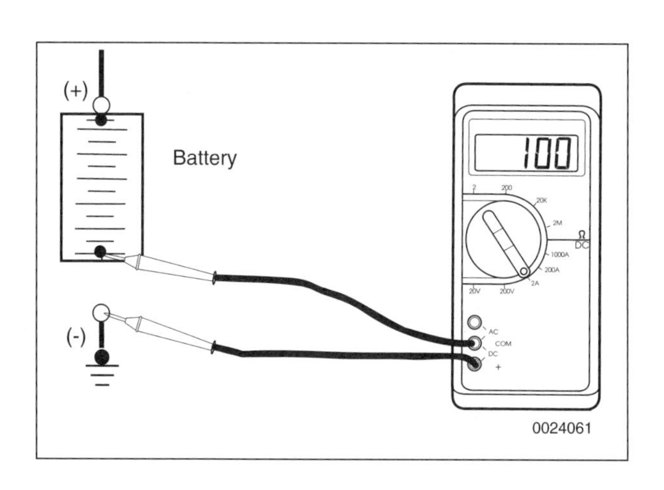

Connect a digital ammeter between battery negative post and negative battery cable and measure current draw.

Static current draw parameters	
Normal	0 - 100 mA
Too high	> 500 mA

– If current draw is 500 mA (0.5 amp) or higher, remove one fuse at a time until the current drops to a normal range. See **EWD Electrical Wiring Diagrams** to help locate wiring or component faults.

Open circuit voltage test

– Load battery with 15 amps for one minute: Switch headlights ON with engine OFF.

– Turn headlights OFF and disconnect battery ground cable.

– Connect accurate digital voltmeter to battery posts and check battery voltage.

– See **Table b** for open-circuit voltage levels and state of charge.

Table b. Open circuit voltage and battery charge	
Open circuit voltage	**State of charge**
12.6 volts or more	Fully charged
12.4 volts	75% charged
12.2 volts	50% charged
12.0 volts	25% charged
11.7 volts or less	Fully discharged

The battery may be in satisfactory condition if open-circuit voltage is at least 12.5 volts.

– If open-circuit voltage is at 12.5 volts or above, but battery still lacks power for starting, perform a load voltage test to determine battery service condition. See **Load voltage, testing** in this repair group.

– If open-circuit voltage is below 12.5 volts, recharge battery. After charging, if battery still fails open circuit voltage test, replace it.

Electrolyte specific gravity, testing

Checking the specific gravity of each battery cells can provide accurate information about the battery condition.

Test electrolyte specific gravity with a battery hydrometer. The hydrometer is a glass or plastic cylinder with a freely moving float inside. When electrolyte is drawn into the cylinder, the level to which the float sinks indicates the specific gravity of the electrolyte. The more dense the concentration of sulfuric acid in the electrolyte, the less the float sinks, resulting in a higher reading and indicating a higher state of charge.

Note that electrolyte temperature affects hydrometer reading. Check electrolyte temperature with a thermometer. Add 0.004 to hydrometer reading for every 6°C (10°F) that electrolyte is above

BentleyPublishers.com
BentleyPublishers.com—All Rights Reserved

27°C (80°F). Subtract 0.004 from reading for every 6°C (10°F) that electrolyte is below 27°C (80°F).Make sure that battery electrolyte temperature is at least 10°C (50°F).

– Load battery with 15 amps for one minute: Switch headlights ON with engine OFF.

– Turn headlights OFF. Remove battery filler caps.

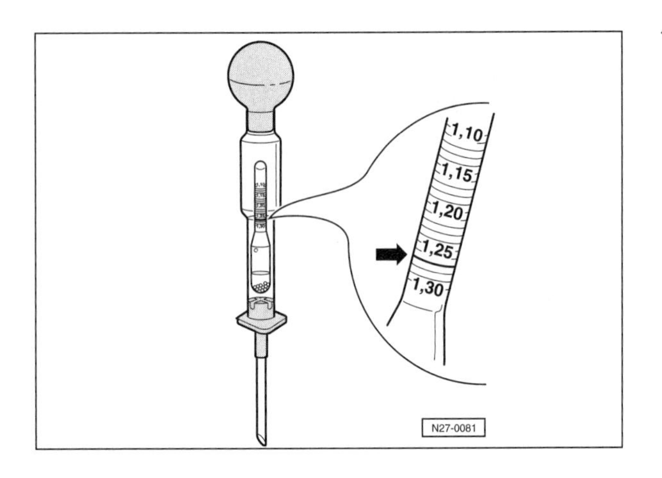

◄ Immerse hydrometer in a cell and extract sufficient electrolyte so that indicator floats free in electrolyte. Read specific gravity and compare to specifications in **Table c**.

Table c. Specific gravity of battery electrolyte at 27°C (80°F)	
Specific gravity	**Charge condition**
1.265	Fully charged
1.225	75% charged
1.190	50% charged
1.155	25% charged
1.120	Fully discharged

– If the average specific gravity of the six cells is 1.225 or higher, the battery is in satisfactory condition.

– If the average specific gravity is 1.225 or higher, but the battery lacks power for starting, determine the battery's service condition with a load voltage test. See **Load voltage, testing** in this repair group.

– If electrolyte density is below 1.225, charge battery. After recharging, if specific gravity varies by more than 0.005 between any two cells, replace battery.

– Install filler caps with O-rings and wipe up spilled electrolyte.

Load voltage, testing

A load voltage battery test is made by connecting a specific resistive load to the battery terminals and then measuring battery voltage.

– Make sure battery is fully charged and at room temperature.

– Disconnect negative (–) battery cable.

– Connect battery load tester and apply specified load (**Table d**, column 3) for 15 seconds and read off battery voltage (**Table d**, column 4). If voltage is below specification, or minimum voltage does not stay at a steady reading, replace battery.

Table d. Battery load current and minimum voltage			
Battery capacity (Ah)	**Cold cranking amps (CCA)**	**Load current (amps)**	**Minimum voltage**
36 Ah	340	100	10.0
40 Ah-49 Ah	220	200	9.2
50 Ah-60 Ah	265-280	200	9.4
61 Ah-80 Ah	300-380	300	9.0
81 Ah-110 Ah	380-500	300	9.5

BentleyPublishers.com
BentleyPublishers.com—All Rights Reserved

Battery, charging

– Recharge a discharged battery using a battery charger. Read and follow the instructions provided by the battery charger's manufacturer.

– Use slow-charging rate (10% of battery capacity) to prevent battery damage caused by overheating.

> ***CAUTION—***
> - *Prolonged battery charging evaporates the electrolyte to a level that can damage the battery.*
> - *Loosen battery cell caps before charging battery but leave them on battery to prevent battery acid splatter.*

Battery, disconnecting and connecting

Make sure the vehicle electrical system is protected by disconnecting the battery negative (–) terminal prior to working on any electrical system components.

> ***CAUTION—***
> - *Do not loosen or remove ground strap from body. Disconnect terminal from battery only.*

 Open engine hood, release battery cover latches (**arrows**) and flip cover up to gain access to battery.

Unbolt and disconnect battery negative (-) cable (**arrow**). Position cable to side where it cannot come in contact with battery negative (-) terminal.

– When reconnecting negative (-) battery terminal, make sure positive (+) terminal is connected first.

Tightening torque	
Battery cable to battery (M6)	9 Nm (80 in-lb)

– After connecting battery terminal:
- Switch ignition ON, then OFF.
- Check for diagnostic trouble codes (DTCs) using VAG scan tool or equivalent.
- Reset electrical consumers. See **Battery reconnection notes** in this repair group.
- Check function of all electrical consumers.

BentleyPublishers.com
BentleyPublishers.com—All Rights Reserved

Battery, removing and installing

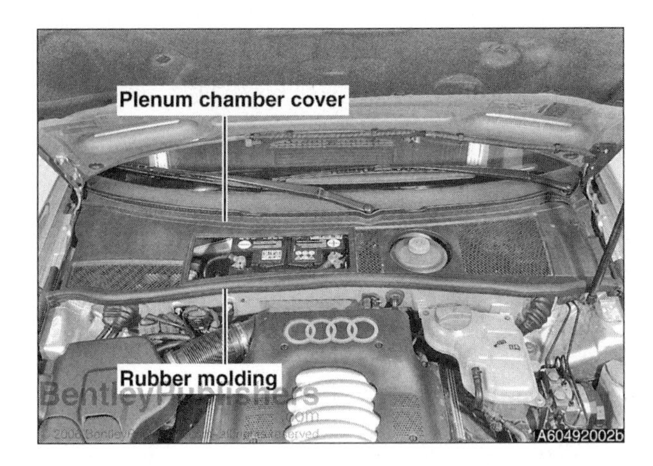

Open engine hood and pull off rubber molding at engine compartment rear bulkhead. Pull plenum chamber cover forward and remove.

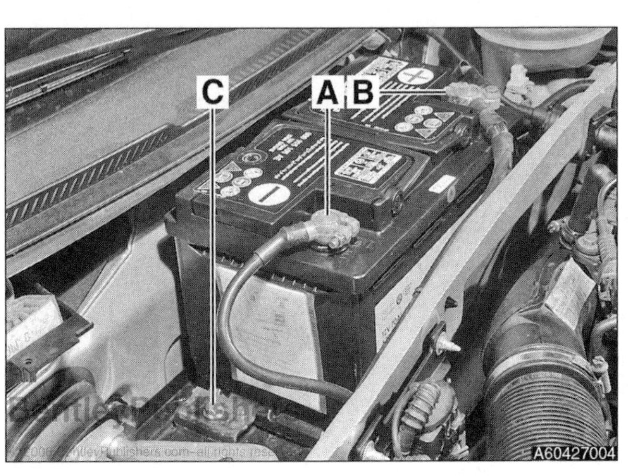

Working at battery:

- Disconnect negative (-) cable (**A**).
- Disconnect positive (+) cable (**B**).
- Remove battery hold-down bracket bolt (**C**).
- Remove hold-down bracket and lift battery out.

> ***CAUTION—***
> - *Disconnect negative (-) cable first and reconnect it last.*

– Installation is reverse of removal. Secure battery firmly, but do not overtighten hold-down bracket bolt.

Tightening torque	
Battery cable to battery (M6)	9 Nm (80 in-lb)
Battery hold-down to plenum chamber (M8)	15 Nm (11 ft-lb)

– Make sure battery venting hoses (if present) are routed correctly to side.

– After connecting battery terminals:

- Switch ignition ON, then OFF.
- Check for diagnostic trouble codes (DTCs) using VAG scan tool or equivalent.
- Reset electrical consumers. See **Battery reconnection notes** in this repair group.
- Check function of all electrical consumers.

STARTER

Starter troubleshooting

A factory-installed anti-theft immobilizer is used on Audi A6 models. This system prevents operation of the starter if a specially coded ignition key is not used. See **96 Interior Lights, Switches, Anti-theft**.

BentleyPublishers.com
BentleyPublishers.com—All Rights Reserved

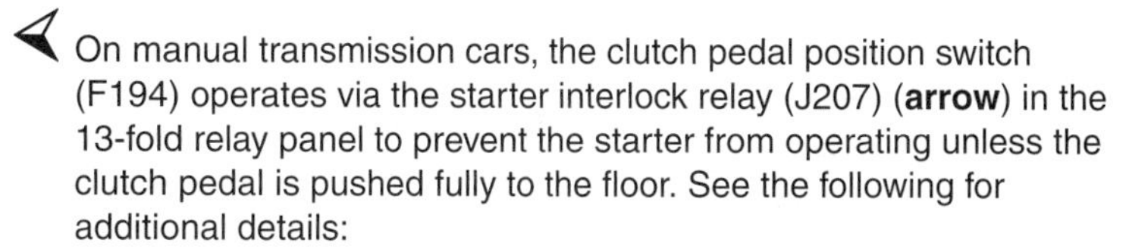

On manual transmission cars, the clutch pedal position switch (F194) operates via the starter interlock relay (J207) (**arrow**) in the 13-fold relay panel to prevent the starter from operating unless the clutch pedal is pushed fully to the floor. See the following for additional details:

- **30 Clutch** for clutch pedal position switch location
- **97 Fuses, Relays, Component Locations** for 13-fold relay panel access information

On automatic transmission cars, the transmission multifunction range switch (F125) signals the immobilizer to prevent the engine from starting in gear positions other than PARK or NEUTRAL.

In models with multitronic (CVT) transmission, the transmission multifunction range switch (F125) signals the park / neutral position (PNP) relay (J226) (**arrow**) in the 13-fold relay panel when the transmission is in PARK or NEUTRAL. The PNP relay then signals the electronic immobilizer to allow the starter to operate. See **97 Fuses, Relays, Component Locations** for 13-fold relay panel access information.

On back of starter, the large wire at terminal 30 is direct battery voltage. The smaller wire at terminal 50 operates the starter solenoid via ignition switch.

– If starter turns engine slowly when ignition is in START position:
  - Check battery state of charge. See **Load voltage, testing** in this repair group.
  - Inspect starter wires, terminals and ground connections for good contact. In particular, make sure ground connections between battery, body and engine are completely clean and tight.
  - If no faults are found, starter may be faulty.

– If starter fails to operate:
  - Check electronic anti-theft immobilizer. Try another ignition key. If no faults can be found, have the immobilizer system checked using VAG scan tool or equivalent.
  - Check clutch pedal operated starter interlock switch or transmission range switch (automatic transmission).

– Check for battery voltage at terminal 50 of starter motor with key in START position.
  - If voltage is not present, check wiring between ignition switch and starter terminal. Check immobilizer system and other inputs that disrupt input to starter. See **EWD Electrical Wiring Diagrams**.

– If voltage is present and no other visible wiring faults can be found, problem is most likely in starter motor.

Starter, removing and installing (V6 engine)

– Disconnect battery ground cable. See **Battery, disconnecting and connecting** in this repair group.

– 2.7 liter turbo engine: Remove alternator. See **Alternator, removing and installing** in this repair group.

BentleyPublishers.com
BentleyPublishers.com—All Rights Reserved

– Raise car and support safely.

> ***WARNING —***
> - *Make sure the car is stable and well supported at all times. Use a professional automotive lift or jack stands designed for the purpose. A floor jack is not adequate support.*

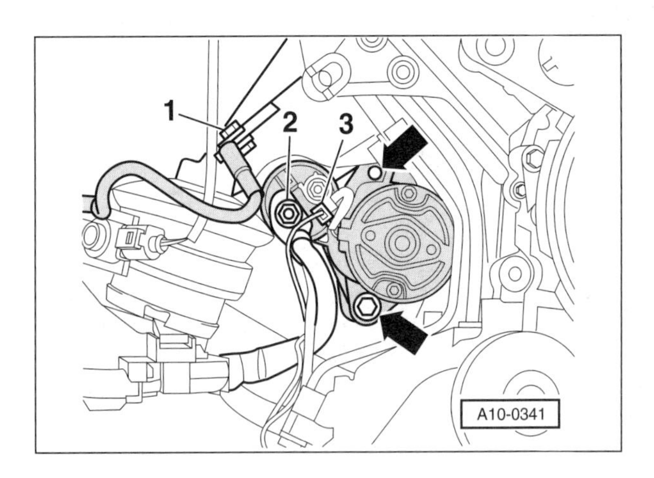

◄ Working at back of starter motor under right cylinder bank:

- Disconnect ground cable (**1**) and starter and solenoid wires (**2**, **3**).
- Working at transmission side of bellhousing, remove starter mounting bolts (**arrows**).
- Lift out starter.

– Installation is reverse of removal.

Tightening torques	
Starter to engine	65 Nm (48 ft-lb)
Starter cable to terminal B+ on starter (M8)	16 Nm (12 ft-lb)

– After connecting battery terminals, reset electrical consumers. See **Battery reconnection notes** in this repair group.

Starter, removing and installing (V8 engine)

– Disconnect battery ground cable. See **Battery, disconnecting and connecting** in this repair group.

– Raise car and support safely.

> ***WARNING —***
> - *Make sure the car is stable and well supported at all times. Use a professional automotive lift or jack stands designed for the purpose. A floor jack is not adequate support.*

– Remove lower engine cover (splash shield). See **03 Maintenance**.

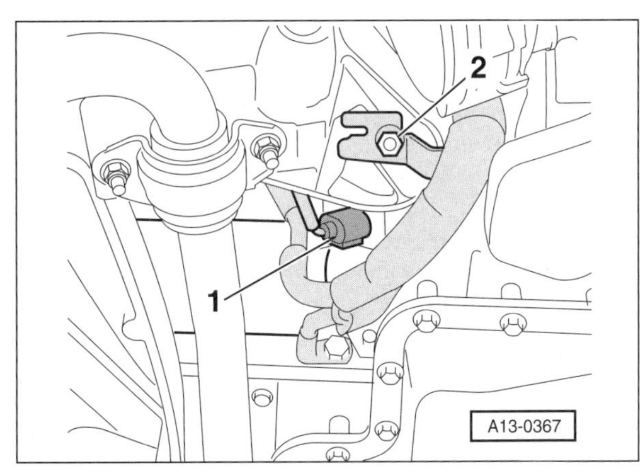

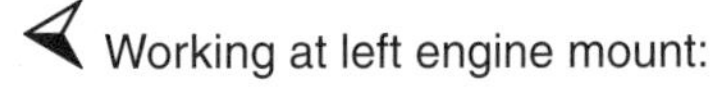

◄ Working at left engine mount:

- Detach engine mount electrical connector (**1**)
- Detach cable retainer (**2**) from engine mount.

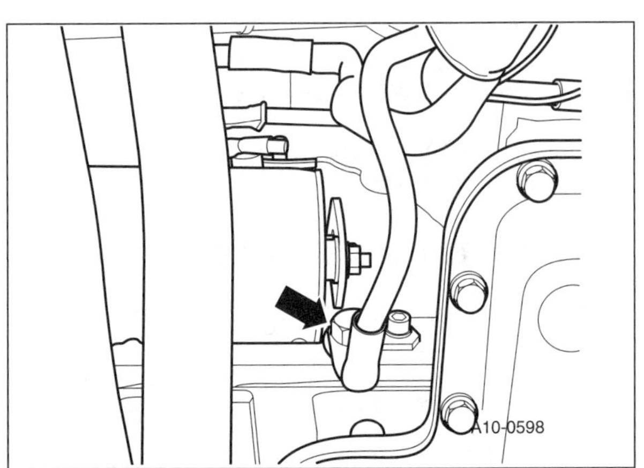

◄ Detach ground cable (**arrow**) at starter housing.

BentleyPublishers.com
BentleyPublishers.com—All Rights Reserved

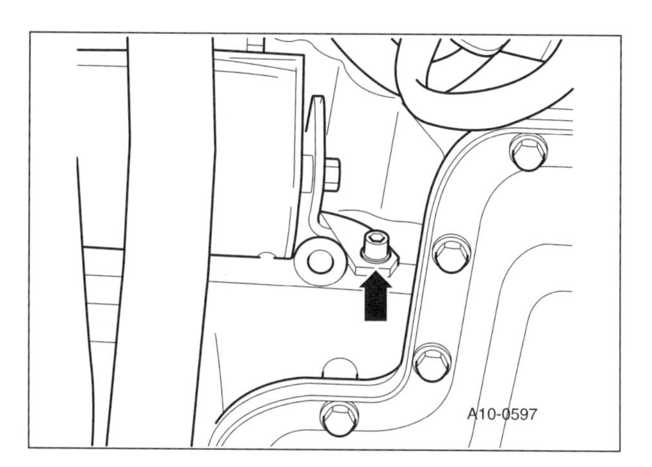

Working at rear of starter motor:

- Remove starter bracket mounting bolt (**arrow**).
- Detach starter and solenoid wires.

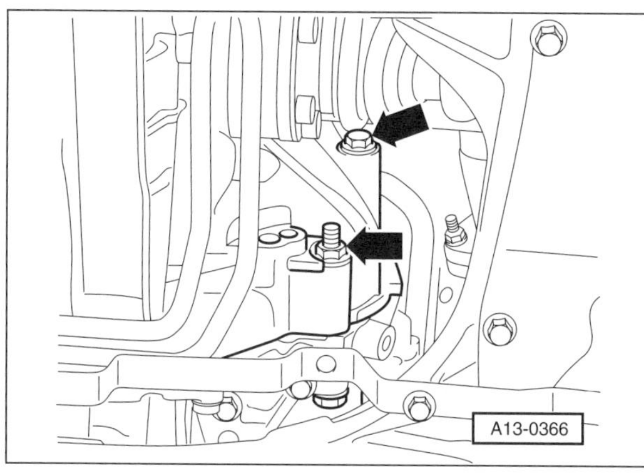

Remove starter mounting bolts (**arrows**).

– Lower starter while turning clockwise.

– Installation is reverse of removal.

Tightening torques	
Cable retainer to engine mount	22 Nm (16 ft-lb)
Starter to engine	65 Nm (48 ft-lb)
Starter bracket to engine	10 Nm (7 ft-lb)
Starter cable to terminal B+ on starter (M8)	16 Nm (12 ft-lb)

– After connecting battery terminals, reset electrical consumers. See **Battery reconnection notes** in this repair group.

ALTERNATOR

Charging system quick check

– Use a digital multimeter to measure voltage across the battery terminals with key OFF and then again with engine running. Compare results to **Table e**.

Table e. Charging system quick check results

Condition	Battery voltage	Probable cause
Key OFF, engine not running	12.6	Battery and charging system normal
	<12.4	Battery discharged
		Battery faulty
		Charging system faulty
Engine running	13.5 - 14.5	Charging system normal
	<13.2	Battery discharged, requires charging
		Charging system faulty
	> 14.8	Charging system faulty

BentleyPublishers.com

BentleyPublishers.com—All Rights Reserved

Alternator, removing and installing

- Disconnect battery ground cable. See **Battery, disconnecting and connecting** in this repair group.
- Raise car and support safely.

> ***WARNING —***
> - *Make sure the car is stable and well supported at all times. Use a professional automotive lift or jack stands designed for the purpose. A floor jack is not adequate support.*

- Remove lower engine cover (splash shield). See **03 Maintenance**.

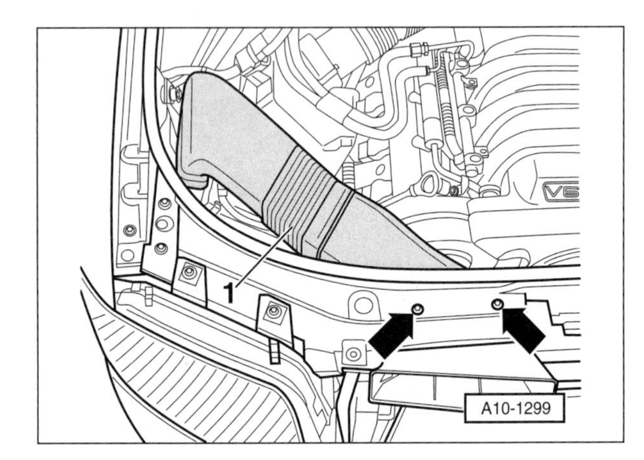

Remove air cleaner duct mounting screws (**arrows**) at lock carrier. Remove air duct (**1**).

- Place lock carrier in service position. See **50 Body–Front**.

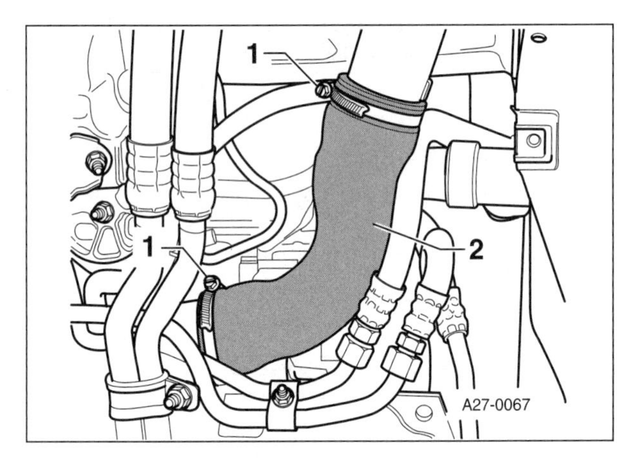

2.7 liter turbocharged engine: Loosen right intercooler hose clamps (**1**) and remove hose (**2**).

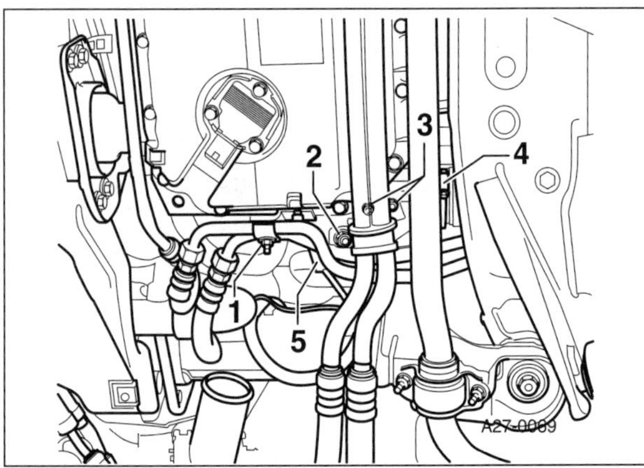

Automatic transmission:

- Loosen ATF lines by loosening bolt (**1**).
- Loosen AC lines by loosening bolt (**2**).
- Remove line (**5**) by removing bolts (**3**) and hose clamp (**4**).

BentleyPublishers.com
BentleyPublishers.com—All Rights Reserved

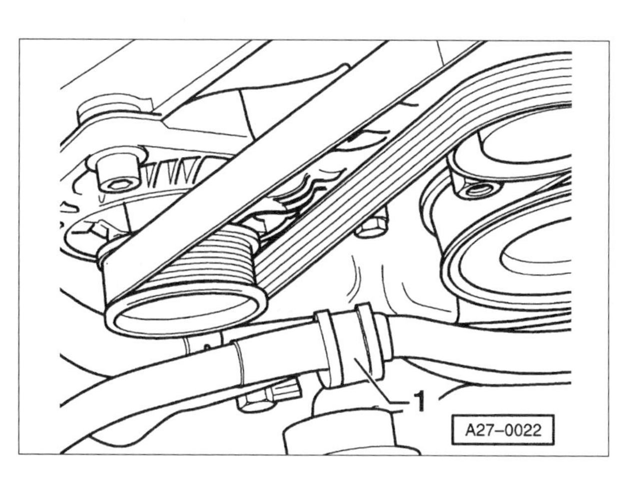

Manual transmission: Remove A/C refrigerant lines mounting bracket (**1**) above torque arm.

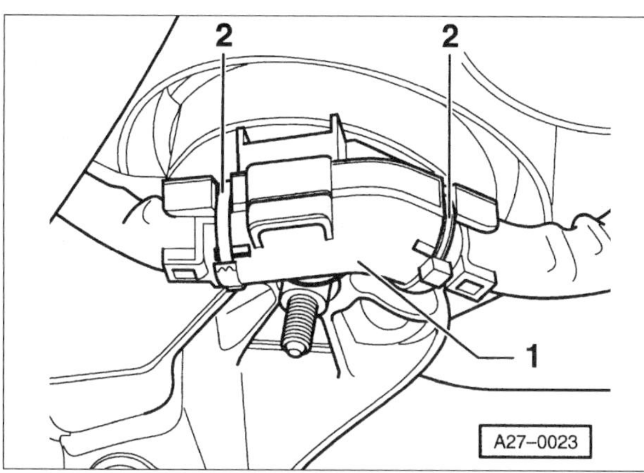

Cut wire-ties (**2**) and unlatch tabs.

- Unhook starter and alternator wires from bracket (**1**).
- Remove engine accessory belt. See **03 Maintenance**. If belt is to be reused, mark direction of rotation prior to removal.

Working under right front of engine, loosen clamp (**arrow**) and remove cooling air duct from alternator.

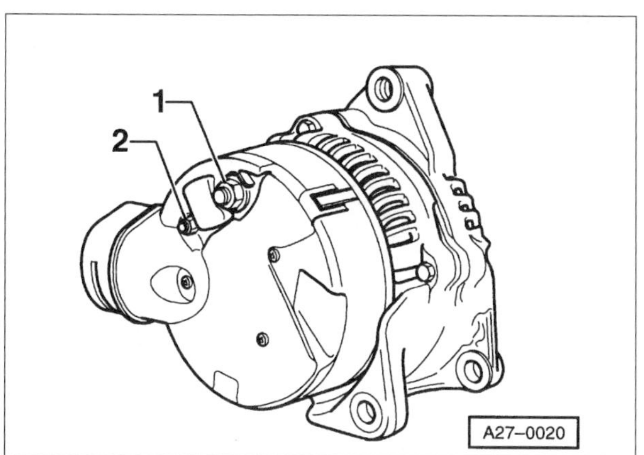

Working at rear of alternator, detach connectors from alternator posts 30 / B+ and D+ (**1**, **2**).

BentleyPublishers.com
BentleyPublishers.com—All Rights Reserved

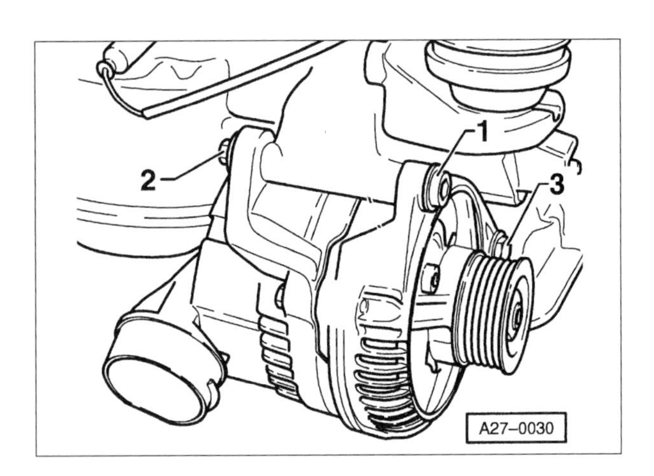

Loosen and remove alternator mounting fasteners:

- Counterhold nut (**2**) to remove top bolt (**1**).
- Loosen and remove lower bolt (**3**).

– Remove alternator from below, taking care to move A/C refrigerant lines and other hoses and wires aside very carefully.

> ***CAUTION—***
> *• Do not bend or kink A/C refrigerant lines.*

– Installation is reverse of removal.

Tightening torques	
Alternator to engine block: • Top bolt (M10) • Lower bolt (M8)	 45 Nm (33 ft lb) 22 Nm (16 ft-lb)
Connector to alternator posts: • 30 / B+ • D+	 16 Nm (12 ft lb) 4 Nm (35 in-lb)

– If reusing accessory belt, install in direction of rotation marked previously.

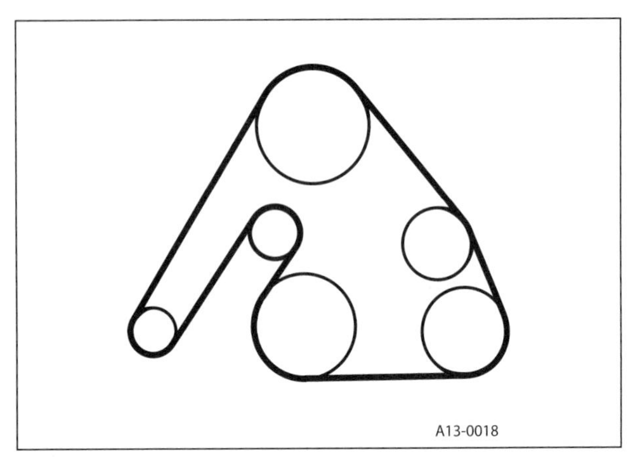

V6 belt routing.

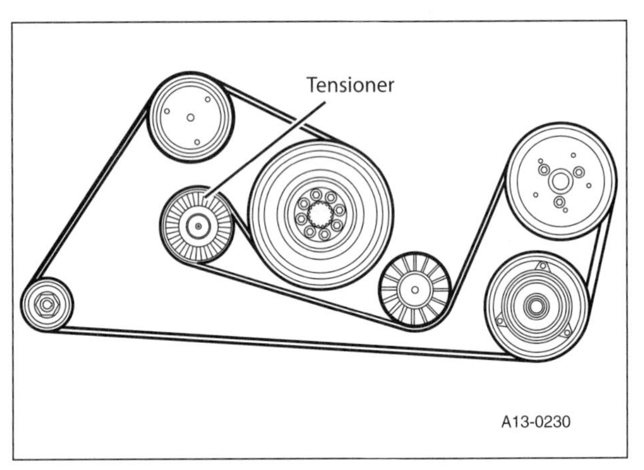

V8 belt routing (*not* allroad quattro).

– Check belt ribs to make sure they fit pulleys correctly.

BentleyPublishers.com
BentleyPublishers.com—All Rights Reserved

28 Ignition System

GENERAL

This repair group provides ignition system information.

See also:

- **03 Maintenance** for spark plug replacement
- **24 Fuel Injection** for Motronic system applications
- **96 Interior Lights, Switches, Anti-theft** for ignition switch replacement
- **EWD Electrical Wiring Diagrams**

Ignition management

Audi A6 ignition is controlled by the Motronic engine management system, which also controls fuel injection and emission control functions.

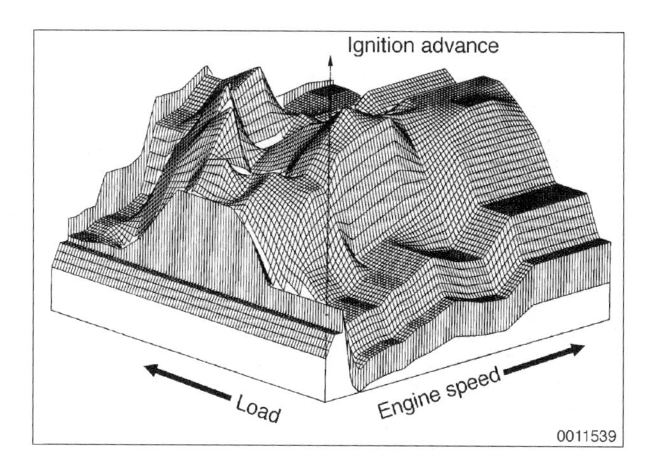

Ignition timing is electronically mapped and not adjustable. A three-dimensional map similar to the illustration is stored in the Motronic engine control module (ECM).

In applying the map, the ECM computes ignition timing based on inputs from the following sensors, in order of importance:

- **Crankshaft position sensor** for timing reference point.
- **Mass airflow sensor** for engine load.
- **Camshaft sensors** for relative camshaft positions.
- **Knock sensors** to modify and correct ignition timing based on changing operating conditions.

BentleyPublishers.com
BentleyPublishers.com—All Rights Reserved

Ignition firing order

V6 engine: Cylinder 1 is at the right front of the engine, followed by cylinders 2 and 3. Cylinders 4, 5 and 6 are on the left side of the engine, from front to back.

V8 engine: Cylinder 1 is at the right front of the engine, followed by cylinders 2, 3 and 4. Cylinders 5, 6, 7 and 8 are on the left side of the engine, from front to back.

Ignition firing order	
V6 engine	1-4-3-6-2-5
V8 engine	1-5-4-8-6-3-7-2

Disabling ignition system

The ignition system operates in a lethal voltage range and also contains sensitive electronic components. To guard against system damage, and for personal and general safety, carry out ignition system service and repair work carefully.

– If the engine must be cranked without starting:
  - Disable fuel system to prevent excess fuel from flooding cylinders.
  - Disable ignition system to prevent discharge of dangerously high voltage.

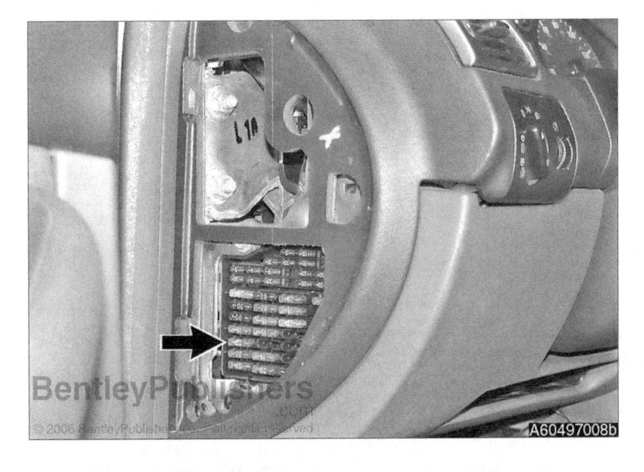

◀ Remove fuel pump fuse (fuse 28, **arrow**) in dashboard fuse panel.

– Remove engine upper cover(s). See **03 Maintenance**.

◀ **2.8 liter engine**: Disconnect ignition coil connector.

BentleyPublishers.com
BentleyPublishers.com—All Rights Reserved

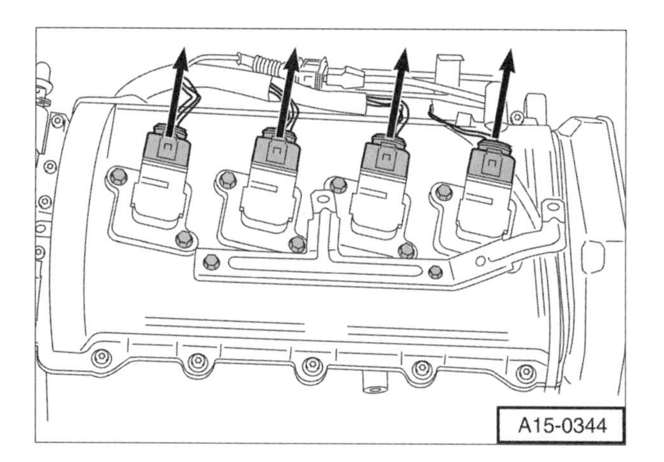

– **2.7 liter, 3.0 liter, V8 engines**: Pull off (**arrows**) each individual ignition coil connector.

On-board diagnostics

Most components and functions of the Motronic system are monitored by on-board diagnostics (OBD II) software in the engine control module (ECM). If faults and malfunctions are detected, one or more diagnostic trouble codes (DTCs) may be set.

In case of an emissions related fault or a malfunction considered catalyst-damaging, the Check Engine light (malfunction indicator light or MIL) or EPC light is illuminated.

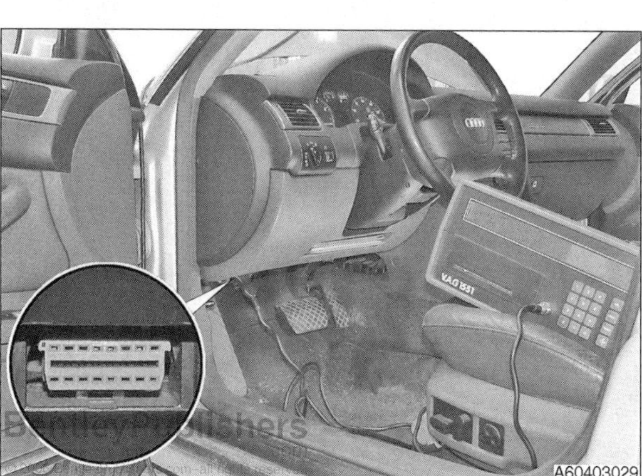

Access fault memory with VAG scan tool or equivalent plugged into the DLC plug (**inset**) under left side dashboard. See **OBD On-Board Diagnostics** for more information.

NOTE—

- *After engine management system tests or repairs, the ECM may recognize a malfunction and store a DTC. Therefore perform the following after ending tests and repairs:*
  -Use VAG scan tool or equivalent to check DTC memory.
  -Erase DTC memory.
  -Reset or erase diagnostic data.
  -Generate readiness code.
- *After engine management system tests or repairs, the engine may start, run for a short period and then cut out because the electronic anti-theft immobilizer disables the ECM. In such cases:*
  -Use VAG scan tool or equivalent to check DTC memory.
  -Adapt ECM.

BentleyPublishers.com
BentleyPublishers.com—All Rights Reserved

Warnings and Cautions

WARNING —
- *Do not touch or disconnect any cables from the coils while the engine is running or being cranked by the starter.*
- *The ignition system produces high voltages that can be fatal. Avoid contact with exposed terminals. Use extreme caution when working on a car with the ignition switched on or the engine running.*
- *Before operating the starter without starting the engine (for example when making a compression test) disable the ignition. See* **Disabling ignition system** *in this repair group.*

CAUTION—
- *Prior to disconnecting the battery, read the battery disconnection precautions given at the front of this manual.*
- *Use a high impedance digital multimeter for voltage and resistance tests.*
- *Use an LED test light in place of an incandescent-type test light.*
- *Do not attempt to disable the ignition by removing the coils from the spark plugs.*
- *Connect and disconnect the ECM, Motronic system wiring or test equipment leads only when the ignition is OFF.*
- *Switch multimeter functions or measurement ranges only with the test probes disconnected.*
- *Do not disconnect the battery while the engine is running.*

IGNITION SYSTEM COMPONENTS

Engine management component location illustrations are in **24 Fuel Injection**. Components specifically associated with ignition are covered below.

Spark plugs

See **03 Maintenance** for spark plug applications and replacement procedure.

Ignition coils

2.8 liter engine: Ignition coils and power stage are ganged at front of engine, under upper engine plastic cover. There is no distributor, distributor cap or ignition rotor.

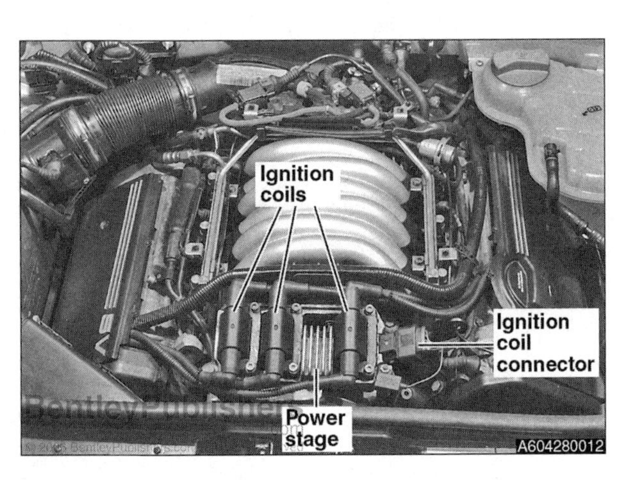

BentleyPublishers.com
BentleyPublishers.com—All Rights Reserved

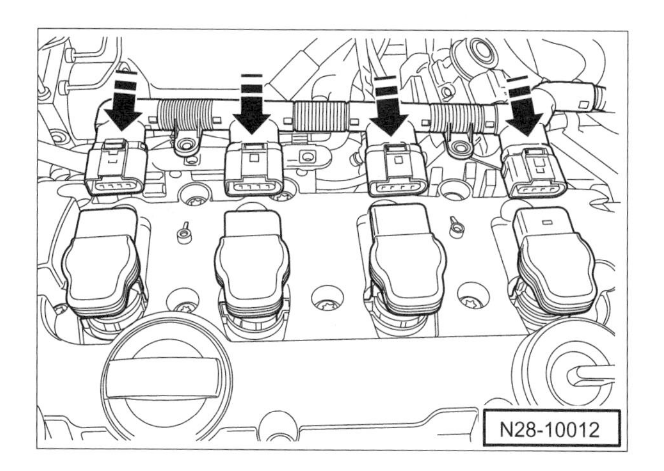

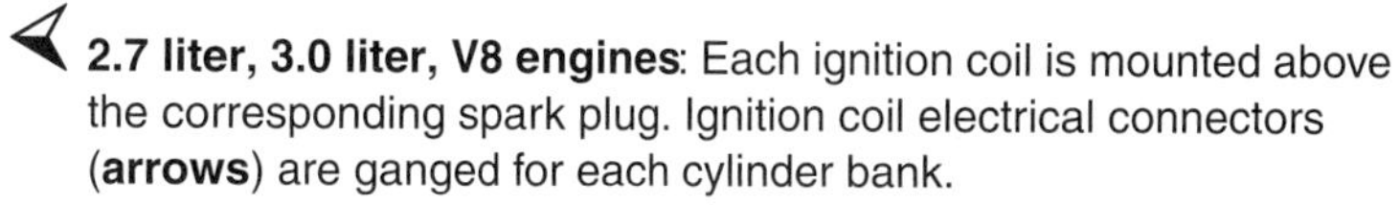

2.7 liter, 3.0 liter, V8 engines: Each ignition coil is mounted above the corresponding spark plug. Ignition coil electrical connectors (**arrows**) are ganged for each cylinder bank.

2.7 liter engine, code APB: Power stages are separate components on air filter housing.

2.7 liter engine, code BEL, 3.0 liter engine, V8 engines: Power stages integrated with ignition coils.

There is no distributor, distributor cap or ignition rotor.

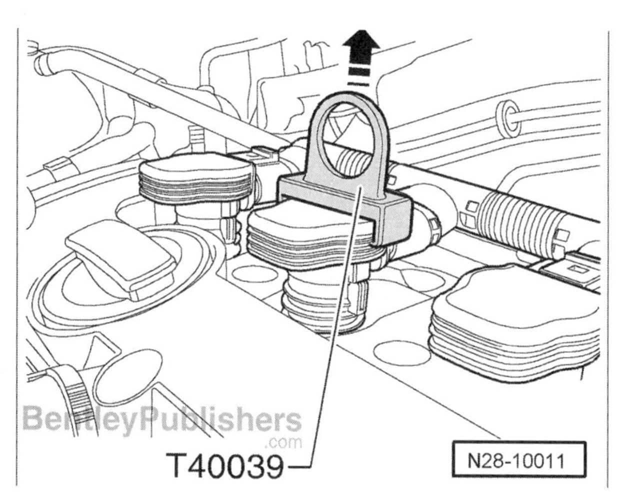

2.7 liter, 3.0 liter, V8 engines: To remove ignition coils, use Audi special tool **T40039**, pull all ignition coils out of spark plug shaft.

– Installation is reverse of removal.
  - Insert all ignition coils loosely into spark plug shaft.
  - Align ignition coils to the connectors and then connect all of the connectors onto ignition coils.
  - Press ignition coils uniformly onto the spark plugs by hand (do not use an impact tool).

Spark plug wires

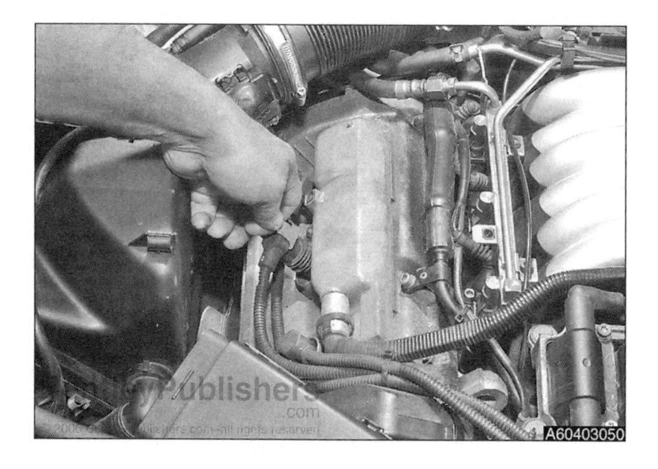

2.8 liter engine: Plug wires are routed from central coil pack to each spark plug.

2.7 liter, 3.0 liter, V8 engines: Spark plug wire for each cylinder is integrated with corresponding ignition coil.

BentleyPublishers.com
BentleyPublishers.com—All Rights Reserved

Camshaft sensors

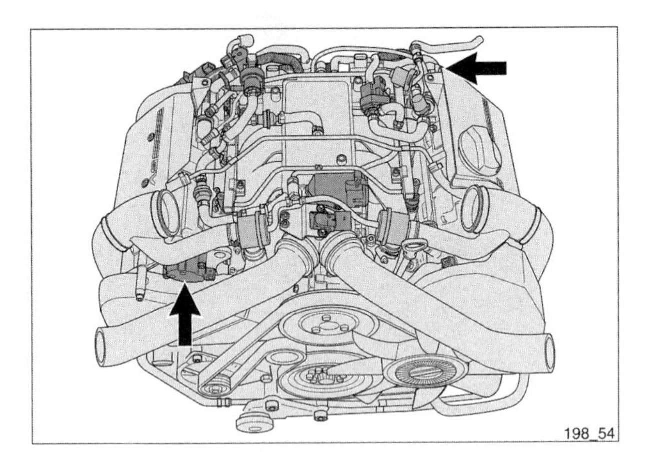

Camshaft sensors (**arrows**) on each camshaft detect camshaft position. This input is used by the ECM to determine when to fire each spark plug.

2.7 liter turbocharged V6 engine illustrated. Other engines are similar.

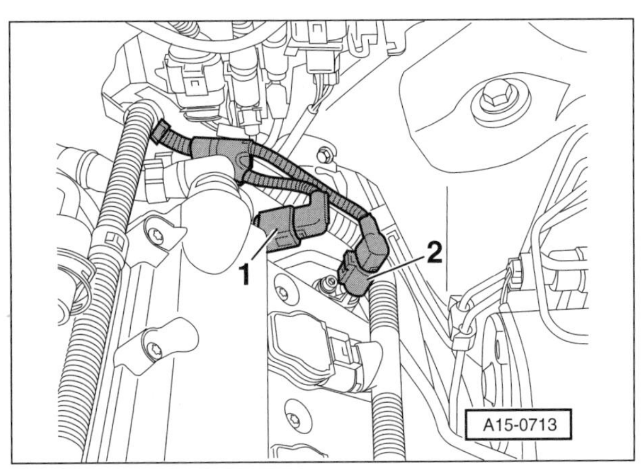

3.0 liter engine: Two camshaft sensors (**1**, **2**) per bank are used.

Knock sensors

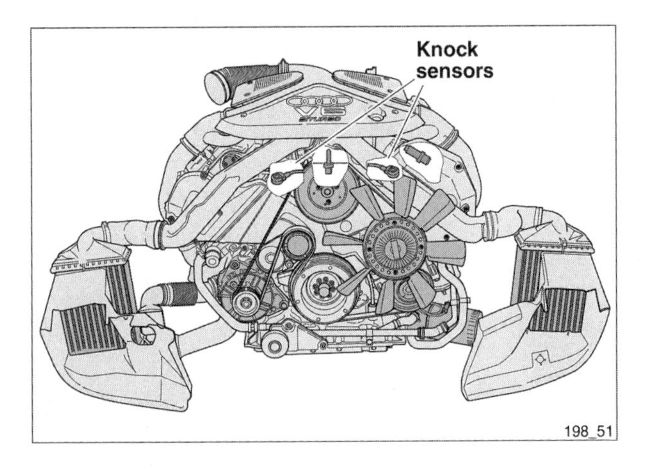

Two or three knock sensors monitor engine knock. The ECM uses signal from knock sensors to optimize ignition timing.

Knock sensors are located under the intake manifold. 2.7 liter turbocharged V6 engine is illustrated. Other engines are similar.

BentleyPublishers.com
BentleyPublishers.com—All Rights Reserved

30 Clutch

GENERAL

This repair group covers clutch hydraulic and mechanical repairs.

Service to the clutch assembly requires removal of the transmission from the vehicle. Special tools and equipment are required to remove the transmission, service the clutch, and bleed the clutch hydraulic system. See **34 Manual transmission** for transmission removal procedure.

Self-adjusting clutch pressure plate

A6 models equipped with 2.7 liter V6 bi-turbo engine feature a Self-Adjusting Clutch (SAC) pressure plate. The SAC clutch pressure plate automatically compensates for clutch disc wear to provide consistent clutch pedal feel. See **Clutch pressure plate (SAC), resetting** in this repair group.

CLUTCH PEDAL

The illustration on the following page shows clutch pedal assembly components for vehicles equipped with 5-speed and 6-speed manual transmissions. When servicing, replace self-locking nuts, circlips and gaskets

NOTE—

- *Audi identifies electrical components by a letter and/or a number in the electrical schematics. See* **EWD Electrical Wiring Diagrams**. *Where appropriate, these electrical identifiers are listed in parentheses as an aid to electrical troubleshooting.*

BentleyPublishers.com
BentleyPublishers.com—All Rights Reserved

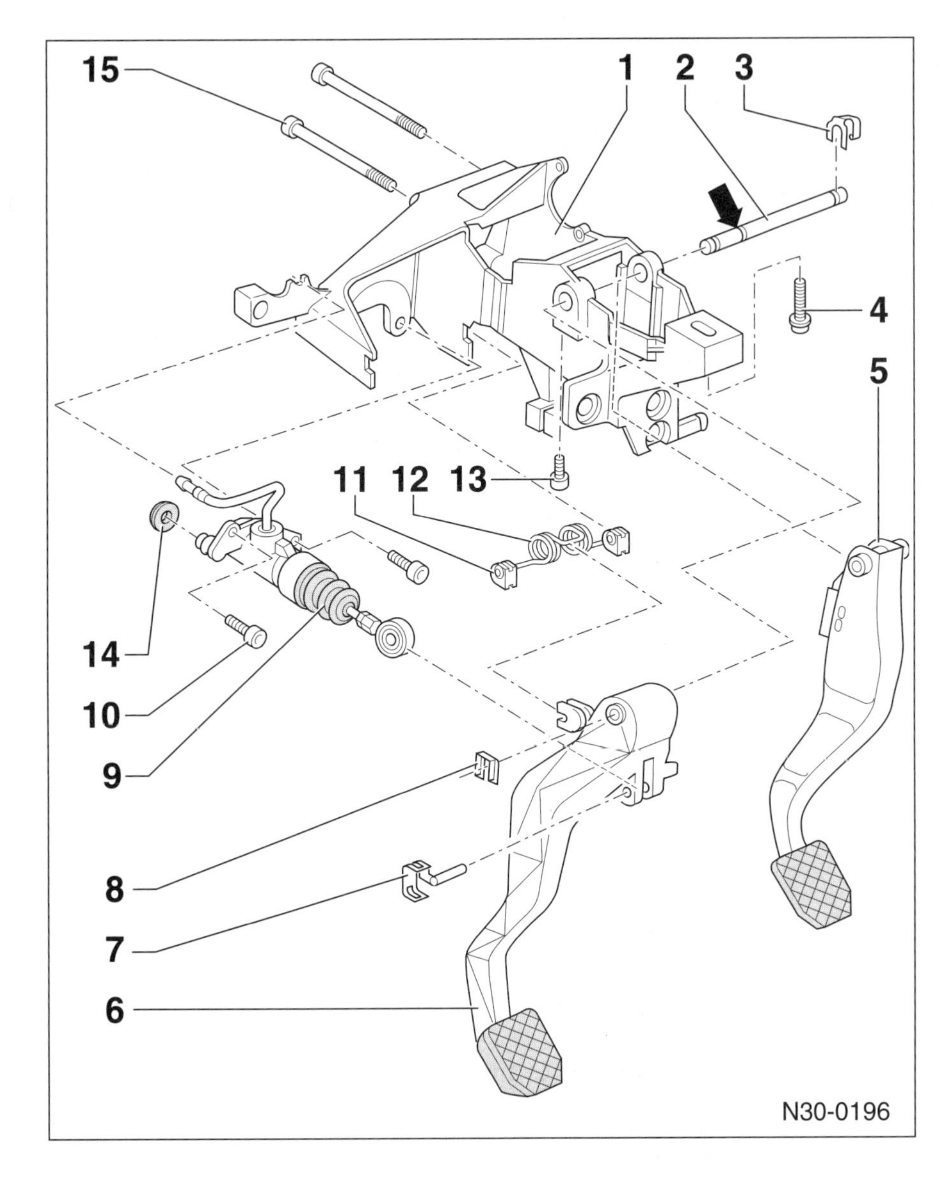

Clutch pedal assembly

1. **Mounting bracket**
2. **Pivot pin**
   - For clutch and brake pedals
   - Install with groove (**arrow**) toward clutch pedal
3. **Locking clip**
4. **Bolt, 25 Nm (18.5 ft-lb)**
5. **Brake pedal**
6. **Clutch pedal**
7. **Pin**
   - Clip to clutch pedal
8. **Locking clip**
9. **Clutch master cylinder**
10. **Bolt, 20 Nm (15 ft-lb)**
11. **Spring mount**
    - Insert in mounting bracket with over-center spring
12. **Over-center spring**
13. **Bolt, 5 Nm (4 ft-lb)**
    - For clutch and brake pedal pivot pin
14. **Seal**
15. **Torx bolt, 25 Nm (18.5 ft-lb)**

BentleyPublishers.com
BentleyPublishers.com—All Rights Reserved

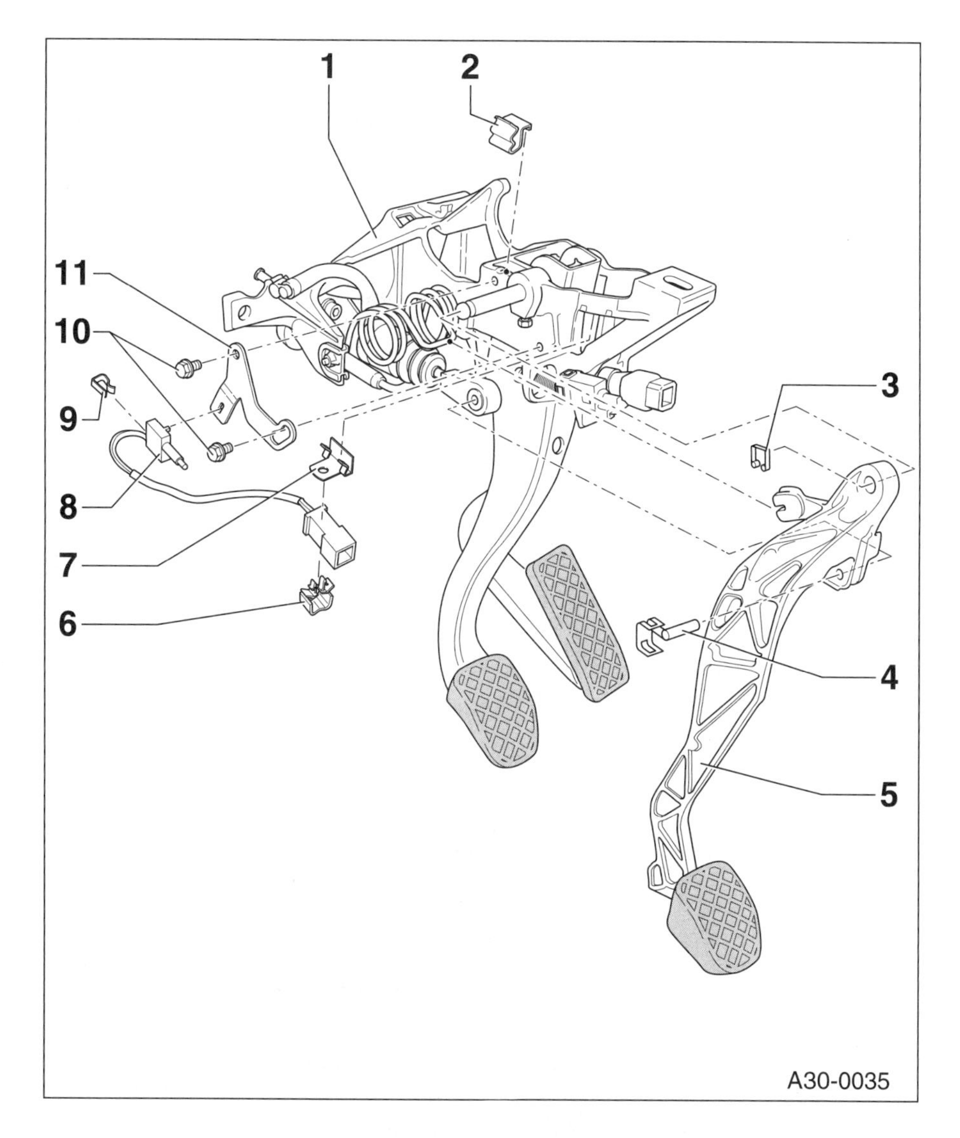

Clutch pedal position switch (F194)

1. **Mounting bracket**
2. **Cable bracket**
3. **Locking clip**
4. **Pin**
   - Clip onto clutch pedal
5. **Clutch pedal**
6. **Clip**
7. **Bracket for harness connector**
8. **Clutch pedal position switch (F194)**
   - See **Clutch pedal position switch, adjusting** in this repair group
9. **Clip**
10. **Screw and washer, 8 Nm (6 ft-lb)**
11. **Securing plate**

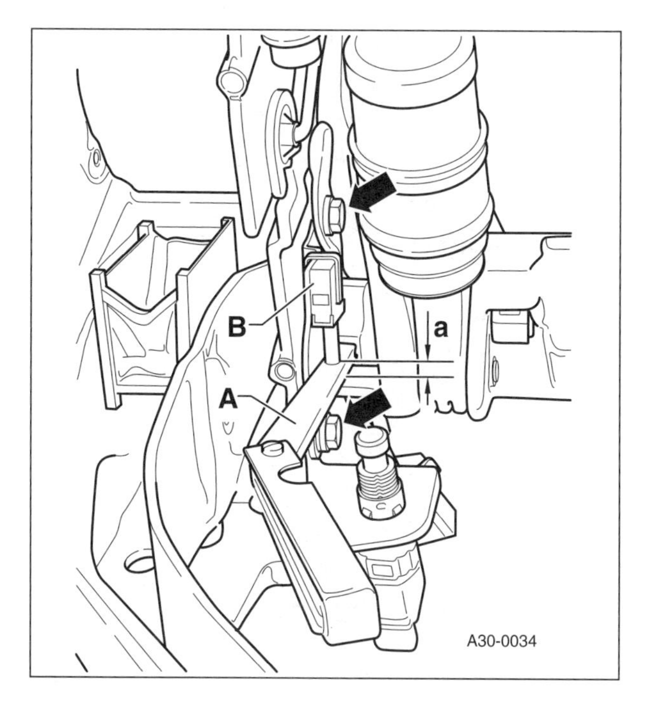

Clutch pedal position switch (F194), adjusting

The clutch pedal position switch (F194) prevents the engine from starting unless the clutch pedal is depressed.

◄ Working above the clutch pedal, loosen 2 bolts (**arrows**).

– Have assistant depress clutch pedal to stop.

– Place feeler gauge (**A**) with selected dimension **a** (3.2 ± 0.2 mm) between operating surface of clutch pedal and switch plunger.

– Swing switch (**B**) toward feeler gauge (**A**) and tighten two mounting bolts (**arrows**).

Tightening torque	
Clutch pedal position switch mounting bracket to pedal assembly bracket	8 Nm (6 ft-lb)

BentleyPublishers.com
BentleyPublishers.com—All Rights Reserved

Clutch Hydraulic System

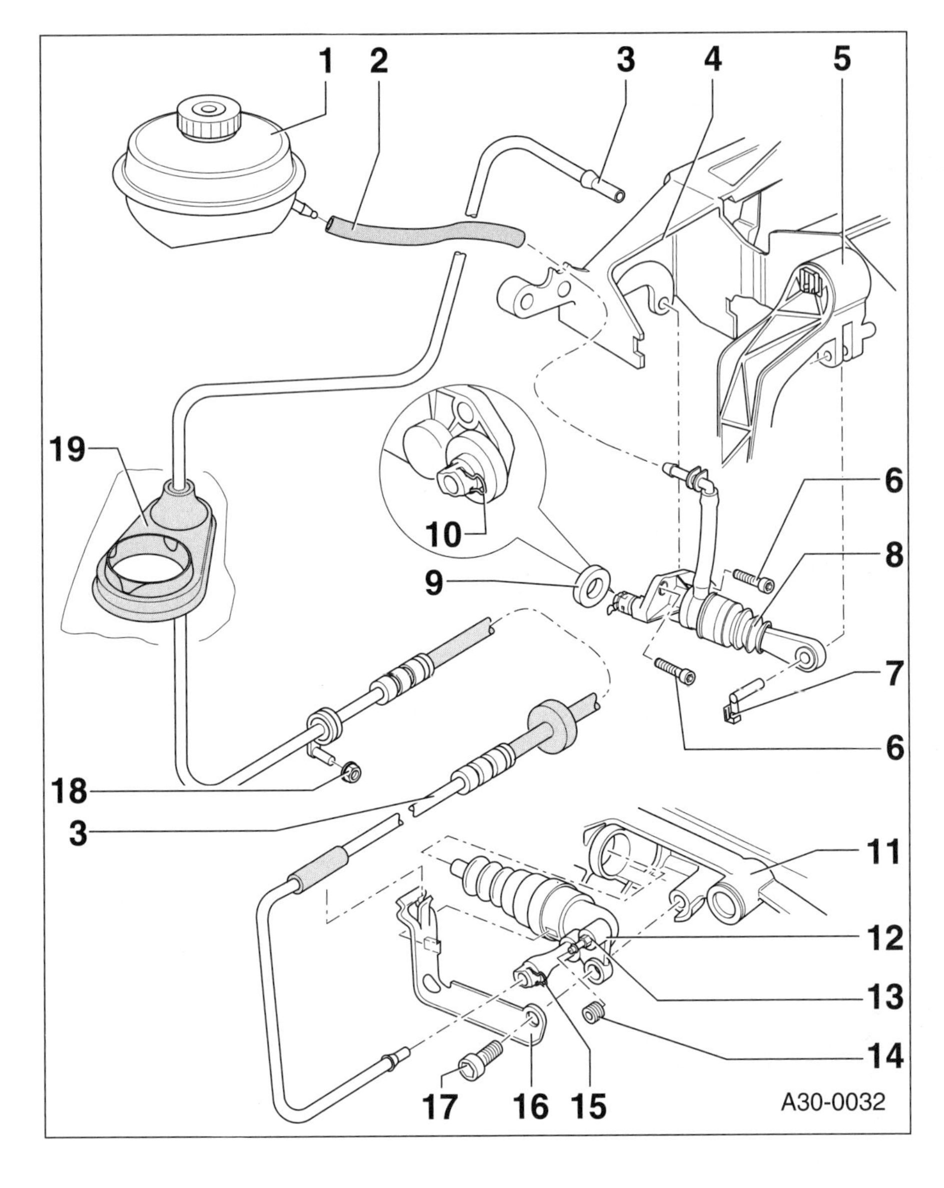

Clutch hydraulic components

1. **Brake fluid reservoir**
2. **Supply hose**
3. **Hose/pipe assembly**
4. **Mounting bracket**
5. **Clutch pedal**
6. **Bolt, 20 Nm (15 ft-lb)**
7. **Pin**
8. **Clutch master cylinder**
   - Do not operate after clutch pedal has been removed
9. **Seal**
10. **Retaining clip**
11. **Transmission**
12. **Clutch slave cylinder**
    - Do not operate after clutch pedal has been removed
    - Brake fluid must not be allowed to get onto the transmission
    - Lightly grease before installing
    - When installing push in until the securing bolt can be fitted.
13. **Bleeder valve/slave cylinder**
    - Tighten to 4.5 Nm (3.3 ft-lb)
    - A broken bleeder valve can be unscrewed with a 3 mm socket
14. **Dust cap**
15. **Retaining clip**
16. **Bracket**
    - Secured to transmission
17. **Bolt, 20 Nm (15 ft-lb)**
18. **Nut**
19. **Seal**

BentleyPublishers.com
BentleyPublishers.com—All Rights Reserved

Master cylinder, removing and installing

– Remove plenum chamber cover.

– Switch ignition OFF and disconnect battery ground (-) strap.

NOTE—

- *Make sure no brake fluid spills into the plenum chamber or on transmission.*
- *When working in footwell, protect carpet from brake fluid using rags.*

Clamp supply hose (**A**) from brake fluid reservoir using special pinch tool 3094. Pull hose off master cylinder and plug hose.

– Remove supply hose grommet from bulkhead.

– Lever out retaining clip (**C**) using a screwdriver and pull pipe (**B**) out slightly.

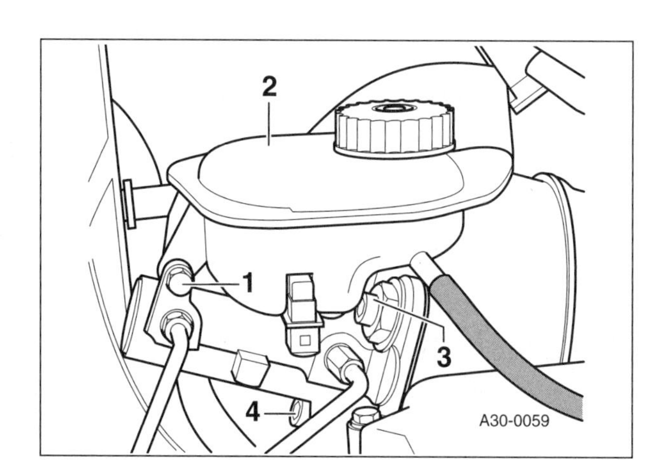

If equipped, remove bolt (**1**) and swing brake fluid container (**2**) to the right.

– Remove Torx bolts (**3** and **4**).

– Working under dash board, remove driver's storage compartment.

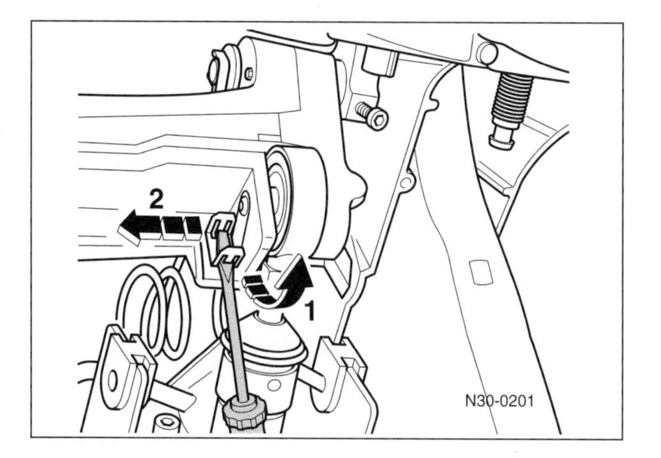

Detach clutch pedal from master cylinder. To do this, unclip pin using a screwdriver (**1**) pull out pin (**2**). Carefully press clutch pedal off.

NOTE—

- *Make sure clutch pedal does not press switch (**1**) out of the retaining clip. This damages the thread of the switch and the switch will have to be replaced.*
- *Switch must not be installed more than once.*

BentleyPublishers.com
BentleyPublishers.com—All Rights Reserved

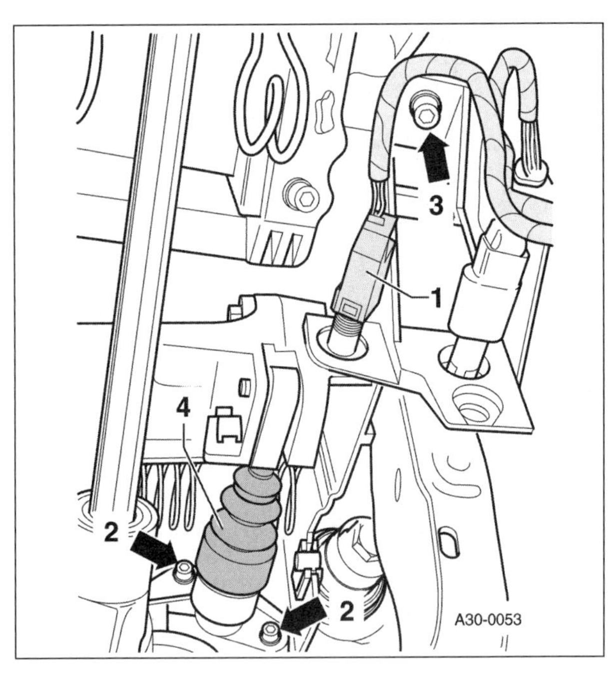

Remove bolts (**2**) for clutch master cylinder and for mounting bracket (**3**).

– Pull complete pedal assembly slightly to rear and remove clutch master cylinder (**4**).

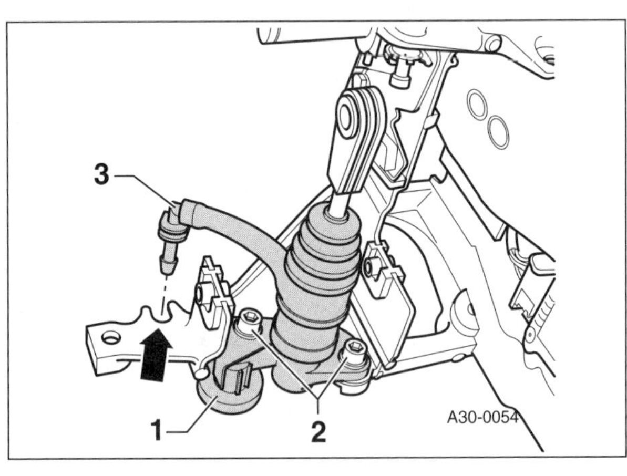

To install, insert clutch master cylinder and tighten bolts (**2**). Make sure seal (**1**) with washer contacts mounting bracket at rear.

Tightening torque	
Master cylinder to mounting bracket	20 Nm (15 ft-lb)

– Press connecting pipe (**3**) into recess (**B**) at mounting bracket.

Slide complete pedal assembly forward, insert bolt (**3**) and tighten by hand. At the same time, have an assistant guide pipe from plenum chamber into clutch master cylinder.

BentleyPublishers.com
BentleyPublishers.com—All Rights Reserved

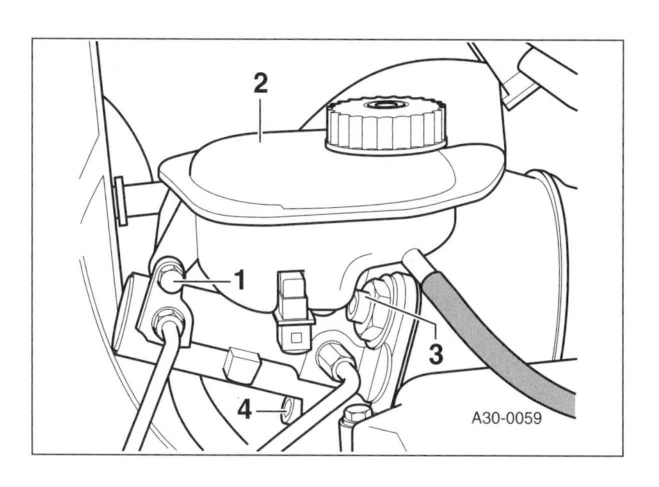

◀ Tighten mounting bolts for brake booster (**3** and **4**).

Tightening torque	
Brake master cylinder to brake booster	25 Nm (18.5 ft-lb)

– Tighten mounting bolt for brake fluid container (**1**)

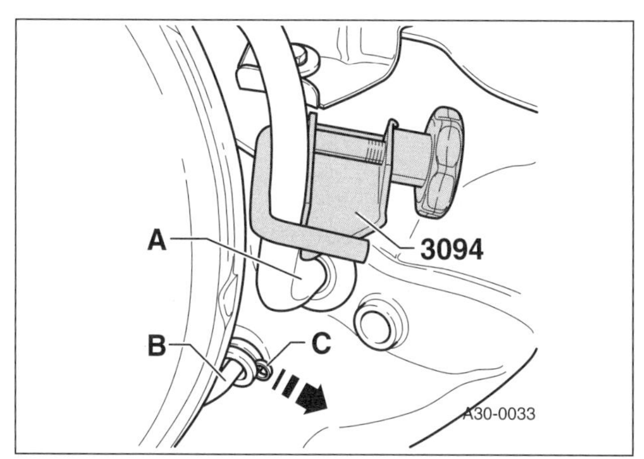

◀ Press in retaining clip (**C**) as far as it will go.

– Push pipe (**B**) into clutch master cylinder until it engages audibly.

– Insert supply hose rubber grommet into place on bulkhead.

– Push in supply hose (**A**) from brake fluid reservoir as far as it will go.

– Remove special tool 3094.

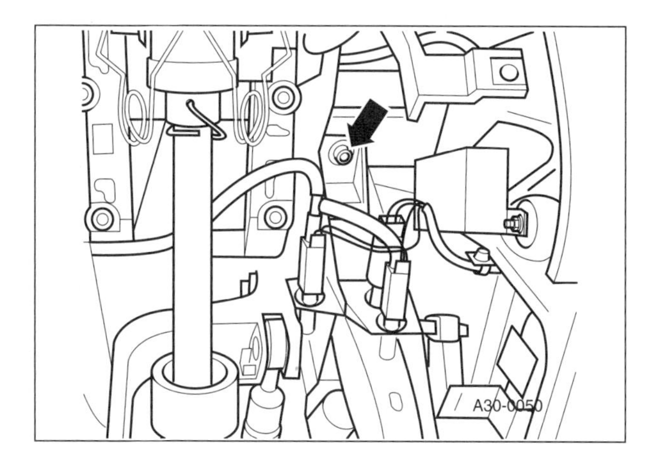

◀ Tighten pedal assembly bolt (**arrow**) to secure pedal assembly to dashboard carrier.

Tightening torque	
Pedal assembly to dashboard carrier	25 Nm (18.5 ft-lb)

– Connect clutch master cylinder to clutch pedal. Engage bolt lock in clutch pedal.

– Bleed clutch system after installing clutch master cylinder. See **Clutch system, bleeding** in this repair group.

BentleyPublishers.com
BentleyPublishers.com—All Rights Reserved

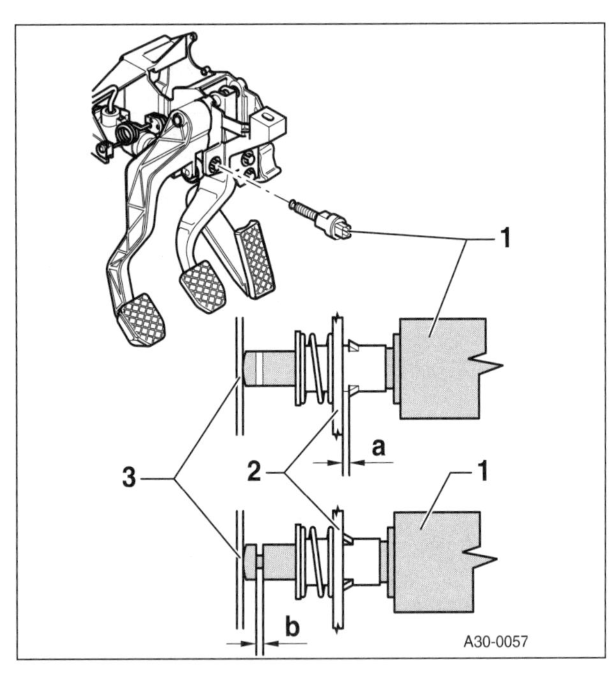

Check adjustment of vent valve (**1**) above clutch pedal. Vent valve must contact clutch pedal (**3**) completely and must operate at full travel.

– Gap measurement **a** between retaining clip and mounting bracket may be a max. of 0.5 mm.

– To adjust, hold clamp and turn vent valve (**1**).

– Install driver's storage compartment.

Clutch slave cylinder, removing and installing

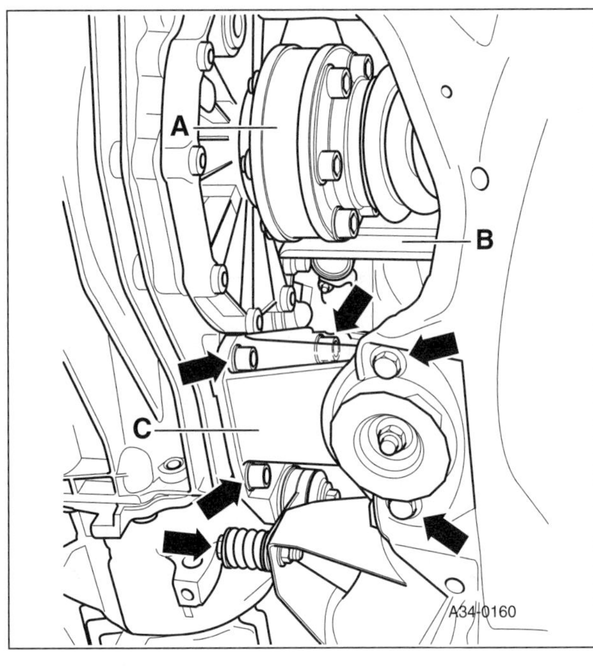

To aid access, unbolt left drive axle (**A**) from transmission and move clear to one side.

– Remove heat shield (**B**) from transmission.

– If necessary, unscrew bolts (**arrows**) and remove left transmission support (**C**) together with transmission mounting.

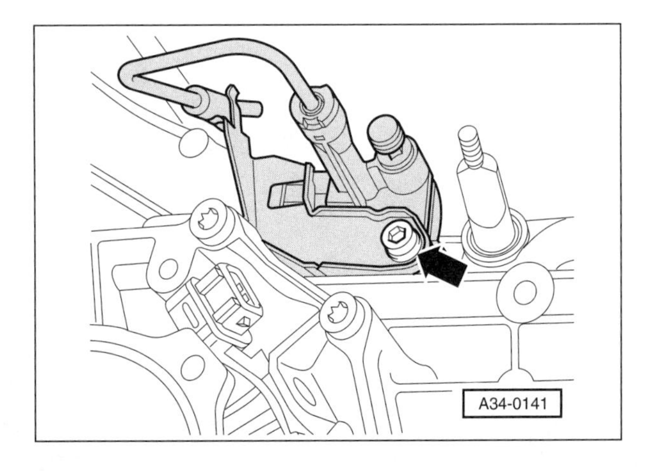

Remove bolt (**arrow**) and take out slave cylinder from the rear.

NOTE —

- *Do not depress clutch pedal after removing slave cylinder.*
- *Do not allow brake fluid to come into contact with transmission. If necessary, clean transmission housing.*

BentleyPublishers.com
BentleyPublishers.com—All Rights Reserved

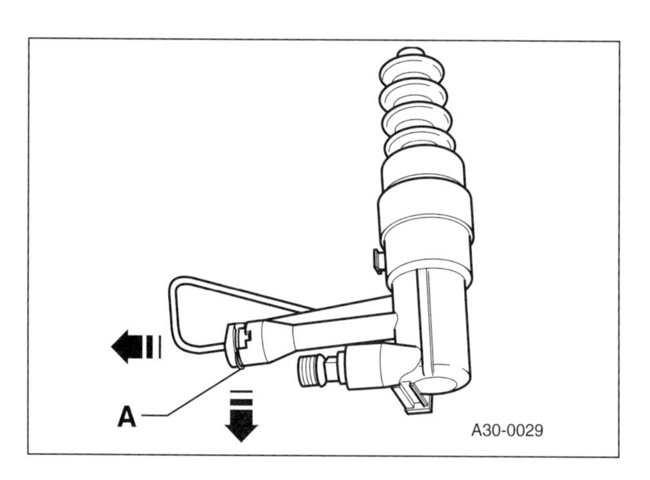

To disconnect pipe, first pry out retaining clip (**A**) with a screwdriver until it disengages audibly. The pipe can then be pulled out.

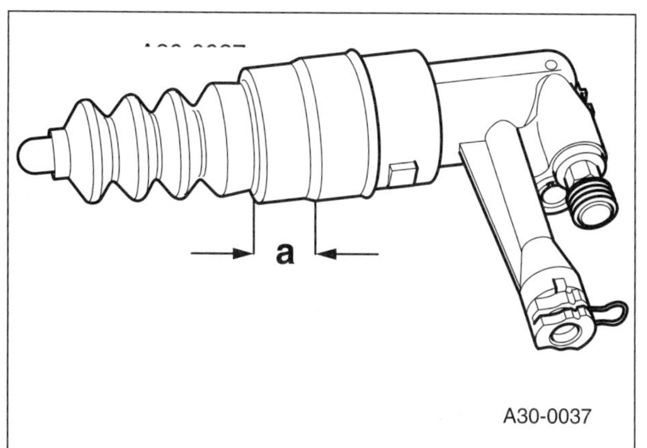

Before mounting the slave cylinder into the transmission housing, coat area **a** of collar with lithium grease.

– Lightly coat tip of push rod using copper grease.

NOTE —

- *To aid installation, engage 6th gear (6-speed transmission) or 4th gear (5-speed transmission) before fitting slave cylinder.*
- *Pre-tension clutch slave cylinder far enough for the securing bolt to be easily inserted.*

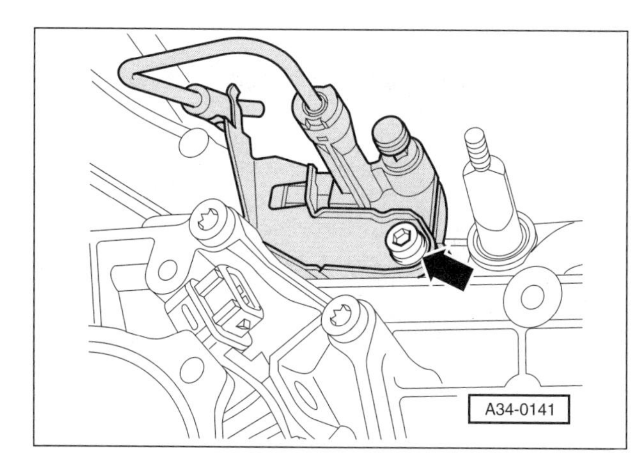

Insert clutch slave cylinder into mounting hole in transmission housing, keeping it in line with direction of operation of push rod, then tighten bolt (**arrow**).

NOTE —

- *If the clutch slave cylinder is inserted off-line there is a danger that the push rod will be guided past the clutch release lever.*
- *Always replace securing bolt.*

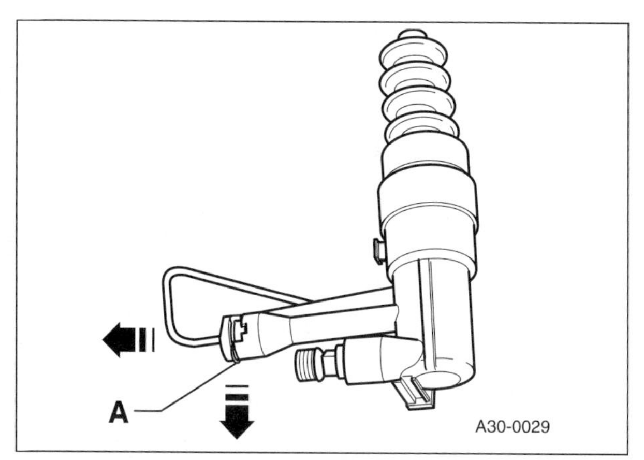

To connect pipe to slave cylinder, press in retaining clip (**A**) as far as it will go. Push pipe into slave cylinder until it engages audibly.

– Reinstall drive axle and heat shield

– Reinstall transmission support and transmission mounting (if removed).

– Bleed clutch system. See **Clutch system, bleeding** in this repair group.

BentleyPublishers.com
BentleyPublishers.com—All Rights Reserved

Tightening torques	
Clutch slave cylinder to transmission	20 Nm (15 ft-lb)
Heat shield for drive axle	25 Nm (18.5 ft-lb)
Drive axle to drive flange: • M8 • M10	 40 Nm (30 ft-lb) 70 Nm (52 ft-lb)
Transmission support to transmission	40 Nm (30 ft-lb)
Transmission mounting to subframe	23 Nm (17 ft-lb)

Clutch system, bleeding

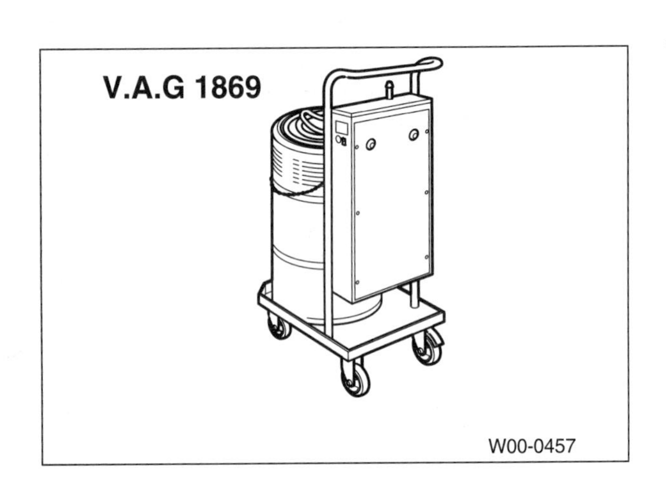

◀ Clutch system bleeding require brake filler/bleeder unit (VAG1869 or equivalent).

NOTE—

- *Open bleeder valve before switching on brake filler/bleeder unit.*
- *Make sure no brake fluid leaks onto transmission.*
- *Bleed clutch system after working on hydraulic clutch mechanism.*
- *Top off brake fluid reservoir to MAX before bleeding clutch system.*

– Pull clutch pedal back to rest position.

– Connect brake filler/bleeder unit (VAG 1869 or equivalent) but do not switch on yet.

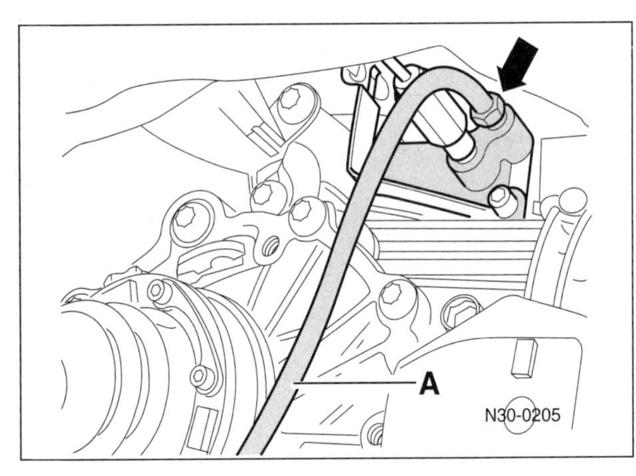

◀ Connect bleeder hose (**A**) to clutch slave cylinder (**arrow**) and open bleeder valve.

– Connect bleeder hose to pressure hose of collector bottle.

– Switch on brake filler/bleeder unit and bleed approx. 100 cc (3 oz) of brake fluid (do not exceed 2.5 bar positive pressure).

– Close bleeder valve.

Tightening torque	
Bleeder valve to clutch slave cylinder	4.5 Nm (3.3 ft-lb)

– Depress clutch pedal several times after completion of bleeding process.

– Bleed system again if necessary.

BentleyPublishers.com
BentleyPublishers.com—All Rights Reserved

CLUTCH ASSEMBLY

NOTE—

- *Clutch replacement requires transmission removal. See* **34 Manual Transmission**.
- *Replace clutch and pressure plate with damaged or loose rivets.*
- *Match clutch plate and pressure plate to engine application.*
- *Clean input shaft splines and hub splines. Remove corrosion and apply a very thin coating of grease (G 000 100 or equivalent) to the splines. Make sure hub moves freely on shaft. Remove excess grease.*
- *Pressure plate has anti-corrosion coating. Clean contact surface before installation.*
- *If clutch was burnt out, thoroughly clean bellhousing, flywheel and parts of the engine facing transmission.*

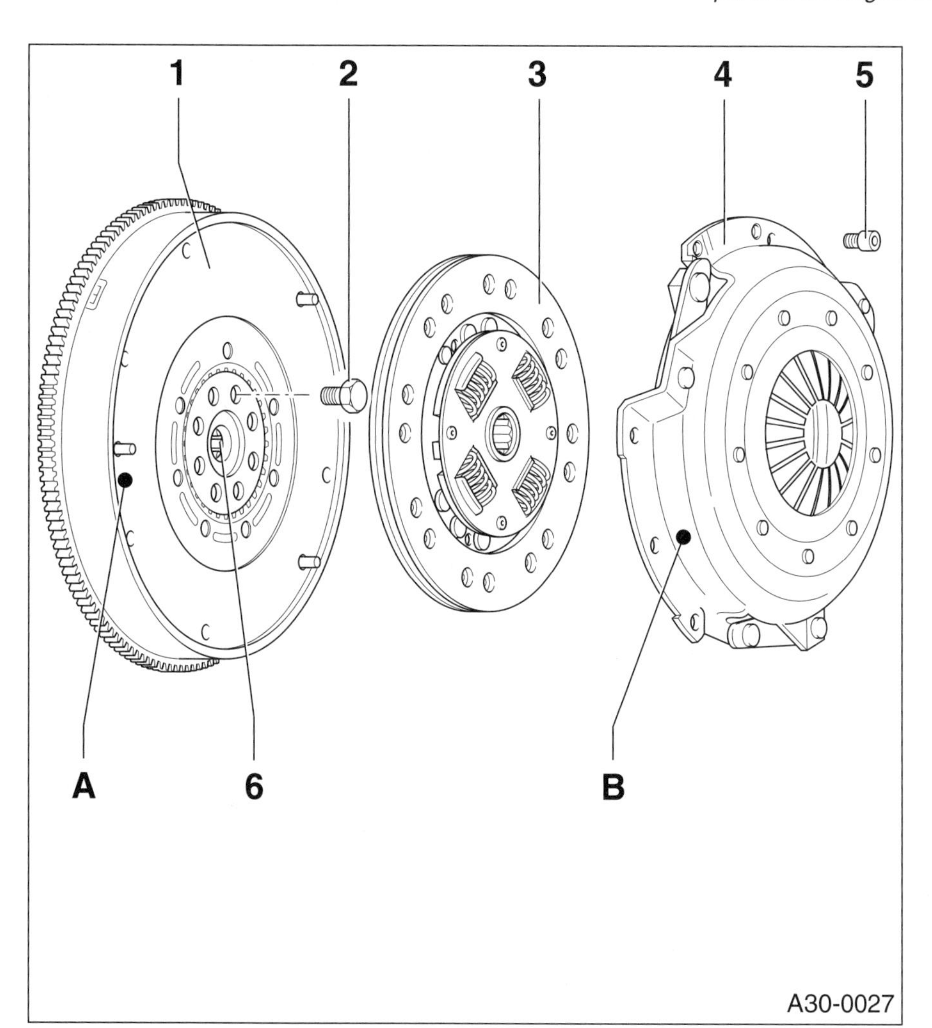

Clutch components (5-speed)

1. **Flywheel**
   - Make sure centering pins are tightly seated
   - Contact surface for clutch lining must be free of scoring, oil and grease
2. **Bolt, 60 Nm (44 ft-lb) + 180° turn**
   - Always replace
3. **Clutch disc**
   - Install spring pack (coil springs) toward pressure plate. Clutch lining must make full contact with flywheel. Marking "Getriebeseite" (if provided) goes toward pressure plate.
   - Lightly grease splines
4. **Pressure plate**
   - Check ends of diaphragm spring
   - Check spring connection and rivets
5. **Bolt, 25 Nm (18 ft-lb)**
   - Loosen and tighten in stages and in diagonal sequence
6. **Needle roller bearing**

NOTE—

- *Mark (**A**) on flywheel must coincide with mark (**B**) on pressure plate.*

BentleyPublishers.com
BentleyPublishers.com—All Rights Reserved

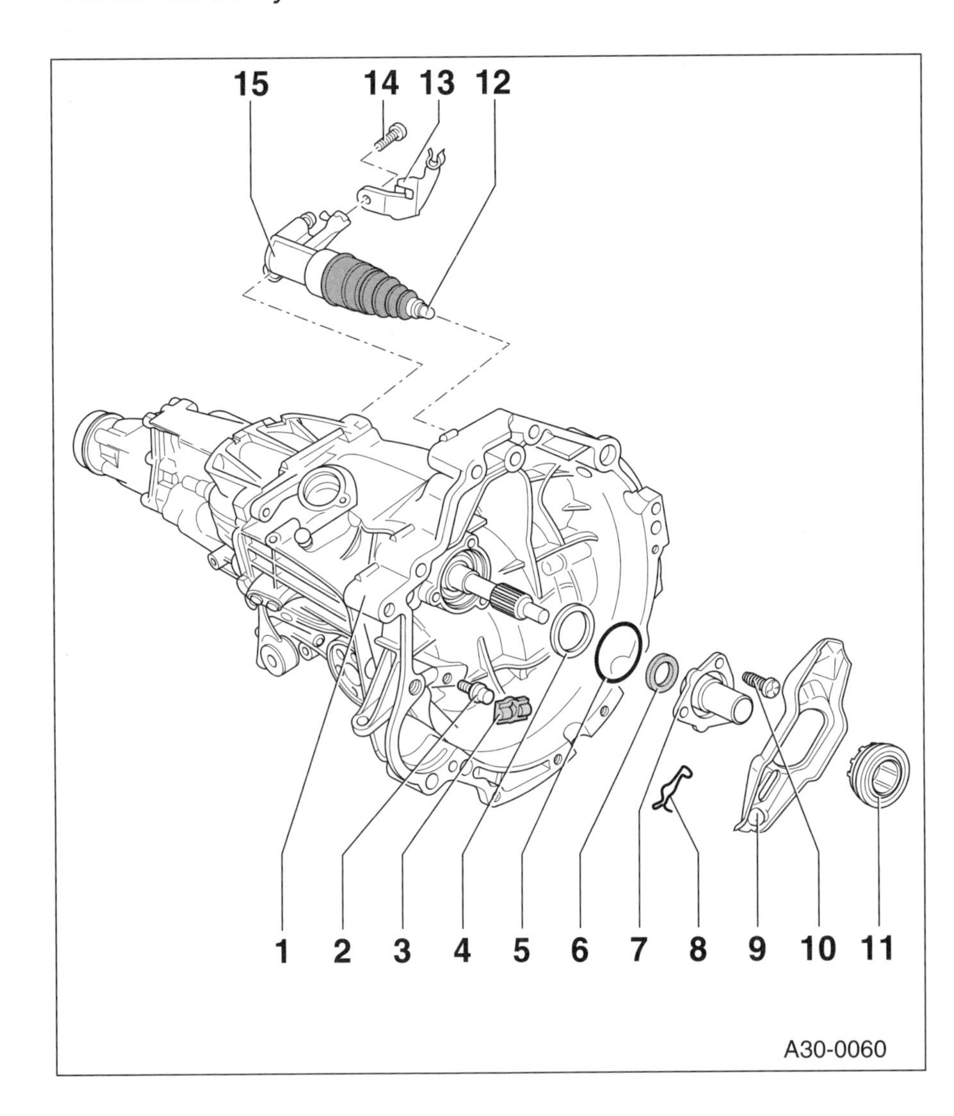

Clutch release mechanism (5-speed)

1. **Transmission**
2. **Ball stud**
   - Tighten to 25 Nm (18.5 ft-lb)
   - Lubricate with MoS2 grease
3. **Intermediate piece**
4. **Dished washer**
   - Smaller diameter (convex side) to guide sleeve
5. **O-ring**
   - Always replace
6. **Input shaft oil seal**
   - Pull out of guide sleeve with oil seal extractor lever VW 681
   - Drive to stop with fitting sleeve VW 192
7. **Guide sleeve**
   - Before removing and installing, cover input shaft splines with a shrink-fit hose to protect oil seal.
8. **Retaining spring**
   - Secure to clutch release lever
9. **Clutch release lever**
   - Before installing, coat contact surface of clutch slave cylinder plunger with a thin layer of copper grease.
10. **Self-locking bolt, 35 Nm (26 ft-lb)**
    - Always replace
11. **Release bearing**
    - Do not wash-out bearing, only wipe
    - Replace noisy bearing
    - Allow retaining tabs on release bearing to engage in release lever
12. **Plunger**
    - Grease end of plunger with copper grease.
13. **Bracket**
    - Secured to transmission along with slave cylinder
14. **Bolt, 20 Nm (15 ft-lb)**
    - Always replace
15. **Clutch slave cylinder**
    - Do not operate clutch pedal after removing
    - Pretension slave cylinder so mounting bolt can be inserted easily. See **Clutch slave cylinder (5-speed), removing and installing**.

BentleyPublishers.com
BentleyPublishers.com—All Rights Reserved

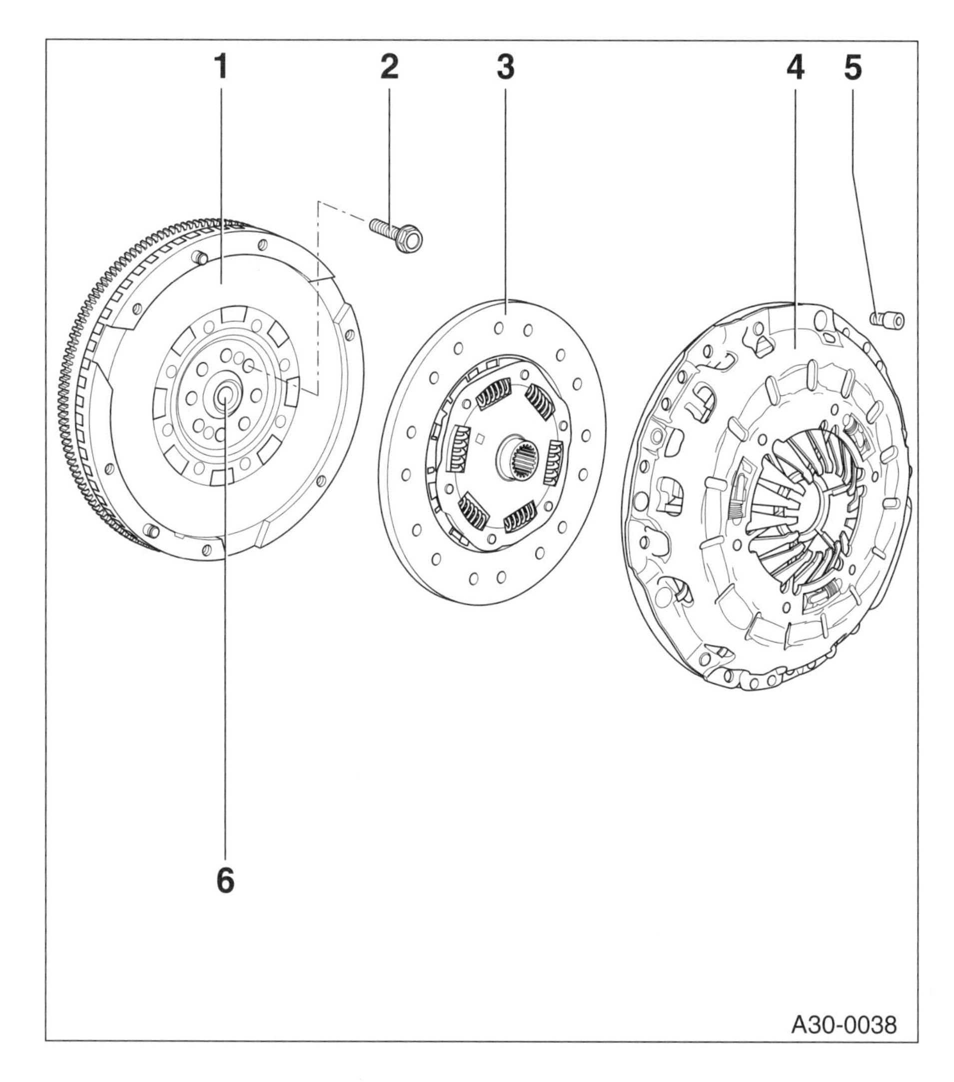

Clutch components (6-speed)

1. **Flywheel**
   - Dual mass flywheel only for 2.7 liter V6 bi-turbo engine
   - Make sure centering pins are tightly seated
   - Contact surface for clutch lining must be free of grooves, oil and grease
2. **Bolt**
   - Vehicle without dual-mass flywheel: tighten to 60 Nm (44 ft-lb) + 90°
   - Vehicle with dual-mass flywheel: tighten to 60 Nm (44 ft-lb) + 180°
   - Always replace
3. **Clutch plate**
   - Install with spring pack (coil springs) or the word "Getriebeseite" towards pressure plate and transmission
   - Do not grease
4. **SAC pressure plate**
   - See **Pressure plate, checking** in this repair group.
   - See **Adjusting ring in SAC clutch pressure plate, resetting** in this repair group.
   - See **Clutch with SAC pressure plate, removing and installing** in this repair group.
5. **Bolt, 22 Nm (16 ft-lb)**
   - Tighten gradually in diagonal sequence and in several stages
6. **Needle roller bearing**

BentleyPublishers.com
BentleyPublishers.com—All Rights Reserved

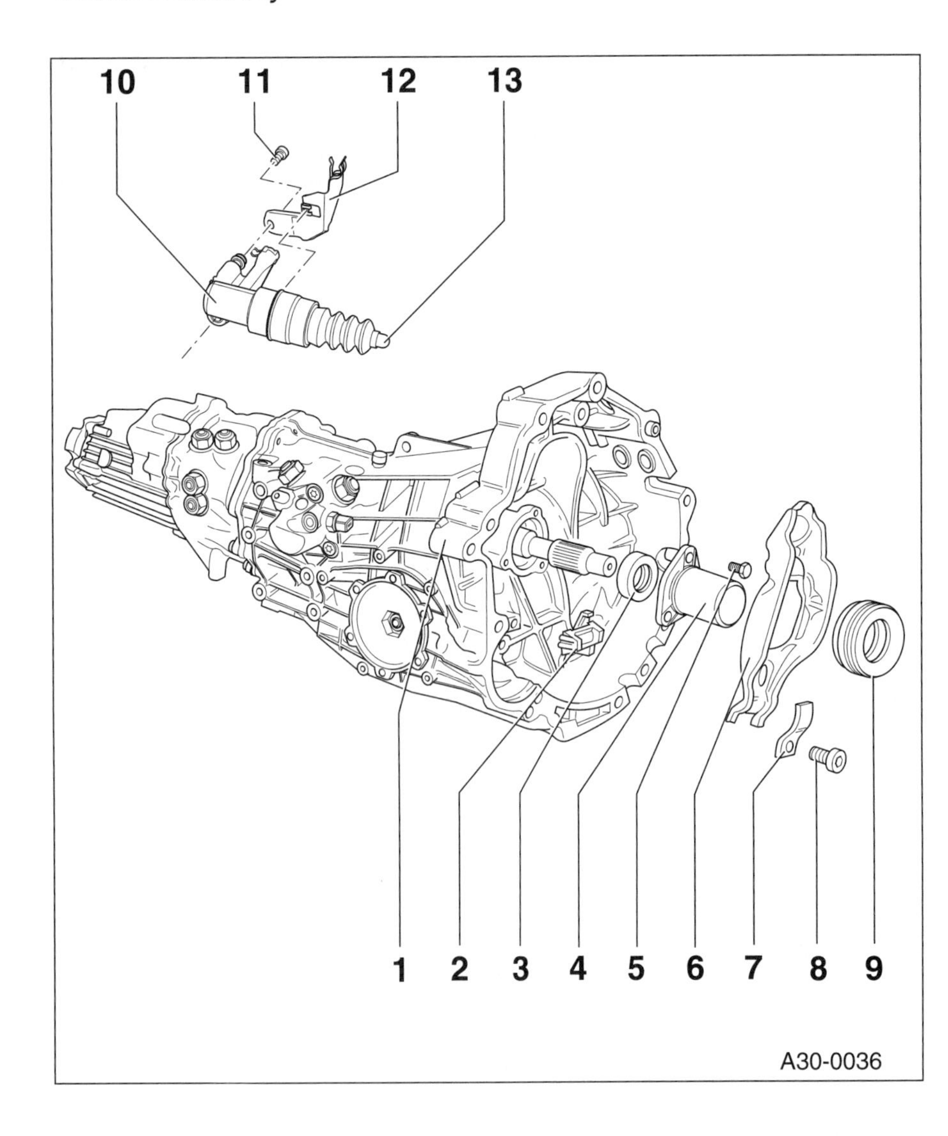

Clutch release mechanism (6-speed)

1. **Transmission**
2. **Intermediate piece**
   - Always replace
3. **Shaft seal**
   - Install depth (factory): 3.5 mm (0.14 in)
   - Install depth (repairs): 4.5 mm (0.18 in)
4. **Guide sleeve**
5. **Bolt, 15 Nm (11 ft-lb)**
6. **Clutch release lever**
   - Engage lever in lugs of intermediate piece
   - Coat contact surface of clutch slave cylinder push rod with a thin layer of copper grease
7. **Leaf spring**
8. **Bolt, 25 Nm (18.5 ft-lb)**
   - Always replace
9. **Release bearing**
   - Do not wash out, wipe clean only
   - Replace noisy bearings
   - Retainer lugs on release bearing must engage in release lever
10. **Clutch slave cylinder**
    - See **Clutch slave cylinder (6-speed), installing** in this repair group.
11. **Bolt, 20 Nm (15 ft-lb)**
    - Always replace
12. **Bracket for hose/pipe assembly**
    - Engage in clutch slave cylinder
13. **Push rod**
    - Coat tip of push rod with copper grease

BentleyPublishers.com
BentleyPublishers.com—All Rights Reserved

Clutch pressure plate, checking

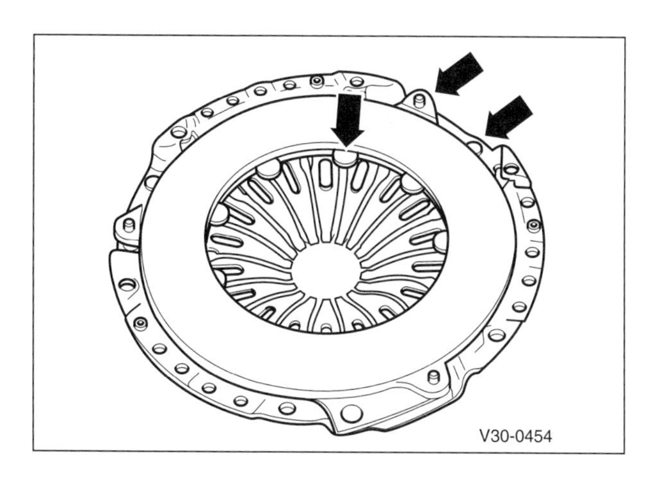

◄ Check spring connection between pressure plate and cover for cracks and make sure rivets are tight.

– Replace clutch with damaged springs or loose rivets (**arrows**).

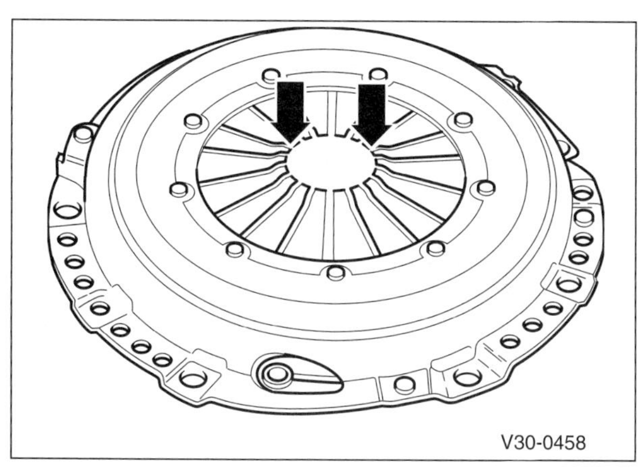

◄ Check ends of the diaphragm spring (**arrows**).

– Wear of diaphragm spring: up to ½ thickness of spring finger.

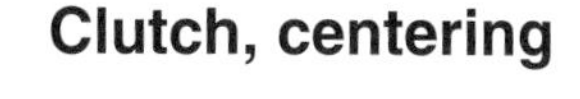

Clutch, centering

NOTE—

- *Models with SAC pressure plate: If necessary, reset SAC pressure plate. See* **Clutch pressure plate (SAC), resetting** *in this repair group.*

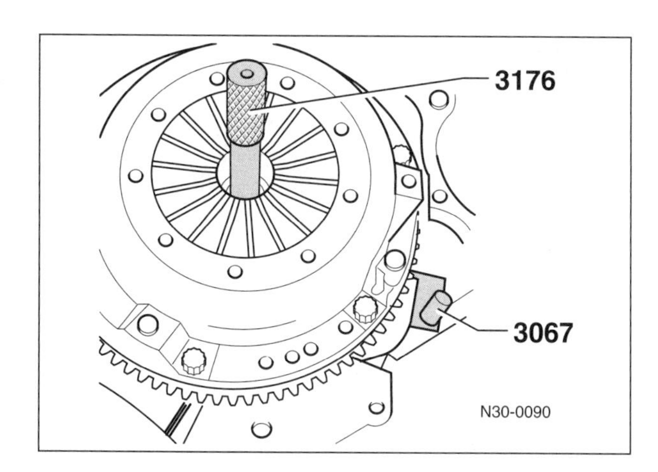

◄ Use clutch tool 3176 to center clutch disc. Use flywheel lock 3067 to remove and install pressure plate.

– Position clutch disc spring pack (coil springs) or marking "Getriebeseite" toward pressure plate and transmission.

– Do not insert mounting bolts until clutch lining and contact surface of pressure plate make full contact with flywheel.

– Loosen and tighten bolts in small stages and in diagonal sequence.

Tightening torque	
Clutch pressure plate to flywheel	25 Nm (18 ft-lb)
Clutch pressure plate (SAC) to flywheel	22 Nm (16 ft-lb)

BentleyPublishers.com
BentleyPublishers.com—All Rights Reserved

Clutch pressure plate (SAC), resetting

NOTE—

- *When fitting a new clutch disc together with a used SAC pressure plate, the adjuster ring in the pressure plate has to be reset by turning it back as far as it will go. If this is not done, the SAC pressure plate will operate with reduced clamping force, which will cause clutch slip and excessive wear (especially of the clutch disc).*
- *If clutch disc is not being replaced, it is not necessary to reset the adjuster ring.*
- *New SAC pressure plates are pre set, and do not have to be reset.*

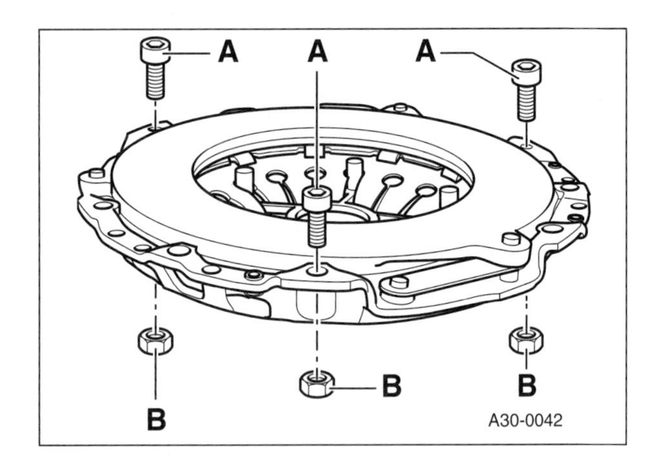

◀ Insert 3 securing bolts (**A**) into pressure plate mounting holes, 120° from each other (1/3 turn), as shown in illustration.

– Screw 3 nuts (**B**) (M8) on bolts (**A**) and tighten nuts slightly.

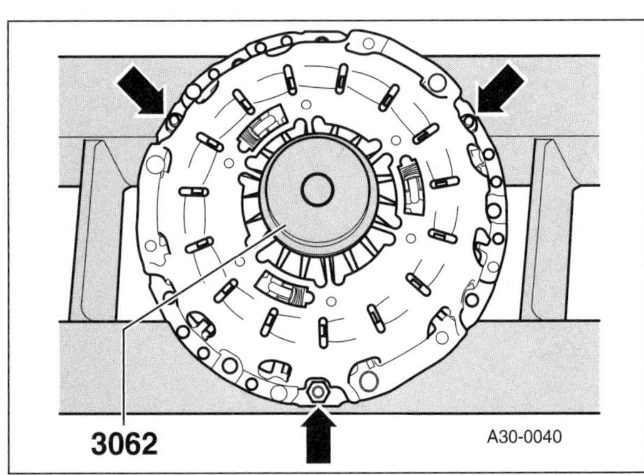

◀ Place SAC pressure plate on press so that only 3 bolt heads (**arrows**) make contact with press plate.

– Position special tool 3062 in center of pressure plate.

NOTE—

- *Do not use force when performing the following steps, otherwise the forks on the adjuster ring can break off.*

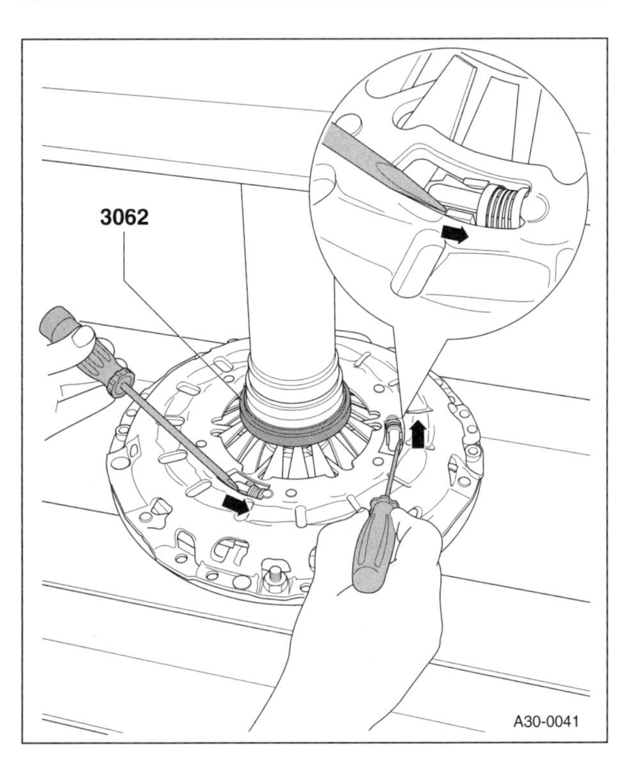

◀ Apply two screwdrivers to forks on adjuster ring. Compress pressure plate until it is just possible to move adjuster ring.

– Using two screwdrivers, turn back adjuster ring evenly in direction of **arrows** until it reaches stop.

– Hold adjuster ring against stop and release press so that adjuster ring is held in this position.

BentleyPublishers.com
BentleyPublishers.com—All Rights Reserved

32 Torque Converter

GENERAL

This repair group covers description and service notes for automatic transmission torque converter. Torque converter service requires removal of the transmission from the vehicle using special tools and equipment. Automatic transmission removal is not covered in this manual.

For additional information, see:

- **37 Automatic Transmission**

Torque converter description

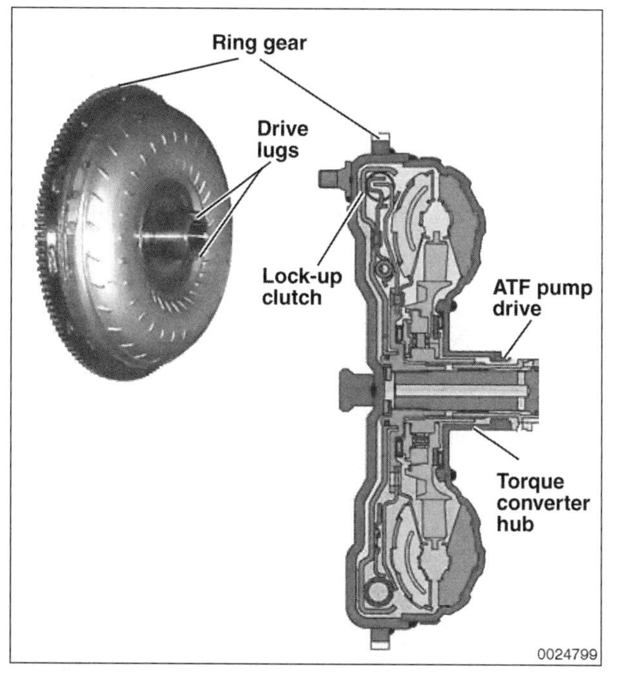

The torque converter is a fluid coupling device connecting the engine and the transmission. It is driven by a steel drive plate attached to the engine crankshaft. In addition to connecting the engine and transmission, the torque converter drives the automatic transmission fluid (ATF) pump within the transmission. The pump delivers ATF under pressure.

Torque converter sections are welded together and then balanced; no internal service is possible.

NOTE—

- *Torque converter with lock-up clutch illustrated. Not all models include this feature.*

Torque converter identification

Different versions of the torque converter are installed depending on engine type. The torque converter is marked with a 3-digit application code on the side facing the engine. For application information, check with an authorized Audi dealer parts department.

BentleyPublishers.com
BentleyPublishers.com—All Rights Reserved

TORQUE CONVERTER SERVICE

Torque converter, removing and installing

– With transmission separated from engine and out of vehicle, remove converter by firmly grasping and carefully pulling straight out.

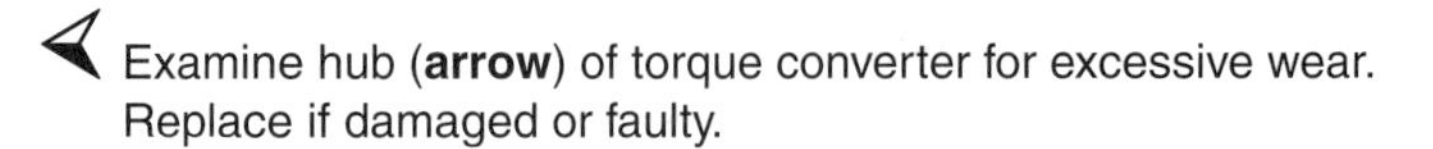

> ***CAUTION—***
> *• Be prepared to catch leaking ATF fluid.*

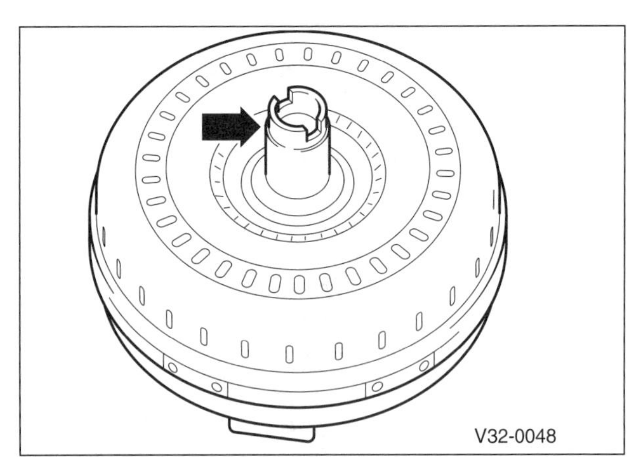

◀ Examine hub (**arrow**) of torque converter for excessive wear. Replace if damaged or faulty.

– Replace oil seal. See **Torque converter oil seal, replacing** in this repair group.

– Push torque converter hub through seal onto transmission shaft as far as first stop.

– Press torque converter into transmission bell housing while rotating it until recess in torque converter hub engages in ATF pump gear shaft and torque converter is felt to slip into place.

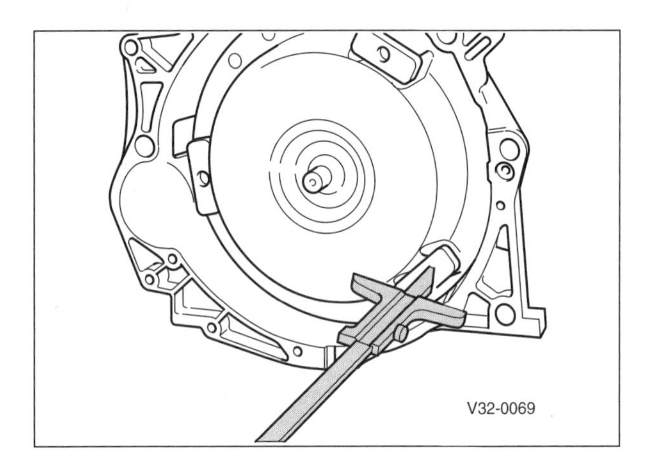

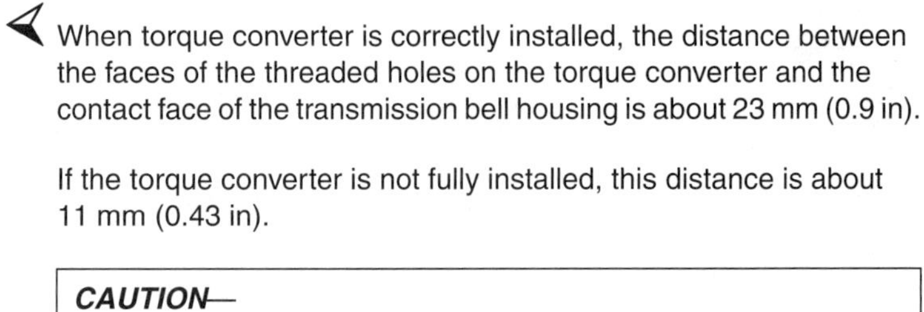

◀ When torque converter is correctly installed, the distance between the faces of the threaded holes on the torque converter and the contact face of the transmission bell housing is about 23 mm (0.9 in).

If the torque converter is not fully installed, this distance is about 11 mm (0.43 in).

> ***CAUTION—***
> *• If the torque converter is not correctly installed, the ATF pump shaft will be destroyed when the transmission is connected to the engine.*

Torque converter, draining

– If torque converter is contaminated by foreign particles, drain torque converter completely. Use VW special tool 1358, 1782 or equivalent.

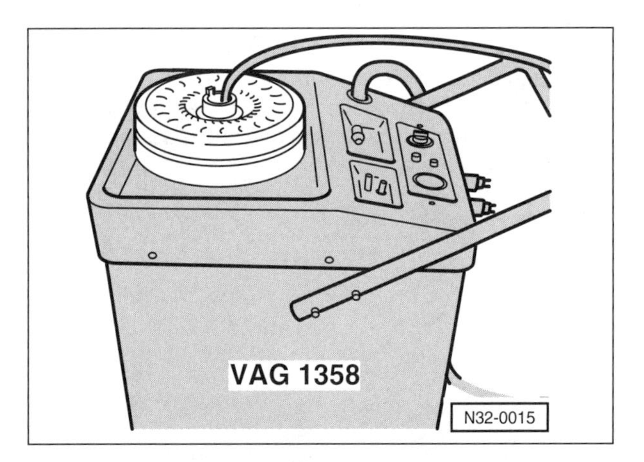

◀ Siphon ATF from torque converter. VAG 1358 A shown with adapter VAG 1358 A/1. Follow manufacturers instructions for operation.

NOTE—
• Be sure to allow sufficient time for fluid to drain from all internal passages of torque converter.

To refill torque converter, see automatic transmission fluid filling instructions in **37 Automatic Transmission**.

BentleyPublishers.com
BentleyPublishers.com—All Rights Reserved

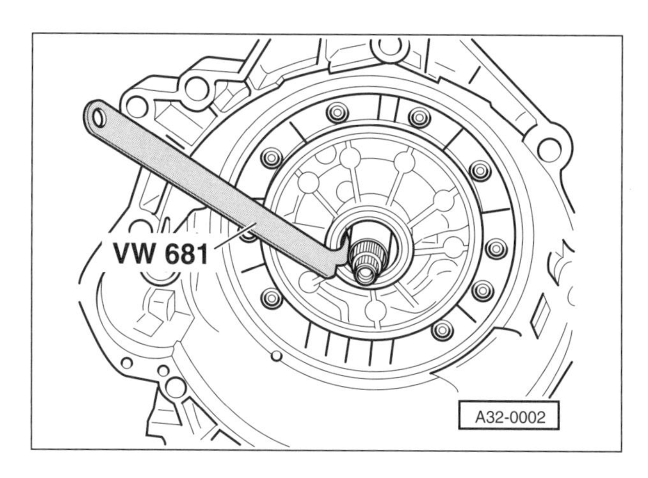

Torque converter oil seal, replacing

Torque converter oil seal is located in the transmission housing front cover at the transmission input shaft.

– With transmission separated from engine and out of vehicle, carefully pull out torque converter.

◄ Front drive models: Insert removal tool VW 681 or equivalent directly behind sealing lip of seal. Make sure contact ring directly behind seal is not damaged.

– Pry oil seal out of transmission housing. Use care to avoid gouging housing with tip of removal tool. Seal is destroyed during removal.

– Install contact ring. Replace contact ring if damaged

– Lightly lubricate outer circumference of seal and sealing lips with ATF.

NOTE—

• Open side of seal points towards transmission.

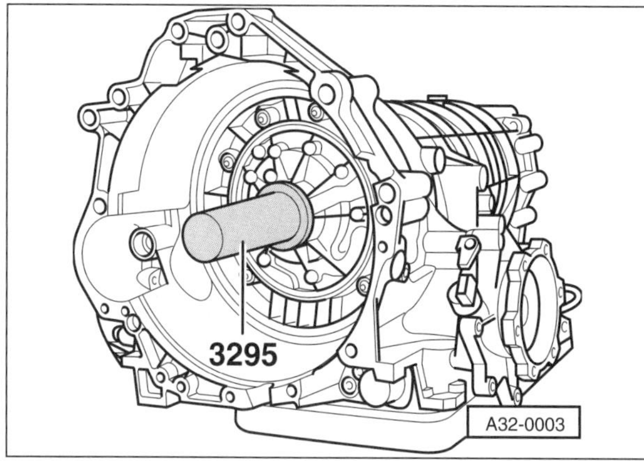

◄ Use drift 3295 to install seal as far as stop on drift.

– Install torque converter. See **Torque converter, removing and installing** in this repair group.

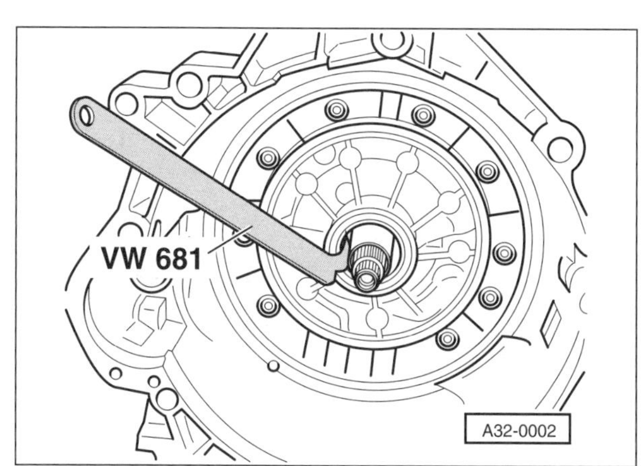

◄ All wheel drive models: Remove circlip in front of seal. Insert removal tool VW 681 or equivalent directly behind sealing lip of seal. Make sure wave washer directly behind seal is not damaged.

– Pry oil seal out of transmission housing. Use care to avoid gouging housing with tip of removal tool. Seal is destroyed during removal.

– Install wave washer. Replace washer if damaged.

– Lightly lubricate outer circumference of seal and sealing lips with ATF.

NOTE—

• Open side of seal points towards transmission.

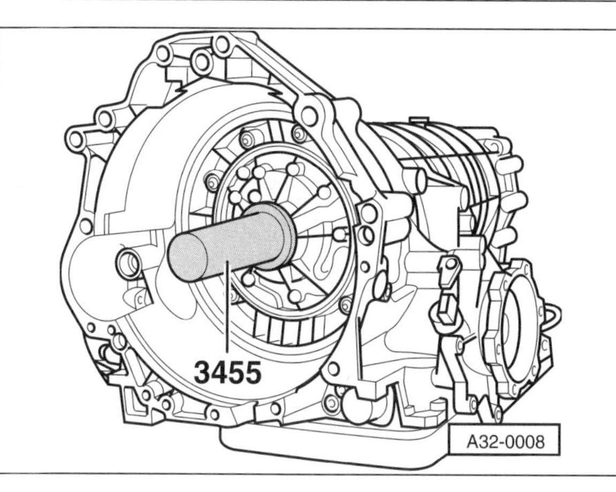

◄ Use drift 3455 to install seal as far as stop on drift.

– Insert circlip. Make sure circlip is seated in bottom of groove.

– Install torque converter. See **Torque converter, removing and installing** in this repair group.

BentleyPublishers.com
BentleyPublishers.com—All Rights Reserved

34 Manual Transmission

General

This repair group covers repair information for cable-operated gear shift mechanisms and removal and installation of the 5-speed or 6-speed manual transmission. Transaxle or transmission teardown, disassembly, or internal repairs are not covered.

For additional information, see the following:

- **03 Maintenance** for engine and transmission identification.
- **30 Clutch** for hydraulic and mechanical clutch repairs
- **40 Front Suspension** for drive axles
- **96 Interior Lights, Switches, Anti-theft** for back up light switch

A6 models were equipped with one of two different quattro (all-wheel drive) manual transmissions.

Transmission	Type
5-speed manual	01A
6-speed manual	01E

Transmissions are identified by type, code, serial number, manufacturer identification code and build date. Transmission identification plates containing some or all of this information are affixed to the transmission housing. See **03 Maintenance** for more information.

> ***CAUTION—***
> - *To avoid damaging plastic interior trim, use a plastic prying tool or a screwdriver with the tip wrapped with masking tape.*

BentleyPublishers.com
BentleyPublishers.com—All Rights Reserved

Transmission and Final Drive Oil

Manual transmission oil

Manual transmission gear box and final drive share a common oil supply. Checking oil level of transmission also checks oil level of final drive (differential).

To remove manual transmission filler plug, use either a 17 mm Allen wrench or a triple-square socket driver, special tool 3357. A combination tool with both 16 mm anti-tamper triple-square (XZN) and 17 mm Allen ends is available as MN2567.

Current specifications for manual transmission oil are shown in **Table a**. If oil container is not marked with appropriate specification, assume that the oil is not suitable.

Table a. Manual transmission oil specifications	
Transmission types 01A, 01E	Synthetic oil SAE 75W90 VAG part no. G 005 000 or G 052 145 S2

CAUTION—
- *Part numbers for transmission lubricants are for reference only. Be sure to check with an authorized Audi parts department for the latest recommendations.*

Approximate refill capacities in **Table b** are listed by transmission type.

Table b. Manual transmission oil capacities (approximate)

Transmission type	Oil capacity in liters (US qt)
01A (5-speed AWD)	2.7 (2.8)
01E (6-speed AWD) • Transmission code EDU • Transmission code EEY or EHS (allroad quattro)	 2.3 (2.4) 3.6 (3.8)

5-speed manual transmission oil level

Make sure vehicle is level during this operation. If front of vehicle is higher than rear, such as when driven up on ramps, oil inside transmission pools toward rear, and level cannot be accurately determined.

Checking level

– Raise vehicle and support safely.

WARNING —
- *Make sure the car is stable and well supported at all times. Use a professional automotive lift or jack stands designed for the purpose. A floor jack is not adequate support.*

BentleyPublishers.com
BentleyPublishers.com—All Rights Reserved

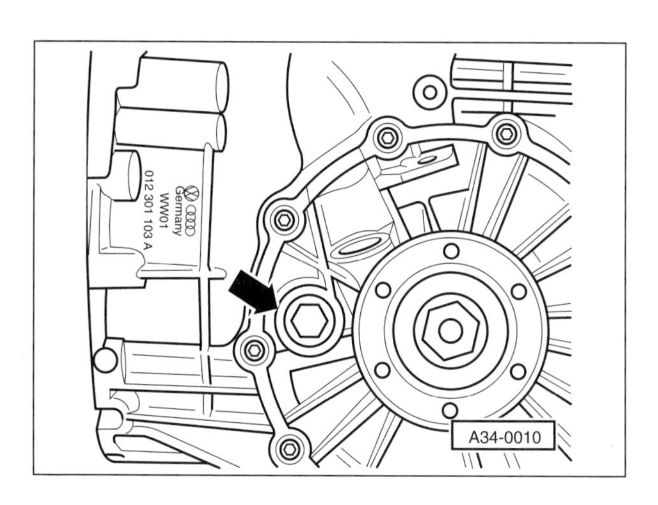

◀ Oil filler plug (**arrow**) is on left of transmission below speedometer sender; it may be concealed by heat shield for drive axle. Depending on transmission version fitted, use either special tool 3357 or 17 mm Allen wrench to loosen and remove oil filler plug.

– Oil level is correct when transmission is filled up to lower edge of filler hole.

Filling

– If oil is low:
  - Place oil drain pan underneath transmission filler plug
  - Top up transmission oil with SAE 75 W 90 (synthetic oil). Make sure oil is filled up to lower edge of filler hole.
  - Allow excess oil to drip out.

– Screw in oil filler plug.

Tightening torque	
Oil filler plug to transmission housing	25 Nm (18 ft-lb)

6-speed manual transmission oil level

When checking 6-speed manual transmission oil level, use a piece of stiff wire (such as a coat-hanger) bent at right angle, as a dipstick.

Make sure vehicle is level during this operation. If front of vehicle is higher than rear, such as when driven up on ramps, oil inside transmission pools toward rear, and level cannot be accurately determined.

Checking level

– Raise vehicle and support safely.

> ***WARNING —***
> - *Make sure the car is stable and well supported at all times. Use a professional automotive lift or jack stands designed for the purpose. A floor jack is not adequate support.*

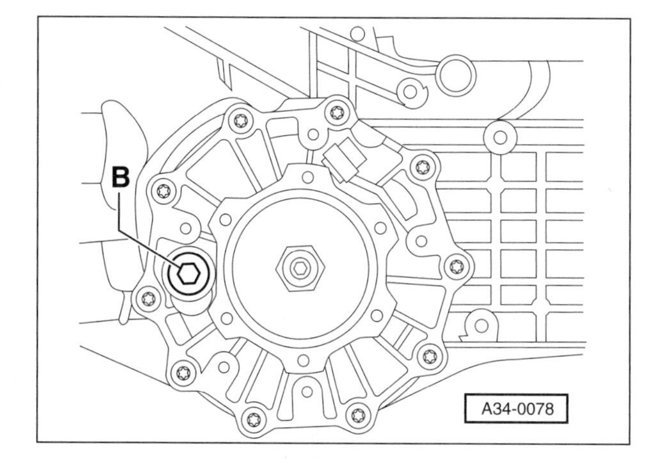

◀ Oil filler plug (**B**) is on left, ahead of drive axle flange. Use either special tool 3357 or 17 mm Allen wrench to loosen and remove oil filler plug.

BentleyPublishers.com
BentleyPublishers.com—All Rights Reserved

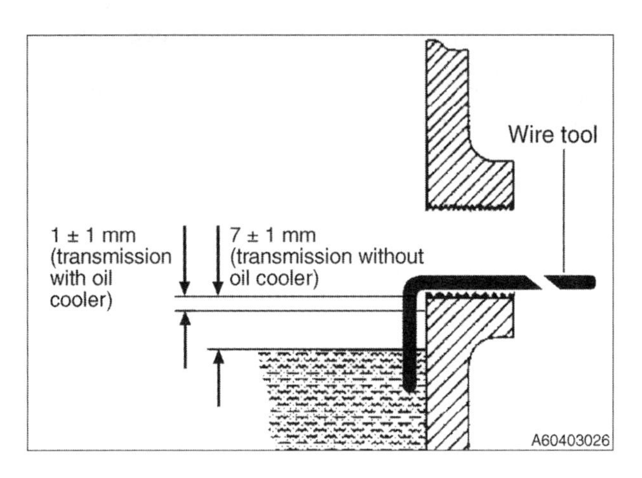

Check transmission oil level using stiff wire hook inserted into fill hole.

6-speed manual transmission oil level guidelines	
Vehicle without transmission oil cooler	7 ± 1 mm (¼ in) below fill hole opening
Vehicle with transmission oil cooler	1 ± 1 mm (1/25 in) below fill hole opening

Filling

– If oil is low, top up transmission oil with SAE 75 W 90 (synthetic oil).

> ***CAUTION—***
> - *Adhere strictly to the oil level guidelines. The transmission is very sensitive to overfilling.*

– Screw in oil filler plug.

Tightening torque	
Oil filler plug to transmission housing	40 Nm (30 ft-lb)

GEAR SELECTOR SERVICE

NOTE—

- *To remove the complete shift linkage, remove the exhaust system. See* **26 Exhaust System**.
- *To disassemble the shift linkage in the installation position, the shift mechanism housing must be lowered.*
- *Lubricate sliding surfaces with polycarbamide grease.*

BentleyPublishers.com
BentleyPublishers.com—All Rights Reserved

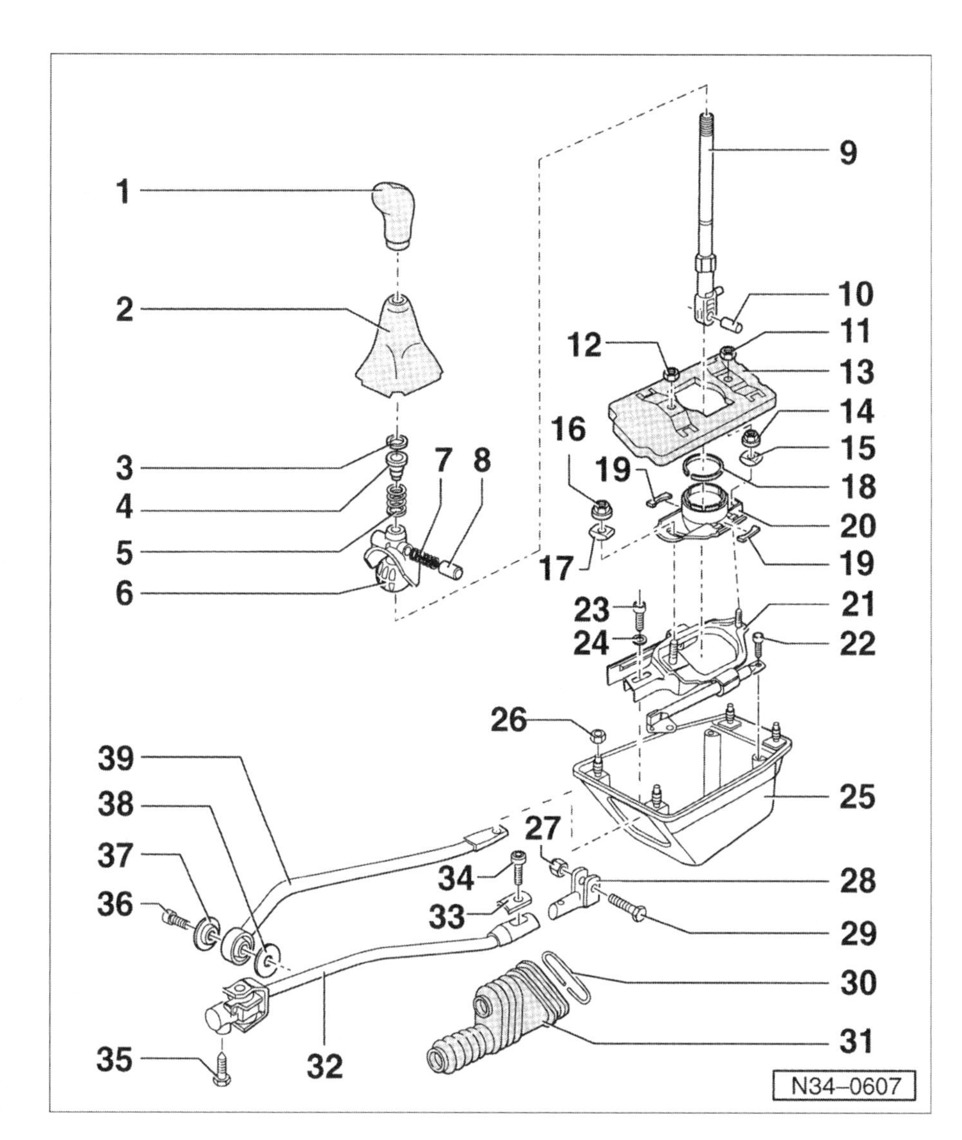

Gear selector mechanism (5-speed)

1. **Shift knob**
2. **Shift boot**
3. **Circlip**
4. **Bushing**
5. **Compression spring**
6. **Ball stop**
7. **Compression spring**
8. **Bushing**
9. **Shift lever**
10. **Spacer tube**
11. **Nut, 8 Nm (ft-lb)**
12. **Nut, 8 Nm (ft-lb)**
13. **Cover**
14. **Nut, 10 Nm (7 ft-lb)**
15. **Connecting piece**
16. **Nut, 23 Nm (17 ft-lb)**
17. **Connecting piece**
18. **Circlip**
19. **Buffer**
20. **Ball housing (upper)**
21. **Ball housing (lower)**
22. **Bolt, 10 Nm (7 ft-lb)**
23. **Bolt, 23 Nm (17 ft-lb)**
24. **Washer**
25. **Shift mechanism housing**
26. **Nut, 10 Nm (7 ft-lb)**
27. **Nut, 10 Nm (7 ft-lb)**
28. **Shift fork**
29. **Bolt**
30. **Tensioning ring**
31. **Boot**
32. **Shift rod**
33. **Clamp**
34. **Bolt, 23 Nm (17 ft-lb)**
35. **Bolt, 23 Nm (17 ft-lb)**
36. **Bolt**
37. **Washer**
38. **Washer**
39. **Pivot rod**

BentleyPublishers.com
BentleyPublishers.com—All Rights Reserved

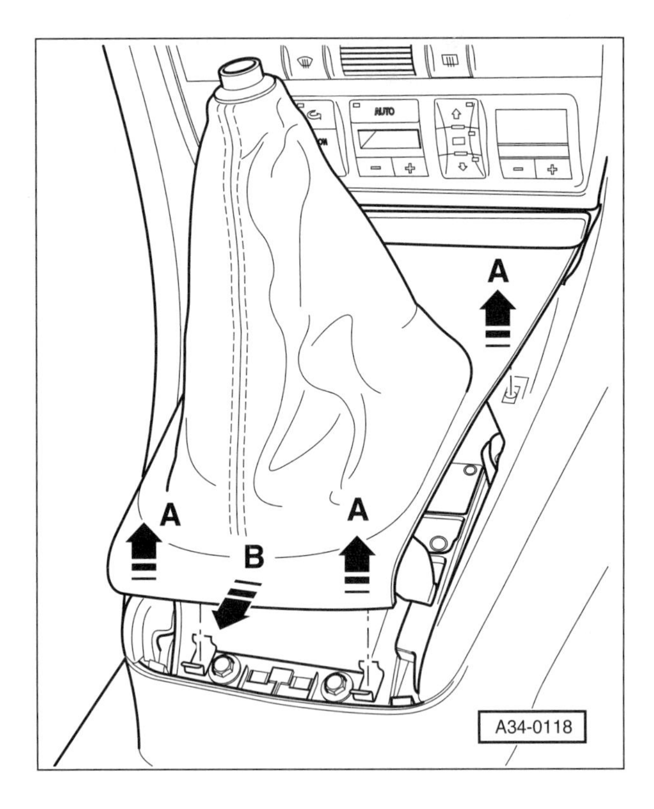

Selector mechanism (5-speed), removing and installing

– Unscrew gear shift knob from shift rod.

NOTE—

- *Cover on gear shift is removed together with cover for center console.*

◄ Use a plastic prying tool to lift up cover for center console slightly (**A**).

– Pull cover slightly toward rear (**B**) and lift off complete cover.

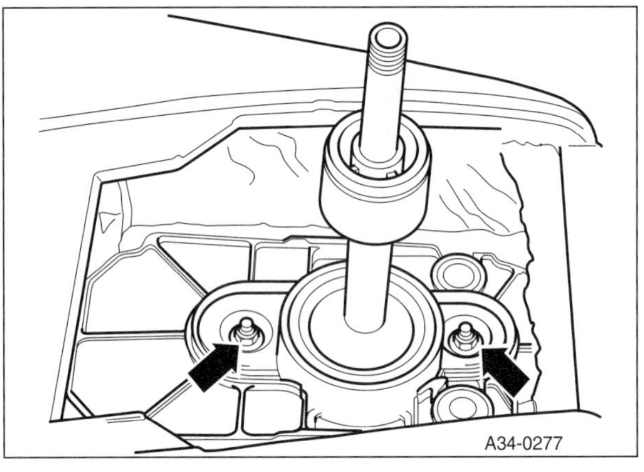

◄ Remove noise insulating cover for gear shift housing (**arrows**).

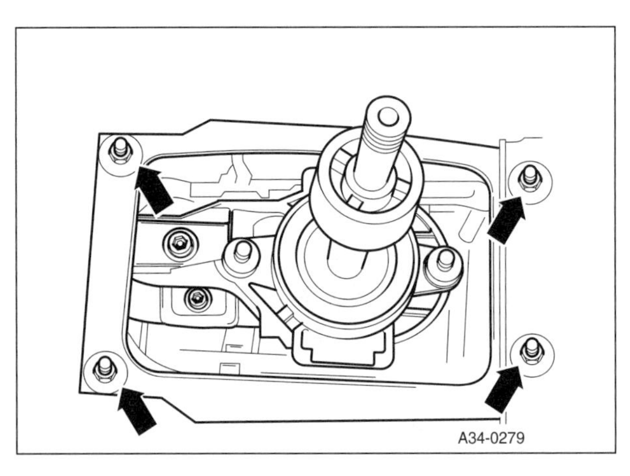

◄ Unscrew nuts securing gear shift housing (**arrows**).

– Separate exhaust system behind catalytic converter. If necessary remove front exhaust system.

– Remove driveshaft.

BentleyPublishers.com
BentleyPublishers.com—All Rights Reserved

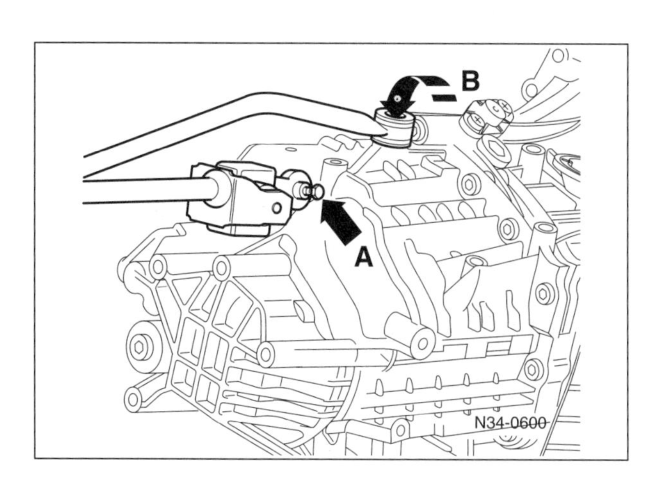

◀ Remove selector rod (**A**).

– Unscrew Allen bolt on push rod (**B**).

– Remove front heat shield above exhaust system.

– Swing gear shift housing with selector rod and push rod down and remove.

Installing

– Install in reverse order.

– Push selector rod on so that the securing bolt fits in the recess in the selector shaft.

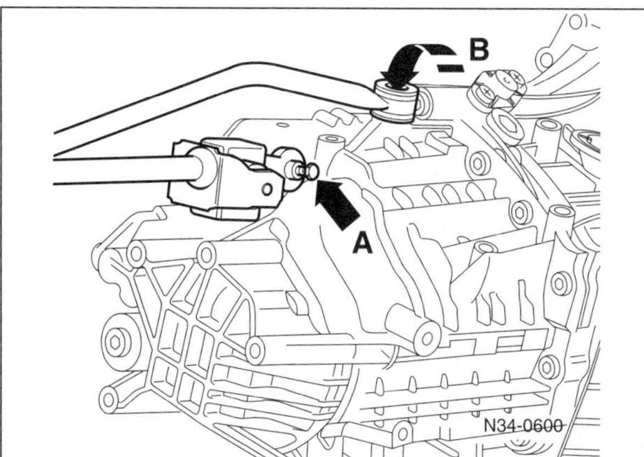

◀ Then secure selector rod (**A**) with bolt.

– Secure push rod to transmission (**B**) with bolt.

– Adjust gear selector mechanism. See **Gear selector mechanism (5-speed), adjusting** in this repair group.

– Install driveshaft and adjust. See **39 Differential and Final Drive**.

Tightening torques	
Gear shift housing to body	10 Nm (7 ft-lb)
Selector rod to transmission	20 Nm (15 ft-lb)
Push rod to transmission	40 Nm (30 ft-lb)

Gear selector mechanism (5-speed), adjusting

– Requirements:
- Components of selector mechanism and linkage must be in good condition.
- Selector mechanism must move freely.
- Transmission, clutch and clutch mechanism must be in good condition.
- Transmission in neutral.

– Unscrew gear shift knob from gear shift.

NOTE—
- *Cover on gear shift is removed together with cover for center console.*

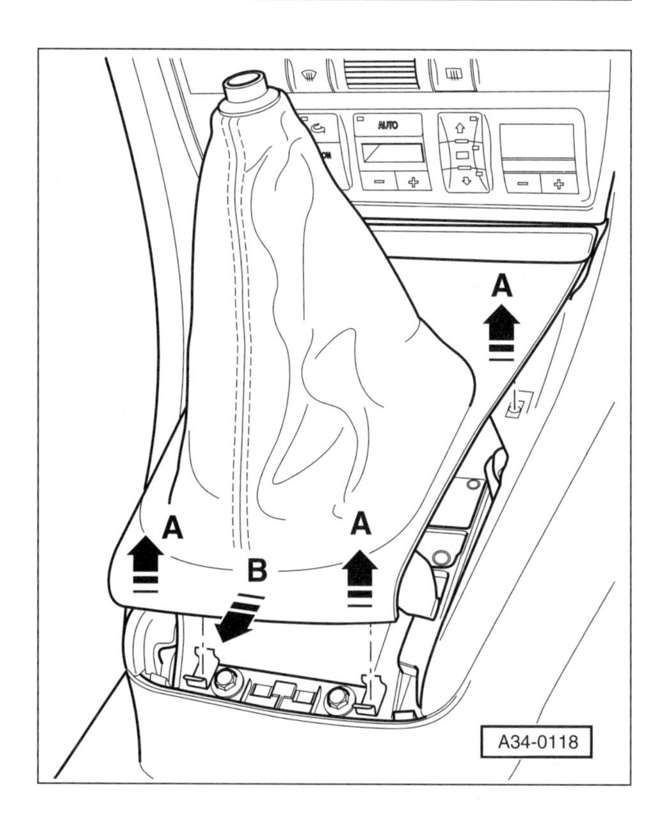

◀ Lift up cover for center console slightly (**A**).

– Pull cover slightly toward the rear (**B**) and then lift off complete cover.

BentleyPublishers.com
BentleyPublishers.com—All Rights Reserved

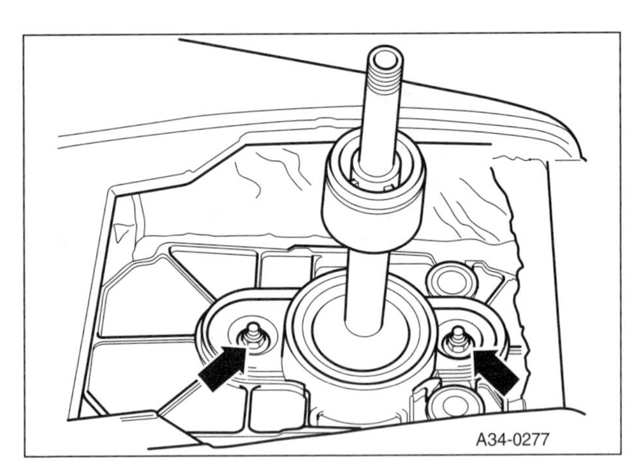

Remove noise insulating cover for gear shift housing (**arrows**).

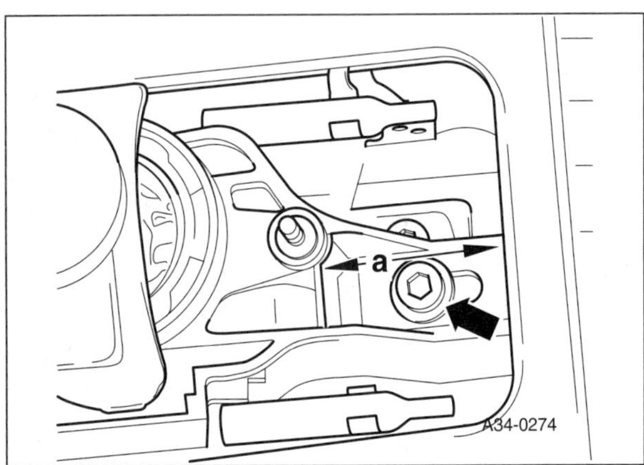

Measure distance between body and rear push rod (in selector mechanism). Distance **a** should equal 41 mm.

– If this is not the case, adjust dimension a as follows:
  - Loosen push rod bolt (**arrow**). Rear push rod (in selector mechanism) should move freely in both directions on slide.
  - Set to distance **a** by moving rear push rod (in selector mechanism).
  - Tighten push rod bolt.

Tightening torque	
Push rod adjusting bolt	25 Nm (18 ft-lb)

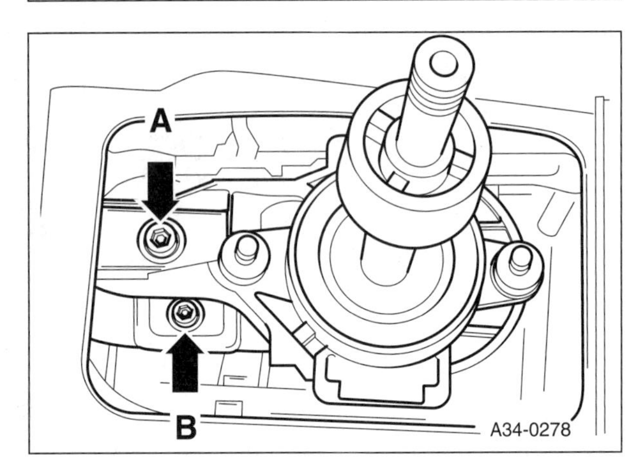

Loosen bolt for shift rod (**arrow**). Shift rod/shift mechanism must move freely.

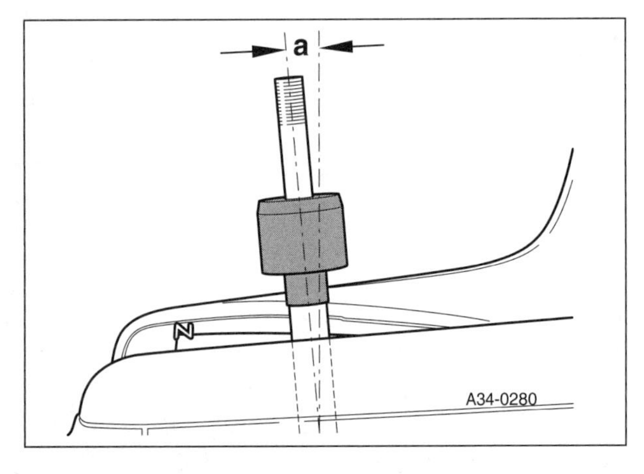

Align shift lever as follows:
  - Shift lever vertical in direction of travel.
  - Shift lever slightly inclined toward rear (**a**).

NOTE—
- *Shift lever is shown from the right side.*

– Hold shift lever in this position.

– Tighten bolt for shift rod.

Tightening torque	
Shift rod adjusting bolt	25 Nm (18 ft-lb)

BentleyPublishers.com
BentleyPublishers.com—All Rights Reserved

NOTE—
- *Make sure position of shift lever does not change when bolt is tightened.*

Gear shift adjustment (5-speed), checking

Make sure gear shift lever rests in the 3rd/4th gear gate when transmission is in neutral.

– Operate clutch.

– Check that all gears can be engaged.

– Check operation of reverse gear lock:
  - The gear shift should return by itself from the 5th/reverse gate into the 3rd/4th gate.
  - It must not be possible to shift directly into reverse gear from 5th gear.
  - It must only be possible to shift into reverse gear when shift lever starts in the neutral gate between 3rd/4th gears.

NOTE—
- *If only 5th gear and reverse gear cannot be engaged, check the 5th and reverse gear locking unit and replace if necessary.*

– Attach covers and gear shift knob.

Selector mechanism (6-speed), removing and installing

– Unscrew shift knob from lever.

NOTE—
- *Shift cover is removed together with cover for center console.*

A34-0118

◀ Use a plastic prying tool to slightly lift cover for center console upward (**A**).

– Pull cover slightly back (**B**) and remove complete cover.

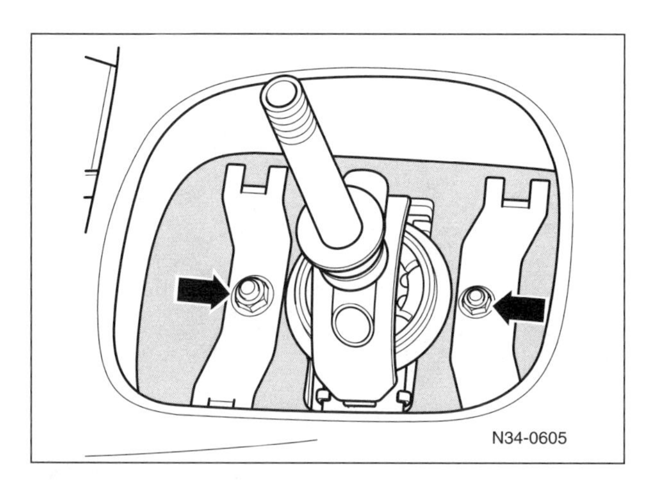

◀ Remove noise insulation for selector mechanism housing (**arrows**).

BentleyPublishers.com
BentleyPublishers.com–All Rights Reserved

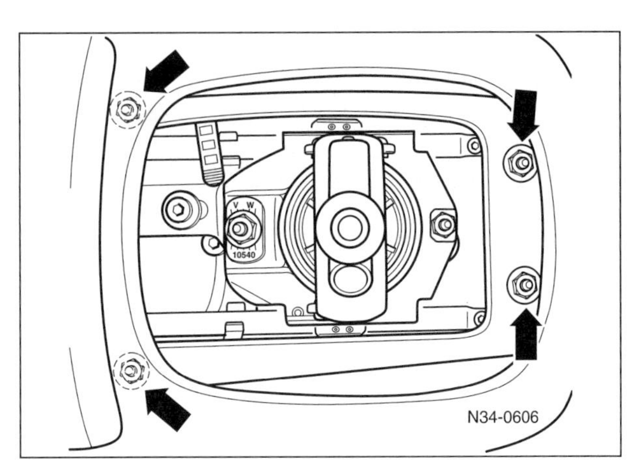

◄ Remove nuts securing selector mechanism housing (**arrows**).

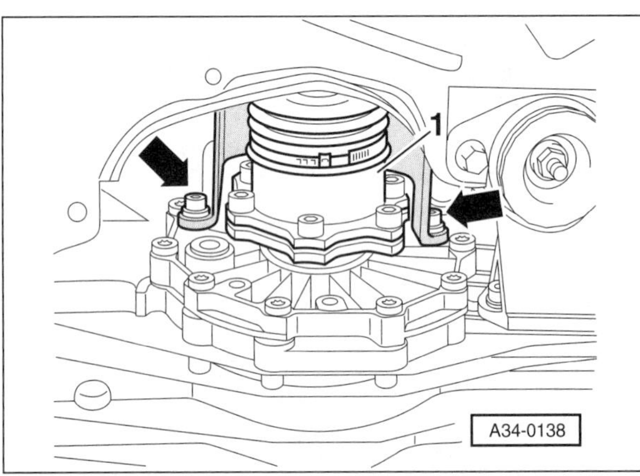

◄ Remove heat shield mounting nuts (**arrows**) for left inner drive axle joint.

– Detach left inner drive axle (**1**). Lift axle towards the front and tie up.

– Remove rear section of exhaust system (rearward of exhaust pipe clamps).

– Remove heat shield above driveshaft.

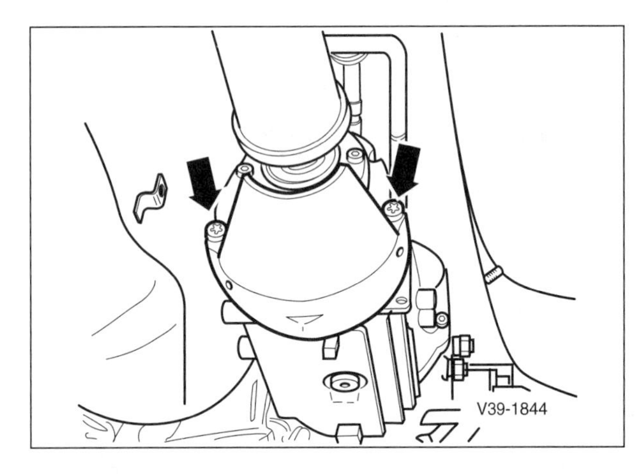

◄ Remove heat shield for driveshaft from differential (**arrows**).

– Remove driveshaft.

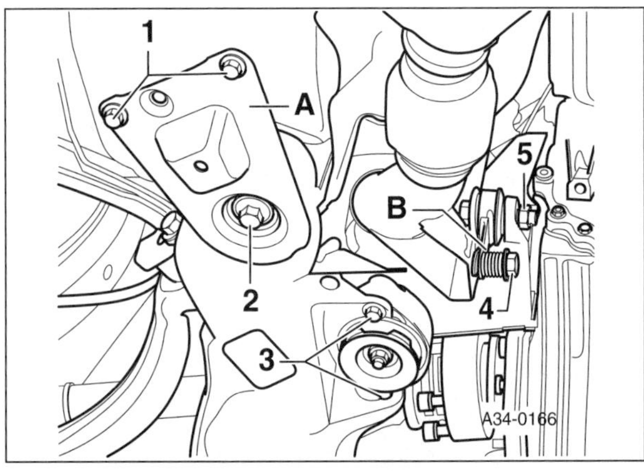

◄ Unscrew bolts (**1** and **2**) on left and right.

BentleyPublishers.com
BentleyPublishers.com—All Rights Reserved

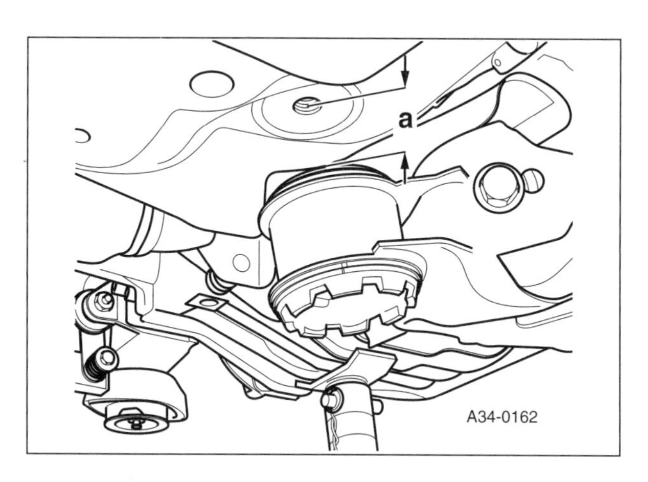

◄ Lower rear of subframe (distance **a**) a maximum of 50 mm (2 in).

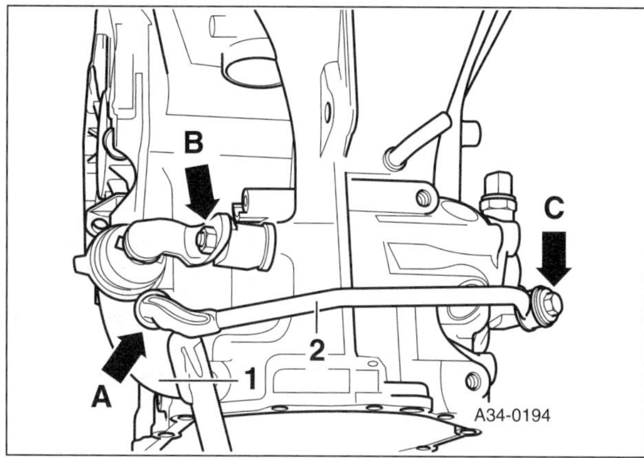

◄ Do not pull ball head (**A**) of connecting rod (**2**) off shift rod (**1**). The ball head is destroyed when pulled off.

– Nut (**B**) and bolt (**C**) must be removed to remove shift rod.

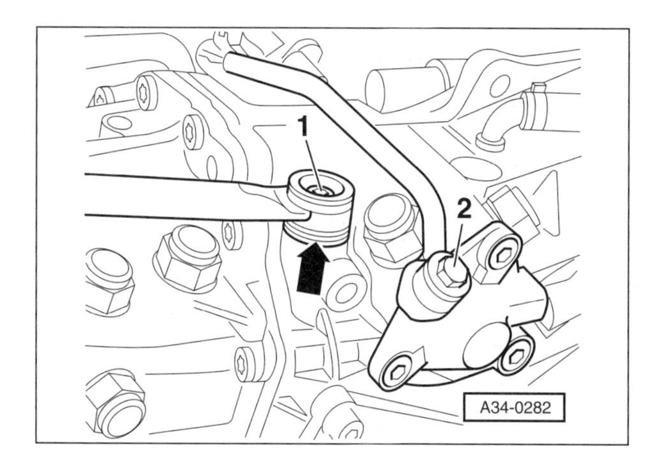

◄ Unbolt connecting rod (**2**) on right side of transmission.

– Remove bolt from push rod (**1**).

NOTE *—*

- *Reuse any washers installed between the pushrod and the transmission (**arrow**).*

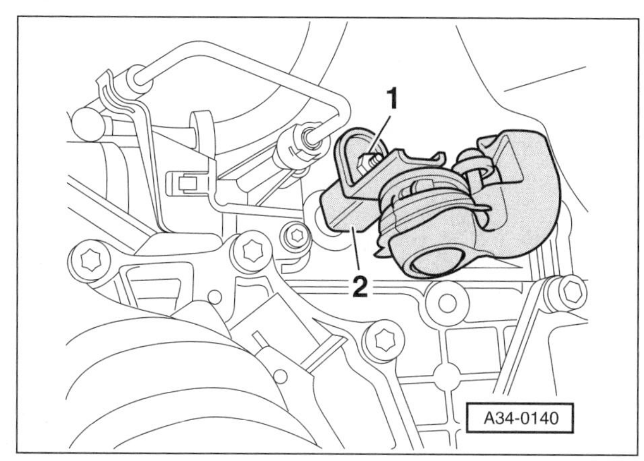

◄ Unscrew nut (**1**) and pull selector rod lever (**2**) off transmission selector shaft.

– Swing gear shift housing with selector rod and push rod down and remove.

BentleyPublishers.com
BentleyPublishers.com—All Rights Reserved

– Installation is carried out in reverse order. Note the following:
  - Bolt on driveshaft.
  - Adjust driveshaft.
  - Adjust gear selector mechanism.
  - Align exhaust system free of stress.

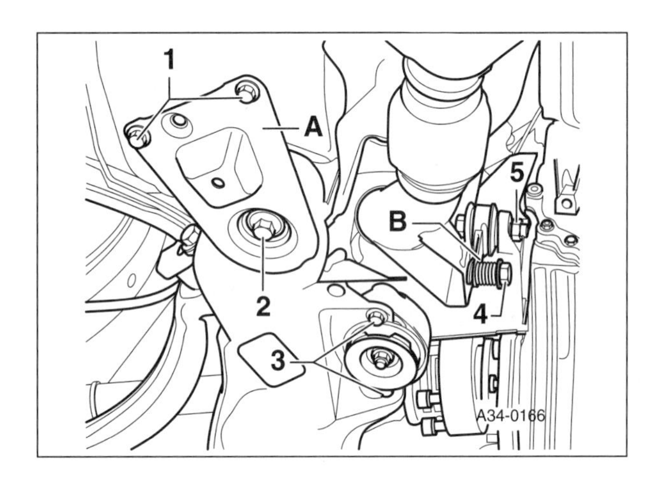

Tightening torques	
Bolt (**1**)	75 Nm (55 ft-lb)
Combination bolt (**2**)	110 Nm (81 ft-lb) + 90° turn
Gear shift housing to body	10 Nm (7 ft-lb)
Selector rod to transmission	25 Nm (19 ft-lb)
Connecting rod to transmission	25 Nm (19 ft-lb)
Push rod to transmission	40 Nm (30 ft-lb)
Drive axle to drive flange: • M8 • M10	 40 Nm (30 ft-lb) 70 Nm (52 ft-lb)
Heat shield for drive axle	25 Nm (19 ft-lb)
Clamp for exhaust pipe	40 Nm (30 ft-lb)

Gear selector mechanism (6-speed), adjusting

– Adjustment requirements:
  - Selector mechanism, operating and relay elements must be in good condition.
  - Selector mechanism must move freely.
  - Transmission, clutch and clutch mechanism must be in good condition.
  - Transmission in neutral.

– Unscrew gear shift knob from gear shift.

NOTE—
- *Shift cover is removed together with the cover for center console.*

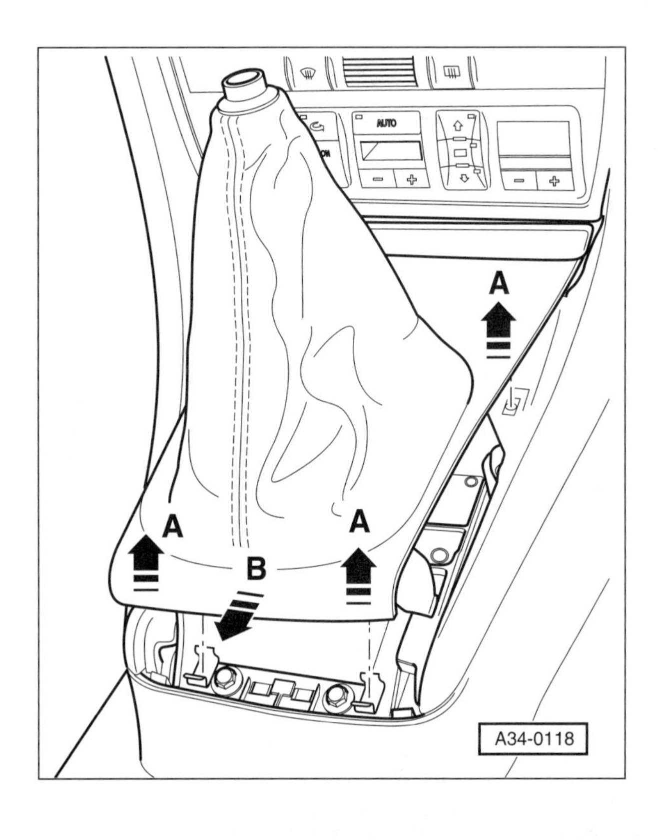

◄ Use a plastic prying tool to slightly lift cover for center console upward (**A**).

– Pull cover slightly back (**B**) and lift off complete cover.

BentleyPublishers.com
BentleyPublishers.com—All Rights Reserved

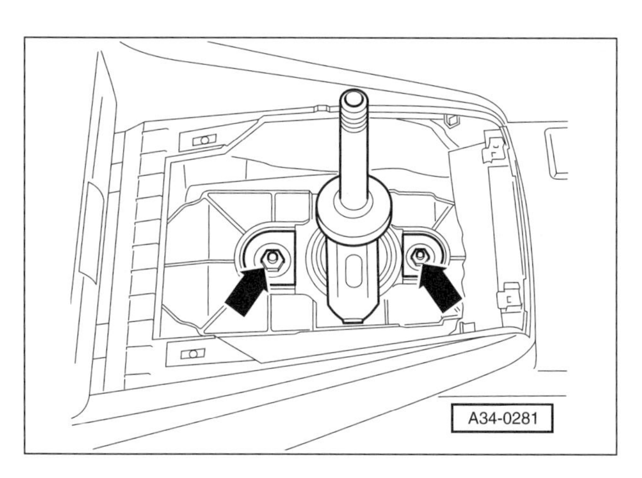

◀ Remove noise insulation for selector mechanism housing (**arrows**).

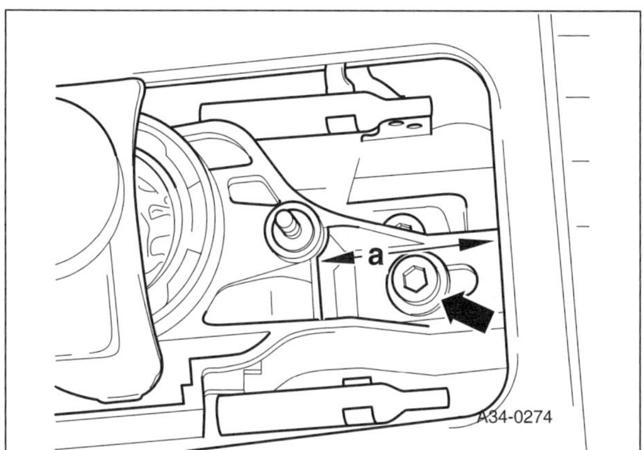

◀ Measure distance between body and rear push rod (in selector mechanism). Distance **a** should equal 43 mm. If not, continue with procedure.

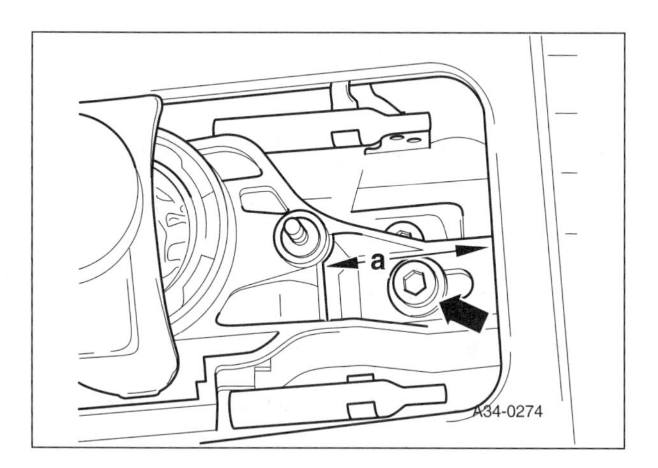

◀ Loosen bolt (**arrow**) for pivot rod.

- Rear pivot rod (in shift control mechanism) must move freely back and forth on sliding piece.

– Adjust measurement **a** by moving pivot rod rear (in shift control mechanism).

– Tighten bolt for pivot rod.

Tightening torque	
Pivot rod adjustment bolt	25 Nm (18 ft-lb)

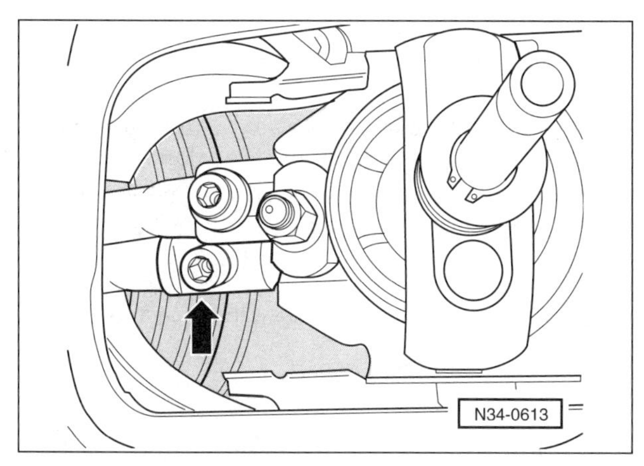

◀ Loosen bolt for selector rod (**arrow**). Connection between selector rod and selector mechanism should move freely.

BentleyPublishers.com
BentleyPublishers.com—All Rights Reserved

Adjust gear shift as follows:

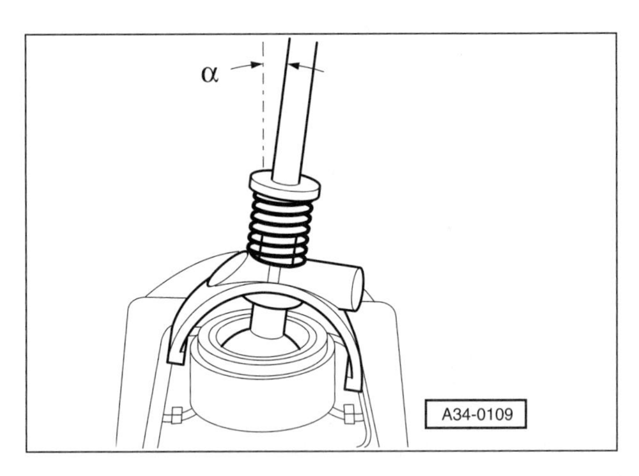

◀ Gear shift vertical, maximum inclination of 3° to right (**a**).

NOTE —

- *Illustration shows gear shift from behind (looking towards front of vehicle)*

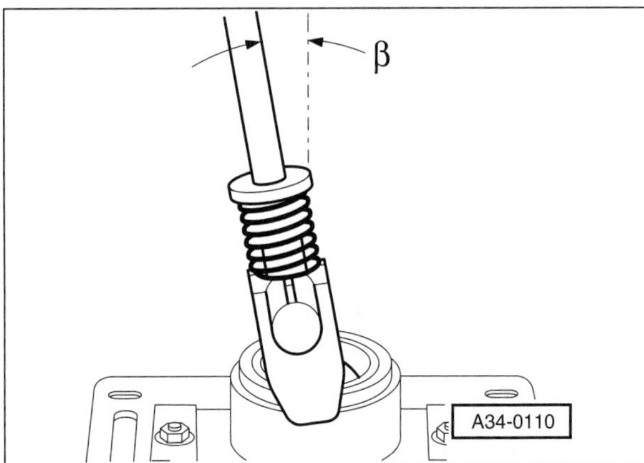

◀ Gear shift inclined slightly backwards (approx. 7°).

NOTE —

- *Illustration shows gear shift from the right.*

– Hold gear shift in this position.

– Tighten selector rod bolt.

Tightening torque	
Selector rod adjustment bolt	25 Nm (18 ft-lb)

NOTE —

- *Gear shift must remain in the same position while bolt is being tightened.*

Gear shift adjustment (6-speed), checking

– Gear shift lever must rest in the 3rd / 4th gear gate when transmission is in neutral.

– Check operation of 1st and 2nd gear stop.

– Engage 2nd gear and push gear shift to the left against the stop.

– Reduce pressure on gear shift until it moves back to pressure point.
  - Spring-back measured at gear shift handle: 3-5 mm.

– Check that all gears can be engaged.

– Check operation of reverse gear lock.
  - It should only be possible to engage reverse gear after pressing the gear shift down to overcome the reverse gear lock.
  - It must be possible to move the gear shift, without pushing and without force, forwards from the reverse gear lock to the 3rd/4th gear plane.

– If the gear shift setting is incorrect, continue with procedure.

BentleyPublishers.com
BentleyPublishers.com—All Rights Reserved

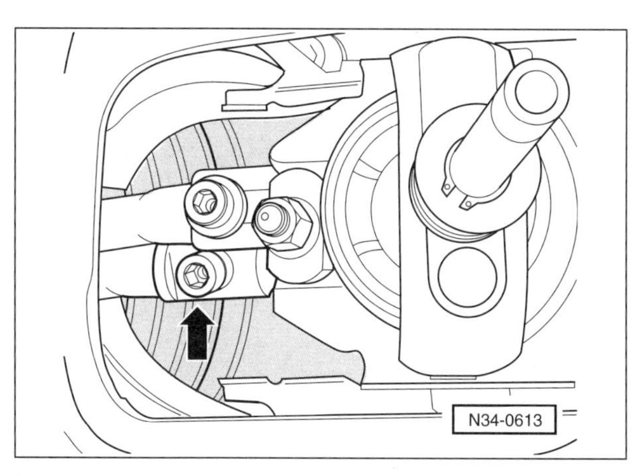

Loosen bolt for selector rod (**arrow**).

NOTE —

- *The angle of forward / backward inclination of the gear shift must not be changed while the following adjustments are being made.*

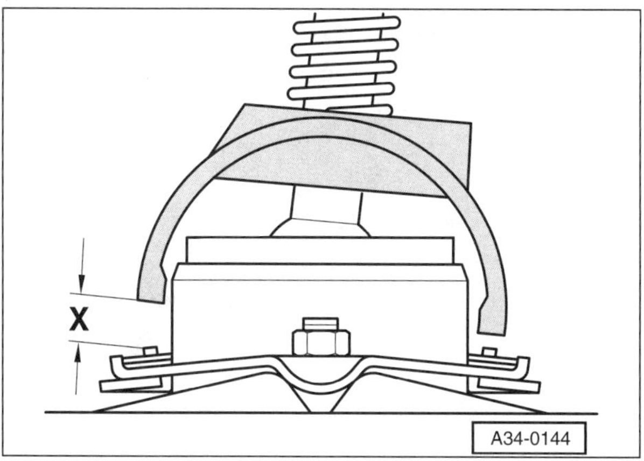

Move gear shift to the left or to the right until distance **x** is 8.5 mm (0.33 in).

– Hold gear shift in this position.

– Tighten selector rod bolt.

– Check gear shift setting again.

– Fit covers and gear shift knob.

Tightening torque	
Selector rod to selector fork	25 Nm (18 ft-lb)

TRANSMISSION SERVICE

Transmission (5-speed), removing and installing

– Special tools:
- Transmission jack V.A.G 1383 A
- Transmission support 3282
- Adjustment plate 3282/10
- Assembling appliance 3139
- Wrench set for oxygen sensors 3337
- Grease G 000 100

– Switch ignition OFF and disconnect battery ground (-) strap.

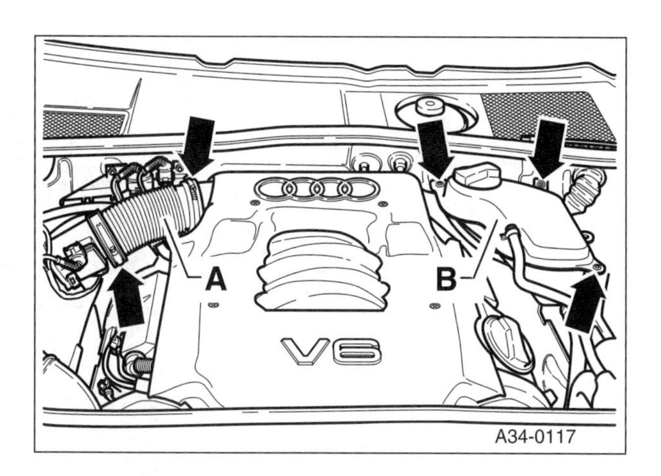

Remove intake hose (**A**).

– Unbolt coolant system expansion tank (**B**) and lay to one side.

– Remove oxygen sensors on left and right exhaust pipes with special tool 3337 and move clear to the side.

– Unscrew engine/transmission securing bolts accessible from above.

BentleyPublishers.com
BentleyPublishers.com—All Rights Reserved

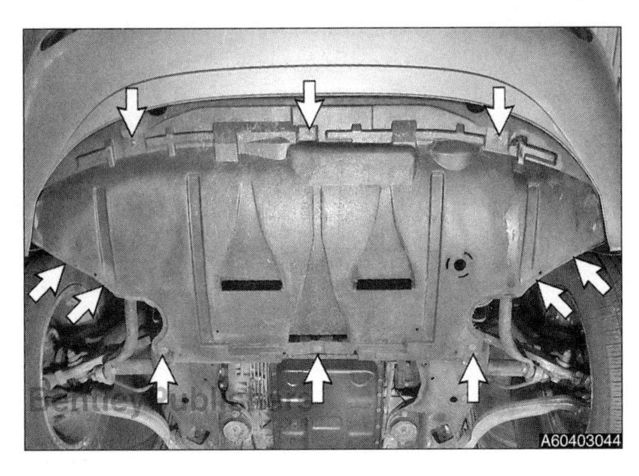

Remove lower engine cover (**arrows**). See **03 Maintenance** for more information.

Unbolt lower engine cover bracket (**arrow**).

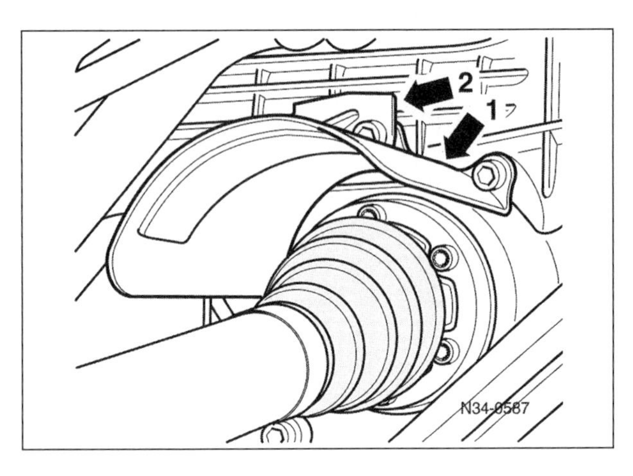

Remove bolts (**1** and **2**) for heat shields above right axle shaft.

- Remove heat shield above left axle shaft.

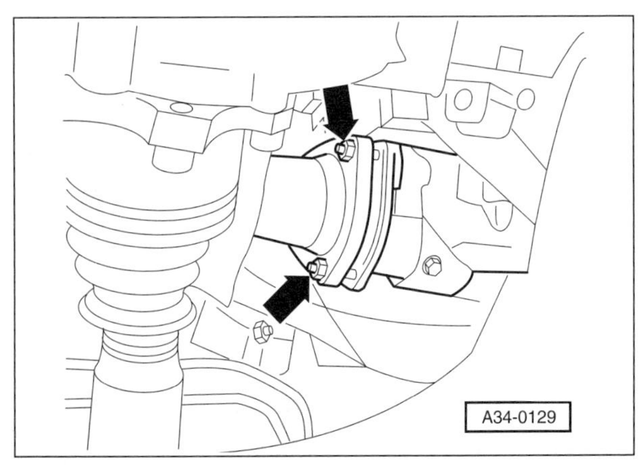

Unbolt front exhaust pipes with catalytic converters from left and right exhaust manifolds (3 nuts on each side) (**arrows**).

- Remove clamping sleeves of exhaust system and remove catalytic converters together with exhaust pipe downward.

BentleyPublishers.com
BentleyPublishers.com—All Rights Reserved

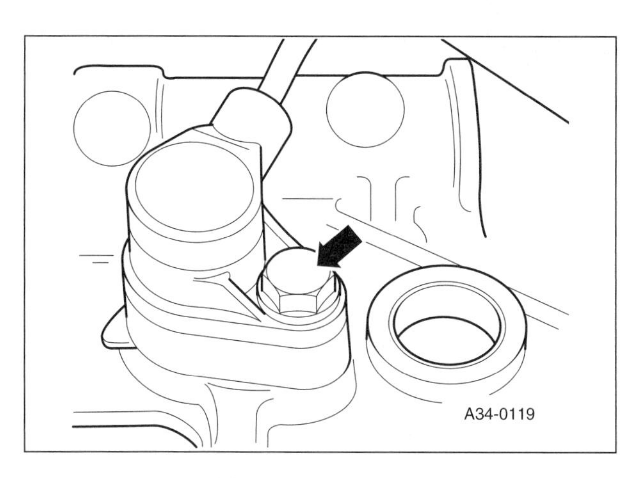

◀ Remove engine speed sender (G28) on left of transmission (**arrow**) and move clear to side.

– Remove drive axles from drive flanges and tie up as high as possible; do not damage protective coating.

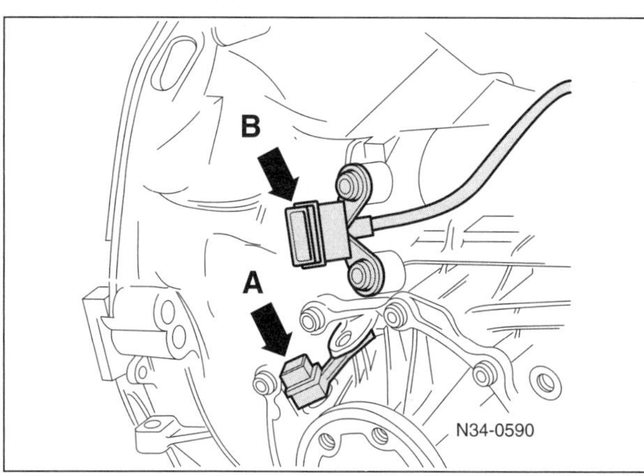

◀ Pull connector off wheel speed sender (**A**).

– Pull off back-up light connector (**B**). Disconnect all other electrical connections and ground wires from transmission and from the engine / transmission securing bolts.

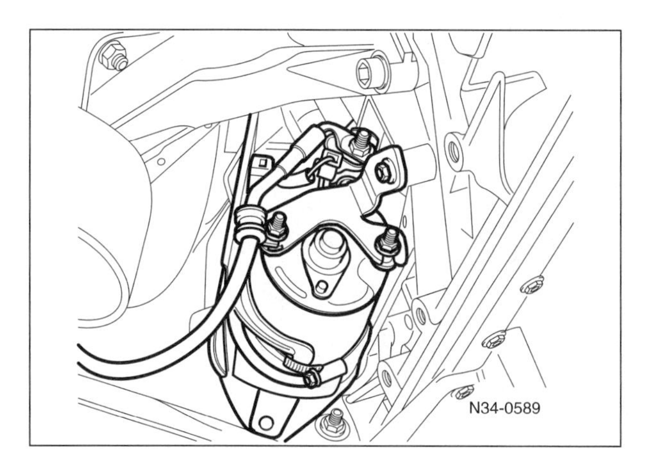

◀ Unbolt starter from engine/transmission and secure as necessary.

NOTE—

- *Starter cables do not have to be disconnected.*

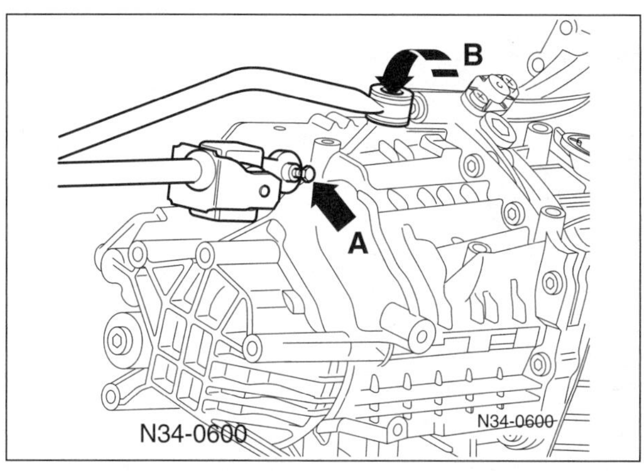

◀ Detach selector rod arrow (**A**).

– Unscrew Allen bolt out of push rod (**B**).

BentleyPublishers.com
BentleyPublishers.com—All Rights Reserved

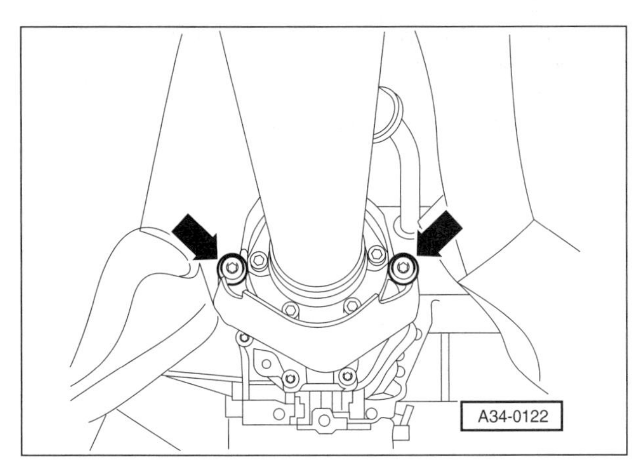

– Unbolt heat shield for driveshaft from differential cover (**arrows**).

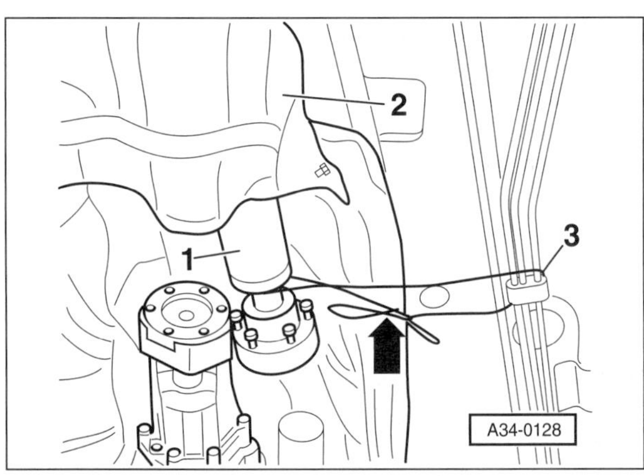

◀ Unbolt driveshaft (**1**) from transmission and rest it on heat shield (**2**).

– Secure driveshaft to fuel pipe bracket (**3**) with wire (**arrow**).

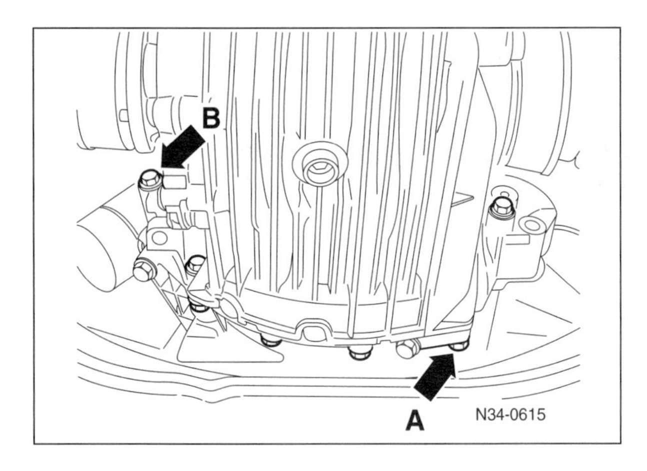

◀ Unscrew bottom engine / transmission securing bolts, except for bolts indicated by arrows (**A** and **B**).

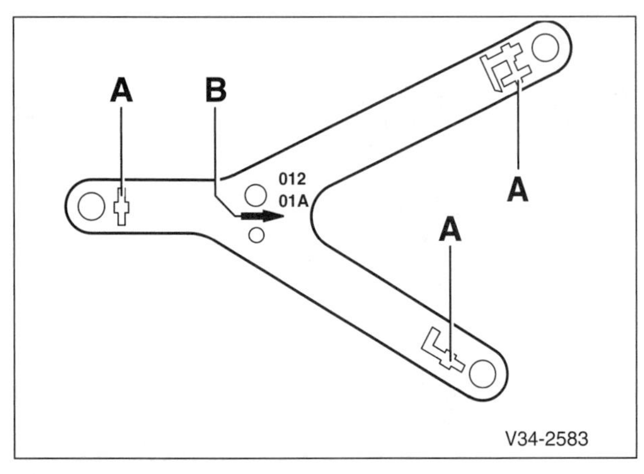

◀ Set up transmission support 3282 for removing manual transmission 01A (all wheel drive) with adjustment plate 3282/10 and attach to transmission jack VAG 1383 A.

NOTE —

- *Attachments (**A**) are shown; arrow (**B**) points in the direction of travel.*
- *Adjustment plate 3282/10 only fits in one position.*

BentleyPublishers.com
BentleyPublishers.com—All Rights Reserved

– Run transmission jack VAG 1383 A with transmission support 3282 in under the transmission and take up the weight of the transmission.

NOTE—

- *If the transmission lift 3282 is not available, the transmission can be removed and installed using transmission lift 1383A and universal mounts VAG1359/2.*

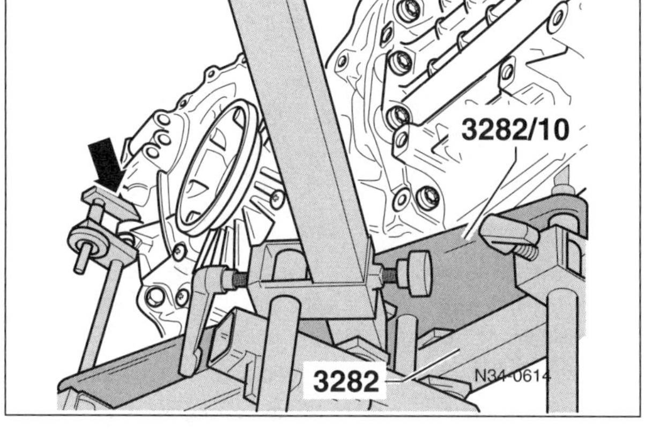

◀ Align adjustment plate parallel to transmission and lock safety support (**arrow**) on transmission.

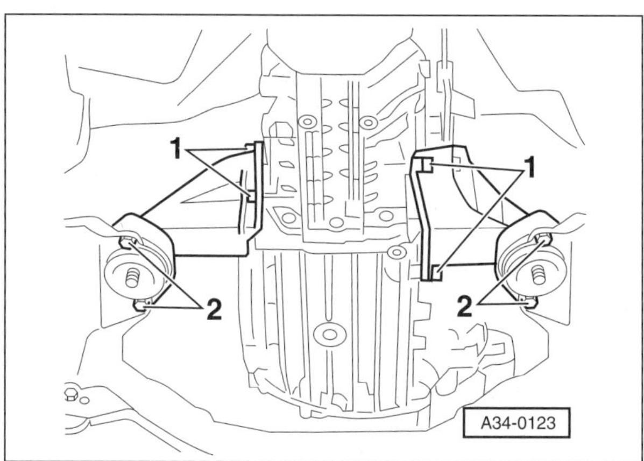

◀ Unbolt right and left transmission supports complete with bonded rubber mountings from transmission and subframe (**1** and **2**).

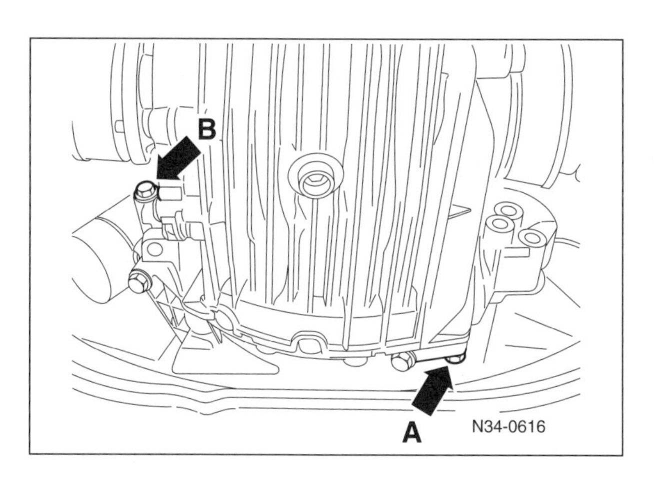

◀ Remove remaining engine / transmission securing bolts (**A** and **B**).

– Press transmission off dowel sleeves and lower carefully with transmission jack VAG 1383 A just far enough for access to clutch slave cylinder.

NOTE—

- *When lowering transmission ensure hydraulic pipe and hose to slave cylinder are not damaged.*

BentleyPublishers.com
BentleyPublishers.com—All Rights Reserved

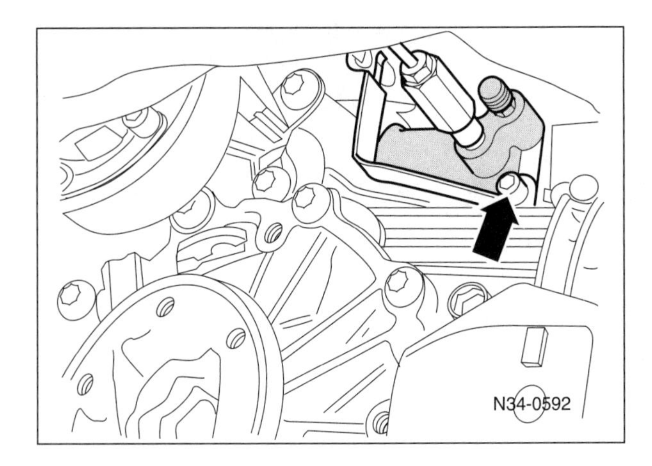

Remove clutch slave cylinder (**arrow**) and secure with wire. Do not disconnect fluid lines.

NOTE—
- *Do not depress clutch pedal after detaching slave cylinder.*

— Lower transmission completely.

NOTE—
- *Remove carefully, watching for hoses, electrical wires or mechanical parts that might become snagged.*

Installing

— Installation is reverse of removal. Remember to:
- Clean input shaft splines and (if reusing clutch disc) hub splines. Remove corrosion and apply only a very thin coating of grease G 000 100 to splines. Do not grease guide sleeve.
- Check clutch release bearing for wear. Replace if necessary.
- Coat slave cylinder push rod tip with thin layer of copper grease.
- Check whether dowel sleeves for aligning engine / transmission are fitted in cylinder block. Install if necessary.
- Use tap to clean threaded holes for mounting clutch slave cylinder to transmission and shift coupling to shift rod.
- Replace self-locking nuts.

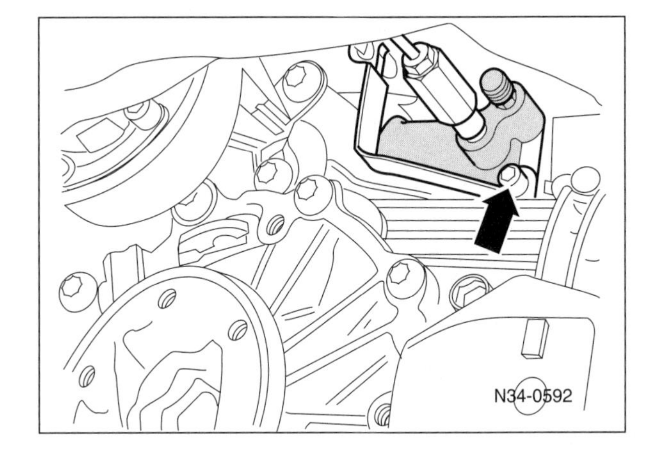

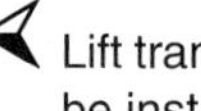

Lift transmission until clutch slave cylinder and fluid line bracket can be installed (**arrow**).

— Fit intermediate plate between transmission and engine on engine dowel sleeves.

— Ensure that intermediate plate is correctly positioned on engine.

— Install lower engine / transmission securing bolts and bolts for starter. Install starter.

— Before installing, use a tap to clean threads in transmission drive axle flange to remove traces of thread locking compound.

— Replace gaskets on axle shafts and on front of driveshaft.

— Bolt on drive axles and drive axle heat shields.

— Bolt on driveshaft heat shields and driveshaft.

— Align exhaust system so it is free of stress.

— Insert oxygen sensors and tighten.

— If equipped, install the engine speed sensor (G28).

— Check adjustment of shift and pivot rod and adjust if necessary.

— Check oil level in transmission.

— After connecting battery, enter anti-theft code for radio.

BentleyPublishers.com
BentleyPublishers.com—All Rights Reserved

– Close electric windows in front doors all the way to their top positions using electric switches. Then operate all electric window switches again for at least one second in the "close" direction to activate automatic one-touch function.

– Set clock to correct time.

Tightening torques

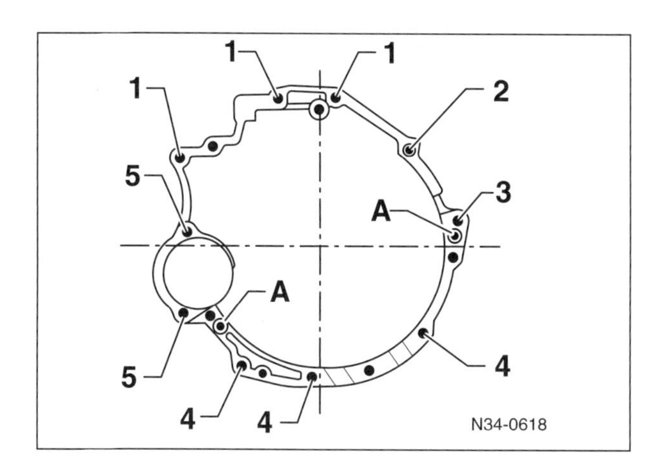

Transmission to engine (6-cylinder)

Item No.	Bolt	Qty.	Torque
1	M 12 x 67	3	65 Nm (48 ft-lb)
2	M 12x 90	1	65 Nm (48 ft-lb)
3	M 12 x 80	1	65 Nm (48 ft-lb)
4	M 10 x 45	3	45 Nm (33 ft-lb)
5 (lower)	M 10 x 135	1	45 Nm (33 ft-lb)
5 (upper)	M 12 x 130	1	65 Nm (48 ft-lb)
A = centering sleeves			

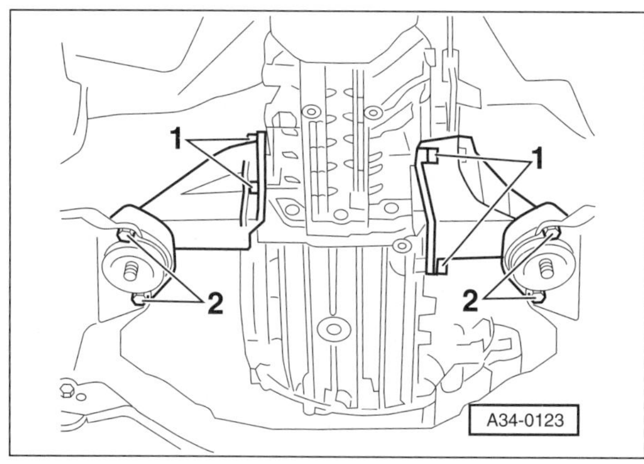

Transmission / engine mounts

Item No.	Bolt	Qty.	Torque
1	M 10 x 35	3	40 Nm (30 ft-lb)
2	M 8x 20	2	20 Nm (15 ft-lb)

Tightening torques	
Axle shaft heat shield to transmission	23 Nm (17 ft-lb)
Bracket for noise insulation to body	10 Nm (7 ft-lb)
Catalytic converter to mounts	25 Nm (19 ft-lb)
Clamp for exhaust pipe	40 Nm (30 ft-lb)
Clutch slave cylinder to transmission	20 Nm (15 ft-lb)
Drive axle to transmission flange: • M8 • M10	 40 Nm (30 ft-lb) 70 Nm (52 ft-lb)
Driveshaft heat shield to differential cover	25 Nm (19 ft-lb)
Driveshaft to transmission	55 Nm (41 ft-lb)
Engine speed sender (G28) to block	10 Nm (7 ft-lb)
Pushrod to transmission	40 Nm (30 ft-lb)
Selector rod to transmission	23 Nm (17 ft-lb)

BentleyPublishers.com
BentleyPublishers.com—All Rights Reserved

Transmission (6-speed), removing and installing

– With ignition switched OFF and disconnect battery ground (-) strap.

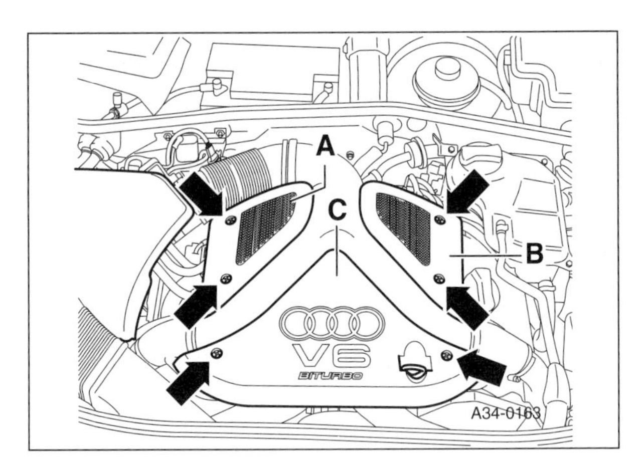

◄ Loosen bolts (**arrows**) and remove engine cover panels (**A**, **B**, and **C**). See **03 Maintenance** for more information.

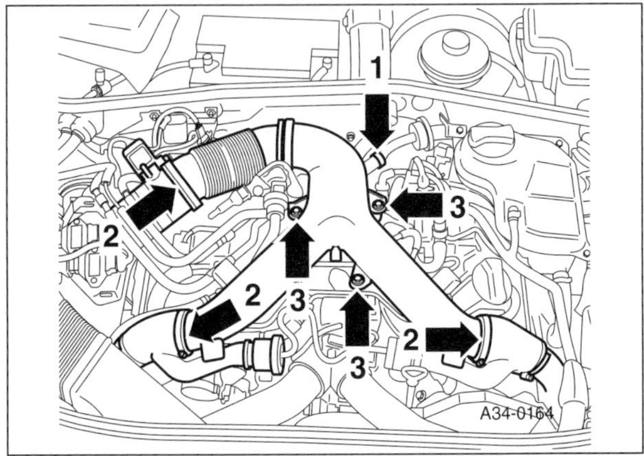

◄ Detach air intake duct. To do this, pull off retainer catch (**1**), loosen clamps (**2**) and unscrew bolts (**3**).

– Remove cooling system expansion reservoir and lay to one side.

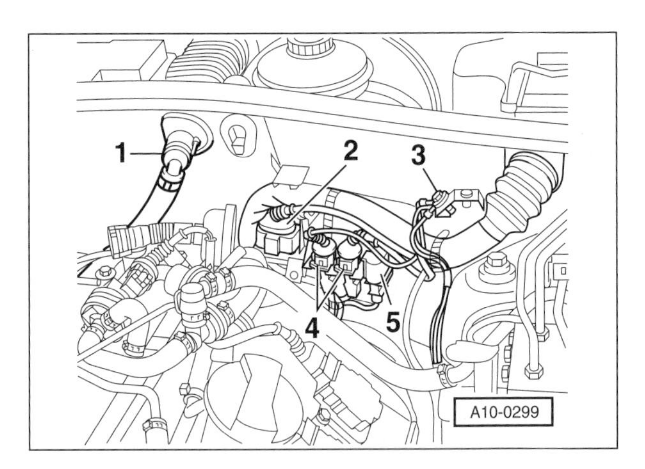

◄ Unplug oxygen sensor connector on left (**2**) and move wire clear.

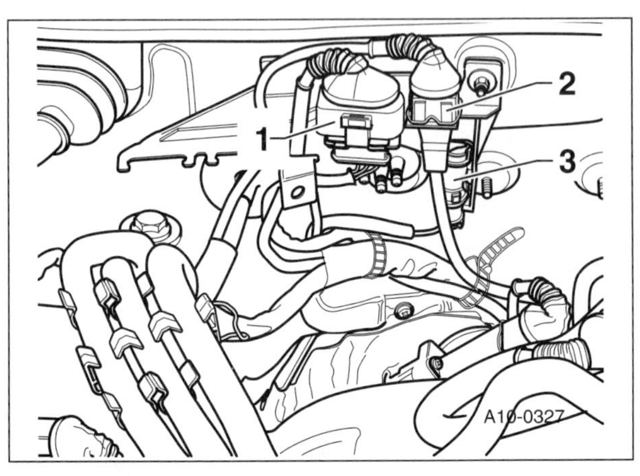

◄ Unplug oxygen sensor connector on right (**1**) and move wire clear.

BentleyPublishers.com
BentleyPublishers.com—All Rights Reserved

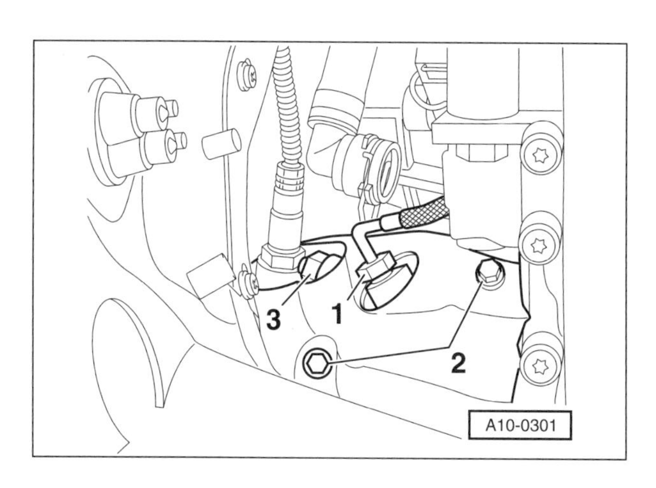

Unscrew two bolts securing heat shield (**2**) on exhaust pipe (right and left).

- Unscrew nuts (**3**) on exhaust pipe (right and left).
- Unscrew engine / transmission mounting bolts accessible from above.

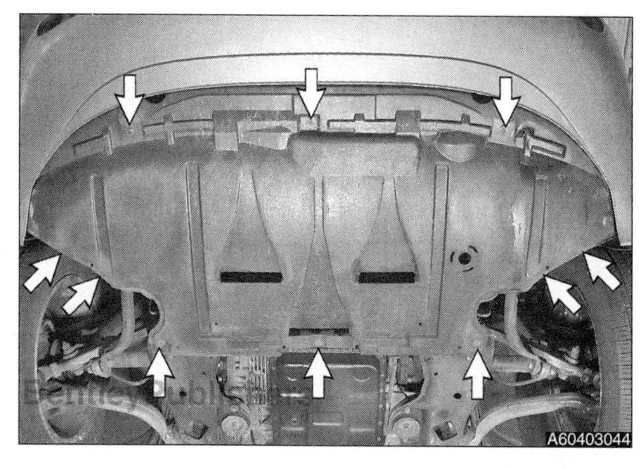

Remove lower engine cover (**arrows**). See **03 Maintenance** for more information.

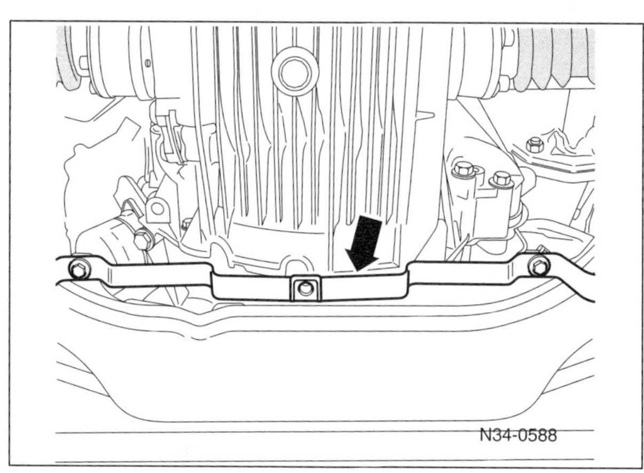

Unbolt lower engine cover bracket (**arrow**).

- Unbolt heat shields from transmission above left and right drive axles.
- Unbolt left and right drive axles from transmission, lift clear and tie aside.

NOTE—

- *Take care not to damage protective coating on drive axles.*

- Loosen exhaust system clamps and push clamps towards rear.

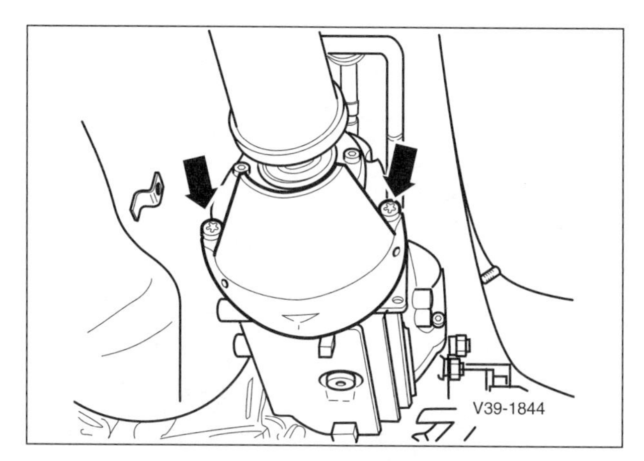

Remove heat shield for driveshaft from cover for differential (**arrows**).

BentleyPublishers.com
BentleyPublishers.com—All Rights Reserved

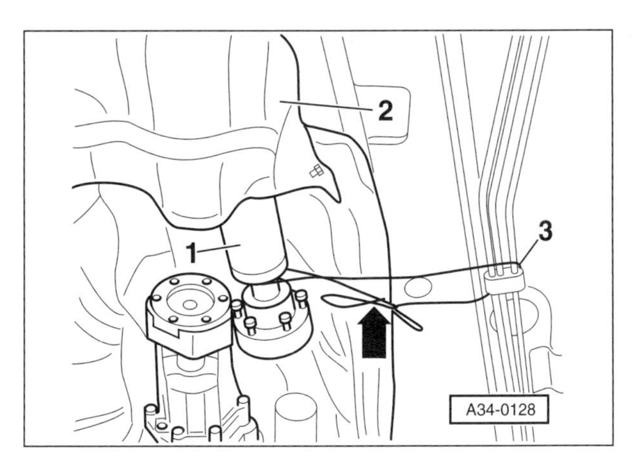

Unbolt driveshaft (**1**) from transmission and rest it on heat shield (**2**).

– Secure driveshaft to fuel pipe bracket (**3**) with wire (**arrow**).

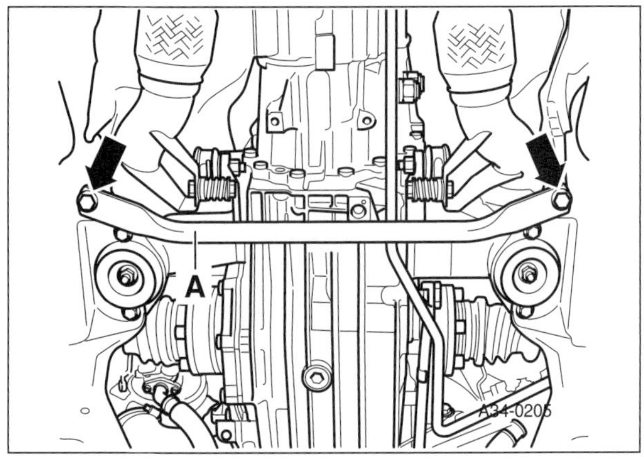

Remove bolts (**arrows**) and remove transverse mount (**A**).

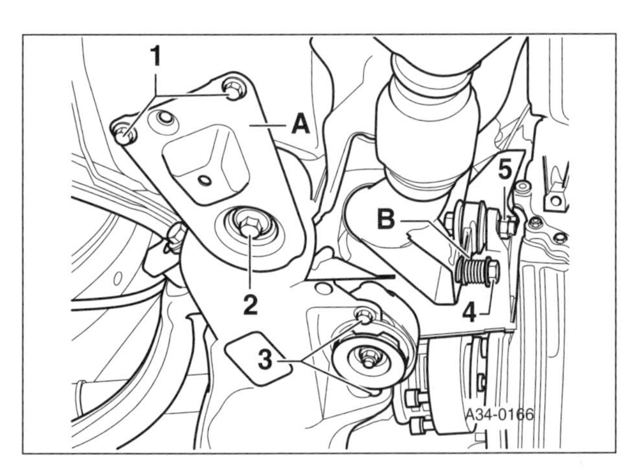

Unscrew and remove bolts (**1** and **2**) on right and left and remove support (**A**).

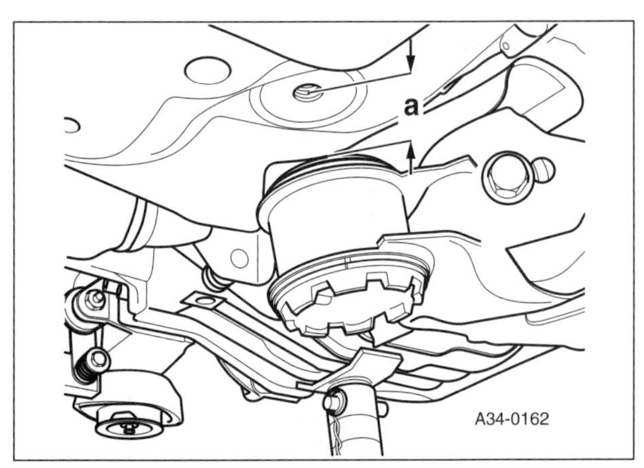

Lower rear of subframe (distance **a**) a maximum of 50 mm (2 in).

BentleyPublishers.com
BentleyPublishers.com—All Rights Reserved

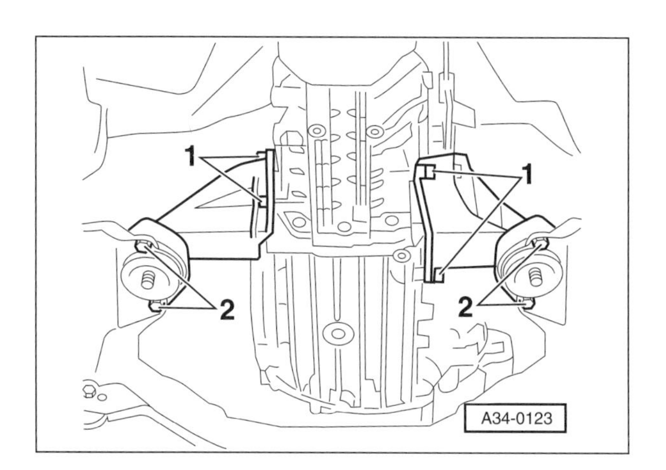

Unscrew bolts (**1** and **2**) and remove side transmission mounts.

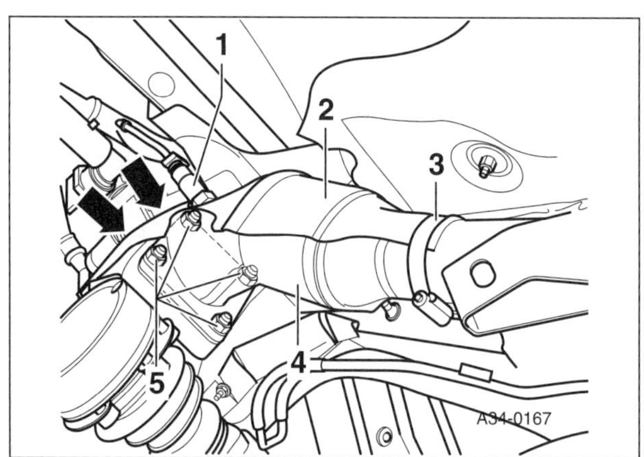

Loosen clamp (**3**), lift heat shield (**2**) slightly and unscrew remaining nuts (**5**).

- Lift heat shield (**2**) and pull out front exhaust pipe together with catalytic converter (**4**).
- Note that oxygen sensor (**1**) and wire have to be guided through opening in heat shield (**2**).

NOTE—

- *Avoid excessive bending of the flexible pipe connection at front exhaust pipe. The angle between the catalytic converter and the front exhaust pipe must not exceed 10°. Otherwise the flexible connection will be damaged.*

Lift transmission to access two lower bolts securing engine to transmission.

- **A** - Axle stand
- **B** - VAG 1383A

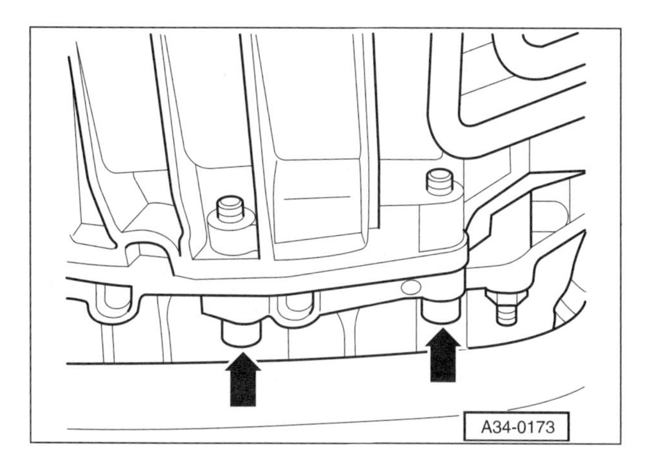

Unscrew two bolts securing transmission to engine (**arrows**).

NOTE—

- *The two M10 x 60 Allen bolts must be replaced by new M10 x 55 bolts (part no. N 104 684 01) so that bolts (**arrows**) can be tightened to the correct torque after installing the transmission.*

- Lower transmission again slowly.

BentleyPublishers.com
BentleyPublishers.com—All Rights Reserved

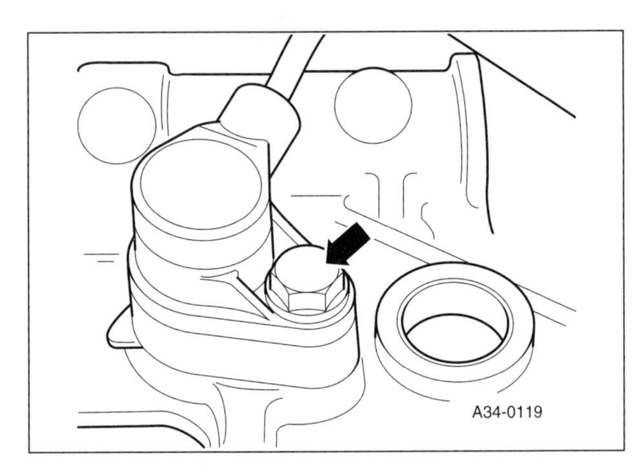

◀ Remove engine speed sender (G28) from left hand side of transmission (**arrow**) and place to one side.

– Pull connector off speedometer sender.

– Pull back-up switch connector off.

– Unbolt starter and push it forward as far as it will go.

NOTE—
- *Do not disconnect starter cables.*

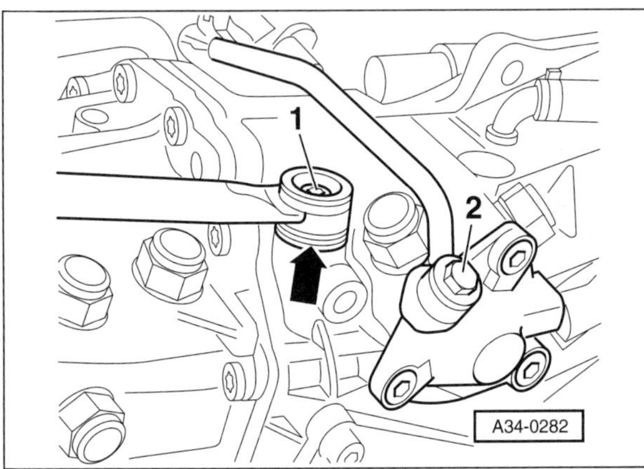

◀ Unbolt connecting rod (**2**) for selector rod on right side of transmission.

– Remove Allen bolt from push rod (**1**).

NOTE—
- *Reuse any washers installed between the pushrod and transmission (**arrow**).*

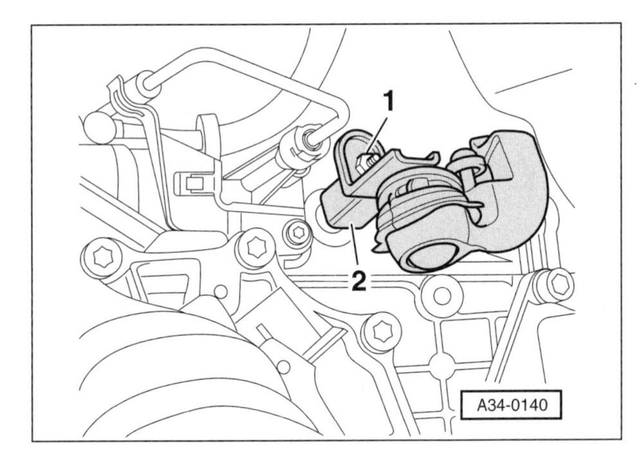

◀ Unscrew nut (**1**) and pull selector rod lever (**2**) off transmission selector shaft.

– Remove all connecting bolts for engine / transmission from below, except for one bolt for securing.

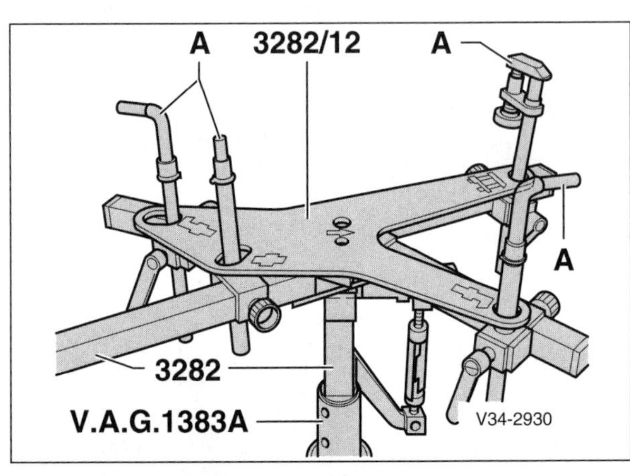

◀ Set up transmission support 3282 with adjustment plate 3282/12 for removal of manual transmission 01E (all-wheel drive) and place on transmission jack VAG 1383 A.

NOTE—
- *The positions for the attachments are shown (**A**).*
- *Adjustment plate 3282/12 can only be fitted in one position.*

BentleyPublishers.com
BentleyPublishers.com—All Rights Reserved

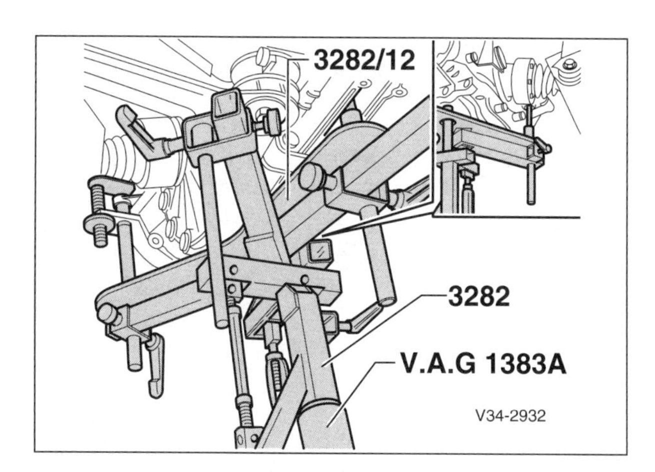

< Support transmission with VAG 1383 and adapter 3282.

NOTE—

- *If adapter 3282 is not available, the transmission can be removed and installed using transmission jack VAG 1383 A.*

– Remove the last remaining engine / transmission securing bolt.

– Press transmission off dowel sleeves and lower carefully just far enough for access to the clutch slave cylinder.

NOTE—

- *When lowering transmission ensure hydraulic line to slave cylinder is not damaged.*

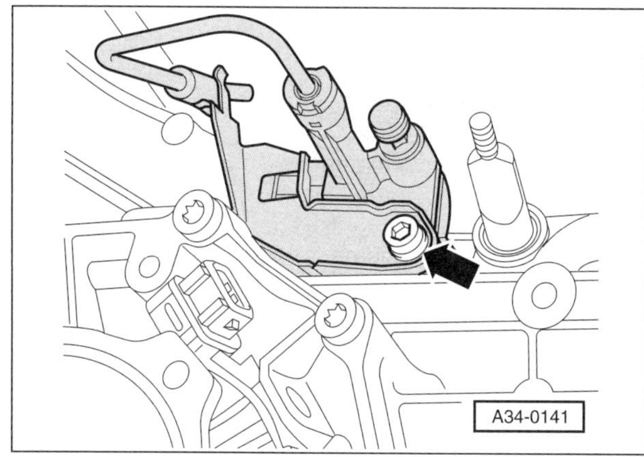

< Remove bolt (**arrow**) and take out slave cylinder from rear. Do not detach hydraulic line.

NOTE—

- *Do not depress clutch pedal after removing slave cylinder.*

– Lower transmission completely.

NOTE—

- *Remove carefully, watching for hoses, electrical wires or mechanical parts that might become snagged.*

– Installation is reverse of removal. Remember to:
  - Replace all seals and gaskets.
  - Replace all self-locking nuts.
  - Check whether dowel sleeves for aligning transmission with engine are in engine flange. Insert if necessary.
  - Clean input shaft splines and (if reusing clutch disc) hub splines. Remove corrosion and apply only a very thin coating of grease G 000 100 to splines. Do not grease guide sleeve.

– Check clutch release bearing for wear and replace if necessary.

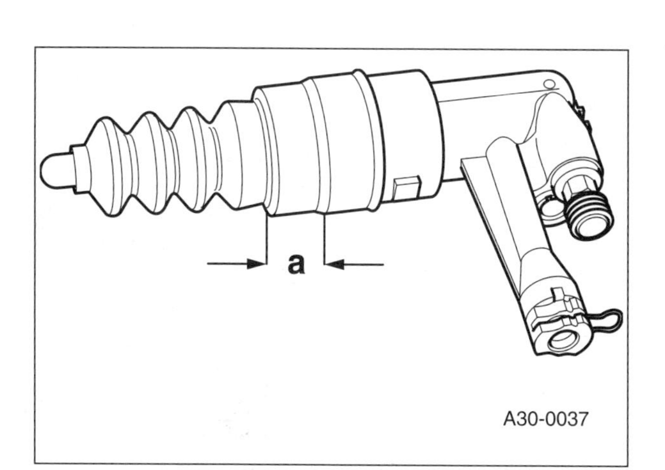

< Coat slave cylinder collar **a** with lithium grease before mounting to transmission housing.

– Coat clutch slave cylinder push rod tip with thin layer of copper grease.

– On vehicle with V6 biturbo engine, fit intermediate plate on dowel sleeves of engine flange.

BentleyPublishers.com
BentleyPublishers.com—All Rights Reserved

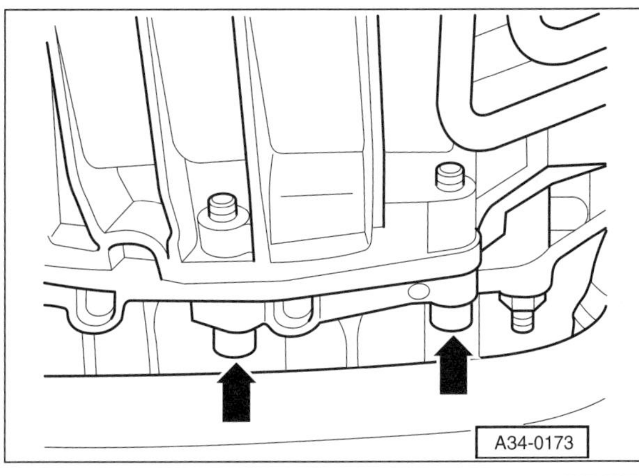

Replace two M10 x 60 Allen bolts with new M10 x 55 bolts (part no. N 104 684 01) so that bolts (**arrows**) can be tightened to correct torque.

– Before installing transmission, tie electrical wiring off to one side so that it cannot be trapped between engine and transmission.

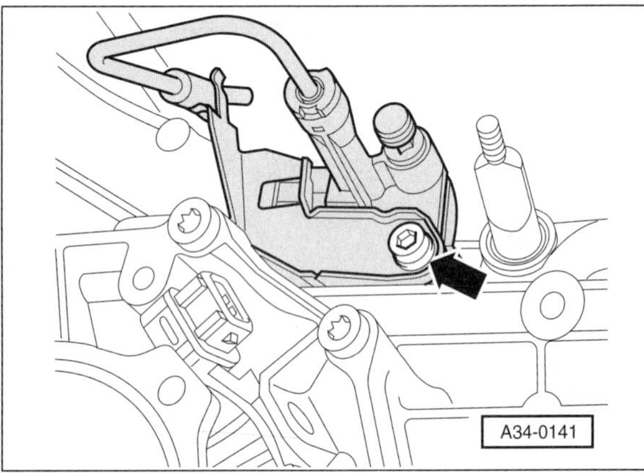

Lift transmission until clutch slave cylinder and hydraulic line bracket can be installed (**arrow**).

– Align exhaust system so it is free of stress.

– Before installing, use a tap to clean threads in transmission drive axle flange to remove traces of thread locking compound.

– Replace gaskets on drive axles and on front of driveshaft.

– Bolt on drive axle heat shields and drive axles.

– Bolt on driveshaft heat shields and driveshaft.

– Check adjustment of selector rod and push rod; readjust if necessary.

– Check oil level in transmission.

– Connect battery and enter anti-theft code for radio.

– Close windows fully using electric window switches. Hold window switches in the "close" position for at least one second to reactivate one-touch function.

– Set clock to correct time.

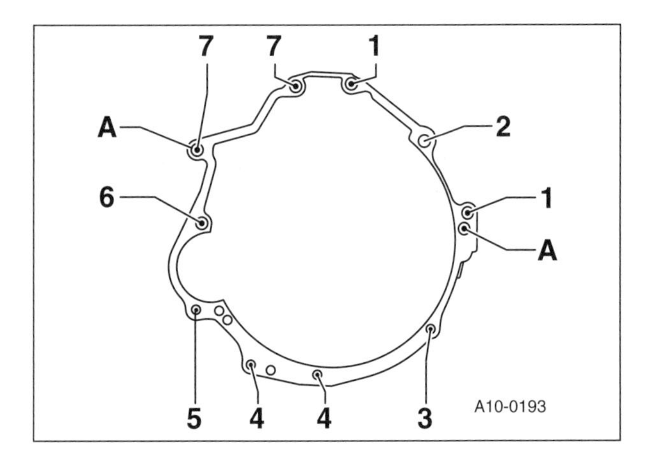

Vehicle with 6-cyl. biturbo engine

Item No.	Bolt	Qty.	Torque
1	M 12 x 90	2	65 Nm (48 ft-lb)
2	M 12 x 100	1	65 Nm (48 ft-lb)
3	M 10 x 60	1	45 Nm (33 ft-lb)
4	M 10 x 60*	2	45 Nm (33 ft-lb)
5	M 10 x 150	1	65 Nm (48 ft-lb)
6	M 12 x 130	1	65 Nm (48 ft-lb)
7	M 12 x 80	2	65 Nm (48 ft-lb)
* Replace two M10 x 60 bolts with new M10 x 55 bolts (part no. 104 684 01) to achieve correct torque specification.			
A = dowel sleeves for aligning			

BentleyPublishers.com
BentleyPublishers.com—All Rights Reserved

Tightening torques	
Bracket for noise insulation to subframe	10 Nm (7 ft-lb)
Catalytic converter to mounting lugs	25 Nm (19 ft-lb)
Clamp for exhaust pipe	40 Nm (30 ft-lb)
Clutch slave cylinder to transmission (always replace)	20 Nm (15 ft-lb)
Connecting rod to transmission	23 Nm (17 ft-lb)
Drive axle to drive flange: • M8 40 • M10 70	 40 Nm (30 ft-lb) 70 Nm (52 ft-lb)
Driveshaft to transmission (always replace)	55 Nm (40 ft-lb)
Engine speed sender (G28) to block	10 Nm (7 ft-lb)
Front exhaust pipe to turbocharger	30 Nm (22 ft-lb)
Heat shield to differential cover	25 Nm (19 ft-lb)
Heat shield to transmission	25 Nm (19 ft-lb)
Heat shield to turbocharger / exhaust pipe	10 Nm (7 ft-lb)
Push rod to transmission	40 Nm (30 ft-lb)
Selector rod to transmission	23 Nm (17 ft-lb)
Transmission support to transmission	40 Nm (30 ft-lb)

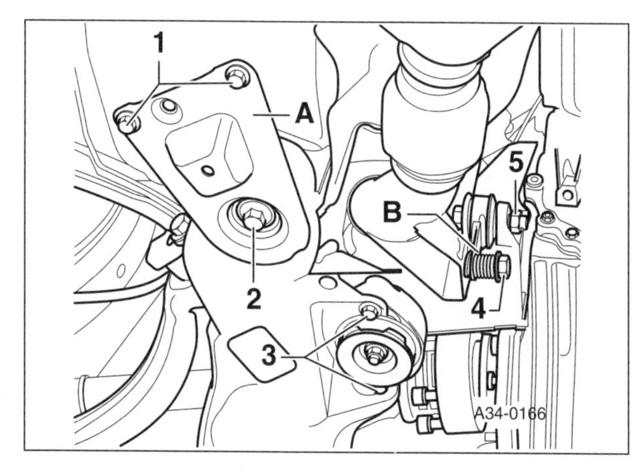

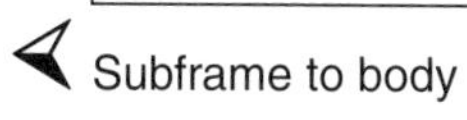
Subframe to body

Tightening torque	
Support (**A**) to body (**1**)	75 Nm (55 ft-lb)
Subframe to body (**2**) (always replace)	115 Nm (85 ft-lb) + 90° turn
Transmission mount to subframe (**3**)	25 Nm (19 ft-lb)

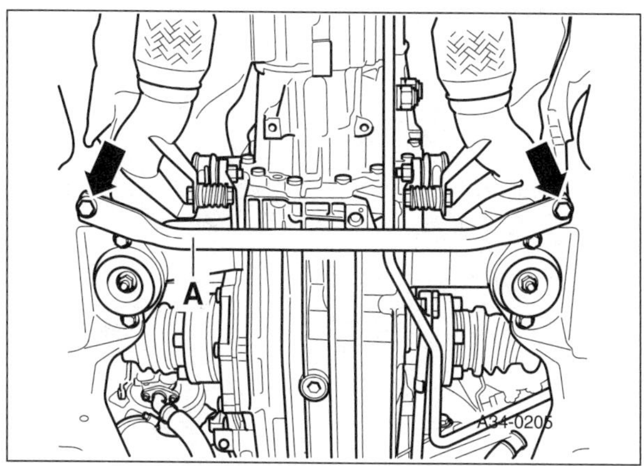

For 6-cylinder biturbo engine only.

Tightening torque	
Cross piece to subframe (always replace) (fit bolt from bottom upwards)	40 Nm (30 ft-lb) + 90°

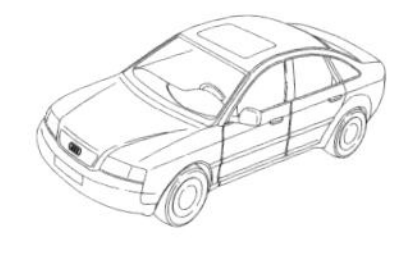

BentleyPublishers.com
BentleyPublishers.com—All Rights Reserved

37 Automatic Transmission

General

This repair group covers external components and adjustments to A6 automatic transmissions. Special tools and equipment are required to remove the automatic transmission. Removal of the automatic transmission is outside the scope of this manual.

A6 models were equipped with one of three different automatic transmission versions, depending on engine type and front-wheel-drive (FWD) or all-wheel-drive (AWD) requirements.

Table a. Automatic transmission codes

Description	Type
5-speed automatic with Tiptronic® AWD	01L
5-speed automatic with Tiptronic® FWD or AWD	01V
Multitronic® or continuously variable transmission (CVT) FWD	01J

Transmissions are identified by type, code, serial number, manufacturer identification code and build date. Transmission identification plates containing some or all of this information are affixed to the transmission housing.

For additional information on identifying engine and transmission, see **03 Maintenance**.

BentleyPublishers.com
BentleyPublishers.com—All Rights Reserved

Back-up light switch function

When automatic transmission is placed in reverse, multifunction transmission range switch (F125) switches back-up lights ON.

When CVT transmission is placed in reverse, transmission control module (J217) indicates gear position to park / neutral position (PNP) relay (J226). PNP relay then switches back-up lights ON.

For more information on back-up lights, see **96 Interior Lights, Switches, Anti-theft**.

TRANSMISSION OIL

Automatic transmission fluid (ATF)

Current specifications for ATF are shown in **Table b**. If fluid container is not marked with appropriate specification, assume that the oil is not suitable.

Table b. ATF specifications	
Transmission types 01L, 01V	VAG ATF part no. G 052 162 A2
Transmission type 01J (Multitronic®, continuously variable transmission or CVT)	CVT ATF part no. G 052 180 A2

CAUTION—
- *ATF used in CVT transmission is different from fluid used in the conventional automatic transmission. Only use ATF specifically intended for CVT transmission.*

Approximate transmission refill capacities in **Table c** are listed by transmission type. Automatic transmission application information is in **Table a** in this repair group.

Table c. ATF capacities (approximate)

	Oil capacity in liters (US qt)	
Transmission type	**New fill**	**Change**
01L	9.8 (10.4)	3.5 - 4.0 (3.7 - 4.2)
01V	9.0 (9.5)	3.5 - 4.0 (3.7 - 4.2)
01J (CVT)	7.5 (7.9)	-

BentleyPublishers.com
BentleyPublishers.com—All Rights Reserved

ATF level

Make sure vehicle is level during this operation. If front of vehicle is higher than rear, such as when driven up on ramps, oil inside transmission pools toward rear, and level cannot be accurately determined.

Checking level

ATF level varies according to temperature as volume increases with a rise in temperature. Therefore, to check ATF level accurately, measure fluid temperature. ATF temperature may be measured using transmission control module on-board diagnostics connected to VAS 5051 or equivalent scan tool.

– Before starting an ATF level check, observe the following:
  - Electrical consumers OFF.
  - A/C system OFF.
  - Transmission not in emergency running ("limp home") mode.
  - ATF temperature not above approximately 30°C (86°F).
  - Vehicle level.

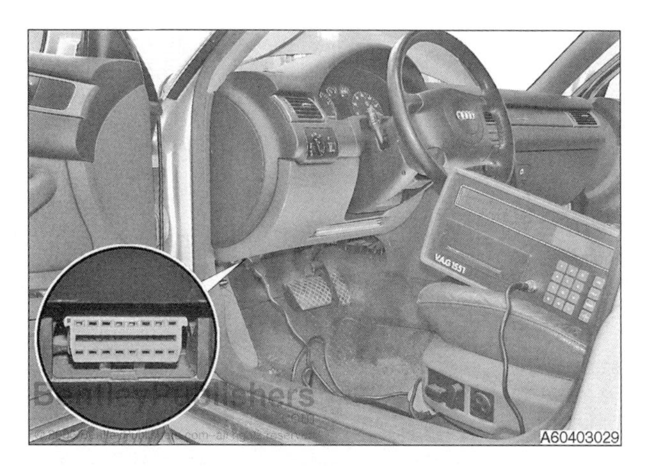

◀ Connect scan tool (VAG 1551 or equivalent) to data link connector (DLC, **inset**), located under driver's knee bar, to left of steering column.

– Start engine with transmission in PARK and read ATF temperature on scan tool.

– Raise vehicle and support safely.

> ***WARNING —***
> - *Make sure the car is stable and well supported at all times. Use a professional automotive lift or jack stands designed for the purpose. A floor jack is not adequate support.*

– Place oil drip pan underneath transmission.

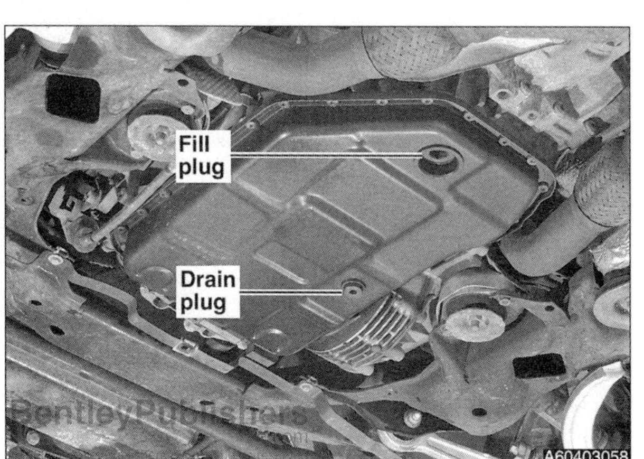

◀ Working underneath vehicle, remove transmission fill plug.

> ***WARNING —***
> - *Hot transmission oil can scald. Wear protective gloves and goggles.*

– ATF level is correct when a little ATF runs out at a temperature between 30°C (86°F) and 45°C (113°F).
  - For tropical climates, allow ATF to drip out until temperature rises to approx. 50°C (122°F).
  - If no ATF runs out when temperature reaches 40°C (104°F), add ATF.

BentleyPublishers.com
BentleyPublishers.com—All Rights Reserved

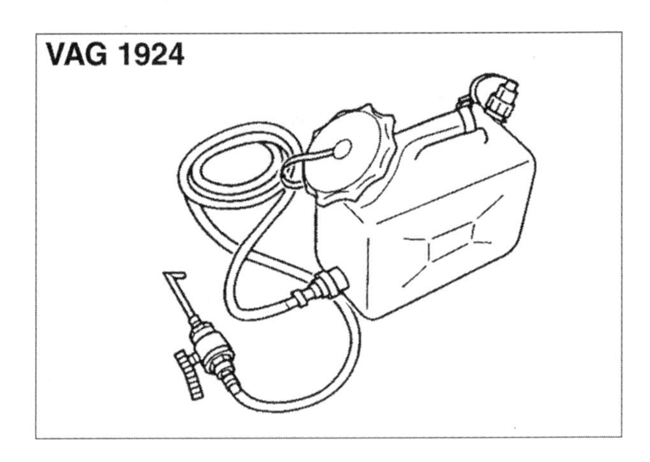

Filling

◀ Use fluid pump with shut-off valve such as VAG 1924 or equivalent to fill transmission.

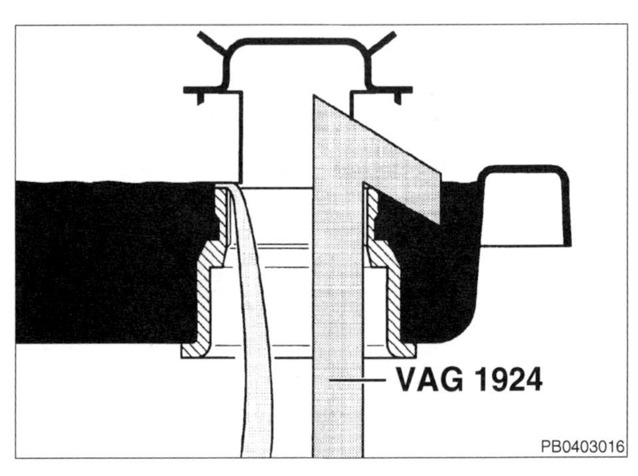

◀ Insert filling hook of VAG 1924 into opening of ATF guard cap in filler.

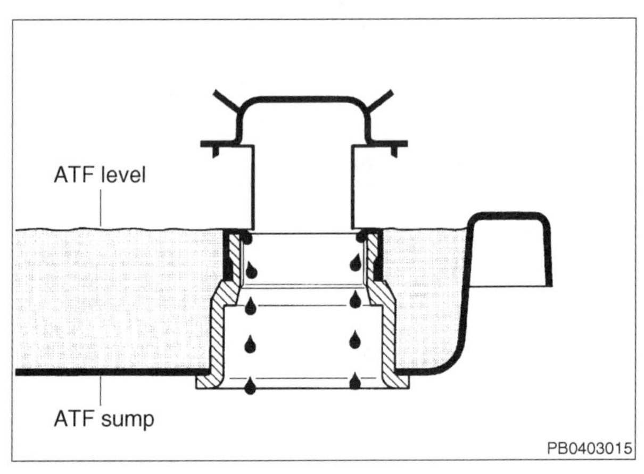

◀ Add ATF until ATF flows from plug opening.

– Replace filler plug. Use new gasket.

Tightening torque	
ATF filler plug to ATF sump (use new gasket)	80 Nm (59 ft-lb)

BentleyPublishers.com
BentleyPublishers.com—All Rights Reserved

CVT ATF level

> ***CAUTION—***
> - *ATF used in 01J CVT (Multitronic®) transmission is different from fluid used in the conventional automatic transmission. Only use ATF specifically intended for CVT transmission,*

Make sure vehicle is level during this operation. If front of vehicle is higher than rear, such as when driven up on ramps, oil inside transmission pools toward rear, and level cannot be accurately determined.

Checking level

ATF level varies according to temperature as volume increases with a rise in temperature. Therefore, to check ATF level accurately, measure fluid temperature. ATF temperature may be measured using transmission control module on-board diagnostics connected to VAS 5051 or equivalent scan tool.

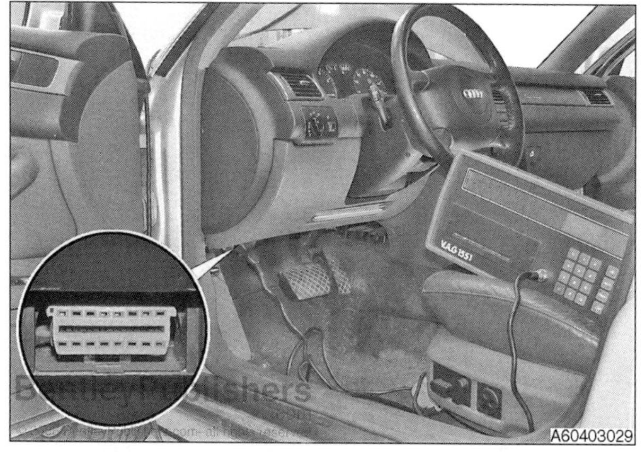

– Before starting an ATF level check, observe the following:
  - Electrical consumers OFF.
  - A/C system OFF.
  - Transmission not in emergency running ("limp home") mode.
  - ATF temperature not above approximately 30°C (86°F).
  - Vehicle level.

◀ Connect scan tool (VAG 1551 or equivalent) to data link connector (DLC, **inset**), located under driver's knee bar, to left of steering column.

– Start engine with transmission in PARK and read ATF temperature on scan tool.

– With engine idling, depress brake pedal and shift through selector lever positions (P, R, N, D). Leave selector in each position approx. 2 seconds.

– Raise vehicle and support safely.

> ***WARNING —***
> - *Make sure the car is stable and well supported at all times. Use a professional automotive lift or jack stands designed for the purpose. A floor jack is not adequate support.*

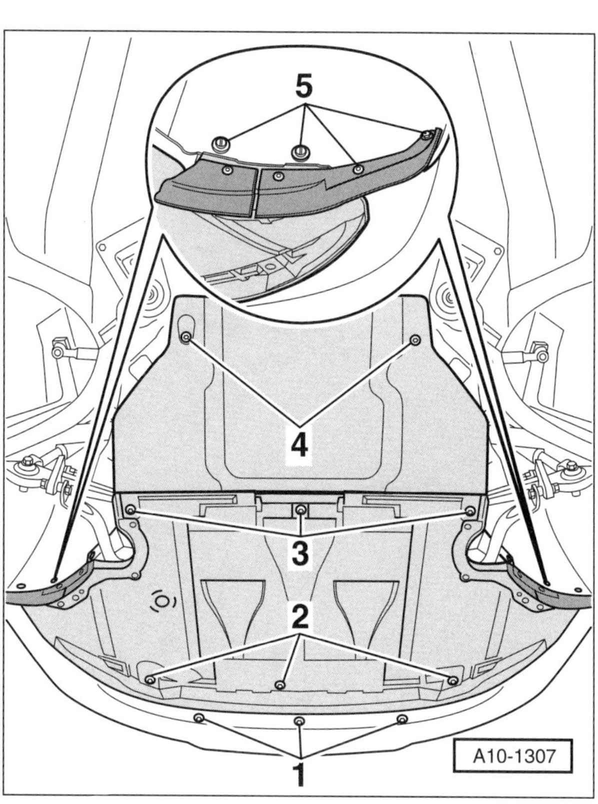

◀ Working underneath vehicle, loosen transmission splash shield mounting fasteners (**3**, **4**). Remove splash shield.

– Place oil drip pan underneath transmission.

BentleyPublishers.com
BentleyPublishers.com—All Rights Reserved

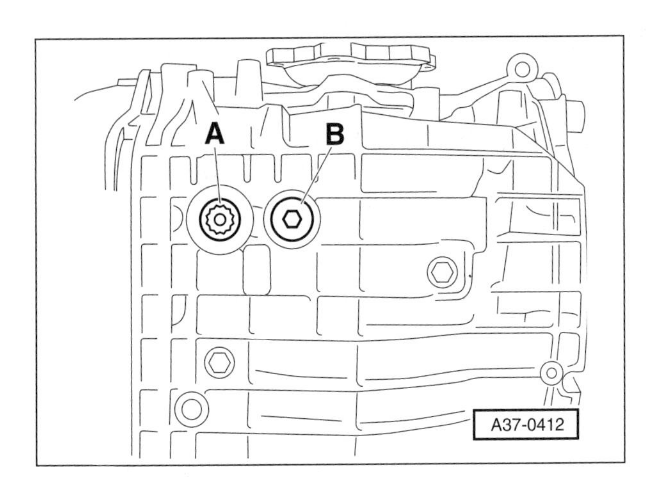

Remove transmission filler plug (**B**).

> ***WARNING —***
> *• Hot transmission oil can scald. Wear protective gloves and goggles.*

– ATF level is correct when a little ATF runs out at a temperature between 35°C (95°F) and 45°C (113°F).

- For tropical climates, allow ATF to drip out until temperature rises to approx. 50°C (122°F).
- If no ATF runs out when temperature reaches 40°C (104°F), add ATF.

Filling

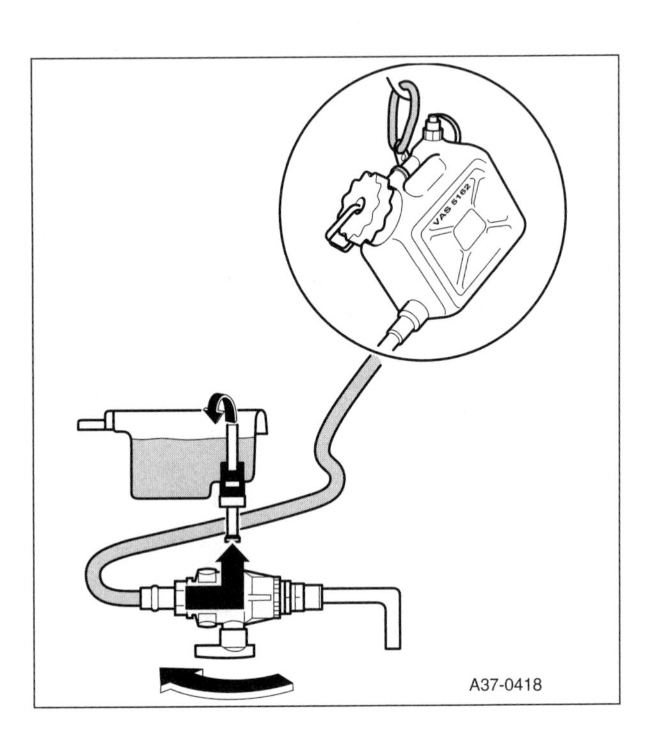

Use fluid pump with shut-off valve such as VAS 5162 or equivalent to fill transmission.

– Add ATF until ATF flows from plug opening.

– Replace filler plug. Use new gasket.

Tightening torque	
ATF filler plug to ATF sump (CVT)	20 Nm (15 ft-lb)

BentleyPublishers.com
BentleyPublishers.com—All Rights Reserved

Shift Mechanism and Components

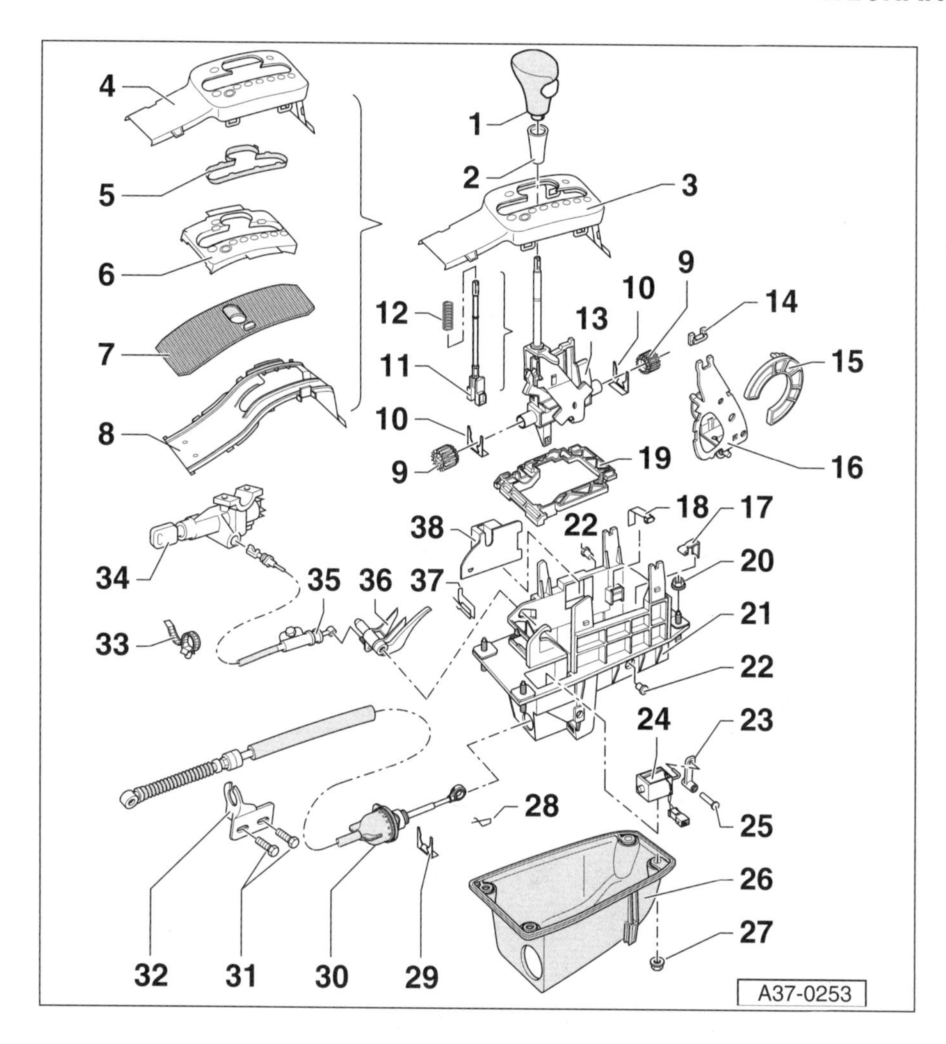

Shift mechanism assembly

1. **Gear selector handle**
    - See **Gear selector handle, removing and installing** in this repair group
2. **Sleeve**
    - Engages into grip for selector handle
3. **Cover**
    - With transmission range display, display for shift lock and Tiptronic
4. **Cover**
5. **Trim piece**
6. **Symbol panel**
    - Circuit board with integrated Tiptronic switch (F189)
7. **Slide cover**
8. **Guide**
9. **Mounting bushing**
10. **Locking clip**
    - Install with angled ends pointing toward inside of mounting bracket.
11. **Pull rod**
12. **Spring**
13. **Selector lever**
14. **Stop buffer**
15. **Detent**
16. **Cable lever**
17. **Spring clip with roller**
18. **Spring clip with roller**
19. **Frame**
    - Clip onto mounting bracket
    - Ribbed side must point up
20. **Nut, 10 Nm (7 ft lb)**
21. **Mounting bracket**
22. **Fulcrum pin**
23. **Locking pawl**
24. **Shift lock solenoid switch (N110)**
25. **Fulcrum pin**
26. **Cover**
    - With seal for mounting bracket
27. **Nut, 10 Nm (7 ft lb)**
28. **Securing clip**
    - Install with angled end forward

BentleyPublishers.com
BentleyPublishers.com—All Rights Reserved

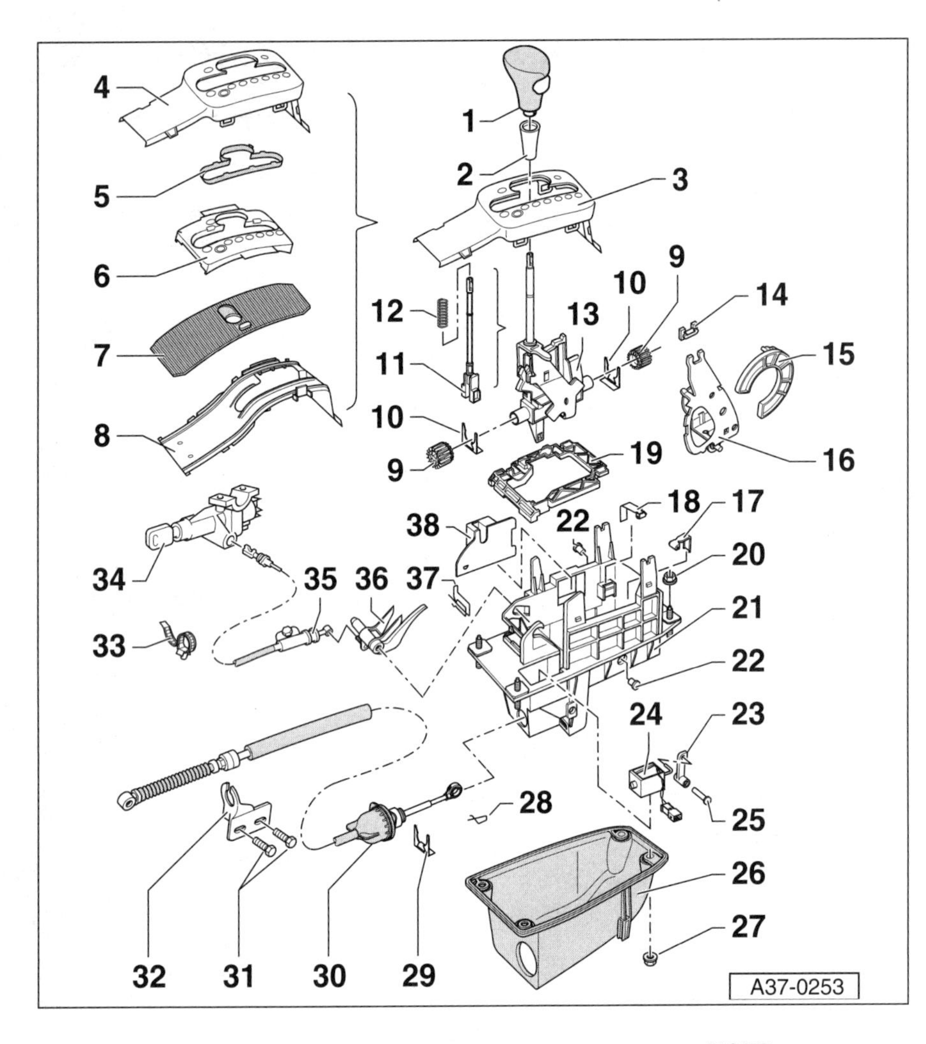

Shift mechanism assembly *(continued)*

29. Locking plate

- Install with angled end toward inside of mounting bracket

30. Selector lever cable

- Do not bend or kink. Lightly lubricate cable eye and ball socket before installing.
- Replace selector cable if rubber sleeves are damaged.
- Make sure rubber sleeves on transmission side are not twisted.

31. Bolt, 23 Nm (17 ft lb)

32. Support bracket

33. Cable tie

34. Ignition / starter switch

35. Ignition interlock cable

- Do not kink

36. Locking lever

- For ignition key removal lock

37. Securing spring

38. Mounting bracket

NOTE—

- *Lubricate bearings and sliding parts with lubricant (part no. G 052 142 A2 or equivalent).*

Shift mechanism, checking

– Without stepping on brake, place selector lever in **P** and switch ignition **ON**:

- The shift lever is locked and cannot be moved from **P** (the shift lock solenoid switch locks the selector lever).

– With brake pedal depressed:

- The shift lock solenoid releases the selector lever. Selecting a driving range must be possible without binding or snagging with the button in the handle depressed.

– Move selector lever slowly from **P** to **R**, **N**, **D**, **4**, **3**, **2** and check whether the driving range positions in instrument cluster display match the console shift lever display.

– Without stepping on brake, place shift lever in **N** with ignition **ON**:

- The shift lever is locked and cannot be moved from **N** with button in lever pressed (the shift lock solenoid switch locks the shift lever).

BentleyPublishers.com
BentleyPublishers.com—All Rights Reserved

– With brake pedal depressed:
  - The shift lock solenoid releases the shift lever. Selecting a driving range must be possible without binding or snagging with the button in the handle depressed.

– With selector lever in **D**, ignition and light switched **ON**.
  - Move selector lever out of driving range **D** into Tiptronic gate.
  - Illumination of **D** symbol on the cover of the shift mechanism must dim and the **+** and **-** symbols must light up.

– Start engine and let idle. Engage parking brake and depress brake pedal:
  - Sift lever position indicator in instrument cluster must change during shifting into Tiptronic-gate from **PRND432** to **54321**.

NOTE—
- *It must not be possible to operate the starter motor in driving ranges* ***2****,* ***3****,* ***4****,* ***D*** *and* ***R****.*
- *For right hand drive (RHD) vehicles, the starter should only be able to be operated with selector lever in* ***P*** *and* ***N*** *with locking button in selector lever handle released.*
- *During vehicle speeds above 5 kph (3 mph) and shifting into* ***N****, the shift lock solenoid must not engage and lock the selector lever. The selector lever can be shifted into a driving range.*
- *When driving slower than 2 kph (1.2 mph) (the car is almost stopped) and shifting into* ***N****, the shift lock solenoid must only engage after about 1 second. The selector lever can only be shifted out of* ***N*** *when the brake pedal is depressed.*

Selector lever handle, removing and installing

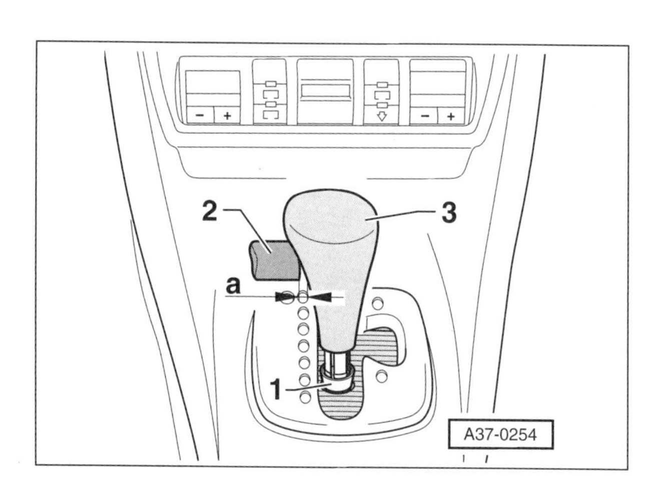

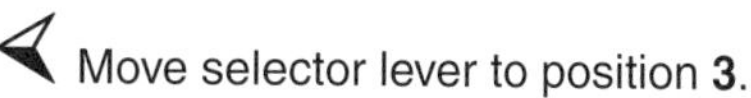

Move selector lever to position **3**.

– Push sleeve (**1**) down to stop.

– Pull button (**2**) out of selector lever handle (**3**) to stop and pull handle upward to remove (dimension **a** = 3 mm).

– To replace handle, move selector lever to position **3**.

– Carefully pull button (**2**) out of handle (**3**) to stop.

– With knob extended, attach handle in driving direction on selector lever to stop.

– Turn handle (**3**) until knob (**2**) is pointing to driver. This causes handle to engage into groove of selector lever.

– Make sure handle engages into circular groove of selector lever.

– Push in knob in handle.

– Pull up sleeve (**1**) until it engages in handle.

BentleyPublishers.com
BentleyPublishers.com—All Rights Reserved

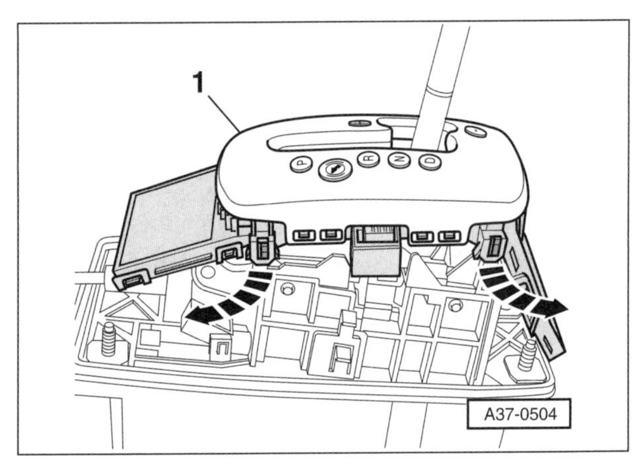

Shift mechanism cover, removing and installing

– Remove center console. See **70 Interior Trim**.

– Remove selector lever handle. See **Selector lever handle, removing and installing** in this repair group.

< Release 4 retaining clips (**arrows**) and lift of cover (**1**) with guide.

– Unplug connector at side of guide.

– Installation is reverse of removal.

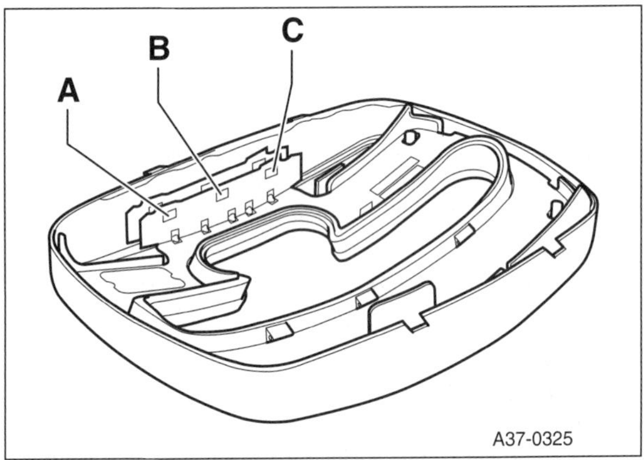

Tiptronic switch (F189), checking

The Tiptronic switch is integrated in the printed circuit of the symbol insert located in the cover of the shift mechanism. See **Tiptronic switch (F189), removing and installing** in this repair group.

It consists of three Hall sensors (**A**, **B** and **C**) that are triggered by a magnet on the slide cover.

- **A** - Sensor for downshift
- **B** - Sensor for Tiptronic recognition
- **C** - Sensor for upshift

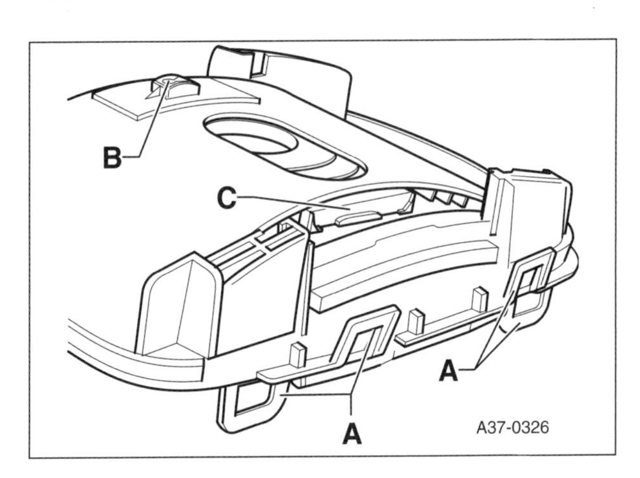

< In case of malfunction first check whether the magnet on slide cover (**C**) is properly mounted. If necessary replace slide cover.

– Check wiring at Tiptronic switch.

– Check steering wheel buttons and their wiring connections.

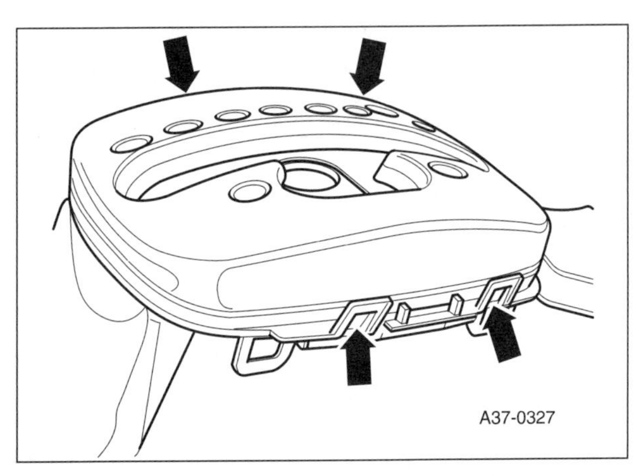

Tiptronic switch (F189), removing and installing

– Remove center console and extension. See **70 Interior Trim**.

– Remove handle for selector lever. See **Selector lever handle, removing and installing** in this repair group.

< Carefully disengage 4 clips (**arrows**) and remove cover upward.

NOTE—

- *If clips are damaged or broken, replace the applicable components.*

BentleyPublishers.com
BentleyPublishers.com—All Rights Reserved

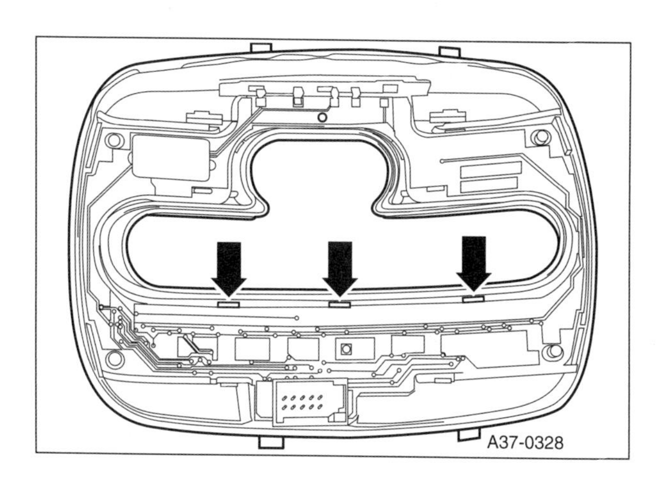

◀ Lift out printed circuit board with small screwdriver over 3 clips (**arrows**).

– Remove printed circuit board with integrated Tiptronic switch.

NOTE—

- *If the printed circuit board is glued into the cover, replace the complete cover with printed circuit board.*

– To install switch, insert printed circuit board with integrated Tiptronic switch into cover of shift mechanism.

– Carefully snap printed circuit board into cover.

NOTE—

- *The 3 clips (**arrows**) must engage visibly above the printed circuit board.*

– Place cover on guide of sliding cover and snap clips in place.

– Install handle for selector lever.

– Install center console and extension.

Shift mechanism, removing and installing

NOTE—

- *Shift mechanism procedure shown is for FWD models. AWD models are similar.*

– Place shift selector lever in **PARK**.

– Remove selector lever handle. See **Selector lever handle, removing and installing** in this repair group.

– Remove shift mechanism cover. See **Shift mechanism cover, removing and installing** in this repair group.

– Disconnect electrical harness connector for shift lock solenoid.

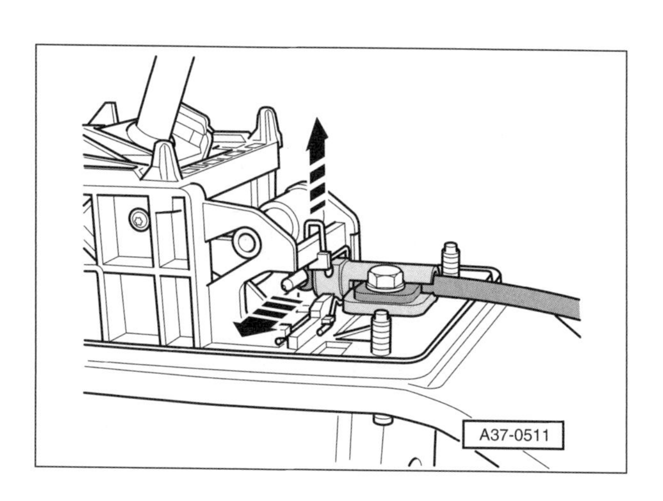

◀ Unhook locking cable (**arrows**).

– Raise vehicle and safely support

WARNING—

- *Make sure the car is stable and well supported at all times. Use a professional automotive lift or jack stands designed for the purpose. A floor jack is not adequate support.*

BentleyPublishers.com
BentleyPublishers.com—All Rights Reserved

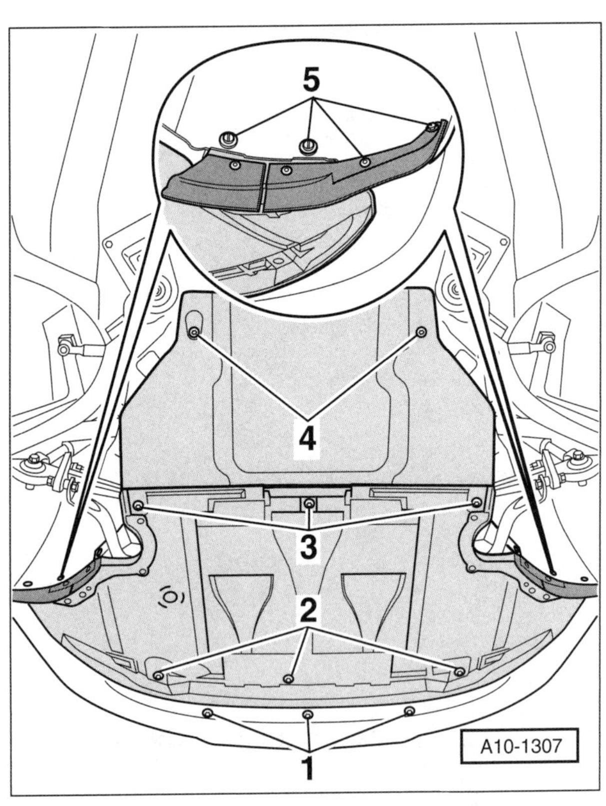

Loosen mounting parts (**3** and **4**) and remove rear part of lower engine cover.

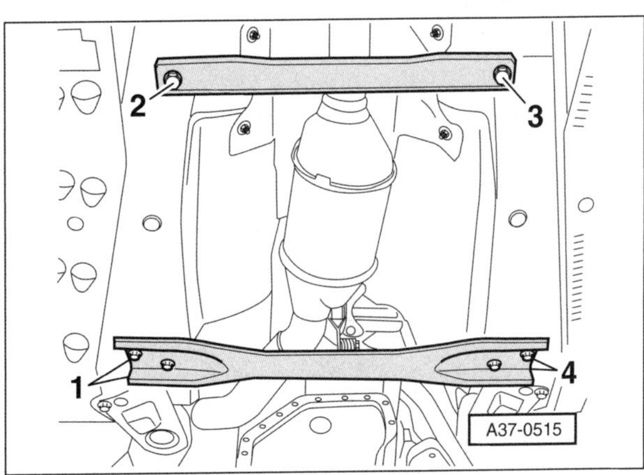

Remove bolts (**1** to **4**) at both floor crossmembers (if applicable).

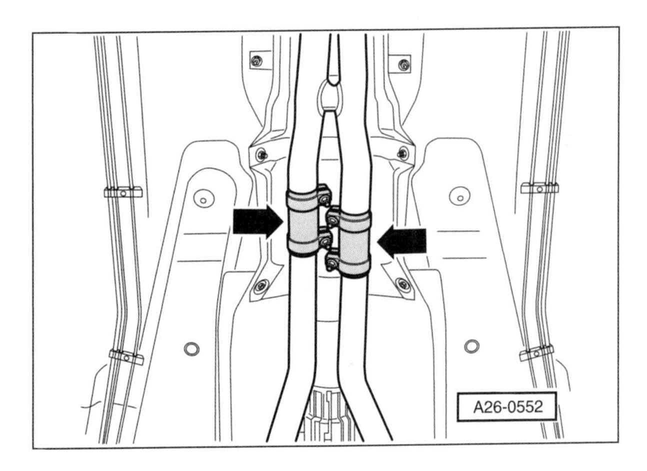

Disconnect exhaust system at double clamps (**arrows**).

NOTE—

- *Flex joint in front exhaust pipe must not be bent more than 10°, otherwise it may be damaged.*

– Remove heat shield for exhaust system beneath shift mechanism.

– AWD models: Disconnect driveshaft from transmission and support from body using stiff wire.

BentleyPublishers.com
BentleyPublishers.com—All Rights Reserved

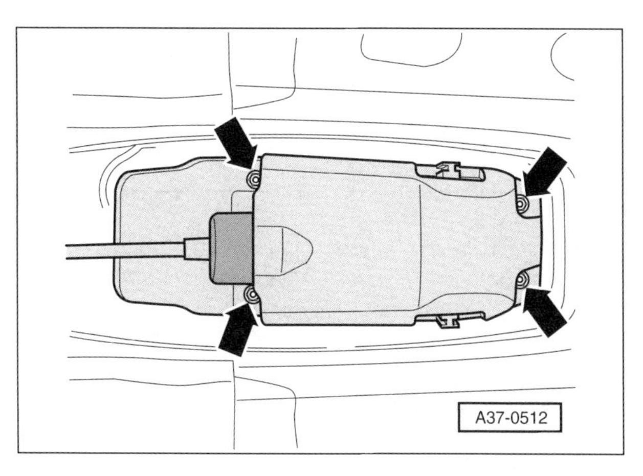

◀ Remove lower cover from shift mechanism (**arrows**).

– Remove selector lever cable boot from cover and slide back.

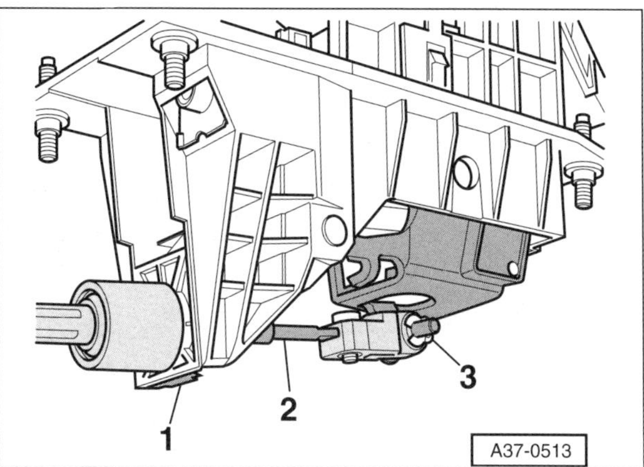

◀ Remove securing plate (**1**) for selector lever cable on shift mechanism downward.

– Remove locking clamp (**3**) from pin.

– Remove selector lever cable (**2**) from pin.

NOTE —

• Do not bend or kink selector lever cable.

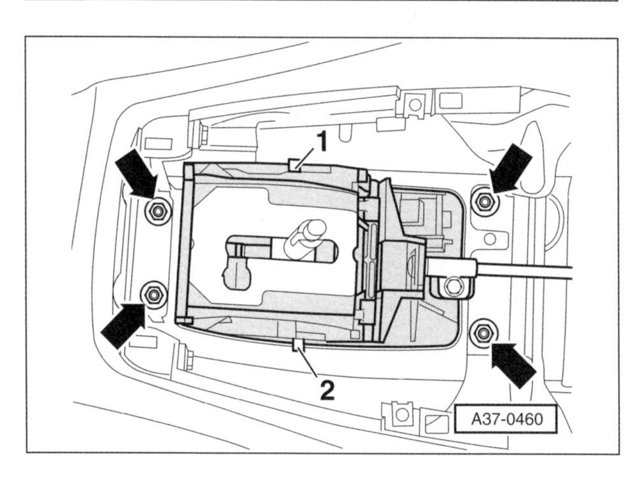

◀ Working from above: Remove nuts (**arrows**) securing mechanism.

– With an assistant supporting shift mechanism from below, press retaining clips (**1** and **2**) inward and shift mechanism downward at the same time.

– Pull out selector lever cable at the same time, so that it does not bend. Remove shift mechanism.

– Installation is reverse of removal. Remember to:

- Install selector lever cable. See **Selector lever cable, removing and installing** in this repair group.
- Install interlock cable. See **Ignition interlock cable, removing and installing** in this repair group.
- Install exhaust system so that it is free from stress.

Tightening torques	
Shift mechanism to body (replace nuts)	10 Nm (7 ft-lb)
Bottom cover to shift mechanism	10 Nm (7 ft-lb)
Floor crossmember to body • M8 • M10	 22 Nm (16 ft-lb) 45 Nm (33 ft-lb)

BentleyPublishers.com
BentleyPublishers.com—All Rights Reserved

Selector lever cable, removing and installing

- Shift selector lever to **P**.
- Raise vehicle and safely support.

> ***WARNING —***
> - *Make sure the car is stable and well supported at all times. Use a professional automotive lift or jack stands designed for the purpose. A floor jack is not adequate support.*

- Disconnect front exhaust system with catalytic converters from rear exhaust system and hang up with wire. 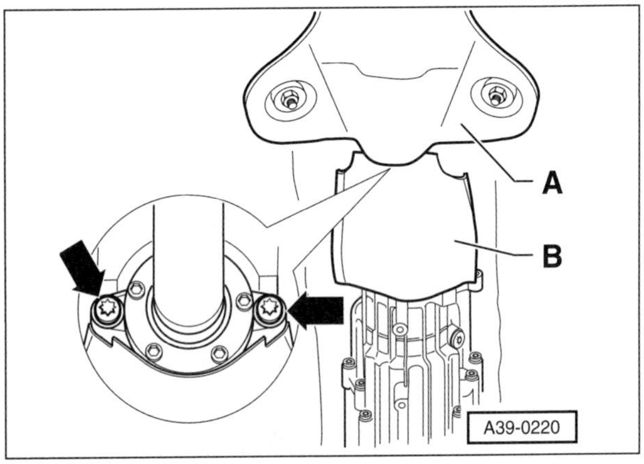
- Remove heat shield (**A**) above driveshaft.
- Remove heat shield (**B**) for driveshaft from differential cover (**arrows**).
- Disconnect driveshaft at transmission and support or hang up with wire.
- Remove rubber sleeve of selector lever cable and push it back.

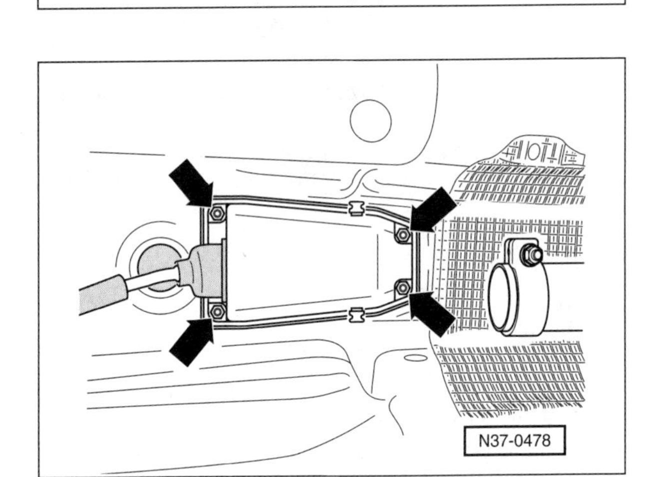

- Remove lower cover from mounting bracket (**arrows**).

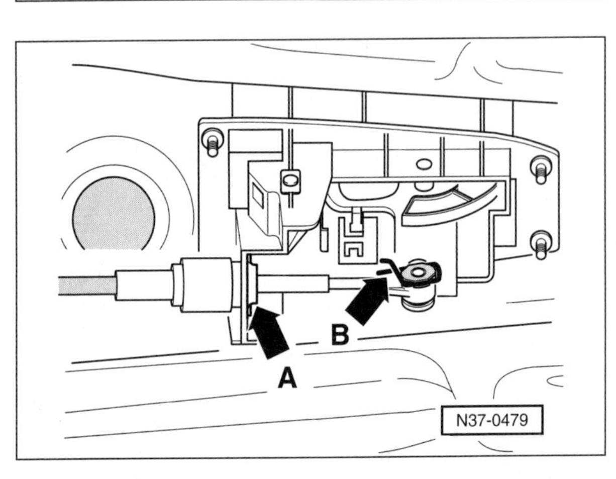

- Press together ends of spring clip (**B**) and remove.
- Pull out lock plate (**A**) for selector lever cable at mounting bracket downward.
- Remove selector lever cable from selector lever.

BentleyPublishers.com
BentleyPublishers.com—All Rights Reserved

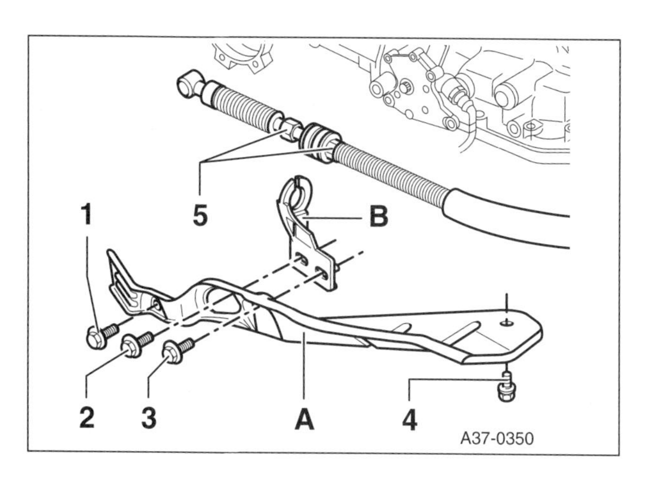

◀ Remove bolts (**1**, **3** and **4**) and remove heat shield (**A**) for selector lever cable.

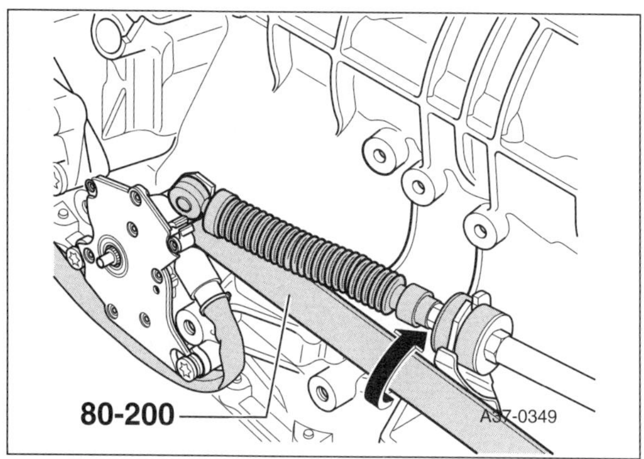

◀ Pry off selector lever cable with special tool 80-200 (or equivalent) from lever and selector shaft (**arrow**)

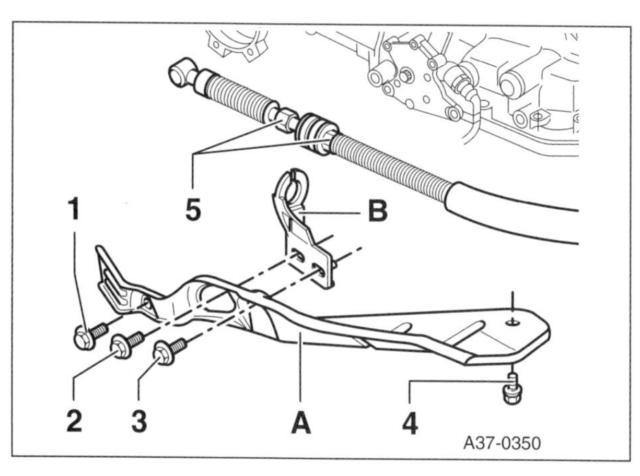

◀ Remove bracket (**B**) from transmission.

– Remove bracket (**B**) from selector lever cable by removing nuts (**5**).

– Pull out selector lever cable from mounting bracket without kinking cable.

– To install, move selector lever, lever and selector shaft in position **P** (parking lock must engage).

NOTE—

- *Slightly lubricate eye and socket before installing.*
- *Do not bend or kink selector lever cable.*

– Guide selector lever cable into mounting bracket.

– Attach selector lever cable to selector lever.

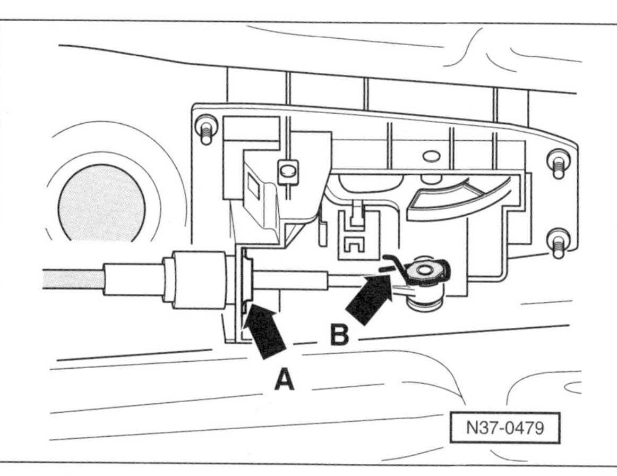

◀ Press ends of spring clamp (**B**) together and secure selector lever cable.

– Install lock plate (**A**) for selector lever cable at mounting bracket.

NOTE—

- *Angled ends of lock plate must point toward end of selector lever cable.*

BentleyPublishers.com
BentleyPublishers.com—All Rights Reserved

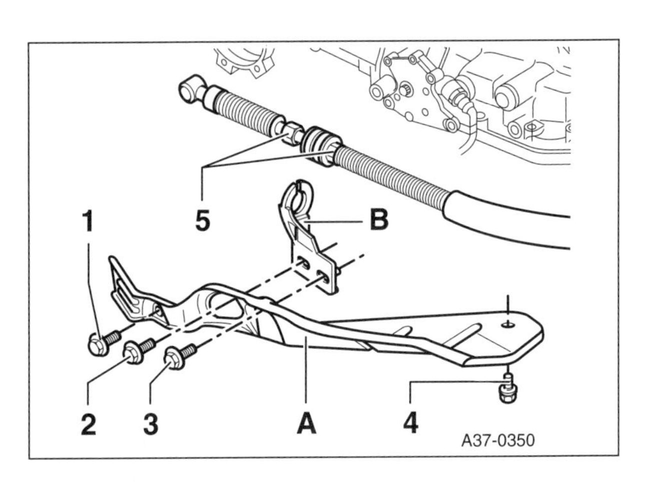

◄ Attach selector lever cable hand-tight to bracket (**B**).

– Attach bracket (**B**) hand-tight to transmission.

– Attach selector lever cable to lever and selector shaft.

– Straighten selector lever cable in bracket (**B**).

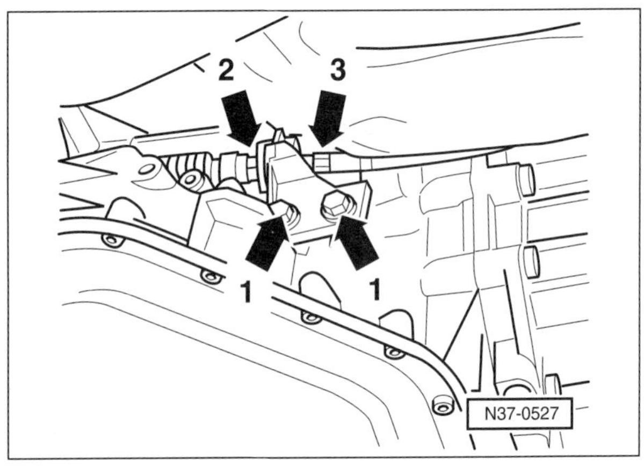

◄ Tighten bracket to transmission (**1**) and selector lever cable to bracket (**2**) with proper torque. Counterhold at hex head (**3**).

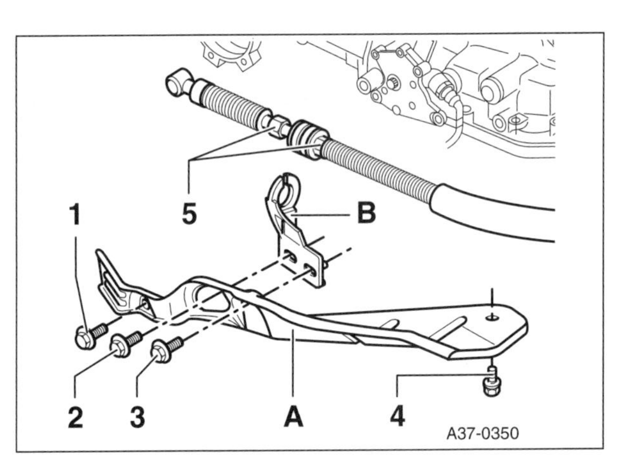

◄ Install heat shield (**A**).

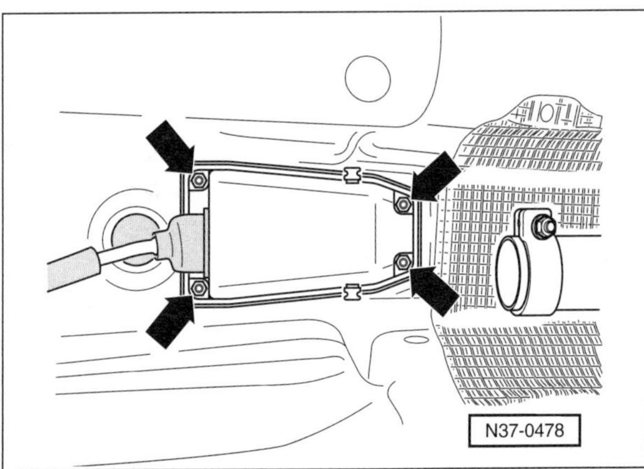

◄ Tighten mounting bracket to body.

– Check adjustment of selector lever cable. See **Selector lever cable, checking and adjusting** in this repair group.

– Reconnect driveshaft to transmission flange.

– Reconnect front and rear exhaust systems.

BentleyPublishers.com
BentleyPublishers.com—All Rights Reserved

Tightening torques	
Bracket to transmission	23 Nm (17 ft-lb)
Heat shield to transmission (M6)	9 Nm (80 in-lb)
Heat shield to transmission (M8)	23 Nm (17 ft-lb)
Mounting bracket to body	10 Nm (7ft lb)
Selector lever cable to bracket	12 Nm (9 ft-lb)
Selector lever mechanism to body	8 Nm (71 in-lb)

Selector lever cable, checking and adjusting

– Shift selector lever to **P**.

– Raise vehicle and safely support.

> ***WARNING —***
> *• Make sure the car is stable and well supported at all times. Use a professional automotive lift or jack stands designed for the purpose. A floor jack is not adequate support.*

– Remove heat shield at selector lever cable.

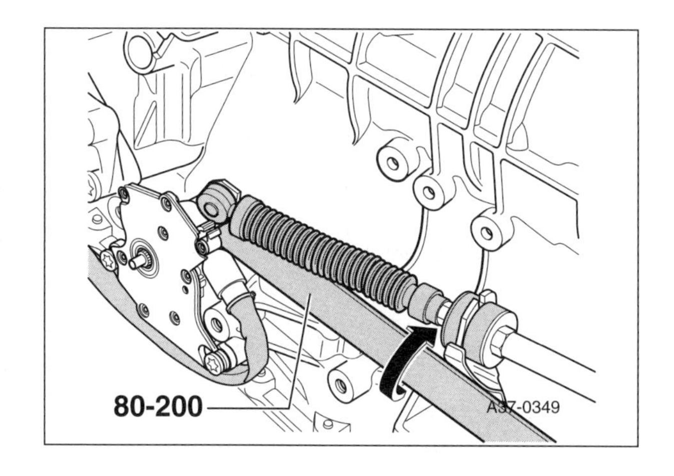

◀ Pry off selector lever cable using lever 80-200 (or equivalent) from lever and selector shaft (**arrow**).

– Move selector lever from **P** to **2**.

- Shift mechanism and selector lever cable must move easily. If necessary replace selector lever cable or repair shift mechanism.

– Shift selector lever to **P**.

- When moving shift lever and selector shaft to **P**, parking lock must engage. Both front wheels must not turn in one direction.
- It must be possible to attach selector lever cable to lever /selector shaft, otherwise adjust selector lever cable.

Adjusting

Selector lever cable must be separated from lever/selector shaft.

– Shift selector lever to **P**.

- When lever/selector shaft are moved to **P**, the parking lock must be engaged and the front wheels must not turn in one direction.

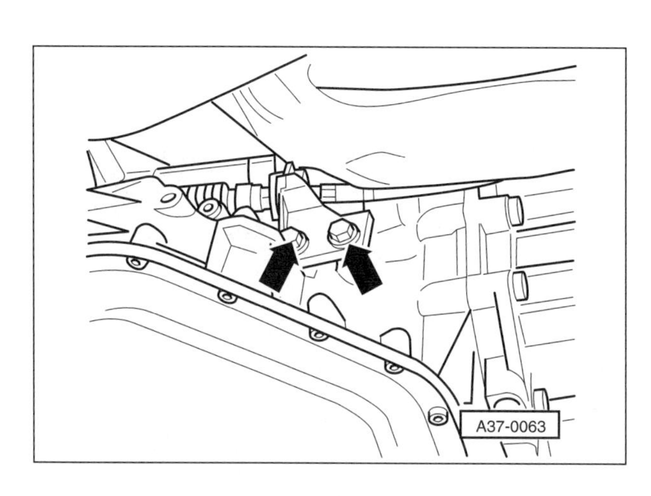

◀ Loosen nuts (**arrow**) for bracket on transmission.

NOTE —
• For illustration purpose the heat shield is not shown.

– Attach selector lever cable on lever/selector shaft.

BentleyPublishers.com
BentleyPublishers.com—All Rights Reserved

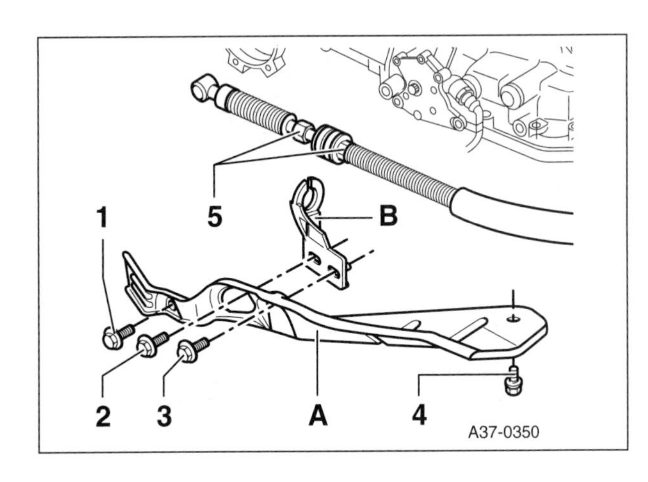

◀ Straighten selector lever cable at bracket (**B**).

– Install heat shield (**A**).

– Tighten bolts (**2** and **3**) at bracket.

– Check shift mechanism. See **Shift mechanism, checking** in this repair group.

Tightening torques	
Bracket to transmission	23 Nm (17 ft-lb)
Heat shield to transmission (M6)	9 Nm (80 in-lb)
Heat shield to transmission (M8)	23 Nm (17 ft-lb)
Selector lever cable to bracket	12 Nm (9 ft-lb)
Selector lever mechanism to body	8 Nm (71 in-lb)

Ignition interlock, checking

The ignition interlock prevents or allows the key and shift mechanism to move depending on their respective positions. Follow the steps below to check the ignition interlock.

– Turn ignition key **ON** then depress brake pedal and hold:
  - Shifting from **P** must be possible without "snagging" when button on shift lever is depressed.
  - Removing the ignition key must not be possible in any driving position except in **P**.

– Move shift lever to **P**.
  - It must be possible to turn Ignition key **OFF** without binding.

– Remove ignition key.
  - It must not be possible to move shift lever from **P**.

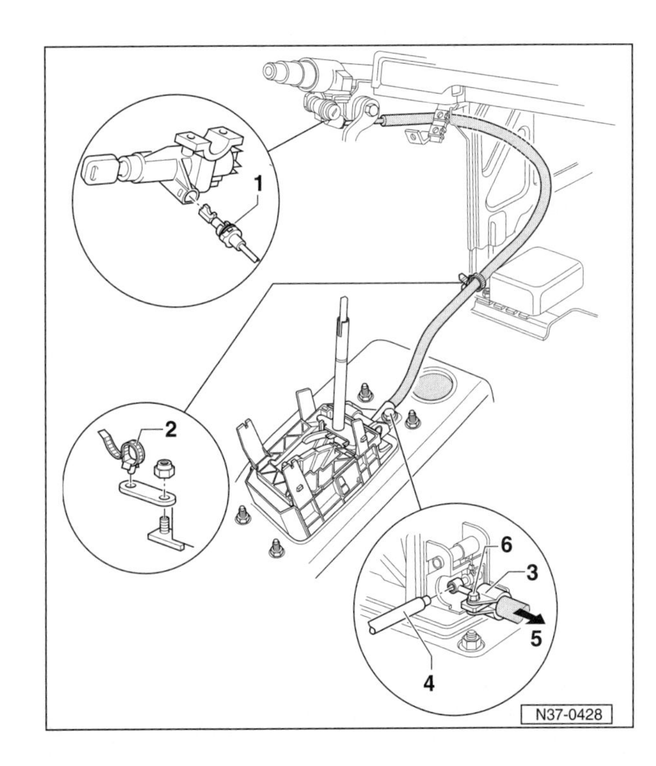

Ignition interlock cable, removing and installing

Removing

– Shift selector lever to **2**.

– Obtain radio anti-theft code.

– Disconnect battery ground connection.

– Remove handle for selector lever. See **Selector lever handle, removing and installing** in this repair group.

– Remove front center console. See **70 Interior Trim**.

– Remove steering wheel. See **48 Steering**.

– Turn ignition key **ON**.

– Shift selector lever to **P**.

BentleyPublishers.com
BentleyPublishers.com—All Rights Reserved

- Lift clip for retainer (**1**) and pull out interlock cable from ignition/starter switch.
- Remove cover with gate (selector lever actuation).
- Unclip interlock cable from securing spring at mounting bracket while slightly lifting securing spring.
- Free cable from cable tie (**2**) and remove.

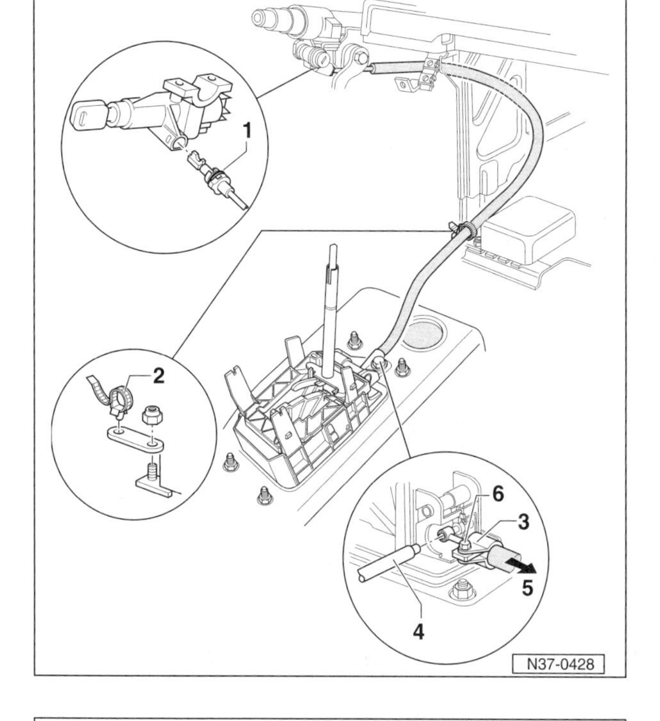

Installing

- Route interlock cable without kinks.
- Turn ignition key **ON**.
- Insert cable in ignition/starter switch.
- Check locking device (**1**) for proper engagement.
- Attach lock cable with cable tie (**2**) to airbag sensor.
- Turn ignition key **OFF**.
- Shift selector lever to **P**.
- Clip in cable in securing spring of mounting bracket.
- Insert cable bracket (3) in shift mechanism and cable eye in lever for cable.

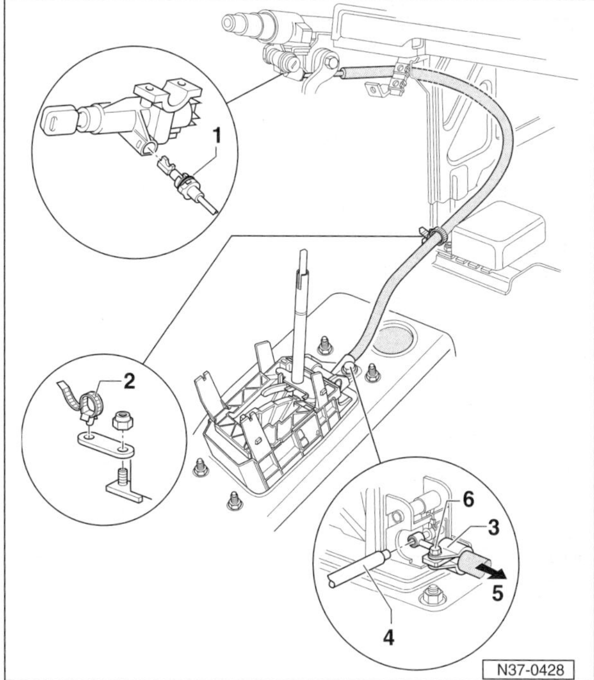

Adjusting

- Shift selector lever to **P**.
- Loosen bolt (**6**).
- Clamp (**3**) must be movable by hand in direction of arrow (**5**).
- Move steering column to lowest position.
- Insert gauge 3352 A (**4**) between lever for interlock cable and cable eye.
- Pull cable in direction of arrow (**5**) and tighten bolt (**6**).

Tightening torque	
Interlock cable clamp bolt	10 Nm (7 ft lb)

- Remove adjustment gauge.
- Check ignition key lock after every adjustment of interlock cable. See **Ignition key lock, checking** in this repair group.
- Install steering wheel.
- Install front center console.

BentleyPublishers.com
BentleyPublishers.com—All Rights Reserved

– Install handle for selector lever.

– Reconnect battery ground cable.

– For vehicles with theft-protected radio, reactivate code.

TRANSMISSION OIL PAN AND COMPONENTS

Transmission oil pan, removing and installing

– Drain automatic transmission fluid (ATF). See **03 Maintenance**.

– Loosen bolts for transmission oil pan in diagonal sequence.

– Remove oil pan.

NOTE—

- *Clean magnet in oil pan (4 magnets). Check that magnets have complete contact with oil pan.*

– Installation is in reverse order of removal. Note the following:

- Replace seals.
- Tighten bolts for oil pan in several steps.
- Refill transmission with ATF.

Tightening torques	
Drain plug to oil pan	12 Nm (9 ft-lb)
Oil pan to transmission (tighten diagonally)	10 Nm (7 ft-lb)

ATF screen, removing and installing

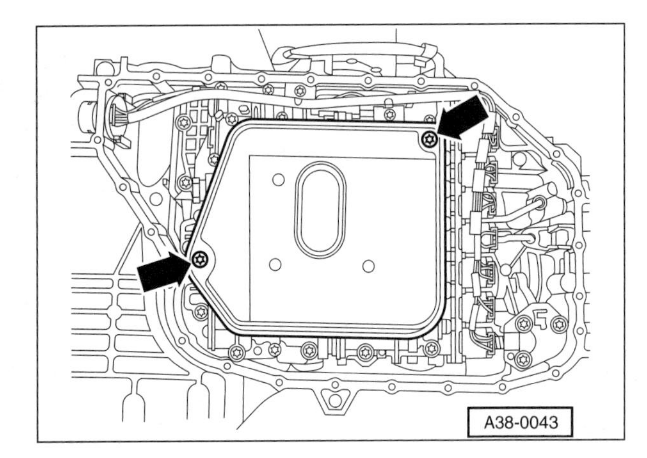

– Remove transmission oil pan. See **Transmission oil pan, removing and installing** in this repair group.

< Remove 2 bolts for ATF screen (**arrow**).

– Remove ATF screen from valve body.

– Apply thin coat of petroleum jelly to intake side of screen gasket.

– Install ATF screen.

– Install oil pan.

– Refill transmission with ATF.

Tightening torques	
ATF screen to valve body	5 Nm (44 in-lb)
Drain plug to oil pan	12 Nm (9 ft-lb)
Oil pan to transmission (tighten diagonally)	10 Nm (7 ft-lb)

BentleyPublishers.com
BentleyPublishers.com—All Rights Reserved

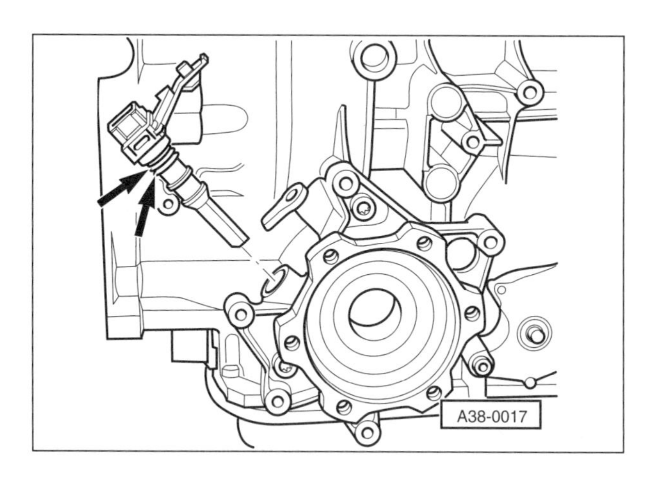

Vehicle speed sensor (G22), removing and installing

◄ Working at transmission, remove connector for vehicle speed sensor (G22).

– Press down clamp for sensor, turn and remove it.

– Installation is reverse of removal. Note the following:
  - Replace both O-rings for sensor (**arrow**).
  - Apply thin coat of petroleum jelly to O-rings.
  - Insert sensor and snap clamp in place.

Transmission rpm sensor (G182), removing and installing

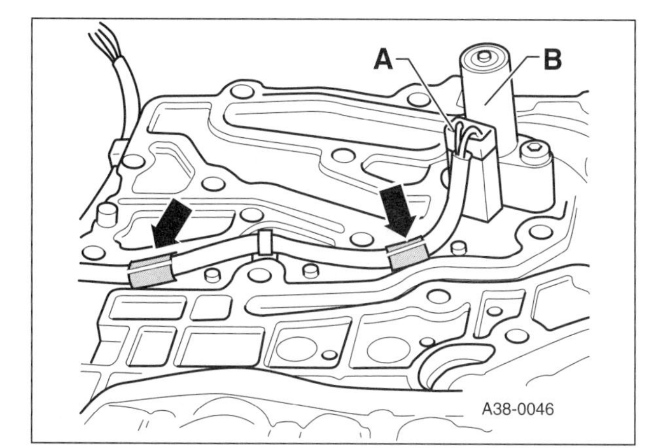

NOTE—
- *Do not reuse a sensor that was dropped (the permanent magnet breaks).*

– Remove valve body.

◄ Disconnect connector (**A**) on back-side of valve body.

– Remove rpm sensor (G182) (**B**) at valve body.

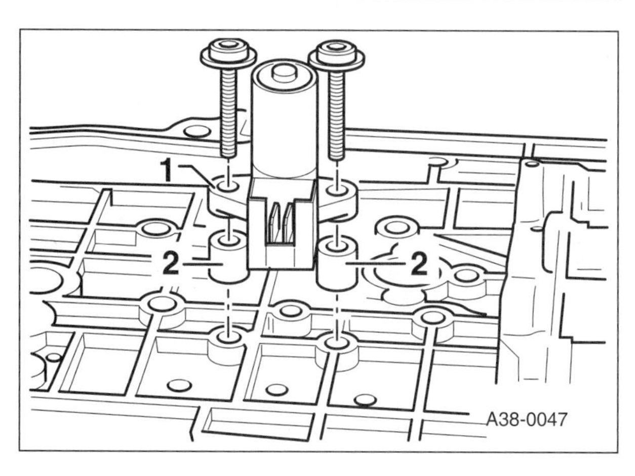

◄ Install rpm sensor (**1**) using spacers (**2**) (length 8.7 mm).

– Reattach electrical connector.

– Install valve body.

NOTE—
- *The sensor side with the connector terminals points toward the valve body.*

– Refill transmission with ATF.

Tightening torques	
ATF drain plug to oil pan	12 Nm (9 ft-lb)
ATF screen to valve body	5 Nm (44 in-lb)
Oil pan to transmission (tighten diagonally)	10 Nm (7 ft-lb)
Transmission rpm sensor to valve body	5 Nm (44 in-lb)
Valve body to transmission (tighten from inside-out)	8 Nm (71 in-lb)

BentleyPublishers.com
BentleyPublishers.com—All Rights Reserved

39 Final Drive, Driveshaft

TABLES

GENERAL

This repair group covers front and rear final drive oil service, and driveshaft removal and adjustment.

CAUTION—
- *Do not bend driveshaft universal joint beyond 25°. Joint may be damaged.*
- *Store and move driveshaft fully extended.*
- *Driveshaft cannot be repaired, only adjusted.*
- *If removing one side of driveshaft, support from body with wire.*
- *Match mark parts before removal. Install parts in previously marked position to maintain driveshaft balance. Imbalance can result in damage to bearings, vibration and noise.*
- *Before replacing driveshaft because of noise / vibrations, check adjustment. See* **Driveshaft, adjusting** *in this repair group.*
- *When disconnecting driveshaft from rear final drive, do not reinstall balance washer between shim and bolt head.*

FINAL DRIVE

Final drive oil

In manual transmission models, the front, center and rear final drives (differentials) are filled with the same oil as the transmission gearbox. In automatic models, the final drive housing(s) are filled with a different oil than ATF.

Current specifications for final drive oil are shown in **Table a**. If oil container is not marked with appropriate specification, assume that the oil is not suitable.

BentleyPublishers.com
BentleyPublishers.com—All Rights Reserved

Table a. Final drive oil specifications	
Manual transmission models	Synthetic oil SAE 75W-90 Audi part no. G 005 000
Automatic transmission models	Synthetic oil SAE 75W-90 Audi part no. G 052 145 S2

Approximate refill capacities for final drive units are listed in **Table b**.

Table b. Final drive oil capacities (approximate)

Transmission type	Final drive location (type)	Oil capacity in liters (US qt)	
		New fill	**Change**
Manual	Front wheel drive	Filled with transmission	
	Quattro (AWD): • Front final drive • Center final drive • Rear final drive, 01R, 01H	 Filled with transmission Filled with transmission 1.5 (1.6)	
Automatic 01L	Quattro (AWD): • Front final drive • Center final drive • Rear final drive, 01R	 1.3 (1.4) 0.88 (0.93) 1.5 (1.6)	
Automatic 01V	Front wheel drive	0.8 (0.85)	
	Quattro (AWD): • Front and center final drive • Rear final drive, 01R	 0.8 (0.85) 1.9 (2.01)	 0.75 (0.8)
Multitronic (CVT) 01J	Front wheel drive	1.3 (1.4)	

Front final drive oil level

Make sure vehicle is level during this operation. If front of vehicle is higher than rear, such as when driven up on ramps, oil inside final drive housing pools toward rear, and level cannot be accurately determined.

– Warm up gear oil in front final drive to approx. 60°C (140°F) by performing a short road test.

– Lift vehicle. See **03 Maintenance**. Make sure vehicle is level.

> ***WARNING —***
> • *Make sure the vehicle is stable and well supported at all times. Use a professional automotive lift or jack stands designed for the purpose. A floor jack is not adequate support.*

– Place suitable drip tray underneath transmission.

BentleyPublishers.com
BentleyPublishers.com—All Rights Reserved

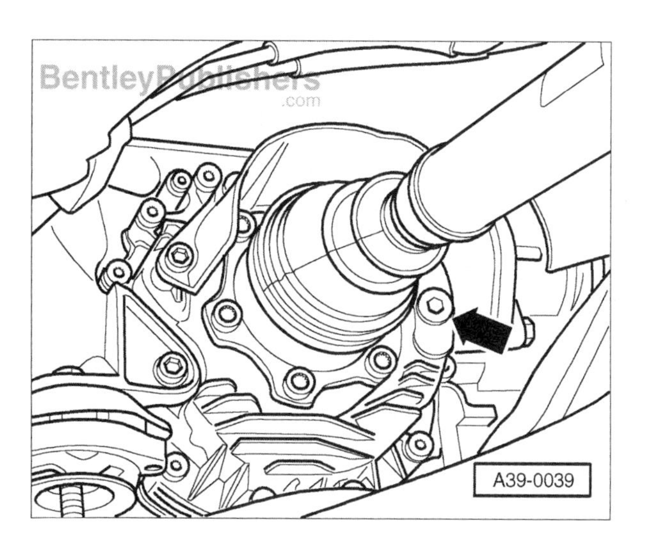

Remove fill plug (**arrow**) using angled Allen wrench.

> ***CAUTION—***
> - *Do not use Allen ball socket 3247. Use of this tool could damage fill plug.*

– Fill front final drive with gear oil slowly and evenly over a period of at least 5 minutes until oil drips from bottom of inspection hole.

NOTE—
- *A filling time of at least 5 minutes is required so that the internal oil level between the differential and oil fill space can equalize.*

– Oil level is correct when front final drive is filled up to lower edge of oil filler hole.

– Let excessive gear oil drain or top up gear oil.

– When the oil level is correct, install new fill plug and seal.

Tightening torque	
Fill plug to final drive housing (replace with new plug and seal)	35 Nm (26 ft-lb)

Rear final drive oil level

Make sure vehicle is level during this operation. If front of vehicle is higher than rear, such as when driven up on ramps, oil inside final drive housing pools toward rear, and level cannot be accurately determined.

Checking level

– Raise vehicle and support safely.

> ***WARNING—***
> - *Make sure the car is stable and well supported at all times. Use a professional automotive lift or jack stands designed for the purpose. A floor jack is not adequate support.*

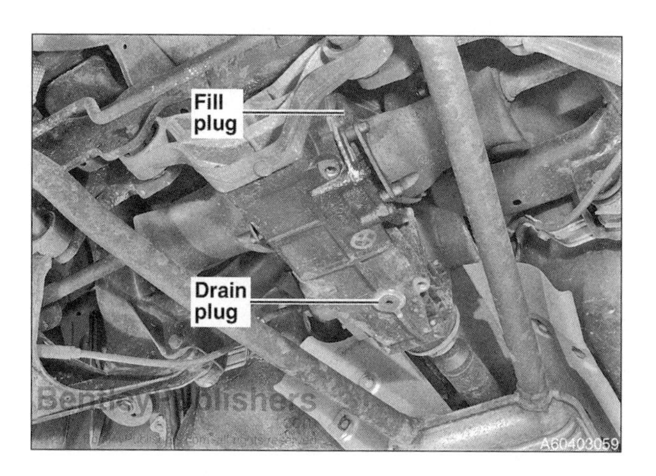

 Use 17 mm Allen wrench to loosen oil fill plug to check rear final drive oil.

– Remove fill plug. Oil level is correct when final drive is filled up to lower edge of filler hole.

BentleyPublishers.com
BentleyPublishers.com—All Rights Reserved

Filling

– To top up oil:
  - Place oil drain pan underneath final drive fill plug
  - Manual transmission: Top up final drive with SAE 75W-90 (synthetic oil), part no. G 005 000.
  - Automatic transmission: Top up final drive with SAE 75W-90 (synthetic oil), part no. G 052 145 S2.
  - Make sure oil is filled up to lower edge of filler hole. Final drive oil capacity is in **Table b**.
  - Allow excess oil to drip out.

– Screw in oil fill plug.

Tightening torque	
Oil fill plug to rear final drive housing	35 Nm (26 ft-lb)

DRIVESHAFT

Driveshaft assembly

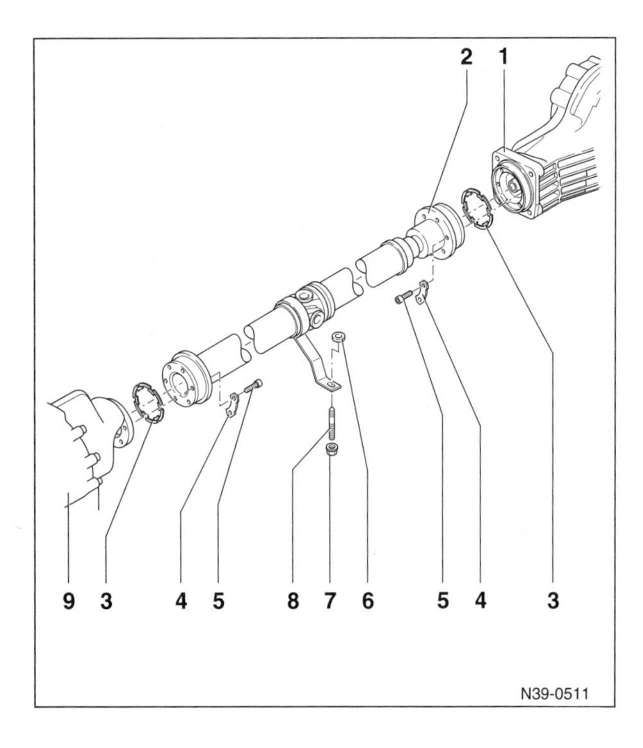

1. Rear final drive
2. Drive shaft
3. Gasket
   - Always replace
4. Backing plate
5. Bolt, 55 Nm (41 ft-lb)
   - Always replace
6. Adjustment shim
7. Nut, 25 Nm (18 ft-lb)
8. Stud bolt, 10 Nm (7 ft-lb)
9. Transmission

Driveshaft, removing and installing

NOTE—
- *Observe driveshaft cautions at the beginning of this repair group.*

– Remove crossmember below exhaust system (if installed).

– Remove rear part of exhaust system behind clamping sleeve(s).

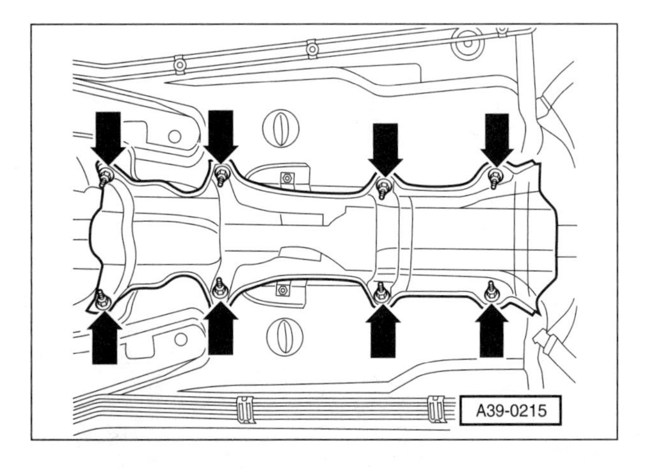

◄ Remove heat shields for driveshaft from body (**arrows**).

BentleyPublishers.com
BentleyPublishers.com—All Rights Reserved

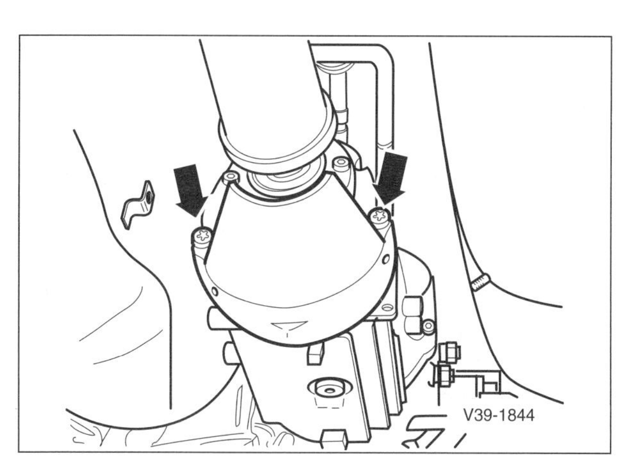

◀ Remove heat shield for driveshaft from differential cover (**arrows**).

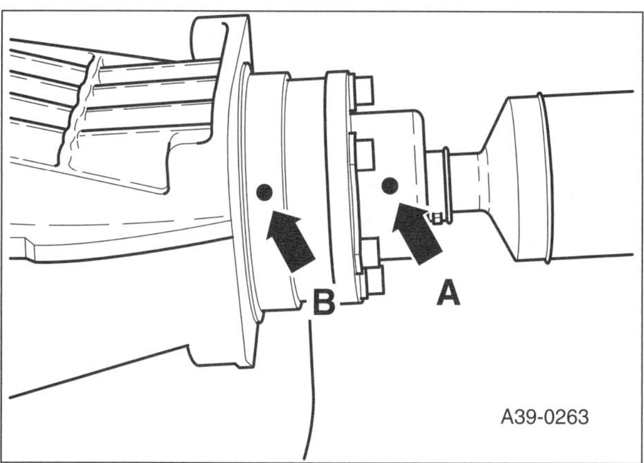

◀ Check for factory marking on driveshaft and at flange/driveshaft at rear final drive. If not found, mark location of driveshaft flange (**A**) to rear final drive (**B**).

– Remove three upper mounting bolts from each drive shaft constant velocity joint.

> ***CAUTION—***
>
> *• Do not remove driveshaft before installing VAG 3139 alignment fixture.*

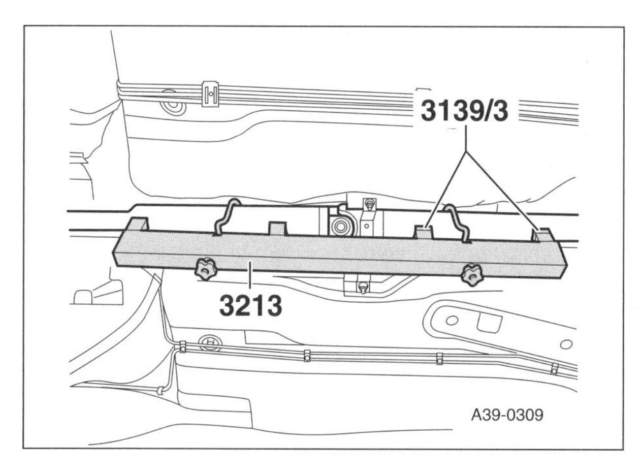

◀ Attach special tool VAG 3139 alignment fixture with 3139/3 spacers and tighten plastic nuts.

– Remove remaining three mounting bolts at front and rear constant velocity joints.

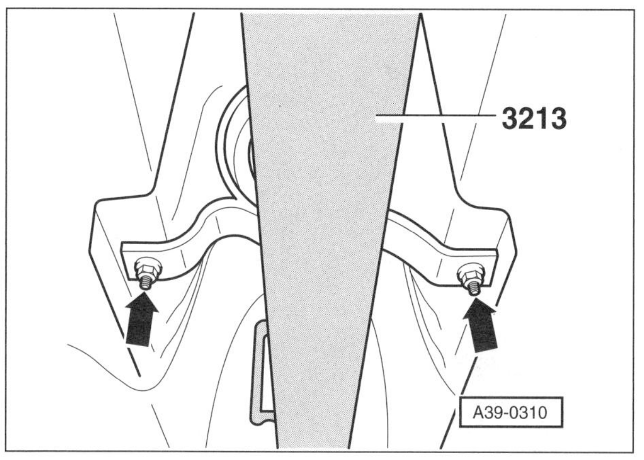

◀ Remove nuts (**arrows**) for intermediate bearing and remove driveshaft with alignment fixture.

NOTE—

• Only store and move driveshaft in fully extended (horizontal) position.

BentleyPublishers.com
BentleyPublishers.com–All Rights Reserved

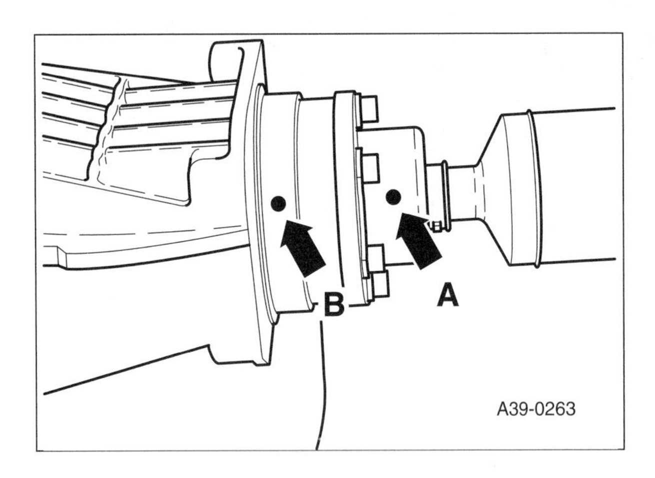

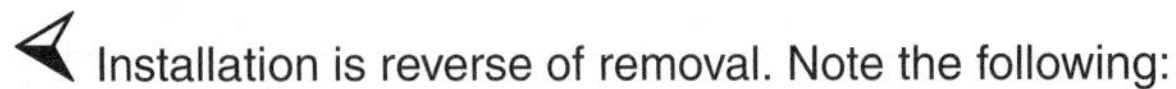

Installation is reverse of removal. Note the following:

- For proper balance, drive shaft flanges (**A**) and rear final drive (**B**) must be installed so that factory marks (or marks made previously) are aligned.
- If new drive shaft is installed and factory marks on rear final drive flange are no longer visible, check and mark radial run-out at flange and align marks on drive shaft to new mark on flange. See **Radial run-out, measuring** in this repair group.
- Replace gaskets on drive flanges (remove protective film and attach gasket to drive flanges).
- After disconnecting driveshaft from rear final drive, do not reinstall additional balance washer (thick washer) between shim and bolt (if equipped).
- Clean thread bores in flanges before installing driveshaft.
- Always replace drive shaft bolts.

– Adjust drive shaft. See **Driveshaft, adjusting** in this repair group.

– Install exhaust system so it is free of stress.

Tightening torques	
Driveshaft to manual transmission	55 Nm (41 ft-lb)
Driveshaft to rear final drive	55 Nm (41 ft-lb)
Front crossmember to body	25 Nm (18 ft-lb)
Heat shield to differential	25 Nm (18 ft-lb)
Intermediate bearing to body	25 Nm (18 ft-lb)
Nuts for clamping sleeve	40 Nm (30 ft-lb)

Driveshaft, adjusting

Driveshaft adjustments must be made precisely. A poorly adjusted drive shaft is usually the cause of vibrations and humming noises.

– Remove crossmember below exhaust system (if installed).

– Remove rear part of exhaust system behind clamping sleeve(s).

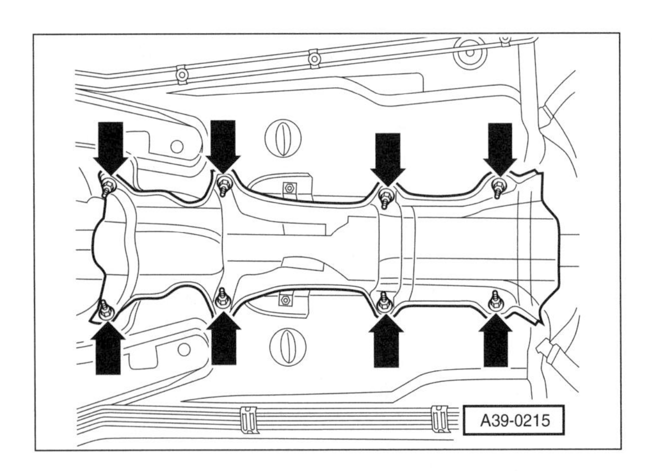

Remove heat shields for driveshaft from body (**arrows**).

BentleyPublishers.com
BentleyPublishers.com—All Rights Reserved

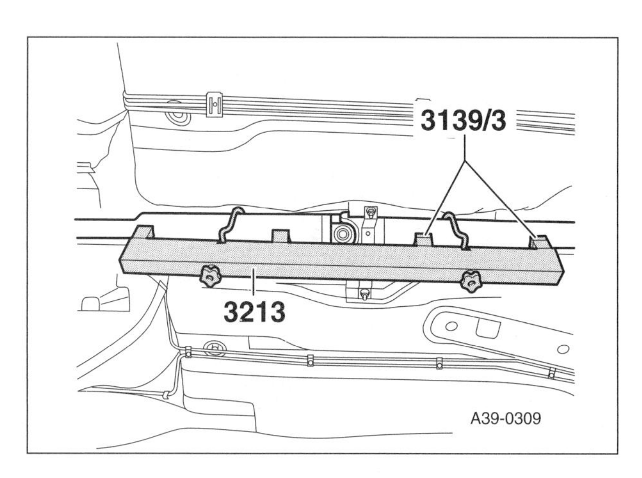

Attach special tool VAG 3213 alignment fixture with 3139/3 spacers and tighten plastic nuts.

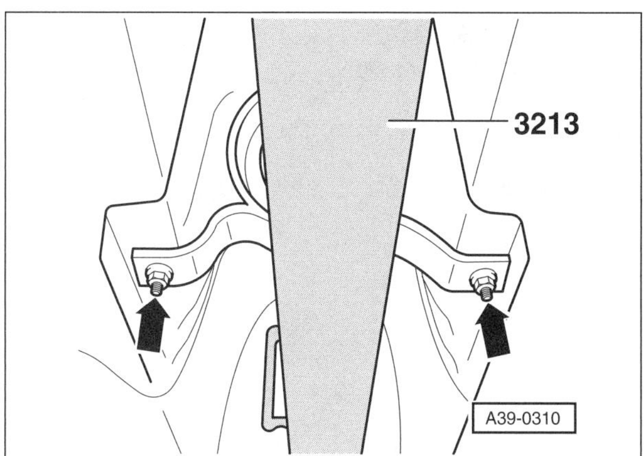

Remove nuts (**arrows**) for intermediate bearing.

NOTE —
- *Do not remove stud bolt.*

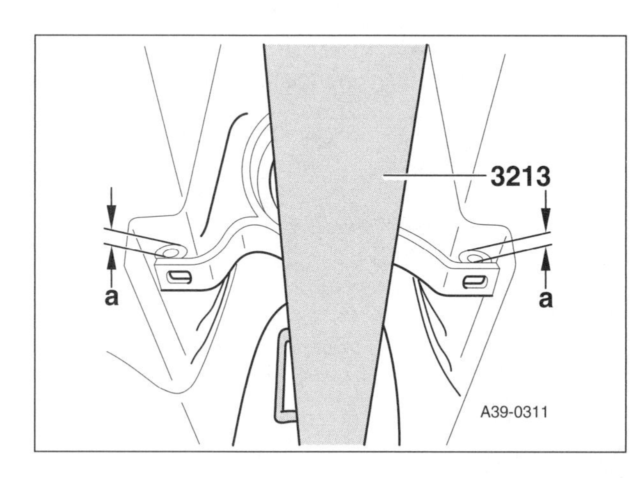

Align intermediate bearing of drive shaft so that distance **a** (left side) is equal to distance **a** (right side).

– Measure distance a to determine adjustment shims according to table below.

Available adjustment shims	
Distance a (mm)	**Shim thickness (mm)**
0 ... 3.0	-
3.1 ... 5.0	2
5.1 ... 7.0	4
7.1 ... 9.0	6
9.1 ... 11.0	8
11.1 ... 13.0	10

BentleyPublishers.com
BentleyPublishers.com—All Rights Reserved

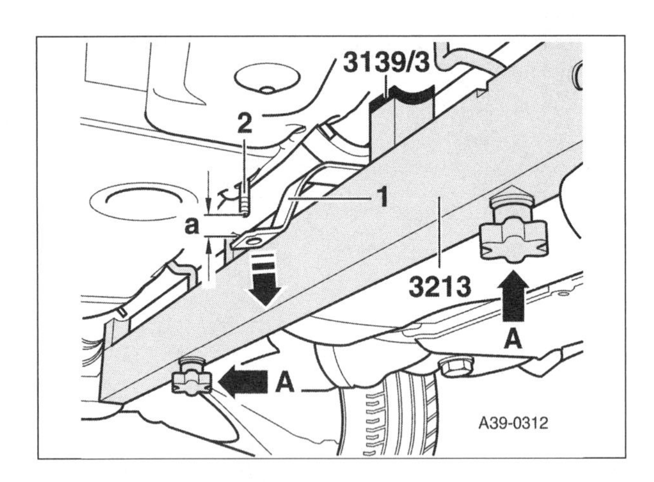

Install adjustment shims as follows:

- Loosen plastic nuts of alignment fixture (**A**) far enough so that bracket for intermediate bearing (**1**) has some clearance (measurement **a**) to stud (**2**).

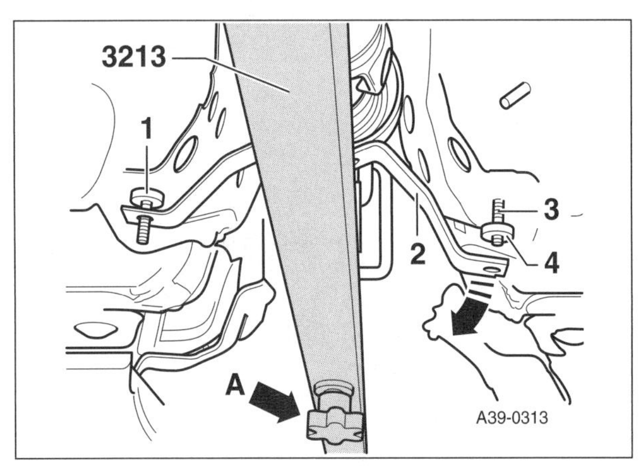

Insert appropriate adjustment shim (**1**) and swing bracket for intermediate bearing (**2**) to other side.

- Place second adjustment shim (**4**) onto stud (**3**).
- Now tighten both plastic nuts of alignment fixture (**A**) again.

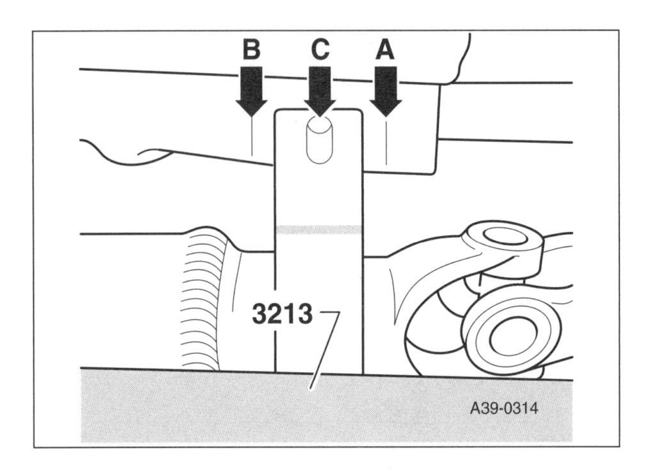

To center driveshaft, slide drive shaft with alignment fixture toward rear to stop and mark location of intermediate bearing (**A**).

- Slide drive shaft with alignment fixture toward front to stop and mark location of intermediate bearing (**B**).
- Center drive shaft so that intermediate bearing (**C**) is centered between markings (**A** and **B**).
- Tighten intermediate bearing in centered position.
- Remove mounting fixture.
- Remainder of installation is reverse of removal.
- Install exhaust system free of stress.

Tightening torque	
Front crossmember to body	25 Nm (18 ft-lb)
Intermediate bearing to body	25 Nm (18 ft-lb)
Nuts for clamping sleeve	40 Nm (30 ft-lb)

BentleyPublishers.com
BentleyPublishers.com—All Rights Reserved

Radial run-out, measuring

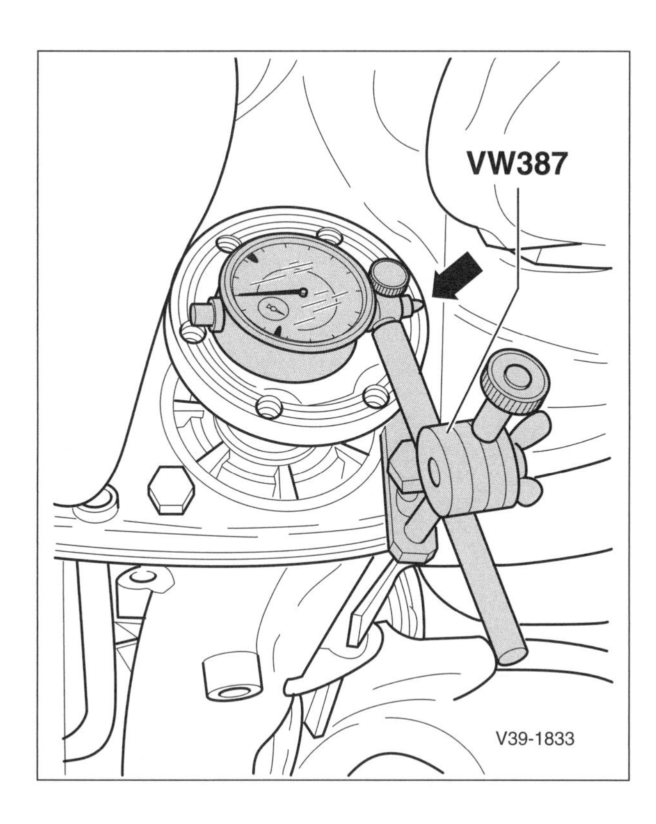

NOTE —

- *Measure radial run-out if rear final drive flange is removed. Make new marks and remove old marks.*
- *Mark location of largest radial run-out if new driveshaft is installed and marks on final drive flange are no longer visible.*
- *Align new mark with mark on driveshaft.*

◀ Attach special tool VAG VW387 dial gauge holder with dial indicator to connection between crossmember and final drive.

– Set dial indicator onto ground diameter of drive shaft flange (**arrow**) and set to 0 with 1 mm preload.

– Turn differential gear via both rear wheels (left and right flanges) at the same time in one direction until drive flange/driveshaft flange has made on complete revolution.

– Mark largest radial run-out on outer edge of flange (equivalent to largest distance from turning axis).

– Remove old mark on flange.

– Installing driveshaft. Match mark on driveshaft to mark on flange. See **Driveshaft, removing and installing** in this repair group.

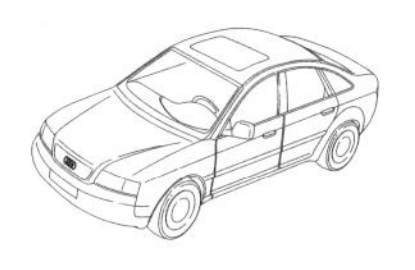

BentleyPublishers.com
BentleyPublishers.com—All Rights Reserved

40 Front Suspension

GENERAL

This section covers repairs to the front suspension and related components. Also see:

- **42 Rear Suspension** for rear suspension servicing and repair information
- **44 Wheels, Tires, Wheel Alignment** for wheel alignment specifications

Special tools, equipment and procedures are required for most front suspension repair and component replacement. On suspension components with bonded rubber bushings, tighten fasteners with vehicle at curb weight position. See **Curb weight position** in this repair group.

Wheel alignment is almost always disturbed when suspension components are removed or replaced. Plan on a wheel alignment after suspension repairs.

Curb weight position

Bonded rubber bushings can only be flexed to a limited extent. Place suspension under load (curb weight position) before tightening suspension arms with bonded rubber bushings.

In this illustration jack stands and wheel ramps are used to place vehicle at curb weight position. Then tighten suspension arms with bonded rubber bushings to torque specification.

BentleyPublishers.com
BentleyPublishers.com—All Rights Reserved

allroad quattro "jack mode"

Prior to jacking or lifting allroad quattro vehicle, set ride height control to "jack mode" to protect suspension shock absorbers and struts.

> ***WARNING —***
> - *Make sure that no one is lying under the vehicle or has head or hands in the wheel housing while the ride height is changing.*

Activating

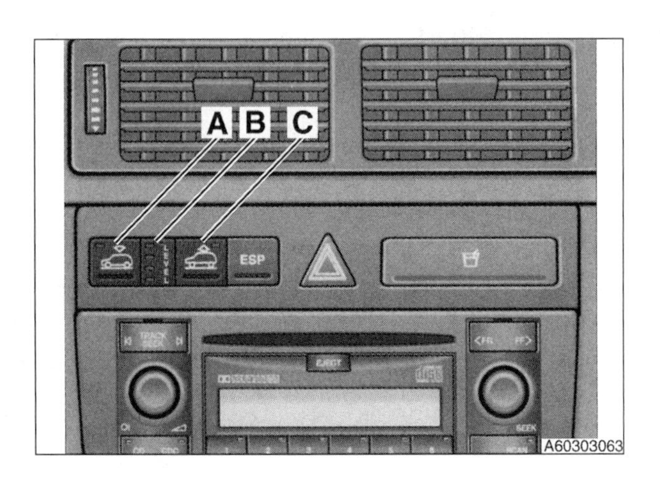

Before lifting vehicle with jack or lift:

- Switch ignition ON.
- Press ride height control buttons **A** and **C** in center dashboard for at least 5 seconds.
- When system is in "jack mode", LEDs on control buttons (**A** and **C**), yellow LED for manual mode on level indicator (**B**) and warning light for level control in instrument cluster all illuminate.

– Switch ignition OFF and lift or jack vehicle.

Deactivating

– Press control buttons **A** and **C** for at least 5 seconds. Warning light in instrument cluster, LED for manual mode and control button LEDs turn OFF.

"Jack mode" is automatically deactivated when vehicle speed exceeds 3 mph (5 kph).

Warnings and Cautions

> ***WARNING —***
> - *Most fasteners are designed to be used only once and become unreliable and may fail when used a second time. This includes, but is not limited to, nuts, bolts, washers, circlips, cotter pins, self-locking nuts and bolts. For replacements, use new parts.*
> - *Do not reinstall bolts and nuts coated with undercoating wax. Clean the threads with a suitable solvent before installing.*
> - *Replace rusted or corroded bolts, nuts and washers even if not specifically indicated.*
> - *Do not attempt to straighten or weld suspension struts, wheel bearing housings, control arms or any other wheel locating or load bearing components of the front suspension.*

> ***CAUTION—***
> - *To avoid wheel bearing damage, do not move vehicle without drive axle(s). If vehicle needs to be moved, install an outer joint in place of the drive axle. Tighten outer joint to 115 Nm (85 ft-lb) for M14 bolt or 190 Nm (140 ft-lb) for M16 bolt.*
> - *Bonded rubber bushings may only be flexed to a limited extent. See* **Curb weight position** *in this repair group.*
> - *Before performing suspension work on allroad quattro, see* **allroad quattro "jack mode"** *in this repair group.*

BentleyPublishers.com
BentleyPublishers.com—All Rights Reserved

FRONT STRUT ASSEMBLY

Front suspension strut, removing and installing

– Working in engine compartment, remove battery cover and plenum cover. See **03 Maintenance**.

– Remove hubcap. For alloy wheels, pull off center cap (use pulling hook in vehicle tools).

– Raise car and support safely.

> ***WARNING —***
> * *Make sure car is stable and well supported at all times. Use a professional automotive lift or jack stands designed for the purpose. A floor jack is not adequate support.*

– Remove front wheels.

NOTE —

* *Vehicle with headlight vertical aim control: Disconnect vehicle height sensor arm from control arm.*
* *In order not to damage lower control arm ball joints, use engine / transmission jack VAG1383-A or equivalent to brace control arms.*

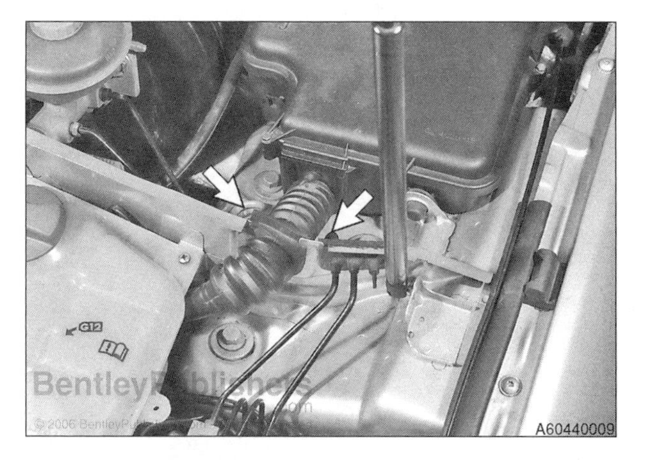

◄ Working in plenum chamber, remove rubber grommets (**arrows**) to gain access to strut upper mounting nuts. Use socket on long extension to remove nuts.

> ***CAUTION—***
> * *Do not bend or damage brake lines.*

– Working under car, pull ABS wheel speed sensor harness out of bracket on brake caliper. Be careful not to damage brake lines.

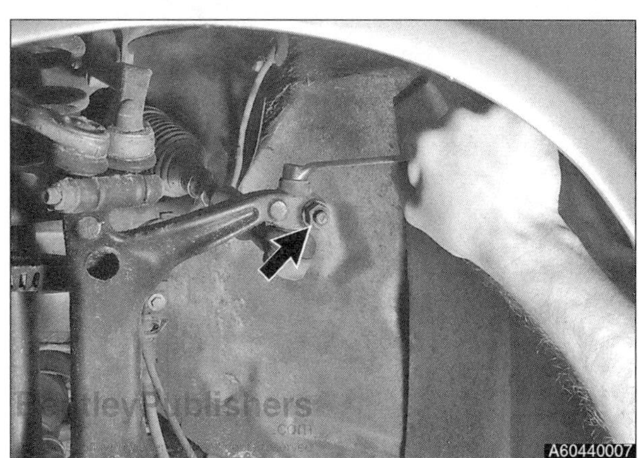

◄ Working at tie rod end:

* Remove nut on tie rod end pinch bolt (**arrow**) and remove bolt.
* Remove bolt from top of tie rod end.

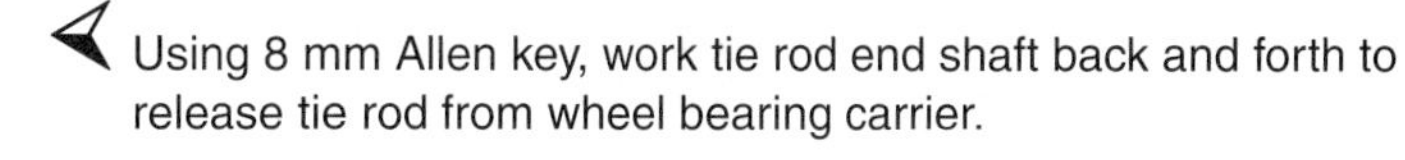

◄ Using 8 mm Allen key, work tie rod end shaft back and forth to release tie rod from wheel bearing carrier.

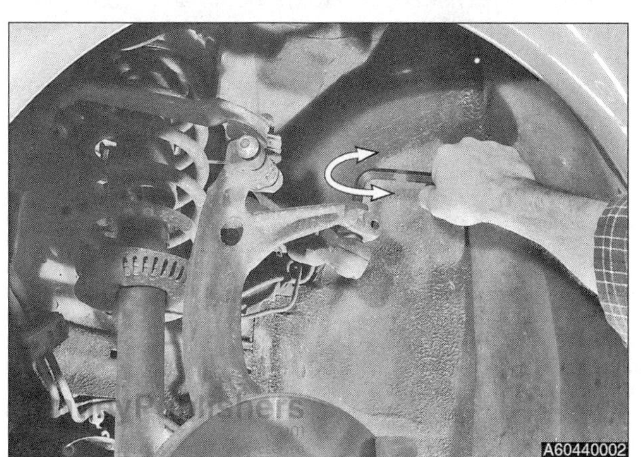

BentleyPublishers.com
BentleyPublishers.com—All Rights Reserved

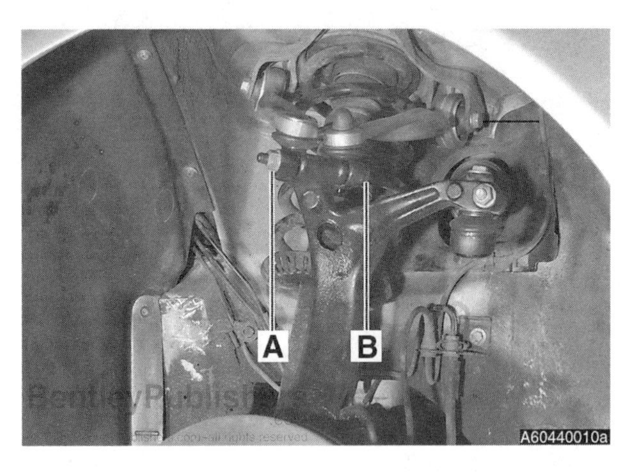

◄ Remove nut (**A**) on front and rear upper control arm pinch bolt and remove pinch bolt (**B**).

– Release upper front and rear control arm outer ends from wheel bearing carrier.

> ***CAUTION—***
> • *Do not use chisel or similar tool to widen slits in wheel bearing housing.*

– Remove nut from lower rear control arm ball joint. Using a suitable ball joint puller, pop control arm end out of wheel bearing carrier. Be careful not to damage CV boot.

NOTE—

• *If reusing control arm, put collar nut back on flush with ball joint stud to protect threads before attaching puller.*

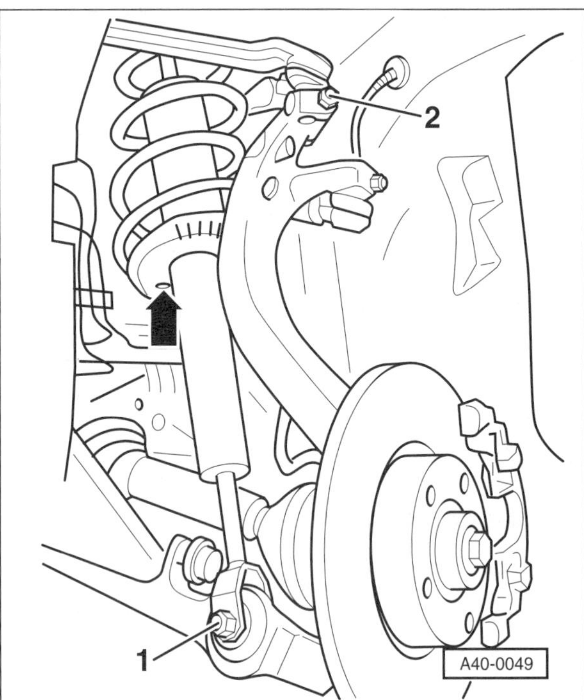

◄ Remove suspension strut lower mount bolt (**1**) and disengage strut from lower front control arm.

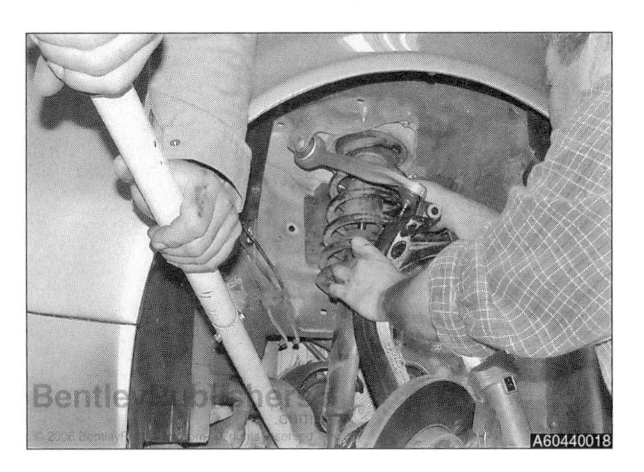

◄ Using a suitable lever arm, pry down on wheel bearing carrier and remove suspension strut upper mount from upper suspension bracket.

– Disassemble strut and coil spring, replacing components as needed. See **Front strut assembly components** and **Front strut assembly details** in this repair group.

BentleyPublishers.com
BentleyPublishers.com—All Rights Reserved

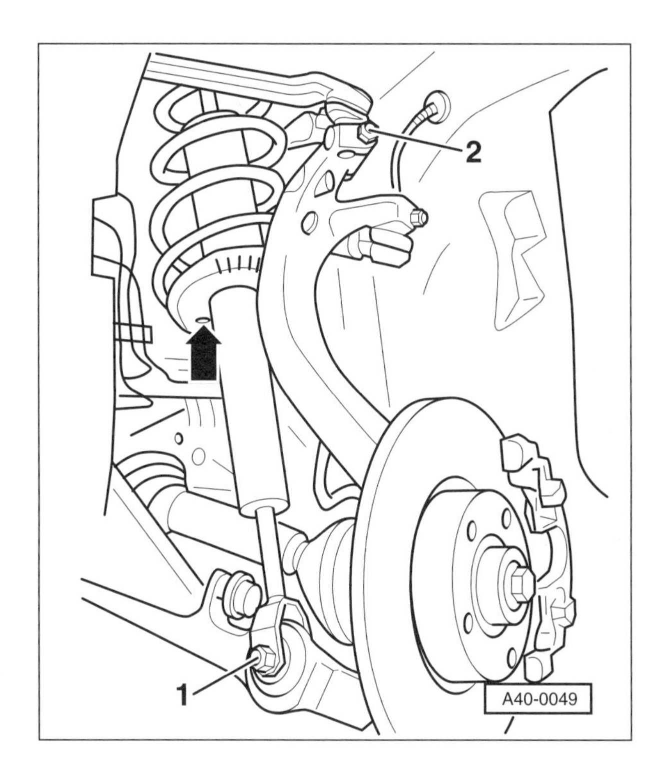

Installation is reverse of removal. Remember to:

- Position suspension strut so that hole (**arrow**) in spring seat faces middle of vehicle. See **Front strut assembly details** in this repair group.
- Guide suspension strut upper mount into position in upper suspension bracket. Install upper mount nuts from above in plenum chamber. Install rubber grommets and plenum chamber cover.
- Bolt suspension strut to lower front control arm. Tighten bolt (**1**) in curb weight position. See **Curb weight position** in this repair group.
- Insert upper control arm ball joints in wheel bearing housing and press down as far as possible. Install new pinch bolt and tighten nut (**2**).
- Reattach lower rear control arm to wheel bearing housing. Use 4 mm Allen wrench to counterhold ball joint stud, if necessary.
- Insert ABS wheel speed sensor harness into bracket on brake caliper.
- Mount wheel and tighten.

Tightening torques	
Lower rear control arm to wheel bearing housing	100 Nm (74 ft-lb)
Suspension strut upper mount to upper suspension bracket	20 Nm (15 ft-lb)
Suspension strut to lower front control arm	90 Nm (66 ft-lb)
Upper control arm to wheel bearing housing	40 Nm (30 ft-lb)
Wheel to wheel bearing hub	120 Nm (89 ft-lb)

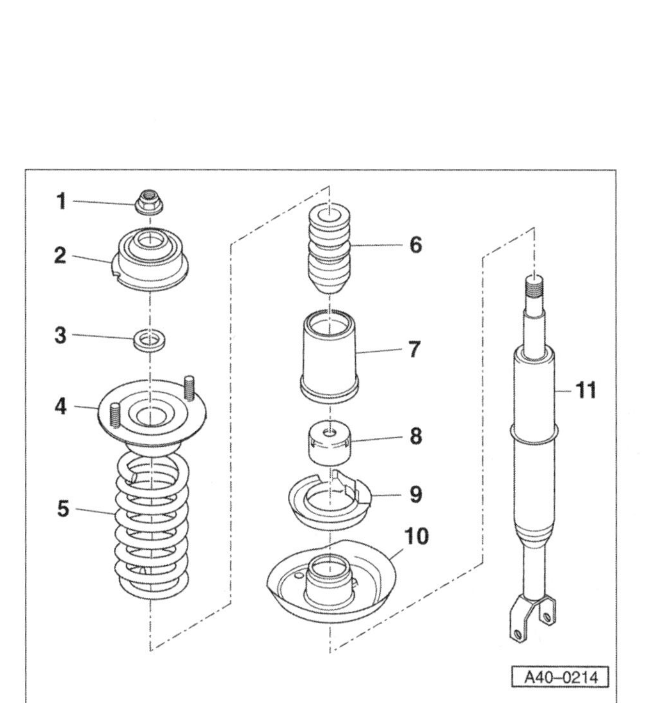

Front strut assembly components

1. Nut, 60 Nm (44 ft-lb)
   - Remove using special tool 3353.
   - Replace.
2. Upper strut mount
3. Washer
4. Upper spring seat: See **Front strut assembly details** in this repair group.
5. Coil spring
   - Inspect surface for pitting and damage.
   - Match color code when replacing. Check vehicle data sticker. See **Front strut assembly details** in this repair group.
6. Bump stop: Insert in upper spring seat.
7. Dust sleeve
8. Protective cap
9. Lower spring buffer
   - Not in all vehicles.
10. Lower spring seat: See **Front strut assembly details** in this repair group.
11. Strut: Match with spring. Check vehicle data sticker. See **Front strut assembly details** in this repair group.

BentleyPublishers.com
BentleyPublishers.com—All Rights Reserved

Front strut assembly details

Special tools, including a coil spring compressor, a special socket and a spring seat angle adjusting gauge, are needed for front strut disassembly and reassembly.

> ***WARNING —***
> - *Do not attempt to disassemble the struts without a spring compressor designed specifically for this job.*
> - *Make sure the spring compressor grabs the spring fully and securely before compressing it.*

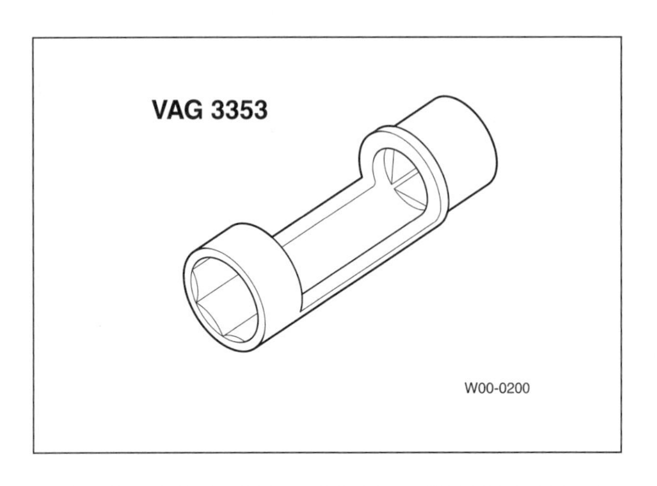

VAG special socket 3353 used in conjunction with Allen key for removing upper strut nut.

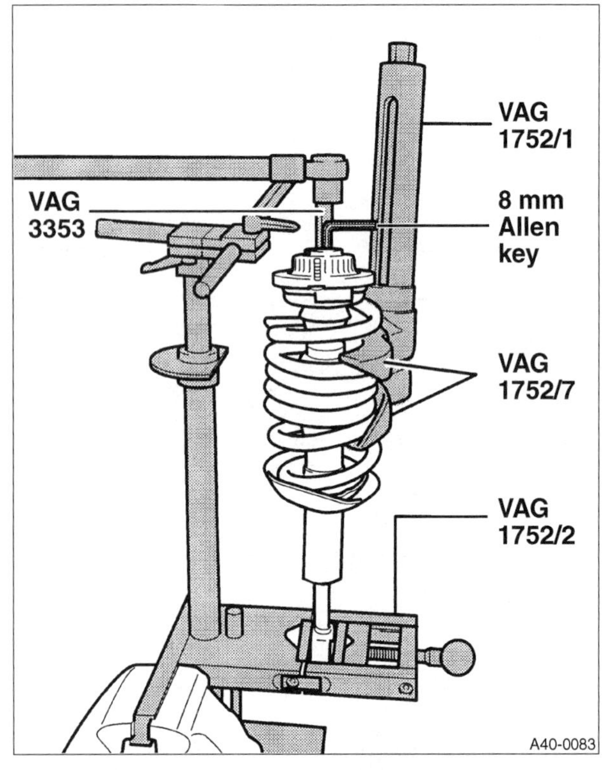

VAG special tools needed for safely compressing coil spring, loosening upper mount and replacing strut.

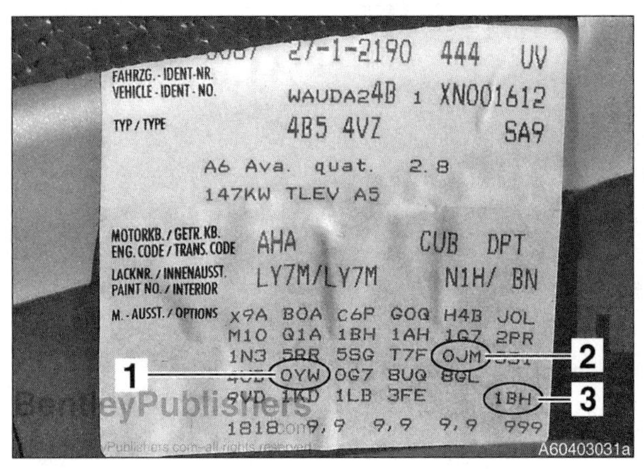

Vehicle data sticker in spare tire well:

1. Rear axle designation
2. Front axle designation
3. Suspension code

Match to strut and coil spring codes found in Audi parts data.

See **44 Wheels, Tires, Alignment** for more information about suspension codes.

BentleyPublishers.com
BentleyPublishers.com—All Rights Reserved

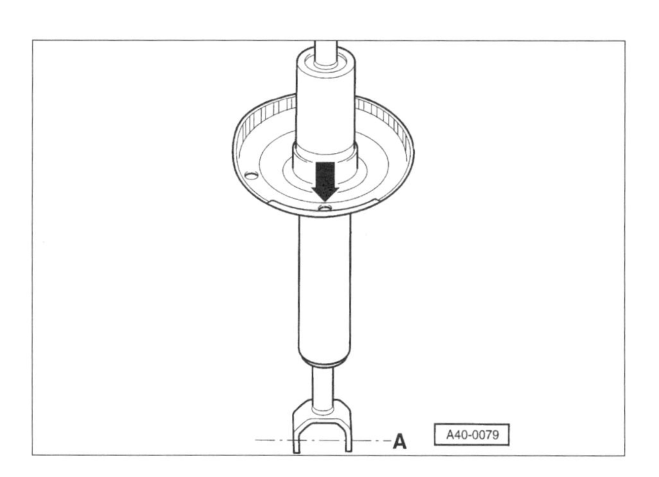

◀ Installed position of lower spring seat: Make sure hole (**arrow**) in seat is at right angles (90° ± 2°) to axis of lower mounting fork (**A**).

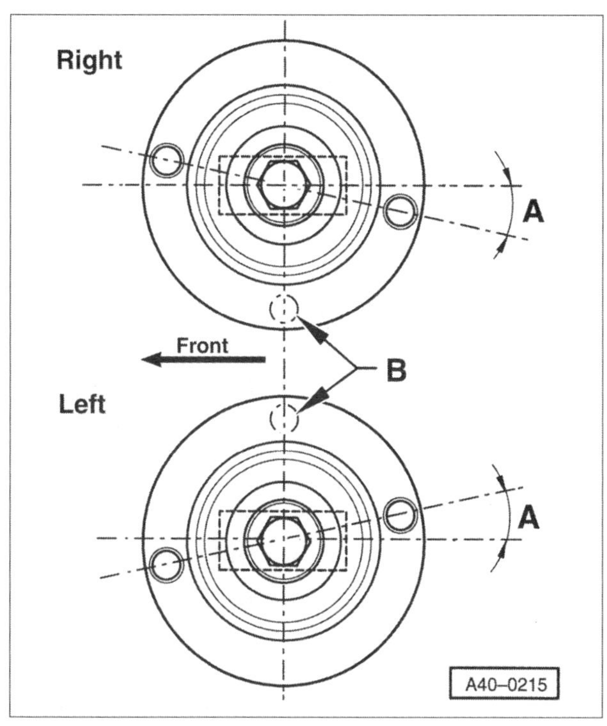

◀ Installation of spring seats:

- Studs in upper spring seats are rotated 11° (**A**) to axis of strut lower mounting fork.
- Holes (**B**) in lower spring seat face middle of vehicle.

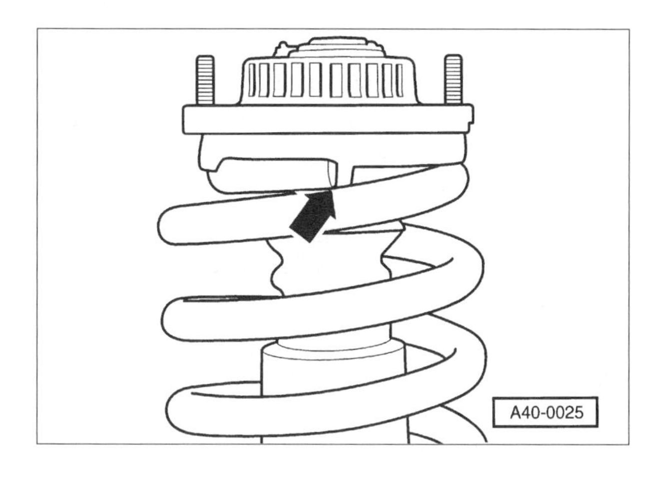

◀ Upper coil spring seat: Make sure upper end of spring (**arrow**) rests against upper spring seat stop.

BentleyPublishers.com
BentleyPublishers.com—All Rights Reserved

FRONT SUSPENSION ARMS

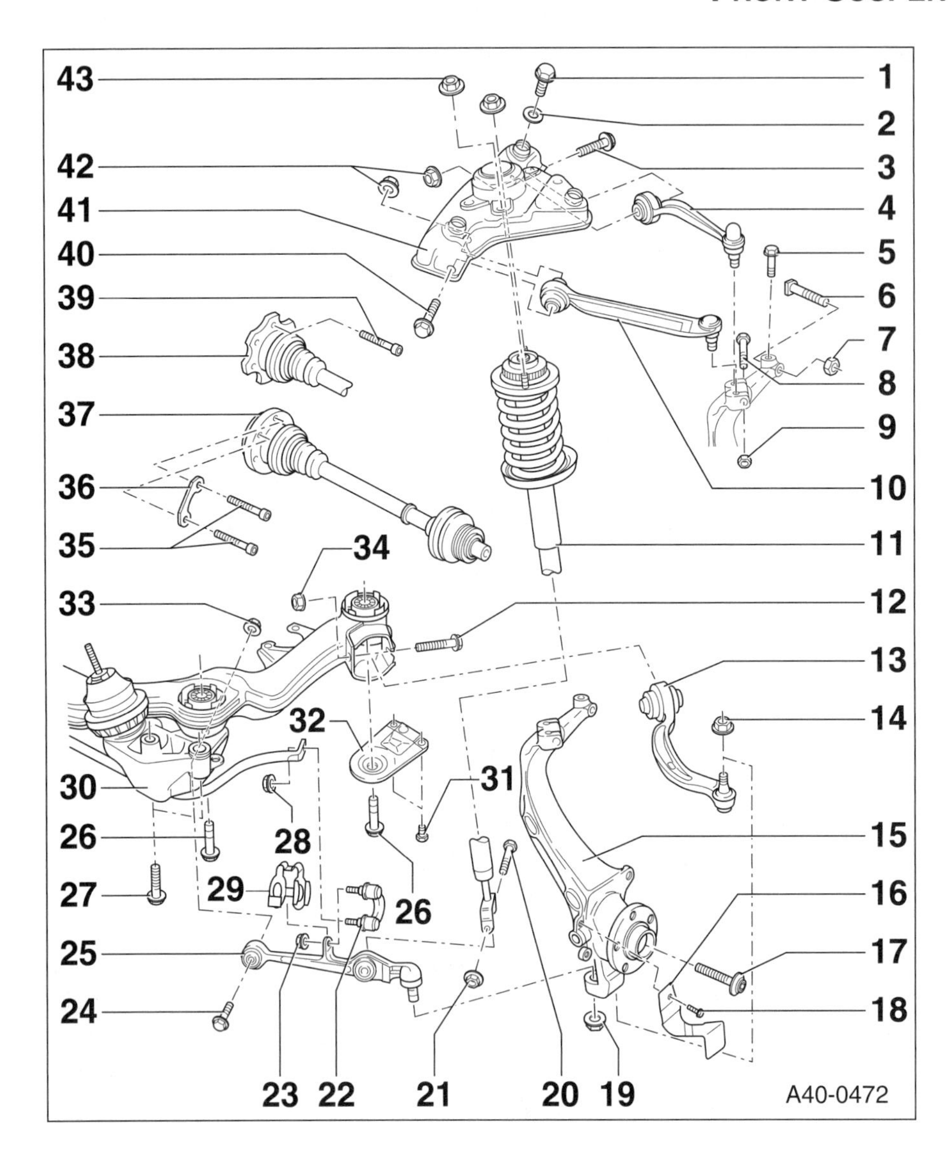
A40-0472

Front suspension components

1. **Bolt M10, 75 Nm (55 ft-lb)**
2. **Washer**
3. **Bolt M10 x 62 mm**
   - Replace.
   - Load suspension when tightening. See **Curb weight position** in this repair group.
4. **Upper rear control arm**
   - Replace bushing
5. **Bolt, 5 Nm (4 ft-lb)**
6. **Bolt**
7. **Self-locking nut**
   - Replace.
   - 50 Nm (37 ft-lb) for steel wheel bearing housing
   - 45 Nm (33 ft-lb) for aluminium wheel bearing housing
8. **Bolt M10 x 100**
9. **Self-locking nut, 40 Nm (30 ft-lb)**
   - Replace.
10. **Upper front control arm**
    - Can be removed together with mounting bracket
11. **Suspension strut**
12. **Bolt M12 x 1.5 x 120 mm**
    - Replace after disassembly.
    - Load suspension when tightening. See **Curb weight position** in this repair group.
13. **Lower rear control arm**
    - Replace arm if fluid-filled bushing is leaky
    - Replace.
14. **Self-locking nut**
    - 100 Nm (74 ft-lb) for steel wheel bearing housing
    - 110 Nm (81 ft-lb) for aluminium wheel bearing housing
15. **Wheel bearing housing**
16. **Backing plate**

BentleyPublishers.com
BentleyPublishers.com—All Rights Reserved

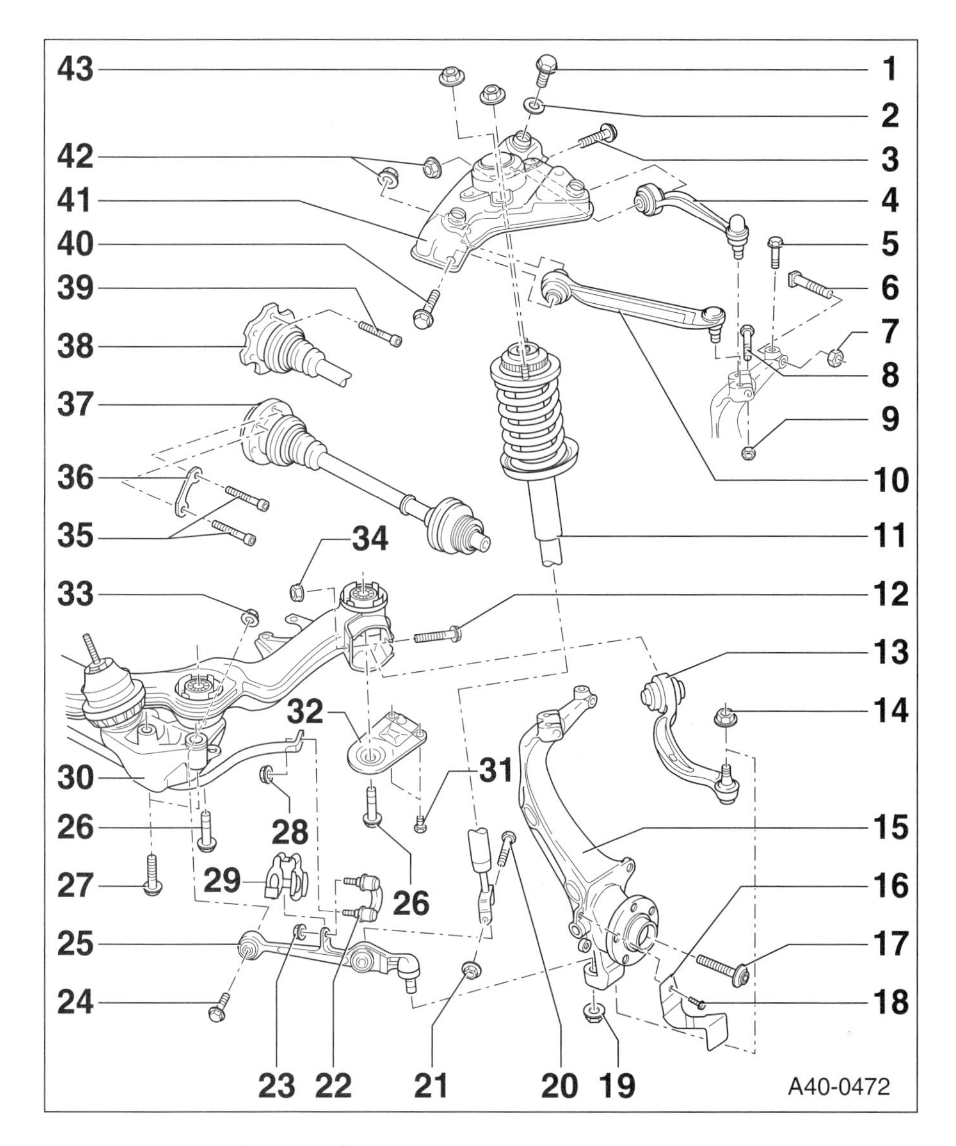

Front suspension components *(continued)*

17. Outer CV joint bolt, M14 or M16

- Replace.
- Tighten in 2 stages. See **Front drive axle, removing and installing** in this repair group.

18. Bolt, 10 Nm (7 ft-lb)

19. Self-locking nut

- 100 Nm (74 ft-lb) for steel wheel bearing housing.
- 120 Nm (89 ft-lb) for aluminium wheel bearing housing.
- Replace.

20. Bolt M12 x 1.5 x 85 mm

21. Self-locking nut, 90 Nm (66 ft-lb)

- Replace.

22. Stabilizer bar link

- Arrow points in direction of travel.

23. Self-locking nut

- Replace after disassembly.
- 40 Nm (30 ft-lb) plus additional 90° (¼ turn).
- Nut has ribs on the bottom, replace with same.

24. Bolt M12 x 1.5 x 100 mm

- Replace.
- Load suspension when tightening. See **Curb weight position** in this repair group.

25. Lower front control arm

26. Bolt M12 x 1.5 x 110

- 110 Nm (81 ft-lb) plus additional 90° (¼ turn).
- Replace.

27. Bolt, 75 Nm (55 ft-lb)

- M10 X 70 mm
- Replace.

28. Self-locking ribbed nut

- 100 Nm (74 ft-lb).
- Replace.

29. Clip

- Inserted in lower front control arm.
- Replace.

30. Subframe

31. Bolt, 25 Nm (18 ft-lb)

- Replace.

BentleyPublishers.com
BentleyPublishers.com—All Rights Reserved

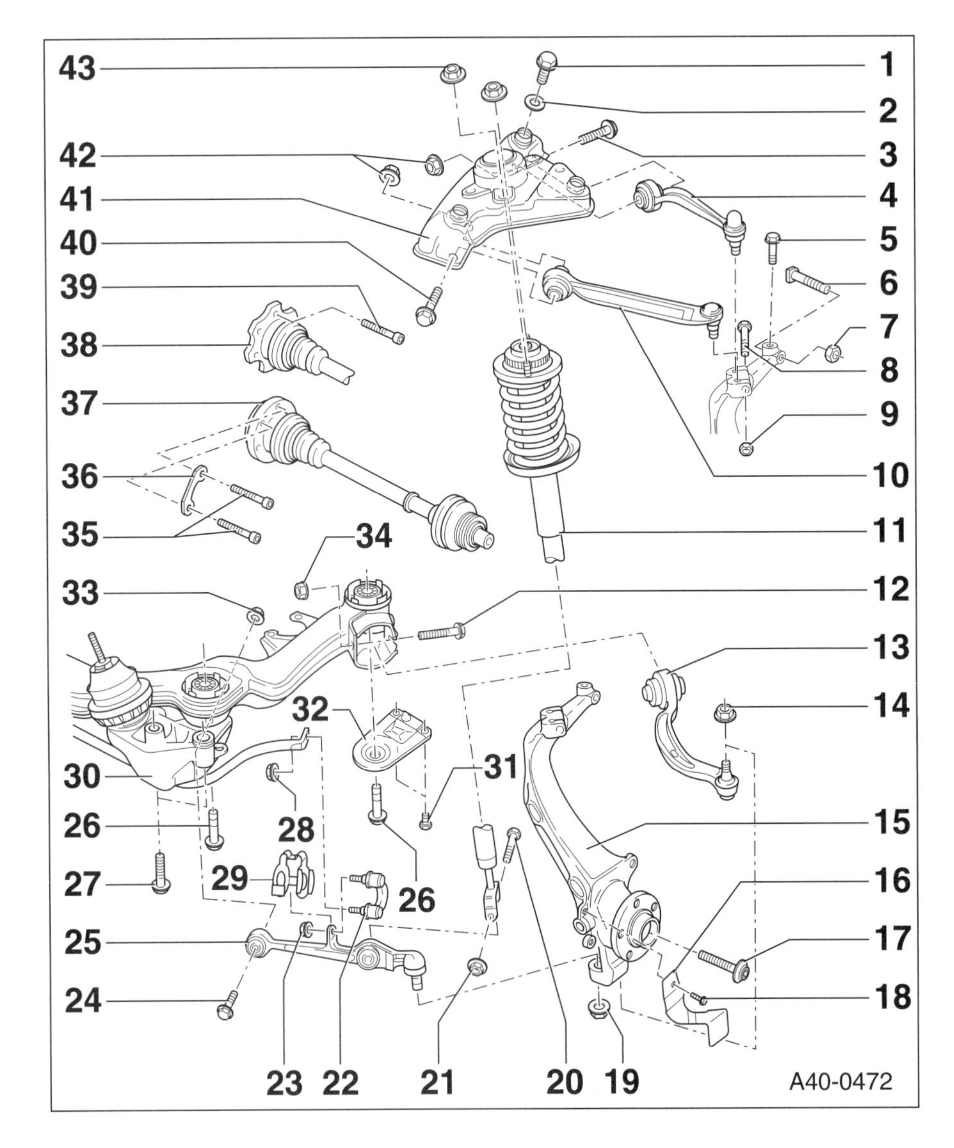

Front suspension components *(continued)*

32. Subframe support

33. Self-locking nut

- 70 Nm (52 ft-lb) plus additional 180° (½ turn).
- Replace.

34. Self-locking nut

- 70 Nm (52 ft-lb) plus additional 180° (½ turn).
- Replace.

35. Bolt

- Vehicle with inner CV joint.
- M8 X 48: 40 Nm (30 ft-lb).
- M10 X 48: 70 Nm (52 ft-lb).

36. Backing plate

37. Drive axle

- See **Front drive axle, removing and installing** in this repair group.

38. Drive axle with triple roller joint

39. Bolt, 70 Nm (52 ft-lb)

- Vehicle with triple roller joint.
- M10 X 20 mm. Bolt M10 x 62
- Replace.
- Load suspension when tightening. See **Curb weight position** in this repair group.

40. Bolt M10 x 62

- Always replace after removal.

41. Suspension upper mounting bracket

42. Self-locking nut

- 50 Nm (37 ft-lb) plus additional 90° (¼ turn).
- Always replace after removal.

43. Self-locking nut with flange, 20 Nm (20 ft-lb)

- Replace after disassembly.

BentleyPublishers.com
BentleyPublishers.com—All Rights Reserved

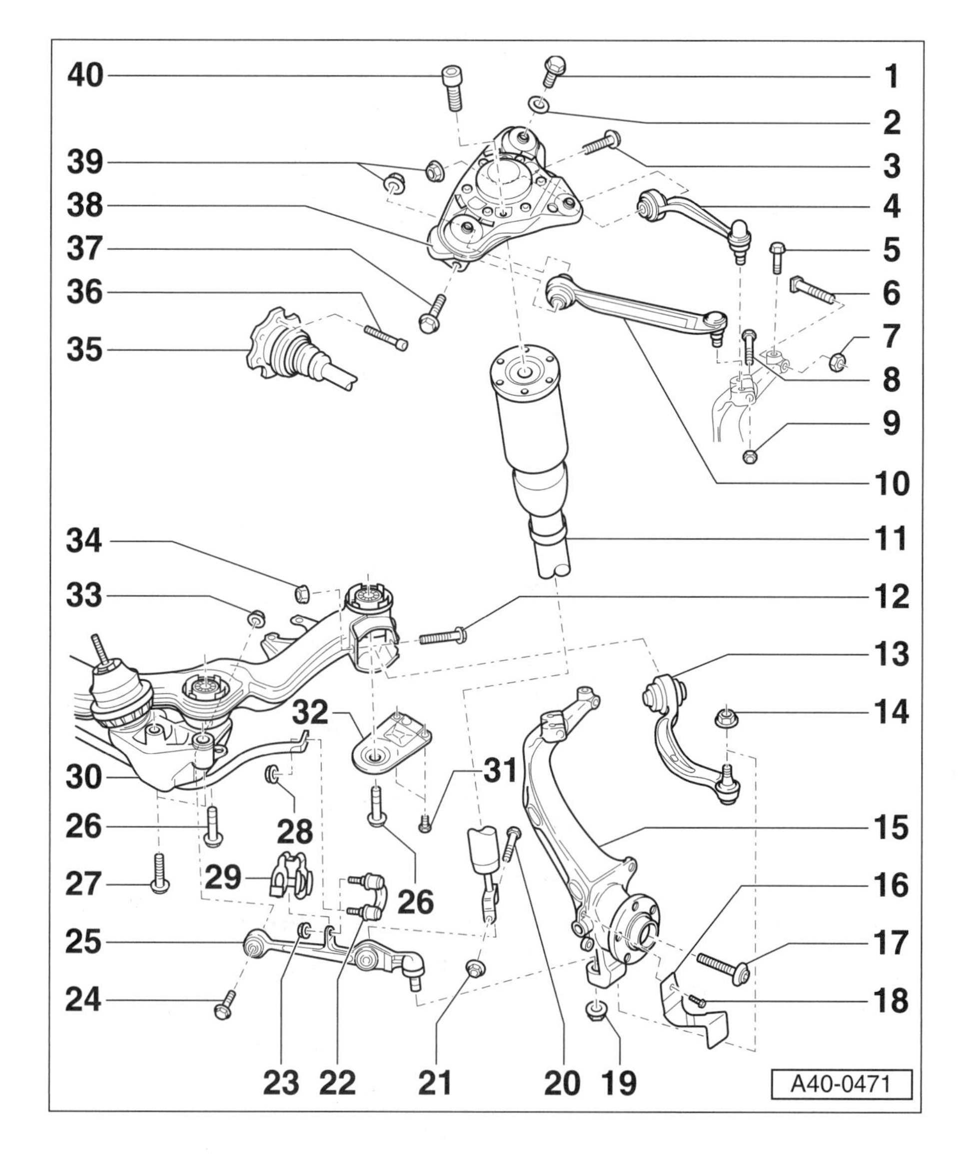

Front suspension components, allroad quattro

1. **Bolt**
   - 45 Nm (33 ft-lb) plus additional 55°
   - Replace.
2. **Washer**
3. **Bolt, M10 x 62 mm**
   - Replace.
   - Load suspension when tightening. See **Curb weight position** in this repair group.
4. **Upper rear control arm**
5. **Bolt, 5 Nm (4 ft-lb)**
6. **Bolt**
7. **Self-locking nut, 50 Nm (37 ft-lb)**
   - Replace.
8. **Bolt M10 x 100 mm**
9. **Self-locking nut, 40 Nm (30 ft-lb)**
   - Replace.
10. **Upper front control arm**
11. **Suspension strut**
12. **Bolt M12 x 1.5 x 120**
    - 90 Nm (66 ft-lb) plus additional 90°
    - Replace.
    - Load suspension when tightening. See **Curb weight position** in this repair group.
13. **Lower rear control arm**
14. **Self-locking nut, 110 Nm (81 ft-lb)**
    - Replace.
15. **Wheel bearing housing**
16. **Backing plate**
17. **Outer CV joint bolt, M16**
    - Replace.
    - Tighten in 2 stages. See **Front drive axle, removing and installing** in this repair group.
18. **Bolt,10 Nm (7 ft-lb)**
19. **Self-locking nut, 110 Nm (81ft-lb)**
    - Replace.
20. **Bolt M12 x 1.5 x 85 mm**
21. **Self-locking nut, 90 Nm (66 ft-lb)**
    - Replace.
22. **Stabilizer bar link**
    - Arrow points in direction of travel

BentleyPublishers.com
BentleyPublishers.com—All Rights Reserved

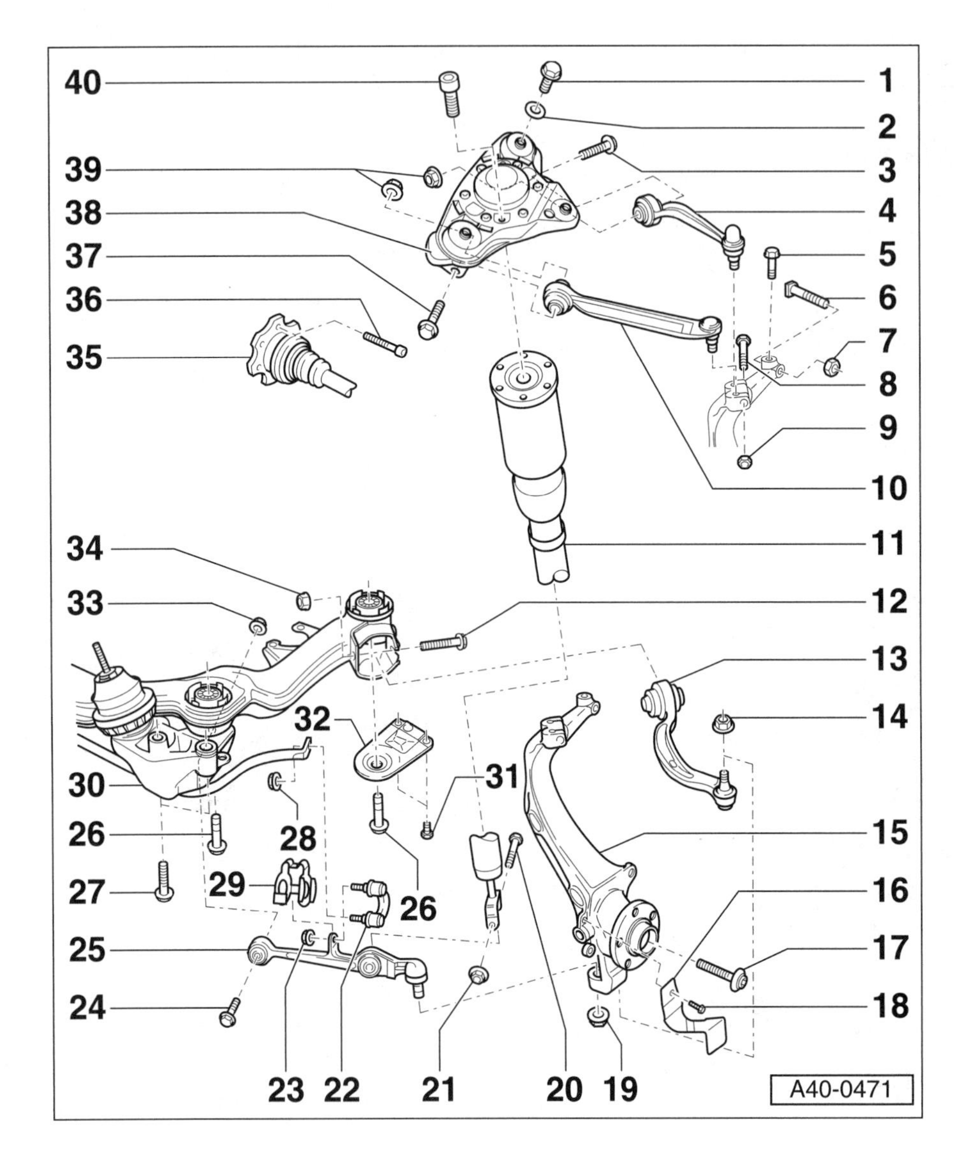

Front suspension components, allroad quattro *(continued)*

23. Self-locking ribbed nut
- Replace after disassembly
- 40 Nm (30 ft-lb) plus additional 90°

24. Bolt M12 x 1.5 x 100 mm
- Replace.
- Load suspension when tightening. See **Curb weight position** in this repair group.

25. Lower front control arm

26. M12 x 1.5 x 140 mm
- 110 Nm (81 ft-lb) plus 90°
- Replace.

27. Bolt, 70 Nm (52 ft-lb)
- M10 X 70 mm
- Replace.

28. Self-locking nut, 60 Nm (44 ft-lb)
- Replace.

29. Clip in lower front control arm
- Replace.

30. Subframe

31. Bolt, 25 Nm (18 ft-lb)

32. Support for subframe

33. Self-locking nut
- 80 Nm (59 ft-lb) plus additional 90°
- Replace.

34. Self-locking nut
- 90 Nm (66 ft-lb) plus additional 90°
- Replace.

35. Drive axle with triple roller joint

36. Bolt M10 X 20 mm
- 70 Nm (52 ft-lb)

37. Bolt M10 x 62 mm
- Replace.
- Load suspension when tightening. See **Curb weight position** in this repair group.

38. Suspension upper mounting bracket
- Replace.

39. Self-locking nut
- 50 Nm (37 ft-lb) plus additional 90°
- Replace.

40. Bolt, 23 Nm (17 ft-lb)
- Replace.

BentleyPublishers.com
BentleyPublishers.com—All Rights Reserved

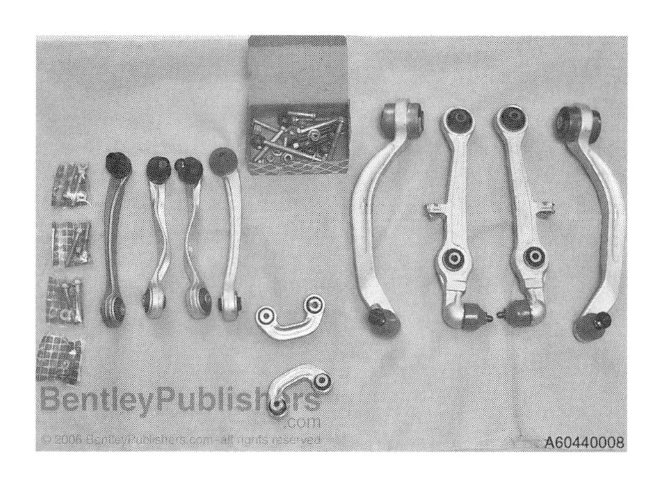

Front suspension control arms

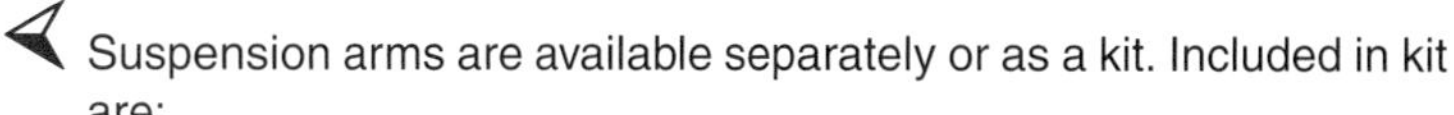

Suspension arms are available separately or as a kit. Included in kit are:

- Upper front control arms
- Upper rear control arms
- Stabilizer bar links
- Lower rear control arms
- Lower front control arms
- Mounting hardware

See **Front suspension components** diagrams in this repair group for suspension arm locations and torque specifications.

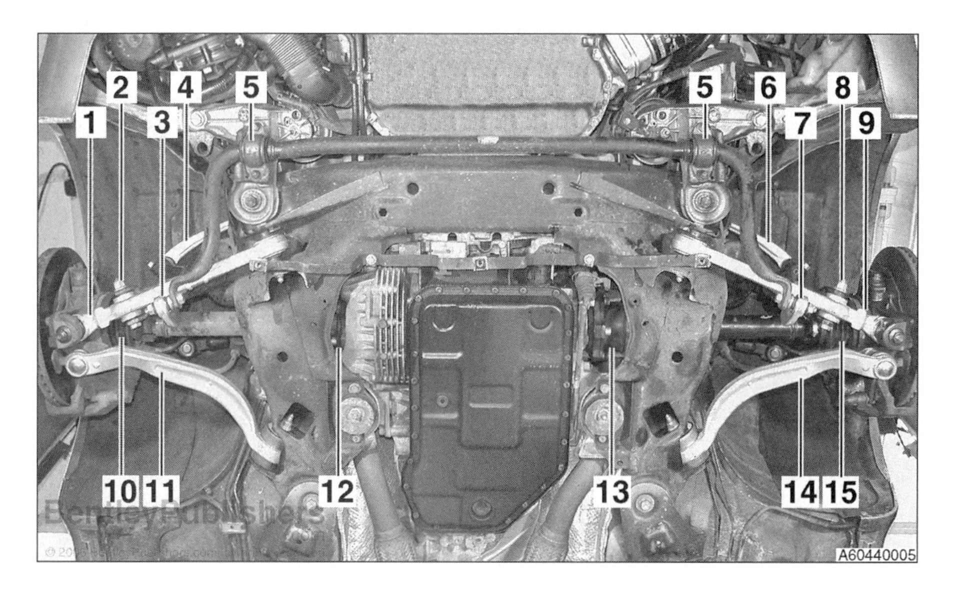

Suspension arm components

1. **Lower front control arm (right)**
2. **Lower strut mount (right)**
3. **Stabilizer bar link (right)**
4. **Upper rear control arm (right)**
5. **Stabilizer bar mounting bracket**
6. **Upper rear control arm (left)**
7. **Stabilizer bar link (left)**
8. **Lower strut mount (left)**
9. **Lower front control arm (left)**
10. **Outer CV boot (right)**
11. **Lower rear control arm (right)**
12. **Axle flange (right)**
13. **Axle flange (left)**
14. **Lower rear control arm (left)**
15. **Outer CV boot (left)**

FRONT STABILIZER BAR

Front stabilizer bar, removing and installing

– Raise car and support safely.

> ***WARNING —***
> - *Make sure car is stable and well supported at all times. Use a professional automotive lift or jack stands designed for the purpose. A floor jack is not adequate support.*

– Remove engine lower cover. See **03 Maintenance**.

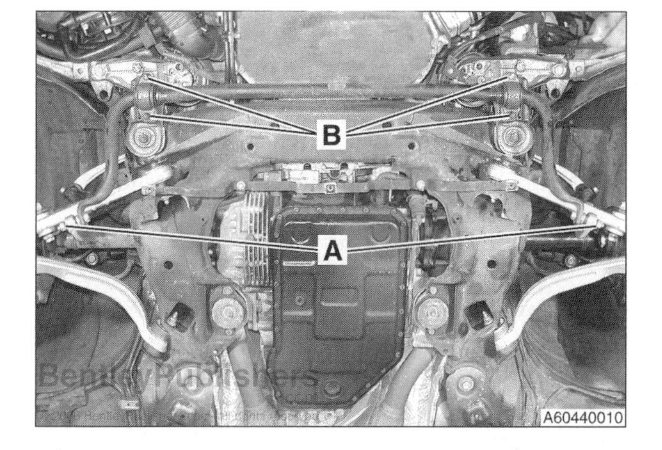

Remove stabilizer link nut (**A**).

– Remove stabilizer link from bar.

– Unscrew mounting bracket nuts (**B**) and remove stabilizer.

BentleyPublishers.com
BentleyPublishers.com—All Rights Reserved

– Installation is reverse of removal. Remember to:
  - Match stabilizer bar to suspension version.
  - Mount bushing and stabilizer bar without grease.
  - Attach stabilizer bar links. See **Front stabilizer bar links** in this repair group.

Tightening torque	
Stabilizer bar bracket to chassis (replace nuts)	25 Nm (18 ft-lb)

Front stabilizer bar links

From model year 1999, stabilizer bar link with bonded rubber bushings replaces unit with ball joints.

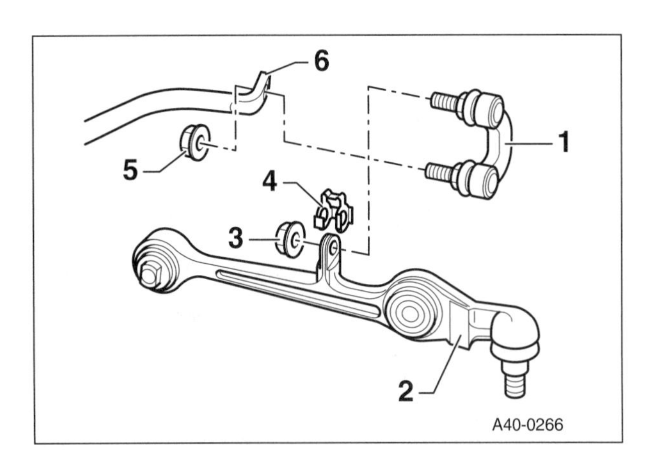

Link with ball joint

1. Link with ball joint
2. Lower front control arm
3. Self-locking nut, 40 Nm (30 ft-lb) plus additional 90°
4. Clip
5. Self-locking nut, 100 Nm (74 ft-lb)
6. Stabilizer bar

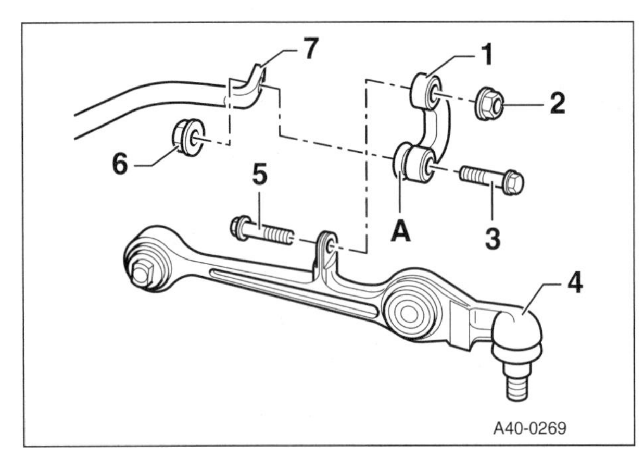

Link with bonded rubber bushings

1. Link with rubber bushing (rubber bushing (**A**) with pressed-on spacer is bolted to stabilizer)
2. Self-locking nut, 40 Nm (30 ft-lb) plus additional 90° turn
3. Bolt
4. Lower front control arm
5. Bolt
6. Self-locking nut, 60 Nm (44 ft-lb)
7. Stabilizer bar

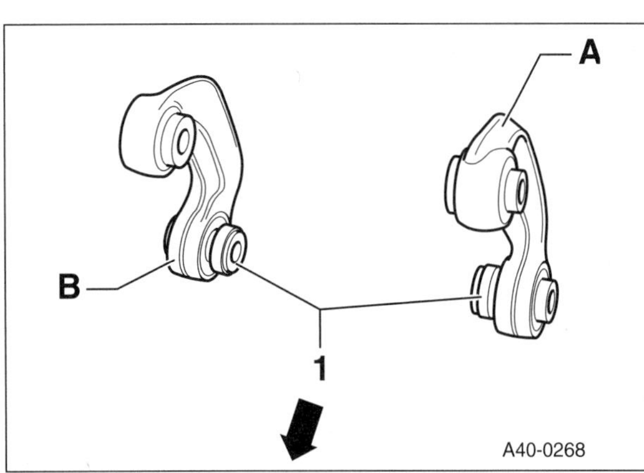

◀ Link (**A**) for left side of vehicle

– Link (**B**) for right side of vehicle

– Arrow on link points in direction of travel.

– Place fixed spacer (**1**) on stabilizer.

> ***CAUTION—***
> - *To avoid damage, bonded rubber bushings may only be flexed to a limited extent. Tighten suspension bushing bolts with vehicle in curb weight position. See* **Curb weight position** *in this repair group.*

BentleyPublishers.com
BentleyPublishers.com—All Rights Reserved

Front Drive Axles

Front drive axle, removing and installing

- Remove hubcap. For alloy wheels pull off center cap (use pulling hook in vehicle tools).
- With vehicle standing on its wheels, loosen outer CV joint bolt a maximum of 90° turn.
- Raise car and support safely.

> ***WARNING —***
> - *Make sure car is stable and well supported at all times. Use a professional automotive lift or jack stands designed for the purpose. A floor jack is not adequate support.*

- Remove wheel.
- Thread in all five wheel bolts again.
- With assistant operating brake, remove outer CV bolt.

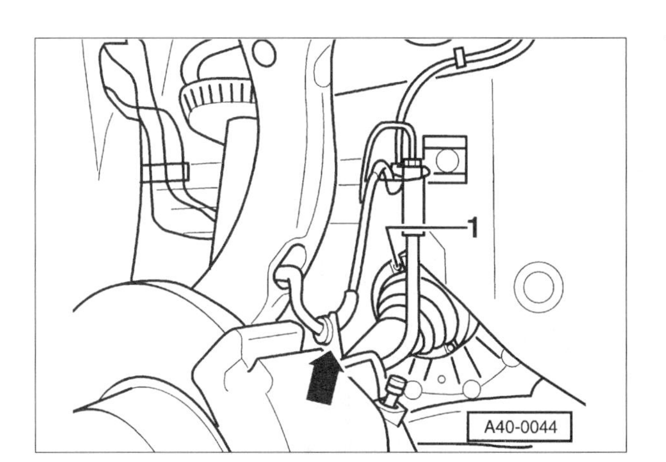

- ◀ Unscrew drive axle bolts (**1**) from transmission flange.
- Pull ABS wheel speed sensor harness out of bracket on brake caliper (**arrow**).
- Pull ABS wheel speed sensor slightly out of wheel bearing housing.

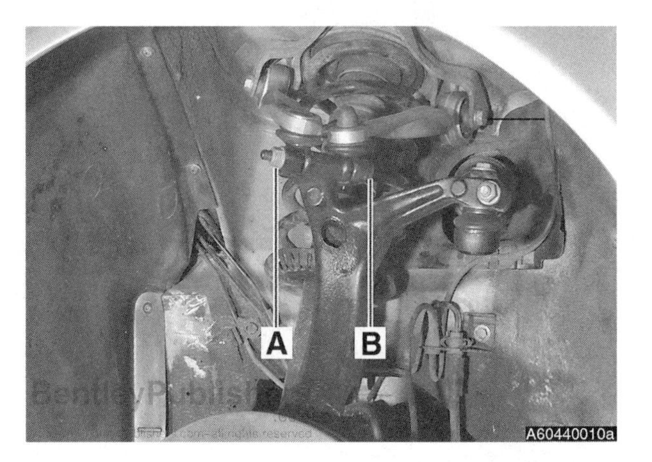

- ◀ Remove nut (**A**) on front and rear upper control arm pinch bolt and remove pinch bolt (**B**).
- Release upper front and rear control arm outer ends from wheel bearing carrier

> ***CAUTION—***
> - *Do not use chisel or similar tool to widen slits in wheel bearing housing.*
> - *Do not remove bolts for outer tie rod end. Wheel alignment may change.*

- Swing wheel bearing carrier away from suspension strut.
- Work outer CV joint out of wheel bearing hub and remove drive axle.
- Repair or replace drive axle as necessary. For axle boot replacement procedure, see **CV joint boot, replacing** in this repair group.
- When installing, insert outer CV joint into wheel bearing hub first.

BentleyPublishers.com
BentleyPublishers.com—All Rights Reserved

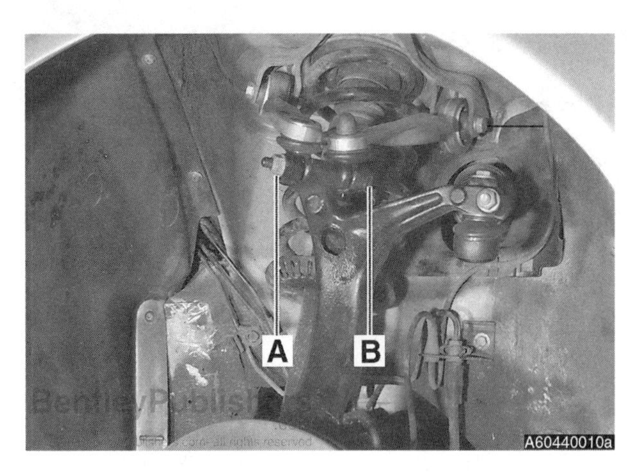

◀ Insert upper control arm ball joints in wheel bearing housing and press down as far as possible. Install new pinch bolt (**B**) and tighten nut (**A**).

– Bolt drive axle to transmission flange.

Tightening torques	
Drive axle to transmission flange: • M8 bolt • M10 bolt	 40 Nm (30 ft-lb) 70 Nm (52 ft-lb)
Upper control arms to wheel bearing housing	40 Nm (30 ft-lb)

– Press ABS wheel speed sensor into wheel bearing housing and insert harness into bracket on brake caliper.

– With an assistant operating brake, tighten outer CV joint bolt to initial tightening torque.

> ***CAUTION—***
> *• Do not allow vehicle to contact the ground when initially tightening outer CV joint bolt.*

Tightening torques	
Drive axle to bearing hub (initial torque) • M14 bolt • M16 bolt	 115 Nm (85 ft-lb) 180 Nm (133 ft-lb)

– Mount wheel and tighten.

Tightening torque	
Wheel lugs to wheel bearing hub	120 Nm (89 ft-lb)

– Tighten bolt for drive axle to final tightening torque with vehicle standing on its wheels.

Tightening torques	
Drive axle to bearing hub (final torque) • M14 bolt • M16 bolt	 115 Nm (85 ft-lb) + 180° 180 Nm (133 ft-lb) + 180°

CV joint boot, replacing

When replacing CV joint boot, use a complete repair kit including new boot, clamping bands, special lubricant and axle circlip.

If CV joints are worn or defective, a complete rebuilt axle shaft is available from an authorized Audi dealer.

– Remove drive axle. See **Front drive axle, removing and installing** in this repair group.

– Place drive axle on workbench in a padded vise.

– Cut off old boot clamps and remove boot.

BentleyPublishers.com
BentleyPublishers.com—All Rights Reserved

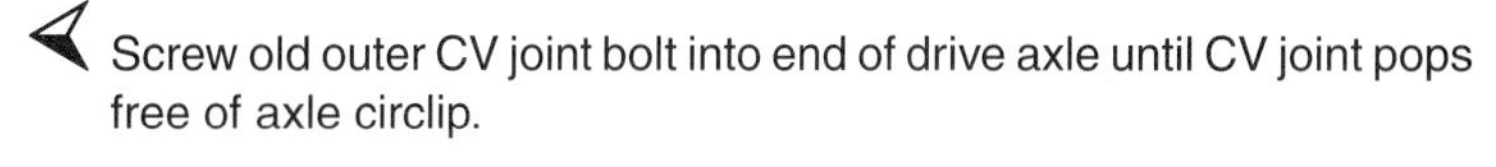

- Screw old outer CV joint bolt into end of drive axle until CV joint pops free of axle circlip.
- Remove CV joint and clean old grease off joints and shaft.
- Inspect CV joint for galling, pitting and other signs of physical damage.

NOTE—

- *Polished surfaces or visible ball tracks alone are not necessarily cause for replacement. Discoloration due to overheating indicates lack of lubricant and the need for a new CV joint.*

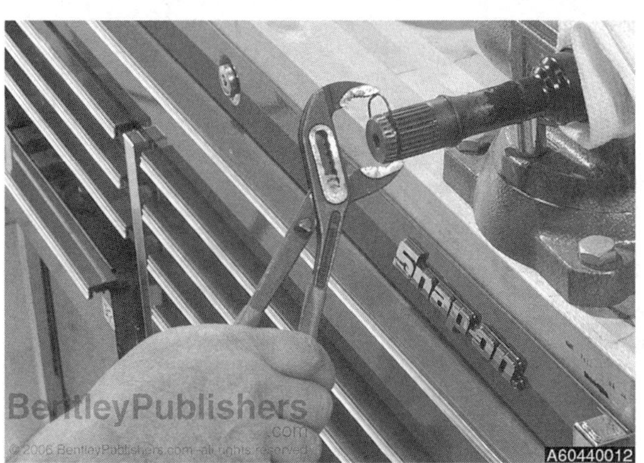

A60440012

- Snap new circlip into groove on end of axle shaft.
- Slide new small clamping band, CV boot and large clamping band on axle shaft.
- Apply special lubricant from CV boot repair kit to CV joint.
- While supporting axle shaft, press inner hub of CV joint on shaft. Make sure circlip snaps securely in place. Use a rubber mallet to drive joint on axle shaft, if necessary.

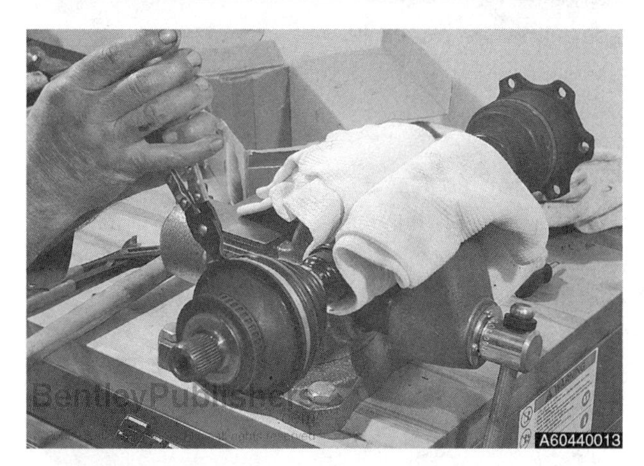

A60440013

- Work large end of CV boot into place on joint. Use band clamp pliers to secure boot clamp.

NOTE—

- *Make sure seal beading of boot fits in grooves on CV joint and axle shaft.*

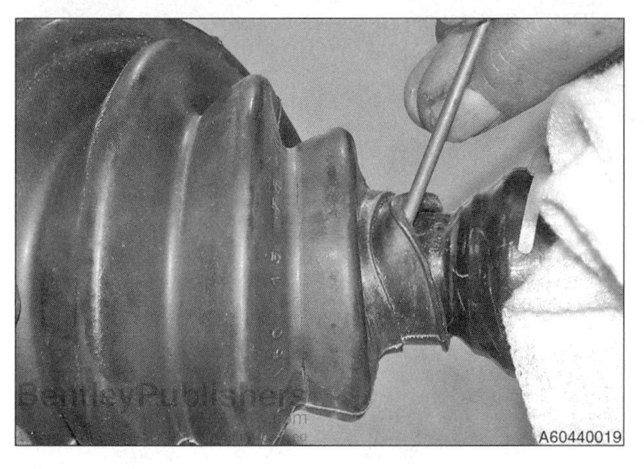

A60440019

- Swivel CV joint as far as it will go, thus compressing axle boot. Use a small screwdriver to "burp" excess air from boot. Be careful not to damage boot.
- Position small end of CV boot in place and tighten small clamping band over boot end.
- Reinstall drive axle using new axle bolt.

BentleyPublishers.com

BentleyPublishers.com—All Rights Reserved

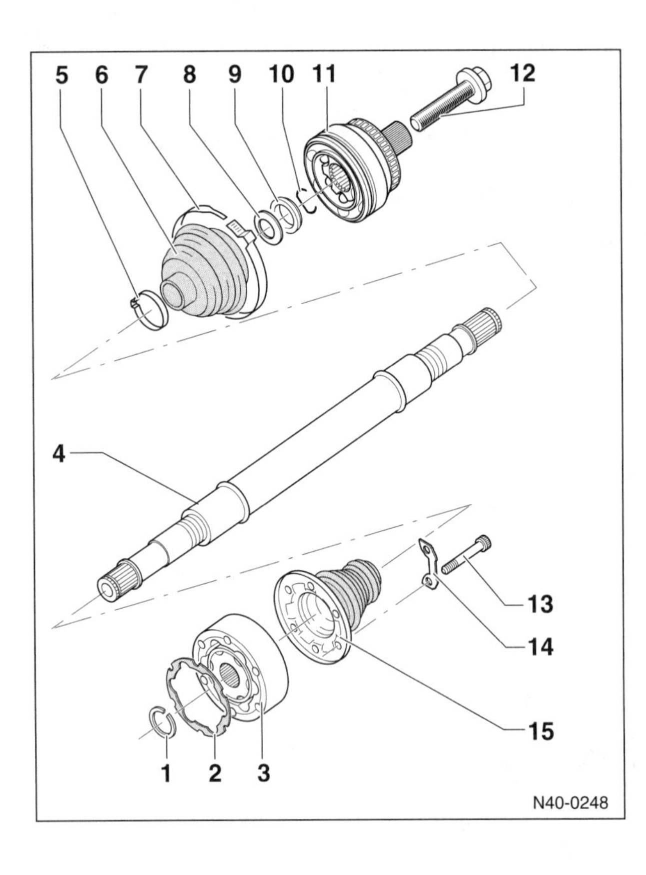

Front drive axle assembly with inner CV joint

1. Circlip
   - Replace.
2. Gasket
   - Replace.
3. Inner CV joint
   - Outside diameter is 100 mm or 108 mm (3.94 in or 4.24 in), depending on engine/transmission combination.
   - Replace only as complete unit.
4. Drive axle shaft
5. Clamp
   - Replace.
6. CV boot
   - Check for tears and scuffing.
   - Before tightening small end clamp, "burp" CV boot.
7. Clamp
   - Replace.
8. Dished washer
   - Large diameter (concave side) rests against thrust washer.
9. Thrust washer
10. Circlip
    - Replace.
    - Insert in groove on shaft.
11. Outer CV joint
    - Replace only as complete unit.
    - Drive on shaft until circlip snaps into place.
12. Bolt
    - Replace.
    - M14 bolt: 115 Nm (85 ft-lb) plus additional 180° (½ turn).
    - M16 bolt: 190 Nm (140 ft-lb) plus additional 180° (½ turn).
13. Bolt
    - M8 x 48 mm: 40 Nm (30 ft-lb).
    - M10 x 48 mm: 70 Nm (52 ft-lb).
14. Reinforcing plate
15. Inner CV joint boot.
    - Check for tears and scuffing.
    - Spread D 454 300 A2 or equivalent sealer on sealing surface before mounting on CV joint.

BentleyPublishers.com
BentleyPublishers.com—All Rights Reserved

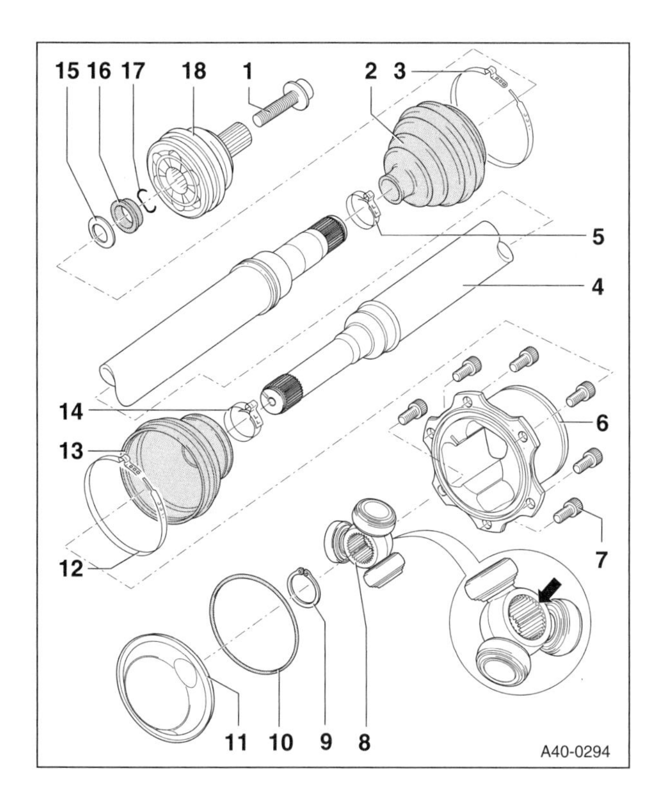

Front drive axle assembly with inner triple roller joint (AAR 2900)

1. Bolt
   - Replace.
   - M16 bolt: 190 Nm (140 ft-lb) plus an additional 180° (½ turn).
2. Outer CV boot
   - Check for tears and scuffing.
   - Before tightening small end hose clamp, "burp" boot.
3. Clamp (matched to boot)
   - Replace.
4. Drive axle
5. Clamp (matched to boot).
   - Replace.
6. Housing
7. Socket-head bolt
   - M 10 x 20 mm: 70 Nm (52 ft-lb).
8. Triple roller
   - Chamfer (**arrow**) faces splines of drive axle.
9. Circlip
   - Replace.
   - Insert in groove on shaft.
10. Sealing ring
    - Replace with new seal supplied in repair kit.
11. Cover
    - Replace.
12. Clamp (matched to boot)
    - Replace.
13. Boot for triple roller joint
14. Clamp
    - Replace.
15. Dished washer
16. Spacer ring (plastic)
17. Circlip
    - Replace.
    - Insert in ring groove of shaft before installation.
18. Outer CV joint
    - Replace only as complete unit.
    - Drive on shaft until circlip snaps in.

BentleyPublishers.com
BentleyPublishers.com—All Rights Reserved

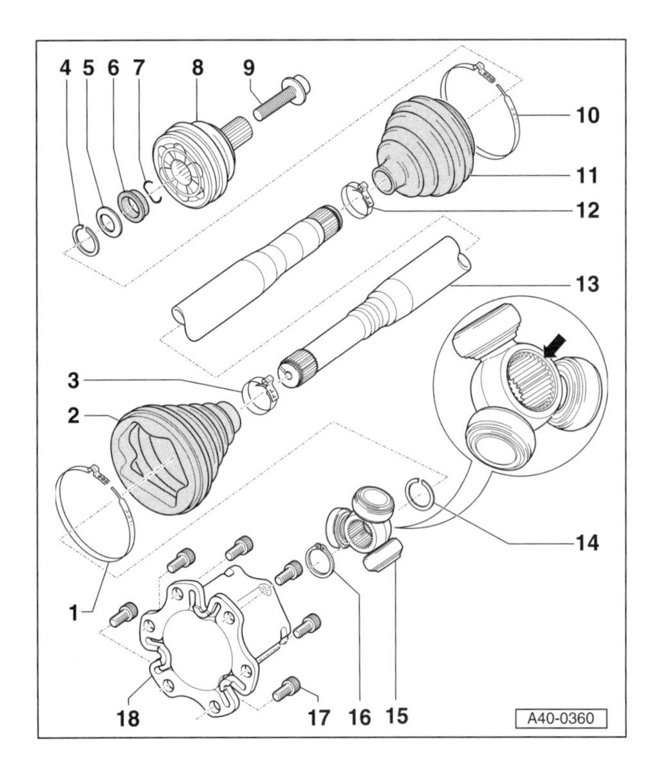

Front drive axle assembly inner triple roller joint (AAR 3300i)

1. Clamp (matched to boot)
   - Replace.
2. CV boot for triple roller joint
3. Clamp
   - Replace.
4. Circlip
5. Dished washer
6. Spacer ring (plastic)
7. Circlip
   - Replace.
   - Insert in ring groove of shaft before installation.
8. Outer CV joint
   - Replace only as complete unit
   - Drive on shaft until circlip snaps in.
9. Bolt
   - Replace.
   - M16 bolt: 190 Nm (140 ft-lb) plus additional 180° (½ turn).
10. Clamp (matched to boot)
    - Replace.
11. CV boot for outer CV joint
    - Check for tears and scuffing.
    - Before tightening small end clamp, "burp" CV boot.
12. Clamp
    - Replace.
13. Drive axle
14. Circlip
15. Triple roller connection
    - Chamfer (**arrow**) faces splines of drive axle.
16. Circlip
    - Replace.
    - Insert in groove on shaft.
17. Socket-head bolt
    - M 10 x 20 mm; 70 Nm (52 ft-lb).
18. Housing

BentleyPublishers.com
BentleyPublishers.com—All Rights Reserved

42 Rear Suspension

General

This section covers repairs to the rear suspension and related components. Also see:

- **40 Front Suspension** for rear suspension servicing and repair information
- **44 Wheels, Tires, Wheel Alignment** for wheel alignment specifications

Special tools, equipment and procedures are required for most rear suspension repair and component replacement. On suspension components with bonded rubber bushings, tighten fasteners with vehicle at curb weight position. See **Curb weight position** in this repair group.

Wheel alignment is almost always disturbed when suspension components are removed or replaced. Plan on a wheel alignment after suspension repairs.

Curb weight position

Bonded rubber bushings can only be flexed to a limited extent. Place suspension under load (curb weight position) before tightening suspension arms with bonded rubber bushings.

In this illustration jack stands and wheel ramps are used to place vehicle at curb weight position. Then tighten suspension arms with bonded rubber bushings to torque specification.

BentleyPublishers.com
BentleyPublishers.com—All Rights Reserved

allroad quattro "jack mode"

Prior to jacking or lifting allroad quattro vehicle, set ride height control to "jack mode" to protect suspension shock absorbers and struts.

> ***WARNING —***
> *• Make sure that no one is lying under the vehicle or has head or hands in the wheel housing while the ride height is changing.*

Activating

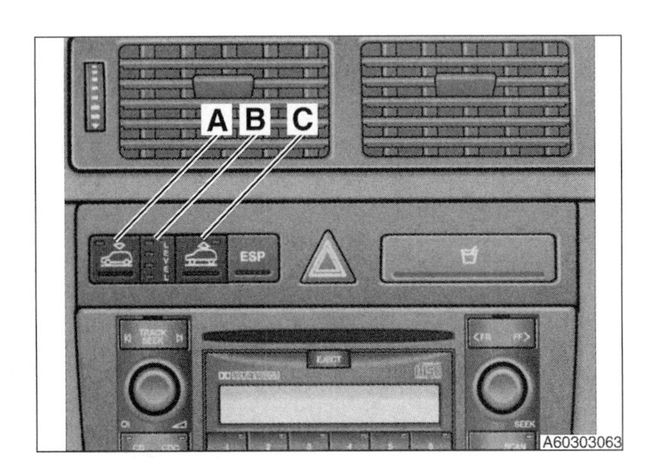

Before lifting vehicle with jack or lift:

- Switch ignition ON.
- Press ride height control buttons **A** and **C** in center dashboard for at least 5 seconds.
- When system is in "jack mode", LEDs on control buttons (**A** and **C**), yellow LED for manual mode on level indicator (**B**) and warning light for level control in instrument cluster all illuminate.

– Switch ignition OFF and lift or jack vehicle.

Deactivating

– Press control buttons **1** and **3** for at least 5 seconds. Warning light in instrument cluster, LED for manual mode and control button LEDs turn OFF.

"Jack mode" is automatically deactivated when vehicle speed exceeds 3 mph (5 kph).

Suspension codes (part numbers)

The spring/shock absorber combination installed in the vehicle is documented by part number on the vehicle data sticker. The vehicle data sticker is located in the spare tire well and in the service booklet.

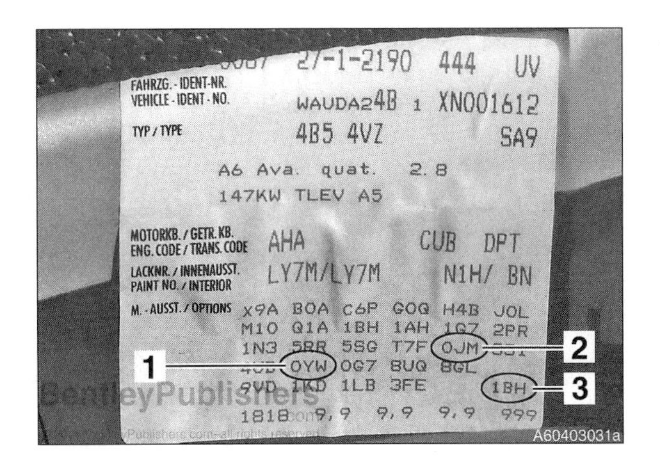

Example of a vehicle data sticker.

1. Rear axle designation
2. Front axle designation
3. Suspension code

Match to strut and coil spring codes found in Audi parts data.

See **44 Wheels, Tires, Alignment** for more information about suspension codes.

BentleyPublishers.com
BentleyPublishers.com—All Rights Reserved

Warnings and Cautions

WARNING—
- *Most fasteners are designed to be used only once and become unreliable and may fail when used a second time. This includes, but is not limited to, nuts, bolts, washers, circlips, cotter pins, self-locking nuts and bolts. For replacements, use new parts.*
- *Do not reinstall bolts and nuts coated with undercoating wax. Clean the threads with a suitable solvent before installing.*
- *Replace rusted or corroded bolts, nuts and washers even if not specifically indicated.*
- *Do not attempt to straighten or weld suspension struts, wheel bearing housings, control arms or any other wheel locating or load bearing components of the front suspension.*

CAUTION—
- *To avoid wheel bearing damage, do not move vehicle without drive axle(s). If vehicle needs to be moved, install an outer joint in place of the drive axle. Tighten outer joint to 115 Nm (85 ft-lb) for M14 bolt or 190 Nm (140 ft-lb) for M16 bolt.*
- *Bonded rubber bushings may only be flexed to a limited extent. See* **Curb weight position** *in this repair group.*
- *Before performing suspension work on allroad quattro, see* **allroad quattro "jack mode"** *in this repair group.*

REAR SUSPENSION

Vehicle level sensor (front-wheel drive)

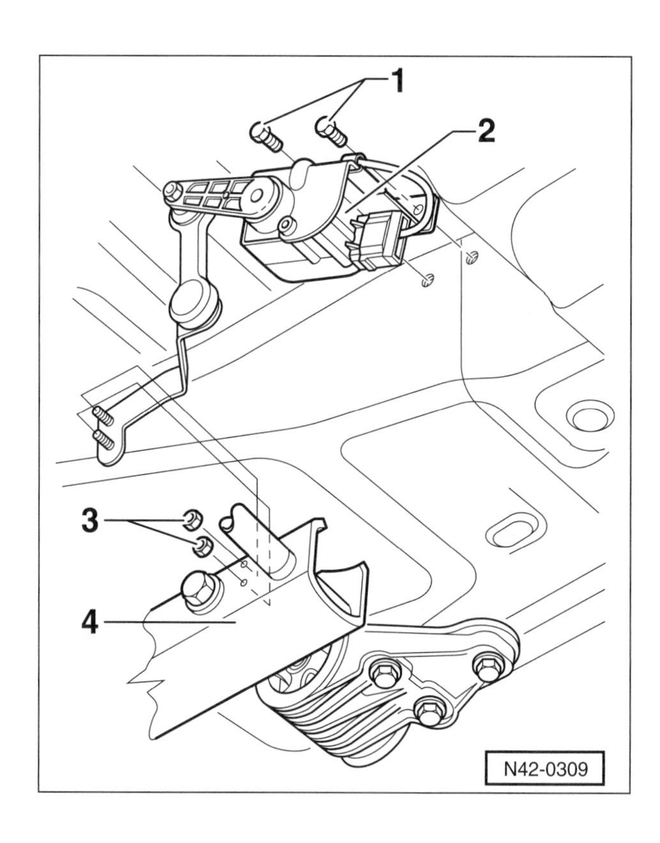

Vehicle with automatic headlight vertical aim control is equipped with vehicle level sensor.

1. Bolt
2. Left rear level control system sensor (G76)
   - Checked via on-board diagnostics (OBD).
   - Do not remove coupling link from ball head.
3. Self-locking nut, 10 Nm (7 ft-lb)
   - Replace
4. Rear axle beam

BentleyPublishers.com
BentleyPublishers.com—All Rights Reserved

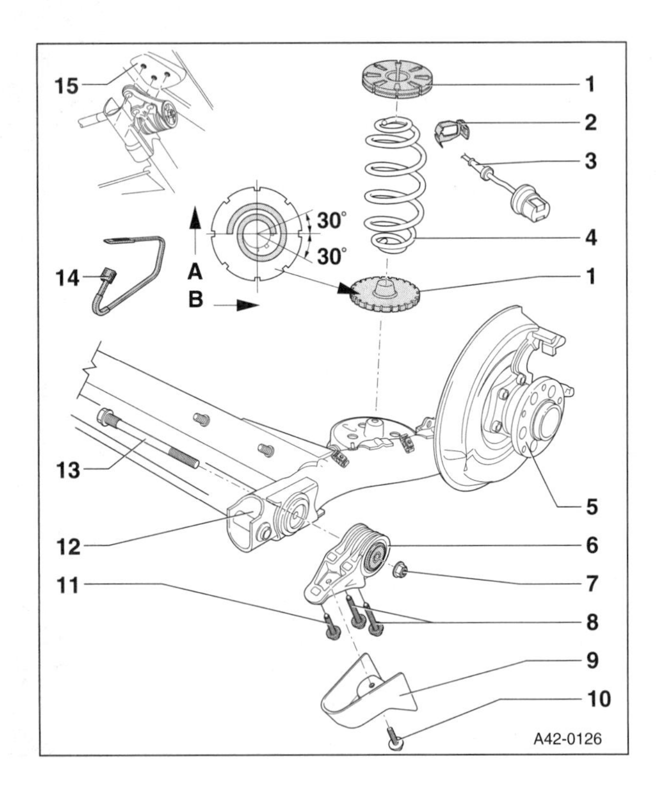

Rear axle components (front-wheel drive)

1. Upper spring seat
2. ABS wheel speed sensor bracket
3. ABS wheel speed sensor
4. Coil spring
   - See **Shock absorber (front wheel drive), removing and installing** in this repair group
   - Make sure surface of spring coil is not damaged
   - Different types of springs are used for different applications
   - Make sure bottom end of coil spring rests in spring base as shown
   - (Arrow **A**) shows direction of travel
   - (Arrow **B**) points right
   - Bottom turns of spring have markings which must point backward during installation
5. Wheel bearing
   - When replacing, also replace ABS wheel speed sensor
6. Mounting bracket
   - After installation check settings and adjust, if necessary
   - Use special care when starting bolts
   - Do not mix cast and forged mounting brackets
7. Nut
   - Always replace
8. Bolt M12 x 1.5 x 90 mm
   - Always replace after disassembly
   - 110 Nm (81 ft-lb) plus additional 90°
9. Stone impact protection plate for mounting bracket
   - Only for rough terrain suspension
10. Bolt, 10 Nm
11. Bolt M12 x 1.5 x 60 mm
   - Always replace
   - Use special care when starting bolts
   - 110 Nm (81 ft-lb) plus additional 90°
12. Axle beam
13. Bolt M14 x 1.5 x 190 mm
   - 120 Nm (120 ft-lb) plus additional 90°
   - Always replace after disassembly
14. Retaining strap
15. Thread in side member
   - Use special care when starting bolts

BentleyPublishers.com
BentleyPublishers.com—All Rights Reserved

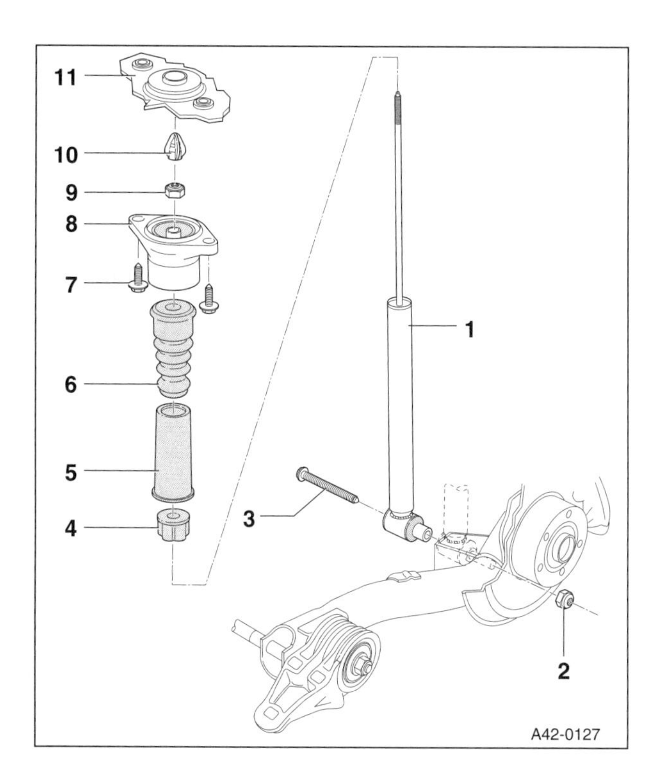

Shock absorber (front-wheel drive)

1. Gas-filled shock absorber
   - Individually replaceable
   - Shock absorbers vary with application.
   - Dispose of properly
   - Check shock absorber for leaks and noises
2. Nut
   - Always replace after disassembly
3. Bolt M10 x 90 mm
   - Always replace after disassembly
   - Only tighten when vehicle is standing on its wheels.
   - 50 Nm (37 ft-lb) plus an additional 90°
4. Protective cap
   - Press on until it stops
5. Protective tube
6. Stop buffer
7. Bolt, 45 Nm (33 ft-lb)
8. Shock absorber upper mount
9. Self-locking nut, 25 Nm (18 ft-lb)
   - Always replace after disassembly
10. Installation cap
    - Only for factory installation, not used for repairs.
11. Thread in wheel housing

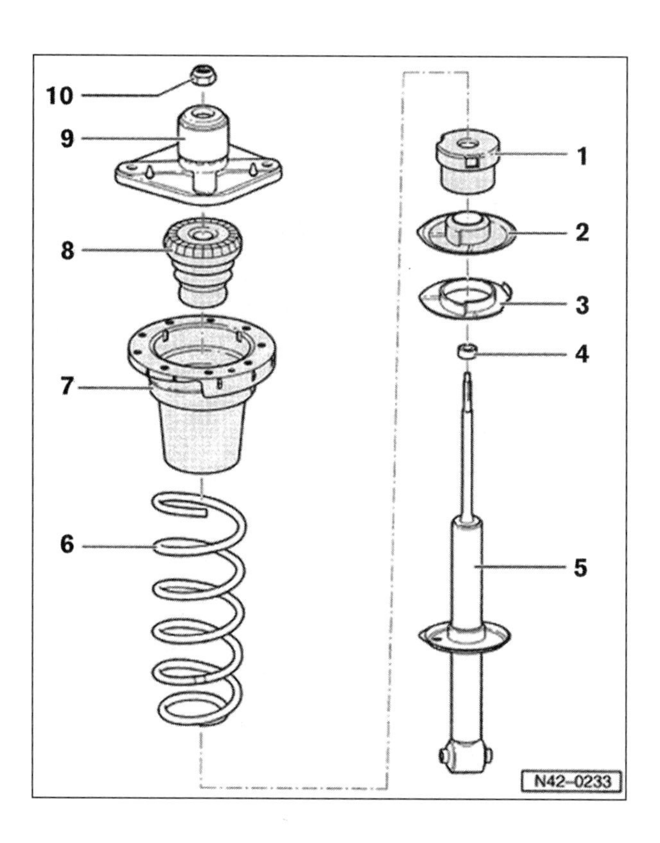

Shock absorber (quattro)

1. Protective cap
   - Preinstalled on shock absorber
2. Rubber insert
   - Preinstalled on shock absorber
3. Spring seat
4. Spacer ring
   - The spacer ring is only present on "Bilstein" shock absorbers. Do not install on shock absorbers of a different make.
5. Gas shock absorber
   - individually replaceable
   - Disposal of properly.
   - Check shock absorber for leaks and noises.
6. Coil spring
   - Check for paint damage, repair if necessary
   - After installation, ends should lie in stop the rubber inserts.
7. Spring support with protective tube
   - Consists of two parts
8. Buffer stop
9. Shock absorber mount
   - Self-locking nuts, 27 Nm
   - Always replace

BentleyPublishers.com
BentleyPublishers.com—All Rights Reserved

Shock absorber (front-wheel drive), removing and installing

Special tools and equipment

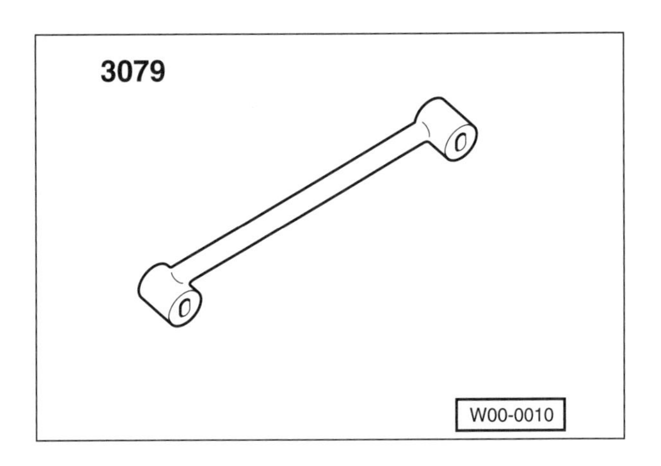

◄ 3079 counter hold tool

– Remove installation cap at top of shock absorber.

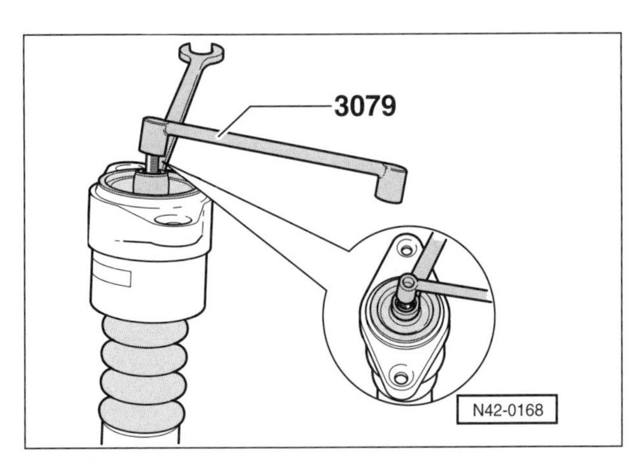

◄ If removing shock absorber from mount, use special tool 3079 to counterhold shock strut when removing upper nut.

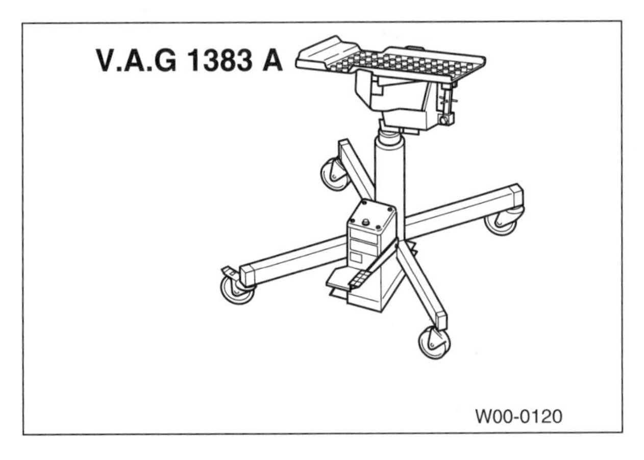

◄ Use VAG 1383 A (engine and transmission jack) and VAG 1359/2 (support bracket) to support rear suspension arm.

BentleyPublishers.com
BentleyPublishers.com—All Rights Reserved

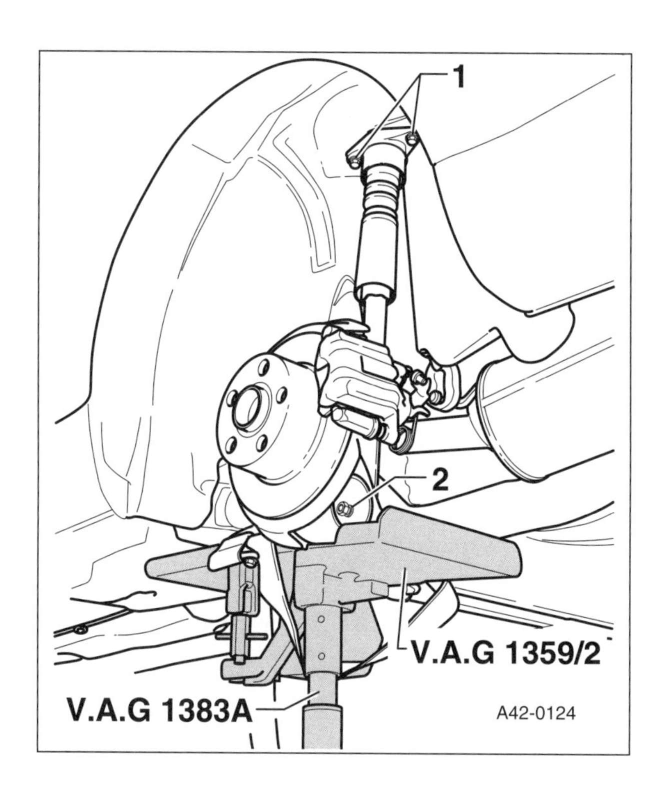

Removing

– Raise vehicle and support safely. Remove wheel.

> ***WARNING—***
> *• Make sure car is stable and well supported at all times. Use a professional automotive lift or jack stands designed for the purpose. A floor jack is not adequate support.*

– Place engine/transmission jack underneath wheel bearing hub and raise to relieve coil spring tension.

◀ Remove bolt (**2**), and lower engine / transmission jack.

– With assistant pulling down rear axle, remove coil spring.

– Unscrew bolts (**1**) at top and remove shock absorber.

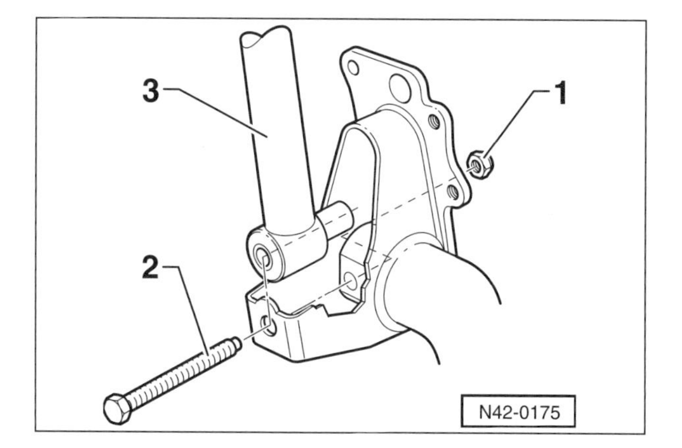

Installing

– Insert shock absorber in body and tighten to 45 Nm (33 ft-lb).

– With assistant pulling down rear axle, install coil spring.

◀ Install lower shock absorber bushing to rear axle bolt.

> ***CAUTION—***
> *• Use new fasteners.*

1. Self-locking nut
2. Bolt
3. Shock absorber

– Tighten lower shock absorber to rear axle bolt with vehicle suspension loaded.

Tightening torque	
Shock absorber to rear axle	50 Nm (37 ft-lb) + 90°

BentleyPublishers.com
BentleyPublishers.com—All Rights Reserved

Shock absorber (quattro), removing and installing

– Raise car and support safely. Remove rear wheel.

> ***WARNING —***
> • *Make sure car is stable and well supported at all times. Use a professional automotive lift or jack stands designed for the purpose. A floor jack is not adequate support.*

◄ Counterhold lower guide pin and remove self-locking bolt from brake caliper housing.

– Remove upper caliper guide bolt and hang caliper from chassis using stiff wire.

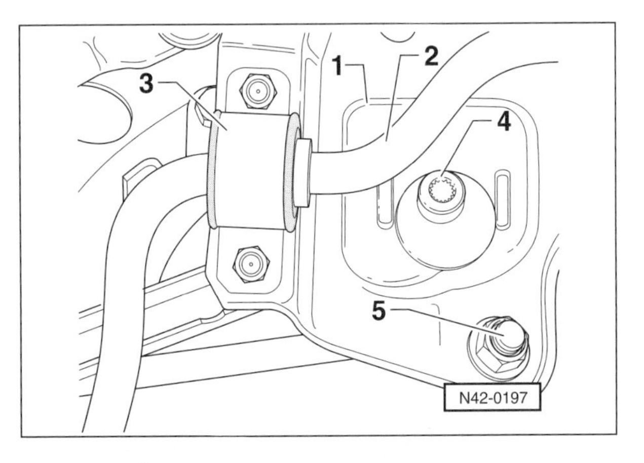

◄ Remove clamp (**3**) for stabilizer bar (**2**).

1. Axle beam
2. Eccentric bolt

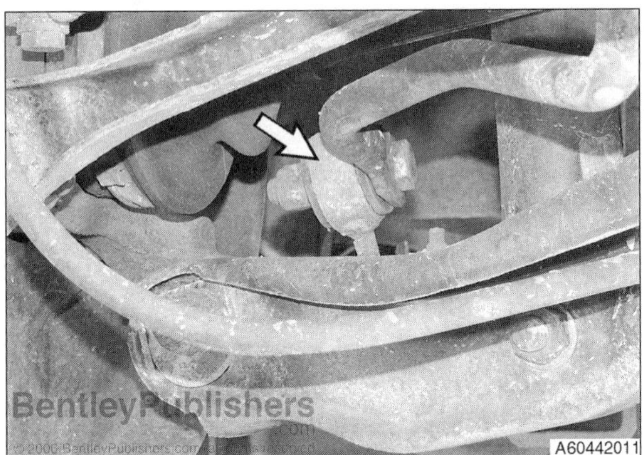

◄ Remove stabilizer bar link (**arrow**) at bar.

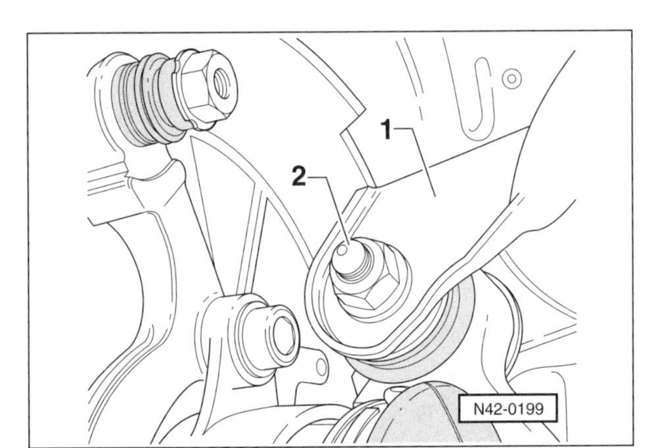

◄ Remove bolt (**2**) for link (**1**).

BentleyPublishers.com
BentleyPublishers.com—All Rights Reserved

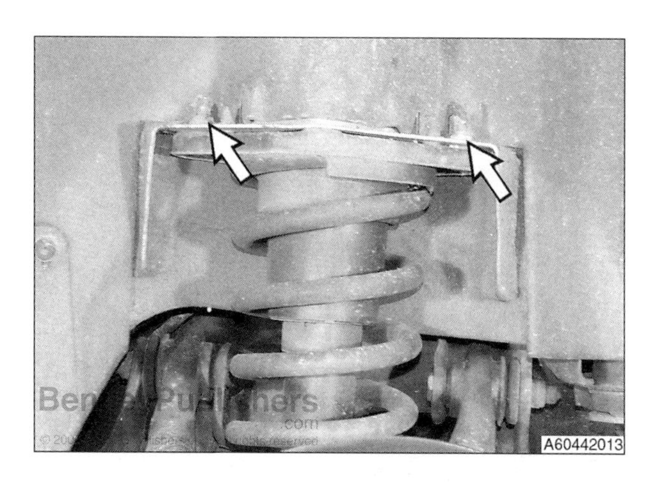

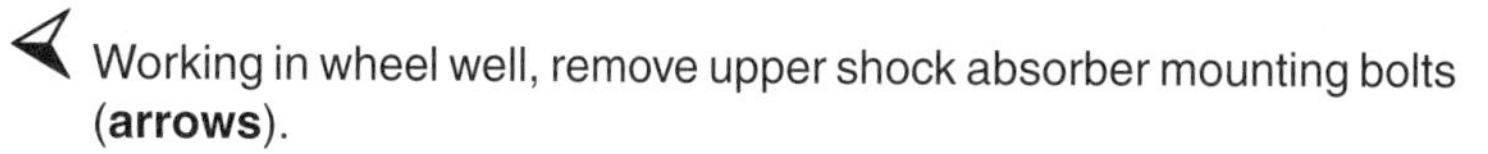

Working in wheel well, remove upper shock absorber mounting bolts (**arrows**).

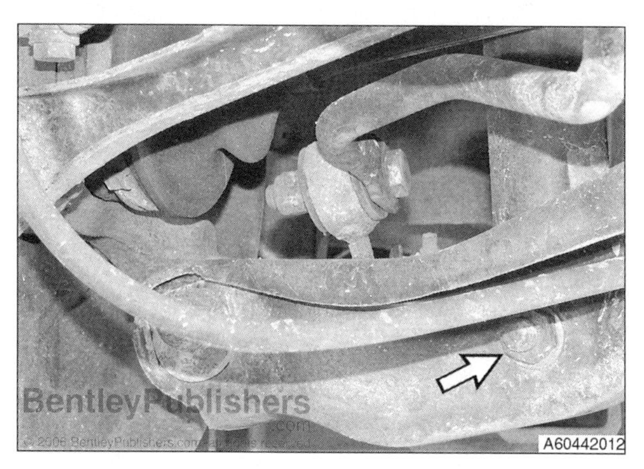

Unscrew lower mounting bolt (**arrow**) for shock absorber and remove shock absorber.

– To remove shock absorber, pull wheel bearing housing down a little.

– Installation is reverse of removal. Remember to tighten suspension bolts with suspension loaded. See **Curb weight position** in this repair group.

Tightening torques	
Stabilizer bar clamp to axle beam	25 Nm (18 ft-lb)
Coupling rod to stabilizer bar	50 Nm (37 ft-lb)
Link to wheel bearing housing	70 Nm (52 ft-lb) + 90° turn
Shock absorber to body	45 Nm (33 ft-lb)
Shock absorber to link (always replace bolts)	70 Nm (52 ft-lb) + 90° turn
Brake caliper to bracket (always replace bolts)	30 Nm (22 ft-lb)

Shock absorber (quattro), disassembling and assembling

The following tools are required for this job:

Spring compressor (Volkswagen spring compressor kit VAG 1752 or equivalent).

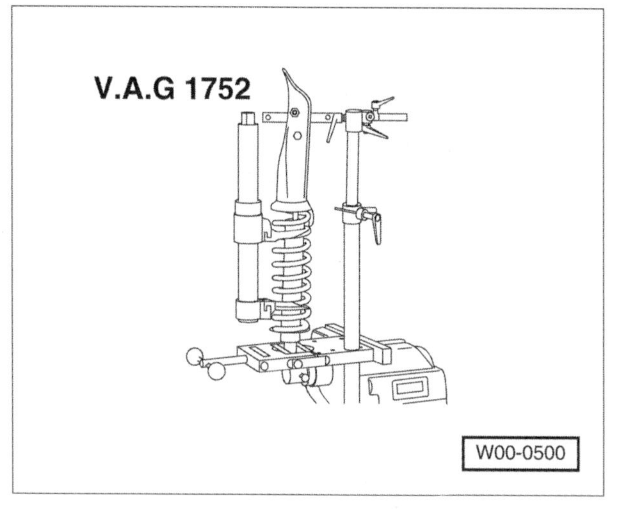

BentleyPublishers.com
BentleyPublishers.com—All Rights Reserved

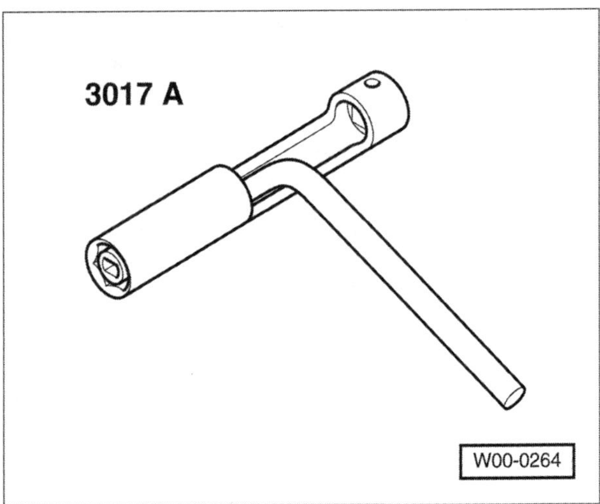

Volkswagen special wrench 3017 A or equivalent.

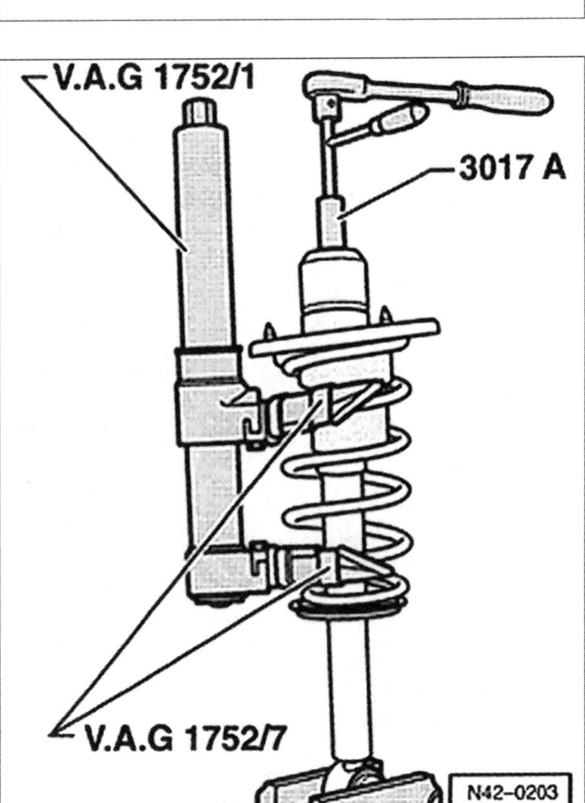

Disassembling

Install spring compressor (VAG1752/1 or equivalent) onto coil spring.

> ***WARNING —***
> *• Make sure spring compressor jaws are fully seated in spring.*

– Pre-load coil spring using Spring Compressor VAG1752/1 far enough until top shock absorber mount is free.

– Remove nut from push rod using Volkswagen special wrench 3017 A or equivalent.

– Remove coil spring and individual parts from shock absorber.

BentleyPublishers.com
BentleyPublishers.com—All Rights Reserved

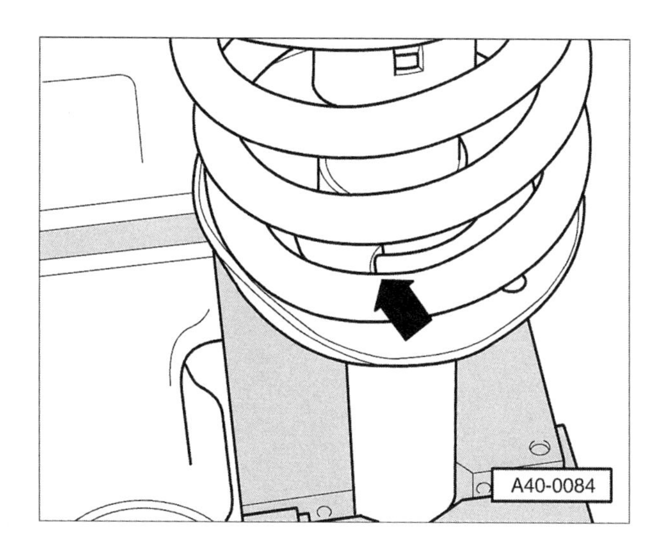

Assembling

Installation is performed in the reverse order of removal, noting the following:

◄ Top and bottom ends of spring must rest against stop of spring support (**arrow**).

- Illustration shows lower spring end. Lower spring end must point toward center of vehicle.

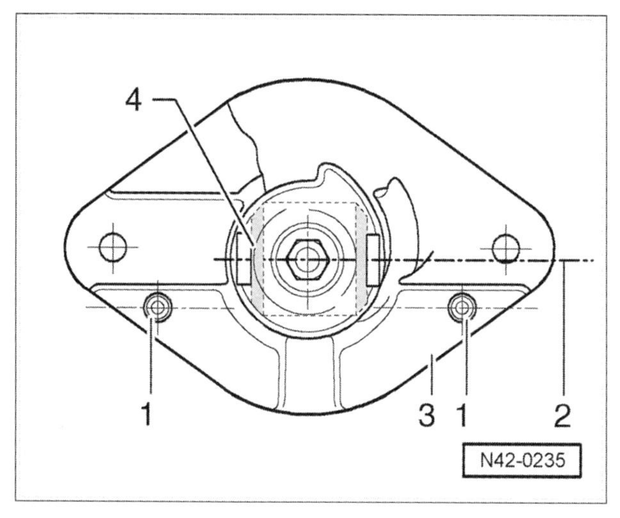

◄ Installed position of shock absorber mount (**3**):

- Mount bolts (**1**) must align with axis (**2**) of bottom shock absorber bushing (**4**).
- Mount bolts (**1**) face outside of vehicle.

Stabilizer bar, removing and installing

– Raise vehicle and support safely. Remove wheel.

> ***WARNING —***
> - *Make sure car is stable and well supported at all times. Use a professional automotive lift or jack stands designed for the purpose. A floor jack is not adequate support.*

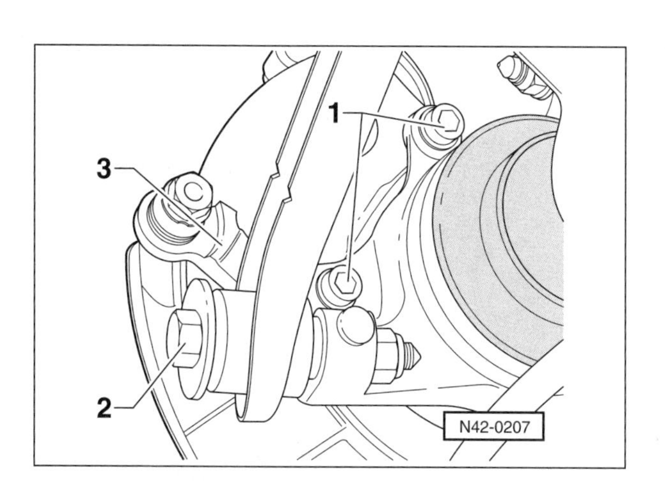

◄ Unscrew attachment screws (**1**) from brake caliper.

BentleyPublishers.com
BentleyPublishers.com—All Rights Reserved

Rear Suspension

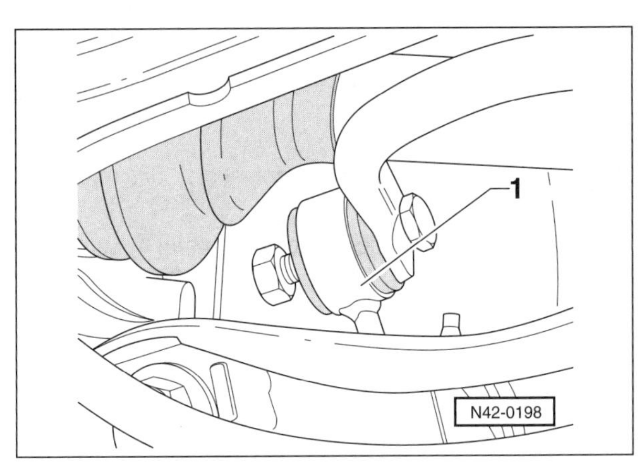

◄ Detach link (**1**) from stabilizer bar.

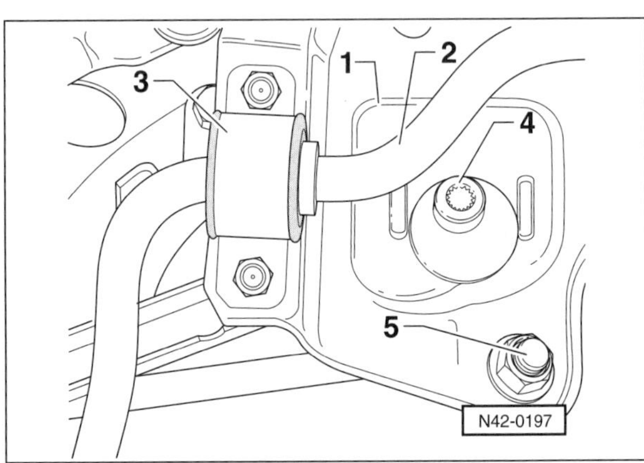

◄ Remove clamp (**3**) for stabilizer (**2**).

1. Axle beam
2. Eccentric bolt

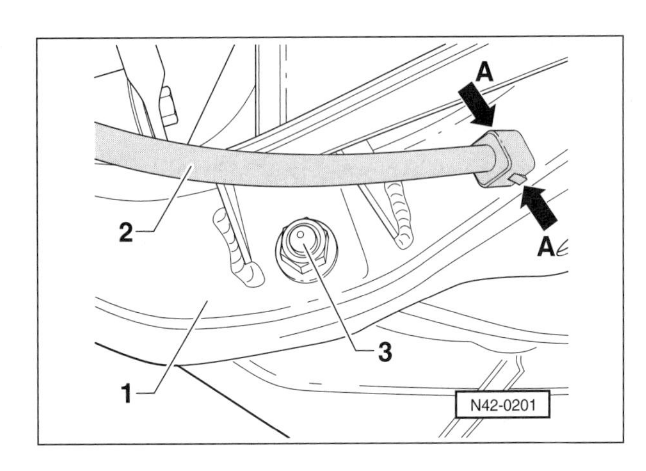

◄ Unscrew bolt (**3**) for shock absorber at lower control arm.

- **1** Control arm
- **2** Parking brake cable
- **3** Bolt for shock absorber
- **A** Holding clip for parking brake cable

– Pull wheel bearing housing downward along with lower control arm.

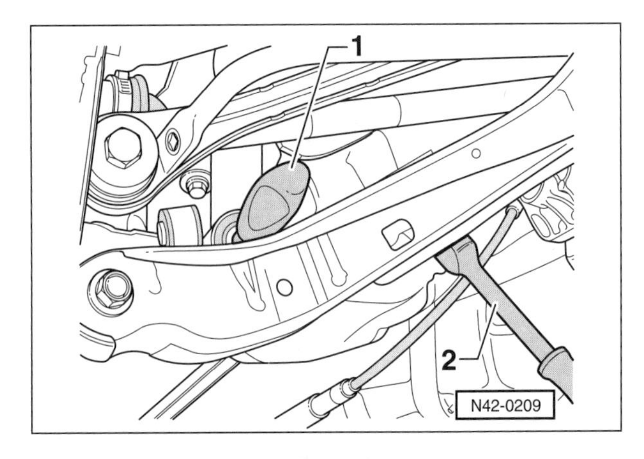

◄ Push drift (**2**) or through hole of shock absorber bushing.

– Unscrew lower hex bolt for coupling link and remove stabilizer bar.

BentleyPublishers.com
BentleyPublishers.com—All Rights Reserved

– Installation is reverse of removal.

Tightening torques	
Shock absorber to control arm (use new bolts)	70 Nm (52 ft-lb) + 90°
Coupling link to wheel bearing	55 Nm (41 ft-lb
Coupling link to stabilizer	50 Nm (37 ft-lb)
Stabilizer bar clamp to axle beam	25 Nm (18 ft-lb)
Brake caliper to wheel bearing housing (steel)	95 Nm (70 ft-lb)
Brake caliper to wheel bearing housing (aluminum)	70 Nm (52 ft-lb) + 90°

REAR DRIVE AXLES

Rear drive axle, removing and installing

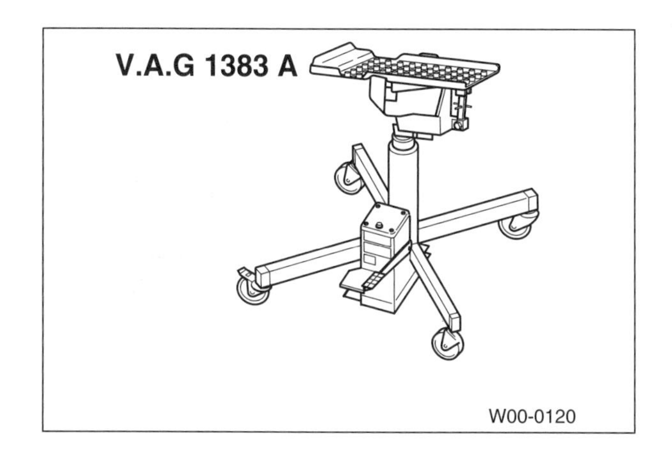

◄ Use VAG 1383 A (engine and transmission jack) to support rear suspension arm during this procedure.

– With vehicle standing on its wheels, loosen axle collar bolt 90°.

> ***WARNING —***
> - *Vehicle must be standing on its wheels when loosening and tightening collar bolt.*
> - *Only loosen collar bolt 90°; otherwise wheel bearing will be damaged*

– Raise car and support safely.

> ***WARNING —***
> - *Make sure the car is stable and well supported at all times. Use a professional automotive lift or jack stands designed for the purpose. A floor jack is not adequate support.*

– Remove wheel.

– Secure brake disc using wheel bolts.

– Remove bolt for drive axle.

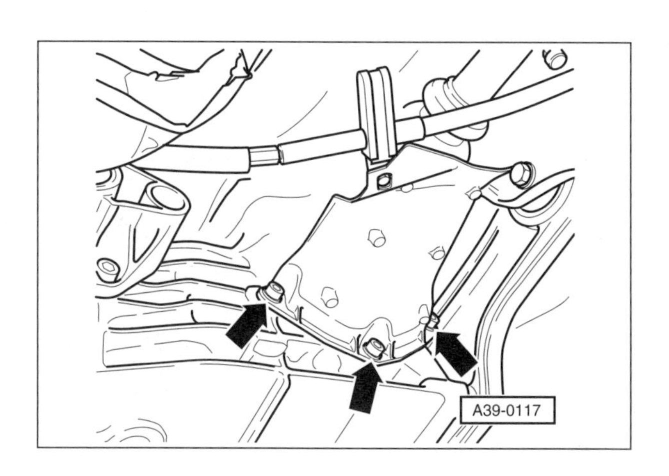

◄ Remove heat shield for drive axle (**arrows**).

– Unscrew drive axle from differential flange.

BentleyPublishers.com
BentleyPublishers.com—All Rights Reserved

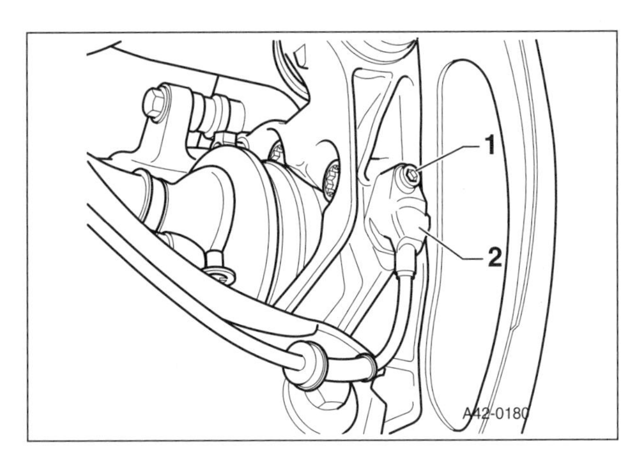

◀ Remove Allen bolt (**1**).

– Remove ABS wheel speed sensor (**2**) from wheel bearing housing.

– Loosen exhaust system in back, and secure to body. See **26 Exhaust System**.

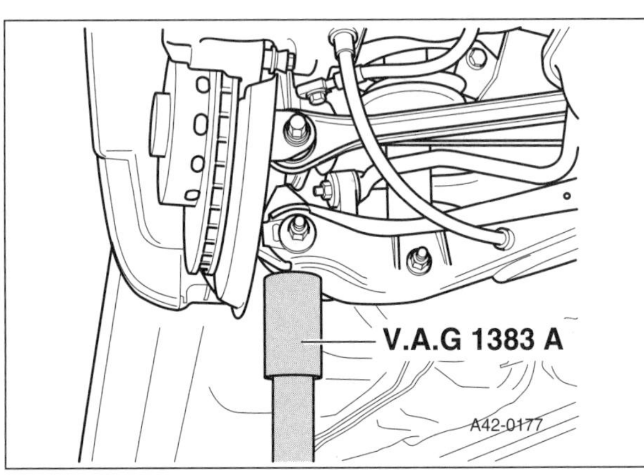

◀ Support lower control arm with VAG1383 A.

– Pound drive axle out of wheel hub with plastic hammer, if necessary.

– Remove drive axle.

– When installing, Place outer CV joint in wheel bearing hub. Install and tighten differential flange bolts.

– With an assistant operating the brake, install and tighten bolt for drive axle to initial tightening torque.

CAUTION—
- *Do not allow vehicle to contact the ground when initially tightening outer CV joint bolt.*

Tightening torques	
Drive axle to bearing hub (initial torque) • M14 (use new bolt) • M16 (use new bolt)	 115 Nm (85 ft-lb) 190 Nm (140 ft-lb)
Drive axle to differential flange: • M8 • M10	 40 Nm (30 ft-lb) 70 Nm (52 ft-lb)
Heat shield to chassis	25 Nm (18 ft-lb)

– Mount wheel and lower vehicle to ground.

Tightening torque	
Wheel lugs to wheel bearing hub	120 Nm (89 ft-lb)

– Tighten drive axle to final torque specification (initial specification plus an additional 180°).

Tightening torques	
Drive axle to bearing hub (final torque) • M14 bolt • M16 bolt	 115 Nm (85 ft-lb) + 180° 190 Nm (140 ft-lb) + 180°

44 Wheels, Tires, Alignment

TABLES

General

This section covers basic tire, wheel, and wheel alignment information. Also covered here are wheel alignment specifications to be used in conjunction with professional alignment tools and measuring equipment.

Wheels and Tires

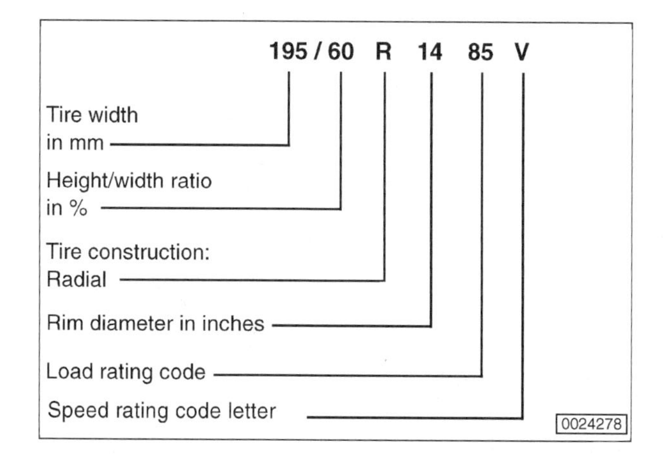

Wheels and tires approved by the manufacturer are matched to the vehicle and contribute largely to the road handling and driving characteristics. Replace tires with tires having the same specifications with regard to size, design, load carrying capacity, speed rating, tread pattern, tread depth, etc. This information can be found on the tire sidewall.

Audi recommends that tires be rotated front to rear, with the tires remaining on the same side of the vehicle. For tire rotation schedules and general tire and wheel service information, see **03 Maintenance**.

– When installing wheel bolts, they should be snugged down in a diagonal pattern and then tightened to final torque using the same pattern.

Tightening torque	
Wheel bolt to wheel hub	120 Nm (90 ft-lb)

BentleyPublishers.com
BentleyPublishers.com—All Rights Reserved

Tire mounting

Models covered by this manual are equipped with alloy wheels. Use tire mounting equipment designed for use with alloy wheels to avoid damage to the wheel and tire.

Alloy wheel valve replacement

On some models with alloy wheels, metal valve stems are installed in place of the conventional rubber valve stems. Replace metal valve stem each time the tire is changed.

Suspension codes (part numbers)

Various suspensions are offered as options. These are identified by part numbers

The suspension version installed is indicated by the factory weight code on the vehicle data sticker. The vehicle data sticker is located in the spare tire well and in the owner's manual.

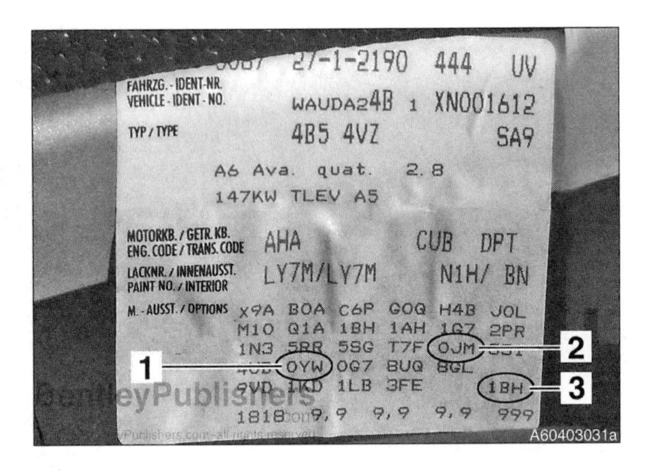

Example of a vehicle data sticker.

1. Rear axle designation
2. Front axle designation
3. Suspension code

The following table shows the meaning of the part numbers which are critical for wheel alignment. (In this example vehicle has USA suspension 1BH.)

- 1BA = standard suspension
- 1BE = sport suspension
- 1BP = rough terrain suspension 1 (curb weight position like 1BA, but with spring limitation)
- 1BT = rough terrain suspension 1= and slight reinforcement (about 7 mm raised position)
- 1BB = rough terrain suspension 2 (about 20 mm raised position)
- 1BG = suspension with level control system
- 1BC = special-purpose vehicles (fire department, medical emergency, police, etc.)
- 1BH = suspension for USA
- 1BD = Audi S6 sport suspension
- 1BJ = suspension for light armor
- 1BY = suspension for allroad (4-wheel air suspension)
- 1BV = sports suspension Audi RS6 quattro
- 2MC = sports suspension Audi RS6 quattro (sport suspension with variable shock absorber)
- 2ME = sports suspension Audi RS6 quattro (super sport suspension with shock absorber)

BentleyPublishers.com
BentleyPublishers.com—All Rights Reserved

Tire Pressure Monitoring System

Some models covered by this manual are equipped with a sophisticated tire pressure monitoring system. Transmitters in each wheel send tire pressure information to a control module located in the left rear of the trunk. If a loss of pressure is detected, a low pressure warning light is displayed on the instrument cluster.

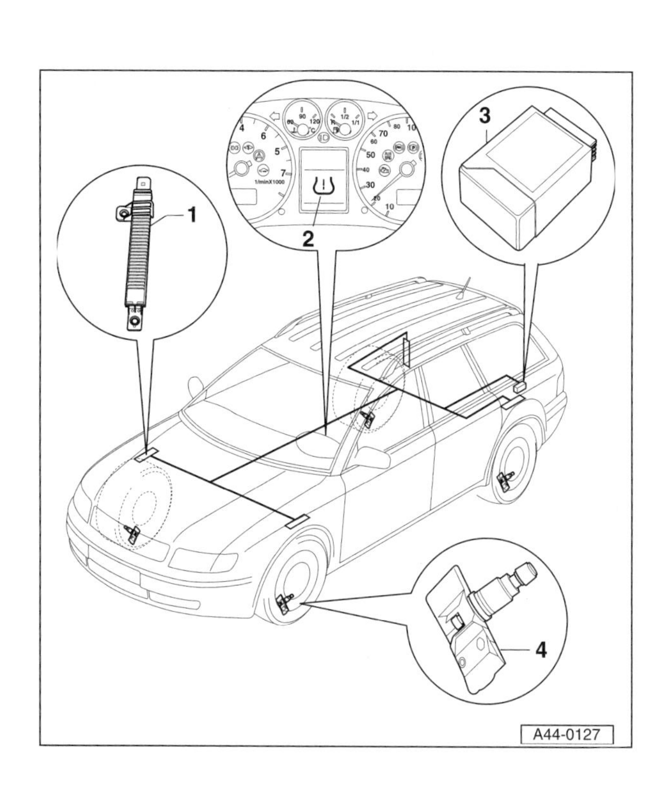

Tire pressure monitoring components

1. Antenna (4x)
2. Display on instrument panel
3. Tire pressure monitoring control module (J502)
   - Left rear of trunk
4. Wheel electronics (4x)

NOTE—

- *If self-locking bolt of the wheel electronics is loosened, replace it.*
- *For safety reasons, damaged wheel electronics or valves must be replaced.*
- *Do not clean wheel electronics using a pressure washer or strong pressurized air stream.*
- *After using tire sealing liquid, the wheel electronics must be replaced, since a possibility of an incorrect measurement exists due to deposits of the liquid at pressure sensor.*

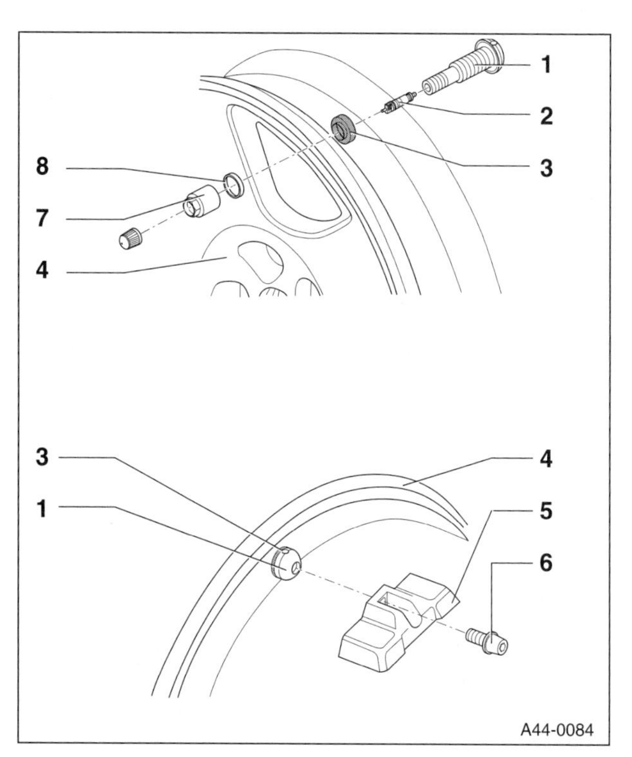

Wheels / tires, assembly overview

1. Metal valve body
   - Supplied complete
   - Replace valve insert at every tire change
2. Valve insert
3. Seal
4. Rim
5. Wheel electronics
   - Battery service life approx. 7 years
   - Remaining service life can be checked with diagnostic tool
   - Must be replaced as a complete unit
6. Micro encapsulated bolt (Torx T20), 4 Nm (3 ft-lb)
7. Union nut, 4 Nm (3 ft-lb)
8. Chamfered washer

BentleyPublishers.com
BentleyPublishers.com—All Rights Reserved

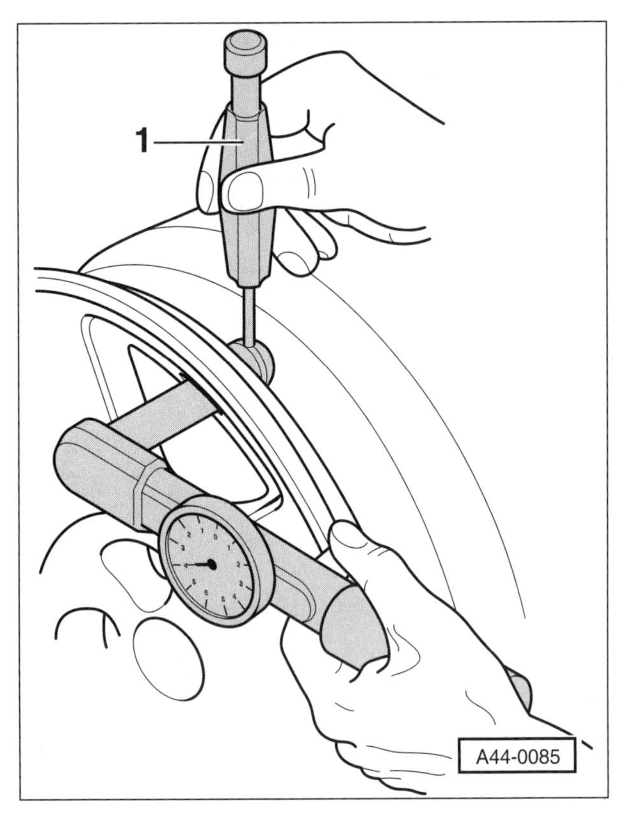

Tire pressure wheel electronics, removing and installing

◀ Install metal valve with rubber seal through rim from inside.

– Set chamfered washer and union nut on outside and tighten by hand.

– Secure against turning, using retainer (**1**) (or 2 mm drill bit).

Tightening torque	
Metal valve to rim	4 Nm (3 ft-lb)

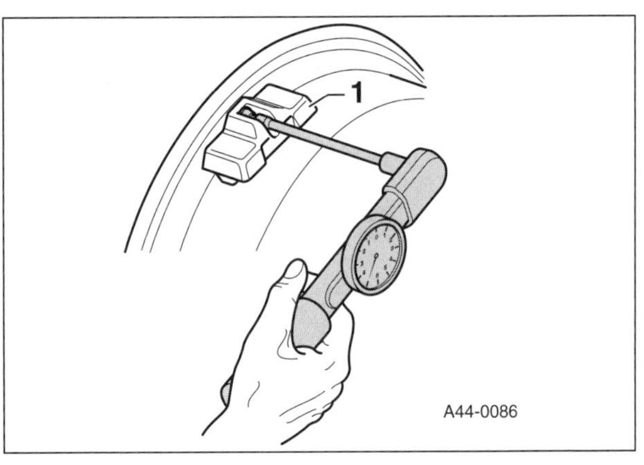

◀ Press wheel electronics (**1**) into position and tighten with self-locking bolt at rear of valve.

Tightening torque	
Wheel electronics to valve	4 Nm (3 ft-lb)

Tire pressure reception antenna, removing and installing

The reception antennas are located behind the wheel housing liners.

– Turn ignition off.

– Remove wheel.

– Remove the wheel housing liner.

– Disconnect harness connector to antenna.

– Remove threaded connection of antenna to bracket.

– Remove antenna.

BentleyPublishers.com
BentleyPublishers.com—All Rights Reserved

Tire pressure monitoring system, setting

After checking inflation pressures and properly inflating tires (including the spare), store the current tire pressures in the monitoring system.

- Switch ignition on and access the start menu from instrument cluster driver information display.
- Move selection arrow to **Set** by rotating turn / push knob.
- Press turn / push knob. Setting menu appears.
- Move selection arrow to **Tire pressure** by rotating the knob.
- Press turn / push knob. Submenu **Tire pressure** appears.
- Move selection arrow to **Store pressures!**.
- Press knob and turn. A check mark appears. System responds with **The current tire pressure will be saved**.
- Move selection arrow to **Back**.
- Press knob to leave tire pressure menu.

NOTE—

- *Reset the tire pressure monitoring system after every change in air pressure and after every wheel change.*

WHEEL ALIGNMENT

Have alignment checked if:

- Vehicle not tracking properly
- Tires wearing unevenly
- Suspension damage suspected
- A repair may have altered the alignment

Accurate wheel alignment is best performed by your Audi dealer or a certified alignment shop. Always include both front and rear axles when checking alignment.

NOTE—

- *Do not align wheels until the vehicle has been driven 1000 - 2000 km (600 - 1200 mi), to allow coil springs to settle.*
- *Vehicle instability can be caused by the wheels having a residual imbalance and/or radial runout which is too great.*
- *Ensure proper attachment and adjustment of measuring equipment; pay attention to manufacturer's operating instructions.*
- *If appropriate, have manufacturer of wheel alignment equipment provide instruction.*
- *Wheel alignment platform and wheel alignment equipment / computer may start to deviate over the course of time from their original position / setting.*
- *Make sure wheel alignment platform and wheel alignment equipment and computer is inspected at least once a year*
- *Treat sensitive alignment equipment carefully and conscientiously.*

BentleyPublishers.com
BentleyPublishers.com—All Rights Reserved

Table a. Wheel alignment data–front-wheel drive					
Front axle	**Suspension (1BA, 1BH, 1BP, 1BC)**	**Sport suspension (1BE)**	**Rough terrain suspension (1BB)**	**Rough terrain suspension (1BT)**	**Level control system (1BG)**
Camber	- 50' ± 25'	- 1° 05' ± 25'	- 35 ± 25'	- 45' ± 25'	- 50' ± 25'
Max permissible difference between sides	30'	30'	30'	30'	30'
Toe per wheel (adjustment value in start position)	+ 10' ± 2'	+ 10' ± 2'	+ 10' ± 2'	+ 10' ± 2'	+ 10' ± 2'
Toe per wheel (control value in start position)	+ 10' ± 5'	+ 10' ± 5'	+ 10' ± 5'	+ 10' ± 5'	+ 10' ± 5'
Toe constant per wheel (setting)	+ 7' ± 2'	+ 7' ± 2'	+ 7' ± 2'	+ 7' ± 2'	+ 7' ± 2'
Toe constant per wheel (control value)	+ 7' ± 7'	+ 7' ± 7'	+ 7' ± 7'	+ 7' ± 7'	+ 7' ± 7'
Toe-out angle at 20° (The angle of the outside wheel is less by this amount. This may be indicated as negative on the alignment computer)	- 1° 30' ± 30'	- 1° 30' ± 30'	- 1° 30' ± 30'	- 1° 30' ± 30'	- 1° 30' ± 30'
Rear axle					
Camber Maximum permissible difference between sides	- 1° 30' ± 20'	- 1° 30' ± 20'	- 1° 30' ± 20'	- 1° 30' ± 20'	- 1° 30' ± 20'
Maximum permissible difference between sides	30'	30'	30'	30'	30'
Overall toe	+ 20' + 15' / - 10'	+ 28 + 15' / - 10'	+ 14 + 15' / - 10'	+ 17 + 15' / - 10'	+ 26 + 15' / - 10'
Maximum permissible deviation from longitudinal axis of vehicle	± 15'	± 15'	± 15'	± 15'	± 15'

BentleyPublishers.com

BentleyPublishers.com—All Rights Reserved

Table b. Wheel alignment data–quattro (V6 engines)						
Front axle	**Suspension (1BA, 1BH, 1BP, 1BC)**	**Sport suspension (1BE)**	**Rough terrain suspension (1BB)**	**Rough terrain suspension (1BR)**	**Rough terrain suspension (1BT, 1BJ)**	**Level control system (1BG)**
Camber	- 50' ± 25'	- 1° 05' ± 25'	- 35 ± 25'	- 40' ± 25'	- 45' ± 25'	- 50' ± 25'
Max permissible difference between sides	30'	30'	30'	30'	30'	30'
Toe per wheel (adjustment value in start position)	+ 10' ± 2'	+ 10' ± 2'	+ 10' ± 2'	+ 10' ± 2'	+ 10' ± 2'	+ 10' ± 2'
Toe per wheel (control value in start position)	+ 10' ± 5'	+ 10' ± 5'	+ 10' ± 5'	+ 10' ± 5'	+ 10' ± 5'	+ 10' ± 5'
Toe constant per wheel (setting)	+ 7' ± 2'	+ 7' ± 2'	+ 7' ± 2'	+ 7' ± 2'	+ 7' ± 2'	+ 7' ± 2'
Toe constant per wheel (control value)	+ 7' ± 7'	+ 7' ± 7'	+ 7' ± 7'	+ 7' ± 7'	+ 7' ± 7'	+ 7' ± 7'
Toe-out angle at 20° (The angle of the outside wheel is less by this amount. This may be indicated as negative on the alignment computer)	- 1° 30' ± 30'	- 1° 30' ± 30'	- 1° 30' ± 30'	- 1° 30' ± 30'	- 1° 30' ± 30'	- 1° 30' ± 30'
Rear axle						
Camber	- 40' ± 30'	- 40' ± 30'	- 40' ± 30'	- 40' ± 30'	- 40' ± 30'	- 40' ± 30'
Maximum permissible difference between sides	30'	30'	30'	30'	30'	30'
Toe per wheel	+ 8' ± 5'	+ 8' ± 5'	+ 8' ± 5'	+ 8' ± 5'	+ 8' ± 5'	+ 8' ± 5'
Maximum permissible deviation from longitudinal axis of vehicle	± 10'	± 10'	± 10'	± 10'	± 10'	± 10'

BentleyPublishers.com
BentleyPublishers.com—All Rights Reserved

Table c. Wheel alignment data–quattro (V8 engines)					
Front axle	**Suspension (1BA, 1BH)**	**Sport suspension (1BE, 1BD)**	**Sport suspension (1BV, 2ME, 2MC)**	**Rough terrain suspension (1BR)**	**Level control system (1BG)**
Camber	- 1° ± 25'	- 1° 05' ± 25'	- 1° 20' ± 20'	- 50' ± 25'	- 1° ± 25'
Max permissible difference between sides	30'	30'	30'	30'	30'
Toe per wheel (adjustment value in start position)	+ 10' ± 2'	+ 10' ± 2'	+ 10' ± 2'	+ 10' ± 2'	+ 10' ± 2'
Toe per wheel (control value in start position)	+ 10' ± 5'	+ 10' ± 5'	+ 10' ± 5'	+ 10' ± 5'	+ 10' ± 5'
Toe constant per wheel (setting)	+ 7' ± 2'	+ 7' ± 2'	+ 7' ± 2'	+ 7' ± 2'	+ 7' ± 2'
Toe constant per wheel (control value)	+ 7' ± 7'	+ 7' ± 7'	+ 7' ± 7'	+ 7' ± 7'	+ 7' ± 7'
Toe-out angle at 20° (The angle of the outside wheel is less by this amount. This may be indicated as negative on the alignment computer)	- 1° 30' ± 30'	- 1° 30' ± 30'	- 1° 30' ± 30'	- 1° 30' ± 30'	- 1° 30' ± 30'
Rear axle					
Camber	- 60' ± 30'	- 60' ± 30'	- 60' ± 30'	- 60' ± 30'	- 1° 10' ± 30'
Maximum permissible difference between sides	30'	30'	30'	30'	30'
Toe per wheel	+ 8' ± 5'	+ 8' ± 5'	+ 8' ± 5'	+ 8' ± 5'	+ 8' ± 5'
Maximum permissible deviation from longitudinal axis of vehicle	± 10'	± 10'	± 10'	± 10'	± 10'

BentleyPublishers.com
BentleyPublishers.com—All Rights Reserved

Table d. Wheel alignment data–allroad quattro	
Front axle	**Suspension (1BY)**
Camber	- 1° ± 25'
Max permissible difference between sides	30'
Toe per wheel (adjustment value in start position)	+ 10' ± 2'
Toe per wheel (control value in start position)	+ 10' ± 5'
Toe constant per wheel (setting)	+ 7' ± 2'
Toe constant per wheel (control value)	+ 7' ± 7'
Toe-out angle at 20° (The angle of the outside wheel is less by this amount. This may be indicated as negative on the alignment computer)	- 1° 25' ± 30'
Rear axle	
Camber	- 60' ± 30'
Maximum permissible difference between sides	30'
Toe per wheel	+ 9' ± 4'
Permissible overall toe	+ 18' ± 8'
Maximum permissible deviation from longitudinal axis of vehicle	± 10'

Test requirements

- Check suspension, wheel bearing, steering and steering linkage for excessive play and damage.
- Tread depth difference of no more than 2 mm on tires per axle.
- Tires inflated to correct pressure.
- Vehicle in curb weight position.
- Fuel tank full.
- Spare tire and vehicle tools installed in appropriate position.
- Container for the windshield/headlight cleaning system is full.
- In vehicles with level control (1BG), switch on ignition before measurement, and wait until vehicle height control process is complete.
- For vehicles with 4-wheel air suspension (1BY), vehicle must be at normal level. Deactivate level control system. Be sure there is sufficient pressure in pressure reservoir. Press up and down buttons in operating unit for level control system simultaneously for more than 5 seconds until LED indicators light up. Indicator in instrument cluster will also light up.
- On vehicles with 4-wheel air suspension (1BY), check specified height between center of wheel and bottom edge of fender (Specification: 402 mm ± 5 mm).

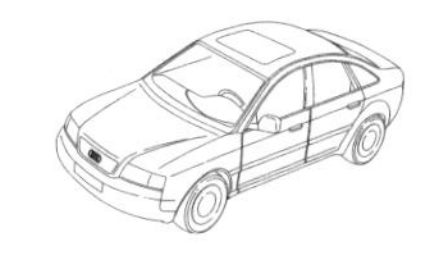

BentleyPublishers.com
BentleyPublishers.com—All Rights Reserved

45 Antilock Brakes (ABS)

GENERAL

This repair group covers antilock brakes (ABS). For related information see:

- **46 Brakes–Mechanical**
- **47 Brakes–Hydraulic**

Warnings and Cautions

WARNING—
- *ABS is a vehicle safety system; appropriate knowledge and special equipment are necessary to work on the system.*
- *If ABS and brake system warning lights illuminate, the ABS system is compromised and wheels may lock up prematurely when braking.*
- *Bleed the ABS system after opening brake hydraulic system.*
- *If an ABS fault is detected by on-board diagnostics, the function of the brake system may be limited and there is a risk of accident.*

BentleyPublishers.com
BentleyPublishers.com—All Rights Reserved

WARNING —

- *The ABS and brake system warning lights are not capable of monitoring all brake system functions; visual inspections are still required for the system.*
- *Do not use silicone-based brake fluids (DOT 5). This fluid is incompatible with the brake system and even the smallest trace can cause severe corrosion.*
- *Absolute cleanliness is required when working on the ABS system. Do not use any products that contain mineral oil such as oil, grease, etc.*
- *Carefully seal or cover opened components if repairs cannot be completed immediately. Sealing plugs are contained in factory supplied repair kit.*
- *Upon completion of repairs, road test the vehicle so that the ABS is felt to engage through the pulsing of the brake pedal.*
- *Do not reuse fasteners that are worn or deformed.*

CAUTION—

- *Before working on the ABS, switch ignition OFF and disconnect the battery ground (-) strap.*
- *For ABS / ESP system to work properly, make sure all four wheels are fitted with identical tires. Any differences in the rolling radius of the tires can cause faults.*
- *Keep brake fluid off painted surfaces; it removes paint.*
- *An illuminated ABS warning light indicates a fault in the system. Check the fault memory with an Audi scan tool or equivalent before disconnecting the battery.*
- *Brake fluid absorbs moisture from the surrounding air. Replace every two years. Use only new brake fluid that complies with Federal Motor Vehicle Safety Standards 116, DOT 4 Super or DOT 4 Plus.*
- *Before doing any electric welding to the vehicle, disconnect the ABS control module.*
- *Do not expose the ABS control module to high heat (as in a drying booth) for a prolonged period. The module may be exposed to a max. temperature of 95°C (203°F) for a short period and to a max. temperature of 85° C (185°F) for 2 hours.*
- *Do not let brake fluid enter wiring harness connectors.*
- *Obtain the anti-theft radio security code (if applicable) before interrupting power to the radio.*
- *On vehicles equipped with Audi Telematics by OnStar®, switch off the emergency (back-up) battery for the telematic/telephone control module prior to disconnecting vehicle battery.*
- *After reconnecting vehicle battery, re-code and check operation of anti-theft radio. Also check operation of clock and power windows according to Owner's manual.*
- *After reconnecting vehicle battery on vehicles equipped with Audi Telematics by OnStar®, switch on the emergency (back-up) battery for the telematic/telephone control module.*

BentleyPublishers.com
BentleyPublishers.com—All Rights Reserved

Antilock brakes (ABS)

A6 vehicles are equipped with antilock braking (ABS). ABS uses electronic control of brakes and throttle to prevent wheels from locking during hard braking, thus helping to increase vehicle direction control and decrease stopping distance.

Basic ABS is supplemented by software and hardware to achieve additional safety features. The electronic components have self diagnostic capabilities.

In this repair group ABS versions are referred to simply as ABS unless a distinction is necessary.

Bosch ABS / ASR 5.3

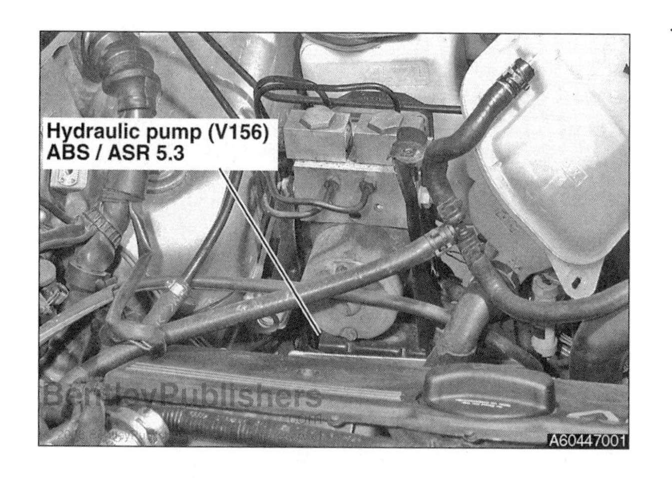

Bosch ABS 5.3 includes the following vehicle stability refinements:

- **Electronic brake pressure distribution (EBD)**. ABS control module regulates brake pressure to eliminate rear wheel lock-up. EBD control ends as soon as ABS control is applied.
- **Electronic differential lock (EDL)**. As an aid to starting on a slick surface, EDL automatically brakes the spinning wheel and shunts driving torque to the wheel with traction. EDL controls slip up to 25 mph (40 kph) in a front-wheel drive vehicle and up to 50 mph (80 kph) in a quattro (all-wheel drive) vehicle.
- **Antislip regulation (ASR)**. On front-wheel drive vehicle, if driving wheels spin during acceleration, ASR regulates wheel spin by reducing engine torque. This is done through retarding ignition timing or cyclically shutting down fuel injectors. ASR is effective across the entire range of speed.

> ***WARNING —***
> - *Bleeding procedure for models with ABS / ASR 5.3 cannot be accomplished without VAG scan tool 1551 / 1552.*

Bosch ABS / ESP 5.7

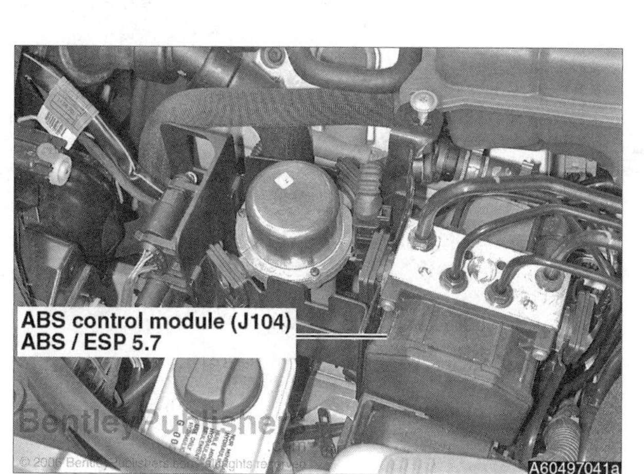

Beginning in 2001, A6 models are equipped with ABS supplemented with electronic stability program (ESP). This traction control system uses acceleration sensors (rotation rate or yaw sensor and lateral acceleration sensor), steering angle sensor and and a hydraulic pump to maintain precise control of the vehicle under difficult traction conditions. ABS / ESP incorporates ABS / ASR features (EDL, EBD) in addition to:

- **Electronic brake control (EBC)**. This feature prevents driven wheels from locking (and skidding) due to engine braking.

BentleyPublishers.com
BentleyPublishers.com—All Rights Reserved

ABS Components

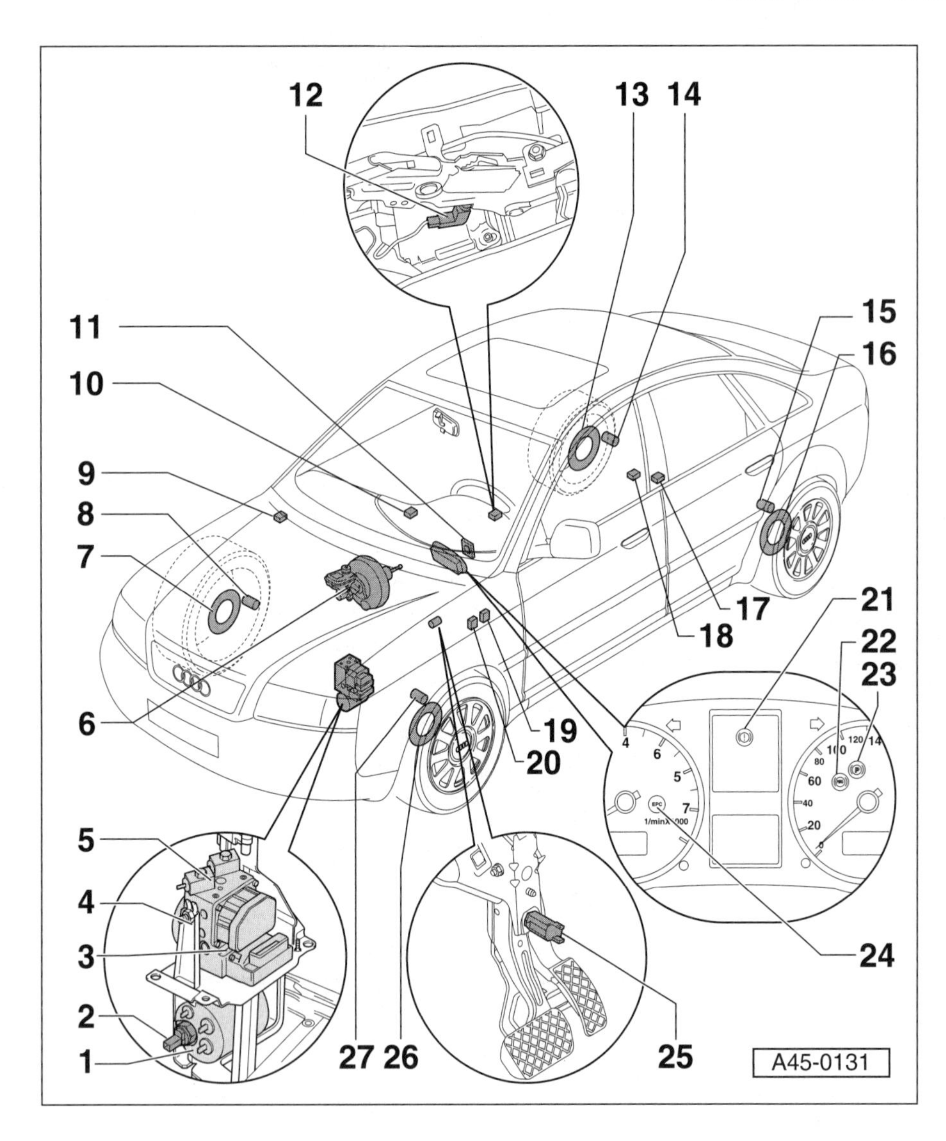

Bosch ABS / ASR 5.3

1. ASR traction control hydraulic pump (V156)
2. Brake pressure sensor 1 (G201) in ESP pump
3. ABS return flow pump relay (J105) for Bosch 5.3
4. ABS solenoid valve relay (J106)
5. ABS hydraulic unit (N55) and ABS hydraulic pump (V64)
6. Brake fluid level warning switch (F34)
7. Right front impulse wheel
8. Right front wheel speed sensor (G45)
9. ABS control module (J104) in passenger footwell
10. Anti-slip control switch (E132) on center console
11. Steering angle sensor (G85) in steering column
12. Parking brake warning light switch (F9) on parking brake lever assembly
13. Right rear impulse wheel
14. Right rear wheel speed sensor (G44)
15. Left rear wheel speed sensor (G46)
16. Left rear impulse wheel
17. Transverse acceleration sensor (G200)
18. Rotation rate sensor (G202)
19. ABS return flow pump relay (J105) on fuse box
20. ABS solenoid valve relay (J106)
21. Warning light for brake system (K118) in instrument cluster
22. ABS warning light (K47) in instrument cluster
23. Parking brake indicator lamp (K14) in instrument cluster
24. Traction control indicator light (K86) in instrument cluster
25. Brake light switch (F)
26. Left front impulse wheel
27. Left front wheel speed sensor (G47)

BentleyPublishers.com
BentleyPublishers.com—All Rights Reserved

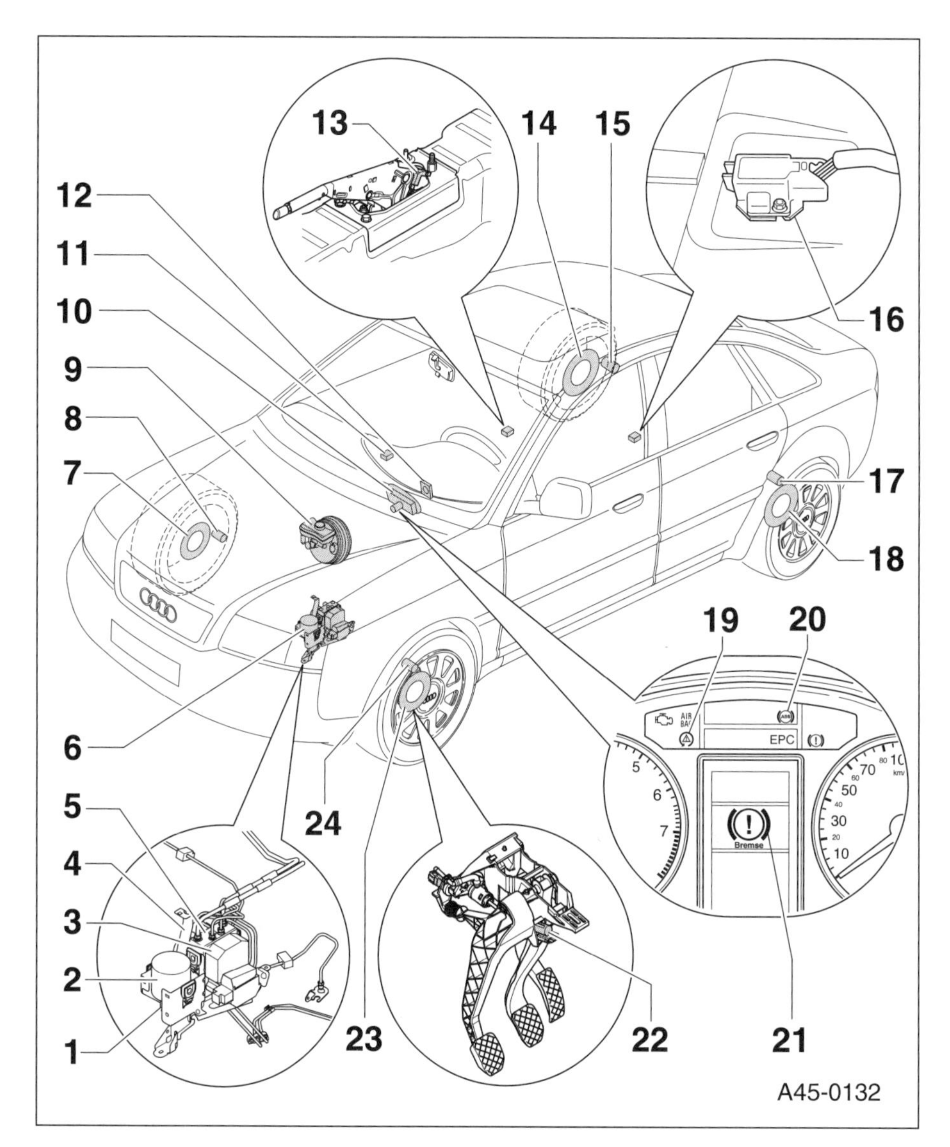

Bosch ABS / ESP 5.7

1. Bracket for hydraulic control unit
2. Vacuum pump for brake servo (engines with automatic transmission)
3. ABS control module (J104)
4. ABS hydraulic unit (N55)
5. Brake pressure sensor 1 (G201) on hydraulic control unit
6. Tightening torque 20 Nm (15 ft-lb)
7. Right front impulse wheel
8. Right front wheel speed sensor (G45)
9. Brake fluid level warning switch (F34)
10. Instrument cluster
11. Anti-slip control switch (E132) on center console
12. Steering angle sensor (G85) for ESP in coil connectorwith slip ring on steering column
    - Test with VAS 5051
13. Parking brake warning light switch (F9) on parking brake lever assembly
14. Right rear impulse wheel
15. Right rear wheel speed sensor (G44)
16. Transverse acceleration sensor (G200) and rotation rate sensor (G202) underneath radio
17. Left rear wheel speed sensor (G46)
18. Left rear impulse wheel
19. Traction control indicator lamp (K86) in instrument cluster
20. ABS warning light (K47) in instrument cluster
21. Warning light for brake system (K118) in instrument cluster
22. Brake light switch (F)
23. Left front impulse wheel
24. Left front wheel speed sensor (G47)

BentleyPublishers.com
BentleyPublishers.com—All Rights Reserved

ABS Electrical Components

ASR control module (J104) (Bosch 5.3), removing and installing

– For vehicles equipped with coded anti-theft radio, note radio code.

– Disconnect battery. See **00 Warnings and Cautions**.

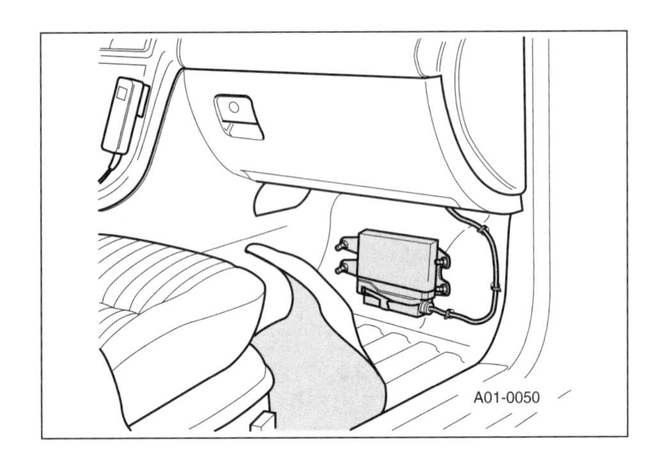

◀ ASR control module (J104) is located under floor covering in front passenger footwell

– Remove bottom right A-pillar trim. See **70 Interior Trim**.

– Loosen screws securing ASR control module (J104).

– Unplug connector from ASR control module (J104).

– Install new control module in reverse sequence.

Tightening torque	
Control module to floor	9 Nm (7 ft-lb)

– Connect battery and enter radio code (if applicable).

Steering angle sensor (G85), removing and installing

– Set wheels to straight ahead position and remove driver's airbag unit and steering wheel. See **69 Seat Belts, Airbags**.

– Extend steering column out and tilt downward as far as it will go.

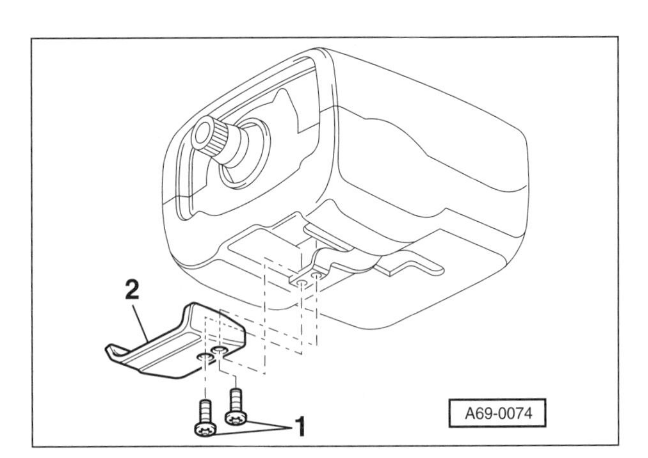

◀ Remove bolts (**1**).

– Remove grip (**2**).

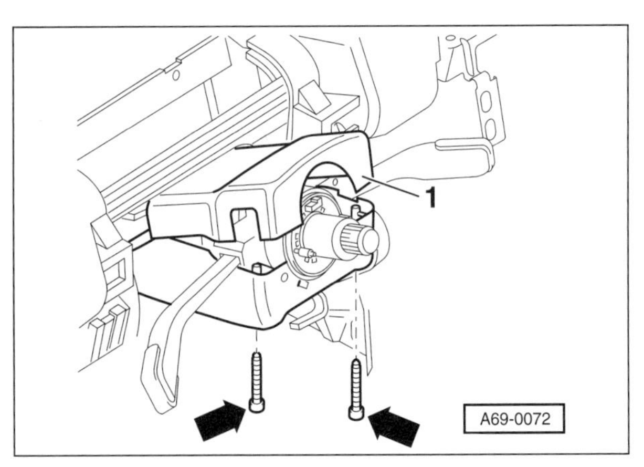

◀ Remove two Phillips-head screws (**arrows**) from column trim.

– Remove upper trim panel from steering column (**1**).

NOTE—

- *When installing, insert upper shell into hooks of lower shell, swivel downward and screw in place.*

BentleyPublishers.com
BentleyPublishers.com—All Rights Reserved

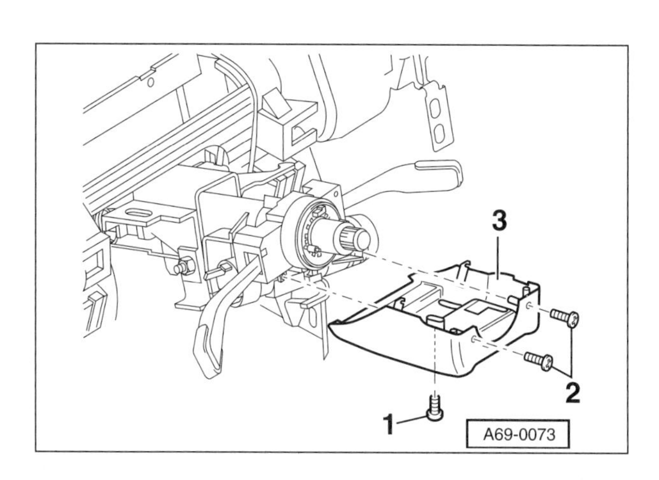

< Remove screws (**2**) from lower trim.

- Remove head bolt (**1**).
- Remove lower trim panel (**3**) from steering column.

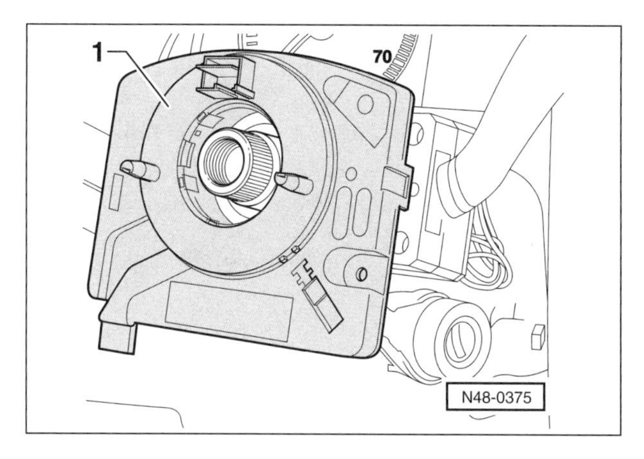

< Steering angle sensor (G85) is installed in housing (**1**) together with coil spring.

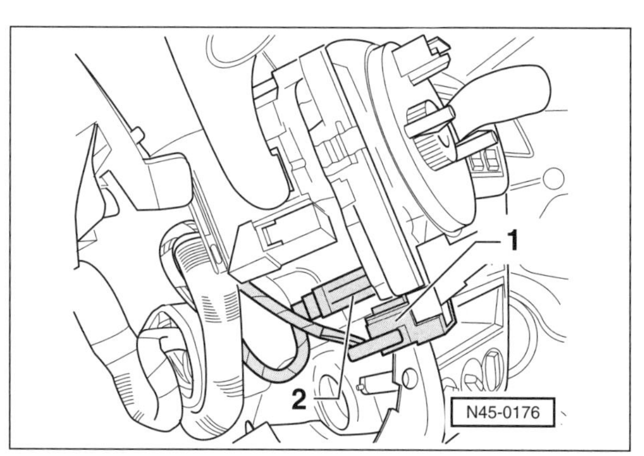

< Remove connectors (**1** and **2**).

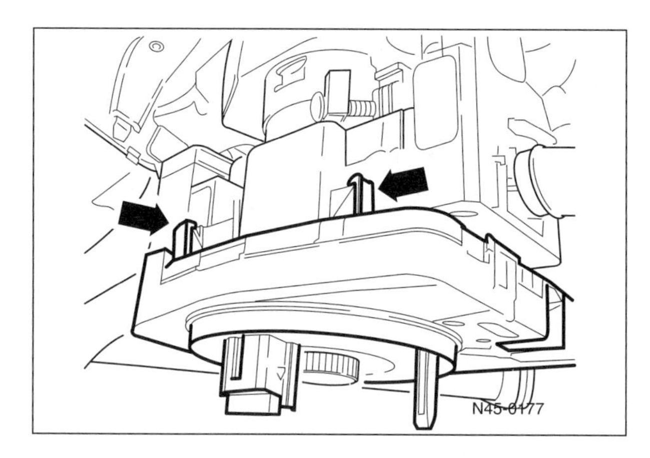

< Carefully lift hooks (**arrows**) and disconnect steering angle sensor.

- When installing the steering angle sensor (**G85**):
  - Place in center position.
  - Mount steering angle sensor until tabs engage.
  - Remove transport protection (if applicable).

BentleyPublishers.com
BentleyPublishers.com—All Rights Reserved

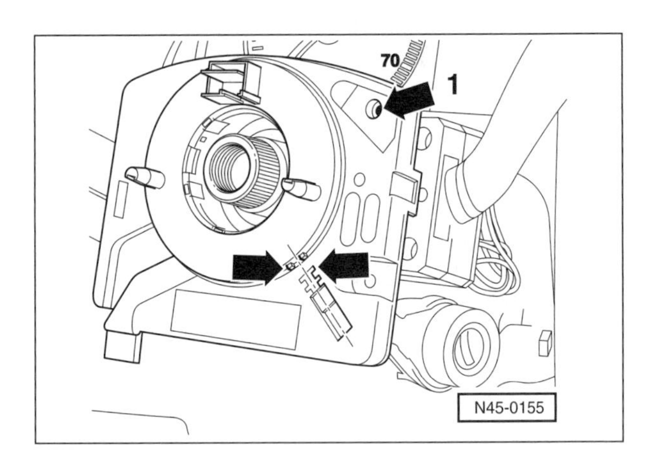

Make sure yellow dot is visible in hole (arrow **1**) and alignment marks (**arrows**) align.

– The remainder of installation is reverse of removal. Remember to:
  - Install airbag unit and steering wheel.
  - Use Audi scan tool or equivalent to carry out basic setting of steering angle sensor (G85).

Transverse acceleration sensor (G200) and Rotation rate sensor (G202) (Bosch 5.3), removing and installing

NOTE—
- *Rotation rate sensor (G202), also called yaw sensor, may be installed in housing with transverse acceleration sensor (G200).*

– For vehicles equipped with coded anti-theft radio, note radio code.

– Remove rear seat. See **72 Seats**.

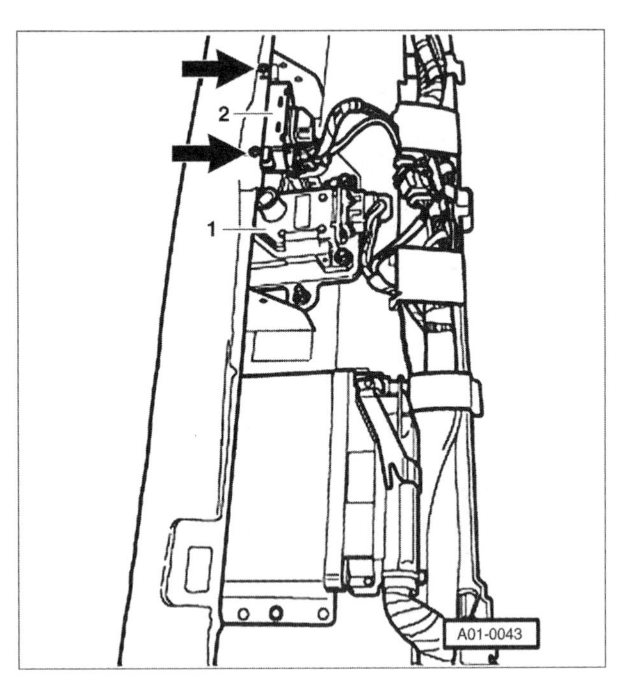

Rotation rate sensor (**1**)(G202) is located underneath the driver's seat. It measures the vehicle's rotation about the vertical axis (yaw rate).

– To remove transverse acceleration sensor (G200), first remove ultrasound sensor control module (**2**)(J347) mounting screws (**arrows**).

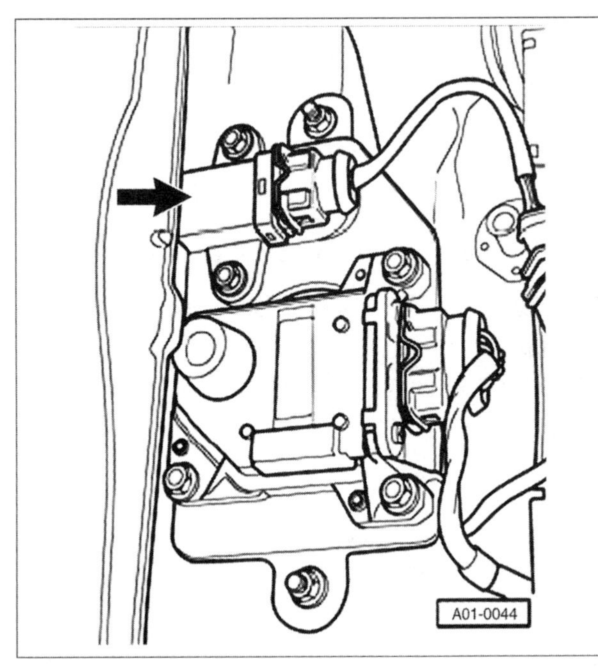

Remove transverse acceleration sensor (G200) following removal of ultrasound sensors control module (J347).

> ***CAUTION—***
> - *Sensor is sensitive to impact.*
> - *If the sensor is dropped by mistake it may not work properly and must not be installed into vehicle. Do not use the sensor again.*

– The remainder of installation is reverse of removal:
  - Install rear seat.
  - Connect battery and enter radio code (if applicable).

BentleyPublishers.com
BentleyPublishers.com—All Rights Reserved

Transverse acceleration sensor (G200) and Rotation rate sensor (G202) (Bosch 5.7), removing and installing

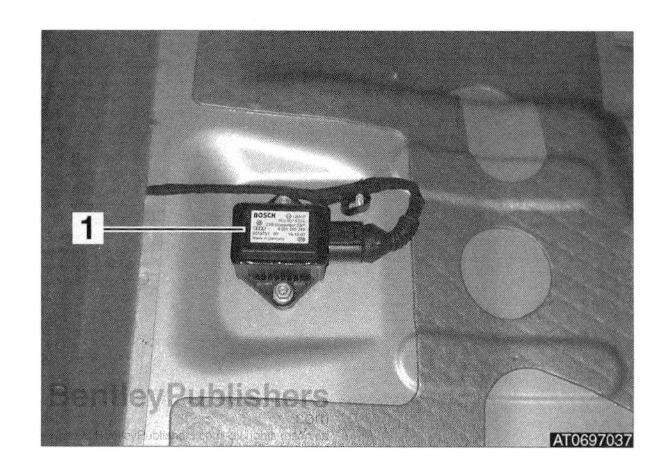

The rotation rate sensor (G202) and the sensor for transverse acceleration are located under the rear seat bench.

NOTE—
- *Rotation rate sensor (G202), also called yaw sensor, may be installed in housing with transverse acceleration sensor (G200).*

– For vehicles equipped with coded anti-theft radio, note radio code.

– Removing rear seat. See **72 Seats**.

– Disconnect connector for sensor.

> *CAUTION—*
> - *The sensor is sensitive to impact.*
> - *If the sensor is dropped by mistake it may not work properly and must not be installed into the vehicle. Do not use the sensor again.*

– The remaining installation steps are carried out in the reverse sequence:
- Install rear seat.
- Connect battery and enter radio code (if applicable).

Brake light switch, removing and installing

Brake light switch with rectangular housing

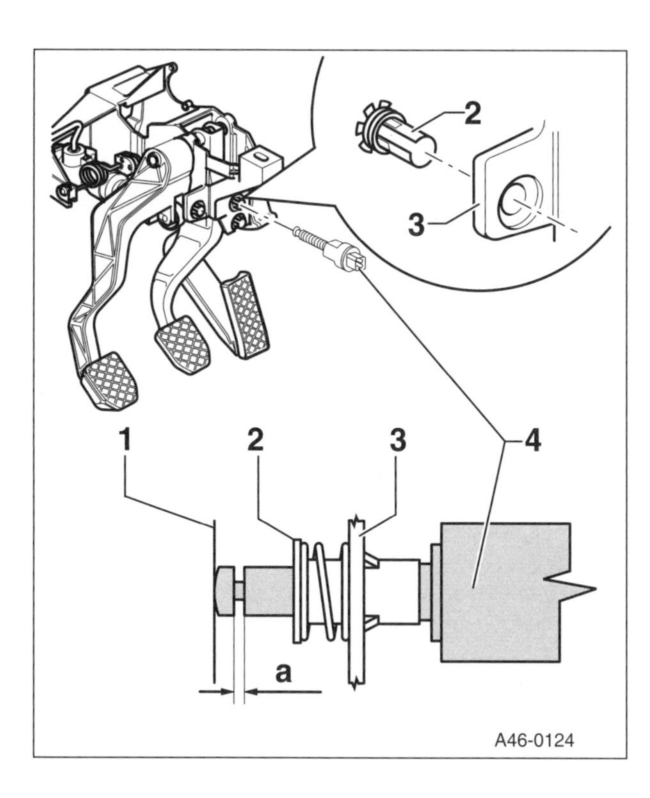

Brake light switch with spring clip mount:
1. Brake pedal impact surface
2. Spring clip
3. Pedal bracket
4. Brake light switch (F)

– Press spring clip (**2**) into pedal bracket opposite direction of travel.

– Guide switch (**4**) into installed spring clip (**2**) and press/thread it in until stop.

– Check dimension **a** using feeler gauge and twist brake light switch to dimension **a** specification if necessary.

NOTE—
- *Dimension **a**: min. 0.5 mm to max. 1.2 mm (0.02 in to 0.05 in)*
- *To ensure a secure fit, the switch may only be installed once.*

Brake light switch with round housing

– Remove driver's storage compartment.

– Unplug electrical connector on switch.

– Turn brake light switch 45° counter clockwise to remove.

BentleyPublishers.com
BentleyPublishers.com—All Rights Reserved

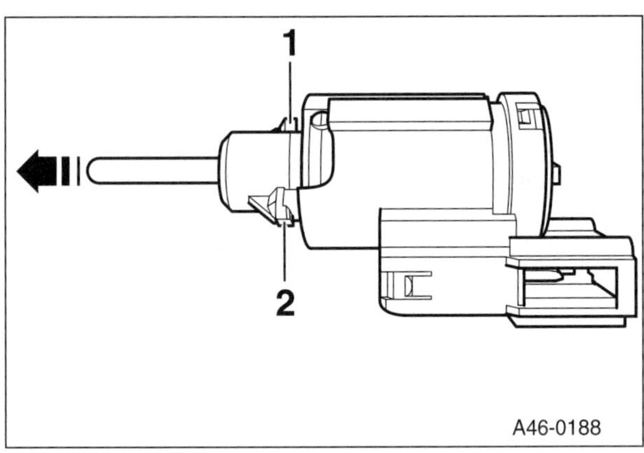

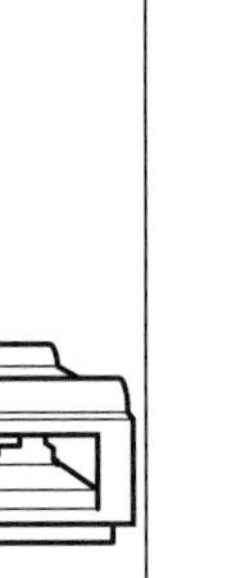

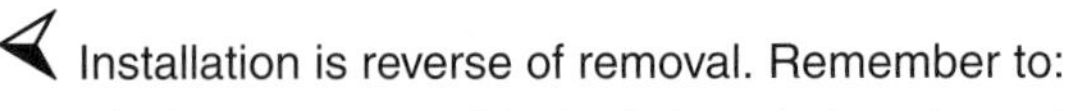

Installation is reverse of removal. Remember to:

- Pull out plunger of brake light switch as far as it will go (**arrow**).
- Insert switch into its mounting so that retaining tabs (**1** and **2**) engage in corresponding slots and secure with 45° turn.
- Do not press brake pedal while installing switch.
- Plunger in switch is adjusted automatically when switch is inserted.
- Press and release brake pedal to check brake light function.

NOTE—

- *To ensure a secure fit, the switch may only be installed once.*

Parking brake switch

The parking brake switch informs the ABS control module (J104) whether or not the parking brake is applied. This information is required because the parking brake influences ABS and traction control functions.

ABS Hydraulic Components

ASR hydraulic control unit (Bosch 5.3), removing and installing

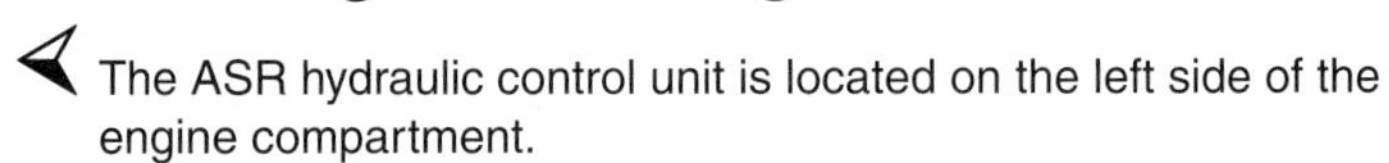

The ASR hydraulic control unit is located on the left side of the engine compartment.

– Vehicles equipped with anti-theft radio, obtain radio code.

– Switch ignition OFF and disconnect battery. See **00 Warnings and Cautions**.

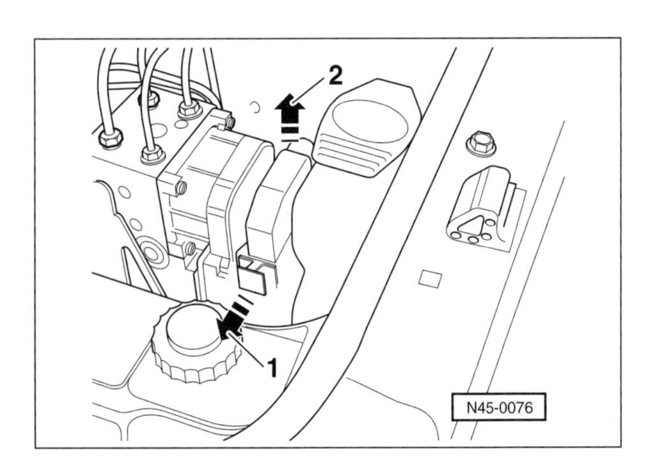

Release connector from control module (**1**) and disconnect by lifting up (**2**).

– Connect bleeder bottle to bleeder screw on left front brake caliper and open bleeder screw.

– Insert brake pedal depressor between brake pedal and driver's seat. Depress brake pedal by at least 60 mm (2.4 in).

NOTE—

- *Do not remove brake pedal depressor until all brake lines have been reconnected to the hydraulic control unit.*

– Close left front bleeder screw.

NOTE—

- *Protect engine compartment from brake fluid spills.*

– Disconnect brake line connections at hydraulic control unit.

> ***CAUTION—***
> - *Do not bend brake lines.*

BentleyPublishers.com
BentleyPublishers.com—All Rights Reserved

– Seal brake lines and threaded holes with plugs from repair kit (part no. 1H0 698 311 A).

– Remove nuts on hydraulic control unit bracket.

– Remove hydraulic control unit.

– When installing, pay particular attention to the following:
  - Only remove sealing plugs on new hydraulic control unit when corresponding brake line is about to be installed.
  - If sealing plugs are removed too early, brake fluid can escape. If this occurs, unit may not be sufficiently filled or adequately bled.
  - Bleed brake system. See **47 Brakes–Hydraulic**.
  - Reconnect battery and enter radio code.

ASR hydraulic pump (Bosch 5.3), removing and installing

◀ The ASR hydraulic pump is located at left of engine compartment below hydraulic unit.

– Vehicles equipped with anti-theft radio, obtain radio code.

– Switch ignition OFF and disconnect battery. See **00 Warnings and Cautions**.

– Raise vehicle and support safely.

> ***WARNING —***
> - *Make sure the car is stable and well supported at all times. Use a professional automotive lift or jack stands designed for the purpose. A floor jack is not adequate support.*

– Remove left front wheel.

– Remove left front wheel housing liner.

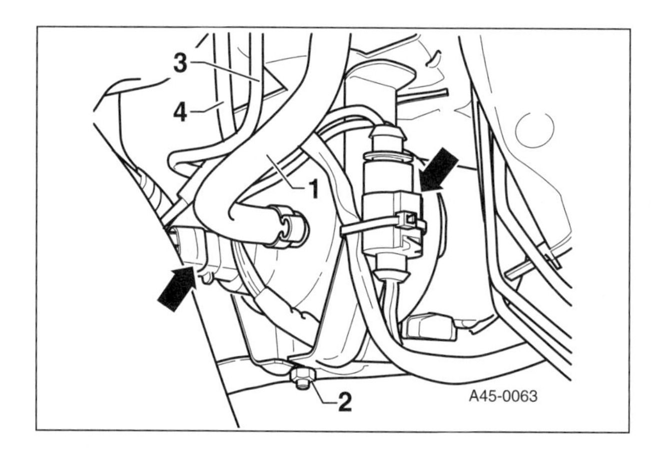

◀ Detach electrical connectors at pump (see **arrows**).

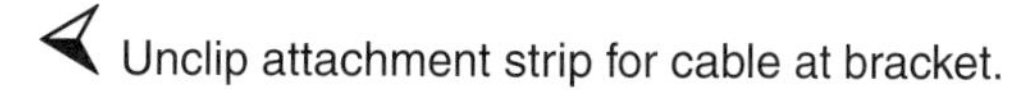

◀ Unclip attachment strip for cable at bracket.

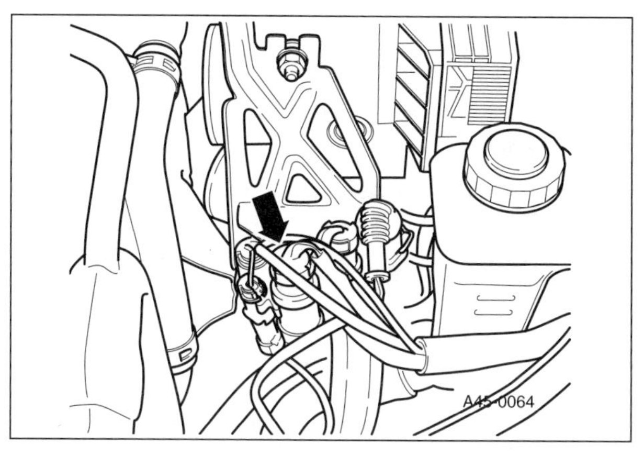

BentleyPublishers.com
BentleyPublishers.com—All Rights Reserved

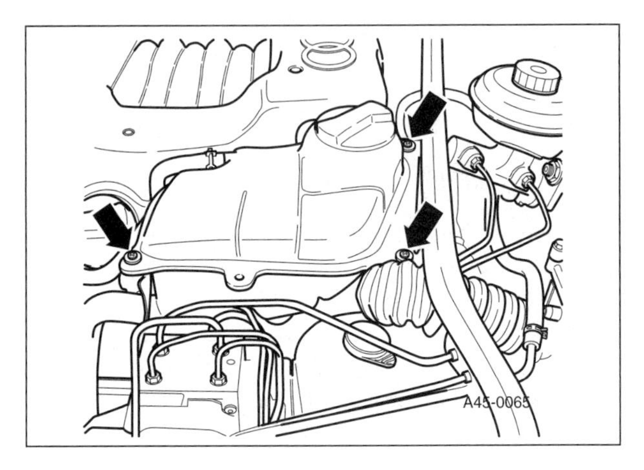

Remove coolant expansion tank mounting bolts (**arrows**) and swing reservoir to side.

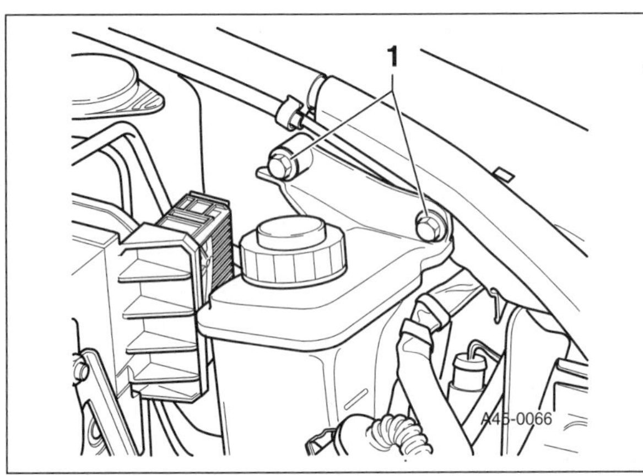

Remove bolts (**1**) for hydraulic fluid reservoir and swing to side.

NOTE—

- *Protect engine compartment from brake fluid spills.*

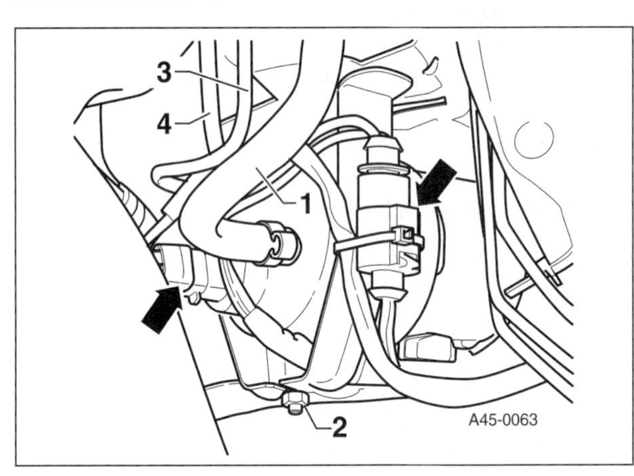

Disengage intake hose on hydraulic unit (**1**) and remove.

- Remove brake lines (**3** and **4**).
- Seal brake lines and threaded holes using plugs from repair kit (part no. 1H0 698 311 A).
- Unclip brake lines from holder.
- Remove nut (**2**).
- Remove bolt from securing bracket.
- Working in engine compartment, lift bracket and hydraulic unit out.
- Pay special attention to the following on installation:
  - Only remove sealing plugs on new hydraulic control unit when corresponding brake line is about to be installed. Otherwise, brake fluid will escape observe paint and corrosion protection measures.
  - Bleed brake system. See **47 Brakes–Hydraulic**.
  - Reconnect battery and enter radio code.

Tightening torque	
Coolant expansion tank to chassis	6 Nm (4 ft-lb)
Hydraulic fluid reservoir to chassis	10 Nm (7 ft-lb)

BentleyPublishers.com
BentleyPublishers.com—All Rights Reserved

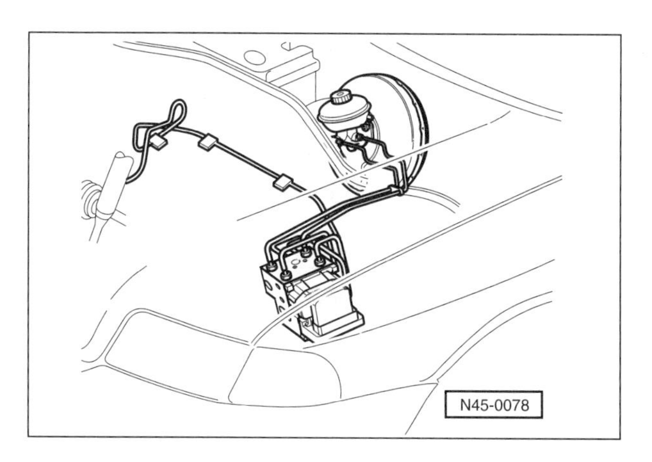

ESP hydraulic control unit (Bosch 5.7), removing and installing

◀ The hydraulic control unit is installed on left side of engine compartment.

– For vehicles equipped with anti-theft radio, note radio code.

– Switch ignition OFF and disconnect battery. See **00 Warnings and Cautions**.

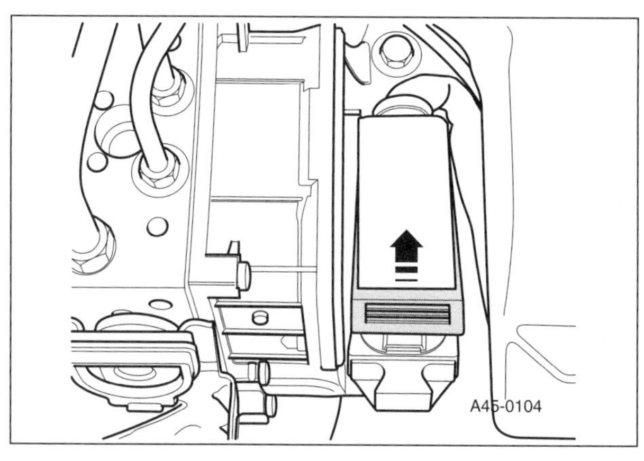

◀ Detach electrical connector from control module in direction of **arrow**.

– Insert brake pedal depressor between brake pedal and driver's seat. Depress brake pedal by at least 60 mm (2.4 in).

NOTE—

- *The brake pedal depressor must remain in position until all the brake lines have been reconnected to the hydraulic unit.*
- *Protect the engine compartment from brake fluid spills.*

– Unscrew brake line connections on hydraulic unit.

> ***CAUTION—***
> - *Do not bend brake lines.*

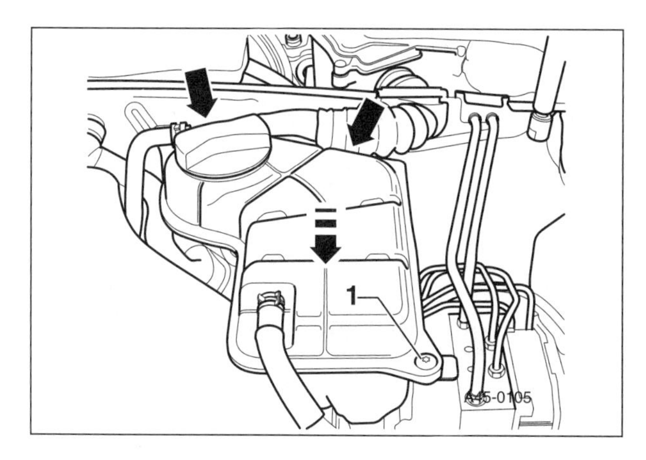

– Seal brake lines and threaded holes with plugs from repair kit (part no. 1H0 698 311 A).

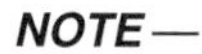

NOTE—

- *Protect the plenum chamber and engine compartment from brake fluid spills.*

◀ Remove bolt (**1**) and pull it out in the direction of **arrow**. Swing coolant expansion tank to side.

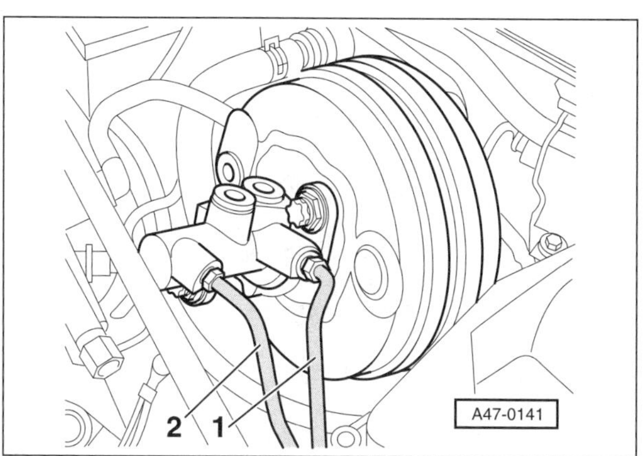

◀ Detach brake lines (**1** and **2**) from master cylinder and seal with dummy plugs from repair kit (part no. 1H0 698 311 A).

BentleyPublishers.com
BentleyPublishers.com—All Rights Reserved

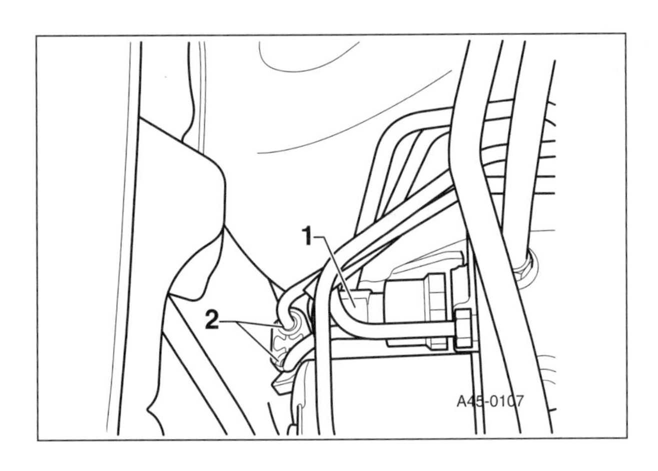

◀ Detach connectors (**1**) and unclip brake lines (**2**) from bracket.

– Remove hydraulic control unit.

– When installing, pay particular attention to the following:
  - Only remove sealing plugs on new hydraulic control unit when corresponding brake line is about to be installed.
  - Bleed brake system. See **47 Brakes–Hydraulic**.
  - Reconnect battery and enter radio code.

Tightening torque	
Coolant expansion tank to chassis	6 Nm (4 ft-lb)

ABS Wheel Speed Sensors

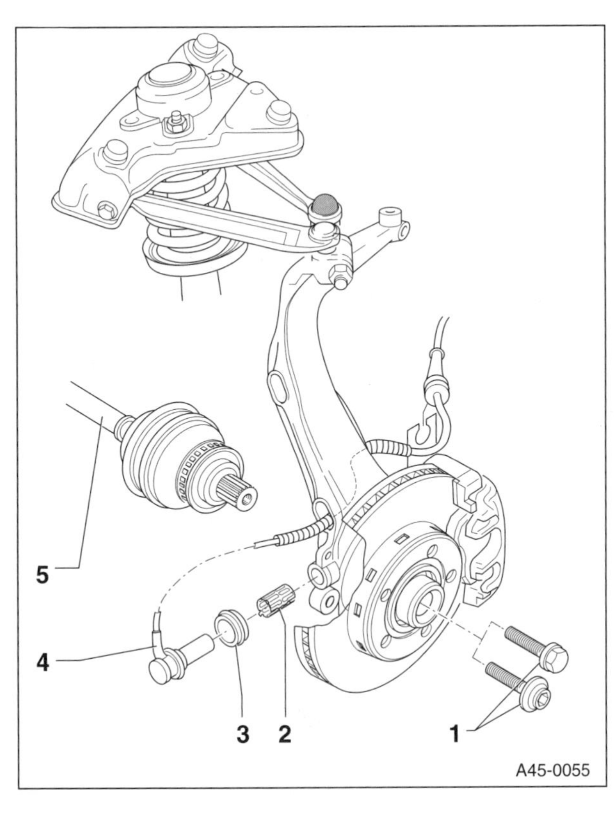

ABS front axle components

1. Drive axle bolt
   - Always replace
2. Clamping sleeve
   - Always replace
   - Before installing, grease hole in the wheel bearing housing with lubricating paste G 000 650
   - Push into wheel bearing housing up to stop
3. Sealing collar
4. ABS wheel speed sensor
   - Install drive axle before installing speed sensor
   - Push into clamping sleeve up to stop
5. Drive axle
   - Inspect impulse wheel for signs of dirt and damage
   - Replace outer CV joint if the impulse wheel is damaged

ABS front wheel speed sensor, removing and installing

– Raise car and support safely.

> ***WARNING —***
> - *Make sure car is stable and well supported at all times. Use a professional automotive lift or jack stands designed for the purpose. A floor jack is not adequate support.*

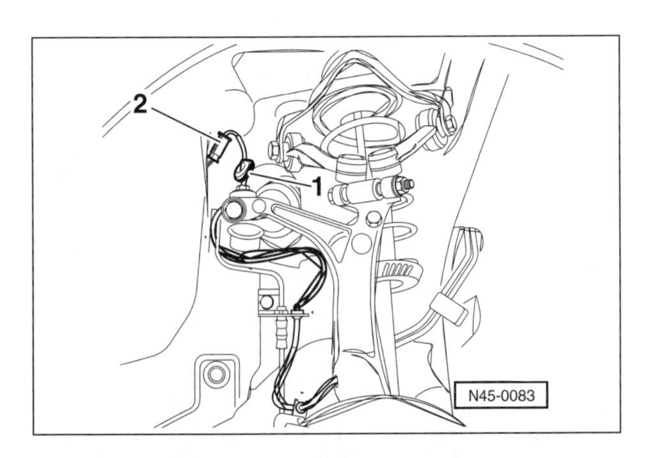

– Remove wheel.

◀ Remove grommet (**1**) from the wheel housing.

– Disconnect ABS wheel speed sensor harness connector (**2**).

– Disconnect ABS wheel speed sensor wiring from retainers (**arrows**).

BentleyPublishers.com
BentleyPublishers.com—All Rights Reserved

– Remove ABS speed sensor from wheel bearing housing.

– When installing pay particular attention to the following points:
- Replace clamping sleeve.
- Before installing speed sensor, install sealing collar.
- Connect ABS wheel speed sensor wiring and fasten grommets to retainers.
- Turn wheels completely to left and right stop positions and check ABS wheel speed sensor wiring clearance.

Front impulse wheel, checking

– Raise car and support safely.

> ***WARNING —***
> - *Make sure car is stable and well supported at all times. Use a professional automotive lift or jack stands designed for the purpose. A floor jack is not adequate support.*

– Remove wheel

– Remove ABS wheel speed sensor from wheel bearing housing.

– Inspect impulse wheel for signs of damage or dirt.

NOTE —
- *If impulse wheel is damaged or dirty remove drive axle. If damaged, replace outer CV joint together with impulse wheel.*

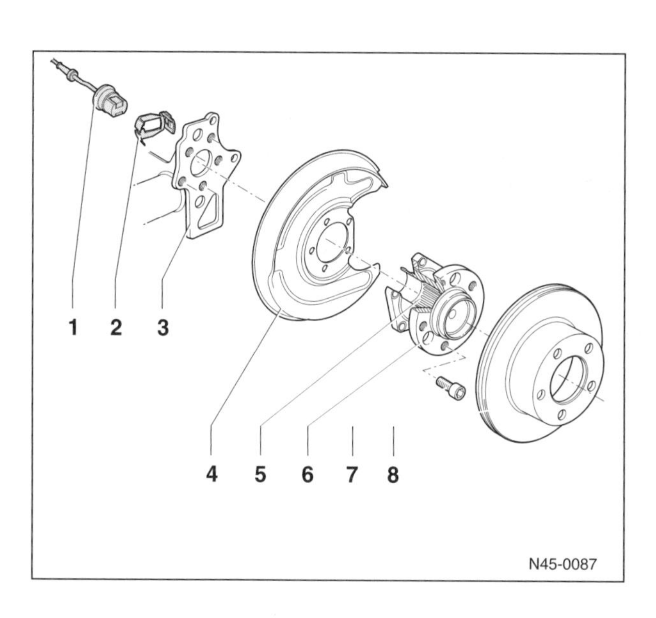

ABS rear axle components (FWD)

1. ABS wheel speed sensor
   - Must be replaced when replacing new wheel bearing/hub unit
   - Always replace O-ring and lubricate with brake cylinder paste (part no. G 000 65)
2. Clip
3. Axle beam
4. Cover plate
5. ABS wheel speed sensor impulse wheel
   - If damaged or faulty replace with wheel bearing/hub unit
6. Wheel bearing/hub unit
7. Allen head bolt, 60 Nm (44 ft-lb)
8. Brake disc

BentleyPublishers.com
BentleyPublishers.com—All Rights Reserved

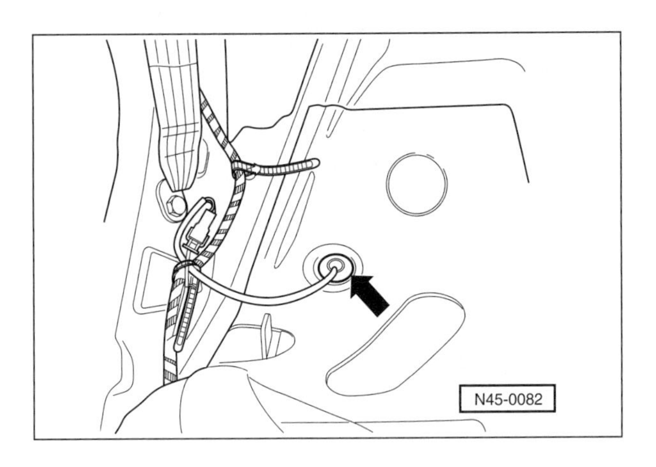

ABS rear wheel speed sensor (FWD), removing and installing

– Lift out rear seat bench.

◄ Disconnect ABS wheel speed sensor harness connector and push out grommet (**arrow**).

– Raise car and support safely.

> ***WARNING*** —
> - *Make sure car is stable and well supported at all times. Use a professional automotive lift or jack stands designed for the purpose. A floor jack is not adequate support.*

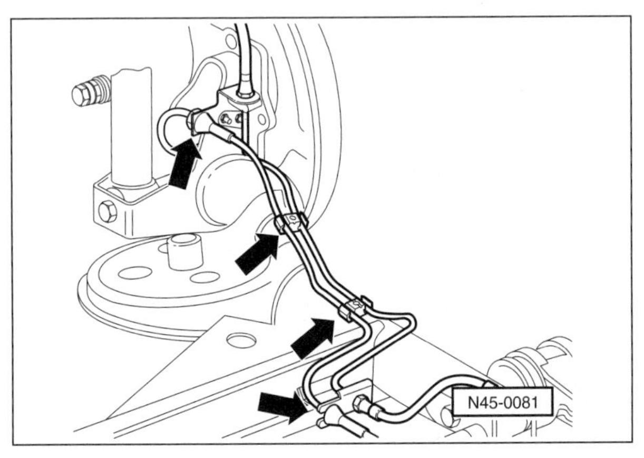

◄ Disconnect ABS wheel speed sensor wiring from retainers (**arrows**).

– Remove retainer for ABS wheel speed sensor.

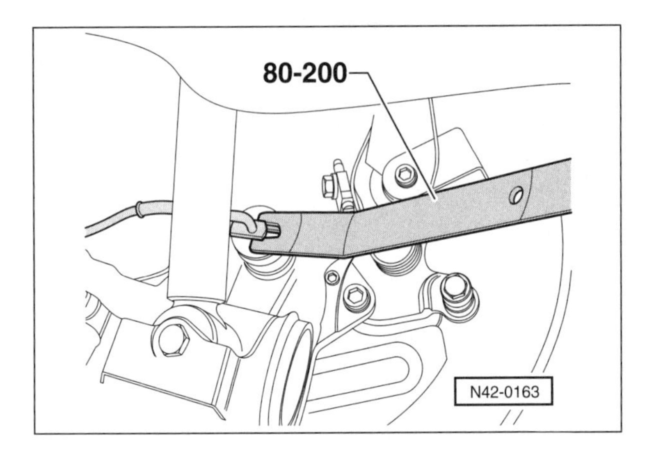

◄ Pry ABS wheel speed sensor from wheel bearing/hub unit using 80-200 pry lever.

– Before inserting ABS wheel speed sensor, lubricate circumference of seal with brake cylinder paste.

BentleyPublishers.com
BentleyPublishers.com—All Rights Reserved

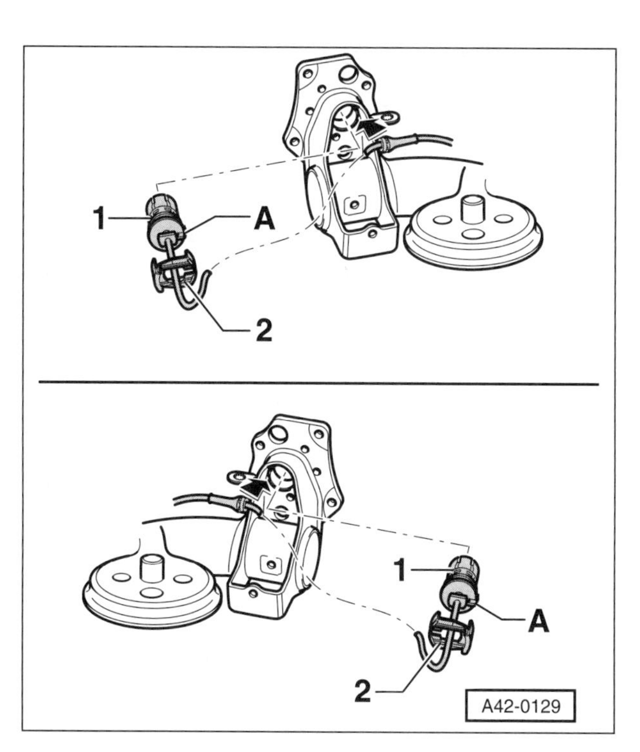

◄ Insert ABS wheel speed sensor (**1**) until stop.

– Locating tab (**A**) faces front on left and rear on right.

– Push ABS wheel speed sensor into wheel bearing housing by hand.

– Insert retainer for ABS wheel speed sensor.

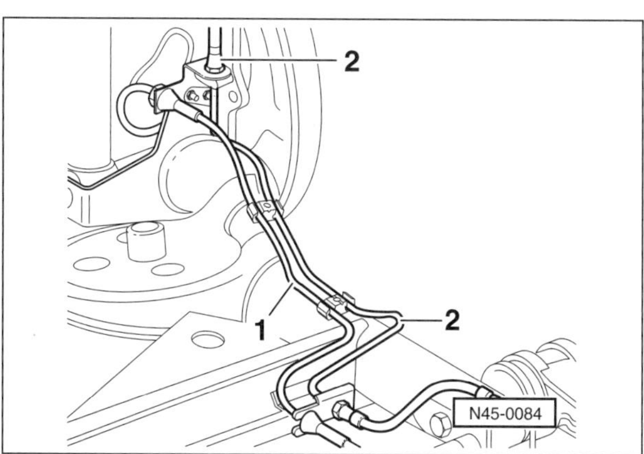

◄ Route ABS wheel speed sensor wiring (**1**) next to brake line (**2**) on both sides as shown.

– On left side, maintain clearance of about 2 cm (0.8 in) between heat shield on the exhaust system and ABS wheel speed sensor wiring.

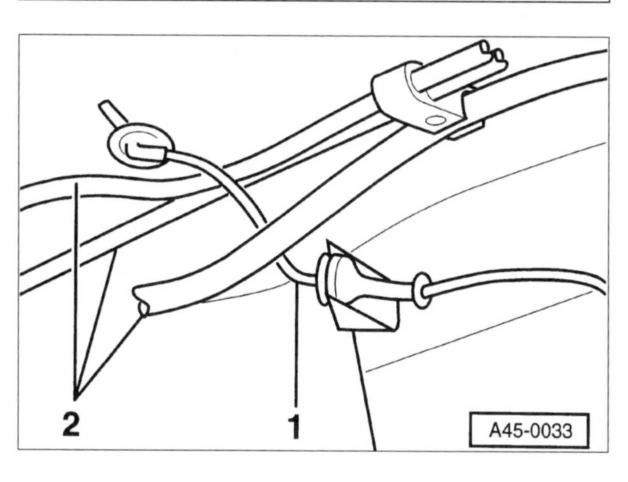

◄ On right side, ABS wheel speed sensor wire (**1**) must be routed between fuel lines (**2**).

BentleyPublishers.com
BentleyPublishers.com—All Rights Reserved

Rear impulse wheel (FWD), checking

– Raise car and support safely.

> ***WARNING —***
> - *Make sure car is stable and well supported at all times. Use a professional automotive lift or jack stands designed for the purpose. A floor jack is not adequate support.*

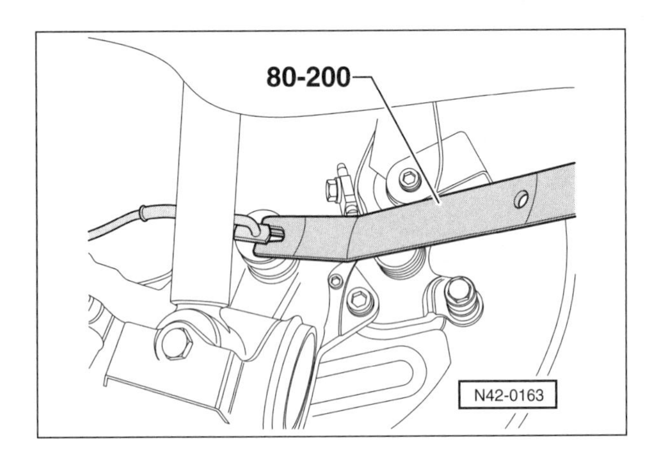

◀ Remove retainer and pry ABS wheel speed sensor from wheel bearing/hub unit using 80-200 pry lever.

– Inspect impulse wheel for dirt and damage through mounting hole of wheel speed sensor.

– If impulse wheel is damaged or dirty, replace together with wheel bearing/hub unit.

– Install ABS wheel speed sensor.

Rear impulse wheel (FWD), removing and installing

– Raise car and support safely.

> ***WARNING —***
> - *Make sure car is stable and well supported at all times. Use a professional automotive lift or jack stands designed for the purpose. A floor jack is not adequate support.*

– Remove rear wheel.

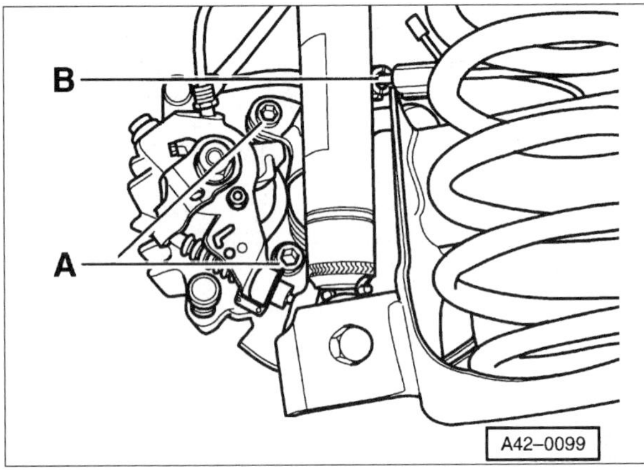

◀ Remove caliper bolts (**A**) and use stiff wire to hang brake caliper.

– Remove retainer for ABS wheel speed sensor.

– Remove brake disc.

– Pry ABS wheel speed sensor from wheel bearing/hub unit using 80-200 pry lever.

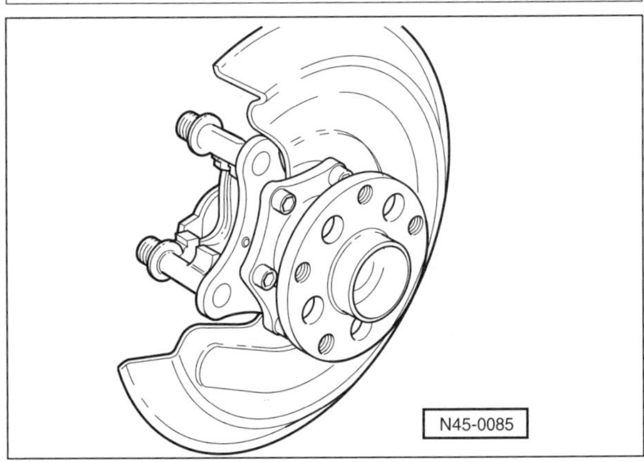

◀ Remove wheel bearing/hub unit mounting bolts and remove wheel bearing/hub unit with backing plate.

– If impulse wheel is damaged or dirty, replace together with wheel bearing/hub unit.

– Install new wheel bearing/hub unit and backing plate to axle beam and tighten.

– Reinstall brake disc.

– Mount brake caliper to axle beam and tighten.

– Install ABS wheel speed sensor.

Tightening torques	
Brake caliper to axle	95 Nm (70 ft-lb)
Wheel bearing/hub to axle beam	60 Nm (44 ft-lb)

BentleyPublishers.com
BentleyPublishers.com—All Rights Reserved

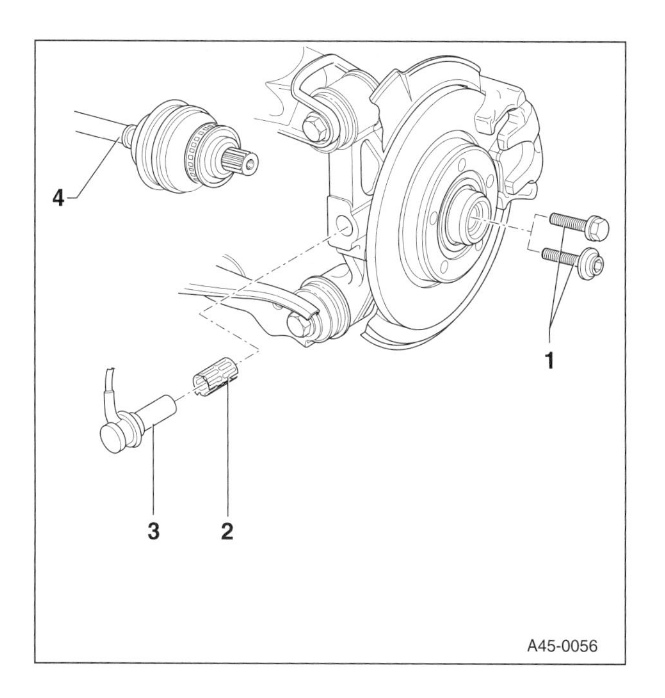

ABS rear axle components (quattro)

1. Drive axle bolt
   - Always replace
2. Clamping sleeve
   - Always replace
   - Before installing coat hole in the wheel bearing housing with brake cylinder paste G 000 650
   - Push into wheel bearing housing up to stop
3. ABS wheel speed sensor
   - Push into clamping sleeve up to stop
4. Drive axle with ABS wheel speed sensor impulse wheel (all wheel drive)
   - Check for dirt and damage.
   - Replace with outer CV joint.
   - Outer CV joint is supplied with impulse wheel.

ABS rear wheel speed sensor (quattro), removing and installing

– Switch ignition off.

– Lift out rear bench.

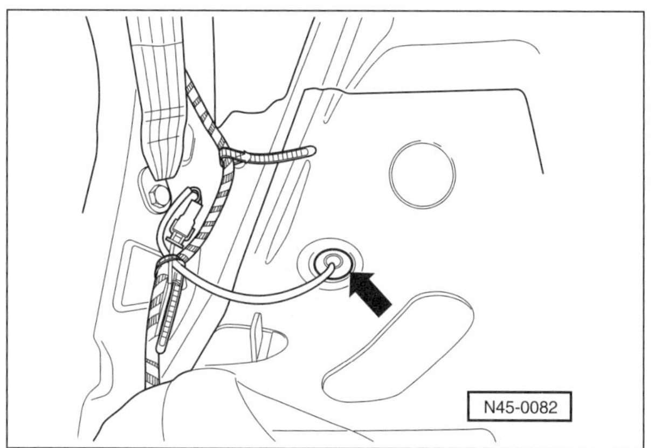

◄ Disconnect ABS wheel speed sensor harness connectors and push out grommet (**arrow**).

– Raise car and support safely.

> ***WARNING*** *—*
> - *Make sure car is stable and well supported at all times. Use a professional automotive lift or jack stands designed for the purpose. A floor jack is not adequate support.*

– Pull ABS wheel speed sensor out of wheel bearing housing.

– Disconnect sensor wiring from retainers. Pull wiring downward together with grommet.

– When installing pay particular attention to the following:
- Replace clamping sleeve.
- Connect ABS wheel speed sensor wiring and snap grommets in retainers.

BentleyPublishers.com
BentleyPublishers.com—All Rights Reserved

Rear impulse wheel (quattro), checking

– Loosen wheel bolts.

– Raise car and support safely.

> ***WARNING—***
> - *Make sure car is stable and well supported at all times. Use a professional automotive lift or jack stands designed for the purpose. A floor jack is not adequate support.*

– Remove wheel.

– Remove ABS wheel speed sensor out of wheel bearing housing.

– Check impulse wheel for damage or dirt.

NOTE—
- *If impulse wheel is damaged or dirty, remove drive axle. If damaged, replace outer CV joint together with impulse wheel.*

BentleyPublishers.com
BentleyPublishers.com—All Rights Reserved

46 Brakes–Mechanical

TABLES

GENERAL

This repair group covers mechanical brake components including front and rear brake pads, calipers, brake light switch and parking brake. For additional brake system information, see:

- **45 Antilock Brakes (ABS)** for ABS related components
- **47 Brakes–Hydraulic** for brake calipers, brake fluid bleeding

Brake discs

Inspect discs for cracks, scoring, glazing and warpage. Replace discs in pairs if either disc is worn below the minimum thickness or if any of the above listed defects are found. For brake disc specifications, see **Table a** and **Table b** in this repair group.

Table a. Front brake caliper and disc specifications

Brake calipers	FN-3 (15")	FN-3 (16")	FNR-G60	HP-2 (16")	Brembo (18")
Production number	1LB, 1LE	1LT, 1LF	1LA, 1LG	1LX, T7Z	1LJ
Brake disc, vented	288 mm	312 mm	321 mm	321 mm	365 mm
Brake disc, thickness	25 mm	25 mm	30 mm	30 mm	34 mm
Brake disc, wear limit	23 mm	23 mm	28 mm	28 mm	32 mm
Brake caliper piston	57 mm	57 mm	60 mm	42.8 mm	28.32 mm
Pad thickness (w/out backing plate)	14 mm	14 mm	14 mm	14 mm	11.4 mm

BentleyPublishers.com
BentleyPublishers.com—All Rights Reserved

Table b. Rear brake caliper and disc specifications				
Brake caliper (Lucas)	**C 38**	**C 43 (16")**	**C 43 (16")**	**C 43 (16")**
Production Number	1KD	1KE	1KE	1KZ
Brake disc, not vented	245 mm	255 mm		
Brake disc, vented			269 mm	355 mm
Brake disc, thickness	10 mm	10 mm	22 mm	22 mm
Brake disc, wear limit	8 mm	8 mm	20 mm	20 mm
Brake caliper, pistons	38.1 / 36.0 mm	42.8 / 40.8 mm	42.8 / 40.8 mm	43 / 41 mm
Pad thickness (w/out backing plate)	12 mm	12 mm	12 mm	12 mm
Wear limit (w/out backing plate)	2 mm	2 mm	2 mm	2 mm

NOTE—
- *Minimum brake disc thickness is also stamped on disc hub.*

Warnings and Cautions

WARNING—
- *If any of the brake warning lights illuminate while driving, check brake system immediately.*
- *Brake fluid is poisonous. Wear safety glasses when working with brake fluid, and wear rubber gloves to prevent brake fluid from entering the bloodstream through cuts or scratches. Do not siphon brake fluid by mouth.*
- *New brake pads require some breaking in. Allow for slightly longer stopping distances for the first 100 to 150 miles of city driving, and avoid hard stops.*
- *Perform brake work under clean conditions with careful attention to specifications and proper working procedures. If you lack the skills, the tools or a clean workplace for servicing the brake system, leave these repairs to an authorized Audi dealer or other qualified shop.*
- *After replacing brake components, depress the brake pedal firmly several times to seat the brakes in their normal operating position. Make sure the pedal is firm and at its normal height. If not, further work is required before driving vehicle.*

CAUTION—
- *Brake fluid damages paint. Immediately wipe up any brake fluid that spills on the vehicle and rinse area with water.*

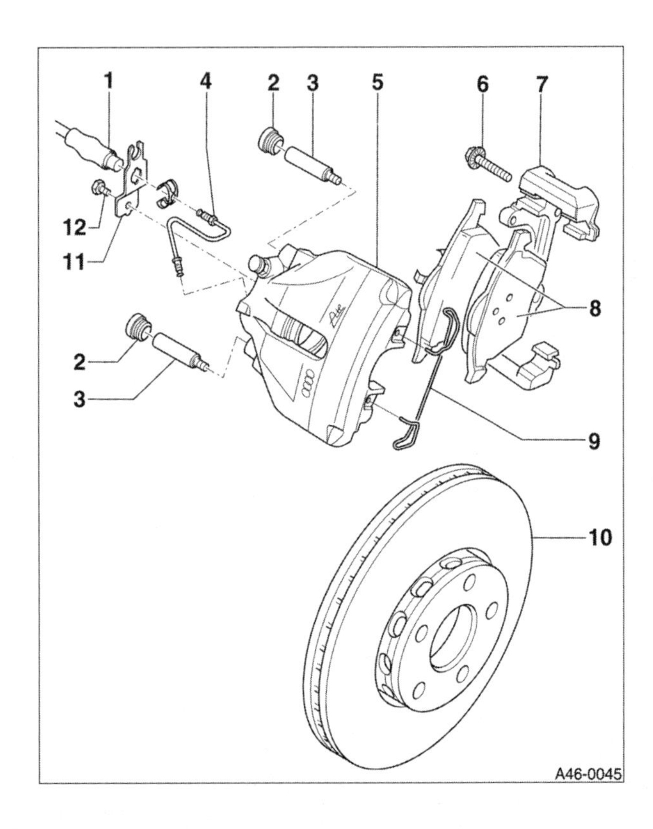

FRONT BRAKES

Front brake assembly (FN-3)

1. Brake hose
   - Do not disconnect when removing brake pads
2. Protective cap
3. Guide bolt, 25 Nm (18 ft-lb)
4. Brake line, 15 Nm (11 ft-lb)
   - Attach to brake caliper.
   - Counterhold brake hose when attaching. Make sure hose is not twisted when installed.

BentleyPublishers.com
BentleyPublishers.com—All Rights Reserved

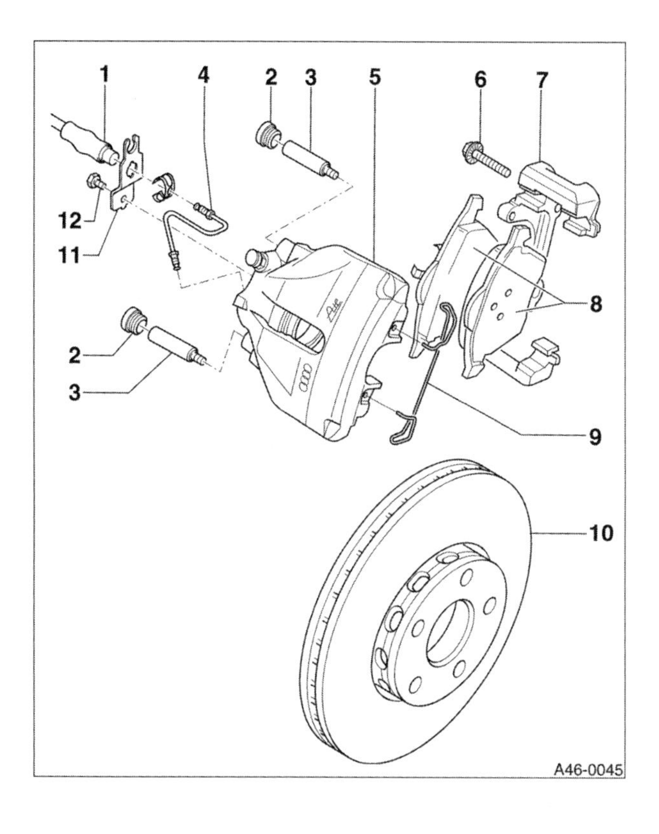

5. Brake caliper
    - Check for free movement by moving from side to side.
    - Use alcohol to clean brake caliper thoroughly of grease and other contaminants before installing new brake pads.
6. Ribbed bolt, tightening torques:
    - M12 x 1.5 x 35, 130 Nm (96 ft-lb)
    - M14 x 1.5 x 35, 200 Nm (148 ft-lb)
    - Clean ribs if reusing. Use thread locking compound.
7. Brake pad carrier
    - Bolt to wheel bearing hub.
8. Brake pads
    - Replace front brake pads as a set.
    - Inner pad has retaining spring.
    - Inner pad (with spring) is marked with arrow pointing in direction of brake disc forward rotation when installed. Noise may result from incorrect installation.
    - Remove adhesive foil on outer brake pads before installing.
9. Retaining spring: Pay attention to correct installation position.
    - Insert in caliper holes.
    - Press retaining spring under pad carrier.

> ***WARNING—***
> - *If spring is improperly installed, brake caliper cannot adjust to compensate for wear on the outer brake pad. This causes brake pedal travel to increase.*

10. Brake disc
    - To replace brake disc, remove brake caliper and pad carrier.
    - Lightly coat contact surfaces between brake disc and wheel bearing hub with polycarbamide grease (VAG part no.G 052 142 A2).
11. Bracket
    - Bolt to brake caliper.
    - Make sure locating tab is seated in caliper.
12. Bolt, 10 Nm (7 ft-lb)

Front brake pads (FN-3), removing and installing

NOTE—
- *If reusing brake pads or brake discs, mark them so they can be installed in their original positions to prevent uneven braking.*

– Raise car and support safely.

> ***WARNING—***
> - *Make sure the car is stable and well supported at all times. Use a professional automotive lift or jack stands designed for the purpose. A floor jack is not adequate support.*

– Remove front wheels.

BentleyPublishers.com
BentleyPublishers.com—All Rights Reserved

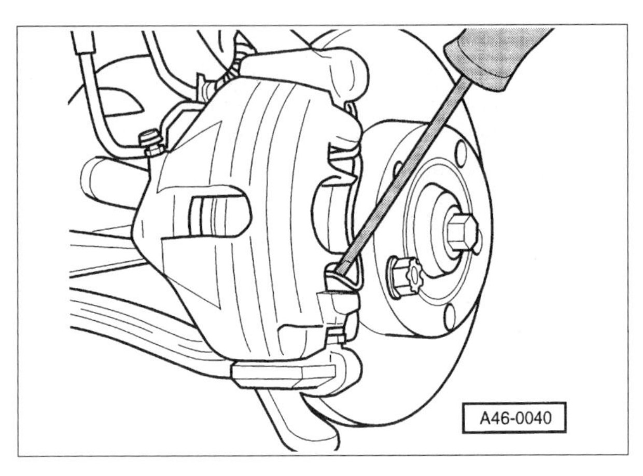

Using screwdriver, pry off brake pad retaining spring from brake caliper housing and remove.

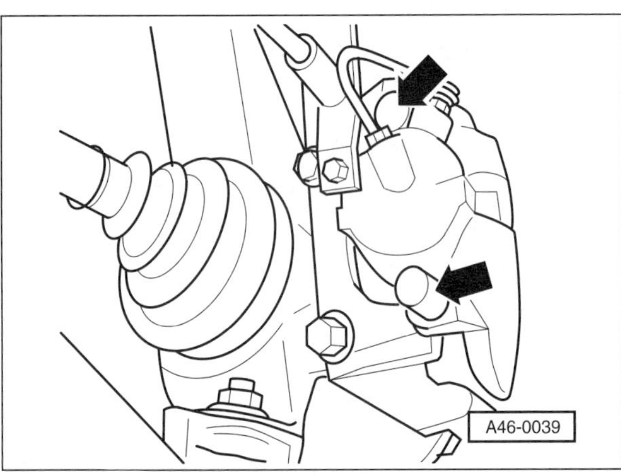

Remove protective caps from caliper guide bolts.

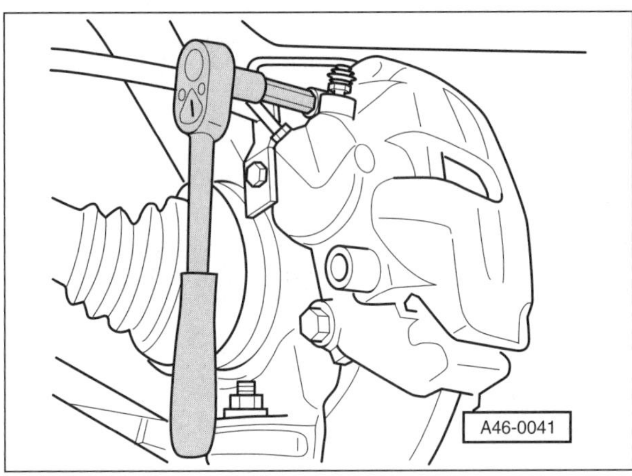

Remove both guide bolts from brake caliper.

- Remove brake caliper and hang from chassis using stiff wire. Do not allow caliper to hang from brake hose.
- Remove brake pads.

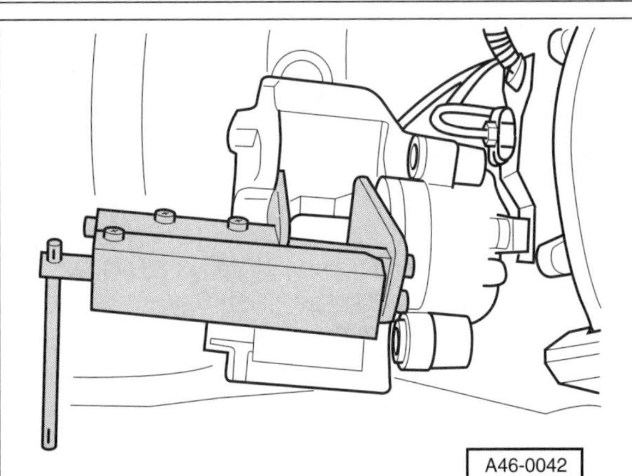

Press piston back into caliper using resetting tool.

> ***CAUTION—***
> - *Before pushing caliper pistons back into caliper, siphon off brake fluid out of brake fluid reservoir. Otherwise, fluid may overflow and cause paint damage.*

NOTE—
- *When siphoning brake fluid, use a bleeder bottle that is used exclusively for brake fluid.*

BentleyPublishers.com
BentleyPublishers.com—All Rights Reserved

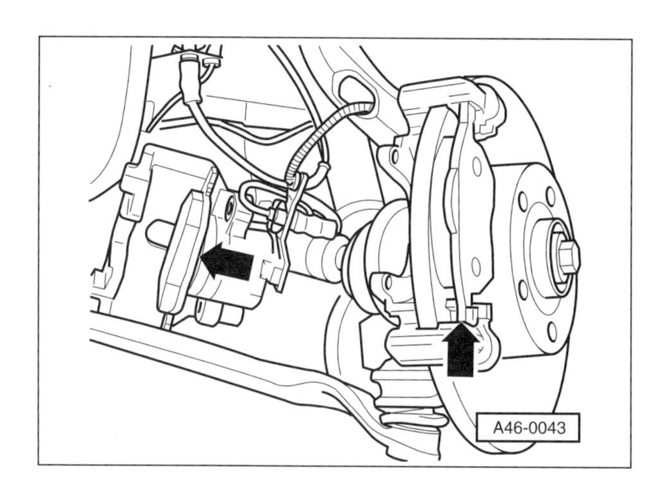

Install complete brake repair kit. Insert brake pad with retaining spring in brake caliper housing. Make sure arrow on inner pad points in direction of forward rotation when installed. Noise may result from incorrect installation.

– Remove protective foil on backing plate of outer brake pad.

– Install outer brake pad on pad carrier.

– Install brake caliper housing in pad carrier using both guide bolts. Install both protective caps.

Tightening torques	
Brake caliper to pad carrier	25 Nm (18 ft-lb)

– Install retaining spring in brake caliper.

– Check caliper for freedom of movement.

– Install wheels.

Tightening torques	
Wheel to hub	120 Nm (89 ft-lb)

– Check brake fluid and top off if necessary.

– Before moving vehicle, depress brake pedal several times firmly to properly seat brake pads in their normal operating position.

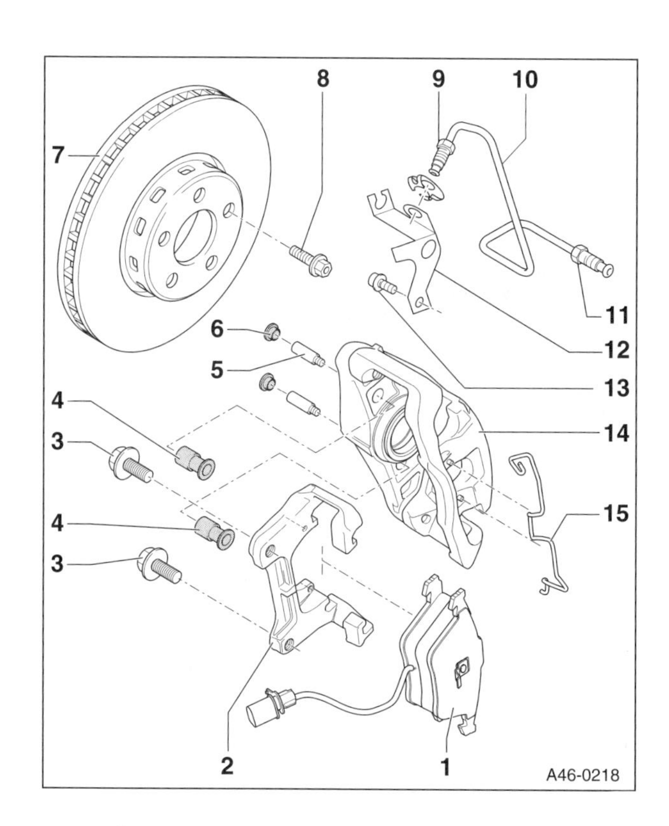

Front brake assembly (FNR-G60)

1. Brake pads
   - Replace front pads as a set.
   - Inner pad (with spring) has wear sensor and harness.
   - Outer brake pad provided with adhesive sheet on backing plate. Remove protective sheet before inserting pad.
2. Brake pad carrier
   - Bolt to wheel bearing housing.
3. M14 x 1.5 x 38 ribbed bolt, 200 Nm (148 ft-lb)
   - Clean ribs if reusing. Use thread locking compound.
4. Bushing
   - Insert into brake caliper housing.
5. Guide bolt, 30 Nm (22 ft-lb)
6. Protective cap
   - Attach wear sensor harness.
7. Brake disc
   - If worn, replace both on front as a set.
   - To replace brake disc, remove brake caliper and pad carrier.
   - Lightly coat contact surfaces between brake disc and wheel bearing hub with polycarbamide grease (VAG part no.G 052 142 A2).
8. Wheel bolts, 120 Nm (89 ft-lb)
9. Brake line connection to brake hose, 15 Nm (11 ft-lb)

BentleyPublishers.com
BentleyPublishers.com—All Rights Reserved

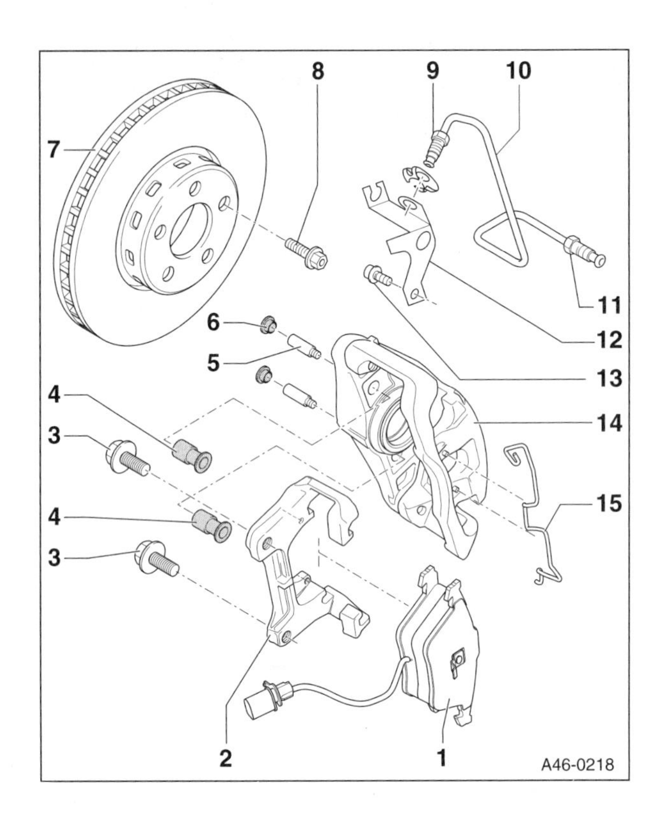

10. Brake line
- Attach to brake caliper.
- Counterhold brake hose when attaching. Make sure brake hose is not twisted when installed.
- Make sure tabs are properly seated in retainer grooves.

11. Brake line to brake caliper, 17 Nm (13 ft-lb)

12. Brake hose bracket
- Bolt to brake caliper housing.
- Attach brake line.
- Make sure locating tab is seated in caliper.

13. M8 x 16 bolt, 25 Nm (18 ft-lb)

14. Brake caliper
- Do not hang by brake hose. Secure to vehicle using wire. Do not twist hose.
- Use alcohol to clean brake caliper thoroughly of grease and other contaminants before installing new brake pads.

15. Retaining spring: Pay attention to correct installation position.
- Engage retaining spring to brake caliper and to outer pad.
- Press retaining spring under pad carrier after it is inserted in caliper holes.

> ***WARNING—***
> - *If spring is improperly installed, brake caliper cannot adjust to compensate for wear on the outer brake pad. This causes brake pedal travel to increase.*

Front brake pads (FNR-G60), removing and installing

– Raise vehicle and support safely.

> ***WARNING—***
> - *Make sure the car is stable and well supported at all times. Use a professional automotive lift or jack stands designed for the purpose. A floor jack is not adequate support.*

– Remove front wheels.

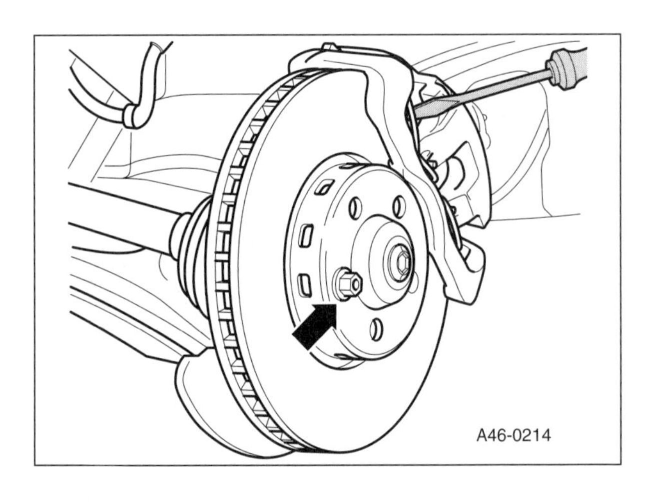

◀ Secure brake disc with a wheel bolt (**arrow**).

– Using screwdriver, pry off brake pad retaining spring from caliper and remove.

– Disconnect brake wear sensor connector. Using a screwdriver, release connector from holder turning connector by 90° at same time. Detach connector and pull upward to unhook wire.

BentleyPublishers.com
BentleyPublishers.com—All Rights Reserved

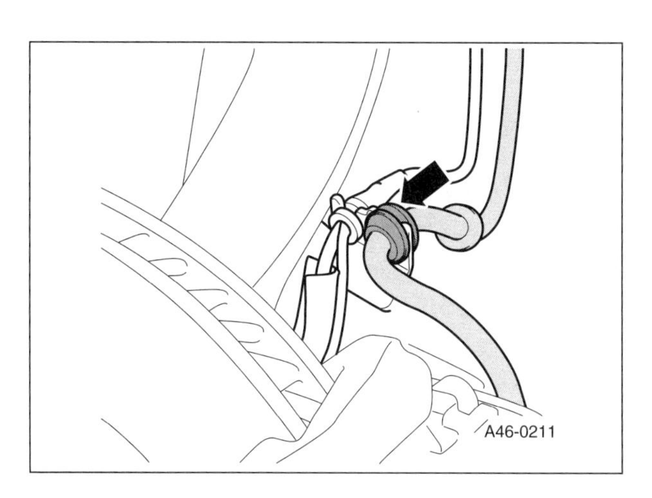

◀ Detach brake hose from bracket (**arrow**).

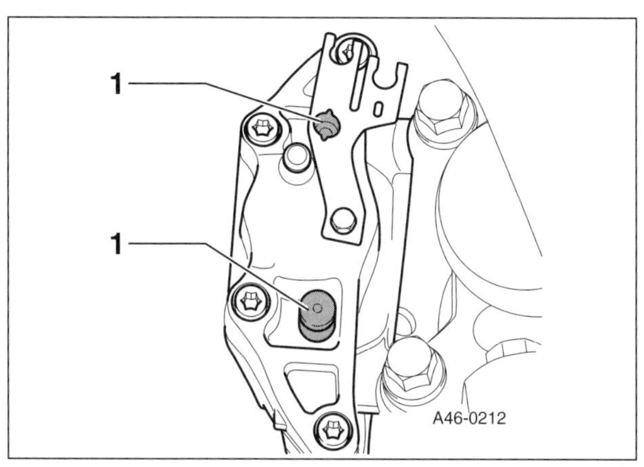

◀ Remove protective caps (**1**) from caliper guide bolts.

NOTE—

- *If necessary, remove brake hose bracket to gain access to upper guide bolt.*

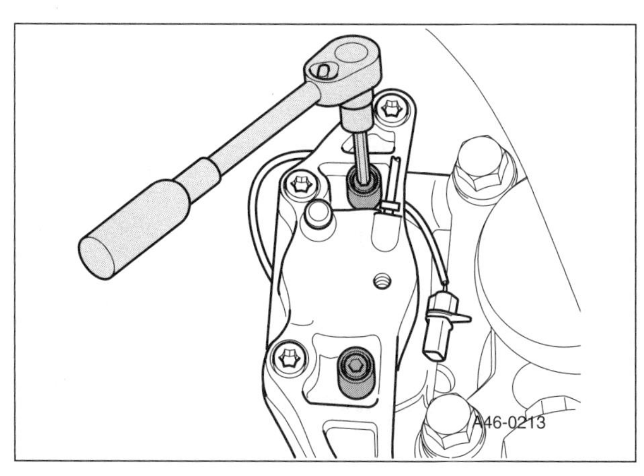

◀ Unscrew both guide bolts and remove from brake caliper.

– Remove brake caliper and hang from chassis using stiff wire. Do not allow caliper to hang from brake hose.

– Remove brake pads.

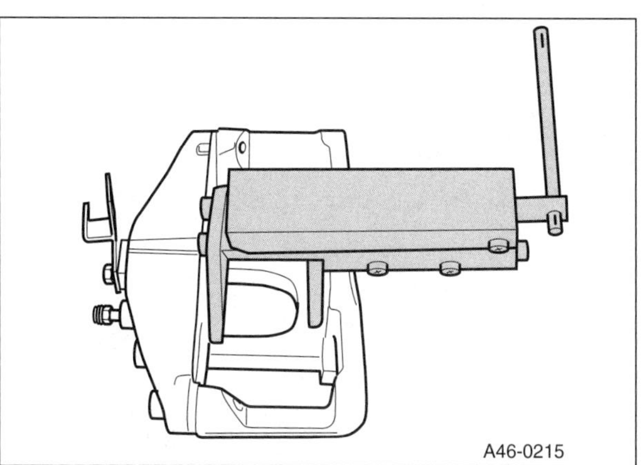

◀ Press piston back into caliper using resetting tool.

CAUTION—

- *Before pushing caliper pistons back into caliper, siphon off brake fluid out of brake fluid reservoir. Otherwise, fluid may overflow and cause paint damage.*

NOTE—

- *When siphoning brake fluid, use a bleeder bottle that is used exclusively for brake fluid.*

BentleyPublishers.com
BentleyPublishers.com—All Rights Reserved

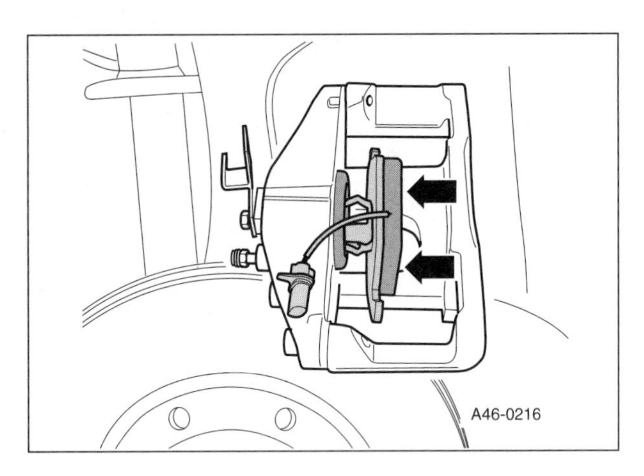

Install complete brake repair kit. Insert inner brake pad with retaining spring and wear sensor into brake caliper piston (**arrows**).

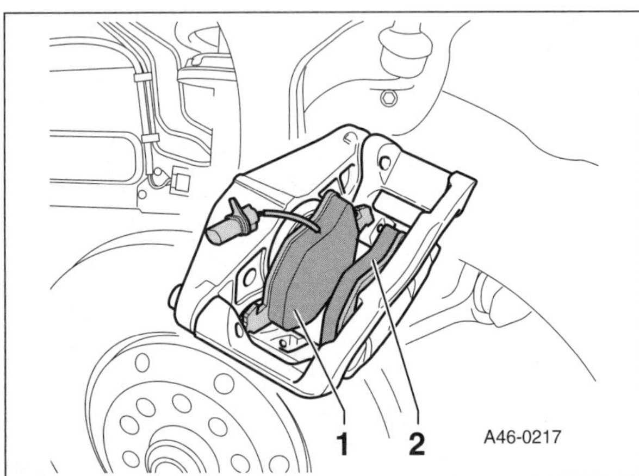

Install outer brake pad (**2**) on brake caliper.

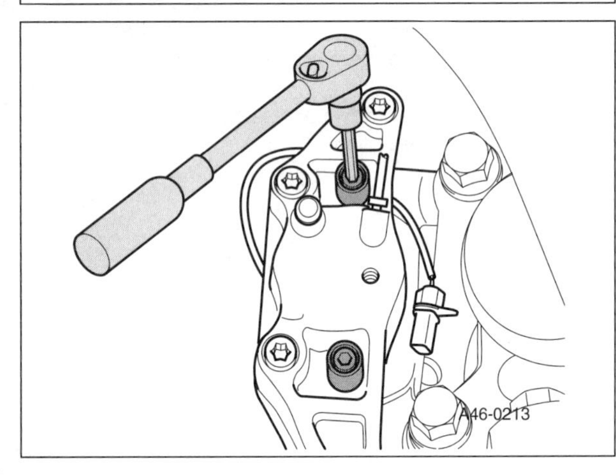

Place caliper over brake pads and secure to pad carrier with guide bolts. Install protective caps.

Tightening torques	
Brake caliper to pad carrier	25 Nm (18 ft-lb)

- Make sure brake hose is not twisted when installed.
- Insert retaining spring into brake caliper housing. Pay attention to correct installation position.
  - Insert in both holes of brake caliper housing.
  - Press retaining spring under pad carrier after it is inserted in both holes.

WARNING —
- *If spring is improperly installed, brake caliper cannot adjust to compensate for wear on the outer brake pad. This causes brake pedal travel to increase.*

- Reconnect brake pad sensor and attach sensor harness to bracket.
- Check caliper for freedom of movement.
- Install wheels.

Tightening torques	
Wheel to hub	120 Nm (89 ft-lb)

BentleyPublishers.com
BentleyPublishers.com—All Rights Reserved

– Check brake fluid and top off if necessary.

– Before moving vehicle, depress brake pedal several times firmly to properly seat brake pads in their normal operating position.

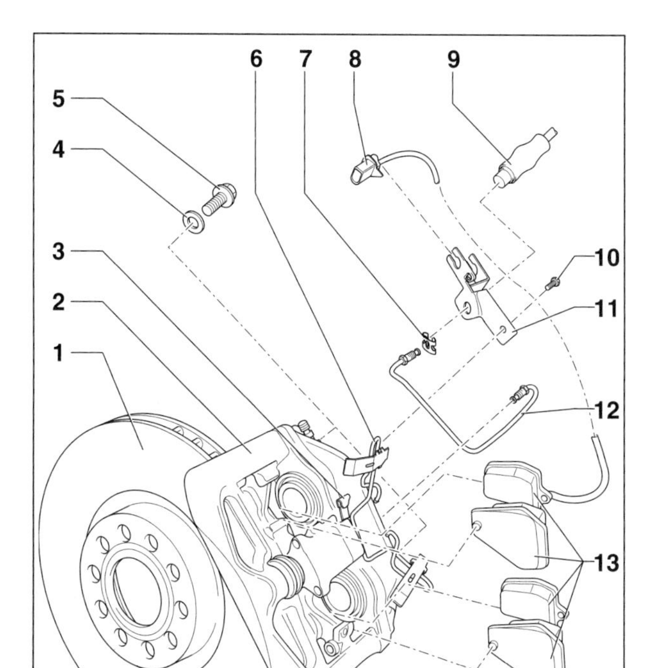

Front brake assembly, allroad (HP-2)

1. Brake disc
   - Check wear limit
   - To replace brake disc, remove brake caliper and pad carrier.
   - Lightly coat contact surfaces between brake disc and wheel bearing hub with polycarbamide grease (VAG part no.G 052 142 A2).
2. Brake caliper
   - If repair work is necessary, do not hang by brake hose. Secure to vehicle using wire
3. Pad sensor harness retainer
4. Ribbed washer
5. Ribbed bolt, 190 Nm (140 ft-lb)
   - Clean ribs if reusing. Use thread locking compound.
6. Retaining spring
   - Slide to side (not removable)
7. Spring clip
8. Pad wear sensor harness connector
9. Brake hose
   - Do not disconnect when removing brake pads
10. Bolt, 25 Nm (18 ft-lb)
11. Pad sensor connector and brake hose bracket
    - Bolt to brake caliper housing.
    - Make sure locating tab is seated in caliper.
    - Insert pad sensor harness connector.
12. Brake line, 15 Nm (11 ft-lb)
13. Brake pads
    - Replace front brake pads as a set.
    - Pads are different on left and right sides.
    - Once wear limit is reached (limit: 3 mm or ⅛ in), warning light in instrument cluster illuminates.
    - Remove adhesive foil from backing plates on outer brake pads before installing.

Front brake pads, allroad (HP-2), removing and installing

NOTE—

- *If reusing brake pads or brake discs, mark them so they can be installed in their original positions to prevent uneven braking.*

– Raise vehicle and support safely.

> *WARNING—*
> - *Make sure the car is stable and well supported at all times. A floor jack is not adequate support.*

BentleyPublishers.com
BentleyPublishers.com—All Rights Reserved

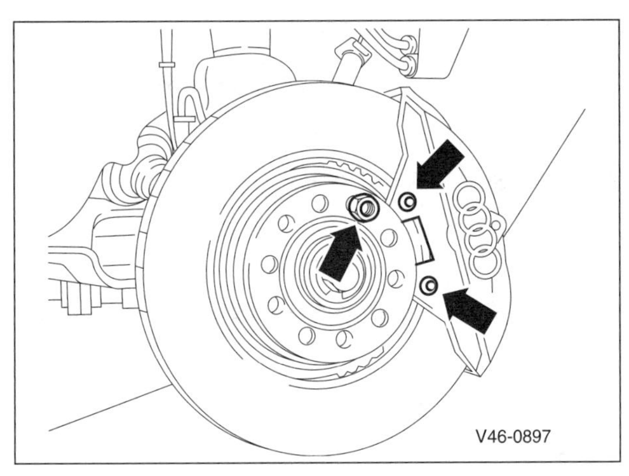

– Remove front wheels.

◄ Secure brake disc using a wheel bolt.

– Disconnect brake pad wear sensor connector.

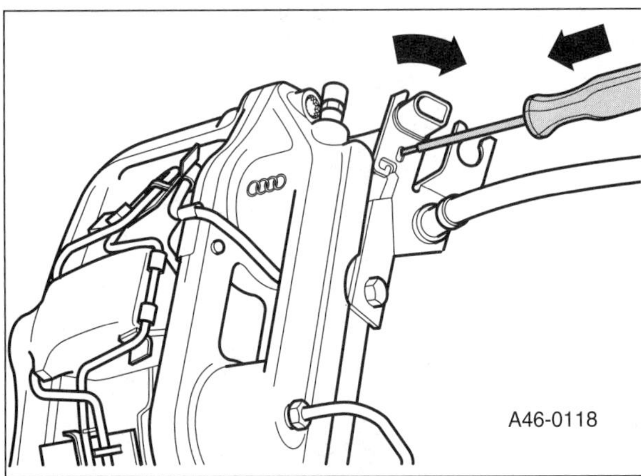

◄ Disengage connector from bracket using a screwdriver and simultaneously turn connector 90°. Lift connector up and out and disengage wire.

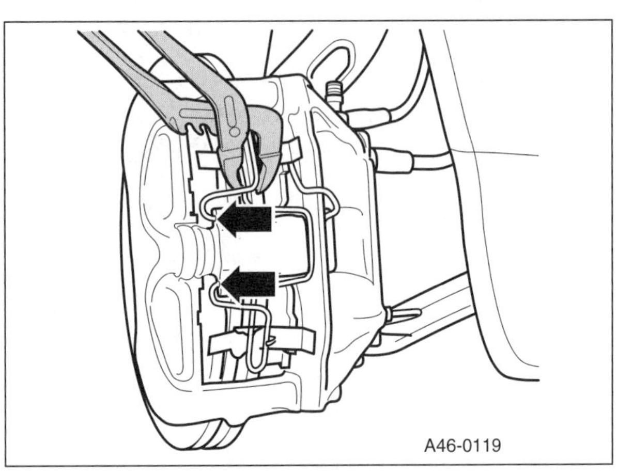

◄ Use pliers to press upper and lower spring tensioner out of locking mechanism (**arrows**). Slide spring aside.

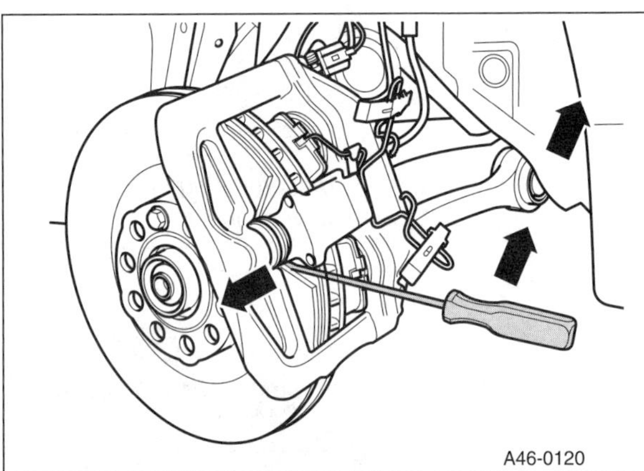

◄ Press piston back into brake caliper housing. To do this, slide floating frame using a screwdriver as shown and press powerfully outward.

CAUTION—

• Before pushing caliper pistons back into caliper, siphon off brake fluid out of brake fluid reservoir. Otherwise, fluid may overflow and cause paint damage.

NOTE—

• When siphoning brake fluid, use a bleeder bottle that is used exclusively for brake fluid.

BentleyPublishers.com
BentleyPublishers.com—All Rights Reserved

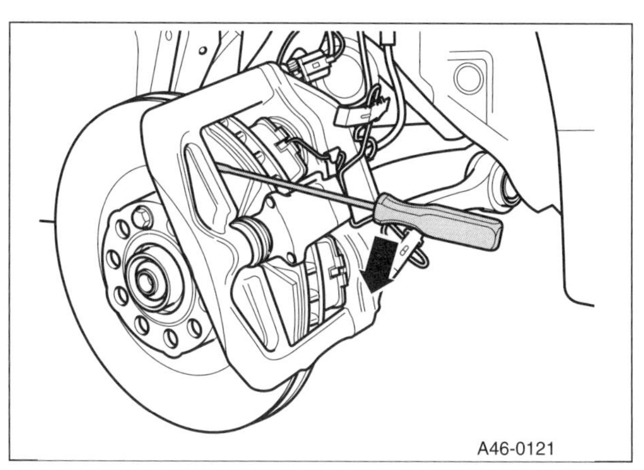

◀ Use screwdriver to press outer brake pads off brake caliper.

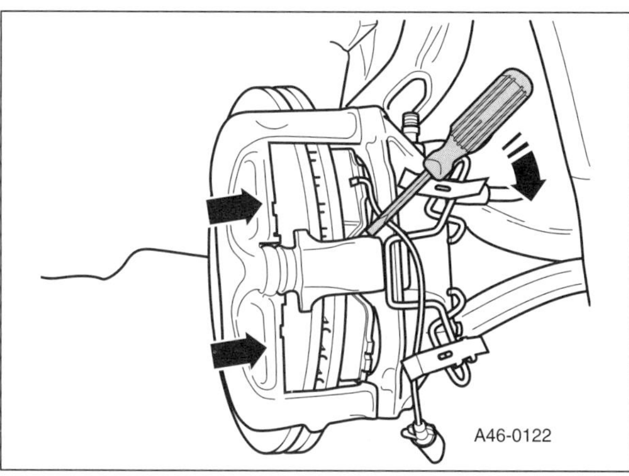

◀ Press floating frame firmly in direction of **arrow** until stop.

- Pry inner brake pads out of caliper pistons. Remove brake caliper housing.
- Use alcohol to clean brake caliper thoroughly of grease and other contaminants before installing new brake pads.

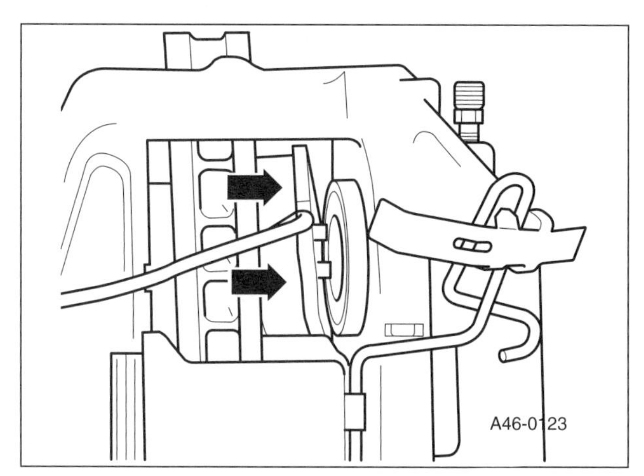

◀ Insert complete brake repair kit. Insert inner brake pads in caliper piston.

> ***CAUTION—***
> *• Do not damage protective piston boots when installing brake pads.*

NOTE —

• Make sure protrusions on backing plates are securely fitted in brake pistons.

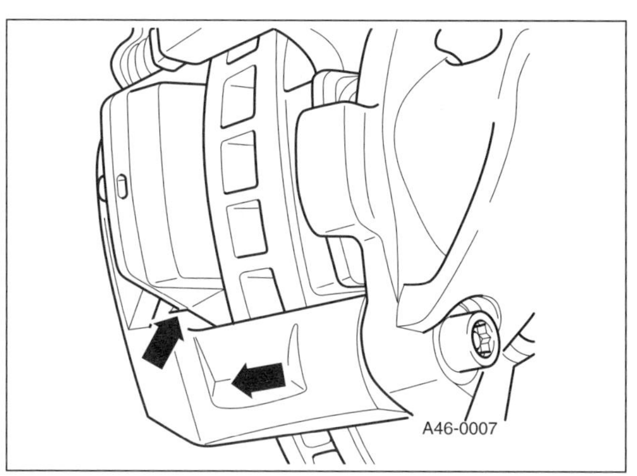

◀ Press floating frame outward until stop.

- Remove protective foil on backing plates of outer brake pads.
- Make sure protrusions on pad backing plates are securely fitted in floating frame.
- Check caliper for freedom of movement.
- Remove wheel bolt from wheel hub.

BentleyPublishers.com
BentleyPublishers.com—All Rights Reserved

– Install wheels.

Tightening torque	
Wheel to wheel hub	120 Nm (89 ft-lb)

NOTE —
- *When installing wheels, make sure rim flange does not rest on brake caliper. This could push brake pads out of position.*

– Check brake fluid and top off if necessary.

– Before moving vehicle, depress brake pedal several times firmly to properly seat brake pads in their normal operating position.

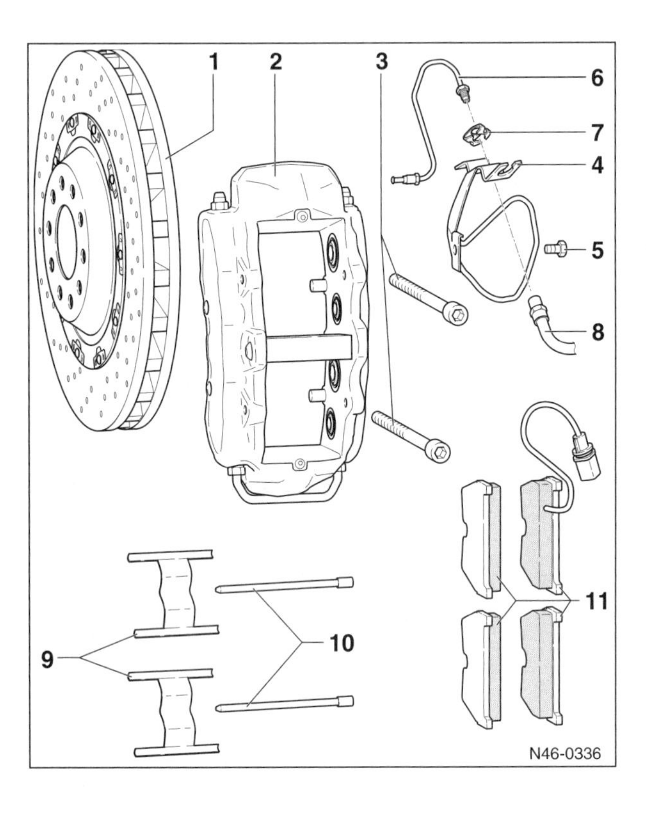

Front brake assembly, RS6 (Brembo)

1. Brake disc
   - Internally ventilated brake disc diameter: 365 mm
   - Brake disc thickness: 34 mm
   - Wear limit: 32 mm
   - Replace front brake pads as a set.
   - To replace brake disc, remove brake caliper and pad carrier.
   - Lightly coat contact surfaces between brake disc and wheel bearing hub with polycarbamide grease (VAG part no.G 052 142 A2).
2. Brake caliper
   - See **Damper bracket, installing** in this repair group.
   - Do not disconnect brake hose when changing brake disc.
3. Replace Allen bolts, 110 Nm (81 ft-lb)
4. Brake hose bracket
   - Bolt to brake caliper housing.
   - Make sure locating tab is seated in caliper.
5. Bolt, 25 Nm (18 ft-lb)
6. Brake line, 15 Nm (11 ft-lb)
   - Attach to brake caliper.
   - Counterhold brake hose when attaching. Make sure brake hose is not twisted when installed.
   - Make sure tabs are properly seated in retainer grooves.
7. Spring clip
8. Brake hose
   - Do not disconnect brake hose when changing brake disc.
9. Pad retaining springs
10. Pad retaining pins
11. Brake pads
    - With wear sensor.
    - Check thickness.

BentleyPublishers.com
BentleyPublishers.com—All Rights Reserved

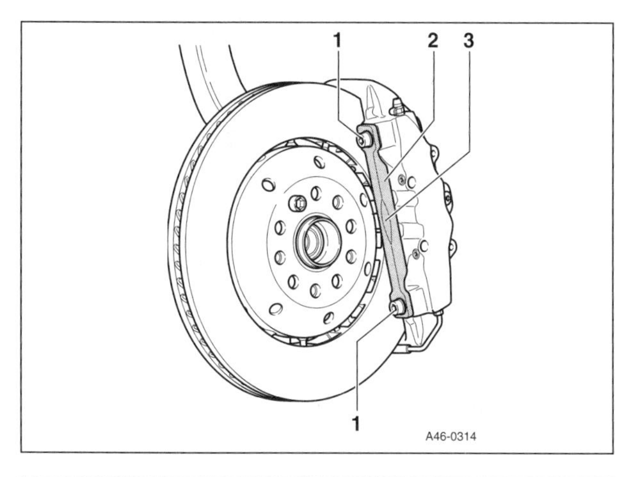

Damper bracket, installing, RS6

When installing damper bracket, check for damage to the surface and paint black if necessary.

Install both damper brackets so the mounting points on brakes make full contact with the contour of damper bracket.

NOTE—

- *The notch on the damper bracket faces toward wheel rim.*

1. Allen bolt M8 x 25, 25 Nm (18 ft-lb)
   - Apply thread locking compound when installing.
2. Damper bracket (note different versions).
3. Notch on damper bracket faces toward wheel rim.

Front brake pads, RS6 (Brembo), removing and installing

– Raise vehicle and support safely.

> ***WARNING—***
> - *Make sure the car is stable and well supported at all times. Use a professional automotive lift or jack stands designed for the purpose. A floor jack is not adequate support.*

– Remove front wheels.

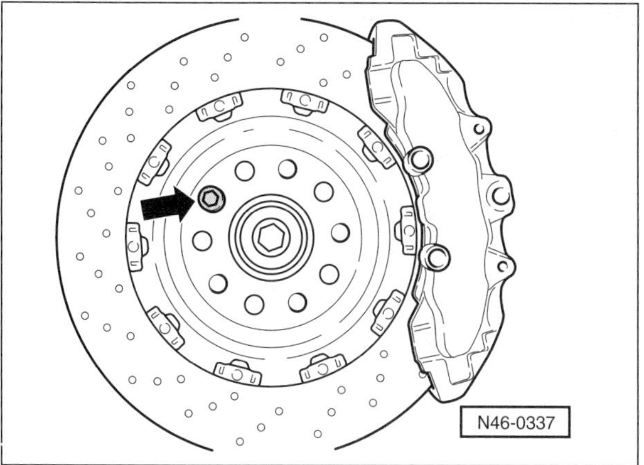

◀ Secure brake disc with a wheel bolt.

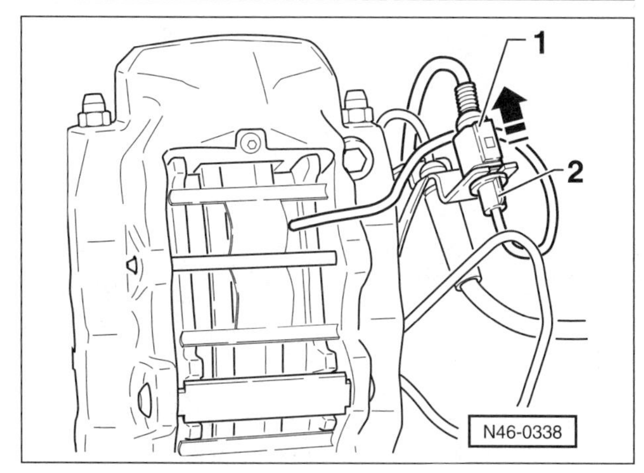

◀ Detach connector (**1**) from brake pad wear sensor.

– Slightly lift up positioning tag on lower part of connector (**2**) and then turn it through 90°. Remove lower half of connector out of holder.

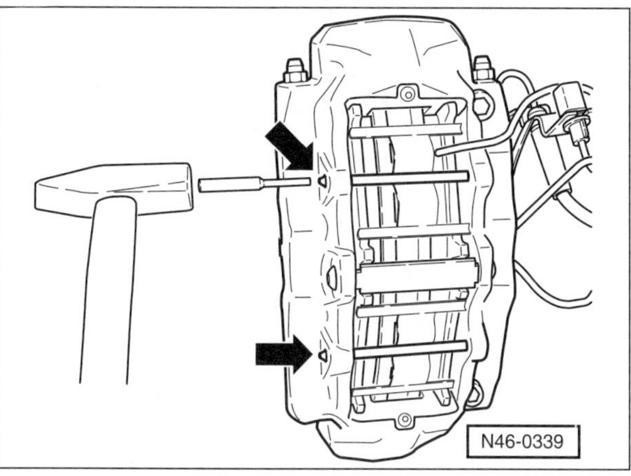

◀ Drive out pad retaining pins (**arrows**).

BentleyPublishers.com
BentleyPublishers.com—All Rights Reserved

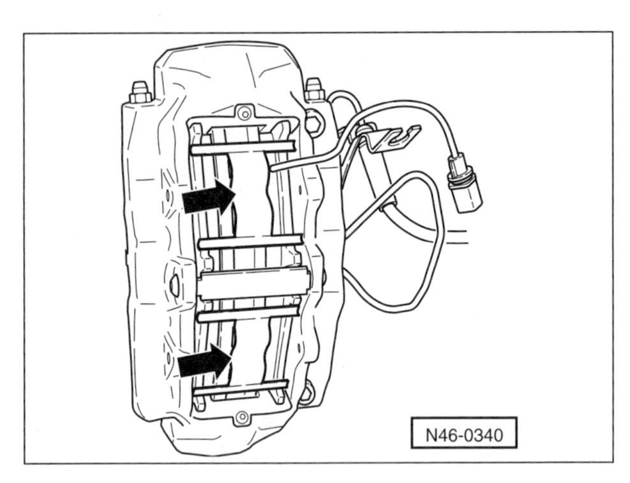

◀ Remove retaining springs (**arrows**).

– Slide brake pads out of brake caliper.

– Press pistons back into caliper housing using resetting tool.

> ***CAUTION—***
> *• Before pushing caliper pistons back into caliper, siphon off brake fluid out of brake fluid reservoir. Otherwise, fluid may overflow and cause paint damage.*

NOTE—
• When siphoning brake fluid, use a bleeder bottle that is used exclusively for brake fluid.

– Insert complete brake repair kit. Insert brake pads into brake caliper housing.

NOTE—
• The brake pads are directional. Note markings.

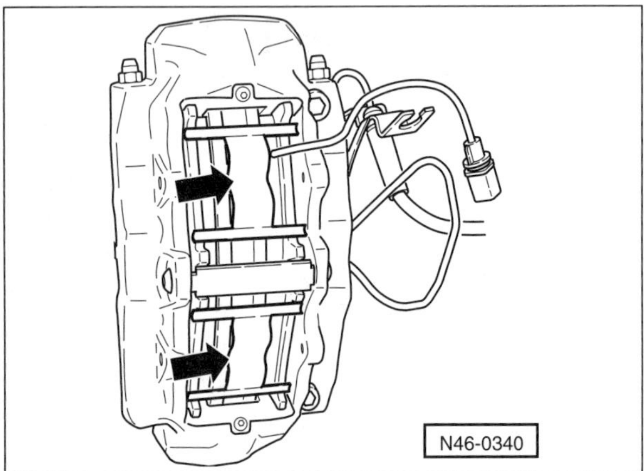

◀ Install retaining springs (**arrows**). Press retaining springs downward and drive in pad retaining pins to stop.

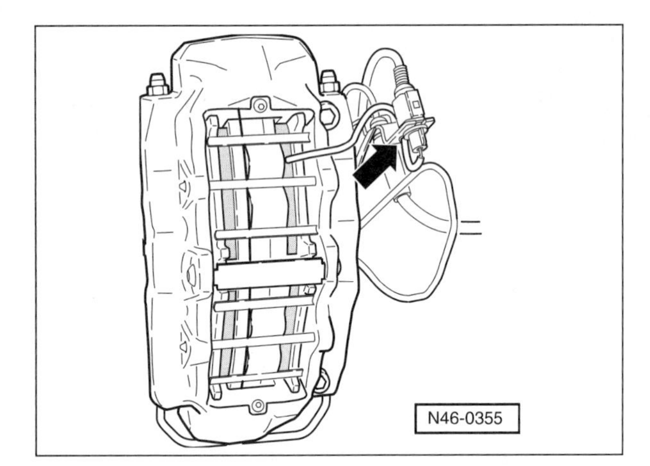

◀ Insert brake pad sensor connector and harness (**arrow**) into bracket at brake caliper. Plug in brake pad sensor connector.

– Check caliper for freedom of movement.

– Install wheels.

Tightening torque	
Wheel to hub	120 Nm (89 ft-lb)

– Check brake fluid and top off if necessary.

– Before moving vehicle, depress brake pedal several times firmly to properly seat brake pads in their normal operating position.

BentleyPublishers.com
BentleyPublishers.com—All Rights Reserved

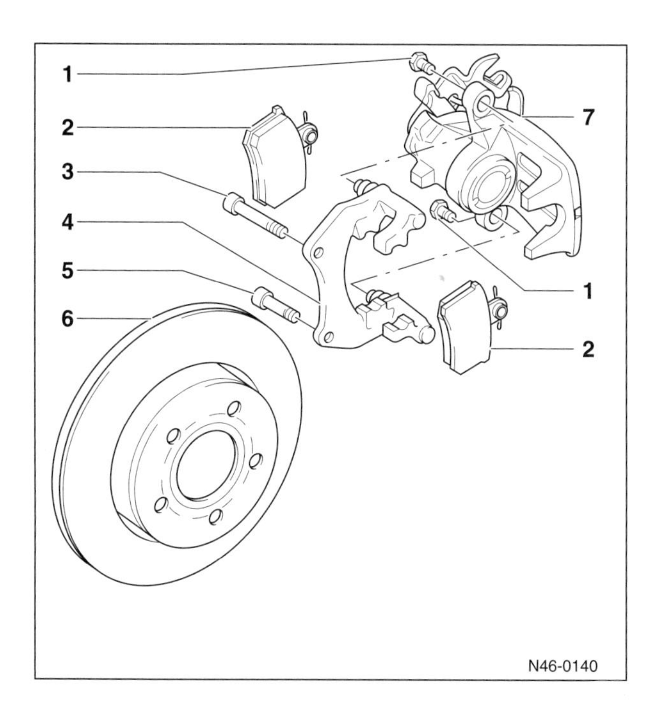

REAR BRAKES

Rear brake assembly, quattro

1. Replace self-locking guide bolt, 35 Nm (26 ft-lb)
   - When loosening and tightening, counterhold guide pin.
2. Brake pads. Replace as a set.
3. Ribbed bolt, 95 Nm (70 ft-lb)
4. Pad carrier with guide bolts and protective cap
   - Supplied with sufficient grease on guide bolts.
   - If protective caps or guide bolts are damaged, install repair kit. Use grease packet supplied to lubricate guide bolts.
5. Ribbed bolt, 95 Nm (70 ft-lb)
6. Brake disc. Replace as a set.
   - To replace brake disc, remove brake caliper and pad carrier.
   - Lightly coat contact surfaces between brake disc and wheel bearing hub with polycarbamide grease (VAG part no.G 052 142 A2).
7. Brake caliper
   - Do not disconnect brake hose when changing brake pads.
   - Brake hose to brake caliper, 45 Nm (33 ft-lb).
   - Make sure brake hose is not twisted when installed.

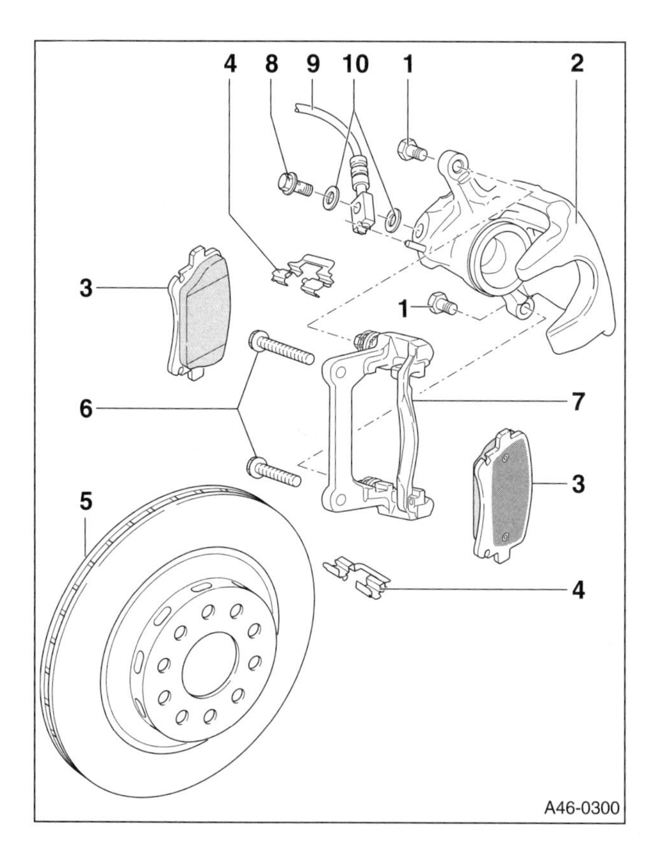

Rear brake assembly, front-wheel drive

1. Replace self-locking guide bolt, 35 Nm (26 ft-lb)
   - When loosening and tightening, counterhold guide pin
2. Brake caliper
   - Unbolt caliper when changing brake pads.
   - Make sure brake hose is not twisted when installed.
3. Brake pads. Replace as a set.
   - Do not disconnect brake hose when changing brake pads.
4. Pad retaining spring
   - Always replace when changing pads.
5. Brake disc. Replace as a set.
   - To replace brake disc, remove brake caliper and pad carrier.
   - Lightly coat contact surfaces between brake disc and wheel bearing hub with polycarbamide grease (VAG part no.G 052 142 A2).
   - Check degree of wear (wear limit)
6. Replace M10 x 1.25 x 20 Allen bolts, 70 Nm (52 ft-lb) plus 90°
7. Pad carrier with guide pin and protective cap
   - Supplied as replacement part assembled with sufficient quantities of grease on guide bolts.
   - If protective caps or guide bolts are damaged, install repair kit. Use grease packet supplied to lubricate guide bolts.
8. Banjo bolt, 40 Nm (30 ft-lb)
9. Brake line with banjo union
   - Make sure brake hose is not twisted when installed.
   - Unbolt from pad carrier when changing pads.
10. Seals

BentleyPublishers.com
BentleyPublishers.com—All Rights Reserved

Rear brake pads and discs, removing and installing

NOTE—

- *If reusing brake pads or brake discs, mark them so they can be installed in their original positions to prevent uneven braking.*

In the following procedure, a 1999 Avant quattro is illustrated. Rear brake pad replacement on a front wheel-drive model is similar.

– Raise vehicle and support safely.

> ***WARNING—***
> - *Make sure the car is stable and well supported at all times. Use a professional automotive lift or jack stands designed for the purpose. A floor jack is not adequate support.*

– Remove rear wheels.

◄ Counterhold lower guide pin and remove self-locking bolt from brake caliper housing.

- If necessary, detach parking brake cable to remove lower guide bolt.
- Similarly, remove upper caliper guide bolt.

– Remove brake caliper and support from chassis using stiff wire. Do not let caliper hang on hose.

◄ Slide brake pads out of brake pad carrier (**arrows**).

– Use alcohol to clean brake caliper thoroughly of grease and other contaminants before installing new brake pads.

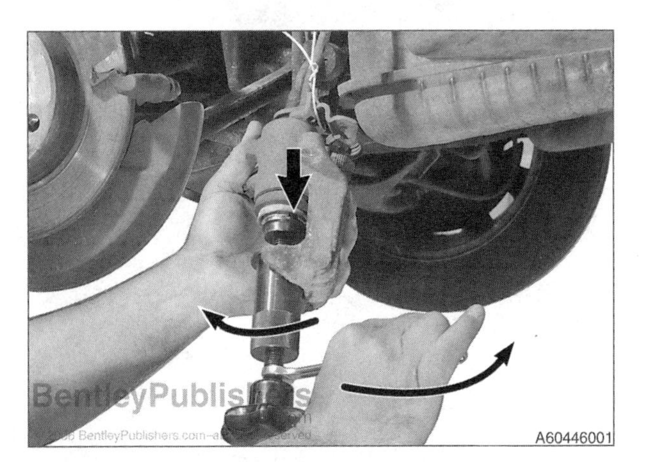

◄ Use VAG special tool 3272 or equivalent to screw caliper piston inward as far as it will go.

- Make sure tab on tool fits caliper piston slot (**arrow**).
- Rotate threaded spindle clockwise and knurled section counterclockwise.

> ***CAUTION—***
> - *Before pushing caliper pistons back into caliper, siphon off brake fluid out of brake fluid reservoir. Otherwise, fluid may overflow and cause paint damage.*

NOTE—

- *When siphoning brake fluid, use a bleeder bottle that is used exclusively for brake fluid.*

BentleyPublishers.com
BentleyPublishers.com—All Rights Reserved

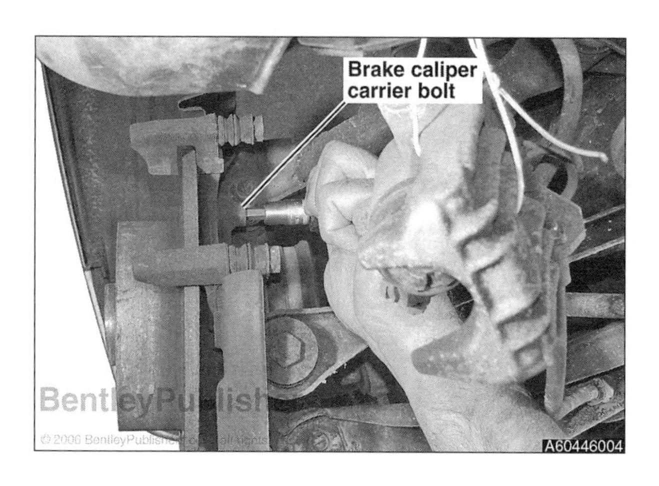

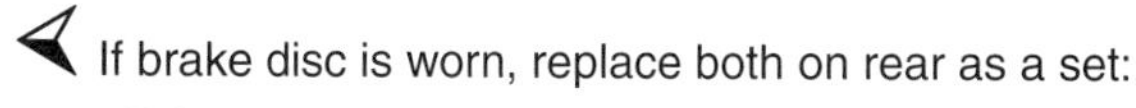

If brake disc is worn, replace both on rear as a set:

- Prior to removing disc, remove brake pad carrier mounting bolts and lift off pad carrier.
- Remove brake disc. Prior to installing new disc, lubricate contact surfaces between disc and wheel hub with polycarbamide grease (VAG part no. G 052 142 A2).
- Use new pad carrier bolts with thread-locking compound.

Tightening torque	
Pad carrier to hub (use new bolts) • quattro, M10 x 1.25 x 20 Allen bolt • Front-wheel drive, ribbed bolt	 95 Nm (70 ft-lb) 70 Nm (52 ft-lb) + 90°

– Install complete brake repair kit. Remove protective foil on brake pad backing plates. Insert brake pads.

– Reinstall brake caliper using new self-locking bolts included in repair kit.

Tightening torque	
Brake caliper to pad carrier	35 Nm (26 ft-lb)

– Check caliper for freedom of movement.

– Install wheels.

Tightening torque	
Wheel to hub	120 Nm (89 ft-lb)

– Check brake fluid and top off if necessary.

– Before moving vehicle, depress brake pedal several times firmly to properly seat brake pads in their normal operating position.

PARKING BRAKE

Parking brake, adjusting

Because the rear brakes are automatically self-adjusting, readjusting the parking brake is usually not necessary. Adjustment is only necessary after replacing parking brake cables, rear brake calipers, rear brake pads or rear brake discs.

Bleed brakes, if necessary, prior to adjusting parking brake. See **47 Brakes–Hydraulic**.

– Depress brake pedal firmly.

– Make sure parking brake lever is released.

– Remove rear air vent from center console.

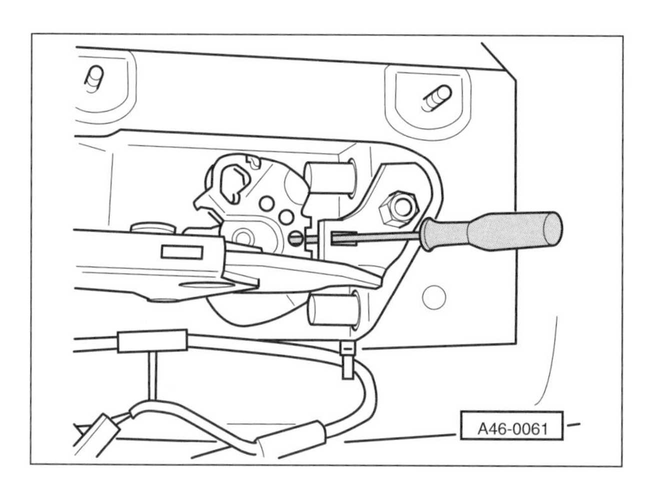

Remove any plastic components on parking brake cable compensator and discard.

BentleyPublishers.com
BentleyPublishers.com–All Rights Reserved

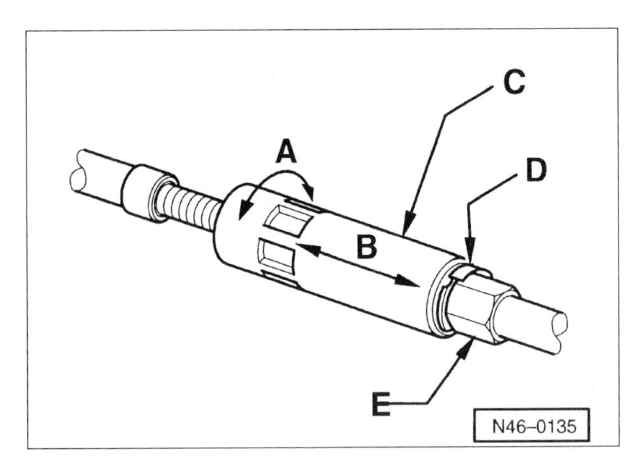

- Use screwdriver to prevent compensator from turning (center console removed in illustration).

NOTE—
- *quattro: parking brake cable adjusting components are located in front of rear lower control arms.*
- *Front-wheel drive: parking brake cable adjusting components are located in the tunnel on the underside of vehicle.*

◄ Remove locking component (**D**).

- Counterhold hex (**E**) using 13 mm open end wrench and turn adjusting nut (**C**) in until stop.
- Slide cable adjuster (**B**) together.
- Unscrew adjusting nut until slot for locking component is visible.
- Install locking component.

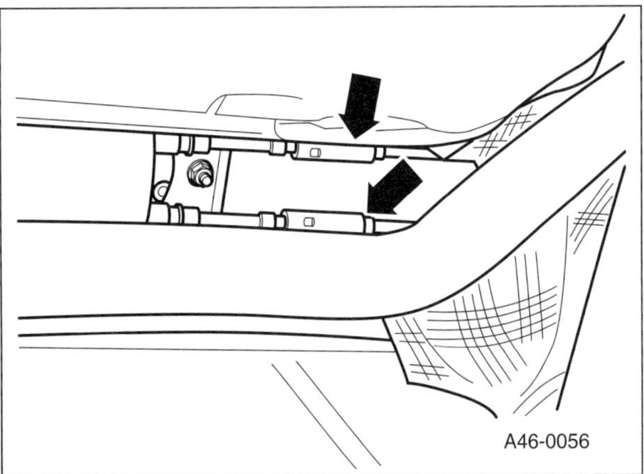

◄ Pull both cable sleeve adjusters apart simultaneously until cables are tensioned. Lever must not lift off at brake caliper during this process.

- Remove screwdriver from compensator and apply parking brake firmly 3 times.

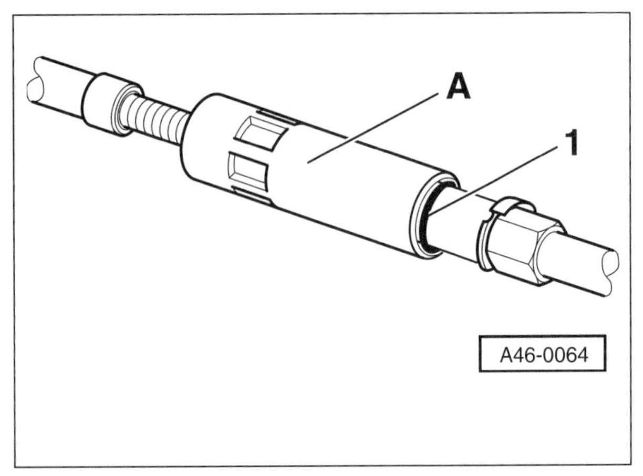

NOTE—
- *Colored O-ring (**1**) must not be visible following successful installation. This ensures that the parking brake cable adjustment mechanism is sufficiently protected against dirt and water spray.*

◄ Check parking brake tension and adjust fine adjustment (**A**) in or out, if necessary.

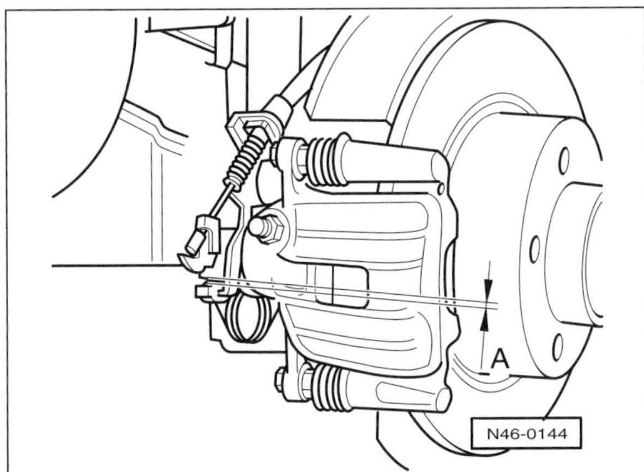

◄ Turn adjustment until lever lifts slightly from brake caliper.

- Note dimension (**A**). Gap should be visible, but not be wider than 1.5 mm (0.06 in).

BentleyPublishers.com
BentleyPublishers.com—All Rights Reserved

Brake Light Switch

> ***CAUTION—***
> *• Make sure brake pedal travel is not restricted by thick floor mats.*

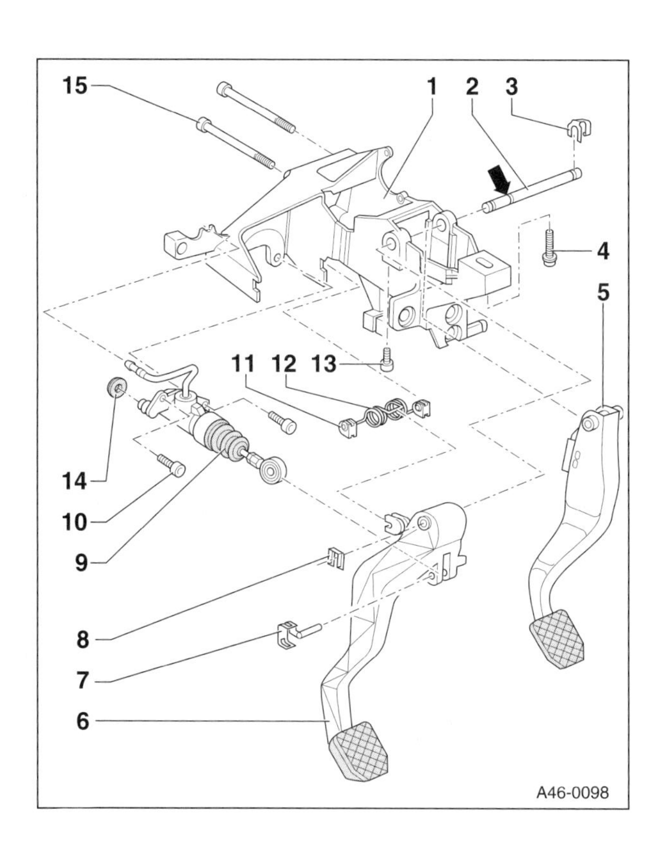

Brake pedal assembly

1. Bracket
2. Pivot pin for clutch and brake pedals
   - Installation position: Groove (**arrow**) faces clutch pedal
3. Locking clip
4. Allen bolt, 25 Nm (18 ft-lb)
5. Brake pedal
6. Clutch pedal
7. Pin
8. Locking clip
9. Clutch master cylinder
10. Allen bolt, 20 Nm (15 ft-lb)
11. Mount
    - Install in mounting bracket with over-center spring.
12. Over-center spring
    - Remove and install together with clutch pedal.
13. Socket-head bolt, 5 Nm (4 ft-lb)
    - Secures pivot pin for clutch and brake pedal.
14. Seal
    - Do not remove from plastic master cylinder.
15. Torx bolt, 25 Nm (18 ft-lb)

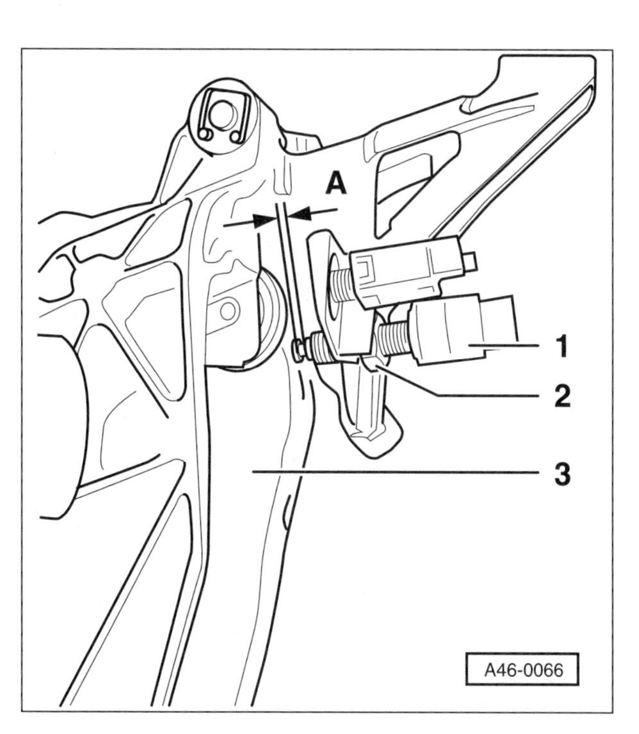

Brake light switch, adjusting

NOTE—

• Before adjusting, make sure pedal cluster is bolted tightly to brake booster and instrument panel carrier.

Threaded brake light switch

1. Brake light switch
2. Nut, M12 x 1.5
3. Brake pedal

NOTE—

*• A spring washer (not shown in illustration) is installed between nut (**2**) and pedal bracket.*

– Turn brake light switch to adjust inward or outward as necessary to obtain dimension **A** = 0.1 - 0.5 mm (0.004 - 0.02 in). Lock in place with nut and tighten nut to 4.5 Nm (3 ft-lb).

– Check dimension **A** again after tightening nut.

BentleyPublishers.com
BentleyPublishers.com—All Rights Reserved

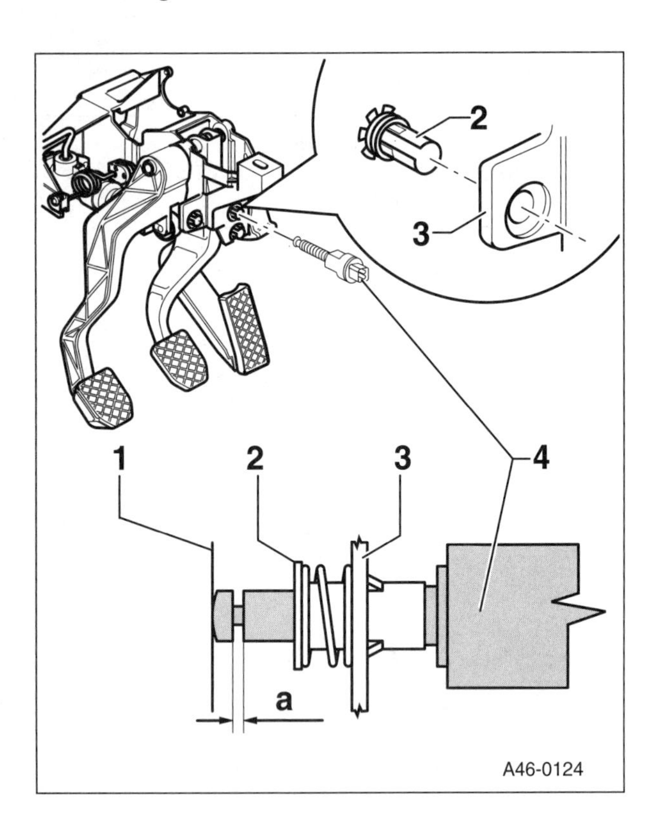

Spring-clip mounted brake light switch

1. Brake pedal contact surface
2. Spring clip (insert in pedal bracket from rear)
3. Pedal bracket
4. Brake light switch (F)
    - Dimension **a**: max. 0.7 mm (0.28 in)

NOTE—
- *In order to fit tightly, do not install switch more than once.*

Adjusting switch

– Guide switch (**4**) into installed clip (**2**) and press in until stop.

NOTE—
- *Do not apply pressure to brake pedal.*
- *Turn back switch if necessary. Observe dimension* **a**.

BentleyPublishers.com
BentleyPublishers.com—All Rights Reserved

47 Brakes–Hydraulic

GENERAL

This repair group covers brake bleeding and replacement procedures for the master cylinder, brake booster and vacuum pump. For additional brake system information, see:

- **03 Maintenance** for brake fluid level check
- **45 Antilock Brakes (ABS)** for ABS related components
- **46 Brakes–Mechanical** for brake pads, discs and parking brake

Models covered by this manual are equipped with power assisted four-wheel disc brakes. A vacuum brake booster provides power assistance to a dual-circuit master cylinder.

Troubleshooting

Brake performance is mainly affected by three factors:

- Level and condition of brake fluid
- Brake system's ability to create and maintain hydraulic pressure
- Condition of friction components

– Air in brake fluid makes brake pedal feel spongy during braking or will increase brake pedal force required to stop. Fluid contaminated by moisture or dirt can corrode components. Inspect brake fluid. If fluid is dirty or murky, or over two years old, flush and replace it.

– Visually check hydraulic system, starting at master cylinder. To check function of master cylinder:
  - Hold brake pedal down hard with car stopped and engine running. Make sure pedal feels and stays solid.
  - If pedal slowly falls to floor, either brake master cylinder is leaking internally, or fluid is escaping from other points.
  - If no leaks can be found, brake master cylinder is faulty; replace it.

BentleyPublishers.com
BentleyPublishers.com—All Rights Reserved

– Check all brake fluid lines and couplings for leaks, kinks, chafing or corrosion.

– Check brake booster by pumping brake pedal approximately 10 times with engine off. Hold pedal down and start engine. Pedal should fall slightly. If not, before suspecting a faulty brake booster:
  - Check for visible faults in vacuum ducting.
  - Check brake booster vacuum pump (if applicable).
  - Check one-way check valve at hose connection on vacuum booster.

Warnings and Cautions

WARNING —
- *If any of the brake warning lights illuminate while driving, check brake system immediately.*
- *Brake fluid is poisonous. Wear safety glasses when working with brake fluid, and wear rubber gloves to prevent brake fluid from entering the bloodstream through cuts or scratches. Do not siphon brake fluid by mouth.*
- *Perform brake work under clean conditions with careful attention to specifications and proper working procedures. If you lack the skills, the tools or a clean workplace for servicing the brake system, leave these repairs to an authorized Audi dealer or other qualified shop.*
- *After replacing brake components, depress the brake pedal firmly several times to seat the brakes in their normal operating position. Make sure the pedal is firm and at its normal height. If not, further work is required before driving vehicle.*
- *Do not mix mineral oil products such as gasoline or engine oil with brake fluid. Mineral oil damages rubber seals in the brake system.*
- *Use only new brake fluid conforming to US standard FMVSS 116 DOT 4. Genuine VW / Audi brake fluid conforms to this specification.*
- *Observe hazardous waste regulations when disposing brake fluid as a hazardous waste.*
- *Do not use DOT 5 (silicone) brake fluid.*

CAUTION —
- *Brake fluid damages paint. Immediately wipe up any brake fluid that spills on the vehicle and rinse area with water.*
- *Brake fluid is hygroscopic (absorbs moisture from the air). Store in an airtight container.*
- *Do not fill brake fluid above* **MAX** *in fluid reservoir.*
- *During final road test, perform at least one ABS-controlled braking operation. Make sure appropriate pulsations are felt at the brake pedal.*

BentleyPublishers.com
BentleyPublishers.com—All Rights Reserved

BLEEDING BRAKES

Brake fluid absorbs moisture from the atmosphere and so is said to be *hygroscopic*. If moisture content of brake fluid is too high, corrosion inside brake system may result. The boiling point of brake fluid is also considerably lowered, resulting in the possible vaporization of brake fluid and reduction in braking force under some conditions.

Brake fluid is also used as the operating fluid for the hydraulic clutch on vehicles with a manual transmission. Fluid is drawn into clutch master cylinder via a small fitting in upper part of reservoir. In case of a hydraulic clutch leak, fluid loss will only be to the level of the fitting. This retains an adequate reserve for the brake system.

Audi specifies use of brake fluid that meets Department of Transportation (DOT) 4 Super specification in addition to Federal Motor Vehicle Safety Standard (FMVSS) 116 and Society of Automotive Engineers (SAE) standard J1703. DOT 4 Super is also known as DOT 4+. Brake fluid meeting this specification is also available in 30 and 50 liter containers for use in pressure bleeders.

Brake fluid specifications	
Audi specification	DOT 4 Super (DOT 4+)
Minimum wet boiling point	329°F (165°C)
Dry boiling point	exceeds 500°F (260°C)

Bleeding brakes is usually done for one of two reasons: either to replace old brake fluid as part of routine maintenance or to expel trapped air in the system that resulted from opening the brake hydraulic system during repairs.

When adding or replacing brake fluid, add new brake fluid from an unopened container. If you are certain no air was introduced into the master cylinder or ABS hydraulic unit, bleed brakes at the calipers using a pressure bleeder.

Because the hydraulic clutch and the brake system share a common fluid and reservoir, be sure to bleed the clutch hydraulic system when replacing the brake fluid.

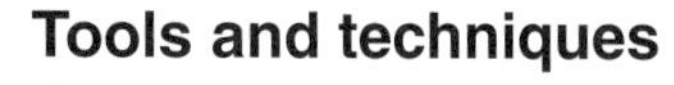

Tools and techniques

To bleed brakes, use special tool VAS 5234 or equivalent pressure bleeder.

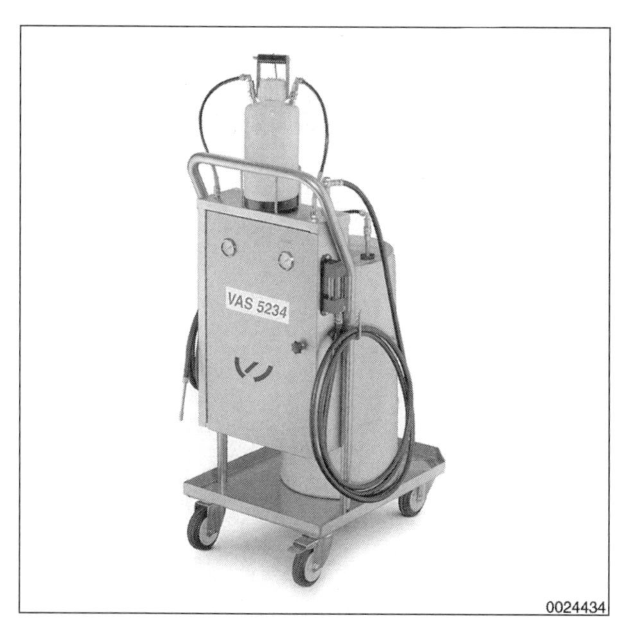

0024434

BentleyPublishers.com
BentleyPublishers.com—All Rights Reserved

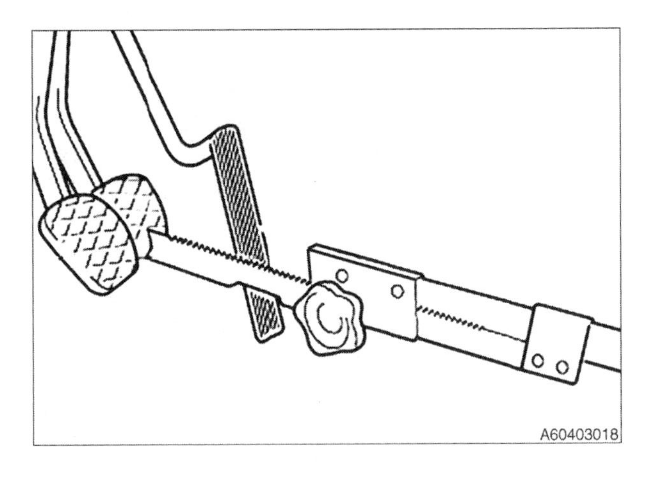

A brake pedal prop is also needed for this procedure.

> ***WARNING —***
> - *The ABS system uses electronic controls and a sophisticated hydraulic unit. Once air enters the hydraulic unit, it cannot be removed using traditional bleeding methods. For this reason, pressure bleed the brakes using VAS 5234 or equivalent pressure bleeder.*
> - *When flushing brake fluid from the system, use extreme care to not let the brake fluid reservoir run dry. If air enters the hydraulic unit, use the VAG 1551 or VAG 1552 scan tool and service tester to bleed the brake system before the vehicle is driven.*

The design of the ABS hydraulic unit requires that brakes be bleed in the following (non-traditional) order:

- Right front
- Left front
- Right rear
- Left rear

ABS versions

For effective brake fluid flushing and bleeding, identify the version of Bosch ABS system installed in vehicle. Bleeding procedure for the two systems differs.

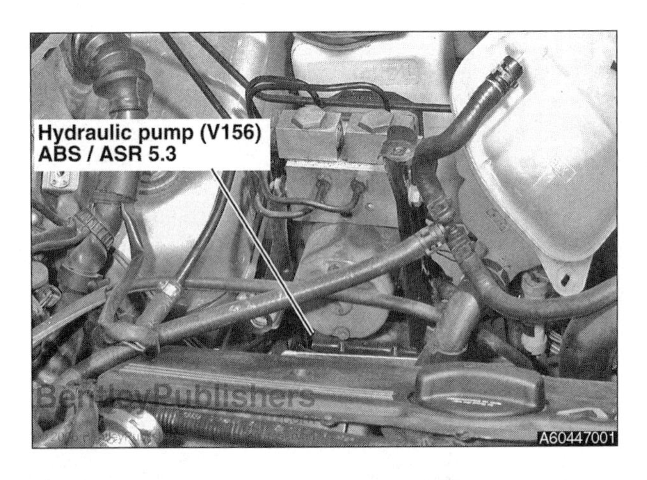

Bosch ABS / ASR 5.3. System is equipped with hydraulic pump (V156), located under ABS hydraulic unit, left side engine compartment. ABS control module (J104) is separate unit under floor covering in right footwell.

> ***WARNING —***
> - *Bleeding procedure for models with ABS / ASR 5.3 cannot be accomplished without VAG 1551 or VAG 1552.*

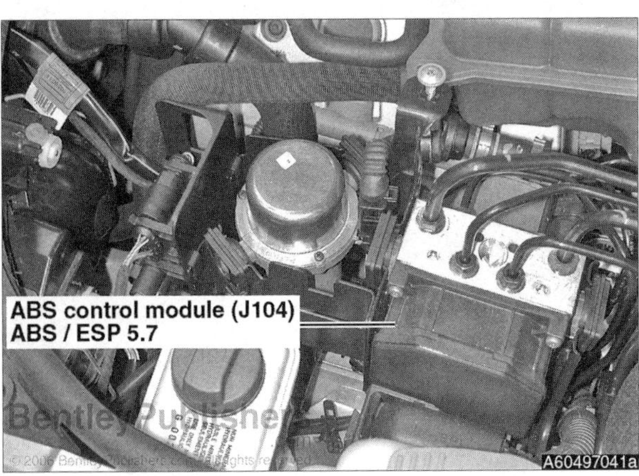

Bosch ABS / ESP 5.7. ABS control module (J104) is combined with ABS hydraulic unit, left side engine compartment. There is no separate hydraulic pump.

BentleyPublishers.com
BentleyPublishers.com—All Rights Reserved

Brake fluid, bleeding or replacing, Bosch ABS / ASR 5.3

In this procedure, brake calipers are first bled using the pressure bleeder. Then the ABS / ASR hydraulic pump is bled using the VAG 1551 or VAG 1552 scan tool. The scan tool activates the hydraulic pump for 10 seconds.

> ***WARNING—***
> - *Bleeding procedure for models with ABS / ASR 5.3 cannot be accomplished without VAG 1551 or VAG 1552.*
> - *Do not allow the brake fluid reservoir to run completely dry at any time during this procedure.*

– Unscrew brake fluid reservoir cap and siphon out as much brake fluid as possible from fluid reservoir.

> ***WARNING—***
> - *Brake fluid is poisonous. Do not siphon fluid by mouth.*

> ***CAUTION—***
> - *Brake fluid damages paint. Immediately wipe up any brake fluid that spills on the vehicle and rinse area with water.*

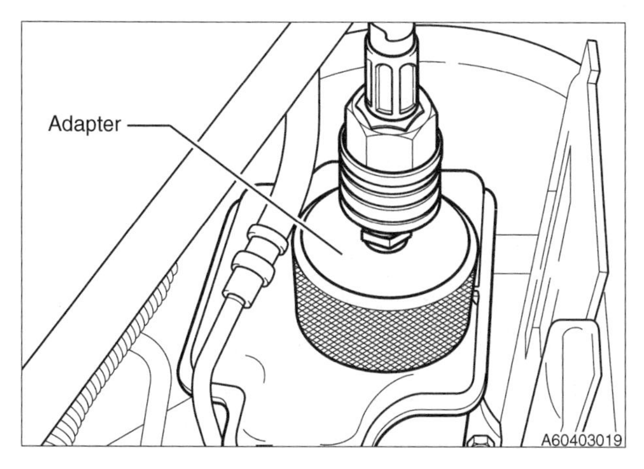

◄ Screw brake bleeder adapter to reservoir. Connect brake bleeder pressure hose to adapter.

NOTE—
- *At least 2 bar (29 psi) of pressure is needed to bleed the ABS hydraulic unit.*

– Install brake pedal prop between driver seat and brake pedal and apply tension.

– Raise car and support safely.

> ***WARNING—***
> - *Make sure the car is stable and well supported at all times. Use a professional automotive lift or jack stands designed for the purpose. A floor jack is not adequate support.*

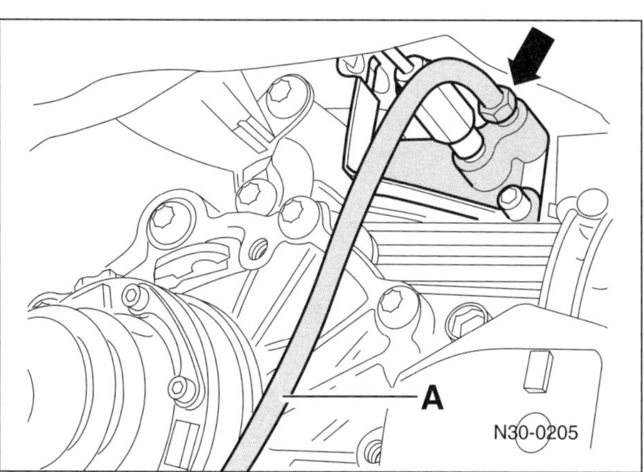

◄ Manual transmission vehicle: Remove clutch slave cylinder bleeder screw cap. Attach brake fluid collecting bottle hose to bleeder screw (**arrow**). Open bleeder and allow approx. 100 cc (4 oz) of fluid to flow out. Close bleeder and reinstall cap.

– Working at each brake caliper, attach brake fluid collecting bottle bleeder hose to bleeder screw. Follow this sequence for bleeding:
- Right front
- Left front
- Right rear
- Left rear

– Open bleeder and allow approx. 250 cc (10 oz) of fluid to flow out of each caliper. Tighten bleeder screws and reinstall caps.

BentleyPublishers.com
BentleyPublishers.com—All Rights Reserved

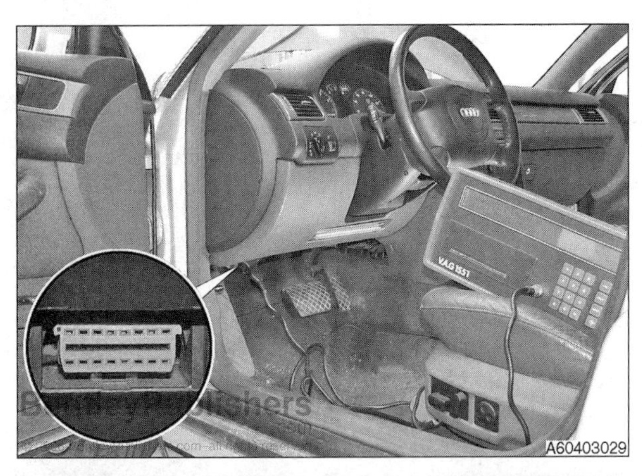

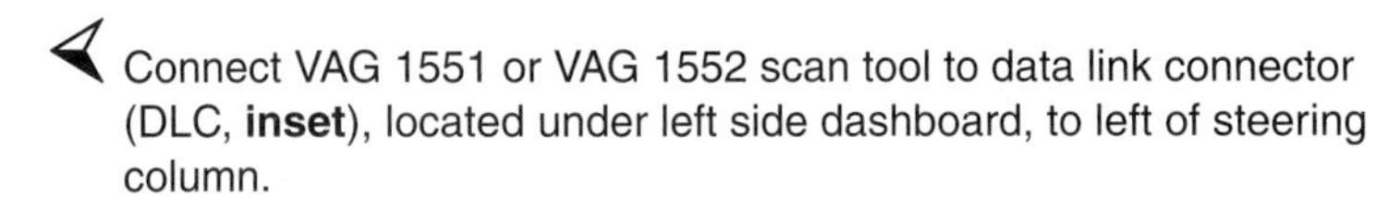

Connect VAG 1551 or VAG 1552 scan tool to data link connector (DLC, **inset**), located under left side dashboard, to left of steering column.

- Perform brake bleeding sequence using scan tool.
- Remove brake bleeding apparatus and pedal prop.

Check brake fluid level and top off if necessary.

- Check brake pedal pressure and free play.

Brake pedal free play	
Maximum pedal play	⅓ of pedal travel

- Manual transmission vehicle: Check clutch operation and clutch pedal free play.

Brake fluid, bleeding or replacing, Bosch ABS / ESP 5.7

Unlike Bosch ABS / ASR version 5.3, ABS / ESP version 5.7 does not incorporate hydraulic pump for traction control (V156). It is not necessary to bleed the system with VAG 1551 or VAG 1552 scan tool unless air enters the ABS hydraulic unit.

> ***WARNING —***
> - *Do not allow the brake fluid reservoir to run completely dry at any time during this procedure.*

- Unscrew brake fluid reservoir cap and siphon out as much brake fluid as possible from fluid reservoir.

> ***WARNING —***
> - *Brake fluid is poisonous. Do not siphon fluid by mouth.*

> ***CAUTION—***
> - *Brake fluid damages paint. Immediately wipe up any brake fluid that spills on the vehicle and rinse area with water.*

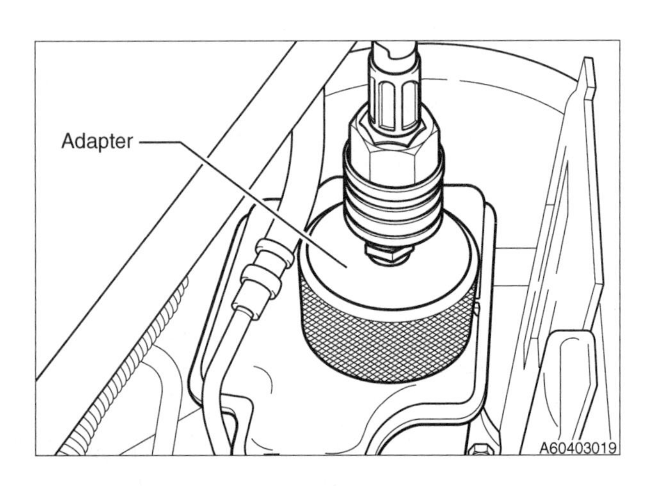

Screw brake bleeder adapter to reservoir. Connect brake bleeder pressure hose to adapter.

NOTE —
- *At least 2 bar (29 psi) of pressure is needed to bleed the ABS hydraulic unit.*

BentleyPublishers.com
BentleyPublishers.com—All Rights Reserved

– Install brake pedal prop between driver seat and brake pedal and apply tension.

– Raise car and support safely.

> ***WARNING—***
> - *Make sure the car is stable and well supported at all times. Use a professional automotive lift or jack stands designed for the purpose. A floor jack is not adequate support.*

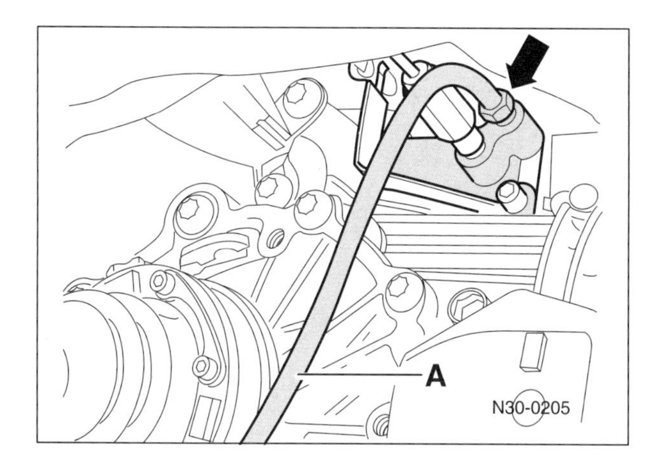

◀ Manual transmission vehicle: Remove clutch slave cylinder bleeder screw cap. Attach brake fluid collecting bottle hose to bleeder screw (**arrow**). Open bleeder and allow approx. 100 cc (4 oz) of fluid to flow out. Tighten bleeder and reinstall cap.

– Working at each brake caliper, attach brake fluid collecting bottle bleeder hose to bleeder screw. Follow this sequence for bleeding:
- Right front
- Left front
- Right rear
- Left rear

– Open bleeder and allow approx. 250 cc (10 oz) of fluid to flow out of each caliper. Tighten bleeders and reinstall caps.

– Remove brake bleeding apparatus and pedal prop.

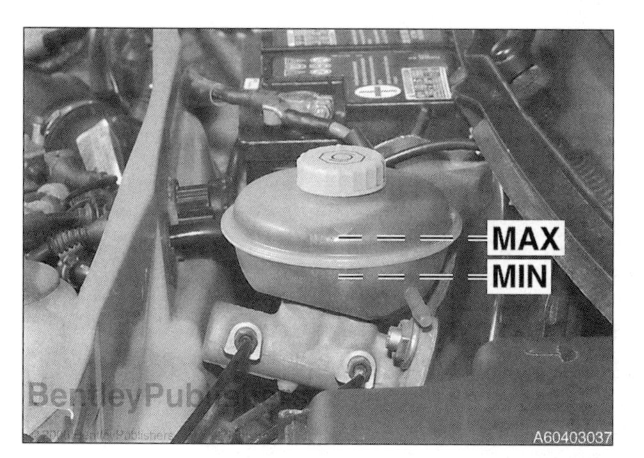

◀ Check brake fluid level and top off if necessary.

– Check brake pedal pressure and free play.

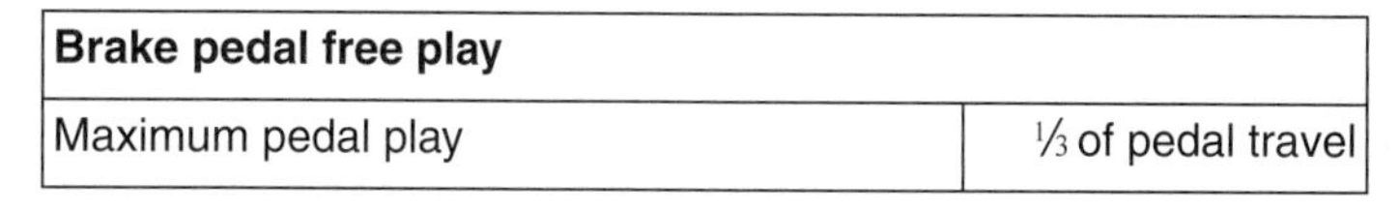

Brake pedal free play	
Maximum pedal play	⅓ of pedal travel

– Manual transmission vehicle: Check clutch operation and clutch pedal free play.

BentleyPublishers.com—All Rights Reserved

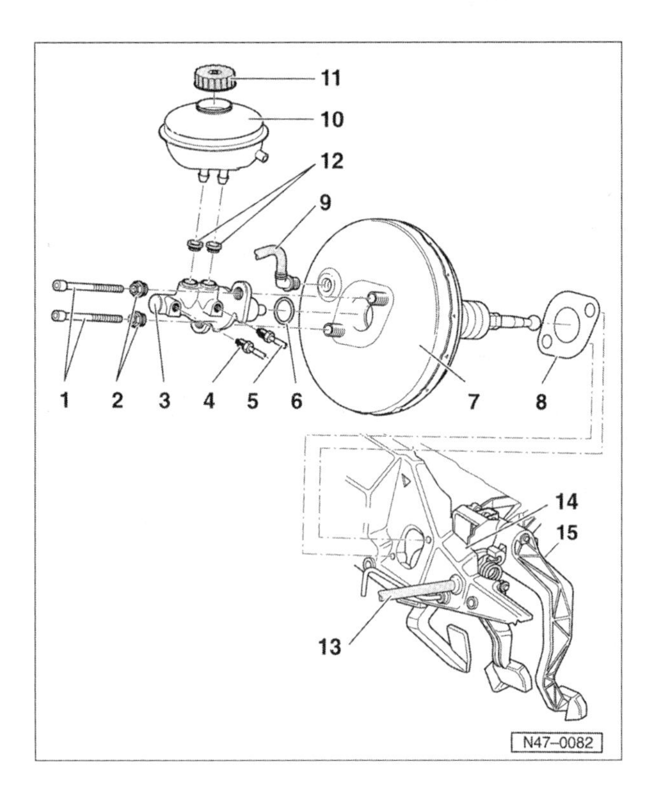

Brake Master Cylinder and Booster

Brake master cylinder and booster assembly

1. T45 Torx, 25 Nm (18 ft-lb)
2. Self-locking nut, 49 Nm (36 ft-lb)
3. Brake master cylinder
4. Brake line, 15 Nm (11 ft-lb)
   • to ABS hydraulic unit
5. Brake line, 15 Nm (11 ft-lb)
   • to ABS hydraulic unit
6. Seal
   • Replace
7. Brake booster
8. Gasket
9. Vacuum hose (with check valve)
   • Insert in brake booster
10. Brake fluid reservoir
11. Cap
12. Sealing plugs
13. Fluid supply hose
   • to Clutch master cylinder
14. Bulkhead
15. Pedal cluster

Brake master cylinder, removing and installing

WARNING *—*
- *The ABS system uses electronic controls and a sophisticated hydraulic unit. Once air enters the hydraulic unit, it cannot be removed using traditional bleeding methods. For this reason, pressure bleed the brakes using VAS 5234 or equivalent pressure bleeder.*
- *When bleeding brake system, use extreme care to not let the brake fluid reservoir run dry. If air enters the hydraulic unit, use the VAG 1551 or VAG 1552 scan tool and service tester to bleed the brake system before the vehicle is driven.*
- *Do not disassemble the brake master cylinder. It is not serviceable.*

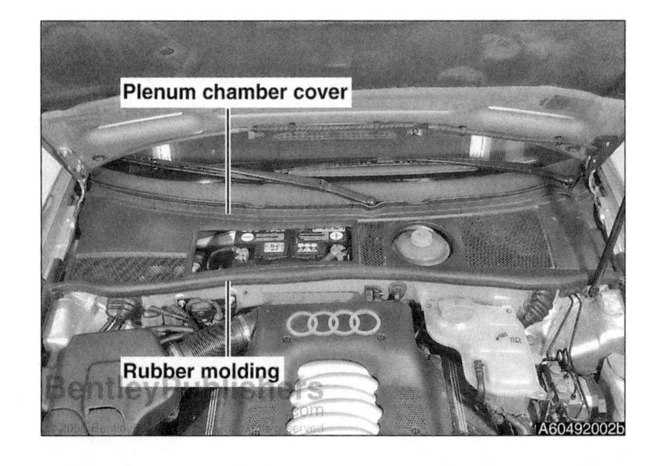

 Pull off rubber molding at engine compartment rear bulkhead. Pull plenum chamber cover forward and remove.

– Unscrew brake fluid reservoir cap and siphon out as much brake fluid as possible from fluid reservoir.

WARNING *—*
- *Brake fluid is poisonous. Do not siphon fluid by mouth.*

CAUTION *—*
- *Brake fluid damages paint. Immediately wipe up any brake fluid that spills on the vehicle and rinse area with water.*

BentleyPublishers.com
BentleyPublishers.com—All Rights Reserved

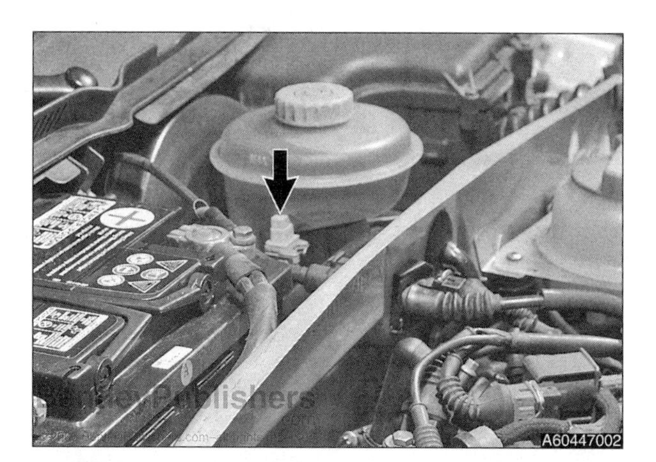

– Manual transmission vehicle: Disconnect clutch master cylinder supply hose and seal with plug.

◄ Disconnect brake fluid level warning switch harness connector (**arrow**).

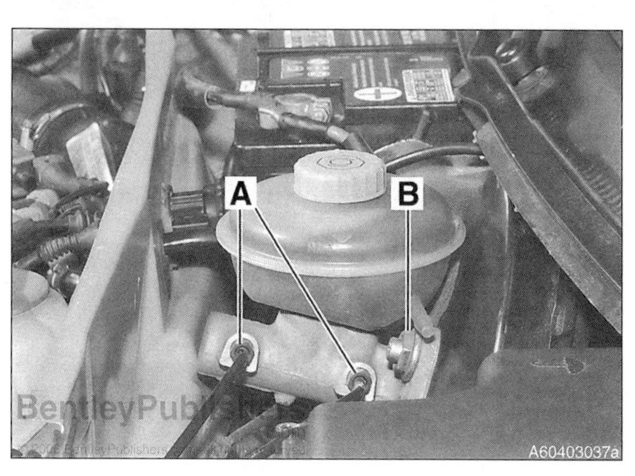

◄ Working below brake fluid reservoir:

- Disconnect brake lines (**A**) at brake master cylinder. Immediately seal brake lines and plug brake fluid ports.
- Remove brake master cylinder M16 mounting nuts (**B**).

– Carefully remove brake master cylinder from brake booster.

– When installing, pay particular attention to the following points:

- When installing brake master cylinder on brake booster, make sure pushrod is correctly located in brake master cylinder.
- Apply slight pressure to brake pedal to move pushrod toward brake master cylinder; this makes it easier to guide brake master cylinder onto pushrod.

– Flush and bleed brake system. See **Bleeding Brakes** in this repair group.

Brake booster, removing and installing

WARNING *—*

- *The ABS system uses electronic controls and a sophisticated hydraulic unit. Once air enters the hydraulic unit, it cannot be removed using traditional bleeding methods. For this reason, pressure bleed the brakes using VAS 5234 or equivalent pressure bleeder.*
- *When bleeding brake system, use extreme care to not let the brake fluid reservoir run dry. If air enters the hydraulic unit, use the VAG 1551 or VAG 1552 scan tool and service tester to bleed the brake system before the vehicle is driven.*

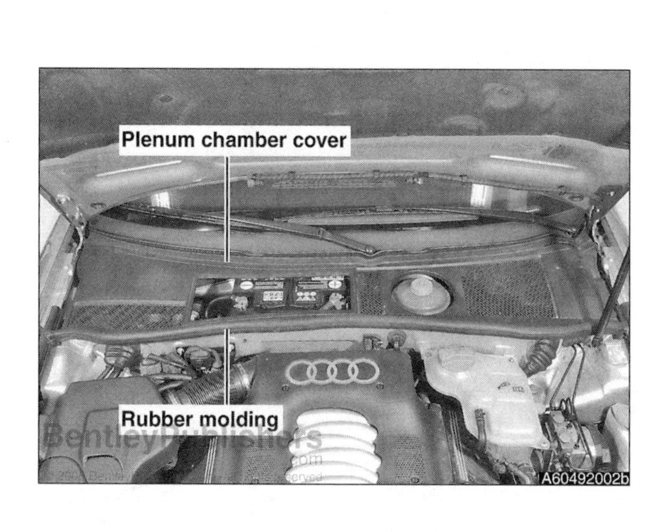

◄ Pull off rubber molding at engine compartment rear bulkhead. Pull plenum chamber cover forward and remove.

– Unscrew brake fluid reservoir cap and siphon out as much brake fluid as possible from fluid reservoir.

WARNING *—*

- *Brake fluid is poisonous. Do not siphon fluid by mouth.*

CAUTION *—*

- *Brake fluid damages paint. Immediately wipe up any brake fluid that spills on the vehicle and rinse area with water.*

BentleyPublishers.com
BentleyPublishers.com—All Rights Reserved

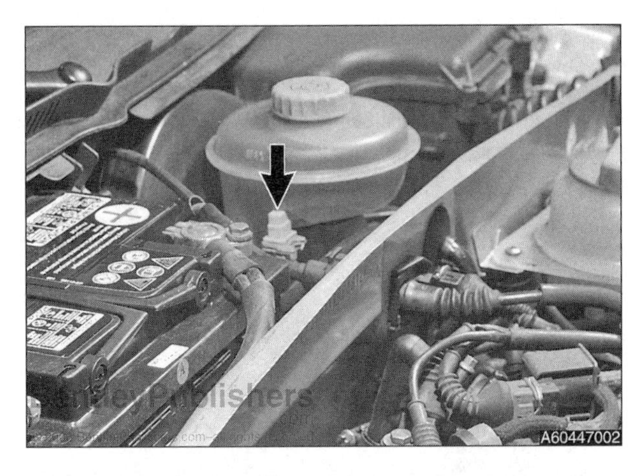

– Manual transmission vehicle: Disconnect clutch master cylinder supply hose and seal with plug.

Disconnect brake fluid level warning switch harness connector (**arrow**).

– Disconnect brake booster vacuum hose.

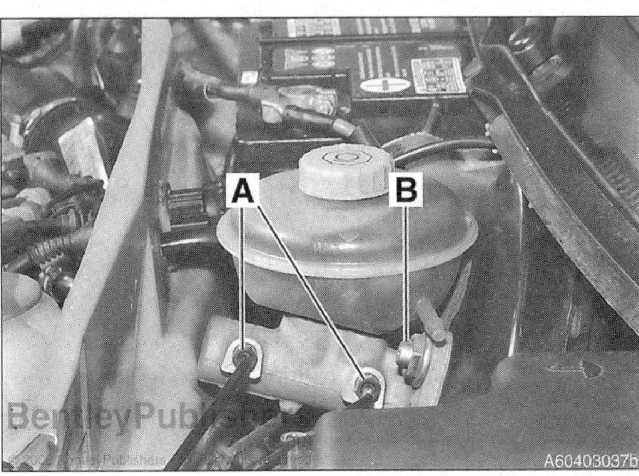

Working below brake fluid reservoir:

- Disconnect brake lines (**A**) at brake master cylinder. Immediately seal brake lines and plug brake fluid ports.
- Remove Torx T45 bolts (**B**) mounting master cylinder and booster assembly to bulkhead.

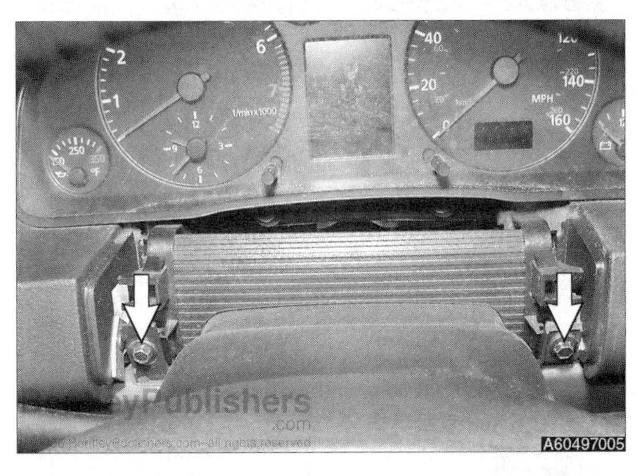

Working in passenger compartment, tilt steering column down and pull off trim under instrument cluster. Remove steering column trim panel fasteners (**arrows**).

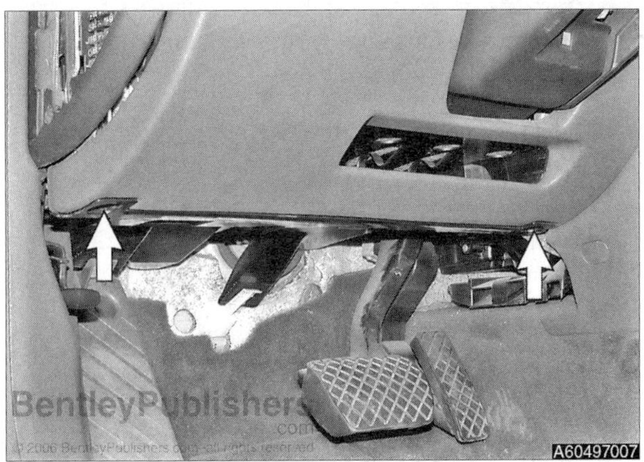

Remove lower trim fasteners (**arrows**) and pull down trim panels under steering column to gain access to pedal cluster.

BentleyPublishers.com
BentleyPublishers.com—All Rights Reserved

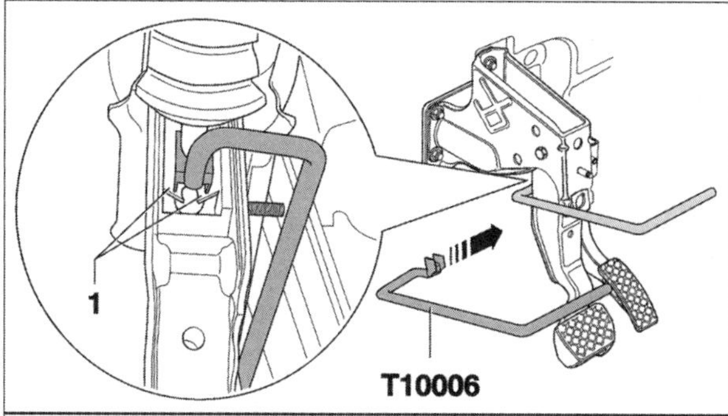

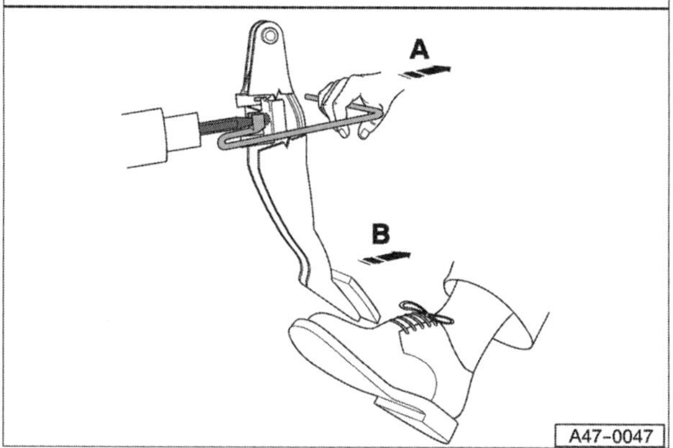

– Working under dashboard above pedals, remove brake light switch.

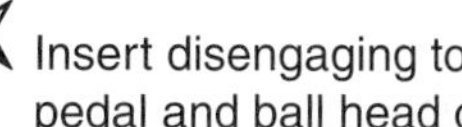

◀ Insert disengaging tool (VAG special tool T10006) between brake pedal and ball head of brake booster pushrod (**1**).

- Pull back on tool and brake pedal simultaneously (**2**).
- Hold brake pedal securely. Pry retaining tabs from ball head by pulling lightly on tool (**arrow A**).
- Keep tension on tool so that retaining tabs do not reengage.
- Pull brake pedal from ball head (**arrow B**).

– Remove brake master cylinder and booster from plenum chamber.

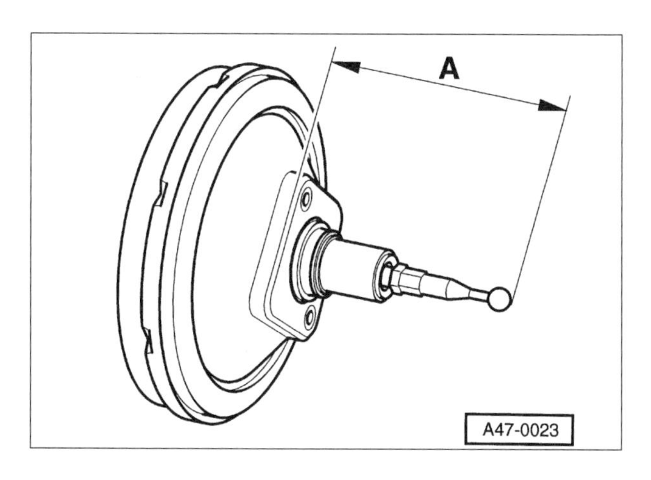

◀ Prior to installation, check and reset brake booster pushrod length.

Brake booster pushrod length	
Measurement **A**: • Left hand drive (LHD) vehicle • Right hand drive (RHD) vehicle	 159 ± 0.5 mm (6.26 ± 0.02 in) 173.2 ± 0.5 mm (6.82 ± 0.02 in)

NOTE—

- *When measuring, make sure ball head is at right angle to surface of brake booster.*
- *Measure from end of ball head to metal base of booster, not to gasket.*

– Installation is reverse of removal. Pay particular attention to the following points:

- Reinstall and adjust brake light switch and brake and clutch pedal cruise control vent valve switches.
- Flush and bleed brake system. See **Bleeding Brakes** in this repair group.

BentleyPublishers.com
BentleyPublishers.com—All Rights Reserved

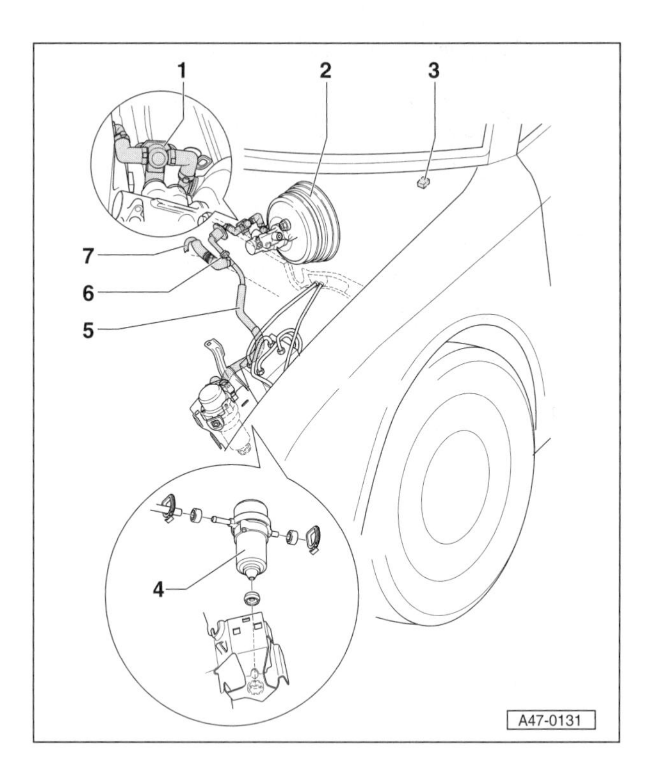

Brake booster vacuum pump assembly

In 3.0 liter V6 models and all V8 models, an electric brake booster vacuum pump is used to supplement intake manifold vacuum.

1. Pressure sensor (G 294)
2. Brake booster
3. Brake booster vacuum pump relay (J569)
4. Brake booster vacuum pump (V192)
5. Vacuum hose
6. Check valve
7. Vacuum hose
   - Connects to engine

Brake booster vacuum pump (V192), removing and installing

If problems with vacuum supply to brake booster are encountered, determine cause by means of a vacuum pump function test and leak test. If a malfunction is found, replace pump.

– Remove protective cover.

◄ Unplug electrical connector (**arrow**).

– Detach vacuum hose from pump.

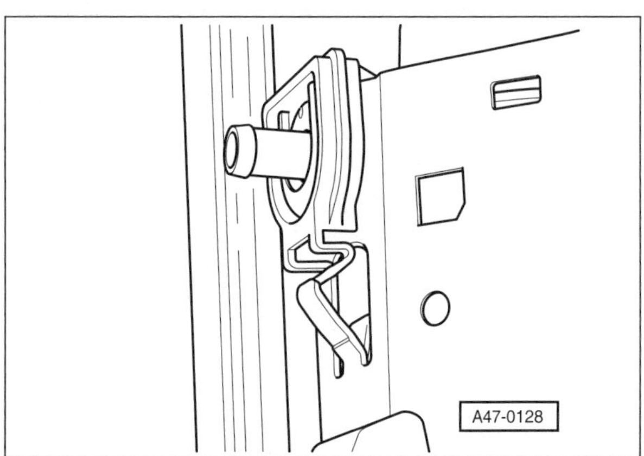

◄ Press clip out of retainer by hand and lift out electric vacuum pump from above.

– Installation is reverse of removal.

BentleyPublishers.com
BentleyPublishers.com—All Rights Reserved

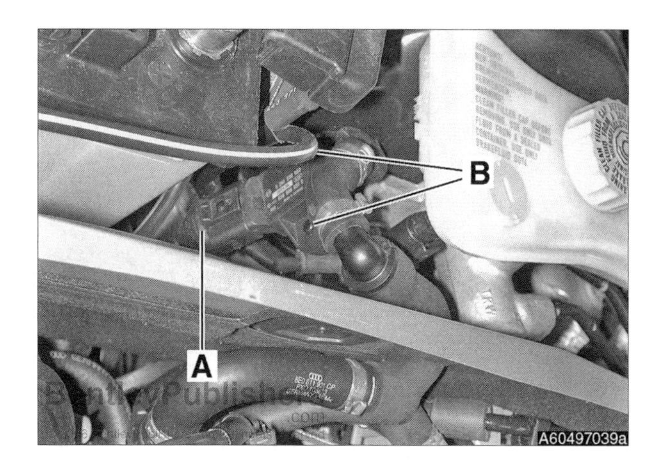

Brake booster vacuum pump pressure sensor (G294), removing and installing

– Remove plenum chamber cover.

< Working to left of battery:

- Pull pressure sensor electrical connector (**A**) downward to disconnect.
- Remove sensor mounting screws (**B**) and remove sensor.

NOTE —

- *The sensor may also be attached by means of clips.*

– Installation is reverse of removal.

Brake booster vacuum pump relay (J569), removing and installing

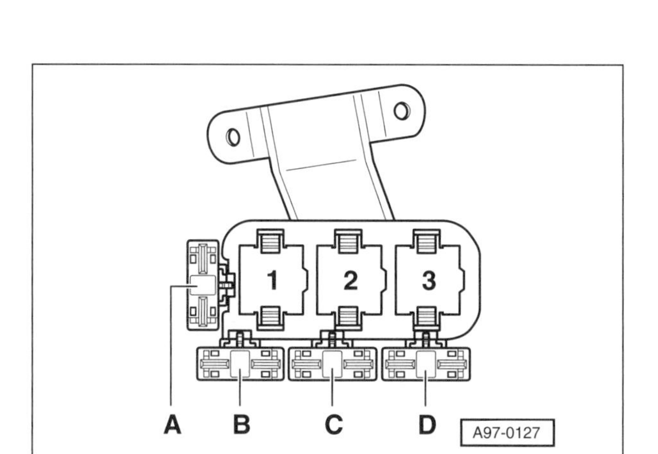

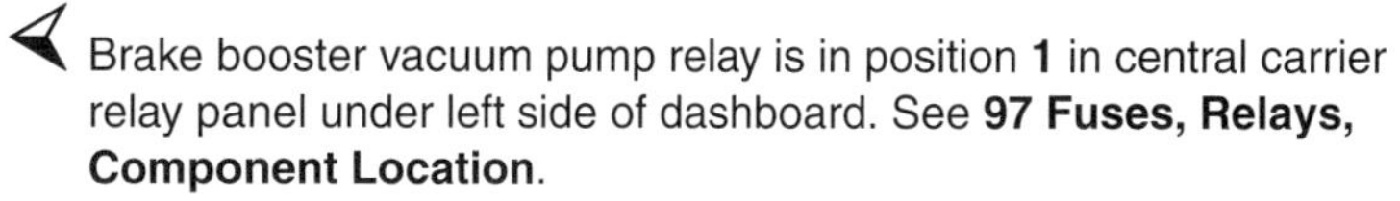

< Brake booster vacuum pump relay is in position **1** in central carrier relay panel under left side of dashboard. See **97 Fuses, Relays, Component Location**.

BentleyPublishers.com
BentleyPublishers.com—All Rights Reserved

48 Steering

GENERAL

This repair group includes a description of the steering system, an overview of steering column and steering system components and steering wheel removal. For more information, see:

- **45 Antilock Brakes (ABS)** for steering angle sensor removal
- **69 Seat Belts, Airbags** for airbag procedures

Power steering fluid

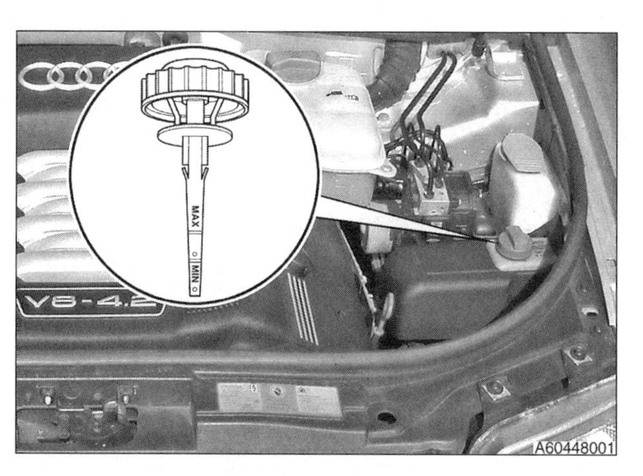

Working in engine compartment, unscrew power steering reservoir cap and wipe off dipstick.

– Fully close and reopen cap. Check level on dipstick. See **03 Maintenance** for more information.

Power steering fluid	
Capacity	0.85L (0.89qt)
Type	Hydraulic oil G002 000 G002 012

> ***CAUTION—***
> *• Do not use automatic transmission fluid (ATF).*

BentleyPublishers.com
BentleyPublishers.com—All Rights Reserved

Warnings

> ***WARNING —***
> - *Do not remove or repair airbag unit unless you have proper training, qualifications and tools to do the work safely.*
> - *Disconnect battery ground strap before performing any work on the airbag system. No waiting time is required after disconnecting the battery. When connecting airbag system to a voltage source, no one should be in the vehicle interior.*
> - *Before handling (touching) the airbag unit, the technician must be electrostatically discharged. This is accomplished by touching grounded metal like e.g. water pipes, heating pipes or metal carriers.*
> - *Always install airbag units in the vehicle, as soon as they are removed from the packaging.*
> - *If installation work is interrupted, immediately return the airbag unit to its original packaging.*
> - *Do not leave an airbag unit unattended.*
> - *Once removed, lay the airbag down with impact absorbing pad facing up.*
> - *Do not install airbag units that have fallen onto a hard surface, or which have signs of damage.*
> - *Airbag units that have not ignited must be returned in original packaging for proper recycling/disposal. Contact your supplier or the manufacturer for more information.*
> - *Ignited airbag units can be disposed of with industrial waste.*

BentleyPublishers.com
BentleyPublishers.com—All Rights Reserved

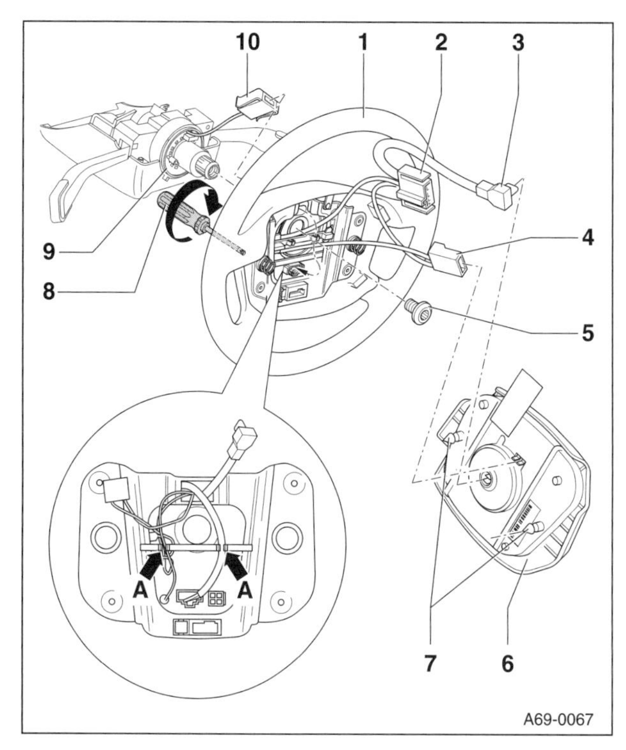

STEERING WHEEL AND COLUMN

Steering wheel, removing and installing

- Remove airbag unit. See **69 Seat Belts, Airbags**.
- Put steering wheel in center position (wheels straight ahead).
- Remove bolt (**5**).
- Disconnect connectors (**2**) and (**10**) and remove wheel.
- When installing, attach steering wheel in center position (wheels straight ahead).
- Attach electrical connectors (**2**) and (**10**). Make sure you hear plugs completely engage.
- Replace bolt (**5**) and tighten.

Tightening torque	
Steering wheel to steering column	50 Nm (37 ft-lb)

- Install airbag unit.

Sport steering wheel, removing and installing

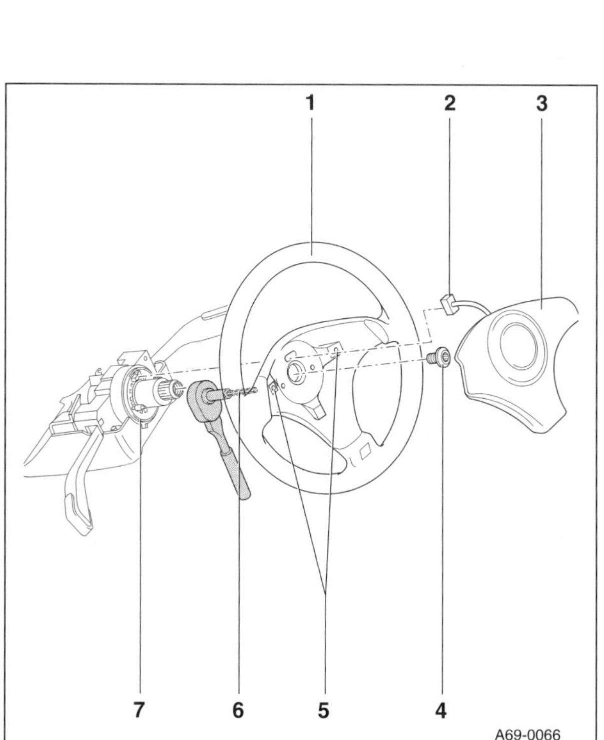

- Remove airbag unit. See **69 Seat Belts, Airbags**.
- Put steering wheel in center position (wheels straight ahead).
- Remove bolt (**4**) and remove wheel.
- When installing, attach steering wheel in center position (wheels straight ahead).
- Replace bolt (**4**) and tighten.

Tightening torque	
Steering wheel to steering column	50 Nm (37 ft-lb)

- Install airbag unit.

BentleyPublishers.com
BentleyPublishers.com—All Rights Reserved

Steering column assembly (electrically adjustable)

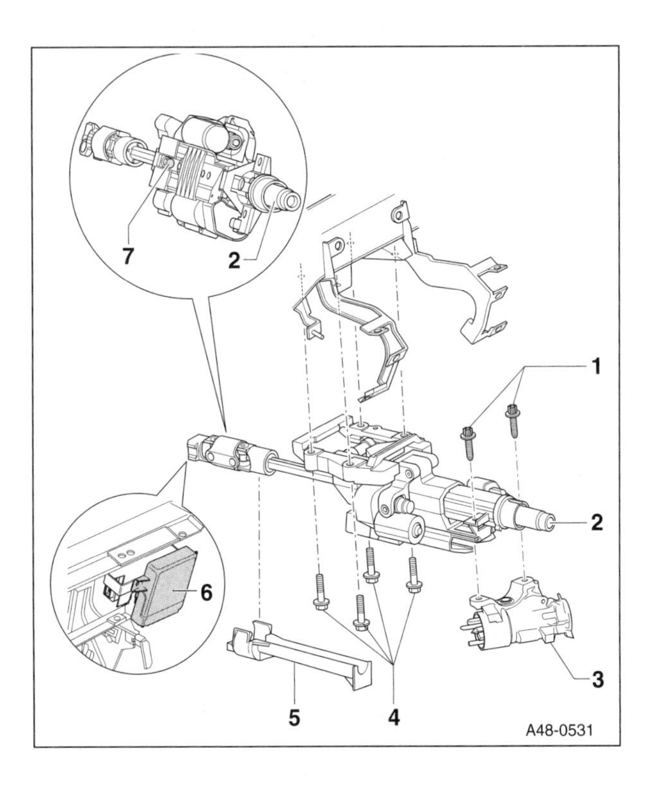

The steering column is only available as a complete unit.

1. Shear bolt, 20 Nm (15 ft-lb)
2. Steering column
3. Ignition/starter switch
4. Bolt, 22 Nm (16 ft-lb)
5. Bracket (used for transport)
6. Steering column height adjustment control module (J352)
    - No self-diagnosis capability.
    - Sliding spline between upper and lower steering column must not be separated under any circumstances.
    - Sliding over a range in excess of 5 cm can lead to steering column damage.
7. Tolerance adjusting screw, 22 Nm (16 ft-lb)
    - Provided with electrically adjustable steering column
    - Tighten 4 bolts for attaching steering column to dashboard brace to torque, before tightening tolerance adjusting screw.
    - Do not apply force to steering column or steering wheel while tightening tolerance adjusting screw.

Steering column stalk switch, removing and installing

WARNING—
- *Before removing steering wheel, place wheels in straight-ahead position.*
- *Observe safety precautions for working on airbags. See* **69 Seat Belts, Airbags**.
- *If these procedures are not observed, the airbag system may not function properly during vehicle operation.*

– Remove airbag unit. See **69 Seat Belts, Airbags**.

– Remove steering wheel. See **Steering Wheel, removing and installing** in this repair group.

– Remove steering angle sensor in vehicles with Electronic Stabilization Program (ESP). See **45 Antilock Brakes** for more information.

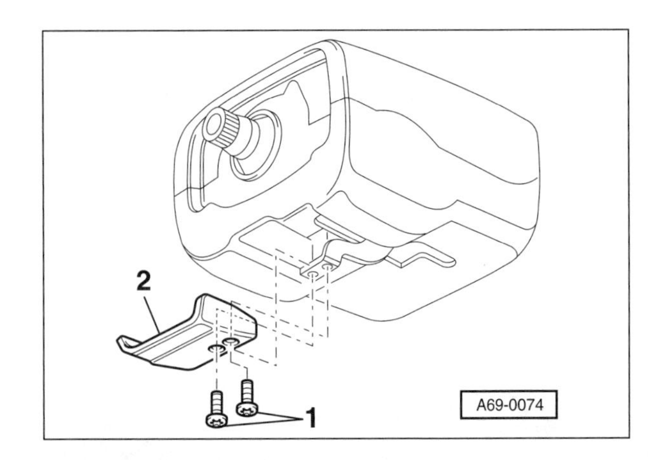

◀ Remove bolt (**1**).

– Remove handle (**2**).

BentleyPublishers.com
BentleyPublishers.com—All Rights Reserved

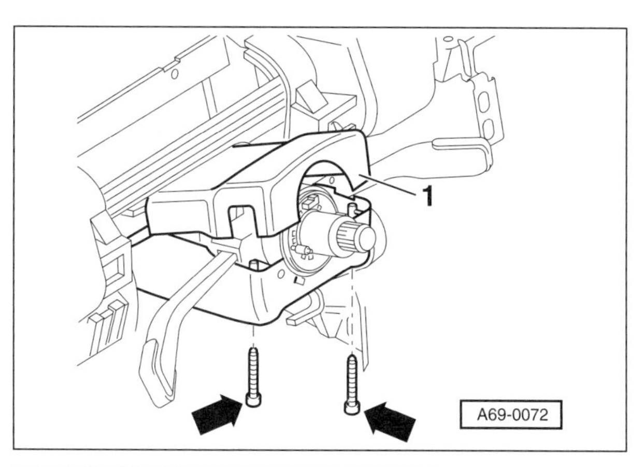

◀ Remove two Phillips head screws (**arrows**).

– Remove upper steering column trim (**1**).

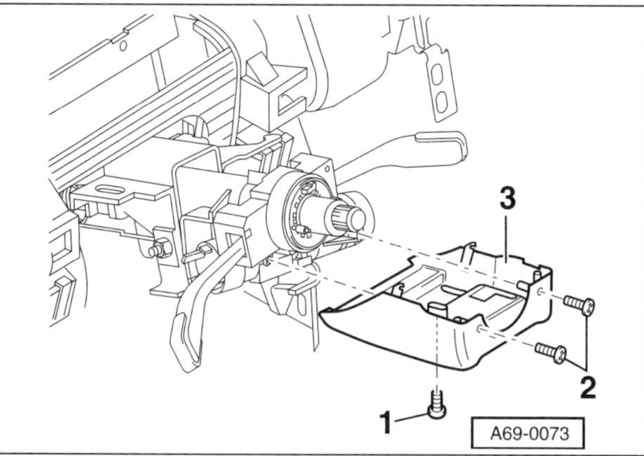

◀ Remove bolts (**2**).

– Remove bolt (**1**).

– Remove lower steering column trim (**3**).

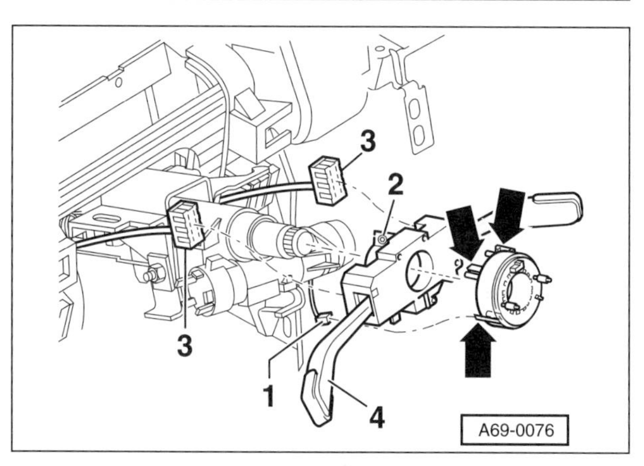

◀ Disconnect connector (**1**).

– Loosen bolt (**2**).

– Disconnect connector (**3**).

– Remove steering column stalk switch (**4**).

Steering column, removing and installing

The steering column is only available as a complete unit. Servicing is not possible.

> ***WARNING —***
> - *Before removing steering wheel, place wheels in straight-ahead position.*
> - *Observe safety precautions for working on airbags. See* **69 Seat Belts, Airbags**.
> - *If these procedures are not observed, the airbag system may not function properly during vehicle operation.*

BentleyPublishers.com
BentleyPublishers.com—All Rights Reserved

– Remove airbag unit. See **69 Seat Belts, Airbags**.

– Remove steering wheel. See **Steering Wheel, removing and installing** in this repair group.

– Remove steering angle sensor in vehicles with Electronic Stabilization Program (ESP). See **45 Antilock Brakes** for more information.

– Remove steering column stalk switch. See Steering column stalk switch, removing and installing in this repair group.

– Disengage steering column adjustment.

A48-0061

◄ Release cable ties and connectors located at steering column.

NOTE—

- *For vehicles with automatic transmission, locking cable must be disengaged.*

– Place selector lever in PARK position.

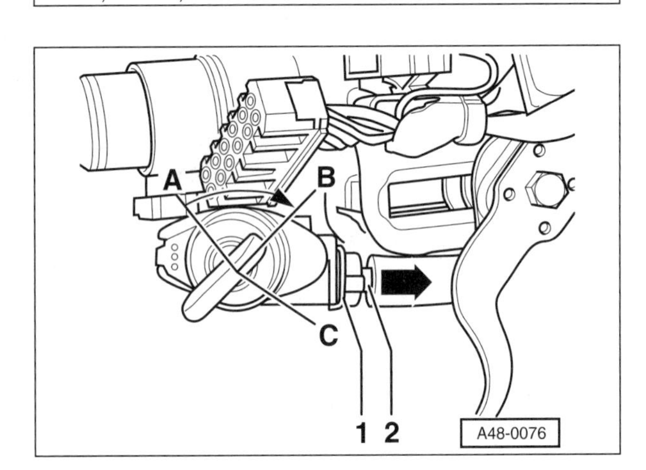

◄ Switch ignition ON (position **B**).

– Lift locking lever (**1**) slightly and pull shift interlock cable (**2**) from ignition lock housing.

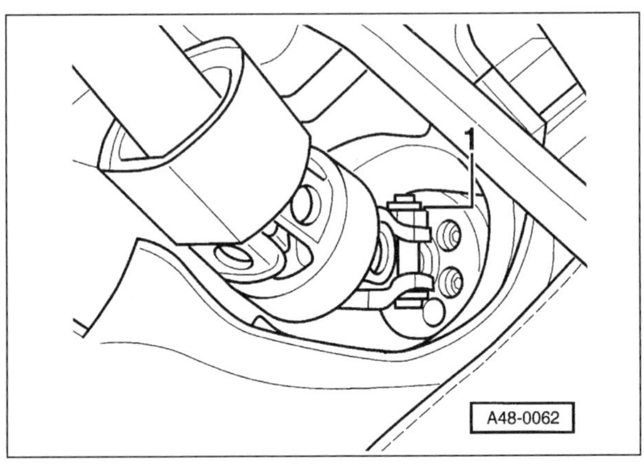

◄ Working at base of steering column, remove nut (**1**) at steering shaft universal joint.

– Rotate eccentric pinch bolt (Torx T50) clockwise to relieve tension, then remove bolt.

– Disengage steering column adjustment.

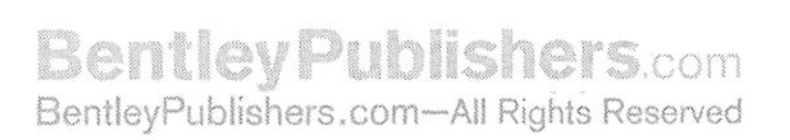
BentleyPublishers.com
BentleyPublishers.com—All Rights Reserved

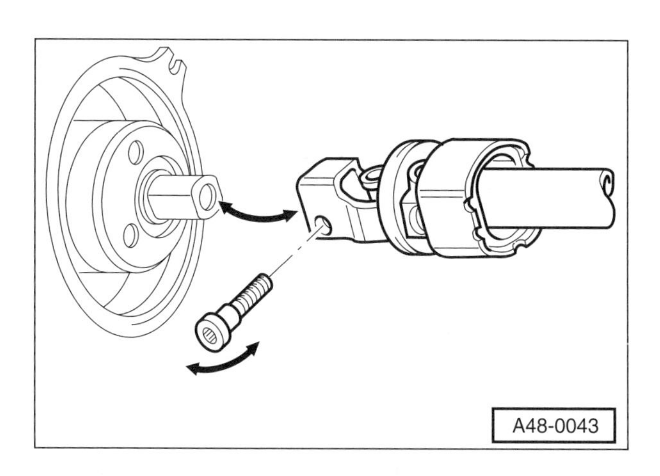
A48-0043

◀ Swing universal joint to side.

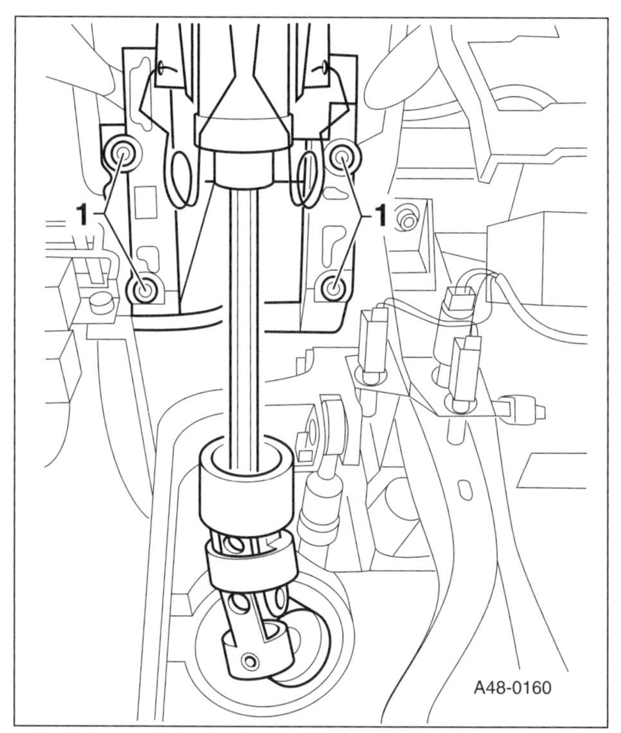

A48-0160

◀ Remove Allen bolts (**1**) for steering column and remove steering column.

– Check steering column for damage.

Installing

– Attach steering column to dashboard brace using 4 Allen bolts (**1**).

– Disengage steering column adjustment.

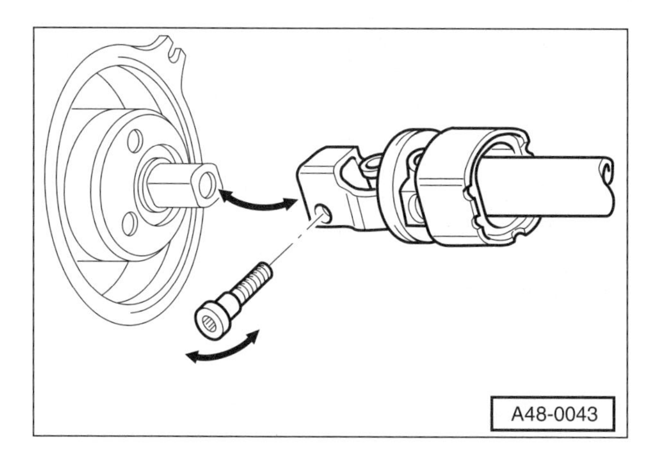
A48-0043

◀ Attach steering shaft universal joint to steering rack pinion gear.

– Insert pinch bolt (Torx T50) through lower part and pretension counterclockwise. Insert nut and tighten to 30 Nm (22 ft-lb).

Tightening torque	
Steering shaft pinch bolt to steering gear	30 Nm (22 ft-lb)

NOTE—

- *For vehicles with automatic transmission, engage shift interlock cable.*

BentleyPublishers.com
BentleyPublishers.com—All Rights Reserved

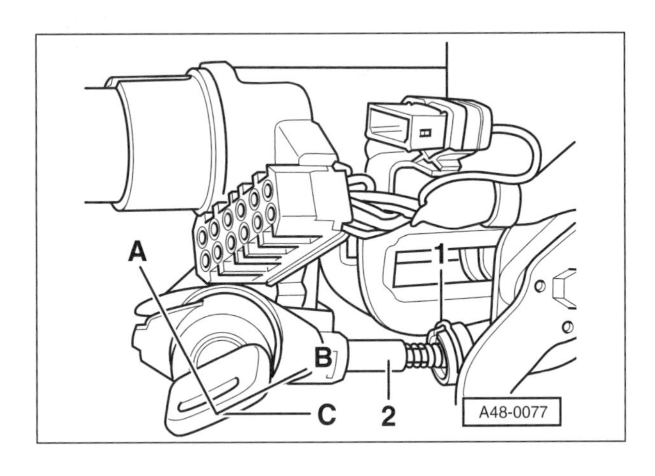

◀ Switch ignition key to ON (position **B**).

– Place selector lever in PARK.

– Insert shift interlock cable (**2**) in ignition lock housing. Make sure locking bracket (**1**) engages.

Shift interlock function check:

The ignition interlock prevents or allows the key and shift mechanism to move depending on their respective positions. Follow the steps below to check the ignition interlock.

– Turn ignition key **ON** then depress brake pedal and hold:
  - Shifting from **P** must be possible without "snagging" when button on shift lever is depressed.
  - Removing the ignition key must not be possible in any driving position except in **P**.

– Move shift lever to **P**.
  - It must be possible to turn Ignition key **OFF** without binding.

– Remove ignition key.
  - It must not be possible to move shift lever from **P**.

– To continue with procedure, install and position steering column stalk trim so that gaps in trim are uniform.

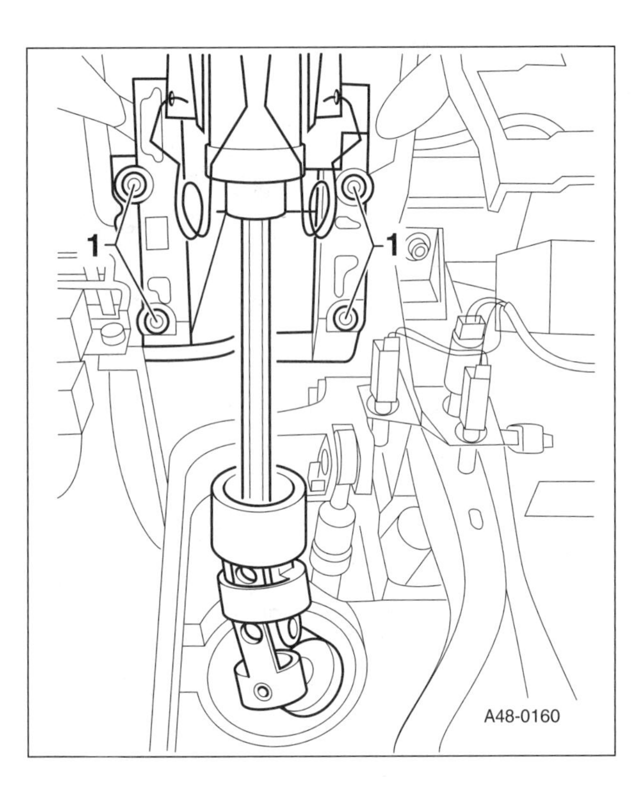

◀ Tighten socket head bolts (**1**) for steering column.

Tightening torque	
Steering column bracket to dashboard brace	25 Nm (18 ft-lb)

Vehicles with electrically adjustable steering column:

– Tighten tolerance adjusting screw.

– Connect wiring of servo motors.

– Use cable tie to attach wires to steering column.

All vehicles (continued)

– Install steering column trim.

– Install steering wheel.

– Install airbag module.

BentleyPublishers.com
BentleyPublishers.com—All Rights Reserved

STEERING RACK

Power steering rack components (1998 models)

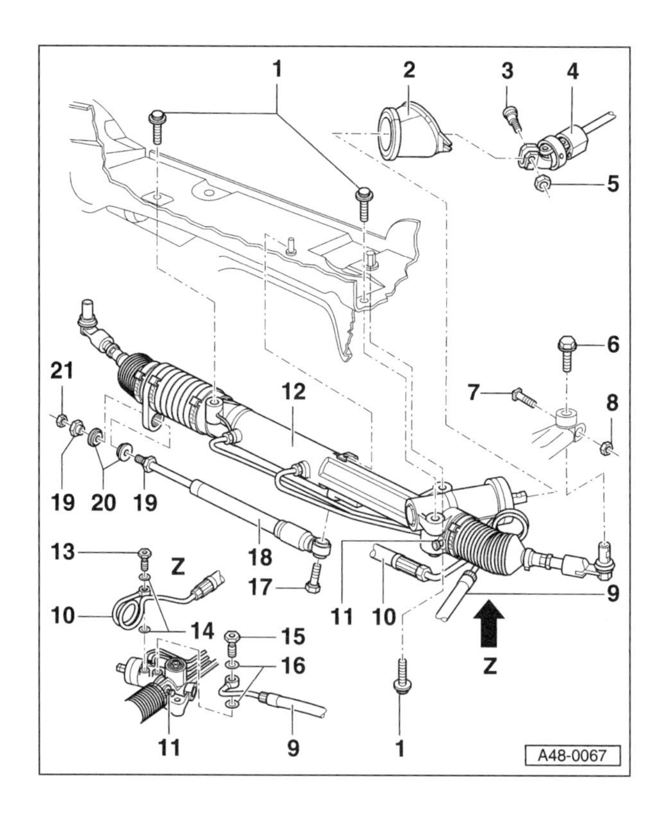

1. Bolt, 67 Nm (49 ft-lb)
2. Boot seal
   - Check for tears and scuffing
3. Eccentric pinch bolt, Torx T50
   - To release tension, turn clockwise
   - To tension, turn counterclockwise
4. Steering column
5. Self-locking nut, 30 Nm (22 ft-lb)
   - Always replace
6. Screw and washer assembly, 7 Nm (5 ft-lb)
   - For adjusting toe-in
7. Bolt
8. Self-locking nut, 45 Nm (33 ft-lb)
   - Always replace
9. Return hose
10. Expansion hose
11. Cylinder bolt, 13 Nm (10 ft-lb)
    - Locking screw for centering steering
12. Power steering rack with tie rods
13. Banjo bolt, 40 Nm (30 ft-lb)
    - With integrated check-valve
14. Seal
    - Always replace
15. Banjo bolt, 47 Nm (35 ft-lb)
16. Seal
    - Always replace
17. Bolt, 35 Nm (26 ft-lb)
18. Steering damper
    - V6 models only
19. Socket
20. Rubber bushing
    - Two-piece
21. Nut, tighten to 10 Nm (7 ft-lb)
    - To loosen and tighten, counterhold hex on steering damper piston rod

BentleyPublishers.com
BentleyPublishers.com—All Rights Reserved

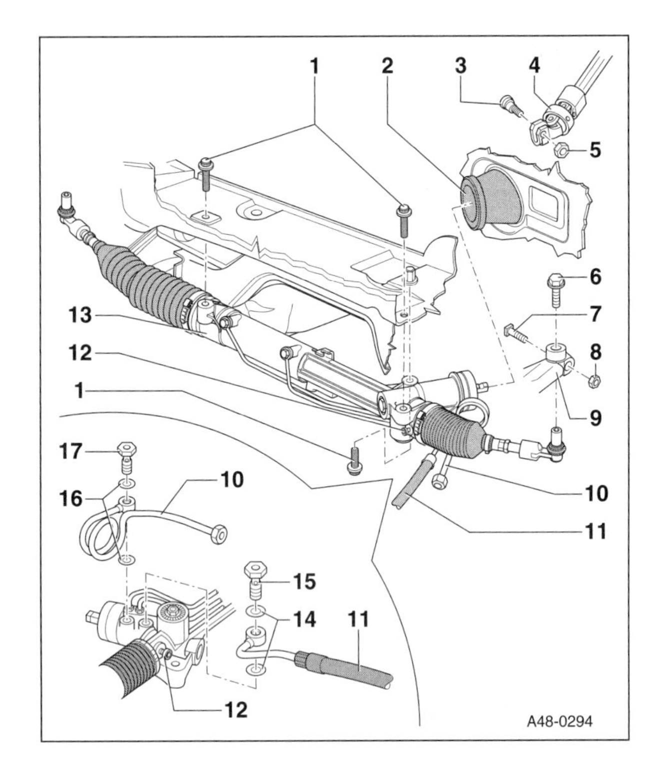

Power steering rack components (1999 - 2004)

1. Bolt, 67 Nm (49 ft-lb)
2. Boot seal
   - Check for tears and scuffing
3. Eccentric pinch bolt, Torx T50
   - To release tension, turn clockwise
   - To tension, turn counterclockwise
4. Steering column
5. Self-locking nut, 30 Nm (22 ft-lb)
   - Always replace
6. Screw and washer assembly, 7 Nm (5 ft-lb)
   - For adjusting toe-in
7. Bolt
8. Self-locking nut, 45 Nm (33 ft-lb)
   - Always replace
9. Steering knuckle from wheel bearing housing
10. Pressure line, 40 Nm (30 ft-lb)
11. Return hose
    - Note installation position at steering rack
12. Cylinder bolt, 13 Nm (10 ft-lb)
    - Locking screw for centering steering
13. Power steering rack with tie rods
14. Seal
    - Always replace
15. Banjo bolt, 47 Nm (35 ft-lb)
16. Seal
    - Always replace
17. Banjo bolt, 40 Nm (30 ft-lb)
    - With integrated check-valve

Tie rod, removing and installing

NOTE—
- *Left and right tie rods are identical.*
- *Tie rod can be removed and installed with steering rack installed.*

– Remove front wheel.

∢ Remove upper bolt (**1**) and pinch bolt (**2**).

– Detach tie rod from steering knuckle.

BentleyPublishers.com—All Rights Reserved

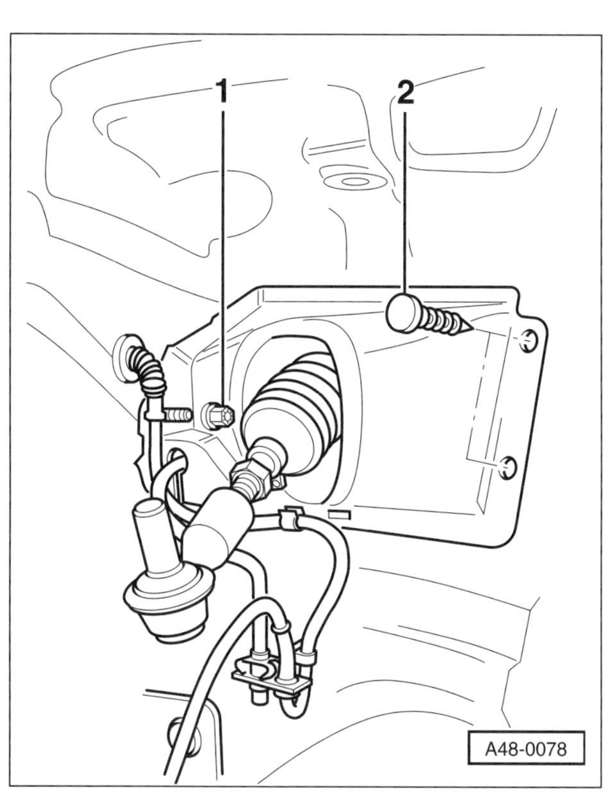

- Unscrew plastic nut (**1**).
- Pry out clip (**2**).
- Remove cover for tie rod.

NOTE —
- *Shown without wheel bearing housing and strut for the sake of illustration.*

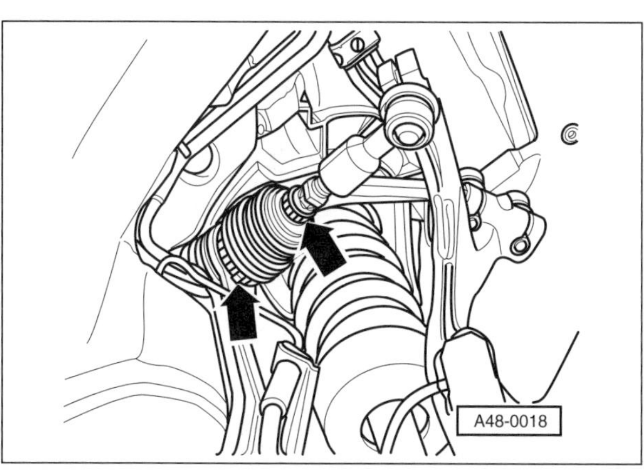

- Cut rack boot clamps.
- Pull boot as far toward outside as possible.

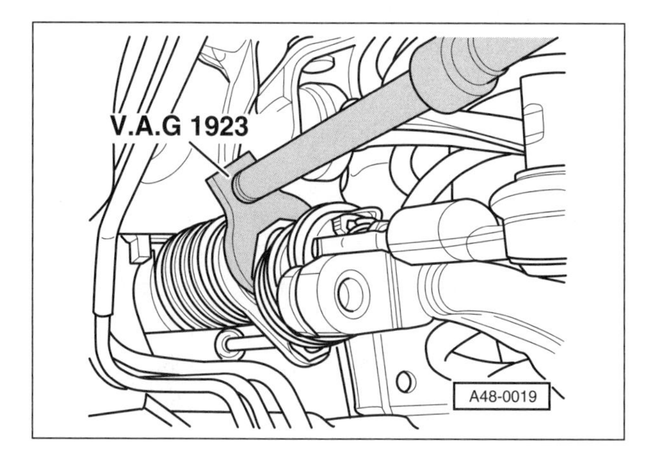

- Remove tie rod using special tool VAG 1923 (or equivalent).
- Replace boot or tie rod depending on extent of damage.
- Attach tie rod to to steering rack.

Tightening torque	
Tie rod to steering rack	100 Nm (74 ft-lb)

- When installing boot:
  - Smaller diameter of boot snaps into groove at tie rod.
  - Do not twist boot.
  - Replace boot clamps.
- Check alignment.

BentleyPublishers.com
BentleyPublishers.com—All Rights Reserved

Steering Pump

Power steering pump (2.8 liter engine), removing and installing

Unclip upper engine cover and remove.

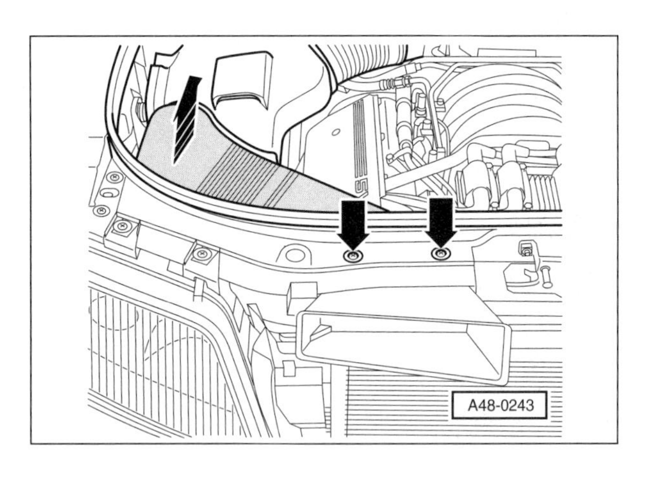

Remove bolts (**arrows**) for air filter duct housing at front of lock carrier and lift air duct out.

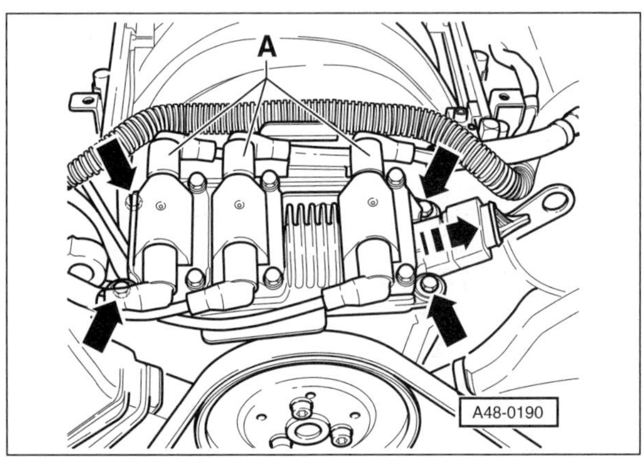

Disconnect 5-pin connector from ignition coils.

- Disconnect three rear connectors (**A**) from ignition coils.
- Remove 4 mounting bolts (**arrows**).
- Swing ignition coil bracket toward rear.

BentleyPublishers.com
BentleyPublishers.com—All Rights Reserved

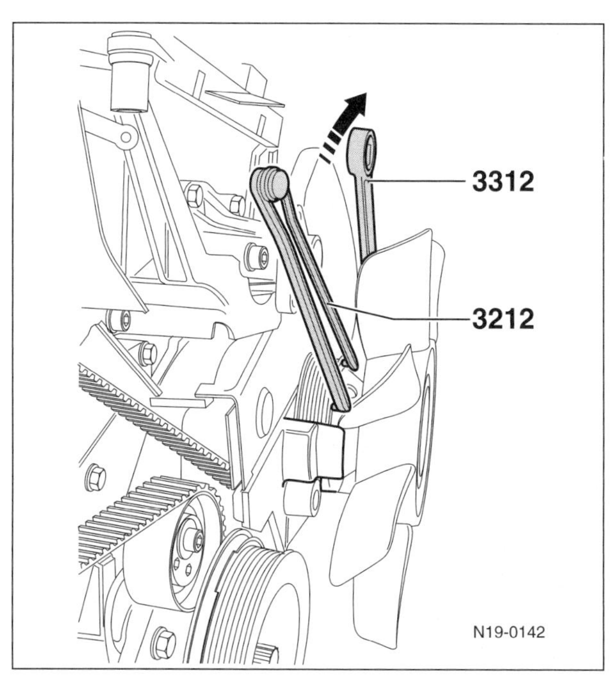

Counterhold pulley for viscous fan using VAG 3212 spanner and remove viscous fan using VAG 3312 open-end spanner (left hand thread).

– Lift viscous fan up and out.

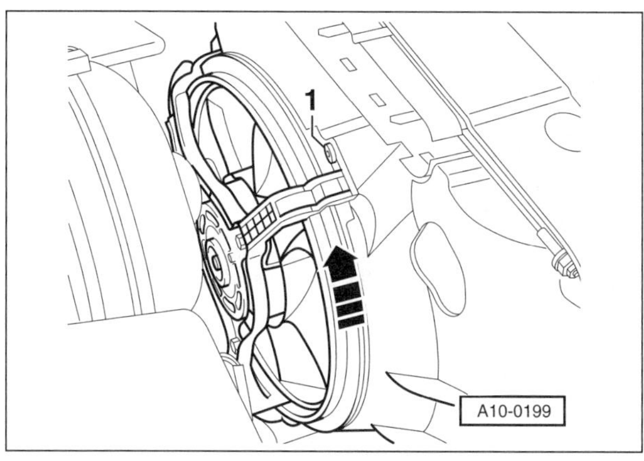

Remove electric fan by first removing bolt (**1**).

– Release electric fan harness clips on viscous fan air guide.

– Rotate electric fan slightly in direction of (**arrow**) and place fan aside to right.

NOTE —

- *Before removing engine accessory belt, mark direction of rotation with chalk or felt-tip marker.*

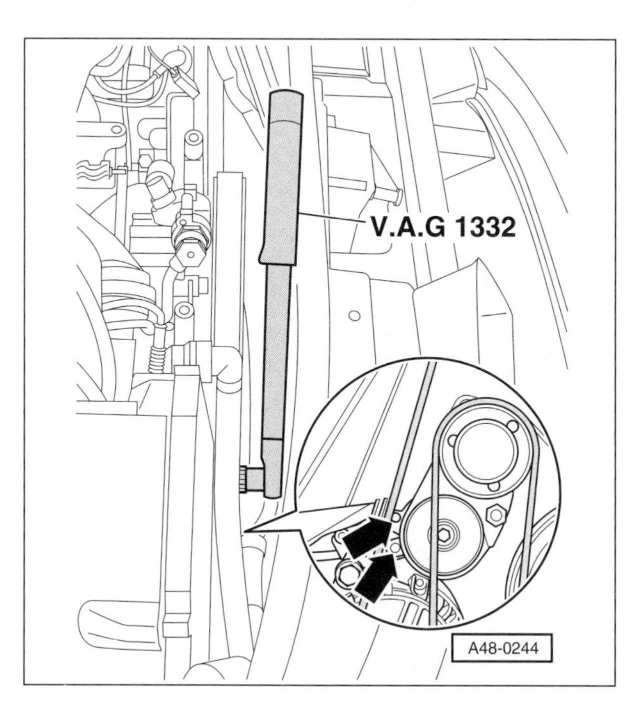

Turn belt tensioner toward right using a long torque wrench and 17 mm socket insert.

– When both bores align (**arrows**), secure using VAG 3204 drift (or equivalent).

– Remove accessory belt from belt pulley.

BentleyPublishers.com
BentleyPublishers.com—All Rights Reserved

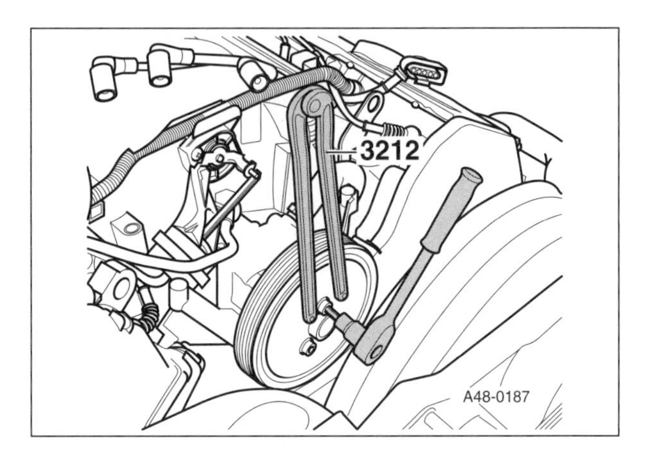

Unbolt belt pulley for power steering pump. Counterhold with VAG 3212 spanner.

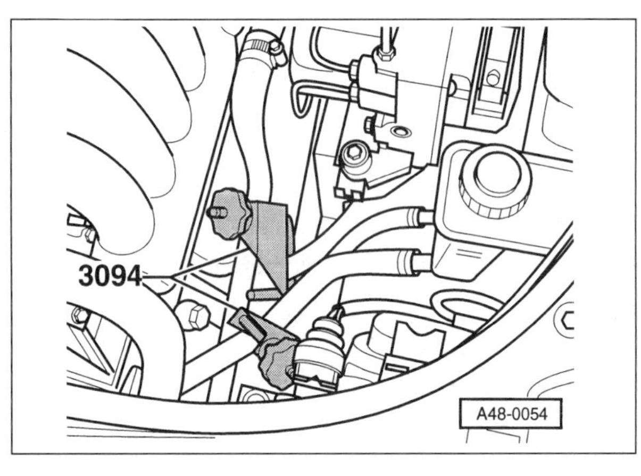

Pinch off intake line and return line using 3094 hose clamps.

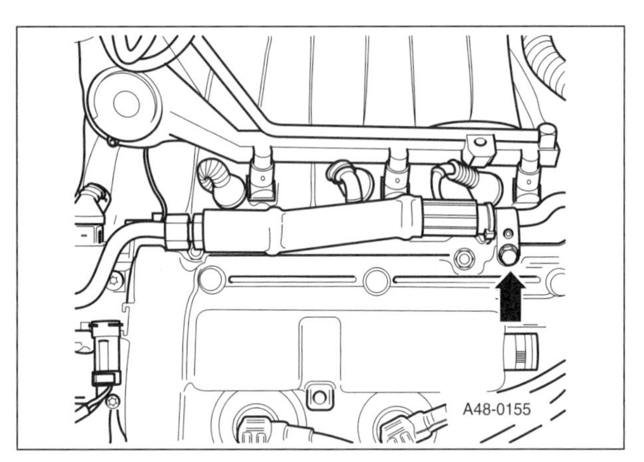

Remove expansion hose bracket bolts (**arrow**) on cylinder head.

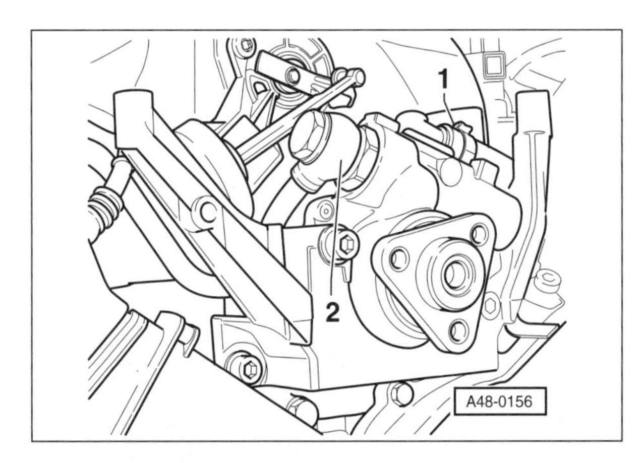

Open clamp (**1**) and remove fluid supply hose.

- Unbolt expansion hose (**2**).

NOTE—

- *Be prepared to catch dripping fluid.*

BentleyPublishers.com
BentleyPublishers.com—All Rights Reserved

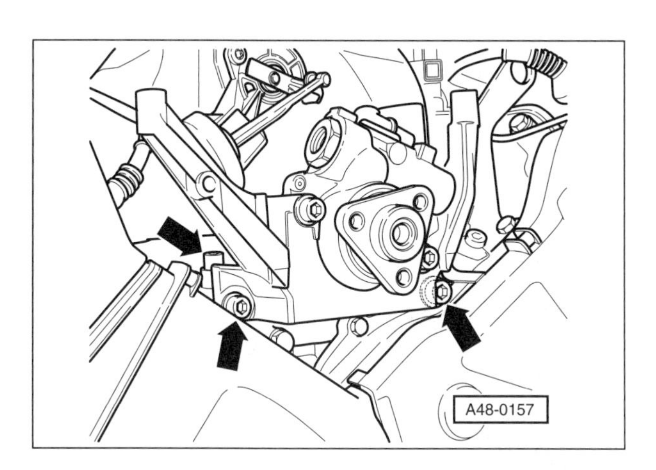

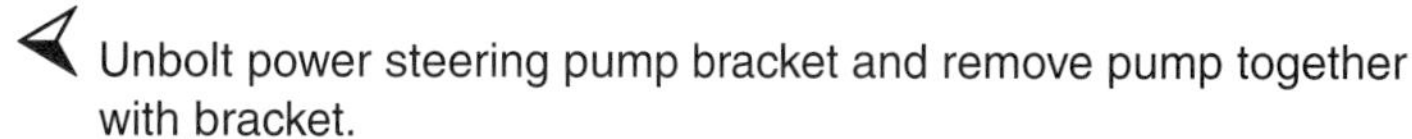

Unbolt power steering pump bracket and remove pump together with bracket.

– Unbolt power steering pump from bracket.

– When installing, pay particular attention to the following points:
  - Before installing a new pump, fill intake side with hydraulic fluid and turn pump by hand until fluid comes out the pump side.
  - Clean any spilled fluid in engine compartment.

Tightening torque	
Power steering pump to bracket	22 Nm (16 ft-lb)

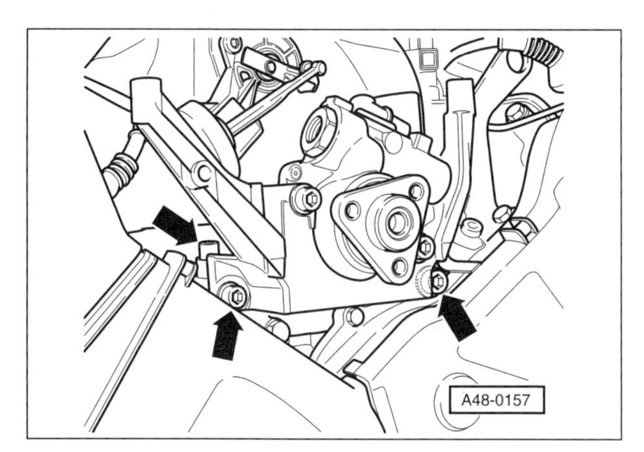

Install pump together with bracket.

– First tighten front two bolts, and then rear two bolts to 22 Nm (16 ft-lb).

– Install new seals on banjo fitting.

Tightening torque	
Power steering pump bracket to engine	22 Nm (16 ft-lb)

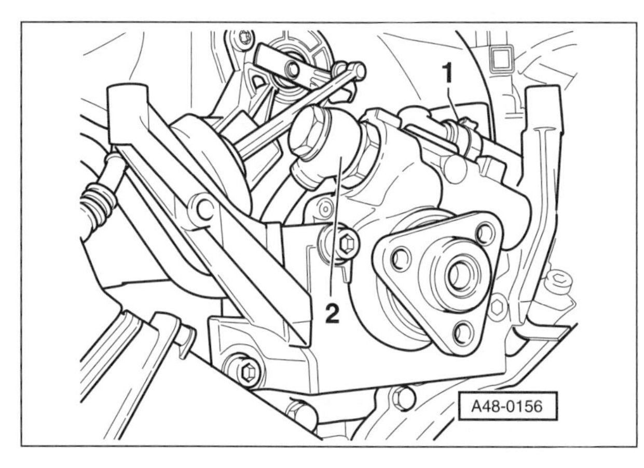

Tighten banjo fitting for expansion hose (**2**).

– Bolt line for expansion hose to cylinder head and tighten to 6 Nm (6 ft-lb).

– Install intake hose (**1**) with marking **P** facing toward pump

– Remove VAG 3094 hose clamps.

Tightening torques	
Power steering fluid supply line to pump	47 Nm (35 ft-lb)
Steering fluid clamp to cylinder head	6 Nm (4.5 ft-lb)

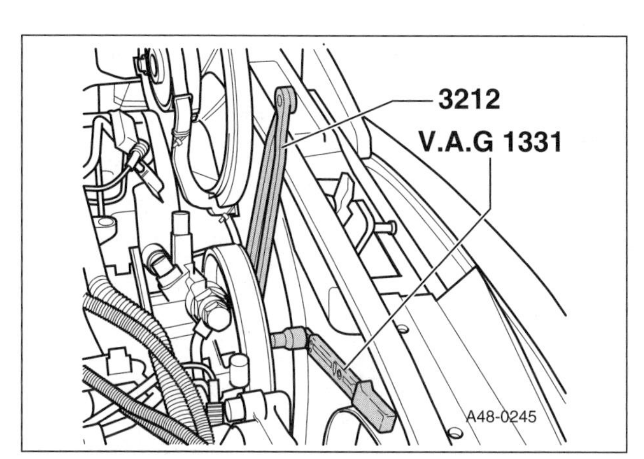

Tighten Allen bolts for power steering pump belt pulley.

Tightening torques	
Steering pump pulley to steering pump	22 Nm (16 ft-lb)

BentleyPublishers.com
BentleyPublishers.com—All Rights Reserved

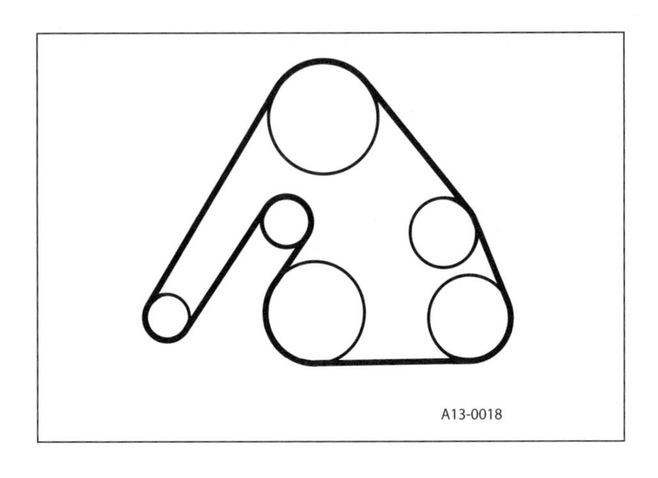

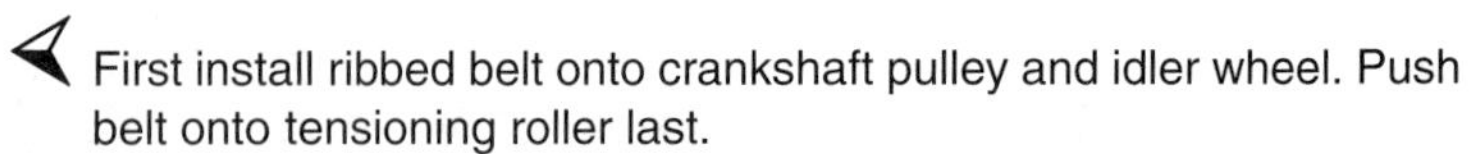

First install ribbed belt onto crankshaft pulley and idler wheel. Push belt onto tensioning roller last.

NOTE—

- *Note previously marked direction of rotation of belt and be sure that it is seated correctly on pulley.*

– Remove VAG 3204 drift.

– First install electric fan and then viscous fan.

Tightening torques	
Viscous fan with VAG 3312 wrench	37 Nm (27 ft-lb)
Viscous fan without VAG 3312 wrench	70 Nm (52 ft-lb)

– Fill and bleed steering system.

– Check steering system for leaks.

Power steering pump (2.7 liter biturbo engine), removing and installing

– Remove front bumper. See **63 Bumpers**.

– Bring front lock carrier into service position. See **50 Body–Front**.

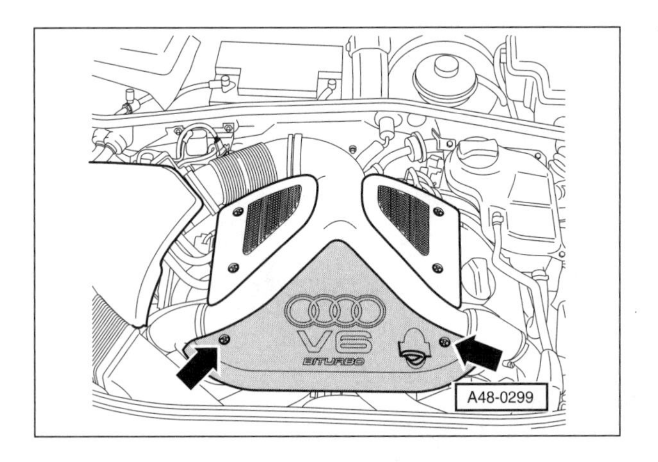

Remove bolts (**arrows**) and remove front engine cover.

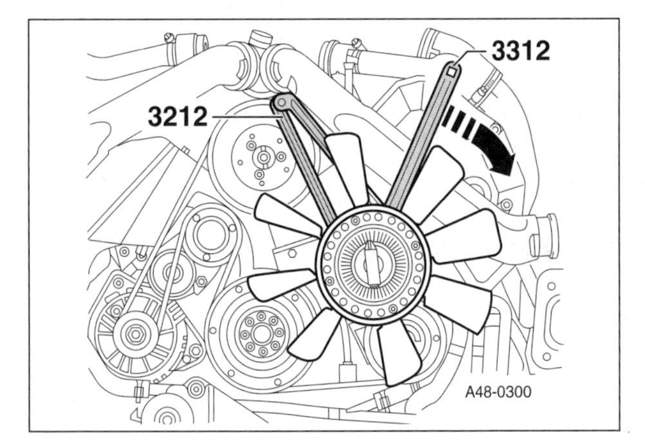

Counterhold viscous fan pulley using VAG 3212 spanner wrench and remove viscous fan using VAG 3312 open-end wrench.

NOTE—

- *The viscous fan has a left-hand thread. Unbolt in direction of* ***arrow****.*

– Lift viscous fan up and out.

NOTE—

- *Before removing engine accessory belt, mark direction of rotation with chalk or felt-tip marker.*

BentleyPublishers.com
BentleyPublishers.com—All Rights Reserved

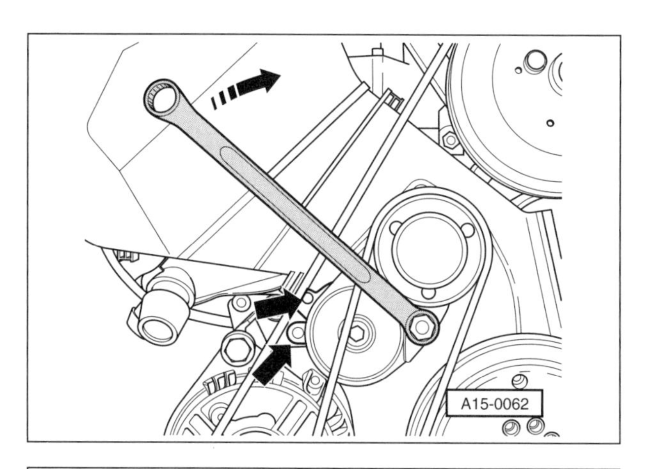

To loosen accessory belt, turn tensioner clockwise using 17 mm box wrench until two holes are aligned (**arrows**). Lock in position using VAG 3204 drift.

– Remove accessory belt from belt pulley.

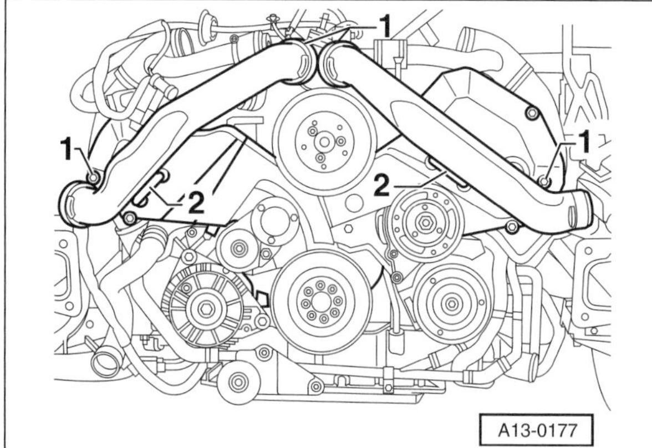

Remove turbo pressure duct mounting bolts (**1**). Remove ducts.

NOTE —

- *Pay attention to retaining strips (**2**).*

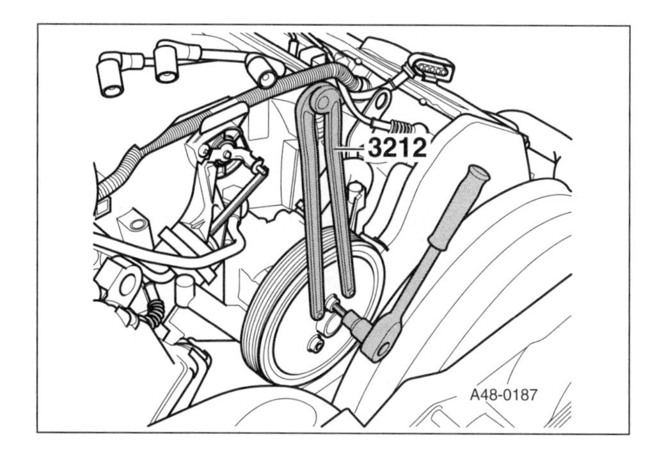

Unbolt belt pulley for power steering pump. Counterhold with VAG 3212 spanner.

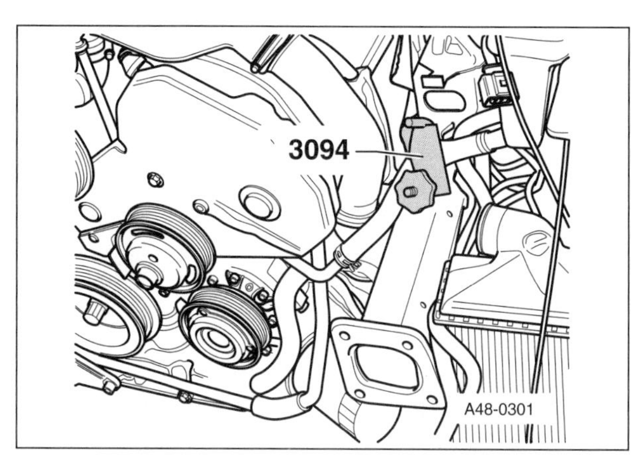

Pinch off intake line using VAG 3094 hose clamp.

BentleyPublishers.com
BentleyPublishers.com—All Rights Reserved

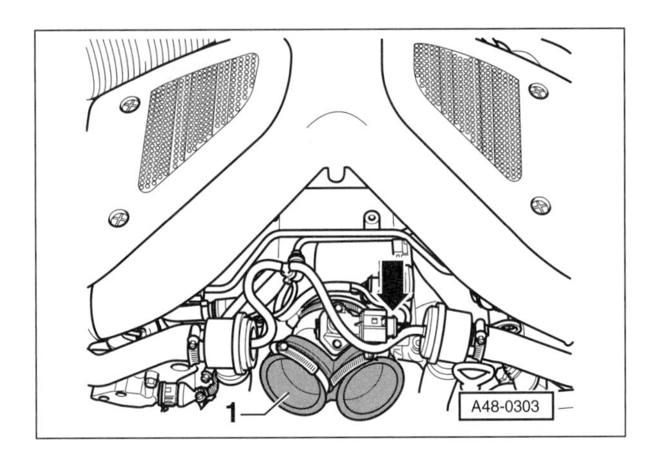

◀ Disconnect harness connector at charge air pressure sensor (G31) (**arrow**).

– Remove rubber shroud (**1**).

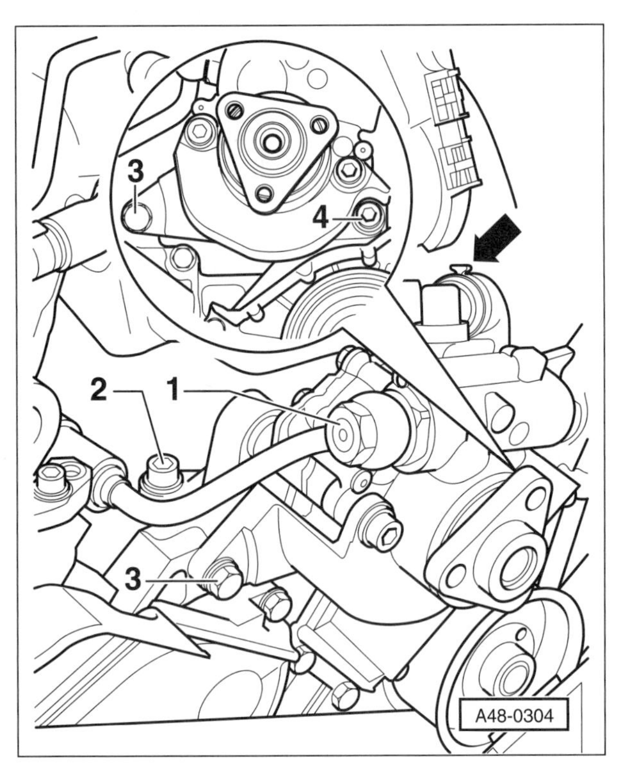

◀ Open clamp (**arrow**) and disconnect intake hose from pump.

– Unbolt pressure line (**1**).

– Remove power steering pump bracket fasteners (**2**, **3** and **4**).

– Remove pump together with bracket.

– Unbolt power steering pump from bracket.

– When installing, pay particular attention to the following points:
  - Before installing a new pump, fill intake side with hydraulic fluid and turn pump by hand until fluid comes out the pump side.
  - Clean any areas in engine compartment covered with fluid.

– Bolt pump to bracket.

– Tighten front two bolts first, and then rear two bolts.

Tightening torque	
Power steering pump to bracket	22 Nm (16 ft-lb)

BentleyPublishers.com
BentleyPublishers.com—All Rights Reserved

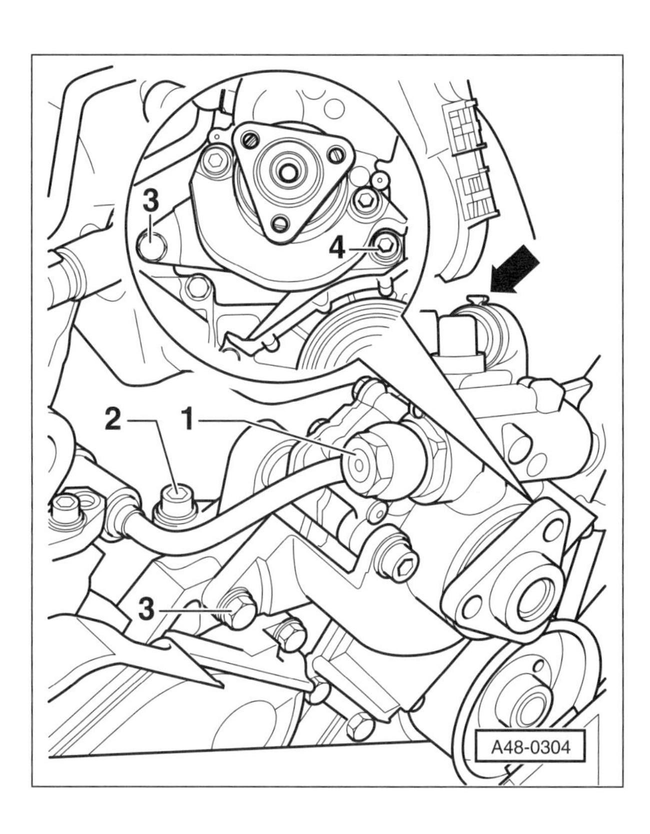

◀ Install pump together with bracket.

– Install new seals on banjo bolt.

– Tighten banjo bolt for pressure line (**1**) to 50 Nm (37 ft-lb).

Tightening torques	
Power steering fluid supply line to pump	50 Nm (37 ft-lb)

– Install intake hose (align mark **P** with flash on pump).

– Remove VAG 3094 hose clamps.

– Tighten Allen bolts for power steering pump belt pulley.

Tightening torques	
Power steering pulley to steering pump	22 Nm (16 ft-lb)

– Install viscous fan.

Tightening torques	
Viscous fan with VAG 3312 wrench	37 Nm (27 ft-lb)
Viscous fan without VAG 3312 wrench	70 Nm (52 ft-lb)

– Install lock carrier. See **03 Maintenance**.

– Install bumper. See **63 Bumpers**.

– Fill and bleed steering system.

– Check steering system for leaks.

Power steering pump (3.0 liter engine), removing and installing

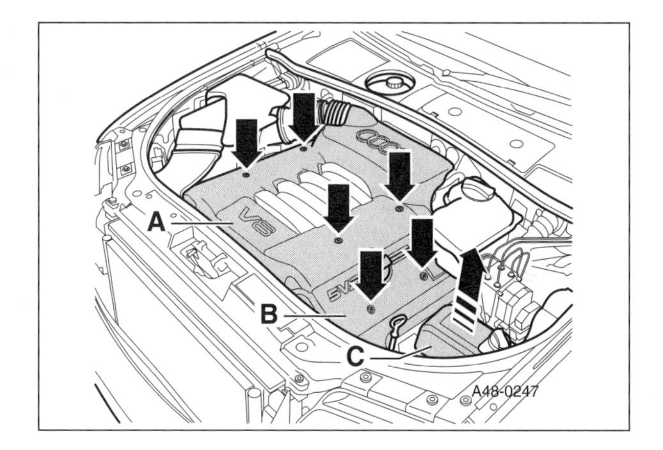

◀ Remove bolts (**arrows**) and remove engine covers (**A**) and (**B**).

– Remove cover (**C**) from power steering fluid reservoir.

BentleyPublishers.com
BentleyPublishers.com—All Rights Reserved

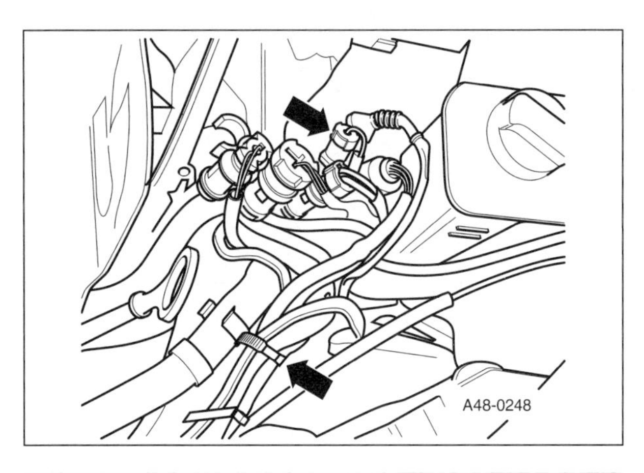

- Disconnect for A/C clutch green connector (upper **arrow**) and remove lower part of connector from bracket.
- Open wire tie (lower **arrow**) on line to A/C compressor and let line hang free.
- Remove right timing belt cover.

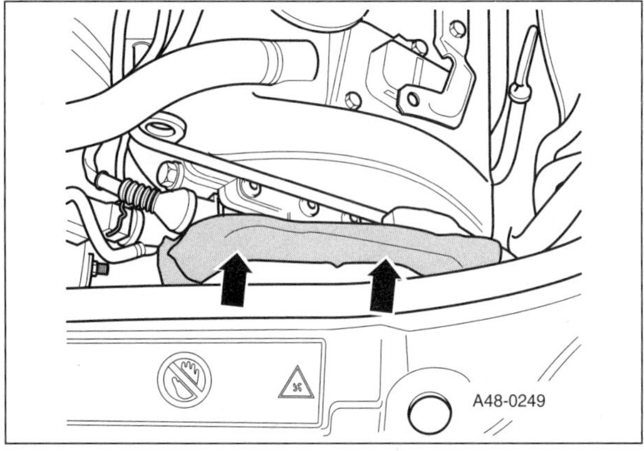

- Cover sharp corners of shroud with tape.

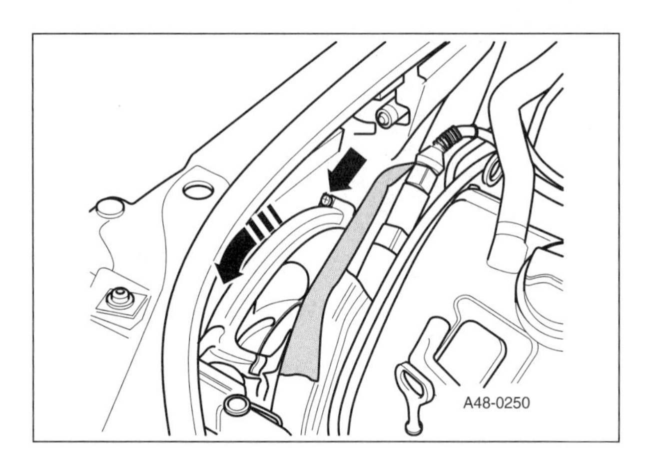

- Remove electric engine cooling fan by removing bolt (solid **arrow**).
- Pull sealing lip away from front end slightly in area near electric fan.
- Rotate electric fan slightly in direction of **arrow** and place fan to side.

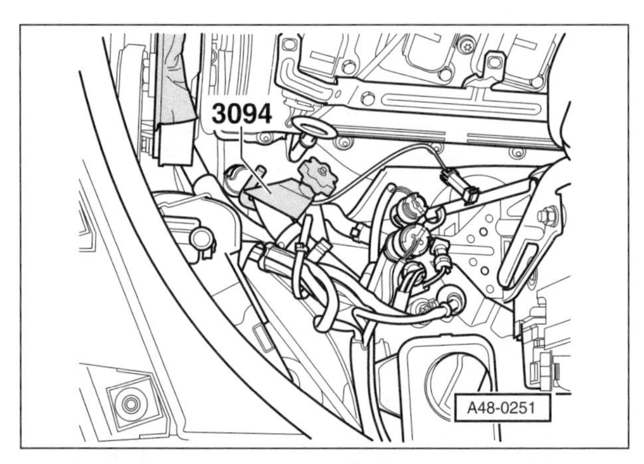

- Pinch off supply line using VAG 3094 hose clamp.

BentleyPublishers.com
BentleyPublishers.com—All Rights Reserved

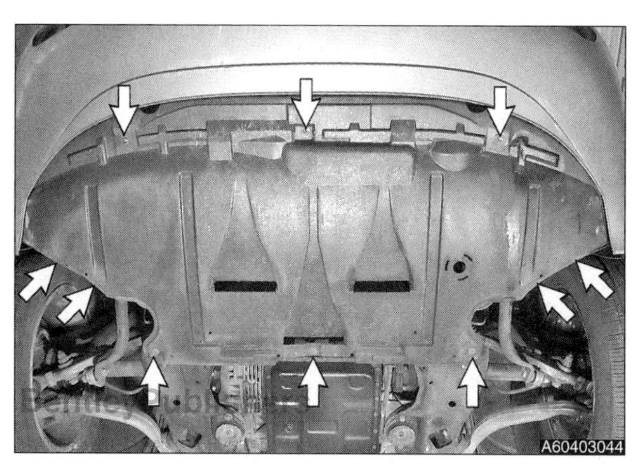

Remove engine lower cover.

NOTE—

- *Before removing engine accessory belt, mark direction of rotation with chalk or felt-tip marker.*

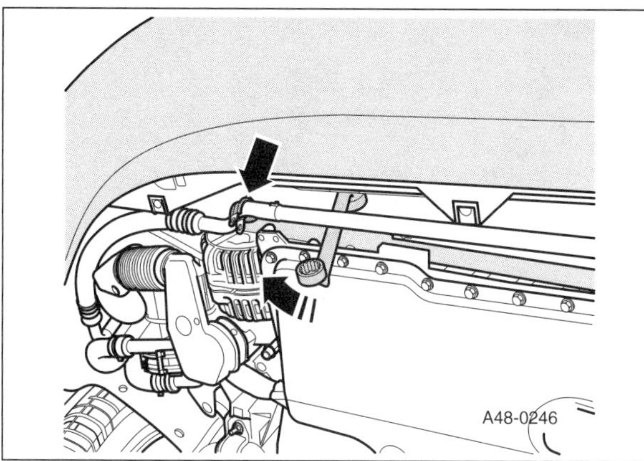

Remove clamp from refrigerant line.

– Move tensioner in direction of **arrow** to release tension on accessory belt.

– Remove accessory belt and release tensioning roller.

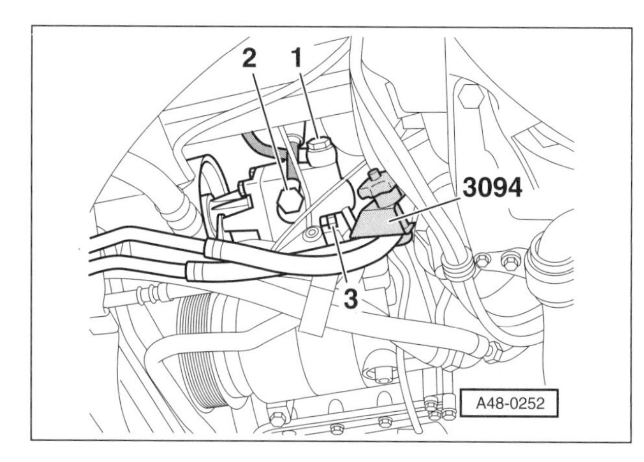

Pinch off return lines using VAG 3094 hose clamp.

– Place pan underneath vehicle to collect steering fluid.

– Unbolt expansion hose (**1**) and intake hose (**2**) from pump.

– Seal connection holes using plugs.

– Remove bolt (**3**) for power steering pump and bracket.

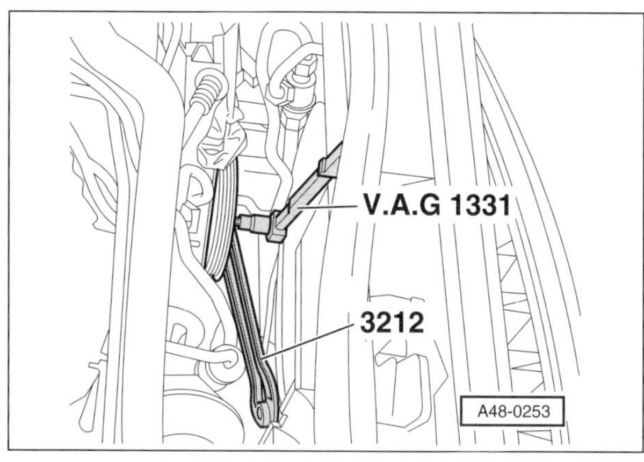

Unbolt belt pulley for power steering pump. Counterhold with VAG 3212 spanner.

BentleyPublishers.com
BentleyPublishers.com—All Rights Reserved

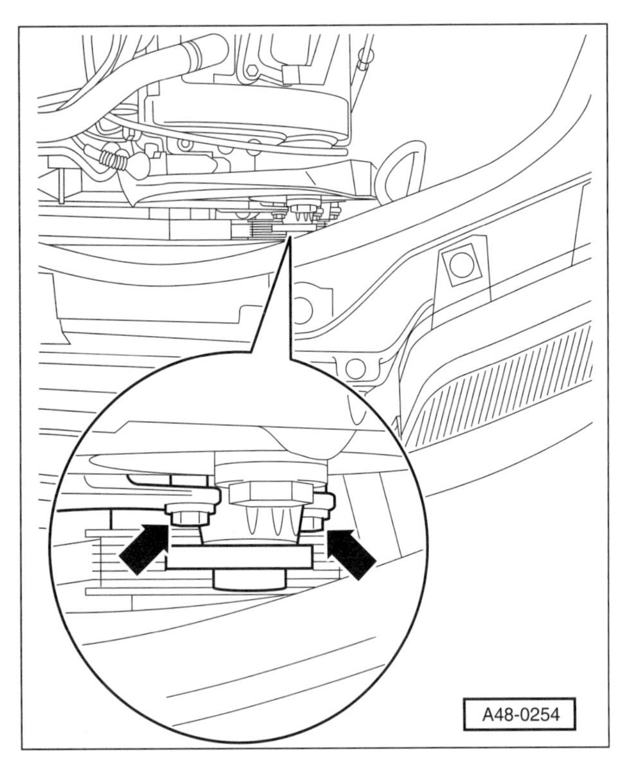

Remove pump mounting bolts (**arrows**). Remove pump downward and out.

– When installing, pay particular attention to the following points:
  - Before installing a new pump, fill intake side with steering fluid and turn pump by hand until fluid comes out the pump side.
  - Clean areas in engine compartment contaminated with fluid.
  - Return expansion and intake hose to installed position.
  - Place pump in bracket from below and hand-tighten all three bolts.
  - Install expansion and intake hoses with new seals, but do not tighten all the way.

– Tighten two front bolts (**arrows**).

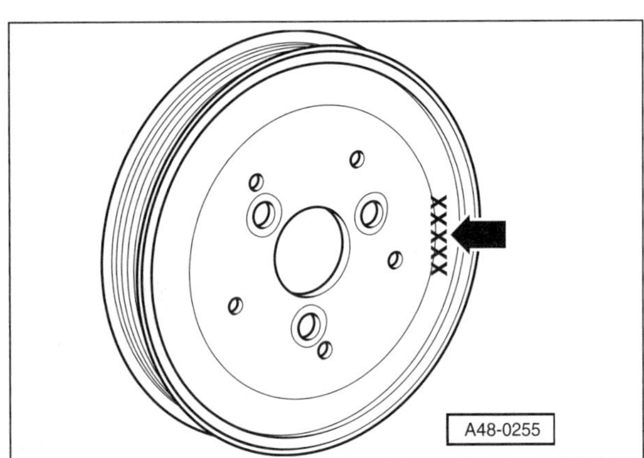

When installing belt pulley, note that front of pulley is marked "front".

BentleyPublishers.com
BentleyPublishers.com—All Rights Reserved

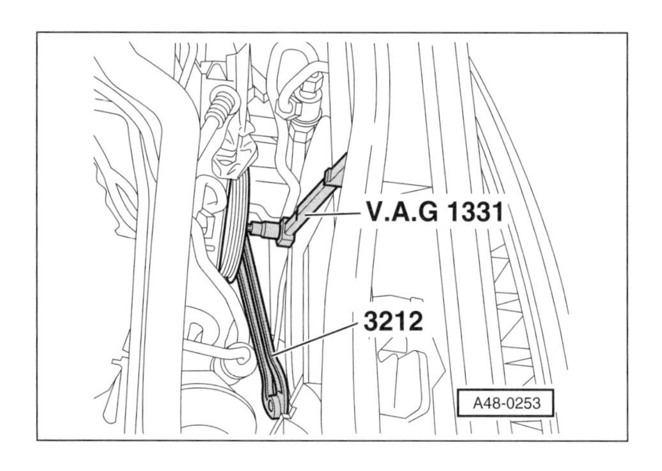

– Bolt on belt pulley for power steering pump. Counterhold with VAG 3212 spanner wrench.

Check and adjust pulley alignment

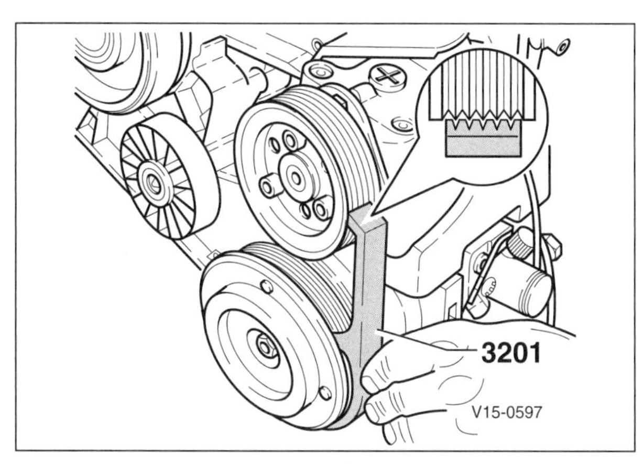

◄ Install VAG 3201 level on A/C compressor pulley.

– Pulley for power steering pump must align with pulley for A/C compressor.

– If the two pulleys do not align, unbolt pulley for power steering pump.

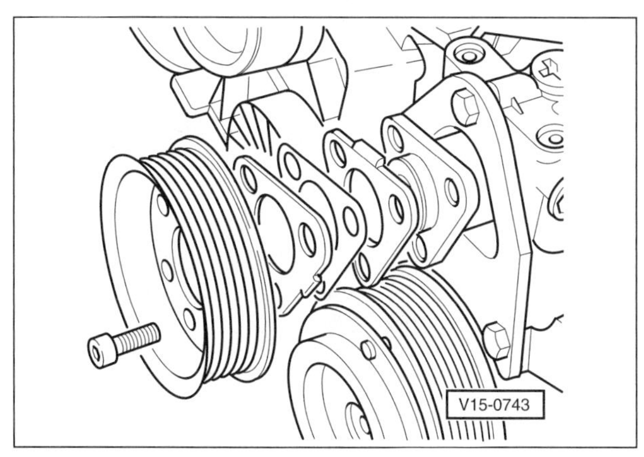

◄ Use shims (available as replacement parts) to adjust pulley with A/C pulley and steering pump pulley until it aligns.

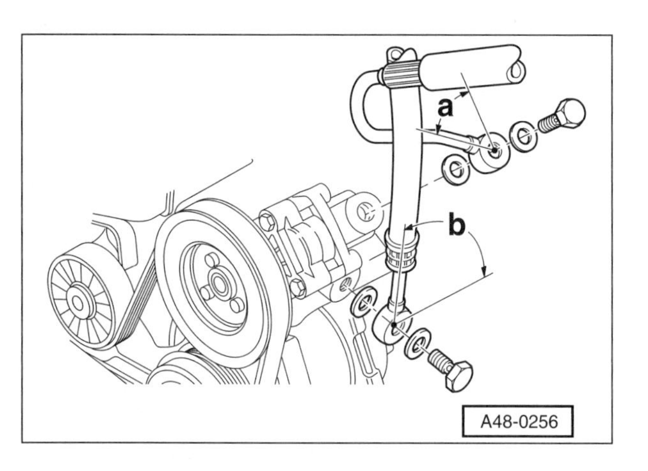

◄ Note installation position of expansion hose. Angle **a** is approximately 57°.

– Align expansion hose and tighten.

– Install electric cooling fan.

BentleyPublishers.com
BentleyPublishers.com—All Rights Reserved

Steering Pump

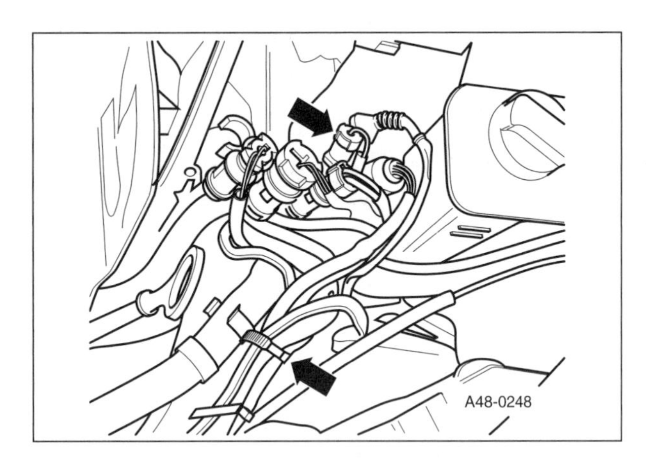

◀ Connect A/C harness and replace wire ties.

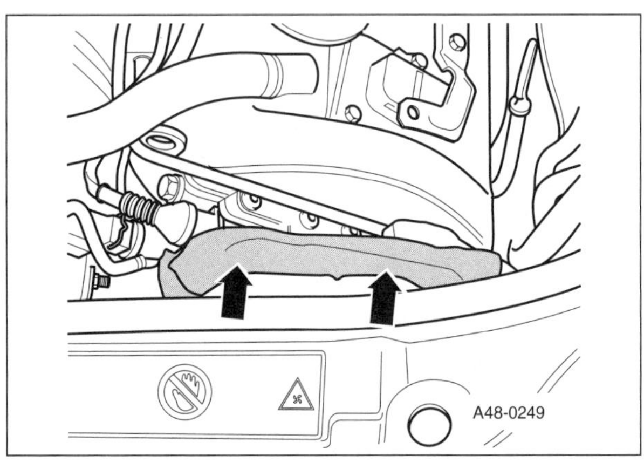

◀ Remove adhesive tape (installed earlier).

– Install covers for timing belt, power steering fluid reservoir and engine.

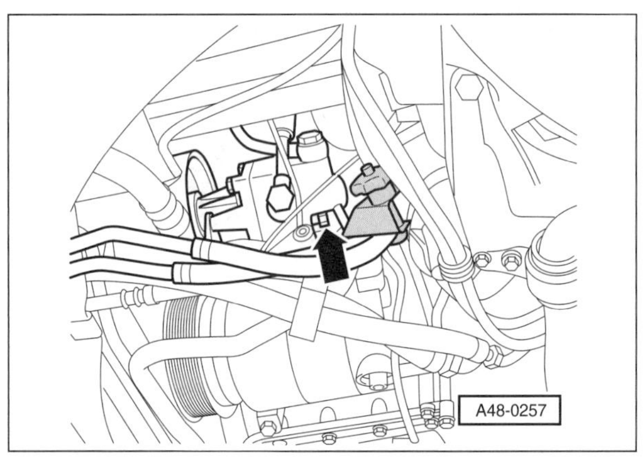

◀ Tighten rear bolt on power steering pump.

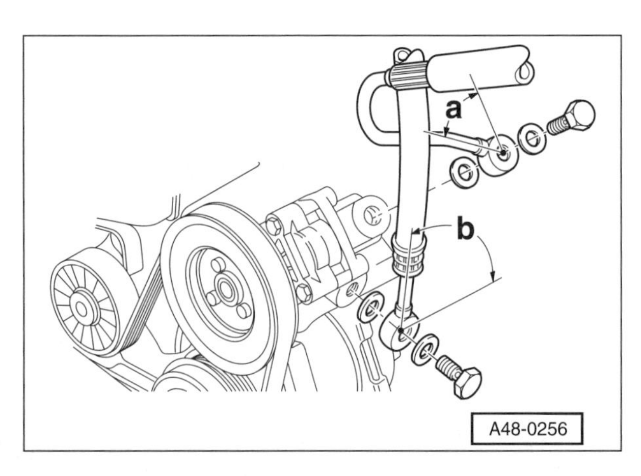

◀ Note installation position of intake line. Angle **b** is approximately 90°.

– Align intake line and tighten.

– Remove VAG 3094 hose clamps.

BentleyPublishers.com
BentleyPublishers.com—All Rights Reserved

Install accessory belt

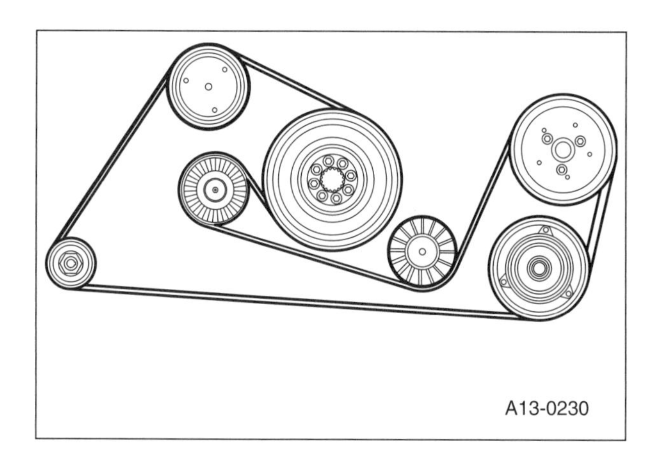

- First install engine accessory belt onto crankshaft pulley and idler wheel. Push belt onto tensioning roller last.

NOTE—

- *Note previously marked direction of rotation of belt and be sure that it is seated correctly on pulley.*

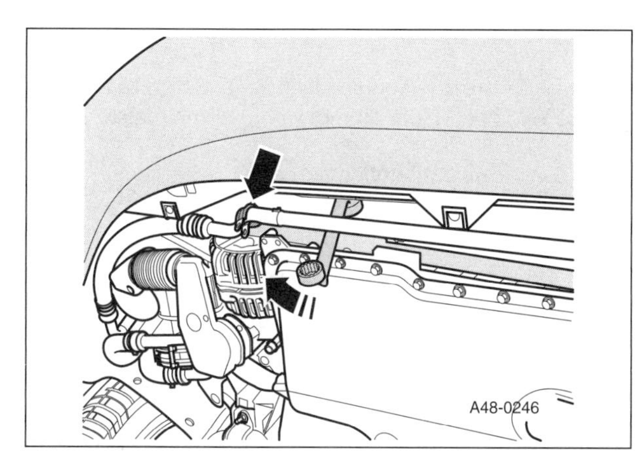

- Swing tensioner in direction of dashed **arrow**.
- Position accessory belt and release tensioning roller.
- Install refrigerant line clamp.
- Fill and bleed steering system.
- Check steering system for leaks.

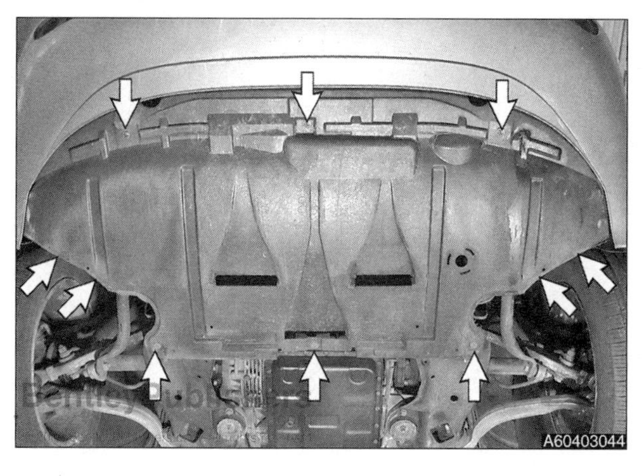

- Install engine lower cover.

BentleyPublishers.com
BentleyPublishers.com—All Rights Reserved

Power steering pump (allroad quattro V8), removing and installing

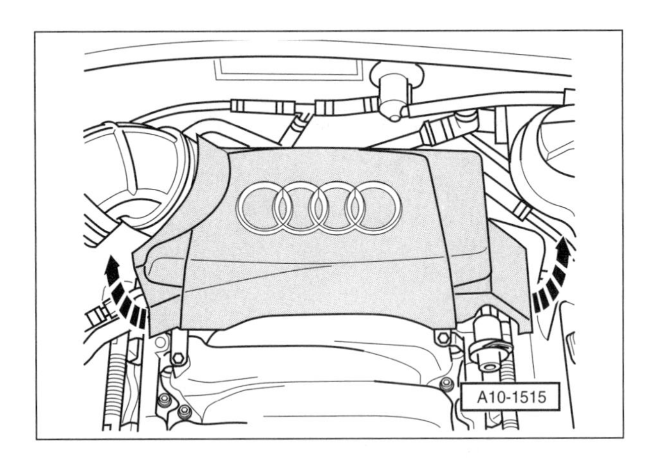

◀ Remove rear engine cover (**arrows**).

– Extract power steering hydraulic fluid from reservoir using VAG 1358 A or VAG 1782.

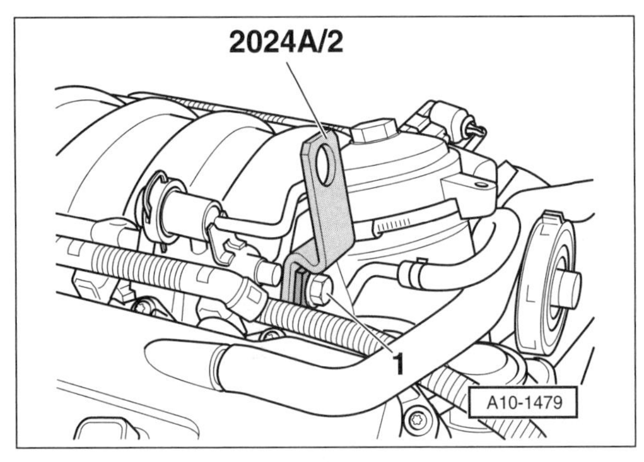

◀ Attach plate 2024 A/2 to rear left engine lifting eye. Screw out bolt at plate and replace with bolt M10 x 15 mm with packing plate.

– Remove rubber seal at wing mounting flanges.

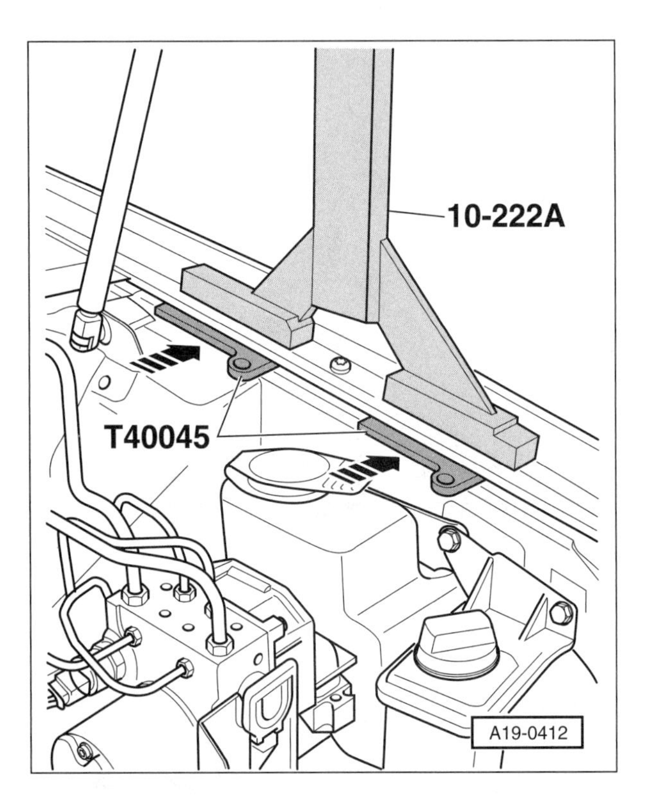

◀ To stop support bar from damaging edges of wing, insert special tool T40045 on either side between wing mounting flange and splash panel beneath.

BentleyPublishers.com
BentleyPublishers.com—All Rights Reserved

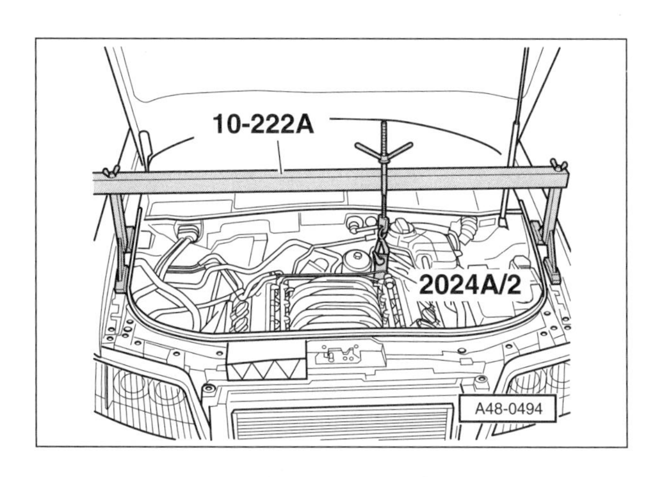

◀ Position support bar VAG 10-222 A on wing mounting flanges.

– Engage plate 2024 A/2 at spindle of support bar.

– Slightly pretension engine with spindle of support bar, but do not lift.

– Remove left front wheel.

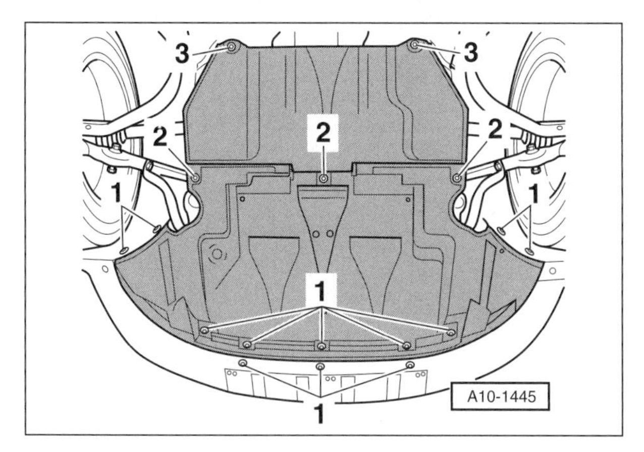

◀ Unfasten quick-release fasteners (**1**, **2** and **3**) and detach lower engine cover (2 sections).

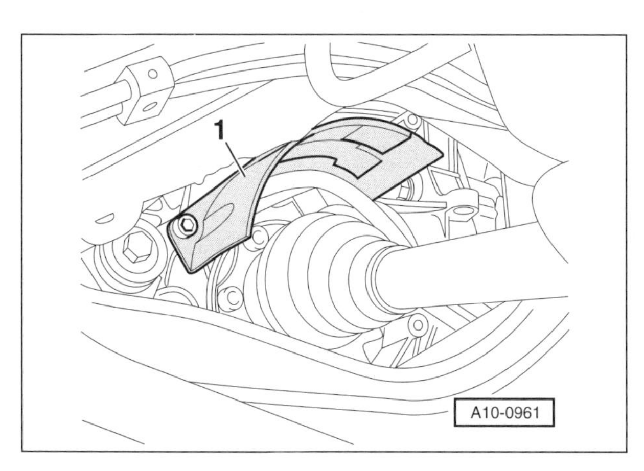

◀ Remove heat shield (**1**) for drive axle on left.

– Remove left drive axle. See **40 Front Suspension**.

BentleyPublishers.com
BentleyPublishers.com—All Rights Reserved

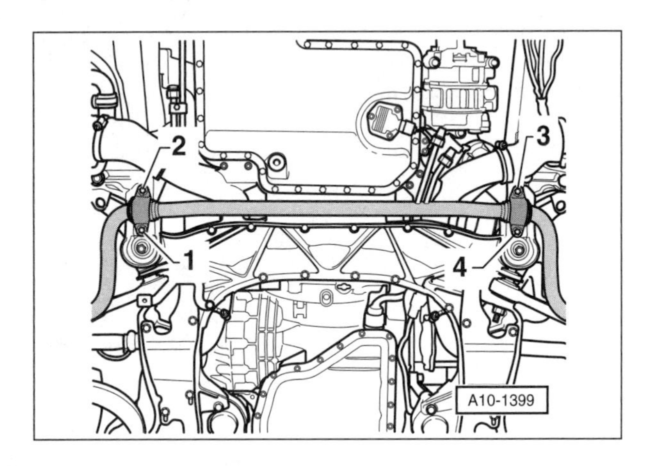

Unscrew nuts (**1** - **4**) for stabilizer bar mounting on left and right.

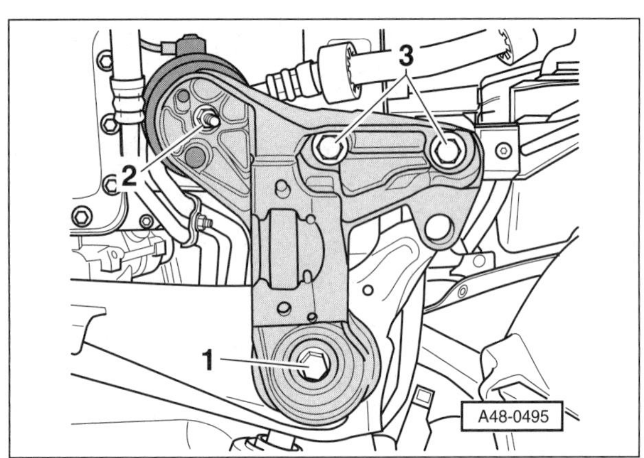

Unscrew nuts (**3**).

- Remove bolts (**1** and **2**) and detach engine mount bracket.

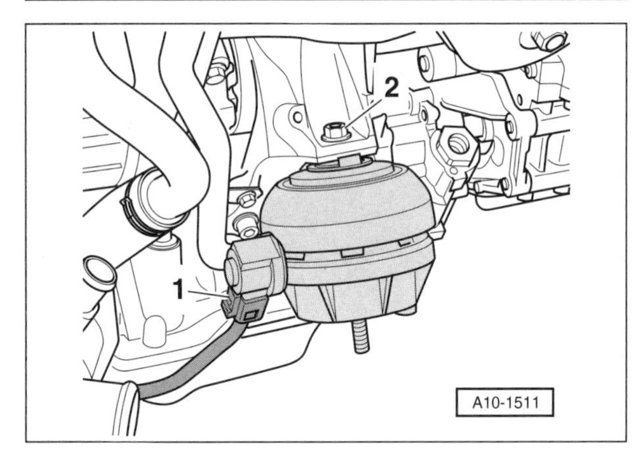

Disconnect connector (**1**) at left engine mounting.

- Unscrew nut (**2**) and detach engine mounting.

NOTE —

- *Illustration shows engine removed.*

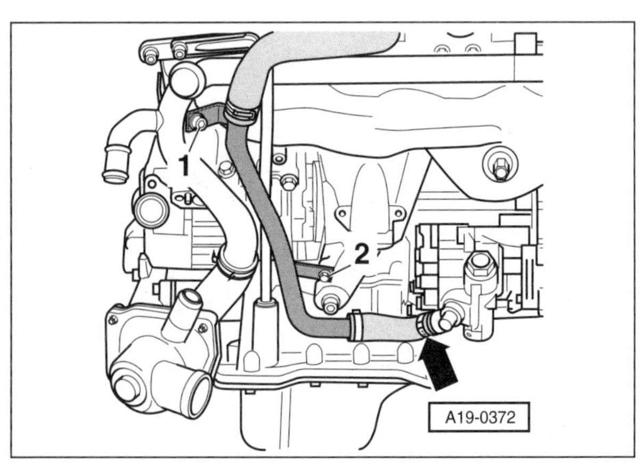

Remove bolts (**1** and **2**).

- Remove return line from power steering pump (**arrow**).

BentleyPublishers.com
BentleyPublishers.com—All Rights Reserved

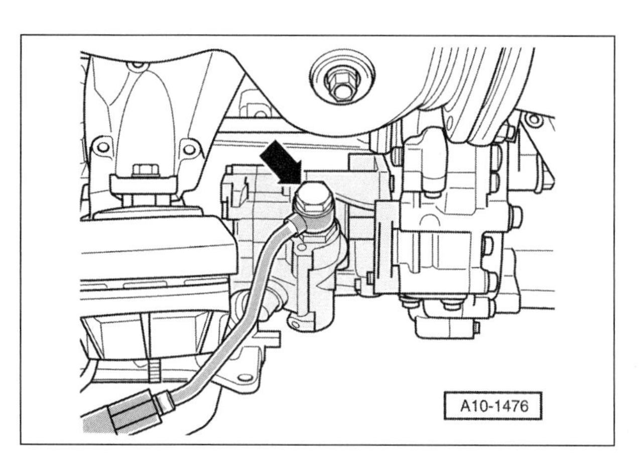

◀ Remove hydraulic pressure line banjo bolt (**arrow**) at power steering pump.

– Set aside hydraulic pressure line.

NOTE—

• Be prepared to catch dripping steering fluid

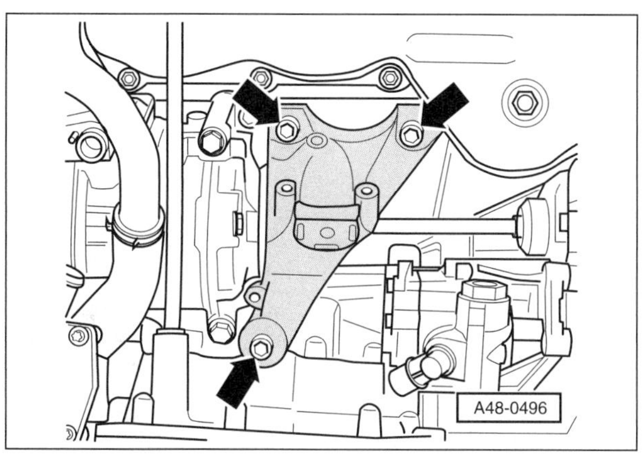

◀ Remove left engine support (**arrows**).

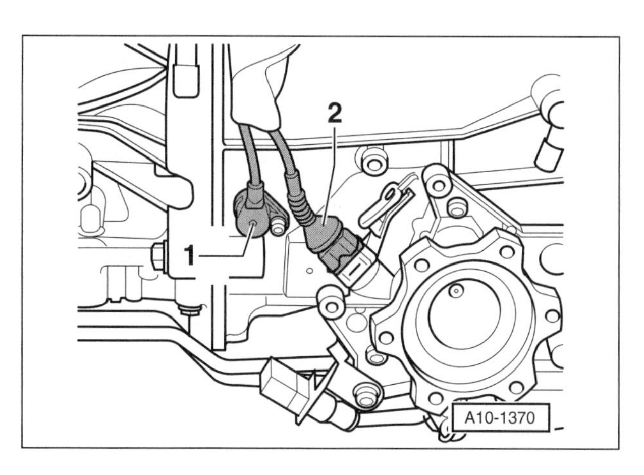

◀ Disconnect connector (**2**) at speedometer vehicle speed sensor (G22).

NOTE—

• Ignore item 1.

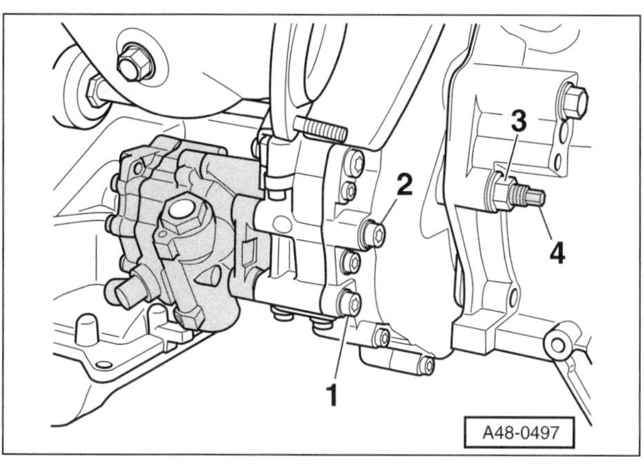

◀ Loosen lock nut (**3**) by several turns.

– Screw out as far as possible stud bolt (**4**).

– Screw out bolts (**1**) and (**2**) and detach power steering pump.

BentleyPublishers.com
BentleyPublishers.com—All Rights Reserved

- Pay particular attention to the following on installation:
  - In engine compartment, clean areas contaminated with fluid.
  - Before installing new power steering pump, pour in hydraulic fluid on intake end and crank by hand until fluid emerges on discharge end.
  - Replace O-ring on drive shaft of power steering pump.
  - Replace oil seals.
  - Secure all hose connections with standard hose clamps.
- Install engine bracket and screw on subframe.
- Install stabilizer bar.
- Install drive axle.
- Fill and bleed power steering system.
- Check power steering system for leaks.

Tightening torques	
Power steering pump to engine	22 Nm (16 ft-lb)
Stud bolt to power steering pump	40 Nm (30 ft-lb)
Lock nut to stud bolt	65 Nm (48 ft-lb)
Engine support to cylinder block	42 Nm (31 ft-lb)
Hydraulic pressure line to pump	47 Nm (31 ft-lb)
Return line to: • engine support • coolant pipe	 10 Nm (7 ft-lb) 10 Nm (7 ft-lb)
Engine mounting to: • engine support • engine bracket	 23 Nm (17 ft-lb) 23 Nm (17 ft-lb)
Drive axle heat shield to transmission	23 Nm (17 ft-lb)

BentleyPublishers.com
BentleyPublishers.com—All Rights Reserved

Power steering pump (RS6 engine), removing and installing

– Bring front lock carrier into service position. See **50 Body–Front**.

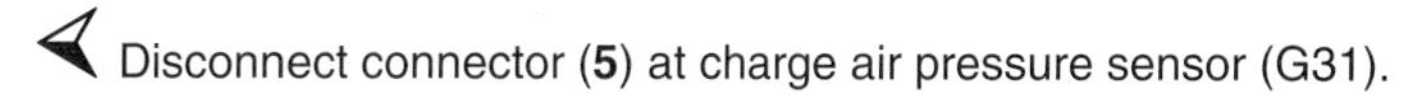

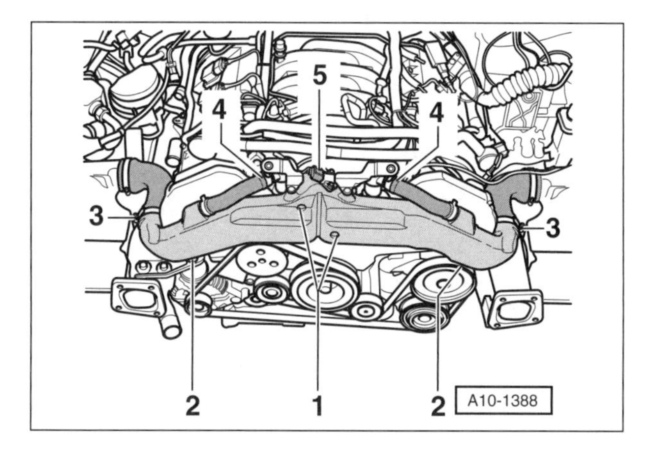

◄ Disconnect connector (**5**) at charge air pressure sensor (G31).

– Unfasten hose clamps (**3**) and (**4**).

– Detach hoses at air recirculation valves.

– Screw out bolts (**1** and **2**) and detach air line.

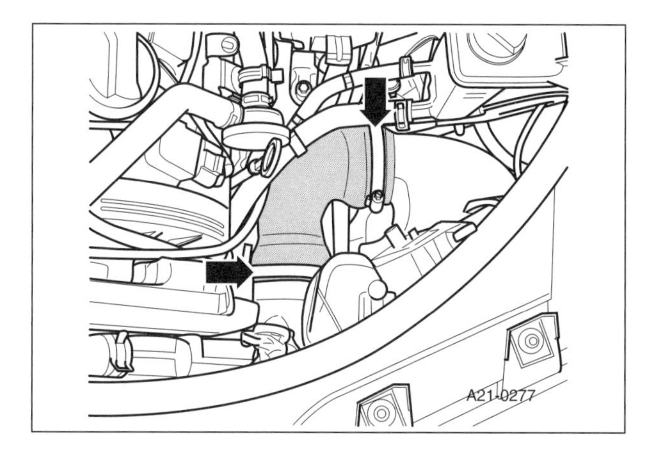

◄ Remove air hose (**arrows**).

NOTE—

• Before removing engine accessory belt, mark direction of rotation with chalk or felt-tip marker.

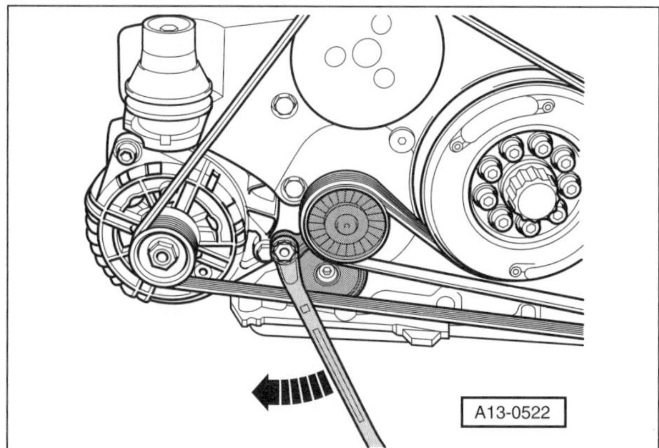

◄ To loosen accessory belt swing tensioner in direction of **arrow**.

BentleyPublishers.com
BentleyPublishers.com—All Rights Reserved

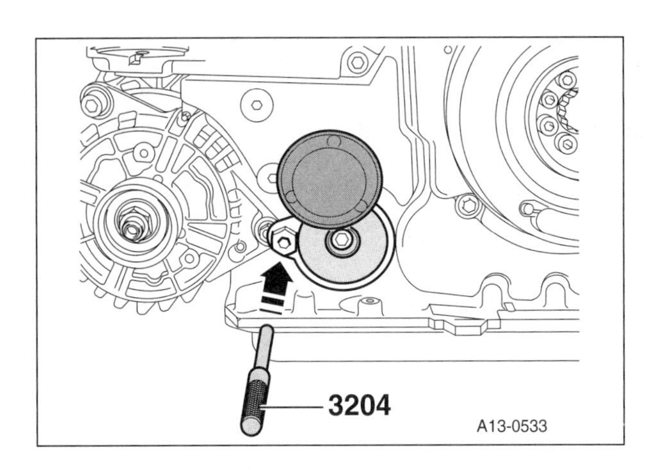

◄ Insert VAG 3204 drift into locating holes (**arrow**) to lock tensioner.

– Detach accessory belt from power steering pump pulley.

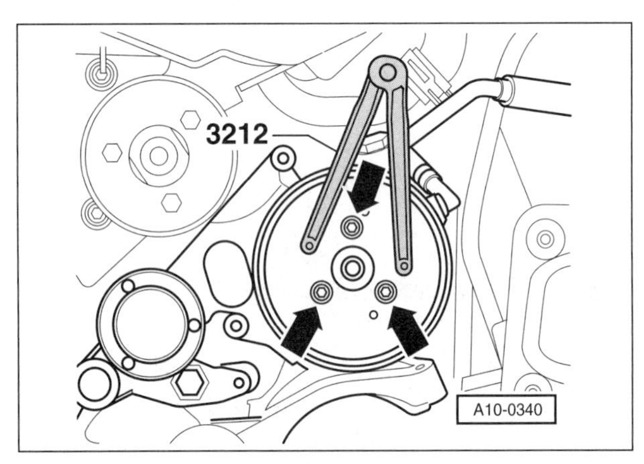

◄ Remove bolts (**arrows**) from power steering pump pulley. Counterhold with VAG 3212 spanner when loosening and tightening bolts.

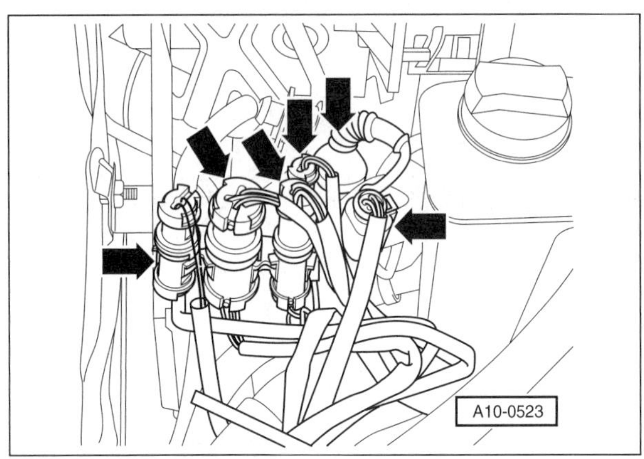

◄ Take connectors (**arrows**) out of holder.

– Remove holder for connectors.

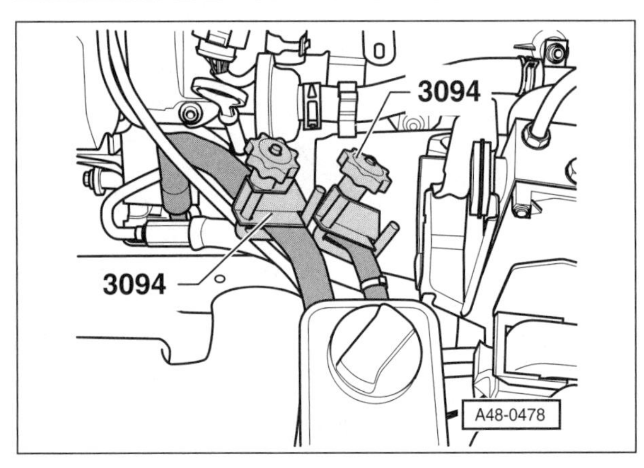

◄ Use VAG 3094 hose clamps to pinch off hydraulic hoses.

NOTE —

- *Be prepared to catch dripping steering fluid.*

BentleyPublishers.com
BentleyPublishers.com—All Rights Reserved

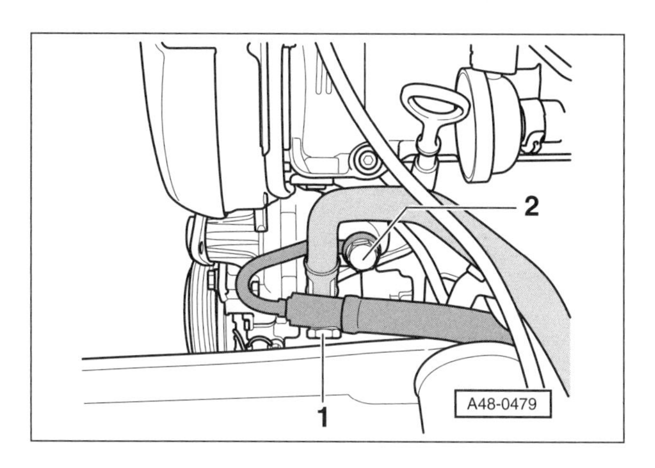

Disconnect intake hose (**1**) and expansion hose (**2**) at power steering pump.

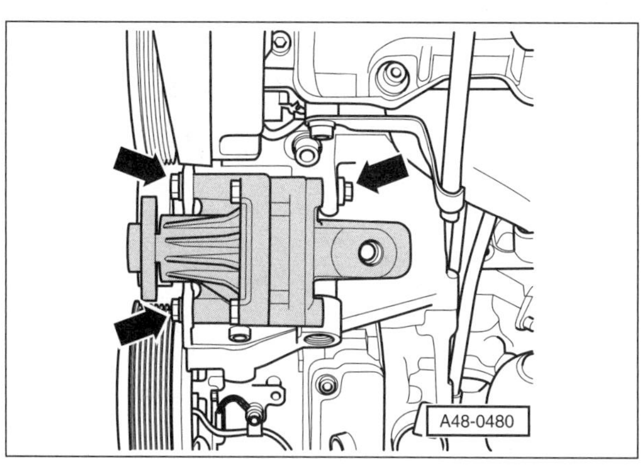

Remove bolts (**arrows**).

– Take out power steering pump.

– Pay particular attention to the following on installation:
  - Before installing a new pump, fill intake side with hydraulic fluid and turn pump by hand until fluid comes out the pump side.
  - Clean any areas in engine compartment contaminated with fluid.
  - Return expansion hose and suction hose to installed position.
  - Replace oil seals and O-rings.
  - Secure all hose connections with standard hose clamps.

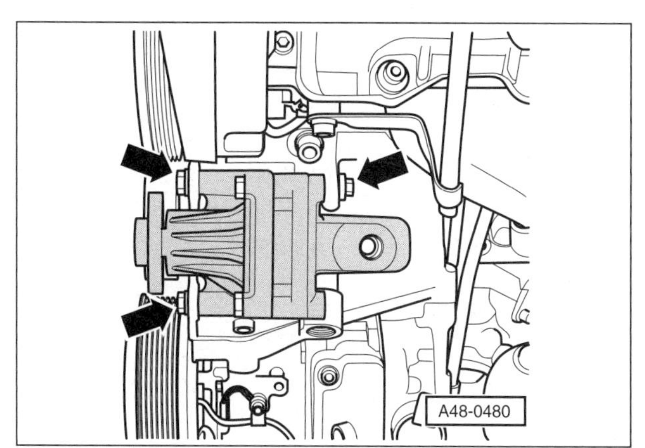

Place pump in bracket and screw in all three bolts (**arrows**).

– Install expansion and intake hoses with new seals, but do not tighten all the way.

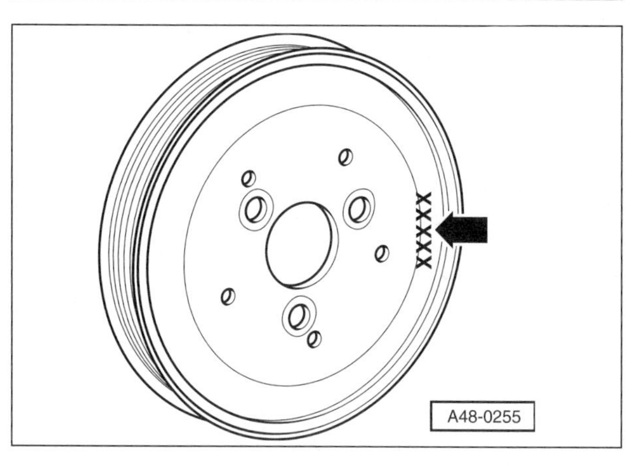

Front of the belt pulley is marked "front".

BentleyPublishers.com
BentleyPublishers.com—All Rights Reserved

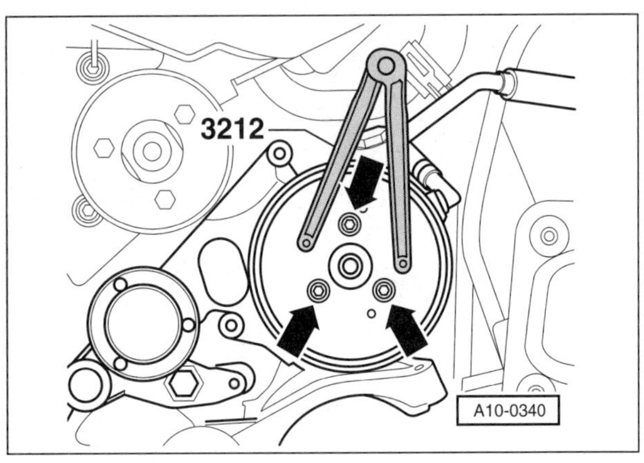

Screw pulley on power steering pump while counterholding with VAG 3212 spanner.

Checking and adjusting pulley alignment

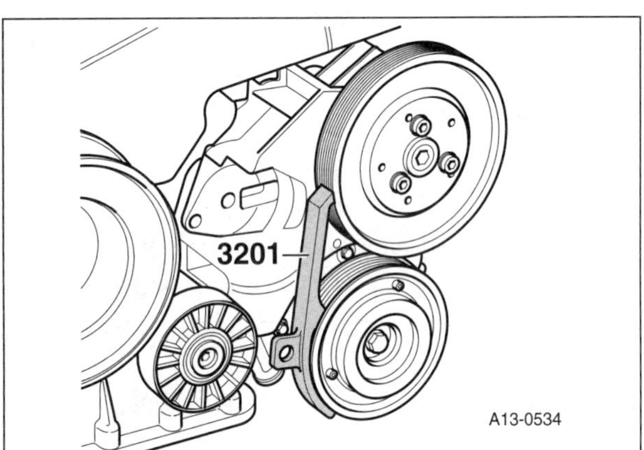

Position alignment gauge VAG 3201 on A/C compressor pulley.

- Ribs of power steering pump pulley must coincide with ribs of A/C compressor pulley.

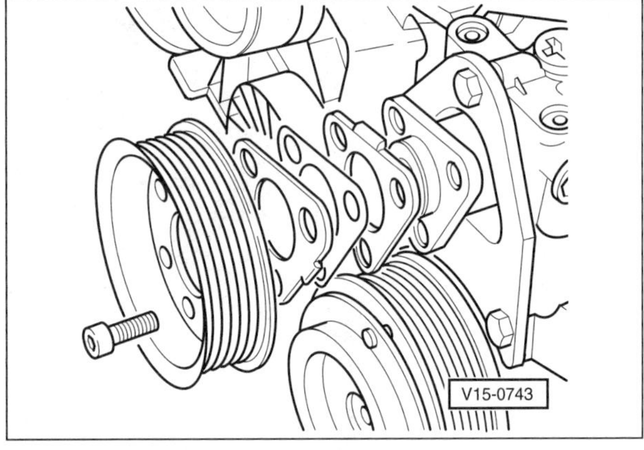

If the two pulleys do not coincide, align by installing shims.

– Check alignment of pulleys again with alignment gauge VAG 3201. Repeat adjustment if necessary.

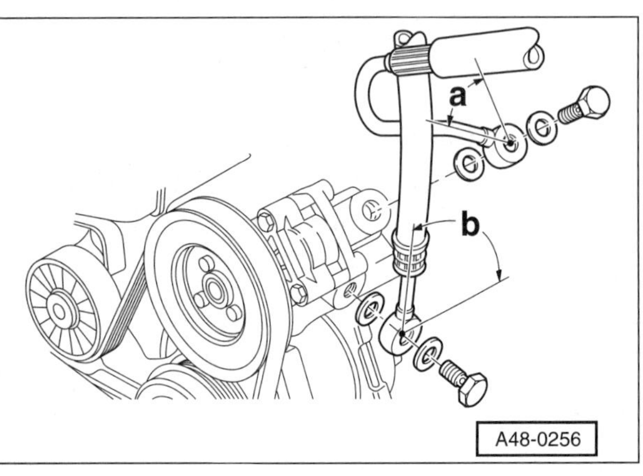

Note installed position of intake line:

- Angle **b** is approximately 90°.
- Align and tighten intake line.

– Note installed position of expansion hose:

- Angle **a** is approximately 75°.
- Align and tighten expansion hose.

– Detach hose clamps VAG 3094.

BentleyPublishers.com
BentleyPublishers.com—All Rights Reserved

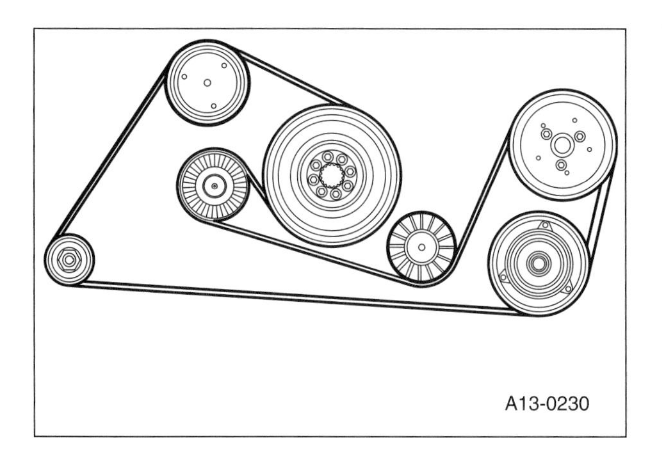

Installing accessory belt

< Position engine accessory belt over crankshaft pulley and other pulleys and then on tensioning roller last.

NOTE—
- *Make sure accessory belt is properly seated on pulleys.*
- *Replace O-ring for front air line.*
- *Before installing hose connections and hoses for charge air system must be free from oil and grease. Never use lubricant.*

– Remainder of installation is reverse of removal. Remember to:
- Check hydraulic fluid level.
- Bleed steering system.
- Check steering system for leaks.
- Check belt routing and start engine.
- Install lock carrier. See **50 Body–Front**.

Tightening torques	
Power steering pump to bracket	22 Nm (16 ft-lb)
Pulley to power steering pump	22 Nm (16 ft-lb)
Hydraulic line to power steering pump: • M16 • M18	 50 Nm (37 ft-lb) 47 Nm (35 ft-lb)
Front air line to engine block	9 Nm (7 ft-lb)
Clamps for air hoses	5 Nm (4 ft-lb)

BentleyPublishers.com
BentleyPublishers.com—All Rights Reserved

50 Body–Front

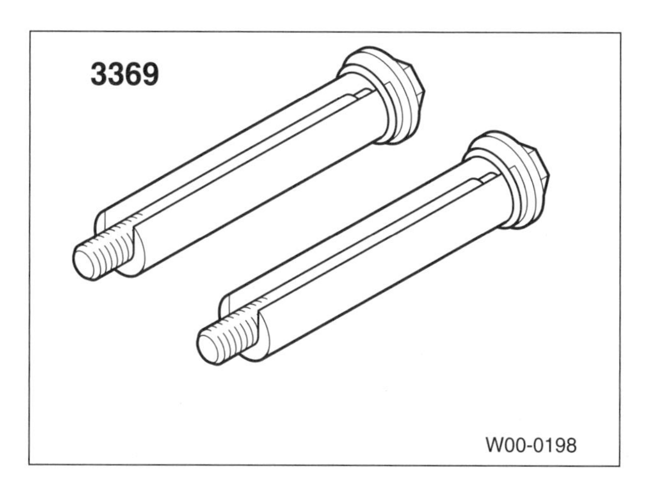

GENERAL

This repair group covers the front lock carrier and front fenders. The lock carrier can be moved to a convenient service position for accessing components located in the front of the engine compartment.

For additional information see:

- **03 Maintenance** for lower engine cover removal.
- **63 Bumpers** for bumper and bumper cover removal.

FRONT LOCK CARRIER

The radiator support frame in front of the engine is referred to as the front lock carrier.

A number of repair procedures require that the lock carrier be unbolted from the frame and moved forward into service position. It is then supported on special tools 3369 (or equivalent).

Remove engine covers as needed to free up and gain access to lock carrier fasteners.

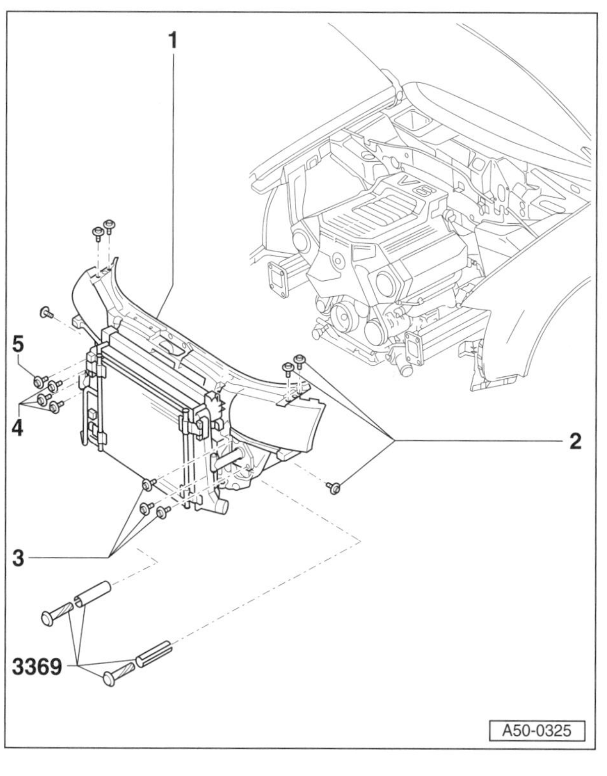

Front lock carrier assembly

1. Lock carrier
2. Torx bolts
   - 10 Nm (7 ft-lb)
3. Bolt and washer assembly
   - 50 Nm (37 ft-lb)
4. Bolt and washer assembly
   - 50 Nm (37 ft-lb)
5. Bolt and washer assembly
   - Remove bolts and install special tools 3369
   - 50 Nm (37 ft-lb)

BentleyPublishers.com
BentleyPublishers.com—All Rights Reserved

Front lock carrier service position

– Raise car and support safely.

> ***WARNING —***
> *• Make sure the car is stable and well supported at all times. Use a professional automotive lift or jack stands designed for the purpose. A floor jack is not adequate support.*

– Remove lower engine cover. See **03 Maintenance**.

– Remove front bumper cover and bumper. See **63 Bumpers**.

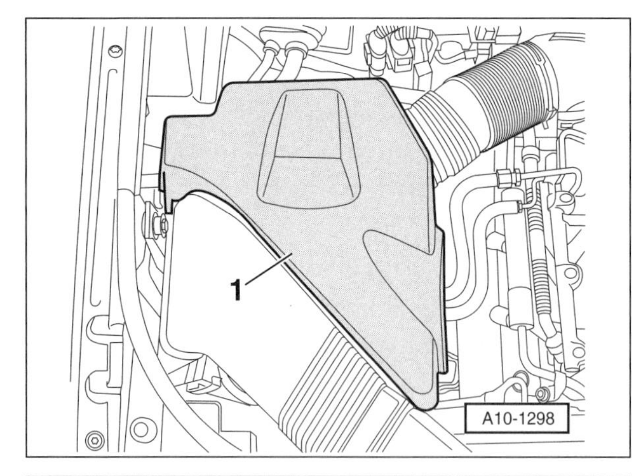

◀ Working in engine compartment right side, remove air filter housing cover (**1**).

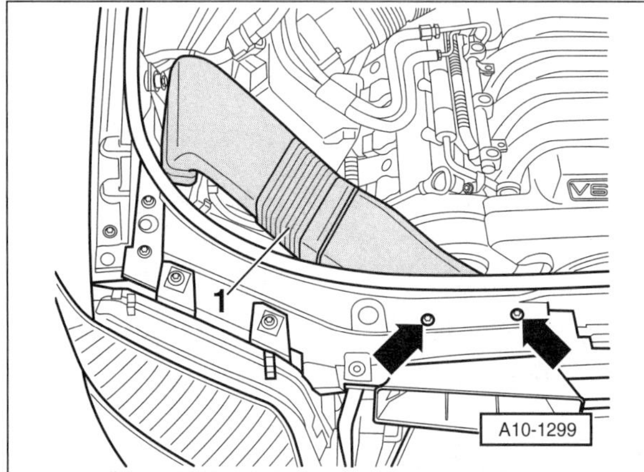

◀ Remove air duct mounting bolts (**arrows**) at lock carrier. Remove air duct (**1**).

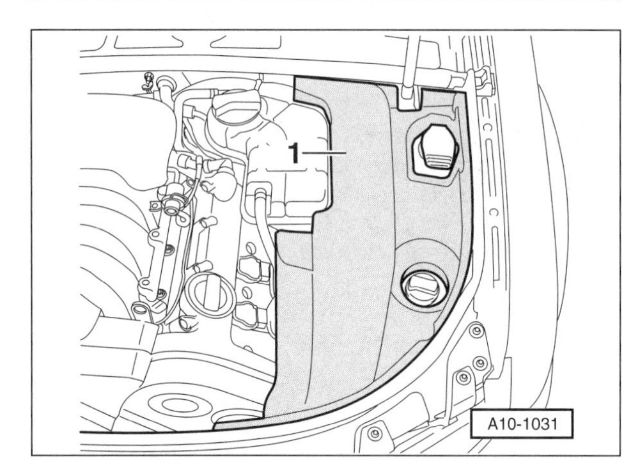

◀ Remove cover (**1**) in engine compartment (left side).

– Pull off hood seal from lock carrier and fender edges.

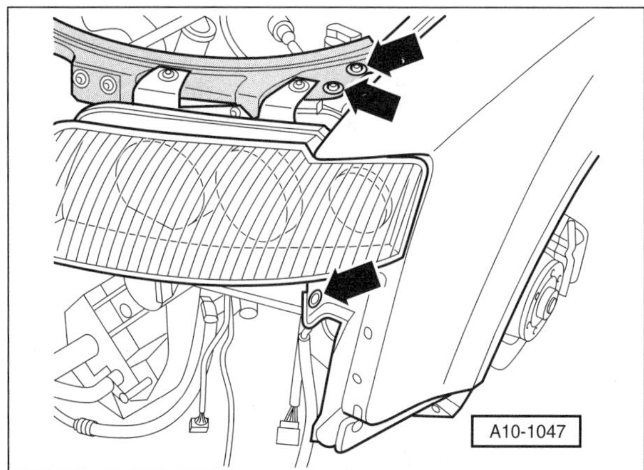

◀ Unscrew bolts (**arrows**) at left and right.

BentleyPublishers.com
BentleyPublishers.com—All Rights Reserved

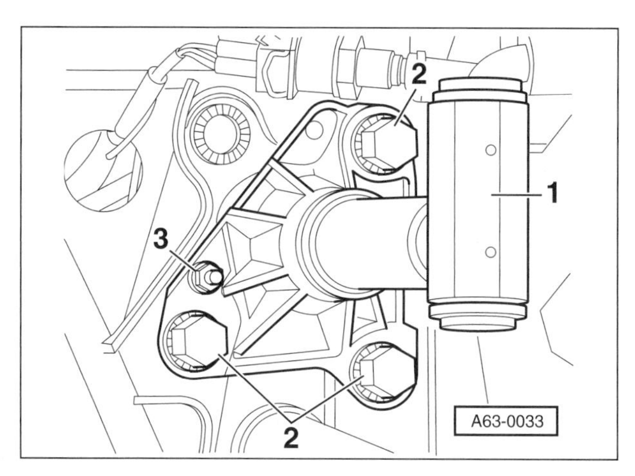

◀ Remove bolts (**2**) at impact absorbers.

NOTE—

- *Support lock carrier until special tool 3369 (shown in following step) is in place.*

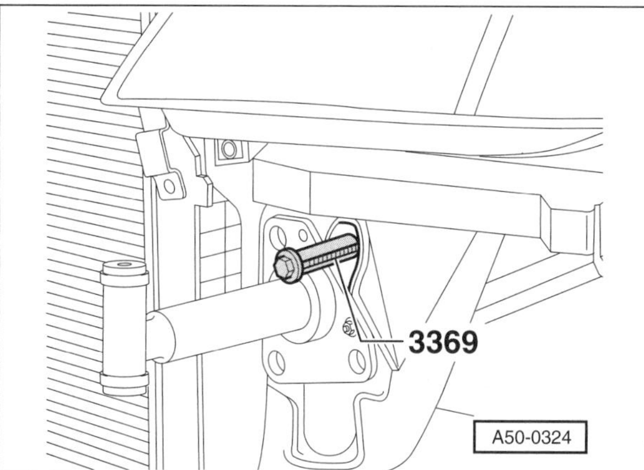

◀ Insert special tool 3369 (or equivalent) as illustrated.

– Pull lock carrier forward until rear hole in lock carrier lines up with front threaded hole in fender flange.

CAUTION—

- *When extending lock carrier, make sure wiring, A/C lines and coolant hoses are not damaged.*

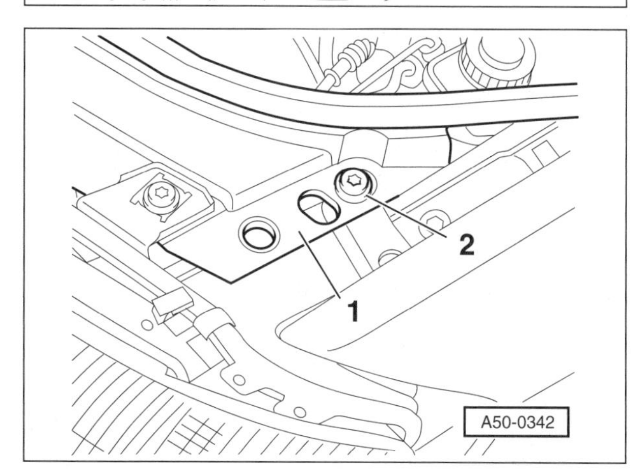

◀ Secure lock carrier in service position using bolts (**2**).

– Returning lock carrier to normal position is reverse of previous steps.

Tightening torques	
Lock carrier to fender	10 Nm (7 ft-lb)
Lock carrier to impact absorber	50 Nm (37 ft-lb)

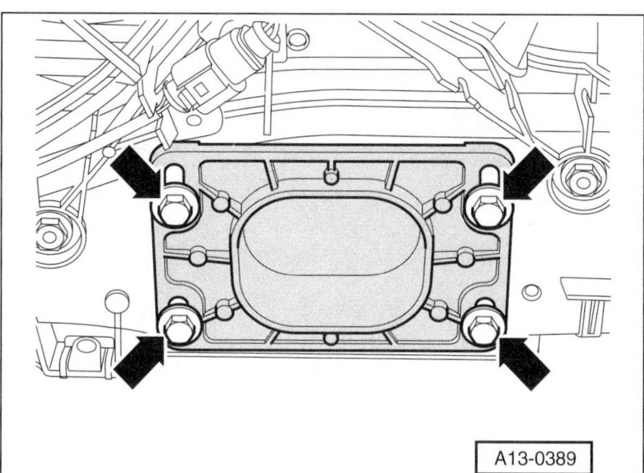

◀ Models with engine torque support, loosen support bolts (**arrows**).

– Allow torque support to rest on rubber buffer and tighten torque support bolts.

– All models: Reinstall bumper and bumper cover.

Tightening torques	
Torque support to lock carrier	28 Nm (21 ft-lb)
Front bumper to fender	10 Nm (7 ft-lb)
Front bumper to impact absorber	50 Nm (37 ft-lb)

BentleyPublishers.com
BentleyPublishers.com—All Rights Reserved

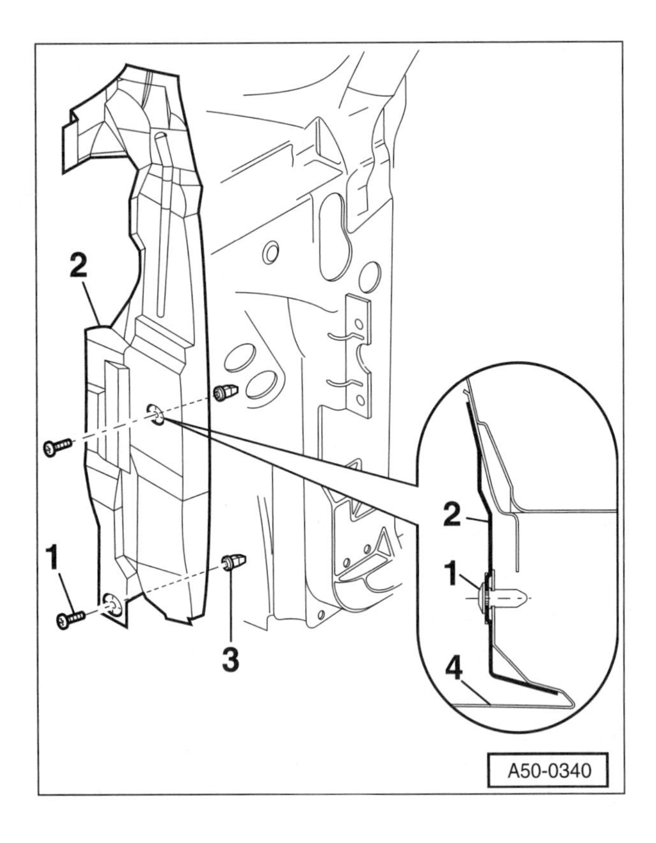

FRONT FENDER

Front fender end plate, removing

- Remove front wheel.
- Remove wheel housing liner.
- Remove bolts (**1**) from expansion clips (**3**).
- Remove plate (**2**) from fender (**4**).
- Installation is reverse of removal.

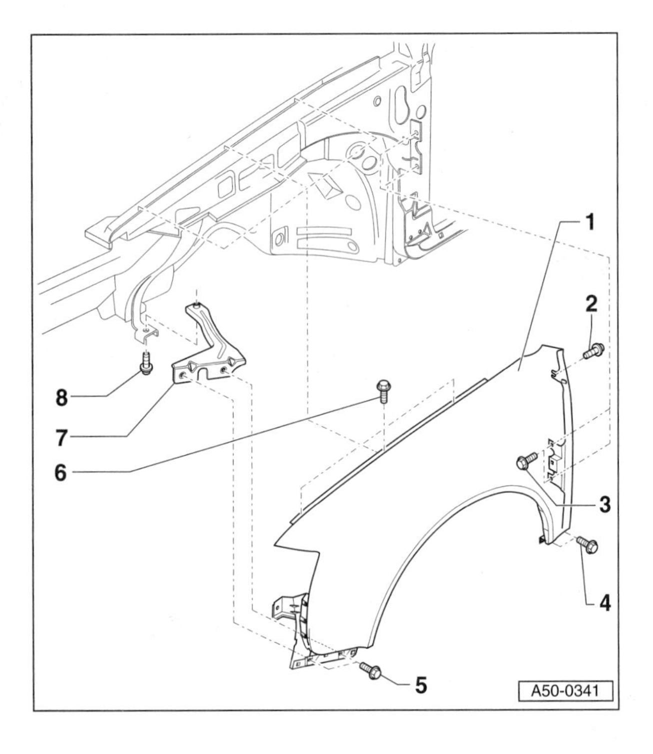

Front fender, removing and installing

- Remove bumper. See **63 Bumpers**.
- Remove front wheel.
- Remove wheel housing liner.
- Remove end plate. See **Front fender end plate, removing and installing** in this repair group.
- Remove lower engine cover. See **03 Maintenance**.
- Remove headlights. See **94 Exterior Lights**
- Disconnect side turn signal harness connector.
- Remove bolts (**2**, **3**, **4**, **5**, and **6**) and remove fender.
- Installation is reverse of removal.

Tightening torques	
Fender bolt to body (**2**, **3**, **4**, **5** and **8**)	10 Nm (7 ft-lb)
Fender bolt to body (**6**)	15 Nm (11 ft-lb)

BentleyPublishers.com
BentleyPublishers.com—All Rights Reserved

55 Hoods and Lids

General

This repair group covers removal and installation of the front hood, trunk lid, tailgate and associated components.

NOTE—

- *Removal of large body components, such as the hood or trunk lid, is best accomplished with a helper to avoid body damage.*

Aluminum body panels

Models covered by this manual include extensive use of aluminum body panels. Corrosion caused by contact between dissimilar metals may result if the wrong fastening components (screws, bolts, nuts, washers, rivets, plugs grommets, adhesives, etc.) are installed.

For this reason, fasteners coated with Dachromet (recognized by its greenish color) and rubber parts, plastic parts and adhesives that are electrically non-conductive are used on aluminum components.

Dachromet coated fasteners are available as replacement parts from your authorized Audi dealer.

NOTE—

- *If there is any doubt about whether old parts are suitable for re-use, always install new replacement parts.*
- *Damage from contact corrosion caused by improper assembly with unsuitable parts is not covered under warranty.*

BentleyPublishers.com
BentleyPublishers.com—All Rights Reserved

FRONT HOOD

Hood, removing and installing

NOTE —

- *The hood is made of aluminum. See* ***Aluminum body panels*** *in this repair group.*
- *Have an assistant support and lift hood during removal.*
- *Before removing hood, mark position of hinges to maintain alignment.*

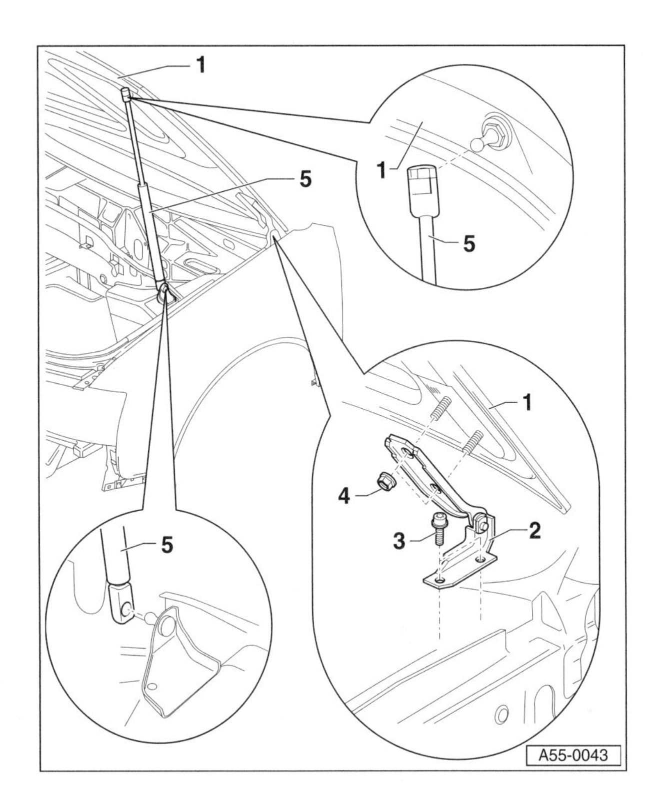

1. Hood
2. Hood hinge
3. Torx bolt, 15 Nm (11 ft-lb)
4. Nut, 15 Nm (11 ft-lb)
5. Gas strut
   - See **Hood strut, removing and installing** in this repair group.

– Disconnect hose to windshield washer system and unclip from hood.

– Remove gas strut (**5**) from hood (**1**). See **Hood strut, removing and installing** in this repair group.

– Remove nuts (**4**) and remove hood.

– Install in reverse order of removal. Perform the following adjustments (if necessary):
- Align hood (**1**) between fenders. See **Hood gap dimensions** in this repair group.
- Adjust height of hood at hood lock. See **Hood lock, removing and installing** in this repair group.
- Adjust height of hood relative to fenders. See **Hood stops, adjusting** in this repair group.

Hood strut, removing and installing

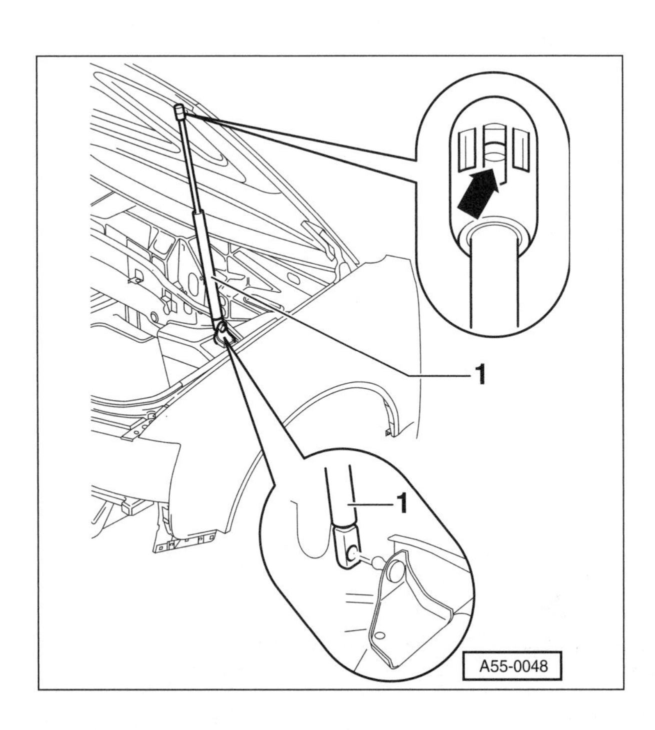

– Prop up and secure hood in open position.

◄ Lift retaining spring (**arrow**) slightly using small screwdriver and pull hood strut (**1**) off ball stud.

– Detach strut at bottom ball stud.

– To install, press strut onto ball studs so that it engages in position.

– Install with tube end of strut facing down.

BentleyPublishers.com
BentleyPublishers.com—All Rights Reserved

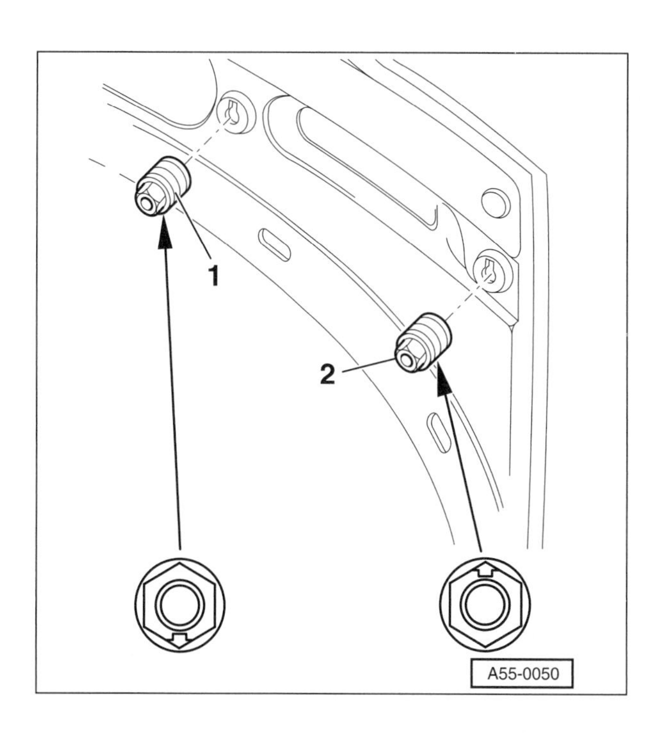

Hood stops, adjusting

NOTE —

- *Use hood stops to adjust hood height.*

Turn hood stops (**1** and **2**) until hood is flush with fenders.

– When hood is closed, hood stops should lightly contact hood at contact area.

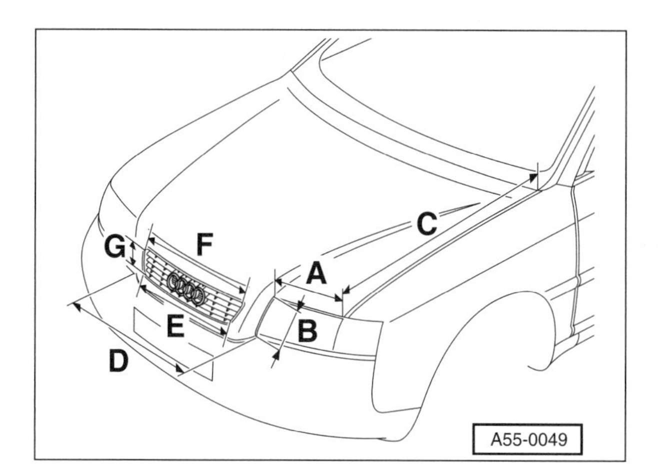

Hood gap dimensions

NOTE —

- *Use 3371 (or equivalent) adjustment gauge to check hood gap.*

Hood gap dimensions	
Dimension	**Specification**
A	5.0 +1/-0 mm (0.197 +0.039/-0 in)
B	5.0 +1/-0 mm (0.197 +0.039/-0 in)
C	3.0 +1/-0 mm (0.118 +0.039/-0 in)
D	5.0 +1/-0 mm (0.197 +0.039/-0 in)
E	2.0 ± 0.2 mm (0.079 ± 0.008 in)
F	2.0 ± 0.2 mm (0.079 ± 0.008 in)
G	2.5 ± 0.2 mm (0.098 ± 0.008 in)

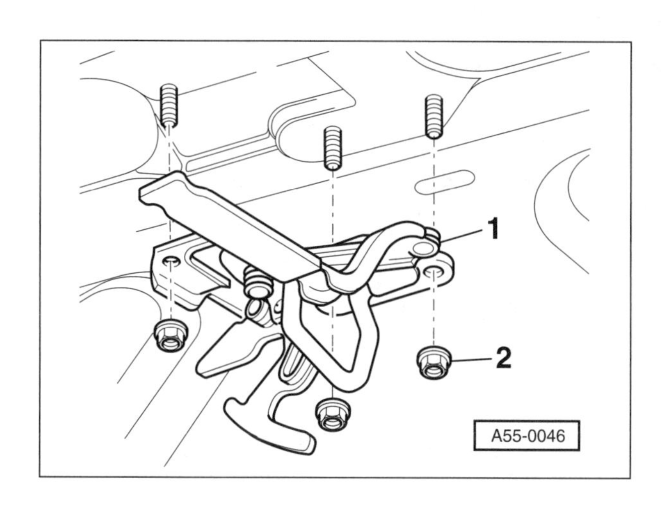

Hood lock, removing and installing

Remove nuts (**2**) and remove hood striker (**1**).

– Disengage hood lock cable. See **Hood lock cable, removing and installing** in this repair group.

Tightening torque	
Hood striker to hood	8 Nm (71 in-lb)

BentleyPublishers.com
BentleyPublishers.com–All Rights Reserved

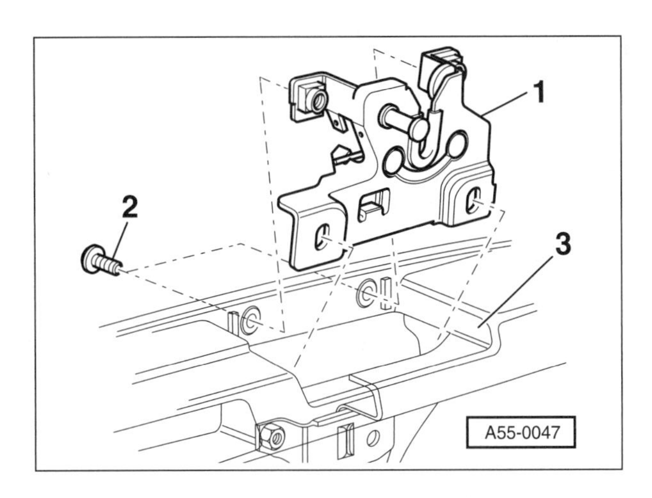

Remove 4 bolts (**2**).

Tightening torque	
Hood lock to lock carrier	10 Nm (7 ft-lb)

– Remove lower section of lock (**1**) from lock carrier (**3**).

– To adjust hood, move lower section of lock up or down as required.

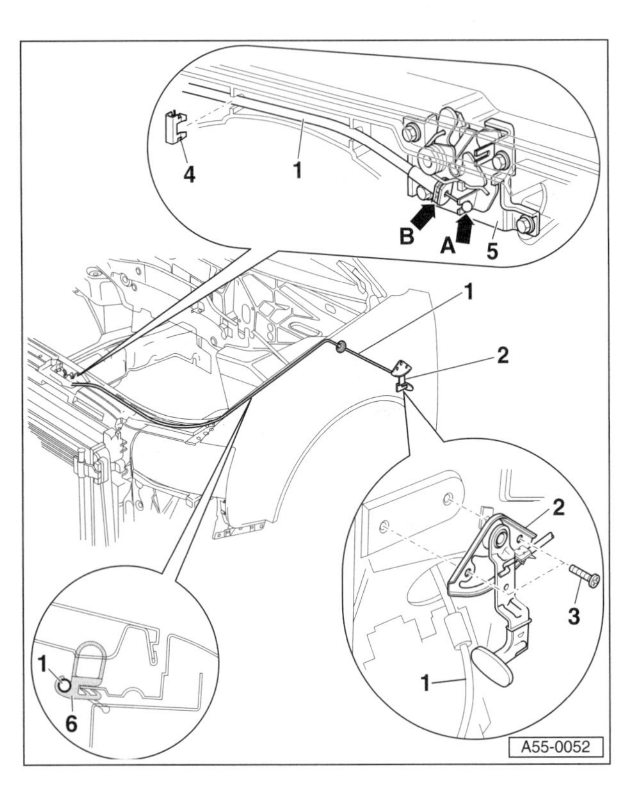

Hood lock cable, removing and installing

1. Hood lock cable
   - Remove storage compartment on driver's side. See **70 Interior Trim**.
   - Remove screws (**3**).
   - Detach cable at support (arrow **B**) and disengage end of cable (arrow **A**) at lower section of hood lock (**5**).
   - Remove securing clip (**4**).
   - Remove cable (**1**) from guide channel in seal (**6**) and pull cable through bulkhead into interior of vehicle.
   - Installation is reverse of removal.
2. Hood lock release
3. Self-tapping screw, 1.5 Nm (13 in-lb)
4. Securing clip
5. Lower section of hood lock
   - Can be moved to adjust hood height.

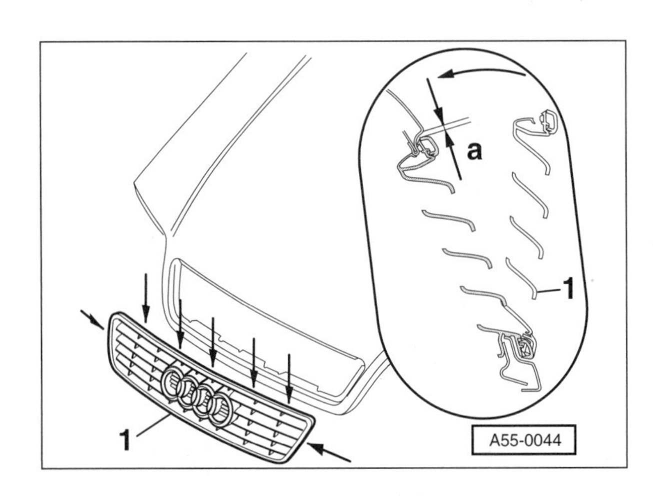

Radiator grill, removing and installing

Release catches (**arrows**).

– Swing out radiator grill at top, then pull off at bottom.

– Install radiator grill at bottom in center (make sure it is properly centered) and pivot it into place in direction shown in inset (**arrow**).

– Dimension **a** = 2 ± 0.5 mm (0.079 ± 0.019 in)

– Make sure radiator grill engages audibly in position.

BentleyPublishers.com
BentleyPublishers.com—All Rights Reserved

Trunk Lid

Trunk lid, removing and installing

NOTE—

- *Have an assistant support the lid during removal.*
- *Before removing trunk lid, mark position of hinges to maintain alignment.*

1. Trunk lid
2. Lid stops (2x)
3. Expanding nuts
4. Self-tapping screws, 1.5 Nm (13 in-lb)
5. Holder
6. Striker plate
7. Nuts, 8 Nm (71 in-lb)

– Remove warning triangle and holder.

– Remove trunk lid lining.

– Disconnect electrical harness connectors.

– Remove trunk lid gas strut. See **Trunk lid strut, removing and installing** in this repair group.

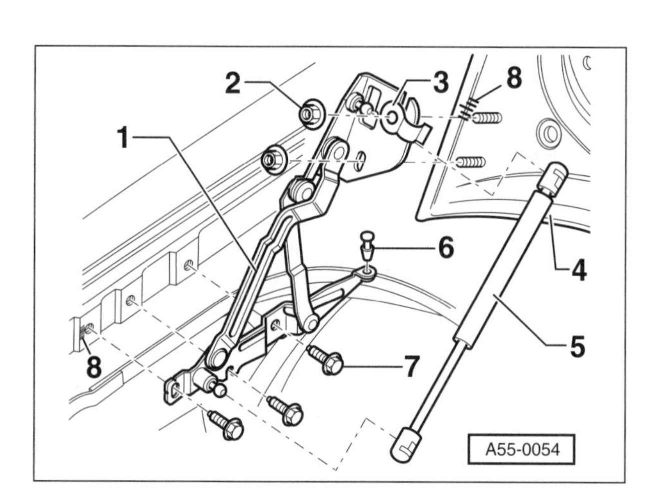

◀ Remove nuts (**2**) at hinge and carefully remove lid.

– Install in reverse order of removal. Perform the following adjustments (if necessary):

- Adjust lid height at trunk lid hinges. See **Trunk lid hinge, removing and installing** in this repair group.
- Adjust lid stops. See **Trunk lid stops, adjusting** in this repair group.
- Adjust striker plate. See **Trunk lid striker plate, adjusting** in this repair group.
- Check that trunk lid gap dimensions are correct. See **Trunk lid gap dimensions** in this repair group.

Trunk lid strut, removing and installing

– Using screwdriver, push back retaining springs slightly at both ends of gas strut and unclip strut.

NOTE—

- *When removing or installing the gas strut, push back the opened trunk lid slightly.*

– To install, clip in gas strut at bottom first, then at top.

– Install with tube end of strut facing up.

BentleyPublishers.com
BentleyPublishers.com—All Rights Reserved

Trunk lid stops, adjusting

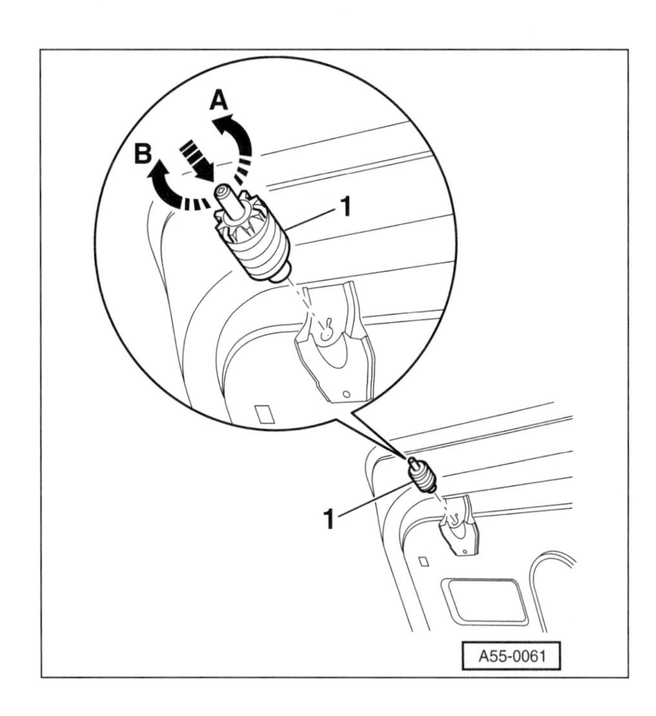

- Open trunk lid.
- Press in stop buffer spring pin in direction of arrow using Phillips-head screwdriver.
- Lock spring pin by turning counterclockwise (arrow **A**).
- Screw in stop buffer (**1**) until fully home.
- Unscrew stop buffer again in stages.
- Adjust so that lid stops make light contact with rear sill trim when trunk lid is closed.
- Unlock spring pin by turning clockwise (arrow **B**).

Trunk lid striker plate, adjusting

- Loosen striker plate nuts. Retighten nuts until finger-tight (striker plate should still be movable).
- Carefully close trunk lid until flush with rear side panels.
- Carefully open trunk lid and tighten striker plate nuts.

Tightening torque	
Trunk lid striker plate to body	8 Nm (71 in-lb)

Trunk lid gap dimensions

NOTE —
- *Use 3371 (or equivalent) adjustment gauge to check trunk lid gap.*

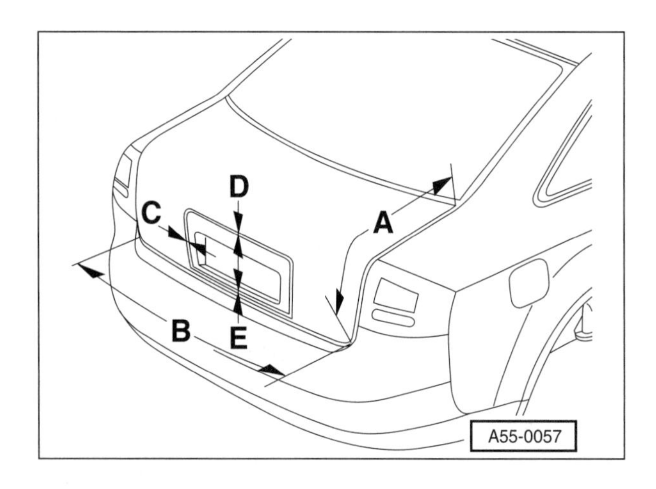

Trunk lid gap dimensions	
Dimension	**Specification**
A	3 +1/-0 mm (0.118 +0.039/-0 in)
B	5 +1/-0 mm (0.197 +0.039/-0 in)
C	1.5 ± 0.3 mm (0.059 ± 0.012 in)
D	Flush with trunk lid
E	1.5 -0.5/+0 mm (0.059 -0.019/+0 in)

BentleyPublishers.com
BentleyPublishers.com—All Rights Reserved

Trunk lid hinge, removing and installing

Have an assistant support the lid during hinge removal.

– Unclip trunk lid hinge cover from top of hinge and remove.

– If necessary, remove trunk lid gas strut. See **Trunk lid strut, removing and installing** in this repair group.

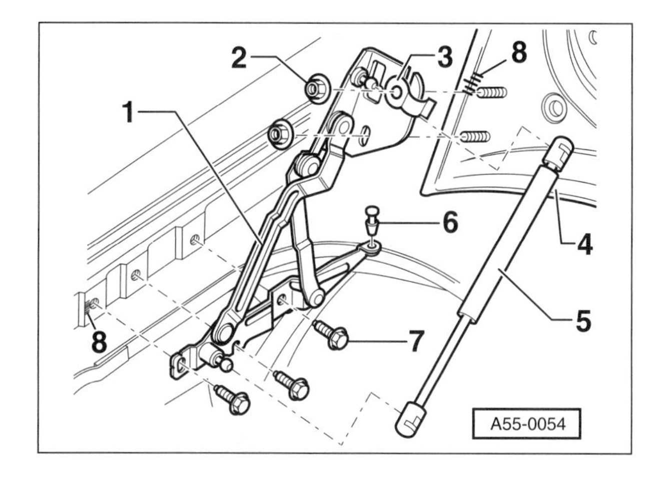

< Loosen top nuts (**2**) on hinge at trunk lid.

– Remove bottom nuts on hinge at trunk lid.

– Lift trunk lid off hinge and support.

– Remove left and right bolts (**7**) on hinge at body.

– Loosen, but do not remove center bolt on hinge at body.

– Remove hinge (**1**).

– When reinstalling hinge and trunk lid check trunk lid gap. See **Trunk lid, removing and installing** in this repair group.

NOTE —

- *For easier adjustment of the trunk lid hinge, use graduations (**8**) marked on the hinge and the trunk lid.*

Tightening torques	
Trunk lid hinge to body	21 Nm (15 ft-lb)
Trunk lid hinge to trunk lid	21 Nm (15 ft-lb)

Trunk lid seal, installing

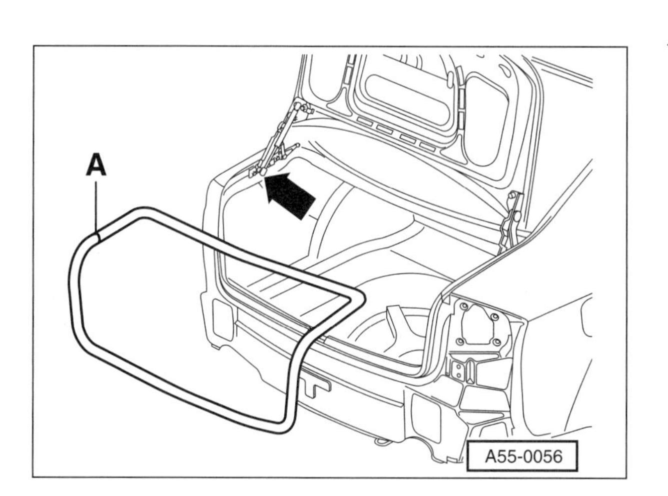

< Align butt joint (**A**) of the trunk lid seal with the ball socket of the left hinge (**arrow**).

– Press seal into place.

BentleyPublishers.com
BentleyPublishers.com—All Rights Reserved

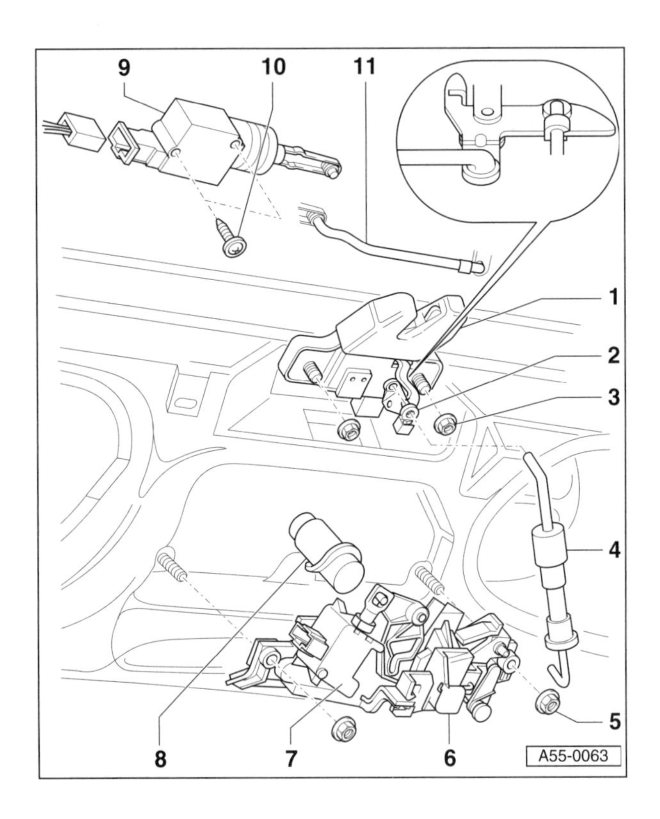

Trunk lid lock, removing and installing

1. Trunk lid lock
   - To remove, swing out hinge arm (**2**) from operating rod (**4**) and unhook from trunk lid lock.
   - Remove nuts (**3**) and remove trunk lid lock.
2. Hinge arm
   - After installing trunk lid lock, swing it over operating rod.
3. Nuts, 8 Nm (71 in-lb)
4. Operating rod
5. Nut, 8 Nm (71 in-lb)
6. Handle mechanism
   - To remove, remove trunk lid trim.
   - Unclip operating rod (**4**) from handle mechanism.
   - Unscrew nuts (**5**) and remove handle mechanism.
7. Trunk lid lock actuator
8. Lock cylinder
   - Remove release handle.
   - Remove operating rod.
   - Remove lock cylinder out of handle mechanism.
9. Trunk lid lock motor
10. Screws (2x)
11. Operating rod

Trunk lid lock motor, removing and installing

– Remove trunk lid lining.

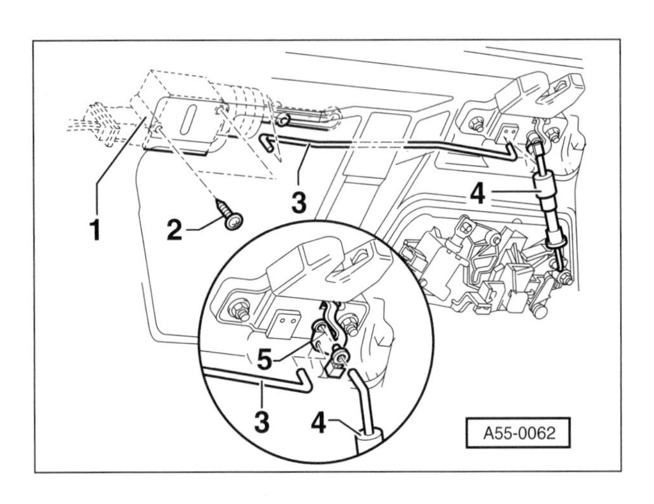

◄ Unclip operating rod (**3**).

– Remove screws (**2**) (2x).

– Disconnect harness connector from motor (**1**).

– Installation is reverse of removal.

NOTE—
- *Do not unclip operating rod (**4**) from fulcrum lever (**5**).*
- *Make sure the harness connector is connected to the motor before installation.*

BentleyPublishers.com
BentleyPublishers.com—All Rights Reserved

TAILGATE (AVANT)

Tailgate, removing and installing

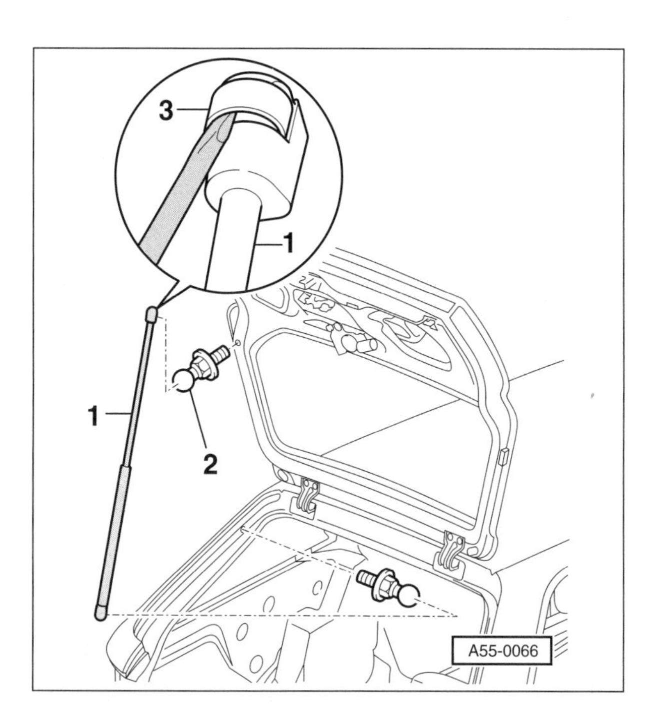

- Remove tailgate trim. See **70 Interior Trim**.
- Disconnect or remove harness connectors and hoses for tailgate lock and rear window washer.
- Prop up and secure tailgate in open position.
- ◀ Using screwdriver, lift upper strut retaining spring (**3**) as shown and remove gas strut (**1**) from upper ball stud (**2**).
- Lift lower strut retaining spring slightly and pull gas strut (**1**) off lower ball stud.

NOTE—
- *Before removing tailgate, mark position of hinges to maintain tailgate alignment.*
- *Have an assistant support the tailgate during removal.*

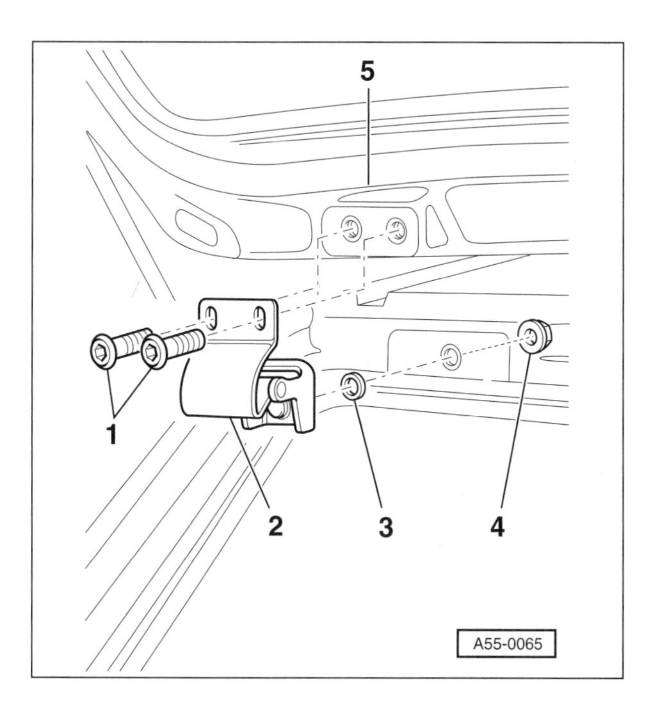

- ◀ Remove bolts (**1**) from hinges (**2**) and remove tailgate.
- To reinstall, align hinges to tailgate using marks made previously and fasten hinges.
- Install tube end of strut towards body.
- Press gas strut onto ball stud so that it engages in position.
- Adjust tailgate alignment (if necessary). See **Tailgate, adjusting** in this repair group.

Tightening torque	
Ball stud to tailgate and body	15 Nm (11 ft-lb)
Hinge bolts to tailgate	21 Nm (15 ft-lb)

BentleyPublishers.com
BentleyPublishers.com—All Rights Reserved

Tailgate, adjusting

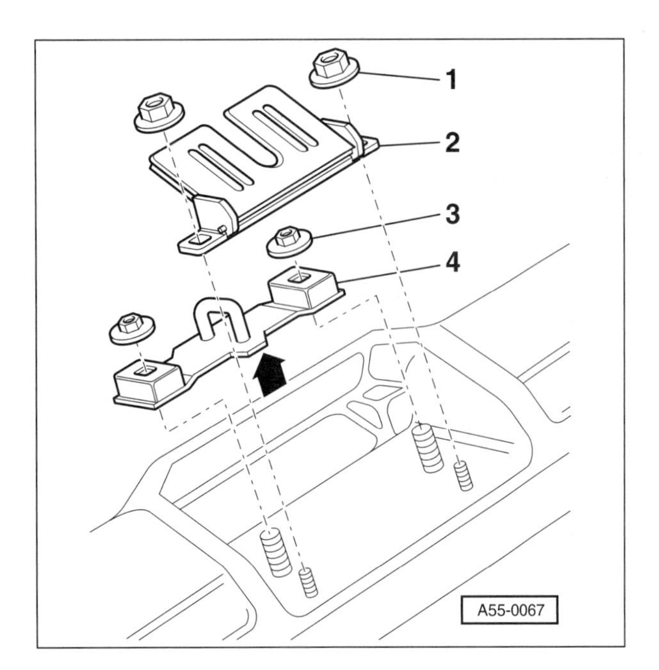

– Remove rear sill trim. See **70 Interior Trim**.

< Unscrew nuts (**1**) and remove cover for panel (**2**).

– Loosen nuts (**3**) until striker (**4**) is moveable.

NOTE—

- *When installing striker, note the correct installation position (the tip (**arrow**) must point toward the back).*
- *The stop on the body side is two-part and slides into basic position when installing the rear hatch.*

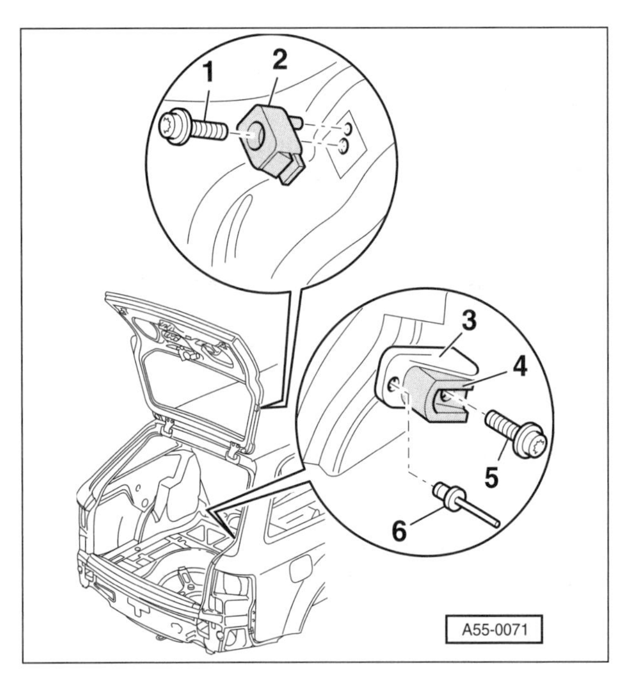

< Loosen screw (**5**) and completely remove upper part (**4**) of stop (**3**).

– Tighten screw (**5**) slightly.

– Close tailgate and press approx. 2 mm (0.079 in.) below basic position.

– Open tailgate and pull out upper part of stop three notches.

– Tighten screw (**5**) to 8 Nm (71 in-lb).

– When installing, fasten stop (**3**) with rivet (**6**).

– When installing stop (**2**), tighten screw (**1**) to 8 Nm (71 in-lb).

BentleyPublishers.com
BentleyPublishers.com—All Rights Reserved

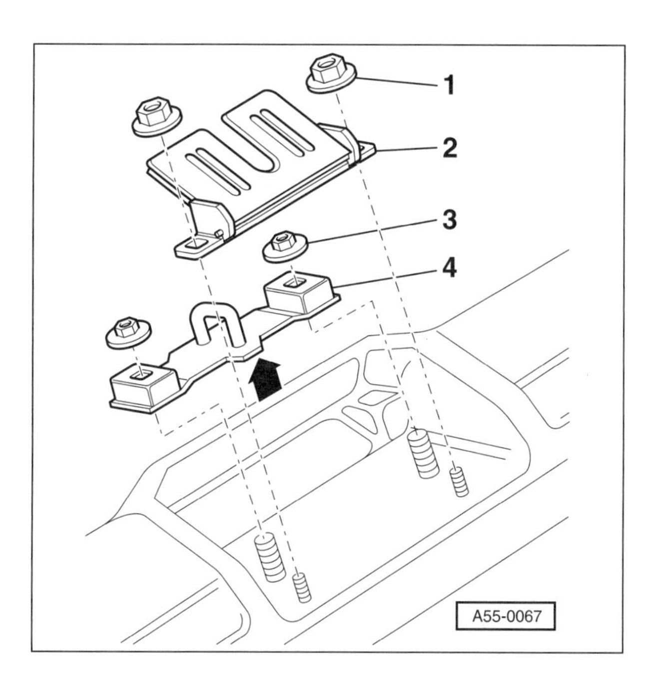

◀ Tighten nuts (**3**) to 21 Nm (15 ft-lb).

– Check tailgate alignment. See **Tailgate gap dimensions** in this repair group.

– Install cover for lock bracket (**2**) and tighten nuts (**1**) to 5 Nm (44 in-lb).

Tailgate hinge, removing and installing

– Remove rear roof trim. See **70 Interior Trim**.

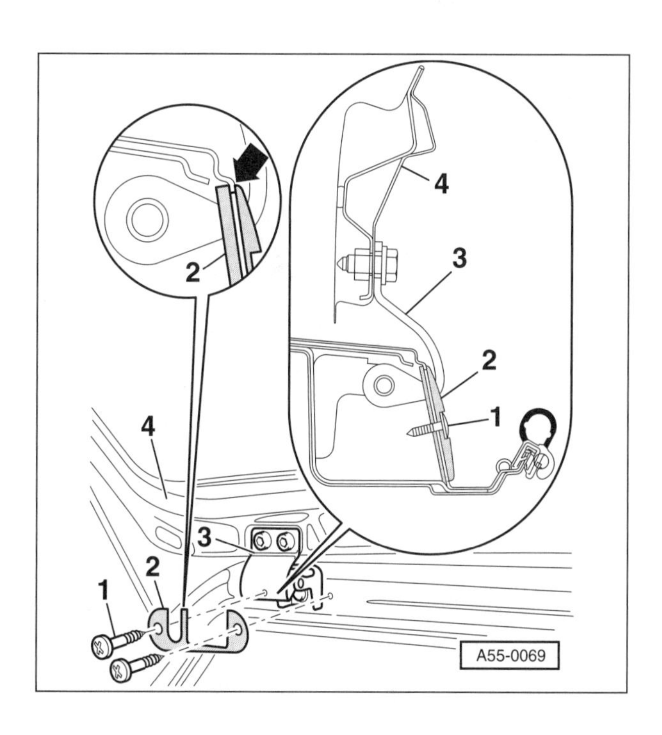

◀ Remove screws (**1**) and remove cover (**2**) for hinge (**3**).

– When installing, the cover must first be centered in the notches (**arrow**) up in the roof flange.

BentleyPublishers.com
BentleyPublishers.com—All Rights Reserved

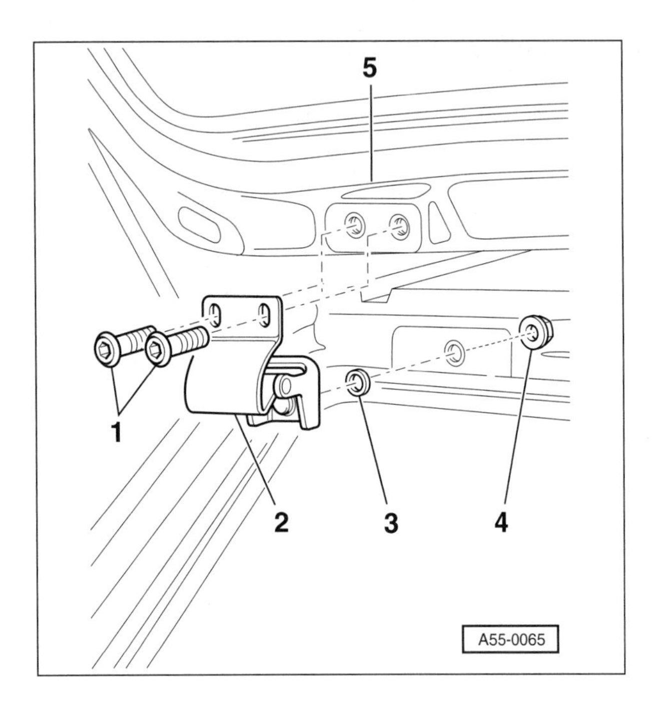

◀ Remove nut (**4**).

– Remove bolts (**1**) and hinge (**2**).

– Installation is reverse of removal. Remember to replace seal (**3**).

Tailgate seal, removing and installing

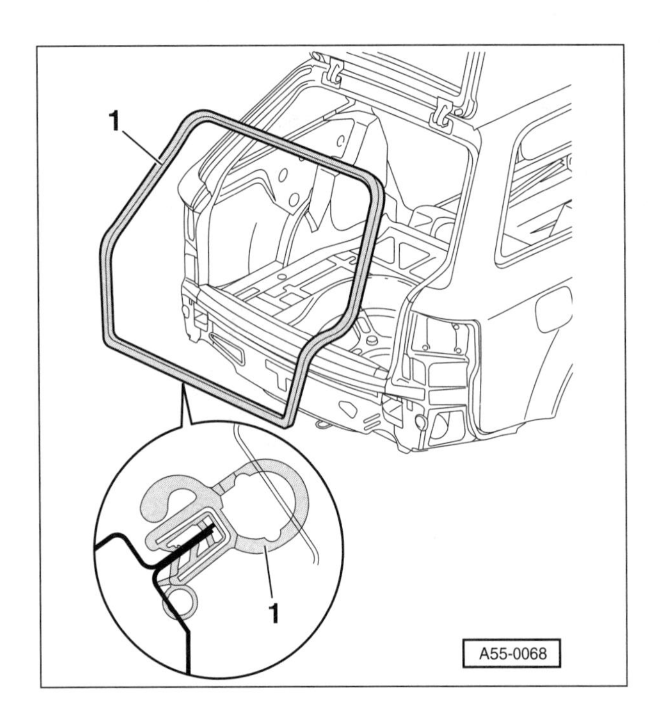

◀ Open tailgate and remove seal.

– When installing, first place seal at left and right in corners.

– Press seal (**1**) onto body as shown.

BentleyPublishers.com
BentleyPublishers.com—All Rights Reserved

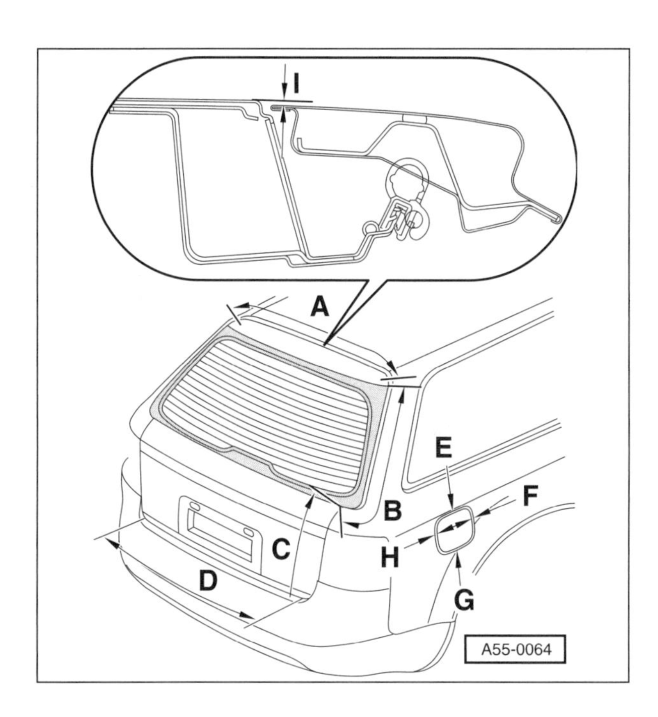

Tailgate gap dimensions

NOTE—
- *Use 3371 (or equivalent) adjustment gauge to check tailgate gap.*

Tailgate gap dimensions	
Dimension	**Specification**
A	4 +1/-0 mm (0.157 +0.039/-0 in)
B	4 +1/-0 mm (0.157 +0.039/-0 in)
C	4 +1/-0 mm (0.157 +0.039/-0 in)
D	5 +1/-0 mm (0.197 +0.039/-0 in)
E	2.5 ± 0.1 mm (0.098 ± 0.004 in)
F	3.3 ± 0.1 mm (0.130 ± 0.004 in)
G	3 ± 0.2 mm (0.118 ± 0.008 in)
H	2.5 ± 0.4 mm (0.098 ± 0.016 in)
I	0 to ± 1 mm (0 to 0.039 in)

Tailgate lock and handle mechanism, removing and installing

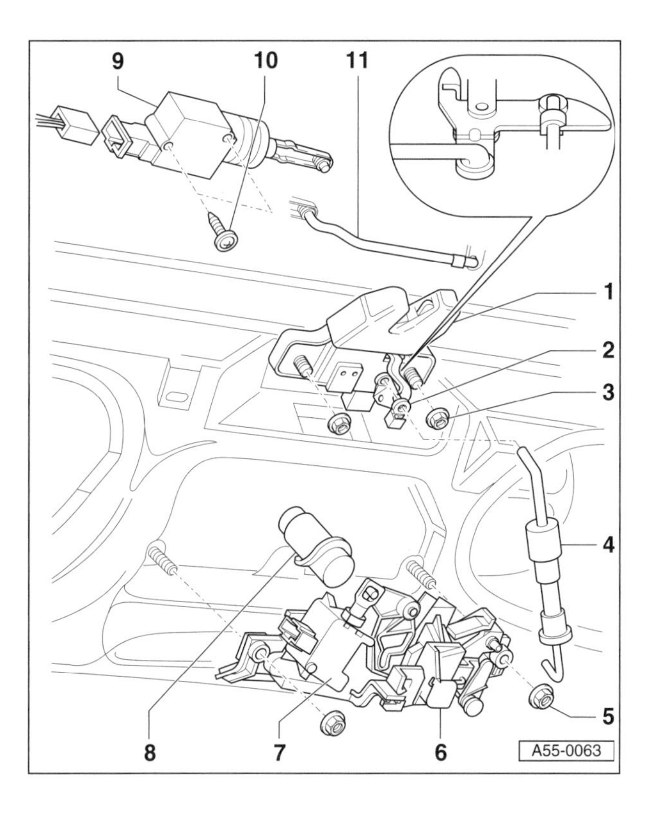

NOTE—
- *The tailgate lock consists of two locks (two stage).*
- *For installation to be correct, the tailgate lock must lock in the second stage (main lock).*

1. Tailgate lock
   - To remove, swing out hinge (**2**) from operating rod (**4**) and unhook from tailgate lock.
   - Remove nuts (**3**) and remove tailgate lock.
2. Hinge arm
   - After installing tailgate lock move operating rod over.
3. Nuts, 8 Nm (71 in-lb)
4. Operating rod
   - Adjust operating rod so there is no tension or play.
5. Nuts, 8 Nm (71 in-lb)
6. Handle mechanism
   - Remove tailgate trim.
   - Unclip operating rod (**4**) from handle mechanism.
   - Remove nuts (**5**) and remove handle mechanism.
7. Actuator element
   - See **Tailgate lock actuator, removing and installing** in this repair group.
8. Lock cylinder
   - Remove handle mechanism.
   - Remove operation operating rod.
   - Remove lock cylinder from handle mechanism.

BentleyPublishers.com
BentleyPublishers.com—All Rights Reserved

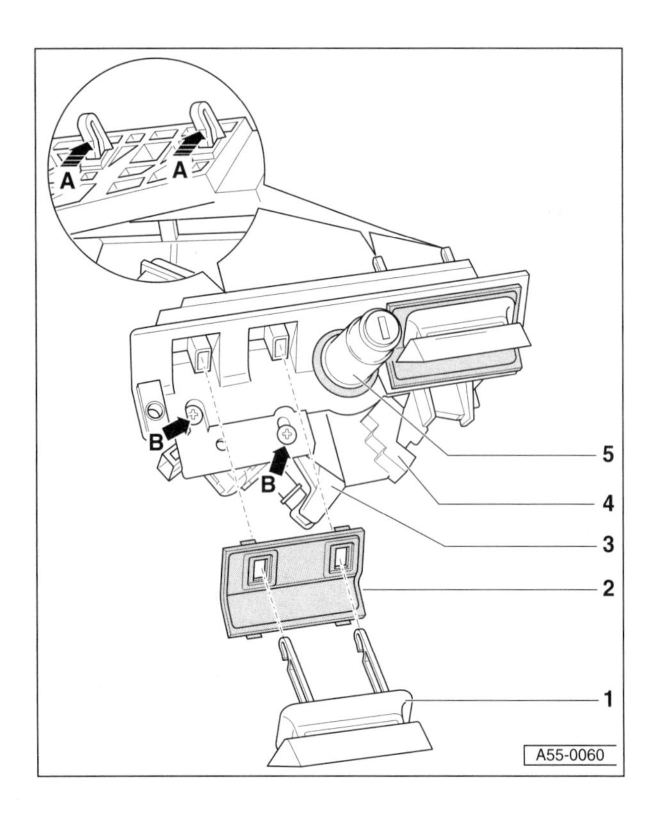

Tailgate lock actuator, removing and installing

1. Release handle
   - Loosen lock and remove handle.
2. Seal
3. Lock actuator
   - Remove tailgate trim and lock system.
   - Using screwdriver, press in and disengage locating hooks on left release handle (**1**) in direction of arrow (**A**).
   - Pull left release handle (**1**) downward and out of handle mechanism (**4**).
   - Remove screws (**B**) and lock actuator.
4. Handle mechanism
   - See **Tailgate lock handle mechanism, removing and installing** in this repair group.
5. Lock cylinder
   - Spray with sealant.

BentleyPublishers.com
BentleyPublishers.com—All Rights Reserved

57 Doors and Locks

GENERAL

This repair group covers the front and rear door assemblies. Each door includes a door component carrier that can be removed without removing the door. Removing the component carrier provides access to the door lock, window regulator, window motor and other components.

For related information see:

- **64 Door Windows** for window glass, regulator and motor service
- **70 Interior Trim** for door panels and trim
- **91 Radio and Communication** for door speakers
- **96 Interior Lights, Switches, Anti-theft** for window switches and central locking

DOORS

Front door service procedures are described in this repair group. Rear door is similar. See the assembly diagrams in this section for slight variations in front and rear door design.

Front door assembly

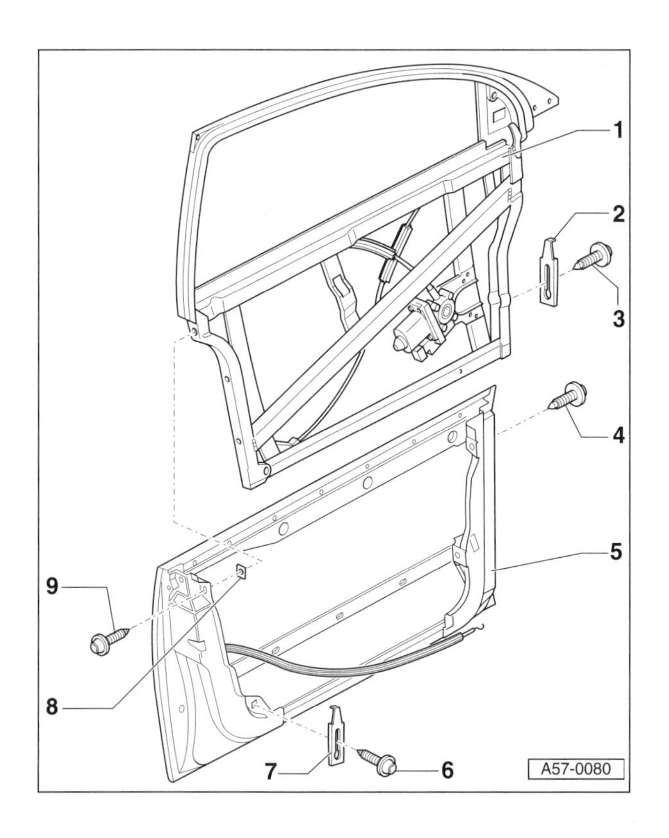

1. Door component carrier
2. Adjusting plate.
3. Bolt w/washer assembly, 30 Nm (22 ft-lb)
4. Bolt w/washer assembly, 30 Nm (22 ft-lb)
5. Front door
6. Bolt w/washer assembly, 30 Nm (22 ft-lb)
7. Adjusting plate
8. Washer
9. Bolt w/washer assembly, 30 Nm (22 ft-lb)

BentleyPublishers.com
BentleyPublishers.com—All Rights Reserved

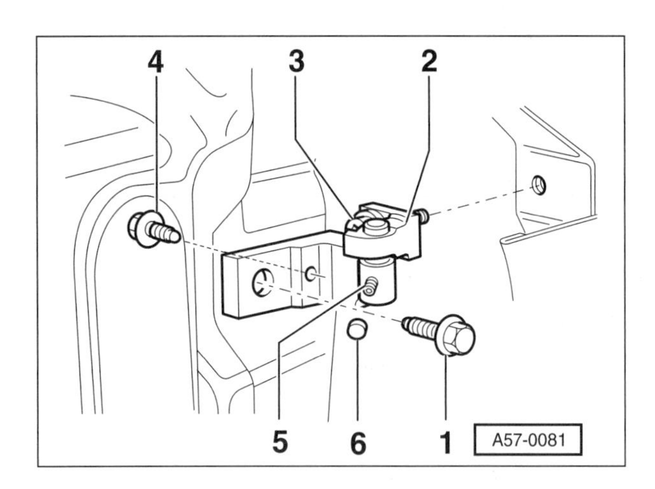

Front door upper hinge

1. Bolt, 30 Nm (22 ft-lb)
   - Install from inside vehicle
2. Upper door hinge
3. Torx bolt 30 Nm (22 ft-lb)
4. Bolt 30 Nm (22 ft-lb)
   - Install from inside vehicle
5. Stud, 23 Nm (17 ft-lb)
6. Cap

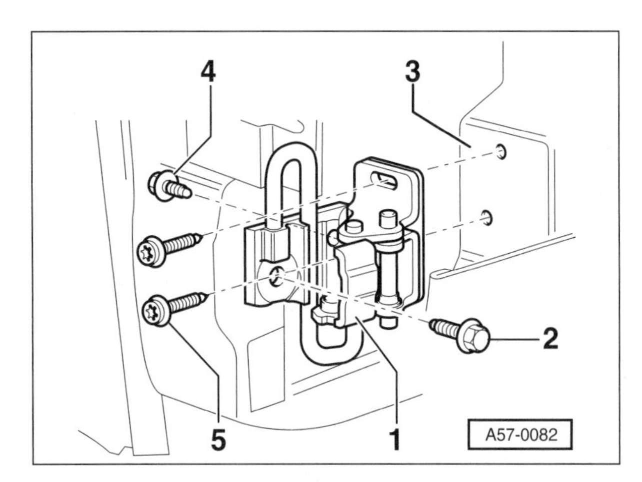

Front door lower hinge

1. Lower door hinge
2. Bolt, 30 Nm (22 ft-lb)
3. Front door
4. Bolt, 30 Nm (22 ft-lb)
5. Torx bolt 30 Nm (22 ft-lb)

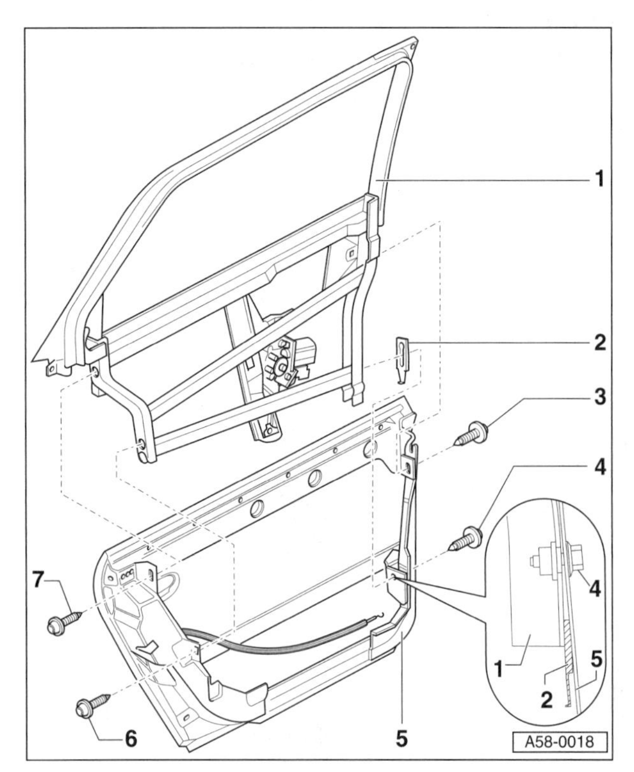

Rear door assembly

1. Door component carrier
2. Adjusting plate
3. Bolt w/washer assembly, 30 Nm (22 ft lb)
4. Bolt w/washer assembly, 30 Nm (22 ft lb)
5. Rear door
6. Bolt w/washer assembly, 30 Nm (22 ft lb)
7. Bolt w/washer assembly, 30 Nm (22 ft lb)

BentleyPublishers.com
BentleyPublishers.com—All Rights Reserved

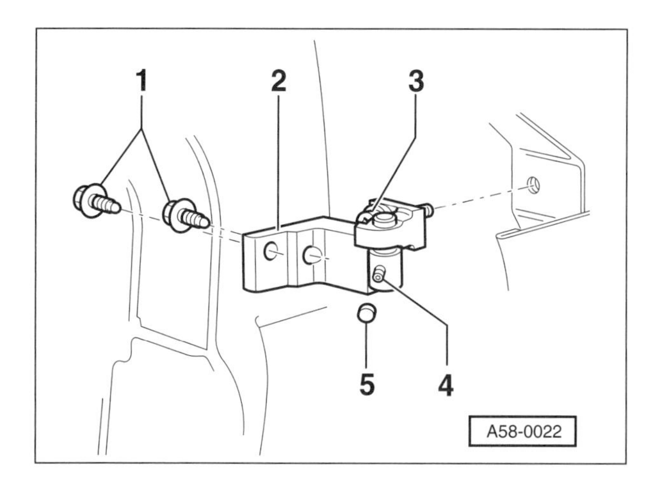

Rear door upper hinge

1. Bolt, 30 Nm (22 ft-lb)
2. Upper door hinge
3. Torx bolt, 30 Nm (22 ft-lb)
4. Stud, 23 Nm (17 ft-lb)
5. Cap

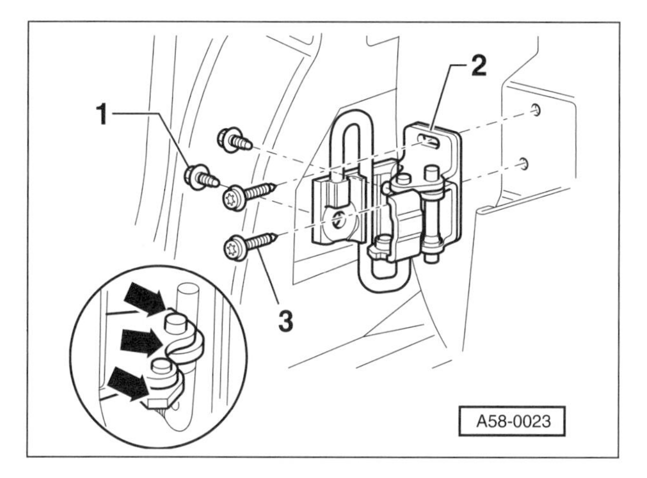

Rear door lower hinge

1. Bolt, 30 Nm (22 ft-lb)
2. Lower door hinge
3. Torx bolt, 30 Nm (22 ft-lb)

NOTE—

*• Lubrication points indicated by **arrows**.*

Door, removing and installing

NOTE—

• Front door removal described, rear door is similar

• To remove component carrier without removing door, see **Door component carrier, removing and installing** *in this repair group.*

– Remove bottom section of A-pillar trim. See **70 Interior Trim**.

– Disconnect harness connectors at connection point on A-pillar.

– Unclip rubber boot between front door and A-pillar and pull wiring harness out from A-pillar.

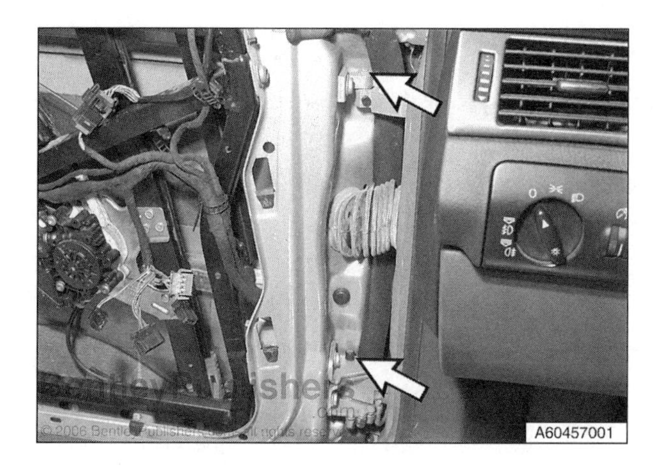

< Remove Torx bolts securing upper and lower hinges to door and remove door with door component carrier.

– Installation is reverse of removal. Remember to check door adjustment. See **Door gap, adjusting** in this repair group.

BentleyPublishers.com
BentleyPublishers.com—All Rights Reserved

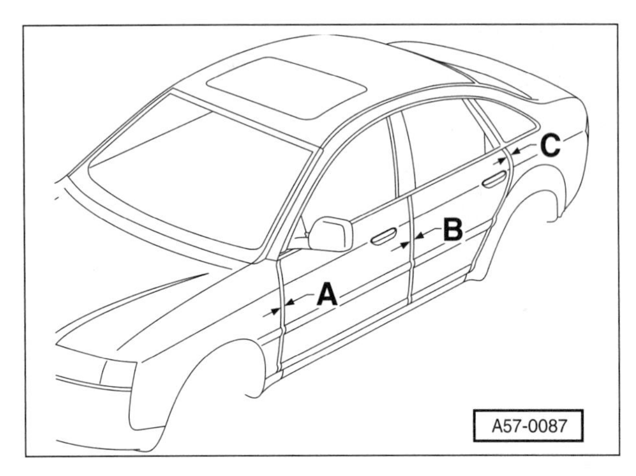

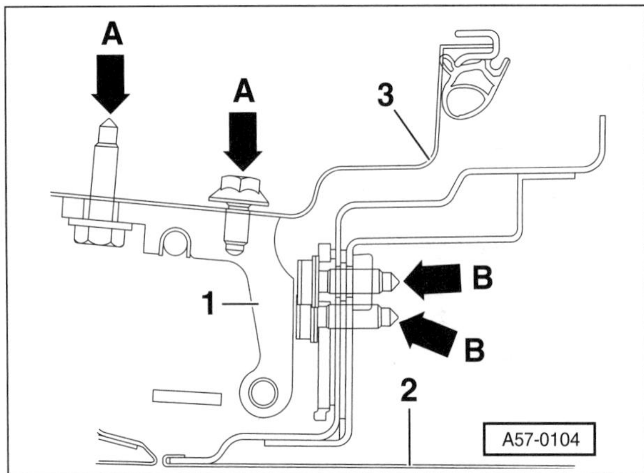

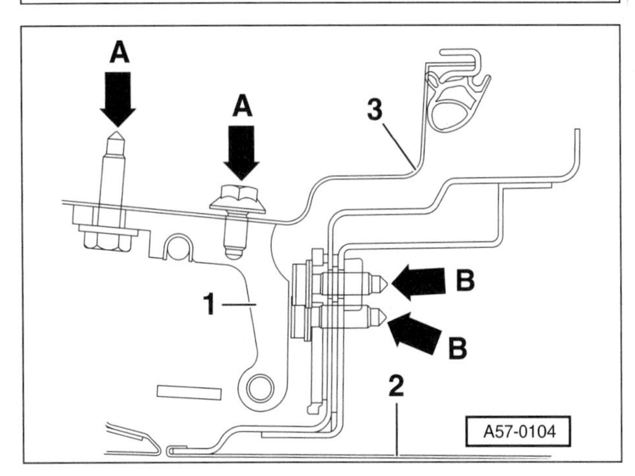

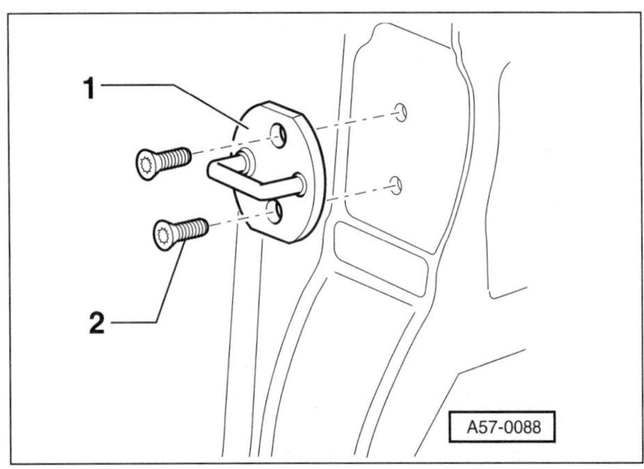

Door gap, adjusting

NOTE—

- *First adjust rear door (if necessary).*
- *Check gap dimensions using 3371 adjustment gauge (or equivalent).*

Door gap dimensions	
Dimension	**Specification**
A	3 +1/-0 mm (0.118 +0.039/-0 in)
B	4.5 ± 0.5 mm (0.177 0.019 in)
C	3.5 ± 0.5 mm (0.138 0.019 in)

Adjusting door longitudinally

1. Door hinge
2. A-pillar
3. Door

– Adjust door longitudinally by loosening hinge bolts on (**A**) at upper and/or lower door hinge and moving door in elongated holes in hinge on A-pillar.

NOTE—

- *Lower door hinge shown.*

Adjusting door flush with contour of body

Adjust door inward or outward by loosening upper and/or lower door hinge bolts (**B**) and moving door within elongated holes in hinge.

– Adjust the rear end of door by altering the position of the striker plate inward or outward.

Tightening torque	
Door hinge to A-pillar	30 Nm (22 ft-lb)
Door hinge to door	30 Nm (22 ft-lb)

Adjusting door at striker plate

Use this adjustment only to align door inward or outward, not vertically.

– Loosen bolts (**2**) and move striker plate (**1**) until door shell is flush with contour of body.

Tightening torque	
Striker plate bolts to body	20 Nm (15 ft-lb)

BentleyPublishers.com
BentleyPublishers.com—All Rights Reserved

Door component carrier, removing and installing

NOTE—

- *Front door shown, rear door is similar.*
- *Door can be removed without removing component carrier. See* **Door, removing and installing** *in this repair group.*
- *Before removing component carrier, mark position with felt-tip pen.*

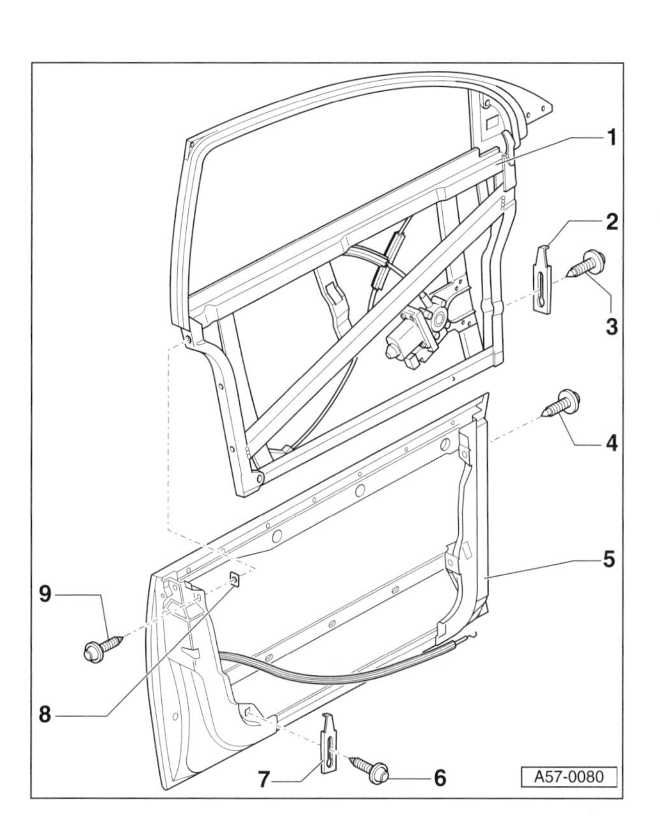

– Remove door trim panel. See **70 Interior trim**.

– Disconnect harness connector and unclip harness from door component carrier.

< Remove component carrier mounting bolts (**3**, **9**, **6**, and **4**). Lift component carrier upward and out of door.

– Before installing component carrier, check door adjustment. See **Door gap, adjusting** in this repair group.

– Insert component carrier into door and align with marks made previously.

– Secure carrier loosely with bolts (**3**, **4**, **6** and **9**) and adjusting plate (**2** and **7**).

– Shut door and check adjustments. See **Door component carrier, adjusting** in this repair group.

– When adjustment is correct, tighten component carrier to door in specified sequence (**3**, **9**, **6**, **4**).

Tightening torque	
Door component carrier to door	30 Nm (22 ft-lb)

Door component carrier, adjusting

– If reinstalling door, adjust door first. See **Door gap, adjusting** in this repair group.

– Remove door trim panel. See **70 Interior Trim**.

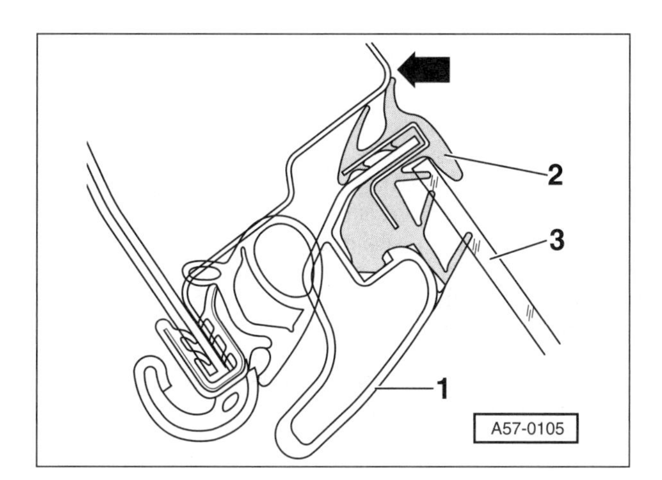

< Align door component carrier by altering position of adjusting plate at bottom of door. Make sure sealing lip (**2**) conforms to contour of A-pillar and roof frame.

– Adjust door component carrier height.

– If height adjustment is correct, loosen securing bolts at bottom of door component carrier and press carrier (**1**) in toward center of vehicle.

– Push carrier in slightly further to compress rubber seal.

– Position adjusting plate at B-pillar as necessary.

– Tighten lower securing bolts to 30 Nm (22 ft-lb).

NOTE—

- *Item* **3** *is the door glass.*

BentleyPublishers.com
BentleyPublishers.com—All Rights Reserved

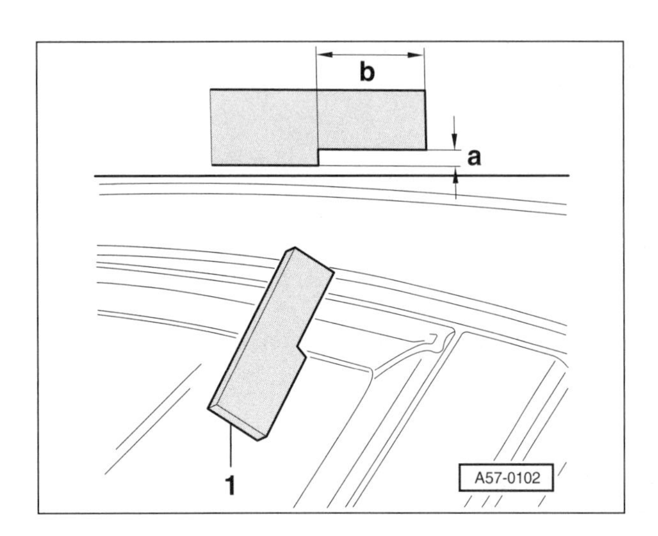

Check adjustment using shop-made distance gauge.

- Make a distance gauge according to dimensions given below using suitable material (e.g. plastic or wood):
  - Dimension a = 3.7 mm (0.146 in)
  - Dimension b = 28.2 mm (1.110 in)
- If door component carrier is correctly adjusted, gauge should contact door window and roof trim molding when applied at a point approx. 100 mm (3.94 in) from B-pillar.

Door seal, removing and installing

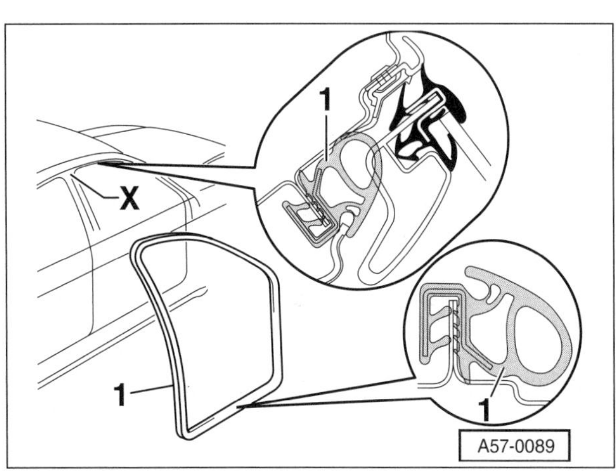

- Remove trim panels at top and bottom of A and B-pillars together with sill trim panel. See **70 Interior Trim**.
- Pull door seal off door flange.

To install, place end of door seal PVC lip onto flange at marking **X** (5 mm, or 0.197 in) near B-pillar.

- Press door seal (**1**) onto door flange all around door opening.

NOTE—

• Front door shown, rear door is similar.

DOOR LOCKS

Central locking service notes

- Lock actuators in the doors are integrated in the lock and have 2 electric motors.
- Lock actuators cannot be replaced separately.
- The first electric motor locks the door and the second locks the interior door mechanism.
- Locked doors can only be opened electrically from inside the car.
- Lock actuators for the fuel filler flap and rear lid each have one electric motor, and can be replaced separately.
- After an accident that activates an airbag, the control module for central locking will unlock any active door locks.
- The control module for central locking is located in a protective box beneath the driver's seat and carpet.

See **96 Interior Lights, Switches, Anti-theft** for locations of central locking and anti-theft alarm components.

BentleyPublishers.com
BentleyPublishers.com—All Rights Reserved

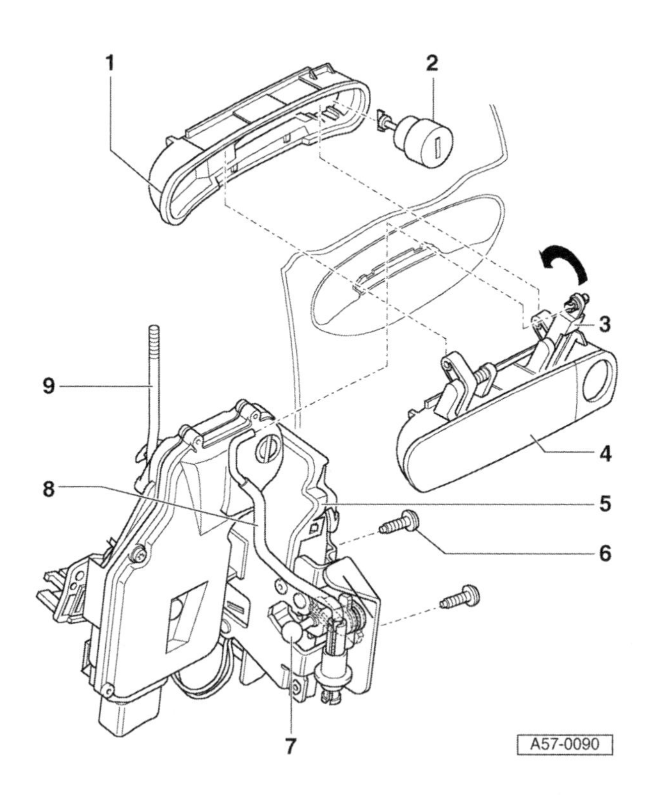

Door handle and door lock assembly

1. Door handle mount
2. Lock cylinder
3. Fulcrum lever
4. Door handle
5. Door lock
6. Bolt (M8), 20 Nm (15 ft-lb)
7. Locating pin
8. Operating rod
9. Operating rod (door lock)

Door lock, removing and installing

– Remove door trim panel. See **70 Interior Trim**.

– Run door window up to top position.

– Disconnect harness connector for central locking at door lock.

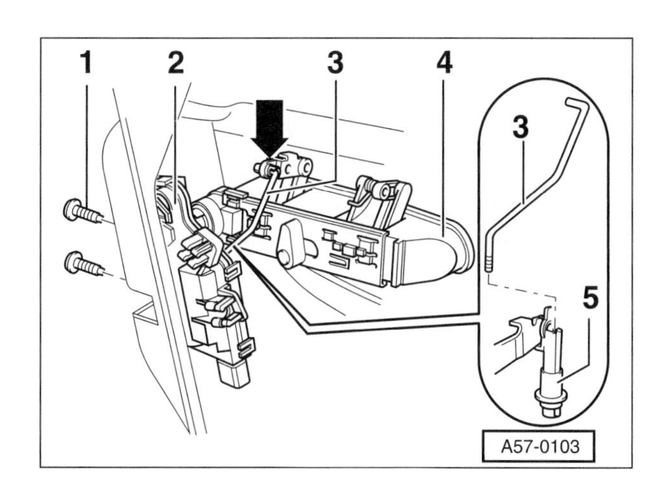

◀ Disengage operating rod (**3**).

– Remove combination bolts (**1**) and remove door lock (**2**) in downward direction, between door component carrier and front door shell.

– Installation is reverse of removal. Note the following:

- When reinstalling door lock, disengage operating rod (**3**) at fulcrum lever clip (**arrow**).
- When installing new door lock, disengage operating rod (**3**) at sliding sleeve (**5**).

BentleyPublishers.com
BentleyPublishers.com—All Rights Reserved

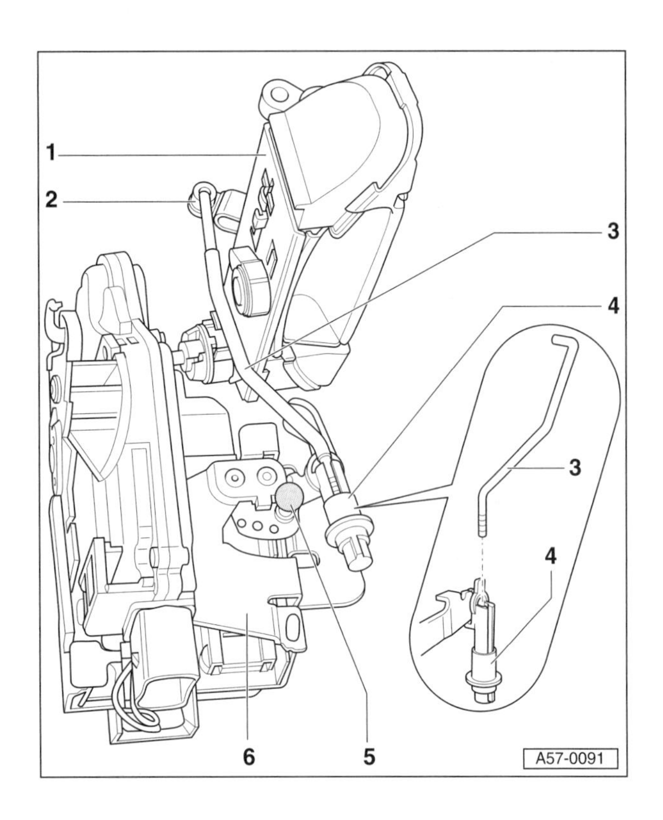

Door handle, removing and installing

1. Door handle
2. Clip
3. Operating rod
4. Sliding sleeve
5. Locating pin
6. Door lock

– Remove door trim panel. See **70 Interior Trim**.

– Run door window up to top position.

– Disengage operating rod at clip (**2**).

– Remove door handle.

– Installation is reverse of removal. Remember to:
  - Attach operating rod to handle at clip (**2**).
  - If installing new door lock, insert operating rod (**3**) and pull out locating pin (**5**) (if equipped).
  - Push up sliding sleeve (**4**) to lock operating rod (**3**) in place.

Fuel flap lock actuator, removing and installing

– Disconnect filler flap lock actuator harness connector.

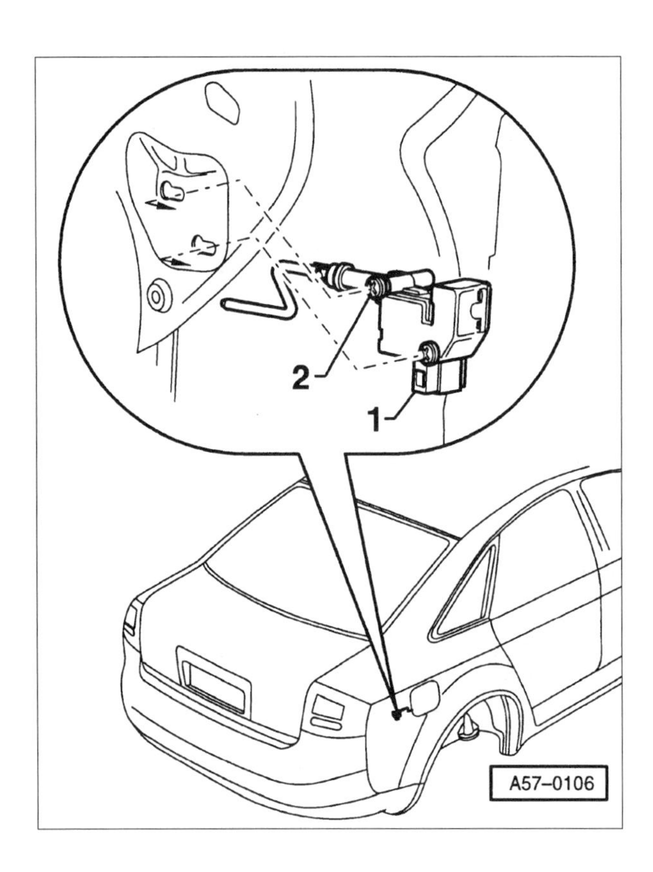

< Loosen screws (**2**). Push back lock actuator (**1**) and remove.

– Installation is reverse of removal.

BentleyPublishers.com
BentleyPublishers.com—All Rights Reserved

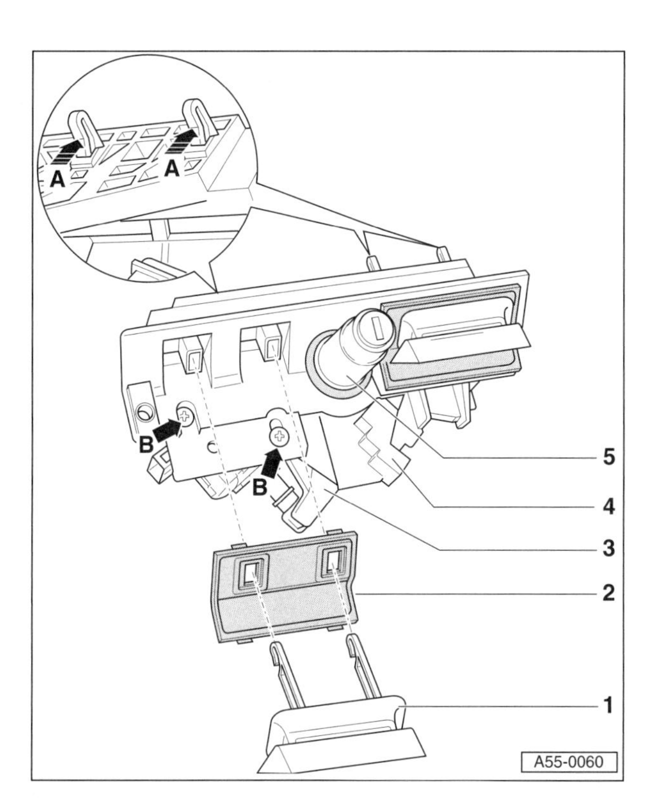

Rear lid lock actuator, removing and installing

1. Release handle
2. Seal
3. Lock actuator
4. Handle mechanism
5. Lock cylinder

– Remove tailgate / trunk lid trim. See **70 Interior Trim**.

– Remove lock. See **55 Hoods and Lids**.

– Using screwdriver, press in and disengage locating hooks on left release handle (**1**) in direction of arrow (**A**).

– Pull left release handle (**1**) downward out of handle mechanism (**4**).

– Remove Phillips-head screws (**B**) and remove lock actuator.

– Installation is reverse of removal.

BentleyPublishers.com
BentleyPublishers.com—All Rights Reserved

60 Sunroof

GENERAL

This repair group covers sunroof repair and adjustment procedures including sunroof drain hose maintenance. The power sunroof is a two-way design that tilts open or slides back in the roof.

Function

The sunroof is opened and closed by switching the ignition ON and turning the rotary sunroof switch to the desired position. Push the rotary switch to tilt the sunroof up, and pull to return it to closed position.

The sunroof can also be opened or closed after turning the ignition OFF, and before opening either front door. Should the sunroof electronics malfunction or the vehicle lose battery power, the sunroof can be closed manually. See **Sunroof emergency closing** in this repair group.

Sunroof motor and switch, accessing

To access the sunroof motor or rotary switch for testing or service, remove the switch panel from the headliner.

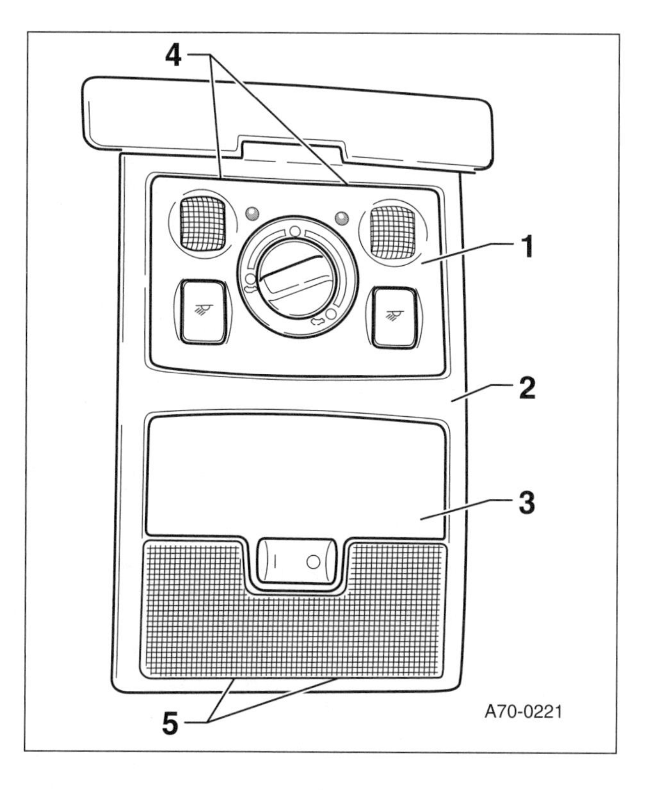

- To access the rotary switch, use a plastic prying tool to pry cover off switch panel (**1**) at front of panel (**4**) and remove switch.

- To access sunroof motor, use a plastic prying tool to pry cover off interior light (**3**) at back of panel (**5**).

BentleyPublishers.com
BentleyPublishers.com—All Rights Reserved

Sunroof emergency closing

If the sunroof will not close electrically, it can be closed manually. An emergency closing tool can be found on the inside of the fuse panel cover.

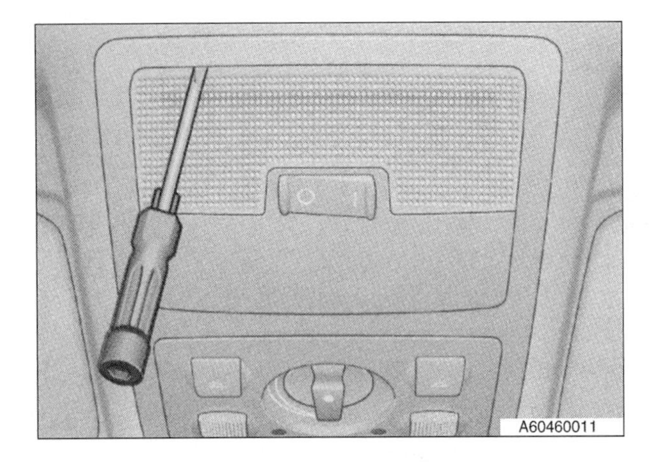

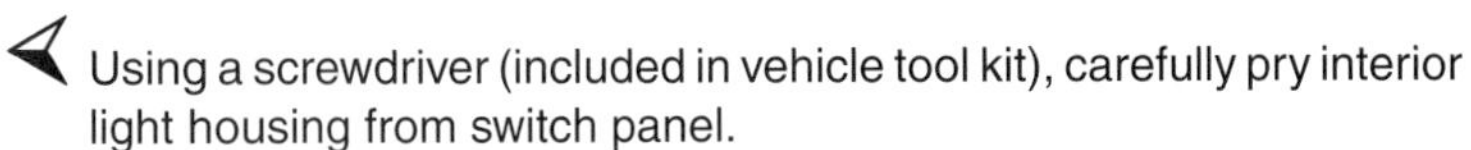

◄ Using a screwdriver (included in vehicle tool kit), carefully pry interior light housing from switch panel.

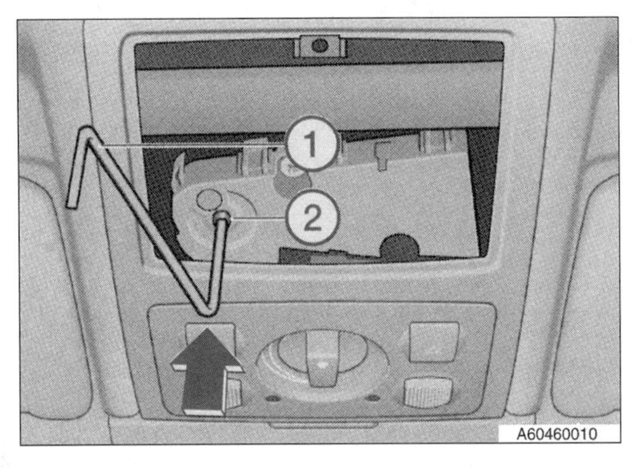

◄ Insert Allen end of closing tool (**1**) into socket on sunroof motor (**2**) as far as it will go. Crank sunroof closed.

SUNROOF SERVICE

Wind deflector, removing and installing

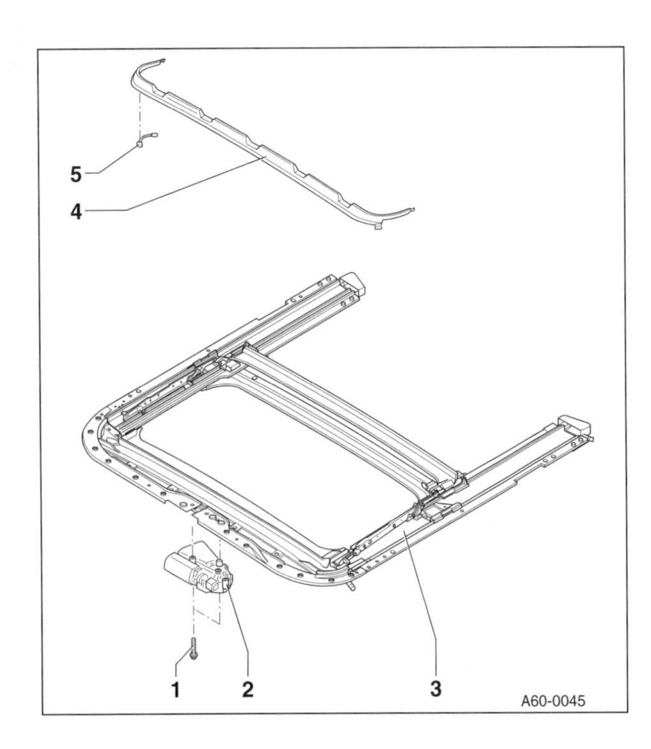

1. Screw w/washer
2. Sunroof motor
3. Frame
4. Wind deflector
5. Wind defector spring

– Slide back sunroof panel to fully open.

BentleyPublishers.com
BentleyPublishers.com—All Rights Reserved

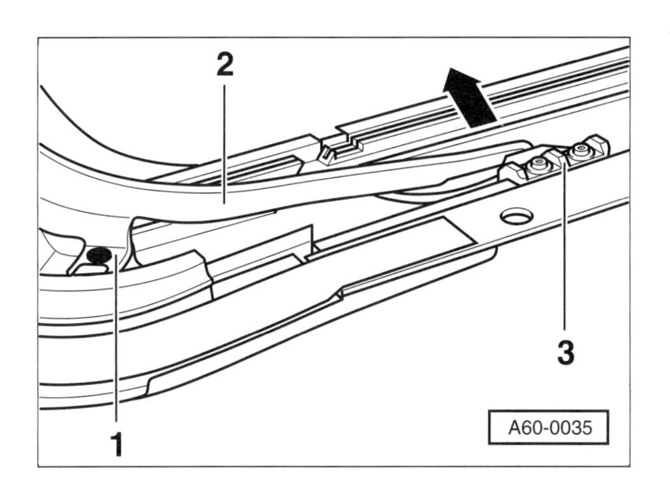

◀ Press wind deflector (**2**) out of holder (**3**) in direction of arrow.

– Pull wind deflector backward and out of roof cut-out.

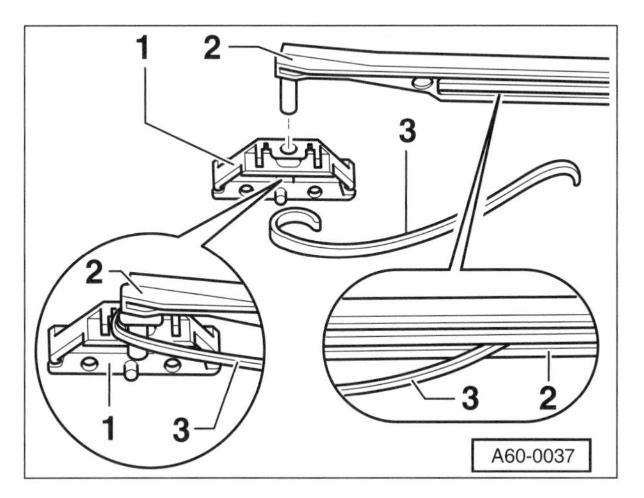

◀ Install wind deflector by hooking spring (**3**) onto retaining pin (**2**).

– Insert other end of spring into groove of wind deflector.

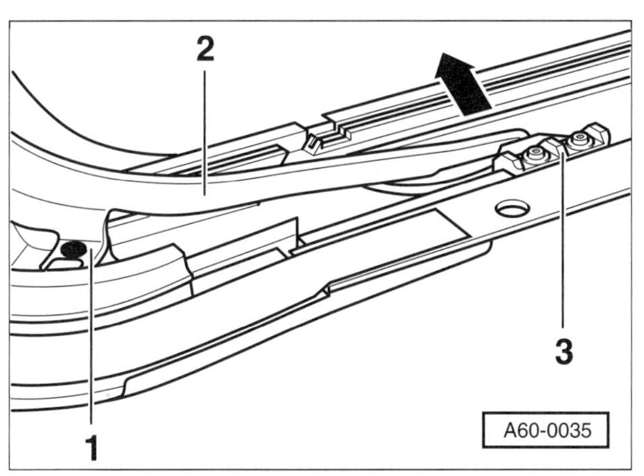

◀ Push stop (**1**) of wind deflector (**2**) forward below roof cut-out.

– Insert retaining pin of wind deflector into holder (**3**).

– On opposite side, push deflector in toward center of vehicle and lock other retaining pin into holder.

BentleyPublishers.com
BentleyPublishers.com—All Rights Reserved

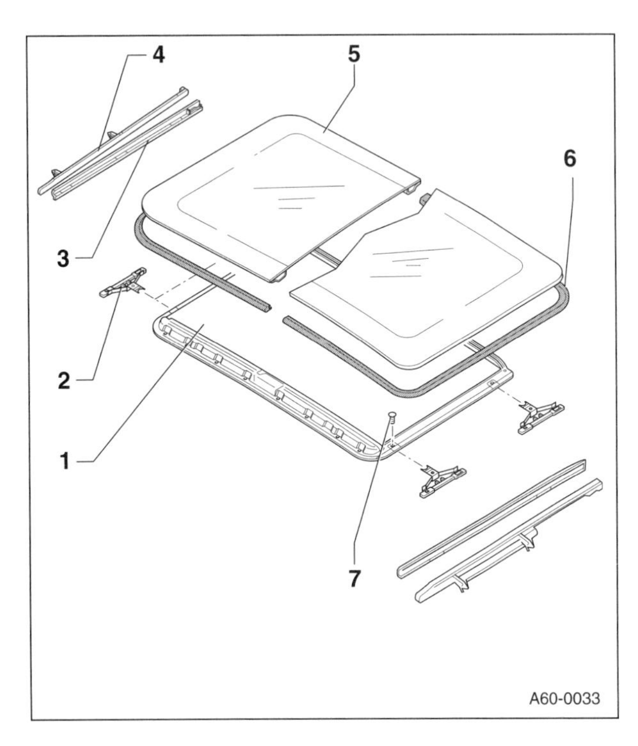

Sunroof panel, removing and installing

1. Sliding headliner
2. Sliding headliner spring
3. Lower trim
4. Upper trim
5. Sunroof panel
6. Sunroof panel sleeve
7. Countersunk screw

– Remove sliding headliner by removing screws (**7**) and front and rear.

– Push front spring (**2**) forward and rear spring backward off sliding headliner and remove springs.

– Pull sliding headliner sideways out of opposite guide rail and lift out.

– Open sunroof to tilt position.

– Remove trim on left and right sides. See **Sunroof trim, removing and installing** in this repair group.

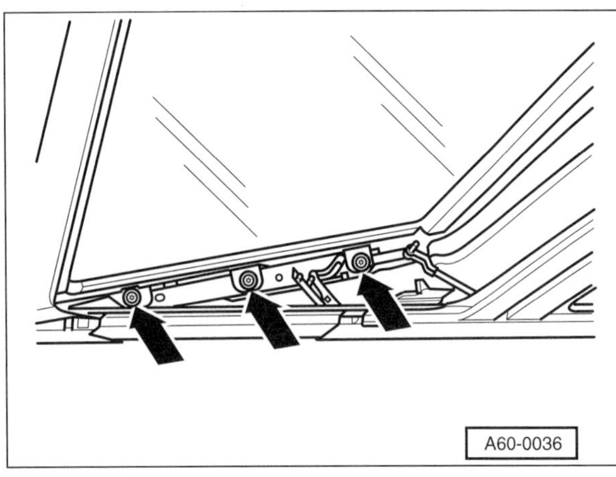

◄ Unscrew Torx bolts (T25) on both sides and lift panel out from top.

– Installation is reverse of removal. Remember to:
  - Make sure sunroof motor and sunroof guide plate are set to zero position. See **Sunroof motor, setting zero position**and **Sunroof guide plate, setting zero position** in this repair group.
  - Adjust sunroof panel height (if necessary). See **Sunroof, adjusting** in this repair group.

Tightening torque	
Sunroof fasteners to sunroof panel	4.5 + 0.5 / -0 Nm (40 + 4.4 / -0 in-lb)

Sunroof trim, removing and installing

– Open sunroof to tilt position.

– Push sliding headliner fully back.

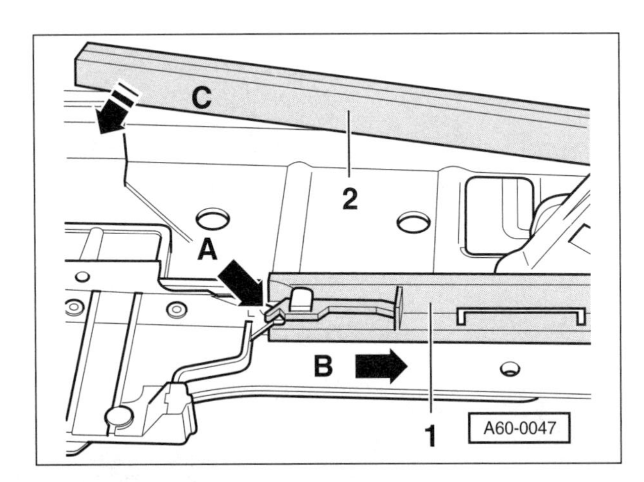

◄ Unclip lower trim (**1**) at rear in direction of arrow (**A**) and slide it forward (**B**).

– Pull lower trim off guide rail toward center of vehicle.

– Unclip upper trim (**2**) at front and center in direction of dashed **arrow**.

BentleyPublishers.com
BentleyPublishers.com—All Rights Reserved

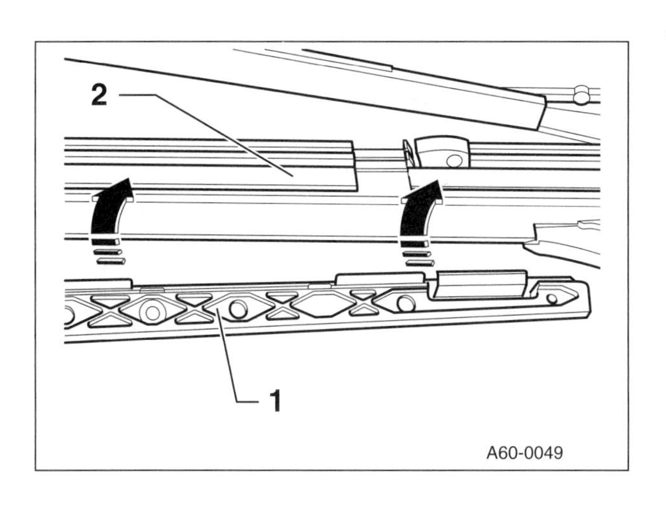

To install, press lower trim (**1**) onto upper edge (**arrows**) of guide rail (**2**).

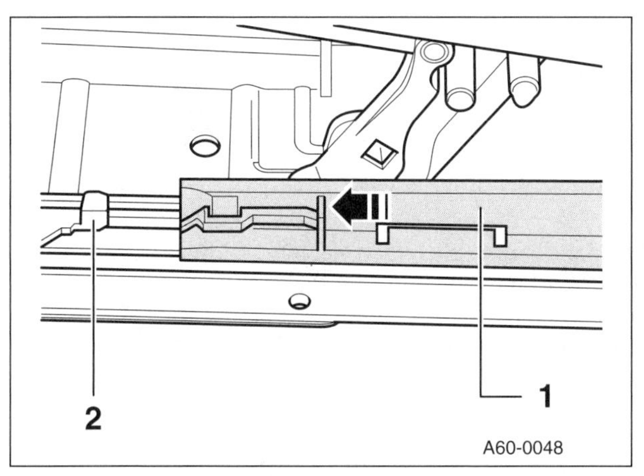

Press lower trim (**1**) backward in direction of **arrow** until locating hook locks into actuator (**2**).

– Press upper trim into guide plate.

Sunroof panel, adjusting

– To avoid wind noise, adjust sunroof panel at front and rear as shown.

– Check height adjustment at points where rounded corners merge with straight parts of roof cut-out at front and rear.

NOTE —

• Lower arrow points toward front.

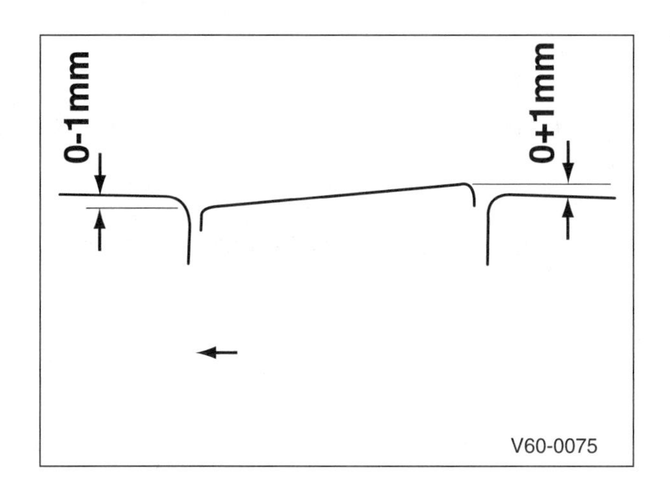

Sunroof panel should not be higher than the roof at any point at the front or lower than the roof at any point at the rear.

– Run sunroof from open position (i.e, not tilted open) to closed position and adjust according to specifications.

BentleyPublishers.com
BentleyPublishers.com—All Rights Reserved

Sunroof panel seal, replacing

– Remove sunroof panel. See **Sunroof panel, removing and installing** in this repair group.

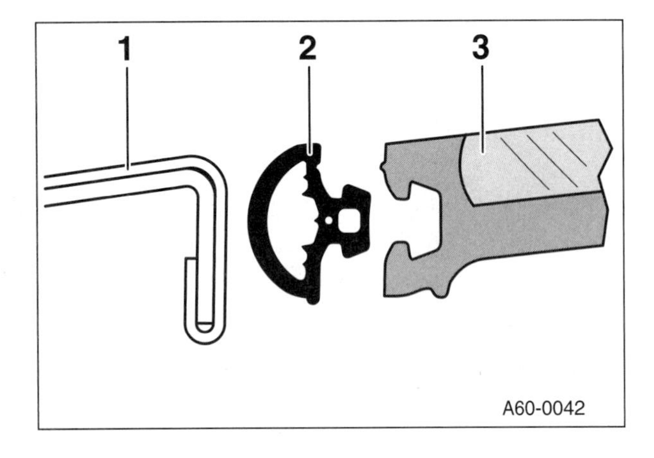

◀ Pull seal (**2**) out of sunroof panel (**3**).

– Press new seal into sunroof panel, starting from center of rear edge of panel.

– To aid installation, lubricate edge of sunroof panel with soapy water.

Sunroof motor, setting zero position

Set sunroof motor zero position before installing motor in frame.

– With sunroof motor removed, attach electrical harness connector to motor and move rotary control to fully closed position.

Sunroof winds to zero position and cuts out automatically.

– Install motor in this position with sunroof closed.

Sunroof guide plate, setting zero position

Set guide plate zero position before installing sunroof panel. Guide plate zero position can be checked using a pocket mirror and a flashlight, with the sunroof panel installed.

Checking zero position

– Remove sunroof trim. See **Sunroof trim, removing and installing** in this repair group.

– Run sunroof panel back and then into closed position.

– Push sliding headliner fully back.

– Shine flashlight on sunroof panel and visually check zero position from the side.

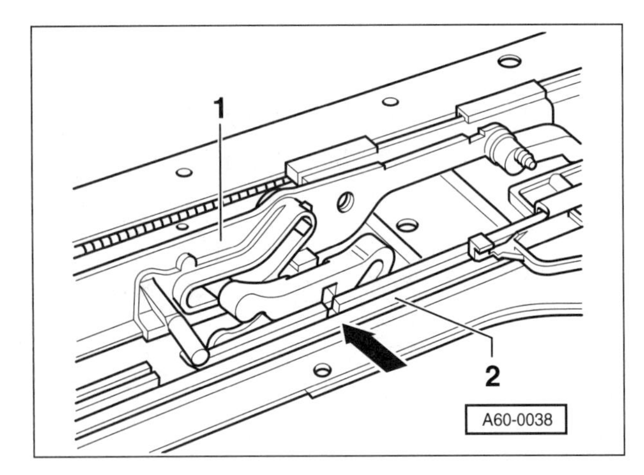

◀ Guide plate zero position is correctly set if the square hole in the tilt mechanism of the guide plate (**1**) is aligned with the recess (**arrow**) in the guide rail (**2**).

BentleyPublishers.com
BentleyPublishers.com—All Rights Reserved

Setting zero position

– Remove sunroof panel. See **Sunroof panel, removing and installing** in this repair group.

– Remove sunroof motor.

– Remove sunroof trim. See **Sunroof trim, removing and installing** in this repair group.

– Slide guide plates (**1**) on both sides from rear to front.

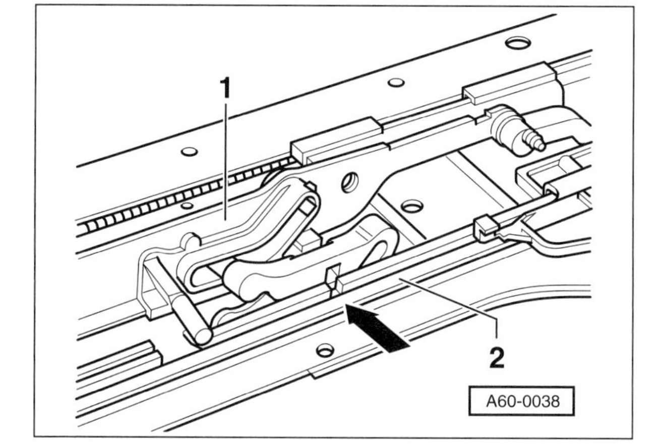

◀ Align square hole in tilt mechanism of guide plate (**1**) with recess (**arrow**) in guide rail (**2**) for sliding headliner.

This is zero position.

– Reinstall sunroof motor.

– Reinstall sunroof panel and trim.

Sunroof drain hoses, cleaning

Sunroof drain hoses should be checked periodically. Blocked drain hoses can allow water to enter the interior of the vehicle. To clean drain hoses a long cable is needed. An old speedometer inner cable, approx. 2300 mm (90 in) long will work for this.

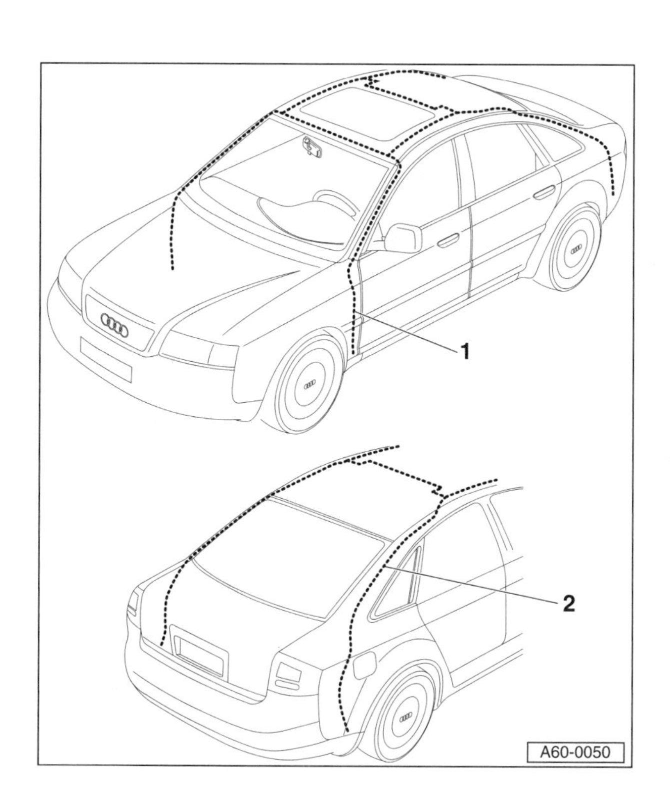

◀ Front water drain hoses (**1**) are routed in the A-pillars and exit between the door and A-pillar. Cleaning is performed from sunroof cut-out.

– Rear water drain hoses (**2**) are routed in the C-pillars and exit behind the bumper cover at side. Remove bumper cover and clean from the lower end of hose. See **63 Bumpers** for more information.

BentleyPublishers.com
BentleyPublishers.com—All Rights Reserved

63 Bumpers

GENERAL

This repair group covers removal and installation of the front and rear bumper components. The bumpers consist of a bumper cover and crossmember bolted to impact absorbers. The front bumper must be removed to extend the lock carrier assembly (radiator support) into service position. See **50 Body–Front** for more information.

FRONT BUMPER

Front bumper assembly

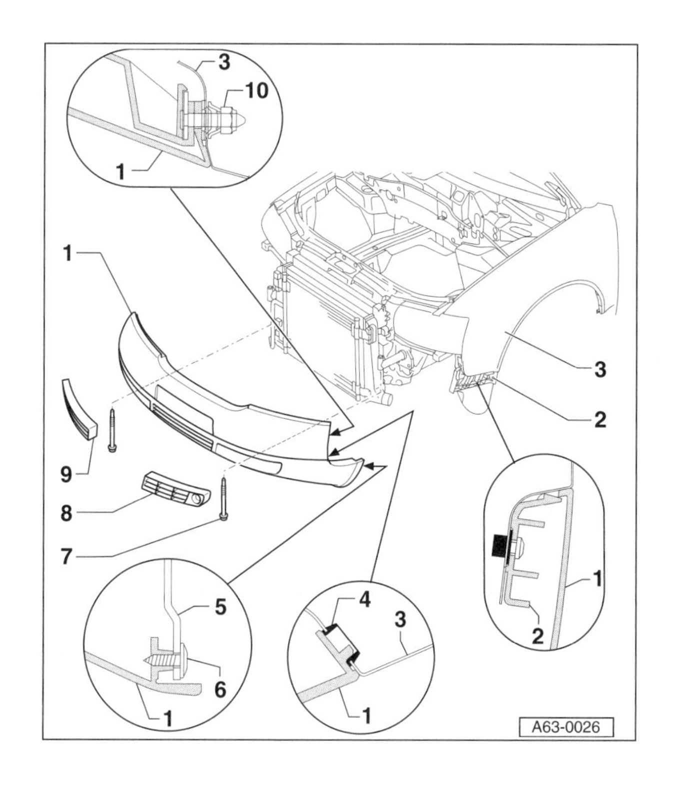

1. Bumper cover
2. Guide
3. Fender
4. Clip
   - Bumper cover tab (**1**) must latch into clip (**4**)
5. Wheelhousing liner
6. Screw w/washer
7. Bolt, 23 Nm (17 ft-lb)
8. Air inlet grill
9. Air inlet grill
10. Flanged nut, 6 Nm (53 in-lb)

BentleyPublishers.com
BentleyPublishers.com–All Rights Reserved

Front bumper, removing and installing

NOTE—
- *Bumper cover is removed along with crossmember.*

– Remove front tires.

– Loosen fasteners securing front half of wheelhousing liners to bumper.

– Remove screw w/washer (**6**) from wheelhousing liner (**5**).

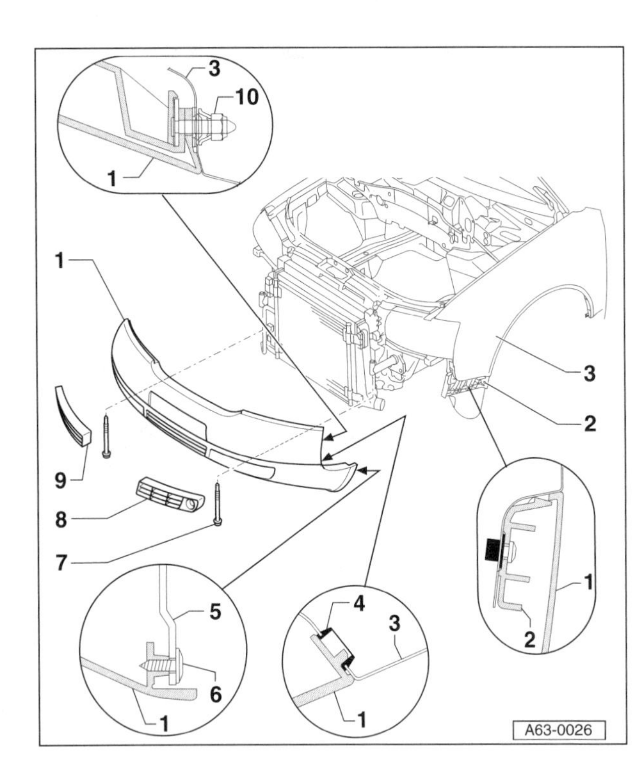

< Unscrew flanged nuts (**10**) inside wheelhousing on both sides.

– Remove bolts (**7**) and remove bumper cover together with crossmember toward front.

NOTE—
- *Before installing the bumper, make sure expanding clips are in place on fenders.*
- *Make sure bumper cover tabs latch into clips on fenders.*

– To install, align bumper cover onto impact absorbers.

– Latch bumper cover (**1**) into guide (**2**).

– Tighten flanged nuts (**10**) in wheelhousing.

– Tighten bolt (**7**) holding bumper cover to impact absorber.

Tightening torque	
Bumper cover to inner fender	6 Nm (53 in-lb)
Bumper cover to impact absorber	23 Nm (17 ft-lb)

Front bumper impact absorber, removing and installing

– Remove bumper. See **Bumper, removing and installing** in this repair group.

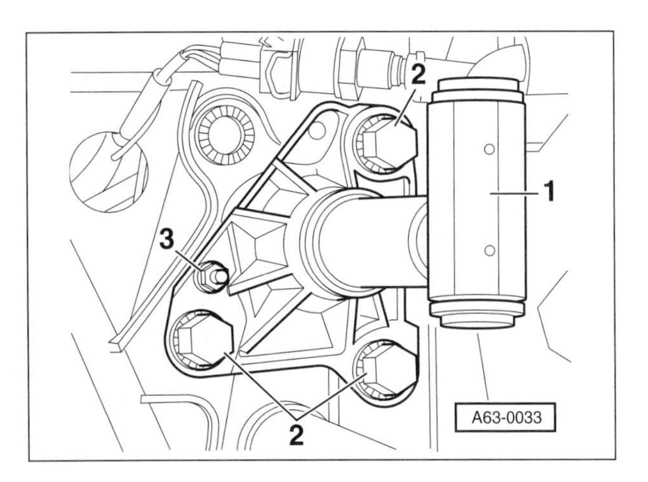

< Remove bolts (**2**) from impact absorber.

– Remove nut (**3**) and remove impact absorber.

– Installation is reverse of removal.

Tightening torques	
Impact absorber retaining nut	4 Nm (35 in-lb)
Impact absorber to body	50 Nm (37 ft-lb)

BentleyPublishers.com
BentleyPublishers.com—All Rights Reserved

Rear Bumper

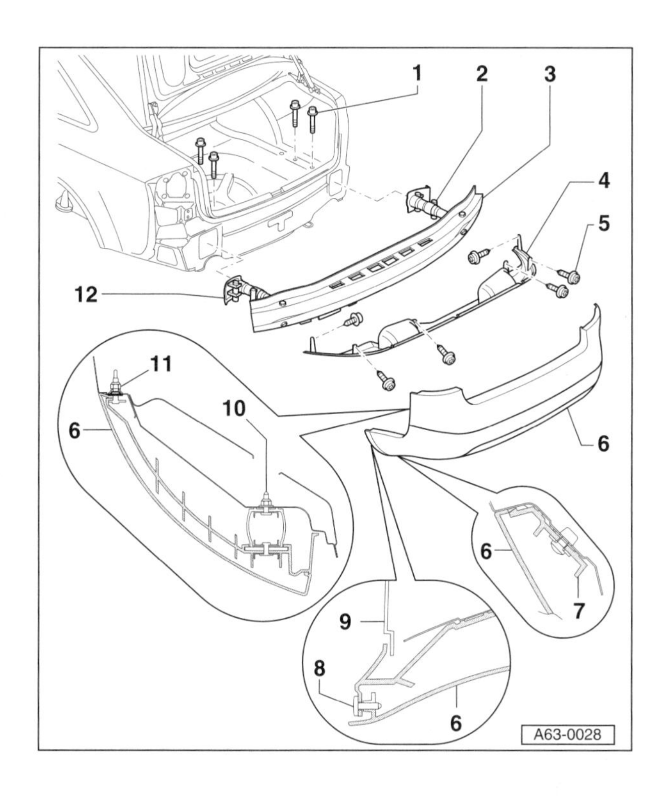

Rear bumper assembly

1. Bolt w/washer, 23 Nm (17 ft-lb)
2. Impact absorber
3. Crossmember
4. Lower back panel
5. Self-tapping screw
6. Bumper cover
7. Guide
8. Screw w/washer
9. Wheelhousing liner
10. Flanged nut (2x), 6 Nm (53 in-lb)
11. Flanged nut (2x), 6 Nm (53 in-lb)
12. Seal

Rear bumper cover, removing and installing

NOTE—

• Bumper cover is removed independent of crossmember.

– Partially remove luggage compartment trim to expose bumper cover fasteners. See **70 Interior Trim**.

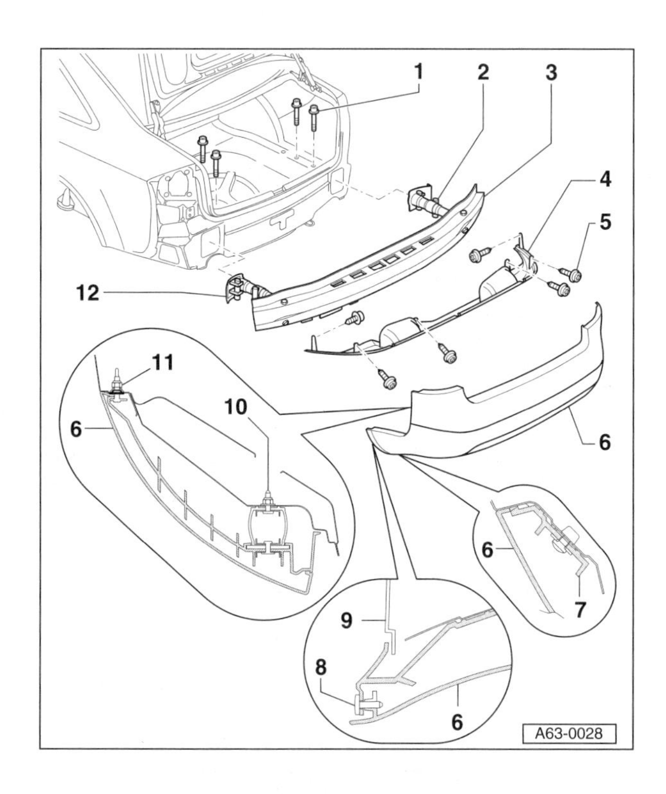

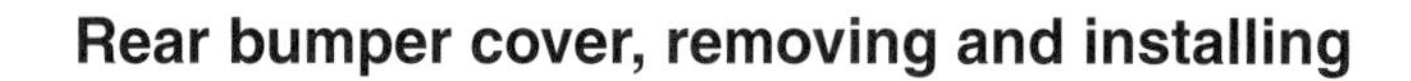

Unscrew flanged nuts (**10**) accessible through luggage compartment.

– Remove flanged nuts (**11**) accessible through trunk under taillight.

– Remove rear bumper cover.

– Installation is reverse of removal.

BentleyPublishers.com
BentleyPublishers.com—All Rights Reserved

Rear crossmember, removing and installing

– Remove bumper cover. See **Rear bumper cover, removing and installing** in this repair group.

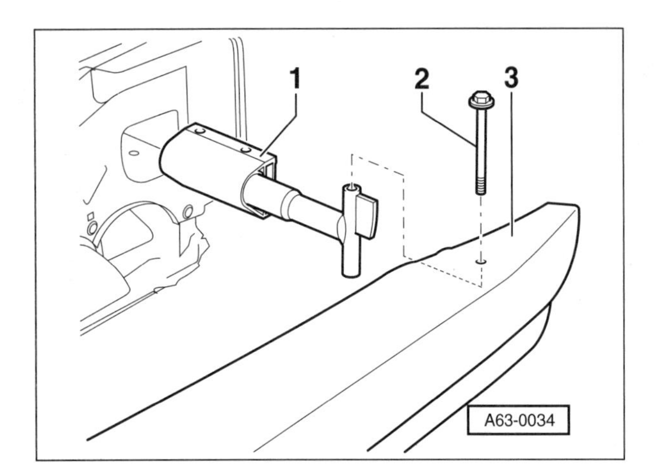

< Remove bolt w/washer (**2**) from each impact absorber (**1**) and remove crossmember (**3**).

– Installation is reverse of removal.

Tightening torque	
Crossmember to impact absorber	23 Nm (17 ft-lb)

BentleyPublishers.com
BentleyPublishers.com—All Rights Reserved

64 Door Windows

GENERAL

This repair group covers replacement of door glass, window regulator and power window motor. Replacing the windshield or rear glass requires special glue and tools and is not covered.

For additional information see:

- **57 Doors and Locks** for door component carrier
- **70 Interior Trim** for interior door panel
- **96 Interior Lights, Switches, Anti-theft** for door window switches

Window motor reinitializing

If the battery is disconnected, the window control module loses its memory of window current position and disables one-touch automatic up / down function. To restore one-touch operation:

– Switch ignition ON.

– Use window switches to raise windows to top.

– Operate each window switch in CLOSE direction for 1 second to reinitialize one-touch operation.

Warnings and Cautions

WARNING—
- *When working on door window, disconnect the harness connector to the window regulator to prevent pinching fingers in the moving window mechanism.*
- *Wear hand and eye protection when working with broken glass.*
- *If a window is broken, vacuum all of the glass bits out of the door cavity. Use a blunt screwdriver to clean out any remaining glass pieces from the window guide.*

CAUTION—
- *To avoid damaging interior trim, use a plastic prying tool or a screwdriver with the tip wrapped with masking tape.*

BentleyPublishers.com
BentleyPublishers.com—All Rights Reserved

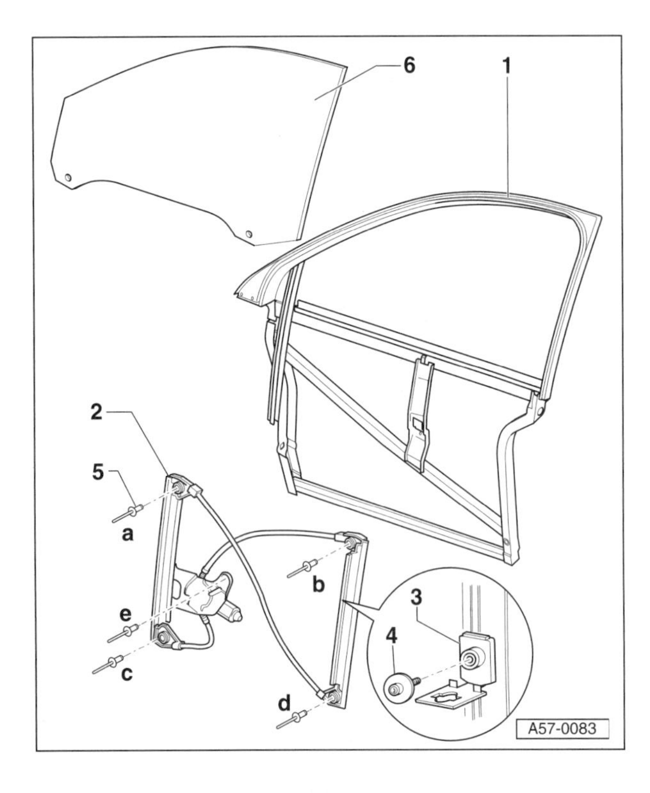

Door Window Glass Service

Front door window assembly

1. Door component carrier
2. Window regulator
3. Mount
4. Torx bolt, 4.5 Nm (40 in-lb)
5. Pop rivet (**a**, **b**, **c**, **d** = 4.8 mm, **e** = 6.0 mm)
6. Door window

Front window regulator, removing and installing

– Remove door trim. See **70 Interior Trim**.

– Remove door component carrier. See **57 Doors and Locks**.

– Drill heads off pop rivets (**5**), knock out rivet stems and detach window regulator.

– Shake door component carrier so that pop rivet stems fall out.

– To install, fasten rivets (**5**) in order (**a**, **b**, **c**, **d** and **e**) using pop rivet pliers (note different rivet diameters).

Front door window, removing and installing

– Run window down to bottom position.

– Remove door trim. See **70 Interior Trim**.

– Remove door component carrier. See **57 Doors and Locks**.

– Mark longitudinal position of window at both mounts (**3**) using felt-tip pen.

– Remove clamp bolts (**4**) and lift window out of mount (**3**).

– To install original window, insert window into window regulator and align with marks made previously.

– Tighten clamp bolts (**4**).

– To install a new window, insert window into guide and push it back toward B-pillar as far as it will go.

– Move window upward as far as stop using window regulator without bolt (**4**).

– Close mount (**3**) and tighten clamp bolts (**4**).

Tightening torque	
Front window clamp bolt to mount	4.5 Nm (40 in-lb)

BentleyPublishers.com
BentleyPublishers.com—All Rights Reserved

Rear door window assembly

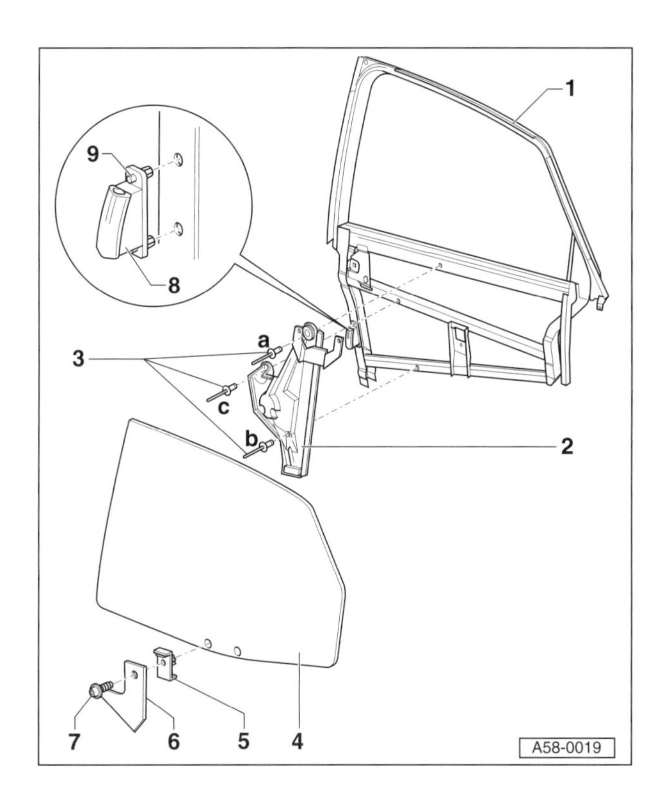

1. Door component carrier
2. Window regulator
3. Pop rivet (**a** = 4.8 mm, **b** and **c** = 6.0 mm)
4. Door window
5. Locating bracket
6. Carrier
7. Bolt, 2.5 Nm (22 in-lb)
8. Buffer
   - After inserting, press in clamping pin (**9**)
9. Clamping pin

Rear window regulator, removing and installing

- Remove door trim. See **70 Interior Trim**.
- Remove door component carrier. See **57 Doors and Locks**.
- Drill heads off pop rivets (**5**), knock out rivet stems and detach window regulator.
- Shake door component carrier so that pop rivet stems fall out.
- To install, install rivets (**5**) in order (**a**, **b**, and **c**) using pop rivet pliers (note different rivets diameters).

Rear door window, removing and installing

- Run window down to bottom position.
- Remove door trim. See **70 Interior Trim**.
- Remove door component carrier. See **57 Doors and Locks**.
- Mark position of window in vertical and longitudinal directions at mount using felt-tip pen.
- Remove clamp bolt (**7**) and lift window out of carrier (**6**).
- To install original window, insert window into window regulator and align with marks made previously.
- Tighten clamp bolt (**7**).
- To install a new window, insert window into guide and push it forward toward B-pillar.
- Make sure sealing lips of window channel strip are not pressed downward.
- Run window up to top stop using window regulator without installing clamp bolt (**7**).
- Tighten bolt clamp bolt (**7**).

Tightening torque	
Rear window clamp bolt to mount	2.5 Nm (22 in-lb)

BentleyPublishers.com
BentleyPublishers.com—All Rights Reserved

Window Motor Service

Front window motor removal shown. Rear window motor removal is similar.

Window motor, removing and installing

- Remove door trim. See **70 Interior Trim**.
- Disconnect electrical connector from window motor.

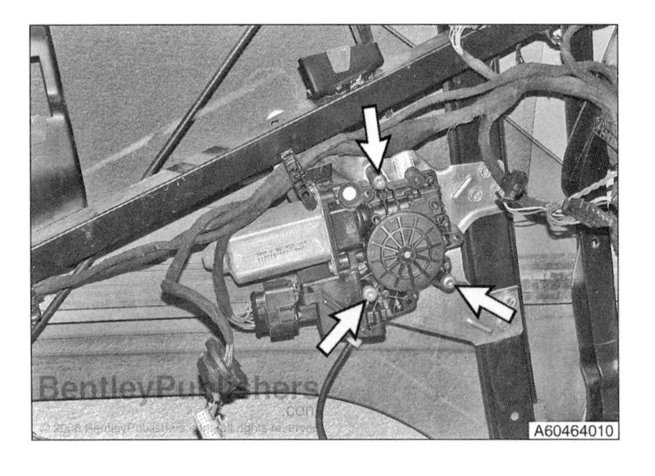

- Remove bolts (**arrows**) and remove window motor.
- Installation is reverse of removal. Note the following:
  - Window regulator motor can be installed or removed with motor and window in any position.
  - After installing motor and wiring, turn ignition ON and OFF.
  - Intitalize window motor. See **Window motor initializing** in this repair group.

Tightening torques	
Window regulator motor to bracket	6.5 Nm (58 in-lb)

BentleyPublishers.com
BentleyPublishers.com—All Rights Reserved

66 Body–Exterior Equipment

GENERAL

This repair group covers wheel housing liners and outside rear view mirrors. For related information see the following repair groups:

- **63 Bumpers**
- **94 Exterior Lights**

WHEEL HOUSING LINERS

Front wheel housing liner assembly

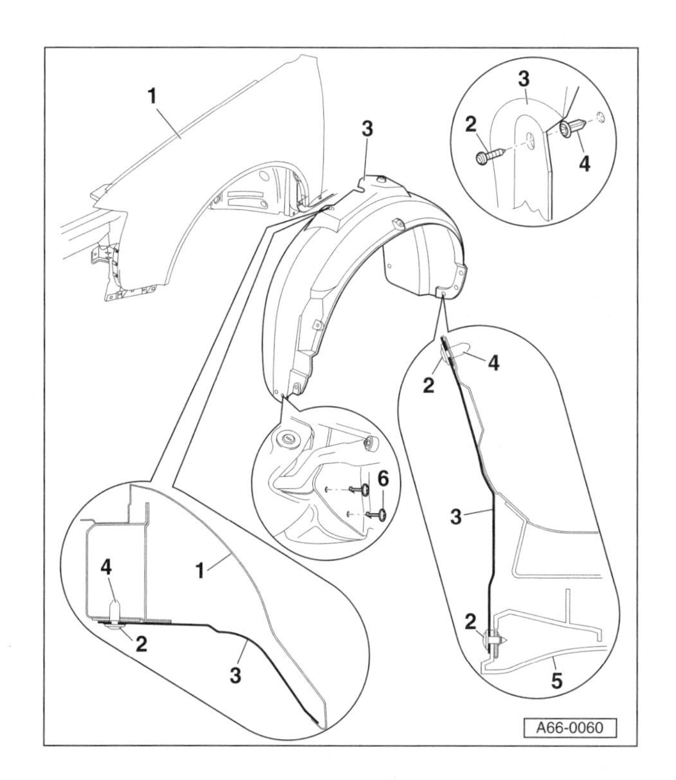

1. Fender
2. Torx bolt (T25), 1.5 Nm (13 in-lb)
3. Wheel housing liner, front
4. Expanding nuts (13x)
5. Sill panel trim

BentleyPublishers.com
BentleyPublishers.com—All Rights Reserved

Front wheel housing liner, removing and installing

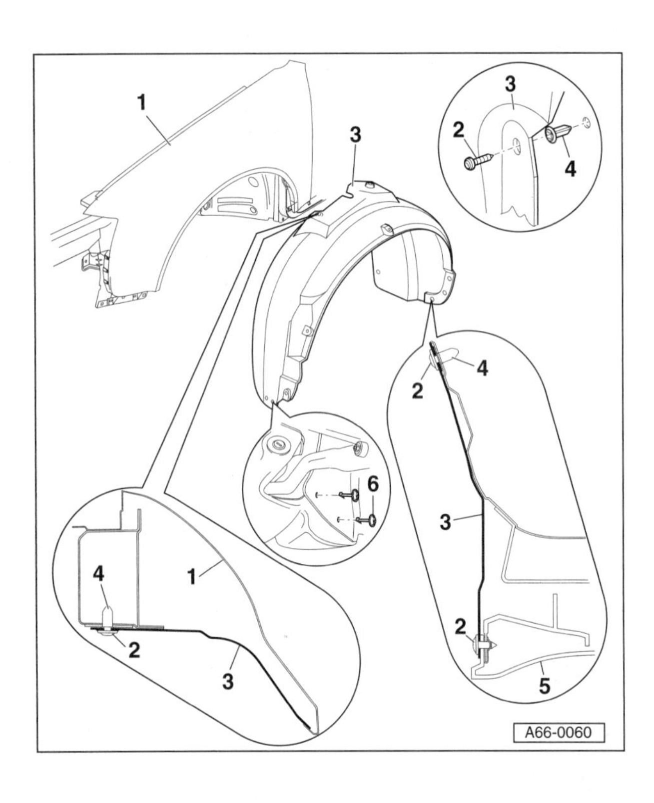

- Raise car and support safely. Remove wheel.

> ***WARNING—***
> *• Make sure care is stable and well supported at all times. Use a professional automotive lift or jack stands designed for the purpose. A floor jack is not adequate support.*

- Release expanding fasteners (**6**).
- Remove Torx bolts (**2**) (3x).
- Disengage wheel housing liner from fender and pull out in downward direction.
- Installation is reverse of removal.

Rear wheel housing liner assembly

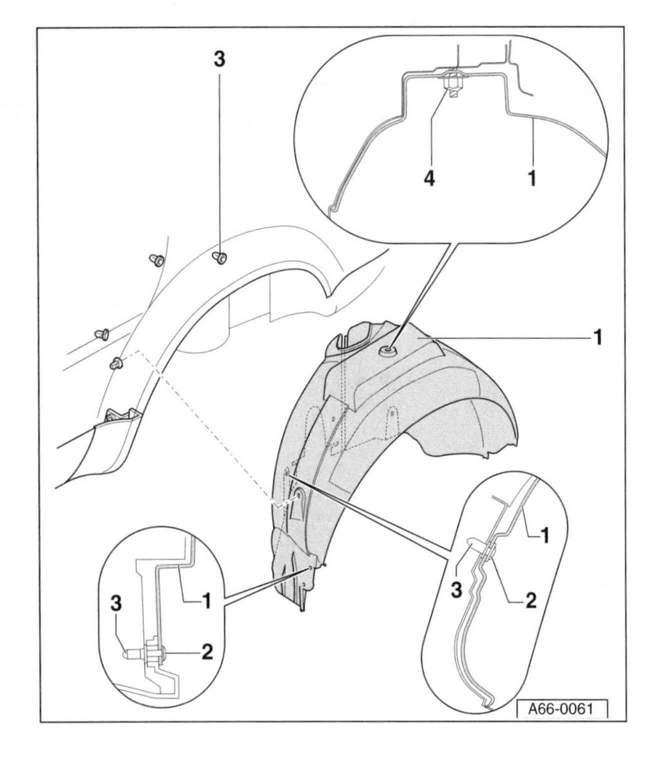

1. Rear wheel housing liner
2. Self-tapping screw, 1.5 Nm (13 in-lb)
3. Expanding nut
4. Cap nut, 1.5 Nm (13 in-lb)

BentleyPublishers.com
BentleyPublishers.com—All Rights Reserved

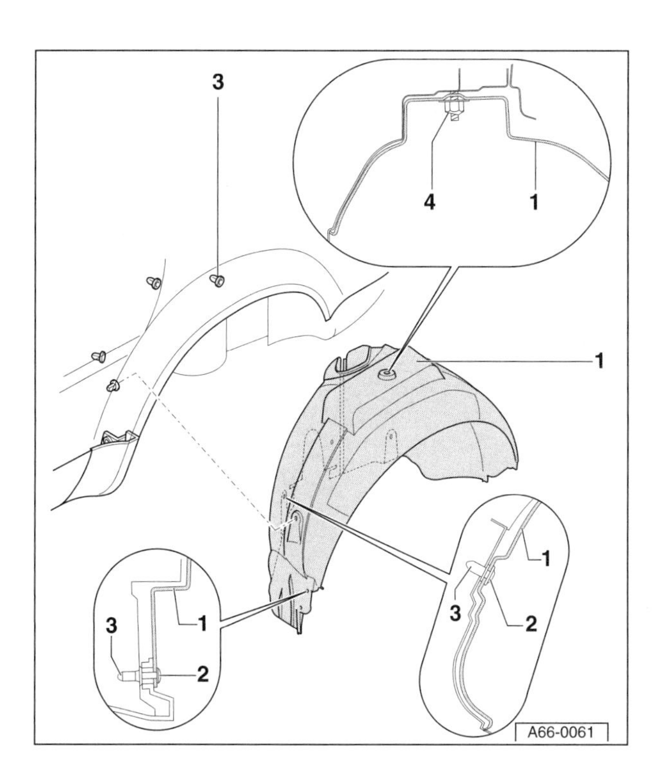

Rear wheel housing liner, removing and installing

– Raise car and support safely. Remove wheel.

> ***WARNING —***
> - *Make sure care is stable and well supported at all times. Use a professional automotive lift or jack stands designed for the purpose. A floor jack is not adequate support.*

– Remove screws (**2**) and cap nut (**4**).

– Pull wheel housing liner out in downward direction and remove.

– Installation is reverse of removal.

Exterior Rear View Mirror

Exterior rear view mirror components and service

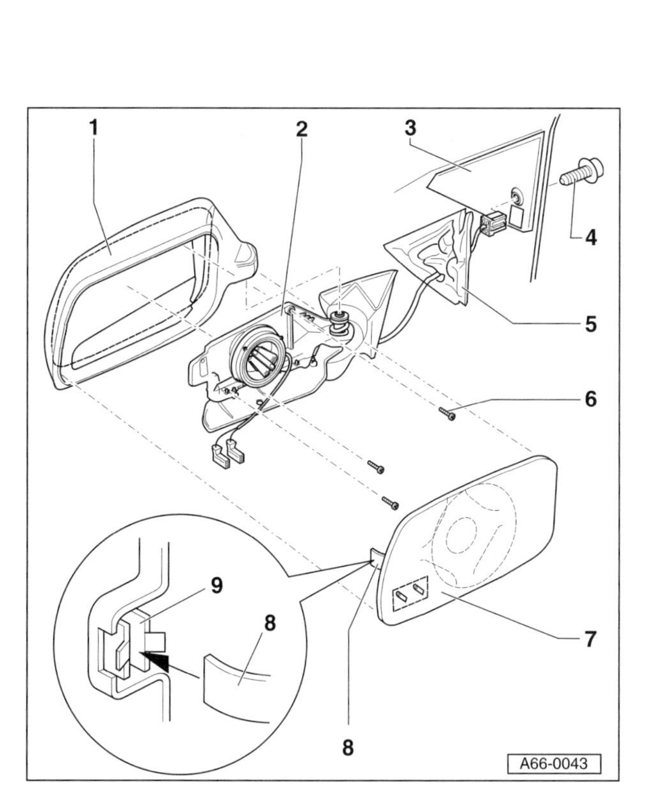

1. Mirror housing
   - Can be removed from door without disassembling complete rear view mirror.
2. Mirror motor
   - Before removing, disassemble rear view mirror from door
   - Carefully pry off mirror glass and disconnect harness connectors for mirror heater. Unscrew Phillips head screws (**6**) and remove motor unit from mirror housing (**1**).
   - On vehicles with seat memory, remove door trim panel before disassembling mirror motor.
3. Door component carrier
4. Socket-head screw, 12 Nm (9 ft-lb)
5. Seal
6. Phillips head screw, 1 Nm (9 in-lb)
7. Mirror glass
   - Use special tool 80-200 lever to unclip mirror glass
   - Protect mirror housing from damage by applying fabric reinforced adhesive tape at top and bottom of housing
   - Press off mirror glass, first at bottom then at top.
   - To install, insert mirror glass into guide stud and friction finger into friction spring and press on.
   - Press only on center of mirror (always use protective gloves).
8. Friction finger
   - When installing mirror glass, friction finger (**8**) must be inserted into friction spring (**9**) in housing
9. Friction spring

BentleyPublishers.com
BentleyPublishers.com—All Rights Reserved

Exterior mirror housing, replacing

– Remove mirror glass. See **Exterior rear view mirror components and service** in this repair group.

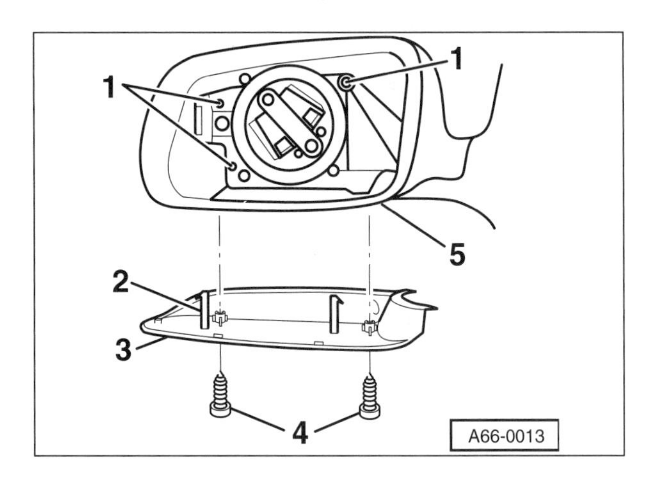

◄ Remove screws (**4**).

– Press back locking lugs (**2**) and detach cover (**3**) downward.

– Remove screws (**1**) at mirror motor.

– Lift off housing (**5**) upward.

Power mirror motor, replacing

– Disconnect battery negative (-) cable.

– Remove mirror glass. See **Exterior rear view mirror components and service** in this repair group.

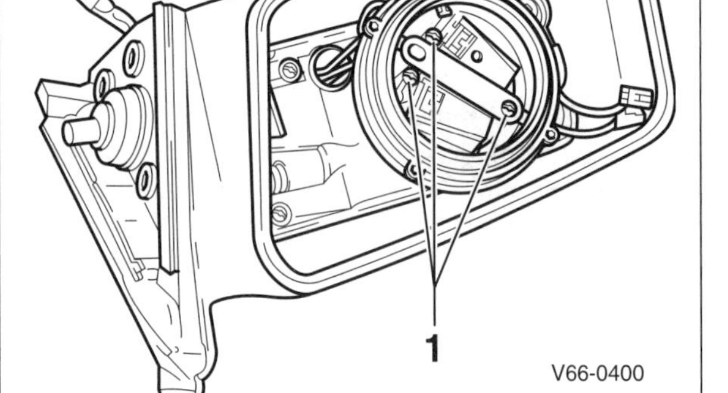

◄ Remove screws (**1**).

– Remove mirror motor harness connectors (if equipped).

NOTE —

- *If mirror motor has soldered wiring, electric leads must be separated individually directly at the housing.*

– Install wiring in connector housing supplied with replacement part.

NOTE —

- *Note configuration of wiring on old motor. See* ***EWD Electrical Wiring Diagrams****.*

BentleyPublishers.com
BentleyPublishers.com—All Rights Reserved

69 Seat Belts, Airbags

GENERAL

This repair group covers emergency tensioning (pyrotechnic) seat belt assemblies and airbag system components. It does not cover airbag system or pyrotechnic seat belt fault diagnosis or repair. Service and repair to these systems requires special test equipment, knowledge and training and should only be carried out by and authorized Audi Dealer. Before starting repairs involving these systems, always read and observe all warnings, cautions and notes in this repair group.

Child seat anchors

Vehicle manufactures are required to have an anchorage point installed in vehicles to facilitate installation of a tether strap for child seats. Lower anchorages were originally designated as "**ISOFIX**" points. Lower anchor points in combination with upper tether anchor points are known as "**L**ower **A**nchor and **T**ether for **CH**ildren" or simply "**LATCH**" points. For information regarding proper installation of child seats booster seats, infant seats, and restraints, always consult vehicle owner's manual or authorized Audi dealer.

Rear side airbags

If special situations require, your dealer can disconnect the rear side airbags upon request. For more information, see your owner's manual or authorized Audi dealer.

BentleyPublishers.com
BentleyPublishers.com—All Rights Reserved

Seat belt inspection

Seat belts should periodically be checked for proper operation, wear and damage. Check points include visual and operational inspection of the following:

- Belt webbing.
- Automatic retraction (stop function).
- Belt latch mechanisms.
- Guide stop and seat belt locking tongue
- Anchoring points and components.

> ***WARNING —***
> - *After an accident, be sure to inspect seatbelts carefully and replace damaged ones.*

Belt webbing

– Pull belt webbing completely off automatic belt retractor.

– Check belt webbing for soiling and wash with mild soap and water solution as needed. Avoid harsh chemical solutions.

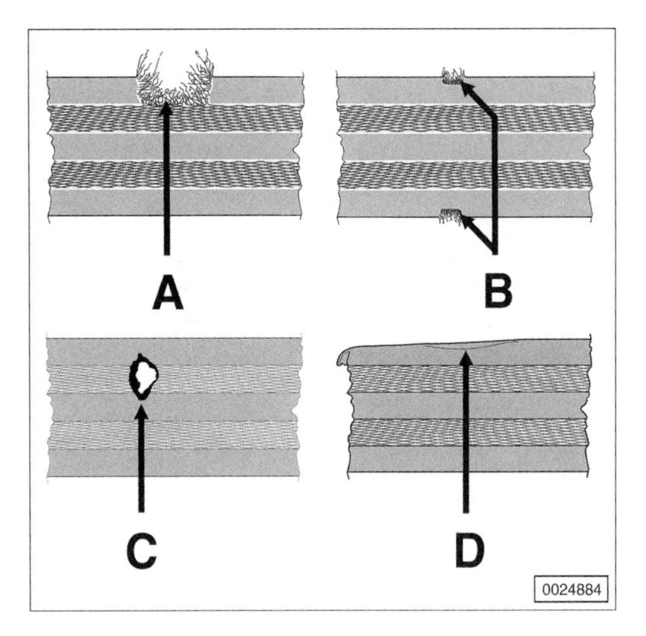

◄ Inspect webbing for damage.

A. Webbing cut, torn or chafed.

B. Webbing loops on belt edges torn.

C. Burned spots from cigarettes or other sources.

D. One side of belt edge deformed, or area of belt edge is wavy or rolled over.

– **A**, **B** - If noted in a post-accident inspection, replace complete belt assembly and height adjuster where applicable.

– **A**, **B**, **C**, **D**, - If noted during routine inspection and vehicle has not been involved in an accident, replace damaged parts as needed.

Automatic retraction (stop function)

Once activated, pyrotechnic seat belts will not retract. These units are no longer functional and require replacement.

The automatic belt retractor has two locking functions triggered by either occupant or vehicle movement.

– Function 1 (occupant movement): Locking is triggered by rapid unrolling of belt from automatic belt retractor reel (belt extraction acceleration).

- To check, pull belt out of automatic belt retractor with firm jerk. Seat belt should lock while being pulled out. If not, replace seat belt and belt latch.

NOTE —
- *If problems are noted during function test 1, check mounting/installation position of automatic retractor. Incorrect installation position can influence operation.*

BentleyPublishers.com
BentleyPublishers.com—All Rights Reserved

Automatic retraction (stop function) continued

– Function 2 (vehicle movement): Locking function triggered by a change in vehicle movement (vehicle-dependent locking function).
  - To check, securely fasten belt and drive vehicle on a clear, level, even surface to a speed of 12 mph (20 kph).
  - Apply brake pedal fully in a panic-style application.
  - Seat belt should lock as vehicle suddenly stops while braking. If not, replace seat belt and belt latch.

> ***WARNING —***
> - *For safety reasons, perform road test in an area without traffic, pedestrians, or obstacles such as curbs, light and sign poles, or parked vehicles.*

Belt latch mechanisms

– Inspect seat belt latch for cracks and fractures.
  - If damaged, replace seat belt and belt latch.

– Check operation of seat belt tongue into latch by inserting until an audible engagement is heard. Pull seat belt webbing forcefully to be sure of proper engagement in lock mechanism.
  - Repeat test at least five (5) times. If seat belt fails to engage even once, replace seat belt and belt latch.

– Check release operation by pressing release button with finger while seat belt is in a relaxed state. Locking tongue must spring out of seat belt lock completely and without assistance.
  - Repeat test at least five (5) times. If seat belt fails to release even once, replace seat belt and belt latch.

> ***WARNING —***
> - *Do not lubricate seat belt tongue, buttons, or release mechanisms to repair noises or stiffness.*

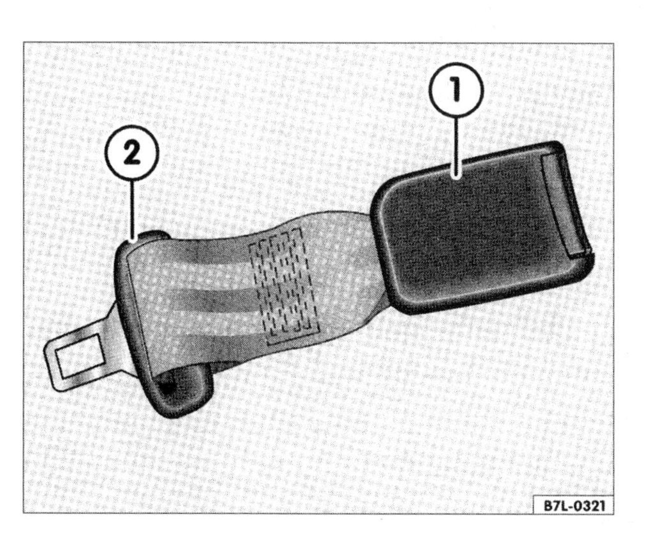

Inspect seat belt extender (if equipped) lock tongue (**2**) and belt latch mechanisms (**1**) for the same conditions found with seat belt latch mechanisms and seat belt lock tongue.

Guides and seat belt lock tongue

– Inspect all plastic coated seat belt guides and plastic coating on seat belt lock tongue. Check for deformation, cracking, tearing and wear.
  - If damaged, replace seat belt and belt latch.

Anchoring points and components

– Visually inspect and functionally test lock strap, height adjuster and anchor points on seats, pillars, and floor. Check for deformation, secure and correct mounting, and proper operation.
  - If damage is found at components, replace seat belt and belt latch.
  - If damage is found on anchoring points, replace components as required.
  - If damage is found from general wear and not due to accident, replace only damaged component(s).

BentleyPublishers.com
BentleyPublishers.com—All Rights Reserved

Warnings

In addition to the warnings listed below, watch for airbag identification tags at various locations around the vehicle interior. The number and placement of tags varies with vehicle equipment.

> ***WARNING —***
> - *Always disconnect the battery ground (-) strap before starting to work on airbag or pyrotechnic seat belt systems. Work can begin immediately after disconnecting battery. No waiting time is needed.*
> - *Never let anyone sit in passenger compartment when the airbag or seat belt system is being worked on, especially when either system is being connected to a power source or the battery.*
> - *Never test, remove, repair or install airbag units or pyrotechnic seat belts or work on either system unless you have proper training, qualifications and tools to do the work safely and correctly.*
> - *Always discharge static electricity from your body before handling or even touching an airbag unit. Place your hands on a suitable metal object that is grounded such as a water or heating pipe or metal frame.*
> - *Never use impact wrenches to remove airbag or pyrotechnic seat belt tensioners.*
> - *Always install airbag unit immediately after removing it from its shipping container.*
> - *If installation work is interrupted, immediately return airbag unit to its original packaging.*
> - *Never leave an airbag unit unattended. If work is interrupted, always return airbag unit to original packaging or shipping container.*
> - *Never place airbag unit on a workbench or other surface with the side that points towards inside of passenger compartment facing down. If airbag unit should deploy, it can cause personal injury. Always make sure that the side of the airbag unit that faces passenger compartment faces up.*
> - *Never install airbag or pyrotechnic seat belt units that have dropped onto a hard surface, or which have signs of damage.*
> - *Never open or attempt to repair airbag or seat belt tensioner components; always use new parts (danger of injury).*
> - *Always replace airbag and seat belt tensioner units which have experienced mechanical damage (dents, cracks).*
> - *Always put airbag units that have not deployed or have not completely deployed in original airbag shipping container and mark outside of container so that others will know that there is still a possibility of airbag deployment.*
> - *Always store and handle airbag units according to all applicable legal requirements.*
> - *Seat belt pyrotechnic ignition charge does not have an expiration date, i.e. it has an infinite shelf life and is maintenance free.*
> - *Never treat a pyrotechnic seat belt tensioner unit with grease, spray lubricants, cleaning products or similar products.*
> - *Do not expose pyrotechnic seat belt tensioner or airbag units to temperatures above 212°F (100°C) even for a few seconds.*
> - *Airbag and pyrotechnic seat belt tensioner storage and transportation are covered under several explosive substance laws. Always be aware and observant of applicable regulations.*

BentleyPublishers.com
BentleyPublishers.com—All Rights Reserved

Seat Belts

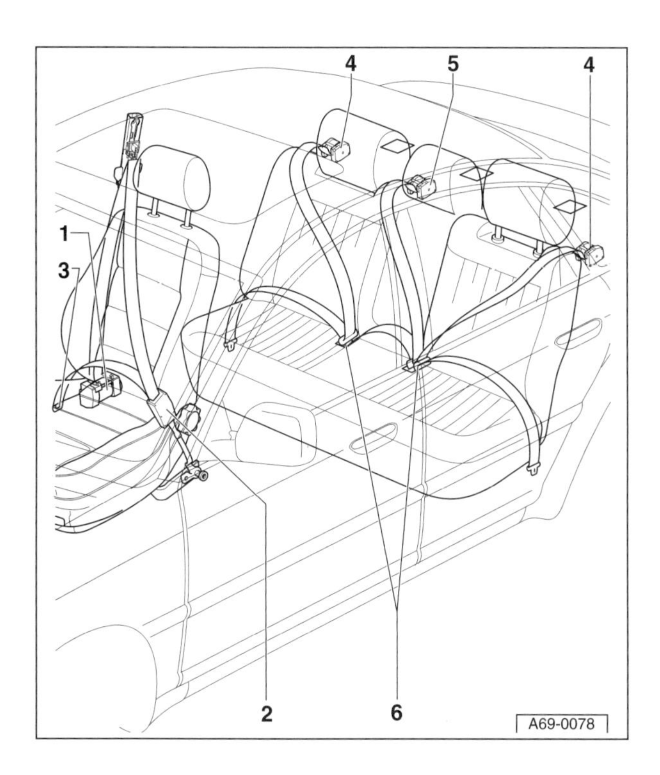

Component overview

1. 3-point seat belt with belt height adjuster and belt tensioner
2. Attachment point for belt latch
3. Outer floor panel attachment point
4. 3-point seat belts, rear left and right
5. 3-point seat belts, rear center
6. Center floor panel attachment points

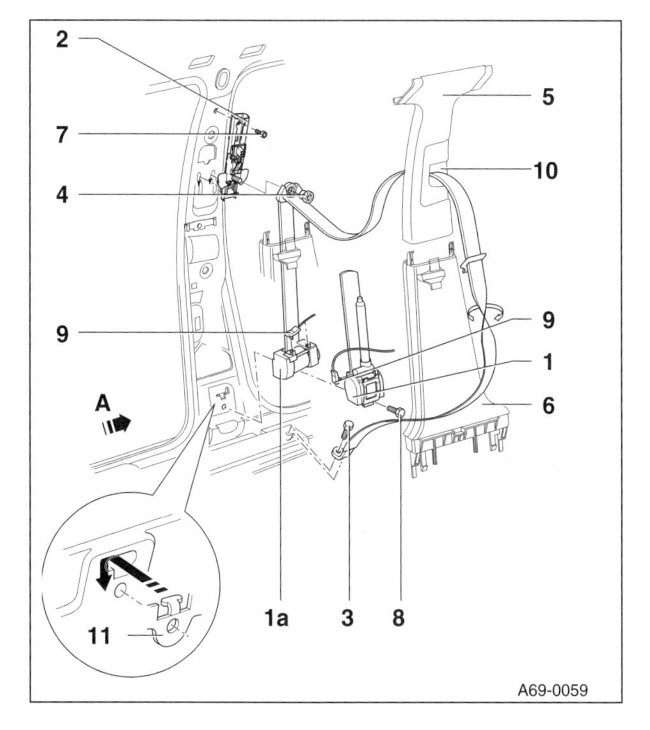

Front seat belt, removing and installing

1. Belt tensioner
2. Belt height adjuster
3. Bolt, 55 Nm (41 ft-lb)
4. Shouldered nut, 55 Nm (41 ft-lb)
5. Upper B-pillar trim
6. Lower B-pillar trim
7. Bolt, 23 Nm (17 ft-lb)
8. Bolt, 55 Nm (41 ft-lb)
9. Belt tensioner connector
10. Button for height adjuster
11. Anti-twist tab

Removing three-point seat belt

– Disconnect battery negative (-) cable.

– Remove sill trim. See **70 Interior Trim**.

– Remove lower belt bolt (**3**).

– Remove upper B-pillar trim (**5**) and lower B-pillar trim (**6**). See **70 Interior Trim**.

– Remove shouldered nut (**4**).

– Disconnect connector (**9**) at belt tensioner.

– Remove bolt (**8**).

– Lift belt tensioner (**1** or **1a**) out of tab (**11**).

BentleyPublishers.com
BentleyPublishers.com—All Rights Reserved

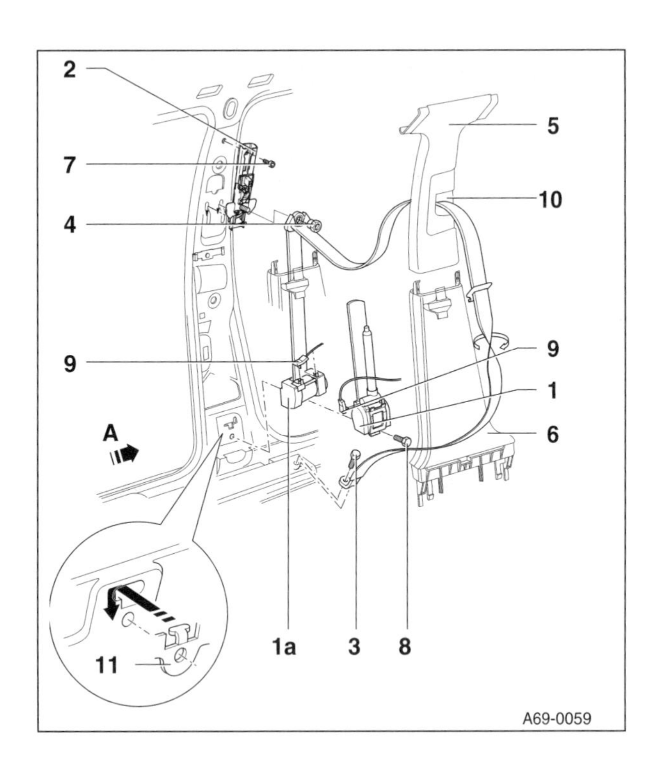

Installing 3-point seat belt:

- Insert belt tensioner (**1** or **1a**) in anti-twist element, as shown in magnified view (**11**).
- Install webbing with a 180° turn (**arrow**) as shown in illustration.

NOTE—

- *Arrow (**A**) points in direction of travel.*

- Remainder of installation is reverse of removal. After installing, make sure 3-point seat belt is not impeded. Check belt-fastened sensor (if equipped) using VAS 5051 or equivalent.

Removing belt height adjuster

- Remove upper B-pillar trim (**5**). See **70 Interior Trim**.
- Remove shouldered nut (**4**) and bolt (**7**).
- Lift out belt height adjuster (**2**).
- Installation is reverse of removal. Check that:
  - Belt operates properly after installing trim.
  - Height adjuster engages audibly in each of 5 possible positions (including top position).
  - Height adjustment fitting returns by itself to up position after being pressed.
  - After tightening shouldered nut, guide fitting automatically settles in correct direction.

NOTE—

- *Do not press button (**10**) when installing.*

Belt latch, removing and installing

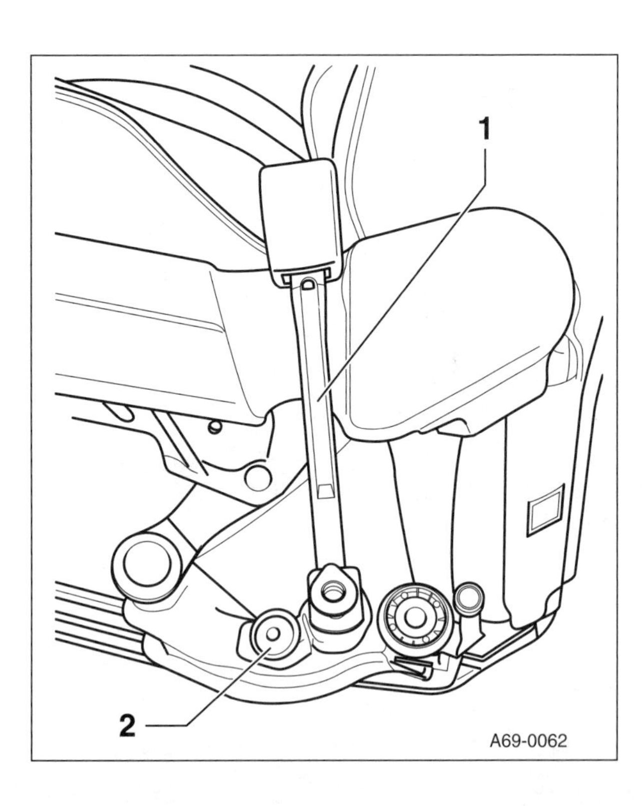

- Remove seat. See **72 Seats**.
- Remove bolt (**2**) from belt latch (**1**).
- Installation is reverse of removal.

NOTE—

- *It is possible that seat occupation recognition and seat belt fastened sensor have been installed between VIN 4BWN 024500 and 4BWN 033300 that are without function.*
- *When replacing a seat belt latch on vehicles in the above mentioned VIN range, install a belt latch without a "belt fastened" detector.*
- *On vehicles with a "belt fastened" detector and belt latch switch (only on drivers side, in models for some countries), remove wires from red connector housing.*

Tightening torque	
Lower belt latch bolt	24 Nm (18 ft-lb)

BentleyPublishers.com
BentleyPublishers.com—All Rights Reserved

Rear seat belt, removing and installing

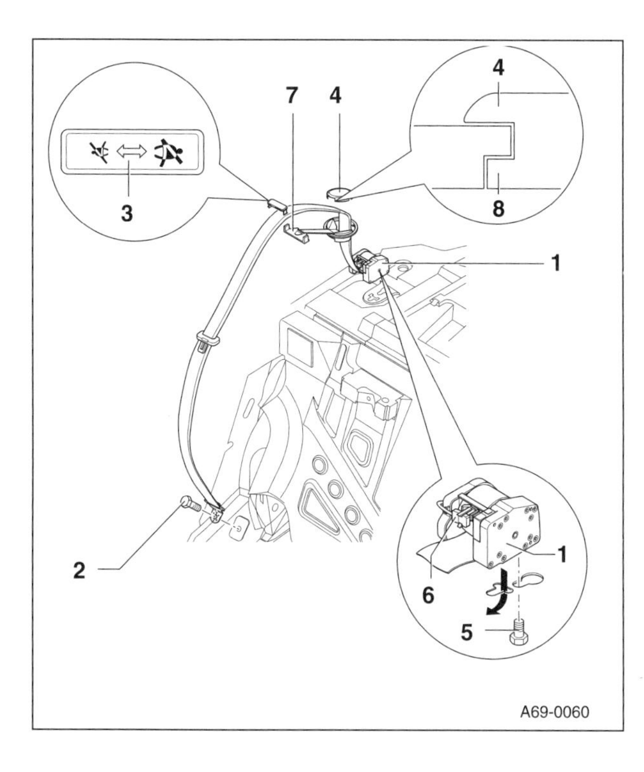

1. Belt tensioner
2. Bolt, 55 Nm (41 ft-lb)
3. Belt guide trim
   • Large symbol must always face center of vehicle.
4. Belt adjuster trim
5. Bolt, 55 Nm (41 ft-lb)
6. Belt tensioner connector
7. Belt adjuster
8. Tab

– Remove rear seat bench. See **72 Seats**.

– Remove bolt (**2**).

– Unclip belt guide trim (**3**).

– Pry off belt adjuster trim (**4**) on left and right at tab (**8**).

– Remove rear shelf (with folding backrest). See **70 Interior Trim**.

– Disconnect connector (**6**) at belt tensioner.

– Open cover for luggage compartment side trim.

– Vehicles with CD player and / or subwoofer or similar: Remove trunk / tailgate side trim. See **70 Interior Trim**.

– Remove bolt (**5**).

– Installation is reverse of removal. Check belt-fastened sensor (if equipped) using VAS 5051 or equivalent.

Rear seat belt adjuster, removing

The rear shelf does not have to be taken out in order to remove the rear seat belt adjuster.

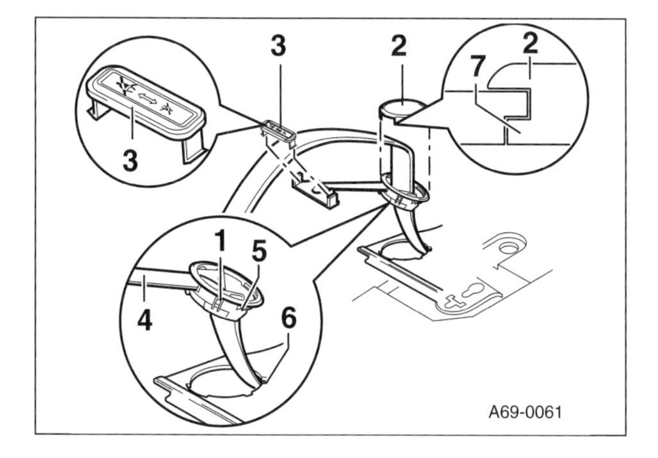

 Unfasten lower attachment point of rear 3-point seat belt. See **Rear seat belt, removing and installing** in this repair group.

– Unclip belt guide trim (**3**).

– Pry off belt adjuster trim (**2**) on left and right at tab (**7**).

– Raise arm of belt adjuster (**4**) with webbing.

BentleyPublishers.com
BentleyPublishers.com—All Rights Reserved

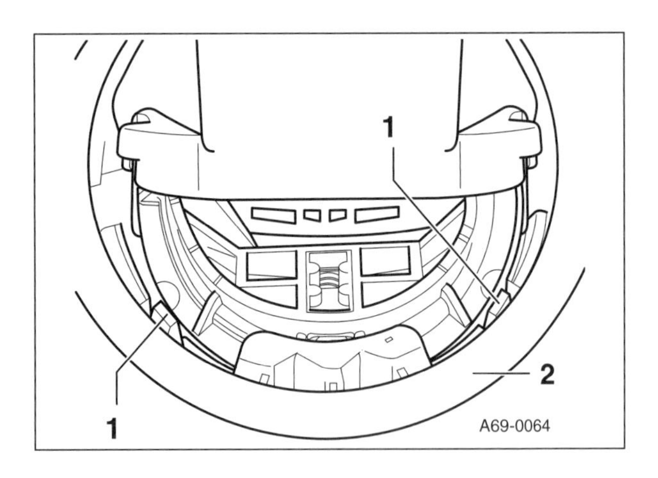

◄ Use screwdriver to press retainer tabs (**1**) inward and at the same time pull belt adjuster (**2**) upward.

– Pull webbing through belt adjuster (**2**).

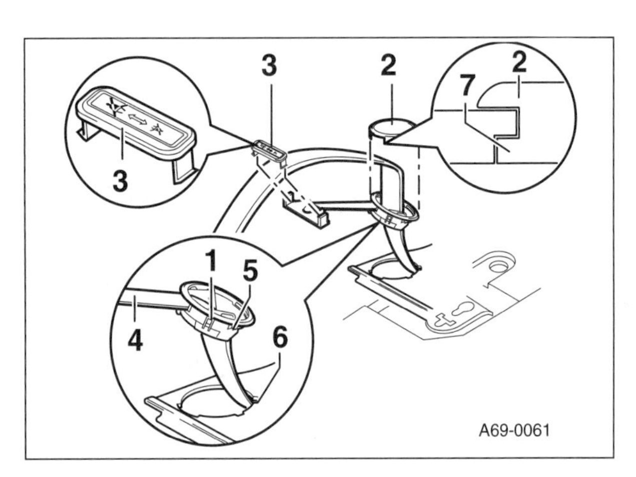

◄ To install, make sure retainers (**1**) engage properly.

– Make sure rear stud engages in recess (**6**).

– Locating element (**5**) must face center of vehicle.

Tightening torque	
Rear belt lower bolt to body	55 Nm (41 ft-lb)

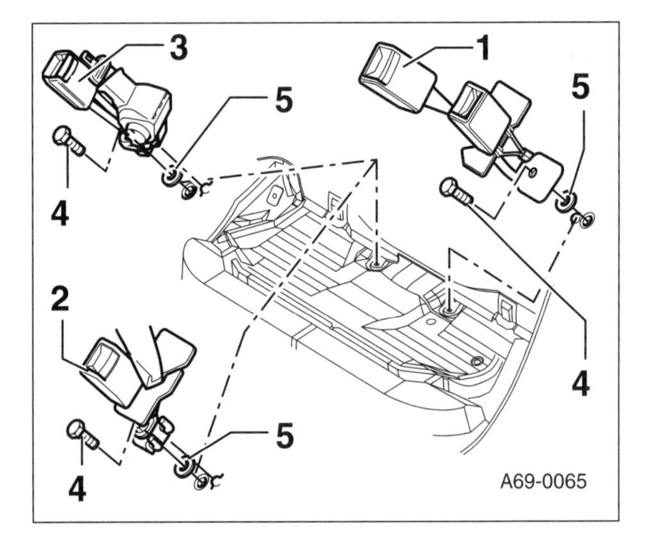

Rear belt floor panel attachment points

1. Latch
2. Lap belt
3. Lap belt with retractor
4. Bolt, 55 Nm (41 ft-lb)
   • Use OEM bolts.
5. Seal
   • Always replace

– Remove rear seat bench. See **72 Seats**.

– Remove bolt (**4**).

– When installing, make sure latch and/or lap belt engage(s) in anti-twist projections.

– Narrow side of latch and/or lap belt must face in direction of travel.

BentleyPublishers.com
BentleyPublishers.com—All Rights Reserved

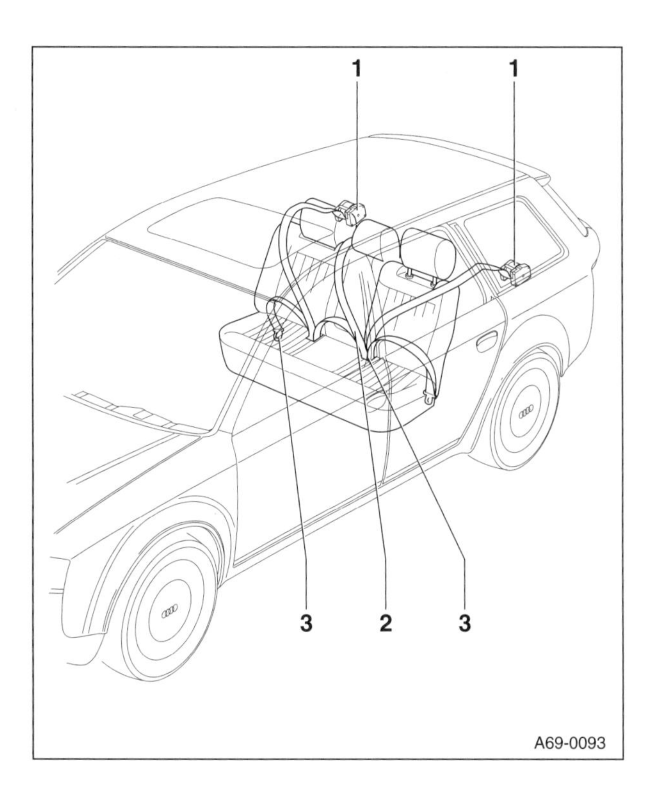

Rear seat belt attachment points (Avant)

1. 3-point seat belts, rear left and right (Avant)
2. 3-point seat belt, rear center
3. Center floor panel attachment points

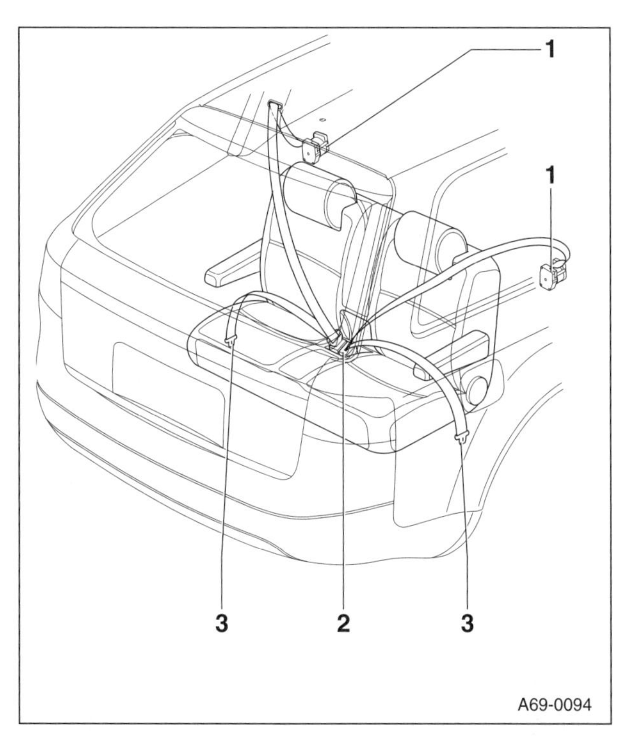

Rear seat belt attachment points (children's bench)

1. Rear 3-point seat belts for children's bench
2. Center rear attachment point for children's bench
3. Outer rear attachment points

BentleyPublishers.com
BentleyPublishers.com—All Rights Reserved

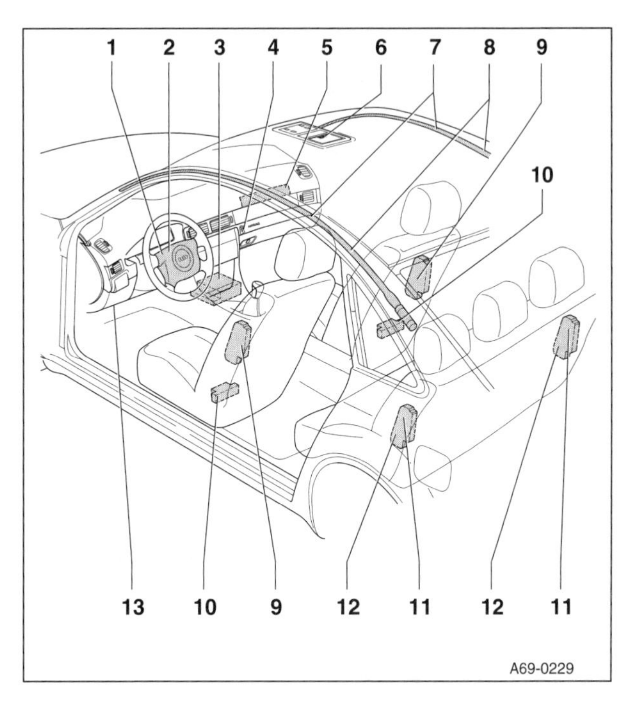

Airbags

Airbag components

1. Airbag unit - drivers side
2. Airbag unit drivers side (sport steering wheel)
3. Airbag control module
4. Not applicable for USA/Canada
5. Passenger airbag unit
6. Not applicable for USA/Canada
7. Side curtain airbag, Sedan
8. Side curtain airbag, Avant
9. Front side airbag
10. Lateral acceleration sensor
11. Rear side airbag (folding backrest)
12. Rear side airbag (fixed backrest)
13. Data link connector (DLC)

Airbag control module (J234), removing and installing

– Disconnect battery negative (-) cable.

– Remove front section of center console. See **70 Interior Trim**.

– Remove connecting pieces for air ducts to rear left and right footwell vents.

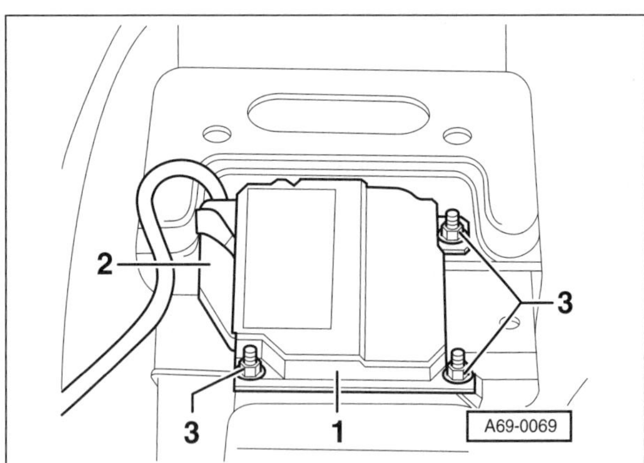

◄ Through production date August 1999:

- Release retainer on connector (**2**).
- Pull connector (**2**) out of control module (**1**).
- Remove nuts (**3**).
- Remove control module

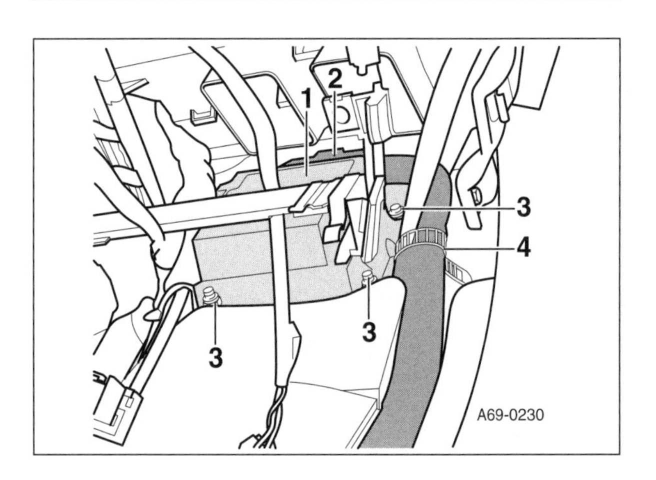

◄ From production date September 1999 on:

- Open cable ties (**4**).
- Remove nuts (**3**).
- Pull out control module (**1**).
- Release retainer on connector (**2**).
- Pull connector (**2**) out of control module (**1**).
- Remove control module

– Installation is reverse of removal. Note the following:

- If control module is replaced, it must be coded.
- If airbag warning light remains on, erase DTC memory using VAS 5051 or equivalent.

Tightening torque	
Airbag control module (J234) to body	6 Nm (53 in-lb)

BentleyPublishers.com
BentleyPublishers.com—All Rights Reserved

Driver's front airbag, removing and installing

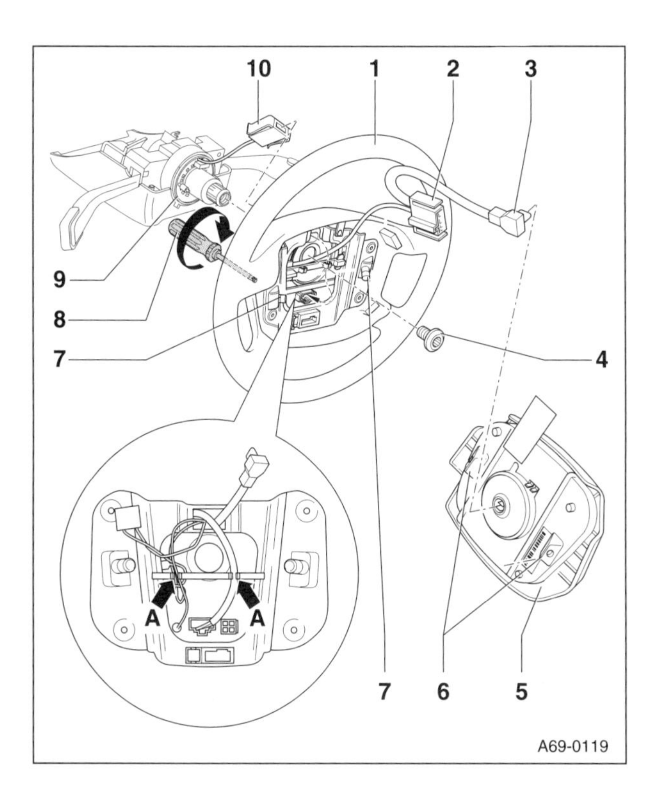

1. Steering wheel
2. Connector for self-canceling ring
3. Connector for airbag unit
4. Bolt (always replace)
   - Allen head, 60 Nm (44 ft-lb)
   - Hex head, 70 Nm (52 ft-lb)
5. Airbag unit
6. Threaded inserts
7. Torx bolts T30 (if equipped), 7 Nm (62 in-lb)
8. Torx wrench T30
9. Self-canceling ring with slip ring
10. Connector for heated steering wheel

– Disconnect battery negative (-) cable.

– Release steering column adjuster. Extend steering column out and upward as far as possible.

– Move steering wheel so that hub and spokes are vertical.

– Release one Torx bolt (**7**).

– Move steering wheel back 1/2 turn and release second Torx bolt (**7**).

– Release electrical connectors from airbag unit.

– Lay airbag down with padded side facing upward.

– Installation is reverse of removal. Remember to:
  - Make sure electrical connectors fully (audibly) engage.
  - Reset vehicle equipment (radio, clock, power windows) as per owner's manual.
  - Re-connect battery and switch ignition ON.
  - If airbag warning light remains on, erase DTC memory using VAS 5051 or equivalent.

> ***WARNING —***
> *• Make sure no one is in vehicle when reconnecting battery.*

BentleyPublishers.com
BentleyPublishers.com—All Rights Reserved

Passenger front airbag, removing and installing

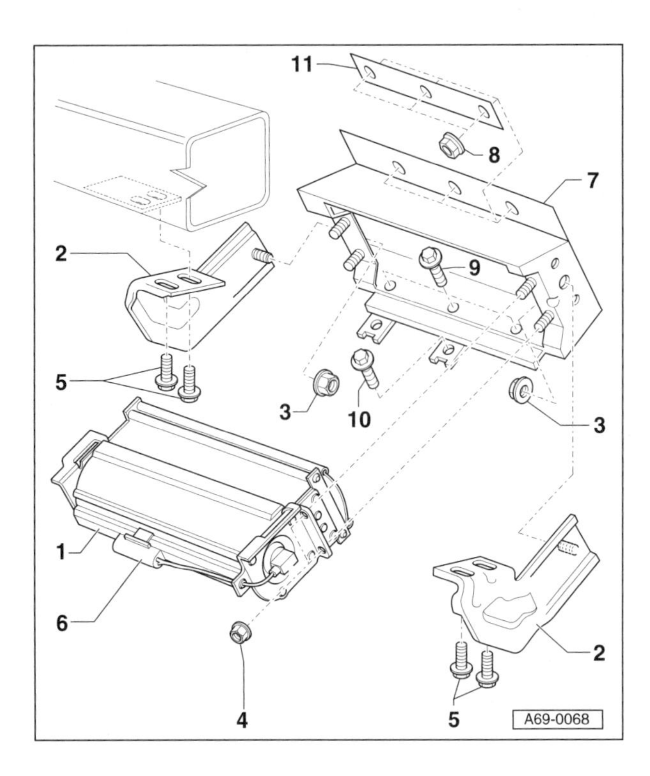

1. Passenger front airbag unit
2. Support
   - Must be replaced if airbag has triggered
3. Nut (2x), 18 Nm (13 ft-lb)
   - Must be replaced if airbag has triggered
4. Nut (4x), 9 Nm (80 in-lb)
   - Must be replaced if airbag has triggered
5. Bolts (4x), 9 Nm (80 in-lb)
   - Must be replaced if airbag has triggered
6. Connector
7. Frame for airbag unit
   - Must be replaced if airbag has triggered
8. Nut (3x), 4.5 Nm (40 in-lb)
9. Bolts (3x), 4.5 Nm (40 in-lb)
10. Bolt (1x), 4.5 Nm (40 in-lb)
11. Backing plate

– Disconnect battery negative (-) cable.

– Remove glove box. See **70 Interior Trim**.

– Disconnect electrical connector (**6**). See **Disconnecting passenger airbag connector** in this repair group.

> ***WARNING —***
> *• Never disconnect red 2-pin connector on front side of airbag.*

– Remove nuts (**4**) and remove airbag unit.

– Set down airbag unit with padded side facing upward.

– Installation is reverse of removal. Remember to:
- Make sure electrical connectors fully (audibly) engage.
- Reset vehicle equipment (radio, clock, power windows) as per owner's manual.
- Reconnect battery and switch ignition ON.
- If airbag warning light remains on, erase DTC memory using VAS 5051 or equivalent.

> ***WARNING —***
> *• Make sure no one is in vehicle when reconnecting battery.*

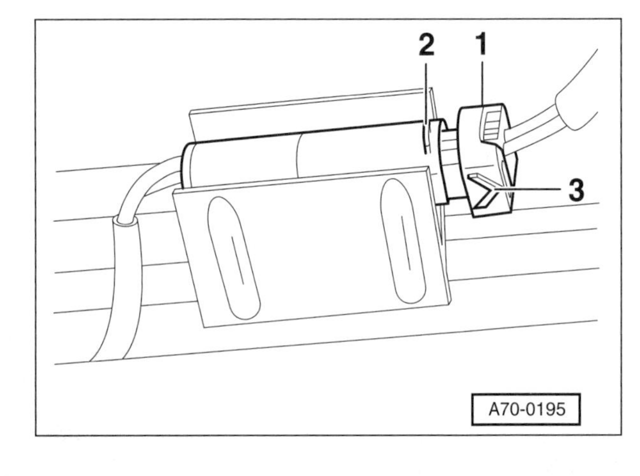

Disconnecting passenger airbag connector

Use screwdriver to lift connector (**1**) for front passenger side airbag over retainer tab (**2**) and pull off connector in direction of (**3**).

BentleyPublishers.com
BentleyPublishers.com—All Rights Reserved

70 Interior Trim

GENERAL

This repair group covers interior equipment and trim panels, including center console removal. For interior lights and accessories see **96 Interior Lights, Switches, Anti-theft**.

WARNING —

- *Cars covered by this manual are equipped with airbags. When servicing these components, disconnect the negative (-) battery terminal. See* **69 Seat Belts, Airbags**.

BentleyPublishers.com
BentleyPublishers.com—All Rights Reserved

> ***CAUTION—***
> - *Prior to disconnecting the battery, acquire radio anti-theft code.*
> - *Read the battery disconnection cautions in* **00 Warnings and Cautions**.

INTERIOR REAR VIEW MIRROR

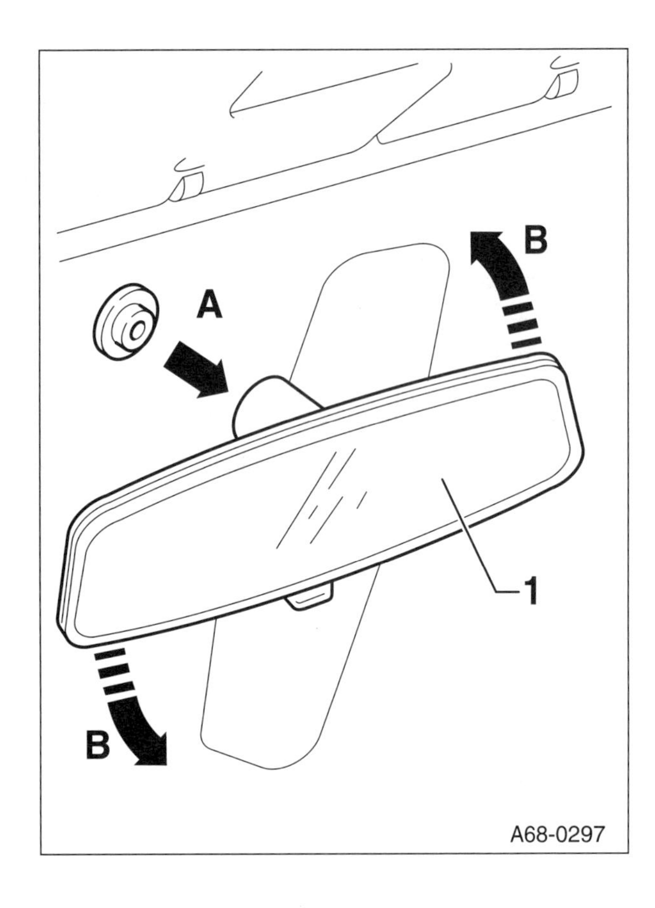

Interior rear view mirror, removing and replacing

- Rotate mirror (**1**) 60° to 90° in direction of arrows (**B**) until locking spring releases and remove mirror.

- To install, place mirror at an angle of 60° to 90° to attachment and turn until locking spring clicks in place.

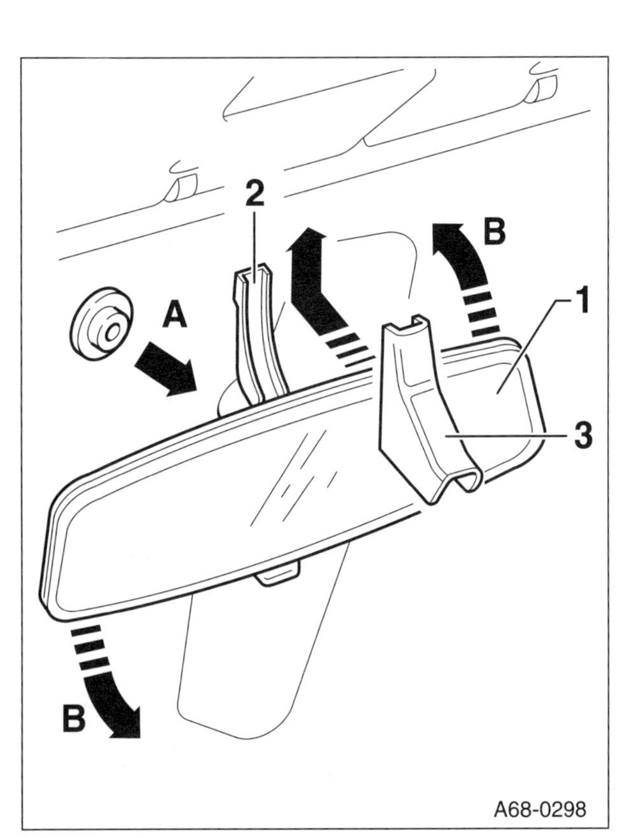

Interior rear view mirror (automatic dimming), removing and replacing

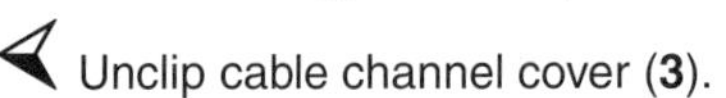

- Unclip cable channel cover (**3**).

- Remove electrical connector from retainer and disconnect.

- Rotate mirror (**1**) 60° to 90° in direction of arrows (**B**) until locking spring releases and remove mirror.

> ***CAUTION—***
> - *Do not use cable channel (2) to turn mirror.*

- To install, place mirror at an angle of 60° to 90° to attachment and turn until locking spring clicks in place.

- Reconnect electrical connector. Place connector in retainer and attach channel cover (**3**).

BentleyPublishers.com
BentleyPublishers.com—All Rights Reserved

Storage Compartments, Covers and Trim

Steering column trim, removing and installing

- Disconnect battery negative (-) cable.
- Remove driver's airbag. See **69 Seat Belts, Airbags**.
- Remove steering wheel. See **48 Steering**.
- Extend steering column out and down.

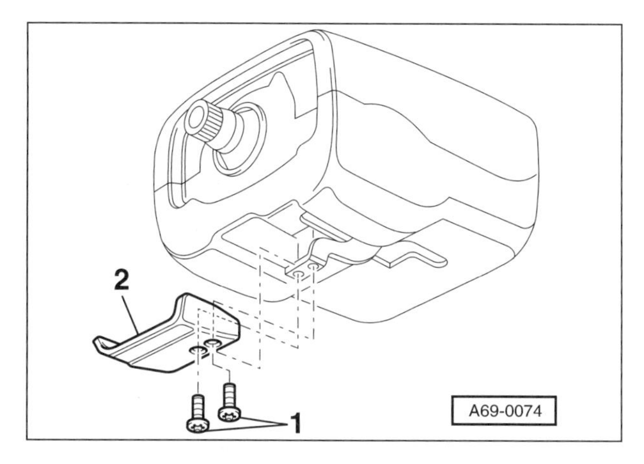

◀ Remove screws (**1**) and remove grip (**2**).

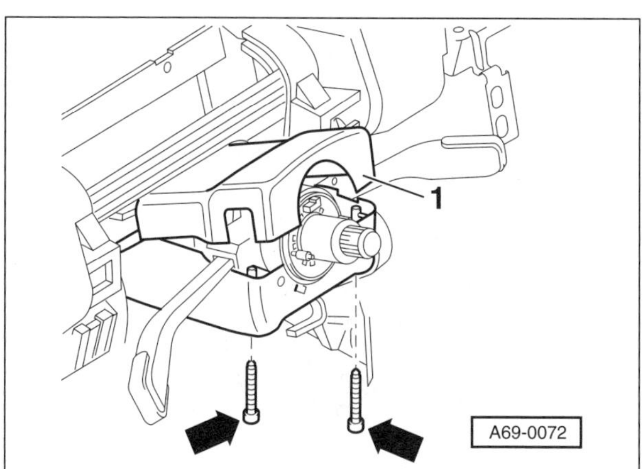

◀ Remove two screws (**arrows**) and remove upper trim piece.

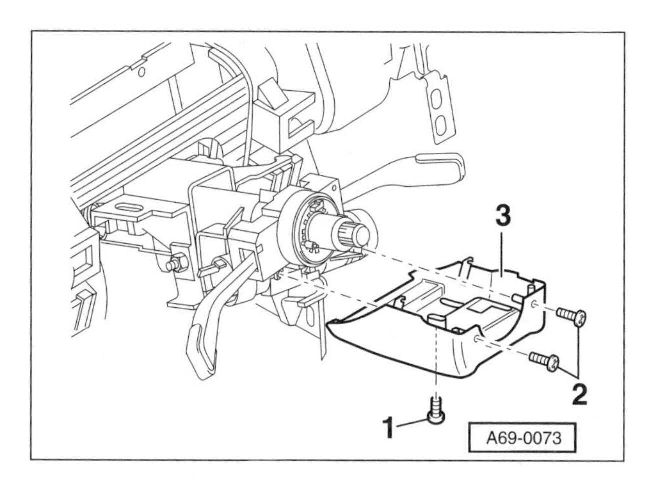

◀ Remove screws (**2**) and bolt (**1**). Remove lower trim piece.

- To install, replace lower section first.
- Insert upper section into hooks on lower section, swivel downward and secure.

BentleyPublishers.com
BentleyPublishers.com—All Rights Reserved

Dashboard end trim, removing and installing

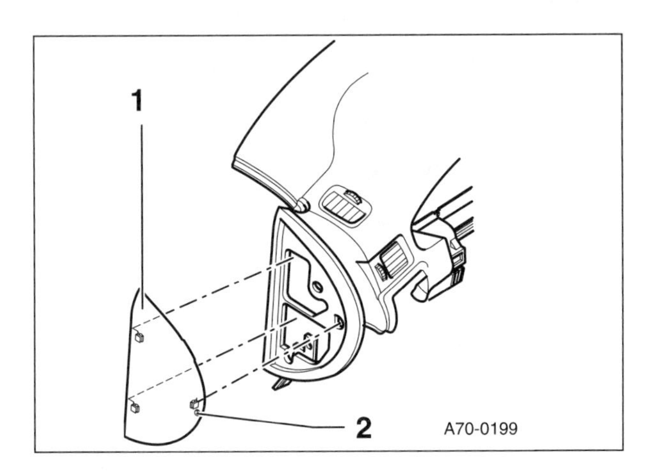

◄ Use a plastic prying tool or screwdriver with tip wrapped with tape to unclip left or right side panel end trim.

– To install insert trim toward front and clip back firmly in place.

Glove compartment, removing and installing

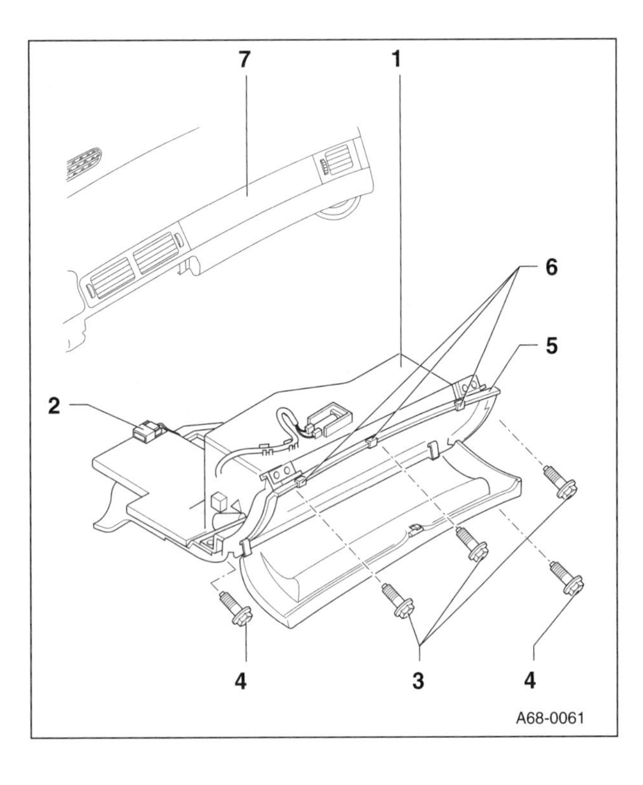

1. Glove box
2. Connector
3. Bolts (3x)
4. Bolts (2x)
5. Strip
6. Rubber studs
7. Dash board

– Open glove box lid and remove bolts (**3**).

– Remove bolts (**4**).

– Remove glove box downward.

– Disconnect electrical connector (**2**).

– To install, position strip (**5**) with rubber studs (**6**) against dash board (**7**) from underneath and install glove box.

BentleyPublishers.com
BentleyPublishers.com—All Rights Reserved

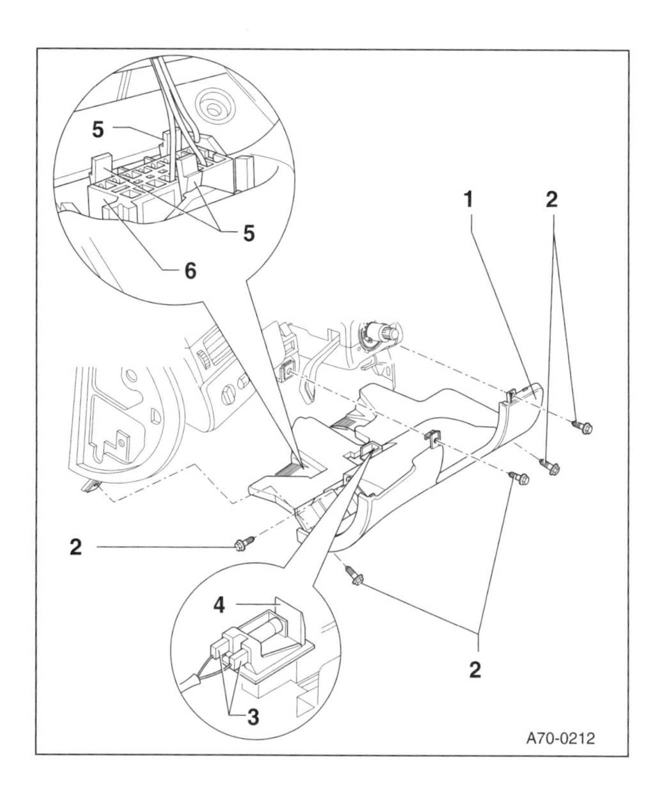

Driver's storage compartment, removing and installing

1. Driver's storage compartment
2. Bolts (5x)
3. Connector
4. Footwell light
5. Tabs
6. Data link connector (DLC)

– Pull out dashboard molding to rear.

NOTE—

• Steering wheel shown removed for clarity.

– Unclip instrument panel end trim on drivers side. See **Instrument panel end trim, removing and installing** in this repair group.

– Remove bolts (**2**) and set down driver's storage compartment.

– Disconnect electrical connector (**3**) from footwell light (**4**).

– Unclip data link connector (DLC) (**6**) by squeezing together tabs (**5**).

– Installation is reverse of removal.

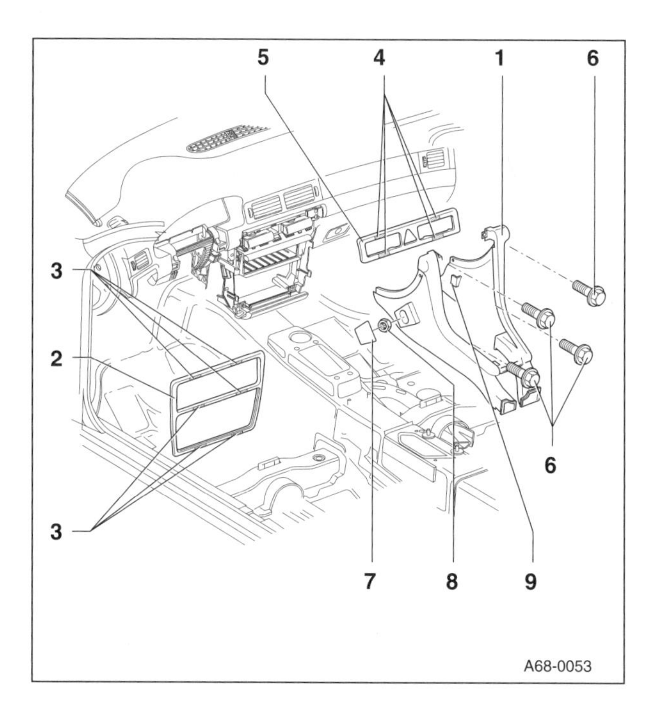

Center console (front section), removing and installing

1. Center console (front section)
2. Trim section
3. Fasteners
4. Fasteners
5. Switch module trim
6. Bolts (4x)
7. Trim cap
8. Nut
9. Retainer

– Remove rear section of center console. See **Center console (rear section), removing and installing** in this repair group.

– Open front ashtray and unclip trim section (**2**) at fasteners (**3**) (Chorus and Concert radios only).

– Remove radio. See **91 Radio and Communication**.

– Remove navigation system controls (if equipped).

– Remove A/C controls. See **87 Heating and Air-conditioning**.

– Remove front ashtray. See **Front ashtray, removing and installing** in this repair group.

– Unclip switch-panel trim (**5**) at fasteners (**4**).

– Remove console bolts (**6**).

BentleyPublishers.com
BentleyPublishers.com—All Rights Reserved

– Pry off trim cap (**7**) on side of console and remove nut (**8**).

– Pull center console on driver's side over securing pin.

– Remove center console to rear.

– Installation is reverse of removal. Make sure retainer (**9**) locks in place.

Center console (rear section), removing and installing

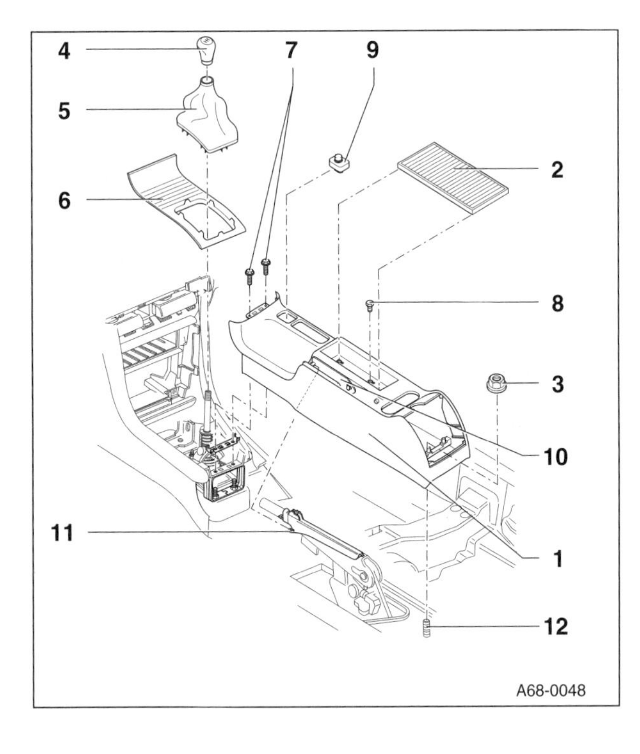

1. Center console (rear section)
2. Mat lining
3. Nut, 10 Nm (7 ft-lb)
4. Shifter knob
   • Twist on / off
5. Protective boot
   • Unclip upward. Make sure clips lock in place when installing
6. Center console insert
7. Screws (2x)
8. Screw (1x)
9. Mirror adjustment switch
10. Parking brake trim
11. Retainer tab
12. Stud

– Remove rear air vent and rear air duct connecting piece. See **Rear vent and duct, removing and installing** in this repair group.

– Remove center armrest (if necessary). See **Center armrest, removing and installing** in this repair group.

– Remove console nut (**3**).

– Remove mat lining (**2**) upward.

– Remove console bolt (**8**).

– Models with manual transmission:
  • Twist off shifter knob (**4**).
  • Unclip protective boot (**5**) upward.

– Unclip center console insert (**6**) upward.

– Remove console bolts (**7**).

– Remove parking brake trim.

– Raise rear section of center console (**1**) at front and detach connector for mirror adjustment switch (**9**).

– Fully apply parking brake.

– Unfasten trim (**10**) from retainer (**11**) on parking brake.

– Raise rear section of center console (**1**) over stud (**12**) and pull center console to rear.

BentleyPublishers.com
BentleyPublishers.com—All Rights Reserved

– Lift out rear section of center console (**1**) forward over parking brake.

– Installation is reverse of removal. Remember to:
  - Install trim (**10**) at retainer (**11**) of parking brake.
  - Make sure air ducts are properly installed when installing rear section of center console (**1**).

Rear vent and duct, removing and installing

– Use or fabricate a suitable tool to remove vent trim (a screwdriver bent 90° at 6 - 10 mm along its length will work).

◄ Pull rear vent (**1**) out center console (rear section).

– Disconnect electrical connectors for cigarette lighter, vent light and rear heated seat (if equipped).

– Remove rear air duct connecting piece (**2**) from air duct (**3**).

– Installation is reverse of removal

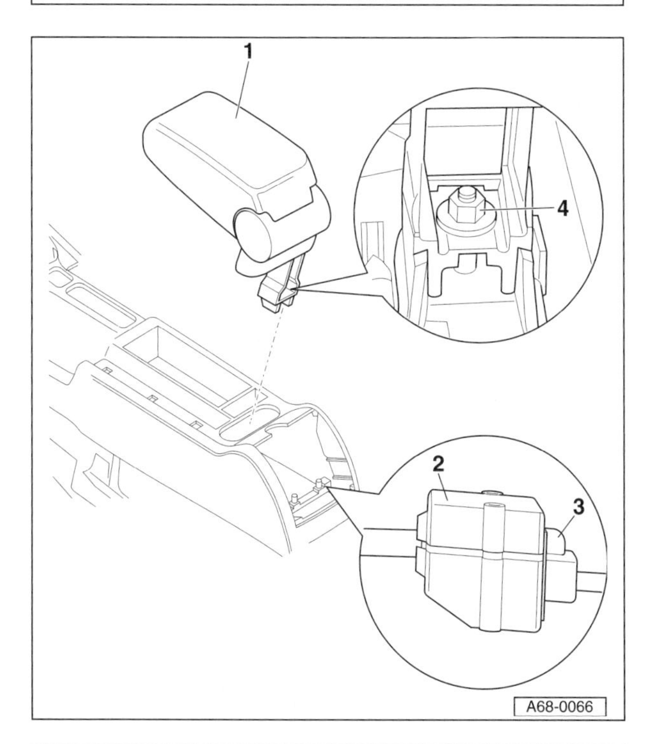

Center armrest, removing and installing

1. Center armrest
2. Connector
3. Retainer tabs
4. Nut, 26 Nm (18 ft-lb)

– Remove rear vent and duct. See **Rear vent and duct, removing and installing** in this repair group.

– Models with phone: Disconnect electrical connector (**2**) by squeezing retaining tabs (**3**).

– Remove nut (**4**) and remove armrest.

– Installation is reverse of removal. Make sure to clamp in connector (**2**) on right side between support bracket and center console.

Front ashtray (Concert radio), removing and installing

– Open radio flap and open front ashtray.

– Starting at bottom, carefully unclip front center console trim.

– Working above radio, carefully unclip switch panel trim.

◄ Remove bolts (**1**).

BentleyPublishers.com
BentleyPublishers.com—All Rights Reserved

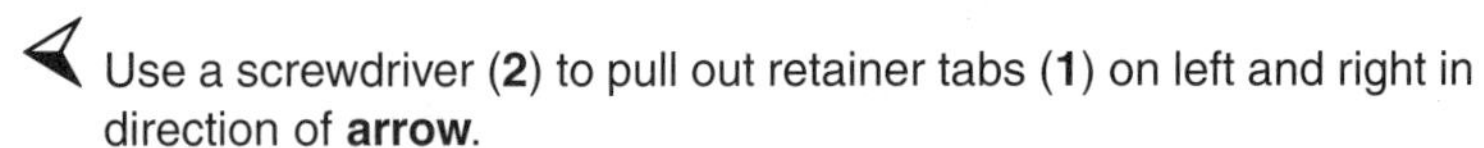

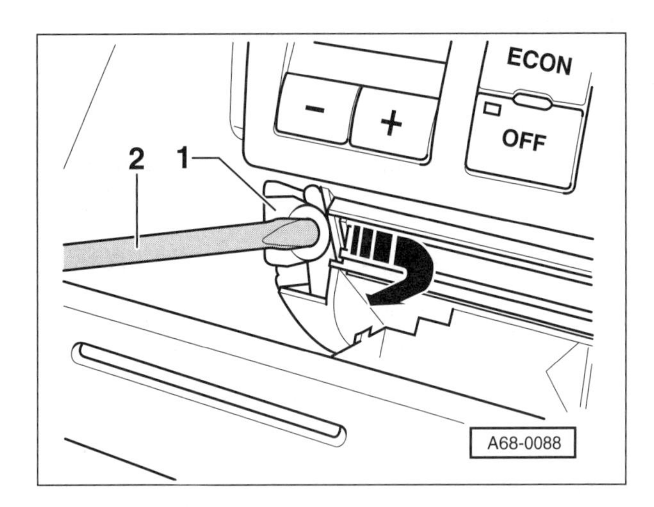

◀ Use a screwdriver (**2**) to pull out retainer tabs (**1**) on left and right in direction of **arrow**.

– Carefully pull A/C control head out slightly toward the rear, and pull out ashtray downward and toward rear.

– Disconnect electrical connections for cigarette lighter and ashtray light. Remove ashtray.

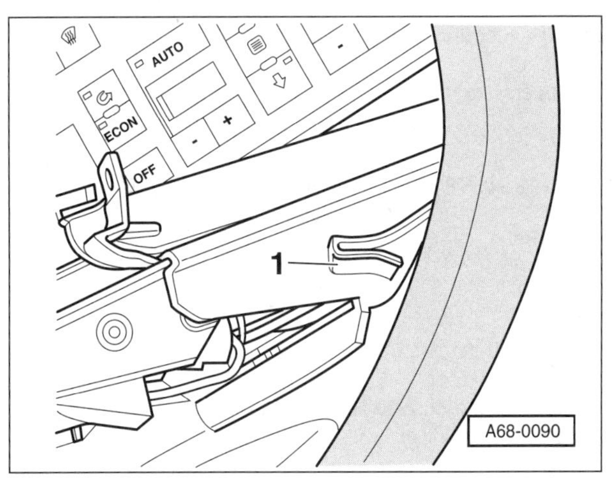

◀ When installing, make sure guide (**1**) engages on left and right in center console.

Front ashtray (Symphony radio), removing and installing

– Remove radio. See **91 Radio and Communications**.

– Remove navigation system controls (if equipped).

– Remove A/C controls and heat controls. See **87 Heating and Air-conditioning**.

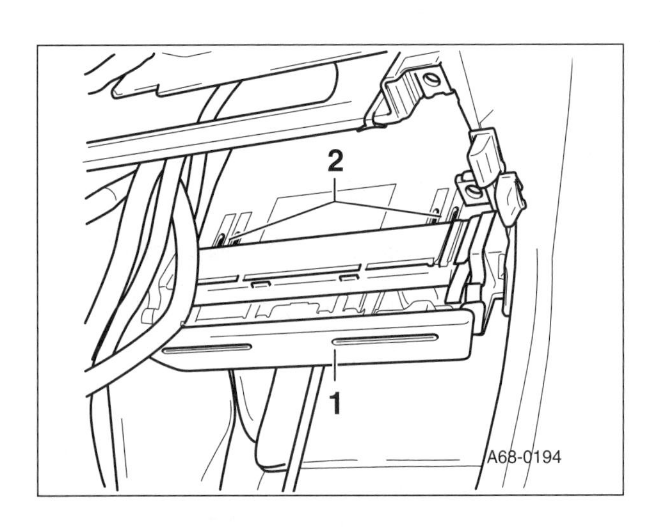

◀ Press retaining tabs (**2**) and pull ashtray out forward.

– Disconnect electrical connections and remove ashtray.

– Installation is reverse of removal. Make sure retaining tabs lock in place.

BentleyPublishers.com
BentleyPublishers.com—All Rights Reserved

Door Trim

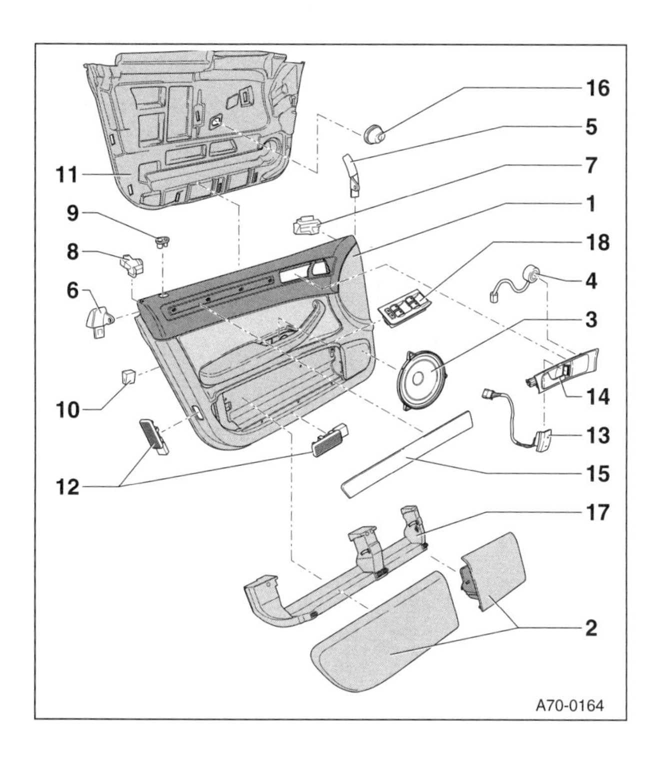

Front door trim components

1. Door trim
2. Hinged pockets
3. Bass loudspeaker
4. Treble speaker
5. Inner connecting piece
6. Outer connecting piece
7. Inner filler piece
8. Outer filler piece
9. Locking knob guide
10. Clip locator
11. Insulation
12. Door courtesy light
13. Central locking switch (drivers side only)
14. Interior door handle
15. Decorative trim
16. Foam seal
17. Mount for hinged pockets
18. Window switch

Front door trim, removing and installing

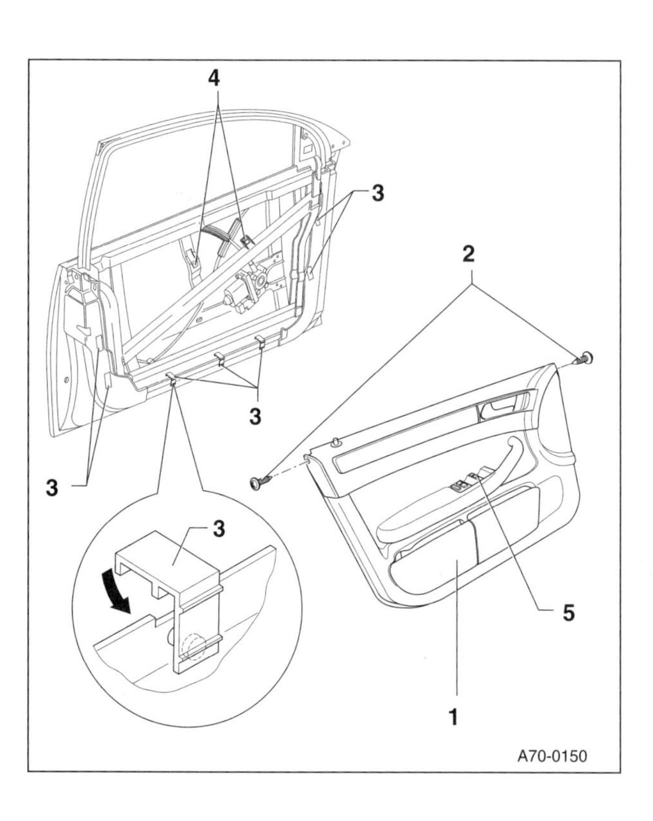

Remove screws (**2**) from door trim (**1**).

BentleyPublishers.com
BentleyPublishers.com—All Rights Reserved

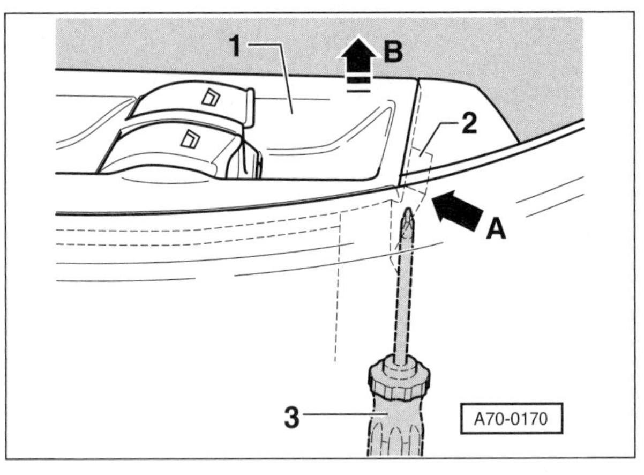

Remove window switch unit by inserting screwdriver (**3**) in hole in armrest. Press locking tab (**2**) in direction of arrow (**A**).

NOTE —

- *On models without hole in armrest, use a plastic prying tool to pry front edge of window switch up slightly and release locking tab.*

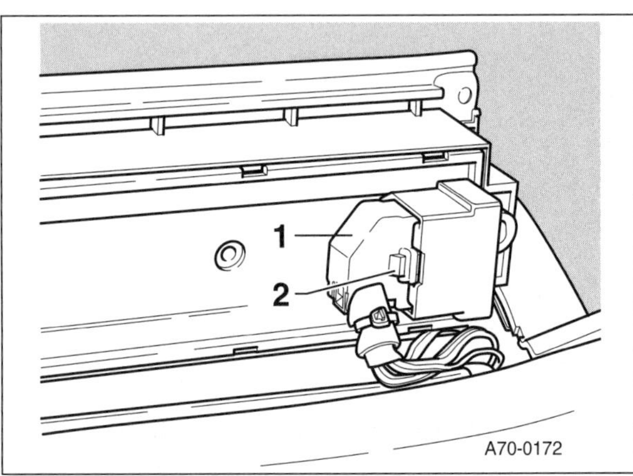

Lift switch unit up. Press retainer tab (**2**) to release electrical connector (**1**).

– Lift door trim about 20 cm (8 in) upward and remove from door.

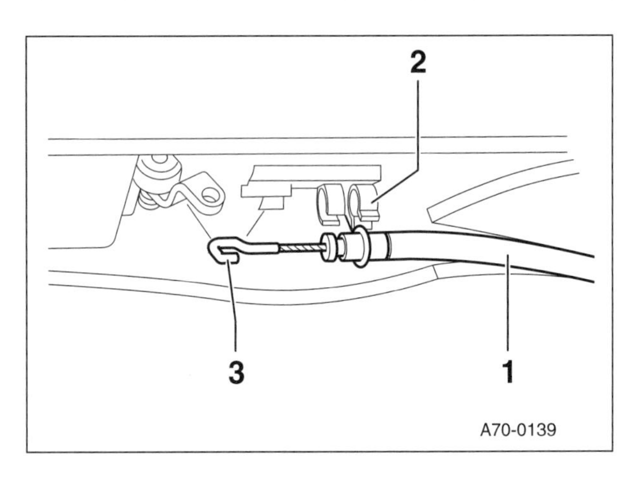

Pull bowden cable (**1**) out of guide (**2**) and detach.

NOTE —

- *When installing, make sure hook (**3**) is facing upward.*

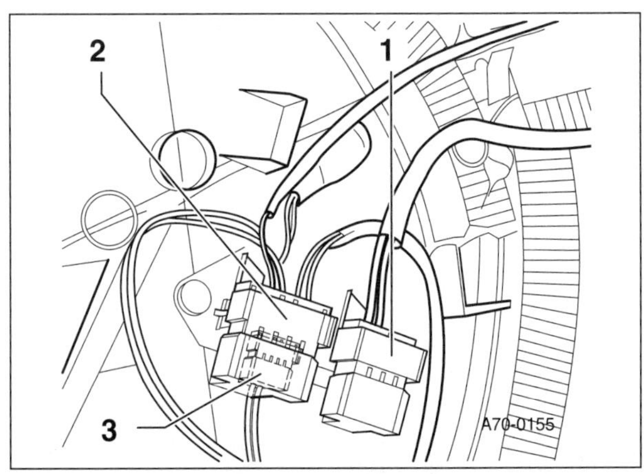

Disconnect electrical connectors for central locking switch (**1**), door warning light (**2**) and radio speaker (**3**).

BentleyPublishers.com
BentleyPublishers.com—All Rights Reserved

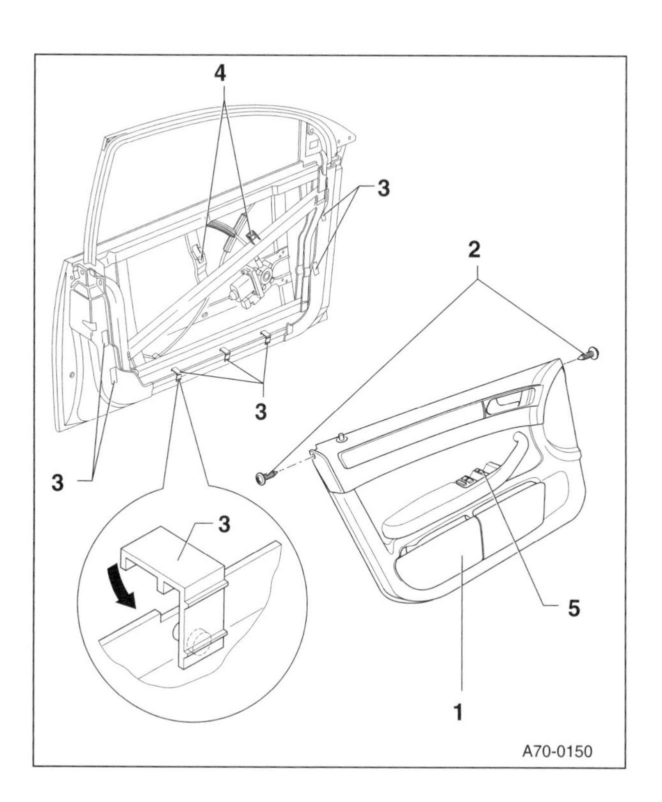

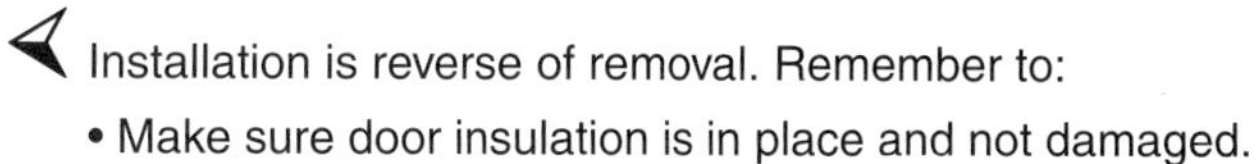
Installation is reverse of removal. Remember to:

- Make sure door insulation is in place and not damaged.
- Press center of door trim on door first.
- Clip locators (**3** and **4**) in place.

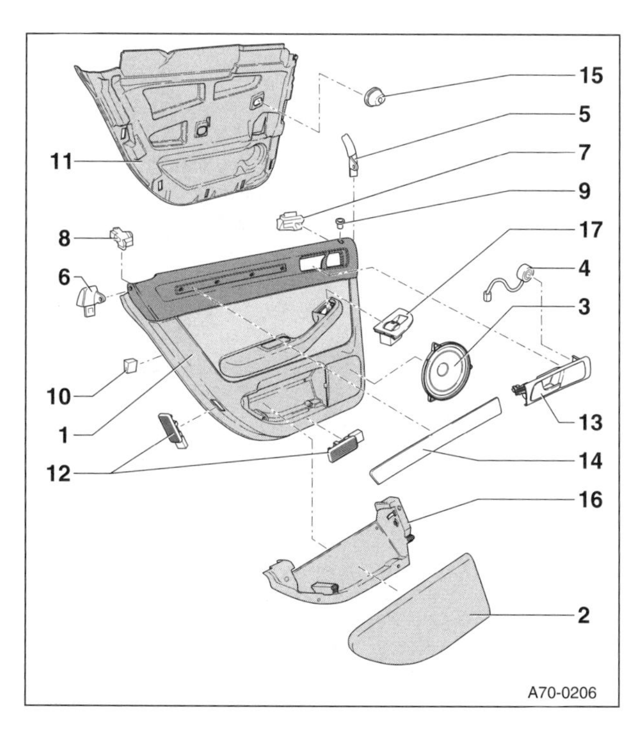

Rear door trim components

1. Rear door trim
2. Hinged pocket
3. Bass speaker
4. Treble speaker
5. Inner connecting piece
6. Outer connecting piece
7. Inner filler piece
8. Outer filler piece
9. Locking knob guide
10. Clip locator
11. Insulation
12. Door courtesy light light
13. Interior door handle
14. Decorative trim
15. Foam seal
16. Mount for hinged pockets
17. Window switch

BentleyPublishers.com
BentleyPublishers.com—All Rights Reserved

Rear door trim, removing and installing

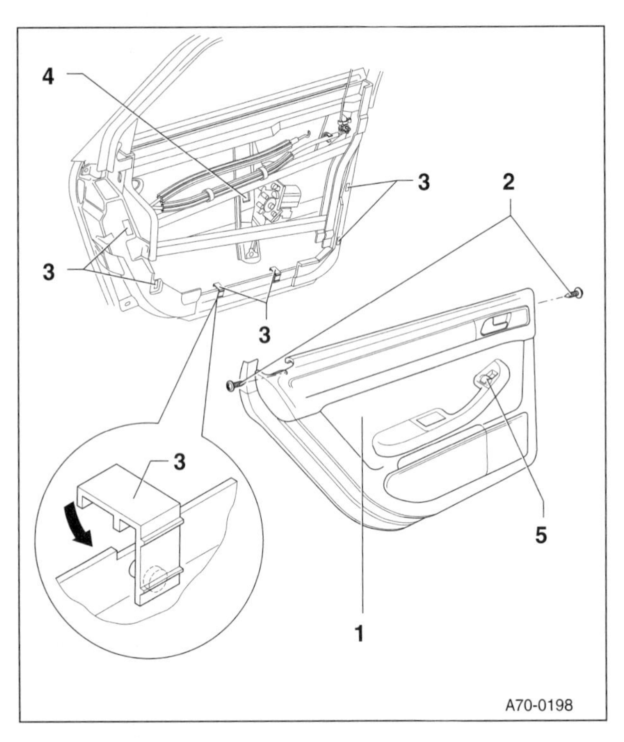

◀ Remove screws (**2**) from door trim (**1**).

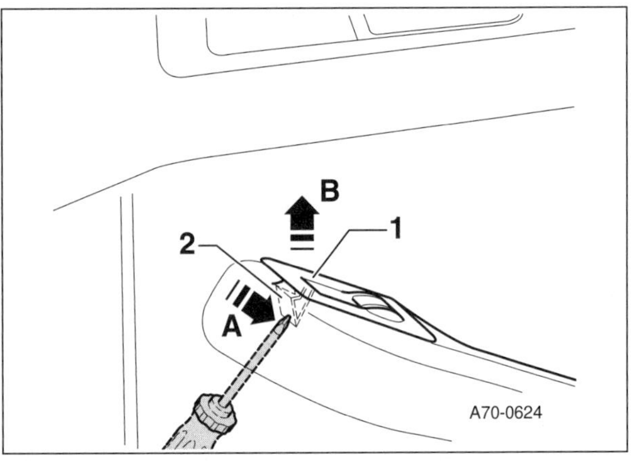

◀ Remove window switch by inserting screwdriver in hole in armrest. Press locking tab (**2**) in direction of arrow (**A**).

NOTE—

- *On models without hole in armrest, use a plastic prying tool to pry front edge of window switch up slightly and release locking tab.*

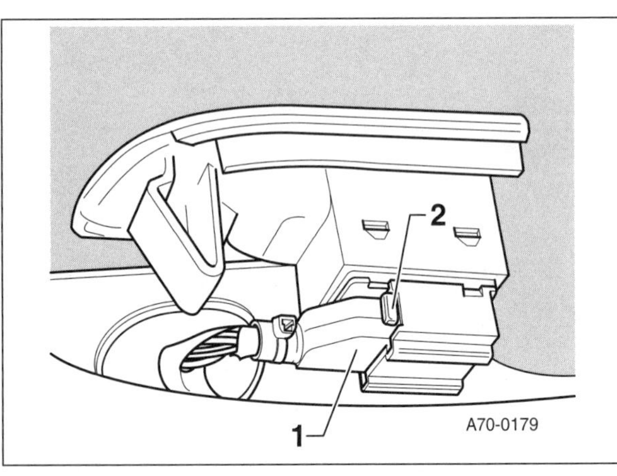

◀ Lift switch unit up. Press retainer tab (**2**) to release electrical connector (**1**).

– Lift door trim about 20 cm (8 in) upward and remove from door.

BentleyPublishers.com
BentleyPublishers.com—All Rights Reserved

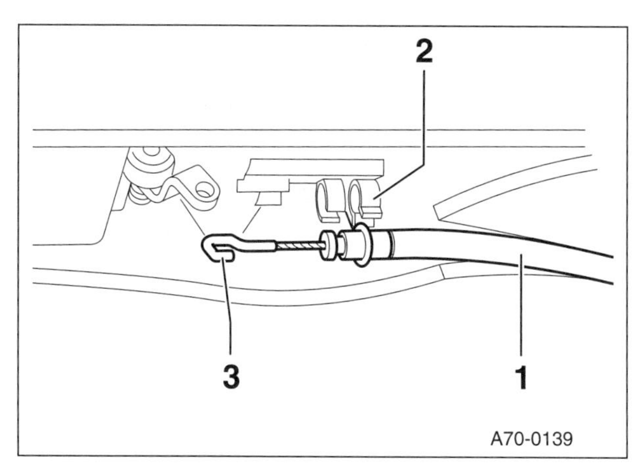

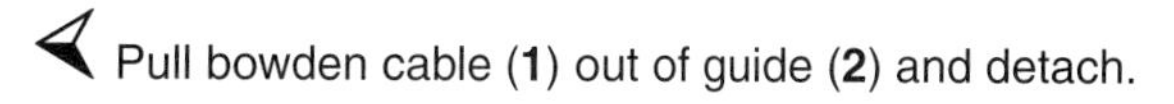

Pull bowden cable (**1**) out of guide (**2**) and detach.

NOTE—

- *When installing, make sure hook (**3**) is facing upward.*

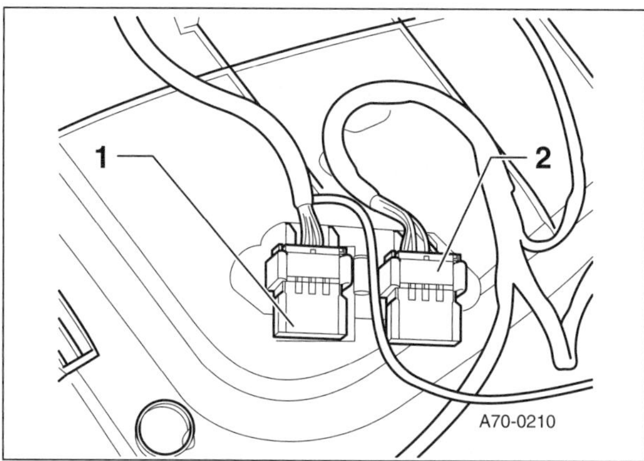

Disconnect electrical connector for door warning light (**2**) and radio speaker (**1**).

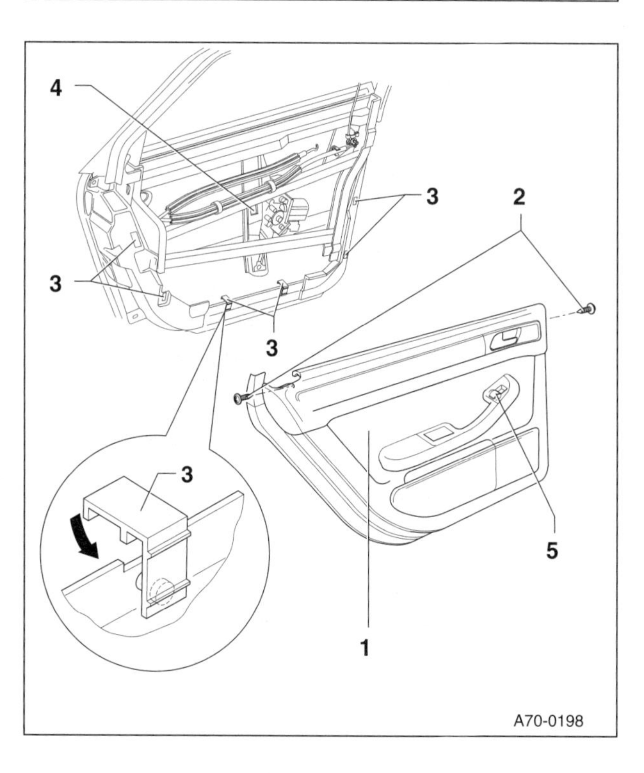

Installation is reverse of removal. Remember to:

- Make sure door insulation is in place and not damaged.
- Press center of door trim on door first.
- Clip locators (**3** and **4**) in place.

BentleyPublishers.com
BentleyPublishers.com—All Rights Reserved

PILLAR AND SIDE TRIM

Upper A-pillar trim, removing and installing

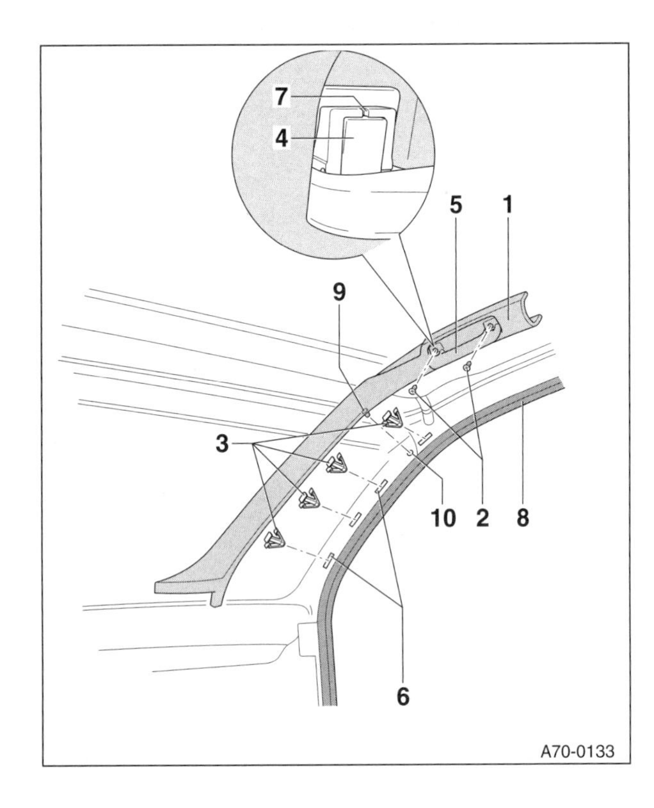

1. Upper A-pillar trim
2. Screws (2x)
3. Clips (4x)
4. Flap for roof grab handle
5. Roof grab handle
6. Mounts
7. Groove
8. Seal
9. Pin
10. Mount

– Pivot down flap (**4**) on roof grab handle (**5**) at groove (**7**).

– Remove screws (**2**).

– Unclip A-pillar trim (**1**), starting at top.

– Pull out trim and rotate upward.

– On models with hands-free telephone function, disconnect connector for hands-free unit.

– Installation is reverse of removal. Remember to:
  - Carefully route wiring and connector for hands-free telephone unit from bottom to top (if equipped).
  - Move sun visor out of the way and insert trim (**1**) with clips (**3**) in mounts (**6**).
  - Insert pin (**9**) in mount (**10**).
  - Clip in trim (**1**) and install door seal over (**A**) pillar trim.

Lower A-pillar trim, removing and installing

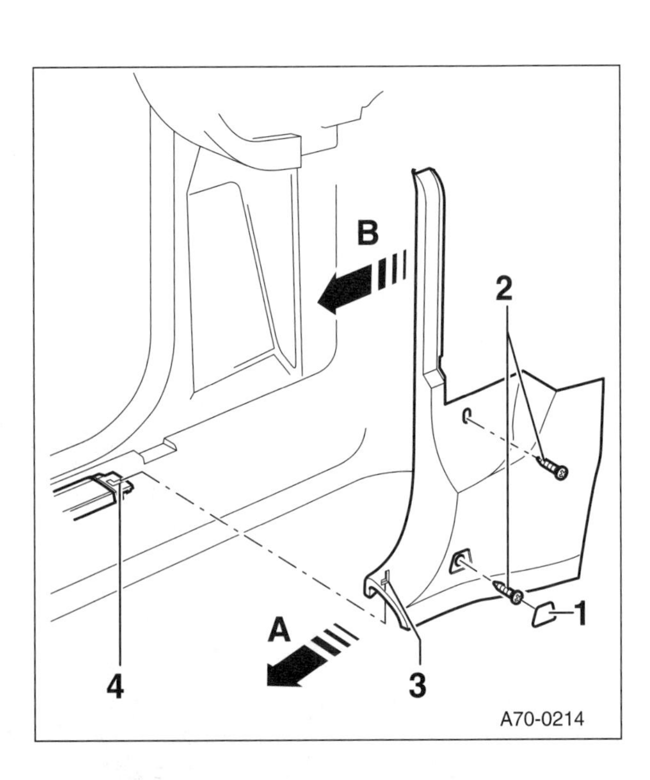

– Pry out cover (**1**) and remove screws (**2**).

◄ Pull lower A-pillar trim in direction of arrow (**A**) until pin (**3**) can be detached from groove (**4**).

– Pull extension of trim in direction of arrow (**B**) out of retainers.

– Installation is reverse of removal.

BentleyPublishers.com
BentleyPublishers.com—All Rights Reserved

Sill trim, removing and installing

– Remove lower A-pillar trim. See **Lower A-pillar trim, removing** in this repair group.

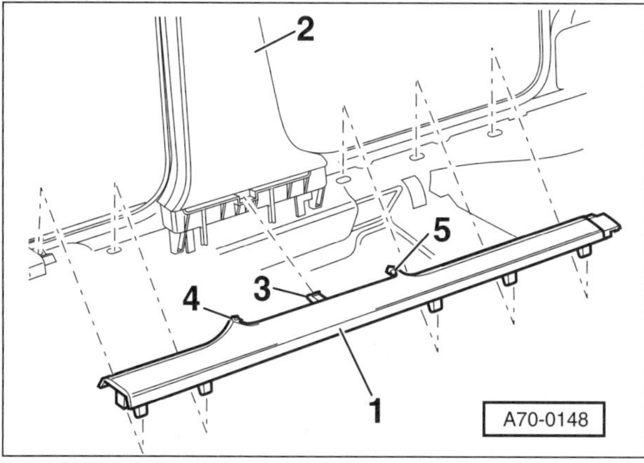

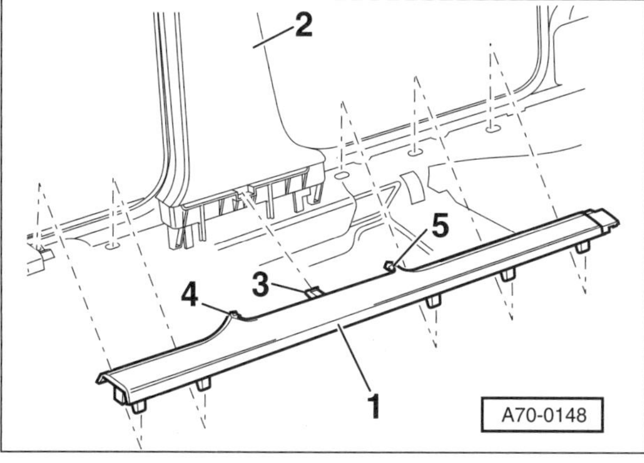

◀ Unclip front sill trim (**1**).

– Detach sill trim (**1**) from lower B-pillar trim (**2**) at retainer (**5**).

– Unclip rear of sill trim (**1**).

– Detach sill panel trim from lower B-pillar trim (**2**) at retainer (**4**).

– Detach sill panel trim (**1**) from lower B-pillar trim (**2**) at retainer (**3**).

– Installation is reverse of removal.

Upper B-pillar trim, removing and installing

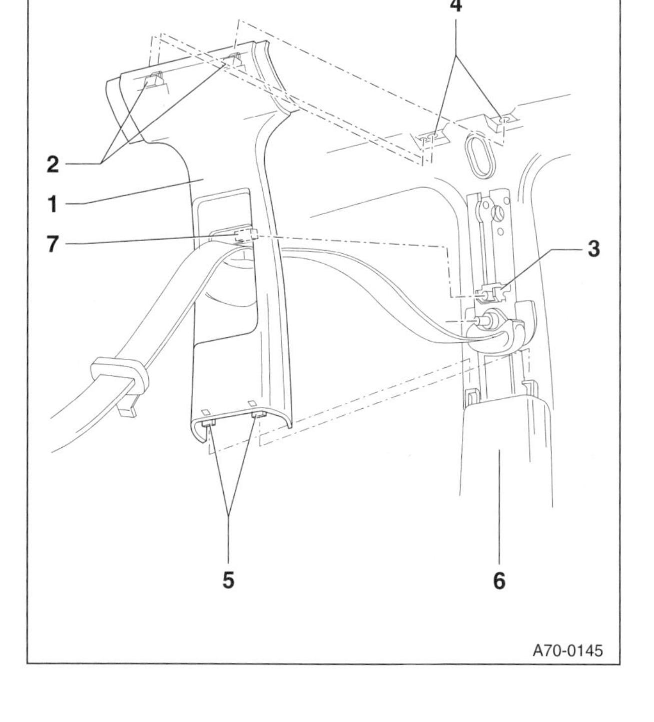

1. Upper B-pillar trim
2. Clip
3. Height adjuster
4. Mounts
5. Retainer tab
6. Lower B-pillar trim
7. Actuating button for height adjuster

– Remove outer seat belt attachment point. See **69 Seat Belts, Airbags**.

– Unfasten roof grab handle at upper A-pillar trim. See **Upper A-pillar trim, removing and installing** in this repair group.

– Unfasten roof grab handle and coat hook at upper C/D-pillar trim. See **Upper C-pillar trim, removing and installing** in this repair group.

– Pull clips (**2**) on trim (**1**) out of mounts (**4**) at the top.

– Detach retainer tabs (**5**) upwards from lower B-pillar trim (**6**).

– Installation is reverse of removal. Remember to:
- Engage retainers (**5**) on trim (**1**) in lower B-pillar trim (**6**).
- Insert actuating button for height adjuster (**7**) in height adjuster (**3**).
- Insert clips (**2**) on trim (**1**) in mounts (**4**).
- Make sure door seal is correctly seated and check operation of height adjuster.
- Check that height adjuster engages audibly in each of 5 possible positions (including top position) and that adjuster button always returns to up position.

NOTE—

• Do not press height adjustment fitting when installing.

BentleyPublishers.com
BentleyPublishers.com—All Rights Reserved

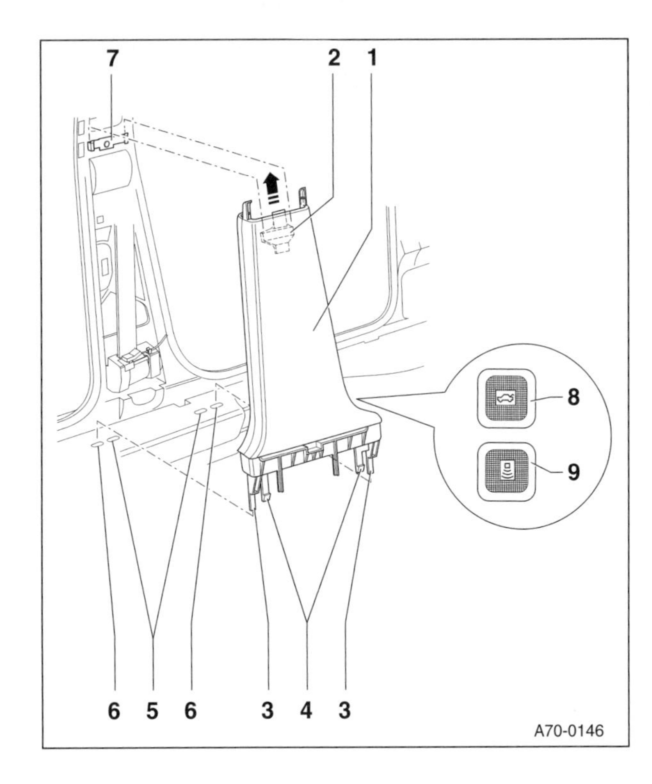

Lower B-pillar trim, removing and installing

1. Lower B-pillar trim
2. Belt guide
3. Guide pins
4. Retainer tabs
5. Mounts
6. Mounts
7. Retainer
8. Push-button switch for trunk lid
9. Push-button switch for interior monitoring

NOTE—

- *The outer seat belt attachment point does not have to be removed when taking out the lower B-pillar trim.*

– Remove upper B-pillar trim. See **Upper B-pillar trim, removing and installing** in this repair group.

– Remove sill trim. See **Sill trim, removing and installing** in this repair group.

– Use screwdriver to press retainer tabs (**4**) outward.

– Pull up lower B-pillar trim in direction of **arrow** until it disengages at belt guide (**2**).

– Disconnect connectors for trunk lid and interior monitoring switches and remove trim.

– Installation is reverse of removal. Remember to:

- Insert belt guide (**2**) from top into retainer (**7**).
- Insert guide pins (**3**) and retainer tabs (**4**) in mounts (**5**) and (**6**).
- Make sure door seal is correctly seated.

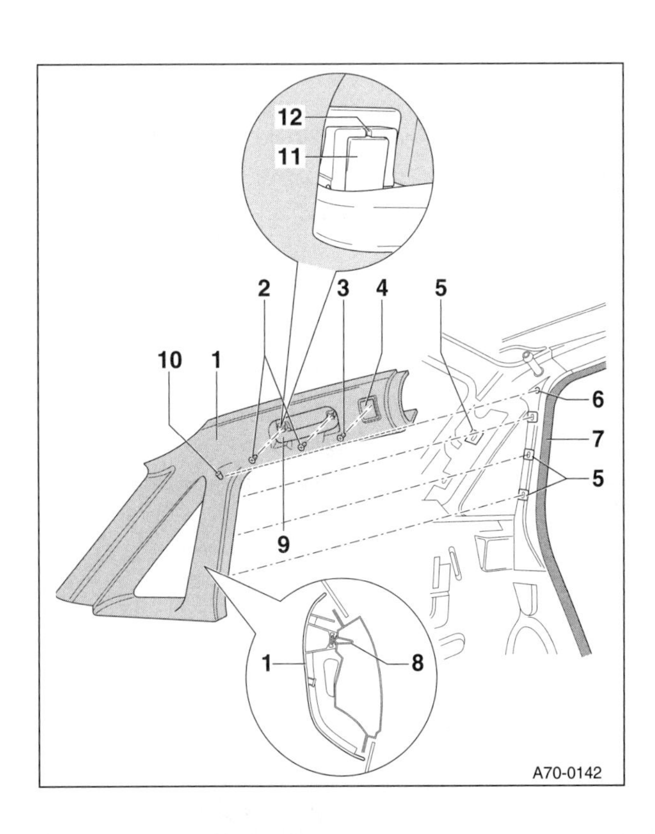

Upper C-pillar trim, removing and installing

1. Upper C-pillar trim
2. Screws (2x)
3. Screw (1x)
4. Coat hook
5. Mounts
6. Mount
7. Seal
8. Clips (7x)
9. Roof grab handle
10. Pin
11. Roof grab handle flap
12. Groove

– Open out coat hook and remove screw (**3**).

– Pivot down flap (**11**) on roof grab handle (**9**) at groove (**12**).

– Remove trim screws (**2**) and unclip C-pillar trim (**1**), starting at top.

BentleyPublishers.com
BentleyPublishers.com—All Rights Reserved

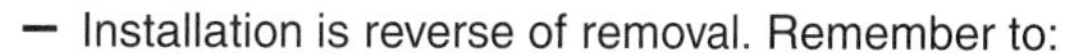

– Installation is reverse of removal. Remember to:
  - Insert trim (**1**) in rear shelf mounts.
  - Insert pin (**10**) on trim (**1**) in mount (**6**).
  - Insert clips (**8**) on trim (**1**) in mounts (**5**).
  - Clip trim (**1**) into place.
  - Make sure door seal is correctly seated.

Lower C-pillar trim, removing and installing

– Remove sill trim. See **Sill trim, removing** in this repair group.

– Remove rear seat bench and rear side padding. See **72 Seats**.

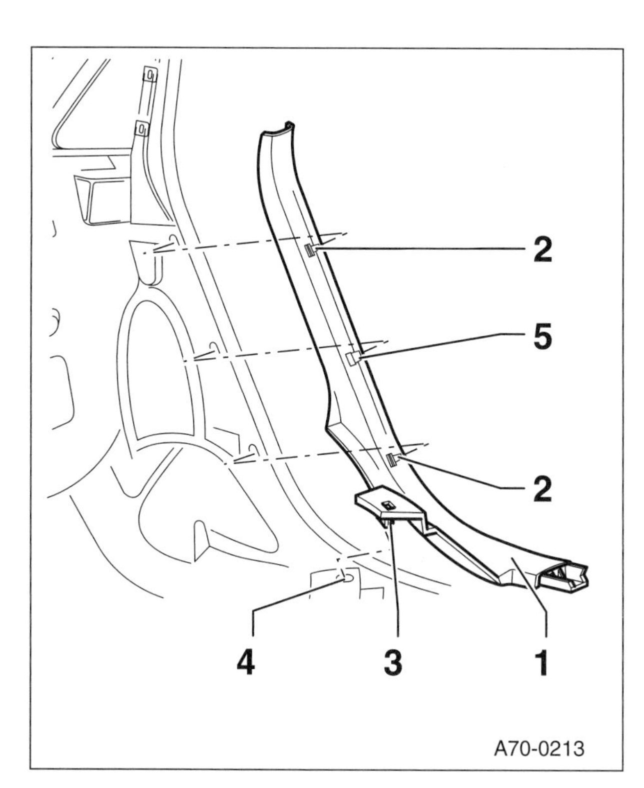

◄ Unclip upper C-pillar trim around lower C-pillar trim.

– Unclip lower C-pillar trim (**1**) at retaining clips (**2**).

– Detach retainer tab (**3**) from mount (**4**) and remove trim.

– Installation is reverse of removal. Remember to align guide pin (**5**).

Upper C-pillar trim (Avant), removing and installing

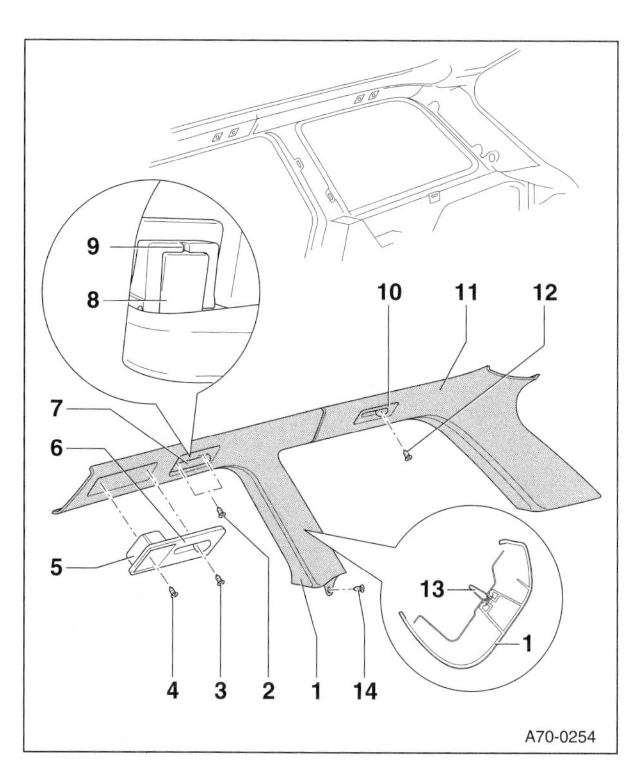

1. C-pillar trim
2. Screws (2x)
3. Screw
4. Bolt
5. Coat hook
6. Front partition grille mount
7. Roof grab handle
8. Roof grab handle flap
9. Groove
10. Rear partition grille mount
11. D-pillar trim
12. Screw
13. Clip
14. Screw

BentleyPublishers.com
BentleyPublishers.com—All Rights Reserved

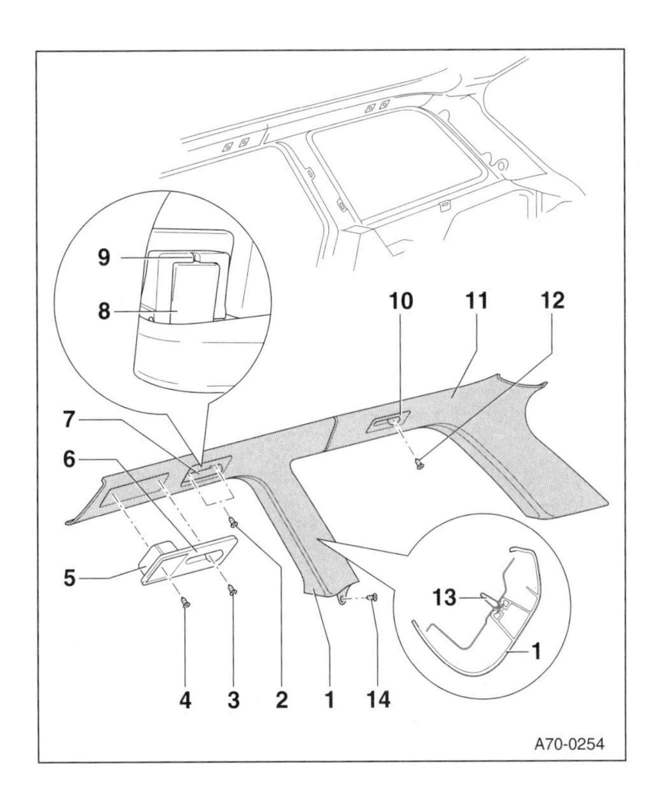

◀ Pull down coat hook (**5**) and remove screw (**4**).

– Remove screw (**3**) from front partition mount (**6**).

– Pull down flap (**8**) on roof grab handle (**7**) at groove (**9**).

– Remove screws (**2**).

– Remove screw (**12**) from rear partition mount (**10**).

– Remove trim plate for seat belt.

– Remove screw (**14**).

– Unclip D-pillar trim (**11**) at C-pillar trim (**1**).

– Carefully pull away D-pillar trim (**11**).

– Unclip C-pillar trim (**1**), working from the top downward.

– Installation is reverse of removal. Remember to:
  - Push C-pillar trim in behind D-pillar trim.
  - Insert clips (**1**) on trim (**13**) into mounts.
  - Clip trim (**1**) into place.

Upper D-pillar trim (Avant), removing and installing

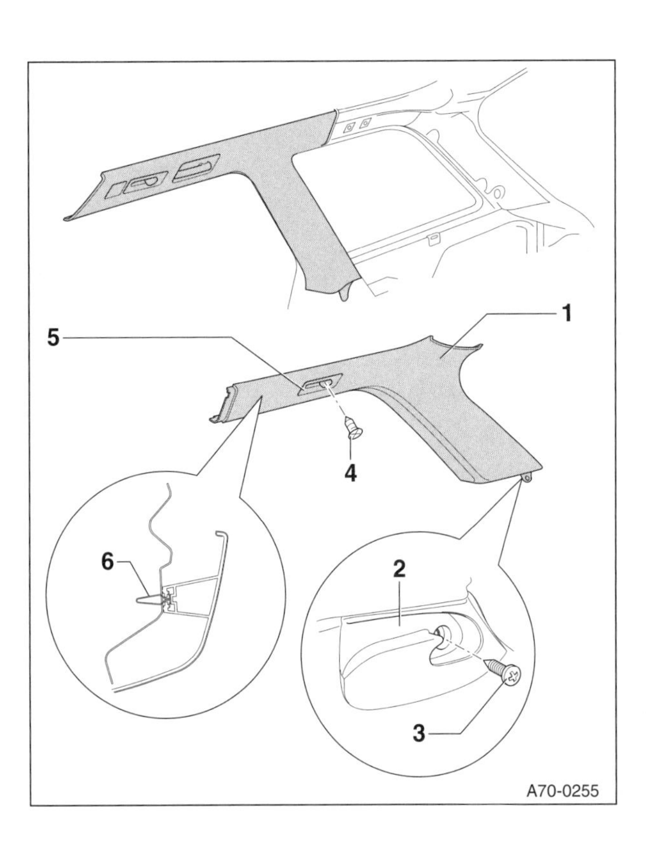

1. D-pillar trim
2. Mount for cargo area cover
3. Screw
4. Screw
5. Rear partition grille mount
6. Clip

– Remove screw (**4**) from rear mount (**5**) for partition grille.

– Remove bolt (**3**).

– Unclip D-pillar trim (**1**) working from the top downward.

– Lift D-pillar trim out of cargo area side trim.

– Remainder of installation is reverse of removal. Remember to:
  - Insert D-pillar trim behind cargo area side trim.
  - Insert clips (**6**) on trim (**1**) in mounts.
  - Clip trim (**1**) into place.

BentleyPublishers.com
BentleyPublishers.com—All Rights Reserved

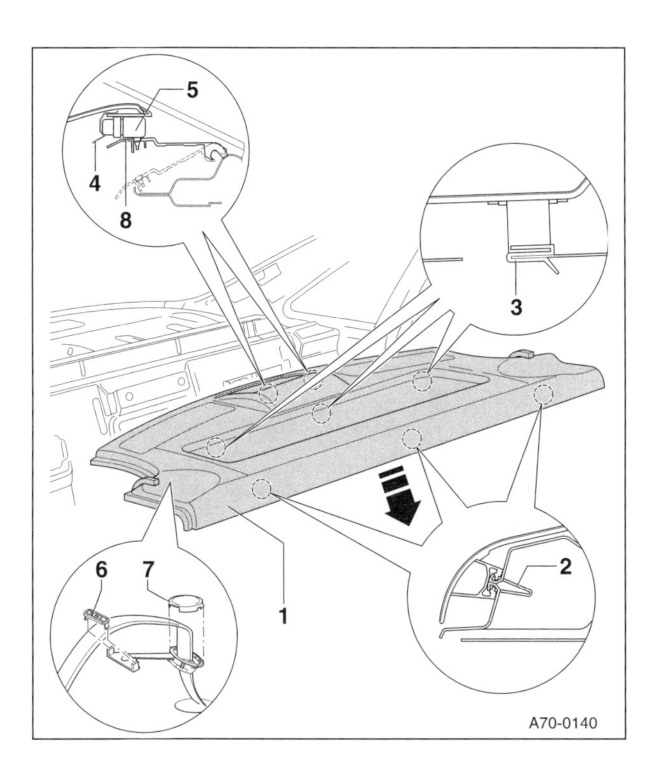

Rear shelf (folding backrest), removing and installing

1. Rear shelf
2. Clip (3x)
3. Clip (3x)
4. Spring
5. Center brake light
6. Guide
7. Trim
8. Flap

– Fold down rear seat backrest.

– Unbolt lower attachment points for rear left and right three-point seat belts. See **69 Seat Belts, Airbags.**

– Remove rear left and right side padding. See **72 Seats**.

– Unclip trim (**7**) and guide (**6**) for rear left and right three-point seat belts.

– Remove upper C-pillar trim on left and right. See **Upper C-pillar trim, removing and installing** in this repair group.

– For vehicles with flap in rear shelf: Open flap (**8**) beneath high-mount brake light (**5**) in luggage compartment.

– For all vehicles: Unclip center brake light (**5**) by unfastening spring (**4**) in luggage compartment.

– On vehicles with electric sun shade, loosen sunshade screws in luggage compartment.

– Pull rear shelf (**1**) forward in direction of arrow and unclip rear shelf at clip (**2**).

– Disconnect electrical connector for sun shade (if equipped).

– Remove rear shelf upward and forward.

– Installation is reverse of removal. Remember to:

• Press down rear shelf slightly until clips (**3**) and springs (**4**) engage.

• Press in on front edge of self to engage clips (**2**).

BentleyPublishers.com
BentleyPublishers.com—All Rights Reserved

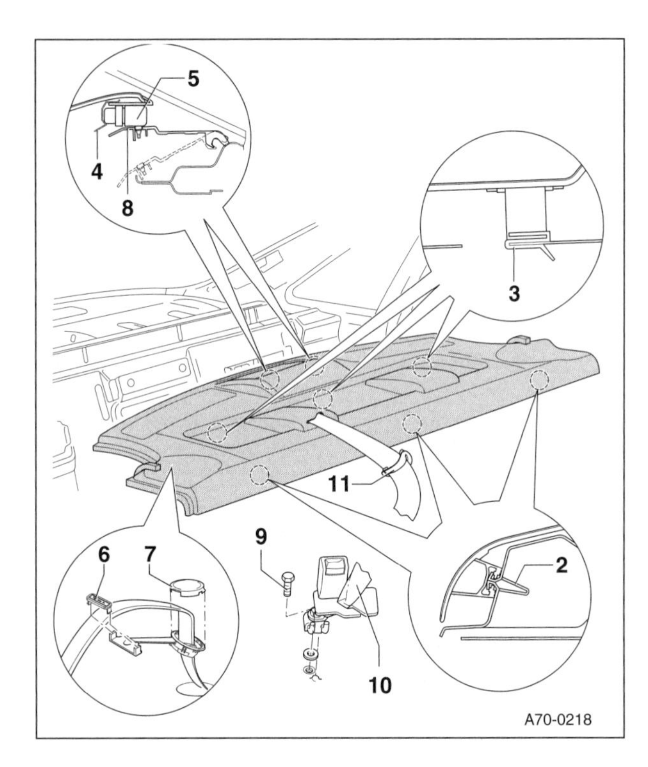

Rear shelf (fixed backrest), removing and installing

1. Rear shelf
2. Clip (3x)
3. Clip (3x)
4. Spring
5. Center brake light
6. Guide
7. Trim
8. Flap
9. Bolt (1x), 55 Nm (41 ft-lb)
10. Webbing
11. Guide

- Unbolt lower attachment points for rear left and right three-point seat belts. See **69 Seat Belts, Airbags**.
- Unbolt lower attachment point (**8**) for center three-point seat belt. See **69 Seat Belts, Airbags**.
- Remove rear seat backrest. See **72 Seats**.
- Unclip trim (**7**) and guide (**6**) for rear left and right three-point seat belts. See **69 Seat Belts, Airbags**.
- Remove upper C-pillar trim on left and right. See **Upper C-pillar trim, removing and installing** in this repair group.
- For vehicles with flap in parcel shelf: Open flap (**8**) beneath high-mount brake light (**5**) in luggage compartment.
- For all vehicles: Unclip center brake light (**5**) by unfastening spring (**4**) in luggage compartment.
- On vehicles with electric sun shade, loosen sunshade screws in luggage compartment
- Pull rear shelf (**1**) forward in direction of arrow and unclip rear shelf at clip (**2**).
- Unclip trim and guide (**10**) for center three-point seat belt from the rear.
- Disconnect electrical connector for sun shade (if equipped).
- Remove rear shelf (**1**) upward and forward.
- Installation is reverse of removal. Remember to:
  - Press down rear shelf slightly until clips (**3**) and springs (**4**) engage.
  - Press in on front edge of shelf to engage clips (**2**).

BentleyPublishers.com
BentleyPublishers.com—All Rights Reserved

Luggage Compartment Trim

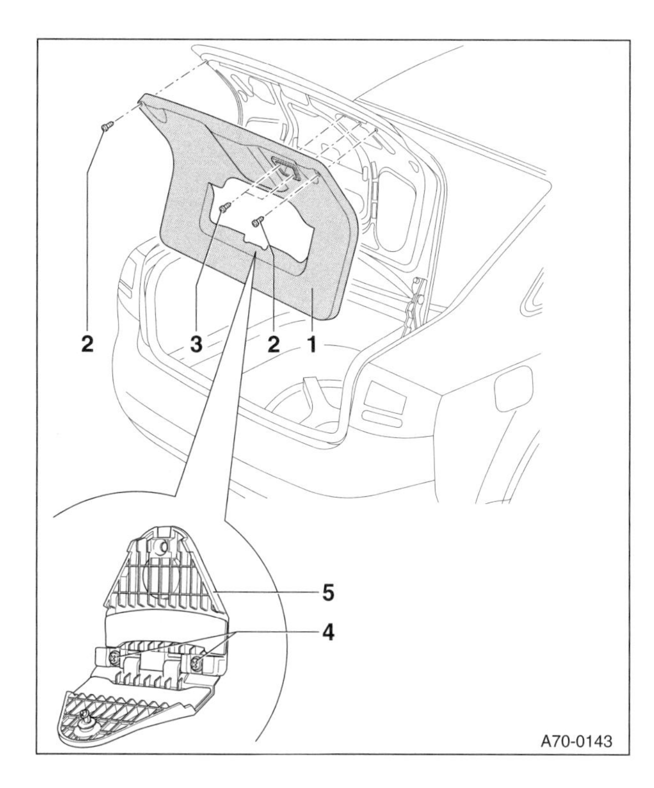

Trunk lid trim, removing and installing

1. Rear lid trim
2. Screws (2x)
3. Screws (2x)
4. Screws (2x)
5. Bracket for hazard warning triangle

– Remove hazard warning triangle.

– Remove screws (**4**) and remove bracket for hazard warning triangle (**5**).

– Remove screws (**2**).

– Vehicle with emergency release: Unclip cover and remove bolt for emergency release handle.

– All vehicles: Remove screws (**3**).

– Unclip trunk lid trim (**1**) from underneath and remove trim.

– Installation is reverse of removal.

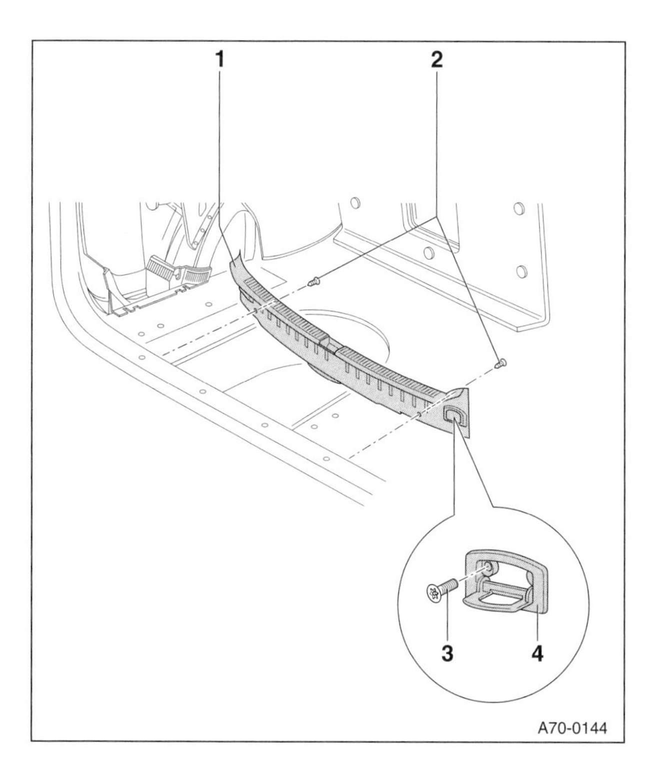

Trunk sill trim, removing and installing

1. Trunk sill trim
2. Screws (2x)
3. Screws (2x)
4. Fastening rings

– Remove trunk light from sill trim (**1**).

– Remove screws (**2**).

– Remove screws (**3**) from fastening rings (**4**) on left and right.

– Unclip sill trim (**1**) upward.

– Guide trunk light through opening in trim.

– Installation is reverse of removal. Make sure seal is correctly seated.

BentleyPublishers.com
BentleyPublishers.com—All Rights Reserved

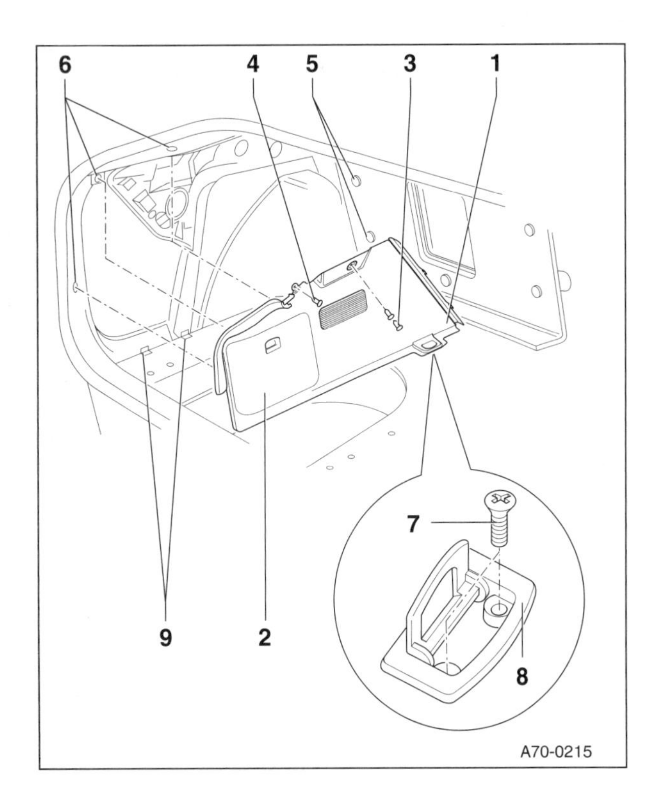

Trunk side trim, removing and installing

1. Trunk side trim
2. Cover
3. Clip
4. Clip
5. Clips (2x)
6. Mounts
7. Bolts (2x)
8. Fastening rings
9. Fastening clip

– Remove trunk floor covering.

– Models with split rear seat: Remove respective backrest. See **72 Seats**.

– Remove trunk sill trim. See **Trunk sill trim, removing and installing** in this repair group.

– Remove screws (**7**) for fastening rings (**8**).

– Unclip front trunk trim clips (**5**).

– Unclip trim clips (**3** and **4**) and remove cover (**2**).

– Unclip remaining clips and pull side trim (**1**) upward out of locating tabs (**9**).

– Pull side trim (**1**) out from behind back panel trim.

– Installation is reverse of removal. Remember to insert side trim behind front trunk trim before replacing clips (**5**).

BentleyPublishers.com
BentleyPublishers.com—All Rights Reserved

Tailgate trim (Avant), removing and installing

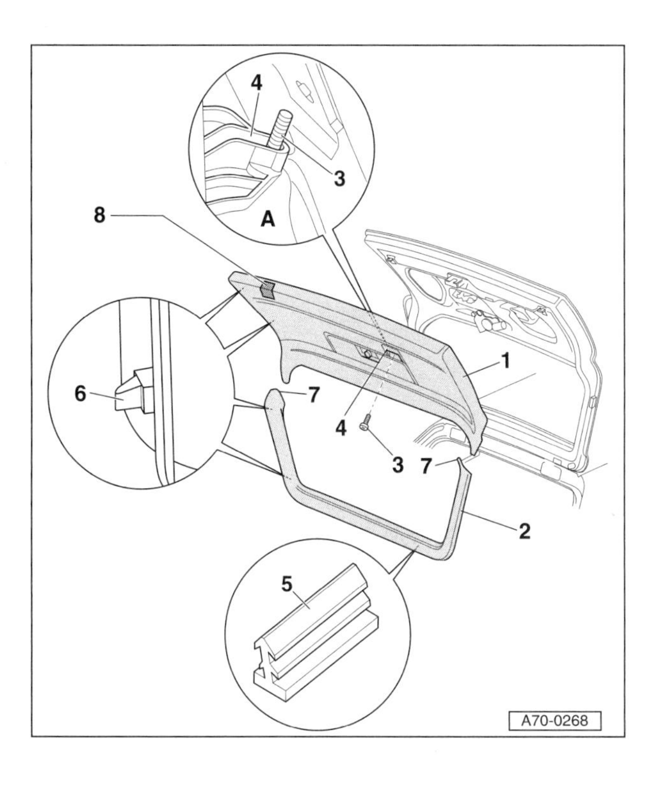

1. Tailgate trim (bottom section)
2. Tailgate trim (top section)
3. Screw
4. Grip molding
5. Rubber buffer (10x)
6. Clip
7. Retainer
8. Light

Tailgate trim bottom section

- Remove bolt (**3**).
- Remove screw behind warning triangle
- Pull bottom section of trim (**1**) away from tailgate slightly.
- Using a plastic trim tool or screwdriver with tip wrapped with tape, release clips (**6**) (8x) on bottom section of trim (**1**). Start at bottom right and proceed counterclockwise.
- Press out light (**3**) from behind trim.
- Disconnect connector for light (**8**).
- To install, locate bottom section of trim (**3**) at rear window first, and clip into place.
- Make sure that screw (**4**) also secures grip molding (**3**) (shown in inset **A**).

Tailgate trim top section

- Remove bottom section of trim (**1**).
- Unclip retainer tabs (**7**) on both sides.
- Release clips (**6**) (7x) on top section of trim (**2**) by pulling sharply (do not use excessive force).
- Before installing top section of trim (**2**), make sure that rubber buffers (**5**) (10x) are installed properly.
- First install top section of trim (**2**) behind side tailgate flanges.
- Clip in upper tailgate trim (**2**) into place behind side flanges on tailgate.

BentleyPublishers.com
BentleyPublishers.com—All Rights Reserved

Tailgate sill trim (Avant), removing and installing

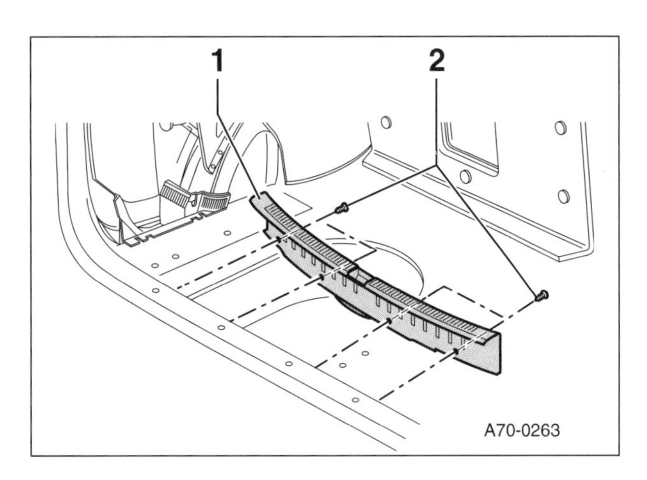

- ◄ Remove screws (**2**) (4x).
- Remove sill trim (**1**).
- Install in reverse order. Make sure tailgate seal is positioned over sill panel trim.

Tailgate side trim (Avant), removing and installing

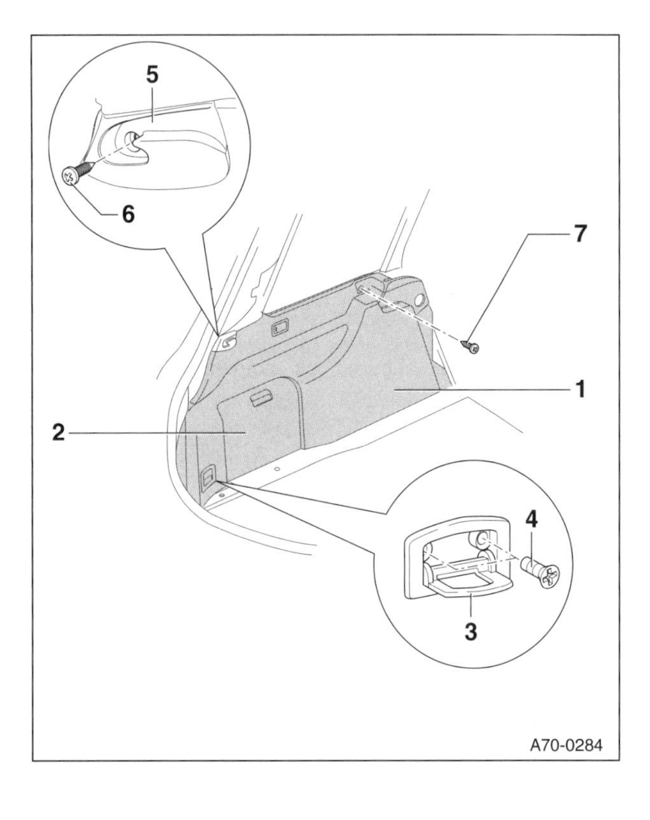

1. Cargo area side trim
2. Cover
3. Fastening rings
4. Screw (2x)
5. Mount for cargo area cover
6. Screw
7. Screw

- Remove cargo area cover.
- Remove rear seat bench and respective side padding. See **72 Seats**.
- Detach seat belt from bottom attachment. Remove belt trim plate. See **69 Seat Belts, Airbags**.
- Remove tailgate sill trim. See **Tailgate sill trim, removing and installing** in this repair group.
- Remove screws (**4**) for fastening rings (**3**).
- Remove storage cover (**2**).
- Remove mounting bracket screw (**6**).
- Remove bolt (**7**).
- Unclip rear trim.
- Pull trim slightly toward front, and disconnect connector for rear electrical socket.
- Installation is reverse of removal. Remember to engage cargo area trim behind side window flange.

BentleyPublishers.com
BentleyPublishers.com—All Rights Reserved

72 Seats

GENERAL

This repair group covers removing and installing the front and rear seats. Both driver and passenger seats may be equipped with airbags mounted in the seat backrest frames. Additionally, airbags may be installed in the rear seat side upholstery. For airbag precautions and handling procedures see **69 Seat Belts, Airbags**.

> ***WARNING—***
> - *Do not install aftermarket upholstery on any vehicle equipped with seat mounted airbags. Non-factory upholstery including "beads'" and "sheepskins" may cause seat mounted airbags to deploy when they are not supposed to; fail to deploy when they should; or deploy in some manner other than designed. Serious injury or death could result.*

FRONT SEATS

Front seat components

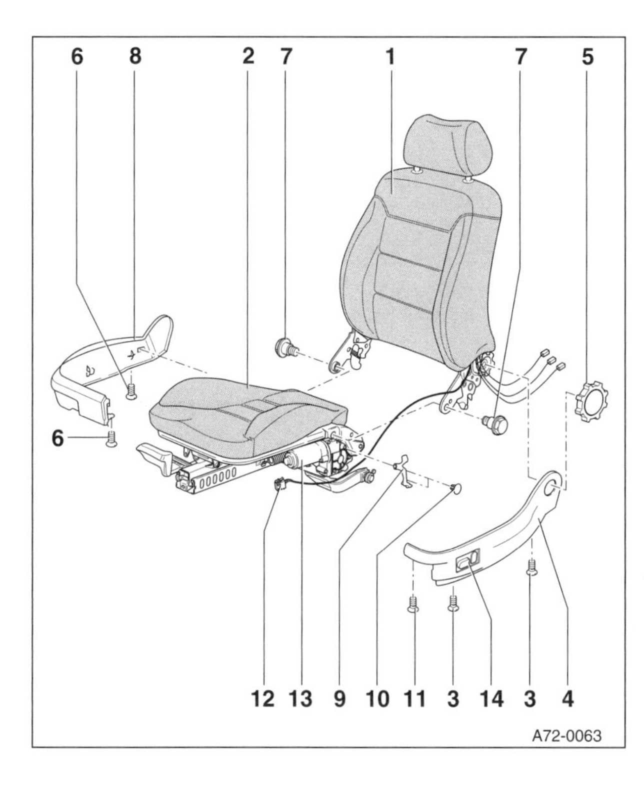

1. Backrest
2. Seat pan
3. Bolts
4. Outer side panel of seat
5. Adjustment knob
6. Bolts
7. Bolts, 20 Nm (15 ft lb)
8. Inner side panel of seat
9. Bracket
10. Expanding pin
11. Screw
12. Connector
13. Motor for seat height adjuster
14. Seat adjustment switch

BentleyPublishers.com
BentleyPublishers.com—All Rights Reserved

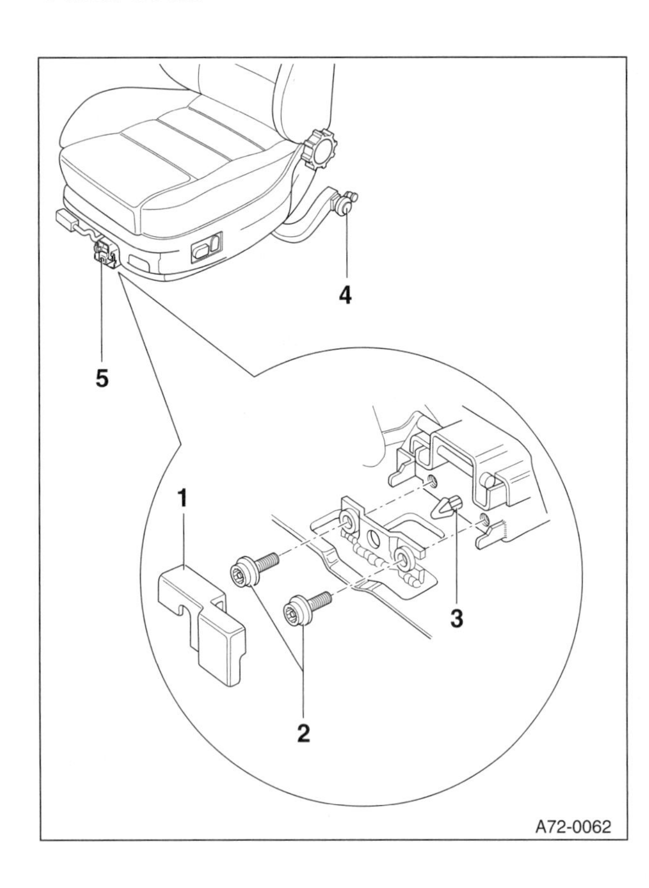

Front seat, removing and installing

1. Cover
2. Bolts, 23 Nm (17 ft-lb)
   - Self-locking
   - Always replace
3. Clip
4. Rear rollers
5. Front sliders
   - Lubricate with multi-purpose grease

NOTE—
- *Removal and installation procedures may vary slightly depending on vehicle equipment.*

– Raise seat to highest position and adjust seat forward.

– Remove trim covers at rear of inner and outer seat rails.

– Adjust seat backward.

– Disconnect battery negative (-) cable.

NOTE—
- *Obtain radio code before disconnecting battery.*

– Remove trim cover (**1**) at front of front seat rail.

– Remove bolts (**2**) and release clip (**3**).

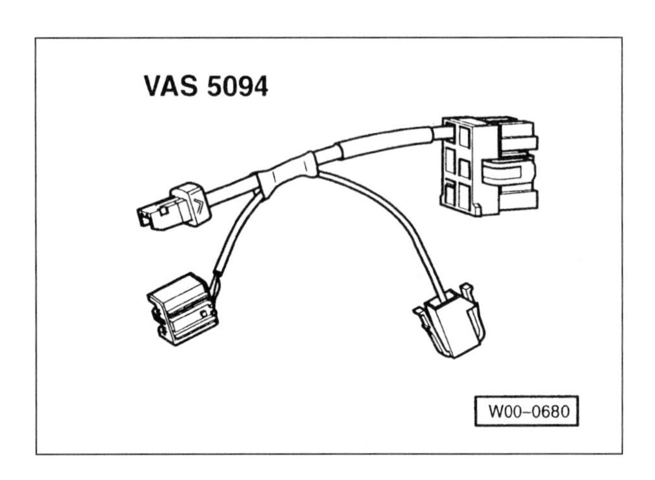

◀ Connect airbag adapter VAS 5094 to wiring harness for side airbag.

> ***WARNING—***
> - *Observe all airbag warnings. See* **69 Seat Belts, Airbags**.
> - *Before disconnecting airbag harness connector, discharge static electricity by briefly grasping chassis or door striker.*
> - *Failure to install VAS 5094 could result in deployment of side airbag.*

– Disconnect remaining electrical connectors.

– Fold back floor covering at rear around guide rail.

– Push seat backward out of guide and remove seat.

– Installation is reverse of removal. Remember to:
  - Remove airbag adapter VAS 5094 and reattach electrical connectors.
  - Replace seat mounting bolts with new.

Tightening torque	
Front seat to seat rail	23 Nm (17 ft-lb)

BentleyPublishers.com
BentleyPublishers.com—All Rights Reserved

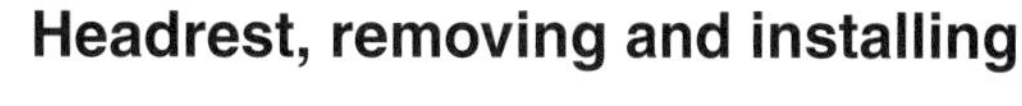

Headrest, removing and installing

Style 1:

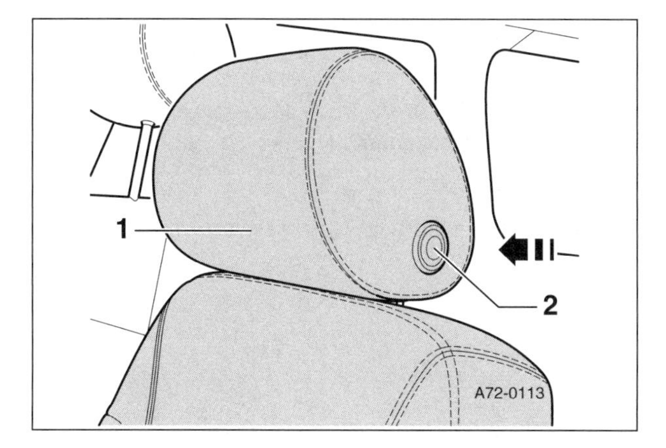

Press release button (**2**) fully and lift headrest up and off seat back.

– To install, press release button (**2**) to first stage and slide headrest down on seat back. Press release button to adjust headrest height.

Style 2:

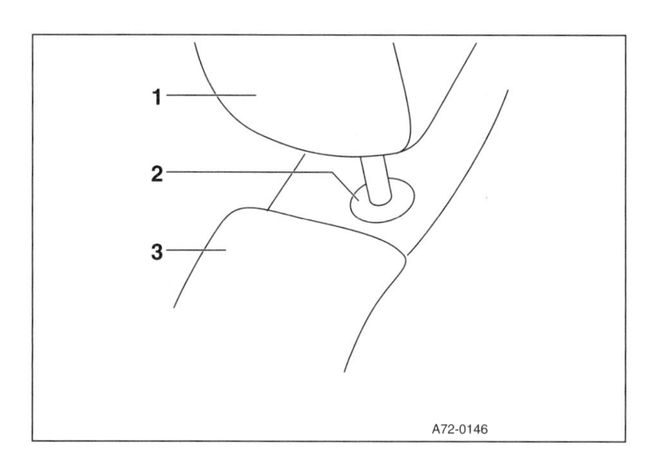

Press down on guide sleeve (**2**) from above.

– Press down slightly on headrest (**1**) to release catch.

– Pull headrest up and off seat back.

– To install, slide head rest down on seat back and press on guide sleeve (**2**) to adjust headrest height.

Seat memory control module, removing and installing

– Remove driver's seat. See **Front seat, removing and installing** in this repair group.

– Working under seat, remove bolts for inner side panel of seat.

– Release clip on inner side panel. Push panel forward and remove. See **Front seat components** in this repair group for more information.

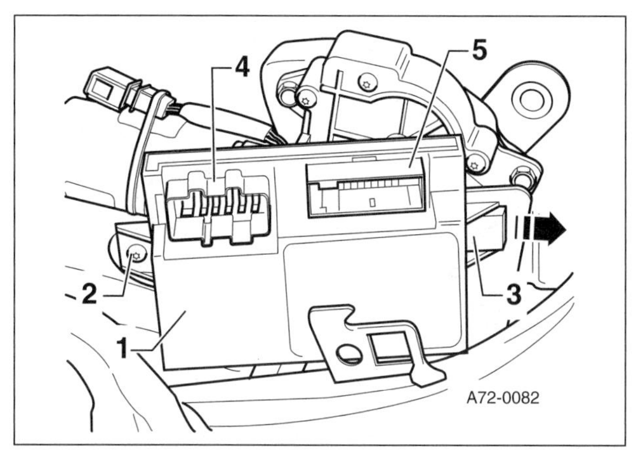

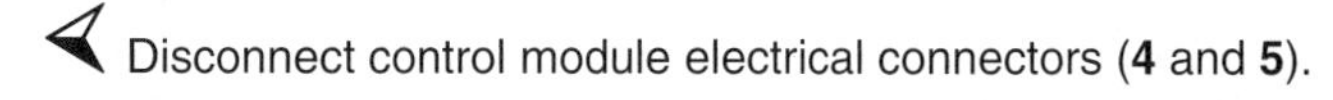

Disconnect control module electrical connectors (**4** and **5**).

– Remove screw (**2**).

– Press clip (**3**) in direction of arrow using a screwdriver and remove control module.

– Installation is reverse of removal.

BentleyPublishers.com
BentleyPublishers.com—All Rights Reserved

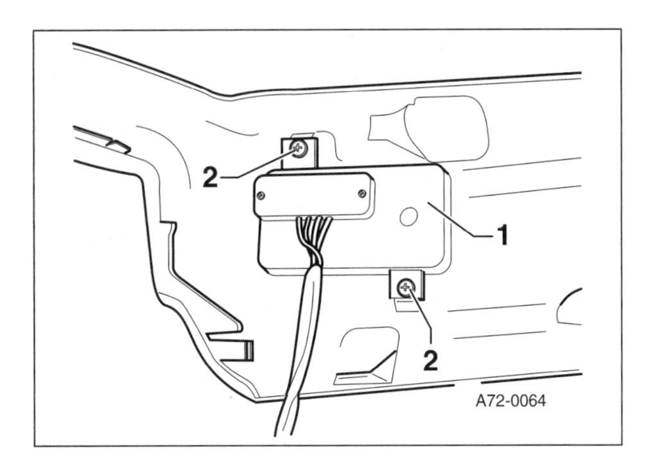

Seat control switch, removing and installing

– Remove driver's seat. See **Front seat, removing and installing** in this repair group.

– Working under seat, remove bolts for outer side panel of seat.

– Release clip on inside of side panel. Push panel forward and remove. See **Front seat components** in this repair group for more information.

◄ Remove screws (2) and remove seat control switch.

– Installation is reverse of removal.

REAR SEATS

Seat bench, removing and installing

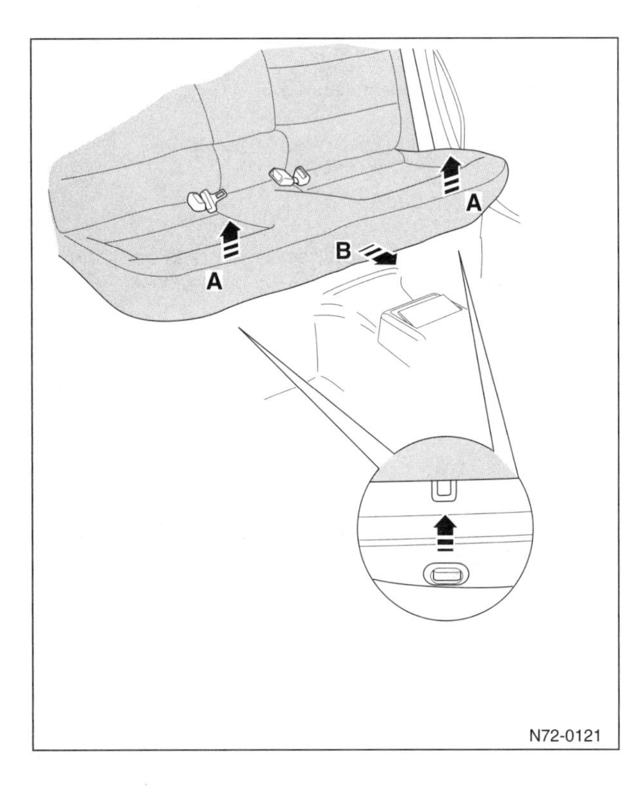

> ***WARNING —***
> - *Observe all airbag warnings. See* **69 Seat Belts, Airbags***.*
> - *Before disconnecting airbag harness connector, discharge static electricity by briefly grasping chassis or door striker.*

– Disconnect battery negative (-) cable.

NOTE —
- *Obtain radio code before disconnecting battery.*

◄ Lift front edge of seat bench in direction of arrows (**A**) and pull forward (**B**).

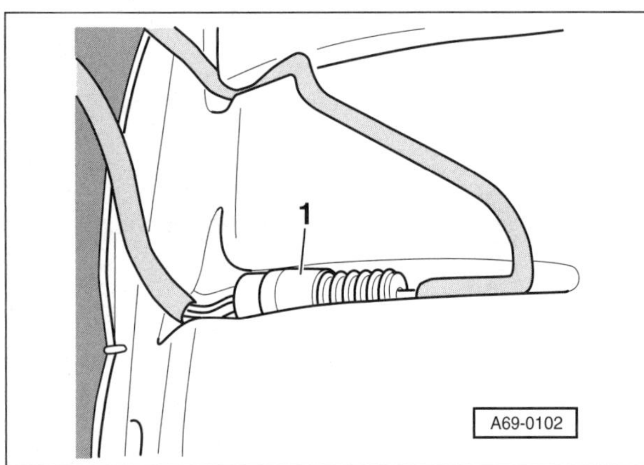

◄ Remove harness connector for side airbag (if equipped) from groove and disconnect.

– Disconnect connector for heated seats (if equipped) and remove seat bench.

BentleyPublishers.com
BentleyPublishers.com—All Rights Reserved

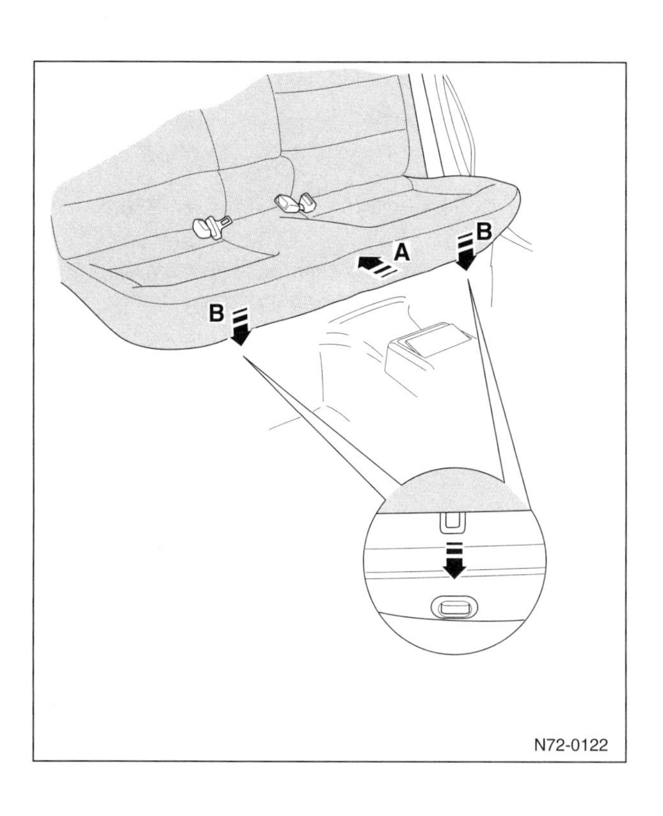

Installation is reverse of removal. Remember to:

- Connect harness connector for side airbags and heated seats (if equipped).
- Slide bench rearward in direction of arrow (**A**).
- Guide anchoring eyelets into plastic sleeves (arrow **B**)
- Avoid damage to plastic sleeves (**inset**).

Backrest (fixed), removing and installing

> ***WARNING —***
> - *Observe all airbag warnings. See* **69 Seat Belts, Airbags**.
> - *Before disconnecting airbag harness connector, discharge static electricity by briefly grasping chassis or door striker.*

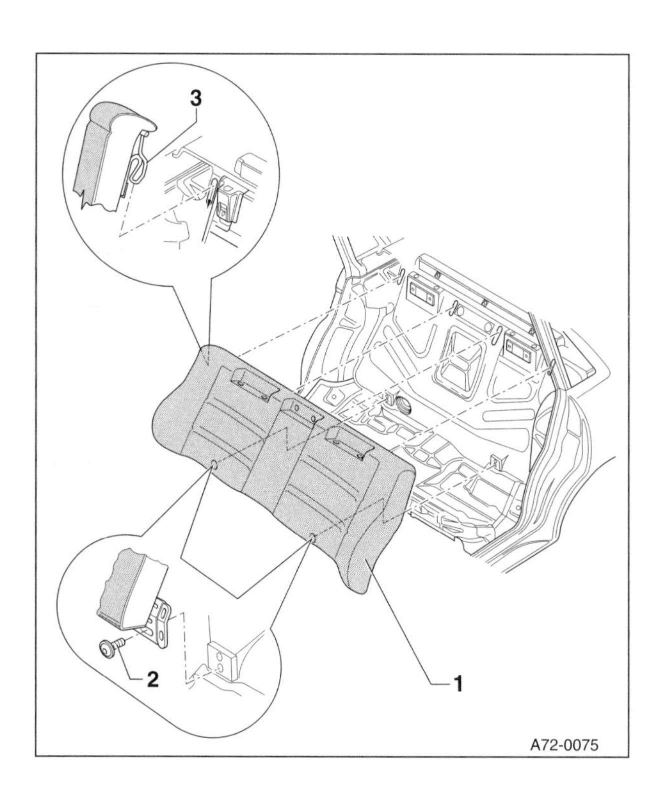

1. Fixed backrest
2. Screws (2x), 8 Nm (71 in-lb)
3. Hooks (4x)

- Remove rear seat bench. See **Seat bench, removing and installing** in this repair group.
- Remove head rest (if necessary). See **Rear headrest, removing and installing** in this repair group.
- Pry out frame for ski bag (if equipped)
- Disconnect electrical connectors for side airbags and heated seats (if equipped).
- Remove screws (**2**) (2x).
- Push backrest (**1**) upward and detach at hooks (**3**) (4x).
- Installation is reverse of removal.

BentleyPublishers.com
BentleyPublishers.com—All Rights Reserved

Backrest (split-folding), removing and installing

> ***WARNING —***
> - *Observe all airbag warnings. See* **69 Seat Belts, Airbags**.
> - *Before disconnecting airbag harness connector, discharge static electricity by briefly grasping chassis or door striker.*

1/3 section

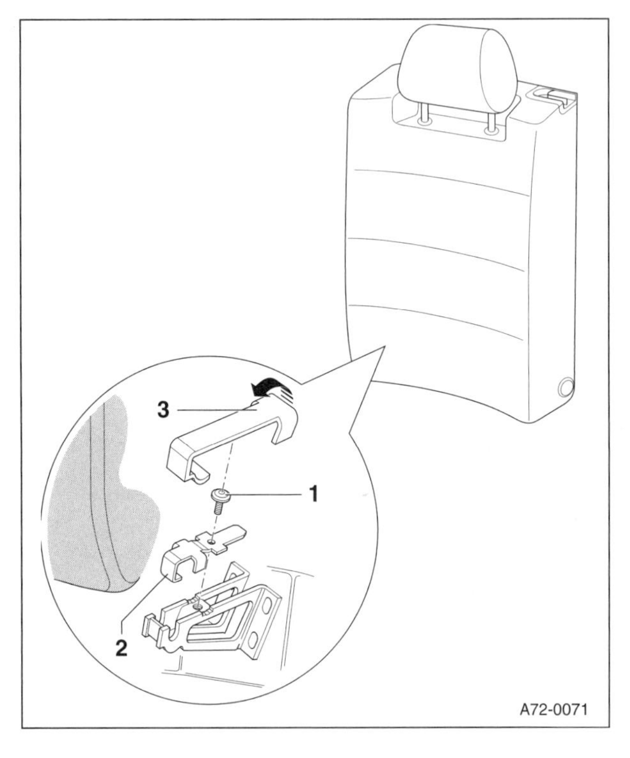

– Fold both backrest sections forward.

◄ Unclip trim piece (**3**) from rear in direction of **arrow**. If necessary, pry out trim clip from rear through hole.

– Remove bolt (**1**) and clamp (**2**).

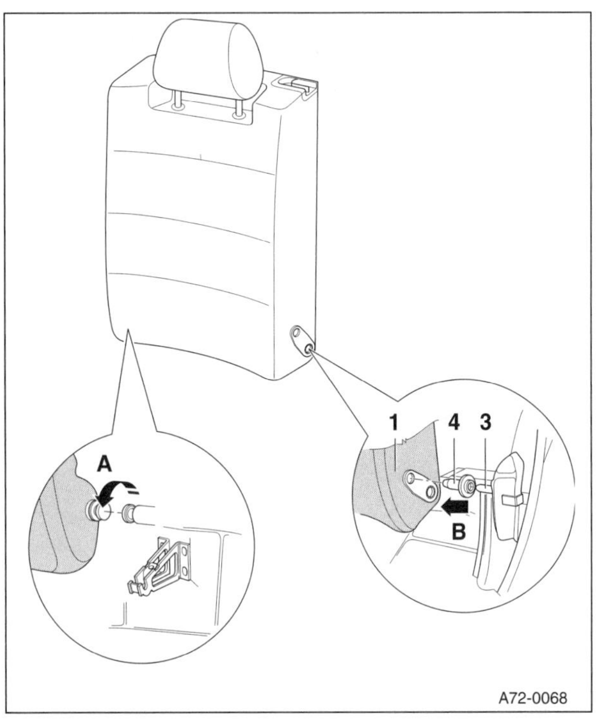

◄ Lift backrest out of center support bracket (**2**) in direction of arrow (**A**) and pull off fitting (**3**) in direction of arrow (**B**).

– Disconnect electrical connectors for heated seats (if equipped).

– Installation is reverse of removal. Remember to reinstall grommet (**4**) on pin (**3**) if it pulled off during removal.

Tightening torque	
Backrest bracket bolt	10 Nm (7 ft-lb)

BentleyPublishers.com
BentleyPublishers.com—All Rights Reserved

2/3 section

– Fold both backrest sections forward.

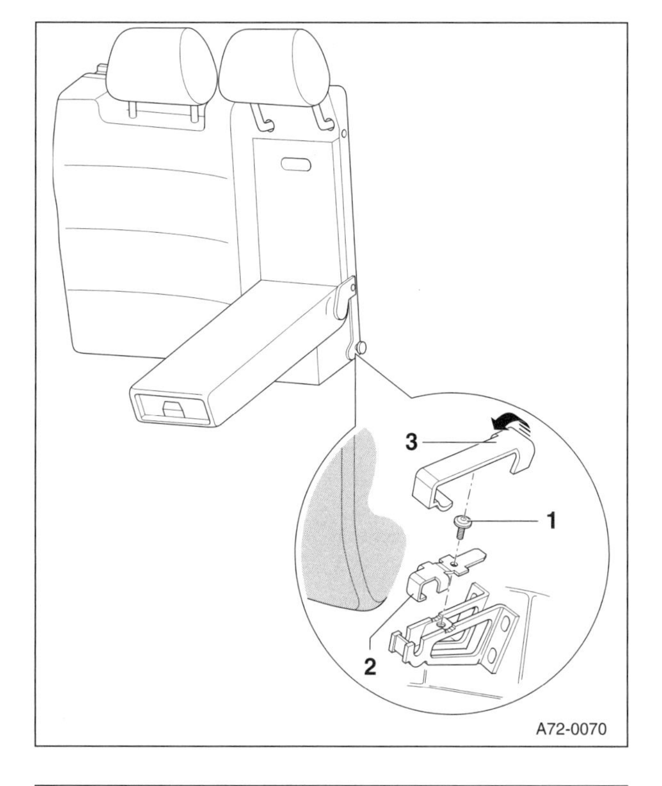

◄ Unclip trim piece (**3**) from rear in direction of **arrow**. If necessary, pry out clip from rear through hole.

– Remove bolt (**1**) and clamp (**2**).

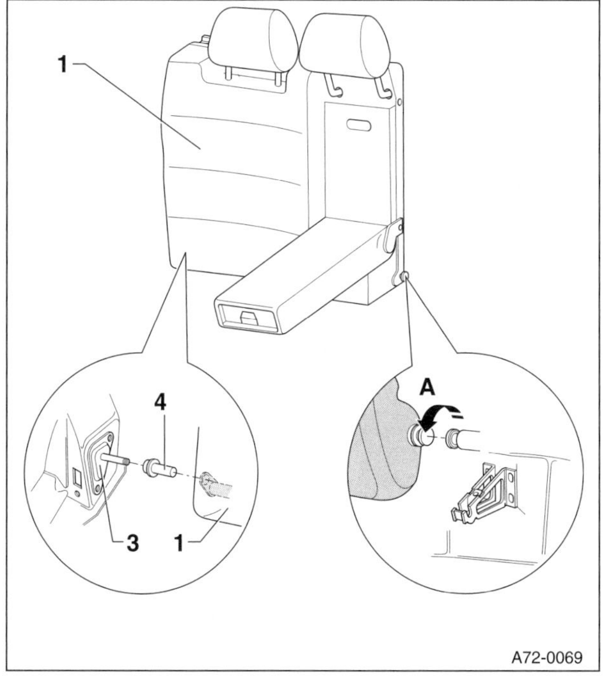

◄ Lift backrest out of center support bracket in direction of arrow (**A**) and pull off fitting (**3**) in direction of arrow (**B**).

– Disconnect electrical connectors for heated seats (if equipped).

– Installation is reverse of removal. Remember to reinstall grommet (**4**) on pin (**3**) if it pulled off of fitting during removal.

Tightening torque	
Backrest bracket bolt	10 Nm (7 ft-lb)

BentleyPublishers.com
BentleyPublishers.com—All Rights Reserved

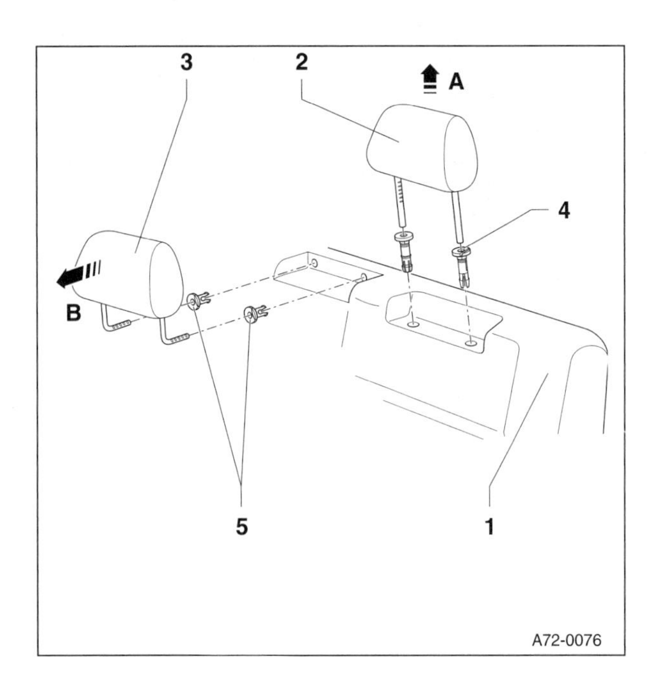

Rear headrest, removing and installing

1. Rear backrest
2. Outer headrest
3. Center headrest
4. Guide piece with release button for outer headrest
5. Guide pieces with release buttons for center headrest

– To remove outer headrest (**2**), press release button (**4**) on left guide piece and pull out in direction of arrow (**A**).

– To remove inner headrest (**3**), press release buttons (**5**) on both guide pieces and pull out in direction of arrow (**B**).

– Installation is reverse of removal. Note that front and rear headrests are not interchangeable.

Rear seat side padding, removing and installing

> ***WARNING—***
> - *Observe all airbag warnings. See* **69 Seat Belts, Airbags**.
> - *Before disconnecting airbag harness connector, discharge static electricity by briefly grasping chassis or door striker.*

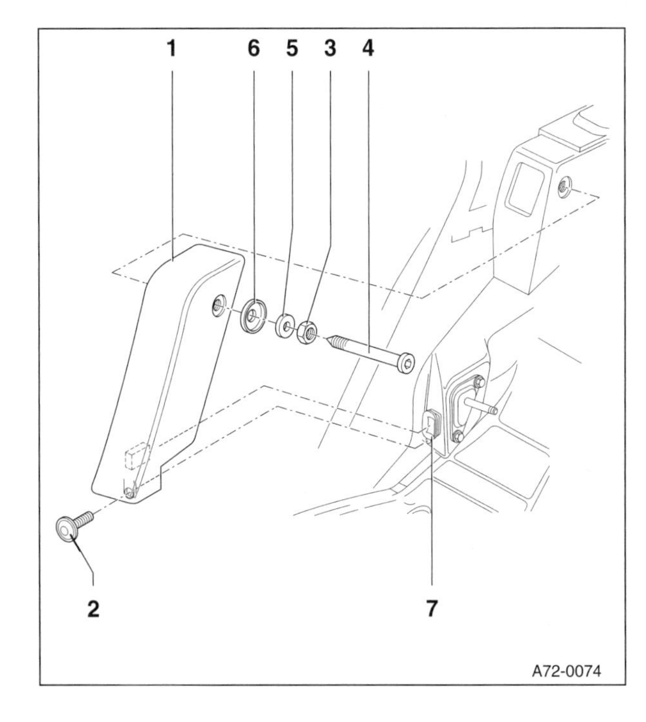

1. Side padding
2. Bolt, 8 Nm (71 in-lb)
3. Nut, 50 Nm (37 ft-lb)
4. Bolt
5. Washer
6. Dished washer
7. Retainer

– Remove rear seat bench. See **Rear seat bench, removing and installing** in this repair group.

– Remove lower bolt (**2**) and on side padding (**1**).

NOTE—
- *Bolt (**2**) is only installed on vehicles with side airbag*

– Disconnect electrical connector for side airbag (if equipped).

– Remove nut (**3**) on striker pin (**4**) together with washer (**5**) and dished washer (**6**).

– Pull side padding out of bottom retainer (**7**).

– Installation is reverse of removal. remember to:
- Press side padding into bottom retainer (**7**).
- Adjust striker pin (**4**) and tighten nut (**3**).

BentleyPublishers.com
BentleyPublishers.com—All Rights Reserved

Children's bench (Avant), removing and installing

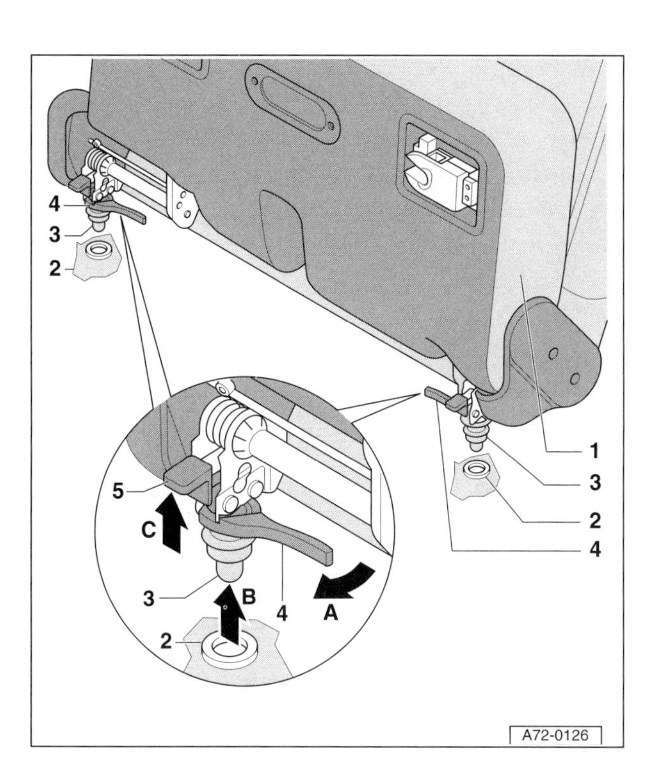

On Avant models, a rear facing children's bench seat in the luggage compartment is an option.

1. Children's bench
2. Mounts
3. Head of locking mechanism
4. Locking lever
5. Securing lever

– Release children's bench using locking levers (**4**).

– Press both securing levers (**5**) upwards in direction of arrow (**C**).

– Guide bench out of mounts (**2**) and remove bench.

– To install, guide locking mechanism heads (**3**) onto mounts (**2**). Make sure heads seat fully.

– Lock bench with locking levers (**4**) in direction of arrow (**A**).

BentleyPublishers.com
BentleyPublishers.com—All Rights Reserved

87 Heating and Air-conditioning

TABLES

General

This repair group covers interior ventilation and air-conditioning. For additional information, see:

- **03 Maintenance** for engine accessory (A/C compressor) drive belt and dust and pollen filter service
- **19 Cooling System** for cooling system fan replacement
- **EWD Electrical Wiring Diagrams**

Some A/C or heater system repairs procedures require that the A/C refrigerant charge be evacuated. A/C system evacuating and recharging procedures are beyond the scope of this manual.

Climate control

A60487001a

The climate control system (A/C and heater) automatically regulates the interior temperature. The left and right sides (driver and passenger sides) are controlled separately. Either side may be adjusted manually.

Setting the system at 75°F (23°C) on both sides turns automatic climate regulation ON.

BentleyPublishers.com
BentleyPublishers.com—All Rights Reserved

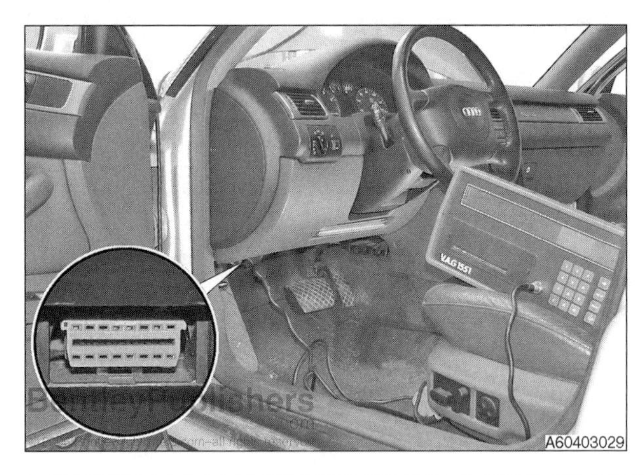

Troubleshooting

If the on-board diagnostics (OBD II) system recognizes malfunctions in the climate control system, it sets diagnostic fault codes (DTCs).

Access fault memory with VAG scan tool or equivalent plugged into DLC plug (**inset**) under left side dashboard.

Table a provides common symptoms and possible corrective actions to aid in diagnosing simple faults in the A/C system.

Table a. Air-conditioning troubleshooting

Symptom	Probable cause	Corrective action
Blower does not work.	Blower fuse blown.	Inspect blower fuse. See **Climate control and engine cooling fuses and relays** in this repair group.
	Blower resistor (power stage) faulty.	Test blower resistor. Replace if necessary.
	Blower motor wiring faulty.	Troubleshoot electrical harness. Repair as necessary.
	Blower motor faulty.	Test motor. Replace if necessary.
Blower does not operate in lower speeds.	Blower resistor (power stage) faulty.	Test blower resistor. Replace if necessary.
	Blower switch faulty.	Replace A/C and heating control panel.
Air-conditioning system does not produce cold air or produces low volume.	A/C compressor not engaging.	Check A/C compressor clutch electrical connector. Repair or reconnect as necessary.
	Refrigerant charge too low or too high.	Test A/C system pressures. Repair as necessary.
	Temperature control flap malfunction.	Test temperature control flap and motor. Repair or replace as necessary.
	Evaporator iced up.	Shut A/C off and allow to thaw.
	Pressure switch faulty.	Test pressure switch and replace if necessary.
Air-conditioned air smells musty.	Mold growing in A/C evaporator vanes.	Apply disinfectant using evaporator cleaning wand.
A/C compressor noisy.	Loose A/C compressor or mounting bracket.	Tighten A/C compressor and mounting bracket bolts.
	Refrigerant lines contacting each other or other components.	Reposition and realign refrigerant lines.
A/C compressor noisy immediately after switching ON or when driving around corners.	A/C system overcharged with refrigerant.	Discharge refrigerant and recharge correctly.
Water sprays out of dashboard vents.	Evaporator drain not functioning.	Check evaporator housing drain hose for correct routing. Make sure it is not kinked.
		Check evaporator housing drain valve.

BentleyPublishers.com
BentleyPublishers.com—All Rights Reserved

Warnings and Cautions

WARNING—

- *Wear hand and eye protection (gloves and goggles) when working around the A/C system. If refrigerant oil comes in contact with your skin or eyes:*
  -Remove contact lenses, if worn.
  -Do not rub skin or eyes.
  -Immediately flush skin or eyes with cool water for 15 minutes.
  -Rush to a doctor or hospital. Do not induce vomiting.
- *Work in a well ventilated area. Switch on building exhaust / ventilation system when working on the A/C system.*
- *If you inhale refrigerant oil, breathe fresh air immediately and then seek medical attention.*
- *Do not expose any component of the A/C system to high temperatures (above 80°C / 176°F). Excessive heat could burst the system.*
- *Keep refrigerant away from open flames. Poisonous gas is produced if it burns. Do not smoke near refrigerant gases for the same reason.*
- *The A/C system is filled with refrigerant gas which is under pressure. Pressurized refrigerant in the presence of oxygen may form a combustible mixture. Do not introduce compressed air into any refrigerant container.*
- *Electric welding near refrigerant hoses causes refrigerant to decompose. Discharge system before welding.*
- *Dispose of refrigerant oil as hazardous waste.*

CAUTION—

- *US law requires that any person who services a motor vehicle air-conditioner is properly trained and certified, and uses approved refrigerant recycling equipment. Technicians must complete an EPA-approved recycling course to be certified.*
- *It is recommended that all A/C service be left to an authorized Audi dealer or other qualified A/C service facility.*
- *State and local governments may have additional requirements regarding A/C servicing. Always comply with state and local laws.*
- *Leak test and repair any A/C system which is known to lose its charge.*
- *Do not top off a partially charged refrigerant system. Discharge system, evacuate and then recharge system.*
- *Do not use R12 refrigerant, refrigerant oils or system components in R134a system. Component damage and system contamination results.*
- *Refrigerant oil (PAG oil) is extremely hygroscopic (water-absorbent). Do not store in open container.*
- *The mixture of refrigerant oil and refrigerant (R134a) attacks some metals and alloys (for example, copper) and breaks down certain hose materials. Use only hoses and pipes that are identified with a green mark (stripe) or the lettering "R134a".*
- *Immediately plug open connections on A/C components and pipes to prevent dirt and moisture contamination.*

BentleyPublishers.com
BentleyPublishers.com—All Rights Reserved

> ***CAUTION—***
> - *Do not steam clean A/C condensers or evaporators. Use only cold water or compressed air.*
> - *Do not reuse refrigerant oil.*
> - *Do not store refrigerant oil in vicinity of flames, heat sources or strongly oxidizing materials.*
> - *In case of fire, use carbon dioxide (CO_2) extinguisher or sprayed water jet on A/C components.*
> - *After refilling or testing refrigerant lines, replace protective caps. They serve as additional seals.*
> - *In winter, run the A/C compressor at least 10 minutes each month to maintain a film of oil on its moving parts and to ensure seal integrity.*
> - *If a new compressor is installed or an old compressor is charged with fresh refrigerant, turn over the compressor by hand approx. 10 revolutions. This ensures that the compressor is not damaged when activated.*

A/C Fluids and Capacities

Refrigerant

The air-conditioning refrigerant in Audi A6 models is R134a, known chemically as *tetrafluoroethane*. As it contains no chlorine, it is considered harmless to atmospheric ozone if accidentally discharged. Nevertheless, strict regulations govern the handling and disposal of automotive refrigerants. Be sure to read **Warnings and Cautions** in this group if working with A/C and heating components.

Refrigerant capacities for different models are shown in **Table b**.

Table b. A/C system refrigerant (R134a) capacities

Engine, production date	Capacity in grams (oz)
V6 engine	
11 / 1997 - 08 / 1998	850 + 50 (30 + 2)
08 / 1998 - 04 / 1999	750 + 50 (26 + 2)
All except 3.0 liter, 04 / 1999 on	650 + 50 (23 + 2)
3.0 liter	550 + 50 (19 + 2)
V8 engine	
to 03 / 1999	550 + 50 (19 + 2)
03 / 1999 - 02 / 2000	650 + 50 (23 + 2)
02 / 2000 on	550 + 50 (19 + 2)

Refrigerant oil

A synthetic oil, *polyalkylene glycol* (PAG), is used to lubricate the A/C compressor. This type of oil is highly hygroscopic (attracts water). Be sure to immediately reseal an opened container after use.

VAG recommends a PAG oil with a specific part number for each type of A/C system used.

BentleyPublishers.com
BentleyPublishers.com—All Rights Reserved

PAG oil recommendation	
Denso compressor	G 052 300 A2
Sanden compressor	G 052 154 A2
Zexel compressor	G 052 154 A2 or G 052 200 A2

A new A/C compressor is filled at the factory with refrigerant oil. If installing a new compressor, recharge with refrigerant using A/C recharging equipment. Do not reuse oil from old compressor.

Refrigerant oil capacities for different models are shown in **Table c**.

Table c. A/C system refrigerant oil (PAG) capacities	
Engine, production date	**Capacity in cc (oz)**
V6 engine	
All except 3.0 liter	250 + 50 (8.5 + 1.7 oz)
3.0 liter	220 + 50 (7.4 + 1.7 oz)
V8 engine	
All with timing belt	250 + 50 (8.5 + 1.7 oz)
allroad quattro (engine code BAS)	220 + 50 (7.4 + 1.7 oz)

Climate Control and Engine Cooling Controls

A/C and heater control panel

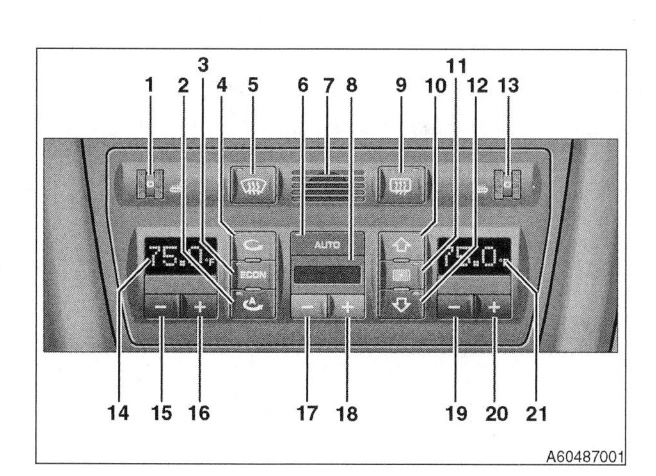

Climate control system automatically regulates interior temperature:

1. Left seat heater control
2. Recirculation OFF
3. A/C compressor OFF
4. Recirculation ON
5. Windshield defroster
6. Automatic setting: 23°C (75°F)
7. Interior air temperature sampling vent
8. Blower fan speed display
9. Rear window defogger
10. A/C and heater air directed to windshield defroster vents
11. A/C and heater air directed to dashboard vents
12. A/C and heater air directed to footwell vents
13. Right seat heater control
14. Left side temperature setting
15. Decrease left side temperature
16. Increase left side temperature
17. Decrease blower fan speed
18. Increase blower fan speed
19. Decrease right side temperature setting
20. Increase right side temperature setting
21. Right side temperature setting

BentleyPublishers.com
BentleyPublishers.com—All Rights Reserved

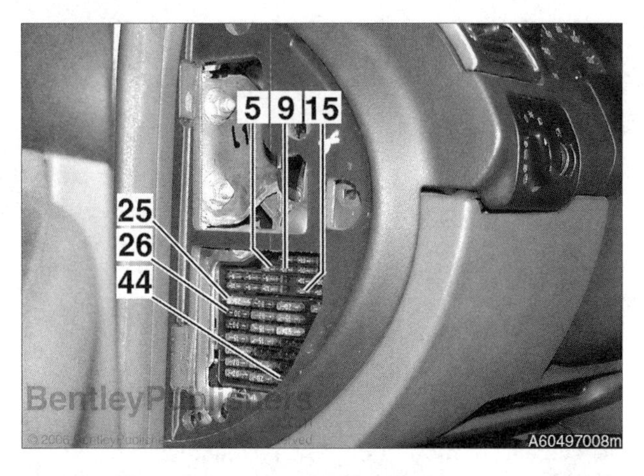

Climate control and engine cooling fuses and relays

Main fuse panel, left end of dashboard.

Table d. Climate control and engine cooling fuses		
Function	**Fuse #**	**Amperage**
Climate control	5	10
Outside mirror heating	9	10
Climate control	15	10
Blower	25	30
Rear window defogger	26	30
Seat heater	44	30

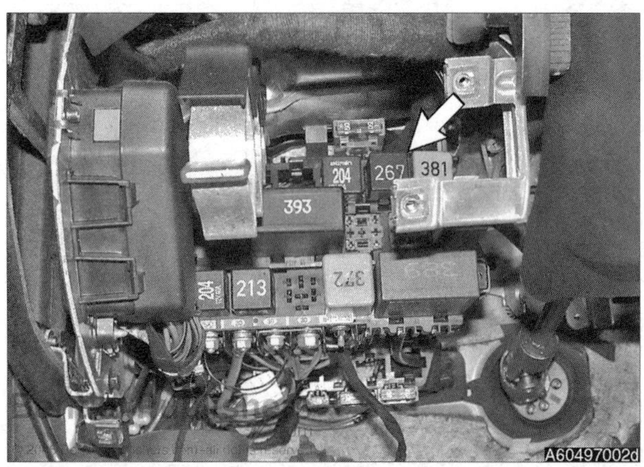

13-fold relay panel, underneath left side dashboard, trim panel removed.

- A/C clutch relay (J44) (**arrow**)

See **97 Fuses, Relays, Component Locations** for access information.

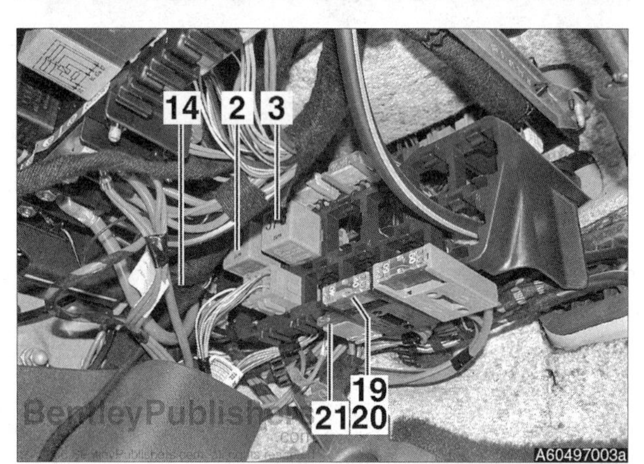

8-fold relay panel, underneath left dashboard, trim panel removed.

- **2**: Engine cooling fan relay, high speed (J101)
- **3**: Engine cooling fan relay (J26)
- **14**: Engine cooling fan fuse, 60A (S26), 2002 - 2004 allroad quattro
- **19** and **20**: Engine cooling fan fuse, 40A (S42)
- **21**: Engine cooling fan control module fuse, 5A (S142)

See **97 Fuses, Relays, Component Locations** for access information.

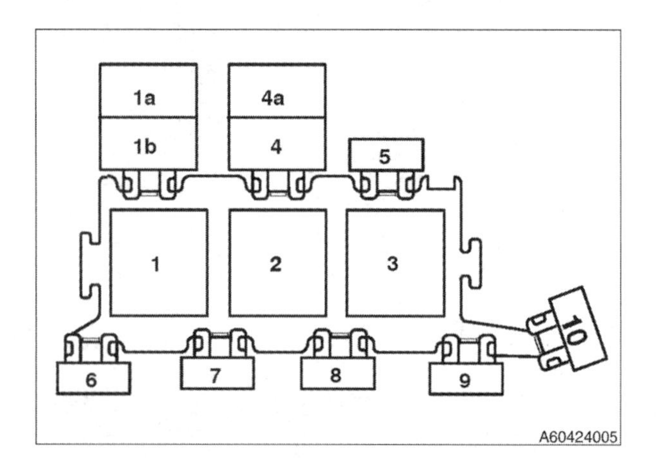

3-fold relay panel in E-box, under plenum chamber cover.

- **1b**: After-run coolant pump relay (J151)

See **97 Fuses, Relays, Component Locations** for access information.

BentleyPublishers.com
BentleyPublishers.com—All Rights Reserved

Engine cooling fan and blower controls

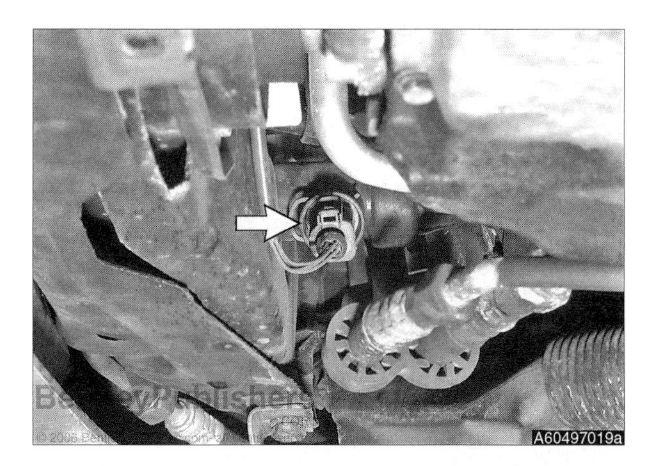

◄ Engine cooling fan switch (**arrow**) at radiator outlet pipe, right lower radiator.

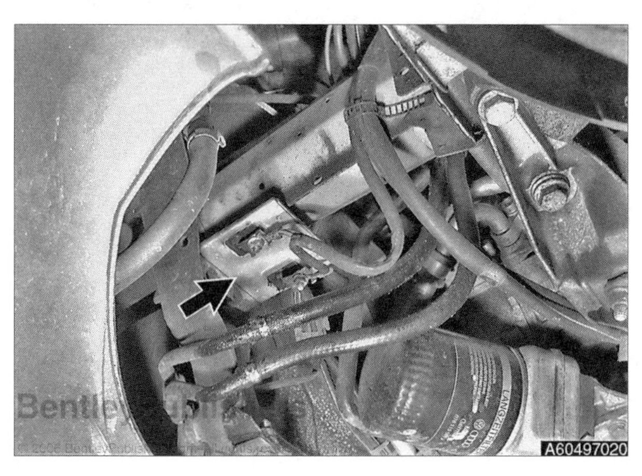

◄ Engine cooling fan series resistor (N39) (**arrow**), underneath left front of engine.

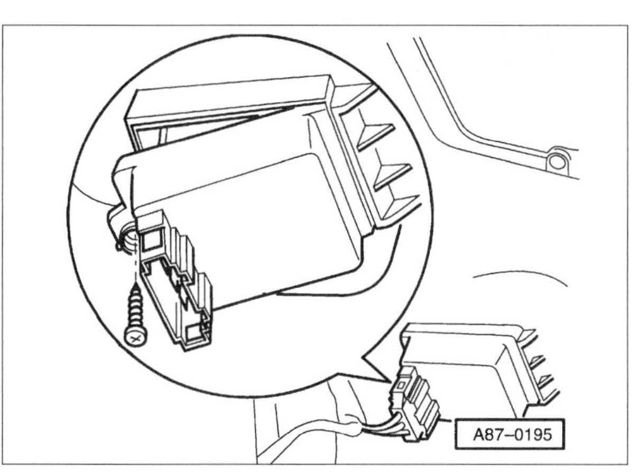

◄ Blower motor control module (J126) (**inset**), behind glove compartment on evaporator housing, left of blower motor.

NOTE —

- *Audi identifies electrical components by a letter and/or a number in the electrical schematics. See* **EWD Electrical Wiring Diagrams**. *These electrical identifiers are listed in parentheses as an aid to electrical troubleshooting.*

BentleyPublishers.com
BentleyPublishers.com—All Rights Reserved

CLIMATE CONTROL OPERATION

A/C system schematics

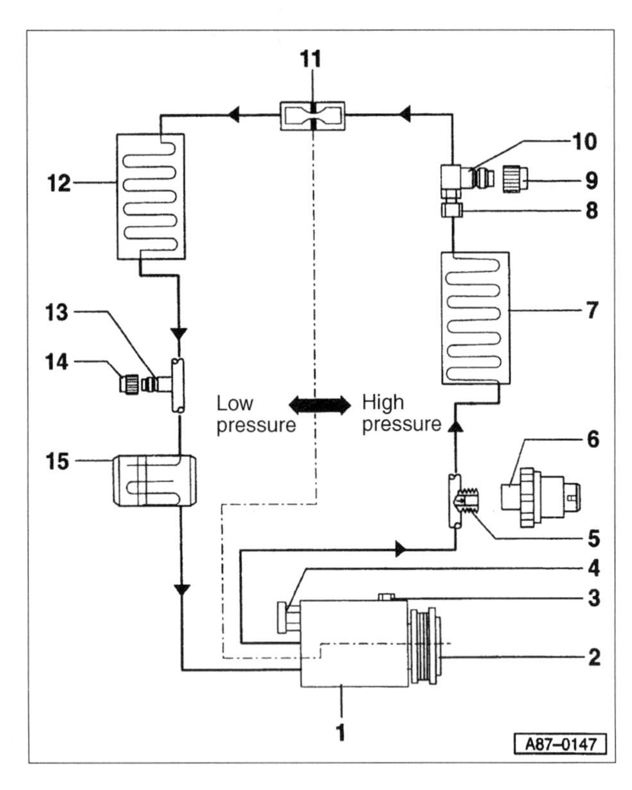

A/C refrigerant circuit, compressor with clutch:

1. A/C compressor
2. Compressor clutch (N25)
3. Refrigerant oil drain plug, 30 Nm (22 ft-lb)
4. High pressure relief valve, 10 Nm (7 ft-lb)
5. Threaded junction
6. A/C pressure switch (F129)
   High pressure sensor (G65)
7. A/C condenser
8. Connector, 15 Nm (11 ft-lb)
9. Sealing cap
10. High pressure service port
11. Restrictor
12. A/C evaporator
13. Low pressure service port
14. Sealing cap
15. Receiver-dryer (accumulator)

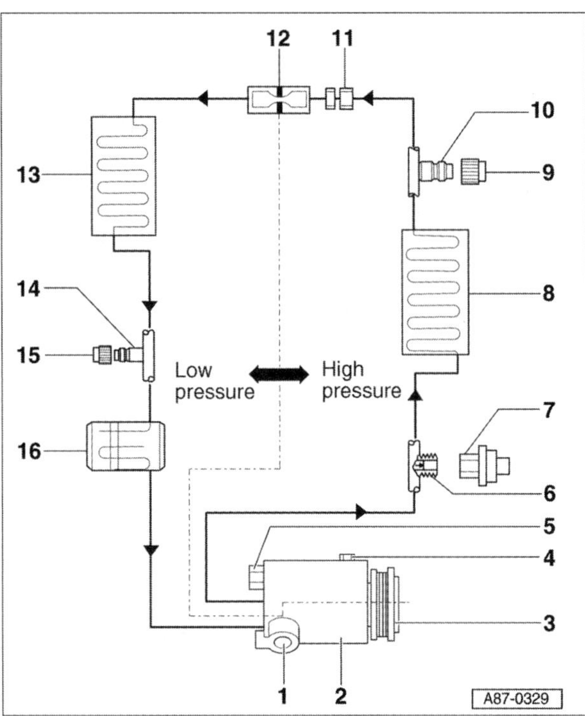

A/C refrigerant circuit, compressor with regulator valve:

1. Compressor regulator valve (N280)
2. A/C compressor
3. Belt pulley
4. Refrigerant oil drain plug, 30 Nm (22 ft-lb)
5. High pressure relief valve, 10 Nm (7 ft-lb)
6. Threaded junction
7. High pressure sensor (G65)
8. A/C condenser
9. Sealing cap
10. High pressure service port
11. Connector
12. Restrictor
13. A/C evaporator
14. Low pressure service port
15. Sealing cap
16. Receiver-dryer (accumulator)

BentleyPublishers.com
BentleyPublishers.com—All Rights Reserved

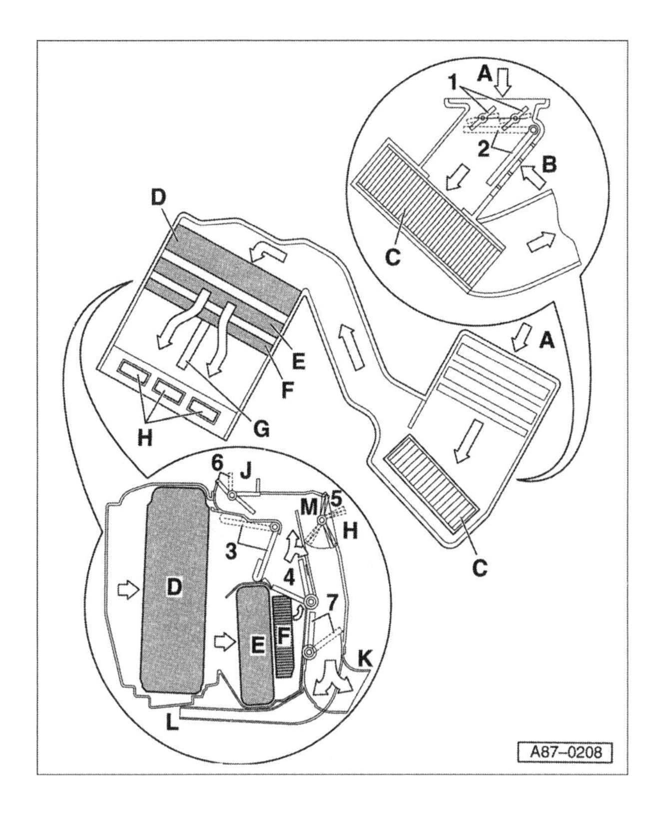

Air distribution

Evaporator housing components and airflow:

A. Dust and pollen filter element in plenum chamber
B. Right front footwell
C. Blower (V2)
D. A/C evaporator
E. Heater core
F. Auxiliary heater element (Z35) (not for US market)
G. Left / right side divider
H. to Dashboard vents
I. - - -
J. to Windshield vents
K. to Rear footwell vents
L. to Footwell vents
M. Wire mesh
1. Air flow flaps
2. Fresh air / recirculated air flap
3. Temperature flap 1
4. Temperature flap 2
5. Center vent flap
6. Defroster flap
7. Footwell flap

CLIMATE CONTROL COMPONENTS

A/C and heating component locations are described in **Table e** and shown in illustrations.

Table e. A/C and heater component locations

Component	Code	Location
A/C and heater control module	J255 J301	A/C and heater control panel in center dashboard
A/C compressor		Left front engine compartment
A/C compressor clutch	N25	On A/C compressor, left front engine compartment
A/C compressor regulator valve	N280	On A/C compressor, left front engine compartment
A/C high pressure sensor	G65	Right front engine compartment (2002 - 2004)
A/C pressure switch	F129	Right side of A/C condenser (1998 - 2001)
Air flow (back pressure) flap motor	V71	Near blower motor fresh air intake
Air flow flap motor position sensor	G113	Part of air flow (back pressure) flap motor
Blower control module	J126	In heating and A/C duct between evaporator housing and blower motor
Blower motor	V2	In evaporator housing behind glove box
Blower motor control module	J126	Under right side dashboard, to left of blower motor
Blower motor resistor	N24	Under right side of dashboard, on blower motor
Center air flow flap motor	V70	Right side of evaporator housing, lowest position
Center air flow flap motor sensor	G112	Part of center air flow flap motor

BentleyPublishers.com
BentleyPublishers.com—All Rights Reserved

Table e. A/C and heater component locations		
Component	**Code**	**Location**
Compressor clutch relay	J44	13-fold relay panel underneath left side dashboard
Defroster flap motor	V107	Right side of evaporator housing, topmost position
Defroster flap motor sensor	G135	Part of defroster flap motor
Evaporator temperature sensor	G263	Center of A/C and heater housing, under right side of dashboard
Footwell temperature sensor	G192	A/C and heater housing, under right side of dashboard
Fresh air intake duct temperature sensor	G89	Behind instrument panel, above blower motor housing, near fresh air intake duct
Interior temperature sensor	G56	On A/C and heater control panel
Interior temperature sensor fan	V42	On A/C and heater control panel
Left footwell vent temperature sensor	G192	Left side of A/C and heater housing, under center of dashboard
Left temperature regulator flap motor	V158	Left side of evaporator housing, above heater core
Left temperature regulator sensor	G220	Part of left temperature regulator flap motor
Left vent temperature sensor	G150	In ducting behind dashboard vents
Recirculation flap motor	V113	In plenum chamber, rear of engine compartment (2002 - 2004)
Right temperature regulator flap motor	V159	Right side of evaporator housing, center position
Right temperature regulator sensor	G221	Part of right temperature regulator flap motor
Right vent temperature sensor	G151	In ducting behind dashboard vents

BentleyPublishers.com
BentleyPublishers.com—All Rights Reserved

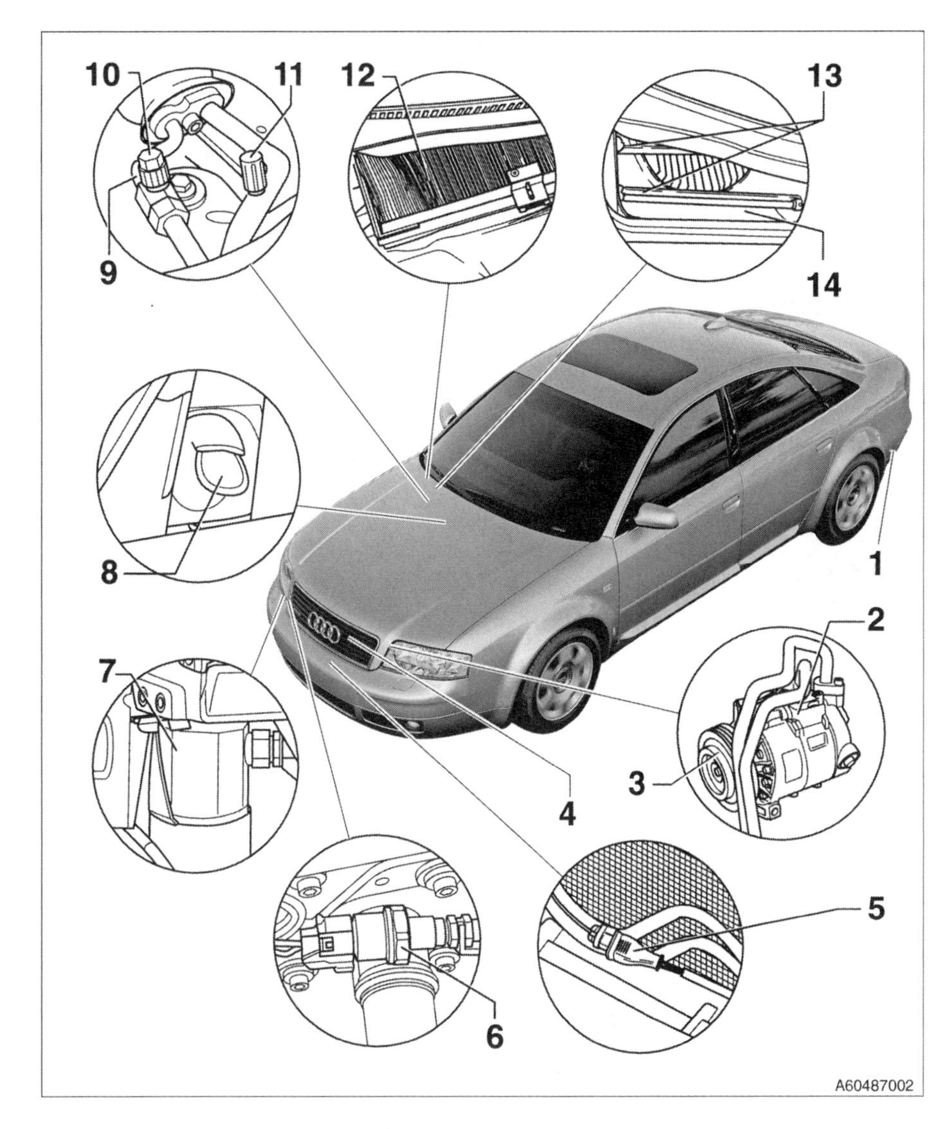

Climate control components in engine compartment

1. **Air extraction vents**
    - Under left and right rear bumper
    - Free air movement through vents necessary for vehicle ventilation
2. **A/C compressor with clutch**
    - Left front of engine
3. **Compressor clutch (N25)**
    - Front of A/C compressor
4. **A/C condenser**
    - In front of radiator
5. **Outside air temperature sensor (G17)**
    - Behind front bumper
6. **A/C pressure switch (F129)**
    - 1998 - 2001
7. **Receiver - dryer (accumulator)**
    - Under right headlight
8. **Evaporator housing drain valve**
    - Underneath vehicle
9. **Restrictor**
    - In refrigerant line
10. **Service connection, high pressure side**
    - Reinstall cap after repairs
11. **Service connection, low pressure side**
    - Reinstall cap after repairs
12. **Dust and pollen filter**
    - See **03 Maintenance** for replacement procedure
13. **Air flow flap motor (V71)**
    - Includes back pressure flap motor position sensor (G113)
14. **Recirculation flap motor (V11)**

BentleyPublishers.com
BentleyPublishers.com—All Rights Reserved

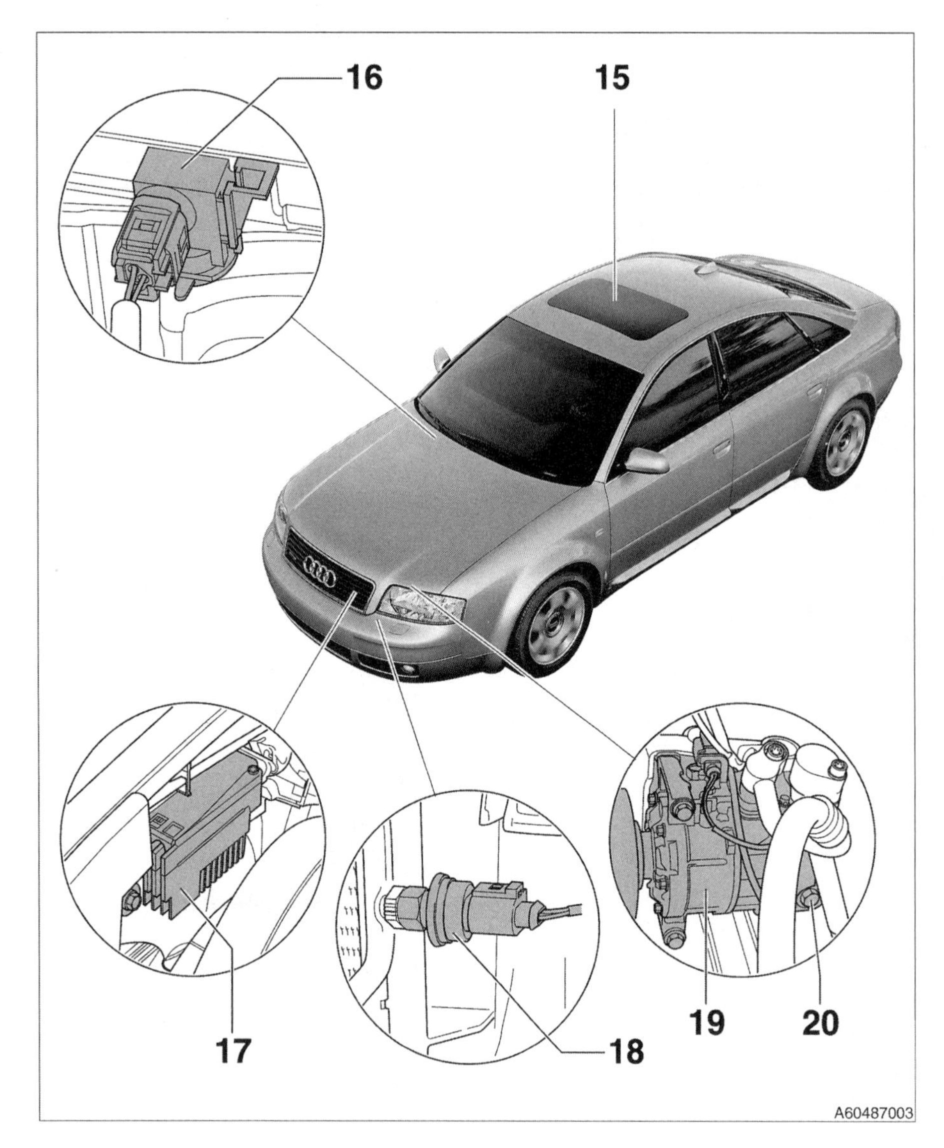

Climate control components in engine compartment
(continued)

15. Sunroof with solar sensors

16. Air quality sensor (G238)
- In dust and pollen filter housing

17. Engine cooling fan control module (J293)
- V6 3.0 liter engine
- In front of engine at cooling fan

18. A/C high pressure sensor (G65)
- 2002 - 2004

19. A/C compressor with regulator valve
- Left front of engine

20. A/C compressor regulator valve (N280)
- Side of A/C compressor

BentleyPublishers.com
BentleyPublishers.com—All Rights Reserved

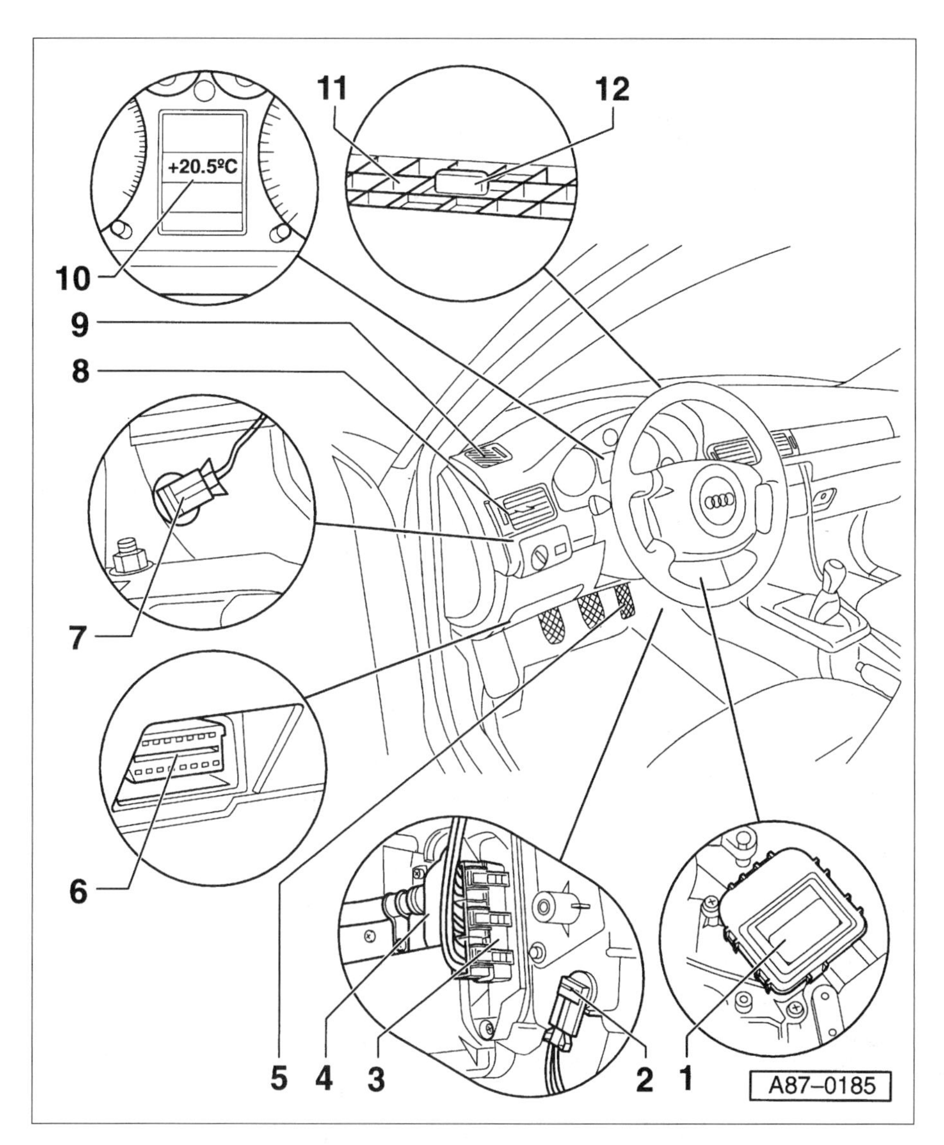

Climate control components in dashboard

1. **Left temperature flap motor (V158)**
   - With temperature sensor (G220)
2. **Footwell temperature sensor (G192)**
3. **Auxiliary heater (Z35) (not for US market)**
4. **Heater core**
5. **A/C shut-off switch**
   - Controlled by ECM
6. **Data link connector (DLC)**
7. **Left vent temperature sensor (G150)**
8. **Left dashboard vent**
9. **Left side window defroster vent**
10. **Outside air temperature display (G106)**
11. **Windshield defroster vent**
12. **Sunlight photo sensor (G107)**

BentleyPublishers.com
BentleyPublishers.com—All Rights Reserved

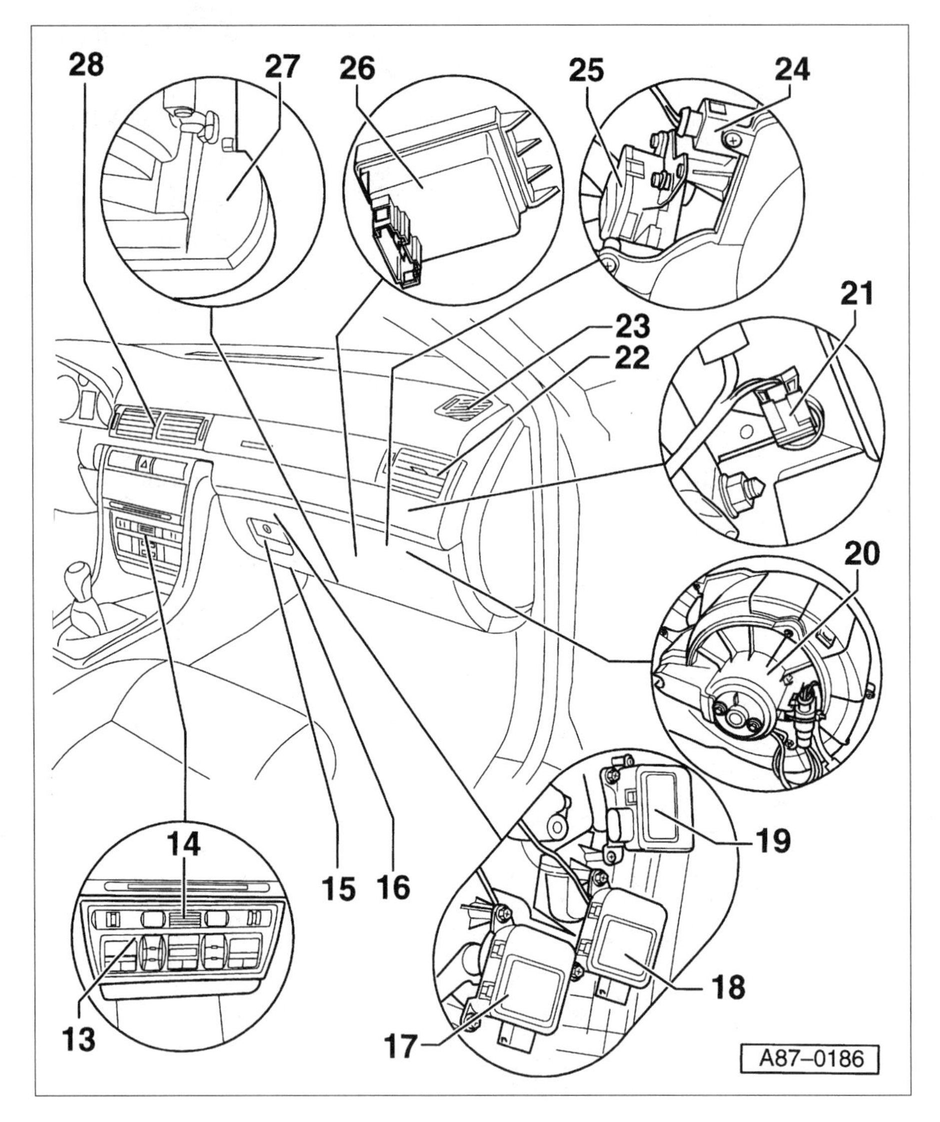

Climate control components in dashboard *(continued)*

13. **A/C control panel (E87)**
14. **Interior temperature sensor (G56) and sensor fan (V42)**
15. **Evaporator housing (A/C and heating unit)**
16. **Footwell vent**
17. **Center vent flap motor (V70)**
    - With temperature sensor (G112)
18. **Right temperature flap motor (V159)**
    - With temperature sensor (G221)
19. **Defroster flap motor (V107)**
    - With temperature sensor (G135)
20. **Blower (V2)**
21. **Right vent temperature sensor (G151)**
22. **Right dashboard vent**
23. **Right side window defroster vent**
24. **Air intake duct temperature sensor (G89)**
25. **Air flow flap motor (V71)**
    - With back pressure flap motor position sensor (G113) or recirculation flap motor position sensor (G143)
    - Up to 2001
26. **Blower control module (J126)**
27. **Evaporator housing drain**
28. **Center dashboard vent**

BentleyPublishers.com
BentleyPublishers.com—All Rights Reserved

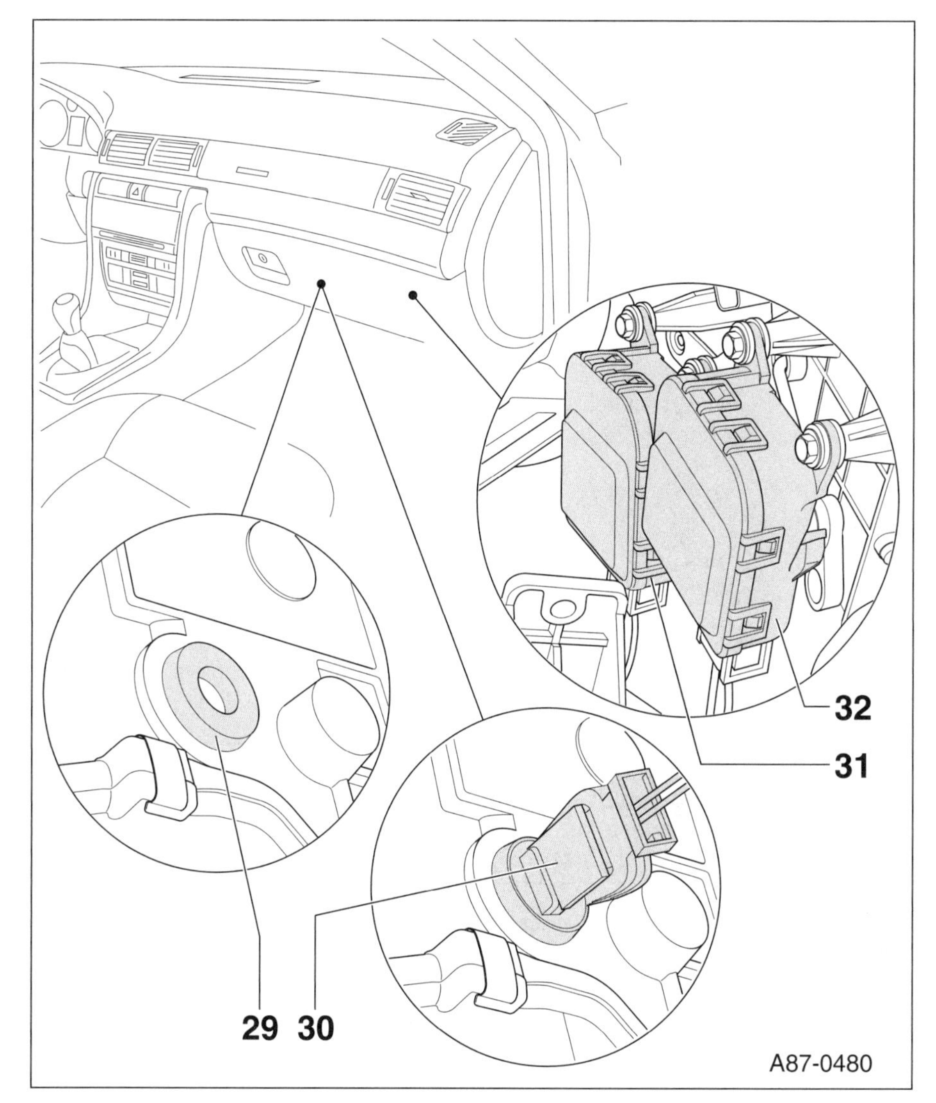

Climate control components in dashboard

29. **Sealing plug**
    - From 2002
30. **Evaporator vent temperature sensor (G263)**
    - From 2002
31. **Recirculation flap motor (V113)**
    - From 2002
32. **Air flow flap motor (V71)**
    - With back pressure flap motor position sensor (G113)
    - From 2002

BentleyPublishers.com
BentleyPublishers.com—All Rights Reserved

Climate Control Repairs

Blower, removing and installing

– Remove glove compartment. See **70 Interior Trim**.

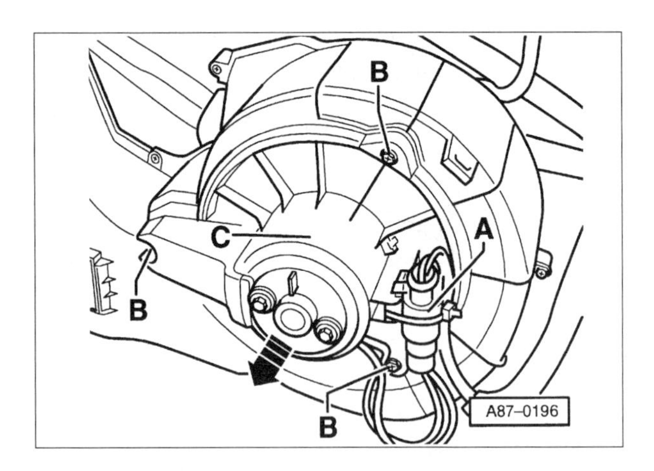

Working under right side dashboard:

- Separate harness connector (**A**).
- Remove blower mounting screws (**B**).
- Pull blower (**C**) out of evaporator housing in direction of **arrow**.

– Prior to installing new blower, carefully compare length of blower wheel blades. Two versions are used, with one approx. 10 mm (⅓ in) shorter.

– Installation is reverse of removal.

A/C compressor with clutch, removing and installing

– Following manufacturer's instructions, connect an approved refrigerant recovery / recycling / recharging unit to A/C system and discharge system.

> ***WARNING —***
> - *Do not attempt to discharge or charge the A/C system without proper equipment and training. Damage to the vehicle and personal injury may result.*

– Raise car and support safely.

> ***WARNING —***
> - *Make sure the car is stable and well supported at all times. Use a professional automotive lift or jack stands designed for the purpose. A floor jack is not adequate support.*

– Remove engine lower cover. See **03 Maintenance**.

– Remove engine accessory belt. See **03 Maintenance**. Mark belt with direction of rotation if it is to be reused.

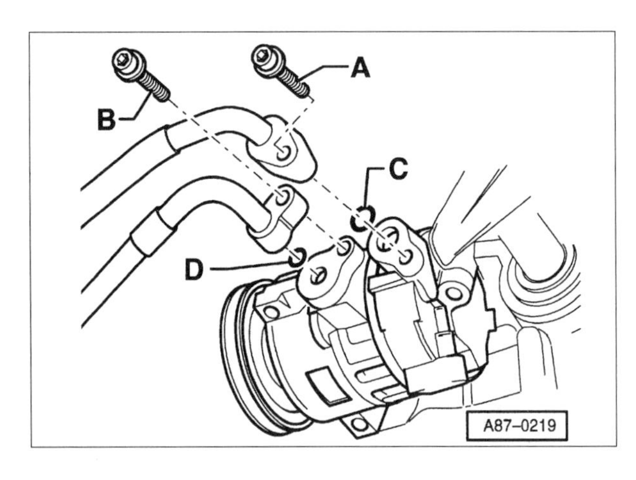

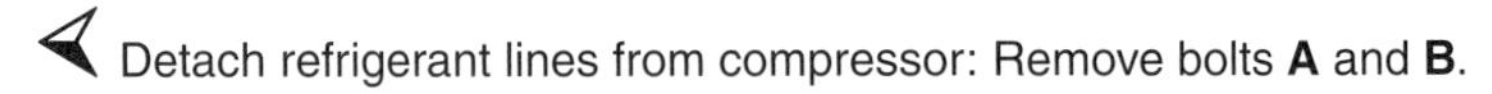

Detach refrigerant lines from compressor: Remove bolts **A** and **B**.

> ***CAUTION—***
> - *Immediately seal off open refrigerant lines and ports.*

– Unbolt compressor from bracket.

– Installation is reverse of removal. Be sure to replace O-ring seals **C** and **D**.

Tightening torque	
Compressor to engine bracket	25 Nm (18 ft-lb)
Refrigerant line to compressor	25 Nm (18 ft-lb)

BentleyPublishers.com
BentleyPublishers.com—All Rights Reserved

– Prior to starting engine, rotate compressor manually approx. 10 revolutions to lubricate internal components.

– Start engine with climate control OFF. Wait until idle is stable. Then switch A/C ON and allow to run at least 10 minutes at idle.

– Recheck accessory belt alignment.

A/C compressor with regulator valve, removing and installing

– Following manufacturer's instructions, connect an approved refrigerant recovery / recycling / recharging unit to A/C system and discharge system.

> *WARNING—*
> • *Do not attempt to discharge or charge the A/C system without proper equipment and training. Damage to the vehicle and personal injury may result.*

– Place lock carrier (radiator support frame) in service position. See **50 Body–Front**.

– Remove oil filter.

– Raise car and support safely.

> *WARNING—*
> • *Make sure the car is stable and well supported at all times. Use a professional automotive lift or jack stands designed for the purpose. A floor jack is not adequate support.*

– Remove engine lower cover. See **03 Maintenance**.

– Remove engine accessory belt. See **03 Maintenance**. Mark belt with direction of rotation if it is to be reused.

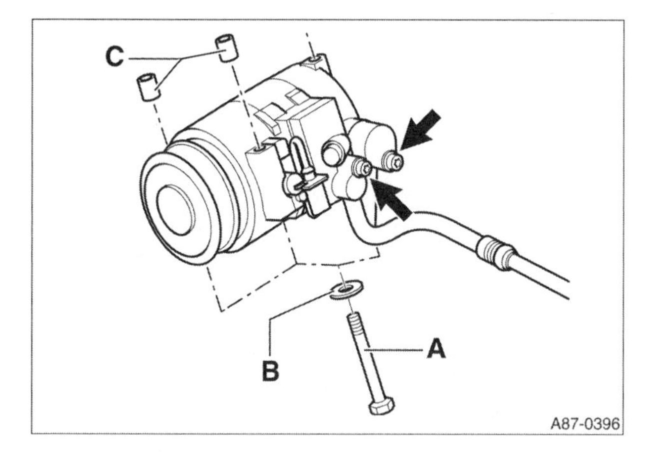

◄ Working at side of compressor, remove regulator valve mounting bolts (**arrows**) from compressor. Tie regulator valve safely to side.

> *CAUTION—*
> • *Immediately seal off open refrigerant lines and ports.*

– Unbolt compressor from engine bracket.

– Installation is reverse of removal.
  - Install one washer (**B**) on each bolt (**A**).
  - Check that bushings (**C**) are seated correctly.

Tightening torque	
Compressor to engine bracket	25 Nm (18 ft-lb)
Regulator valve to compressor	25 Nm (18 ft-lb)

– Check refrigerant lines for correct placement.

– Prior to starting engine, rotate compressor manually approx. 10 revolutions to lubricate internal components.

BentleyPublishers.com
BentleyPublishers.com—All Rights Reserved

– Start engine with climate control OFF. Wait until idle is stable. Then switch A/C ON and allow to run at least 10 minutes at idle.

– Recheck accessory belt alignment.

Heater core, removing and installing

> ***CAUTION—***
> *• To avoid personal injury, be sure the engine is cold before opening the cooling system.*

– Vehicle with power seats: Move seats all the way back.

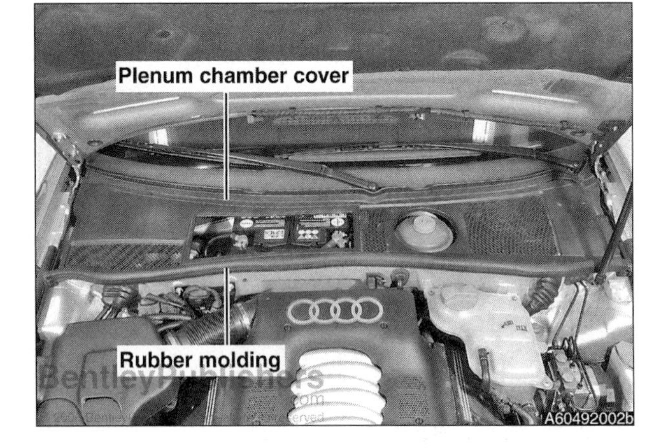

◀ Open engine hood and pull off rubber molding at engine compartment rear bulkhead. Pull plenum chamber cover forward and remove.

– Disconnect battery cables and remove battery. See **27 Battery, Alternator, Starter**.

> ***CAUTION—***
> *• Be sure to have the radio anti-theft code before disconnecting the battery.*

– 1998 model: Remove complete windshield wiper assembly. See **92 Wiper and Washers**.

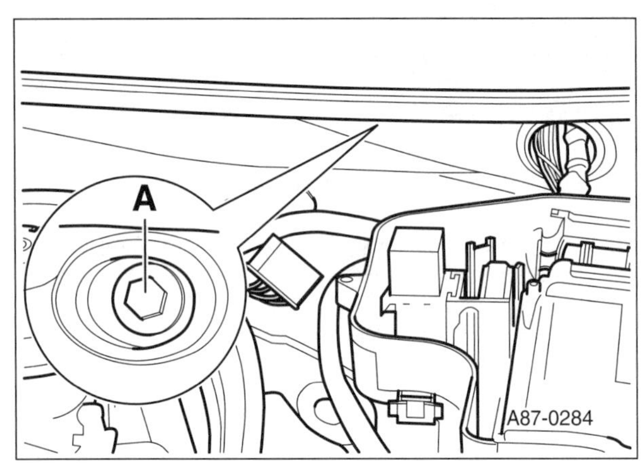

◀ Working in plenum chamber:

• Remove insulating grommet.

• Loosen screw (**A**) approx. 10 mm (1/8 in) or remove.

– Loosen coolant reservoir cap to release cooling system pressure.

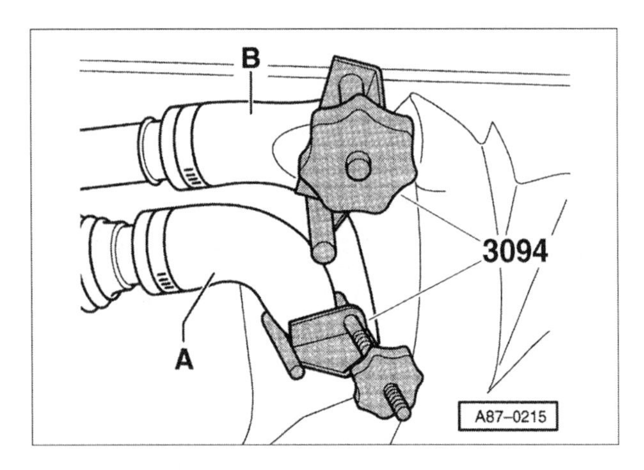

◀ Use coolant hose pinch-off clamps (special tool VAG 3094 or equivalent) to clamp off heater core hoses (**A** and **B**). Disconnect both hoses.

NOTE—

• Mark the hoses so that they can be reattached correctly:
*Hose **A**: Coolant supply from cylinder head*
*Hose **B**: Coolant return to coolant pump*

BentleyPublishers.com
BentleyPublishers.com—All Rights Reserved

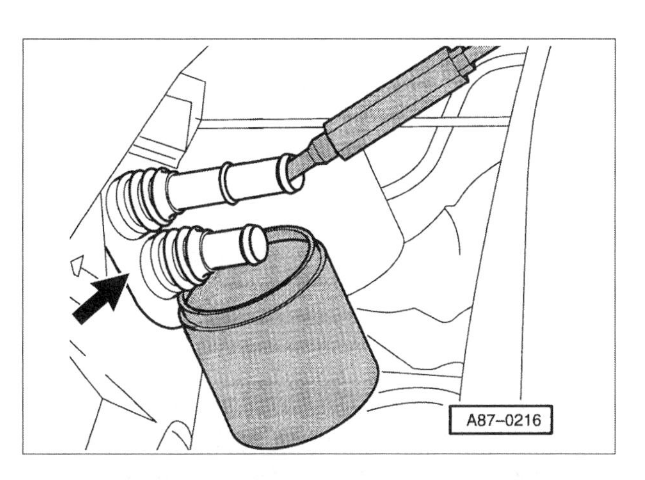

◄ Place container under heater core connection and use compressed air to blow out coolant.

- Remove rubber grommet (**arrow**).

◄ Working in passenger compartment, tilt steering column down and pull off trim under instrument cluster. Remove steering column trim panel fasteners (**arrows**).

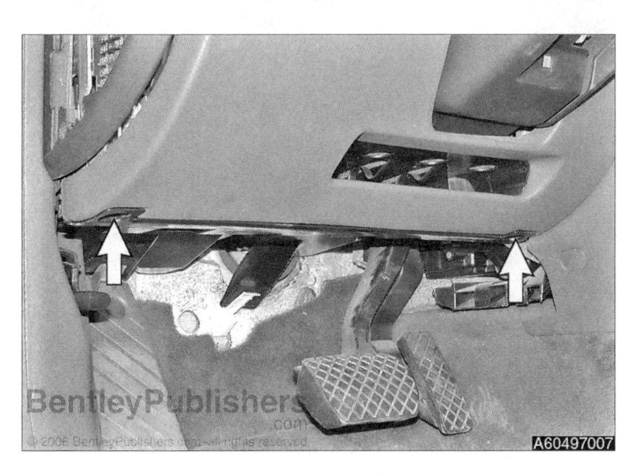

◄ Remove lower trim fasteners (**arrows**) and pull down trim panels under steering column to gain access to left side of evaporator housing.

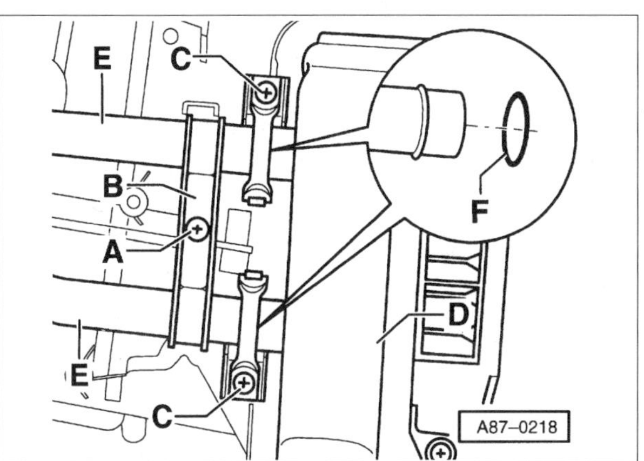

◄ Working at left side of evaporator housing underneath dashboard:

- Unscrew screw (**A**) and remove bracket (**B**).
- Loosen clamps (**C**).
- Pull heater core (**D**) out of evaporator housing slightly.
- Pull coolant lines (**E**) off heater core.
- Remove heater core.

CAUTION—
• Be prepared to catch dripping coolant.

BentleyPublishers.com
BentleyPublishers.com—All Rights Reserved

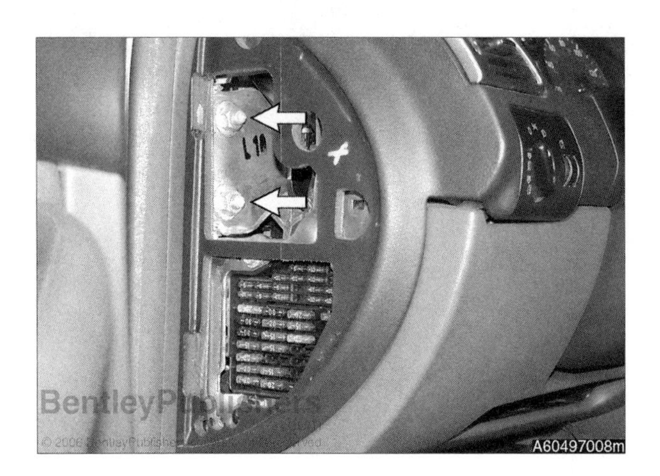

If brake pedal mounting bracket interfered with heater core removal:

- Remove dashboard end covers on left and right.
- Loosen dashboard cross brace mounting fasteners (**arrows**) on left and right. This allows dashboard and pedals to be moved sufficiently out of the way for heater core removal.

> ***CAUTION—***
> - *The pedal assembly in 1998 models may interfere severely with heater core removal. In that case, the entire dashboard may need to be removed.*

– Installation is reverse of removal. Bear in mind:

- Replace heater core hose clamps and sealing O-rings.
- Install hose clamps so that they do not contact and rattle against evaporator housing.
- Fill and bleed cooling system. See **19 Cooling System**. Check system for leaks.

BentleyPublishers.com
BentleyPublishers.com—All Rights Reserved

9 Electrical System–General

GENERAL

This repair group presents a brief description of the principal parts of the electrical system, including the bus systems. Also covered here are basic electrical system troubleshooting tips.

Audi A6 models are equipped with a 12-volt, direct current (DC), negative-ground electrical system. The voltage regulator maintains the voltage in the system at approximately 12 volts. All circuits are grounded by direct or indirect connection to the negative (-) terminal of the battery.

All electrical circuits except those required for starting and operating the engine are protected by fuses. For a full listing of fuses and fuse locations, see **97 Fuses, Relays, Component Locations**.

Also see:

- **27 Battery, Alternator, Starter**
- **57 Doors and Locks**
- **64 Door Windows**
- **96 Interior Lights, Switches, Anti-theft**
- **EWD Electrical Wiring Diagrams**

Warnings and Cautions

Please read the following before doing any work on your electrical system.

WARNING—

- *Airbags and pyrotechnic seat belt tensioners utilize explosive devices. Handle with extreme care. Refer to the warnings and cautions in* **69 Seat Belts, Airbags**.
- *The ignition system of the car operates at lethal voltages. If you have a weak heart or wear a pacemaker, do not expose yourself to ignition system electric currents. Take extra precautions when working on the ignition system or when servicing the engine while it is running or the key is ON. See* **28 Ignition System** *for additional ignition system warnings and cautions.*
- *Keep hands, clothing and other objects clear of the electric radiator cooling fan when working on a warm engine. The fan may start at any time, even when the ignition is switched OFF.*

BentleyPublishers.com
BentleyPublishers.com—All Rights Reserved

> ***CAUTION—***
> - *Do not disconnect the battery with the engine running.*
> - *Switch the ignition OFF and remove the negative (-) battery cable before removing any electrical components. Connect and disconnect electrical connectors and ignition test equipment leads only while the ignition is switched OFF.*
> - *Relay and fuse positions are subject to change and may vary from car to car. If questions arise, an authorized Audi dealer is the best source for the most accurate and up-to-date information.*
> - *Use a digital multimeter for electrical tests. Switch the multimeter to the appropriate function and range before making test connections.*
> - *Many control modules are static sensitive. Static discharge damages them permanently. Handle the modules using proper static prevention equipment and techniques.*
> - *To avoid damaging harness connectors or relay panel sockets, use jumper wires with flat-blade connectors that are the same size as the connector or relay terminals.*
> - *Do not try to start the engine of a car which has been heated above 176°F (80°C) (for example, in a paint drying booth). Allow it to cool to normal temperature.*
> - *Disconnect the battery before doing any electric welding on the car. Do not wash the engine while it is running, or any time the ignition is ON.*
> - *Choose test equipment carefully. Use a digital multimeter with at least 10 MΩ input impedance or an LED test light. An analog meter (swing-needle) or a test light with a normal incandescent bulb may draw enough current to damage sensitive electronic components.*
> - *Do not use an ohmmeter to measure resistance on solid state components such as control modules.*
> - *Disconnect the battery before making resistance (ohm) measurements on a circuit.*

Bus Systems

Audi A6 vehicles are electrically complex. Many vehicle systems and subsystems are interconnected or integrated. In addition, the requirements of second generation on-board diagnostics (OBD II) are such that there are many more circuits and wires in the vehicle than ever before. The components must exchange large volumes of data with one another in order to perform their various functions.

To handle this complexity effectively, the A6 makes extensive use of control module networking using controller area network (CAN) bus systems. Signals are shared digitally among several control modules on a CAN-bus, eliminating the need for separate connections for each pair of control modules. The use of bus communication for controls and accessories reduces wiring complexity and improves system response time.

The CAN-bus consists of a pair of wires twisted together to ensure reliable communication among modules without conflicts or interference. The two wires are referred to as CAN-high and CAN-low.

BentleyPublishers.com
BentleyPublishers.com—All Rights Reserved

Data transfer over a the CAN-bus is similar to a telephone conference. A control module on the bus transmits a stream of data which other modules receive at the same time. Each control module then decides to use or ignore this data.

The benefits of the bus method of data transfer are as follows:

- As data and programs are modified and extended, only software modifications are necessary.
- Continuous verification of transmitted data leads to low error rate.
- Sensors and signal wires can be simplified or eliminated due to the transmittal of multiplexed digital data.
- Control modules transfer data at a high rate.
- Control module sizes and connector sizes are smaller.
- CAN-bus conforms to international standards. This facilitates data interchange between control modules of different manufacture.

As of 2000, the A6 CAN-bus systems consisted of the following:

- Drivetrain CAN-bus: 500k baud transfer rate
- Convenience CAN-bus: 100k baud transfer rate
- Infotainment CAN-bus: 100k baud transfer rate

Drivetrain CAN-bus

1. ABS / ESP control module (J104)
2. Engine control module (ECM) (J271)
3. Instrument cluster control module (J285)
4. Steering angle sensor (G85)
5. Automatic transmission control module (J217)

Electrical Troubleshooting

Four things are required for current to flow in any electrical circuit:

- Voltage source
- Wires or connections to transport voltage
- Load or device that uses electricity
- Connection to ground

Most problems can be found using a digital multimeter (volt / ohm / ammeter) to check the following:

- Voltage supply
- Breaks in the wiring (infinite resistance / no continuity)
- A path to ground that completes the circuit

Electric current is logical in its flow, always moving from the voltage source toward ground. Electrical faults can usually be located through a process of elimination. When troubleshooting a complex circuit, separate the circuit into smaller parts. General tests outlined below may be helpful in finding electrical problems. The information is most helpful when used with wiring diagrams.

Be sure to analyze the problem. Use wiring diagrams to determine the most likely cause. Get an understanding of how the circuit works by following the circuit from ground back to the power source.

When making test connections at connectors and components, use care to avoid spreading or damaging the connectors or terminals.

BentleyPublishers.com
BentleyPublishers.com—All Rights Reserved

Some tests may require jumper wires to bypass components or connections in the wiring harness. When connecting jumper wires, use blade connectors at the wire ends that match the size of the terminal being tested. The small internal contacts are easily spread apart, and this can cause intermittent or faulty connections that can lead to more problems.

Voltage and voltage drop

Wires, connectors, and switches that carry current are designed with very low resistance so that current flows with a minimum loss of voltage. A voltage drop is caused by higher than normal resistance in a circuit. This additional resistance actually decreases or stops the flow of current. Excessive voltage drop can be noticed by problems ranging from dim headlights to sluggish wipers. Some common sources of voltage drops are corroded or dirty switches, dirty or corroded connections or contacts, and loose or corroded ground wires and ground connections.

A voltage drop test is a good test to perform if current is flowing through the circuit but the circuit is not operating correctly. A voltage drop test helps pinpoint a corroded ground strap or a faulty switch. Normally, there should be less than 1 volt drop across most wires or closed switches. A voltage drop across a connector or short cable should not exceed 0.5 volt.

A voltage drop test is generally more accurate than a simple resistance check because the resistances involved are often too small to measure with most ohmmeters. For example, a resistance as small as 0.02 Ω would results in a 3 volt drop in a typical 150 amp starter circuit. (150 amps x 0.02 Ω = 3 volts).

Keep in mind that voltage with ignition key ON and voltage with engine running are not the same. With ignition ON and engine OFF (battery voltage), voltage should be approximately 12.6 volts. With engine running (charging voltage), voltage should be approximately 14.0 volts. Measure voltage at battery with ignition ON and then with engine running to get exact measurements.

Voltage, measuring

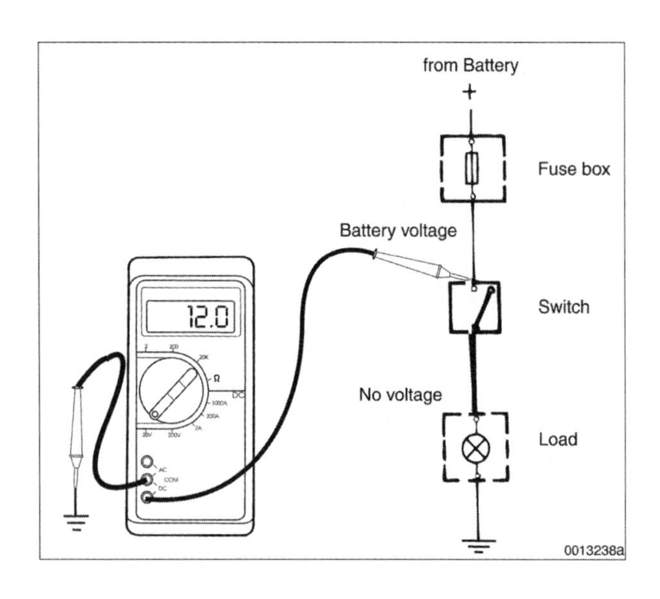

- Connect digital multimeter negative lead to a reliable ground point on car.

NOTE—
- *The negative (-) battery terminal is always a good ground point.*

- Connect digital multimeter positive lead to point in circuit you wish to measure.

- Voltage reading should not deviate more than 1 volt from voltage at battery. If voltage drop is more than this, check for a corroded connector or loose ground wire.

BentleyPublishers.com
BentleyPublishers.com—All Rights Reserved

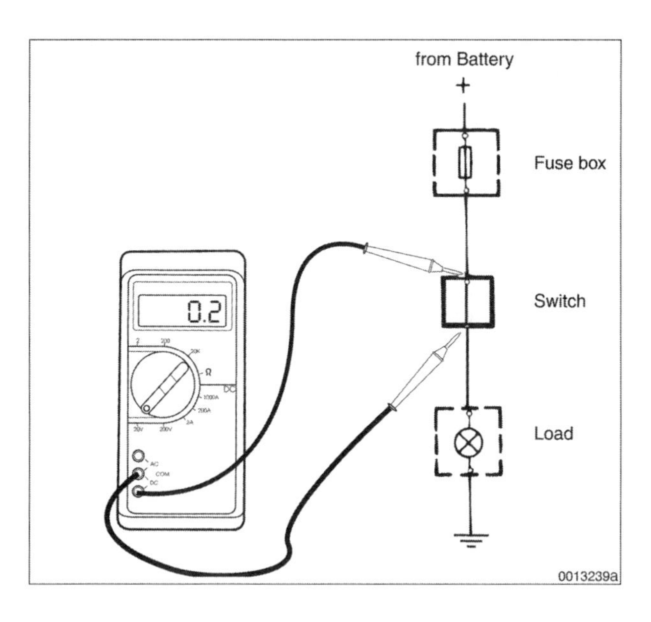

Voltage drop, testing

Check voltage drop only when there is a load on the circuit, such as when operating the starter motor or turning on the headlights. Use a digital multimeter to ensure accurate readings.

- Connect digital multimeter positive lead to positive (+) battery terminal or a positive power supply close to battery source.
- Connect digital multimeter negative lead to other end of cable or switch being tested.
- With power on and circuit working, meter shows voltage drop (difference between two points). This value should not exceed 1 volt.
- Maximum voltage drop in an automotive circuit, as recommended by the Society of Automotive Engineers (SAE), is as follows:
  - 0 volt for small wire connections
  - 0.1 volt for high current connections
  - 0.2 volt for high current cables
  - 0.3 volt for switch or solenoid contacts
- On longer wires or cables, the drop may be slightly higher. In any case, a voltage drop of more than 1.0 volt usually indicates a problem.

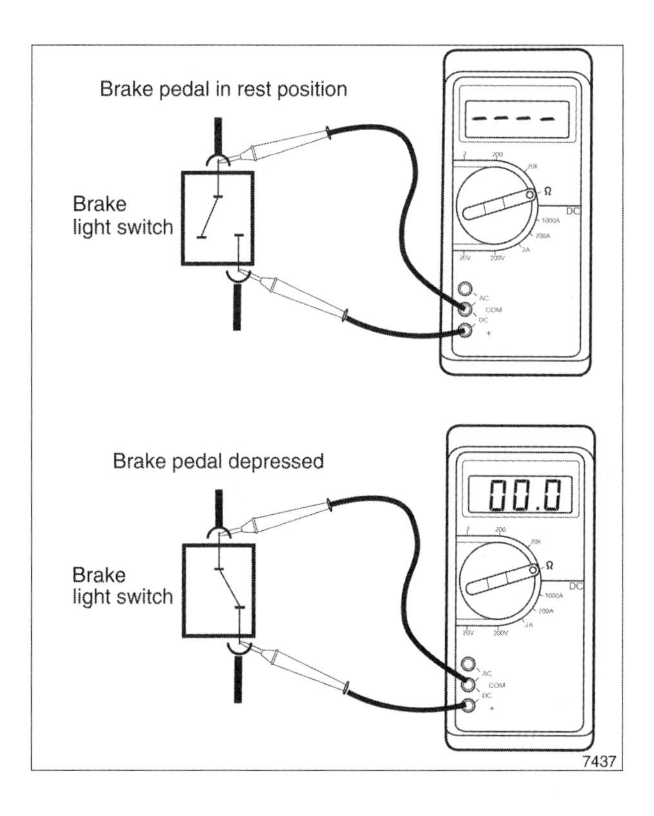

Continuity, checking

Use continuity test to check a circuit or switch. Because most automotive circuits are designed to have little or no resistance, a circuit or part of a circuit can be easily checked for faults using an ohmmeter. An open circuit or a circuit with high resistance does not allow current to flow. A circuit with little or no resistance allows current to flow easily.

When checking continuity, turn ignition OFF. On circuits that are powered at all times, disconnect battery. Using the appropriate wiring diagram, test circuit for faulty connections, wires, switches, relays and engine sensors by checking for continuity.

- Example: Test brake light switch for continuity:
  - With brake pedal in rest position (switch open) there is no continuity (infinite Ω).
  - With pedal depressed (switch closed) there is continuity (0 Ω).

Short circuits, testing

Short circuits are exactly what the name implies. Current in the circuit takes a shorter path than it was designed to take. The most common short that causes problems is a short to ground where the insulation on a positive (+) wire wears away and the metal wire is exposed. When the wire rubs against a metal part of the car or other ground source, the circuit is shorted to ground. If the exposed wire is live (positive battery voltage), a fuse blows and the circuit may be damaged.

BentleyPublishers.com
BentleyPublishers.com—All Rights Reserved

Short circuits are often difficult to locate and may vary in nature. Short circuits can be found using a logical approach based on knowledge of the current path.

Use a digital multimeter to locate short circuits.

> ***CAUTION—***
> - *In circuits protected with high rating fuses (25 amp and greater), wires or circuit components may be damaged before the fuse blows. Check for wiring damage before replacing fuses of this rating. Also, check for correct fuse rating.*

Testing with ohmmeter

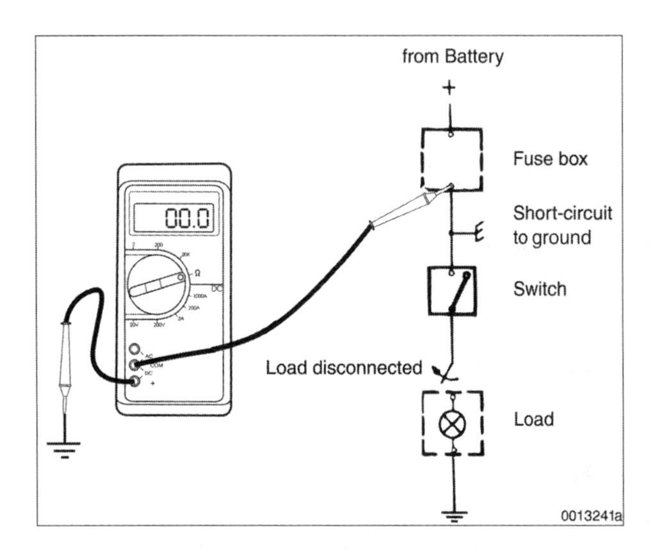

- Remove blown fuse from circuit and disconnect cables from battery. Disconnect harness connector from circuit load or consumer.
- ◄ Using an ohmmeter, connect one test lead to load side of fuse terminal (terminal leading to circuit) and other test lead to ground.
- If there is continuity to ground, there is a short to ground.
- If there is no continuity, work from wire harness nearest to fuse / relay panel and move or wiggle wires while observing meter. Continue to move down harness until meter displays a reading. This is the location of the short to ground.
- Visually inspect wire harness at this point for any faults. If no faults are visible, carefully slice open harness cover or wire insulation for further inspection. Repair any faults found.

Testing with voltmeter

- Remove blown fuse from circuit. Disconnect harness connector from circuit load or consumer.

> ***NOTE—***
> - *Most fuses power more than one consumer. Be sure all consumers are disconnected when checking for a short circuit.*

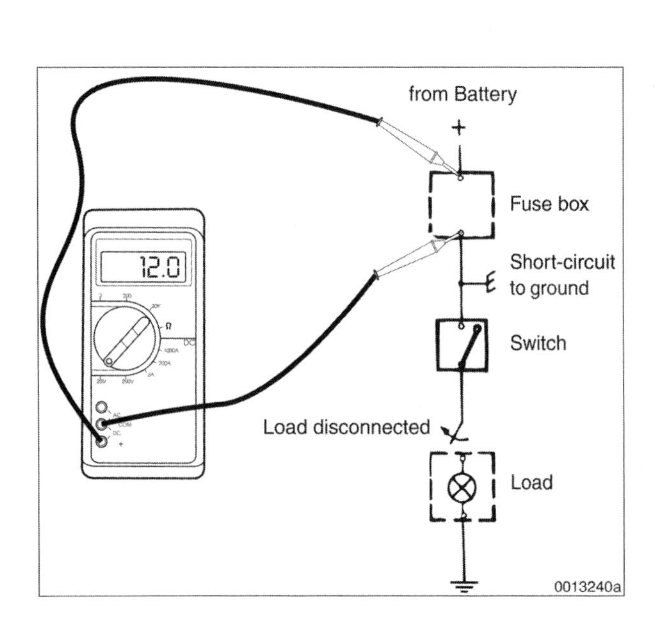

- ◄ Using a digital multimeter, connect test leads across fuse terminals. Make sure power is present in circuit. If necessary switch ignition ON.
- If voltage is present at voltmeter, there is a short to ground.
- If voltage is not present, work from wire harness nearest to fuse / relay panel and move or wiggle wires while observing meter. Continue to move down harness until meter displays a reading. This is the location of the short to ground.

Visually inspect wire harness at this point for any faults. If no faults are visible, carefully slice open harness cover or wire insulation for further inspection. Repair any faults found.

BentleyPublishers.com
BentleyPublishers.com—All Rights Reserved

90 Instruments

GENERAL

This repair group covers the instrument cluster and warning lights.

See **91 Radio and Communication** for information about the navigation system.

Instrument cluster

The instrument cluster consists of

- Engine oil temperature gauge
- Tachometer with clock
- Engine coolant temperature gauge
- Fuel gauge
- Multifunction indicator (MFI) screen
- Electronic speedometer, odometer, trip odometer
- Voltmeter

Warning and indicator lights in the speedometer and tachometer faces are LEDs which cannot be serviced separately. An LED malfunction requires instrument cluster replacement.

The instrument cluster supplements various warning light displays with acoustic warnings.

BentleyPublishers.com
BentleyPublishers.com—All Rights Reserved

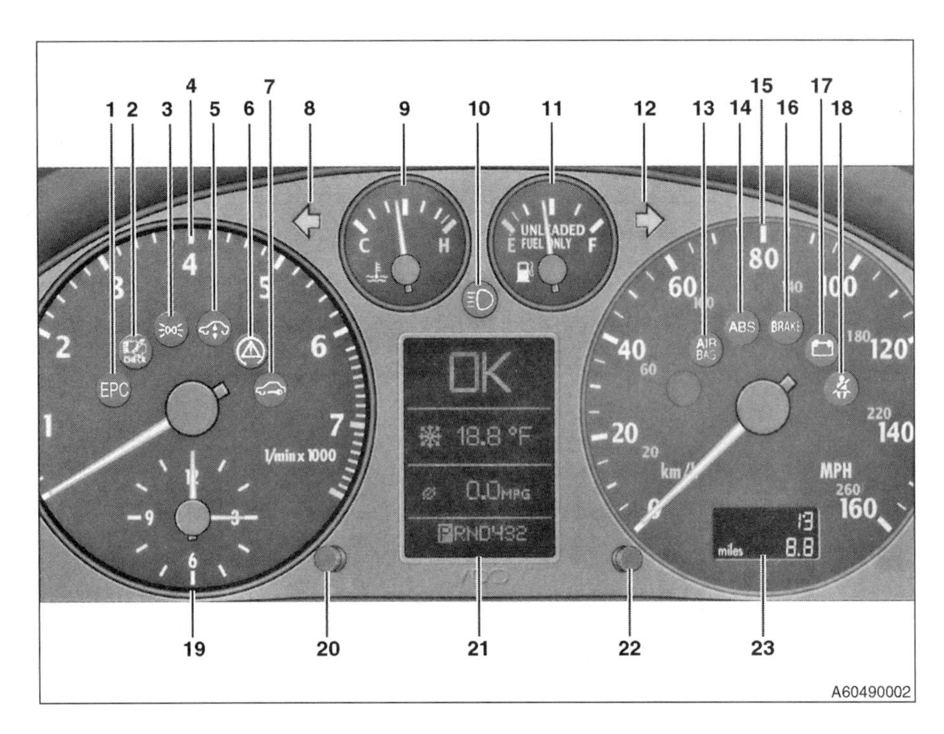

Instrument cluster and warning lights

1. Electronic throttle (EPC)
2. Malfunction indicator light (MIL)
3. Headlights and side marker lights on
4. Tachometer
5. Self-leveling suspension malfunction (allroad quattro)
6. Anti-slip regulation (ASR) or electronic stabilization program (ESP)
7. Electronic immobilizer on
8. Left turn signal indicator
9. Engine coolant temperature gauge
10. High beams on
11. Fuel gauge
12. Right turn signal indicator
13. Airbags malfunction
14. Anti-lock braking (ABS) malfunction
15. Speedometer
16. Brake fluid low or parking brake on
17. Charging system malfunction
18. Seat belts not buckled
19. Clock
20. Left setting button:
    - Clock
    - Auto check system
    - Trip odometer
21. Driver information display
22. Right setting button:
    - Service interval
    - Trip odometer
23. Odometer and service indicator

NOTE—

- *Warning light placement and other functions vary slightly among models.*

BentleyPublishers.com
BentleyPublishers.com—All Rights Reserved

Driver information

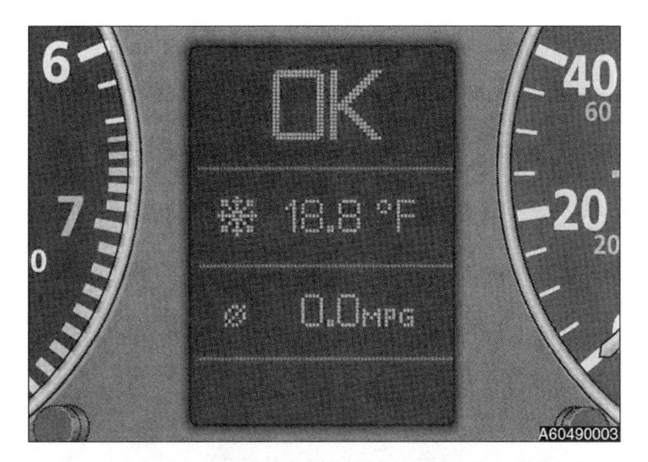

The driver information display in the center of the instrument cluster provides useful driver information and service reminders, including:

- Service interval
- Auto-check system
- Outside air temperature
- Trip computer
- Automatic transmission gear shift position
- Radio frequency
- Door open indicator
- Speed warning

If equipped, the driver information display doubles as the navigation display. See **91 Radio, Communication**.

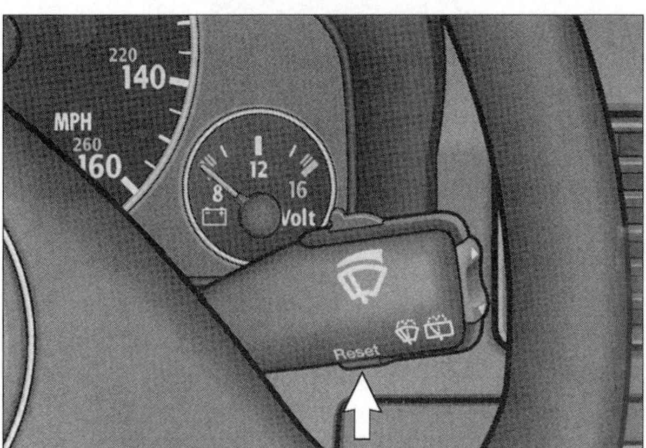

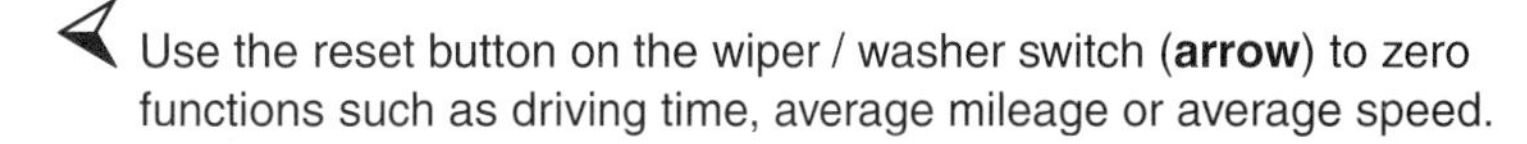

Use the reset button on the wiper / washer switch (**arrow**) to zero functions such as driving time, average mileage or average speed.

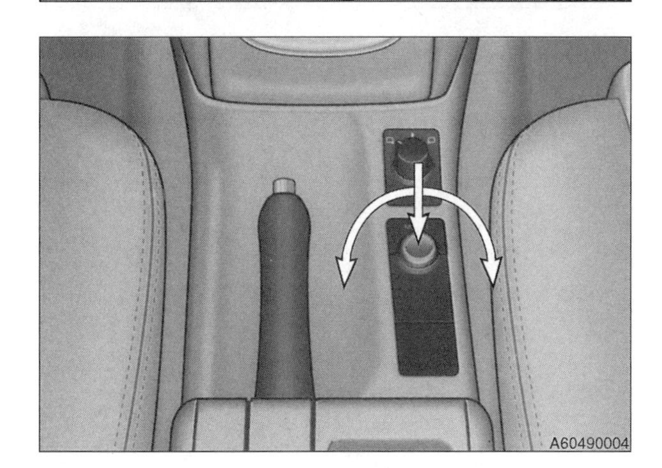

2002 - 2004: Select or reset driver information display using Menu button on center console, to the right of the parking brake handle.

On-board diagnostics

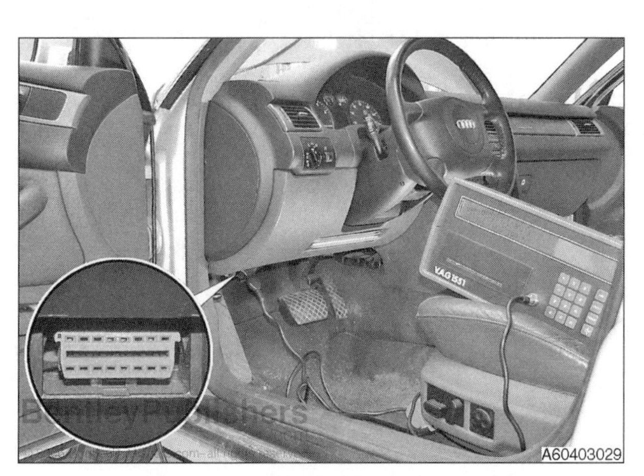

If the on-board diagnostics (OBD II) system recognizes instrument cluster malfunctions relating to odometer or speedometer calibration, the trip odometer displays dEF.

Before performing troubleshooting or inspection, access fault memory with VAG scan tool or equivalent plugged into DLC plug (**inset**) under left side dashboard.

BentleyPublishers.com
BentleyPublishers.com—All Rights Reserved

Cautions

> ***CAUTION—***
> - *Switch the ignition OFF and remove the negative (-) battery cable before removing any electrical components.*
> - *Connect and disconnect electrical connectors and ignition test equipment leads only while the ignition is OFF.*
> - *Only use a digital multimeter for electrical tests.*
> - *To avoid damaging plastic interior trim, use a plastic prying tool or a screwdriver with the tip wrapped with masking tape.*

Instrument Cluster Repairs

Instrument cluster, removing and installing

It is not necessary to remove steering wheel to remove instrument cluster.

– If replacing cluster with new, note odometer reading on old cluster.

– Adjust steering wheel backward and down.

– Switch ignition OFF. Disconnect battery negative cable.

> ***CAUTION—***
> - *Prior to disconnecting the battery, read the battery disconnection precautions in* **27 Battery, Alternator, Starter**.

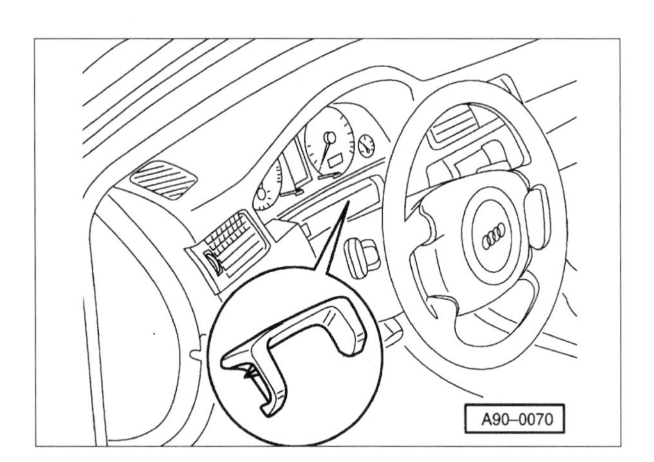

 Press in sides of instrument cluster lower trim (**insert**) to disengage from dashboard. Lift out trim.

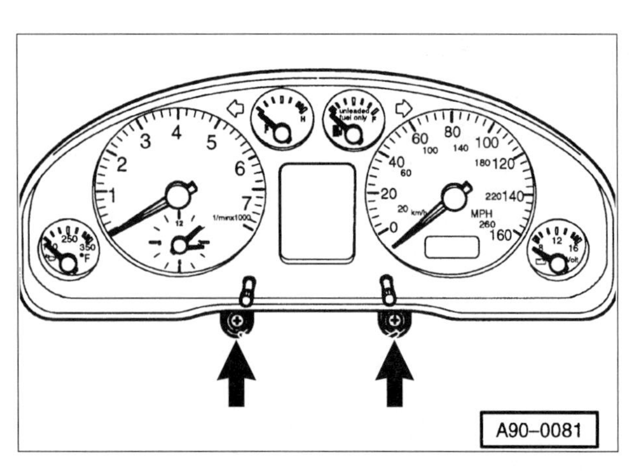

Remove cluster mounting screws (**arrows**)

– Tilt instrument cluster rearward and cut wire ties at rear. Unclip connectors securing latches and detach connectors.

– Installation is reverse of removal. If new cluster is installed, use VAG scan tool or equivalent for coding and adaptation functions:
  - Code new cluster to vehicle version and equipment level.
  - Adapt odometer reading.
  - Adapt electronic immobilizer functions to engine control module (ECM).

BentleyPublishers.com
BentleyPublishers.com—All Rights Reserved

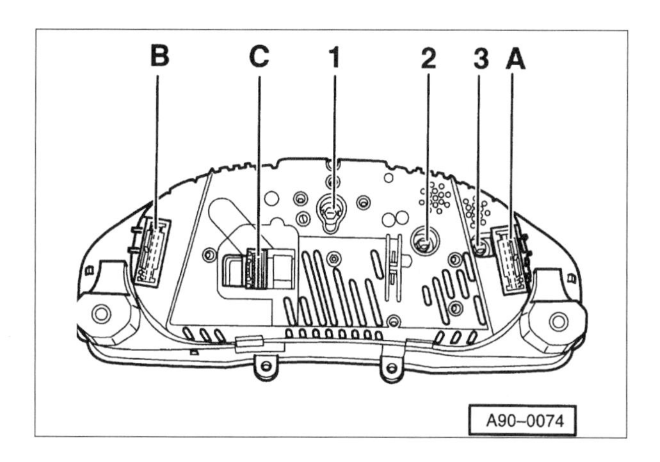

Instrument cluster details

◄ 1998 - 1999: Lights **1**, **2**, **3** are 1.2 W replaceable bulbs:

1. High beam indicator
2. ASR / ESP indicator
3. Malfunction indicator light (MIL)

A. 32-pin connector, blue, basic functions
B. 32-pin connector, green, additional functions
C. 20-pin connector, red, driver information screen

– All other indicator lights are LEDs. In case of failure, replace instrument cluster.

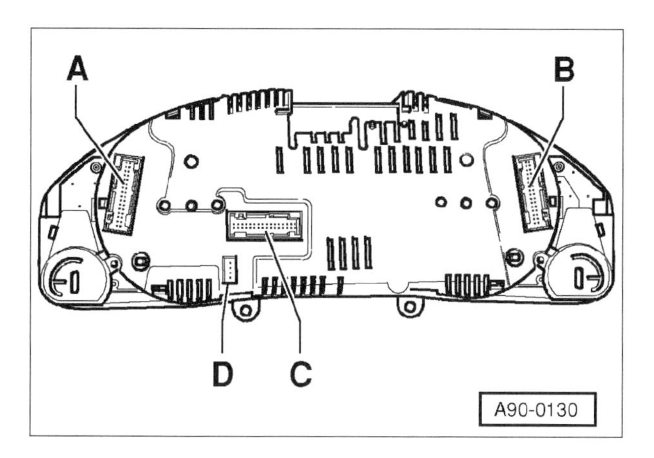

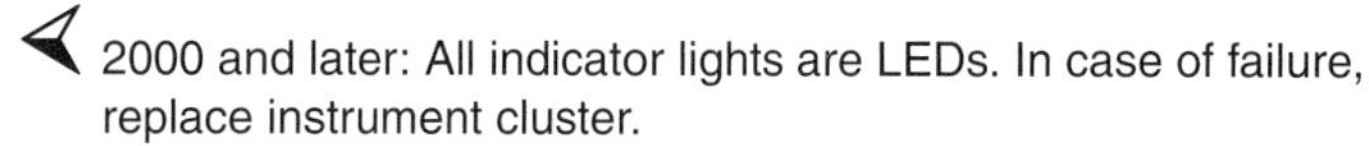

◄ 2000 and later: All indicator lights are LEDs. In case of failure, replace instrument cluster.

A. 32-pin connector, green, basic functions
B. 32-pin connector, blue, additional functions
C. 32-pin connector, grey, driver information screen
D. 4-pin connector, black, remote clock (not connected in US)

BentleyPublishers.com
BentleyPublishers.com—All Rights Reserved

91 Radio and Communication

Concert radio

A60491001

Symphony radio

A60491002

GENERAL

This repair group covers Audi A6 factory installed sound systems. Also described are the optional telephone, Telematics, navigation system and multifunction steering wheel.

The basic and premium Audi A6 sound systems center on the Concert radio and the Symphony radio. The radio is coded to the vehicle using VAG scan tool or equivalent. Coding is based on other equipment installed such as telephone, CD changer, navigation or amplifier.

If you plan to install an aftermarket system, see **Cautions** in this repair group.

See also **EWD Electrical Wiring Diagrams**.

BentleyPublishers.com
BentleyPublishers.com—All Rights Reserved

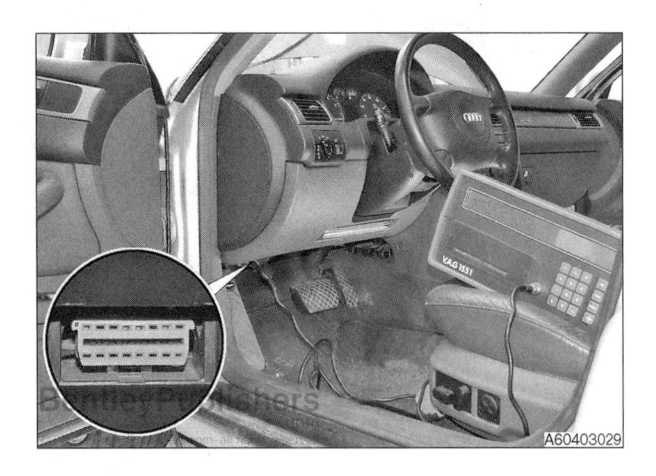

On-board diagnostics

If the on-board diagnostics (OBD II) system recognizes malfunctions in the radio or communication systems, it sets diagnostic fault codes (DTCs).

Access fault memory with VAG scan tool or equivalent plugged into DLC plug (**inset**) under left side dashboard.

Cautions

Read the following cautions and the additional warnings and cautions in **97 Fuses, Relays, Component Locations**before doing any work on your electrical system.

CAUTION—

- *The radio is wired to the vehicle alarm system. If the alarm system is armed, removing the radio activates the alarm even if the proper tools are used. Do not attempt to remove the radio without disarming the alarm.*
- *The factory installed connectors are designed for the genuine Audi radio. If installing a different radio, remember that it may not fit properly into the space provided, the electrical connections may not be compatible and different terminals may be needed.*
- *The factory installed radio is electronically linked to the instrument cluster, ignition, alarm and data link circuits.*
- *Prior to disconnecting the battery, read the battery disconnection precautions in* **27 Battery, Alternator, Starter**.
- *Switch the ignition OFF and remove the negative (-) battery cable before removing any electrical components.*
- *Connect and disconnect electrical connectors and ignition test equipment leads only while the ignition is OFF.*
- *Only use a digital multimeter for electrical tests.*
- *To avoid damaging plastic interior trim, use a plastic prying tool or a screwdriver with the tip wrapped with masking tape.*

BentleyPublishers.com
BentleyPublishers.com—All Rights Reserved

RADIO

Audi A6 sound system components and options are as follows:

- **Basic** 140 watt ten speaker system
- **Premium** 200 watt ten speaker system
- **Concert** in-dash radio
- **Symphony** double-height (double-DIN) radio
- **Bass, midrange and treble speakers** in doors
- **Subwoofer** behind trunk or luggage compartment right side trim
- **Bose® amplifier** behind trunk or luggage compartment right side trim
- **CD changer** behind trunk or luggage compartment left side trim
- **Satellite radio** behind trunk or luggage compartment left side storage box
- **Antenna**
  -Sedan: Rear window
  -Avant: Side rear windows
- **Antenna amplifiers**
  -Sedan: Behind left and right D-pillar trim
  -Avant: Below left and right luggage compartment side trim
- **Antenna selection control module**
  -Sedan: Underneath rear parcel shelf trim
  -Avant: Under headliner near roof antenna
- **Satellite radio antenna** on roof

Radio power supply

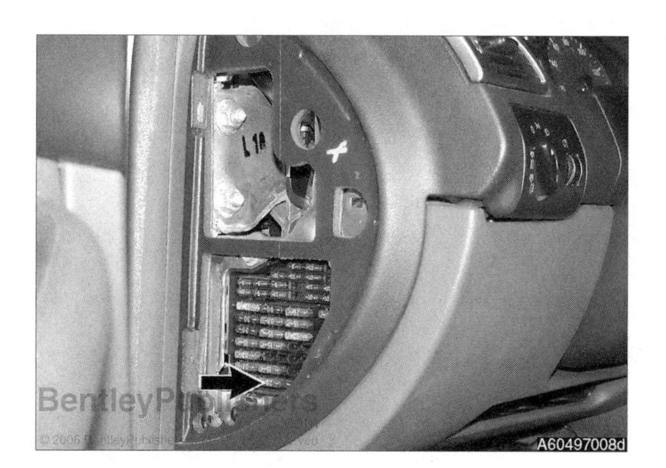

In main fuse panel, left end of dashboard.

Sound system fuse rating	
Fuse 37	20 A

BentleyPublishers.com
BentleyPublishers.com—All Rights Reserved

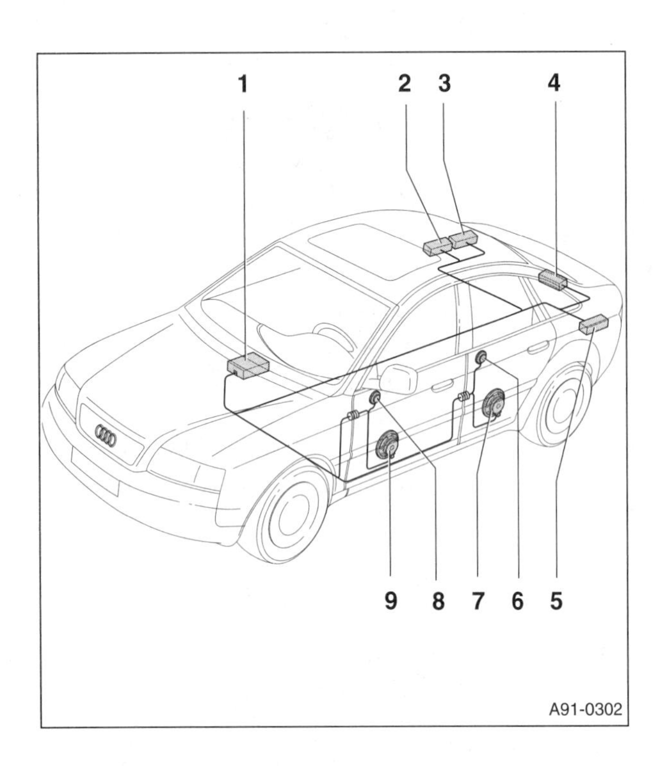

Sound system components, sedan (1998 - 2001)

1. Radio (R) in center dashboard
2. Subwoofer (R100) in right rear trunk behind trim
3. Amplifier (R12) in right rear trunk behind trim, Bose®
4. Antenna amplifier behind left D-pillar trim
5. CD changer (R41)
6. Rear treble / midrange speakers (R34, R35) in rear door upper trim
7. Rear bass speakers (R15, R17) in rear door lower trim
8. Front treble / midrange speakers (R26, R27) in front door upper trim
9. Front bass speakers (R21, R23) in front door lower trim

NOTE—
- *Audi identifies electrical components by a letter and/or a number in the electrical schematics. See* **EWD Electrical Wiring Diagrams**. *These electrical identifiers are listed in parentheses as an aid to electrical troubleshooting.*
- *Avant (station wagon) sound system layout and components are similar.*

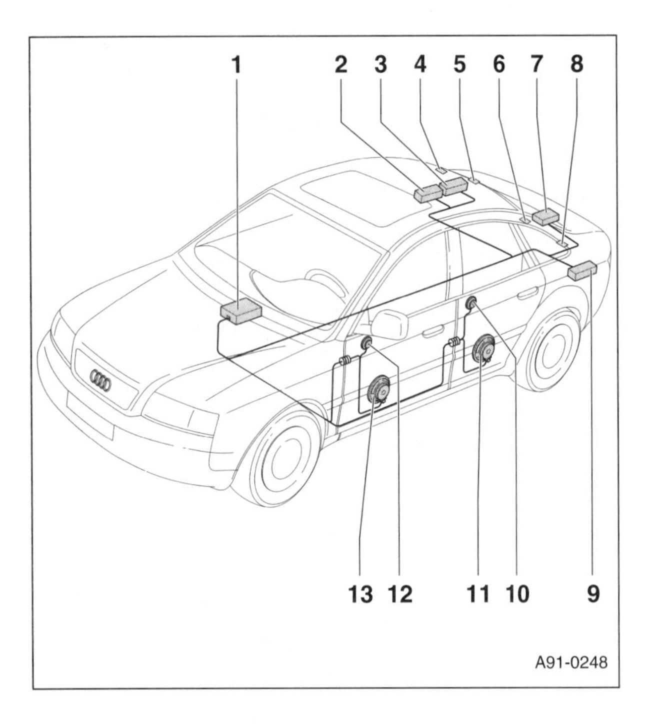

Sound system components, sedan (2002 - 2004)

1. Radio (R) in center dashboard
2. Subwoofer (R100) in right rear trunk behind trim
3. Amplifier (R12) in right rear trunk behind trim, Bose®
4. Antenna amplifier (R82) behind right upper D-pillar trim
5. Antenna amplifier (R83) behind right lower D-pillar trim
6. Antenna amplifier (R85) behind left upper D-pillar trim
7. Antenna selection control module (J515) underneath left side parcel shelf
8. Antenna amplifier (R84) behind lower D-pillar trim
9. CD changer (R41)
10. Rear treble / midrange speakers (R34, R35) in rear door upper trim
11. Rear bass speakers (R15, R17) in rear door lower trim
12. Front treble / midrange speakers (R26, R27) in front door upper trim
13. Front bass speakers (R21, R23) in front door lower trim

NOTE—
- *Avant (station wagon) sound system layout and components are similar.*

BentleyPublishers.com
BentleyPublishers.com—All Rights Reserved

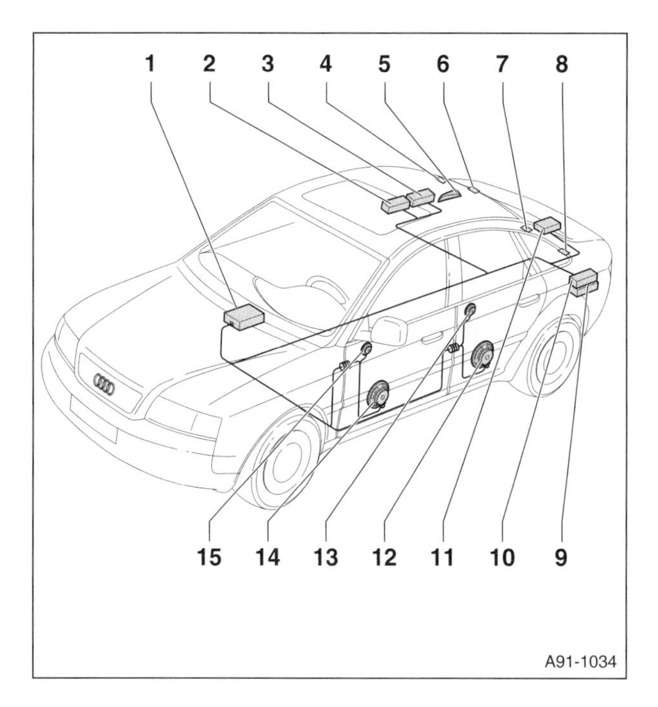

Sound system with satellite radio (SDARS), sedan

1. Radio (R) in center dashboard
2. Subwoofer (R100) in right rear trunk behind trim
3. Amplifier (R12) in right rear trunk behind trim, Bose®
4. Antenna amplifier (R82) behind right upper D-pillar trim
5. Satellite antenna (R182) on roof
6. Antenna amplifier (R83) behind right lower D-pillar trim
7. Antenna amplifier (R85) behind left upper D-pillar trim
8. Antenna amplifier (R84) behind left lower D-pillar trim
9. CD changer (R41)
10. Satellite tuner (R146)
11. Antenna selection control module (J515) underneath left side parcel shelf
12. Rear bass speakers (R15, R17) in rear door lower trim
13. Rear treble / midrange speakers (R34, R35) in rear door upper trim
14. Front bass speakers (R21, R23) in front door lower trim
15. Front treble / midrange speakers (R26, R27) in front door upper trim

NOTE—
- *Avant (station wagon) sound system layout and components are similar.*

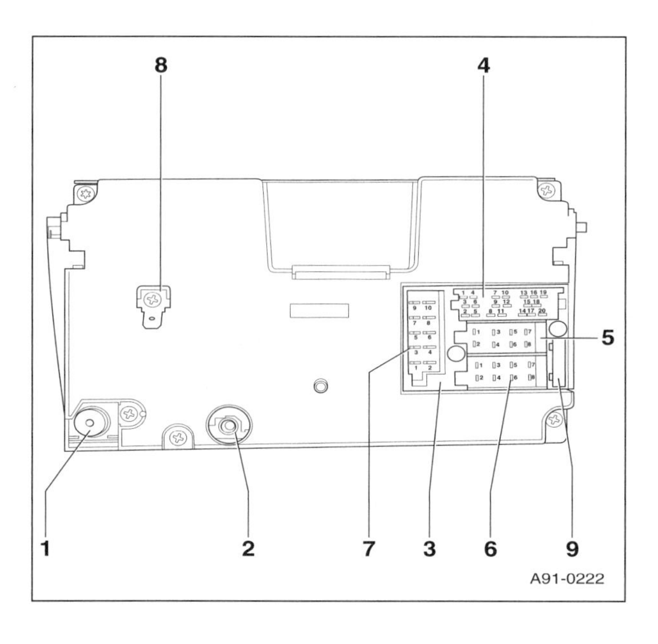

Radio connections

At rear of radio:

1. Antenna connection socket for radio (HF wire)
2. ZF-wire
3. Radio switch block
4. Connector 1, 20-pin, black
5. Connector 2, 8-pin, brown
6. Connector 3, 8-pin, black
7. Connector 4, 10-pin, red
8. Anti-theft alarm system ground (1998 - 2001)
9. Fuse

BentleyPublishers.com
BentleyPublishers.com—All Rights Reserved

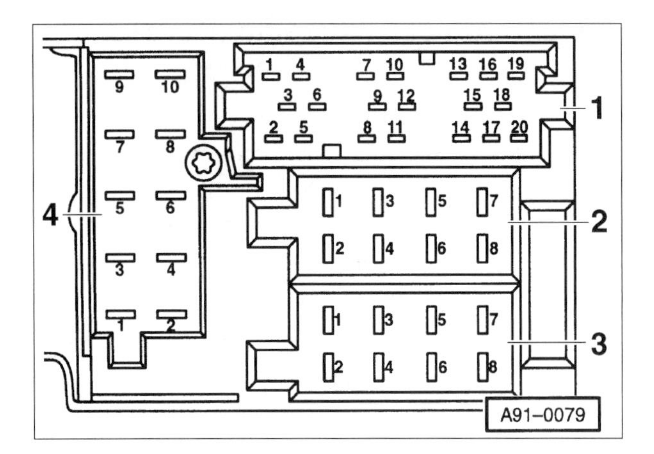

Radio pin assignments, basic

Connector 1, 20-pin, black

1. Line output, left rear
2. Line output, right rear
3. Line output, ground
4. Line output, left front, Bose®
5. Line output, right front, Bose®
6. Switched power for speakers
   Switched power for amplifier, Bose®
7. CAN-bus high (infotainment), 2002 - 2004
8. Radio clock, 1998 - 2001
9. Radio frequency data signal, 1998 - 2001
10. Radio frequency display enable signal, 1998 - 2001
    CD ground, 2002 - 2004
11. Steering wheel remote control, 1998 - 2001
12. CAN-bus low (infotainment), 2002 - 2004
13. CD display signal in
14. CD display signal out
15. CD clock
16. Power supply (B+), CD changer
17. Switched power supply, CD changer
18. Ground, CD changer
19. Line output, left, CD changer
20. Line output, right, CD changer

Connector 2, 8-pin, brown

1. Right front speaker signal
2. Right front speaker ground
3. Left front speaker signal
4. Left front speaker ground

Connector 3, 8-pin, black

1. GALA (vehicle speed signal), 1998 - 2001
   Bose® coding, 2002 - 2004
2. Telephone muting, 1998 - 2001
   Anti-theft ground, 2002 - 2004
3. K or L line, on-board diagnostics
4. Radio switching connection to ignition switch, 1998 - 2001
5. Automatic antenna control, 1998 - 2001
6. Switch illumination 1998 - 2001
7. Power supply (B+)
8. Ground

BentleyPublishers.com
BentleyPublishers.com—All Rights Reserved

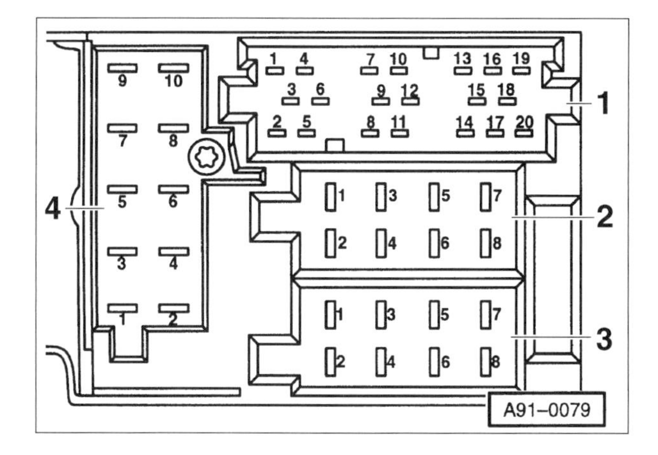

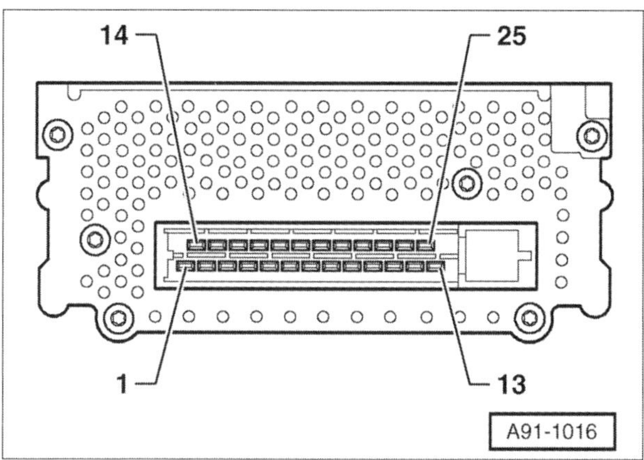

Connector 4, 10 pin, red

1. Telephone muting
2. Terminal 15, navigation only, 1998 - 2001
3. Telephone (NF+), without navigation
4. Telephone (NF-), without navigation
5. Navigation (NF+)
6. Navigation (NF-)
7. Navigation control wire
8. Not used
9. Display illumination, terminal 58d, 1998 - 2001
10. CD changer ground, 1998 - 2001

Bose® amplifier pin assignments

1. Not used
2. Not used
3. Not used
4. Not used
5. Not used
6. Line input from radio, ground
7. Line input from radio, right rear
8. Line input from radio, left rear
9. Line input from radio, right front
10. Line input from radio, left front
11. Not used
12. Not used
13. Power (B+)
14. Not used
15. Not used
16. Left front speaker ground
17. Left front speaker power
18. Right front speaker power
19. Right front speaker power
20. Left rear speaker power
21. Left rear speaker ground
22. Right rear speaker power
23. Right rear speaker ground
24. Switched power

BentleyPublishers.com
BentleyPublishers.com—All Rights Reserved

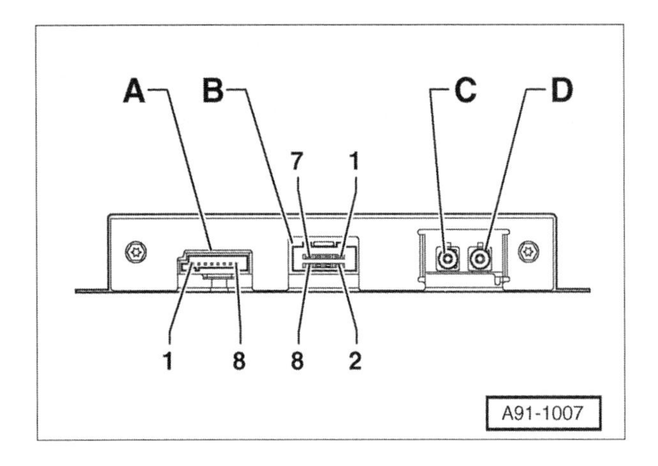

Satellite radio pin assignments

A. Connector, 8-pin:
1. CAN-bus high (infotainment)
2. CAN-bus low (infotainment)
3. CD in, left
4. CD in, right
5. Ground NF (radio)
6. Left channel NF
7. Right channel NF
8. Ground (CD)
B. Connector, 8-pin:
1. Power
2. Ground
C. Satellite antenna, brown
D. Satellite antenna, green

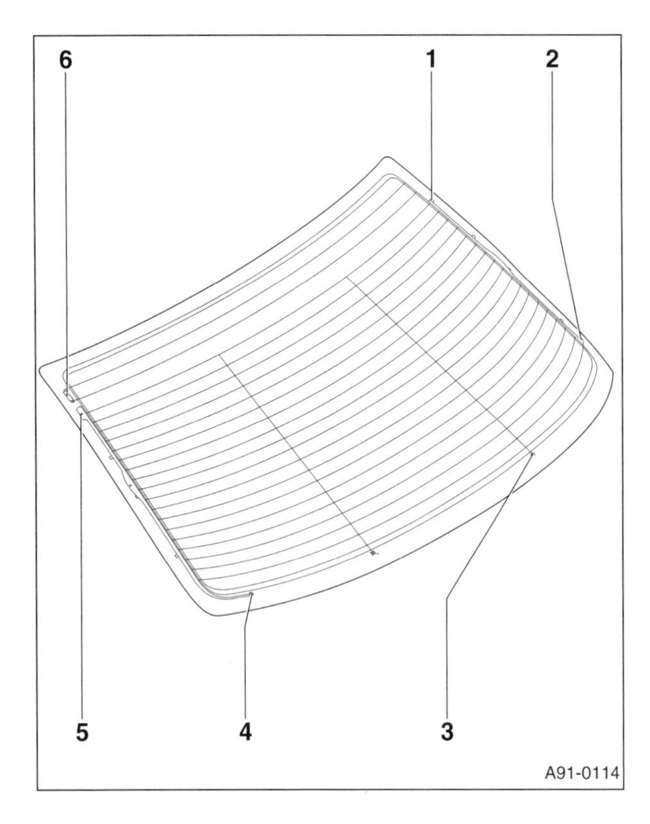

Rear window antenna connections (sedan)

1. FM antenna 1
2. FM antenna 2
3. Auxiliary heater antenna (if applicable)
4. FM antenna 3
5. FM antenna 4
6. AM antenna

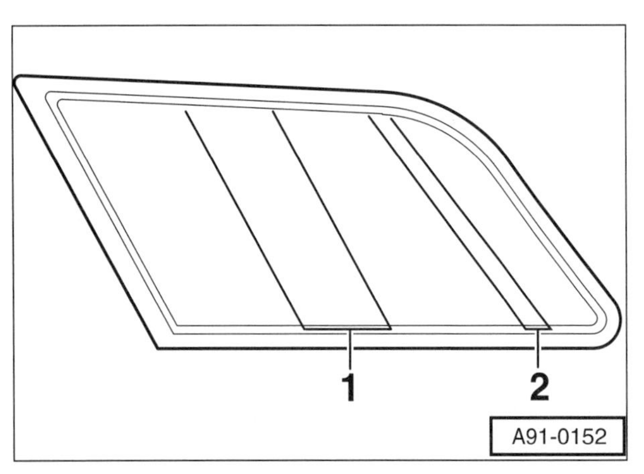

Side window antenna connection (Avant)

On Avant (station wagon), FM antennas 1 - 3 are in the side rear windows:

1. FM antenna 1 on right rear window
1. FM antenna 2 on left rear window
2. FM antenna 3 on left rear window

FM antenna 4 and AM antenna are on roof.

If equipped with satellite radio (SDARS) with antenna on roof, the following side window configuration is used:

1. FM antenna 1 on right rear window
2. FM antenna 2 on right rear window
1. FM antenna 4, AM antenna on left rear window
2. FM antenna 3 on left rear window

BentleyPublishers.com
BentleyPublishers.com—All Rights Reserved

Radio, removing and installing

- Obtain radio anti-theft security code.
- Switch ignition OFF. Disconnect battery negative cable.

> ***CAUTION—***
> • *Prior to disconnecting the battery, read the battery disconnection precautions in* **27 Battery, Alternator, Starter**.

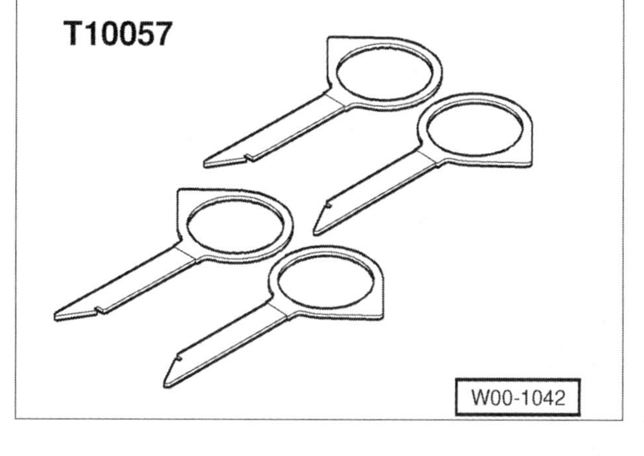

◀ Use VAG special tools for radio removal:

- VAG T10057 consists of 4 prying hooks for Symphony radio removal.
- VAG 3344 consists of 2 prying hooks for Concert radio removal

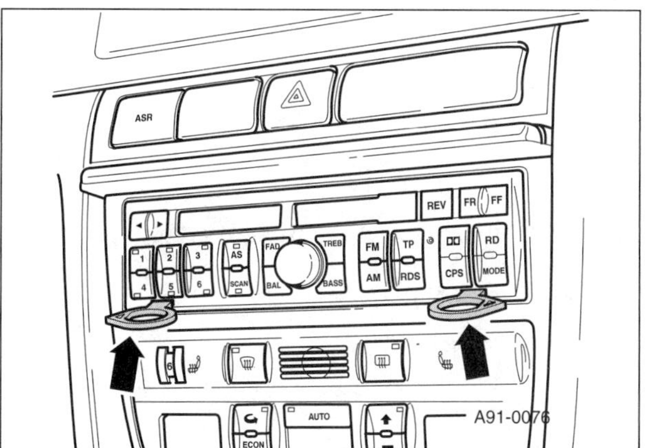

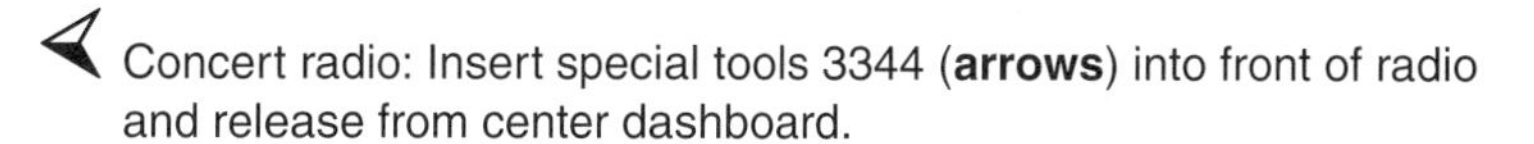

◀ Concert radio: Insert special tools 3344 (**arrows**) into front of radio and release from center dashboard.

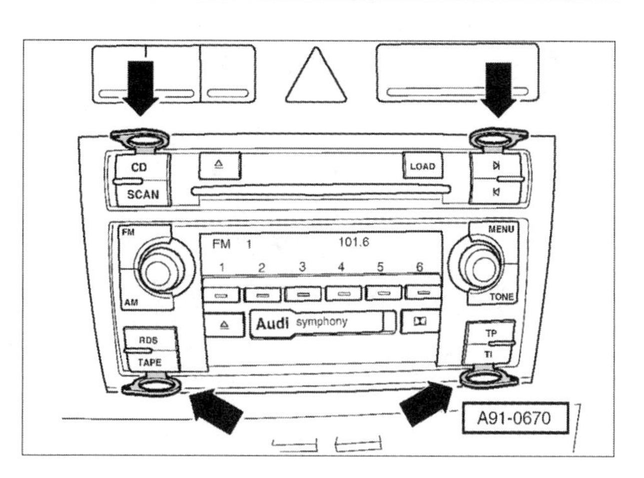

◀ Symphony radio: Insert special tools T10057 (**arrows**) into upper and lower slots in front of radio and release from center dashboard.

- Pull radio out and detach connectors and antenna cable.
- Prior to reinstalling radio, remove special tools from panel.
- Attach electrical connections and antenna.
- Slide radio into center dashboard frame until it engages firmly.
- Reconnect battery and activate anti-theft coding.

BentleyPublishers.com
BentleyPublishers.com—All Rights Reserved

TELEPHONE

The telephone is an option supplied in two versions:

- **Hands-free telephone**, 1998 - 2001
- **Digital cellular telephone**, 2002 - 2004, used in conjunction with Telematics

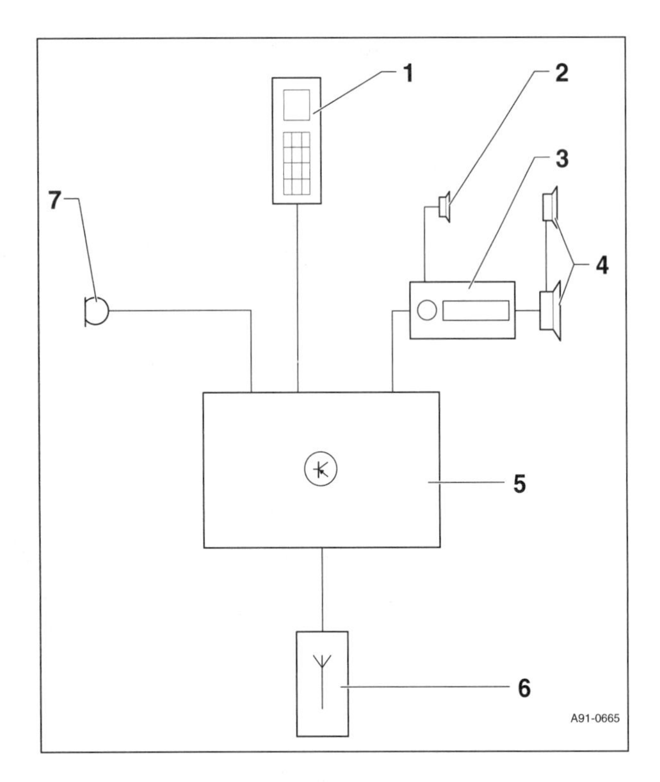

Hands-free telephone components

1. Telephone handset (R37)
2. Telephone-navigation speaker (R39) behind left B-pillar lower trim (Bose® system)
3. Radio (R)
4. Front speakers behind front door panels
5. Telephone transceiver (R36)
   -1998: Under rear parcel shelf
   -1999 - 2004: in trunk or luggage compartment left side storage compartment
6. Antenna
   -Sedan: In rear window
   -Avant: On roof
7. Telephone microphone (R38)
   -1998 - 1999: In left A-pillar
   -2000 - 2004: In front interior light module

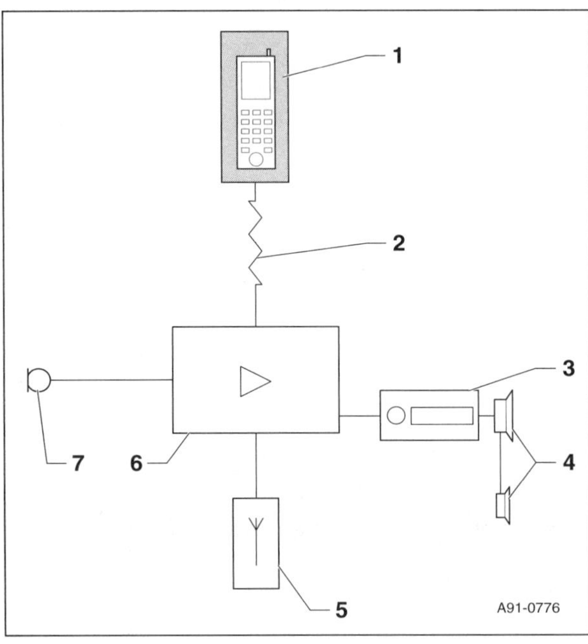

Digital cellular telephone components

1. Telephone handset docking station (R54)
2. Spiral cord
3. Radio (R)
4. Front speakers behind front door panels
5. Telephone antenna (R65) in rear bumper
6. Telematics-telephone control module (J526) in right front footwell
7. Telephone microphone (R38)

BentleyPublishers.com
BentleyPublishers.com—All Rights Reserved

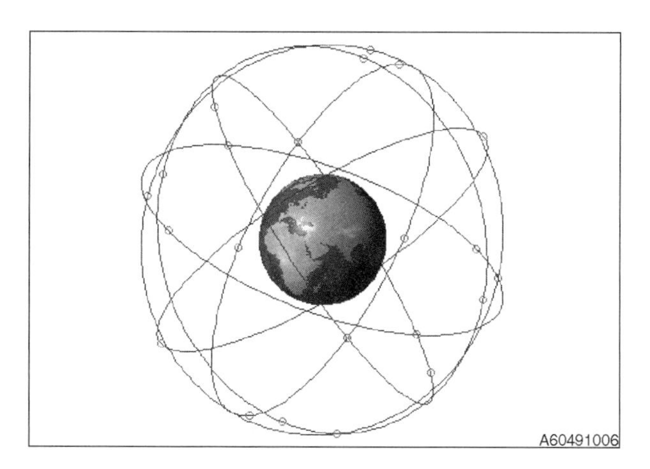

Navigation

Vehicle navigation systems utilize the signal from 24 global positioning system (GPS) satellites.

The navigation system is available as an option in Audi A6 vehicles. Two versions of the system are used;

- **Navigation III**, 1998 - 2001:
  -Display in instrument cluster driver information screen
  -Data entries via multifunction switch on center console
  -Control module with CD reader in trunk or luggage compartment
  -Navigation antenna on roof
- **Navigation IV**, 2002 - 2004, has similar components but offers the following additional functions and features:
  -Dynamic route guidance
  -Integration with Audi Telematics
  -Compass
  -Help function

The navigation system is integrated with the vehicle CAN-bus network. The dynamic route guidance feature in Navigation IV enables local traffic report information to be accessed from either the radio or the telephone.

The navigation system is self-diagnosing. DTCs are stored in the navigation control module.

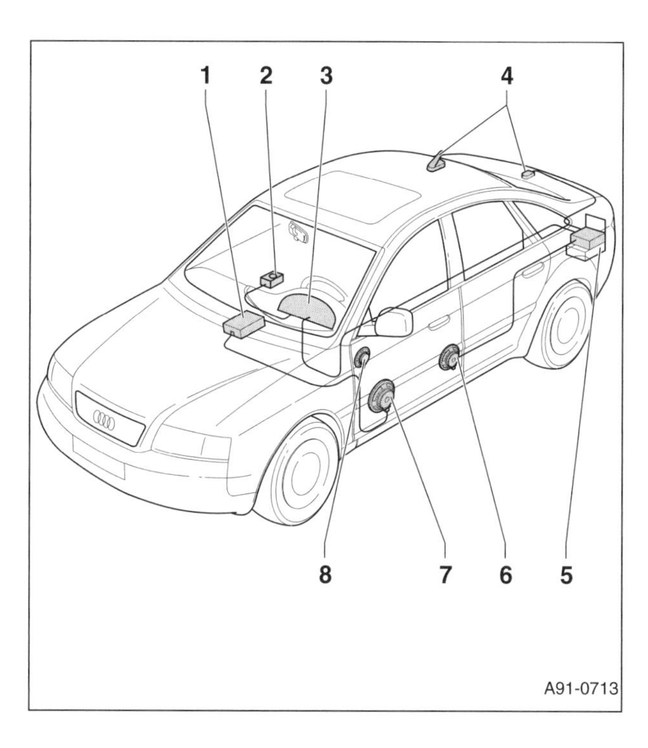

Navigation system components

1. Radio (R) in center dashboard
2. Function selector switch (E272) in center console
3. Instrument cluster driver information display
4. Navigation antenna (R50)
   Radio-telephone-navigation antenna (R52)
   1998 - 2001: On trunk lid
   2002 - 2004: On roof
5. Navigation control module with CD reader (J401) in trunk or luggage compartment left side storage compartment
6. Telephone-navigation speaker (R39) in B-pillar trim (vehicle with Bose® sound system)
7. Bass speaker in front door lower trim
8. Tweeter / midrange speaker in front door upper trim

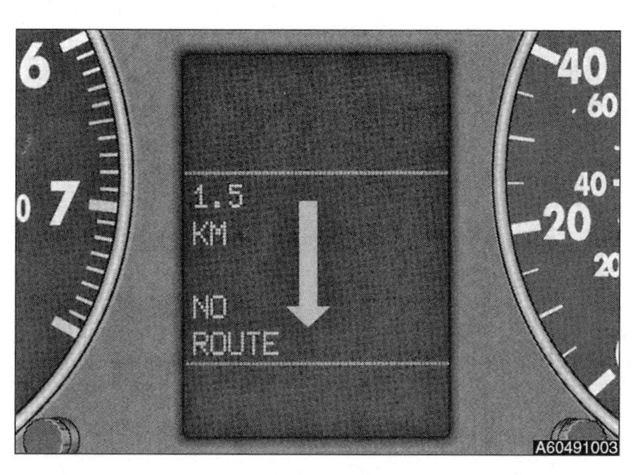

Navigation display

The driver information display at center of instrument cluster serves as navigation display.

BentleyPublishers.com
BentleyPublishers.com—All Rights Reserved

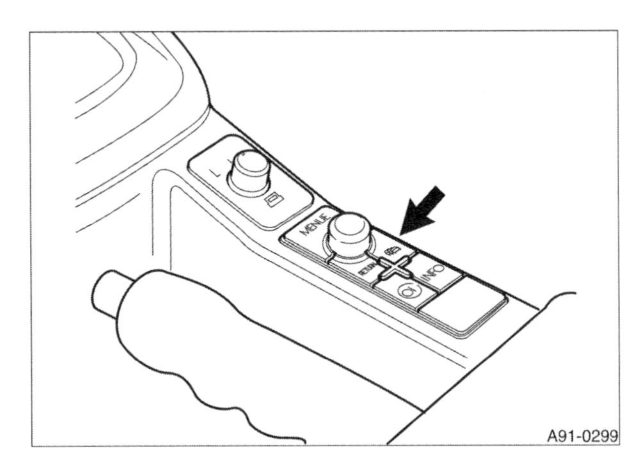

Navigation function selector

 The function selector switch (**arrow**) is on center console, to right of parking brake handle.

Navigation control module

Navigation control module with CD reader (**arrow**) is in storage compartment at left side of trunk or luggage compartment.

TELEMATICS

Telematics is an optional system that uses cellular, internet and global positioning system (GPS) technologies to provide the driver with safety, security and convenience services.

The Telematics-telephone control module receives GPS data, decodes it, and transmits vehicle location to the service provider OnStar® call center via cellular network. OnStar® service plans may provide route directions, concierge services or emergency vehicle repair services.

If the airbag control module detects a crash, vehicle emergency flashers are activated and doors are unlocked via door control modules. In addition, the Telematics system automatically places an emergency call to the OnStar® call center.

An emergency (back-up) battery for the Telematics control module is used in case the vehicle battery goes dead or becomes disconnected.

BentleyPublishers.com
BentleyPublishers.com—All Rights Reserved

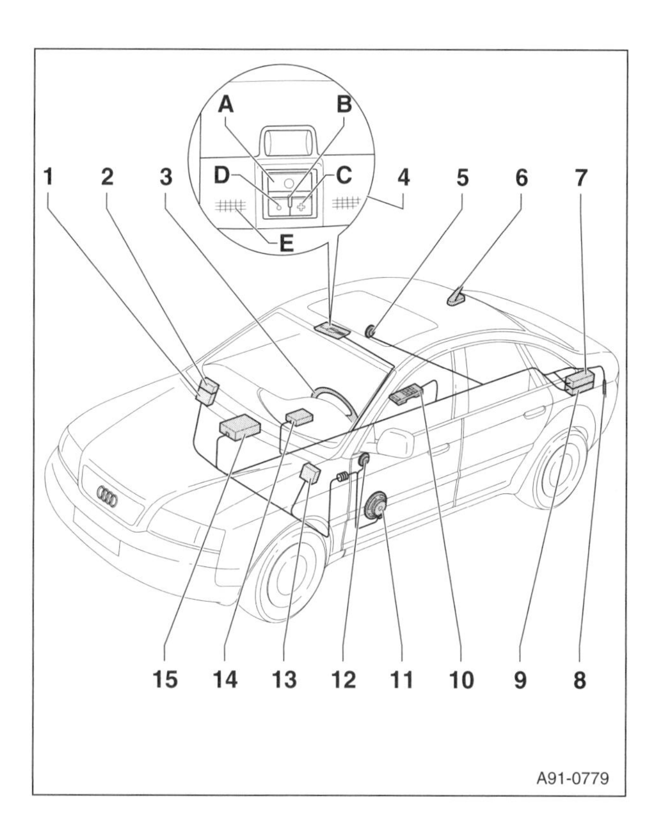

Telematics components

1. Emergency battery (A16) on Telematics-telephone control module in right front footwell
2. Telematics-telephone control module (J526) in right front footwell
3. Multifunction steering wheel
4. Front interior light module with Telematics control head (E264)

A. OnStar® button

B. System status LED

C. Emergency button

D. Communication button

E. Telephone microphone (R38)

5. Telematics speaker (R91) behind right B-pillar lower trim
6. Navigation-Telematics antenna (R50, R90) on roof
7. Antenna splitter (R87) in trunk or luggage compartment left storage compartment
8. Telephone antenna (R65) in rear bumper
9. Navigation control module with CD reader (J401) in trunk or luggage compartment left side storage compartment
10. Cellular telephone (R54) in center console (optional with Telematics)
11. Front bass speakers (R21, R23) behind front door lower trim
12. Front midrange speakers (R26, R27) behind front door upper trim
13. Multifunction steering wheel control module (J453)
14. Instrument cluster
15. Radio

MULTIFUNCTION STEERING WHEEL

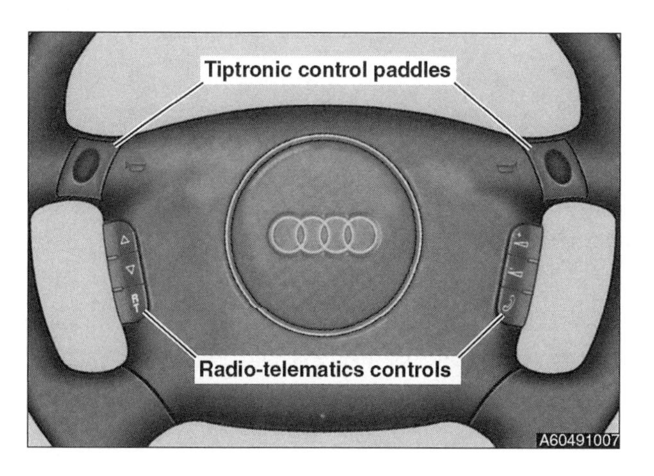

Switches and buttons integrated into the optional multifunction steering wheel allow the driver to operate the radio, telephone and Tiptronic controls without taking hands off the steering wheel. In addition, the steering wheel is electrically heated.

Integrated electronics in the steering wheel regulate steering wheel heating and transmit switch operations to the multifunction steering wheel control module via vehicle CAN-bus network. The control module controls radio, navigation, telephone, instrument cluster, steering wheel heating and Tiptronic functions.

> ***WARNING —***
> - *The steering wheel houses the driver airbag. An airbag is an explosive device and part of the vehicle safety system. Prior to working on the steering wheel or its components, read the airbag precautions in* **69 Seat Belts, Airbags**.

BentleyPublishers.com
BentleyPublishers.com—All Rights Reserved

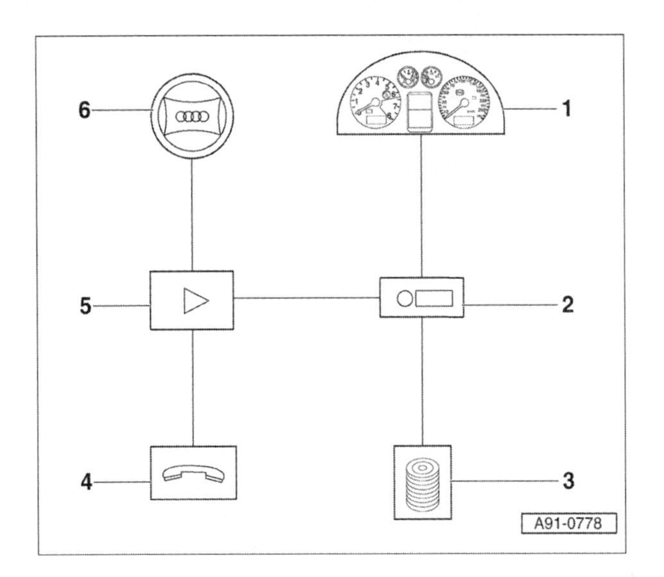

Multifunction steering wheel, 1998 - 2001

1. Instrument cluster
2. Radio (R)
3. CD changer (R41) in trunk or luggage compartment left storage compartment
4. Telephone transceiver (R36)
5. Multifunction steering wheel control module (J453) in 13-fold relay panel, underneath left side dashboard
6. Multifunction steering wheel

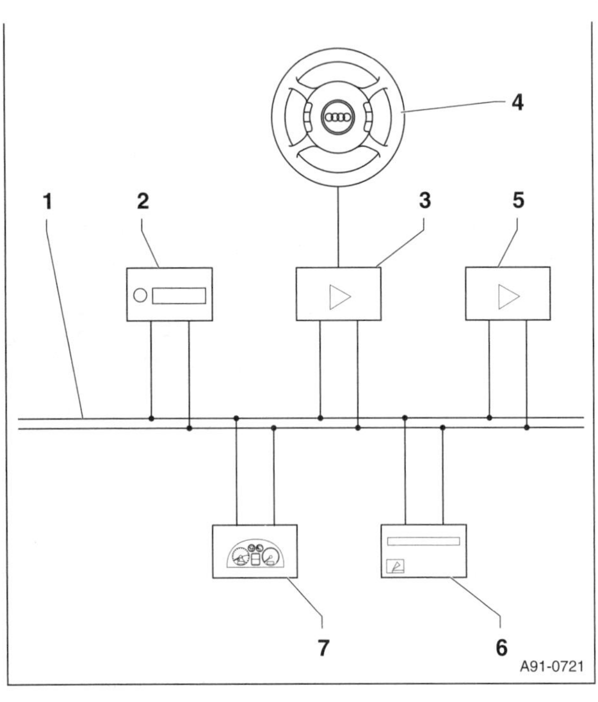

Multifunction steering wheel, 2002 - 2004

1. CAN-bus (infotainment)
2. Radio (R)
3. Multifunction steering wheel control module (J453) in 13-fold relay panel, underneath left side dashboard
4. Multifunction steering wheel
5. Telematics-telephone control module (J526) in right front footwell
6. Navigation control module with CD reader (J401) in trunk or luggage compartment left side storage compartment
7. Instrument cluster

Multifunction steering wheel control module (J453)

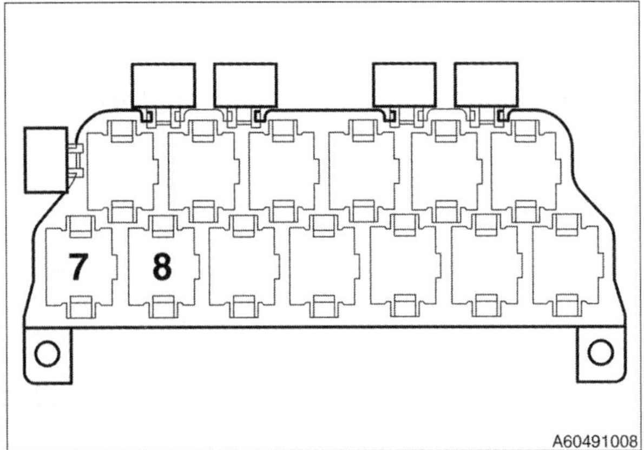

The multifunction steering wheel control module is a double-relay in positions 7 and 8 of the 13-fold relay panel under left side dashboard.

BentleyPublishers.com
BentleyPublishers.com—All Rights Reserved

92 Wipers and Washers

GENERAL

This repair group covers wiper blade service and wiper assembly repairs. Also covered are the windshield, rear window and headlight washer assemblies.

Also see:

- **48 Steering** for steering column stalk switch (wiper and washer switch) replacement
- **97 Fuses, Relays, Component Locations**
- **EWD Electrical Wiring Diagrams**

Cautions

CAUTION—
- *Switch the ignition OFF and disconnect the negative (-) battery cable before removing any electrical components.*
- *Prior to disconnecting the battery, read the battery disconnection precautions in* **00 Warnings and Cautions**.
- *Prior to disconnecting the battery:*
  -Obtain radio code.
  -Record radio presets.
- *Only use a digital multimeter for electrical tests.*
- *Do not operate the windshield wipers with the engine compartment lid open. The wiper arms may scrape.*

BentleyPublishers.com
BentleyPublishers.com—All Rights Reserved

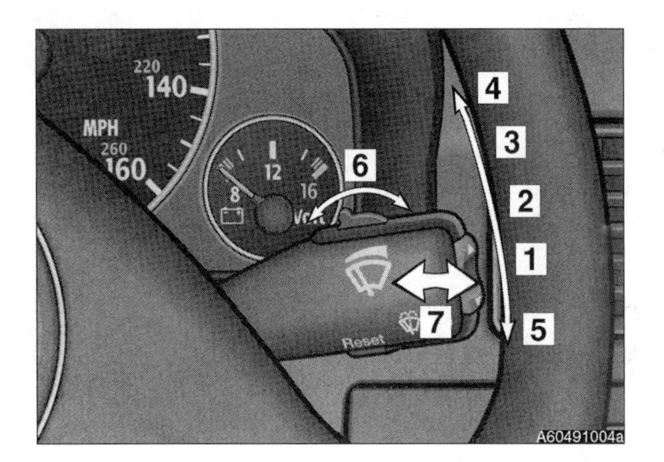

A60491004a

Wiper and Washer Controls

Wiper and washer switch

Switch positions are as follows:

1. Wipers parked
2. Intermittent wipe
3. Slow wipe
4. Fast wipe
5. One-tap wipe
6. Intermittent wiper speed adjuster
7. Pull back: Automatic windshield wiper and washer function
   Push forward: Automatic rear window wiper and washer

Wiper and washer stalk switch replacement is covered in **48 Steering**.

Wiper and washer system fuses

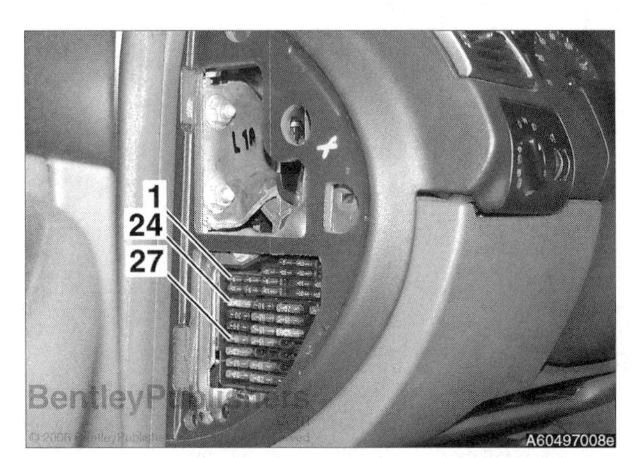

A60497008e

In main fuse panel, left end of dashboard.

Wiper and washer fuses		
Protected circuit	**Fuse #**	**Amperage**
Washer nozzle heater	1	5
Front wiper / washer	24	25
Rear wiper / washer	27	15

Wiper and washer system relay

A60497002e

In micro central electric panel, underneath left side dashboard.

NOTE—

- *To gain access to relays, see* **97 Fuses, Relays, Component Locations**.

Wiper and washer system relay	
Intermittent wiper / washer relay (J31)	Relay positions 5 and 6, micro central electric panel

BentleyPublishers.com
BentleyPublishers.com—All Rights Reserved

WIPER ARMS AND BLADES

Wiper blade cleaning problems

Common problems with the windshield wipers include streaking or sheeting, water drops after wiping, and blade chatter.

Streaking is usually caused when wiper blades are coated with road film or car wash wax.

– Clean blades using soapy water. If cleaning does not cure problem, replace blades.

Drops that remain behind after wiping are usually caused by oil, road film, or diesel exhaust residue on the glass.

– Use an alcohol or ammonia solution or other non-abrasive cleaner to clean windshield.

Chatter may be caused by dirty or worn blades, or by wiper arms that are out of alignment.

– Clean blades and windshield as described above.

– Bend wiper arm so that there is even pressure along blade, and so that blade is perpendicular to windshield at rest.

– If problems persist, replace blades and wiper arms.

Wiper blade, replacing

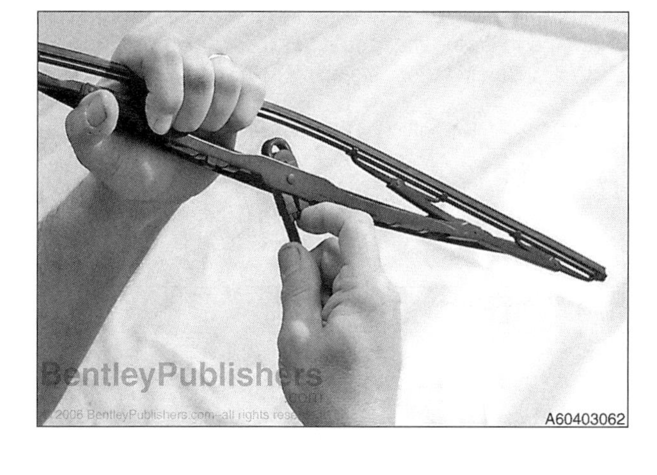

– Fold wiper arm away from windshield.

< Remove wiper blade:

• Rotate blade on arm approximately ¼ turn.

• Squeeze plastic retainer and slide blade off arm.

> ***CAUTION—***
> *• Do not allow wiper arm to snap back against windshield*

– Installation is reverse of removal.

Wiper arm, removing and installing

– Make sure wipers are in PARK position and ignition switch is OFF. Open engine hood.

< Gently pry off plastic trim plugs (**arrows**) from base of wiper arm.

– Remove wiper arm mounting nuts.

– On each side, mark position of wiper arm on shaft to aid installation.

– Remove wiper arms.

– Installation is reverse of removal. Tighten mounting nuts after adjusting position of arms.

NOTE—
• The right wiper arm is longer.

BentleyPublishers.com
BentleyPublishers.com—All Rights Reserved

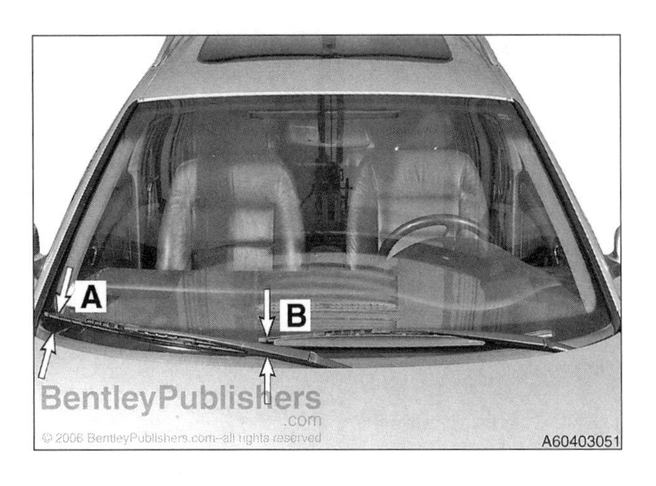

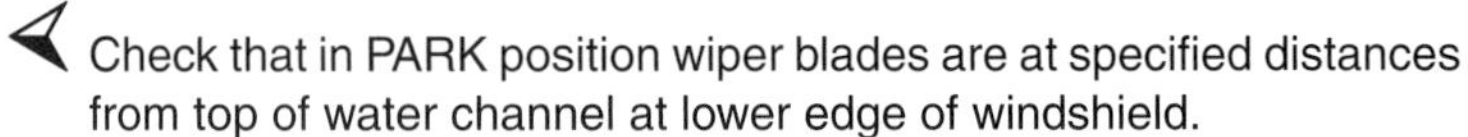

Check that in PARK position wiper blades are at specified distances from top of water channel at lower edge of windshield.

Wiper arm position at PARK	
Distance from wiper blade tip to lower edge of windshield: • Right wiper (**A**) • Left wiper (**B**)	 approx. 12 mm (½ in) approx. 25 mm (1 in)

– If necessary, loosen wiper arm mounting nut, adjust arm and retighten mounting nut.

Tightening torque	
Wiper arm to shaft (M8)	16 Nm (12 ft-lb)

Wiper blade contact angle, adjusting

Possible causes for wiper blades skipping or chattering are as follows:

- Windshield is scratched.
- Wiper blade rubber has become detached from arm or is split.
- Wiper arm or blade is loose or bent.
- Wiper blades are coated with wax or road grime.

If wiper blades are skipping or chattering and none of the causes listed apply, use the following procedure to adjust angle of wiper arms relative to windshield before installing new wiper blades.

– Move wiper arms to PARK position. Remove wiper blades.

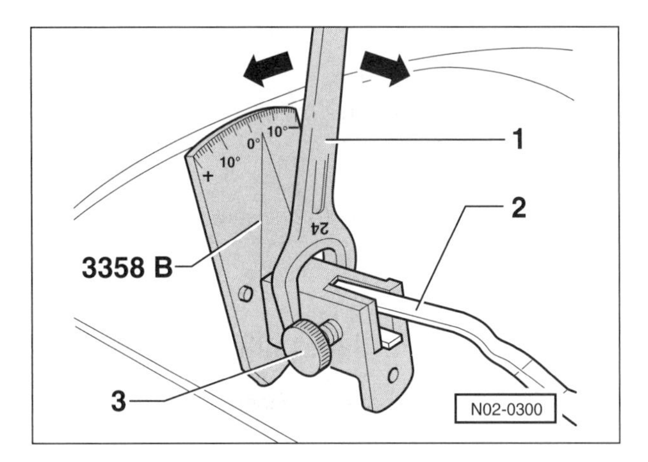

Insert wiper arm (**2**) into windshield wiper adjuster tool (Audi special tool 3358 B) and secure with locking screw (**3**). Compare angle with specifications.

Wiper arm angle specifications	
Left wiper arm	-9°
Right wiper arm	-5°

– If necessary, use 24 mm wrench (**1**) to adjust angle to specification.

BentleyPublishers.com
BentleyPublishers.com—All Rights Reserved

WINDSHIELD WIPER ASSEMBLY REPAIRS

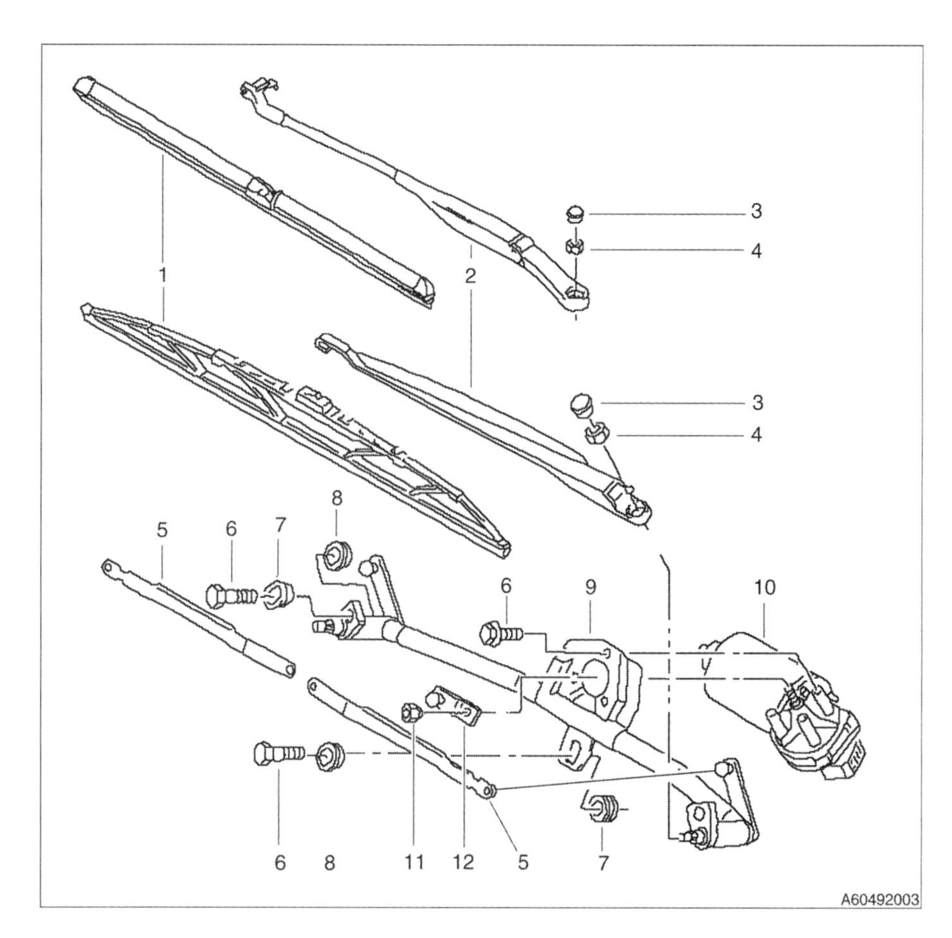

Windshield wiper assembly

1. **Wiper blade**
2. **Wiper arm**
3. **Trim cap**
4. **Nut, 16 Nm (12 ft-lb)**
5. **Link**
6. **M6 bolt**
7. **Grommet**
8. **Spacer bushing**
9. **Wiper assembly frame**
10. **Wiper motor**
11. **M8 self-locking nut**
12. **Crank**

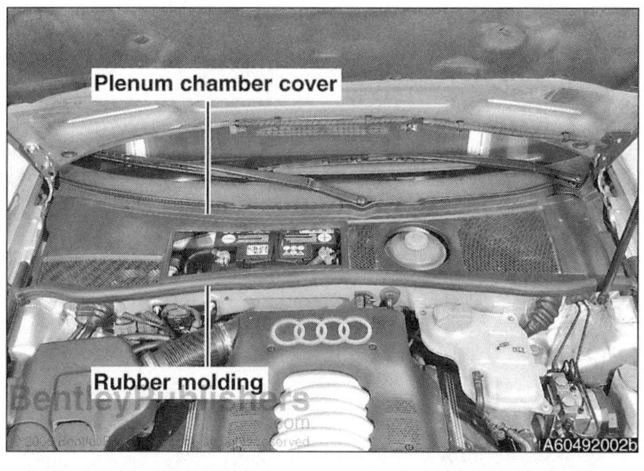

Windshield wiper assembly, removing and installing

– Make sure wipers are in PARK position and ignition switch is OFF. Open engine hood.

< Pull off rubber molding at engine compartment rear bulkhead. Pull plenum chamber cover forward and remove.

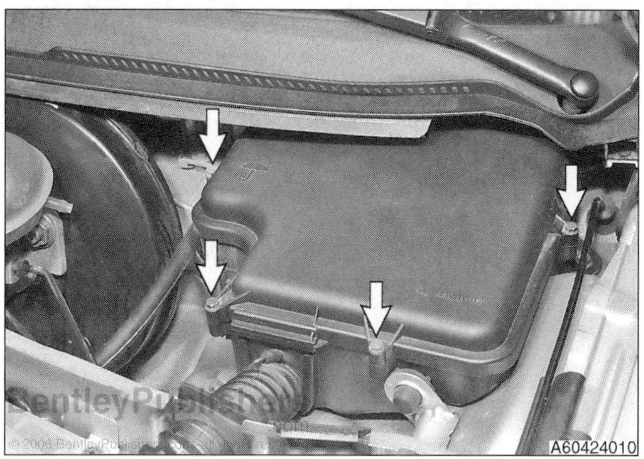

< Remove electronics box (E-box) cover retaining bolts (**arrows**). Lift off cover.

– Remove wiper arms. See **Wiper arm, removing and installing** in this repair group.

BentleyPublishers.com
BentleyPublishers.com—All Rights Reserved

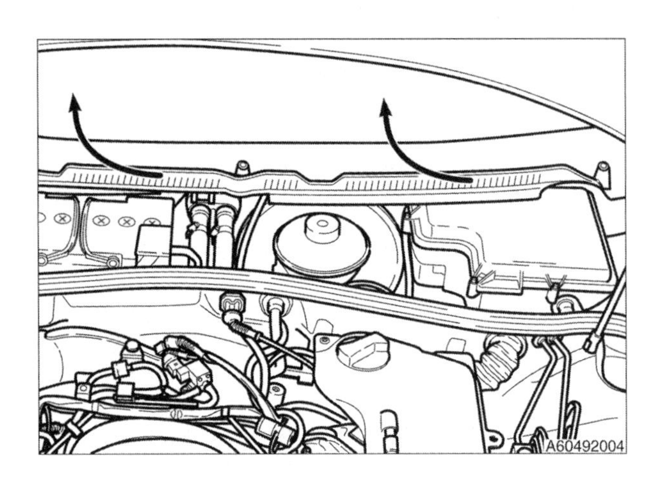

Pull out 3 retaining clips and carefully lift out (**arrows**) plastic cowl cover at base of windshield.

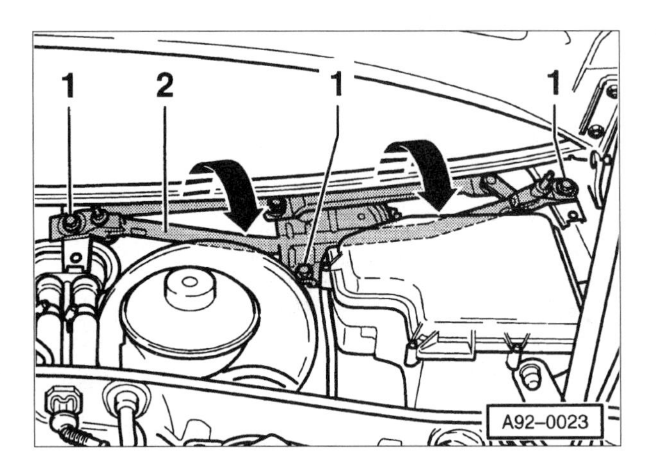

Working inside plenum chamber:

- Remove wiper assembly mounting bolts (**1**).
- Tilt (**arrows**) wiper assembly (**2**) forward and slide to right to remove from plenum chamber.
- Detach electrical connector at wiper motor.

– Installation is reverse of removal.

> ***CAUTION—***
> - *Before reinstalling wiper arms, switch ignition ON, make sure wiper motor is in PARK position, then switch ignition OFF.*

Tightening torque	
Wiper arm to shaft (M8)	16 Nm (12 ft-lb)
Wiper assembly to plenum chamber	8 Nm (6 ft-lb)

– Adjust wiper arms to correct position. See **Wiper arm, removing and installing** in this repair group.

Windshield wiper motor, removing and installing

– Remove wiper assembly. See **Windshield wiper assembly, removing and installing** in this repair group.

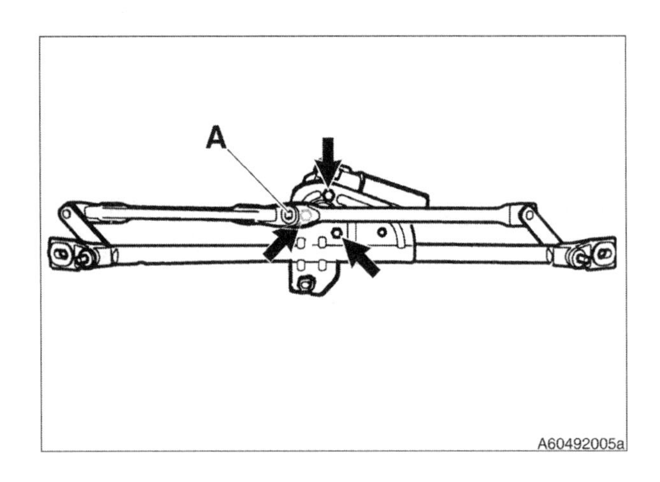

Place wiper assembly on work bench:

- Detach wiper links at wiper crank ball joint (**A**).
- Remove wiper motor mounting bolts (**arrows**) and lift motor off frame.

BentleyPublishers.com
BentleyPublishers.com—All Rights Reserved

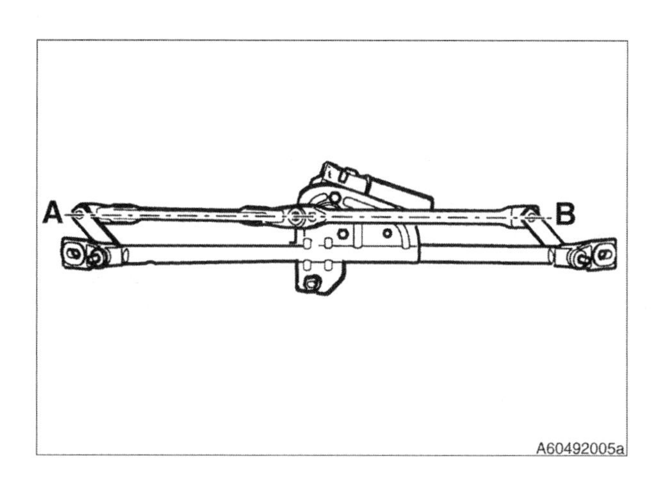

Installation is reverse of removal. Make sure:

- With motor in PARK, wiper crank is parallel to wiper frame.
- When wiper links are attached to crank, they lie in a straight line (**AB**) aligned with wiper crank.

– Reinstall wiper assembly in vehicle. See **Windshield wiper assembly, removing and installing** in this repair group.

Rear Wiper and Washer Assembly

On Avant (station wagon) models, the rear window wiper is controlled by the wiper switch. Washer fluid is supplied from the same reservoir as the windshield washers. See **Washer fluid reservoir and motor, removing and installing**in this repair group.

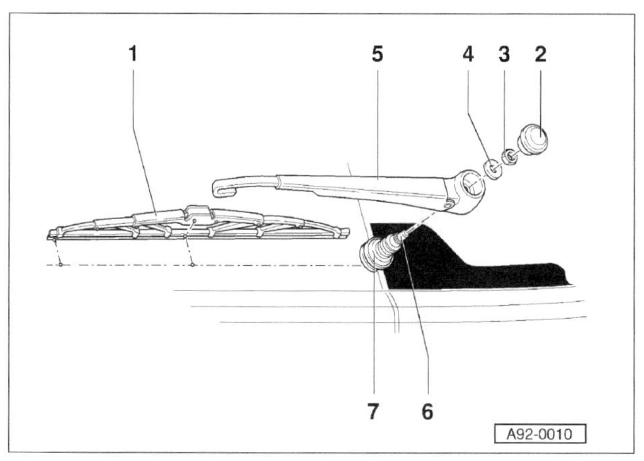

Rear wiper

1. Wiper blade
2. Cap
3. M8 nut, 16 Nm (12 ft-lb)
4. Spring washer
5. Wiper arm
6. Washer jet
7. Wiper shaft

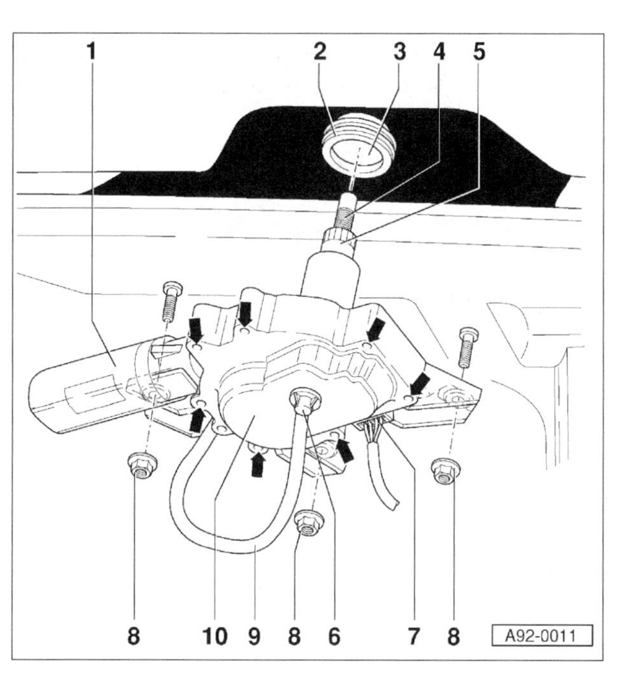

Rear wiper assembly

1. Wiper motor
2. Seal
3. Hole in rear window
4. Washer jet
5. Wiper shaft
6. Washer fluid hose connector
7. Harness connector
8. M6 mounting nut
9. Washer fluid hose
10. Wiper motor cover

Washers

The windshield washers, rear window washer (Avant models) and headlight washers are supplied from a fluid reservoir under the left front fender.

When the headlights are ON, the headlight washers operate each time the windshield washer is operated.

BentleyPublishers.com
BentleyPublishers.com—All Rights Reserved

In Avant models, a single bidirectional washer motor supplies fluid to the windshield and headlight washers when activated in one direction and to the rear window when activated in the opposite direction.

Washer fluid reservoir and motor, removing and installing

◀ The fluid reservoir is a one piece plastic bottle on the left side of the engine compartment. It extends into the left wheel housing.

– Prior to removal, siphon out as much fluid as possible.

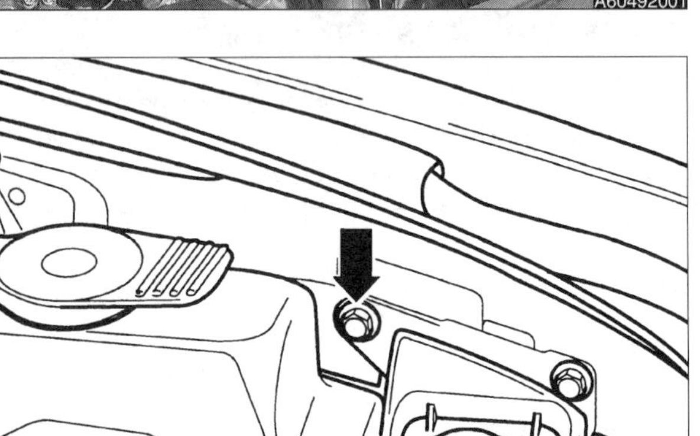

◀ Remove washer reservoir upper mounting bolt (**arrow**).

– Raise left front of vehicle and support safely.

> ***WARNING —***
> • *Make sure the car is stable and well supported at all times. Use a professional automotive lift or jack stands designed for the purpose. A floor jack is not adequate support.*

– Remove left front wheel housing liner. See **66 Exterior Equipment**.

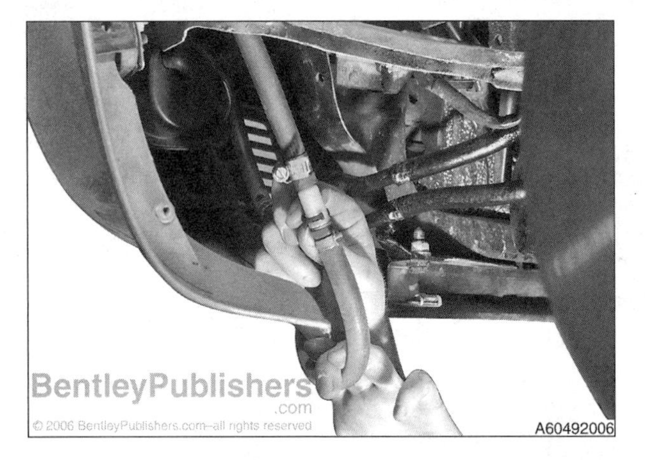

◀ Place 2-gallon bucket underneath left fender. Pull off locking clip and detach fluid hose. Allow fluid to drain into bucket.

◀ Working underneath left wheel housing:

• Detach fluid level sensor harness connector.

• Remove reservoir mounting bolts (**arrows**).

– Lower reservoir carefully. Reach up in front of reservoir and pull washer fluid pump out of front of reservoir.

> ***CAUTION—***
> • *Be prepared to catch spilling washer fluid.*

– Installation is reverse of removal. Be sure to replace washer motor sealing grommet in reservoir.

BentleyPublishers.com
BentleyPublishers.com—All Rights Reserved

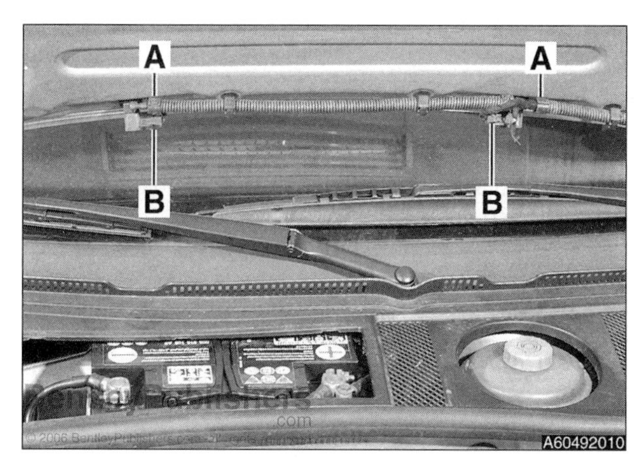

Windshield washer nozzle, removing and installing

– Open engine hood.

◄ Working at underside of rear of hood:

- Detach fluid hoses (**A**) from nozzles.
- Detach harness connectors (**B**) from washer nozzle heaters.

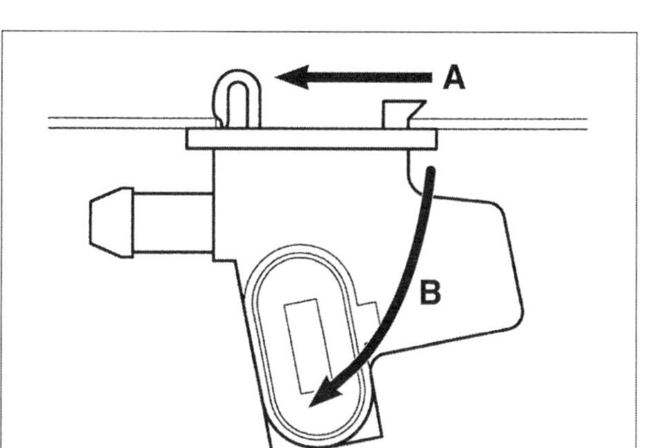

◄ Press nozzle forward (**A**) and tilt (**B**) out of engine hood.

– Installation is reverse of removal.

Windshield washer nozzles, aiming

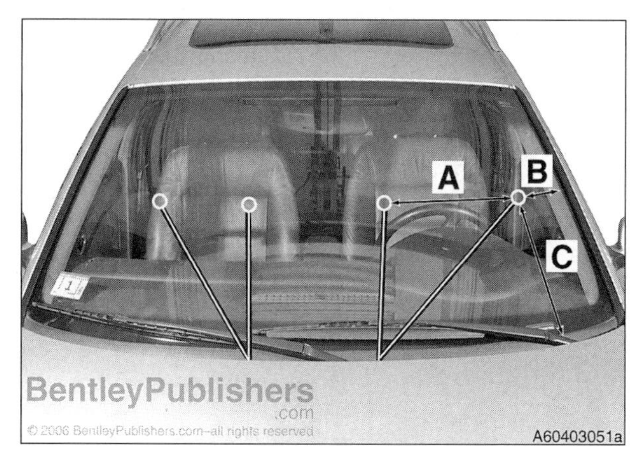

Washer fluid sprays at a slightly different angle when the car is stationary. The specified adjustment points take this into account.

◄ Mark spots on windshield:

- **A** = 400 mm (15.75 in)
- **B**= 190 mm (7.5 in) (distance to windshield side molding)
- **C** = 420 mm (16.5 in) (distance to top of cowl cover)

– Using special tool VAG 3125 (washer nozzle aiming tool or a safety pin), adjust nozzles to correct settings.

– If spray stream is not uniform or cannot be adjusted to specified dimensions, replace nozzle.

Rear window washer nozzle, removing and installing

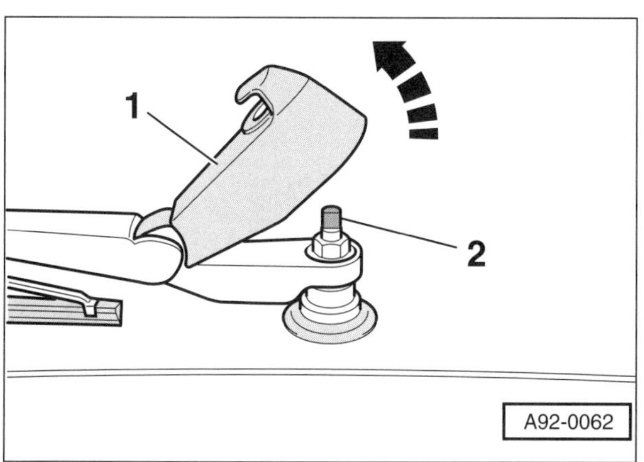

◄ Working at rear window wiper:

- Lift up washer nozzle cover (**1**).
- Pull out spray nozzle (**2**) and insert new.

BentleyPublishers.com
BentleyPublishers.com—All Rights Reserved

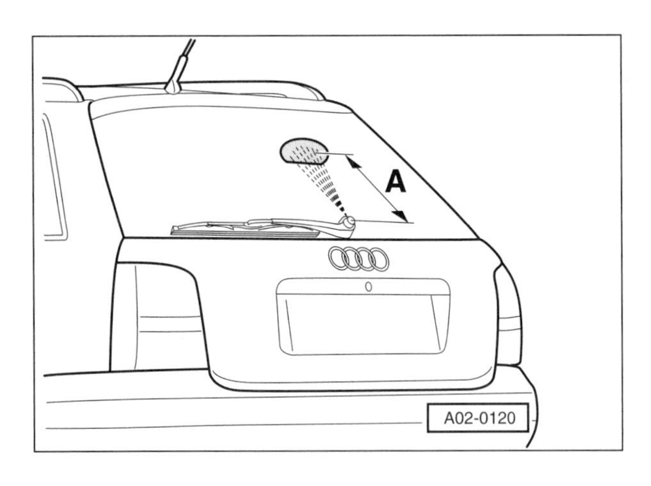

Rear window washer nozzle, aiming

Mark spot on rear window:

- **A** = 280 mm (11 in) above wiper shaft.

– Using special tool VAG 3125 (washer nozzle aiming tool or a safety pin), adjust nozzle to correct setting.

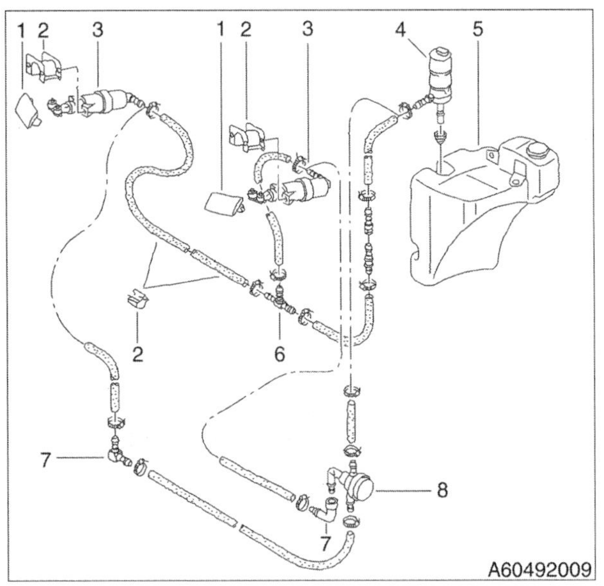

Headlight washer system

1. Cap
2. Bracket
3. Washer nozzle
4. Washer pump
5. Washer fluid reservoir
6. T-fitting
7. Elbow fitting
8. Check valve

Headlight washer nozzles aim

Headlight washer nozzles, installed in the front bumper, are aimed correctly by the manufacturer and do not need to be adjusted.

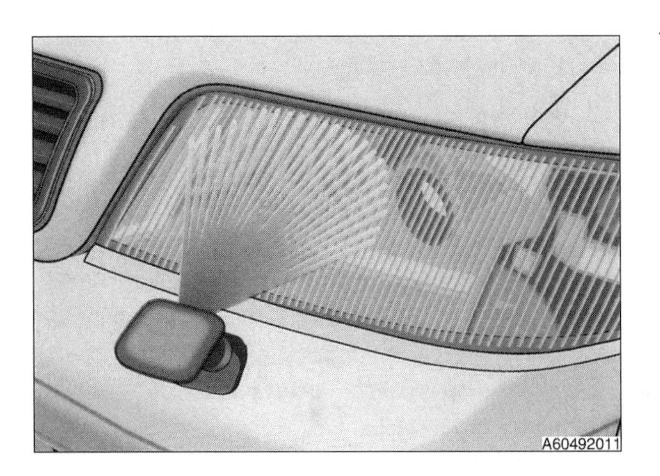

BentleyPublishers.com
BentleyPublishers.com—All Rights Reserved

94 Exterior Lights

TABLE

General

This repair group covers exterior lighting repairs and bulb replacement.

For headlight beam adjustment, see **03 Maintenance**.

Also see:

- **96 Interior Lights, Switches, Anti-theft**
- **97 Fuses, Relays, Component Locations**
- **EWD Electrical Wiring Diagrams**

Headlights

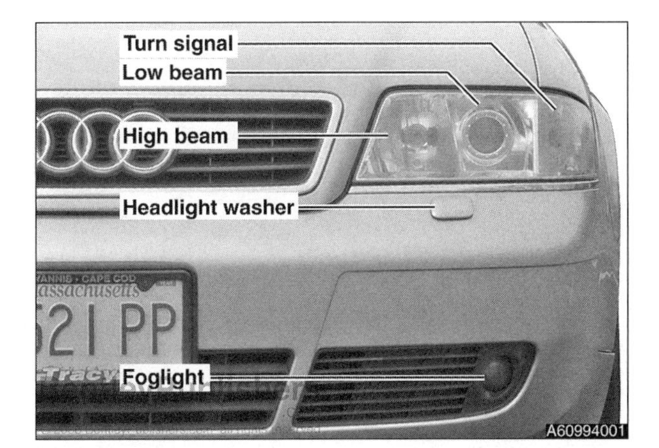

The headlights and front turn signals are combined in one unit on each side. The headlight lens is made of impact-resistant flexible plastic coated with varnish, which also makes it scratch resistant.

High beams use a free-space reflector, producing a precisely formed cone of light. H7 halogen bulbs are used for extra brightness.

BentleyPublishers.com
BentleyPublishers.com—All Rights Reserved

Automatic headlight beam adjustment

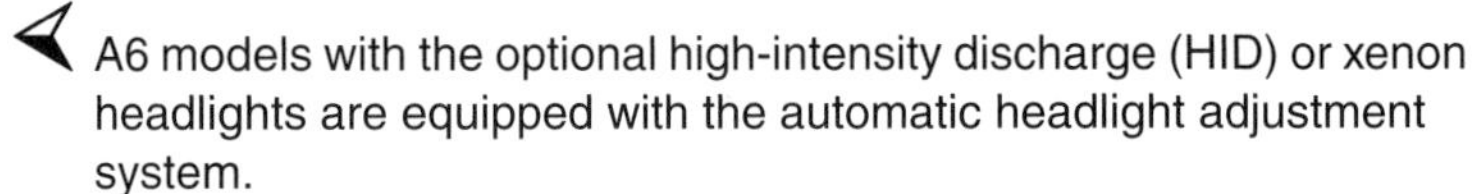

A6 models with the optional high-intensity discharge (HID) or xenon headlights are equipped with the automatic headlight adjustment system.

1. Headlight adjuster stepper motor in each headlight assembly
2. Headlight adjuster control module
   -1998 - 2000: Behind trim in left side trunk or luggage compartment
   -2001 - 2004: Behind glove compartment
3. Front suspension level sensor at left front control arm
4. Rear suspension level sensor at left rear control arm

Taillights

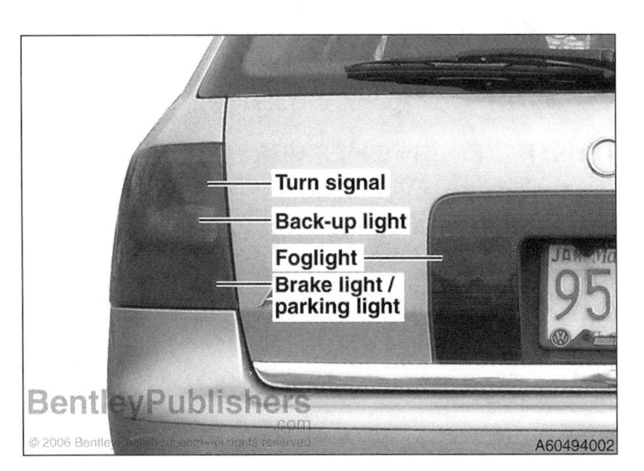

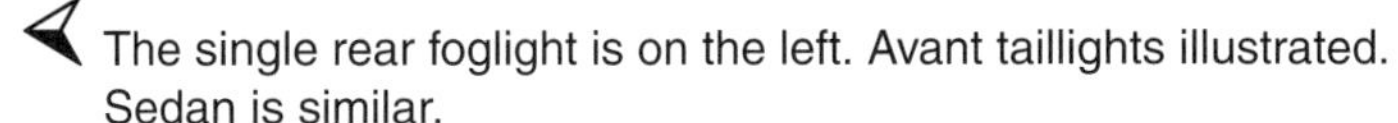

The single rear foglight is on the left. Avant taillights illustrated. Sedan is similar.

Exterior bulb applications

For convenience, exterior bulb applications for A6 models are listed in **Table a**.

Table a. A6 exterior bulb applications

Function	Type	Wattage
Back-up light	P21W	21
Brake light • Center (sedan) • Corners (with parking light)	 W5/1.2 P21/5W	 2.3 21 / 5
Foglight • Front • Rear	 H3 P21W	 55 21
Headlight • High beam • Low beam • Xenon	 H7 H1 D2S	 55 55 35
License plate light	L	5
Parking light • Front (with turn signal) • Rear (with brake light)	 P25-2-1 (amber) P21/5W	 21 / 5 21 / 5
Turn signal • Front (with parking light) • Rear • Side	 P25-2-1 (amber) P21W WY5W	 21 / 5 21 / 5 5

BentleyPublishers.com
BentleyPublishers.com—All Rights Reserved

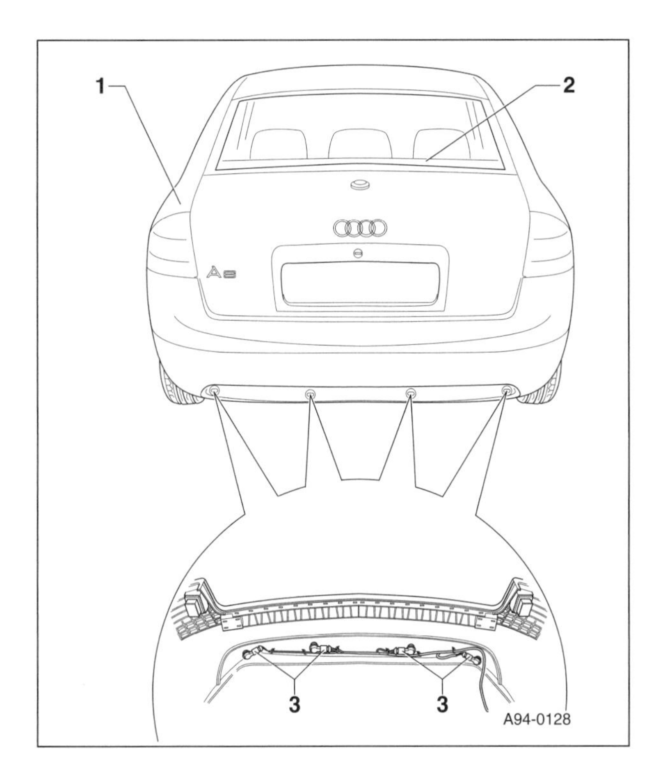

Parking aid

The optional parking aid system helps the driver during parking maneuvers by giving audible and visual signals when the vehicle approaches obstructions in the rear.

Parking aid system components are as follows:

1. Parking aid control module (J446) behind left trunk trim
2. Parking aid warning buzzer (H15) on rear parcel shelf
3. Parking aid sensors in rear bumper cover

Warnings and Cautions

WARNING —

- *Xenon (high intensity discharge or HID) bulbs operate at high voltages. When working on xenon headlight components (control module, ignition unit, gas discharge bulb, drive motor), be sure to switch lights OFF.*
- *Some light bulbs, especially headlight bulbs, can become hot enough to cause injury. Allow bulbs to cool before working on the lighting system.*

CAUTION —

- *Switch the ignition OFF and disconnect the negative (-) battery cable before removing any electrical components.*
- *Prior to disconnecting the battery, read the battery disconnection precautions in* **27 Battery, Alternator, Starter**.
- *Only use a digital multimeter for electrical tests.*
- *Do not touch the glass of bulbs with bare skin. If necessary wipe bulb using a clean cloth dampened with rubbing alcohol.*
- *Use only original equipment replacement bulbs. Non-original bulbs may cause false failure readings on the dashboard display.*
- *Use only specified bulbs. A bulb with higher wattage may cause damage to bulb housing.*
- *To avoid marring car paint or trim, work with a plastic prying tool.*

NOTE —

- *Xenon (HID) components are often marked with a yellow high voltage sticker.*

BentleyPublishers.com
BentleyPublishers.com—All Rights Reserved

Lighting Controls

In this manual, light switches are covered as follows:

- Brake light switch in **46 Brakes–Mechanical**
- Headlight flasher switch in **48 Steering**
- Light switch, emergency flasher switch, back-up light switch in **96 Interior Lights, Switches, Anti-theft**.

Lighting fuses

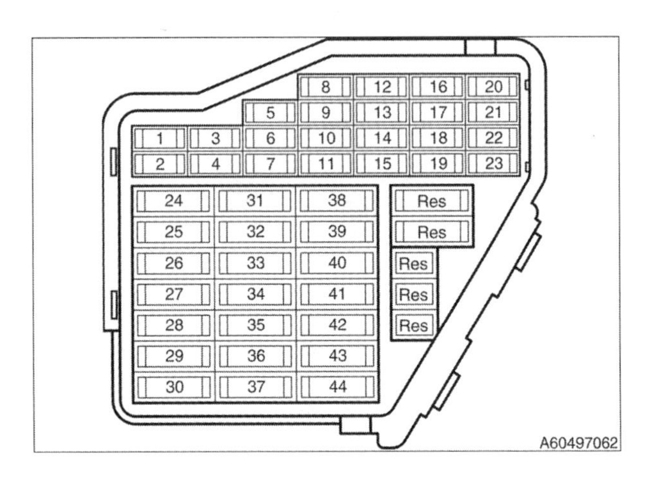

In main fuse panel, left end of dashboard.

Table b. Lighting fuses		
Protected circuit	**Fuse #**	**Amperage**
Back-up lights	31	15
Brake lights	13	10
Emergency flasher	39	15
Foglights	36	15
Headlights • Automatic adjustment • High beam left • High beam right • Low beam left • Low beam right	 10 19 18 21 20	 5 10 10 15 15
License plate lights	4	5
Taillight, parking light • Left • Right	 23 22	 5 5
Turn signals	2	10

Lighting system relays

13-fold relay panel underneath left side dashboard:

1. Relay positions 9 - 10: Headlight relay (J123)
2. Relay position 6: Foglight relay (J5)

Headlight controls (xenon)

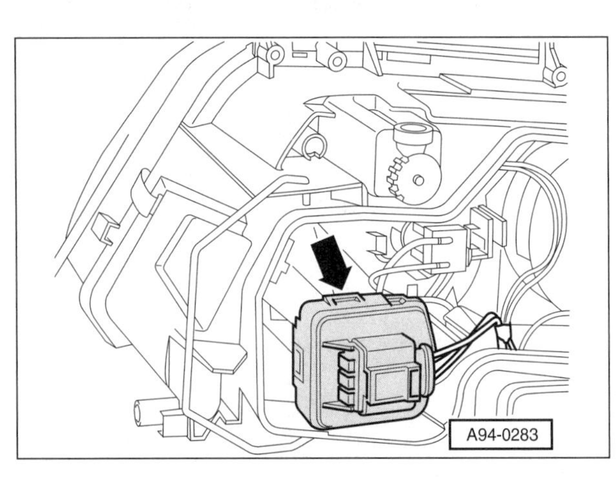

Xenon (HID) headlight control module (**arrow**) at rear of headlight assembly (static headlight aim adjusting system).

BentleyPublishers.com
BentleyPublishers.com–All Rights Reserved

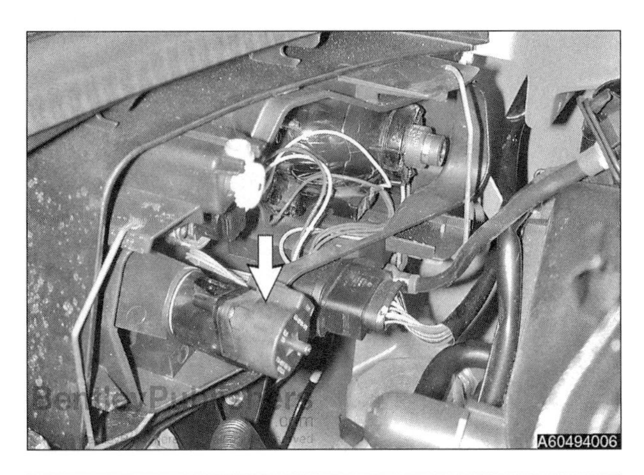

Xenon (HID) headlight adjustment stepper motor (**arrow**) at rear of headlight assembly (automatic headlight aim adjusting system).

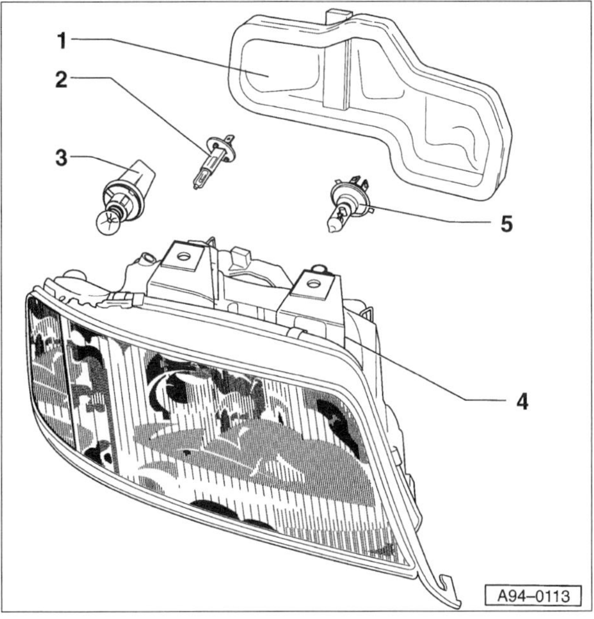

HEADLIGHT REPAIRS

Headlight assembly

1. Headlight housing rear cover
2. Low beam bulb
3. Turn signal bulb
4. Headlight housing
5. High beam bulb

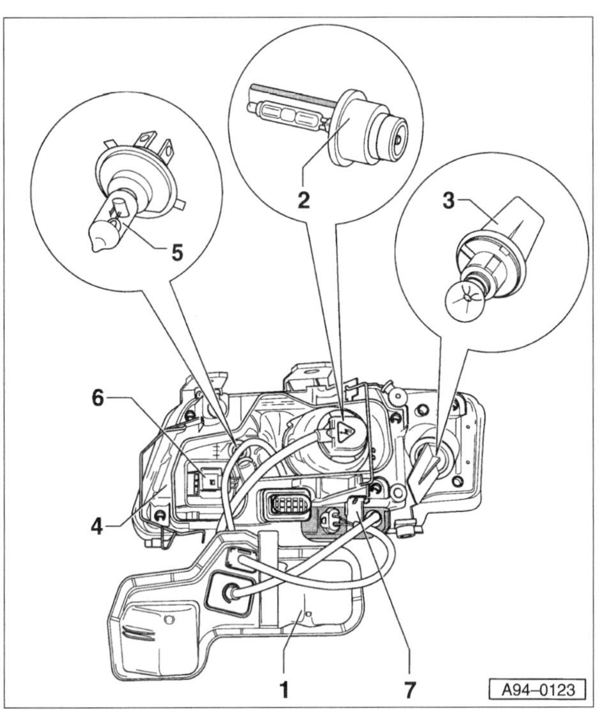

Xenon (HID) headlight assembly

1. Headlight housing rear cover
2. Low beam xenon bulb
3. Turn signal bulb
4. Headlight housing
5. High beam bulb
6. Headlight aim adjusting motor (static)
7. Igniter for xenon bulb

BentleyPublishers.com
BentleyPublishers.com—All Rights Reserved

Headlight, removing and installing

– Remove front bumper. See **63 Bumpers**.

– Xenon (HID) headlight: Disconnect negative (–) battery cable and cover battery terminal to keep cable from accidentally contacting terminal.

> ***CAUTION—***
> *• Be sure to have the radio anti-theft code before disconnecting the battery.*

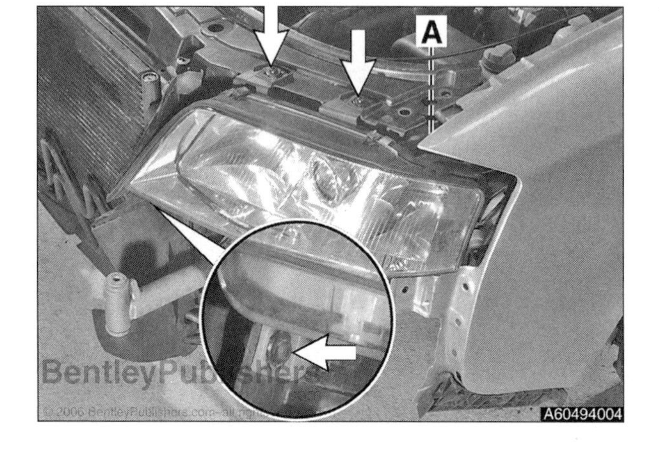

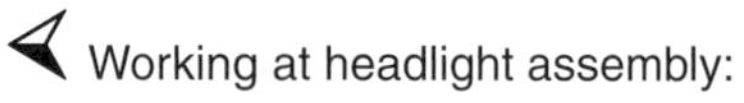
Working at headlight assembly:

- Remove 3 T30 Torx screws (**arrows**). Use 8 inch (200 mm) Torx driver to remove lower inner screw (**insert**).
- Use VAG special too 3249 (ball-head Allen wrench) to reach behind headlight and loosen outer lower screw (**A**).
- Disconnect headlight electrical connector.

– Swivel headlight assembly toward center of car and remove.

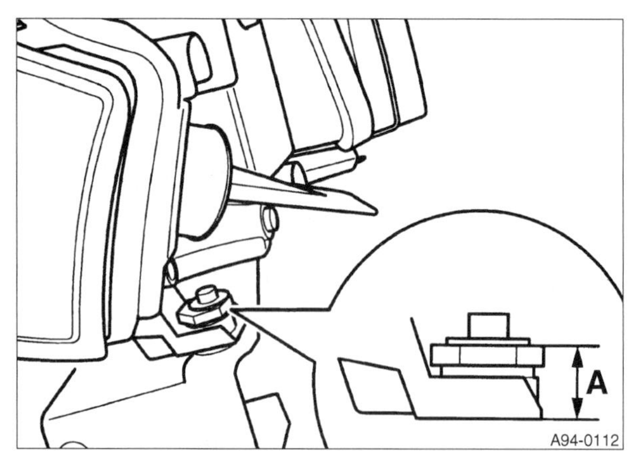

Prior to installing new headlight, measure distance **A** at plastic bolt on old headlight.

– Adjust plastic bolt on new headlight accordingly.

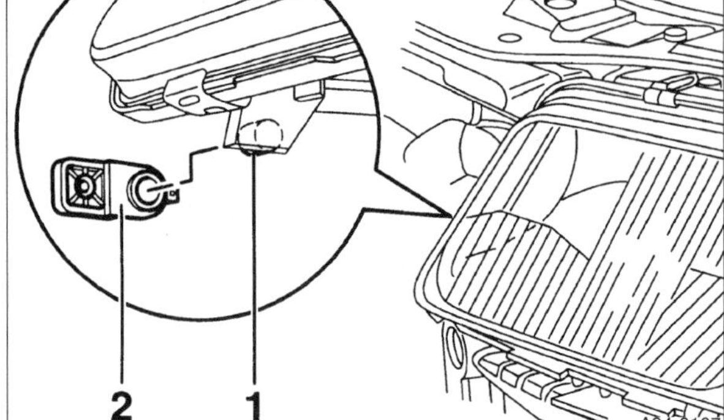

When installing new headlight, first engage lug **1** in guide **2**.

– Remainder of installation is reverse of removal. Position headlight with even gap to surrounding body panels.

Tightening torque	
Headlight to lock carrier (T30 bolts)	6 Nm (53 in-lb)

– Adjust headlight aim. See **03 Maintenance**.

BentleyPublishers.com
BentleyPublishers.com—All Rights Reserved

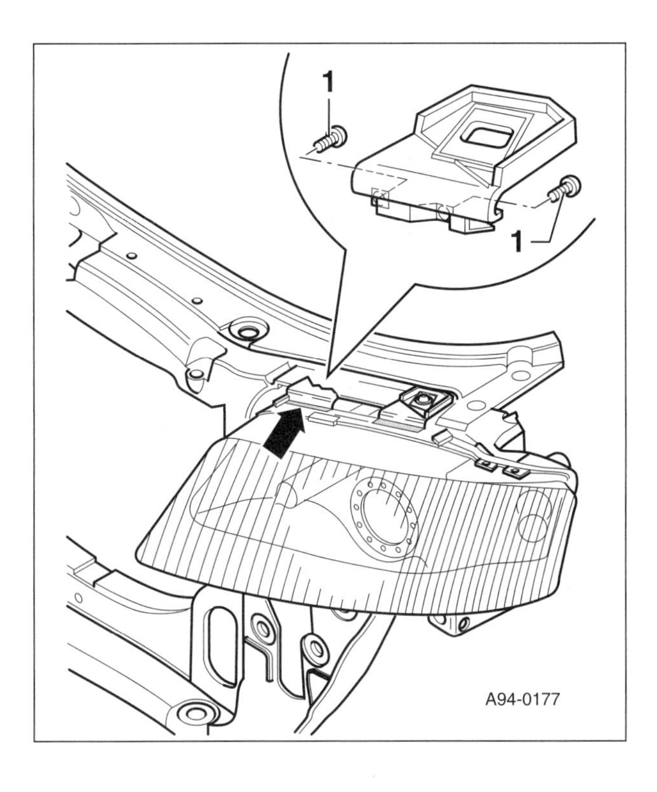

Headlight housing, repairing

A repair kit is available for broken headlight mounting tabs.

Headlight tab repair kits	
Left headlight	Part no. 4B0 998 121
Right headlight	Part no. 4B0 998 122

NOTE—

- *VAG part numbers given were correct at publication time but may change over time. Consult an authorized Audi dealer parts department for current part information.*

– Remove headlight with broken mounting tab. See **Headlight, removing and installing** in this repair group. Remove pieces of broken mount from headlight and from lock carrier.

◄ Place repair kit mount at edge of headlight housing (**arrow**). Secure with 2 screws (**1**) from behind.

– Reinstall headlight.

Headlight bulb, halogen, removing and installing

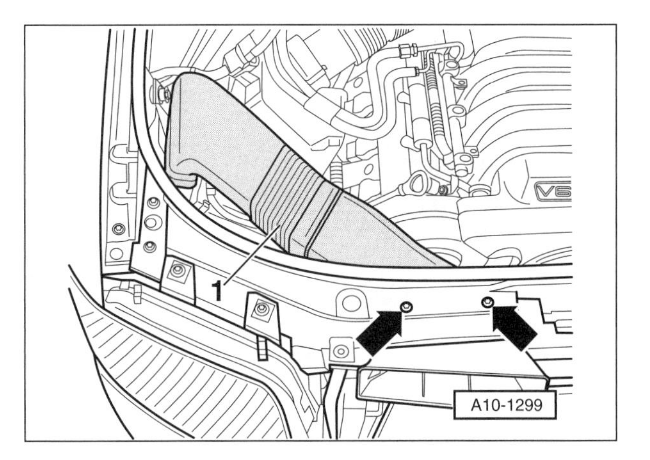

◄ Right headlight: Remove air filter housing cover. Then remove air duct mounting bolts (**arrows**) at lock carrier (radiator support frame). Remove air duct (**1**).

– Remove headlight housing cover.

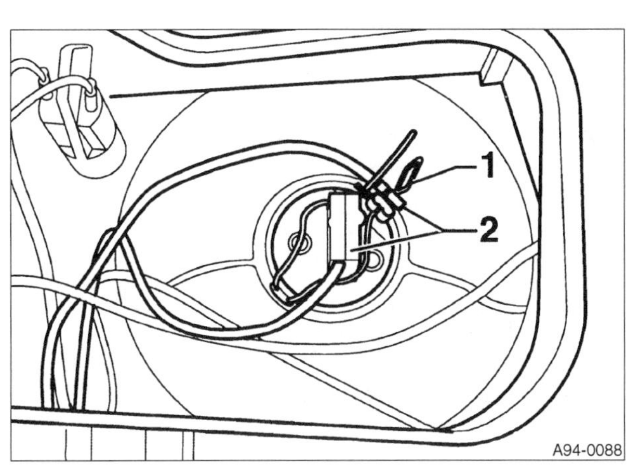

◄ Low beam:

- Detach harness connector (**1**).
- Remove wire clip (**2**) and take bulb out of housing.

BentleyPublishers.com
BentleyPublishers.com—All Rights Reserved

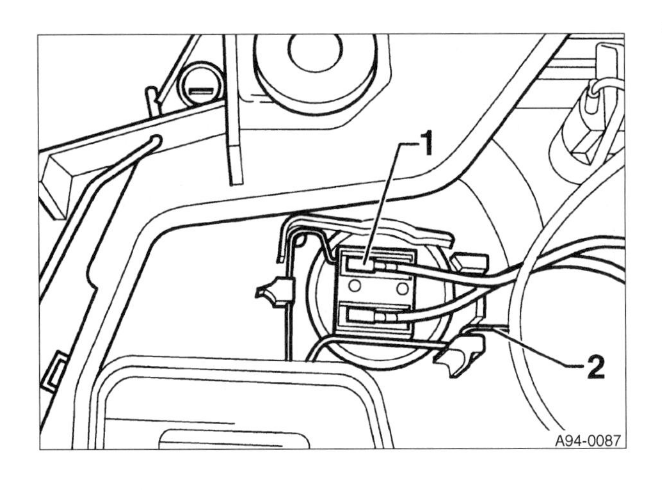

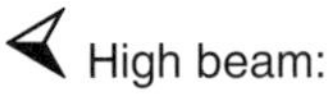 High beam:

- Detach harness connector (**1**).
- Remove wire clip (**2**) and take bulb out of housing.

– Insert new bulb, lining up tabs on bulb with notches in headlight housing.

Bulb applications	
Headlight high beam	H7 55 watt
Headlight low beam	H1 55 watt

> ***CAUTION—***
> *• Do not touch bulb glass with bare skin.*

– Reinstall wire clip, reattach harness connector and install cover.

Xenon (HID) bulb, removing and installing

– Disconnect negative (–) battery cable and cover battery terminal to keep cable from accidentally contacting terminal.

> ***CAUTION—***
> *• Be sure to have the radio anti-theft code before disconnecting the battery.*

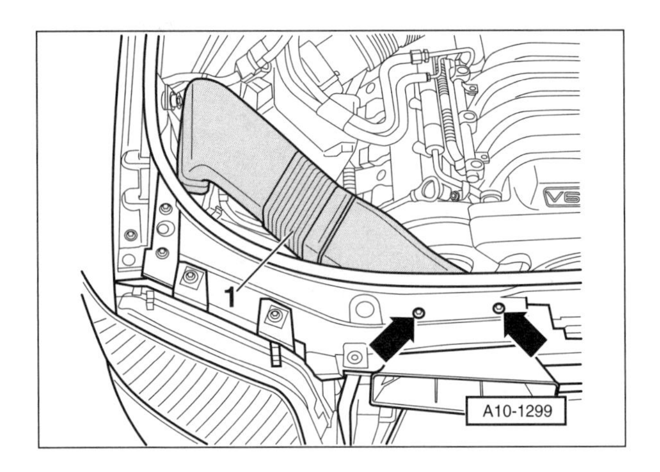

 Right headlight: Remove air filter housing cover. Then remove air duct mounting bolts (**arrows**) at lock carrier (radiator support frame). Remove air duct (**1**).

– Remove headlight housing cover.

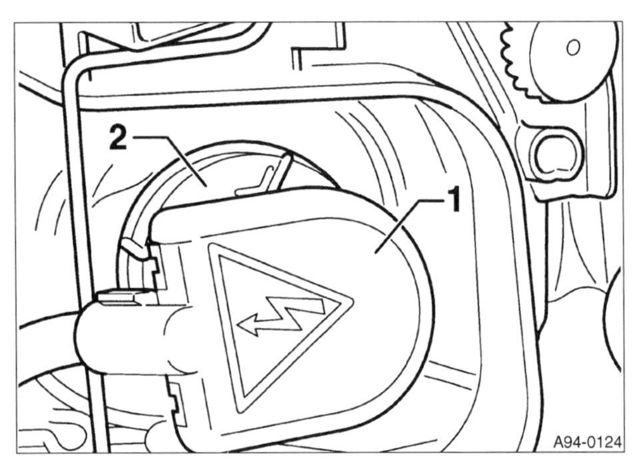

Working at rear of headlight:

- Turn low beam harness connector (**1**) counterclockwise to detach.
- Rotate bulb retainer ring (**2**) counterclockwise and remove.
- Remove bulb and insert new.

Bulb application	
Xenon headlight	D2S 35 watt

> ***CAUTION—***
> *• Do not touch bulb glass with bare skin.*

BentleyPublishers.com
BentleyPublishers.com—All Rights Reserved

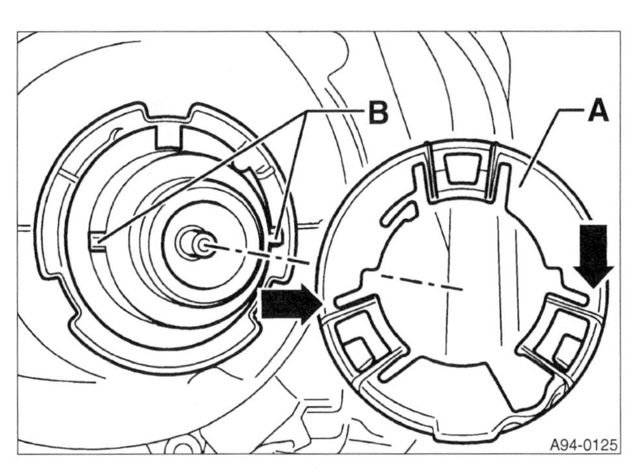

Insert retainer ring (**A**) on bulb lugs (**B**), lining up with notches (**arrows**). Rotate retainer clockwise to secure.

– Reattach harness connector and install cover.

FRONT TURN SIGNAL REPAIRS

Front turn signal bulb, removing and installing

Front turn signal bulb also functions as parking light.

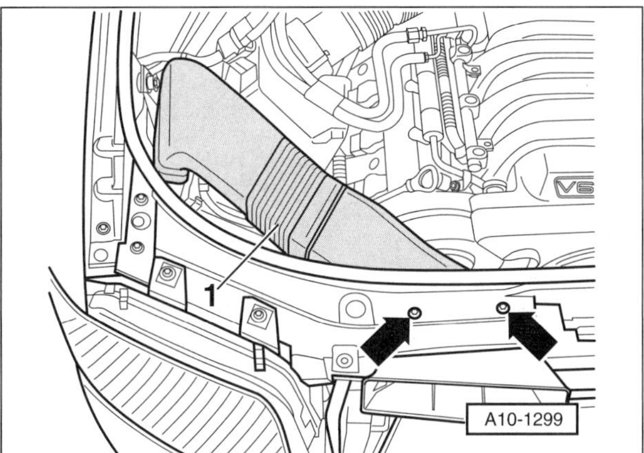

Right front turn signal: Remove air filter housing cover. Then remove air duct mounting bolts (**arrows**) at lock carrier (radiator support frame). Remove air duct (**1**).

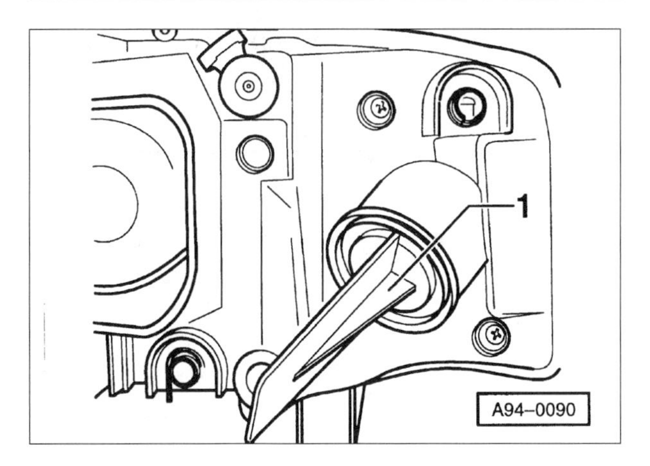

Working at outboard corner of headlight assembly, rotate turn signal bulb socket (**1**) counterclockwise to remove.

– Rotate bulb to remove from socket. Replace with new and reassemble socket.

Bulb application	
Front turn signal and parking (amber)	P25-2-1 21 / 5 watt

> ***CAUTION—***
> *• Do not touch bulb glass with bare skin.*

Side turn signal bulb, removing and installing

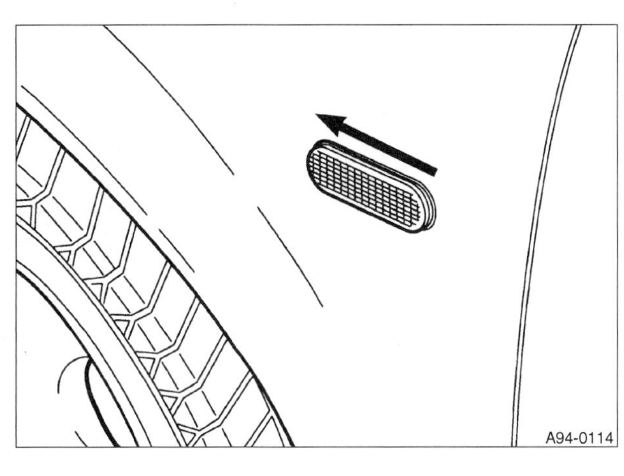

Push side turn signal forward (**arrow**) against spring clip and remove from front fender.

– Detach electrical harness. Remove bulb and replace.

Bulb application	
Side turn signal (amber)	WY5W 5 watt

> ***CAUTION—***
> *• Do not touch bulb glass with bare skin.*

– Installation is reverse of removal.

BentleyPublishers.com
BentleyPublishers.com–All Rights Reserved

FOGLIGHT REPAIRS

Front foglight, removing and installing

- Insert screwdriver in lower grille slot. Pry gently on lock tab (**arrow**), then pull grille off.

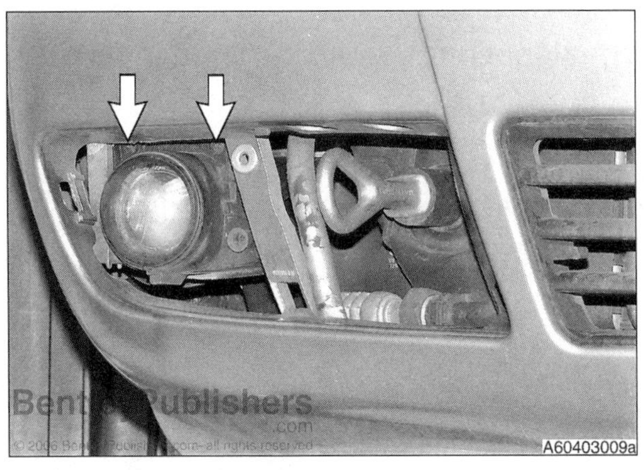

- Remove foglight mounting screws (**arrows**).
- Pull foglight assembly forward and detach electrical harness.
- Installation is reverse of removal. Adjust foglight beam. See **Front foglight, adjusting** in this repair group.

Front foglight bulb, removing and installing

Foglight with cover screwed on

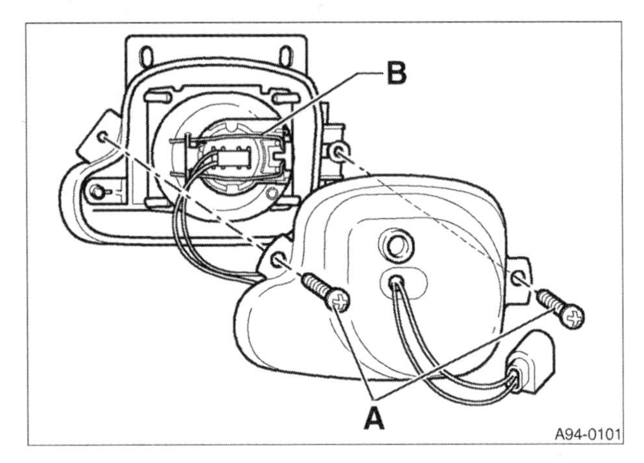

- Remove foglight. See **Front foglight, removing and installing** in this repair group.
- Remove 2 Phillips screws (**A**) from foglight assembly cover. Pull cover back.
- Disengage wire (**B**) clip and remove bulb.

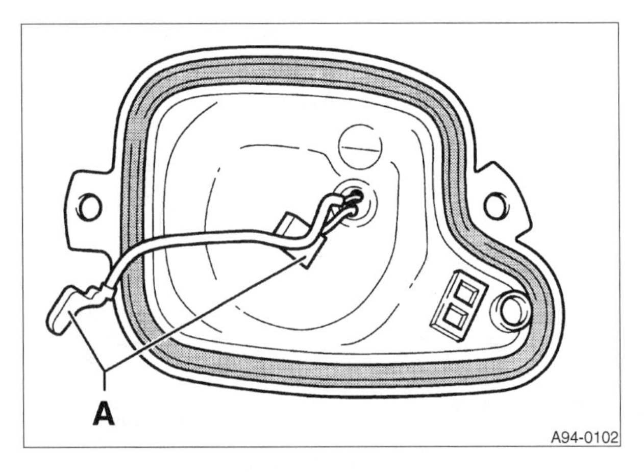

- Detach electrical connectors (**A**) from bulb holder in housing cover.

Bulb application	
Front foglight	H3 55 watt

> ***CAUTION—***
> • *Do not touch bulb glass with bare skin.*

- Installation is reverse of removal.

BentleyPublishers.com
BentleyPublishers.com—All Rights Reserved

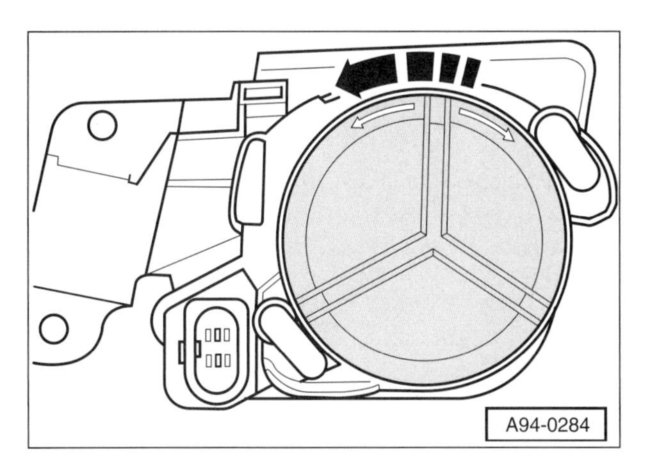

Foglight with bayonet mounted cover

– Remove foglight. See **Front foglight, removing and installing** in this repair group.

◄ Rotate cover counterclockwise (**arrow**) to remove.

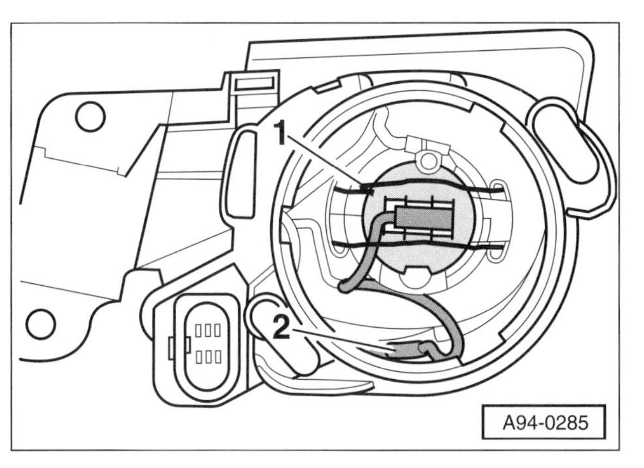

◄ Disengage wire (**1**) clip and remove bulb.

– Detach electrical connectors (**2**) from bulb holder in housing cover.

Bulb application	
Front foglight	H3 55 watt

> ***CAUTION—***
> *• Do not touch bulb glass with bare skin.*

– Installation is reverse of removal.

Front foglight, adjusting

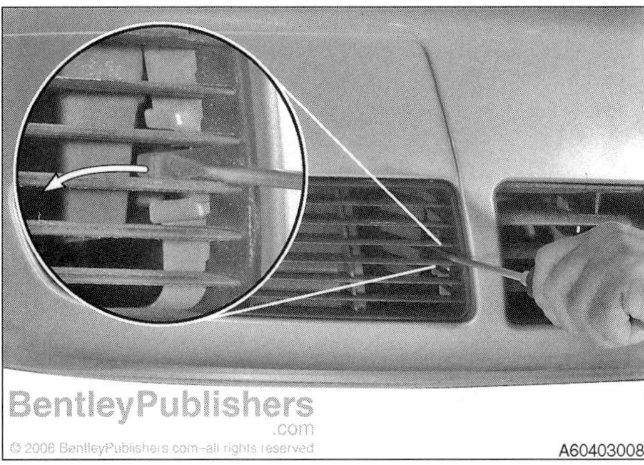

◄ Insert screwdriver in lower grille slot. Pry gently on lock tab (**arrow**), then pull grille off.

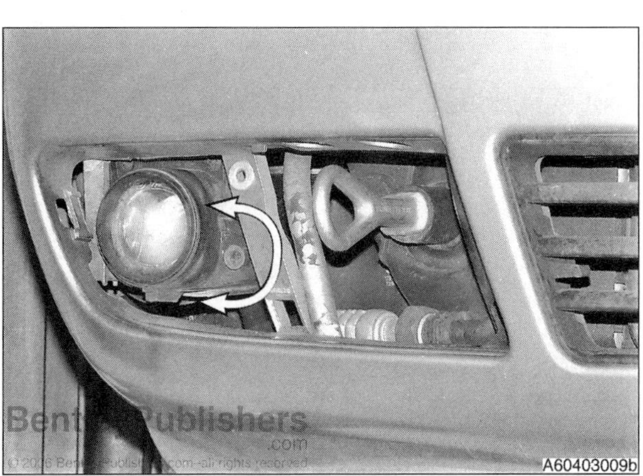

◄ Use adjustment screw to adjust foglight beam.

- Counterclockwise lowers beam.
- Clockwise raises beam.

It is not possible to adjust foglight beam horizontally.

BentleyPublishers.com
BentleyPublishers.com—All Rights Reserved

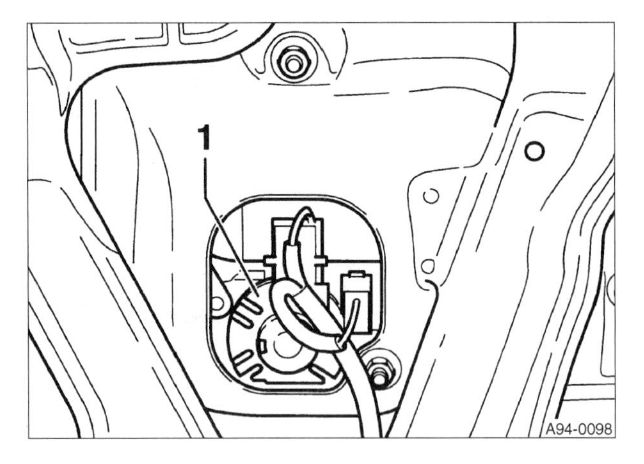

Rear foglight bulb, replacing

– Open trunk or luggage compartment. Remove trunk lid or rear deck lid trim.

< Rotate rear foglight bulb holder (**1**) clockwise and pull out. Replace rear foglight bulb.

Bulb applications	
Rear foglight	P21W 21 watt

> ***CAUTION—***
> *• Do not touch bulb glass with bare skin.*

– Installation is reverse of removal.

TAILLIGHT REPAIRS

Taillight assembly

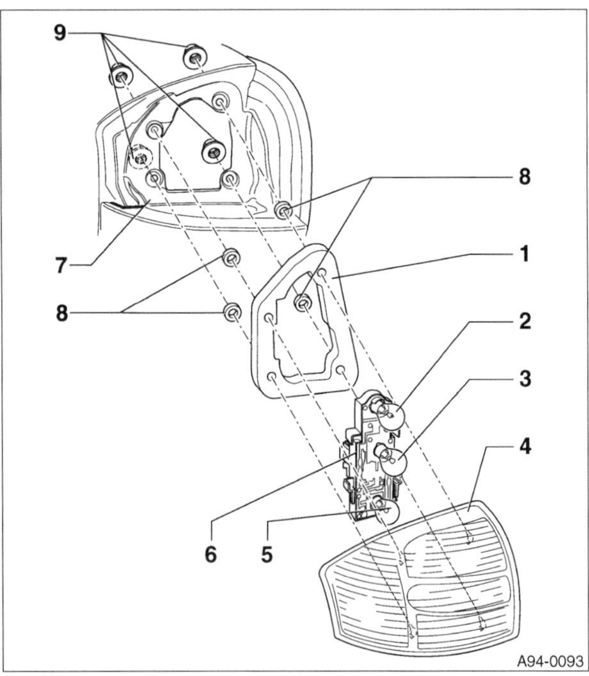

1. Gasket
2. Turn signal bulb
3. Back-up light bulb
4. Tail lens
   • Sedan version illustrated
   • Avant (station wagon) version has additional lateral cover
5. Brake light / parking light bulb
6. Bulb carrier
7. Taillight housing
8. Spacers
9. Mounting nuts, 4 Nm (35 in-ft)

Taillight bulb, replacing

– Open trunk or luggage compartment. Remove side storage compartment cover.

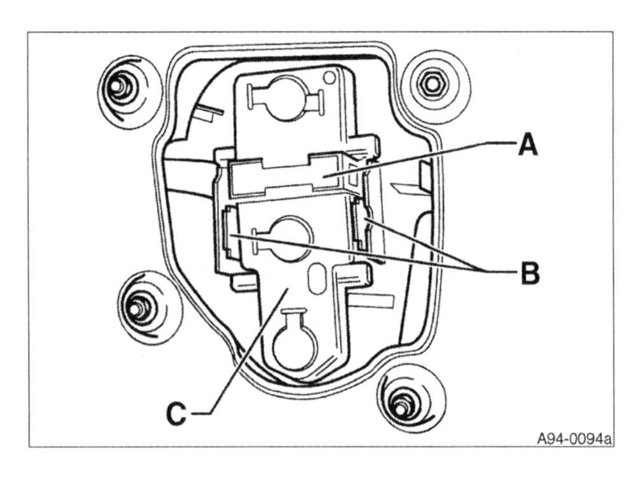

< Working at rear of taillight inside luggage compartment:
- Detach electrical harness (**A**).
- Press mounting bracket tabs (**B**) together and remove bulb carrier (**C**).
- Replace defective bulbs as necessary.

Taillight bulb applications	
Back-up light	P21W 21 watt
Brake light / parking light	P21/5W 21 / 5 watt
Turn signal	P21W 21 watt

> ***CAUTION—***
> *• Do not touch bulb glass with bare skin.*

– Installation is reverse of removal.

BentleyPublishers.com
BentleyPublishers.com—All Rights Reserved

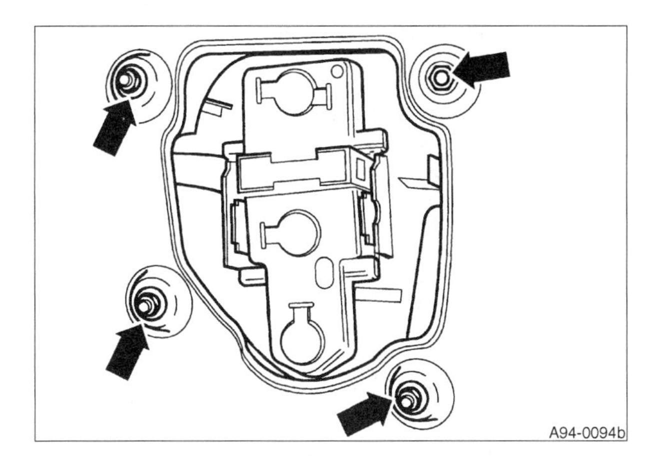

Taillight lens, removing and installing

– Open trunk or luggage compartment. Remove bulb carrier. See **Taillight bulb, replacing** in this repair group.

> ***CAUTION—***
> *• Do not touch bulb glass with bare skin.*

< Continuing to work inside luggage compartment, remove taillight lens mounting nuts (**arrows**).

– Remove taillight lens from body.

– Installation is reverse of removal, noting the following:

- Be sure to seat tail lens moisture seal correctly.
- Align tail lens to body contours when tightening mounting nuts.

Tightening torque	
Taillight lens to body	4 Nm (35 in-ft)

Center Brake Light Repairs

Center brake light, removing and installing (sedan)

– Disconnect negative (–) battery cable and cover battery terminal to keep cable from accidentally contacting terminal.

> ***CAUTION—***
> *• Be sure to have the radio anti-theft code before disconnecting the battery.*

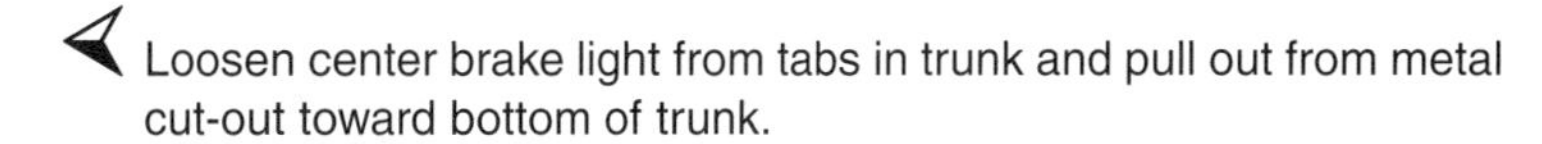

< Loosen center brake light from tabs in trunk and pull out from metal cut-out toward bottom of trunk.

– Detach electrical connection and remove bulb holder from trunk.

< Carefully lever off plastic cover from bulb carrier. Replace bulbs as necessary.

Bulb applications	
Center brake light bulbs (qty. 10)	W5/1.2 2.3 watt

> ***CAUTION—***
> *• Do not touch bulb glass with bare skin.*

– Installation is reverse of removal.

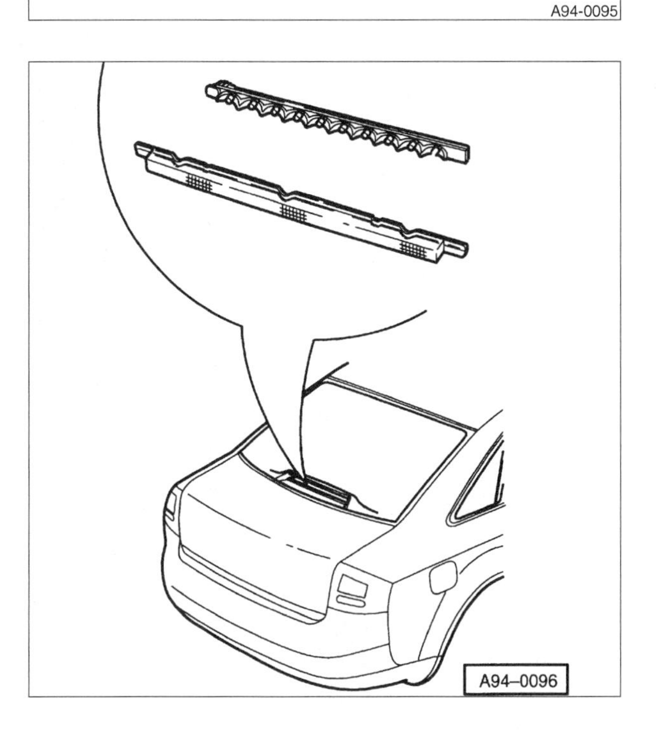

BentleyPublishers.com
BentleyPublishers.com—All Rights Reserved

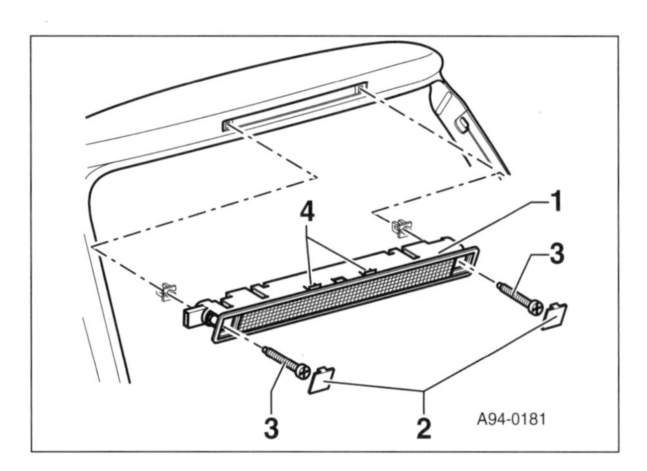

Center brake light, removing and installing (Avant)

– Disconnect negative (–) battery cable and cover battery terminal to keep cable from accidentally contacting terminal.

> ***CAUTION—***
> • *Be sure to have the radio anti-theft code before disconnecting the battery.*

◄ Working at top of rear deck:

- Remove trim caps (**2**) at ends of brake light.
- Remove mounting screws (**3**).
- Lift out center brake light and detach harness connector.

– If defective, replace entire socket. There are no separate bulbs.

– Installation is reverse of removal. Make sure socket seals securely into rear deck opening.

LICENSE PLATE LIGHT REPAIRS

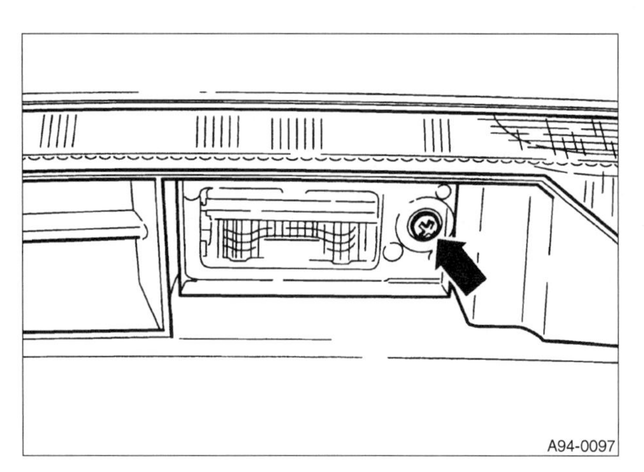

License plate light bulb, replacing

◄ Working in trunk or luggage compartment handle recess, remove screw (**arrow**) from license plate light socket.

– Pry out socket and replace bulb.

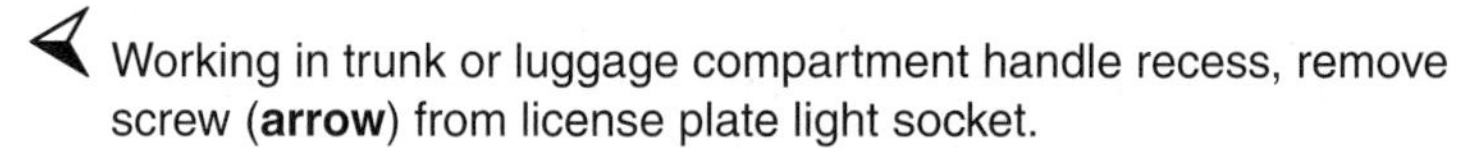

Bulb applications	
License plate light	L 5 watt

> ***CAUTION—***
> • *Do not touch bulb glass with bare skin.*

– Installation is reverse of removal. Make sure socket seals securely into handle recess.

BentleyPublishers.com
BentleyPublishers.com—All Rights Reserved

96 Interior Lights, Switches, Anti-theft

TABLE

General

This repair group covers interior light repairs, switch replacement, horns, electronic immobilizer and central locking. Also included is information about the optional anti-theft alarm and interior motion monitoring system.

For additional information, see:

- **97 Fuses, Relays, Component Locations**
- **EWD Electrical Wiring Diagrams**

BentleyPublishers.com
BentleyPublishers.com—All Rights Reserved

Cautions

> ***CAUTION—***
> - *Switch the ignition OFF and disconnect the negative (-) battery cable before removing any electrical components.*
> - *Prior to disconnecting the battery, read the battery disconnection precautions in* **27 Battery, Alternator, Starter**.
> - *Only use a digital multimeter for electrical tests.*
> - *To avoid marring paint or trim, work with a plastic prying tool or wrap screwdriver tip with tape.*

Interior bulb applications

For convenience, interior bulb applications for A6 models are listed in **Table a**.

Table a. A6 interior bulb applications		
Function	**Type**	**Wattage**
Dome light	K	10
Door courtesy light	W	5
Footwell	L	5
Glove compartment	L	5
Reading light	L	5
Trunk or luggage compartment	L	5

Interior Lights

Warnings and Cautions

> ***WARNING—***
> - *Some light bulbs can become hot enough to cause injury. Allow bulbs to cool before working on the lighting system.*

> ***CAUTION—***
> - *Do not touch bulb glass with bare skin. Dirt and skin oils may cause the bulb to fail prematurely. If necessary wipe bulb using clean cloth dampened with rubbing alcohol.*
> - *Use only original equipment replacement bulbs. Non-original bulbs may cause false failure readings on the dashboard display.*
> - *Use only specified bulbs. A bulb with higher wattage may cause damage to bulb housing.*

Interior lights power supply

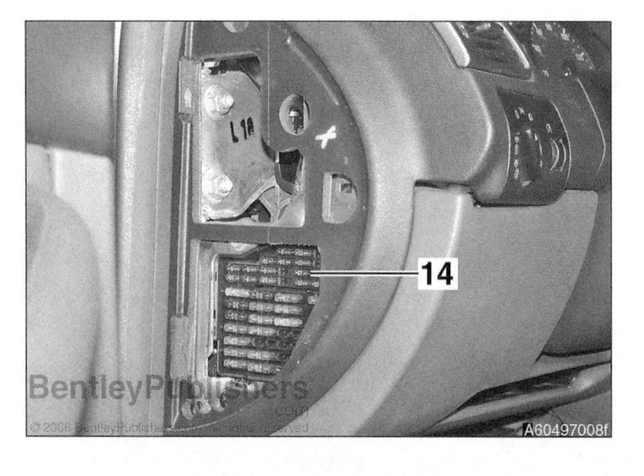

 In main fuse panel, left end of dashboard.

Interior lights fuse rating	
Fuse 14	10 A

BentleyPublishers.com
BentleyPublishers.com—All Rights Reserved

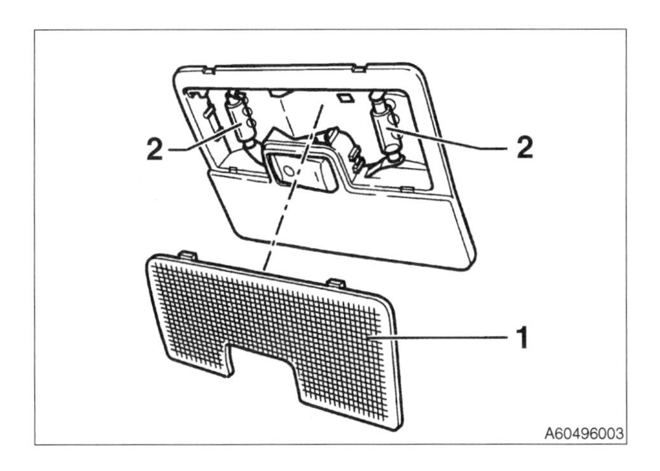

Front interior light bulbs, servicing

Front dome light bulb, replacing

< Using plastic prying tool:

- Carefully pry off dome light lens (**1**).
- Pry out and replace dome light bulbs (**2**) as necessary.

Bulb application	
Rear dome light	K 10 watt

> ***CAUTION—***
> *• Do not touch bulb glass with bare skin.*

– Installation is reverse of removal.

Front reading light bulb, replacing

– Disconnect negative (–) battery cable and cover battery terminal to keep cable from accidentally contacting terminal.

> ***CAUTION—***
> *• Be sure to have the radio anti-theft code before disconnecting the battery.*

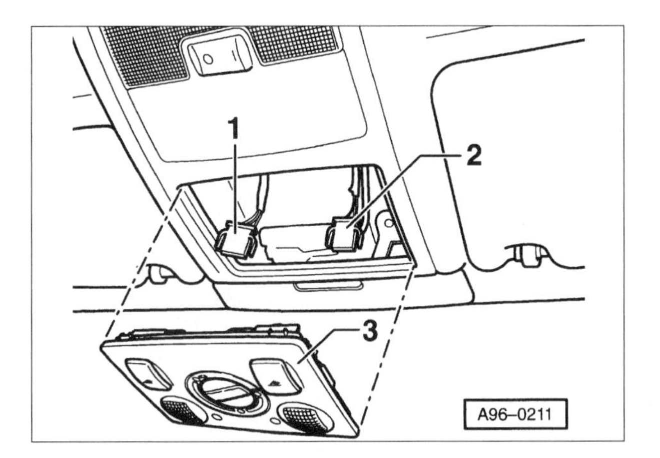

 Using plastic prying tool:

- Carefully pry dome light assembly (**3**) out of head liner.
- Disconnect electrical connectors (**1**, **2**).

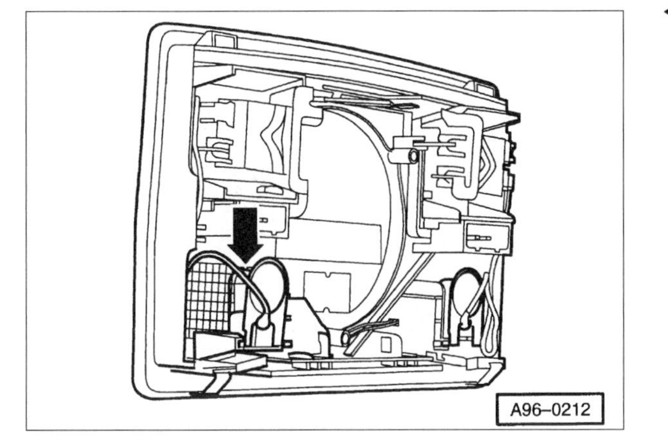

< Remove reading light bulb (**arrow**) from back of light assembly and replace with new.

Bulb application	
Rear reading light	L 5 watt

> ***CAUTION—***
> *• Do not touch bulb glass with bare skin.*

– Reattach electrical connector.

– When installing assembly into head liner, hook in one side and clip in at opposite side.

BentleyPublishers.com
BentleyPublishers.com—All Rights Reserved

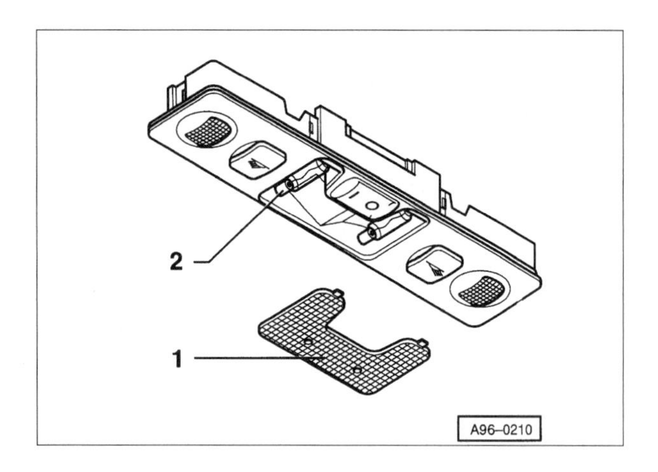

Rear interior light bulbs, servicing

Rear dome light bulb, replacing

 Using plastic prying tool or flat screwdriver tip wrapped with masking tape:

- Carefully pry off dome light lens (**1**).
- Pry out and replace dome light bulbs (**2**) as necessary.

Bulb application	
Rear dome light	K 10 watt

CAUTION—
- *Do not touch bulb glass with bare skin.*

– Installation is reverse of removal.

Rear reading light bulb, replacing

– Disconnect negative (–) battery cable and cover battery terminal to keep cable from accidentally contacting terminal.

CAUTION—
- *Be sure to have the radio anti-theft code before disconnecting the battery.*

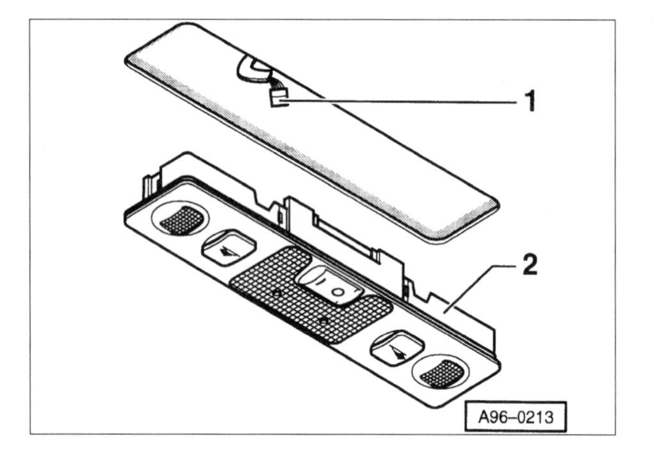

 Using plastic prying tool or flat screwdriver tip wrapped with masking tape:

- Carefully pry dome light assembly (**2**) out of head liner.
- Disconnect electrical connector (**1**).

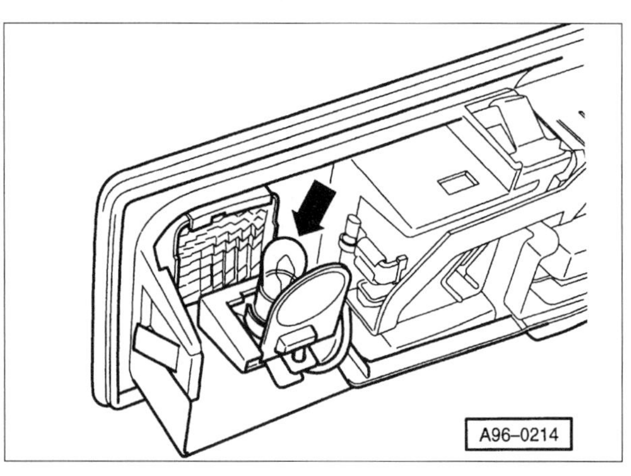

Remove reading light bulb (**arrow**) from back of light assembly and replace with new.

Bulb application	
Rear reading light	L 5 watt

CAUTION—
- *Do not touch bulb glass with bare skin.*

– Reattach electrical connector.

– When installing assembly into head liner, hook in one side and clip in at opposite side.

BentleyPublishers.com
BentleyPublishers.com—All Rights Reserved

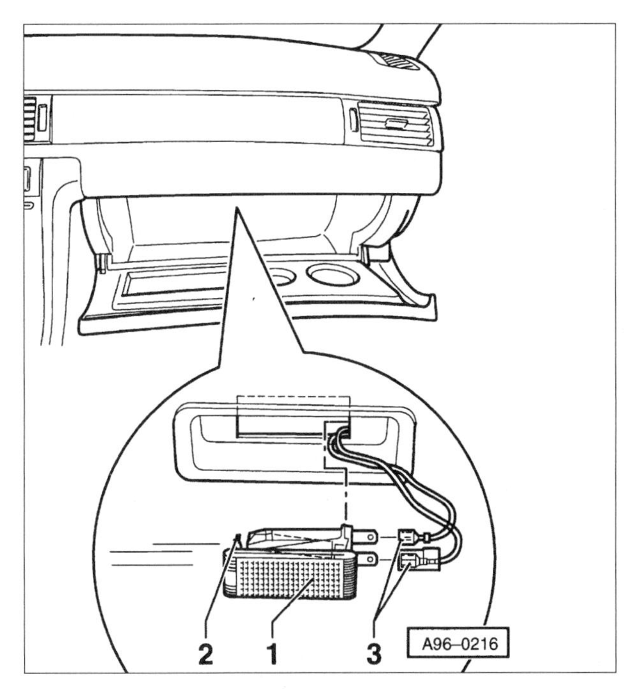

Glove compartment light bulb, replacing

◀ Using plastic prying tool or flat screwdriver tip wrapped with masking tape:

- Carefully push in retaining tab (**2**) and pry out glove compartment light socket (**1**).
- Detach electrical connectors (**3**).

– Remove bulb from back of socket and replace with new.

Bulb application	
Glove compartment light	L 5 watt

> ***CAUTION—***
> - *Do not touch bulb glass with bare skin.*

– Reattach electrical connectors.

– When installing socket, hook in one side and clip in at opposite side.

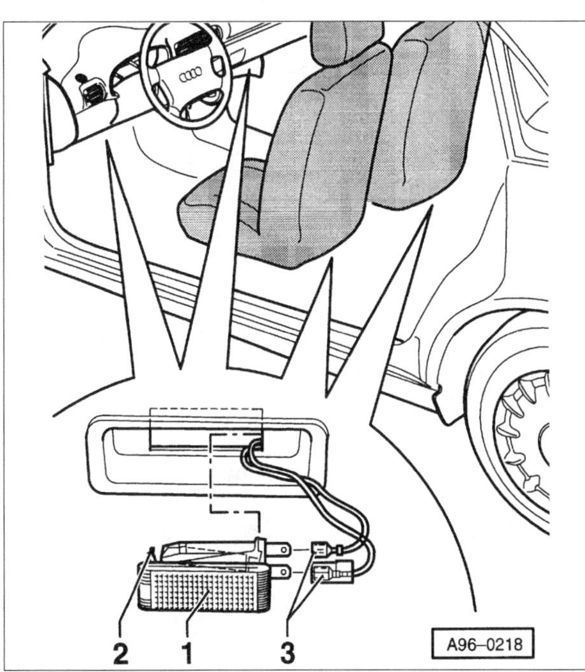

Footwell light bulb, replacing

◀ Using plastic prying tool or flat screwdriver tip wrapped with masking tape:

- Carefully push in retaining tab (**2**) and pry out footwell light socket (**1**).
- Detach electrical connectors (**3**).

– Remove bulb from back of socket and replace with new.

Bulb application	
Footwell light	L 5 watt

> ***CAUTION—***
> - *Do not touch bulb glass with bare skin.*

– Reattach electrical connectors.

– When installing socket, hook in one side and clip in at opposite side.

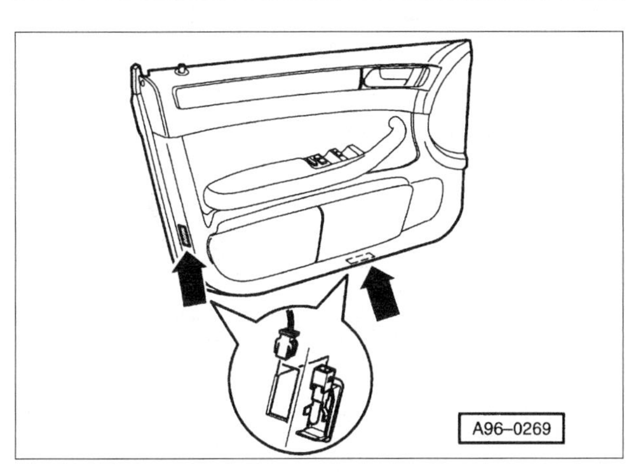

Door courtesy light bulb, replacing

◀ Using plastic prying tool or flat screwdriver tip wrapped with masking tape:

- Carefully push in retaining tab and pry out courtesy light socket (**arrow**).
- Detach electrical connectors.

BentleyPublishers.com
BentleyPublishers.com—All Rights Reserved

– Remove bulb from back of socket and replace with new.

Bulb application	
Door courtesy light	W 5 watt

> ***CAUTION—***
> *• Do not touch bulb glass with bare skin.*

– Installation is reverse of removal.

Trunk light bulb, replacing

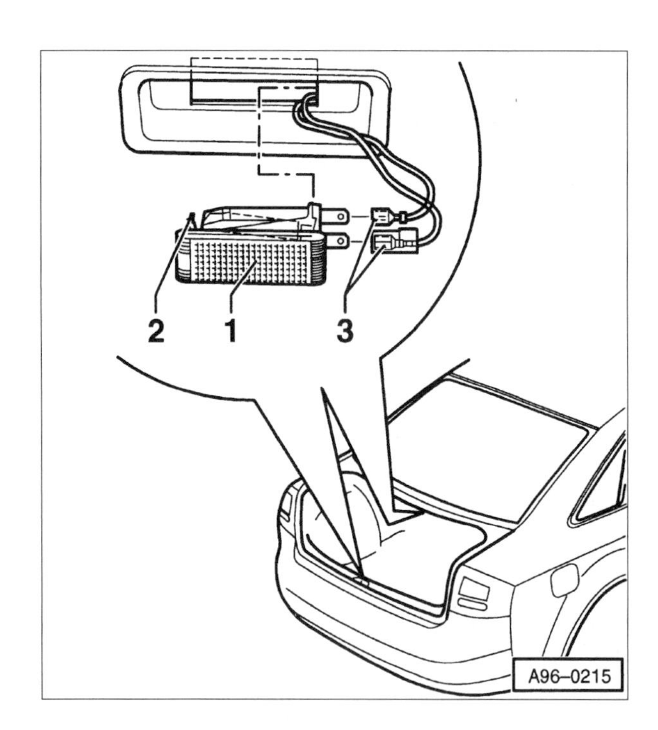

Using plastic prying tool or flat screwdriver tip wrapped with masking tape:

- Carefully push in retaining tab (**2**) and pry out trunk light socket (**1**).
- Detach electrical connectors (**3**).

– Remove bulb from back of socket and replace with new.

Bulb application	
Trunk light	L 5 watt

> ***CAUTION—***
> *• Do not touch bulb glass with bare skin.*

– Reattach electrical connectors.

– When installing socket, hook in one side and clip in at opposite side.

NOTE —
• Avant luggage compartment light bulb replacement is similar.

BACK-UP LIGHTS

Back-up light bulbs are covered in **94 Exterior Lights**.

Back-up lights power supply

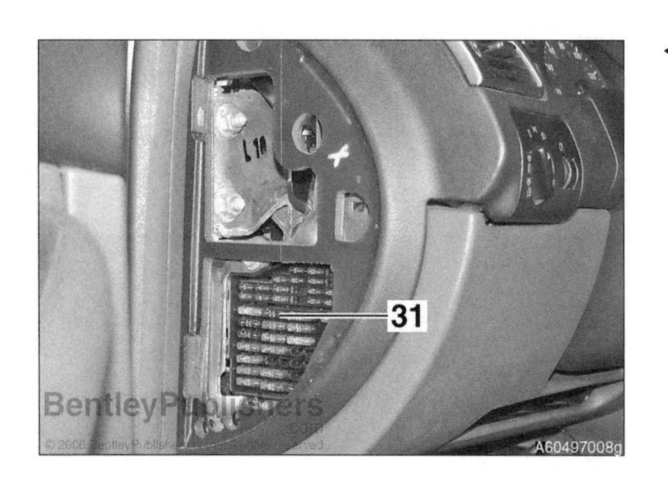

In main fuse panel, left end of dashboard.

Back-up lights fuse rating	
Fuse 31	15 A

BentleyPublishers.com
BentleyPublishers.com—All Rights Reserved

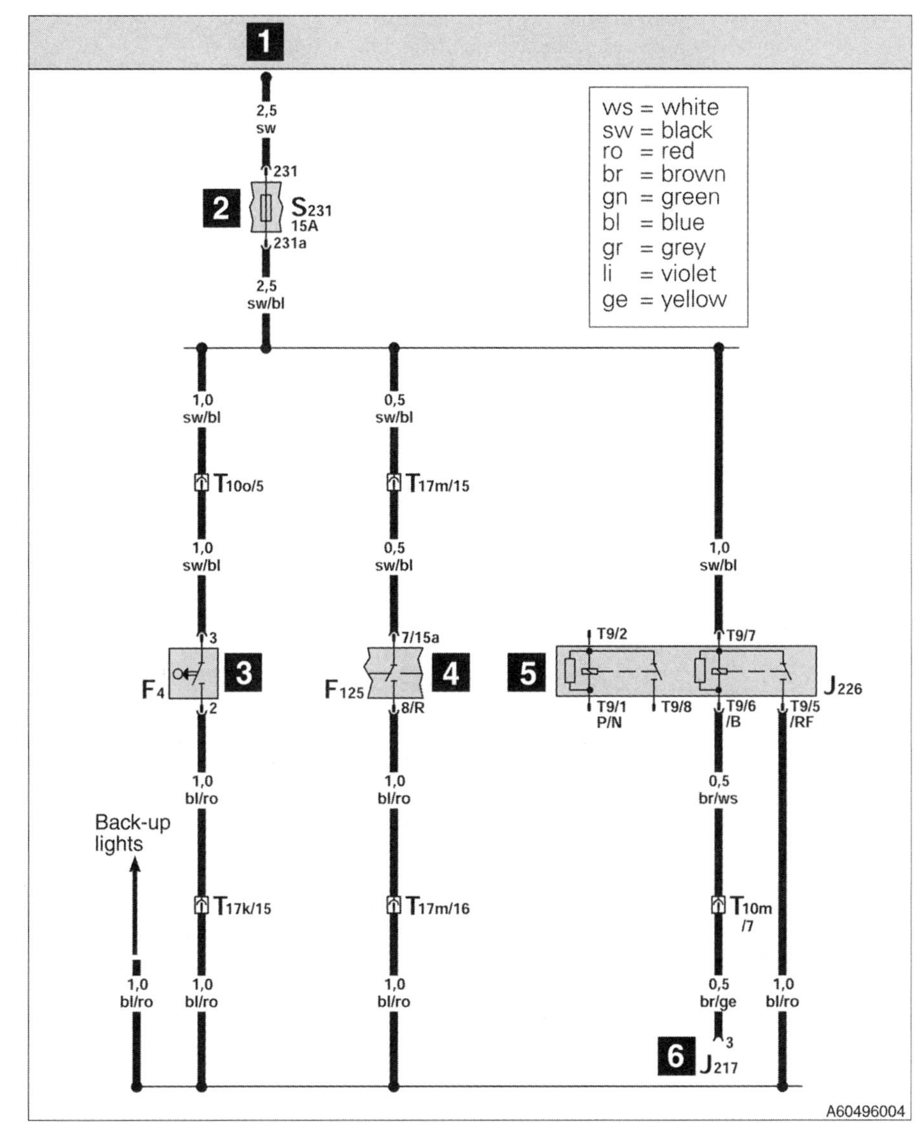

Back-up lights circuits

1. **Power supply**
   - Terminal 15, switched power in dashboard wiring harness.
2. **Fuse 31, 15 A**
   - In main fuse panel, left end of dashboard. See **Back-up lights power supply** in this repair group.
3. **Back-up light switch (F4)**
   - Manual transmission model.
   - In transmission case. See **Back-up light switch, manual transmission** in this repair group.
4. **Multifunction transmission range switch (F125)**
   - Automatic transmission model.
   - Switch assembly on side of transmission, underneath vehicle.
5. **Park / neutral position (PNP) relay (J226)**
   - Multitronic (CVT) transmission model.
   - In 13-fold relay panel underneath left side dashboard. See **Back-up light switch function, multitronic transmission (CVT)** in this repair group.
6. **Transmission control module (J217)**
   - Multitronic (CVT) transmission model.
   - Inside transmission at rear of case, underneath end cover.

NOTE—

- *Audi identifies electrical components by a letter and/or a number in the electrical schematics. See* **EWD Electrical Wiring Diagrams**. *These electrical identifiers are listed in parentheses as an aid to electrical troubleshooting.*

Back-up light switch, manual transmission

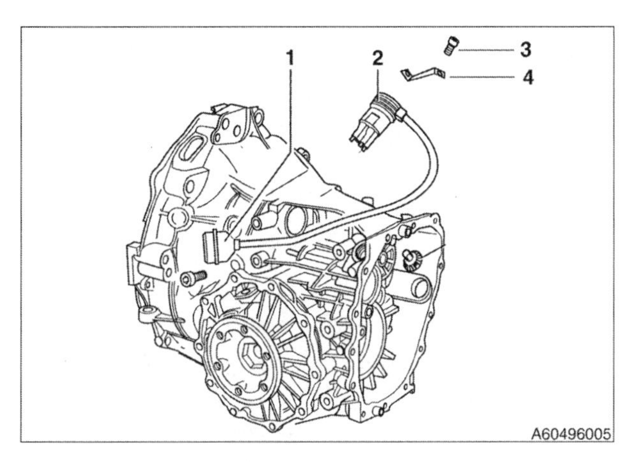

5-speed manual transmission:

1. Harness connector above left drive axle flange.
2. Gear position sender at top right of transmission housing.
   - Includes back-up light switch (F4).
3. Bolt, 25 Nm (18 ft-lb).
4. Hold-down bracket.

BentleyPublishers.com
BentleyPublishers.com—All Rights Reserved

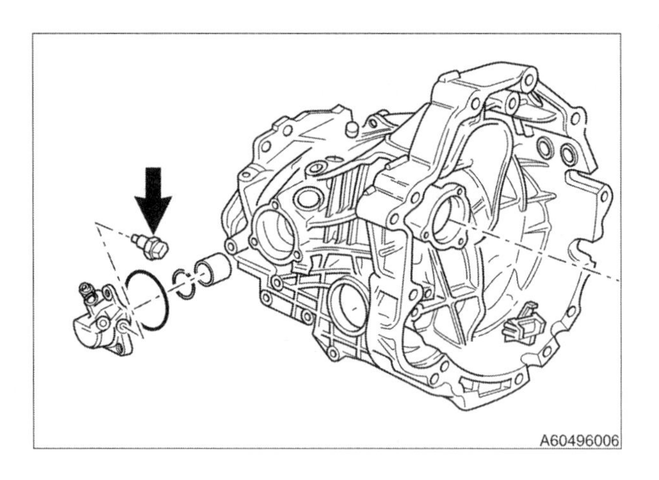

6-speed manual transmission: Backup light switch (F4)(**arrow**) on right side of transmission at selector shaft housing.

- Tighten to 20 Nm (15 ft-lb)

Back-up light switch function, automatic transmission

When automatic transmission is placed in REVERSE, multifunction transmission range switch (F125) switches back-up lights ON.

Back-up light switch function, multitronic transmission (CVT)

When CVT transmission is placed in reverse, transmission control module (J217), under cover in rear of transmission, indicates gear position to park / neutral position (PNP) relay (J226) (**arrow**) in 13-fold relay panel under left side dashboard.

– PNP relay switches back-up lights ON.

Dashboard and Center Console Switches

Light switch, removing and installing

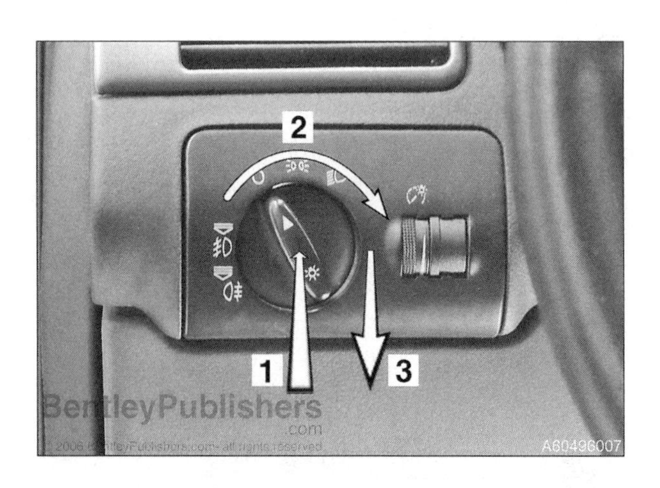

Working at light switch in dashboard:

- Push in switch knob (**1**) and at same time turn it clockwise (**2**).
- Hold in this position and pull switch housing out of dashboard (**3**).
- Detach electrical connectors from switch.

BentleyPublishers.com
BentleyPublishers.com—All Rights Reserved

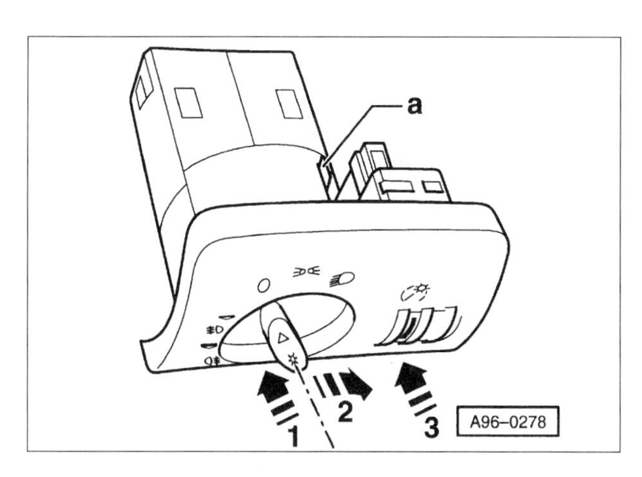

– To reinstall, reattach electrical connectors.

◀ Push switch knob in direction of arrow (**1**) and at same time turn it clockwise (**2**) to engage retaining catch (**a**).

– With switch in this position, push switch housing into recess in dashboard (**3**) until it clicks into place.

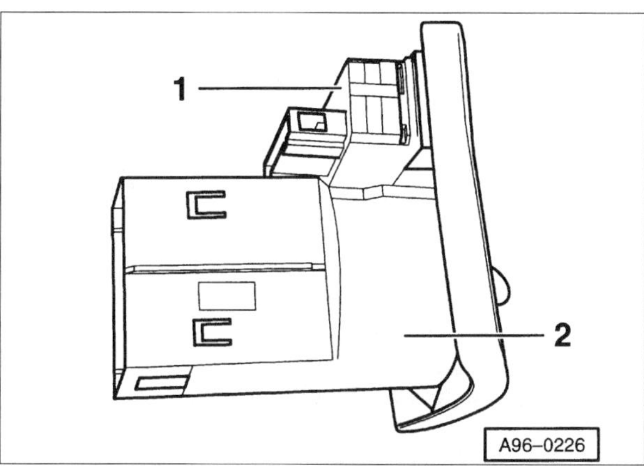

Instrument light dimmer switch, removing and installing

The instrument light dimmer is part of the light switch, but can be replaced separately.

– Remove light switch. See **Light switch, removing and installing** in this repair group.

◀ Use thin flat screwdriver to pry off dimmer switch (**1**) and separate from light switch housing (**2**).

– Installation is reverse of removal

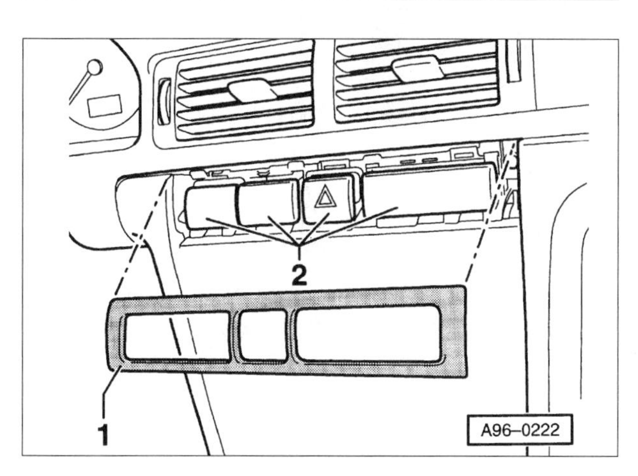

Center dashboard switches, removing and installing

◀ Using plastic prying tool or flat screwdriver tip wrapped with tape, carefully pry off center dashboard trim (**1**).

– Pull out switch (**2**) all the way and detach electrical connector.

– Installation is reverse of removal.

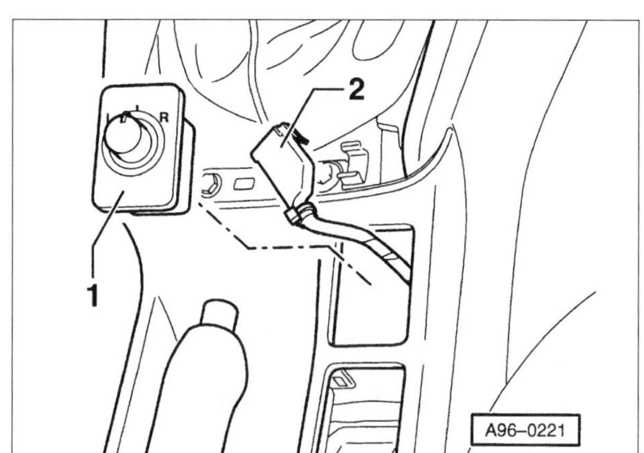

Mirror adjustment switch, removing and installing

– Using plastic prying tool or flat screwdriver tip wrapped with tape, carefully pry off center console insert and shifter boot.

◀ Reach under mirror switch (**1**) with fingers or screwdriver to push switch up and out of console. Detach electrical harness (**2**).

– Installation is reverse of removal. Check for missing clips at front and rear of mirror switch

BentleyPublishers.com
BentleyPublishers.com—All Rights Reserved

Steering Column Switches

The steering column contains the following switches:

- Ignition key cylinder and ignition electrical switch, covered in this repair group.
- Steering column stalk switches, covered in **48 Steering**:
  - -Turn signal and headlight flasher switch.
  - -Cruise control switch.
  - -Wiper and washer switch.
  - -MFI controller.
- Steering angle sensor, covered in **48 Steering**.

Steering column switch replacement requires removal of the driver airbag and steering wheel. Be sure to read airbag warnings and cautions in **48 Steering** and **69 Seat Belts, Airbags** prior to working on steering column switches.

Ignition key cylinder, removing and installing

– Remove driver airbag, steering wheel and steering column upper and lower trim. Remove steering column stalk switches. See **48 Steering**.

> ***WARNING —***
> - *The airbag is inflated by an explosive device. Handled improperly or without adequate safeguards, the airbag system can be very dangerous. See* **69 Seat Belts, Airbags**.

Vehicle with automatic transmission: Disengage interlock cable:

- Move transmission selector lever to P.
- Switch ignition key ON (position **B**).
- Lift ignition interlock lever (**1**) slightly and pull cable (**2**) out of steering lock housing.

NOTE —
- *For the following step, do not attempt to use the main vehicle key with the keyless entry radio transmitter.*

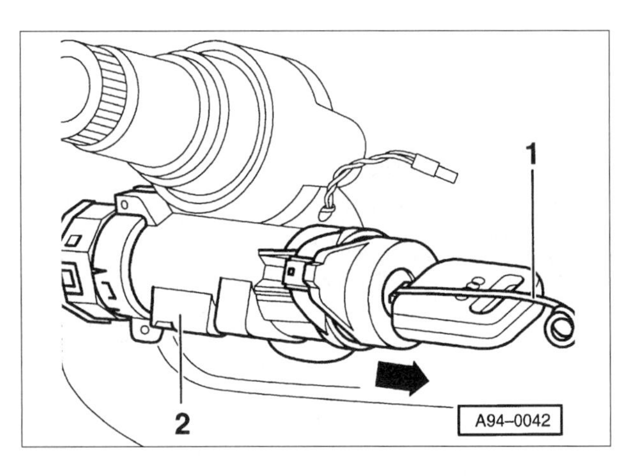

Using spare key or valet key:

- Insert key into ignition lock and switch ignition ON.
- Disconnect immobilizer induction coil harness.
- Insert steel wire (**1**) approx. 1.5 mm (0.06 in) in diameter as far as it will go in bore next to ignition key. A large paper clip is suitable.
- Pull ignition cylinder out of steering lock housing (**2**) in direction of **arrow**.

– When installing:

- Insert spare key or valet key into ignition cylinder and turn to ON.
- Insert steel wire into bore next to ignition key. Push wire in as far as it will go.
- Push lock cylinder together with ignition key all the way into steering lock housing.
- Pull wire out and push in lock cylinder until catch engages audibly.
- Reinstall steering column switches, steering wheel and airbag.

BentleyPublishers.com
BentleyPublishers.com—All Rights Reserved

Ignition electrical switch, removing and installing

– Remove driver airbag, steering wheel and steering column upper and lower trim. Remove steering column stalk switches. See **48 Steering**.

> ***WARNING —***
> - *The airbag is inflated by an explosive device. Handled improperly or without adequate safeguards, the airbag system can be very dangerous. See* **69 Seat Belts, Airbags**.

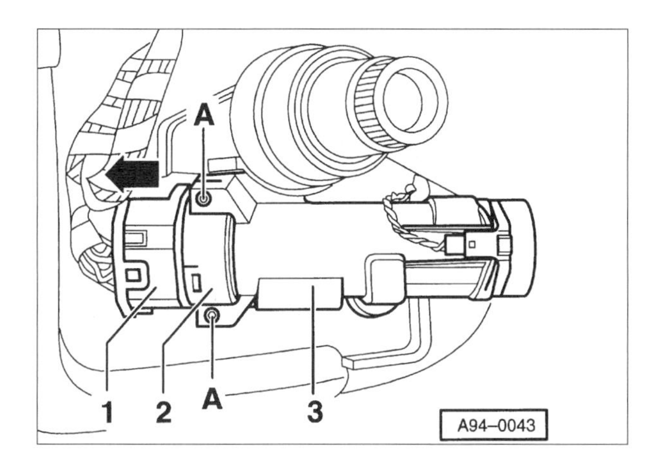

◀ Disconnect electrical connector (**1**) at ignition electrical switch (**2**).

– Scrape off sealing lacquer from two retaining screws (**A**). Loosen screws slightly.

– Pull electrical switch (**2**) out of steering lock housing (**3**), in direction of **arrow**.

– When installing:
- Place ignition electrical switch and ignition cylinder in ON position.
- Slide electrical switch and tighten two retaining screws on steering lock housing.
- Seal screw heads with lacquer.
- Reinstall steering column switches, steering wheel and airbag.

Ignition electrical switch pin assignments

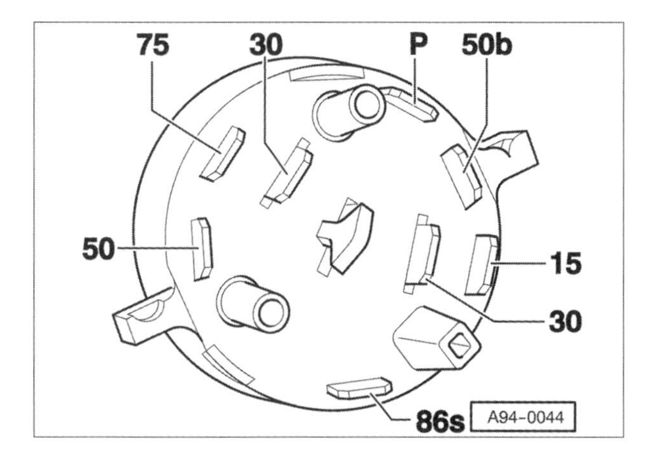

15: Ignition ON.

30: Battery power.

50 and **50B**: Power to starter solenoid.

75: Power to foglights, load reduction relay.

86s: Power to accessories.

P: Park position.

Window Switches

Left door window switch assembly, removing and installing

◀ Using plastic prying tool or flat screwdriver tip wrapped with masking tape, gently pry up (**arrow**) on front edge of switch assembly.

BentleyPublishers.com
BentleyPublishers.com—All Rights Reserved

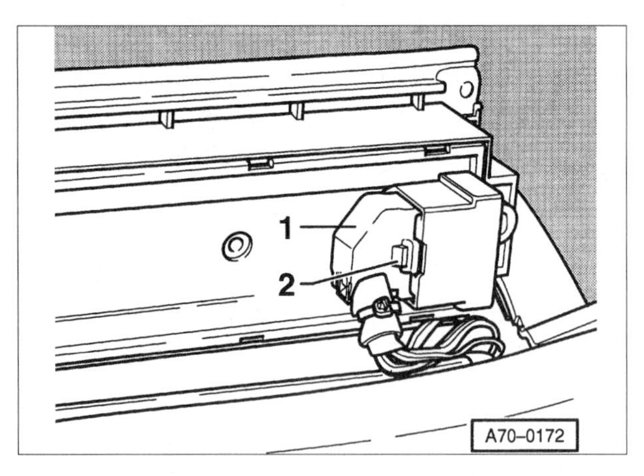

Turn switch assembly over and press on retainer tab (**2**) to detach harness connector (**1**).

– Installation is reverse of removal.

Right door window or rear window switch assembly, removing and installing

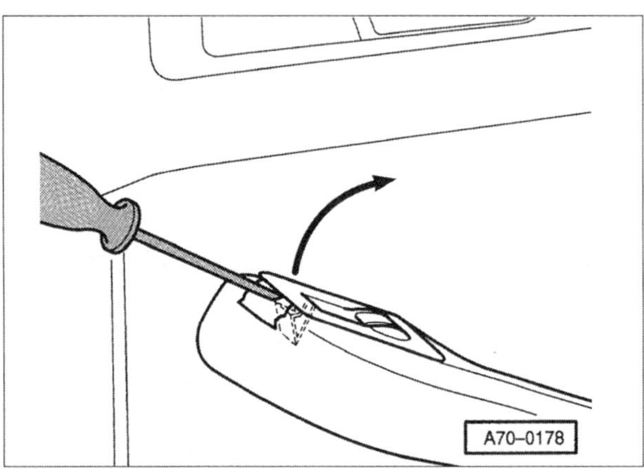

Using plastic prying tool or flat screwdriver tip wrapped with masking tape, gently pry up on front edge of switch assembly (**1**).

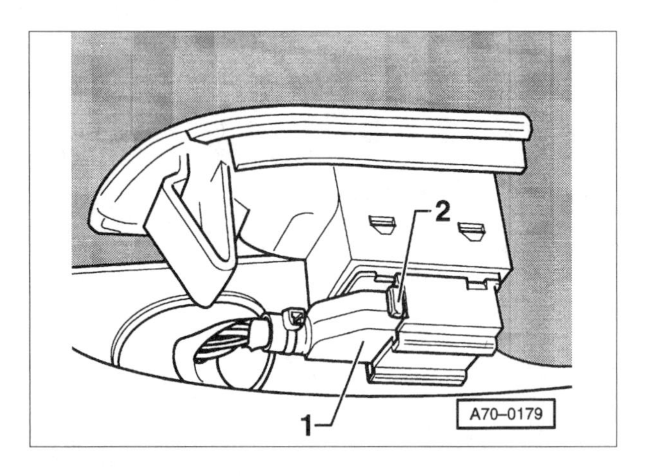

Turn switch assembly over and press on retainer tab (**2**) to detach harness connector (**1**).

– Installation is reverse of removal.

Horns

One horn is installed behind each end of the front bumper.

Horn, removing and installing

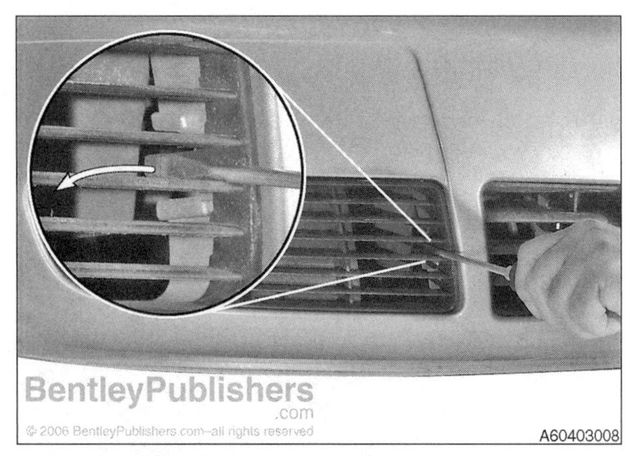

Insert screwdriver in lower grille slot. Pry gently on lock tab (**arrow**), then pull grille off.

BentleyPublishers.com
BentleyPublishers.com—All Rights Reserved

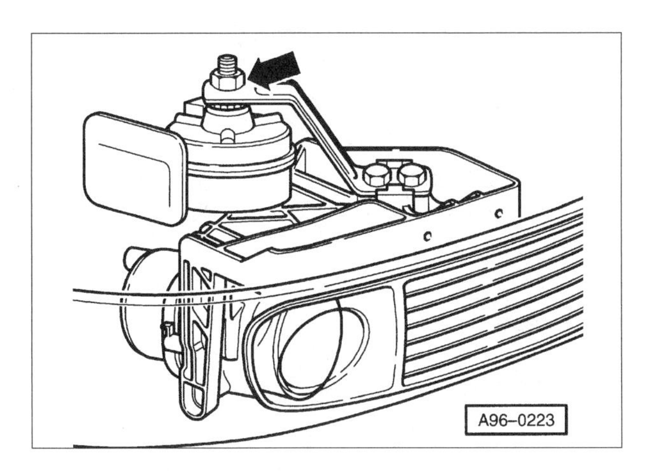

- ◀ Remove horn mounting nut (**arrow**).
- – Disconnect harness connector and lift out horn.
- – Installation is reverse of removal.

ANTI-THEFT SYSTEM

Electronic immobilizer system version 3 is standard on models covered by this manual. Anti-theft alarm is available as an option.

The ultrasonic interior monitor, supplied in conjunction with the optional anti-theft alarm, triggers the anti-theft alarm if an attempt is made to break into the vehicle.

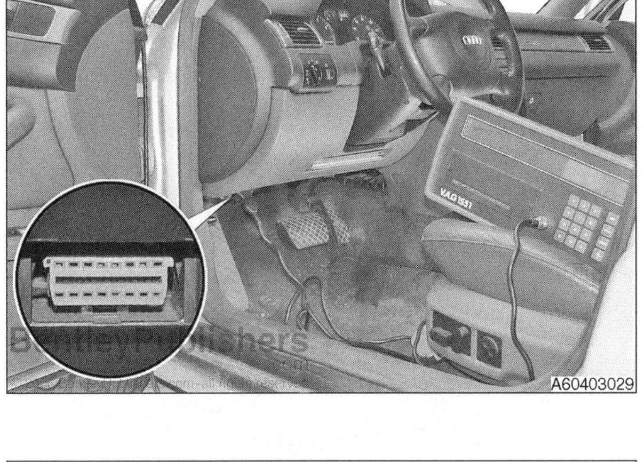
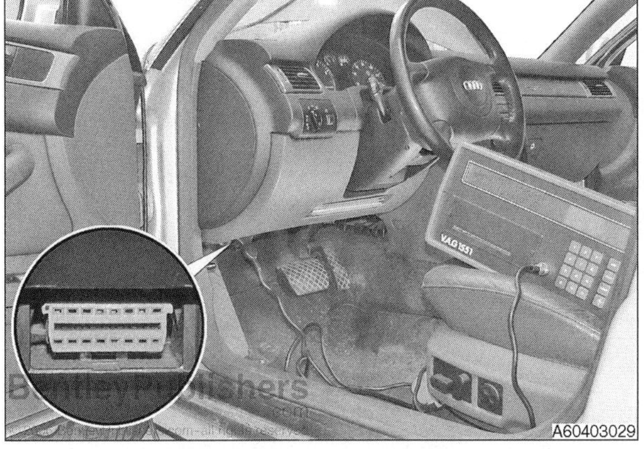

On-board diagnostics

If the on-board diagnostics (OBD II) system recognizes malfunctions in the immobilizer, anti-theft or central locking systems, it sets diagnostic fault codes (DTCs).

◀ Access fault memory with VAG scan tool or equivalent plugged into DLC plug (**inset**) under left side dashboard.

Electronic immobilizer

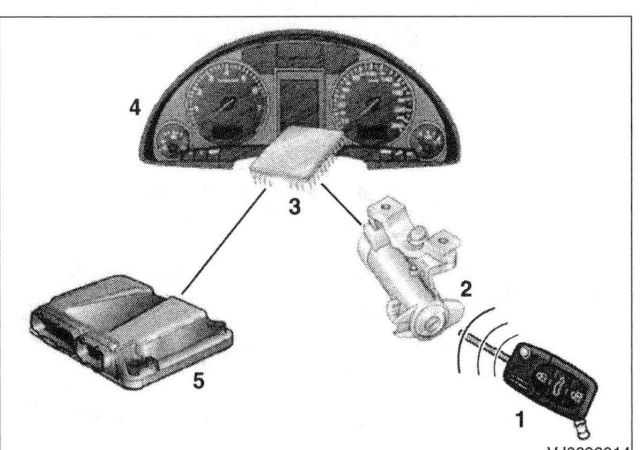

◀ Electronic immobilizer components are as follows:

1. Ignition key with wireless transponder
2. Ignition lock cylinder with immobilizer induction coil
3. Immobilizer control module integrated with **4**
4. Instrument cluster
5. Engine control module (ECM)

Electronic immobilization functions as follows:

- With vehicle key inserted into ignition switch, signal from ignition cylinder induction coil energizes key transponder.
- Key transponder sends unique signal to induction coil.
- Key signal is relayed to immobilizer control module in instrument cluster.
- If control module recognizes key as valid, signal is relayed to engine control module (ECM) via CAN-bus network and vehicle can be started.
- If key is not recognized as valid, starter and engine management system are blocked: Vehicle cannot be started.

BentleyPublishers.com
BentleyPublishers.com—All Rights Reserved

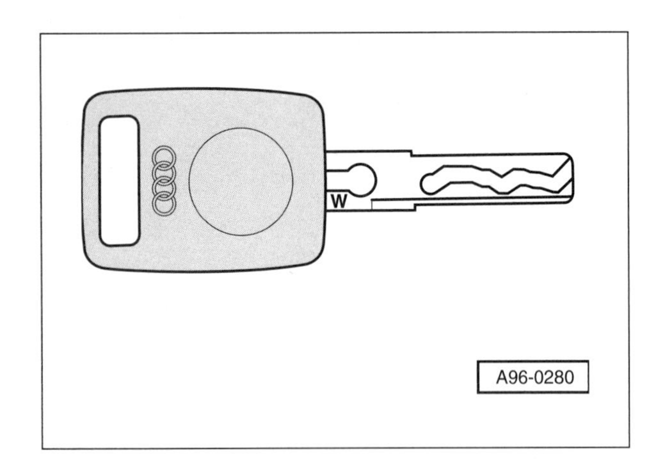

Ignition keys are inscribed with a W indicating that the key is capable of being coded to electronic immobilizer version 3. Only keys with this basic code can be adapted to the vehicle.

The immobilizer control module (J362) is integrated with the instrument cluster control module. In case of anti-theft immobilizer control module failure, replace the entire instrument cluster. See **90 Instruments**.

The immobilizer induction coil (D2) is integrated with the ignition key cylinder and cannot be replaced separately. In case of induction coil malfunction, replace the ignition cylinder. See **Ignition key cylinder, removing and installing** in this repair group.

The key transponder is integrated with the vehicle key. In case of transponder malfunction, replace the complete key. Once a key is replaced, adapt all the vehicle keys to the vehicle using the VAG scan tool or equivalent. If only one key is adapted, the remaining keys will not start the vehicle.

The immobilizer warning light illuminates briefly (max. 3 seconds) and then goes out when an authorized vehicle key is used. If an unauthorized vehicle key is used or when a system malfunction occurs, the warning light blinks constantly during ignition ON.

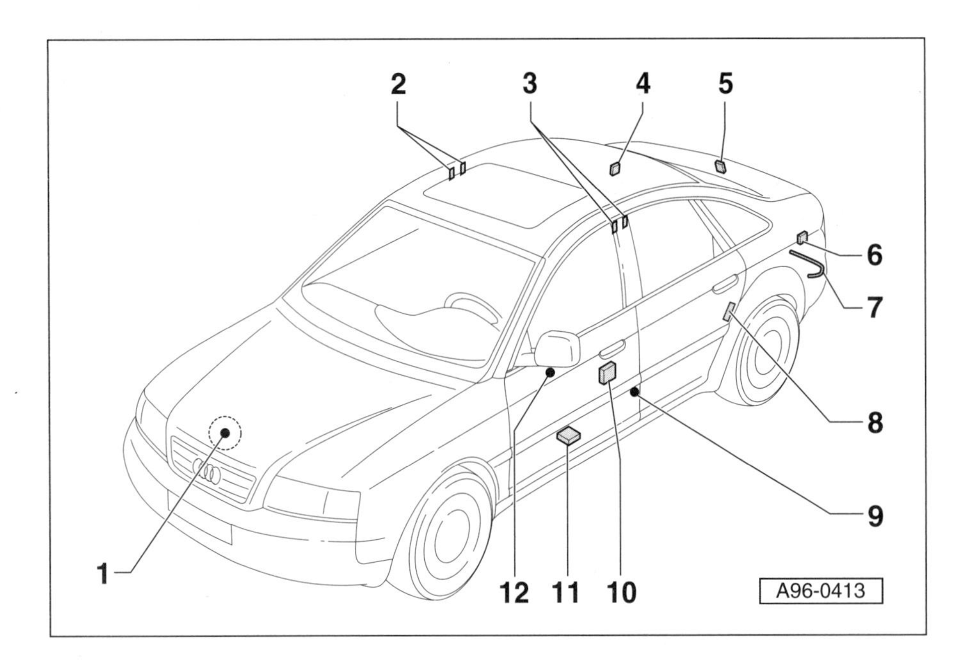

Anti-theft and central locking components

1. **Hood alarm switch (F120)**
2. **Ultrasound interior motion sensor, right (G170)**
3. **Ultrasound interior motion sensor, left (G171)**
4. **Fuel tank filler flap lock actuator (V155)**
5. **Trunk lid lock actuator (F124, F206, F218, V53, V139)**
6. **Alarm horn (H8)**
7. **Alarm system and keyless entry antenna (R47)**
8. **Interior motion sensor control module (J347)**
9. **Interior motion sensor disable switch (E183)**
10. **Door lock actuator (F2, F131, F241, F243, V56, V161)**
11. **Central locking and anti-theft control module (J429)**
12. **Central locking master switch (E150)**
13. **Avant: Glass breakage sensors (G183, G184) below rear side glass (not in this illustration)**

NOTE —

- *2000 model is illustrated. Other models are similar.*

BentleyPublishers.com
BentleyPublishers.com—All Rights Reserved

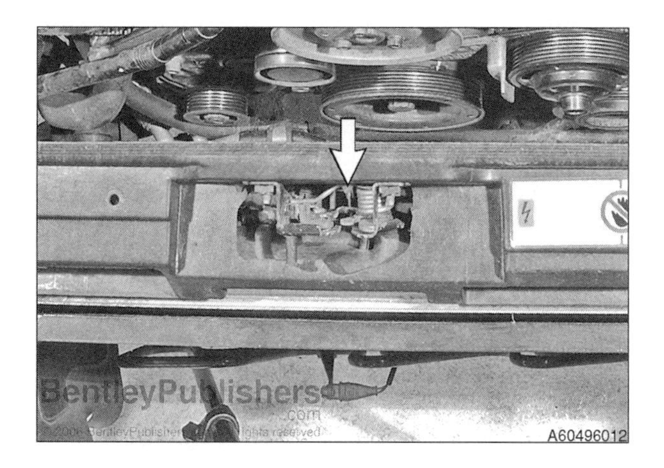

Hood alarm switch (F120)

In hood lock mechanism on front lock carrier (radiator support frame)

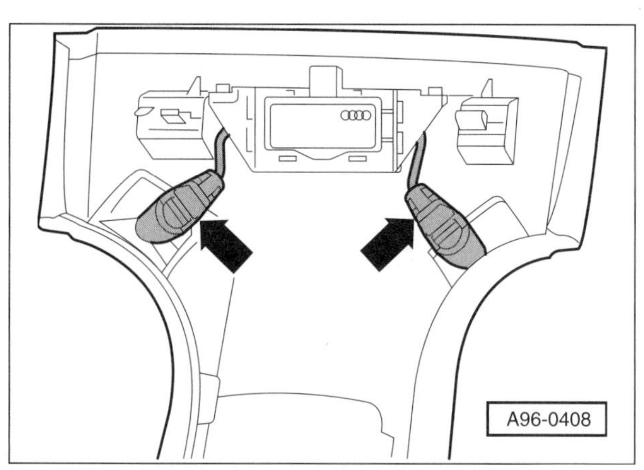

Ultrasound interior motion sensor (G170, G171)

At head liner above left and right B-pillar

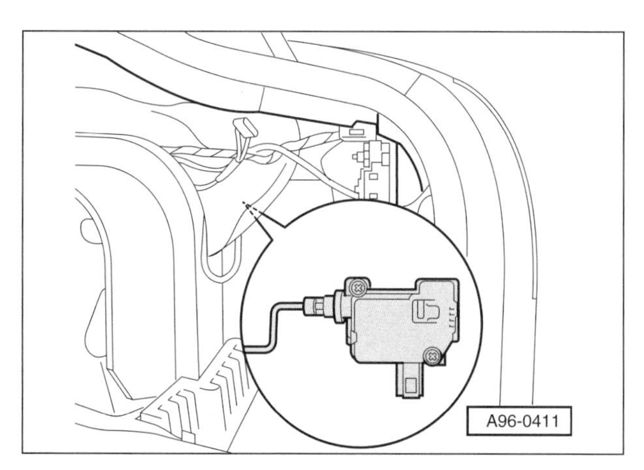

Fuel tank filler flap lock actuator (V155)

In luggage compartment behind right rear side panel trim.

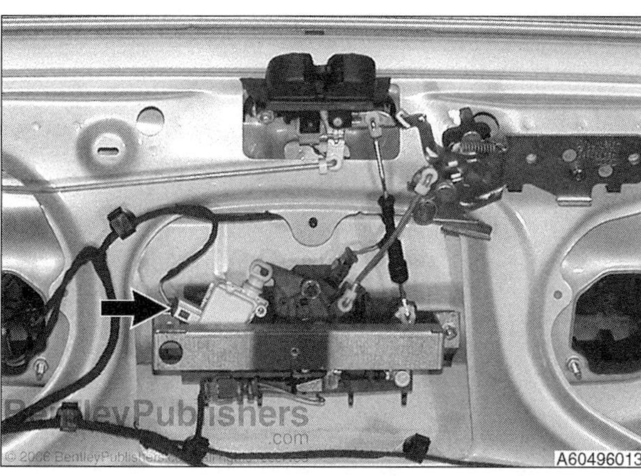

Luggage compartment lid lock actuator (F124, F206, F218, V53, V139)

Behind trim (Avant tailgate lock illustrated)

BentleyPublishers.com
BentleyPublishers.com—All Rights Reserved

Alarm horn (H8)

⮜ Sedan: In trunk behind left storage compartment cover

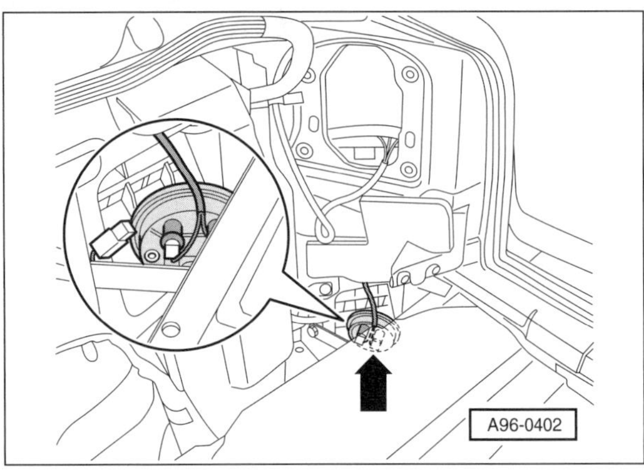

⮜ Avant: In luggage compartment behind left storage compartment cover

Alarm system and keyless entry antenna (R47)

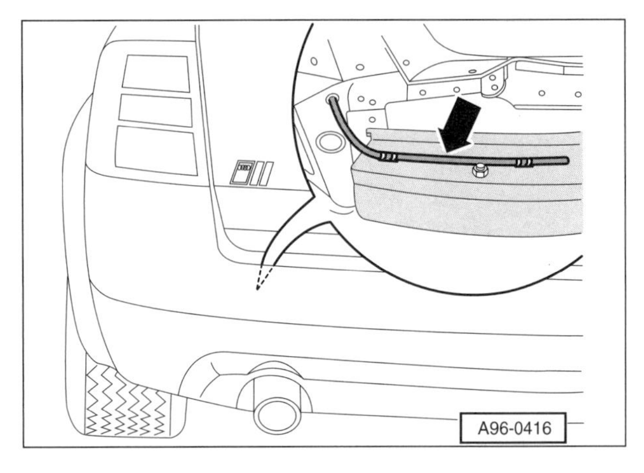

⮜ Sedan: Left side of rear bumper cover

- Avant: Right side of rear bumper cover

Interior motion sensor control module (J347)

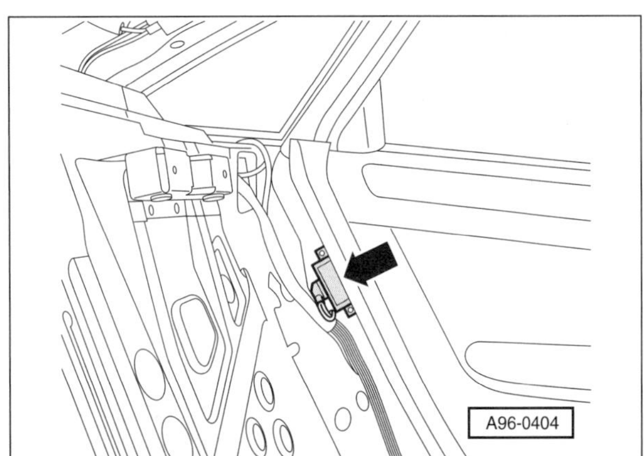

⮜ At lower C-pillar, behind trim

Interior motion sensor control module actuates warning lights next to door lock buttons.

BentleyPublishers.com
BentleyPublishers.com—All Rights Reserved

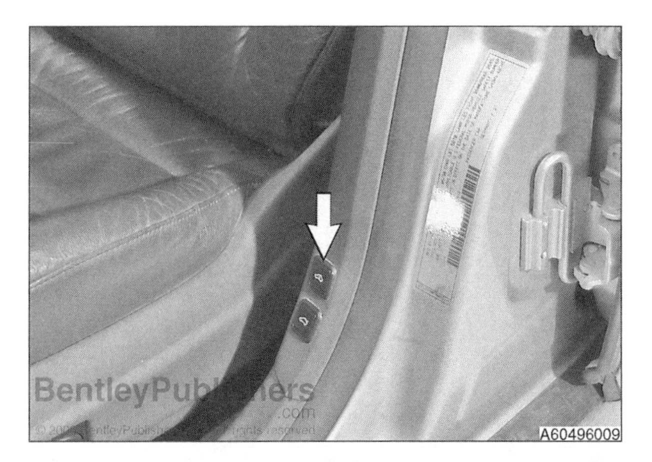

Interior motion sensor disable switch (E183)

◄ On left lower B-pillar

When locking up vehicle, deactivate interior monitor manually by pressing interior monitor switch (**arrow**) if, for example, a dog is to remain in vehicle.

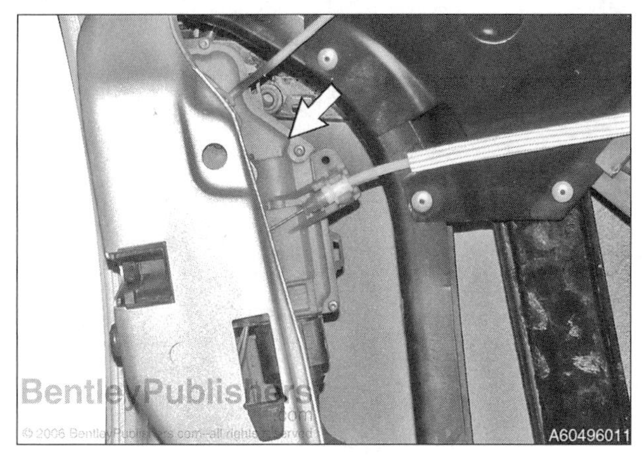

Door lock actuator (F2, F131, F241, F243, V56, V161)

◄ In door cavity behind door trim panel

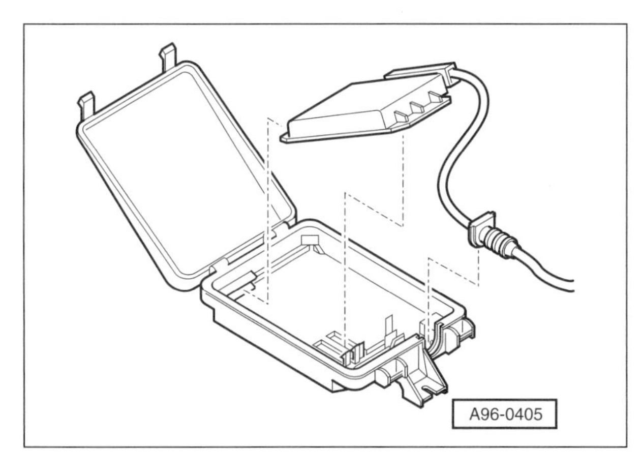

Central locking and anti-theft control module (J429)

◄ In protective case underneath driver seat

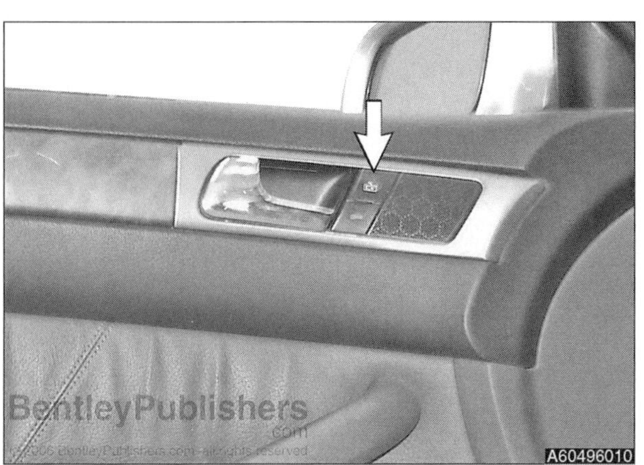

Central locking master switch (E150)

◄ Top front of driver door

BentleyPublishers.com
BentleyPublishers.com—All Rights Reserved

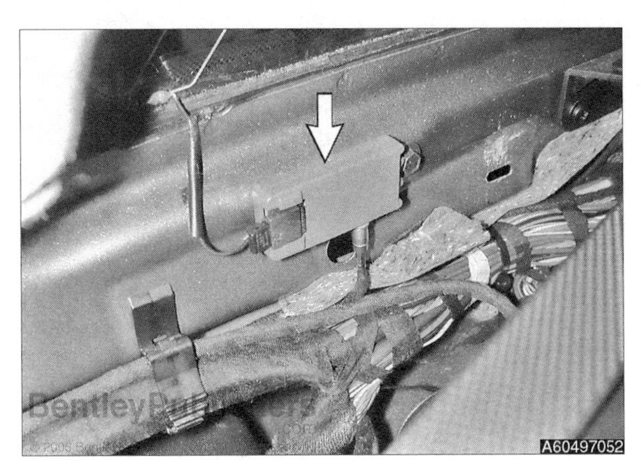

Glass breakage sensors (G183, G184)

< Avant: Below rear side glass, behind trim

BentleyPublishers.com
BentleyPublishers.com—All Rights Reserved

97 Fuses, Relays, Component Locations

TABLES

GENERAL

This repair group covers fuses, relays, ground locations and electrical component locations, primarily via photos or illustrations.

The complex nature of the Audi A6 electrical systems requires a large number of fuses and electrical components. Locating the correct component in this array of equipment is an important first step in electrical diagnosis. Electrical equipment and accessories installed vary depending on model and model year. Always confirm that the proper electrical component is correctly identified.

Almost all electrical circuits are protected by fuses. For a full listing of fuses and fuse locations, see **Fuse and Relay Panels** in this repair group.

Investigating and correcting ground problems often solves mysterious and difficult to trace electrical problems. For ground information, see **Grounds** in this repair group.

Also see:

- **27 Battery, Starter, Alternator**
- **87 Heating and Air-conditioning** for climate control flap motors and temperature sensors
- **9 Electrical System–General** for information on bus systems
- **91 Radio, Communication** for sound system, telephone and navigation component locations
- **EWD Electrical Wiring Diagrams**

BentleyPublishers.com
BentleyPublishers.com—All Rights Reserved

Warnings and Cautions

Please read the following before doing any work on your electrical system.

WARNING —

- *Airbags utilize explosive devices. Handle with extreme care. Refer to the warnings and cautions in* **69 Seat Belts, Airbags**.
- *The ignition system of the car operates at lethal voltages. If you have a weak heart or wear a pacemaker, do not expose yourself to ignition system electric currents. Take extra precautions when working on the ignition system or when servicing the engine while it is running or the key is ON. See* **28 Ignition System** *for additional ignition system warnings and cautions.*

CAUTION—

- *Do not disconnect the battery with the engine running.*
- *Switch the ignition OFF and remove the negative (-) battery cable before removing any electrical components. Connect and disconnect electrical connectors and ignition test equipment leads only while the ignition is switched OFF.*
- *Relay and fuse positions are subject to change and may vary from car to car. If questions arise, an authorized Audi dealer is the best source for the most accurate and up-to-date information.*
- *Use a digital multimeter for electrical tests. Switch the multimeter to the appropriate function and range before making test connections.*
- *Many control modules are static sensitive. Static discharge damages them permanently. Handle the modules using proper static prevention equipment and techniques.*
- *To avoid damaging harness connectors or relay panel sockets, use jumper wires with flat-blade connectors that are the same size as the connector or relay terminals.*
- *Do not try to start the engine of a car which has been heated above 176°F (80°C) (for example, in a paint drying booth). Allow it to cool to normal temperature.*
- *Disconnect the battery before doing any electric welding on the car.*
- *Do not wash the engine while it is running, or any time the ignition is ON.*
- *Choose test equipment carefully. Use a digital multimeter with at least 10 MΩ input impedance, or an LED test light. An analog meter (swing-needle) or a test light with a normal incandescent bulb may draw enough current to damage sensitive electronic components.*
- *Do not use an ohmmeter to measure resistance on solid state components such as control modules.*
- *Disconnect the battery before making resistance (ohm) measurements on a circuit.*

BentleyPublishers.com
BentleyPublishers.com—All Rights Reserved

ELECTRICAL COMPONENT LOCATIONS

Table a is a cross-referenced listing of electrical components for cars covered by this manual.

Audi identifies electrical components by a letter and/or a number in electrical schematics. These codes, listed in column 2, are also used in **EWD Electrical Wiring Diagrams**.

Figure numbers in column 4 and page numbers in column 5 refer to photographs and illustrations immediately following the table.

NOTE—
- *Every component is not installed in every car.*
- *Due to changes in production, component locations may vary from what is illustrated. Consult your Audi dealer for the latest information.*
- *For information about bus systems and associated control modules, see* **9 Electrical System–General**.

Table a. Electrical component locations

Component	Code	Location	Fig. #	Pg. #
13-fold relay panel		Underneath left side dashboard	**Fig. 48** **Fig. 49**	97-27 97-28
3-fold relay panel		E-box, left side of plenum chamber under plastic cover	**Fig. 44** **Fig. 45**	97-24 97-24
8-fold relay panel		Underneath left side dashboard	**Fig. 48** **Fig. 51**	97-27 97-29
A/C and heater blower motor	V2	In evaporator housing behind glove compartment	**Fig. 30** **Fig. 31**	97-20 97-20
A/C and heater blower motor control module	J126	Under right side dashboard, to left of blower motor on evaporator housing		
A/C and heater blower motor resistor	N24	On blower motor		
A/C and heater control module	J255 J301	A/C and heater control panel in center dashboard		
A/C and heater left side flap motors and temperature sensors		Left side of evaporator housing *see also* **87 Heating and Air-conditioning**	**Fig. 29**	97-20
A/C and heater right side flap motors and temperature sensors		Right side of evaporator housing *see also* **87 Heating and Air-conditioning**	**Fig. 31**	97-20
A/C compressor clutch	N25	On A/C compressor, left front engine compartment		
A/C compressor clutch relay	J44	13-fold relay panel underneath left side dashboard	**Fig. 49**	97-28
A/C compressor regulator valve, 2002 - 2004	N280	On A/C compressor, left front engine compartment		
A/C high pressure sensor, 2002 - 2004	G65	Right front engine compartment		
A/C pressure switch, 1998 - 2001	F129	Right side of A/C condenser		
ABS		*see also* ESP entries		
ABS control module, Bosch 5.3	J104	Right footwell under floor covering		
ABS control module, Bosch 5.7	J104	Left side of engine compartment	**Fig. 12**	97-14
ABS hydraulic pump, Bosch 5.3	V156	Under ABS hydraulic unit, left side of engine compartment	**Fig. 1**	97-11
ABS hydraulic unit	N55	Left side of engine compartment	**Fig. 1** **Fig. 12**	97-11 97-14
ABS return flow pump	V39	Part of ABS hydraulic unit, engine compartment left front	**Fig. 12**	97-14
ABS return flow pump relay	J105	8-fold relay panel underneath left side dashboard	**Fig. 52**	97-29
ABS solenoid valve relay	J106	8-fold relay panel underneath left side dashboard	**Fig. 52**	97-29

BentleyPublishers.com
BentleyPublishers.com—All Rights Reserved

Electrical Component Locations

Table a. Electrical component locations				
Component	**Code**	**Location**	**Fig. #**	**Pg. #**
ABS wheel speed sensor	G44 - G47	Integral with wheel bearing assembly (front wheel drive) Mounted at back of wheel bearing housing (quattro)	**Fig. 22**	97-16
Accelerator pedal module		Base of accelerator pedal	**Fig. 29**	97-20
Accelerator pedal position sensor 2	G185	Base of accelerator pedal	**Fig. 29**	97-20
After-run coolant pump	V51	Under vehicle, behind left side of oil pan	**Fig. 15**	97-15
After-run coolant pump relay	J151	3-fold relay panel in E-box, left side of plenum chamber under plastic cover	**Fig. 45**	97-24
Air quality sensor	G238	Under plenum cover, attached to dust and pollen filter housing	**Fig. 11**	97-14
Airbag connector, left front		Under steering column trim covers		
Airbag connector, right front		Underneath right dashboard, above glove compartment		
Airbag control module	J234	On center tunnel under front of center console		
Airbag crash sensor, front side airbag	G179 G180	Under front seat and carpet, on rear of floor pan brace		
Airbag crash sensor, rear side airbag	G256 G257	On rear door sill		
Airbag igniter, right, left	N131	Base of B-pillar behind trim	**Fig. 34**	97-21
Airbag spiral spring	F138	Steering column		
Airbag, driver		Steering wheel		
Airbag, passenger		Right side dashboard	**Fig. 30**	97-20
Airbag, side		Outer seat bolster		
Alarm horn	H8	Left side luggage compartment, behind trim	**Fig. 39**	97-22
Alarm system		*see* Central locking		
Alternator (generator)	C	Right front of engine compartment, on engine	**Fig. 16**	97-15
Amplifier, audio		Right side luggage compartment, behind trim *see* **91 Radio and Communication**		
Antenna amplifier	R82	Behind D-pillar trim		
Antenna selection control module	J515	Under parcel shelf, left side		
Antenna, alarm and central locking	R47	Avant: Rear bumper, right side Sedan: Rear bumper, left side *see also* **96 Interior Lights, Switches, Anti-theft**		
Antenna, navigation / telephone / auxiliary heater	R66 R172	Roof *see also* **91 Radio and Communication**		
Antenna, radio and telephone	R51	Left rear fender		
Antenna, sound system	R11 R50	Roof (Avant) Trunk lid (sedan) *see also* **91 Radio and Communication**		
Antenna, tire pressure check	R59 R60 R61 R62	Behind wheel housing liner, above wheel (allroad)		
Anti-theft control module		*see* Central locking control module		
ASR		*see* ABS *see also* ESP		
Audio components		*see* **91 Radio and Communication**		
Automatic headlight vertical aim control system module		*see* Headlight aim control module		
Automatic transmission control module, 5-speed automatic transmission	J217	Inside protective box in floor cavity underneath front passenger seat and carpet, in front of floor pan brace		

BentleyPublishers.com
BentleyPublishers.com—All Rights Reserved

Table a. Electrical component locations				
Component	**Code**	**Location**	**Fig. #**	**Pg. #**
Automatic transmission control module, Multitronic® transmission (CVT)	J217	Inside transmission, at rear of case		
Automatic transmission gear position switch		*see* Automatic transmission multifunction range switch		
Automatic transmission multifunction range switch	F125	Left side of transmission housing	**Fig. 21**	97-16
Automatic transmission selector lever switch		At shifter		
Auxiliary coolant pump		*see* After-run coolant pump		
Back-up light switch, automatic transmission		*see* Automatic transmission multifunction switch		
Back-up light switch, manual transmission	F4	Top of transmission		
Battery	A	Rear of engine compartment in plenum chamber	**Fig. 1**	97-11
Blower motor		*see* A/C and heater blower motor		
Blower motor resistor		*see* A/C and heater blower motor resistor		
Brake booster pressure sensor	G294	Under plenum cover, directly in front of brake booster	**Fig. 5**	97-12
Brake booster vacuum pump	V192	Engine compartment, left side	**Fig. 12**	97-14
Brake booster vacuum pump fuse, allroad	S279	13-fold relay panel underneath left side dashboard	**Fig. 49**	97-28
Brake booster vacuum pump fuse, except allroad	S279	Central carrier relay panel under left side dashboard	**Fig. 55**	97-30
Brake booster vacuum pump relay	G569	Central carrier relay panel under left side dashboard	**Fig. 55**	97-30
Brake fluid level warning switch	F34	Brake fluid reservoir in plenum chamber	**Fig. 1**	97-11
Brake light switch	F	On pedal cluster above brake vacuum vent switch		
Brake vacuum vent switch	F47	On pedal cluster, below brake light switch		
Broken window glass sensor, left rear	G183	Under left rear window behind trim	**Fig. 41**	97-23
Broken window glass sensor, right rear	G184	Under right rear window behind trim		
Camshaft adjustment valve 1	N205	Rear of right cylinder head (bank 1)		
Camshaft adjustment valve 2	N208	Front of left cylinder head (bank 2)		
Camshaft sensor, left	G163	Rear of left cylinder head (bank 2)		
Camshaft sensor, right	G40	Front of right cylinder head (bank 1)	**Fig. 2** **Fig. 8**	97-11 97-13
CD changer	R41	Left side luggage compartment in storage compartment *see also* **91 Radio and Communication**	**Fig. 39**	97-22
Central carrier relay panel		Underneath left side dashboard	**Fig. 55**	97-30
Central locking control module	J429	In protective box underneath driver seat and carpet, in front of floor pan brace		
Charge air pressure sensor, biturbo	G31	In air intake boot, on intake manifold, front	**Fig. 8**	97-13
Connector station, A-pillar		Behind A-pillar trim in kick panels, left and right	**Fig. 26** **Fig. 27**	97-18 97-19
Coolant		*see* Engine coolant		
Cooling fan		*see* Engine cooling fan		
Crankshaft sensor		*see* Engine speed (rpm) sensor		
Crash sensor		*see* Airbag crash sensor		
Cruise control module	J213	Behind glove compartment, near blower motor housing	**Fig. 30**	97-20
Cruise control stalk switch	E45	On steering column, left		
Cruise control vacuum pump	V18	Underneath ABS control module, left front wheel housing	**Fig. 19**	97-16
Data link connector (DLC)		Under left side dashboard	**Fig. 28**	97-19
Diagnostic connector		*see* Data link connector (DLC)		

BentleyPublishers.com
BentleyPublishers.com—All Rights Reserved

Electrical Component Locations

Table a. Electrical component locations				
Component	**Code**	**Location**	**Fig. #**	**Pg. #**
Door contact switch	F2 F3 F10 F11	Door lock motor	**Fig. 32**	97-20
Door courtesy light		Door edge	**Fig. 33**	97-21
Door lock motor	V56 V57 V97 V115	In door lock	**Fig. 32**	97-20
E-box (electronics box)		Left side plenum chamber, left rear engine compartment	**Fig. 1** **Fig. 44**	97-11 97-24
E-box connector station		In E-box, left side of plenum chamber under plastic cover	**Fig. 44** **Fig. 46**	97-24 97-25
ECM		*see* Engine control module (ECM)		
Electronic immobilizer induction coil		Integrated with ignition key cylinder		
Electronic power control (EPC)		*see* Electronic throttle drive		
Electronic throttle drive	G186	Top rear of engine, behind intake manifold *see* Throttle valve control module *see also* Throttle angle sensor *see also* Throttle position sensor		
Electronics box		*see* E-box		
Emergency flasher	E3	Center of dashboard		
Engine control module (ECM)	J220	In E-box, left side of plenum chamber under plastic cover	**Fig. 6** **Fig. 44**	97-12 97-24
Engine control module (ECM) fuse, 2001 - 2004	S102	3-fold relay panel in E-box, left side of plenum chamber under plastic cover	**Fig. 45**	97-24
Engine control module (ECM) power supply relay, 2001 - 2004	J271	3-fold relay panel in E-box, left side of plenum chamber under plastic cover	**Fig. 45**	97-24
Engine coolant level warning switch	F66	Bottom of coolant reservoir expansion tank	**Fig. 1**	97-11
Engine coolant pump relay		*see* After-run coolant pump relay		
Engine coolant temperature (ECT) sensor	G2 G62	In coolant pipe behind intake manifold		
Engine cooling fan	V7 V177	In front of engine at radiator		
Engine cooling fan control module, 1998 - 2001	J293	Left lower engine compartment on frame rail		
Engine cooling fan control module, 2002 - 2004	J293	In front of engine compartment attached to fan		
Engine cooling fan relay	J26	8-fold relay panel underneath left side dashboard	**Fig. 52**	97-29
Engine cooling fan relay, high speed	J101	8-fold relay panel underneath left side dashboard	**Fig. 52**	97-29
Engine cooling fan series resistor	N39	Left front, underneath longitudinal chassis member	**Fig. 18**	97-15
Engine cooling fan switch	F54 F18	On radiator, lower right	**Fig. 14**	97-14
Engine oil temperature sensor	G8	Left front of cylinder block, above oil filter housing		
Engine speed (rpm) sensor	G28	Left transmission bellhousing near drive flange		
EPC		*see* Electronic throttle drive		
ESP		*see also* ABS entries		
ESP sensor unit		Underneath rear seat	**Fig. 35**	97-21
ESP switch	E256	Center console		
EVAP purge valve		*see* Fuel tank evaporative control canister purge regulator valve		
Evaporator temperature sensor		*see* **87 Heating and Air-conditioning**		
Exhaust flap valve	N321	On exhaust system between muffler and tailpipe tips	**Fig. 25**	97-17

BentleyPublishers.com
BentleyPublishers.com—All Rights Reserved

Table a. Electrical component locations				
Component	**Code**	**Location**	**Fig. #**	**Pg. #**
Exhaust temperature sensor 1, 2 (2.7 liter engine)	G235 G236	In turbocharger outlet pipe, right, left (2.7 liter engine)	**Fig. 10**	97-13
Exhaust temperature sensor modules (2.7 liter engine)		Center top of engine, right of intake manifold	**Fig. 9**	97-13
Flap motor		*see* **87 Heating and Air-conditioning**		
Flasher		*see* Emergency flasher		
Foglight		Front bumper under left and right headlights	**Fig. 18**	97-15
Foglight relay	J5	13-fold relay panel underneath left side dashboard	**Fig. 49**	97-28
Fuel injectors		Top of intake manifold, front of engine		
Fuel level sensor, center	G237	Top of fuel tank, access through luggage compartment (quattro)	**Fig. 37**	97-22
Fuel level sensor, left	G169	Top of fuel tank, left	**Fig. 36**	97-21
Fuel level sensor, right	G	Top of fuel tank, right	**Fig. 36**	97-21
Fuel pump control module fuse, RS6	S81	Central carrier relay panel under left side dashboard	**Fig. 55**	97-30
Fuel pump control module, RS6	J538	Top right side of cylinder head		
Fuel pump relay	J17	Micro central electric panel under left side dashboard	**Fig. 50**	97-28
Fuel pump, all models except RS6	G6	In fuel tank, access under rear seat bench	**Fig. 36**	97-21
Fuel pump, RS6		Underneath right rear of vehicle, ahead of fuel tank		
Fuel tank evaporative control canister purge regulator valve, 2.7 liter engine	N80	Right side of intake manifold	**Fig. 9**	97-13
Fuel tank evaporative control canister purge regulator valve, 2.8 liter engine	N80	Right side of engine compartment	**Fig. 1**	97-11
Fuel tank filler door lock motor	V155	Right side of luggage compartment		
Fuel tank filler neck ground terminal		Underneath right rear wheel housing liner	**Fig. 24**	97-17
Fuel tank leak detection pump	V144	Back of left rear wheel housing, behind liner	**Fig. 23**	97-17
Fuse panel S1		3-fold relay panel in E-box, left side of plenum chamber under plastic cover	**Fig. 44** **Fig. 45**	97-24 97-24
Fuse panel S2		Main fuse panel, left end of dashboard	**Fig. 47**	97-25
Fuses		*see* **Fuse and Relay Panels** in this repair group		
Generator		*see* Alternator		
GPS		*see* **91 Radio and Communication**		
Grounds		*see* **Grounds** in this repair group	**Fig. 56**	97-30
Headlight aim adjustment motor	V48 V49	In headlight housing, below headlight	**Fig. 7**	97-13
Headlight aim control module, 1998 - 2000	J431	Left side luggage compartment, behind trim		
Headlight aim control module, 2001 - 2004	J431	Behind glove compartment on bracket	**Fig. 31**	97-20
Headlight bulb control module	J343 J344	In headlight housing, below headlight		
Headlight washer pump	V11	Washer fluid reservoir under left front fender	**Fig. 19**	97-16
Heater core		In evaporator housing, access from left side	**Fig. 29**	97-20
High intensity discharge (HID) headlight		*see* Headlight entries		
High pressure sensor		*see* A/C high pressure sensor		
Hood alarm switch	F120	Part of hood latch assembly	**Fig. 1**	97-11
Horn button		Steering wheel		
Horn relay	J4	Micro central electric panel under left side dashboard	**Fig. 50**	97-28
Horn, high tone	H2	Behind front bumper, left side	**Fig. 17**	97-15
Horn, low tone	H7	Behind front bumper, right side	**Fig. 18**	97-15
Hydraulic pump relay, 2001 - 2004	J555	Micro central electric panel under left side dashboard	**Fig. 50**	97-28

BentleyPublishers.com
BentleyPublishers.com—All Rights Reserved

Electrical Component Locations

Table a. Electrical component locations

Component	Code	Location	Fig. #	Pg. #
Ignition coil pack, 2.8 liter engine code AHA		Front of engine compartment in front of intake manifold	**Fig. 1**	97-11
Ignition coils power output stage, 2.7 liter V6 engine code APB		Air filter housing		
Ignition coils power output stage, 2.8 liter V6 engine code AHA	N122	Front of engine compartment in front of intake manifold	**Fig. 1**	97-11
Ignition coils power output stage, all except 2.7 liter engine code APB, 2.8 liter engine code AHA		Integrated with ignition coils		
Ignition coils, all except 2.8 liter engine code AHA		Above each spark plug, top of cylinder head covers		
Immobilizer		*see* Electronic immobilizer *see also* Central locking		
Induction coil		*see* Electronic immobilizer induction coil		
Instrument cluster combination processor	J218	Behind instrument cluster		
Instrument cluster control module	J285	Instrument cluster		
Intake air temperature (IAT) sensor	G42	Mounted on intake air ducting, rear of intake manifold		
Intake manifold change-over valve, 2.8 liter engine	N156	Left rear intake manifold above intake air duct	**Fig. 1**	97-11
Intake manifold change-over valve, 3.0 liter engine	N261	Front of intake manifold	**Fig. 13**	97-14
Kickdown switch	F8	In engine compartment on bulkhead	**Fig. 1**	97-11
Knock sensors	G61 G66	Near right cylinder head underneath intake manifold Near left cylinder head underneath intake manifold		
Lamp control module	J123	13-fold relay panel underneath left side dashboard	**Fig. 49**	97-28
Leak detection pump		*see* Fuel tank leak detection pump		
Level control system accumulator valve	N311	Under vehicle, rear, center, above stabilizer bar (allroad)	**Fig. 42**	97-23
Level control system compressor motor	V66	Under vehicle, rear, driver side	**Fig. 43**	97-23
Level control system compressor relay	J403	8-fold relay panel underneath left side dashboard	**Fig. 52**	97-29
Level control system control module	J197	In luggage compartment, left rear	**Fig. 41**	97-23
Level control system sensors	G76 G77 G78 G84	In front wheel housing In left rear wheel housing	**Fig. 40**	97-22
Level control system solenoid valve	N111	Under vehicle, rear, driver side	**Fig. 43**	97-23
Light switch	E1	Left of steering wheel on dashboard		
Lighter		In center console below radio		
Load reduction relay	J59	Micro central electric panel under left side dashboard	**Fig. 50**	97-28
Manifold absolute pressure (MAP) sensor	G71	Biturbo only		
Mass air flow sensor	G70	In intake air duct near top of air cleaner housing	**Fig. 1**	97-11
Micro central electric panel		Underneath left side dashboard	**Fig. 48** **Fig. 50**	97-27 97-28
Mirror fold away control module	J351	13-fold relay panel underneath left side dashboard	**Fig. 49**	97-28
Motronic control module		*see* Engine control module (ECM)		
Motronic main relay		*see* Engine control module power supply relay		
Multifunction steering wheel control module	J453	13-fold relay panel underneath left side dashboard	**Fig. 49**	97-28
Navigation control module		Left side luggage compartment in storage compartment *see also* **91 Radio and Communication**	**Fig. 39**	97-22
Neutral safety switch		*see* Park / neutral position (PNP) relay		
OBD II plug		*see* Data link connector (DLC)		
Oil level thermal sensor	G266	Under engine on oil pan	**Fig. 16**	97-15

BentleyPublishers.com
BentleyPublishers.com—All Rights Reserved

Table a. Electrical component locations				
Component	**Code**	**Location**	**Fig. #**	**Pg. #**
Oil pressure sensor	F1	On oil filter housing, underneath left front of engine		
Outside air temperature sensor	G17	In front of A/C condenser		
Outside rear view mirror		*see* Mirror		
Oxygen sensor, post-catalyst	G130 G131	Behind catalytic converter	**Fig. 20**	97-16
Oxygen sensor, precatalyst	G39 G108	Before catalytic converter in front exhaust pipe	**Fig. 3**	97-12
Park / neutral position (PNP) relay, Multitronic® transmission (CVT)	J226	13-fold relay panel underneath left side dashboard	**Fig. 49**	97-28
Parking aid control module	J446	Left side luggage compartment, behind trim	**Fig. 39**	97-22
Parking aid sensors	G204 G205 G334 G335	In rear bumper		
Parking aid warning buzzer, Avant	H15	Left side luggage compartment, behind trim		
Parking aid warning buzzer, sedan	H15	Rear shelf		
Parking brake warning light switch	F9	At parking brake lever mechanism beneath center console		
Pressure switch		*see* A/C high pressure switch		
Radiator outlet hose coolant temperature sensor		*see* Engine coolant temperature (ECT) sensor		
Radio	R	Center dashboard *see also* **91 Radio and Communication**		
Rear window shade control module	J262	Left side luggage compartment, behind trim		
Rear window shade motor fuse	S100	Micro central electric panel under left side dashboard	**Fig. 50**	97-28
Rear window wiper motor	V12	Under center of tailgate trim	**Fig. 38**	97-22
Recirculating pump		*see* After-run coolant pump		
Relays		*see* **Fuse and Relay Panels** in this repair group		
Remote rear lid unlock motor	V151	Under deck lid trim, center		
Rotation rate sensor (yaw sensor)	G202	Under rear seat bench, center	**Fig. 35**	97-21
Satellite radio	R146	In left side of luggage compartment *see also* **91 Radio and Communication**		
Seat adjustment circuit breaker	S44 S80	8-fold relay panel	**Fig. 54**	97-29
Seat belt lock microswitch		Seat belt buckle on seat		
Seat belt tensioner igniter	N153 N154	Behind lower B-pillar trim	**Fig. 34**	97-21
Seat memory control module, left	J136	Under driver seat		
Seatbelt reel microswitch, left	F145	Behind left B-pillar lower trim		
Seatbelt reel microswitch, right	F146	Behind right B-pillar lower trim	**Fig. 34**	97-21
Secondary air injection pump	V101	Underneath right front of engine	**Fig. 16** **Fig. 17**	97-15 97-15
Secondary air injection pump fuse, 1998	S130	13-fold relay panel underneath left side dashboard	**Fig. 49**	97-28
Secondary air injection pump fuse, 1999 - 2004	S130	3-fold relay panel in E-box, left side of plenum chamber under plastic cover	**Fig. 45**	97-24
Secondary air injection pump relay	J299	3-fold relay panel in E-box, left side of plenum chamber under plastic cover	**Fig. 45**	97-24
Secondary air injection valve	N112	Back of intake manifold above intake air duct	**Fig. 1**	97-11
Selector lever light relay, 1998		13-fold relay panel underneath left side dashboard	**Fig. 49**	97-28
Servotronic control module	J236	13-fold relay panel underneath left side dashboard	**Fig. 49**	97-28
Shiftlock solenoid	N110	Part of gear selector mechanism underneath center console		

BentleyPublishers.com
BentleyPublishers.com—All Rights Reserved

Electrical Component Locations

Table a. Electrical component locations

Component	Code	Location	Fig. #	Pg. #
Solar cell separation relay	J309	13-fold relay panel underneath left side dashboard	**Fig. 49**	97-28
Solar operation control module	J355	On sunroof inner surface		
Sound system		*see* **91 Radio and Communication**		
Speakers		*see* **91 Radio and Communication**		
Speedometer vehicle speed sensor	G22	On left side of transmission near drive axle flange		
Starter	B	Right side of engine compartment connected to transmission		
Starter interlock relay, manual transmission	J207	13-fold relay panel underneath left side dashboard	**Fig. 49**	97-28
Steering angle sensor	G85	Steering column		
Steering column adjustment motor fuse	S275	Micro central electric panel under left side dashboard	**Fig. 50**	97-28
Stop light switch		*see* Brake light switch		
Sunlight sensor	G107	Top center of dashboard		
Sunroof solar sensor control module	J355	Above headliner on sunroof panel		
Suspension strut valve	N148 -N151	Under vehicle, center rear, above stabilizer bar	**Fig. 42**	97-23
Tailgate lock motor	V53	Under center of tailgate trim	**Fig. 38**	97-22
Tailgate remote unlock motor	V151	Under center of tailgate trim	**Fig. 38**	97-22
Telephone control module	J412	Below center console rear, near parking brake lever *see also* **91 Radio and Communication**		
Telephone transceiver	R36	Right front footwell		
Throttle angle sensor	G187 G188	Integrated with throttle valve control module (J338)	**Fig. 4** **Fig. 8**	97-12 97-13
Throttle position sensor	G79	Base of accelerator pedal	**Fig. 29**	97-20
Throttle valve control module	J338	On throttle body at back of intake manifold	**Fig. 4** **Fig. 8**	97-12 97-13
Tiptronic gear selector		At shift selector housing		
Tiptronic switch	F189	Left side of gear selector assembly		
Tire pressure monitoring control module	J502	Left side luggage compartment, behind trim	**Fig. 39** **Fig. 41**	97-22 97-23
Tire pressure sensor	G222 -G226	On wheel rim near valve stem		
Traction control		*see* ESP		
Transfer fuel pump, RS6		In fuel tank		
Transmission control module		*see* Automatic transmission control module		
Transverse acceleration sensor	G200	Under rear seat bench, center	**Fig. 35**	97-21
Trunk lid lock motor	V53	Under center of trunk lid trim		
Trunk light		In trunk trim		
Trunk remote unlock motor	V151	Under center of trunk lid trim		
Turn signal / high beam stalk switch	E2,E4	On steering column		
Turn signal flasher	E3	Emergency flasher		
Ultra-sound sensor control module	J347	Below C-pillar on inside left rear door jamb, behind rear seat side trim		
Washer fluid level sensor	G33	Washer fluid reservoir, behind right front wheel housing liner	**Fig. 19**	97-16
Waste gate bypass regulator valve, 2.7 liter engine	N75	Center of intake manifold, right of air intake duct	**Fig. 9**	97-13
Wheel speed sensor		*see* ABS wheel speed sensor		
Window control module	J295 - J298	Integrated with window motor and window regulator assembly	**Fig. 32**	97-20
Window motor		In door	**Fig. 32**	97-20
Window motor switch		In door	**Fig. 32**	97-20

BentleyPublishers.com
BentleyPublishers.com—All Rights Reserved

Table a. Electrical component locations				
Component	**Code**	**Location**	**Fig. #**	**Pg. #**
Windshield washer fluid pump	V5	Washer fluid reservoir, behind left front wheel housing liner		
Windshield washer nozzle, heated		Under engine hood		
Windshield wiper motor	V	Under left cowl in plenum chamber		
Wiper / washer intermittent relay	J31	Micro central electric panel under left side dashboard	**Fig. 50**	97-28
Wiper / washer stalk switch		On steering column		
Wiper motor, rear window (Avant)	V12	In tailgate underneath trim panel	**Fig. 38**	97-22
Woofer		*see* Speaker		
Yaw sensor		*see* Rotation rate sensor (yaw sensor)		

Engine compartment components

Component locations pictures, unless otherwise noted, are of a 1999 A6 Avant quattro with 2.8 liter engine code AHA. Components and locations vary depending on year and model.

Fig. 1 1999 2.8 liter engine compartment, upper cover and plenum chamber cover removed

1. **Fuel tank evaporative control canister purge regulator valve (N80)**
2. **Dust and pollen filter**
3. **Mass air flow (MAF) sensor (G70)**
4. **Oxygen sensor harness connectors**
5. **Battery (A)**
6. **Secondary air injection valve (N112)**
7. **Intake air temperature (IAT) sensor (G42)**
8. **Intake manifold change-over (tuning) valve (N156)**
9. **Transmission kick-down switch (F8)**
10. **Brake fluid level warning switch (F34)**
11. **Fuel pressure regulator**
12. **Electronics box (E-box)**
13. **Power output stage (N122) Ignition coils (N, N128, N158)**
14. **Hood alarm switch (F120)**
15. **Engine coolant level switch (F66)**
16. **Left camshaft adjustment valve (N208)**
17. **ABS hydraulic pump (V156)) ABS hydraulic unit (N55)**

Fig. 2 Right front of engine

Right upper timing belt cover removed

1. Right camshaft sensor (G40)

BentleyPublishers.com
BentleyPublishers.com—All Rights Reserved

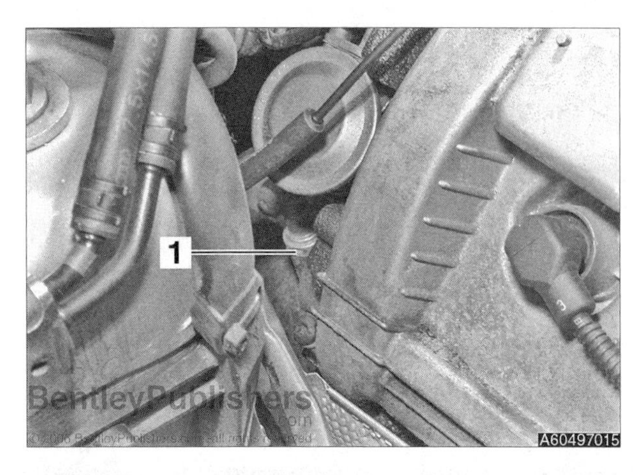

Fig. 3 Right rear of engine

Intake air duct and engine cover removed

1. Precatalyst oxygen sensor 1 (G39)

Fig. 4 Rear of engine

Intake air duct and upper engine cover removed

1. Throttle valve control module (J338)
   Throttle angle sensor (G187, G188)
   Throttle position sensor (G79)

Fig. 5 Plenum chamber, left of battery, 3.0 liter V6, all V8 engines

Plenum chamber cover removed

1. Brake booster vacuum pressure sensor (G294)

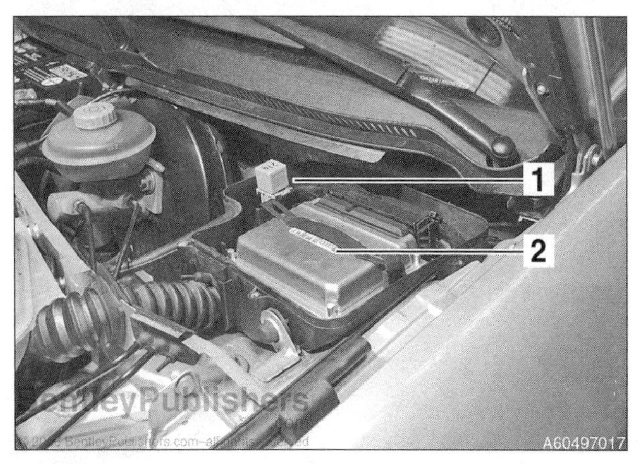

Fig. 6 Plenum chamber, left side

Plenum chamber and E-box covers removed

1. Secondary air injection relay (J299)
2. Engine control module (J220)

The electronics box (E-box) in the left rear of the engine compartment is protected by a plastic cover. It contains 3-fold relay panel and S1 fuse panel. For fuse ratings and relay applications, see **Fuse and Relay Panels** in this repair group.

BentleyPublishers.com
BentleyPublishers.com—All Rights Reserved

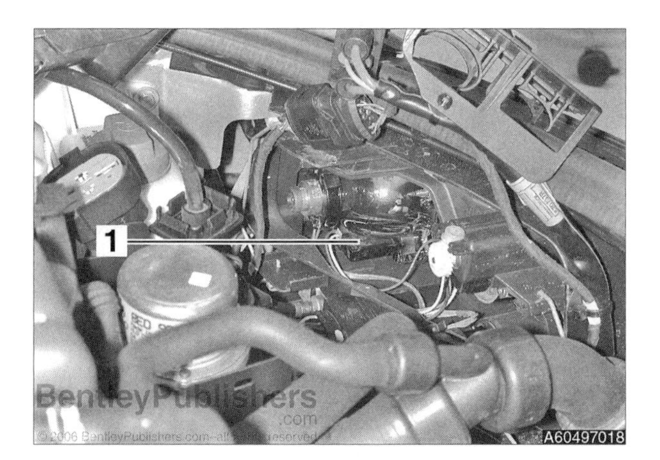

Fig. 7 Left headlight, xenon (HID)

◄ Headlight cover removed

1. Left headlight beam adjusting motor (V48)

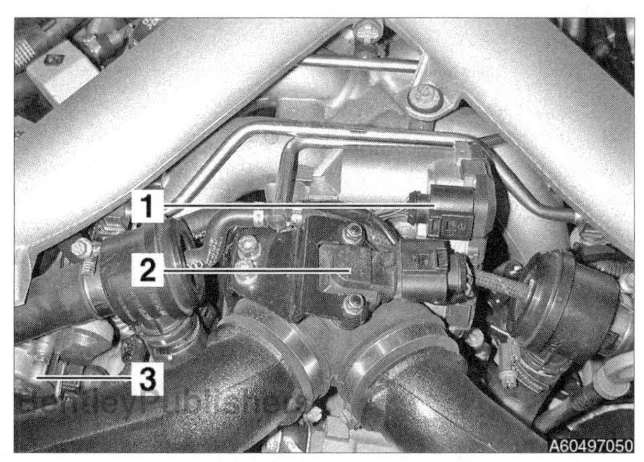

Fig. 8 Top of engine, 2.7 liter biturbo V6 engine

◄ Upper engine cover removed

1. Throttle valve control module (J338)
   Throttle drive (G186)
   Throttle valve angle sensors (G187, G188)
2. Charge air pressure sensor (G31)
3. Right camshaft sensor (G40)

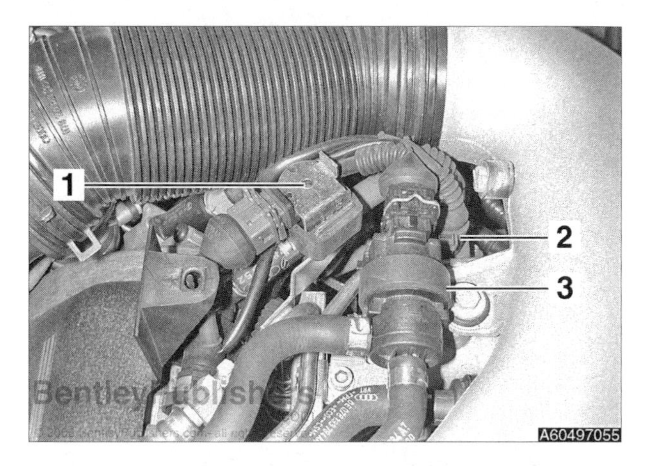

Fig. 9 Top right of engine, 2.7 liter biturbo V6 engine

◄ Top of intake manifold, upper engine cover removed

1. Waste gate bypass regulator valve (N75)
2. Right exhaust temperature sensor module (G235)
3. Fuel tank evaporative control canister purge regulator valve (N80)

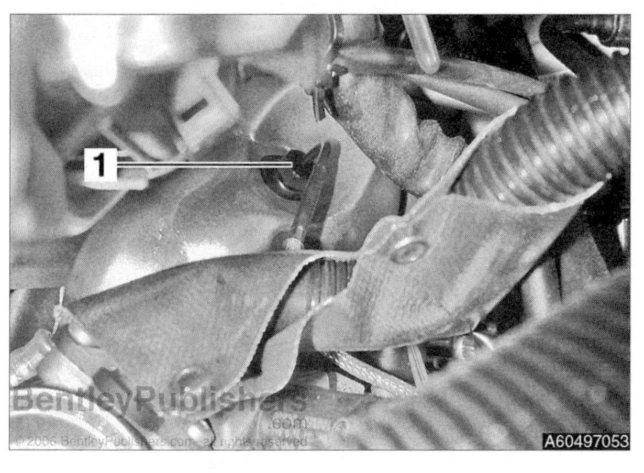

Fig. 10 Rear of engine, 2.7 liter biturbo V6 engine

◄ Right turbocharger outlet pipe

1. Right exhaust temperature sensor (G235)

BentleyPublishers.com
BentleyPublishers.com—All Rights Reserved

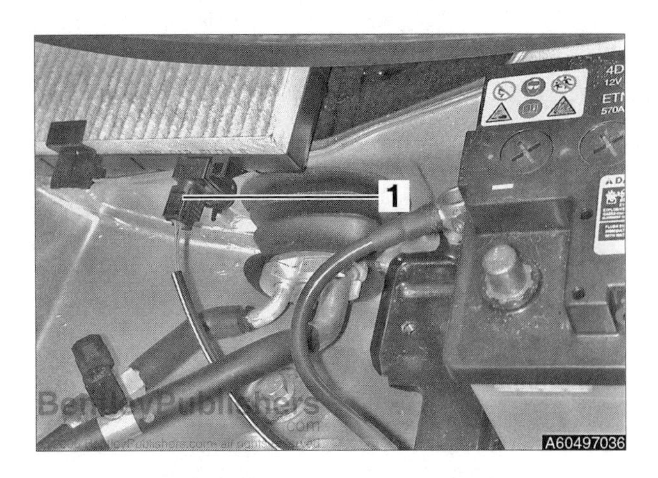

Fig. 11 Plenum chamber, right of battery, 3.0 liter V6 engine

◄ Plenum chamber cover removed

1. Air quality sensor (G238)

Fig. 12 Engine compartment, left side, 3.0 liter V6 engine

◄ Left side engine cover removed

1. Brake booster vacuum pump (V192)
2. ABS hydraulic unit (N55)
3. ABS control module (J104)
4. ABS return flow pump (V39)

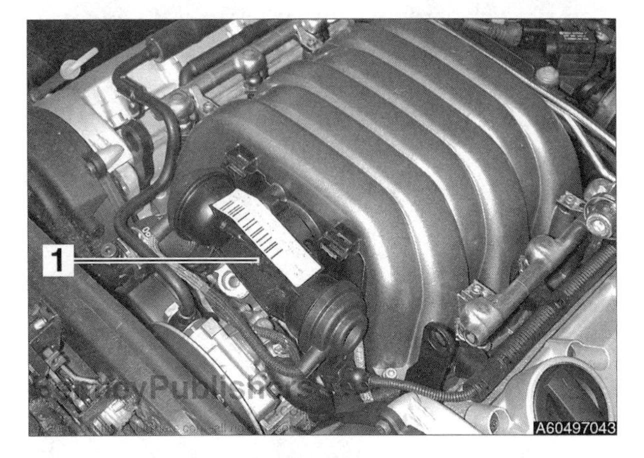

Fig. 13 Top of engine, 3.0 liter V6 engine

◄ Upper engine cover removed

1. Intake manifold change-over (tuning) valve (N261)

Components underneath vehicle

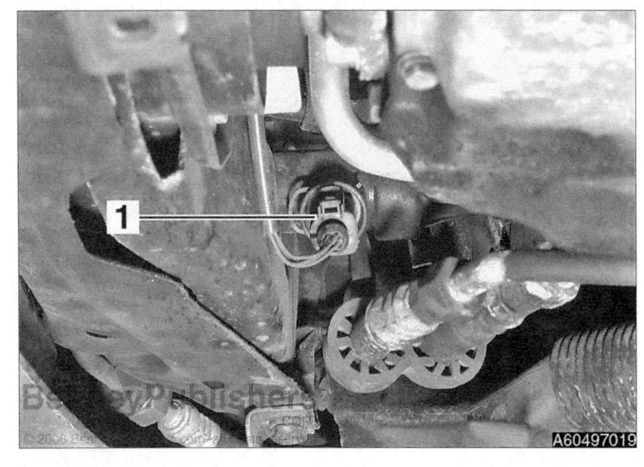

Fig. 14 Underneath right side of engine

◄ Radiator outlet pipe, engine lower cover removed

1. Engine cooling fan switch (F54, F18)

BentleyPublishers.com
BentleyPublishers.com—All Rights Reserved

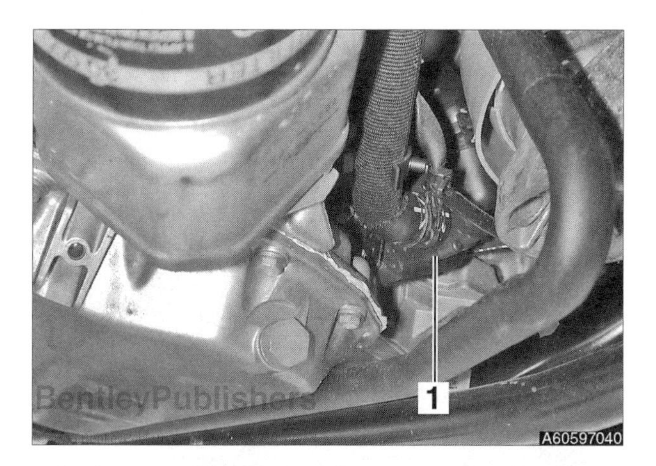

Fig. 15 Underneath left side of engine

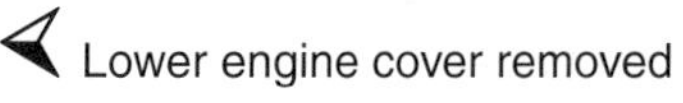

Lower engine cover removed

1. After-run coolant pump (V51)

Fig. 16 Underneath center of engine

Lower sound absorber panel removed, 3.0 liter engine

1. Alternator (C)
2. Secondary air injection pump motor (V101)
3. Oil level thermal sensor (G266)

Fig. 17 Underneath right front of vehicle

Lower engine cover removed

1. Secondary air injection pump (V101)
2. Low tone horn (H7)

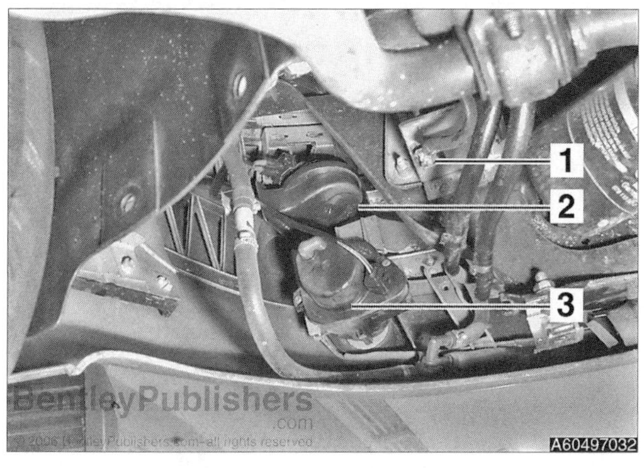

Fig. 18 Underneath left front of vehicle

Lower engine cover removed

1. Engine cooling fan series resistance (N39)
2. High tone horn (H2)
3. Left foglight

BentleyPublishers.com
BentleyPublishers.com—All Rights Reserved

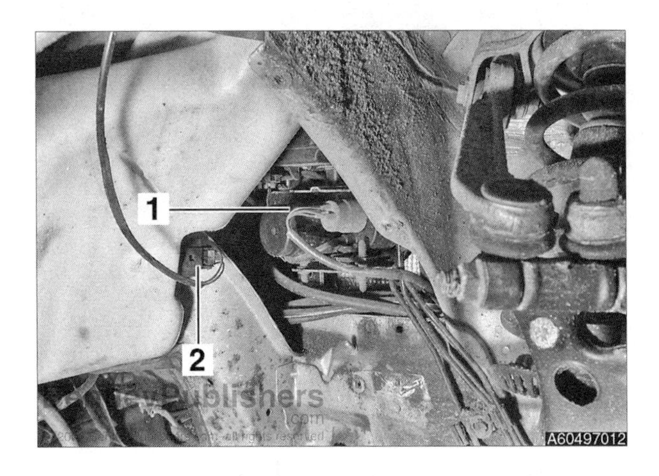

Fig. 19 Left front wheel housing

◀ Wheel housing liner removed.

1. Cruise control vacuum pump (V18)
2. Washer fluid level sensor (G33)

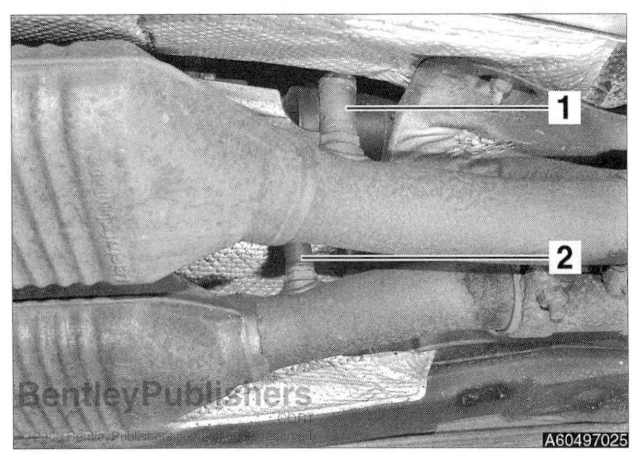

Fig. 20 Underneath middle of vehicle

◀ Rear of catalytic converter, in exhaust system

1. Post-catalyst oxygen sensor 2 (G131)
2. Post-catalyst oxygen sensor 1 (G130)

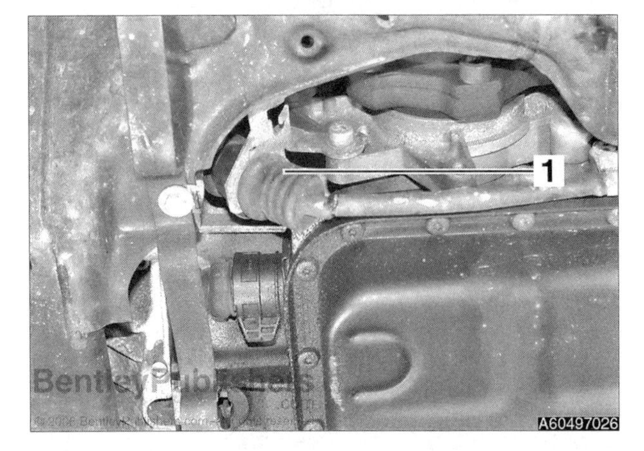

Fig. 21 Underneath automatic transmission

◀ Lower sound absorber panel removed

1. Multifunction transmission range switch (F125)

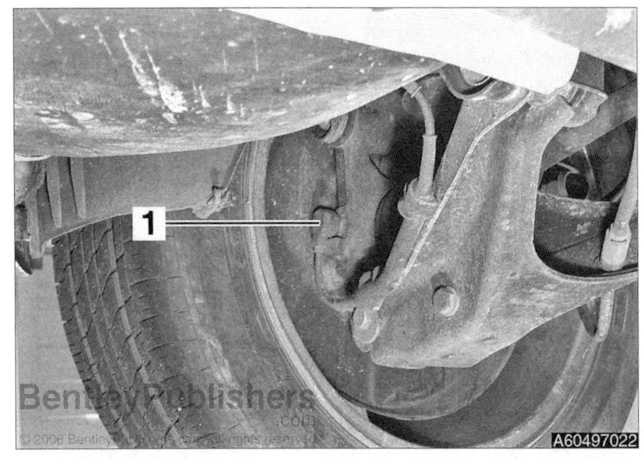

Fig. 22 Underneath right rear of vehicle

◀ Right rear wheel hub

1. Right rear ABS wheel speed sensor (G46)

BentleyPublishers.com
BentleyPublishers.com—All Rights Reserved

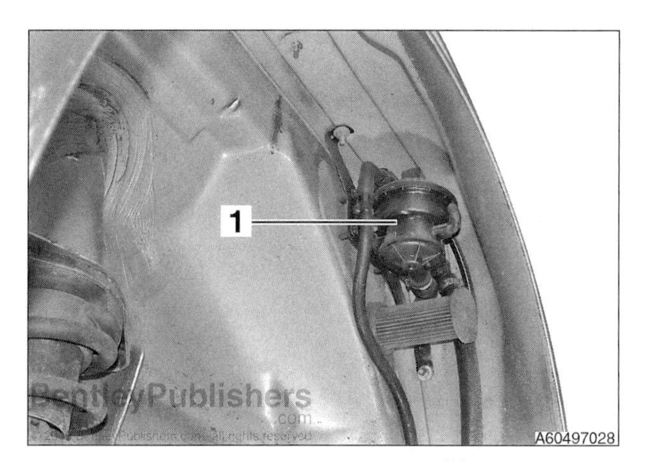

Fig. 23 Left rear wheel housing

Wheel housing liner removed

1. Fuel tank leak detection pump (144)

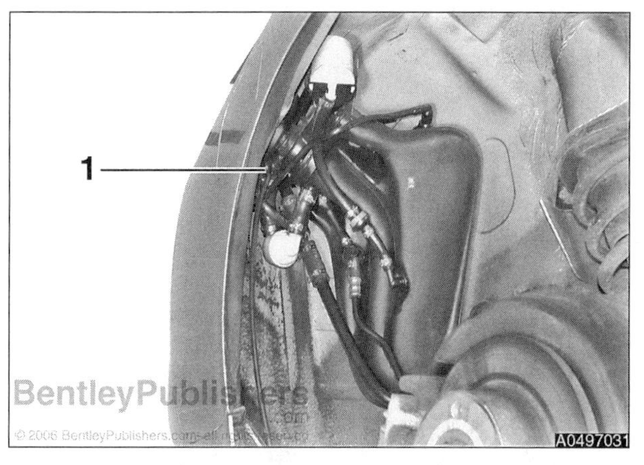

Fig. 24 Right rear wheel housing

Wheel housing liner removed

1. Fuel tank filler neck ground

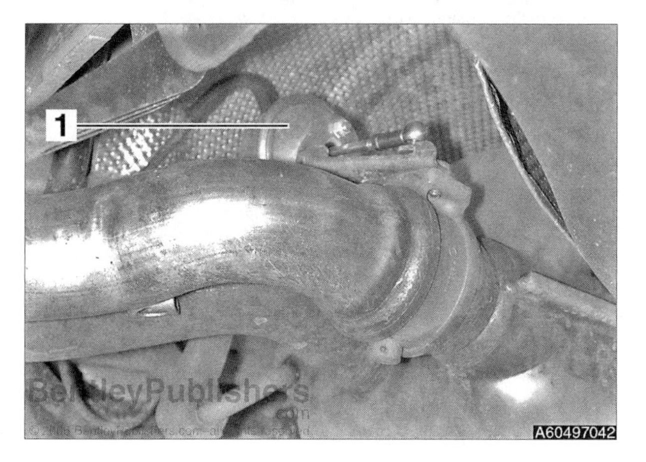

Fig. 25 Above tailpipe, 3.0 liter engine

In exhaust pipe

1. Exhaust flap valve (N321)

BentleyPublishers.com
BentleyPublishers.com—All Rights Reserved

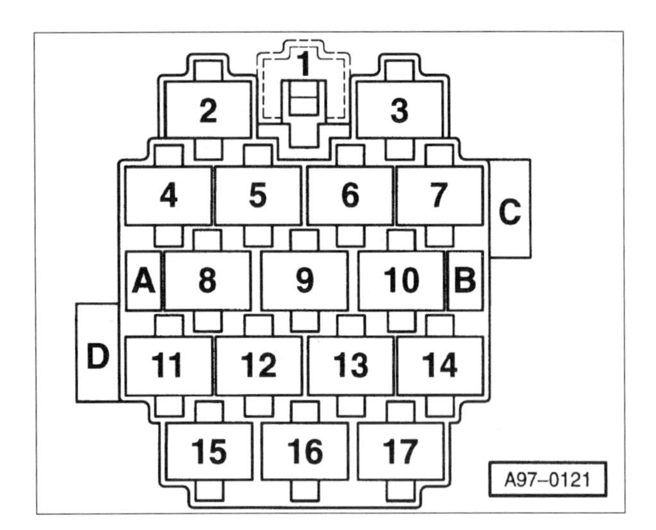

Interior components

Fig. 26 Left A-pillar connector station

Behind left kick panel trim

Table b. Left A-pillar connector station plugs			
No.	**Color**	**From wiring harness**	**To wiring harness**
1	Black	Dashboard	Mirror adjustment
2	Light brown	Central locking, anti-theft	Door lock, anti-theft
3	Red	Power window	Power window in door
4	Black	Cruise control	Dashboard
5	Blue	Speaker	Speaker
6	Grey	Power window	Power window
7	Blue	Rear of vehicle	Dashboard
8	Brown	Rear of vehicle	Dashboard
9	Red	Xenon headlight, left	Dashboard
10	Violet	Xenon headlight, left	Dashboard
11	Blue	Rear of vehicle	Dashboard
12	White	Seats, seat belt warning	Dashboard
13	Brown	Power window, sunroof	Dashboard
14	Grey	Engine cooling fan, A/C compressor	Dashboard
15	Orange	ABS	Dashboard
16	Violet	Dome light	Dashboard
17	Black	Wiper motor	Dashboard
A	Black	Rear of vehicle	Dashboard
B	Black	Washer	Dashboard
C	Blue	Dashboard	Temperature gauge
D	Yellow	Airbag, driver	Dashboard

BentleyPublishers.com
BentleyPublishers.com—All Rights Reserved

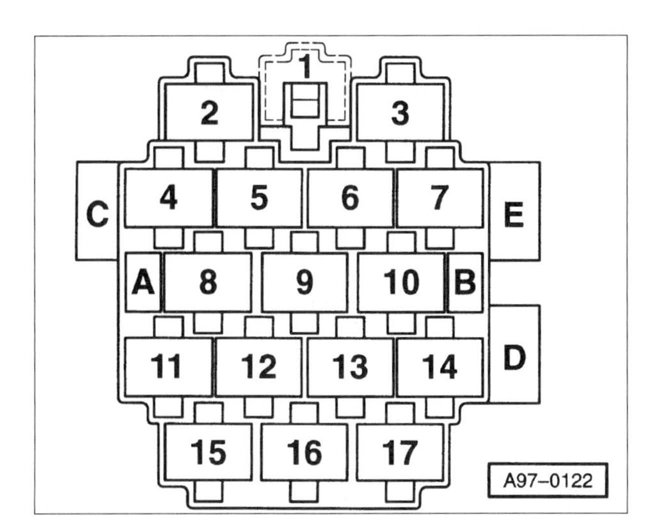

Fig. 27 Right A-pillar connector station

Behind right kick panel trim

Table c. Right A-pillar connector station plugs			
No.	**Color**	**From wiring harness**	**To wiring harness**
1	Black	Dashboard	Mirror adjustment
2	Red	Power window	Power window in door
3	Yellow	Central locking, anti-theft	Door lock, anti-theft
4	White	Parking aid	Dashboard
5	Blue	Headlight aim, xenon bulb	Dashboard
6	Green	Speaker	Speaker
7	Black	Central locking, anti-theft	Dashboard
8	Violet	Xenon headlight, right	Dashboard
9	red	Xenon headlight, right	Dashboard
10	Blue	Automatic transmission	Dashboard
11	Grey	ABS	Dashboard
12	Yellow	Airbag	Dashboard
13	Black	Reading lights	Dashboard
14	Grey	Automatic transmission	Dashboard
15	Yellow	Daytime running lights	Dashboard
16	Brown	Leak detection pump, automatic day/night mirror	Dashboard
17	Orange	Central locking, anti-theft	Dashboard
A		Open	
B		Open	
C		Open	
D	White	Seat belt tensioner, passenger airbag	Dashboard
E		Open	

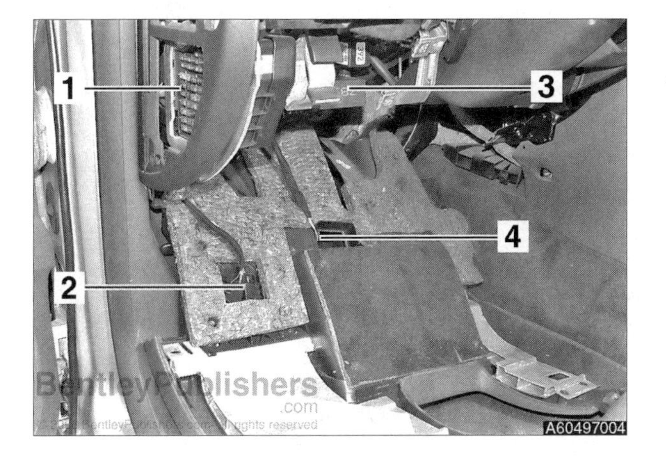

BentleyPublishers.com

Fig. 28 Under dashboard, left side

Lower left dashboard trim panel pulled down (back side shown)

1. Main fuse panel
2. Data link connector (DLC)
3. 13-fold relay panel
   Micro central electric panel relay panel
4. Footwell light

BentleyPublishers.com
BentleyPublishers.com—All Rights Reserved

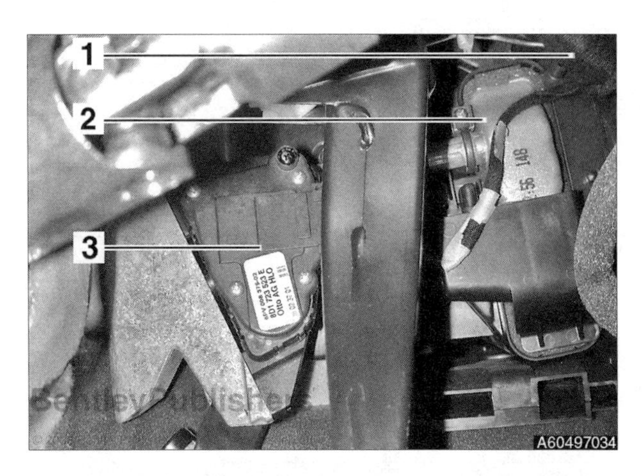

Fig. 29 Left footwell

◀ Lower left dashboard trim panel removed

1. Left temperature regulator flap motor (V158) and sensor (G220)
2. Heater core
3. Accelerator pedal module:
   Throttle position sensor (G79)
   Accelerator pedal position sensor 2 (G185)

Fig. 30 Under dashboard, right side

◀ Glove compartment removed

1. Passenger airbag ignitor (N131)
2. Blower motor (V2)

Fig. 31 Under dashboard, right side

◀ Components behind glove compartment

1. Right temperature regulator flap motor (V159) and sensor (G221)
2. Center air flow flap motor (V70) and sensor (G112)
3. Defroster flap motor (V107) and sensor (G135)
4. Cruise control module (J213)
5. Headlight aim control module (J431)
6. Blower motor (V2)

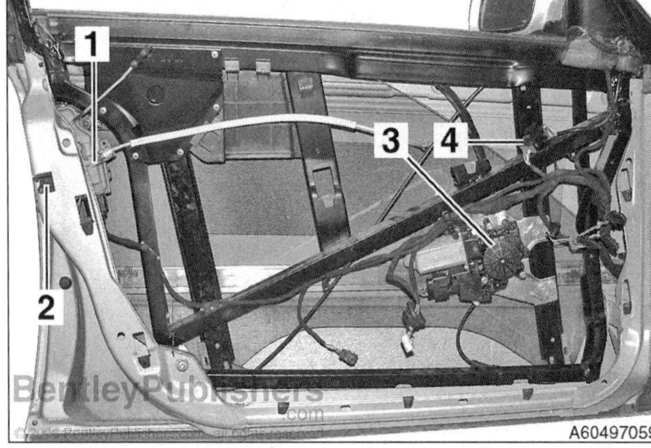

Fig. 32 Driver's door

◀ Door panel removed

1. Door lock motor (V56)
   Door contact switch (F3)
2. Door lock
3. Window motor (V14)
   Window control module (J295)
4. Outside mirror electrical connector

BentleyPublishers.com
BentleyPublishers.com—All Rights Reserved

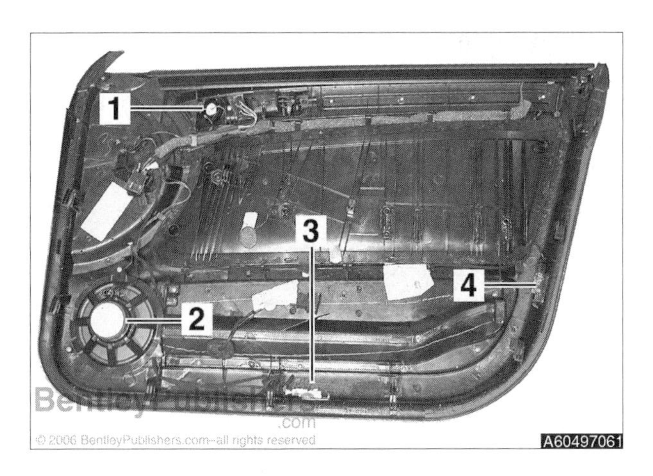

Fig. 33 Driver's door panel

Backside of panel removed from door

1. High frequency speaker (tweeter) (R34)
2. Midrange speaker (R21)
3. Door courtesy light
4. Door warning light

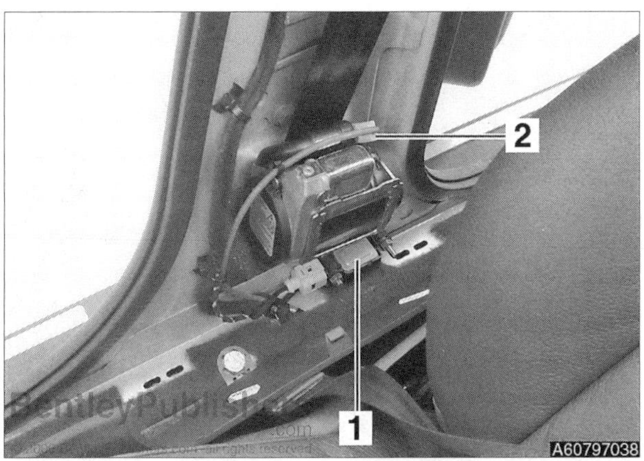

Fig. 34 Lower B-pillar

B-pillar trim removed

1. Seat belt reel microswitch (F146)
2. Seat belt tensioner igniter (N153)

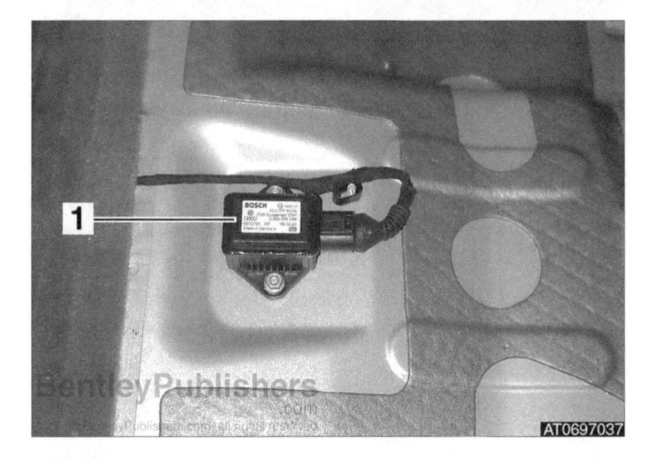

Fig. 35 Under rear seat, 2000 - 2004 sedan

Rear seat cushion removed

1. ESP sensor:
   Transverse acceleration (G200)
   Rotation rate (yaw) sensor (G202)

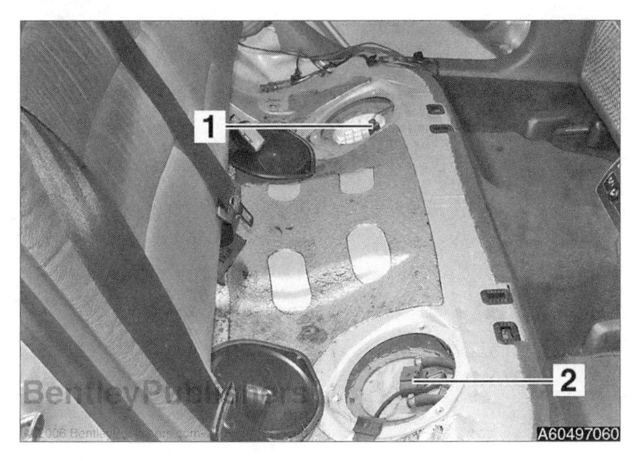

Fig. 36 Under rear seat

Rear seat cushion and fuel tank access covers removed

1. Fuel level sensor, left (G169)
2. Fuel level sensor, right (G)
   Fuel pump (G6)

BentleyPublishers.com
BentleyPublishers.com—All Rights Reserved

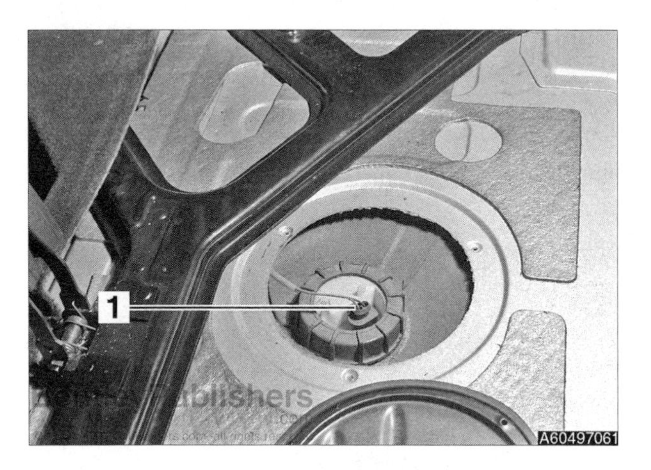

Fig. 37 Luggage compartment, Avant

Behind rear seats, rear seat folded forward, luggage compartment carpet removed, fuel tank access cover removed

1. Fuel level sensor, center (G237) (quattro only)

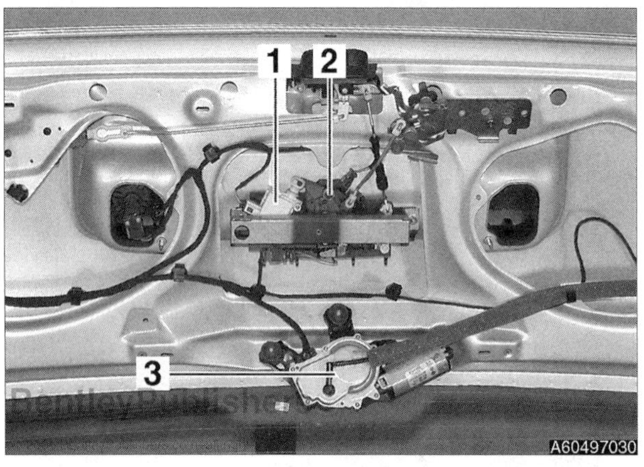

Fig. 38 Tailgate, Avant

Tailgate trim removed

1. Tailgate remote unlock motor (V151)
2. Tailgate lock motor (V53)
3. Rear window wiper motor (V12)

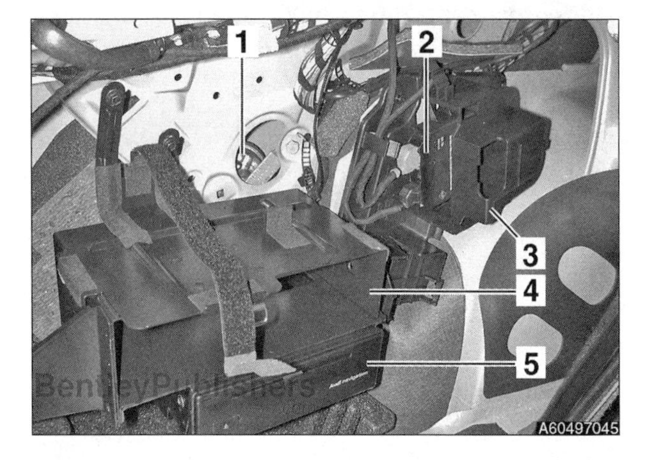

Fig. 39 Left side luggage compartment, sedan

Left storage cover removed

1. Alarm horn (H8)
2. Tire pressure monitoring control module (J502)
3. Parking aid control module (J446)
4. CD changer (R41)
5. Navigation control module

allroad quattro components

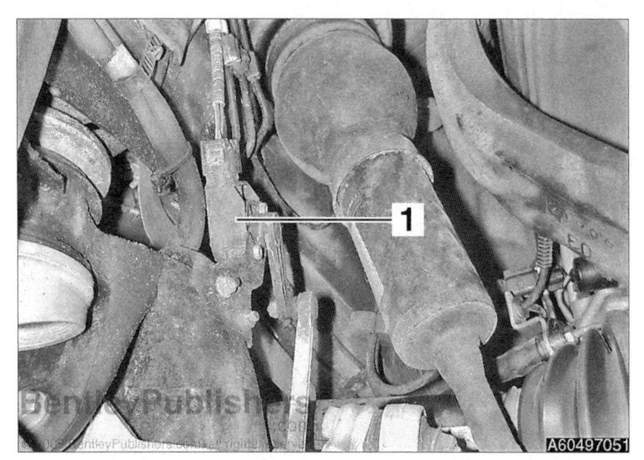

Fig. 40 Left front wheel housing

Wheel housing liner removed

1. Left front level control system sensor (G78)

NOTE—

- *Level control sensors for the other wheels are as follows:*
  -Right front: G84
  -Left rear: G76
  -Right rear: G77

BentleyPublishers.com
BentleyPublishers.com—All Rights Reserved

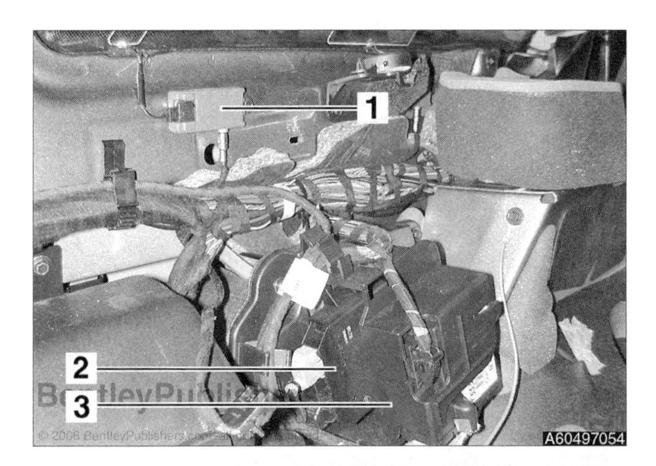

Fig. 41 Left side luggage compartment

Left storage cover and trim removed

1. Broken window glass sensor (G183, G184)
2. Tire pressure monitoring control module (J502)
3. Level control system control module (J197)

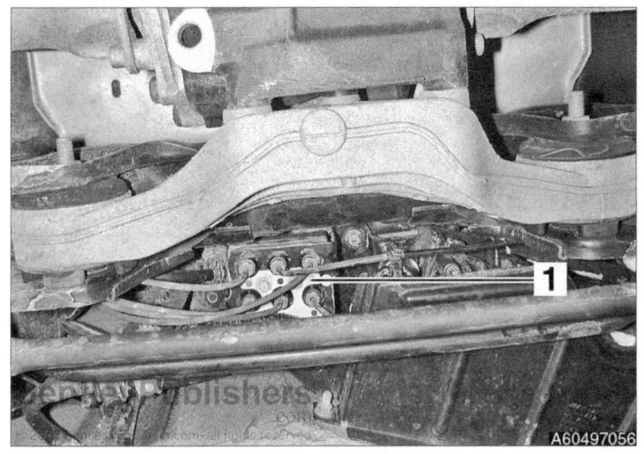

Fig. 42 Rear underside of vehicle

Above stabilizer bar

1. Suspension strut valve (N148, N149, N150, N151)
   Level control system accumulator valve (N311)

Fig. 43 Left rear underside of vehicle

Between exhaust system and spare tire well

1. Level control system compressor motor (V66)
   Level control system solenoid valve (N111)

Fuse and Relay Panels

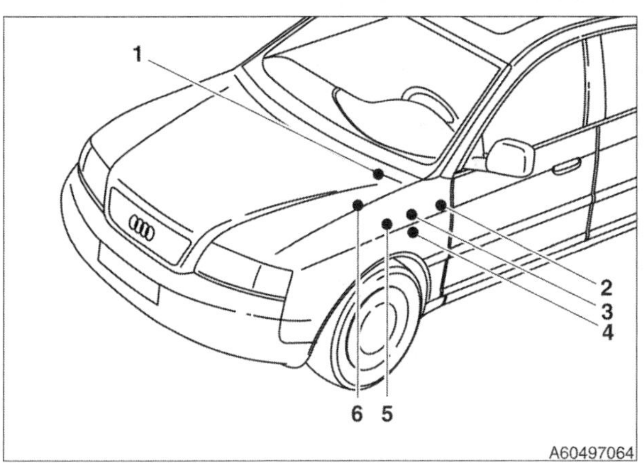

The majority of A6 fuses and relays are in the locations shown in the illustration. See **Table d** for names and locations.

BentleyPublishers.com
BentleyPublishers.com—All Rights Reserved

Table d. A6 fuse and relay panels			
No.	**Panel**	**Location**	**See section in this repair group**
1	3-fold relay panel, fuse panel S1	Left side plenum chamber, rear of engine compartment	**E-box (electronics box) components**
2	Main fuse panel, fuse panel S2	Left end of dashboard	**Main fuse panel, fuse panel S2**
3	13-fold relay panel	Under left side dashboard	**Under-dash relay panels**
4	Micro central electric panel	Under left side dashboard	**Under-dash relay panels**
5	8-fold relay panel	Under left side dashboard	**Under-dash relay panels**
6	Central carrier relay panel	Under left side dashboard	**Under-dash relay panels**

E-box (electronics box) components

– Open engine hood and remove plenum chamber cover.

< Remove E-box cover retaining bolts (**arrows**). Lift off cover.

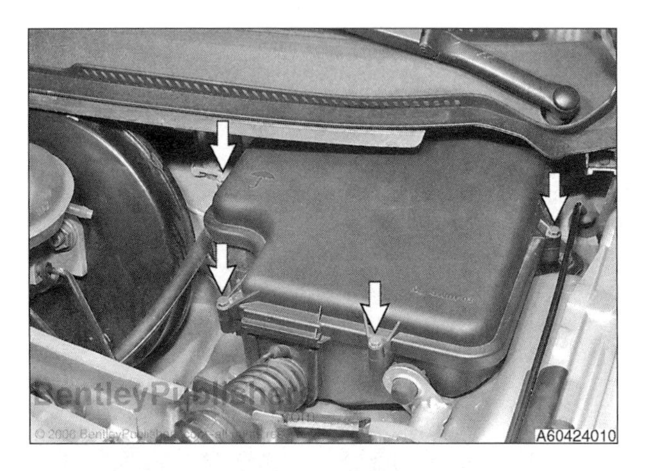

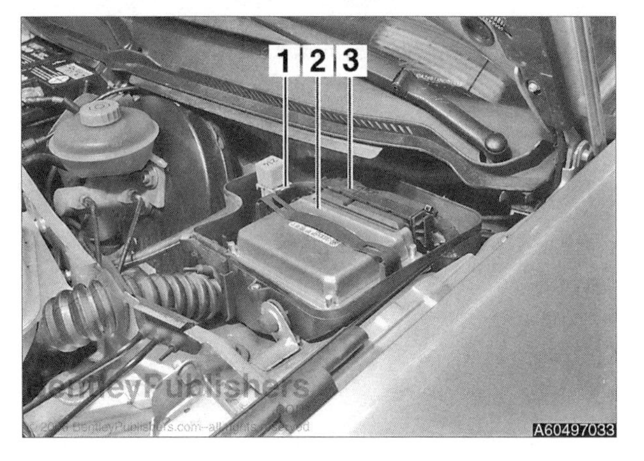

Fig. 44 E-box

< Plenum chamber cover and E-box cover removed

1. 3-fold relay panel
   Fuse panel S1
2. Engine control module (ECM)
3. E-box connector station

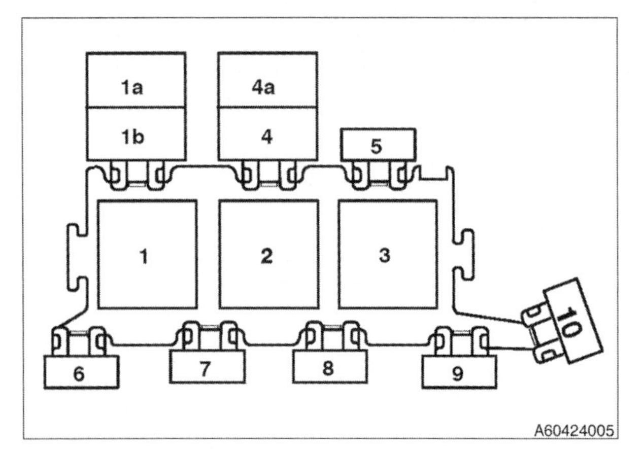

Fig. 45 3-fold relay panel, fuse panel S1

< Relay and fuse configuration varies by year and model

Table e. 3-fold relay panel fuses and relays		
Fuse	**Amps**	**Protected circuit**
1		Open
1a		Open
1b		After-run coolant pump relay (J151)
2		Secondary air injection pump relay (J299)
3		ECM power relay (J271)
4		Open
4a		Open

BentleyPublishers.com
BentleyPublishers.com—All Rights Reserved

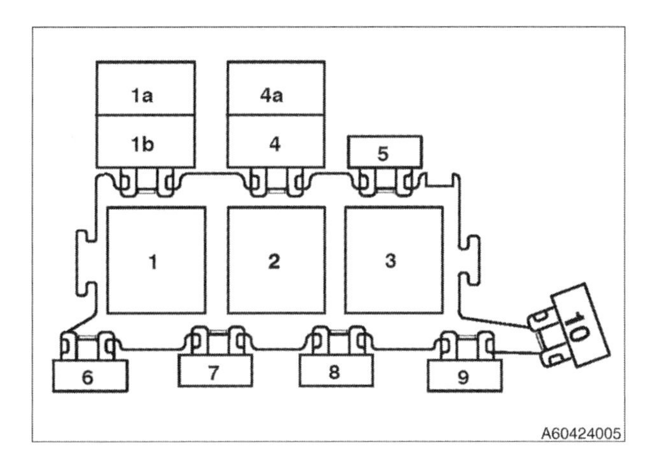

Table e. 3-fold relay panel fuses and relays		
Fuse	**Amps**	**Protected circuit**
5	15	ECM fuse (S102)
6		Open
7	40	Secondary air injection pump fuse (S130)
8		Open
9		Open
10		Open

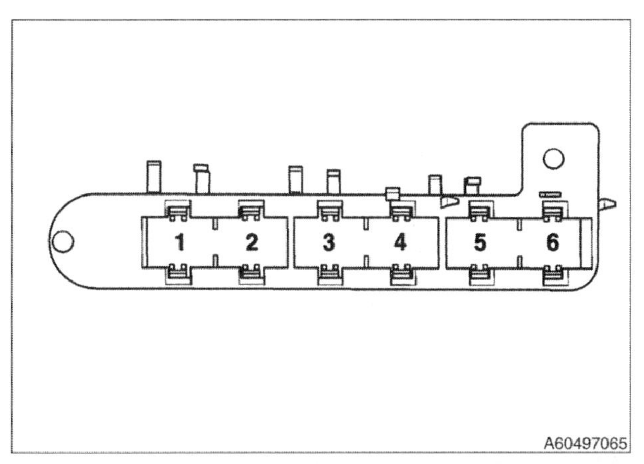

Fig. 46 E-box connector station

Table f. E-box connector station plugs			
No.	**Color**	**From wiring harness**	**To wiring harness**
1	White	Dashboard	Engine
2	Blue	Dashboard	Engine / transmission
3	Orange	Dashboard	Engine
4	Red	Dashboard	Engine
5	Brown	Dashboard	Engine/ transmission
6	Black	Dashboard	Engine compartment

Main fuse panel, fuse panel S2

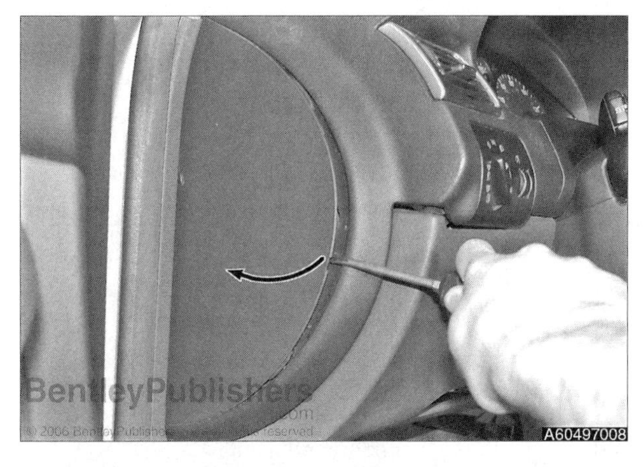

Gently pry open cover at left end of dashboard to gain access to main fuse panel.

Fig. 47 Dashboard fuse panel

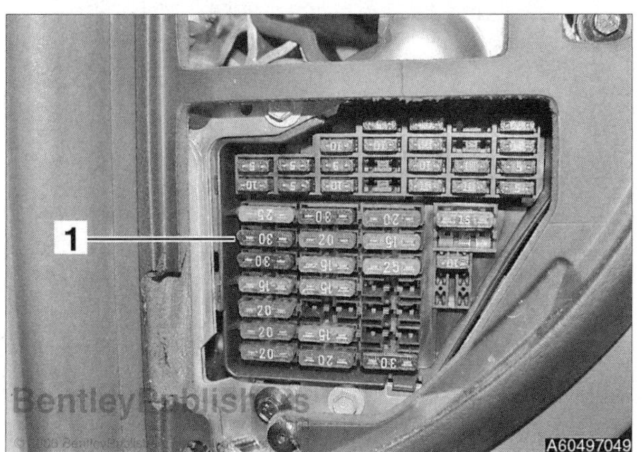

Left end of dashboard, under cover.

1. Fuse panel S2

See **Table g** for fuse ratings and applications. Be sure to replace a bad fuse with one of equivalent rating.

NOTE—

- *Starting with fuse 23, fuses in the main fuse panel are identified with an additional S2 in the wiring diagrams.*
  -Example: Fuse 23 = S223

BentleyPublishers.com
BentleyPublishers.com—All Rights Reserved

Fuse and Relay Panels

Table g. Main fuse panel

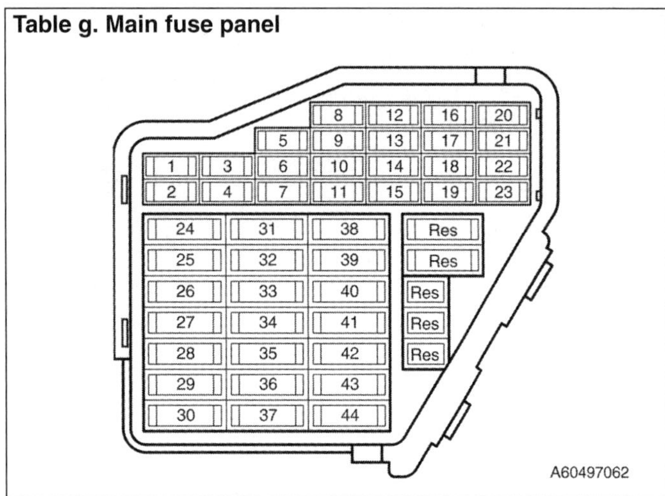

Fuse	Amps	Protected circuit
1	5	Washer nozzle heater
2	10	Turn signal
3	5	Illumination (glove compartment, climate control, instrument cluster)
4	5	License plate lights
5	10	Instrument cluster Climate control Outside mirror Lamp control module Seat heating Rear window shade Telephone
6	5	Central locking
7	10	ABS
8	5	Telephone
9	10	Mirror heating
10	5	Automatic headlight aim control
11	5	Cruise control
12	10	Supply voltage diagnostic
13	10	Brake lights
14	10	Interior lights Reading lights Central locking Vanity mirror light Seat memory
15	10	Instrument cluster Automatic transmission Climate control Mirror memory
16	5	Steering angle sensor
17	10	Level control system
18	10	Headlight high beam, right
19	10	Headlight high beam, left
20	15	Headlight low beam, right Headlight adjusting
21	15	Headlight low beam, left Headlight adjusting
22	5	Tail and parking light, right

Table g. Main fuse panel

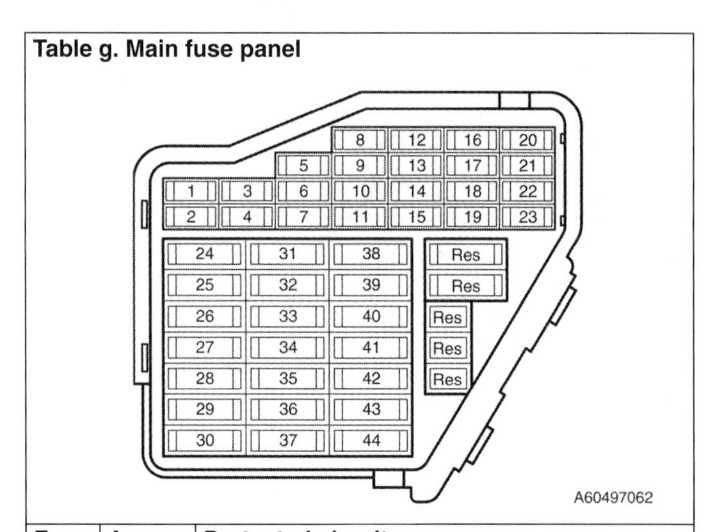

Fuse	Amps	Protected circuit
23	5	Tail and parking light, left
24	25	Wiper and washer system
25	30	Blower, climate control
26	30	Rear window defogger
27	15	Rear window wiper and washer Heated steering wheel
28	15	Fuel pump (1998 - 2000)
28	20	Fuel pump (2001 - 2004)
29	20	Engine control module (ECM) (1998 - 2000)
29	30	Engine control module (ECM) (2001 - 2004)
30	20	Sunroof
31	15	Automatic transmission Back-up lights Cruise control Automatic day / night interior mirror Data link connector
32	20	Engine control module
33	15	Cigarette lighter
34	15	Engine control module, fuel injectors
35		Open
36	15	Foglights
37	20	Sound system
38	15	Luggage compartment light Central locking system (1998 - 2000)
38	20	Luggage compartment light Central locking system (2001 - 2004)
39	15	Emergency flasher
40	25	Horn
41		Open
42	25	ABS / ESP
43		Open
44	30	Seat heater

Under-dash relay panels

Tilt steering column down and pull off trim under instrument cluster. Remove steering column trim panel fasteners (**arrows**).

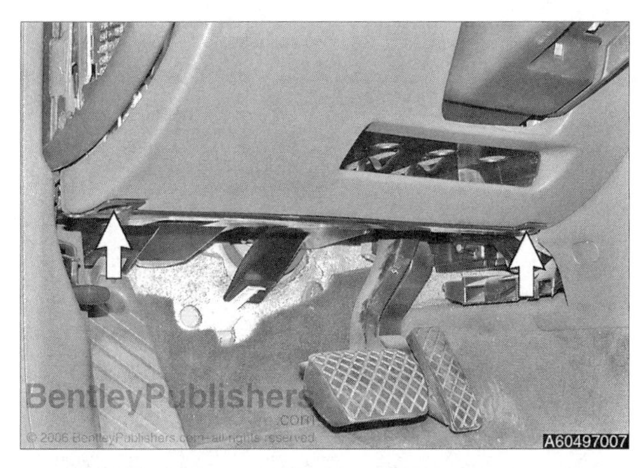

Remove lower trim fasteners (**arrows**) and pull down trim panels under steering column to gain access to fuse and relay panels under dashboard.

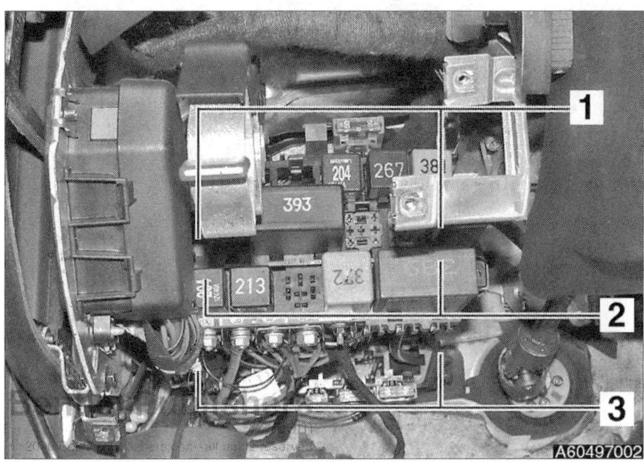

Fig. 48 Underneath left side of dashboard

Lower steering column trim panel removed

1. 13-fold relay panel
2. Micro central electric panel
3. 8 fold relay panel

BentleyPublishers.com
BentleyPublishers.com—All Rights Reserved

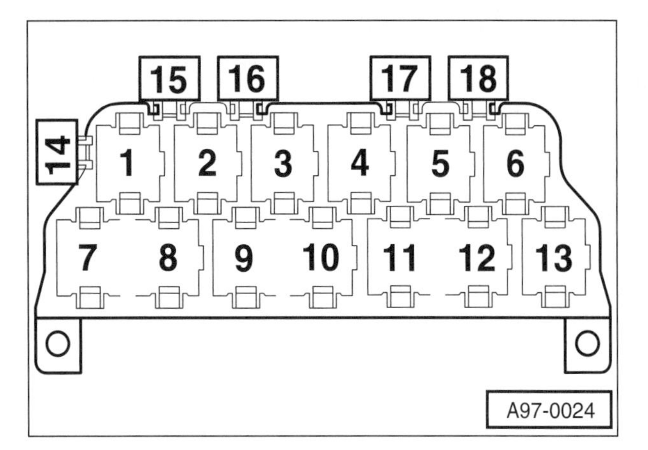

Fig. 49 13-fold relay panel

Underneath left side dashboard, trim panel removed

Table h. 13-fold relay panel fuses and relays		
Pos	**Amps**	**Circuit or component**
1		Selector lever light relay (1998)
2		Open
3		Solar cell separation relay (J309)
4		Starter interlock relay (J207), manual transmission Park / neutral position (PNP) relay (J226), Multitronic® transmission (CVT)
5		A/C clutch relay (J44)
6		Foglight relay (J5)
7		Multi-function steering wheel control module (J453)
8		
9		Lamp control module (J123)
10		
11		Mirror fold away control module (J351)
12		
13		Servotronic control module (J236)
14	15	Brake booster vacuum pump fuse (S279) (allroad)
15		Open
16		Open
17	50	Secondary air pump fuse (S130) (1998)
18		Open

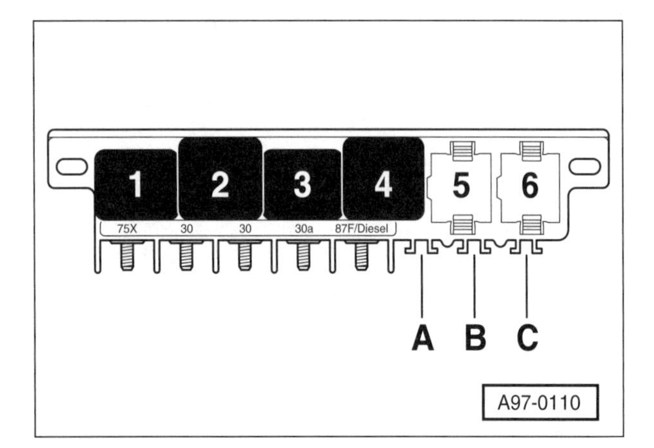

Fig. 50 Micro central electric panel

Underneath left side dashboard, trim panel removed

Table i. Micro central electric panel fuses and relays		
Pos	**Amps**	**Circuit or component**
1		Horn relay (J4)
2		Load reduction relay (J59)
3		Hydraulic pump relay (J555) (2001 - 2004)
4		Fuel pump relay (J17)
5		Wiper / washer intermittent relay (J31)
6		
A	20	Steering column adjustment motor fuse (S275)
B	10	Rear window shade motor fuse (S100)
C		Open

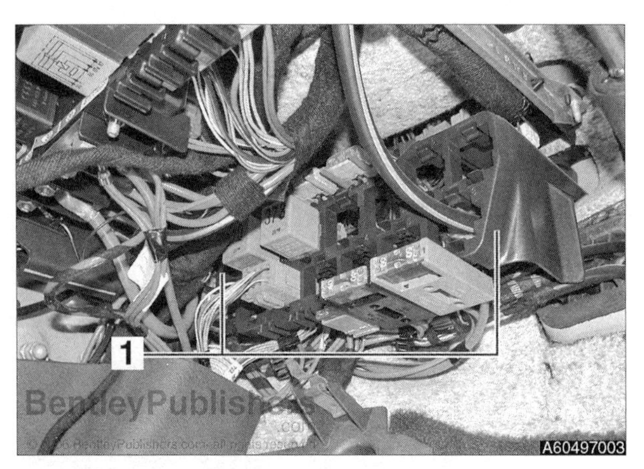

Fig. 51 8-fold relay panel

Underneath left side dashboard, trim panel removed

1. 8-fold relay and fuse panel

BentleyPublishers.com
BentleyPublishers.com—All Rights Reserved

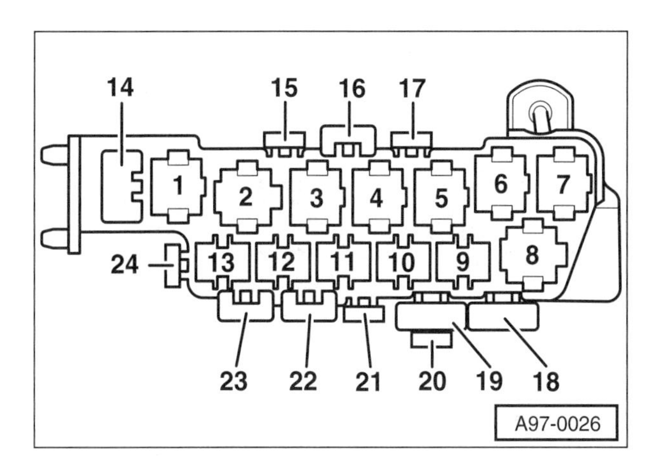

Fig. 52 8-fold relay panel relays

Table j. 8-fold relay panel relays	
Relay	**Circuit**
1	ABS solenoid valve relay (J106) (2002 - 2004)
2	Engine cooling fan relay, high speed (J101)
3	Engine cooling fan relay (J26)
4	ABS solenoid valve relay (J106) (2001)
5	Open
6	Level control system compressor relay (J403) (2001 - 2004)
7	ABS return flow pump relay (J105) (2001 - 2004)
8	Open

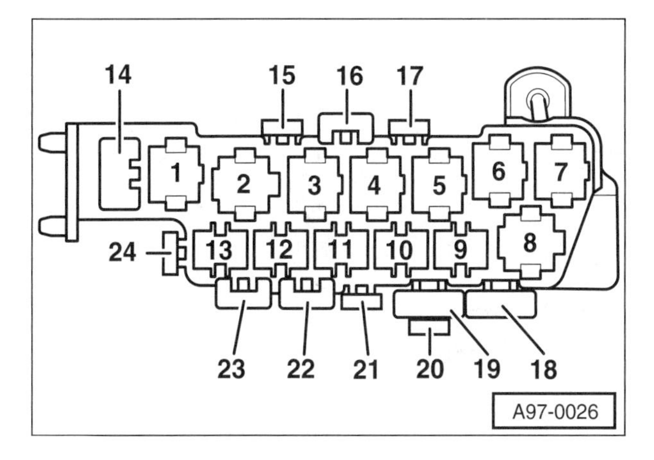

Fig. 53 8-fold relay panel connectors

Table k. 8-fold relay panel connectors			
No.	**Color**	**From wiring harness**	**To wiring harness**
9		Open	
10	Grey	Level control system	Dashboard
11	Black	ABS / ESP	Dashboard
12	Violet	Rear of vehicle	Central locking, anti-theft
13		Open	

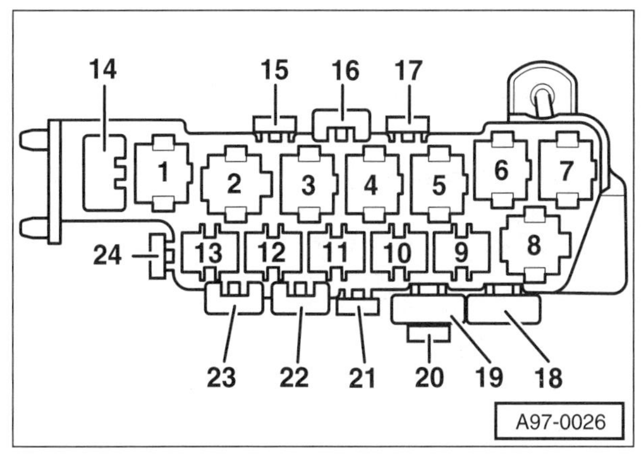

Fig. 54 8-fold relay panel fuses

Table l. 8-fold relay panel fuses		
Fuse	**Amps**	**Protected circuit**
14		Driver seat heater circuit breaker 1 (S44) (1998 - 2001)
14	60	Engine cooling fan (S42) (2002 - 2004 allroad)
15	20	Luggage compartment auxiliary power socket (S184)
16		Front window circuit breaker (S37)
17		Rear window circuit breaker (S43)
18	50	ABS control module (S123)
19	40	Engine cooling fan (S42) (1998 - 2000)
19		Open
20	40	Engine cooling fan (S42)
21	5	Engine cooling fan control module fuse (S142)
22		Driver seat heater circuit breaker 1 (S44) (2002 -2004)
23	40	Level control system (S110)
24		Passenger seat heater circuit breaker (S80)

BentleyPublishers.com
BentleyPublishers.com—All Rights Reserved

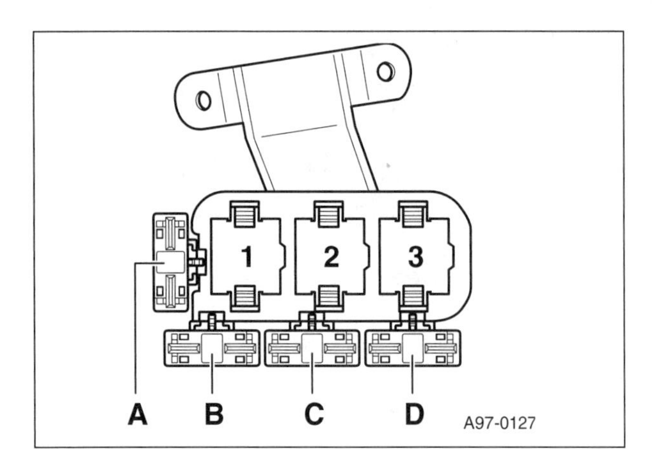

Fig. 55 Central carrier relay panel

 Underneath left side dashboard, attached to dashboard center support, trim panel removed

Table m. Central carrier relay panel fuses and relays		
Pos	**Amps**	**Circuit**
1		Brake booster vacuum pump relay (J569)
2		Open
3		Open
A		Open
B		Open
C	15	Brake booster vacuum pump fuse (S279)
D	15	Fuel pump control module fuse (S81) (RS6)

Grounds

Grounds are distributed throughout the vehicle body. Many are found under the interior carpets or behind trim panels. Several components grounds are often ganged. Ground positions vary among models. Lugs and connectors attached to ground are susceptible to damage and corrosion. Clean or renew as necessary.

Fig. 56 Ground locations

 Major grounding locations throughout the vehicle body:

1. Right side engine compartment
2. Right lower A-pillar
3. Battery to body ground cable
4. Right lower B-pillar
5. Right lower D-pillar
6. Near right taillight
7. Left side luggage compartment
8. Left lower A-pillar
9. ABS hydraulic unit ground
10. Left side engine compartment
11. Engine to body ground strap
12. Near ignition coil pack

NOTE—

- *In wiring diagrams, ground locations are described following a circled number.*

-Example:  *– Ground connection on left A-pillar lower part*

BentleyPublishers.com
BentleyPublishers.com—All Rights Reserved

EWD Electrical Wiring Diagrams

GENERAL

This section contains selected wiring diagrams for vehicles covered by this manual. These diagrams are provided by Audi of America, Inc. The publisher of this manual cannot vouch for the accuracy of these diagrams. Additional wiring diagrams can be found in the *Official Factory Repair Manual* available on CD-ROM or via web subscription at **www.BentleyPublishers.com**.

For a comprehensive list of fuses, relays and electrical component locations, see **97 Fuses, Relays, Component Locations**.

BentleyPublishers.com—All Rights Reserved

Warnings and Cautions

Please read the following before doing any work on your electrical system.

WARNING—

- *Airbags and pyrotechnic seat belt tensioners utilize explosive devices. Handle with extreme care. Refer to the warnings and cautions in* **69 Seat Belts, Airbags**.
- *The ignition system of the car operates at lethal voltages. If you have a weak heart or wear a pacemaker, do not expose yourself to the ignition system electric currents. Take extra precautions when working on the ignition system or when servicing the engine while it is running or the key is ON. See* **28 Ignition System** *for additional ignition system warnings and cautions.*
- *Keep hands, clothing and other objects clear of the electric radiator cooling fan when working on a warm engine. The fan may start at any time, even when the ignition is switched OFF.*

CAUTION—

- *Do not disconnect the battery with the engine running.*
- *Switch the ignition OFF and remove the negative (-) battery cable before removing any electrical components. Connect and disconnect electrical connectors and ignition test equipment leads only while the ignition is switched OFF.*
- *Relay and fuse positions are subject to change and may vary from car to car. If questions arise, an authorized Audi dealer is the best source for the most accurate and up-to-date information.*
- *Use a digital multimeter for electrical tests. Switch the multimeter to the appropriate function and range before making test connections.*
- *Many control modules are static sensitive. Static discharge damages them permanently. Handle the modules using proper static prevention equipment and techniques.*
- *To avoid damaging harness connectors or relay panel sockets, use jumper wires with flat-blade connectors that are the same size as the connector or relay terminals.*
- *Do not try to start the engine of a car which has been heated above 176°F (80°C) (for example, in a paint drying booth). Allow it to cool to normal temperature.*
- *Disconnect the battery before doing any electric welding on the car.*
- *Do not wash the engine while it is running, or any time the ignition is ON.*
- *Choose test equipment carefully. Use a digital multimeter with at least 10 MΩ input impedance, or an LED test light. An analog meter (swing-needle) or a test light with a normal incandescent bulb may draw enough current to damage sensitive electronic components.*
- *Do not use an ohmmeter to measure resistance on solid state components such as control modules.*
- *Disconnect the battery before making resistance (ohm) measurements on a circuit.*

BentleyPublishers.com—All Rights Reserved

Wiring diagram organization

Audi wiring diagrams in this manual are organized as follows:

- **Standard equipment** covers
  Grounds
  Power supply
  Alternator (generator) and starter
  Exterior lights
  ABS
  Wipers and washers
  Seats and mirrors
- **Central locking system**
- **Daytime running lights** including foglights
- **Engine management**
- **Headlights and headlight washers**
- **Instruments and controls** covers switches and instrument panel on dashboard
- **Sound system**
- **Windows**

Terminal (circuit) identification

Several circuits in the vehicle's electrical system are identified with a number or letter designation. These circuits are universally identified in many wiring diagrams and may also be used to identify switch connector terminals.

Terminal	Identification
1	Ignition coil/distributor low voltage (i.e. RPM signal)
15	Switched battery positive (B+) from ignition switch
30	Battery positive (B+), hot at all times
31	Ground (–)
31b	Switched ground (–)
50	Starter control; switched B+ from ignition switch
58	Switched parking light, taillight, B+ from light switch
S (SU)	Key in ignition, switched B+ from ignition switch
X	Switched B+ from load reduction relay

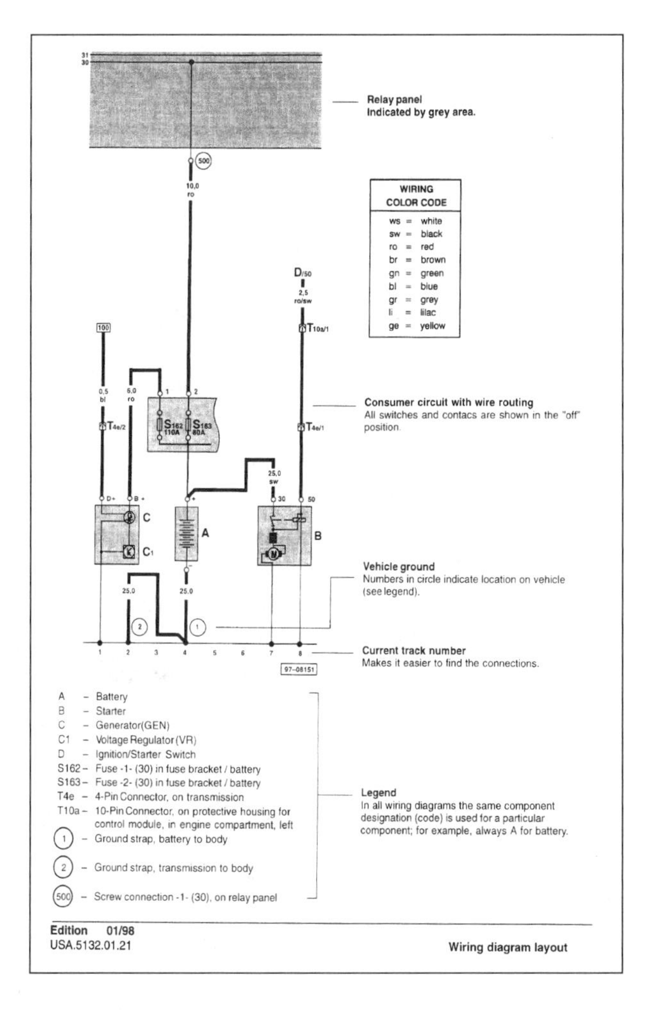

How to read wiring diagrams

 This diagram shows how a typical wiring diagram page is laid out.

Details of wiring diagrams are explained on the next 3 pages.

BentleyPublishers.com—All Rights Reserved

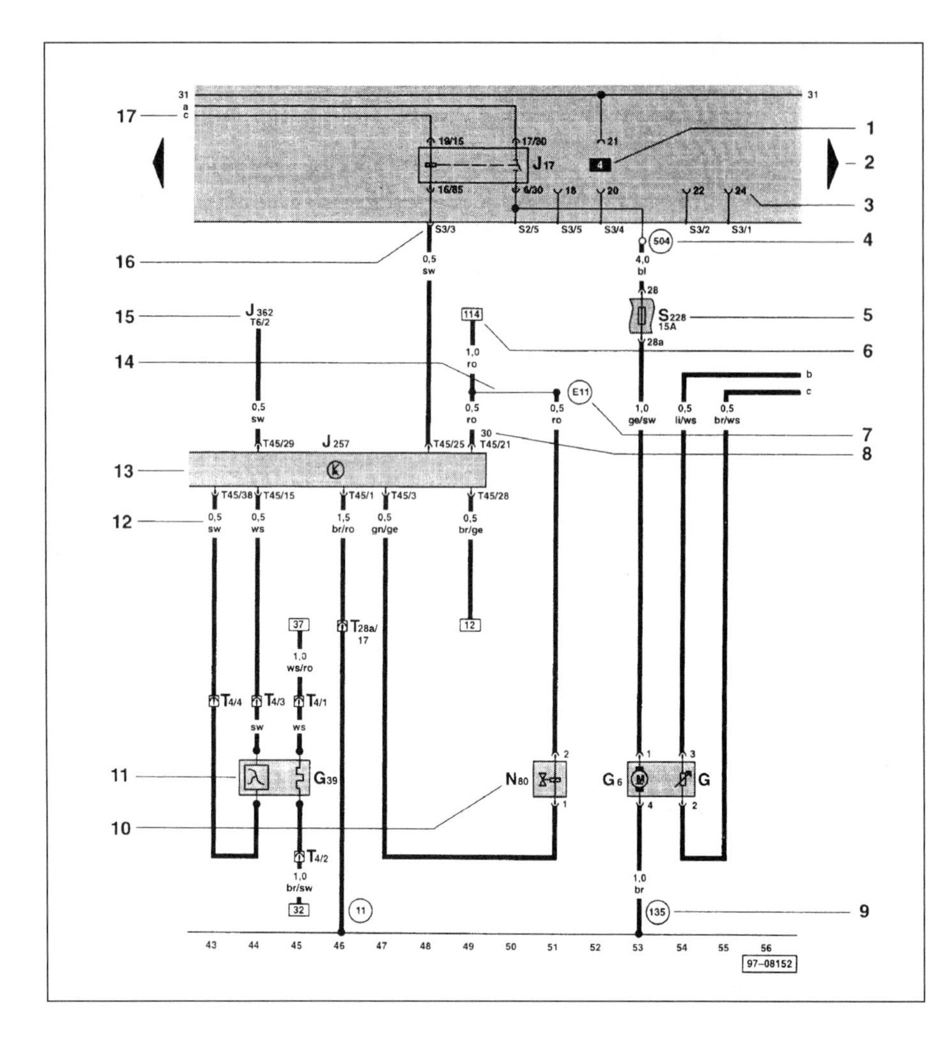

Wiring diagram details

1. **Relay location number**
    - Indicates location on relay panel.
2. **Arrow**
    - Indicates wiring circuit is continued on the previous and/or next page.
3. **Connection designation - relay control module on relay panel**
    - Shows individual terminals in a multi-point connector.
    - For example: terminal 24 indicated.
4. **Diagram of threaded pin on relay panel**
    - White circle shows detachable connection.
5. **Fuse designation**
    - For example: S228 = fuse number 228, 15 amps, in fuse holder.
6. **Reference of wire continuation (current track number)**
    - Number in frame indicates current track where wire is continued on another wiring diagram page.
7. **Wire connection designation in wiring harness**
    - Location of wire connections are indicated in legend.
8. **Terminal designation**
    - Designation which appears on actual component and/or terminal number of a multi-point connector.
9. **Ground connection designation in wire harness**
    - Locations of ground connections are indicated in legend.
10. **Component designation**
    - Use legend to identify component code.
11. **Component symbols**
    - See **Symbols used in wiring diagrams** in this repair group.
12. **Wire cross-section size (in mm^2) and wire colors**
    - Abbreviations are explained in color chart beside wiring diagram.
13. **Component symbol with open drawing side**
    - Indicated component is continued on another wiring diagram page.

BentleyPublishers.com—All Rights Reserved

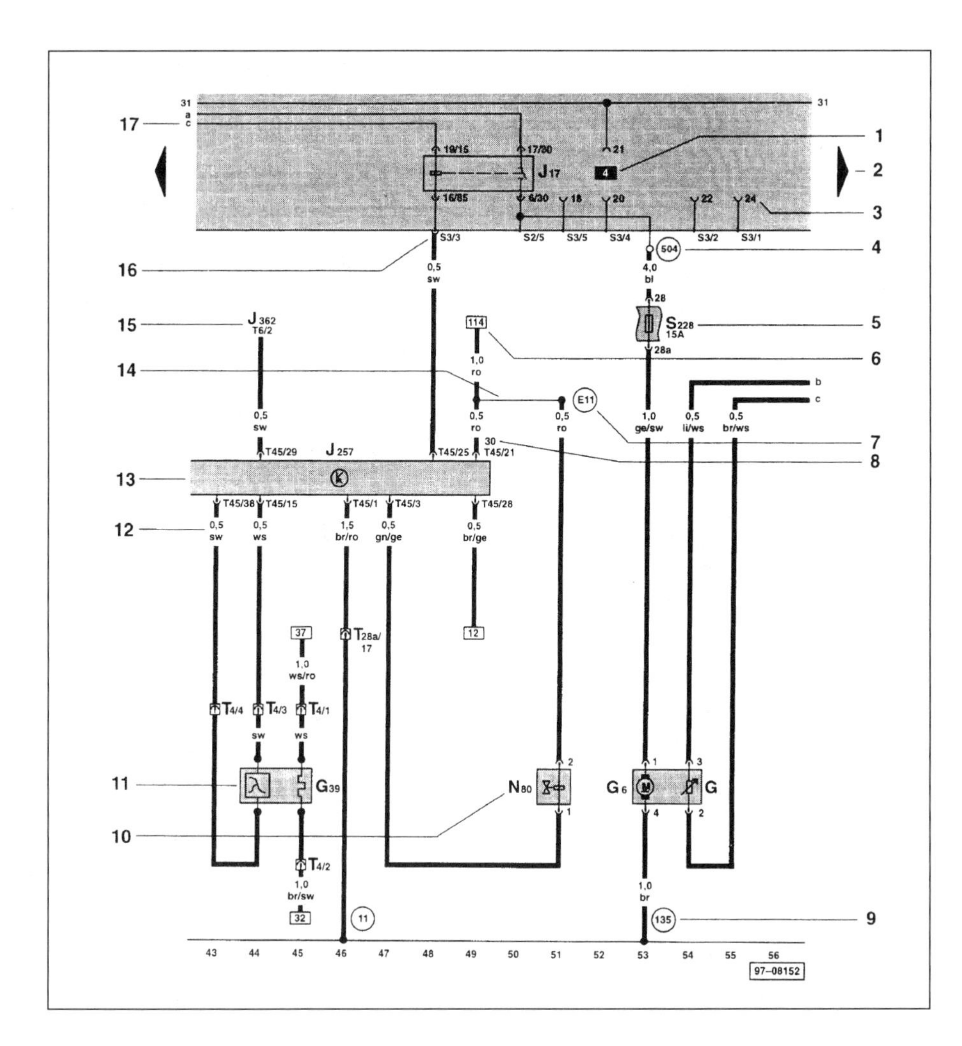

14. Internal connections (thin lines)

- These connections are not wires.
- Internal connections are current carrying and are shown to allow tracing of current flow inside components or wiring harnesses.

15. Reference of continuation of wire to component

- For example: Anti-theft immobilizer (J362) control module on 6-pin connector, terminal 2.

16. Relay panel connectors

- Shows wiring of multi-point or single connectors on relay panel.
- For example: S3/3 = multi-point connector S3, terminal 3.

17. Reference of internal connection continuation

- Letters indicate where connection continues on previous and/or next page.

BentleyPublishers.com—All Rights Reserved

Symbols used in wiring diagrams

Various symbols identified below are used in the wiring diagrams.

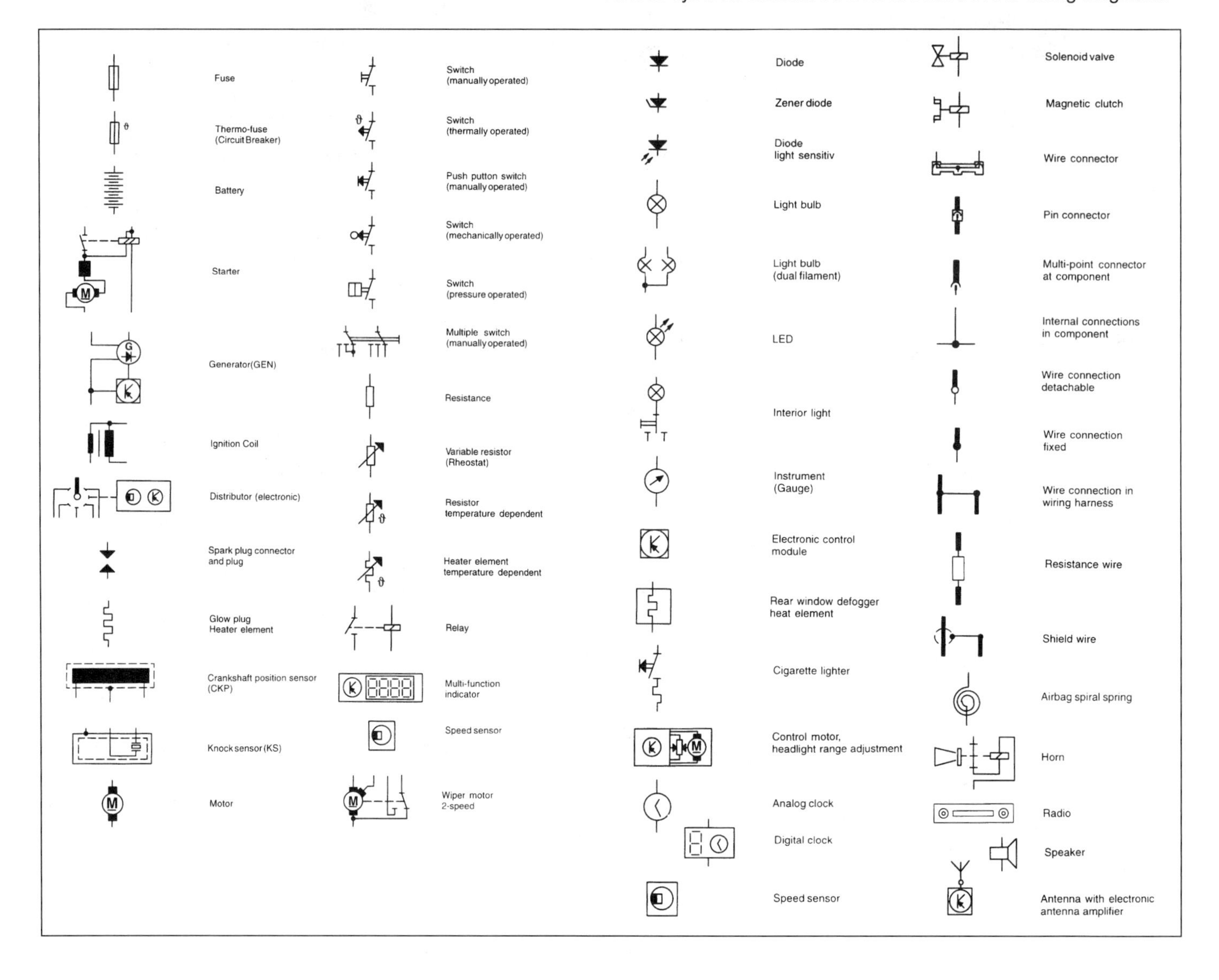

BentleyPublishers.com—All Rights Reserved

ELECTRICAL WIRING DIAGRAM INDEX

BentleyPublishers.com—All Rights Reserved

Electrical Wiring Diagram Index

BentleyPublishers.com—All Rights Reserved

BentleyPublishers.com—All Rights Reserved

Electrical Wiring Diagram Index

BentleyPublishers.com—All Rights Reserved

BentleyPublishers.com—All Rights Reserved

BentleyPublishers.com—All Rights Reserved

BentleyPublishers.com—All Rights Reserved

BentleyPublishers.com—All Rights Reserved

BentleyPublishers.com—All Rights Reserved

BentleyPublishers.com—All Rights Reserved

BentleyPublishers.com—All Rights Reserved

BentleyPublishers.com—All Rights Reserved

BentleyPublishers.com—All Rights Reserved

Electrical Wiring Diagram Index

BentleyPublishers.com—All Rights Reserved

BentleyPublishers.com—All Rights Reserved

Electrical Wiring Diagram Index

BentleyPublishers.com—All Rights Reserved

BentleyPublishers.com—All Rights Reserved

BentleyPublishers.com—All Rights Reserved

BentleyPublishers.com—All Rights Reserved

(2,8 l - Injection Engine, 6-Cylinder), Code AHA

1998 m. y.

Fuse Panel

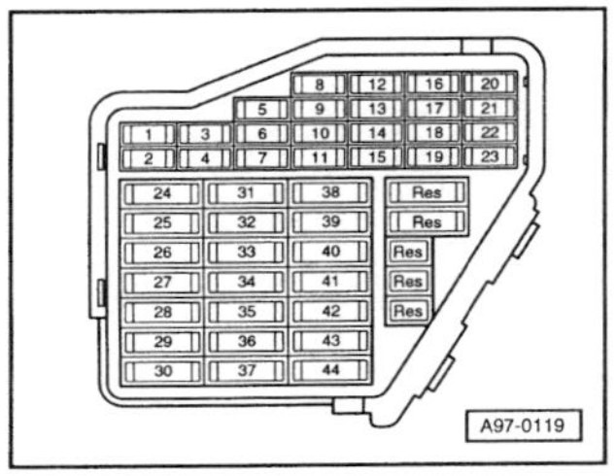

Fuse Colors:

30 A - Green
25 A - White
20 A - Yellow
15 A - Blue
10 A - Red
5 A - Beige

Starting with fuse position 23, fuses in the fuse holder are identified with 223 in the wiring diagram.

Micro Central Electric Panel

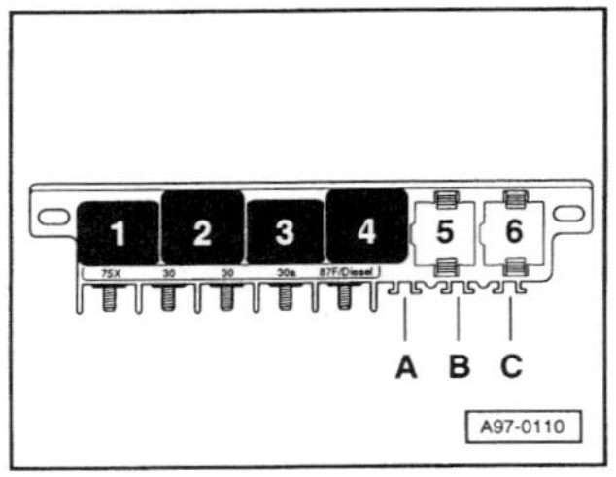

Relay Location:

- **1** - Dual Horn Relay, J4
- **2** - Load Reduction Relay, J59
- **4** - Fuel Pump (FP) Relay, J17
- **5** - Wiper/Washer Intermittent Relay, J31
- **6** - Wiper/Washer Intermittent Relay, J31

13 - Fold Relay Panel

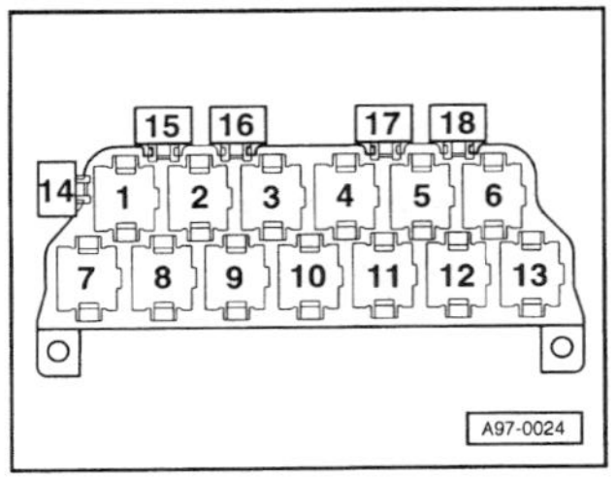

Relay Location:

- **4** - Starting Interlock Relay, J207
- **6** - Fog Light Relay, J5
- **17** - Fuse for secondary air pump, S130

8 - Fold Relay Panel

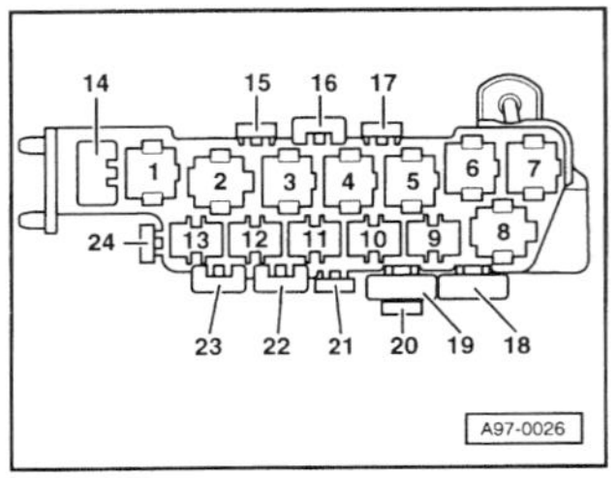

Relay Location:

- **2** - Second Speed Coolant Fan Control (FC) Relay, J101
- **3** - Coolant Fan control (FC) Relay, J26
- **20** - Coolant Fan Fuse, S42
- **21** - Control module fuse for coolant fan, S142

3 - Fold Relay Panel in E-Box plenum chamber

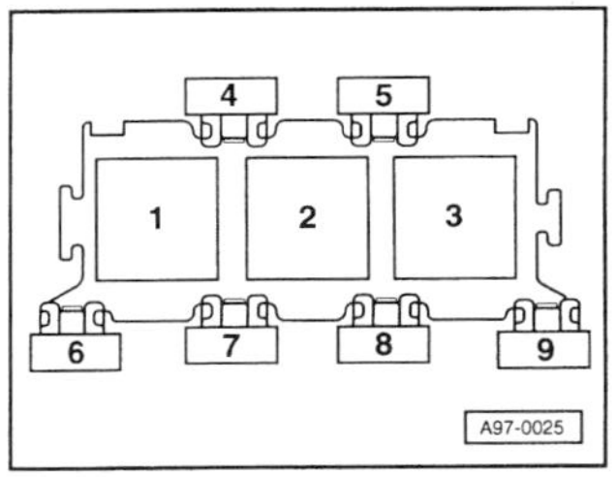

Relay Location:

- **2** - Secondary Air Injection (AIR) Pump Relay, J299

BentleyPublishers.com—All Rights Reserved

Engine management, 2.8 liter engine, code AHA (1998)

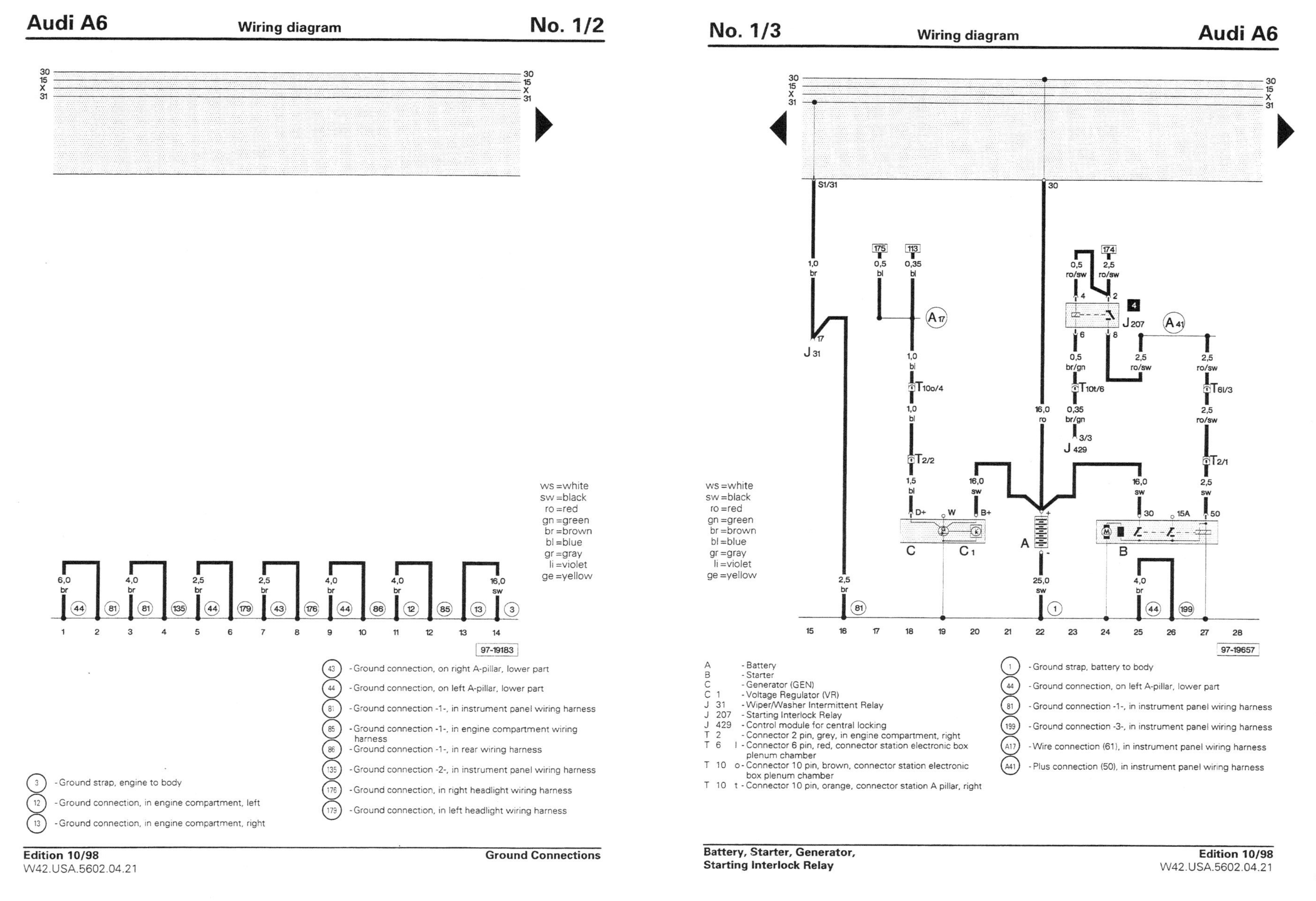

BentleyPublishers.com–All Rights Reserved

Engine management, 2.8 liter engine, code AHA (1998)

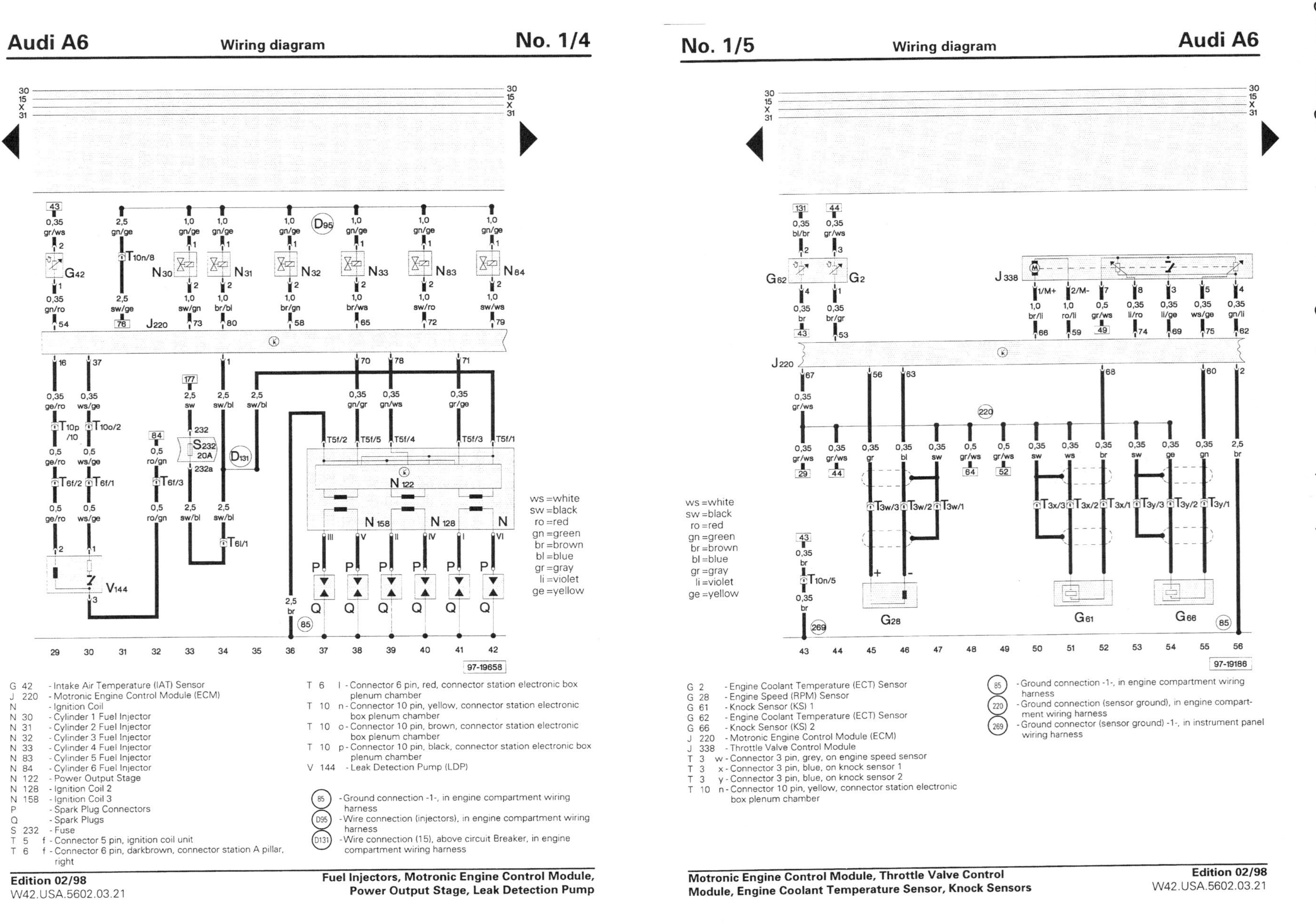

Audi A6
Wiring diagram
No. 1/4
G 42 - Intake Air Temperature (IAT) Sensor
J 220 - Motronic Engine Control Module (ECM)
N - Ignition Coil
N 30 - Cylinder 1 Fuel Injector
N 31 - Cylinder 2 Fuel Injector
N 32 - Cylinder 3 Fuel Injector
N 33 - Cylinder 4 Fuel Injector
N 83 - Cylinder 5 Fuel Injector
N 84 - Cylinder 6 Fuel Injector
N 122 - Power Output Stage
N 128 - Ignition Coil 2
N 158 - Ignition Coil 3
P - Spark Plug Connectors
Q - Spark Plugs
S 232 - Fuse
T 5 f - Connector 5 pin, ignition coil unit
T 6 f - Connector 6 pin, darkbrown, connector station A pillar, right
T 6 l - Connector 6 pin, red, connector station electronic box plenum chamber
T 10 n - Connector 10 pin, yellow, connector station electronic box plenum chamber
T 10 o - Connector 10 pin, brown, connector station electronic box plenum chamber
T 10 p - Connector 10 pin, black, connector station electronic box plenum chamber
V 144 - Leak Detection Pump (LDP)
85 - Ground connection -1-, in engine compartment wiring harness
D95 - Wire connection (injectors), in engine compartment wiring harness
D131 - Wire connection (15), above circuit Breaker, in engine compartment wiring harness
ws = white
sw = black
ro = red
gn = green
br = brown
bl = blue
gr = gray
li = violet
ge = yellow
97-19658
Edition 02/98
W42.USA.5602.03.21
Fuel Injectors, Motronic Engine Control Module, Power Output Stage, Leak Detection Pump
No. 1/5
Wiring diagram
Audi A6
G 2 - Engine Coolant Temperature (ECT) Sensor
G 28 - Engine Speed (RPM) Sensor
G 61 - Knock Sensor (KS) 1
G 62 - Engine Coolant Temperature (ECT) Sensor
G 66 - Knock Sensor (KS) 2
J 220 - Motronic Engine Control Module (ECM)
J 338 - Throttle Valve Control Module
T 3 w - Connector 3 pin, grey, on engine speed sensor
T 3 x - Connector 3 pin, blue, on knock sensor 1
T 3 y - Connector 3 pin, blue, on knock sensor 2
T 10 n - Connector 10 pin, yellow, connector station electronic box plenum chamber
85 - Ground connection -1-, in engine compartment wiring harness
220 - Ground connection (sensor ground), in engine compartment wiring harness
269 - Ground connector (sensor ground) -1-, in instrument panel wiring harness
97-19186
Motronic Engine Control Module, Throttle Valve Control Module, Engine Coolant Temperature Sensor, Knock Sensors
Edition 02/98
W42.USA.5602.03.21

BentleyPublishers.com—All Rights Reserved

Engine management, 2.8 liter engine, code AHA (1998)

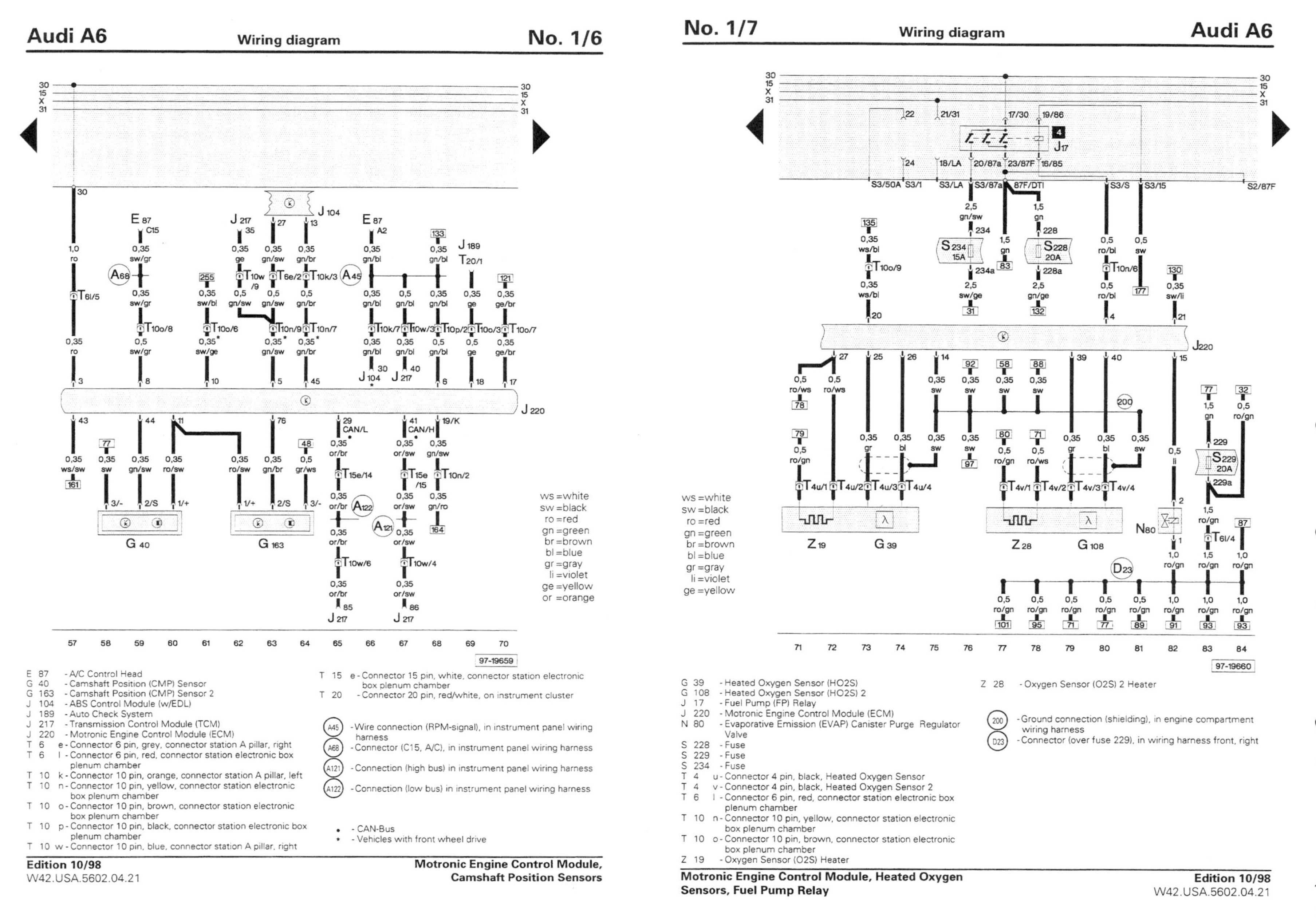

Audi A6
Wiring diagram
No. 1/6
ws =white
sw =black
ro =red
gn =green
br =brown
bl =blue
gr =gray
li =violet
ge =yellow
or =orange
97-19659
E 87 - A/C Control Head
G 40 - Camshaft Position (CMP) Sensor
G 163 - Camshaft Position (CMP) Sensor 2
J 104 - ABS Control Module (w/EDL)
J 189 - Auto Check System
J 217 - Transmission Control Module (TCM)
J 220 - Motronic Engine Control Module (ECM)
T 6 e - Connector 6 pin, grey, connector station A pillar, right
T 6 l - Connector 6 pin, red, connector station electronic box plenum chamber
T 10 k - Connector 10 pin, orange, connector station A pillar, left
T 10 n - Connector 10 pin, yellow, connector station electronic box plenum chamber
T 10 o - Connector 10 pin, brown, connector station electronic box plenum chamber
T 10 p - Connector 10 pin, black, connector station electronic box plenum chamber
T 10 w - Connector 10 pin, blue, connector station A pillar, right
T 15 e - Connector 15 pin, white, connector station electronic box plenum chamber
T 20 - Connector 20 pin, red/white, on instrument cluster
A45 - Wire connection (RPM-signal), in instrument panel wiring harness
A68 - Connector (C15, A/C), in instrument panel wiring harness
A121 - Connection (high bus) in instrument panel wiring harness
A122 - Connection (low bus) in instrument panel wiring harness
* - CAN-Bus
* - Vehicles with front wheel drive
Edition 10/98
W42.USA.5602.04.21
Motronic Engine Control Module, Camshaft Position Sensors
No. 1/7
Wiring diagram
Audi A6
ws =white
sw =black
ro =red
gn =green
br =brown
bl =blue
gr =gray
li =violet
ge =yellow
97-19660
G 39 - Heated Oxygen Sensor (HO2S)
G 108 - Heated Oxygen Sensor (HO2S) 2
J 17 - Fuel Pump (FP) Relay
J 220 - Motronic Engine Control Module (ECM)
N 80 - Evaporative Emission (EVAP) Canister Purge Regulator Valve
S 228 - Fuse
S 229 - Fuse
S 234 - Fuse
T 4 u - Connector 4 pin, black, Heated Oxygen Sensor
T 4 v - Connector 4 pin, black, Heated Oxygen Sensor 2
T 6 l - Connector 6 pin, red, connector station electronic box plenum chamber
T 10 n - Connector 10 pin, yellow, connector station electronic box plenum chamber
T 10 o - Connector 10 pin, brown, connector station electronic box plenum chamber
Z 19 - Oxygen Sensor (O2S) Heater
Z 28 - Oxygen Sensor (O2S) 2 Heater
200 - Ground connection (shielding), in engine compartment wiring harness
D23 - Connector (over fuse 229), in wiring harness front, right
Motronic Engine Control Module, Heated Oxygen Sensors, Fuel Pump Relay
Edition 10/98
W42.USA.5602.04.21

BentleyPublishers.com—All Rights Reserved

Engine management, 2.8 liter engine, code AHA (1998)

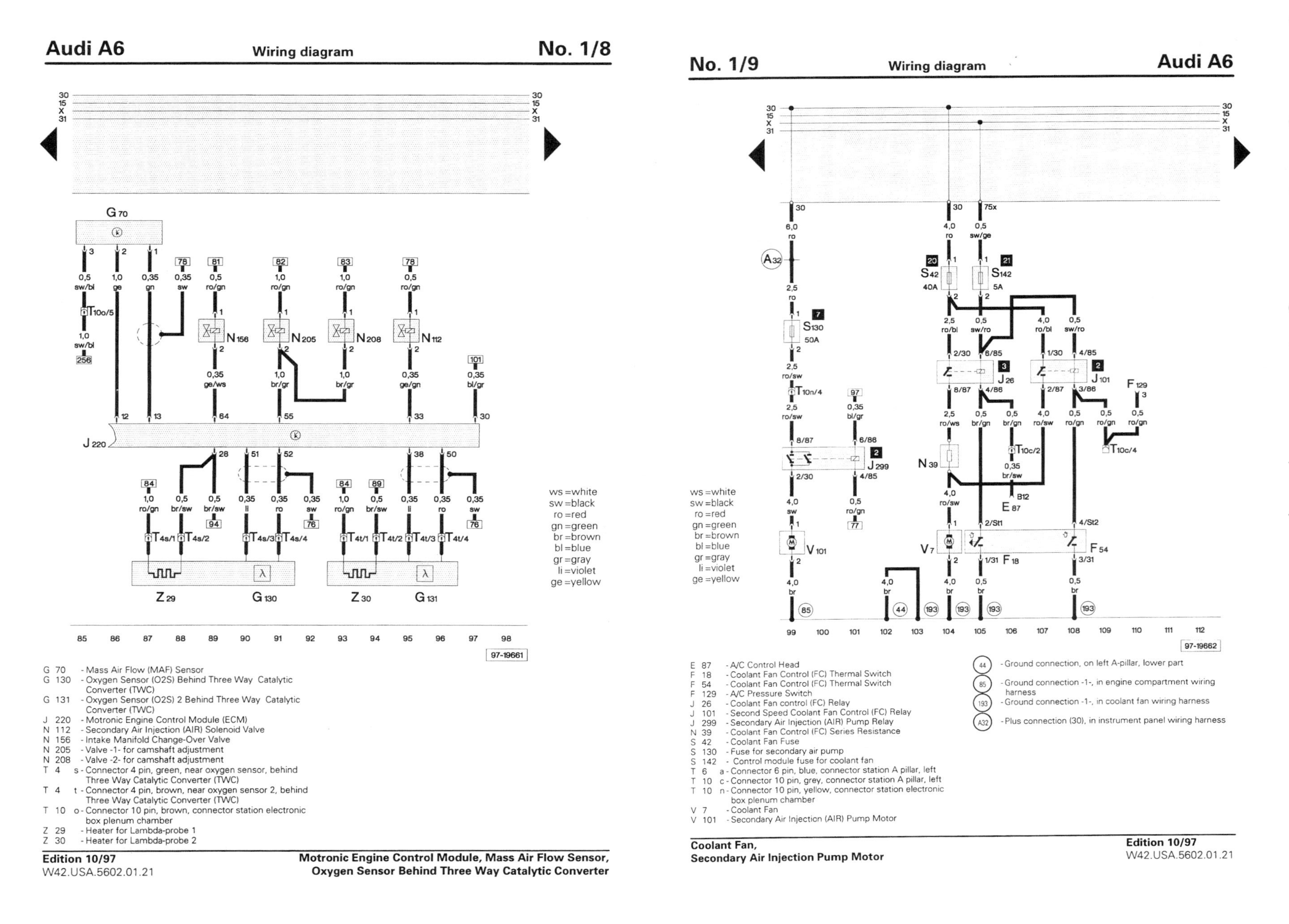

BentleyPublishers.com—All Rights Reserved

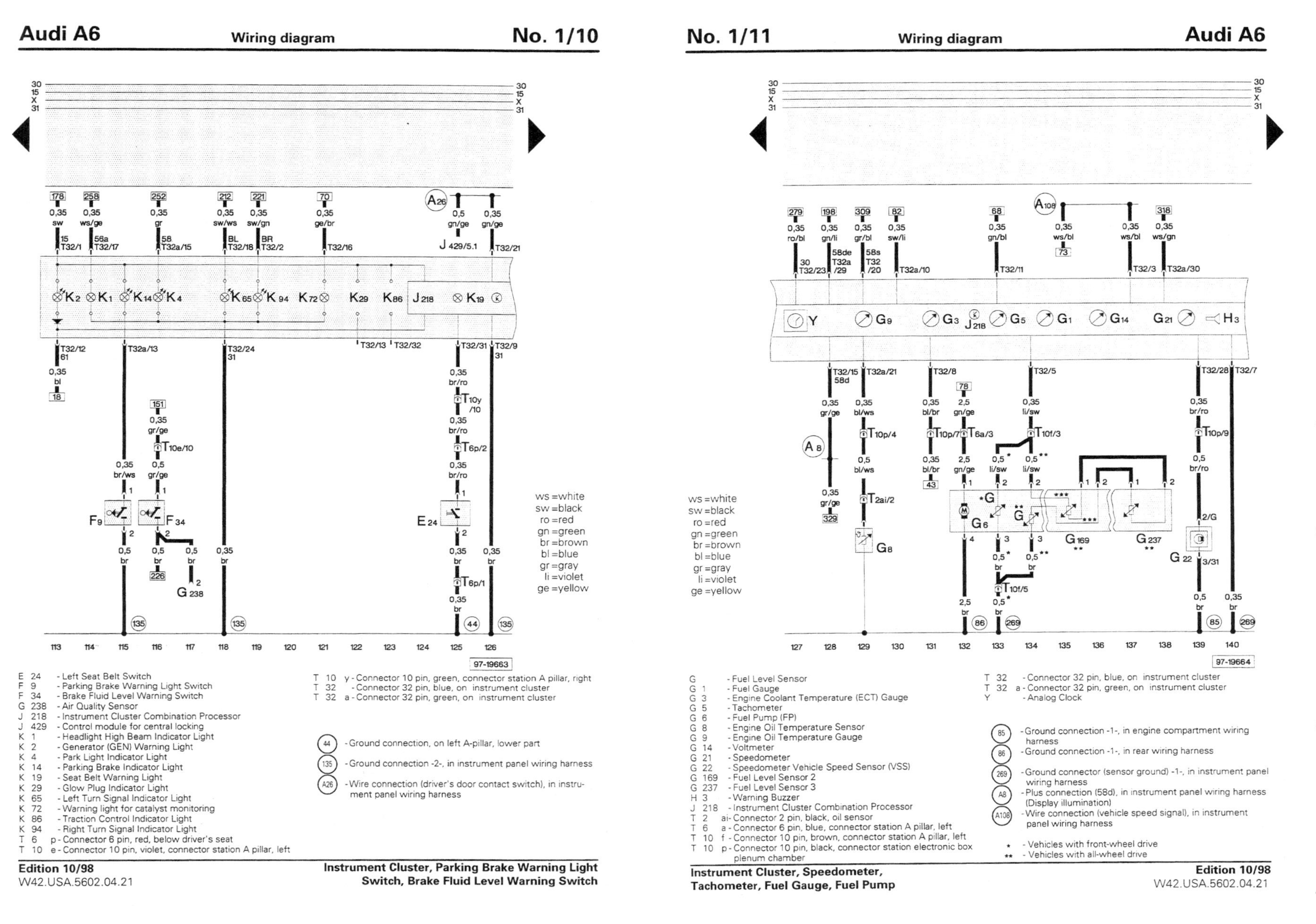

Audi A6
Wiring diagram
No. 1/10
ws =white
sw =black
ro =red
gn =green
br =brown
bl =blue
gr =gray
li =violet
ge =yellow
97-19663
E 24 - Left Seat Belt Switch
F 9 - Parking Brake Warning Light Switch
F 34 - Brake Fluid Level Warning Switch
G 238 - Air Quality Sensor
J 218 - Instrument Cluster Combination Processor
J 429 - Control module for central locking
K 1 - Headlight High Beam Indicator Light
K 2 - Generator (GEN) Warning Light
K 4 - Park Light Indicator Light
K 14 - Parking Brake Indicator Light
K 19 - Seat Belt Warning Light
K 29 - Glow Plug Indicator Light
K 65 - Left Turn Signal Indicator Light
K 72 - Warning light for catalyst monitoring
K 86 - Traction Control Indicator Light
K 94 - Right Turn Signal Indicator Light
T 6 p - Connector 6 pin, red, below driver's seat
T 10 e - Connector 10 pin, violet, connector station A pillar, left
T 10 y - Connector 10 pin, green, connector station A pillar, right
T 32 - Connector 32 pin, blue, on instrument cluster
T 32 a - Connector 32 pin, green, on instrument cluster
44 - Ground connection, on left A-pillar, lower part
135 - Ground connection -2-, in instrument panel wiring harness
A26 - Wire connection (driver's door contact switch), in instrument panel wiring harness
Edition 10/98
W42.USA.5602.04.21
Instrument Cluster, Parking Brake Warning Light Switch, Brake Fluid Level Warning Switch
No. 1/11
Wiring diagram
Audi A6
ws =white
sw =black
ro =red
gn =green
br =brown
bl =blue
gr =gray
li =violet
ge =yellow
97-19664
G - Fuel Level Sensor
G 1 - Fuel Gauge
G 3 - Engine Coolant Temperature (ECT) Gauge
G 5 - Tachometer
G 6 - Fuel Pump (FP)
G 8 - Engine Oil Temperature Sensor
G 9 - Engine Oil Temperature Gauge
G 14 - Voltmeter
G 21 - Speedometer
G 22 - Speedometer Vehicle Speed Sensor (VSS)
G 169 - Fuel Level Sensor 2
G 237 - Fuel Level Sensor 3
H 3 - Warning Buzzer
J 218 - Instrument Cluster Combination Processor
T 2 ai - Connector 2 pin, black, oil sensor
T 6 a - Connector 6 pin, blue, connector station A pillar, left
T 10 f - Connector 10 pin, brown, connector station A pillar, left
T 10 p - Connector 10 pin, black, connector station electronic box plenum chamber
T 32 - Connector 32 pin, blue, on instrument cluster
T 32 a - Connector 32 pin, green, on instrument cluster
Y - Analog Clock
85 - Ground connection -1-, in engine compartment wiring harness
86 - Ground connection -1-, in rear wiring harness
269 - Ground connector (sensor ground) -1-, in instrument panel wiring harness
A8 - Plus connection (58d), in instrument panel wiring harness (Display illumination)
A108 - Wire connection (vehicle speed signal), in instrument panel wiring harness
* - Vehicles with front-wheel drive
** - Vehicles with all-wheel drive
Instrument Cluster, Speedometer, Tachometer, Fuel Gauge, Fuel Pump
Edition 10/98
W42.USA.5602.04.21

BentleyPublishers.com—All Rights Reserved

Engine management, 2.8 liter engine, code AHA (1998)

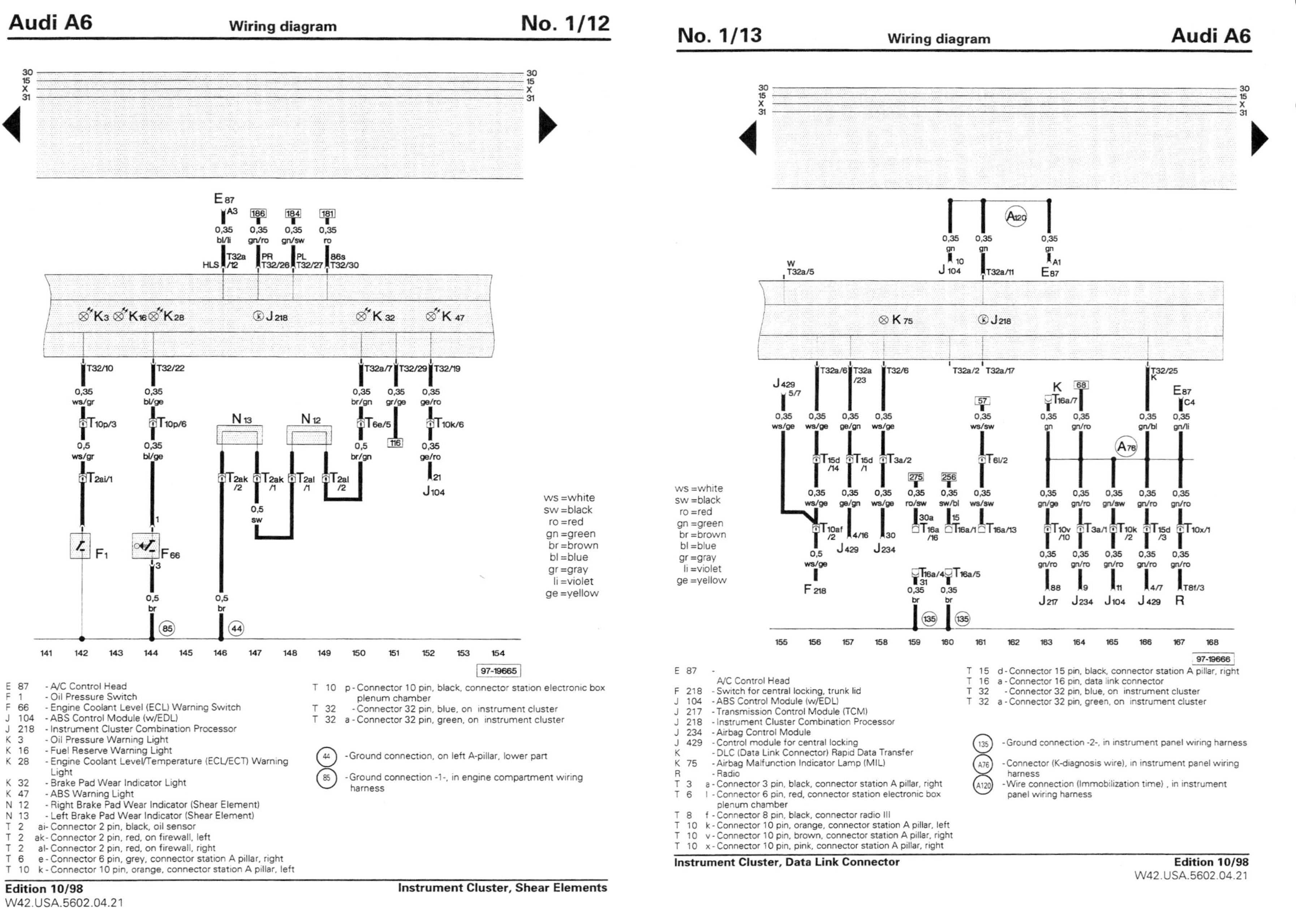

Audi A6
Wiring diagram
No. 1/12
E 87 - A/C Control Head
F 1 - Oil Pressure Switch
F 66 - Engine Coolant Level (ECL) Warning Switch
J 104 - ABS Control Module (w/EDL)
J 218 - Instrument Cluster Combination Processor
K 3 - Oil Pressure Warning Light
K 16 - Fuel Reserve Warning Light
K 28 - Engine Coolant Level/Temperature (ECL/ECT) Warning Light
K 32 - Brake Pad Wear Indicator Light
K 47 - ABS Warning Light
N 12 - Right Brake Pad Wear Indicator (Shear Element)
N 13 - Left Brake Pad Wear Indicator (Shear Element)
T 2 ai- Connector 2 pin, black, oil sensor
T 2 ak- Connector 2 pin, red, on firewall, left
T 2 al- Connector 2 pin, red, on firewall, right
T 6 e - Connector 6 pin, grey, connector station A pillar, right
T 10 k - Connector 10 pin, orange, connector station A pillar, left
T 10 p - Connector 10 pin, black, connector station electronic box plenum chamber
T 32 - Connector 32 pin, blue, on instrument cluster
T 32 a - Connector 32 pin, green, on instrument cluster
44 - Ground connection, on left A-pillar, lower part
85 - Ground connection -1-, in engine compartment wiring harness
ws = white
sw = black
ro = red
gn = green
br = brown
bl = blue
gr = gray
li = violet
ge = yellow
97-19665
Edition 10/98
W42.USA.5602.04.21
Instrument Cluster, Shear Elements
No. 1/13
Wiring diagram
Audi A6
E 87 - A/C Control Head
F 218 - Switch for central locking, trunk lid
J 104 - ABS Control Module (w/EDL)
J 217 - Transmission Control Module (TCM)
J 218 - Instrument Cluster Combination Processor
J 234 - Airbag Control Module
J 429 - Control module for central locking
K - DLC (Data Link Connector) Rapid Data Transfer
K 75 - Airbag Malfunction Indicator Lamp (MIL)
R - Radio
T 3 a - Connector 3 pin, black, connector station A pillar, right
T 6 l - Connector 6 pin, red, connector station electronic box plenum chamber
T 8 f - Connector 8 pin, black, connector radio III
T 10 k - Connector 10 pin, orange, connector station A pillar, left
T 10 v - Connector 10 pin, brown, connector station A pillar, right
T 10 x - Connector 10 pin, pink, connector station A pillar, right
T 15 d - Connector 15 pin, black, connector station A pillar, right
T 16 a - Connector 16 pin, data link connector
T 32 - Connector 32 pin, blue, on instrument cluster
T 32 a - Connector 32 pin, green, on instrument cluster
135 - Ground connection -2-, in instrument panel wiring harness
A76 - Connector (K-diagnosis wire), in instrument panel wiring harness
A120 - Wire connection (Immobilization time), in instrument panel wiring harness
97-19666
Instrument Cluster, Data Link Connector
Edition 10/98
W42.USA.5602.04.21

BentleyPublishers.com—All Rights Reserved

Engine management, 2.8 liter engine, code AHA (1998)

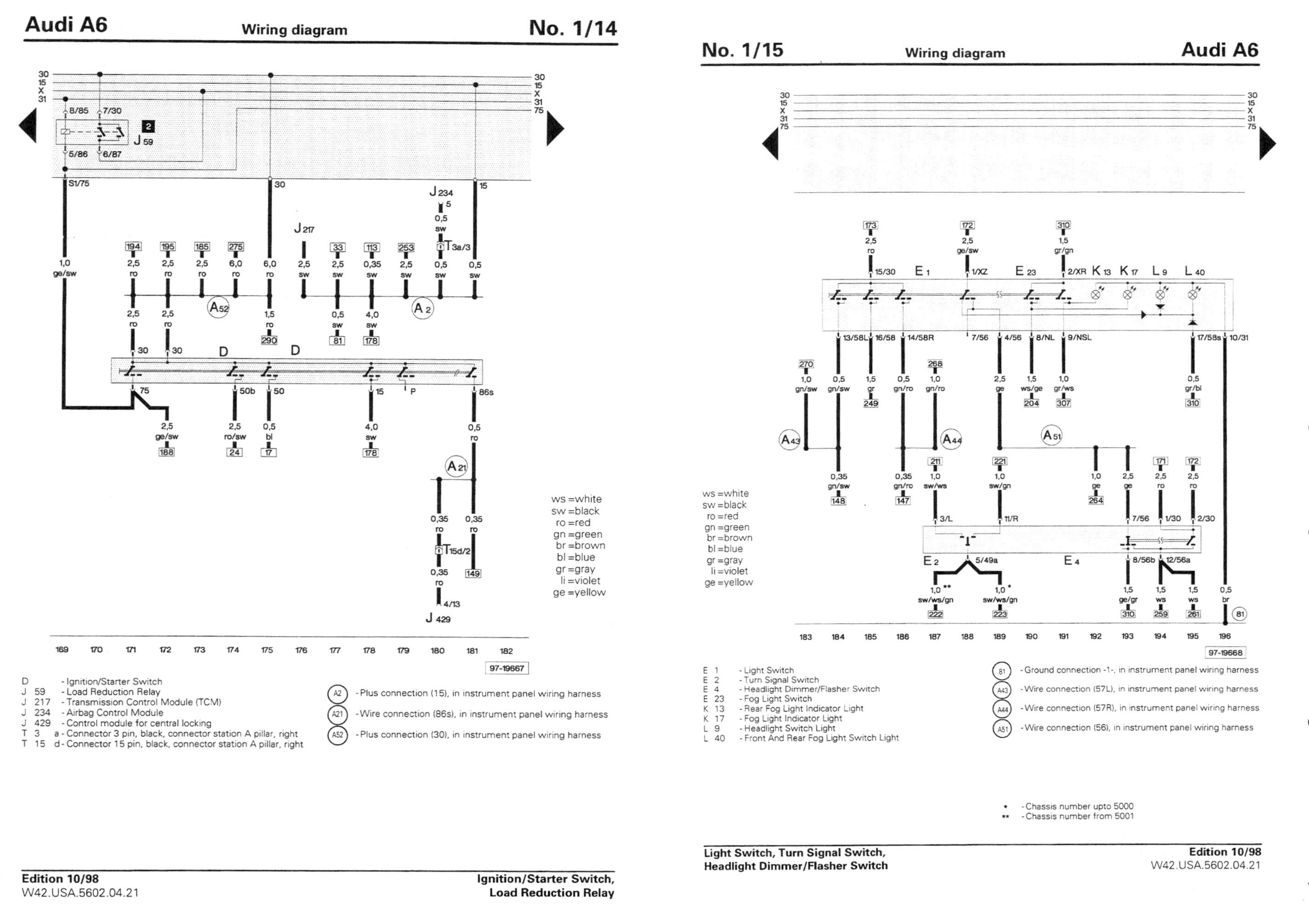

BentleyPublishers.com—All Rights Reserved

Engine management, 2.8 liter engine, code AHA (1998)

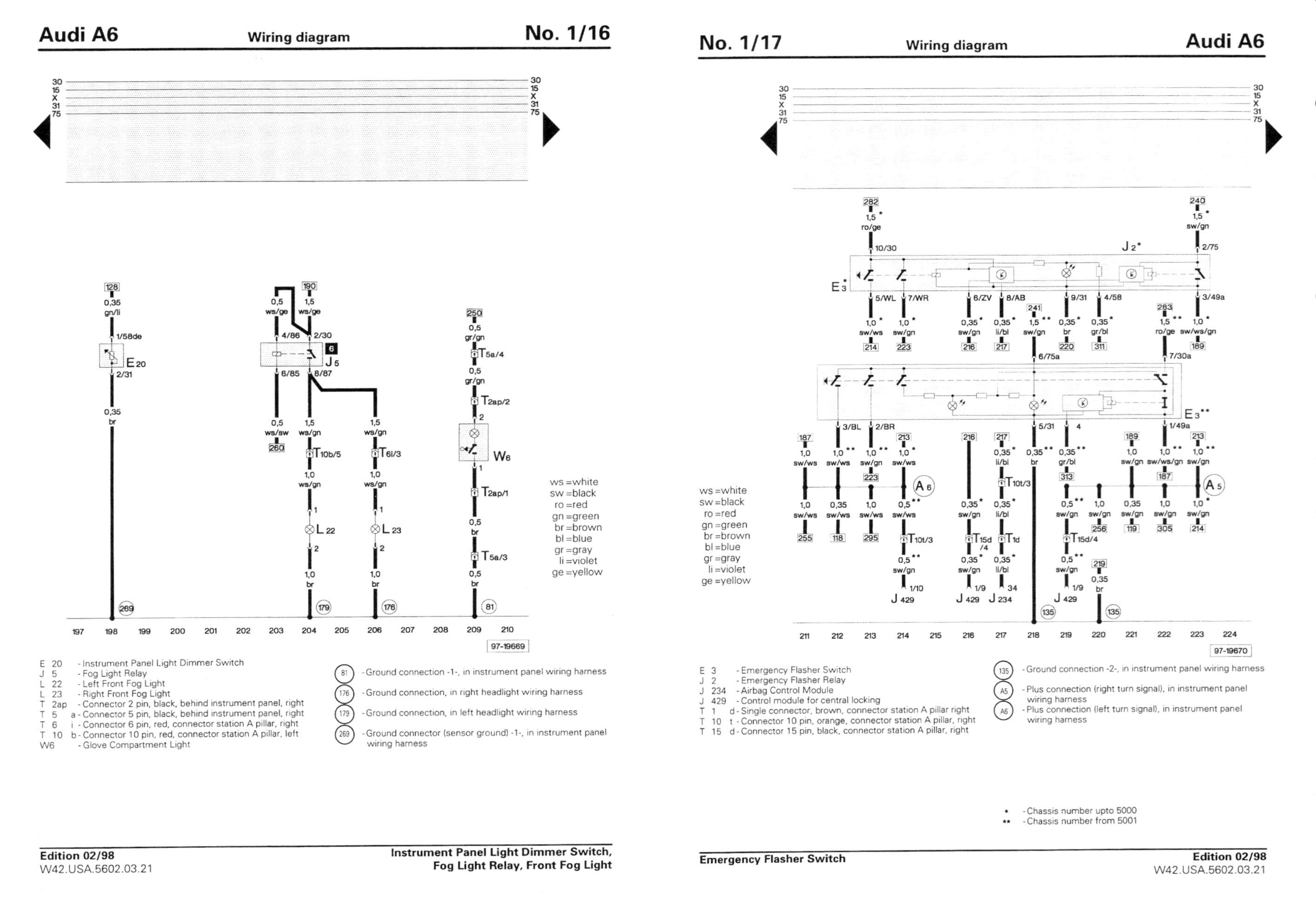

BentleyPublishers.com—All Rights Reserved

Engine management, 2.8 liter engine, code AHA (1998)

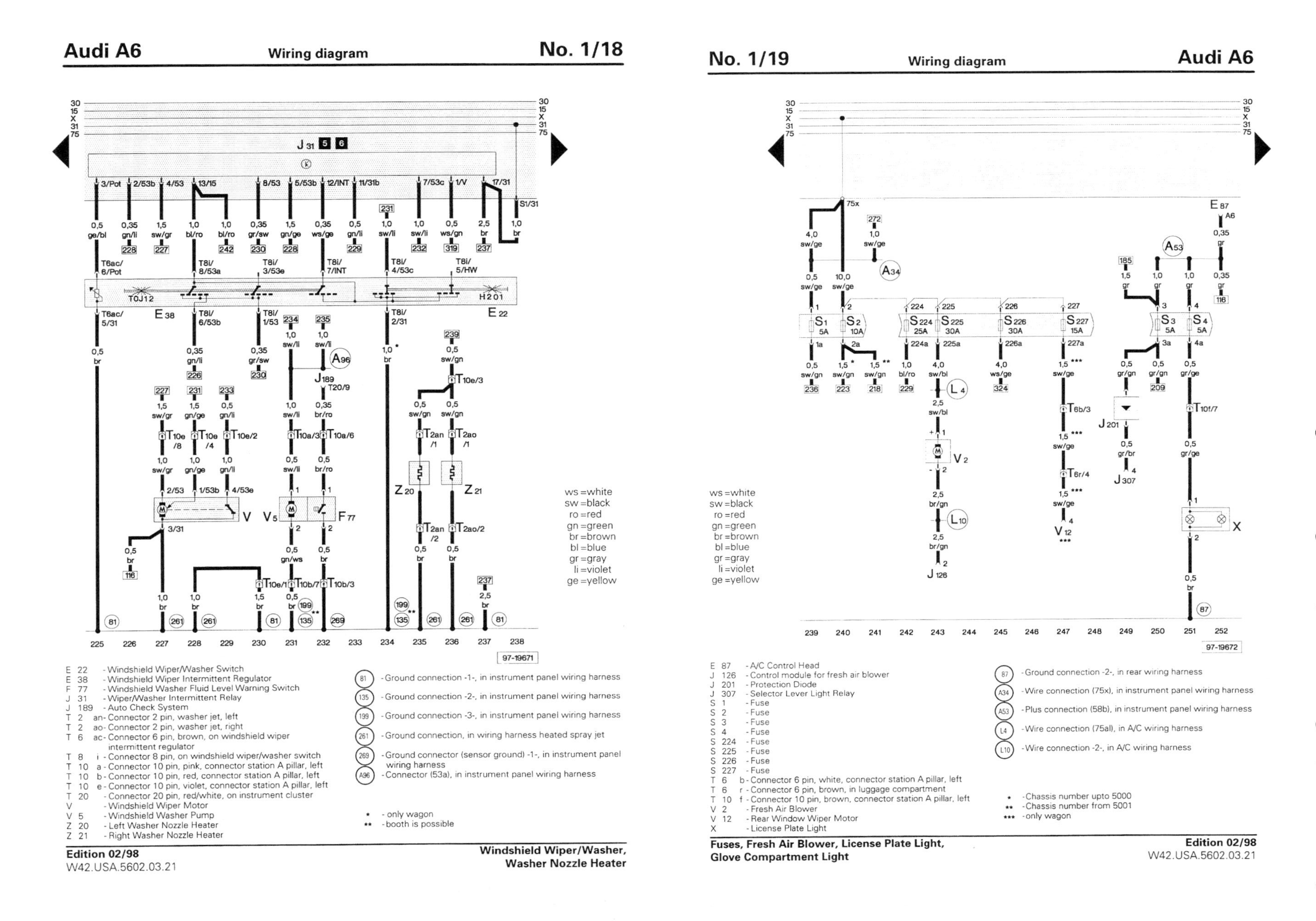

Audi A6
Wiring diagram
No. 1/18
J 31
ws =white
sw =black
ro =red
gn =green
br =brown
bl =blue
gr =gray
li =violet
ge =yellow
97-19671
E 22 - Windshield Wiper/Washer Switch
E 38 - Windshield Wiper Intermittent Regulator
F 77 - Windshield Washer Fluid Level Warning Switch
J 31 - Wiper/Washer Intermittent Relay
J 189 - Auto Check System
T 2 an - Connector 2 pin, washer jet, left
T 2 ao - Connector 2 pin, washer jet, right
T 6 ac - Connector 6 pin, brown, on windshield wiper intermittent regulator
T 8 i - Connector 8 pin, on windshield wiper/washer switch
T 10 a - Connector 10 pin, pink, connector station A pillar, left
T 10 b - Connector 10 pin, red, connector station A pillar, left
T 10 e - Connector 10 pin, violet, connector station A pillar, left
T 20 - Connector 20 pin, red/white, on instrument cluster
V - Windshield Wiper Motor
V 5 - Windshield Washer Pump
Z 20 - Left Washer Nozzle Heater
Z 21 - Right Washer Nozzle Heater
81 - Ground connection -1-, in instrument panel wiring harness
135 - Ground connection -2-, in instrument panel wiring harness
199 - Ground connection -3-, in instrument panel wiring harness
261 - Ground connection, in wiring harness heated spray jet
269 - Ground connector (sensor ground) -1-, in instrument panel wiring harness
A96 - Connector (53a), in instrument panel wiring harness
* - only wagon
** - booth is possible
Edition 02/98
W42.USA.5602.03.21
Windshield Wiper/Washer, Washer Nozzle Heater
No. 1/19
Wiring diagram
Audi A6
ws =white
sw =black
ro =red
gn =green
br =brown
bl =blue
gr =gray
li =violet
ge =yellow
97-19672
E 87 - A/C Control Head
J 126 - Control module for fresh air blower
J 201 - Protection Diode
J 307 - Selector Lever Light Relay
S 1 - Fuse
S 2 - Fuse
S 3 - Fuse
S 4 - Fuse
S 224 - Fuse
S 225 - Fuse
S 226 - Fuse
S 227 - Fuse
T 6 b - Connector 6 pin, white, connector station A pillar, left
T 6 r - Connector 6 pin, brown, in luggage compartment
T 10 f - Connector 10 pin, brown, connector station A pillar, left
V 2 - Fresh Air Blower
V 12 - Rear Window Wiper Motor
X - License Plate Light
87 - Ground connection -2-, in rear wiring harness
A34 - Wire connection (75x), in instrument panel wiring harness
A53 - Plus connection (58b), in instrument panel wiring harness
L4 - Wire connection (75al), in A/C wiring harness
L10 - Wire connection -2-, in A/C wiring harness
* - Chassis number upto 5000
** - Chassis number from 5001
*** - only wagon
Fuses, Fresh Air Blower, License Plate Light, Glove Compartment Light
Edition 02/98
W42.USA.5602.03.21

BentleyPublishers.com—All Rights Reserved

Engine management, 2.8 liter engine, code AHA (1998)

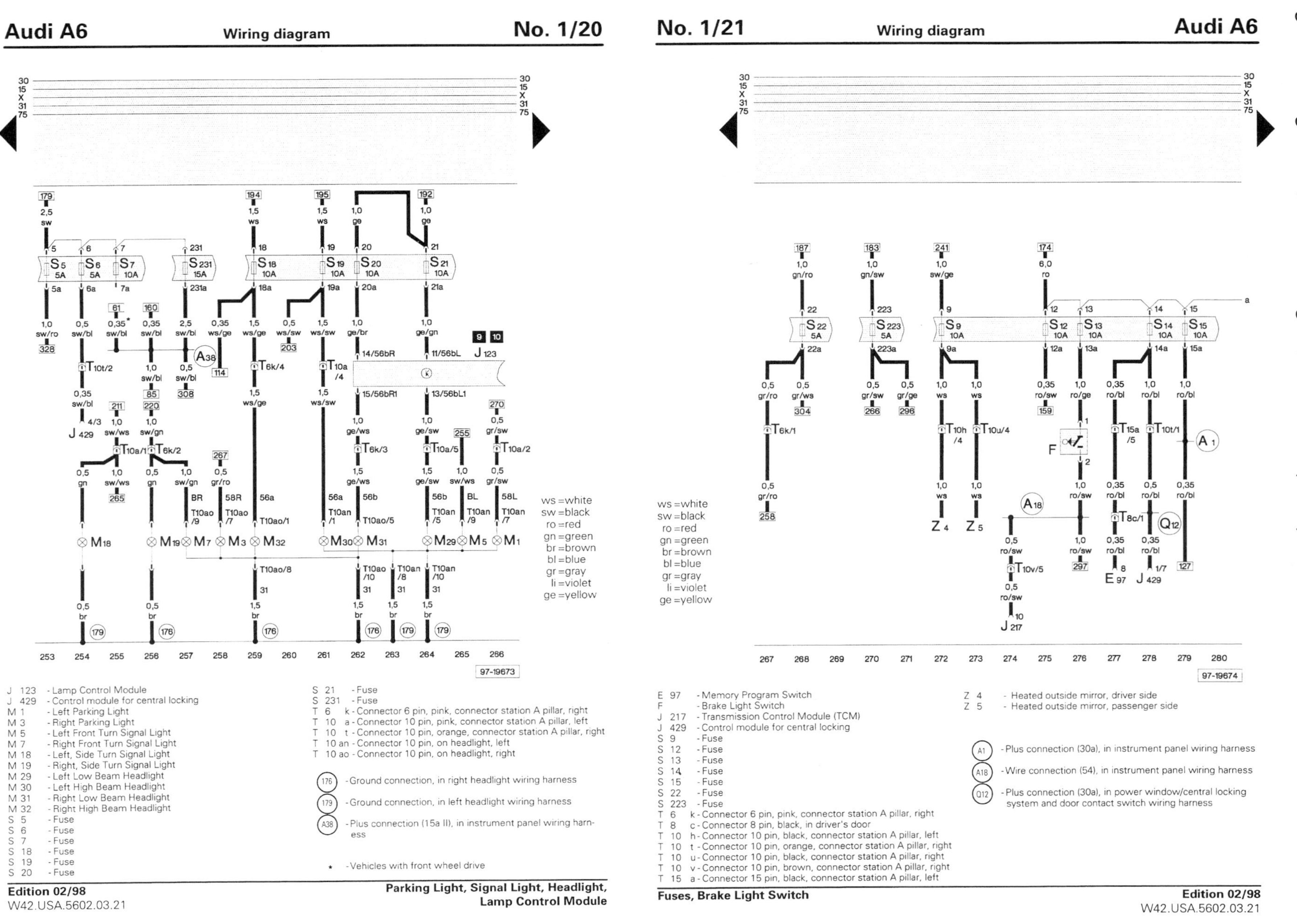

Audi A6
Wiring diagram
No. 1/20
ws = white
sw = black
ro = red
gn = green
br = brown
bl = blue
gr = gray
li = violet
ge = yellow
97-19673
J 123 - Lamp Control Module
J 429 - Control module for central locking
M 1 - Left Parking Light
M 3 - Right Parking Light
M 5 - Left Front Turn Signal Light
M 7 - Right Front Turn Signal Light
M 18 - Left, Side Turn Signal Light
M 19 - Right, Side Turn Signal Light
M 29 - Left Low Beam Headlight
M 30 - Left High Beam Headlight
M 31 - Right Low Beam Headlight
M 32 - Right High Beam Headlight
S 5 - Fuse
S 6 - Fuse
S 7 - Fuse
S 18 - Fuse
S 19 - Fuse
S 20 - Fuse
S 21 - Fuse
S 231 - Fuse
T 6 k - Connector 6 pin, pink, connector station A pillar, right
T 10 a - Connector 10 pin, pink, connector station A pillar, left
T 10 t - Connector 10 pin, orange, connector station A pillar, right
T 10 an - Connector 10 pin, on headlight, left
T 10 ao - Connector 10 pin, on headlight, right
176 - Ground connection, in right headlight wiring harness
179 - Ground connection, in left headlight wiring harness
A38 - Plus connection (15a II), in instrument panel wiring harness
* - Vehicles with front wheel drive
Edition 02/98
W42.USA.5602.03.21
Parking Light, Signal Light, Headlight, Lamp Control Module
No. 1/21
Wiring diagram
Audi A6
97-19674
E 97 - Memory Program Switch
F - Brake Light Switch
J 217 - Transmission Control Module (TCM)
J 429 - Control module for central locking
S 9 - Fuse
S 12 - Fuse
S 13 - Fuse
S 14 - Fuse
S 15 - Fuse
S 22 - Fuse
S 223 - Fuse
T 6 k - Connector 6 pin, pink, connector station A pillar, right
T 8 c - Connector 8 pin, black, in driver's door
T 10 h - Connector 10 pin, black, connector station A pillar, left
T 10 t - Connector 10 pin, orange, connector station A pillar, right
T 10 u - Connector 10 pin, black, connector station A pillar, right
T 10 v - Connector 10 pin, brown, connector station A pillar, right
T 15 a - Connector 15 pin, black, connector station A pillar, left
Z 4 - Heated outside mirror, driver side
Z 5 - Heated outside mirror, passenger side
A1 - Plus connection (30a), in instrument panel wiring harness
A18 - Wire connection (54), in instrument panel wiring harness
Q12 - Plus connection (30a), in power window/central locking system and door contact switch wiring harness
Fuses, Brake Light Switch
Edition 02/98
W42.USA.5602.03.21

BentleyPublishers.com—All Rights Reserved

Engine management, 2.8 liter engine, code AHA (1998)

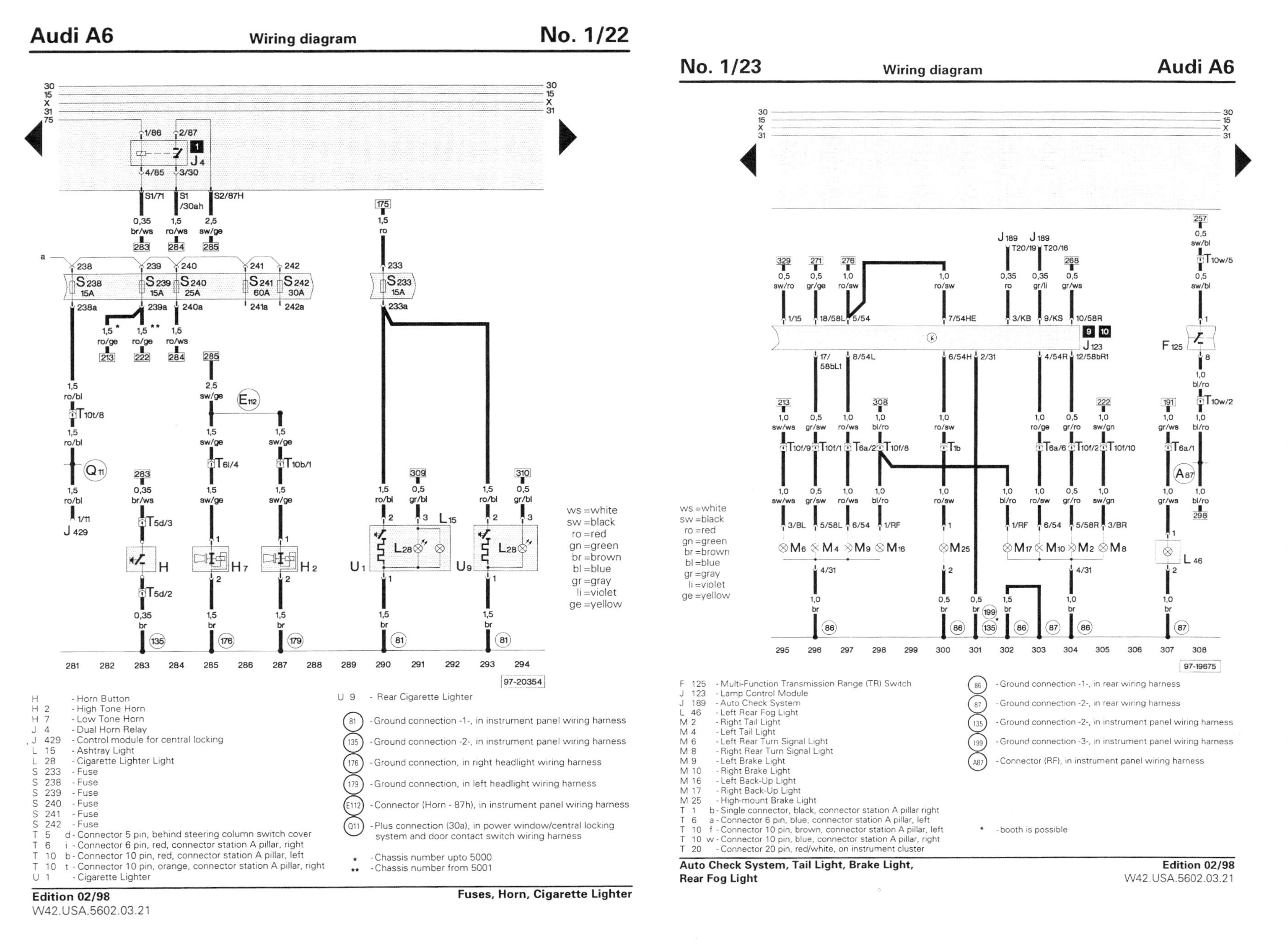

BentleyPublishers.com—All Rights Reserved

Engine management, 2.8 liter engine, code AHA (1998)

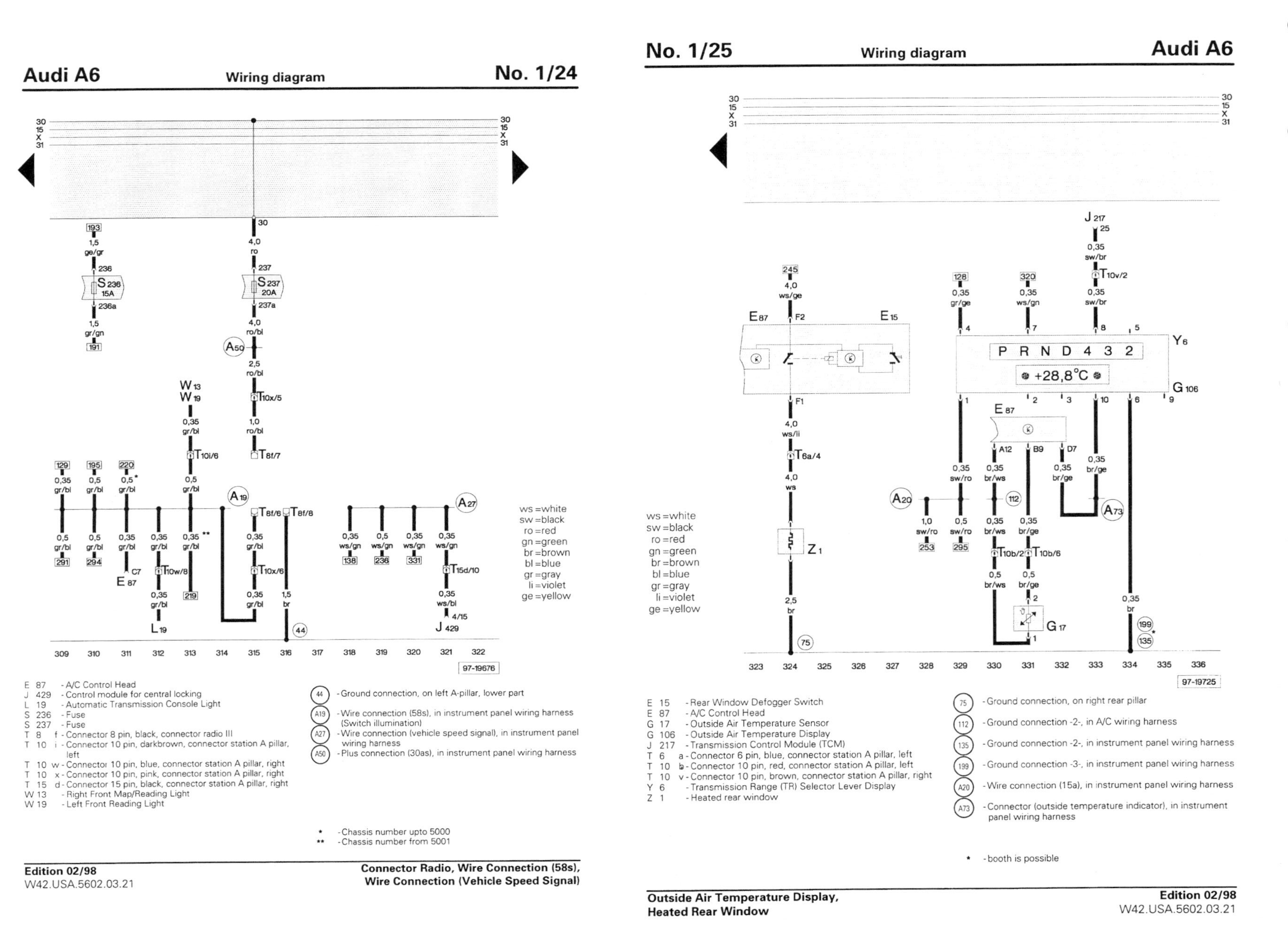

BentleyPublishers.com—All Rights Reserved

Driver's Information System

1998 m. y.

Fuse Panel

A97-0119

Fuse Colors:

30 A - Green
25 A - White
20 A - Yellow
15 A - Blue
10 A - Red
5 A - Beige

Starting with fuse position 23, fuses in the fuse holder are identified with 223 in the wiring diagram.

13 - Fold Relay Panel

A97-0024

Relay Location:

9 - Lamp Control Module, J123

10 - Lamp Control Module, J123

ws = white
sw = black
ro = red
gn = green
br = brown
bl = blue
gr = gray
li = violet
ge = yellow

97-19732

E 91 - Board Computer Function Selector Switch
E 92 - Board Computer Reset Button
F 77 - Windshield Washer Fluid Level Warning Switch
J 189 - Auto Check System
J 220 - Motronic Engine Control Module (ECM)
R - Radio
T 6 x - Connector 6 pin, green, connector radio I - 2
T 10 a - Connector 10 pin, pink, connector station A pillar, left
T 10 b - Connector 10 pin, red, connector station A pillar, left
T 10 o - Connector 10 pin, brown, connector station electronic box plenum chamber
T 10 x - Connector 10 pin, pink, connector station A pillar, right
T 20 - Connector 20 pin, red/white, on instrument cluster

81 - Ground connection -1-, in instrument panel wiring harness
135 - Ground connection -2-, in instrument panel wiring harness
199 - Ground connection -3-, in instrument panel wiring harness
269 - Ground connector (sensor ground) -1-, in instrument panel wiring harness

* - booth is possible

Auto Check System, Board Computer Funtion Selector Switch, Windshield Washer Fluid Level Warning Switch

BentleyPublishers.com—All Rights Reserved

Instruments and Controls (1998)

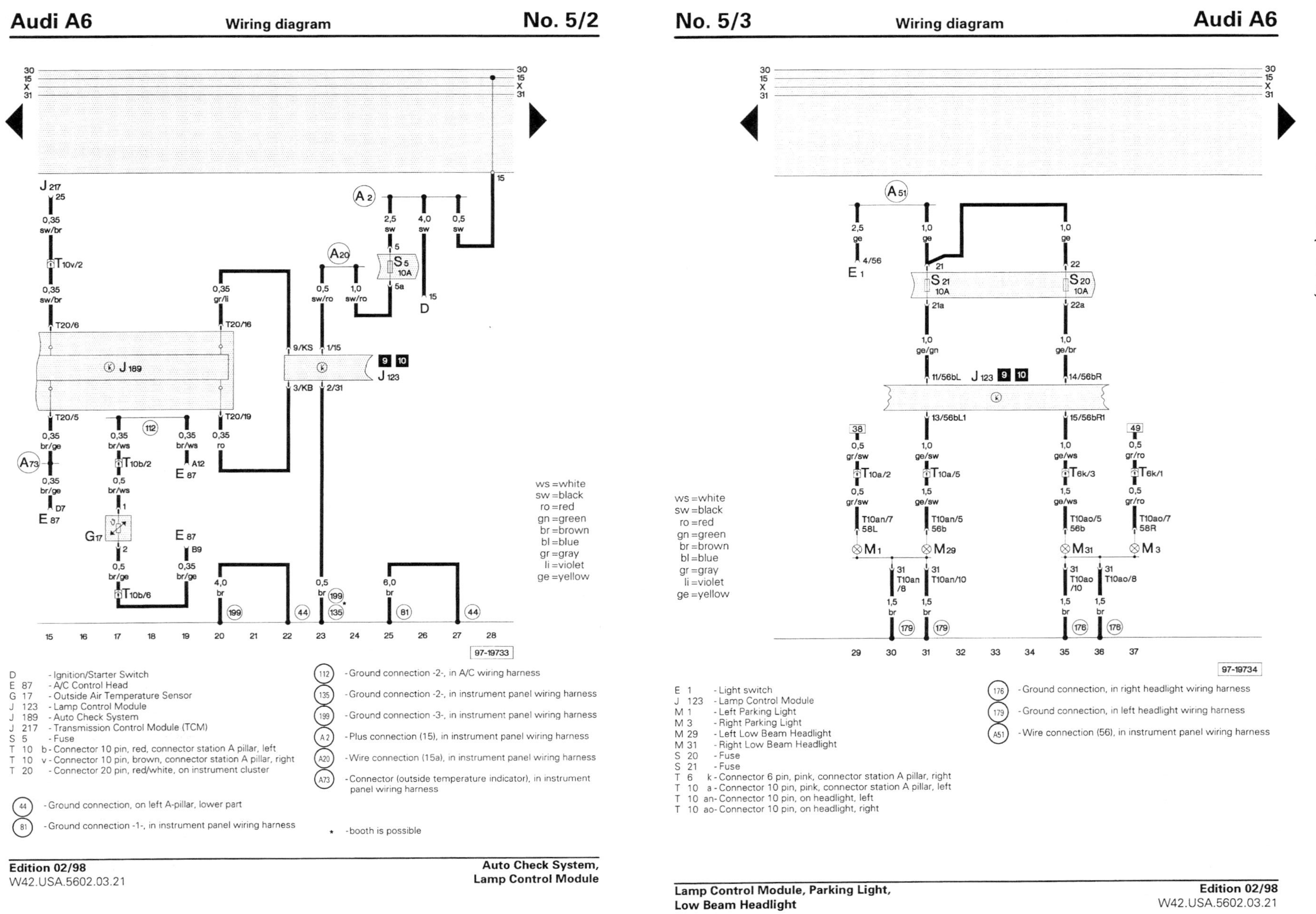

BentleyPublishers.com—All Rights Reserved

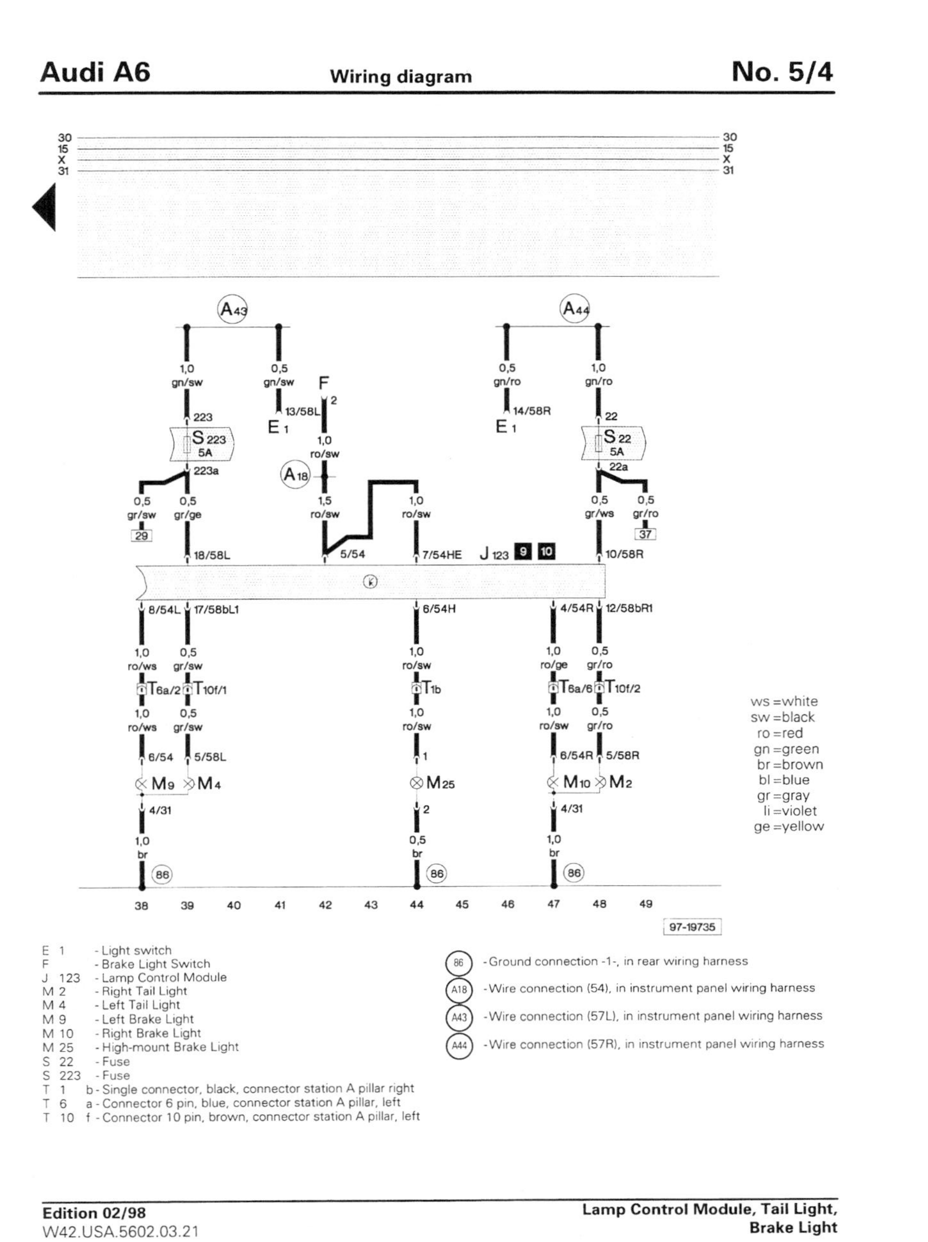

THIS SPACE INTENTIONALLY LEFT BLANK

BentleyPublishers.com—All Rights Reserved

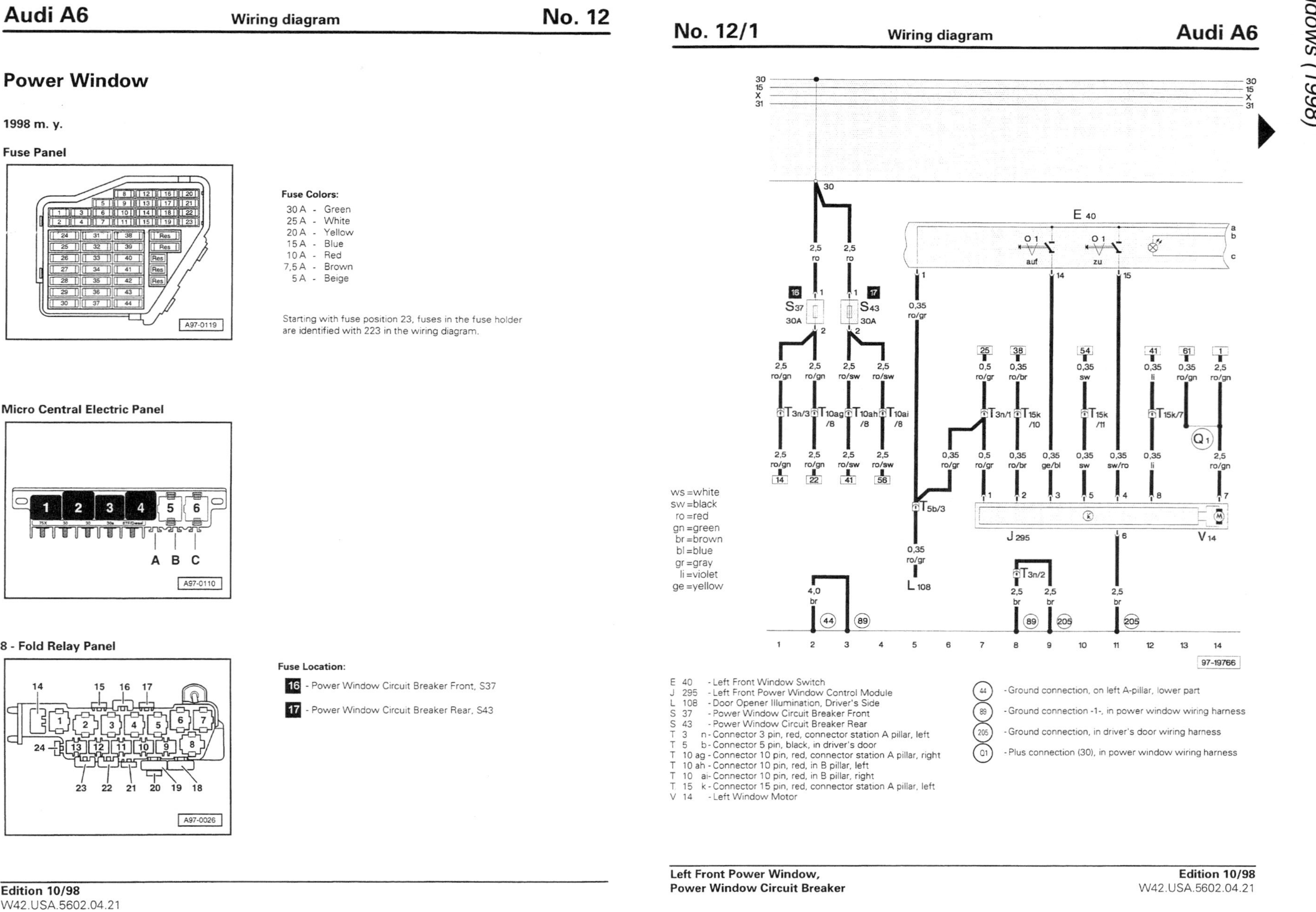

BentleyPublishers.com—All Rights Reserved

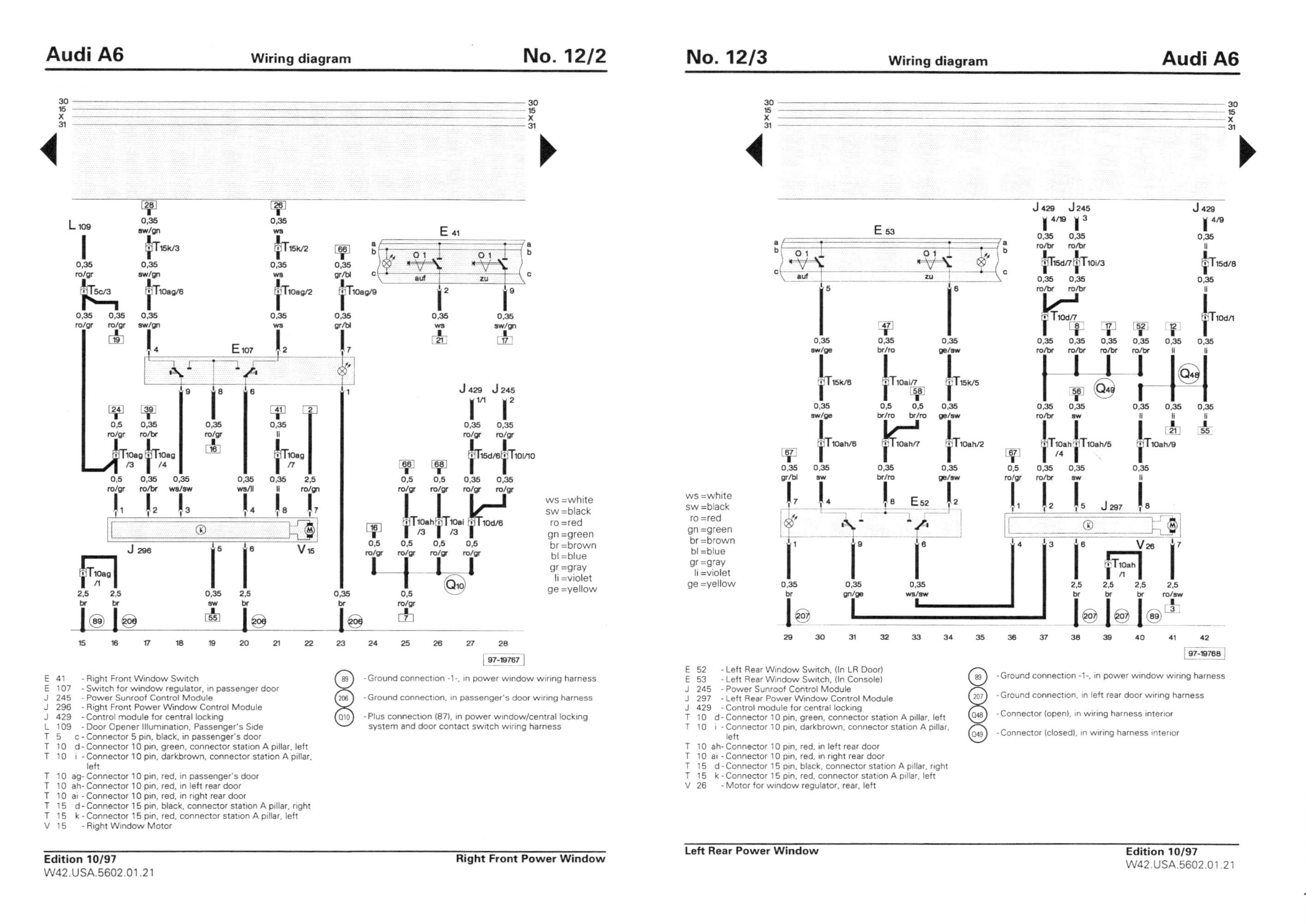

BentleyPublishers.com—All Rights Reserved

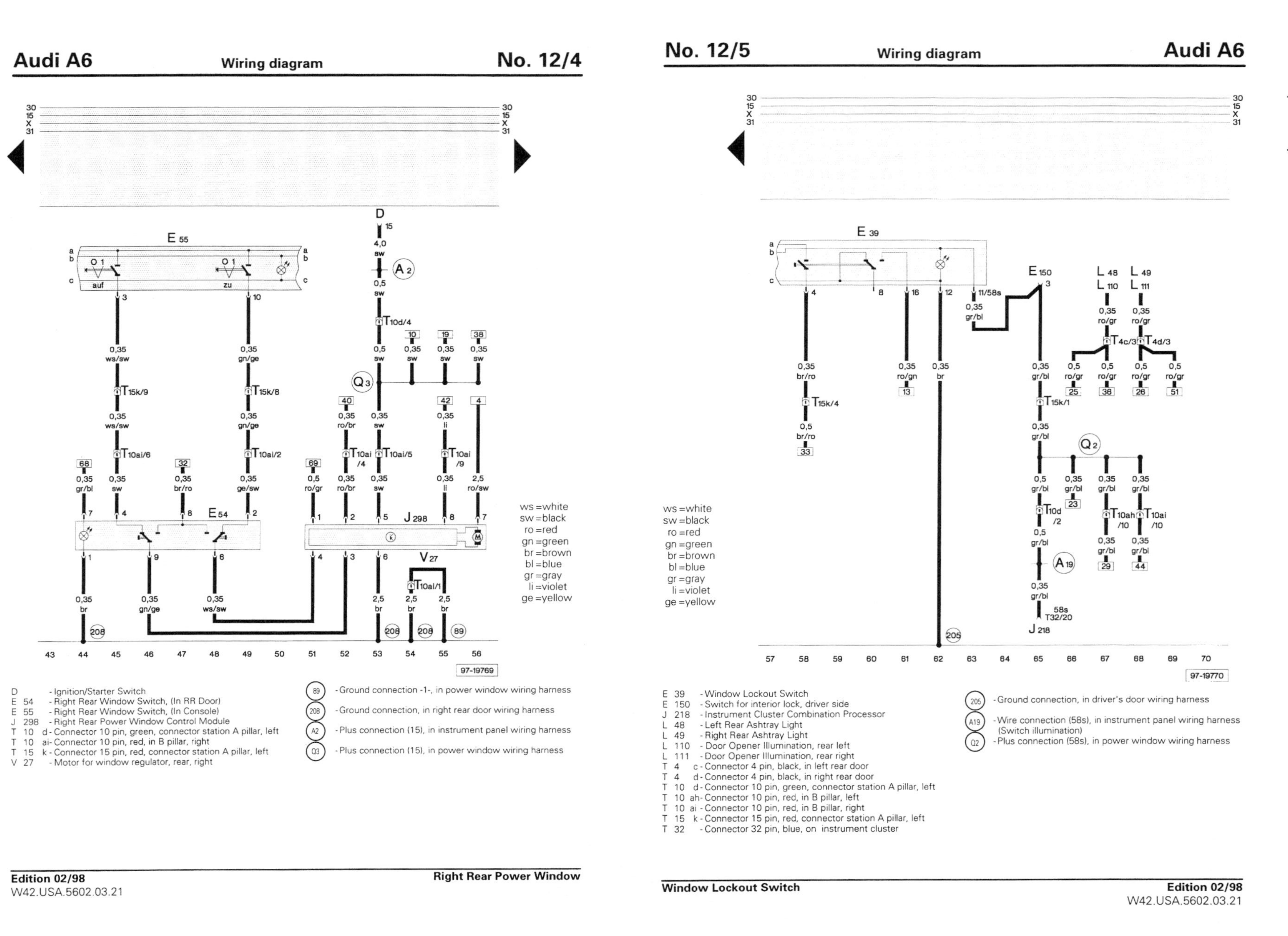

BentleyPublishers.com—All Rights Reserved

Radio Concert, Bose with and without CD Changer

1998 m. y.

Fuse Panel

A97-0119

Fuse Colors:

30 A - Green
25 A - White
20 A - Yellow
15 A - Blue
10 A - Red
7,5 A - Brown
5 A - Beige

Starting with fuse position 23, fuses in the fuse holder are identified with 223 in the wiring diagram.

Micro Central Electric Panel

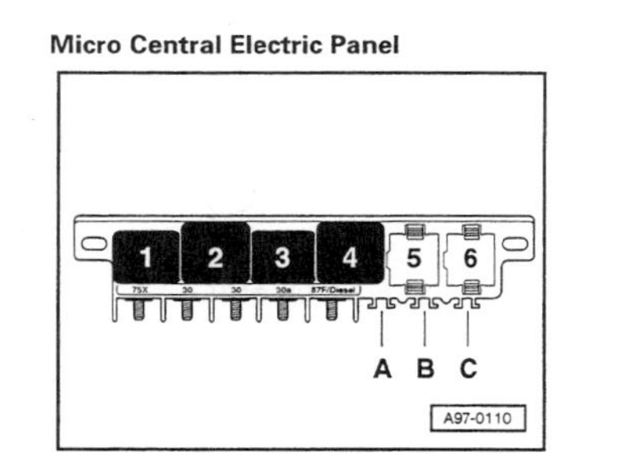

ws = white
sw = black
ro = red
gn = green
br = brown
bl = blue
gr = gray
li = violet
ge = yellow
or = orange

97-50637

J 218 - Instrument Cluster Combination Processor
J 429 - Control module for central locking
R - Radio
R 41 - CD Changer Unit
T 1 e - Single connector, on Radio
T 2 ar - Connector 2 pin, black, behind instrument panel, center
T 8 f - Connector 8 pin, black, connector radio III
T 8 g - Connector 8 pin, blue, connector radio I - 3
T 10 x - Connector 10 pin, pink, connector station A pillar, right
T 10 z - Connector 10 pin, red, connector radio IV
T 15 d - Connector 15 pin, black, connector station A pillar, right
T 32 - Connector 32 pin, blue, on instrument cluster

(133) - Ground connection, in speaker wiring harness
(A19) - Wire connection (58s), in instrument panel wiring harness (Switch illumination)
(A50) - Plus connection (30as), in instrument panel wiring harness
(V1) - Plus connection (30), in rear speaker wiring harness
(V14) - Wire connection (shielding), in CD-changer wiring harness

* - Vehicles with CD Changer
** - Automatic Antenna

BentleyPublishers.com—All Rights Reserved

Sound System, Bose (1998)

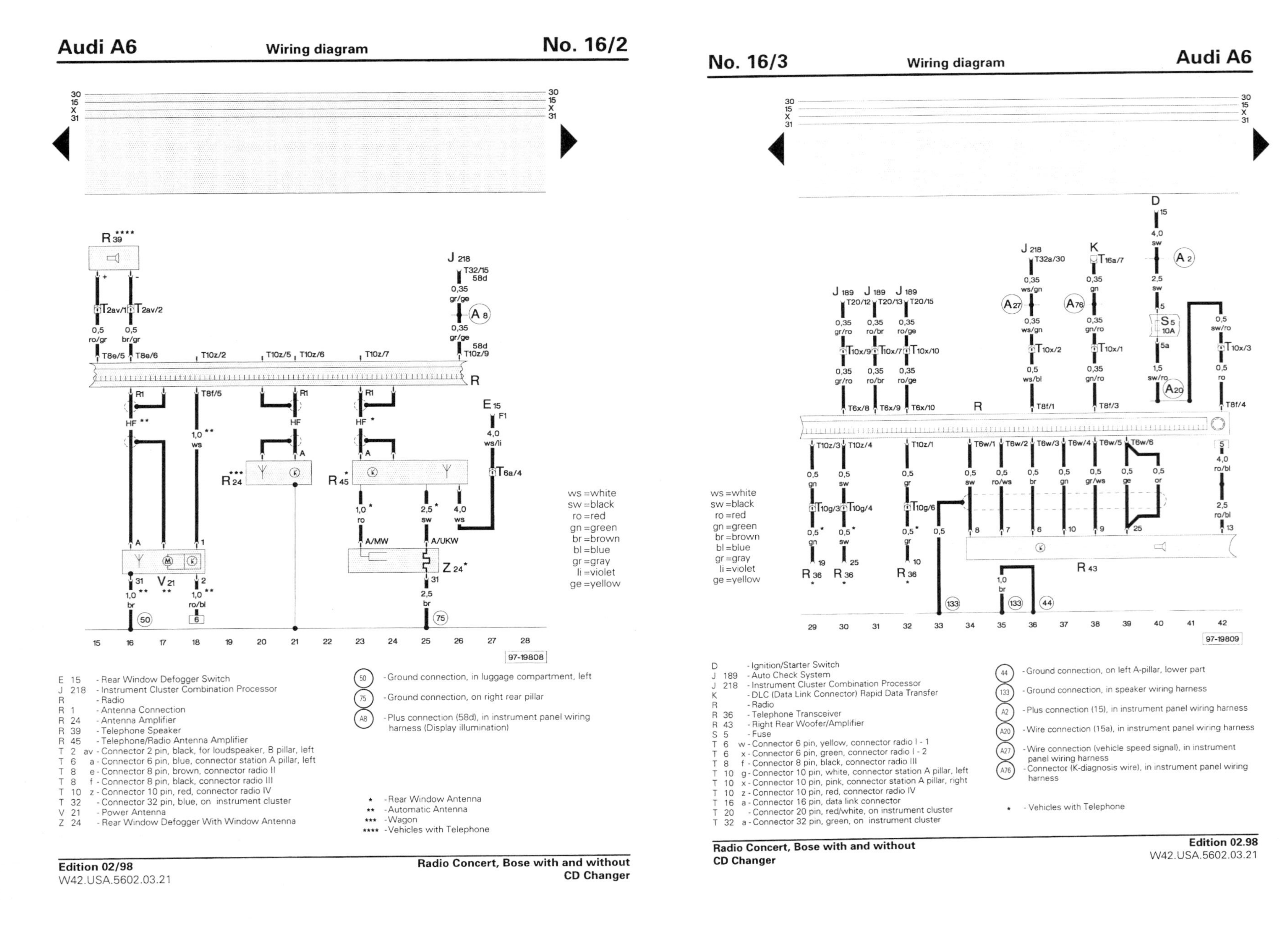

Audi A6
Wiring diagram
No. 16/2
E 15 - Rear Window Defogger Switch
J 218 - Instrument Cluster Combination Processor
R - Radio
R 1 - Antenna Connection
R 24 - Antenna Amplifier
R 39 - Telephone Speaker
R 45 - Telephone/Radio Antenna Amplifier
T 2 av - Connector 2 pin, black, for loudspeaker, B pillar, left
T 6 a - Connector 6 pin, blue, connector station A pillar, left
T 8 e - Connector 8 pin, brown, connector radio II
T 8 f - Connector 8 pin, black, connector radio III
T 10 z - Connector 10 pin, red, connector radio IV
T 32 - Connector 32 pin, blue, on instrument cluster
V 21 - Power Antenna
Z 24 - Rear Window Defogger With Window Antenna
50 - Ground connection, in luggage compartment, left
75 - Ground connection, on right rear pillar
A8 - Plus connection (58d), in instrument panel wiring harness (Display illumination)
* - Rear Window Antenna
** - Automatic Antenna
*** - Wagon
**** - Vehicles with Telephone
ws =white
sw =black
ro =red
gn =green
br =brown
bl =blue
gr =gray
li =violet
ge =yellow
97-19808
Edition 02/98
W42.USA.5602.03.21
Radio Concert, Bose with and without CD Changer
No. 16/3
Wiring diagram
Audi A6
D - Ignition/Starter Switch
J 189 - Auto Check System
J 218 - Instrument Cluster Combination Processor
K - DLC (Data Link Connector) Rapid Data Transfer
R - Radio
R 36 - Telephone Transceiver
R 43 - Right Rear Woofer/Amplifier
S 5 - Fuse
T 6 w - Connector 6 pin, yellow, connector radio I - 1
T 6 x - Connector 6 pin, green, connector radio I - 2
T 8 f - Connector 8 pin, black, connector radio III
T 10 g - Connector 10 pin, white, connector station A pillar, left
T 10 x - Connector 10 pin, pink, connector station A pillar, right
T 10 z - Connector 10 pin, red, connector radio IV
T 16 a - Connector 16 pin, data link connector
T 20 - Connector 20 pin, red/white, on instrument cluster
T 32 a - Connector 32 pin, green, on instrument cluster
44 - Ground connection, on left A-pillar, lower part
133 - Ground connection, in speaker wiring harness
A2 - Plus connection (15), in instrument panel wiring harness
A20 - Wire connection (15a), in instrument panel wiring harness
A27 - Wire connection (vehicle speed signal), in instrument panel wiring harness
A76 - Connector (K-diagnosis wire), in instrument panel wiring harness
* - Vehicles with Telephone
ws =white
sw =black
ro =red
gn =green
br =brown
bl =blue
gr =gray
li =violet
ge =yellow
97-19809
Radio Concert, Bose with and without CD Changer
Edition 02.98
W42.USA.5602.03.21

BentleyPublishers.com—All Rights Reserved

Audi A6 Wiring diagram **No. 16/4**

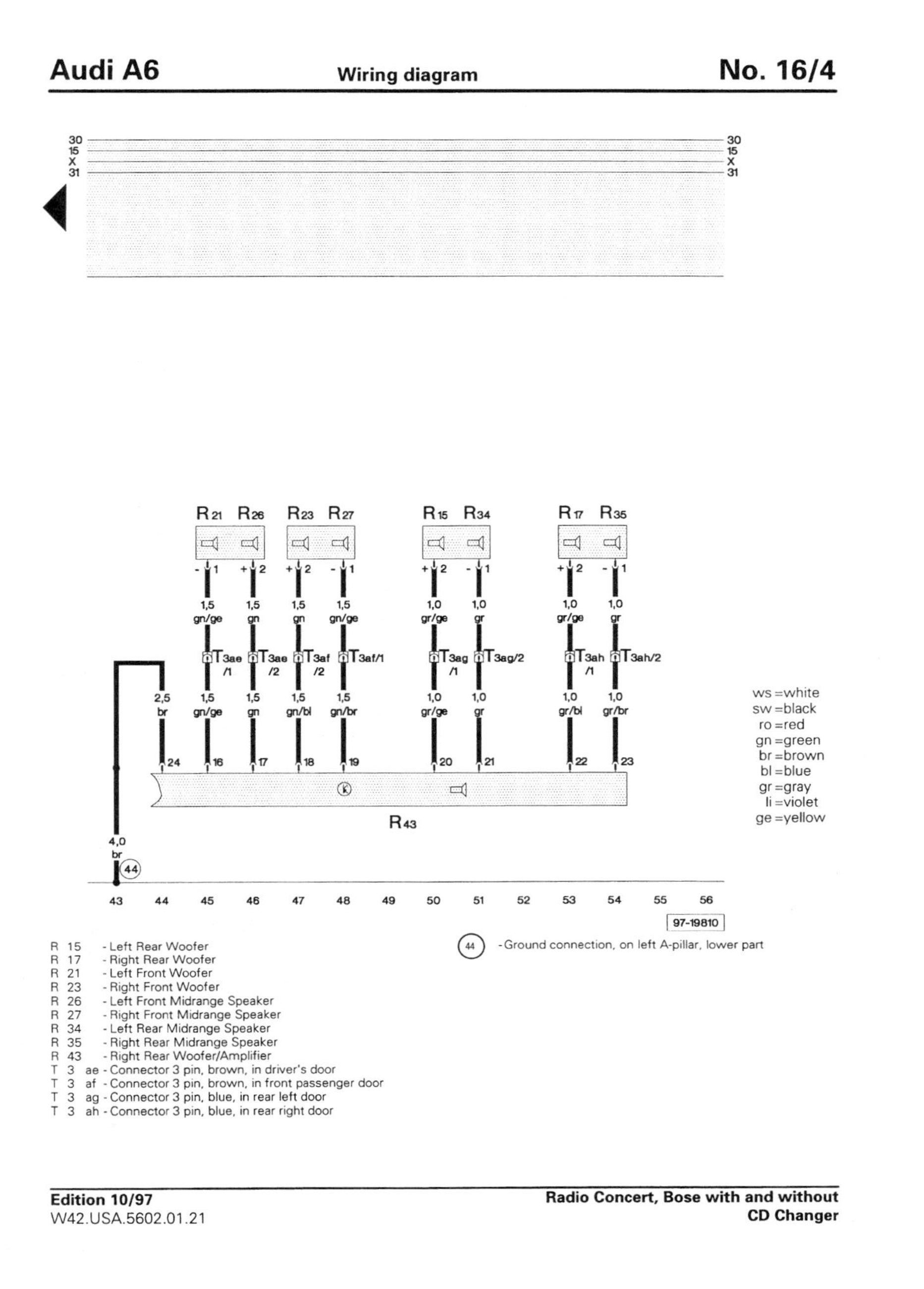

Edition 10/97
W42.USA.5602.01.21

Radio Concert, Bose with and without CD Changer

THIS SPACE INTENTIONALLY LEFT BLANK

BentleyPublishers.com—All Rights Reserved

Audi A6 — Wiring diagram — No. 17

Radio-Stereo, Concert

1998 m. y.

Fuse Panel

A97-0119

Fuse Colors:

- 30 A - Green
- 25 A - White
- 20 A - Yellow
- 15 A - Blue
- 10 A - Red
- 5 A - Beige

Starting with fuse position 23, fuses in the fuse holder are identified with 223 in the wiring diagram.

Micro Central Electric Panel

A97-0110

Edition 02/98
W42.USA.5602.03.21

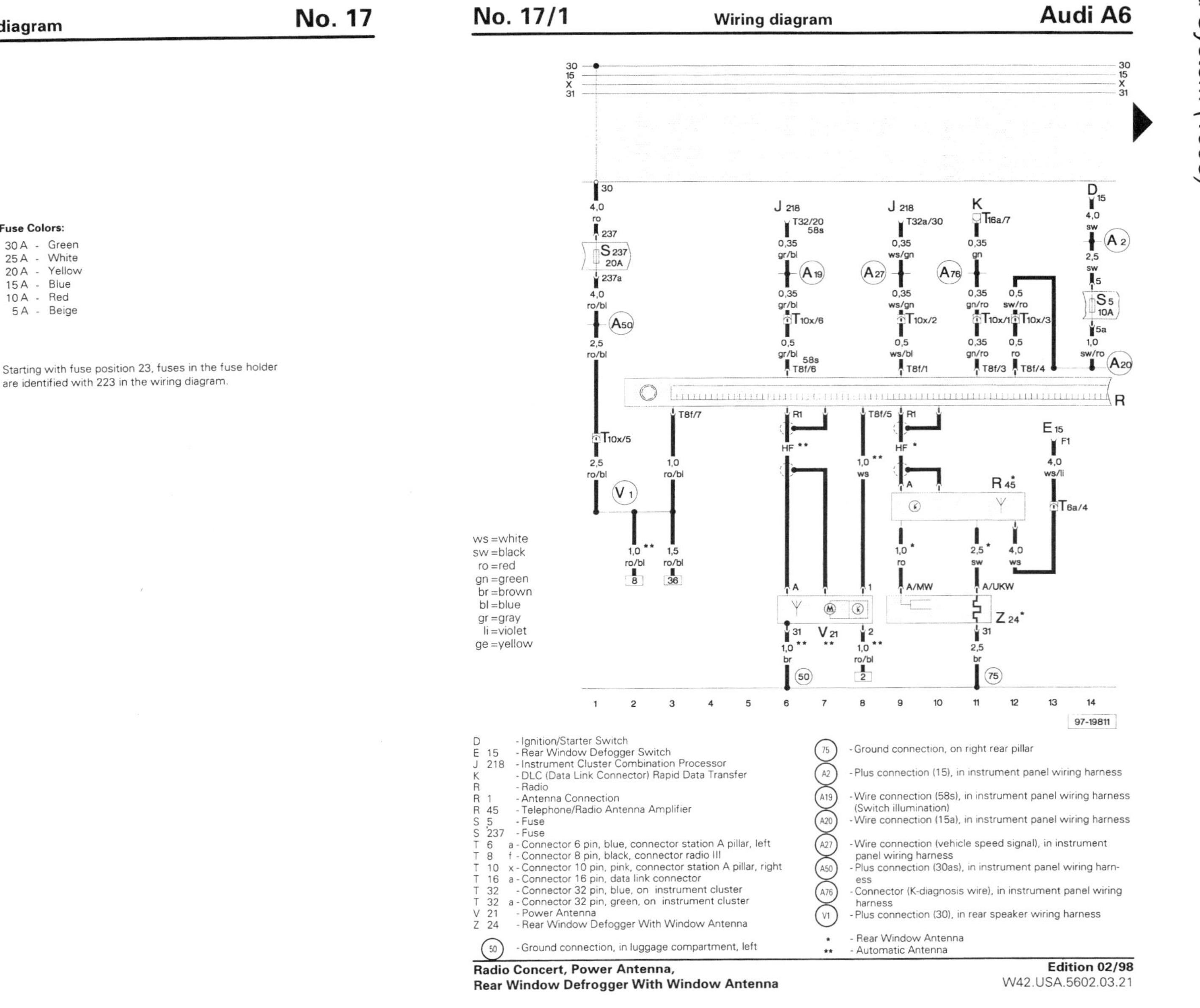

BentleyPublishers.com—All Rights Reserved

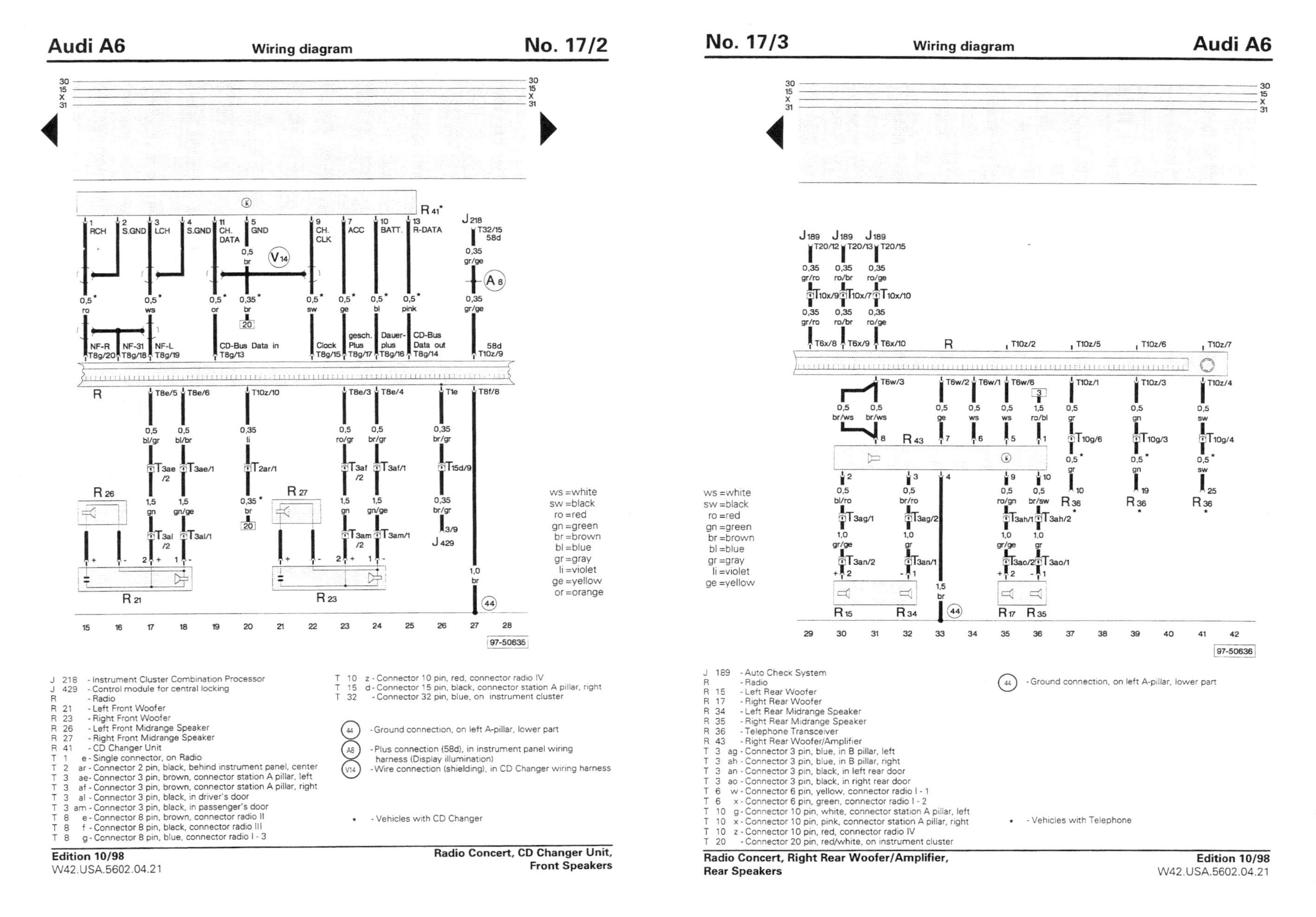

BentleyPublishers.com—All Rights Reserved

Headlight Washer

1998 m. y.

Fuse Panel

A97-0119

Fuse Colors:

30 A - Green
25 A - White
20 A - Yellow
15 A - Blue
10 A - Red
5 A - Beige

Starting with fuse position 23, fuses in the fuse holder are identified with 223 in the wiring diagram.

Micro Central Electric Panel

A97-0110

Relay Location:

5 - Wiper/Washer Intermittent Relay, J31

6 - Wiper/Washer Intermittent Relay, J31

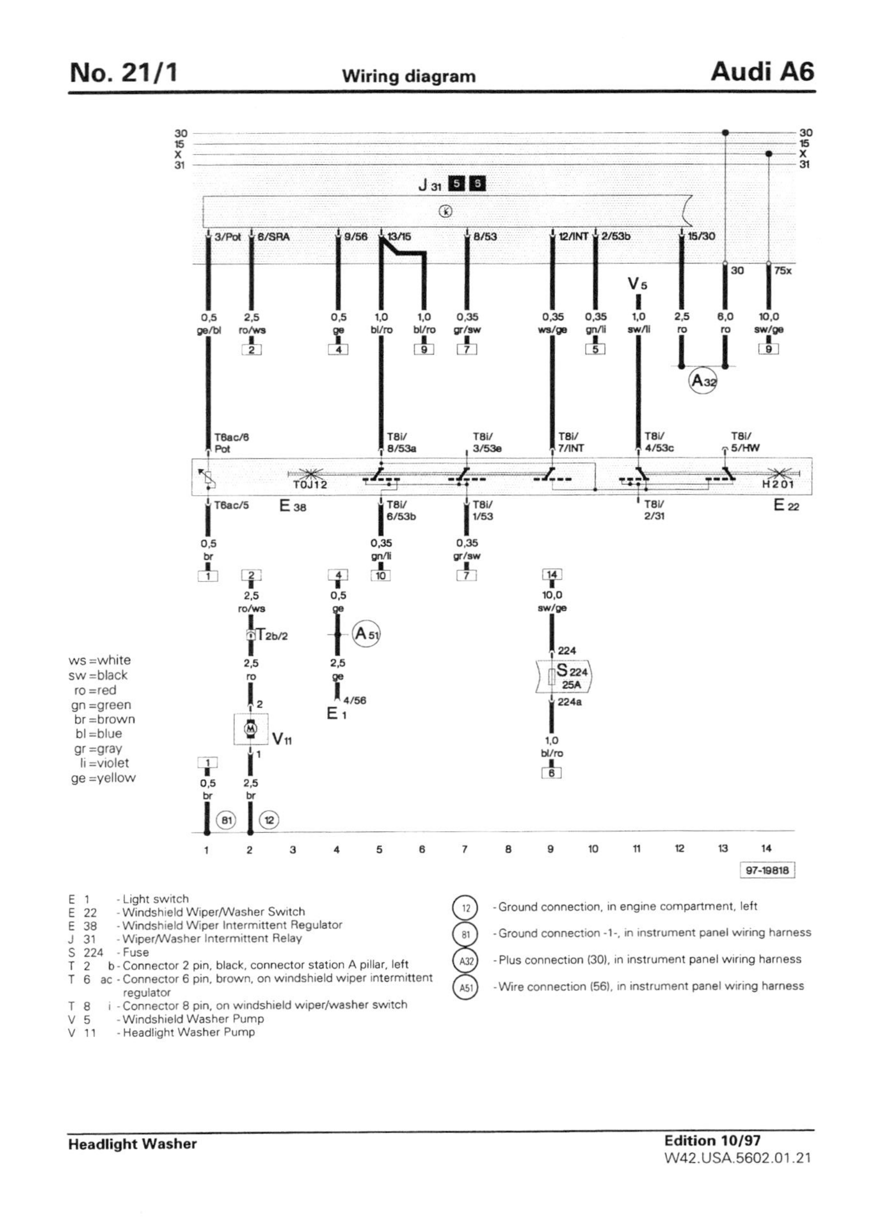

ws = white
sw = black
ro = red
gn = green
br = brown
bl = blue
gr = gray
li = violet
ge = yellow

E 1 - Light switch
E 22 - Windshield Wiper/Washer Switch
E 38 - Windshield Wiper Intermittent Regulator
J 31 - Wiper/Washer Intermittent Relay
S 224 - Fuse
T 2 b - Connector 2 pin, black, connector station A pillar, left
T 6 ac - Connector 6 pin, brown, on windshield wiper intermittent regulator
T 8 i - Connector 8 pin, on windshield wiper/washer switch
V 5 - Windshield Washer Pump
V 11 - Headlight Washer Pump

(12) - Ground connection, in engine compartment, left
(81) - Ground connection -1-, in instrument panel wiring harness
(A32) - Plus connection (30), in instrument panel wiring harness
(A51) - Wire connection (56), in instrument panel wiring harness

BentleyPublishers.com—All Rights Reserved

Audi A6 Wiring diagram **No. 22**

Headlight with Hight Intensity Gas discharge Lamps and automatic Headlight Beam Adjusting

1998 m. y.

Fuse Panel

A97-0119

Fuse Colors:

30 A - Green
25 A - White
20 A - Yellow
15 A - Blue
10 A - Red
7,5 A - Brown
5 A - Beige

Starting with fuse position 23, fuses in the fuse holder are identified with 223 in the wiring diagram.

Micro Central Electric Panel

A B C
A97-0110

Relay Location:

2 - Load Reduction Relay, J59

13 - Fold Relay Panel

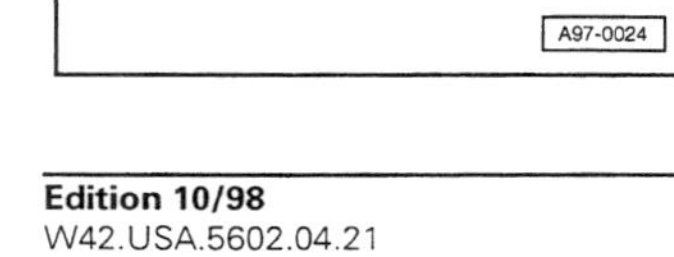

A97-0024

Relay Location:

9 - Lamp Control Module, J123

10 - Lamp Control Module, J123

Edition 10/98
W42.USA.5602.04.21

No. 22/1 Wiring diagram **Audi A6**

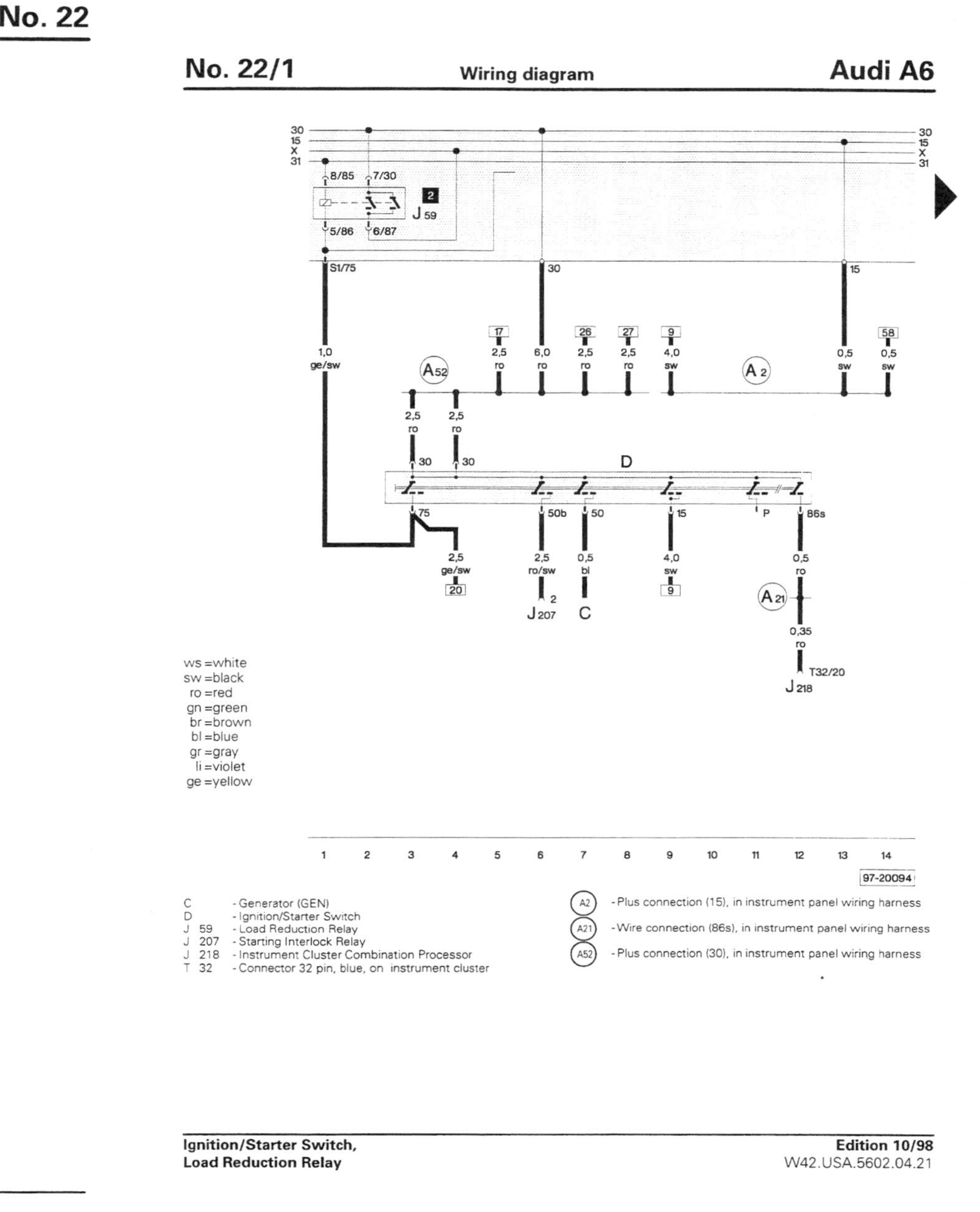

ws = white
sw = black
ro = red
gn = green
br = brown
bl = blue
gr = gray
li = violet
ge = yellow

C - Generator (GEN)
D - Ignition/Starter Switch
J 59 - Load Reduction Relay
J 207 - Starting Interlock Relay
J 218 - Instrument Cluster Combination Processor
T 32 - Connector 32 pin, blue, on instrument cluster

A2 - Plus connection (15), in instrument panel wiring harness
A21 - Wire connection (86s), in instrument panel wiring harness
A52 - Plus connection (30), in instrument panel wiring harness

Ignition/Starter Switch, Load Reduction Relay

Edition 10/98
W42.USA.5602.04.21

BentleyPublishers.com—All Rights Reserved

Headlights, Xenon (HID) (1998)

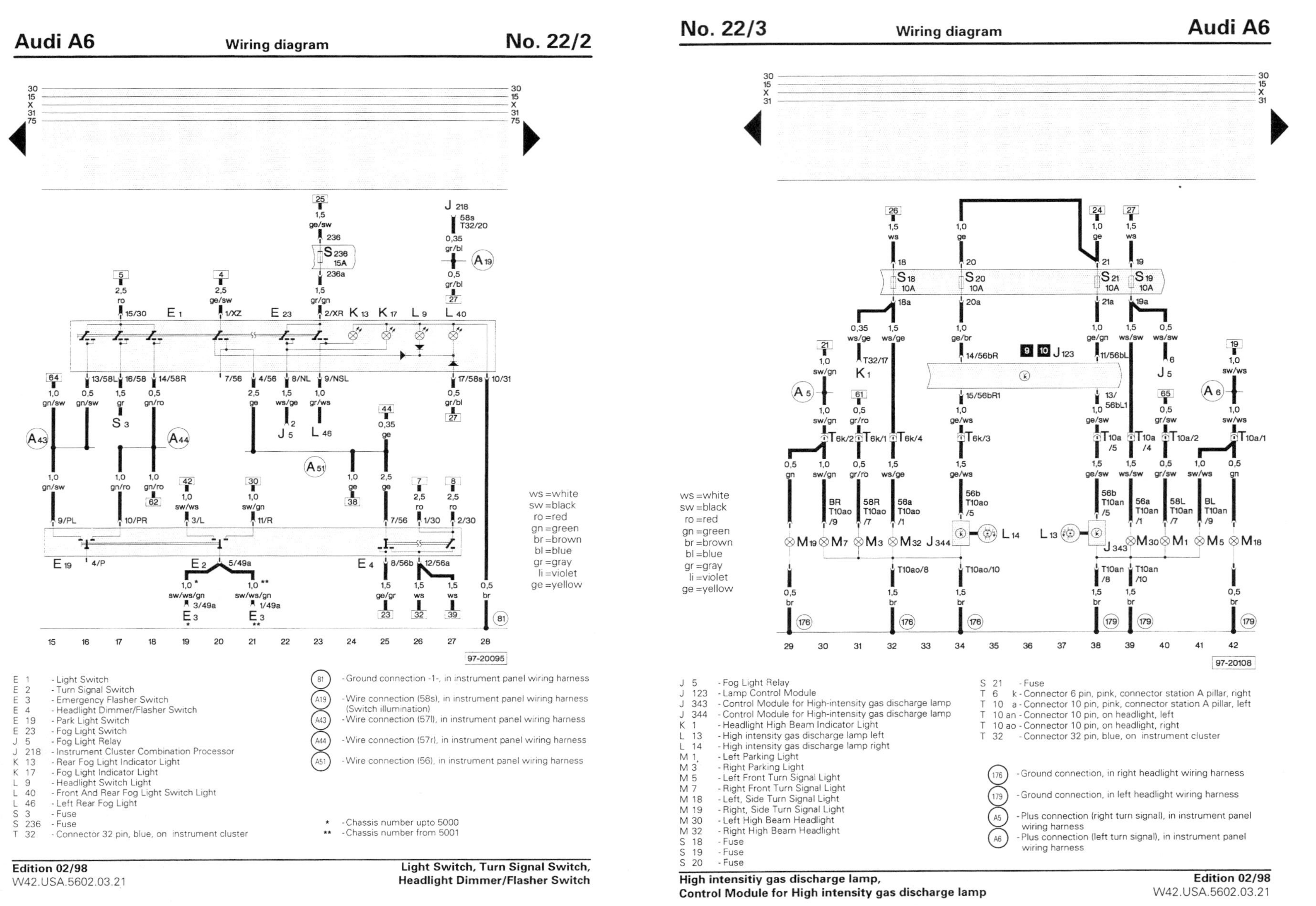

BentleyPublishers.com—All Rights Reserved

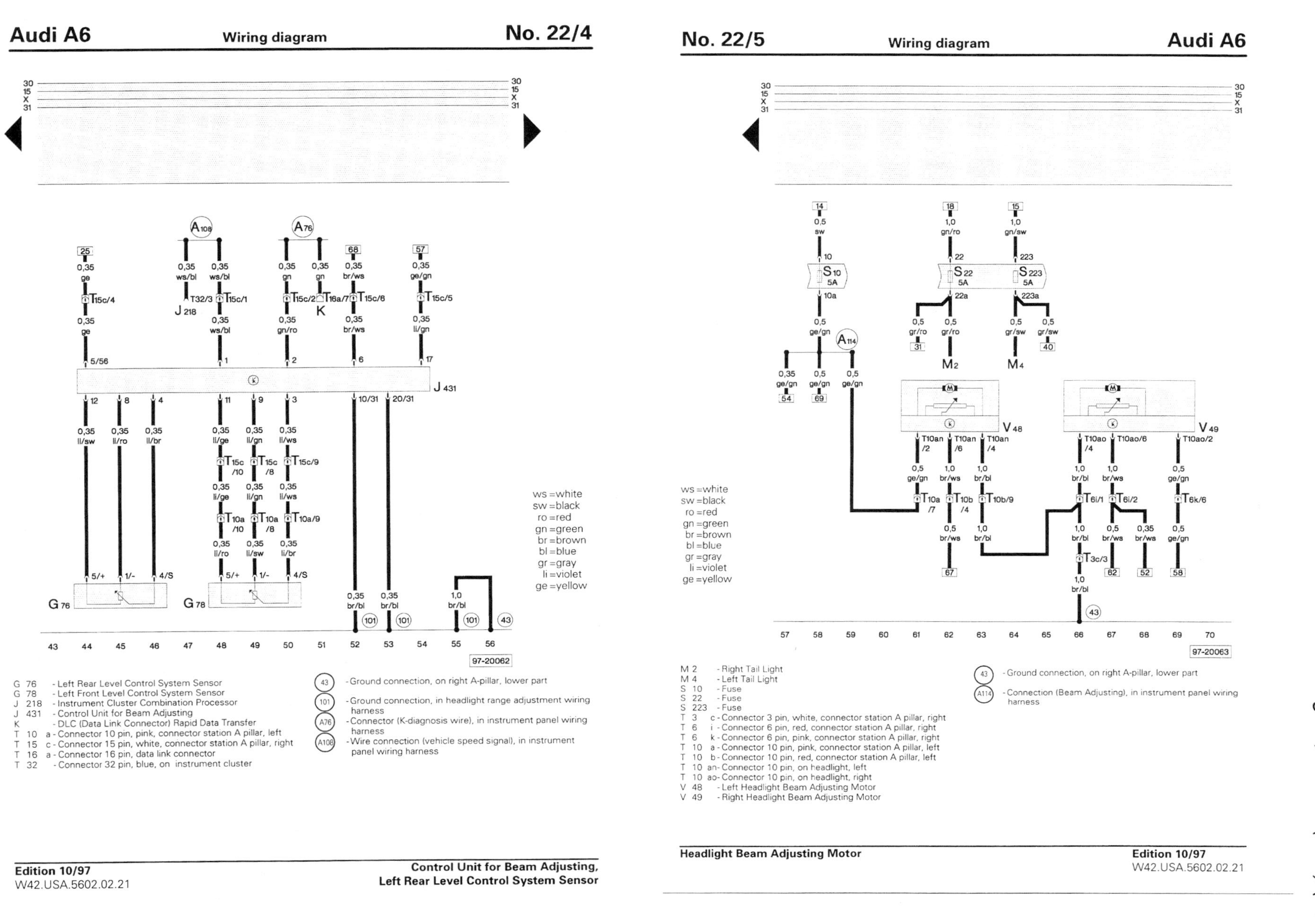

BentleyPublishers.com—All Rights Reserved

Central Locking System/Anti Theft Alarm System/ Ultra-Sound Interior Monitor

1999 m. y.

Fuse Panel

Fuse Colors:

30 A - Green
25 A - White
20 A - Yellow
15 A - Blue
10 A - Red
7,5 A - Brown
5 A - Beige

Starting with fuse position 23, fuses in the fuse holder are identified with 223 in the wiring diagram.

13 - Fold Relay Panel

Relay Location:

4 - Starting Interlock Relay, J207

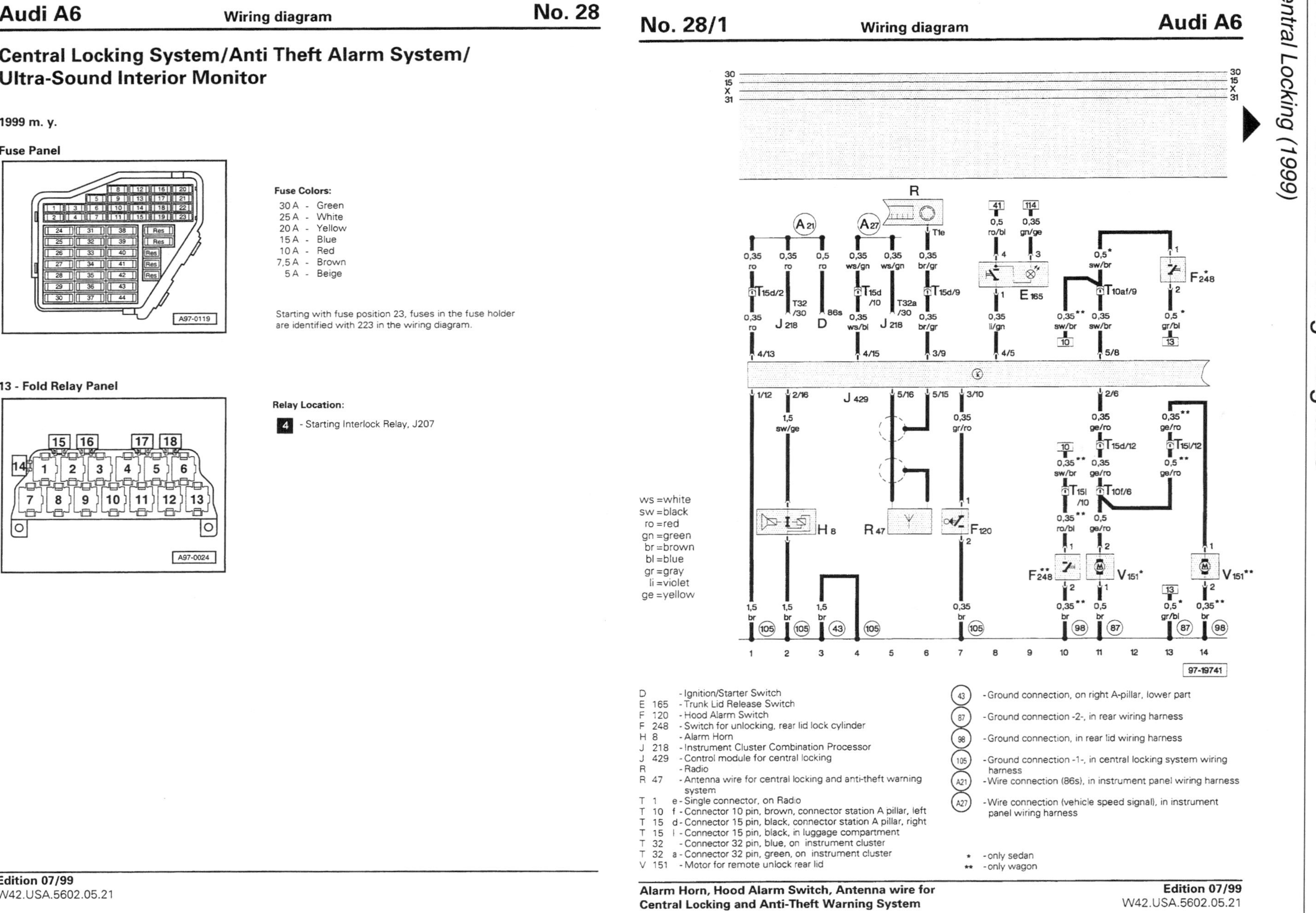

ws = white
sw = black
ro = red
gn = green
br = brown
bl = blue
gr = gray
li = violet
ge = yellow

D - Ignition/Starter Switch
E 165 - Trunk Lid Release Switch
F 120 - Hood Alarm Switch
F 248 - Switch for unlocking, rear lid lock cylinder
H 8 - Alarm Horn
J 218 - Instrument Cluster Combination Processor
J 429 - Control module for central locking
R - Radio
R 47 - Antenna wire for central locking and anti-theft warning system
T 1 e - Single connector, on Radio
T 10 f - Connector 10 pin, brown, connector station A pillar, left
T 15 d - Connector 15 pin, black, connector station A pillar, right
T 15 l - Connector 15 pin, black, in luggage compartment
T 32 - Connector 32 pin, blue, on instrument cluster
T 32 a - Connector 32 pin, green, on instrument cluster
V 151 - Motor for remote unlock rear lid

(43) - Ground connection, on right A-pillar, lower part
(87) - Ground connection -2-, in rear wiring harness
(98) - Ground connection, in rear lid wiring harness
(105) - Ground connection -1-, in central locking system wiring harness
(A21) - Wire connection (86s), in instrument panel wiring harness
(A27) - Wire connection (vehicle speed signal), in instrument panel wiring harness

* - only sedan
** - only wagon

Alarm Horn, Hood Alarm Switch, Antenna wire for Central Locking and Anti-Theft Warning System

BentleyPublishers.com—All Rights Reserved

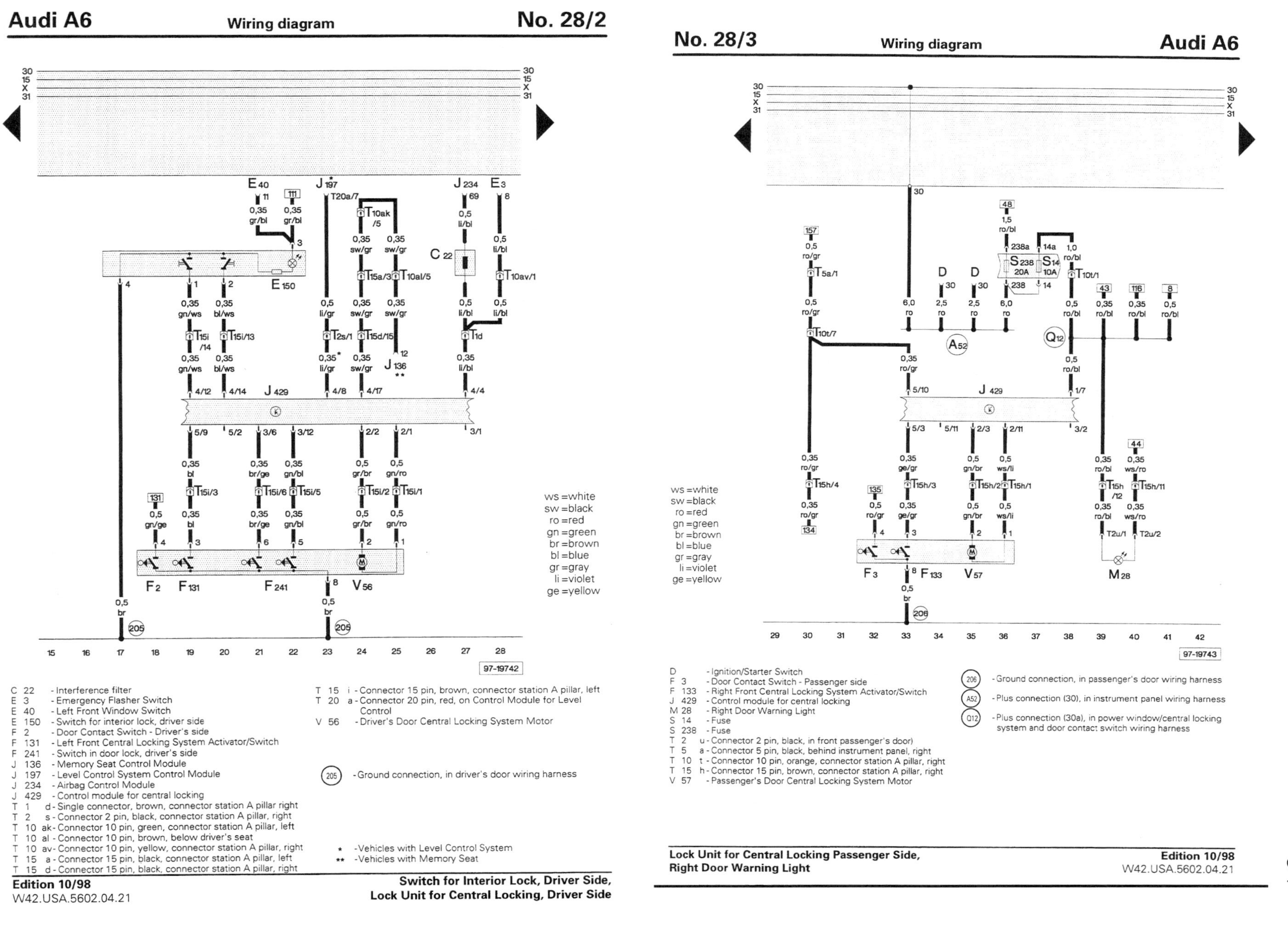

Audi A6
Wiring diagram
No. 28/2
ws = white
sw = black
ro = red
gn = green
br = brown
bl = blue
gr = gray
li = violet
ge = yellow
97-19742
C 22 - Interference filter
E 3 - Emergency Flasher Switch
E 40 - Left Front Window Switch
E 150 - Switch for interior lock, driver side
F 2 - Door Contact Switch - Driver's side
F 131 - Left Front Central Locking System Activator/Switch
F 241 - Switch in door lock, driver's side
J 136 - Memory Seat Control Module
J 197 - Level Control System Control Module
J 234 - Airbag Control Module
J 429 - Control module for central locking
T 1 d - Single connector, brown, connector station A pillar right
T 2 s - Connector 2 pin, black, connector station A pillar, right
T 10 ak - Connector 10 pin, green, connector station A pillar, left
T 10 al - Connector 10 pin, brown, below driver's seat
T 10 av - Connector 10 pin, yellow, connector station A pillar, right
T 15 a - Connector 15 pin, black, connector station A pillar, left
T 15 d - Connector 15 pin, black, connector station A pillar, right
T 15 i - Connector 15 pin, brown, connector station A pillar, left
T 20 a - Connector 20 pin, red, on Control Module for Level Control
V 56 - Driver's Door Central Locking System Motor
205 - Ground connection, in driver's door wiring harness
* - Vehicles with Level Control System
** - Vehicles with Memory Seat
Edition 10/98
W42.USA.5602.04.21
Switch for Interior Lock, Driver Side,
Lock Unit for Central Locking, Driver Side
No. 28/3
Wiring diagram
Audi A6
ws = white
sw = black
ro = red
gn = green
br = brown
bl = blue
gr = gray
li = violet
ge = yellow
97-19743
D - Ignition/Starter Switch
F 3 - Door Contact Switch - Passenger side
F 133 - Right Front Central Locking System Activator/Switch
J 429 - Control module for central locking
M 28 - Right Door Warning Light
S 14 - Fuse
S 238 - Fuse
T 2 u - Connector 2 pin, black, in front passenger's door
T 5 a - Connector 5 pin, black, behind instrument panel, right
T 10 t - Connector 10 pin, orange, connector station A pillar, right
T 15 h - Connector 15 pin, brown, connector station A pillar, right
V 57 - Passenger's Door Central Locking System Motor
206 - Ground connection, in passenger's door wiring harness
A52 - Plus connection (30), in instrument panel wiring harness
Q12 - Plus connection (30a), in power window/central locking system and door contact switch wiring harness
Lock Unit for Central Locking Passenger Side,
Right Door Warning Light
Edition 10/98
W42.USA.5602.04.21

BentleyPublishers.com—All Rights Reserved

Central Locking (1999)

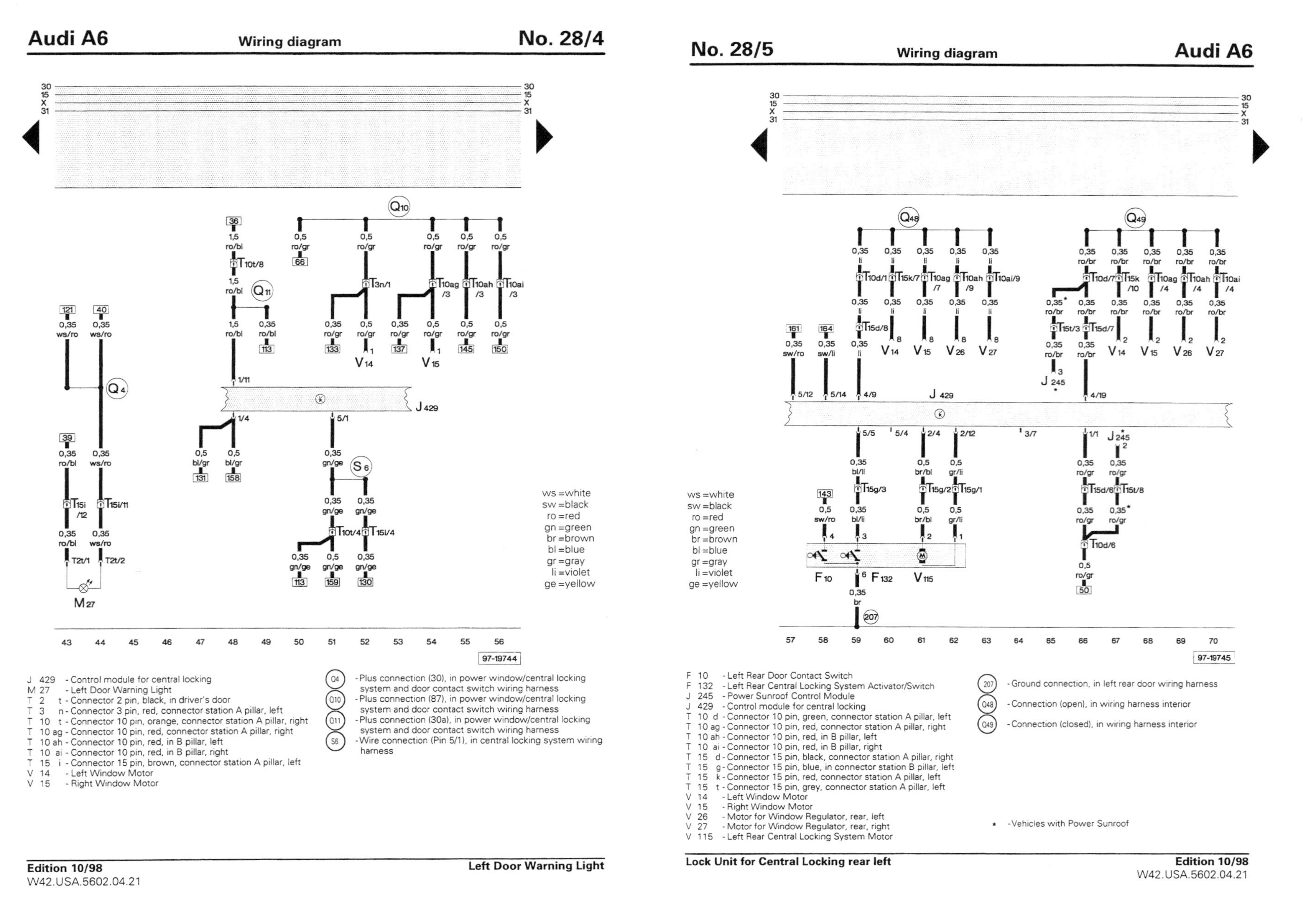

BentleyPublishers.com—All Rights Reserved

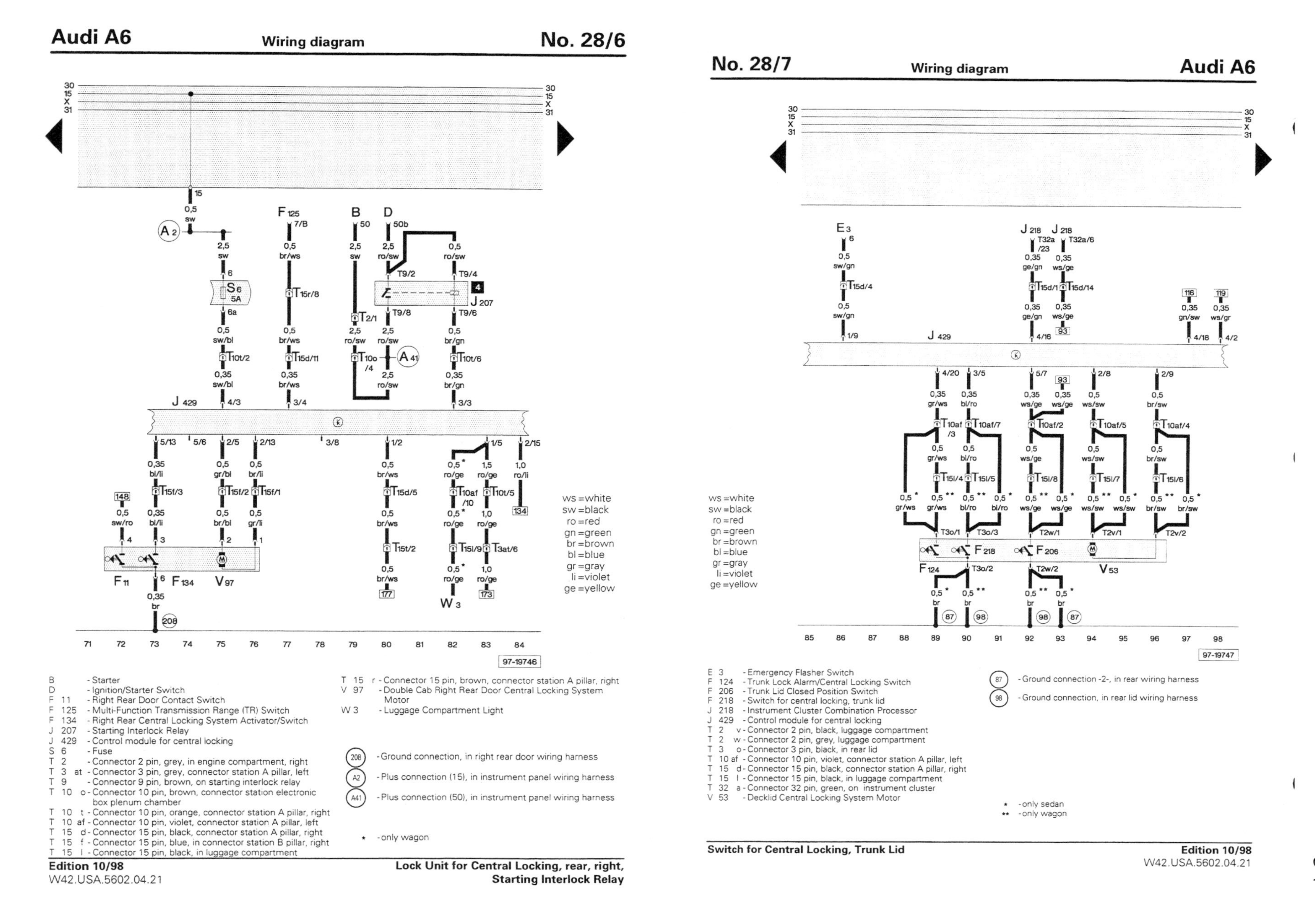

BentleyPublishers.com—All Rights Reserved

Central Locking (1999)

Audi A6 — Wiring diagram — No. 28/8

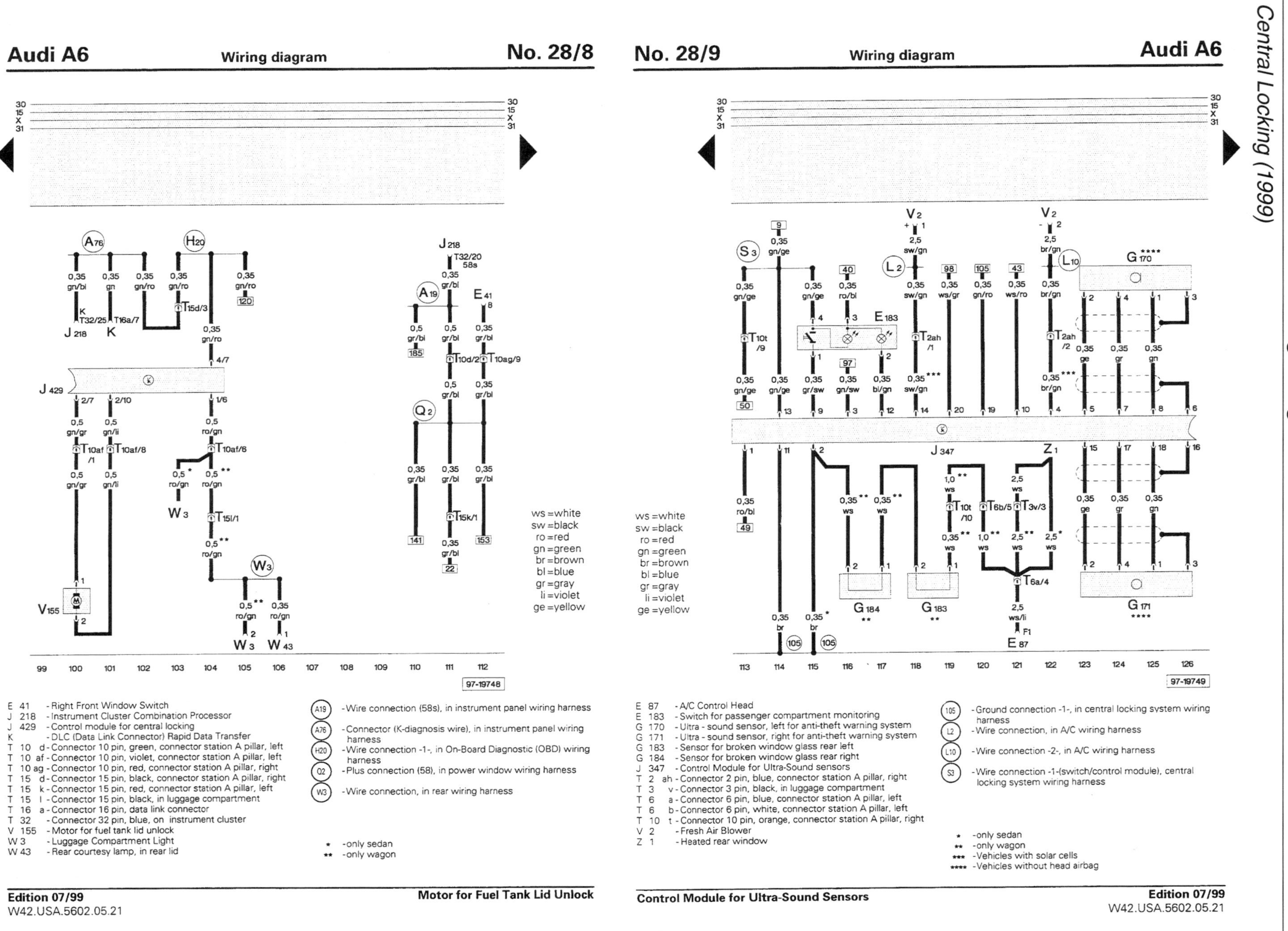

E 41 - Right Front Window Switch
J 218 - Instrument Cluster Combination Processor
J 429 - Control module for central locking
K - DLC (Data Link Connector) Rapid Data Transfer
T 10 d - Connector 10 pin, green, connector station A pillar, left
T 10 af - Connector 10 pin, violet, connector station A pillar, left
T 10 ag - Connector 10 pin, red, connector station A pillar, right
T 15 d - Connector 15 pin, black, connector station A pillar, right
T 15 k - Connector 15 pin, red, connector station A pillar, left
T 15 l - Connector 15 pin, black, in luggage compartment
T 16 a - Connector 16 pin, data link connector
T 32 - Connector 32 pin, blue, on instrument cluster
V 155 - Motor for fuel tank lid unlock
W 3 - Luggage Compartment Light
W 43 - Rear courtesy lamp, in rear lid

(A19) - Wire connection (58s), in instrument panel wiring harness
(A76) - Connector (K-diagnosis wire), in instrument panel wiring harness
(H20) - Wire connection -1-, in On-Board Diagnostic (OBD) wiring harness
(Q2) - Plus connection (58), in power window wiring harness
(W3) - Wire connection, in rear wiring harness

* - only sedan
** - only wagon

ws = white, sw = black, ro = red, gn = green, br = brown, bl = blue, gr = gray, li = violet, ge = yellow

97-19748

Edition 07/99 W42.USA.5602.05.21 — **Motor for Fuel Tank Lid Unlock**

No. 28/9 — Wiring diagram — Audi A6

E 87 - A/C Control Head
E 183 - Switch for passenger compartment monitoring
G 170 - Ultra - sound sensor, left for anti-theft warning system
G 171 - Ultra - sound sensor, right for anti-theft warning system
G 183 - Sensor for broken window glass rear left
G 184 - Sensor for broken window glass rear right
J 347 - Control Module for Ultra-Sound sensors
T 2 ah - Connector 2 pin, blue, connector station A pillar, right
T 3 v - Connector 3 pin, black, in luggage compartment
T 6 a - Connector 6 pin, blue, connector station A pillar, left
T 6 b - Connector 6 pin, white, connector station A pillar, left
T 10 t - Connector 10 pin, orange, connector station A pillar, right
V 2 - Fresh Air Blower
Z 1 - Heated rear window

(105) - Ground connection -1-, in central locking system wiring harness
(L2) - Wire connection, in A/C wiring harness
(L10) - Wire connection -2-, in A/C wiring harness
(S3) - Wire connection -1-(switch/control module), central locking system wiring harness

* - only sedan
** - only wagon
*** - Vehicles with solar cells
**** - Vehicles without head airbag

ws = white, sw = black, ro = red, gn = green, br = brown, bl = blue, gr = gray, li = violet, ge = yellow

97-19749

Control Module for Ultra-Sound Sensors — **Edition 07/99** W42.USA.5602.05.21

BentleyPublishers.com—All Rights Reserved

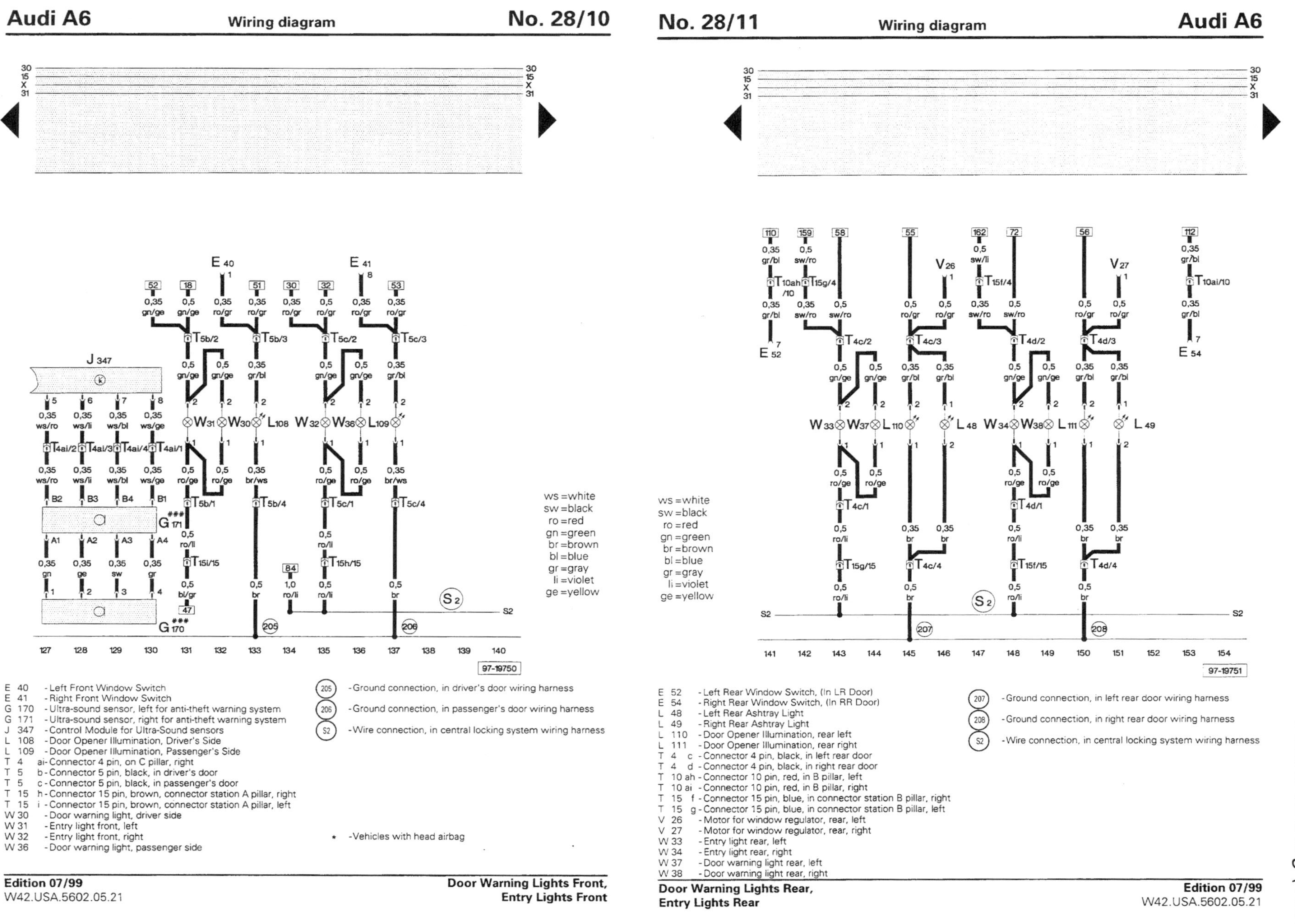

BentleyPublishers.com—All Rights Reserved

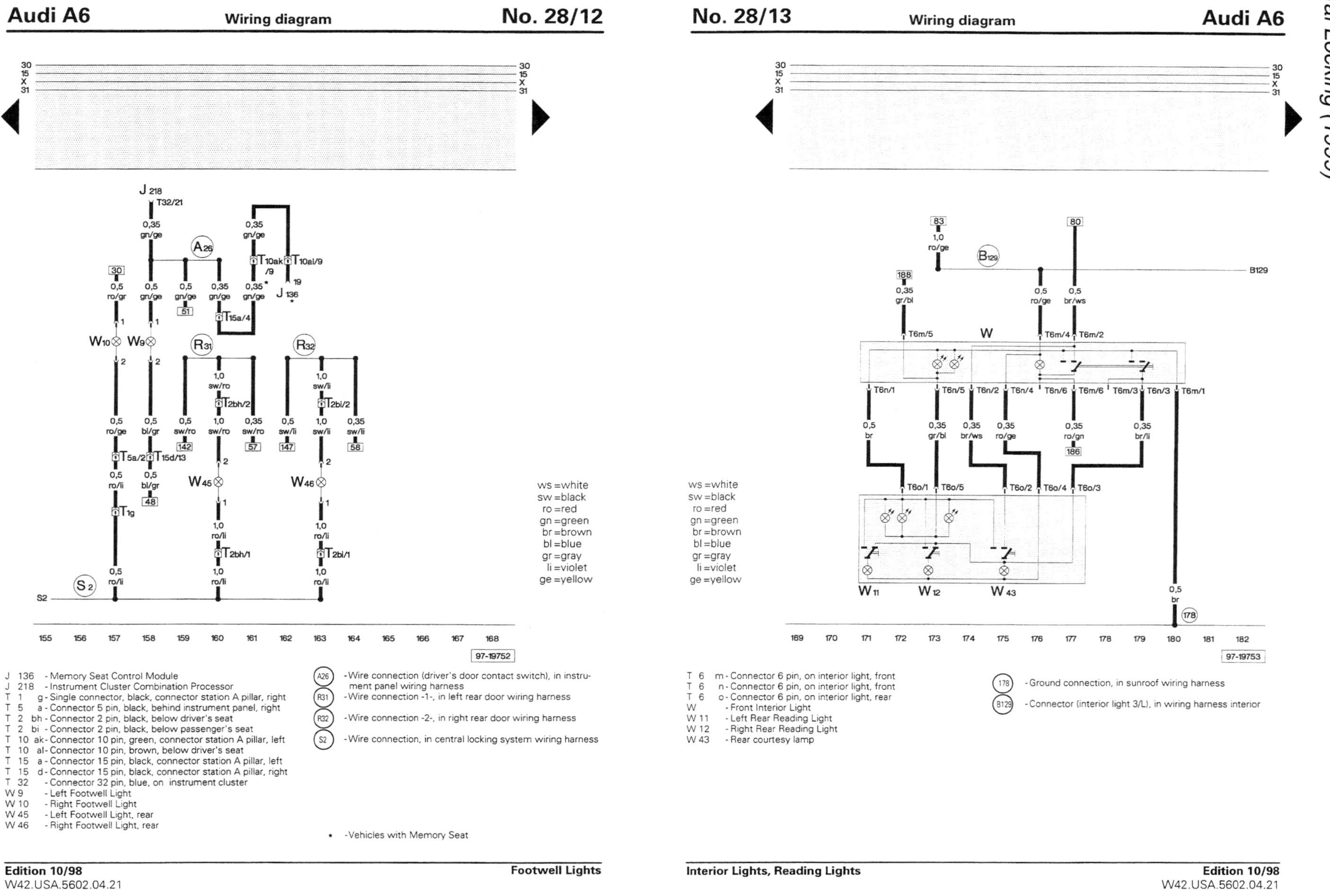

Audi A6
Wiring diagram
No. 28/12
ws =white
sw =black
ro =red
gn =green
br =brown
bl =blue
gr =gray
li =violet
ge =yellow
97-19752
J 136 - Memory Seat Control Module
J 218 - Instrument Cluster Combination Processor
T 1 g - Single connector, black, connector station A pillar, right
T 5 a - Connector 5 pin, black, behind instrument panel, right
T 2 bh - Connector 2 pin, black, below driver's seat
T 2 bi - Connector 2 pin, black, below passenger's seat
T 10 ak - Connector 10 pin, green, connector station A pillar, left
T 10 al - Connector 10 pin, brown, below driver's seat
T 15 a - Connector 15 pin, black, connector station A pillar, left
T 15 d - Connector 15 pin, black, connector station A pillar, right
T 32 - Connector 32 pin, blue, on instrument cluster
W 9 - Left Footwell Light
W 10 - Right Footwell Light
W 45 - Left Footwell Light, rear
W 46 - Right Footwell Light, rear
A26 - Wire connection (driver's door contact switch), in instrument panel wiring harness
R31 - Wire connection -1-, in left rear door wiring harness
R32 - Wire connection -2-, in right rear door wiring harness
S2 - Wire connection, in central locking system wiring harness
* - Vehicles with Memory Seat
Edition 10/98
W42.USA.5602.04.21
Footwell Lights
No. 28/13
Wiring diagram
Audi A6
ws =white
sw =black
ro =red
gn =green
br =brown
bl =blue
gr =gray
li =violet
ge =yellow
97-19753
T 6 m - Connector 6 pin, on interior light, front
T 6 n - Connector 6 pin, on interior light, front
T 6 o - Connector 6 pin, on interior light, rear
W - Front Interior Light
W 11 - Left Rear Reading Light
W 12 - Right Rear Reading Light
W 43 - Rear courtesy lamp
178 - Ground connection, in sunroof wiring harness
B129 - Connector (interior light 3/L), in wiring harness interior
Interior Lights, Reading Lights
Edition 10/98
W42.USA.5602.04.21

BentleyPublishers.com—All Rights Reserved

THIS SPACE INTENTIONALLY

LEFT BLANK

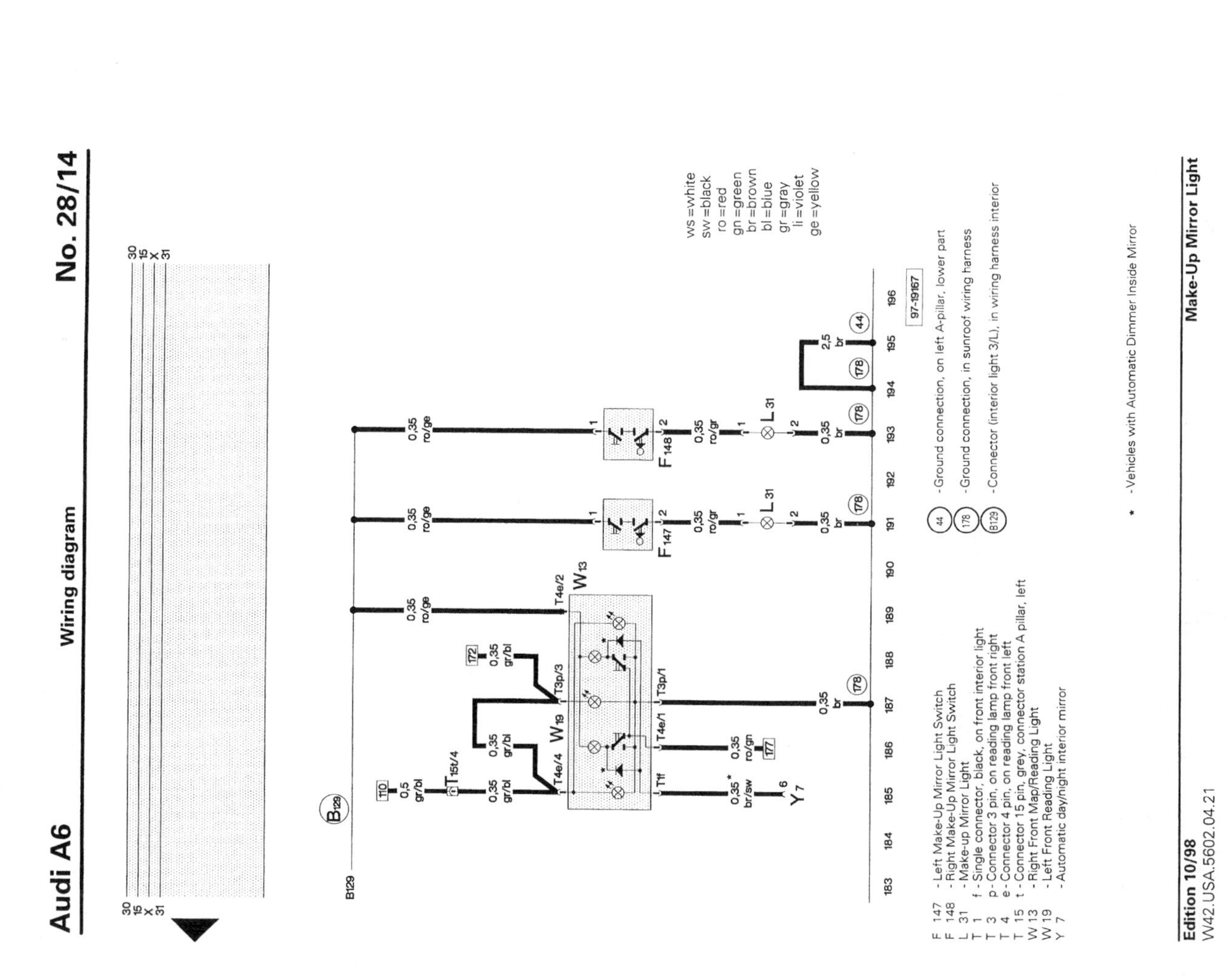

BentleyPublishers.com—All Rights Reserved

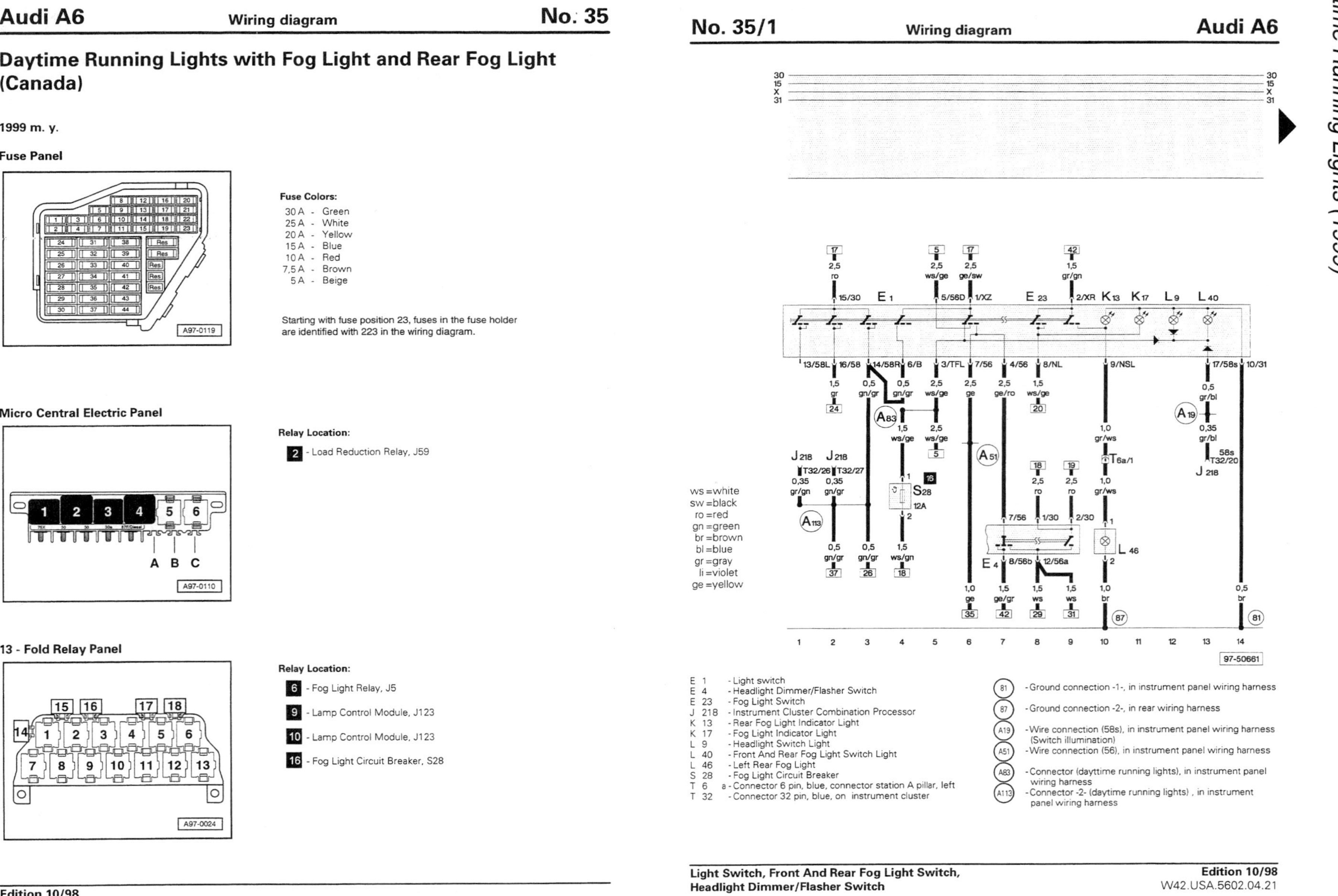

BentleyPublishers.com—All Rights Reserved

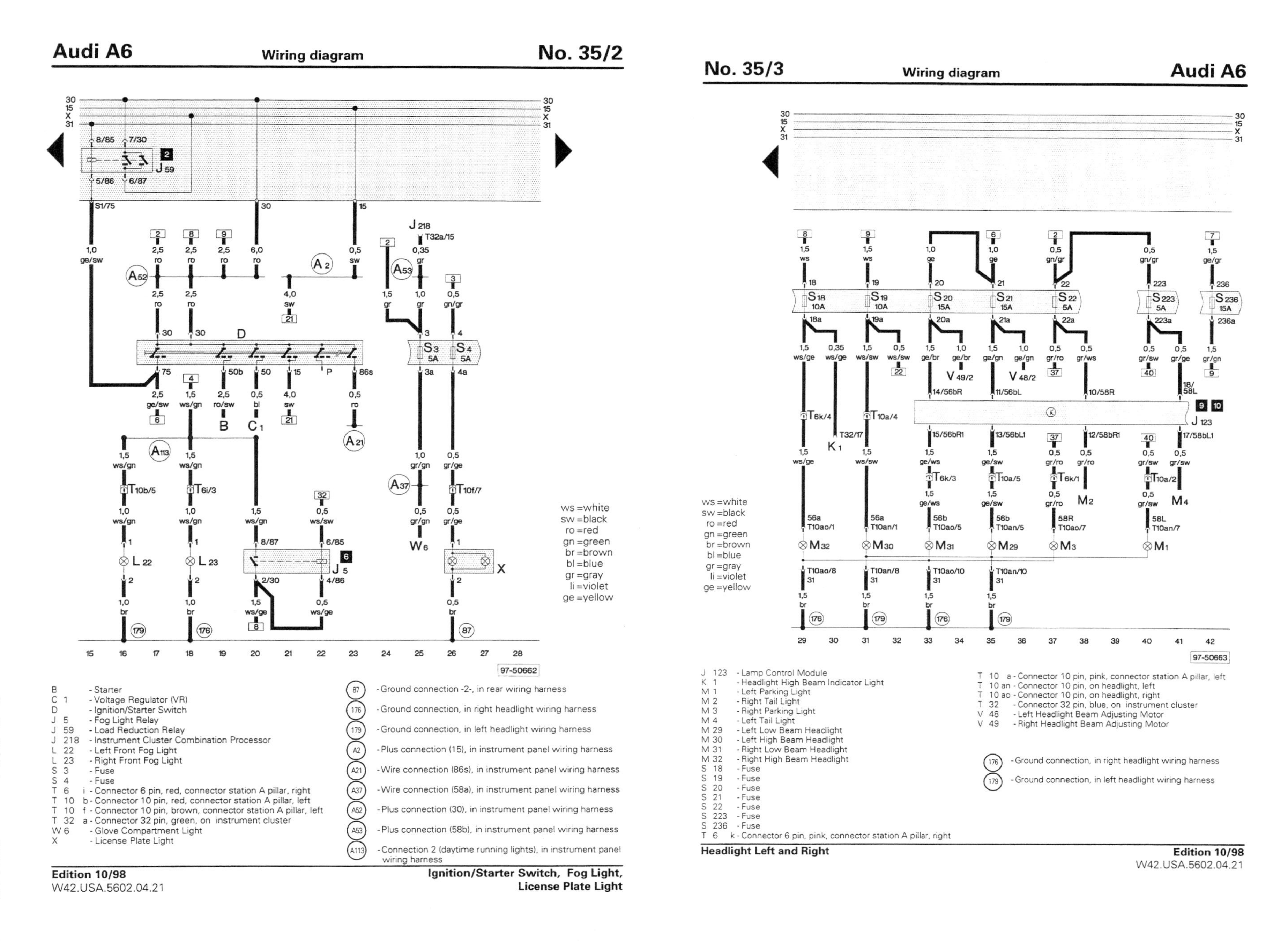

BentleyPublishers.com—All Rights Reserved

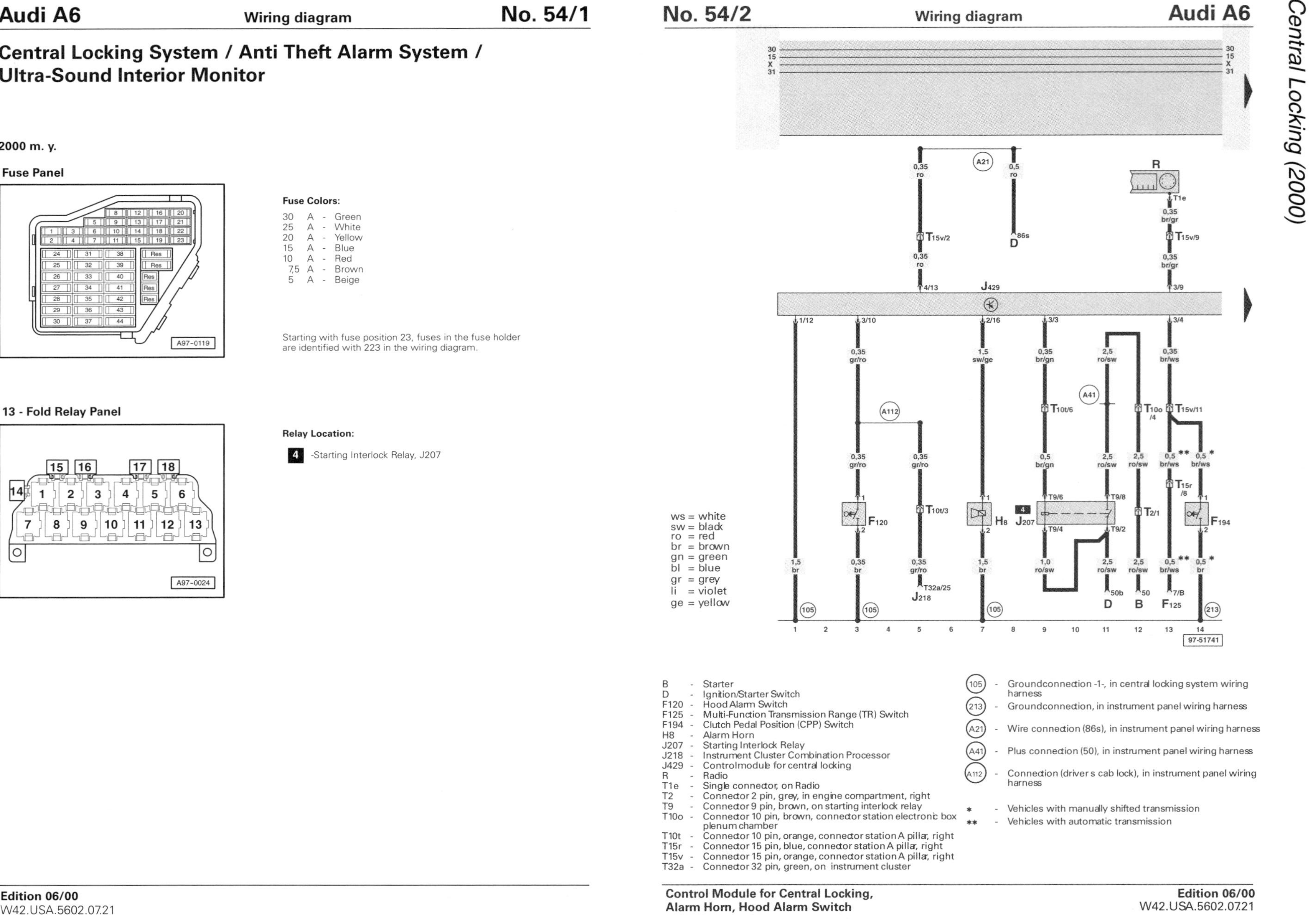

BentleyPublishers.com—All Rights Reserved

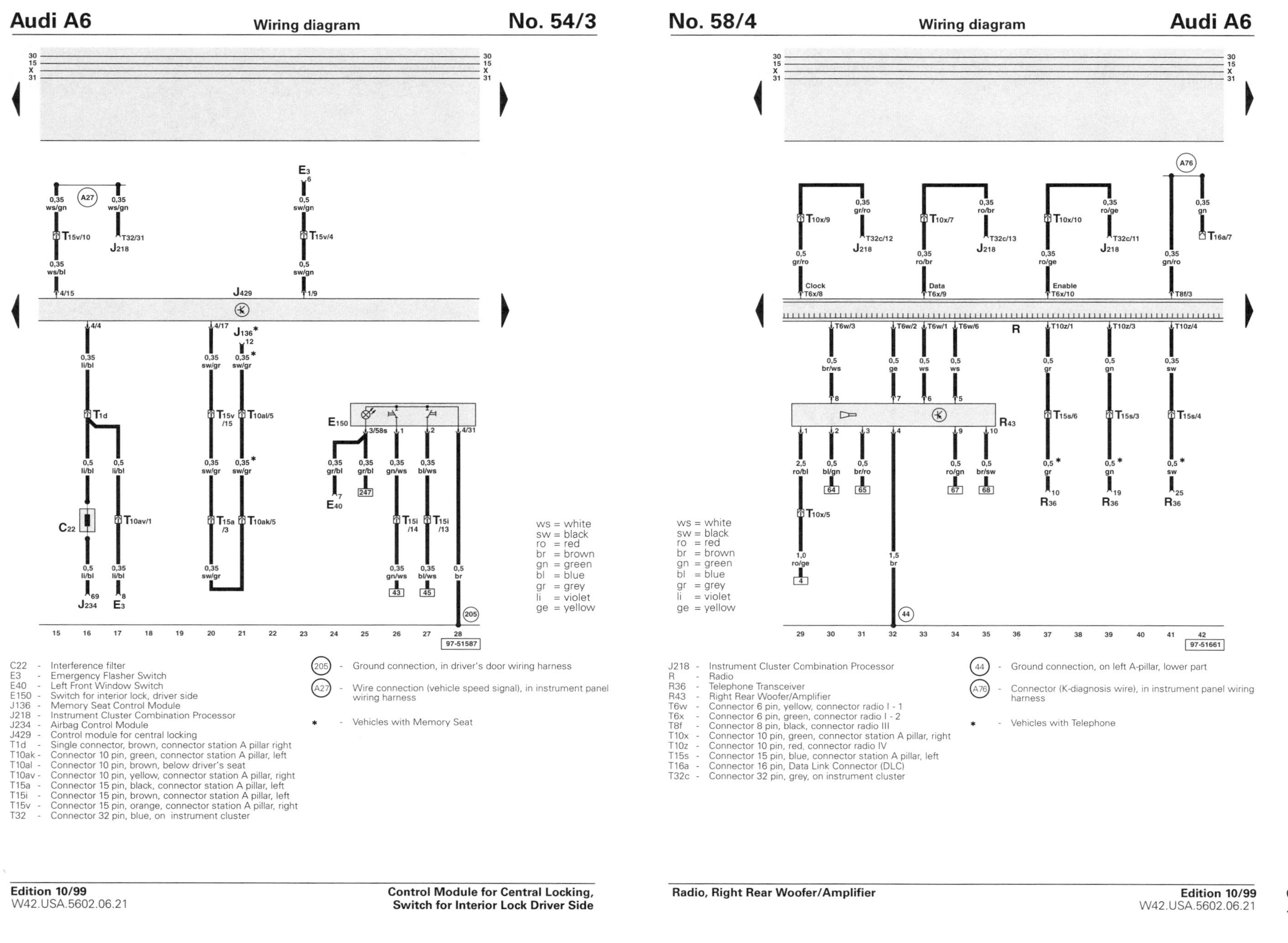

BentleyPublishers.com—All Rights Reserved

Central Locking (2000)

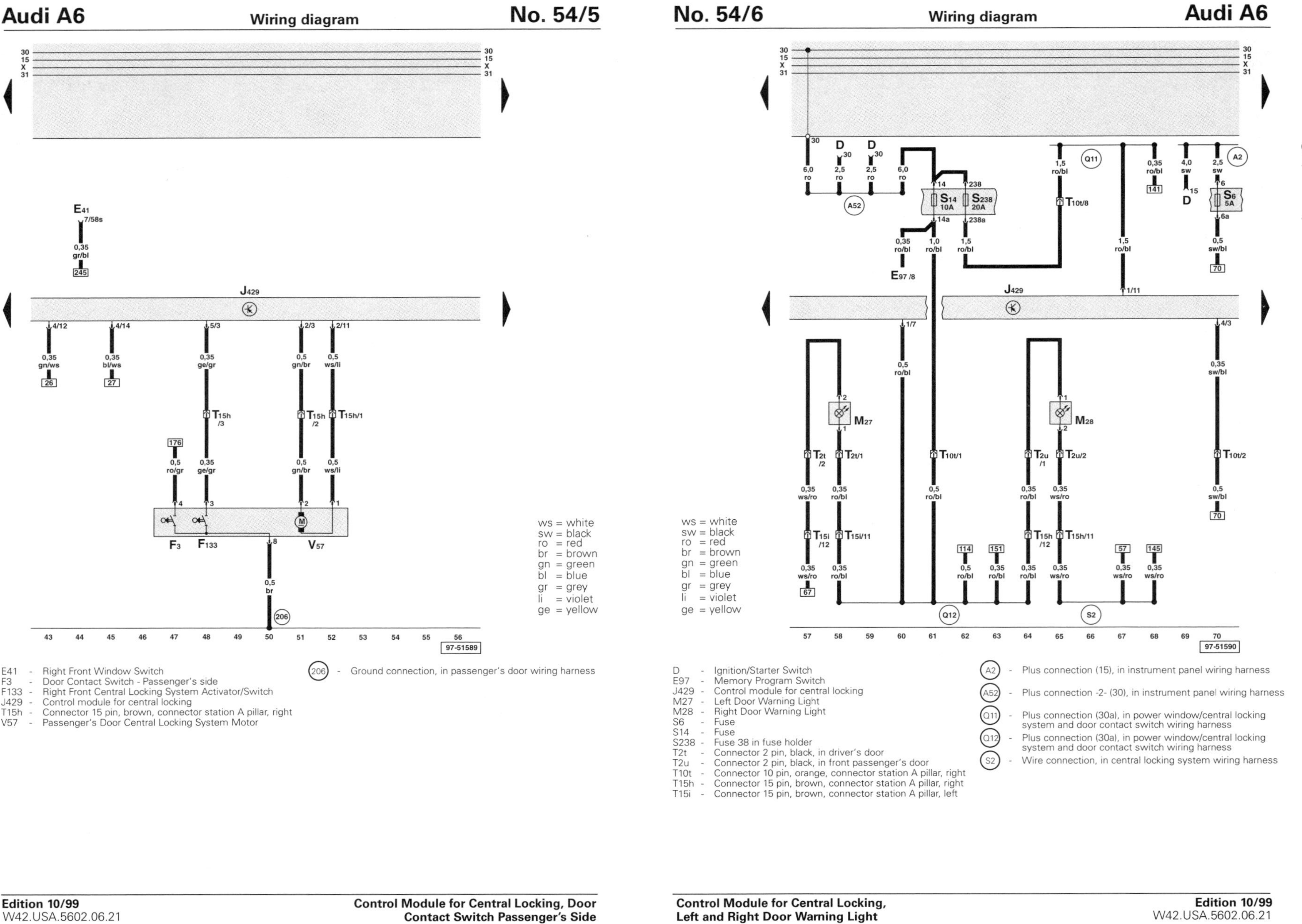

BentleyPublishers.com—All Rights Reserved

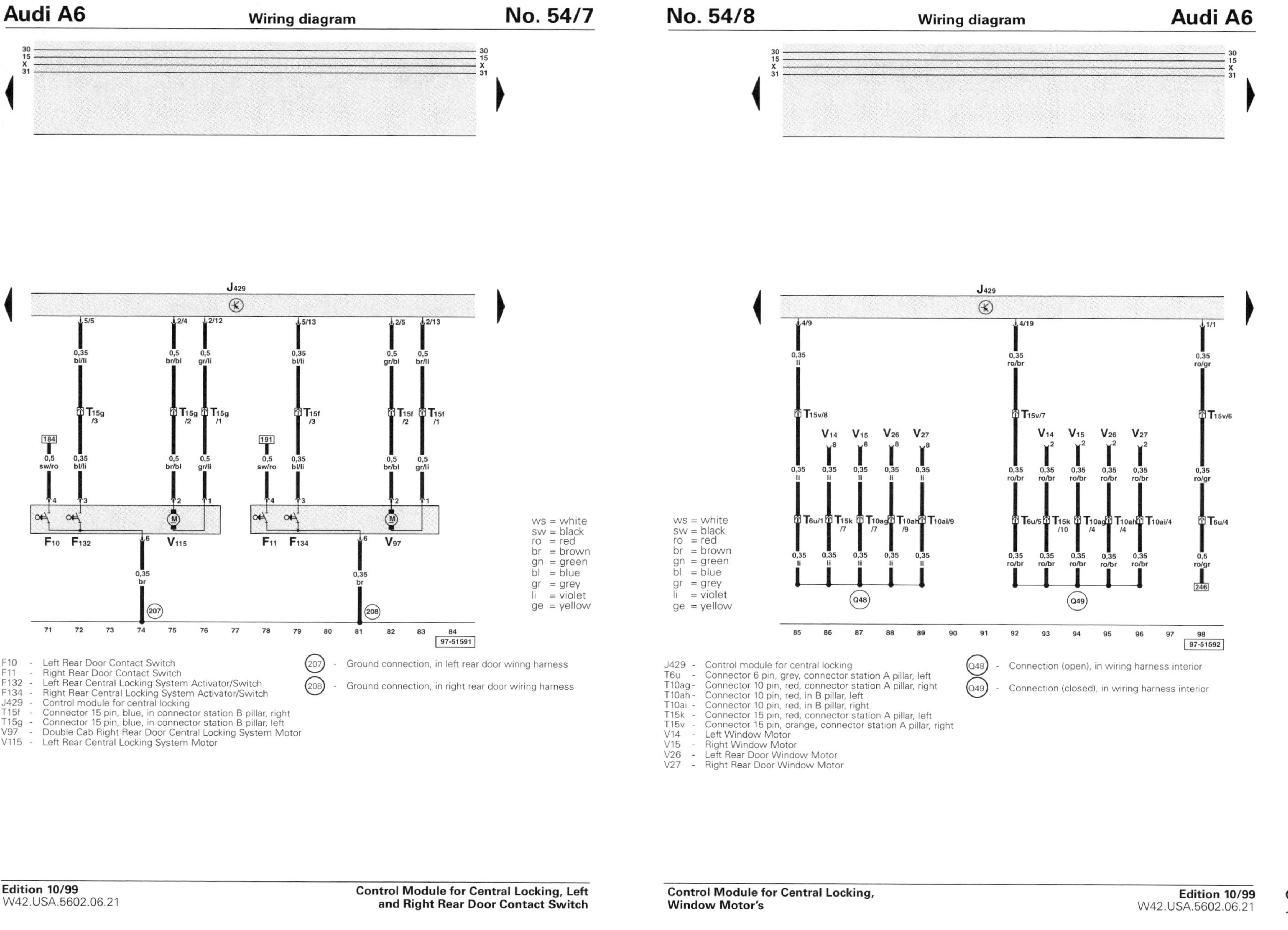

BentleyPublishers.com—All Rights Reserved

Central Locking (2000)

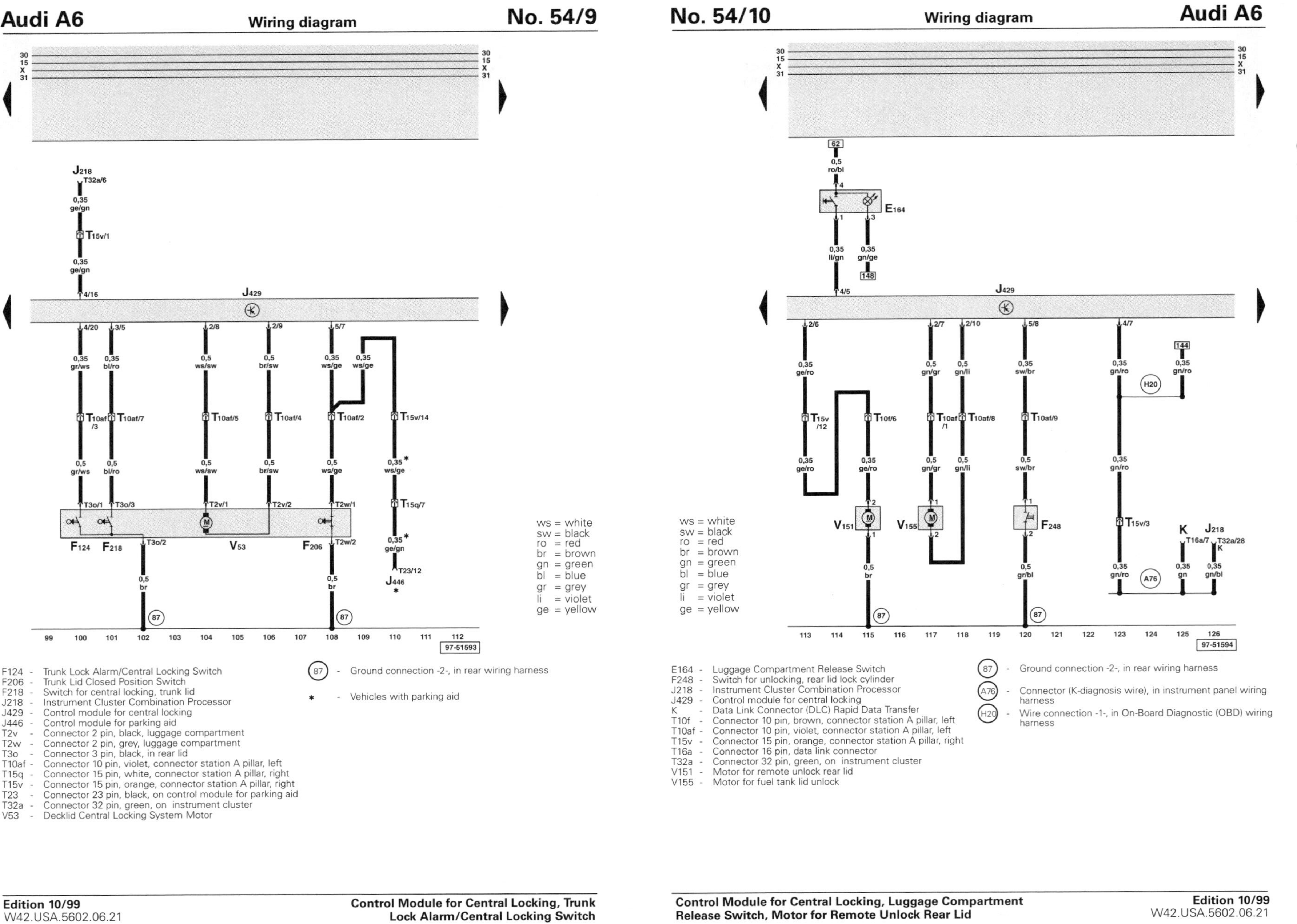

BentleyPublishers.com—All Rights Reserved

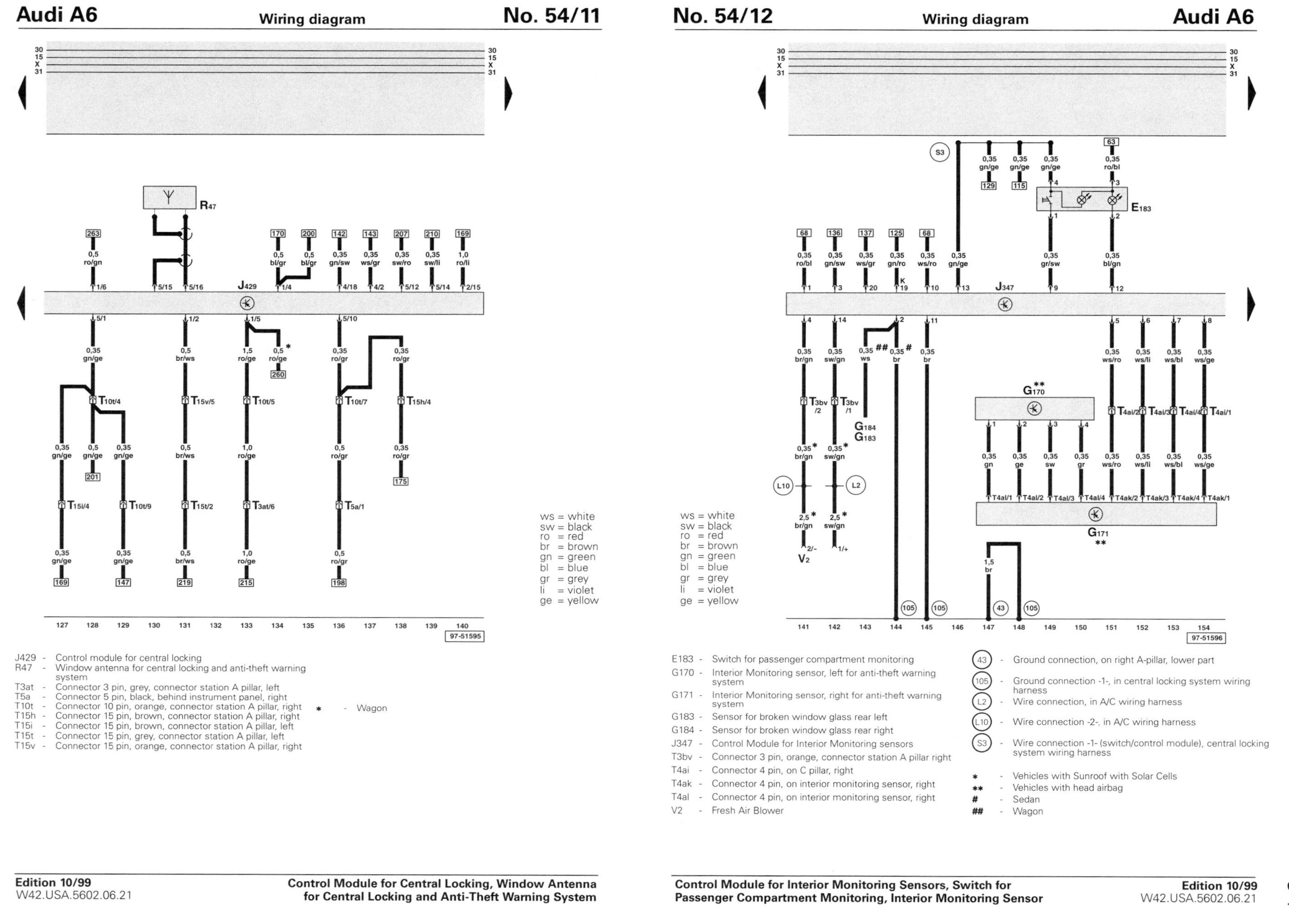

BentleyPublishers.com—All Rights Reserved

Central Locking (2000)

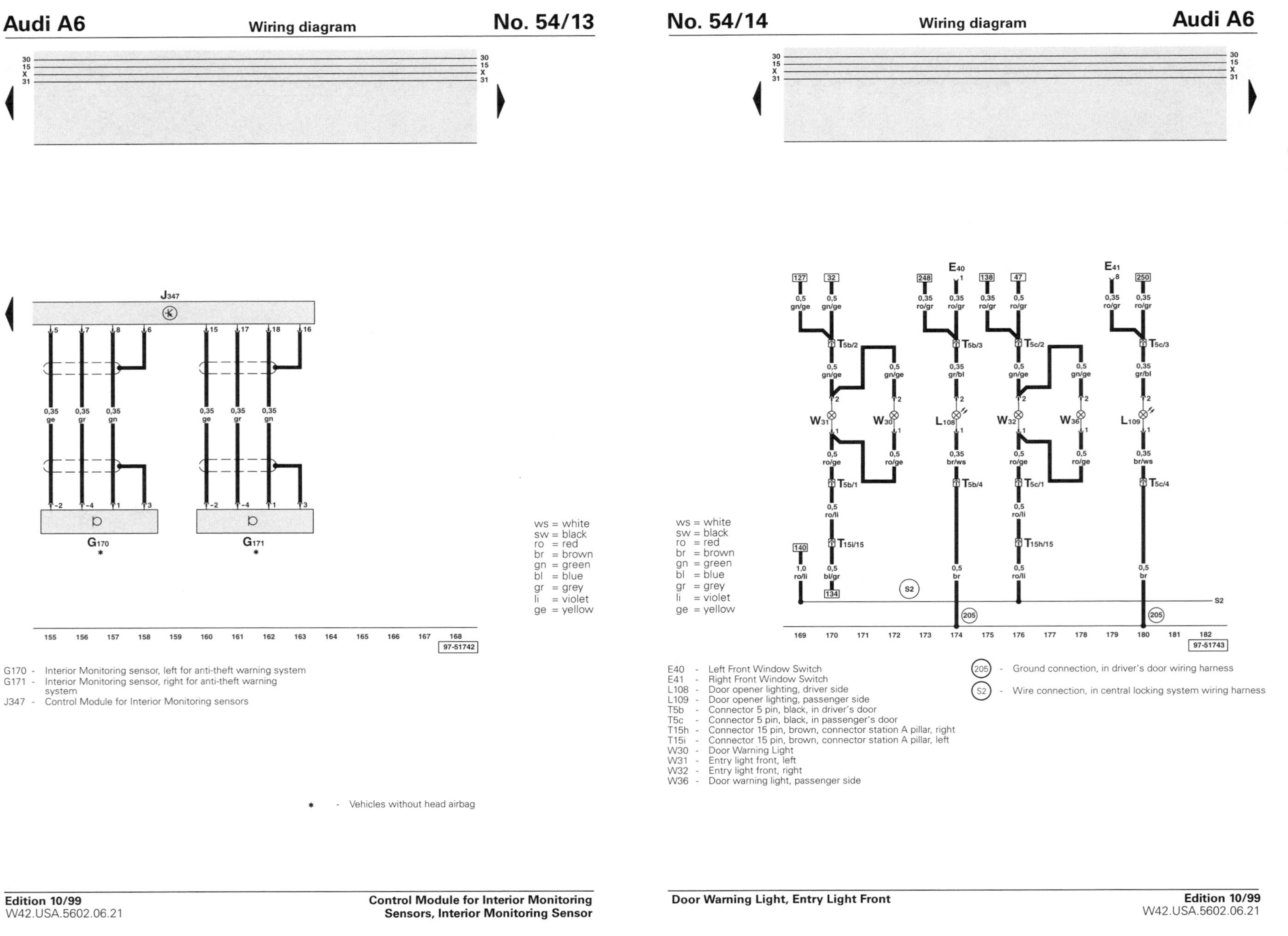

BentleyPublishers.com—All Rights Reserved

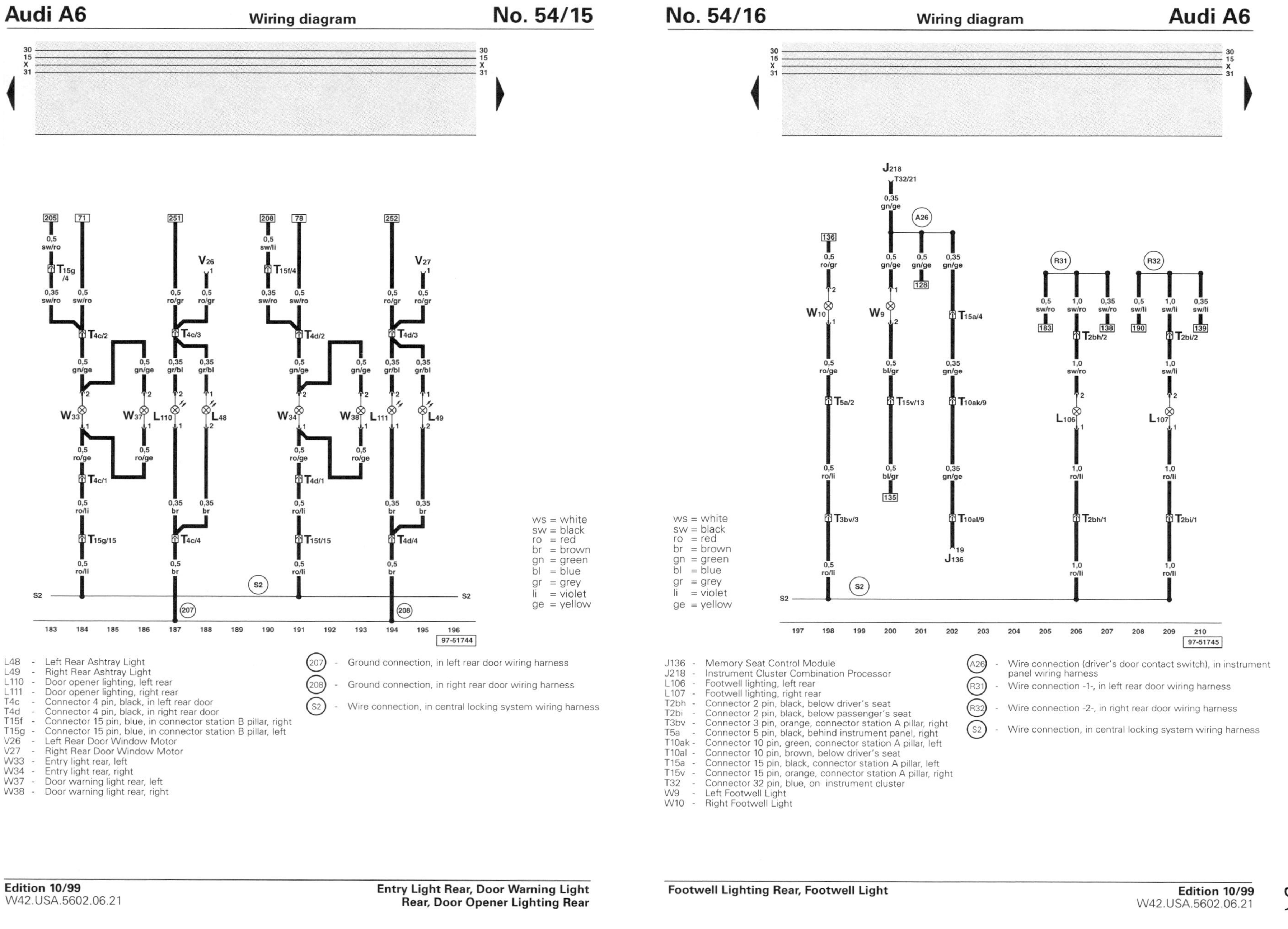

BentleyPublishers.com—All Rights Reserved

Central Locking (2000)

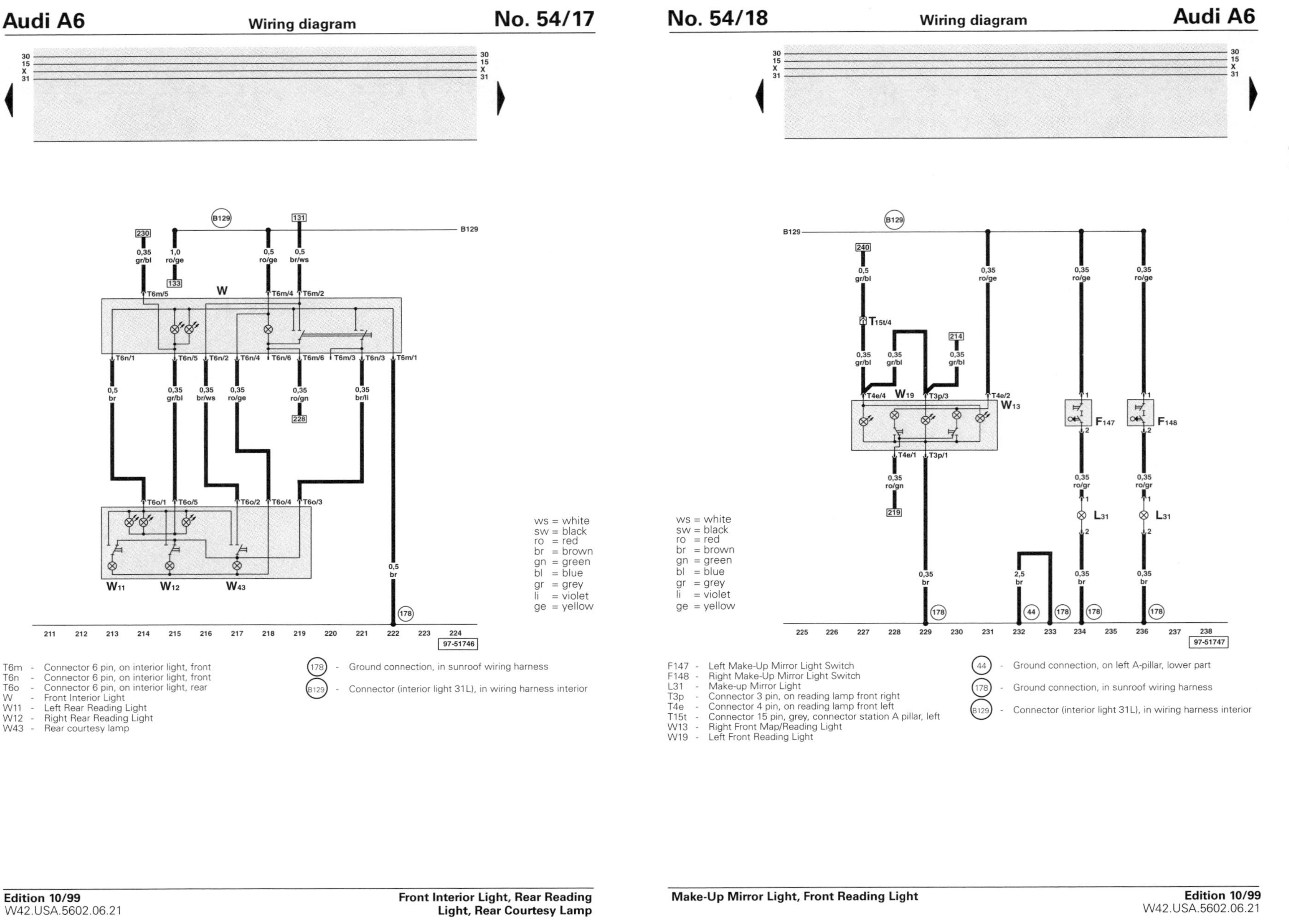

BentleyPublishers.com—All Rights Reserved

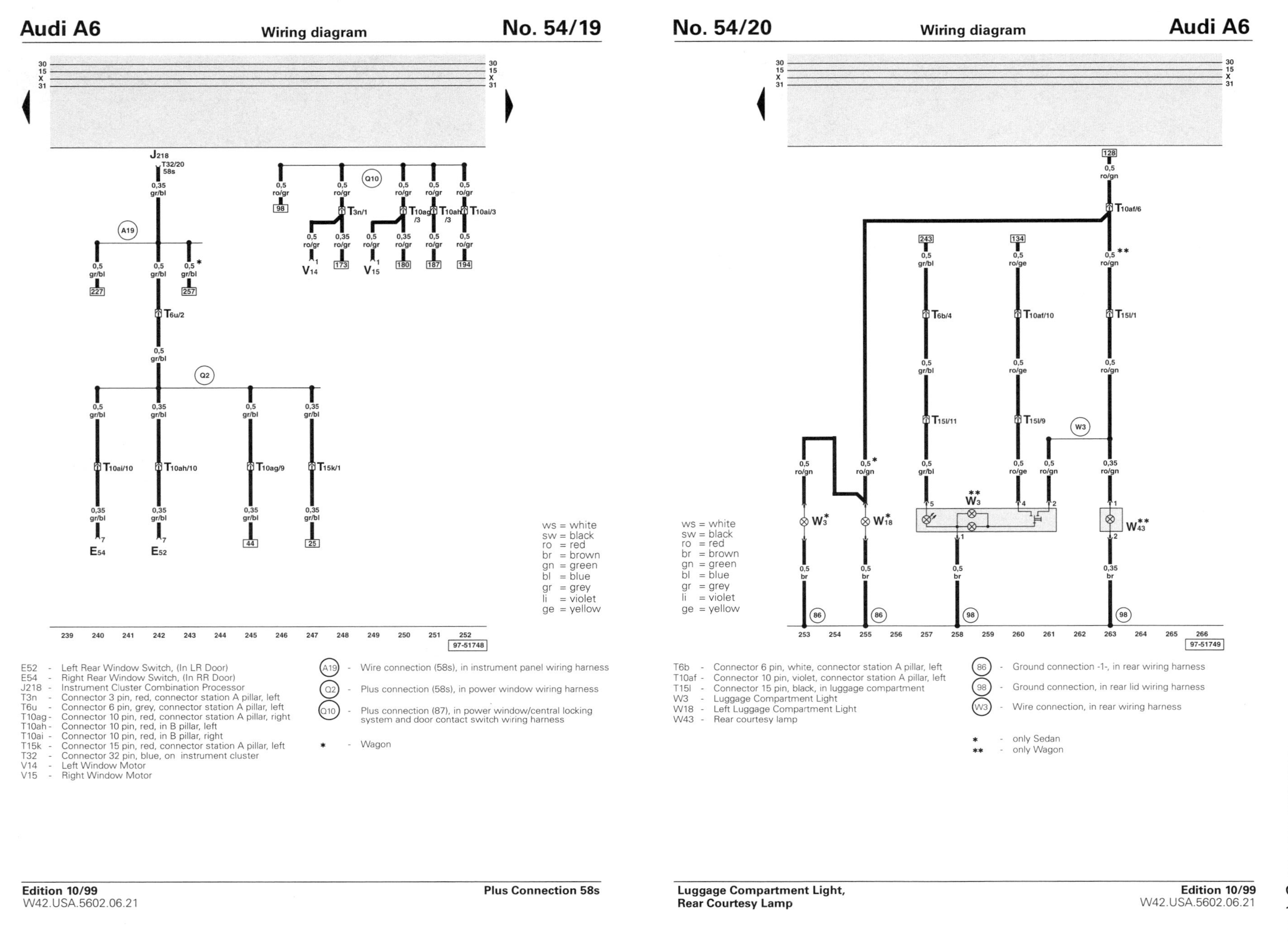

BentleyPublishers.com—All Rights Reserved

Radio-Stereo, Concert with and without CD-Changer

2000 m. y.

Fuse Panel

A97-0119

Fuse Colors:

30 A - Green
25 A - White
20 A - Yellow
15 A - Blue
10 A - Red
7,5 A - Brown
5 A - Beige

Starting with fuse position 23, fuses in the fuse holder are identified with 223 in the wiring diagram.

Micro Central Electric Panel

A B C

A97-0110

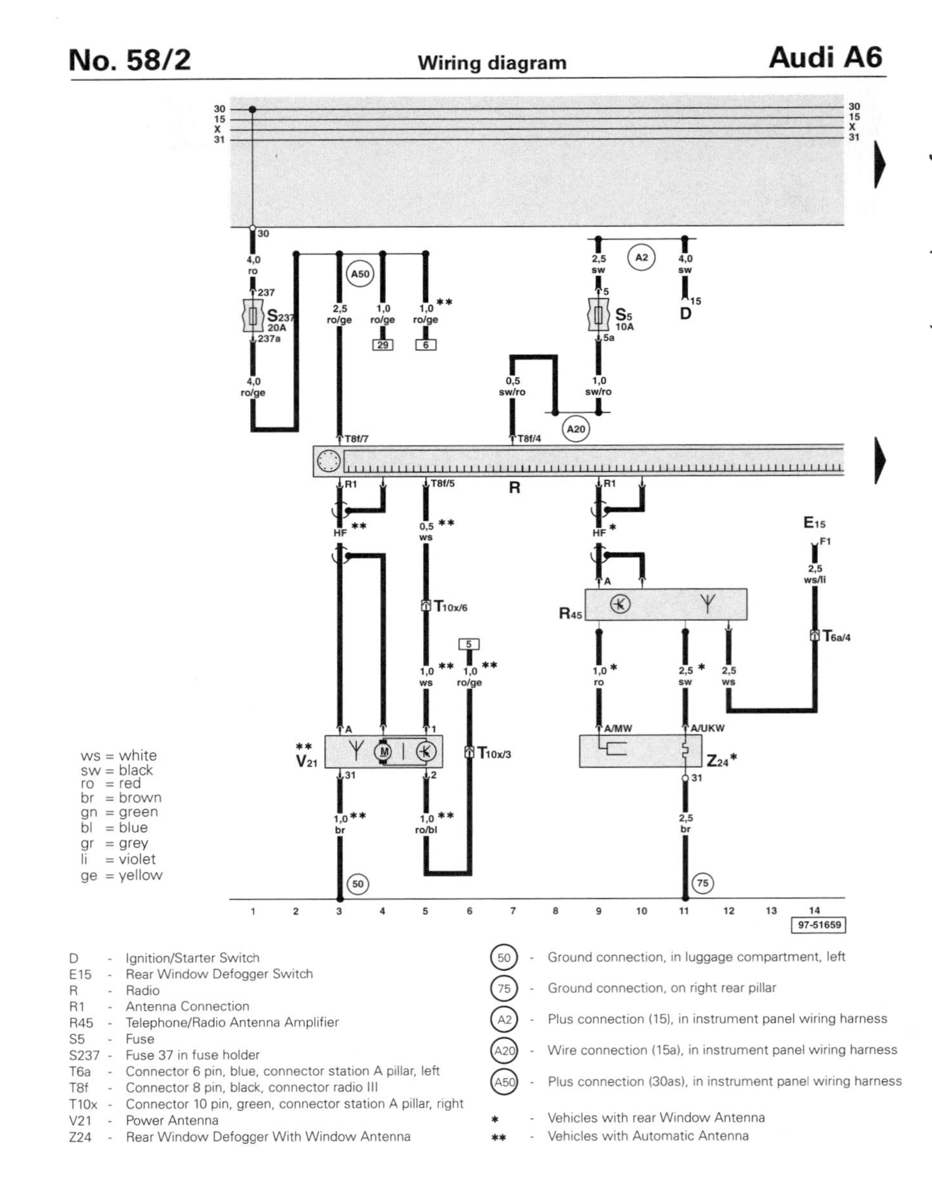

ws = white
sw = black
ro = red
br = brown
gn = green
bl = blue
gr = grey
li = violet
ge = yellow

D - Ignition/Starter Switch
E15 - Rear Window Defogger Switch
R - Radio
R1 - Antenna Connection
R45 - Telephone/Radio Antenna Amplifier
S5 - Fuse
S237 - Fuse 37 in fuse holder
T6a - Connector 6 pin, blue, connector station A pillar, left
T8f - Connector 8 pin, black, connector radio III
T10x - Connector 10 pin, green, connector station A pillar, right
V21 - Power Antenna
Z24 - Rear Window Defogger With Window Antenna

(50) - Ground connection, in luggage compartment, left
(75) - Ground connection, on right rear pillar
(A2) - Plus connection (15), in instrument panel wiring harness
(A20) - Wire connection (15a), in instrument panel wiring harness
(A50) - Plus connection (30as), in instrument panel wiring harness
* - Vehicles with rear Window Antenna
** - Vehicles with Automatic Antenna

Radio, Rear Window Defogger with Window Antenna

BentleyPublishers.com—All Rights Reserved

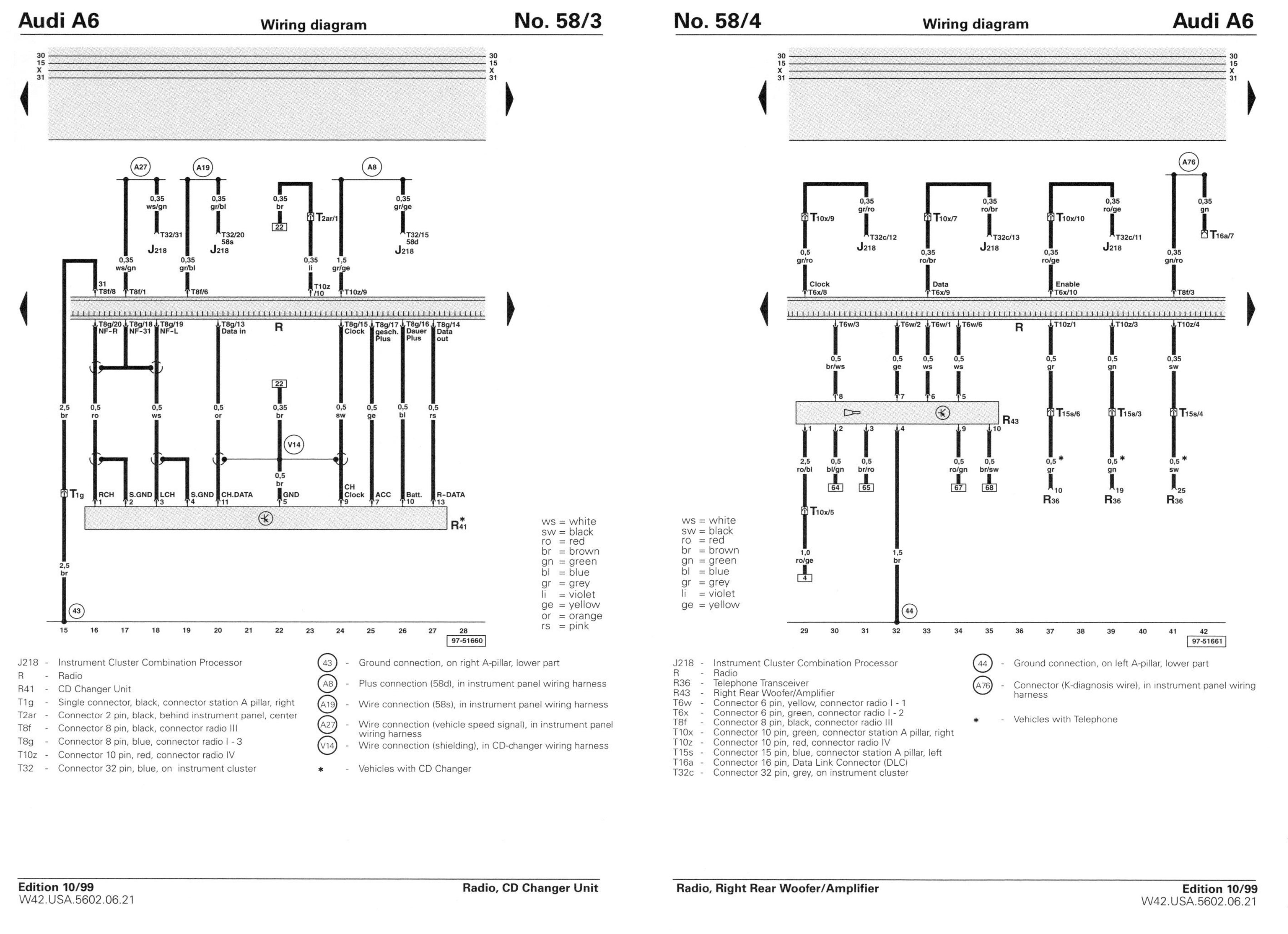

BentleyPublishers.com—All Rights Reserved

Sound System (2000)

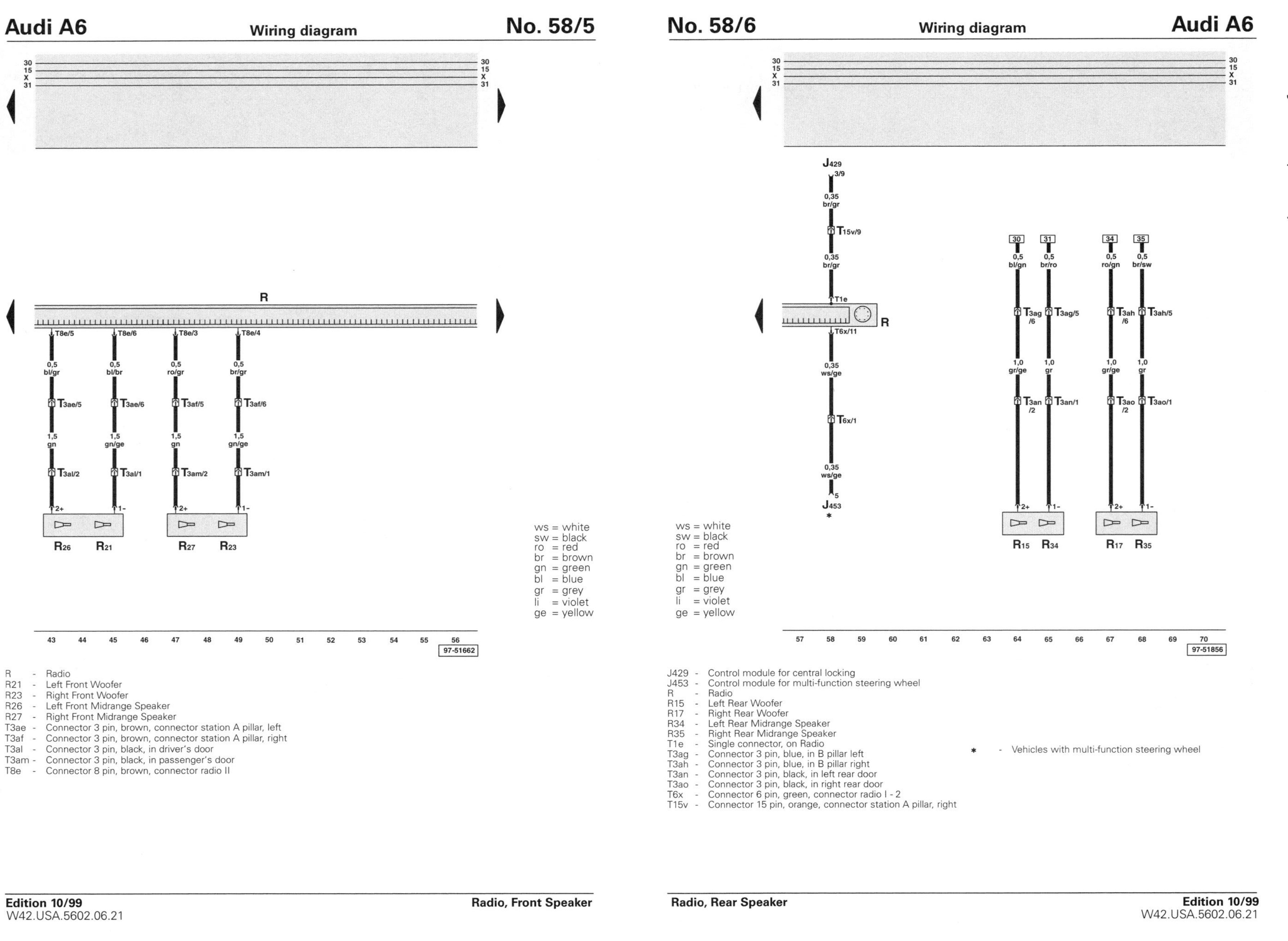

BentleyPublishers.com—All Rights Reserved

Radio Concert, Symphony, Bose with and without CD Changer

2000 m. y.

Fuse Panel

A97-0119

Fuse Colors:

30 A - Green
25 A - White
20 A - Yellow
15 A - Blue
10 A - Red
7,5 A - Brown
5 A - Beige

Starting with fuse position 23, fuses in the fuse holder are identified with 223 in the wiring diagram.

Micro Central Electric Panel

A97-0110

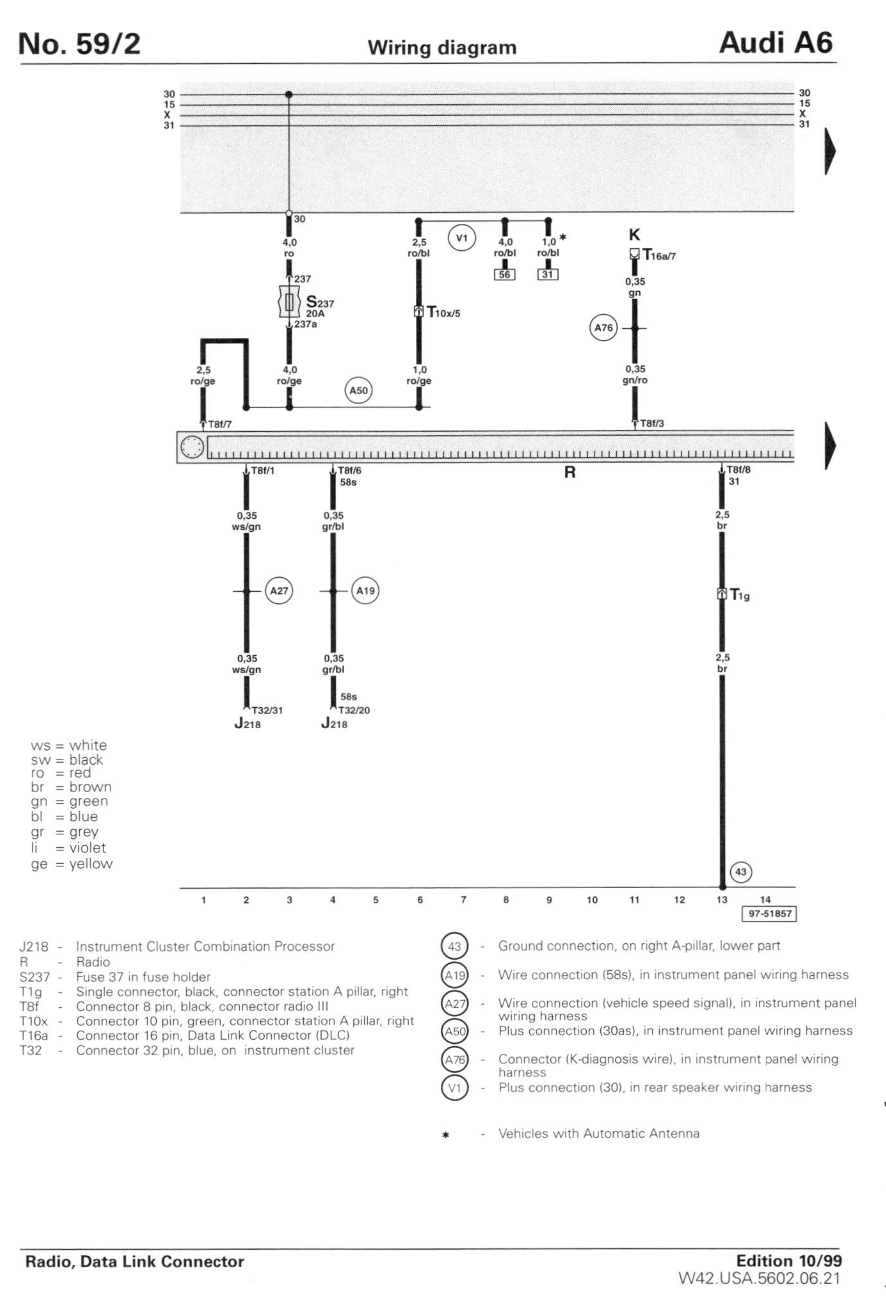

ws = white
sw = black
ro = red
br = brown
gn = green
bl = blue
gr = grey
li = violet
ge = yellow

J218 - Instrument Cluster Combination Processor
R - Radio
S237 - Fuse 37 in fuse holder
T1g - Single connector, black, connector station A pillar, right
T8f - Connector 8 pin, black, connector radio III
T10x - Connector 10 pin, green, connector station A pillar, right
T16a - Connector 16 pin, Data Link Connector (DLC)
T32 - Connector 32 pin, blue, on instrument cluster

(43) - Ground connection, on right A-pillar, lower part
(A19) - Wire connection (58s), in instrument panel wiring harness
(A27) - Wire connection (vehicle speed signal), in instrument panel wiring harness
(A50) - Plus connection (30as), in instrument panel wiring harness
(A76) - Connector (K-diagnosis wire), in instrument panel wiring harness
(V1) - Plus connection (30), in rear speaker wiring harness

* - Vehicles with Automatic Antenna

Radio, Data Link Connector

BentleyPublishers.com—All Rights Reserved

Sound System, Bose (2000)

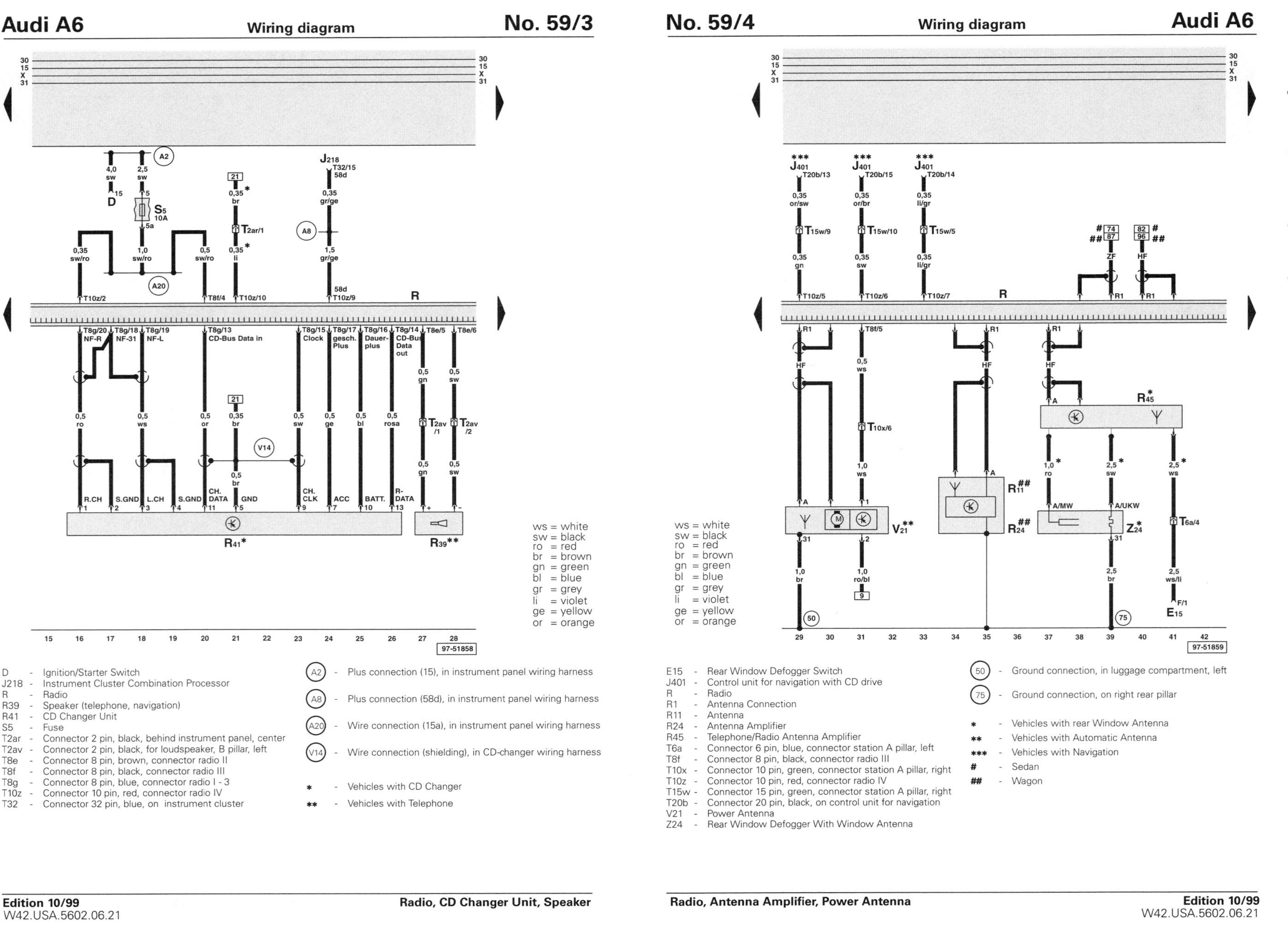

BentleyPublishers.com—All Rights Reserved

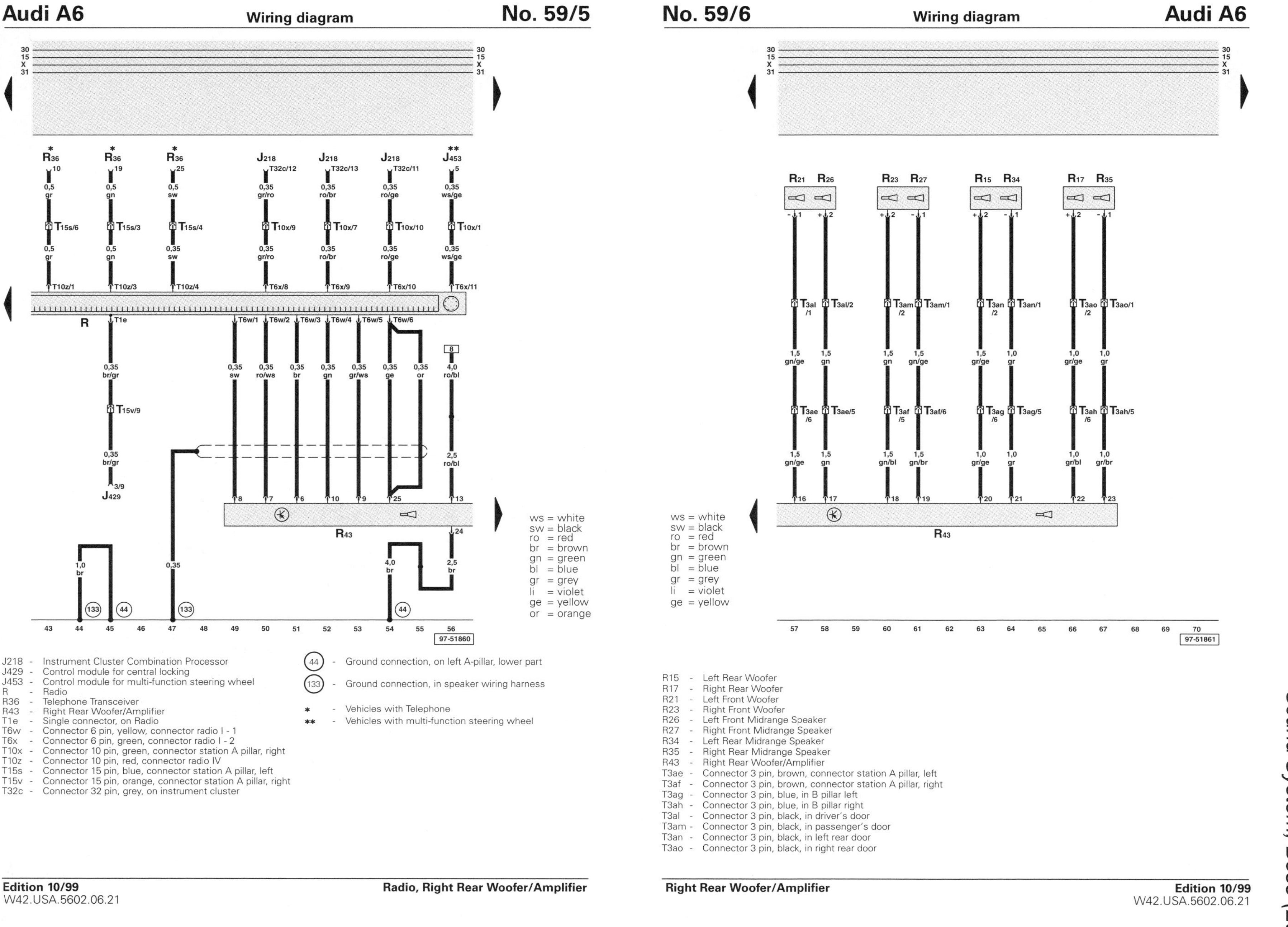

BentleyPublishers.com—All Rights Reserved

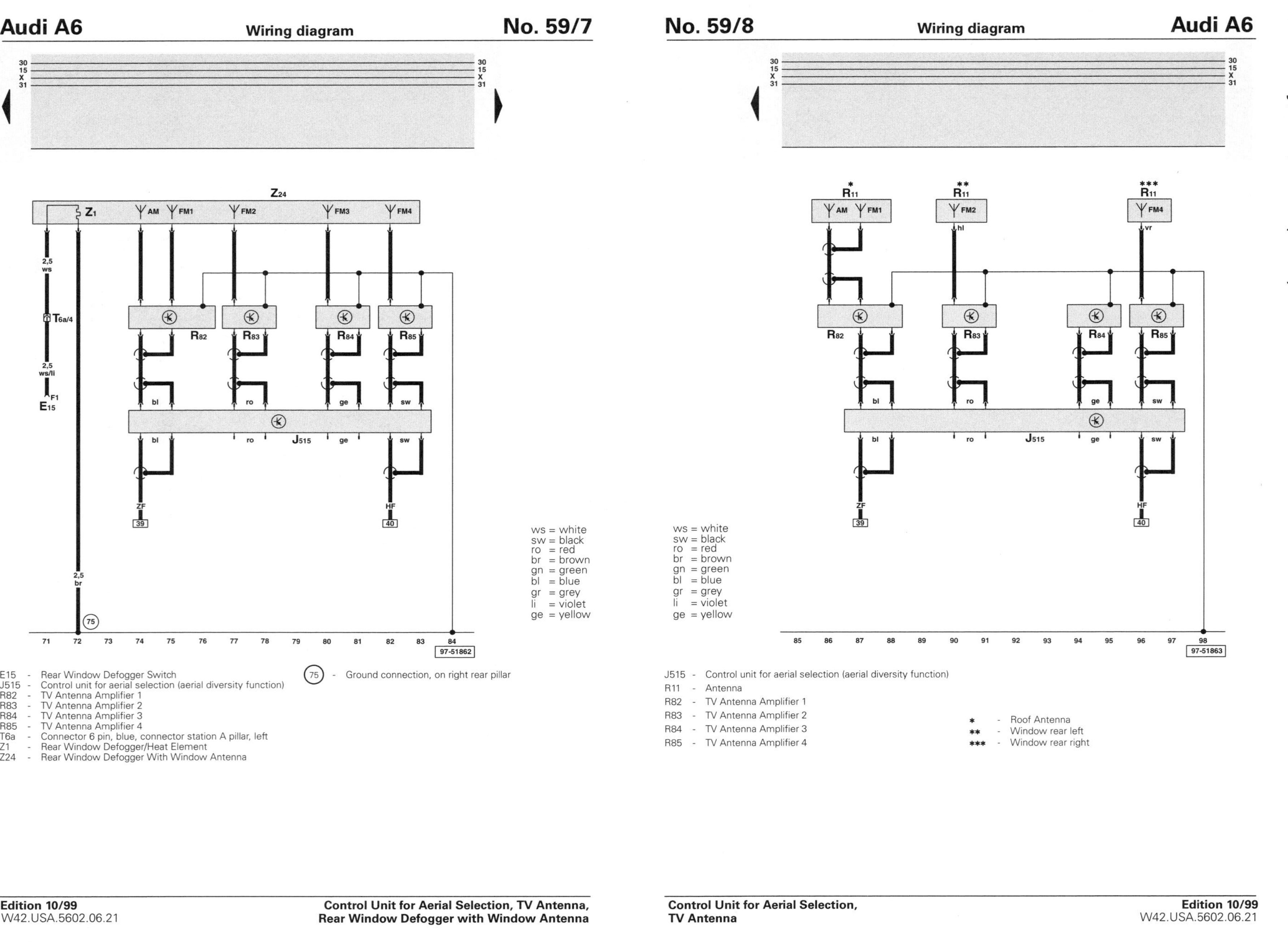

BentleyPublishers.com—All Rights Reserved

Driver's Information System

2000 m. y.

Fuse Panel

Fuse Colors:

30 A - Green
25 A - White
20 A - Yellow
15 A - Blue
10 A - Red
7,5 A - Brown
5 A - Beige

Starting with fuse position 23, fuses in the fuse holder are identified with 223 in the wiring diagram.

13 - Fold Relay Panel

Relay Location:

9 -Lamp Control Module, J123

10 -Lamp Control Module, J123

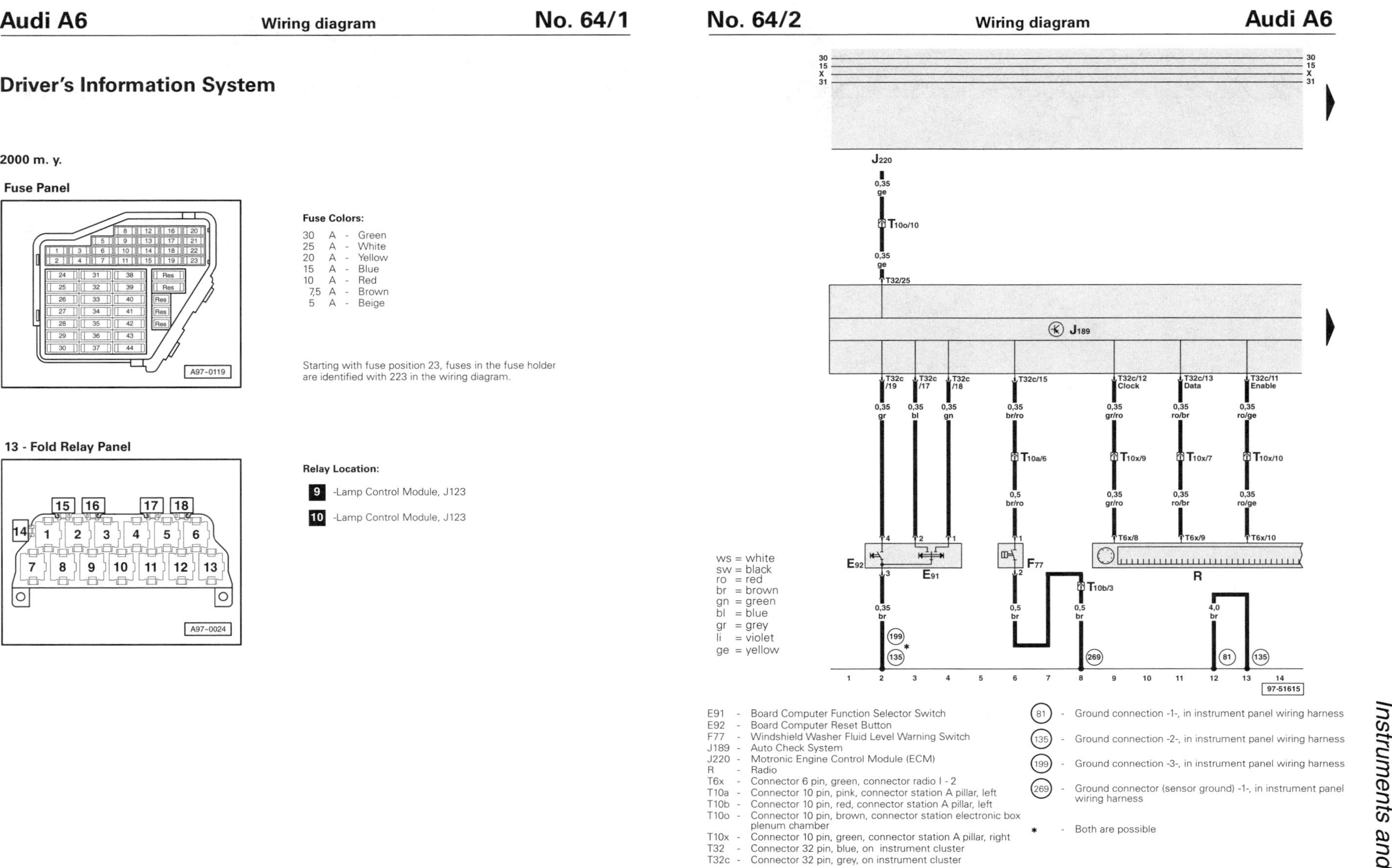

ws = white
sw = black
ro = red
br = brown
gn = green
bl = blue
gr = grey
li = violet
ge = yellow

E91 - Board Computer Function Selector Switch
E92 - Board Computer Reset Button
F77 - Windshield Washer Fluid Level Warning Switch
J189 - Auto Check System
J220 - Motronic Engine Control Module (ECM)
R - Radio
T6x - Connector 6 pin, green, connector radio I - 2
T10a - Connector 10 pin, pink, connector station A pillar, left
T10b - Connector 10 pin, red, connector station A pillar, left
T10o - Connector 10 pin, brown, connector station electronic box plenum chamber
T10x - Connector 10 pin, green, connector station A pillar, right
T32 - Connector 32 pin, blue, on instrument cluster
T32c - Connector 32 pin, grey, on instrument cluster

(81) - Ground connection -1-, in instrument panel wiring harness
(135) - Ground connection -2-, in instrument panel wiring harness
(199) - Ground connection -3-, in instrument panel wiring harness
(269) - Ground connector (sensor ground) -1-, in instrument panel wiring harness
* - Both are possible

Board Computer Function Selector Switch, Windshield Washer Fluid Level Warning Switch

BentleyPublishers.com—All Rights Reserved

Instruments and Controls (2000)

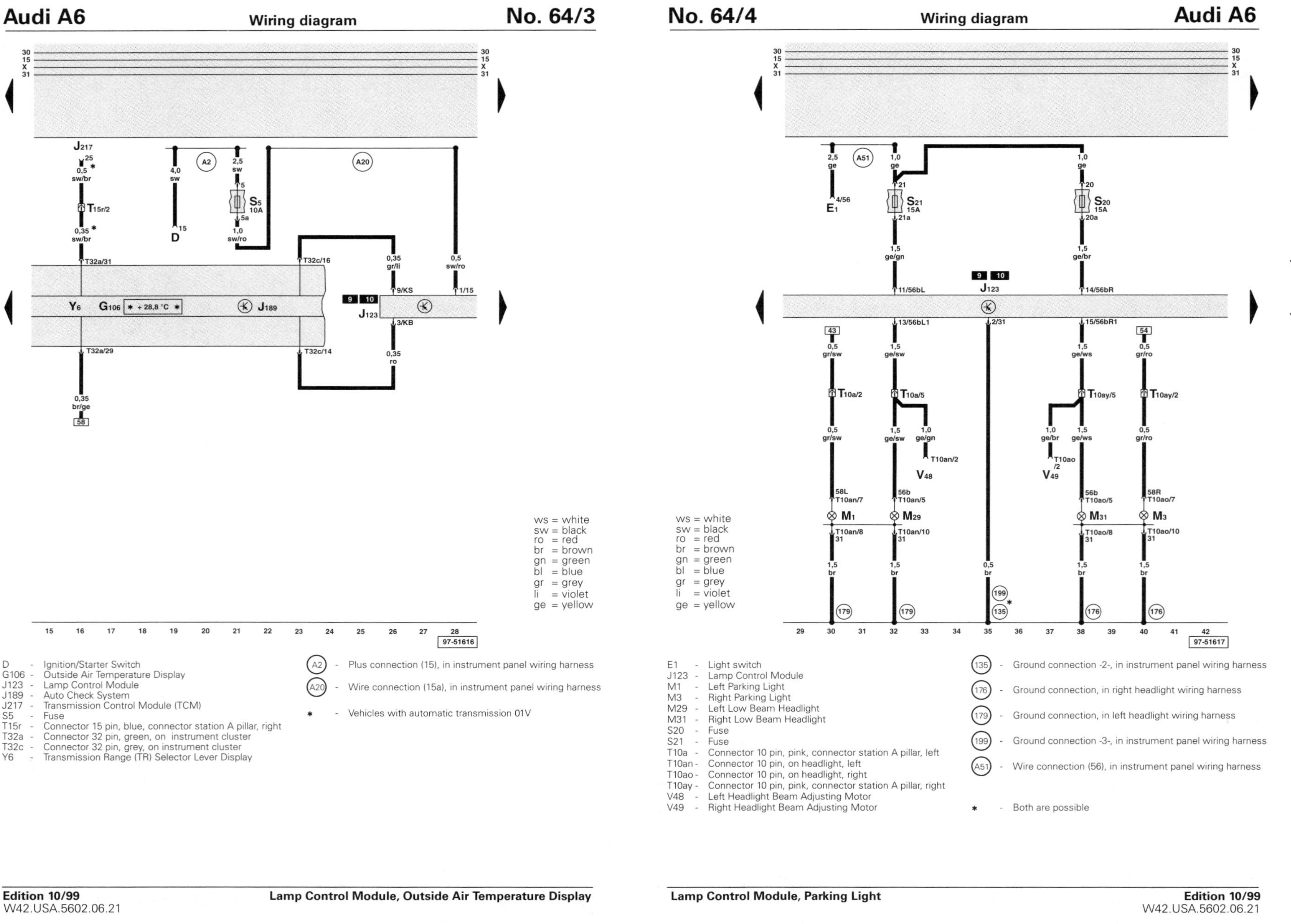

BentleyPublishers.com—All Rights Reserved

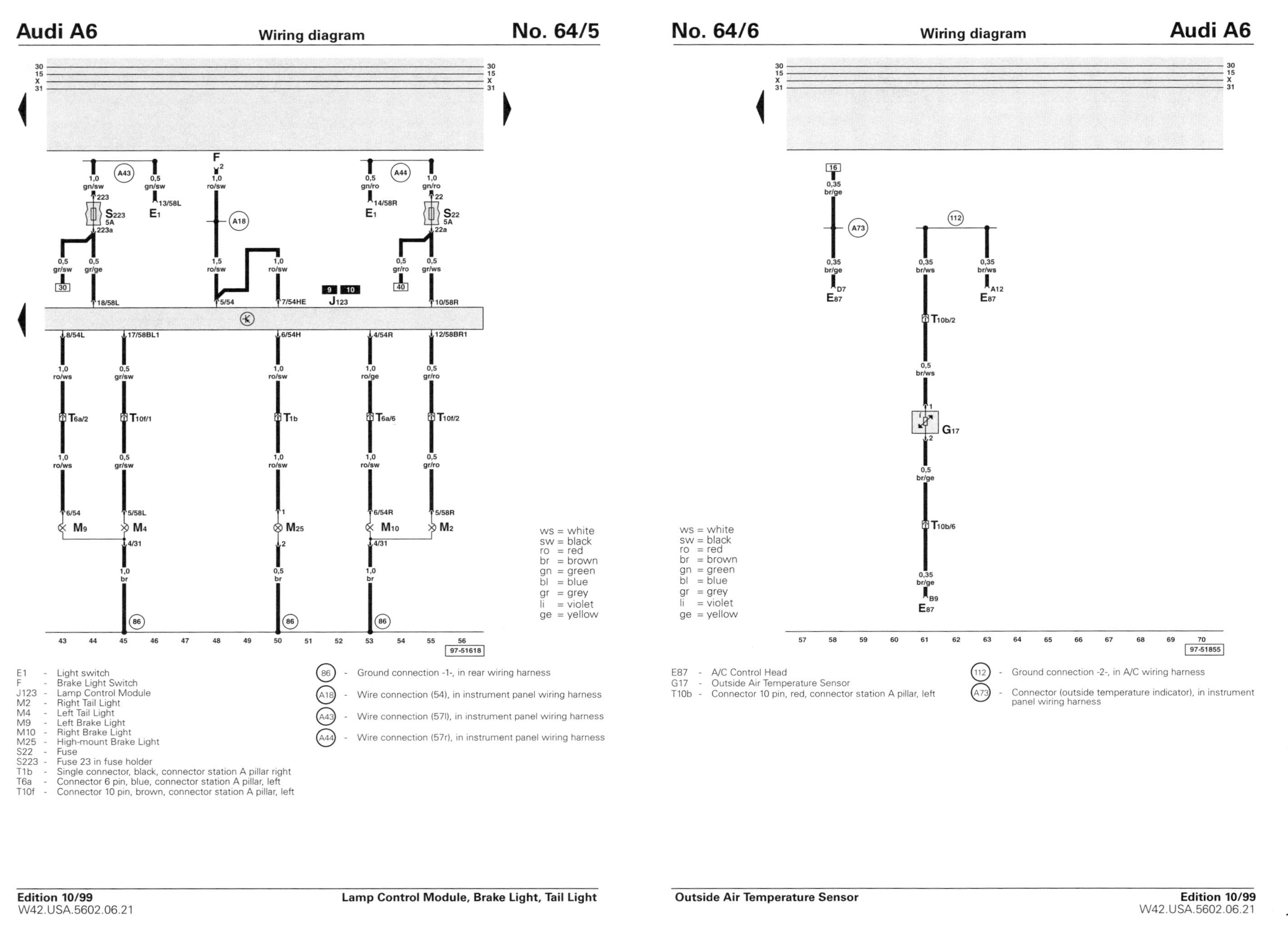

BentleyPublishers.com—All Rights Reserved

Daytime Running Lights with Fog Light and Rear Fog Light (Canada)

2000 m. y.

Fuse Panel

A97-0119

Fuse Colors:

30 A - Green
25 A - White
20 A - Yellow
15 A - Blue
10 A - Red
7,5 A - Brown
5 A - Beige

Starting with fuse position 23, fuses in the fuse holder are identified with 223 in the wiring diagram.

Micro Central Electric Panel

A B C

A97-0110

Relay Location:

2 -Load Reduction Relay, J59

13 - Fold Relay Panel

A97-0024

Relay Location:

6 -Fog Light Relay, J5
9 -Lamp Control Module, J123
10 -Lamp Control Module, J123
17 -Fog Light Circuit Breaker, S28

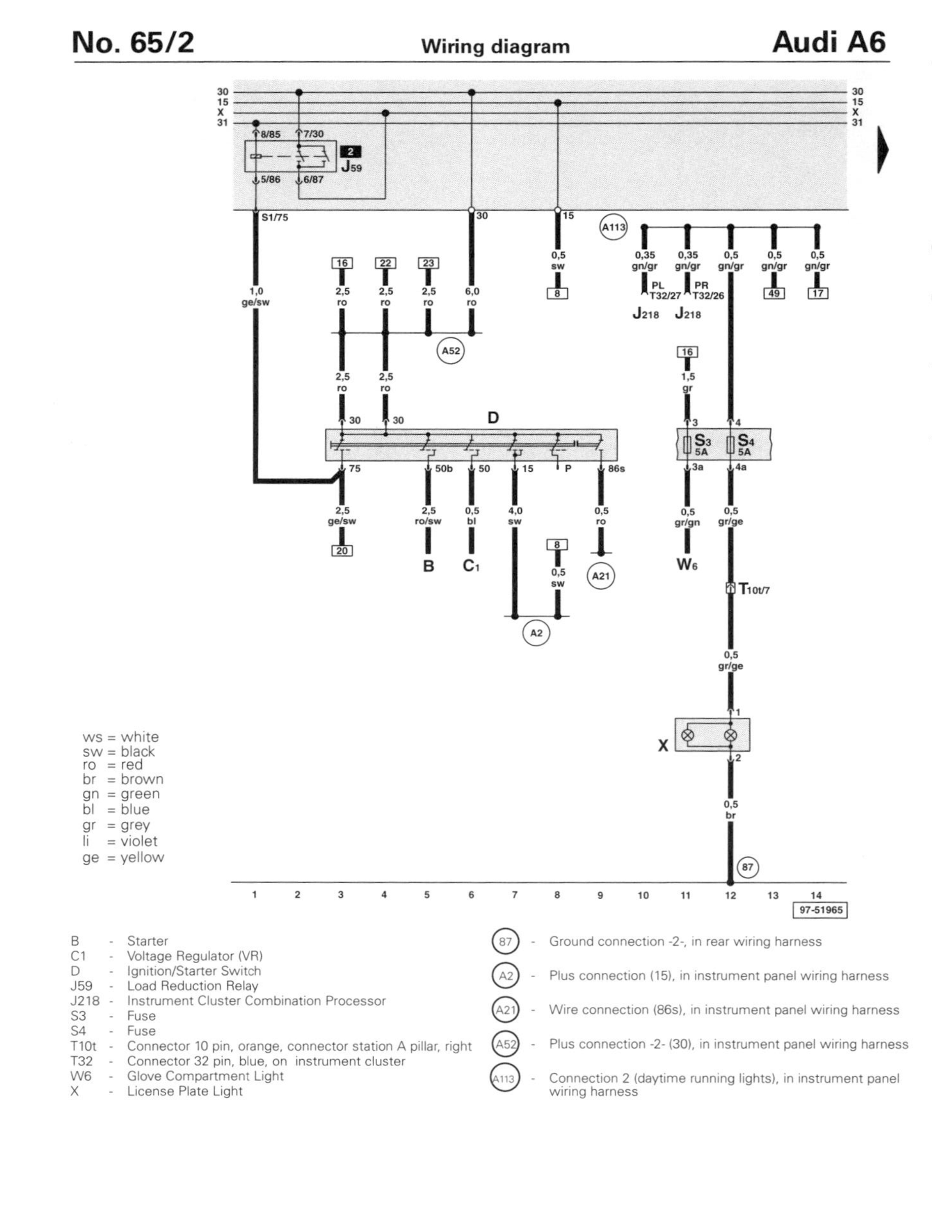

ws = white
sw = black
ro = red
br = brown
gn = green
bl = blue
gr = grey
li = violet
ge = yellow

B - Starter
C1 - Voltage Regulator (VR)
D - Ignition/Starter Switch
J59 - Load Reduction Relay
J218 - Instrument Cluster Combination Processor
S3 - Fuse
S4 - Fuse
T10t - Connector 10 pin, orange, connector station A pillar, right
T32 - Connector 32 pin, blue, on instrument cluster
W6 - Glove Compartment Light
X - License Plate Light

(87) - Ground connection -2-, in rear wiring harness
(A2) - Plus connection (15), in instrument panel wiring harness
(A21) - Wire connection (86s), in instrument panel wiring harness
(A52) - Plus connection -2- (30), in instrument panel wiring harness
(A113) - Connection 2 (daytime running lights), in instrument panel wiring harness

Ignition/Starter Switch, License Plate Light, Load Reduction Relay

BentleyPublishers.com—All Rights Reserved

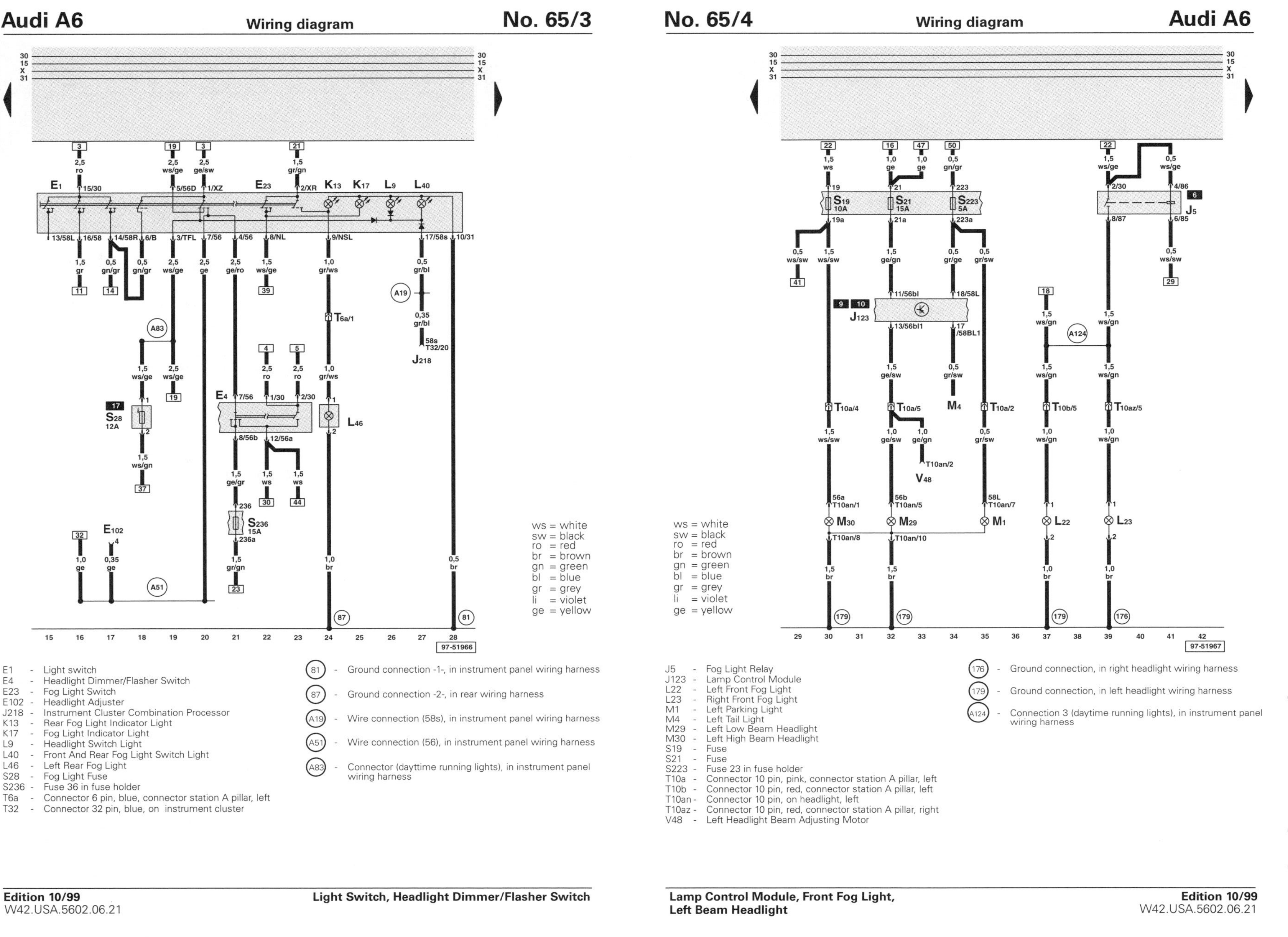

BentleyPublishers.com—All Rights Reserved

THIS SPACE INTENTIONALLY
LEFT BLANK

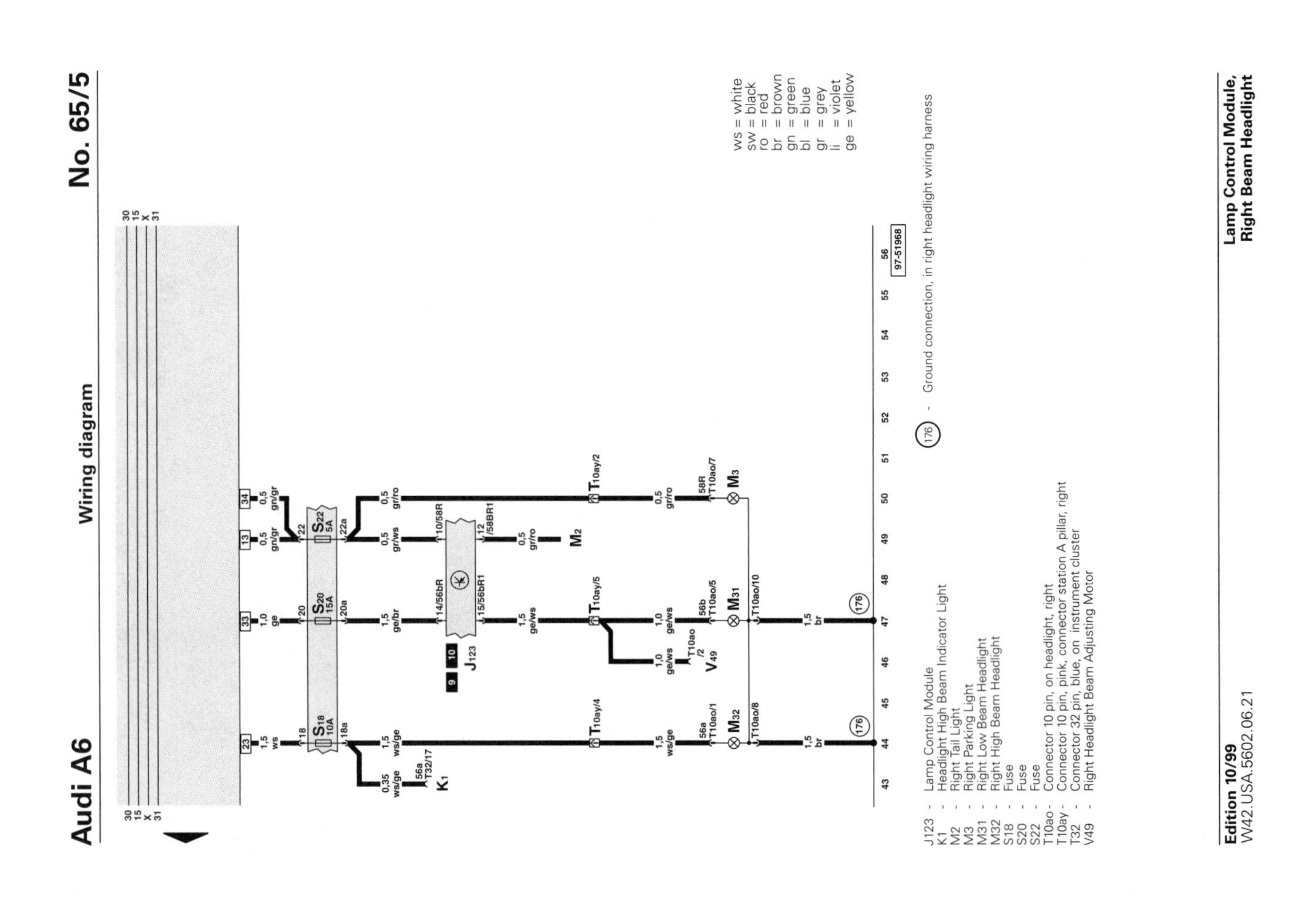

BentleyPublishers.com—All Rights Reserved

Headlight with High Intensity Gas discharge Lamps and automatic Headlight Beam Adjusting

2000 m. y.

Fuse Panel

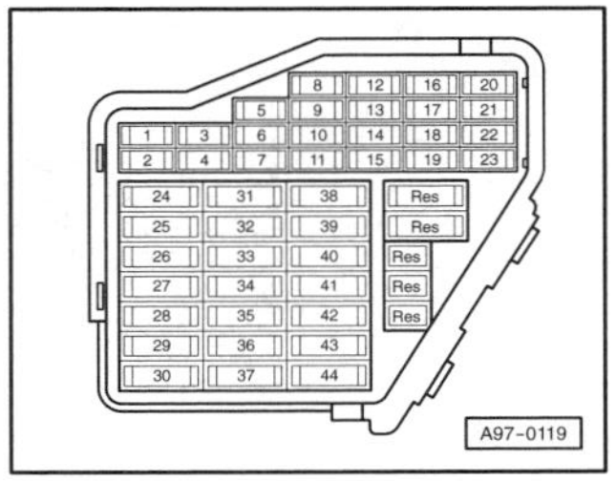

A97-0119

Fuse Colors:

30 A - Green
25 A - White
20 A - Yellow
15 A - Blue
10 A - Red
7,5 A - Brown
5 A - Beige

Starting with fuse position 23, fuses in the fuse holder are identified with 223 in the wiring diagram.

Micro Central Electric Panel

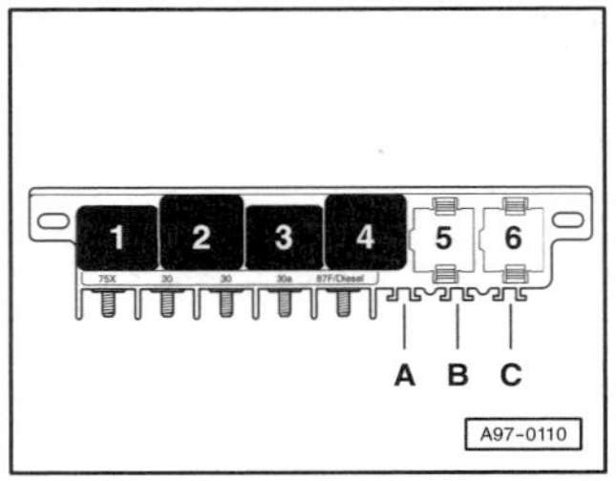

A97-0110

Relay Location:

2 -Load Reduction Relay, J59

13 - Fold Relay Panel

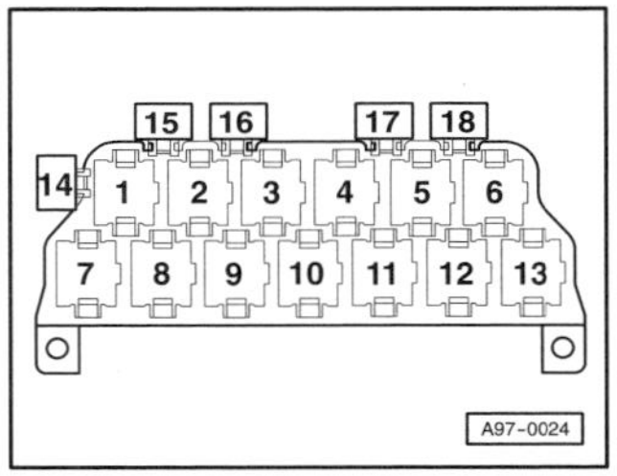

A97-0024

Relay Location:

9 -Lamp Control Module, J123

10 -Lamp Control Module, J123

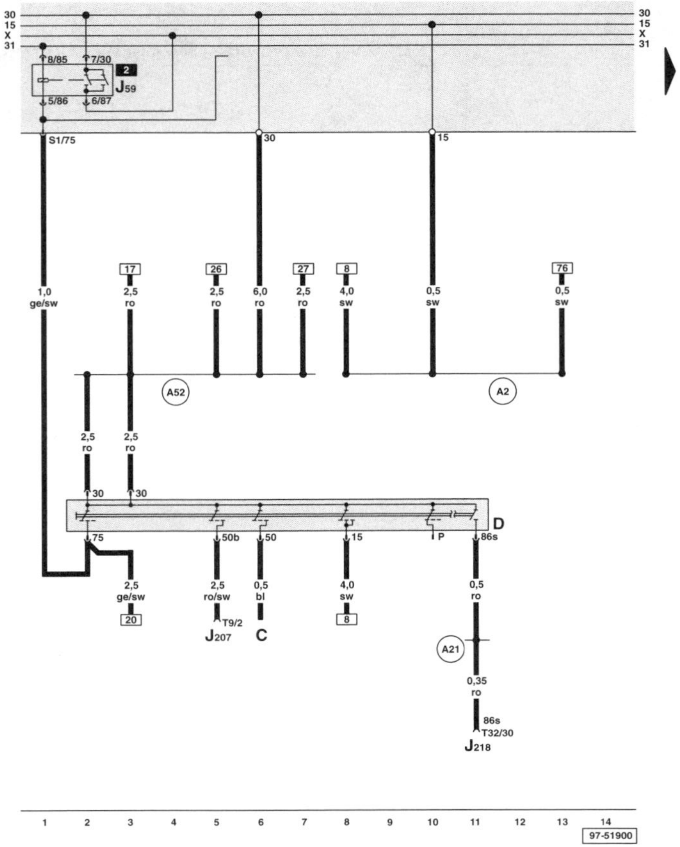

ws = white
sw = black
ro = red
br = brown
gn = green
bl = blue
gr = grey
li = violet
ge = yellow

97-51900

C - Generator (GEN)
D - Ignition/Starter Switch
J59 - Load Reduction Relay
J207 - Starting Interlock Relay
J218 - Instrument Cluster Combination Processor
T9 - Connector 9 pin, brown, on starting interlock relay
T32 - Connector 32 pin, blue, on instrument cluster

A2 - Plus connection (15), in instrument panel wiring harness
A21 - Wire connection (86s), in instrument panel wiring harness
A52 - Plus connection -2- (30), in instrument panel wiring harness

Ignition/Starter Switch

BentleyPublishers.com—All Rights Reserved

Headlights, Xenon (HID) (2000)

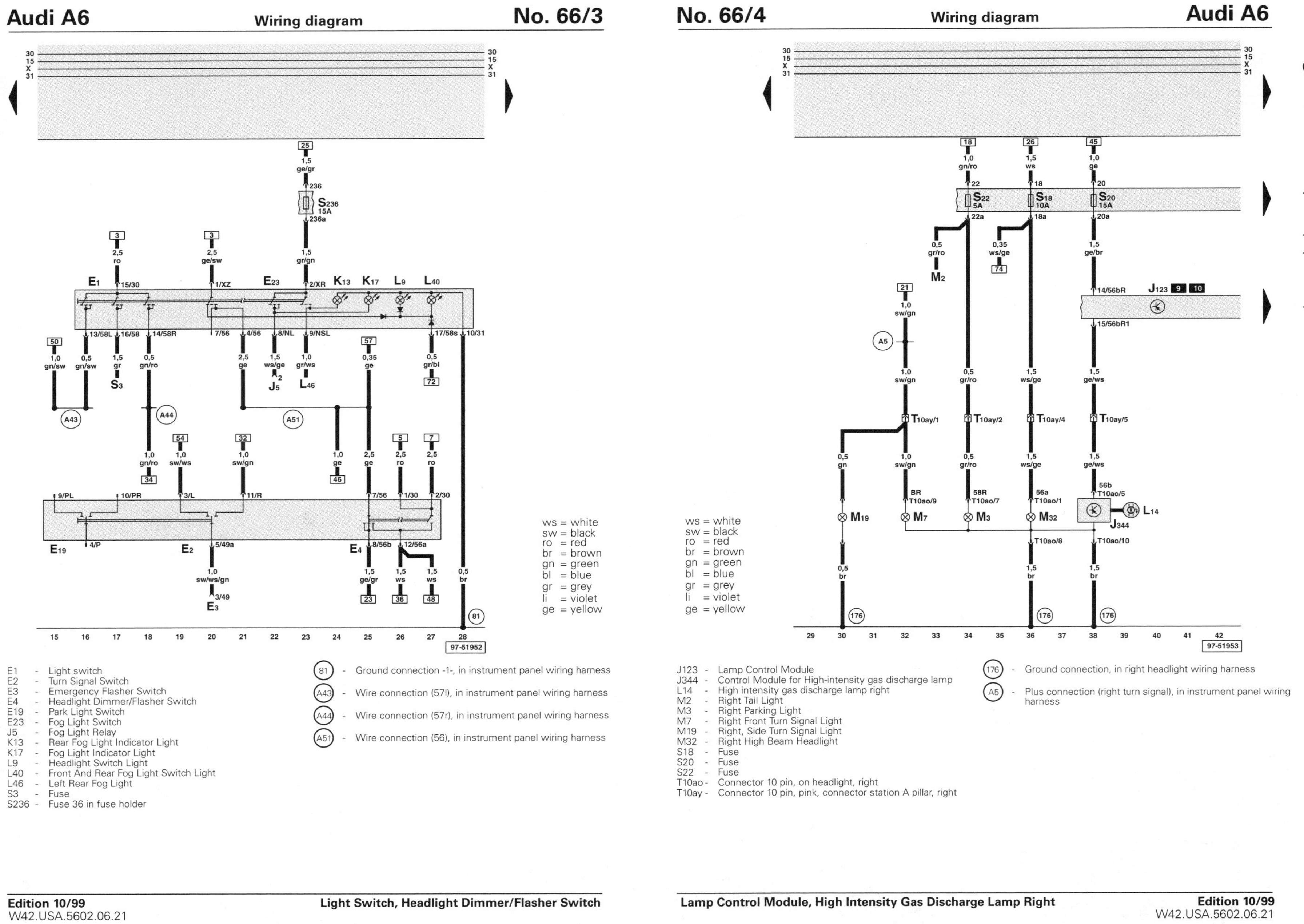

BentleyPublishers.com—All Rights Reserved

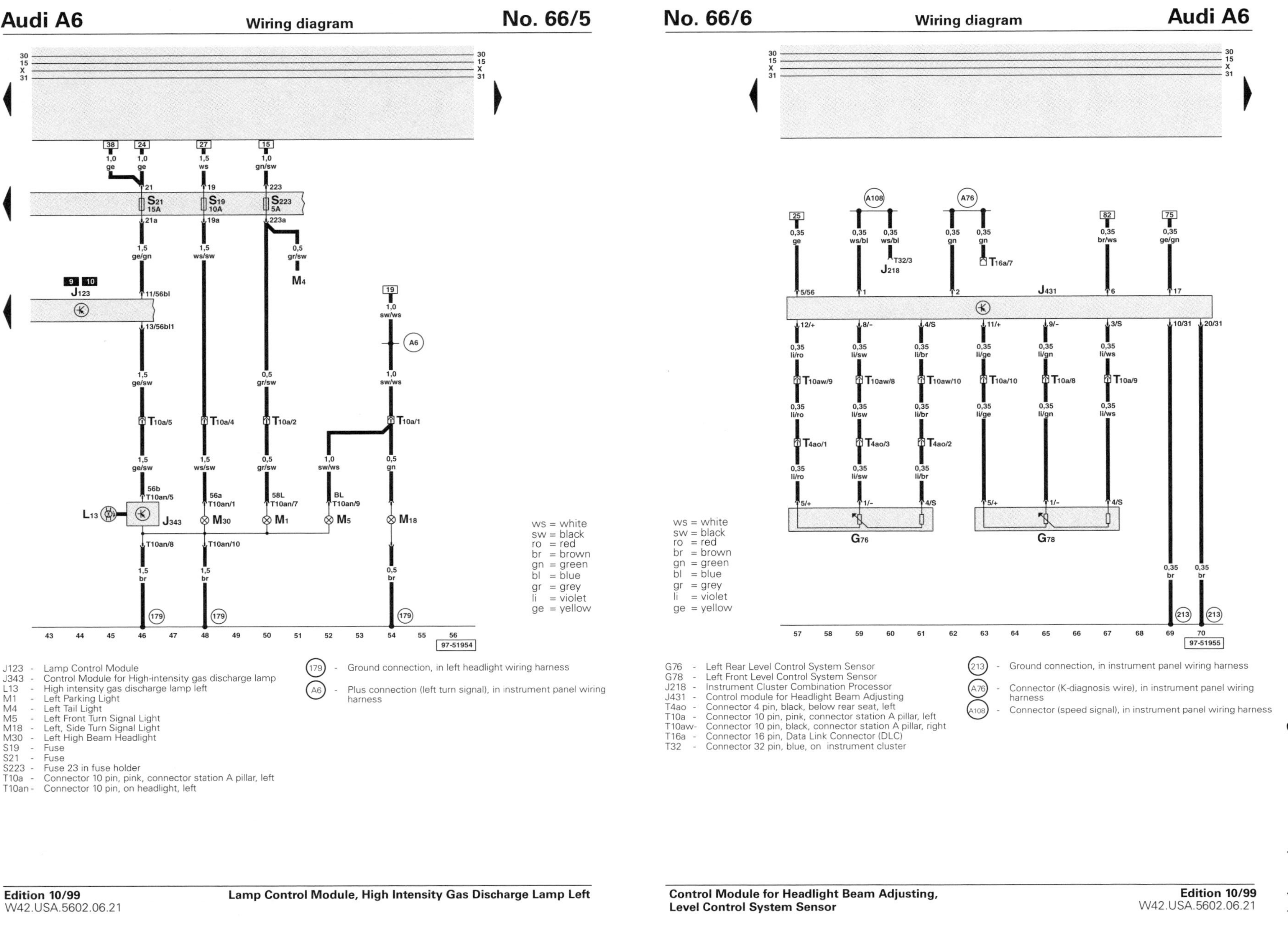

BentleyPublishers.com—All Rights Reserved

Audi A6 Wiring diagram **No. 66/7**

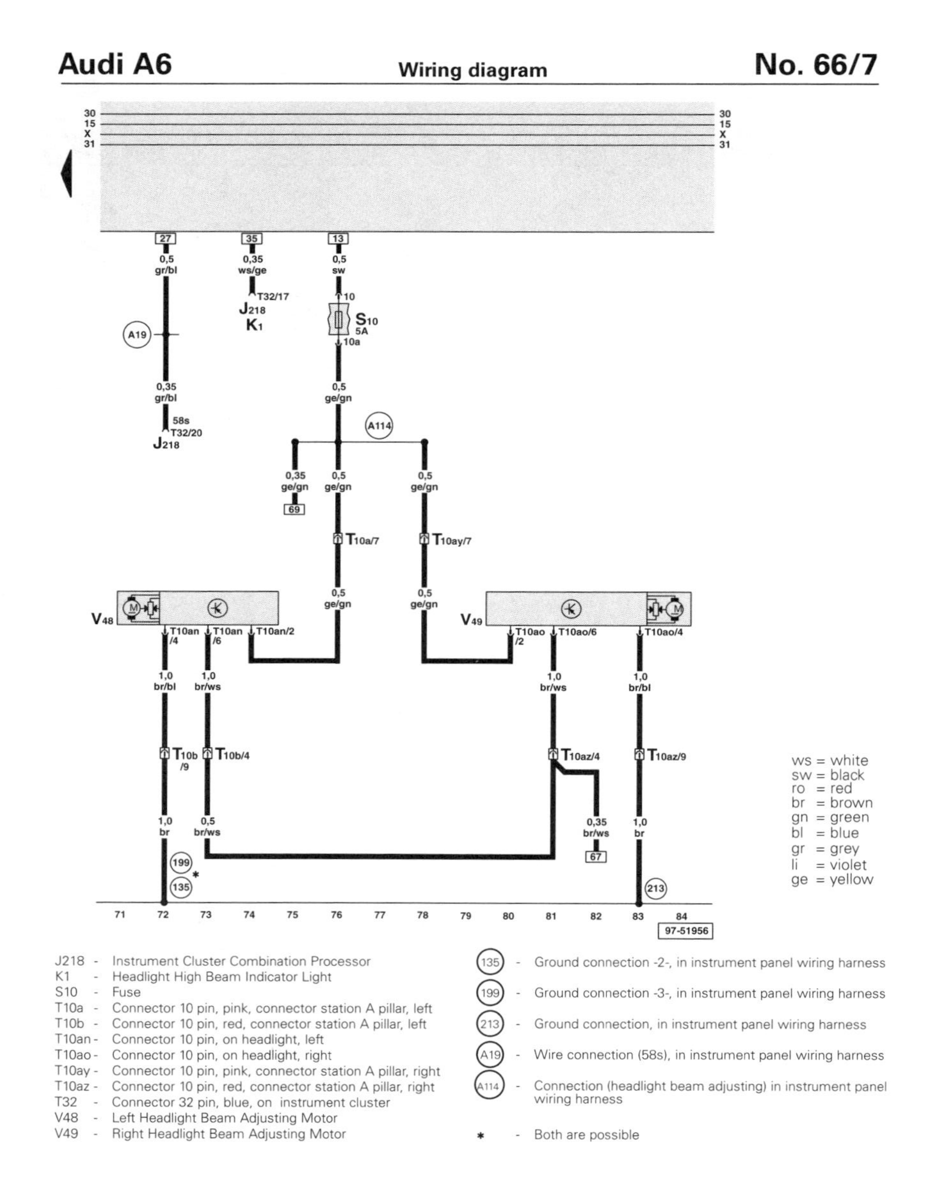

J218 - Instrument Cluster Combination Processor
K1 - Headlight High Beam Indicator Light
S10 - Fuse
T10a - Connector 10 pin, pink, connector station A pillar, left
T10b - Connector 10 pin, red, connector station A pillar, left
T10an - Connector 10 pin, on headlight, left
T10ao - Connector 10 pin, on headlight, right
T10ay - Connector 10 pin, pink, connector station A pillar, right
T10az - Connector 10 pin, red, connector station A pillar, right
T32 - Connector 32 pin, blue, on instrument cluster
V48 - Left Headlight Beam Adjusting Motor
V49 - Right Headlight Beam Adjusting Motor

(135) - Ground connection -2-, in instrument panel wiring harness
(199) - Ground connection -3-, in instrument panel wiring harness
(213) - Ground connection, in instrument panel wiring harness
(A19) - Wire connection (58s), in instrument panel wiring harness
(A114) - Connection (headlight beam adjusting) in instrument panel wiring harness
* - Both are possible

Edition 10/99
W42.USA.5602.06.21

Left Headlight Beam Adjusting Motor,
Right Headlight Beam Adjusting Motor

THIS SPACE INTENTIONALLY

LEFT BLANK

BentleyPublishers.com—All Rights Reserved

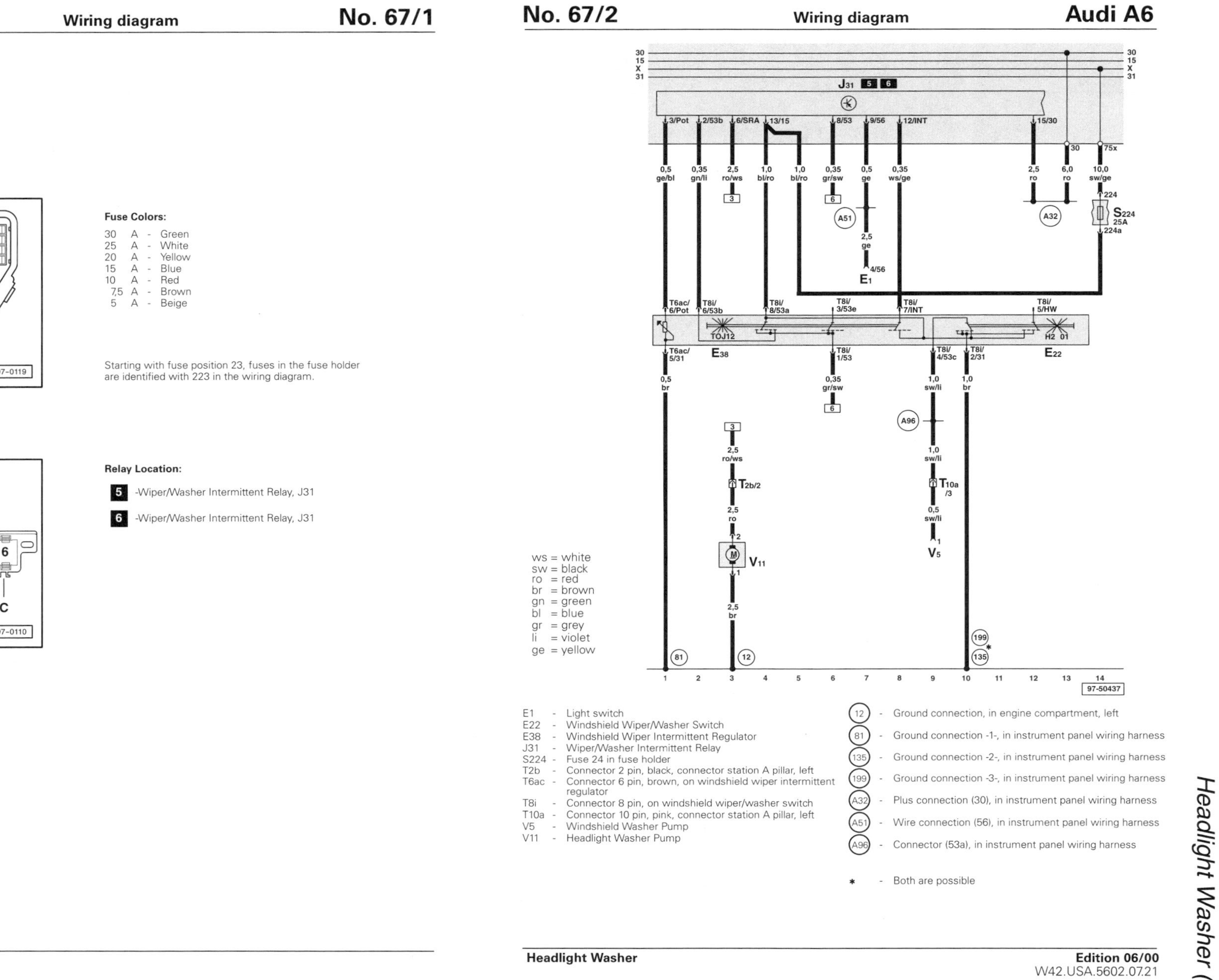

BentleyPublishers.com—All Rights Reserved

Power Windows

2000 m. y.

Fuse Panel

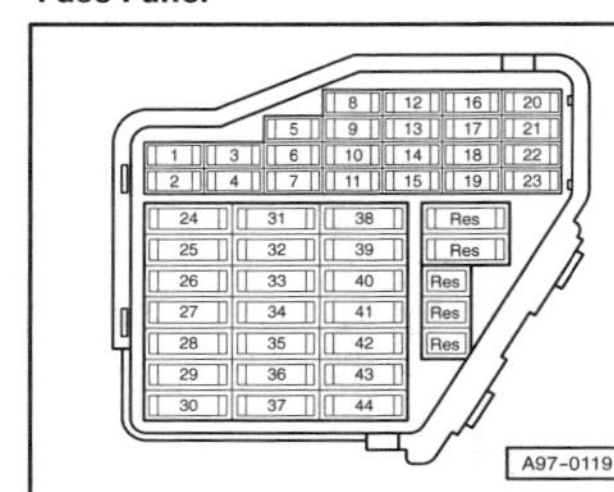

Fuse Colors:

30 A - Green
25 A - White
20 A - Yellow
15 A - Blue
10 A - Red
7,5 A - Brown
5 A - Beige

Starting with fuse position 23, fuses in the fuse holder are identified with 223 in the wiring diagram.

Micro Central Electric Panel

A B C
A97-0110

8 - Fold Relay Panel

A97-0026

Fuse Location:

16 -Power Window Circuit Breaker Front, S37

17 -Power Window Circuit Breaker Rear, S43

ws = white
sw = black
ro = red
br = brown
gn = green
bl = blue
gr = grey
li = violet
ge = yellow

E40 - Left Front Window Switch
J295 - Left Front Power Window Control Module
L108 - Door opener lighting, driver side
S37 - Power Window Circuit Breaker Front
S43 - Power Window Circuit Breaker Rear
T3n - Connector 3 pin, red, connector station A pillar, left
T5b - Connector 5 pin, black, in driver's door
T10ag - Connector 10 pin, red, connector station A pillar, right
T10ah - Connector 10 pin, red, in B pillar, left
T10ai - Connector 10 pin, red, in B pillar, right
T15k - Connector 15 pin, red, connector station A pillar, left
V14 - Left Window Motor

(44) - Ground connection, on left A-pillar, lower part

(89) - Ground connection -1-, in power window wiring harness

(205) - Ground connection, in driver's door wiring harness

(Q1) - Plus connection (30), in power window wiring harness

97-51873

Left Front Power Window Control Module,
Left Front Window Switch

BentleyPublishers.com—All Rights Reserved

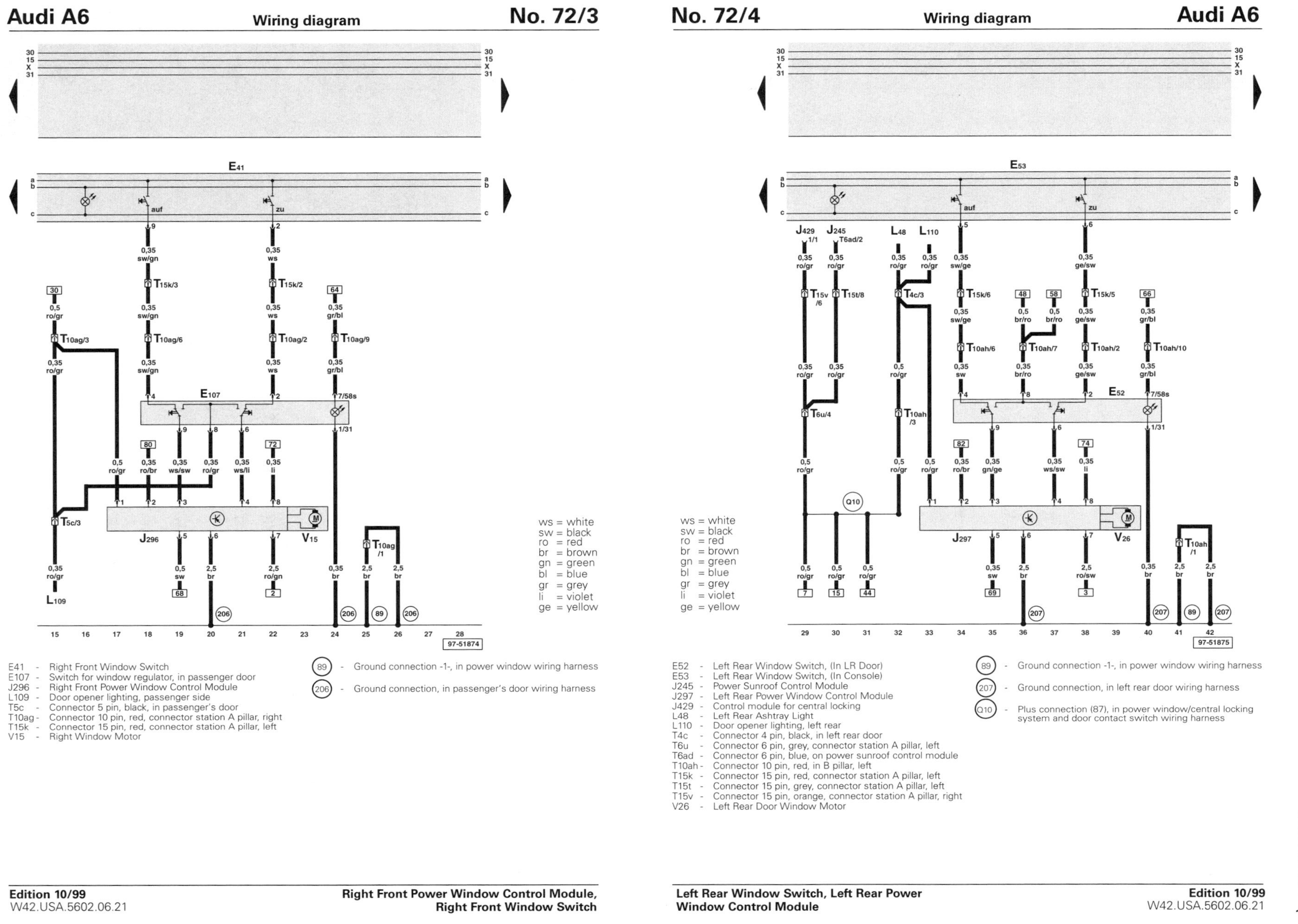

BentleyPublishers.com—All Rights Reserved

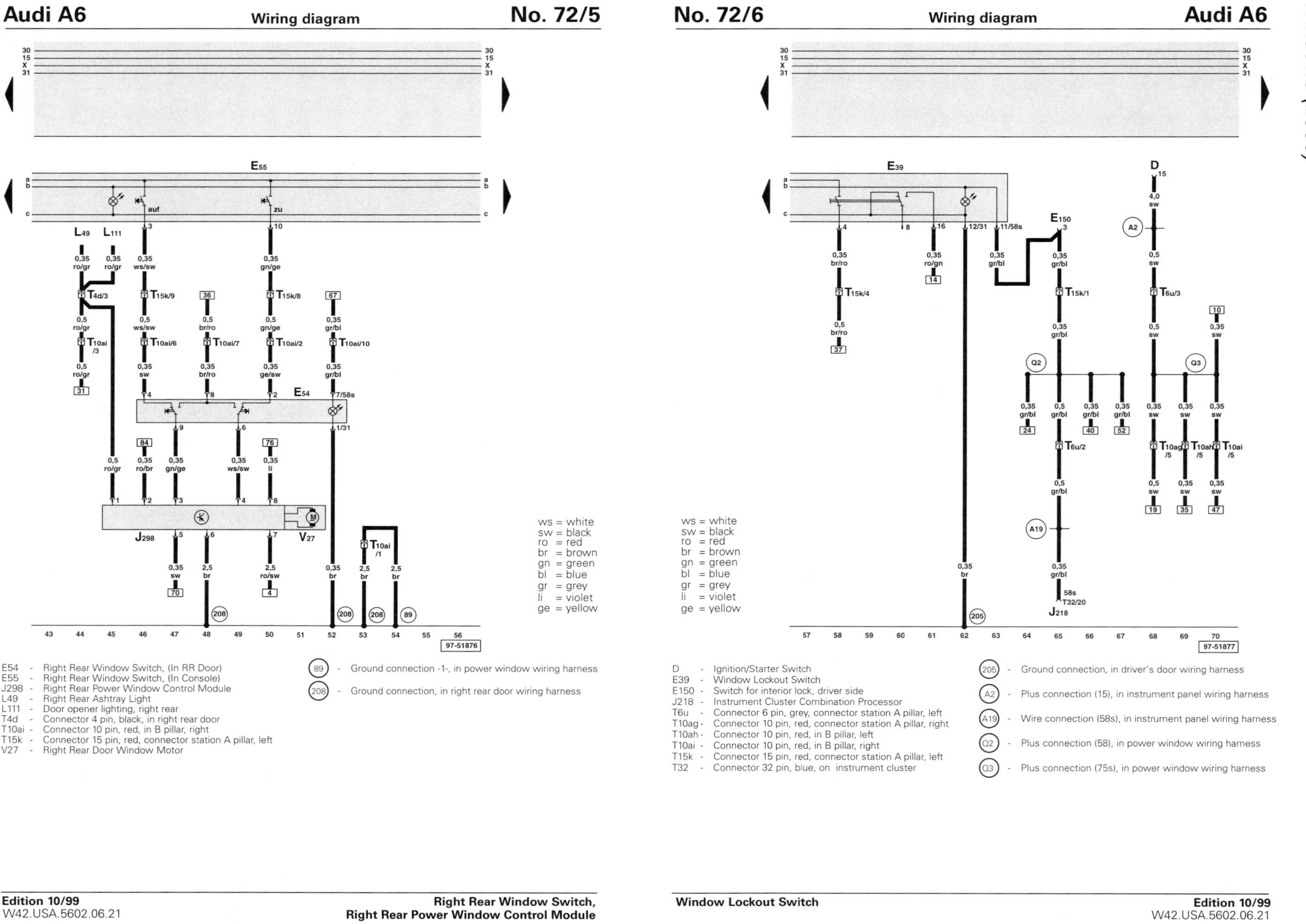

BentleyPublishers.com—All Rights Reserved

Audi A6 **Wiring diagram** **No. 72/7**

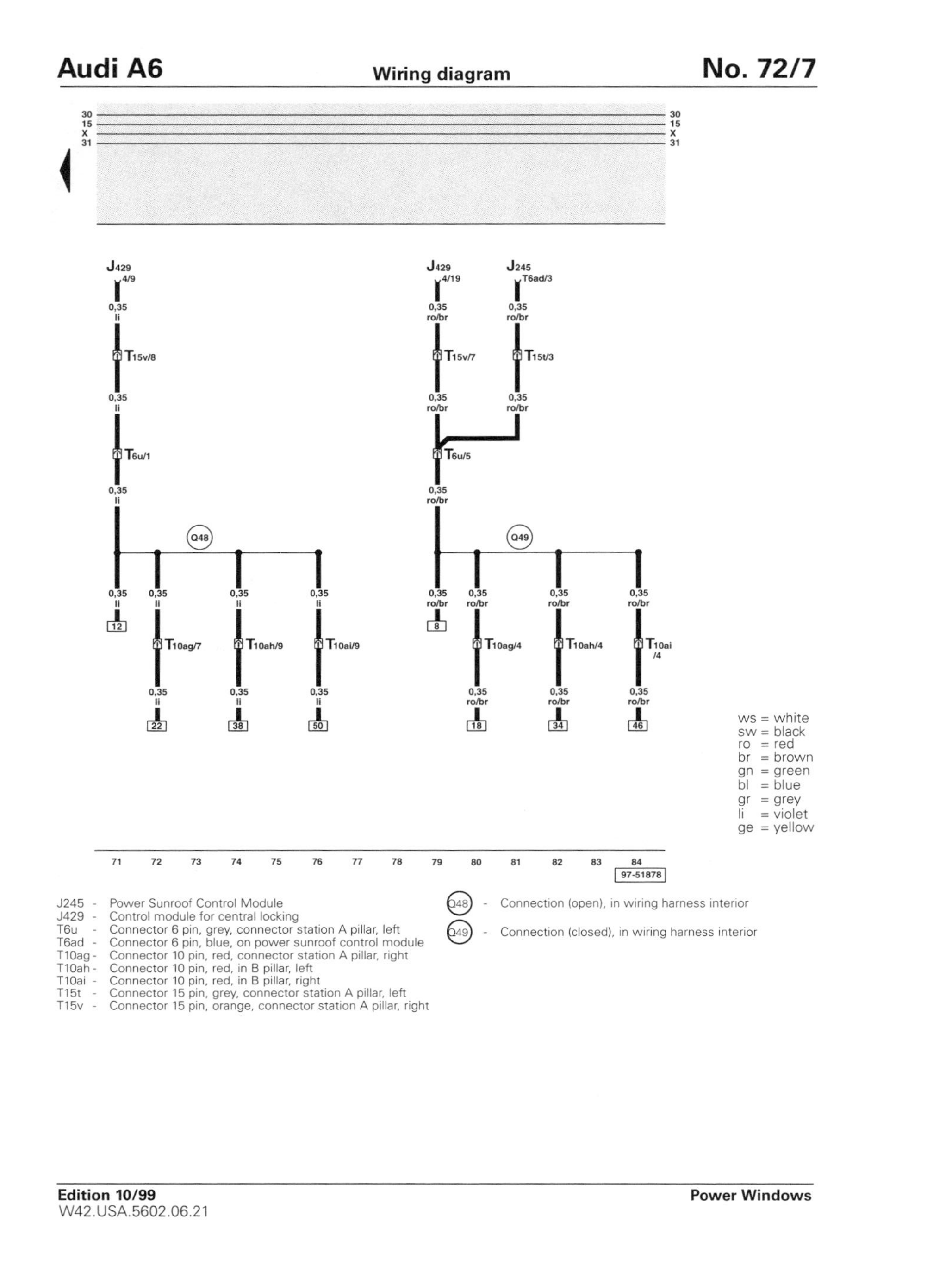

ws = white
sw = black
ro = red
br = brown
gn = green
bl = blue
gr = grey
li = violet
ge = yellow

J245 - Power Sunroof Control Module
J429 - Control module for central locking
T6u - Connector 6 pin, grey, connector station A pillar, left
T6ad - Connector 6 pin, blue, on power sunroof control module
T10ag - Connector 10 pin, red, connector station A pillar, right
T10ah - Connector 10 pin, red, in B pillar, left
T10ai - Connector 10 pin, red, in B pillar, right
T15t - Connector 15 pin, grey, connector station A pillar, left
T15v - Connector 15 pin, orange, connector station A pillar, right

Q48 - Connection (open), in wiring harness interior
Q49 - Connection (closed), in wiring harness interior

THIS SPACE INTENTIONALLY LEFT BLANK

Edition 10/99 **Power Windows**
W42.USA.5602.06.21

BentleyPublishers.com—All Rights Reserved

Power Sunroof with Automatic Preselection

2000 m. y.

Fuse Panel

A97-0119

Fuse Colors:

30 A - Green
25 A - White
20 A - Yellow
15 A - Blue
10 A - Red
7,5 A - Brown
5 A - Beige

Starting with fuse position 23, fuses in the fuse holder are identified with 223 in the wiring diagram.

Micro Central Electric Panel

A B C

A97-0110

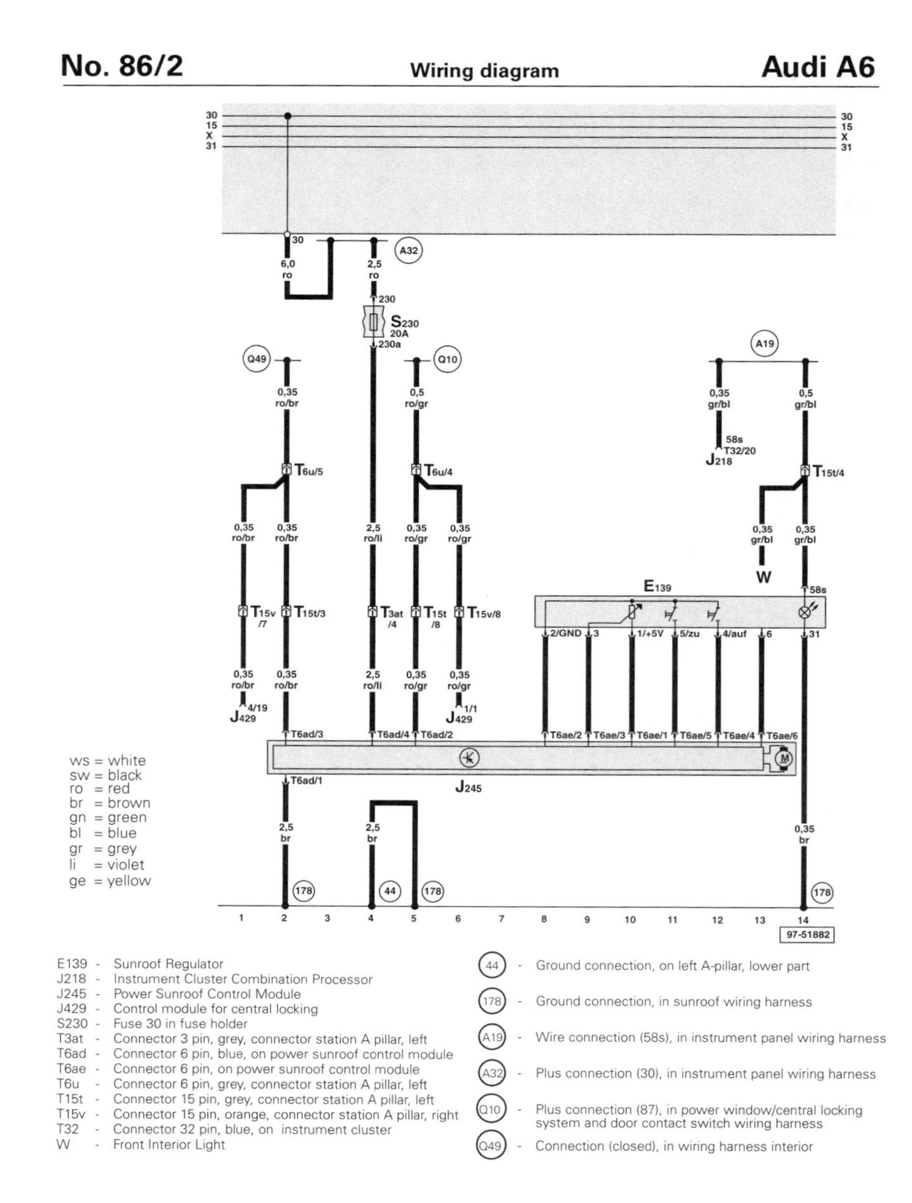

ws = white
sw = black
ro = red
br = brown
gn = green
bl = blue
gr = grey
li = violet
ge = yellow

E139 - Sunroof Regulator
J218 - Instrument Cluster Combination Processor
J245 - Power Sunroof Control Module
J429 - Control module for central locking
S230 - Fuse 30 in fuse holder
T3at - Connector 3 pin, grey, connector station A pillar, left
T6ad - Connector 6 pin, blue, on power sunroof control module
T6ae - Connector 6 pin, on power sunroof control module
T6u - Connector 6 pin, grey, connector station A pillar, left
T15t - Connector 15 pin, grey, connector station A pillar, left
T15v - Connector 15 pin, orange, connector station A pillar, right
T32 - Connector 32 pin, blue, on instrument cluster
W - Front Interior Light

(44) - Ground connection, on left A-pillar, lower part
(178) - Ground connection, in sunroof wiring harness
(A19) - Wire connection (58s), in instrument panel wiring harness
(A32) - Plus connection (30), in instrument panel wiring harness
(Q10) - Plus connection (87), in power window/central locking system and door contact switch wiring harness
(Q49) - Connection (closed), in wiring harness interior

BentleyPublishers.com—All Rights Reserved

Standard Equipment

2001 m. y.

Fuse Panel

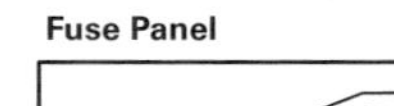
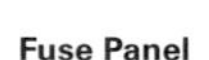
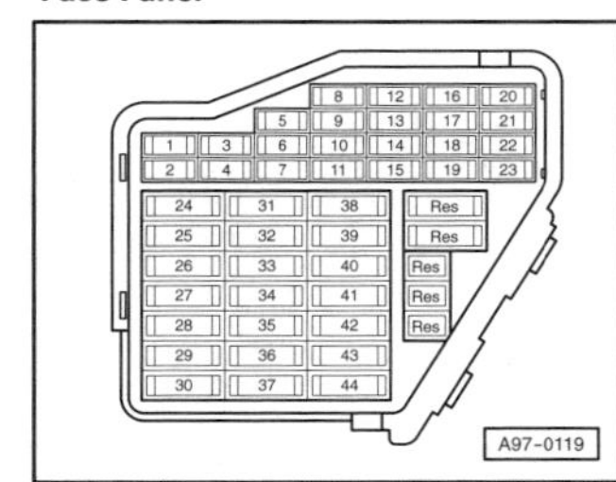

Fuse Colors:

30 A - Green
25 A - White
20 A - Yellow
15 A - Blue
10 A - Red
7.5 A - Brown
5 A - Beige

Starting with fuse position 23, fuses in the fuse holder are identified with 223 in the wiring diagram.

Micro Central Electric Panel

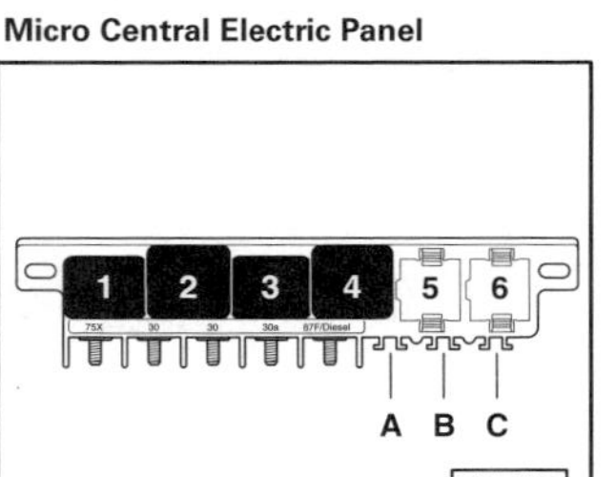

Relay Location:

- **1** - Dual Horn Relay, J4
- **2** - Load Reduction Relay, J59
- **4** - Fuel Pump (FP) Relay, J17
- **5** - Wiper/Washer Intermittent Relay, J31
- **6** - Wiper/Washer Intermittent Relay, J31

13 - Fold Relay Panel

A97-0024

Relay Location:

- **4** - Starting Interlock Relay, J207
- **6** - Fog Light Relay, J5

8 - Fold Relay Panel

A97-0026

Fuse Location:

- **15** - Fuse for Luggage Compartment Socket, S184

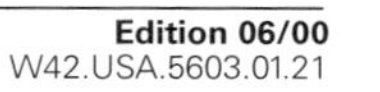

BentleyPublishers.com—All Rights Reserved

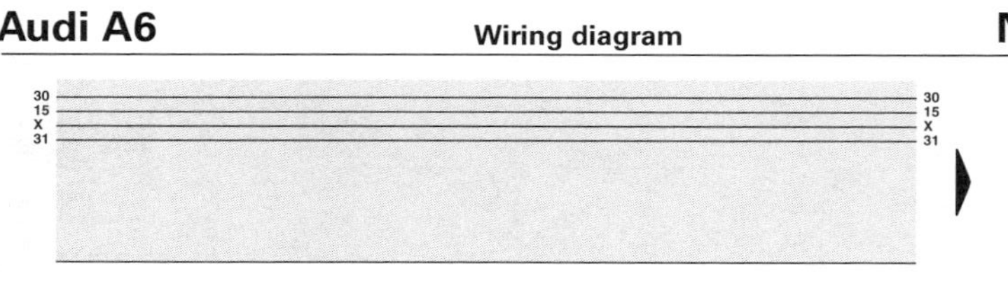

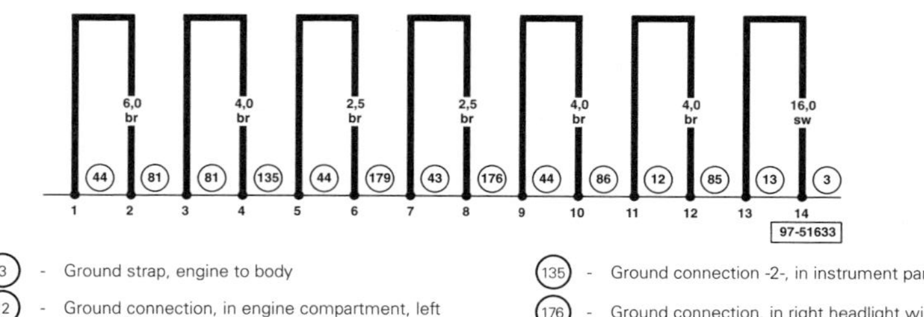

ws = white
sw = black
ro = red
br = brown
gn = green
bl = blue
gr = grey
li = violet
ge = yellow

- (3) - Ground strap, engine to body
- (12) - Ground connection, in engine compartment, left
- (13) - Ground connection, in engine compartment, right
- (43) - Ground connection, on right A-pillar, lower part
- (44) - Ground connection, on left A-pillar, lower part
- (81) - Ground connection -1-, in instrument panel wiring harness
- (85) - Ground connection -1-, in engine compartment wiring harness
- (86) - Ground connection -1-, in rear wiring harness
- (135) - Ground connection -2-, in instrument panel wiring harness
- (176) - Ground connection, in right headlight wiring harness
- (179) - Ground connection, in left headlight wiring harness

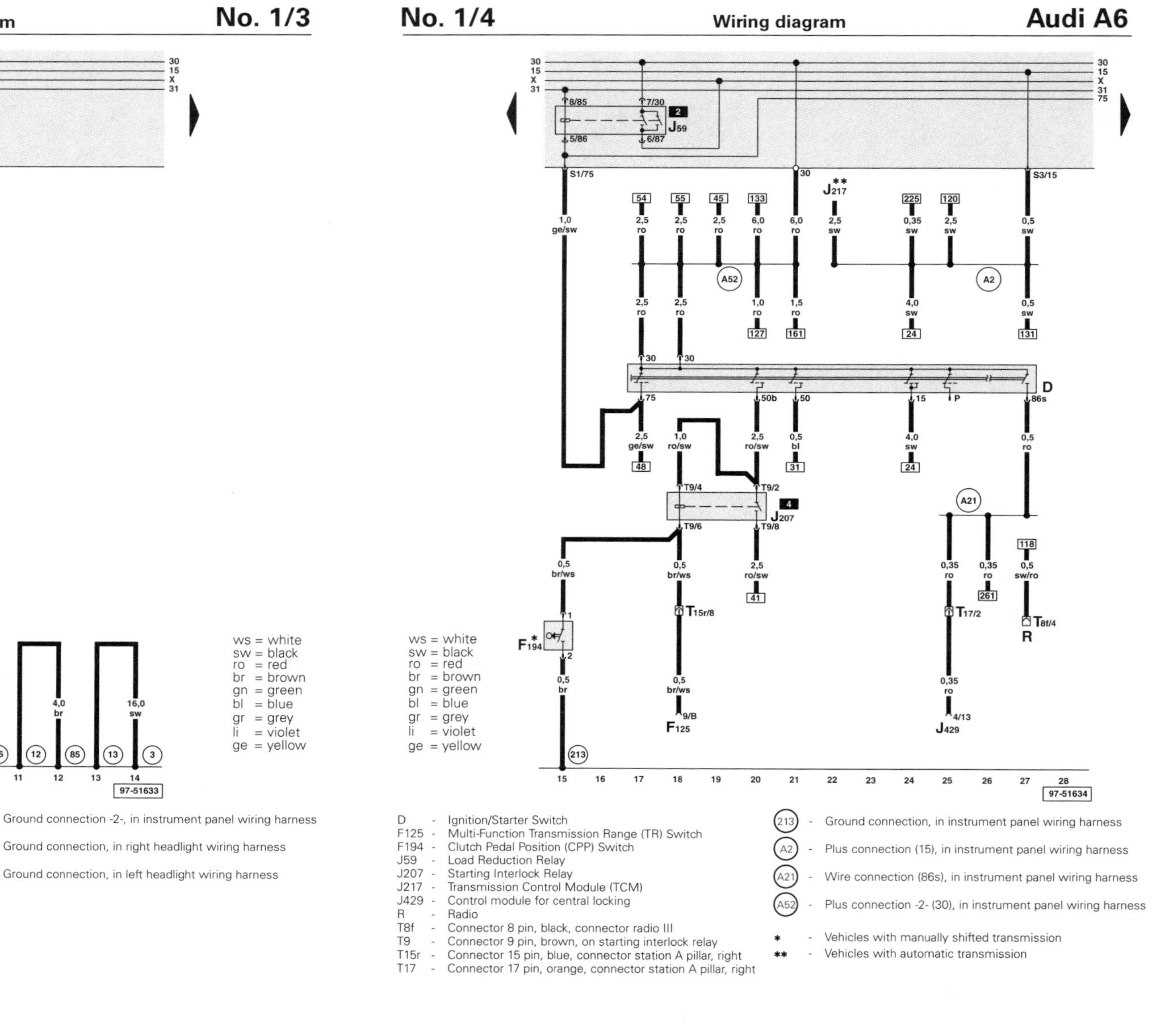

ws = white
sw = black
ro = red
br = brown
gn = green
bl = blue
gr = grey
li = violet
ge = yellow

- D - Ignition/Starter Switch
- F125 - Multi-Function Transmission Range (TR) Switch
- F194 - Clutch Pedal Position (CPP) Switch
- J59 - Load Reduction Relay
- J207 - Starting Interlock Relay
- J217 - Transmission Control Module (TCM)
- J429 - Control module for central locking
- R - Radio
- T8f - Connector 8 pin, black, connector radio III
- T9 - Connector 9 pin, brown, on starting interlock relay
- T15r - Connector 15 pin, blue, connector station A pillar, right
- T17 - Connector 17 pin, orange, connector station A pillar, right

- (213) - Ground connection, in instrument panel wiring harness
- (A2) - Plus connection (15), in instrument panel wiring harness
- (A21) - Wire connection (86s), in instrument panel wiring harness
- (A52) - Plus connection -2- (30), in instrument panel wiring harness

- * - Vehicles with manually shifted transmission
- ** - Vehicles with automatic transmission

BentleyPublishers.com—All Rights Reserved

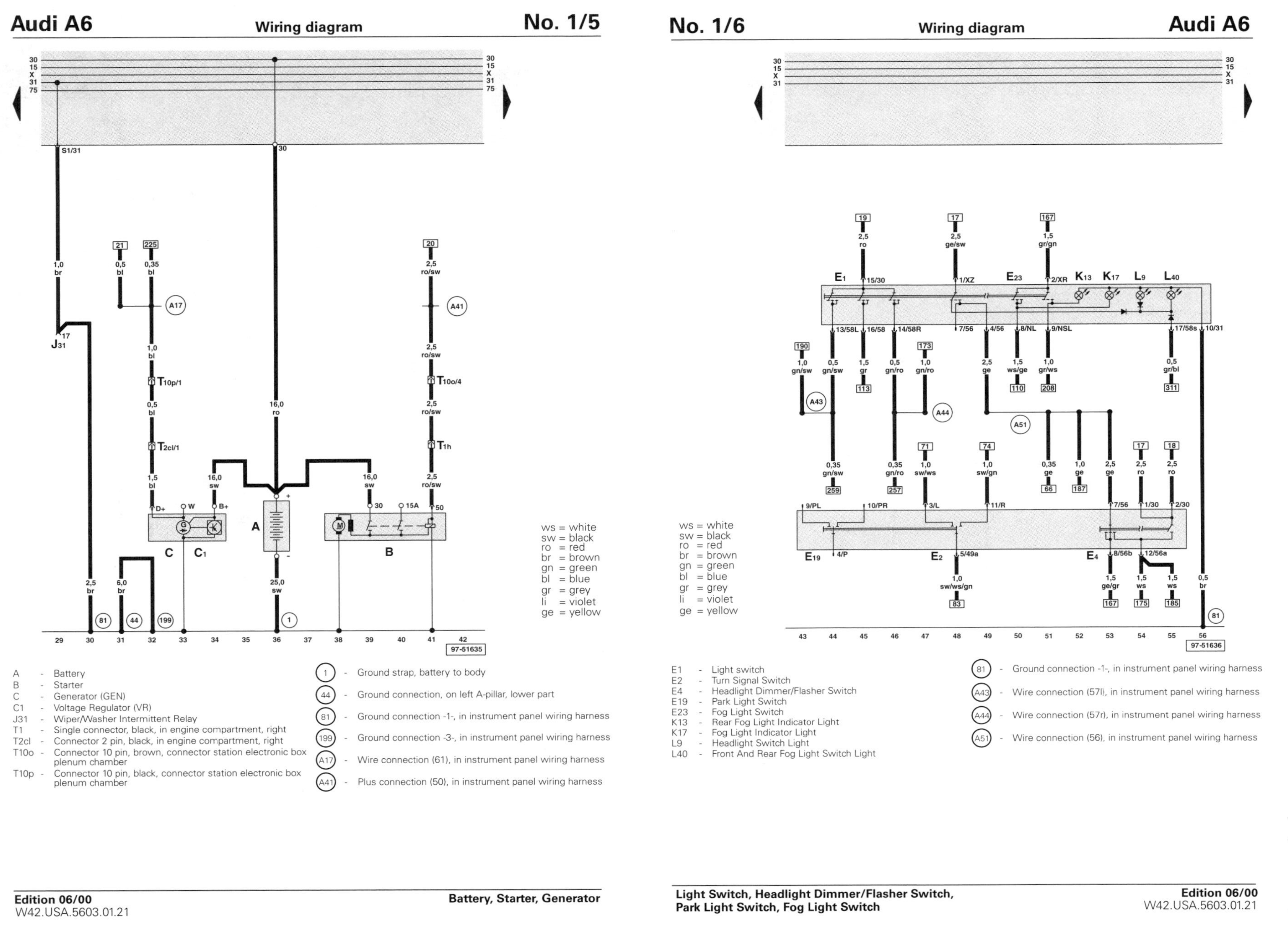

BentleyPublishers.com—All Rights Reserved

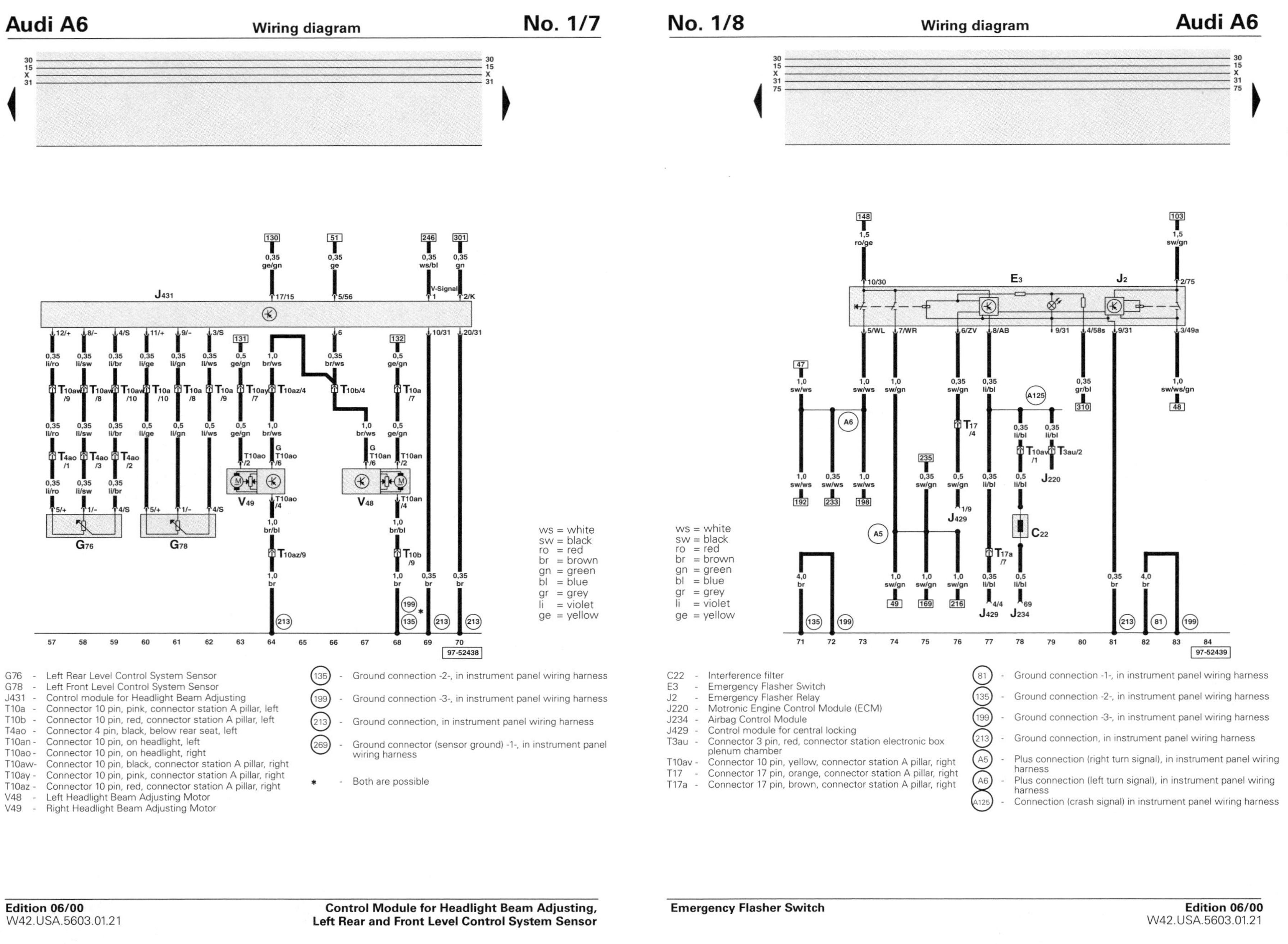

Audi A6 — Wiring diagram — No. 1/7

ws = white
sw = black
ro = red
br = brown
gn = green
bl = blue
gr = grey
li = violet
ge = yellow

G76 - Left Rear Level Control System Sensor
G78 - Left Front Level Control System Sensor
J431 - Control module for Headlight Beam Adjusting
T10a - Connector 10 pin, pink, connector station A pillar, left
T10b - Connector 10 pin, red, connector station A pillar, left
T4ao - Connector 4 pin, black, below rear seat, left
T10an - Connector 10 pin, on headlight, left
T10ao - Connector 10 pin, on headlight, right
T10aw - Connector 10 pin, black, connector station A pillar, right
T10ay - Connector 10 pin, pink, connector station A pillar, right
T10az - Connector 10 pin, red, connector station A pillar, right
V48 - Left Headlight Beam Adjusting Motor
V49 - Right Headlight Beam Adjusting Motor

(135) - Ground connection -2-, in instrument panel wiring harness
(199) - Ground connection -3-, in instrument panel wiring harness
(213) - Ground connection, in instrument panel wiring harness
(269) - Ground connector (sensor ground) -1-, in instrument panel wiring harness
* - Both are possible

Edition 06/00
W42.USA.5603.01.21

Control Module for Headlight Beam Adjusting, Left Rear and Front Level Control System Sensor

No. 1/8 — Wiring diagram — Audi A6

ws = white
sw = black
ro = red
br = brown
gn = green
bl = blue
gr = grey
li = violet
ge = yellow

C22 - Interference filter
E3 - Emergency Flasher Switch
J2 - Emergency Flasher Relay
J220 - Motronic Engine Control Module (ECM)
J234 - Airbag Control Module
J429 - Control module for central locking
T3au - Connector 3 pin, red, connector station electronic box plenum chamber
T10av - Connector 10 pin, yellow, connector station A pillar, right
T17 - Connector 17 pin, orange, connector station A pillar, right
T17a - Connector 17 pin, brown, connector station A pillar, right

(81) - Ground connection -1-, in instrument panel wiring harness
(135) - Ground connection -2-, in instrument panel wiring harness
(199) - Ground connection -3-, in instrument panel wiring harness
(213) - Ground connection, in instrument panel wiring harness
(A5) - Plus connection (right turn signal), in instrument panel wiring harness
(A6) - Plus connection (left turn signal), in instrument panel wiring harness
(A125) - Connection (crash signal) in instrument panel wiring harness

Emergency Flasher Switch

Edition 06/00
W42.USA.5603.01.21

BentleyPublishers.com—All Rights Reserved

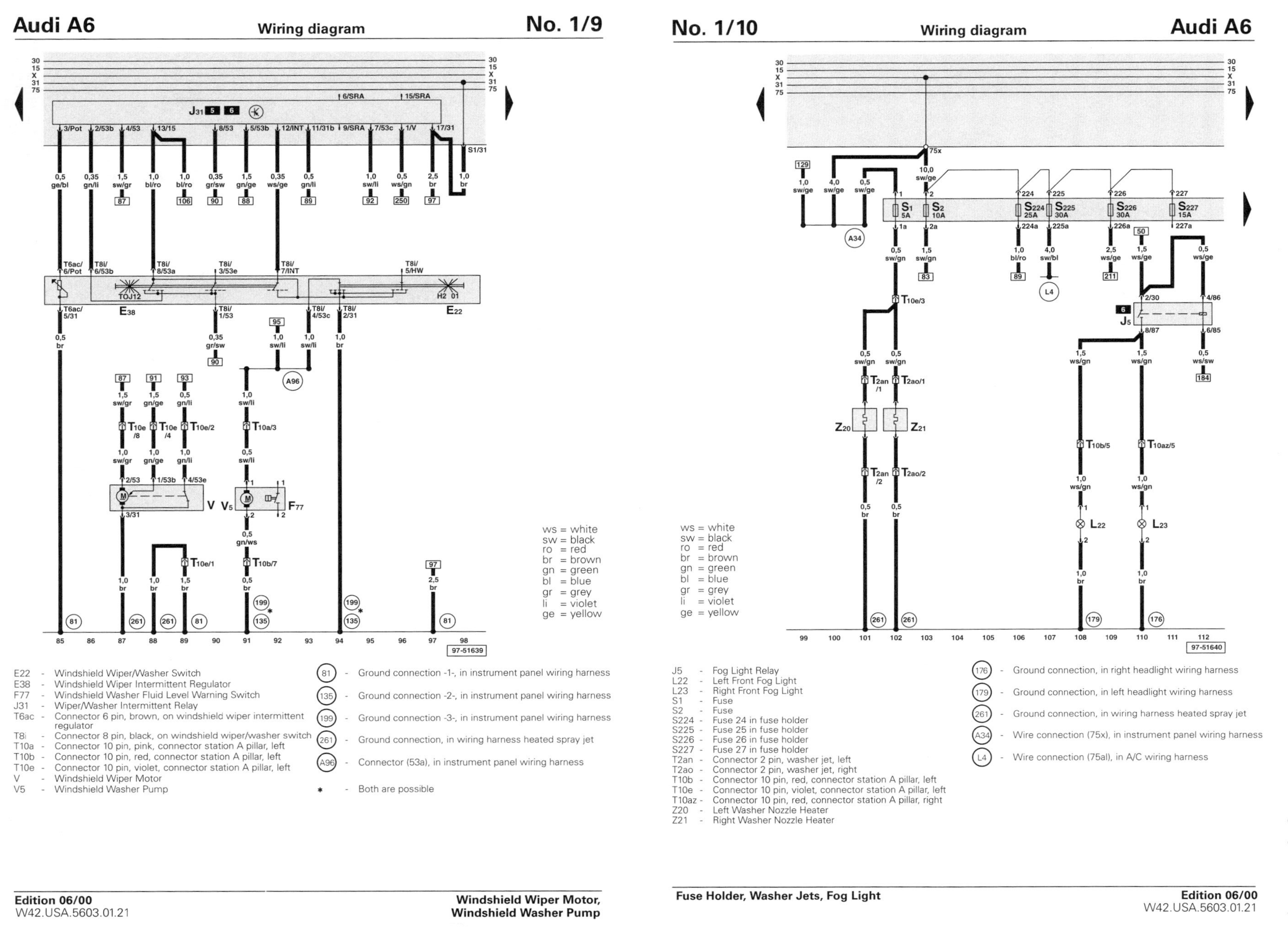

BentleyPublishers.com—All Rights Reserved

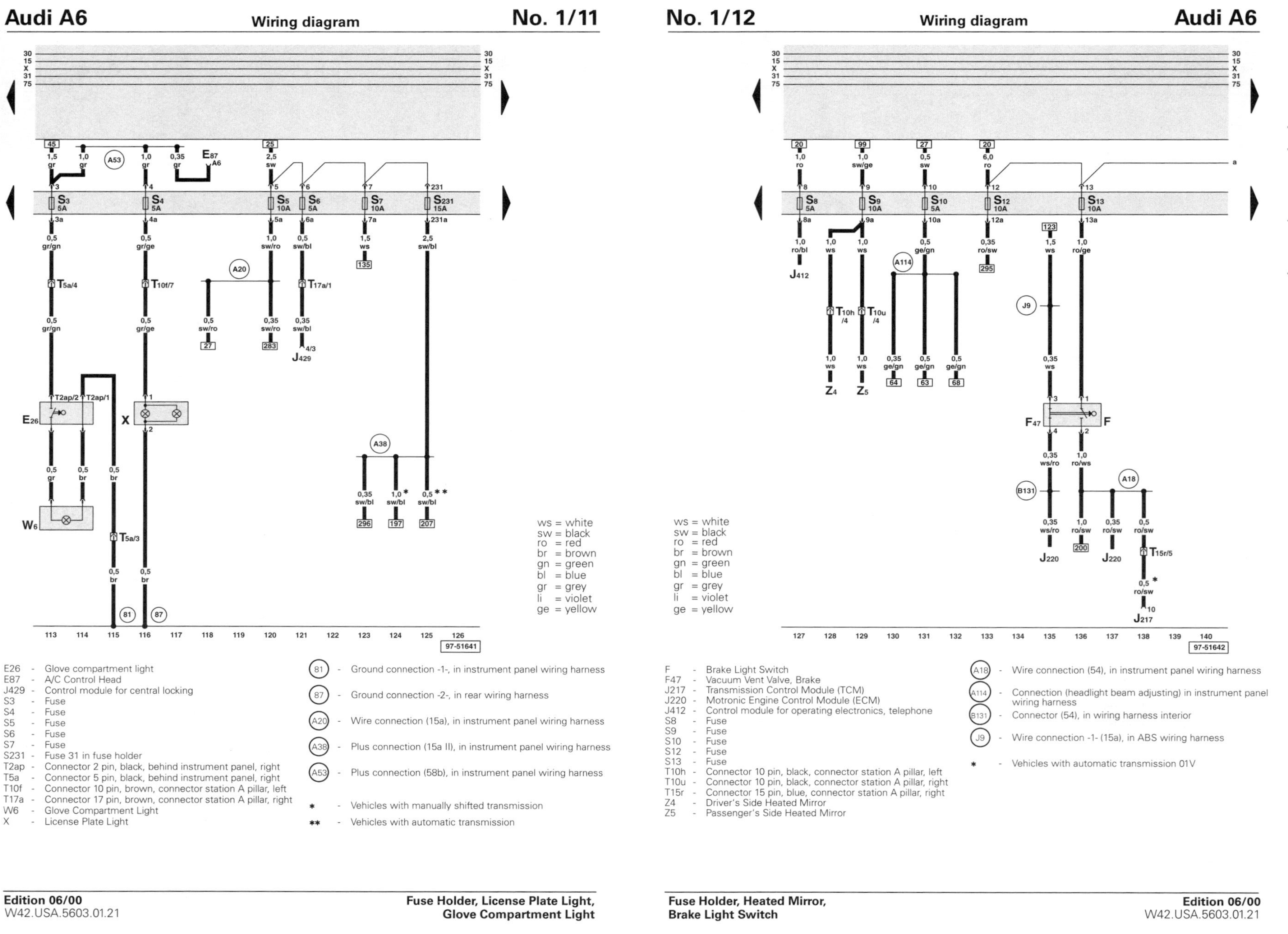

BentleyPublishers.com—All Rights Reserved

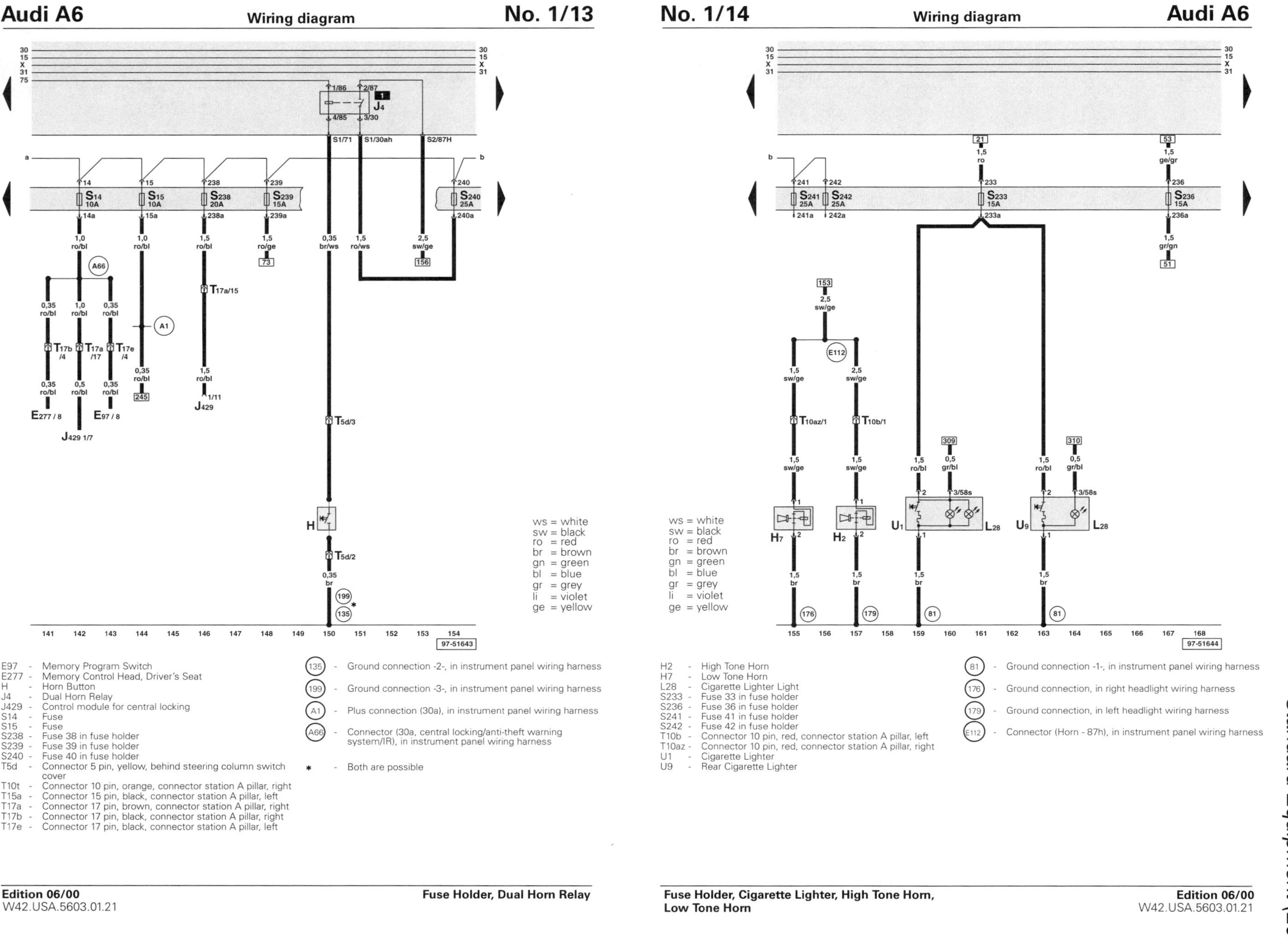

BentleyPublishers.com—All Rights Reserved

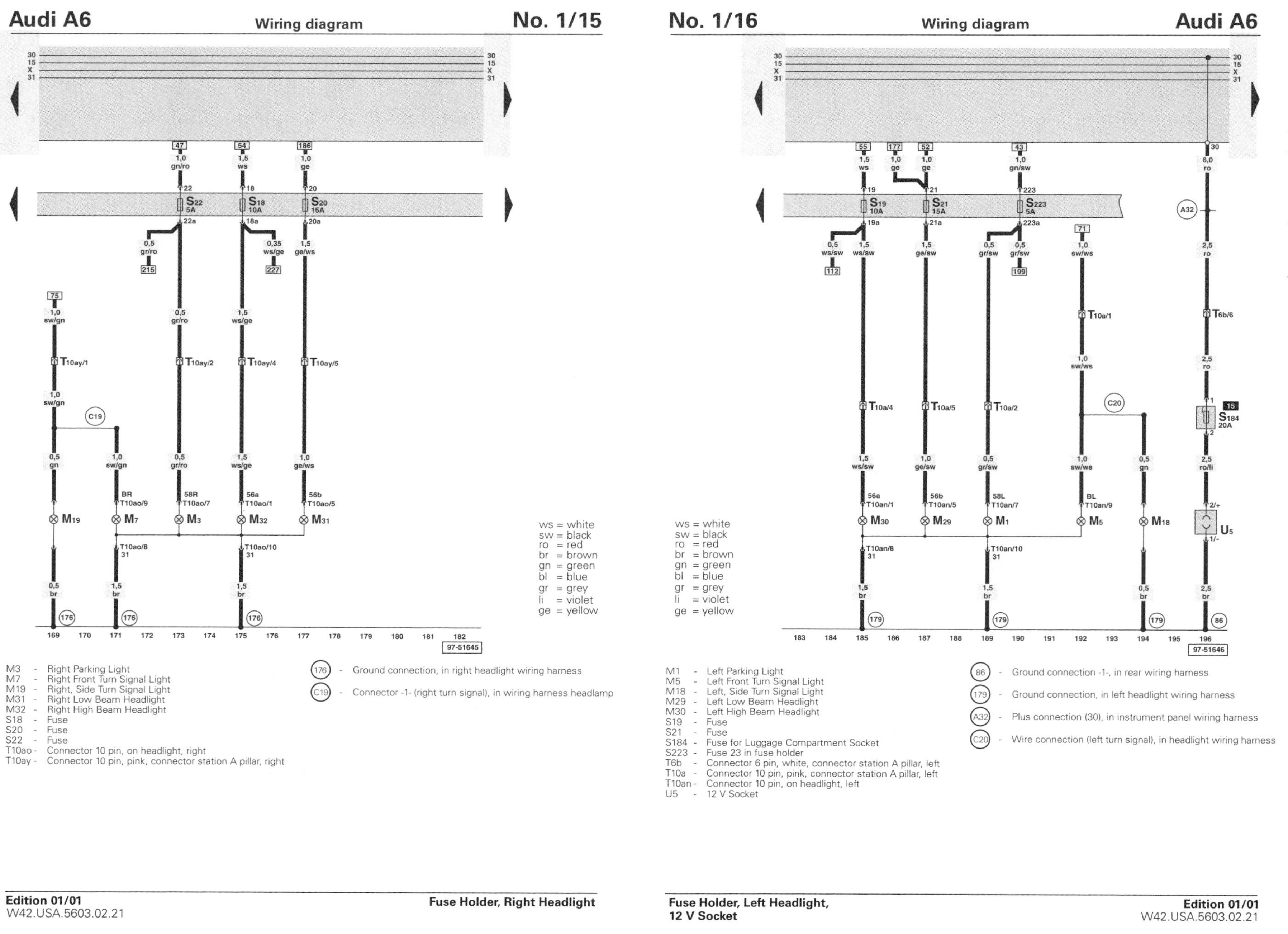

BentleyPublishers.com—All Rights Reserved

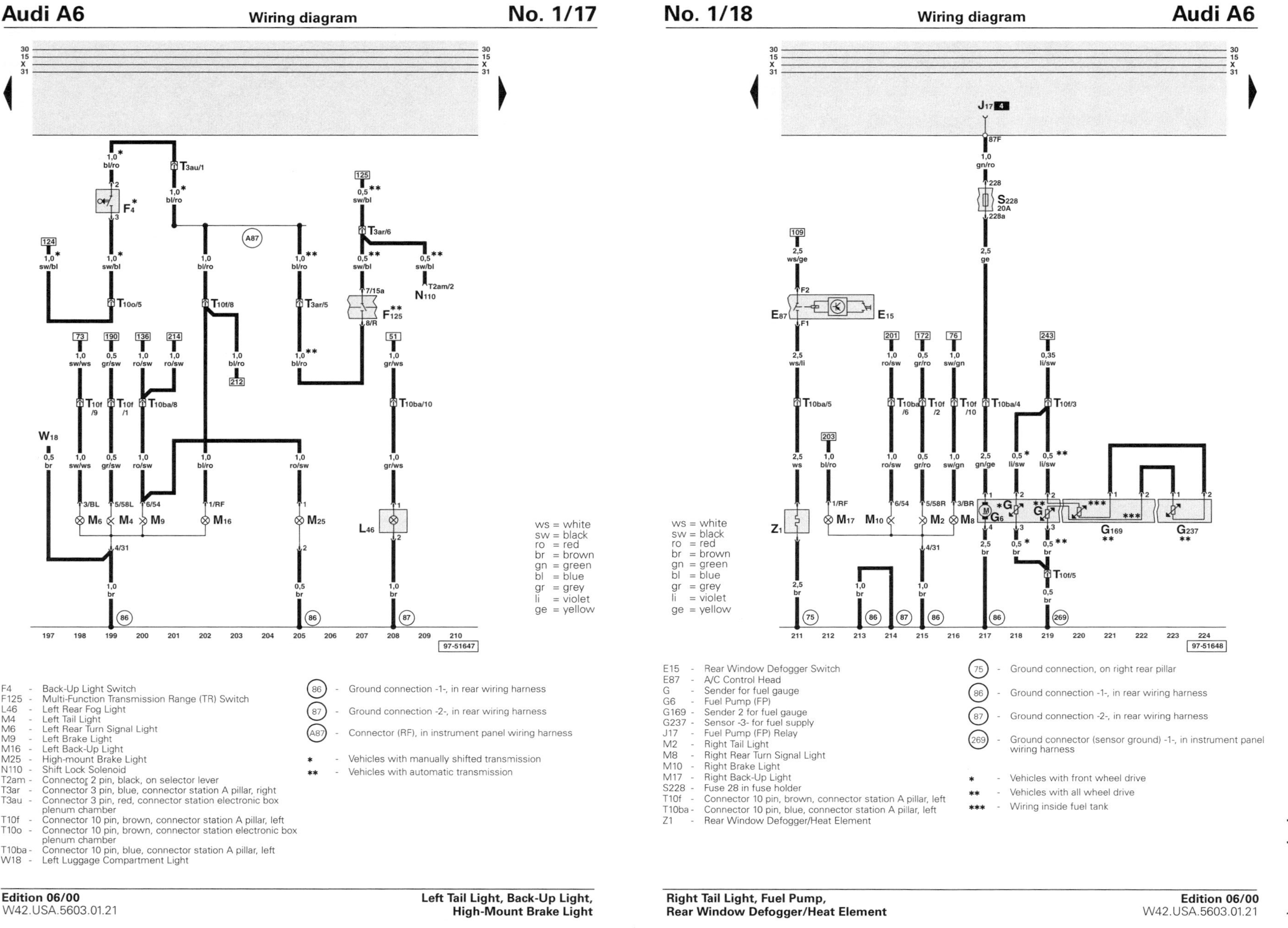

BentleyPublishers.com—All Rights Reserved

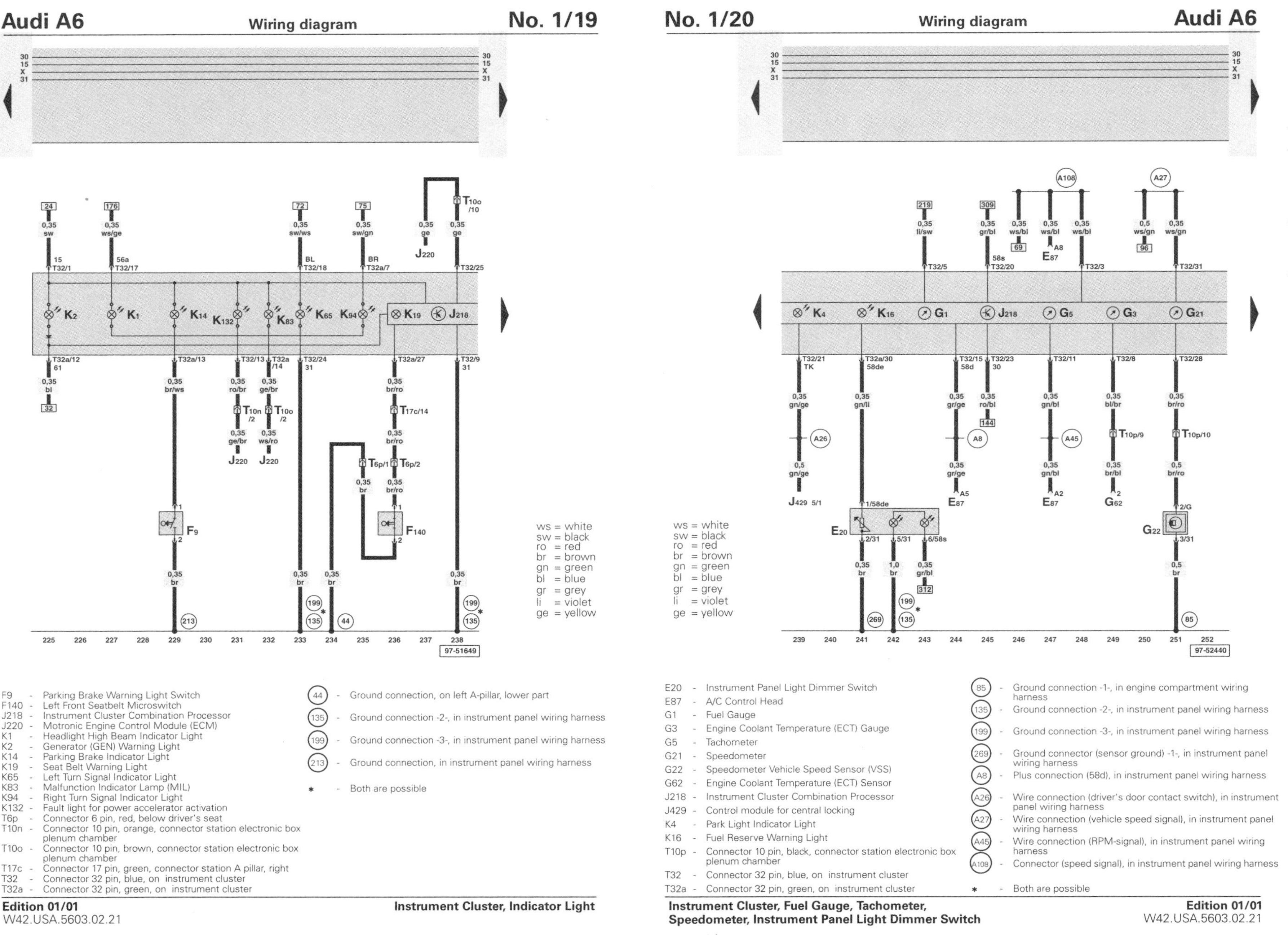

BentleyPublishers.com—All Rights Reserved

Audi A6 — Wiring diagram — No. 1/21

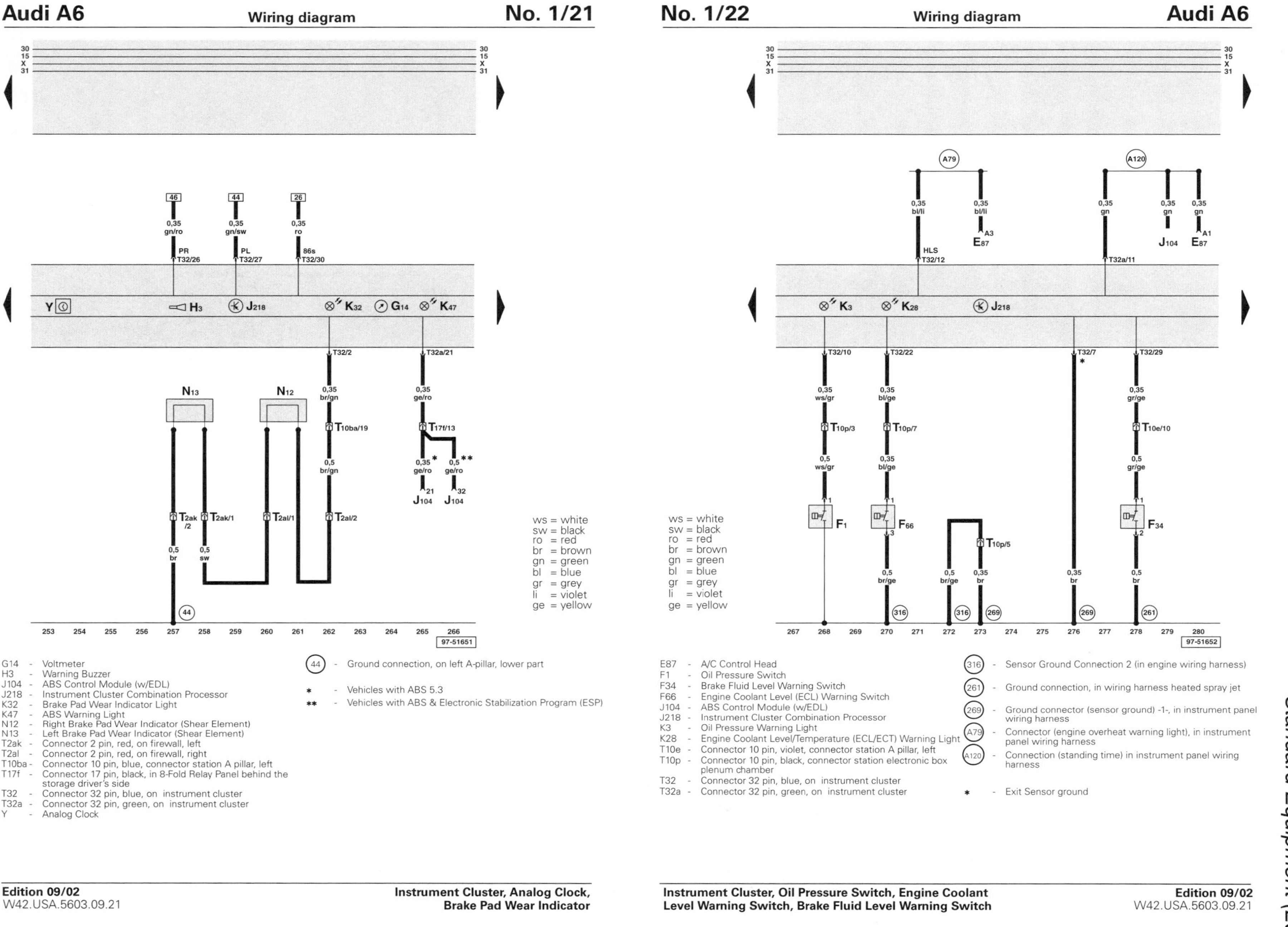

ws = white
sw = black
ro = red
br = brown
gn = green
bl = blue
gr = grey
li = violet
ge = yellow

G14 - Voltmeter
H3 - Warning Buzzer
J104 - ABS Control Module (w/EDL)
J218 - Instrument Cluster Combination Processor
K32 - Brake Pad Wear Indicator Light
K47 - ABS Warning Light
N12 - Right Brake Pad Wear Indicator (Shear Element)
N13 - Left Brake Pad Wear Indicator (Shear Element)
T2ak - Connector 2 pin, red, on firewall, left
T2al - Connector 2 pin, red, on firewall, right
T10ba - Connector 10 pin, blue, connector station A pillar, left
T17f - Connector 17 pin, black, in 8-Fold Relay Panel behind the storage driver's side
T32 - Connector 32 pin, blue, on instrument cluster
T32a - Connector 32 pin, green, on instrument cluster
Y - Analog Clock

(44) - Ground connection, on left A-pillar, lower part

* - Vehicles with ABS 5.3

** - Vehicles with ABS & Electronic Stabilization Program (ESP)

Edition 09/02
W42.USA.5603.09.21

Instrument Cluster, Analog Clock, Brake Pad Wear Indicator

No. 1/22 — Wiring diagram — Audi A6

ws = white
sw = black
ro = red
br = brown
gn = green
bl = blue
gr = grey
li = violet
ge = yellow

E87 - A/C Control Head
F1 - Oil Pressure Switch
F34 - Brake Fluid Level Warning Switch
F66 - Engine Coolant Level (ECL) Warning Switch
J104 - ABS Control Module (w/EDL)
J218 - Instrument Cluster Combination Processor
K3 - Oil Pressure Warning Light
K28 - Engine Coolant Level/Temperature (ECL/ECT) Warning Light
T10e - Connector 10 pin, violet, connector station A pillar, left
T10p - Connector 10 pin, black, connector station electronic box plenum chamber
T32 - Connector 32 pin, blue, on instrument cluster
T32a - Connector 32 pin, green, on instrument cluster

(316) - Sensor Ground Connection 2 (in engine wiring harness)
(261) - Ground connection, in wiring harness heated spray jet
(269) - Ground connector (sensor ground) -1-, in instrument panel wiring harness
(A79) - Connector (engine overheat warning light), in instrument panel wiring harness
(A120) - Connection (standing time) in instrument panel wiring harness

* - Exit Sensor ground

Instrument Cluster, Oil Pressure Switch, Engine Coolant Level Warning Switch, Brake Fluid Level Warning Switch

Edition 09/02
W42.USA.5603.09.21

BentleyPublishers.com—All Rights Reserved

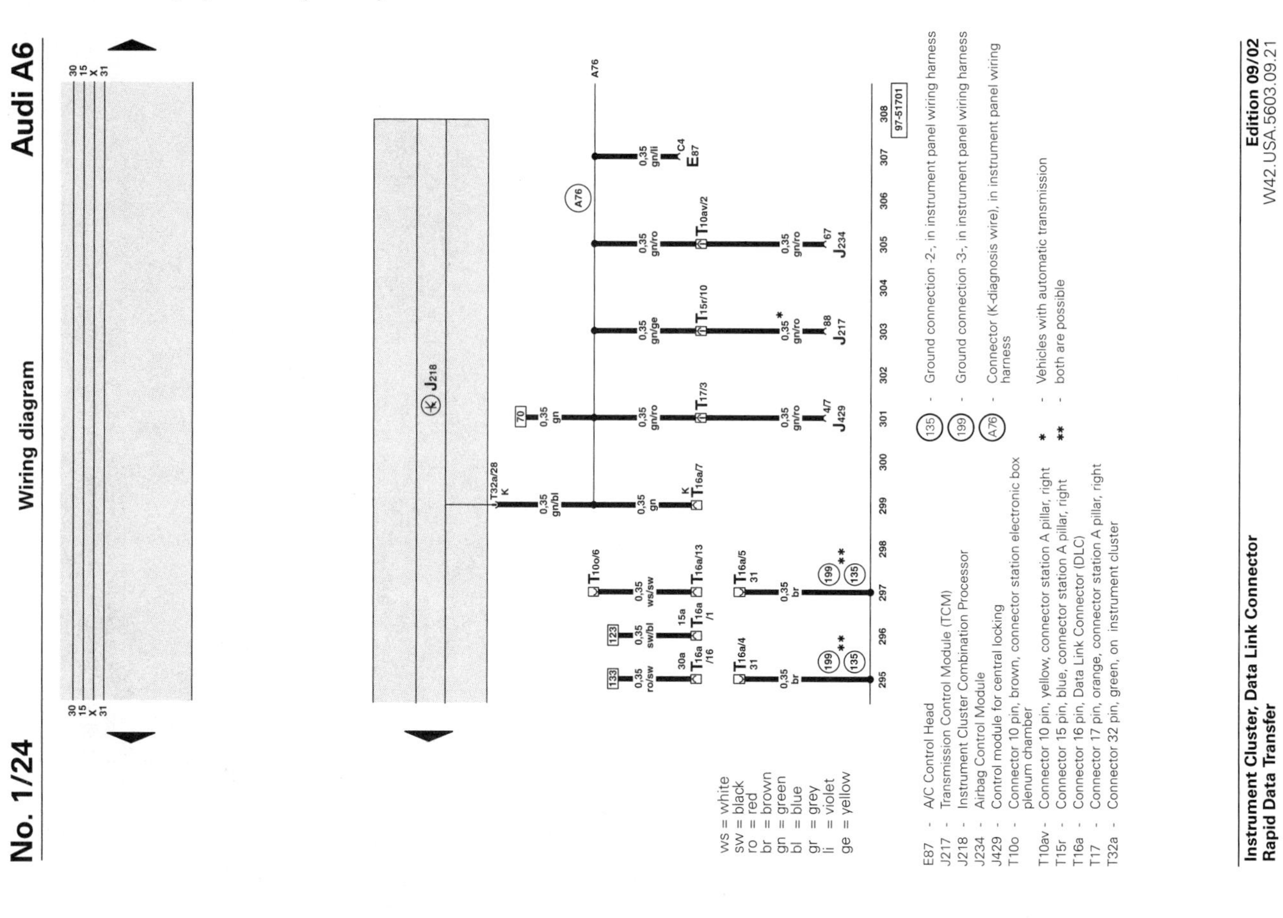

ws = white
sw = black
ro = red
br = brown
gn = green
bl = blue
gr = grey
li = violet
ge = yellow

E87 - A/C Control Head
J217 - Transmission Control Module (TCM)
J218 - Instrument Cluster Combination Processor
J234 - Airbag Control Module
J429 - Control module for central locking
T10o - Connector 10 pin, brown, connector station electronic box plenum chamber
T10av - Connector 10 pin, yellow, connector station A pillar, right
T15r - Connector 15 pin, blue, connector station A pillar, right
T16a - Connector 16 pin, Data Link Connector (DLC)
T17 - Connector 17 pin, orange, connector station A pillar, right
T32a - Connector 32 pin, green, on instrument cluster

(135) - Ground connection -2-, in instrument panel wiring harness
(199) - Ground connection -3-, in instrument panel wiring harness
(A76) - Connector (K-diagnosis wire), in instrument panel wiring harness
* - Vehicles with automatic transmission
** - both are possible

Instrument Cluster, Data Link Connector
Rapid Data Transfer

Edition 09/02
W42.USA.5603.09.21

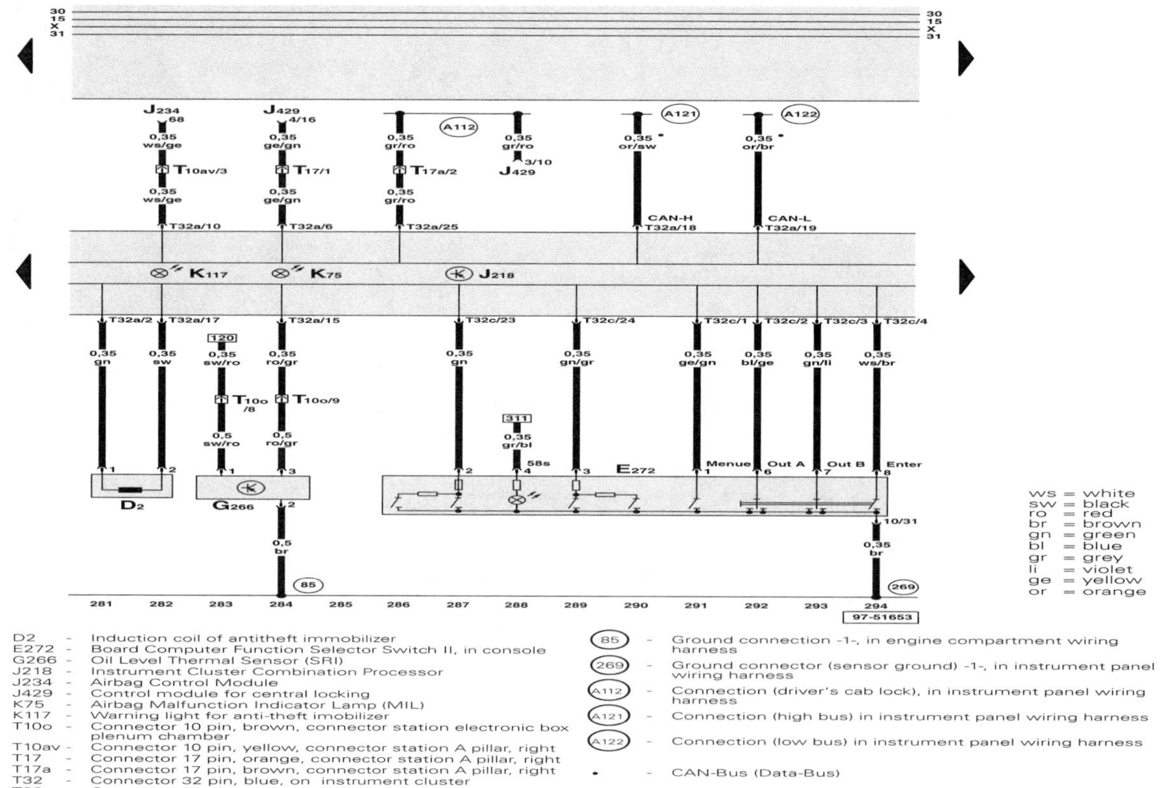

ws = white
sw = black
ro = red
br = brown
gn = green
bl = blue
gr = grey
li = violet
ge = yellow
or = orange

D2 - Induction coil of antitheft immobilizer
E272 - Board Computer Function Selector Switch II, in console
G266 - Oil Level Thermal Sensor (SRI)
J218 - Instrument Cluster Combination Processor
J234 - Airbag Control Module
J429 - Control module for central locking
K75 - Airbag Malfunction Indicator Lamp (MIL)
K117 - Warning light for anti-theft immobilizer
T10o - Connector 10 pin, brown, connector station electronic box plenum chamber
T10av - Connector 10 pin, yellow, connector station A pillar, right
T17 - Connector 17 pin, orange, connector station A pillar, right
T17a - Connector 17 pin, brown, connector station A pillar, right
T32 - Connector 32 pin, blue, on instrument cluster
T32a - Connector 32 pin, green, on instrument cluster
T32c - Connector 32 pin, grey, on instrument cluster

(85) - Ground connection -1-, in engine compartment wiring harness
(269) - Ground connector (sensor ground) -1-, in instrument panel wiring harness
(A112) - Connection (driver's cab lock), in instrument panel wiring harness
(A121) - Connection (high bus) in instrument panel wiring harness
(A122) - Connection (low bus) in instrument panel wiring harness
• - CAN-Bus (Data-Bus)

BentleyPublishers.com—All Rights Reserved

Audi A6 **Wiring diagram** **No. 1/25**

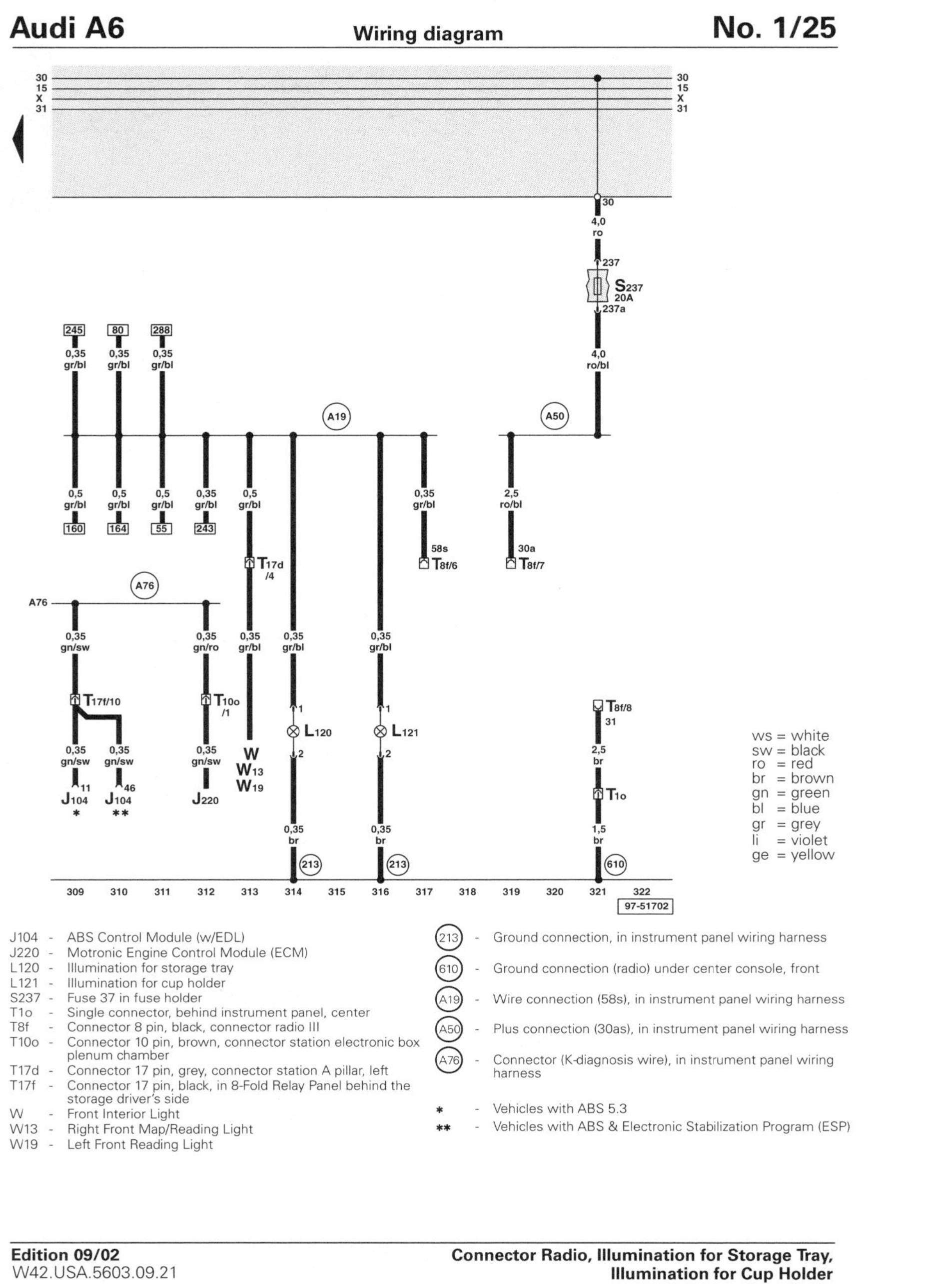

J104 - ABS Control Module (w/EDL)
J220 - Motronic Engine Control Module (ECM)
L120 - Illumination for storage tray
L121 - Illumination for cup holder
S237 - Fuse 37 in fuse holder
T1o - Single connector, behind instrument panel, center
T8f - Connector 8 pin, black, connector radio III
T10o - Connector 10 pin, brown, connector station electronic box plenum chamber
T17d - Connector 17 pin, grey, connector station A pillar, left
T17f - Connector 17 pin, black, in 8-Fold Relay Panel behind the storage driver's side
W - Front Interior Light
W13 - Right Front Map/Reading Light
W19 - Left Front Reading Light

(213) - Ground connection, in instrument panel wiring harness
(610) - Ground connection (radio) under center console, front
(A19) - Wire connection (58s), in instrument panel wiring harness
(A50) - Plus connection (30as), in instrument panel wiring harness
(A76) - Connector (K-diagnosis wire), in instrument panel wiring harness
* - Vehicles with ABS 5.3
** - Vehicles with ABS & Electronic Stabilization Program (ESP)

Edition 09/02
W42.USA.5603.09.21

Connector Radio, Illumination for Storage Tray, Illumination for Cup Holder

THIS SPACE INTENTIONALLY LEFT BLANK

BentleyPublishers.com—All Rights Reserved

(2,7 l - Injection Engine, 6 Cylinder), Code APB

2001 m. y.

Fuse Panel

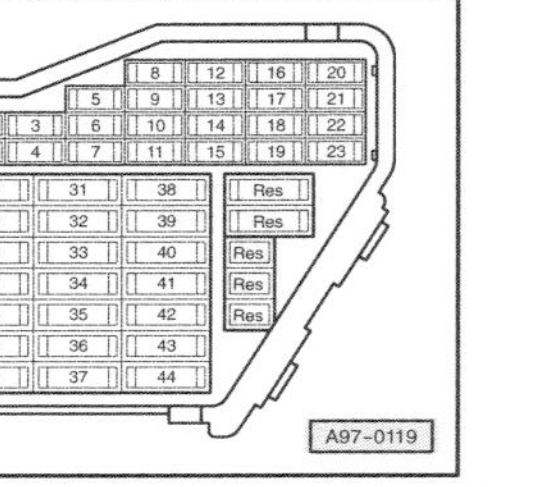
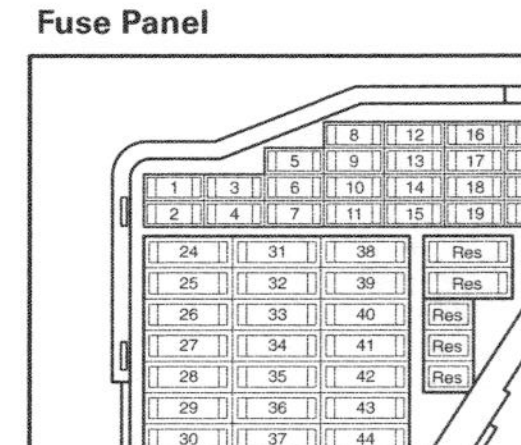

Fuse Colors:

30 A - Green
25 A - White
20 A - Yellow
15 A - Blue
10 A - Red
7,5 A - Brown
5 A - Beige

Starting with fuse position 23, fuses in the fuse holder are identified with 223 in the wiring diagram.

Micro Central Electric Panel

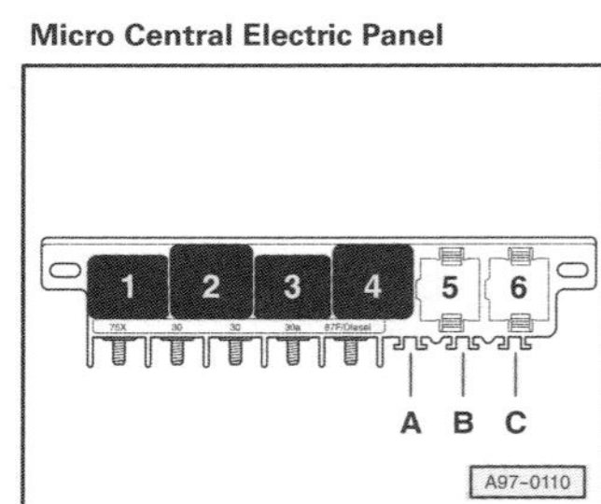

Relay Location:

4 -Fuel Pump (FP) Relay , J17

13 - Fold Relay Panel

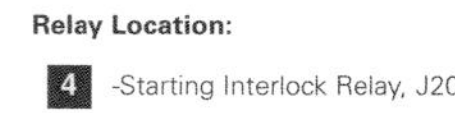

Relay Location:

4 -Starting Interlock Relay, J207

8 - Fold Relay Panel

Fuse Location:

20 -Coolant Fan Fuse, S42

21 -Control module fuse for coolant fan, S142

3 - Fold Relay Panel in E-Box plenum chamber

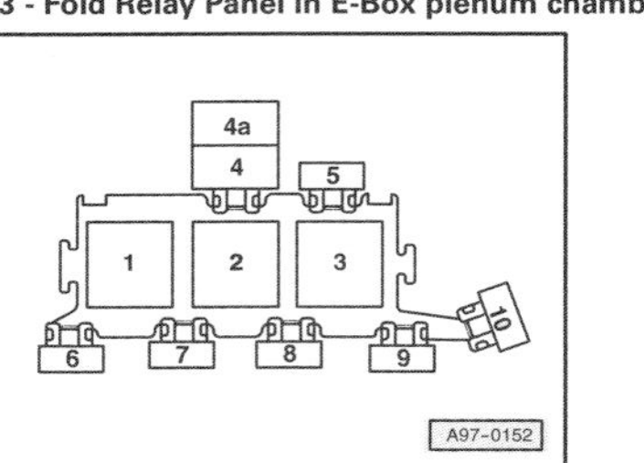

Relay Location:

2 -Secondary Air Injection (AIR) Pump Relay, J299

7 -Fuse for secondary air pump , S130

Relay Panel behind Instrument Panel left, on central girder

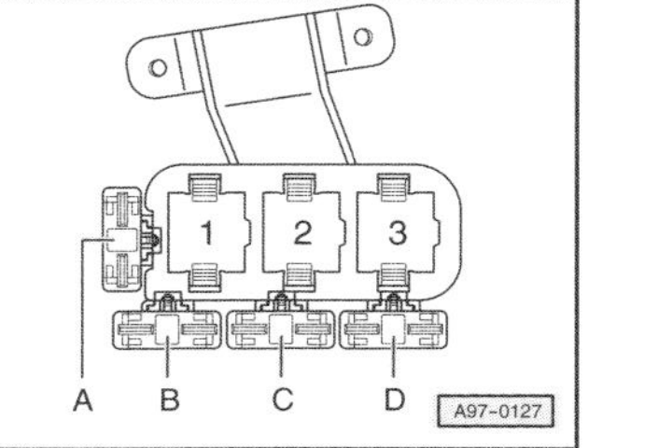

Relay Location:

1 -Brake Booster Relay, J569

C -Fuse for Hydraulic Pump Relay, S279

BentleyPublishers.com—All Rights Reserved

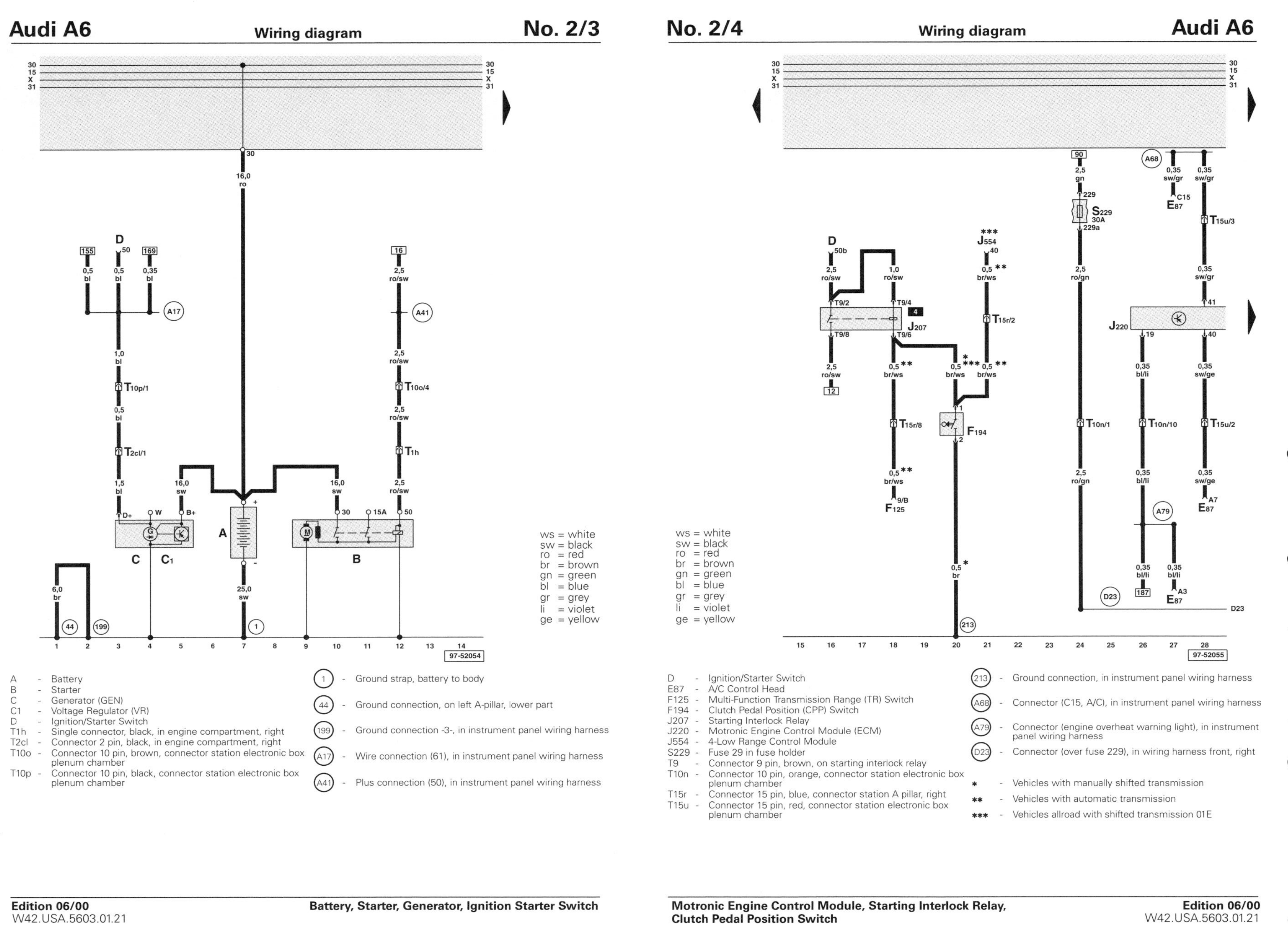

BentleyPublishers.com—All Rights Reserved

Engine Management, 2.7 Liter, Engine Code APB (2001)

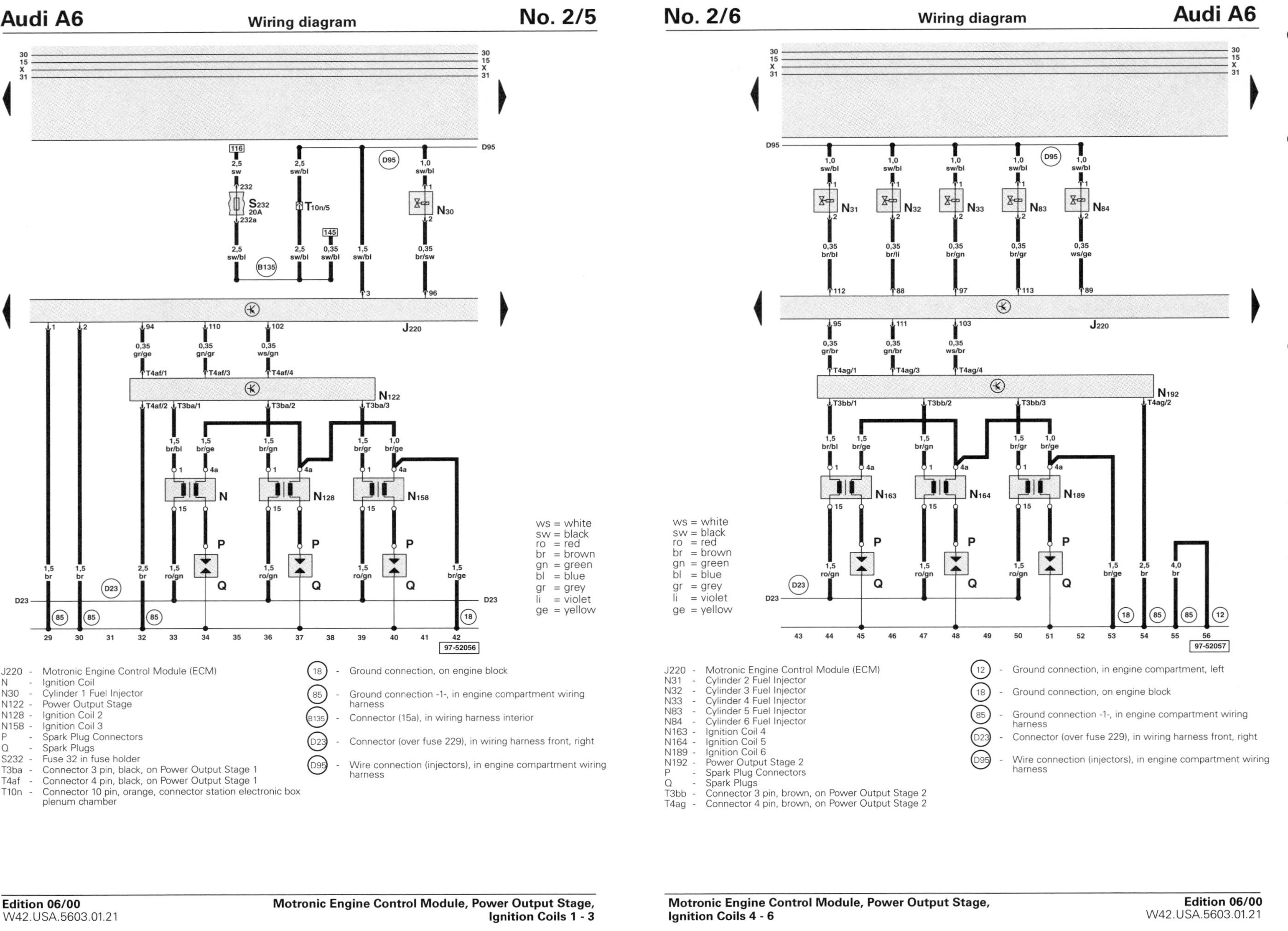

BentleyPublishers.com—All Rights Reserved

Engine Management, 2.7 Liter, Engine Code APB (2001)

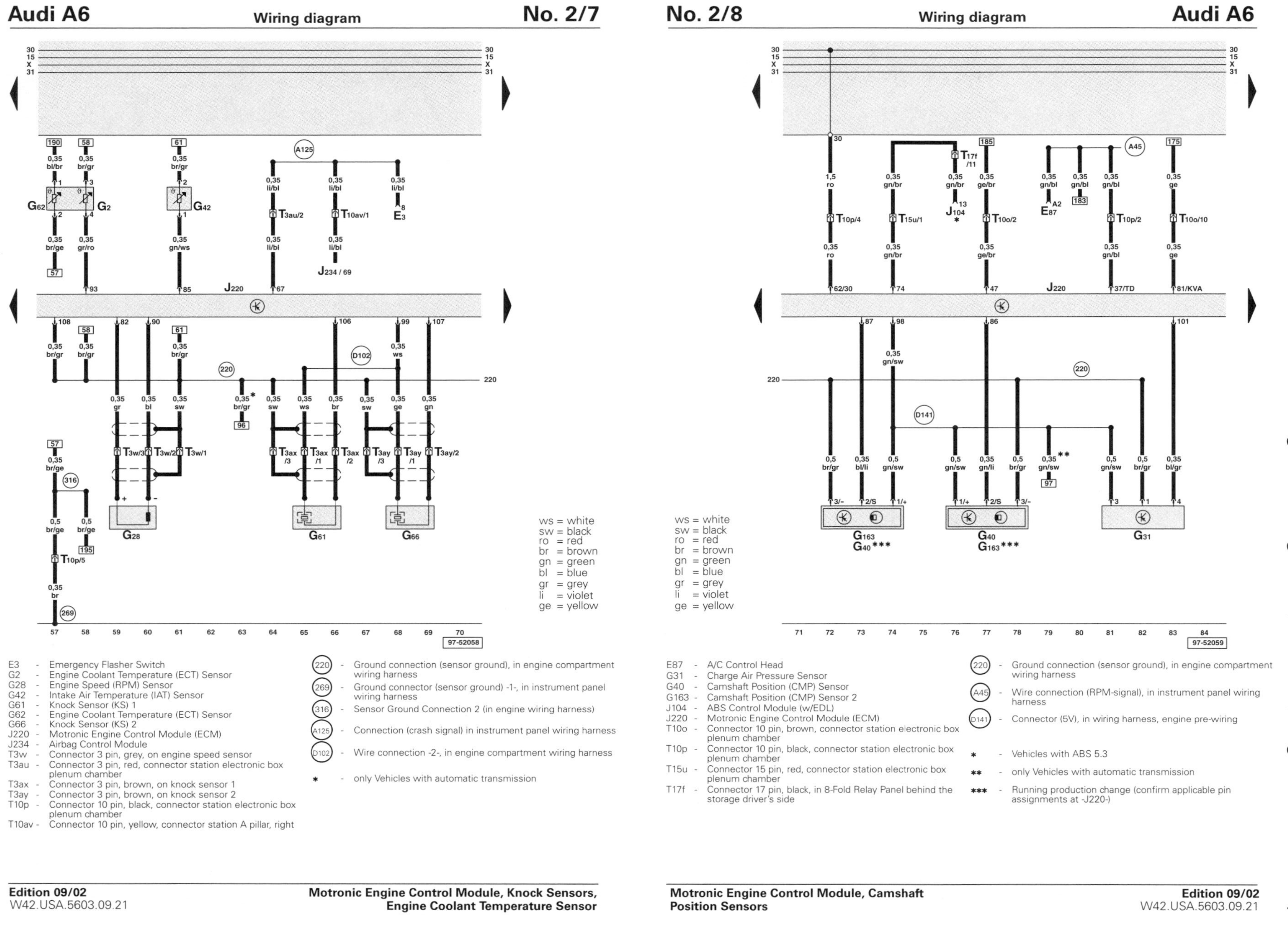

BentleyPublishers.com—All Rights Reserved

Engine Management, 2.7 Liter, Engine Code APB (2001)

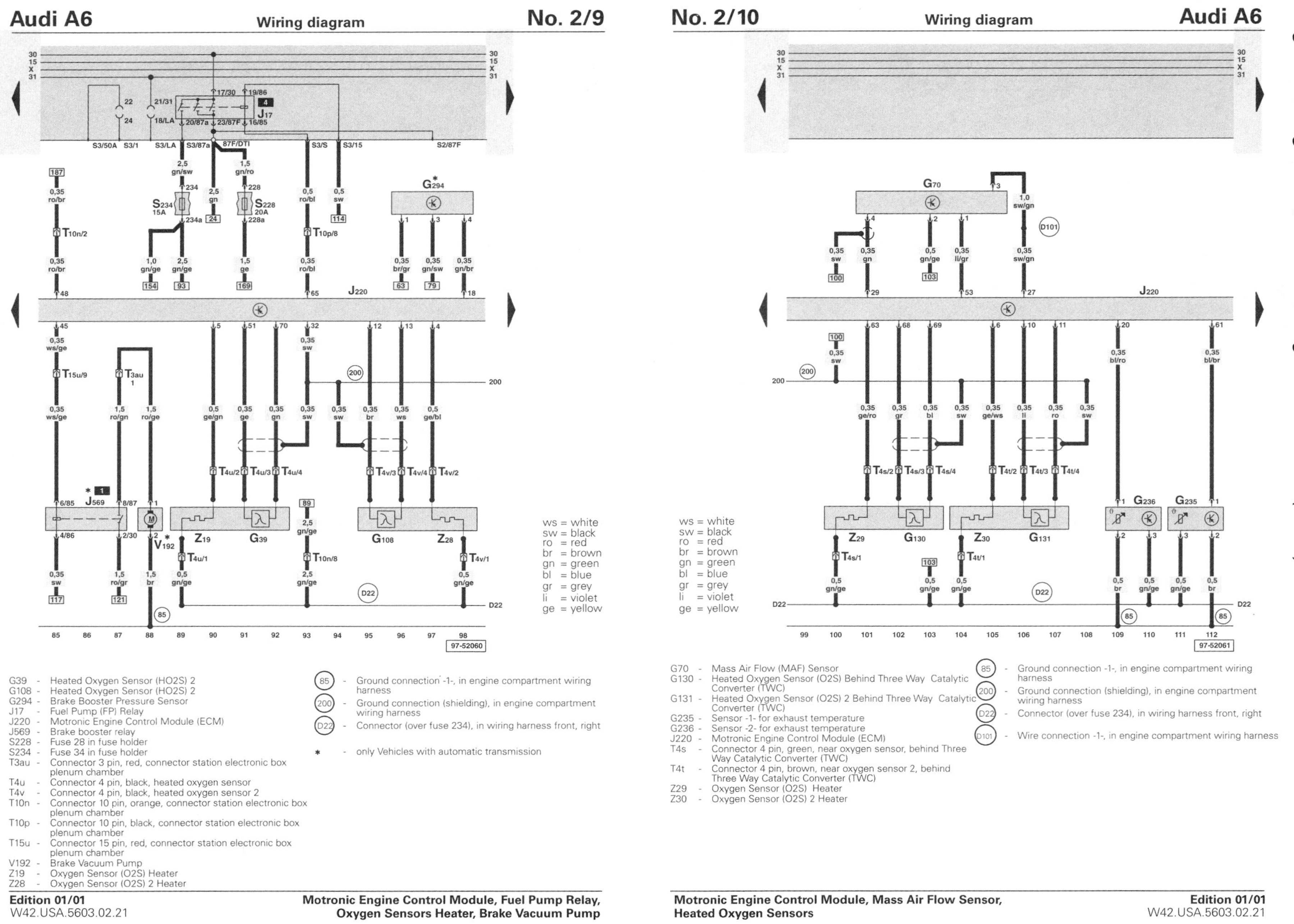

BentleyPublishers.com—All Rights Reserved

Engine Management, 2.7 Liter, Engine Code APB (2001)

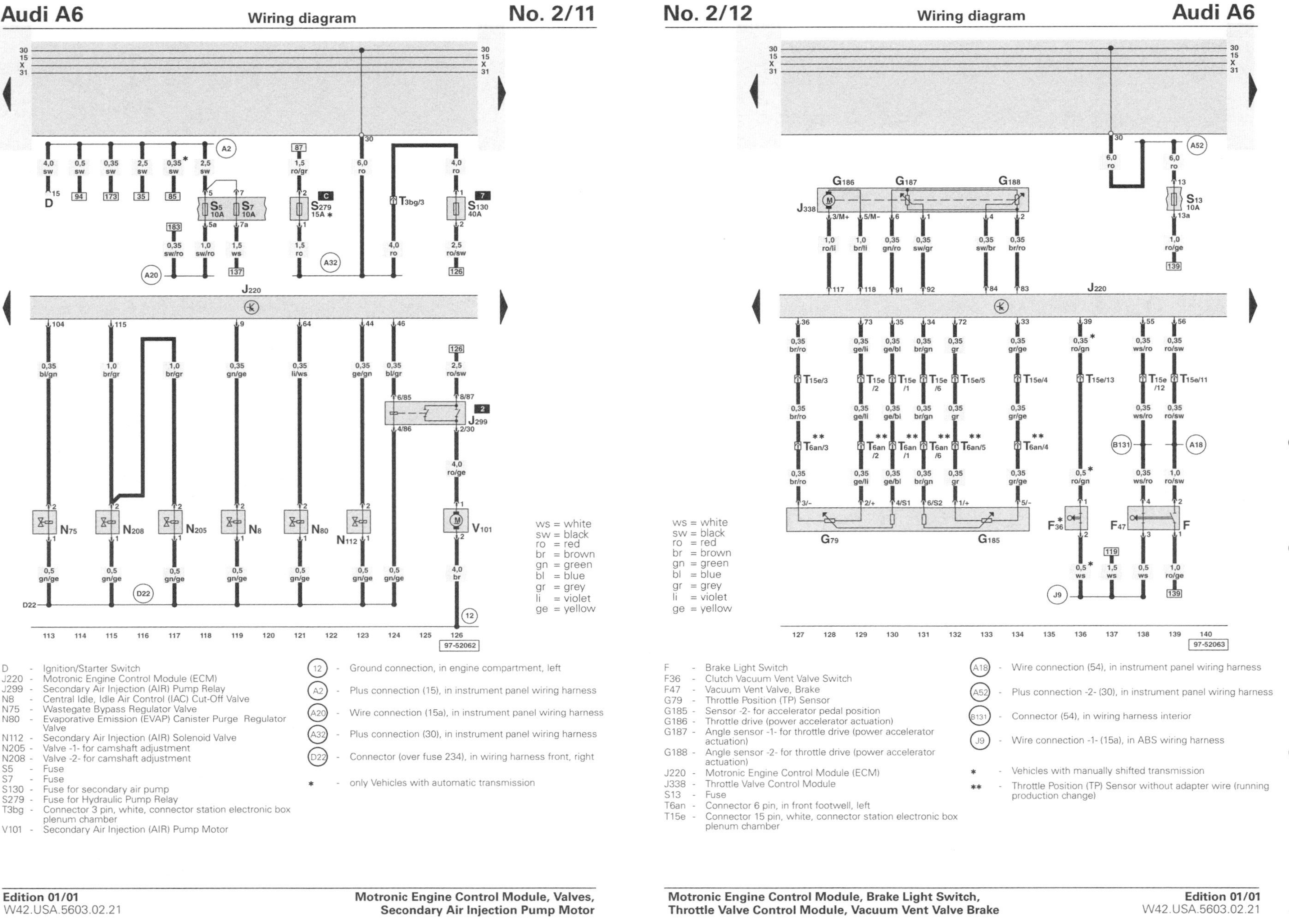

BentleyPublishers.com—All Rights Reserved

Engine Management, 2.7 Liter, Engine Code APB (2001)

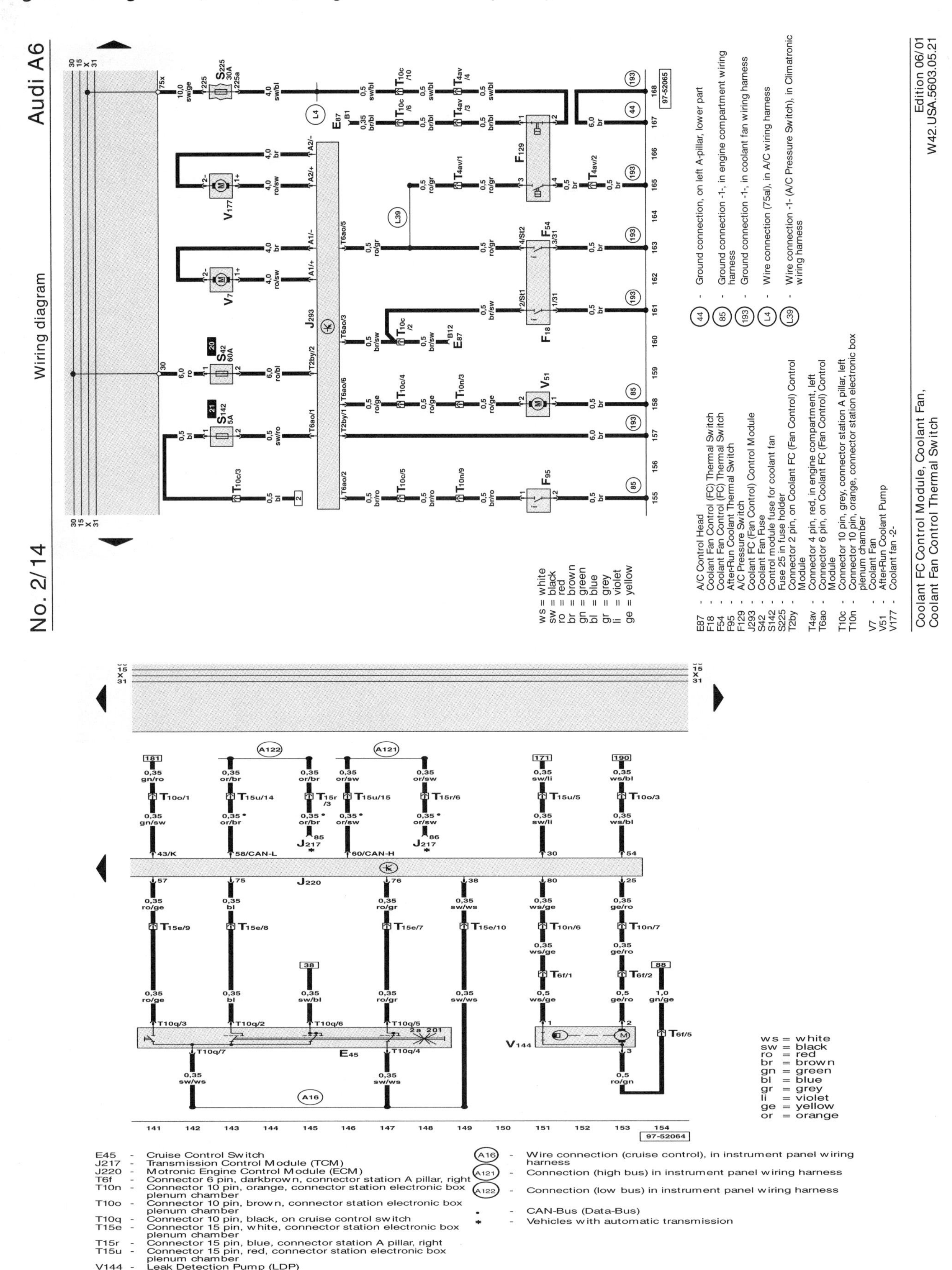

BentleyPublishers.com—All Rights Reserved

Engine Management, 2.7 Liter, Engine Code APB (2001)

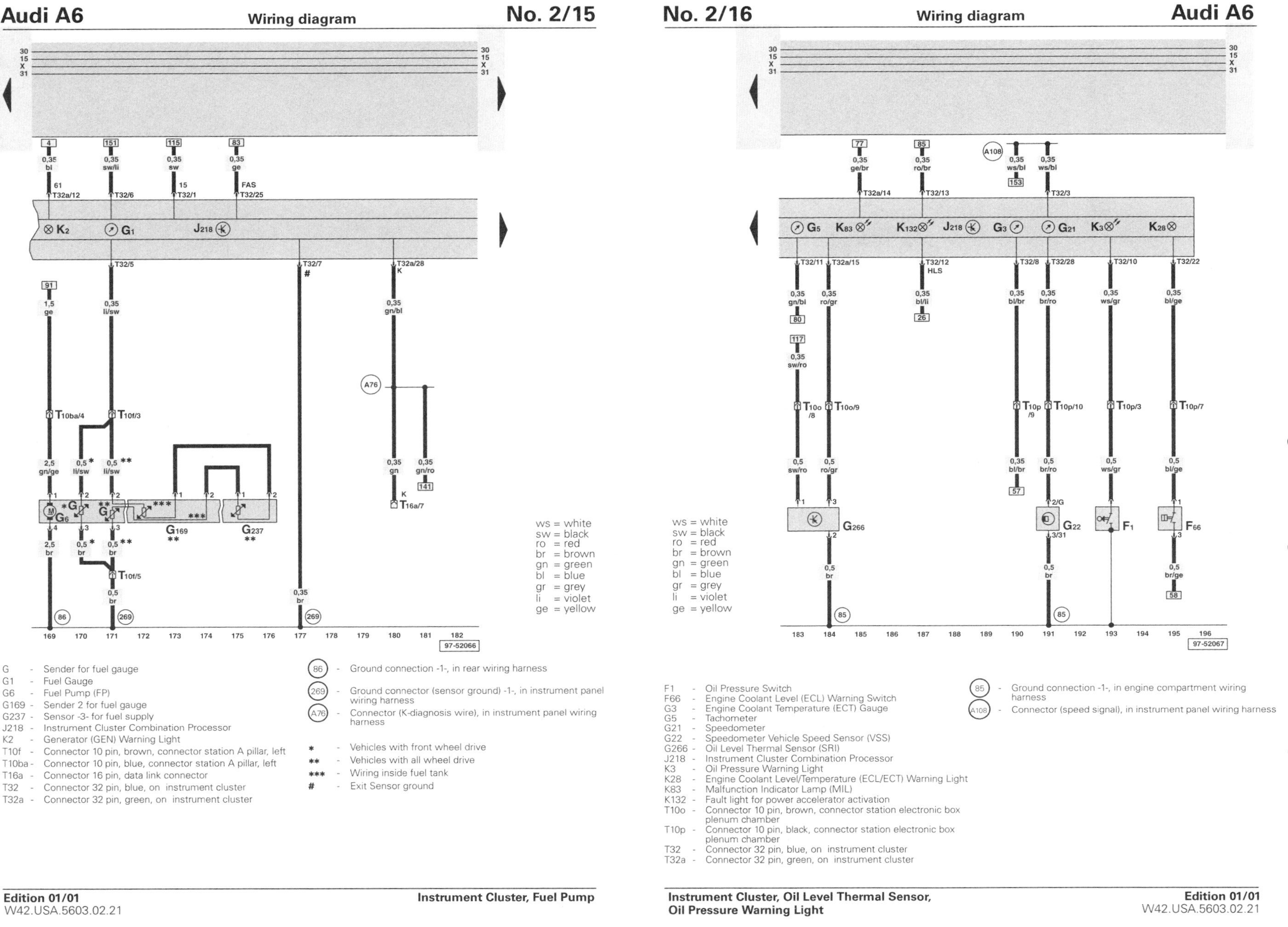

BentleyPublishers.com—All Rights Reserved

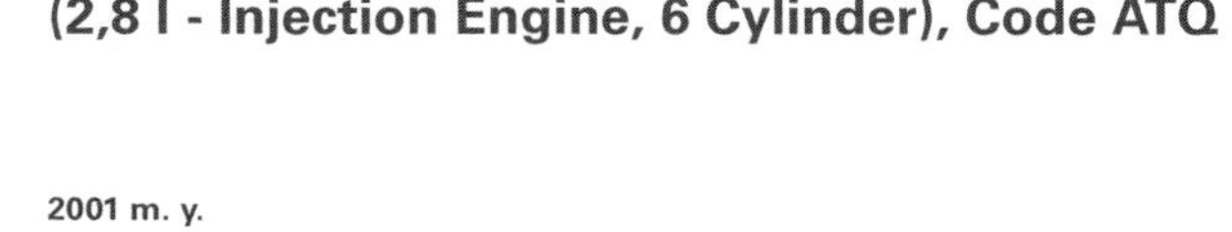

(2,8 l - Injection Engine, 6 Cylinder), Code ATQ

2001 m. y.

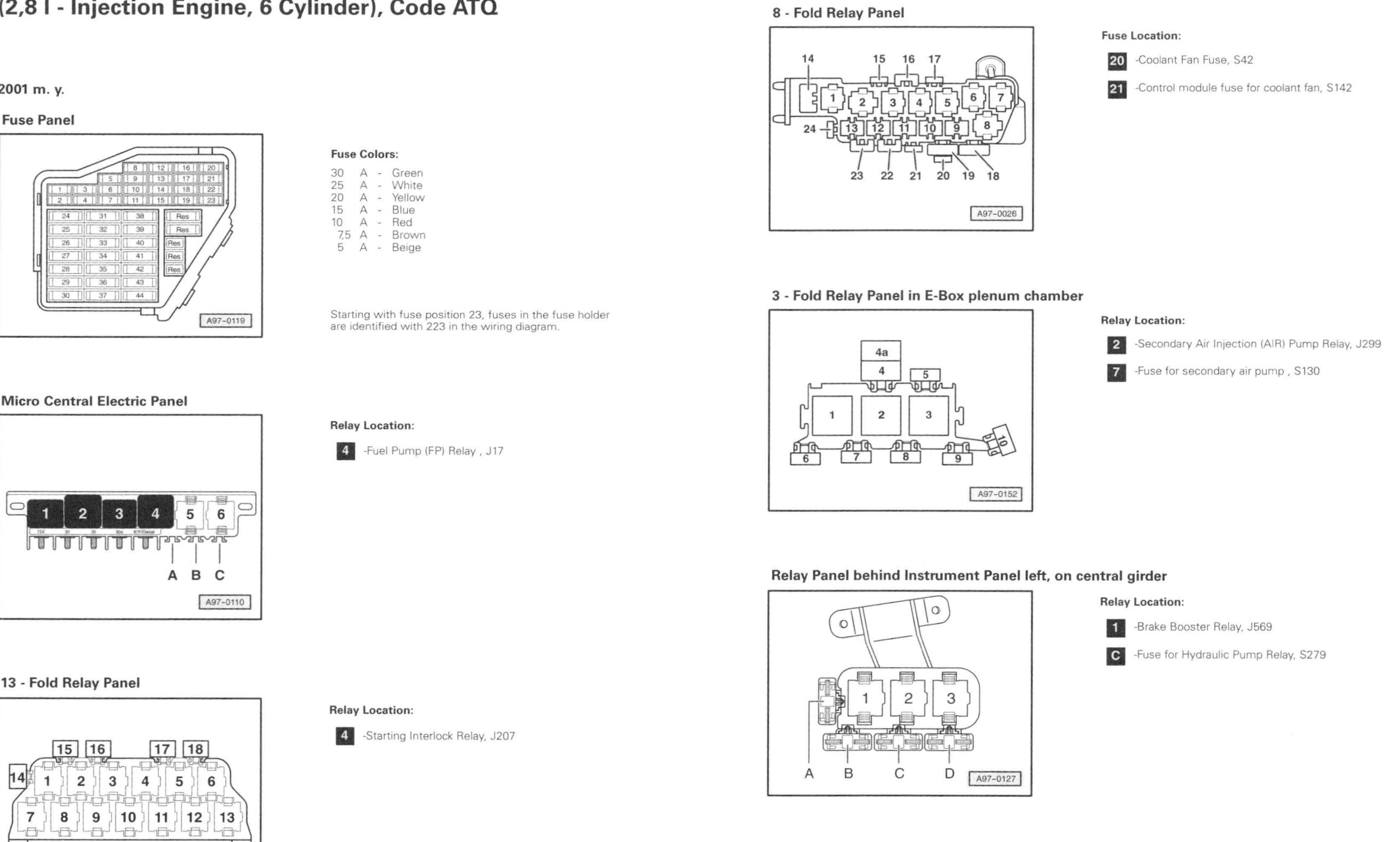

Fuse Panel

Fuse Colors:

30 A - Green
25 A - White
20 A - Yellow
15 A - Blue
10 A - Red
7,5 A - Brown
5 A - Beige

Starting with fuse position 23, fuses in the fuse holder are identified with 223 in the wiring diagram.

Micro Central Electric Panel

Relay Location:

4 -Fuel Pump (FP) Relay , J17

13 - Fold Relay Panel

Relay Location:

4 -Starting Interlock Relay, J207

8 - Fold Relay Panel

Fuse Location:

20 -Coolant Fan Fuse, S42

21 -Control module fuse for coolant fan, S142

3 - Fold Relay Panel in E-Box plenum chamber

Relay Location:

2 -Secondary Air Injection (AIR) Pump Relay, J299

7 -Fuse for secondary air pump , S130

Relay Panel behind Instrument Panel left, on central girder

Relay Location:

1 -Brake Booster Relay, J569

C -Fuse for Hydraulic Pump Relay, S279

BentleyPublishers.com—All Rights Reserved

Engine Management, 2.8 Liter, Engine Code ATQ

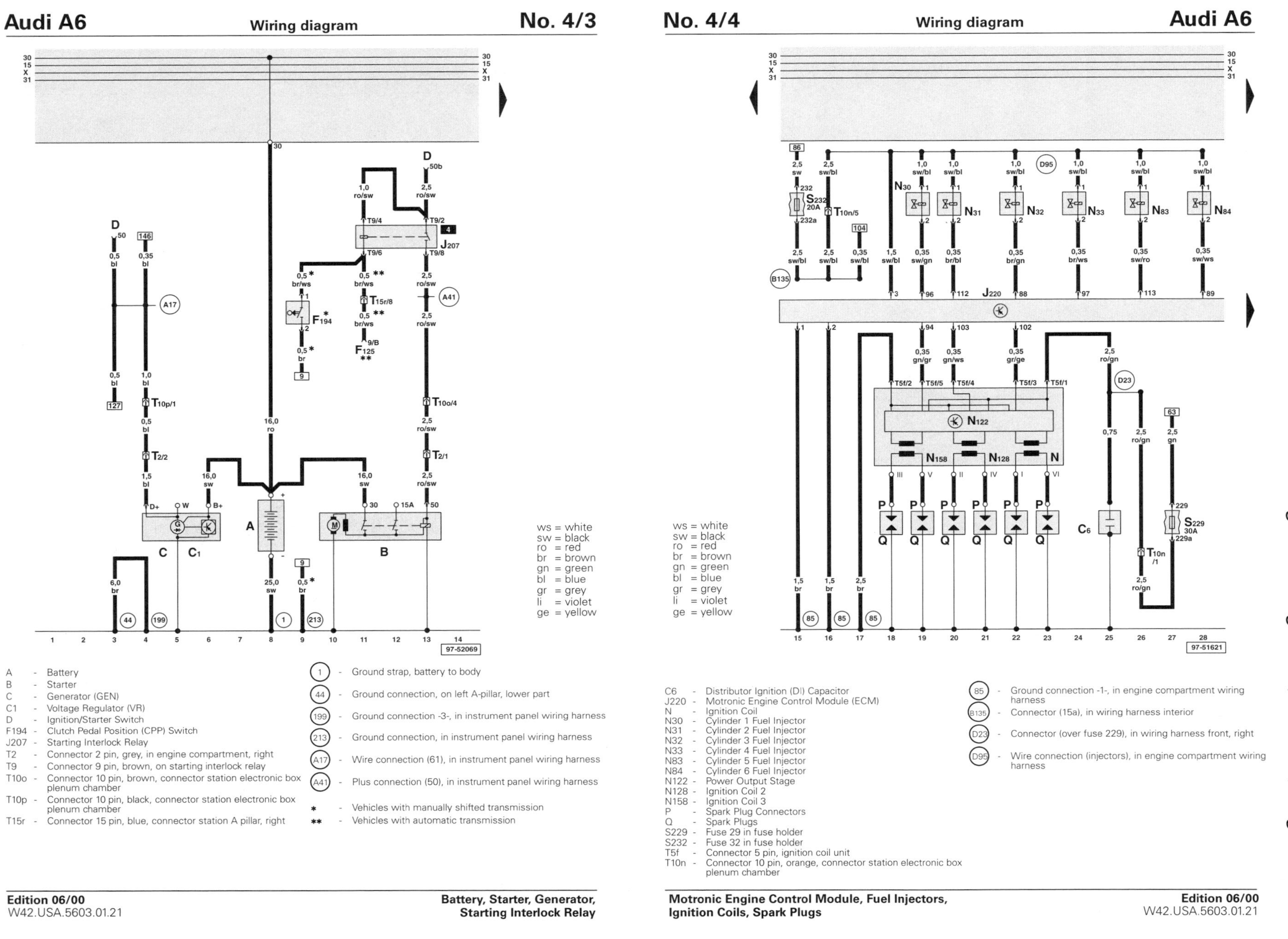

BentleyPublishers.com—All Rights Reserved

Engine Management, 2.8 Liter, Engine Code ATQ

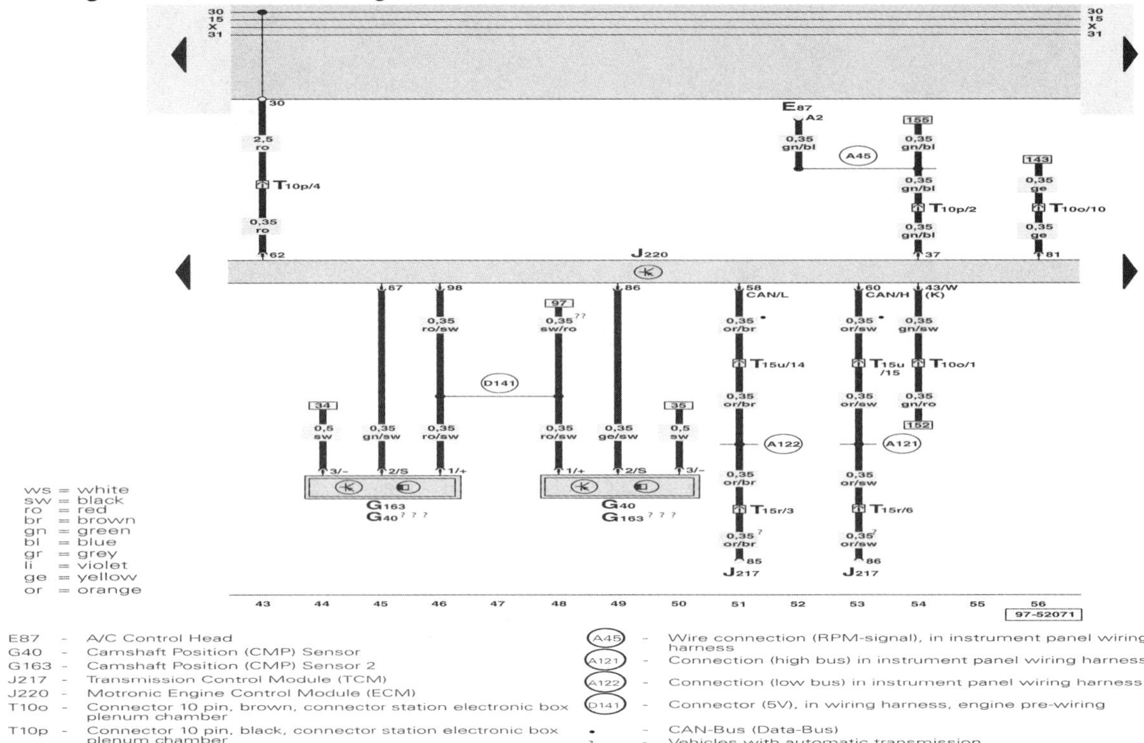

E87 - A/C Control Head
G40 - Camshaft Position (CMP) Sensor
G163 - Camshaft Position (CMP) Sensor 2
J217 - Transmission Control Module (TCM)
J220 - Motronic Engine Control Module (ECM)
T10o - Connector 10 pin, brown, connector station electronic box plenum chamber
T10p - Connector 10 pin, black, connector station electronic box plenum chamber
T15r - Connector 15 pin, blue, connector station A pillar, right
T15u - Connector 15 pin, red, connector station electronic box plenum chamber

(A45) - Wire connection (RPM-signal), in instrument panel wiring harness
(A121) - Connection (high bus) in instrument panel wiring harness
(A122) - Connection (low bus) in instrument panel wiring harness
(D141) - Connector (5V), in wiring harness, engine pre-wiring
• - CAN-Bus (Data-Bus)
? - Vehicles with automatic transmission
?? - only Vehicles with automatic transmission
??? - Running production change (confirm applicable pin assignments at -J220-)

Motronic Engine Control Module, Transmission Control Module, Camshaft Position Sensor

Edition 01/01
W42.USA.5603.02.21

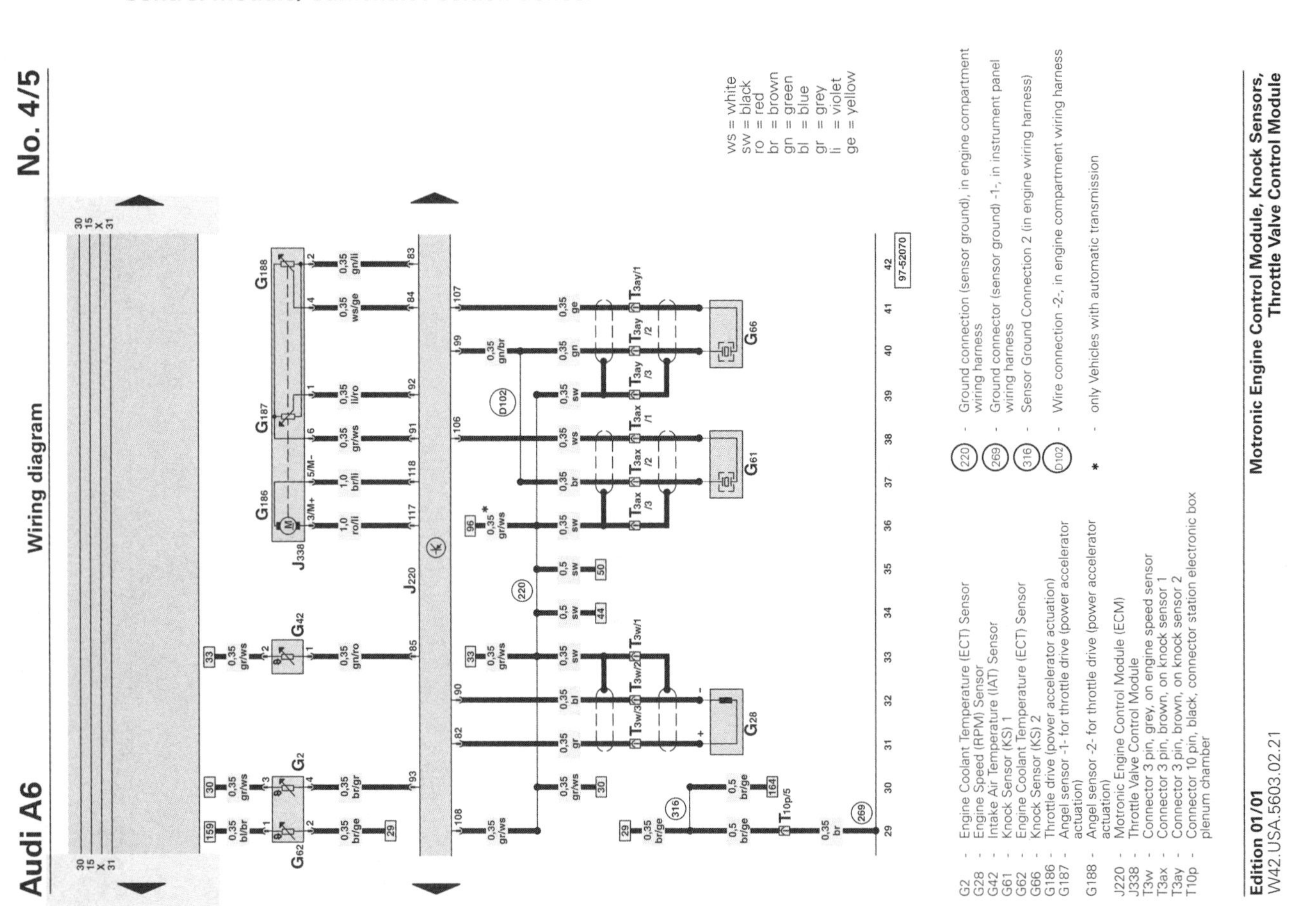

BentleyPublishers.com—All Rights Reserved

Engine Management, 2.8 Liter, Engine Code ATQ

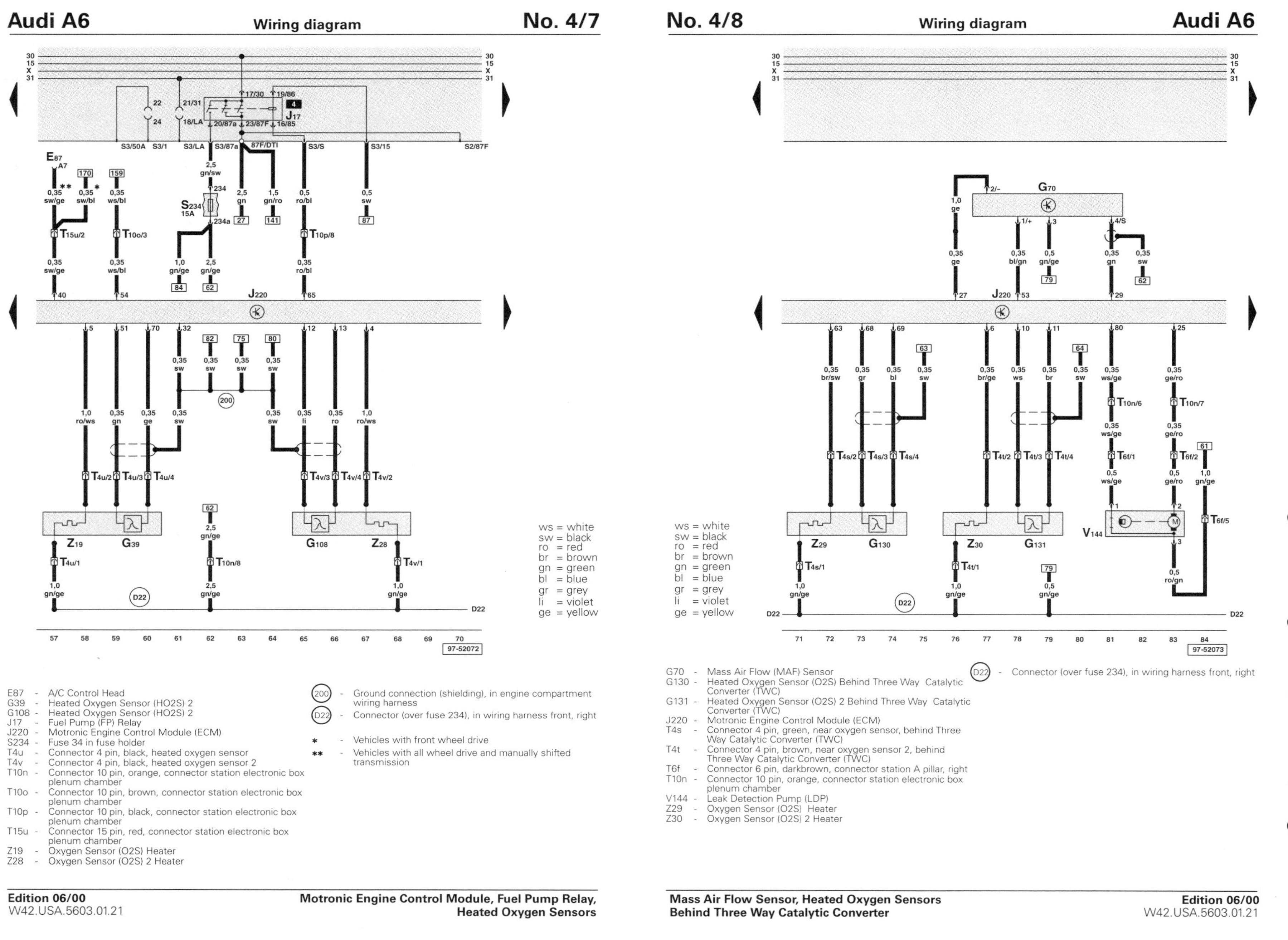

BentleyPublishers.com—All Rights Reserved

Engine Management, 2.8 Liter, Engine Code ATQ

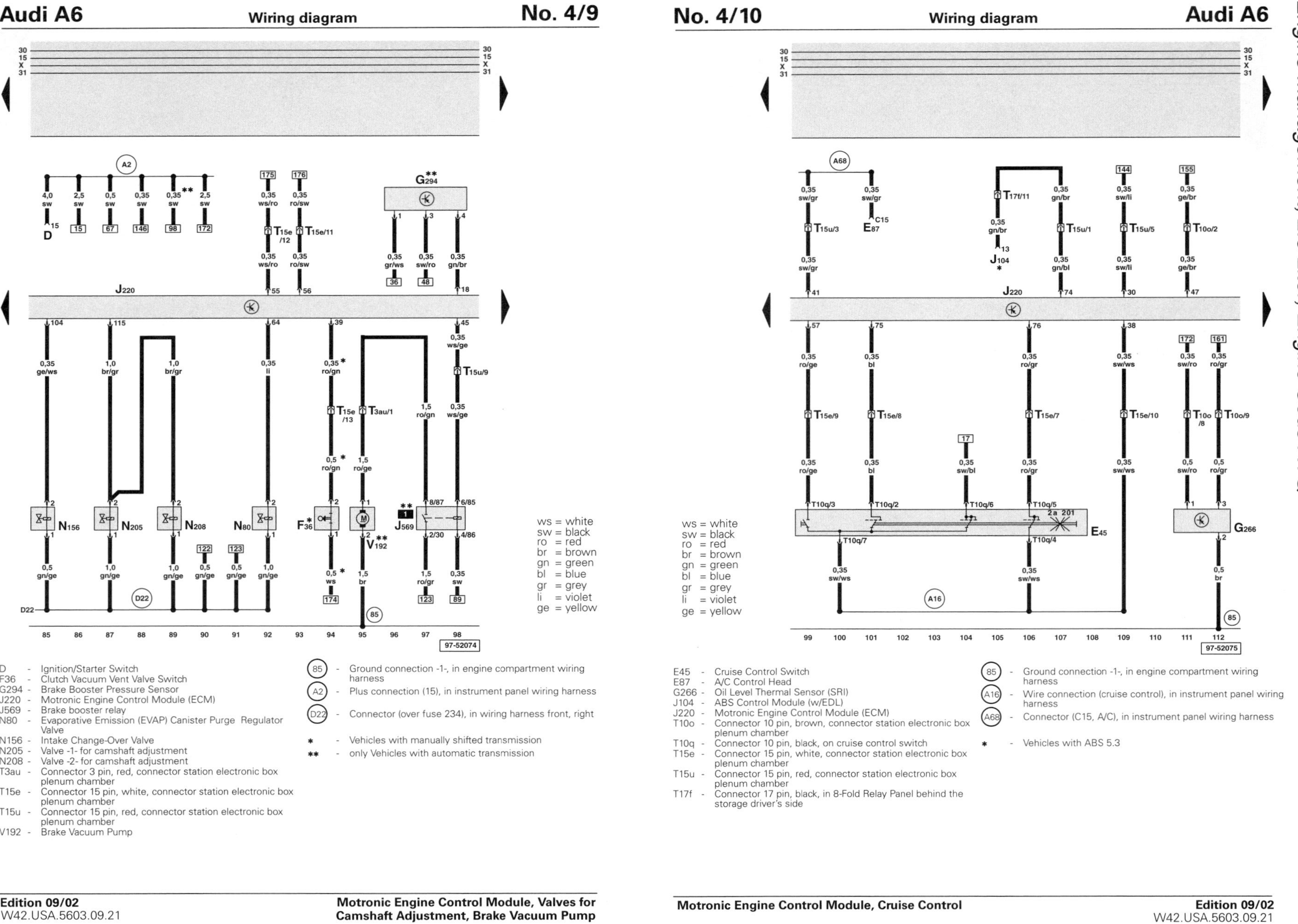

BentleyPublishers.com—All Rights Reserved

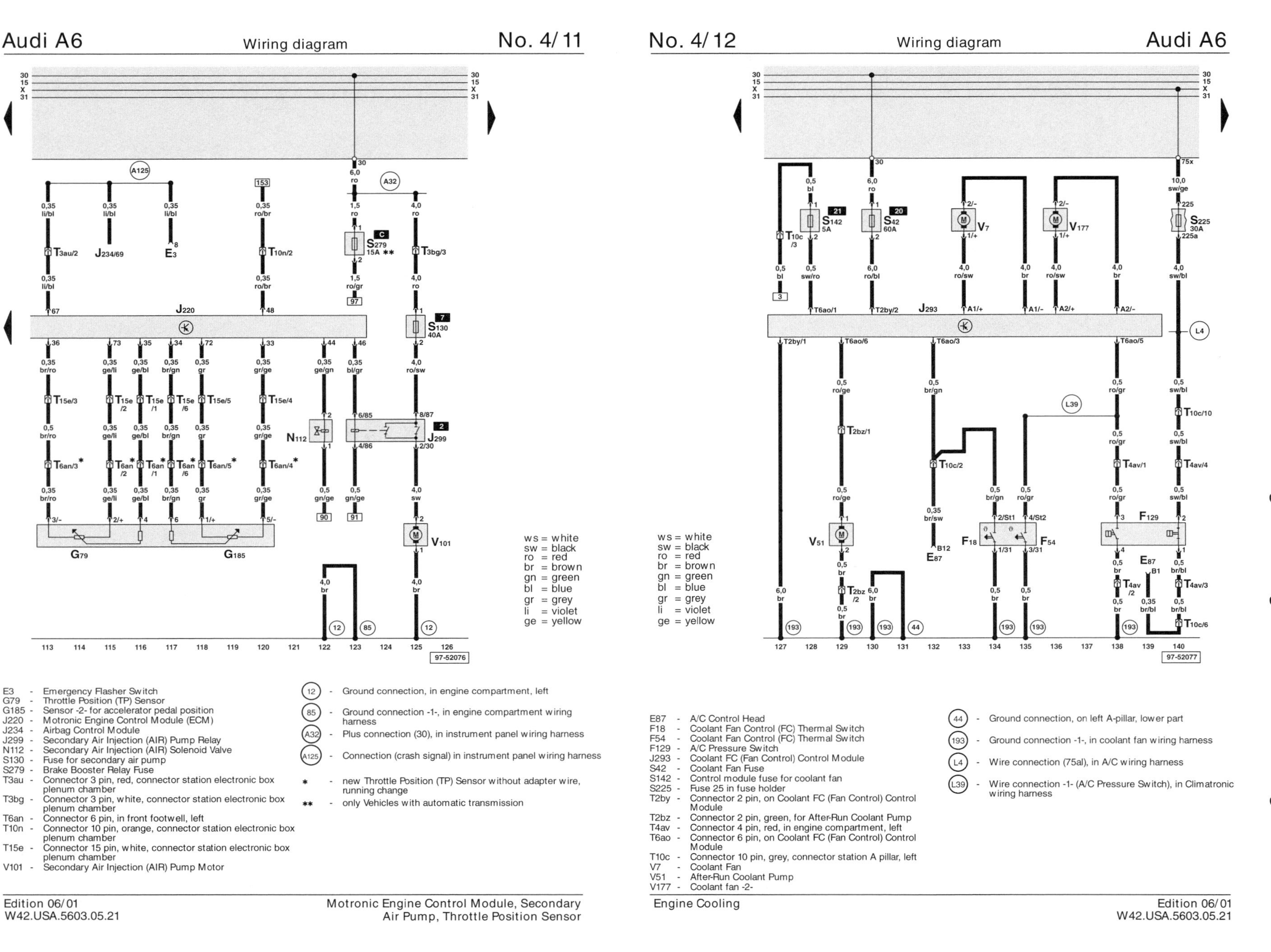

Audi A6
Wiring diagram
No. 4/11
ws = white
sw = black
ro = red
br = brown
gn = green
bl = blue
gr = grey
li = violet
ge = yellow
E3 - Emergency Flasher Switch
G79 - Throttle Position (TP) Sensor
G185 - Sensor -2- for accelerator pedal position
J220 - Motronic Engine Control Module (ECM)
J234 - Airbag Control Module
J299 - Secondary Air Injection (AIR) Pump Relay
N112 - Secondary Air Injection (AIR) Solenoid Valve
S130 - Fuse for secondary air pump
S279 - Brake Booster Relay Fuse
T3au - Connector 3 pin, red, connector station electronic box plenum chamber
T3bg - Connector 3 pin, white, connector station electronic box plenum chamber
T6an - Connector 6 pin, in front footwell, left
T10n - Connector 10 pin, orange, connector station electronic box plenum chamber
T15e - Connector 15 pin, white, connector station electronic box plenum chamber
V101 - Secondary Air Injection (AIR) Pump Motor
12 - Ground connection, in engine compartment, left
85 - Ground connection -1-, in engine compartment wiring harness
A32 - Plus connection (30), in instrument panel wiring harness
A125 - Connection (crash signal) in instrument panel wiring harness
* - new Throttle Position (TP) Sensor without adapter wire, running change
** - only Vehicles with automatic transmission
Edition 06/01
W42.USA.5603.05.21
Motronic Engine Control Module, Secondary Air Pump, Throttle Position Sensor
No. 4/12
Wiring diagram
Audi A6
ws = white
sw = black
ro = red
br = brown
gn = green
bl = blue
gr = grey
li = violet
ge = yellow
E87 - A/C Control Head
F18 - Coolant Fan Control (FC) Thermal Switch
F54 - Coolant Fan Control (FC) Thermal Switch
F129 - A/C Pressure Switch
J293 - Coolant FC (Fan Control) Control Module
S42 - Coolant Fan Fuse
S142 - Control module fuse for coolant fan
S225 - Fuse 25 in fuse holder
T2by - Connector 2 pin, on Coolant FC (Fan Control) Control Module
T2bz - Connector 2 pin, green, for After-Run Coolant Pump
T4av - Connector 4 pin, red, in engine compartment, left
T6ao - Connector 6 pin, on Coolant FC (Fan Control) Control Module
T10c - Connector 10 pin, grey, connector station A pillar, left
V7 - Coolant Fan
V51 - After-Run Coolant Pump
V177 - Coolant fan -2-
44 - Ground connection, on left A-pillar, lower part
193 - Ground connection -1-, in coolant fan wiring harness
L4 - Wire connection (75al), in A/C wiring harness
L39 - Wire connection -1- (A/C Pressure Switch), in Climatronic wiring harness
Engine Cooling
Edition 06/01
W42.USA.5603.05.21

BentleyPublishers.com—All Rights Reserved

Engine Management, 2.8 Liter, Engine Code ATQ

Audi A6 — Wiring diagram — No. 4/13

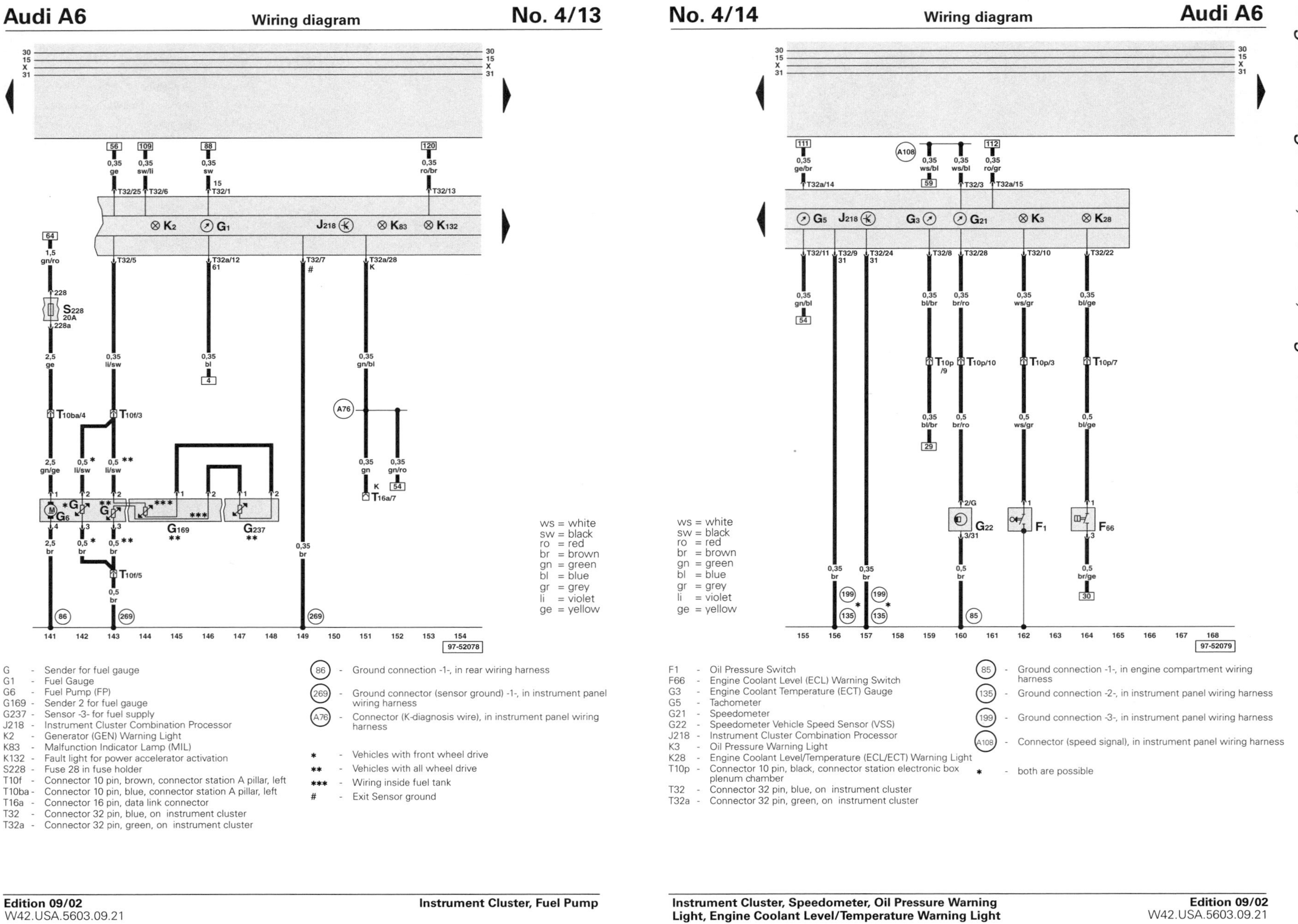

G - Sender for fuel gauge
G1 - Fuel Gauge
G6 - Fuel Pump (FP)
G169 - Sender 2 for fuel gauge
G237 - Sensor -3- for fuel supply
J218 - Instrument Cluster Combination Processor
K2 - Generator (GEN) Warning Light
K83 - Malfunction Indicator Lamp (MIL)
K132 - Fault light for power accelerator activation
S228 - Fuse 28 in fuse holder
T10f - Connector 10 pin, brown, connector station A pillar, left
T10ba - Connector 10 pin, blue, connector station A pillar, left
T16a - Connector 16 pin, data link connector
T32 - Connector 32 pin, blue, on instrument cluster
T32a - Connector 32 pin, green, on instrument cluster

(86) - Ground connection -1-, in rear wiring harness
(269) - Ground connector (sensor ground) -1-, in instrument panel wiring harness
(A76) - Connector (K-diagnosis wire), in instrument panel wiring harness

* - Vehicles with front wheel drive
** - Vehicles with all wheel drive
*** - Wiring inside fuel tank
\# - Exit Sensor ground

ws = white
sw = black
ro = red
br = brown
gn = green
bl = blue
gr = grey
li = violet
ge = yellow

Edition 09/02
W42.USA.5603.09.21

Instrument Cluster, Fuel Pump

No. 4/14 — Wiring diagram — Audi A6

F1 - Oil Pressure Switch
F66 - Engine Coolant Level (ECL) Warning Switch
G3 - Engine Coolant Temperature (ECT) Gauge
G5 - Tachometer
G21 - Speedometer
G22 - Speedometer Vehicle Speed Sensor (VSS)
J218 - Instrument Cluster Combination Processor
K3 - Oil Pressure Warning Light
K28 - Engine Coolant Level/Temperature (ECL/ECT) Warning Light
T10p - Connector 10 pin, black, connector station electronic box plenum chamber
T32 - Connector 32 pin, blue, on instrument cluster
T32a - Connector 32 pin, green, on instrument cluster

(85) - Ground connection -1-, in engine compartment wiring harness
(135) - Ground connection -2-, in instrument panel wiring harness
(199) - Ground connection -3-, in instrument panel wiring harness
(A108) - Connector (speed signal), in instrument panel wiring harness

* - both are possible

ws = white
sw = black
ro = red
br = brown
gn = green
bl = blue
gr = grey
li = violet
ge = yellow

Instrument Cluster, Speedometer, Oil Pressure Warning Light, Engine Coolant Level/Temperature Warning Light

Edition 09/02
W42.USA.5603.09.21

BentleyPublishers.com—All Rights Reserved

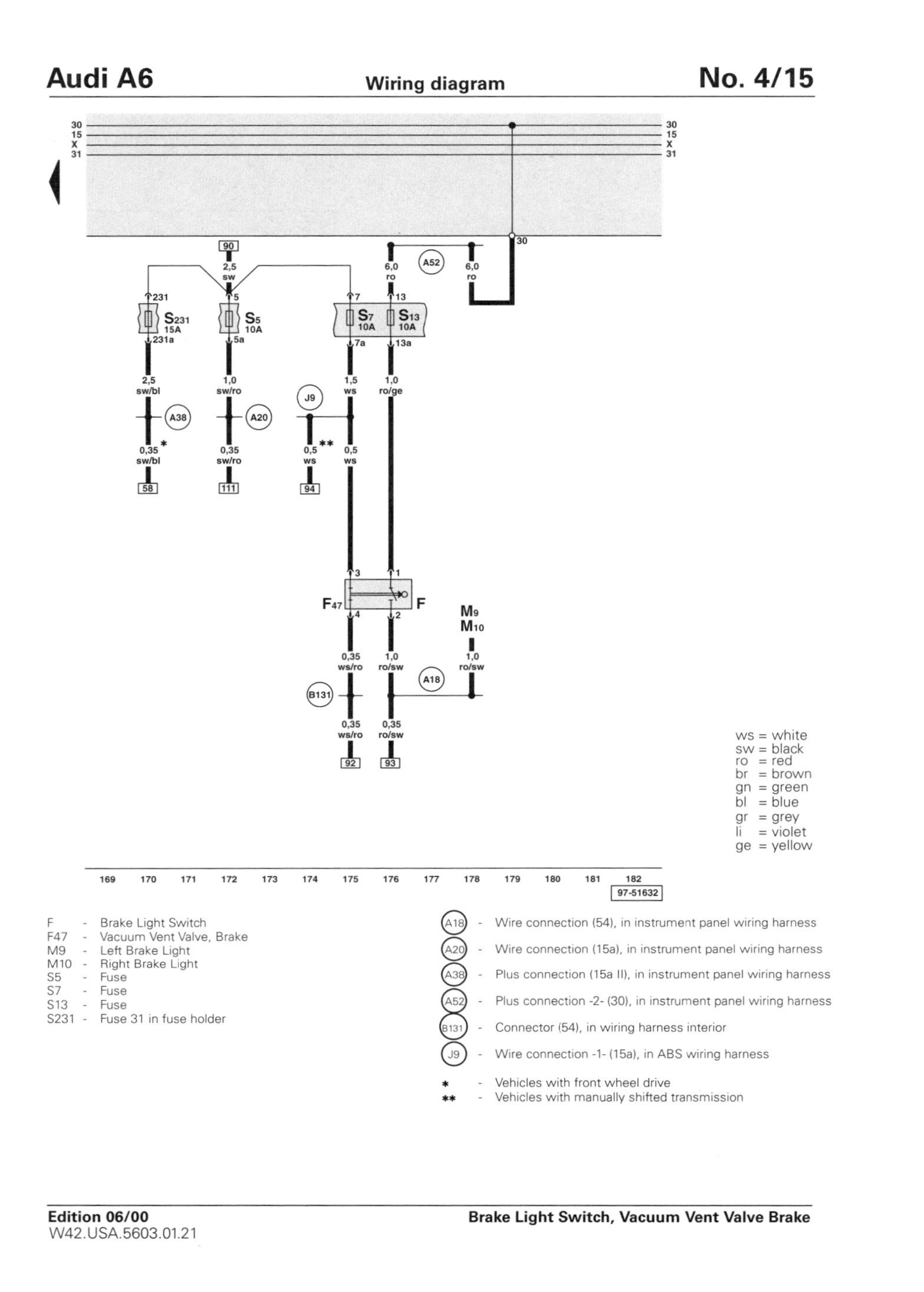

THIS SPACE INTENTIONALLY LEFT BLANK

BentleyPublishers.com—All Rights Reserved

(4,2 l - Injection Engine, 8 Cylinder), Code ART

2001 m. y.

Fuse Panel

Fuse Colors:

30 A - Green
25 A - White
20 A - Yellow
15 A - Blue
10 A - Red
7,5 A - Brown
5 A - Beige

Starting with fuse position 23, fuses in the fuse holder are identified with 223 in the wiring diagram.

Micro Central Electric Panel

Relay Location:

4 -Fuel Pump (FP) Relay , J17

5 -Wiper/Washer Intermittent Relay, J31

6 -Wiper/Washer Intermittent Relay, J31

13 - Fold Relay Panel

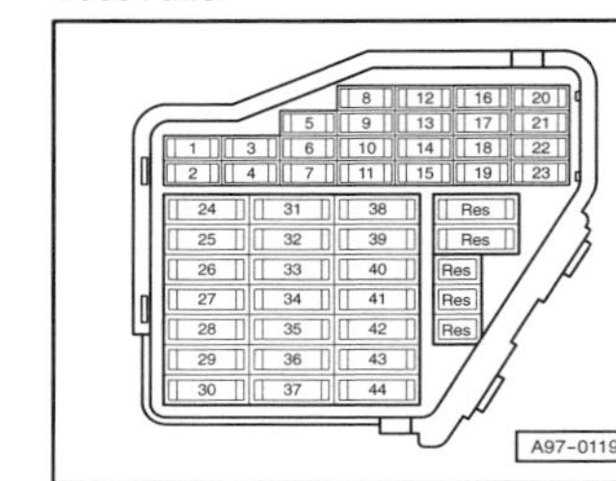

Relay Location:

4 -Starting Interlock Relay, J207

8 - Fold Relay Panel

Fuse Location:

20 -Coolant Fan Fuse, S42

21 -Control module fuse for coolant fan, S142

3 - Fold Relay Panel in E-Box plenum chamber

Relay Location:

2 -Secondary Air Injection (AIR) Pump Relay, J299

3 -Motronic Engine Control Module (ECM) Power Supply Relay, J271

7 -Fuse for secondary air pump , S130

BentleyPublishers.com—All Rights Reserved

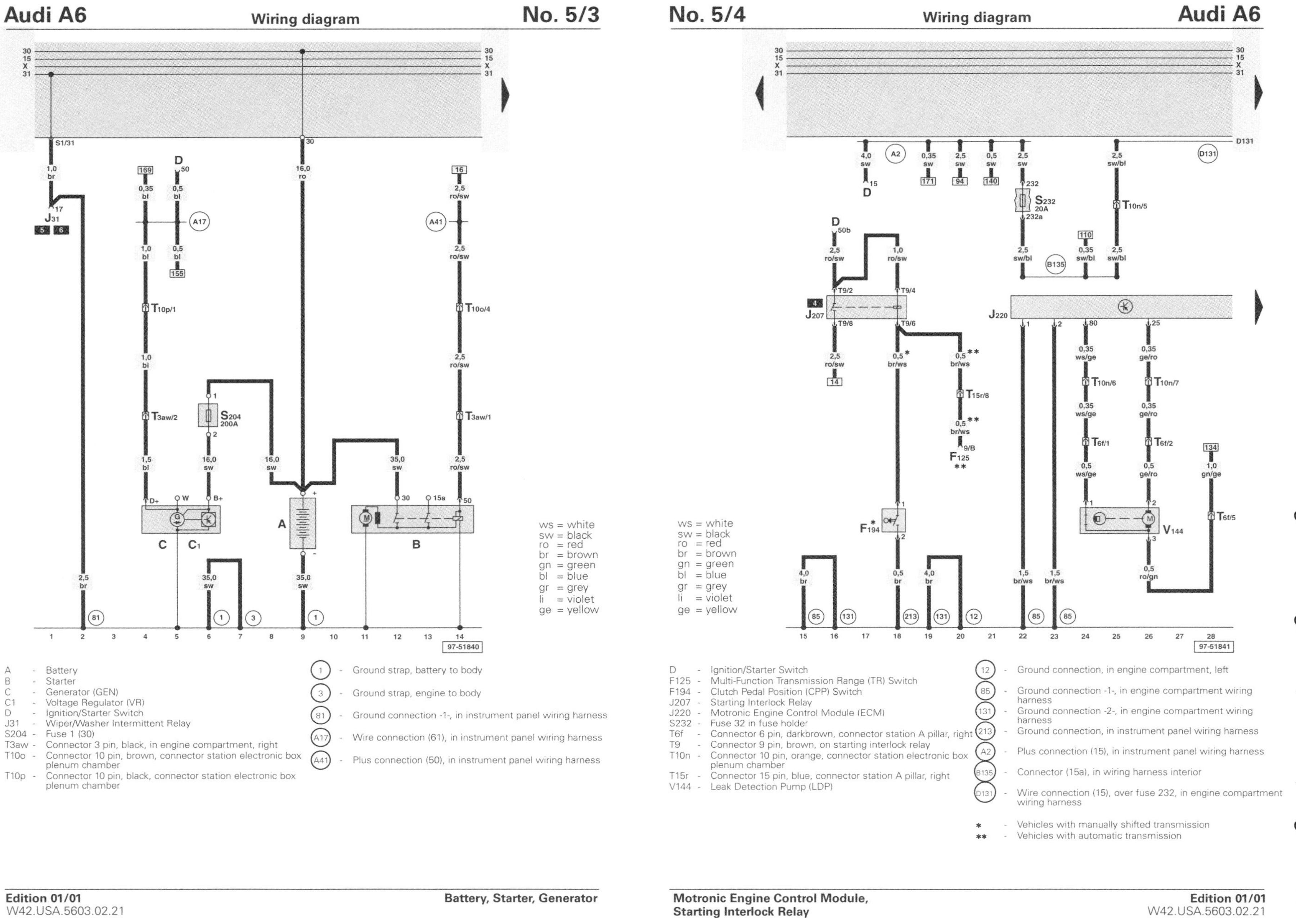

BentleyPublishers.com—All Rights Reserved

Engine Management, 4.2 Liter, Engine Code ART

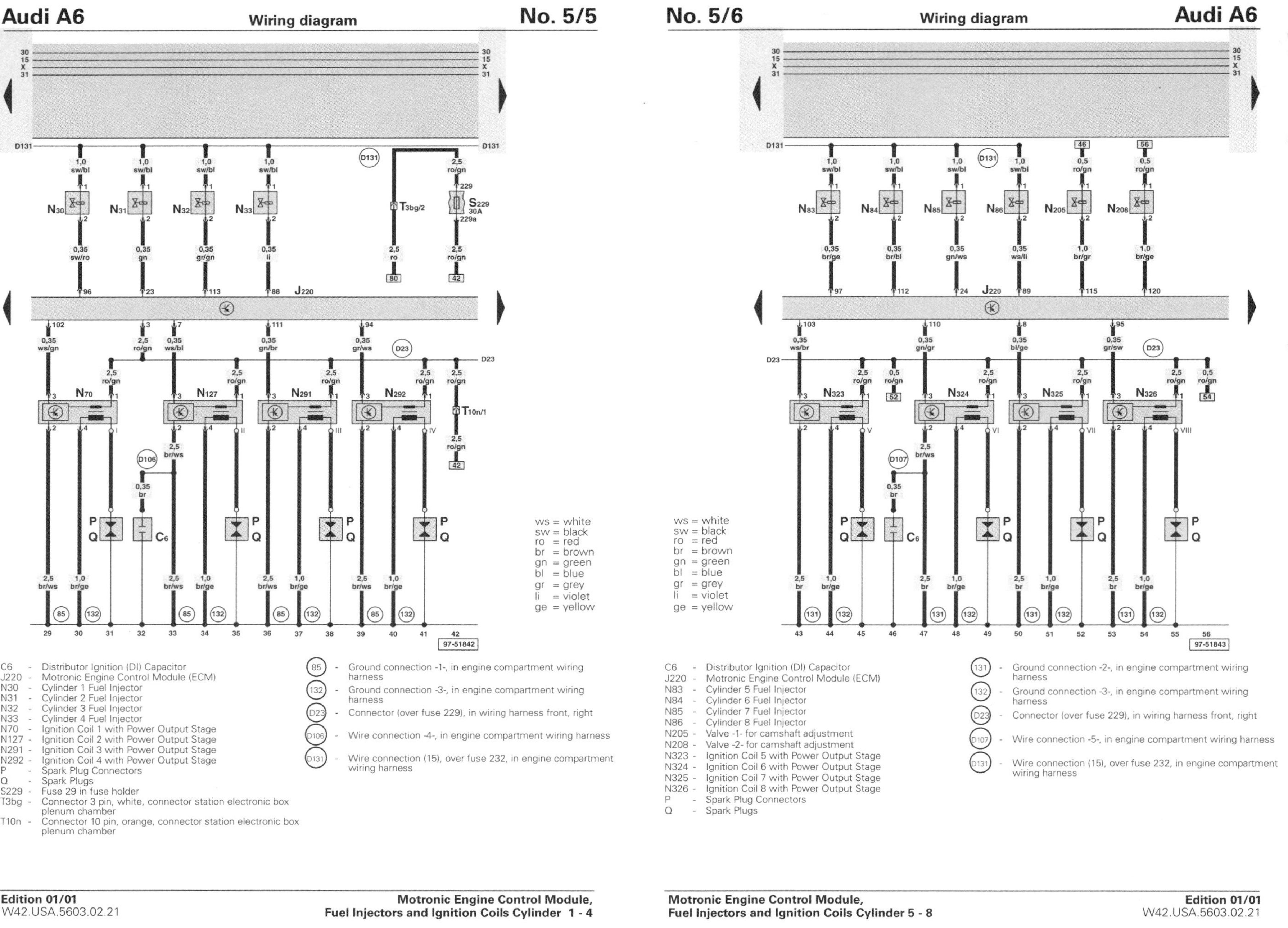

BentleyPublishers.com—All Rights Reserved

Engine Management, 4.2 Liter, Engine Code ART

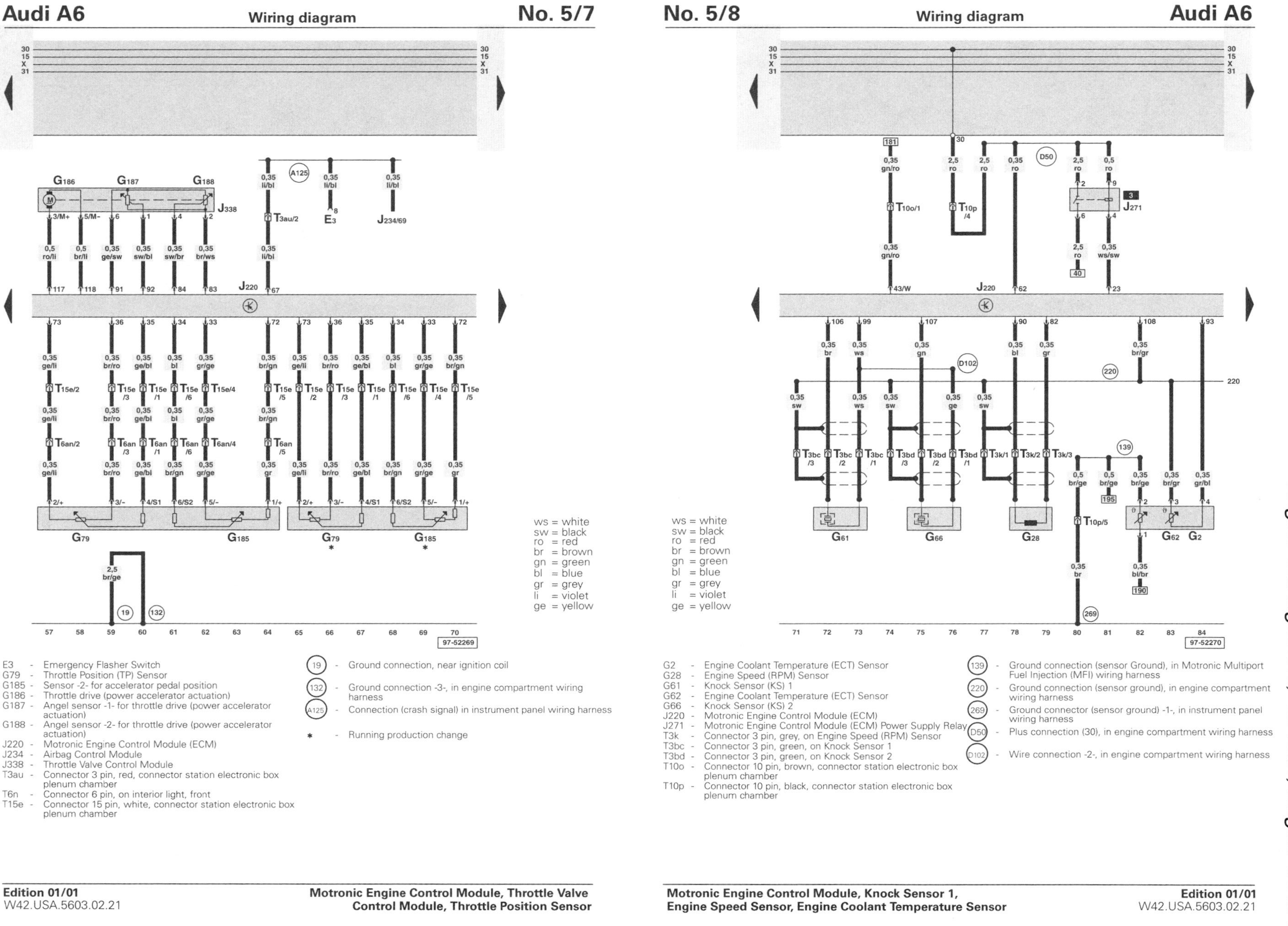

BentleyPublishers.com—All Rights Reserved

Engine Management, 4.2 Liter, Engine Code ART

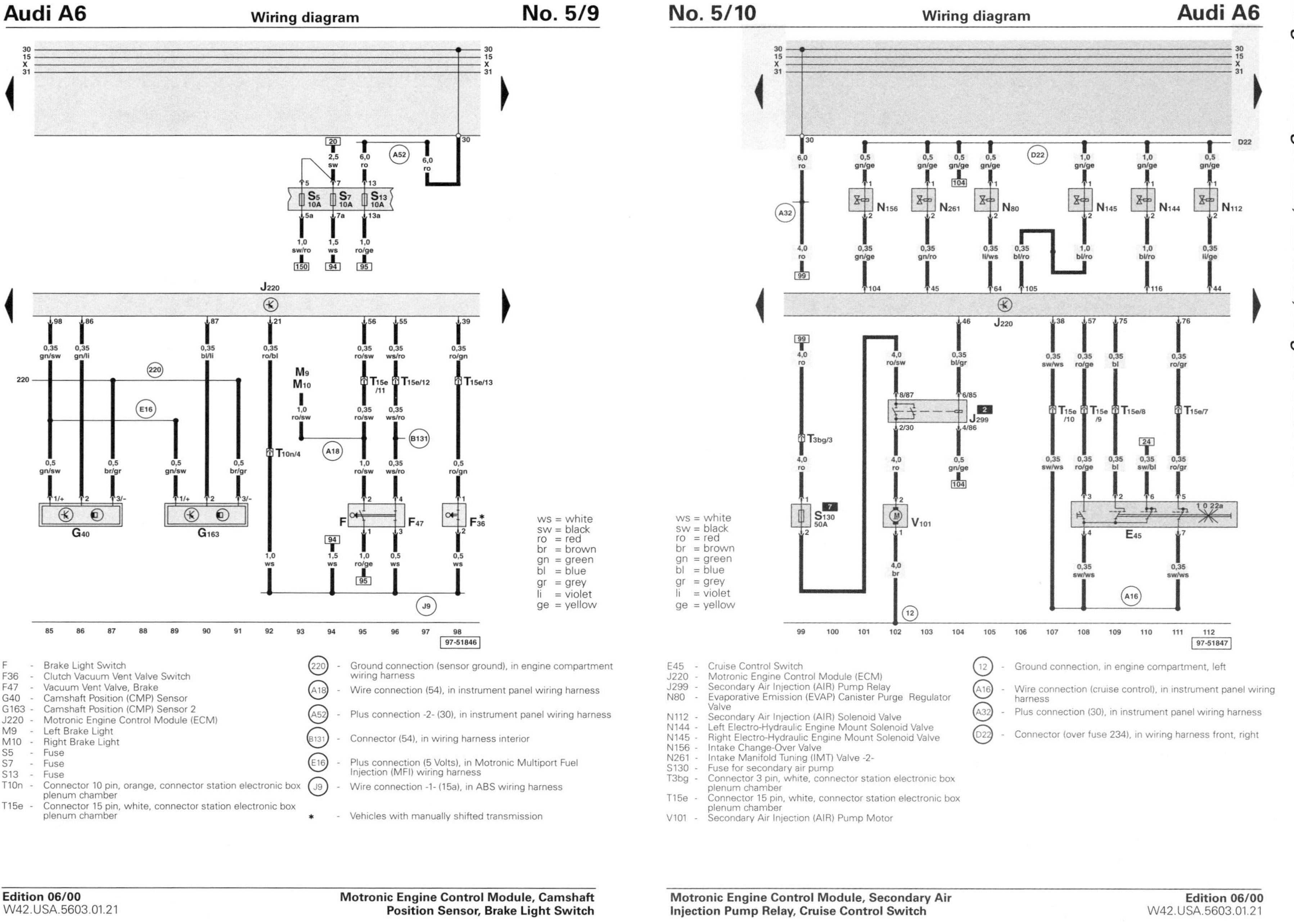

Audi A6
Wiring diagram
No. 5/9
ws = white
sw = black
ro = red
br = brown
gn = green
bl = blue
gr = grey
li = violet
ge = yellow
97-51846
F - Brake Light Switch
F36 - Clutch Vacuum Vent Valve Switch
F47 - Vacuum Vent Valve, Brake
G40 - Camshaft Position (CMP) Sensor
G163 - Camshaft Position (CMP) Sensor 2
J220 - Motronic Engine Control Module (ECM)
M9 - Left Brake Light
M10 - Right Brake Light
S5 - Fuse
S7 - Fuse
S13 - Fuse
T10n - Connector 10 pin, orange, connector station electronic box plenum chamber
T15e - Connector 15 pin, white, connector station electronic box plenum chamber
220 - Ground connection (sensor ground), in engine compartment wiring harness
A18 - Wire connection (54), in instrument panel wiring harness
A52 - Plus connection -2- (30), in instrument panel wiring harness
B131 - Connector (54), in wiring harness interior
E16 - Plus connection (5 Volts), in Motronic Multiport Fuel Injection (MFI) wiring harness
J9 - Wire connection -1- (15a), in ABS wiring harness
* - Vehicles with manually shifted transmission
Edition 06/00
W42.USA.5603.01.21
Motronic Engine Control Module, Camshaft Position Sensor, Brake Light Switch
No. 5/10
Wiring diagram
Audi A6
ws = white
sw = black
ro = red
br = brown
gn = green
bl = blue
gr = grey
li = violet
ge = yellow
97-51847
E45 - Cruise Control Switch
J220 - Motronic Engine Control Module (ECM)
J299 - Secondary Air Injection (AIR) Pump Relay
N80 - Evaporative Emission (EVAP) Canister Purge Regulator Valve
N112 - Secondary Air Injection (AIR) Solenoid Valve
N144 - Left Electro-Hydraulic Engine Mount Solenoid Valve
N145 - Right Electro-Hydraulic Engine Mount Solenoid Valve
N156 - Intake Change-Over Valve
N261 - Intake Manifold Tuning (IMT) Valve -2-
S130 - Fuse for secondary air pump
T3bg - Connector 3 pin, white, connector station electronic box plenum chamber
T15e - Connector 15 pin, white, connector station electronic box plenum chamber
V101 - Secondary Air Injection (AIR) Pump Motor
12 - Ground connection, in engine compartment, left
A16 - Wire connection (cruise control), in instrument panel wiring harness
A32 - Plus connection (30), in instrument panel wiring harness
D22 - Connector (over fuse 234), in wiring harness front, right
Motronic Engine Control Module, Secondary Air Injection Pump Relay, Cruise Control Switch
Edition 06/00
W42.USA.5603.01.21

BentleyPublishers.com—All Rights Reserved

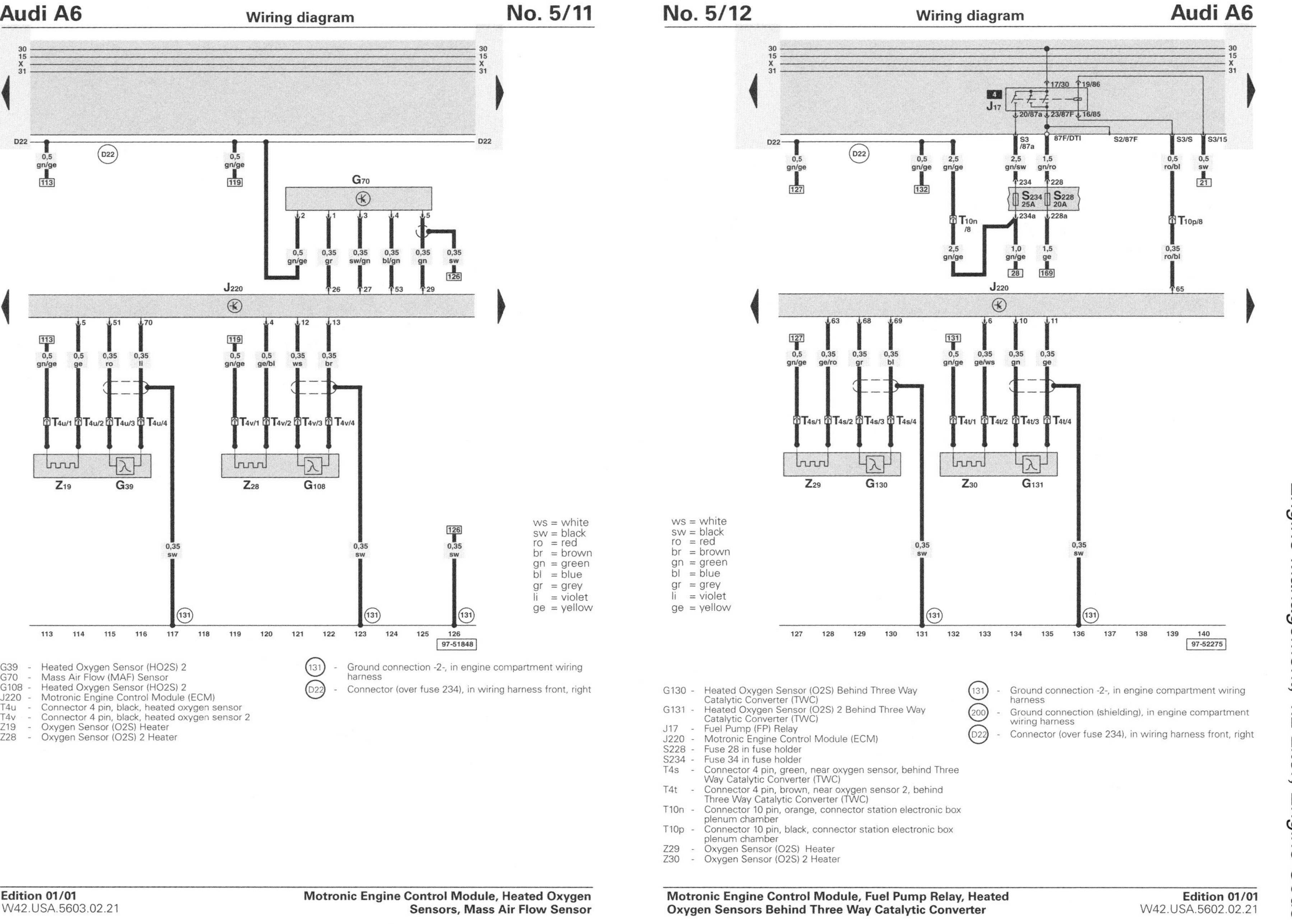

BentleyPublishers.com—All Rights Reserved

Engine Management, 4.2 Liter, Engine Code ART

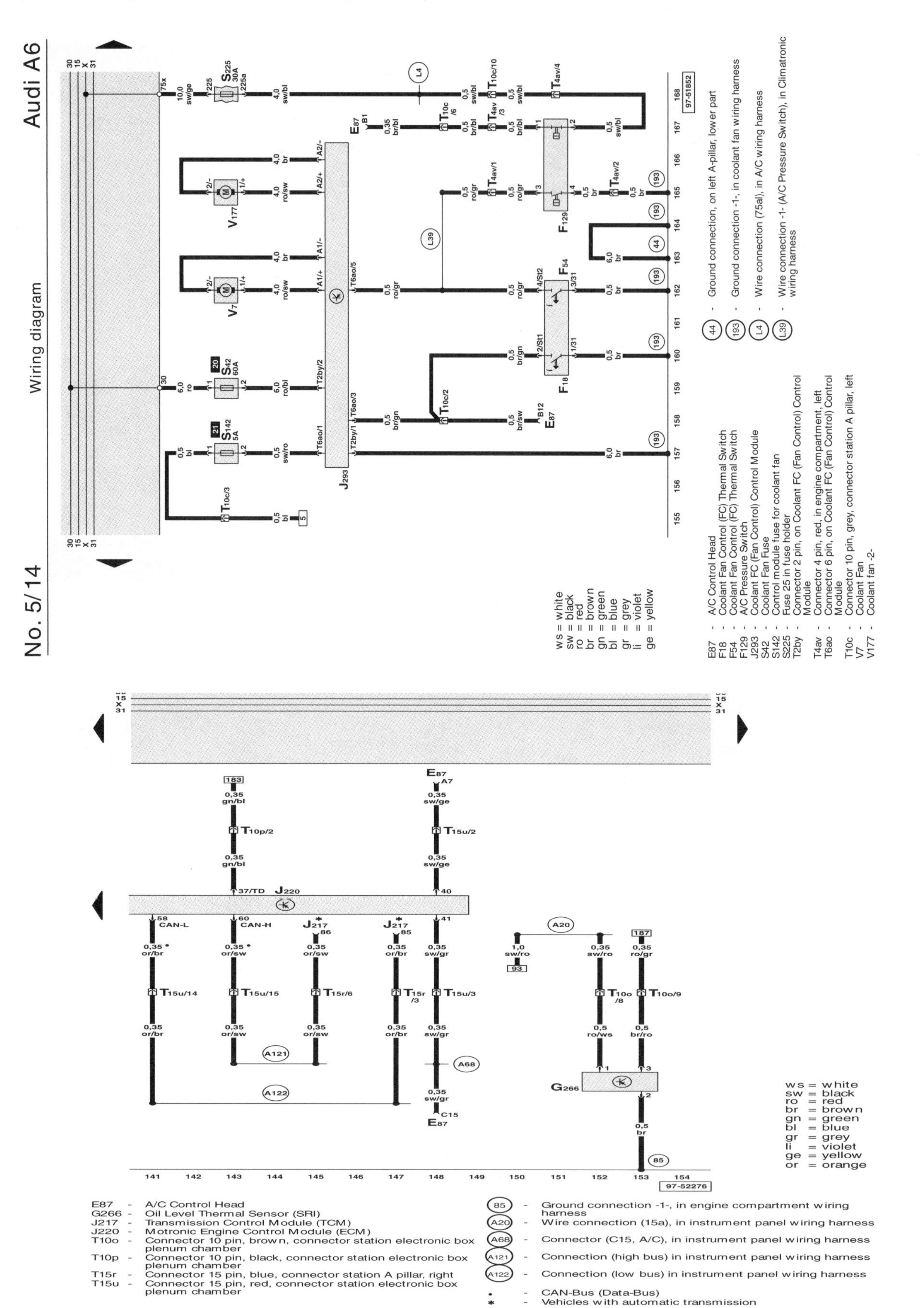

BentleyPublishers.com—All Rights Reserved

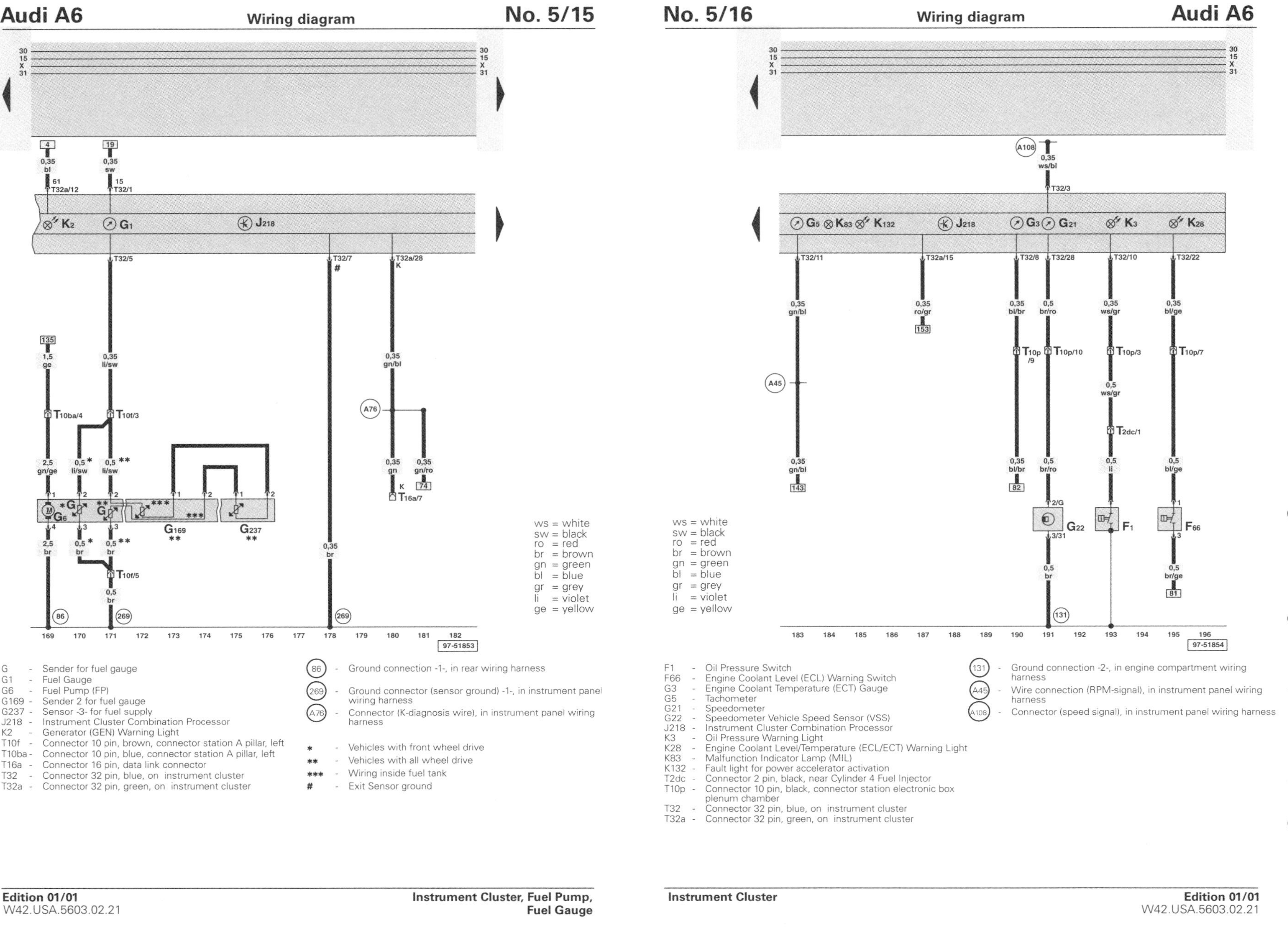

BentleyPublishers.com—All Rights Reserved

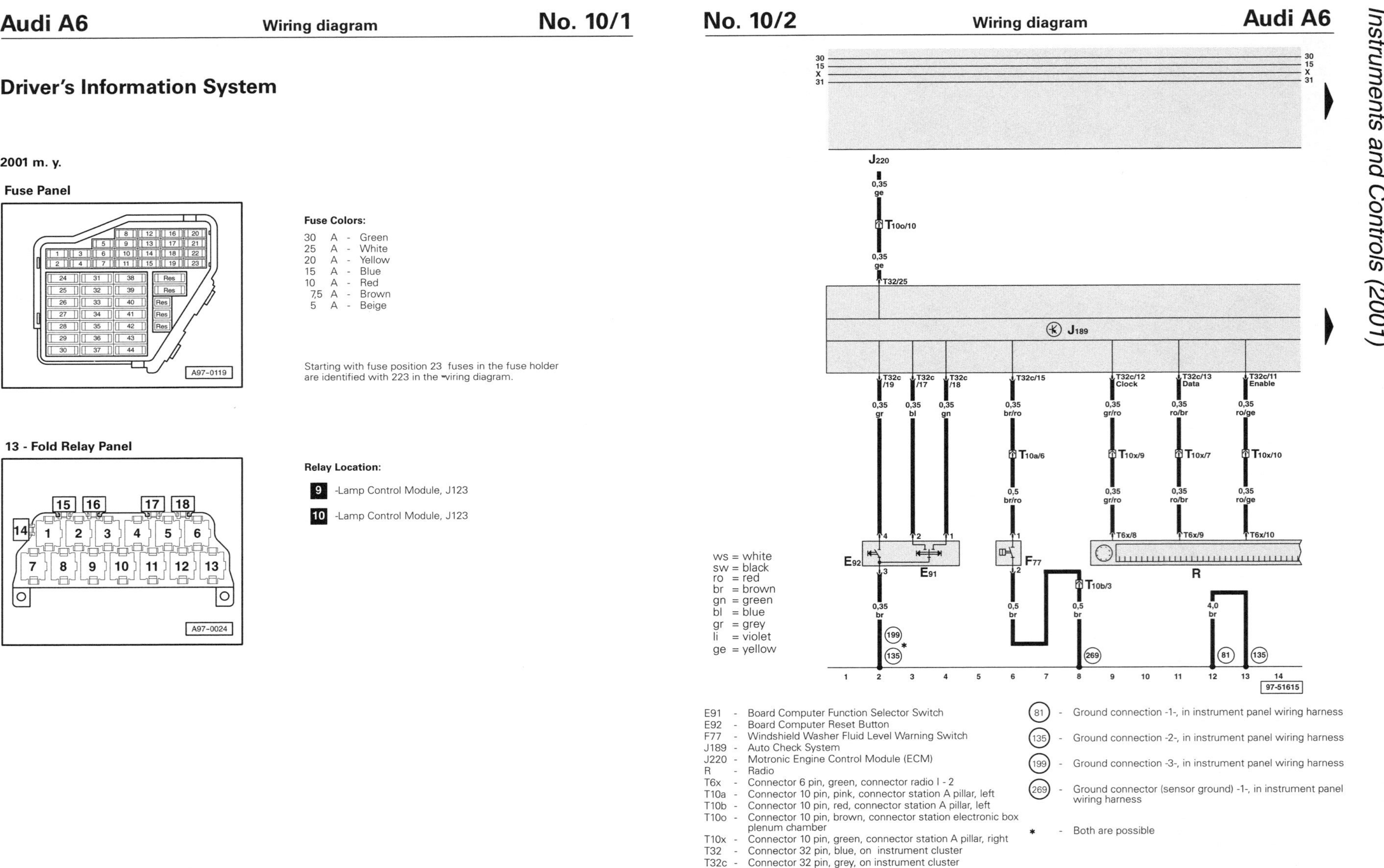

BentleyPublishers.com—All Rights Reserved

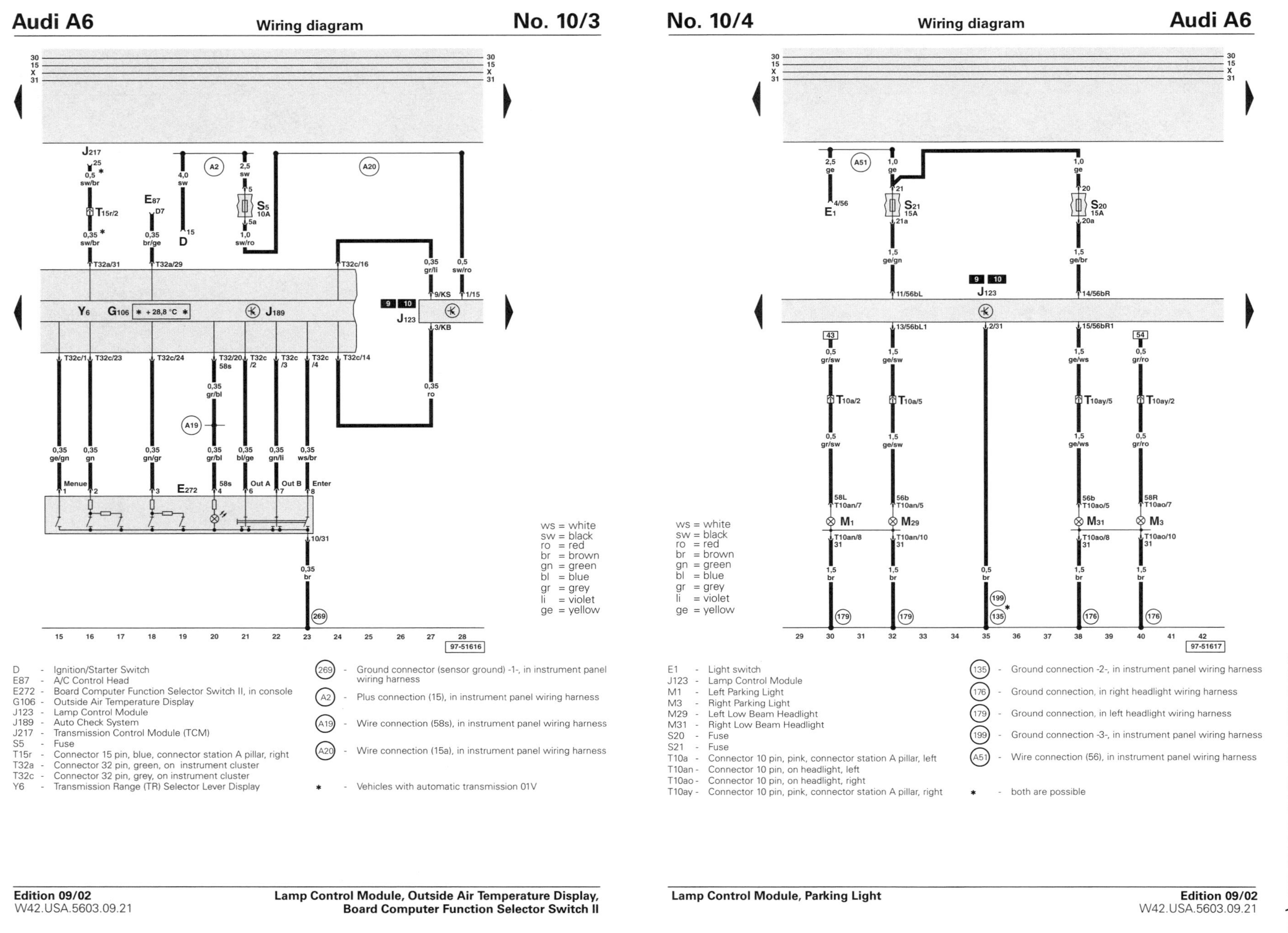

BentleyPublishers.com—All Rights Reserved

Instruments and Controls (2001)

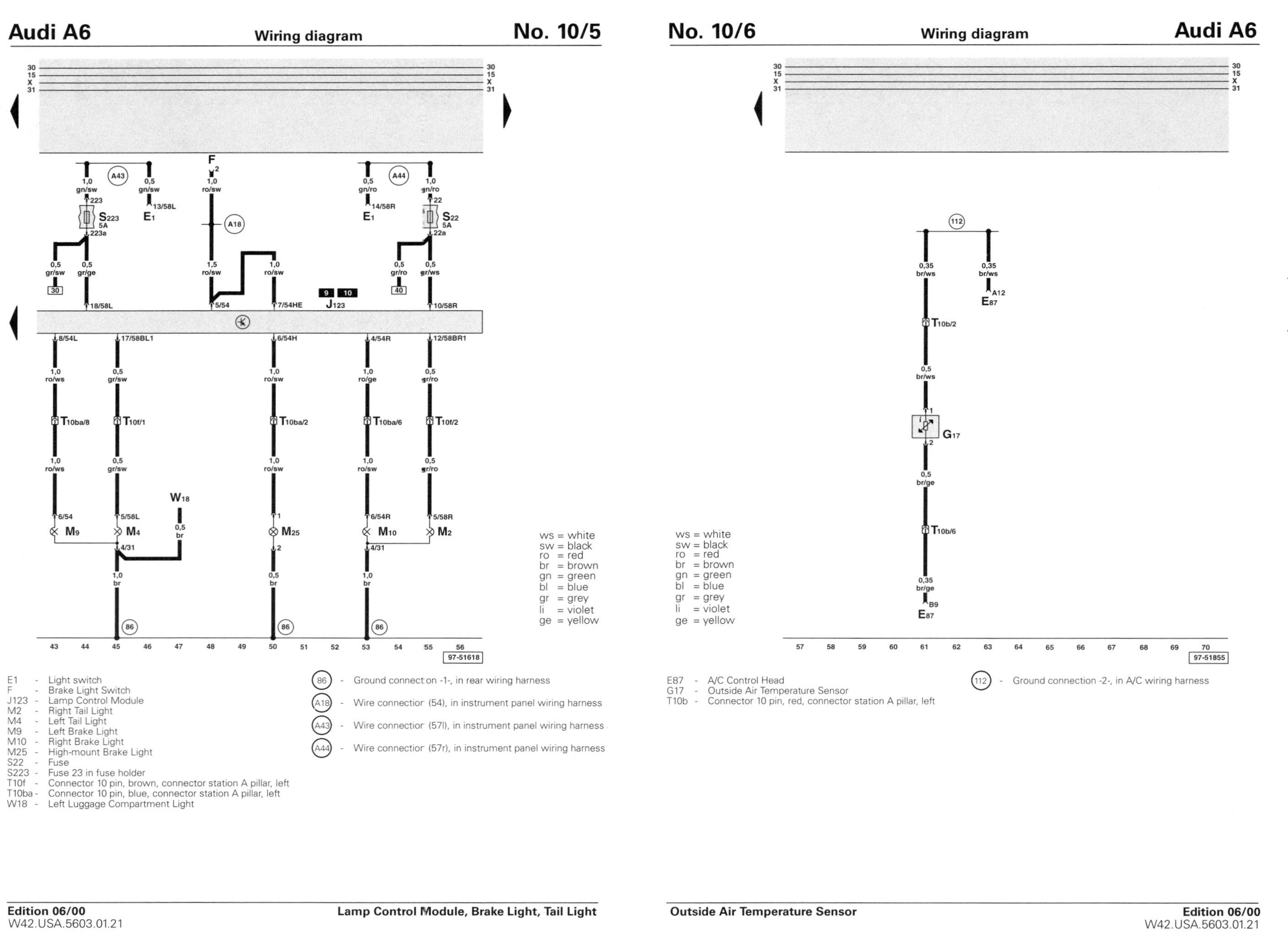

BentleyPublishers.com—All Rights Reserved

Daytime Running Lights with Fog Light and Rear Fog Light (Canada)

2001 m. y.

Fuse Panel

Fuse Colors:

30 A - Green
25 A - White
20 A - Yellow
15 A - Blue
10 A - Red
7,5 A - Brown
5 A - Beige

Starting with fuse position 23, fuses in the fuse holder are identified with 223 in the wiring diagram.

Micro Central Electric Panel

Relay Location:

2 -Load Reduction Relay, J59

13 - Fold Relay Panel

Relay Location:

6 -Fog Light Relay, J5
9 -Lamp Control Module, J123
10 -Lamp Control Module, J123
17 -Fog Light Circuit Breaker, S28

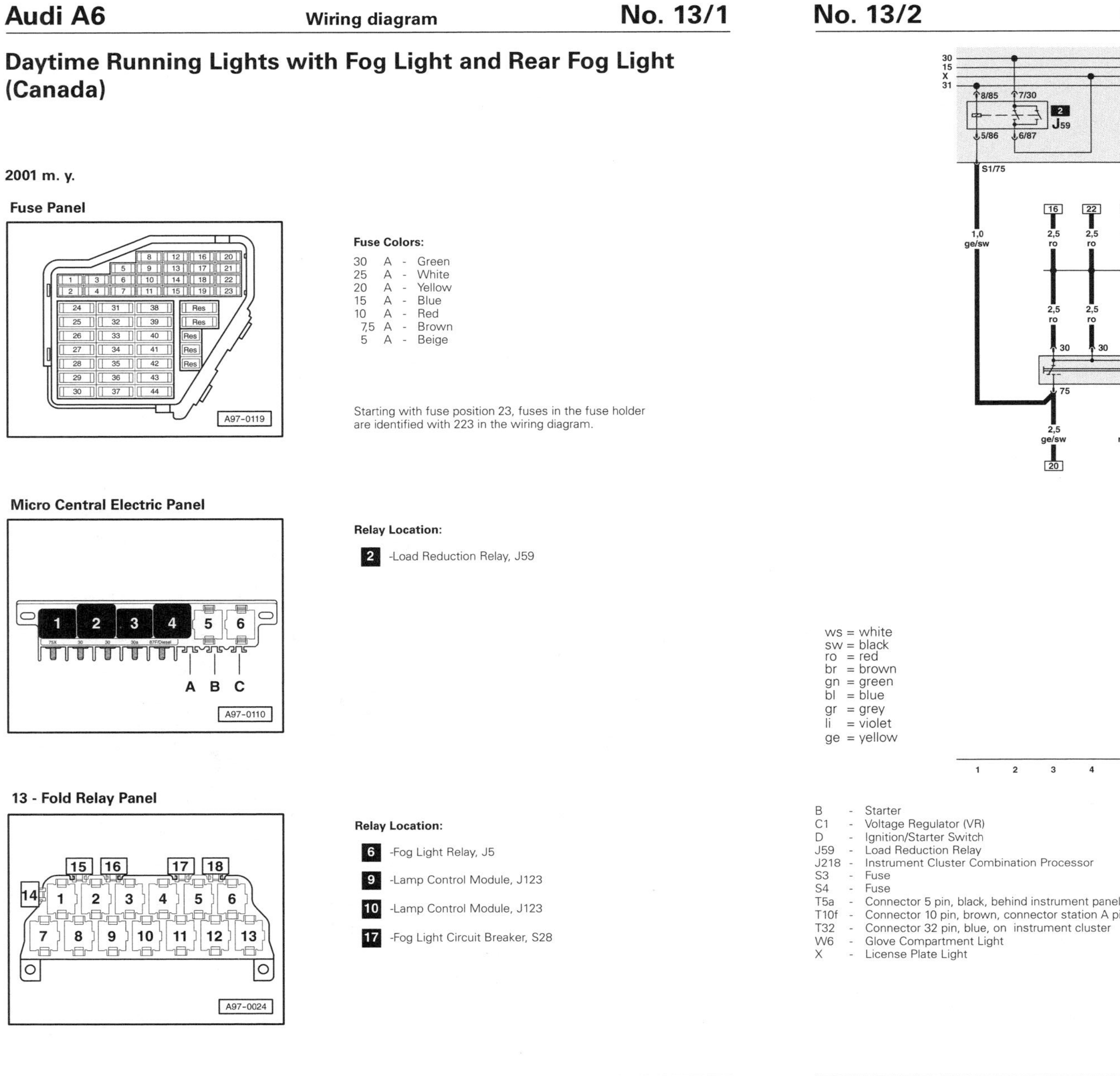

ws = white
sw = black
ro = red
br = brown
gn = green
bl = blue
gr = grey
li = violet
ge = yellow

B - Starter
C1 - Voltage Regulator (VR)
D - Ignition/Starter Switch
J59 - Load Reduction Relay
J218 - Instrument Cluster Combination Processor
S3 - Fuse
S4 - Fuse
T5a - Connector 5 pin, black, behind instrument panel, right
T10f - Connector 10 pin, brown, connector station A pillar, left
T32 - Connector 32 pin, blue, on instrument cluster
W6 - Glove Compartment Light
X - License Plate Light

(87) - Ground connection -2-, in rear wiring harness
(A2) - Plus connection (15), in instrument panel wiring harness
(A21) - Wire connection (86s), in instrument panel wiring harness
(A52) - Plus connection -2- (30), in instrument panel wiring harness
(A113) - Connection 2 (daytime running lights), in instrument panel wiring harness

Ignition/Starter Switch, License Plate Light, Load Reduction Relay

BentleyPublishers.com—All Rights Reserved

Daytime Running Lights (2001)

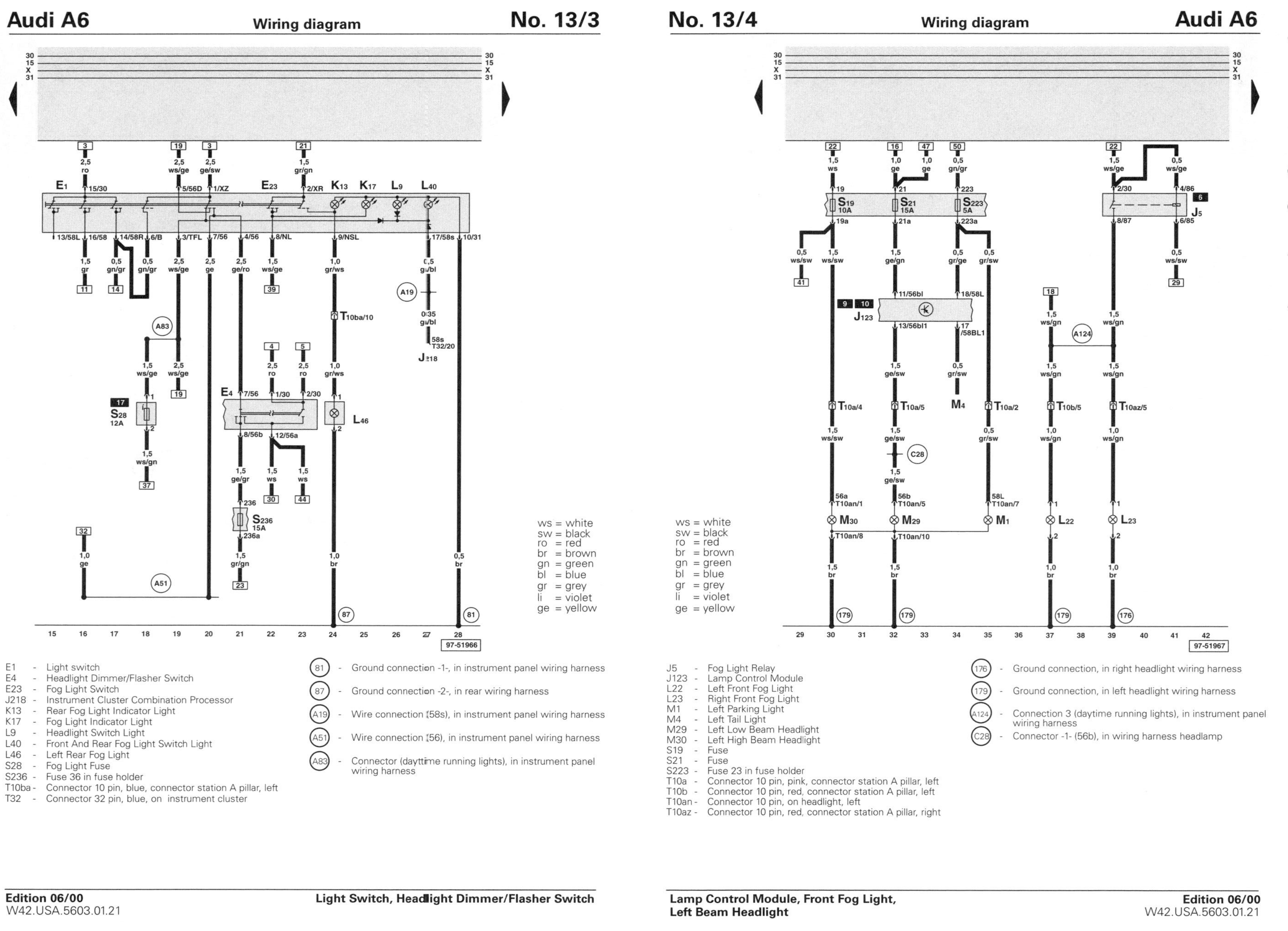

Audi A6
Wiring diagram
No. 13/3
ws = white
sw = black
ro = red
br = brown
gn = green
bl = blue
gr = grey
li = violet
ge = yellow
E1 - Light switch
E4 - Headlight Dimmer/Flasher Switch
E23 - Fog Light Switch
J218 - Instrument Cluster Combination Processor
K13 - Rear Fog Light Indicator Light
K17 - Fog Light Indicator Light
L9 - Headlight Switch Light
L40 - Front And Rear Fog Light Switch Light
L46 - Left Rear Fog Light
S28 - Fog Light Fuse
S236 - Fuse 36 in fuse holder
T10ba - Connector 10 pin, blue, connector station A pillar, left
T32 - Connector 32 pin, blue, on instrument cluster
81 - Ground connection -1-, in instrument panel wiring harness
87 - Ground connection -2-, in rear wiring harness
A19 - Wire connection (58s), in instrument panel wiring harness
A51 - Wire connection (56), in instrument panel wiring harness
A83 - Connector (daytime running lights), in instrument panel wiring harness
97-51966
Light Switch, Headlight Dimmer/Flasher Switch
Edition 06/00
W42.USA.5603.01.21
No. 13/4
Wiring diagram
Audi A6
ws = white
sw = black
ro = red
br = brown
gn = green
bl = blue
gr = grey
li = violet
ge = yellow
J5 - Fog Light Relay
J123 - Lamp Control Module
L22 - Left Front Fog Light
L23 - Right Front Fog Light
M1 - Left Parking Light
M4 - Left Tail Light
M29 - Left Low Beam Headlight
M30 - Left High Beam Headlight
S19 - Fuse
S21 - Fuse
S223 - Fuse 23 in fuse holder
T10a - Connector 10 pin, pink, connector station A pillar, left
T10b - Connector 10 pin, red, connector station A pillar, left
T10an - Connector 10 pin, on headlight, left
T10az - Connector 10 pin, red, connector station A pillar, right
176 - Ground connection, in right headlight wiring harness
179 - Ground connection, in left headlight wiring harness
A124 - Connection 3 (daytime running lights), in instrument panel wiring harness
C28 - Connector -1- (56b), in wiring harness headlamp
97-51967
Lamp Control Module, Front Fog Light, Left Beam Headlight
Edition 06/00
W42.USA.5603.01.21

BentleyPublishers.com—All Rights Reserved

Audi A6 **Wiring diagram** **No. 13/5**

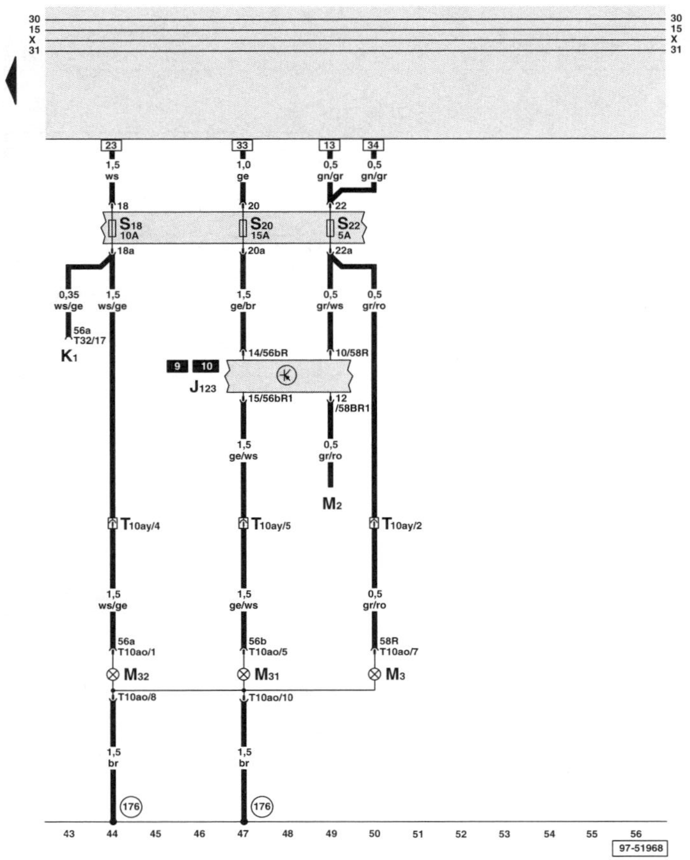

ws = white
sw = black
ro = red
br = brown
gn = green
bl = blue
gr = grey
li = violet
ge = yellow

J123 - Lamp Control Module
K1 - Headlight High Beam Indicator Light
M2 - Right Tail Light
M3 - Right Parking Light
M31 - Right Low Beam Headlight
M32 - Right High Beam Headlight
S18 - Fuse
S20 - Fuse
S22 - Fuse
T10ao - Connector 10 pin, on headlight, right
T10ay - Connector 10 pin, pink, connector station A pillar, right
T32 - Connector 32 pin, blue, on instrument cluster

(176) - Ground connection, in right headlight wiring harness

Edition 06/00
W42.USA.5603.01.21

Lamp Control Module, Right Beam Headlight

THIS SPACE INTENTIONALLY LEFT BLANK

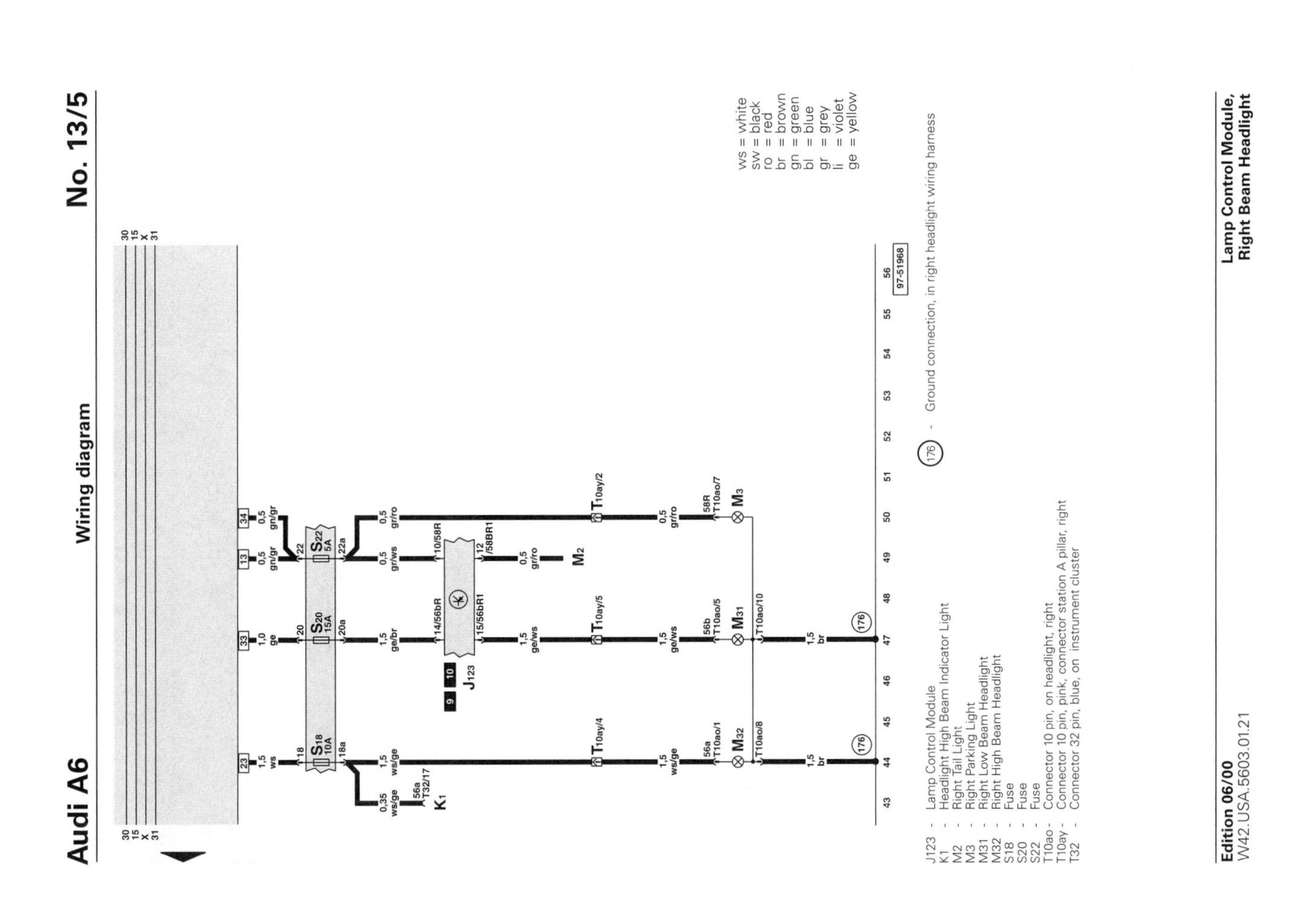

BentleyPublishers.com—All Rights Reserved

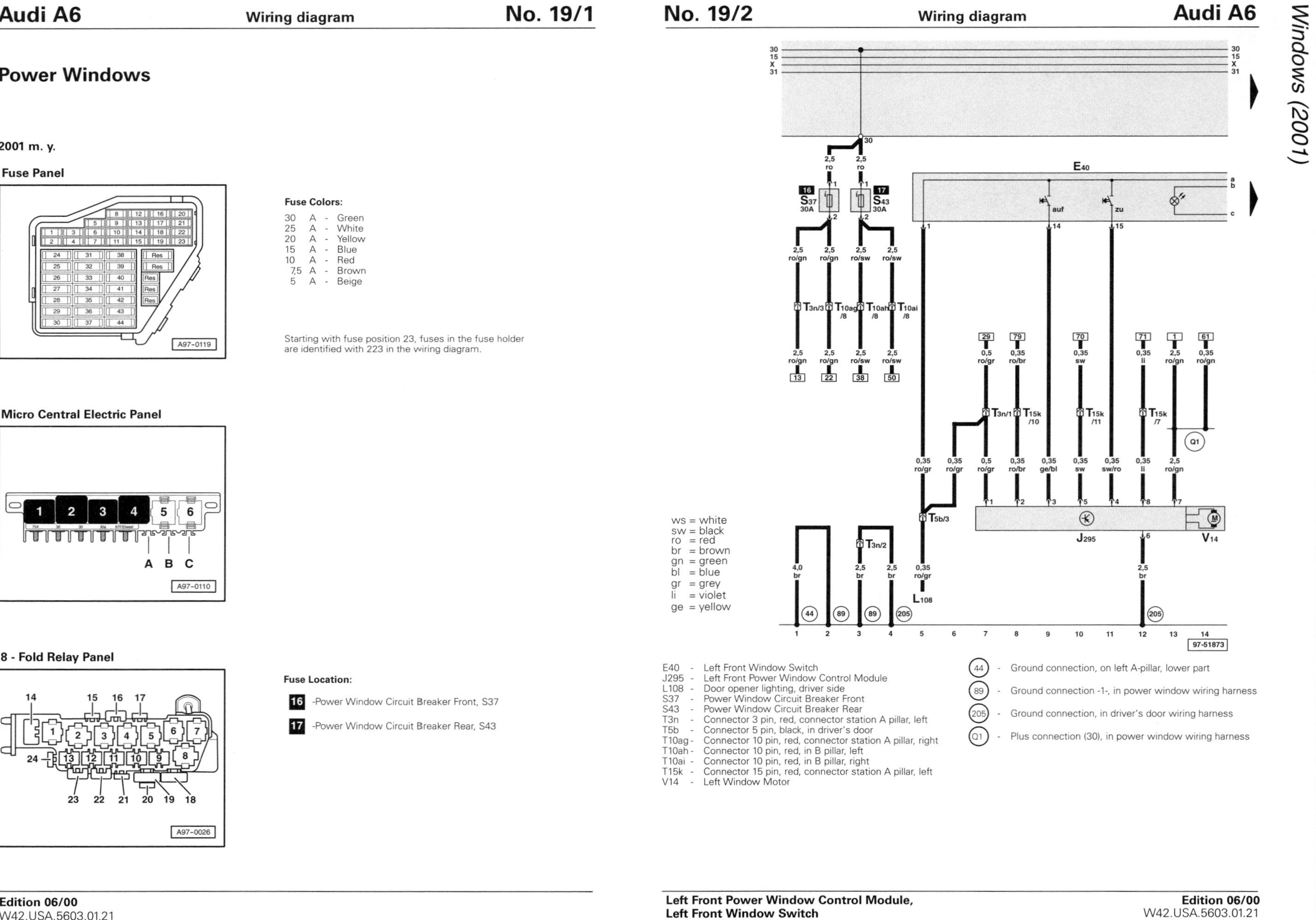

BentleyPublishers.com—All Rights Reserved

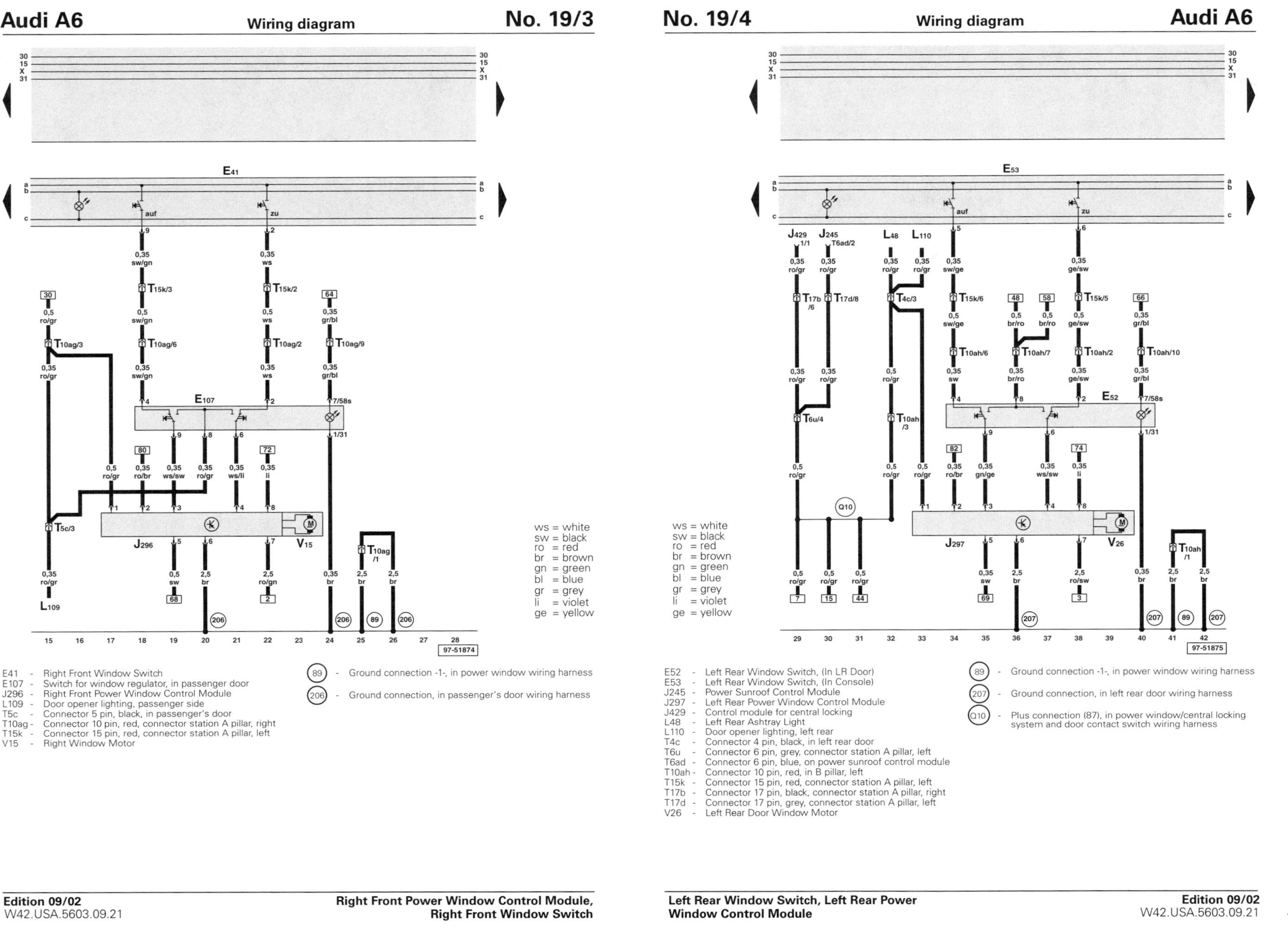

Audi A6 — Wiring diagram — No. 19/3

E41 - Right Front Window Switch
E107 - Switch for window regulator, in passenger door
J296 - Right Front Power Window Control Module
L109 - Door opener lighting, passenger side
T5c - Connector 5 pin, black, in passenger's door
T10ag - Connector 10 pin, red, connector station A pillar, right
T15k - Connector 15 pin, red, connector station A pillar, left
V15 - Right Window Motor

(89) - Ground connection -1-, in power window wiring harness
(206) - Ground connection, in passenger's door wiring harness

Edition 09/02
W42.USA.5603.09.21

Right Front Power Window Control Module, Right Front Window Switch

No. 19/4 — Wiring diagram — Audi A6

E52 - Left Rear Window Switch, (In LR Door)
E53 - Left Rear Window Switch, (In Console)
J245 - Power Sunroof Control Module
J297 - Left Rear Power Window Control Module
J429 - Control module for central locking
L48 - Left Rear Ashtray Light
L110 - Door opener lighting, left rear
T4c - Connector 4 pin, black, in left rear door
T6u - Connector 6 pin, grey, connector station A pillar, left
T6ad - Connector 6 pin, blue, on power sunroof control module
T10ah - Connector 10 pin, red, in B pillar, left
T15k - Connector 15 pin, red, connector station A pillar, left
T17b - Connector 17 pin, black, connector station A pillar, right
T17d - Connector 17 pin, grey, connector station A pillar, left
V26 - Left Rear Door Window Motor

(89) - Ground connection -1-, in power window wiring harness
(207) - Ground connection, in left rear door wiring harness
(Q10) - Plus connection (87), in power window/central locking system and door contact switch wiring harness

Left Rear Window Switch, Left Rear Power Window Control Module

Edition 09/02
W42.USA.5603.09.21

BentleyPublishers.com—All Rights Reserved

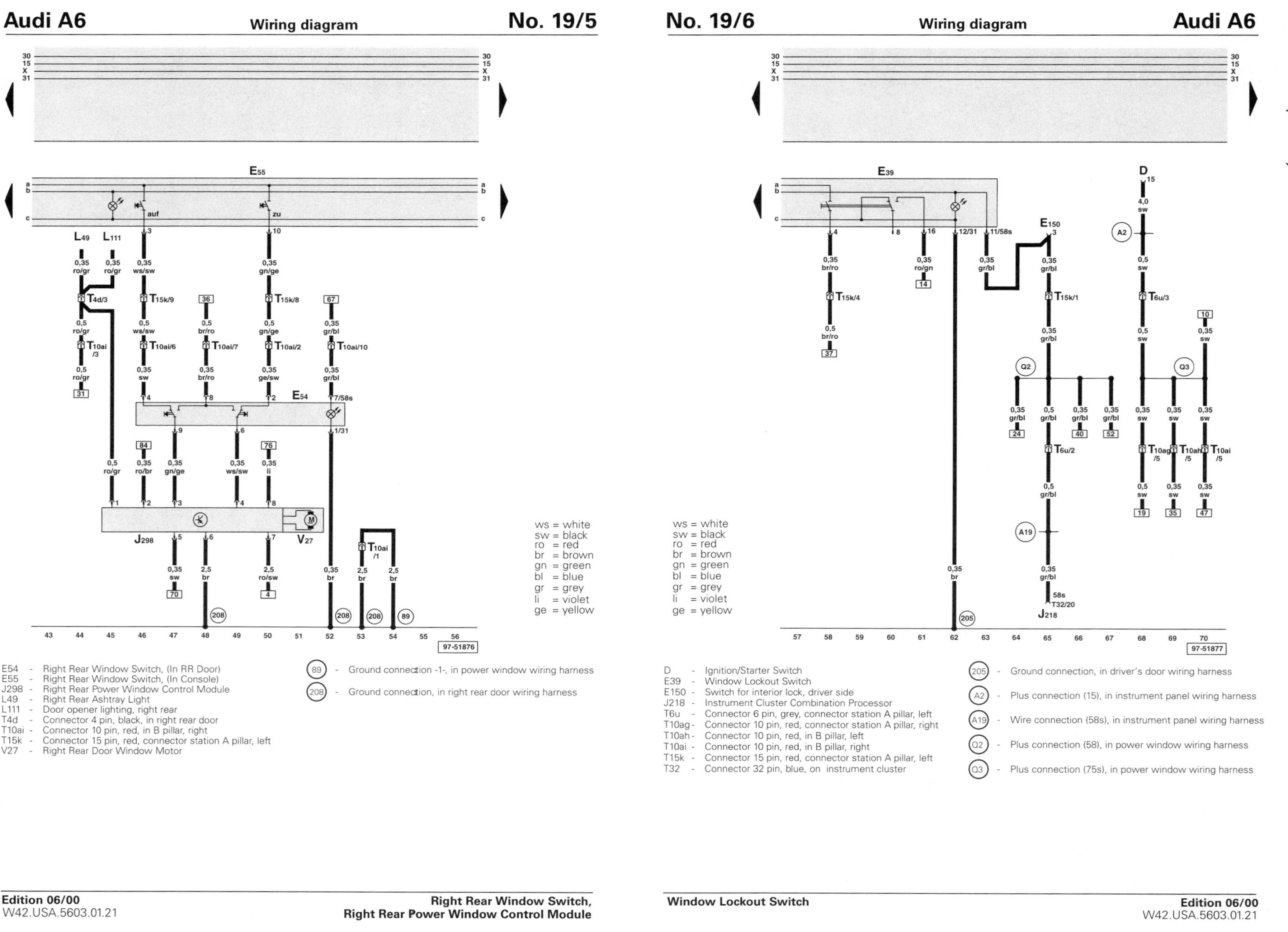

BentleyPublishers.com—All Rights Reserved

THIS SPACE INTENTIONALLY LEFT BLANK

Audi A6 — Wiring diagram — No. 19/7

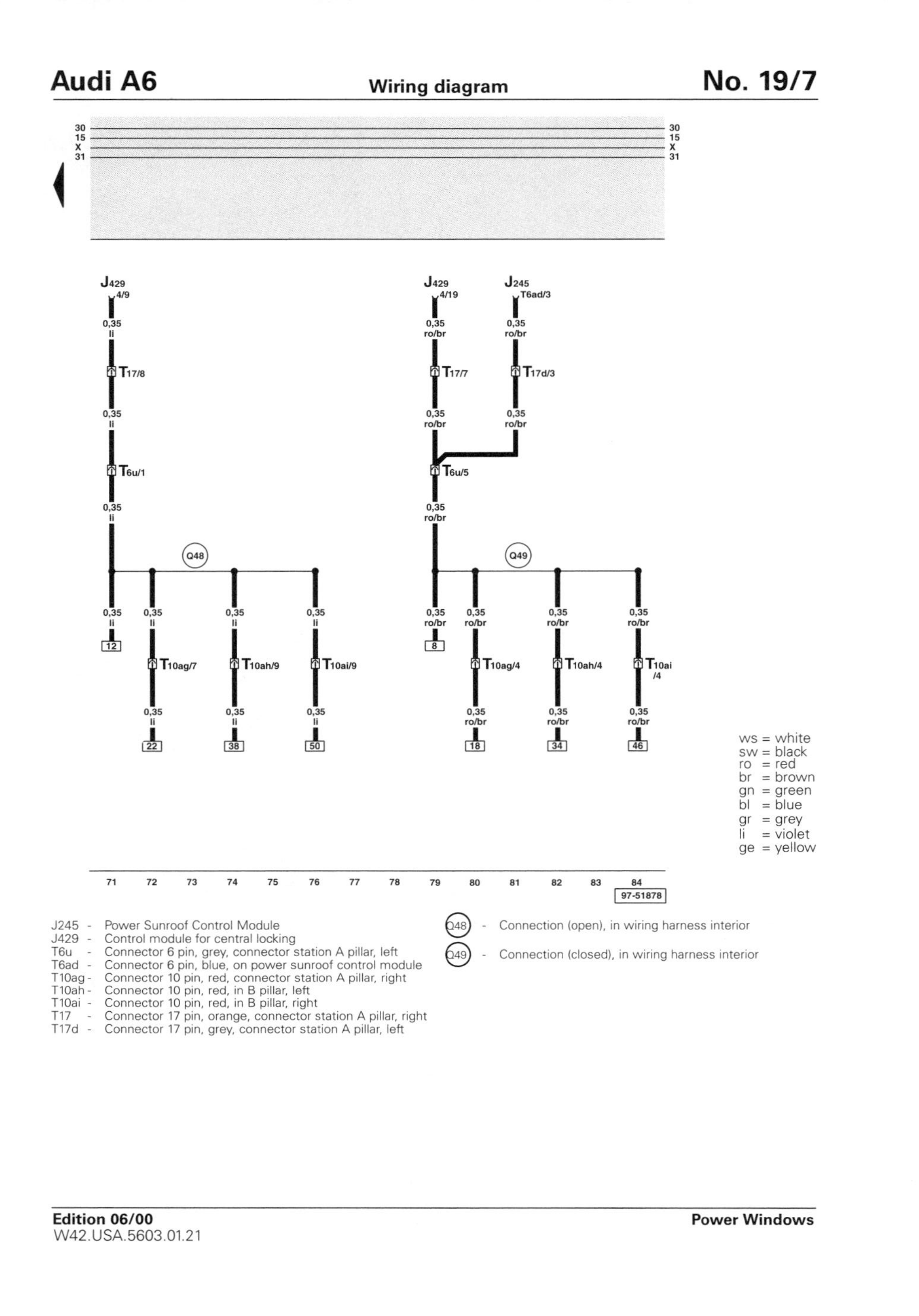

ws = white
sw = black
ro = red
br = brown
gn = green
bl = blue
gr = grey
li = violet
ge = yellow

J245 - Power Sunroof Control Module
J429 - Control module for central locking
T6u - Connector 6 pin, grey, connector station A pillar, left
T6ad - Connector 6 pin, blue, on power sunroof control module
T10ag - Connector 10 pin, red, connector station A pillar, right
T10ah - Connector 10 pin, red, in B pillar, left
T10ai - Connector 10 pin, red, in B pillar, right
T17 - Connector 17 pin, orange, connector station A pillar, right
T17d - Connector 17 pin, grey, connector station A pillar, left

Q48 - Connection (open), in wiring harness interior
Q49 - Connection (closed), in wiring harness interior

BentleyPublishers.com—All Rights Reserved

Heated Seats

2001 m. y.

Fuse Panel

A97-0119

Fuse Colors:

30 A - Green
25 A - White
20 A - Yellow
15 A - Blue
10 A - Red
7,5 A - Brown
5 A - Beige

Starting with fuse position 23, fuses in the fuse holder are identified with 223 in the wiring diagram.

Micro Central Electric Panel

A B C

A97-0110

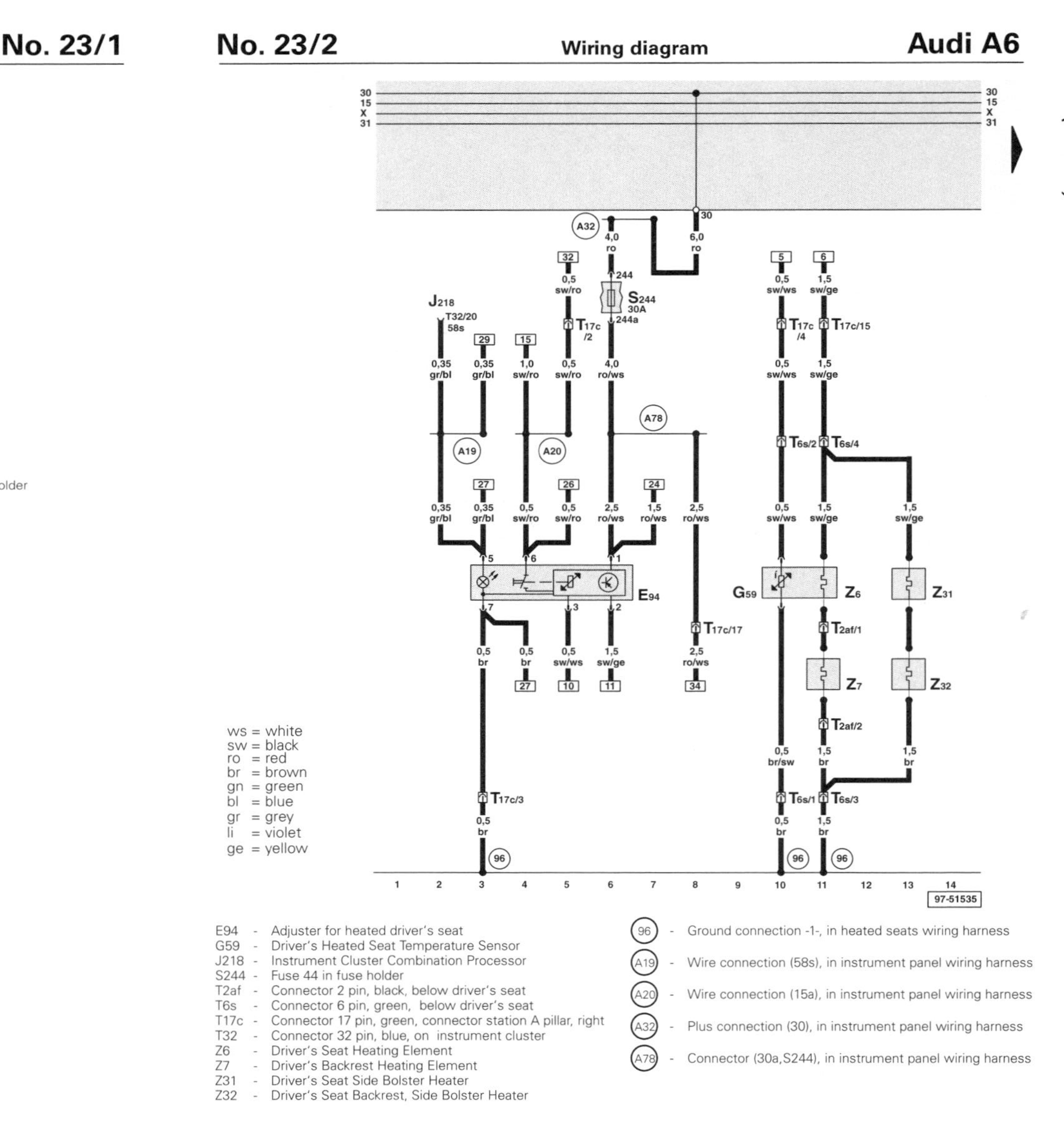

ws = white
sw = black
ro = red
br = brown
gn = green
bl = blue
gr = grey
li = violet
ge = yellow

E94 - Adjuster for heated driver's seat
G59 - Driver's Heated Seat Temperature Sensor
J218 - Instrument Cluster Combination Processor
S244 - Fuse 44 in fuse holder
T2af - Connector 2 pin, black, below driver's seat
T6s - Connector 6 pin, green, below driver's seat
T17c - Connector 17 pin, green, connector station A pillar, right
T32 - Connector 32 pin, blue, on instrument cluster
Z6 - Driver's Seat Heating Element
Z7 - Driver's Backrest Heating Element
Z31 - Driver's Seat Side Bolster Heater
Z32 - Driver's Seat Backrest, Side Bolster Heater

(96) - Ground connection -1-, in heated seats wiring harness
(A19) - Wire connection (58s), in instrument panel wiring harness
(A20) - Wire connection (15a), in instrument panel wiring harness
(A32) - Plus connection (30), in instrument panel wiring harness
(A78) - Connector (30a,S244), in instrument panel wiring harness

Adjuster for Heated Driver's Seat,
Driver's Backrest Heating Element

BentleyPublishers.com—All Rights Reserved

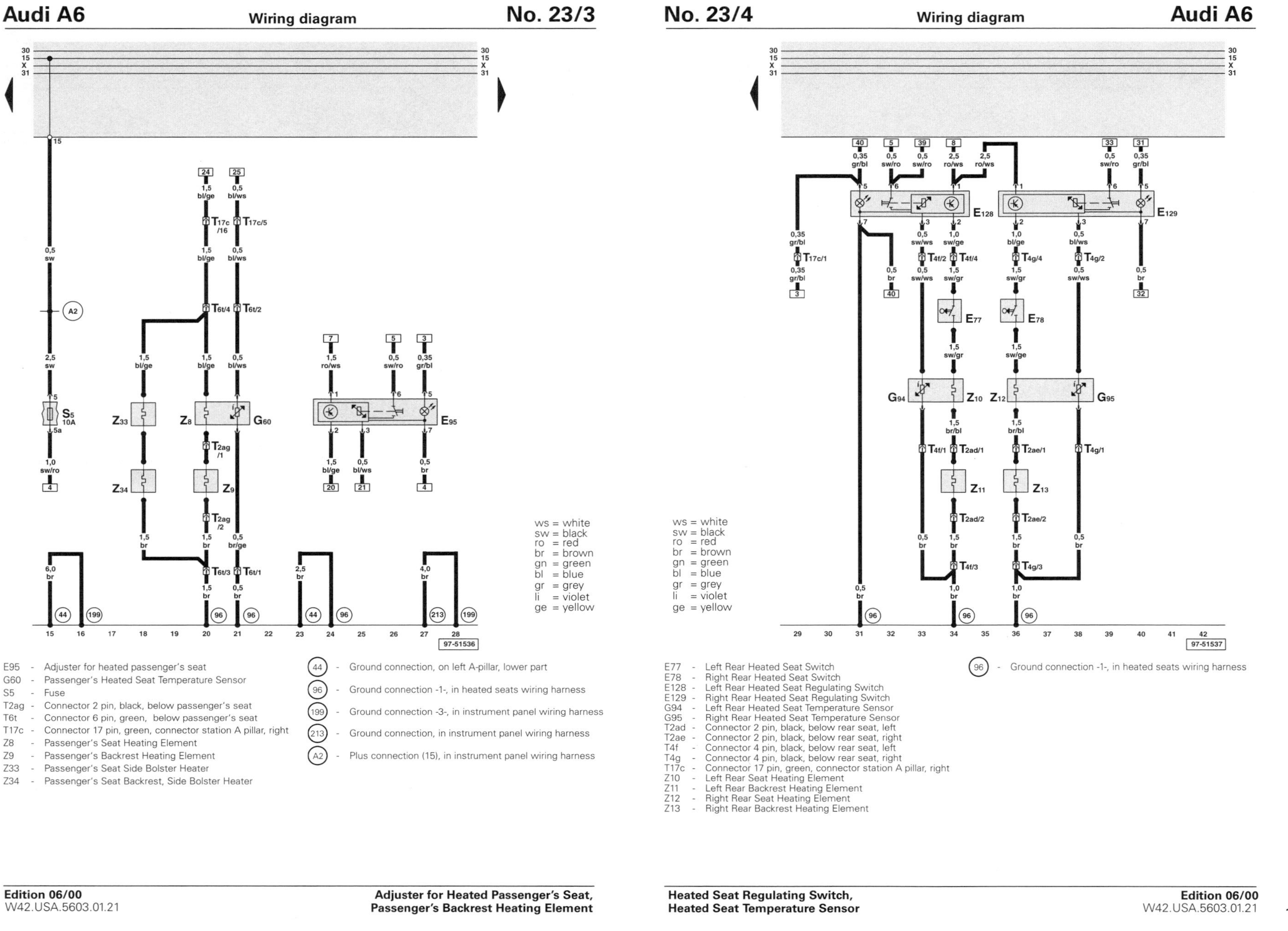

BentleyPublishers.com—All Rights Reserved

Headlight with High Intensity Gas discharge Lamps and automatic dynamic Headlight Beam Adjusting

2001 m. y.

Fuse Panel

A97-0119

Fuse Colors:

30 A - Green
25 A - White
20 A - Yellow
15 A - Blue
10 A - Red
7,5 A - Brown
5 A - Beige

Starting with fuse position 23, fuses in the fuse holder are identified with 223 in the wiring diagram.

Micro Central Electric Panel

A97-0110

Relay Location:

2 -Load Reduction Relay, J59

13 - Fold Relay Panel

A97-0024

Relay Location:

9 -Lamp Control Module, J123

10 -Lamp Control Module, J123

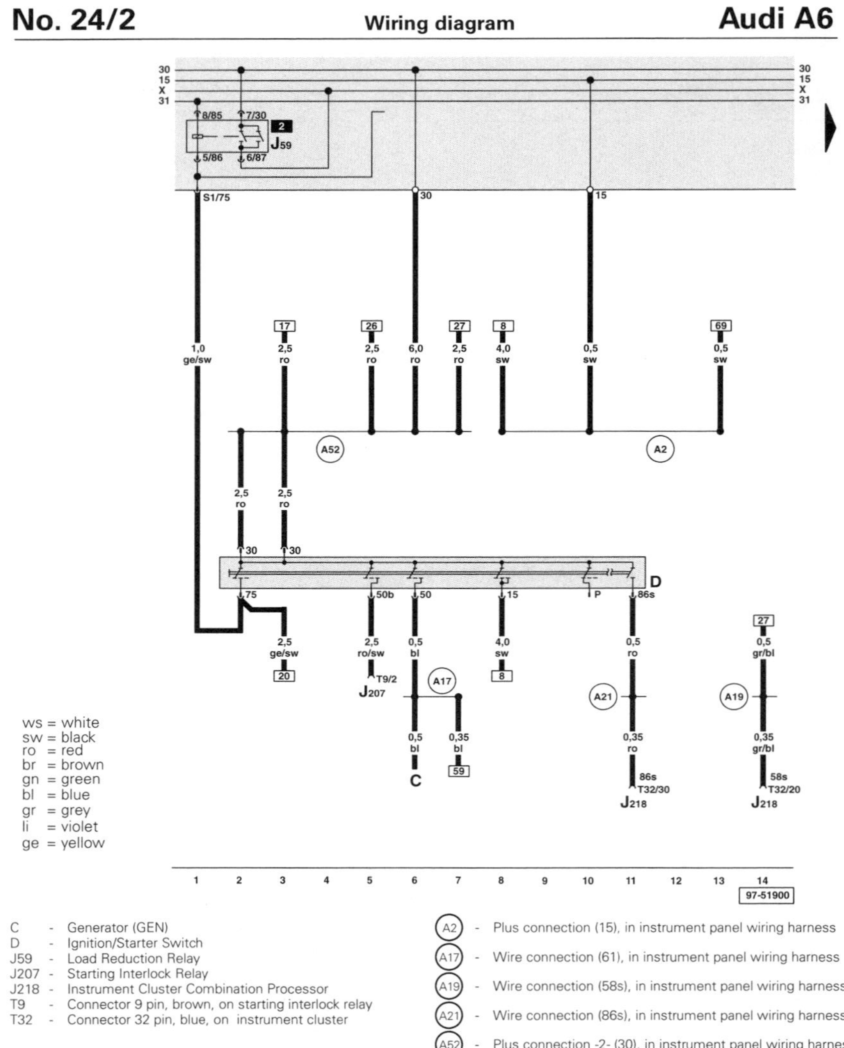

ws = white
sw = black
ro = red
br = brown
gn = green
bl = blue
gr = grey
li = violet
ge = yellow

C - Generator (GEN)
D - Ignition/Starter Switch
J59 - Load Reduction Relay
J207 - Starting Interlock Relay
J218 - Instrument Cluster Combination Processor
T9 - Connector 9 pin, brown, on starting interlock relay
T32 - Connector 32 pin, blue, on instrument cluster

A2 - Plus connection (15), in instrument panel wiring harness
A17 - Wire connection (61), in instrument panel wiring harness
A19 - Wire connection (58s), in instrument panel wiring harness
A21 - Wire connection (86s), in instrument panel wiring harness
A52 - Plus connection -2- (30), in instrument panel wiring harness

Ignition/Starter Switch

BentleyPublishers.com—All Rights Reserved

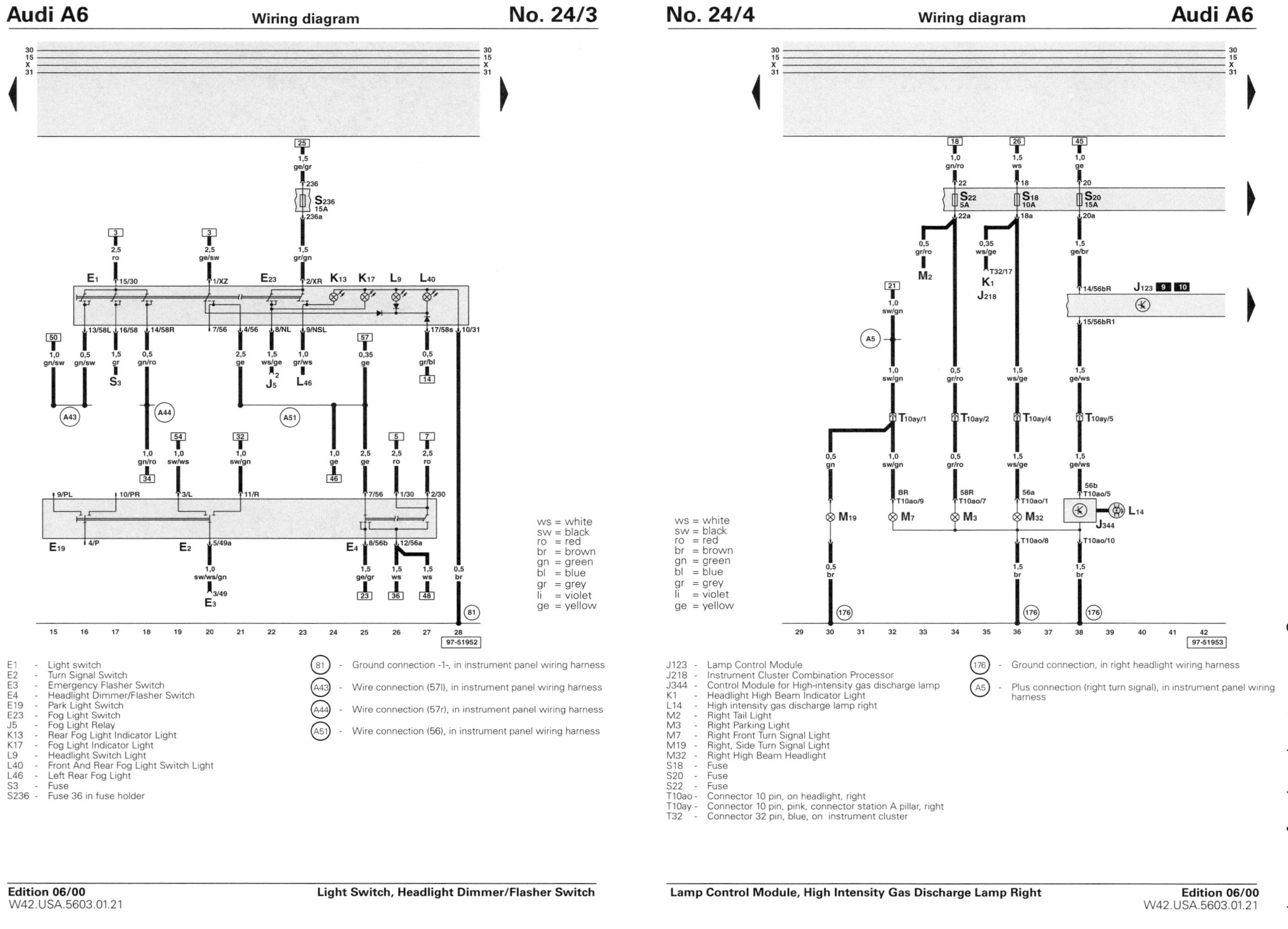

BentleyPublishers.com—All Rights Reserved

Headlights, Xenon (HID), Dynamic (2001)

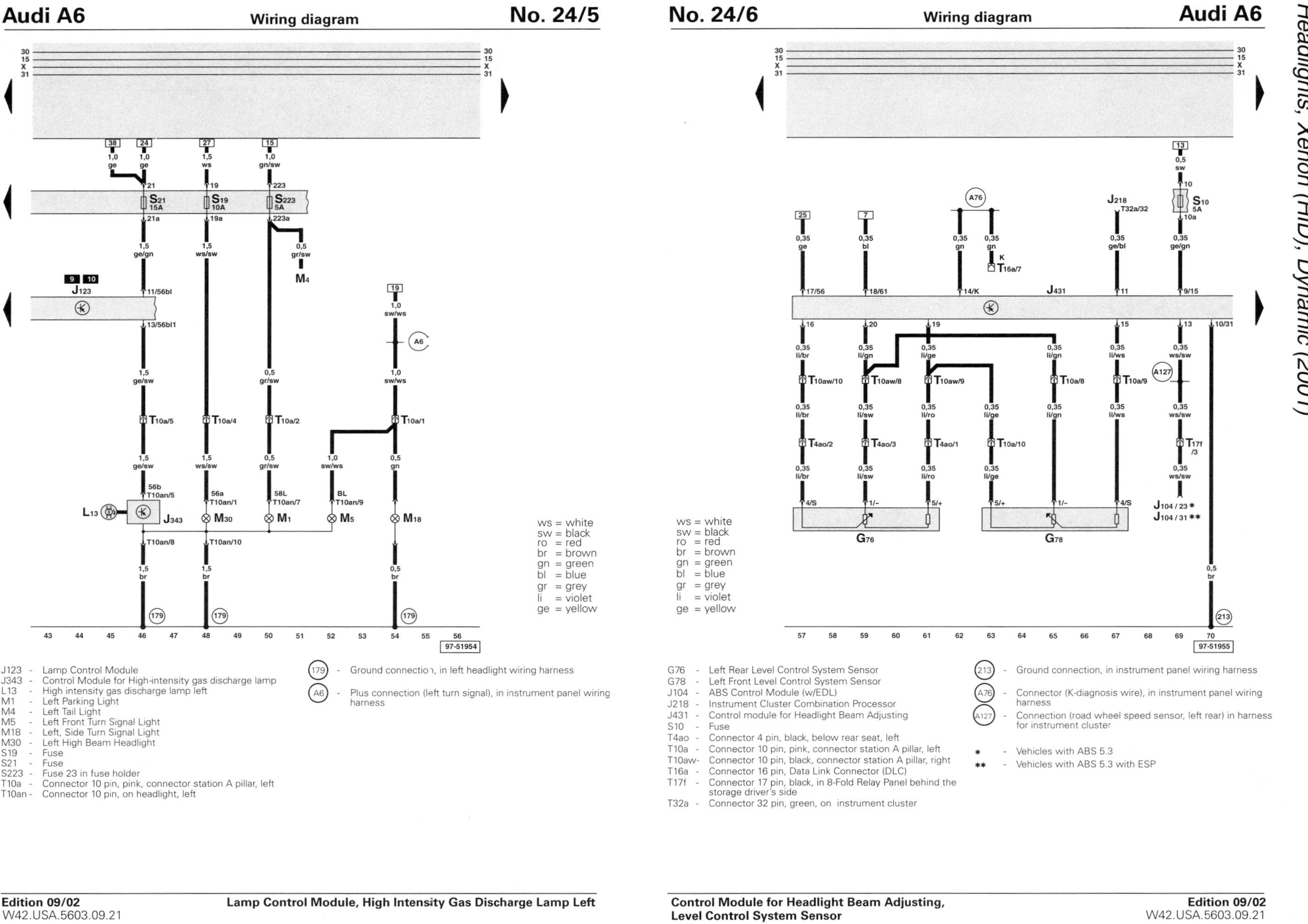

BentleyPublishers.com—All Rights Reserved

Audi A6 **Wiring diagram** **No. 24/7**

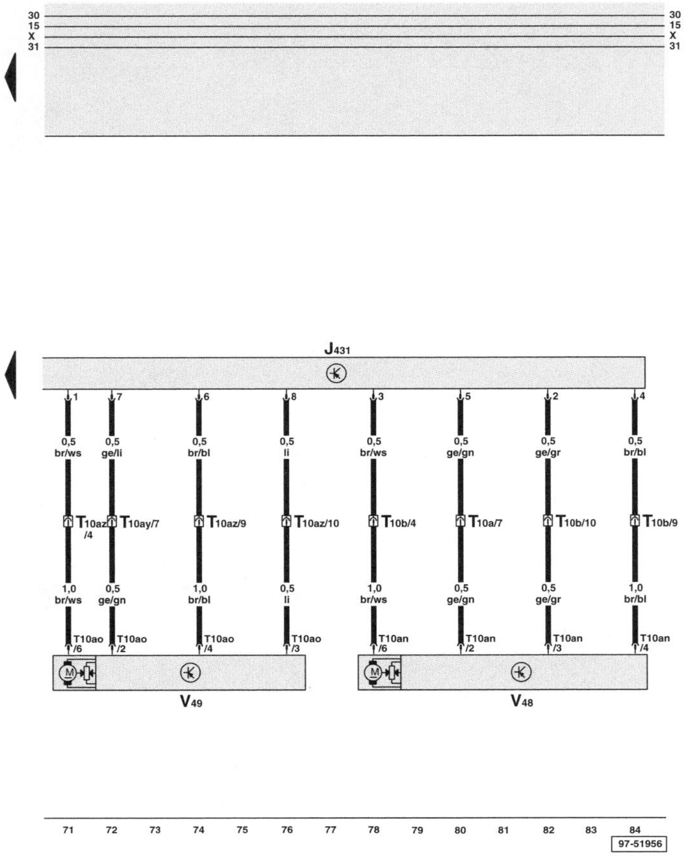

ws = white
sw = black
ro = red
br = brown
gn = green
bl = blue
gr = grey
li = violet
ge = yellow

J431 - Control module for Headlight Beam Adjusting
T10a - Connector 10 pin, pink, connector station A pillar, left
T10b - Connector 10 pin, red, connector station A pillar, left
T10an - Connector 10 pin, on headlight, left
T10ao - Connector 10 pin, on headlight, right
T10ay - Connector 10 pin, pink, connector station A pillar, right
T10az - Connector 10 pin, red, connector station A pillar, right
V48 - Left Headlight Beam Adjusting Motor
V49 - Right Headlight Beam Adjusting Motor

Edition 06/00
W42.USA.5603.01.21

Left Headlight Beam Adjusting Motor,
Right Headlight Beam Adjusting Motor

THIS SPACE INTENTIONALLY LEFT BLANK

BentleyPublishers.com—All Rights Reserved

Radio-Stereo, Concert with and without CD-Changer

2001 m. y.

Fuse Panel

A97-0119

Fuse Colors:

30 A - Green
25 A - White
20 A - Yellow
15 A - Blue
10 A - Red
7,5 A - Brown
5 A - Beige

Starting with fuse position 23, fuses in the fuse holder are identified with 223 in the wiring diagram.

Micro Central Electric Panel

A B C

A97-0110

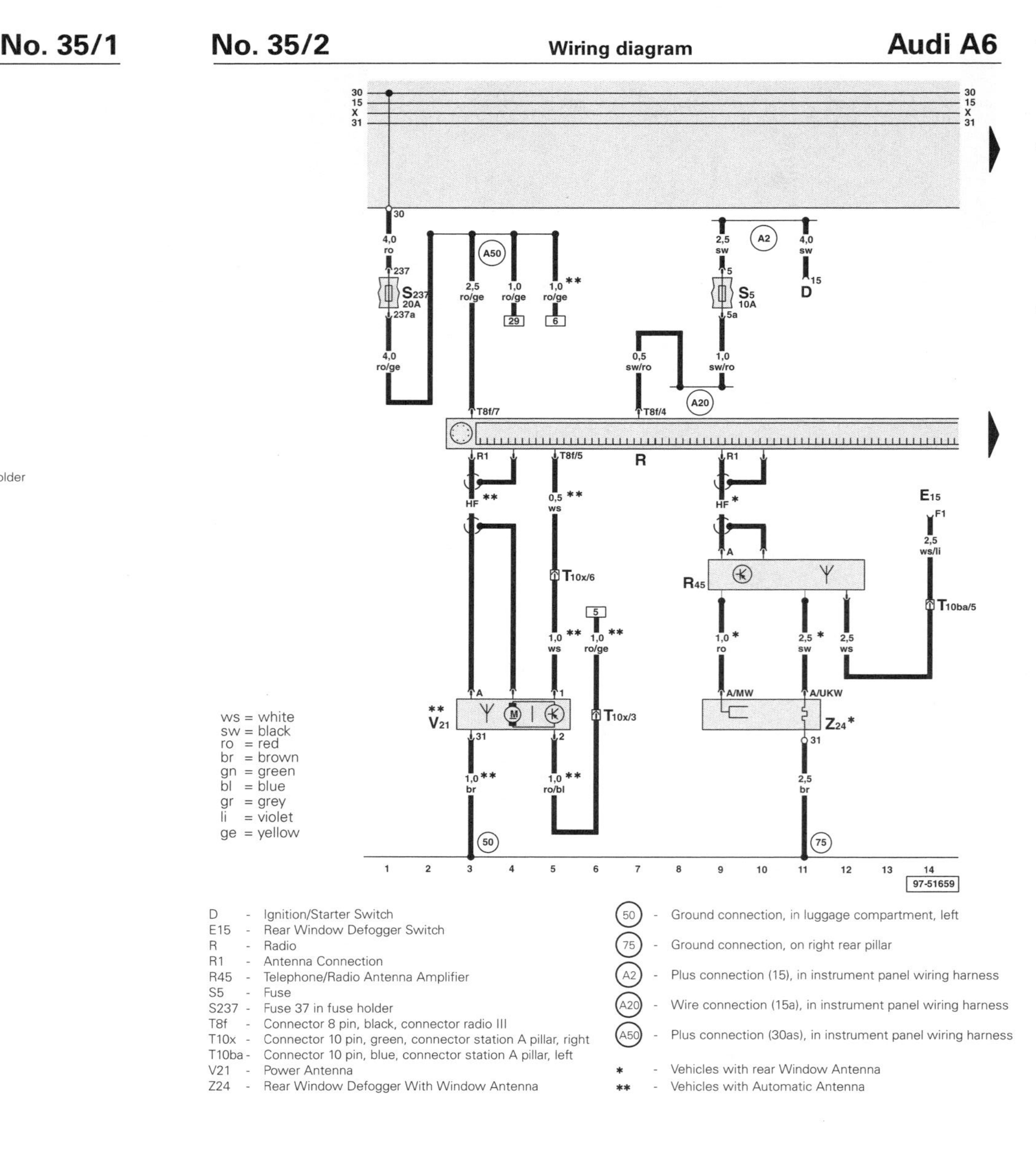

ws = white
sw = black
ro = red
br = brown
gn = green
bl = blue
gr = grey
li = violet
ge = yellow

D - Ignition/Starter Switch
E15 - Rear Window Defogger Switch
R - Radio
R1 - Antenna Connection
R45 - Telephone/Radio Antenna Amplifier
S5 - Fuse
S237 - Fuse 37 in fuse holder
T8f - Connector 8 pin, black, connector radio III
T10x - Connector 10 pin, green, connector station A pillar, right
T10ba - Connector 10 pin, blue, connector station A pillar, left
V21 - Power Antenna
Z24 - Rear Window Defogger With Window Antenna

(50) - Ground connection, in luggage compartment, left
(75) - Ground connection, on right rear pillar
(A2) - Plus connection (15), in instrument panel wiring harness
(A20) - Wire connection (15a), in instrument panel wiring harness
(A50) - Plus connection (30as), in instrument panel wiring harness

* - Vehicles with rear Window Antenna
** - Vehicles with Automatic Antenna

Radio, Rear Window Defogger with Window Antenna

BentleyPublishers.com—All Rights Reserved

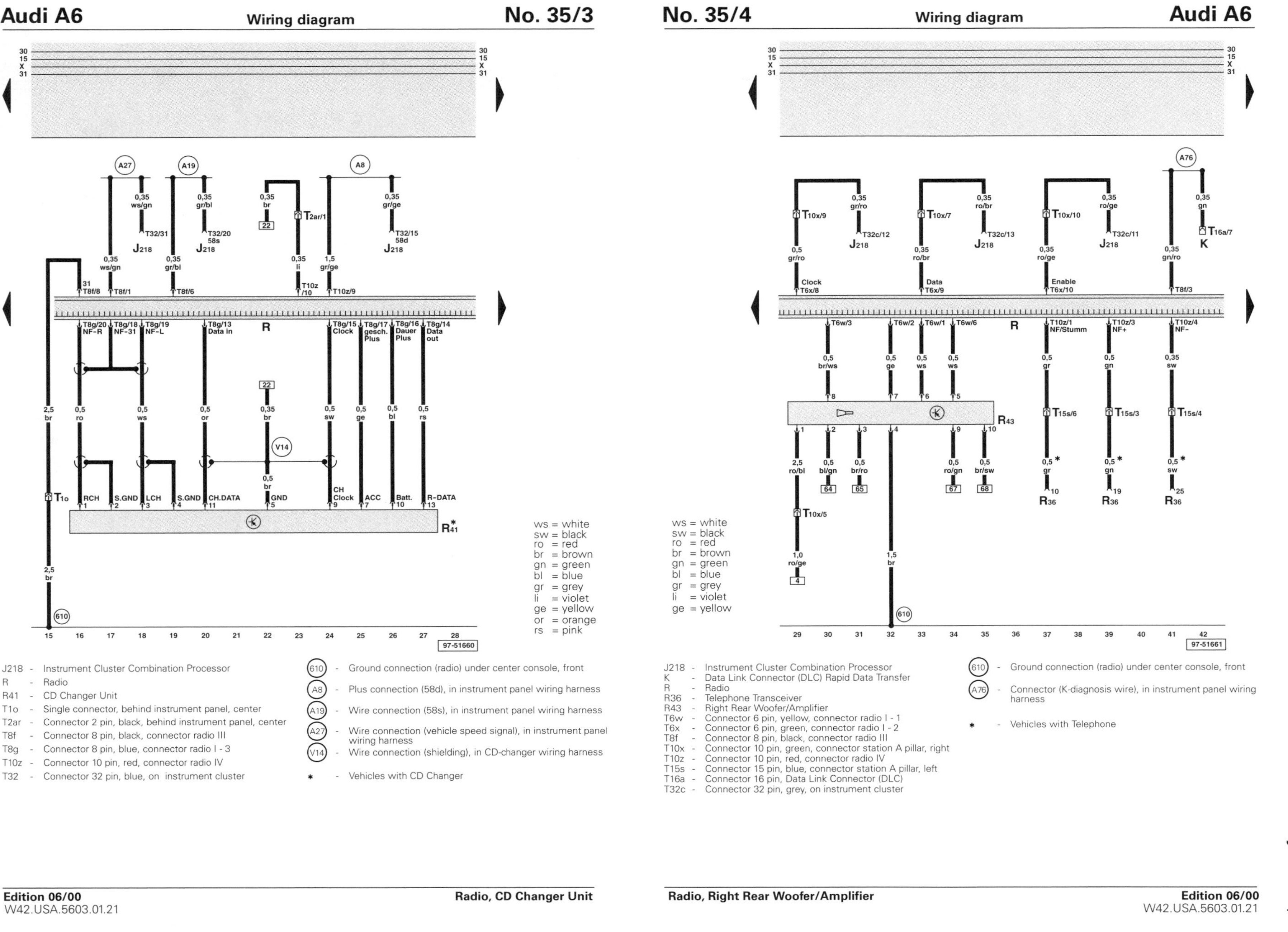

BentleyPublishers.com—All Rights Reserved

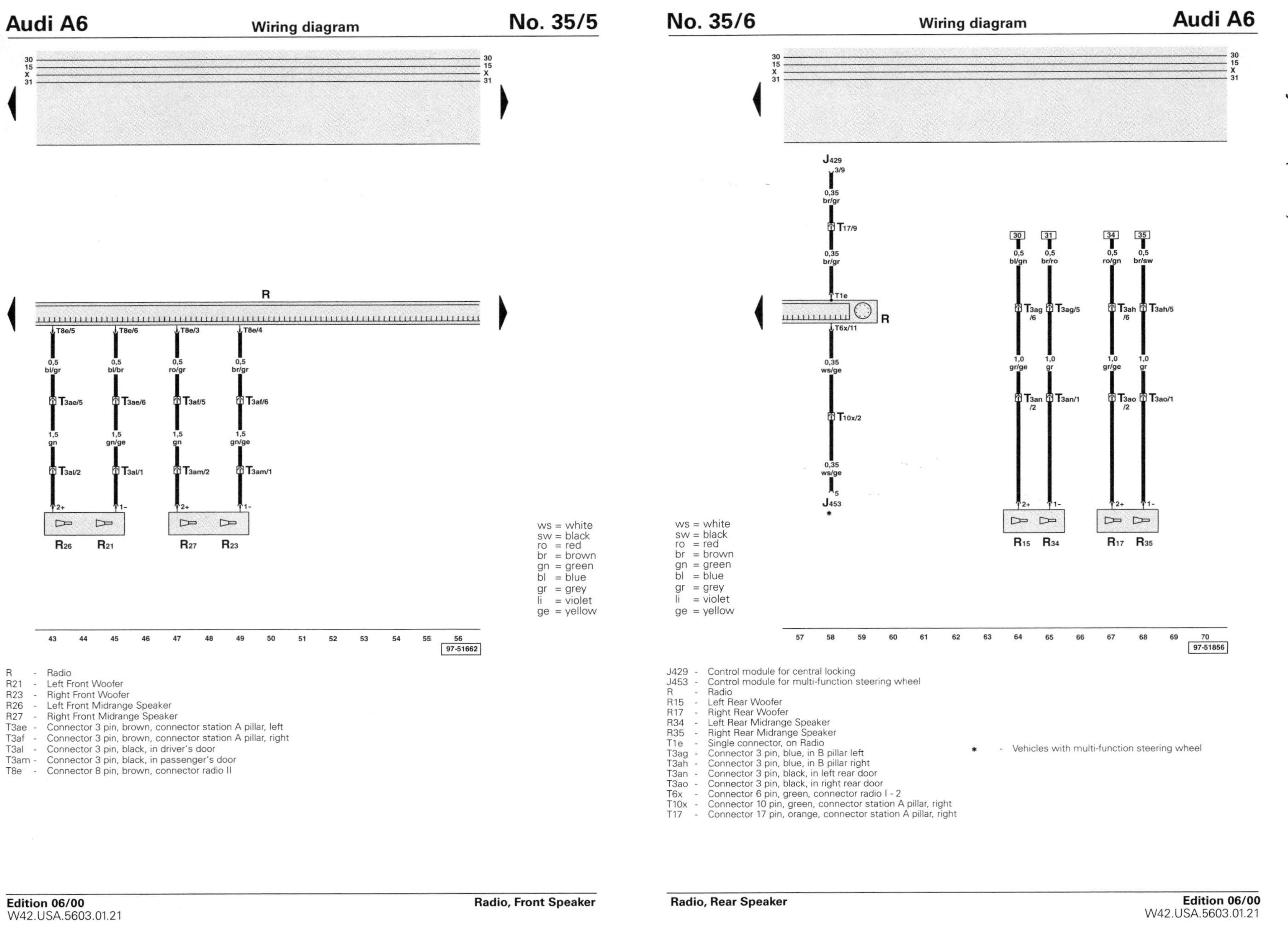

BentleyPublishers.com—All Rights Reserved

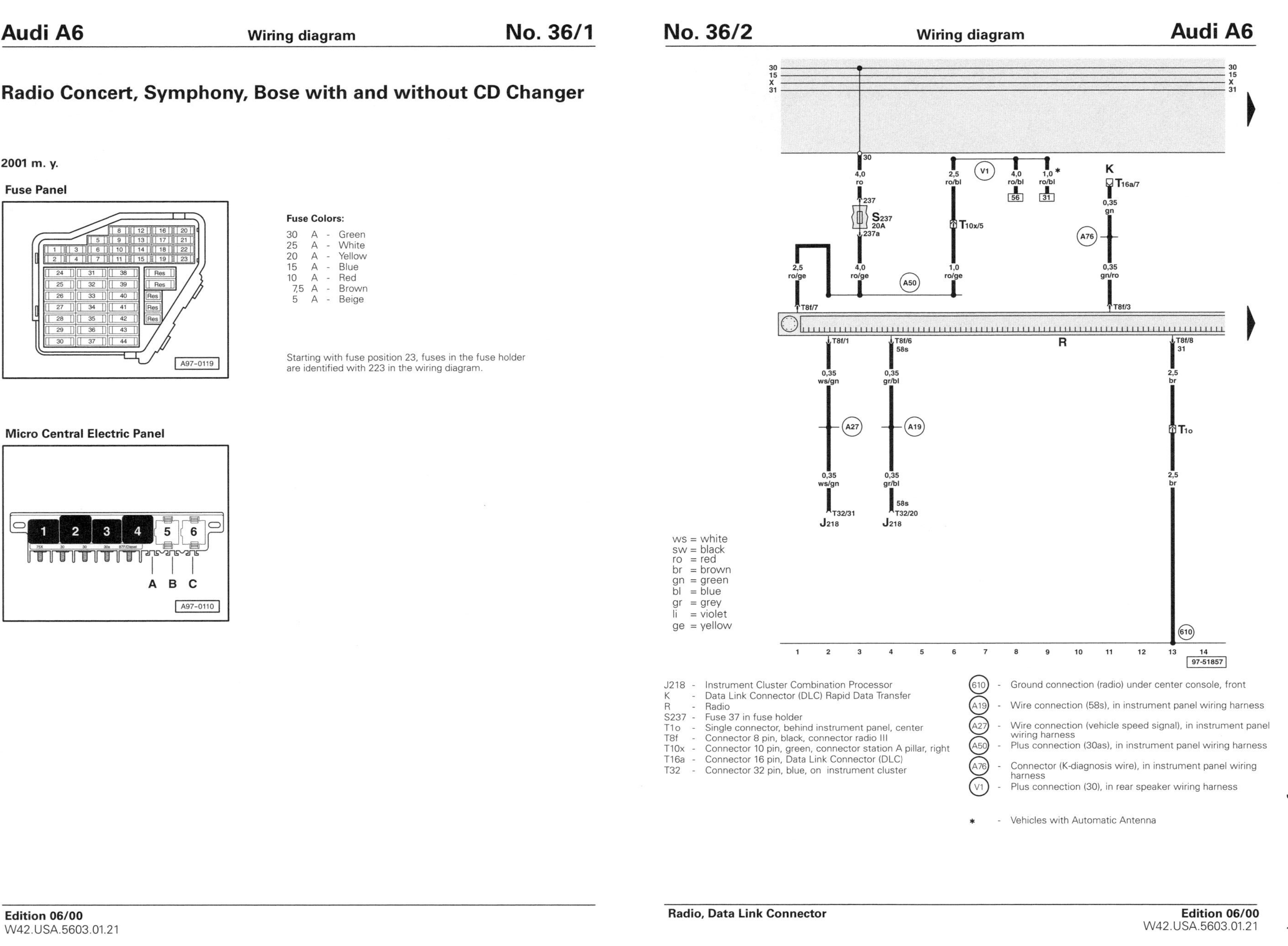

BentleyPublishers.com—All Rights Reserved

Sound System, Bose (2001)

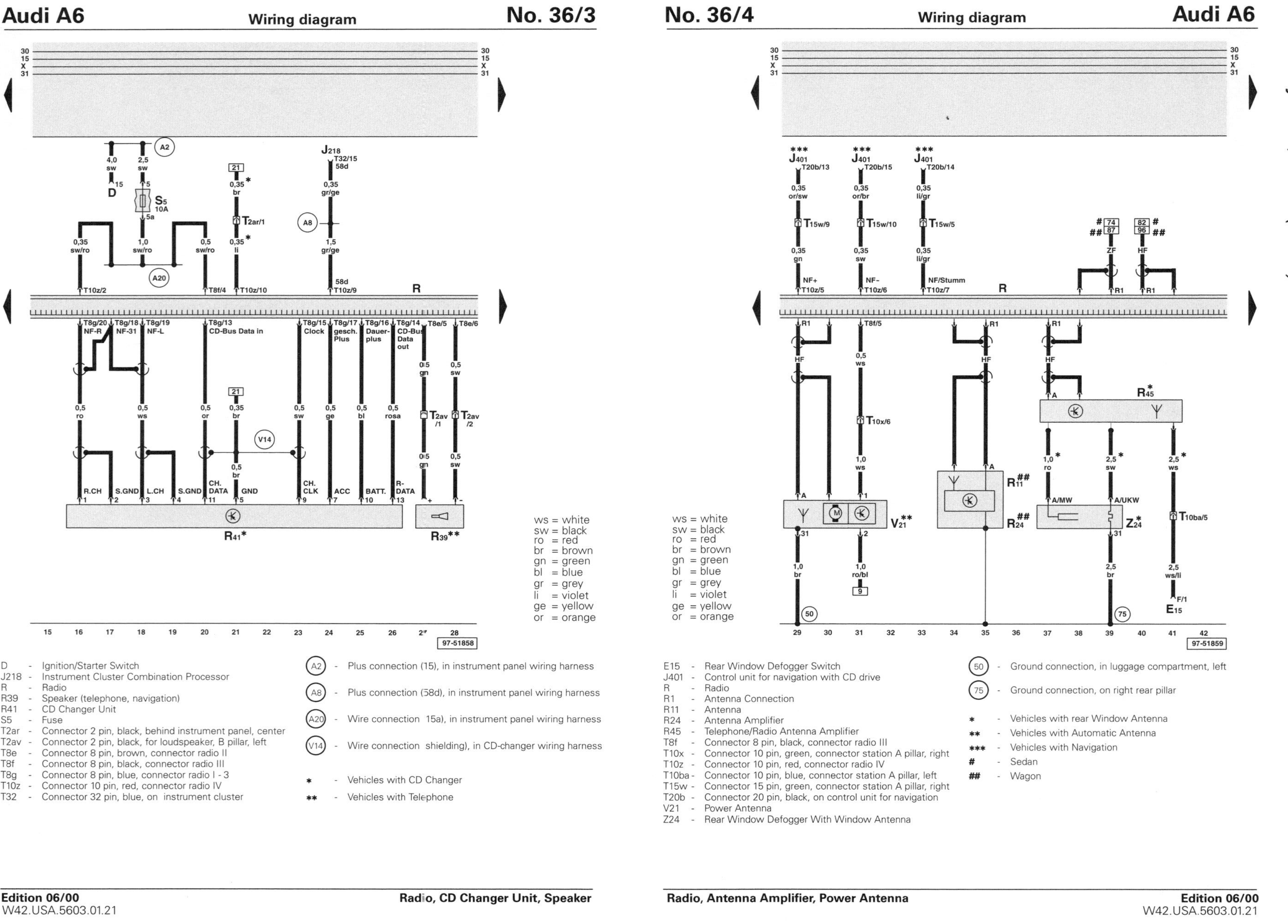

BentleyPublishers.com—All Rights Reserved

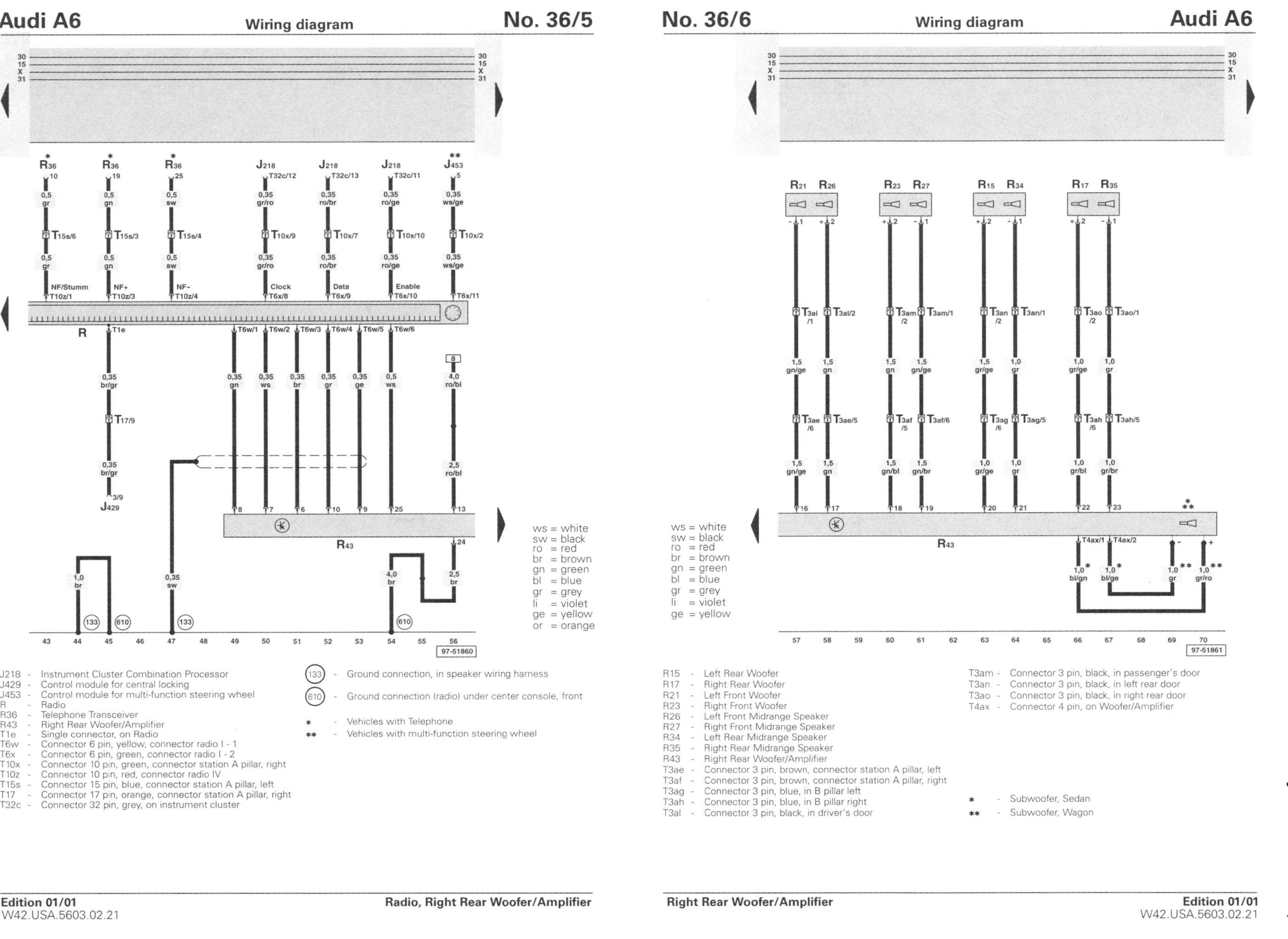

BentleyPublishers.com—All Rights Reserved

Sound System, Bose (2001)

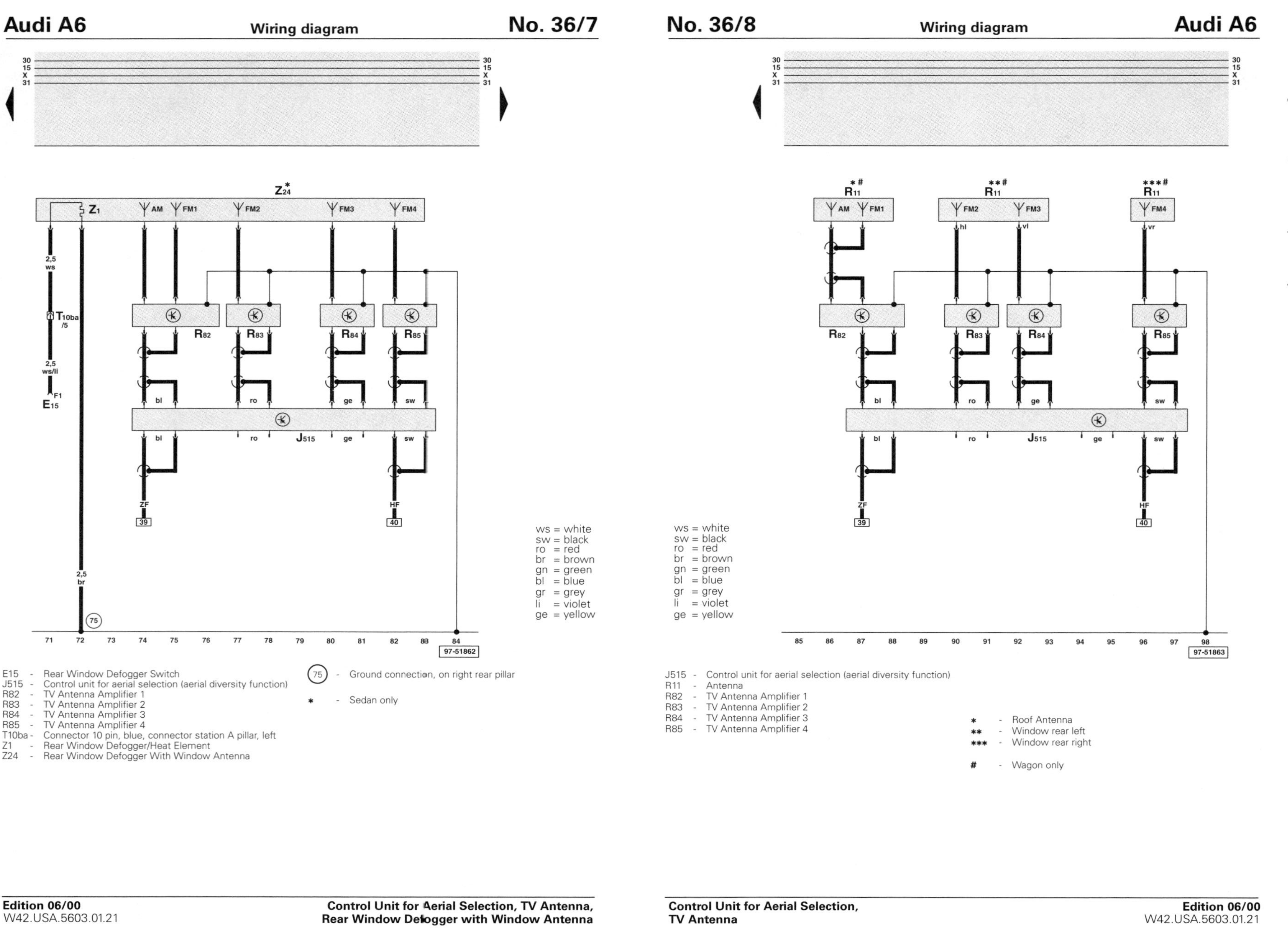

BentleyPublishers.com—All Rights Reserved

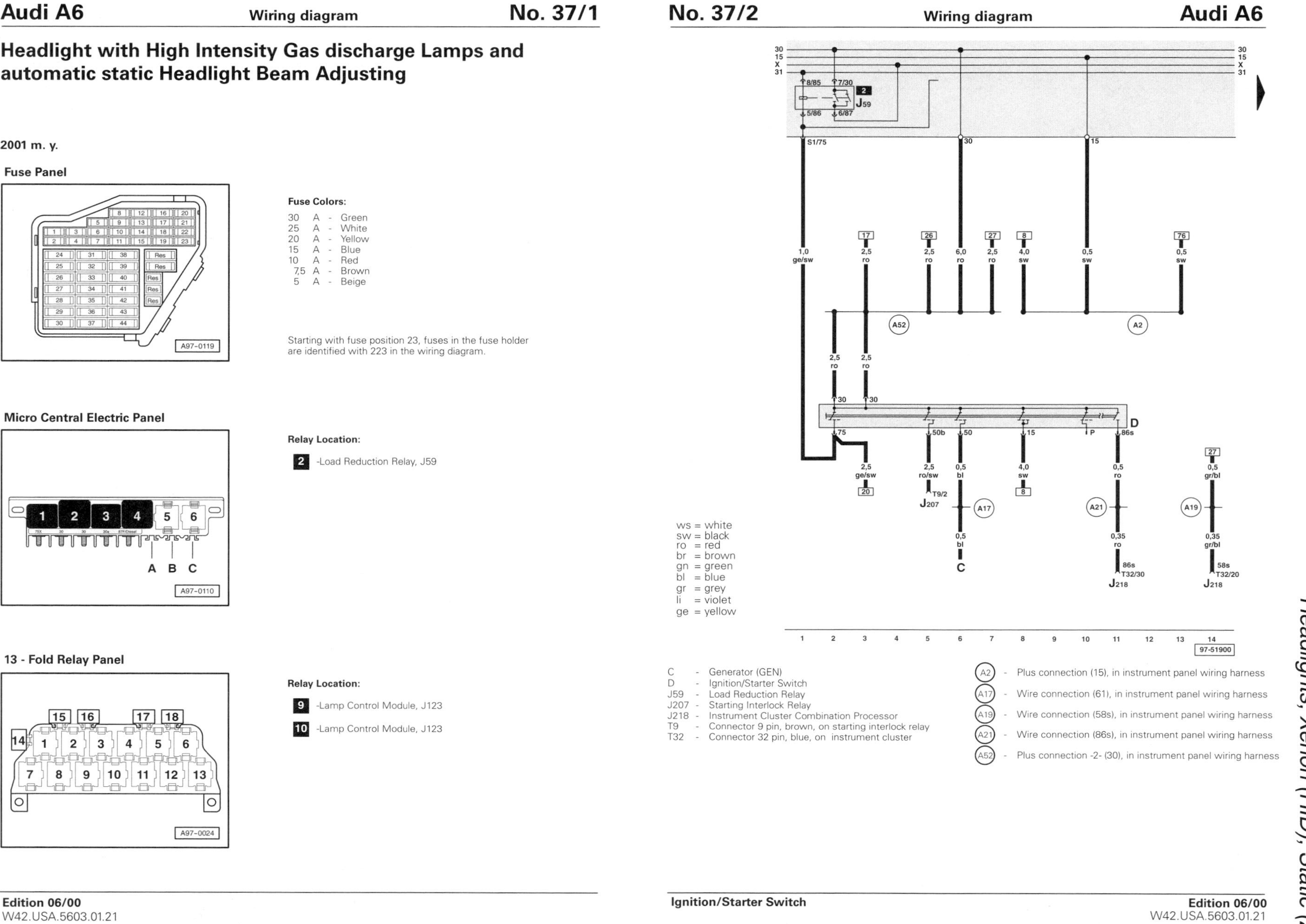

Audi A6 Wiring diagram **No. 37/1**

Headlight with High Intensity Gas discharge Lamps and automatic static Headlight Beam Adjusting

2001 m. y.

Fuse Panel

Fuse Colors:

30 A - Green
25 A - White
20 A - Yellow
15 A - Blue
10 A - Red
7,5 A - Brown
5 A - Beige

Starting with fuse position 23, fuses in the fuse holder are identified with 223 in the wiring diagram.

Micro Central Electric Panel

Relay Location:

2 -Load Reduction Relay, J59

13 - Fold Relay Panel

Relay Location:

9 -Lamp Control Module, J123

10 -Lamp Control Module, J123

Edition 06/00
W42.USA.5603.01.21

No. 37/2 Wiring diagram **Audi A6**

ws = white
sw = black
ro = red
br = brown
gn = green
bl = blue
gr = grey
li = violet
ge = yellow

C - Generator (GEN)
D - Ignition/Starter Switch
J59 - Load Reduction Relay
J207 - Starting Interlock Relay
J218 - Instrument Cluster Combination Processor
T9 - Connector 9 pin, brown, on starting interlock relay
T32 - Connector 32 pin, blue, on instrument cluster

A2 - Plus connection (15), in instrument panel wiring harness
A17 - Wire connection (61), in instrument panel wiring harness
A19 - Wire connection (58s), in instrument panel wiring harness
A21 - Wire connection (86s), in instrument panel wiring harness
A52 - Plus connection -2- (30), in instrument panel wiring harness

Ignition/Starter Switch

Edition 06/00
W42.USA.5603.01.21

BentleyPublishers.com—All Rights Reserved

Headlights, Xenon (HID), Static (2001)

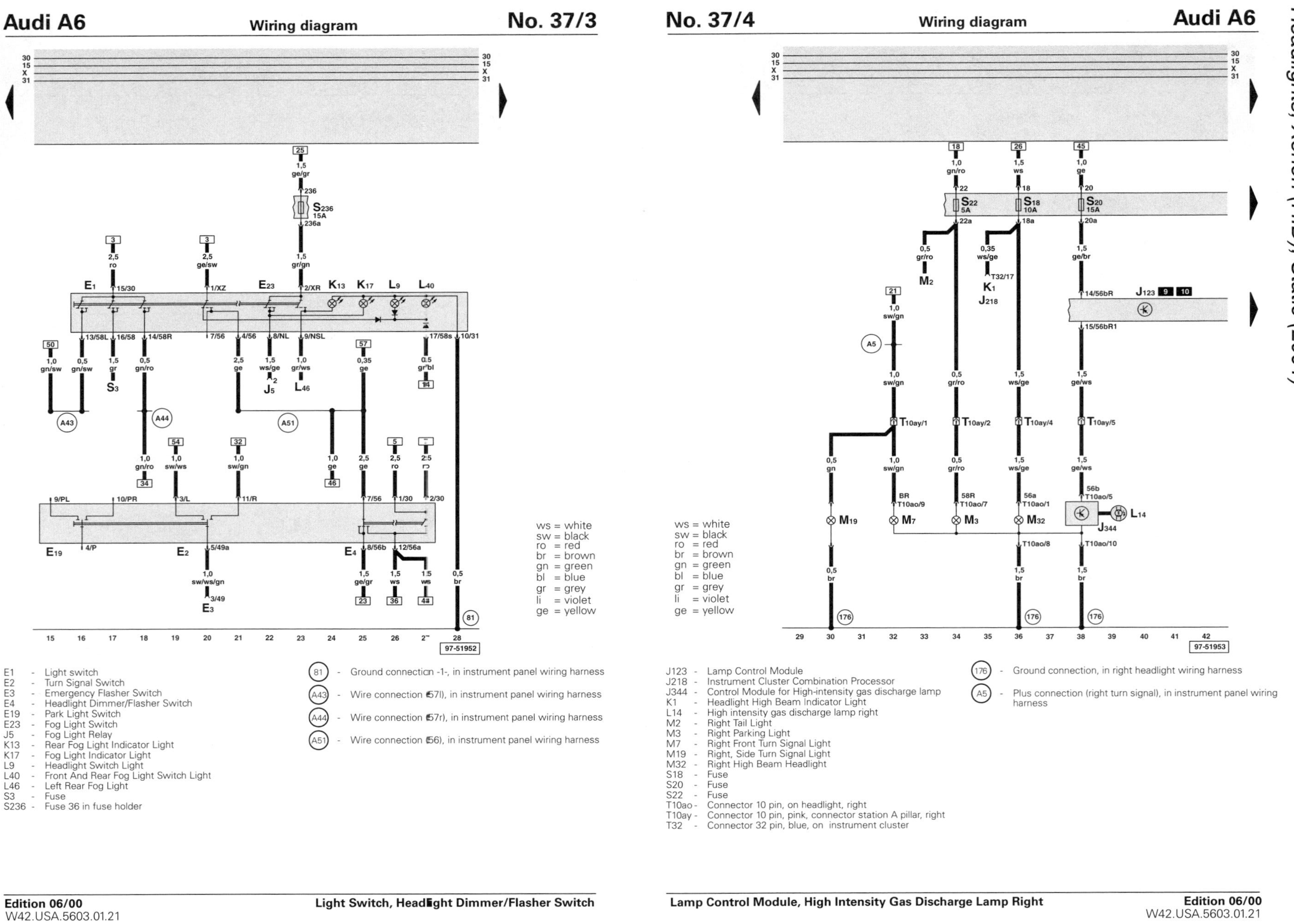

BentleyPublishers.com—All Rights Reserved

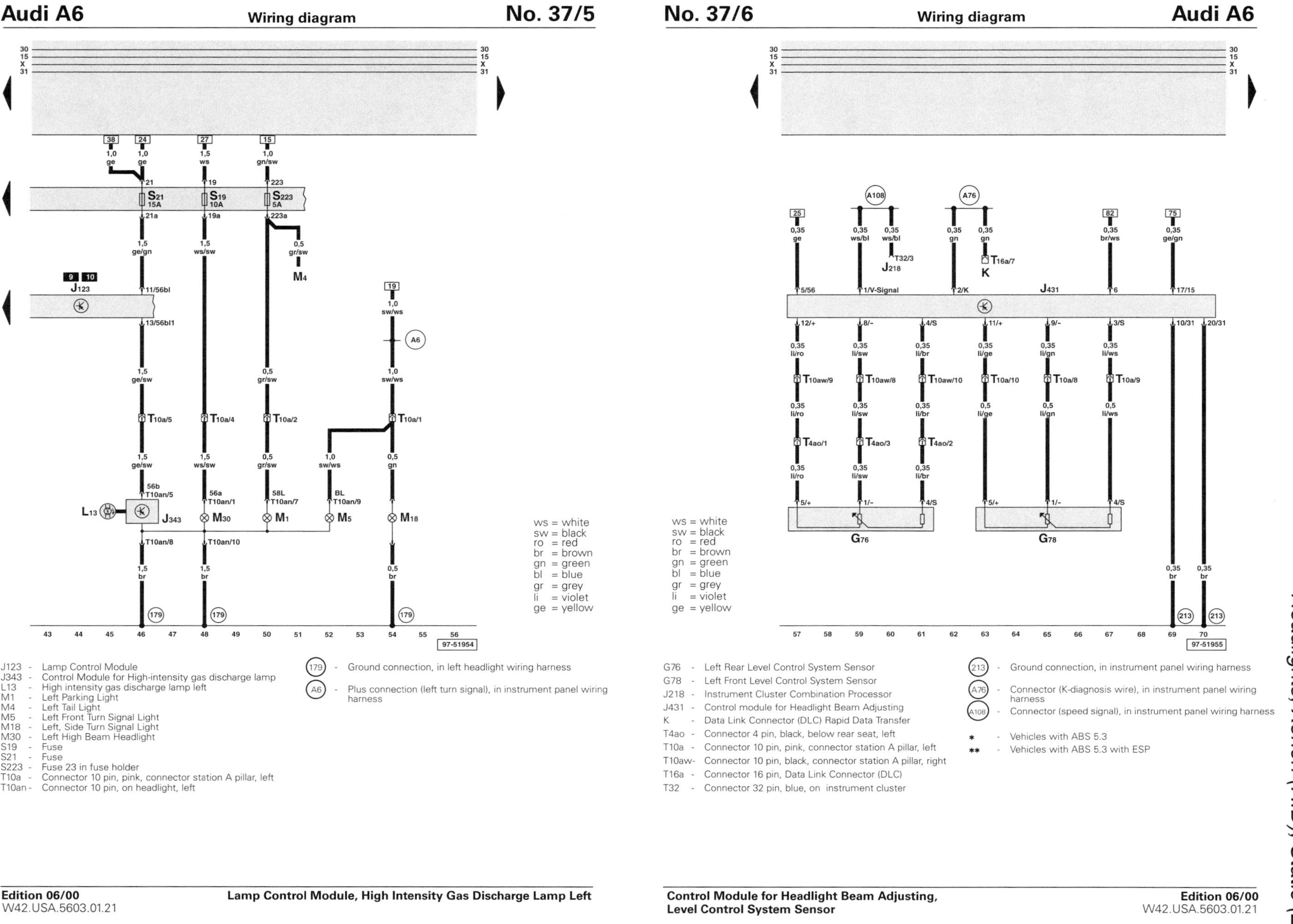

BentleyPublishers.com—All Rights Reserved

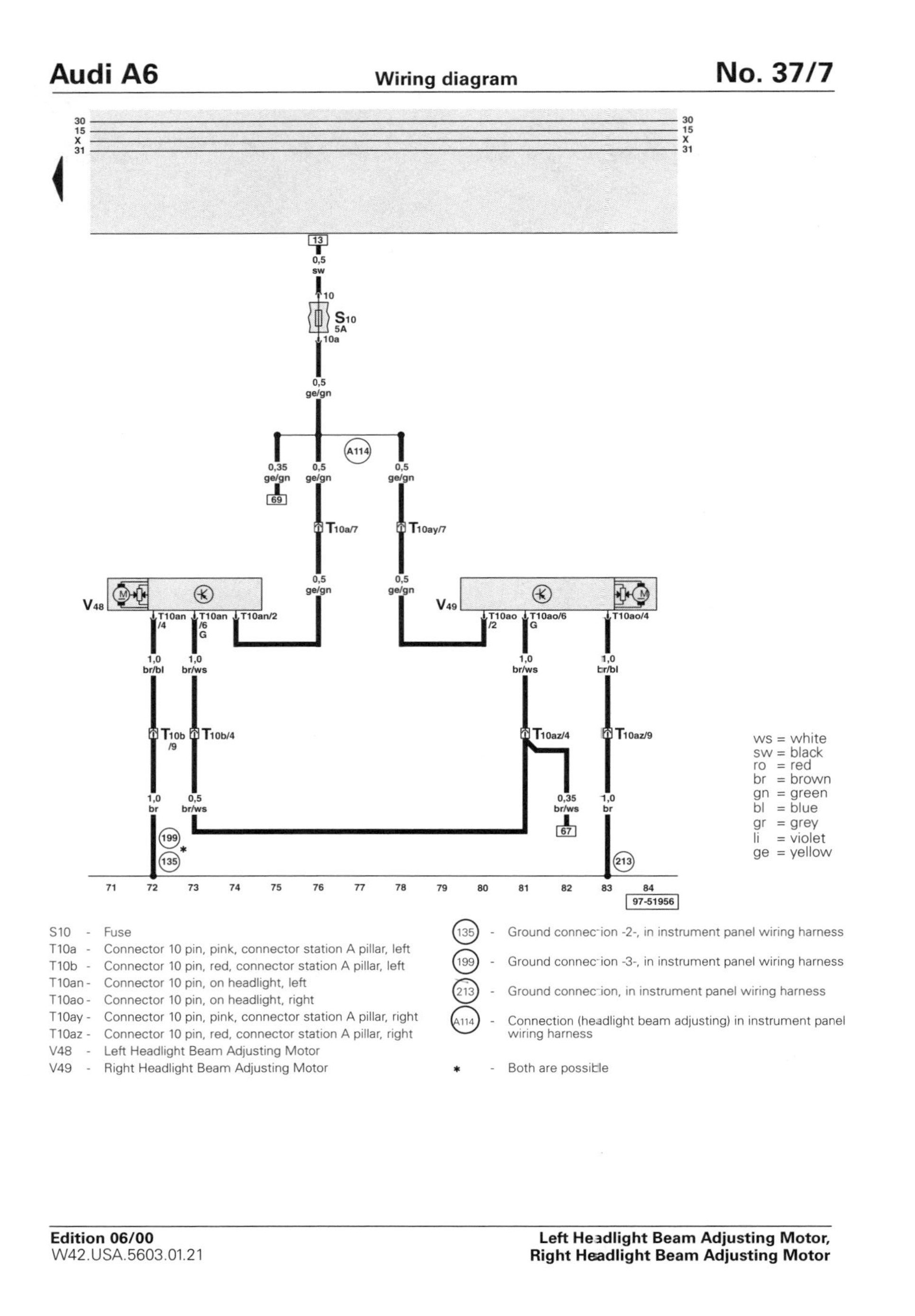

THIS SPACE INTENTIONALLY LEFT BLANK

BentleyPublishers.com—All Rights Reserved

Standard Equipment

2002 m. y.

Fuse Panel

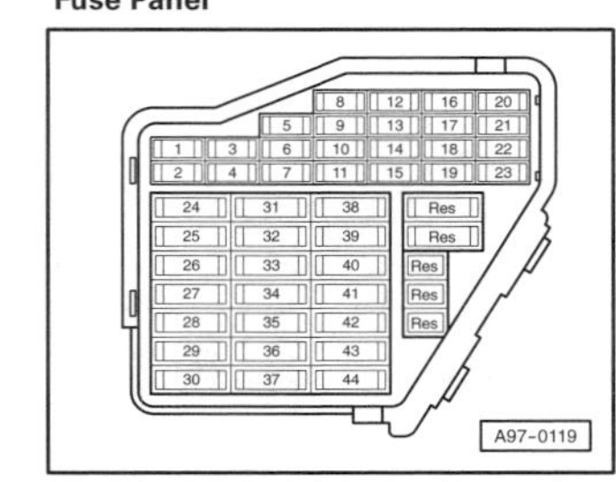

Fuse Colors:

30 A - Green
25 A - White
20 A - Yellow
15 A - Blue
10 A - Red
7,5 A - Brown
5 A - Beige

Starting with fuse position 23, fuses in the fuse holder are identified with 223 in the wiring diagram.

Micro Central Electric Panel

Relay Location:

- **1** - Dual Horn Relay, J4
- **2** - Load Reduction Relay, J59
- **4** - Fuel Pump (FP) Relay, J17
- **5** - Wiper/Washer Intermittent Relay, J31
- **6** - Wiper/Washer Intermittent Relay, J31

13 - Fold Relay Panel

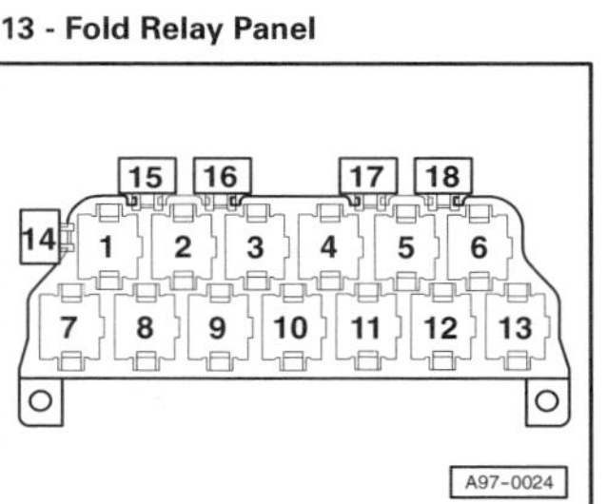

Relay Location:

- **4** - Starting Interlock Relay, J207
- **6** - Fog Light Relay, J5

8 - Fold Relay Panel

Fuse Location:

- **15** - Fuse for Luggage Compartment Socket, S184

BentleyPublishers.com—All Rights Reserved

Standard Equipment (2002)

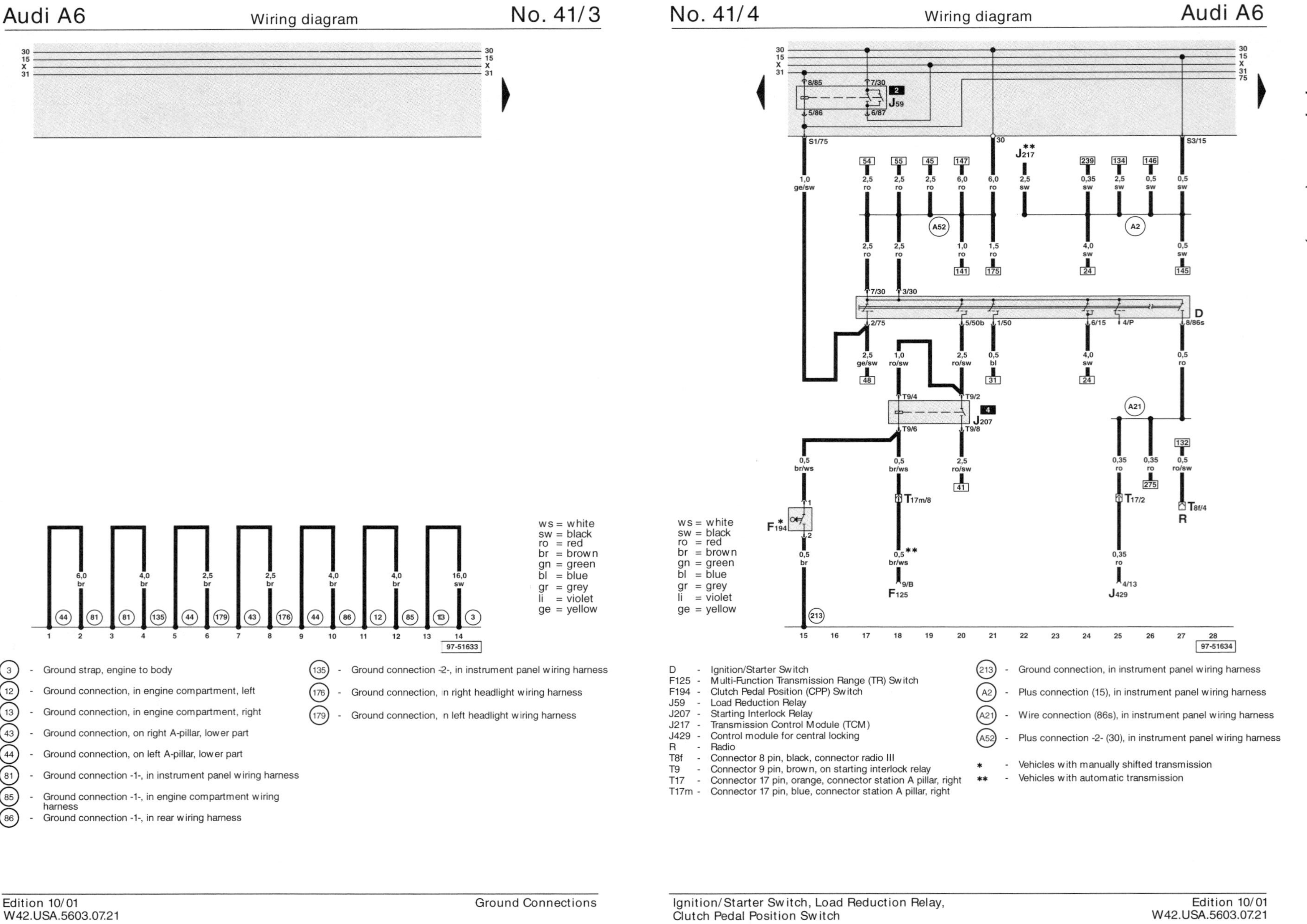

BentleyPublishers.com—All Rights Reserved

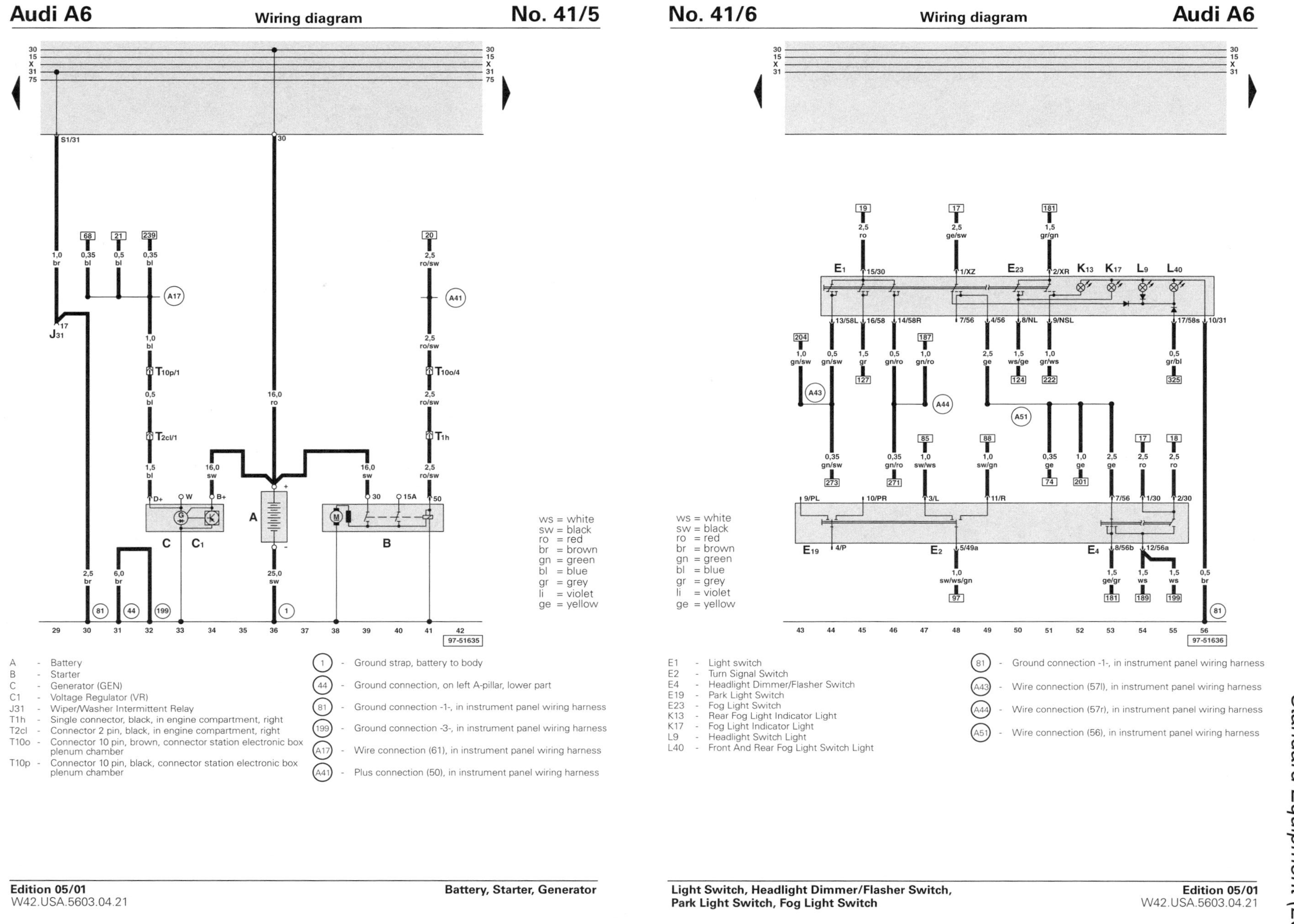

Audi A6
Wiring diagram
No. 41/5
ws = white
sw = black
ro = red
br = brown
gn = green
bl = blue
gr = grey
li = violet
ge = yellow
A - Battery
B - Starter
C - Generator (GEN)
C1 - Voltage Regulator (VR)
J31 - Wiper/Washer Intermittent Relay
T1h - Single connector, black, in engine compartment, right
T2cl - Connector 2 pin, black, in engine compartment, right
T10o - Connector 10 pin, brown, connector station electronic box plenum chamber
T10p - Connector 10 pin, black, connector station electronic box plenum chamber
1 - Ground strap, battery to body
44 - Ground connection, on left A-pillar, lower part
81 - Ground connection -1-, in instrument panel wiring harness
199 - Ground connection -3-, in instrument panel wiring harness
A17 - Wire connection (61), in instrument panel wiring harness
A41 - Plus connection (50), in instrument panel wiring harness
97-51635
Edition 05/01
W42.USA.5603.04.21
Battery, Starter, Generator
No. 41/6
Wiring diagram
Audi A6
ws = white
sw = black
ro = red
br = brown
gn = green
bl = blue
gr = grey
li = violet
ge = yellow
E1 - Light switch
E2 - Turn Signal Switch
E4 - Headlight Dimmer/Flasher Switch
E19 - Park Light Switch
E23 - Fog Light Switch
K13 - Rear Fog Light Indicator Light
K17 - Fog Light Indicator Light
L9 - Headlight Switch Light
L40 - Front And Rear Fog Light Switch Light
81 - Ground connection -1-, in instrument panel wiring harness
A43 - Wire connection (57l), in instrument panel wiring harness
A44 - Wire connection (57r), in instrument panel wiring harness
A51 - Wire connection (56), in instrument panel wiring harness
97-51636
Light Switch, Headlight Dimmer/Flasher Switch, Park Light Switch, Fog Light Switch
Edition 05/01
W42.USA.5603.04.21

BentleyPublishers.com—All Rights Reserved

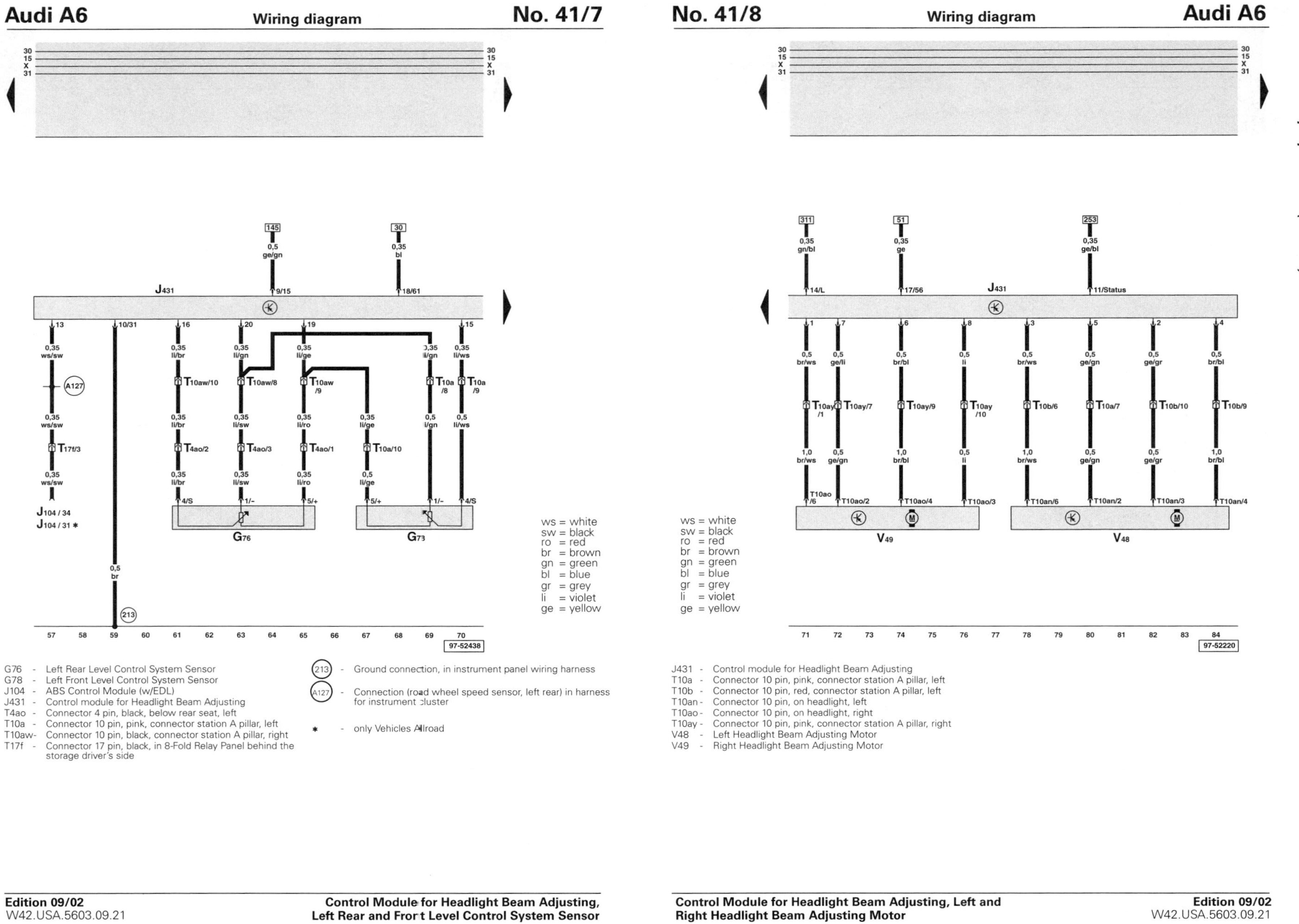

BentleyPublishers.com—All Rights Reserved

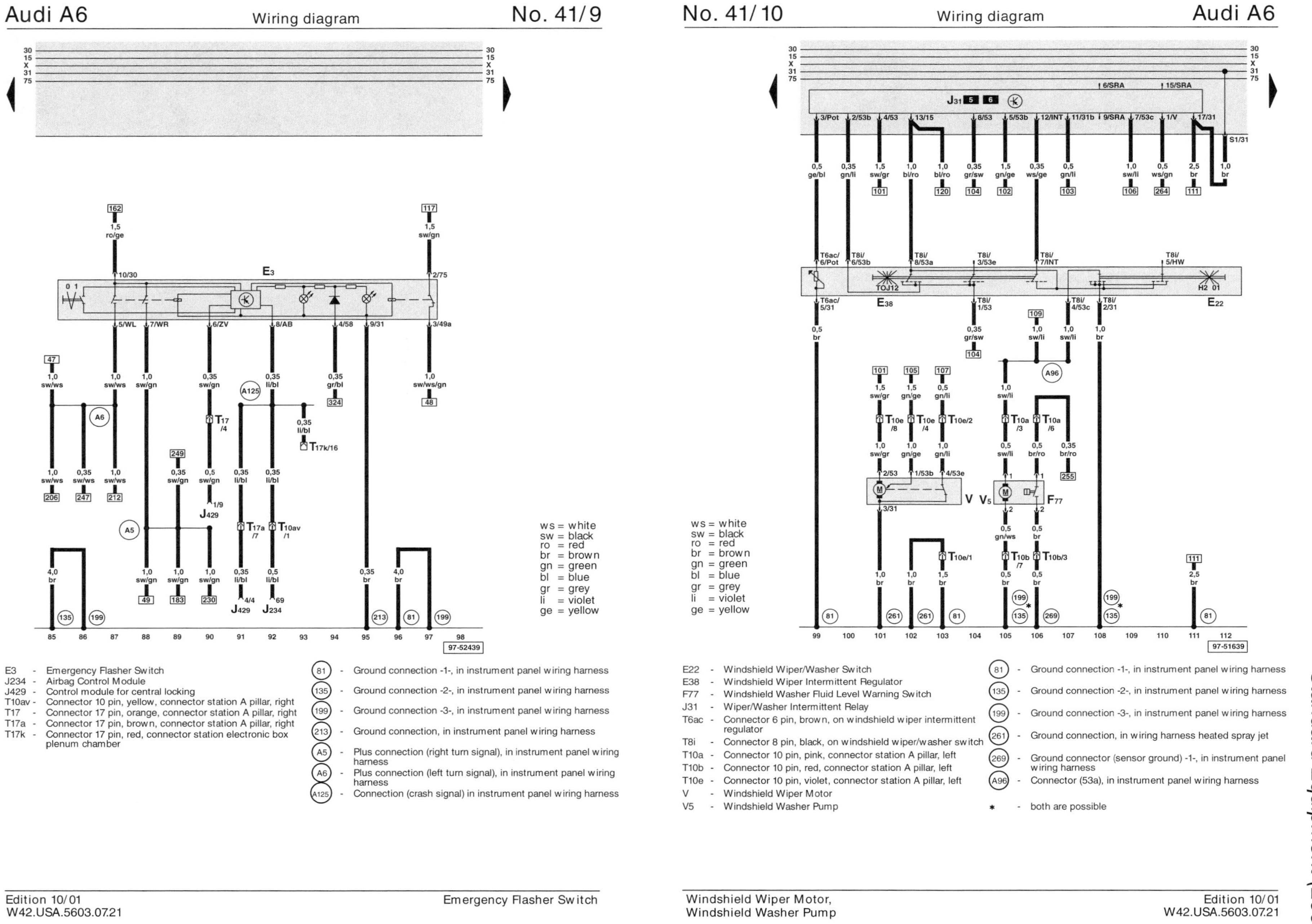

BentleyPublishers.com—All Rights Reserved

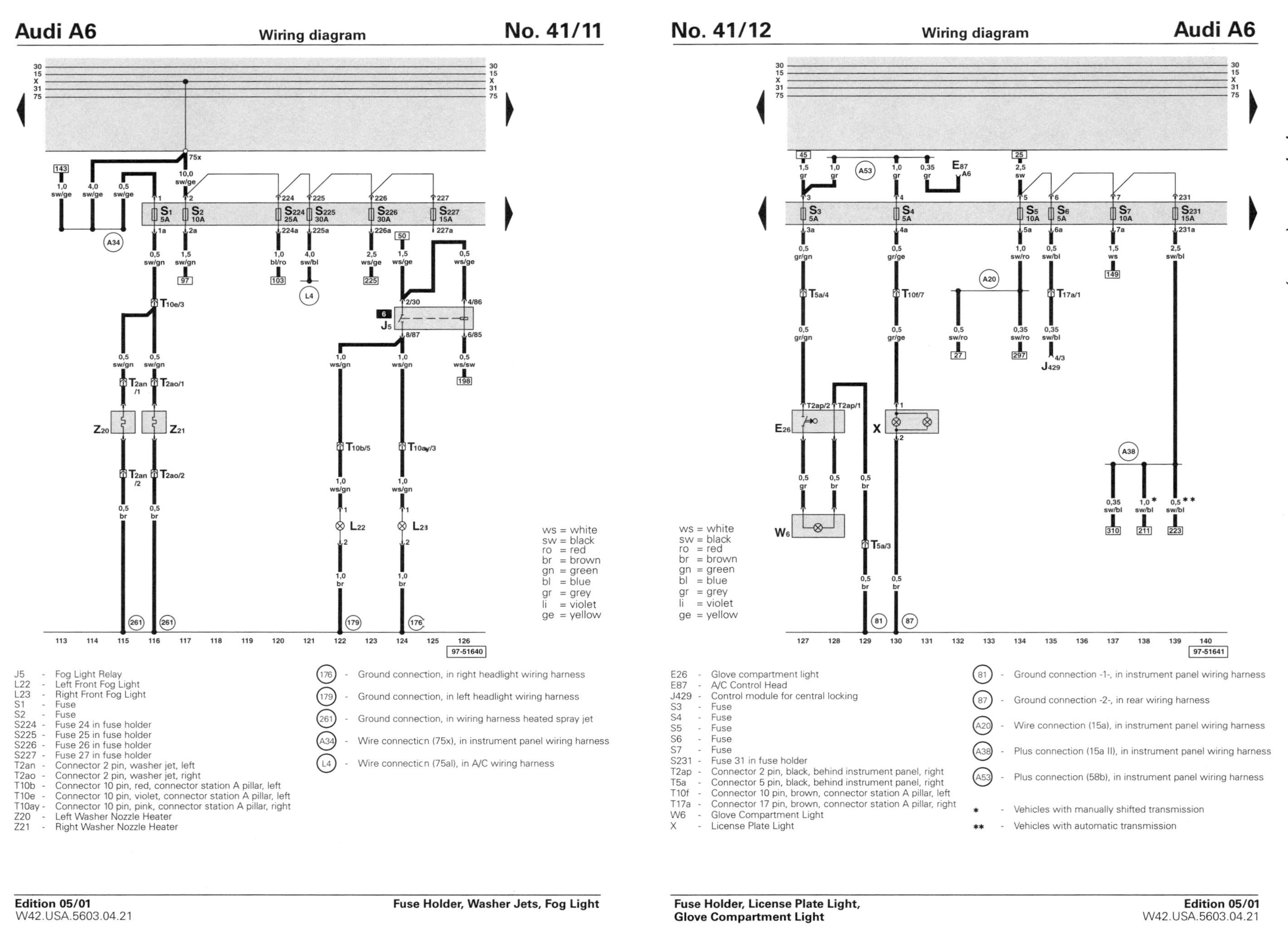

Audi A6
Wiring diagram
No. 41/11
ws = white
sw = black
ro = red
br = brown
gn = green
bl = blue
gr = grey
li = violet
ge = yellow
97-51640
J5 - Fog Light Relay
L22 - Left Front Fog Light
L23 - Right Front Fog Light
S1 - Fuse
S2 - Fuse
S224 - Fuse 24 in fuse holder
S225 - Fuse 25 in fuse holder
S226 - Fuse 26 in fuse holder
S227 - Fuse 27 in fuse holder
T2an - Connector 2 pin, washer jet, left
T2ao - Connector 2 pin, washer jet, right
T10b - Connector 10 pin, red, connector station A pillar, left
T10e - Connector 10 pin, violet, connector station A pillar, left
T10ay - Connector 10 pin, pink, connector station A pillar, right
Z20 - Left Washer Nozzle Heater
Z21 - Right Washer Nozzle Heater
176 - Ground connection, in right headlight wiring harness
179 - Ground connection, in left headlight wiring harness
261 - Ground connection, in wiring harness heated spray jet
A34 - Wire connection (75x), in instrument panel wiring harness
L4 - Wire connection (75al), in A/C wiring harness
Edition 05/01
W42.USA.5603.04.21
Fuse Holder, Washer Jets, Fog Light
No. 41/12
Wiring diagram
Audi A6
ws = white
sw = black
ro = red
br = brown
gn = green
bl = blue
gr = grey
li = violet
ge = yellow
97-51641
E26 - Glove compartment light
E87 - A/C Control Head
J429 - Control module for central locking
S3 - Fuse
S4 - Fuse
S5 - Fuse
S6 - Fuse
S7 - Fuse
S231 - Fuse 31 in fuse holder
T2ap - Connector 2 pin, black, behind instrument panel, right
T5a - Connector 5 pin, black, behind instrument panel, right
T10f - Connector 10 pin, brown, connector station A pillar, left
T17a - Connector 17 pin, brown, connector station A pillar, right
W6 - Glove Compartment Light
X - License Plate Light
81 - Ground connection -1-, in instrument panel wiring harness
87 - Ground connection -2-, in rear wiring harness
A20 - Wire connection (15a), in instrument panel wiring harness
A38 - Plus connection (15a II), in instrument panel wiring harness
A53 - Plus connection (58b), in instrument panel wiring harness
* - Vehicles with manually shifted transmission
** - Vehicles with automatic transmission
Fuse Holder, License Plate Light, Glove Compartment Light
Edition 05/01
W42.USA.5603.04.21

BentleyPublishers.com—All Rights Reserved

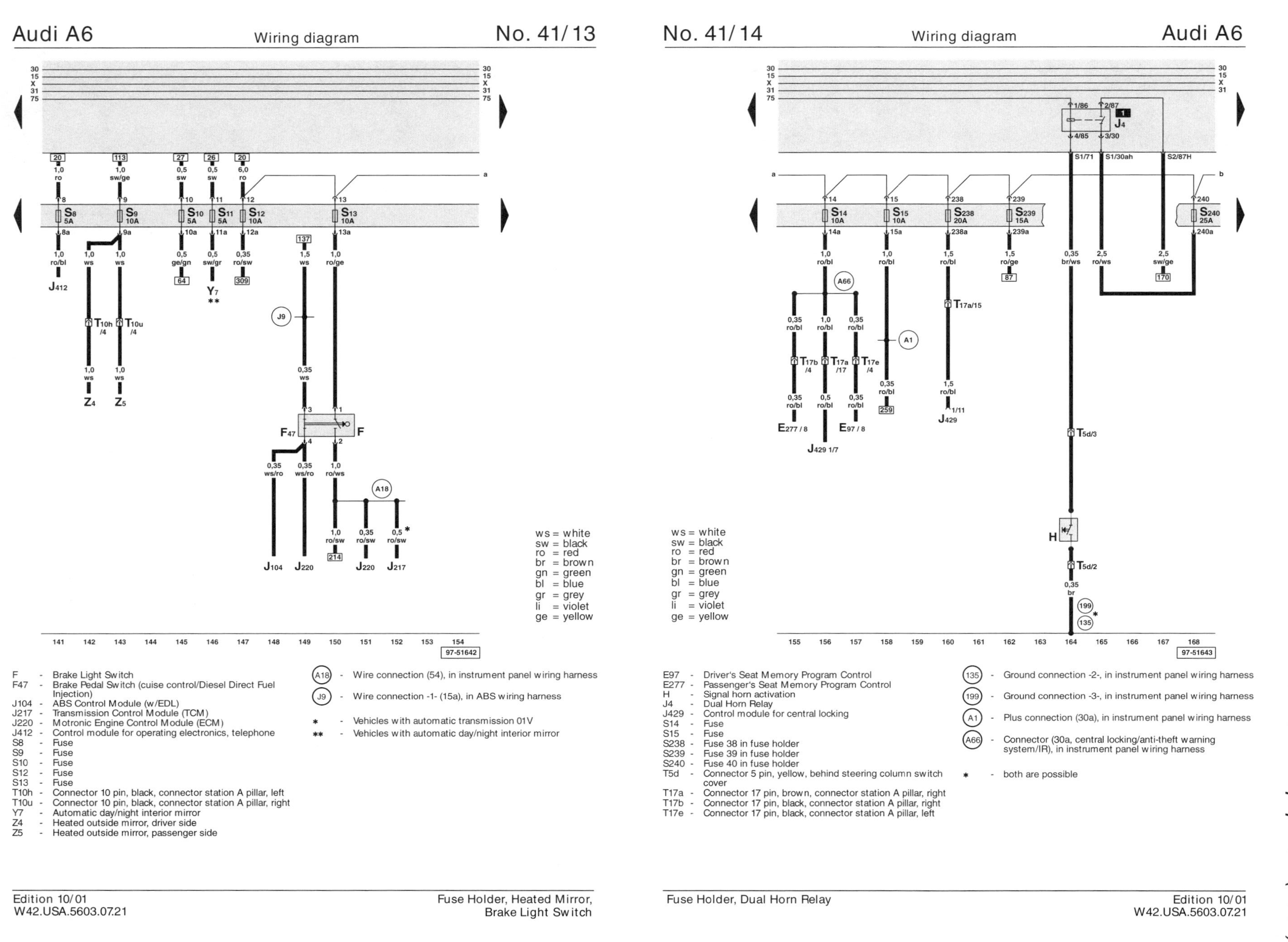

BentleyPublishers.com—All Rights Reserved

Standard Equipment (2002)

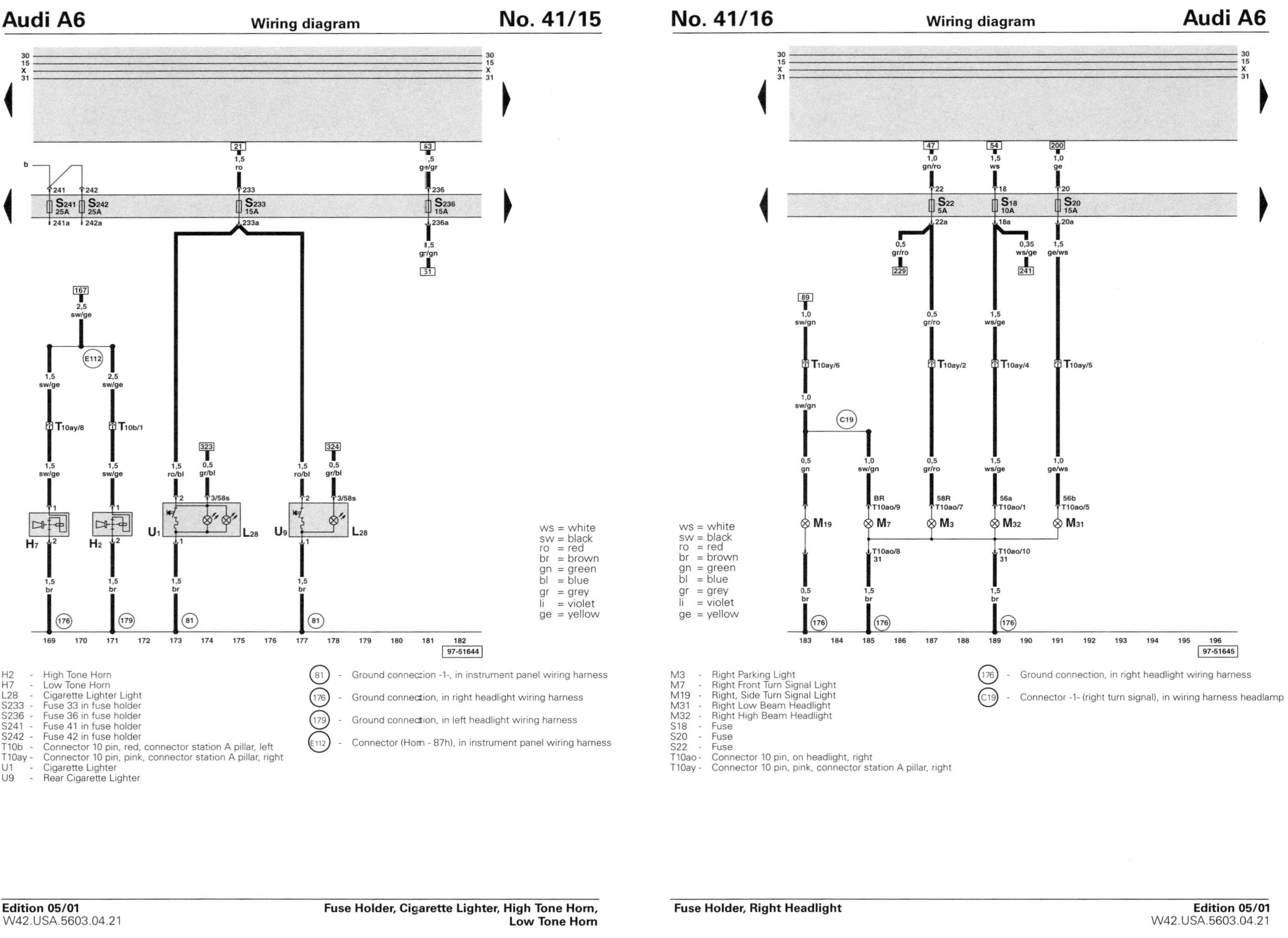

BentleyPublishers.com—All Rights Reserved

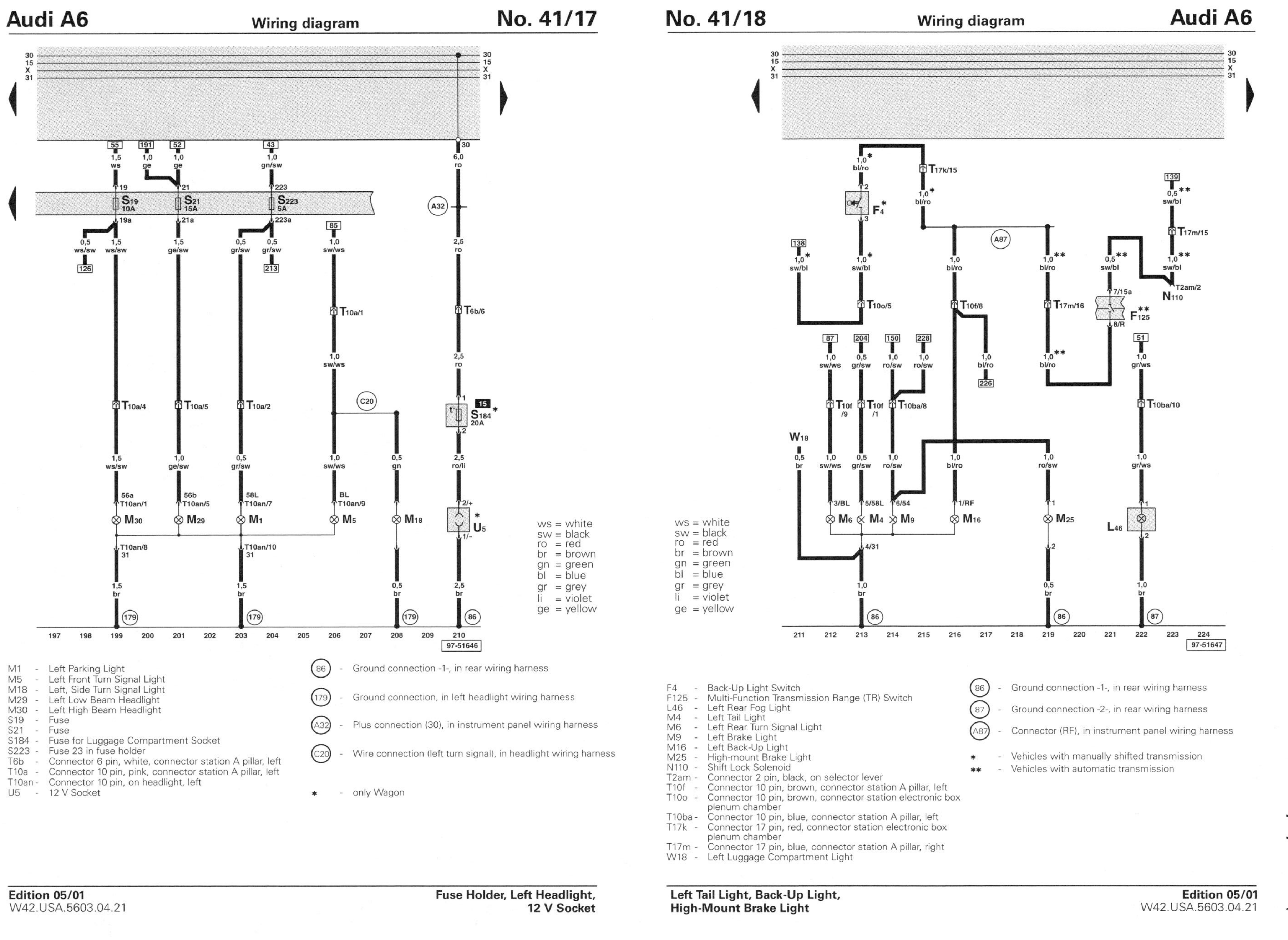

Audi A6
Wiring diagram
No. 41/17
ws = white
sw = black
ro = red
br = brown
gn = green
bl = blue
gr = grey
li = violet
ge = yellow
M1 - Left Parking Light
M5 - Left Front Turn Signal Light
M18 - Left, Side Turn Signal Light
M29 - Left Low Beam Headlight
M30 - Left High Beam Headlight
S19 - Fuse
S21 - Fuse
S184 - Fuse for Luggage Compartment Socket
S223 - Fuse 23 in fuse holder
T6b - Connector 6 pin, white, connector station A pillar, left
T10a - Connector 10 pin, pink, connector station A pillar, left
T10an - Connector 10 pin, on headlight, left
U5 - 12 V Socket
86 - Ground connection -1-, in rear wiring harness
179 - Ground connection, in left headlight wiring harness
A32 - Plus connection (30), in instrument panel wiring harness
C20 - Wire connection (left turn signal), in headlight wiring harness
* - only Wagon
Edition 05/01
W42.USA.5603.04.21
Fuse Holder, Left Headlight, 12 V Socket
No. 41/18
Wiring diagram
Audi A6
ws = white
sw = black
ro = red
br = brown
gn = green
bl = blue
gr = grey
li = violet
ge = yellow
F4 - Back-Up Light Switch
F125 - Multi-Function Transmission Range (TR) Switch
L46 - Left Rear Fog Light
M4 - Left Tail Light
M6 - Left Rear Turn Signal Light
M9 - Left Brake Light
M16 - Left Back-Up Light
M25 - High-mount Brake Light
N110 - Shift Lock Solenoid
T2am - Connector 2 pin, black, on selector lever
T10f - Connector 10 pin, brown, connector station A pillar, left
T10o - Connector 10 pin, brown, connector station electronic box plenum chamber
T10ba - Connector 10 pin, blue, connector station A pillar, left
T17k - Connector 17 pin, red, connector station electronic box plenum chamber
T17m - Connector 17 pin, blue, connector station A pillar, right
W18 - Left Luggage Compartment Light
86 - Ground connection -1-, in rear wiring harness
87 - Ground connection -2-, in rear wiring harness
A87 - Connector (RF), in instrument panel wiring harness
* - Vehicles with manually shifted transmission
** - Vehicles with automatic transmission
Left Tail Light, Back-Up Light, High-Mount Brake Light
Edition 05/01
W42.USA.5603.04.21

BentleyPublishers.com—All Rights Reserved

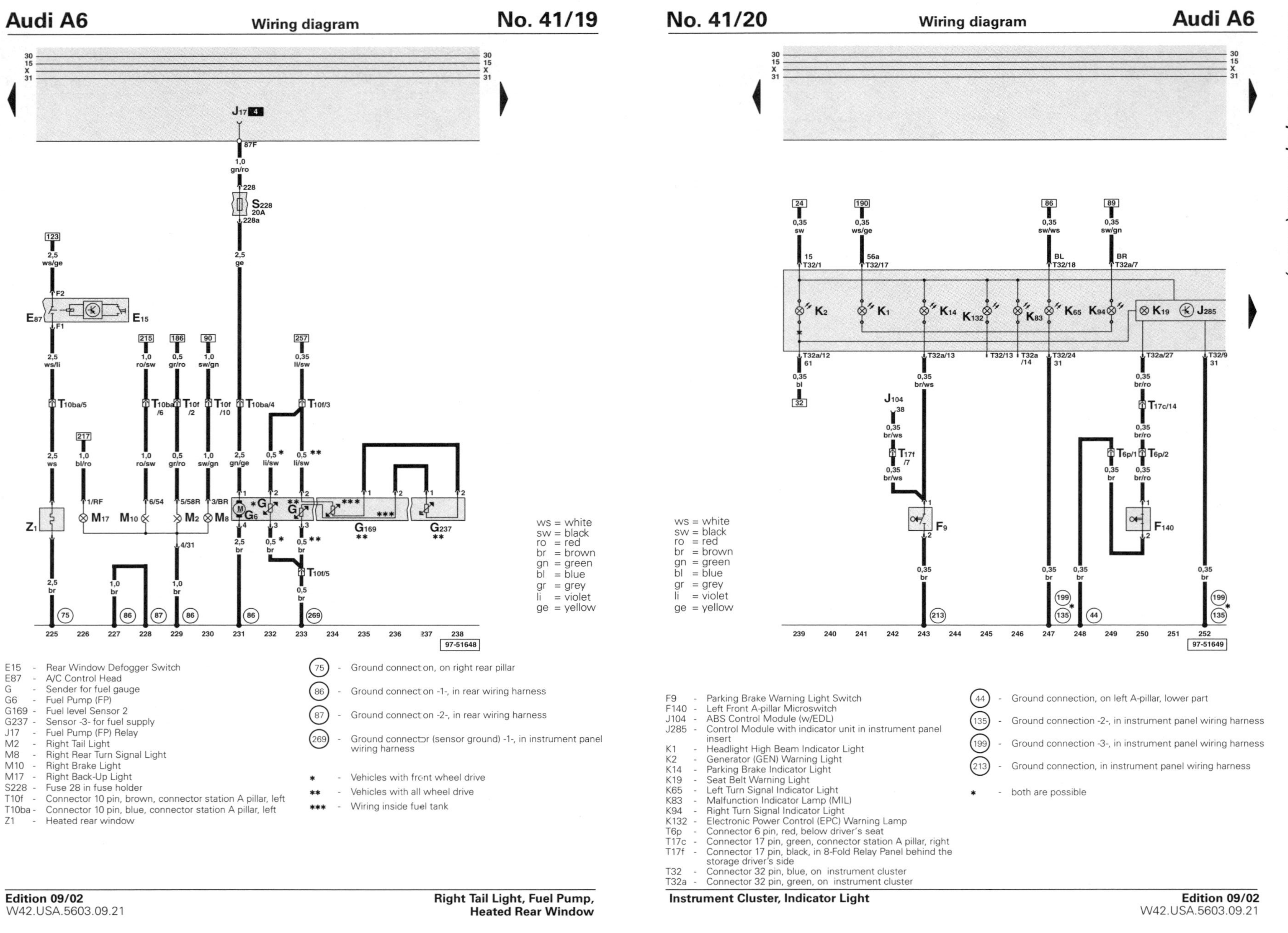

BentleyPublishers.com—All Rights Reserved

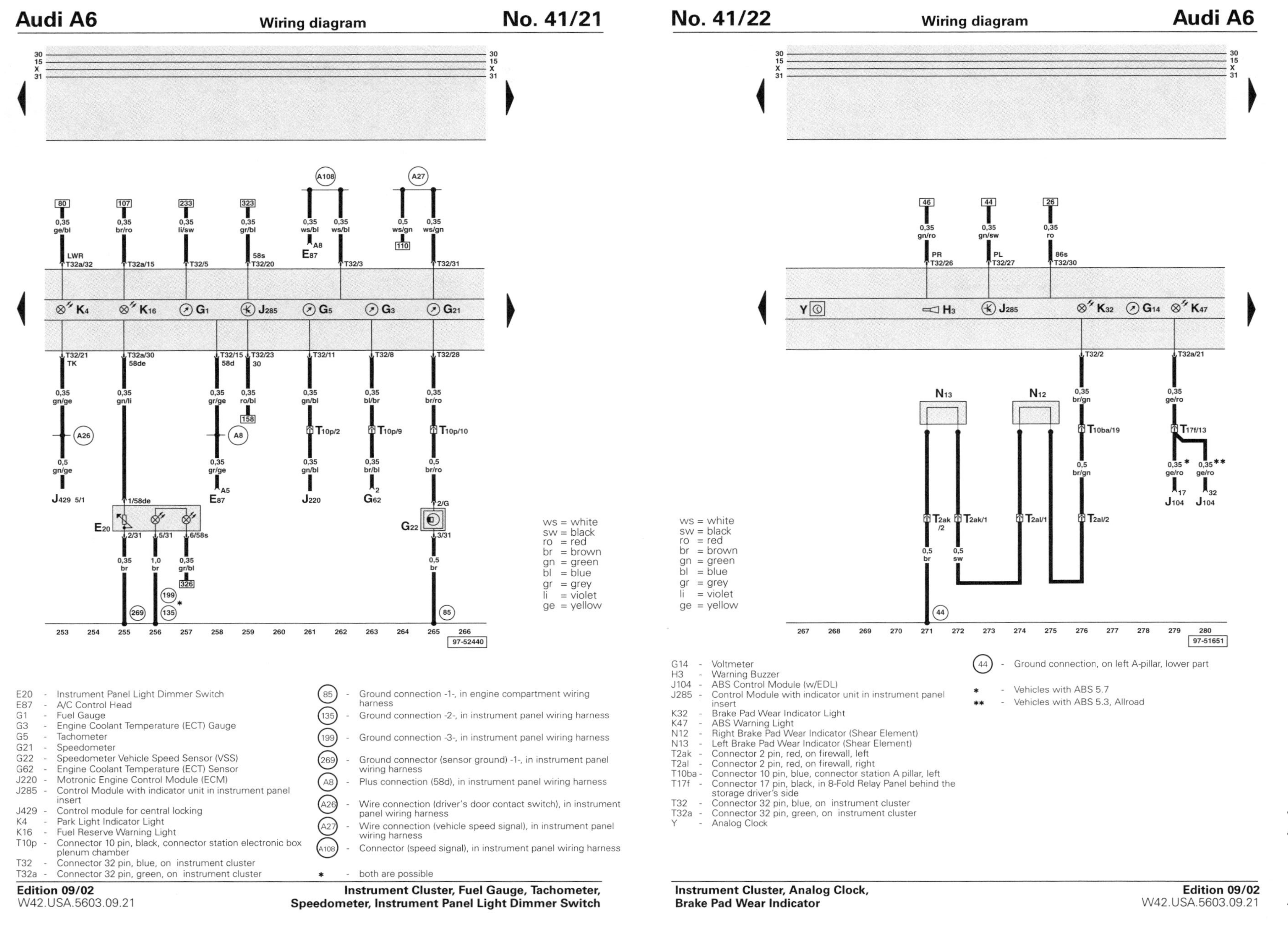

BentleyPublishers.com—All Rights Reserved

Standard Equipment (2002)

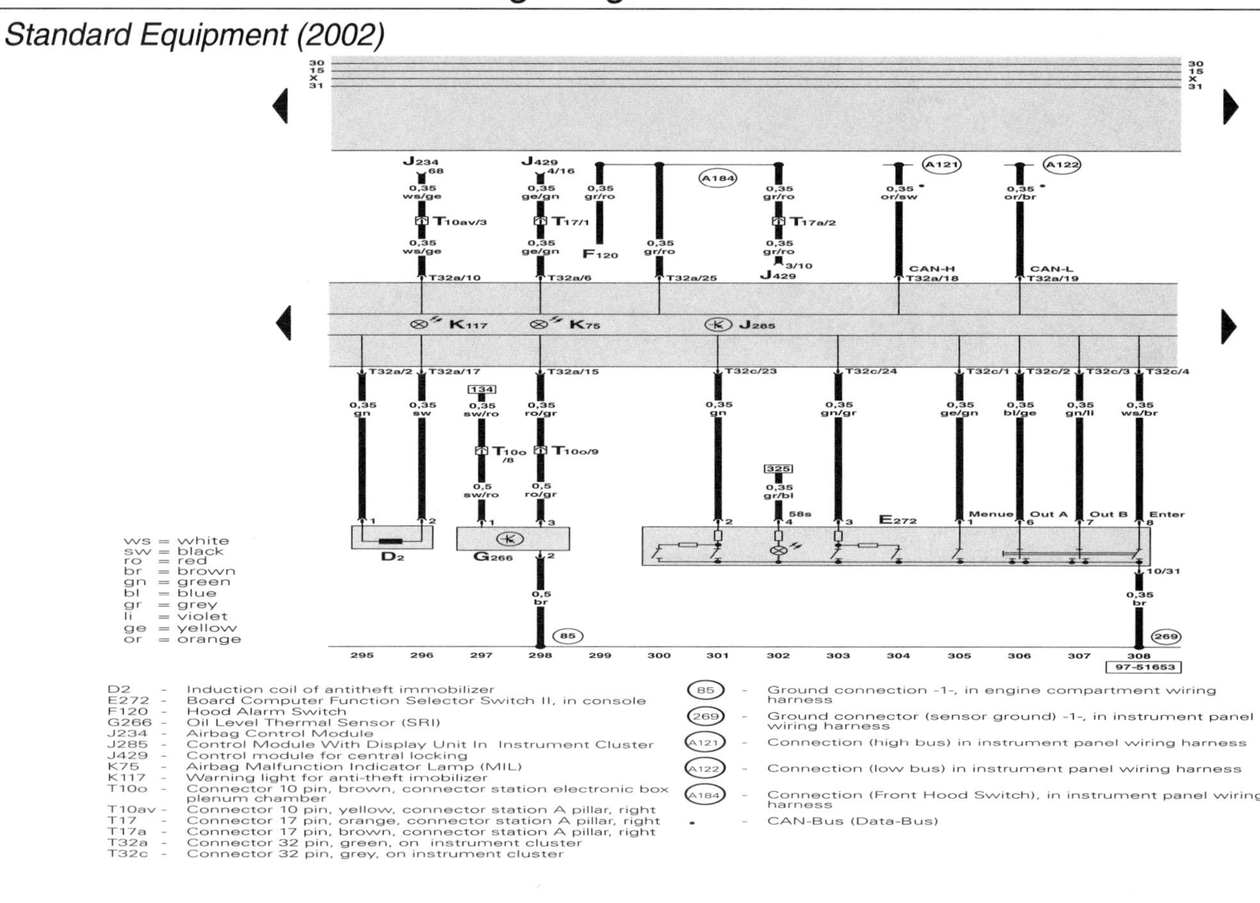

Instrument Cluster, Oil Level Thermal Sensor, Board Computer Function Selector Switch II

Edition 05/01
W42.USA.5603.04.21

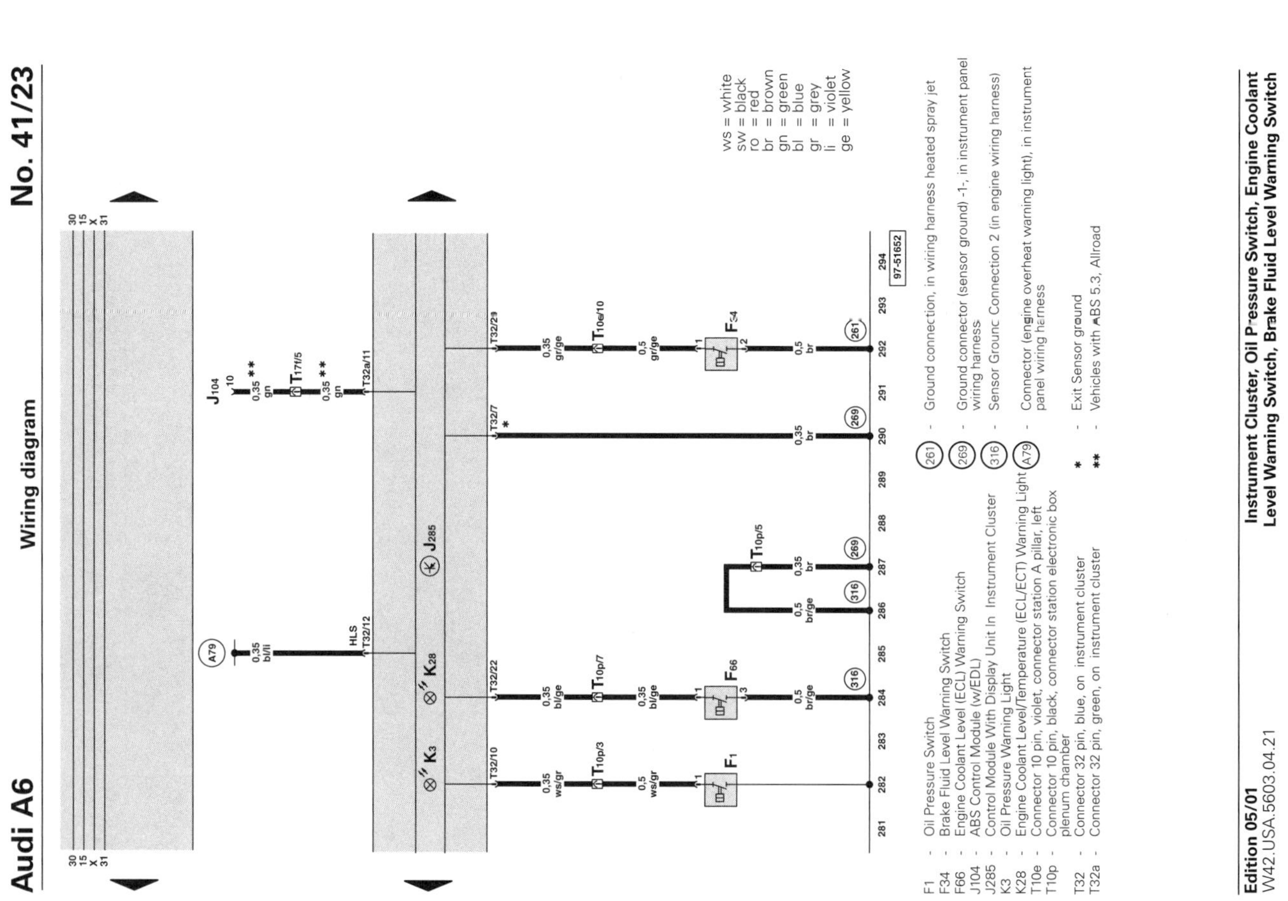

BentleyPublishers.com—All Rights Reserved

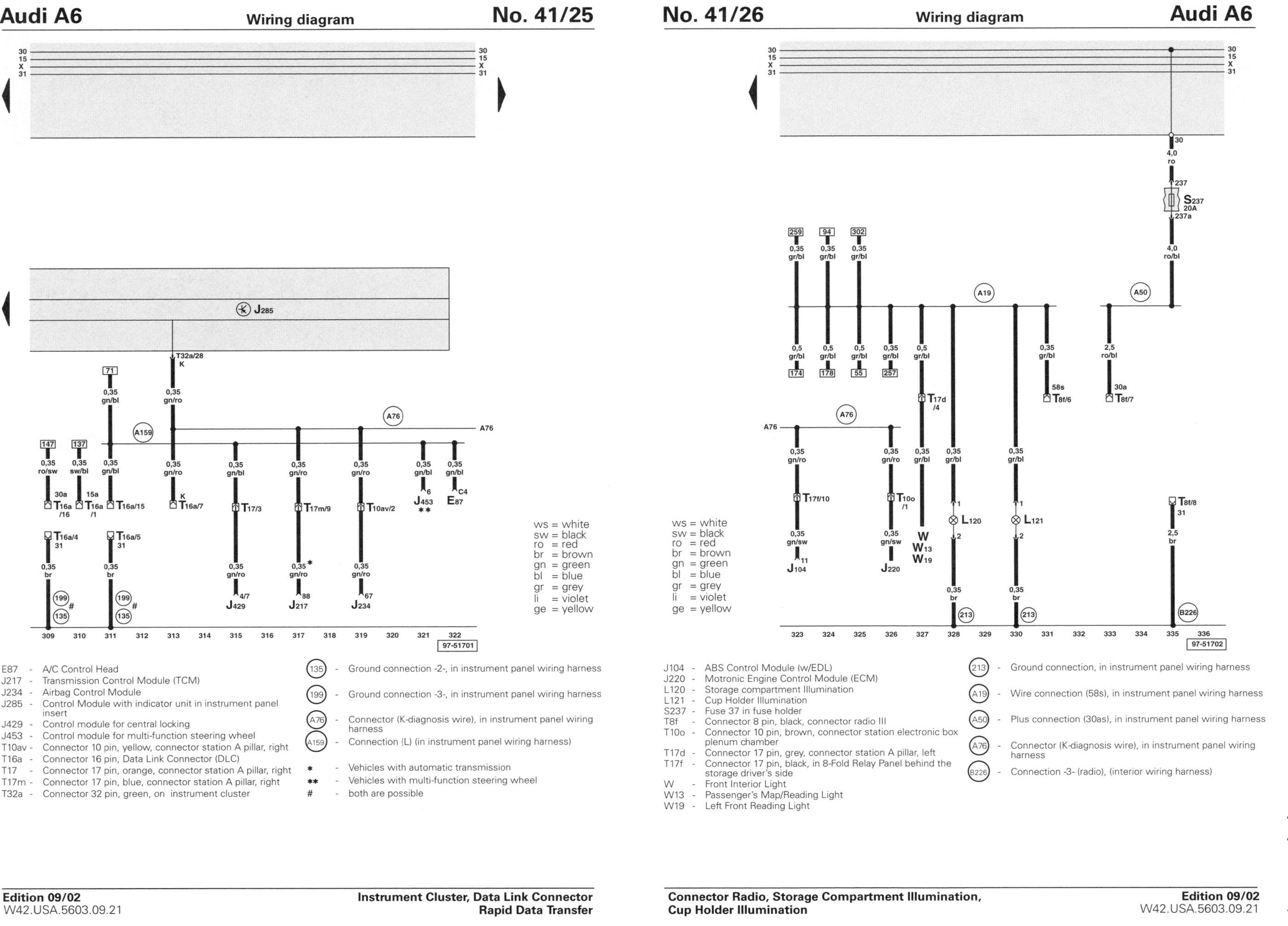

BentleyPublishers.com—All Rights Reserved

(3,0 l - Injection Engine, 6 Cylinder), Code AVK

2002 m. y.

Fuse Panel

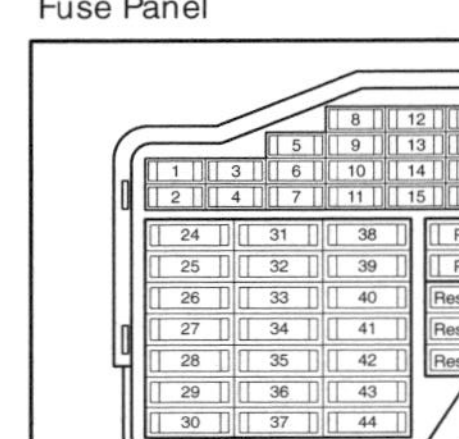

Fuse Colors:

30 A - Green
25 A - White
20 A - Yellow
15 A - Blue
10 A - Red
7,5 A - Brown
5 A - Beige

Starting with fuse position 23, fuses in the fuse holder are identified with 223 in the wiring diagram.

Micro Central Electric Panel

A97-0110

Relay Location:

4 -Fuel Pump (FP) Relay , J17

13 - Fold Relay Panel

A97-0024

Relay Location:

4 -Starting Interlock Relay, J207

8 - Fold Relay Panel

A97-0026

Relay Location:

20 -Coolant Fan Fuse, S42

21 -Control module fuse for coolant fan,S142

3 - Fold Relay Panel in E-Box plenum chamber

A97-0327

Relay Location:

1b -Coolant Circulation Pump Relay, J151

2 -Secondary Air Injection (AIR) Pump Relay, J299

3 -Motronic Engine Control Module (ECM) Power Supply Relay, J271

5 -Engine Electronic Fuse, S282

7 -Fuse for secondary air pump , S130

Relay Panel behind Instrument Panel left, on central girder

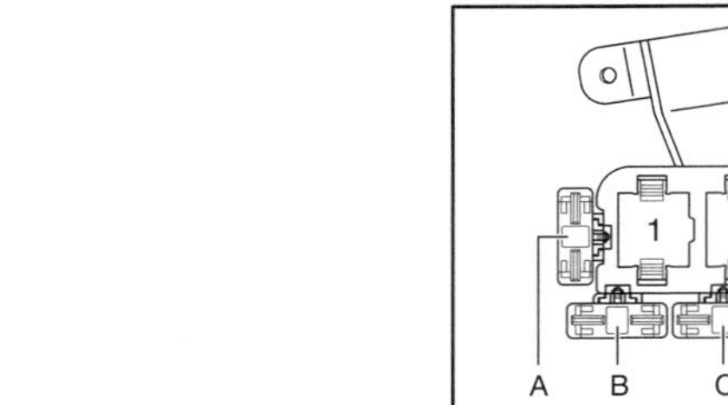

Relay Location:

1 -Brake Booster Relay, J569

C -Fuse for Hydraulic Pump Relay, S279

BentleyPublishers.com—All Rights Reserved

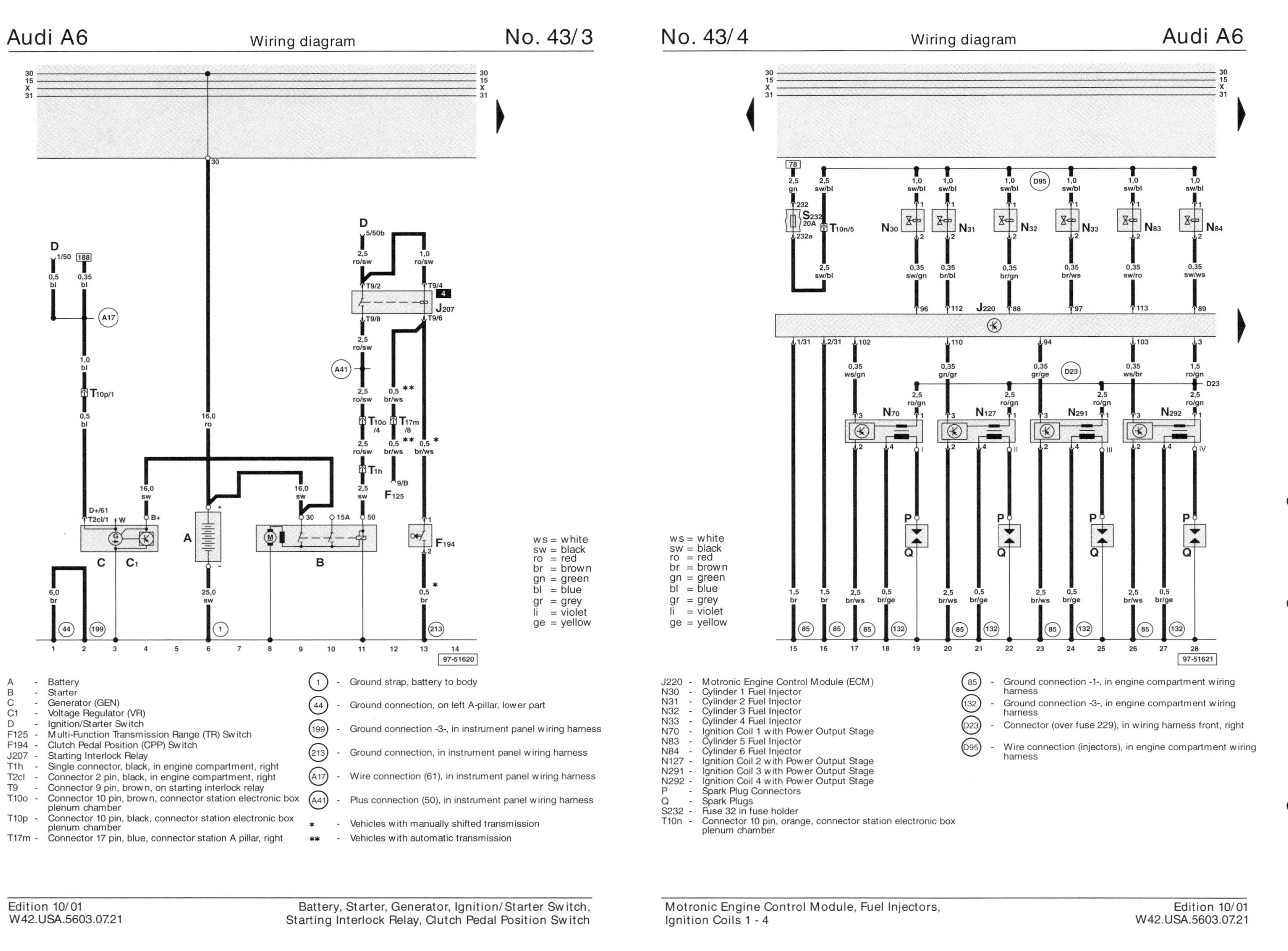

BentleyPublishers.com—All Rights Reserved

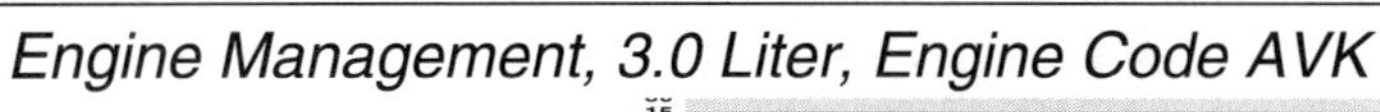

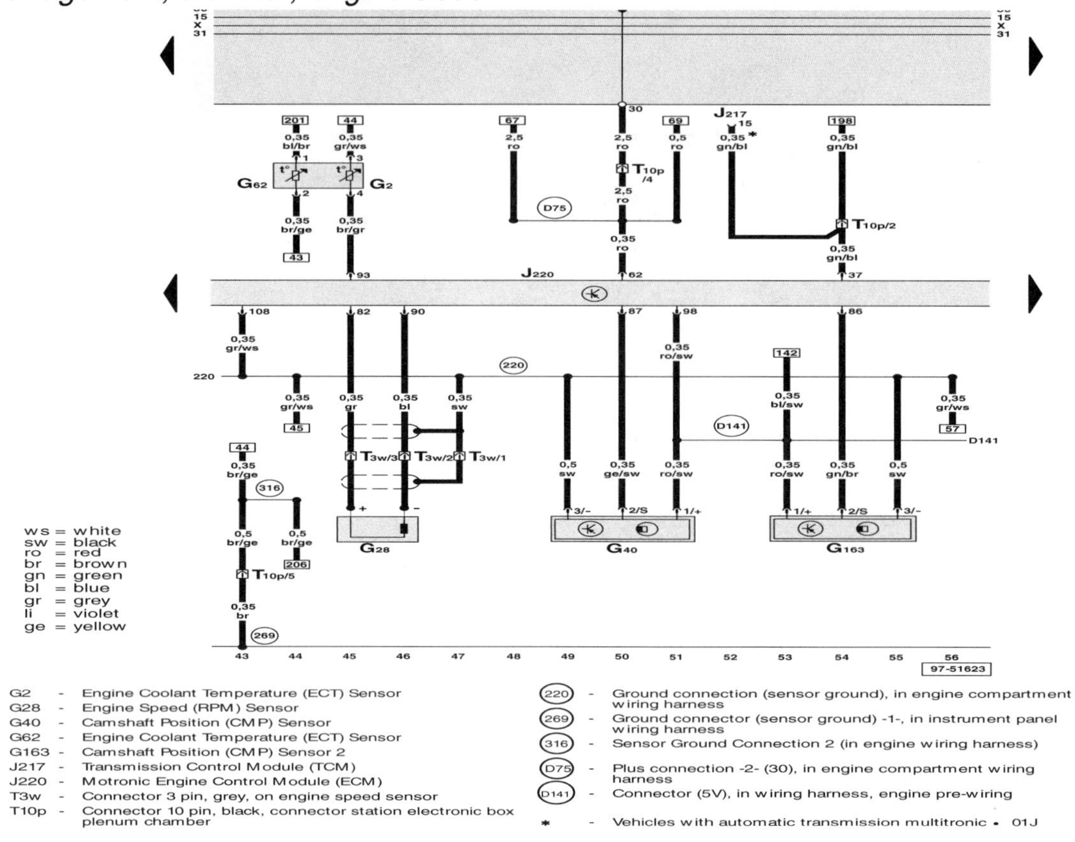

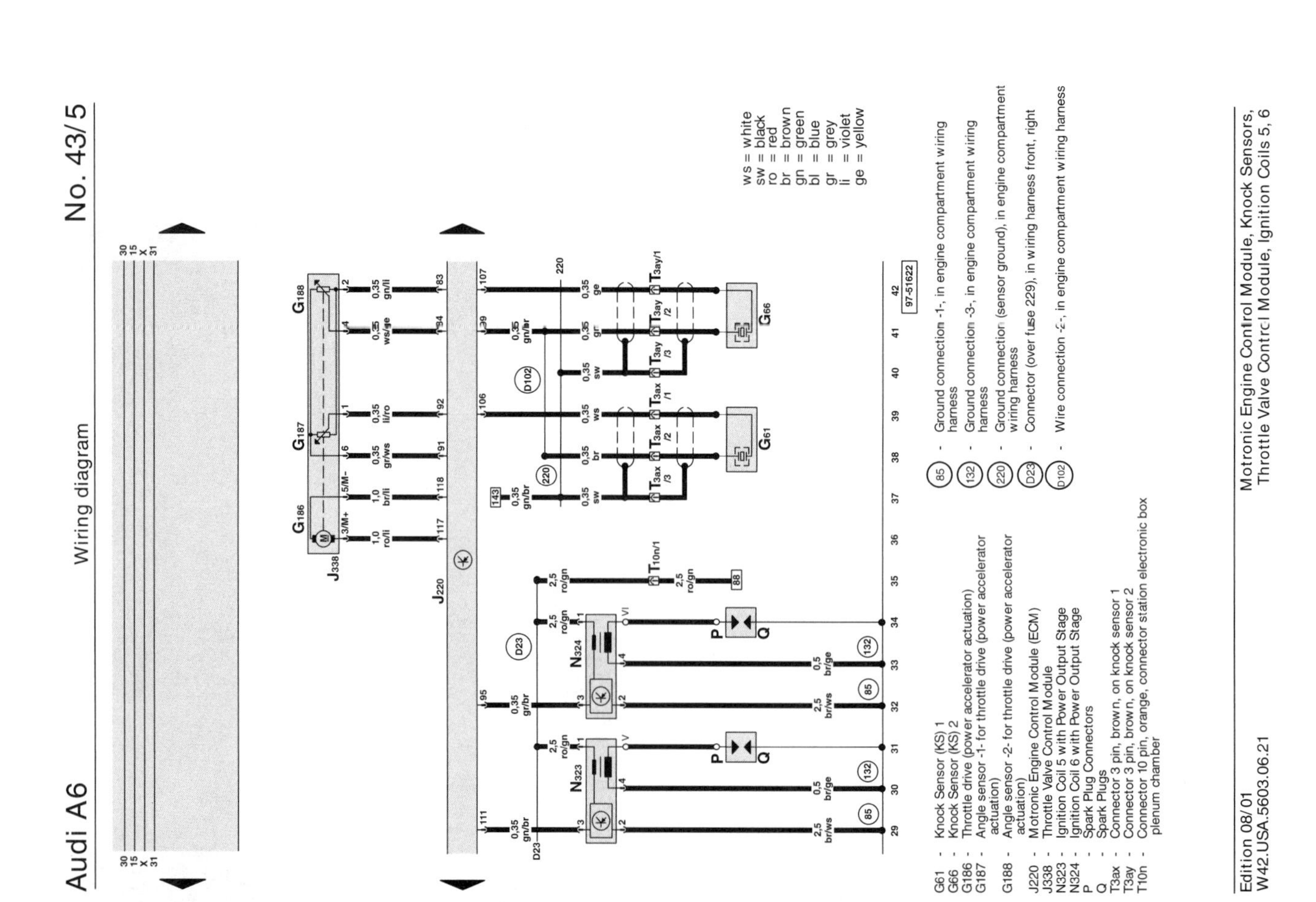

Audi A6
Wiring diagram
No. 43/5
ws = white
sw = black
ro = red
br = brown
gn = green
bl = blue
gr = grey
li = violet
ge = yellow
G61 - Knock Sensor (KS) 1
G66 - Knock Sensor (KS) 2
G186 - Throttle drive (power accelerator actuation)
G187 - Angle sensor -1- for throttle drive (power accelerator actuation)
G188 - Angle sensor -2- for throttle drive (power accelerator actuation)
J220 - Motronic Engine Control Module (ECM)
J338 - Throttle Valve Control Module
N323 - Ignition Coil 5 with Power Output Stage
N324 - Ignition Coil 6 with Power Output Stage
P - Spark Plug Connectors
Q - Spark Plugs
T3ax - Connector 3 pin, brown, on knock sensor 1
T3ay - Connector 3 pin, brown, on knock sensor 2
T10n - Connector 10 pin, orange, connector station electronic box plenum chamber
85 - Ground connection -1-, in engine compartment wiring harness
132 - Ground connection -3-, in engine compartment wiring harness
220 - Ground connection (sensor ground), in engine compartment wiring harness
D23 - Connector (over fuse 229), in wiring harness front, right
D102 - Wire connection -2-, in engine compartment wiring harness
97-51622
Edition 08/01
W42.USA.5603.06.21
Motronic Engine Control Module, Knock Sensors, Throttle Valve Control Module, Ignition Coils 5, 6

BentleyPublishers.com—All Rights Reserved

Audi A6

Wiring diagram

No. 43/8

ws = white
sw = black
ro = red
br = brown
gn = green
bl = blue
gr = grey
li = violet
ge = yellow

- G39 - Heated Oxygen Sensor (HO2S) 2
- G108 - Heated Oxygen Sensor (HO2S) 2
- J17 - Fuel Pump (FP) Relay
- J220 - Motronic Engine Control Module (ECM)
- S234 - Fuse 34 in fuse holder
- T6as - Connector 6 pin, black, heated oxygen sensor
- T6av - Connector 6 pin, black, heated oxygen sensor 2
- T10n - Connector 10 pin, orange, connector station electronic box plenum chamber
- T10p - Connector 10 pin, black, connector station electronic box plenum chamber
- Z19 - Oxygen Sensor (O2S) Heater
- Z28 - Oxygen Sensor (O2S) 2 Heater

- (200) - Ground connection (shielding), in engine compartment wiring harness
- (D22) - Connector (over fuse 234), in wiring harness front, right

Motronic Engine Control Module, Fuel Pump Relay, Heated Oxygen Sensors

Edition 08/01
W42.USA.5603.06.21

97-51624

97-53299

ws = white
sw = black
ro = red
br = brown
gn = green
bl = blue
gr = grey
li = violet
ge = yellow

- G300 - Camshaft Position Sensor 3
- G301 - Camshaft Position Sensor 4
- J217 - Transmission Control Module (TCM)
- J220 - Motronic Engine Control Module (ECM)
- J271 - Motronic Engine Control Module (ECM) Power Supply Relay
- S199 - Fuse 1 (15)
- S282 - Engine Electronics Fuse

- (200) - Ground connection (shielding), in engine compartment wiring harness
- (D52) - Plus connection (15a), in engine compartment wiring harness
- (D141) - Connector (5V), in wiring harness, engine pre-wiring
- (D173) - Connection (87) (in engine compartment wiring harness)
- * - Vehicles without automatic transmission multitronic • 01J
- ** - Vehicles with automatic transmission multitronic • 01J

BentleyPublishers.com—All Rights Reserved

Engine Management, 3.0 Liter, Engine Code AVK

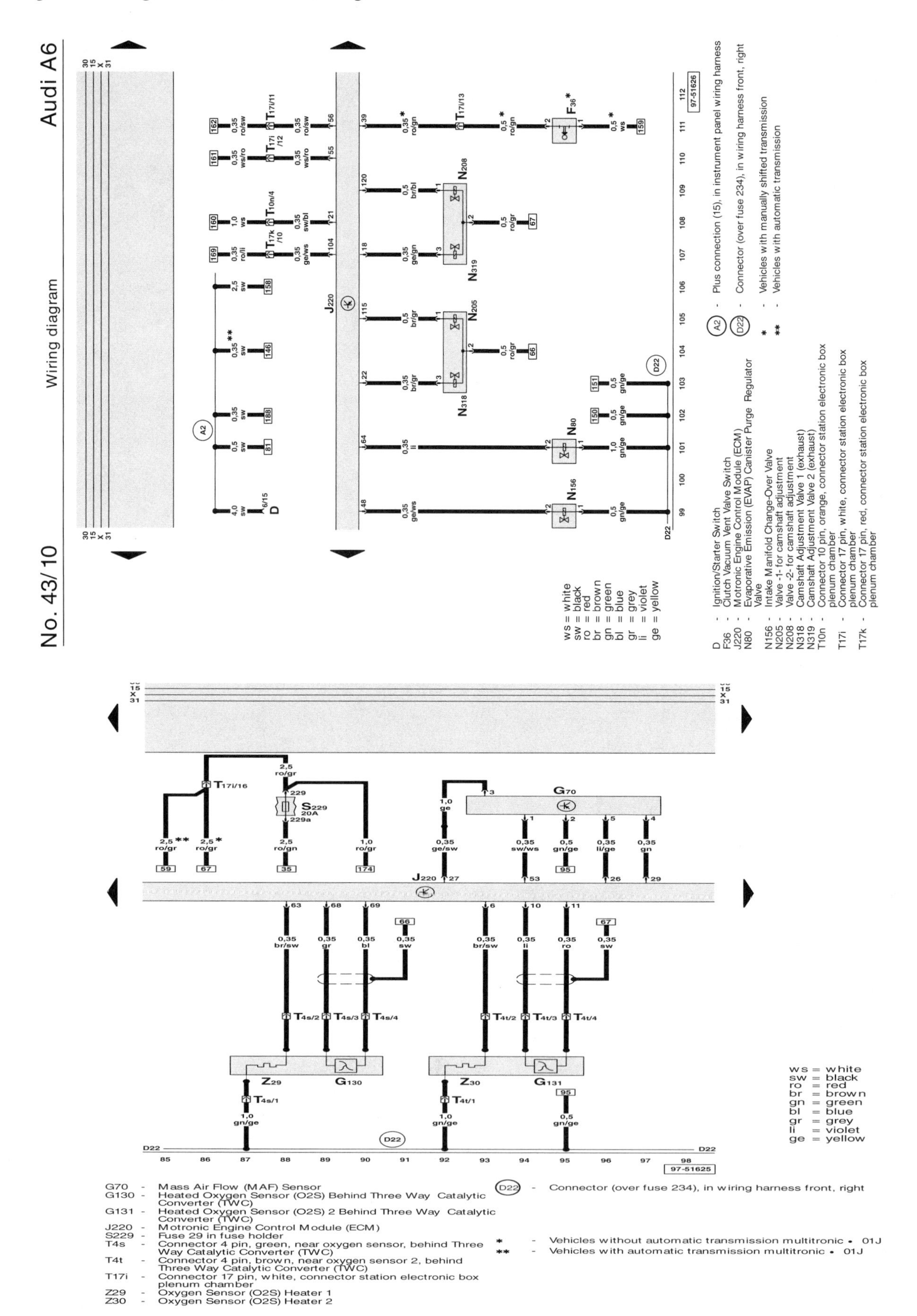

BentleyPublishers.com—All Rights Reserved

Engine Management, 3.0 Liter, Engine Code AVK

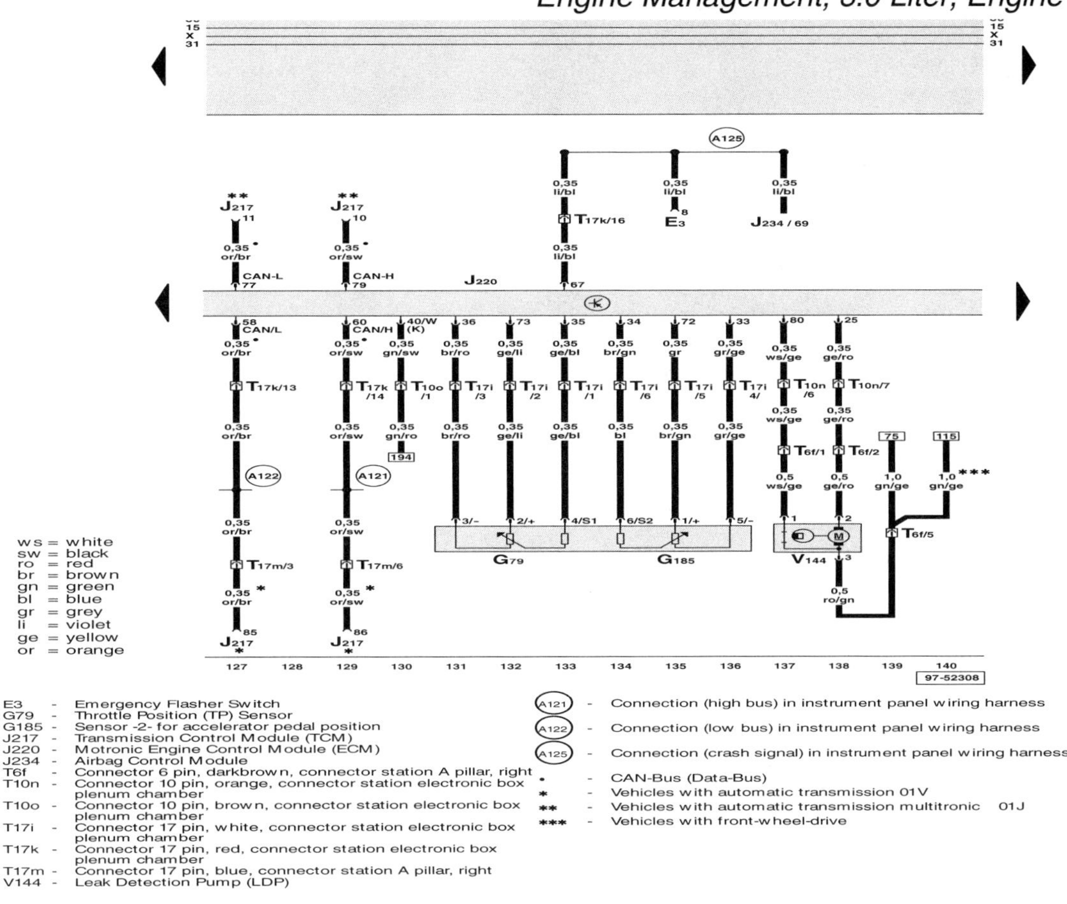

E3 - Emergency Flasher Switch
G79 - Throttle Position (TP) Sensor
G185 - Sensor -2- for accelerator pedal position
J217 - Transmission Control Module (TCM)
J220 - Motronic Engine Control Module (ECM)
J234 - Airbag Control Module
T6f - Connector 6 pin, darkbrown, connector station A pillar, right
T10n - Connector 10 pin, orange, connector station electronic box plenum chamber
T10o - Connector 10 pin, brown, connector station electronic box plenum chamber
T17i - Connector 17 pin, white, connector station electronic box plenum chamber
T17k - Connector 17 pin, red, connector station electronic box plenum chamber
T17m - Connector 17 pin, blue, connector station A pillar, right
V144 - Leak Detection Pump (LDP)

(A121) - Connection (high bus) in instrument panel wiring harness
(A122) - Connection (low bus) in instrument panel wiring harness
(A125) - Connection (crash signal) in instrument panel wiring harness
• - CAN-Bus (Data-Bus)
* - Vehicles with automatic transmission 01V
** - Vehicles with automatic transmission multitronic 01J
*** - Vehicles with front-wheel-drive

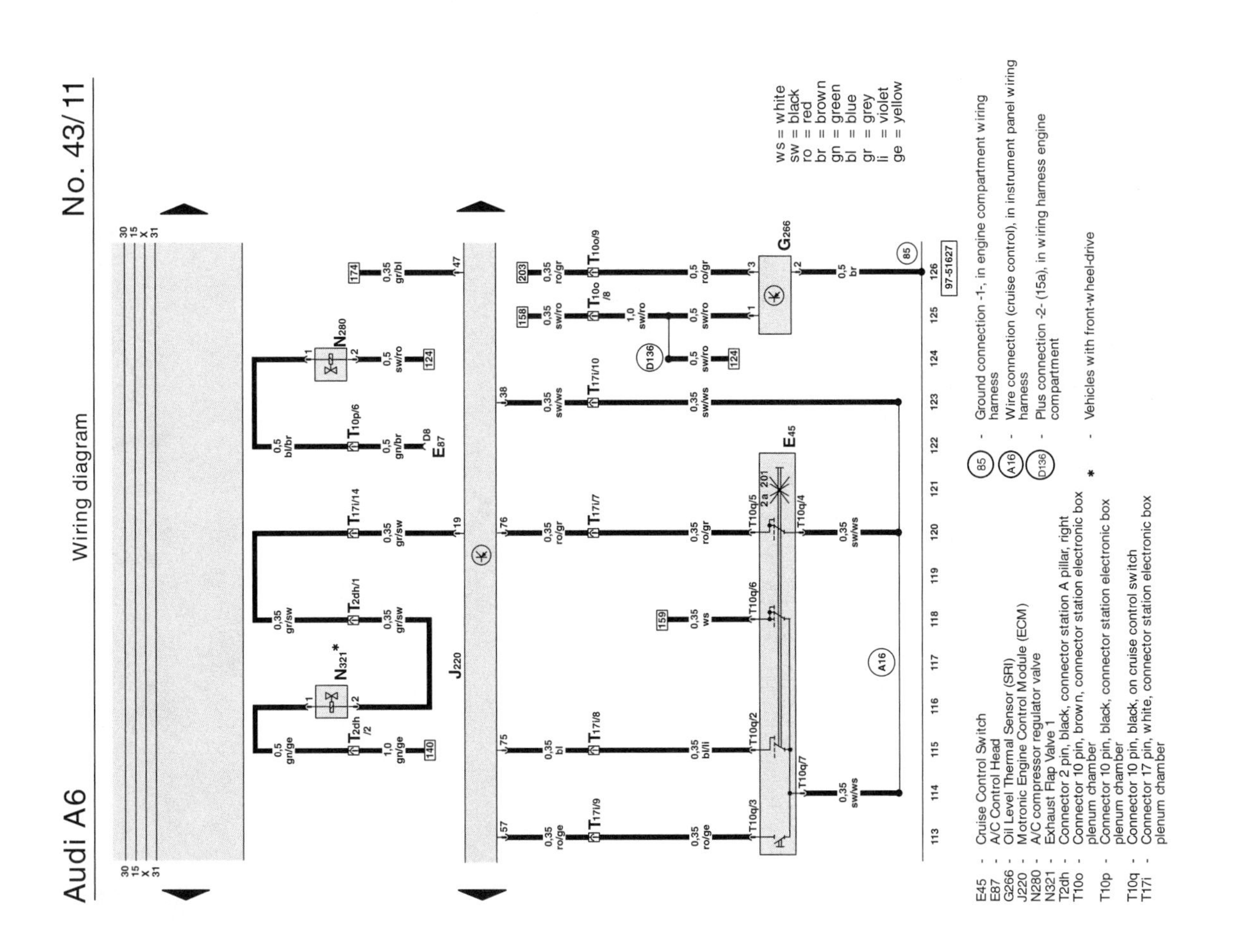

BentleyPublishers.com—All Rights Reserved

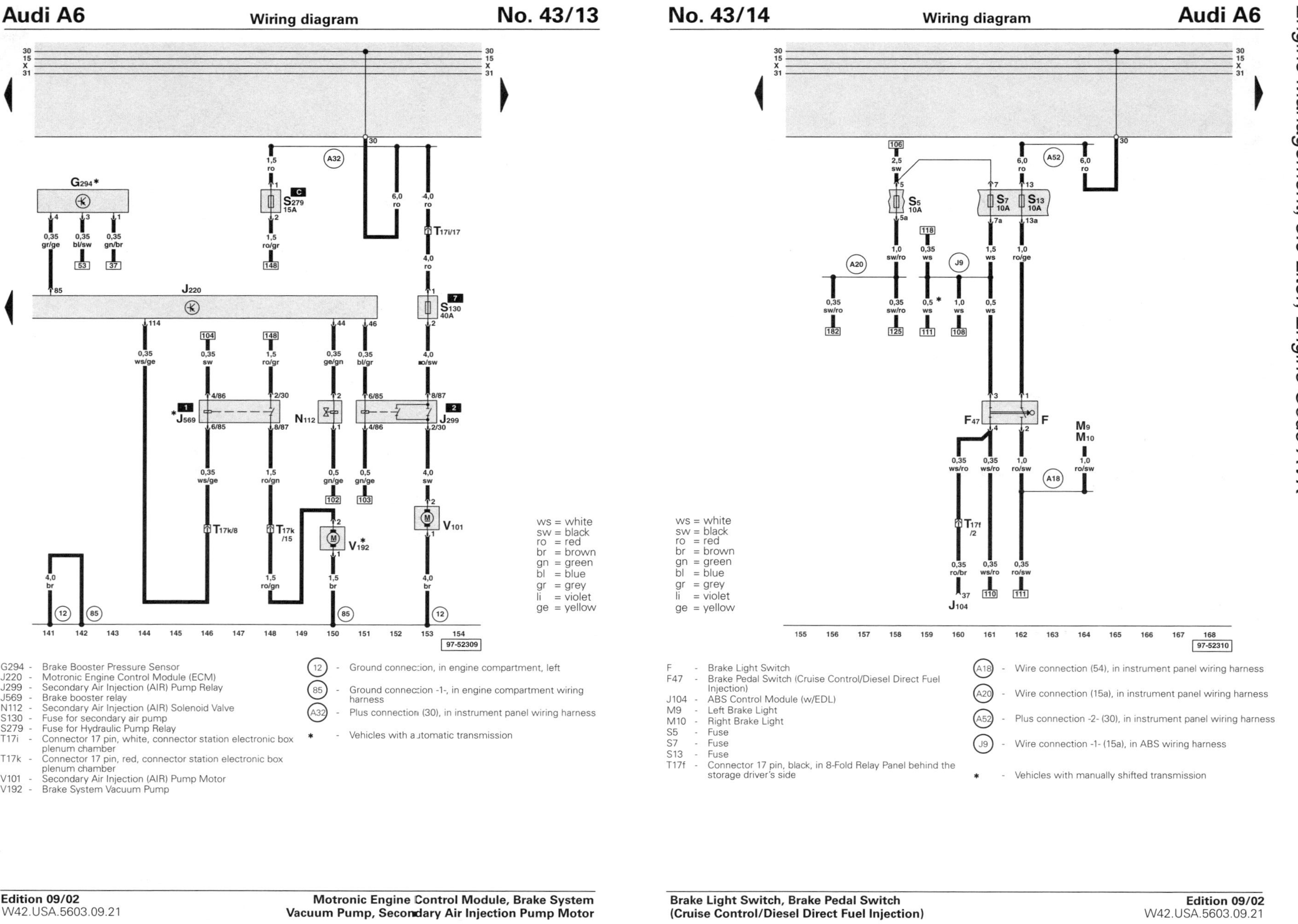

BentleyPublishers.com—All Rights Reserved

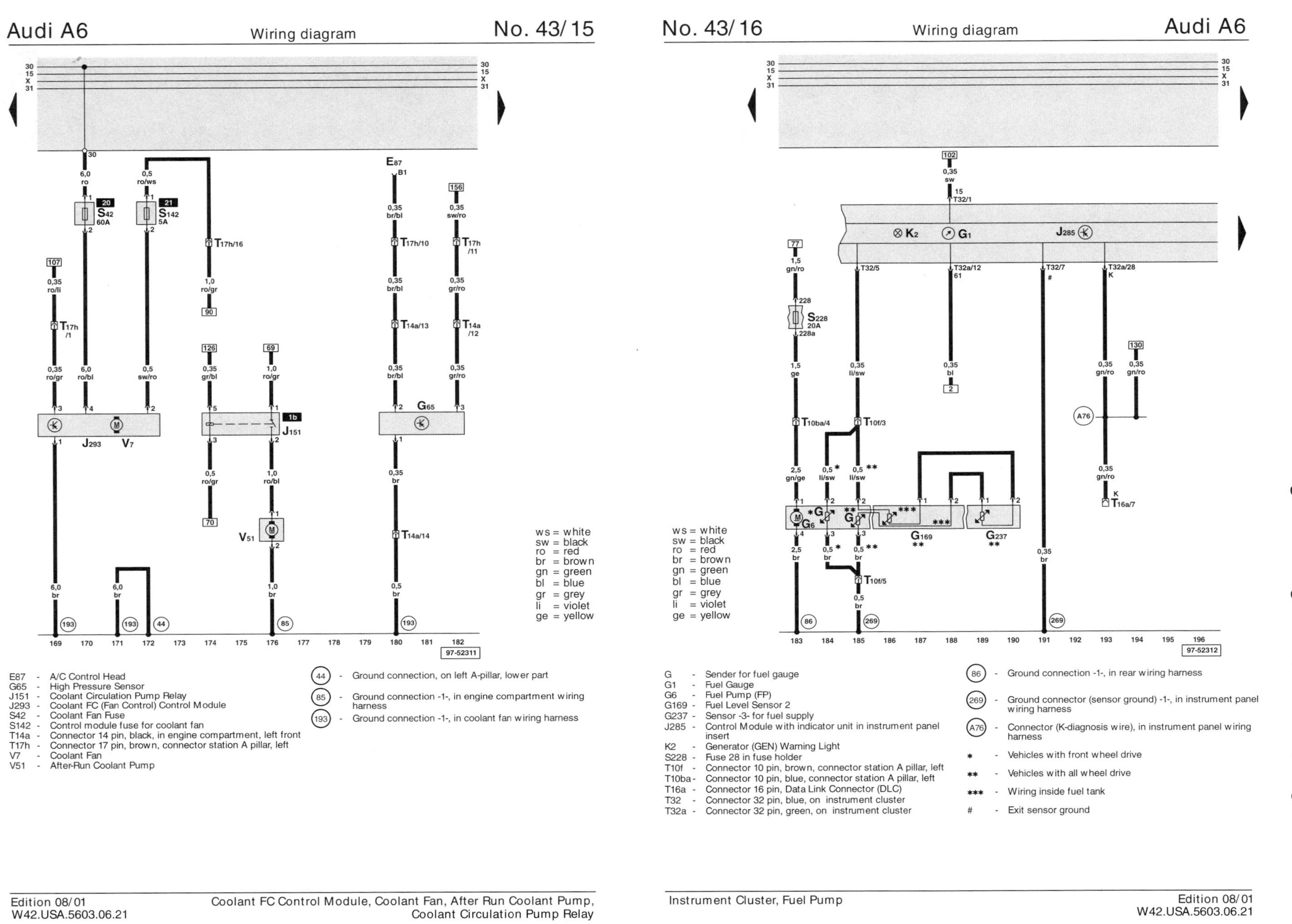

BentleyPublishers.com—All Rights Reserved

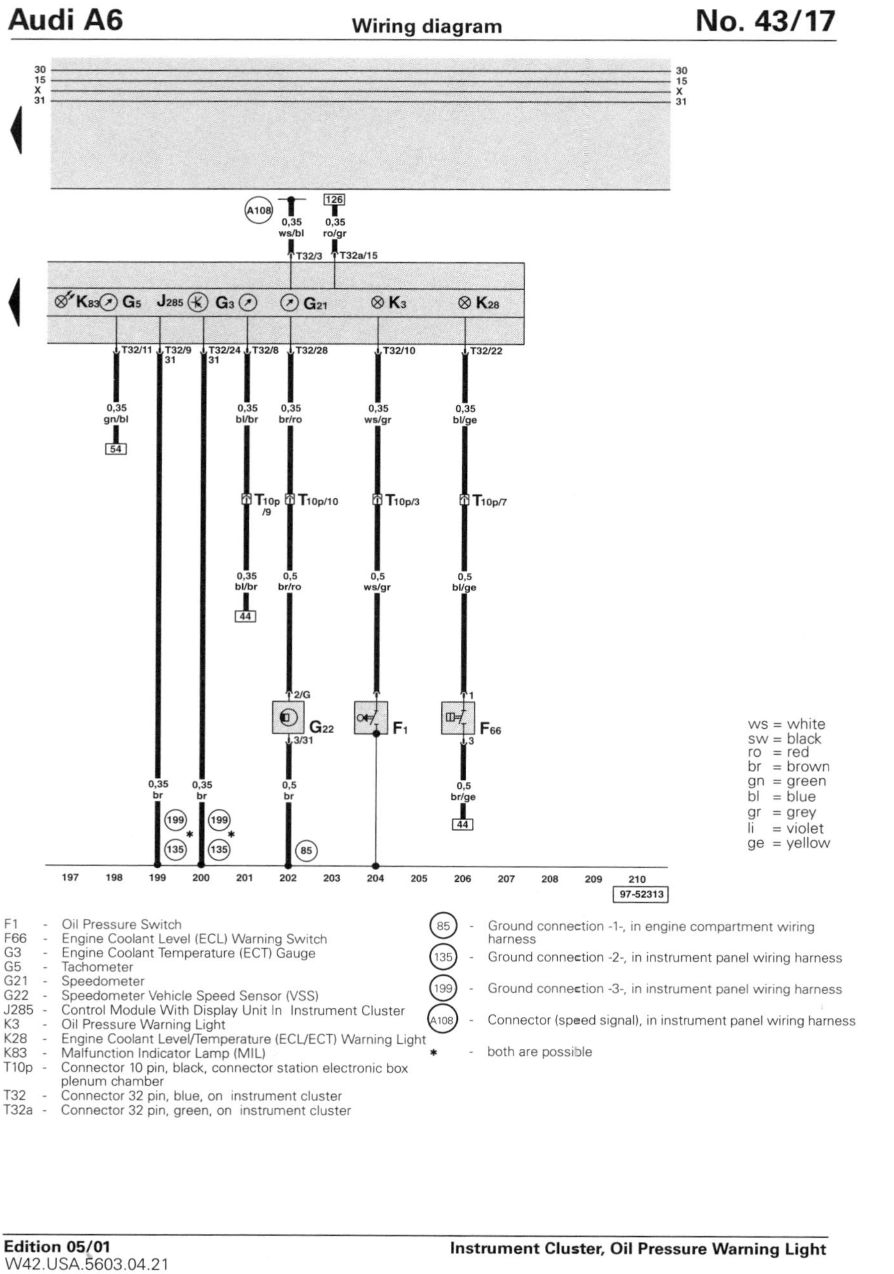

THIS SPACE INTENTIONALLY LEFT BLANK

BentleyPublishers.com—All Rights Reserved

(4,2 l - Injection Engine, 8 Cylinder), Code AWN, BBD

2002 m. y.

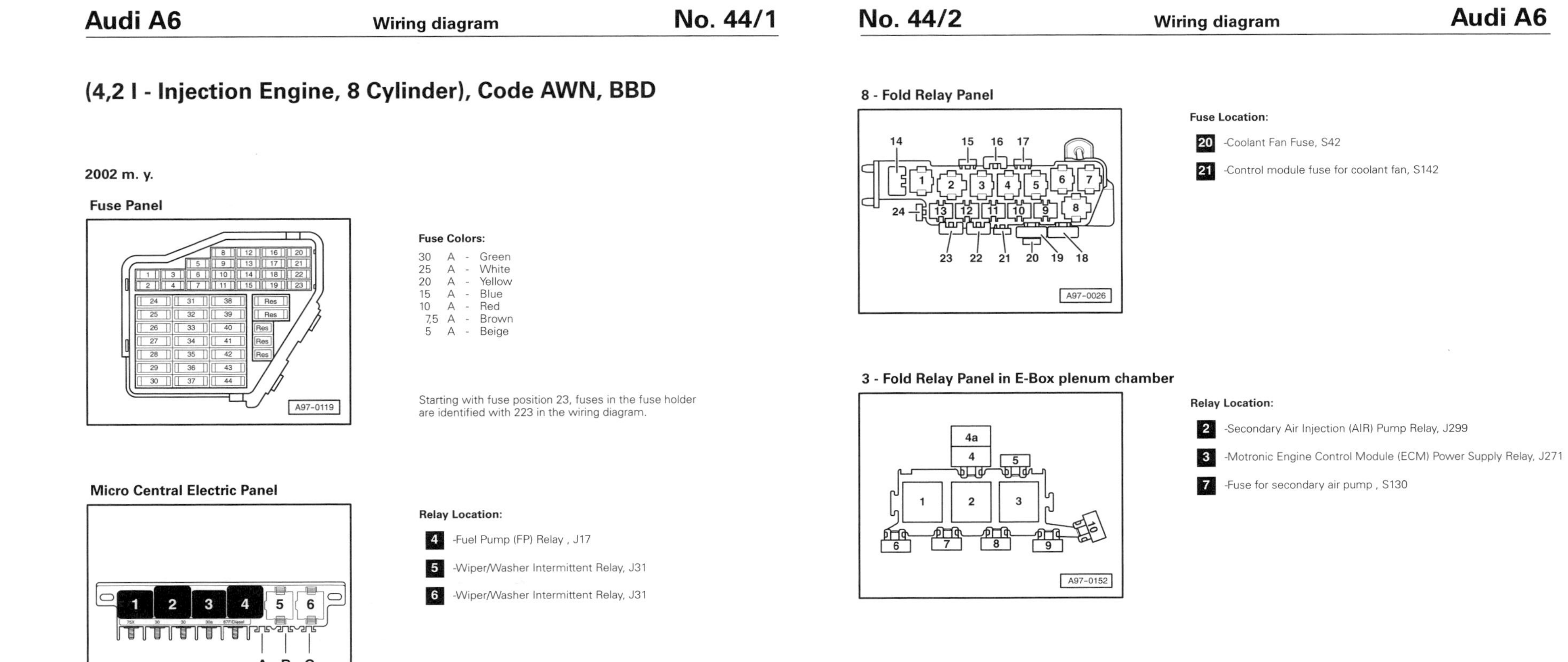

Fuse Colors:

30 A - Green
25 A - White
20 A - Yellow
15 A - Blue
10 A - Red
7,5 A - Brown
5 A - Beige

Starting with fuse position 23, fuses in the fuse holder are identified with 223 in the wiring diagram.

Micro Central Electric Panel – Relay Location:

- **4** -Fuel Pump (FP) Relay , J17
- **5** -Wiper/Washer Intermittent Relay, J31
- **6** -Wiper/Washer Intermittent Relay, J31

13 - Fold Relay Panel – Relay Location:

- **4** -Starting Interlock Relay, J207

8 - Fold Relay Panel – Fuse Location:

- **20** -Coolant Fan Fuse, S42
- **21** -Control module fuse for coolant fan, S142

3 - Fold Relay Panel in E-Box plenum chamber – Relay Location:

- **2** -Secondary Air Injection (AIR) Pump Relay, J299
- **3** -Motronic Engine Control Module (ECM) Power Supply Relay, J271
- **7** -Fuse for secondary air pump , S130

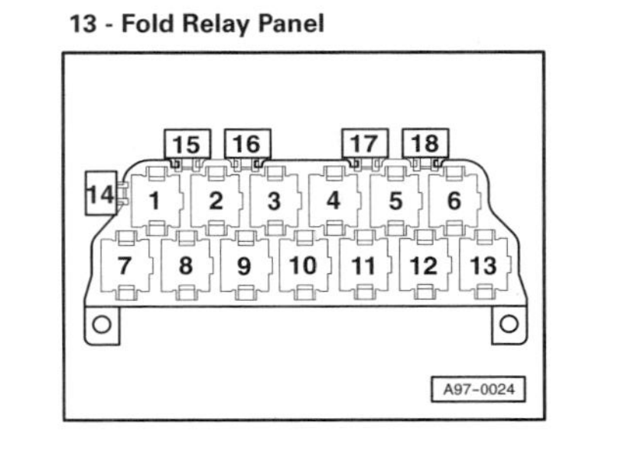

BentleyPublishers.com—All Rights Reserved

Engine Management, 4.2 Liter, Engine Codes AWN, BBD

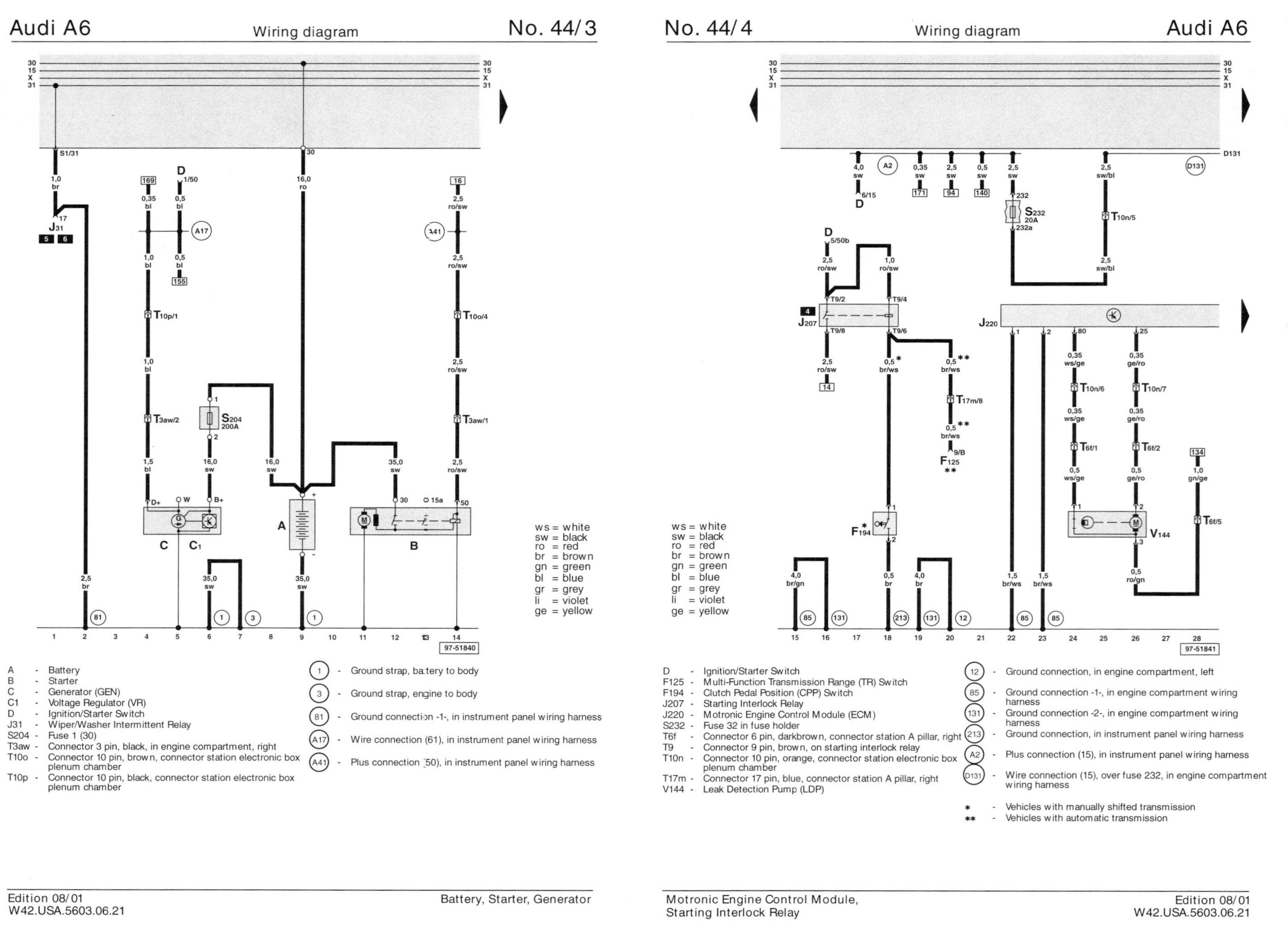

BentleyPublishers.com—All Rights Reserved

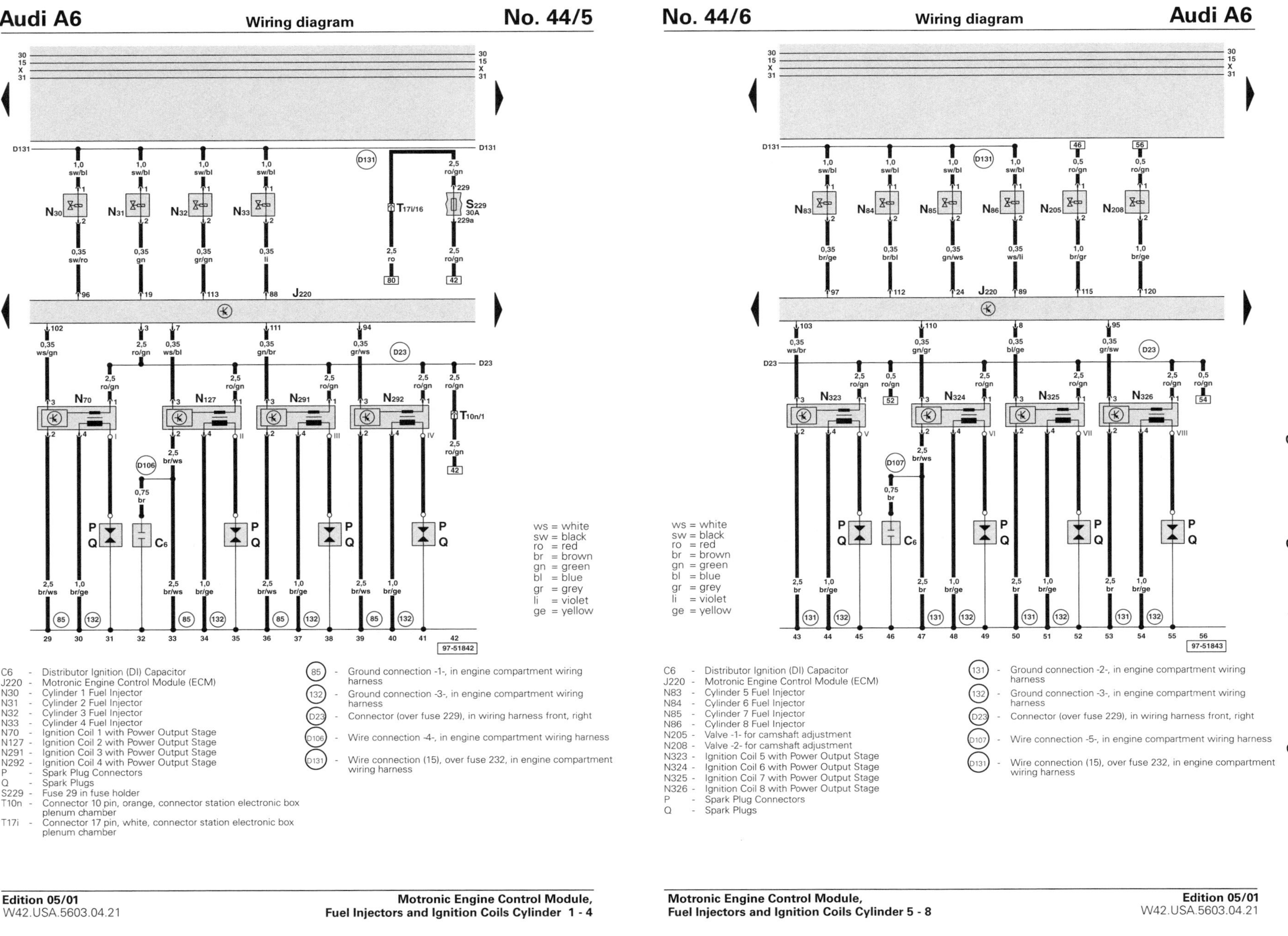

BentleyPublishers.com—All Rights Reserved

Engine Management, 4.2 Liter, Engine Codes AWN, BBD

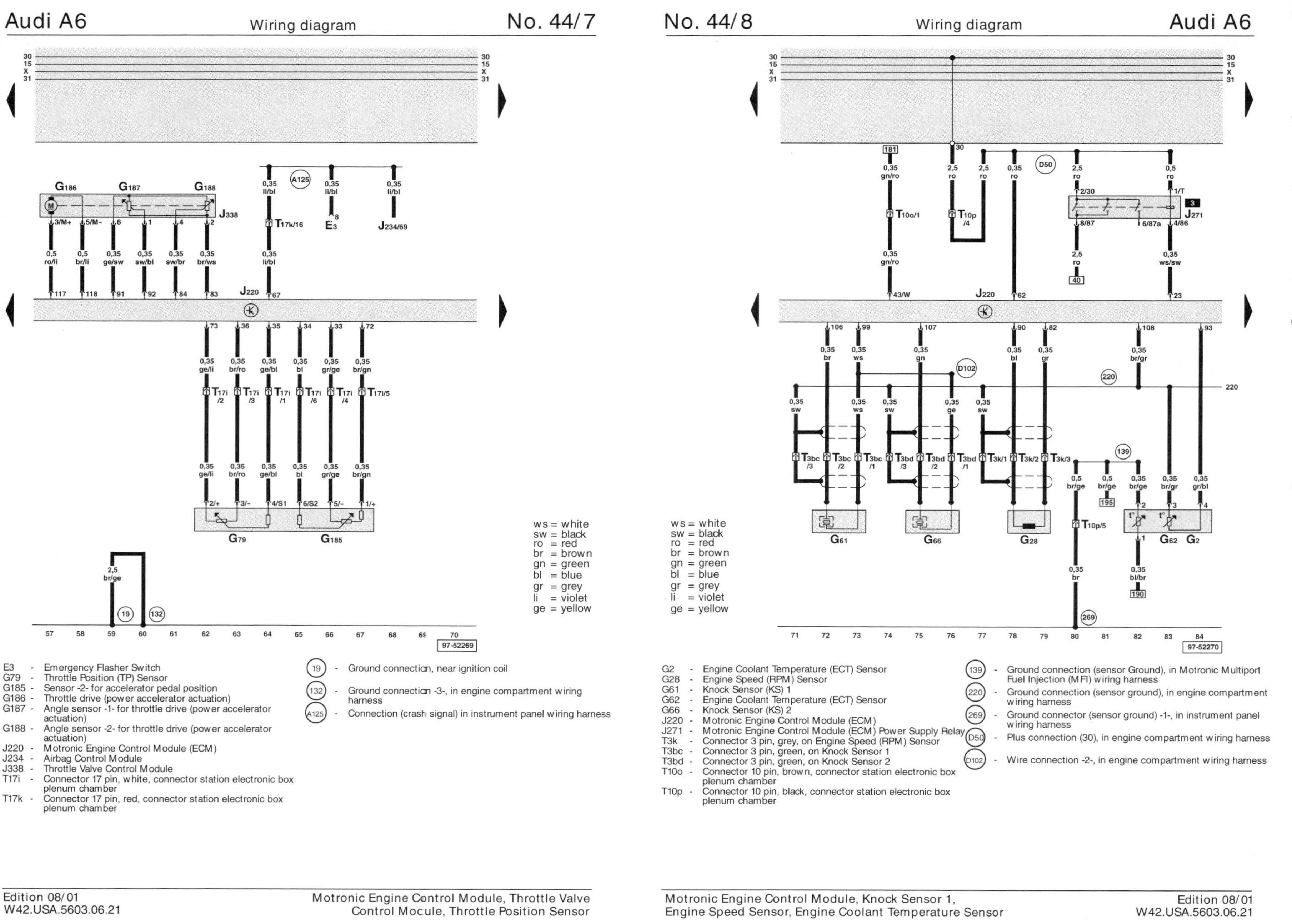

BentleyPublishers.com—All Rights Reserved

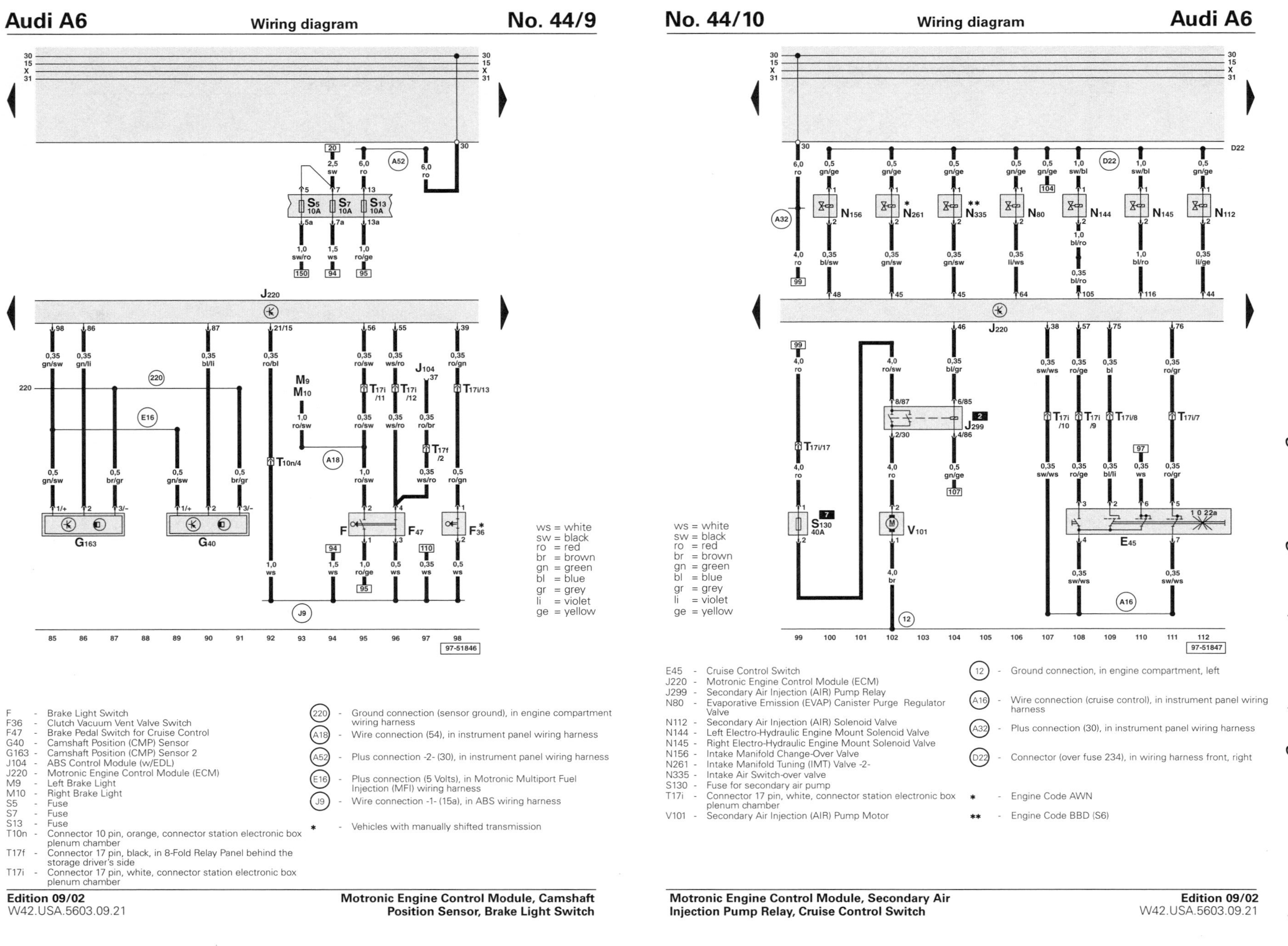

BentleyPublishers.com—All Rights Reserved

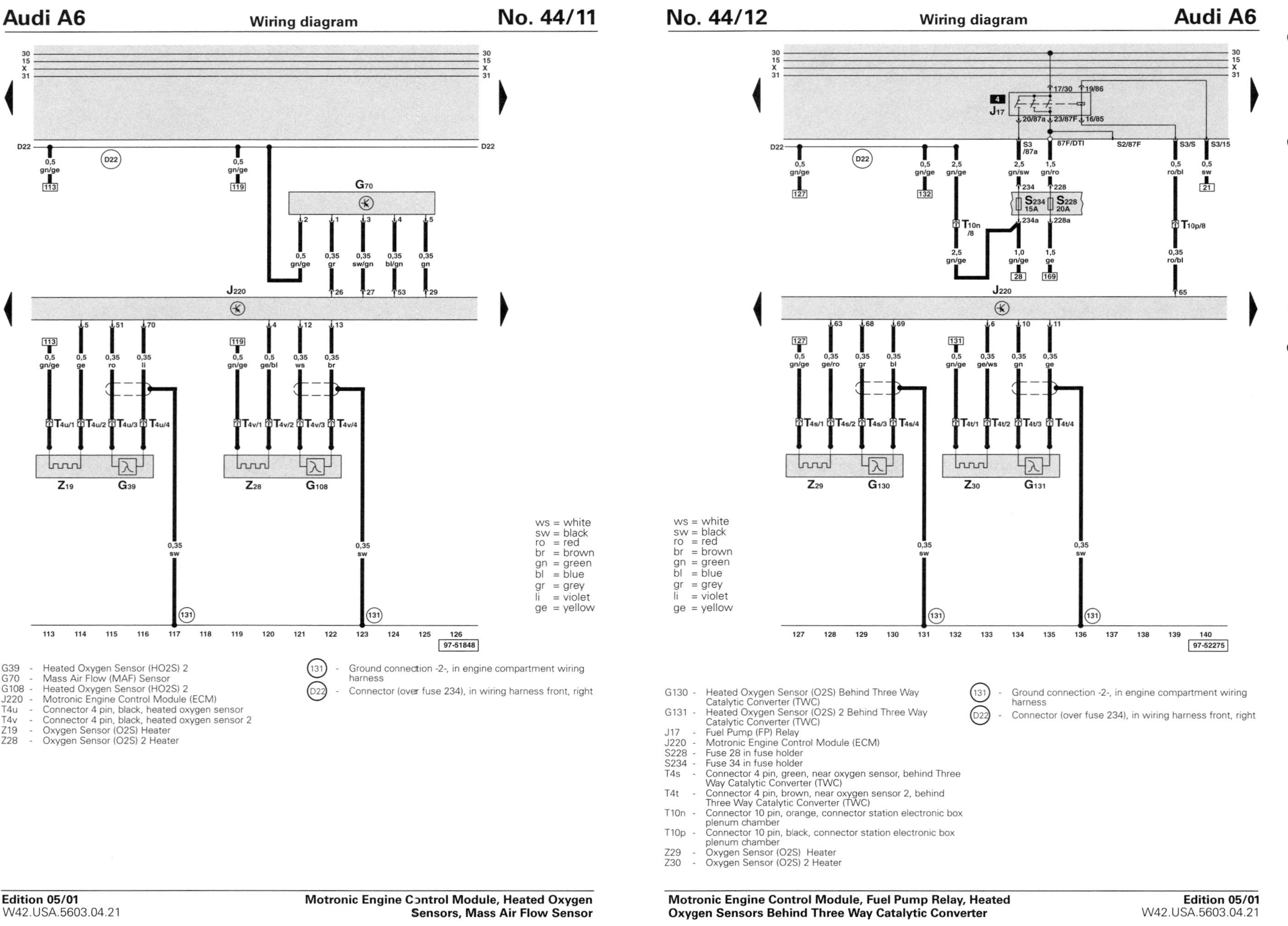

BentleyPublishers.com—All Rights Reserved

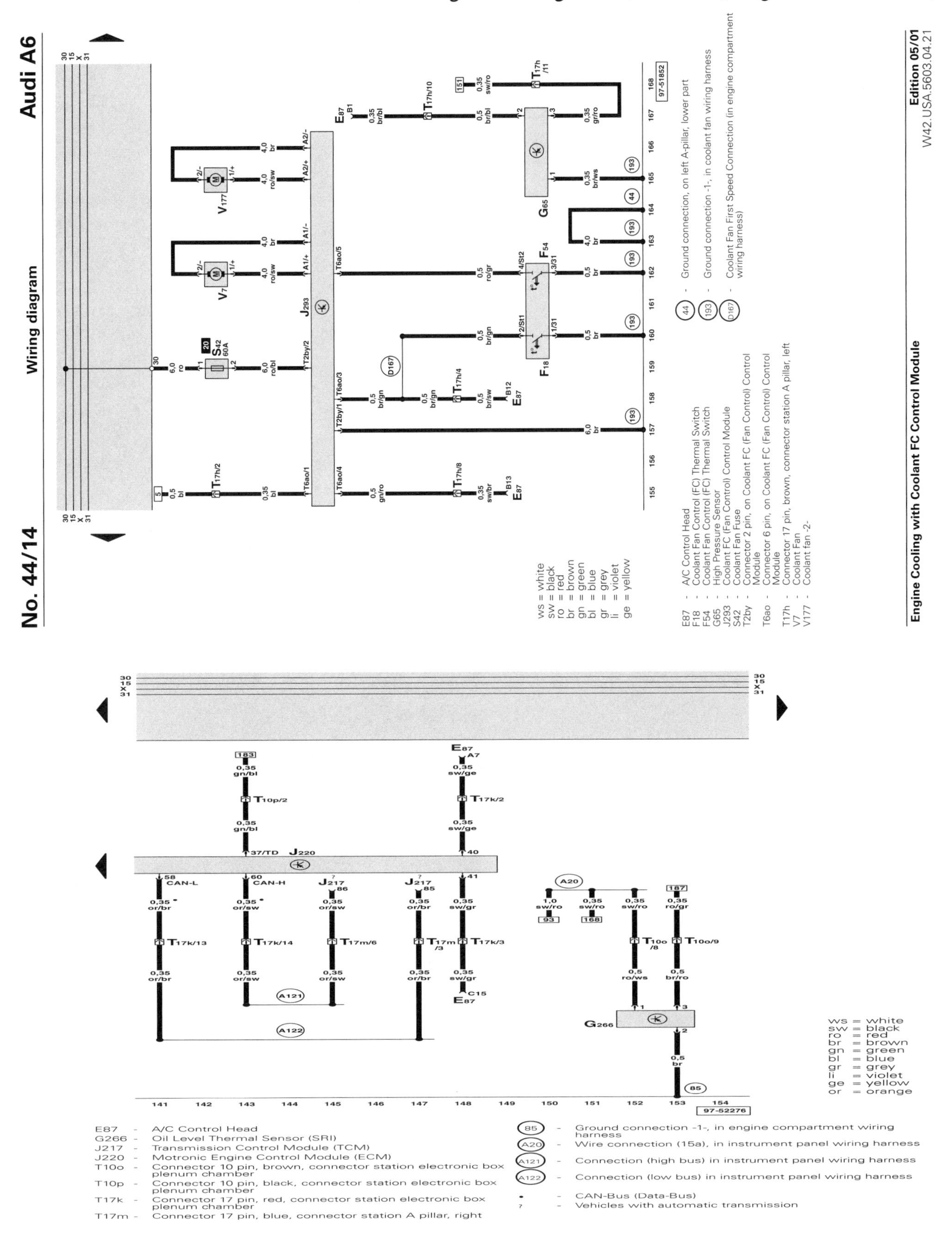

BentleyPublishers.com—All Rights Reserved

Engine Management, 4.2 Liter, Engine Codes AWN, BBD

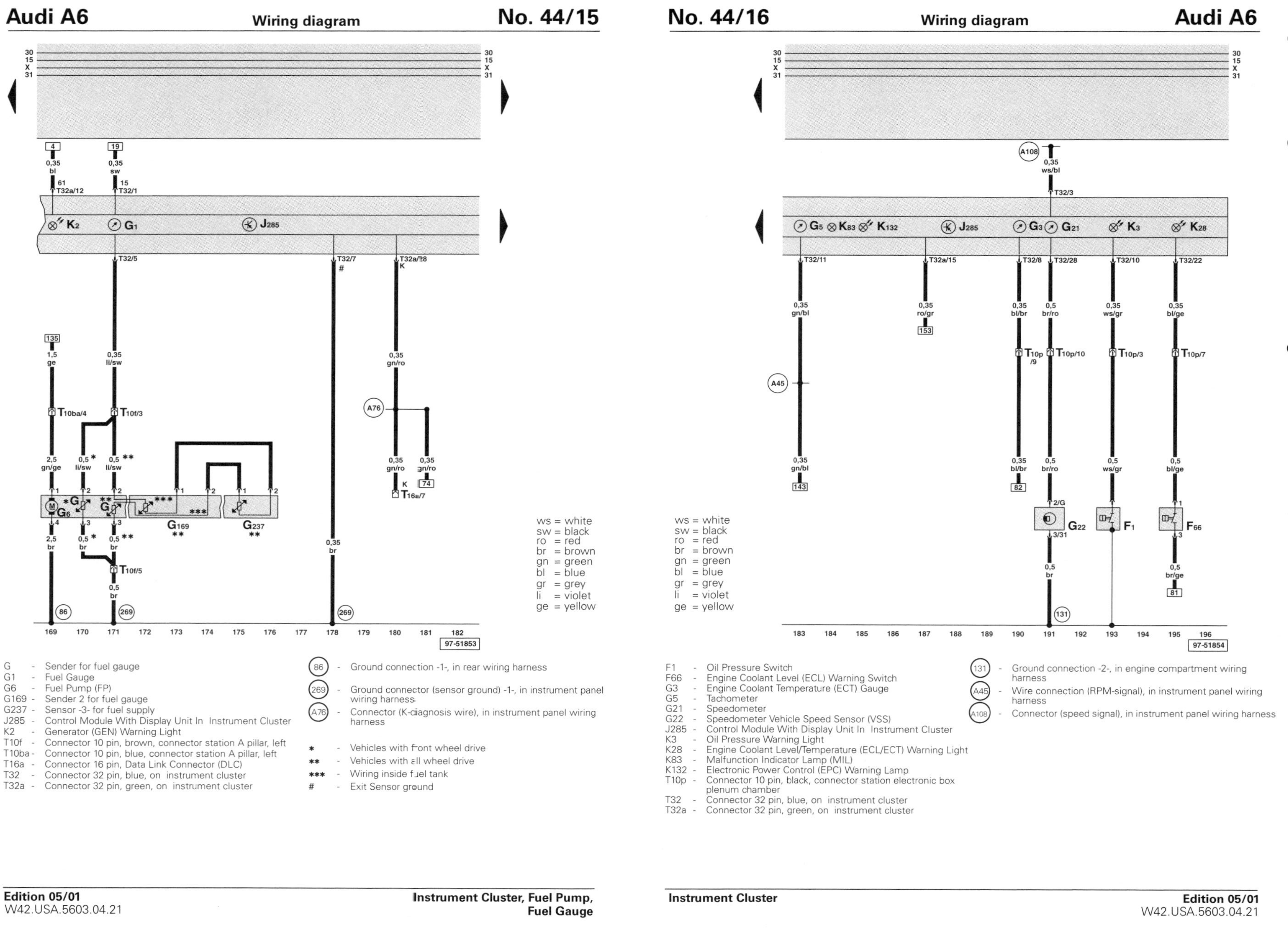

BentleyPublishers.com—All Rights Reserved

Central Locking System / Anti Theft Alarm System / Ultra-Sound Interior Monitor

2002 m. y.

Fuse Panel

A97-0119

Fuse Colors:

30	A	-	Green
25	A	-	White
20	A	-	Yellow
15	A	-	Blue
10	A	-	Red
7,5	A	-	Brown
5	A	-	Beige

Starting with fuse position 23, fuses in the fuse holder are identified with 223 in the wiring diagram.

13 - Fold Relay Panel

A97-0024

Relay Location:

4 -Starting Interlock Relay, J207

ws = white
sw = black
ro = red
br = brown
gn = green
bl = blue
gr = grey
li = violet
ge = yellow

97-52437

- B - Starter
- D - Ignition/Starter Switch
- F120 - Hood Alarm Switch
- F125 - Multi-Function Transmission Range (TR) Switch
- F194 - Clutch Pedal Position (CPP) Switch
- H8 - Alarm Horn
- J207 - Starting Interlock Relay
- J285 - Control Module with indicator unit in instrument panel insert
- J429 - Control module for central locking
- R - Radio
- T1h - Single connector, black, in engine compartment, right
- T8f - Connector 8 pin, black, connector radio III
- T9 - Connector 9 pin, brown, on starting interlock relay
- T10o - Connector 10 pin, brown, connector station electronic box plenum chamber
- T14a - Connector 14 pin, black, in engine compartment, left front
- T17 - Connector 17 pin, orange, connector station A pillar, right
- T17a - Connector 17 pin, brown, connector station A pillar, right
- T17h - Connector 17 pin, brown, connector station A pillar, left
- T17m - Connector 17 pin, blue, connector station A pillar, right
- T32a - Connector 32 pin, green, on instrument cluster
- (44) - Ground connection, on left A-pillar, lower part
- (105) - Ground connection -1-, in central locking system wiring harness
- (213) - Ground connection, in instrument panel wiring harness
- (A21) - Wire connection (86s), in instrument panel wiring harness
- (A41) - Plus connection (50), in instrument panel wiring harness
- (A184) - Front Hood Contact Switch Connection (in instrument panel wiring harness)
- * - Vehicles with manually shifted transmission
- ** - Vehicles with automatic transmission

Control Module for Central Locking, Alarm Horn, Hood Alarm Switch

BentleyPublishers.com—All Rights Reserved

Central Locking (2002)

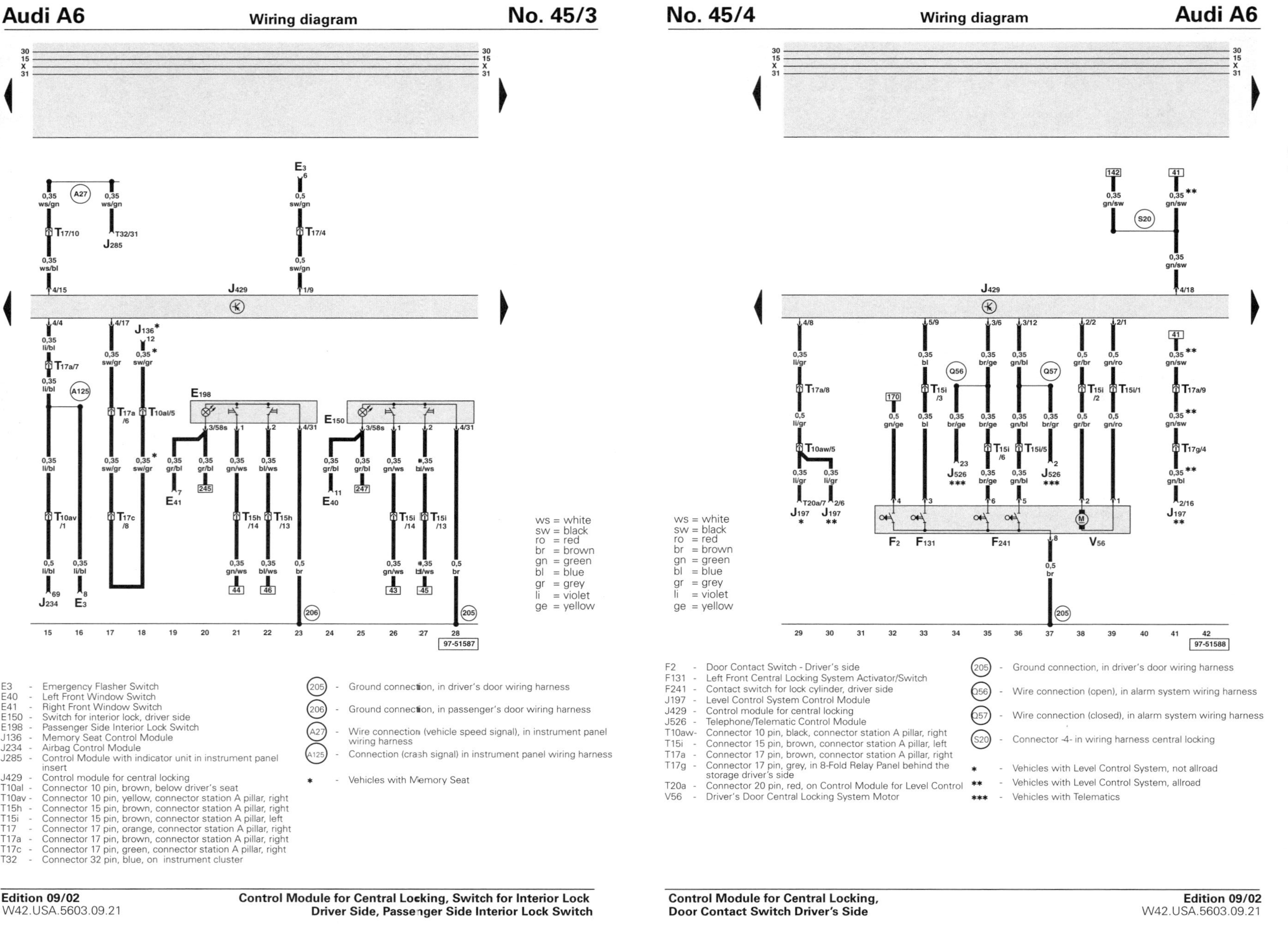

BentleyPublishers.com—All Rights Reserved

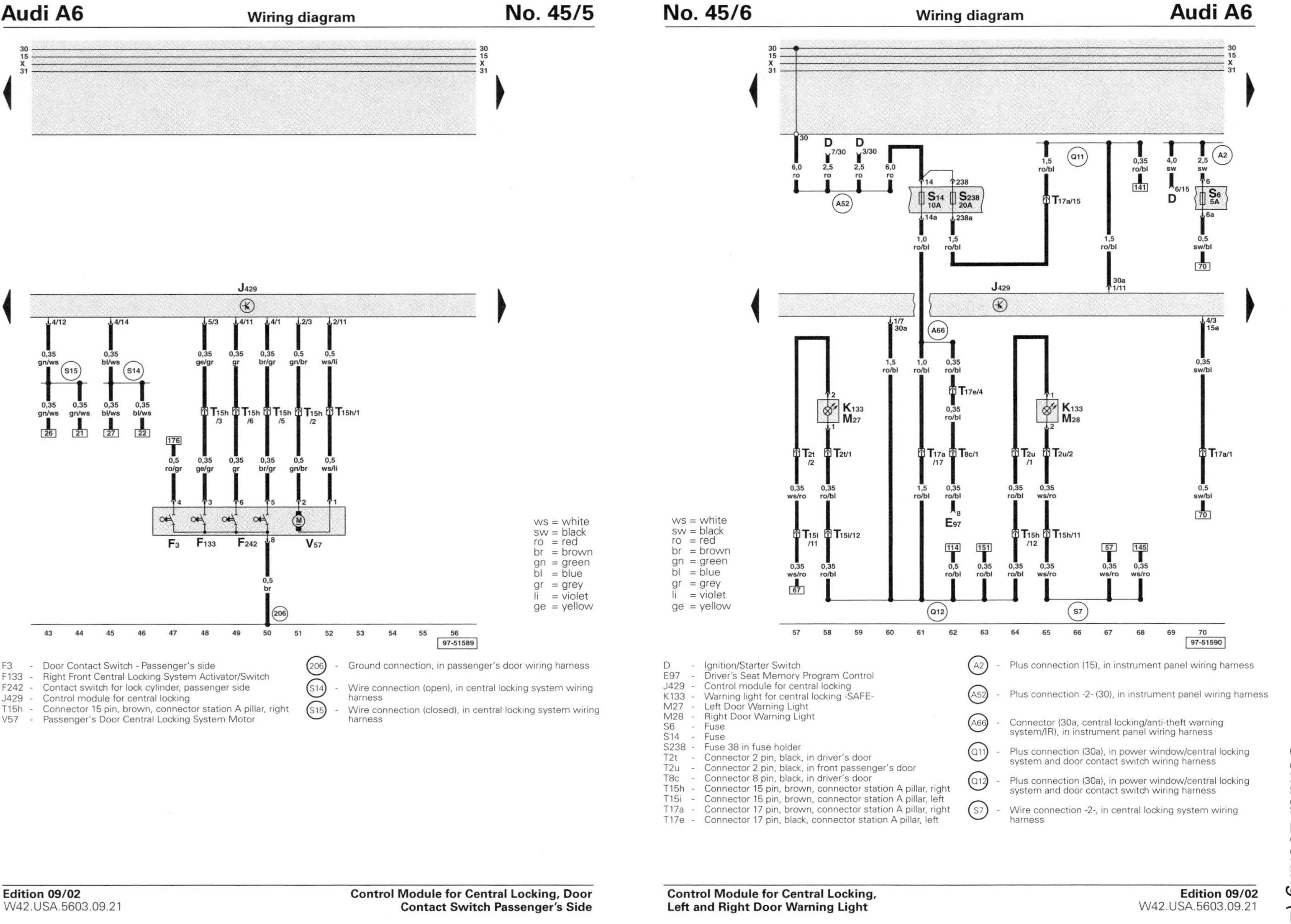

F3 - Door Contact Switch - Passenger's side
F133 - Right Front Central Locking System Activator/Switch
F242 - Contact switch for lock cylinder, passenger side
J429 - Control module for central locking
T15h - Connector 15 pin, brown, connector station A pillar, right
V57 - Passenger's Door Central Locking System Motor

(206) - Ground connection, in passenger's door wiring harness
(S14) - Wire connection (open), in central locking system wiring harness
(S15) - Wire connection (closed), in central locking system wiring harness

Edition 09/02
W42.USA.5603.09.21

Control Module for Central Locking, Door Contact Switch Passenger's Side

D - Ignition/Starter Switch
E97 - Driver's Seat Memory Program Control
J429 - Control module for central locking
K133 - Warning light for central locking -SAFE-
M27 - Left Door Warning Light
M28 - Right Door Warning Light
S6 - Fuse
S14 - Fuse
S238 - Fuse 38 in fuse holder
T2t - Connector 2 pin, black, in driver's door
T2u - Connector 2 pin, black, in front passenger's door
T8c - Connector 8 pin, black, in driver's door
T15h - Connector 15 pin, brown, connector station A pillar, right
T15i - Connector 15 pin, brown, connector station A pillar, left
T17a - Connector 17 pin, brown, connector station A pillar, right
T17e - Connector 17 pin, black, connector station A pillar, left

(A2) - Plus connection (15), in instrument panel wiring harness
(A52) - Plus connection -2- (30), in instrument panel wiring harness
(A66) - Connector (30a, central locking/anti-theft warning system/IR), in instrument panel wiring harness
(Q11) - Plus connection (30a), in power window/central locking system and door contact switch wiring harness
(Q12) - Plus connection (30a), in power window/central locking system and door contact switch wiring harness
(S7) - Wire connection -2-, in central locking system wiring harness

Control Module for Central Locking, Left and Right Door Warning Light

Edition 09/02
W42.USA.5603.09.21

BentleyPublishers.com—All Rights Reserved

Central Locking (2002)

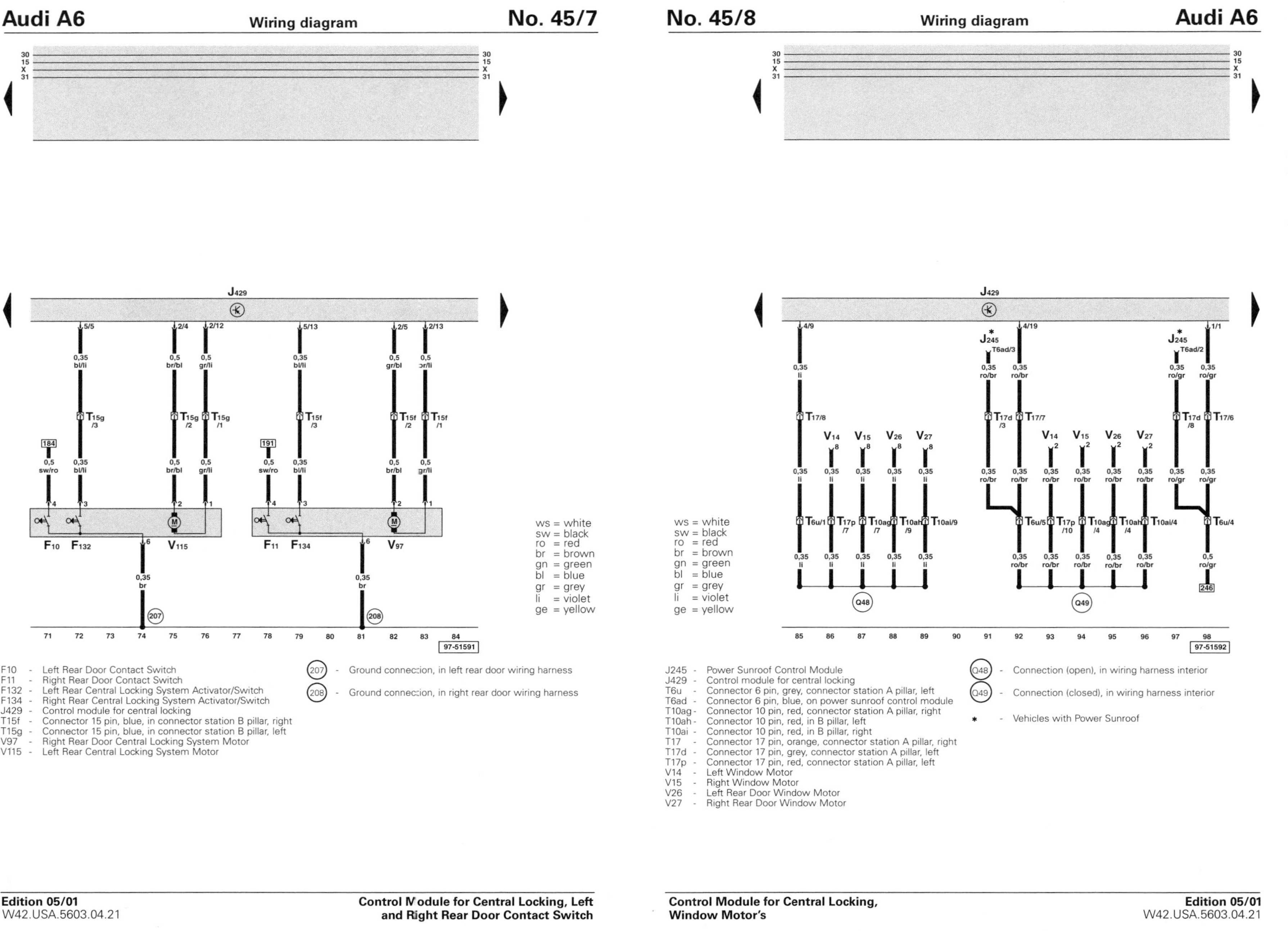

ws = white
sw = black
ro = red
br = brown
gn = green
bl = blue
gr = grey
li = violet
ge = yellow

F10 - Left Rear Door Contact Switch
F11 - Right Rear Door Contact Switch
F132 - Left Rear Central Locking System Activator/Switch
F134 - Right Rear Central Locking System Activator/Switch
J429 - Control module for central locking
T15f - Connector 15 pin, blue, in connector station B pillar, right
T15g - Connector 15 pin, blue, in connector station B pillar, left
V97 - Right Rear Door Central Locking System Motor
V115 - Left Rear Central Locking System Motor

(207) - Ground connection, in left rear door wiring harness
(208) - Ground connection, in right rear door wiring harness

Edition 05/01
W42.USA.5603.04.21

Control Module for Central Locking, Left and Right Rear Door Contact Switch

ws = white
sw = black
ro = red
br = brown
gn = green
bl = blue
gr = grey
li = violet
ge = yellow

J245 - Power Sunroof Control Module
J429 - Control module for central locking
T6u - Connector 6 pin, grey, connector station A pillar, left
T6ad - Connector 6 pin, blue, on power sunroof control module
T10ag - Connector 10 pin, red, connector station A pillar, right
T10ah - Connector 10 pin, red, in B pillar, left
T10ai - Connector 10 pin, red, in B pillar, right
T17 - Connector 17 pin, orange, connector station A pillar, right
T17d - Connector 17 pin, grey, connector station A pillar, left
T17p - Connector 17 pin, red, connector station A pillar, left
V14 - Left Window Motor
V15 - Right Window Motor
V26 - Left Rear Door Window Motor
V27 - Right Rear Door Window Motor

(Q48) - Connection (open), in wiring harness interior
(Q49) - Connection (closed), in wiring harness interior
* - Vehicles with Power Sunroof

Control Module for Central Locking, Window Motor's

Edition 05/01
W42.USA.5603.04.21

BentleyPublishers.com—All Rights Reserved

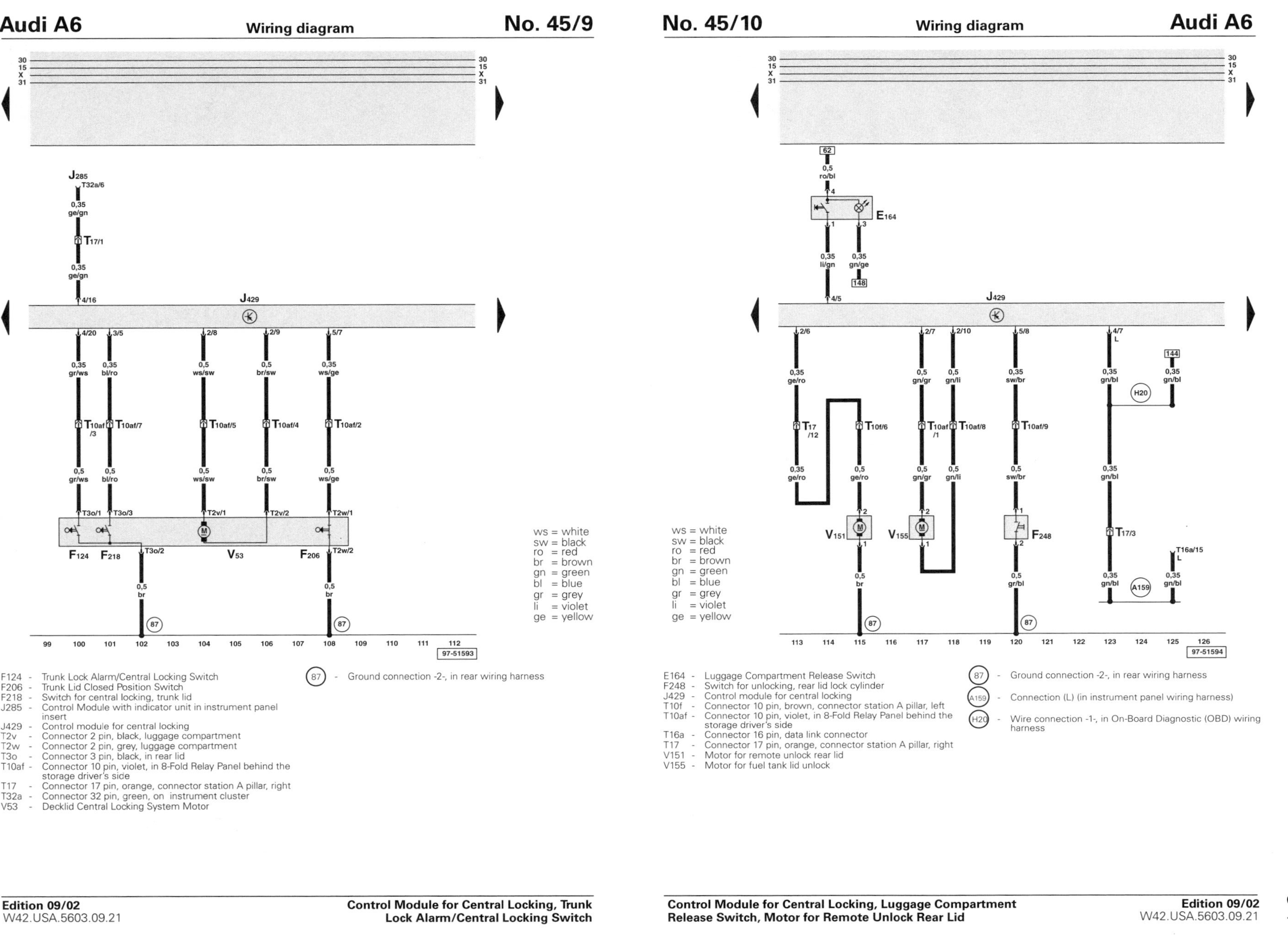

BentleyPublishers.com—All Rights Reserved

Central Locking (2002)

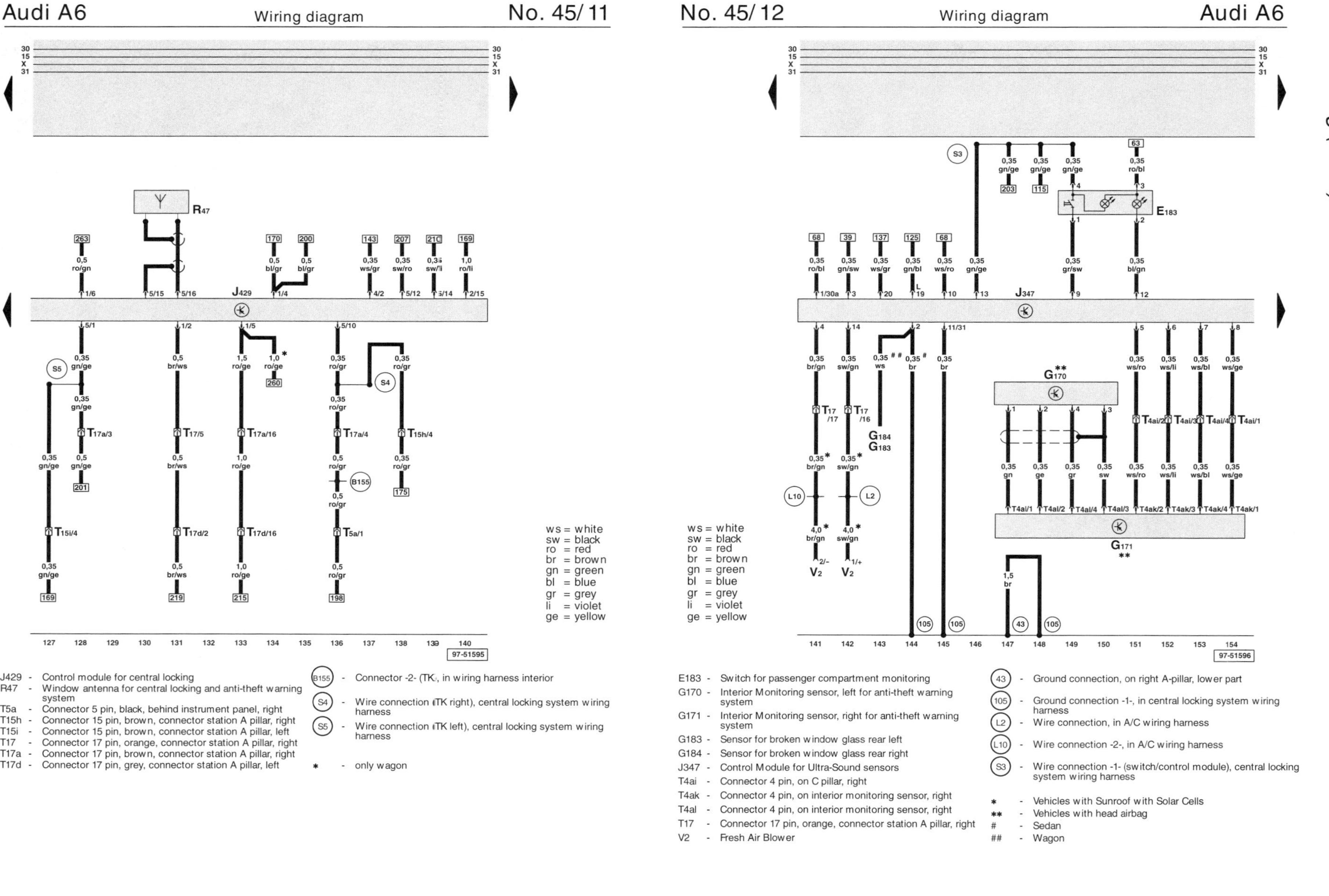

BentleyPublishers.com—All Rights Reserved

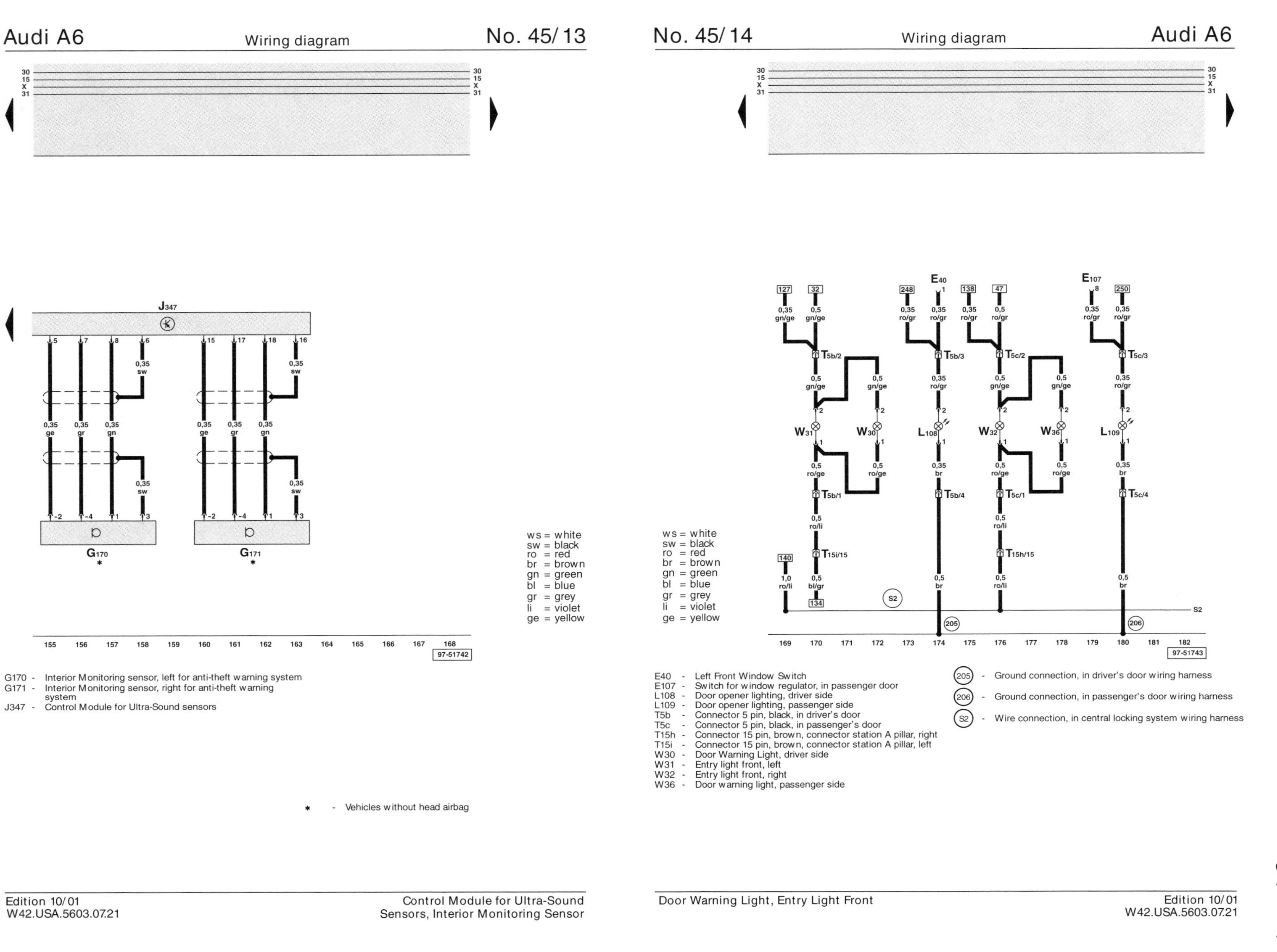

BentleyPublishers.com—All Rights Reserved

Central Locking (2002)

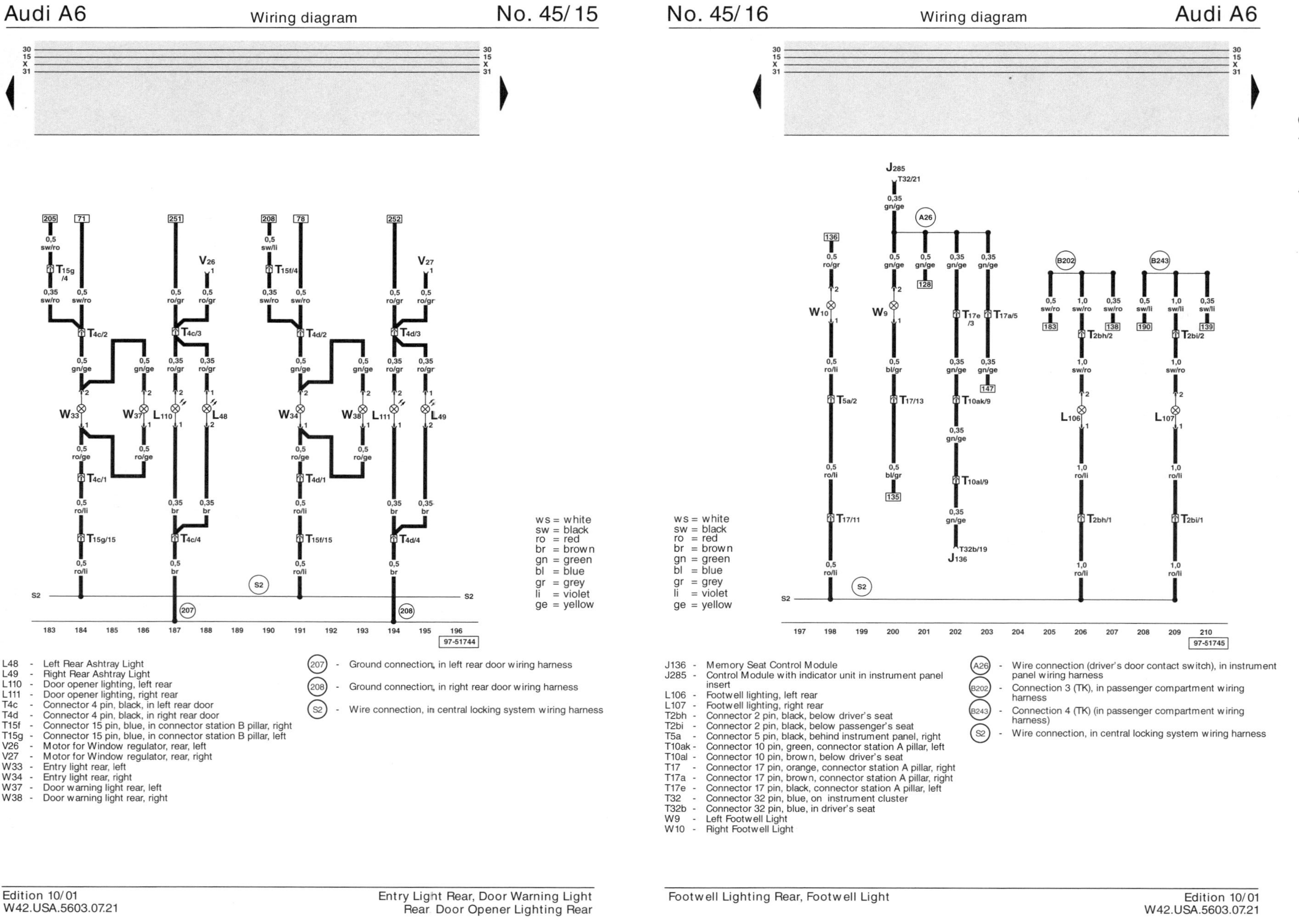

BentleyPublishers.com—All Rights Reserved

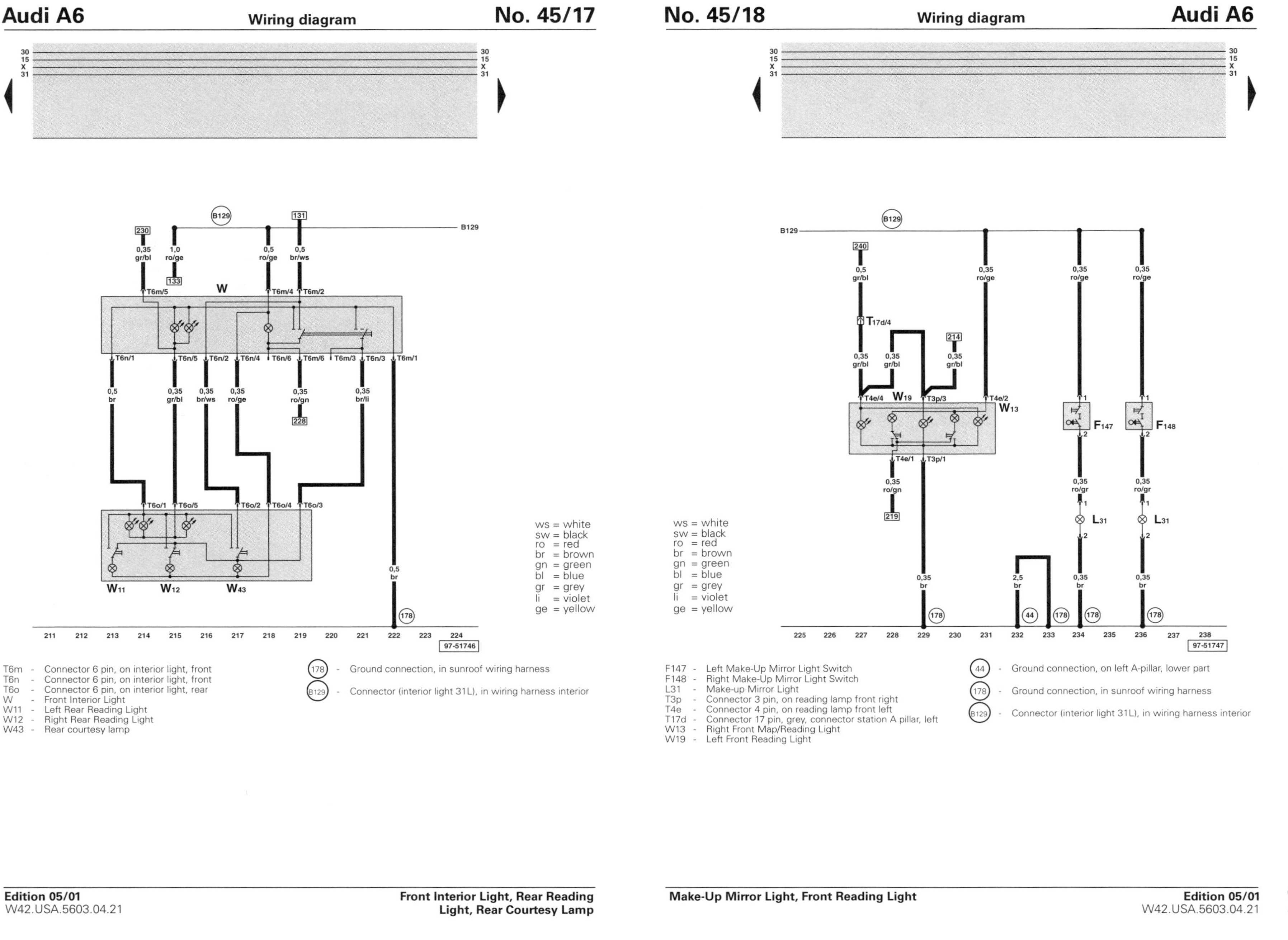

BentleyPublishers.com—All Rights Reserved

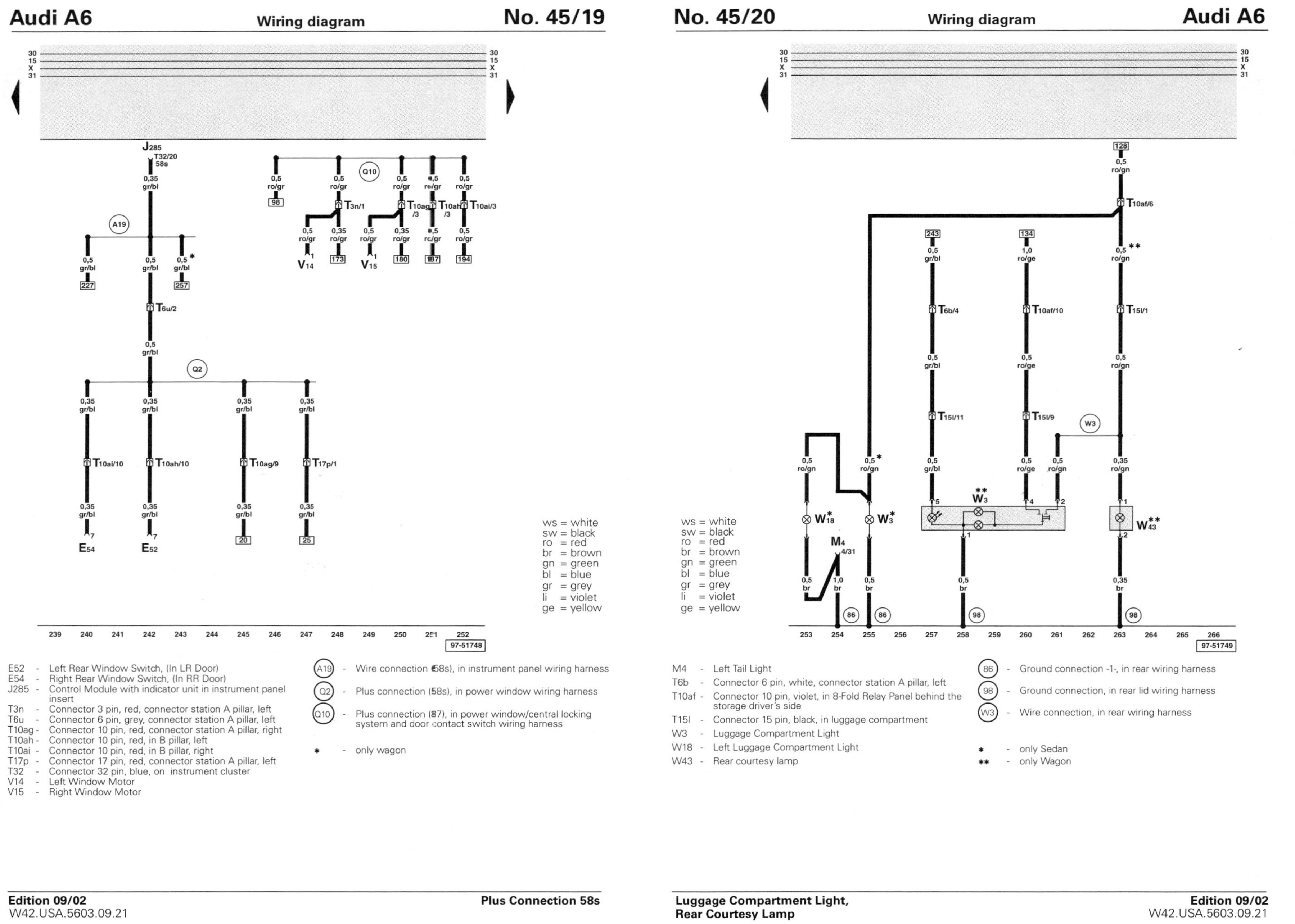

BentleyPublishers.com—All Rights Reserved

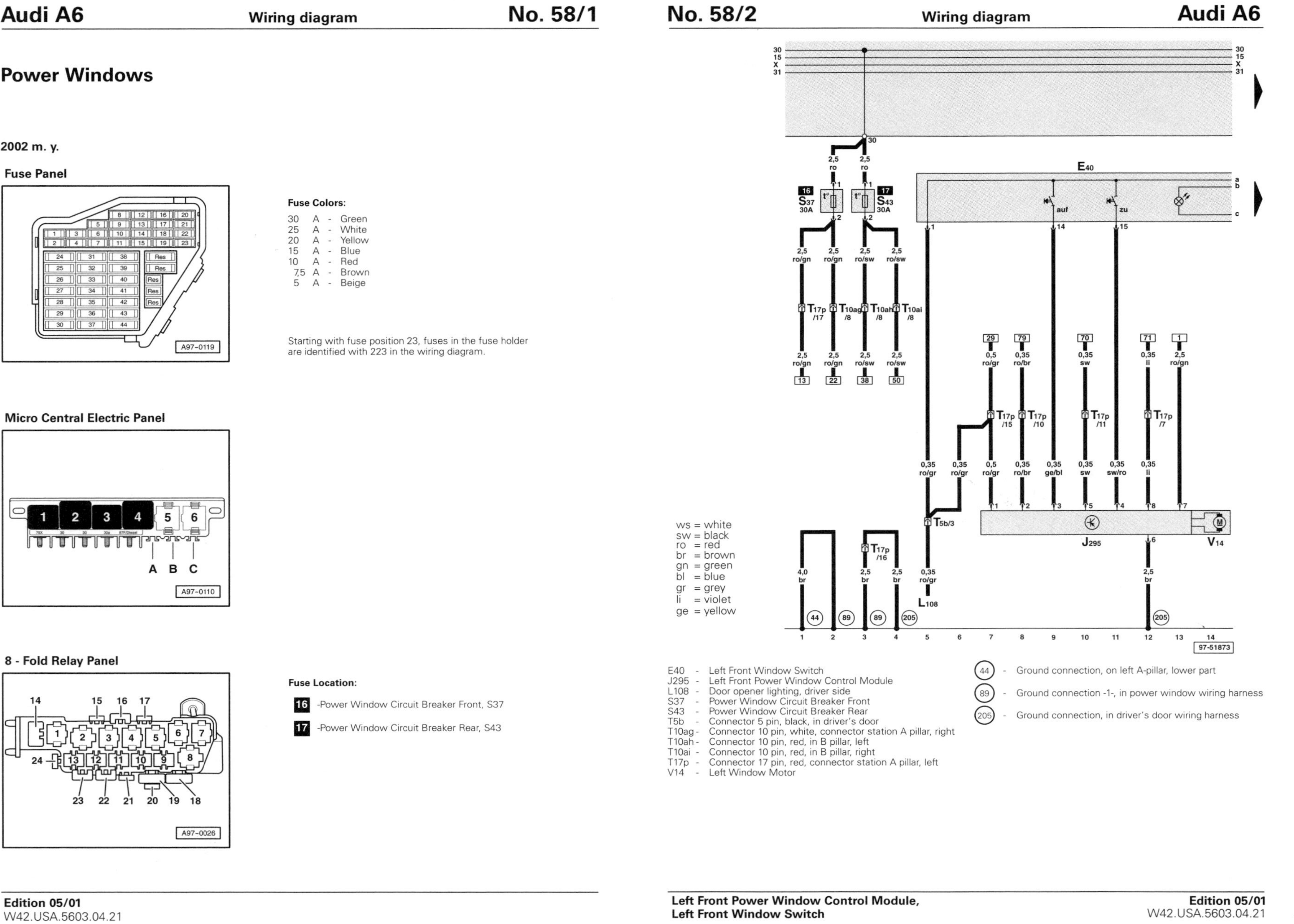

BentleyPublishers.com—All Rights Reserved

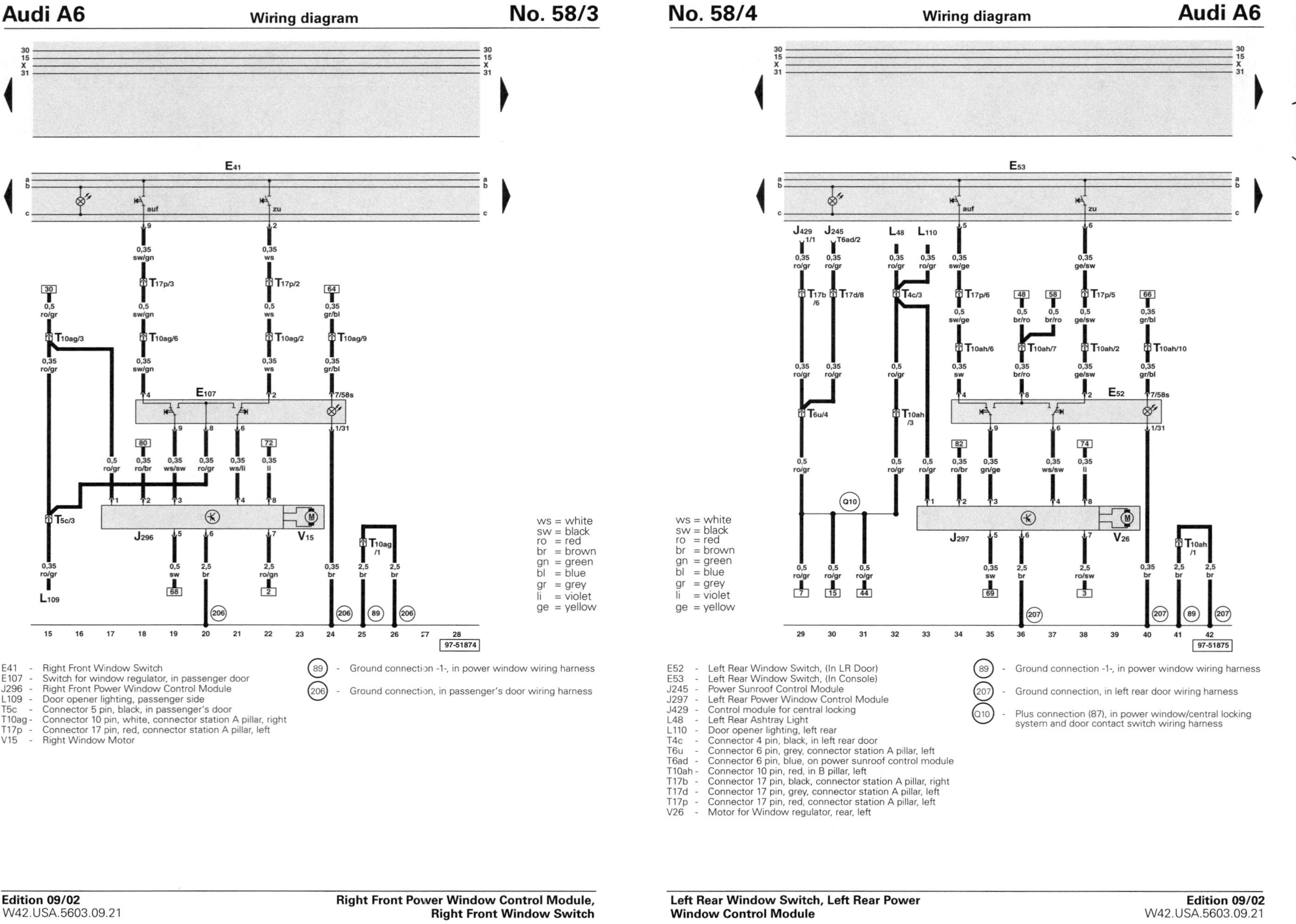

BentleyPublishers.com—All Rights Reserved

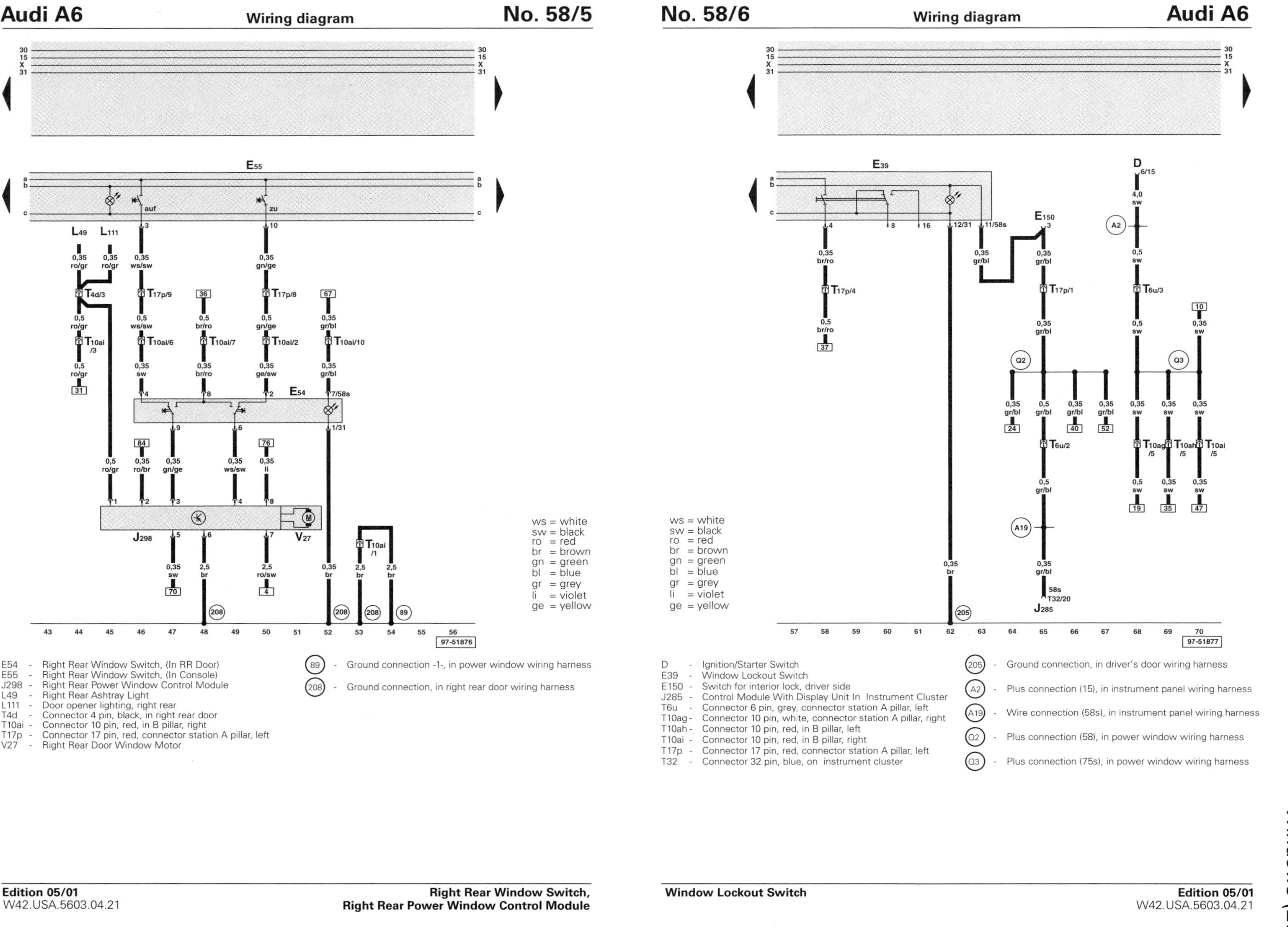

BentleyPublishers.com—All Rights Reserved

THIS SPACE INTENTIONALLY LEFT BLANK

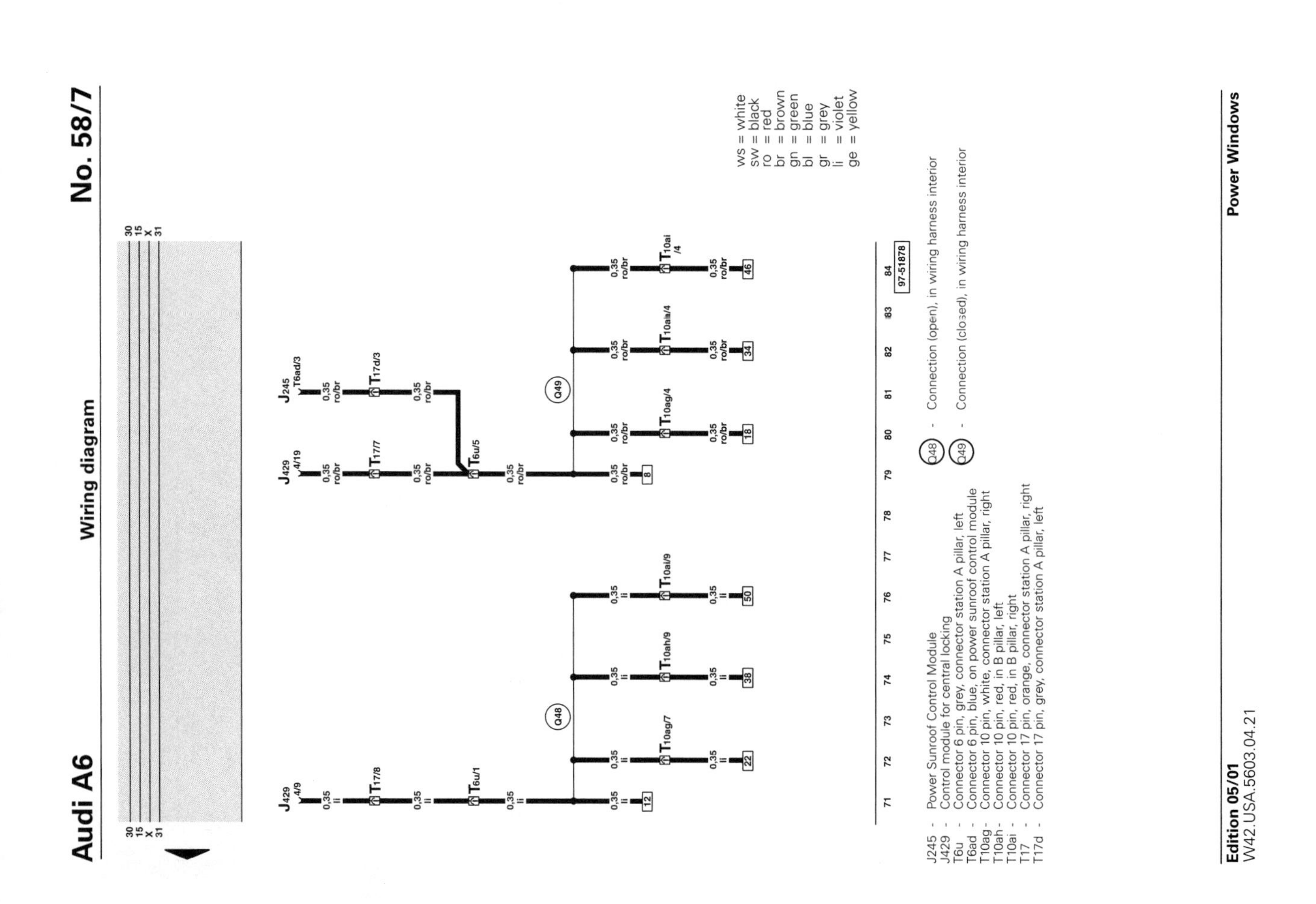

BentleyPublishers.com—All Rights Reserved

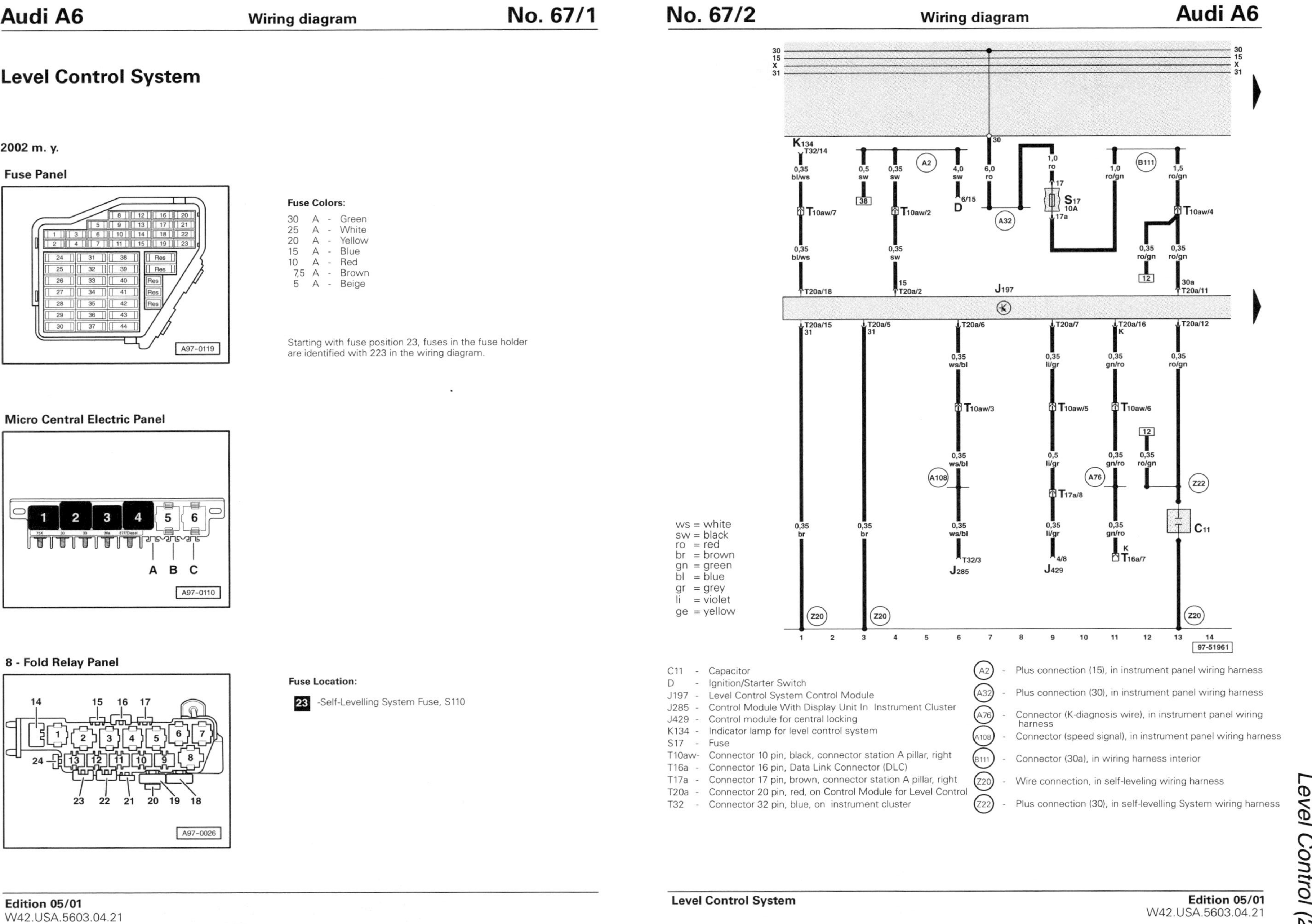

BentleyPublishers.com—All Rights Reserved

Audi A6 Wiring diagram **No. 67/3**

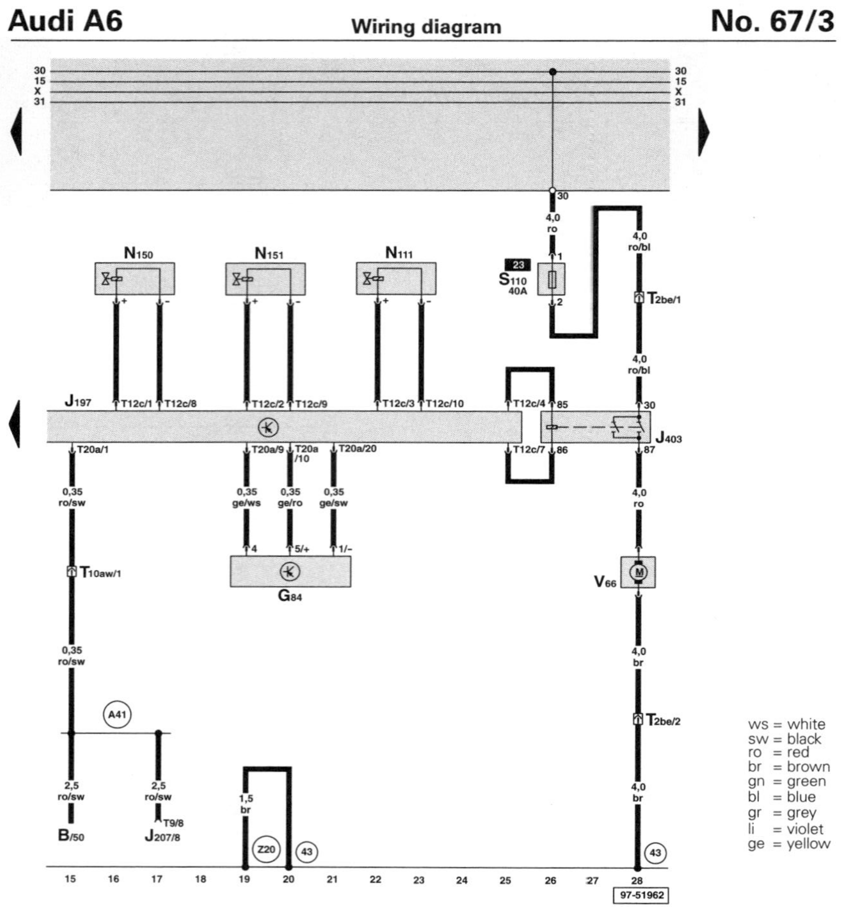

ws = white
sw = black
ro = red
br = brown
gn = green
bl = blue
gr = grey
li = violet
ge = yellow

- B - Starter
- G84 - Level Control System Sensor (ESP)
- J197 - Level Control System Control Module
- J207 - Starting Interlock Relay
- J403 - Relay for compressor level control system
- N111 - Solenoid for level control system
- N150 - Left Rear Suspension Strut Valve
- N151 - Right Rear Suspension Strut Valve
- S110 - Self-Levelling System Fuse
- T2be - Connector 2 pin, near Control Module for Level Control
- T9 - Connector 9 pin, brown, on starting interlock relay
- T10aw- Connector 10 pin, black, connector station A pillar, right
- T12c - Connector 12 pin, red, on Control Module for Level Control
- T20a - Connector 20 pin, red, on Control Module for Level Control
- V66 - Level Control System Compressor Motor

- (43) - Ground connection, on right A-pillar, lower part
- (A41) - Plus connection (50), in instrument panel wiring harness
- (Z20) - Wire connection, in self-leveling wiring harness

Edition 09/02
W42.USA.5603.09.21

Level Control System Control Module, Level Control System Sensor (ESP)

No. 67/4 Wiring diagram **Audi A6**

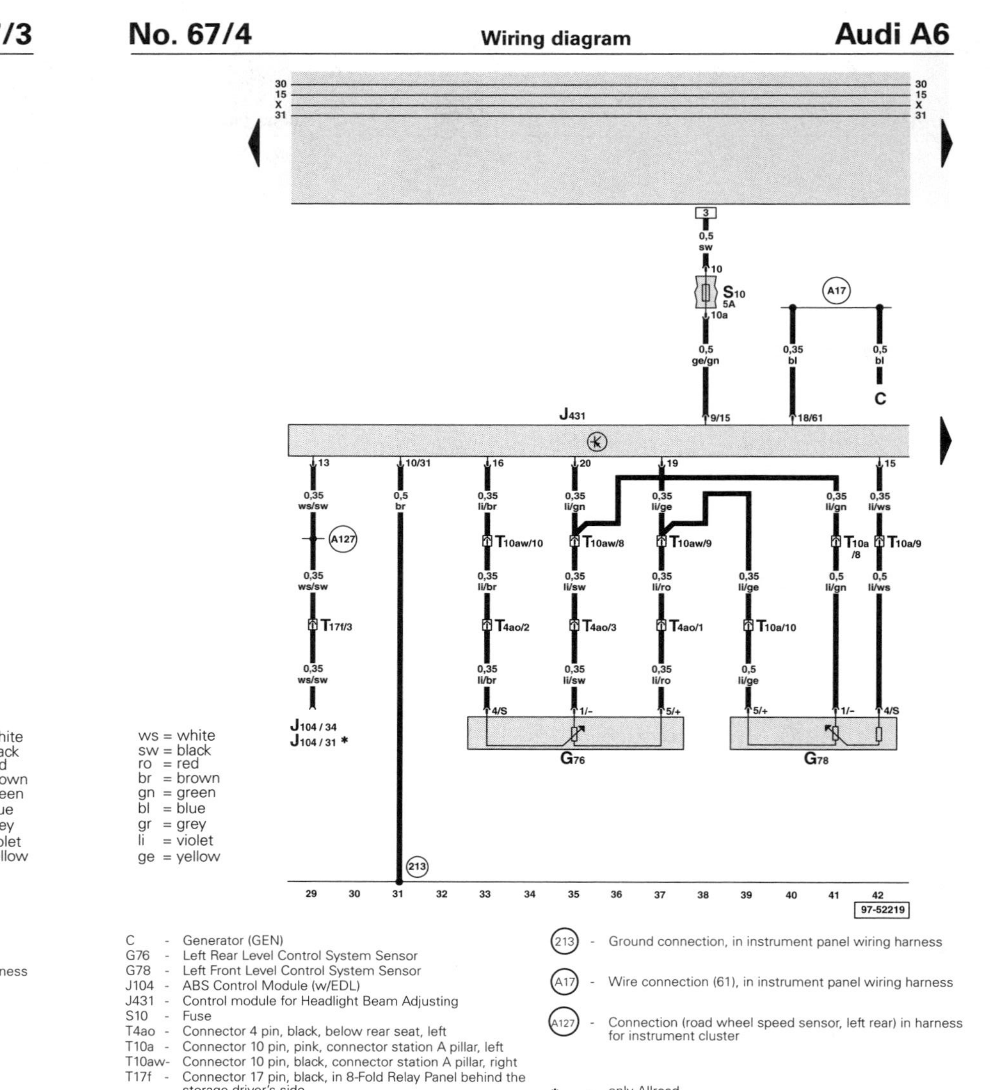

ws = white
sw = black
ro = red
br = brown
gn = green
bl = blue
gr = grey
li = violet
ge = yellow

- C - Generator (GEN)
- G76 - Left Rear Level Control System Sensor
- G78 - Left Front Level Control System Sensor
- J104 - ABS Control Module (w/EDL)
- J431 - Control module for Headlight Beam Adjusting
- S10 - Fuse
- T4ao - Connector 4 pin, black, below rear seat, left
- T10a - Connector 10 pin, pink, connector station A pillar, left
- T10aw- Connector 10 pin, black, connector station A pillar, right
- T17f - Connector 17 pin, black, in 8-Fold Relay Panel behind the storage driver's side

- (213) - Ground connection, in instrument panel wiring harness
- (A17) - Wire connection (61), in instrument panel wiring harness
- (A127) - Connection (road wheel speed sensor, left rear) in harness for instrument cluster
- * - only Allroad

Control Module for Headlight Beam Adjusting, Left Front and Rear Level Control System Sensor

Edition 09/02
W42.USA.5603.09.21

BentleyPublishers.com—All Rights Reserved

Audi A6 **Wiring diagram** **No. 67/5**

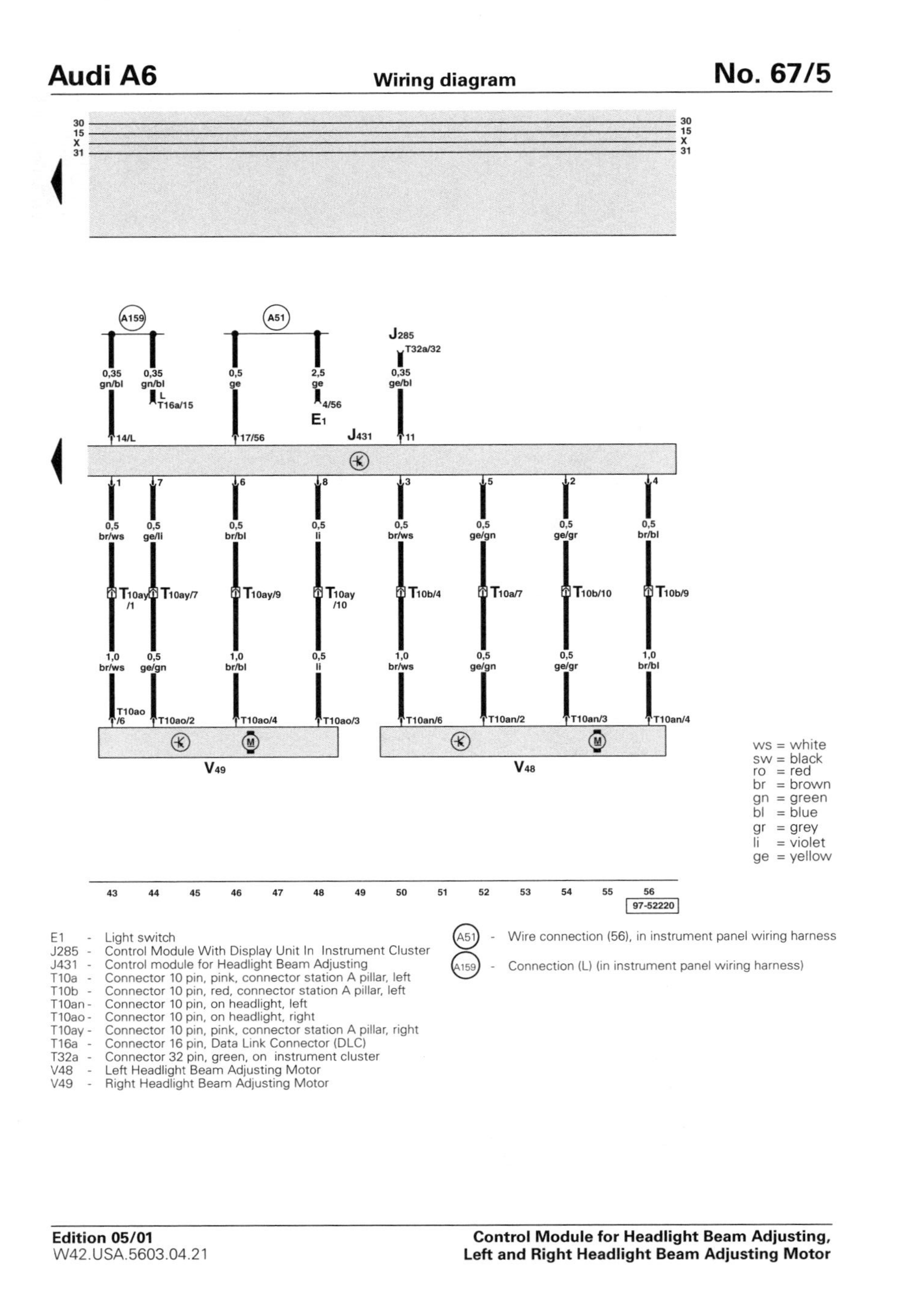

ws = white
sw = black
ro = red
br = brown
gn = green
bl = blue
gr = grey
li = violet
ge = yellow

E1 - Light switch
J285 - Control Module With Display Unit In Instrument Cluster
J431 - Control module for Headlight Beam Adjusting
T10a - Connector 10 pin, pink, connector station A pillar, left
T10b - Connector 10 pin, red, connector station A pillar, left
T10an - Connector 10 pin, on headlight, left
T10ao - Connector 10 pin, on headlight, right
T10ay - Connector 10 pin, pink, connector station A pillar, right
T16a - Connector 16 pin, Data Link Connector (DLC)
T32a - Connector 32 pin, green, on instrument cluster
V48 - Left Headlight Beam Adjusting Motor
V49 - Right Headlight Beam Adjusting Motor

(A51) - Wire connection (56), in instrument panel wiring harness
(A159) - Connection (L) (in instrument panel wiring harness)

Edition 05/01
W42.USA.5603.04.21

Control Module for Headlight Beam Adjusting, Left and Right Headlight Beam Adjusting Motor

THIS SPACE INTENTIONALLY LEFT BLANK

BentleyPublishers.com—All Rights Reserved

Audi A6 **Wiring Diagram** **No. 1/1**

Standard Equipment

2003 m.y.

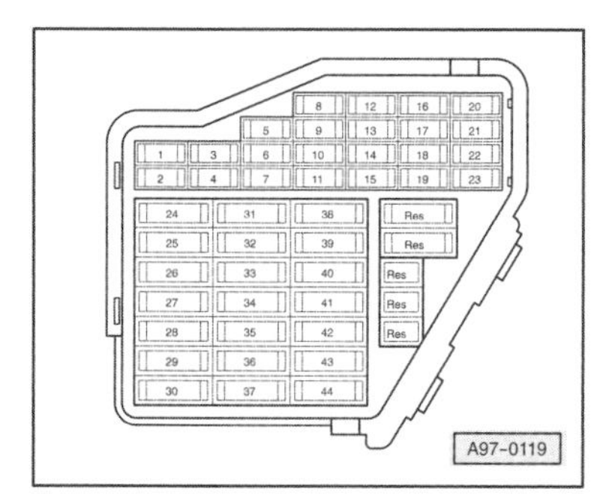

Fuse Panel

Fuse Colors:

30 A - Green
25 A - White
20 A - Yellow
15 A - Blue
10 A - Red
7,5 A - Brown
5 A - Beige
1 A - Black

Starting with fuse position 23, fuses in the fuse holder are identified with 223 in the wiring diagram.

Audi A6 **Wiring Diagram** **No. 1/2**

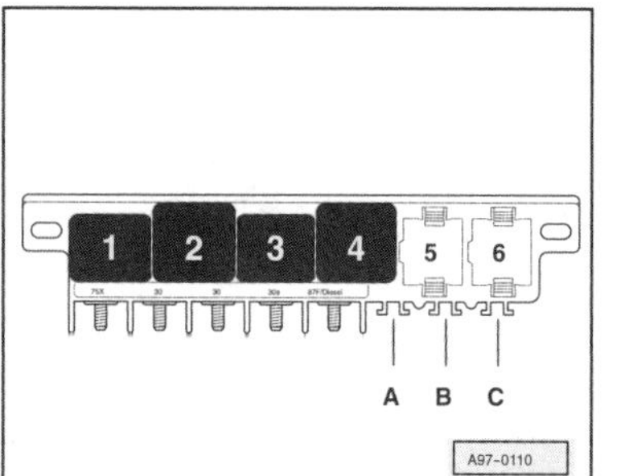

Micro Central Electric Panel

Relay Location:

1 - Dual Horn Relay, J4
2 - Load Reduction Relay, J59
4 - Fuel Pump (FP) Relay, J17
5 - Wiper/Washer Intermittent Relay, J31
6 - Wiper/Washer Intermittent Relay, J31

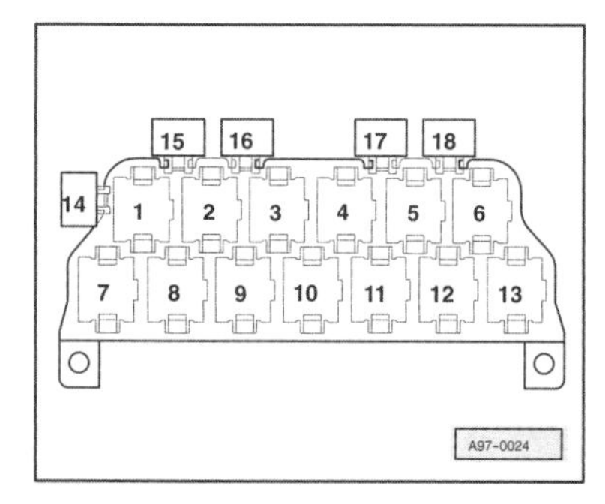

13 - Fold Relay Panel

Relay Location:

4 - Starting Interlock Relay, J207
6 - Fog Light Relay, J5

BentleyPublishers.com—All Rights Reserved

Audi A6

Wiring Diagram **No. 1/3**

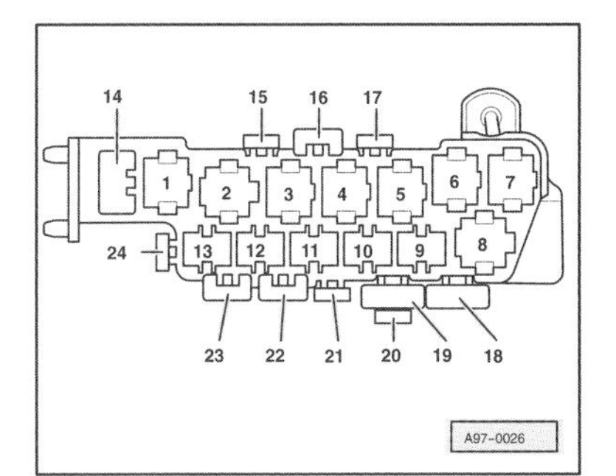

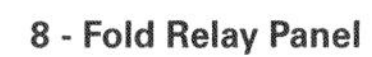

8 - Fold Relay Panel

Fuse Location:

15 - Fuse for Luggage Compartment Socket, S184

Audi A6

Wiring Diagram **No. 1/4**

Ground Connections

(3) - Ground strap, engine to body

(12) - Ground connection, in engine compartment, left

(13) - Ground connection, in engine compartment, right

(43) - Ground connection, on right A-pillar, lower part

(44) - Ground connection, on left A-pillar, lower part

(81) - Ground connection -1-, in instrument panel wiring harness

(85) - Ground connection -1-, in engine compartment wiring harness

(86) - Ground connection -1-, in rear wiring harness

(176) - Ground connection, in right headlight wiring harness

(179) - Ground connection, in left headlight wiring harness

(213) - Ground connection, in instrument panel wiring harness

* - Vehicles without Trailer Socket

** - Vehicles with Trailer Socket

ws = white
sw = black
ro = red
br = brown
gn = green
bl = blue
gr = grey
li = violet
ge = yellow
or = orange

97-51633

BentleyPublishers.com—All Rights Reserved

Standard Equipment (2003)

Audi A6

Wiring Diagram

No. 1/5

Ignition/Starter Switch, Load Reduction Relay, Clutch Pedal Position Switch

D - Ignition/Starter Switch
F125 - Multi-Function Transmission Range (TR) Switch
F194 - Clutch Pedal Position (CPP) Switch
J59 - Load Reduction Relay
J207 - Starting Interlock Relay
J217 - Transmission Control Module (TCM)
J429 - Control module for central locking
T9 - Connector 9 pin, brown, on starting interlock relay
T17 - Connector 17 pin, orange, connector station A pillar, right
T17m - Connector 17 pin, blue, connector station A pillar, right

(213) - Ground connection, in instrument panel wiring harness
(A2) - Plus connection (15), in instrument panel wiring harness
(A21) - Wire connection (86s), in instrument panel wiring harness
(A52) - Plus connection -2- (30), in instrument panel wiring harness

* - Vehicles with manual transmission
** - Vehicles with automatic transmission

ws = white
sw = black
ro = red
br = brown
gn = green
bl = blue
gr = grey
li = violet
ge = yellow
or = orange

97-51634

USA.5604.03.21

Edition 05/03

Audi A6

Wiring Diagram

No. 1/6

Battery, Starter, Generator

A - Battery
B - Starter
C - Generator (GEN)
C1 - Voltage Regulator (VR)
J31 - Wiper/Washer Intermittent Relay
S205 - Fuse 2 (30)
T1h - Single connector, black, in engine compartment, right
T2cl - Connector 2 pin, black, in engine compartment, right
T10o - Connector 10 pin, brown, connector station electronic box plenum chamber
T10p - Connector 10 pin, black, connector station electronic box plenum chamber

(1) - Ground strap, battery to body
(44) - Ground connection, on left A-pillar, lower part
(81) - Ground connection -1-, in instrument panel wiring harness
(199) - Ground connection -3-, in instrument panel wiring harness
(A17) - Wire connection (61), in instrument panel wiring harness
(A41) - Plus connection (50), in instrument panel wiring harness

ws = white
sw = black
ro = red
br = brown
gn = green
bl = blue
gr = grey
li = violet
ge = yellow
or = orange

97-51635

USA.5604.03.21

Edition 05/03

BentleyPublishers.com—All Rights Reserved

Audi A6 — Wiring Diagram — No. 1/7

Light Switch, Headlight Dimmer/Flasher Switch, Park Light Switch, Fog Light Switch

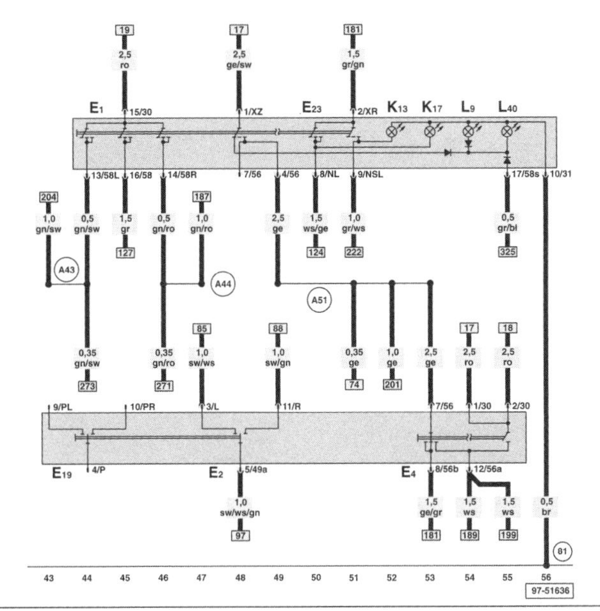

- E1 - Light switch
- E2 - Turn Signal Switch
- E4 - Headlight Dimmer/Flasher Switch
- E19 - Park Light Switch
- E23 - Fog Light Switch
- K13 - Rear Fog Light Indicator Light
- K17 - Fog Light Indicator Light
- L9 - Headlight Switch Light
- L40 - Front And Rear Fog Light Switch Light

- (81) - Ground connection -1-, in instrument panel wiring harness
- (A43) - Wire connection (57l), in instrument panel wiring harness
- (A44) - Wire connection (57r), in instrument panel wiring harness
- (A51) - Wire connection (56), in instrument panel wiring harness

ws = white
sw = black
ro = red
br = brown
gn = green
bl = blue
gr = grey
li = violet
ge = yellow
or = orange

USA.5604.03.21 Edition 05/03

Audi A6 — Wiring Diagram — No. 1/8

Control Module for Headlight Beam Adjusting, Left Rear and Front Level Control System Sensor

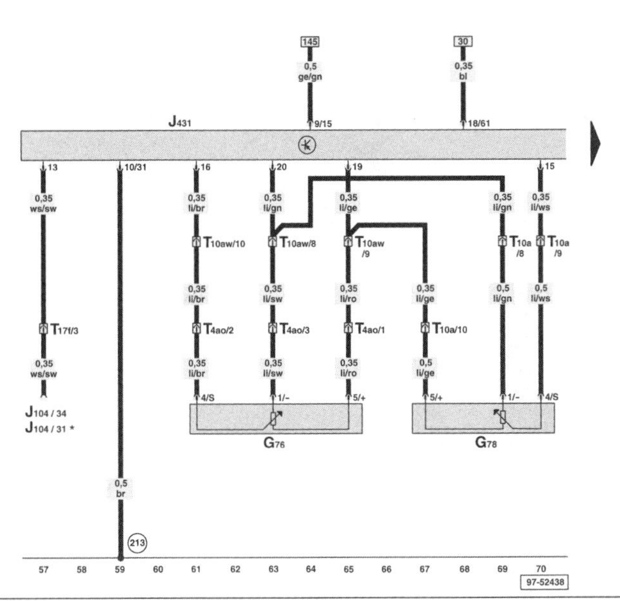

- G76 - Left Rear Level Control System Sensor
- G78 - Left Front Level Control System Sensor
- J104 - ABS Control Module (w/EDL)
- J431 - Control module for Headlight Beam Adjusting
- T4ao - Connector 4 pin, black, below rear seat, left
- T10a - Connector 10 pin, pink, connector station A pillar, left
- T10aw - Connector 10 pin, black, connector station A pillar, right
- T17f - Connector 17 pin, black, in 8-Fold Relay Panel behind the storage driver's side

- (213) - Ground connection, in instrument panel wiring harness

- * - allroad only

ws = white
sw = black
ro = red
br = brown
gn = green
bl = blue
gr = grey
li = violet
ge = yellow
or = orange

USA.5604.03.21 Edition 05/03

BentleyPublishers.com—All Rights Reserved

Audi A6 — Wiring Diagram — No. 1/9

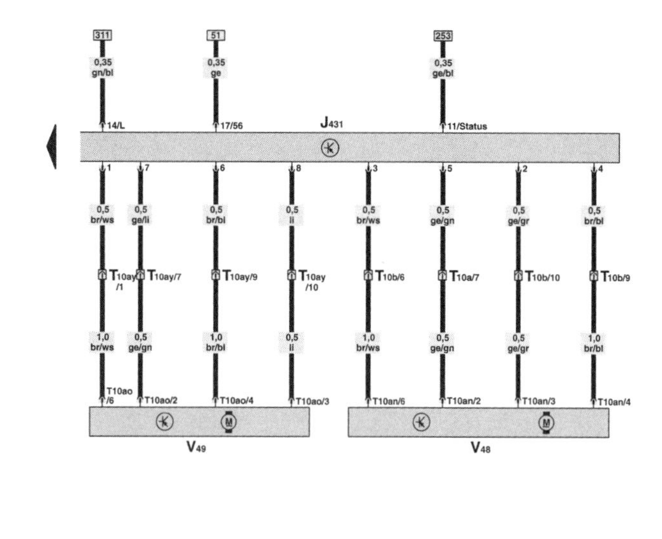

71 72 73 74 75 76 77 78 79 80 81 82 83 84

97-52220

Control Module for Headlight Beam Adjusting, Left and Right Headlight Beam Adjusting Motor

- J431 - Control module for Headlight Beam Adjusting
- T10a - Connector 10 pin, pink, connector station A pillar, left
- T10b - Connector 10 pin, red, connector station A pillar, left
- T10an - Connector 10 pin, on headlight, left
- T10ao - Connector 10 pin, on headlight, right
- T10ay - Connector 10 pin, pink, connector station A pillar, right
- V48 - Left Headlight Beam Adjusting Motor
- V49 - Right Headlight Beam Adjusting Motor

ws = white
sw = black
ro = red
br = brown
gn = green
bl = blue
gr = grey
li = violet
ge = yellow
or = orange

USA.5604.03.21 Edition 05/03

Audi A6 — Wiring Diagram — No. 1/10

30
15
X
31
75

162
1,5 ro/ge
117
1,5 sw/gn
10/30
E3
2/75
5/WL 7/WR 6/ZV 8/AB 4/58 9/31 3/49a
47
324
48
T17/4
T17k/16
249
206 247 212
J429
A5
A6
A125
T17a/7
T10av/1
49 183 230
J429 J234
135 199 213

85 86 87 88 89 90 91 92 93 94 95 96 97 98

97-52439

Emergency Flasher Switch

- E3 - Emergency Flasher Switch
- J234 - Airbag Control Module
- J429 - Control module for central locking
- T10av - Connector 10 pin, yellow, connector station A pillar, right
- T17 - Connector 17 pin, orange, connector station A pillar, right
- T17a - Connector 17 pin, brown, connector station A pillar, right
- T17k - Connector 17 pin, red, connector station electronic box plenum chamber

- (135) - Ground connection -2-, in instrument panel wiring harness
- (199) - Ground connection -3-, in instrument panel wiring harness
- (213) - Ground connection, in instrument panel wiring harness
- (A5) - Plus connection (right turn signal), in instrument panel wiring harness
- (A6) - Plus connection (left turn signal), in instrument panel wiring harness
- (A125) - Connection (crash signal) in instrument panel wiring harness

ws = white
sw = black
ro = red
br = brown
gn = green
bl = blue
gr = grey
li = violet
ge = yellow
or = orange

USA.5604.03.21 Edition 05/03

BentleyPublishers.com—All Rights Reserved

Audi A6 — Wiring Diagram — No. 1/11

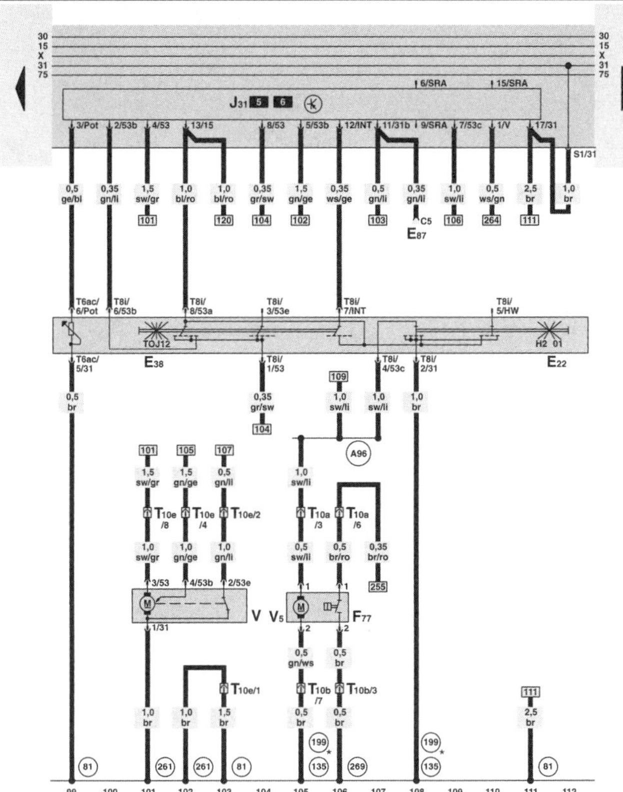

Windshield Wiper Motor, Windshield Washer Pump

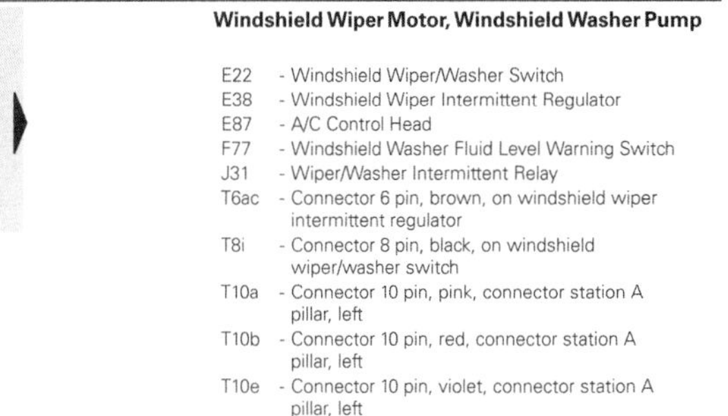

- E22 - Windshield Wiper/Washer Switch
- E38 - Windshield Wiper Intermittent Regulator
- E87 - A/C Control Head
- F77 - Windshield Washer Fluid Level Warning Switch
- J31 - Wiper/Washer Intermittent Relay
- T6ac - Connector 6 pin, brown, on windshield wiper intermittent regulator
- T8i - Connector 8 pin, black, on windshield wiper/washer switch
- T10a - Connector 10 pin, pink, connector station A pillar, left
- T10b - Connector 10 pin, red, connector station A pillar, left
- T10e - Connector 10 pin, violet, connector station A pillar, left
- V - Windshield Wiper Motor
- V5 - Windshield Washer Pump
- (81) - Ground connection -1-, in instrument panel wiring harness
- (135) - Ground connection -2-, in instrument panel wiring harness
- (199) - Ground connection -3-, in instrument panel wiring harness
- (261) - Ground connection, in wiring harness heated spray jet
- (269) - Ground connector (sensor ground) -1-, in instrument panel wiring harness
- (A96) - Connector (53a), in instrument panel wiring harness
- * - both are possible

ws = white
sw = black
ro = red
br = brown
gn = green
bl = blue
gr = grey
li = violet
ge = yellow
or = orange

Audi A6 — Wiring Diagram — No. 1/12

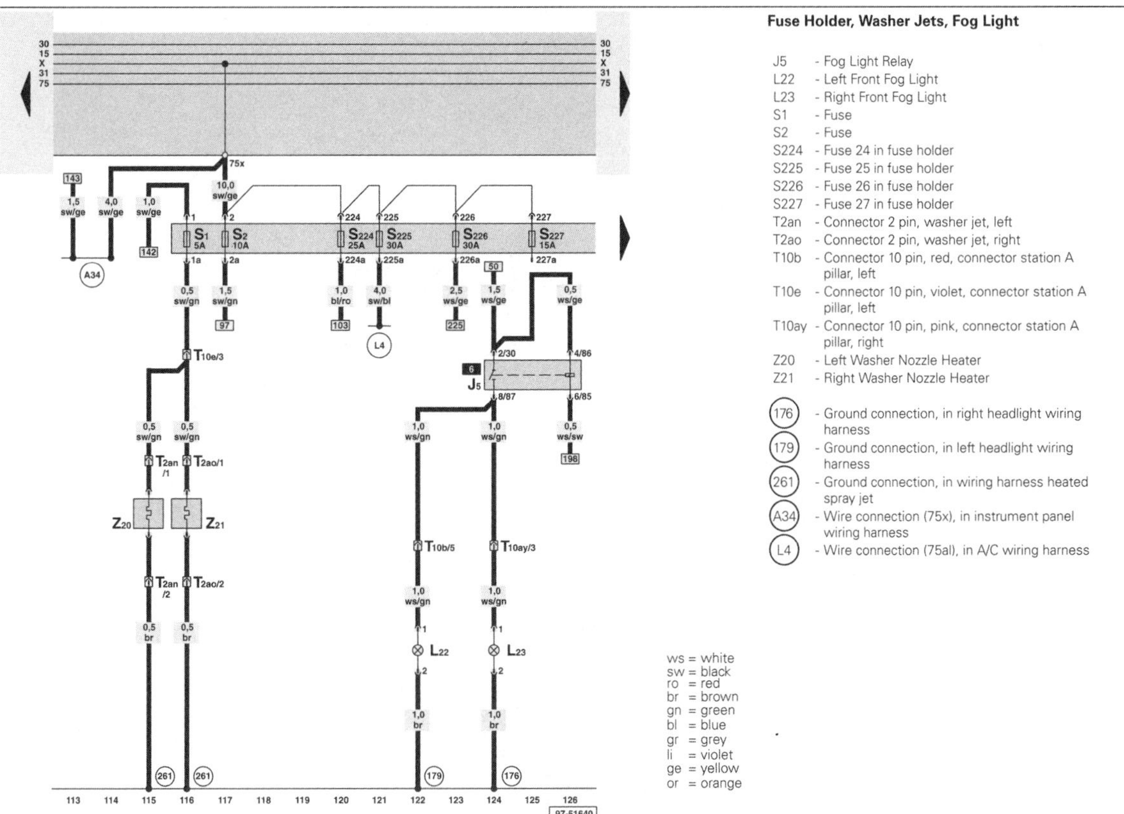

Fuse Holder, Washer Jets, Fog Light

- J5 - Fog Light Relay
- L22 - Left Front Fog Light
- L23 - Right Front Fog Light
- S1 - Fuse
- S2 - Fuse
- S224 - Fuse 24 in fuse holder
- S225 - Fuse 25 in fuse holder
- S226 - Fuse 26 in fuse holder
- S227 - Fuse 27 in fuse holder
- T2an - Connector 2 pin, washer jet, left
- T2ao - Connector 2 pin, washer jet, right
- T10b - Connector 10 pin, red, connector station A pillar, left
- T10e - Connector 10 pin, violet, connector station A pillar, left
- T10ay - Connector 10 pin, pink, connector station A pillar, right
- Z20 - Left Washer Nozzle Heater
- Z21 - Right Washer Nozzle Heater
- (176) - Ground connection, in right headlight wiring harness
- (179) - Ground connection, in left headlight wiring harness
- (261) - Ground connection, in wiring harness heated spray jet
- (A34) - Wire connection (75x), in instrument panel wiring harness
- (L4) - Wire connection (75al), in A/C wiring harness

ws = white
sw = black
ro = red
br = brown
gn = green
bl = blue
gr = grey
li = violet
ge = yellow
or = orange

BentleyPublishers.com—All Rights Reserved

Audi A6

Wiring Diagram

No. 1/13

Fuse Holder, License Plate Light, Glove Compartment Light

- E26 - Glove compartment light
- E87 - A/C Control Head
- J429 - Control module for central locking
- S3 - Fuse
- S4 - Fuse
- S5 - Fuse
- S6 - Fuse
- S7 - Fuse
- S231 - Fuse 31 in fuse holder
- T2ap - Connector 2 pin, black, behind instrument panel, right
- T5a - Connector 5 pin, black, behind instrument panel, right
- T10f - Connector 10 pin, brown, connector station A pillar, left
- T17a - Connector 17 pin, brown, connector station A pillar, right
- W6 - Glove Compartment Light
- X - License Plate Light
- (81) - Ground connection -1-, in instrument panel wiring harness
- (87) - Ground connection -2-, in rear wiring harness
- (A20) - Wire connection (15a), in instrument panel wiring harness
- (A38) - Plus connection -2- (15a), in instrument panel wiring harness
- (A53) - Plus connection (58), in instrument panel wiring harness
- * - Vehicles with manual transmission
- ** - Vehicles with automatic transmission

ws = white
sw = black
ro = red
br = brown
gn = green
bl = blue
gr = grey
li = violet
ge = yellow
or = orange

127 128 129 130 131 132 133 134 135 136 137 138 139 140

97-51641

USA.5604.03.21 Edition 05/03

Audi A6

Wiring Diagram

No. 1/14

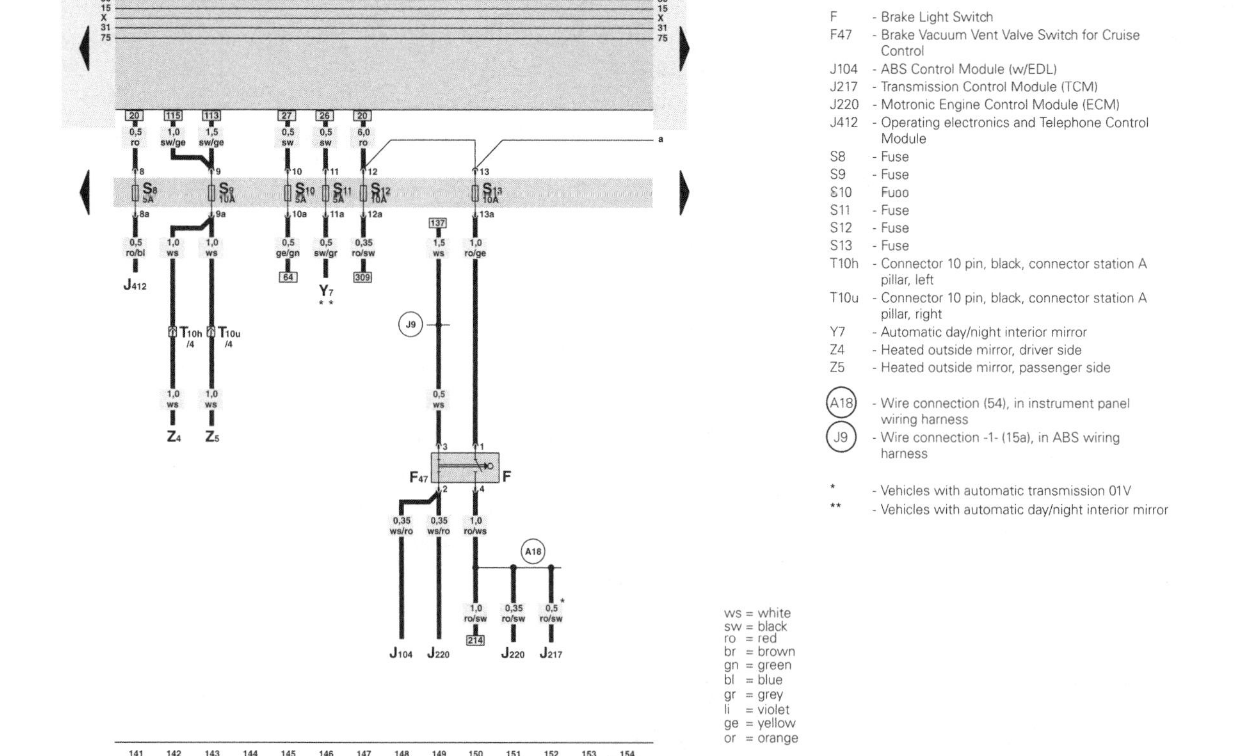

Fuse Holder, Heated Mirror, Brake Light Switch

- F - Brake Light Switch
- F47 - Brake Vacuum Vent Valve Switch for Cruise Control
- J104 - ABS Control Module (w/EDL)
- J217 - Transmission Control Module (TCM)
- J220 - Motronic Engine Control Module (ECM)
- J412 - Operating electronics and Telephone Control Module
- S8 - Fuse
- S9 - Fuse
- S10 - Fuse
- S11 - Fuse
- S12 - Fuse
- S13 - Fuse
- T10h - Connector 10 pin, black, connector station A pillar, left
- T10u - Connector 10 pin, black, connector station A pillar, right
- Y7 - Automatic day/night interior mirror
- Z4 - Heated outside mirror, driver side
- Z5 - Heated outside mirror, passenger side
- (A18) - Wire connection (54), in instrument panel wiring harness
- (J9) - Wire connection -1- (15a), in ABS wiring harness
- * - Vehicles with automatic transmission 01V
- ** - Vehicles with automatic day/night interior mirror

ws = white
sw = black
ro = red
br = brown
gn = green
bl = blue
gr = grey
li = violet
ge = yellow
or = orange

141 142 143 144 145 146 147 148 149 150 151 152 153 154

97-51642

USA.5604.03.21 Edition 05/03

BentleyPublishers.com—All Rights Reserved

Audi A6

Wiring Diagram

No. 1/15

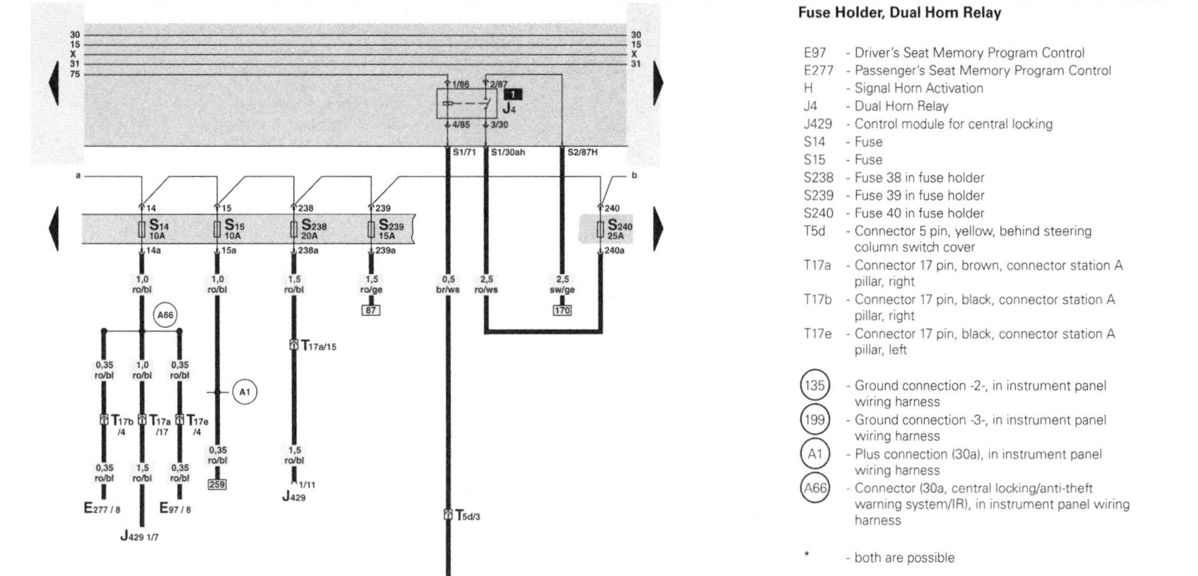

Fuse Holder, Dual Horn Relay

- E97 - Driver's Seat Memory Program Control
- E277 - Passenger's Seat Memory Program Control
- H - Signal Horn Activation
- J4 - Dual Horn Relay
- J429 - Control module for central locking
- S14 - Fuse
- S15 - Fuse
- S238 - Fuse 38 in fuse holder
- S239 - Fuse 39 in fuse holder
- S240 - Fuse 40 in fuse holder
- T5d - Connector 5 pin, yellow, behind steering column switch cover
- T17a - Connector 17 pin, brown, connector station A pillar, right
- T17b - Connector 17 pin, black, connector station A pillar, right
- T17e - Connector 17 pin, black, connector station A pillar, left

- (135) - Ground connection -2-, in instrument panel wiring harness
- (199) - Ground connection -3-, in instrument panel wiring harness
- (A1) - Plus connection (30a), in instrument panel wiring harness
- (A66) - Connector (30a, central locking/anti-theft warning system/IR), in instrument panel wiring harness

- * - both are possible

ws = white
sw = black
ro = red
br = brown
gn = green
bl = blue
gr = grey
li = violet
ge = yellow
or = orange

Audi A6

Wiring Diagram

No. 1/16

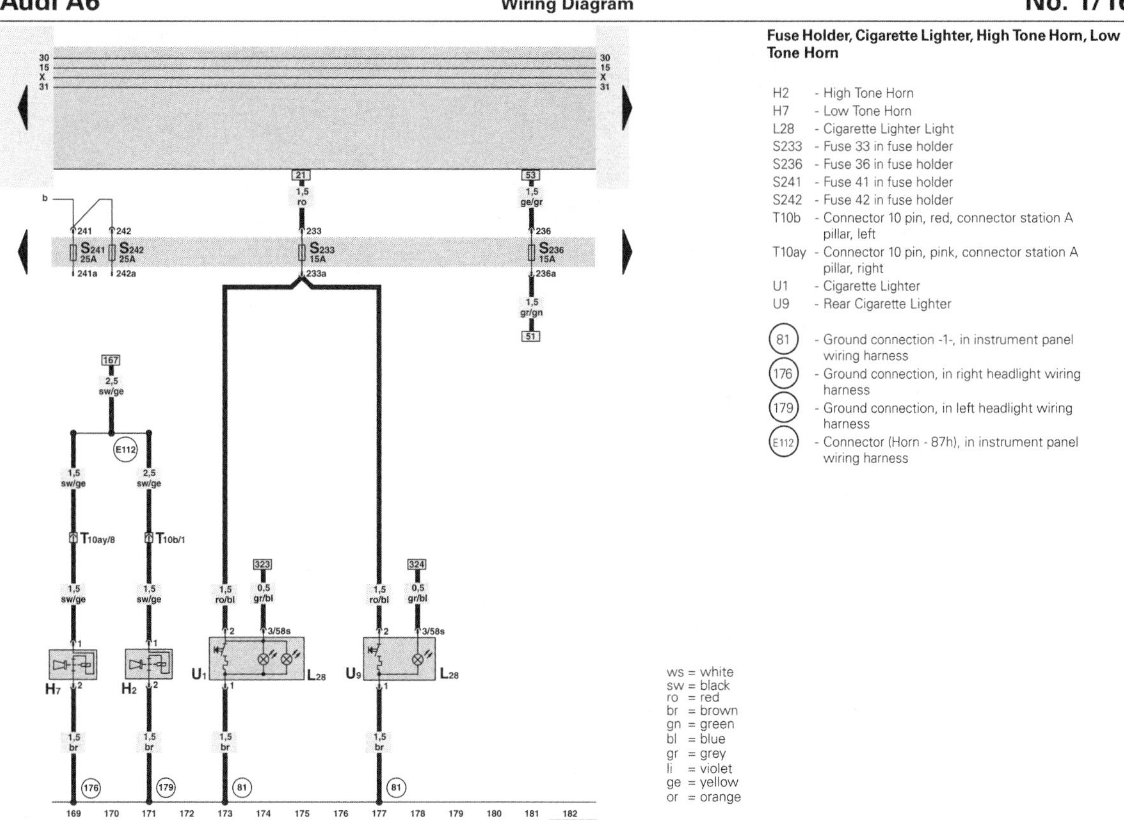

Fuse Holder, Cigarette Lighter, High Tone Horn, Low Tone Horn

- H2 - High Tone Horn
- H7 - Low Tone Horn
- L28 - Cigarette Lighter Light
- S233 - Fuse 33 in fuse holder
- S236 - Fuse 36 in fuse holder
- S241 - Fuse 41 in fuse holder
- S242 - Fuse 42 in fuse holder
- T10b - Connector 10 pin, red, connector station A pillar, left
- T10ay - Connector 10 pin, pink, connector station A pillar, right
- U1 - Cigarette Lighter
- U9 - Rear Cigarette Lighter

- (81) - Ground connection -1-, in instrument panel wiring harness
- (176) - Ground connection, in right headlight wiring harness
- (179) - Ground connection, in left headlight wiring harness
- (E112) - Connector (Horn - 87h), in instrument panel wiring harness

ws = white
sw = black
ro = red
br = brown
gn = green
bl = blue
gr = grey
li = violet
ge = yellow
or = orange

BentleyPublishers.com—All Rights Reserved

Audi A6 — Wiring Diagram — No. 1/17

Fuse Holder, Right Parking Light, Right Turn Signal Light, Right Low and High Beam Headlight

M3 - Right Parking Light
M7 - Right Front Turn Signal Light
M19 - Right, Side Turn Signal Light
M31 - Right Low Beam Headlight
M32 - Right High Beam Headlight
S18 - Fuse
S20 - Fuse
S22 - Fuse
T10ao - Connector 10 pin, on headlight, right
T10ay - Connector 10 pin, pink, connector station A pillar, right

(176) - Ground connection, in right headlight wiring harness
(C19) - Connector -1- (right turn signal), in wiring harness headlamp
(C29) - Connector -2- (56b), in wiring harness headlamp

ws = white
sw = black
ro = red
br = brown
gn = green
bl = blue
gr = grey
li = violet
ge = yellow
or = orange

97-51645

USA.5604.03.21 — Edition 05/03

Audi A6 — Wiring Diagram — No. 1/18

Fuse Holder, Left Low and High Beam Headlight, Left Turn Signal Light, Left Parking Light, 12 V Sok-ket

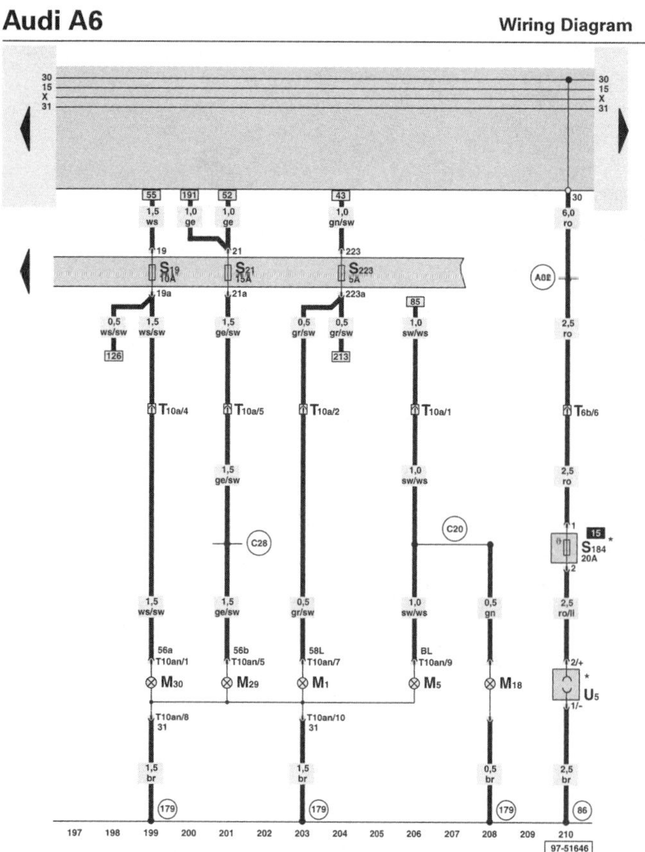

M1 - Left Parking Light
M5 - Left Front Turn Signal Light
M18 - Left, Side Turn Signal Light
M29 - Left Low Beam Headlight
M30 - Left High Beam Headlight
S19 - Fuse
S21 - Fuse
S184 - Power Outlet Fuse 1
S223 - Fuse 23 in fuse holder
T6b - Connector 6 pin, white, connector station A pillar, left
T10a - Connector 10 pin, pink, connector station A pillar, left
T10an - Connector 10 pin, on headlight, left
U5 - 12 V Socket

(86) - Ground connection -1-, in rear wiring harness
(179) - Ground connection, in left headlight wiring harness
(A32) - Plus connection (30), in instrument panel wiring harness
(C20) - Wire connection (left turn signal), in headlight wiring harness
(C28) - Connector -1- (56b), in wiring harness headlamp

* - Wagon only

ws = white
sw = black
ro = red
br = brown
gn = green
bl = blue
gr = grey
li = violet
ge = yellow
or = orange

USA.5604.03.21 — Edition 05/03

BentleyPublishers.com—All Rights Reserved

Audi A6 — Wiring Diagram — No. 1/19

Left Tail Light, Back-Up Light, High Mount Brake Light

- F4 - Back-Up Light Switch
- F125 - Multi-Function Transmission Range (TR) Switch
- L46 - Left Rear Fog Light
- M4 - Left Tail Light
- M6 - Left Rear Turn Signal Light
- M9 - Left Brake Light
- M16 - Left Back-Up Light
- M25 - High-mount Brake Light
- N110 - Shift Lock Solenoid
- T2am - Connector 2 pin, black, on selector lever
- T10f - Connector 10 pin, brown, connector station A pillar, left
- T10o - Connector 10 pin, brown, connector station electronic box plenum chamber
- T10ba - Connector 10 pin, blue, connector station A pillar, left
- T17k - Connector 17 pin, red, connector station electronic box plenum chamber
- T17m - Connector 17 pin, blue, connector station A pillar, right
- W18 - Luggage Compartment light, left
- (86) - Ground connection -1-, in rear wiring harness
- (87) - Ground connection -2-, in rear wiring harness
- (A87) - Connector (RF), in instrument panel wiring harness
- * - Vehicles with manual transmission
- ** - Vehicles with automatic transmission

ws = white
sw = black
ro = red
br = brown
gn = green
bl = blue
gr = grey
li = violet
ge = yellow
or = orange

USA.5604.03.21 Edition 05/03

Audi A6 — Wiring Diagram — No. 1/20

Right Tail Light, Fuel Pump, Heated Rear Window

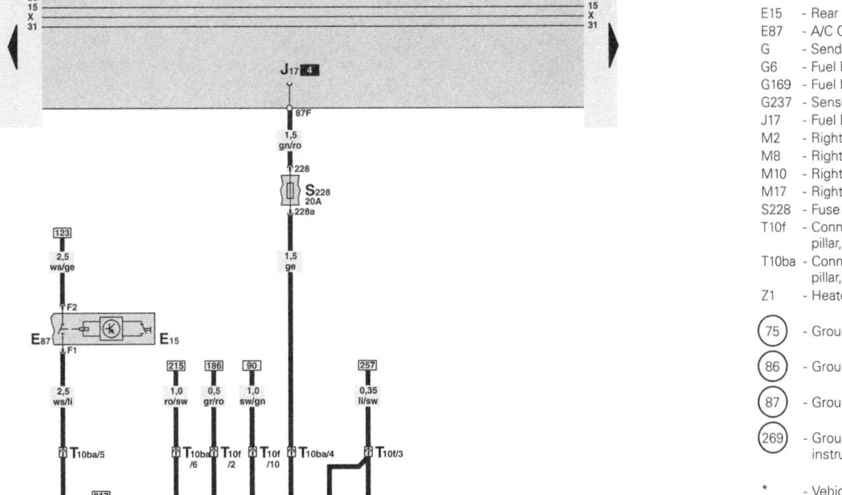

- E15 - Rear Window Defogger Switch
- E87 - A/C Control Head
- G - Sender for fuel gauge
- G6 - Fuel Pump (FP)
- G169 - Fuel Level Sensor 2
- G237 - Sensor -3- for fuel supply
- J17 - Fuel Pump (FP) Relay
- M2 - Right Tail Light
- M8 - Right Rear Turn Signal Light
- M10 - Right Brake Light
- M17 - Right Back-Up Light
- S228 - Fuse 28 in fuse holder
- T10f - Connector 10 pin, brown, connector station A pillar, left
- T10ba - Connector 10 pin, blue, connector station A pillar, left
- Z1 - Heated rear window
- (75) - Ground connection, on right rear pillar
- (86) - Ground connection -1-, in rear wiring harness
- (87) - Ground connection -2-, in rear wiring harness
- (269) - Ground connector (sensor ground) -1-, in instrument panel wiring harness
- * - Vehicles with front wheel drive
- ** - Vehicles with all wheel drive
- *** - Wiring inside fuel tank

ws = white
sw = black
ro = red
br = brown
gn = green
bl = blue
gr = grey
li = violet
ge = yellow
or = orange

USA.5604.03.21 Edition 05/03

BentleyPublishers.com—All Rights Reserved

Audi A6

Wiring Diagram

No. 1/21

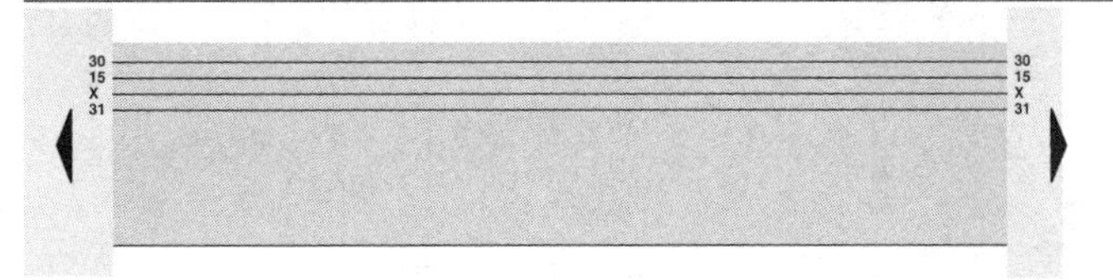

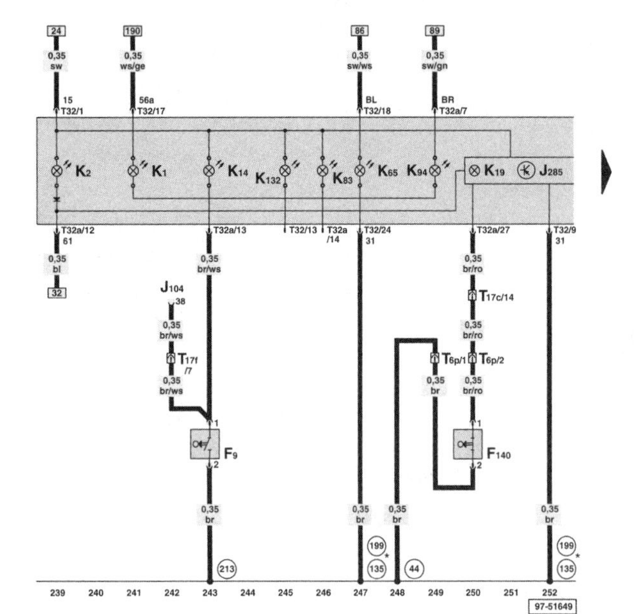

Instrument Cluster, Indicator Light

- F9 - Parking Brake Warning Light Switch
- F140 - Left Front Seatbelt Microswitch
- J104 - ABS Control Module (w/EDL)
- J285 - Control Module with indicator unit in instrument panel insert
- K1 - Headlight High Beam Indicator Light
- K2 - Generator (GEN) Warning Light
- K14 - Parking Brake Indicator Light
- K19 - Seat Belt Warning Light
- K65 - Left Turn Signal Indicator Light
- K83 - Malfunction Indicator Lamp (MIL)
- K94 - Right Turn Signal Indicator Light
- K132 - Electronic Power Control (EPC) Warning Lamp
- T6p - Connector 6 pin, red, below driver's seat
- T17c - Connector 17 pin, green, connector station A pillar, right
- T17f - Connector 17 pin, black, in 8-Fold Relay Panel behind the storage driver's side
- T32 - Connector 32 pin, blue, on instrument cluster
- T32a - Connector 32 pin, green, on instrument cluster
- (44) - Ground connection, on left A-pillar, lower part
- (135) - Ground connection -2-, in instrument panel wiring harness
- (199) - Ground connection -3-, in instrument panel wiring harness
- (213) - Ground connection, in instrument panel wiring harness
- * - both are possible

ws = white
sw = black
ro = red
br = brown
gn = green
bl = blue
gr = grey
li = violet
ge = yellow
or = orange

USA.5604.03.21 Edition 05/03

Audi A6

Wiring Diagram

No. 1/22

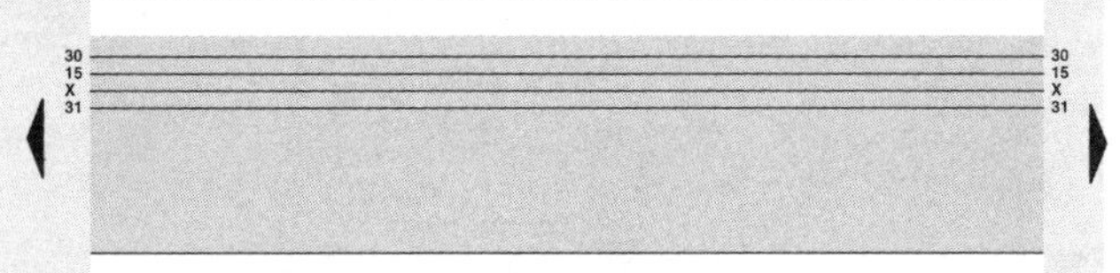

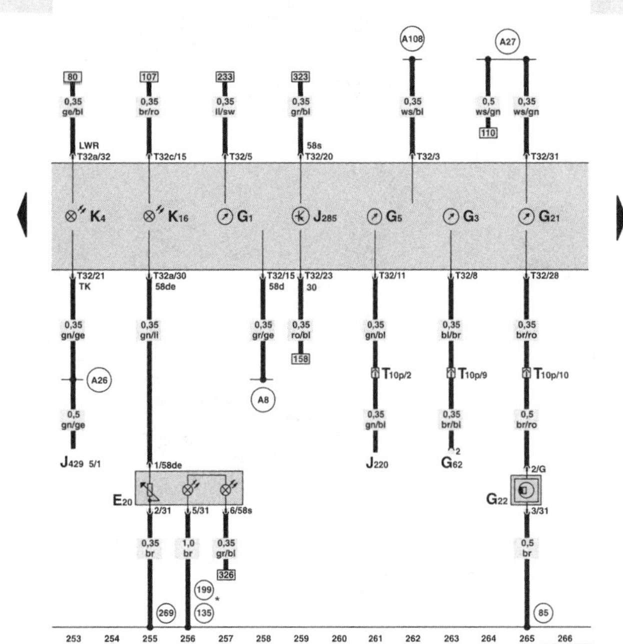

Instrument Cluster, Fuel Gauge, Tachometer, Speedometer, Instrument Panel Light Dimmer Switch

- E20 - Instrument Panel Light Dimmer Switch
- G1 - Fuel Gauge
- G3 - Engine Coolant Temperature (ECT) Gauge
- G5 - Tachometer
- G21 - Speedometer
- G22 - Speedometer Vehicle Speed Sensor (VSS)
- G62 - Engine Coolant Temperature (ECT) Sensor
- J220 - Motronic Engine Control Module (ECM)
- J285 - Control Module with indicator unit in instrument panel insert
- J429 - Control module for central locking
- K4 - Park Light Indicator Light
- K16 - Fuel Reserve Warning Light
- T10p - Connector 10 pin, black, connector station electronic box plenum chamber
- T32 - Connector 32 pin, blue, on instrument cluster
- T32a - Connector 32 pin, green, on instrument cluster
- T32c - Connector 32 pin, grey, on instrument cluster
- (85) - Ground connection -1-, in engine compartment wiring harness
- (135) - Ground connection -2-, in instrument panel wiring harness
- (199) - Ground connection -3-, in instrument panel wiring harness
- (269) - Ground connector (sensor ground) -1-, in instrument panel wiring harness
- (A8) - Plus connection (58d), in instrument panel wiring harness
- (A26) - Wire connection (driver's door contact switch), in instrument panel wiring harness
- (A27) - Wire connection (vehicle speed signal), in instrument panel wiring harness
- (A108) - Connector (speed signal), in instrument panel wiring harness
- * - both are possible

ws = white
sw = black
ro = red
br = brown
gn = green
bl = blue
gr = grey
li = violet
ge = yellow
or = orange

USA.5604.03.21 Edition 05/03

BentleyPublishers.com—All Rights Reserved

Audi A6

Wiring Diagram

No. 1/23

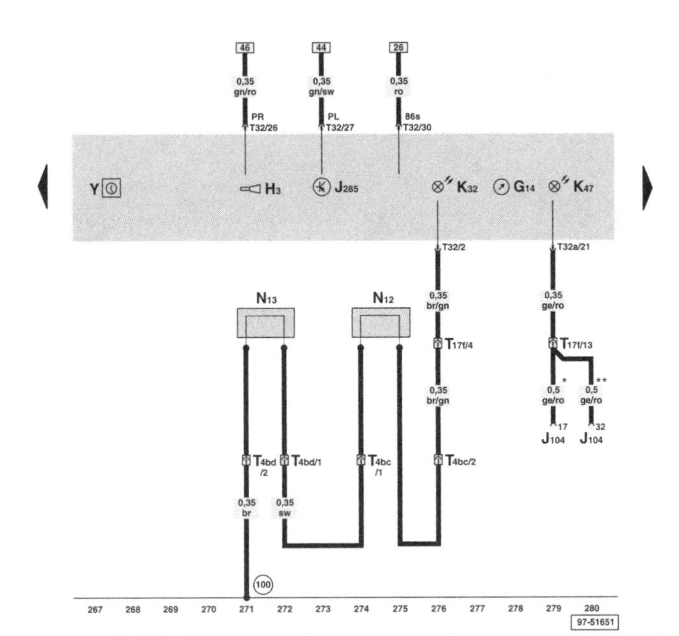

Instrument Cluster, Analog Clock, Brake Pad Wear Indicator

- G14 - Voltmeter
- H3 - Warning Buzzer
- J104 - ABS Control Module (w/EDL)
- J285 - Control Module with indicator unit in instrument panel insert
- K32 - Brake Pad Wear Indicator Light
- K47 - ABS Warning Light
- N12 - Right Brake Pad Wear Indicator (Shear Element)
- N13 - Left Brake Pad Wear Indicator (Shear Element)
- T4bc - Connector 4 pin, black, for Left Front ABS Wheel Speed Sensor
- T4bd - Connector 4 pin, black, for Right Front ABS Wheel Speed Sensor
- T17f - Connector 17 pin, black, in 8-Fold Relay Panel behind the storage driver's side
- T32 - Connector 32 pin, blue, on instrument cluster
- T32a - Connector 32 pin, green, on instrument cluster
- Y - Analog Clock

- (100) - Ground connection -1-, in ABS wiring harness

- * - Vehicles with ABS 5.7
- ** - Vehicles with ABS 5.3, allroad

ws = white
sw = black
ro = red
br = brown
gn = green
bl = blue
gr = grey
li = violet
ge = yellow
or = orange

USA.5604.03.21

Edition 05/03

Audi A6

Wiring Diagram

No. 1/24

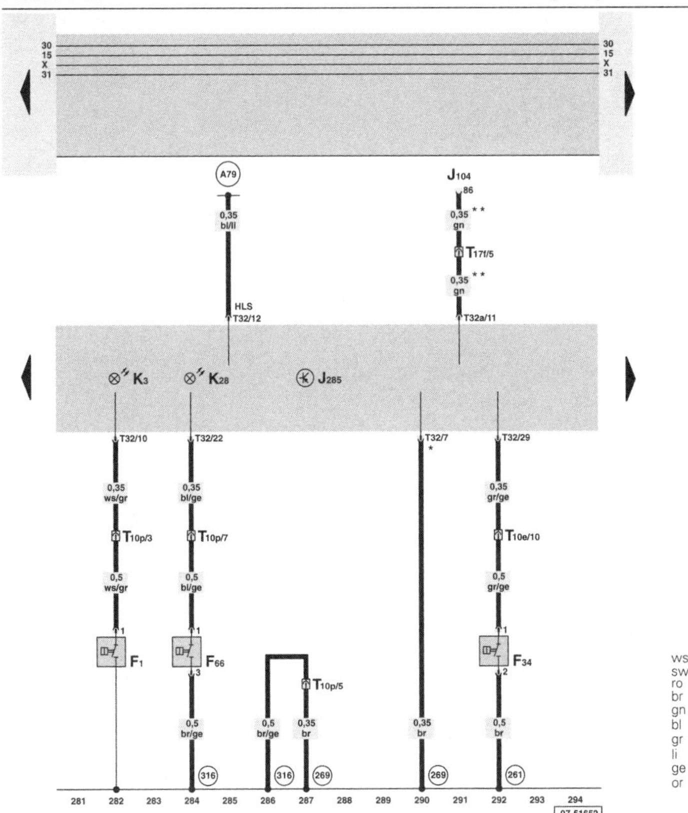

Instrument Cluster, Oil Pressure Switch, Engine Coolant Level Warning Switch, Brake Fluid Level Warning Switch

- F1 - Oil Pressure Switch
- F34 - Brake Fluid Level Warning Switch
- F66 - Engine Coolant Level (ECL) Warning Switch
- J104 - ABS Control Module (w/EDL)
- J285 - Control Module with indicator unit in instrument panel insert
- K3 - Oil Pressure Warning Light
- K28 - Engine Coolant Level/Temperature (ECL/ECT) Warning Light
- T10e - Connector 10 pin, violet, connector station A pillar, left
- T10p - Connector 10 pin, black, connector station electronic box plenum chamber
- T17f - Connector 17 pin, black, in 8-Fold Relay Panel behind the storage driver's side
- T32 - Connector 32 pin, blue, on instrument cluster
- T32a - Connector 32 pin, green, on instrument cluster

- (261) - Ground connection, in wiring harness heated spray jet
- (269) - Ground connector (sensor ground) -1-, in instrument panel wiring harness
- (316) - Sensor Ground Connection 2 (in engine wiring harness)
- (A79) - Connector (engine overheat warning light), in instrument panel wiring harness

- * - Exit Sensor Ground
- ** - Vehicles with ABS 5.3, allroad

ws = white
sw = black
ro = red
br = brown
gn = green
bl = blue
gr = grey
li = violet
ge = yellow
or = orange

USA.5604.03.21

Edition 05/03

BentleyPublishers.com—All Rights Reserved

Audi A6 — Wiring Diagram — No. 1/25

Instrument Cluster, Oil Level Thermal Sensor, Function Selector Switch II

- D2 - Induction coil of antitheft immobilizer
- E272 - Board Computer Function Selector Switch II, in console
- F120 - Hood Alarm Switch
- G266 - Oil Level Thermal Sensor (SRI)
- J234 - Airbag Control Module
- J285 - Control Module with indicator unit in instrument panel insert
- J429 - Control module for central locking
- K75 - Airbag Malfunction Indicator Lamp (MIL)
- K117 - Warning light for anti-theft immobilizer sensor
- T10o - Connector 10 pin, brown, connector station electronic box plenum chamber
- T10av - Connector 10 pin, yellow, connector station A pillar, right
- T17 - Connector 17 pin, orange, connector station A pillar, right
- T17a - Connector 17 pin, brown, connector station A pillar, right
- T32a - Connector 32 pin, green, on instrument cluster
- T32c - Connector 32 pin, grey, on instrument cluster

- (85) - Ground connection -1-, in engine compartment wiring harness
- (269) - Ground connector (sensor ground) -1-, in instrument panel wiring harness
- (A121) - Connection (high bus) in instrument panel wiring harness
- (A122) - Connection (low bus) in instrument panel wiring harness
- (A184) - Front Hood Contact Switch Connection (in instrument panel wiring harness)

- • - CAN-Bus (Data-Bus)

ws = white
sw = black
ro = red
br = brown
gn = green
bl = blue
gr = grey
li = violet
ge = yellow
or = orange

295 296 297 298 299 300 301 302 303 304 305 306 307 308

97-51653

USA.5604.03.21 — Edition 05/03

Audi A6 — Wiring Diagram — No. 1/26

Instrument Cluster, Data Link Connector (DLC)

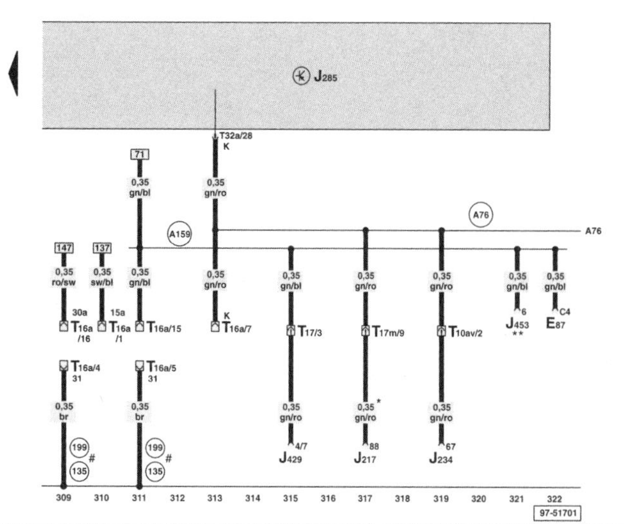

- E87 - A/C Control Head
- J217 - Transmission Control Module (TCM)
- J234 - Airbag Control Module
- J285 - Control Module with indicator unit in instrument panel insert
- J429 - Control module for central locking
- J453 - Control module for multi-function steering wheel
- T10av - Connector 10 pin, yellow, connector station A pillar, right
- T16a - Connector 16 pin, Data Link Connector (DLC)
- T17 - Connector 17 pin, orange, connector station A pillar, right
- T17m - Connector 17 pin, blue, connector station A pillar, right
- T32a - Connector 32 pin, green, on instrument cluster

- (135) - Ground connection -2-, in instrument panel wiring harness
- (199) - Ground connection -3-, in instrument panel wiring harness
- (A76) - Connector (K-diagnosis wire), in instrument panel wiring harness
- (A159) - Connection (L) (in instrument panel wiring harness)

- * - Vehicles with automatic transmission
- ** - Vehicles with multi-function steering wheel
- # - both are possible

ws = white
sw = black
ro = red
br = brown
gn = green
bl = blue
gr = grey
li = violet
ge = yellow
or = orange

309 310 311 312 313 314 315 316 317 318 319 320 321 322

97-51701

USA.5604.03.21 — Edition 05/03

BentleyPublishers.com—All Rights Reserved

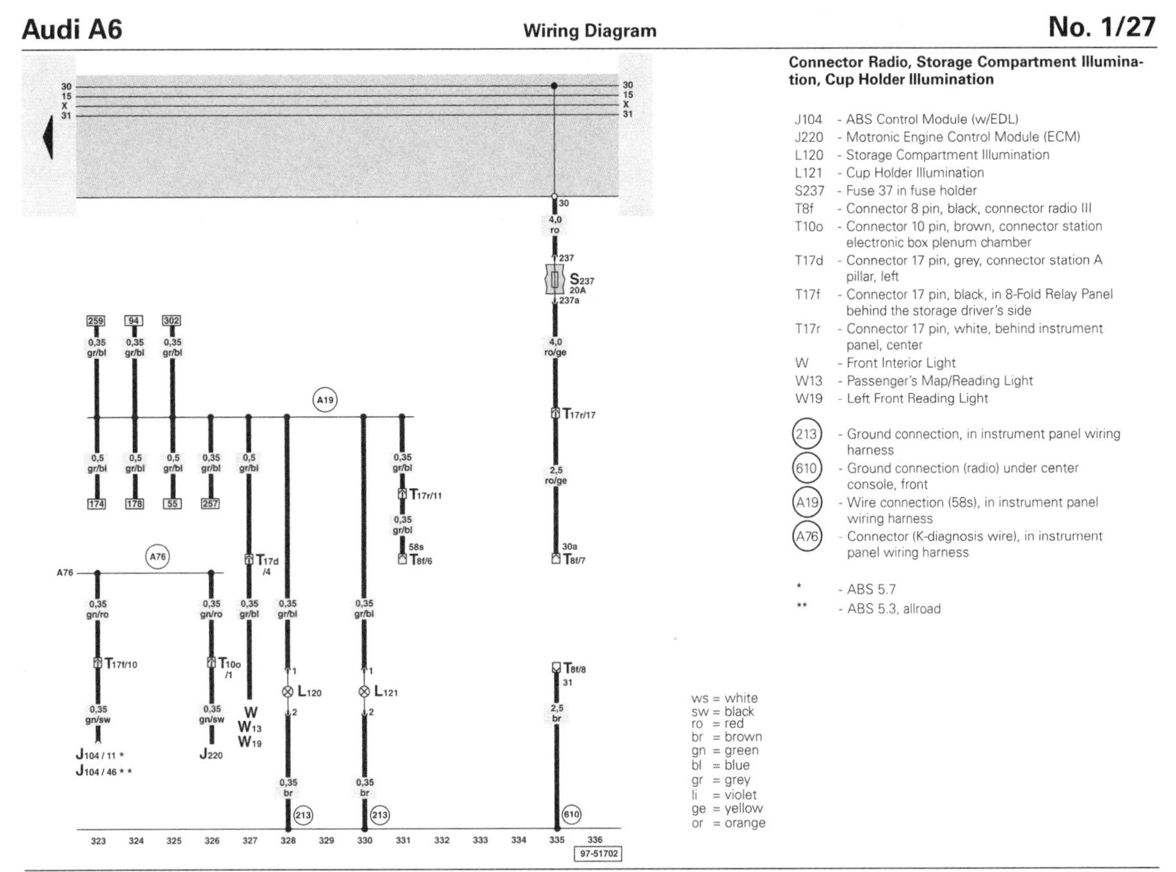

THIS SPACE INTENTIONALLY

LEFT BLANK

BentleyPublishers.com—All Rights Reserved

(2,7 l - Injection Engine, 6 Cylinder), Code APB, BEL

2003 m.y.

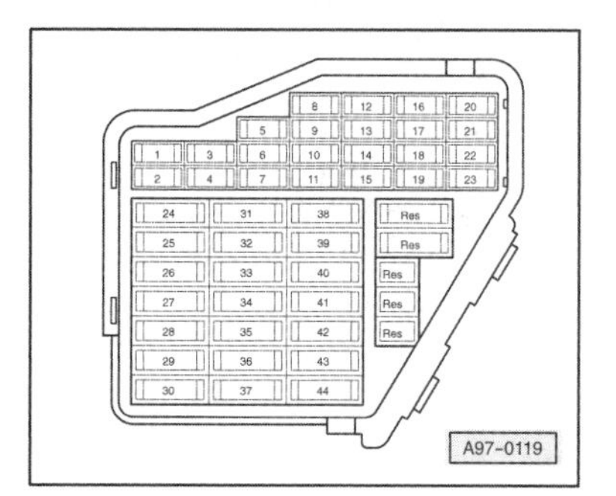

Fuse Panel

Fuse Colors:

30 A - Green
25 A - White
20 A - Yellow
15 A - Blue
10 A - Red
7,5 A - Brown
5 A - Beige
1 A - Black

Starting with fuse position 23, fuses in the fuse holder are identified with 223 in the wiring diagram.

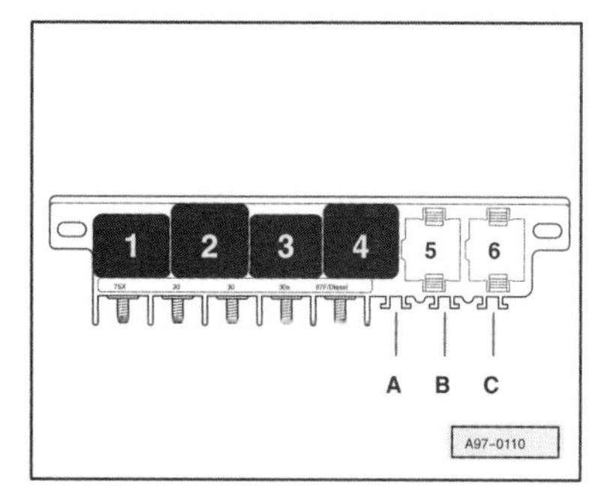

Micro Central Electric Panel

Relay Location:

4 - Fuel Pump (FP) Relay, J17

13 - Fold Relay Panel

Relay Location:

4 - Starting Interlock Relay, J207

BentleyPublishers.com—All Rights Reserved

Audi A6 **Wiring Diagram** **No. 2/3**

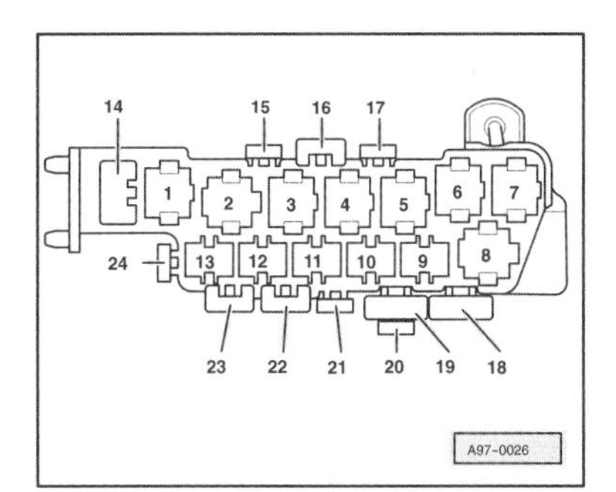

8 - Fold Relay Panel

Fuse Location:

19 - Coolant Fan Fuse, S42

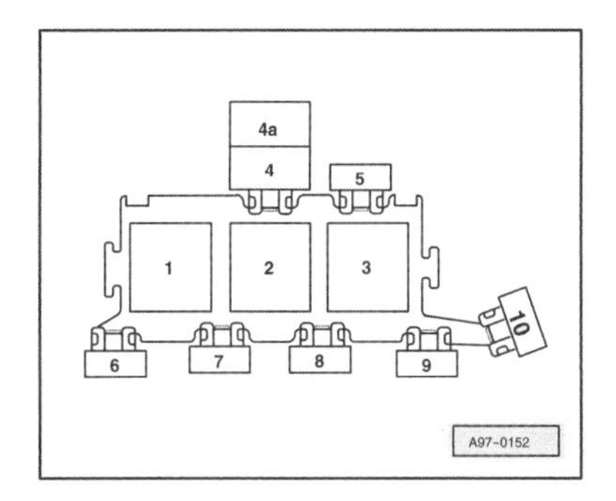

3 - Fold Relay Panel in E-Box plenum chamber

Relay Location:

2 - Secondary Air Injection (AIR) Pump Relay, J299

3 - Motronic Engine Control Module (ECM) Power Supply Relay, J271

5 - Engine Electronics Fuse, S282

7 - Fuse for Secondary Air Pump, S130

Audi A6 **Wiring Diagram** **No. 2/4**

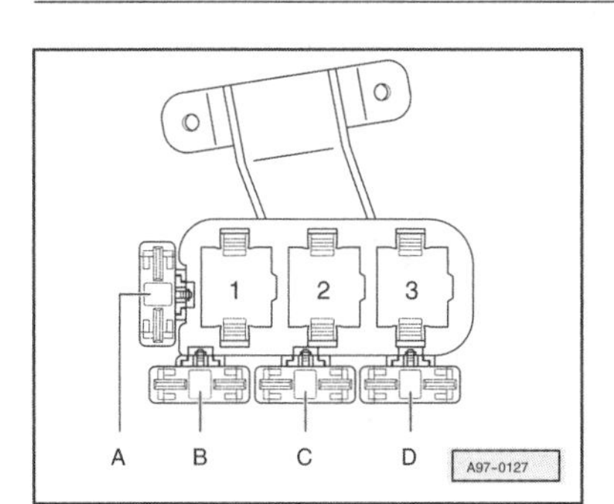

Relay Panel behind Instrument Panel left, on central girder

Relay Location:

1 - Brake Booster Relay, J569

C - Fuse for Hydraulic Pump Relay, S279

BentleyPublishers.com—All Rights Reserved

Engine Management, 2.7 Liter, Engine Codes APB, BEL (2003)

Audi A6 **Wiring Diagram** **No. 2/5**

Battery, Starter, Generator, Ignition/Starter Switch

A - Battery
B - Starter
C - Generator (GEN)
C1 - Voltage Regulator (VR)
D - Ignition/Starter Switch
S205 - Fuse 2 (30)
T1h - Single connector, black, in engine compartment, right
T2cl - Connector 2 pin, black, in engine compartment, right
T10o - Connector 10 pin, brown, connector station electronic box plenum chamber
T10p - Connector 10 pin, black, connector station electronic box plenum chamber

(1) - Ground strap, battery to body
(44) - Ground Connection, on left A-pillar, lower part
(199) - Ground Connection 3, in instrument panel wiring harness
(A17) - Connection (61), in instrument panel wiring harness
(A41) - Plus Connection (50), in instrument panel wiring harness

ws = white
sw = black
ro = red
br = brown
gn = green
bl = blue
gr = grey
li = violet
ge = yellow
or = orange

97-52054

USA.5604.05.21 Edition 07/04

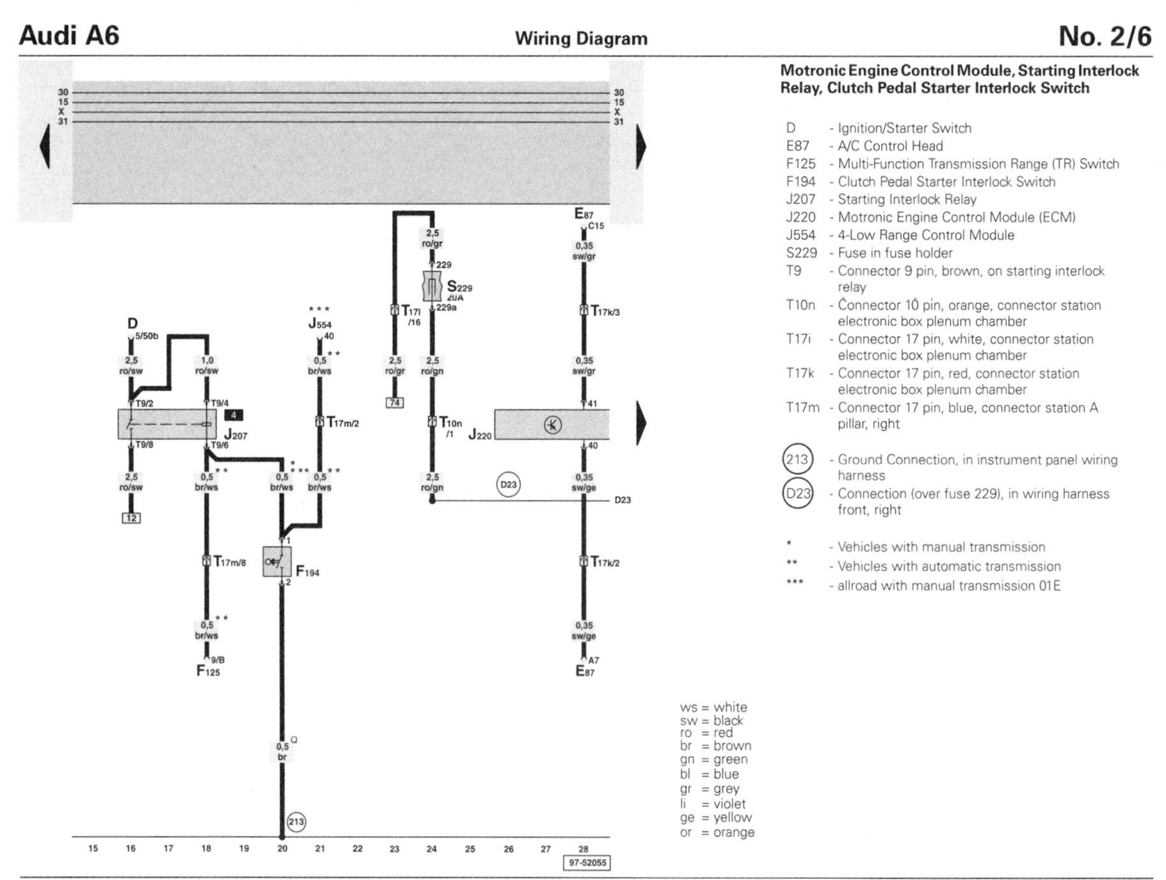

Audi A6
Wiring Diagram
No. 2/6
Motronic Engine Control Module, Starting Interlock Relay, Clutch Pedal Starter Interlock Switch
D - Ignition/Starter Switch
E87 - A/C Control Head
F125 - Multi-Function Transmission Range (TR) Switch
F194 - Clutch Pedal Starter Interlock Switch
J207 - Starting Interlock Relay
J220 - Motronic Engine Control Module (ECM)
J554 - 4-Low Range Control Module
S229 - Fuse in fuse holder
T9 - Connector 9 pin, brown, on starting interlock relay
T10n - Connector 10 pin, orange, connector station electronic box plenum chamber
T17i - Connector 17 pin, white, connector station electronic box plenum chamber
T17k - Connector 17 pin, red, connector station electronic box plenum chamber
T17m - Connector 17 pin, blue, connector station A pillar, right
(213) - Ground Connection, in instrument panel wiring harness
(D23) - Connection (over fuse 229), in wiring harness front, right
* - Vehicles with manual transmission
** - Vehicles with automatic transmission
*** - allroad with manual transmission 01E
ws = white
sw = black
ro = red
br = brown
gn = green
bl = blue
gr = grey
li = violet
ge = yellow
or = orange
97-52055

USA.5604.05.21 Edition 07/04

BentleyPublishers.com—All Rights Reserved

Engine Management, 2.7 Liter, Engine Codes APB, BEL (2003)

Audi A6

Wiring Diagram

No. 2/7

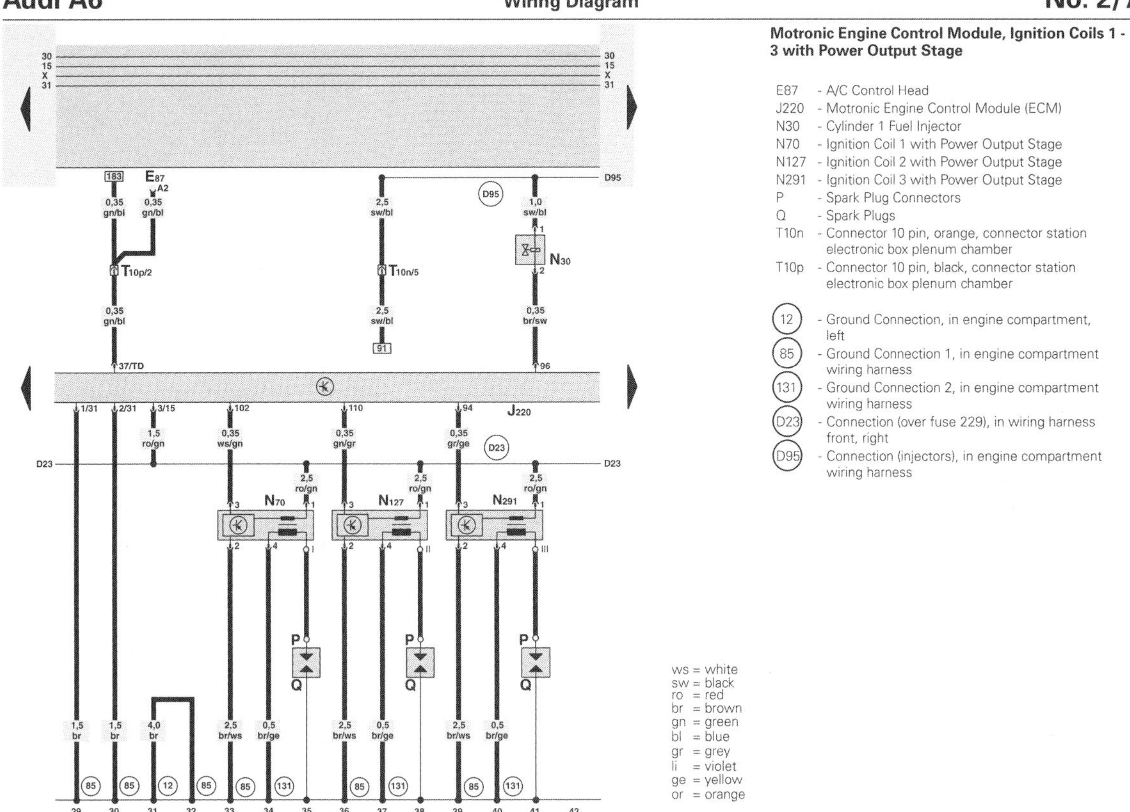

Motronic Engine Control Module, Ignition Coils 1 - 3 with Power Output Stage

- E87 - A/C Control Head
- J220 - Motronic Engine Control Module (ECM)
- N30 - Cylinder 1 Fuel Injector
- N70 - Ignition Coil 1 with Power Output Stage
- N127 - Ignition Coil 2 with Power Output Stage
- N291 - Ignition Coil 3 with Power Output Stage
- P - Spark Plug Connectors
- Q - Spark Plugs
- T10n - Connector 10 pin, orange, connector station electronic box plenum chamber
- T10p - Connector 10 pin, black, connector station electronic box plenum chamber

- (12) - Ground Connection, in engine compartment, left
- (85) - Ground Connection 1, in engine compartment wiring harness
- (131) - Ground Connection 2, in engine compartment wiring harness
- (D23) - Connection (over fuse 229), in wiring harness front, right
- (D95) - Connection (injectors), in engine compartment wiring harness

ws = white
sw = black
ro = red
br = brown
gn = green
bl = blue
gr = grey
li = violet
ge = yellow
or = orange

USA.5604.05.21 Edition 07/04

Audi A6

Wiring Diagram

No. 2/8

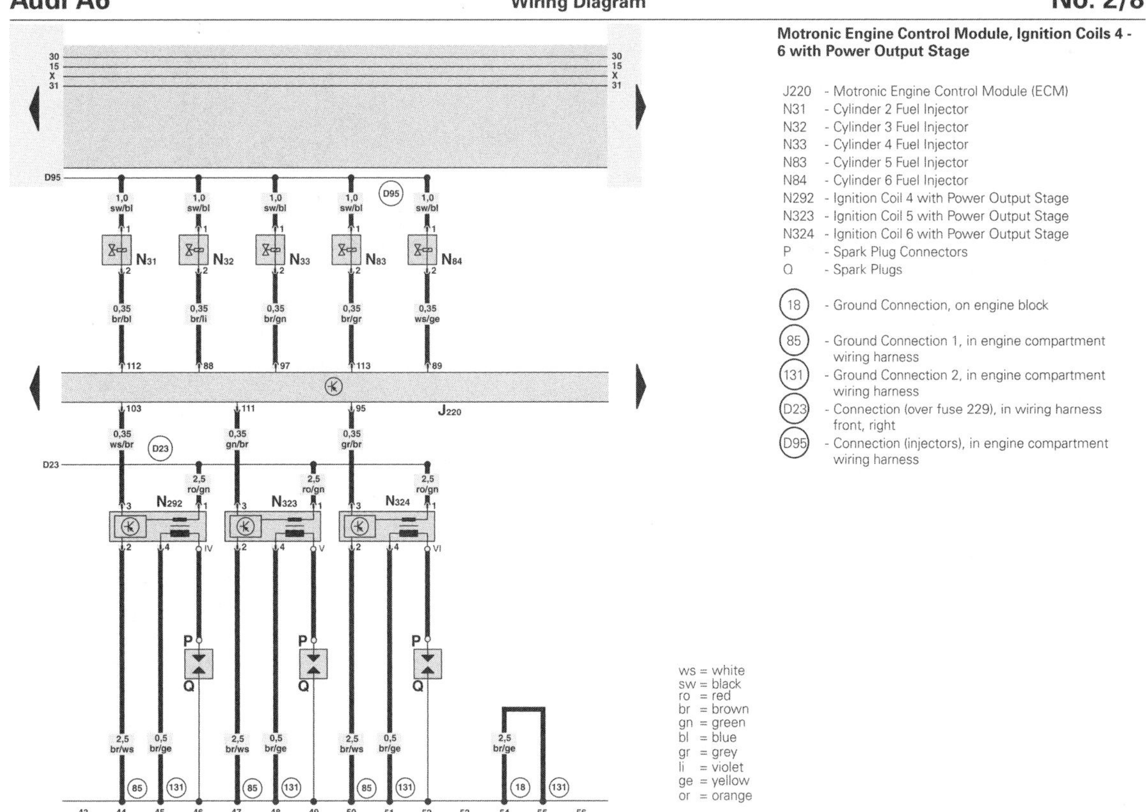

Motronic Engine Control Module, Ignition Coils 4 - 6 with Power Output Stage

- J220 - Motronic Engine Control Module (ECM)
- N31 - Cylinder 2 Fuel Injector
- N32 - Cylinder 3 Fuel Injector
- N33 - Cylinder 4 Fuel Injector
- N83 - Cylinder 5 Fuel Injector
- N84 - Cylinder 6 Fuel Injector
- N292 - Ignition Coil 4 with Power Output Stage
- N323 - Ignition Coil 5 with Power Output Stage
- N324 - Ignition Coil 6 with Power Output Stage
- P - Spark Plug Connectors
- Q - Spark Plugs

- (18) - Ground Connection, on engine block
- (85) - Ground Connection 1, in engine compartment wiring harness
- (131) - Ground Connection 2, in engine compartment wiring harness
- (D23) - Connection (over fuse 229), in wiring harness front, right
- (D95) - Connection (injectors), in engine compartment wiring harness

ws = white
sw = black
ro = red
br = brown
gn = green
bl = blue
gr = grey
li = violet
ge = yellow
or = orange

USA.5604.05.21 Edition 07/04

BentleyPublishers.com—All Rights Reserved

Engine Management, 2.7 Liter, Engine Codes APB, BEL (2003)

Audi A6 — Wiring Diagram — No. 2/9

Motronic Engine Control Module, Knock Sensors, Engine Coolant Temperature Sensor

- E3 - Emergency Flasher Switch
- G2 - Engine Coolant Temperature (ECT) Sensor
- G28 - Engine Speed (RPM) Sensor
- G42 - Intake Air Temperature (IAT) Sensor
- G61 - Knock Sensor (KS) 1
- G62 - Engine Coolant Temperature (ECT) Sensor
- G66 - Knock Sensor (KS) 2
- J220 - Motronic Engine Control Module (ECM)
- J234 - Airbag Control Module
- T3w - Connector 3 pin, grey, on engine speed sensor
- T3ax - Connector 3 pin, brown, on knock sensor 1
- T3ay - Connector 3 pin, brown, on knock sensor 2
- T10p - Connector 10 pin, black, connector station electronic box plenum chamber
- T10av - Connector 10 pin, yellow, connector station A pillar, right
- T17k - Connector 17 pin, red, connector station electronic box plenum chamber

- (220) - Ground Connection (sensor ground), in engine compartment wiring harness
- (269) - Ground connector (sensor ground) 1, in instrument panel wiring harness
- (316) - Sensor Ground Connection 2 (in engine wiring harness)
- (A125) - Connection (crash signal) in instrument panel wiring harness
- (D102) - Connection 2, in engine compartment wiring harness

* - Vehicles with automatic transmission

ws = white
sw = black
ro = red
br = brown
gn = green
bl = blue
gr = grey
li = violet
ge = yellow
or = orange

97-52058

USA.5604.05.21 — Edition 07/04

Audi A6 — Wiring Diagram — No. 2/10

Motronic Engine Control Module, Camshaft Position Sensors, Valves for Camshaft Adjustment

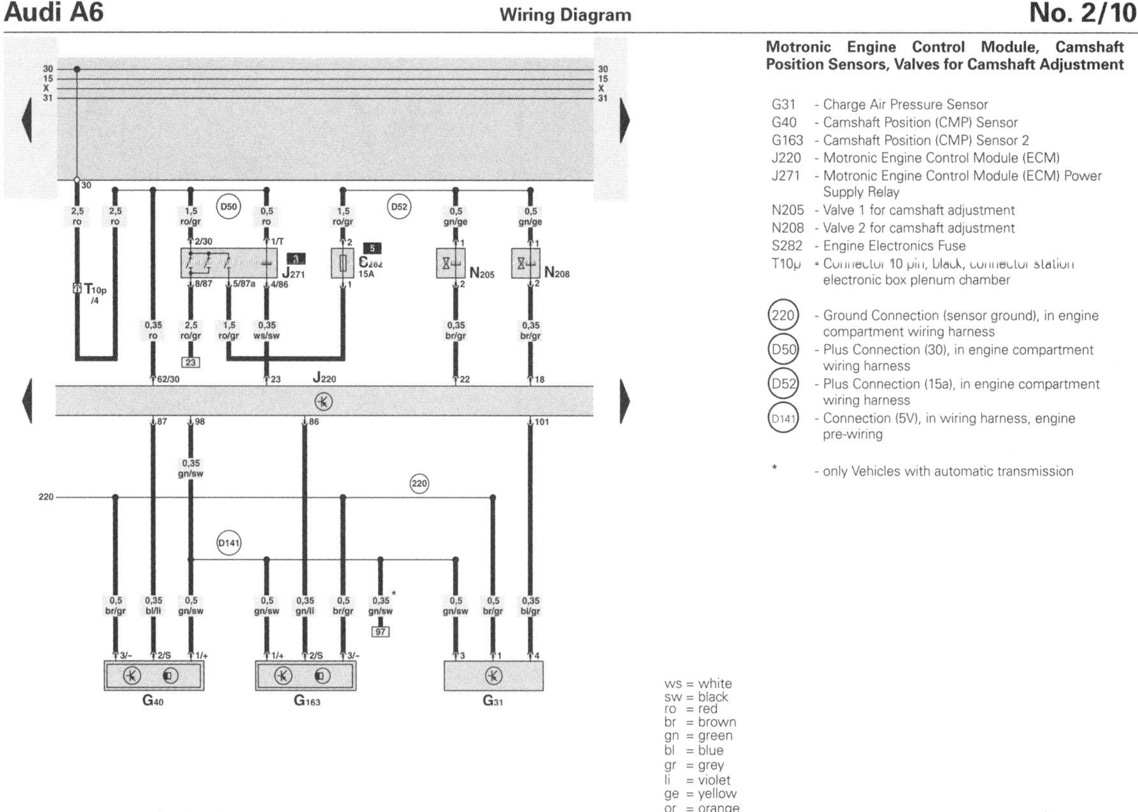

- G31 - Charge Air Pressure Sensor
- G40 - Camshaft Position (CMP) Sensor
- G163 - Camshaft Position (CMP) Sensor 2
- J220 - Motronic Engine Control Module (ECM)
- J271 - Motronic Engine Control Module (ECM) Power Supply Relay
- N205 - Valve 1 for camshaft adjustment
- N208 - Valve 2 for camshaft adjustment
- S282 - Engine Electronics Fuse
- T10p - Connector 10 pin, black, connector station electronic box plenum chamber

- (220) - Ground Connection (sensor ground), in engine compartment wiring harness
- (D50) - Plus Connection (30), in engine compartment wiring harness
- (D52) - Plus Connection (15a), in engine compartment wiring harness
- (D141) - Connection (5V), in wiring harness, engine pre-wiring

* - only Vehicles with automatic transmission

ws = white
sw = black
ro = red
br = brown
gn = green
bl = blue
gr = grey
li = violet
ge = yellow
or = orange

97-52059

USA.5604.05.21 — Edition 07/04

BentleyPublishers.com—All Rights Reserved

Audi A6 — Wiring Diagram — No. 2/11

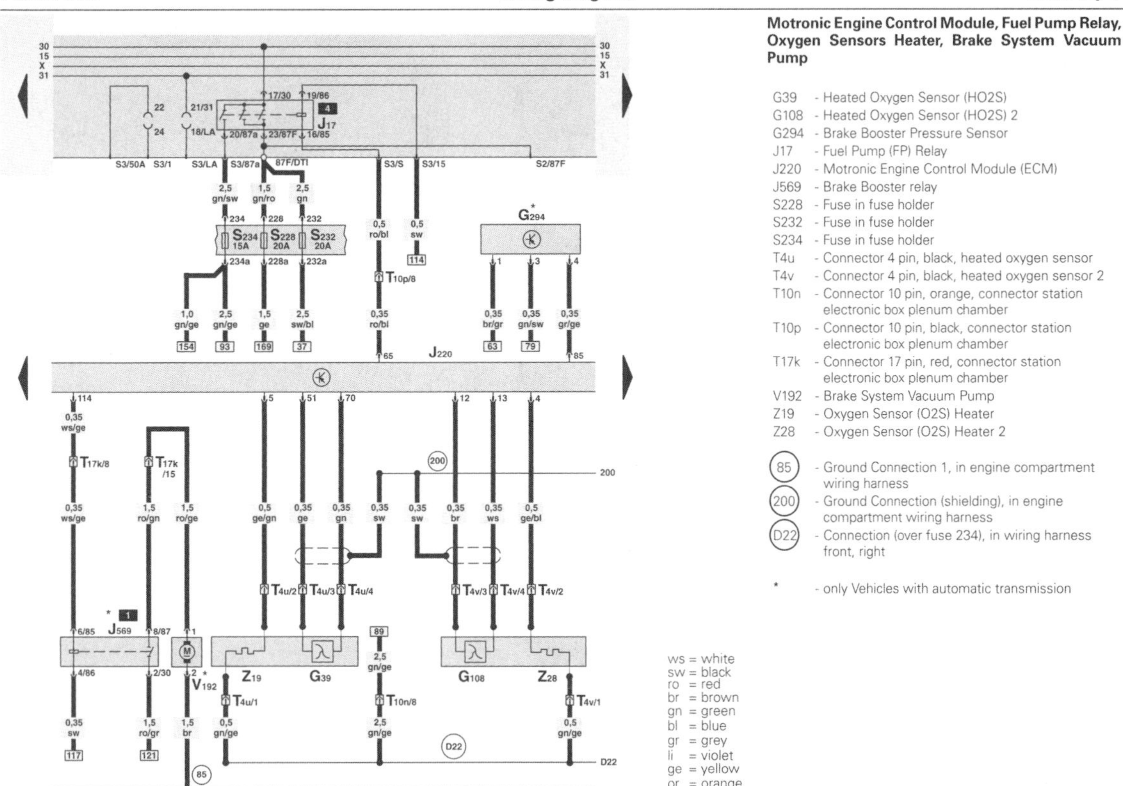

Motronic Engine Control Module, Fuel Pump Relay, Oxygen Sensors Heater, Brake System Vacuum Pump

- G39 - Heated Oxygen Sensor (HO2S)
- G108 - Heated Oxygen Sensor (HO2S) 2
- G294 - Brake Booster Pressure Sensor
- J17 - Fuel Pump (FP) Relay
- J220 - Motronic Engine Control Module (ECM)
- J569 - Brake Booster relay
- S228 - Fuse in fuse holder
- S232 - Fuse in fuse holder
- S234 - Fuse in fuse holder
- T4u - Connector 4 pin, black, heated oxygen sensor
- T4v - Connector 4 pin, black, heated oxygen sensor 2
- T10n - Connector 10 pin, orange, connector station electronic box plenum chamber
- T10p - Connector 10 pin, black, connector station electronic box plenum chamber
- T17k - Connector 17 pin, red, connector station electronic box plenum chamber
- V192 - Brake System Vacuum Pump
- Z19 - Oxygen Sensor (O2S) Heater
- Z28 - Oxygen Sensor (O2S) Heater 2

- (85) - Ground Connection 1, in engine compartment wiring harness
- (200) - Ground Connection (shielding), in engine compartment wiring harness
- (D22) - Connection (over fuse 234), in wiring harness front, right

* - only Vehicles with automatic transmission

ws = white
sw = black
ro = red
br = brown
gn = green
bl = blue
gr = grey
li = violet
ge = yellow
or = orange

97-52060

Audi A6 — Wiring Diagram — No. 2/12

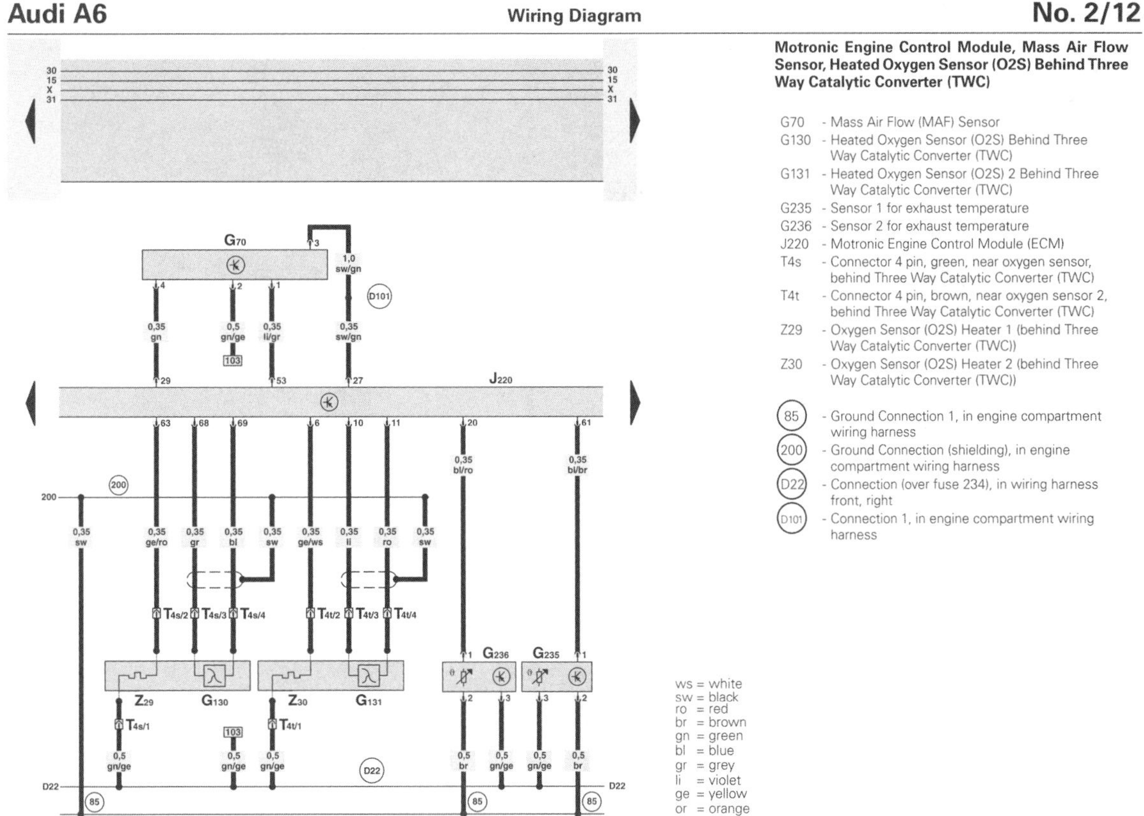

Motronic Engine Control Module, Mass Air Flow Sensor, Heated Oxygen Sensor (O2S) Behind Three Way Catalytic Converter (TWC)

- G70 - Mass Air Flow (MAF) Sensor
- G130 - Heated Oxygen Sensor (O2S) Behind Three Way Catalytic Converter (TWC)
- G131 - Heated Oxygen Sensor (O2S) 2 Behind Three Way Catalytic Converter (TWC)
- G235 - Sensor 1 for exhaust temperature
- G236 - Sensor 2 for exhaust temperature
- J220 - Motronic Engine Control Module (ECM)
- T4s - Connector 4 pin, green, near oxygen sensor, behind Three Way Catalytic Converter (TWC)
- T4t - Connector 4 pin, brown, near oxygen sensor 2, behind Three Way Catalytic Converter (TWC)
- Z29 - Oxygen Sensor (O2S) Heater 1 (behind Three Way Catalytic Converter (TWC))
- Z30 - Oxygen Sensor (O2S) Heater 2 (behind Three Way Catalytic Converter (TWC))

- (85) - Ground Connection 1, in engine compartment wiring harness
- (200) - Ground Connection (shielding), in engine compartment wiring harness
- (D22) - Connection (over fuse 234), in wiring harness front, right
- (D101) - Connection 1, in engine compartment wiring harness

ws = white
sw = black
ro = red
br = brown
gn = green
bl = blue
gr = grey
li = violet
ge = yellow
or = orange

97-52061

BentleyPublishers.com—All Rights Reserved

Engine Management, 2.7 Liter, Engine Codes APB, BEL (2003)

Audi A6 — Wiring Diagram — No. 2/13

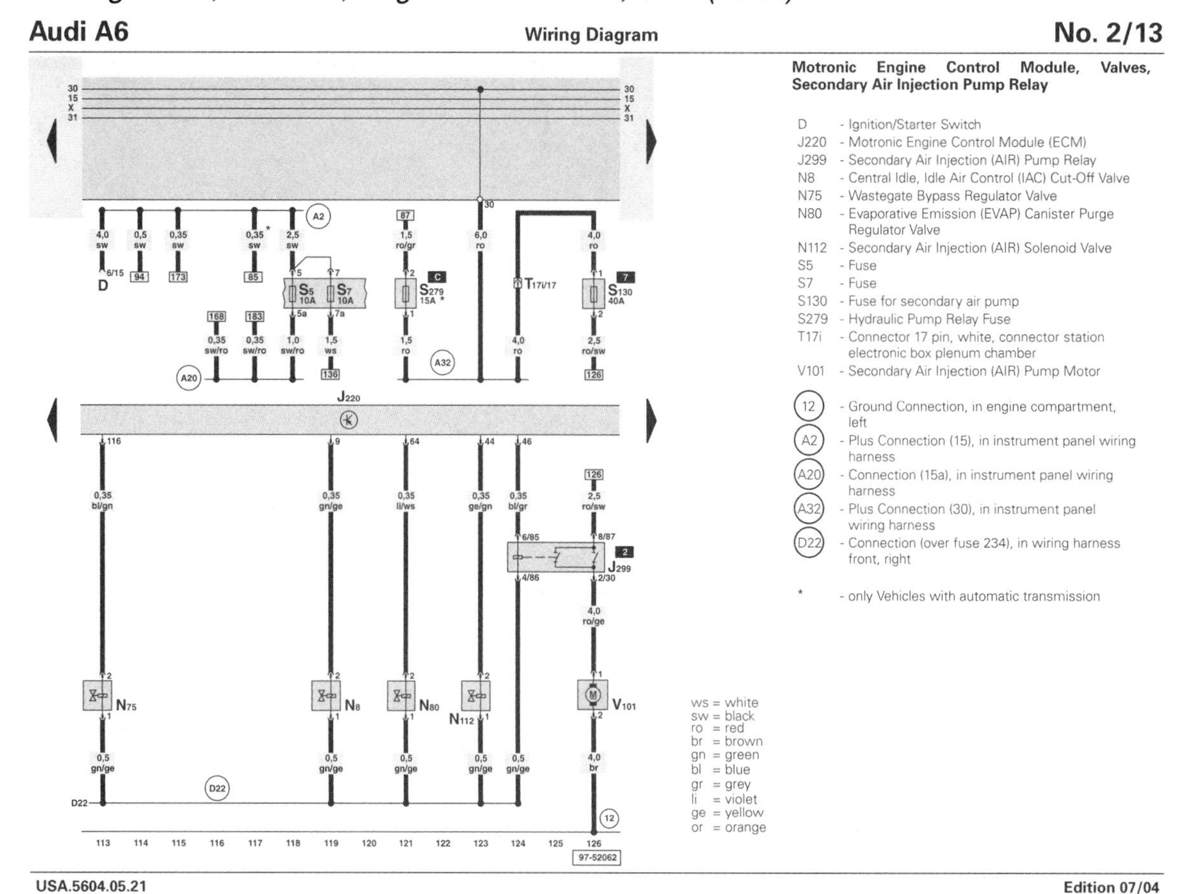

Motronic Engine Control Module, Valves, Secondary Air Injection Pump Relay

- D - Ignition/Starter Switch
- J220 - Motronic Engine Control Module (ECM)
- J299 - Secondary Air Injection (AIR) Pump Relay
- N8 - Central Idle, Idle Air Control (IAC) Cut-Off Valve
- N75 - Wastegate Bypass Regulator Valve
- N80 - Evaporative Emission (EVAP) Canister Purge Regulator Valve
- N112 - Secondary Air Injection (AIR) Solenoid Valve
- S5 - Fuse
- S7 - Fuse
- S130 - Fuse for secondary air pump
- S279 - Hydraulic Pump Relay Fuse
- T17i - Connector 17 pin, white, connector station electronic box plenum chamber
- V101 - Secondary Air Injection (AIR) Pump Motor

- (12) - Ground Connection, in engine compartment, left
- (A2) - Plus Connection (15), in instrument panel wiring harness
- (A20) - Connection (15a), in instrument panel wiring harness
- (A32) - Plus Connection (30), in instrument panel wiring harness
- (D22) - Connection (over fuse 234), in wiring harness front, right

* - only Vehicles with automatic transmission

ws = white
sw = black
ro = red
br = brown
gn = green
bl = blue
gr = grey
li = violet
ge = yellow
or = orange

USA.5604.05.21 — Edition 07/04

Audi A6 — Wiring Diagram — No. 2/14

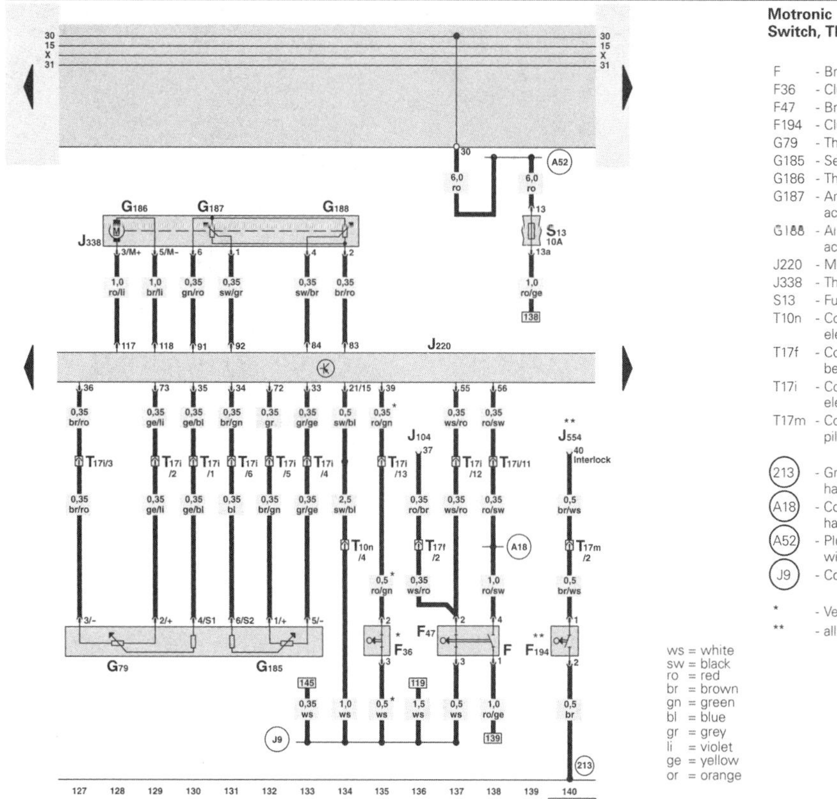

Motronic Engine Control Module, Brake Light Switch, Throttle Valve Control Module

- F - Brake Light Switch
- F36 - Clutch Pedal Switch
- F47 - Brake Pedal Switch
- F194 - Clutch Pedal Starter Interlock Switch
- G79 - Throttle Position (TP) Sensor
- G185 - Sensor 2 for accelerator pedal position
- G186 - Throttle drive (power accelerator actuation)
- G187 - Angle sensor 1 for throttle drive (power accelerator actuation)
- G188 - Angle sensor 2 for throttle drive (power accelerator actuation)
- J220 - Motronic Engine Control Module (ECM)
- J338 - Throttle Valve Control Module
- S13 - Fuse
- T10n - Connector 10 pin, orange, connector station electronic box plenum chamber
- T17f - Connector 17 pin, black, in 8-Fold Relay Panel behind the storage driver's side
- T17i - Connector 17 pin, white, connector station electronic box plenum chamber
- T17m - Connector 17 pin, blue, connector station A pillar, right

- (213) - Ground Connection, in instrument panel wiring harness
- (A18) - Connection (54), in instrument panel wiring harness
- (A52) - Plus Connection 2 (30), in instrument panel wiring harness
- (J9) - Connection 1 (15a), in ABS wiring harness

* - Vehicles with manual transmission
** - allroad with manual transmission 01E

ws = white
sw = black
ro = red
br = brown
gn = green
bl = blue
gr = grey
li = violet
ge = yellow
or = orange

USA.5604.05.21 — Edition 07/04

BentleyPublishers.com—All Rights Reserved

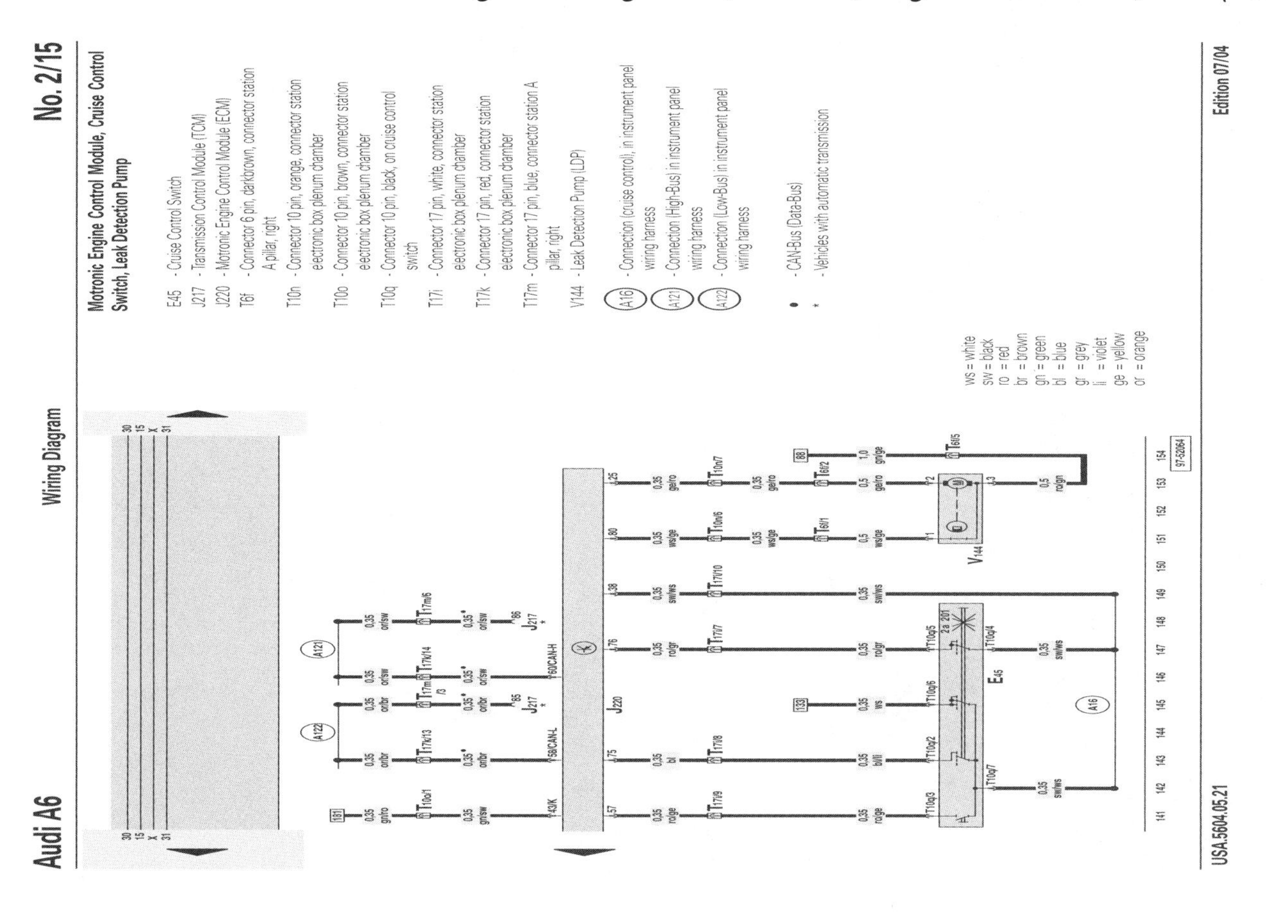

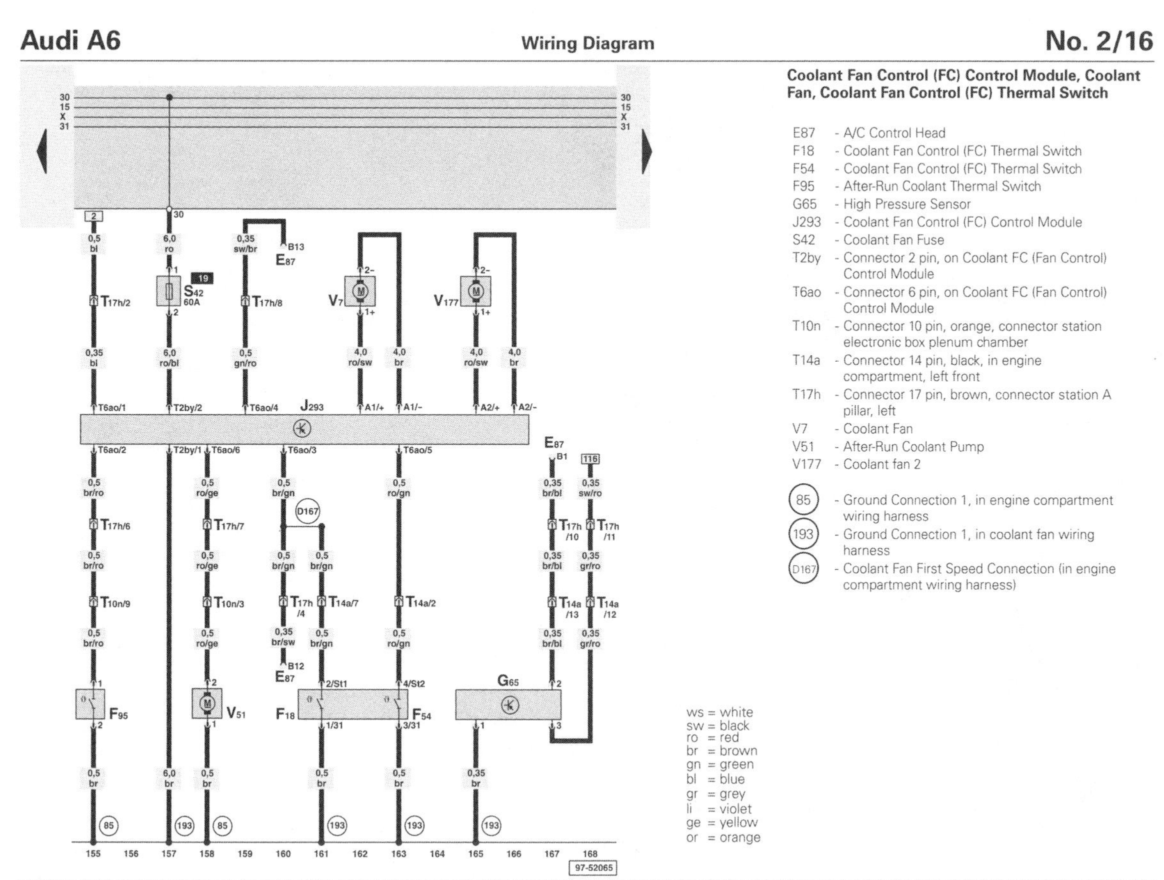

Audi A6
Wiring Diagram
No. 2/16
Coolant Fan Control (FC) Control Module, Coolant Fan, Coolant Fan Control (FC) Thermal Switch
E87 - A/C Control Head
F18 - Coolant Fan Control (FC) Thermal Switch
F54 - Coolant Fan Control (FC) Thermal Switch
F95 - After-Run Coolant Thermal Switch
G65 - High Pressure Sensor
J293 - Coolant Fan Control (FC) Control Module
S42 - Coolant Fan Fuse
T2by - Connector 2 pin, on Coolant FC (Fan Control) Control Module
T6ao - Connector 6 pin, on Coolant FC (Fan Control) Control Module
T10n - Connector 10 pin, orange, connector station electronic box plenum chamber
T14a - Connector 14 pin, black, in engine compartment, left front
T17h - Connector 17 pin, brown, connector station A pillar, left
V7 - Coolant Fan
V51 - After-Run Coolant Pump
V177 - Coolant fan 2
85 - Ground Connection 1, in engine compartment wiring harness
193 - Ground Connection 1, in coolant fan wiring harness
D167 - Coolant Fan First Speed Connection (in engine compartment wiring harness)
ws = white
sw = black
ro = red
br = brown
gn = green
bl = blue
gr = grey
li = violet
ge = yellow
or = orange
97-52065

BentleyPublishers.com—All Rights Reserved

Engine Management, 2.7 Liter, Engine Codes APB, BEL (2003)

Audi A6 — Wiring Diagram — **No. 2/17**

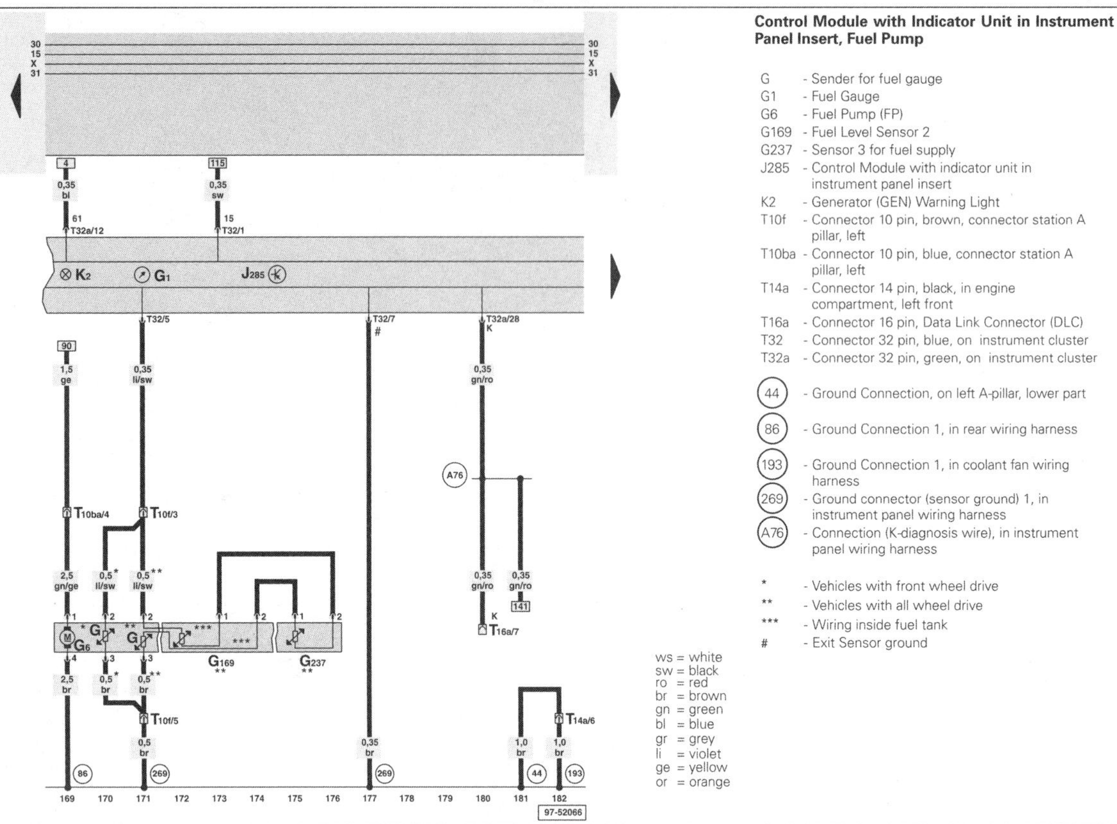

Control Module with Indicator Unit in Instrument Panel Insert, Fuel Pump

- G - Sender for fuel gauge
- G1 - Fuel Gauge
- G6 - Fuel Pump (FP)
- G169 - Fuel Level Sensor 2
- G237 - Sensor 3 for fuel supply
- J285 - Control Module with indicator unit in instrument panel insert
- K2 - Generator (GEN) Warning Light
- T10f - Connector 10 pin, brown, connector station A pillar, left
- T10ba - Connector 10 pin, blue, connector station A pillar, left
- T14a - Connector 14 pin, black, in engine compartment, left front
- T16a - Connector 16 pin, Data Link Connector (DLC)
- T32 - Connector 32 pin, blue, on instrument cluster
- T32a - Connector 32 pin, green, on instrument cluster
- (44) - Ground Connection, on left A-pillar, lower part
- (86) - Ground Connection 1, in rear wiring harness
- (193) - Ground Connection 1, in coolant fan wiring harness
- (269) - Ground connector (sensor ground) 1, in instrument panel wiring harness
- (A76) - Connection (K-diagnosis wire), in instrument panel wiring harness
- * - Vehicles with front wheel drive
- ** - Vehicles with all wheel drive
- *** - Wiring inside fuel tank
- # - Exit Sensor ground

ws = white
sw = black
ro = red
br = brown
gn = green
bl = blue
gr = grey
li = violet
ge = yellow
or = orange

USA.5604.05.21 — **Edition 07/04**

Audi A6 — Wiring Diagram — **No. 2/18**

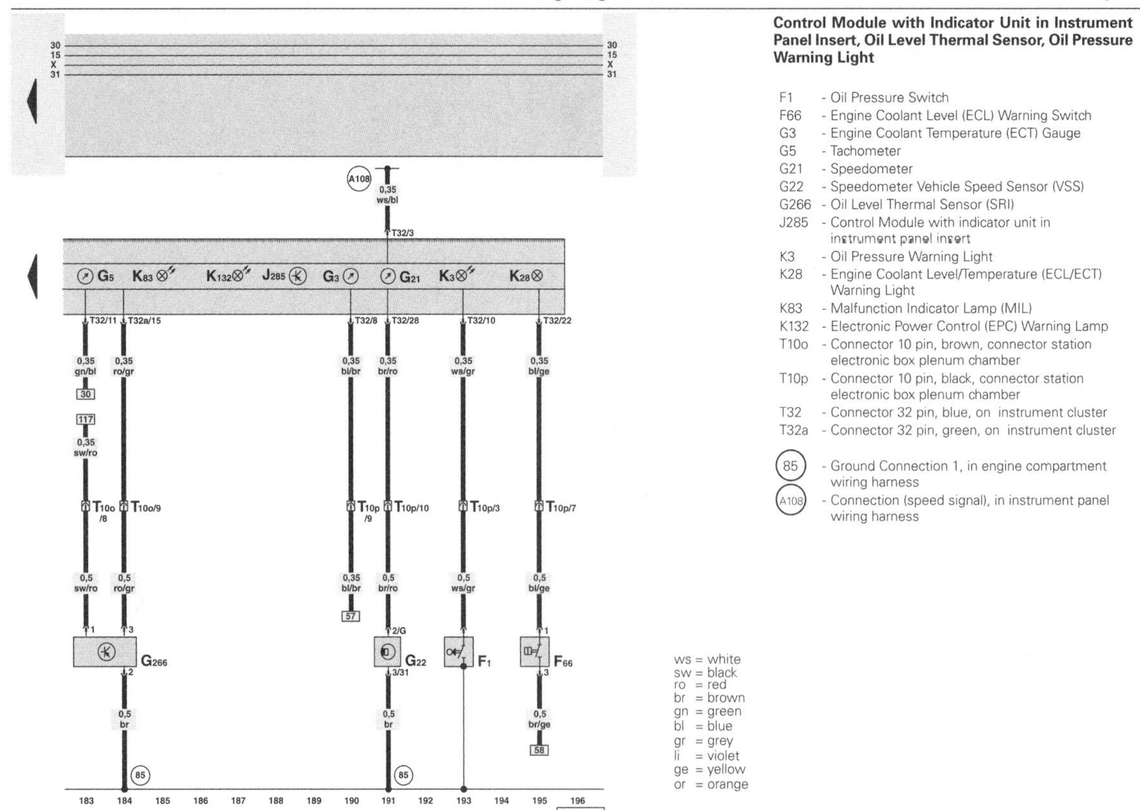

Control Module with Indicator Unit in Instrument Panel Insert, Oil Level Thermal Sensor, Oil Pressure Warning Light

- F1 - Oil Pressure Switch
- F66 - Engine Coolant Level (ECL) Warning Switch
- G3 - Engine Coolant Temperature (ECT) Gauge
- G5 - Tachometer
- G21 - Speedometer
- G22 - Speedometer Vehicle Speed Sensor (VSS)
- G266 - Oil Level Thermal Sensor (SRI)
- J285 - Control Module with indicator unit in instrument panel insert
- K3 - Oil Pressure Warning Light
- K28 - Engine Coolant Level/Temperature (ECL/ECT) Warning Light
- K83 - Malfunction Indicator Lamp (MIL)
- K132 - Electronic Power Control (EPC) Warning Lamp
- T10o - Connector 10 pin, brown, connector station electronic box plenum chamber
- T10p - Connector 10 pin, black, connector station electronic box plenum chamber
- T32 - Connector 32 pin, blue, on instrument cluster
- T32a - Connector 32 pin, green, on instrument cluster
- (85) - Ground Connection 1, in engine compartment wiring harness
- (A108) - Connection (speed signal), in instrument panel wiring harness

ws = white
sw = black
ro = red
br = brown
gn = green
bl = blue
gr = grey
li = violet
ge = yellow
or = orange

USA.5604.05.21 — **Edition 07/04**

BentleyPublishers.com—All Rights Reserved

(4,2 l - Injection Engine, 8 Cylinder), Code BAS - allroad

2003 m. y.

Fuse Panel

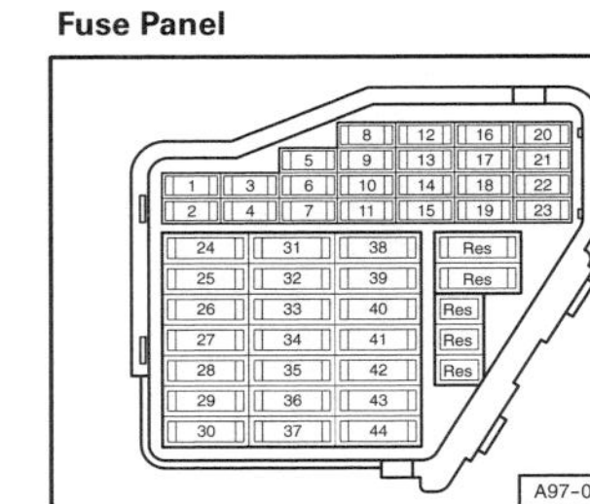

Fuse Colors:

30 A - Green
25 A - White
20 A - Yellow
15 A - Blue
10 A - Red
7,5 A - Brown
5 A - Beige

Starting with fuse position 23, fuses in the fuse holder are identified with 223 in the wiring diagram.

Micro Central Electric Panel

Relay Location:

- **4** -Fuel Pump (FP) Relay, J17
- **5** -Wiper/Washer Intermittent Relay, J31
- **6** -Wiper/Washer Intermittent Relay, J31

13 - Fold Relay Panel

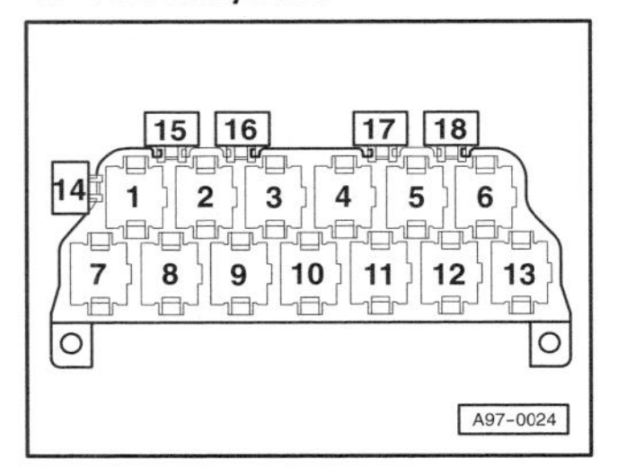

Relay Location:

- **4** -Starting Interlock Relay, J207

8 - Fold Relay Panel

Relay Location:

- **19** -Coolant Fan Fuse, S42
- **21** -Fuse for Fan for coolant 1st and 2nd stage, S146
- **22** -Second Speed Coolant Fan Fuse, S104

Relay Panel behind Instrument panel left, on central girder

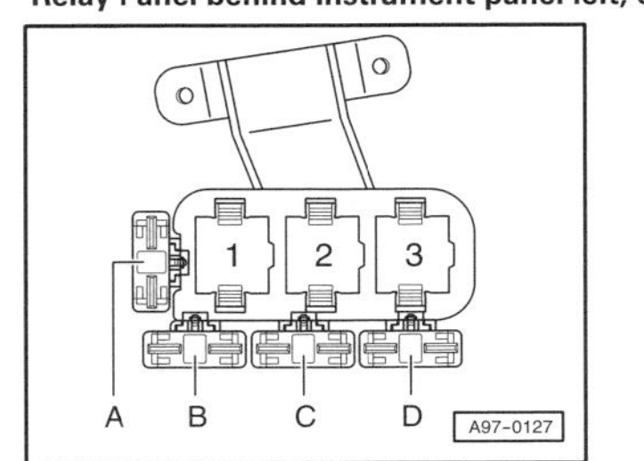

Relay Location:

- **1** -Brake Booster Relay, J569
- **C** -Fuse for Hydraulic Pump Relay, S279

3 - Fold Relay Panel in E-Box plenum chamber

Relay Location:

- **1a** -Coolant Fan Control (FC) Relay, J26
- **1b** -Coolant Circulation Pump Relay, J151
- **2** -Secondary Air Injection (AIR) Pump Relay, J299
- **3** -Motronic Engine Control Module (ECM) Power Supply Relay, J271
- **5** -Engine Electronics Fuse, S282
- **7** -Fuse for secondary air pump, S130
- **9** -Third Speed Coolant Fan Fuse, S94
- **10** -Control module fuse for coolant fan, S142

BentleyPublishers.com—All Rights Reserved

Engine Management, 4.2 Liter, Engine Code BAS (allroad quattro 2003)

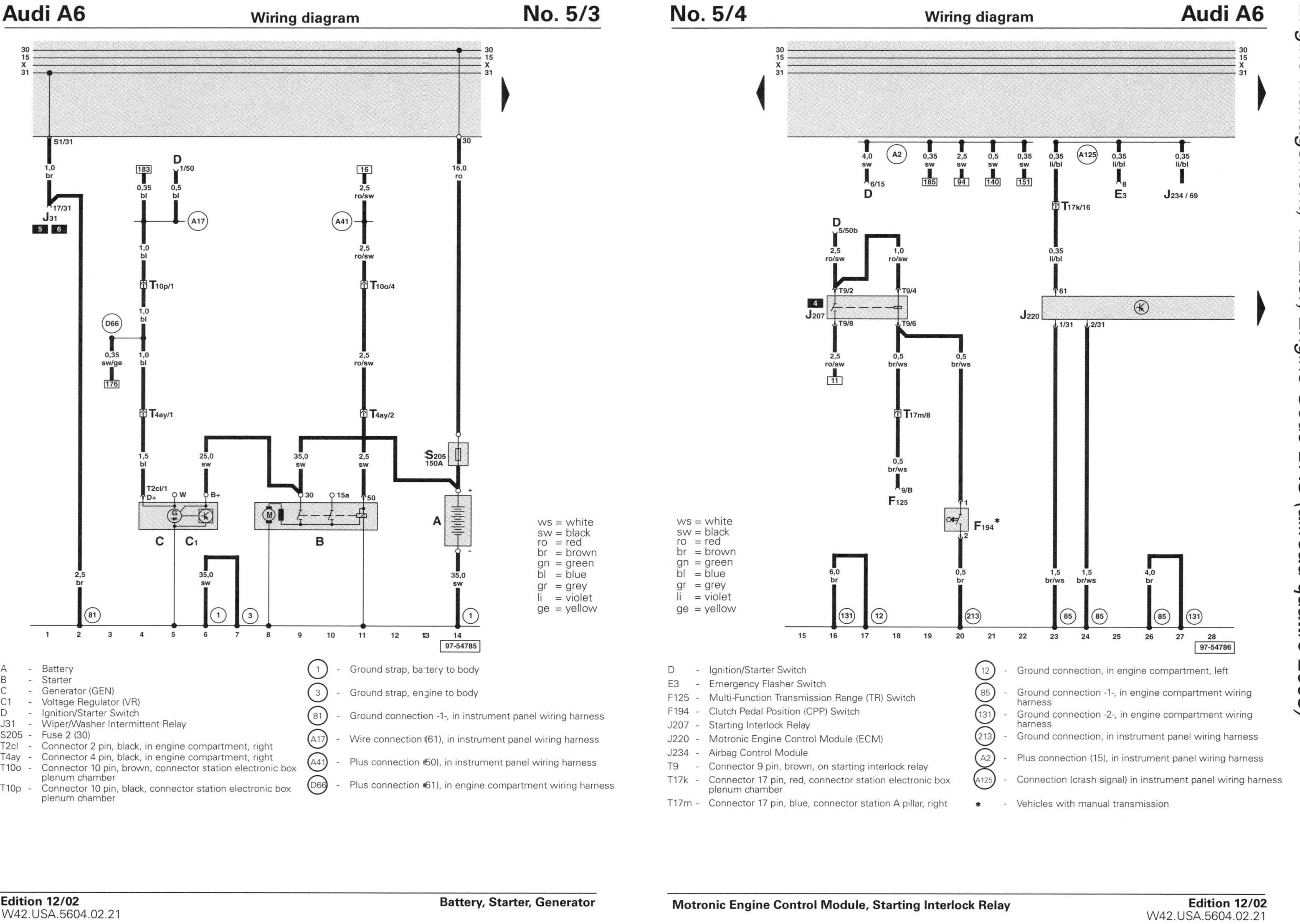

BentleyPublishers.com—All Rights Reserved

Engine Management, 4.2 Liter, Engine Code BAS (allroad quattro 2003)

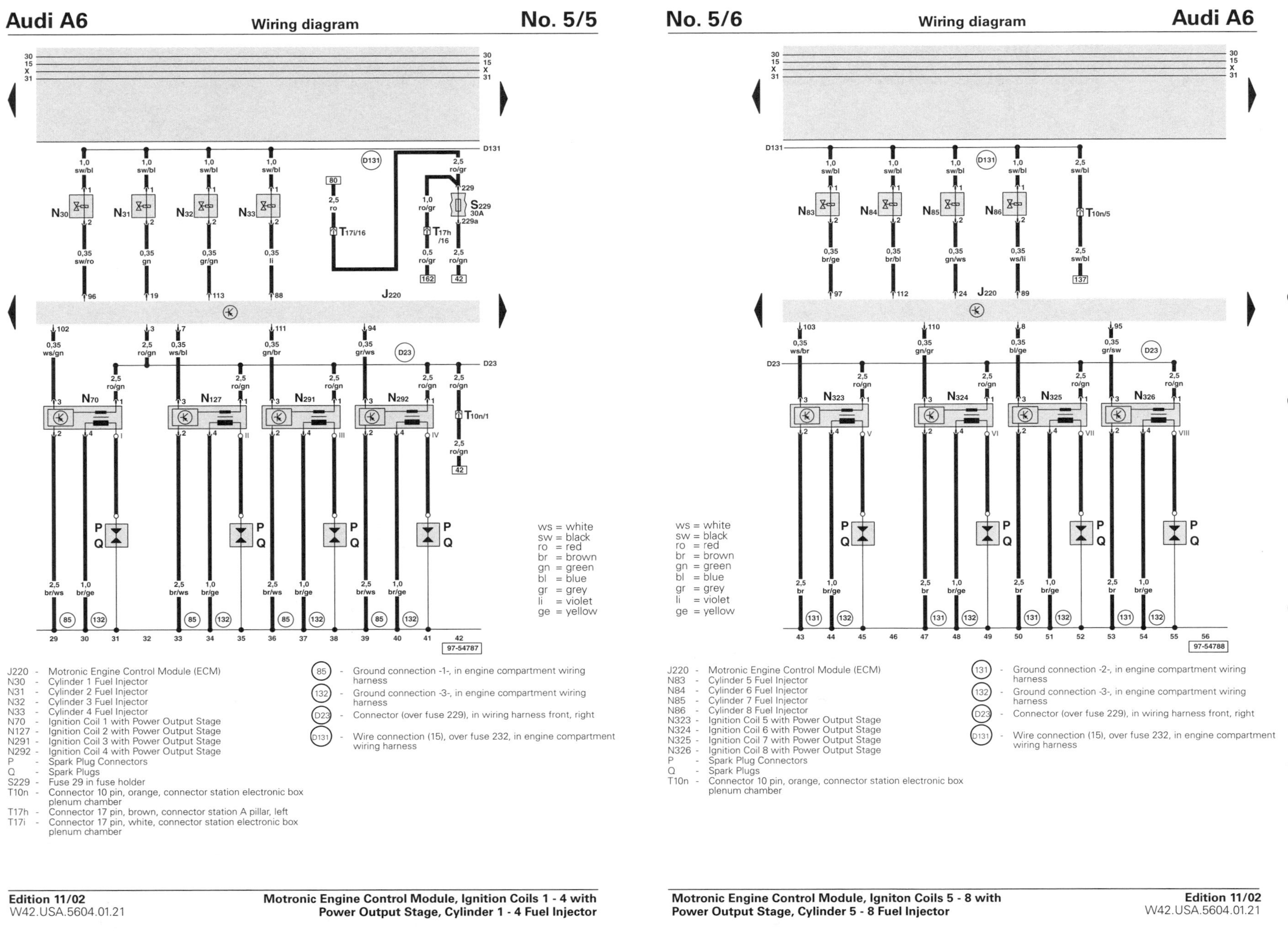

BentleyPublishers.com—All Rights Reserved

Engine Management, 4.2 Liter, Engine Code BAS (allroad quattro 2003)

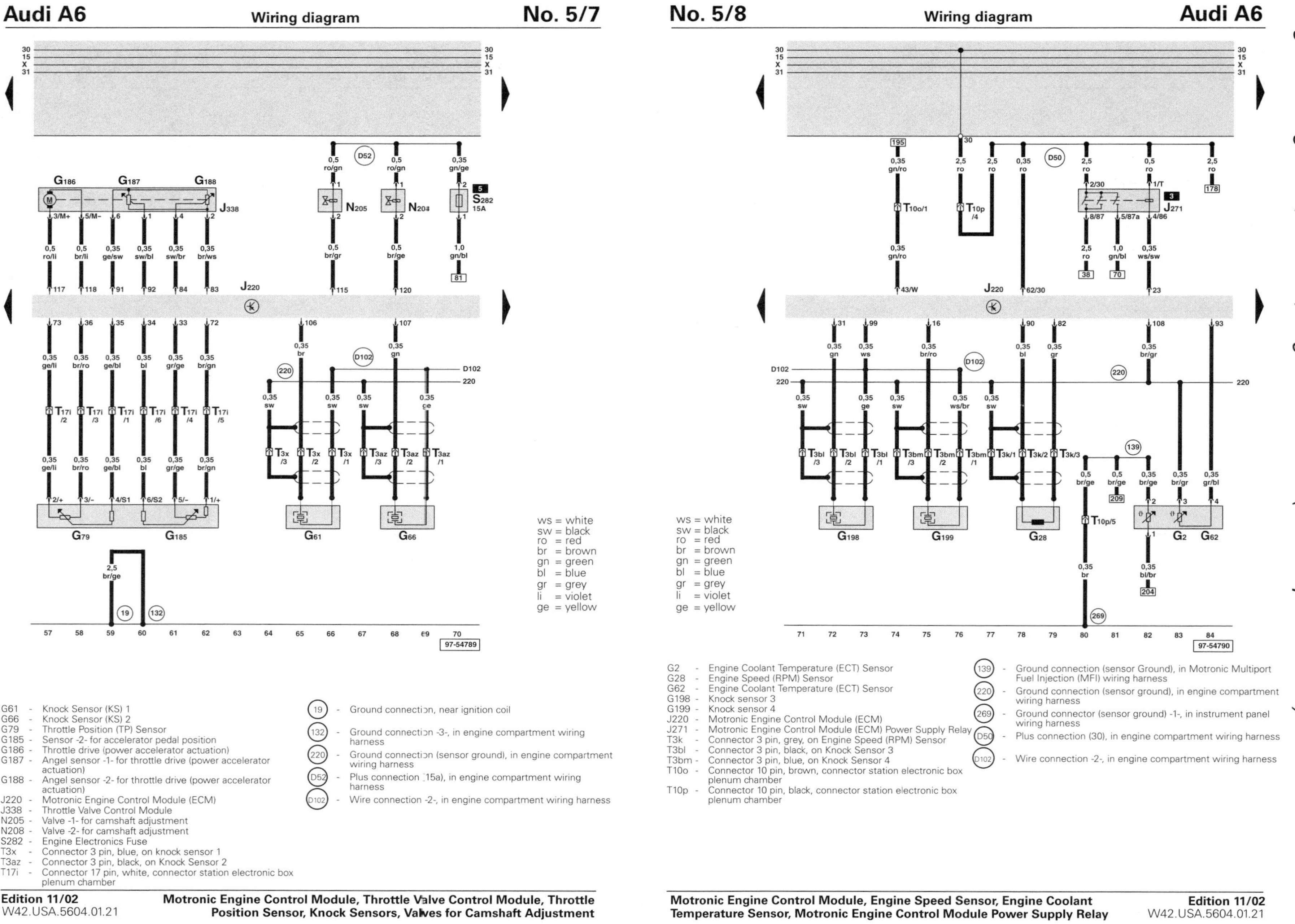

BentleyPublishers.com—All Rights Reserved

Engine Management, 4.2 Liter, Engine Code BAS (allroad quattro 2003)

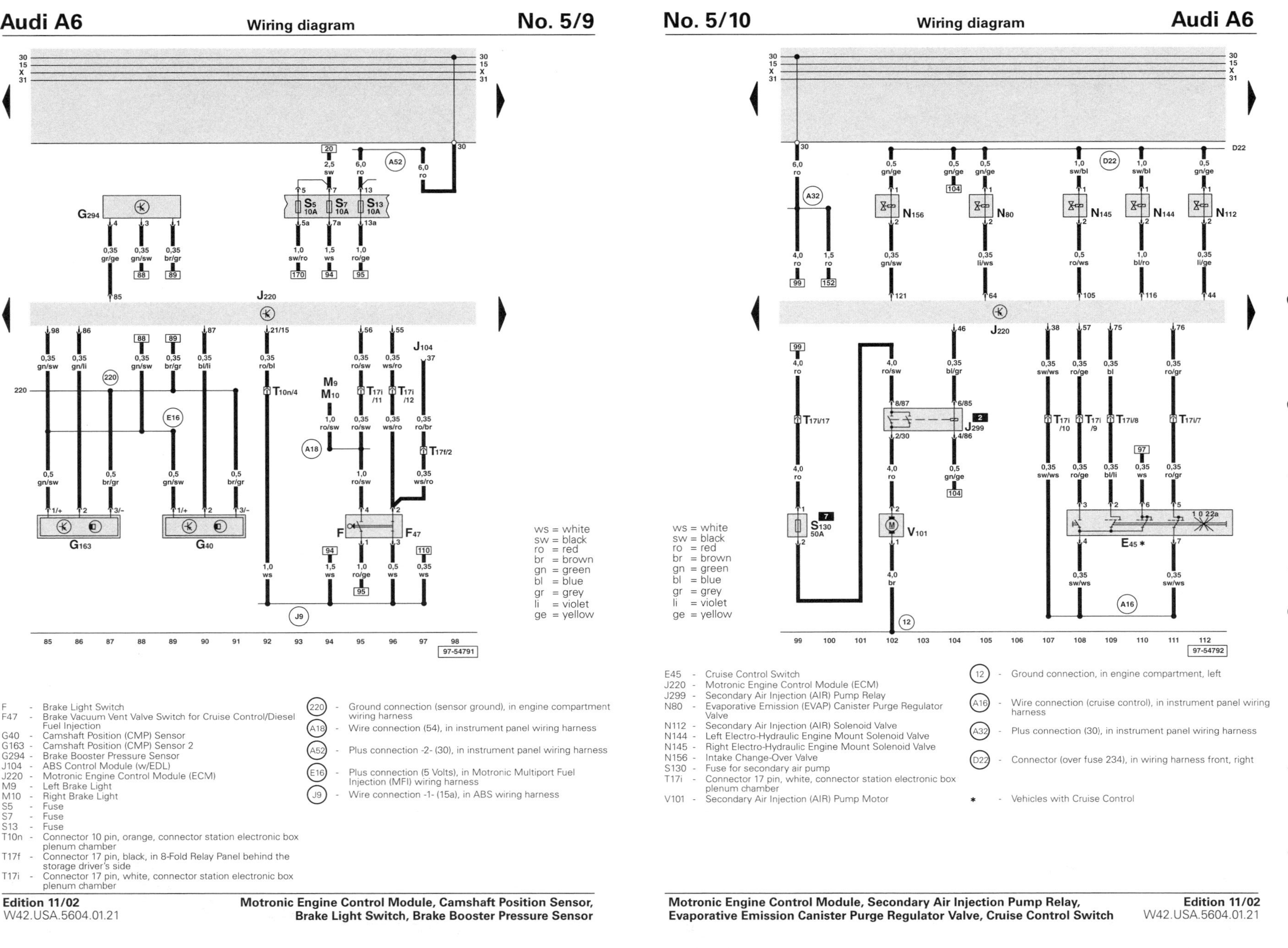

BentleyPublishers.com—All Rights Reserved

Engine Management, 4.2 Liter, Engine Code BAS (allroad quattro 2003)

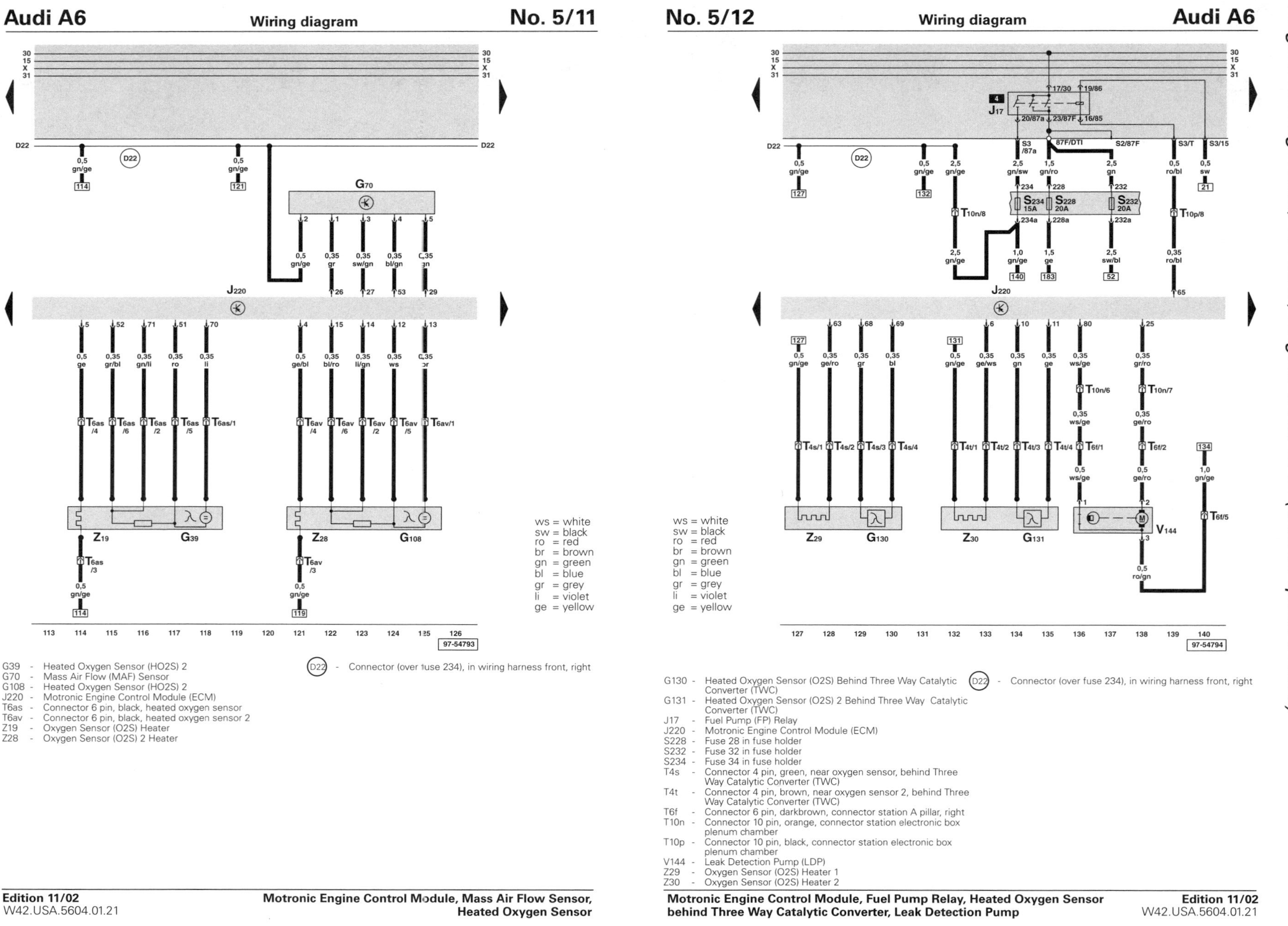

BentleyPublishers.com—All Rights Reserved

Engine Management, 4.2 Liter, Engine Code BAS (allroad quattro 2003)

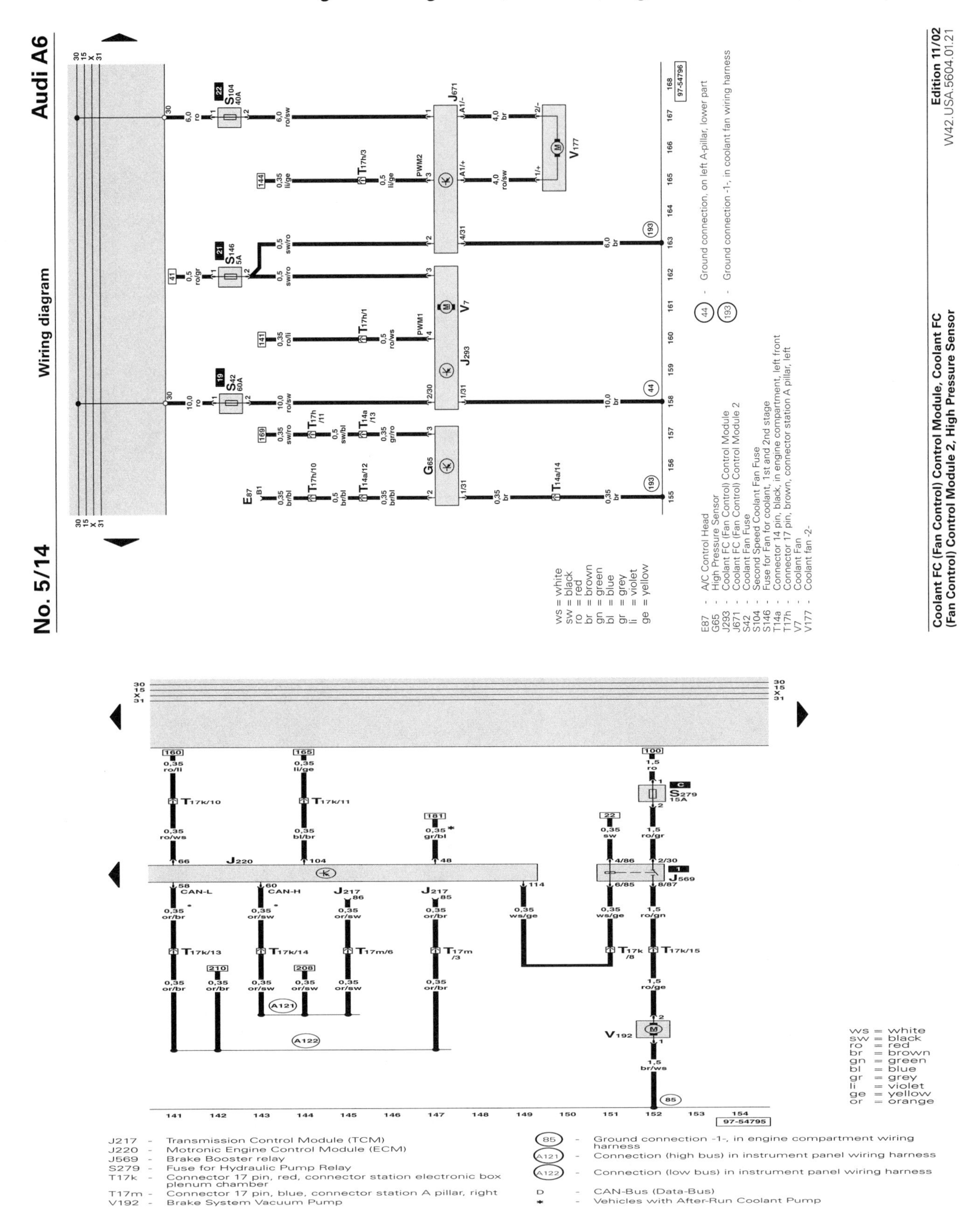

BentleyPublishers.com—All Rights Reserved

Audi A6 — Wiring diagram — No. 5/15

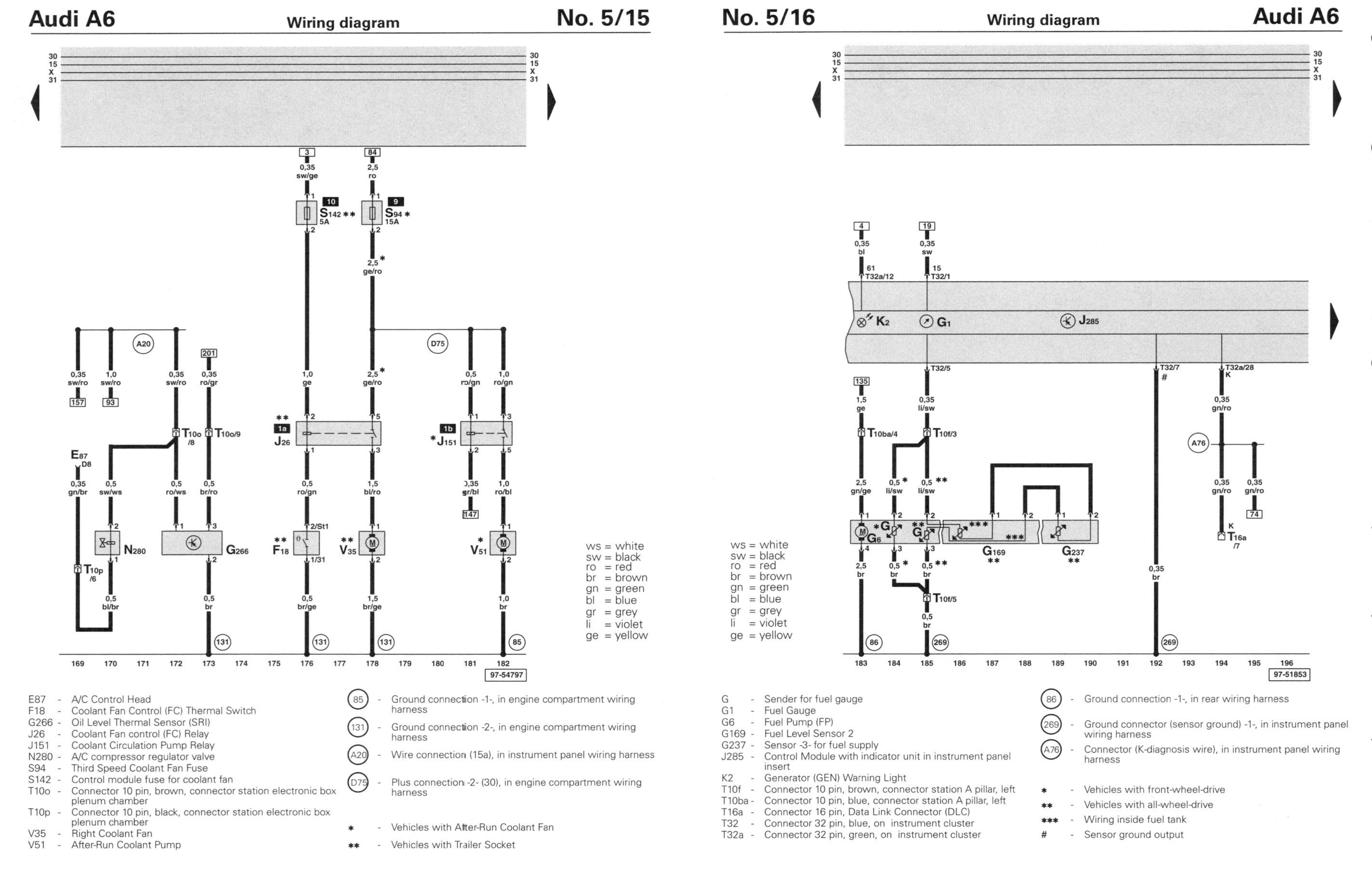

ws = white
sw = black
ro = red
br = brown
gn = green
bl = blue
gr = grey
li = violet
ge = yellow

E87 - A/C Control Head
F18 - Coolant Fan Control (FC) Thermal Switch
G266 - Oil Level Thermal Sensor (SRI)
J26 - Coolant Fan control (FC) Relay
J151 - Coolant Circulation Pump Relay
N280 - A/C compressor regulator valve
S94 - Third Speed Coolant Fan Fuse
S142 - Control module fuse for coolant fan
T10o - Connector 10 pin, brown, connector station electronic box plenum chamber
T10p - Connector 10 pin, black, connector station electronic box plenum chamber
V35 - Right Coolant Fan
V51 - After-Run Coolant Pump

(85) - Ground connection -1-, in engine compartment wiring harness
(131) - Ground connection -2-, in engine compartment wiring harness
(A20) - Wire connection (15a), in instrument panel wiring harness
(D75) - Plus connection -2- (30), in engine compartment wiring harness

* - Vehicles with After-Run Coolant Fan
** - Vehicles with Trailer Socket

Edition 11/02
W42.USA.5604.01.21

Coolant Fan Control (FC) Relay, Coolant Circulation Pump Relay, Right Coolant Fan, After-Run Coolant Pump, Oil Level Thermal Sensor

No. 5/16 — Wiring diagram — Audi A6

ws = white
sw = black
ro = red
br = brown
gn = green
bl = blue
gr = grey
li = violet
ge = yellow

G - Sender for fuel gauge
G1 - Fuel Gauge
G6 - Fuel Pump (FP)
G169 - Fuel Level Sensor 2
G237 - Sensor -3- for fuel supply
J285 - Control Module with indicator unit in instrument panel insert
K2 - Generator (GEN) Warning Light
T10f - Connector 10 pin, brown, connector station A pillar, left
T10ba - Connector 10 pin, blue, connector station A pillar, left
T16a - Connector 16 pin, Data Link Connector (DLC)
T32 - Connector 32 pin, blue, on instrument cluster
T32a - Connector 32 pin, green, on instrument cluster

(86) - Ground connection -1-, in rear wiring harness
(269) - Ground connector (sensor ground) -1-, in instrument panel wiring harness
(A76) - Connector (K-diagnosis wire), in instrument panel wiring harness

* - Vehicles with front-wheel-drive
** - Vehicles with all-wheel-drive
*** - Wiring inside fuel tank
\# - Sensor ground output

Instrument Panel, Fuel Pump, Fuel Gauge

Edition 11/02
W42.USA.5604.01.21

BentleyPublishers.com—All Rights Reserved

Engine Management, 4.2 Liter, Engine Code BAS (allroad quattro 2003)

THIS SPACE INTENTIONALLY LEFT BLANK

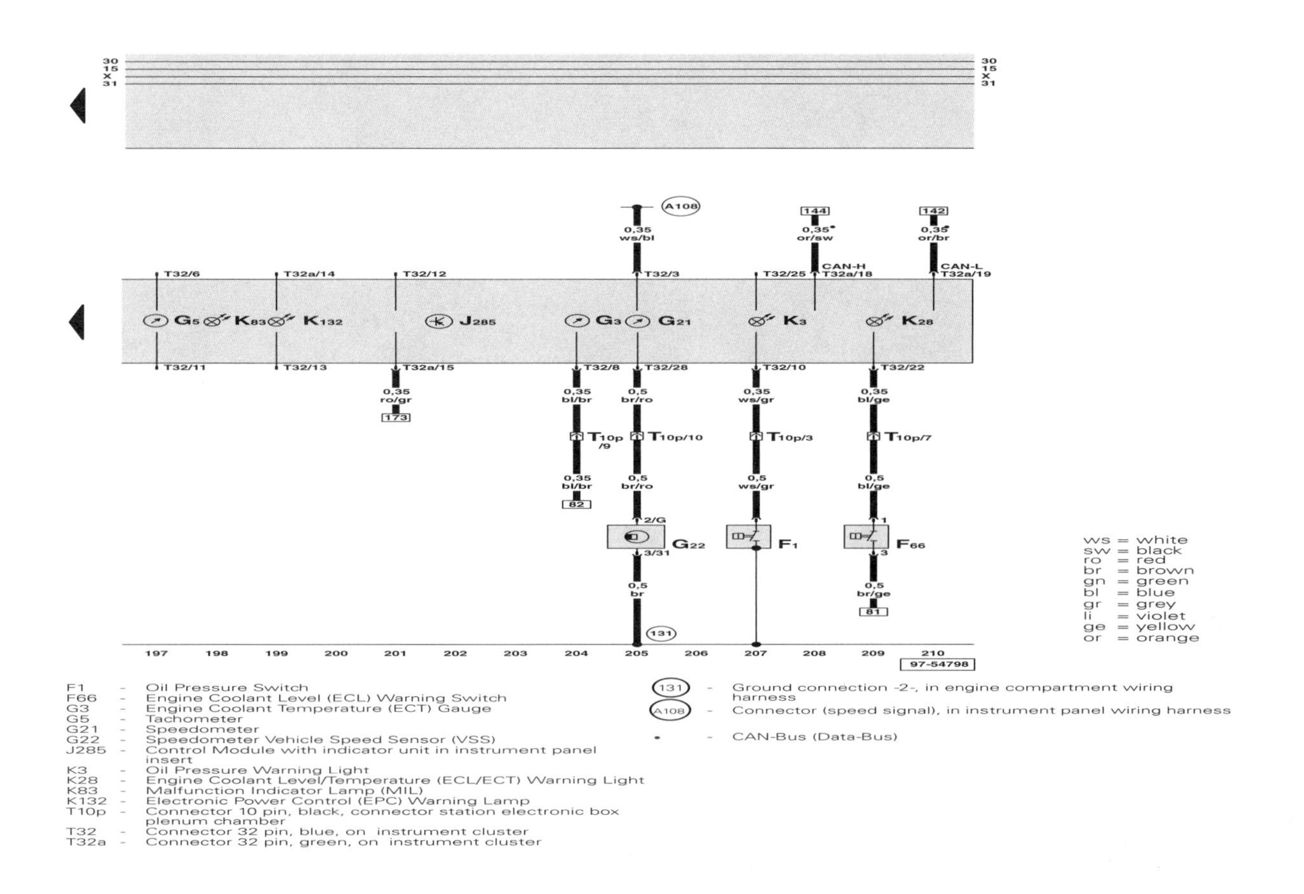

BentleyPublishers.com—All Rights Reserved

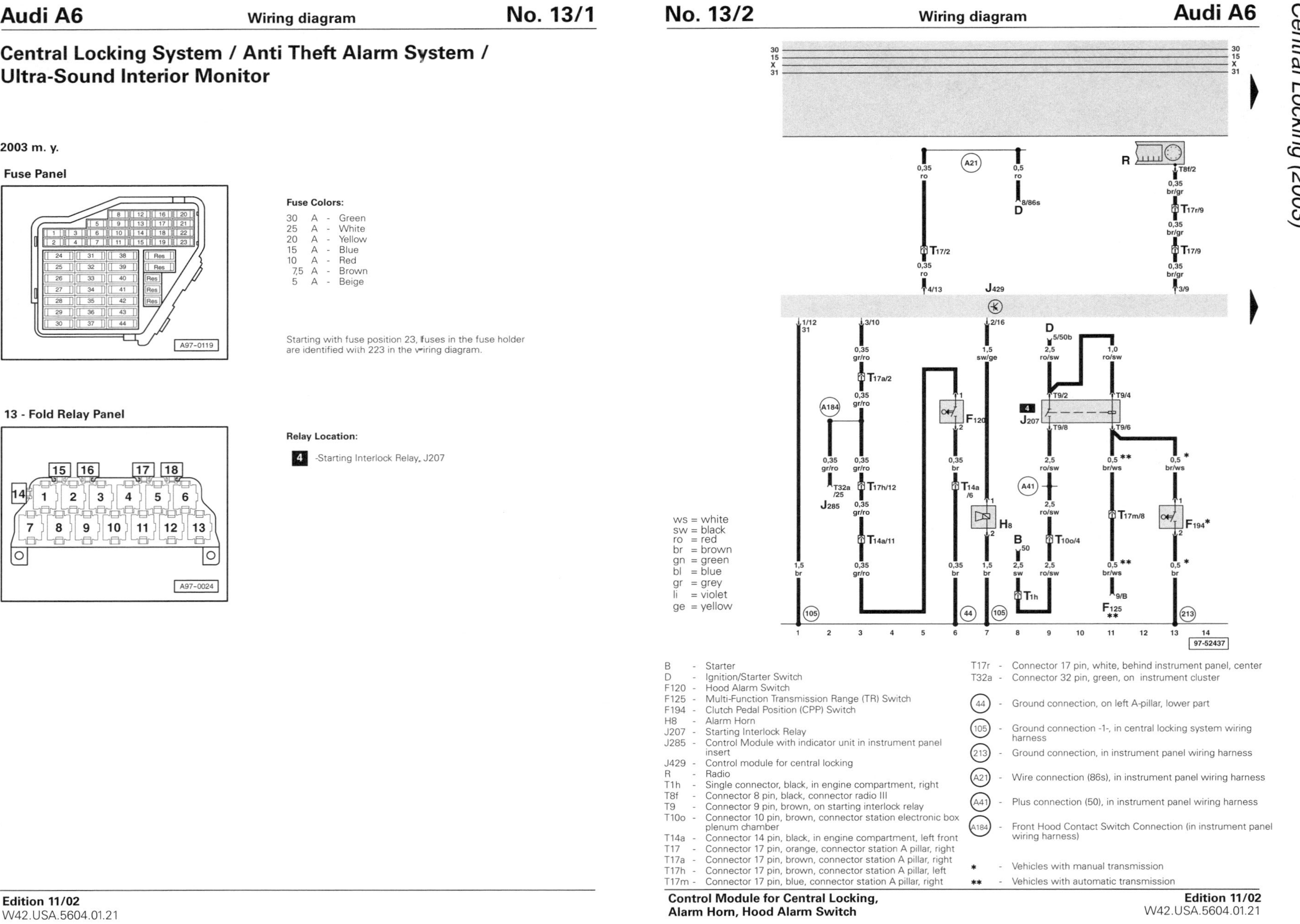

BentleyPublishers.com—All Rights Reserved

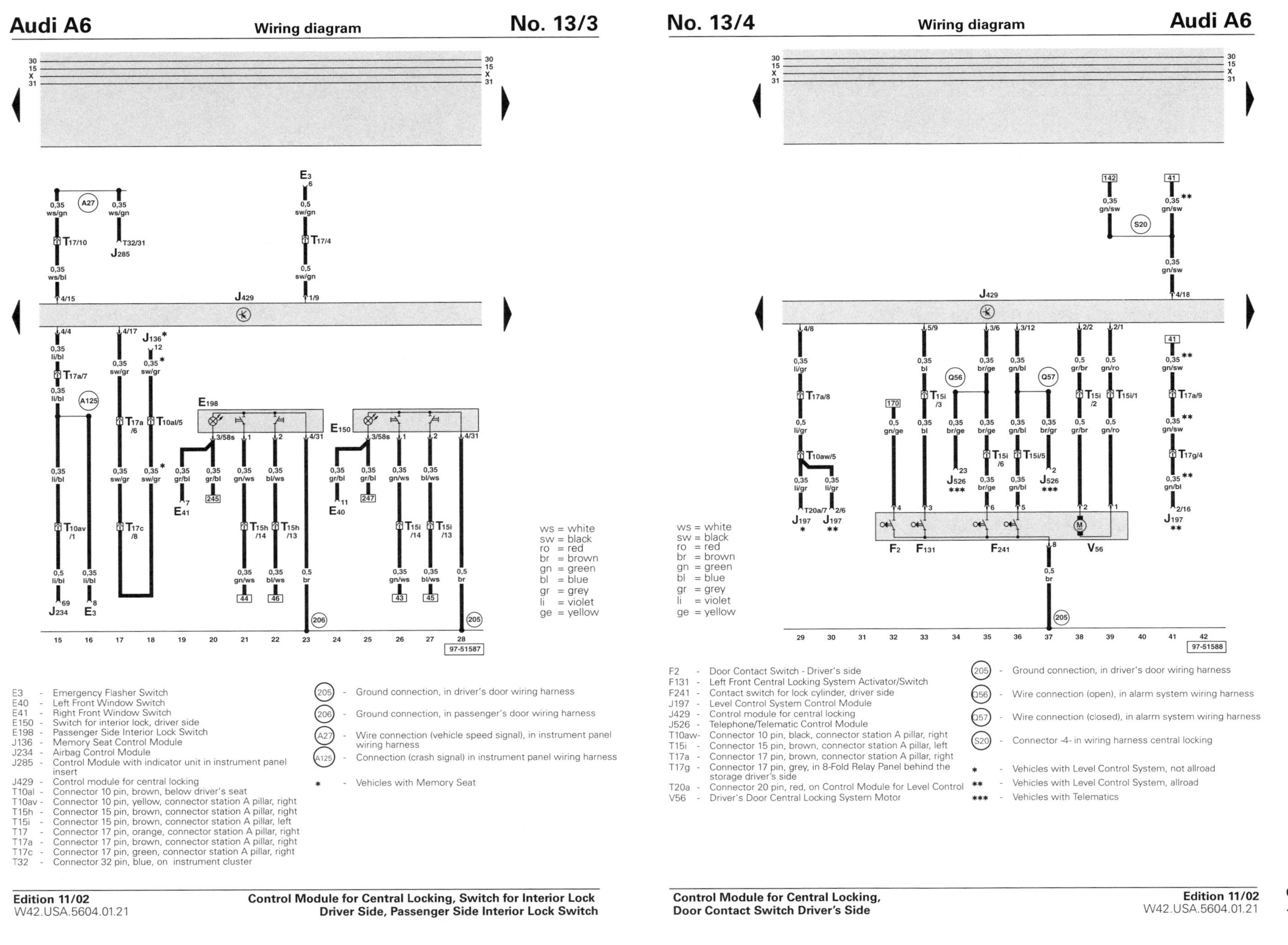

BentleyPublishers.com—All Rights Reserved

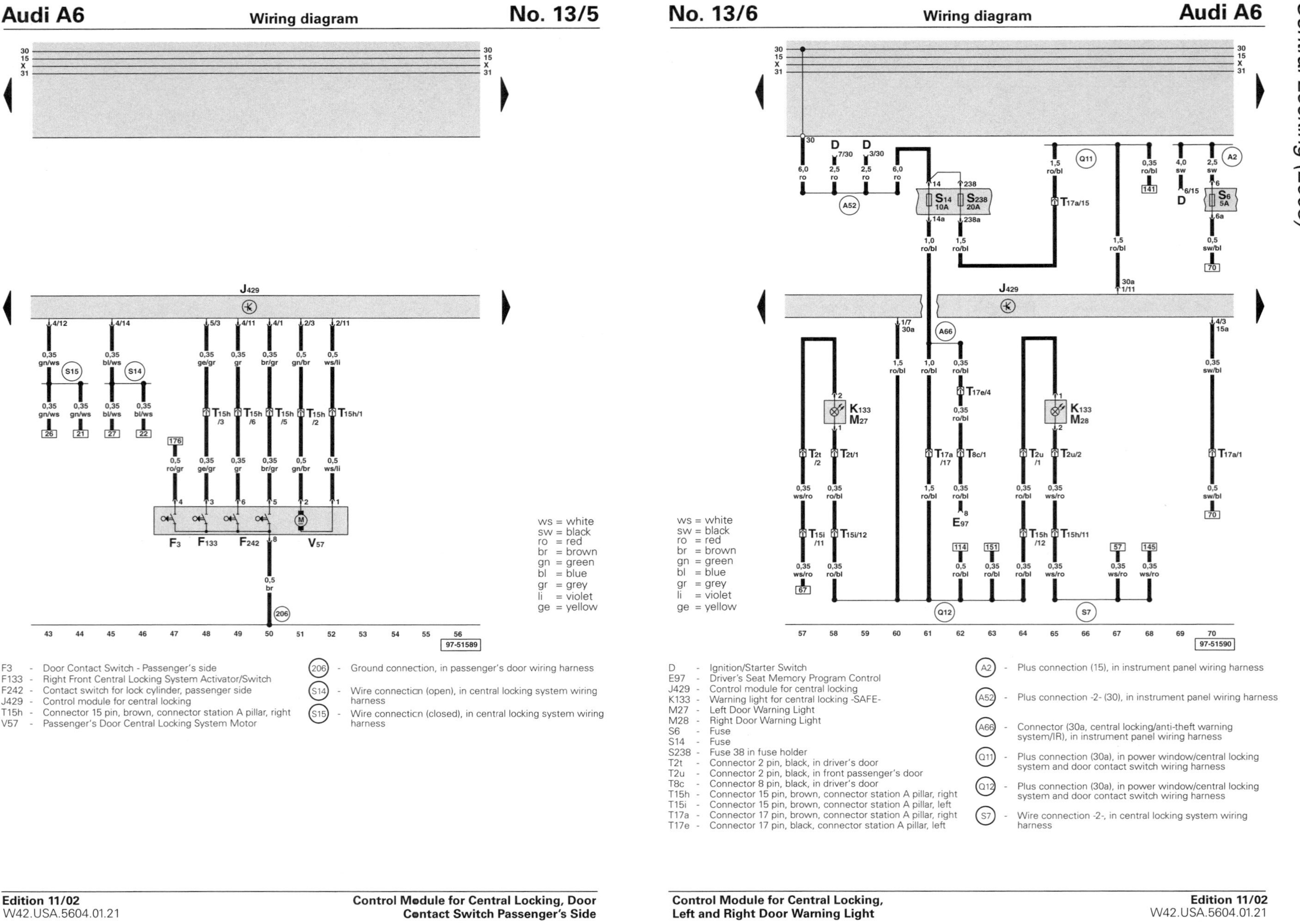

BentleyPublishers.com—All Rights Reserved

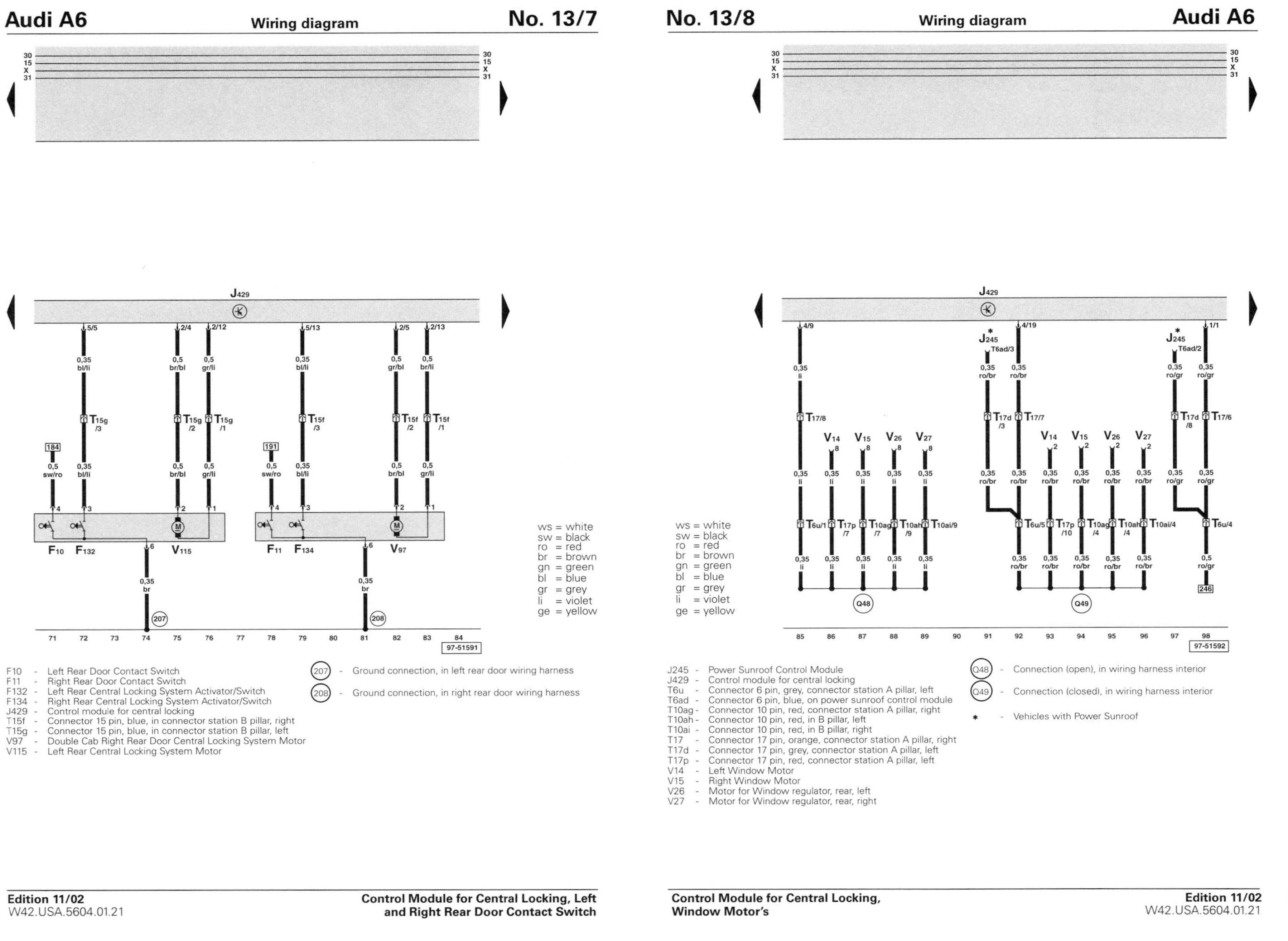

BentleyPublishers.com—All Rights Reserved

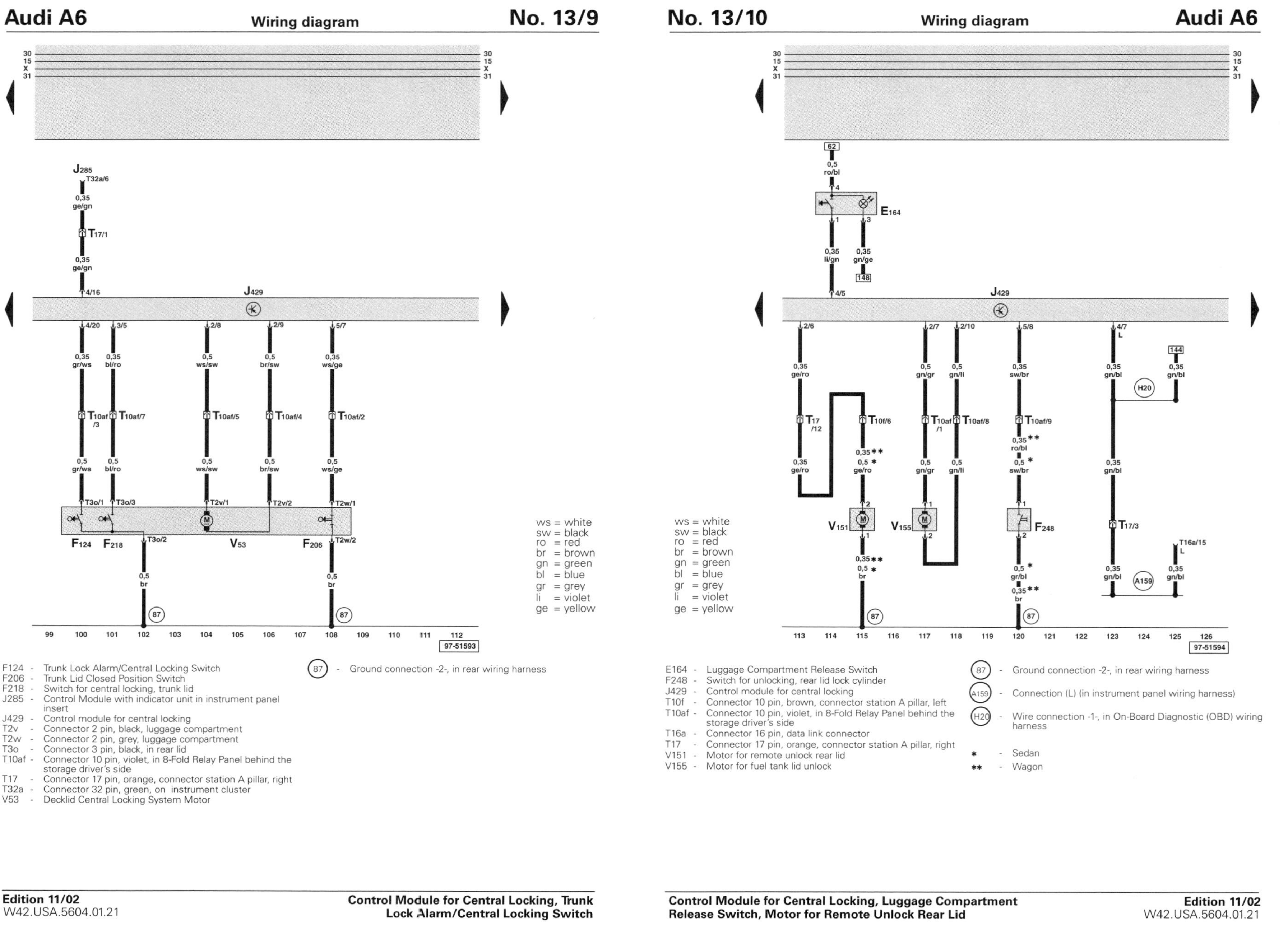

BentleyPublishers.com—All Rights Reserved

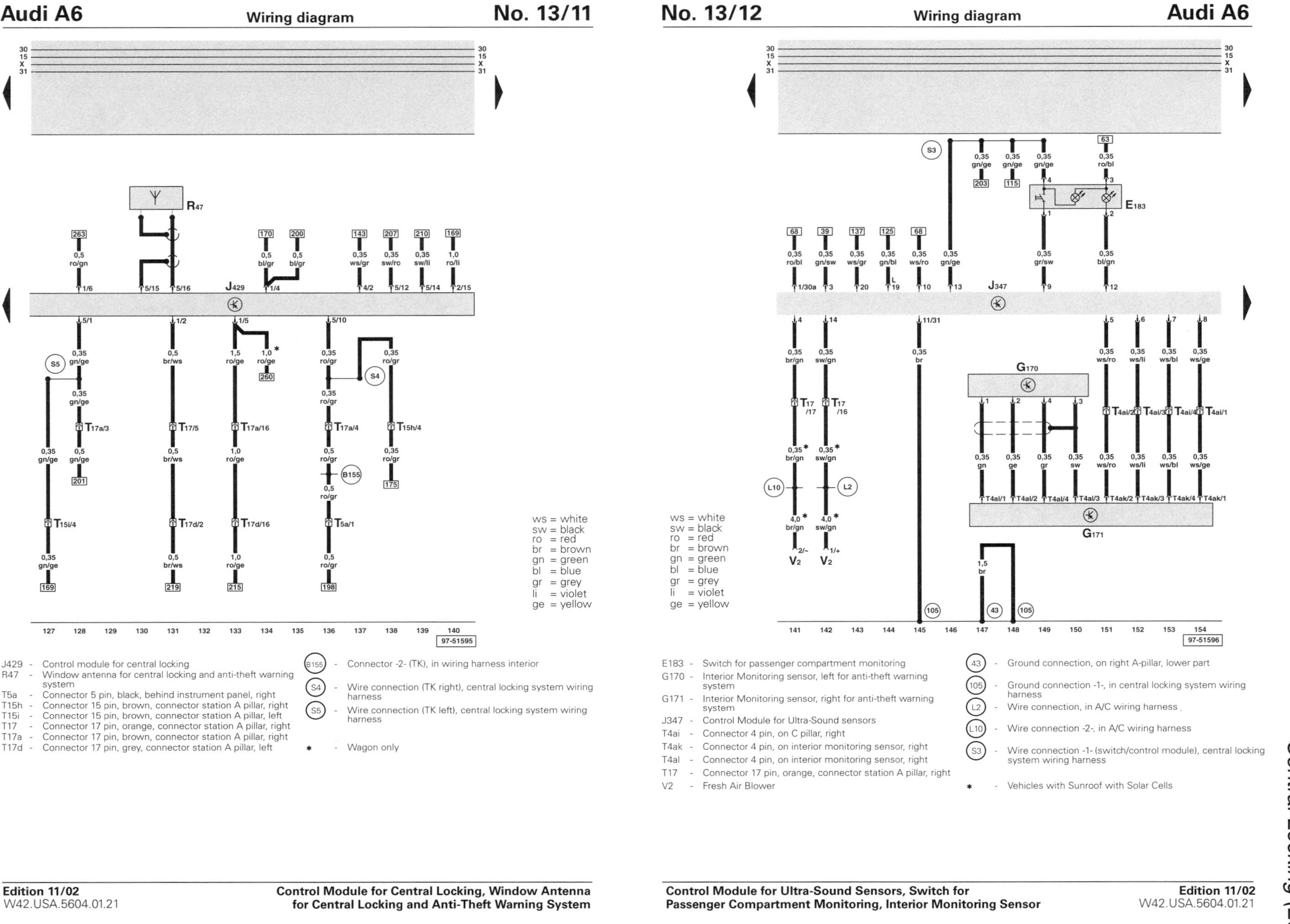

BentleyPublishers.com—All Rights Reserved

Audi A6 Wiring diagram No. 13/13

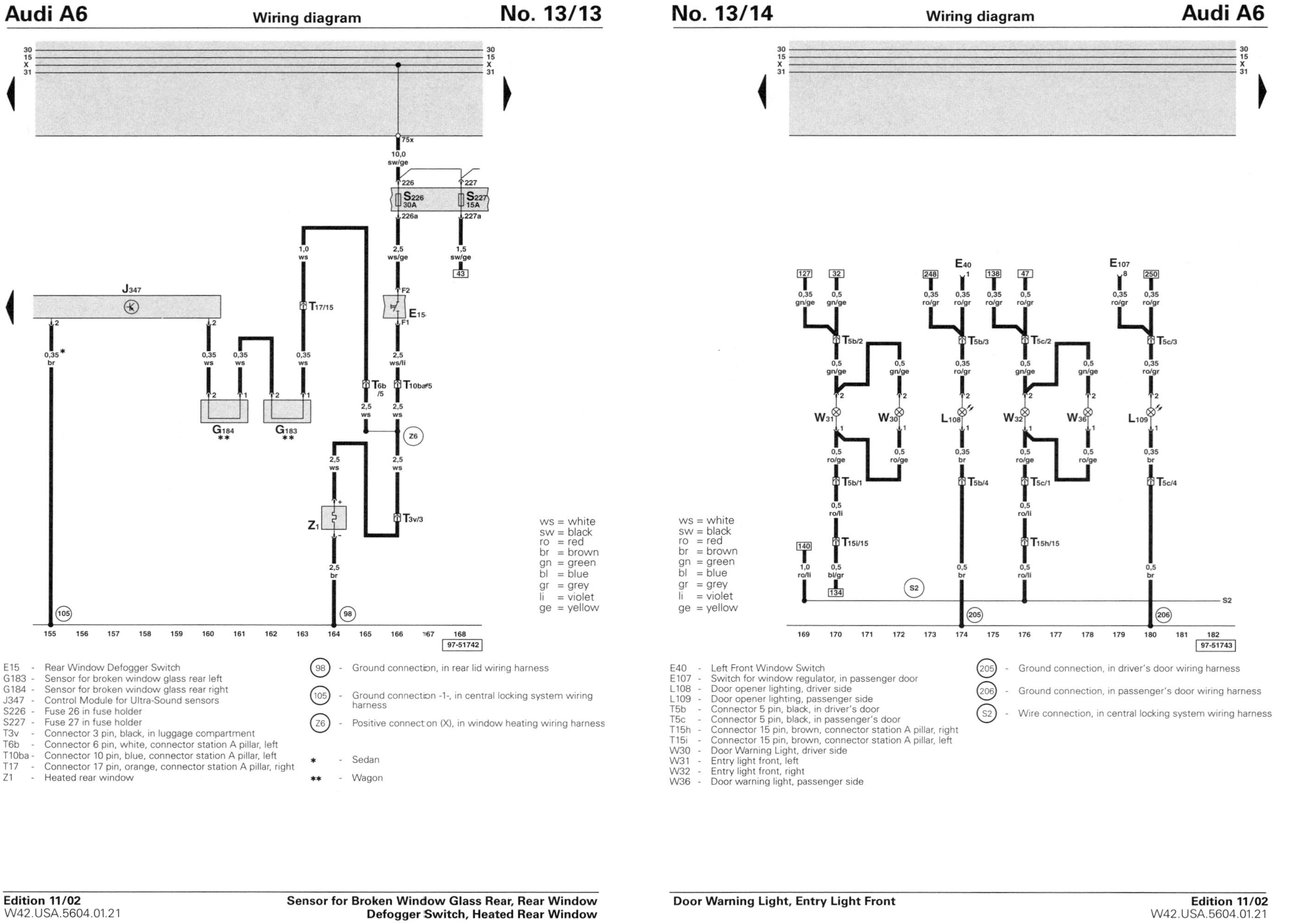

Central Locking (2003)

BentleyPublishers.com—All Rights Reserved

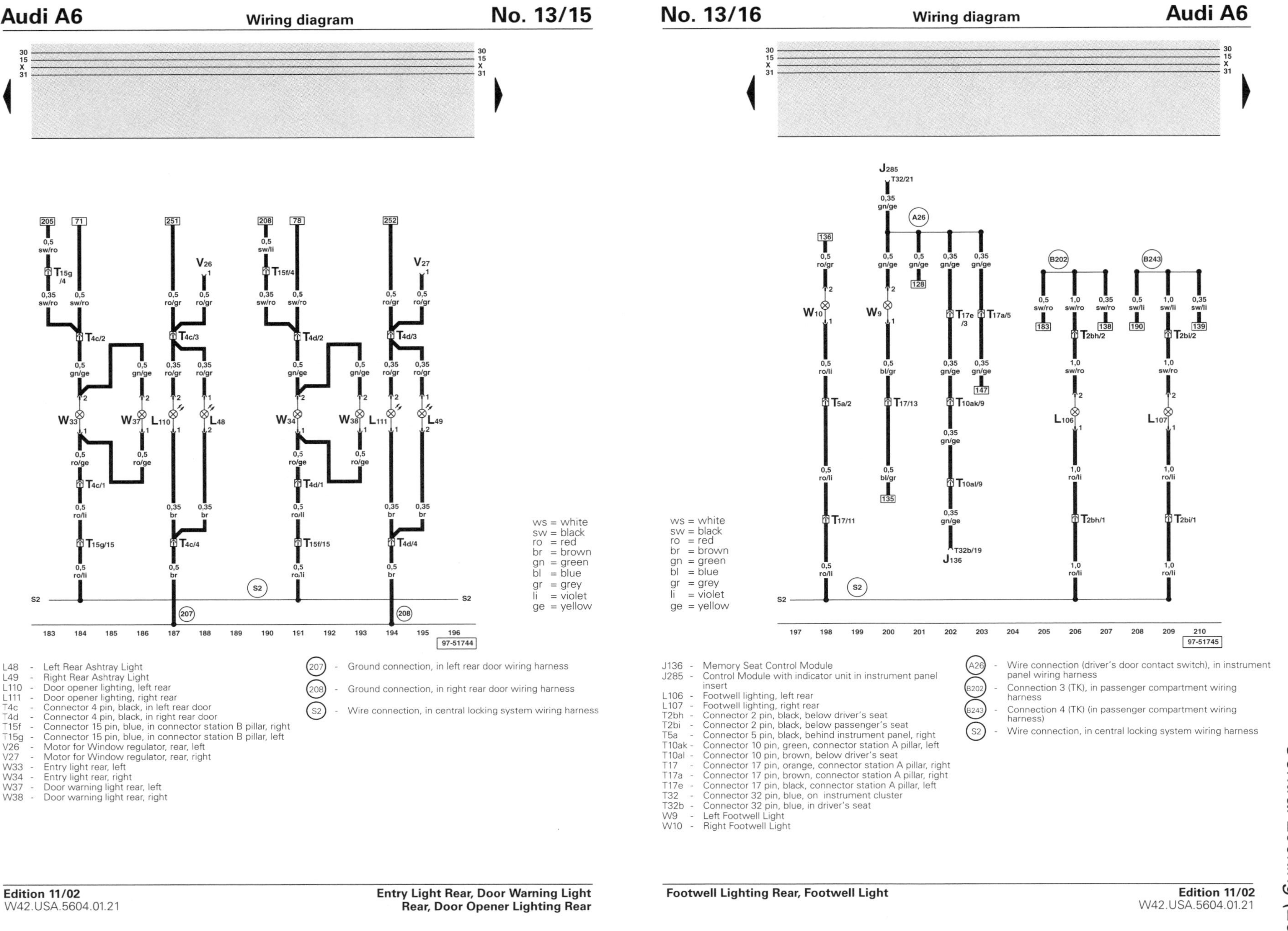

BentleyPublishers.com—All Rights Reserved

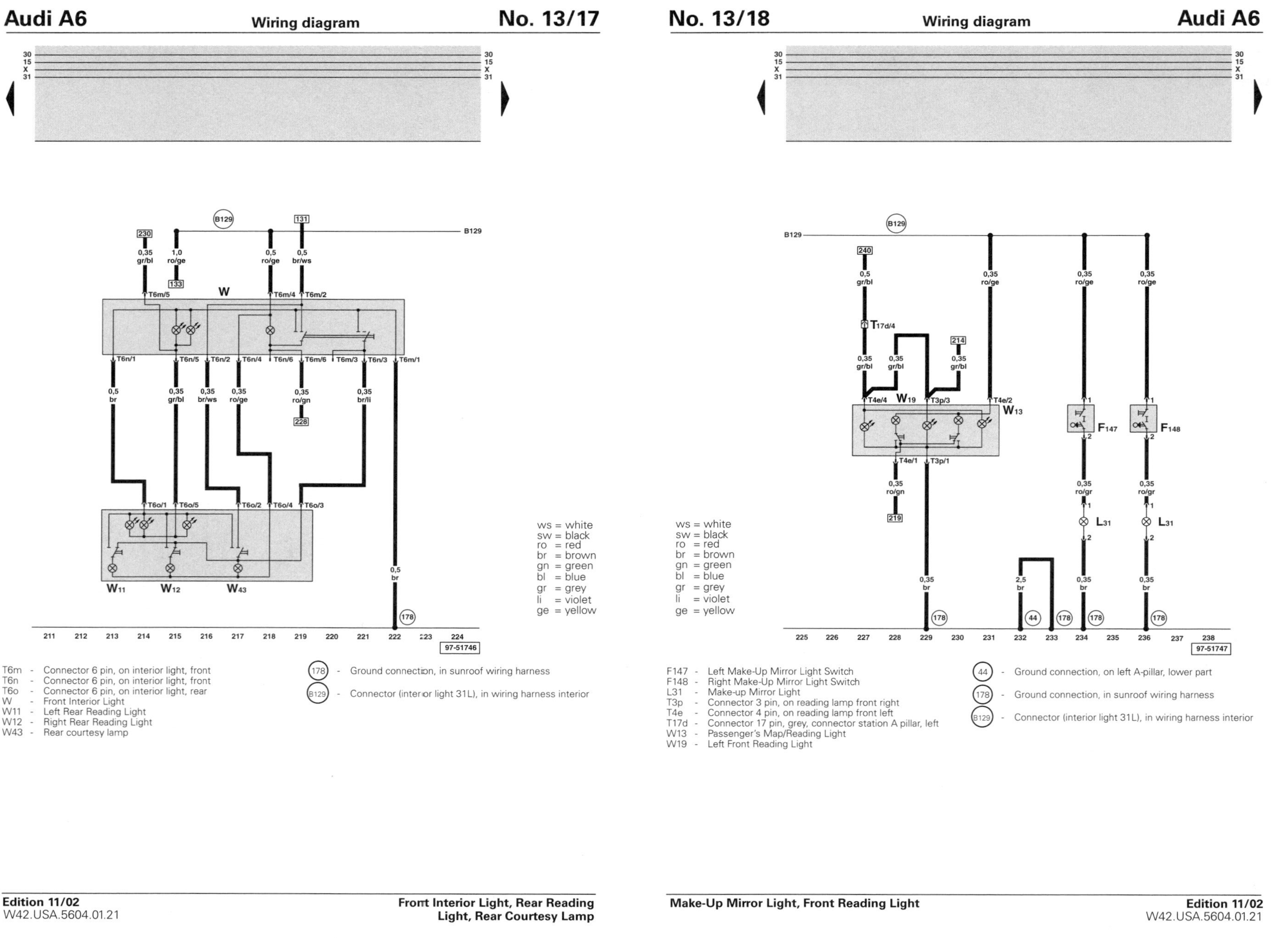

BentleyPublishers.com—All Rights Reserved

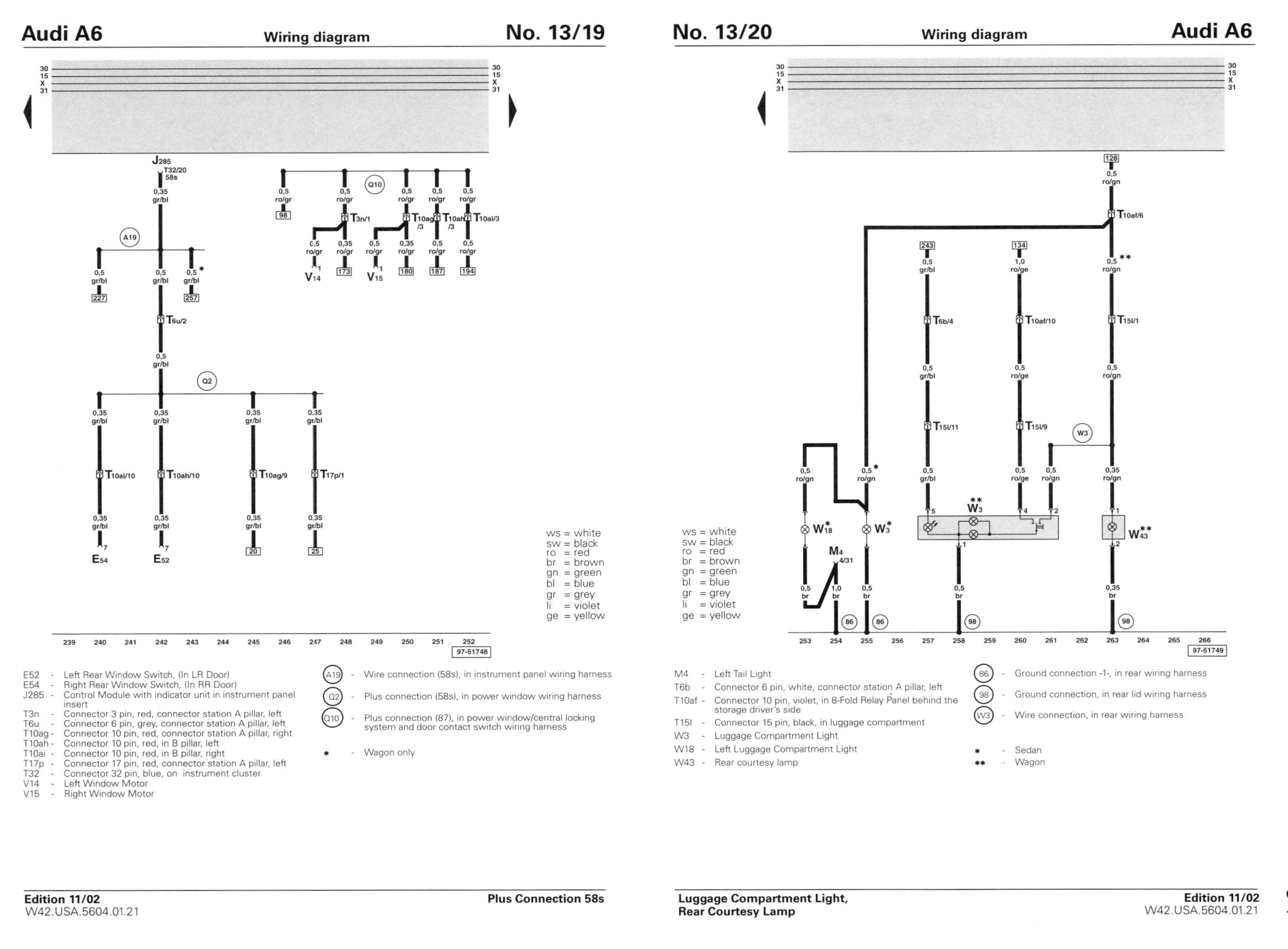

BentleyPublishers.com—All Rights Reserved

Daytime Running Lights with Fog Light and Rear Fog Light (Canada)

2003 m. y.

Fuse Panel

A97-0119

Fuse Colors:

30 A - Green
25 A - White
20 A - Yellow
15 A - Blue
10 A - Red
7,5 A - Brown
5 A - Beige

Starting with fuse position 23, fuses in the fuse holder are identified with 223 in the wiring diagram.

Micro Central Electric Panel

A B C

A97-0110

Relay Location:

2 -Load Reduction Relay, J59

13 - Fold Relay Panel

A97-0024

Relay Location:

6 -Fog Light Relay, J5

9 -Lamp Control Module, J123

10 -Lamp Control Module, J123

17 -Fog Light Circuit Breaker, S28

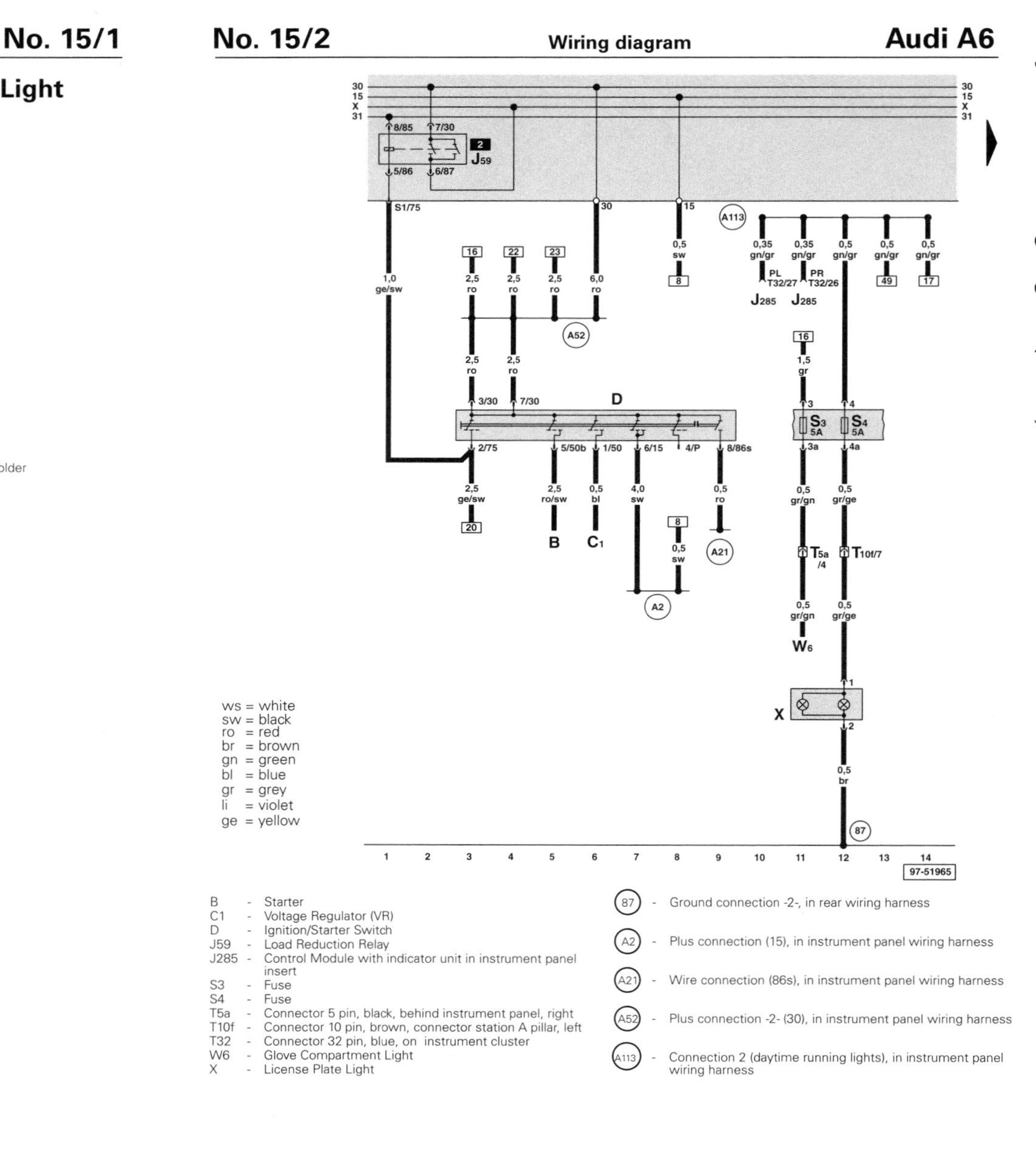

ws = white
sw = black
ro = red
br = brown
gn = green
bl = blue
gr = grey
li = violet
ge = yellow

B - Starter
C1 - Voltage Regulator (VR)
D - Ignition/Starter Switch
J59 - Load Reduction Relay
J285 - Control Module with indicator unit in instrument panel insert
S3 - Fuse
S4 - Fuse
T5a - Connector 5 pin, black, behind instrument panel, right
T10f - Connector 10 pin, brown, connector station A pillar, left
T32 - Connector 32 pin, blue, on instrument cluster
W6 - Glove Compartment Light
X - License Plate Light

(87) - Ground connection -2-, in rear wiring harness

(A2) - Plus connection (15), in instrument panel wiring harness

(A21) - Wire connection (86s), in instrument panel wiring harness

(A52) - Plus connection -2- (30), in instrument panel wiring harness

(A113) - Connection 2 (daytime running lights), in instrument panel wiring harness

Ignition/Starter Switch, License Plate Light, Load Reduction Relay

BentleyPublishers.com—All Rights Reserved

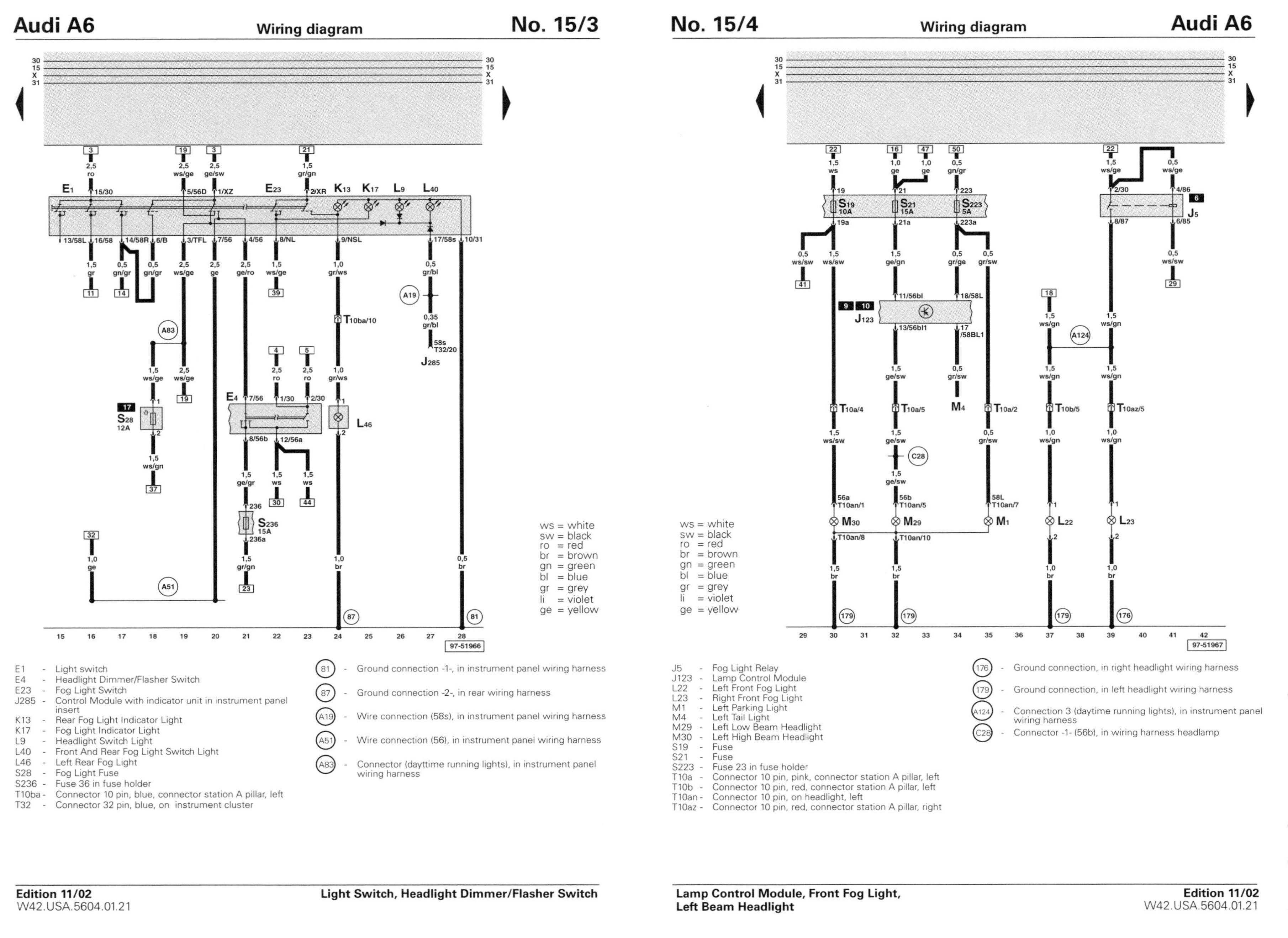

BentleyPublishers.com—All Rights Reserved

Daytime Running Lights (2003)

THIS SPACE INTENTIONALLY LEFT BLANK

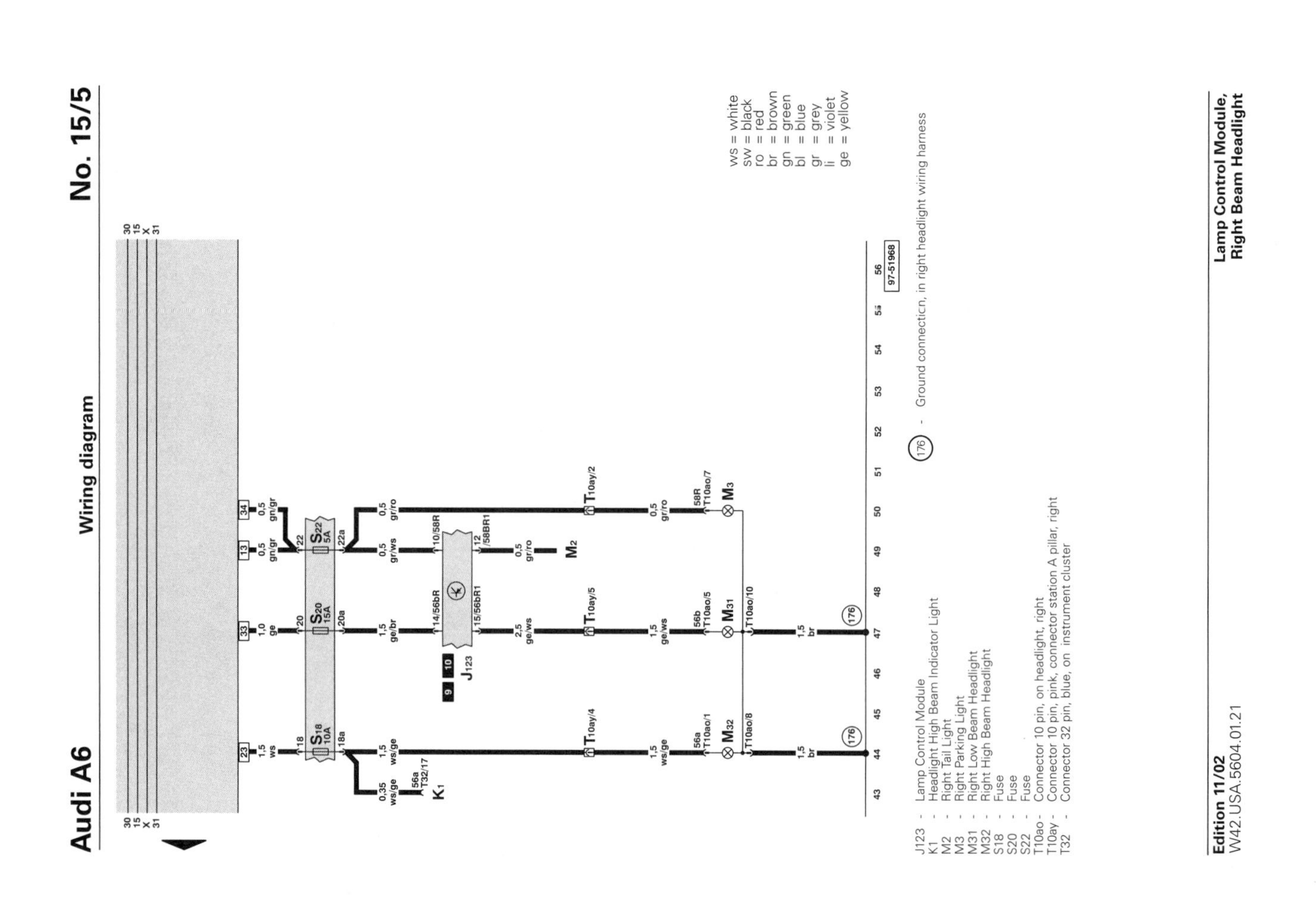

BentleyPublishers.com—All Rights Reserved

Audi A6 **Wiring Diagram** **No. 16/1**

Driver's Information System

2003 m.y.

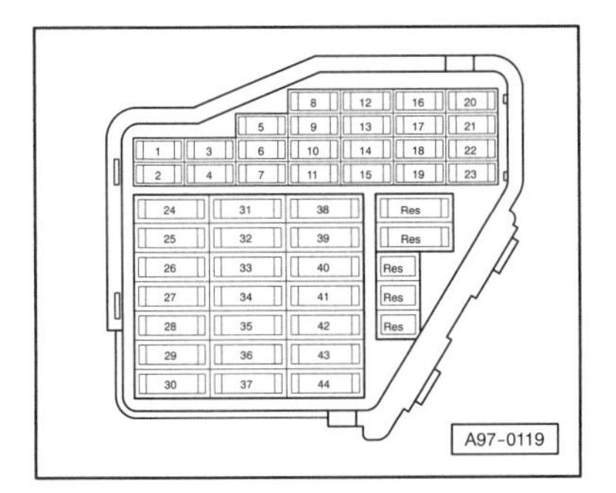

Fuse Panel

Fuse Colors:
- 30 A - Green
- 25 A - White
- 20 A - Yellow
- 15 A - Blue
- 10 A - Red
- 7,5 A - Brown
- 5 A - Beige
- 1 A - Black

Starting with fuse position 23, fuses in the fuse holder are identified with 223 in the wiring diagram.

Audi A6 **Wiring Diagram** **No. 16/2**

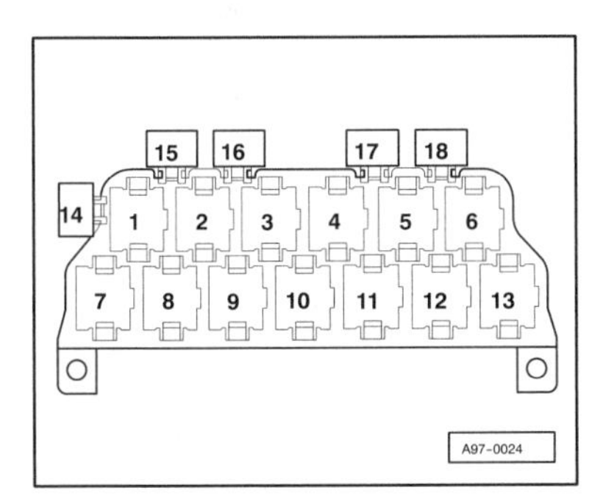

13 - Fold Relay Panel

Relay Location:
- 9 - Lamp Control Module, J123
- 10 - Lamp Control Module, J123

BentleyPublishers.com—All Rights Reserved

Audi A6

Wiring Diagram

No. 16/3

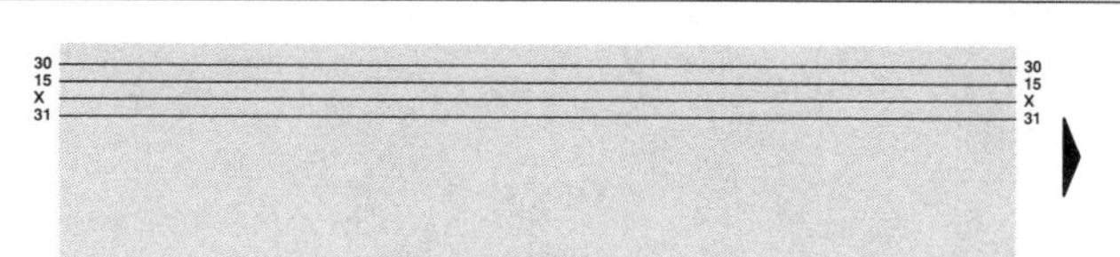

Board Computer Function Selector Switch, Windshield Washer Fluid Level Warning Switch

- E91 - Board Computer Function Selector Switch
- E92 - Board Computer Reset Button
- F77 - Windshield Washer Fluid Level Warning Switch
- J189 - Auto Check System
- J285 - Control Module with indicator unit in instrument panel insert
- T10a - Connector 10 pin, pink, connector station A pillar, left
- T10b - Connector 10 pin, red, connector station A pillar, left
- T32c - Connector 32 pin, grey, on instrument cluster

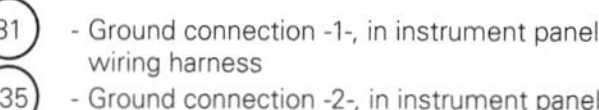

- (81) - Ground connection -1-, in instrument panel wiring harness
- (135) - Ground connection -2-, in instrument panel wiring harness
- (199) - Ground connection -3-, in instrument panel wiring harness
- (269) - Ground connector (sensor ground) -1-, in instrument panel wiring harness
- * - both are possible

ws = white
sw = black
ro = red
br = brown
gn = green
bl = blue
gr = grey
li = violet
ge = yellow
or = orange

97-51615

USA.5604.03.21 **Edition 05/03**

Audi A6

Wiring Diagram

No. 16/4

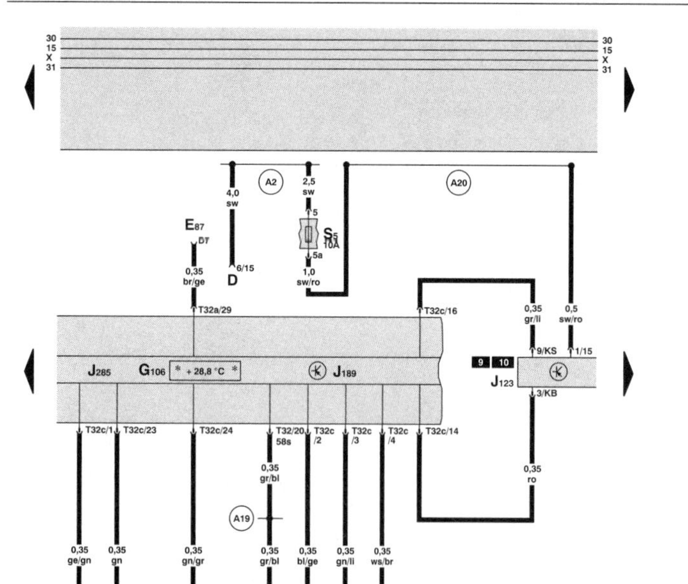

Lamp Control Module, Outside Air Temperature Display, Board Computer Function Selector Switch II

- D - Ignition/Starter Switch
- E87 - A/C Control Head
- E272 - Board Computer Function Selector Switch II, in console
- G106 - Outside Air Temperature Display
- J123 - Lamp Control Module
- J189 - Auto Check System
- J285 - Control Module with indicator unit in instrument panel insert
- S5 - Fuse
- T32 - Connector 32 pin, blue, on instrument cluster
- T32a - Connector 32 pin, green, on instrument cluster
- T32c - Connector 32 pin, grey, on instrument cluster
- (269) - Ground connector (sensor ground) -1-, in instrument panel wiring harness
- (A2) - Plus connection (15), in instrument panel wiring harness
- (A19) - Wire connection (58s), in instrument panel wiring harness
- (A20) - Wire connection (15a), in instrument panel wiring harness

ws = white
sw = black
ro = red
br = brown
gn = green
bl = blue
gr = grey
li = violet
ge = yellow
or = orange

97-51616

USA.5604.03.21 **Edition 05/03**

BentleyPublishers.com—All Rights Reserved

Audi A6 — Wiring Diagram — No. 16/5

Lamp Control Module, Parking Light, Low Beam Headlight

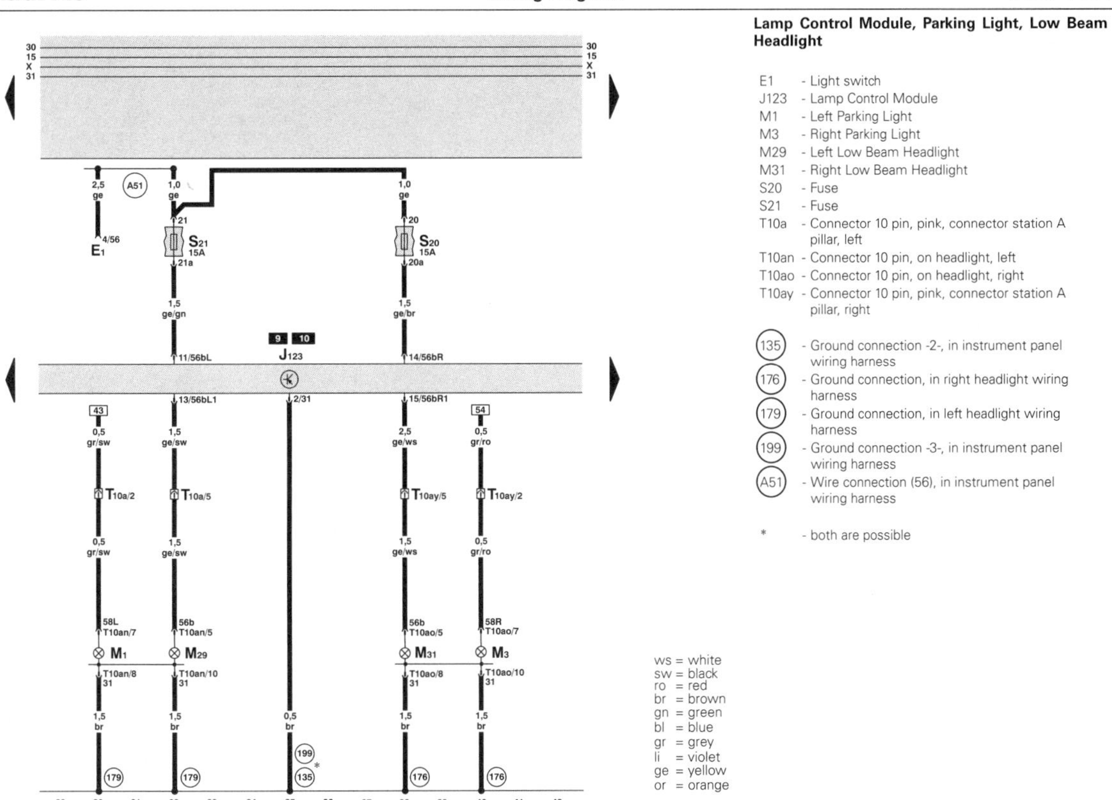

- E1 - Light switch
- J123 - Lamp Control Module
- M1 - Left Parking Light
- M3 - Right Parking Light
- M29 - Left Low Beam Headlight
- M31 - Right Low Beam Headlight
- S20 - Fuse
- S21 - Fuse
- T10a - Connector 10 pin, pink, connector station A pillar, left
- T10an - Connector 10 pin, on headlight, left
- T10ao - Connector 10 pin, on headlight, right
- T10ay - Connector 10 pin, pink, connector station A pillar, right

- (135) - Ground connection -2-, in instrument panel wiring harness
- (176) - Ground connection, in right headlight wiring harness
- (179) - Ground connection, in left headlight wiring harness
- (199) - Ground connection -3-, in instrument panel wiring harness
- (A51) - Wire connection (56), in instrument panel wiring harness

* - both are possible

ws = white
sw = black
ro = red
br = brown
gn = green
bl = blue
gr = grey
li = violet
ge = yellow
or = orange

USA.5604.03.21 Edition 05/03

Audi A6 — Wiring Diagram — No. 16/6

Lamp Control Module, Brake Light, Tail Light

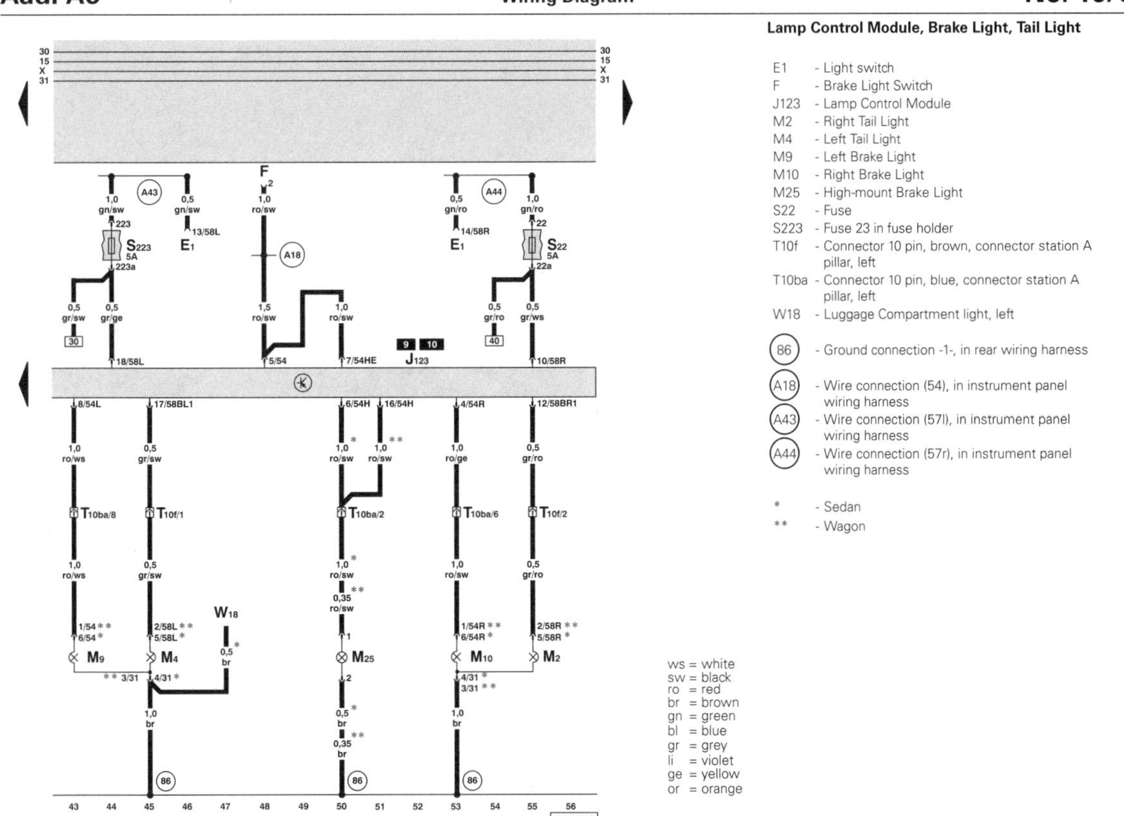

- E1 - Light switch
- F - Brake Light Switch
- J123 - Lamp Control Module
- M2 - Right Tail Light
- M4 - Left Tail Light
- M9 - Left Brake Light
- M10 - Right Brake Light
- M25 - High-mount Brake Light
- S22 - Fuse
- S223 - Fuse 23 in fuse holder
- T10f - Connector 10 pin, brown, connector station A pillar, left
- T10ba - Connector 10 pin, blue, connector station A pillar, left
- W18 - Luggage Compartment light, left

- (86) - Ground connection -1-, in rear wiring harness
- (A18) - Wire connection (54), in instrument panel wiring harness
- (A43) - Wire connection (57l), in instrument panel wiring harness
- (A44) - Wire connection (57r), in instrument panel wiring harness

* - Sedan
** - Wagon

ws = white
sw = black
ro = red
br = brown
gn = green
bl = blue
gr = grey
li = violet
ge = yellow
or = orange

USA.5604.03.21 Edition 05/03

BentleyPublishers.com—All Rights Reserved

Audi A6 Wiring Diagram **No. 16/7**

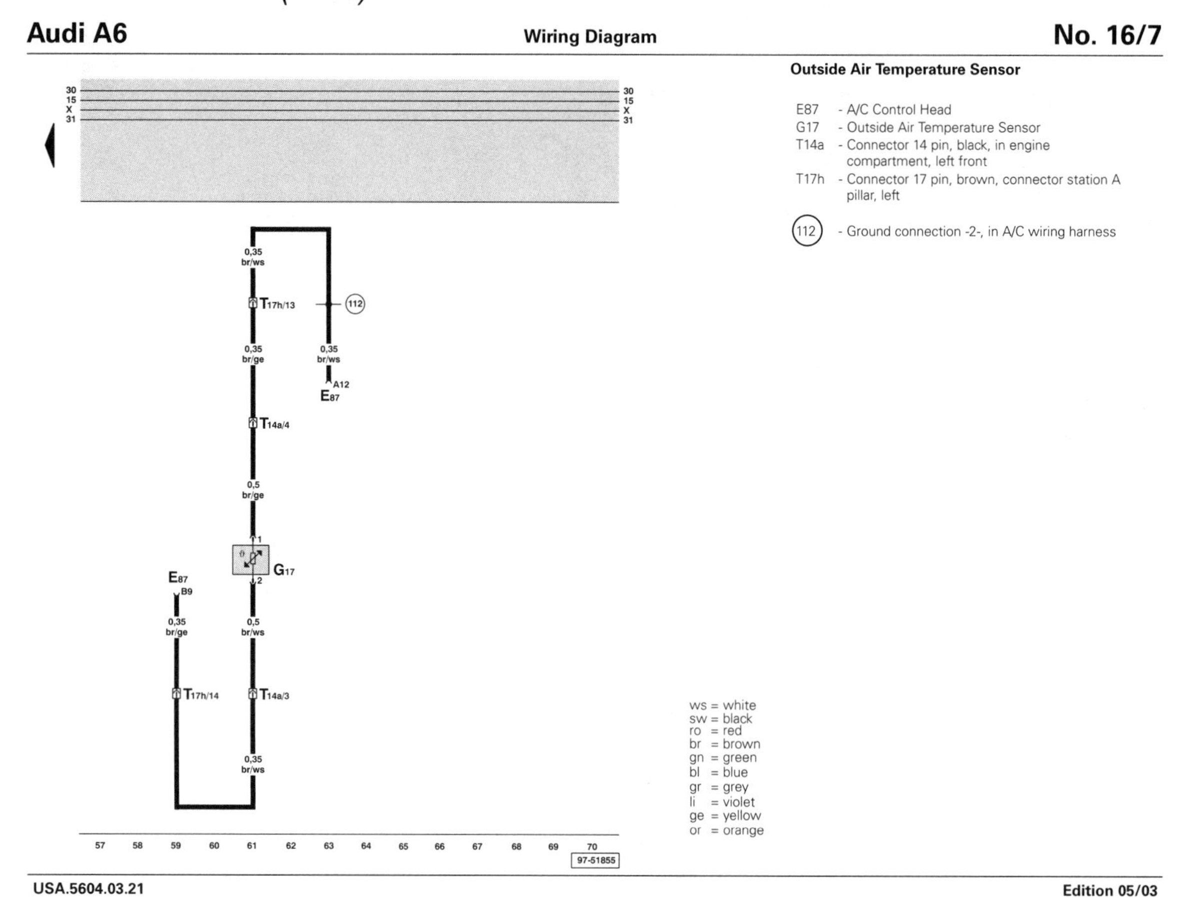

Outside Air Temperature Sensor

E87 - A/C Control Head
G17 - Outside Air Temperature Sensor
T14a - Connector 14 pin, black, in engine compartment, left front
T17h - Connector 17 pin, brown, connector station A pillar, left

(112) - Ground connection -2-, in A/C wiring harness

USA.5604.03.21 **Edition 05/03**

THIS SPACE INTENTIONALLY LEFT BLANK

BentleyPublishers.com—All Rights Reserved

Audi A6 Wiring diagram **No. 18/1**

Headlights with High Intensity Gas Discharge (HID) Lamps and Automatic Dynamic Headlight Beam Adjusting

2003 m. y.

Fuse Panel

A97-0119

Fuse Colors:

30 A - Green
25 A - White
20 A - Yellow
15 A - Blue
10 A - Red
7,5 A - Brown
5 A - Beige

Starting with fuse position 23, fuses in the fuse holder are identified with 223 in the wiring diagram.

Micro Central Electric Panel

A97-0110

Relay Location:

2 -Load Reduction Relay, J59

13 - Fold Relay Panel

A97-0024

Relay Location:

9 -Lamp Control Module, J123

10 -Lamp Control Module, J123

Edition 11/02
W42.USA.5604.01.21

No. 18/2 Wiring diagram **Audi A6**

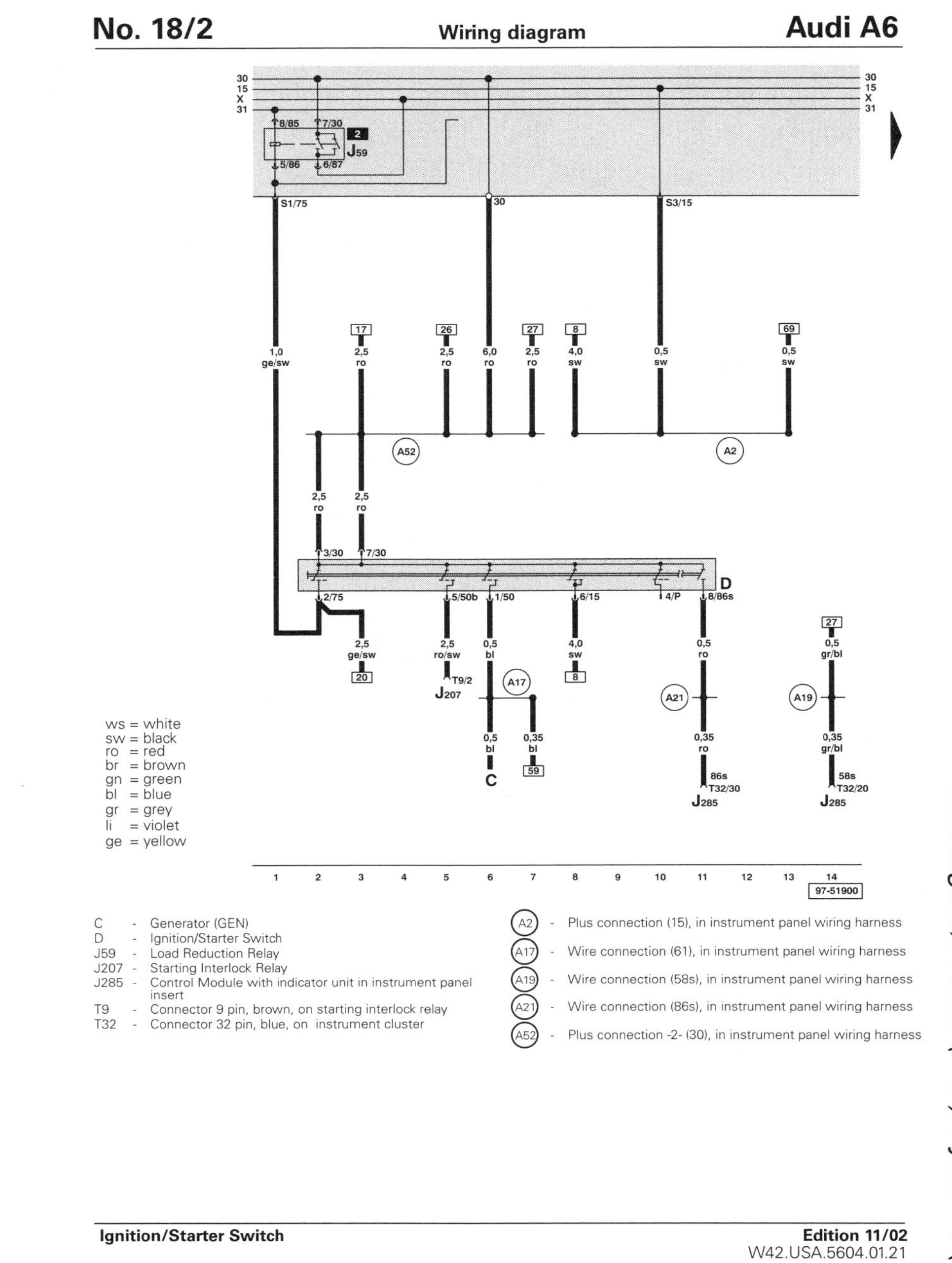

ws = white
sw = black
ro = red
br = brown
gn = green
bl = blue
gr = grey
li = violet
ge = yellow

C - Generator (GEN)
D - Ignition/Starter Switch
J59 - Load Reduction Relay
J207 - Starting Interlock Relay
J285 - Control Module with indicator unit in instrument panel insert
T9 - Connector 9 pin, brown, on starting interlock relay
T32 - Connector 32 pin, blue, on instrument cluster

A2 - Plus connection (15), in instrument panel wiring harness
A17 - Wire connection (61), in instrument panel wiring harness
A19 - Wire connection (58s), in instrument panel wiring harness
A21 - Wire connection (86s), in instrument panel wiring harness
A52 - Plus connection -2- (30), in instrument panel wiring harness

Ignition/Starter Switch

Edition 11/02
W42.USA.5604.01.21

BentleyPublishers.com—All Rights Reserved

Headlights, Xenon (HID) Dynamic (2003)

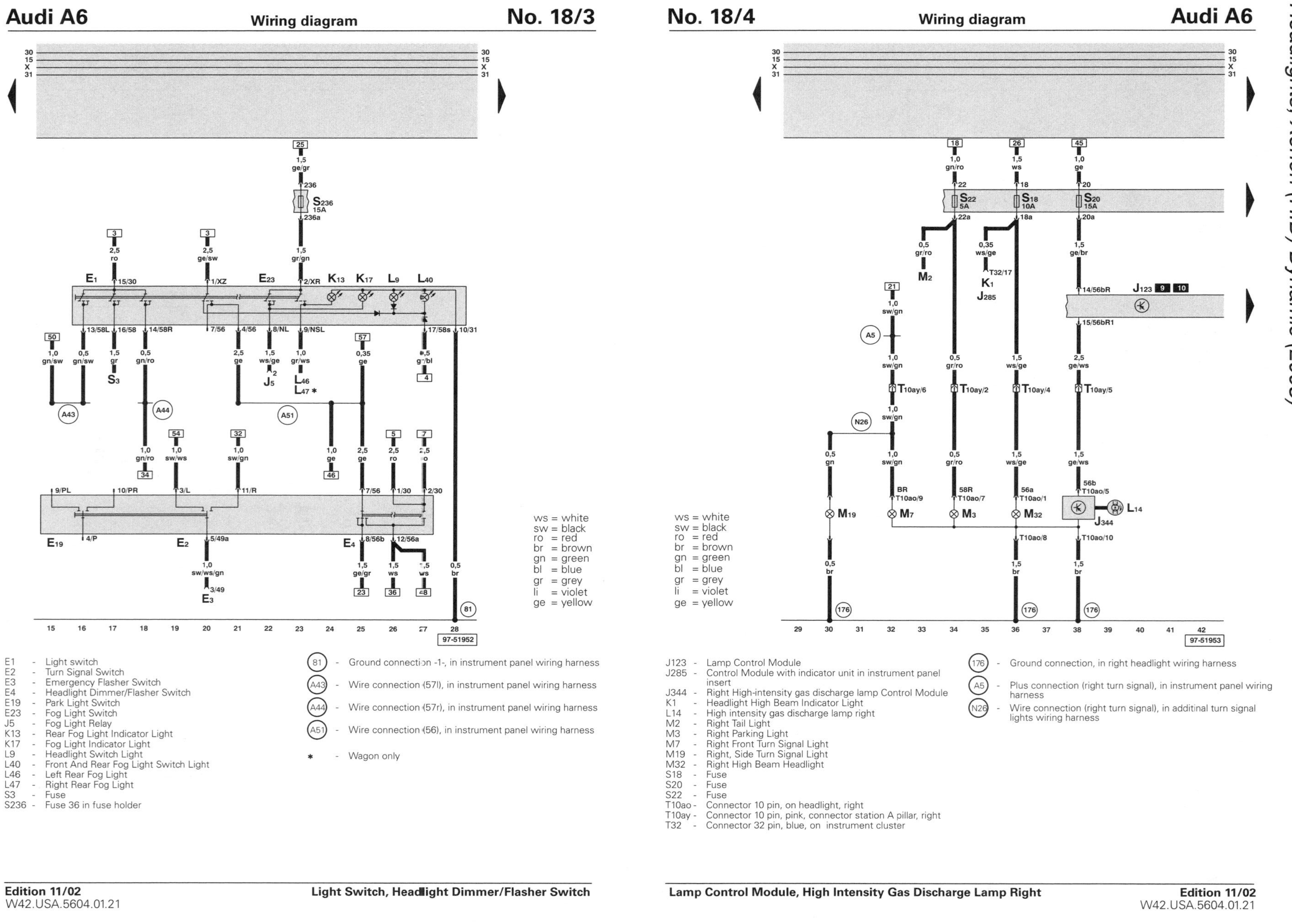

BentleyPublishers.com—All Rights Reserved

Audi A6 — Wiring diagram — No. 18/5

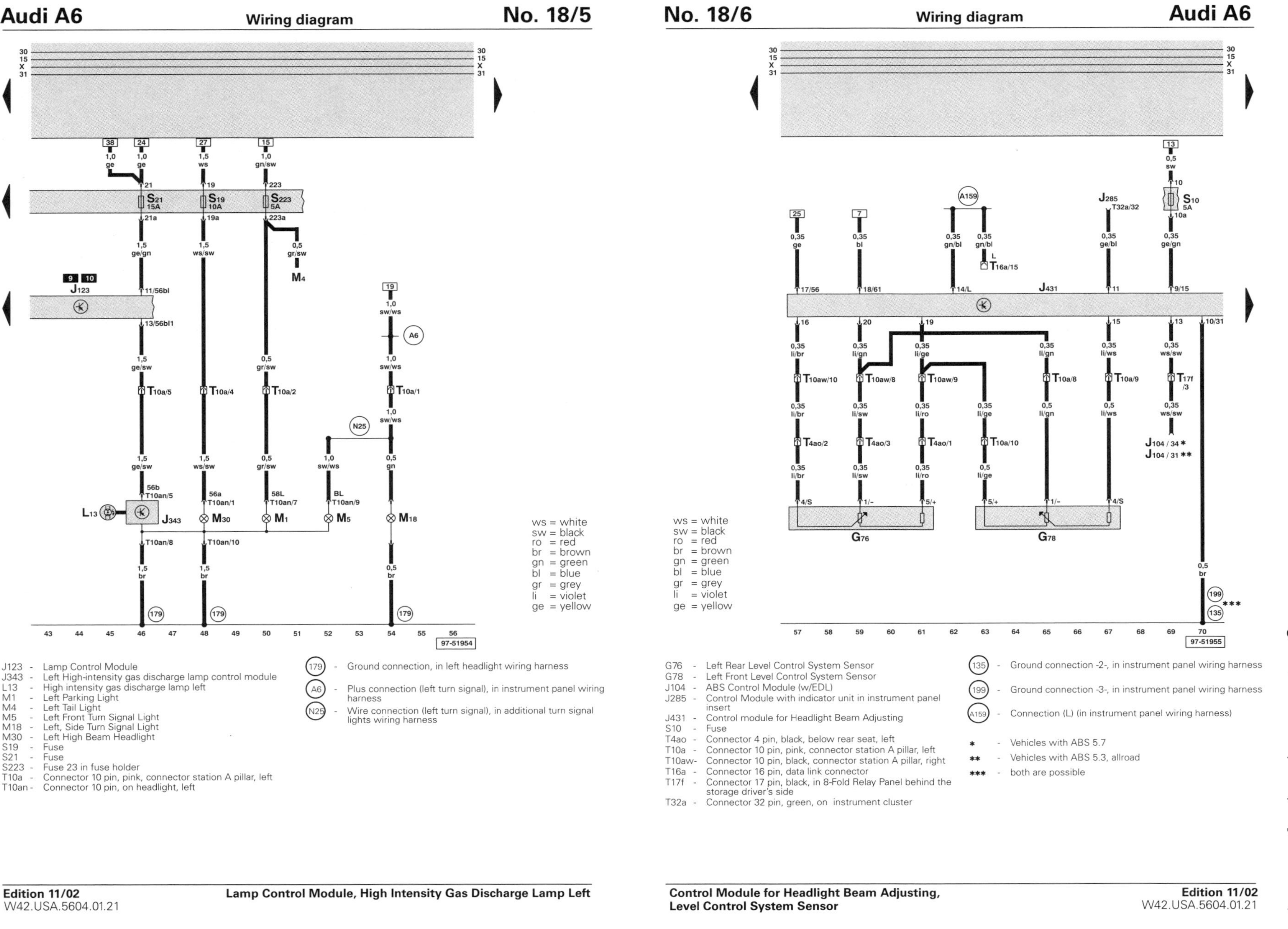

ws = white
sw = black
ro = red
br = brown
gn = green
bl = blue
gr = grey
li = violet
ge = yellow

J123 - Lamp Control Module
J343 - Left High-intensity gas discharge lamp control module
L13 - High intensity gas discharge lamp left
M1 - Left Parking Light
M4 - Left Tail Light
M5 - Left Front Turn Signal Light
M18 - Left, Side Turn Signal Light
M30 - Left High Beam Headlight
S19 - Fuse
S21 - Fuse
S223 - Fuse 23 in fuse holder
T10a - Connector 10 pin, pink, connector station A pillar, left
T10an - Connector 10 pin, on headlight, left

(179) - Ground connection, in left headlight wiring harness
(A6) - Plus connection (left turn signal), in instrument panel wiring harness
(N25) - Wire connection (left turn signal), in additional turn signal lights wiring harness

Edition 11/02
W42.USA.5604.01.21

Lamp Control Module, High Intensity Gas Discharge Lamp Left

No. 18/6 — Wiring diagram — Audi A6

ws = white
sw = black
ro = red
br = brown
gn = green
bl = blue
gr = grey
li = violet
ge = yellow

G76 - Left Rear Level Control System Sensor
G78 - Left Front Level Control System Sensor
J104 - ABS Control Module (w/EDL)
J285 - Control Module with indicator unit in instrument panel insert
J431 - Control module for Headlight Beam Adjusting
S10 - Fuse
T4ao - Connector 4 pin, black, below rear seat, left
T10a - Connector 10 pin, pink, connector station A pillar, left
T10aw- Connector 10 pin, black, connector station A pillar, right
T16a - Connector 16 pin, data link connector
T17f - Connector 17 pin, black, in 8-Fold Relay Panel behind the storage driver's side
T32a - Connector 32 pin, green, on instrument cluster

(135) - Ground connection -2-, in instrument panel wiring harness
(199) - Ground connection -3-, in instrument panel wiring harness
(A159) - Connection (L) (in instrument panel wiring harness)

* - Vehicles with ABS 5.7
** - Vehicles with ABS 5.3, allroad
*** - both are possible

Control Module for Headlight Beam Adjusting, Level Control System Sensor

Edition 11/02
W42.USA.5604.01.21

BentleyPublishers.com—All Rights Reserved

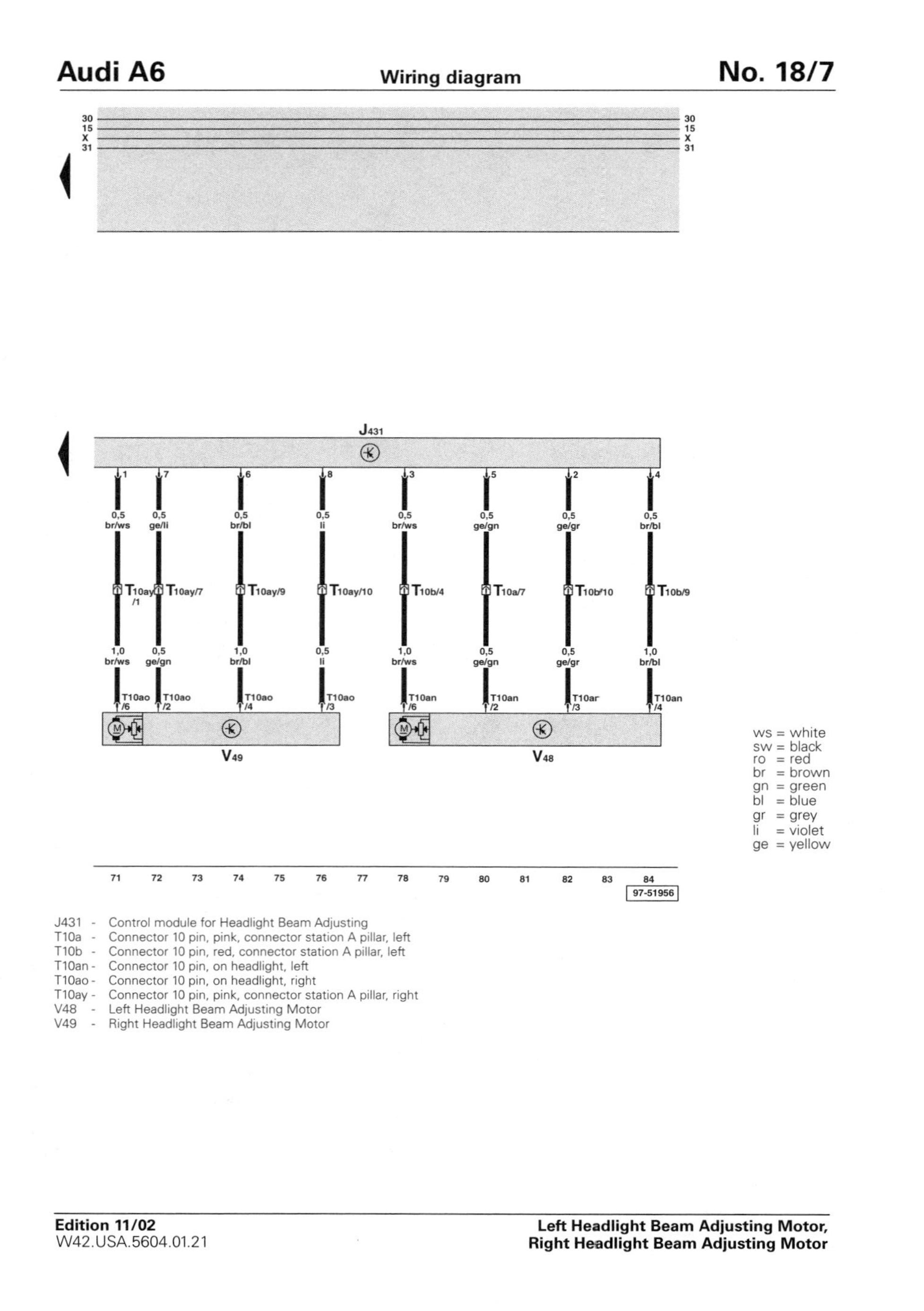

THIS SPACE INTENTIONALLY LEFT BLANK

BentleyPublishers.com—All Rights Reserved

Audi A6 Wiring Diagram **No. 24/1**

Radio Concert I, Concert II, Symphony, Bose with and without CD Changer

2003 m.y.

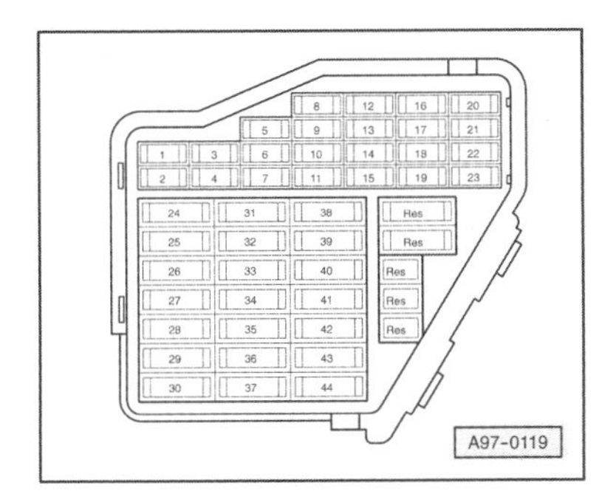

Fuse Panel

Fuse Colors:
- 30 A - Green
- 25 A - White
- 20 A - Yellow
- 15 A - Blue
- 10 A - Red
- 7,5 A - Brown
- 5 A - Beige
- 1 A - Black

Starting with fuse position 23, fuses in the fuse holder are identified with 223 in the wiring diagram.

USA.5604.05.21 Edition 07/04

Audi A6 Wiring Diagram **No. 24/2**

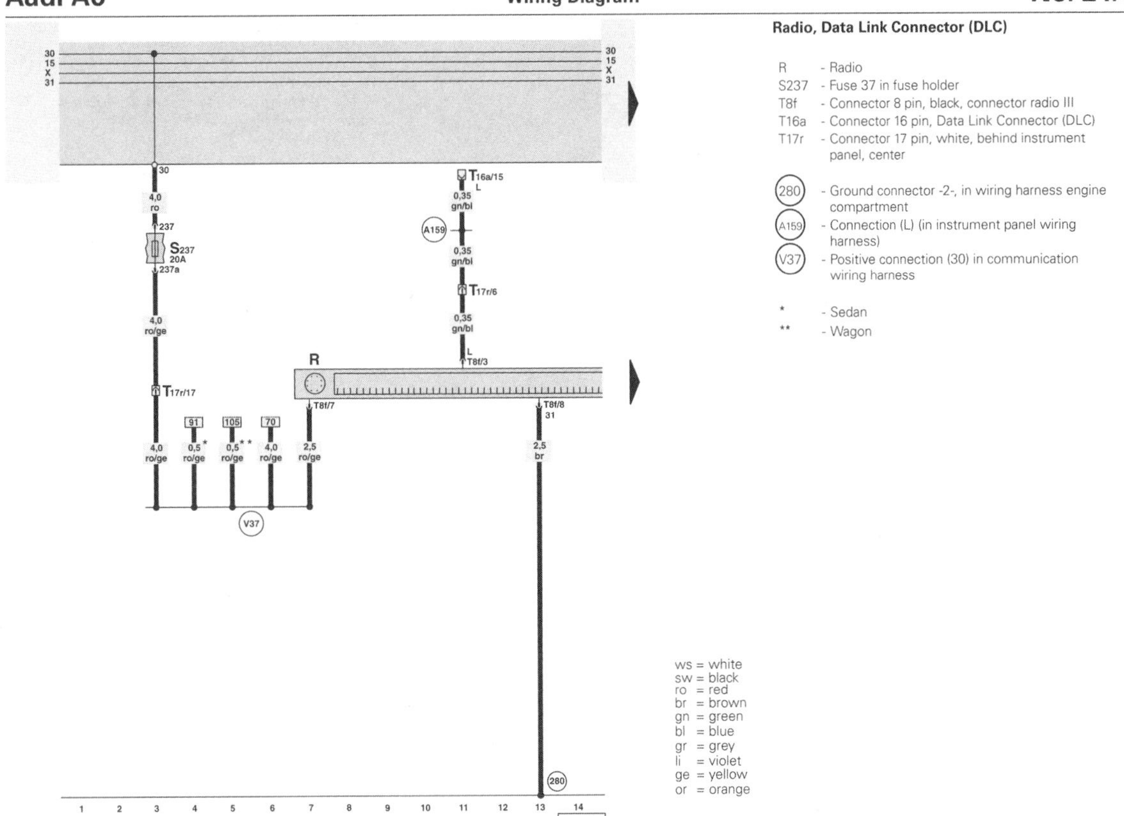

Radio, Data Link Connector (DLC)

- R - Radio
- S237 - Fuse 37 in fuse holder
- T8f - Connector 8 pin, black, connector radio III
- T16a - Connector 16 pin, Data Link Connector (DLC)
- T17r - Connector 17 pin, white, behind instrument panel, center

- (280) - Ground connector -2-, in wiring harness engine compartment
- (A159) - Connection (L) (in instrument panel wiring harness)
- (V37) - Positive connection (30) in communication wiring harness

- * - Sedan
- ** - Wagon

ws = white
sw = black
ro = red
br = brown
gn = green
bl = blue
gr = grey
li = violet
ge = yellow
or = orange

USA.5604.03.21 Edition 05/03

BentleyPublishers.com—All Rights Reserved

Sound System, Bose (2003)

Audi A6 **Wiring Diagram** **No. 24/3**

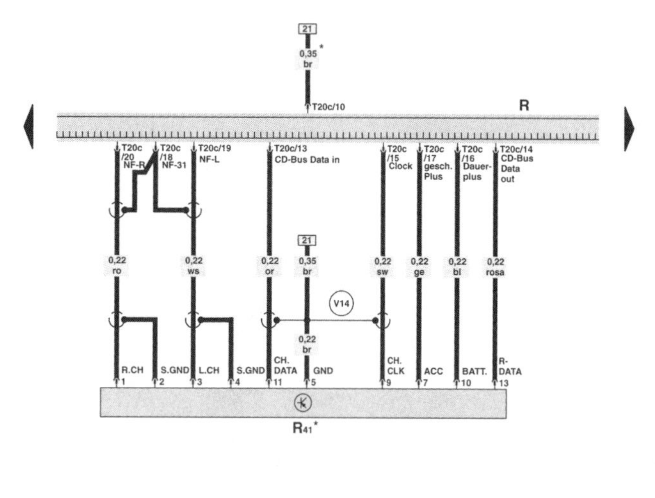

Radio, CD Changer Unit

R - Radio
R41 - CD Changer Unit
T20c - Connector 20 pin, black, connector radio I

 - Wire connection (shielding), in CD-changer wiring harness

* - Vehicles with CD Changer

ws = white
sw = black
ro = red
br = brown
gn = green
bl = blue
gr = grey
li = violet
ge = yellow
or = orange

USA.5604.03.21 Edition 05/03

Audi A6 **Wiring Diagram** **No. 24/4**

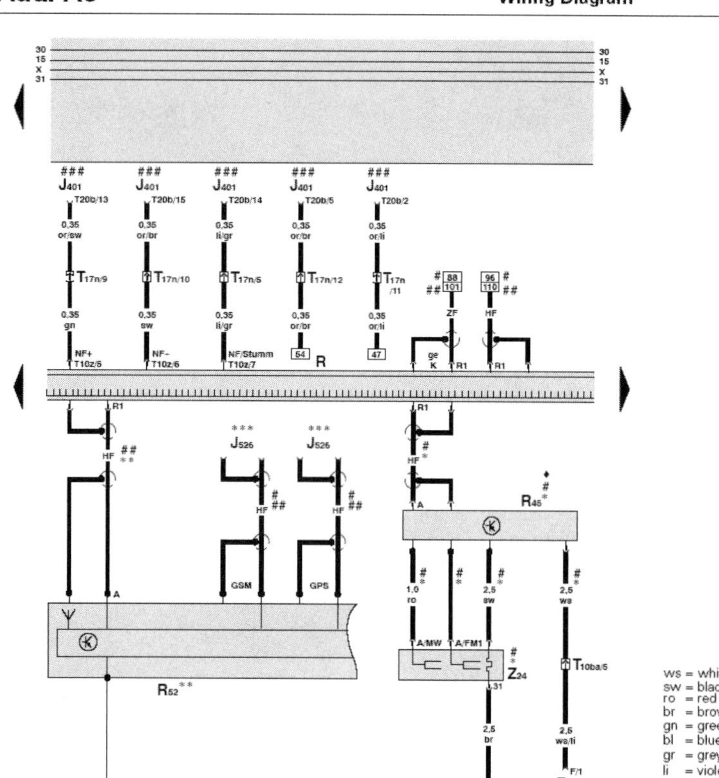

Radio, Telephone/Radio Antenna Amplifier, Antenna for Radio/Telephone/Navigation (GPS)

E15 - Rear Window Defogger Switch
J401 - Control unit for navigation with CD-mechanism
J526 - Telephone/Telematic Control Module
R - Radio
R1 - Antenna Connection
R45 - Telephone/Radio Antenna Amplifier
R52 - Antenna for radio/telephone/navigation (GPS)
T10z - Connector 10 pin, red, connector radio IV
T10ba - Connector 10 pin, blue, connector station A pillar, left
T17n - Connector 17 pin, brown, connector station A pillar, right
T20b - Connector 20 pin, black, on control unit for navigation
Z24 - Rear Window Defogger With Window Antenna

(75) - Ground connection, on right rear pillar

* - Rear Window Antenna
** - Roof Antenna
*** - Vehicles with Telematics
# - Sedan
- Wagon
- Vehicles with Navigation
♦ - only Radio Concert I

ws = white
sw = black
ro = red
br = brown
gn = green
bl = blue
gr = grey
li = violet
ge = yellow
or = orange

USA.5604.05.21 Edition 07/04

BentleyPublishers.com—All Rights Reserved

Audi A6

Wiring Diagram

No. 24/5

CAN-Bus Connection

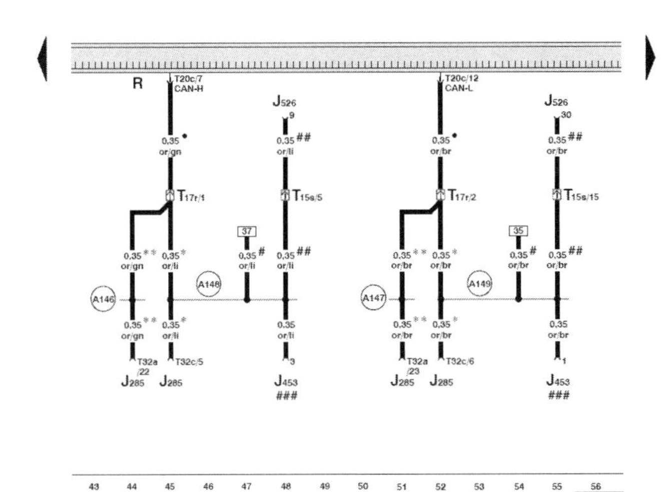

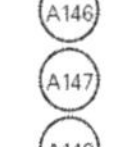

- J285 - Control Module with indicator unit in instrument panel insert
- J453 - Control module for multi-function steering wheel
- J526 - Telephone/Telematic Control Module
- R - Radio
- T15s - Connector 15 pin, blue, connector station A pillar, left
- T17r - Connector 17 pin, white, behind instrument panel, center
- T20c - Connector 20 pin, black, connector radio I
- T32a - Connector 32 pin, green, on instrument cluster
- T32c - Connector 32 pin, grey, on instrument cluster

- (A146) - Connection (high bus, comfort), in instrument panel wiring harness
- (A147) - Connection (low bus, comfort), in instrument panel wiring harness
- (A148) - Connection (high bus, navigation), in instrument panel wiring harness
- (A149) - Connection (low bus, navigation), in instrument panel wiring harness

- • - CAN-Bus (Data-Bus)
- * - Vehicles with Auto Check System
- ** - Vehicles without Auto Check System
- # - Vehicles with Navigation
- ## - Vehicles with Telematics, Telephone
- ### - Vehicles with multi-function steering wheel

ws = white
sw = black
ro = red
br = brown
gn = green
bl = blue
gr = grey
li = violet
ge = yellow
or = orange

USA.5604.03.21

Edition 05/03

Audi A6

Wiring Diagram

No. 24/6

Radio, Right Rear Woofer/Amplifier

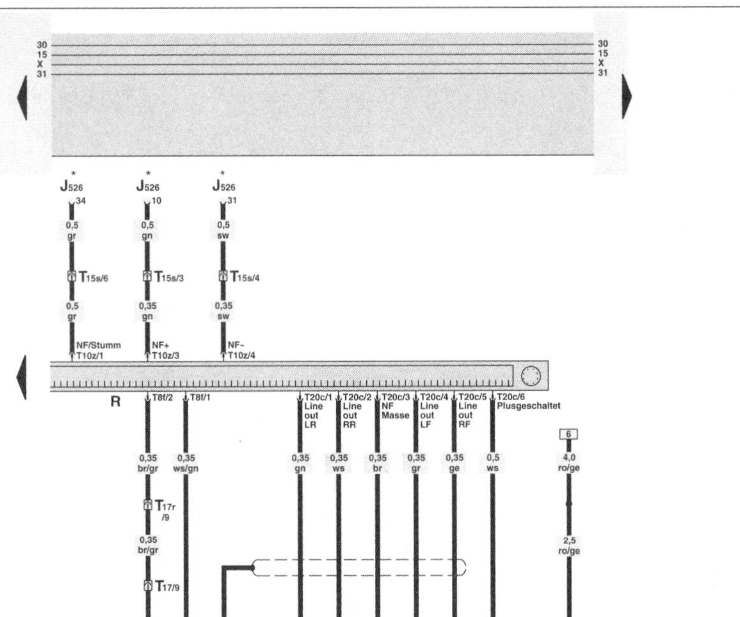

- J429 - Control module for central locking
- J526 - Telephone/Telematic Control Module
- R - Radio
- R43 - Right Rear Woofer/Amplifier
- T8f - Connector 8 pin, black, connector radio III
- T10z - Connector 10 pin, red, connector radio IV
- T15s - Connector 15 pin, blue, connector station A pillar, left
- T17 - Connector 17 pin, orange, connector station A pillar, right
- T17r - Connector 17 pin, white, behind instrument panel, center
- T20c - Connector 20 pin, black, connector radio I

- (280) - Ground connector -2-, in wiring harness engine compartment
- (610) - Ground connection (radio) under center console, front

- * - Vehicles with Telematics, Telephone

ws = white
sw = black
ro = red
br = brown
gn = green
bl = blue
gr = grey
li = violet
ge = yellow
or = orange

USA.5604.03.21

Edition 05/03

BentleyPublishers.com—All Rights Reserved

Sound System, Bose (2003)

Audi A6

Wiring Diagram

No. 24/7

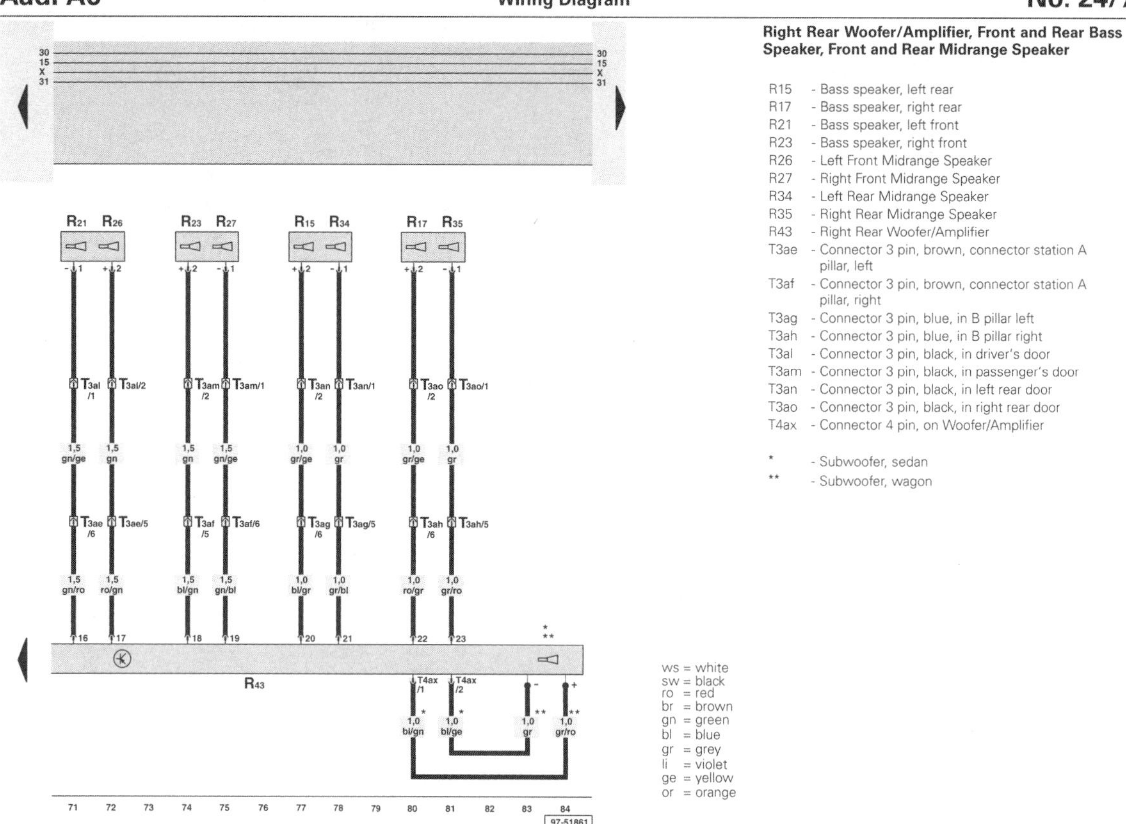

Right Rear Woofer/Amplifier, Front and Rear Bass Speaker, Front and Rear Midrange Speaker

R15 - Bass speaker, left rear
R17 - Bass speaker, right rear
R21 - Bass speaker, left front
R23 - Bass speaker, right front
R26 - Left Front Midrange Speaker
R27 - Right Front Midrange Speaker
R34 - Left Rear Midrange Speaker
R35 - Right Rear Midrange Speaker
R43 - Right Rear Woofer/Amplifier
T3ae - Connector 3 pin, brown, connector station A pillar, left
T3af - Connector 3 pin, brown, connector station A pillar, right
T3ag - Connector 3 pin, blue, in B pillar left
T3ah - Connector 3 pin, blue, in B pillar right
T3al - Connector 3 pin, black, in driver's door
T3am - Connector 3 pin, black, in passenger's door
T3an - Connector 3 pin, black, in left rear door
T3ao - Connector 3 pin, black, in right rear door
T4ax - Connector 4 pin, on Woofer/Amplifier

* - Subwoofer, sedan
** - Subwoofer, wagon

ws = white
sw = black
ro = red
br = brown
gn = green
bl = blue
gr = grey
li = violet
ge = yellow
or = orange

USA.5604.03.21

Edition 05/03

Audi A6

Wiring Diagram

No. 24/8

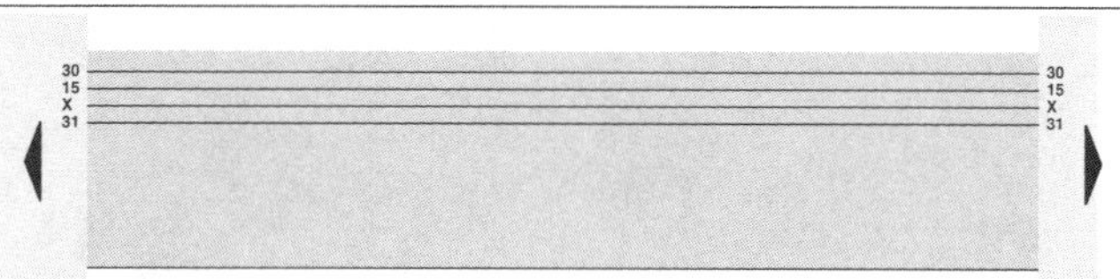

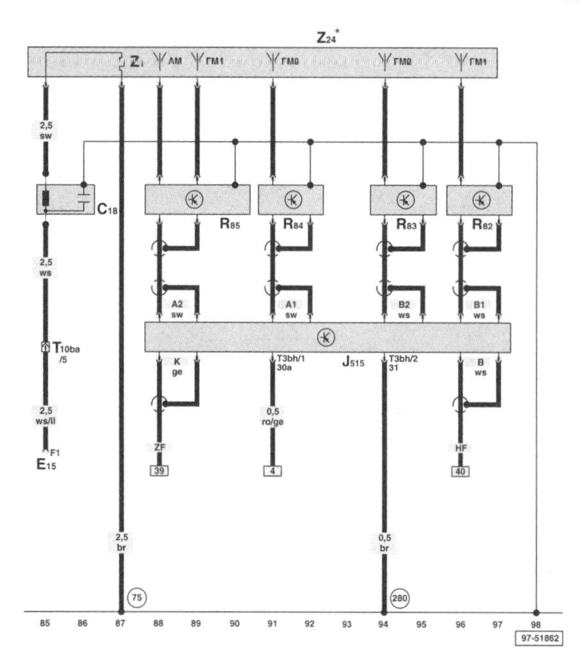

Antennas for Radio Concert II, Symphony, Antenna Selection Control Module, TV Antenna Amplifier, Rear Window Defogger with Window Antenna

C18 - Windshield Antenna Suppression Filter
E15 - Rear Window Defogger Switch
J515 - Antenna Selection Control Module
R82 - TV Antenna Amplifier 1
R83 - TV Antenna Amplifier 2
R84 - TV Antenna Amplifier 3
R85 - TV Antenna Amplifier 4
T3bh - Connector 3 pin, black, on Control unit for aerial selection (aerial diversity function)
T10ba - Connector 10 pin, blue, connector station A pillar, left
Z1 - Heated rear window
Z24 - Rear Window Defogger With Window Antenna

(75) - Ground connection, on right rear pillar

(280) - Ground connector -2-, in wiring harness engine compartment

* - Sedan

ws = white
sw = black
ro = red
br = brown
gn = green
bl = blue
gr = grey
li = violet
ge = yellow
or = orange

USA.5604.05.21

Edition 07/04

BentleyPublishers.com—All Rights Reserved

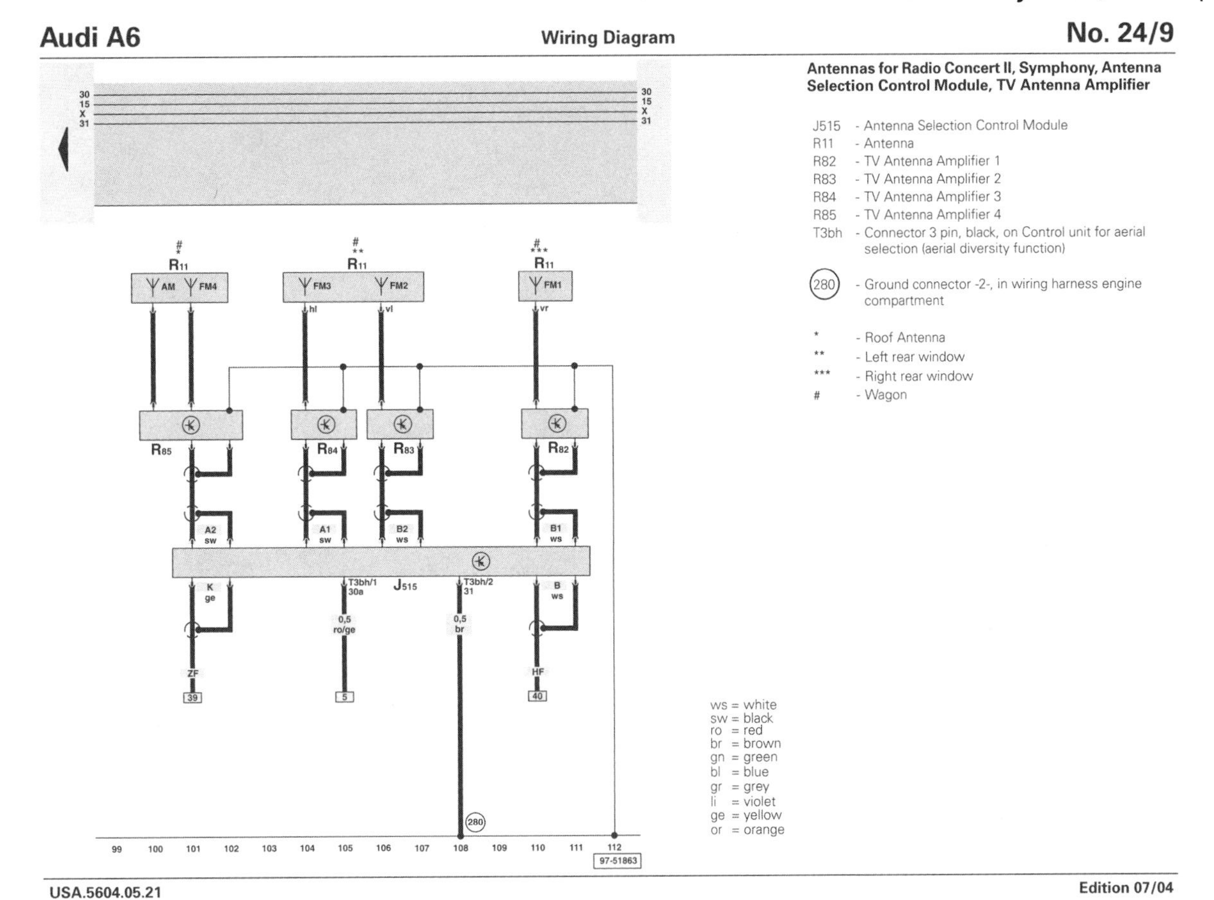
Audi A6
Wiring Diagram
No. 24/9
Antennas for Radio Concert II, Symphony, Antenna Selection Control Module, TV Antenna Amplifier
J515 - Antenna Selection Control Module
R11 - Antenna
R82 - TV Antenna Amplifier 1
R83 - TV Antenna Amplifier 2
R84 - TV Antenna Amplifier 3
R85 - TV Antenna Amplifier 4
T3bh - Connector 3 pin, black, on Control unit for aerial selection (aerial diversity function)
280 - Ground connector -2-, in wiring harness engine compartment
* - Roof Antenna
** - Left rear window
*** - Right rear window
# - Wagon
30
15
X
31
R11
AM
FM4
FM3
FM2
FM1
hl
vl
vr
R85
R84
R83
R82
A2 sw
A1 sw
B2 ws
B1 ws
K ge
T3bh/1 30a
J515
T3bh/2 31
B ws
0,5 ro/ge
0,5 br
ZF
HF
39
5
40
280
ws = white
sw = black
ro = red
br = brown
gn = green
bl = blue
gr = grey
li = violet
ge = yellow
or = orange
99 100 101 102 103 104 105 106 107 108 109 110 111 112
97-51863
USA.5604.05.21
Edition 07/04

THIS SPACE INTENTIONALLY

LEFT BLANK

BentleyPublishers.com—All Rights Reserved

Audi A6 Wiring Diagram **No. 28/1**

(4,2 l - Injection Engine, 8 Cylinder), Code BCY - RS6

2003 m.y.

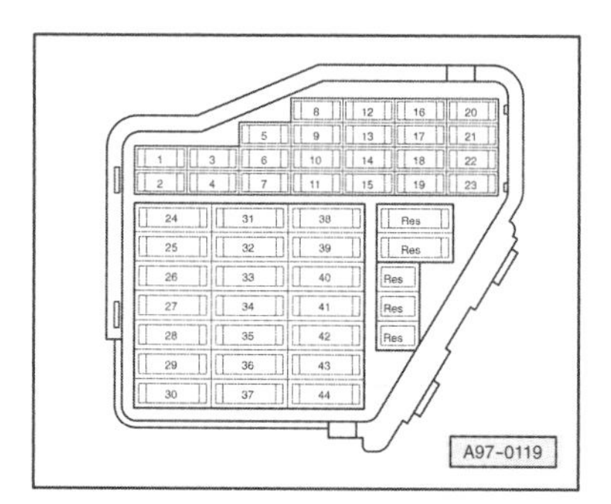

Fuse Panel

Fuse Colors:

30 A - Green
25 A - White
20 A - Yellow
15 A - Blue
10 A - Red
7,5 A - Brown
5 A - Beige
1 A - Black

Starting with fuse position 23, fuses in the fuse holder are identified with 223 in the wiring diagram.

Audi A6 Wiring Diagram **No. 28/2**

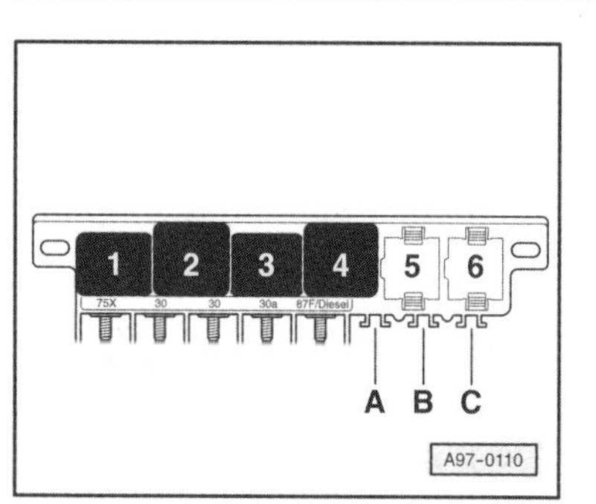

Micro Central Electric Panel

Relay Location:

4 - Fuel Pump (FP) Relay, J17
5 - Wiper/Washer Intermittent Relay, J31
6 - Wiper/Washer Intermittent Relay, J31

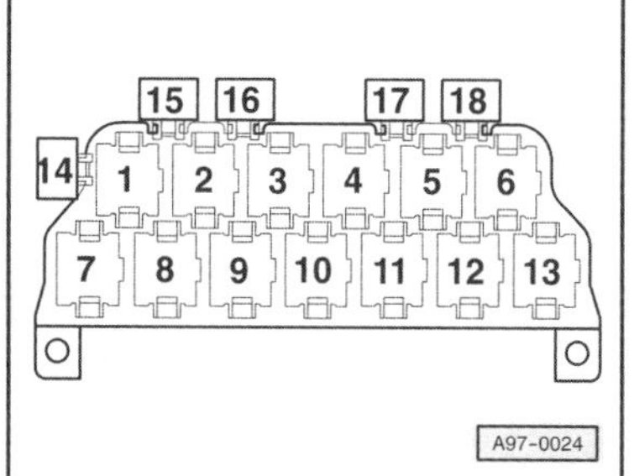

13 - Fold Relay Panel

Relay Location:

4 - Starting Interlock Relay, J207

BentleyPublishers.com—All Rights Reserved

Audi A6

Wiring Diagram **No. 28/3**

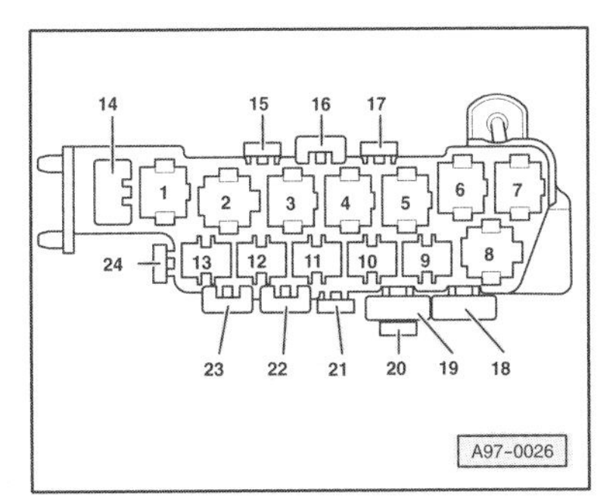

8 - Fold Relay Panel

Relay Location:

19 - Coolant Fan Fuse, S42

21 - Control Module Fuse for Coolant Fan, S142

22 - Second Speed Coolant Fan Fuse, S104

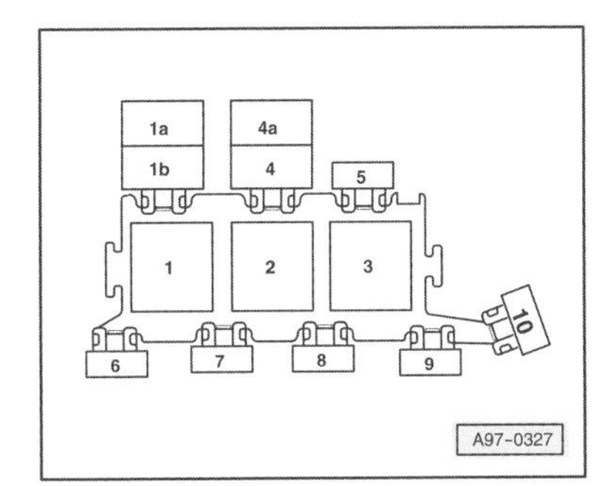

3 - Fold Relay Panel in E-Box plenum chamber

Relay Location:

1 - Secondary Air Injection (AIR) Pump Relay, J299

3 - Motronic Engine Control Module (ECM) Power Supply Relay, J271

4 - Coolant Circulation Pump Relay, J151

7 - Fuse for Secondary Air Pump, S130

9 - Engine Coolant Pump Circuit Breaker, S78

Audi A6

Wiring Diagram **No. 28/4**

Relay Panel behind Instrument Panel left, on central girder

Relay Location:

1 - Brake Booster Relay, J569

C - Hydraulic Pump Relay Fuse, S279

D - Fuel Pump (FP) Fuse, S81

BentleyPublishers.com—All Rights Reserved

Engine Management, 4.2 Liter, Engine Code BCY (RS6 2003)

Audi A6 — Wiring Diagram — No. 28/5

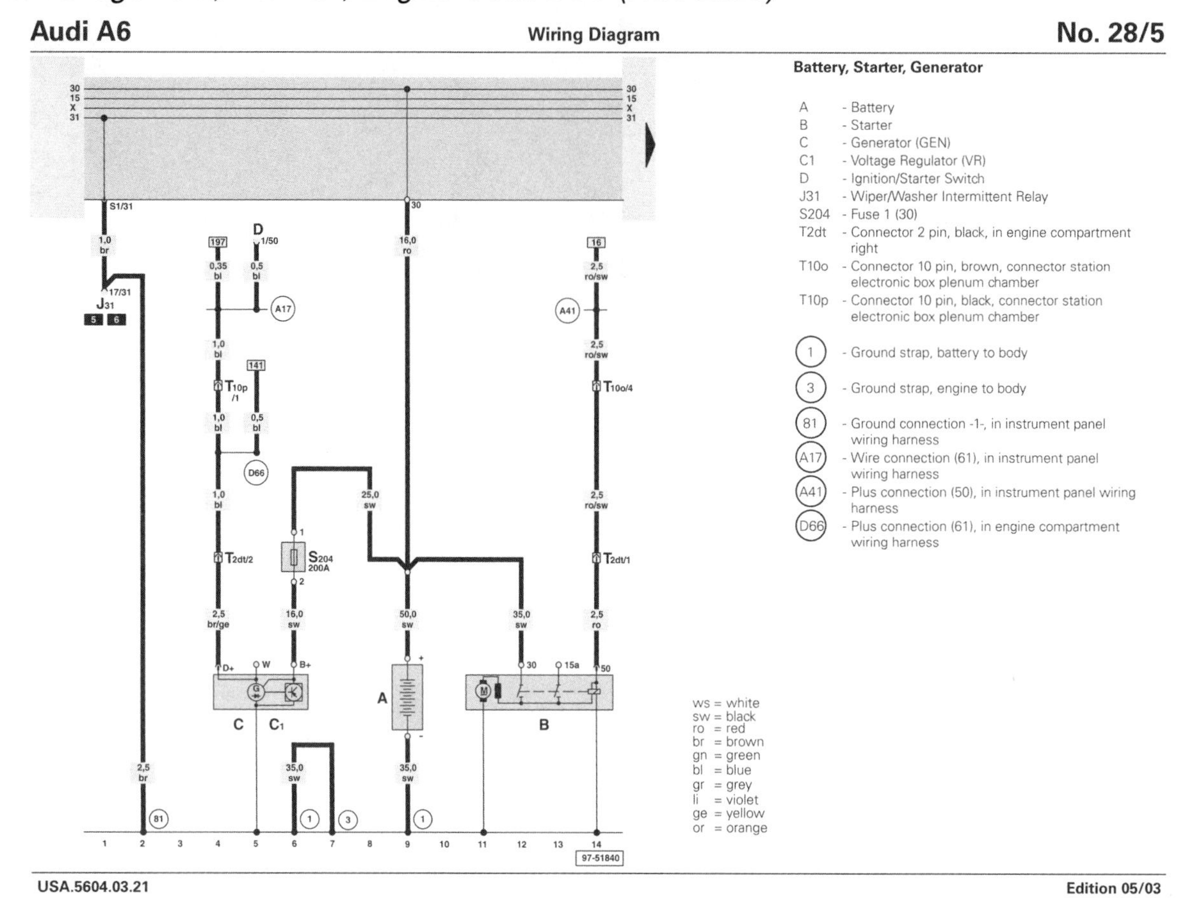

Battery, Starter, Generator

- A - Battery
- B - Starter
- C - Generator (GEN)
- C1 - Voltage Regulator (VR)
- D - Ignition/Starter Switch
- J31 - Wiper/Washer Intermittent Relay
- S204 - Fuse 1 (30)
- T2dt - Connector 2 pin, black, in engine compartment right
- T10o - Connector 10 pin, brown, connector station electronic box plenum chamber
- T10p - Connector 10 pin, black, connector station electronic box plenum chamber
- (1) - Ground strap, battery to body
- (3) - Ground strap, engine to body
- (81) - Ground connection -1-, in instrument panel wiring harness
- (A17) - Wire connection (61), in instrument panel wiring harness
- (A41) - Plus connection (50), in instrument panel wiring harness
- (D66) - Plus connection (61), in engine compartment wiring harness

ws = white
sw = black
ro = red
br = brown
gn = green
bl = blue
gr = grey
li = violet
ge = yellow
or = orange

USA.5604.03.21 — Edition 05/03

Audi A6 — Wiring Diagram — No. 28/6

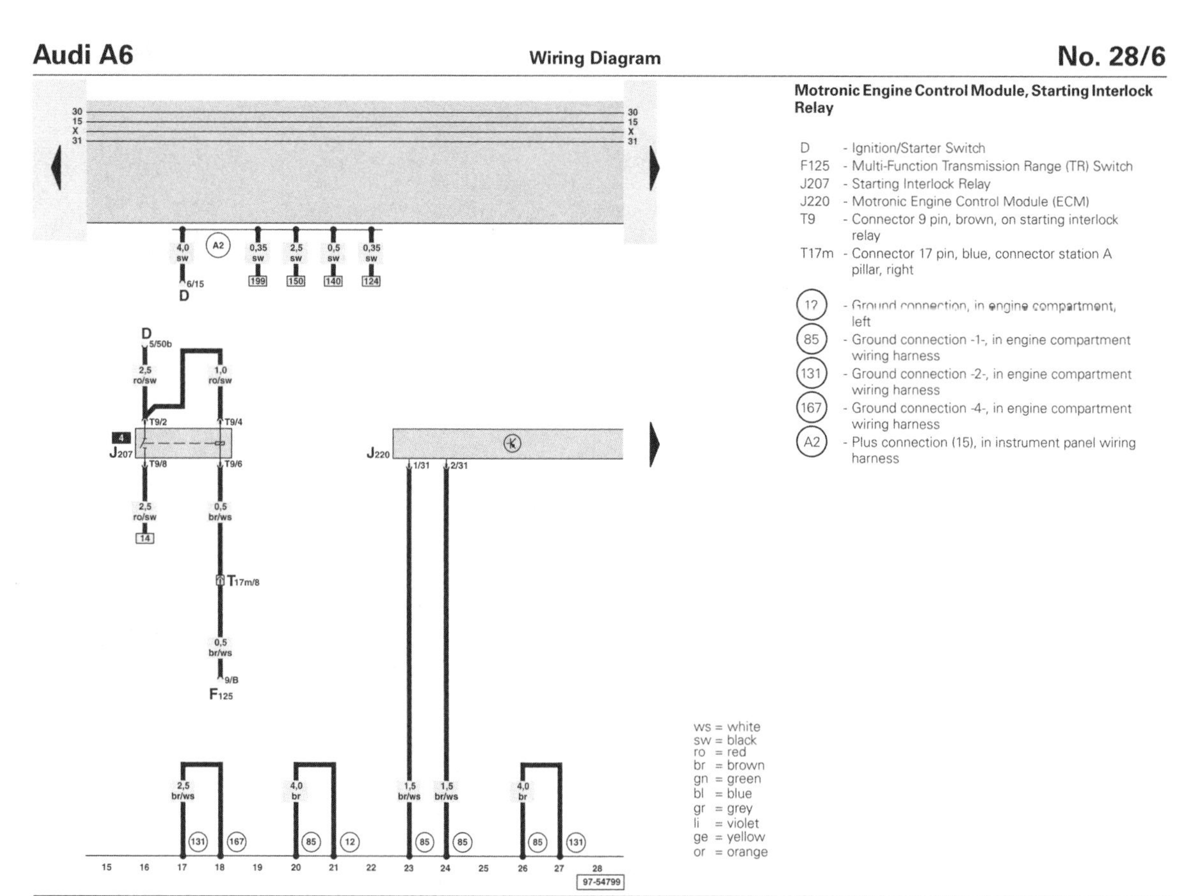

Motronic Engine Control Module, Starting Interlock Relay

- D - Ignition/Starter Switch
- F125 - Multi-Function Transmission Range (TR) Switch
- J207 - Starting Interlock Relay
- J220 - Motronic Engine Control Module (ECM)
- T9 - Connector 9 pin, brown, on starting interlock relay
- T17m - Connector 17 pin, blue, connector station A pillar, right
- (12) - Ground connection, in engine compartment, left
- (85) - Ground connection -1-, in engine compartment wiring harness
- (131) - Ground connection -2-, in engine compartment wiring harness
- (167) - Ground connection -4-, in engine compartment wiring harness
- (A2) - Plus connection (15), in instrument panel wiring harness

ws = white
sw = black
ro = red
br = brown
gn = green
bl = blue
gr = grey
li = violet
ge = yellow
or = orange

USA.5604.03.21 — Edition 05/03

BentleyPublishers.com—All Rights Reserved

Audi A6 — Wiring Diagram — No. 28/7

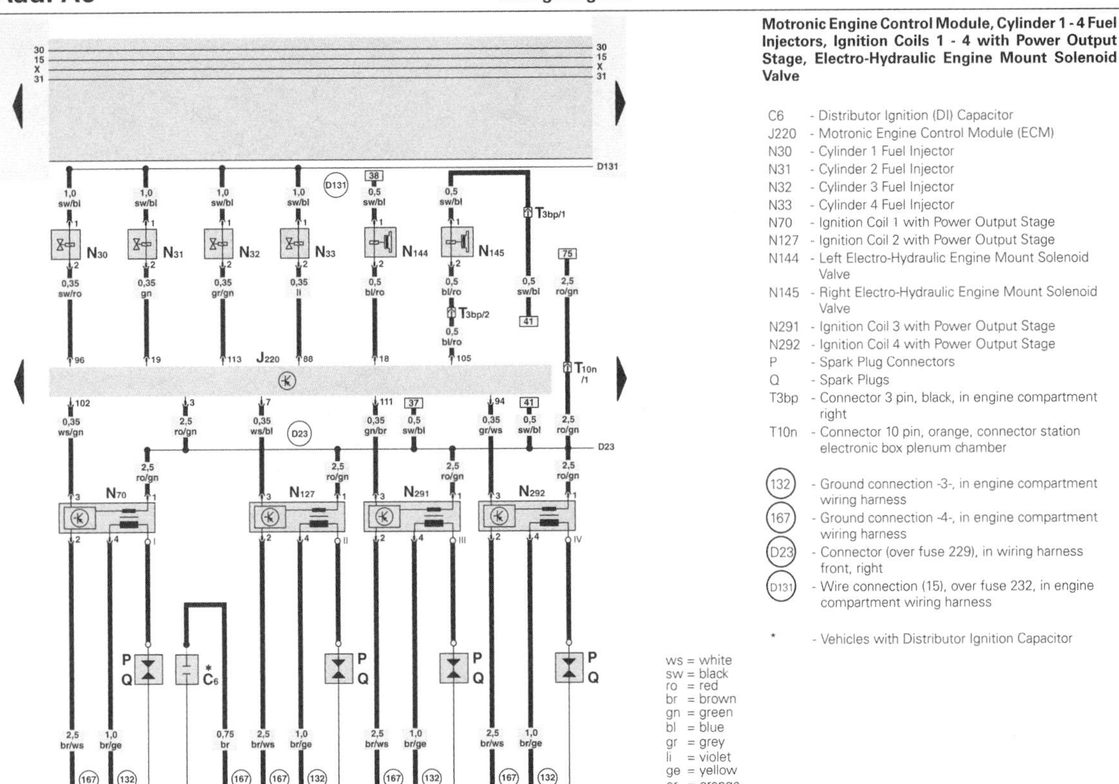

Motronic Engine Control Module, Cylinder 1 - 4 Fuel Injectors, Ignition Coils 1 - 4 with Power Output Stage, Electro-Hydraulic Engine Mount Solenoid Valve

- C6 - Distributor Ignition (DI) Capacitor
- J220 - Motronic Engine Control Module (ECM)
- N30 - Cylinder 1 Fuel Injector
- N31 - Cylinder 2 Fuel Injector
- N32 - Cylinder 3 Fuel Injector
- N33 - Cylinder 4 Fuel Injector
- N70 - Ignition Coil 1 with Power Output Stage
- N127 - Ignition Coil 2 with Power Output Stage
- N144 - Left Electro-Hydraulic Engine Mount Solenoid Valve
- N145 - Right Electro-Hydraulic Engine Mount Solenoid Valve
- N291 - Ignition Coil 3 with Power Output Stage
- N292 - Ignition Coil 4 with Power Output Stage
- P - Spark Plug Connectors
- Q - Spark Plugs
- T3bp - Connector 3 pin, black, in engine compartment right
- T10n - Connector 10 pin, orange, connector station electronic box plenum chamber

- (132) - Ground connection -3-, in engine compartment wiring harness
- (167) - Ground connection -4-, in engine compartment wiring harness
- (D23) - Connector (over fuse 229), in wiring harness front, right
- (D131) - Wire connection (15), over fuse 232, in engine compartment wiring harness

- * - Vehicles with Distributor Ignition Capacitor

ws = white
sw = black
ro = red
br = brown
gn = green
bl = blue
gr = grey
li = violet
ge = yellow
or = orange

USA.5604.03.21 — Edition 05/03

Audi A6 — Wiring Diagram — No. 28/8

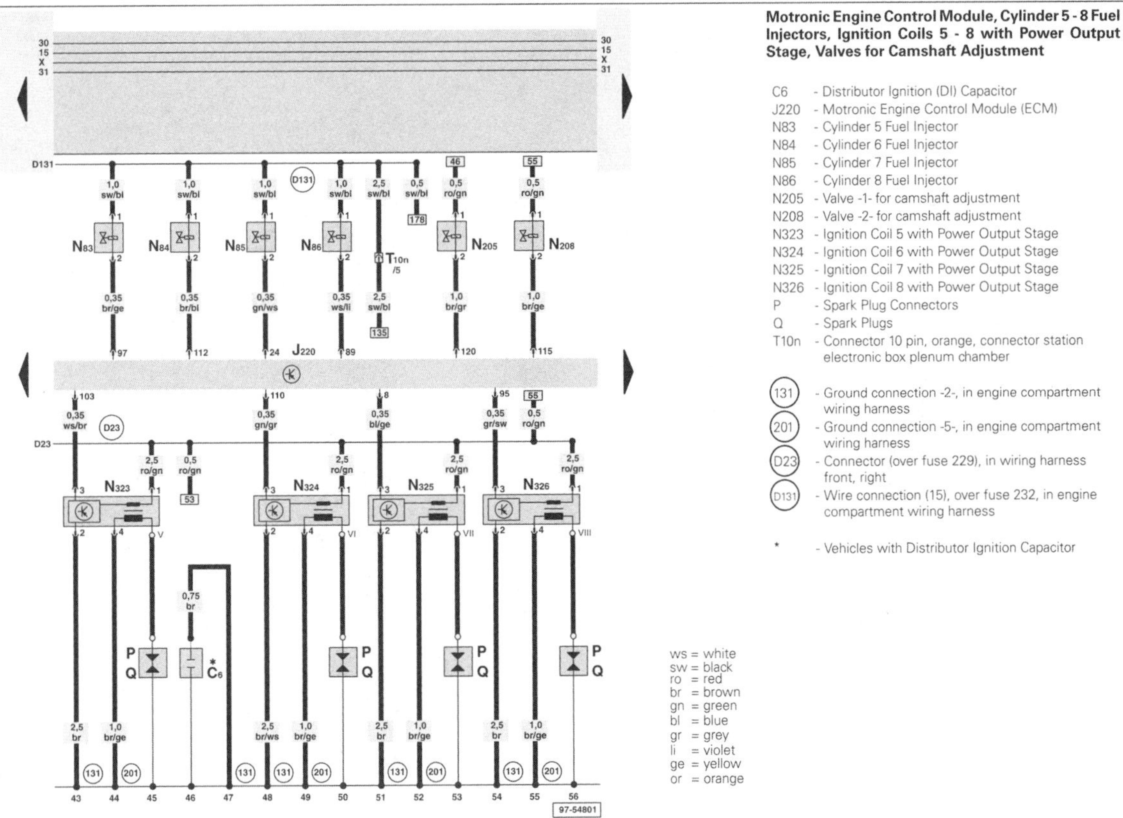

Motronic Engine Control Module, Cylinder 5 - 8 Fuel Injectors, Ignition Coils 5 - 8 with Power Output Stage, Valves for Camshaft Adjustment

- C6 - Distributor Ignition (DI) Capacitor
- J220 - Motronic Engine Control Module (ECM)
- N83 - Cylinder 5 Fuel Injector
- N84 - Cylinder 6 Fuel Injector
- N85 - Cylinder 7 Fuel Injector
- N86 - Cylinder 8 Fuel Injector
- N205 - Valve -1- for camshaft adjustment
- N208 - Valve -2- for camshaft adjustment
- N323 - Ignition Coil 5 with Power Output Stage
- N324 - Ignition Coil 6 with Power Output Stage
- N325 - Ignition Coil 7 with Power Output Stage
- N326 - Ignition Coil 8 with Power Output Stage
- P - Spark Plug Connectors
- Q - Spark Plugs
- T10n - Connector 10 pin, orange, connector station electronic box plenum chamber

- (131) - Ground connection -2-, in engine compartment wiring harness
- (201) - Ground connection -5-, in engine compartment wiring harness
- (D23) - Connector (over fuse 229), in wiring harness front, right
- (D131) - Wire connection (15), over fuse 232, in engine compartment wiring harness

- * - Vehicles with Distributor Ignition Capacitor

ws = white
sw = black
ro = red
br = brown
gn = green
bl = blue
gr = grey
li = violet
ge = yellow
or = orange

USA.5604.03.21 — Edition 05/03

BentleyPublishers.com—All Rights Reserved

Engine Management, 4.2 Liter, Engine Code BCY (RS6 2003)

Audi A6 Wiring Diagram **No. 28/9**

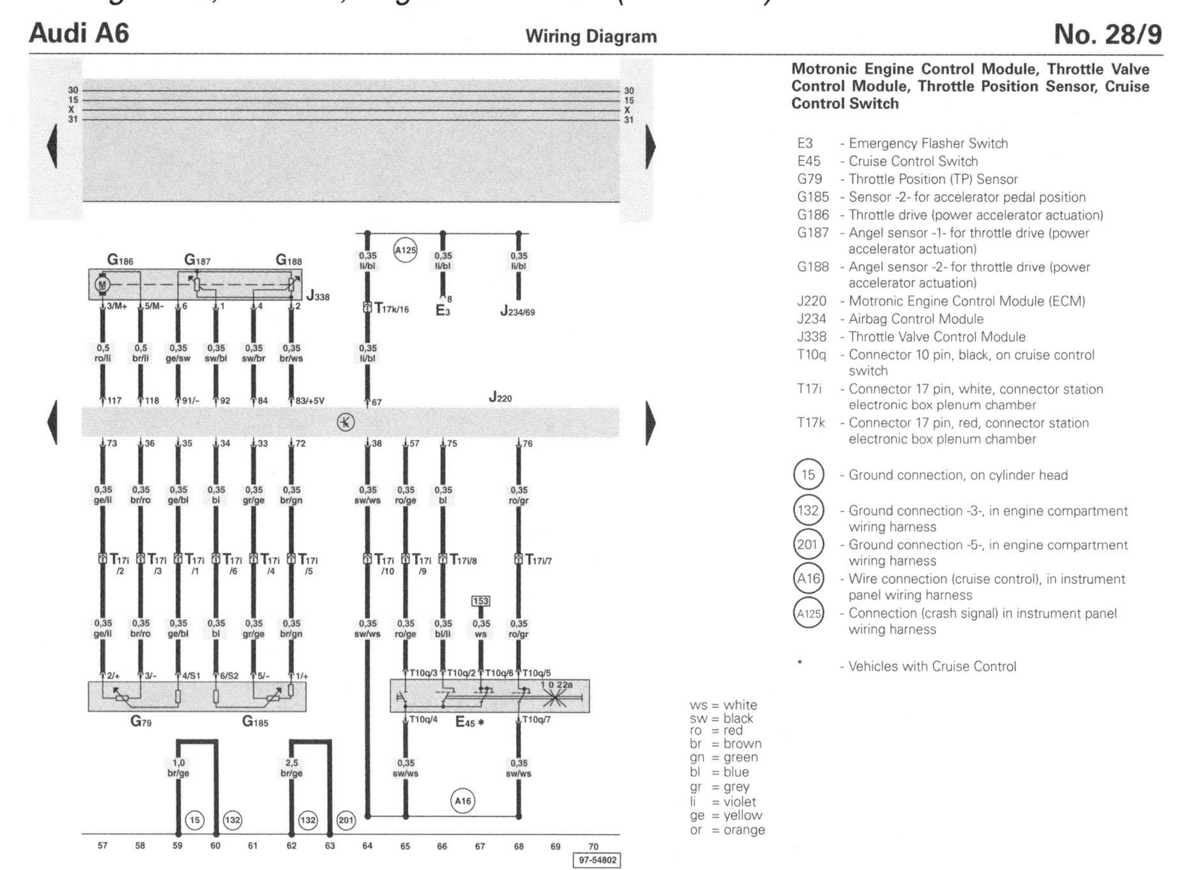

Motronic Engine Control Module, Throttle Valve Control Module, Throttle Position Sensor, Cruise Control Switch

- E3 - Emergency Flasher Switch
- E45 - Cruise Control Switch
- G79 - Throttle Position (TP) Sensor
- G185 - Sensor -2- for accelerator pedal position
- G186 - Throttle drive (power accelerator actuation)
- G187 - Angel sensor -1- for throttle drive (power accelerator actuation)
- G188 - Angel sensor -2- for throttle drive (power accelerator actuation)
- J220 - Motronic Engine Control Module (ECM)
- J234 - Airbag Control Module
- J338 - Throttle Valve Control Module
- T10q - Connector 10 pin, black, on cruise control switch
- T17i - Connector 17 pin, white, connector station electronic box plenum chamber
- T17k - Connector 17 pin, red, connector station electronic box plenum chamber

- (15) - Ground connection, on cylinder head
- (132) - Ground connection -3-, in engine compartment wiring harness
- (201) - Ground connection -5-, in engine compartment wiring harness
- (A16) - Wire connection (cruise control), in instrument panel wiring harness
- (A125) - Connection (crash signal) in instrument panel wiring harness

* - Vehicles with Cruise Control

ws = white
sw = black
ro = red
br = brown
gn = green
bl = blue
gr = grey
li = violet
ge = yellow
or = orange

USA.5604.03.21 Edition 05/03

Audi A6 Wiring Diagram **No. 28/10**

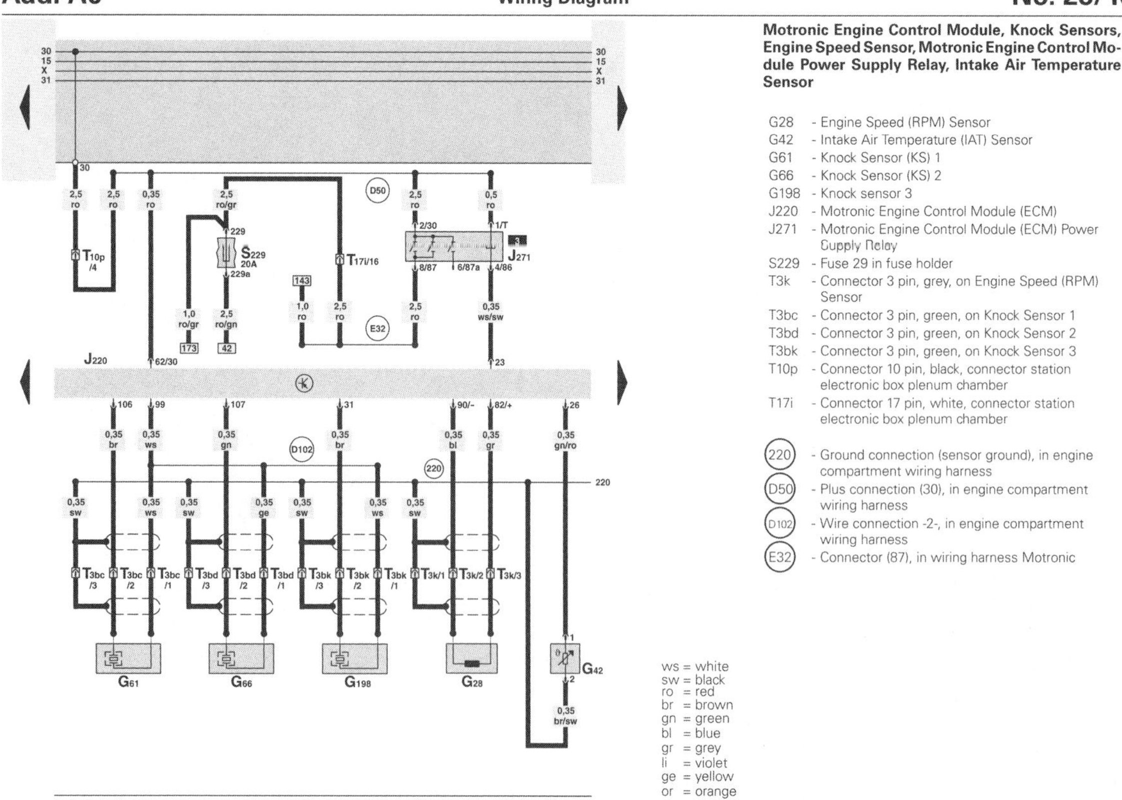

Motronic Engine Control Module, Knock Sensors, Engine Speed Sensor, Motronic Engine Control Module Power Supply Relay, Intake Air Temperature Sensor

- G28 - Engine Speed (RPM) Sensor
- G42 - Intake Air Temperature (IAT) Sensor
- G61 - Knock Sensor (KS) 1
- G66 - Knock Sensor (KS) 2
- G198 - Knock sensor 3
- J220 - Motronic Engine Control Module (ECM)
- J271 - Motronic Engine Control Module (ECM) Power Supply Relay
- S229 - Fuse 29 in fuse holder
- T3k - Connector 3 pin, grey, on Engine Speed (RPM) Sensor
- T3bc - Connector 3 pin, green, on Knock Sensor 1
- T3bd - Connector 3 pin, green, on Knock Sensor 2
- T3bk - Connector 3 pin, green, on Knock Sensor 3
- T10p - Connector 10 pin, black, connector station electronic box plenum chamber
- T17i - Connector 17 pin, white, connector station electronic box plenum chamber

- (220) - Ground connection (sensor ground), in engine compartment wiring harness
- (D50) - Plus connection (30), in engine compartment wiring harness
- (D102) - Wire connection -2-, in engine compartment wiring harness
- (E32) - Connector (87), in wiring harness Motronic

ws = white
sw = black
ro = red
br = brown
gn = green
bl = blue
gr = grey
li = violet
ge = yellow
or = orange

USA.5604.03.21 Edition 05/03

BentleyPublishers.com—All Rights Reserved

Audi A6 **Wiring Diagram** **No. 28/11**

Motronic Engine Control Module, Camshaft Position Sensor, Engine Coolant Temperature Sensor, Charge Air Pressure Sensor, Brake Booster Pressure Sensor

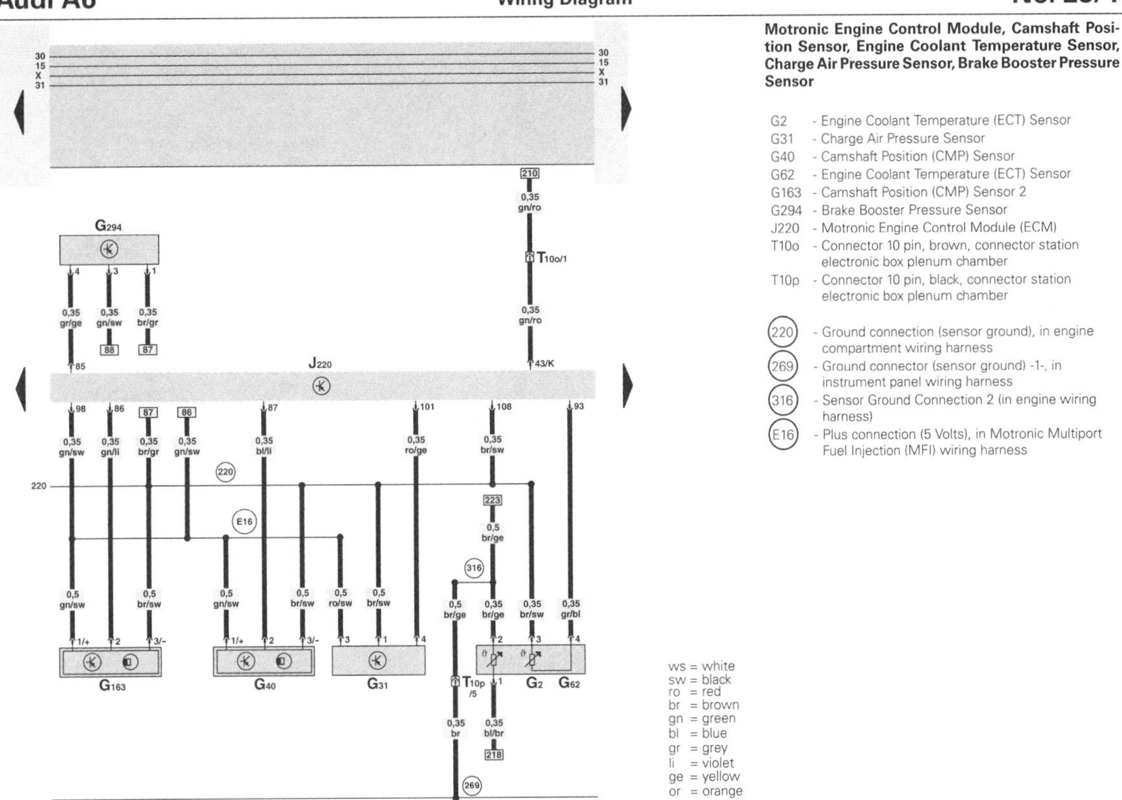

- G2 - Engine Coolant Temperature (ECT) Sensor
- G31 - Charge Air Pressure Sensor
- G40 - Camshaft Position (CMP) Sensor
- G62 - Engine Coolant Temperature (ECT) Sensor
- G163 - Camshaft Position (CMP) Sensor 2
- G294 - Brake Booster Pressure Sensor
- J220 - Motronic Engine Control Module (ECM)
- T10o - Connector 10 pin, brown, connector station electronic box plenum chamber
- T10p - Connector 10 pin, black, connector station electronic box plenum chamber

- (220) - Ground connection (sensor ground), in engine compartment wiring harness
- (269) - Ground connector (sensor ground) -1-, in instrument panel wiring harness
- (316) - Sensor Ground Connection 2 (in engine wiring harness)
- (E16) - Plus connection (5 Volts), in Motronic Multiport Fuel Injection (MFI) wiring harness

ws = white
sw = black
ro = red
br = brown
gn = green
bl = blue
gr = grey
li = violet
ge = yellow
or = orange

USA.5604.03.21 Edition 05/03

Audi A6 **Wiring Diagram** **No. 28/12**

Motronic Engine Control Module, Secondary Air Injection (AIR) Pump Relay, Evaporative Emission Canister Purge Regulator Valve, Mass Air Flow Sensor

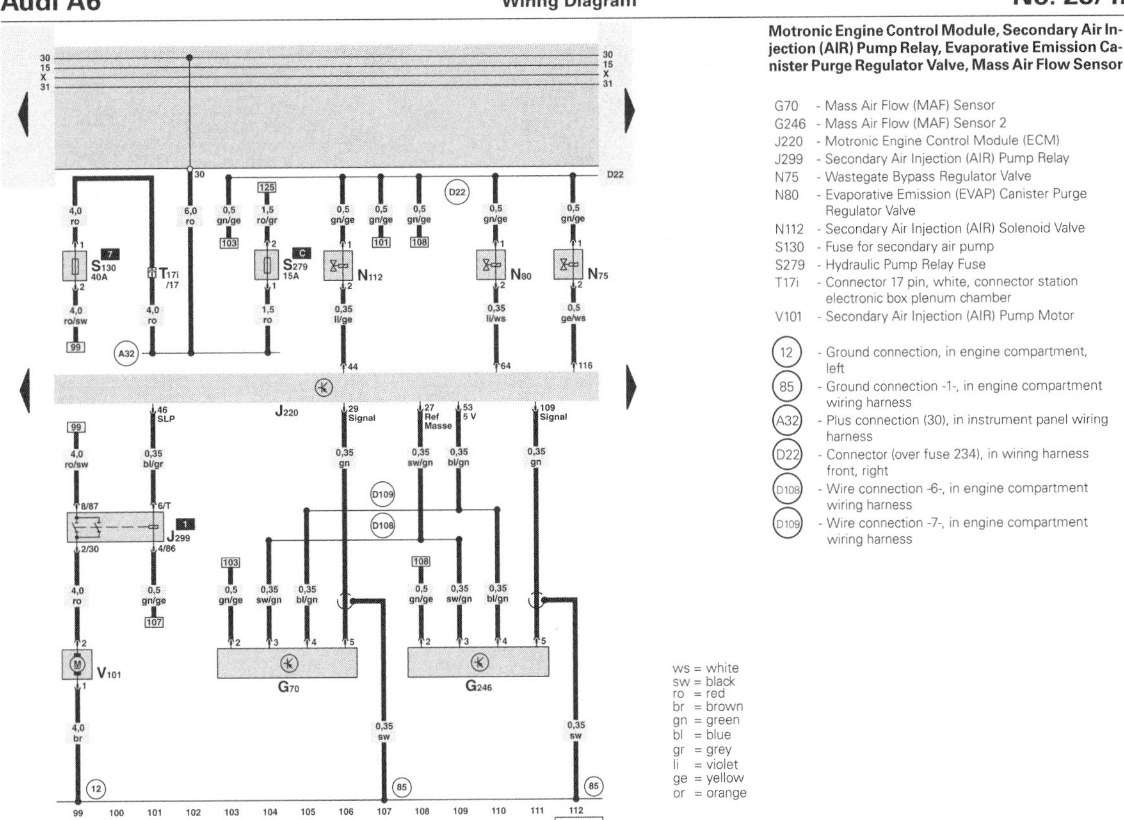

- G70 - Mass Air Flow (MAF) Sensor
- G246 - Mass Air Flow (MAF) Sensor 2
- J220 - Motronic Engine Control Module (ECM)
- J299 - Secondary Air Injection (AIR) Pump Relay
- N75 - Wastegate Bypass Regulator Valve
- N80 - Evaporative Emission (EVAP) Canister Purge Regulator Valve
- N112 - Secondary Air Injection (AIR) Solenoid Valve
- S130 - Fuse for secondary air pump
- S279 - Hydraulic Pump Relay Fuse
- T17i - Connector 17 pin, white, connector station electronic box plenum chamber
- V101 - Secondary Air Injection (AIR) Pump Motor

- (12) - Ground connection, in engine compartment, left
- (85) - Ground connection -1-, in engine compartment wiring harness
- (A32) - Plus connection (30), in instrument panel wiring harness
- (D22) - Connector (over fuse 234), in wiring harness front, right
- (D108) - Wire connection -6-, in engine compartment wiring harness
- (D109) - Wire connection -7-, in engine compartment wiring harness

ws = white
sw = black
ro = red
br = brown
gn = green
bl = blue
gr = grey
li = violet
ge = yellow
or = orange

USA.5604.03.21 Edition 05/03

BentleyPublishers.com—All Rights Reserved

Engine Management, 4.2 Liter, Engine Code BCY (RS6 2003)

Audi A6 — **Wiring Diagram** — **No. 28/13**

Motronic Engine Control Module, Heated Oxygen Sensors, Sensors for Exhaust Temperature, Recirculating Valve for Turbocharger, Brake Booster Relay

- G39 - Heated Oxygen Sensor (HO2S)
- G108 - Heated Oxygen Sensor (HO2S) 2
- G235 - Sensor -1- for exhaust temperature
- G236 - Sensor -2- for exhaust temperature
- J220 - Motronic Engine Control Module (ECM)
- J569 - Brake Booster relay
- N249 - Recirculating valve for turbocharger
- T2do - Connector 2 pin, black, for Recirculating valve for turbocharger
- T2dp - Connector 2 pin, green, on Recirculating valve for turbocharger
- T4u - Connector 4 pin, black, heated oxygen sensor
- T4v - Connector 4 pin, black, heated oxygen sensor 2
- T17k - Connector 17 pin, red, connector station electronic box plenum chamber
- V192 - Brake System Vacuum Pump
- Z19 - Oxygen Sensor (O2S) Heater
- Z28 - Oxygen Sensor (O2S) 2 Heater

- (85) - Ground connection -1-, in engine compartment wiring harness
- (D22) - Connector (over fuse 234), in wiring harness front, right
- (D110) - Wire connection -8-, in engine compartment wiring harness

ws = white
sw = black
ro = red
br = brown
gn = green
bl = blue
gr = grey
li = violet
ge = yellow
or = orange

97-54806

USA.5604.03.21 — Edition 05/03

Audi A6 — **Wiring Diagram** — **No. 28/14**

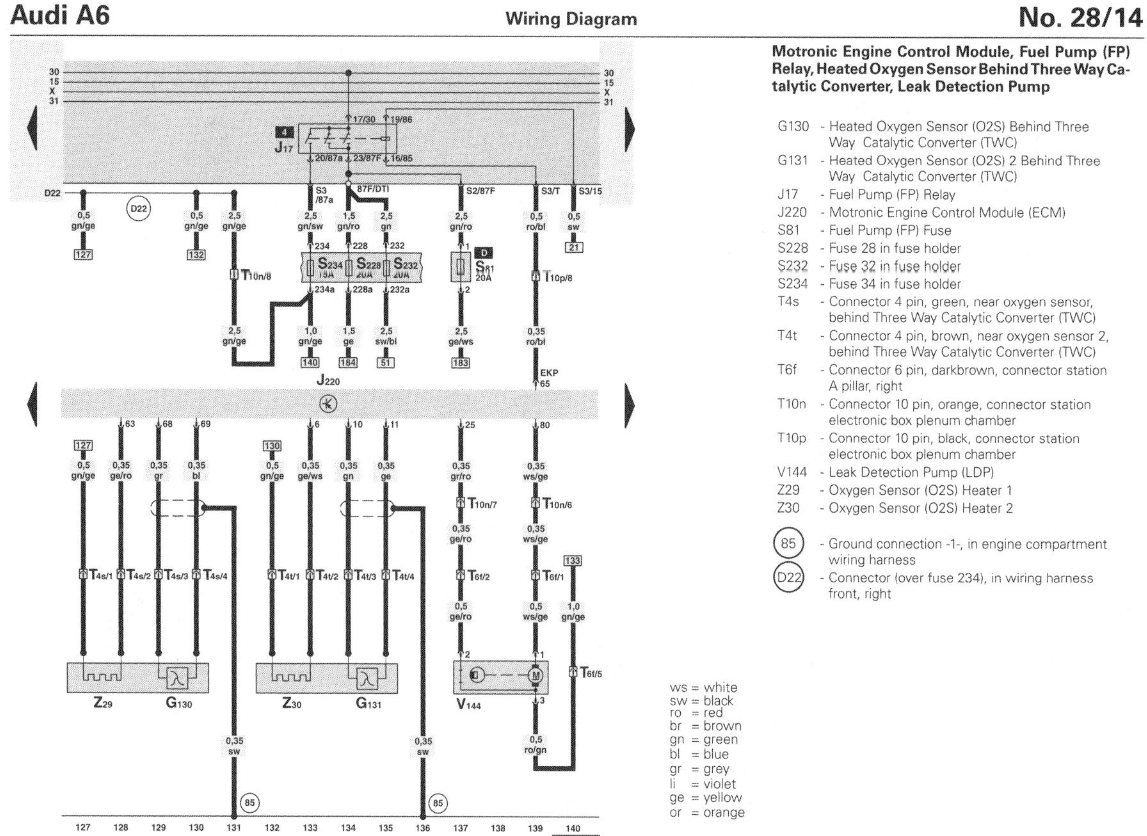

Motronic Engine Control Module, Fuel Pump (FP) Relay, Heated Oxygen Sensor Behind Three Way Catalytic Converter, Leak Detection Pump

- G130 - Heated Oxygen Sensor (O2S) Behind Three Way Catalytic Converter (TWC)
- G131 - Heated Oxygen Sensor (O2S) 2 Behind Three Way Catalytic Converter (TWC)
- J17 - Fuel Pump (FP) Relay
- J220 - Motronic Engine Control Module (ECM)
- S81 - Fuel Pump (FP) Fuse
- S228 - Fuse 28 in fuse holder
- S232 - Fuse 32 in fuse holder
- S234 - Fuse 34 in fuse holder
- T4s - Connector 4 pin, green, near oxygen sensor, behind Three Way Catalytic Converter (TWC)
- T4t - Connector 4 pin, brown, near oxygen sensor 2, behind Three Way Catalytic Converter (TWC)
- T6f - Connector 6 pin, darkbrown, connector station A pillar, right
- T10n - Connector 10 pin, orange, connector station electronic box plenum chamber
- T10p - Connector 10 pin, black, connector station electronic box plenum chamber
- V144 - Leak Detection Pump (LDP)
- Z29 - Oxygen Sensor (O2S) Heater 1
- Z30 - Oxygen Sensor (O2S) Heater 2

- (85) - Ground connection -1-, in engine compartment wiring harness
- (D22) - Connector (over fuse 234), in wiring harness front, right

ws = white
sw = black
ro = red
br = brown
gn = green
bl = blue
gr = grey
li = violet
ge = yellow
or = orange

97-54807

USA.5604.03.21 — Edition 05/03

BentleyPublishers.com—All Rights Reserved

Engine Management, 4.2 Liter, Engine Code BCY (RS6 2003)

Audi A6 **Wiring Diagram** **No. 28/15**

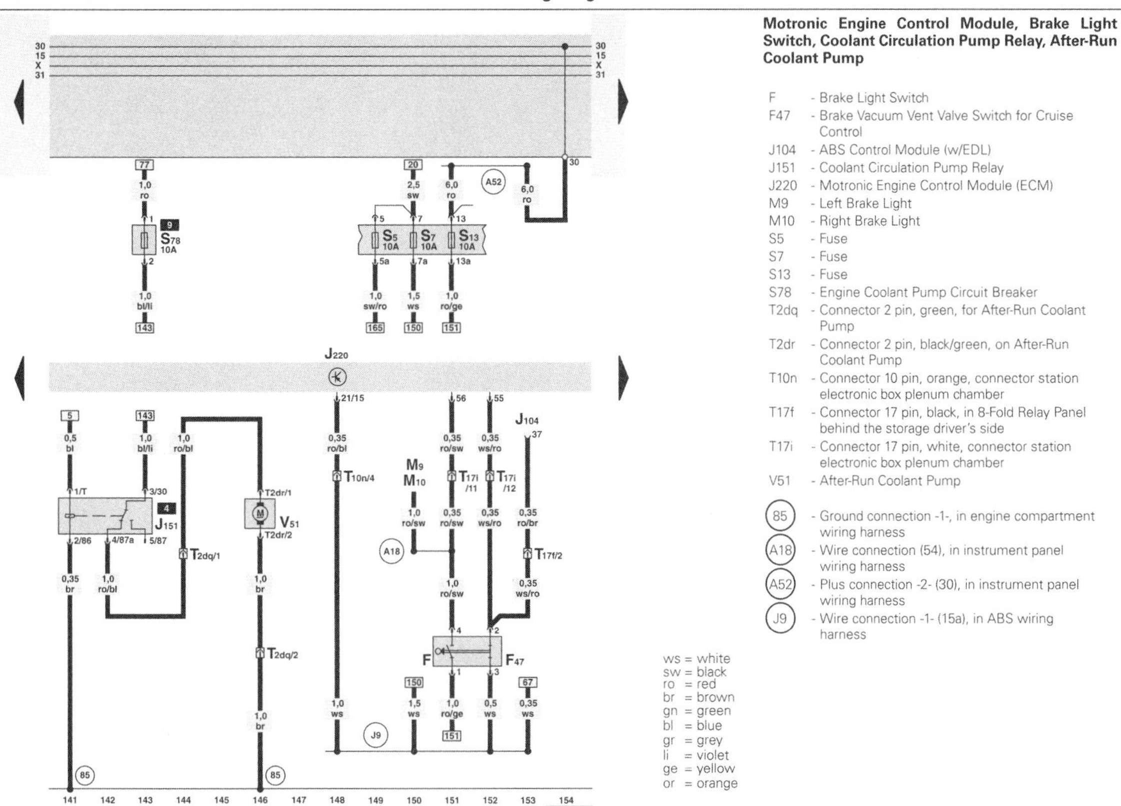

Motronic Engine Control Module, Brake Light Switch, Coolant Circulation Pump Relay, After-Run Coolant Pump

- F - Brake Light Switch
- F47 - Brake Vacuum Vent Valve Switch for Cruise Control
- J104 - ABS Control Module (w/EDL)
- J151 - Coolant Circulation Pump Relay
- J220 - Motronic Engine Control Module (ECM)
- M9 - Left Brake Light
- M10 - Right Brake Light
- S5 - Fuse
- S7 - Fuse
- S13 - Fuse
- S78 - Engine Coolant Pump Circuit Breaker
- T2dq - Connector 2 pin, green, for After-Run Coolant Pump
- T2dr - Connector 2 pin, black/green, on After-Run Coolant Pump
- T10n - Connector 10 pin, orange, connector station electronic box plenum chamber
- T17f - Connector 17 pin, black, in 8-Fold Relay Panel behind the storage driver's side
- T17i - Connector 17 pin, white, connector station electronic box plenum chamber
- V51 - After-Run Coolant Pump

- (85) - Ground connection -1-, in engine compartment wiring harness
- (A18) - Wire connection (54), in instrument panel wiring harness
- (A52) - Plus connection -2- (30), in instrument panel wiring harness
- (J9) - Wire connection -1- (15a), in ABS wiring harness

ws = white
sw = black
ro = red
br = brown
gn = green
bl = blue
gr = grey
li = violet
ge = yellow
or = orange

USA.5604.03.21 Edition 05/03

Audi A6 **Wiring Diagram** **No. 28/16**

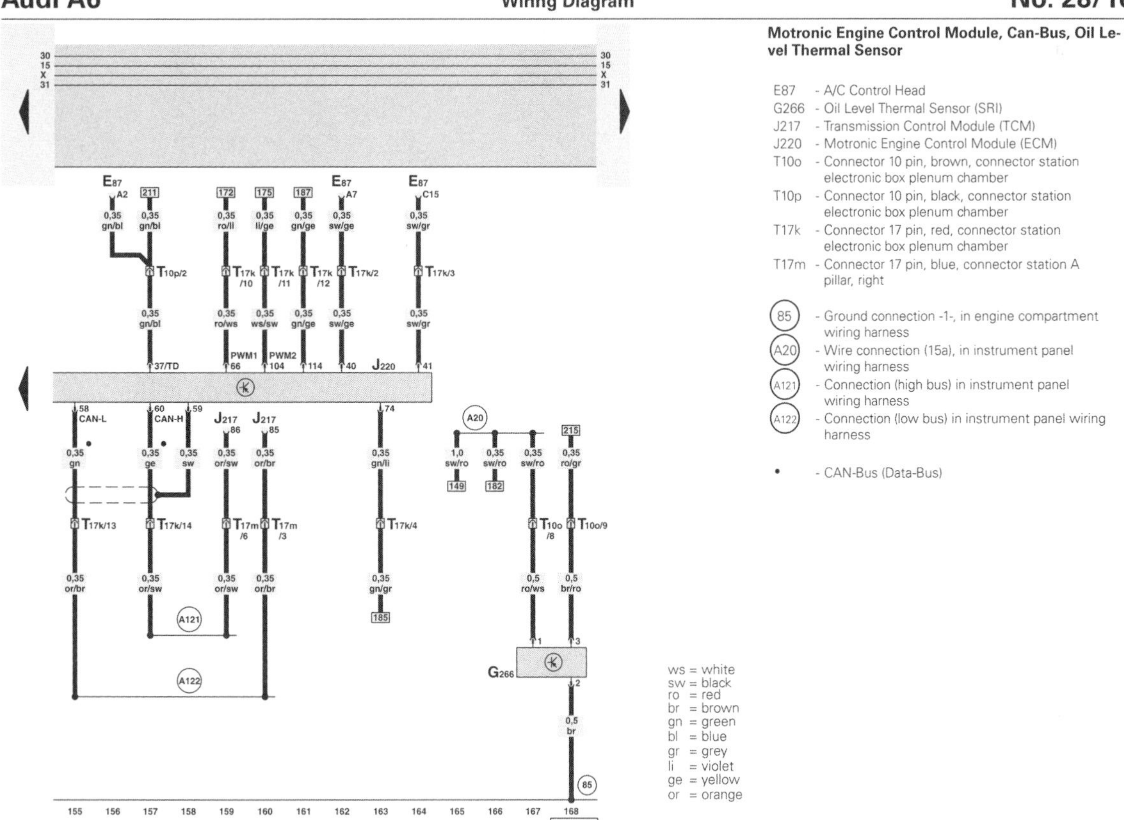

Motronic Engine Control Module, Can-Bus, Oil Level Thermal Sensor

- E87 - A/C Control Head
- G266 - Oil Level Thermal Sensor (SRI)
- J217 - Transmission Control Module (TCM)
- J220 - Motronic Engine Control Module (ECM)
- T10o - Connector 10 pin, brown, connector station electronic box plenum chamber
- T10p - Connector 10 pin, black, connector station electronic box plenum chamber
- T17k - Connector 17 pin, red, connector station electronic box plenum chamber
- T17m - Connector 17 pin, blue, connector station A pillar, right

- (85) - Ground connection -1-, in engine compartment wiring harness
- (A20) - Wire connection (15a), in instrument panel wiring harness
- (A121) - Connection (high bus) in instrument panel wiring harness
- (A122) - Connection (low bus) in instrument panel wiring harness

- • - CAN-Bus (Data-Bus)

ws = white
sw = black
ro = red
br = brown
gn = green
bl = blue
gr = grey
li = violet
ge = yellow
or = orange

USA.5604.03.21 Edition 05/03

BentleyPublishers.com—All Rights Reserved

Engine Management, 4.2 Liter, Engine Code BCY (RS6 2003)

Audi A6 — Wiring Diagram — No. 28/17

Coolant FC (Fan Control) Control Module, Coolant FC (Fan Control) Control Module 2, High Pressure Sensor, Valve for Coolant Circulation, Coolant Fan

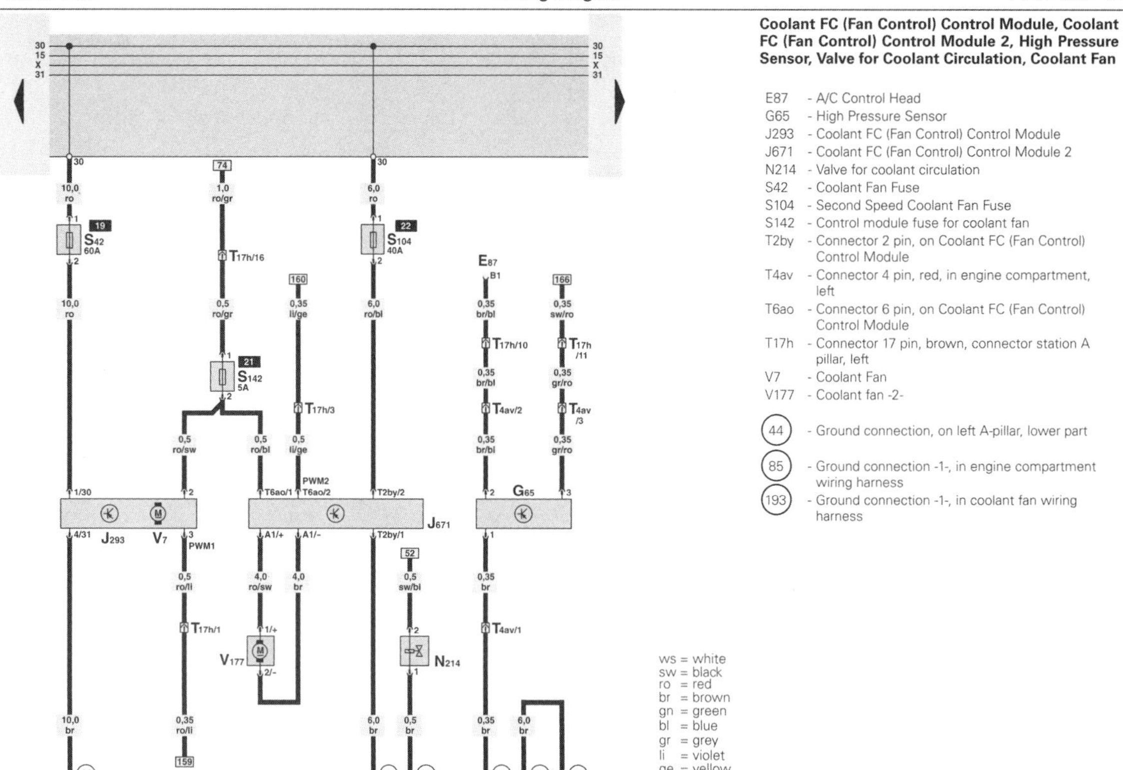

- E87 - A/C Control Head
- G65 - High Pressure Sensor
- J293 - Coolant FC (Fan Control) Control Module
- J671 - Coolant FC (Fan Control) Control Module 2
- N214 - Valve for coolant circulation
- S42 - Coolant Fan Fuse
- S104 - Second Speed Coolant Fan Fuse
- S142 - Control module fuse for coolant fan
- T2by - Connector 2 pin, on Coolant FC (Fan Control) Control Module
- T4av - Connector 4 pin, red, in engine compartment, left
- T6ao - Connector 6 pin, on Coolant FC (Fan Control) Control Module
- T17h - Connector 17 pin, brown, connector station A pillar, left
- V7 - Coolant Fan
- V177 - Coolant fan -2-
- (44) - Ground connection, on left A-pillar, lower part
- (85) - Ground connection -1-, in engine compartment wiring harness
- (193) - Ground connection -1-, in coolant fan wiring harness

ws = white
sw = black
ro = red
br = brown
gn = green
bl = blue
gr = grey
li = violet
ge = yellow
or = orange

USA.5604.03.21 — Edition 05/03

Audi A6 — Wiring Diagram — No. 28/18

Fuel Pump (FP) Control Module, Fuel Pump (FP), Sender for Fuel Gauge

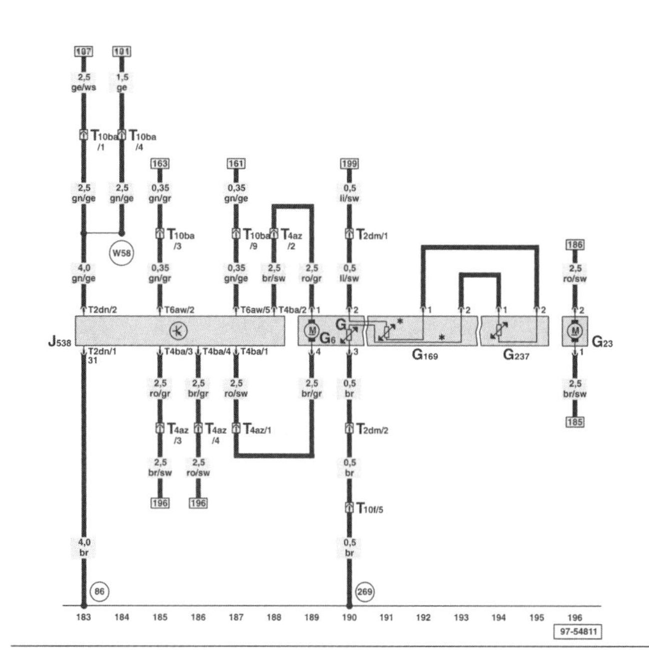

- G - Sender for fuel gauge
- G6 - Fuel Pump (FP)
- G23 - Transfer Fuel Pump (FP)
- G169 - Fuel Level Sensor 2
- G237 - Sensor -3- for fuel supply
- J538 - Fuel Pump (FP) Control Module
- T2dm - Connector 2 pin, below rear seat
- T2dn - Connector 2 pin, on Fuel Pump (FP) Control Module
- T4az - Connector 4 pin, below rear seat
- T4ba - Connector 4 pin, on Fuel Pump (FP) Control Module
- T6aw - Connector 6 pin, on Fuel Pump (FP) Control Module
- T10f - Connector 10 pin, brown, connector station A pillar, left
- T10ba - Connector 10 pin, blue, connector station A pillar, left
- (86) - Ground connection -1-, in rear wiring harness
- (269) - Ground connector (sensor ground) -1-, in instrument panel wiring harness
- (W58) - Connection (87) (in rear wiring harness)
- * - Wiring inside Fuel Tank

ws = white
sw = black
ro = red
br = brown
gn = green
bl = blue
gr = grey
li = violet
ge = yellow
or = orange

USA.5604.03.21 — Edition 05/03

BentleyPublishers.com—All Rights Reserved

Audi A6 — Wiring Diagram — No. 28/19

Control Module with Indicator Unit in Instrument Panel Insert, Fuel Gauge

- G1 - Fuel Gauge
- J285 - Control Module with indicator unit in instrument panel insert
- K2 - Generator (GEN) Warning Light
- T10f - Connector 10 pin, brown, connector station A pillar, left
- T16a - Connector 16 pin, Data Link Connector (DLC)
- T32 - Connector 32 pin, blue, on instrument cluster
- T32a - Connector 32 pin, green, on instrument cluster
- (269) - Ground connector (sensor ground) -1-, in instrument panel wiring harness
- (A76) - Connector (K-diagnosis wire), in instrument panel wiring harness
- # - Sensor Ground Output

ws = white
sw = black
ro = red
br = brown
gn = green
bl = blue
gr = grey
li = violet
ge = yellow
or = orange

197 198 199 200 201 202 203 204 205 206 207 208 209 210

97-54812

Audi A6 — Wiring Diagram — No. 28/20

Control Module with Indicator Unit in Instrument Panel Insert, Oil Pressure Switch, Engine Coolant Level Warning Switch, Speedometer Vehicle Speed Sensor (VSS)

- F1 - Oil Pressure Switch
- F66 - Engine Coolant Level (ECL) Warning Switch
- G3 - Engine Coolant Temperature (ECT) Gauge
- G5 - Tachometer
- G21 - Speedometer
- G22 - Speedometer Vehicle Speed Sensor (VSS)
- J285 - Control Module with indicator unit in instrument panel insert
- K3 - Oil Pressure Warning Light
- K28 - Engine Coolant Level/Temperature (ECL/ECT) Warning Light
- K83 - Malfunction Indicator Lamp (MIL)
- K132 - Electronic Power Control (EPC) Warning Lamp
- T2dv - Connector 2 pin, black, for Speedometer Vehicle Speed Sensor (VSS)
- T3bp - Connector 3 pin, black, in engine compartment right
- T3br - Connector 3 pin, on Speedometer Vehicle Speed Sensor (VSS)
- T10p - Connector 10 pin, black, connector station electronic box plenum chamber
- T32 - Connector 32 pin, blue, on instrument cluster
- T32a - Connector 32 pin, green, on instrument cluster
- (85) - Ground connection -1-, in engine compartment wiring harness
- (A108) - Connector (speed signal), in instrument panel wiring harness

ws = white
sw = black
ro = red
br = brown
gn = green
bl = blue
gr = grey
li = violet
ge = yellow
or = orange

211 212 213 214 215 216 217 218 219 220 221 222 223 224

97-51854

BentleyPublishers.com—All Rights Reserved

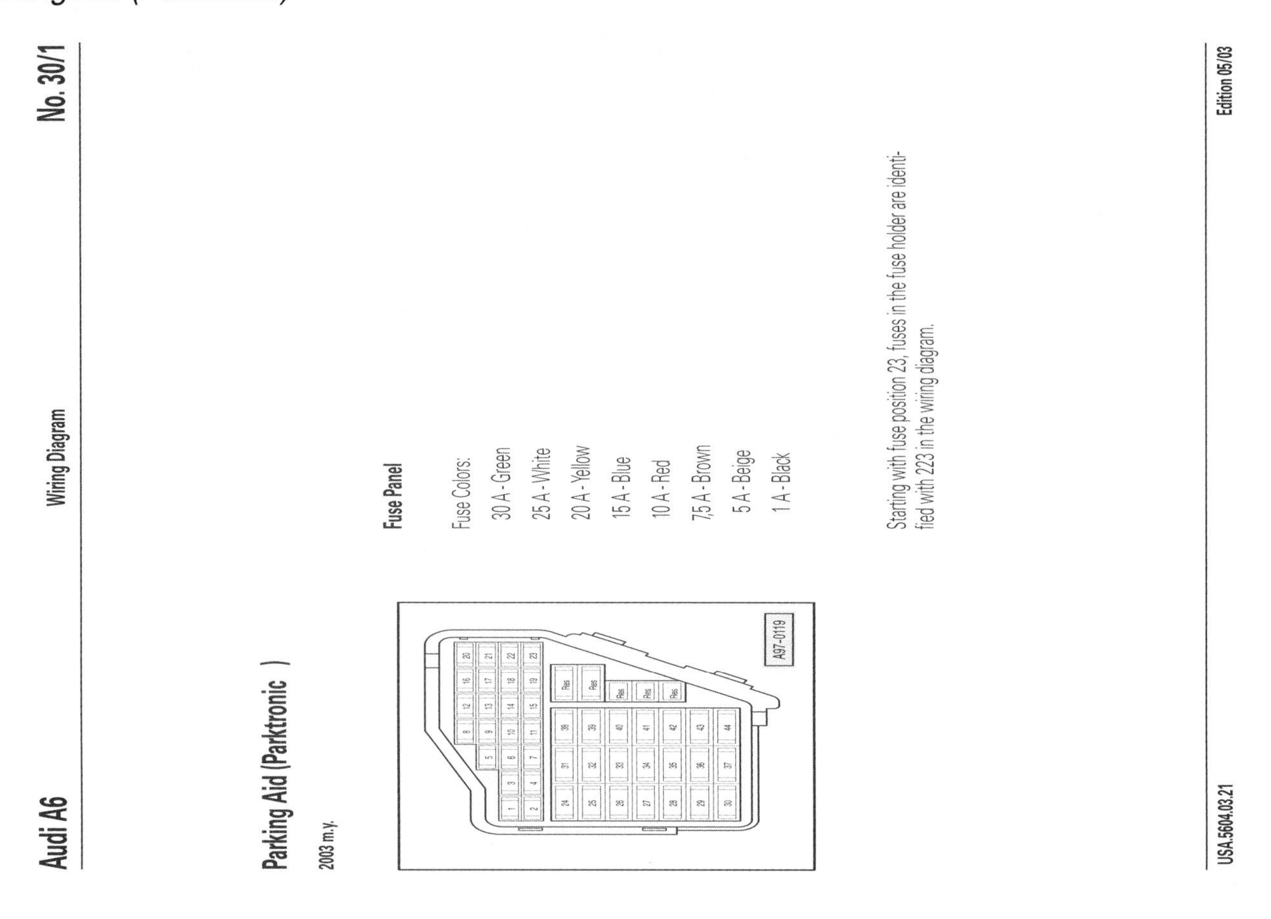

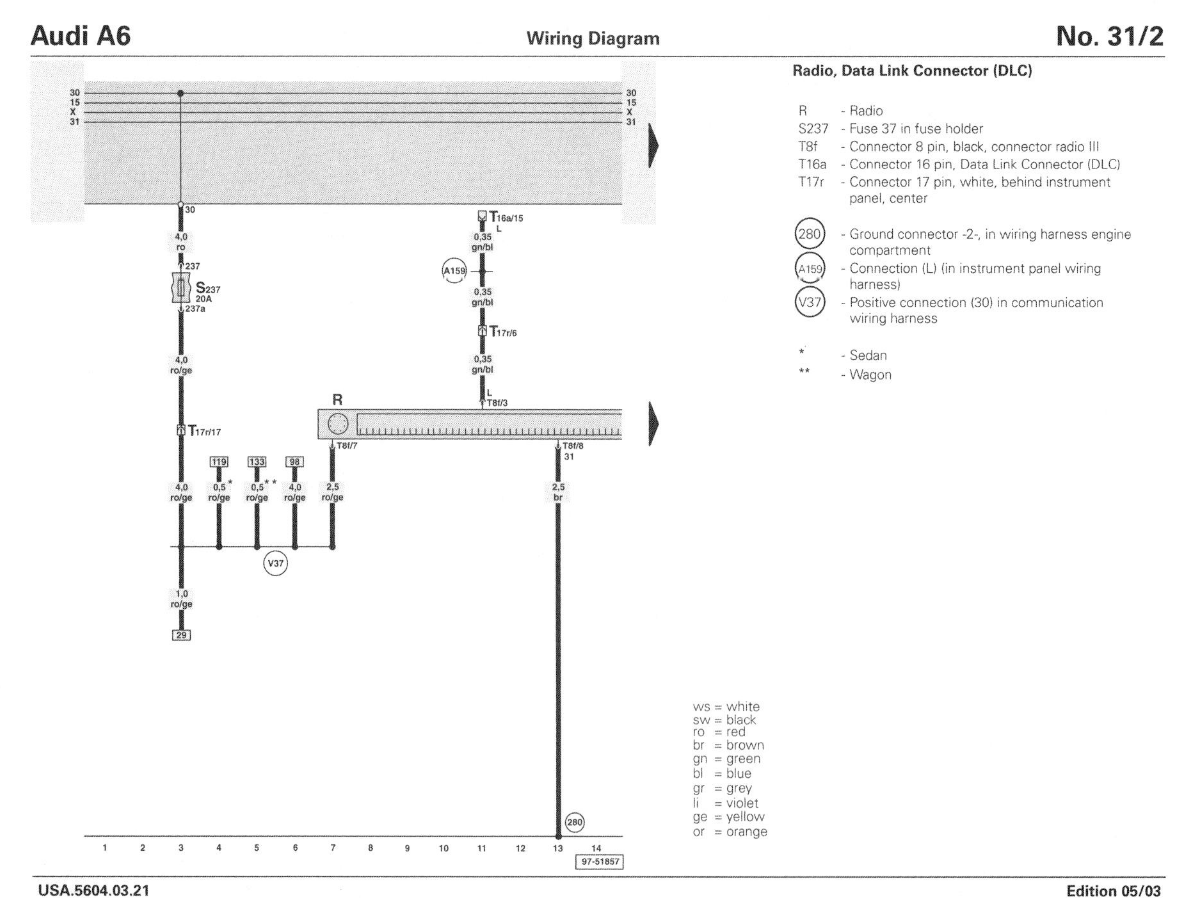

BentleyPublishers.com—All Rights Reserved

Audi A6 **Wiring Diagram** **No. 30/3**

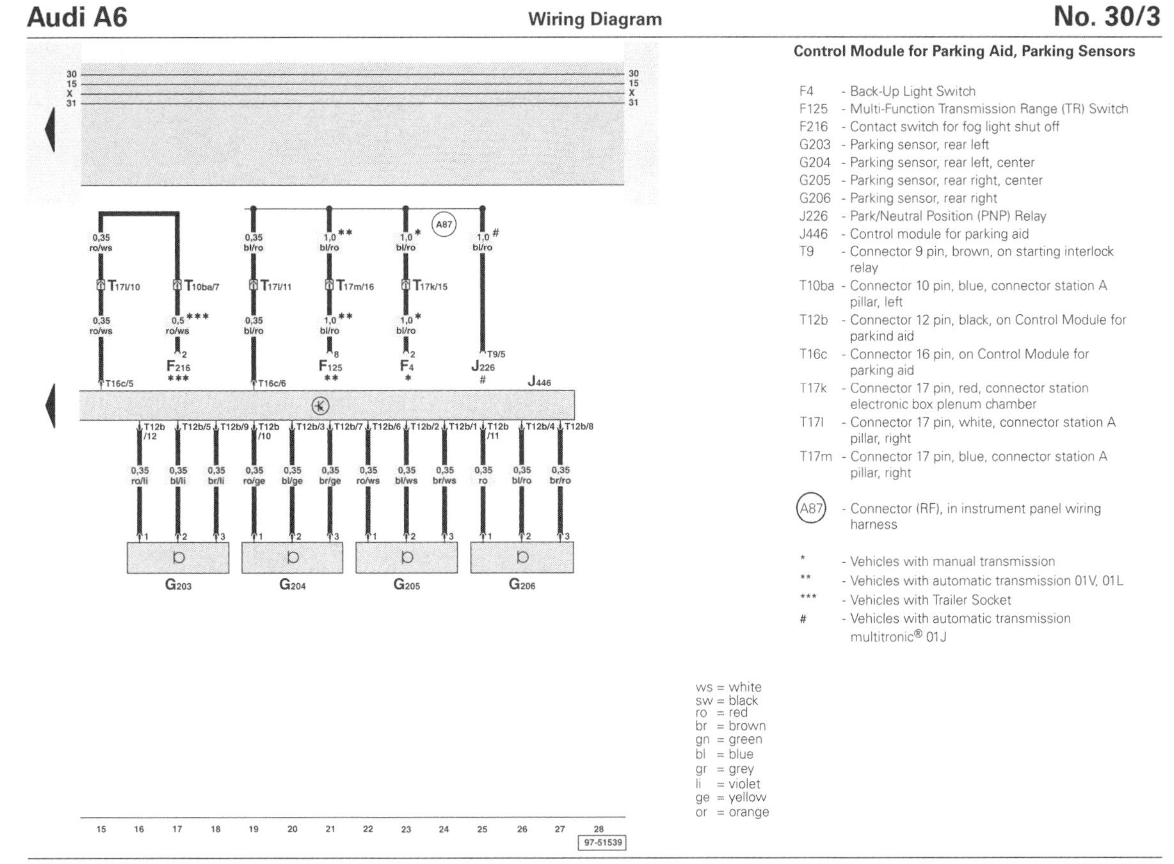

Control Module for Parking Aid, Parking Sensors

F4 - Back-Up Light Switch
F125 - Multi-Function Transmission Range (TR) Switch
F216 - Contact switch for fog light shut off
G203 - Parking sensor, rear left
G204 - Parking sensor, rear left, center
G205 - Parking sensor, rear right, center
G206 - Parking sensor, rear right
J226 - Park/Neutral Position (PNP) Relay
J446 - Control module for parking aid
T9 - Connector 9 pin, brown, on starting interlock relay
T10ba - Connector 10 pin, blue, connector station A pillar, left
T12b - Connector 12 pin, black, on Control Module for parkind aid
T16c - Connector 16 pin, on Control Module for parking aid
T17k - Connector 17 pin, red, connector station electronic box plenum chamber
T17l - Connector 17 pin, white, connector station A pillar, right
T17m - Connector 17 pin, blue, connector station A pillar, right

(A87) - Connector (RF), in instrument panel wiring harness

* - Vehicles with manual transmission
** - Vehicles with automatic transmission 01V, 01L
*** - Vehicles with Trailer Socket
# - Vehicles with automatic transmission multitronic® 01J

ws = white
sw = black
ro = red
br = brown
gn = green
bl = blue
gr = grey
li = violet
ge = yellow
or = orange

THIS SPACE INTENTIONALLY LEFT BLANK

BentleyPublishers.com—All Rights Reserved

Audi A6 Wiring Diagram **No. 31/1**

Radio, Satellite Radio, Bose with and without CD Changer

2003 m.y.

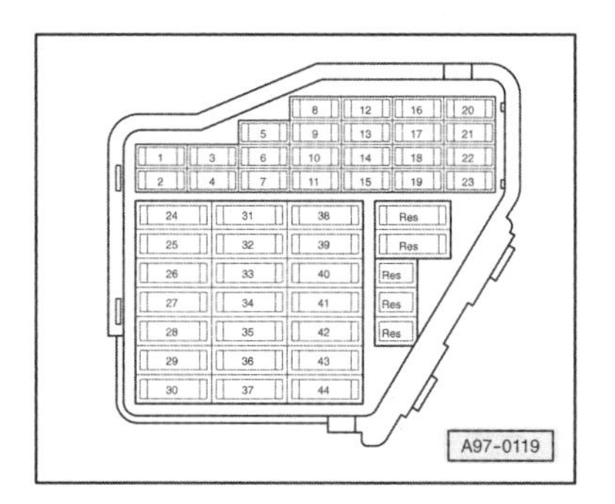

Fuse Panel

Fuse Colors:
- 30 A - Green
- 25 A - White
- 20 A - Yellow
- 15 A - Blue
- 10 A - Red
- 7,5 A - Brown
- 5 A - Beige
- 1 A - Black

Starting with fuse position 23, fuses in the fuse holder are identified with 223 in the wiring diagram.

USA.5604.03.21 Edition 05/03

Audi A6 Wiring Diagram **No. 31/2**

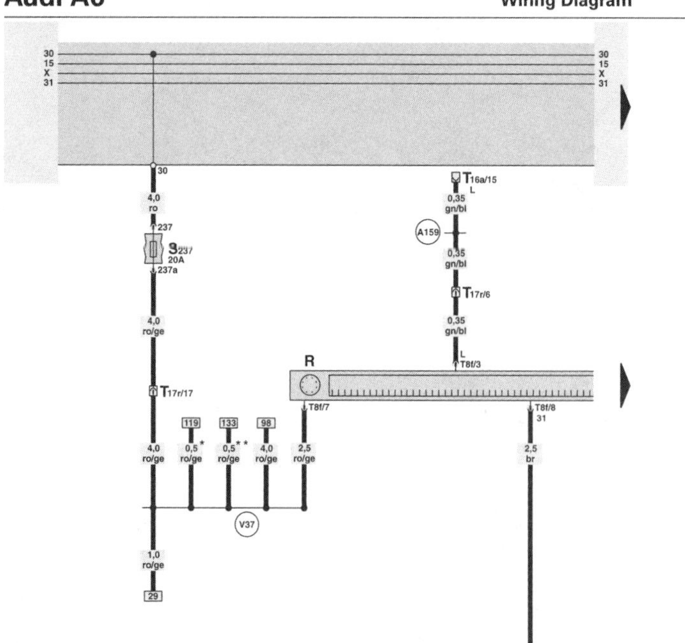

Radio, Data Link Connector (DLC)

- R - Radio
- S237 - Fuse 37 in fuse holder
- T8f - Connector 8 pin, black, connector radio III
- T16a - Connector 16 pin, Data Link Connector (DLC)
- T17r - Connector 17 pin, white, behind instrument panel, center
- (280) - Ground connector -2-, in wiring harness engine compartment
- (A159) - Connection (L) (in instrument panel wiring harness)
- (V37) - Positive connection (30) in communication wiring harness
- * - Sedan
- ** - Wagon

ws = white
sw = black
ro = red
br = brown
gn = green
bl = blue
gr = grey
li = violet
ge = yellow
or = orange

USA.5604.03.21 Edition 05/03

BentleyPublishers.com—All Rights Reserved

Audi A6

Wiring Diagram

No. 31/3

Radio, CD Changer Unit

R - Radio
R41 - CD Changer Unit
T20c - Connector 20 pin, black, connector radio I

(V14) - Wire connection (shielding), in CD-changer wiring harness

* - Vehicles with CD Changer Unit

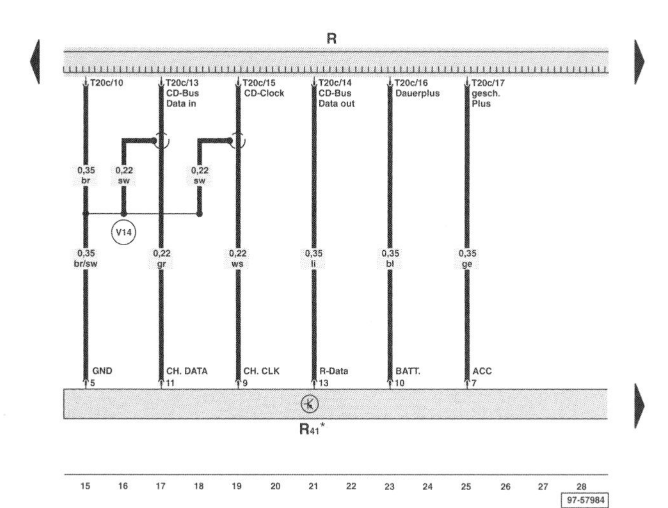

ws = white
sw = black
ro = red
br = brown
gn = green
bl = blue
gr = grey
li = violet
ge = yellow
or = orange

Audi A6

Wiring Diagram

No. 31/4

Radio, Satellite Radio, CD Changer Unit

R - Radio
R41 - CD Changer Unit
R146 - Satellite Radio
T8r - Connector 8 pin, black, on Satellite Radio
T8s - Connector 8 pin, black, on Satellite Radio
T20c - Connector 20 pin, black, connector radio I

(280) - Ground connector -2-, in wiring harness engine compartment
(V45) - Connection 2 (shielding) (in CD changer wiring harness)
(V50) - Low Frequency - Connection (in communication wiring harness)

* - Vehicles without Satellite Radio to bridge with counter plug
** - Vehicles with CD Changer Unit

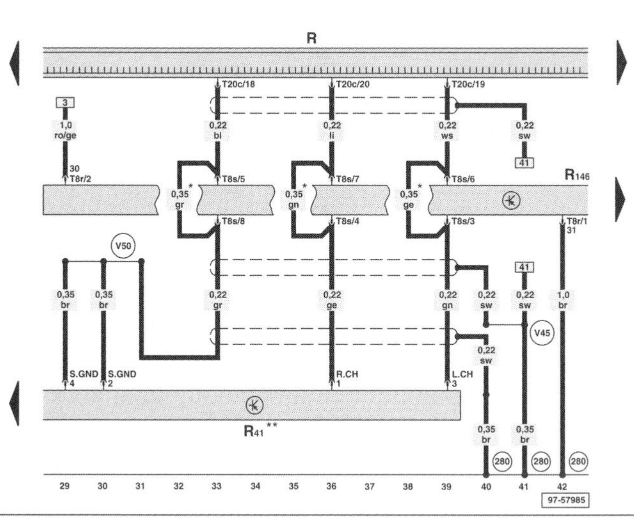

ws = white
sw = black
ro = red
br = brown
gn = green
bl = blue
gr = grey
li = violet
ge = yellow
or = orange

BentleyPublishers.com–All Rights Reserved

Audi A6

Wiring Diagram

No. 31/5

Radio, Satellite Radio, CAN-Bus Connection

J285 - Control Module with indicator unit in instrument panel insert
J526 - Telephone/Telematic Control Module
R - Radio
R146 - Satellite Radio
T8s - Connector 8 pin, black, on Satellite Radio
T15s - Connector 15 pin, blue, connector station A pillar, left
T17r - Connector 17 pin, white, behind instrument panel, center
T20c - Connector 20 pin, black, connector radio I
T32c - Connector 32 pin, grey, on instrument cluster

A148 - Connection (high bus, navigation), in instrument panel wiring harness
A149 - Connection (low bus, navigation), in instrument panel wiring harness
V41 - High-Bus Connection (in communications wiring harness)
V42 - Low-Bus Connection (in communications wiring harness)

• - CAN-Bus (Data-Bus)
* - Vehicles with Telematics, Telephone
** - Vehicles with Navigation System

ws = white
sw = black
ro = red
br = brown
gn = green
bl = blue
gr = grey
li = violet
ge = yellow
or = orange

97-57986

USA.5604.03.21

Edition 05/03

Audi A6

Wiring Diagram

No. 31/6

Radio, Satellite Radio, Antennas for Sedan

J401 - Control unit for navigation with CD-mechanism
J526 - Telephone/Telematic Control Module
R - Radio
R50 - Antenna for navigation system (GPS)
R52 - Antenna for radio/telephone/navigation (GPS)
R110 - GPS Antenna Splitter
R146 - Satellite Radio

* - Vehicles with Telematics
** - Vehicles with Navigation System
*** - Roof Antenna Sedan

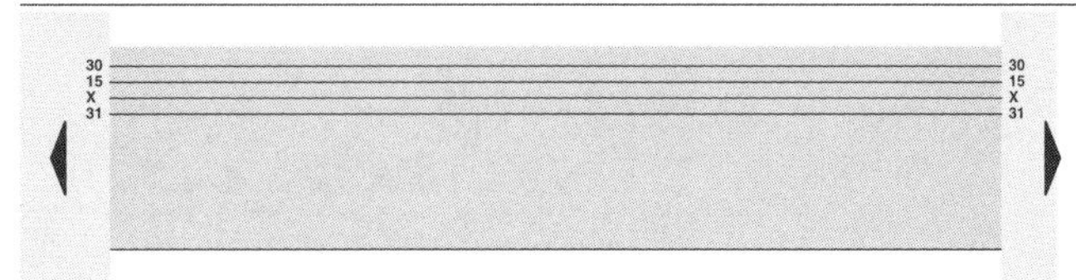

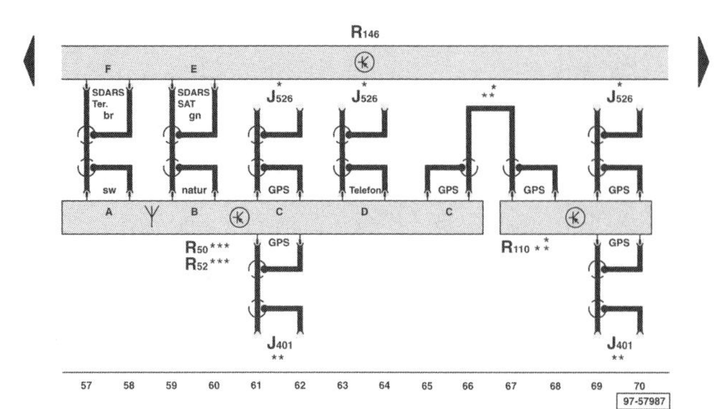

ws = white
sw = black
ro = red
br = brown
gn = green
bl = blue
gr = grey
li = violet
ge = yellow
or = orange

97-57987

USA.5604.03.21

Edition 05/03

BentleyPublishers.com—All Rights Reserved

Audi A6 — Wiring Diagram — No. 31/7

Radio, Satellite Radio, Antennas for Wagon

- J401 - Control unit for navigation with CD-mechanism
- J526 - Telephone/Telematic Control Module
- R - Radio
- R50 - Antenna for navigation system (GPS)
- R52 - Antenna for radio/telephone/navigation (GPS)
- R110 - GPS Antenna Splitter
- R146 - Satellite Radio
- T10z - Connector 10 pin, red, connector radio IV
- T17n - Connector 17 pin, brown, connector station A pillar, right
- T20b - Connector 20 pin, black, on control unit for navigation

- * - Vehicles with Telematics
- ** - Vehicles with Navigation System
- *** - Roof Antenna Wagon

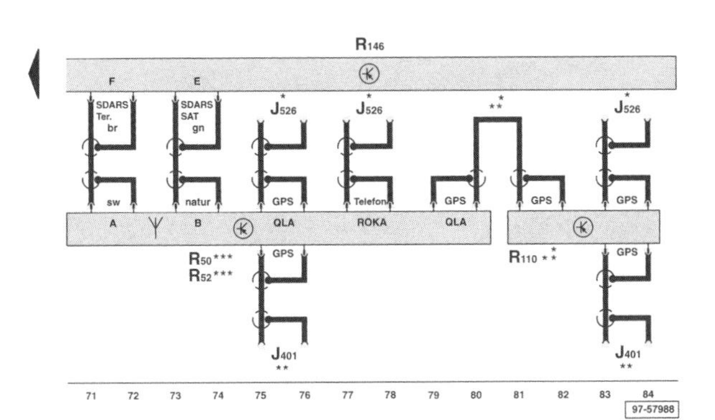

ws = white
sw = black
ro = red
br = brown
gn = green
bl = blue
gr = grey
li = violet
ge = yellow
or = orange

Audi A6 — Wiring Diagram — No. 31/8

Radio, Right Rear Woofer/Amplifier

- J429 - Control module for central locking
- J526 - Telephone/Telematic Control Module
- R - Radio
- R43 - Right Rear Woofer/Amplifier
- T8f - Connector 8 pin, black, connector radio III
- T10z - Connector 10 pin, red, connector radio IV
- T15s - Connector 15 pin, blue, connector station A pillar, left
- T17 - Connector 17 pin, orange, connector station A pillar, right
- T17r - Connector 17 pin, white, behind instrument panel, center
- T20c - Connector 20 pin, black, connector radio I
- (280) - Ground connector -2-, in wiring harness engine compartment
- (610) - Ground connection (radio) under center console, front

- * - Vehicles with Telematics, Telephone
- # - Sedan
- ## - Wagon

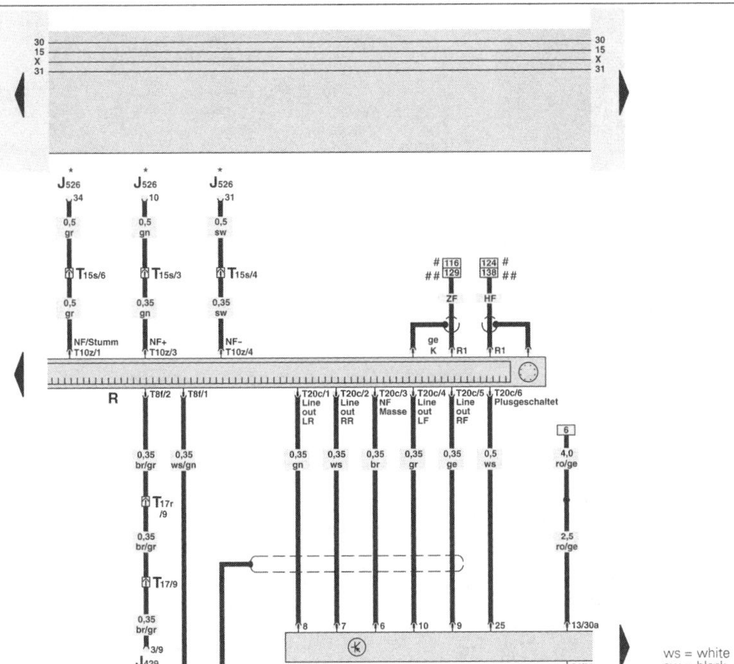

ws = white
sw = black
ro = red
br = brown
gn = green
bl = blue
gr = grey
li = violet
ge = yellow
or = orange

BentleyPublishers.com—All Rights Reserved

Audi A6

Wiring Diagram

No. 31/9

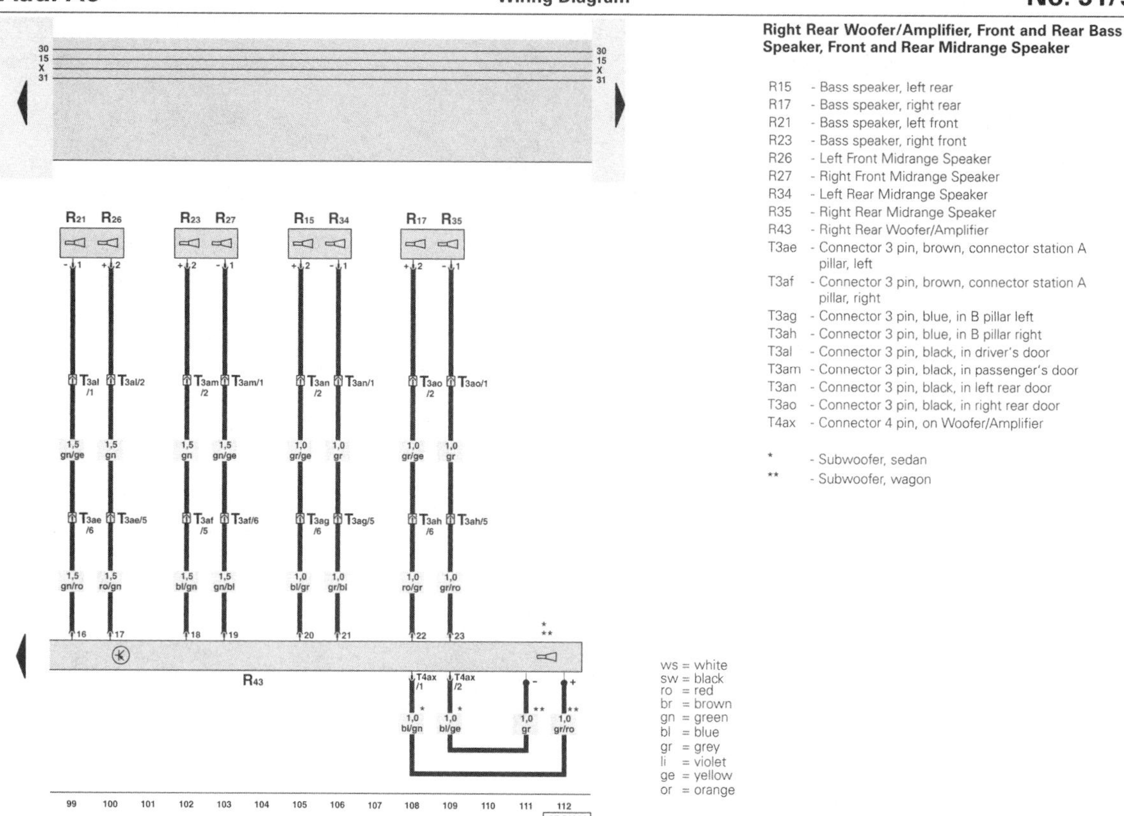

Right Rear Woofer/Amplifier, Front and Rear Bass Speaker, Front and Rear Midrange Speaker

- R15 - Bass speaker, left rear
- R17 - Bass speaker, right rear
- R21 - Bass speaker, left front
- R23 - Bass speaker, right front
- R26 - Left Front Midrange Speaker
- R27 - Right Front Midrange Speaker
- R34 - Left Rear Midrange Speaker
- R35 - Right Rear Midrange Speaker
- R43 - Right Rear Woofer/Amplifier
- T3ae - Connector 3 pin, brown, connector station A pillar, left
- T3af - Connector 3 pin, brown, connector station A pillar, right
- T3ag - Connector 3 pin, blue, in B pillar left
- T3ah - Connector 3 pin, blue, in B pillar right
- T3al - Connector 3 pin, black, in driver's door
- T3am - Connector 3 pin, black, in passenger's door
- T3an - Connector 3 pin, black, in left rear door
- T3ao - Connector 3 pin, black, in right rear door
- T4ax - Connector 4 pin, on Woofer/Amplifier

* - Subwoofer, sedan

** - Subwoofer, wagon

USA.5604.03.21 Edition 05/03

Audi A6

Wiring Diagram

No. 31/10

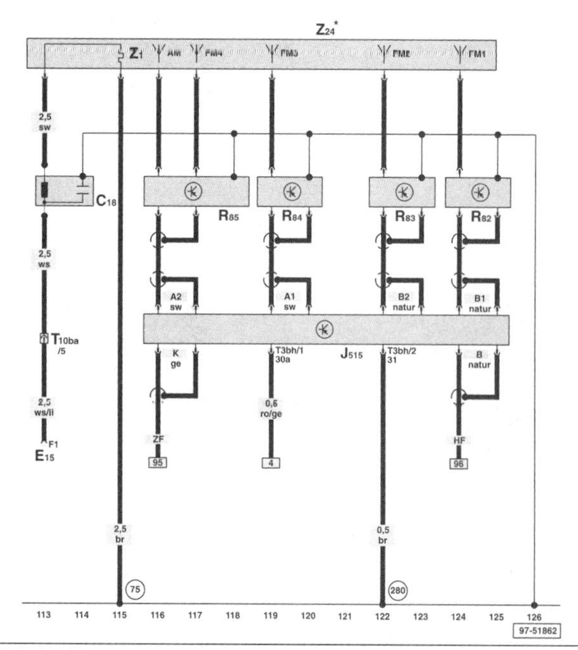

ws = white
sw = black
ro = red
br = brown
gn = green
bl = blue
gr = grey
li = violet
ge = yellow
or = orange

Antenna Selection Control Module, TV Antenna Amplifier, Rear Window Defogger with Window Antenna

- C18 - Windshield Antenna Suppression Filter
- E15 - Rear Window Defogger Switch
- J515 - Antenna Selection Control Module
- R82 - TV Antenna Amplifier 1
- R83 - TV Antenna Amplifier 2
- R84 - TV Antenna Amplifier 3
- R85 - TV Antenna Amplifier 4
- T3bh - Connector 3 pin, black, on Control unit for aerial selection (aerial diversity function)
- T10ba - Connector 10 pin, blue, connector station A pillar, left
- Z1 - Heated rear window
- Z24 - Rear Window Defogger With Window Antenna
- (75) - Ground connection, on right rear pillar
- (280) - Ground connector -2-, in wiring harness engine compartment

* - Sedan

USA.5604.03.21 Edition 05/03

BentleyPublishers.com—All Rights Reserved

Audi A6 **Wiring Diagram** **No. 31/11**

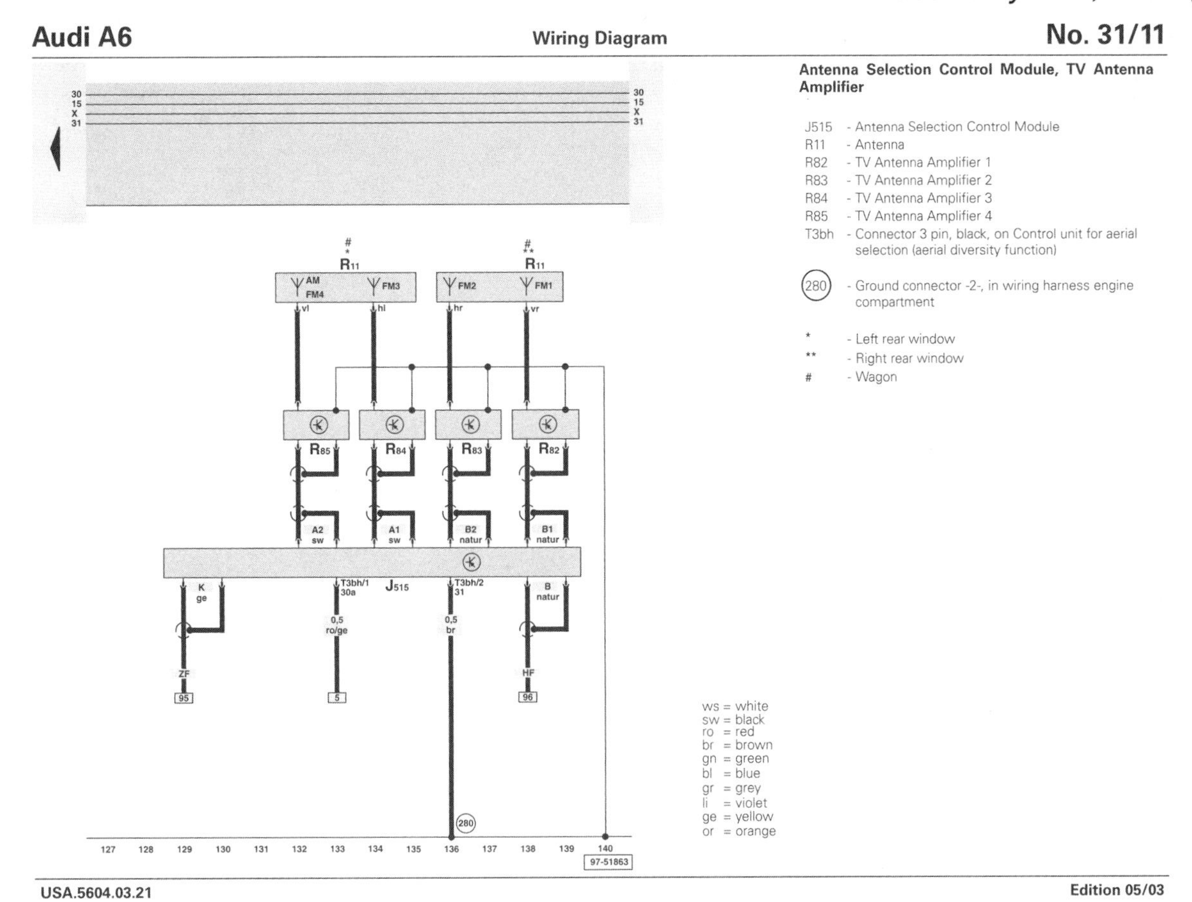

Antenna Selection Control Module, TV Antenna Amplifier

J515 - Antenna Selection Control Module
R11 - Antenna
R82 - TV Antenna Amplifier 1
R83 - TV Antenna Amplifier 2
R84 - TV Antenna Amplifier 3
R85 - TV Antenna Amplifier 4
T3bh - Connector 3 pin, black, on Control unit for aerial selection (aerial diversity function)

(280) - Ground connector -2-, in wiring harness engine compartment

* - Left rear window
** - Right rear window
# - Wagon

ws = white
sw = black
ro = red
br = brown
gn = green
bl = blue
gr = grey
li = violet
ge = yellow
or = orange

THIS SPACE INTENTIONALLY

LEFT BLANK

BentleyPublishers.com—All Rights Reserved

Audi A6 **Wiring Diagram** **No. 33/1**

CD Changer Unit

2004 m.y.

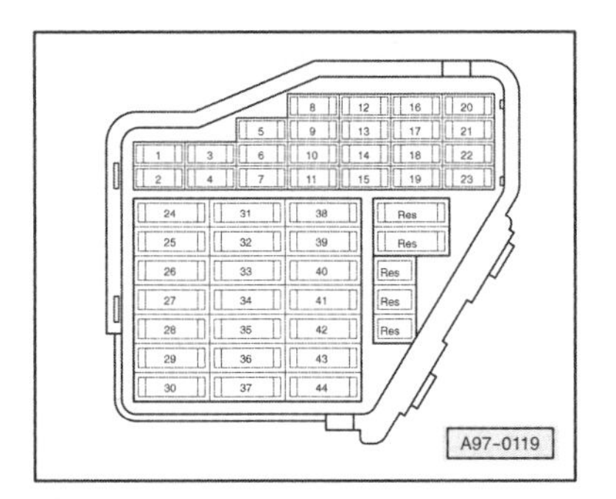

Fuse Panel

Fuse Colors:
- 30 A - Green
- 25 A - White
- 20 A - Yellow
- 15 A - Blue
- 10 A - Red
- 7,5 A - Brown
- 5 A - Beige
- 1 A - Black

Starting with fuse position 23, fuses in the fuse holder are identified with 223 in the wiring diagram.

USA.5604.05.21 Edition 07/04

Audi A6 **Wiring Diagram** **No. 33/2**

CD Changer Unit, Radio

- R - Radio
- R41 - CD Changer Unit
- T12r - Connector 12 pin, black, on CD Changer Unit
- T20c - Connector 20 pin, black, connector radio I
- (309) - Ground Connection (shielding) (in CD changer wiring harness)

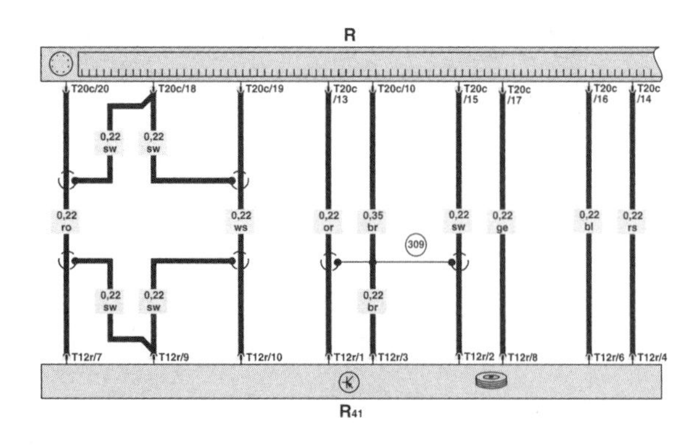

ws = white
sw = black
ro = red
br = brown
gn = green
bl = blue
gr = grey
li = violet
ge = yellow
or = orange

USA.5604.05.21 Edition 07/04

BentleyPublishers.com—All Rights Reserved

OBD On-Board Diagnostics

TABLES

GENERAL

This section gives a brief overview of on-board diagnostics (OBD) and a list of diagnostic trouble codes (DTCs) applicable to the powertrain.

While most DTCs can be accessed with a generic scan tool, full and complete vehicle system diagnostic information can only be obtained using a VW / Audi scan tool such as the VAS 5051 / 5052 and derivatives, or an equivalent aftermarket scan tool.

ON-BOARD DIAGNOSTICS

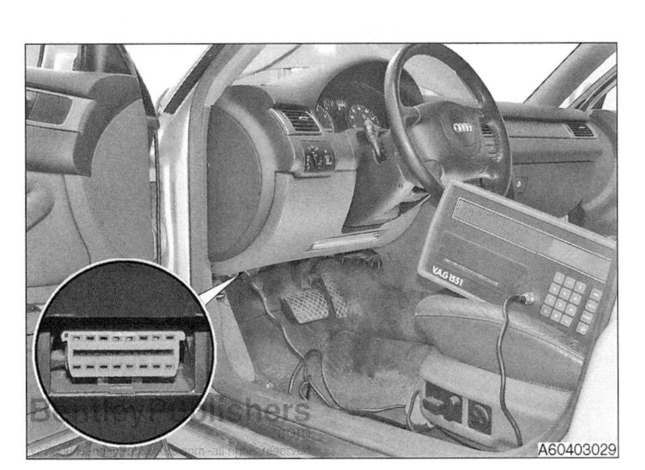

Vehicles covered by this manual are equipped with on-board diagnostics. Diagnosis system access is via the data link connector (DLC) (**inset**) under left side dashboard, to left of steering solumn.

The DLC uses the industry standard SAE 16-pin connector. The connector is recessed into the panel under the dashboard and does not use a cover.

OBD II

Second generation on-board diagnostics (OBD II) is a system mandated by federal regulations for all cars sold in the United States since 1996. OBD II is integrated with VW / Audi factory system diagnostics.

OBD II, programmed into the engine control module (ECM) monitors operation and function of engine management and automatic transmission to insure compliance with specified exhaust and evaporative emission levels. When the OBD II system recognizes a malfunction (or fault) a DTC is stored in the ECM and the malfunction indicator light (MIL) in the instrument cluster is illuminated. This alerts the driver of a problem and the need to have the system checked.

Each DTC is assigned two codes. The first is a numerical code assigned by VW / Audi (VAG code). The second is referred to as a P-code or U-code and follows a structure required by law and defined by the Society of Automotive Engineers (SAE). DTCs follow

BentleyPublishers.com
BentleyPublishers.com–All Rights Reserved

a standard format. This standard uses a letter to designate the system and four numbers or letters to further identify and detail the malfunction as listed below.

First digit structure is as follows:
- **Pxxxx**: Powertrain
- **Uxxxx**: Vehicle network communications
- **Bxxxx**: Body
- **Cxxxx**: Chassis

Second digit structure is:
- **P0xxx**: Government required codes
- **P1xxx**: Manufacturer codes for additional emission system function; not required but reported to government

Third digit structure is:
- **Px1xx**: Air and fuel measurement
- **Px2xx**: Air and fuel measurement
- **Px3xx**: Ignition system
- **Px4xx**: Additional emission controls
- **Px5xx**: Speed and idle regulation
- **Px6xx**: Computer and output signals
- **Px7xx**: Transmission
- **Px8xx**: Transmission
- **Px9xx**: Control modules, input and output signals

The fourth and fifth digits denote the individual components and systems. In some cases, the fifth digit can be a letter.

The digit structure shown above has P as the first digit. However, letters B, C, and U could be used where applicable. Only P and U codes are currently used in vehicles covered by this manual.

Table a in this repair group contains P-codes and VAG codes with brief descriptions of each.

Scan tool

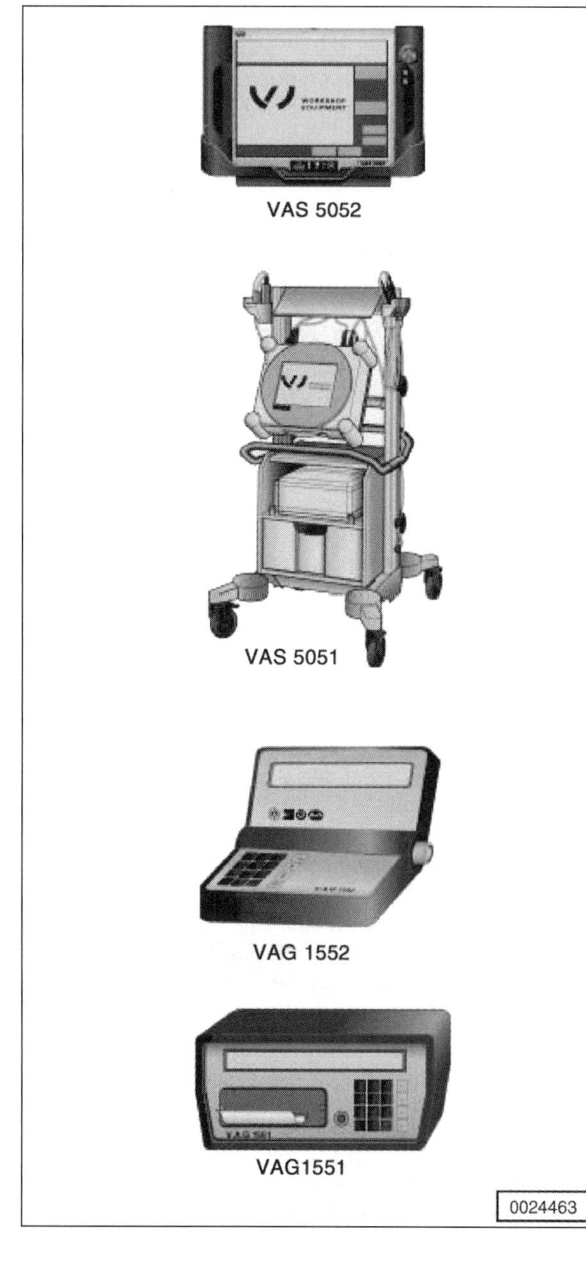

Sampling of VW / Audi scan tools. VAS 5051 is superseded by VAS 5051B. Note that VAG 1551 and VAG 1552 do not have the ability to access all of the systems on the CAN-bus.

BentleyPublishers.com
BentleyPublishers.com—All Rights Reserved

Diagnostic scan tool and special tool suppliers

The following suppliers offer diagnostic scan tools or computerized scan tool programs. Some of them also offer special tools and equipment for specialized repair work.

Assenmacher Specialty Tools, Inc.
6440 Odell Place
Boulder, CO 80301
(800) 525-2943
www.asttool.com

Baum Tools Unlimited, Inc.
PO Box 5867
Sarasota, FL 34277
(800) 848-6657
www.baumtools.com

Equipment Solutions
P.O. Box 1450
Kenosha, WI 53141-1450
(800) 892-9650

Metalnerd
5113 Crestview Drive, Suite B
Greensburg, PA, 15601 USA
(412) 601-4270
www.metalnerd.com

Ross-Tech
920 South Broad Street
Lansdale, PA 19446
215-361-8942
www.ross-tech.com

Samstag Sales
115 Main St. N., Suite 216
Carthage, TN 37030
(615) 735-3388
www.samstagsales.com

Shade Tree Software
4186 Culebra Ct.
Boulder, CO 80301
(303) 449-1664
(303) 940-2468
www.shadetreesoftware.com

ZDMAK Tools
P.O. Box 5100
Sarasota, FL 34277
(877) 938-6657
www.zdmak.com

Zelenda Automotive, Inc.
66-02 Austin St.
Forest Hills, NY 11374
(888) 892-8348
www.zelenda.com

DIAGNOSTIC TROUBLE CODES

Table a contains a list of available powertrain scan tool codes for VW / Audi vehicles available at the time of publication. Not all codes apply to all vehicles.

Table a. Diagnostic trouble codes (DTCs or P-codes)

VAG code	SAE P-code	Description
16394	P0010	-A- Camshaft Pos. Actuator Circ. Bank 1 Malfunction
16395	P0020	-A- Camshaft Pos. Actuator Circ. Bank 2 Malfunction
16449	P0065	Air Assisted Injector Control Range/Performance
16450	P0066	Air Assisted Injector Control Low Input/Short to ground
16451	P0067	Air Assisted Injector Control Input/Short to B+
16485	P0101	Mass or Volume Air Flow Circ Range/Performance
16486	P0102	Mass or Volume Air Flow Circ Low Input
16487	P0103	Mass or Volume Air Flow Circ High Input

BentleyPublishers.com
BentleyPublishers.com—All Rights Reserved

Table a. Diagnostic trouble codes (DTCs or P-codes)

VAG code	SAE P-code	Description
16489	P0105	Manifold Abs.Pressure or Bar.Pressure Voltage supply
16490	P0106	Manifold Abs.Pressure or Bar.Pressure Range/Performance
16491	P0107	Manifold Abs.Pressure or Bar.Pressure Low Input
16492	P0108	Manifold Abs.Pressure or Bar.Pressure High Input
16496	P0112	Intake Air Temp.Circ Low Input
16497	P0113	Intake Air Temp.Circ High Input
16500	P0116	Engine Coolant Temp.Circ Range/Performance
16501	P0117	Engine Coolant Temp.Circ Low Input
16502	P0118	Engine Coolant Temp.Circ High Input
16504	P0120	Throttle/Pedal Pos.Sensor A Circ Malfunction
16505	P0121	Throttle/Pedal Pos.Sensor A Circ Range/Performance
16506	P0122	Throttle/Pedal Pos.Sensor A Circ Low Input
16507	P0123	Throttle/Pedal Pos.Sensor A Circ High Input
16509	P0125	Insufficient Coolant Temp.for Closed Loop Fuel Control
16512	P0128	Coolant Thermostat/Valve Temperature below control range
16514	P0130	O2 Sensor Circ.,Bank1-Sensor1 Malfunction
16515	P0131	O2 Sensor Circ.,Bank1-Sensor1 Low Voltage
16516	P0132	O2 Sensor Circ.,Bank1-Sensor1 High Voltage
16517	P0133	O2 Sensor Circ.,Bank1-Sensor1 Slow Response
16518	P0134	O2 Sensor Circ.,Bank1-Sensor1 No Activity Detected
16519	P0135	O2 Sensor Heater Circ.,Bank1-Sensor1 Malfunction
16520	P0136	O2 Sensor Circ.,Bank1-Sensor2 Malfunction
16521	P0137	O2 Sensor Circ.,Bank1-Sensor2 Low Voltage
16522	P0138	O2 Sensor Circ.,Bank1-Sensor2 High Voltage
16523	P0139	O2 Sensor Circ.,Bank1-Sensor2 Slow Response
16524	P0140	O2 Sensor Circ.,Bank1-Sensor2 No Activity Detected
16525	P0141	O2 Sensor Heater Circ.,Bank1-Sensor2 Malfunction
16534	P0150	O2 Sensor Circ.,Bank2-Sensor1 Malfunction
16535	P0151	O2 Sensor Circ.,Bank2-Sensor1 Low Voltage
16536	P0152	O2 Sensor Circ.,Bank2-Sensor1 High Voltage
16537	P0153	O2 Sensor Circ.,Bank2-Sensor1 Slow Response
16538	P0154	O2 Sensor Circ.,Bank2-Sensor1 No Activity Detected
16539	P0155	O2 Sensor Heater Circ.,Bank2-Sensor1 Malfunction
16540	P0156	O2 Sensor Circ.,Bank2-Sensor2 Malfunction
16541	P0157	O2 Sensor Circ.,Bank2-Sensor2 Low Voltage
16542	P0158	O2 Sensor Circ.,Bank2-Sensor2 High Voltage
16543	P0159	O2 Sensor Circ.,Bank2-Sensor2 Slow Response

BentleyPublishers.com
BentleyPublishers.com—All Rights Reserved

Table a. Diagnostic trouble codes (DTCs or P-codes)		
VAG code	**SAE P-code**	**Description**
16544	P0160	O2 Sensor Circ.,Bank2-Sensor2 No Activity Detected
16545	P0161	O2 Sensor Heater Circ.,Bank2-Sensor2 Malfunction
16554	P0170	Fuel Trim,Bank1 Malfunction
16555	P0171	Fuel Trim,Bank1 System too Lean
16556	P0172	Fuel Trim,Bank1 System too Rich
16557	P0173	Fuel Trim,Bank2 Malfunction
16558	P0174	Fuel Trim,Bank2 System too Lean
16559	P0175	Fuel Trim,Bank2 System too Rich
16566	P0182	Fuel temperature sender-G81 Short to ground
16567	P0183	Fuel temperature sender-G81 Interruption/Short to B+
16581	P0197	Engine Oil Temperature Circuit Low Input
16582	P0198	Engine Oil Temperature Circuit High Input
16585	P0201	Cyl.1, Injector Circuit Fault in electrical circuit
16586	P0202	Cyl.2, Injector Circuit Fault in electrical circuit
16587	P0203	Cyl.3, Injector Circuit Fault in electrical circuit
16588	P0204	Cyl.4, Injector Circuit Fault in electrical circuit
16589	P0205	Cyl.5 Injector Circuit Fault in electrical circuit
16590	P0206	Cyl.6 Injector Circuit Fault in electrical circuit
16591	P0207	Cyl.7 Injector Circuit Fault in electrical circuit
16592	P0208	Cyl.8 Injector Circuit Fault in electrical circuit
16599	P0215	Engine Shut-Off Solenoid Malfunction
16600	P0216	Injector/Injection Timing Control Malfunction
16603	P0219	Engine Overspeed Condition
16605	P0221	Throttle Pos. Sensor -B- Circuit Range/Performance
16606	P0222	Throttle Pos. Sensor -B- Circuit Low Input
16607	P0223	Throttle Pos. Sensor -B- Circuit High Input
16609	P0225	Throttle Pos. Sensor -C- Circuit Voltage supply
16610	P0226	Throttle Pos. Sensor -C- Circuit Range/Performance
16611	P0227	Throttle Pos. Sensor -C- Circuit Low Input
16612	P0228	Throttle Pos. Sensor -C- Circuit High Input
16614	P0230	Fuel Pump Primary Circuit Fault in electrical circuit
16618	P0234	Turbocharger Overboost Condition Control limit exceeded
16619	P0235	Turbocharger Boost Sensor (A) Circ Control limit not reached
16620	P0236	Turbocharger Boost Sensor (A) Circ Range/Performance
16621	P0237	Turbocharger Boost Sensor (A) Circ Low Input
16622	P0238	Turbocharger Boost Sensor (A) Circ High Input
16627	P0243	Turbocharger Wastegate Solenoid (A) Open/Short Circuit to Ground
16629	P0245	Turbocharger Wastegate Solenoid (A) Low Input/Short to ground

BentleyPublishers.com
BentleyPublishers.com—All Rights Reserved

Table a. Diagnostic trouble codes (DTCs or P-codes)		
VAG code	SAE P-code	Description
16630	P0246	Turbocharger Wastegate Solenoid (A) High Input/Short to B+
16636	P0252	Injection Pump Metering Control (A) Range/Performance
16645	P0261	Cyl.1 Injector Circuit Low Input/Short to ground
16646	P0262	Cyl.1 Injector Circuit High Input/Short to B+
16648	P0264	Cyl.2 Injector Circuit Low Input/Short to ground
16649	P0265	Cyl.2 Injector Circuit High Input/Short to B+
16651	P0267	Cyl.3 Injector Circuit Low Input/Short to ground
16652	P0268	Cyl.3 Injector Circuit High Input/Short to B+
16654	P0270	Cyl.4 Injector Circuit Low Input/Short to ground
16655	P0271	Cyl.4 Injector Circuit High Input/Short to B+
16657	P0273	Cyl.5 Injector Circuit Low Input/Short to ground
16658	P0274	Cyl.5 Injector Circuit High Input/Short to B+
16660	P0276	Cyl.6 Injector Circuit Low Input/Short to ground
16661	P0277	Cyl.6 Injector Circuit High Input/Short to B+
16663	P0279	Cyl.7 Injector Circuit Low Input/Short to ground
16664	P0280	Cyl.7 Injector Circuit High Input/Short to B+
16666	P0282	Cyl.8 Injector Circuit Low Input/Short to ground
16667	P0283	Cyl.8 Injector Circuit High Input/Short to B+
16684	P0300	Random/Multiple Cylinder Misfire Detected
16685	P0301	Cyl.1 Misfire Detected
16686	P0302	Cyl.2 Misfire Detected
16687	P0303	Cyl.3 Misfire Detected
16688	P0304	Cyl.4 Misfire Detected
16689	P0305	Cyl.5 Misfire Detected
16690	P0306	Cyl.6 Misfire Detected
16691	P0307	Cyl.7 Misfire Detected
16692	P0308	Cyl.8 Misfire Detected
16697	P0313	Misfire Detected Low Fuel Level
16698	P0314	Single Cylinder Misfire
16705	P0321	Ign./Distributor Eng.Speed Inp.Circ Range/Performance
16706	P0322	Ign./Distributor Eng.Speed Inp.Circ No Signal
16709	P0325	Knock Sensor 1 Circuit Electrical Fault in Circuit
16710	P0326	Knock Sensor 1 Circuit Range/Performance
16711	P0327	Knock Sensor 1 Circ Low Input
16712	P0328	Knock Sensor 1 Circ High Input
16716	P0332	Knock Sensor 2 Circ Low Input
16717	P0333	Knock Sensor 2 Circ High Input

BentleyPublishers.com
BentleyPublishers.com—All Rights Reserved

Table a. Diagnostic trouble codes (DTCs or P-codes)		
VAG code	**SAE P-code**	**Description**
16719	P0335	Crankshaft Pos. Sensor (A) Circ Malfunction
16720	P0336	Crankshaft Pos. Sensor (A) Circ Range/Performance/Missing tooth
16721	P0337	Crankshaft Pos.Sensor (A) Circ Low Input
16724	P0340	Camshaft Pos. Sensor (A) Circ Incorrect allocation
16725	P0341	Camshaft Pos.Sensor Circ Range/Performance
16726	P0342	Camshaft Pos.Sensor Circ Low Input
16727	P0343	Camshaft Pos.Sensor Circ High Input
16735	P0351	Ignition Coil (A) Cyl.1 Prim./Sec. Circ Malfunction
16736	P0352	Ignition Coil (B) Cyl.2 Prim./Sec. Circ Malfunction
16737	P0353	Ignition Coil (C) Cyl.3 Prim./Sec. Circ Malfunction
16738	P0354	Ignition Coil (D) Cyl.4 Prim./Sec. Circ Malfunction
16739	P0355	Ignition Coil (E) Cyl.5 Prim./Sec. Circ Malfunction
16740	P0356	Ignition Coil (F) Cyl.6 Prim./Sec. Circ Malfunction
16741	P0357	Ignition Coil (G) Cyl.7 Prim./Sec. Circ Malfunction
16742	P0358	Ignition Coil (H) Cyl.8 Prim./Sec. Circ Malfunction
16764	P0380	Glow Plug/Heater Circuit (A) Electrical Fault in Circuit
16784	P0400	Exhaust Gas Recirc.Flow Malfunction
16785	P0401	Exhaust Gas Recirc.Flow Insufficient Detected
16786	P0402	Exhaust Gas Recirc.Flow Excessive Detected
16787	P0403	Exhaust Gas Recirc. Contr. Circ Malfunction
16788	P0404	Exhaust Gas Recirc. Contr. Circ Range/Performance
16789	P0405	Exhaust Gas Recirc. Sensor (A) Circ Low Input
16790	P0406	Exhaust Gas Recirc. Sensor (A) Circ High Input
16791	P0407	Exhaust Gas Recirc. Sensor (B) Circ Low Input
16792	P0408	Exhaust Gas Recirc. Sensor (B) Circ High Input
16794	P0410	Sec.Air Inj.Sys Malfunction
16795	P0411	Sec.Air Inj.Sys. Incorrect Flow Detected
16796	P0412	Sec.Air Inj.Sys.Switching Valve A Circ Malfunction
16802	P0418	Sec. Air Inj. Sys. Relay (A) Contr. Circ Malfunction
16804	P0420	Catalyst System,Bank1 Efficiency Below Threshold
16806	P0422	Main Catalyst,Bank1 Below Threshold
16811	P0427	Catalyst Temperature Sensor, Bank 1 Low Input/Short to ground
16812	P0428	Catalyst Temperature Sensor, Bank 1 High Input/Open/Short Circuit to B+
16816	P0432	Main Catalyst,Bank2 Efficiency Below Threshold
16820	P0436	Catalyst Temperature Sensor, Bank 2 Range/Performance
16821	P0437	Catalyst Temperature Sensor, Bank 2 Low Input/Short to ground
16822	P0438	Catalyst Temperature Sensor, Bank 2 High Input/Open/Short Circuit to B+

BentleyPublishers.com
BentleyPublishers.com—All Rights Reserved

Table a. Diagnostic trouble codes (DTCs or P-codes)		
VAG code	**SAE P-code**	**Description**
16824	P0440	EVAP Emission Contr.Sys. Malfunction
16825	P0441	EVAP Emission Contr.Sys.Incorrect Purge Flow
16826	P0442	EVAP Emission Contr.Sys.(Small Leak) Leak Detected
16827	P0443	EVAP Emiss. Contr. Sys. Purge Valve Circ Electrical Fault in Circuit
16836	P0452	EVAP Emission Contr.Sys.Press.Sensor Low Input
16837	P0453	EVAP Emission Contr.Sys.Press.Sensor High Input
16839	P0455	EVAP Emission Contr.Sys.(Gross Leak) Leak Detected
16845	P0461	Fuel Level Sensor Circ Range/Performance
16846	P0462	Fuel Level Sensor Circuit Low Input
16847	P0463	Fuel Level Sensor Circuit High Input
16885	P0501	Vehicle Speed Sensor Range/Performance
16887	P0503	Vehicle Speed Sensor Intermittent/Erratic/High Input
16889	P0505	Idle Control System Malfunction
16890	P0506	Idle Control System RPM Lower than Expected
16891	P0507	Idle Control System Higher than Expected
16894	P0510	Closed Throttle Pos.Switch Malfunction
16915	P0531	A/C Refrigerant Pressure Sensor Circuit Range/Performance
16916	P0532	A/C Refrigerant Pressure Sensor Circuit Low Input
16917	P0533	A/C Refrigerant Pressure Sensor Circuit High Input
16935	P0551	Power Steering Pressure Sensor Circuit Range/Performance
16944	P0560	System Voltage Malfunction
16946	P0562	System Voltage Low Voltage
16947	P0563	System Voltage High Voltage
16952	P0568	Cruise Control Set Signal Incorrect Signal
16955	P0571	Cruise/Brake Switch (A) Circ Malfunction
16984	P0600	Serial Comm. Link (Data Bus) Message Missing
16985	P0601	Internal Contr.Module Memory Check Sum Error
16986	P0602	Control Module Programming Error/Malfunction
16987	P0603	Internal Contr.Module (KAM) Error
16988	P0604	Internal Contr.Module Random Access Memory (RAM) Error
16989	P0605	Internal Contr.Module ROM Test Error
16990	P0606	ECM/PCM Processor
17026	P0642	Knock Control Module Malfunction
17029	P0645	A/C Clutch Relay Control Circuit
17034	P0650	MIL Control Circuit Electrical Fault in Circuit
17038	P0654	Engine RPM Output Circuit Electrical Fault in Circuit
17040	P0656	Fuel Level Output Circuit Electrical Fault in Circuit

BentleyPublishers.com
BentleyPublishers.com—All Rights Reserved

Table a. Diagnostic trouble codes (DTCs or P-codes)

VAG code	SAE P-code	Description
17084	P0700	Transm.Contr.System Malfunction
17086	P0702	Transm.Contr.System Electrical
17087	P0703	Torque Converter/Brake Switch B Circ Malfunction
17089	P0705	Transm.Range Sensor Circ.(PRNDL Inop.) Malfunction
17090	P0706	Transm.Range Sensor Circ Range/Performance
17091	P0707	Transm.Range Sensor Circ Low Input
17092	P0708	Transm.Range Sensor Circ High Input
17094	P0710	Transm.Fluid Temp.Sensor Circ. Malfunction
17095	P0711	Transm.Fluid Temp.Sensor Circ. Range/Performance
17096	P0712	Transm.Fluid Temp.Sensor Circ. Low Input
17097	P0713	Transm.Fluid Temp.Sensor Circ. High Input
17099	P0715	Input Turbine/Speed Sensor Circ. Malfunction
17100	P0716	Input Turbine/Speed Sensor Circ. Range/Performance
17101	P0717	Input Turbine/Speed Sensor Circ. No Signal
17105	P0721	Output Speed Sensor Circ Range/Performance
17106	P0722	Output Speed Sensor Circ No Signal
17109	P0725	Engine Speed Inp.Circ. Malfunction
17110	P0726	Engine Speed Inp.Circ. Range/Performance
17111	P0727	Engine Speed Inp.Circ. No Signal
17114	P0730	Gear Incorrect Ratio
17115	P0731	Gear 1 Incorrect Ratio
17116	P0732	Gear 2 Incorrect Ratio
17117	P0733	Gear 3 Incorrect Ratio
17118	P0734	Gear 4 Incorrect Ratio
17119	P0735	Gear 5 Incorrect Ratio
17124	P0740	Torque Converter Clutch Circ Malfunction
17125	P0741	Torque Converter Clutch Circ Performance or Stuck Off
17132	P0748	Pressure Contr.Solenoid Electrical
17134	P0750	Shift Solenoid A malfunction
17135	P0751	Shift Solenoid A Performance or Stuck Off
17136	P0752	Shift Solenoid A Stuck On
17137	P0753	Shift Solenoid A Electrical
17140	P0756	Shift Solenoid B Performance or Stuck Off
17141	P0757	Shift Solenoid B Stuck On
17142	P0758	Shift Solenoid B Electrical
17145	P0761	Shift Solenoid C Performance or Stuck Off
17146	P0762	Shift Solenoid C Stuck On

BentleyPublishers.com
BentleyPublishers.com—All Rights Reserved

Table a. Diagnostic trouble codes (DTCs or P-codes)

VAG code	SAE P-code	Description
17147	P0763	Shift Solenoid C Electrical
17152	P0768	Shift Solenoid D Electrical
17157	P0773	Shift Solenoid E Electrical
17174	P0790	Normal/Performance Switch Circ Malfunction
17509	P1101	O2 Sensor Circ.,Bank1-Sensor1Voltage too Low/Air Leak
17510	P1102	O2 Sensor Heating Circ.,Bank1-Sensor1 Short to B+
17511	P1103	O2 Sensor Heating Circ.,Bank1-Sensor1 Output too Low
17512	P1104	Bank1-Sensor2 Voltage too Low/Air Leak
17513	P1105	O2 Sensor Heating Circ.,Bank1-Sensor2 Short to B+
17514	P1106	O2 Sensor Circ.,Bank2-Sensor1 Voltage too Low/Air Leak
17515	P1107	O2 Sensor Heating Circ.,Bank2-Sensor1 Short to B+
17516	P1108	O2 Sensor Heating Circ.,Bank2-Sensor1 Output too Low
17517	P1109	O2 Sensor Circ.,Bank2-Sensor2 Voltage too Low/Air Leak
17518	P1110	O2 Sensor Heating Circ.,Bank2-Sensor2 Short to B+
17519	P1111	O2 Control (Bank 1) System too lean
17520	P1112	O2 Control (Bank 1) System too rich
17521	P1113	Bank1-Sensor1 Internal Resistance too High
17522	P1114	Bank1-Sensor2 Internal Resistant too High
17523	P1115	O2 Sensor Heater Circ.,Bank1-Sensor1 Short to Ground
17524	P1116	O2 Sensor Heater Circ.,Bank1-Sensor1 Open
17525	P1117	O2 Sensor Heater Circ.,Bank1-Sensor2 Short to Ground
17526	P1118	O2 Sensor Heater Circ.,Bank1-Sensor2 Open
17527	P1119	O2 Sensor Heater Circ.,Bank2-Sensor1 Short to Ground
17528	P1120	O2 Sensor Heater Circ.,Bank2-Sensor1 Open
17529	P1121	O2 Sensor Heater Circ.,Bank2-Sensor2 Short to Ground
17530	P1122	O2 Sensor Heater Circ.,Bank2-Sensor2 Open
17531	P1123	Long Term Fuel Trim Add.Air.,Bank1 System too Rich
17532	P1124	Long Term Fuel Trim Add.Air.,Bank1 System too Lean
17533	P1125	Long Term Fuel Trim Add.Air.,Bank2 System too Rich
17534	P1126	Long Term Fuel Trim Add.Air.,Bank2 System too Lean
17535	P1127	Long Term Fuel Trim mult.,Bank1 System too Rich
17536	P1128	Long Term Fuel Trim mult.,Bank1 System too Lean
17537	P1129	Long Term Fuel Trim mult.,Bank2 System too Rich
17538	P1130	Long Term Fuel Trim mult.,Bank2 System too Lean
17539	P1131	Bank2-Sensor1 Internal Resistance too High
17540	P1132	O2 Sensor Heating Circ.,Bank1+2-Sensor1 Short to B+
17541	P1133	O2 Sensor Heating Circ.,Bank1+2-Sensor1 Electrical Malfunction

BentleyPublishers.com
BentleyPublishers.com—All Rights Reserved

Table a. Diagnostic trouble codes (DTCs or P-codes)		
VAG code	**SAE P-code**	**Description**
17542	P1134	O2 Sensor Heating Circ.,Bank1+2-Sensor2 Short to B+
17543	P1135	O2 Sensor Heating Circ.,Bank1+2-Sensor2 Electrical Malfunction
17544	P1136	Long Term Fuel Trim Add.Fuel,Bank1 System too Lean
17545	P1137	Long Term Fuel Trim Add.Fuel,Bank1 System too Rich
17546	P1138	Long Term Fuel Trim Add.Fuel,Bank2 System too Lean
17547	P1139	Long Term Fuel Trim Add.Fuel,Bank2 System too Rich
17548	P1140	Bank2-Sensor2 Internal Resistance too High
17549	P1141	Load Calculation Cross Check Range/Performance
17550	P1142	Load Calculation Cross Check Lower Limit Exceeded
17551	P1143	Load Calculation Cross Check Upper Limit Exceeded
17552	P1144	Mass or Volume Air Flow Circ Open/Short to Ground
17553	P1145	Mass or Volume Air Flow Circ Short to B+
17554	P1146	Mass or Volume Air Flow Circ Supply Malfunction
17555	P1147	O2 Control (Bank 2) System too lean
17556	P1148	O2 Control (Bank 2) System too rich
17557	P1149	O2 Control (Bank 1) Out of range
17558	P1150	O2 Control (Bank 2) Out of range
17559	P1151	Bank1, Long Term Fuel Trim, Range 1 Leanness Lower Limit Exceeded
17560	P1152	Bank1, Long Term Fuel Trim, Range 2 Leanness Lower Limit Exceeded
17562	P1154	Manifold Switch Over Malfunction
17563	P1155	Manifold Abs.Pressure Sensor Circ. Short to B+
17564	P1156	Manifold Abs.Pressure Sensor Circ. Open/Short to Ground
17565	P1157	Manifold Abs.Pressure Sensor Circ. Power Supply Malfunction
17566	P1158	Manifold Abs.Pressure Sensor Circ. Range/Performance
17568	P1160	Manifold Temp.Sensor Circ. Short to Ground
17569	P1161	Manifold Temp.Sensor Circ. Open/Short to B+
17570	P1162	Fuel Temp.Sensor Circ. Short to Ground
17571	P1163	Fuel Temp.Sensor Circ. Open/Short to B+
17572	P1164	Fuel Temperature Sensor Range/Performance/Incorrect Signal
17573	P1165	Bank1, Long Term Fuel Trim, Range 1 Rich Limit Exceeded
17574	P1166	Bank1, Long Term Fuel Trim, Range 2 Rich Limit Exceeded
17579	P1171	Throttle Actuation Potentiometer Sign.2 Range/Performance
17580	P1172	Throttle Actuation Potentiometer Sign.2 Signal too Low
17581	P1173	Throttle Actuation Potentiometer Sign.2 Signal too High
17582	P1174	Fuel Trim, Bank 1 Different injection times
17584	P1176	O2 Correction Behind Catalyst,B1 Limit Attained
17585	P1177	O2 Correction Behind Catalyst,B2 Limit Attained

BentleyPublishers.com
BentleyPublishers.com—All Rights Reserved

Table a. Diagnostic trouble codes (DTCs or P-codes)

VAG code	SAE P-code	Description
17586	P1178	Linear 02 Sensor / Pump Current Open Circuit
17587	P1179	Linear 02 Sensor / Pump Current Short to ground
17588	P1180	Linear 02 Sensor / Pump Current Short to B+
17589	P1181	Linear 02 Sensor / Reference Voltage Open Circuit
17590	P1182	Linear 02 Sensor / Reference Voltage Short to ground
17591	P1183	Linear 02 Sensor / Reference Voltage Short to B+
17592	P1184	Linear 02 Sensor / Common Ground Wire Open Circuit
17593	P1185	Linear 02 Sensor / Common Ground Wire Short to ground
17594	P1186	Linear 02 Sensor / Common Ground Wire Short to B+
17595	P1187	Linear 02 Sensor / Compens. Resistor Open Circuit
17596	P1188	Linear 02 Sensor / Compens. Resistor Short to ground
17597	P1189	Linear 02 Sensor / Compens. Resistor Short to B+
17598	P1190	Linear 02 Sensor / Reference Voltage Incorrect Signal
17604	P1196	O2 Sensor Heater Circ.,Bank1-Sensor1 Electrical Malfunction
17605	P1197	O2 Sensor Heater Circ.,Bank2-Sensor1 Electrical Malfunction
17606	P1198	O2 Sensor Heater Circ.,Bank1-Sensor2 Electrical Malfunction
17607	P1199	O2 Sensor Heater Circ.,Bank2-Sensor2 Electrical Malfunction
17609	P1201	Cyl.1-Fuel Inj.Circ. Electrical Malfunction
17610	P1202	Cyl.2-Fuel Inj.Circ. Electrical Malfunction
17611	P1203	Cyl.3-Fuel Inj.Circ. Electrical Malfunction
17612	P1204	Cyl.4-Fuel Inj.Circ. Electrical Malfunction
17613	P1205	Cyl.5-Fuel Inj.Circ. Electrical Malfunction
17614	P1206	Cyl.6-Fuel Inj.Circ. Electrical Malfunction
17615	P1207	Cyl.7-Fuel Inj.Circ. Electrical Malfunction
17616	P1208	Cyl.8-Fuel Inj.Circ. Electrical Malfunction
17617	P1209	Intake valves for cylinder shut-off Short circuit to ground
17618	P1210	Intake valves for cylinder shut-off Short to B+
17619	P1211	Intake valves for cylinder shut-off Open circuit
17621	P1213	Cyl.1-Fuel Inj.Circ. Short to B+
17622	P1214	Cyl.2-Fuel Inj.Circ. Short to B+
17623	P1215	Cyl.3-Fuel Inj.Circ. Short to B+
17624	P1216	Cyl.4-Fuel Inj.Circ. Short to B+
17625	P1217	Cyl.5-Fuel Inj.Circ. Short to B+
17626	P1218	Cyl.6-Fuel Inj.Circ. Short to B+
17627	P1219	Cyl.7-Fuel Inj.Circ. Short to B+
17628	P1220	Cyl.8-Fuel Inj.Circ. Short to B+
17629	P1221	Cylinder shut-off exhaust valves Short circuit to ground

BentleyPublishers.com
BentleyPublishers.com—All Rights Reserved

Table a. Diagnostic trouble codes (DTCs or P-codes)		
VAG code	**SAE P-code**	**Description**
17630	P1222	Cylinder shut-off exhaust valves Short to B+
17631	P1223	Cylinder shut-off exhaust valves Open circuit
17633	P1225	Cyl.1-Fuel Inj.Circ. Short to Ground
17634	P1226	Cyl.2-Fuel Inj.Circ. Short to Ground
17635	P1227	Cyl.3-Fuel Inj.Circ. Short to Ground
17636	P1228	Cyl.4-Fuel Inj.Circ. Short to Ground
17637	P1229	Cyl.5-Fuel Inj.Circ. Short to Ground
17638	P1230	Cyl.6-Fuel Inj.Circ. Short to Ground
17639	P1231	Cyl.7-Fuel Inj.Circ. Short to Ground
17640	P1232	Cyl.8-Fuel Inj.Circ. Short to Ground
17645	P1237	Cyl.1-Fuel Inj.Circ. Open Circ.
17646	P1238	Cyl.2-Fuel Inj.Circ. Open Circ.
17647	P1239	Cyl.3-Fuel Inj.Circ. Open Circ.
17648	P1240	Cyl.4-Fuel Inj.Circ. Open Circ.
17649	P1241	Cyl.5-Fuel Inj.Circ. Open Circ.
17650	P1242	Cyl.6-Fuel Inj.Circ. Open Circ.
17651	P1243	Cyl.7-Fuel Inj.Circ. Open Circ.
17652	P1244	Cyl.8-Fuel Inj.Circ. Open Circ.
17653	P1245	Needle Lift Sensor Circ. Short to Ground
17654	P1246	Needle Lift Sensor Circ. Range/Performance
17655	P1247	Needle Lift Sensor Circ. Open/Short to B+
17656	P1248	Injection Start Control Deviation
17657	P1249	Fuel consumption signal Electrical Fault in Circuit
17658	P1250	Fuel Level Too Low
17659	P1251	Start of Injection Solenoid Circ Short to B+
17660	P1252	Start of Injection Solenoid Circ Open/Short to Ground
17661	P1253	Fuel consumption signal Short to ground
17662	P1254	Fuel consumption signal Short to B+
17663	P1255	Engine Coolant Temp.Circ Short to Ground
17664	P1256	Engine Coolant Temp.Circ Open/Short to B+
17665	P1257	Engine Coolant System Valve Open
17666	P1258	Engine Coolant System Valve Short to B+
17667	P1259	Engine Coolant System Valve Short to Ground
17688	P1280	Fuel Inj.Air Contr.Valve Circ. Flow too Low
17691	P1283	Fuel Inj.Air Contr.Valve Circ. Electrical Malfunction
17692	P1284	Fuel Inj.Air Contr.Valve Circ. Open
17693	P1285	Fuel Inj.Air Contr.Valve Circ. Short to Ground

BentleyPublishers.com
BentleyPublishers.com—All Rights Reserved

Table a. Diagnostic trouble codes (DTCs or P-codes)

VAG code	SAE P-code	Description
17694	P1286	Fuel Inj.Air Contr.Valve Circ. Short to B+
17695	P1287	Turbocharger bypass valve open
17696	P1288	Turbocharger bypass valve short to B+
17697	P1289	Turbocharger bypass valve short to ground
17704	P1296	Cooling system malfunction
17705	P1297	Connection turbocharger - throttle valve pressure hose
17708	P1300	Misfire detected Reason: Fuel level too low
17721	P1319	Knock Sensor 1 Circ. Short to Ground
17728	P1320	Knock Sensor 2 Circ. Short to Ground
17729	P1321	Knock Sensor 3 Circ. Low Input
17730	P1322	Knock Sensor 3 Circ. High Input
17731	P1323	Knock Sensor 4 Circ. Low Input
17732	P1324	Knock Sensor 4 Circ. High Input
17733	P1325	Cyl.1-Knock Contr. Limit Attained
17734	P1326	Cyl.2-Knock Contr. Limit Attained
17735	P1327	Cyl.3-Knock Contr. Limit Attained
17736	P1328	Cyl.4-Knock Contr. Limit Attained
17737	P1329	Cyl.5-Knock Contr. Limit Attained
17738	P1330	Cyl.6-Knock Contr. Limit Attained
17739	P1331	Cyl.7-Knock Contr. Limit Attained
17740	P1332	Cyl.8-Knock Contr. Limit Attained
17743	P1335	Engine Torque Monitoring 2 Control Limit Exceeded
17744	P1336	Engine Torque Monitoring Adaptation at limit
17745	P1337	Camshaft Pos.Sensor,Bank1 Short to Ground
17746	P1338	Camshaft Pos.Sensor,Bank1 Open Circ./Short to B+
17747	P1339	Crankshaft Pos./Engine Speed Sensor Cross Connected
17748	P1340	Crankshaft-/Camshaft Pos.Sens.Signals Out of Sequence
17749	P1341	Ignition Coil Power Output Stage 1 Short to Ground
17750	P1342	Ignition Coil Power Output Stage 1 Short to B+
17751	P1343	Ignition Coil Power Output Stage 2 Short to Ground
17752	P1344	Ignition Coil Power Output Stage 2 Short to B+
17753	P1345	Ignition Coil Power Output Stage 3 Short to Ground
17754	P1346	Ignition Coil Power Output Stage 3 Short to B+
17755	P1347	Bank2,Crankshaft-/Camshaft os.Sens.Sign. Out of Sequence
17756	P1348	Ignition Coil Power Output Stage 1 Open Circuit
17757	P1349	Ignition Coil Power Output Stage 2 Open Circuit
17758	P1350	Ignition Coil Power Output Stage 3 Open Circuit

BentleyPublishers.com
BentleyPublishers.com—All Rights Reserved

Table a. Diagnostic trouble codes (DTCs or P-codes)		
VAG code	**SAE P-code**	**Description**
17762	P1354	Modulation Piston Displ.Sensor Circ. Malfunction
17763	P1355	Cyl. 1, ignition circuit Open Circuit
17764	P1356	Cyl. 1, ignition circuit Short to B+
17765	P1357	Cyl. 1, ignition circuit Short to ground
17766	P1358	Cyl. 2, ignition circuit Open Circuit
17767	P1359	Cyl. 2, ignition circuit Short Circuit to B+
17768	P1360	Cyl. 2, ignition circuit Short Circuit to Ground
17769	P1361	Cyl. 3, ignition circuit Open Circuit
17770	P1362	Cyl. 3, ignition circuit Short Circuit to B+
17771	P1363	Cyl. 3, ignition circuit Short Circuit to ground
17772	P1364	Cyl. 4 ignition circuit Open Circuit
17773	P1365	Cyl. 4 ignition circuit Short circuit to B+
17774	P1366	Cyl. 4 ignition circuit Short circuit to ground
17775	P1367	Cyl. 5, ignition circuit Open Circuit
17776	P1368	Cyl. 5, ignition circuit Short Circuit to B+
17777	P1369	Cyl. 5, ignition circuit short to ground
17778	P1370	Cyl. 6, ignition circuit Open Circuit
17779	P1371	Cyl. 6, ignition circuit Short Circuit to B+
17780	P1372	Cyl. 6, ignition circuit short to ground
17781	P1373	Cyl. 7, ignition circuit Open Circuit
17782	P1374	Cyl. 7, ignition circuit Short Circuit to B+
17783	P1375	Cyl. 7, ignition circuit short to ground
17784	P1376	Cyl. 8, ignition circuit Open Circuit
17785	P1377	Cyl. 8, ignition circuit Short Circuit to B+
17786	P1378	Cyl. 8, ignition circuit short to ground
17794	P1386	Internal Control Module Knock Control Circ.Error
17795	P1387	Internal Contr. Module altitude sensor error
17796	P1388	Internal Contr. Module drive by wire error
17799	P1391	Camshaft Pos.Sensor,Bank2 Short to Ground
17800	P1392	Camshaft Pos.Sensor,Bank2 Open Circ./Short to B+
17801	P1393	Ignition Coil Power Output Stage 1 Electrical Malfunction
17802	P1394	Ignition Coil Power Output Stage 2 Electrical Malfunction
17803	P1395	Ignition Coil Power Output Stage 3 Electrical Malfunction
17804	P1396	Engine Speed Sensor Missing Tooth
17805	P1397	Engine speed wheel Adaptation limit reached
17806	P1398	Engine RPM signal, TD Short to ground
17807	P1399	Engine RPM signal, TD Short Circuit to B+

BentleyPublishers.com
BentleyPublishers.com—All Rights Reserved

Table a. Diagnostic trouble codes (DTCs or P-codes)

VAG code	SAE P-code	Description
17808	P1400	EGR Valve Circ Electrical Malfunction
17809	P1401	EGR Valve Circ Short to Ground
17810	P1402	EGR Valve Circ Short to B+
17811	P1403	EGR Flow Deviation
17812	P1404	EGR Flow Basic Setting not carried out
17814	P1406	EGR Temp.Sensor Range/Performance
17815	P1407	EGR Temp.Sensor Signal too Low
17816	P1408	EGR Temp.Sensor Signal too High
17817	P1409	Tank Ventilation Valve Circ. Electrical Malfunction
17818	P1410	Tank Ventilation Valve Circ. Short to B+
17819	P1411	Sec.Air Inj.Sys.,Bank2 Flow too Flow
17820	P1412	EGR Different.Pressure Sensor Signal too Low
17821	P1413	EGR Different.Pressure Sensor Signal too High
17822	P1414	Sec.Air Inj.Sys.,Bank2 Leak Detected
17825	P1417	Fuel Level Sensor Circ Signal too Low
17826	P1418	Fuel Level Sensor Circ Signal too High
17828	P1420	Sec.Air Inj.Valve Circ Electrical Malfunction
17829	P1421	Sec.Air Inj.Valve Circ Short to Ground
17830	P1422	Sec.Air Inj.Sys.Contr.Valve Circ Short to B+
17831	P1423	Sec.Air Inj.Sys.,Bank1 Flow too Low
17832	P1424	Sec.Air Inj.Sys.,Bank1 Leak Detected
17833	P1425	Tank Vent.Valve Short to Ground
17834	P1426	Tank Vent.Valve Open
17840	P1432	Sec.Air Inj.Valve Open
17841	P1433	Sec.Air Inj.Sys.Pump Relay Circ. open
17842	P1434	Sec.Air Inj.Sys.Pump Relay Circ. Short to B+
17843	P1435	Sec.Air Inj.Sys.Pump Relay Circ. Short to ground
17844	P1436	Sec.Air Inj.Sys.Pump Relay Circ. Electrical Malfunction
17847	P1439	EGR Potentiometer Error in Basic Setting
17848	P1440	EGR Valve Power Stage Open
17849	P1441	EGR Valve Circ Open/Short to Ground
17850	P1442	EGR Valve Position Sensor Signal too high
17851	P1443	EGR Valve Position Sensor Signal too low
17852	P1444	EGR Valve Position Sensor range/performance
17853	P1445	Catalyst Temp.Sensor 2 Circ. Range/Performance
17854	P1446	Catalyst Temp.Circ Short to Ground
17855	P1447	Catalyst Temp.Circ Open/Short to B+

BentleyPublishers.com
BentleyPublishers.com—All Rights Reserved

Table a. Diagnostic trouble codes (DTCs or P-codes)		
VAG code	SAE P-code	Description
17856	P1448	Catalyst Temp.Sensor 2 Circ. Short to Ground
17857	P1449	Catalyst Temp.Sensor 2 Circ. Open/Short to B+
17858	P1450	Sec.Air Inj.Sys.Circ Short to B+
17859	P1451	Sec.Air Inj.Sys.Circ Short to Ground
17860	P1452	Sec.Air Inj.Sys. Open Circ.
17861	P1453	Exhaust gas temperature sensor 1 open/short to B+
17862	P1454	Exhaust gas temperature sensor short 1 to ground
17863	P1455	Exhaust gas temperature sensor 1 range/performance
17864	P1456	Exhaust gas temperature control bank 1 limit attained
17865	P1457	Exhaust gas temperature sensor 2 open/short to B+
17866	P1458	Exhaust gas temperature sensor 2 short to ground
17867	P1459	Exhaust gas temperature sensor 2 range/performance
17868	P1460	Exhaust gas temperature control bank 2 limit attained
17869	P1461	Exhaust gas temperature control bank 1 Range/Performance
17870	P1462	Exhaust gas temperature control bank 2 Range/Performance
17873	P1465	Additive Pump Short Circuit to B+
17874	P1466	Additive Pump Open/Short to Ground
17875	P1467	EVAP Canister Purge Solenoid Valve Short Circuit to B+
17876	P1468	EVAP Canister Purge Solenoid Valve Short Circuit to Ground
17877	P1469	EVAP Canister Purge Solenoid Valve Open Circuit
17878	P1470	EVAP Emission Contr.LDP Circ Electrical Malfunction
17879	P1471	EVAP Emission Contr.LDP Circ Short to B+
17880	P1472	EVAP Emission Contr.LDP Circ Short to Ground
17881	P1473	EVAP Emission Contr.LDP Circ Open Circ.
17882	P1474	EVAP Canister Purge Solenoid Valve electrical malfunction
17883	P1475	EVAP Emission Contr.LDP Circ Malfunction/Signal Circ.Open
17884	P1476	EVAP Emission Contr.LDP Circ Malfunction/Insufficient Vacuum
17885	P1477	EVAP Emission Contr.LDP Circ Malfunction
17886	P1478	EVAP Emission Contr.LDP Circ Clamped Tube Detected
17908	P1500	Fuel Pump Relay Circ. Electrical Malfunction
17909	P1501	Fuel Pump Relay Circ. Short to Ground
17910	P1502	Fuel Pump Relay Circ. Short to B+
17911	P1503	Load signal from Alternator Term. DF Range/performance/Incorrect Signal
17912	P1504	Intake Air Sys.Bypass Leak Detected
17913	P1505	Closed Throttle Pos. Does Not Close/Open Circ
17914	P1506	Closed Throttle Pos.Switch Does Not Open/Short to Ground
17915	P1507	Idle Sys.Learned Value Lower Limit Attained

BentleyPublishers.com
BentleyPublishers.com—All Rights Reserved

Table a. Diagnostic trouble codes (DTCs or P-codes)		
VAG code	**SAE P-code**	**Description**
17916	P1508	Idle Sys.Learned Value Upper Limit Attained
17917	P1509	Idle Air Control Circ. Electrical Malfunction
17918	P1510	Idle Air Control Circ. Short to B+
17919	P1511	Intake Manifold Changeover Valve circuit electrical malfunction
17920	P1512	Intake Manifold Changeover Valve circuit Short to B+
17921	P1513	Intake Manifold Changeover Valve2 circuit Short to B+
17922	P1514	Intake Manifold Changeover Valve2 circuit Short to ground
17923	P1515	Intake Manifold Changeover Valve circuit Short to Ground
17924	P1516	Intake Manifold Changeover Valve circuit Open
17925	P1517	Main Relay Circ. Electrical Malfunction
17926	P1518	Main Relay Circ. Short to B+
17927	P1519	Intake Camshaft Contr.,Bank1 Malfunction
17928	P1520	Intake Manifold Changeover Valve2 circuit Open
17929	P1521	Intake Manifold Changeover Valve2 circuit electrical malfunction
17930	P1522	Intake Camshaft Contr.,Bank2 Malfunction
17931	P1523	Crash Signal from Airbag Control Unit range/performance
17933	P1525	Intake Camshaft Contr.Circ.,Bank1 Electrical Malfunction
17934	P1526	Intake Camshaft Contr.Circ.,Bank1 Short to B+
17935	P1527	Intake Camshaft Contr.Circ.,Bank1 Short to Ground
17936	P1528	Intake Camshaft Contr.Circ.,Bank1 Open
17937	P1529	Camshaft Control Circuit Short to B+
17938	P1530	Camshaft Control Circuit Short to ground
17939	P1531	Camshaft Control Circuit open
17941	P1533	Intake Camshaft Contr.Circ.,Bank2 Electrical Malfunction
17942	P1534	Intake Camshaft Contr.Circ.,Bank2 Short to B+
17943	P1535	Intake Camshaft Contr.Circ.,Bank2 Short to Ground
17944	P1536	Intake Camshaft Contr.Circ.,Bank2 Open
17945	P1537	Engine Shutoff Solenoid Malfunction
17946	P1538	Engine Shutoff Solenoid Open/Short to Ground
17947	P1539	Clutch Vacuum Vent Valve Switch Incorrect signal
17948	P1540	Vehicle Speed Sensor High Input
17949	P1541	Fuel Pump Relay Circ Open
17950	P1542	Throttle Actuation Potentiometer Range/Performance
17951	P1543	Throttle Actuation Potentiometer Signal too Low
17952	P1544	Throttle Actuation Potentiometer Signal too High
17953	P1545	Throttle Pos.Contr Malfunction
17954	P1546	Boost Pressure Contr.Valve Short to B+

BentleyPublishers.com
BentleyPublishers.com—All Rights Reserved

Table a. Diagnostic trouble codes (DTCs or P-codes)

VAG code	SAE P-code	Description
17955	P1547	Boost Pressure Contr.Valve Short to Ground
17956	P1548	Boost Pressure Contr.Valve Open
17957	P1549	Boost Pressure Contr.Valve Short to Ground
17958	P1550	Charge Pressure Deviation
17959	P1551	Barometric Pressure Sensor Circ. Short to B+
17960	P1552	Barometric Pressure Sensor Circ. Open/Short to Ground
17961	P1553	Barometric/manifold pressure signal ratio out of range
17962	P1554	Idle Speed Contr.Throttle Pos. Basic Setting Conditions not met
17963	P1555	Charge Pressure Upper Limit exceeded
17964	P1556	Charge Pressure Contr. Negative Deviation
17965	P1557	Charge Pressure Contr. Positive Deviation
17966	P1558	Throttle Actuator Electrical Malfunction
17967	P1559	Idle Speed Contr.Throttle Pos. Adaptation Malfunction
17968	P1560	Maximum Engine Speed Exceeded
17969	P1561	Quantity Adjuster Deviation
17970	P1562	Quantity Adjuster Upper Limit Attained
17971	P1563	Quantity Adjuster Lower Limit Attained
17972	P1564	Idle Speed Contr.Throttle Pos. Low Voltage During Adaptation
17973	P1565	Idle Speed Control Throttle Position lower limit not attained
17974	P1566	Load signal from A/C compressor range/performance
17975	P1567	Load signal from A/C compressor no signal
17976	P1568	Idle Speed Contr.Throttle Pos. mechanical Malfunction
17977	P1569	Cruise control switch Incorrect signal
17978	P1570	Contr.Module Locked
17979	P1571	Left Eng. Mount Solenoid Valve Short to B+
17980	P1572	Left Eng. Mount Solenoid Valve Short to ground
17981	P1573	Left Eng. Mount Solenoid Valve Open circuit
17982	P1574	Left Eng. Mount Solenoid Valve Electrical fault in circuit
17983	P1575	Right Eng. Mount Solenoid Valve Short to B+
17984	P1576	Right Eng. Mount Solenoid Valve Short to ground
17985	P1577	Right Eng. Mount Solenoid Valve Open circuit
17986	P1578	Right Eng. Mount Solenoid Valve Electrical fault in circuit
17987	P1579	Idle Speed Contr.Throttle Pos. Adaptation not started
17988	P1580	Throttle Actuator B1 Malfunction
17989	P1581	Idle Speed Contr.Throttle Pos. Basic Setting Not Carried Out
17990	P1582	Idle Adaptation at Limit
17991	P1583	Transmission mount valves Short to B+

BentleyPublishers.com
BentleyPublishers.com—All Rights Reserved

Table a. Diagnostic trouble codes (DTCs or P-codes)

VAG code	SAE P-code	Description
17992	P1584	Transmission mount valves Short to ground
17993	P1585	Transmission mount valves Open circuit
17994	P1586	Engine mount solenoid valves Short to B+
17995	P1587	Engine mount solenoid valves Short to ground
17996	P1588	Engine mount solenoid valves Open circuit
18008	P1600	Power Supply (B+) Terminal 15 Low Voltage
18010	P1602	Power Supply (B+) Terminal 30 Low Voltage
18011	P1603	Internal Control Module Malfunction
18012	P1604	Internal Control Module Driver Error
18013	P1605	Rough Road/Acceleration Sensor Electrical Malfunction
18014	P1606	Rough Road Spec Engine Torque ABS-ECU Electrical Malfunction
18015	P1607	Vehicle speed signal Error message from instrument cluster
18016	P1608	Steering angle signal Error message from steering angle sensor
18017	P1609	Crash shut-down activated
18019	P1611	MIL Call-up Circ./Transm.Contr.Module Short to Ground
18020	P1612	Electronic Control Module Incorrect Coding
18021	P1613	MIL Call-up Circ Open/Short to B+
18022	P1614	MIL Call-up Circ./Transm.Contr.Module Range/Performance
18023	P1615	Engine Oil Temperature Sensor Circuit range/performance
18024	P1616	Glow Plug/Heater Indicator Circ. Short to B+
18025	P1617	Glow Plug/Heater Indicator Circ. Open/Short to Ground
18026	P1618	Glow Plug/Heater Relay Circ. Short to B+
18027	P1619	Glow Plug/Heater Relay Circ. Open/Short to Ground
18028	P1620	Engine coolant temperature signal open/short to B+
18029	P1621	Engine coolant temperature signal short to ground
18030	P1622	Engine coolant temperature signal range/performance
18031	P1623	Data Bus Powertrain No Communication
18032	P1624	MIL Request Sign.active
18033	P1625	Data-Bus Powertrain Implausible Message from Transm.Contr.
18034	P1626	Data-Bus Powertrain Missing Message from Transm.Contr.
18035	P1627	Data-Bus Powertrain missing message from fuel injection pump
18036	P1628	Data-Bus Powertrain missing message from steering sensor
18037	P1629	Data-Bus Powertrain missing message from distance control
18038	P1630	Accelera.Pedal Pos.Sensor 1 Signal too Low
18039	P1631	Accelera.Pedal Pos.Sensor 1 Signal too High
18040	P1632	Accelera.Pedal Pos.Sensor 1 Power Supply Malfunction
18041	P1633	Accelera.Pedal Pos.Sensor 2 Signal too Low

BentleyPublishers.com
BentleyPublishers.com—All Rights Reserved

Table a. Diagnostic trouble codes (DTCs or P-codes)

VAG code	SAE P-code	Description
18042	P1634	Accelera.Pedal Pos.Sensor 2 Signal too High
18043	P1635	Data Bus Powertrain missing message f.air condition control
18044	P1636	Data Bus Powertrain missing message from Airbag control
18045	P1637	Data Bus Powertrain missing message f.central electr.control
18046	P1638	Data Bus Powertrain missing message from clutch control
18047	P1639	Accelera.Pedal Pos.Sensor 1+2 Range/Performance
18048	P1640	Internal Contr.Module (EEPROM) Error
18049	P1641	Please check DTC Memory of Air Condition ECU
18050	P1642	Please check DTC Memory of Airbag ECU
18051	P1643	Please check DTC Memory of central electric ECU
18052	P1644	Please check DTC Memory of clutch ECU
18053	P1645	Data Bus Powertrain missing message f.all wheel drive contr.
18054	P1646	Please Check DTC Memory of all wheel drive ECU
18055	P1647	Please check coding of ECUs in Data Bus Powertrain
18056	P1648	Data Bus Powertrain Malfunction
18057	P1649	Data Bus Powertrain Missing message from ABS Control Module
18058	P1650	Data Bus Powertrain Missing message fr.instrument panel ECU
18059	P1651	Data Bus Powertrain missing messages
18060	P1652	Please check DTC Memory of transmission ECU
18061	P1653	Please check DTC Memory of ABS Control Module
18062	P1654	Please check DTC Memory of control panel ECU
18063	P1655	Please check DTC Memory of ADR Control Module
18064	P1656	A/C clutch relay circuit short to ground
18065	P1657	A/C clutch relay circuit short to B+
18066	P1658	Data Bus Powertrain Incorrect signal from ADR Control Module
18084	P1676	Drive by Wire-MIL Circ. Electrical Malfunction
18085	P1677	Drive by Wire-MIL Circ. Short to B+
18086	P1678	Drive by Wire-MIL Circ. Short to Ground
18087	P1679	Drive by Wire-MIL Circ. Open
18089	P1681	Contr.Unit Programming, Programming not Finished
18092	P1684	Contr.Unit Programming Communication Error
18094	P1686	Contr.Unit Error Programming Error
18098	P1690	Malfunction Indication Light Malfunction
18099	P1691	Malfunction Indication Light Open
18100	P1692	Malfunction Indication Light Short to Ground
18101	P1693	Malfunction Indication Light Short to B+
18102	P1694	Malfunction Indication Light Open/Short to Ground

BentleyPublishers.com
BentleyPublishers.com—All Rights Reserved

Table a. Diagnostic trouble codes (DTCs or P-codes)

VAG code	SAE P-code	Description
18112	P1704	Kick Down Switch Malfunction
18113	P1705	Gear/Ratio Monitoring Adaptation limit reached
18119	P1711	Wheel Speed Signal 1 Range/Performance
18124	P1716	Wheel Speed Signal 2 Range/Performance
18129	P1721	Wheel Speed Signal 3 Range/Performance
18131	P1723	Starter Interlock Circ. Open
18132	P1724	Starter Interlock Circ. Short to Ground
18134	P1726	Wheel Speed Signal 4 Range/Performance
18136	P1728	Different Wheel Speed Signals Range/Performance
18137	P1729	Starter Interlock Circ. Short to B+
18141	P1733	Tiptronic Switch Down Circ. Short to Ground
18147	P1739	Tiptronic Switch up Circ. Short to Ground
18148	P1740	Clutch temperature control
18149	P1741	Clutch pressure adaptation at limit
18150	P1742	Clutch torque adaptation at limit
18151	P1743	Clutch slip control signal too high
18152	P1744	Tiptronic Switch Recognition Circ. Short to Ground
18153	P1745	Transm.Contr.Unit Relay Short to B+
18154	P1746	Transm.Contr.Unit Relay Malfunction
18155	P1747	Transm.Contr.Unit Relay Open/Short to Ground
18156	P1748	Transm.Contr.Unit Self-Check
18157	P1749	Transm.Contr.Unit Incorrect Coded
18158	P1750	Power Supply Voltage Low Voltage
18159	P1751	Power Supply Voltage I ligh Voltage
18160	P1752	Power Supply Malfunction
18168	P1760	Shift Lock Malfunction
18169	P1761	Shift Lock Short to Ground
18170	P1762	Shift Lock Short to B+
18171	P1763	Shift Lock Open
18172	P1764	Transmission temperature control
18173	P1765	Hydraulic Pressure Sensor 2 adaptation at limit
18174	P1766	Throttle Angle Signal Stuck Off
18175	P1767	Throttle Angle Signal Stuck On
18176	P1768	Hydraulic Pressure Sensor 2 Too High
18177	P1769	Hydraulic Pressure Sensor 2 Too Low
18178	P1770	Load Signal Range/Performance
18179	P1771	Load Signal Stuck Off

BentleyPublishers.com
BentleyPublishers.com—All Rights Reserved

Table a. Diagnostic trouble codes (DTCs or P-codes)		
VAG code	**SAE P-code**	**Description**
18180	P1772	Load Signal Stuck On
18181	P1773	Hydraulic Pressure Sensor 1 Too High
18182	P1774	Hydraulic Pressure Sensor 1 Too Low
18183	P1775	Hydraulic Pressure Sensor 1 adaptation at limit
18184	P1776	Hydraulic Pressure Sensor 1 range/performance
18185	P1777	Hydraulic Pressure Sensor 2 range/performance
18186	P1778	Solenoid EV7 Electrical Malfunction
18189	P1781	Engine Torque Reduction Open/Short to Ground
18190	P1782	Engine Torque Reduction Short to B+
18192	P1784	Shift up/down Wire Open/Short to Ground
18193	P1785	Shift up/down Wire Short to B+
18194	P1786	Reversing Light Circ. Open
18195	P1787	Reversing Light Circ. Short to Ground
18196	P1788	Reversing Light Circ. Short to B+
18197	P1789	Idle Speed Intervention Circ. Error Message from Engine Contr.
18198	P1790	Transmission Range Display Circ. Open
18199	P1791	Transmission Range Display Circ. Short to Ground
18200	P1792	Transmission Range Display Circ. Short to B+
18201	P1793	Output Speed Sensor 2 Circ. No Signal
18203	P1795	Vehicle Speed Signal Circ. Open
18204	P1796	Vehicle Speed Signal Circ. Short to Ground
18205	P1797	Vehicle Speed Signal Circ. Short to B+
18206	P1798	Output Speed Sensor 2 Circ. Range/Performance
18207	P1799	Output Speed Sensor 2 Circ. RPM too High
18221	P1813	Pressure Contr.Solenoid 1 Electrical
18222	P1814	Pressure Contr.Solenoid 1 Open/Short to Ground
18223	P1815	Pressure Contr.Solenoid 1 Short to B+
18226	P1818	Pressure Contr.Solenoid 2 Electrical
18227	P1819	Pressure Contr.Solenoid 2 Open/Short to Ground
18228	P1820	Pressure Contr.Solenoid 2 Short to B+
18231	P1823	Pressure Contr.Solenoid 3 Electrical
18232	P1824	Pressure Contr.Solenoid 3 Open/Short to Ground
18233	P1825	Pressure Contr.Solenoid 3 Short to B+
18236	P1828	Pressure Contr.Solenoid 4 Electrical
18237	P1829	Pressure Contr.Solenoid 4 Open/Short to Ground
18238	P1830	Pressure Contr.Solenoid 4 Short to B+
18242	P1834	Pressure Contr.Solenoid 5 Open/Short to Ground

BentleyPublishers.com
BentleyPublishers.com—All Rights Reserved

Table a. Diagnostic trouble codes (DTCs or P-codes)

VAG code	SAE P-code	Description
18243	P1835	Pressure Contr.Solenoid 5 Short to B+
18249	P1841	Engine/Transmission Control Modules Versions do not match
18250	P1842	Please check DTC Memory of instrument panel ECU
18251	P1843	Please check DTC Memory of ADR Control Module
18252	P1844	Please check DTC Memory of central electric control ECU
18255	P1847	Please check DTC Memory of brake system ECU
18256	P1848	Please check DTC Memory of engine ECU
18257	P1849	Please check DTC Memory of transmission ECU
18258	P1850	Data-Bus Powertrain Missing Message from Engine Contr.
18259	P1851	Data-Bus Powertrain Missing Message from Brake Contr.
18260	P1852	Data-Bus Powertrain Implausible Message from Engine Contr.
18261	P1853	Data-Bus Powertrain Implausible Message from Brake Contr.
18262	P1854	Data-Bus Powertrain Hardware Defective
18263	P1855	Data-Bus Powertrain Software version Contr.
18264	P1856	Throttle/Pedal Pos.Sensor A Circ. Error Message from Engine Contr.
18265	P1857	Load Signal Error Message from Engine Contr.
18266	P1858	Engine Speed Input Circ. Error Message from Engine Contr.
18267	P1859	Brake Switch Circ. Error Message from Engine Contr.
18268	P1860	Kick Down Switch Error Message from Engine Contr.
18269	P1861	Throttle Position (TP) sensor Error Message from ECM
18270	P1862	Data Bus Powertrain Missing message from instr. panel ECU
18271	P1863	Data Bus Powertrain Missing Message from St. Angle Sensor
18272	P1864	Data Bus Powertrain Missing message from ADR control module
18273	P1865	Data Bus Powertrain Missing message from central electronics
18274	P1866	Data Bus Powertrain Missing messages

BentleyPublishers.com
BentleyPublishers.com—All Rights Reserved

VAG 1551/1552 OPERATING OVERVIEW FLOW CHART

MODE MENU

1 - Rapid Data =>
2 - Blink Code
3 - Self Test
4 - Dealership Code
5 - LT - Diagnosis

ADDRESS WORDS MENU (V9)

01 - Engine Electronics =>
11 - Engine Electronics II
41 - Diesel Pump Electronics
51 - Electric Drive
61 - Battery Control
71 - Battery Charger
02 - Transmission Electronics
12 - Clutch Electronics
22 - 4 Wheel Drive Electronics
03 - Brake Electronics
13 - Distance Control
33 - OBD II
14 - Suspension Electronics
24 - Anti - Slip Control
34 - Level Control
44 - Steering Assist
54 - Rear Spoiler
15 - Airbag
25 - Anti-Theft Immobilizer Sensor
35 - Central Locking
45 - Interior Monitor
55 - Headlamp Vert. Aim Control
65 - Tire Pressure Monitoring
75 - Emergency Call Module
16 - Steering Wheel Electronics
26 - Automatic Roof Control
36 - Seat Adjustment Driver's Side
46 - Comfort System (AC CM)
56 - Radio
66 - Seat & Mirror Adjustment
76 - Parking Aid
17 - Instrument Cluster
37 - Navigation System
47 - Sound System
57 - Airbag (Cabriolet only)
67 - Language Control
08 - A/C & Heating Electronics
18 - Auxiliary Heater
09 - Electronic Control Module
19 - Diagnostic Interface for Data Bus
29 - Left Light Control
39 - Right Light Control
49 - Auto Light Switch
00 - Automatic Test Sequence

FUNCTIONS MENU

01 - Check Control Module Versions
02 - Check DTC Memory
03 - Output Diagnostic Test Mode
04 - Basic Setting
  Display Group xxx =>
05 - Erase DTC Memory
06 - End Output
07 - Code Control Module
08 - Read Measuring Value Block
  Display Group xxx =>
09 - Read Individual Measuring Values
  Channel xx
10 - Adaptation
  Channel xx
11 - Log-In Procedure
15 - Readiness Code

DISPLAY GROUP FORMAT

Display Group 000 (only)
xx xx xx xx xx xx xx xx xx xx

Display Group 001 (and higher)
xxxxx xxxxx xxxxx xxxxx

BentleyPublishers.com
BentleyPublishers.com—All Rights Reserved

WARNING

Your common sense, good judgement and general alertness are crucial to safe and successful service work. Before attempting any work on your Audi, be sure to read **00 Warnings and Cautions** *and the copyright page at the front of the manual. Review these warnings and cautions each time you prepare to work on your car. Please also read any warnings and cautions that accompany the procedures in the manual.*

A

B

BentleyPublishers.com
BentleyPublishers.com—All Rights Reserved

BentleyPublishers.com
BentleyPublishers.com—All Rights Reserved

WARNING

Your common sense, good judgement and general alertness are crucial to safe and successful service work. Before attempting any work on your Audi, be sure to read **00 Warnings and Cautions** *and the copyright page at the front of the manual. Review these warnings and cautions each time you prepare to work on your car. Please also read any warnings and cautions that accompany the procedures in the manual.*

E

BentleyPublishers.com
BentleyPublishers.com—All Rights Reserved

BentleyPublishers.com
BentleyPublishers.com—All Rights Reserved

WARNING

Your common sense, good judgement and general alertness are crucial to safe and successful service work. Before attempting any work on your Audi, be sure to read **00 Warnings and Cautions** *and the copyright page at the front of the manual. Review these warnings and cautions each time you prepare to work on your car. Please also read any warnings and cautions that accompany the procedures in the manual.*

I

J

K

L

M

BentleyPublishers.com
BentleyPublishers.com—All Rights Reserved

BentleyPublishers.com
BentleyPublishers.com—All Rights Reserved

WARNING

Your common sense, good judgement and general alertness are crucial to safe and successful service work. Before attempting any work on your Audi, be sure to read **00 Warnings and Cautions** *and the copyright page at the front of the manual. Review these warnings and cautions each time you prepare to work on your car. Please also read any warnings and cautions that accompany the procedures in the manual.*

BentleyPublishers.com
BentleyPublishers.com—All Rights Reserved

V

W

X

BentleyPublishers.com
BentleyPublishers.com—All Rights Reserved

Selected Books and Repair Information From Bentley Publishers

Motorsports

Alex Zanardi: My Sweetest Victory *Alex Zanardi and Gianluca Gasparini*
ISBN 978-0-8376-1249-2

The Unfair Advantage *Mark Donohue and Paul van Valkenburgh*
ISBN 978-0-8376-0069-7

Equations of Motion - Adventure, Risk and Innovation
William F. Milliken
ISBN 978-0-8376-1570-7

Engineering

Bosch Fuel Injection and Engine Management *Charles O. Probst, SAE*
ISBN 978-0-8376-0300-1

Maximum Boost: Designing, Testing, and Installing Turbocharger Systems
Corky Bell ISBN 978-0-8376-0160-1

Supercharged! Design, Testing and Installation of Supercharger Systems
Corky Bell ISBN 978-0-8376-0168-7

Scientific Design of Exhaust and Intake Systems *Phillip H. Smith & John C. Morrison* ISBN 978-0-8376-0309-4

Audi

Audi A4 (B6, B7) Service Manual: 2002-2008, 1.8L Turbo, 2.0L Turbo, 3.0L, 3.2L including Avant and Cabriolet
Bentley Publishers
ISBN 978-0-8376-1574-5

Audi A4 Service Manual: 1996-2001, 1.8L Turbo, 2.8L, including Avant and quattro *Bentley Publishers*
ISBN 978-0-8376-1675-9

Audi A6 Service Manual: 1998-2004
Bentley Publishers
ISBN 978-0-8376-1670-4

Audi TT Service Manual: 2000-2006, 1.8L turbo, 3.2 L, including Roadster and quattro *Bentley Publishers*
ISBN 978-0-8376-1625-4

Audi A6, S6: 2005-2007 Repair Manual on DVD-ROM *Audi of America*
ISBN 978-0-8376-1362-8

BMW

Memoirs of a Hack Mechanic
Bob Siegel ISBN 978-0-8376-1720-6

BMW 5 Series (E60, E61) Service Manual: 2004-2010 *Bentley Publishers*
ISBN 978-0-8376-1689-6

BMW X5 (53) Service Manual: 2000-2006 *Bentley Publishers*
ISBN 978-0-8376-1643-8

BMW Z3 (E36/7) Service Manual: 1996-2002 *Bentley Publishers*
ISBN 978-0-8376-1617-9

BMW 3 Series (E30) Service Manual: 1984-1990 *Bentley Publishers*
ISBN 978-0-8376-1647-6

BMW 3 Series (E36) Service Manual: 1992-1998 *Bentley Publishers*
ISBN 978-0-8376-0326-1

BMW 3 Series (E46) Service Manual: 1999-2005 *Bentley Publishers*
ISBN 978-0-8376-1657-5

BMW 3 Series (E90, E91, E92, E93) Service Manual: 2006-2010
Bentley Publishers
ISBN 978-0-8376-1685-8

BMW 5 Series (E39) Service Manual: 1997-2003 *Bentley Publishers*
ISBN 978-0-8376-1672-8

BMW 7 Series (E38) Service Manual: 1995-2001 *Bentley Publishers*
ISBN 978-0-8376-1618-6

Porsche

Porsche 911 (993) Service Manual 1995-1998 *Bentley Publishers*
ISBN 978-0-8376-1719-0

Porsche 911 Service Manual 1999-2005
Bentley Publishers
ISBN 978-0-8376-1710-7

Porsche Boxster Service Manual: 1997-2004 *Bentley Publishers*
ISBN 978-0-8376-1645-2

Porsche 911 Carrera Service Manual: 1984-1989 *Bentley Publishers*
ISBN 978-0-8376-1696-4

Porsche: Excellence Was Expected
Karl Ludvigsen ISBN 978-0-8376-0235-6

Porsche — Origin of the Species *Karl Ludvigsen* ISBN 978-0-8376-1331-4

Volkswagen

Battle for the Beetle *Karl Ludvigsen*
ISBN 978-0-8376-1695-7

Volkswagen Rabbit, GTI Service Manual: 2006-2009 *Bentley Publishers*
ISBN 978-0-8376-1664-3

Volkswagen Jetta, Golf, GTI Service Manual: 1999-2005 *Bentley Publishers*
ISBN 978-0-8376-1678-0

Volkswagen Jetta, Golf, GTI: 1993-1999, Cabrio: 1995-2002 Service Manual *Bentley Publishers*
ISBN 978-0-8376-0366-7

Volkswagen GTI, Golf, Jetta Service Manual: 1985-1992 *Bentley Publishers*
ISBN 978-0-8376-1637-7

Volkswagen Corrado Repair Manual: 1990-1994 *Bentley Publishers*
ISBN 978-0-8376-1699-5

Volkswagen Jetta: 2005-2009 Repair Manual on DVD-ROM
Volkswagen of America
ISBN 978-0-8376-1360-4

Volkswagen Passat, Passat Wagon: 1998-2005 Service Manual
Bentley Publishers
ISBN 978-0-8376-1669-8

MINI Repair Manuals

MINI Cooper Service Manual: 2007-2011 *Bentley Publishers*
ISBN 978-0-8376-1671-1

MINI Cooper Service Manual: 2002-2006 *Bentley Publishers*
ISBN 978-0-8376-1639-1

MINI Cooper Diagnosis Without Guesswork: 2002-2006
Bentley Publishers
ISBN 978-0-8376-1571-4

Mercedes-Benz

Mercedes-Benz C-Class (W202) Service Manual 1994-2000
Bentley Publishers
ISBN 978-0-8376-1692-6

Mercedes Benz E-Class (W124) Owner's Bible: 1986-1995
Bentley Publishers
ISBN 978-0-8376-0230-1

Mercedes-Benz Technical Companion
Staff of The Star and members of Mercedes-Benz Club of America
ISBN 978-0-8376-1033-7

B BentleyPublishers™ .com

Automotive Reference

Bentley Publishers has published service manuals and automobile books since 1950. For more information, please contact Bentley Publishers at 1734 Massachusetts Avenue, Cambridge, MA 02138, visit our web site at **www.BentleyPublishers.com**